Element	Definition and Example
labi-	lip: *levator labii superioris*
lacri-	tears: *nasolacrimal*
later-	side: *lateral*
leuk-	white: *leukocyte*
lip-	fat: *lipid*
-logy	science of: *morphology*
-lysis	solution, dissolve: *hemolysis*
macro-	large, great: *macrophage*
mal-	bad, abnormal, disorder: *malignant*
medi-	middle: *medial*
mega-	great, large: *megakaryocyte*
meso-	middle or moderate: *mesoderm*
meta-	after, beyond: *metatarsal*
micro-	small: *microtome*
mito-	thread: *mitosis*
mono-	alone, one, single: *monocyte*
mons	mountain: *mons pubis*
morph-	form, shape: *morphology*
multi-	many, much: *multinuclear*
myo-	muscle: *myology*
narc-	numbness, stupor: *narcotic*
necro-	corpse, dead: *necrosis*
neo-	new, young: *neonatal*
nephro-	kidney: *nephritis*
neuro-	nerve: *neurolemma*
noto-	back: *notochord*
ob-	against, toward, in front of: *obturator*
oc-	against: *occlusion*
-oid	resembling, likeness: *sigmoid*
oligo-	few, small: *oligodendrocyte*
-oma	tumor: *lymphoma*
oo-	egg: *oocyte*
or-	mouth: *oral*
orchi-	testicles: *cryptorchidism*
ortho-	straight, normal: *orthopnea*
-ory	pertaining to: *sensory*
-ose	full of: *adipose*
osteo-	bone: *osteoblast*
oto-	ear: *otolith*
ovo-	egg: *ovum*
par-	give birth to, bear: *parturition*
para-	near, beyond, beside: *paranasal*
path-	disease, that which undergoes sickness: *pathology*
-pathy	abnormality, disease: *neuropathy*
ped-	children: *pediatrician*
pen-	need, lack: *penicillin*
-penia	deficiency: *thrombocytopenia*
per-	through: *percutaneous*
peri-	near, around: *pericardium*
phag-	to eat: *phagocyte*
-phil	have an affinity for: *neutrophil*
phlebo-	vein: *phlebitis*
-phobe	abnormal fear, dread: *hydrophobia*
-plasty	reconstruction of: *rhinoplasty*
platy-	flat, side: *platysma*
-plegia	stroke, paralysis: *paraplegia*
-pnea	to breathe: *apnea*
penumato-	breathing: *pneumonia*
pod-	foot: *podiatry*

Element	Definition and Example
-poiesis	formation of: *hemopoiesis*
poly-	many, much: *polyploid*
post-	after, behind: *postnatal*
pre-	before in time or place: *prenatal*
prim-	first: *primitive*
pro-	before in time or place: *prosect*
proct-	anus: *proctology*
pseudo-	false: *pseudostratified*
psycho-	mental: *psychology*
pyo-	pus: *pyoculture*
quad-	fourfold: *quadriceps femoris*
re-	back, again: *repolarization*
rect-	straight: *lateral rectus*
reno-	kidney: *renal*
rete-	network: *retina*
retro-	backward, behind: *retroperitoneal*
rhin-	nose: *rhinitis*
-rrhage	excessive flow: *hemorrhage*
-rrhea	flow or discharge: *diarrhea*
sanguin-	blood: *sanguiferous*
sarc-	flesh: *sarcoplasm*
-scope	instrument for examining a part: *stethoscope*
-sect	cut: *dissect*
semi-	half: *semilunar*
serrate-	saw-edged: *serratus anterior*
-sis	process or action: *dialysis*
steno-	narrow: *stenohaline*
-stomy	surgical opening: *tracheostomy*
sub-	under, beneath, below: *subcutaneous*
super-	above, beyond, upper: *superficial*
supra-	above, over: *suprarenal*
syn- (sym-)	together, joined, with: *synapse*
tachy-	swift, rapid: *tachycardia*
tele-	far: *telencephalon*
tens-	stretch: *tensor fascia lata*
tetra-	four: *tetrad*
therm-	heat: *thermogram*
thorac-	chest: *thoracic cavity*
thrombo-	lump, clot: *thrombocyte*
-tomy	cut: *appendectomy*
tox-	poison: *toxic*
tract-	draw, drag: *traction*
trans-	across, over: *transfuse*
tri-	three: *trigone*
trich-	hair: *trichology*
-trophy	a state relating to nutrition: *hypertrophy*
-tropic	turning toward, changing: *gonadotropic*
ultra-	beyond, excess: *ultrasonic*
uni-	one: *unicellular*
-uria	urine: *polyuria*
uro-	urine, urinary organs or tract: *uroscope*
vas-	vessel: *vasoconstriction*
vermi-	worm: *vermiform*
viscer-	organ: *visceral*
vit-	life: *vitamin*
zoo-	animal: *zoology*
zygo-	union, join: *zygote*

Concepts of Human Anatomy and Physiology

Concepts of Human Anatomy and Physiology

Second Edition

Kent M. Van De Graaff
Brigham Young University

Stuart Ira Fox
Pierce College

wcb

Wm. C. Brown Publishers
Dubuque, Iowa

Book Team

Editor *Edward G. Jaffe*
Developmental Editor *Carol Mills*
Designer *K. Wayne Harms*
Production Editor *Ann Fuerste*
Art Editor *Barbara J. Grantham*
Photo Research Editor *Mary Roussel*
Permissions Editor *Vicki Krug*
Visuals Processor *Joyce E. Watters*

wcb group

Chairman of the Board *Wm. C. Brown*
President and Chief Executive Officer *Mark C. Falb*

wcb

Wm. C. Brown Publishers, College Division

President *G. Franklin Lewis*
Vice President, Editor-in-Chief *George Wm. Bergquist*
Vice President, Director of Production *Beverly Kolz*
Vice President, National Sales Manager *Bob McLaughlin*
Director of Marketing *Thomas E. Doran*
Marketing Communications Manager *Edward Bartell*
Marketing Information Systems Manager *Craig S. Marty*
Executive Editor *Edward G. Jaffe*
Manager of Visuals and Design *Faye M. Schilling*
Production Editorial Manager *Colleen A. Yonda*
Production Editorial Manager *Julie A. Kennedy*
Publishing Services Manager *Karen J. Slaght*

Cover design and interior design by Julie E. Anderson.

Cover collage by Joan Hall, 1988.

Cover collage photo of fetus, © Lennart Nilsson, A CHILD IS BORN, Dell Publishing Company, New York. Cover collage photos of the eye and tastebuds, © Lennart Nilsson, BEHOLD MAN, Little, Brown and Company, Boston.

The credits section for this book begins on page 1066, and is considered an extension of the copyright page.

Library of Congress Catalog Card Number: 88–72157

ISBN 0–697–05675–9

Printed in the United States of America by Wm. C. Brown Publishers
2460 Kerper Boulevard, Dubuque, IA 52001

10 9 8 7 6 5 4 3

Cover concept

Although there are many concepts in human anatomy and physiology, probably none is so important as homeostasis, implied in the "body landscape" collage depicted on this cover. The body is a composite of dynamic organ-systems. Maintaining homeostasis between the many body organs requires intricate feedback mechanisms provided by specific molecules transported in the blood. In theory, each cell can chemically communicate with any other body cell through the myriad blood vessels. With the exception of the reproductive system, all the body systems provide maintenance and sustenance roles necessary to maintain homeostasis. The reproductive system provides a mechanism for propagation and for nurturing the offspring, thus insuring the continuity of the human species.

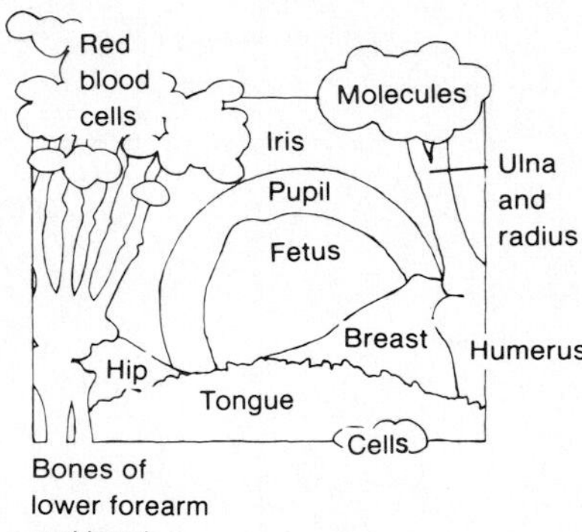

Reviewers

We were very fortunate to have a knowledgeable and hardworking panel of reviewers, whose forthright criticisms and helpful suggestions added immeasurably to the quality of the final text. The review panel for the second edition included:

John C. Conroy
University of Winnipeg

Esta R. Schwarz
Bucks County Community College

Enid Farrand
Harford Community College

Louis D. Trombetta
St. John's University

Jeffry W. Gerst
North Dakota State University

David Saltzman
Santa Fe Community College

Thomas J. Nowak
British Columbia Institute of Technology

Robert L. Azen
Cypress College

Matthew F. Sak
Salem State College

Reviewers of the First Edition

Donald S. Kisiel
Suffolk County Community College

Werner Wieland
Mary Washington College

Donna Van Wynsberghe
University of Wisconsin, Milwaukee

David E. Grosland
Iowa Central Community College

William C. Kleinelp, Jr.
Middlesex County College

Roberta M. O'Dell-Smith
University of New Orleans

Michael P. Schenk
University of Texas Health Science Center, Dallas

This book could not have been written without the enduring patience and support of our wives, Karen Van De Graaff and Ellen Fox, to whom this book is gratefully dedicated.

Brief Contents

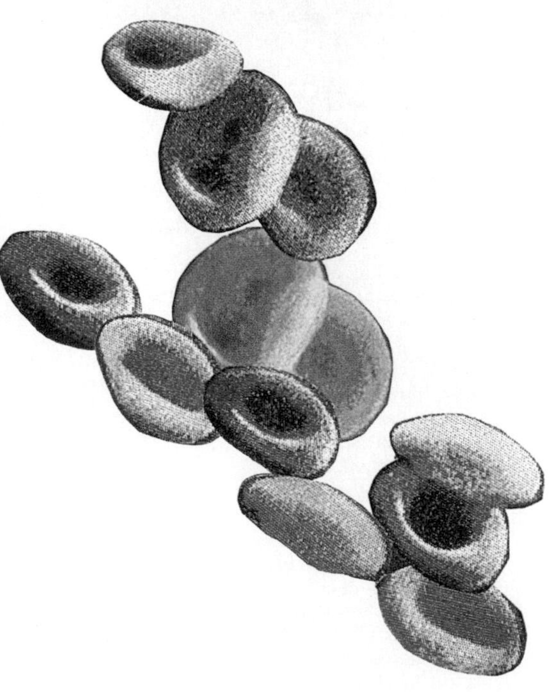

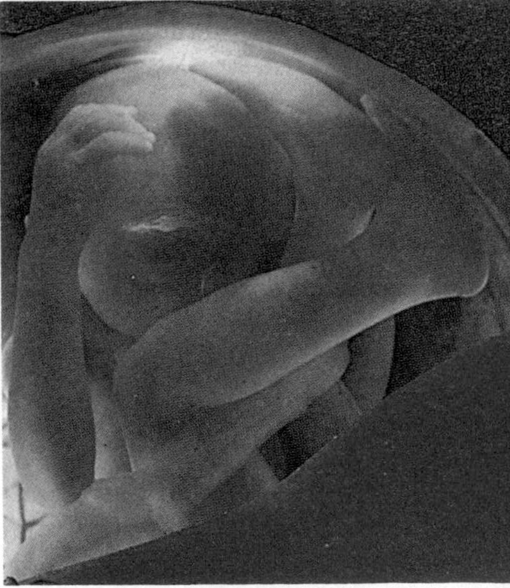

EXPANDED CONTENTS

Unit 3 Integration and Control Systems of the Human Body 396

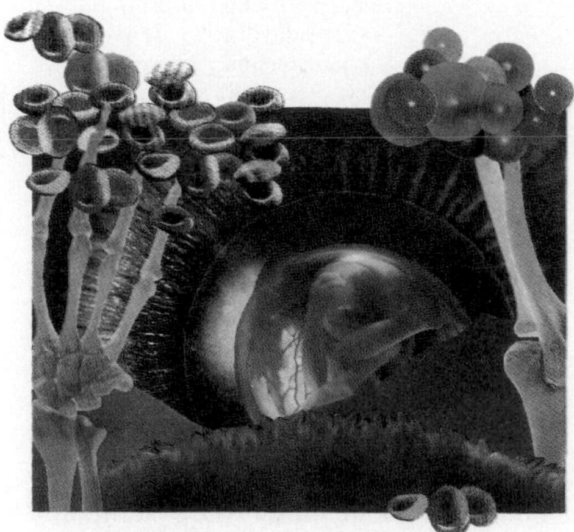

Unit 5 Continuance of the Human Species 922

Preface

The second edition of *Concepts of Human Anatomy and Physiology,* like the first edition, organizes the facts of anatomy and physiology around unifying concepts and learning objectives. Complete descriptions and clear explanations, many accurate and attractive illustrations, and numerous pedagogical devices help students learn the material in each conceptual unit. The way in which the material is presented promotes understanding, not simply memorization, of the information required in the anatomy and physiology course. The organization of the text also enables instructors to tailor the required text readings to their course needs.

This textbook is designed for students who do not have extensive science backgrounds but who plan to enter health and other careers that require extensive knowledge of anatomy and physiology. Therefore, the chapters in the first unit present basic chemical, cellular, biological, and anatomical concepts. The chapters in the remaining four units then present in detail the anatomy and physiology of organs and systems. Furthermore, descriptions of the frontiers of knowledge, as well as explanations of the experimental basis for the information presented, heighten the excitement of learning. Also, the second edition is scientifically current, containing many new and revised concepts that reflect ongoing research efforts.

Organization

The thirty chapters in this text are grouped into five units.

Unit 1: Orientation and Organization of the Human Body. Introductory material containing information about the history of anatomy and physiology, the scientific method, and anatomical concepts and terminology is presented in chapter 1. Chapters 2–7 provide necessary background information on chemistry, cell structure and function, and tissues. Students are shown how this basic knowledge will apply to concepts studied later in the course. Chapter 7 emphasizes the fundamental importance of homeostasis in understanding body function.

Unit 2: Support and Movement of the Human Body. The structures and functions of the integument and musculoskeletal systems are described in chapters 8–13.

Unit 3: Integration and Control Systems of the Human Body. The nervous system, endocrine system, and sensory organs are described in chapters 14–19.

Unit 4: Regulation and Maintenance of the Human Body. In chapters 20–26, the structure and function of the circulatory, immune, respiratory, urinary, and digestive systems are discussed within the overall theme of homeostasis. Since these systems are of great clinical importance, their basic anatomy and physiology is explained in sufficient detail to allow students to understand important pathological processes. The clinical applications presented in these chapters, in turn, reinforce the basic concepts presented.

Unit 5: Continuance of the Human Species. Chapters 27–30 describe the male and female reproductive systems, prenatal development, and postnatal development and aging. The important concepts of these chapters are presented in sufficient depth to promote understanding.

Second Edition

The narrative and illustrations in each chapter have been revised in response to reviewers' and users' comments as well as to new scientific findings. Therefore, students will find the text very readable and visually appealing, and instructors will find that it is scientifically current and accurate. Some of the changes in the second edition include, but are not limited to, the following.

1. All chapters have been reorganized to some extent to improve the flow of information and to enhance understanding.
2. The outstanding illustration program has been improved and now includes many new pieces, more full color, careful attention to placement near text reference, and the color-coded learning system initiated in the first edition.
3. In the second edition the skeletal system is covered in two separate chapters: one on the axial skeleton, and one on the appendicular skeleton.
4. Mechanisms of muscle contraction have been updated, and a thorough discussion of the causes of muscle fatigue has been added to chapter 12.

5. Memory and language function has been added to the discussion of the central nervous system (chapter 15).
6. Recent knowledge about the transport and cellular uptake of cholesterol, and the etiology of atherosclerosis, appears in chapters 19 and 25.
7. Blood composition and clotting mechanisms are discussed in chapter 20, and the cardiovascular system topics are now covered in two, instead of three, chapters.
8. The subject of atrial natriuretic factor has been added to chapters 20 and 24.
9. Discussion of AIDS has been updated, and the topics of macrophage-leukocyte interactions and interleukin-2 have been added to chapter 22.
10. New findings that relate to sex determination are discussed in chapter 27.
11. Selected readings are presented in an appendix, where they are grouped by topic, instead of at the ends of the chapters.

Pedagogical Aids

This text is organized to maximize students' ability to locate information, associate facts in a conceptual framework, and test their mastery of the material. The wide variety of pedagogical devices within the text further assist students. These devices are described in detail under the heading Student Aids on page xx.

Supplementary Materials

Supplementary materials accompany the text and are designed to aid students in their learning activities and to help instructors plan coursework and presentations. These supplementary materials include:

1. A *Laboratory Manual* to accompany *Concepts of Human Anatomy and Physiology,* second edition, by Stuart I. Fox and Kent M. Van De Graaff. The laboratory manual includes information required for each laboratory exercise and requires that students read relevant background information in the textbook before attempting and completing the laboratory exercises. It provides the laboratory experience required to support the lecture portion of the course and includes exercises that have been classroom-tested for a number of years.
2. The *Instructor's Manual for the Laboratory Manual* provides the answers to the questions contained in the laboratory reports in the *Laboratory Manual.*
3. A *Student Study Guide* to accompany *Concepts of Human Anatomy and Physiology,* second edition, by Kent M. Van De Graaff. This study guide aids students in their mastery of the material presented in the course and helps them test their knowledge by taking sample quizzes, answering clinical questions, working crossword puzzles, and adding labels to figures.
4. An updated set of *70 color slides,* which depict important pathological conditions, helps support information given in the text and is free to adopters of this textbook. A narrative describing each slide is provided in the *Instructor's Manual.*
5. A set of *100 acetate transparencies* is available to instructors who adopt this text. The transparencies are made from selected illustrations in the text and have been chosen for their value in reinforcing lecture presentations.
6. An *Instructor's Manual and Test Item File* by Kerry Openshaw provides instructional support in the use of the textbook. It also contains a test item file with approximately fifty items for each chapter to aid instructors in constructing examinations.
7. A *Computer Review of Human Anatomy and Physiology* by S. Scott Zimmerman, Thomas V. Davis, and Kent M. Van De Graaff is contained in a thirteen-disk set designed for Apple® II- or IBM®-compatible computers. This computer disk set provides a graphic and innovative way to learn and review human anatomy and physiology.
8. *Study Cards for Anatomy and Physiology* by Kent M. Van De Graaff, R. Ward Rhees, and Christopher H. Creek. This boxed set of more than three hundred 3-by-5-inch cards presents a well-organized and illustrated synopsis of the structure and function of the human body. Clinical information is presented as it applies to specific body organs and systems or physiological processes. The Study Cards present a quick and effective way for students to review human anatomy and physiology.
9. *Anatomy and Physiology Study Notes* by Kent M. Van De Graaff, R. Ward Rhees, and Christopher H. Creek. This spiral-bound book contains a well-organized and illustrated synopsis of the structure and function of the human body. Clinical information is presented as it applies to specific body organs and systems or physiological processes. The Study Notes present a quick and effective way for students to review human anatomy and physiology.
10. **wcb** *QuizPak,* a student self-testing program that operates on an Apple® IIe or IIc, IBM® PC, or Macintosh® microcomputer, is available to instructors who adopt this text.
11. **wcb** *TestPak,* a free, computerized testing service for generating examinations, is available to instructors who adopt this text.

Student Aids

The organization and pedagogical devices of this text are designed to help you learn anatomy and physiology. Don't just read this text as you would read a novel. Interact with it. Use the pedagogical tools provided. For example, check off the objectives as they are achieved, and write out the flowcharts and essays asked for at the ends of the sections. The more active you are in your studies, the better you will learn and the more enjoyable the study will become.

The following information about the organization and pedagogical devices in the text will help you use this book to the best advantage.

Unit Introduction

The chapters of this text are grouped into five units, each beginning with an introduction that succinctly summarizes the concepts presented in the unit and lists the chapters in the unit. Read these unit introductions for a general orientation to the topics presented in the units and to learn the common characteristics of the chapters that allow them to be grouped together as a unit.

Chapter Introductions

The opening page of each chapter contains an overview, in outline form, of the contents of the chapter. The "concepts" of each major heading in the chapter are also listed here. Read these before beginning a chapter to gain a perspective on the material to be covered. This page will also be a useful reference for you afterward, when studying for exams.

Concepts and Objectives

One of the unique attributes of this text is the way in which major headings are introduced. Each major heading begins with a succinct summary of the primary organizing concept in that section, and a list of learning objectives. Read the concepts and the objectives before starting the section, and then check back as you read the section to assure that you have fulfilled the learning objectives.

Study Activities

Each major heading in the chapter ends with a box containing study activities: pictures and flowcharts to draw, essays to answer, and other activities. The more you actually perform these activities, the better you will understand the material presented in the sections. Write these out, rather than just think about them.

Understanding Terminology

Where each technical term first appears in the narrative, it is set off by boldface or italic type and followed by a phonetic pronunciation in parentheses. You should pause in your readings when these appear, and learn the correct pronunciation of the words (you will not be able to remember a term if you can't pronounce it). Then go back and reread the sentence for understanding.

Word Derivations

The derivation of some of the terms in the text are provided in footnotes at the bottom of the page. These are often interesting in themselves, and if you know how a word was derived, it becomes more meaningful and is easier to remember.

Boxed Clinical Commentaries

Following a discussion of a basic concept of anatomy or physiology, you may find a colored box of text. These contain short discussions of the clinical or practical applications of the information preceding the boxes. You will find it enjoyable, as well as instructive, to see how your newly acquired basic science knowledge is applied to clinical and practical problems.

Illustrations and Tables

This text contains abundant tables and illustrations to support the concepts presented. Carefully studying the tables will help you to understand the text more completely, and the summary tables will be useful when you review for examinations. Though many of the figures are visually beautiful, they are constructed with one primary purpose—to illustrate concepts presented in the text. Therefore, analyze and try to understand each figure as it is referenced in the text. This will probably require rereading of sentences, but it is the only way to derive maximum benefit from the information presented.

Chapter Summaries

At the end of each chapter the material is summarized for you in outline form. This outline summary is organized by major headings followed by the essence of textual information. Read the summary after studying the chapter to be sure that you have not missed any points, and use the chapter summaries to help you review for examinations.

Review Activities

Following each chapter summary is a section called Review Activities. These sections contain objective questions (with the answers in Appendix A) and essay questions. Be sure to take these self-quizzes in a "closed book" fashion after studying the chapters, and then correct your answers using the appendix. Review the information that relates to any missed questions. These practice exams will help you to anticipate the types of questions that could be on real exams, and they provide a "reality check" so you can be sure you learned the required information.

Appendixes

The appendixes, following chapter 30, contain valuable reference information. Appendix A (page 1037) contains the answers to the objective questions in the review activities of each chapter. Appendix B (page 1038) lists the important scientific journals that publish current research relative to anatomy and physiology. Appendix C (page 1039) provides selected readings that will help you to learn more about particular topics of interest. Appendix D (page 1044) lists the medical and pharmacological abbreviations used in clinical practice. Appendix E (page 1045) lists common clinical laboratory tests performed on human blood and urine.

Glossary

The glossary provides pronunciations and definitions of the more important scientific and clinical terms used in the text. Whenever you encounter an unfamiliar term or would like additional information about a term, look it up in the glossary.

Index

The index at the end of the book allows you to locate the pages on which specific terms and concepts are discussed. There are numerous cross-references that aid in the location of referenced subjects. Use the index as frequently as needed; it is particularly useful when you are studying for exams and as a reference source in your later coursework.

Acknowledgments

The second edition of *Concepts of Human Anatomy and Physiology* was greatly improved by comments that the authors have received from the many users of the first edition. Though there are too many to acknowledge individually, we are deeply grateful to each one. As in the past, our colleagues at our respective institutions have been very supportive and helpful. In particular, we would like to thank professors Richard W. Heninger, Lawrence H. Thouin, R. Ward Rhees, James Rikel, and William M. Hess.

We also wish to thank friends who have assisted in specific ways. Drs. J. Phillip Freestone, Douglas W. Hacking, and Charles H. Stewart provided professional advice. Dr. Brent C. Chandler provided many of the X rays used in the text. The photographs of surface anatomy were taken by Dr. Sheril D. Burton. Drs. James N. Jones and Paul Urie assisted in updating the clinical information. Trent D. Rasmussen, Connie J. Erdmann, and Pamela Barker assisted in various ways in the preparation of the ancillaries. Sincere gratitude is extended to Kathlyn J. Loveridge for her meticulous and dedicated typing of portions of the manuscript.

Quality illustrations for this text were provided by a number of talented artists. We are grateful for their tremendous contributions.

The editorial and production staffs at Wm. C. Brown have inspired, guided, and shaped this enormous project. We owe a large debt of gratitude to Executive Editor Ed Jaffe, Developmental Editor Carol Mills, Production Editor Ann Fuerste, and many other talented people at Wm. C. Brown Publishers.

Kent Marshall Van De Graaff

Stuart Fox

Concepts of Human Anatomy and Physiology

UNIT 1

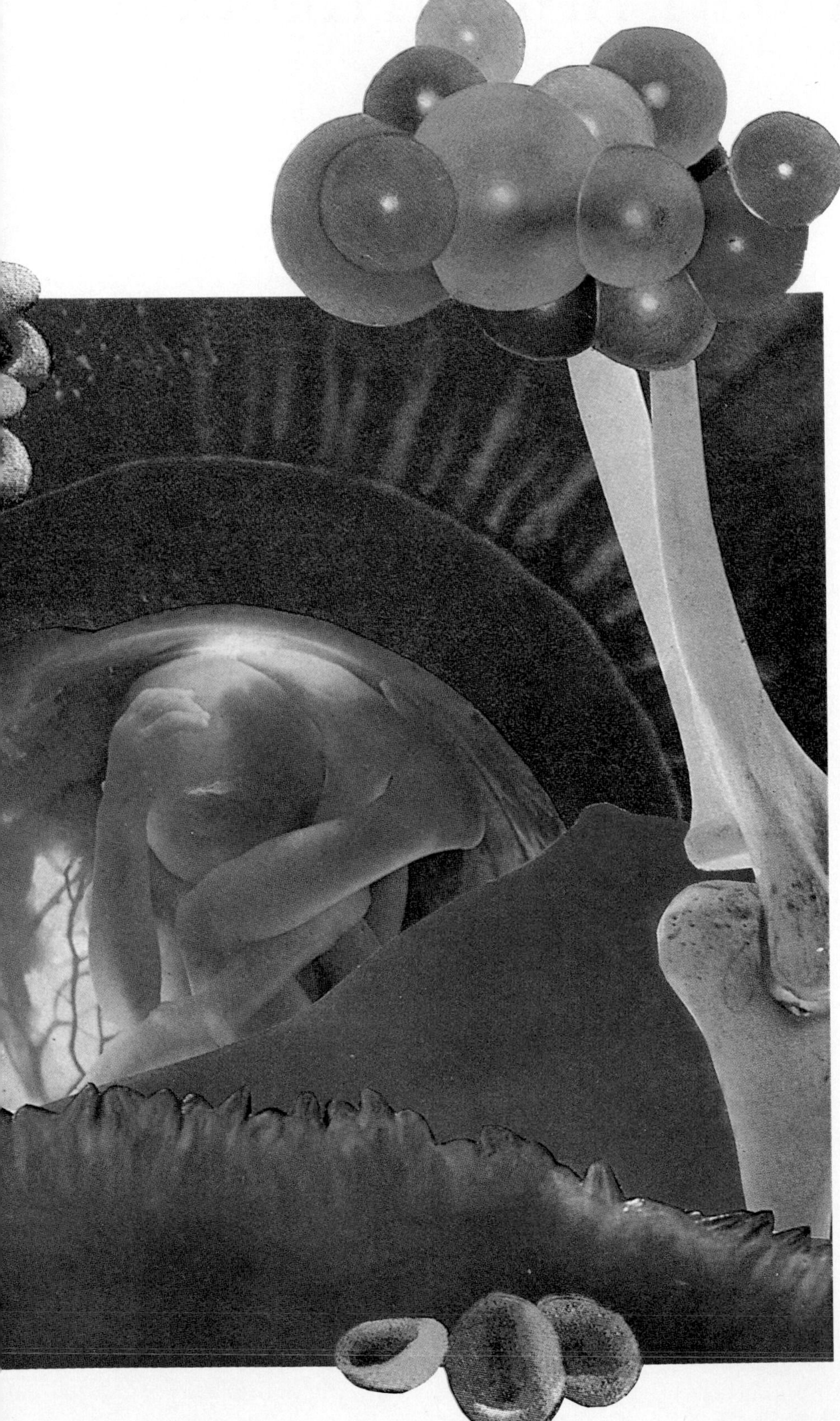

The chapters of Unit I introduce the sciences of anatomy and physiology by explaining the basic organization and principal processes of the human body. The fundamental terminology for describing the structure and function of the body is presented. The various levels of body composition and organization, from the chemical and cellular to the tissue (histological) and organ systems, are defined and described. The theme of homeostasis is explained clearly to set the stage for understanding the body systems, which are discussed in the chapters that follow.

ORIENTATION AND ORGANIZATION OF THE HUMAN BODY

1 **INTRODUCTION TO ANATOMY AND PHYSIOLOGY**
Anatomy and physiology are integrated and dynamic sciences that have a precise, descriptive vocabulary of Greek and Latin derivation. The cells of the human body are highly organized into tissue and organ systems, functioning together to maintain homeostasis.

2 **CHEMICAL COMPOSITION OF THE BODY**
The body is composed of atoms chemically bonded together to form molecules. The principal molecules in the body are proteins, nucleic acids, carbohydrates, and lipids.

3 **CELL STRUCTURE AND GENETIC REGULATION**
Cellular structure and function depend upon the organelles contained within a cell and the controlling influence of the DNA within a cell's nucleus.

4 **ENZYMES, ENERGY, AND METABOLISM**
Cellular metabolism and the release of energy is dependent upon the appropriate extracellular environment, the availability of the proper nutrients to the cell, the removal of wastes, a constant optimal temperature, and the activity of enzymes.

5 **MEMBRANE TRANSPORT AND THE MEMBRANE POTENTIAL**
The cell membrane is dynamic in selectively permitting the passage of materials through passive and active transport. The distribution of molecules and ions results in a potential difference, or voltage, across the membrane.

6 **HISTOLOGY**
Tissues are classified, on the basis of their cellular composition and histological appearance, as epithelial, connective, muscle, or nervous tissue; each type occurs in specific organs and performs specialized functions.

7 **HOMEOSTASIS AND CONTROL MECHANISMS**
Homeostasis in the body's organ systems is maintained by numerous and precise negative feedback mechanisms involving the actions of the endocrine and nervous systems.

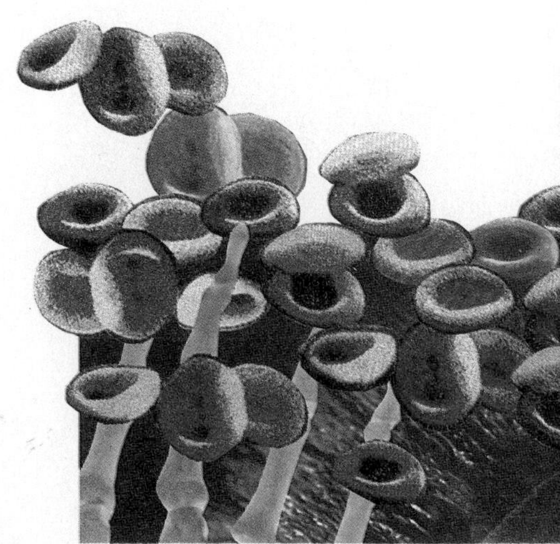

1

Introduction to Anatomy and Physiology

Outline

Concepts

Anatomy and physiology are integrated sciences.

Anatomy and physiology are dynamic sciences that have long, exciting heritages and currently provide the foundation for medical, biochemical, developmental, cytogenetic, and biomechanical research.

Humans are biological organisms belonging to the phylum Chordata within the kingdom Animalia, and to the family Hominidae within the class Mammalia and the order Primates.

The maintenance of the life of an organism depends upon the presence of certain vital factors for its existence.

Structural and functional levels of organization exist in the human body, with each of its parts contributing to the total organism.

As sciences, anatomy and physiology have descriptive terminologies that must be mastered by students who would understand and be conversant in these subjects.

The human body is divided into regions and specific localized areas, which can be identified on the surface. Each region contains internal organs, the location of which are anatomically, physiologically, and clinically important.

For functional and protective purposes, the viscera is compartmentalized and supported in specific body cavities by connective and epithelial membranes.

All of the descriptive planes of reference and terms of direction used in anatomy are standardized because of their reference to the body in anatomical position.

Measurements are the human way of expressing relative lengths, masses, temperatures, and volumes of substances, which are important because they bring uniformity and consistency to scientific communication.

Definitions of the Sciences

Anatomy and physiology are integrated sciences.

Objective 1. Define anatomy and physiology.
Objective 2. Explain the concept of complementarity of organization between structure and function and the importance of knowing this concept in studying human anatomy and physiology.

Human anatomy and physiology are sciences concerned with the structure and function of the human body. The term **anatomy** *(an-nat'o-me)* is derived from a Greek word meaning "to cut up"; in the past, the word *anatomize* was more commonly used than the word *dissect*. The dissection of a human **cadaver** *(kah-dav'er)* has been the basis for understanding the structure of the human body for a long time. **Physiology** *(fiz"e-ol'o-je)* is also derived from a Greek word meaning "the study of nature"—the "nature" of an organism is its function. Physiology is a dynamic science that attempts to explain physical and chemical processes occurring in the body. Much of the knowledge of physiology is gained through experimentation.

Anatomy and physiology are both subdivisions of the science of *biology*, the study of living organisms. Frequently, anatomy and physiology are studied as separate disciplines, in which case learning the anatomy of the body precedes learning its physiology. Anatomical structure and function are complementary, however, and the simultaneous study of both facilitates learning about the body. The anatomy of every structure of the body is adapted for performing a function or perhaps several functions. Natural selection eliminates organisms with inappropriate structures and functions and determines which favorable structures will be passed from one generation to the next.

1. Explain the statement that anatomy is a science of observation, and physiology is a science of experimentation and observation.
2. Use an example of a specific disease (such as lung cancer, which alters the anatomy of the lung) to show complementarity of organization between structure and function. Include in the answer how diagnosis (detection) of the disease is frequently determined by altered body functions.

anatomy: Gk. *ana*, up; *tome*, a cutting
cadaver: L. *cadere*, to fall
physiology: Gk. *physis*, nature; *logos*, study
biology: Gk. *bios*, life; *logos*, study

Historical Perspective

Anatomy and physiology are dynamic sciences that have long, exciting heritages and currently provide the foundation for medical, biochemical, developmental, cytogenetic, and biomechanical research.

Objective 3. Discuss key historical events in the science of anatomy and physiology.
Objective 4. Relate how the scientific method is currently used in physiological experiments.
Objective 5. Explain why an understanding of anatomy and physiology is important and personally applicable.
Objective 6. List ways a person may remain informed of anatomical and physiological research, and discuss why this is important.

The sciences of human anatomy and physiology have had rich, long, and frequently troubled heritages. The histories of the sciences of human anatomy and physiology parallel that of medicine. In fact, an interest in the structure and function of the body has often resulted from the desire of the medical profession to explain a body dysfunction. Various religions, on the other hand, have stifled the study of human anatomy and physiology through their restrictions on human dissections and their religious explanations of diseases and debilitations.

People have always had an innate interest in their own structure, function, and physical capabilities. The Greeks esteemed physical athletic competition and expressed the beauty of the body in their art. Many of the great masters of the Renaissance portrayed human figures in their art. Indeed, many of these artists were excellent anatomists because of their attention to detail. Such an artistic genius was Michelangelo, who was preoccupied with the detail and beauty of the human form and captured it in sculpture with the *David* (fig. 1.1) and in paintings such as those executed in the Sistine Chapel.

Even Shakespeare was awed by the structure of the human body as he wrote, "What a piece of work is a man! How noble in reason! How infinite in faculty! In form and moving how express and admirable! In action how like an angel! In apprehension how like a god! The beauty of the world! The paragon of animals" (*Hamlet* 2.2.315–319).

As a science, anatomy preceded that of physiology. It was necessary to describe the structures of the body before their functions could be determined. In addition, technology in the physical sciences, like physics and chemistry, had to make sufficient progress so that physiological experiments could be conducted. Much current physiological research is ongoing as new technology is continuously being developed.

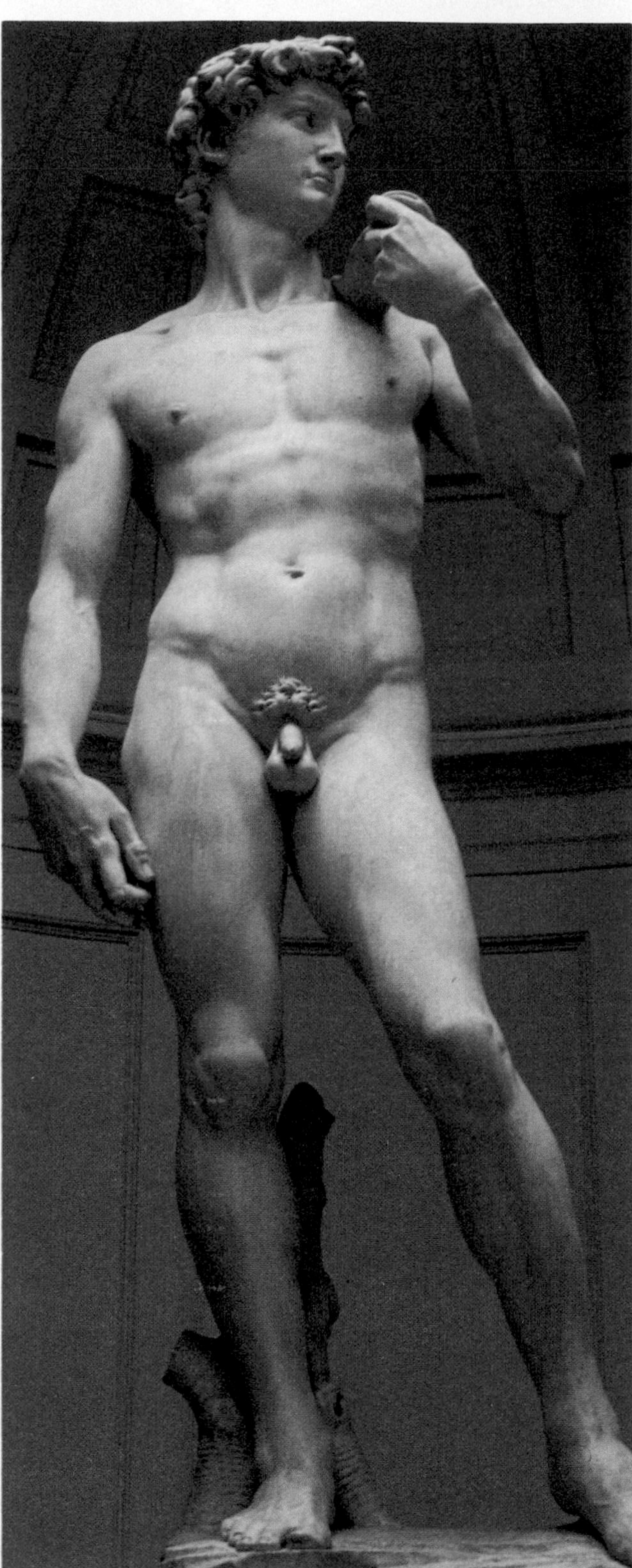

Figure 1.1. Michelangelo completed the 17-foot-tall *David* in 1504, sculptured from a single block of white, unflawed Carrara marble. This masterpiece captures the physical beauty of the human body in an expression of art.

Figure 1.2. A fourteenth-century painting of the famous Greek physician Hippocrates. Hippocrates is referred to as the "father of modern medicine," and his creed is immortalized as the Hippocratic oath.

Historical Synopsis of Human Anatomy and Physiology

Grecian Period. Anatomy and physiology were first widely accepted as sciences in ancient Greece. The writings of several Greek philosophers had a tremendous impact on future scientific thinking. *Hippocrates* (460–337 B.C.) was the famous Grecian physician (fig. 1.2) who is regarded as the "father of modern medicine" because of the sound principles of medical practice that he established. Perhaps the greatest contribution of Hippocrates was that he attributed diseases to natural causes rather than to the displeasure of the gods. His application of logic and reason to medicine was the beginning of observational medicine.

Hippocrates probably had only a limited exposure to human dissections but was well disciplined in the popular humoral theory of body organization. Four body humors were recognized, and each was associated with a particular body organ: sanguine with the liver; choler, or yellow bile, with the gallbladder; phlegm with the lungs; and melancholic, or black bile, with the spleen. A healthy person was thought to have a balance of the four humors.

physician: Gk. *physikos,* natural
sanguine: L. *sanguis,* bloody
choler: Gk. *chole,* bile
phlegm: Gk. *phlegm,* inflammation
melancholic: Gk. *melan,* black; *chole,* bile

Figure 1.3. This Roman copy of a Greek sculpture is believed to be of Aristotle, the famous Grecian philosopher.

The concept of humors has long since been discarded, but it dominated medical thought for over two thousand years.

Aristotle (384–332 B.C.), a pupil of Plato, made careful investigations of all kinds of animals—he even included references to humans—and established a type of scientific method in obtaining data (fig. 1.3). He wrote the first known account of embryology, in which he described the development of the heart in a chick embryo. His best-known zoological works are *History of Animals, Parts of Animals,* and *Generation of Animals.*

In spite of his tremendous accomplishments, Aristotle perpetuated some erroneous views of anatomy. For example, the doctrine of the humors formed the boundaries of Aristotle's thought. Plato had described the brain as the seat of feeling and thought, but Aristotle disagreed. He stated that the heart was the seat of intelligence and that the function of the brain, which was bathed in fluid, was to cool the blood that was pumped from the heart.

The Grecian scientist *Erasistratus* (about 300 B.C.) was more interested in body functions than structure and is, therefore, frequently referred to as the "father of physiology." Erasistratus authored a book on the causes of diseases, in which he included observations on the heart, vessels, brain, and cranial nerves. He noted the toxic effects of snake venom on various visceral organs and described changes in the liver resulting from various diseases. Although some of the writings of Erasistratus were very scientific, other concepts of his were primitive and mystical. For instance, he thought that the cranial nerves carried animal spirits and that muscles contracted because of distention by spirits.

humor: L. *humor,* fluid

Both *Erasistratus* and another Grecian philosopher, *Herophilus,* were greatly criticized later in history for the performance of *vivisection (viv"i-sek'shun).* Celsus (about A.D. 30) and Tertullian (about A.D. 200) were particularly critical of the practice of vivisection. Herophilus was described as a butcher of men who had dissected as many as six hundred living persons, some of them as public demonstrations.

Roman Era. In many respects, the Roman Empire stifled scientific advancements and set the stage for the Dark Ages that were to follow. The interest and emphasis of science shifted from the theoretical to the practical during this time. Few dissections of cadavers were performed other than at *autopsies* in attempts to determine the cause of death in criminal cases. Medicine was not preventive but was almost solely limited to the treatment of soldiers injured in battle. Later in Roman history, laws were established that indicated the influence of the church on medical practice. According to Roman law, for example, no deceased pregnant women could be buried without prior removal of the fetus from the womb so that it could be baptized.

Claudius Galen (A.D. 130–201) was perhaps the best physician since Hippocrates. He was certainly the most influential writer of all times on medical subjects. For nearly fifteen hundred years, the writings of Galen were the unquestionable authority on anatomy and medical treatment. Galen probably dissected no more than two or three human cadavers during his career, of necessity limiting his anatomical descriptions to animal dissections. He compiled nearly five hundred medical papers (of which eighty-three have been preserved) from earlier works as well as his personal studies. Galen believed in the humors of the body and perpetuated this concept; he also gave authoritative explanations of nearly all body functions.

Galen's works contain many errors, primarily because of his desire to present definitive answers and his interpretation of data from nonhuman animals. He did, however, provide some astute and accurate anatomical details that are still regarded as classics. He proved to be an experimentalist, demonstrating that the heart of a pig would continue to beat when the spinal nerve was transected so that nerve impulses could not reach the heart. He showed that the squealing of a pig stopped when the

vivisect: L. *vivus,* living; *sectus,* to cut
autopsy: Gk. *autopsia,* seeing with one's own eyes

Figure 1.4. Plates from *De Humani Corporis Fabrica,* which Vesalius completed at the age of twenty-eight. This book, published in 1543, revolutionized the sciences of anatomy and physiology.

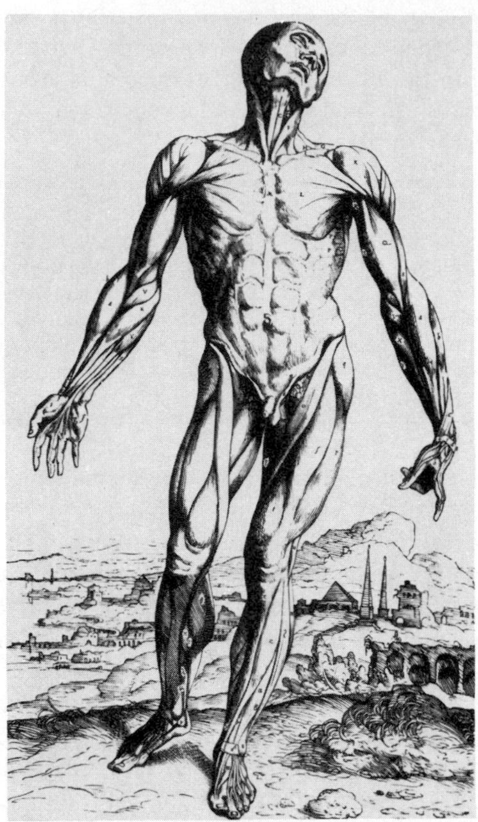

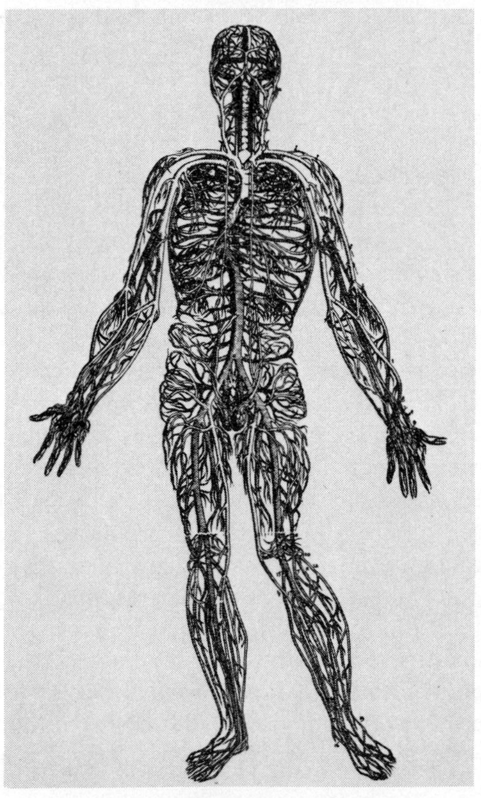

particular nerve that innervated its vocal cords was cut. He also proved that arteries contained blood rather than pneuma.

Middle Ages. The Middle Ages (Dark Ages) came with the fall of the Roman Empire in A.D. 476 and lasted nearly one thousand years. Dissections of cadavers were totally prohibited during this period, and molesting a corpse was a criminal act frequently punishable by burning at the stake. If mysterious deaths occurred, examinations by inspection and palpation were acceptable. During the plague epidemic in the sixth century, however, a few *necropsies (nek'rop-se)* and dissections were performed in hopes of determining the cause of this dreaded disease.

Renaissance. The period of time known as the Renaissance was characterized by a rebirth of science. It lasted from the fourteenth through the sixteenth century and was a transitional period from the Middle Ages to the modern age of science. The development of movable type about 1450 revolutionized the production of printed books and helped to usher in the Renaissance.

With the increased interest in anatomy and physiology during the Renaissance, obtaining cadavers for dissection became a serious problem. Medical students regularly practiced grave robbing until finally an official decree was issued that permitted the bodies of executed criminals to be used as specimens.

The major advancement in anatomy during the Renaissance came from the artistic and scientific ability of *Andreas Vesalius* (1514–64). Vesalius apparently had enormous energy and ambitions. He completed the masterpiece of his life, *De Humani Corporis Fabrica,* by the time he was twenty-eight years old. The various body systems and individual organs were beautifully illustrated and described in the *Fabrica* (fig. 1.4). The magnitude of this work and the impact it had on science earned Vesalius the title of "father of modern anatomy." His book was especially important in that it boldly challenged hundreds of Galen's erroneous teachings. Vesalius wrote of his surprise at finding numerous anatomical errors that were taught as fact, and he refused to accept Galen's explanations on faith. Bitter controversies ensued between Vesalius and the traditional Galenic anatomists, including Vesalius's former teacher, Sylvius. Vesalius became so incensed by the relentless attack that he destroyed much of his later unpublished work and ceased his dissections.

Seventeenth and Eighteenth Centuries. The two most outstanding contributions to anatomy and physiology of the seventeenth and eighteenth centuries were the explanation of blood flow and the development and use of the microscope.

---

epidemic: Gk. *epi,* upon; *demos,* people
necropsy: Gk. *nekros,* corpse; *opsy,* view

Figure 1.5. The English physician William Harvey demonstrated with experiments in 1628 that blood circulates and does not flow back and forth through the same vessels.

In 1628, the English physician *William Harvey* (1578–1657) published his outstanding work *On the Movement of the Heart and Blood in Animals.* This important research established brilliant proof of the continuous circulation of blood within contained vessels, and the technique of investigation presented in this publication is still regarded as a classic example of the scientific method for conducting research (fig. 1.5). Harvey is frequently regarded as the "father of modern physiology." Like Vesalius, Harvey was severely criticized for his departure from Galenic philosophy. The controversy over the circulation of the blood raged for twenty years until other anatomists finally repeated Harvey's experiments and added information.

Antony van Leeuwenhoek (1632–1723) was a Dutch lens grinder who so improved the microscope that he achieved a magnification of 270 times. His many contributions included developing techniques for tissue examination and describing blood cells, spermatozoa, and the striped appearance of skeletal muscle.

The development of the microscope added an entirely new dimension to anatomy and physiology and eventually led to explanations of basic body functions. In addition, the improved microscope was invaluable for understanding the etiologies (causes) of diseases and discovering cures for many of them.

Nineteenth and Twentieth Centuries. The major contribution in the nineteenth century was the formulation of the cell theory and the implications it had for a clearer understanding of the structure and functioning of the body. Once the microscope was invented, it was merely a matter of time before cells were discovered and described.

Johannes Müller (1801–58) is noted for applying the sciences of physics, chemistry, and psychology to the study of the human body. With this increased dimensional breadth to the subject, anatomy and physiology became comparative sciences.

The study of anatomy and physiology during the twentieth century became specialized, and the research more detailed and complex. In response to the increased technology and depths of understanding, new disciplines and specialities have appeared in the sciences of anatomy and physiology to categorize and use the new knowledge (fig. 1.6).

The Scientific Method

The Nobel laureates for anatomy and physiology research are listed on the inside of the back cover of this text. In addition to having an intense interest in biology and humanity, these distinguished researchers all employed the processes of scientific inquiry in conducting their investigations and making discoveries. Although there may be considerable creativity within the boundaries of these investigative processes, scientific experiments progress in well-defined, orderly ways. Such a disciplined approach for the acquisition of truth is referred to as the *scientific method.* Simply stated, the scientific method depends on a number of processes of gaining information in orderly ways. Actually, the scientific method may be used in searching for answers to questions encountered in everyday life as well as in methodical research.

All of the information in this text has been gained by the application of the scientific method. Although many different techniques are involved in the scientific method, all share three attributes: (1) confidence that the natural world, including ourselves, is ultimately explainable in terms we can understand; (2) descriptions and explanations of the natural world that are honestly based on observations and that could be modified or refuted by other observations; and (3) humility, that is, the willingness to accept the fact that we could be wrong. If further study should yield conclusions that refute all or part of an idea, the idea must be accordingly modified. In short, the scientific method is based on a confidence in our rational ability, honesty, and humility. Practicing scientists may not always display these attributes, but the validity of the large body of scientific knowledge that has been accumulated—as shown by the technological applications and the predictive value of scientific hypotheses—are ample testimony to the fact that the scientific method works.

Figure 1.6. Some of the subdivisions or specialities of human anatomy and physiology.

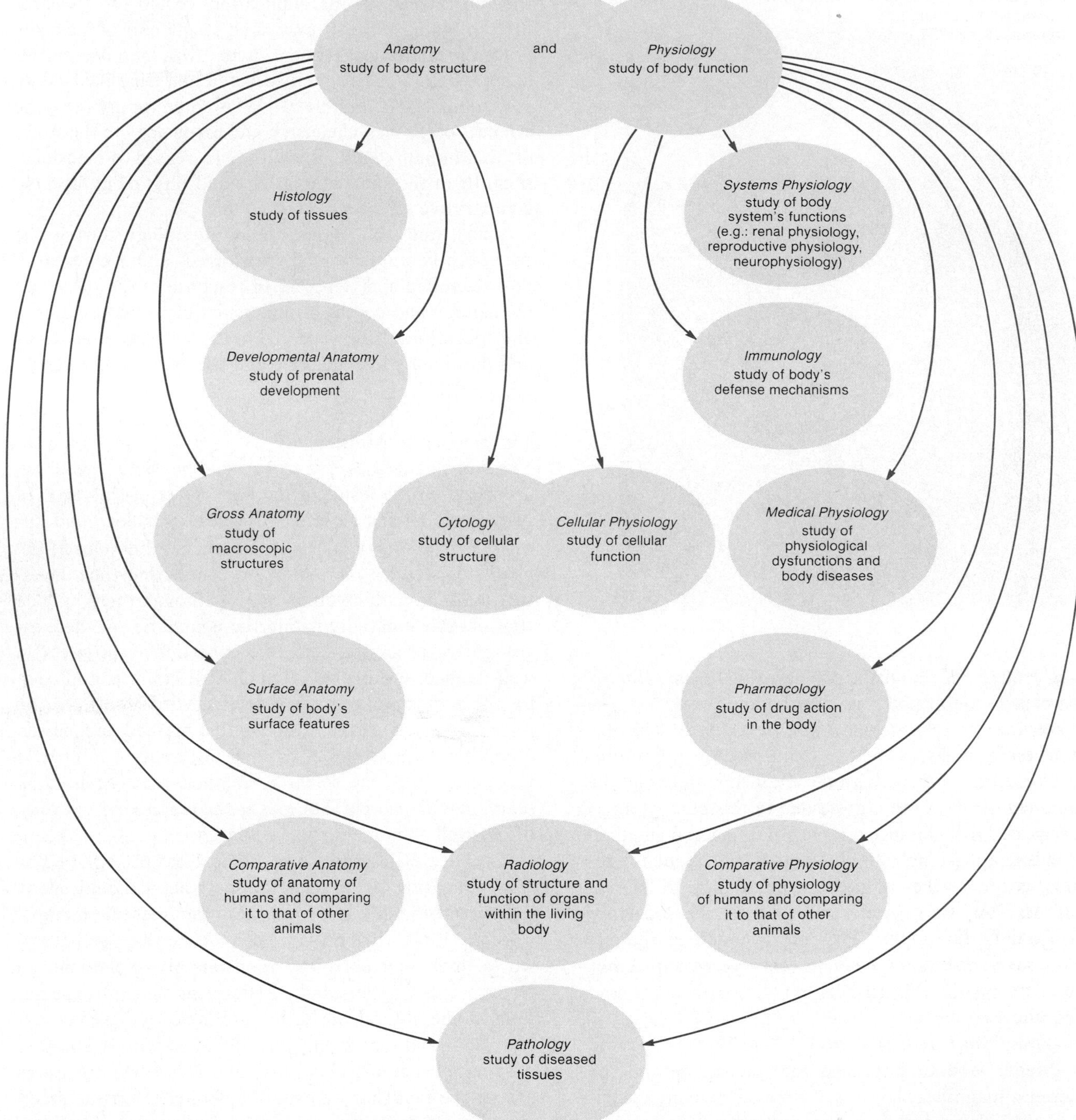

For an example of the scientific method, suppose the resting pulse rate of one athlete and one sedentary person is measured, and the athlete has a lower rate. If the scientific method is followed, it can be stated that the athlete had a lower resting pulse than the sedentary person at the time the measurements were made, but no generalization is scientifically justified by these data. If these measurements are repeated at different times and the differences remain, a scientific hypothesis can be constructed stating that athletes have lower resting pulse rates than sedentary people. This hypothesis is scientific because it is *testable;* the pulse rates of 100 athletes and 100 sedentary people can be measured to see if statistically significant differences are obtained. If they are, one is justified in saying that athletes, on the average, have lower resting pulse rates than sedentary people, *based on obtained data.* An investigator must still be open to the fact that the hypothesis

Figure 1.7. A positron emission tomography (PET) scan of the brain following an injection of radioactive glucose. The left side of the brain, concerned with cognitive activity, and the frontal lobes, concerned with planning and organization of behavior, are seen as active areas in this PET scan.

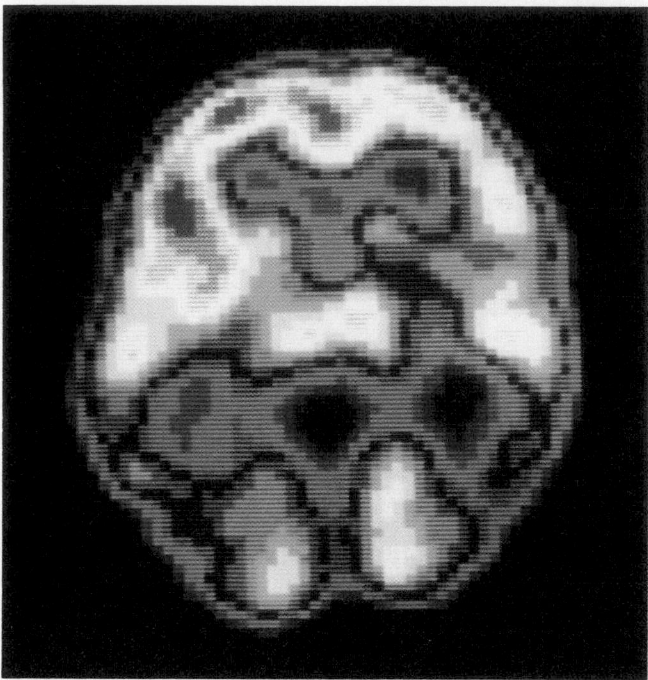

could be wrong (the measurement techniques may have been biased, or the sample of people may not have been representative of the general population), or that there could be alternative explanations for the results. Before the discovery becomes generally accepted, other scientists have to consistently replicate the results because scientific theories are based on *reproducible data.*

It is quite possible that when others attempt to replicate previous experiments, their results will be slightly different. They may then construct scientific hypotheses that the differences in resting pulse rate also depend on factors such as the nature of the exercise performed by the athlete or on nutrition or on genetic influences. When other scientists attempt to test these hypotheses, they will likely encounter new problems, requiring new explanatory hypotheses, which must be tested by additional experiments.

In this way, a large body of highly specialized information is gradually accumulated and a more generalized explanation can be formulated. This explanation will almost always be different from preconceived notions. People who follow the scientific method will then appropriately modify their concepts, realizing that their new ideas will probably have to be changed again in the future as additional experiments are performed.

New concepts that emerge from the scientific method are not always welcome; people often prefer comfortable, older ideas, particularly when these are part of a large, cherished belief structure. In this situation, the knowledge gained by the scientific method may be rejected, and the scientists involved vilified, persecuted, or even murdered. Because the open communication of knowledge is a scientific tradition, however, those ideas that are closest to the truth have survived and eventually supplanted their competitors in the minds of most educated people.

Anatomy and Physiology Today

As depicted in figure 1.6, anatomy and physiology form the bases for numerous scientific and medical disciplines. Anatomy and physiology are dynamic, applied sciences, which are changing constantly as discoveries are made. Innovative techniques and ideas for conducting research provide data for formulating hypotheses in attempts to understand physiological processes or etiologies of diseases. A positron emission tomography (PET) scan, for example, is used to quantify the relative metabolic activity within different parts of an organ, such as the brain (fig. 1.7).

With such a proliferation of scientific knowledge, the question is, "How does one keep up in the dynamic subject of anatomy and physiology?" The answer is different for the scientist and the student. Being on the forefront of discovery, the scientist must specialize and stay current in a narrow realm of inquiry. The scientist collaborates with professional colleagues, conducts research, presents findings at scientific meetings, and publishes new discoveries in scientific journals (see appendix B).

In order for a student of anatomy and physiology to keep up, it is first necessary to establish a solid and broad understanding of the sciences. This text as used in an introductory course is designed to provide a foundation for the student. A student must learn sufficient facts so that concepts can be understood and applied in a cognitive fashion. Not only should students of anatomy and physiology be able to describe a body structure and know how it functions, but they should also be able to predict the dysfunctions if certain processes were not to occur.

An objective of this text is to enable students to become educated and conversant in anatomy and physiology. An excellent way for students to keep up with anatomy and physiology during and after the formal course is completed is to subscribe to and read scientific magazines such as *Science, Scientific American, Discover,* or *Science News.* These publications and others have many articles on scientific discoveries written in an understandable manner. There are many specialities of anatomy and physiology that are undergoing a rapid proliferation of exciting knowledge. It is most important and interesting to become and stay informed so that you may be an educated contributor to society.

Figure 1.8. A schematic diagram of the front part of the embryo of a chordate.

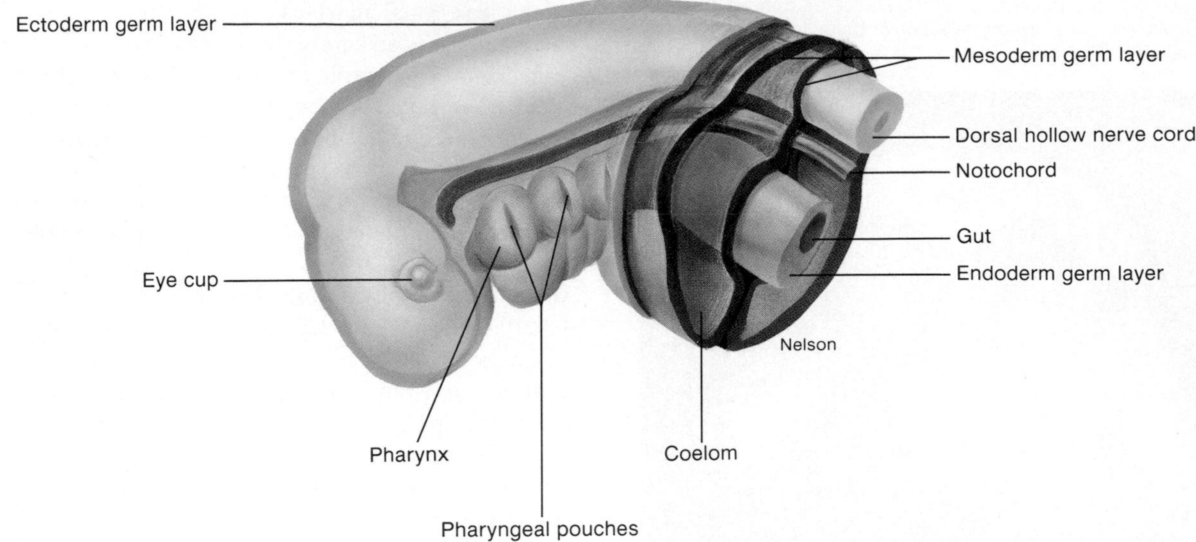

1. Define *vivisection, necropsy,* and *autopsy.*
2. What effect did religion have on the early interest in anatomy?
3. Why is Erasistratus regarded as the "father of physiology" and Vesalius the "father of anatomy"?
4. Explain the importance of experimentation and careful observation in gaining new knowledge. Why must a scientific experiment be performed many times before it is reported as fact?
5. How can a student stay current in the sciences of anatomy and physiology? Why is it important to do so?

Classification and Characteristics of Humans

Humans are biological organisms belonging to the phylum Chordata within the kingdom Animalia, and to the family Hominidae within the class Mammalia and the order Primates.

Objective 7. List the taxonomic classification of humans.
Objective 8. List the characteristics that identify humans as chordates and as mammals.
Objective 9. Describe the characteristics that humans have but other primates do not.

The classification, or taxonomic, scheme has been established by biologists to organize the structural and evolutionary relationships of living organisms. Each category of classification is referred to as a *taxon.* The highest taxon is the kingdom, and the most specific taxon is the species. Humans are species belonging to the **animal kingdom.** *Phylogeny (fi-loj'ĕny)* is the origin and evolutionary development of animal species.

Phylum Chordata

Human beings belong to the phylum Chordata *(fi'lum kor-dah'tah)* along with fish, amphibians, reptiles, birds, and other mammals. All chordates have three structures in common: a **notochord** *(no'to-kord),* a **dorsal hollow nerve cord,** and **pharyngeal pouches** *(fah-rin'je-al)* (fig. 1.8). These chordate characteristics are well expressed during the embryonic stage of development, and to a certain extent are present in an adult. The notochord is a flexible rod of tissue that extends the length of the back of an embryo. A portion of the notochord persists in the adult as the **nucleus pulposus,** located within each **intervertebral disc** (fig. 1.9). The dorsal hollow nerve cord is positioned above the notochord and develops into the **brain** and **spinal cord,** which are highly functional as the **central nervous system** in the adult. Pharyngeal pouches form gill openings in fish and some amphibians. In other chordates, such as humans, embryonic pharyngeal pouches develop, but only one of the pouches persists, becoming the **auditory (eustachian) canal** *(u-sta'ke-an),* a connection between the middle ear and **pharynx** *(far'ingks)* (throat area).

Class Mammalia

Mammals are chordate animals with hair and mammary glands. Hair is a thermoregulatory protective covering for most mammals, and mammary glands serve for suckling the young (fig. 1.10). Other characteristics of mammals include three ear ossicles (bones), fleshy external ear (auricle), heterodont dentition (differently shaped teeth), squamosal-dentary jaw articulation (a joint between lower

heterodont: Gk. *heteros,* other; *odontos,* tooth

Figure 1.9. A midsagittal section through vertebrae to show the intervertebral discs and nuclei pulposi.

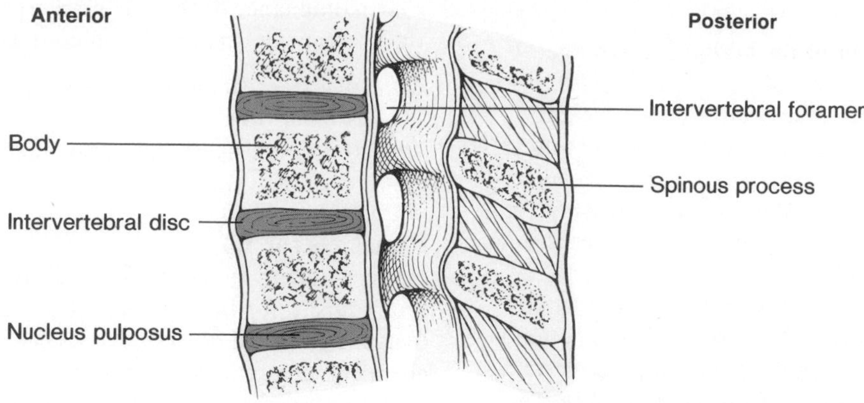

Figure 1.10. Characteristics of mammals—the class to which humans belong.

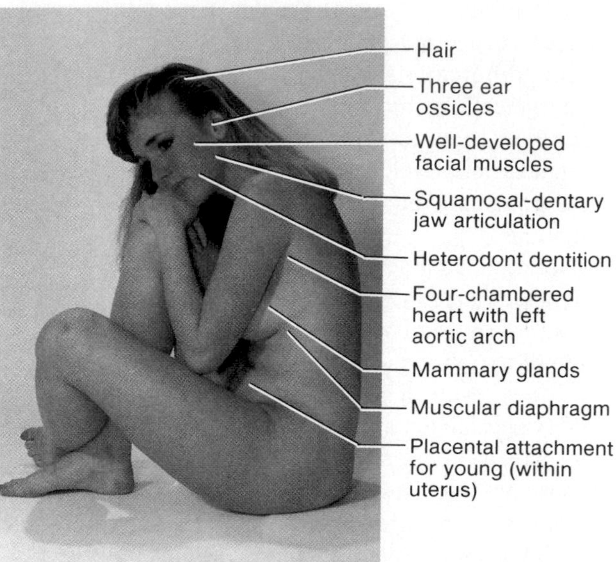

Table 1.1 Classification scheme of human beings

Taxon	Designated grouping	Characteristics
Kingdom	Animalia	Eucaryotic cells that lack walls, plastids, and photosynthetic pigments
Phylum	Chordata	Dorsal hollow nerve cord; notochord; pharyngeal pouches
Subphylum	Vertebrata	Vertebral column
Class	Mammalia	Mammary glands; hair
Order	Primates	Well-developed brain; prehensile hands
Family	Hominidae	Large cerebrum; bipedal locomotion
Genus	*Homo*	Flattened face; prominent chin and nose with inferiorly positioned nostrils
Species	*sapiens*	Largest cerebrum

From Kent M. Van De Graaff, *Human Anatomy,* 2d ed. Copyright © 1988 Wm. C. Brown Publishers, Dubuque, Iowa. All Rights Reserved. Reprinted by permission.

jaw and skull), usually seven cervical vertebrae, an attached placenta *(plah-sen'tah),* well-developed facial muscles, a muscular diaphragm, and a four-chambered heart with a left aortic arch.

Order Primates

There are several subdivisions of closely related groupings of mammals, called orders. Humans, along with lemurs, monkeys, and great apes, belong to the order called Primates. Members of this order have prehensile hands, digits modified for grasping, and relatively large, well-developed brains.

placenta: L. *placenta,* flat cake
Primates: L. *primas,* first
prehensile: L. *prehensus,* to grasp

Family Hominidae

Humans are the sole living members of the family Hominidae. *Homo sapiens* is included within this family, to which all the varieties or ethnic groups of humans belong. Each "racial group" has distinguishing features that have been established in isolated populations over thousands of years. The taxonomic classification of humans is presented in table 1.1.

Characteristics of Humans

Human beings have a few anatomical characteristics that are so specialized that they are diagnostic in separating them from other animals and even from other closely related mammals. Humans also have characteristics that are equally well developed in other animals, but when these function with the human brain, they provide remarkable capabilities. The anatomical characteristics of humans include the following.

1. **Size and development of the brain.** The average human brain weighs between 1,350 and 1,400 g. This gives humans a large brain-to-body-weight ratio. But more important is the development of portions of the brain. Certain extremely specialized regions and structures within the brain account for emotion, thought, reasoning, memory, and precise, coordinated movement.
2. **Style of locomotion.** Because humans stand and walk on two appendages, their style of locomotion is said to be bipedal. Upright posture imposes certain other diagnostic structural features such as the sigmoid (S-shaped) curvature of the spine, the anatomy of the hip and thighs, and arched feet. Some of these features may cause clinical problems in older individuals.
3. **Opposable thumb.** The human thumb joint is structurally adapted for tremendous versatility in grasping objects. Most primates have opposable thumbs.
4. **Vocal structures.** Humans, like no other animals, have developed articulated speech. The anatomical structure of the vocal organs and the well-developed brain have made this possible.
5. **Stereoscopic vision.** Although this characteristic is well developed in several other animals, it is also keen in humans. Human eyes are directed forward so that when focused upon an object, it is viewed from two angles. Stereoscopic vision gives depth perception, or a three-dimensional image.

Humans also differ from other animals in the number and arrangement of vertebrae (vertebral formula), the kinds and number of teeth (tooth formula), well-developed facial muscles, and the structural specializations of various body organs.

1. What is a chordate? Why are humans considered members of the phylum Chordata?
2. Why are humans designated as mammals and primates? What characteristics distinguish humans from other primates?
3. Which of the characteristics of humans are adaptive for social organization?

Maintenance of Life

The maintenance of the life of an organism depends upon the presence of certain vital factors for its existence.

Objective 10. List and describe the principal requisites for life.

Objective 11. Discuss the physical needs of an organism to sustain life.

Requisites for Life

Humans, being living organisms, have several traits that are requisite for life. Some of the requisites for life are dynamic chemical processes that occur continuously. Such chemical interactions and reactions are referred to as *metabolism (mĕ-tab'o-lizm)*. Other requisites for life are organizational; that is, they relate to the structural complexity of an organism. The requisites for life include the following.

1. **Development.** It is during the developmental process, called **morphogenesis** *(mor''fo-jen'ĭ-sis)* that undifferentiated (unspecialized) tissues undergo cellular differentiation, resulting in organs that are structurally and functionally distinct.
2. **Organization.** Each body structure is genetically determined and adapted to carry out specific functions.
3. **Adaptation.** Just as an organism must have a tolerance range in which it can be functional, structures in the body have tolerance ranges for each of the physical needs. Narrow levels of tolerance result in a structure or organism being highly specialized and less adaptable than those that have wide levels of tolerance.
4. **Responsiveness.** Responsiveness is the ability of an organism (or a cell) to monitor internal or external conditions and respond to changes that might be threatening. A white blood cell, for example, will respond to toxins from bacteria and move toward the pathogens in an attempt to dispose of them. Pulling a hand away from a hot object demonstrates responsiveness, as does drinking water to satisfy thirst.
5. **Movement.** It is the movement either within cells or of the cells themselves that results in the pumping of blood, respiratory movements, and even changes in an organism's position from one place to another as muscle fibers are stimulated to contract. Even nerve impulses result from the movements of ions across cell membranes.

Table 1.2 Physical needs for life

Physical needs	Examples of physiological effects
Water	Comprises about 60% of body weight and is the medium in which most chemical reactions in the body take place; essential for body homeostasis
Oxygen	Comprises 20% of normal air and is used in the body in releasing energy from food substances
Food	Nutrient-containing substances that can be broken down into chemicals used for energy, building new compounds, or regulating body metabolic processes
Heat	A form of energy expressed as temperature; the body can tolerate a wide range in environmental temperature but maintains a narrow range in body temperature; many physiological mechanisms insure a consistent internal body temperature
Pressure	External, such as atmospheric pressure, important for respiration; internal, such as blood pressure, important for filtration of blood in kidneys
Protection	Clothing and shelter create artificial environments for our bodies

6. **Reproduction.** Reproduction refers to both cellular replication, important for body growth and repair, and to sexual reproduction when parents propagate (participate in having offspring) a new individual.
7. **Growth.** Growth refers to processes of both the cells and the organism. After cellular replication, it is important for the new (daughter) cells to grow and mature in order to be functional. Likewise, there is a genetically determined growth process of an organism when body structures reflect metabolic needs.
8. **Respiration.** Respiration refers to all of the processes that permit the exchange of oxygen and carbon dioxide between the body cells and the external environment and to the utilization of oxygen and production of carbon dioxide by the tissue cells.
9. **Digestion and absorption.** Digestion and absorption is the process by which complex food substances are chemically altered into simpler forms that pass through membranes lining the digestive tract into the body fluids.
10. **Circulation.** A complex, multicellular organism must have the means for all of its living cells to receive continuous sustenance; this is provided by the circulation of body fluids, which transports nutrients and oxygen to cells and removes metabolic wastes.
11. **Synthesis.** Synthesis *(sin'thĕ-sis)* is the process by which simple molecules are united to form more complex molecules, such as DNA, RNA, proteins, and fats.
12. **Secretion.** Secretions are specific chemical compounds synthesized (produced) within specialized cells. Various cellular secretions are necessary for intercellular communication (such as the secretion of hormones) or for other purposes. Digestive enzymes, for example, are secreted into the digestive tract to break down complex food molecules so that absorption can occur.
13. **Excretion.** Excretion is the removal of wastes that are produced as the result of cellular metabolism. These wastes are frequently toxic to the cell and must be eliminated from the cellular environment as well as from the body of a multicellular organism.
14. **Survival instinct.** An innate survival instinct may be essential for an organism if it is to avoid death long enough to propagate. A survival instinct is, therefore, not only important for an individual, but it is also important for the perpetuation of a species population.

Physical Conditions Required for Life

Humans live in an environment conducive to life. Not only is the body adapted to take advantage of environmental parameters, but it has become dependent upon the consistency of these factors for survival. Life is fragile and is strictly dependent on the presence of particular environmental conditions. These are summarized in table 1.2.

1. Which of the requisites for life and the physical needs of life are necessary for metabolism?
2. Distinguish between secretion and excretion.
3. Why is maintaining a constant body temperature important for physiological mechanisms?

Body Organization

Structural and functional levels of organization exist in the human body, with each of its parts contributing to the total organism.

Objective 12. Identify the structures of a cell, tissue, organ, and system, and explain the relationships among these structures as they constitute an organism.

Objective 13. Explain the general function of each system.

Figure 1.11. The levels of structural organization and complexity within the human body.

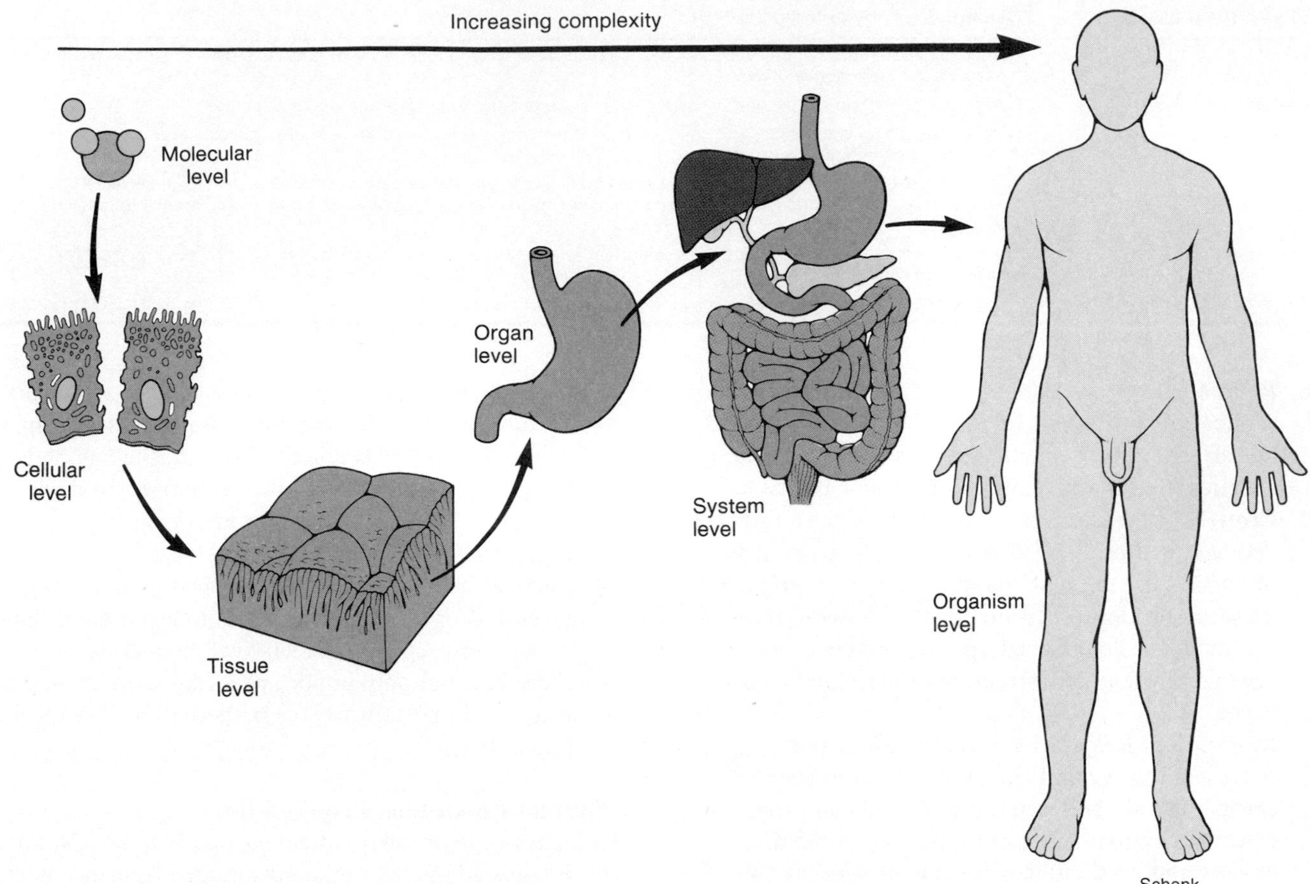

Cellular Level

The **cell** is the basic structural and functional component of life. Humans are multicellular organisms *(or'gah-nizm)* composed of between 60 and 100 trillion cells. It is at the cellular level (fig. 1.11) that the vital functions of life are carried on, such as metabolism, growth, irritability and adaptability, repair, and reproduction. All cells are composed of chemical substances called **protoplasm** *(pro'to-plaz''m)*. Certain protoplasmic compounds are arranged into small functional structures called **organelles** *(or''gah-nelz)*. Each organelle carries out a specific function within the cell. The nucleus, mitochondria, and endoplasmic reticulum are all organelles.

The body contains many distinct kinds of cells, each specialized to perform specific functions. Examples of specialized cells are bone cells, muscle cells, fat cells, blood cells, and nerve cells. Each of these cell types has a unique structure directly related to its function.

cell: L. *cella,* small room
protoplasm: Gk. *protos,* first; *plassein,* to mold

Tissue Level

Tissues are layers or aggregations of similar cells that perform specific functions. An example of a tissue is muscle, which functions to contract and produce movement within the body. The outer layer of skin is a tissue because it is composed of similar cells that are bound together and function to protect and contain body contents.

Organ Level

An **organ** is an aggregate of two or more tissues integrated to perform a particular function. Organs occur throughout the body and vary greatly in size and function. Examples of organs are the heart, spleen, pancreas, ovary, skin, and even any bone within the body. Each organ usually has one or more primary tissues and several secondary tissues. In the stomach, for example, the inside epithelial lining is considered the primary tissue because the basic functions of secretion and absorption occur within this layer. The secondary tissues of the stomach are the supporting connective tissue and the vascular, nervous, and muscle tissues.

tissue: Fr. *tissu,* woven; from L. *texo,* to weave
organ: Gk. *organon,* instrument

System Level

The **systems** of the body constitute the next level of structural organization. A body system consists of various organs that have similar or related functions. Examples of systems are the circulatory system, nervous system, digestive system, and endocrine system. Certain organs may serve several systems. The pancreas, for example, functions with both the endocrine and digestive systems. All of the systems of the body are interrelated and function together, constituting the total organism.

A *systemic (sis-tem'ik) approach* to studying anatomy and physiology emphasizes the functional relationships of various organs within a system. For example, the functional role of the digestive system can be better understood if all of the organs within that system are studied together. Another approach to anatomy, the *regional approach,* has merit in professional schools because the structural relationships of portions of several systems can be observed simultaneously. Dissections of cadavers are usually conducted on a regional basis. Trauma or injury usually affects a region of the body, whereas a disease that affects a region may also involve an entire system.

This text uses a systemic approach to anatomy and physiology. In the chapters that follow, you will become acquainted system by system with the structural and functional aspects of the entire body. An overview of the structure and function of each of the body systems is presented in figure 1.12.

---

1. Construct a diagram to illustrate the levels of structural organization that characterize the body. Which of these levels are microscopic?
2. Why is the skin considered an organ?
3. Which body systems control the functioning of the others; which are supportive of the organism; and which serve a transportive role?

Anatomical and Physiological Terminology

As sciences, anatomy and physiology have descriptive terminologies that must be mastered by students who would understand and be conversant in these subjects.

Objective 14. Explain how anatomical and physiological terms are derived.

Objective 15. Define what is meant by prefixes and suffixes.

Anatomy and physiology are descriptive sciences. Analyzing anatomical and physiological terminology can be a rewarding experience in itself as one learns something of the character of antiquity. Not only is an understanding of the roots of words of academic interest, but a familiarity with technical terms reinforces the learning process. The majority of scientific terms are of Greek or Latin derivation. Some of the more recent terms are of German and French derivation. Unfortunately, more recent anatomical and medical terms have been coined in honor of various anatomists or physicians. Such terms have no descriptive basis, cannot be associated with anything, and must simply be memorized.

Many Greek and Latin terms were coined more than two thousand years ago. It is exciting to decipher the meanings of these terms and gain a glimpse into our medical heritage. Many terms referred to common plants or animals. Thus, the term *vermis* means "worm," *cochlea (kok'le-ah),* "snail shell," *cancer,* "crab," *uvula,* "grape," and even *muscle* comes from the Latin *musculus,* "mouse." Other terms reveal the warlike environment of the Greek and Latin era. *Thyroid,* for example, means "shield," *xiphos (zi'fos),* "sword," and *thorax,* "breastplate." *Sella* means "saddle," and *stapes (sta'pēz)* means "stirrup." Various tools or instruments were referred to in early anatomy. The malleus and anvil resemble miniatures of a blacksmith's implements, and tympanum refers to a drum.

You will encounter many new terms throughout your study of anatomy and physiology. You can learn these terms more easily by understanding the prefixes and suffixes of the new words. Use the glossary of prefixes and suffixes (on the inside of the front cover) as an aid in learning new terms. Pronouncing these terms as you learn them will also aid your retention.

Learning the material presented in the remainder of this chapter establishes a basic foundation for anatomy and physiology, as well as for all medical and paramedical fields. Anatomy is a very precise science because there is a universally accepted and used reference language for describing body parts and locations. That language will now be introduced.

---

1. Explain the statement that "anatomy and physiology are descriptive sciences."
2. Refer to the glossary of prefixes and suffixes listed on the inside of the back cover to decipher the terms: *blastocoel, hypodermic, dermatitis,* and *orchiectomy.*

---

system: Gk. *systema,* being together

Figure 1.12. Various systems of the human body.

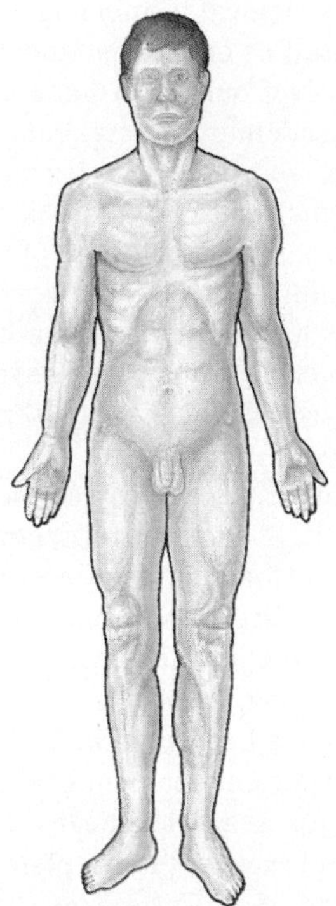
Integumentary system
Function: external support and protection of body

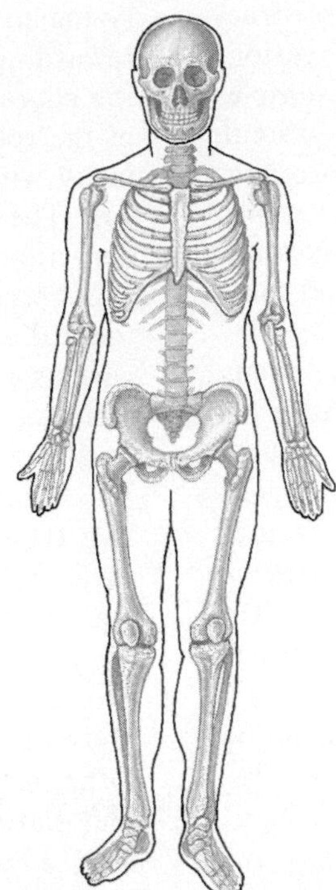
Skeletal system
Function: internal support and flexible framework for body movement; production of blood cells

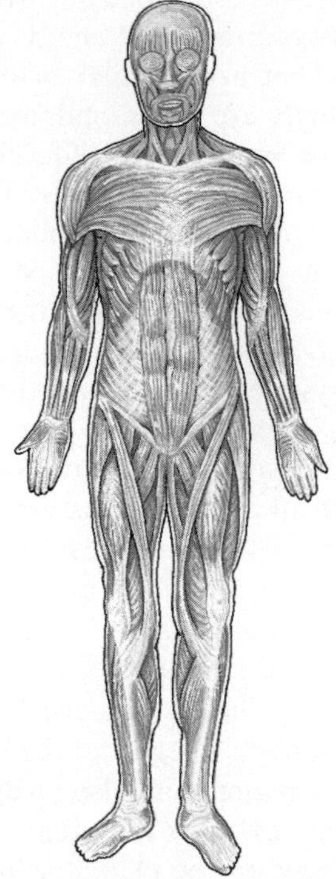
Muscular system
Function: body movement; production of body heat

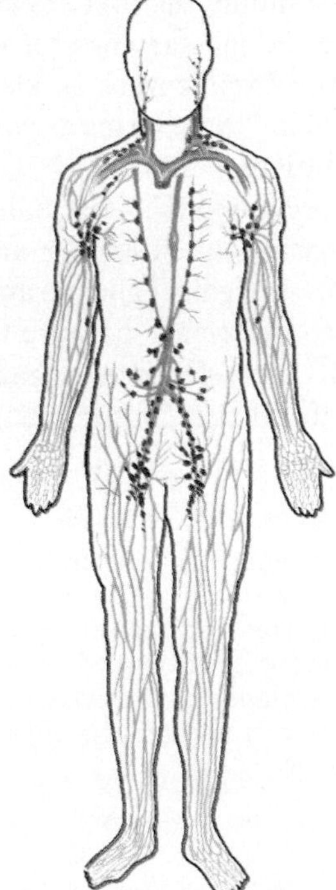
Lymphatic system
Function: body immunity; absorption of fats; drainage of tissue fluid

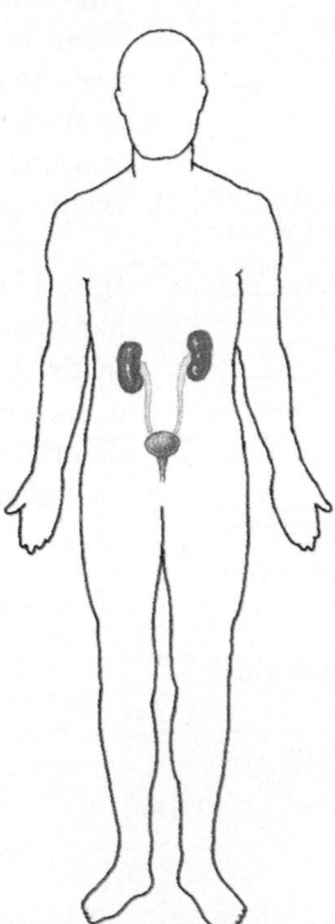
Urinary system
Function: filtration of blood; maintenance of volume and chemical composition of the blood

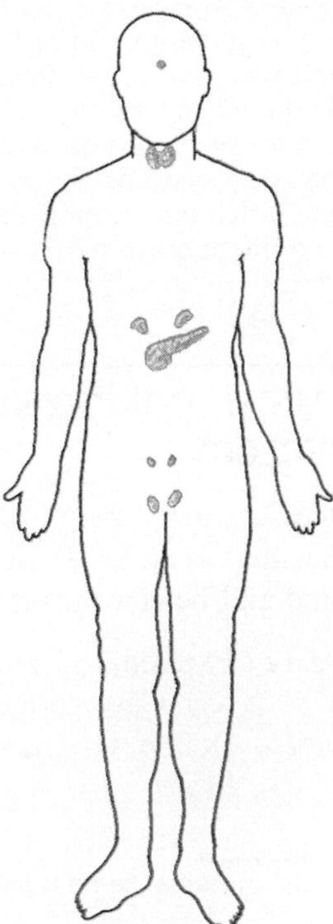
Endocrine system
Function: secretion of hormones for chemical regulation

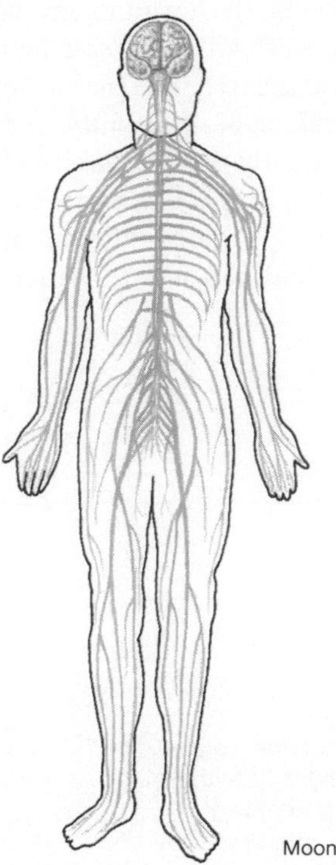

Nervous system
Function: regulation of all body activities: learning and memory

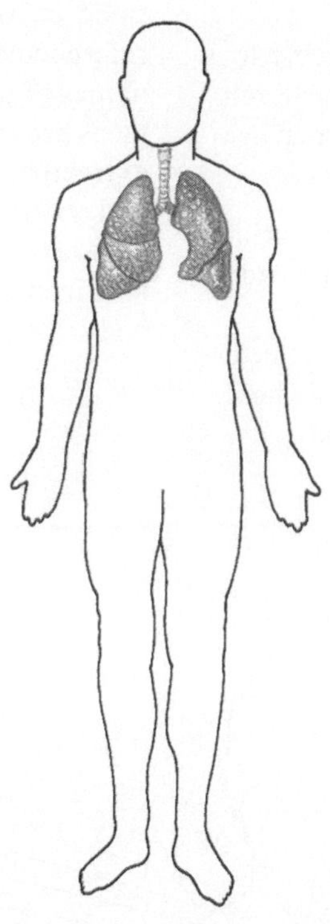

Respiratory system
Function: gaseous exchange between external environment and blood

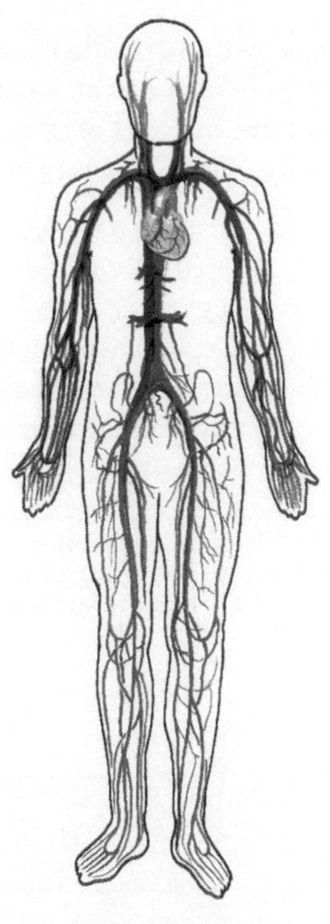

Circulatory system
Function: transport of life-sustaining materials to body cells; removal of metabolic wastes from cells

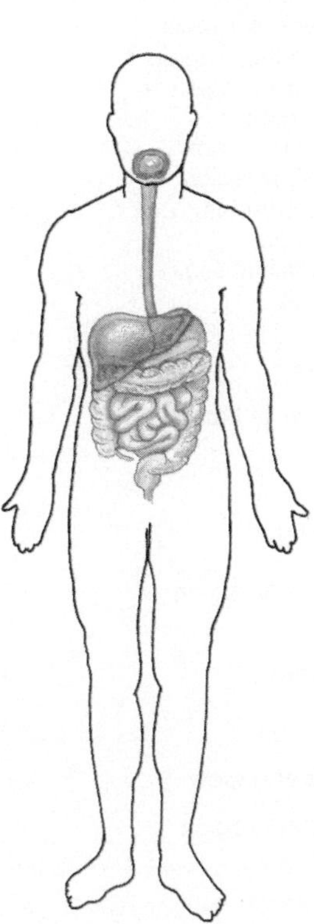

Digestive system
Function: breakdown and absorption of food materials

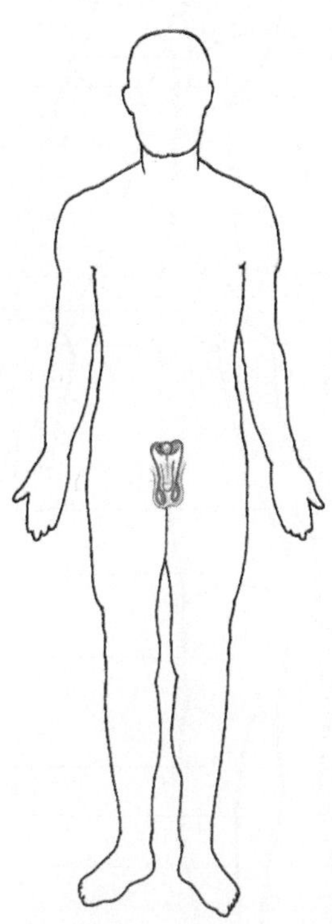

Male reproductive system
Function: production of male sex cells (sperm); transfer of sperm to reproductive system of female

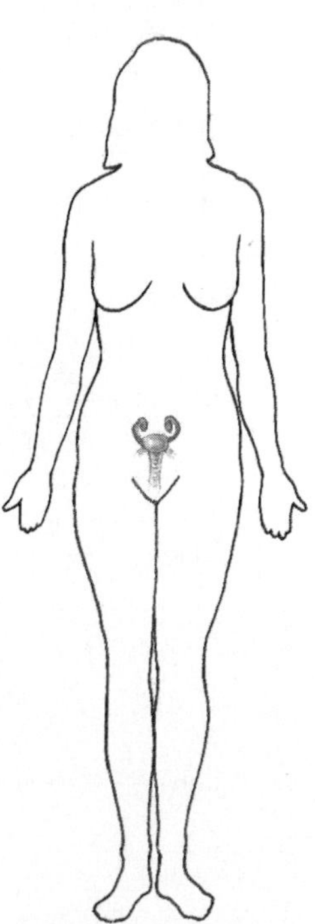

Female reproductive system
Function: production of female sex cells (ova); receptacle of sperm from male; site for fertilization of ovum, implantation, and development of embryo and fetus; delivery of fetus

Body Regions

The human body is divided into regions and specific localized areas, which can be identified on the surface. Each region contains internal organs, the location of which are anatomically, physiologically, and clinically important.

Objective 16. List the regions of the body and the principal localized areas comprising each region.

Objective 17. Explain why it is important to be able to describe the body areas and regions where each major internal organ is located.

The body is divided into several regions that can be identified on the surface. Learning the terminology used in reference to these regions now will help you learn the names of underlying structures later. The major body regions are the **head, neck, trunk, upper extremity,** and **lower extremity.** The trunk is frequently divided into the thorax and abdomen. Table 1.3 and figures 1.13 and 1.14 present a basic outline of the major body regions and some specific localized areas that can be identified on the surface of the body.

Figure 1.13. Surface regions of the body in anterior view. The generalized body regions are shaded and shown on the figure's right side, and the more specific body regions are shown on its left side.

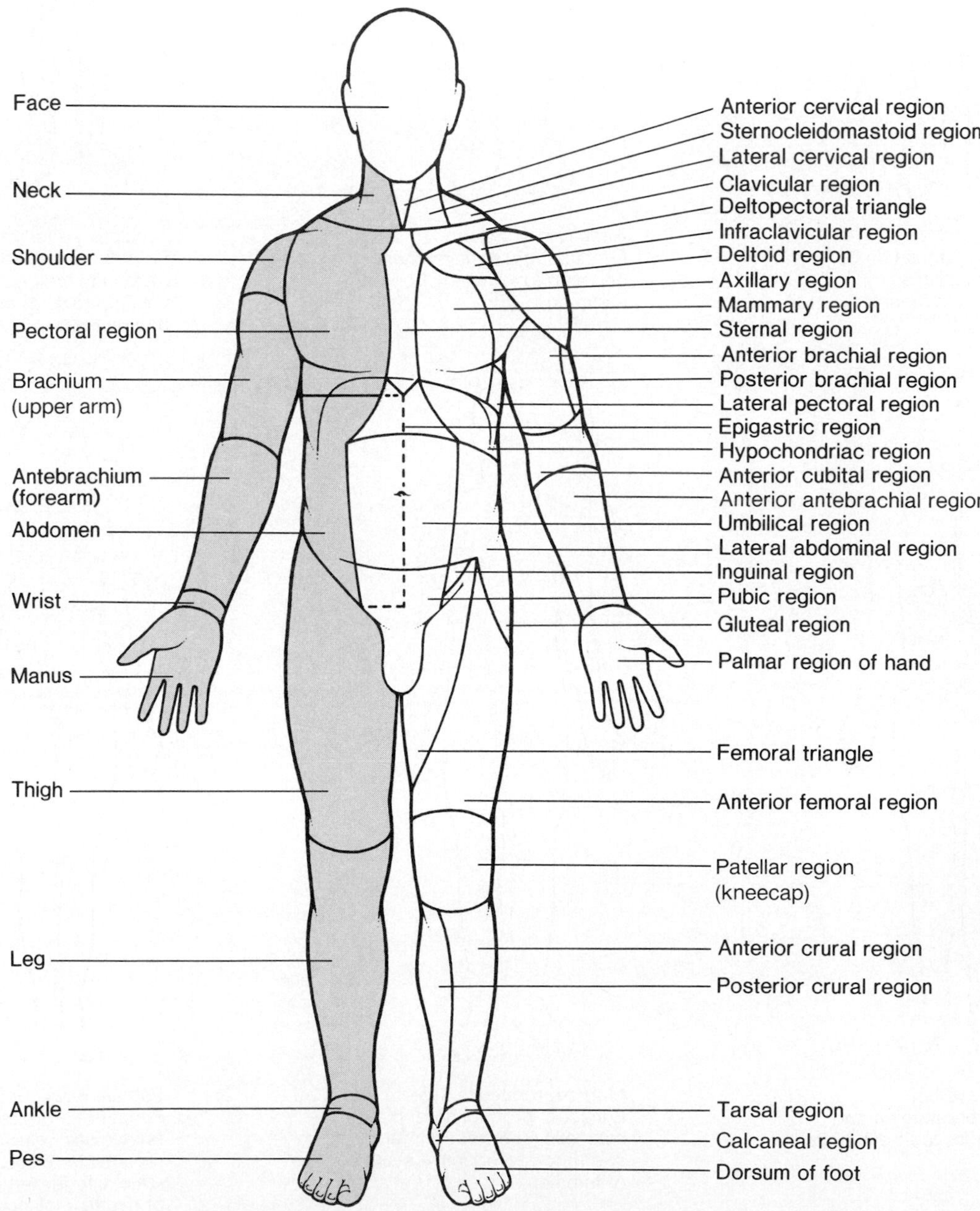

Table 1.3 Body regions and localized areas

Major region	Localized area	Major region	Localized area
I. Head (caput)	Cranium Face (facies)		Antebrachium (forearm) Manus (hand) Wrist
II. Neck (cervix)	Anterior neck Posterior neck (nuchal region)		Palm Dorsum of hand
III. Trunk (truncus)	Back (dorsum) Thorax (pectus) Abdomen (venter) Pelvis	V. Lower extremity	Buttock Thigh (femur) Knee Leg (crus)
IV. Upper extremity	Shoulder (omos) Axilla Brachium (upper arm) Elbow		Ankle Pes (foot) Heel Sole (plantar surface) Dorsum of foot

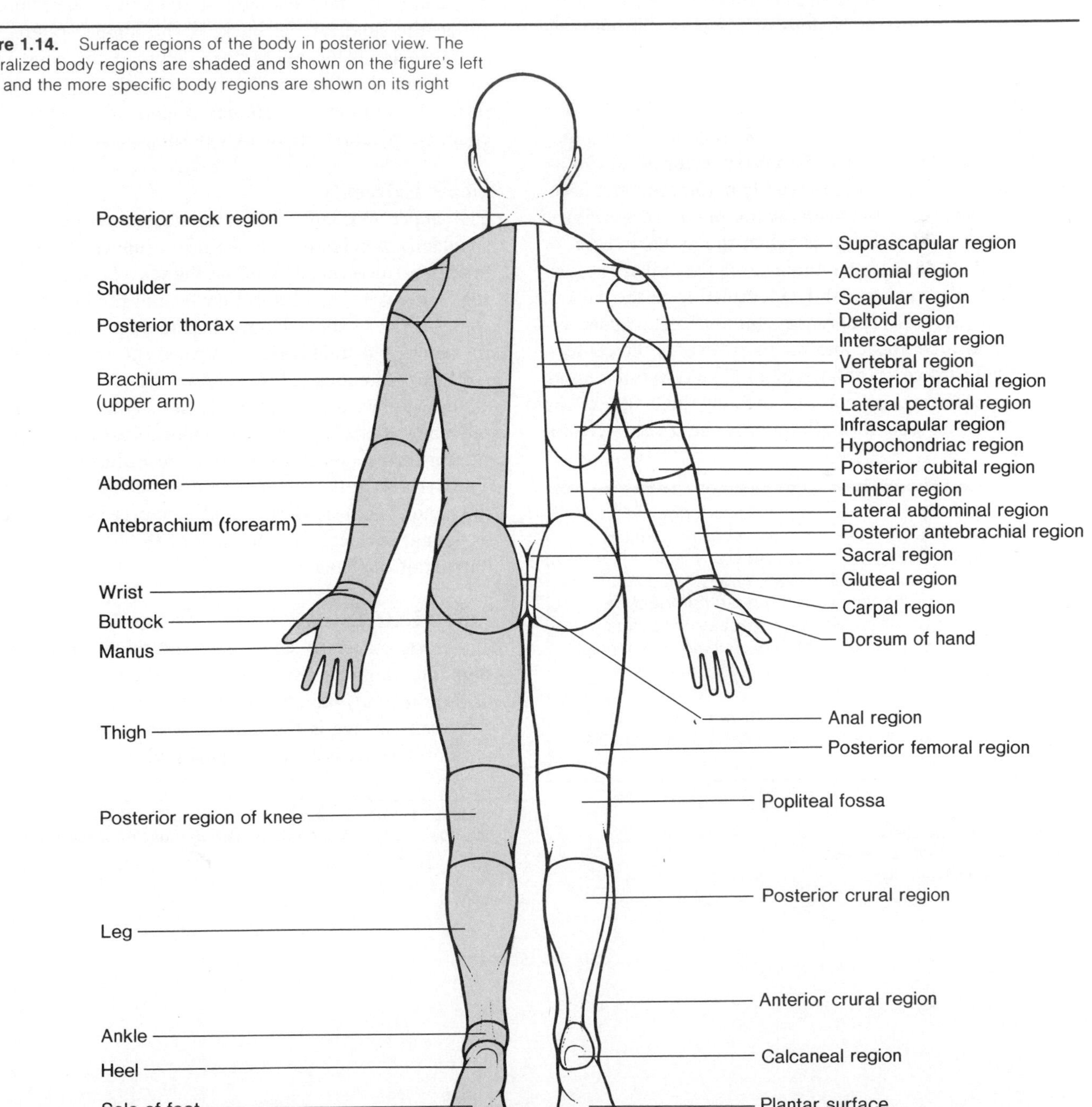

Figure 1.14. Surface regions of the body in posterior view. The generalized body regions are shaded and shown on the figure's left side, and the more specific body regions are shown on its right side.

Head

The **head,** or **caput** *(kap'ut),* is divided into a **facial region,** which includes the eyes, nose, and mouth, and a **cranium** *(kra'ne-um),* or **cranial region,** which covers and supports the brain. Further regionalization of the head is important for identifying specific glands, muscles, nerves, or vessels located under the skin. The identifying names for detailed surface regions are based on associated organs—such as the orbital (eye), nasal (nose), oral (mouth), mental (chin), and auricular (ear) regions—or underlying bones—such as the frontal, zygomatic, temporal, parietal, and occipital regions.

Neck

The **neck,** referred to as the **cervix** *(ser'viks),* or **cervical region,** supports the head and permits it to move. As with the head, detailed subdivisions of the neck can be identified, and these are fully discussed in chapter 13.

Thorax

The **thorax,** or **thoracic** *(tho-ras'ik)* **region,** is commonly referred to as the chest. The **mammary region** of the thorax surrounds the nipple and in sexually mature females is enlarged as the breast. Between the mammary regions is the **sternal region.** The armpit is called the **axillary fossa,** or simply **axilla,** and the surrounding area the **axillary region.** Paired **scapular regions** can be identified from the back of the thorax (fig. 1.14). **Suprascapular** and **interscapular regions** are located above and medial (towards the center), respectively, to the scapular regions. The **vertebral region,** following the vertebral column, extends the length of the back. On either side of the thorax are the **lateral pectoral regions.**

The heart and lungs are contained within the thoracic cavity. Easily identified surface landmarks greatly facilitate a physical examination of them. A physician must know, for example, where the valves of the heart can best be detected and where to listen for respiratory sounds. The axilla becomes important in examining for infected lymph nodes. When fitting a patient for crutches, a physician will instruct the patient to avoid supporting the weight of the body on the axillary area because of the possibility of damaging the underlying nerves and vessels.

thorax: L. *thorax,* chest
mammary: L. *mamma,* breast
axillary: L. *axilla,* armpit

Abdomen

The **abdomen** is located below the thorax. The **navel,** or **umbilicus,** is an obvious landmark on the front and center of the abdomen. The abdomen has been divided into nine regions in order to describe the location of internal organs. Figure 1.15 diagrams the subdivisions of the abdomen, and table 1.4 identifies the internal organs located within these regions. Subdividing the abdomen into four quadrants (fig. 1.16) is a common clinical practice so that various pains and conditions can be more easily located.

The **pelvic** *(pel''vik)* **region** forms the lower portion of the abdomen. Within the pelvic region is the **pubic area,** which is covered with pubic hair in sexually mature persons. The **perineum** *(per''ĭ-ne'um)* (fig. 1.17) is the region containing the external sex organs and the anal opening. The center of the back side of the abdomen, commonly called the small of the back, is the **lumbar region.** The **sacral region** is located farther down, at the point where the vertebral column terminates. The large hip muscles form the **buttock,** or **gluteal region.** This region is a common injection site of hypodermic needles.

Upper Extremity

The **upper extremity** is anatomically divided into the **shoulder, brachium** *(bra'ke-um)* **(upper arm), antebrachium (forearm),** and **manus (hand).** The shoulder is the region between the pectoral girdle and the brachium that contains the shoulder joint. The shoulder is referred to as the **omos,** or **deltoid region.** Between the arm and forearm is a flexible joint called the **elbow.** The surface area of the elbow is known as the **cubital region.** The front surface of the elbow is also known as the **cubital fossa,** an important site for intravenous injections or the withdrawal of blood. The wrist is the flexible junction between the forearm and the hand. The front of the hand is referred to as the **palm,** or **palmar surface,** and the back of the hand is called the **dorsum of the hand.**

Lower Extremity

The **lower extremity** consists of the **thigh, knee, leg,** and **foot.** The thigh is commonly called the **upper leg,** or **femoral** *(fem'o-ral)* **region.** The knee joint has two surfaces: the front surface is the **patellar region,** or kneecap; the back of the knee is called the **popliteal** *(pop''lĭ-te'al)* **fossa.**

cubital: L. *cubitis,* elbow
popliteal: L. *poples,* ham (hamstring muscle) of the knee

Figure 1.15. The abdomen is subdivided into nine regions. The vertical planes are positioned just medial to the nipples; the upper horizontal plane is positioned at the level of the rib cage; and the lower horizontal plane is even with the upper border of the hipbones.

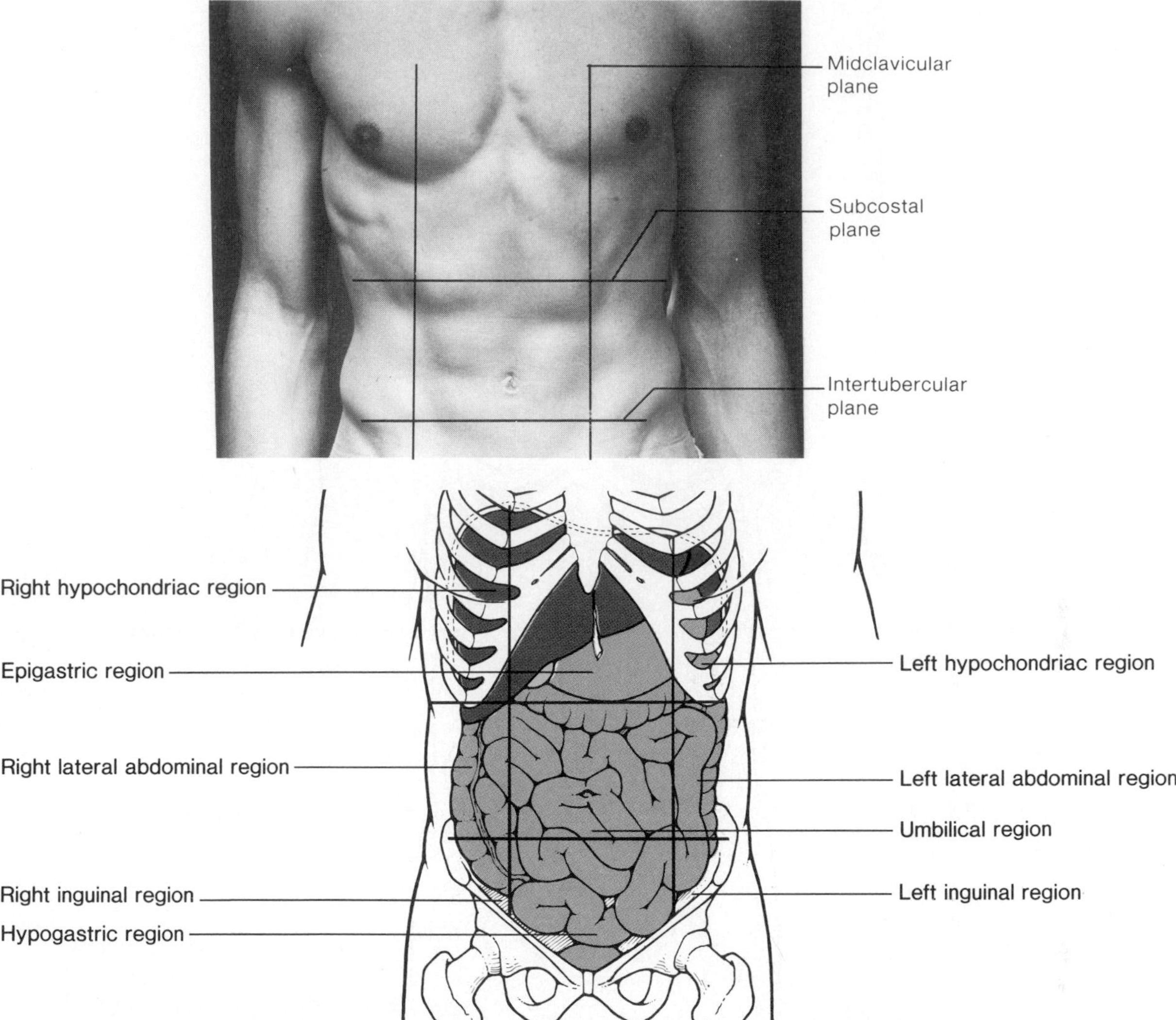

Table 1.4 Regions of the abdomen and pelvis

Region	Location	Internal organs
Right hypochondriac	Right, upper one-third of abdomen	Gallbladder; portions of liver and right kidney
Epigastric	Upper, median abdomen	Portions of liver, stomach, pancreas, and duodenum
Left hypochondriac	Left, upper one-third of abdomen	Spleen; splenic flexure of colon; portions of left kidney and small intestine
Right lateral	Right, lateral one-third of abdomen	Cecum; ascending colon; hepatic flexure; portions of right kidney and small intestine
Umbilical	Center of abdomen	Jejunum; ileum; portions of duodenum, colon, kidneys, and major abdominal vessels
Left lateral	Left, lateral one-third of abdomen	Descending colon; portions of left kidney and small intestine
Right inguinal	Right, lower one-third of abdomen	Appendix; portions of cecum and small intestine
Pubic (hypogastric)	Lower, center one-third of abdomen	Urinary bladder; portions of small intestine and sigmoid colon
Left inguinal	Left, lower one-third of abdomen	Portions of small intestine and descending and sigmoid colon

Figure 1.16. A clinical subdivision of the abdomen into four quadrants.

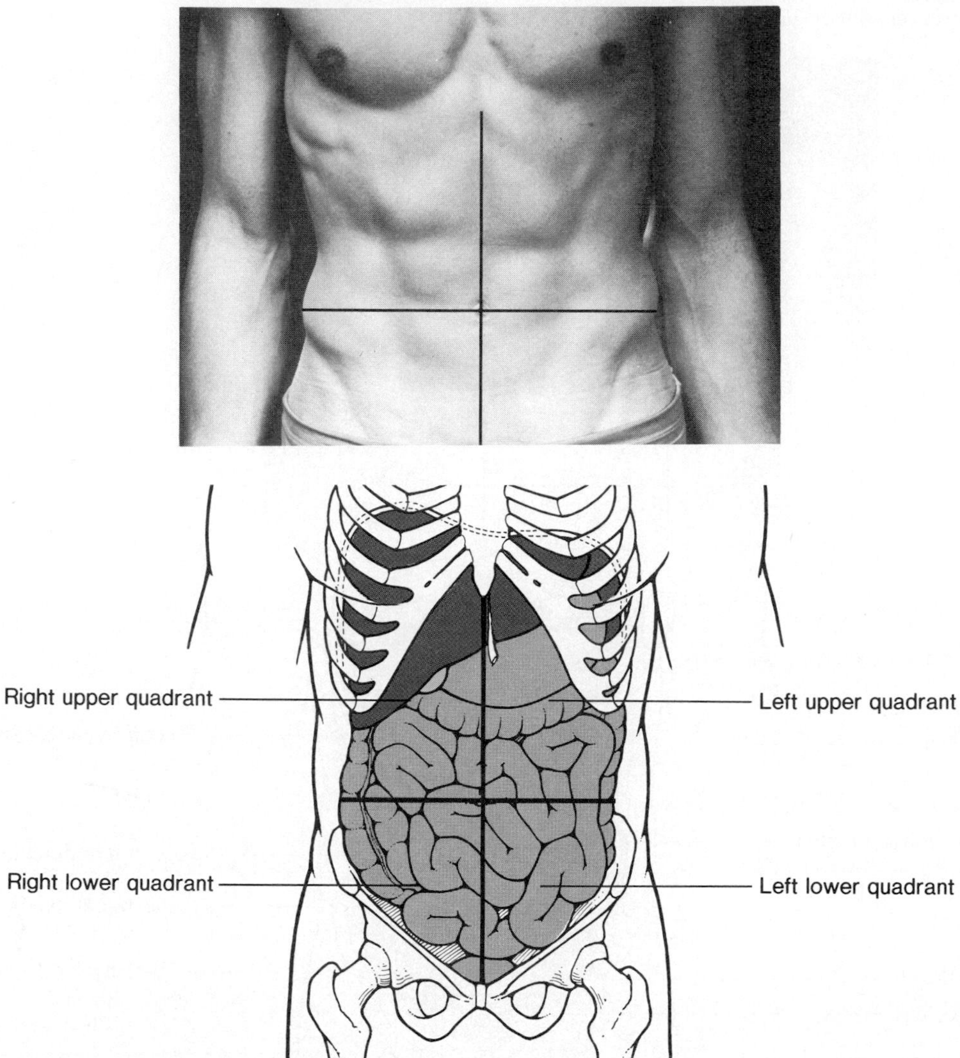

The leg has anterior and posterior crural regions (see fig. 1.13). The **shin** is a prominent bony ridge extending longitudinally along the anterior crural region, and the **calf** is the thickened muscular mass of the posterior crural region. The **ankle** is the junction between the leg and the foot. The **heel** is the back of the foot, and the **sole** of the foot is referred to as the **plantar surface.** The **dorsum of the foot** is the top surface.

1. Using yourself as a model, identify the various body regions that are depicted in figures 1.13 and 1.14. Which of these regions have surface landmarks that help distinguish their boundaries?
2. In which regions of the body are intravenous injections made?
3. What is the distinction between the pelvic, pubic, and perineal regions?
4. Identify the joint between each of the following regions: the brachium and antebrachium, the pectoral girdle and brachium, the leg and foot, the antebrachium and hand, the thigh and leg.
5. Explain how a knowledge of the body regions has practical medical application.

Figure 1.17. A superficial view of the (*a*) male perineum, and the (*b*) female perineum. The perineum consists of a urogenital triangle and an anal triangle.

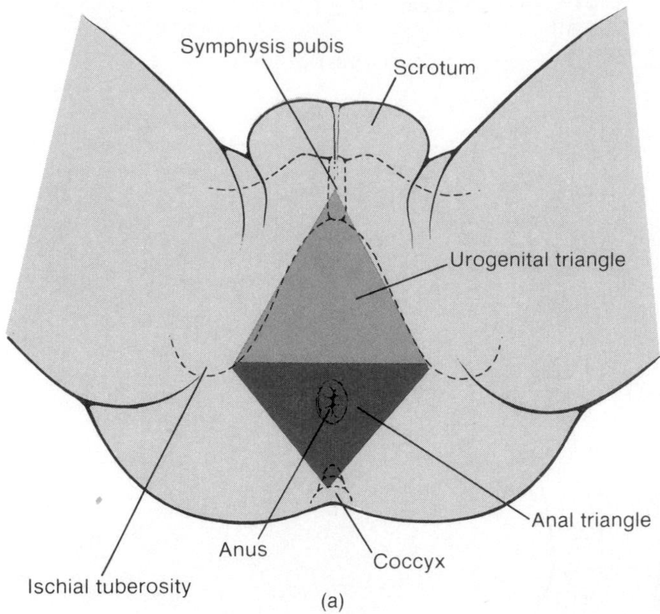

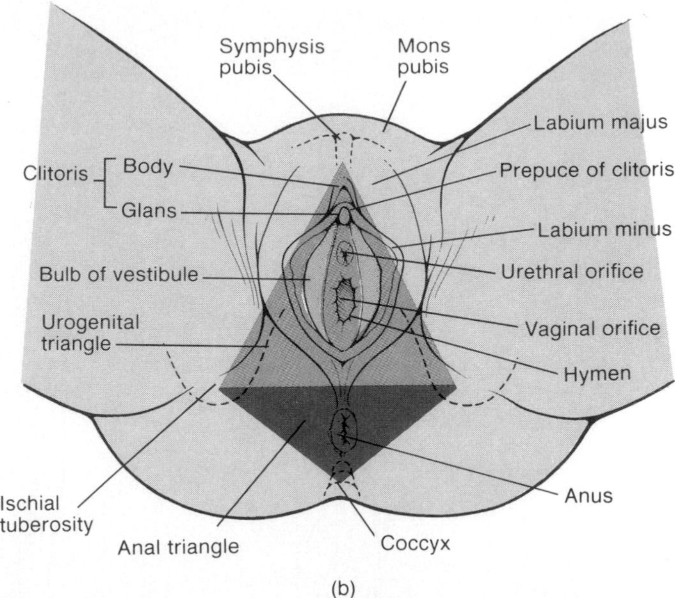

Body Cavities and Membranes

For functional and protective purposes, the viscera are compartmentalized and supported in specific body cavities by connective and epithelial membranes.

Objective 18. Identify the various body cavities and the organs found in each.

Objective 19. Discuss the types and functions of the various body membranes.

Body Cavities

Body cavities are confined spaces within the body. They contain organs that are protected, compartmentalized, and supported by associated membranes. There are two principal body cavities: the **dorsal body cavity** and the larger **ventral body cavity.** The dorsal body cavity contains the brain and the spinal cord.

During development, the ventral cavity forms from a cavity called the **coelom** *(see'lom)*. The coelom is a body cavity within the trunk, which is lined with a membrane that secretes a lubricating fluid. As development progresses, the coelom is partitioned by the muscular diaphragm into an upper **thoracic cavity,** or chest cavity, and a lower **abdominopelvic cavity** (figs. 1.18, 1.19). Organs within the coelom are collectively called **viscera,** or **visceral** *(vis'er-al)* **organs** (fig. 1.20). Within the thoracic cavity are two **pleural** *(ploo'ral)* **cavities** for the right and left lungs and a **pericardial** *(per''i-kar'de-al)* **cavity** containing the heart. The area between the two lungs is known as the **mediastinum** *(me''de-ah-sti'num)*.

The abdominopelvic cavity consists of an upper **abdominal cavity** and a lower **pelvic cavity.** The abdominal cavity contains the stomach, small intestine, large intestine, liver, gallbladder, pancreas, spleen, and kidneys. The pelvic cavity is occupied by the terminal portion of the large intestine, the urinary bladder, and certain reproductive organs (the uterus, uterine tubes, and ovaries in the female; the seminal vesicles and prostate gland in the male).

Body cavities serve to confine organs and systems that have related functions. The major portion of the nervous system occupies the dorsal cavity; the principal organs of the respiratory and circulatory systems, the thoracic cavity; the primary organs of digestion, the abdominal cavity; and the reproductive organs, the pelvic cavity. Not only do these cavities house and support various body organs, they also effectively compartmentalize them so that infections and diseases cannot spread from one compartment to another. For example, pleurisy of one lung membrane does not usually spread to the other, and an injury to the thoracic cavity will usually cause only one lung to collapse rather than both.

coelom: Gk. *koiloma,* cavity

Figure 1.18. A midsagittal (median) section showing the body cavities.

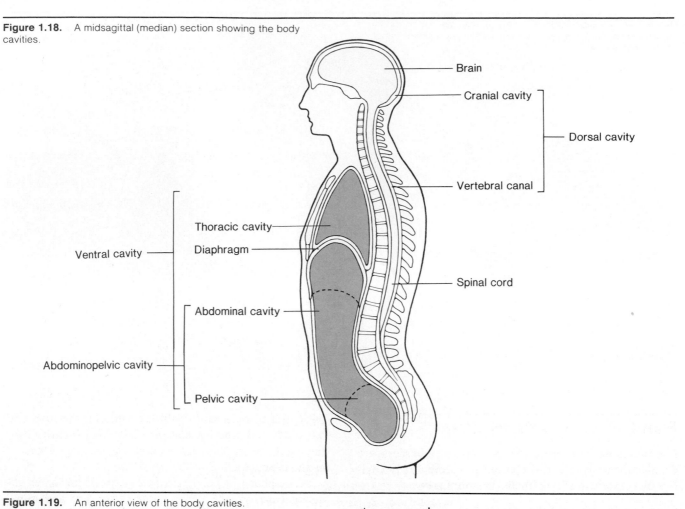

Brain
Cranial cavity
Dorsal cavity
Vertebral canal
Thoracic cavity
Diaphragm
Ventral cavity
Spinal cord
Abdominal cavity
Abdominopelvic cavity
Pelvic cavity

Figure 1.19. An anterior view of the body cavities.

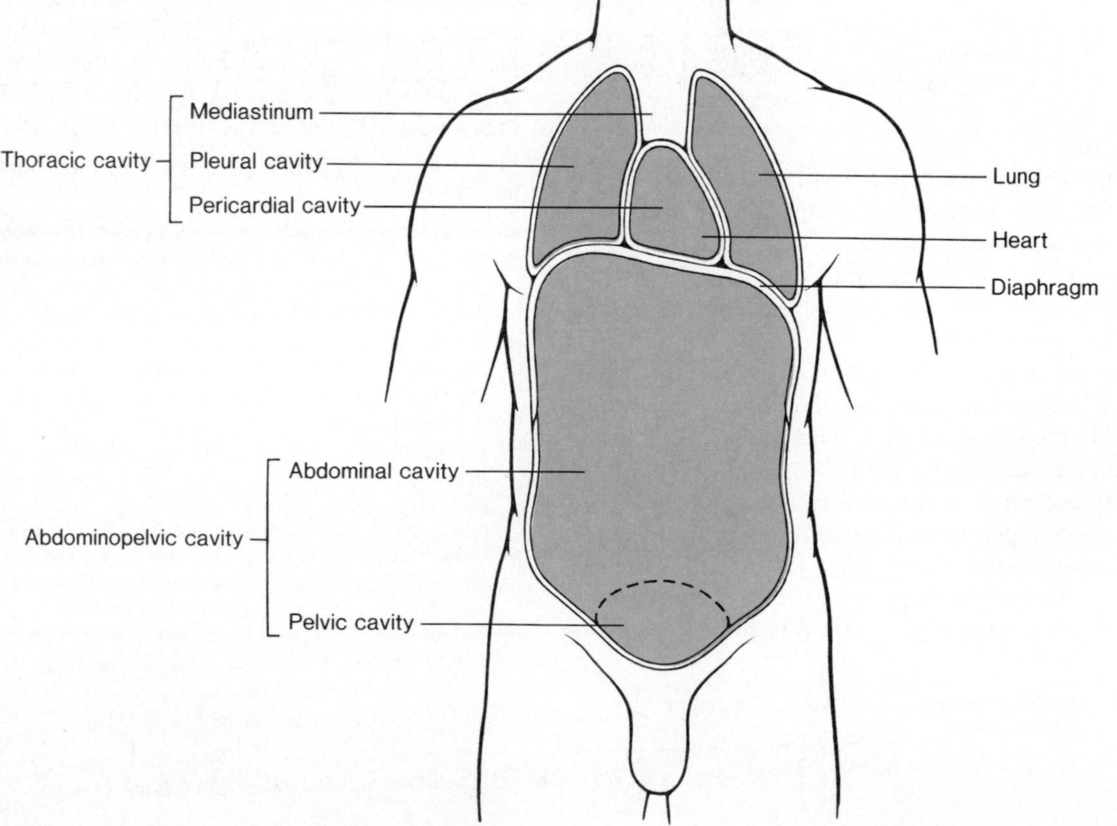

Figure 1.20. Visceral organs of the abdominal cavity and supporting serous membranes.

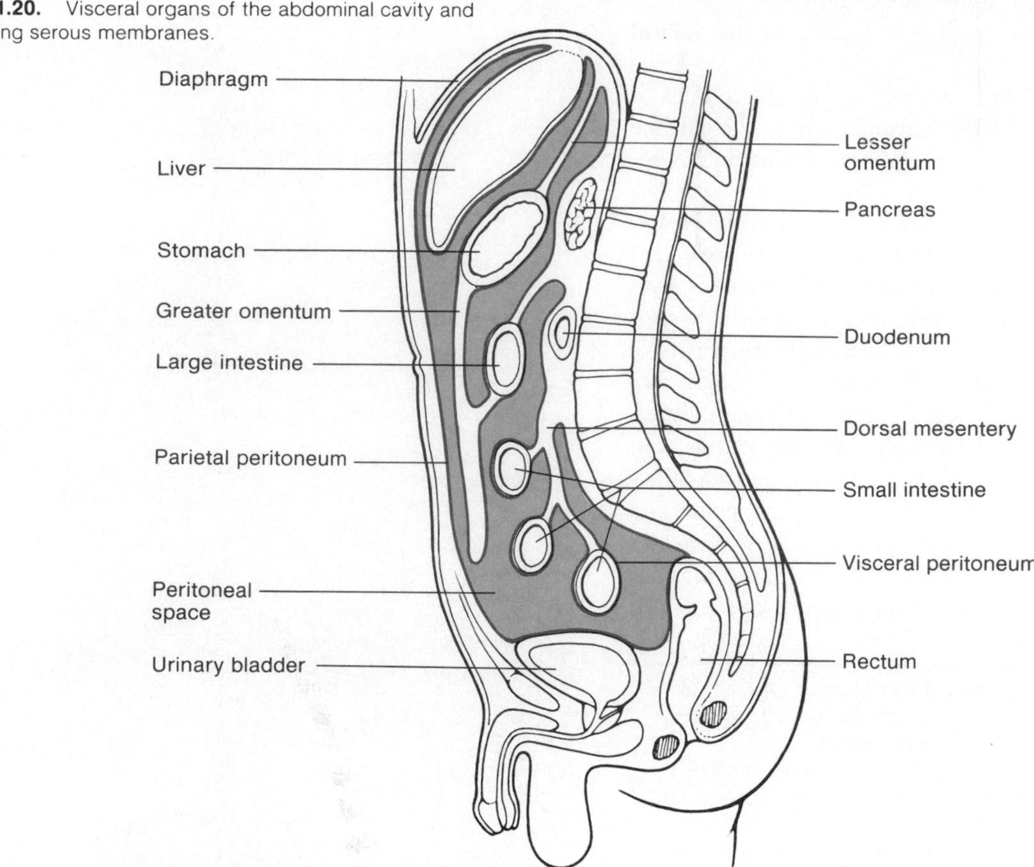

Figure 1.21. Cavities within the head.

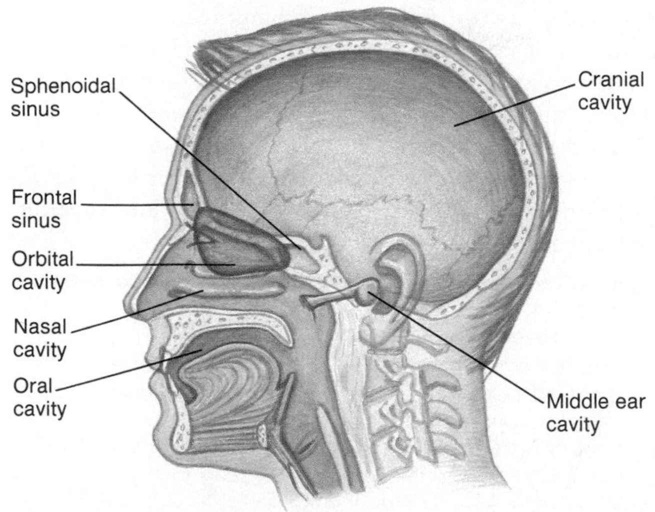

In addition to the large ventral and dorsal cavities, there are several smaller cavities within the head. The **oral,** or **buccal** *(buk'al),* **cavity** functions primarily with digestion and secondarily with respiration. It contains the teeth and tongue. The **nasal cavity,** which is part of the respiratory system, has two chambers created by a nasal septum. There are two **orbital cavities,** each of which houses an eyeball and its associated muscles, vessels, and nerves. Likewise, there are two **middle ear cavities** that contain ear ossicles (bones) and function with hearing. Figure 1.21 shows the location of the cavities within the head.

Body Membranes

Body membranes are composed of thin layers of connective and epithelial tissue, which cover, separate, and support visceral organs and line body cavities. There are two basic types of body membranes: **mucous** *(mu'kus)* **membranes** and **serous** *(se'rus)* **membranes.**

Mucous membranes secrete a thick, viscid substance called *mucus.* Generally, mucus lubricates or protects the associated organs where it is secreted. Mucous membranes line various cavities and tubes that enter or exit from the body, such as the oral and nasal cavities and the tubes of the respiratory, reproductive, urinary, and digestive systems.

Serous membranes line the thoracic and abdominopelvic cavities and cover visceral organs, secreting a watery lubricant called *serous fluid.* **Pleurae** are serous membranes associated with the lungs. Each pleura (pleura of right lung and pleura of left lung) has two parts. The **visceral pleura** adheres to the outer surface of the lung, while the **parietal** *(pah-ri'ĕ-tal)* **pleura** lines the thoracic walls and the thoracic surface of the diaphragm. The moistened space between the two pleurae is known as the **pleural cavity.**

Pericardial membranes are the serous membranes of the heart. A thin **visceral pericardium** covers the surface of the heart, and a thicker **parietal pericardium** is the durable covering that surrounds the heart. The space between these two membranes is called the **pericardial cavity.**

Serous membranes of the abdominal cavity are called **peritoneal** *(per''i-to-ne'al)* **membranes.** The **parietal peritoneum** lines the abdominal wall, and the **visceral peritoneum** covers the visceral organs. The **peritoneal cavity** is the potential space within the abdominopelvic cavity between the parietal and visceral peritoneal membranes. Certain organs, such as the kidneys, adrenal glands, and a portion of the pancreas, which are within the abdominal cavity, are positioned behind the parietal peritoneum and therefore are said to be **retroperitoneal. Mesenteries** *(mes'en-ter''ez)* are double folds of peritoneum that connect the parietal to the visceral peritoneum (see fig. 1.20).

1. Describe the divisions and boundaries of the ventral body cavity, and list the major organs contained within each division of the ventral body cavity.
2. Distinguish between mucous and serous membranes. List the specific serous membranes of the thoracic and abdominopelvic cavities.
3. Explain the importance of separate and distinct body cavities.

Figure 1.22. Planes of reference through the body.

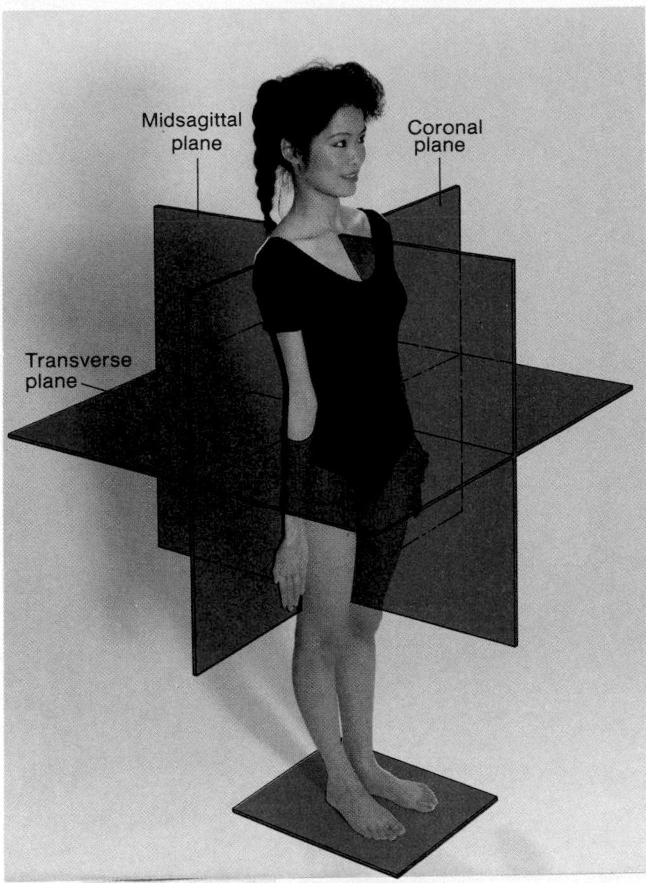

Planes of Reference and Descriptive Terminology

All of the descriptive planes of reference and terms of direction used in anatomy are standardized because of their reference to the body in anatomical position.

Objective 20. Identify the planes of reference used to locate structures within the body.
Objective 21. Describe the anatomical position.
Objective 22. Define and be able to use properly the descriptive and directional terms that have been designated to refer to body structures.

Planes of Reference

In order to visualize and study the structural arrangements of various organs, the body may be sectioned (cut) and diagramed according to planes of reference. Three fundamental planes, **midsagittal** *(mid-saj'i-tal),* **coronal,** and **transverse,** are frequently used to depict structural arrangement (fig. 1.22).

A *midsagittal* plane passes lengthwise through the midplane of the body, dividing it into right and left halves. *Sagittal* planes also extend vertically and divide the body into unequal right and left portions. *Coronal,* or *frontal,* planes pass lengthwise and divide the body into front and back portions. *Transverse* planes, also called *horizontal,* or *cross-sectional,* planes, divide the body into superior (upper) and inferior (lower) portions.

The value of the computerized tomographic X-ray (CT) scan is that it displays an image along a transverse plane similar to that which could otherwise be obtained only in an actual section through the body. Prior to the development of this X-ray technique, conventional X-ray images were on a vertical plane, and the dimensions of body irregularities were difficult, if not impossible, to ascertain.

peritoneum: Gk. *peritonaion,* stretched over
coronal: L. *corona,* crown

Figure 1.23. In the anatomical position, the body is erect, the feet are parallel, the eyes directed forward, the arms to the side with the palms directed forward: (*a*) anterior view; (*b*) lateral view; and (*c*) posterior view.

(a)

(b)

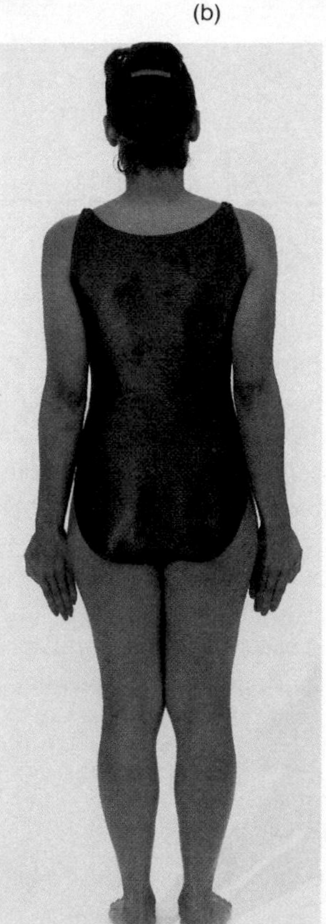

(c)

Descriptive Terminology

Anatomical Position. All terms of direction that describe the relationship of one body part to another are made in reference to the **anatomical position.** In the anatomical position, the body is erect; the feet are parallel to one another and flat on the floor, the eyes are directed forward, and the arms are at the sides of the body with the palms of the hands turned forward (fig. 1.23).

Directional Terms. Directional terms are used to locate the position of structures, surfaces, and regions of the body. These terms are always relative to the specimen positioned in the anatomical position. A summary of directional terms is presented in table 1.5.

Clinical Procedures. Certain clinical procedures are important in determining anatomical structure and function in a living individual. The more common of these are as follows.

1. **Palpation.** Applying the fingers with firm pressure to the surface of the body to detect surface landmarks, lumps, tender spots, or pulsations.
2. **Percussion.** Tapping sharply on various locations on the thorax or abdomen to detect resonating vibrations in determining fluid concentrations and organ densities.
3. **Auscultation.** Listening to the sounds that various organs make as they perform their functions (breathing sounds, heartbeats, digestive sounds, etc.).
4. **Reflex response.** Reflex responses are used to determine the condition of parts of the nervous system and some associated organs. One test of a reflex mechanism involves tapping a predetermined tendon with a reflex hammer and observing the response.

---

1. Explain why a transverse plane through the body is important in studying regional anatomy, whereas a transverse plane through an organ is more important in studying specific systems.
2. What does it mean that directional terms are relative and must be used in reference to a body structure or a body in anatomical position?
3. Write a list of statements, similar to the examples in table 1.5, that correctly expresses the directional terms used to describe the relative positions of various body structures.

Table 1.5 Directional terms for the human body

Term	Definition	Example
Superior (cranial, cephalic)	Toward the head; toward the top	The thorax is superior to the abdomen.
Inferior (caudal)	Away from the head; toward the bottom	The legs are inferior to the trunk.
Anterior (ventral)	Toward the front	The navel is on the anterior side of the body.
Posterior (dorsal)	Toward the back	The kidneys are posterior to the intestine.
Medial	Toward the midline of the body	The heart is medial to the lungs.
Lateral	Toward the side of the body	The ears are on the lateral sides of the head.
Internal (deep)	Away from the surface of the body	The brain is internal to the cranium.
External (superficial)	Toward the surface of the body	The skin is external to the muscles.
Proximal	Toward the main mass of the body	The knee is proximal to the foot.
Distal	Away from the main mass of the body	The hand is distal to the elbow.
Visceral	Related to internal organs	The lungs are covered by a thin membrane called the visceral pleura.
Parietal	Related to the body walls	The parietal pleura is the inside lining of the thoracic cavity.

Table 1.6 Units of length

Metric unit	Equivalent in meters	Apothecaries' equivalent (U.S. equivalent)
1 kilometer (km) kilo = 1,000	1,000 m	3,280.84 ft; 0.62 mi 1 mi = 1.61 km
1 hectometer (hm) hecto = 100	100 m	328 ft
1 dekameter (dam) deka = 10	10 m	32.8 ft
1 meter (m) Standard unit of length	—	39.37 in; 3.28 ft; 1.09 yd
1 decimeter (dm) deci = 1/10	0.1 m	3.94 in
1 centimeter (cm) centi = 1/100	0.01 m	0.394 in 1 in = 2.54 cm
1 millimeter (mm) milli = 1/1,000	0.001 m	0.0394 in
1 micrometer (μm) micro = 1/1,000,000	0.000,000,1 m	3.94×10^{-5} in
1 nanometer (nm) nano = 1/1,000,000,000	0.000,000,001 m	3.94×10^{-8} in
1 angstrom (Å)	0.000,000,000,1 m	3.94×10^{-9} in

Units of Measurements

Measurements are the human way of expressing the relative lengths, masses, temperatures, and volumes of substances, which are important because they bring uniformity and consistency to scientific communication.

Objective 23. Define the frequently used metric units of length, mass, and volume, and their equivalents.

Units of measurements are used frequently in anatomy and physiology to describe body structures and how they function. The speed of an impulse, the pressure of blood, the weight of an organ, and the temperature of the body are examples of measurements important in understanding physiological events as well as in determining a deviation from what is considered normal. Such data are also of clinical importance in determining the amounts of medication to be administered.

The length, mass, and volume units of measurement presented in this text are given in metric form followed in parentheses by the apothecary (U.S.) equivalent. The temperatures presented are in Celsius, followed in parentheses by the Fahrenheit equivalent. Tables 1.6, 1.7, 1.8, and 1.9 will help you become acquainted with the metric units of measurement and their apothecary equivalents and with common Celsius and Fahrenheit temperature conversions. Reference to these tables will probably be necessary as you continue your study of anatomy and physiology.

metric: Fr. *metron,* measure
apothecary: Gr. *apotheke,* storehouse
Celsius: from Anders Celsius, Swedish astronomer, 1701–44
Fahrenheit: from Gabriel D. Fahrenheit, German physicist, 1686–1736

Table 1.7 Units of mass

Metric unit	Equivalent in grams	Apothecaries' equivalent
1 kilogram (kg)	1,000 g	2.205 lb
1 hectogram (hg)	100 g	
1 dekagram (dag)	10 g	
1 gram (g)	—	1 lb = 453.6 g 1 oz = 28.35 g
1 decigram (dg)	0.1 g	
1 centigram (cg)	0.01 g	
1 milligram (mg)	0.001 g	
1 microgram (μg)	0.000,001 g	

Table 1.8 Units of volume

Metric unit	Metric equivalent	Apothecaries' equivalent
1 liter (l)	1,000 ml	33.81 fl oz or 1.057 qt 1 qt = 946 ml
1 milliliter (ml)	0.001 liter	0.0338 fl oz; 1 fl oz = 30 ml 1 teaspoon = 5 ml
1 cubic centimeter (cm^3)	0.999972 ml	0.0338 fl oz

Table 1.9 Units of temperature in °Celsius and °Fahrenheit

°C	°F	°C	°F
35.0	95.0	37.8	100.0
35.1	95.2	37.9	100.2
35.2	95.4	38.0	100.4
35.3	95.6	38.1	100.6
35.4	95.8	38.2	100.8
35.5	96.0	38.3	101.0
35.7	96.2	38.4	101.2
35.8	96.4	38.6	101.4
35.9	96.6	38.7	101.6
36.0	96.8	38.8	101.8
36.1	97.0	38.9	102.0
36.2	97.2	39.0	102.2
36.3	97.4	39.1	102.4
36.4	97.6	39.2	102.6
36.6	97.8	39.3	102.8
36.7	98.0	39.4	103.0
36.8	98.2	39.6	103.2
36.9	98.4	39.7	103.4
37.0	98.6	39.8	103.6
37.1	98.8	39.9	103.8
37.2	99.0	40.0	104.0
37.3	99.2	40.1	104.2
37.4	99.4	40.2	104.4
37.6	99.6	40.3	104.6
37.7	99.8	40.4	104.8
		40.6	105.0

To convert °C to °F
Multiply °C by 9/5 and add 32.
________ °C × 9/5 + 32 = ________ °F

To convert °F to °C
Subtract 32 from °F and multiply by 5/9.
________ °F − 32 × 5/9 = ________ °C

1. Describe a situation that involves measurements of the human body.
2. Referring to tables 1.6, 1.7, 1.8, and 1.9, solve the following problems:
   (a) A body cell measures 100 μm in diameter. Express this measurement in nanometers.
   (b) A person weighs 150 lb. What is the weight in kilograms?
   (c) An experiment is conducted at 150°F. Express this temperature in degrees Celsius.
   (d) If a patient excreted 1,350 ml of urine in a day, how many liters would this be?

Chapter Summary

I. Definitions of the Sciences
   A. Dissections and descriptions of human cadavers are an important aspect of anatomy.
   B. Physiology is a dynamic science that explains physical and chemical processes of the body.

II. Historical Perspective
   A. The history of anatomy and physiology parallels that of medicine; Hippocrates, the famous Grecian physician, is regarded as the "father of modern medicine."
   B. Aristotle established a type of scientific method in obtaining data.
   C. Erasistratus is frequently referred to as the "father of physiology."
   D. There was a de-emphasis of theoretical data during the Roman era. The writings of the Roman physician Galen, however, became the unquestionable authority on anatomy and medical treatment for nearly fifteen hundred years.
   E. The development of movable type (A.D. 1450) and the monumental work of the "father of modern anatomy," Vesalius (1514–64), accelerated an interest and research in anatomy and physiology during the Renaissance.
   F. Two major contributions of the seventeenth and eighteenth centuries were an explanation of blood flow by Harvey and the development and utilization of the microscope.
   G. Cellular biology became established as a science separate from anatomy and physiology during the nineteenth century.
   H. During the twentieth century many specialities have developed within anatomy and physiology and medicine.
   I. The scientific method is a disciplined approach for the acquisition of truth. It includes observation, gathering data, and formulating a hypothesis, which may become accepted as a theory or scientific principle.

III. Classification and Characteristics of Humans
   A. Humans are biological organisms belonging to the phylum Chordata within the kingdom Animalia, and to the family Hominidae within the class Mammalia and the order Primates.
   B. Humans belong to the phylum Chordata because of the presence of a notochord, a dorsal hollow nerve cord, and pharyngeal pouches during the embryonic stage of development.
   C. Some of the characteristics of humans include the size and development of the brain, bipedal locomotion, an opposable thumb, vocal structures, and stereoscopic vision.

IV. Maintenance of Life
   A. Requisites for life include development, organization, adaptation, responsiveness, movement, reproduction, growth, respiration, digestion and absorption, circulation, synthesis, secretion, excretion, and survival instinct.
   B. Some physical needs include water, oxygen, food, heat, pressure, and protection.

V. Body Organization
   A. The cell is the structural and functional component of life.
   B. Tissues are aggregations of similar cells that perform specific functions.
   C. An organ is an aggregate of two or more tissues that performs specific functions.
   D. Body systems consist of various organs that have similar or interrelated functions.

VI. Anatomical and Physiological Terminology
   A. As sciences, anatomy and physiology have descriptive terminologies that must be mastered by students who would understand and be conversant in these subjects.
   B. The majority of anatomical and physiological terms are of Greek or Latin derivation.

VII. Body Regions
   A. The human body is divided into regions and specialized localized areas, which can be identified on the surface. Each region contains internal organs, the location of which are anatomically, physiologically, and clinically important.
   B. The abdomen is divided into four quadrants and nine regions. The pelvic region forms the lower portion of the abdomen.

VIII. Body Cavities and Membranes
   A. For functional and protective purposes, the viscera are compartmentalized and supported in specific body cavities by connective and epithelial membranes.
   B. The body has two principal types of membranes: mucous membranes that secrete protective mucus; and serous membranes that line the ventral cavities and cover visceral organs.
   C. There are three categories of serous membranes: pleural membranes, pericardial membranes, and peritoneal membranes.

IX. Planes of Reference and Descriptive Terminology
   A. In the anatomical position, the subject stands erect and faces forward with arms at the sides and palms turned forward.
   B. Directional terms are used to describe the location of one body part with respect to another part.

X. Units of Measurements
   A. Measurements are the human way of expressing relative lengths, masses, temperatures, and volumes of substances.
   B. Measurements are important because they bring uniformity and consistency to scientific communication.

Review Activities

Objective Questions

1. Which of the following men would be more likely to disagree with the concept of body humors?
   (a) Galen
   (b) Hippocrates
   (c) Vesalius
   (d) Aristotle
2. The most important contribution of William Harvey was his research on the
   (a) continuous circulation of blood
   (b) microscopic structure of spermatozoa
   (c) detailed structure of kidney
   (d) striped appearance of skeletal muscle
3. The taxonomic scheme from specific to general is
   (a) species, class, order, phylum
   (b) genus, family, kingdom, phylum
   (c) species, family, class, kingdom
   (d) genus, phylum, class, kingdom
4. Which of the following is *not* a principal chordate characteristic?
   (a) dorsal hollow nerve cord
   (b) distinct head, thorax, and abdomen
   (c) notochord
   (d) pharyngeal pouches
5. The cubital fossa is located in the
   (a) thorax
   (b) upper extremity
   (c) abdomen
   (d) lower extremity
6. Which of the following is *not* a fundamental plane?
   (a) coronal
   (b) transverse
   (c) vertical
   (d) midsagittal
7. The abdominal region superior to the umbilical region that contains most of the stomach is the
   (a) hypochondriac region
   (b) epigastric region
   (c) diaphragmatic region
   (d) inguinal region
8. Regarding serous membranes, which of the following word pairs is *incorrect*?
   (a) visceral pleura—lung
   (b) parietal peritoneum—body wall
   (c) mesentery—heart
   (d) parietal pleura—body wall
   (e) visceral peritoneum—intestines
9. In the anatomical position the
   (a) arms are extended away from the body
   (b) palms of the hands face posteriorly
   (c) body is erect and palms face anteriorly
   (d) body is in a fetal position
10. Listening to sounds that functioning visceral organs make is
   (a) percussion
   (b) palpation
   (c) audiotation
   (d) auscultation

Essay Questions

1. What is meant by the humoral theory of body organization? Which great anatomists were influenced by this theory? When did the humoral theory cease to be an influence upon the anatomical and physiological investigation of the body?
2. Discuss the impact that Galen had on the advancement of anatomy and physiology and medicine. What ideological circumstances permitted the philosophies of Galen to survive for such a long period?
3. What role did the development of the microscope play in the advancement of the sciences of anatomy and physiology and medicine? What specialities of anatomical and physiological study have arisen since the introduction of the microscope?
4. Describe the scientific method, and relate how it is of immense importance in anatomical and physiological research.
5. What is a serous membrane? Diagram the serous membranes of the thoracic cavity, showing the subcompartments of the pericardial cavity, pleural cavities, and mediastinum.

2

Chemical Composition of the Body

Outline

Concepts

The anatomical structures and physiological processes of the body are based on the properties and interactions of atoms, ions, and molecules.

Carbohydrates as a group share a characteristic ratio of carbon, hydrogen, and oxygen atoms, and they are divided into subgroups depending on the number of simple sugars that each molecule contains. Lipids are a related but distinct group of molecules that are also subdivided into smaller categories. Carbohydrates and lipids are the major sources of energy in the body.

Proteins are large molecules composed of amino acid subunits. The complex and diverse structures of different proteins allow these molecules to perform a wide variety of functions in the body. Nucleic acids are also very large molecules. They are composed of nucleotide subunits, which provide the basis for genetic control of cells and the body.

Atoms, Ions, and Molecules

The anatomical structures and physiological processes of the body are based on the properties and interactions of atoms, ions, and molecules.

Objective 1. Describe the structure of an atom, and explain the meaning of the terms *atomic mass* and *orbital.*

Objective 2. Explain how different types of chemical bonds are formed, and describe the strengths of different chemical bonds.

Objective 3. Define the terms *acid, base, pH,* and *ion.*

Objective 4. Discuss the structure and chemical properties of water, and explain why some molecules are hydrophilic and others are hydrophobic.

Objective 5. Describe the structure of some organic molecules, and explain the nature of different functional groups.

Water is the most plentiful chemical substance (molecule) in the body and constitutes 65% to 75% of the total weight of an average adult. Of this amount, 30% to 40% is within the body cells (in the *intracellular compartment*); the remainder is between the body cells (in the *extracellular compartment*), including the blood and tissue fluids. Dissolved or suspended in this water are many organic molecules (carbon-containing molecules such as carbohydrates, lipids, proteins, and nucleic acids) and inorganic molecules and ions (atoms with a net charge). Before describing the structure and function of organic molecules within the body, some basic chemical concepts, terminology, and symbols will be introduced.

Atoms

Atoms are much too small to be seen individually, even with the most powerful microscope. Through the efforts of generations of scientists, however, the structure of atoms is now well understood. At the center of an atom is its *nucleus.* The nucleus contains two types of particles: *protons,* which have a positive charge, and *neutrons,* which are uncharged. The mass of a proton is approximately equal to the mass of a neutron, and the sum of the number of protons and neutrons in an atom is equal to the **atomic mass** of the atom. For example, an atom of carbon, which contains six protons and six neutrons, has an atomic mass of 12.

The number of protons in an atom is given as its **atomic number.** Carbon has six protons and thus has an atomic number of six. Outside the positively charged nucleus are negatively charged *electrons.* Since the number of electrons in an atom is equal to the number of protons, atoms have a net charge of zero.

Although it is often convenient to think of electrons as orbiting the nucleus like planets orbiting the sun, this older view of atomic structure is no longer believed to be correct. A given electron can occupy any position in a certain volume of space surrounding the nucleus. The outer boundary of this volume of space is called the *orbital* of the electron. The orbital is like an energy "shell," or barrier, beyond which the electron usually does not pass.

There are potentially several such orbitals around a nucleus, with each successive orbital being farther from the nucleus. The first orbital, closest to the nucleus, can contain only two electrons. If an atom has more than two electrons (as do all atoms except hydrogen and helium), the additional electrons must occupy orbitals that are farther removed from the nucleus. The second orbital can

Table 2.1 Atoms commonly present in organic molecules

Atom	Symbol	Atomic number	Atomic mass	Orbital 1	Orbital 2	Orbital 3	Number of chemical bonds
Hydrogen	H	1	1	1	0	0	1
Carbon	C	6	12	2	4	0	4
Nitrogen	N	7	14	2	5	0	3
Oxygen	O	8	16	2	6	0	2
Phosphorous	P	15	31	2	8	5	5
Sulfur	S	16	32	2	8	6	2

contain a maximum of eight electrons, and the third can have a maximum of eighteen. Thus, the orbitals are filled from the innermost outward. Carbon, with six electrons, has two electrons in its first orbital and four electrons in its second orbital (fig. 2.1).

It is always the electrons in the outermost orbital, if this orbital is incomplete, that participate in chemical reactions and form chemical bonds. These outermost electrons are known as the *valence electrons* of the atom.

Isotopes. A particular atom with a given number of protons in its nucleus may exist in several forms that differ from each other in their number of neutrons. The atomic number of these forms is thus the same, but their atomic mass is different. These different forms are called **isotopes.** All of the isotopic forms of a given atom are included in the term **chemical element.** The element hydrogen, for example, has three isotopes. The most common of these has a nucleus consisting of only one proton. Another isotope of hydrogen (called *deuterium*) has one proton and one neutron in the nucleus, whereas the third isotope (*tritium*) has one proton and two neutrons. Tritium is a radioactive isotope that is commonly used in physiological research and in many clinical laboratory procedures.

Chemical Bonds, Molecules, and Ionic Compounds

Molecules are formed when two or more atoms share their outer valence electrons. Such sharing of electrons produces **chemical bonds** (fig. 2.2). The number of bonds that each atom can have is determined by the number of electrons needed to make the outermost orbital complete. Hydrogen, for example, must obtain only one more electron—and can thus form only one chemical bond—to complete the first orbital of two electrons. Carbon, in contrast, must obtain four more electrons—and can thus form four chemical bonds—to complete the second orbital of eight electrons (fig. 2.3).

Figure 2.1. Diagrams of the hydrogen and carbon atoms. The electron orbitals on the left are represented by dots indicating probable positions of the electrons. The orbitals on the right are represented by concentric circles.

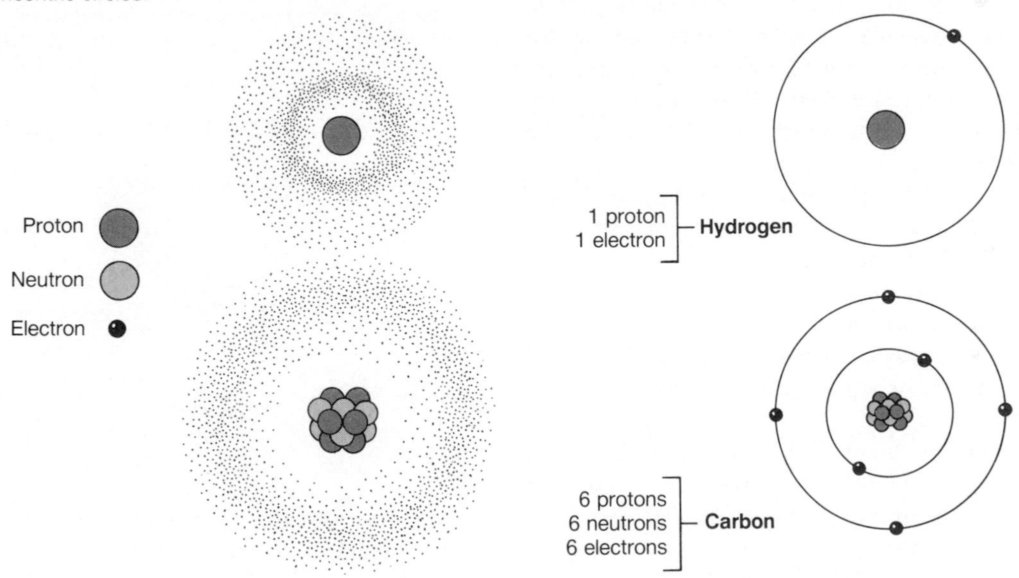

Figure 2.2. The hydrogen molecule, showing the covalent bonds between hydrogen atoms formed by the equal sharing of electrons.

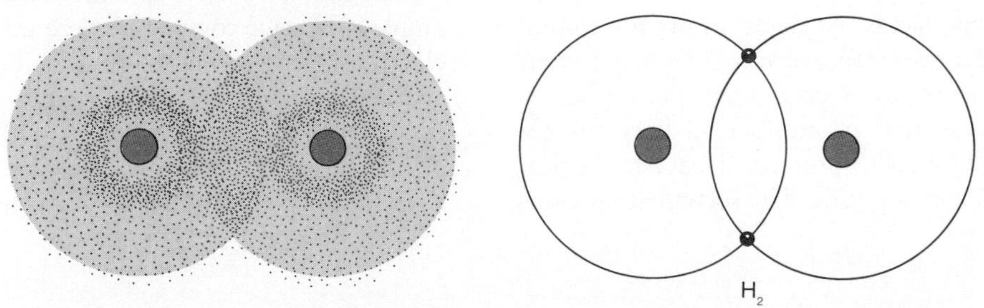

Figure 2.3. The molecules methane and ammonia represented in three different ways. Notice that a bond between two atoms consists of a pair of shared electrons (the electrons from the outer orbital of each atom).

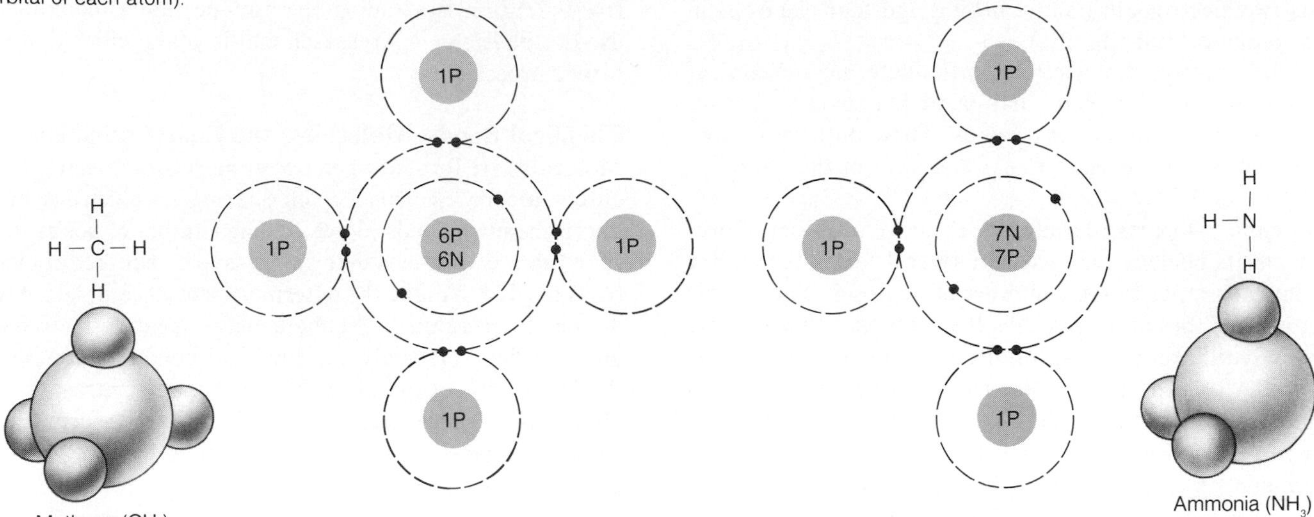

Covalent Bonds. In **covalent bonds,** atoms share electrons. Covalent bonds that are formed between identical atoms, as in oxygen gas (O_2) and hydrogen gas (H_2), are the strongest because their electrons are equally shared. Since the electrons are equally distributed between the two atoms, these molecules are said to be *nonpolar* and are bonded by *nonpolar covalent bonds.* When covalent bonds are formed between two different atoms, however, the electrons may be pulled more toward one atom than the other; they are unequally shared. The side of the molecule toward which the electrons are pulled is electrically more negative than the other side. Such a molecule is said to be *polar* (has positive and negative "poles"), and is bonded by a *polar covalent bond.* Atoms of oxygen, nitrogen, and phosphorous have a particularly strong tendency to pull electrons toward themselves when they bond with other atoms.

Water is the most abundant molecule in the body and serves as the solvent of body fluids. Water is a good solvent because it is polar; the oxygen atom pulls its electrons from the two hydrogens toward its side of the water molecule, so that the oxygen side is more negatively charged than the hydrogen side of the molecule (fig. 2.4). The significance of the polar nature of water in its function as a solvent is discussed in the next section.

Ionic Bonds. In **ionic bonds** the electrons are not shared at all. Instead, one or more valence electrons from one atom are completely transferred to a second atom. The first atom thus loses electrons, so that its number of electrons becomes less than its number of protons; it becomes a positively charged **ion** *(i'on)* or *cation.* The second atom gains more electrons than it has protons and becomes a negatively charged ion, or *anion.* (A positively charged ion is called a cation because it moves toward the negative pole, or cathode, of an electric field; a negatively charged anion moves toward the positive pole, or anode.) The cation and anion attract each other to form an **ionic compound.**

Figure 2.4. A model of a water molecule showing its polar nature. Notice that the oxygen side of the molecule is negative, whereas the hydrogen side is positive. Polar covalent bonds are weaker than nonpolar covalent bonds. As a result, some water molecules ionize to form OH^- (hydroxyl ion) and H^+ (hydrogen ion). The H^+ combines with water molecules to form hydronium (H_3O^+) ions (not shown).

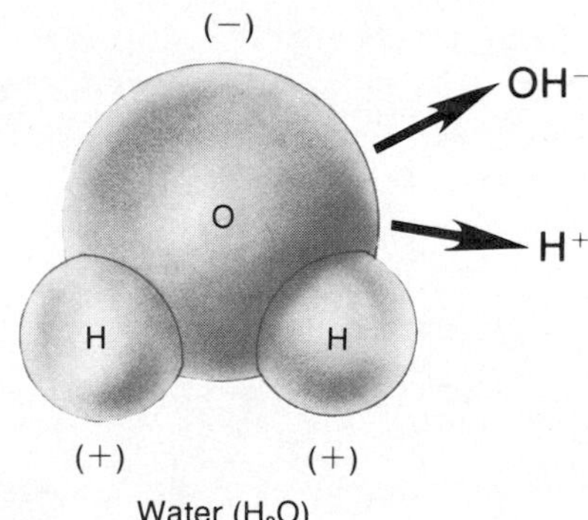

Common table salt, sodium chloride (NaCl), is an example of an ionic compound. Sodium, with a total of eleven electrons, has two in its first orbital, eight in its second orbital, and only one in its third orbital. Chlorine, conversely, is only one electron short of completing its outer orbital of eight electrons. The lone electron in sodium's outer orbital is attracted to chlorine's outer orbital. This

Figure 2.5. The dissociation of sodium and chlorine to produce sodium and chlorine ions. The positive sodium and the negative chloride ions attract each other to produce the ionic compound sodium chloride (NaCl).

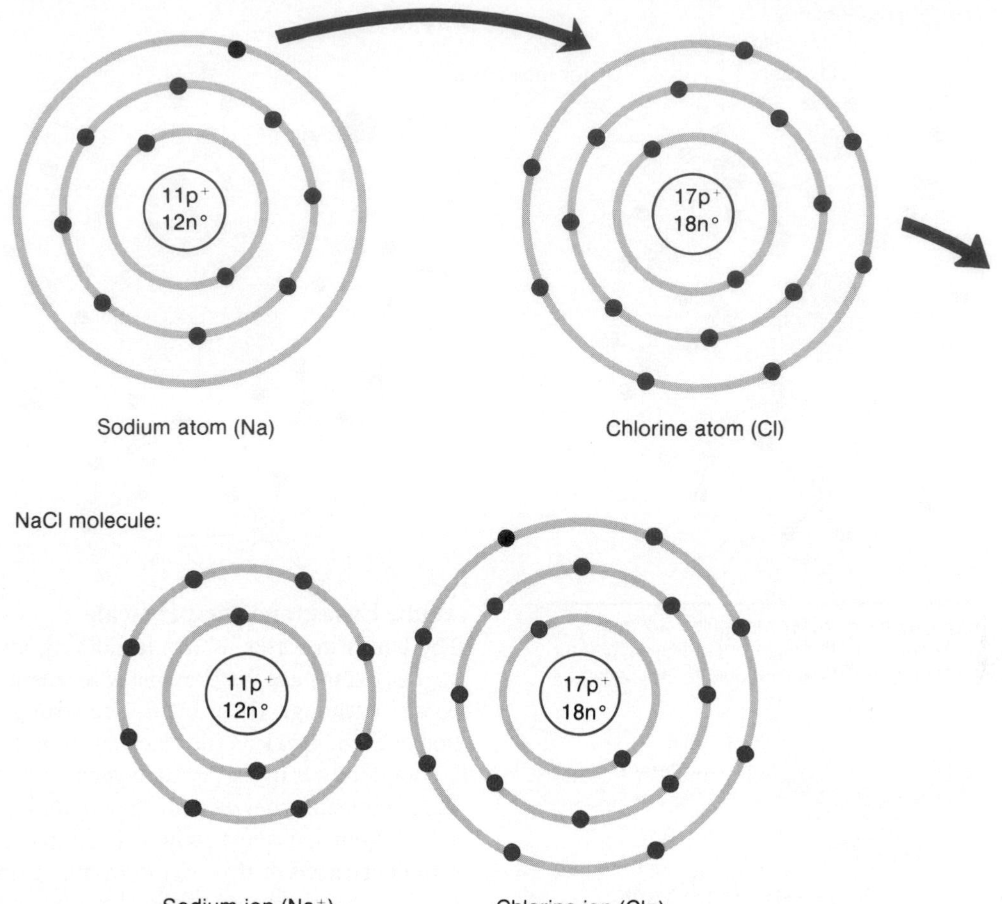

creates a chloride ion (represented as Cl^-) and a sodium ion (Na^+). Although table salt is shown as NaCl, it is actually composed of Na^+Cl^- (fig. 2.5).

Ionic bonds are weaker than polar covalent bonds, and therefore ionic compounds easily dissociate when dissolved in water to yield their separate ions. Dissociation of NaCl, for example, yields Na^+ and Cl^-. Each of these ions attracts polar water molecules; the negative ends of water molecules are attracted to the Na^+, and the positive ends are attracted to the Cl^- (fig. 2.6). The water molecules that surround these ions in turn attract other molecules of water to form *hydration spheres* around each ion.

The formation of hydration spheres makes an ion or a molecule soluble in water. Glucose, amino acids, and many other organic molecules are water-soluble because hydration spheres can form around atoms of oxygen, nitrogen, and phosphorous, which are joined by polar covalent bonds to other atoms in the molecule. Such molecules are said to be **hydrophilic** *(hi"dro-fil'ik)*. In contrast, molecules composed primarily of nonpolar covalent bonds, such as the hydrocarbon chains of fat molecules, have few charges and thus cannot form hydration spheres. They are insoluble in water, and in fact they actually avoid water; they are **hydrophobic.**

Hydrogen Bonds. **Hydrogen bonds** are very weak bonds that help to stabilize the delicate folding and bending of long organic molecules such as proteins. When hydrogen forms a polar covalent bond with an atom of oxygen or nitrogen, the hydrogen gains a slight positive charge as the electron is pulled toward the electronegative atom. Since this hydrogen has a slight positive charge, it will have a weak attraction for a second electronegative atom that may be located near it. This weak attraction is called a *hydrogen bond.* Hydrogen bonds are usually shown with dotted lines (fig. 2.7) to distinguish them from strong covalent bonds, which are shown with solid lines.

hydrophilic: Gk. *hydor,* water; *philos,* fond

hydrophobic: Gk. *hydor,* water; *phobos,* fear

Figure 2.6. The negatively charged oxygen-ends of water molecules are attracted to the positively charged Na^+, whereas the positively charged hydrogen-ends of water molecules are attracted to the negatively charged Cl^-. Other water molecules are attracted to this first concentric layer of water, forming hydration spheres around the sodium and chloride ions.

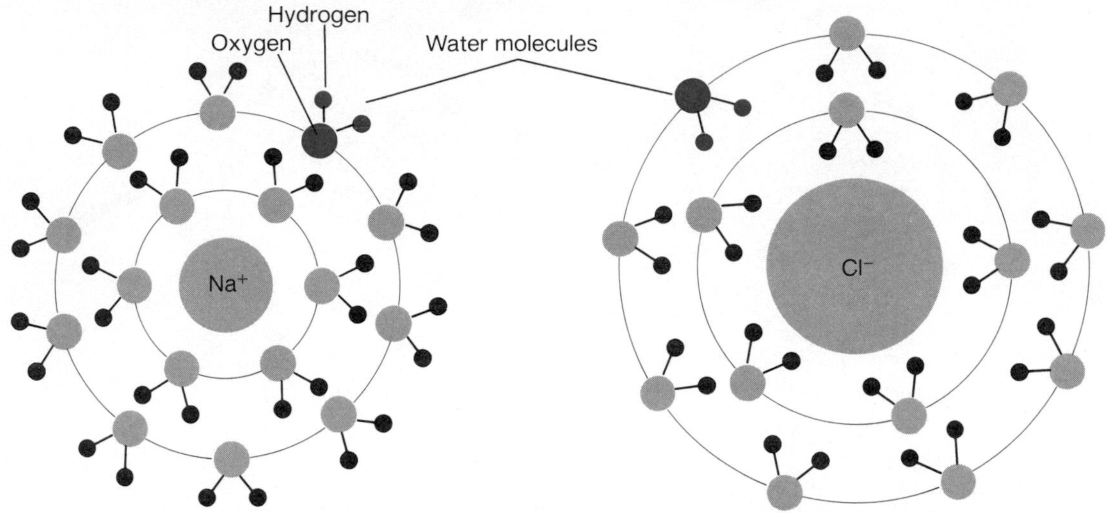

Figure 2.7. The oxygen atoms of water molecules are weakly joined together by the attraction of the electronegative oxygen for the positively charged hydrogen. These weak bonds are called *hydrogen bonds*.

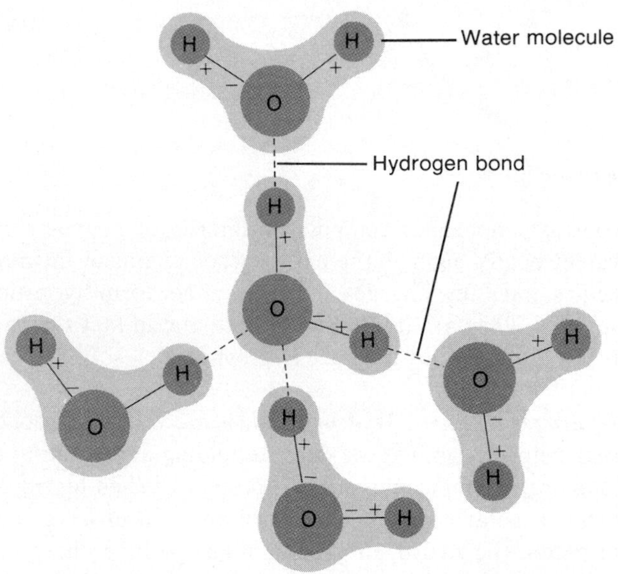

Hydrogen bonds can be formed within the folds of a long molecule, such as a protein (discussed in a later section). They can also be formed between adjacent water molecules (fig. 2.7). The hydrogen bonding between water molecules is responsible for many of the physical properties of water, including its *surface tension* and its ability to be pulled as a column through narrow channels in a process called *capillary action*.

Acids, Bases, and the pH Scale

The bonds in water molecules joining hydrogen atoms to oxygen atoms are, as previously discussed, polar covalent bonds. Although these bonds are strong, a small proportion of them break as the electron from the hydrogen atom is completely transferred to oxygen. When this occurs, the water molecule ionizes to form a *hydroxyl ion* (OH^-) and a hydrogen ion (H^+), which is simply a free proton. A proton released in this way does not remain free for long, because it is attracted to the electrons of oxygen atoms in water molecules. This forms a *hydronium ion,* shown by the formula H_3O^+. For the sake of clarity in the following discussion, however, H^+ will be used to indicate the cation resulting from the ionization of water.

Ionization of water molecules thus produces equal amounts of OH^- and H^+ (which becomes hydronium ions). Since only a small proportion of water molecules ionize, the concentration of H^+ (hydronium ions) and OH^- are each only equal to 10^{-7} molar. (The term *molar* is a unit of concentration, described in chapter 5; for hydrogen, one molar equals one gram per liter.) A solution with this H^+ concentration, which is produced by the ionization of water molecules in which the H^+ concentration equals the OH^- concentration, is said to be **neutral.**

A solution that contains a higher H^+ concentration than that of water is called *acidic (a-sid'ik),* and a solution with a lower H^+ concentration is called *basic.* An **acid** is defined as a molecule that can release protons (H^+) to a solution; it is a "proton donor." A **base** is a negatively

Table 2.2 Common acids and bases			
Acid	**Symbol**	**Base**	**Symbol**
Hydrochloric acid	HCl	Sodium hydroxide	$NaOH$
Phosphoric acid	H_3PO_4	Potassium hydroxide	KOH
Nitric acid	HNO_3	Calcium hydroxide	$Ca(OH)_2$
Sulfuric acid	H_2SO_4	Ammonium hydroxide	NH_4OH
Carbonic acid	H_2CO_3		

From Stuart Ira Fox, *A Laboratory Guide to Human Physiology: Concepts and Clinical Applications,* 4th ed.

charged ion (anion), or a molecule that ionizes to produce the anion, which can combine with H^+ and thus remove the H^+ from solution; it is a "proton acceptor." Most strong bases release OH^- into a solution, which combines with H^+ to form water and which thus lowers the H^+ concentration. Examples of common acids and bases are shown in table 2.2.

pH. The H^+ concentration of a solution is usually indicated in pH units on a pH scale that runs from 0 to 14. The pH number is equal to the logarithm of one over the H^+ concentration:

$$pH = \log \frac{1}{[H^+]}$$

where $[H^+]$ = molar H^+ concentration.

Pure water has an H^+ concentration of 10^{-7} molar at 25°C, and thus it has a pH of 7 (neutral). Because of the logarithmic relationship, a solution with ten times the hydrogen ion concentration (10^{-6} M) has a pH of 6, whereas a solution with one-tenth the H^+ concentration (10^{-8} M) has a pH of 8. The pH number is easier to write than the molar H^+ concentration, but it is admittedly confusing because it is *inversely related* to the H^+ concentration: a solution with a higher H^+ concentration has a lower pH number; one with a lower H^+ concentration has a higher pH number. A strong acid with a high H^+ concentration of 10^{-2} molar, for example, has a pH of 2, whereas a solution with only 10^{-10} molar H^+ has a pH of 10. **Acidic solutions,** therefore, have a pH of less than 7 (that of pure water), whereas **basic solutions** have a pH of between 7 and 14 (table 2.3).

Buffers. A *buffer* is a system of molecules and ions that acts to prevent changes in H^+ concentration, thus serving to stabilize the pH of a solution. In blood plasma, for example, the pH is stabilized by the following reversible reaction involving the bicarbonate ion (HCO_3^-) and carbonic acid (H_2CO_3):

$$HCO_3^- + H^+ \rightleftharpoons H_2CO_3.$$

Table 2.3 The pH scale			
	H^+ concentration (molar)	**pH**	**OH^- concentration (molar)**
	1.0	0	10^{-14}
	0.1	1	10^{-13}
	0.01	2	10^{-12}
Acids	0.001	3	10^{-11}
	0.0001	4	10^{-10}
	10^{-5}	5	10^{-9}
	10^{-6}	6	10^{-8}
Neutral	10^{-7}	7	10^{-7}
	10^{-8}	8	10^{-6}
	10^{-9}	9	10^{-5}
	10^{-10}	10	0.0001
Bases	10^{-11}	11	0.001
	10^{-12}	12	0.01
	10^{-13}	13	0.1
	10^{-14}	14	1.0

From Stuart Ira Fox, *Human Physiology,* 2d ed.

The double arrows indicate that the reaction could go either to the right or to the left; the net direction depends on the concentration of molecules and ions on each side. If an acid (such as lactic acid) should release H^+ into the solution, for example, the following reaction would be promoted:

$$HCO_3^- + H^+ \longrightarrow H_2CO_3.$$

The above reaction serves to decrease the effect of added H^+ on the pH of the blood. Bicarbonate, in fact, is the major buffer of the blood. Under the opposite condition, when the concentration of free H^+ in the blood is falling, the reaction previously described can be reversed:

$$H_2CO_3 \longrightarrow H^+ + HCO_3^-.$$

The dissociation of carbonic acid yields free H^+, which helps to prevent an increase in pH. Bicarbonate ions and carbonic acid thus act as a *buffer pair* to prevent either decreases or increases in pH, respectively. This buffering action normally maintains the blood pH at a very stable 7.40 ± 0.05.

Organic Molecules

Organic molecules are those that contain the atom *carbon.* Since the carbon atom has four electrons in its outer orbital, it must share four additional electrons by covalent bonding with other atoms to fill its outer orbital with eight electrons. The unique bonding requirements of carbon enable it to join with other carbon atoms to form chains and rings, while still allowing the carbons to bond with hydrogen and other atoms.

Figure 2.8. Two carbon atoms joined by a single covalent bond (*left*) or by a double covalent bond (*right*). In both cases each carbon atom shares four pairs of electrons (has four bonds) to complete the eight electrons required to fill its outer orbital.

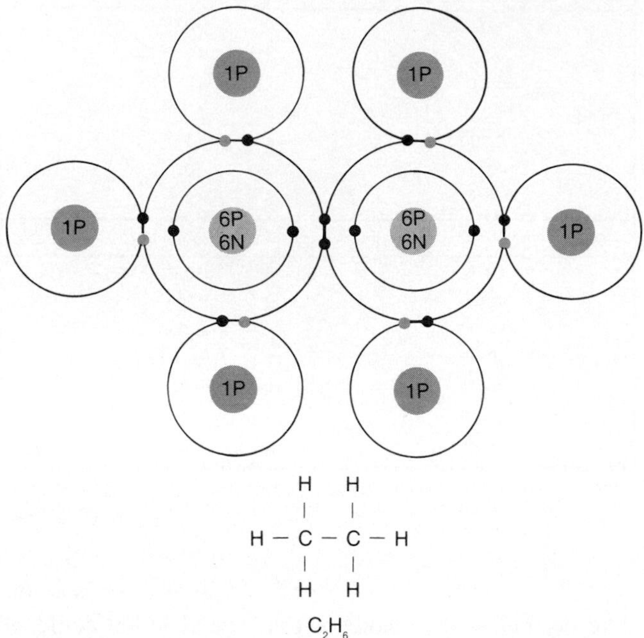

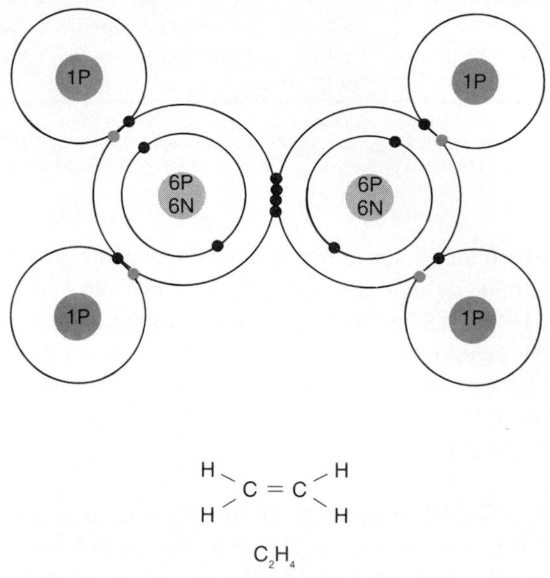

Figure 2.9. Hydrocarbons that are (*a*) linear, (*b*) cyclic, and (*c*) aromatic rings. Notice that the alternating double bonds in an aromatic ring can be represented by a circle.

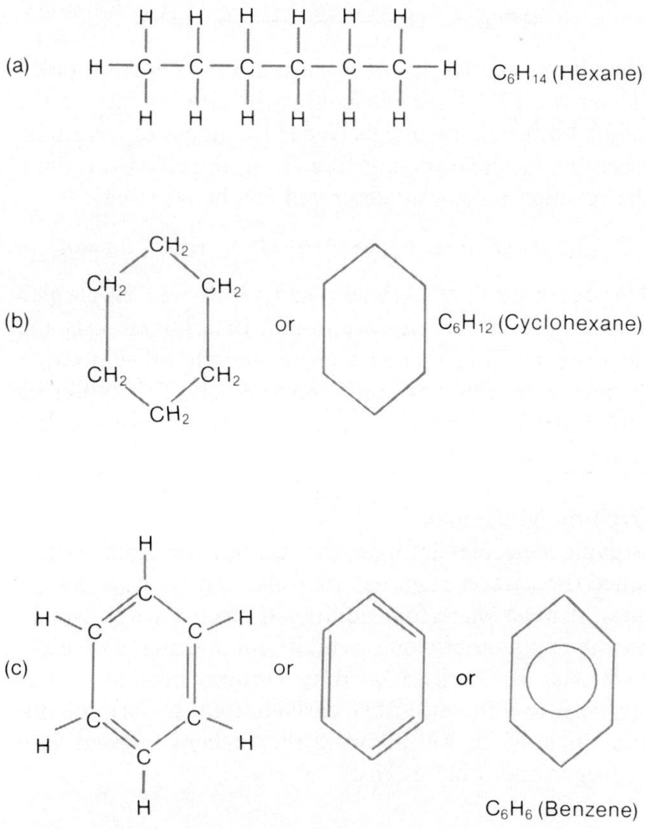

(a) H—C—C—C—C—C—C—H (each carbon bonded to H above and H below) C_6H_{14} (Hexane)

(b) CH_2 ring (six CH_2 groups) or [hexagon] C_6H_{12} (Cyclohexane)

(c) H—C ring with alternating double bonds (six C, each bonded to H) or [hexagon with alternating double bonds] or [hexagon with circle] C_6H_6 (Benzene)

Most organic molecules in the body contain hydrocarbon chains and rings as well as other atoms bonded to carbon. Two adjacent carbon atoms in a chain or ring may share one or two pairs of electrons. If the two carbon atoms share one pair of electrons, they are said to have a *single covalent bond;* this leaves each carbon atom free to bond to as many as three other atoms. If the two carbon atoms share two pairs of electrons, they have a *double covalent bond,* and each carbon atom can only bond to a maximum of two additional atoms (fig. 2.8).

The ends of some hydrocarbons are joined together to form rings. In the shorthand structural formulas of these molecules, the carbon atoms are not shown but are understood to be located at the corners of the ring. Some of these cyclic molecules have a double bond between two adjacent carbon atoms. Benzene and related molecules are shown as a six-sided ring with alternating double bonds. Such compounds are called *aromatic*. Since all of the carbons in an aromatic ring are equivalent, double bonds can be shown between any two adjacent carbons in the ring. Alternatively, this may be illustrated with a circle in the center of the ring (fig. 2.9).

The hydrocarbon chain or ring of many organic molecules provides a relatively inactive molecular "backbone," to which more reactive groups of atoms are attached. Known as *functional groups* of the molecule, these reactive groups usually contain atoms of oxygen, nitrogen, phosphorous, or sulfur and are, in large part, responsible for the unique chemical properties of the molecule (fig. 2.10).

Figure 2.10. Various functional groups of organic molecules.

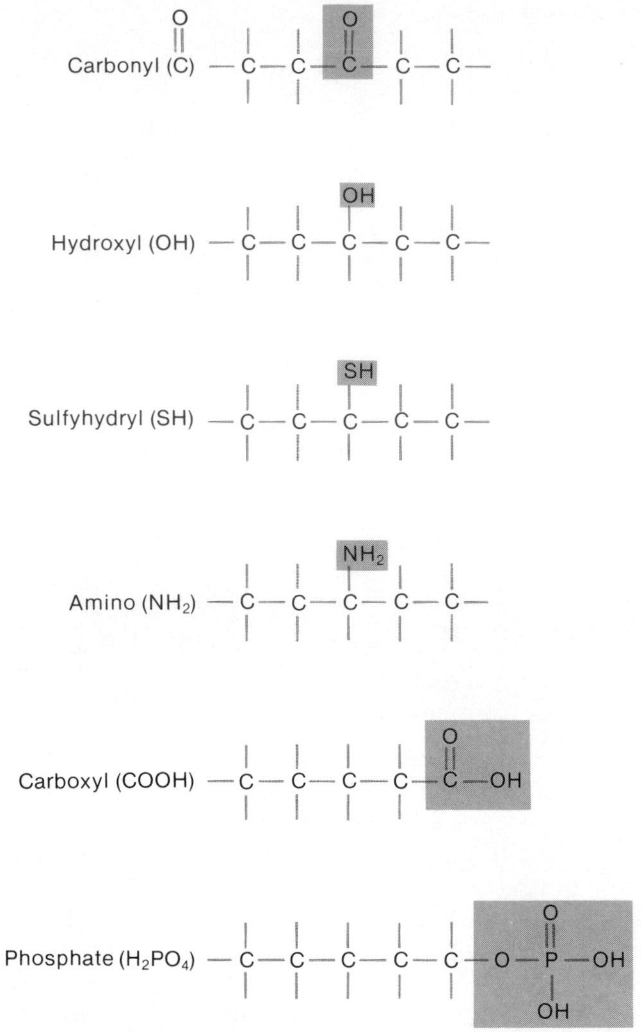

Carbonyl (C)

Hydroxyl (OH)

Sulfyhydryl (SH)

Amino (NH_2)

Carboxyl (COOH)

Phosphate (H_2PO_4)

Classes of organic molecules can be named according to their functional groups. *Ketones (ke'tōnz),* for example, have a carbonyl group within the carbon chain. An organic molecule is an *alcohol* if it has a hydroxyl group at one end of the chain. All *organic acids* (such as acetic acid, citric acids, and others) have a *carboxyl (kar-bok'sil)* group (fig. 2.11).

Stereoisomers. Two molecules may have exactly the same atoms arranged in exactly the same sequence, yet may differ with respect to the spatial orientation of key functional groups. Such molecules are called *stereoisomers* of each other. The isomers that have a key functional group represented on the right side of the molecule are called **D-isomers** (for *dextro,* or right-handed). Molecules that are represented by structures showing functional groups on the left side are called **L-isomers** (for *levo,* or left-handed).

Figure 2.11. Categories of organic molecules based on functional groups.

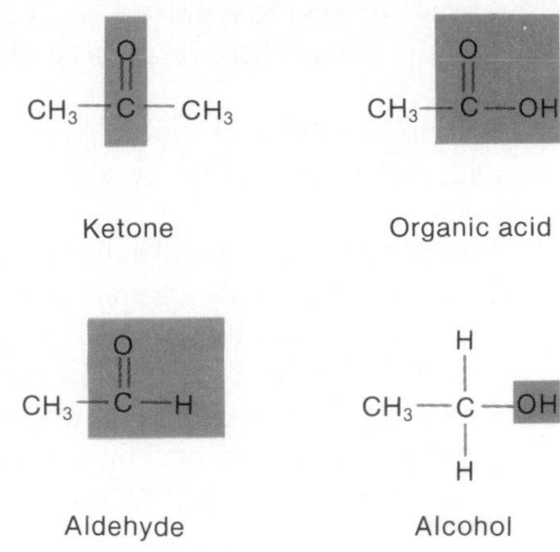

The two stereoisomers are mirror images of each other—they cannot be superimposed. These subtle differences in structure are extremely important biologically, because enzymes—which interact with such molecules in a stereo-specific way in chemical reactions—cannot combine with the "wrong" stereoisomer. The enzymes of all cells (human and others) can only combine with L-amino acids and D-sugars, for example. The opposite stereoisomers (D-amino acids and L-sugars) cannot be used by the body.

1. Describe the structure of an atom, and define the terms *atomic mass* and *atomic number.* Explain why different atoms are able to form characteristic numbers of chemical bonds.
2. Describe the nature of nonpolar and polar covalent bonds, ionic bonds, and hydrogen bonds. Explain the reasons why ions and polar molecules are soluble in water.
3. Define the terms *acidic, basic, acid,* and *base.* Define *pH,* and describe the relationship between pH and the H^+ concentration of a solution.
4. Explain how carbon atoms can bond to each other and to atoms of hydrogen, oxygen, and nitrogen. Describe some of the functional groups to be found in organic molecules, and explain the significance of these groups.

Carbohydrates and Lipids

Carbohydrates as a group share a characteristic ratio of carbon, hydrogen, and oxygen atoms, and they are divided into subgroups depending on the number of simple sugars that each molecule contains. Lipids are a related but distinct group of molecules that are also subdivided into smaller categories. Carbohydrates and lipids are the major sources of energy in the body.

Objective 6. List the different subcategories of carbohydrates, and give examples of each.

Objective 7. Describe dehydration synthesis and hydrolysis reactions using the different subcategories of carbohydrates as examples. Identify where these reactions occur in the body.

Objective 8. Identify the different subcategories of lipids and explain why they are all classified as lipids.

Objective 9. Describe the structure of saturated and unsaturated fatty acids. Explain how triglycerides are formed by dehydration synthesis and broken down by hydrolysis reactions.

Objective 10. Describe the structure of phospholipids and prostaglandins, and explain the functions of these molecules in the body.

Carbohydrates and lipids are similar in many ways. Both groups of molecules consist primarily of the atoms carbon, hydrogen, and oxygen, and both serve as major sources of energy in the body (comprising most of the calories consumed in food). Carbohydrates and lipids differ, however, in some important aspects of their chemical structures and physical properties. Such differences significantly affect the functions of these molecules in the body.

Carbohydrates

Carbohydrates are organic molecules that contain carbon, hydrogen, and oxygen in the ratio described by their name—*carbo* (carbon) and *hydrate* (water, H_2O). The general formula of a carbohydrate molecule is thus CH_2O; the molecule contains twice the number of hydrogen atoms as it contains carbon or oxygen atoms.

Monosaccharides, Disaccharides, and Polysaccharides. Carbohydrates include simple sugars, or **monosaccharides** *(mon″o-sak′ah-rīdz),* and longer molecules that contain a number of monosaccharides joined together. The suffix *-ose* denotes a sugar molecule; the term *hexose,* for example, refers to a six-carbon monosaccharide with the formula $C_6H_{12}O_6$. This formula is adequate for some purposes, but it does not distinguish between related hexose sugars, which are *structural isomers* of each other. The structural isomers glucose, fructose, and galactose, for example, are monosaccharides that have the same ratio of atoms arranged in slightly different ways (fig. 2.12).

Two monosaccharides can be joined covalently to form a **disaccharide** *(di-sak′ah-rīd),* or double sugar. Common disaccharides include table sugar, or *sucrose* (composed of glucose and fructose), milk sugar, or *lactose* (composed of glucose and galactose), and malt sugar, or *maltose* (composed of two glucose molecules). When many monosaccharides are joined together, the resulting molecule is called a **polysaccharide.** *Starch,* for example, is a polysaccharide found in many plants and is formed by the bonding together of thousands of glucose subunits. Animal starch (**glycogen**) *(gli′co-gen),* found in the liver and muscles, likewise consists of repeating glucose molecules but differs from plant starch in that it is more highly branched (fig. 2.13).

Figure 2.12. The structural formulas of three hexose sugars—(*a*) glucose, (*b*) galactose, and (*c*) fructose. All three have the same ratio of atoms—$C_6H_{12}O_6$.

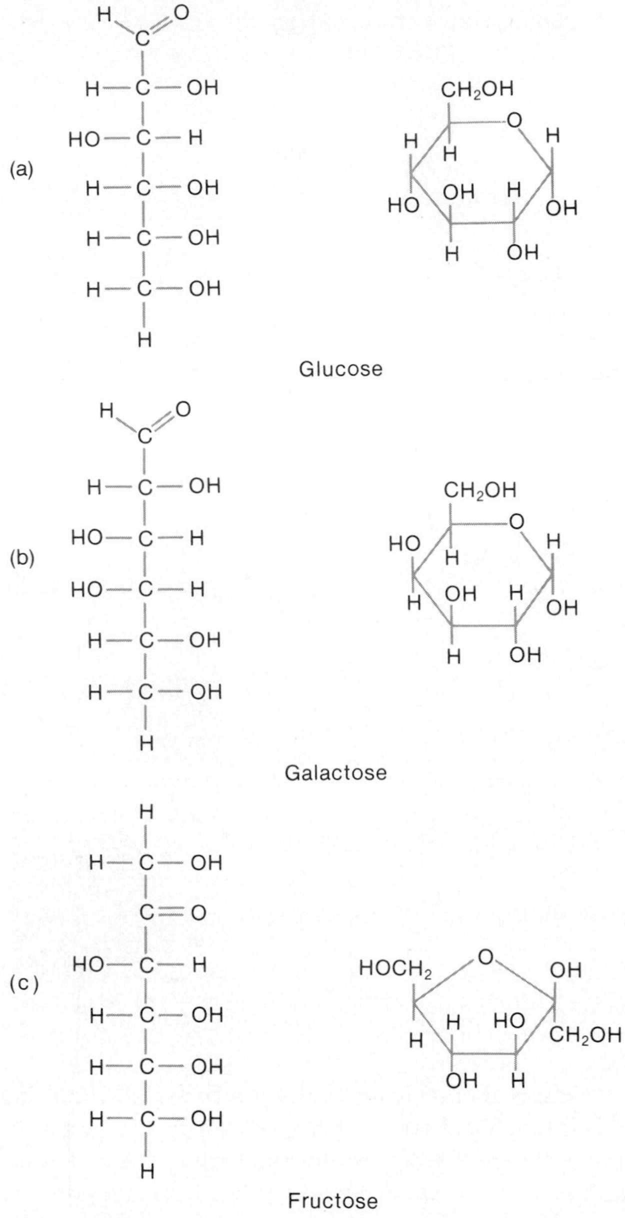

(a) Glucose

(b) Galactose

(c) Fructose

Dehydration Synthesis and Hydrolysis. In the formation of disaccharides and polysaccharides, the separate subunits (monosaccharides) are bonded together covalently by a type of reaction called **dehydration synthesis,** or **condensation.** In this reaction, which requires the participation of specific enzymes (chapter 4), a hydrogen atom is removed from one monosaccharide and a hydroxyl group (OH) is removed from another. As a covalent bond is formed between the two monosaccharides water (H_2O) is produced. Dehydration synthesis reactions are illustrated in figure 2.14.

monosaccharide: Gk. *monos,* single; *sakcharon,* sugar

Figure 2.13. Glycogen is a polysaccharide composed of glucose subunits joined together to form a large, highly branched molecule.

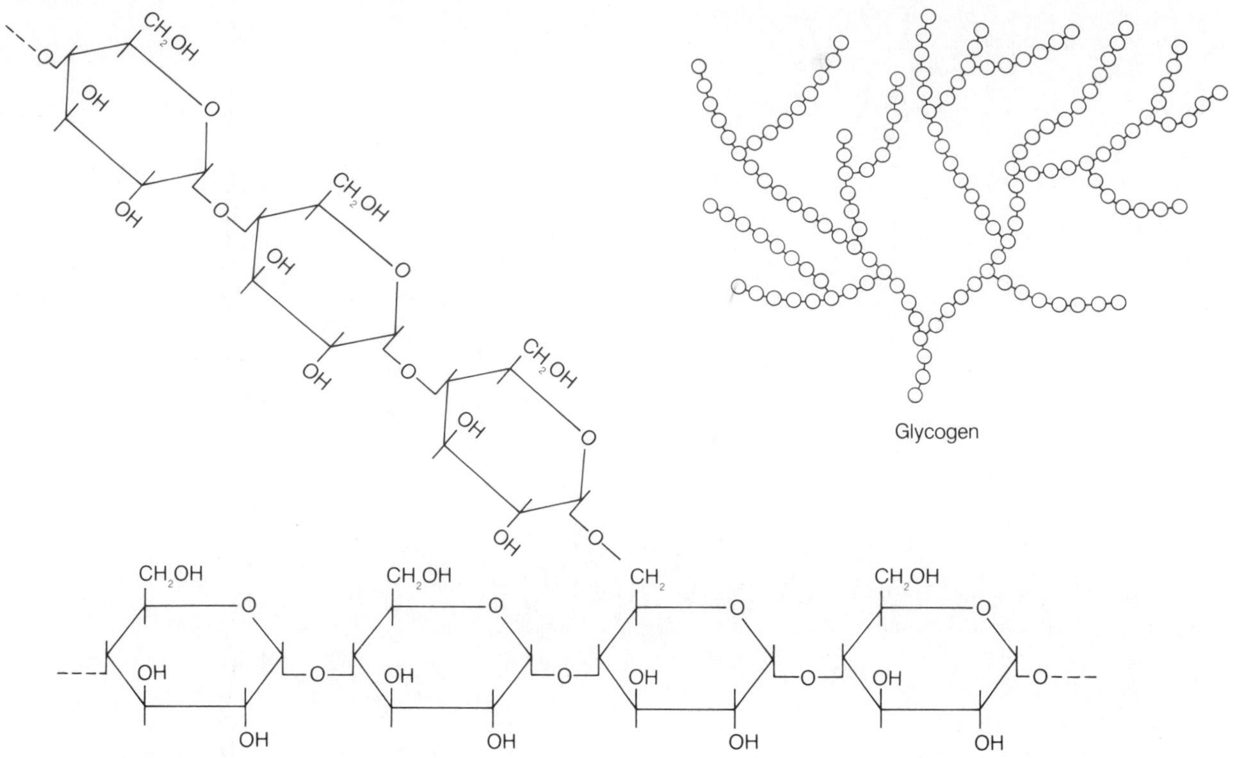

Figure 2.14. Dehydration synthesis of two disaccharides, (*a*) maltose and (*b*) sucrose. Notice that as the disaccharides are formed, a molecule of water is produced.

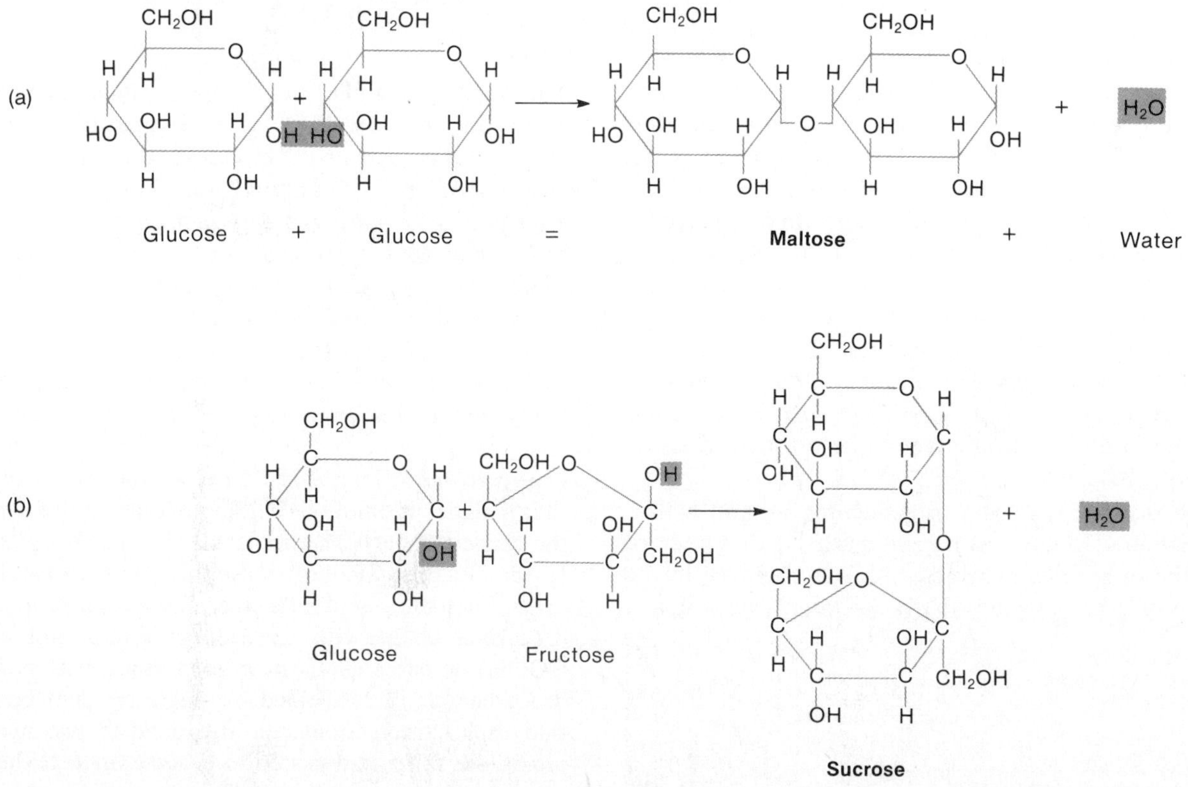

Figure 2.15. The hydrolysis of starch (*a*) into disaccharides (maltose) and (*b*) into monosaccharides (glucose). Notice that as the covalent bond between the subunits breaks, a molecule of water is split. In this way the hydrogen and hydroxyl from water is added to the ends of the released subunits.

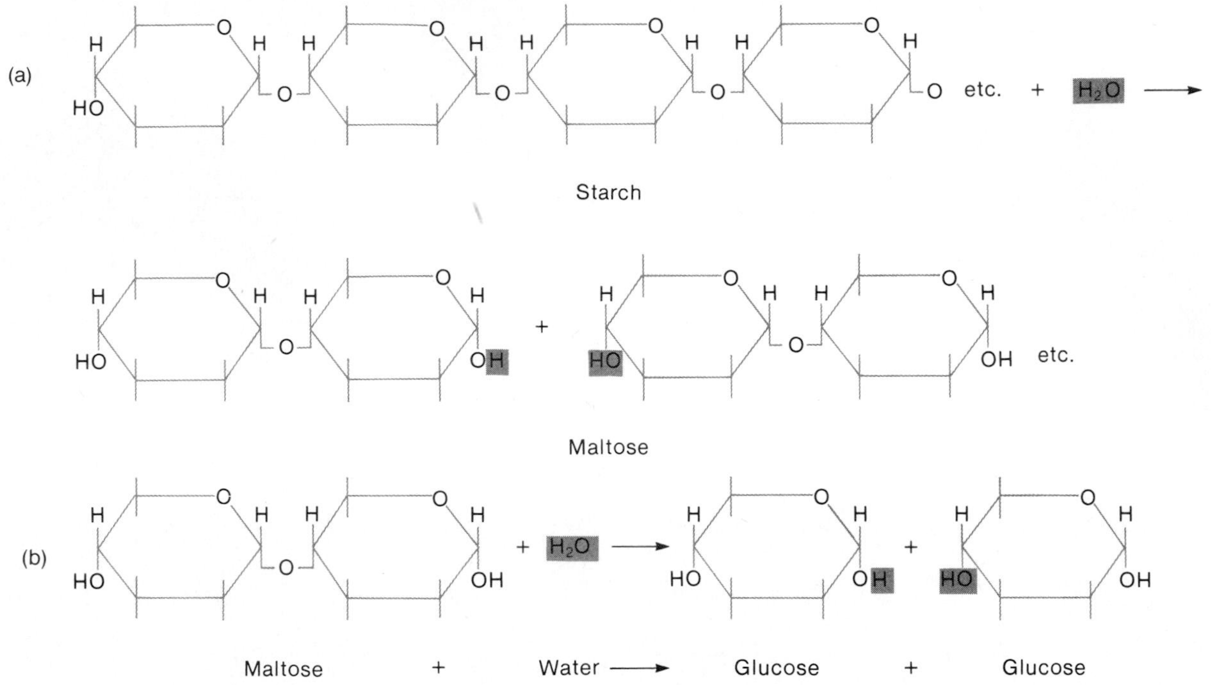

(a) Starch + H_2O → Maltose + Maltose etc.

(b) Maltose + Water → Glucose + Glucose

When a person eats disaccharides and polysaccharides, or when the stored glycogen in the liver and muscles is to be used by tissue cells, the covalent bonds that join monosaccharides into disaccharides and polysaccharides must be broken. These *digestion* reactions occur by means of **hydrolysis** *(hi-drol'ĭ-sis)*. Hydrolysis is the reverse of dehydration synthesis. A water molecule is split, as implied by the word *hydrolysis,* and the resulting hydrogen atom is added to one of the free glucose molecules as the hydroxyl group is added to the other (fig. 2.15).

When a potato is eaten, the starch within it is hydrolyzed into separate glucose molecules within the intestine. This glucose is absorbed into the blood and carried to the tissues. Some tissue cells may use this glucose for energy. Liver and muscles, however, can store excess glucose in the form of glycogen by dehydration synthesis reactions in these cells. During fasting or prolonged exercise, the liver can add glucose to the blood through hydrolysis of its stored glycogen.

Dehydration synthesis, or condensation, and hydrolysis reactions do not occur spontaneously; they require the action of specific enzymes. Similar reactions, in the presence of other enzymes, build and break down lipids, proteins, and nucleic acids. In general, therefore, hydrolysis reactions digest molecules into their subunits, and dehydration synthesis reactions build larger molecules by the bonding together of their subunits.

hydrolysis: Gk. *hydor,* water; *lysis,* dissolution, or break

Lipids

The category of molecules known as **lipids** includes several types of molecules that differ greatly in chemical structure. These diverse molecules are all in the lipid category by virtue of a common physical property—they are all *insoluble in polar solvents* such as water. This is because lipids consist primarily of hydrocarbon chains and rings, which are nonpolar and, thus, hydrophobic. Although lipids are insoluble in water, they can be dissolved in nonpolar solvents such as ether, benzene, and related compounds. So that they can travel in the aqueous plasma, lipids are attached to proteins in the form of lipoproteins.

Triglycerides. Triglycerides are a subcategory of lipids that includes fat and oil. These molecules are formed by the condensation of one molecule of *glycerol (glis'er-ol)* (a three-carbon alcohol) with three molecules of *fatty acids.* Each fatty acid molecule consists of a nonpolar hydrocarbon chain with a carboxyl group (abbreviated COOH) on one end. If the carbon atoms within the hydrocarbon chain are joined by single covalent bonds, so that each carbon atom can also bond to two hydrogen atoms, the fatty acid is said to be *saturated.* If there are

Figure 2.16. Structural formulas for (*a*) saturated and (*b*) unsaturated fatty acids.

(a)

Palmitic acid,
a saturated fatty acid

(b)

Linolenic acid,
an unsaturated fatty acid

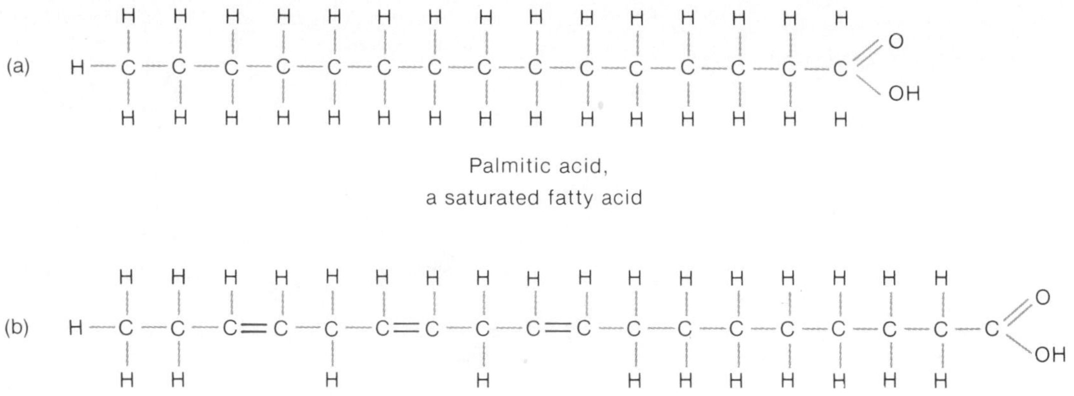

Figure 2.17. Dehydration synthesis of a triglyceride molecule from a glycerol and three fatty acids. A molecule of water is produced as an ester bond forms between each fatty acid and the glycerol. Sawtooth lines represent carbon chains, which are symbolized by an *R*.

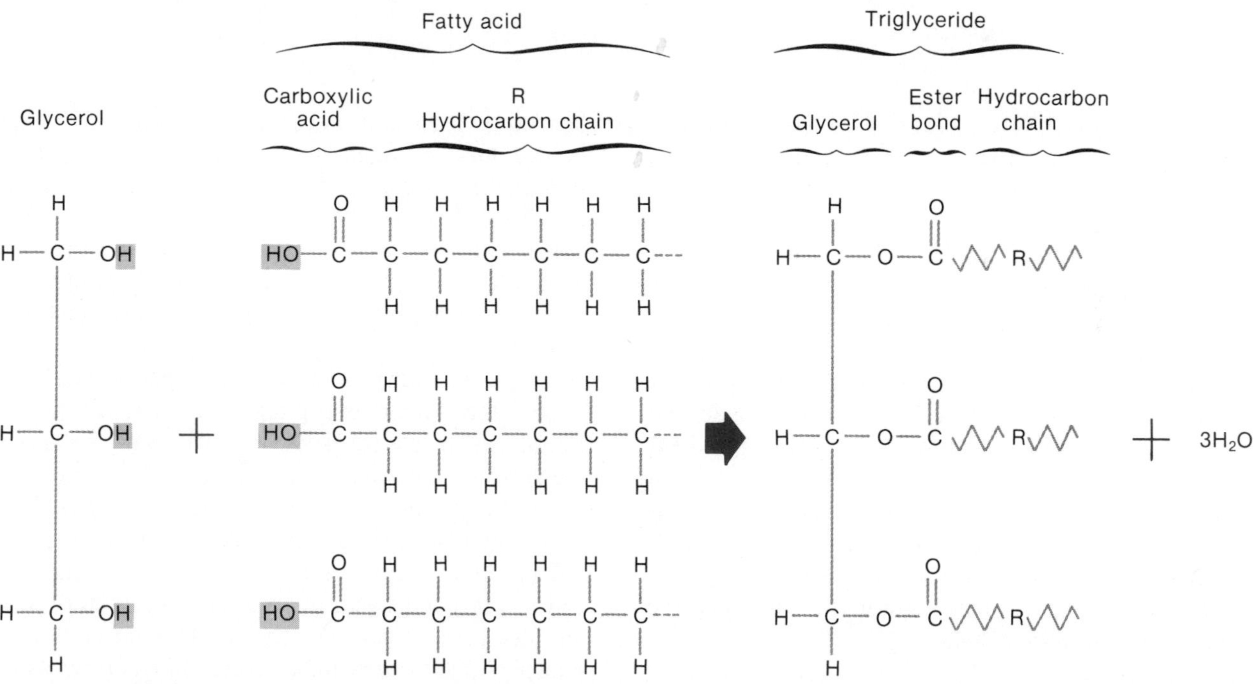

a number of double covalent bonds within the hydrocarbon chain, so that each carbon atom can bond to only one hydrogen atom, the fatty acid is said to be *unsaturated*. Triglycerides that contain saturated fatty acids are called **saturated fats;** those that contain unsaturated fatty acids are **unsaturated fats** (fig. 2.16).

Within the adipose cells of the body, triglycerides are formed as the carboxylic acid ends of fatty acid molecules condense with the hydroxyl groups of a glycerol molecule (fig. 2.17). Since the hydrogen atoms from the carboxyl ends of fatty acid molecules form water molecules during

Figure 2.18. The structure of lecithin, a typical phospholipid (*above*), and its more simplified representation (*below*).

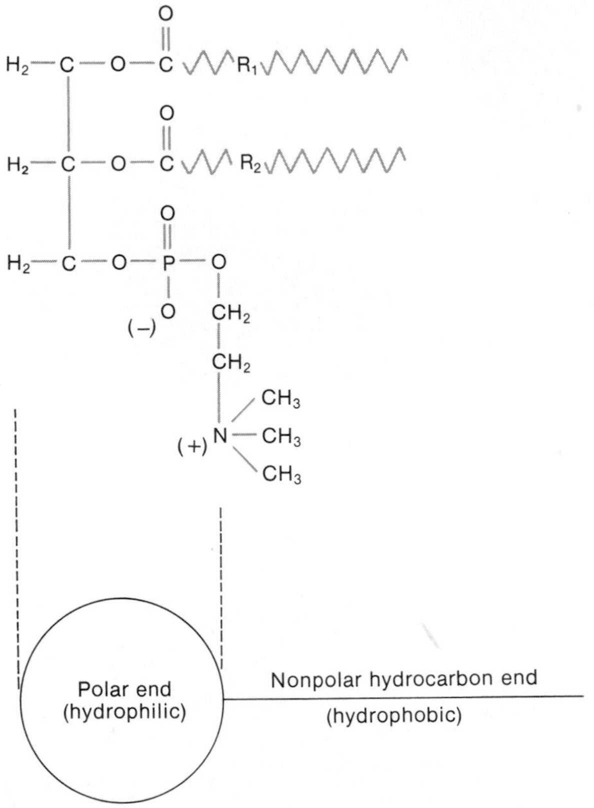

O
H_2—C—O—C R₁
O
H_2—C—O—C R₂
O
H_2—C—O—P—O
(−) O CH_2
CH_2
(+) N — CH_3, CH_3, CH_3

Polar end (hydrophilic)
Nonpolar hydrocarbon end (hydrophobic)

Figure 2.19. The formation of a micelle structure by phospholipids such as lecithin. The straight lines represent the hydrophobic fatty acid parts of the molecule, and the circles represent the polar phosphate part of the molecule. The detailed structure of lecithin is shown in one part of the micelle.

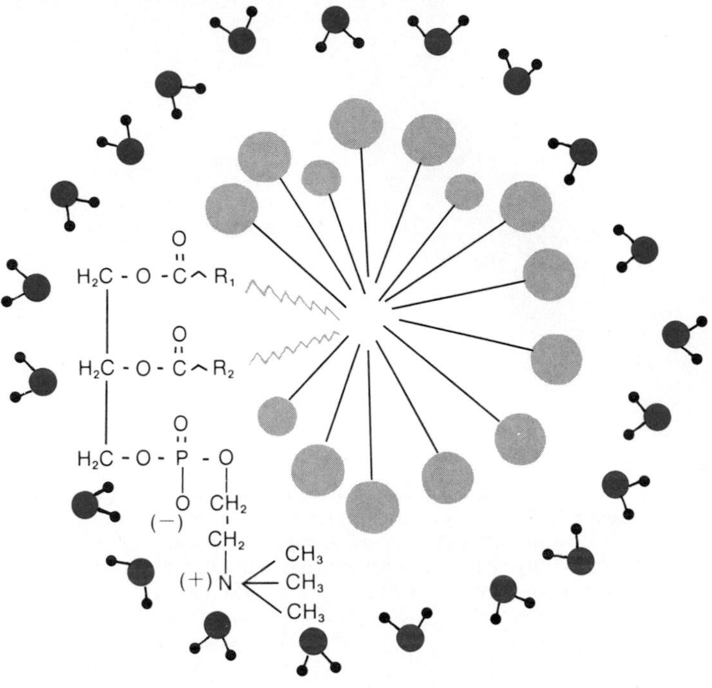

dehydration synthesis, fatty acids that are combined with glycerol can no longer release H^+ and function as acids. For this reason, triglycerides are described as *neutral fats.*

Ketone Bodies. Hydrolysis of triglycerides within adipose tissue releases *free fatty acids* into the blood. Free fatty acids can be used as an immediate source of energy by many organs; they can also be converted by the liver into derivatives called *ketone bodies.* These include four-carbon-long acidic molecules (acetoacetic acid and β-hydroxybutyric acid) and acetone (the active ingredient in nail polish remover). A rapid breakdown of fat, such as occurs during dieting and in uncontrolled diabetes mellitus, results in elevated levels of ketone bodies in the blood, a condition called **ketosis.** If there are sufficient amounts of ketone bodies in the blood to lower the blood pH, the condition is called **ketoacidosis.**

Phospholipids. The class of lipids known as *phospholipids* contains a number of different categories, which have in common the fact that they are lipids that contain a phosphate group. The most common type of phospholipid molecule has this structure: the three-carbon alcohol molecule glycerol is attached to two fatty acid molecules; the third carbon atom of the glycerol molecule is attached to a phosphate group, and the phosphate group in turn is bonded to other molecules. If the phosphate group is attached to a nitrogen-containing choline molecule, the phospholipid molecule thus formed is known as *lecithin* (fig. 2.18). Figure 2.18 shows a simple way of illustrating the structure of a phospholipid. The parts of the molecule capable of ionizing and thus becoming charged are shown as a circle, whereas the nonpolar parts of the molecule are represented by lines.

Since the nonpolar ends of phospholipids are hydrophobic, they tend to group together when mixed in water; this allows the hydrophilic parts (which are polar) to face the surrounding water molecules (fig. 2.19). Such aggregates of molecules are called **micelles** *(mi-selz').* The dual nature of phospholipid molecules (part polar, part nonpolar) allows them to form the major component of the cell membrane as well as to alter the structure of water and decrease its surface tension. This latter function of phospholipids, which makes them *surfactants* (surface-active-agents), prevents collapse of the lungs.

Steroids. The structure of steroid molecules is quite different from that of triglycerides or phospholipids, and yet steroids are still included in the lipid category of molecules because they are nonpolar and insoluble in water.

Figure 2.20. Cholesterol and some steroid hormones derived from cholesterol. The carbon atoms in cholesterol are numbered according to the standard sequence for steroids.

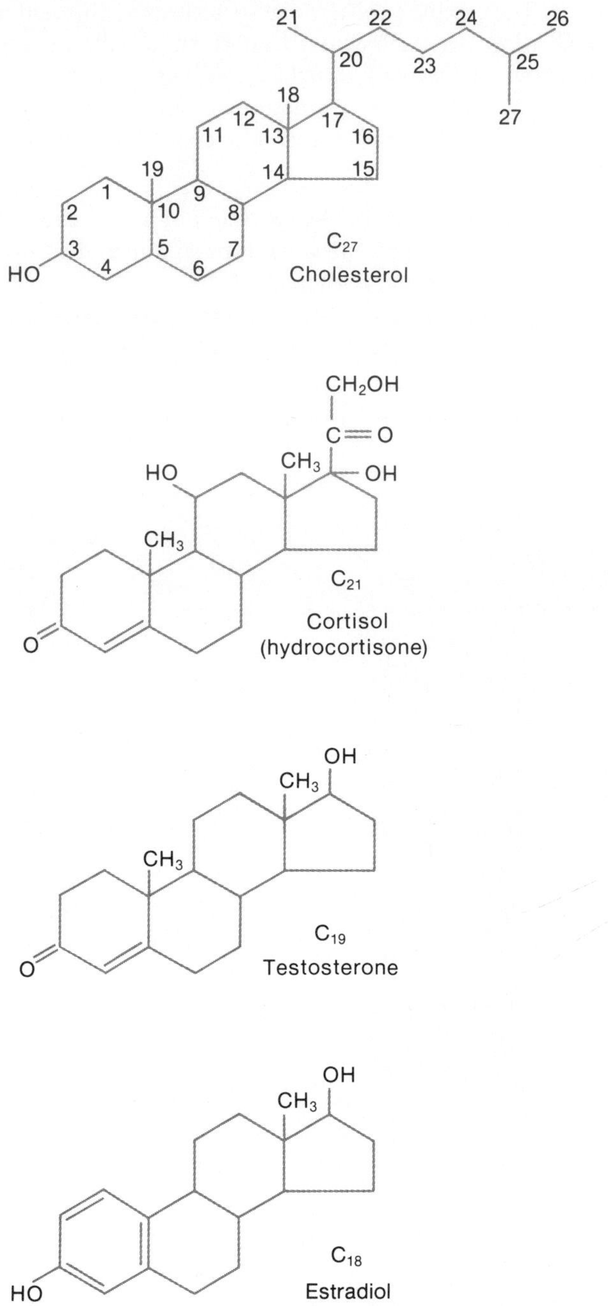

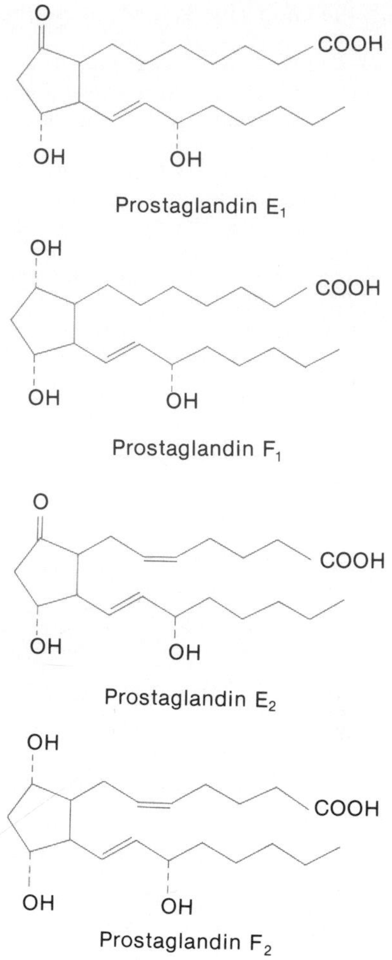

All steroid molecules have the same basic structure; three six-carbon rings are joined to one five-carbon ring (fig. 2.20). However, different kinds of steroids have different functional groups attached to this basic structure, and they vary in the number and position of the double covalent bonds between the carbon atoms in the rings.

Figure 2.21. Structural formulas of some prostaglandins.

Cholesterol is an important molecule in the body because it serves as the precursor (parent molecule) for the steroid hormones produced by the gonads and adrenal cortex. The testes and ovaries (collectively called the *gonads*) secrete **sex steroids,** which include estradiol and progesterone from the ovaries and testosterone from the testes. The adrenal cortex secretes the **corticosteroids,** including hydrocortisone and aldosterone, among others.

Prostaglandins. Prostaglandins are a type of fatty acid (with a cyclic hydrocarbon group), which have a variety of regulatory functions. Although their name is derived from the fact that they were originally found in the semen as a secretion of the prostate gland, it has since been shown that they are produced by and active in almost all tissues. Prostaglandins are implicated in the regulation of blood vessel diameter, ovulation, uterine contraction during labor, inflammation reactions, blood clotting, and many other functions. Some of the different types of prostaglandins are shown in figure 2.21. Because many of the regulatory functions of prostaglandins are associated with the actions of hormones, prostaglandins are discussed in more detail, together with the endocrine system, in chapter 19.

1. Describe the common structure of all carbohydrates, and distinguish between monosaccharides, disaccharides, and polysaccharides.
2. Using dehydration synthesis and hydrolysis reactions, explain how disaccharides and monosaccharides can be interconverted and how triglycerides can be formed and broken down.
3. Explain what a lipid is, and describe the different subcategories of lipids.
4. Relate the functions of phospholipids to their structure, and describe the significance of the prostaglandins.

Proteins and Nucleic Acids

Proteins are large molecules composed of amino acid subunits. The complex and diverse structures of different proteins allow these molecules to perform a wide variety of functions in the body. Nucleic acids are also very large molecules. They are composed of nucleotide subunits, which provide the basis for genetic control of cells and the body.

Objective 11. Describe the structure of amino acids, and explain how one type of amino acid differs from another.

Objective 12. Describe the primary, secondary, tertiary, and quaternary structures of proteins.

Figure 2.22. Representative amino acids, showing different types of functional (*R*) groups.

Functional group

R
H
N—C—C
H
O
OH
H

Amino group Carboxylic acid group

Nonpolar amino acids

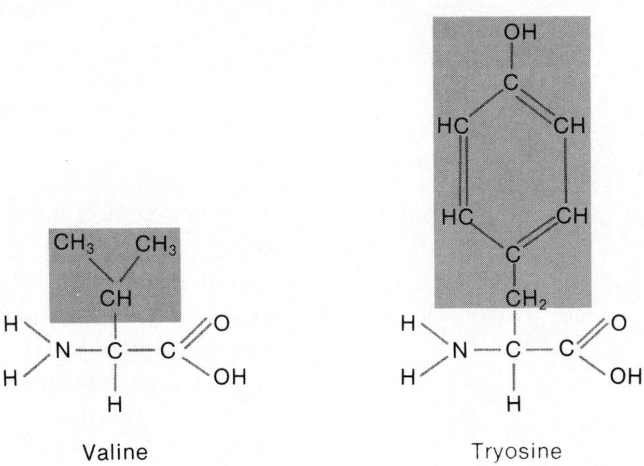

Polar amino acids

Basic Sulfur-containing Acidic

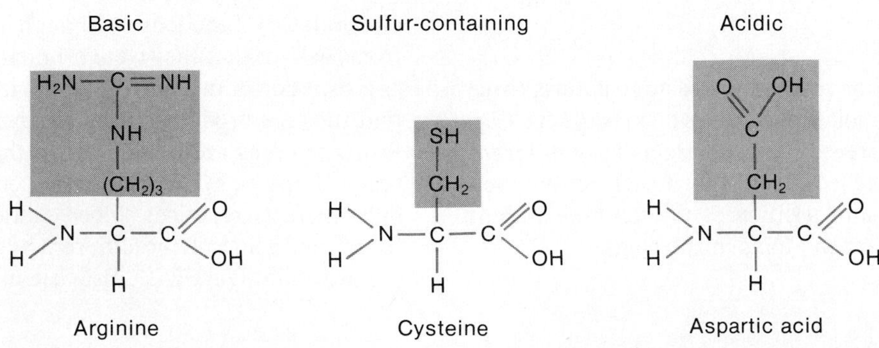

Objective 13. List some of the functions of different proteins, and explain the significance of the great diversity of protein structure.

Objective 14. Describe the structure of DNA and explain the principle of complementary base pairing.

Objective 15. Describe the structure of RNA, list the different types of RNA, and explain how RNA differs from DNA.

Proteins *(pro'te-in)* and nucleic acids are giant molecules, or **macromolecules.** The DNA contains the genetic code; RNA and protein provide the means by which this code can be expressed within the cells of the body. In order to understand these genetic control mechanisms (chapter 3), a knowledge of the structure of proteins and nucleic acids is required.

Proteins

Proteins consist of long chains of subunits called **amino acids** *(ah-me'no as'idz).* As the name implies, each amino acid contains an *amino group* (NH_2) on one end of the molecule and a carboxyl group (COOH) on the other end. There are approximately twenty different amino acids in the body, with different structures and chemical properties. These differences are due to differences in the *functional groups* of these amino acids, which is abbreviated as *R* in the general formula for an amino acid (fig. 2.22). The *R* symbol actually stands for the word *residue,* but it can be thought of as indicating the "rest of the molecule."

When amino acids are joined by dehydration synthesis, the hydrogen from the amino end of one amino acid combines with the hydroxyl group of the carboxylic acid end of another amino acid. As a covalent bond is formed between the two amino acids, water is produced (fig. 2.23). The bond between adjacent amino acids is called a **peptide bond,** and the compound formed is called a *peptide.* When many amino acids are joined in this way, a chain of amino acids, or **polypeptide,** is produced.

The lengths of polypeptide chains vary greatly. A hormone called *thyrotrophin-releasing hormone,* for example, is only three amino acids long, whereas *myosin (mi'o-sin)* (a muscle protein) contains about forty-five hundred amino acids. When the length of a polypeptide chain becomes very long (greater than about a hundred amino acids), the molecule is called a **protein.**

Protein Structure. The structure of a protein can be described at four different levels. At the first level, the sequence of amino acids in the protein, called the **primary structure** of the protein, is described. Each type of protein has a different primary structure. All of the billions of *copies* of a given type of protein in the body, however, have the same structure, because the structure of a given protein is coded by the genes. The primary structure of a protein is illustrated in figure 2.24.

Weak interactions (such as hydrogen bonds) between functional (*R*) groups of amino acids in nearby positions in the polypeptide chain cause this chain to twist into a *helix.* The extent and location of the helical structure is different for each protein because of differences in amino acid composition. A description of the helical structure of a protein is termed its **secondary structure** (fig. 2.24).

Most polypeptide chains bend and fold on themselves to produce complex, three-dimensional shapes, called the **tertiary structure** of the proteins. Each type of protein has its own characteristic tertiary structure. This is because the folding and bending of the polypeptide chain is produced by chemical interactions between particular amino acids that are located in different regions of the chain.

Figure 2.23. The formation of peptide bonds by dehydration synthesis reactions between amino acids.

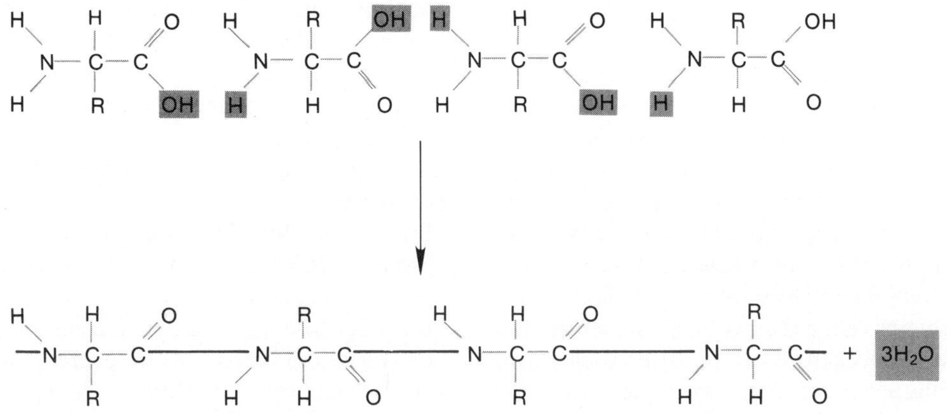

+ $3H_2O$

Figure 2.24. A polypeptide chain, showing (*a*) its primary structure and (*b*) secondary structure.

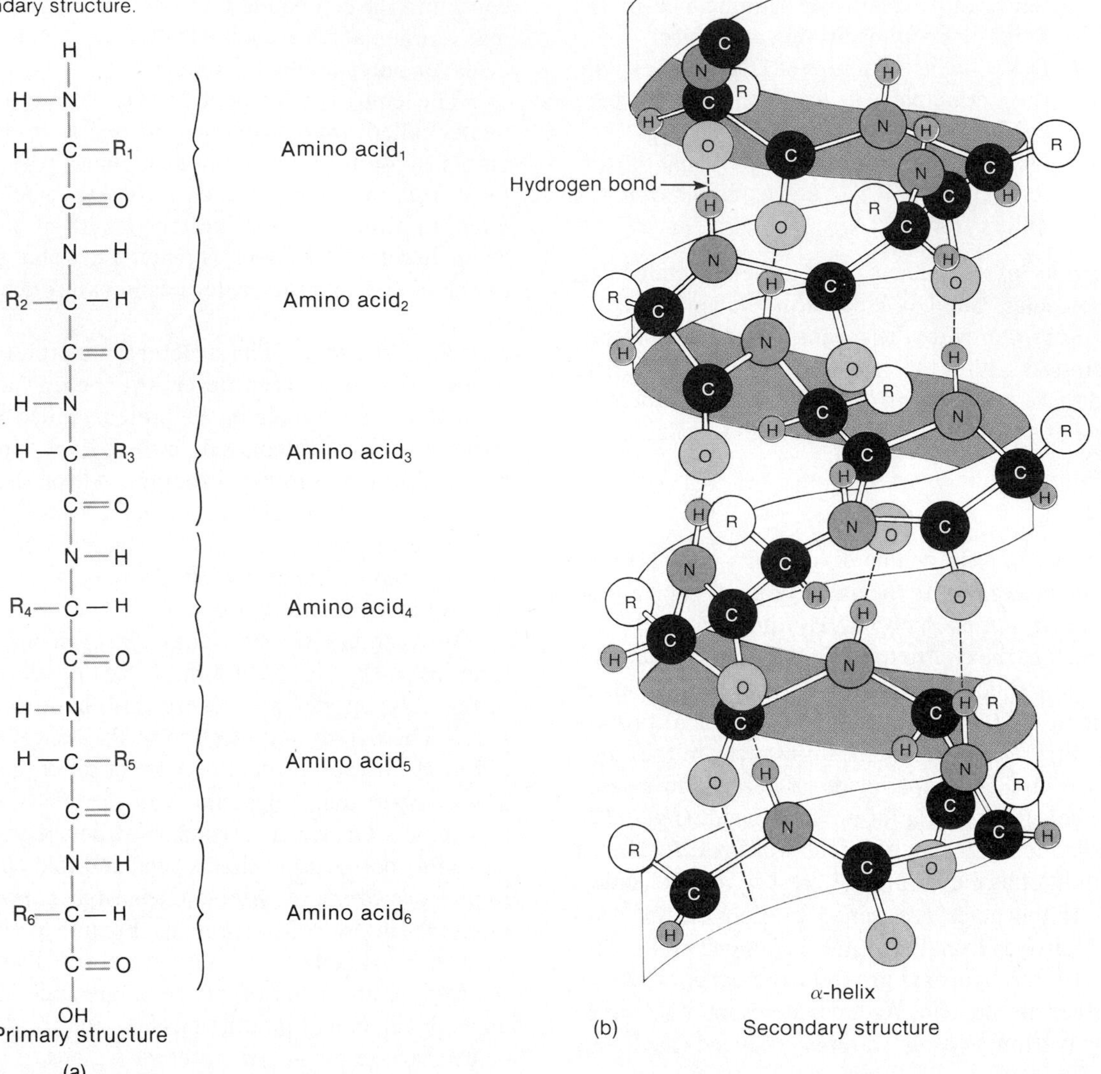

Most of the tertiary structure of proteins is formed and stabilized by weak chemical interactions (such as hydrogen bonds) between widely spaced amino acids. The tertiary structure of some proteins, however, is made more stable by covalent bonds between sulfur atoms (called *disulfide bonds* and abbreviated S-S) in the functional (*R*) groups of amino acids known as cysteines (fig. 2.25). These strong, covalent bonds are the exception. Since most of the tertiary structure is stabilized by weak bonds, this structure can easily be disrupted by high temperature or by changes in pH. Irreversible changes in the tertiary structure of proteins produced by this means are referred to as *denaturation (de''na-chu-ra'shun)* of the proteins.

Denatured proteins retain their primary structure (the peptide bonds are not broken) but have altered chemical properties. Cooking a pot roast, for example, alters the texture of the meat proteins—it doesn't result in an amino acid soup. Denaturation is most dramatically demonstrated by frying an egg. Egg-albumin proteins are soluble in their native state, in which they form the clear, viscous fluid of a raw egg. When denatured by cooking, these proteins change shape, cross-bond with each other, and by this means form an insoluble white precipitate—the egg white.

Some proteins (such as hemoglobin and insulin) are composed of a number of polypeptide chains covalently bonded together. This is the **quaternary structure** of these proteins. Insulin, for example, is composed of two polypeptide chains, one that is twenty-one amino acids long, the other that is thirty amino acids long. Hemoglobin (the protein in red blood cells that carries oxygen) is composed of four separate polypeptide chains. The composition of various body proteins is shown in table 2.4.

Figure 2.25. The tertiary structure of a protein. (*a*) interactions between functional (*R*) groups of amino acids result in (*b*) the formation of complex three-dimensional shapes of proteins.

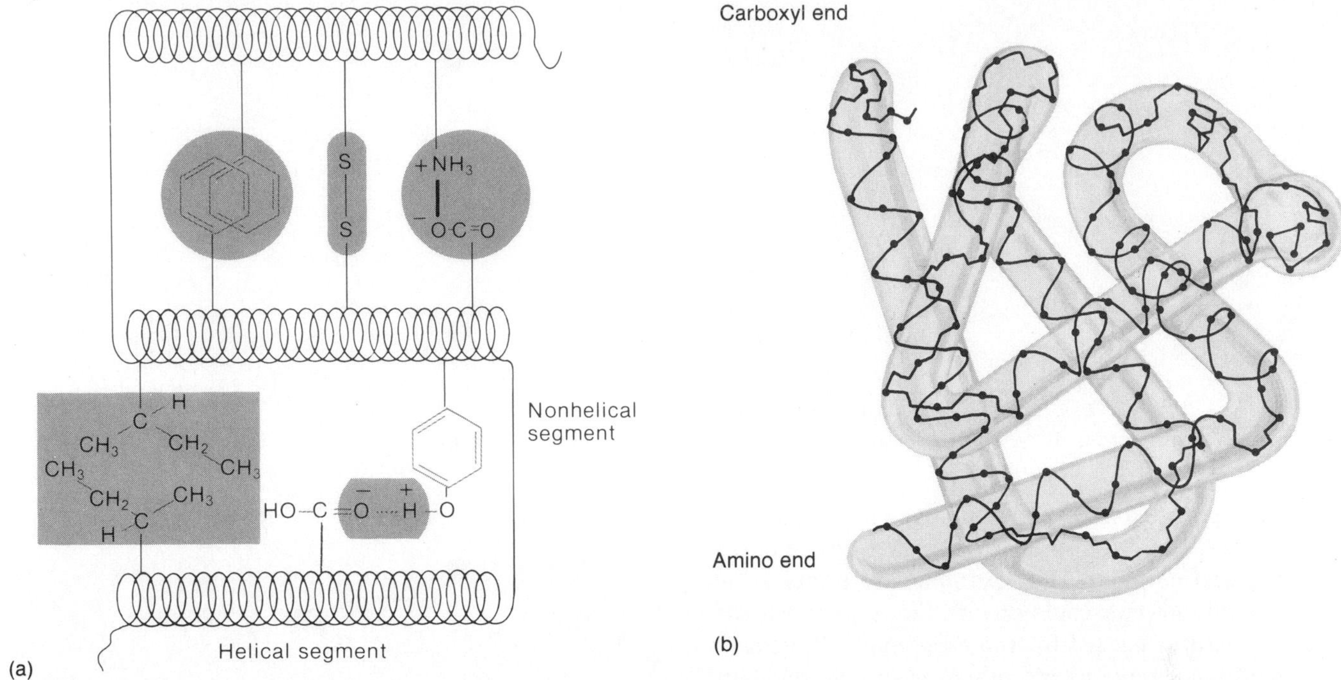

Table 2.4 Composition of selected proteins in the body			
Protein	**Number of polypeptide chains**	**Nonprotein component**	**Function**
Hemoglobin	4	Heme pigment	Carries oxygen in the blood
Myoglobin	2	Heme pigment	Stores oxygen in muscle
Insulin	2	None	Hormone-regulating metabolism
Luteinizing hormone	1	Carbohydrate	Hormone that stimulates gonads
Fibrinogen	1	Carbohydrate	Involved in blood clotting
Mucin	1	Carbohydrate	Forms mucus
Blood group proteins	1	Carbohydrate	Produces blood types
Lipoproteins	1	Lipids	Transports lipids in blood

Many proteins in the body are normally found combined, or *conjugated,* with other types of molecules. **Glycoproteins** are proteins conjugated with carbohydrates. Examples of such molecules include certain hormones and some proteins found in the cell membrane. **Lipoproteins** are proteins conjugated with lipids. These are found in cell membranes and in the plasma (the fluid portion of the blood). Proteins conjugated with pigment molecules are **chromoproteins.** These include hemoglobin, which transports oxygen in red blood cells, and the cytochromes, which are needed for oxygen utilization and energy production within cells.

Functions of Proteins. Because of their tremendous structural diversity, proteins can serve a wider variety of functions than any other type of molecule in the body. Many proteins, for example, contribute significantly to the structure of different tissues and in this way play a passive role in the functions of these tissues. Examples of such *structural proteins* include collagen (fig. 2.26) and keratin. **Collagen** is a fibrous protein that provides tensile strength to connective tissues, such as tendons and ligaments. **Keratin** is found in the outer layer of dead cells in the epidermis, where it serves to prevent water loss through the skin.

chromoprotein: Gk. *chroma,* color; *proteios,* of the first quality

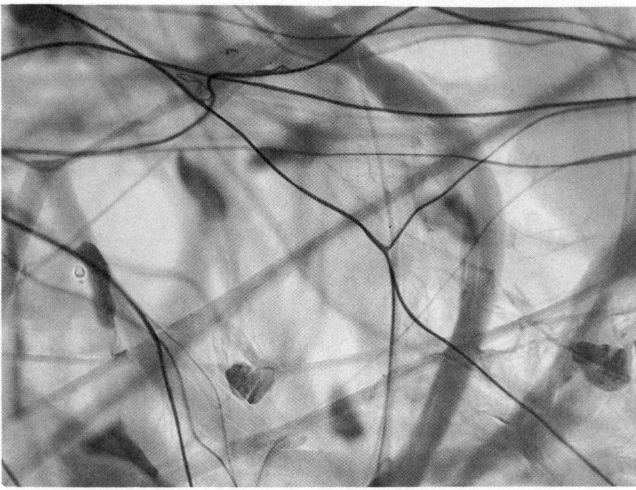

Figure 2.26. A photomicrograph of collagen fibers.

Figure 2.27. The general structure of a nucleotide and the formation of sugar-phosphate bonds between nucleotides to form a polymer.

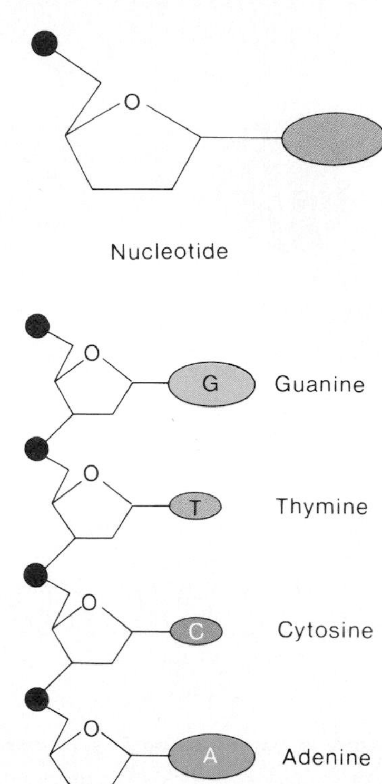

Many proteins serve a more active role in the body where specialized structure and function are required. *Enzymes* and *antibodies,* for example, are proteins—no other type of molecule could provide the vast array of different structures needed for these functions. Proteins in cell membranes serve as *receptors* for specific regulator molecules (such as hormones) and as *carriers* that transport specific molecules across the membrane. Proteins provide the diversity of shape and chemical properties for the specificity required by these functions.

Nucleic Acids

Nucleic acids include the macromolecules of **DNA** (**deoxyribonucleic acid**) and **RNA** (**ribonucleic acid**), which are critically important in genetic regulation, and the subunits from which these molecules are formed. These subunits are known as *nucleotides.*

Nucleotides are used as subunits in the formation of long polynucleotide chains. The nucleotides themselves, however, are composed of subunits. Each nucleotide is composed of three parts—a five-carbon sugar, a phosphate group bonded to one end of the sugar, and a *nitrogenous (ni-trahj'ĕ-nus) base* bonded to the other end of the sugar (fig. 2.27). The nucleotide bases are cyclic nitrogen-containing molecules with either one ring of carbons (the *pyrimidines*) *(pi-rim'i-dēnz)* or two rings (the *purines*).

Deoxyribonucleic Acid (DNA). The structure of DNA serves as the basis for the genetic code. One might, therefore, expect DNA to have an extremely complex structure. Actually, although DNA is the largest molecule in the cell, it has a simpler structure than that of most proteins. This simplicity of structure deceived some of the early scientists into believing that the protein content of chromosomes, rather than their DNA content, provided the basis for the genetic code.

Sugar molecules in the nucleotides of DNA are a type of pentose (five-carbon) sugar called **deoxyribose** *(de-ok''se-ri'bōs)* (hence the name for this nucleic acid). Each deoxyribose sugar can be covalently bonded to one of four possible bases. These bases include the two purines (adenine and guanine) and the two pyrimidines (cytosine and thymine). There are thus four different types of nucleotides that can be used to produce the long DNA chains.

When nucleotides combine to form a chain, the phosphate group of one condenses with the deoxyribose sugar of another nucleotide. This forms a sugar-phosphate chain as water is removed in dehydration synthesis. Since the nitrogenous bases are attached to the sugar molecules, the sugar-phosphate chain looks like a "backbone" from which the bases project. Each of these bases can form hydrogen bonds with other bases, which are in turn joined to a different chain of nucleotides. Such hydrogen bonding between bases thus produces a *double-stranded* DNA molecule; the two strands are like a staircase, with the paired bases as steps (fig. 2.28).

Actually, the two chains of DNA twist about each other to form a **double helix**—the molecule is like a spiral staircase (fig. 2.29). It has been shown that the number of purine bases in DNA is equal to the number of pyrimidine bases. The reason for this is explained by the law of **complementary base pairing;** adenine can pair only with

Figure 2.28. The four nitrogenous bases in deoxyribonucleic acid (DNA). Notice that hydrogen bonds can form between guanine and cytosine and between thymine and adenine.

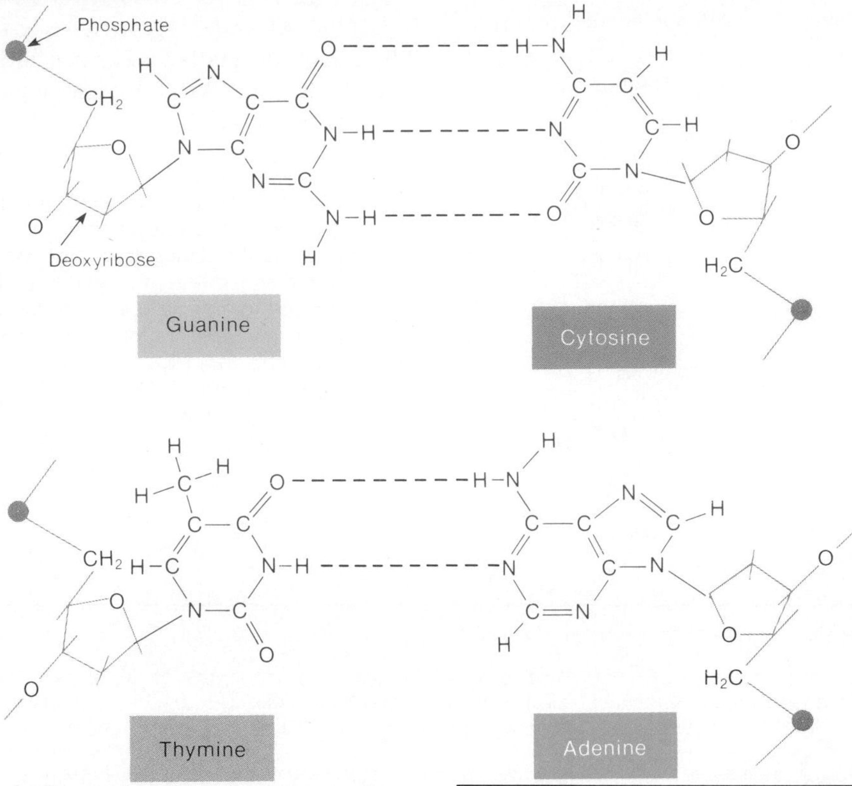

thymine (through two hydrogen bonds), whereas guanine can pair only with cytosine (through three hydrogen bonds). Knowing this rule, we could predict the base sequence of one DNA strand if we knew the sequence of bases in the complementary strand.

Although we can predict which base is opposite a given base in DNA, we cannot predict which bases are above or below that position within a single polynucleotide chain. Although there are only four bases, the number of possible base sequences along a stretch of several thousand nucleotides (the length of a gene) is almost infinite. Despite the almost infinite possible variety of sequences, almost all of the billions of copies of a particular gene in a person are identical.

Ribonucleic Acid (RNA). The genetic information contained in DNA functions to direct the activities of the cell through its production of another type of nucleic acid—*RNA (ribonucleic acid).* Like DNA, RNA consists of long chains of nucleotides joined together by sugar-phosphate bonds. Nucleotides in RNA, however, differ from those in DNA (fig. 2.30) in three ways: (1) a **ribonucleotide** *(ri''bo-nu'kle-o-tīd)* contains the sugar *ribose* (instead of deoxyribose); (2) the base *uracil (yûr'ah-sil)* is found in place of thymine; and (3) RNA is composed of a single polynucleotide strand (it is not double stranded like DNA).

Figure 2.29. The double helix structure of DNA.

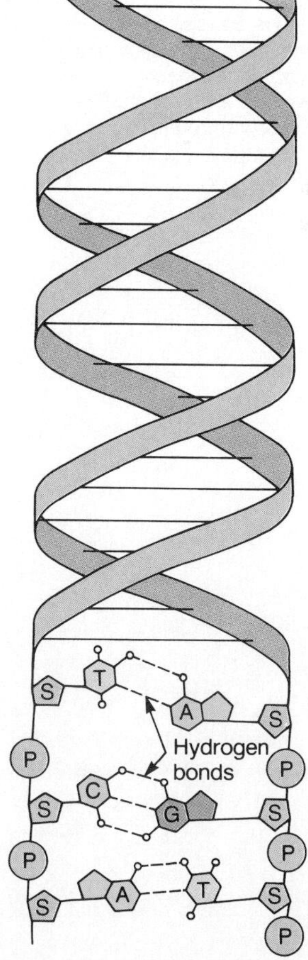

Figure 2.30. Differences between the nucleotides and sugars in DNA and RNA.

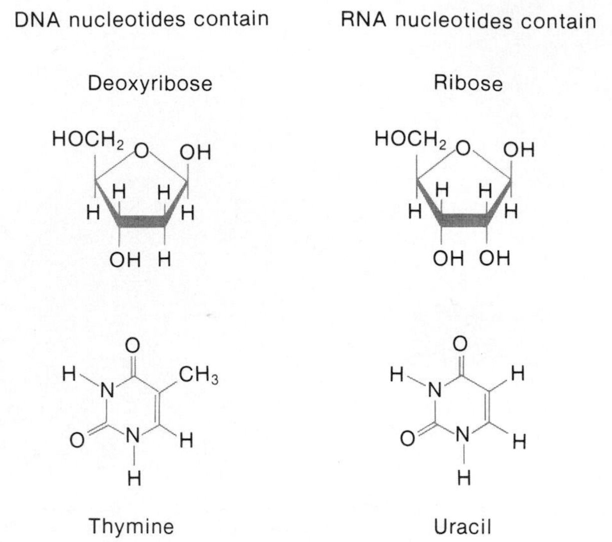

DNA nucleotides contain — RNA nucleotides contain

Deoxyribose — Ribose

Thymine — Uracil

There are three types of RNA molecules that function in the cytoplasm of cells. These are *messenger RNA (mRNA), transfer RNA (tRNA),* and *ribosomal RNA (rRNA)*. All three types are made within the cell nucleus by using information contained in DNA as a guide. The synthesis of RNA and its role in gene expression are discussed in chapter 3.

1. Write the general formula for an amino acid, and describe how amino acids differ from each other.
2. Describe the structure of proteins, list some of the functions of different proteins, and explain how this structure grants specificity to the actions of proteins.
3. Define the law of complementary base pairing, and explain how this law allows us to predict the sequence of bases on one DNA strand if we know the sequence of bases on the other strand.
4. Compare the structure of DNA with that of RNA.

Chapter Summary

I. Atoms, Ions, and Molecules
   A. Covalent bonds are formed by atoms that share electrons; these are the strongest type of chemical bond.
   B. Ionic bonds are formed by atoms that transfer electrons; these weak bonds join atoms together in an ionic compound.
   C. When hydrogen is bonded to an electronegative atom, it gains a slight positive charge and is weakly attracted to another electronegative atom; this weak attraction is a hydrogen bond.
   D. Acids donate hydrogen ions to solution, whereas bases lower the hydrogen ion concentration of a solution.
   E. Organic molecules contain atoms of carbon joined together by covalent bonds; atoms of nitrogen, oxygen, phosphorous, or sulfur may be present as specific functional groups in the organic molecule.

II. Carbohydrates and Lipids
   A. Carbohydrates contain carbon, hydrogen, and oxygen, usually in the ratio of 1:2:1.
   B. Lipids are organic molecules that are insoluble in polar solvents such as water.

III. Proteins and Nucleic Acids
   A. Proteins are composed of long chains of amino acids bonded together by covalent peptide bonds.
   B. Nucleic acids include the macromolecules DNA and RNA, and their nucleotide subunits.

Review Activities

Objective Questions

1. Which of the following statements about atoms is *true*?
   (a) They have more protons than electrons.
   (b) They have more electrons than protons.
   (c) They are electrically neutral.
   (d) They have as many neutrons as they have electrons.
2. The bond between oxygen and hydrogen in a water molecule is a(n)
   (a) hydrogen bond
   (b) polar covalent bond
   (c) nonpolar covalent bond
   (d) ionic bond
3. Which of the following is a nonpolar covalent bond?
   (a) the bond between two carbons
   (b) the bond between sodium and chloride
   (c) the bond between two water molecules
   (d) the bond between nitrogen and hydrogen.
4. Solution A has a pH of 2, and solution B has a pH of 10. Which of the following statements about these solutions is *true*?
   (a) Solution A has a higher H^+ concentration than solution B.
   (b) Solution B is basic.
   (c) Solution A is acidic.
   (d) All of these are true.
5. Glucose is a
   (a) disaccharide
   (b) polysaccharide
   (c) monosaccharide
   (d) phospholipid
6. Digestion reactions occur by means of
   (a) dehydration synthesis
   (b) hydrolysis
7. Carbohydrates are stored in the liver and muscles in the form of
   (a) glucose
   (b) triglycerides
   (c) glycogen
   (d) cholesterol
8. Lecithin is a
   (a) carbohydrate
   (b) protein
   (c) steroid
   (d) phospholipid
9. Which of the following lipids have regulatory roles in the body?
   (a) steroids
   (b) prostaglandins
   (c) triglycerides
   (d) both a and b
   (e) both b and c
10. The tertiary structure of a protein is *directly* determined by
    (a) the genes
    (b) the primary structure of the protein
    (c) enzymes that "mold" the shape of the protein
    (d) the position of peptide bonds
11. If four bases in one DNA strand are A (adenine), G (guanine), C (cytosine), and T (thymine), the complementary bases in the opposite strand will be
    (a) T,C,G,A
    (b) C,G,A,T
    (c) A,G,C,T
    (d) U,C,G,A
12. Which of the following statements about RNA is *true*?
    (a) It is made in the nucleus.
    (b) It contains the base uracil.
    (c) It is double stranded.
    (d) Both a and b are true.
    (e) Both b and c are true.

Essay Questions

1. Compare and contrast nonpolar covalent bonds, polar covalent bonds, and ionic bonds.
2. Give the definition of an acid and base, and explain how these influence the pH of a solution.
3. Using dehydration synthesis and hydrolysis reactions, explain the relationships between starch in an ingested potato, liver glycogen, and blood glucose.
4. "All fats are lipids, but not all lipids are fats." Explain why this statement is true.
5. Explain the relationship between the primary structure of a protein and its secondary and tertiary structures.

3 Cell Structure and Genetic Regulation

Outline

Concepts

The cell is the basic unit of structure and function in the body. The structure and function of cells is dependent upon the specific membranes and organelles characteristic of each type of cell.

Many of the functions of a cell that are carried out in the cytoplasmic compartment are performed by specific organelles. Among these are the microtubules and microfilaments of the cytoskeleton; the lysosomes, which contain digestive enzymes; and the mitochondria, where most of the cellular energy is produced.

The nucleus contains threads of chromatin, which is a combination of DNA and protein. The base sequence of DNA comprises the genetic code. Genes serve as templates for the production of RNA molecules, which can leave the nucleus through the large pores in the nuclear membrane.

After the genetic code is transcribed into the base sequence of messenger RNA, it is translated into the structure of proteins. Translation of the language of nucleotide bases into the language of amino acids requires transfer RNA, specific enzymes, and ribosomes. When the cells produce protein for secretion, additional organelles—the endoplasmic reticulum, Golgi apparatus, and secretory vesicles—are required.

When a cell is going to divide, each of the two strands of DNA acts as a template for the formation of a new, complementary strand. Organs grow and repair themselves through a type of cell division known as mitosis, in which the two daughter cells receive the same DNA as the parent cell. Gametes, which contain only half the number of chromosomes as the parent cell, are formed by meiosis.

Cell Membrane and Associated Structures

The cell is the basic unit of structure and function in the body. The structure and function of cells is dependent upon the specific membranes and organelles characteristic of each type of cell.

Objective 1. List the parts of a generalized cell.
Objective 2. Describe the structure of the cell membrane.
Objective 3. Differentiate between cilia and flagella.
Objective 4. Explain the processes of endocytosis and exocytosis.

As the basic functional unit of the body, each cell is a highly organized molecular factory. Cells come in a great variety of shapes and sizes (although most are too small to be clearly seen with the unaided eye). This great variation, which is also apparent in the subcellular structures within different cells, reflects the variation of functions of different cells in the body. All cells, however, share certain characteristics—such as the fact that they are surrounded by a cell membrane—and most cells possess the structures listed in table 3.1. Thus, although no single cell can be considered "typical," the general structure of cells can be shown with a single illustration (fig. 3.1).

For descriptive purposes, a cell can be divided into three principal parts.

1. **Cell (plasma) membrane.** The outer, differentially permeable cell membrane gives form to the cell and separates the cell's internal structures from the extracellular environment.
2. **Cytoplasm** and **organelles** *(or''gah-nelz')*. The cytoplasm is the aqueous content of a cell between the nucleus and the cell membrane. Organelles are minute structures within the cytoplasm of a cell that are concerned with specific functions.
3. **Nucleus.** The nucleus is a large, generally spheroid body within a cell that contains the DNA, or genetic material, of a cell.

Cell Membrane

Because both the intracellular and extracellular environments (or "compartments") are aqueous, a barrier must be present to prevent the loss of cellular molecules such as enzymes, nucleotides, and others that are water soluble. Since this barrier cannot itself be composed of water-soluble molecules, it makes sense that the cell membrane is composed of lipids.

aqueous: L. *aqua*, water

Table 3.1 Structure and function of cellular components

Component	Structure	Function
Cell (plasma) membrane	Membrane composed of phospholipid and protein molecules	Gives form to cell and controls passage of materials in and out of cell
Cytoplasm	Fluid, jellylike substance in which organelles are suspended	Serves as matrix substance in which chemical reactions occur
Endoplasmic reticulum	System of interconnected membrane-forming canals and tubules	Smooth endoplasmic reticulum metabolizes nonpolar compounds and stores Ca^{++} in striated muscle cells; rough endoplasmic reticulum assists in protein synthesis
Ribosomes	Granular particles composed of protein and RNA	Synthesize proteins
Golgi apparatus	Cluster of flattened, membranous sacs	Synthesizes carbohydrates and packages molecules for secretion; secretes lipids and glycoproteins
Mitochondria	Membranous sacs with folded inner partitions	Release energy from food molecules and transform energy into usable ATP
Lysosomes	Membranous sacs	Digest foreign molecules and worn and damaged cells
Peroxisomes	Spherical membranous vesicles	Contain certain enzymes; form hydrogen peroxide
Centrosome	Nonmembranous mass of two rodlike centrioles	Helps organize spindle fibers and distribute chromosomes during mitosis
Vacuoles	Membranous sacs	Store and excrete various substances within the cytoplasm
Fibrils and microtubules	Thin, hollow tubes	Support cytoplasm and transport materials within the cytoplasm
Cilia and flagella	Minute cytoplasmic extensions from cell	Move particles along surface of cell or move cell
Nuclear membrane	Membrane surrounding nucleus, composed of protein and lipid molecules	Supports nucleus and controls passage of materials between nucleus and cytoplasm
Nucleolus	Dense, nonmembranous mass composed of protein and RNA molecules	Forms ribosomes
Chromatin	Fibrous strands composed of protein and DNA molecules	Controls cellular activity for carrying on life processes

Figure 3.1. A generalized cell and the principal organelles.

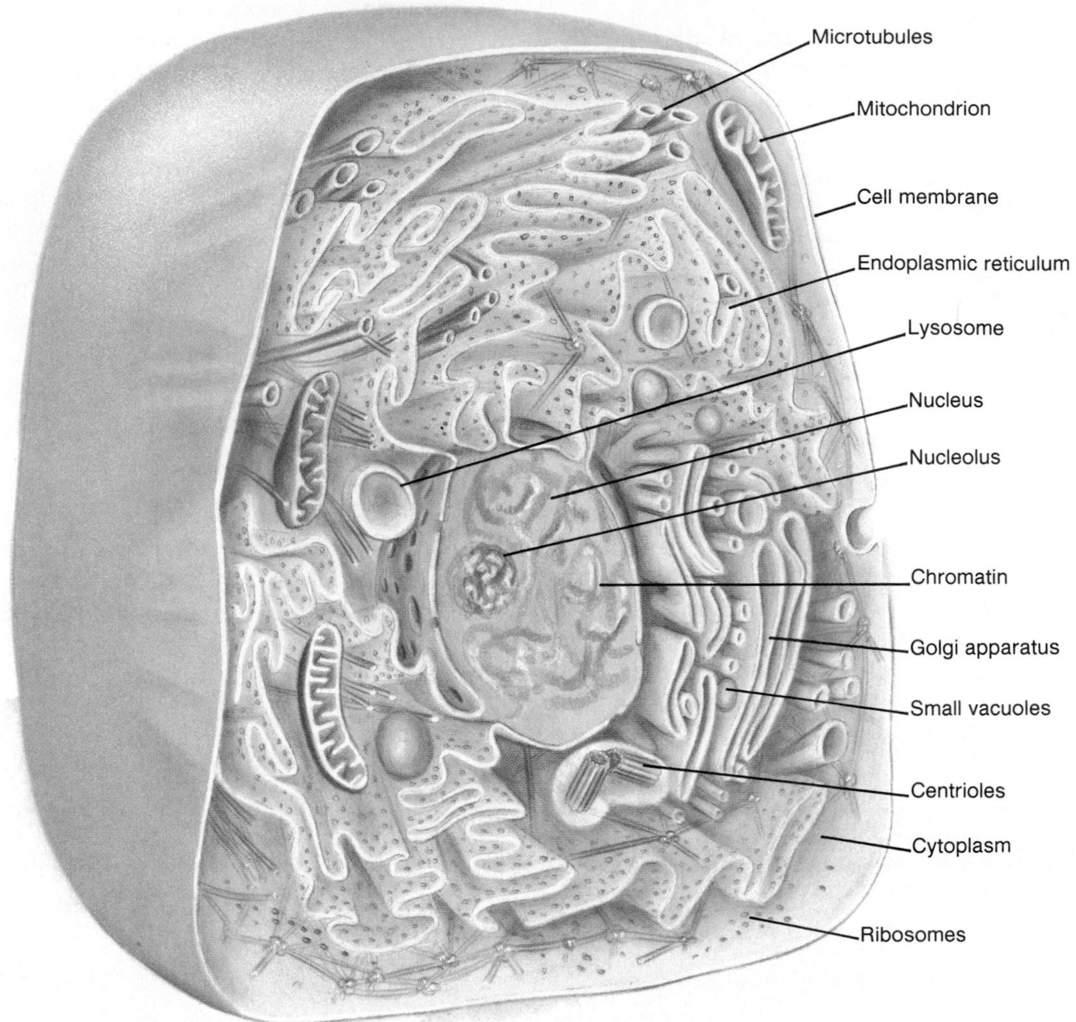

The **cell membrane** (also called the **plasma membrane,** or **plasmalemma**), and indeed all of the membranes surrounding organelles within the cell, are composed primarily of phospholipids and proteins. Phospholipids, as described in chapter 2, are polar on the end that contains the phosphate group and nonpolar (and hydrophobic) throughout the rest of the molecule. Since there is an aqueous environment on each side of the membrane, the hydrophobic parts of the molecules "huddle together" in the center of the membrane, leaving the polar ends exposed to water on both surfaces. This results in the formation of a double layer of phospholipids in the cell membrane.

The hydrophobic core of the membrane restricts the passage of water and water-soluble molecules and ions. Certain of these polar compounds, however, do pass through the membrane. The specialized functions and selective transport properties of the membrane are believed to be due to its protein content. Some proteins are found partially submerged on each side of the membrane; other proteins completely span the membrane from one side to the other. Since the membrane is not solid—phospholipids and proteins are free to move laterally—the proteins within the phospholipid "sea" are not uniformly distributed, but rather present a mosaic pattern. This structure is known as the **fluid-mosaic model** of membrane structure (fig. 3.2).

The proteins found in the cell membrane serve a variety of functions, including (1) structural support; (2) transport of molecules across the membrane; (3) enzymatic control of chemical reactions at the cell surface; (4) receptors for hormones and other regulatory molecules that arrive at the outer surface of the membrane; and (5) cellular "markers" (antigens), which identify the blood and tissue type.

Figure 3.2. The fluid-mosaic model of the cell membrane. The membrane consists of a double layer of phospholipids, with the phosphates (*shown by spheres*) oriented outward and the hydrophobic hydrocarbons (*wavy lines*) oriented toward the center. Proteins may completely or partially span the membrane. Carbohydrates are attached to the outer surface.

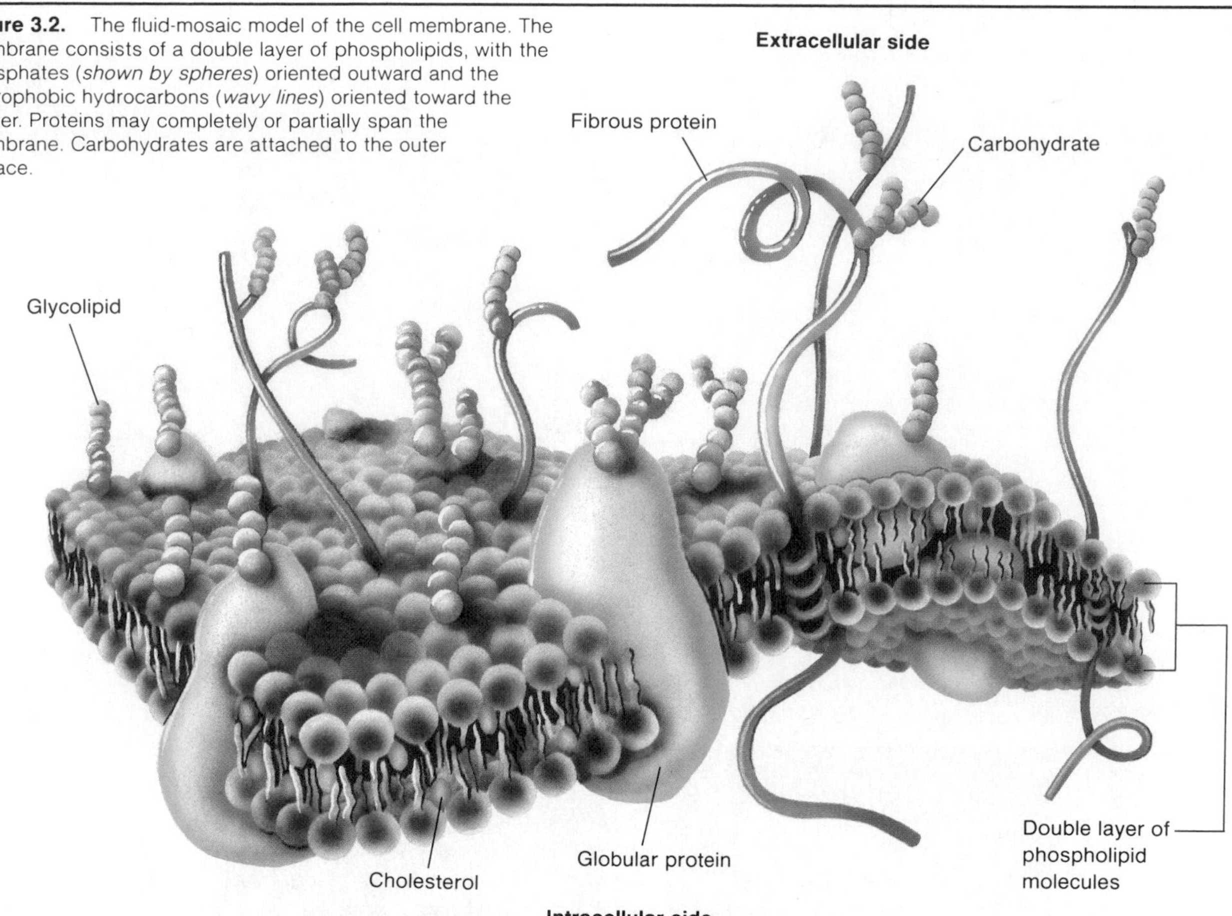

The cell membranes of all higher organisms contain cholesterol. The ratio of cholesterol to phospholipids, as well as the ratio of different types of phospholipids, determines the flexibility of the cell membrane. When there is an inherited defect in these ratios, the flexibility of the cell may be reduced. This could result, for example, in the inability of red blood cells to flex at the middle when passing through narrow blood channels (thereby causing the occlusion of these small vessels).

In addition to lipids and proteins, the cell membrane also contains carbohydrates, which are primarily attached to the outer surface of the membrane as glycoproteins and glycolipids. These surface carbohydrates have many negative charges and, as a result, affect the interaction of regulatory molecules with the membrane. The negative charges at the surface also affect interactions between cells—they help keep red blood cells apart, for example. Stripping the carbohydrates from the outer red blood cell surface results in their more rapid destruction by the liver, spleen, and bone marrow.

Cellular Movements

Some body cells—including certain white blood cells and macrophages in connective tissues and microglial cells in the brain—are able to move like an amoeba (a single-celled animal). This "amoeboid" movement is performed by the extension of parts of the cytoplasm to form *pseudopods (soo'do-pods)*, which attach to a substrate and pull the cell along.

Cilia and Flagella. **Cilia** *(sil'e-ah)* are tiny hairlike structures that protrude from the cell and, like the coordinated action of oarsmen in a boat, stroke in unison. Cilia in the human body are found on the apical surface (the surface facing the lumen, or cavity) of stationary epithelial cells in the respiratory and female genital tracts. In the respiratory system, the cilia transport strands of mucus, which are then conveyed by ciliary action to a region (the pharynx) where the mucus can either be swallowed or expectorated. In the female genital tract, ciliary movements in the epithelial lining draw the egg (ovum) into the uterine tube and move it toward the uterus.

Sperm are the only cells in the human body that have **flagella** *(fla-jel'ah)*. The flagellum is a single, whiplike structure that propels the sperm through its environment.

pseudopod: Gk. *pseudes,* false; *pod,* foot
cilia: L. *cili,* small hair
flagellum: L. *flagrum,* whip

Figure 3.3. Electron micrographs of cilia, showing (*a*) longitudinal and (*b*) cross sections. Notice the characteristic "9 + 2" arrangement of microtubules in the cross sections.

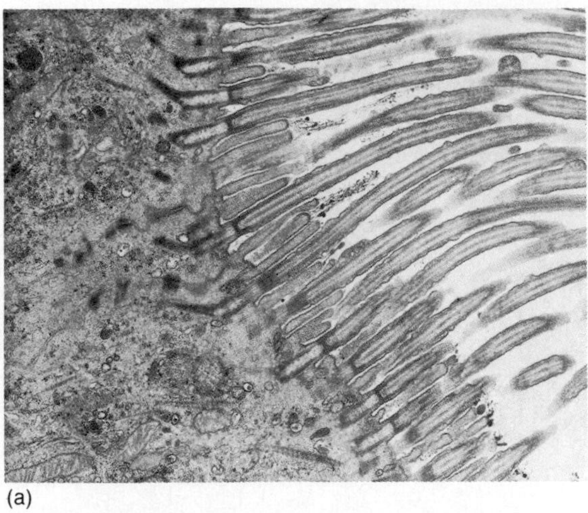

(a)

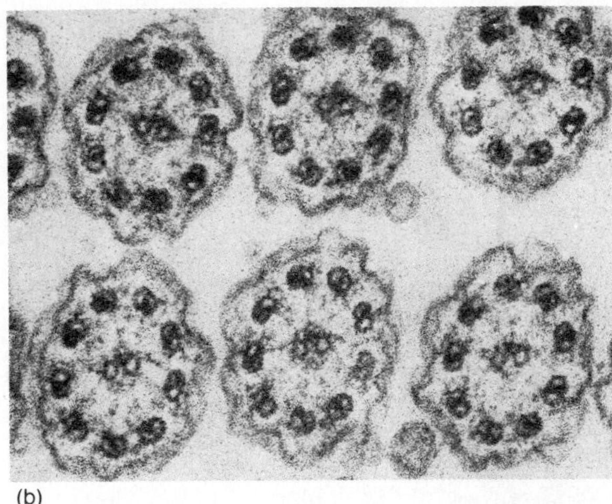

(b)

Figure 3.4. Scanning electron micrographs of phagocytosis, showing the formation of pseudopods and the entrapment of the prey within a food vacuole.

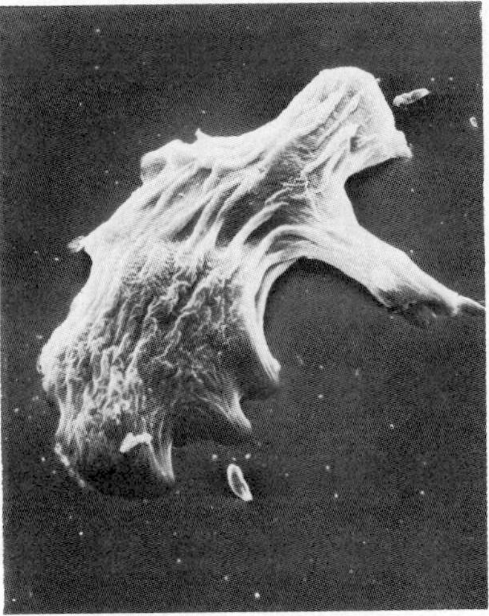

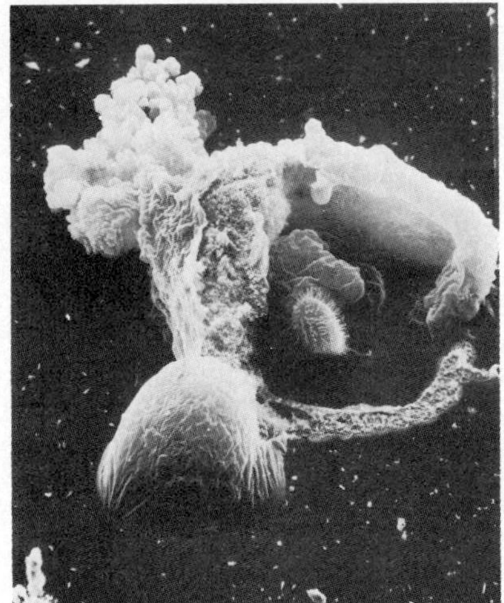

Both cilia and flagella are composed of *microtubules* (described in a later section) arranged in a characteristic way. Nine pairs of microtubules in the periphery of a cilium or a flagellum surround a single pair of microtubules in the center (fig. 3.3).

Endocytosis and Exocytosis

Regions of the cell membrane can invaginate (move inward to form a pouch) and pinch off to produce a membrane-enclosed body within the cytoplasm. This process removes regions of cell membrane as it brings part of the extracellular environment into the cell. The processes by which part of the extracellular environment is brought into a cell by invagination of the cell membrane are called **endocytosis.** There are three types of endocytosis: phagocytosis, pinocytosis, and receptor-mediated endocytosis.

endocytosis: Gk. *endo,* within; *kytos,* hollow body

Phagocytosis and Pinocytosis. Cells that move by amoeboid motion (such as white blood cells)—as well as liver cells, which are not capable of amoeboid movement—use pseudopods to surround and engulf particles of organic matter (such as bacteria). This process is a type of cellular "eating" called **phagocytosis** *(fag''o-si-to'sis),* which serves to protect the body from invading microorganisms and to remove extracellular debris.

Phagocytic cells surround their victim with pseudopods, which join together and fuse (fig. 3.4). After the inner membrane of the pseudopods forms a continuous membranous barrier around the ingested particle, it pinches off from the cell membrane so that the ingested particle is contained in a *food vacuole* within the cell. The

phagocytosis: Gk. *phagein,* to eat; *kytos,* hollow body
vacuole: L. *vacuus,* empty

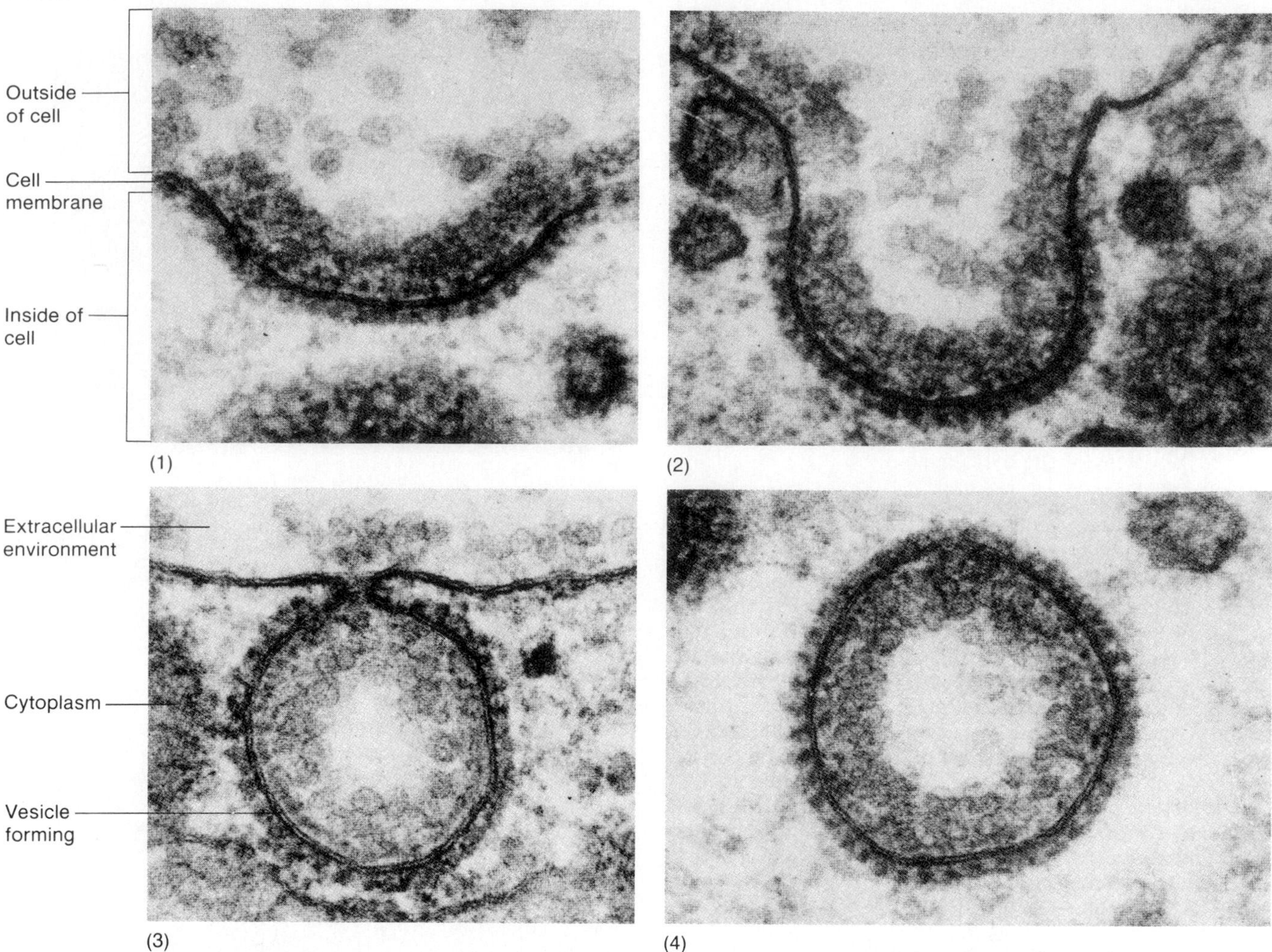

Figure 3.5. Stages (1–4) of endocytosis, in which specific bonding of extracellular particles to membrane receptor proteins is believed to occur.

ingested particle will subsequently be digested by enzymes contained in a different organelle (the lysosome).

Pinocytosis *(pin''o-si-to'sis)* is a related process performed by many cells. Instead of forming pseudopods, the cell membrane invaginates to produce a deep, narrow furrow. The membrane near the surface of this furrow then fuses, and a small vacuole containing extracellular fluid is pinched off and enters the cell.

Receptor-Mediated Endocytosis. This type of endocytosis involves the smallest area of cell membrane, and it occurs only in response to specific molecules in the extracellular environment. In receptor-mediated endocytosis, the interaction of very specific molecules in the extracellular fluid with specific membrane receptor proteins causes the membrane to invaginate, fuse, and pinch off to form a *vesicle*—a small vacuole (fig. 3.5). Vesicles formed in this way contain extracellular fluid and molecules that could not have passed by other means into the cell. Cholesterol attached to specific proteins, for example, is taken up into artery cells by receptor-mediated endocytosis. (This is in part responsible for atherosclerosis, as described in chapter 20.)

pinocytosis: Gk. *pinein,* to drink; *kytos,* hollow body

Exocytosis. Proteins and other molecules produced within the cell that are destined for export (secretion) are packaged inside of the cell within vesicles by an organelle known as the **Golgi apparatus.** In the process of **exocytosis,** these secretory vesicles fuse with the cell membrane and release their contents into the extracellular environment (see fig. 3.19). This process adds new membrane material, which replaces that which was lost from the cell membrane during endocytosis.

Endocytosis and exocytosis account for only part of the two-way traffic between the intracellular and extracellular compartments. Most of this traffic is due to membrane transport processes, the movement of molecules and ions through the cell membrane (chapter 5).

exocytosis: Gk. *exo,* outside; *kytos,* hollow body

1. Draw the fluid-mosaic model of the cell membrane, and describe the structure of the membrane.
2. Describe the structure of cilia and flagella.
3. Draw a figure showing phagocytosis; explain the events that occur during receptor-mediated endocytosis.
4. Explain the significance of exocytosis.

Cytoplasm and Its Organelles

Many of the functions of a cell that are carried out in the cytoplasmic compartment are performed by specific organelles. Among these are the microtubules and microfilaments of the cytoskeleton; the lysosomes, which contain digestive enzymes; and the mitochondria, where most of the cellular energy is produced.

Objective 5. Describe the structure and some of the functions of the cytoskeleton.

Objective 6. Describe the structure and explain the functional significance of lysosomes.

Objective 7. Describe the structure and function of mitochondria, and explain what is meant by "mitochondrial inheritance."

Objective 8. Describe the structure and functions of the endoplasmic reticulum.

Cytoplasm and Cytoskeleton

The jellylike matrix within a cell (exclusive of that within the nucleus) is known as **cytoplasm** *(si'to-plazm).* When viewed in a microscope without special techniques, the cytoplasm appears to be uniform and unstructured. According to recent evidence, however, the cytoplasm is not a homogenous solution; it is, rather, a highly organized structure in which protein fibers—in the form of *microtubules* and *microfilaments*—are arranged in a complex latticework. These can be seen by fluorescence microscopy with the aid of antibodies against the proteins that comprise these structures (fig. 3.6). The interconnected microfilaments and microtubules are believed to provide structural organization for cytoplasmic enzymes and support for various organelles.

The latticework of microfilaments and microtubules is thus said to function as a **cytoskeleton** (fig. 3.7). The structure of this "skeleton" is not rigid; it has been shown to be capable of quite rapid reorganization. Contractile proteins—including actin and myosin, which are responsible for muscle contraction—may be able to shorten the length of some microfilaments. The cytoskeleton may thus represent the cellular "musculature." Microtubules, for example, form the *spindle apparatus* that pulls chromosomes away from each other in cell division; they also form the central parts of cilia and flagella. Recent evidence suggests that the rapid movement of organelles within nerve fibers is also dependent upon microtubules, which can move different organelles in opposite directions at the same time.

Figure 3.6. An immunofluorescence photograph of microtubules forming the cytoskeleton of a cell. Microtubules are visualized with the aid of antibodies against tubulin, the major protein component of the microtubules.

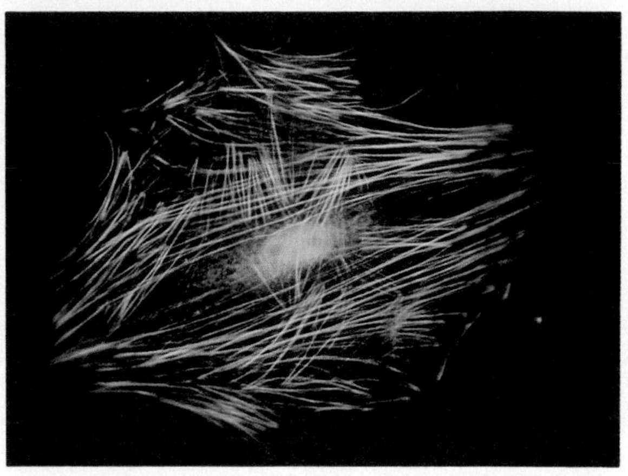

Figure 3.7. A diagram of a proposed structure of the cytoskeleton. (From "The Ground Substance of the Living Cell," by Keith Porter and Jonathon Tucker. Copyright © 1981 by Scientific American, Inc. All rights reserved.)

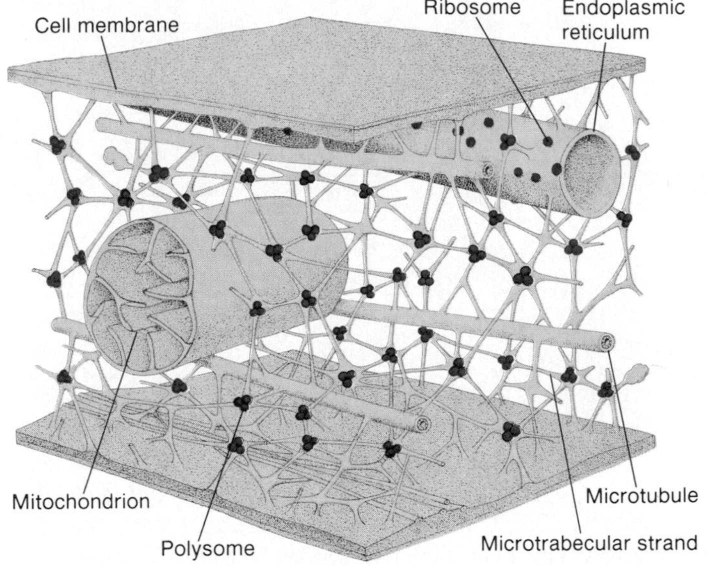

Lysosomes

After a phagocytic cell has engulfed the proteins, polysaccharides, and lipids present in a particle of "food" (such as a bacterium), these molecules are still kept isolated from the cytoplasm by the membranes surrounding the food vacuole. The large molecules of proteins, polysaccharides, and lipids must first be digested into their smaller subunits (amino acids, monosaccharides, and so on) before they can cross the vacuole membrane and enter the cytoplasm.

Figure 3.8. Electron micrograph showing primary lysosomes (Lys_1) and secondary lysosomes (Lys_2). Mitochondria (Mi), Golgi apparatus (GA), and the nuclear envelope (NE) are also seen.

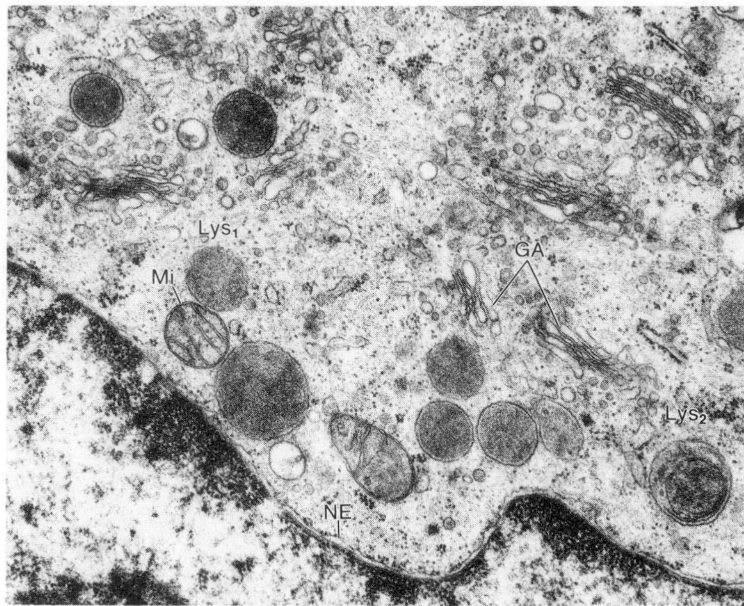

The digestive enzymes of a cell are isolated from the cytoplasm and concentrated within membrane-bound organelles called **lysosomes** (fig. 3.8). A *primary lysosome (li'so-sōm)* may fuse with a food vacuole (or with another cellular organelle) to form a *secondary lysosome.* In this way the products of phagocytosis and worn-out organelles can be digested. A lysosome that contains partially digested remnants of other organelles is called a *secondary lysosome.* A lysosome that contains undigested wastes is called a *residual body.* Residual bodies may eliminate their wastes by exocytosis, or the wastes may accumulate within the cell as the cell ages.

Partly digested membranes of various organelles and other cellular debris are often observed within secondary lysosomes. This is a result of *autophagy,* a process that destroys worn-out organelles so that they can be continuously replaced. Lysosomes are thus aptly referred to as the "digestive system" of the cell.

Lysosomes have also been called "suicide bags," because a break in their membranes would release their digestive enzymes and thus destroy the cell. This happens normally as part of *programmed cell death,* in which the destruction of tissues is part of embryological development. It also occurs in white blood cells during an inflammation reaction.

autophagy: Gk. *autos,* self; *phagein,* to eat

Most, if not all, molecules in the cell have a limited life span. They are continuously destroyed and must be continuously replaced. Glycogen and some complex lipids in the brain, for example, are digested normally at a particular rate by lysosomes. If a person, because of some genetic defect, does not have the proper amount of these lysosomal enzymes, the resulting abnormal accumulation of glycogen and lipids could destroy the tissues.

Mitochondria

All cells in the body, with the exception of mature red blood cells, have a hundred to a few thousand organelles called **mitochondria** *(mi''to-kon'dre-ah).* Mitochondria serve as sites for the production of most of the cellular energy (chapter 4). For this reason, mitochondria are sometimes called the "powerhouses" of the cell.

Mitochondria vary in size and shape, but all have the same basic structure (fig. 3.9). Each is surrounded by an *outer membrane* that is separated by a narrow space from an *inner membrane.* The inner membrane has many folds, called *cristae,* that extend into the central area (or *matrix*) of the mitochondrion. The cristae and the matrix provide different compartments in the mitochondrion and have different roles in the generation of cellular energy.

mitochondrion: Gk. *mitos,* a thread; *chondros,* grain

Figure 3.9. A mitochondrion (*a*). The outer membrane and the infoldings of the inner membrane—the cristae—are clearly seen. The fluid in the center is the matrix. The structure of a mitochondrion is illustrated in (*b*).

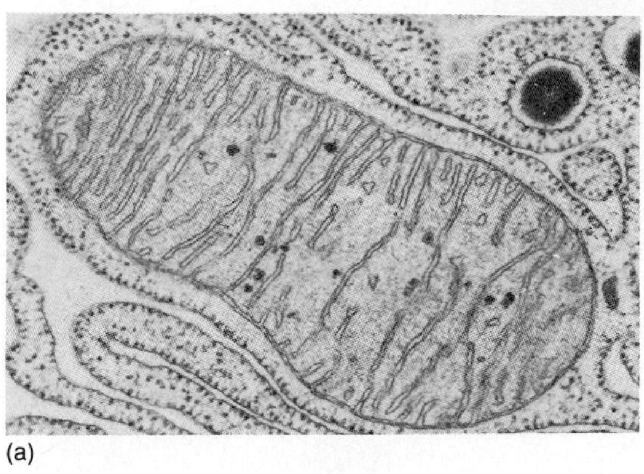

(a)

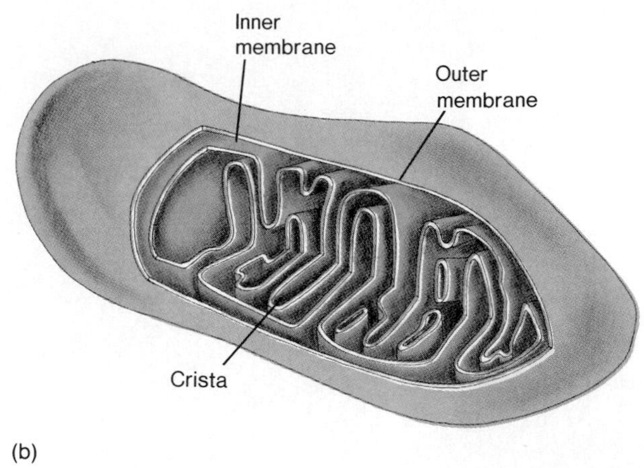

(b)

Figure 3.10. (*a*) an electron micrograph of endoplasmic reticulum magnified about 100,000 times. (*b*) rough endoplasmic reticulum has ribosomes attached to its surface, whereas (*c*) smooth endoplasmic reticulum lacks ribosomes.

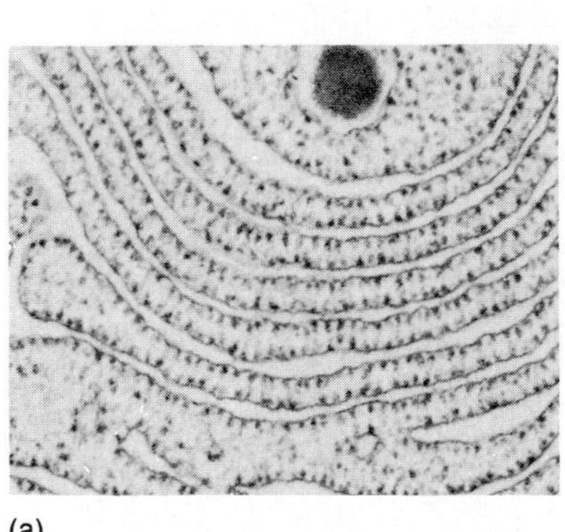

(a)

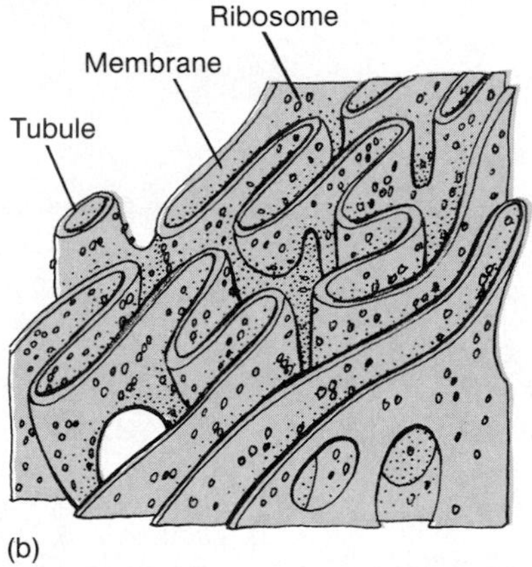

(b)

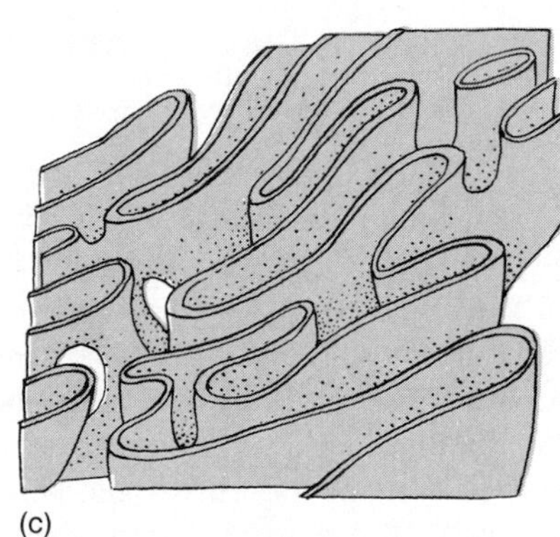

(c)

An ovum (egg cell) contains mitochondria; the head of a sperm contains none. Therefore, all of the mitochondria in a fertilized egg are derived from the mother. The mitochondrial DNA replicates itself and the mitochondria divide, so that all of the mitochondria in the fertilized ovum (zygote) and the cells derived from the fertilized ovum during embryonic and fetal development are genetically identical to those in the original ovum. This provides a unique form of inheritance that is passed only from mother to child. A rare cause of blindness—*Leber's hereditary optic neuropathy*—and perhaps some genetically based neuromuscular disorders, are believed to be inherited in this manner.

Mitochondria are able to migrate through the cytoplasm of a cell, and it is believed that they are able to reproduce themselves. Indeed, mitochondria contain their own DNA! This is a more primitive form of DNA than that found within the cell nucleus. For these and other reasons, many scientists believe that mitochondria evolved from separate organisms, related to bacteria, which entered animal cells and remained in a state of symbiosis.

Endoplasmic Reticulum

Most cells contain a system of membranes known as the endoplasmic reticulum *(en''do-plas'mic rĕ-tik'u-lum),* of which there are two types: (1) a **rough,** or **granular, endoplasmic reticulum;** and (2) a **smooth endoplasmic reticulum** (fig. 3.10). A rough endoplasmic reticulum contains ribosomes—small structures of RNA and protein, described in a later section—on its surface, and a smooth endoplasmic reticulum does not contain ribosomes.

The smooth endoplasmic reticulum is used for a variety of purposes in different cells. It serves as a site for enzyme reactions in steroid hormone production and inactivation, for example, and is the part of the cell where many drugs are inactivated. The smooth endoplasmic reticulum also serves to store Ca^{++} in skeletal muscle cells.

Leber's hereditary optic neuropathy: from Theodor Leber, German ophthalmologist, 1840–1917

symbiosis: Gk. *syn,* with; *bios,* life

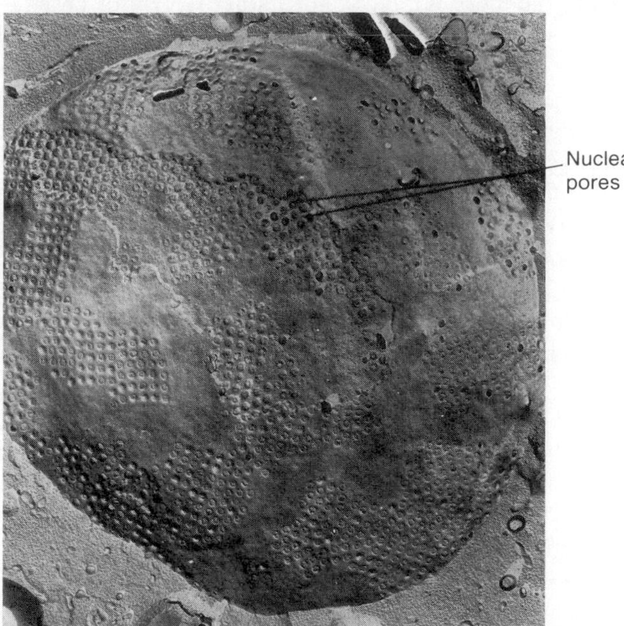

Figure 3.11. An electron micrograph of a freeze-fractured nuclear envelope, showing the nuclear pores.

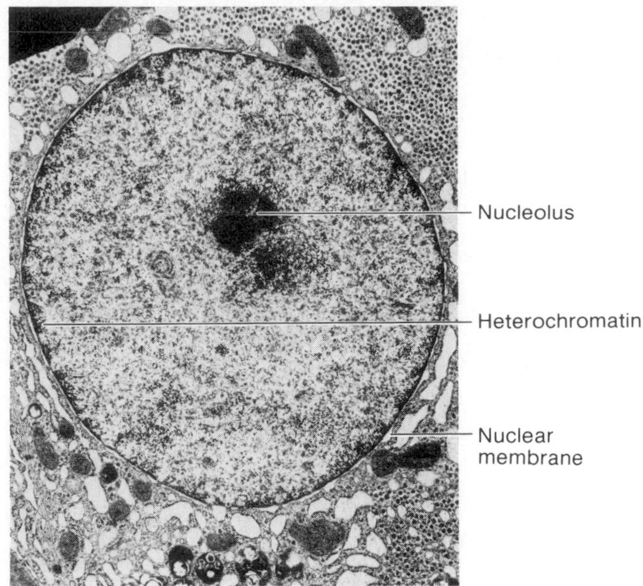

Figure 3.12. The nucleus of a liver cell, showing the nuclear membrane, heterochromatin, and nucleolus.

The rough endoplasmic reticulum is found in cells that are active secretors of proteins, such as those of exocrine and endocrine glands.

Details of the structure and function of the rough endoplasmic reticulum and its associated ribosomes, and of another organelle called the Golgi apparatus, will be described in the section on protein synthesis. The structure of centrioles and the spindle apparatus, which are organelles involved in DNA replication and cell division, also are described in a later section.

1. Explain why microtubules and microfilaments can be thought of as the skeleton and musculature of a cell.
2. Describe the contents of lysosomes and explain the significance of autophagy.
3. Describe the structure and function of mitochondria.
4. Explain how mitochondria can provide a genetic inheritance that is derived from only the mother.
5. Distinguish between the structure and function of a rough versus a smooth endoplasmic reticulum.

Cell Nucleus

The nucleus contains threads of chromatin, which is a combination of DNA and protein. The base sequence of DNA comprises the genetic code. Genes serve as templates for the production of RNA molecules, which can leave the nucleus through the large pores in the nuclear membrane.

Objective 9. Describe the structure of the nuclear membrane and the chromatin.

Objective 10. Explain how DNA directs the synthesis of RNA, and explain why this process is called genetic transcription.

Objective 11. List the different types of RNA, and describe their functions.

Most cells in the body have a single nucleus, although some—such as skeletal muscle cells—are multinucleate. The nucleus is surrounded by a *nuclear envelope* composed of an inner and an outer membrane. These two membranes fuse together to form thin sacs with openings called *nuclear pores* (figs. 3.11, 3.12). These pores allow RNA to exit the nucleus (where it is formed) and enter the cytoplasm but prevent DNA from leaving the nucleus.

Chromatin

Many granulated threads in the nuclear fluid can be seen with an electron microscope. These threads are called **chromatin** *(kro'mah-tin)* and consist of a combination of DNA and protein. There are two forms of chromatin. Thin, extended chromatin—or *euchromatin (u-kro'mah-tin)*—appears to be the active form of DNA in a nondividing cell. Regions of condensed, "blotchy"-appearing chromatin, known as *heterochromatin,* are believed to contain inactive DNA.

One or more dark areas within each nucleus can be seen. These regions, which are not surrounded by membranes, are called **nucleoli** *(nu-kle'o-li).* The DNA within nucleoli contains genes that code for the production of ribosomal RNA (rRNA), an essential component of ribosomes.

chromatin: Gk. *chroma,* color

Figure 3.13. The synthesis of ribosomal RNA on DNA in the nucleolus.

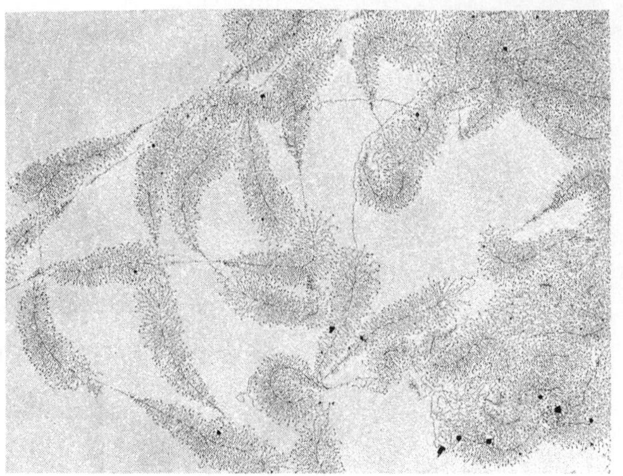

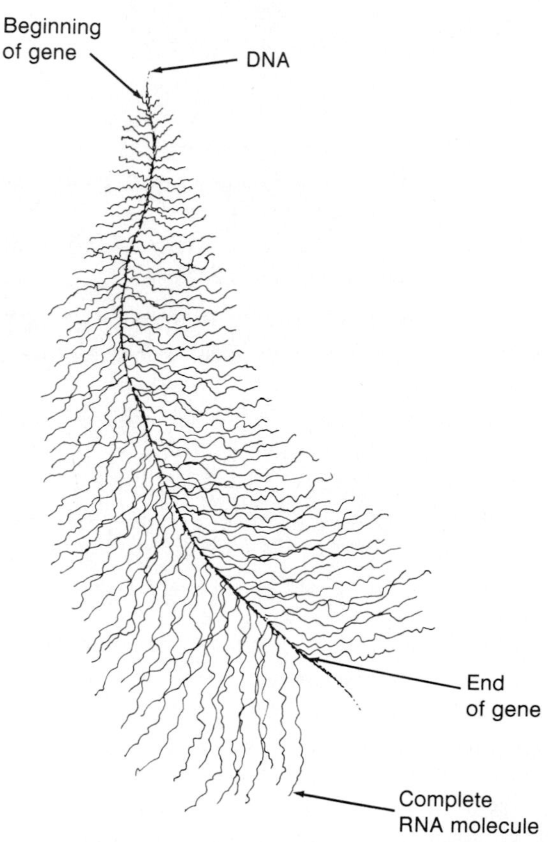

Genetic Transcription—RNA Synthesis

The thin, extended euchromatin is the "working" form of DNA; the more familiar short, stubby form of chromosomes seen during cell division are inactive packages of DNA. The genes do not become active until the chromosomes unravel. Active DNA directs the metabolism of the cell indirectly through its regulation of RNA and protein synthesis.

One gene codes for one polypeptide chain. Each gene is a stretch of DNA that is several thousand nucleotide pairs long. The DNA in a human cell contains three to four billion base pairs—enough to code for at least three million proteins. Since the average human cell contains less than this amount (30,000 to 150,000 different proteins), it follows that only a fraction of the DNA in each cell is used to code for proteins. The remainder of the DNA may be inactive, may be redundant, and may serve to regulate those regions that do code for proteins.

In order for the genetic code to be translated into the synthesis of specific proteins, the DNA code must first be transcribed into an RNA code (fig. 3.13). This is accomplished by DNA-directed RNA synthesis, or **genetic transcription.**

In RNA synthesis, the enzyme *RNA polymerase* breaks the weak hydrogen bonds between paired DNA bases. This does not occur throughout the length of DNA, but only in the regions that are to be transcribed (there are base sequences that code for "start" and "stop"). Double-stranded DNA, therefore, separates in these regions so that the freed bases can pair with the complementary RNA nucleotide bases.

This pairing of bases, like that which occurs in DNA replication (described in a later section), follows the law of complementary base pairing: guanine bonds with cytosine (and vice versa), and adenine bonds with uracil (because uracil in RNA is equivalent to thymine in DNA). Unlike DNA replication, however, only *one* of the two freed strands of DNA serves as a guide for RNA synthesis.

Types of RNA. Four types of RNA are produced within the nucleus by genetic transcription: (1) **precursor messenger RNA (pre-mRNA),** which is altered within the nucleus to form mRNA; (2) **messenger RNA (mRNA),** which contains the code for the synthesis of specific proteins; (3) **transfer RNA (tRNA),** which is needed for decoding the genetic message contained in mRNA; and (4) **ribosomal RNA (rRNA),** which forms part of the structure of ribosomes. The DNA that codes for rRNA synthesis is located in the part of the nucleus called the nucleolus. The DNA that codes for pre-mRNA and tRNA synthesis is located elsewhere in the nucleus.

Figure 3.14. An electron micrograph of polyribosomes. An RNA strand (*arrow*) joins the ribosomes together.

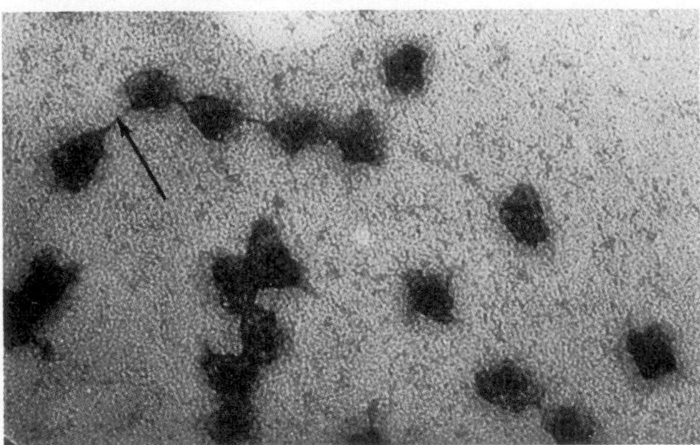

In bacteria, where the pioneering studies of gene function were performed, each gene codes for a finished mRNA that is capable of coding for the synthesis of a protein. In the cells of higher organisms, however, a pre-mRNA is produced within the nucleus that must first be modified before it can enter the cytoplasm as mRNA and direct protein synthesis.

Precursor mRNA is much larger than mRNA. This large size is not due to excess bases on the ends of the molecule that must be trimmed to make mRNA. Rather, the excess bases are *between* regions of bases that will be part of the finished mRNA. Pre-mRNA is converted to mRNA by removal of the stretches of excess bases—called *introns*—which split up the bases that contribute to the mRNA code. After such cutting and splicing, the mRNA is further modified by the addition of characteristic bases at each end. These *post-transcriptional modifications* produce the form of mRNA that enters the cytoplasm.

1. Explain the difference between euchromatin and heterochromatin.
2. Explain how a region of DNA is used to produce precursor mRNA, and how this is modified to produce mRNA.
3. Explain the significance of RNA synthesis in terms of genetic expression.

Protein Synthesis and Secretion

After the genetic code is transcribed into the base sequence of messenger RNA, it is translated into the structure of proteins. Translation of the language of nucleotide bases into the language of amino acids requires transfer RNA, specific enzymes, and ribosomes. When the cells produce proteins for secretion, additional organelles—the endoplasmic reticulum, Golgi apparatus, and secretory vesicles—are required.

Table 3.2 Selected DNA base triplets and mRNA codons

DNA triplet	RNA codon	Amino acid
TAC	AUG	"Start"
ATC	UAG	"Stop"
AAA	UUU	Phenylalanine
AGG	UCC	Serine
ACA	UGU	Cysteine
GGG	CCC	Proline
GAA	CUU	Leucine
GCT	CGA	Arginine
TTT	AAA	Lysine
TGC	ACG	Tyrosine
CCG	GGC	Glycine
CTC	GAG	Aspartic acid

From Stuart Ira Fox, *Human Physiology,* 2d ed.

Objective 12. Describe the functions of mRNA, tRNA, and rRNA.

Objective 13. Explain how codons and anticodons function in protein synthesis.

Objective 14. Describe the functions of the rough endoplasmic reticulum and the Golgi apparatus, and explain how secretory proteins are modified and packaged by these organelles.

When mRNA enters the cytoplasm, it attaches to ribosomes, which are seen in the electron microscope as numerous small particles. **Ribosomes** *(ri'bo-sōm)* are composed of rRNA and protein, arranged to form two subunits of unequal size. The mRNA passes through a number of ribosomes to form a "string of pearls" structure called a *polyribosome* (or *polysome,* for short), shown in figure 3.14. Association of mRNA with ribosomes is needed for **genetic translation**—the production of specific proteins according to the code contained in the mRNA base sequence.

Each mRNA molecule contains several hundred or more nucleotides, arranged in the sequence determined by complementary base pairing with DNA during genetic transcription (RNA synthesis). Every three bases, or *base triplet,* is a code word—called a **codon**—for a specific amino acid. Sample codons and their amino acid "translations" are shown in table 3.2 and in figure 3.15. As mRNA moves through the ribosome, the sequence of codons is translated into a sequence of specific amino acids within a growing polypeptide chain.

Figure 3.15. The genetic code is first transcribed into base triplets (codons) in mRNA and then translated into a specific sequence of amino acids in a protein.

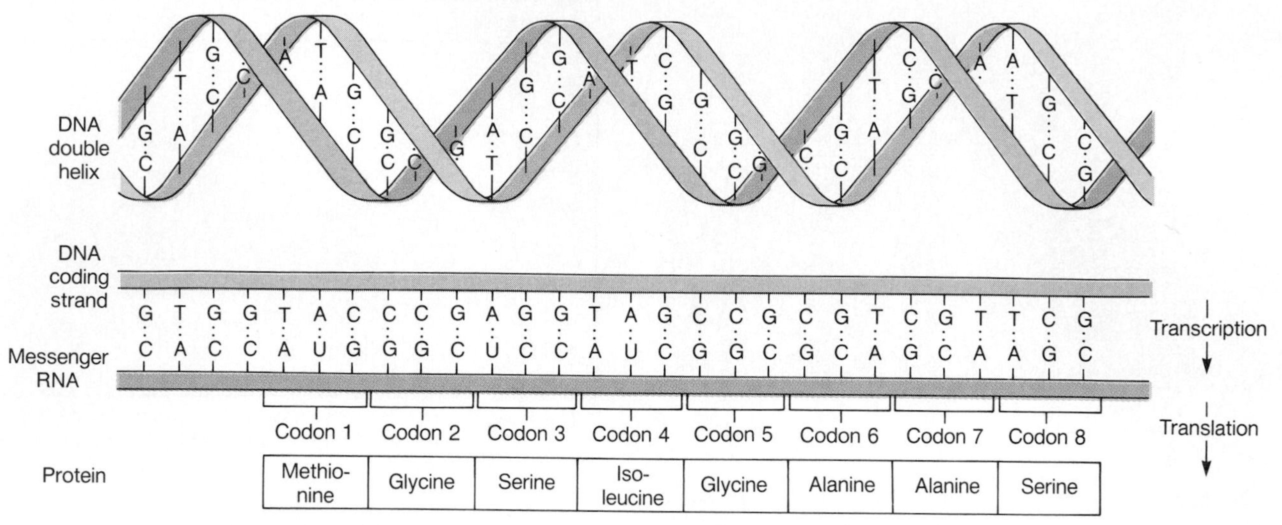

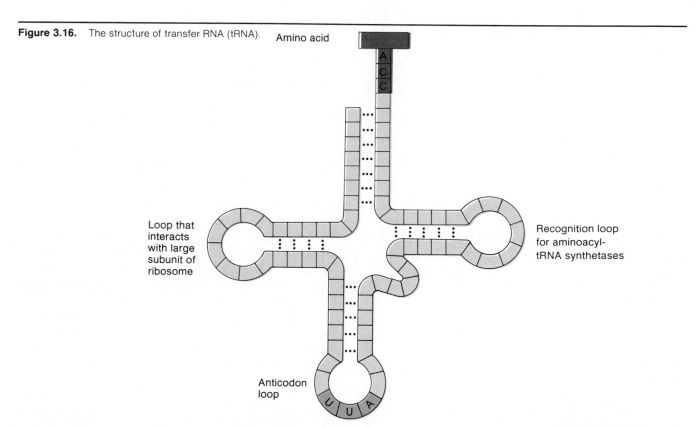

Figure 3.16. The structure of transfer RNA (tRNA).

Amino acid

Loop that interacts with large subunit of ribosome

Recognition loop for aminoacyl-tRNA synthetases

Anticodon loop

Transfer RNA

Translation of the codons is accomplished by tRNA and particular enzymes. Each tRNA molecule, like mRNA and rRNA, is single stranded. Although tRNA is single stranded, it bends on itself to form three loop regions (fig. 3.16). One of these loops contains the anticodon—three nucleotides that are complementary to a specific codon in mRNA.

Enzymes in the cell cytoplasm called *aminoacyl-tRNA synthetase* enzymes join specific amino acids to the ends of tRNA, so that a tRNA with a given anticodon is always bonded to one specific amino acid. These enzymes don't actually "read" the anticodon. Rather, each tRNA with a different anticodon has a different "recognition loop" (fig. 3.16) that is detected by these enzymes.

Figure 3.17. The translation of messenger RNA (mRNA). As the anticodon of each new aminoacyl-tRNA bonds to a codon of the mRNA, new amino acids are joined to the growing tip of the polypeptide chain.

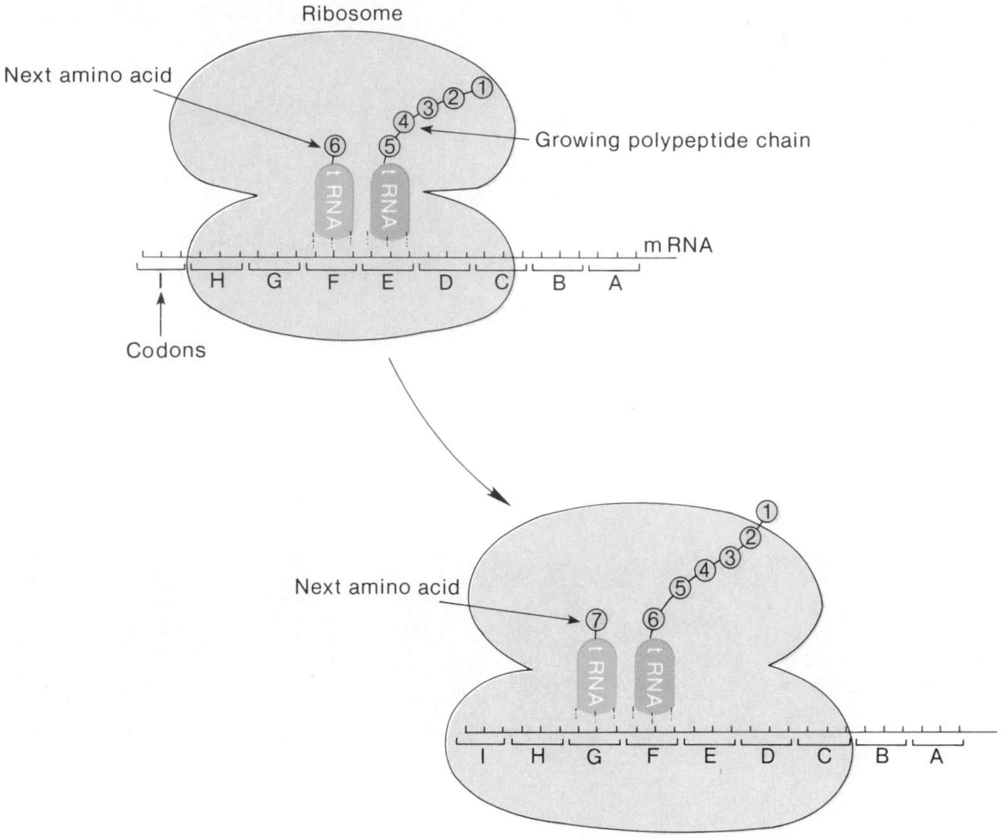

Formation of a Polypeptide

The anticodons of tRNA bond to the codons of mRNA as the mRNA moves through the ribosome. Since each tRNA molecule carries a specific amino acid, the joining together of these amino acids by peptide bonds creates a polypeptide whose amino acid sequence has been determined by the sequence of codons in mRNA.

When the first and second tRNA bring the first and second amino acids together and a peptide bond forms between them, the first amino acid detaches from its tRNA so that a dipeptide is linked by the second amino acid to the second tRNA. When the third tRNA bonds to the third codon, the third amino acid forms a peptide bond with the second amino acid (which detaches from the second tRNA); a tripeptide is then attached by the third amino acid to the third tRNA. The polypeptide chain thus grows as new amino acids are added to its growing tip (fig. 3.17).

As the polypeptide chain grows in length, interactions between its amino acids cause the chain to twist into a helix (secondary structure) and to fold and bend on itself (tertiary structure). At the end of this process, the new protein detaches from the tRNA as the last amino acid is added. Many proteins are further modified after they are formed; these modifications occur in the rough endoplasmic reticulum and Golgi apparatus.

Rough Endoplasmic Reticulum

Proteins that are to be used within the cell are produced in polyribosomes that are free in the cytoplasm. If the protein is a secretory product of the cell, however, it is made by mRNA-ribosome complexes located in the rough endoplasmic reticulum. The membranes of this system enclose fluid-filled cisternae (spaces), which the newly formed proteins may enter.

When proteins that are destined for secretion are produced, the first thirty or so amino acids are primarily hydrophobic. This *leader sequence* is attracted to the lipid component of the membranes of the endoplasmic reticulum. As the polypeptide chain elongates, it is "injected" into the cisterna within the endoplasmic reticulum. The leader sequence is, in a sense, an "address" that directs secretory proteins into the endoplasmic reticulum. Once the proteins are in the cisterna, the leader sequence is removed so the protein cannot reenter the cytoplasm (fig. 3.18).

The processing of the hormone insulin can serve as an example of the post-translational changes that occur within the endoplasmic reticulum. Insulin enters the cisterna as a single polypeptide composed of 109 amino acids. The first twenty-three amino acids serve as a leader sequence and are quickly removed. The remaining chain folds within

Figure 3.18. A protein destined for secretion begins with a *leader sequence* that enables it to be inserted into the endoplasmic reticulum. Once it has been inserted, the leader sequence is removed and carbohydrate is added to the protein.

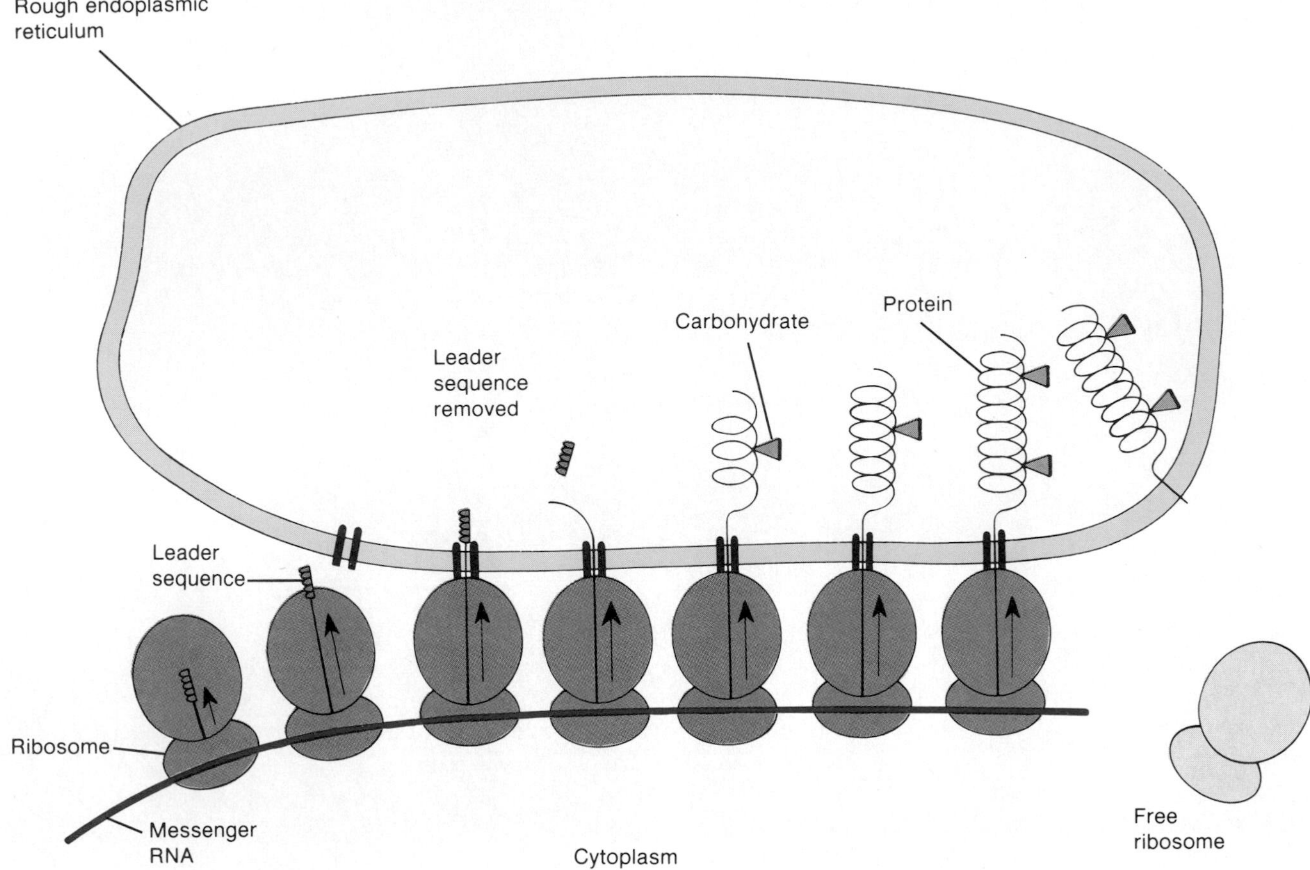

the cisterna so that the first and last amino acids in the polypeptide are brought close together. The central region is then enzymatically removed, producing two chains—one that is twenty-one amino acids long, the other that is thirty amino acids long—which are subsequently joined together by disulfide bonds. This is the form of insulin that is normally secreted from the cell.

Golgi Apparatus

Secretory proteins do not remain trapped within the endoplasmic reticulum; they are transported to another organelle within the cell—the **Golgi** *(gol'je)* **apparatus.** This organelle serves two interrelated functions: (1) further modifications of proteins (such as addition of carbohydrates) occur in the Golgi apparatus; and (2) different types of proteins are separated according to their function and destination. For example, proteins that are to be secreted are separated from those that will be incorporated into the cell membrane and from those that will be introduced into lysosomes.

Golgi: from Camillo Golgi, Italian histologist, 1843–1926

The Golgi apparatus consists of several flattened sacs that enclose cisternae. These cisternae, in turn, communicate with those of the endoplasmic reticulum. This communication may be due to a direct connection between the two organelles, or it may occur by means of vesicles that pass from the endoplasmic reticulum to the Golgi apparatus. Proteins that enter the cisternae of the Golgi apparatus, in turn, are "packaged" within vesicles that bud off from the sacs of the Golgi apparatus (fig. 3.19).

The Golgi apparatus and the rough endoplasmic reticulum contain enzymes that modify the structure of proteins. These changes, combined with the events that occur during genetic transcription and translation, provide numerous possible sites for the regulation of genetic expression. Table 3.3 summarizes the processes involved in genetic expression and possible sites for its regulation.

1. Explain how mRNA, rRNA, and tRNA function during the process of protein synthesis.
2. Describe the rough endoplasmic reticulum, and explain how the processing of secretory proteins differs from the processing of proteins that remain within the cell.
3. Describe the structure of the Golgi apparatus, and explain its functions.

Figure 3.19. (*a*) an electron micrograph of a Golgi apparatus. Notice the formation of vesicles at the ends of some of the flattened sacs. (*b*) an illustration of the processing of proteins by the rough endoplasmic reticulum and Golgi apparatus. (From "The Compartmental Organization of the Golgi Apparatus," by James E. Rothman. Copyright © 1985 by Scientific American, Inc. All Rights Reserved.)

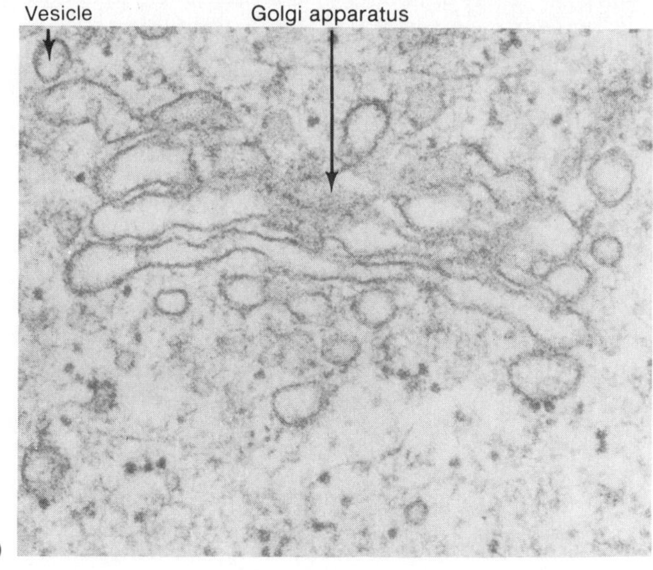

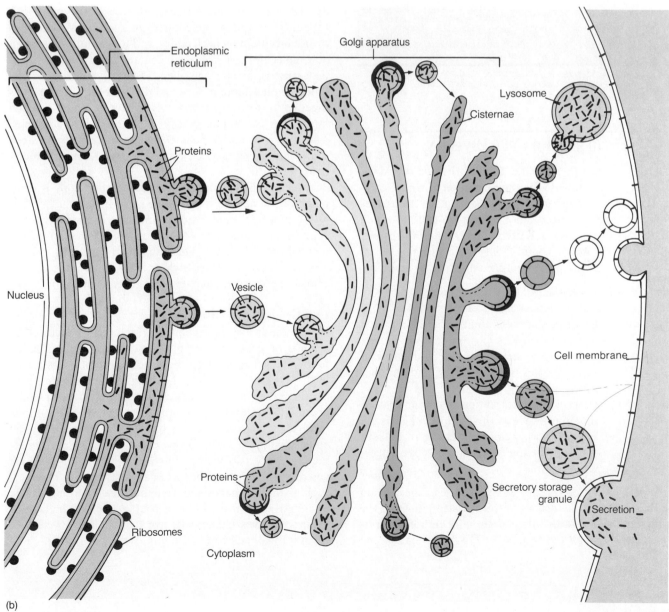

Table 3.3 Stages in genetic expression and possible sites for its regulation

Stage of gene expression	Possible sites for regulation
RNA synthesis	Duplication of genes; cutting and splicing of genes to different positions on a chromosome
	Association of histone and nonhistone proteins with chromatin
Nuclear processing of RNA	Cutting and splicing of nuclear RNA
	Capping of messenger RNA with 7-methyl guanosine and addition of about two hundred adenylate nucleotides
Translation of mRNA	Availability of specific tRNA molecules
Post-translational changes in protein structure	Cleavage of parent protein into biologically active fragments
	Chemical modification of amino acids
	Association of polypeptide chains together to form quaternary structure
	"Addressing" of protein for secretion by leader sequence and insertion into rough endoplasmic reticulum
	Addition of carbohydrate to secreted protein

DNA Synthesis and Cell Division

When a cell is going to divide, each of the two strands of DNA acts as a template for the formation of a new, complementary strand. Organs grow and repair themselves through a type of cell division known as mitosis, in which the two daughter cells receive the same DNA as the parent cell. Gametes, which contain only half the number of chromosomes as the parent cell, are formed by meiosis.

Objective 15. Explain how DNA replicates itself and why this process is called semiconservative.

Objective 16. Describe the events that occur during each phase of the cell cycle.

Objective 17. Describe the structure and significance of the centrioles, spindle fibers, and chromatids.

Objective 18. Describe the different phases of mitosis and the significance of this type of cell division.

Objective 19. Explain how meiosis differs from mitosis, and indicate the function of meiotic cell division.

Genetic information is required for the life of the cell and for the ability of the cell to perform its functions in the body. Each cell obtains this genetic information from its parent cell through the process of DNA replication and cell division. DNA is the only type of molecule in the body capable of replicating itself, and mechanisms exist within the dividing cell to insure that the duplicate copies of DNA are properly distributed to the daughter cells.

DNA Replication

When a cell is going to divide, each DNA molecule replicates itself, and each of the identical DNA copies thus produced is distributed to the two daughter cells. Replication of DNA requires the action of a specific enzyme known as *DNA polymerase (pol″ah-mer-āz″)*. This enzyme moves along the DNA molecule, breaking the weak hydrogen bonds between complementary bases as it travels. As a result, the bases of each of the two DNA strands become free to bond to new complementary bases (which are part of nucleotides) that are available within the surrounding environment.

Because of the rules of complementary base pairing, the bases of each original strand will bond to the appropriate free nucleotides: adenine bases pair with thymine-containing nucleotides; guanine bases pair with cytosine-containing nucleotides, and so on. In this way, two new molecules of DNA, each containing two complementary strands, are formed. The DNA polymerase enzyme links the phosphate groups and deoxyribose sugar groups together to form a second polynucleotide chain in each DNA that is complementary to the first DNA strands. Thus two new double-helix DNA molecules are produced that contain the same base sequence as the parent molecule (fig. 3.20).

When DNA replicates, therefore, each copy is composed of one new strand and one strand from the original DNA molecule. Replication is said to be *semiconservative* (half of the original DNA is "conserved" in each of the new DNA molecules). Through this mechanism, the sequence of bases in DNA—which is the basis of the genetic code—is preserved from one cell generation to the next.

Cell Growth and Division

Unlike the life of an organism, which can be pictured as a linear progression from birth to death, the life of a cell follows a cyclical pattern. Each cell is produced as a part of its "parent" cell; when the daughter cell divides, it in turn becomes two new cells. In a sense, then, each cell is potentially immortal as long as its progeny can continue to divide. Some cells in the body divide frequently; the epidermis of the skin, for example, is renewed approximately every two weeks, and the stomach lining is renewed about every two or three days. Other cells, however, such as mature nerve and skeletal muscle, do not divide at all. All cells in the body, of course, live only as long as the person lives (some cells live longer than others, but eventually all cells die when vital functions cease).

Figure 3.20. The replication of DNA. Each new double helix is composed of one old and one new strand. The base sequences of each of the new molecules is identical to that of the parent DNA because of complementary base pairing.

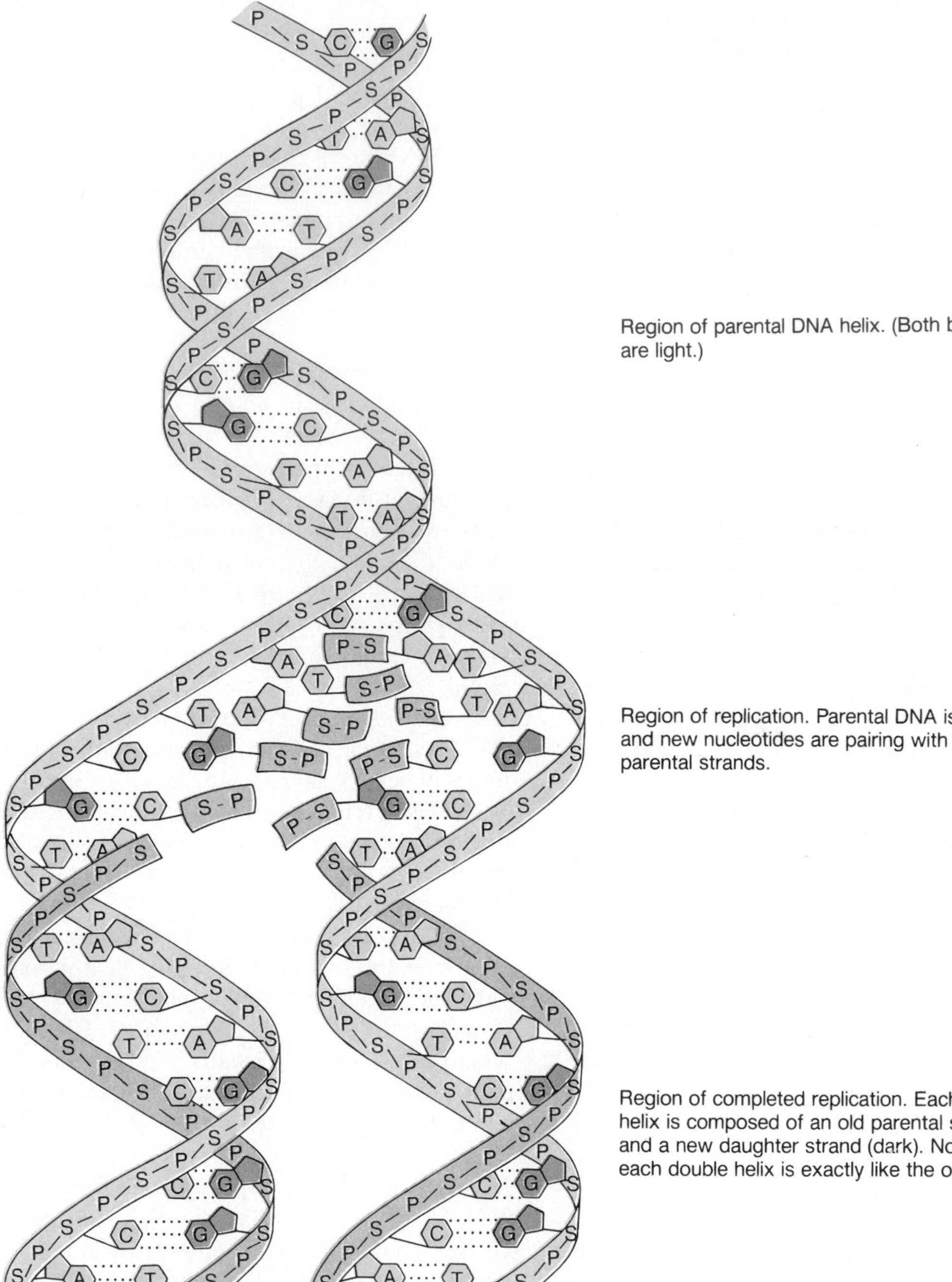

Region of parental DNA helix. (Both backbones are light.)
Region of replication. Parental DNA is unzipped, and new nucleotides are pairing with those in parental strands.
Region of completed replication. Each double helix is composed of an old parental strand (light) and a new daughter strand (dark). Notice that each double helix is exactly like the other one.

Figure 3.21. The life cycle of a cell.

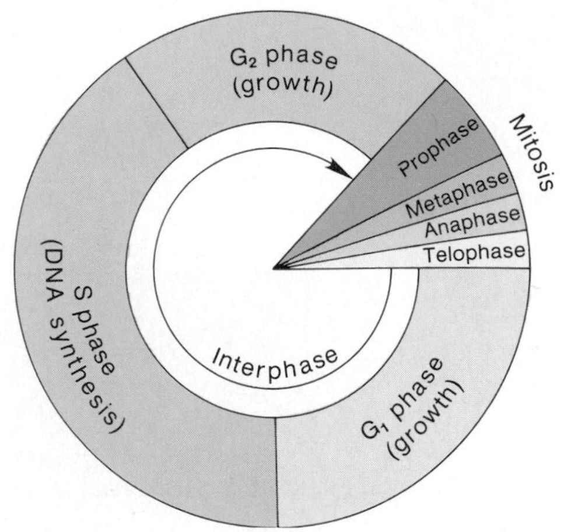

The Cell Cycle. The nondividing cell is in a part of its life cycle known as **interphase** (fig. 3.21), which is subdivided into G_1, S, and G_2 phases, as will be described. The chromosomes are in their extended form (as euchromatin), and their genes actively direct the synthesis of RNA. Through their direction of RNA synthesis, genes control the metabolism of the cell. During this time the cell may be growing, so this part of interphase is known as the *G_1 phase* (*G* stands for *growth*). Although sometimes described as "resting," cells in the G_1 phase perform the physiological functions characteristic of the tissue in which they are found.

If a cell is going to divide, it replicates its DNA in a part of interphase known as the *S phase* (*S* stands for *synthesis*). The mechanisms that cause transformation of a cell from the G_1 to the S phase are not known. Some experiments suggest that two nucleotides—cyclic adenosine monophosphate (cAMP) and cyclic guanosine monophosphate (cGMP)—may regulate this transformation. An increase in the intracellular concentration of cAMP and a decrease in cGMP can stimulate cell division. Reverse changes in the concentration of the nucleotides inhibit cell division. Besides helping to regulate cell division, these cyclic nucleotides have a wide variety of other regulatory functions in the cell (as described in later chapters).

Once DNA has replicated, the chromatin condenses to form short, thick, rodlike structures. This is the more familiar form of chromosomes because they are easily seen in the ordinary (light) microscope. Remember that this form of the chromatin represents a "packaged" state of DNA—not the extended, threadlike form that is active in directing the metabolism of the cell during the G_1 phase.

Centrioles. **Centrioles** *(sen'tre-ōlz)* are small, rodlike structures located near the nucleus. There are two centrioles, positioned at right angles to each other, within a spherical nonmembranous mass called a **centrosome.** Each centriole is composed of nine evenly spaced bundles of microtubules, with three microtubules per bundle (fig. 3.22).

Centrosomes are found only in those cells that are capable of division. During the process of cell division, the centrioles take up positions on opposite sides of the nucleus. The centrioles are then involved in the production of, and are attached to, **spindle fibers,** which are also composed of microtubules. The spindle fibers and centrioles help to pull the duplicated chromosomes to opposite poles of the cell during cell division. Cells that lack centrioles, such as mature muscle and nerve cells, cannot divide.

Mitosis. Following the S phase of the cell cycle, each chromosome consists of two strands called **chromatids** *(kro'mah-tidz),* which are joined together by a *centromere.* The two chromatids within a chromosome contain identical DNA base sequences because each is produced by the semiconservative replication of DNA. Each chromatid, therefore, contains a complete double helix DNA molecule that is a copy of the single DNA molecule existing prior to replication. Each chromatid will become a separate chromosome once cell division has been completed.

Following a second resting phase (*G_2*), which is usually shorter than G_1, the cell proceeds through the various stages of cell division or **mitosis** *(mi-to'sis)* (the *M phase* of the cell cycle). In mitosis, the chromosomes line up single file along the equator of the cell. Spindle fibers from the centrioles form and attach to the centromere of each chromosome (fig. 3.23).

When the spindle fibers shorten, the centromeres split apart and the two chromatids in each chromosome are pulled to opposite poles. Each pole therefore gets one copy of each of the forty-six chromosomes. Division of the cytoplasm (cytokinesis) results in the production of two daughter cells, which are genetically identical to each other and to the original parent cell.

Hypertrophy and Hyperplasia. The growth of an individual from a fertilized egg into an adult involves an increase in cell number and an increase in cell size. Growth due to an increase in cell number results from mitotic cell division and is termed **hyperplasia** *(hi''per-pla'ze-ah).* Growth of a tissue or organ due to an increase in cell size is termed **hypertrophy** *(hi-per'tro-fe).*

Most growth is due to hyperplasia. A callous on the palm of the hand, for example, involves thickening of the skin by hyperplasia due to frequent abrasion. An increase in skeletal muscle size as a result of exercise, in contrast, is generally believed to be produced by hypertrophy.

mitosis: Gk. *mitos,* thread

Figure 3.22. The centrioles. (*a*) micrograph of the two centrioles in a centrosome (14,200×). (*b*) a diagram showing that the centrioles are positioned at right angles to one another.

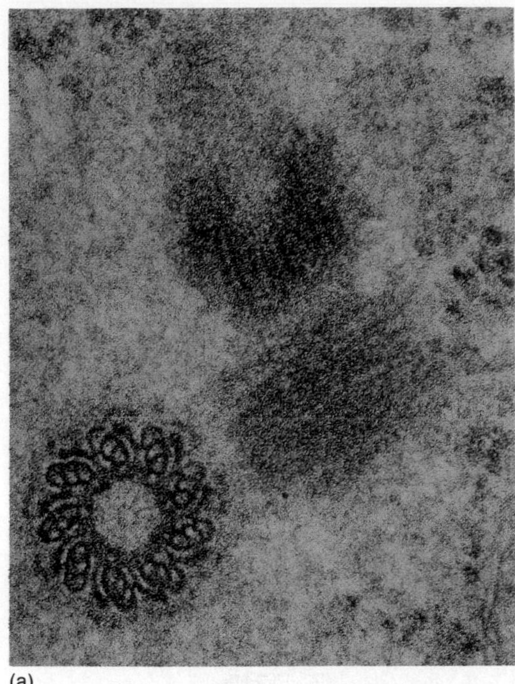
(a)

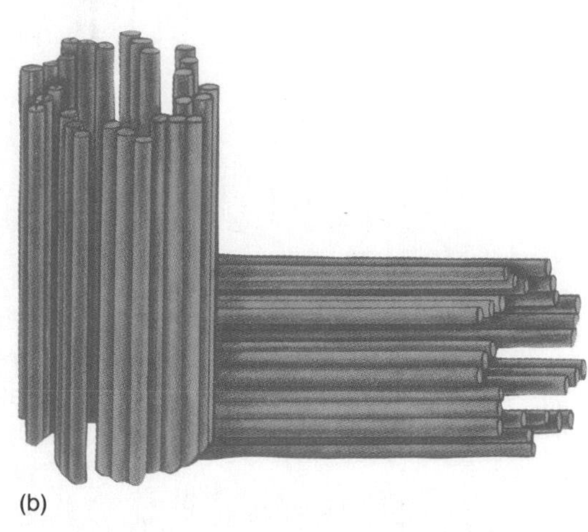
(b)

> Skeletal muscle and cardiac (heart) muscle can only grow by hypertrophy. When this occurs in skeletal muscles in response to an increased workload, during weight training, for example, it is called *compensatory hypertrophy.* The heart muscle may also demonstrate compensatory hypertrophy when its workload increases in response, for example, to hypertension (high blood pressure). The opposite of hypertrophy is *atrophy,* in which the cells become smaller than normal. This may result from the disuse of skeletal muscles (during prolonged bed rest, for example), various diseases, and advanced age.

Meiosis. When a cell is going to divide, either by mitosis or meiosis, the DNA is replicated (forming chromatids) and the chromosomes become shorter and thicker. At this point the chromosomes can be photographed, and similar-looking chromosomes can be matched into pairs.

The matched pairs of chromosomes are called **homologous** *(ho-mol'ah-gus)* **chromosomes.** One member of each homologous pair is derived from a chromosome inherited from the father, and the other member is a copy of one of the chromosomes inherited from the mother. Homologous chromosomes do not have identical DNA base sequences; one member of the pair may code for blue eyes, for example, and the other for brown eyes. There are twenty-two homologous pairs of *autosomal (aw''to-so'mal) chromosomes* and one pair of *sex chromosomes,* described as X and Y. Females have two X chromosomes, whereas males have one X and one Y chromosome (fig. 3.24). In meiosis, as will be described, these homologous chromosomes become joined together side-by-side.

A special type of cell division occurs in the gonads (ovaries and testes) of sexually mature people. This type of cell division, known as **meiosis** *(mi-o'sis),* is used to produce gametes (ova and sperm). Since sperm and ova have half the number of chromosomes (23) as the parent cell (46), meiosis is also known as **reduction division.** There are two stages of meiosis. In the *first meiotic division,* homologous chromosomes pair side-by-side along the equator of the cell (in contrast to single file in mitosis). Since each chromosome at this stage contains two identical chromatids, the paired chromosomes are known as a *tetrad.* Exchanges of genetic information can occur between paired homologous chromosomes, as illustrated in figure 3.25. This process, known as *crossing-over,* results in **genetic recombination.** Genetic recombination results in the production of gametes that are genetically unlike each other and unlike the parent that produced them.

meiosis: Gk. *meioun,* lessen

tetrad: Gk. *tetras,* group of four

Figure 3.23. The stages of mitosis.

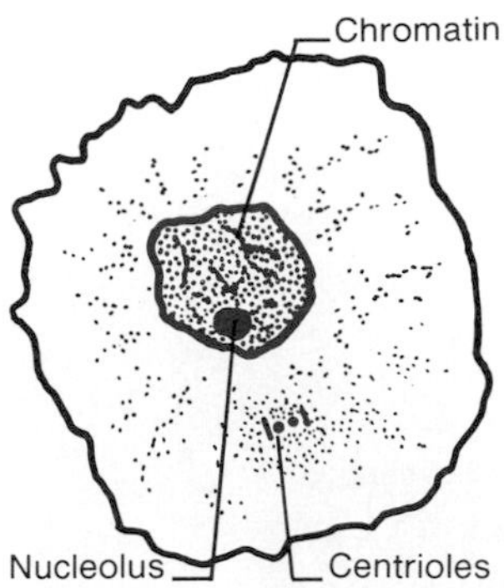

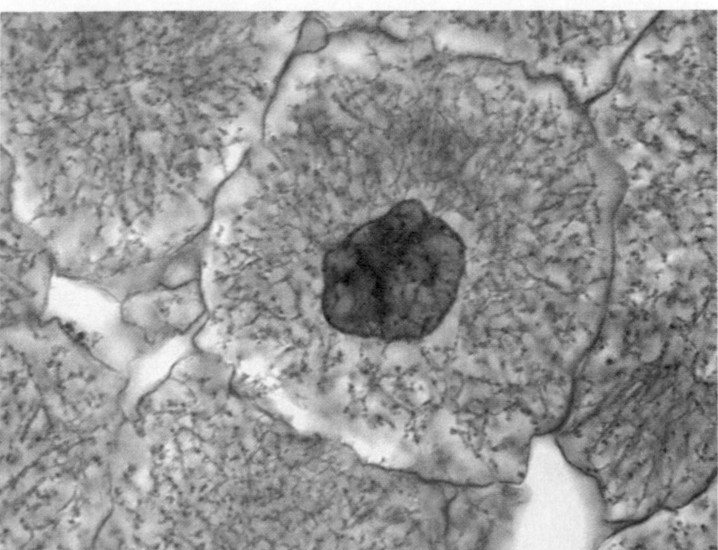

(a) Interphase

The chromosomes are in an extended form and seen as chromatin in the electron microscope. The nucleus is visible.

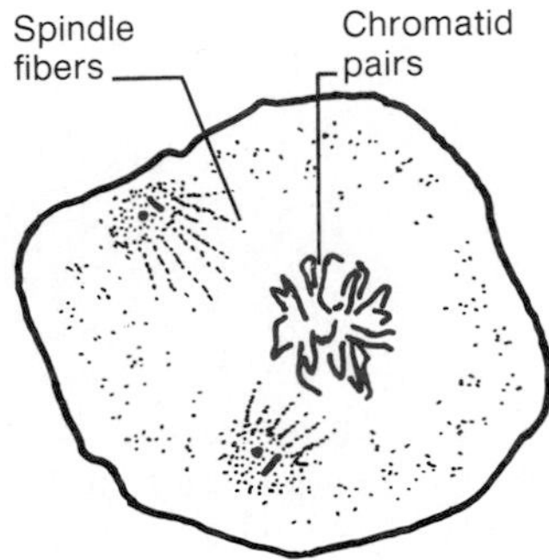

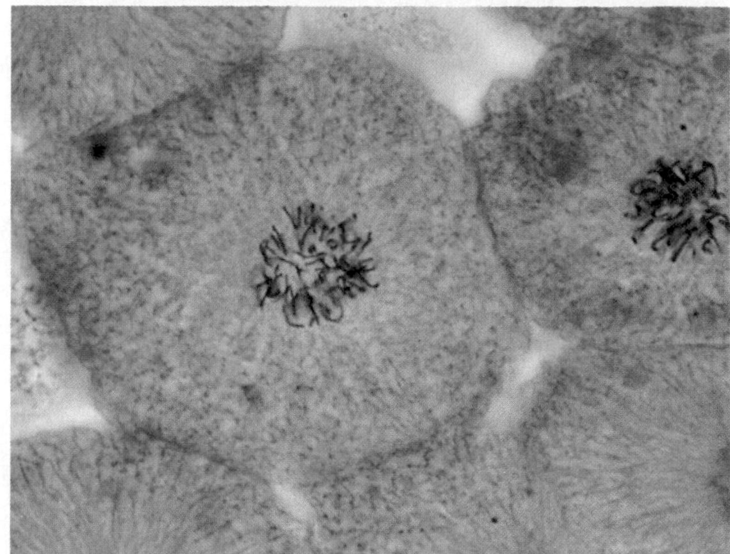

(b) Prophase

The chromosomes are seen and observed to consist of two chromatids joined together by a centromere. The centrioles move apart towards opposite poles of the cell. Spindle fibers are produced and extend from each centriole. The nuclear membrane starts to disappear. The nucleolus is no longer visible.

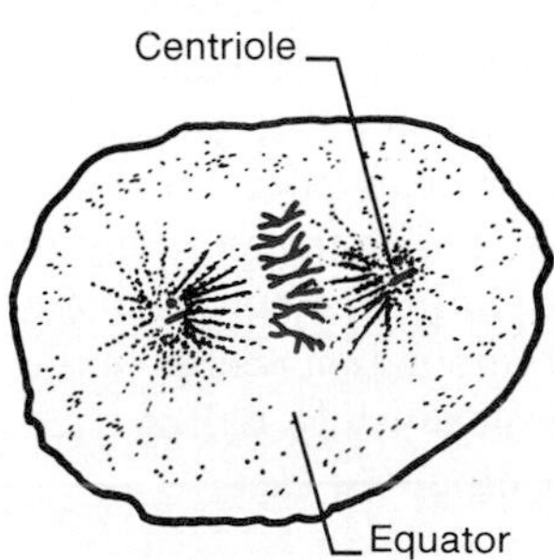

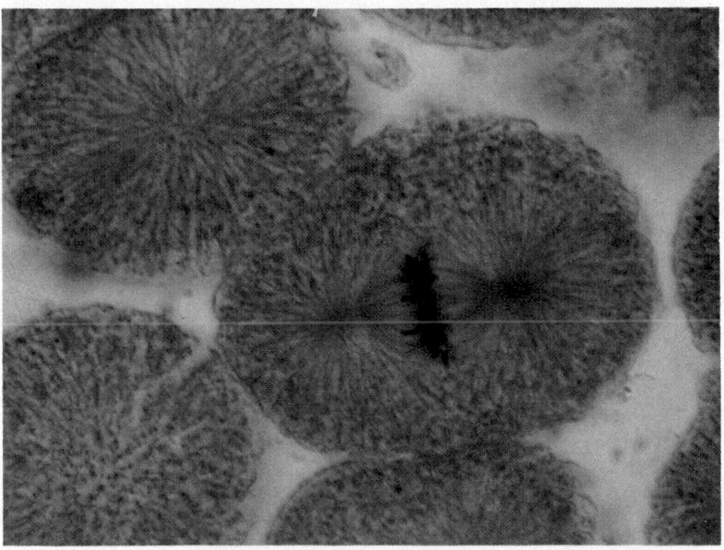

(c) Metaphase

The chromosomes are lined up at the equator of the cell. The spindle fibers from each centriole are attached to the centromeres of the chromosomes. The nuclear membrane has disappeared.

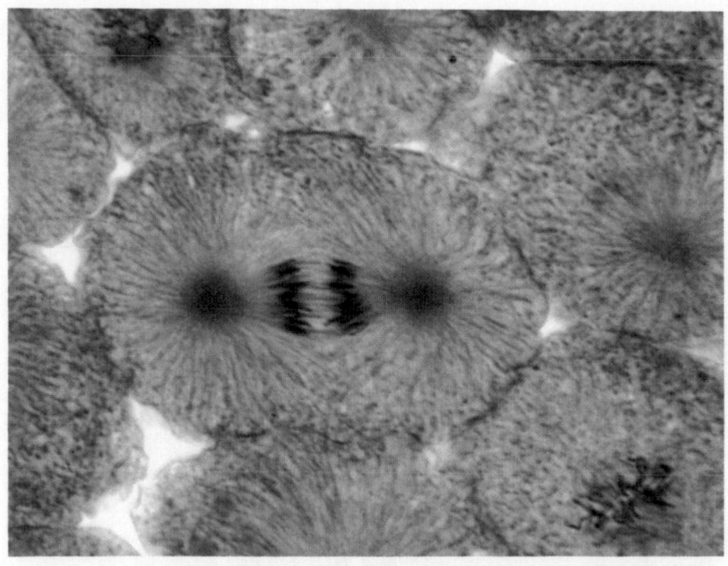

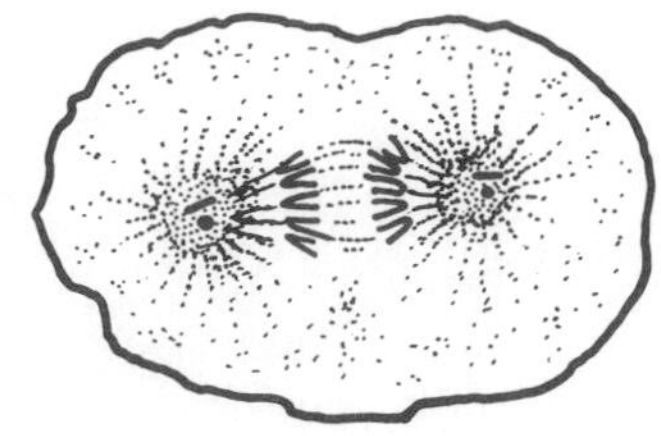

(d) Anaphase

The centromeres split, and the sister chromatids separate as each is pulled to an opposite pole.

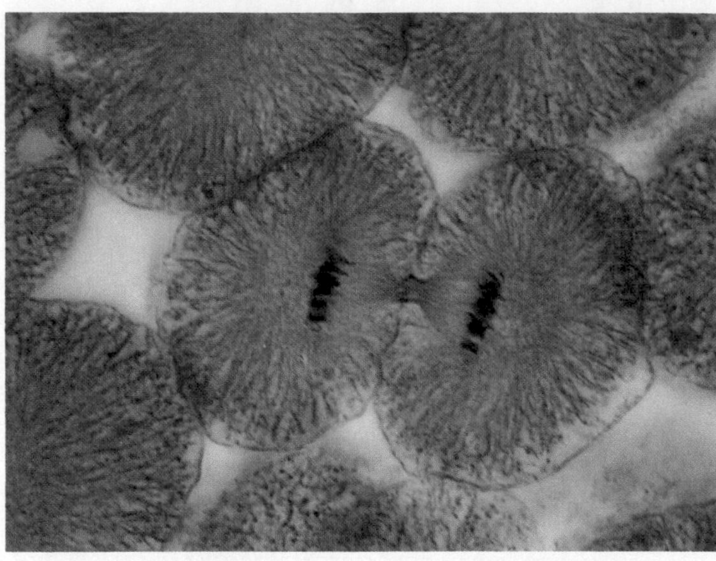

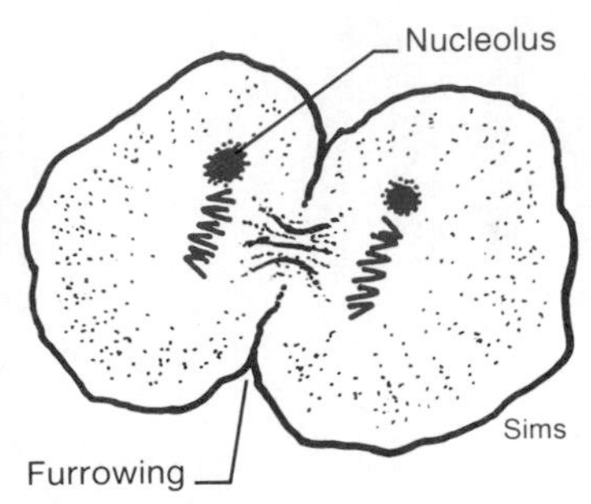

(e) Telophase

The chromosomes become longer, thinner, and less distinct.
New nuclear membranes form.
The nucleolus reappears.
Cell division (cytokinesis) is nearly complete.

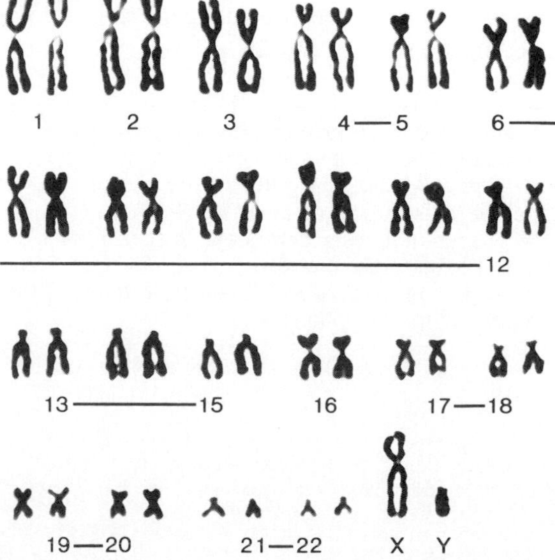

Figure 3.24. A photograph of homologous pairs of chromosomes from a human male cell at the metaphase of mitosis (homologous chromosomes have been paired and numbered according to convention).

Figure 3.25. (*a*) genetic variation results from the crossing-over of tetrads, which occurs during the first meiotic prophase. (*b*) a diagram of the pairing of homologous chromosomes of a single tetrad: (*1*) before crossing-over; (*2*) chromatids crossing-over; and (*3*) results of crossing-over.

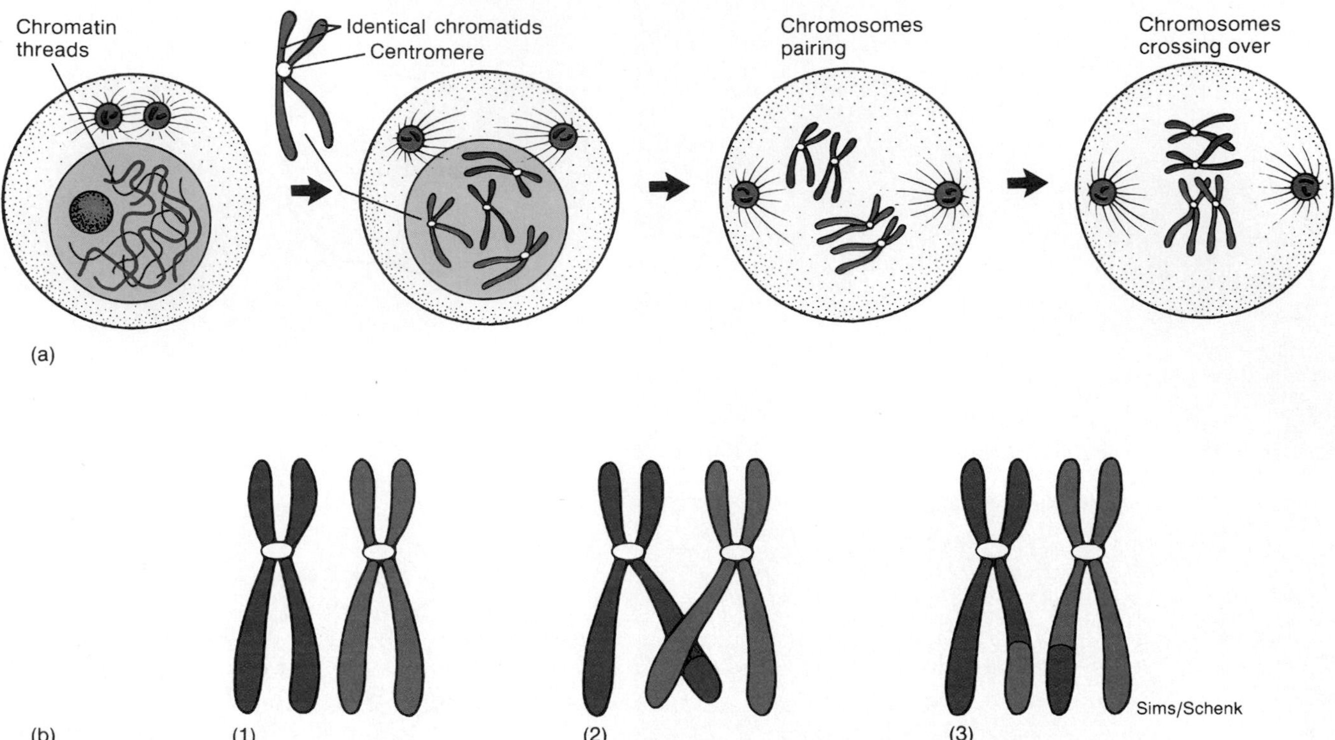

The stages of the first meiotic division are labeled prophase I, metaphase I, anaphase I, and telophase I. At the end of the first meiotic division, the two daughter cells contain only one of each pair of homologous chromosomes; they have twenty-three chromosomes, although each chromosome contains duplicate chromatids. A *second meiotic division* follows, in which the identical chromatids are pulled apart and distributed to different cells (the gametes). The stages of this division are labeled prophase II, metaphase II, anaphase II, and telophase II. The details of meiosis and the production of sperm and ova will be discussed in chapters 27 and 28.

1. Draw a simplified illustration of the semiconservative replication of DNA, using stick figures and two colors.
2. Describe the cell cycle, using the proper symbols to indicate the different stages of the cycle.
3. List the phases of mitosis, and briefly describe the events that occur in each phase.
4. Distinguish between mitosis and meiosis in terms of their final result and their functional significance.

Clinical Considerations

Functions of Cellular Organelles

Lysosomes. Lysosomes are important organelles in the function of phagocytic cells. Organs such as the liver, spleen, and lymph nodes have phagocytes that are immobile, or "fixed," within the organs. These are sometimes referred to as the *reticuloendothelial system.* Among other functions, the reticuloendothelial system helps to remove old red blood cells and bacteria from blood and to inactivate drugs and remove them from the blood. Lysosomes also participate in *autophagy,* and by this process help to prevent the accumulation of damaged organelles and stored molecules. Genetic diseases that result in the defective production of lysosomal enzymes may produce abnormal accumulations of glycogen or lipids within particular cells. Examples of such diseases include *Tay-Sachs disease* and *Gaucher's disease.*

Endoplasmic Reticulum. The smooth endoplasmic reticulum in liver cells and other cells contains enzymes used for the inactivation of steroid hormones and many drugs. This inactivation is generally achieved by reactions that convert these compounds to more water soluble and less active forms, which can be more easily excreted by the kidneys. When people take certain drugs, such as alcohol and phenobarbital, for a long period of time, a larger dose of these compounds is required to produce a given effect. This phenomenon, called *tolerance,* is accompanied by an increase in the smooth endoplasmic reticulum and thus an increase in the enzymes charged with inactivation of these drugs.

Cell Growth and Reproduction

Mitosis and Aging. Certain types of cells can be removed from the body and grown in nutrient solutions (outside the body, or *in vitro*). Under these artificial conditions the potential longevity of different cell lines can be studied. For unknown reasons, normal connective tissue cells (called fibroblasts) stop dividing *in vitro* after about forty to seventy population doublings. Cells that become transformed into cancer, however, apparently do not age and continue dividing indefinitely in culture. It is ironic that these potentially immortal cells may commit suicide by killing their host.

in vitro: L. *in vitro,* in a glass

Table 3.4 Times for cell cycles, measured in hours, in selected normal tissues and tumors

Tissue	Cell-cycle time
	hr
Colon, epithelium of crypt cells	39
Rectum, epithelium of crypt cells	48
Bone-marrow precursor cells	18
Bronchus, epithelial cells	220
Carcinoma of stomach	72
Acute myeloblastic leukemia	80–84
Chronic myeloid leukemia	120
Carcinoma of bronchus	196–260

Reprinted by permission of *The New England Journal of Medicine,* vol. 304, p. 453. Copyright © 1981 Massachusetts Medical Society, Waltham, MA.

The Cell Cycle and Cancer. Mature nerve and muscle cells do not replicate at all; neurons are thus particularly susceptible to damage by alcohol and other drugs. Epithelial cells, in contrast, have very rapid cell cycles that help to replace the continuous loss of cells. *Cancers* have rapid rates of cell division but not necessarily more rapid than normal tissue (table 3.4). The fast growth of some cancers is thus not due simply to a rapid rate of cell division but is rather due to the fact that the rate of cell division is much greater than the rate of cell death. Anticancer agents that act by killing dividing cells can thus have undesirable effects on healthy tissues.

Chapter Summary

I. Cell Membrane and Associated Structures
   A. The structure of the cell, or plasma, membrane is described by a fluid-mosaic model.
   B. Some cells move by extending pseudopods; cilia and flagella protrude from the cell membrane of some specialized cells.
   C. Phagocytosis, pinocytosis, and receptor-mediated endocytosis involve inward-moving invaginations of the cell membrane that brings part of the extracellular environment into the cell.
   D. Exocytosis replaces membrane area with that which was lost during endocytosis and provides a mechanism for the secretion of cellular products.

II. Cytoplasm and Its Organelles
   A. Microfilaments and microtubules produce a cytoskeleton, which aids movements of organelles within the cell.
   B. Lysosomes contain digestive enzymes and are responsible for the elimination of structures and molecules within the cell and for the digestion of the contents of phagocytic food vacuoles.
   C. Mitochondria contain an inner and outer membrane and serve as the major sites for energy production in the cell.
   D. The rough endoplasmic reticulum is covered with ribosomes and is involved in protein synthesis; the smooth endoplasmic reticulum is the site for many enzymatic reactions and, in skeletal muscle cells, serves to store Ca^{++}.

III. Cell Nucleus
   A. The nucleus is surrounded by a nuclear membrane and contains chromatin, which consists of DNA and protein.
   B. Active euchromatin directs the synthesis of RNA; this process is dependent upon RNA polymerase and follows the law of complementary base pairing.
   C. Four types of RNA are made in the nucleus: ribosomal RNA, transfer RNA, precursor messenger RNA, and messenger RNA.

IV. Protein Synthesis and Secretion
   A. Messenger RNA enters the cytoplasm and attaches to ribosomes.
   B. Each transfer RNA bonds to a specific amino acid, so that the base triplet in the anticodon loop of tRNA will come to code for a particular amino acid.

C. The anticodons of tRNA pair with base triplets, called codons, in mRNA; in this way, the sequence of bases in mRNA is translated into a specific sequence of amino acids in the growing polypeptide.
D. Proteins destined for secretion are produced in ribosomes attached to the rough endoplasmic reticulum and enter the cisternae of this organelle.
E. Secretory proteins move from the rough endoplasmic reticulum to the Golgi apparatus, where they are further modified and packaged into vesicles.

V. DNA Synthesis and Cell Division
   A. Replication of DNA is semiconservative; each DNA strand serves as a template for the production of a new strand.
   B. During the G_1 phase of the cell cycle, the cell may grow as DNA directs the synthesis of RNA.
   C. During the S phase of the cell cycle, DNA replicates itself.
   D. After a brief rest (G_2 phase), the cell begins mitosis (M phase).
   E. In mitosis, spindle fibers attach to chromatids, containing duplicated copies of the DNA in each chromosome, and pull the chromatids apart.
   F. Each of the two chromatids from each chromosome go to opposite poles of the cell; this is followed by division of the cytoplasm.
   G. In meiosis there are two cell divisions, which result in the production of gametes that have half the number of chromosomes as the original parent cell.

Review Activities

Objective Questions

1. According to the fluid-mosaic model of the cell membrane,
   (a) protein and phospholipids form a regular, repeating structure
   (b) the membrane is a rigid structure
   (c) phospholipids form a double layer, with the polar parts facing each other
   (d) proteins are free to move within a double layer of phospholipids
2. After the DNA molecule has replicated itself, the duplicate strands are called
   (a) homologous chromosomes
   (b) chromatids
   (c) centromeres
   (d) spindle fibers
3. Nerve and skeletal muscle cells in the adult, which do not divide, remain in the
   (a) G_1 phase
   (b) S phase
   (c) G_2 phase
   (d) M phase
4. The phase of mitosis in which the chromosomes line up at the equator of the cell is called
   (a) interphase
   (b) prophase
   (c) metaphase
   (d) anaphase
   (e) telophase
5. The phase of mitosis in which the chromatids separate is called
   (a) interphase
   (b) prophase
   (c) metaphase
   (d) anaphase
   (e) telophase
6. The RNA nucleotide base that pairs with adenine in DNA is
   (a) thymine
   (b) uracil
   (c) guanine
   (d) cytosine
7. Which of the following statements about RNA is *true?*
   (a) It is made in the nucleus.
   (b) It is double stranded.
   (c) It contains the sugar deoxyribose.
   (d) It is a complementary copy of the entire DNA molecule.
8. Which of the following statements about mRNA is *false?*
   (a) It is produced as a larger pre-mRNA.
   (b) It forms associations with ribosomes.
   (c) Its base triplets are called anticodons.
   (d) It codes for the synthesis of specific proteins.
9. The organelle that combines proteins with carbohydrates and packages them within vesicles for secretion is the
   (a) Golgi apparatus
   (b) rough endoplasmic reticulum
   (c) smooth endoplasmic reticulum
   (d) ribosomes
10. The organelles that contain digestive enzymes are the
   (a) mitochondria
   (b) lysosomes
   (c) endoplasmic reticulum
   (d) Golgi apparatus

Essay Questions

1. The cell membrane is an extremely dynamic structure. Using examples, explain why this statement is true.
2. Explain how one DNA molecule serves as a template for the formation of another DNA and why DNA synthesis is said to be semiconservative.
3. What is the genetic code, and how does it affect the structure and function of the body?
4. Why may tRNA be considered the "interpreter" of the genetic code?
5. Compare the processing of cellular proteins with that of proteins that are secreted by a cell.

4

Enzymes, Energy, and Metabolism

Outline

Concepts

Enzymes are complex and diverse protein catalysts that are very specific in their individual functions.

The rate of an enzyme-catalyzed reaction depends on numerous factors. Variations in these factors control the rate of progress along particular metabolic pathways and thus help to regulate cellular metabolism.

Central to life processes are chemical reactions that are coupled, so that the energy released by one reaction is incorporated into the products of another reaction. The transformation of energy in living systems is largely based on reactions that produce and destroy molecules of ATP and on oxidation-reduction reactions.

In cellular respiration, chemical reactions liberate energy, some of which is used to produce ATP. The complete combustion of a molecule requires the presence of oxygen; some energy can be obtained, however, in the absence of oxygen. This process of anaerobic respiration results in the production of lactic acid from glucose.

Aerobic respiration is equivalent to combustion in terms of its final products (CO_2 and H_2O) and in terms of the total amount of energy liberated. In aerobic respiration, however, the energy is released in small, enzymatically controlled oxidation reactions, and a portion (38%–40%) of the energy released is captured in the high energy bonds of ATP.

Energy can be derived by the cellular respiration of lipids and proteins, using the same aerobic pathway as for the metabolism of pyruvic acid. Pyruvic acid and the Krebs cycle acids also serve as common intermediates in the interconversion of glucose, lipids, and amino acids.

Enzymes as Catalysts

Enzymes are complex and diverse protein catalysts that are very specific in their individual functions.

Objective 1. Define what is meant by the "energy of activation" of a reaction, and explain how this is affected by the action of a catalyst.

Objective 2. Explain how the structure of enzyme proteins allows these molecules to function as catalysts.

The ability of yeast cells to make alcohol from glucose (a process called *fermentation*) had been known since antiquity, yet no chemist by the mid-nineteenth century could duplicate the trick in the absence of living yeast. Also, yeast and other living cells could perform a vast array of chemical reactions at body temperature that could not be duplicated in the chemical laboratory without adding a substantial amount of heat energy. These observations led many mid-nineteenth-century scientists to believe that chemical reactions in living cells were aided by a "vital force" that operated beyond the laws of the physical world. This "vitalist" concept was squashed along with the yeast cells when a pioneering biochemist, Eduard Buchner, demonstrated that juice obtained from yeast could ferment glucose to alcohol. The yeast juice was not alive—evidently some chemicals in the cells were responsible for fermentation. Buchner didn't know what these chemicals were, so he simply named them **enzymes.**

In more recent times, biochemists have demonstrated that *enzymes are proteins* and that enzymes act as *biological catalysts.* (This description is somewhat modified by recent evidence that RNA may also have limited, but important, catalytic ability.) A catalyst is a chemical that (1) increases the rate of a reaction, (2) is not itself changed at the end of the reaction, and (3) does not change the nature of the reaction or its final result. The same reaction would have occurred to the same degree in the absence of the catalyst, but it would have progressed at a much slower rate.

In order for a given reaction to occur, the reactants must have sufficient energy. The amount of energy required for a reaction to proceed is called the **energy of activation.** In a large population of molecules, only a small fraction will possess sufficient energy for a reaction. Adding heat will raise the energy level of all the reactant molecules, thus increasing the fraction of the population that has the activation energy. Heat would make reactions go faster, but it would produce undesirable side effects in cells. Catalysts make the reaction go faster at lower temperatures by *lowering the activation energy* required so that a larger fraction of the population of reactant molecules has sufficient energy to participate in the reaction (fig. 4.1).

Since a small fraction of the reactants has the activation energy required for the reaction even in the absence of a catalyst, the reaction could theoretically occur spontaneously at a slow rate. This rate, however, would be much too slow for the needs of a cell. So, from a biological standpoint, the presence or absence of a specific enzyme catalyst acts as a switch—the reaction will occur if the enzyme is present and will not occur if the enzyme is absent.

Mechanism of Enzyme Action

The ability of enzymes to lower the activation energy of a reaction is a result of their structure. Enzyme proteins are very large molecules with complex, highly ordered, three-dimensional shapes produced by chemical interactions between their amino acids. Each type of enzyme protein has characteristic ridges, grooves, and pockets that are lined with specific amino acids. The particular pockets that are active in catalyzing a reaction are called the *active sites* of the enzyme.

The reactant molecules, which are the *substrates* of the enzyme, have shapes that allow them to fit into the active sites. The fit may not be perfect at first, but a perfect fit may be induced as the substrate gradually slips into the active site. This induced fit, together with temporary bonds that form between the substrate and the amino acids lining the active sites of the enzyme, weaken the existing bonds within the substrate molecules, which allows them to be more easily broken. New bonds are more easily formed as substrates are brought close together in the proper orientation. The *enzyme-substrate complex,* formed temporarily in the course of the reaction, then dissociates to yield *products* and the free, unaltered enzyme. This model of how enzymes work is known as the **lock-and-key model** of enzyme activity (fig. 4.2).

By this mechanism, enzymes lower the activation energy needed and greatly increase the rate of reactions. Hydrogen peroxide (H_2O_2), for example, will slowly decompose by itself if left uncapped for a few days. The addition of *catalase (kat'ah-lās),* an enzyme found in many tissues such as blood and the liver, will make hydrogen peroxide break down into water and oxygen gas a trillion times faster. The ability of enzymes to increase the rate of reactions can be measured by their **turnover number,** which is the number of substrate molecules that the enzyme can convert into products per minute (table 4.1).

fermentation: L. *fermentum,* leaven
Buchner, Eduard: German chemist, 1860–1917
enzyme: Gk. *en,* in; *zyme,* yeast

Figure 4.1. A comparison of a noncatalyzed reaction with a catalyzed reaction. Upper figures compare the proportion of reactant molecules that have sufficient activation energy to participate in the reaction (shown as the green portion of the curves). This proportion is increased in the enzyme-catalyzed reaction, because enzymes lower the activation energy required for the reaction (shown as a barrier on top of an energy "hill" in the bottom figures). Reactants that can overcome this barrier are able to participate in the reaction, as shown by arrows pointing to the bottom of the energy hill.

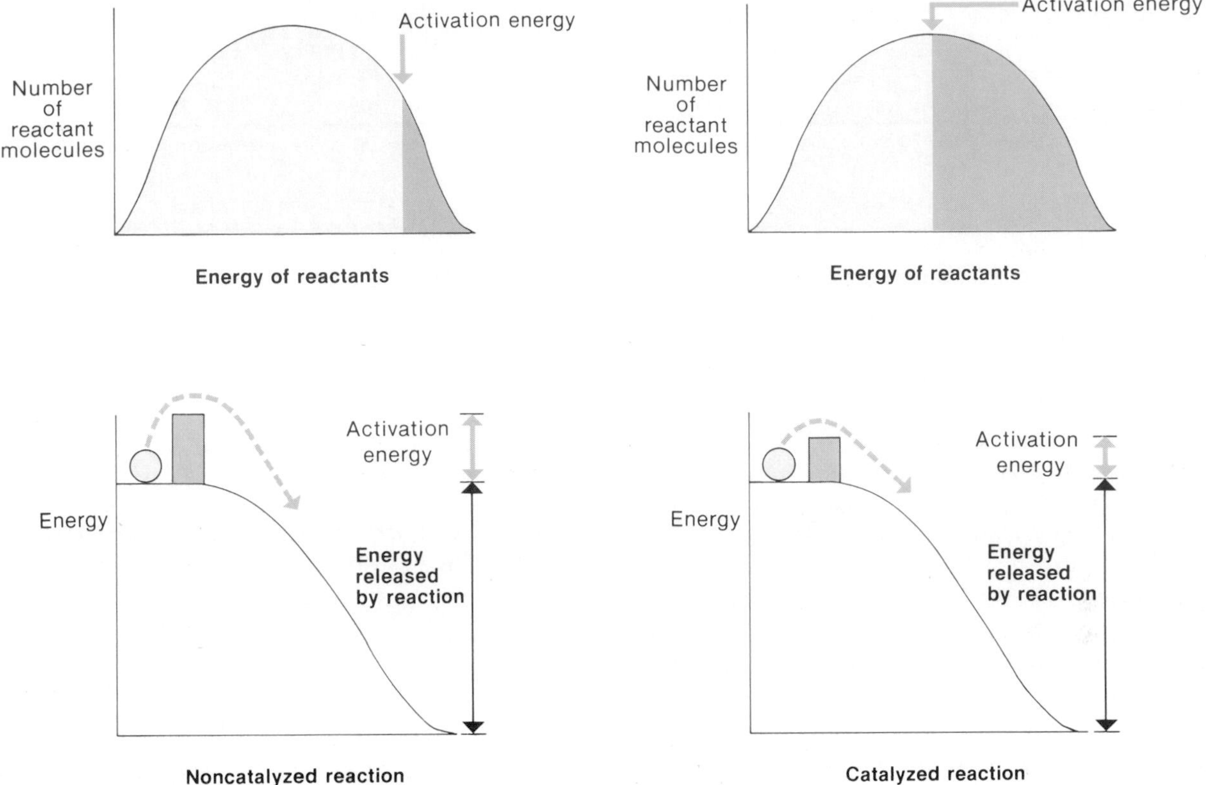

Figure 4.2. The lock-and-key model of enzyme action.

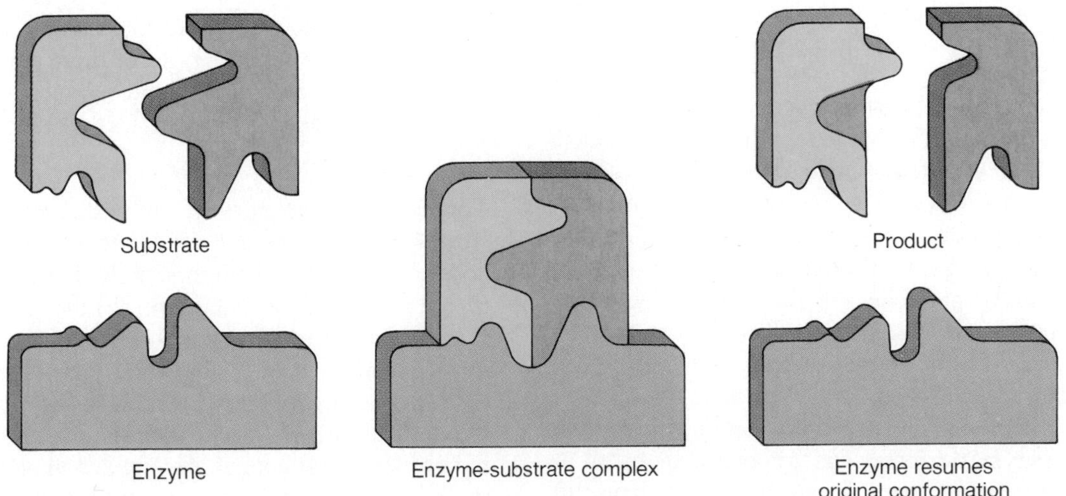

Table 4.1 Some examples of the names of enzymes, the reactions they catalyze, and their turnover numbers (molecules of substrate converted to product per minute under optimum conditions)

Enzyme	Reaction catalyzed	Turnover number
Catalase	$2\ H_2O_2 \rightarrow 2\ H_2O + O_2$	600×10^6
Carbonic anhydrase	$H_2CO_3 \rightarrow H_2O + CO_2$	3.6×10^6
Amylase	starch + $H_2O \rightarrow$ maltose	1.0×10^5
Lactate dehydrogenase	lactic acid $\rightarrow$ pyruvic acid + H_2	6.0×10^4
Ribonuclease	RNA + $H_2O \rightarrow$ ribonucleotides	600

Naming of Enzymes

Although an international committee has established a uniform naming system for enzymes, the names that are in common use do not follow a completely consistent pattern. With the exception of the digestive enzymes that were discovered first (including pepsin and trypsin), all enzyme names end with the suffix *-ase.* Classes of enzymes are named according to their job category. *Hydrolases (hi'dro-lās),* for example, promote hydrolysis reactions. Other enzyme categories include *phosphatases,* which catalyze the removal of phosphate groups, *synthetases (sin'the-tās),* which catalyze dehydration synthesis reactions, and *dehydrogenases (de-hi'dro-jen-ās),* which remove hydrogen atoms from their substrates. Enzymes called *isomerases (i-som'er-ās)* rearrange atoms within their substrate molecules to form structural isomers. (Examples of such structural isomers include glucose and fructose.)

The names of many enzymes specify both the substrate of the enzyme and the activity ("job category") of the enzyme. Lactic acid dehydrogenase, for example, removes hydrogens from lactic acid. Since enzymes are very specific as to their substrates and activity, the concentration of a specific enzyme in a sample of fluid can be measured relatively easily. This is usually done by measuring the rate of conversion of the enzyme's substrates into products under specified conditions.

Enzymes that do exactly the same job (that catalyze the same reaction) in different organs have the same name, since the name describes the activity of the enzyme. Different organs, however, may make slightly different "models" of the enzyme that differ in one or a few amino acids. These different models of the same enzyme are called **isoenzymes** *(i''so-en'zīmz).* The differences in structure do not affect the active sites (otherwise they would not catalyze the same reaction), but they do alter the structure of the enzymes at other locations, so that the different isoenzymatic forms can be separated by standard biochemical procedures. These techniques are useful in the diagnosis of diseases, as described at the end of this chapter.

isoenzymes: Gk. *isos,* equal; *en,* in; *zyme,* yeast

1. Define the "energy of activation" of a reaction; explain how it is affected by a catalyst and how the structure of an enzyme makes it function as a catalyst.
2. Draw a labeled picture of the lock-and-key model of enzyme action, and write a formula for this reaction.
3. Explain how enzymes are named.

Control of Enzyme Activity

The rate of an enzyme-catalyzed reaction depends on numerous factors. Variations in these factors control the rate of progress along particular metabolic pathways and thus help to regulate cellular metabolism.

Objective 3. Describe the effects that changes in pH and temperature have on the rate of enzyme reactions.

Objective 4. Describe the effects that changes in the concentrations of enzymes, substrates, cofactors, and coenzymes have on the rate of enzyme reactions.

Objective 5. Describe a reversible reaction, and explain how the law of mass action can determine the relative rates of these reactions.

Objective 6. Explain how end-product inhibition can affect the direction of a branched metabolic pathway.

The activity of an enzyme, as measured by the rate at which its substrates are converted to products, is influenced by a variety of factors, including (1) the temperature and pH of the solution; (2) the concentration of cofactors and coenzymes *(ko-en'zīm),* which are needed by many enzymes as "helpers" for their catalytic activity; (3) the concentration of enzyme and substrate molecules in the solution; and (4) the stimulatory and inhibitory effects of some products of enzyme action on the activity of the enzymes that helped form these molecules.

Figure 4.3. The effect of temperature on enzyme activity, as measured by the rate of the enzyme-catalyzed reaction under standardized conditions.

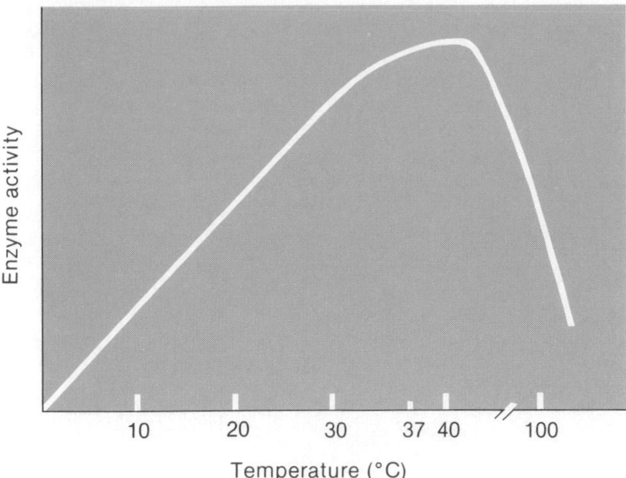

Figure 4.4. The effect of pH on the activity of three digestive enzymes.

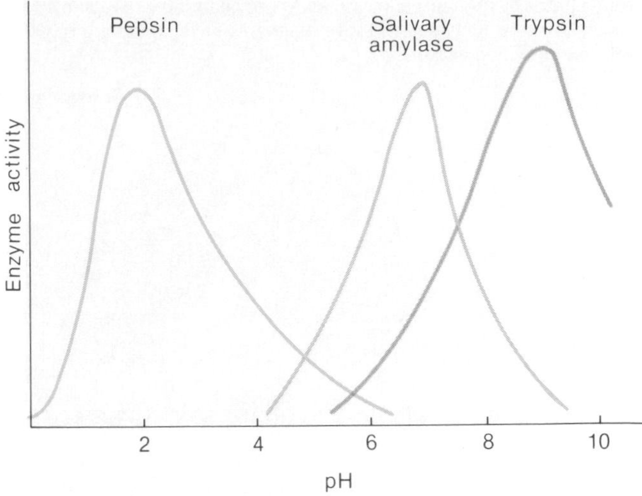

Effects of Temperature and pH

An increase in temperature, as previously described, will increase the rate of non-enzyme-catalyzed reactions because a larger number of reactant molecules will have the activation energy required. A similar relationship between temperature and reaction rate occurs in enzyme-catalyzed reactions. At a temperature of 0°C the reaction rate is unmeasurably slow. As the temperature is raised above 0°C, the reaction rate increases but only up to a point. At a few degrees above body temperature (which is 37°C) the reaction rate reaches a plateau; further increases in temperature actually *decrease* the rate of the reaction (fig. 4.3). This is because the tertiary structure of enzymes becomes altered at higher temperatures.

A similar relationship is observed when the rate of an enzymatic reaction is measured at different pH values. Each enzyme characteristically has its peak activity in a very narrow pH range, which is the **pH optimum** for the enzyme. If the pH is changed from this optimum, the reaction rate decreases (fig. 4.4). This decreased enzyme activity is due to changes in the conformation of the enzyme and in the charges of the R groups of the amino acids lining the active sites.

The pH optimum of an enzyme usually reflects the pH of the body fluid in which the enzyme is found. The acidic pH optimum of the protein-digesting enzyme *pepsin,* for example, allows it to be active in the strong hydrochloric acid of gastric juice. Similarly, the neutral pH optimum of *salivary amylase (am'ĭ-lās)* and the alkaline pH optimum of *trypsin (trip'sin)* in pancreatic juice allow these enzymes to digest starch and protein, respectively, in other parts of the digestive tract.

Table 4.2 The pH optima of selected enzymes

Enzyme	Reaction catalyzed	pH optimum
Pepsin (stomach)	Digestion of protein	2.0
Acid phosphatase (prostate)	Removal of phosphate group	5.5
Salivary amylase (saliva)	Digestion of starch	6.8
Lipase (pancreatic juice)	Digestion of fat	7.0
Alkaline phosphatase (bone)	Removal of phosphate group	9.0
Trypsin (pancreatic juice)	Digestion of protein	9.5
Monoamine oxidase (nerve endings)	Removal of amine group from norepinephrine	9.8

From Stuart Ira Fox, *Human Physiology,* 2d ed.

Although the pH of other body fluids shows less variation than the fluids of the digestive tract, significant differences exist between the pH optima of different enzymes found throughout the body (see table 4.2). Some of these differences can be exploited for diagnostic purposes. Disease of the prostate, for example, may be associated with elevated blood levels of a prostatic phosphatase with an acidic pH optimum (descriptively called acid phosphatase). Bone disease, on the other hand, may be associated with elevated blood levels of alkaline phosphatase, which has a higher pH optimum than the similar enzyme released from the diseased prostate.

Figure 4.5. The roles of cofactors in enzyme function. In (*a*) the cofactor changes the conformation of the active site, allowing a better fit between the enzyme and its substrates. In (*b*) the cofactor participates in the temporary bonding between the active site and the substrates. Only one of these mechanisms is used by a given enzyme and its cofactor.

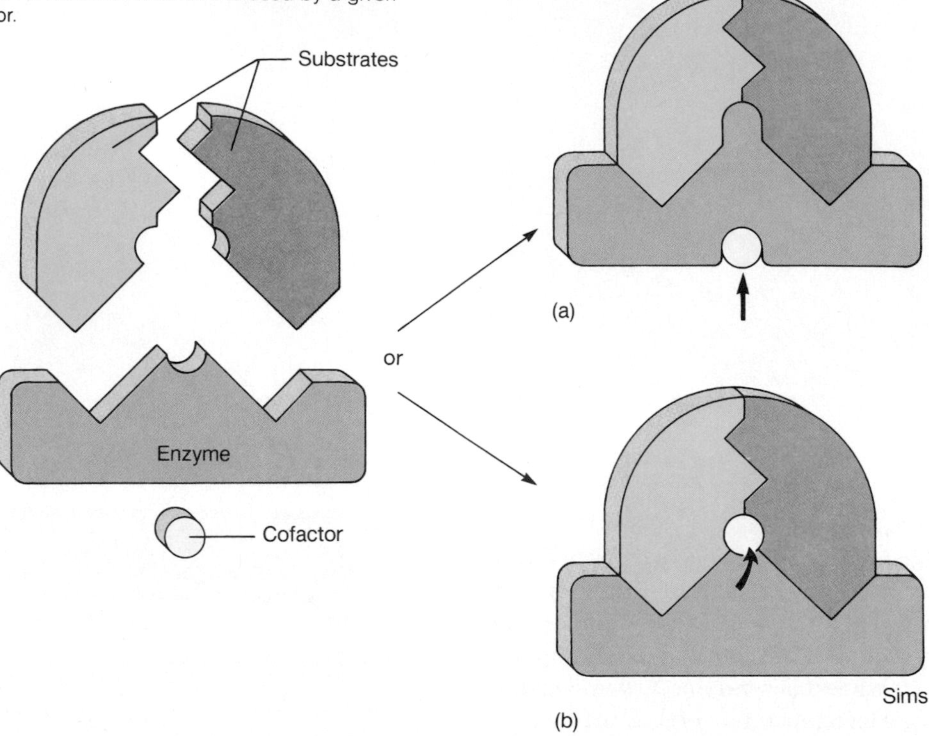

Cofactors and Coenzymes

Many enzymes are completely inactive when they are isolated in a pure state. Evidently some of the ions and smaller organic molecules that were removed in the purification procedure are needed for enzyme activity. These ions and smaller organic molecules are called *cofactors* and *coenzymes (ko-en'zim).*

Cofactors are metal ions such as Ca^{++}, Mg^{++}, Mn^{++}, Cu^{++}, and Zn^{++}. Some enzymes with a cofactor requirement do not have a properly shaped active site in the absence of the cofactor. In these enzymes, the attachment of cofactors causes a conformational change in the protein that allows it to combine with its substrate. The cofactors of other enzymes participate in the temporary bonds between the enzyme and its substrate when the enzyme-substrate complex is formed (fig. 4.5).

Coenzymes are organic molecules that are derived from water-soluble vitamins, such as niacin and riboflavin. Coenzymes participate in enzyme-catalyzed reactions by transporting hydrogen atoms and small molecules from one enzyme to another. Examples of the actions of cofactors and coenzymes in specific reactions will be given in the context of their roles in cellular metabolism later in this chapter.

Figure 4.6. The effect of substrate concentration on the reaction rate of an enzyme-catalyzed reaction. When the reaction rate is maximal, the enzyme is said to be *saturated.*

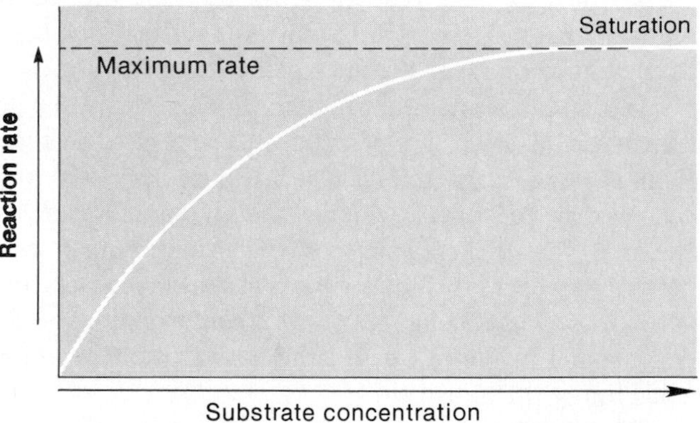

Substrate Concentration and Reversible Reactions

The rate at which an enzymatic reaction converts substrates into products depends on the enzyme concentration and on the concentration of substrates. When the enzyme concentration is at a given level, the rate of product formation will increase as the substrate concentration increases. Eventually, however, a point will be reached where additional increases in substrate concentration do not result in comparable increases in reaction rate. When the relationship between substrate concentration and reaction rate reaches a plateau, the enzyme is said to be *saturated.*

Figure 4.7. A metabolic pathway, where the product of one enzyme becomes the substrate of the next in a multi-enzyme system.

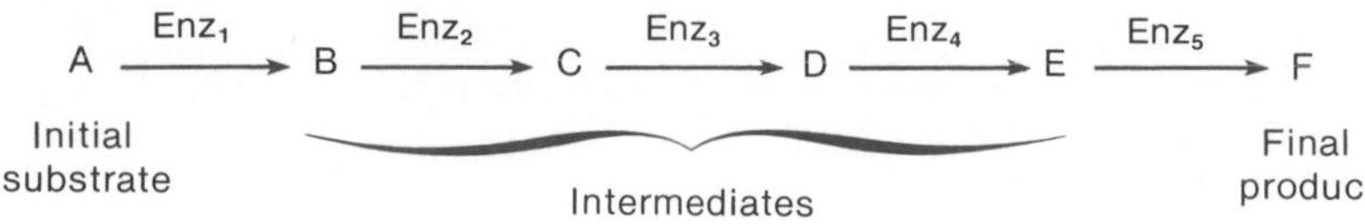

Figure 4.8. A branched metabolic pathway.

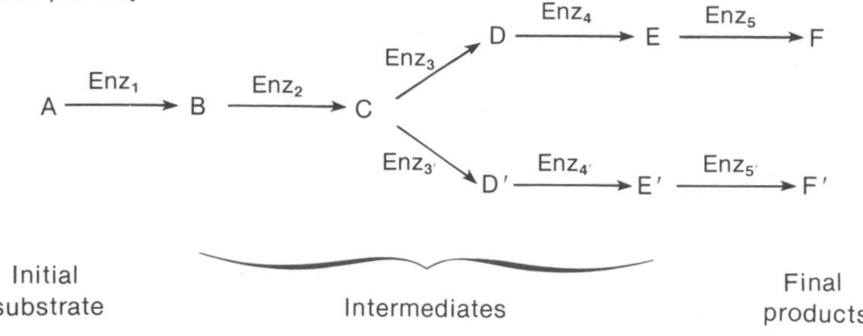

If one thinks of enzymes as workers and substrates as jobs, there is 100% employment when the enzyme is saturated; further availability of jobs (substrate) cannot further increase employment (conversion of substrate to product). This is illustrated in figure 4.6.

Some enzymatic reactions within a cell are reversible, with both the forward and backward reactions catalyzed by the same enzyme. The enzyme *carbonic anhydrase,* for example, is named because it can catalyze the following reaction:

$$H_2CO_3 \rightarrow H_2O + CO_2.$$

The same enzyme, however, can also catalyze the reverse reaction:

$$H_2O + CO_2 \rightarrow H_2CO_3.$$

The two reactions can be more conveniently illustrated by a single equation:

$$H_2O + CO_2 \rightleftarrows H_2CO_3.$$

The direction of the reversible reaction depends, in part, on the relative concentrations of the molecules to the left and right of the arrows. If the concentration of CO_2 is very high (as it is in the tissues), the reaction will be driven to the right. If the concentration of CO_2 is low and that of H_2CO_3 is high (as it is in the lungs), the reaction will be driven to the left. The ability of reversible reactions to be driven from the side of the equation where the concentration is higher to the side where the concentration is lower is known as the **law of mass action.**

Although some enzymatic reactions are not directly reversible, the net effects of the reactions can be reversed by the action of different enzymes. The enzymes that convert glucose to pyruvic acid, for example, are different from those that reverse the pathway and produce glucose from pyruvic acid. Likewise, the formation and breakdown of glycogen (a polymer of glucose) are catalyzed by different enzymes.

Metabolic Pathways

The many thousands of different types of enzymatic reactions within a cell do not occur independently of each other. They are, rather, all linked together by intricate webs of interrelationships, the total pattern of which constitutes cellular metabolism. A part of this web that begins with an *initial substrate,* progresses through a number of *intermediates,* and ends with a *final product* is known as a **metabolic pathway.**

The enzymes in a metabolic pathway cooperate in a manner analogous to workers on an assembly line: each contributes a small part to the final product. In this process, the product of one enzyme in the line becomes the substrate of the next enzyme, and so on (fig. 4.7).

Few metabolic pathways are completely linear. Most are branched so that one intermediate at the branch point can serve as a substrate for two different enzymes. Two different products that serve as intermediates of two divergent pathways can thus be formed (fig. 4.8).

End-Product Inhibition. The activities of enzymes at the branch points of metabolic pathways are often regulated by a process called **end-product inhibition.** In this process, one of the final products of a divergent pathway inhibits the branch point enzyme that began the path toward the production of this inhibitor. This inhibition prevents that final product from accumulating excessively and results in a shift toward the final product of the alternate divergent pathway (fig. 4.9).

Figure 4.9. End-product inhibition in a branched metabolic pathway. Inhibition is shown by a dotted arrow and a negative sign.

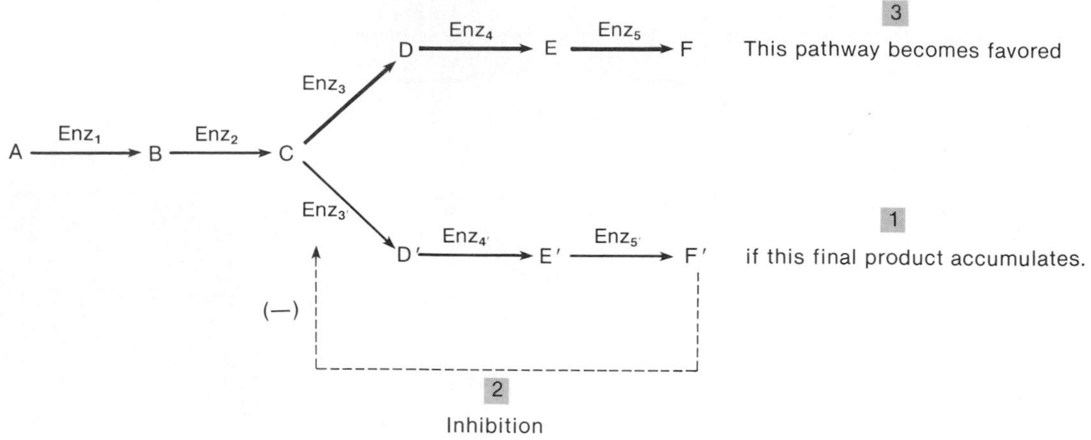

Figure 4.10. The effects of an inborn error of metabolism on a branched metabolic pathway.

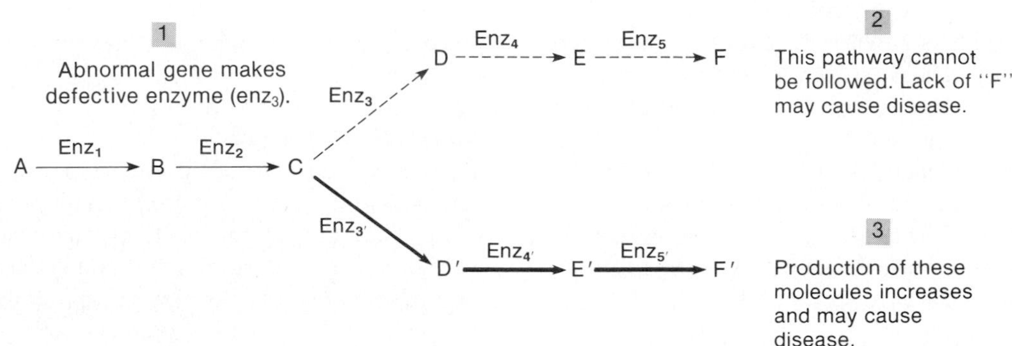

The mechanism by which a final product inhibits an earlier enzymatic step in its pathway is known as **allosteric** *(al″o-ster′ik)* **inhibition.** The allosteric inhibitor combines with a part of the enzyme located away from the active site, causing the active site to change shape so that it can no longer combine properly with its substrate.

Inborn Errors of Metabolism. Each enzyme in a metabolic pathway is coded by a different gene. An inherited defect in one of these genes may result in a disease known as an "inborn error of metabolism." In these diseases there is an *increased* amount of intermediates formed *prior* to the defective step, and a *decrease* in the intermediates and final products formed *after* the defective enzymatic step. Diseases may result from lack of the normal end-product, or from accumulations to toxic levels of intermediates or their alternate derivatives. If the defective enzyme is active at a step after a branch point in a pathway, the intermediates and final products of the divergent pathway will increase as a result of the block in the alternate pathway (fig. 4.10). Specific examples of inborn errors of metabolism are discussed at the end of this chapter.

allosteric: Gk. *allos,* other; *stereos,* position

1. Draw graphs to represent the effects of changes in temperatures, pH, enzyme and substrate concentration, cofactors, and coenzymes on the rate of enzymatic reactions. Explain the mechanisms responsible for the appearance of these graphs.
2. Draw a flowchart of a metabolic pathway (using arrows and letters such as *A, B, C,* etc.) with one branch point.
3. Describe a reversible reaction, and explain how the law of mass action affects this reaction.
4. Define end-product inhibition, and use your diagram of a branched metabolic pathway to explain how this process will affect the concentration of different intermediates.
5. Suppose, due to an inborn error of metabolism, that the enzyme that catalyzes the third reaction in your pathway (question no. 2) is defective. Describe the effects this would have on the concentrations of the intermediates in your pathway.

Bioenergetics

Central to life processes are chemical reactions that are coupled, so that the energy released by one reaction is incorporated into the products of another reaction. The transformation of energy in living systems is largely based on reactions that produce and destroy molecules of ATP and on oxidation-reduction reactions.

Objective 7. Describe the first and second laws of thermodynamics, and use these laws to explain why some molecules have more chemical bond energy than other molecules.

Objective 8. Describe how the coupling of energy-releasing and energy-requiring reactions is used to transform energy in living cells, and—in this context—explain the function of ATP in cells.

Objective 9. Explain how oxidation-reduction reactions involve the transfer of hydrogen from one molecule to another.

Bioenergetics refers to the flow of energy in living systems. Organisms maintain their highly ordered structure and life-sustaining activities through the constant expenditure of energy obtained ultimately from the environment. The energy flow in living systems obeys the first and second laws of a branch of physics known as *thermodynamics*.

According to the **first law of thermodynamics,** energy can be transformed, but it cannot be created or destroyed. This is sometimes called the *law of conservation of energy.* As a result of energy transformations, according to the **second law of thermodynamics,** the universe and its parts (including living systems) have increased amounts of *entropy (en'tro-pe).* Entropy is related to the degree of disorganization of a system. Only energy that is in an organized state—called *free energy*—can be used to do work. Thus, since entropy increases in every energy transformation, the amount of free energy available to do work decreases. As a result of the increased entropy described by the second law, systems tend to go from states of higher to states of lower free energy.

Matter is a form of energy (from Einstein's $E = mc^2$). Atoms that are organized into complex organic molecules, such as glucose, have more free energy (less entropy) than six separate molecules each of carbon dioxide and water. Therefore, in order to convert carbon dioxide and water to glucose, energy must be added. Plants perform this feat using energy from the sun in the process of *photosynthesis* (fig. 4.11).

Endergonic and Exergonic Reactions

Chemical reactions that require the input of energy are known as **endergonic** *(end''er-gon'ik)* **reactions.** Since energy is added to make these reactions "go," the products of endergonic reactions must contain more free energy than the reactants. A portion of the energy added, in other words, is contained within the product molecules. This follows from the fact that energy cannot be created or destroyed (first law of thermodynamics) and from the fact that a more organized state of matter contains more free energy (less entropy) than a less organized state (as described by the second law).

The fact that glucose contains more free energy than carbon dioxide and water can be easily proven by the combustion of glucose to CO_2 and H_2O. This reaction releases energy in the form of heat. Reactions that convert molecules with more free energy to molecules with less—and, therefore, release energy as they proceed—are called **exergonic reactions.**

As illustrated in figure 4.12, the amount of energy released by an exergonic reaction is the same whether the energy is released in a single combustion reaction or in the

Figure 4.11. A simplified diagram of photosynthesis. Some of the sun's radiant energy is captured by the plants and used to produce glucose from carbon dioxide and water. As the product of this endergonic reaction, glucose has a higher free energy content than the initial reactants.

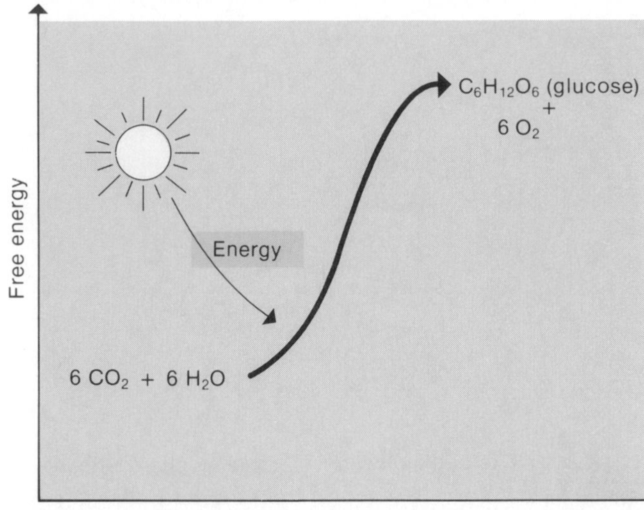

Figure 4.12. Since glucose contains more energy than carbon dioxide and water, the combustion of glucose is an exergonic reaction. The same amount of energy is released if the glucose is broken down stepwise within the cell.

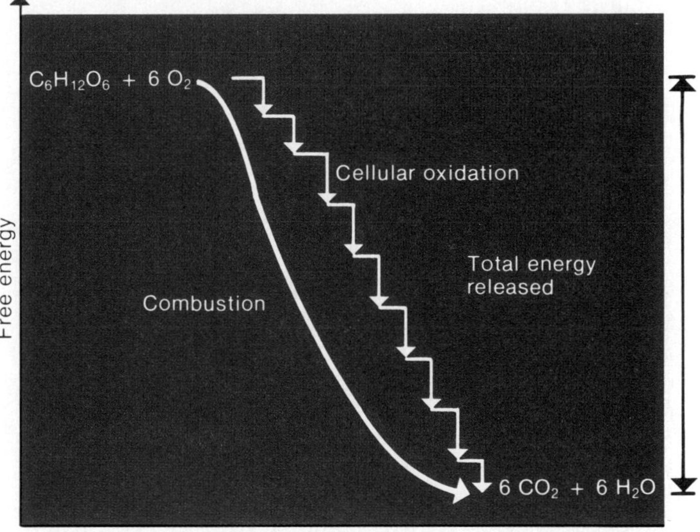

bioenergetics: Gk. *bios,* life; *energeia,* work
thermodynamics: Gk. *therme,* heat; *dynamis,* force
entropy: Gk. *entropia,* turning toward
photosynthesis: Gk. *phos,* light; *synthesis,* a putting together
endergonic: Gk. *endon,* within; *ergon,* work

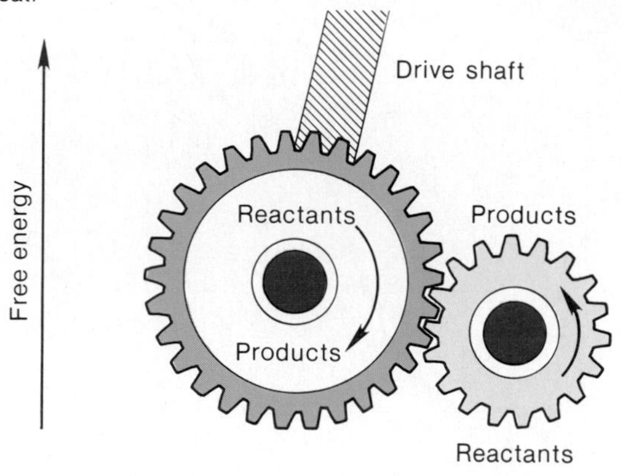

Figure 4.13. A model of the coupling of exergonic and endergonic reactions. The drive shaft (representing the energy of activation) turns the exergonic gear, which turns the endergonic gear. The reactants of the exergonic reaction (represented by the larger gear) have more free energy than the products of the endergonic reaction because the coupling is not 100% efficient—some energy is lost as heat.

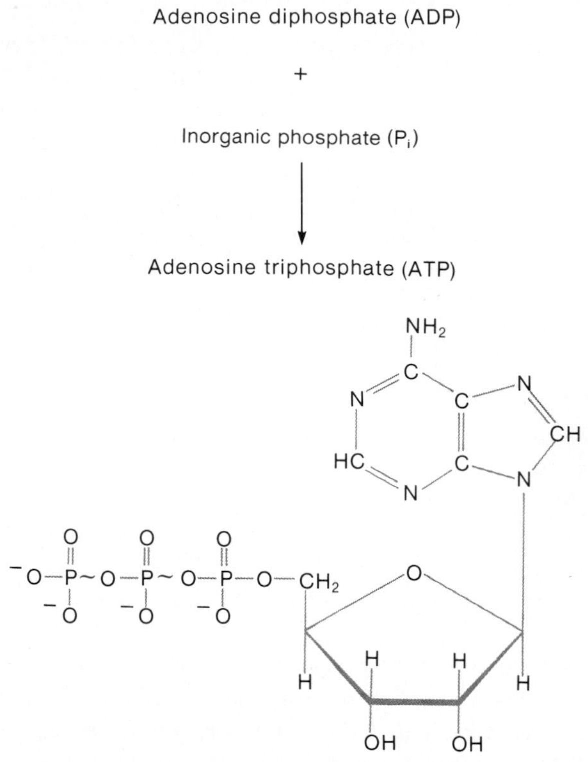

many small, enzymatically controlled steps that occur in tissue cells. The energy that the body obtains from the consumption of particular foods, therefore, can be measured as the amount of heat energy released when these foods are combusted.

Heat is measured in units called *calories*. One calorie is defined as the amount of heat required to raise the temperature of one cubic centimeter of water one degree Celsius. The caloric value of food is usually indicated in kilocalories (one kilocalorie equals a thousand calories), which is commonly expressed with a capital letter—*C*alories.

Coupled Reactions: ATP

In order to remain alive a cell must maintain its highly organized, low entropy state at the expense of free energy in its environment. Accordingly, the cell contains many enzymes that catalyze exergonic reactions, using substrates that come ultimately from the environment. The energy released by these exergonic reactions is used to drive the energy-requiring processes (endergonic reactions) in the cell. Since the cell cannot use heat energy to drive energy-requiring processes, chemical bond energy that is released in the exergonic reactions must be directly transferred to chemical bond energy in the products of endergonic reactions. Energy-liberating reactions are thus *coupled* to energy-requiring reactions. This relationship is like two meshed gears; the turning of one (the energy-releasing, exergonic gear) causes turning of the other (the energy-requiring, endergonic gear—fig. 4.13).

calorie: L. *calor,* heat

Figure 4.14. The formation and structure of adenosine triphosphate (ATP).

Adenosine diphosphate (ADP)

+

Inorganic phosphate (P_i)

↓

Adenosine triphosphate (ATP)

The energy released by most exergonic reactions in the cell is used, either directly or indirectly, to drive *one* endergonic reaction (fig. 4.14): the formation of **adenosine triphosphate** *(ah-den'o-sēn tri-fos'fat)* (**ATP**) from adenosine diphosphate and inorganic phosphate (abbreviated P_i).

The formation of ATP requires the input of a fairly large amount of energy. Since this energy must be conserved (first law of thermodynamics), the bond that is produced by joining P_i to ADP must contain a part of this energy. Thus, when enzymes reverse this reaction and convert ATP to ADP and P_i, a large amount of energy is released. Energy released from the breakdown of ATP is used to power the energy-requiring processes in all cells. As the **universal energy carrier,** ATP serves to couple more efficiently the energy released by the breakdown of food molecules to the energy required by the diverse endergonic processes in the cell (fig. 4.15).

Coupled Reactions: Oxidation-Reduction

When an atom or a molecule gains electrons, it is said to become **reduced;** when it loses electrons, it is said to become **oxidized.** Reduction and oxidation are always coupled reactions: an atom or a molecule cannot become oxidized unless it donates electrons to another, which therefore becomes reduced. The atom or molecule that donates electrons to another is a *reducing agent,* and the one that accepts electrons from another is an *oxidizing agent.* It

Figure 4.15. A model of ATP as the universal energy carrier of the cell. Exergonic reactions are shown as gears with arrows going down (reactions produce a decrease in free energy); endergonic reactions are shown as gears with arrows going up (reactions produce an increase in free energy).

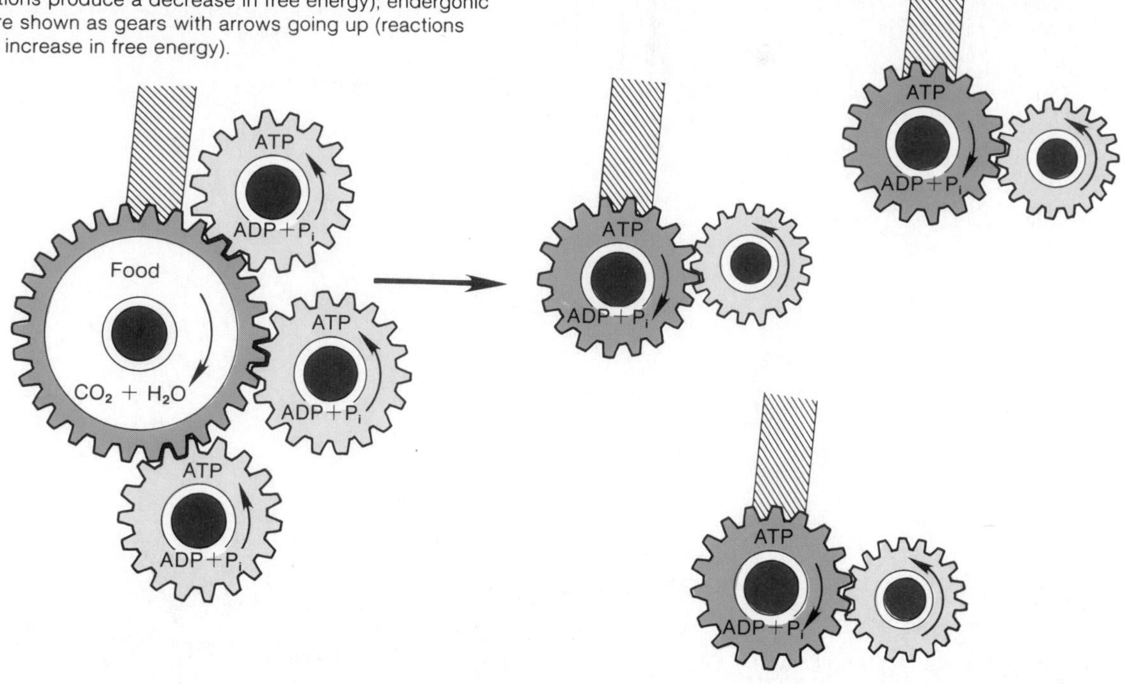

Figure 4.16. NAD_{ox} becomes reduced by the addition of two electrons from hydrogen atoms removed from an organic molecule (*X*). In this reaction, NAD acts as an oxidizing agent. In another cellular location, NAD_{red} can donate these two electrons to a different organic molecule (*Y*). The molecule *Y* is thus reduced by this reaction in which NAD serves as a reducing agent.

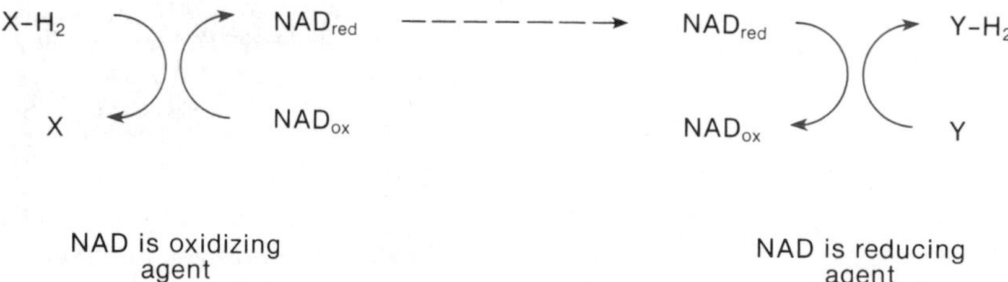

NAD is oxidizing agent

NAD is reducing agent

should be noted that an atom or a molecule may function as an oxidizing agent in one reaction and as a reducing agent in another reaction; it may gain electrons from one atom or molecule and pass them on to another in a series of coupled oxidation-reduction reactions—like a bucket brigade.

Notice that the term *oxidation* does not imply that oxygen participates in the reaction. This term is derived from the fact that oxygen has a great tendency to accept electrons, that is, to act as a strong oxidizing agent. This property of oxygen is exploited by cells; oxygen acts as the final electron acceptor in a chain of oxidation-reduction reactions that provides energy for ATP production.

Oxidation-reduction reactions in cells often involve the transfer of hydrogen atoms rather than of free electrons. Since a hydrogen atom contains one electron (and one proton in the nucleus) a molecule that loses hydrogen becomes oxidized, and one that gains hydrogen becomes reduced. In many oxidation-reduction reactions, pairs of electrons—either as free electrons or as a pair of hydrogen atoms—are transferred from the reducing agent to the oxidizing agent.

Two molecules that serve important roles in the transfer of hydrogens are **nicotinamide** *(nik''o-tin'ah-mīd)* **adenine dinucleotide (NAD),** which is derived from the vitamin niacin (vitamin B_3), and **flavin adenine dinucleotide (FAD),** which is derived from the vitamin riboflavin (vitamin B_2). These molecules are *hydrogen carriers* because they accept hydrogens (becoming reduced) in one cellular location and donate hydrogens (becoming oxidized in the process) at a different cellular location (fig. 4.16).

Figure 4.17. Structures of (*a*) the oxidized form of NAD (nicotinamide adenine dinucleotide) and (*b*) the reduced form of FAD (flavin adenine dinucleotide). Note the two additional hydrogen atoms (shown in color) that reduce FAD.

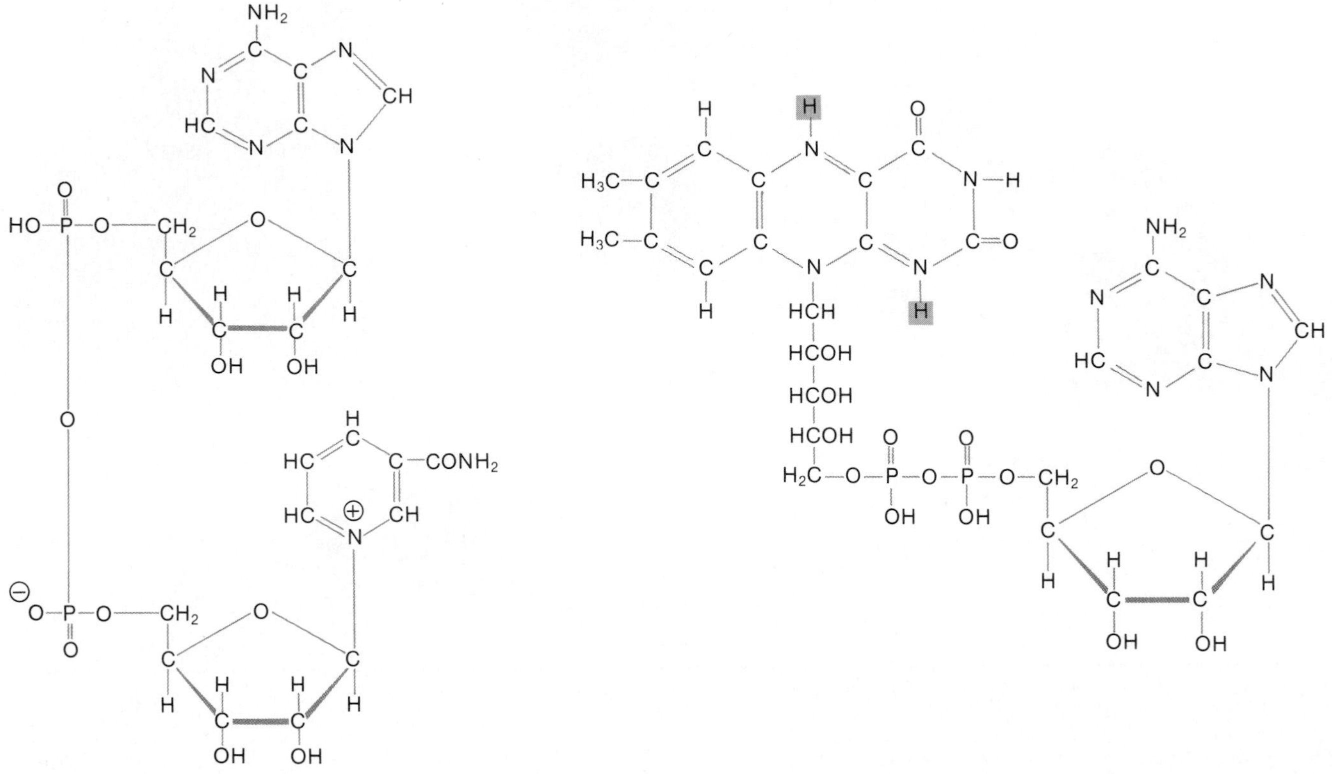

(a)

(b)

Each FAD can accept two electrons and can bind two protons. The reduced form of FAD can, therefore, combine with the equivalent of two hydrogen atoms and can be written as $FADH_2$. Each NAD can also accept two electrons but can only bind one proton. The reduced form of NAD may, therefore, be shown as $NADH + H^+$ (the H^+ represents a free proton). In order to represent these two hydrogen carriers in a way that is both consistent and correct, the oxidized forms will subsequently be shown as NAD_{ox} and FAD_{ox}, and the reduced forms will be shown as NAD_{red} and FAD_{red} (fig. 4.17).

1. Describe the first and second laws of thermodynamics, and use these laws to explain why the chemical bonds in glucose represent a source of potential energy and how cells can obtain this energy.
2. Define the terms *exergonic reaction* and *endergonic reaction*. Use these terms to explain the function of ATP in cells.
3. Using the symbols X-H_2 and Y, draw a coupled oxidation-reduction reaction. Identify the molecule that is reduced and the one that is oxidized, and tell which one is the reducing agent and which is the oxidizing agent.
4. Describe the functions of NAD, FAD, and oxygen (in terms of oxidation-reduction reactions), and explain the meaning of the symbols NAD_{red} and FAD_{red}.

Glycolysis and Anaerobic Respiration

In cellular respiration, chemical reactions liberate energy, some of which is used to produce ATP. The complete combustion of a molecule requires the presence of oxygen; some energy can be obtained, however, in the absence of oxygen. This process of anaerobic respiration results in the production of lactic acid from glucose.

Objective 10. Define glycolysis, and indicate the initial substrate and final products of this metabolic pathway.

Objective 11. Describe the pathway of anaerobic respiration, and explain how lactic acid is produced.

Objective 12. Define *gluconeogenesis*, and explain the significance of this process following exercise.

All of the reactions in the body that involve energy transformation are collectively termed **metabolism** *(me-tab'o-lizm)*. Metabolism may be divided into two categories: *anabolism (ah-nab'o-lizm)* and *catabolism (ca-tab'o-lizm)*. Catabolic reactions release energy, usually by the

metabolism: Gk. *metabole*, change
anabolism: Gk. *anabole*, a raising up
catabolism: Gk. *katabole*, a casting down

Figure 4.18. The energy expenditure and gain in glycolysis. Notice that there is a "net profit" of 2 ATP and 2 NAD_{red} per glucose molecule in glycolysis. Molecules listed by number are (*1*) fructose-1, 6-diphosphate; (*2*) 1, 3-diphosphoglyceric acid; and (*3*) 3-phosphoglyceric acid (see fig. 4.19).

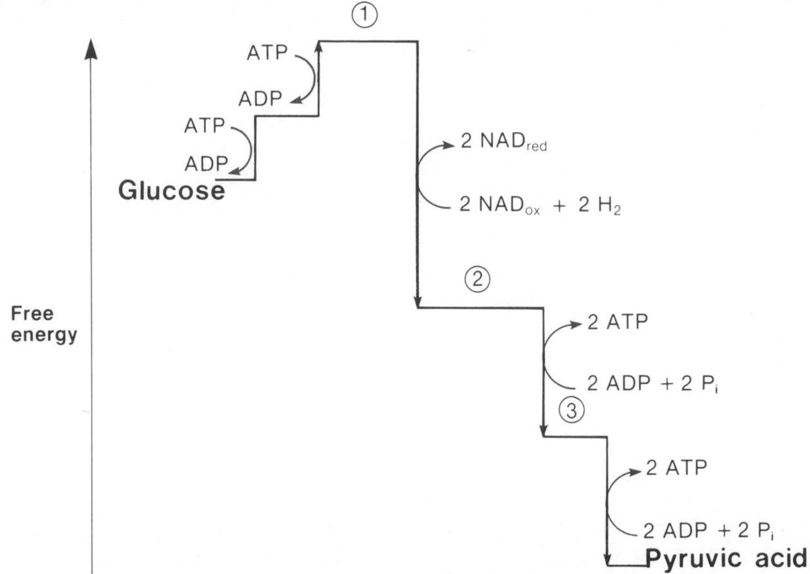

breakdown of larger organic molecules into smaller molecules. Anabolic reactions require the input of energy and include the synthesis of large, energy-storage molecules such as glycogen, fat, and protein.

The catabolic reactions that break down glucose, fatty acids, and amino acids serve as the primary sources of energy for the cellular synthesis of ATP. These metabolic pathways are known collectively as *cellular respiration.* When oxygen serves as the final electron acceptor, these processes are called **aerobic** *(a-er-o'bik)* **cell respiration.** The final products of aerobic respiration are carbon dioxide, water, and energy (a part of which is trapped in the chemical bonds of ATP). The overall equation for aerobic respiration, therefore, is identical to the equation that describes combustion (fuel + $O_2 \rightarrow CO_2 + H_2O$ + energy).

Notice that the term *respiration* refers to chemical reactions that liberate energy for the production of ATP. The oxygen used in aerobic respiration by tissue cells is obtained from the blood; the blood, in turn, becomes oxygenated in the lungs by the process of breathing. Breathing (also called *ventilation* or *external respiration*) is thus needed for, but is different from, aerobic respiration.

Unlike combustion, the conversion of glucose to carbon dioxide and water within the cells occurs in small, enzymatically catalyzed steps. Oxygen is only used at the last step (this will be described in a later section). Since a small amount of the chemical bond energy of glucose is released at early steps in the metabolic pathway, some cells in the body can obtain energy for ATP production in the temporary absence of oxygen. This process is called **anaerobic** *(an''a-er-o'bik)* **respiration.**

anaerobic: Gk. *an,* private; *aer,* air; *bios,* life

Glycolysis

Both the anaerobic and the aerobic respiration of glucose begin with a metabolic pathway known as **glycolysis** *(gli-kol'ĭ-sis).* In glycolysis, glucose—a six-carbon (hexose) sugar—is converted into two molecules of *pyruvic (pi-roo'vik) acid.* Each pyruvic acid molecule contains three carbons, three oxygens, and four hydrogens. The number of carbon and oxygen atoms in glucose—$C_6H_{12}O_6$—can thus be accounted for in the two pyruvic acid molecules. Since the two pyruvic acids together account for only eight hydrogens, however, it is clear that four hydrogen atoms are removed from the intermediates in glycolysis. These hydrogen atoms are used to reduce two molecules of NAD_{ox} to two molecules of NAD_{red}.

Glycolysis is exergonic, and a portion of the energy that is released is used to drive the endergonic reaction ADP + $P_i \rightarrow$ ATP. At the end of the glycolytic pathway there is a net gain of two ATP per glucose molecule, as indicated in the overall equation for glycolysis:

$$\text{Glucose} + 2NAD_{ox} + 2ADP + 2P_i \rightarrow$$
$$2 \text{ pyruvic acid} + 2NAD_{red} + 2ATP.$$

Although the overall equation for glycolysis is exergonic, glucose must be "activated" at the beginning of the pathway before energy can be obtained. This activation requires the addition of two phosphate groups derived from two molecules of ATP. Energy from the reaction ATP $\rightarrow$ ADP + P_i is therefore consumed at the beginning of glycolysis. This is shown as an "up-staircase" in figure 4.18. At later steps in glycolysis, however, four molecules of ATP are produced (and two molecules of NAD are reduced) as energy is liberated (the "down-staircase"

glycolysis: Gk. *glyco,* sugar; *lysis,* breaking

Figure 4.19. In glycolysis, one glucose molecule is converted into two pyruvic acid molecules in nine separate steps. Since two pyruvic acids are produced from one glucose, the products are multiplied by two in the pathway shown. In addition to two pyruvic acids, these products include two molecules of NAD_{red} and four molecules of ATP. Since two ATP molecules were used at the beginning of glycolysis, however, the net gain is two ATP per glucose.

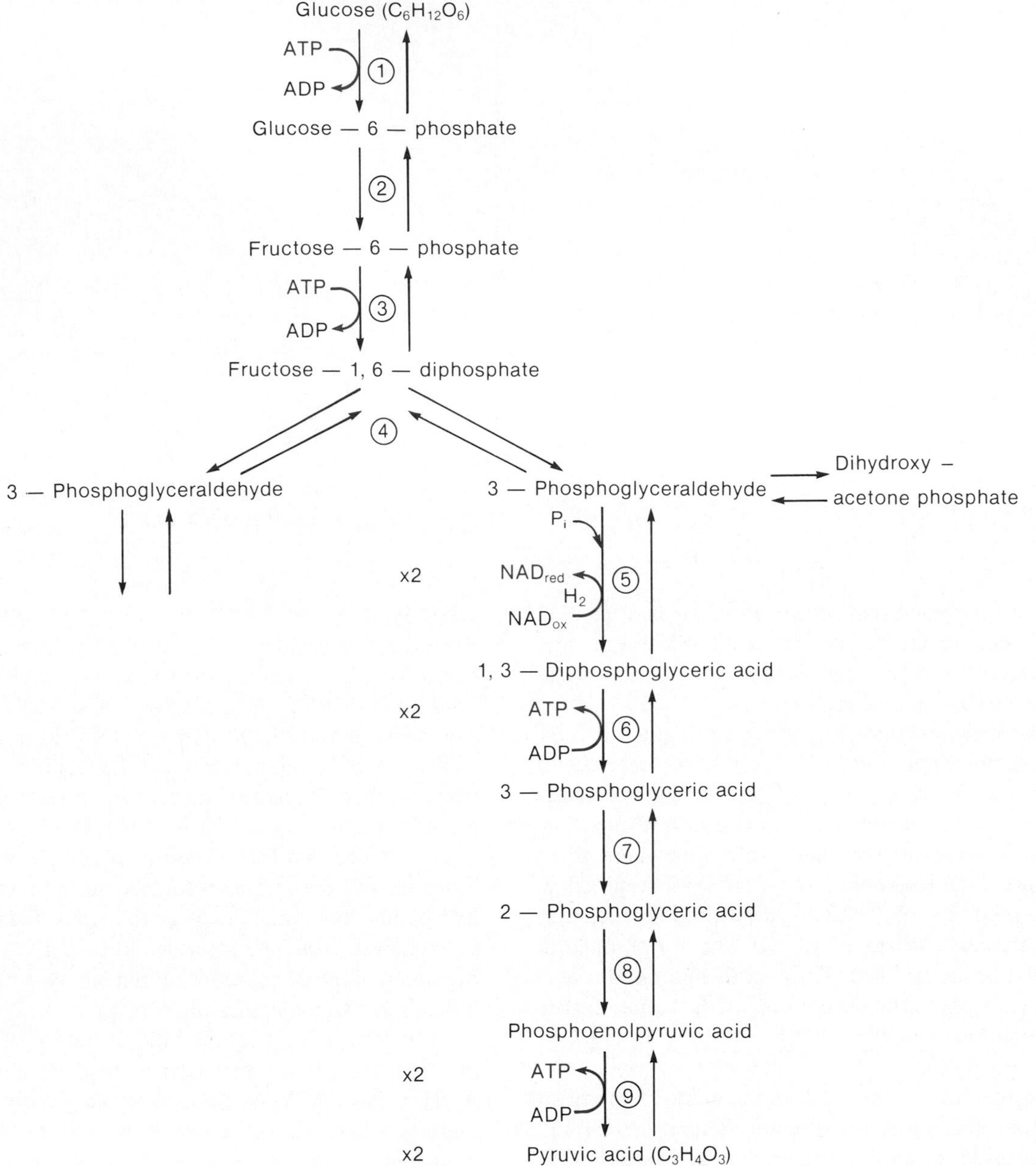

in figure 4.18). The two molecules of ATP used in the beginning, therefore, represent an energy investment; the net gain of two ATP and two NAD_{red} by the end of the pathway represent an energy "profit."

The overall equation for glycolysis obscures the fact that this is a metabolic pathway consisting of nine separate steps. The individual steps in this pathway are shown in figure 4.19, and the enzymes that catalyze these steps are listed in table 4.3.

Anaerobic Respiration

In order for glycolysis to continue, adequate amounts of NAD_{ox} must be available to accept hydrogen atoms. The NAD_{red} that is produced in glycolysis, therefore, must become oxidized by donating its electrons to another molecule. (In aerobic respiration this other molecule is located in the mitochondria and ultimately passes its electrons to oxygen.)

When oxygen is *not* available in sufficient amounts, the NAD_{red} produced in glycolysis are oxidized in the cytoplasm by donating their electrons to pyruvic acid. This results in the formation of NAD_{ox} and the addition of two

Figure 4.20. The addition of two hydrogen atoms (colored boxes) from reduced NAD to pyruvic acid produces lactic acid and oxidized NAD. This reaction is catalyzed by lactate dehydrogenase (LDH).

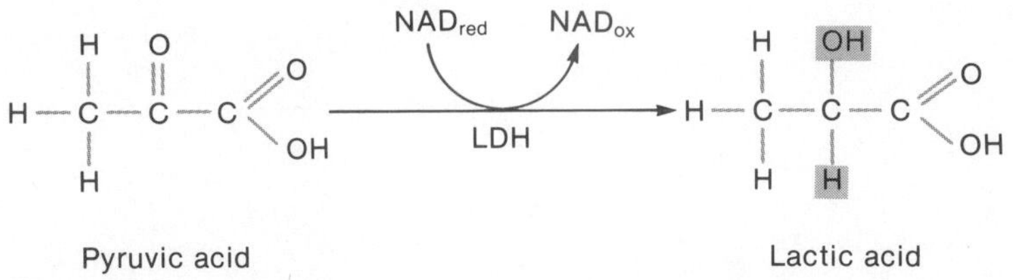

$$\text{Pyruvic acid } (CH_3COCOOH) \xrightarrow[\text{LDH}]{NAD_{red} \rightarrow NAD_{ox}} \text{Lactic acid } (CH_3CHOHCOOH)$$

Table 4.3 Enzymes, cofactors, and coenzymes required for glycolysis

Step	Enzyme	Coenzyme or cofactor	Comments
1	Hexokinase	Mg^{++}	Liver enzyme catalyzes the phosphorylation of glucose, fructose, or mannose
2	Hexose phosphate isomerase	—	Interconverts glucose and fructose
3	Phosphofructokinase	Mg^{++}, ATP	Allosteric enzyme that is inhibited by high ATP
4	Aldolase	—	Splits hexose sugar into two three-carbon compounds
5	Phosphoglyceraldehyde dehydrogenase	NAD	Adds inorganic phosphate and oxidizes aldehyde to acid as NAD is reduced by removal of two hydrogens
6	Phosphoglycerate kinase	Mg^{++}	Two molecules of ATP formed at this step
7	Phosphoglyceromutase	—	Phosphate group transferred to different carbon
8	Enolase	Mg^{++}, Mn^{++}	Catalyzes molecular rearrangement
9	Pyruvate kinase	Mg^{++}, K^{+}	Two molecules of ATP formed at this step

hydrogen atoms to pyruvic acid, which is thus reduced. This addition of two hydrogen atoms to pyruvic acid produces *lactic acid* (fig. 4.20). This metabolic pathway, by which glucose is converted through pyruvic acid to lactic acid, is called **anaerobic respiration.**

Anaerobic respiration yields a net gain of two ATP (produced by glycolysis) per glucose molecule. A cell can survive anaerobically as long as it can produce sufficient energy for its needs in this way and as long as lactic acid concentrations do not become excessive. Some tissues are better adapted to anaerobic respiration than others—skeletal muscles survive better than cardiac muscle, which in turn can survive under anaerobic conditions longer than can the brain.

Except for red blood cells, which can only respire anaerobically (thus sparing the oxygen they carry), anaerobic respiration provides only a temporary sustenance for tissues that have energy requirements in excess of their aerobic ability. Anaerobic respiration can only occur for a limited period of time (longer for skeletal muscles; shorter for the heart, and shortest for the brain) when the *ratio of oxygen supply to oxygen need* (related to the concentration of NAD_{red}) falls below a critical level. Anaerobic respiration is, in a sense, an emergency procedure that provides some ATP until the emergency (oxygen deficiency) has passed. It should be noted that this is a normal, daily occurrence in skeletal muscles during exercise (described later in this chapter), but it represents a pathological condition when it occurs in the heart.

Ischemia refers to inadequate blood flow to an organ, such that the rate of oxygen delivery is insufficient to maintain aerobic respiration. Inadequate blood flow to the heart, or *myocardial ischemia,* may occur if the coronary blood flow is occluded by atherosclerosis, a blood clot, or by an artery spasm. People with myocardial ischemia often experience *angina pectoris,* severe pain in the chest and left (and sometimes the right) arm area. This pain is associated with increased blood levels of lactic acid, which are produced by anaerobic respiration by the heart muscle. The degree of ischemia and angina can be decreased by vasodilator drugs such as nitroglycerin and amyl nitrite, which improve blood flow to the heart and also decrease the work of the heart by dilating peripheral blood vessels.

Gluconeogenesis and the Cori Cycle. Some of the lactic acid produced by exercising skeletal muscles is delivered by the blood to the liver. Lactic acid dehydrogenase within liver cells can convert lactic acid to pyruvic acid, as NAD_{ox} is reduced to NAD_{red}. Unlike most other organs, the liver contains the enzymes needed to convert pyruvic acid to glucose-6-phosphate. This is essentially the reverse of glycolysis. Glucose-6-phosphate in liver cells can be used as an intermediate for glycogen synthesis, or it can be converted to free glucose, which is secreted into the blood.

Figure 4.21. The Cori cycle. The direction of reversible reactions that occurs in the Cori cycle is shown by solid arrows; the sequence of steps is indicated by numbers 1 through 9.

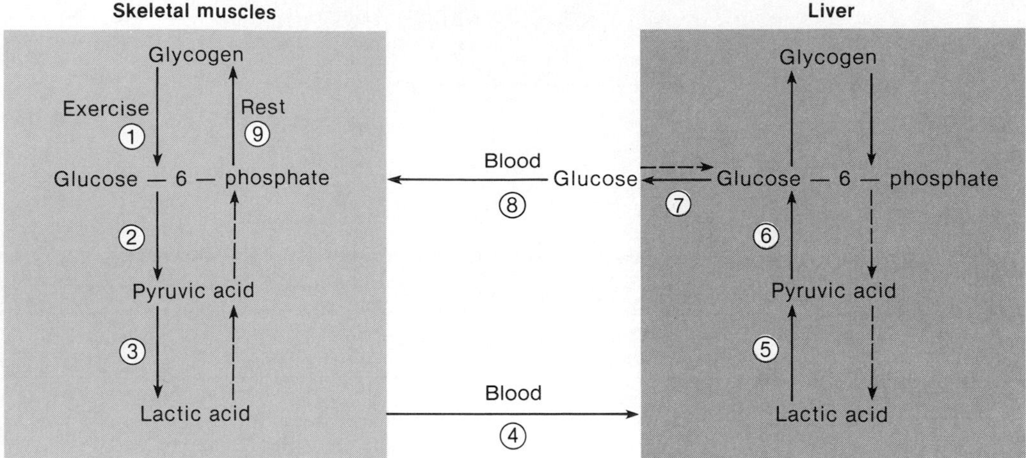

The conversion of noncarbohydrate molecules (lactic acid, amino acids, and glycerol) through pyruvic acid into glucose is called **gluconeogenesis** *(gloo''ko-ne''o-jen'ĕ-sis)*. In starvation and in prolonged exercise, when glycogen stores are depleted, the formation of new glucose in this way becomes the only means for maintaining constant blood-sugar levels. Under these conditions, gluconeogenesis in the liver is the only way that adequate blood glucose levels can be maintained to prevent brain death.

During exercise some of the lactic acid produced by skeletal muscles may be transformed to blood glucose through gluconeogenesis in the liver. This new glucose can serve as an energy source during exercise and can be used after exercise in order to replenish the depleted muscle glycogen. This two-way traffic between skeletal muscles and the liver is called the **Cori cycle** (fig. 4.21). Recent evidence also suggests that some muscle fibers may be able to convert lactic acid to glucose. Through the Cori cycle, gluconeogenesis in the liver (together perhaps with gluconeogenesis in some muscle fibers) allows depleted skeletal muscle glycogen to be restored within forty-eight hours after exercise.

1. Define the term *glycolysis* in terms of its initial substrates and products. Explain why there is a net gain of two ATP in this process.
2. Define the term *anaerobic respiration* in terms of its initial substrates and final products. Explain the significance of lactic acid formation at the end of this process.
3. Explain when and why anaerobic respiration occurs during exercise.
4. Define the term *gluconeogenesis,* and explain how this process functions to restore the glycogen stores of skeletal muscles following exercise.

Cori: from Carl F. Cori, U.S. biochemist, 1896-1984

AEROBIC RESPIRATION

Aerobic respiration is equivalent to combustion in terms of its final products (CO_2 and H_2O) and in terms of the total amount of energy liberated. In aerobic respiration, however, the energy is released in small, enzymatically controlled oxidation reactions, and a portion (38%–40%) of the energy released is captured in the high energy bonds of ATP.

Objective 13. Describe the fate of pyruvic acid in aerobic respiration.
Objective 14. Describe the functional significance of the Krebs cycle, and list its products.
Objective 15. Describe the electron transport system and oxidative phosphorylation; explain the role of oxygen in aerobic respiration.
Objective 16. Explain how the metabolism of glycogen and fat are regulated by end-product inhibition.

The aerobic respiration of glucose begins with glycolysis. Glycolysis in both anaerobic and aerobic respiration results in the production of two molecules of pyruvic acid, two molecules of ATP, and two molecules of NAD_{red} per glucose. In aerobic respiration, however, the electrons in NAD_{red} are *not* donated to pyruvic acid and lactic acid is not formed, as happens in anaerobic respiration.

The enzymes that catalyze glycolysis are located in the cell cytoplasm. In aerobic respiration, pyruvic acid leaves the cell cytoplasm and enters the interior of mitochondria. Once pyruvic acid is inside a mitochondrion, carbon dioxide is enzymatically removed from each three-carbon-long pyruvic acid to form a two-carbon-long organic acid—acetic acid. The enzyme that catalyzes this reaction combines the acetic acid with a coenzyme (derived from the vitamin pantothenic acid) called *coenzyme A*. The combination thus produced is called **acetyl** *(as'ĕ-til)* **coenzyme A,** abbreviated **acetyl CoA** (fig. 4.22).

Figure 4.22. The formation of acetyl coenzyme A in aerobic respiration.

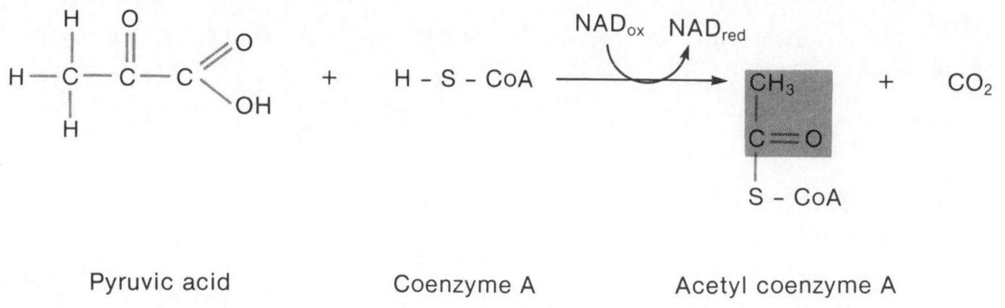

$$\underset{\text{Pyruvic acid}}{\mathrm{H{-}C(H)(H){-}C(=O){-}C(=O)OH}} + \underset{\text{Coenzyme A}}{\mathrm{H{-}S{-}CoA}} \xrightarrow{\mathrm{NAD_{ox}} \rightarrow \mathrm{NAD_{red}}} \underset{\text{Acetyl coenzyme A}}{\mathrm{CH_3{-}C(=O){-}S{-}CoA}} + \mathrm{CO_2}$$

Figure 4.23. A simplified diagram of the Krebs cycle showing how the original four-carbon-long oxaloacetic acid is regenerated at the end of the cyclic pathway. Only the numbers of carbon atoms in the Krebs cycle intermediates are shown; the numbers of hydrogens and oxygens are not accounted for in this simplified scheme.

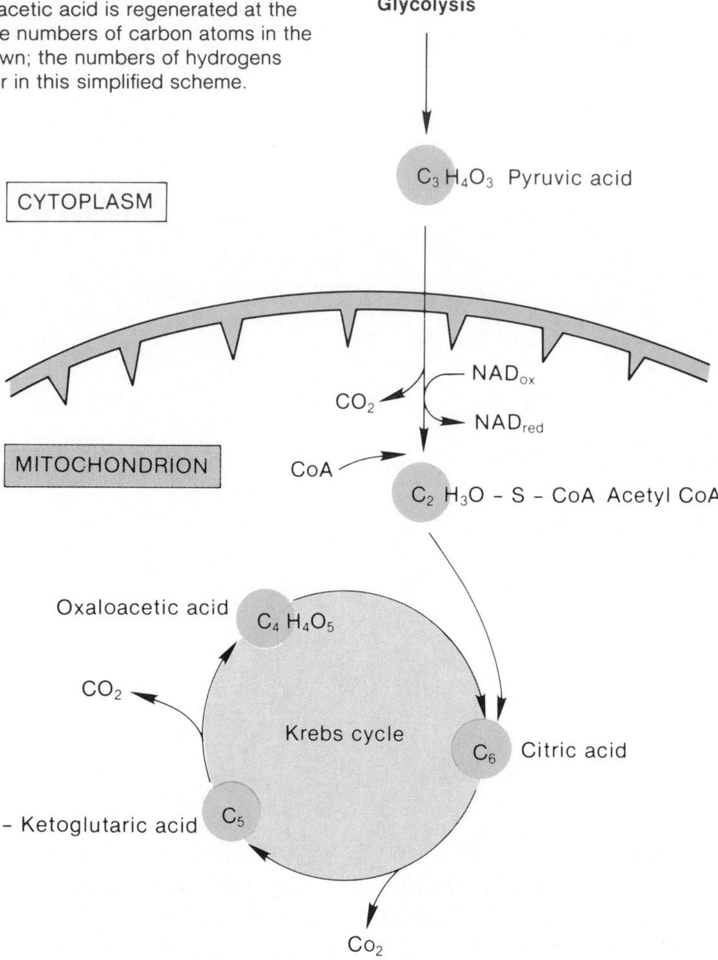

Since glycolysis converts one glucose molecule to two molecules of pyruvic acid, two molecules of acetyl CoA and two molecules of CO_2 are derived from each glucose. The acetyl CoA molecules serve as substrates for mitochondrial enzymes in the aerobic pathway; the carbon dioxide is a waste product in this process, which is carried by the blood to the lungs for elimination. It should be noted that the oxygen in CO_2 is derived from pyruvic acid, not from oxygen gas.

The Krebs Cycle

Once acetyl CoA is formed, the acetic acid subunit (two carbons long) is combined with oxaloacetic acid (four carbons long) to form a molecule of citric acid (six carbons long). Coenzyme A acts only as a transporter of acetic acid from one enzyme to another (similar to the transport of hydrogen by NAD). The formation of citric acid begins a cyclic metabolic pathway known as the **citric acid cycle,** or **TCA cycle** (for tricarboxylic acid; citric acid has three carboxylic acid groups). Most commonly, however, this cyclic pathway is named after its principal discoverer, Sir Hans A. Krebs, and is called the **Krebs cycle.** A simplified illustration of this pathway is shown in figure 4.23.

Through a series of reactions involving the elimination of two carbons and four oxygens (as two CO_2 molecules) and the removal of hydrogens, citric acid is eventually converted to oxaloacetic acid, which completes the cyclic metabolic pathway. In this process the following

Figure 4.24. The Krebs cycle.

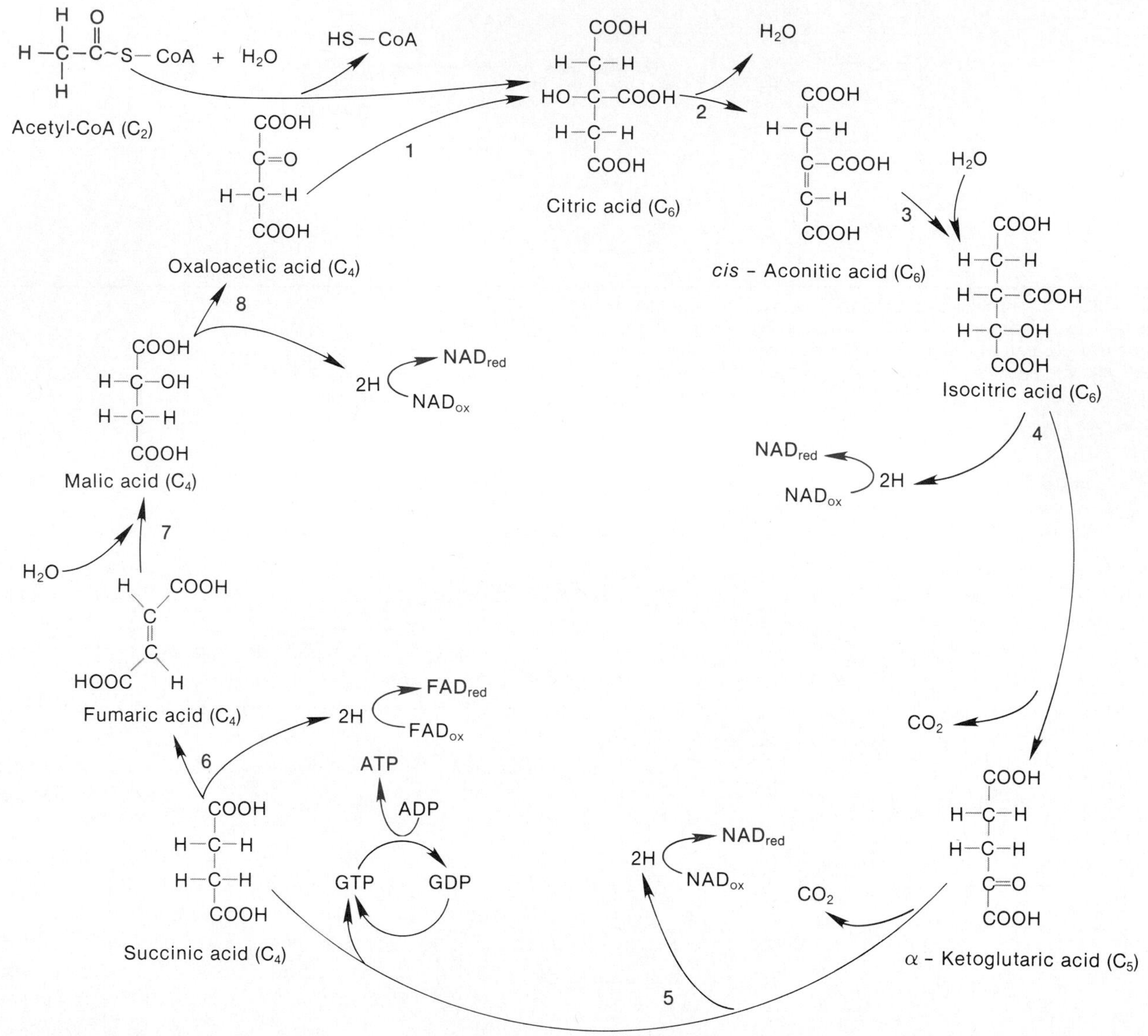

occur: (1) one guanosine triphosphate (GTP) is produced (step 5 of figure 4.24), which is converted to ATP; (2) three molecules of NAD are reduced (steps 4, 5, and 8 of figure 4.24); (3) one molecule of FAD is reduced (step 6). The production of reduced NAD and FAD by each "turn" of the Krebs cycle is far more significant, in terms of energy production, than the single GTP (converted to ATP) produced directly by the cycle.

Electron Transport and Oxidative Phosphorylation

Built into the foldings, or cristae, of the inner mitochondrial membrane are a series of molecules that serve in **electron transport** during aerobic respiration. This electron transport chain of molecules consists of a flavoprotein (derived from the vitamin riboflavin), coenzyme Q (derived from vitamin E), and a group of iron-containing pigments called *cytochromes (si'to-krōmz).* The last of these cytochromes is an *ATPase enzyme* that produces ATP from ADP and P_i, using energy derived from electron transport. This subunit protrudes from the cristae (under some conditions) like a lollipop (fig. 4.25).

In aerobic respiration, NAD_{red} and FAD_{red} become oxidized by transferring their pairs of electrons to the electron transport system of the cristae. In this way the oxidized forms of NAD and FAD are regenerated and able to continue to "shuttle" electrons from the Krebs cycle to the electron transport chain. The first molecule of the electron transport chain, in turn, becomes reduced when it accepts the electron pair from NAD_{red}. When the cytochromes receive a pair of electrons, two ferric ions (Fe^{+++}) become reduced to two ferrous ions (Fe^{++}). Notice that the gain of an electron is indicated by the reduction of the number of positive charges.

Figure 4.25. The ATPase at the end of the electron transport system is attached to the cristae by a stalk. This unit produces ATP, using energy from electron transport (a process called oxidative phosphorylation).

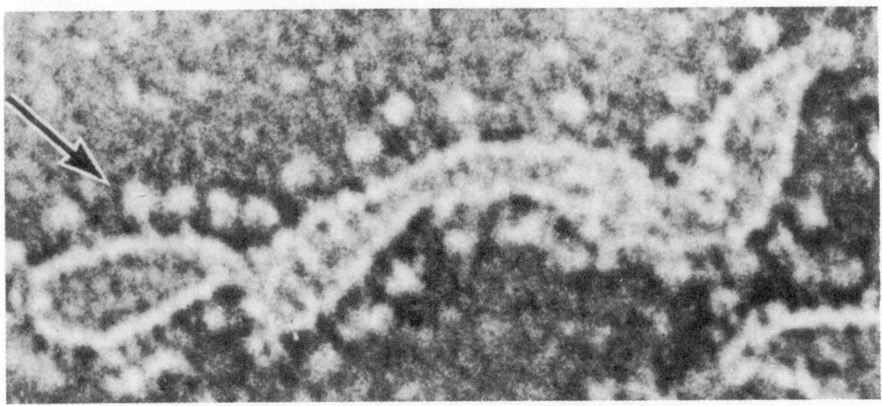

The electron transport chain thus acts as an oxidizing agent for NAD and FAD. Each element in the chain, however, also functions as a reducing agent; one reduced cytochrome transfers its electron pair to the next cytochrome in the chain. In this way, the iron ions in each cytochrome alternately become reduced (to ferrous ions) and oxidized (to ferric ions)—like a "ferrous wheel."

The passage of electrons along the electron transport chain is an exergonic process; the reduced form of the last cytochrome contains less free energy than the first. Energy released by electron transport is used (by a process whose description is beyond the scope of this book) to "drive" the reaction in which ADP is "phosphorylated"—by the addition of P_i—to ATP. The production of ATP in this way is appropriately termed **oxidative phosphorylation** *(fos''for-ĭ-la'shun)* (fig. 4.26).

Function of Oxygen. If the last cytochrome remained in a reduced state, it would be unable to accept more electrons. Electron transport would then progress only to the next-to-last cytochrome. This process would continue until all of the elements of the electron transport chain remained in the reduced state. If this occurred, NAD_{red} and FAD_{red} could not become oxidized by donating their electrons to the cytochrome chain, and through inhibition of Krebs cycle enzymes, no more NAD_{red} and FAD_{red} could be produced in the mitochondria (the Krebs cycle would stop). Respiration would then become anaerobic.

Oxygen, from the air we breathe, allows electron transport to continue by functioning as the **final electron acceptor** of the electron transport chain. This oxidizes the last cytochrome so that electron transport and oxidative phosphorylation can continue. At the very last step of aerobic respiration, therefore, oxygen becomes reduced by the two electrons that were passed to the chain from NAD_{red} and FAD_{red}. This reduced oxygen binds two protons, and a molecule of water (H_2O) is formed.

ATP Balance Sheet

Each time the Krebs cycle turns, three molecules of NAD are reduced by electrons from three pairs of hydrogens removed from Krebs cycle intermediates. Each NAD_{red} donates a pair of electrons to the electron transport chain. Transport of this pair of electrons to oxygen generates energy for the production of three molecules of ATP through oxidative phosphorylation. Electrons from FAD_{red} enter the electron transport chain "down the line" from where the first ATP is produced. Each pair of electrons from FAD_{red}, therefore, produces only two molecules of ATP from oxidative phosphorylation.

The three NAD_{red} produced per turn of the Krebs cycle result in the production of nine ATP molecules. The single FAD_{red} per turn of the Krebs cycle results in the production of two ATP. Together with the single ATP made directly by the Krebs cycle (from GTP), each turn of the Krebs cycle, therefore, yields a total of twelve ATP molecules. Since one molecule of glucose produces two pyruvic acids and thus two turns of the Krebs cycle, a total of twenty-four ATP molecules are produced by a single molecule of glucose through the Krebs cycle and oxidative phosphorylation.

The conversion of pyruvic acid to acetyl CoA also involves the reduction of one NAD. Since two pyruvic acids are produced per glucose, and since each NAD_{red} yields three ATP by oxidative phosphorylation, a total of twenty-four plus six, or thirty, ATP molecules are made in the mitochondrion from the steps that occur after pyruvic acid formation.

Two molecules of NAD_{red} are produced in the cytoplasm during glycolysis (conversion of glucose to pyruvic acid). These NAD_{red} cannot directly enter the mitochondria; instead, they donate their electrons to other molecules that "shuttle" these electrons into the mitochondria. Depending on which shuttle is used, either two or three

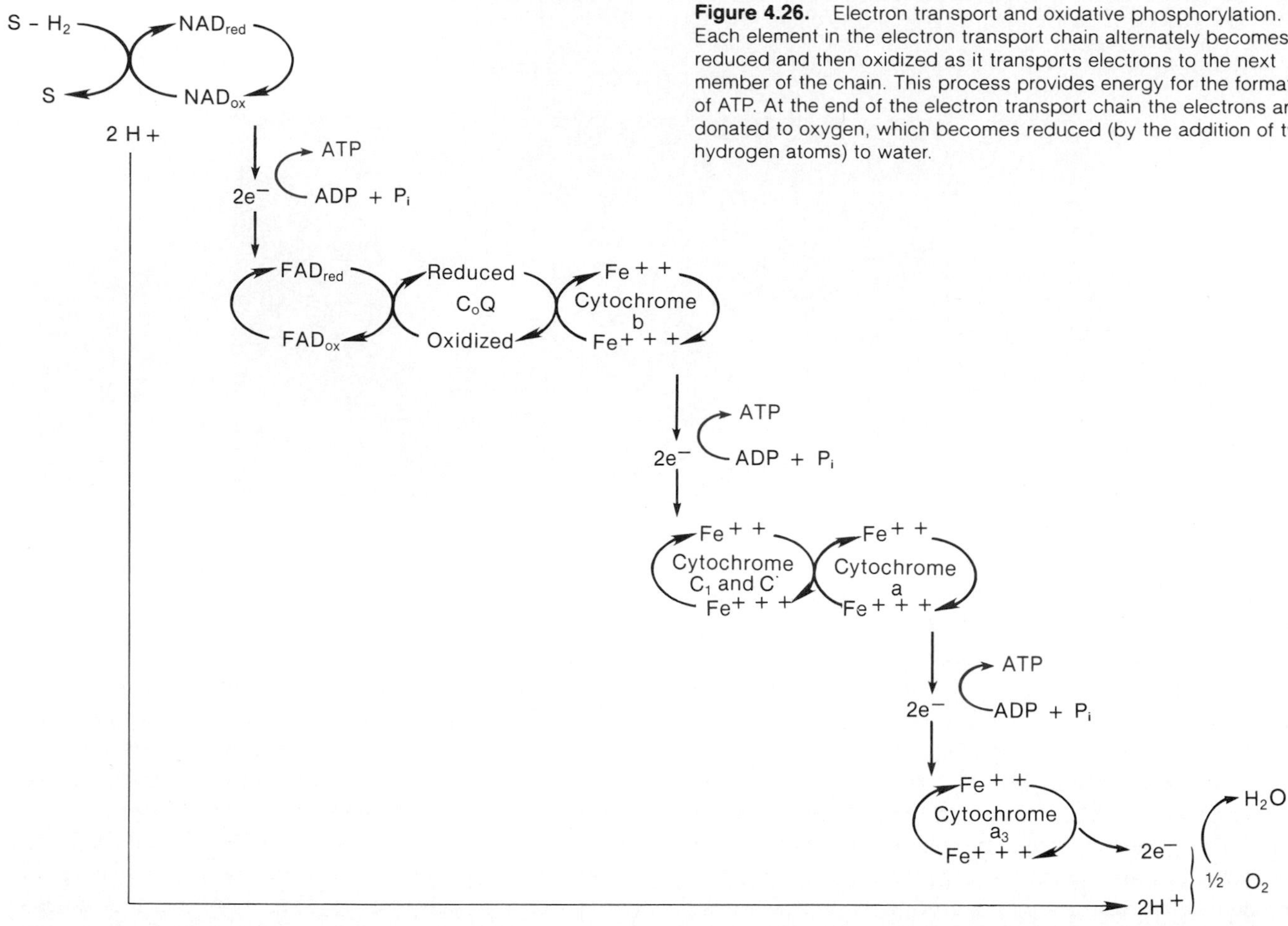

Figure 4.26. Electron transport and oxidative phosphorylation. Each element in the electron transport chain alternately becomes reduced and then oxidized as it transports electrons to the next member of the chain. This process provides energy for the formation of ATP. At the end of the electron transport chain the electrons are donated to oxygen, which becomes reduced (by the addition of two hydrogen atoms) to water.

Table 4.4 Maximum ATP yield per glucose molecule respired anaerobically and aerobically

Phase of respiration	High-energy products	ATP from oxidative phosphorylation	ATP subtotal
Glycolysis (glucose to pyruvic acid)	2 ATP	—	2 (total if anaerobic)
	2 NAD_{red}	6	8 (if aerobic)
Pyruvic acid to acetyl CoA	1 NAD_{red} (X2)	6	14
Krebs cycle	1 ATP (X2)	—	16
	3 NAD_{red} (X2)	18	34
	1 FAD_{red} (X2)	4	38
		Total (aerobic)	38 ATP

ATP molecules can be produced from each pair of these cytoplasmic electrons through oxidative phosphorylation. A total of four or six ATP molecules are thus produced. Added to the thirty ATP previously mentioned, this brings the total to thirty-four or thirty-six (depending on which shuttle is used for the cytoplasmic electrons). Together with the two molecules of ATP produced directly by glycolysis, a grand total of thirty-six to thirty-eight ATP are produced by the aerobic respiration of glucose (table 4.4).

Glycogenesis and Glycogenolysis

Many tissues, particularly the liver, skeletal muscles, and heart, store carbohydrates in the form of glycogen. The formation of glycogen from glucose is called **glycogenesis.** In this process, glucose is converted to glucose-6-phosphate by utilization of the terminal phosphate group of ATP. Glucose-6-phosphate is then converted into its isomer, glucose-1-phosphate. Finally, the enzyme *glycogen synthetase* removes these phosphate groups as it polymerizes glucose to form glycogen.

glycogenesis: Gk. *glyco,* sugar; *genesis,* production

Figure 4.27. Blood glucose that enters tissue cells is rapidly converted to glucose-6-phosphate. This intermediate can be metabolized for energy in glycolysis or can be converted to glycogen (*1*), a process called *glycogenesis.* Glycogen represents a storage form of carbohydrates, which can be used as a source for new glucose-6-phosphate (*2*), in a process called *glycogenolysis.* The liver contains an enzyme that can remove the phosphate from glucose-6-phosphate; liver glycogen thus serves as a source for new blood glucose.

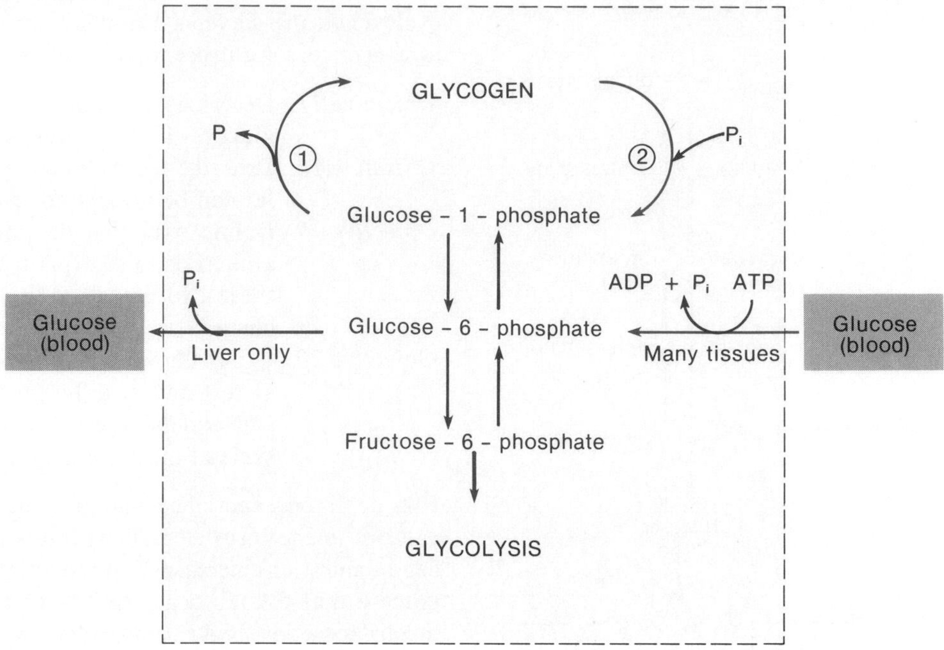

The reverse reactions are similar. The enzyme *glycogen phosphorylase* catalyzes the breakdown of glycogen to glucose-1-phosphate (the phosphates are derived from inorganic phosphate, not from ATP). Glucose-1-phosphate is then converted to glucose-6-phosphate. The conversion of glycogen to glucose-6-phosphate is called **glycogenolysis** *(gli''ko-jĕ-nol'ĭ-sis).* In most tissues, glucose-6-phosphate can then be used to resynthesize glycogen, or it can be respired for energy (through glycolysis). Glucose-6-phosphate *cannot* be released from the cells into the blood because organic molecules with phosphate groups cannot cross cell membranes.

Unlike skeletal muscles, the liver contains an enzyme—known as *glucose-6-phosphatase*—that can remove the phosphate groups and produce free glucose (fig. 4.27). This free glucose can then be transported through the cell membrane; the liver, therefore, can secrete glucose into the blood, whereas skeletal muscles cannot. Liver glycogen can thus supply blood glucose for use by other organs, including exercising skeletal muscles that have depleted much of their own stored glycogen.

Control of Carbohydrate Metabolism

When oxygen is not available in sufficient amounts, the activity of the Krebs cycle enzymes decreases, and glycolysis (and anaerobic respiration) increases. This switching from aerobic to anaerobic respiration is accomplished largely by the allosteric inhibition of enzymes. Key glycolytic enzymes are inhibited by high ATP and stimulated by high ADP levels. When the Krebs cycle is inhibited (by lack of sufficient NAD_{ox} during oxygen deficiency), the decreased ATP concentrations that result stimulate glycolysis. This increased rate of glycolysis under anaerobic conditions is called the *Pasteur effect.*

Cells do not store extra energy in the form of extra ATP. When cellular ATP concentrations rise, because more energy (from food) is available than can be immediately used, high ATP concentrations inhibit glycolysis. Under conditions of high cellular ATP concentrations, when glycolysis is inhibited, glucose is instead converted into glycogen.

High ATP concentrations also inhibit Krebs cycle enzymes, preventing acetyl CoA from joining oxaloacetic acid and thereby forming citric acid. The inhibition of Krebs cycle enzymes results in the increased availability of acetyl CoA, phosphoglyceraldehyde, and dihydroxyacetone phosphate for alternate pathways. These intermediates are instead channeled into pathways leading to triglyceride (fat) production. When food intake occurs at a faster rate than energy consumption, therefore, the excess energy (calories) is stored in the form of glycogen and fat. These pathways are summarized in figure 4.28.

Krebs: from Hans A. Krebs, German biochemist, 1900–1981

Pasteur: from Louis Pasteur, French chemist and bacteriologist, 1822–1895

Figure 4.28. The conversion of glucose into glycogen and fat due to the allosteric inhibition of respiratory enzymes when the cell has adequate amounts of ATP. Favored pathways are indicated by heavy arrows.

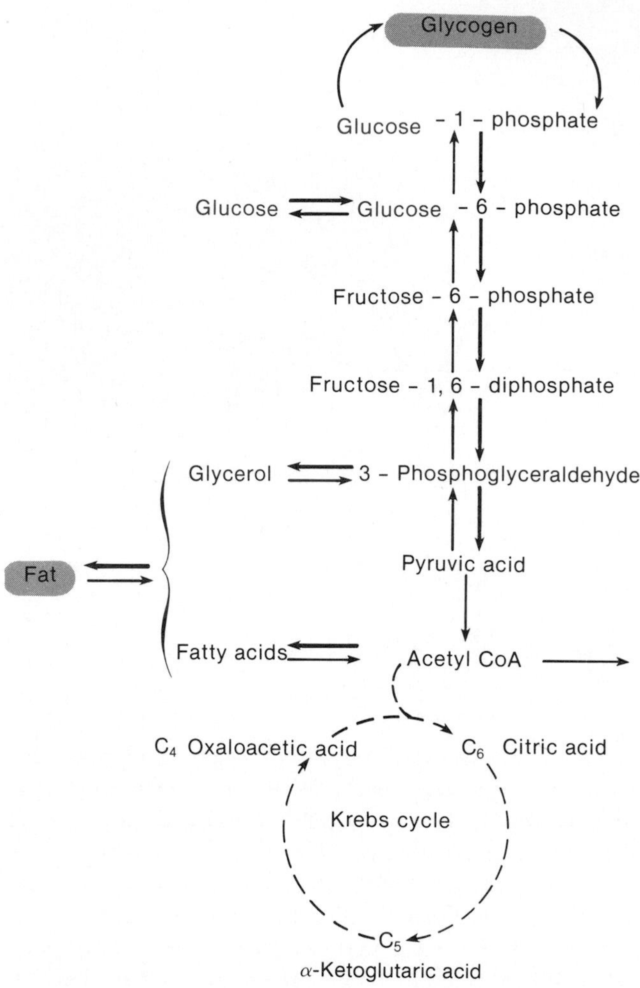

1. Describe the fate of pyruvic acid in aerobic respiration, and explain how this differs from its fate in anaerobic respiration.
2. Draw a simplified Krebs cycle, using C_2 for acetic acid, C_4 for oxaloacetic acid, C_5 for alpha-ketoglutaric acid, and C_6 for citric acid. List the high-energy products that are produced at each turn of the Krebs cycle.
3. Using a diagram, show how electrons from NAD_{red} and FAD_{red} are passed by the cytochromes. Represent the oxidized and reduced forms of the cytochromes with Fe^{+++} and Fe^{++}, respectively.
4. Explain the function of oxygen in the body and how a lack of sufficient oxygen in an organ can make respiration become anaerobic.
5. Explain how the end-product inhibition by ATP controls the rate of cell respiration and the rate of the formation of glycogen and fat.

Metabolism of Lipids and Proteins

Energy can be derived by the cellular respiration of lipids and proteins, using the same aerobic pathway as for the metabolism of pyruvic acid. Pyruvic acid and the Krebs cycle acids also serve as common intermediates in the interconversion of glucose, lipids, and amino acids.

Objective 17. Describe the metabolic pathway by which glucose can be converted into fat.
Objective 18. Describe the metabolic pathways by which fat can be utilized for energy.
Objective 19. Define oxidative deamination and transamination, and explain the significance of these two processes in amino acid metabolism.
Objective 20. Explain how skeletal muscles obtain energy at rest and at different stages of exercise, and explain the roles of the liver during skeletal muscle exercise.

It is common experience that the ingestion of excessive calories in the form of carbohydrates (cakes, ice cream, candy, and so on) increases fat production. The rise in blood glucose that follows carbohydrate-rich meals stimulates insulin secretion, and this hormone, in turn, promotes the entry of blood glucose into adipose cells. Increased availability of glucose within adipose cells, under conditions of high-insulin secretion, promotes the conversion of glucose to fat (fig. 4.28). The hormonal control of carbohydrate and fat metabolism is discussed in chapter 26.

Lipid Metabolism

Glucose can be converted to fat in the following manner. Some of the phosphoglyceraldehyde and dihydroxyacetone phosphate produced by glycolysis from glucose can be converted to glycerol. Most of the glycolytic intermediates, however, are used to form pyruvic acid, which in turn is converted to acetyl CoA. The two-carbon acetic acid subunits of acetyl CoA can then be used to produce a variety of lipids, including steroids such as cholesterol as well as ketone bodies and fatty acids (fig. 4.29). In the formation of fatty acids, a number of acetic acid subunits are joined together to form the fatty acid chain; three of these chains can then condense with glycerol to form a triglyceride molecule.

Formation of fat, or *lipogenesis (lip″o-jen′ĭ-sis)* occurs in a number of tissues—primarily in adipose tissue and in the liver—when blood glucose is elevated. Fat represents the major form of energy storage in the body. In a nonobese 70-kilogram man, 80%–85% of the body's energy is stored as fat, which amounts to about 140,000 Calories.

Figure 4.29. Divergent metabolic pathways for acetyl coenzyme A.

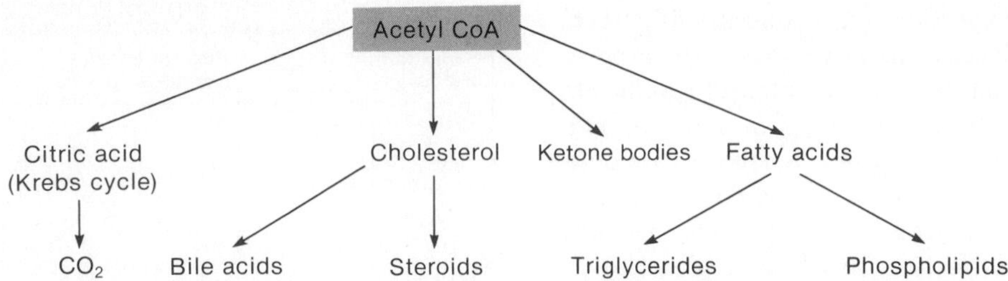

Stored glycogen, in contrast, accounts for less than 2,000 Calories, most of which (about 350 g) is stored in skeletal muscles and is available only for the muscle's use. The liver contains about 80–90 grams of glycogen, which can be converted to glucose and used by other organs. Protein accounts for about 15%–20% of the stored calories in the body, but this is usually not used extensively as an energy source because that would involve the loss of muscle mass.

Breakdown of Fat (Lipolysis). When fat stored in adipose tissue is going to be used as an energy source, lipase enzymes hydrolyze triglycerides into *glycerol* and *free fatty acids,* a process called **lipolysis** *(lī-pol'ĭ-sis)*. These molecules (primarily the free fatty acids) serve as *blood-borne energy carriers* that can be used by the liver, skeletal muscles, and other organs for aerobic respiration.

A few organs can utilize glycerol for energy, by virtue of an enzyme that converts glycerol to phosphoglyceraldehyde. Free fatty acids, however, serve as the major energy source derived from triglycerides. Most fatty acids consist of a long hydrocarbon chain with a carboxylic acid group (COOH) at one end. In a process known as **β-oxidation** (β is the Greek letter *beta*), enzymes remove two-carbon acetic acid molecules from the acid end of a fatty acid. This results in the formation of acetyl CoA, as the third carbon from the end becomes oxidized to produce a new carboxylic acid group. The fatty acid chain is thus decreased in length by two carbons. The process of β-oxidation continues until the entire fatty acid molecule is converted to acetyl CoA (fig. 4.30).

A sixteen-carbon-long fatty acid, for example, yields eight acetyl CoA molecules. Each of these can enter a Krebs cycle and produce twelve ATP per turn of the cycle, so that eight times twelve, or ninety-six, ATP are produced. In addition, each time an acetyl CoA is formed and the end-carbon of the fatty acid chain is oxidized, one NAD_{red} and one FAD_{red} are produced. Oxidative phosphorylation produces three ATP per NAD_{red} and two ATP per FAD_{red}. For a sixteen-carbon-long fatty acid, these five ATP molecules would be formed seven times (producing five times seven, or thirty-five, ATP). Not counting the single ATP used to start β-oxidation (fig. 4.30), this fatty acid could yield a grand total of 35 + 96, or 131, ATP molecules!

Figure 4.30. β-oxidation of a fatty acid. After the attachment of coenzyme A to the carboxylic acid group (step *1*), a pair of hydrogens is removed from the fatty acid and used to reduce one molecule of FAD (step *2*). When this electron pair is donated to the cytochrome chain, two ATP are produced. The addition of a hydroxyl group from water (step *3*), followed by the oxidation of the β-carbon (step *4*), results in the production of three ATP from the electron pair donated by reduced NAD. The bond between the α and β carbons in the fatty acid is broken (step *5*), releasing acetyl coenzyme A and a fatty acid chain that is two carbons shorter than the original. With the addition of a new coenzyme A to the shorter fatty acid the process begins again (step *2*), as acetyl CoA enters the Krebs cycle and generates twelve ATP.

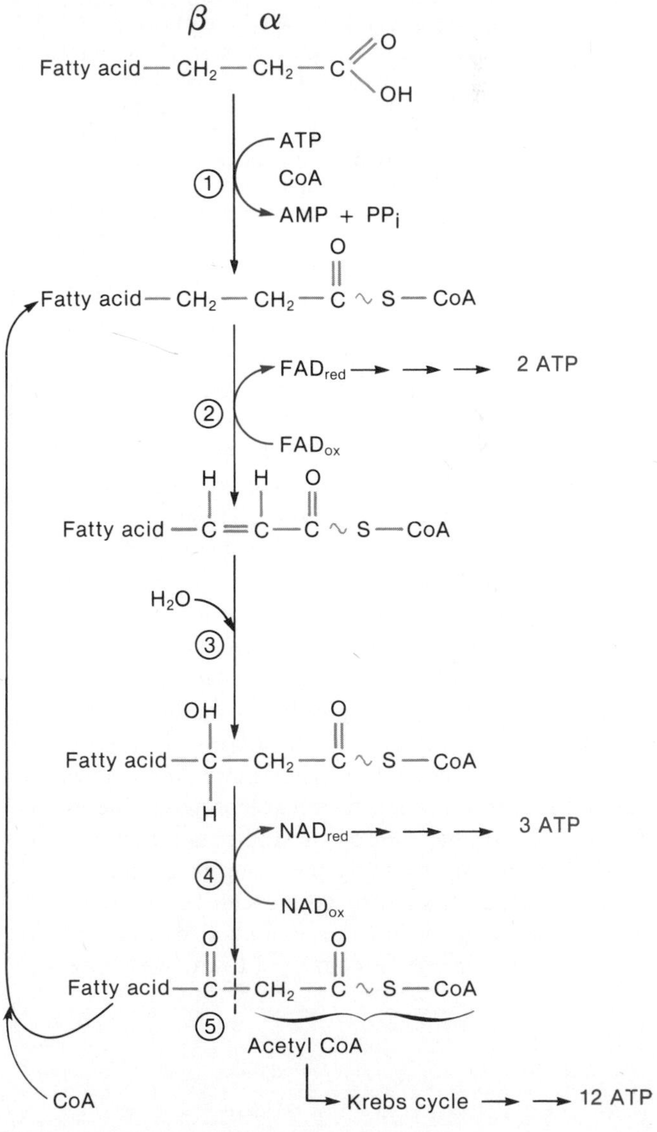

β α
Fatty acid — CH_2 — CH_2 — C(=O)OH
① ATP, CoA → AMP + PP_i
Fatty acid — CH_2 — CH_2 — C(=O) ∿ S — CoA
② FAD_{ox} → FAD_{red} → → → 2 ATP
Fatty acid — C(H) = C(H) — C(=O) ∿ S — CoA
③ H_2O
Fatty acid — C(OH)(H) — CH_2 — C(=O) ∿ S — CoA
④ NAD_{ox} → NAD_{red} → → → 3 ATP
Fatty acid — C(=O) ⑤ CH_2 — C(=O) ∿ S — CoA
Acetyl CoA → Krebs cycle → → 12 ATP
CoA

Ketone Bodies. There is a continuous turnover of triglycerides in adipose tissue, even when a person is not losing weight. New triglycerides are produced, while others are hydrolyzed into glycerol and fatty acids. This turnover insures that the blood normally contains a sufficient amount of fatty acids for aerobic respiration by skeletal muscles, the liver, and other organs. When the rate of lipolysis exceeds the rate of fatty acid utilization—as it may do in starvation, dieting, and in diabetes mellitus—the blood concentrations of fatty acids increase.

If the liver cells contain sufficient amounts of ATP, so that further production of ATP is not needed, some of the acetyl CoA derived from fatty acids is channeled into an alternate pathway. This pathway involves the conversion of two molecules of acetyl CoA into four-carbon-long acidic derivatives, *acetoacetic acid* and *β-hydroxybutyric (hi-drok''se-byu-ti'rik) acid.* Together with *acetone (as'ĕ-tōn),* which is a three-carbon-long derivative of acetoacetic acid, these products are known as **ketone bodies.**

Ketone bodies, which can be used for energy by many organs, are normally found in the blood. Under conditions of fasting or of diabetes mellitus, however, the increased liberation of free fatty acids from adipose tissue results in an elevated production of ketone bodies in the liver. The secretion of abnormally high amounts of ketone bodies into the blood produces *ketosis,* which is one of the signs of fasting or an uncontrolled diabetic state.

Amino Acid Metabolism

Nitrogen is ingested primarily as amino acids and excreted mainly as urea in the urine. In childhood, the amount of nitrogen excreted may be less than the amount ingested because amino acids are incorporated into proteins during growth. Growing children are thus said to be in a state of *positive nitrogen balance.* People who are starving or suffering from prolonged wasting diseases, in contrast, are in a state of *negative nitrogen balance;* they excrete more nitrogen than they ingest because their tissue proteins are breaking down.

Healthy adults maintain a state of nitrogen balance, in which the amount of nitrogen excreted is equal to the amount ingested. This does not imply that the amino acids ingested are unnecessary; on the contrary, they are needed to replace the approximately 400 grams of protein that are "turned over" each day. When more amino acids are ingested than are needed to replace proteins, the excess amino acids are not stored as additional protein (one cannot build muscles simply by eating large amounts of protein). Rather, the amine groups can be removed, and the "carbon skeletons" of the organic acids that are left can be used for energy or converted to carbohydrate and fat.

Table 4.5 The essential, semiessential, and nonessential amino acids

Essential amino acids	Semiessential amino acid	Nonessential amino acids
Lysine	Arginine	Aspartic acid
Tryptophan		Glutamic acid
Phenylalanine		Proline
Threonine		Glycine
Valine		Serine
Methionine		Alanine
Leucine		Cysteine
Isoleucine		
Histidine (children)		

From Stuart Ira Fox, *Human Physiology,* 2d ed.

Transamination. An adequate amount of all twenty amino acids is required to build proteins for growth and for replacement of the proteins that are turned over. Fortunately, only eight (in adults) or nine (in children) amino acids cannot be produced by the body and so must be obtained in the diet; these are the *essential amino acids* (table 4.5). Arginine is a "semiessential" amino acid because it can be produced by transamination, but not in sufficient quantities to classify it as a nonessential amino acid. The remaining amino acids are "nonessential" only in the sense that the body can produce them if it is given a sufficient amount of the essential amino acids and of carbohydrates.

Pyruvic acid and the Krebs cycle acids are collectively termed *keto acids* because they have a ketone group; these should not be confused with the ketone bodies (derived from acetyl CoA). Keto acids can be converted to amino acids by the addition of an amine (NH_2) group. This amine group is usually obtained by "cannibalizing" another amino acid; in this process, a new amino acid is formed as the one that was cannibalized is converted to a new keto acid. This type of reaction, in which the amine group is moved across from one amino acid to form another, is called **transamination.**

Each transamination reaction is catalyzed by a specific enzyme (a transaminase) that requires vitamin B_6 (pyridoxine) as a coenzyme. The amine group from glutamic acid, for example, may be transferred to either pyruvic acid or oxaloacetic acid. The former reaction is catalyzed by the enzyme *glutamate pyruvate transaminase* (GPT), and the latter reaction is catalyzed by *glutamate oxaloacetate (oks''ah-lo-as'ĕ-tāt) transaminase* (GOT). The addition of an amine group to pyruvic acid produces the amino acid alanine; the addition of an amine group to oxaloacetic acid produces the amino acid known as aspartic acid (fig. 4.31).

Oxidative Deamination. As shown in figure 4.32, glutamic acid can be formed through transamination by the combination of an amine group with α-ketoglutaric acid.

Figure 4.31. The formation of the amino acids aspartic acid and alanine using glutamic acid as the amine donor in transamination. (GOT = glutamate oxaloacetate transaminase; GPT = glutamate pyruvate transaminase.) The shaded areas show the parts of the molecules that are changed by transamination reactions.

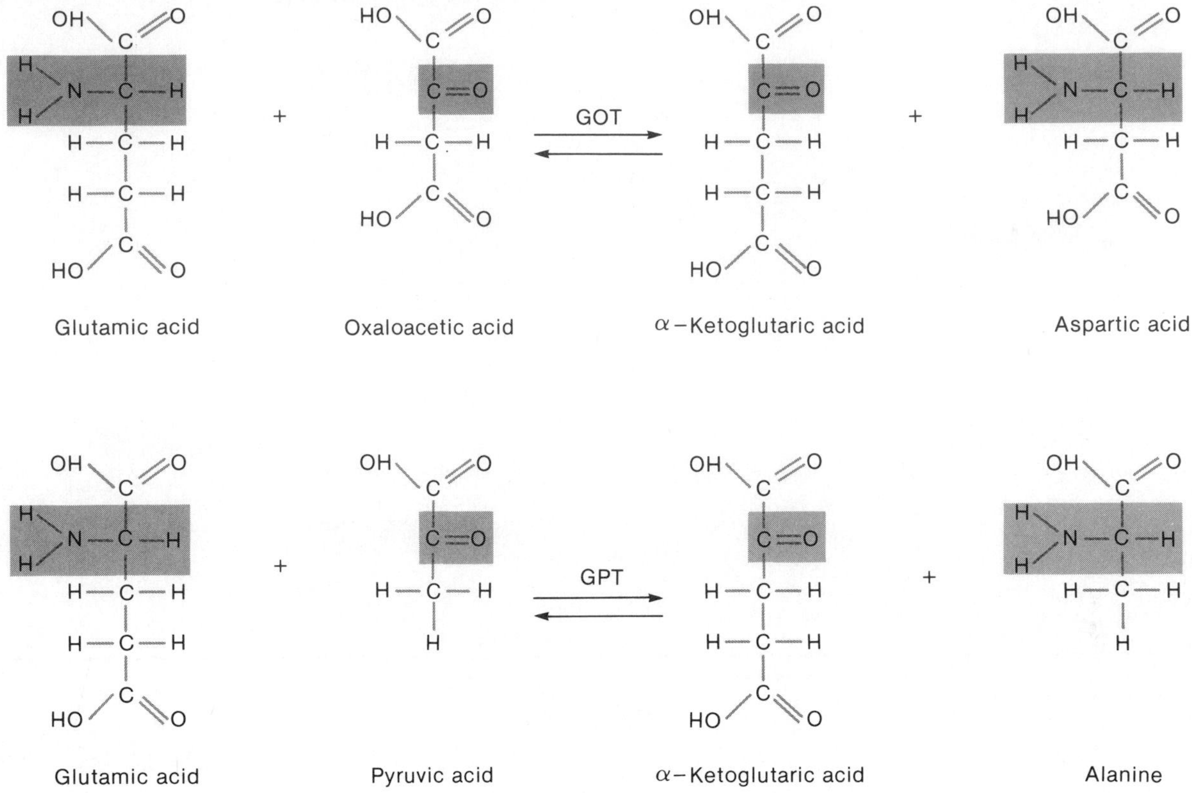

Glutamic acid + Oxaloacetic acid ⇌ (GOT) α-Ketoglutaric acid + Aspartic acid

Glutamic acid + Pyruvic acid ⇌ (GPT) α-Ketoglutaric acid + Alanine

Figure 4.32. Oxidative deamination. Glutamic acid is converted to α-ketoglutaric acid as it donates its amine group to a metabolic pathway that results in the formation of urea.

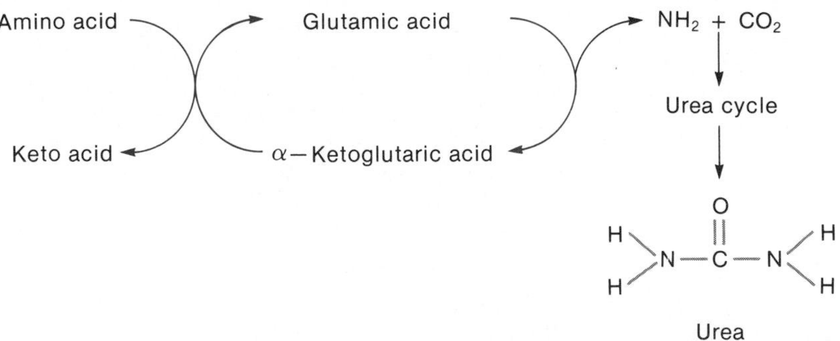

Glutamic acid is also produced in the liver from the ammonia that is generated by intestinal bacteria and carried to the liver in the hepatic portal vein. Since free ammonia is very toxic, its removal from the blood and incorporation into glutamic acid is an important function of the healthy liver.

If there are more amino acids than are needed for protein synthesis, the amine group from glutamic acid may be removed and excreted as *urea* in the urine (fig. 4.32). The metabolic pathway that removes amine groups from amino acids—leaving a keto acid and ammonia (which is converted to urea)—is known as **oxidative deamination** *(de-am''ĭ-na'shun).*

A number of amino acids can be converted into glutamic acid by transamination. Since glutamic acid can donate amine groups to urea (through deamination), it serves as a "funnel" through which other amino acids can be used to produce keto acids (pyruvic acid and Krebs cycle acids). These keto acids may then be used in the Krebs cycle as a source of energy (fig. 4.33).

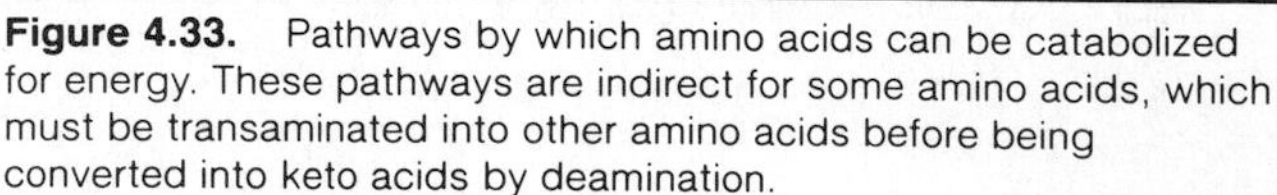

Figure 4.33. Pathways by which amino acids can be catabolized for energy. These pathways are indirect for some amino acids, which must be transaminated into other amino acids before being converted into keto acids by deamination.

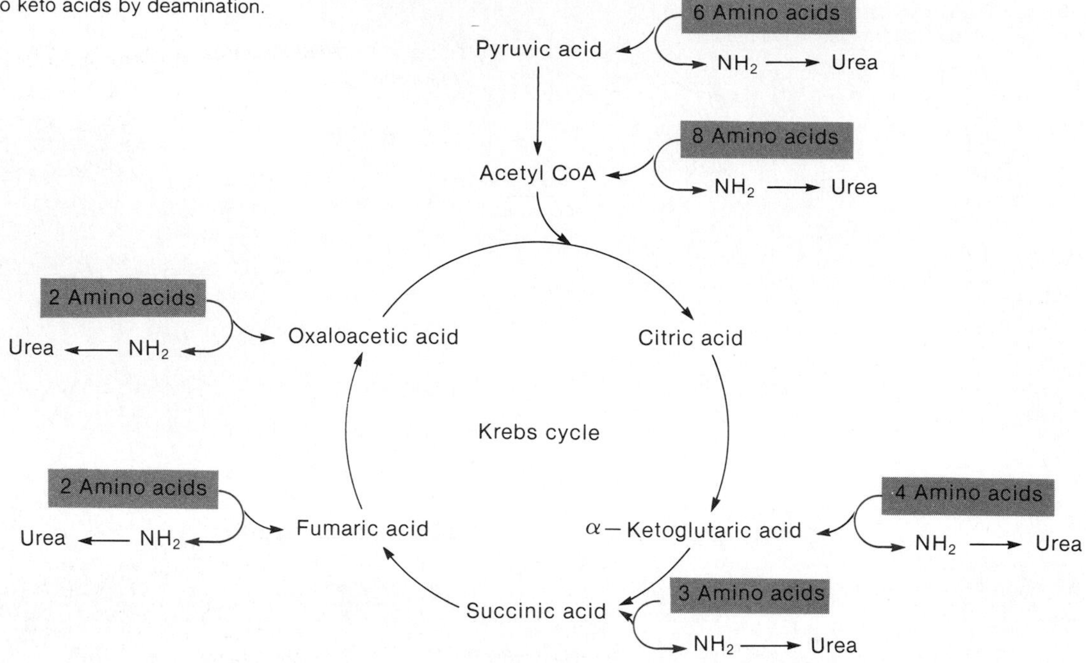

Figure 4.34. Simplified metabolic pathways showing how glycogen, fat, and protein can be interconverted.

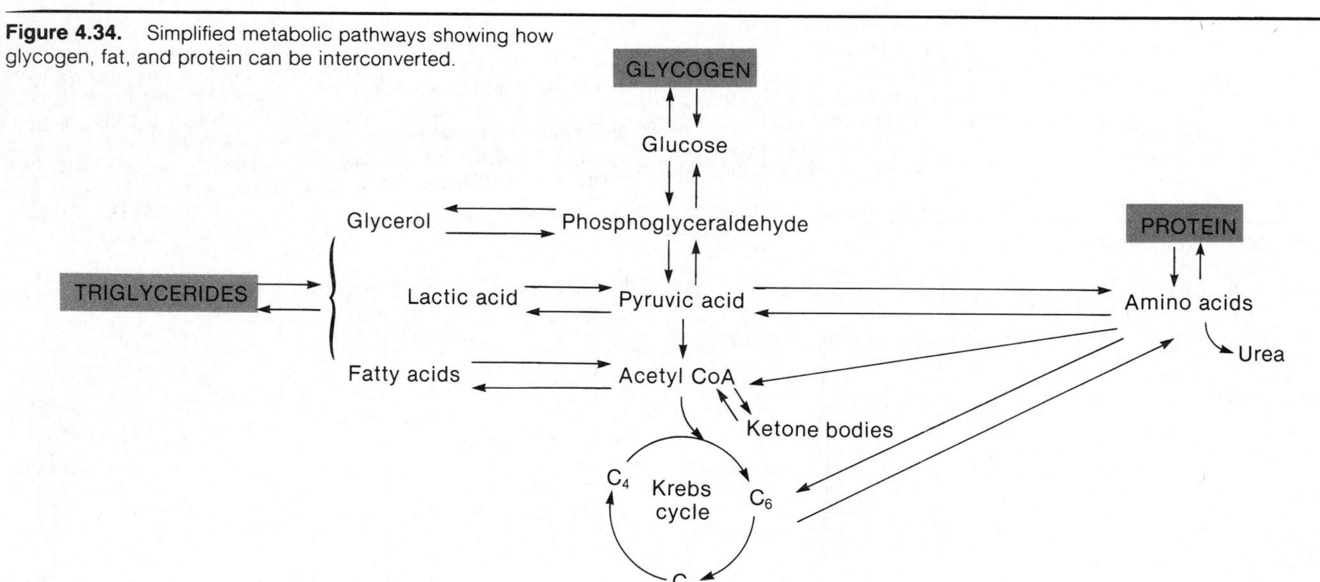

Depending on the amino acid that is deaminated, the keto acid left over may be either pyruvic acid or one of the Krebs cycle acids. These can be respired for energy, converted via pyruvic acid to glucose (amino acids are very important in the production of glucose through *gluconeogenesis*), or converted to fat. The possible interrelationships between amino acids, carbohydrates, and fat are illustrated in figure 4.34.

Uses of Different Energy Sources

The blood serves as a common trough from which all the cells in the body are fed. If all cells used the same energy source, such as glucose, this source would quickly be depleted, and cellular starvation would occur. This is normally prevented because the blood contains different energy sources: glucose and ketone bodies from the liver, fatty acids from adipose tissue, and lactic acid and amino acids from muscles. Some organs preferentially use one energy source more than the others, so that each energy source is "spared" for organs with strict energy needs.

Figure 4.35. The relative contributions of anaerobic and aerobic respiration to the total energy in a well-trained person performing at maximal effort.

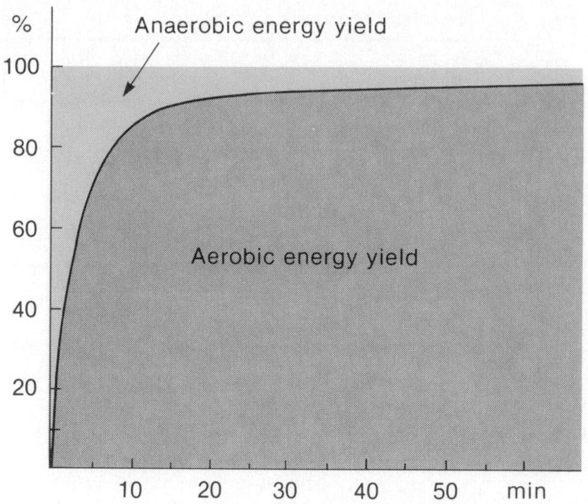

Under normal conditions, for example, the brain uses blood glucose as its major energy source. This glucose is supplied primarily by the liver, which secretes about 150 milligrams of glucose into the blood per minute. About 75% of this glucose is supplied by glycogenolysis, and 25% is produced in the liver by gluconeogenesis, using amino acids, lactic acids, and glycerol as initial substrates. Many organs spare glucose by using fatty acids, ketone bodies, and lactic acid as energy sources.

Metabolism of Skeletal Muscles

The metabolism of skeletal muscles, during rest and during exercise, can provide a summary and a practical example of some of the general principles described in this chapter. Skeletal muscle metabolism is of great significance because of the large contribution of skeletal muscles to total body oxygen requirements, particularly during exercise, and because knowledge of muscle metabolism can be helpful to people interested in exercise and physical fitness.

Anaerobic and Aerobic Respiration. Skeletal muscles respire anaerobically for the first forty-five to ninety seconds of moderate-to-heavy exercise. This is because the energy requirement (need for ATP) increases faster than the rate of oxygen supply by the cardiopulmonary system. If exercise is moderate and the person is in good physical condition, aerobic respiration contributes the major portion of the skeletal muscle energy requirements following the first two minutes of exercise (fig. 4.35).

The maximum rate of oxygen consumption (by aerobic respiration) in the body is called the **maximal oxygen uptake.** The maximal oxygen uptake in a given person is determined primarily by the person's age, size, and sex. It is about 15%–20% higher for males than for females and highest at age twenty for both sexes. Some world-class athletes have maximal oxygen uptakes that are twice the average for their age and sex. This appears to be due largely to genetic factors, but training can increase this value by about 20%.

When the oxygen uptake during exercise is about 60% of the maximum, anaerobic respiration contributes a small share of the energy required, and some lactic acid is produced. The extent of anaerobic respiration and lactic acid production subsequently increases proportionately as more strenuous exercise results in the attainment of higher percentages of the maximal oxygen uptake. Since lactic acid contributes to muscle fatigue (described in chapter 12), people with higher maximal oxygen uptakes can achieve higher levels of exercise before fatigue than people with lower maximal oxygen uptakes.

When a person stops exercising, the rate of oxygen uptake does not immediately go back to pre-exercise levels; it returns slowly (the person continues to breathe heavily for some time afterwards). This extra oxygen is used to repay the **oxygen debt** incurred during exercise. The oxygen debt includes oxygen that was withdrawn from savings deposits—hemoglobin in blood and myoglobin in muscle (chapter 12), the extra oxygen required for metabolism by tissues warmed during exercise, and the oxygen needed for the metabolism of lactic acid, which was produced during anaerobic respiration. Under aerobic conditions lactic acid can be converted to pyruvic acid, which can serve as a substrate for aerobic respiration (as in the heart) or a source of new glucose (in the liver).

Sources of Energy during Rest and Exercise. Skeletal muscles at rest use fatty acids as their major energy source. Metabolism changes at the beginning of exercise, when glucose derived from stored muscle glycogen becomes the prime source of energy. At this time the energy contribution from blood glucose is still low.

After exercise has continued for ten to forty minutes, however, the metabolic pattern changes as muscle glycogen becomes depleted. The contribution of blood glucose to the energy requirements of exercising muscles increases to about 40%. This increased demand for blood glucose is met by the liver, which increases its rate of glycogenolysis two to five times over resting levels.

By forty minutes of exercise, the liver has added about 18 grams of glucose to the blood—roughly 25% of its stored glycogen. Liver glycogenolysis continues beyond this time, but much of the burden is removed by the increased use of fatty acids as an energy source by the exercising muscles. Since some consumption of blood glucose continues, however, the blood glucose concentration is slightly decreased by ninety minutes of exercise. True hypoglycemia (low blood glucose), however, is rare, which is fortunate because the glucose requirements of the brain do not change with exercise.

Table 4.6 Relative energy yield from different sources for skeletal muscles during different phases of exercise

Phase	Free fatty acids	Glucose from muscle glycogen	Blood glucose	Liver glucose output (%)	
				From glycogenolysis	From gluconeogenesis
Rest	+ + +	−	+	75 %	25 %
5–10 min	−	+ + +	+	Decreases	Increases
10–40 min	+	+ +	+ +	Decreases	Increases
40–90 min	+ +	+	+ + +	Decreases	Increases
By 4 hours	+ + +	−	+ +	55 %	45 %

As glycogen stored in the liver becomes depleted during prolonged exercise, the contribution of gluconeogenesis to liver glycogen output increases from 25% (at rest) to 45%. By four hours of exercise, the liver contains only about 25% of its original glycogen stores, and the muscles rely increasingly on free fatty acids as an energy source (table 4.6).

During and immediately following exercise, the lactic acid produced by skeletal muscles can be used by the heart as an energy source and by the liver as a substrate for gluconeogenesis (this is part of the Cori cycle). A recovery period of about forty-eight hours is needed before the muscle glycogen that was depleted during prolonged exercise is completely restored.

1. Construct a flowchart to show the metabolic pathway by which glucose can be converted to fat. Indicate only the major intermediates involved (not all of the steps of glycolysis).
2. Define the terms *lipolysis* and *β-oxidation,* and explain in general terms how fat can be used for energy.
3. Describe transamination and deamination, and explain their functional significance.
4. List five blood-borne energy carriers, and explain in general terms how these are used as sources of energy.
5. Define the terms *glycogenolysis* and *gluconeogenesis,* and explain how these processes help to maintain constant blood glucose concentrations during exercise.

Table 4.7 Examples of the diagnostic value of some enzymes found in plasma

Enzyme	Diseases associated with abnormal plasma enzyme concentrations
Alkaline phosphatase	Obstructive jaundice, Paget's disease (osteitis deformans), carcinoma of bone
Acid phosphatase	Benign hypertrophy of prostate, cancer of prostate
Amylase	Pancreatitus, perforated peptic ulcer
Aldolase	Muscular dystrophy
Creatine kinase (or creatine phosphokinase-CPK)	Muscular dystrophy, myocardial infarction
Lactate dehydrogenase (LDH)	Myocardial infarction, liver disease, renal disease, pernicious anemia
Transaminases (GOT and GPT)	Myocardial infarction, hepatitis, muscular dystrophy

Table 4.8 Some clinical uses of isoenzyme measurements

Enzyme	Isoenzyme form number	Disease associated with abnormal elevation of isoenzyme
Creatine phosphokinase (CPK)	1	Muscular dystrophy
	2	Myocardial infarction
Lactic acid dehydrogenase (LDH)	1	Myocardial infarction
	5	Liver disease (such as hepatitis)

Clinical Considerations

Clinical Enzyme Measurements

Assays of Enzymes in Plasma. When tissues become damaged due to diseases, some of the dead cells disintegrate and release their enzymes into the blood. Most of these enzymes are not normally active in the blood, because of the absence of their specific substrates, but their enzymatic activity can be measured in a test tube by the addition of the appropriate substrates to samples of plasma. Such measurements are clinically useful, because abnormally high plasma concentrations of particular enzymes are characteristic of certain diseases (table 4.7).

Isoenzymes. Isoenzymes can be separated and identified by standardized laboratory techniques. Since different isoenzyme forms catalyze the same reaction, they share the same name; the different forms are thus identified by numbers. Since the isoenzyme form is generally characteristic of a particular organ, elevations in the plasma concentration of one specific isoenzyme can be used diagnostically in the detection of disease states of a particular organ (table 4.8).

Table 4.9 Selected examples of inborn errors in the metabolism of amino acids, carbohydrates, and lipids

Metabolic defect	Disease	Abnormality	Clinical result
Amino acid metabolism	Phenylketonuria (PKU)	Increase in phenylalanine	Mental retardation, epilepsy
	Albinism	Lack of melanin	Susceptibility to skin cancer
	Maple-syrup disease	Increase in leucine, isoleucine, and valine	Degeneration of brain, early death
	Homocystinuria	Accumulation of homocystine	Mental retardation, eye problems
Carbohydrate metabolism	Lactose intolerance	Lactose not utilized	Diarrhea
	Glucose-6-phosphatase deficiency (Gierke's disease)	Accumulation of glycogen in liver	Liver enlargement, hypoglycemia
	Glycogen phosphorylase deficiency (McArdle syndrome)	Accumulation of glycogen in muscle	Muscle fatigue and pain
Lipid metabolism	Gaucher's disease	Lipid accumulation (glucocerebroside)	Liver and spleen enlargement, brain degeneration
	Tay-Sachs disease	Lipid accumulation (ganglioside G_{M2})	Brain degeneration, death by age 5
	Hypercholestremia	High blood cholesterol	Atherosclerosis of coronary and large arteries

Figure 4.36. Metabolic pathways for the degradation of the amino acid phenylalanine. Defective enzyme$_1$ produces phenylketonuria (PKU), defective enzyme$_2$ produces alcaptonuria (not a clinically significant condition), and defective enzyme$_3$ produces albinism.

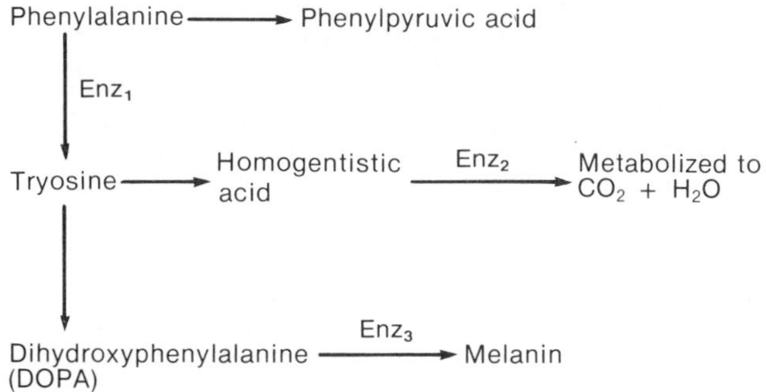

Metabolic Disturbances

Phenylketonuria (PKU). The branched metabolic pathway that begins with phenylalanine as the initial substrate is subject to a number of inborn errors of metabolism (fig. 4.36). When the enzyme that converts this amino acid to tyrosine is defective, the final products of a divergent pathway accumulate and can be detected in the blood and urine. This disease, *phenylketonuria (fen''el-ke''to-nu're-ah) (PKU),* can result in severe mental retardation and a shortened life span. Although no inborn error of metabolism is common, PKU occurs with enough frequency and is so easy to detect that all newborn babies are routinely tested for this defect. If this disease is detected early, brain damage can be prevented by placing the child on an artificial diet low in the amino acid phenylalanine.

Albinism and Other Defects. One of the conversion products of phenylalanine is a molecule called *DOPA,* which is an acronym for dihydroxyphenylalanine (fig. 4.36). DOPA is a precursor of the pigment molecule melanin, which gives skin, eyes, and hair their characteristic colors. An inherited defect in the enzyme that catalyzes the formation of melanin from DOPA results in an *albino (al-bi'no).* Besides PKU and albinism, there are a large number of other inherited defects of amino acid metabolism, as well as inborn errors in the metabolism of carbohydrates and lipids (table 4.9).

Endocrine Disorders and Metabolism. Since metabolism is regulated largely by hormones (chemicals secreted by endocrine glands into the blood), endocrine diseases can produce metabolic disorders. *Diabetes mellitus*—characterized by high blood glucose and the presence of glucose in the urine—for example, results from the inadequate secretion or action of the hormone insulin. This disease may also be associated with the excessive production of ketone bodies, which can alter blood pH and produce *ketoacidosis.* Abnormally low blood glucose, *hypoglycemia,* may be produced by excessive insulin secretion. Other metabolic disorders can result from diseases of the pituitary, thyroid, and adrenal glands, as described in chapter 25.

Chapter Summary

I. Enzymes as Catalysts
  A. Enzymes are biological catalysts.
  B. Enzymes are proteins.
II. Control of Enzyme Activity
  A. The activity of an enzyme is affected by a variety of factors.
  B. Metabolic pathways involve a number of enzyme-catalyzed reactions.
III. Bioenergetics
  A. The flow of energy in the cell is called bioenergetics.
  B. Oxidation-reduction reactions have several characteristics.
IV. Glycolysis and Anaerobic Respiration
  A. Glycolysis refers to the conversion of glucose to two molecules of pyruvic acid.
  B. When respiration is anaerobic, reduced NAD is oxidized by pyruvic acid, which accepts two hydrogen atoms and is thereby reduced to lactic acid.
V. Aerobic Respiration
  A. The Krebs cycle begins when coenzyme A donates acetic acid to an enzyme reaction that adds it to oxaloacetic acid to form citric acid.
  B. Reduced NAD and FAD donate their electrons to an electron transport chain of molecules located in the cristae.
  C. Thirty-six to thirty-eight ATP are produced by the aerobic respiration of one glucose molecule; of these, two are produced in the cytoplasm by glycolysis, and the remainder are produced in the mitochondria.
  D. The formation of glycogen from glucose is called glycogenesis, and the breakdown of glycogen is called glycogenolysis.
  E. Carbohydrate metabolism is influenced by the availability of oxygen and by a negative feedback effect of ATP on glycolysis and the Krebs cycle.
VI. Metabolism of Lipids and Proteins
  A. In lipolysis, triglycerides yield glycerol and fatty acids.
  B. Amino acids can serve as sources of energy.
  C. Each organ uses certain blood-borne energy carriers as its preferred energy source.

Review Activities

Objective Questions

1. Which of the following statements about enzymes is *true?*
   (a) All proteins are enzymes.
   (b) All enzymes are proteins.
   (c) Enzymes are changed by the reactions they catalyze.
   (d) The active sites of enzymes have little specificity for substrates.
2. Which of the following statements about enzyme-catalyzed reactions is *true?*
   (a) The rate of reaction is independent of temperature.
   (b) The rate of all enzyme-catalyzed reactions is decreased when the pH is lowered from 7 to 2.
   (c) The rate of reaction is independent of substrate concentration.
   (d) Under given conditions of substrate concentration, pH, and temperature, the rate of product formation varies directly with enzyme concentration, up to a maximum, at which point the rate cannot be further increased.
3. Which of the following statements about lactate dehydrogenase is *true?*
   (a) It is a protein.
   (b) It oxidizes lactic acid.
   (c) It reduces another molecule (pyruvic acid).
   (d) All of these are true.
4. In an inborn error of metabolism
   (a) a genetic change results in the production of a defective enzyme
   (b) intermediates produced before the defective step accumulate
   (c) alternate pathways are taken by intermediates at branch points located before the defective step
   (d) All of these are true.
5. Which of the following represents an *endergonic* reaction?
   (a) ADP + $P_i \rightarrow$ ATP
   (b) ATP $\rightarrow$ ADP + P_i
   (c) glucose + $O_2 \rightarrow CO_2 + H_2O$
   (d) $CO_2 + H_2O \rightarrow$ glucose
   (e) both *a* and *d*
   (f) both *b* and *c*
6. Which of the following statements about ATP is *true?*
   (a) The bond joining ADP and the third phosphate is a high-energy bond.
   (b) The formation of ATP is coupled to energy-liberating reactions.
   (c) The conversion of ATP to ADP and P_i provides energy for biosynthesis, cell movement, and other cellular processes that require energy.
   (d) ATP is the "universal energy carrier" of cells.
   (e) All of these are true.
7. When oxygen is combined with two hydrogens to make water,
   (a) oxygen is reduced
   (b) the molecule that donated the hydrogens becomes oxidized
   (c) oxygen acts as a reducing agent
   (d) both *a* and *b* apply
   (e) both *a* and *c* apply
8. The net gain of ATP per glucose molecule in anaerobic respiration is ______; the net gain in aerobic respiration is ______.
   (a) 2;4
   (b) 2;38
   (c) 38;2
   (d) 24;30
9. In anaerobic respiration, the oxidizing agent for NAD_{red} (that is, the molecule that removes electrons from NAD_{red}) is
   (a) pyruvic acid
   (b) lactic acid
   (c) citric acid
   (d) oxygen
10. When organs respire anaerobically, there is an increased blood concentration of
   (a) oxygen
   (b) glucose
   (c) lactic acid
   (d) ATP
11. The conversion of lactic acid to pyruvic acid occurs in
   (a) anaerobic respiration
   (b) the heart, where lactic acid is aerobically respired
   (c) the liver, where lactic acid can be converted to glucose
   (d) both *a* and *b*
   (e) both *b* and *c*
12. The oxygen in the air we breathe
   (a) functions as the final electron acceptor of the electron transport chain
   (b) combines with hydrogen to form water
   (c) combines with carbon to form CO_2
   (d) both *a* and *b*
   (e) both *a* and *c*
13. In terms of the number of ATP molecules directly produced, the major energy-yielding process in the cell is
   (a) glycolysis
   (b) the Krebs cycle
   (c) oxidative phosphorylation
   (d) gluconeogenesis
14. Ketone bodies are derived from
   (a) fatty acids
   (b) glycerol
   (c) glucose
   (d) amino acids
15. The conversion of glucose-6-phosphate to free glucose, which can be secreted into the blood, occurs in
   (a) the liver
   (b) the skeletal muscles
   (c) the brain
   (d) all of these

16. The formation of glucose from pyruvic acid that is derived from lactic acid, amino acids, or glycerol is called
    (a) glycogenesis
    (b) glycogenolysis
    (c) glycolysis
    (d) gluconeogenesis
17. Which of the following organs has an almost absolute requirement for blood glucose as its energy source? The
    (a) liver
    (b) brain
    (c) skeletal muscles
    (d) heart
18. When amino acids are used as an energy source,
    (a) oxidative deamination occurs
    (b) pyruvic acid or one of the Krebs cycle acids (keto acids) is formed
    (c) urea is produced
    (d) all of the above occur

Essay Questions

1. Explain the relationship between the chemical structure and the function of an enzyme, and describe how various conditions may alter both the structure and the function of an enzyme.
2. Explain how end-product inhibition represents a form of negative feedback regulation.
3. Explain the advantages and disadvantages of anaerobic respiration.
4. What purpose is served by the formation of lactic acid during anaerobic respiration? How is this purpose achieved during aerobic respiration?
5. The poison cyanide blocks the transfer of electrons from the last cytochrome to oxygen. Describe the effect of this poison on oxidative phosphorylation and on the Krebs cycle, and explain why this poison is deadly.
6. Describe the metabolic pathway by which glucose can be converted into fat, and explain how end-product inhibition by ATP can favor this pathway.
7. Describe the metabolic pathway by which fat can be used as a source of energy, and explain why the metabolism of fatty acids can yield more ATP than the metabolism of glucose.
8. Explain how energy is obtained from the metabolism of amino acids. Why does a starving person have high concentrations of urea in the blood?
9. Explain why the liver is the only organ able to secrete glucose into the blood, and describe the possible sources of this hepatic glucose.
10. Describe the metabolism of glucose and fatty acids by resting and exercising skeletal muscles, and explain the functional significance of fatty acid metabolism by muscles.

5

Membrane Transport and the Membrane Potential

Outline

Concepts

Net diffusion of a molecule or ion through a cell membrane always occurs in the direction of its lower concentration. Nonpolar molecules can penetrate the phospholipid layers, and small inorganic ions can pass through channels in the membrane. Water molecules can also diffuse through cell membranes in a process known as osmosis.

Molecules such as glucose, amino acids, and other organic molecules of comparable size are transported across cell membranes by special protein carriers that are specific and that can become saturated. Carrier-mediated transport that involves a net movement down a concentration gradient, and which therefore is passive, is called facilitated diffusion. Carrier-mediated transport that occurs against a concentration gradient and that requires metabolic energy is called active transport.

Large, negatively charged molecules are concentrated within the cell by their inability to pass through the cell membrane. This creates a polarity across the membrane, with the inside of the cell negatively charged in comparison to the outside of the cell. This polarity is maintained and increased by the activity of the Na^+/K^+ pumps. The magnitude of the difference in charge across the membrane is the membrane potential, measured in millivolts.

Simple Diffusion and Osmosis

Net diffusion of a molecule or ion through a cell membrane always occurs in the direction of its lower concentration. Nonpolar molecules can penetrate the phospholipid layers, and small inorganic ions can pass through channels in the membrane. Water molecules can also diffuse through cell membranes in a process known as osmosis.

Objective 1. Describe how molecules such as O_2, and ions such as K^+, are able to penetrate a cell membrane, and explain how the direction of the net diffusion of these molecules and ions is determined.

Objective 2. Explain why diffusion of molecules and ions through a membrane is called passive transport, and describe the factors that affect the rate of this transport.

Objective 3. Define the term *osmosis,* and describe how the direction of osmosis across a semipermeable membrane is determined.

Objective 4. Describe how the osmolality and osmotic pressure of two solutions are related to their tonicity when these solutions are separated from each other by a semipermeable membrane.

Objective 5. Explain the mechanisms that help to maintain a constant plasma osmolality.

The cell (plasma) membrane separates the intracellular environment from the extracellular environment. Proteins, nucleotides, and other molecules needed for the structure and function of the cell cannot penetrate, or "permeate," the membrane. The cell membrane is, however, **selectively permeable** *(per'me-ah-b'l)* to certain molecules and many ions; this allows two-way traffic in nutrients and wastes needed to sustain metabolism and provides electrical currents created by the movements of ions through the membrane.

The mechanisms involved in the transport of molecules and ions through the cell membrane may be divided into two categories: (1) transport that requires the action of specific *carrier proteins* in the membrane (*carrier-mediated transport*); and (2) transport through the membrane that is not carrier-mediated. Carrier-mediated transport includes *facilitated diffusion* and *active transport;* non-carrier-mediated transport consists of the *simple diffusion* of ions, lipid-soluble molecules, and water through the membrane. The diffusion of water (solvent) through a membrane is called *osmosis.*

Diffusion—whether it requires the presence of carriers or not and whether it involves the movement of solute or solvent through a membrane—is driven by the thermal energy present in the transported substances. Metabolic energy is not required for these transport processes, which,

Figure 5.1. Net diffusion occurs when there is a concentration difference (or concentration gradient) between two regions of a solution (*a*) provided that the membrane separating these regions is permeable to the diffusing substance. Diffusion tends to equalize the concentration of these solutions (*b*) and thus to abolish the concentration differences.

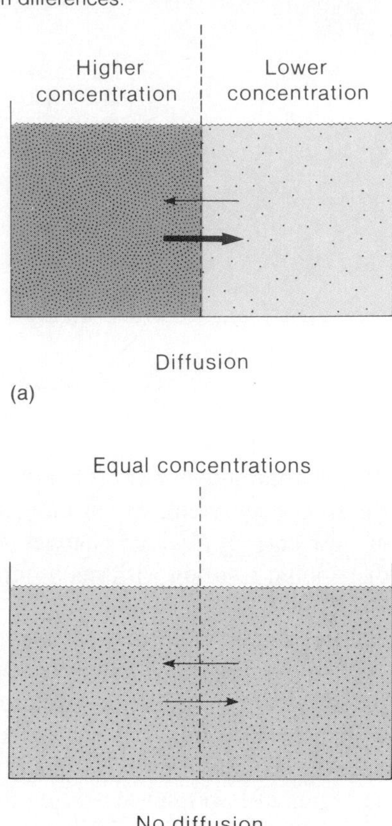

indeed, can occur across artificial membranes or across the membranes of dead cells. Transport through a membrane by means of diffusion is, for these reasons, known as *passive transport.* Active transport, in contrast, requires an active cellular metabolism, because this form of membrane transport does not occur without the energy supplied by the conversion of ATP to ADP and P_i. As will be described in a later section, metabolic energy is needed in active transport because this process moves molecules and ions "uphill," from regions of lower to regions of higher concentrations.

Diffusion

Molecules in a gas and molecules and ions dissolved in a solution are in a constant state of random motion as a result of their thermal (heat) energy. This random motion, called **diffusion,** tends to make the gas or solution evenly mixed, or diffusely spread out, within a given volume. Whenever a *concentration difference,* or *concentration gradient,* exists between two parts of a solution, therefore, random molecular motion tends to abolish the gradient and to make the molecules uniformly distributed (fig. 5.1). In terms of

diffusion: L. *dif-fero,* to carry; *fundo,* to pour in different directions

Figure 5.2. Gas exchange between the intracellular and extracellular compartments occurs by diffusion. The regions of higher concentration are represented by the larger symbols.

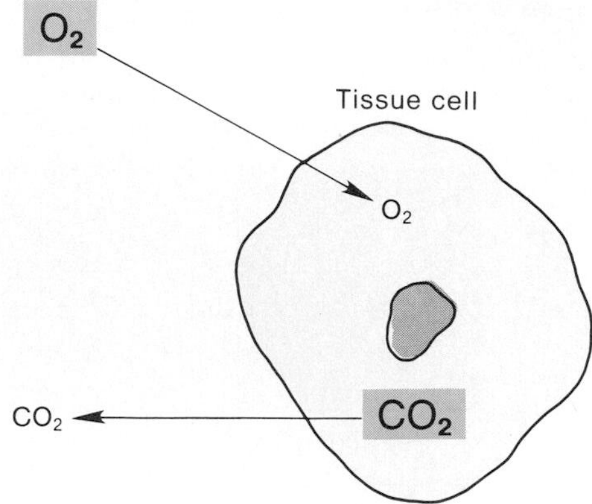

the second law of thermodynamics (chapter 4), the concentration difference represents an unstable state of high organization (low entropy), which changes to produce a uniformly distributed solution with maximum disorganization (high entropy).

As a result of random molecular motion, molecules in the part of the solution with a higher concentration will enter the area of lower concentration. Molecules will also move in the opposite direction, but not as frequently. As a result, there will be a *net movement* from the region of higher to the region of lower concentration until the concentration difference is abolished. This net movement is called **net diffusion.** Net diffusion is a physical process that occurs whenever there is a concentration difference; when the concentration difference exists across a membrane, diffusion becomes a type of membrane transport.

Normal functioning kidneys remove waste products from the blood. After the blood is filtered through pores in capillary walls—which are large enough to permit the passage of wastes and other molecules—the molecules needed by the body are reabsorbed back into the blood. The wastes generally remain in the filtrate and are excreted in the urine. When the kidneys do not function properly, waste molecules can be removed from the blood artificially by a process called *dialysis (di-al'i-sis).* Dialysis refers to the process of removing particular molecules from a solution by having them pass, by means of diffusion, through an artificial porous membrane. Since the pores in this dialysis membrane are large enough to permit the passage of some molecules but too small to permit the passage of others (the plasma proteins), small waste molecules can be removed from the blood by this technique.

entropy: Gk. *entropia,* a turning toward

Diffusion through the Cell Membrane

Since the cell membrane consists primarily of a double layer of phospholipids, molecules that are nonpolar and thus lipid-soluble can easily pass from one side of the membrane to the other. The cell membrane, in other words, does not present a barrier to the diffusion of nonpolar molecules such as oxygen gas (O_2) or steroid hormones. Small organic molecules that have polar covalent bonds but are uncharged, such as CO_2 (as well as ethanol and urea), are also able to penetrate the double-lipid layers. Net diffusion of these molecules can thus easily occur between the intracellular and extracellular compartments when concentration gradients are present.

The oxygen concentration is relatively high, for example, in the extracellular fluid because oxygen is carried from the lungs to the body tissues by the blood. Since oxygen is converted to water in aerobic cell respiration, the oxygen concentration within the cells is lower than in the extracellular fluid. The concentration gradient for carbon dioxide is in the opposite direction because cells produce CO_2. *Gas exchange* thus occurs by diffusion between the tissue cells and their extracellular environments (fig. 5.2).

Although water is not lipid-soluble, water molecules can diffuse through the cell membrane because of their small size and lack of net charge. The passage of water through the membrane may also be aided by the dipolar nature of water molecules (chapter 2), which permits interactions between the water molecules and the negative charges of the membrane phospholipids. The net diffusion of water molecules across the membrane (called *osmosis*) can thus occur when the solution on one side of the membrane is more dilute (has a higher water concentration) than on the other side of the membrane.

Larger polar molecules, such as glucose, cannot pass through the double phospholipid layers of the membrane and thus require special *carrier proteins* in the membrane for transport (described later). The phospholipid portion of the membrane is similarly impermeable to charged inorganic ions, such as Na^+ and K^+. Passage of these ions through the cell membrane may be permitted by tiny **ion channels** through the membrane that are too small to be seen even with an electron microscope. Many scientists believe that these channels are provided by some of the *integral proteins* that span the thickness of the membrane (fig. 5.3).

Rate of Diffusion

The rate of diffusion, measured by the number of diffusing molecules passing through the membrane per unit time, depends on (1) the magnitude of the concentration difference across the membrane (the "steepness" of the concentration gradient); (2) the permeability of the membrane to the diffusing substances; and (3) the surface area of the membrane through which the substances are diffusing.

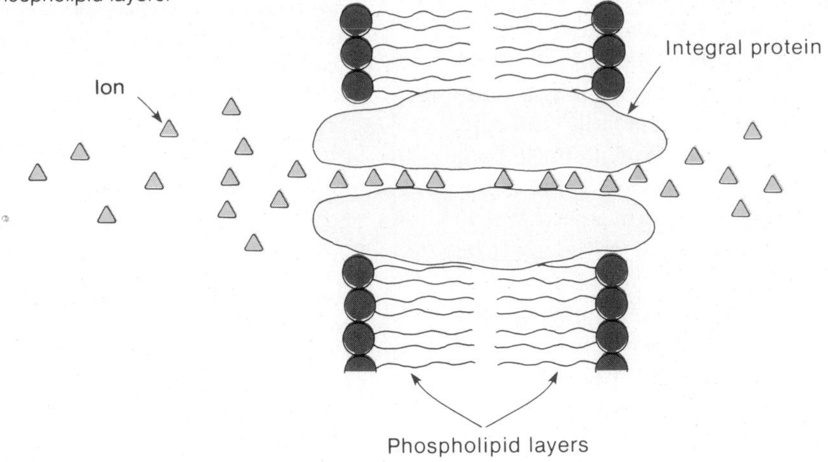

Figure 5.3. Inorganic ions (such as Na^+ and K^+) may penetrate the membrane through pores within integral proteins that span the thickness of the double phospholipid layers.

Figure 5.4. Microvilli (*MV*) in the small intestine, as seen with the transmission (*a*) and scanning (*b*) electron microscopes. (From: *Tissues and Organs: A Text Atlas of Scanning Electron Microscopy* by R. G. Kessel and R. Kardon. W. H. Freeman and Company. © 1979.)

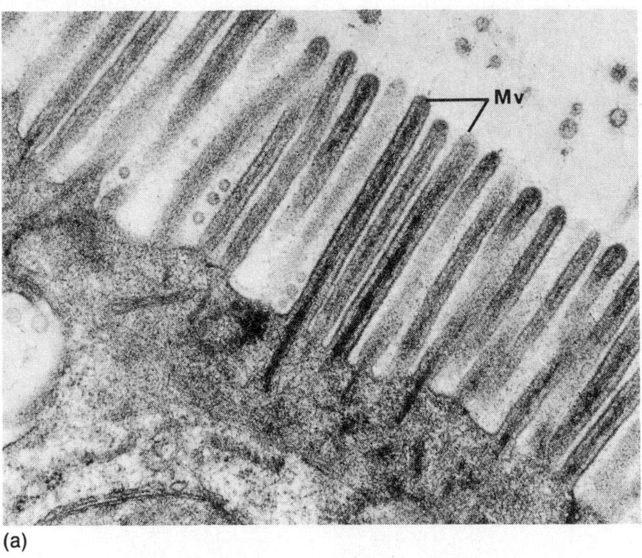

(a)

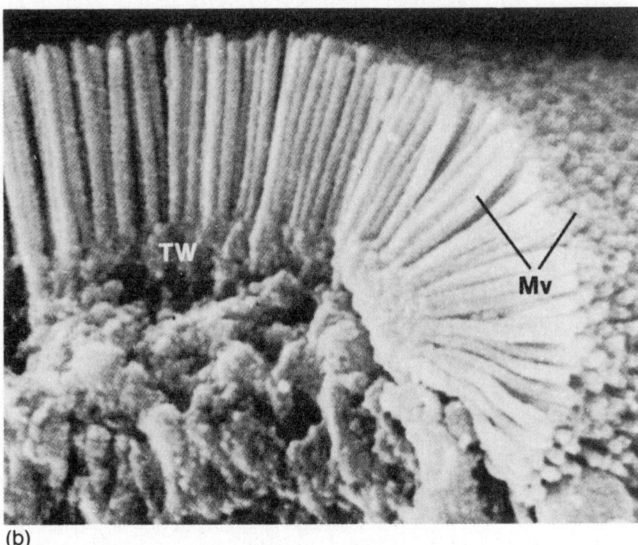

(b)

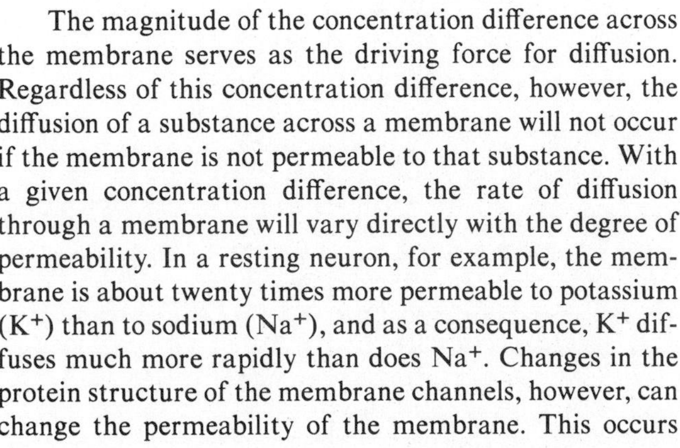

The magnitude of the concentration difference across the membrane serves as the driving force for diffusion. Regardless of this concentration difference, however, the diffusion of a substance across a membrane will not occur if the membrane is not permeable to that substance. With a given concentration difference, the rate of diffusion through a membrane will vary directly with the degree of permeability. In a resting neuron, for example, the membrane is about twenty times more permeable to potassium (K^+) than to sodium (Na^+), and as a consequence, K^+ diffuses much more rapidly than does Na^+. Changes in the protein structure of the membrane channels, however, can change the permeability of the membrane. This occurs during the production of a nerve impulse, when specific stimulation opens Na^+ channels temporarily and allows a faster diffusion rate for Na^+ than for K^+.

In areas of the body that are specialized for rapid diffusion, the surface area of the cell membranes may be increased by numerous folds. The rapid passage of the products of digestion across the epithelial membranes in the intestine, for example, is aided by such structural adaptations. The surface area of the apical membranes (the part facing the lumen) in the intestine is increased by many tiny folds that form fingerlike projections called **microvilli** *(mi″kro-vil′i)* (fig. 5.4). Similar microvilli are also found in the kidney tubule epithelium, which must reabsorb various molecules that are filtered out of the blood.

---

microvillus: Gk. *mikros,* small; L. *villus,* shaggy hair

Osmosis

Suppose that a cylinder is divided into two equal compartments by a membrane partition that can freely move and that one compartment initially contains 180 g/L (grams per liter) of glucose and the other compartment contains 360 g/L of glucose. If the membrane is permeable to glucose, glucose will diffuse from the 360 g/L compartment to the 180 g/L compartment until both compartments contain 270 g/L of glucose. If the membrane is not permeable to glucose but is permeable to water, the same result (270 g/L solutions on both sides of the membrane) will be achieved by the diffusion of water (osmosis). As water diffuses from the 180 g/L compartment to the 360 g/L compartment, the former solution would become more concentrated as the latter becomes more dilute. This is accompanied by volume changes, as illustrated in figure 5.5.

Osmosis *(oz-mo'sis)* refers to the net diffusion of water (the solvent) across a membrane. In order for this to occur, the membrane must be *semipermeable*; that is, it must be more permeable to water molecules than to solutes. Like the diffusion of solute molecules, the diffusion of water occurs when the water is more concentrated on one side of the membrane than on the other side; that is, when one solution is more dilute than the other (fig. 5.6). The more dilute solution has a higher concentration of water molecules because the less dilute solution contains more solute molecules, which reduces the water concentration. The higher solute concentration also reduces the "activity," or the freedom of movement, of water molecules more than does the lower solute concentration. The principles of osmosis apply to the diffusion of any molecule, but the terminology is backward because the term *concentration* is usually used to refer to the density of solute rather than solvent molecules.

osmosis: Gk. *osmos,* a thrust

Figure 5.5. A movable semipermeable membrane (permeable to water but not glucose) separates two solutions of different glucose concentration (*a*). As a result, water moves by osmosis into the solution of greater concentration until (*b*) the volume changes equalize the concentrations on both sides of the membrane.

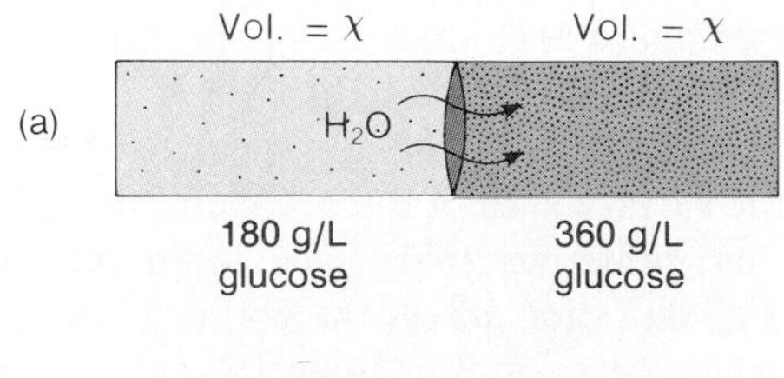

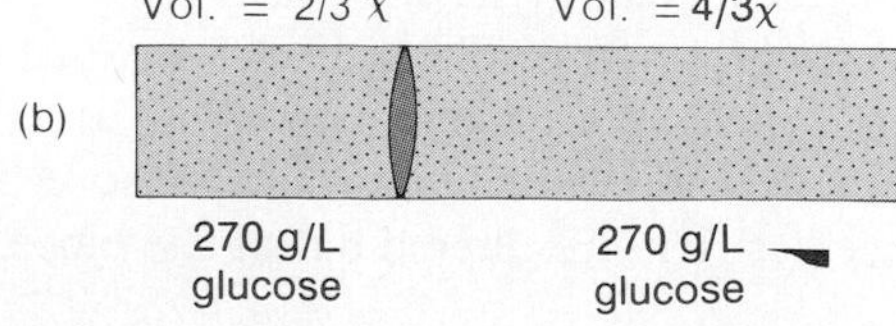

In order for osmosis to occur between two solutions, the two solutions must have different concentrations, and the membrane must be relatively impermeable to the solutes that produce the differences in concentration. Such impermeable solutes are said to be *osmotically active.* Water, for example, returns from tissue fluid to blood capillaries because the protein concentration of blood plasma is higher than the protein concentration of tissue fluid. The plasma proteins, in this case, are osmotically active. This occurs because the plasma proteins, in contrast to other plasma solutes, cannot pass from the capillaries into the tissue fluid. When clinicians want to expand a patient's blood volume (to raise the blood pressure), therefore, they give intravenous infusions of an albumin solution or of plasma (which contains albumin and other proteins). If a person has an abnormally low concentration of plasma proteins, as may occur in liver disease (cirrhosis, for example), fluid may accumulate in the tissues and produce *edema.*

Osmotic Pressure. Osmosis, and the movement of the membrane partition, could be prevented by an opposing force. If one compartment contained 180 g/L of glucose and the other compartment contained pure water, the osmosis of water into the glucose solution could be prevented by pushing against the membrane with a certain force (equal to 22.4 atmospheres pressure in this example). This is illustrated in figure 5.7.

The force that would have to be exerted to prevent osmosis in this situation is the **osmotic** *(oz-mot'ik)* **pressure** of the solution. This backward measurement indicates how strongly the solution "draws" water into it by

Figure 5.6. A model of osmosis, or the net movement of water from the solution of lesser solute concentration to the solution of greater solute concentration.

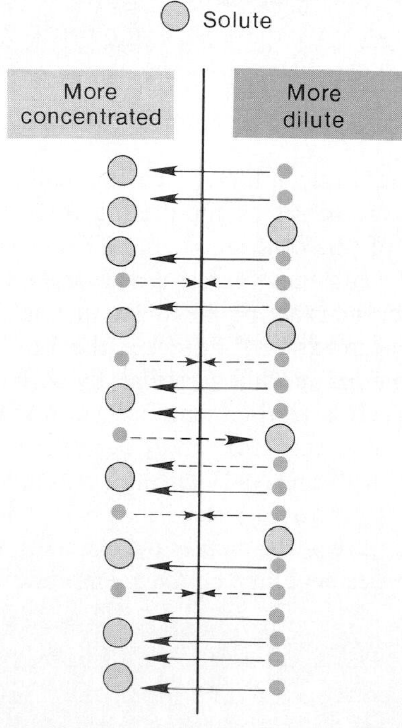

osmosis. The greater the solute concentration of a solution, the greater its osmotic pressure. Pure water, therefore, has an osmotic pressure of zero, and a 360 g/L glucose solution has twice the osmotic pressure of a 180 g/L glucose solution.

Molarity and Molality. Glucose is a monosaccharide with a molecular weight of 180 (the sum of its atomic weights). Sucrose is a disaccharide of glucose and fructose, which have molecular weights of 180 each. When glucose and fructose join together by dehydration synthesis to form a sucrose, a molecule of water (molecular weight = 18) is split off. Therefore, sucrose has a molecular weight of 342 (the sum of 180 + 180 − 18); each sucrose molecule weighs 342/180 times as much as each glucose molecule. It follows that 342 grams of sucrose must contain the same number of molecules as 180 grams of glucose.

Notice that the molecular weight in grams of any compound must contain the same number of molecules as the gram molecular weight of any other compound. This unit of weight is called a *mole,* and it always contains 6.02×10^{23} molecules (**Avogadro's number**). One mole of solute dissolved in water to make one liter of solution is described as a **one-molar** solution (abbreviated 1.0 M). Although this unit of measurement is commonly used in chemistry, it is not completely desirable in discussions of osmosis because the exact ratio of solute to water is not specified. More water, for example, is needed to make a 1.0 M NaCl solution (where a mole of NaCl weighs 58.5 grams) than is needed to make a 1.0 M glucose solution, because 180 grams of glucose take up more volume than 58.5 grams of salt.

Since the ratio of solute to water molecules is of critical importance in osmosis, a more desirable measurement of concentration is **molality.** In a one-molal solution (abbreviated 1.0 m), one mole of solute (180 grams of glucose, for example) is dissolved in one kilogram of water (equal to one liter at 4°C). A 1.0 m NaCl solution and a 1.0 m glucose solution both contain a mole of solute dissolved in exactly the same amount of water (fig. 5.8).

Avogadro: from Amadeo Avogadro, Italian physicist, 1766–1856

Figure 5.7. If a semipermeable membrane separates pure water from a 180 g/L glucose solution, water tends to move by osmosis into the glucose solution, thus creating a hydrostatic pressure that pushes the membrane to the left and expands the volume of the glucose solution. The amount of pressure that must be applied to just counteract this volume change is equal to the osmotic pressure of the glucose solution.

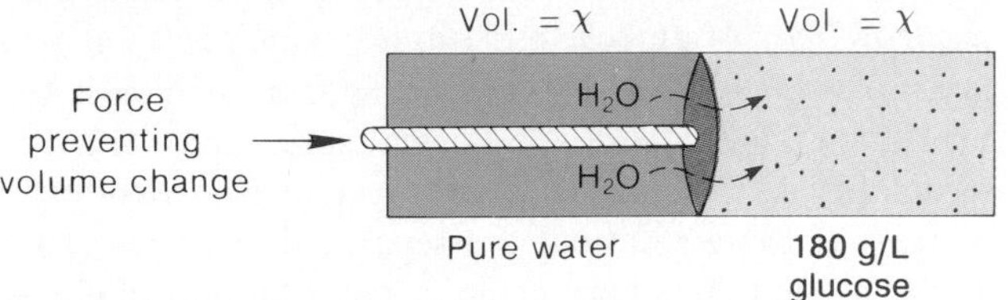

Figure 5.8. Diagram illustrating the difference between a one molar (1.0 M) and a one molal (1.0 m) glucose solution.

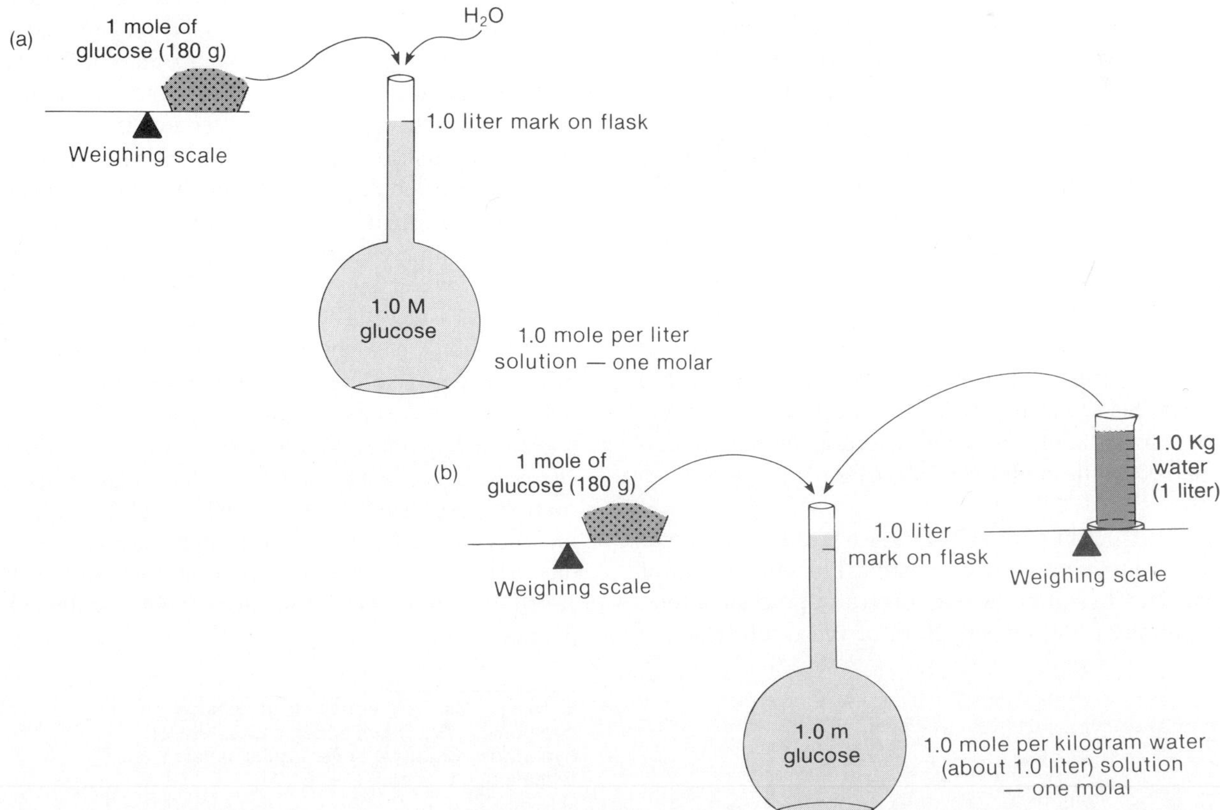

Figure 5.9. The osmolality (Osm) of a solution is equal to the sum of the molalities of each solute in the solution. If a semipermeable membrane separates two solutions with equal osmolalities, no osmosis will occur.

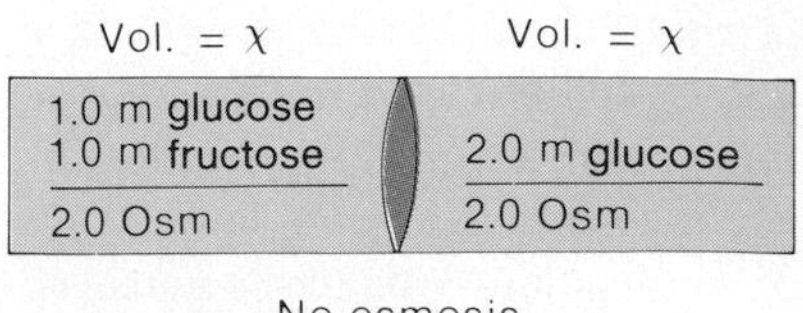

Figure 5.10. If a semipermeable membrane (permeable to water but not to glucose, Na^+, or Cl^-) separates a 1.0 m glucose solution from a 1.0 m NaCl solution (*a*), water will move by osmosis into the NaCl solution. This is because NaCl can ionize to yield one molal Na^+ plus one molal Cl^-. After osmosis (*b*), the total concentration, or osmolality, of the two solutions is equal.

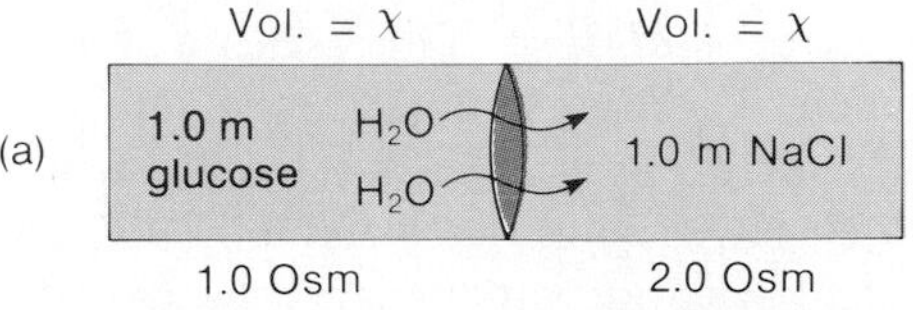

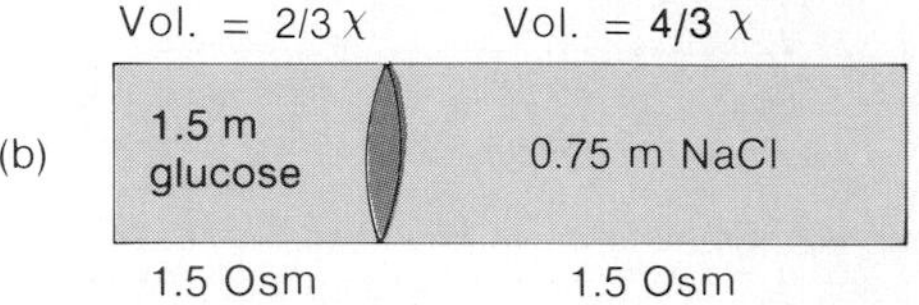

Osmolality. If 180 grams of glucose and 180 grams of fructose were dissolved in the same kilogram of water, the osmotic pressure of the solution would be the same as that of a 360 g/L glucose solution. Osmotic pressure depends on the ratio of solute to solvent, *not* on the chemical nature of the solute molecules. The expression for the total molality of a solution is **osmolality (Osm).** Thus, the solution of 1.0 m glucose plus 1.0 m fructose has a total molality, or *osmolality (oz″mo-lal′i-te),* of 2.0 osmol/L (abbreviated 2.0 Osm). This is the same as the 360 g/L glucose solution, which is 2.0 m and 2.0 Osm (fig. 5.9).

Unlike glucose, fructose, and sucrose, electrolytes such as NaCl ionize when they dissolve in water. One molecule of NaCl dissolved in water yields two ions (Na^+ and Cl^-); one mole of NaCl ionizes to form one mole of Na^+ and one mole of Cl^-. Thus a 1.0 m NaCl solution has a total concentration of 2.0 Osm. The effect of this on osmosis is illustrated in figure 5.10.

Measurement of Osmolality. Plasma and other biological fluids contain many organic molecules and electrolytes. The osmolality of such complex solutions can only be estimated by calculations. Fortunately, however, there is a relatively simple method for measuring osmolality. This method is based on the fact that the freezing point of a solution, like its osmotic pressure, is affected by the total concentration of the solution and not by the chemical nature of the solute.

One mole of solute depresses the freezing point of water by $-1.86°C$. Accordingly, a 1.0 m glucose solution freezes at a temperature of $-1.86°C$, and a 1.0 m NaCl solution freezes at a temperature of $2 \times -1.86 = -3.72°C$, because of ionization. The *freezing point depression* is, thus, a measure of the osmolality. Since plasma freezes at about $-0.56°C$, its osmolality is equal to 0.56/1.86, or 0.3 Osm, which is more commonly indicated as 300 milliosmolal (or 300 mOsm).

Tonicity. A 0.3 m glucose solution has the same osmolality and osmotic pressure as plasma. The same is true of a 0.15 m NaCl solution, which ionizes to produce a total concentration of 300 mOsm. Both of these solutions are used clinically as intravenous infusions, labeled *5.0% dextrose* (5 g of glucose per 100 ml, which is 0.3 m) and *normal saline* (0.9 g of NaCl per 100 ml, which is 0.15 m). Since 5% dextrose and normal saline have the same osmolality as plasma, they are said to be *isosmotic* to plasma. If these solutions are separated from plasma by a membrane that is permeable to water but not to glucose or NaCl, osmosis will not occur. In this case the solutions are said to be **isotonic** to plasma.

Red blood cells placed in an isotonic solution will neither gain nor lose water. It should be noted that a solution may be isosmotic but not isotonic; such is the case whenever the solute in the isosmotic solution can freely penetrate the membrane. A 0.3 m urea solution, for example, is isosmotic but not isotonic because the cell membrane is permeable to urea. When red blood cells are placed in a 0.3 m urea solution, the urea diffuses into the cells until its concentration on both sides of the cell membranes becomes equal. Meanwhile, the solutes within the cells that cannot exit, and that are therefore osmotically active, cause osmosis of water into the cells. Red blood cells placed in 0.3 m urea will thus eventually burst.

Solutions that have a lower total concentration of osmotically active solutes and a lower osmotic pressure than plasma are said to be **hypotonic** to plasma. Red blood cells placed in hypotonic solutions gain water and may burst (*hemolysis*). When red blood cells are placed in a **hypertonic** solution (such as sea water), which has a higher osmolality and osmotic pressure than plasma, they shrink due to the osmosis of water out of the cells. In this process, called *crenation,* the cell surface becomes scalloped in appearance (fig. 5.11).

isotonic: Gk. *isos,* equal; *tonos,* tension
hypotonic: Gk. *hypo,* under; *tonus,* tension
hypertonic: Gk. *hyper,* over; *tonus,* tension
crenation: L. *crena,* a notch

Figure 5.11. A scanning electron micrograph of normal and crenated red blood cells.

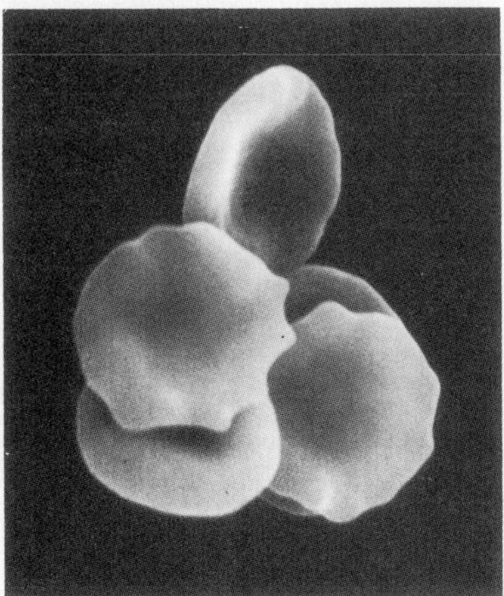

Fluids delivered intravenously must be isotonic to blood in order to maintain the correct osmotic pressure and prevent cells from either expanding or shrinking due to the gain or loss of water. Common fluids used for this purpose are normal saline (approximately 0.9 g NaCl per 100 ml water) and 5% dextrose (5 g glucose per 100 ml water). These solutions have about the same osmolality as normal plasma (approximately 300 mOsm). Another isotonic solution frequently used in hospitals is *Ringer's lactate,* which contains glucose and lactic acid in addition to a number of different salts. Isotonic solutions are also used in heart-lung machines, which take the place of the heart and lungs during open-heart surgery.

Regulation of Blood Osmolality

The osmolality of the blood plasma is normally maintained within very narrow limits by a variety of regulatory mechanisms. When a person becomes dehydrated, for example, the blood becomes more concentrated as the total blood volume is reduced. The increased blood osmolality and osmotic pressure stimulates *osmoreceptors,* which are neurons located in a part of the brain called the hypothalamus.

As a result of increased osmoreceptor stimulation, the person becomes thirsty and drinks, if water is available. Along with increased water intake, a person who is dehydrated excretes a lower volume of urine. This occurs as a result of the following sequence of events: (1) increased plasma osmolality stimulates osmoreceptors in the hypothalamus of the brain; (2) the osmoreceptors stimulate the posterior pituitary gland, by means of a tract of nerve fibers, to secrete **antidiuretic** *(an''tī-di''u-ret'ik)* **hormone (ADH);** (3) ADH acts on the kidneys to promote water retention; so that (4) a lower volume of urine is excreted.

Ringer: from Sidney Ringer, English physiologist, 1835–1910

Figure 5.12. An increase in plasma osmolality (increased concentration and osmotic pressure) due to dehydration stimulates thirst and increased ADH secretion. These effects cause the person to drink more and urinate less. The blood volume, as a result, is increased while the plasma osmolality is decreased. These effects help bring the blood volume back to the normal range and complete the negative feedback (indicated by a negative sign).

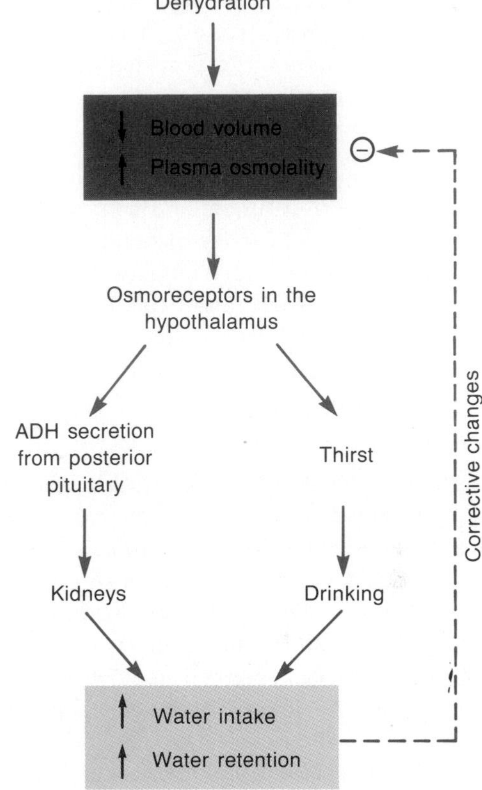

A person who is dehydrated, therefore, drinks more and urinates less. This is represented in figure 5.12, in which the dashed line and arrow to the negative sign represents a correction of the initial deviations from normal conditions. This type of a corrective mechanism is known as a *negative feedback* loop and is discussed in more detail in chapter 7.

1. Define simple diffusion, and describe three factors that influence the diffusion rate.
2. Define the term *osmosis,* and describe the conditions required for it to occur; define the terms *osmolality* and *osmotic pressure.*
3. Define the terms *isotonic, hypotonic,* and *hypertonic,* and explain why hospitals use 5% dextrose and normal saline as intravenous infusions.
4. Explain how changes in the osmolality of plasma are detected and corrected by the body.

Carrier-Mediated Transport

Molecules such as glucose, amino acids, and other organic molecules of comparable size are transported across cell membranes by special protein carriers that are specific and that can become saturated. Carrier-mediated transport that involves a net movement down a concentration gradient, and which therefore is passive, is called facilitated diffusion. Carrier-mediated transport that occurs against a concentration gradient and that requires metabolic energy is called active transport.

Objective 6. Distinguish between simple diffusion and facilitated diffusion, and describe the characteristics of carrier-mediated transport.

Objective 7. Distinguish between facilitated diffusion and active transport, and explain why the term "active transport" is used.

Objective 8. Describe how co-transport occurs, and explain why this is considered to be a type of active transport.

In order to sustain metabolism, cells must be able to take up glucose, amino acids, and other organic molecules from the extracellular environment. Molecules such as these, however, are too large and polar to pass through a lipid barrier by a process of simple diffusion. The transport of glucose, amino acids, and some other molecules is mediated by **protein carriers** within the membrane. Although such carriers cannot be directly observed, their presence has been inferred by the observation that this transport has characteristics in common with enzyme activity. These characteristics include (1) *specificity,* (2) *competition,* and (3) *saturation.*

Like enzyme proteins, protein carriers interact with only specific molecules. Glucose carriers, for example, can only interact with glucose and not with closely related monosaccharides. As a further example of specificity, particular carriers for amino acids transport some types of amino acids but not others. Two amino acids that are transported by the same carrier compete with each other, so that the rate of transport of each when together is lower than it would be if each amino acid were present alone (fig. 5.13).

As the concentration of a transported molecule is increased, its rate of transport will also be increased—but only up to a maximum. Beyond this rate, called the *transport maximum* (or T_m), further increases in concentration do not further increase the transport rate. This indicates that the carriers have become saturated (fig. 5.13).

As an example of saturation, imagine a bus stop that is serviced once per hour by a bus that can hold a maximum of forty people (its "transport maximum"). If ten people wait at the bus stop, ten will be transported per hour. If twenty people wait at the bus stop, twenty will be transported per hour. This linear relationship will hold up to a maximum of forty people; if eighty people are at the bus stop, the transport rate will still be forty per hour.

Figure 5.13. Carrier-mediated transport displays the characteristics of saturation (illustrated by the *transport maximum*) and competition. Molecules *X* and *Y* compete for the same carrier, so that when they are present together the rate of transport of each is less than when either is present separately.

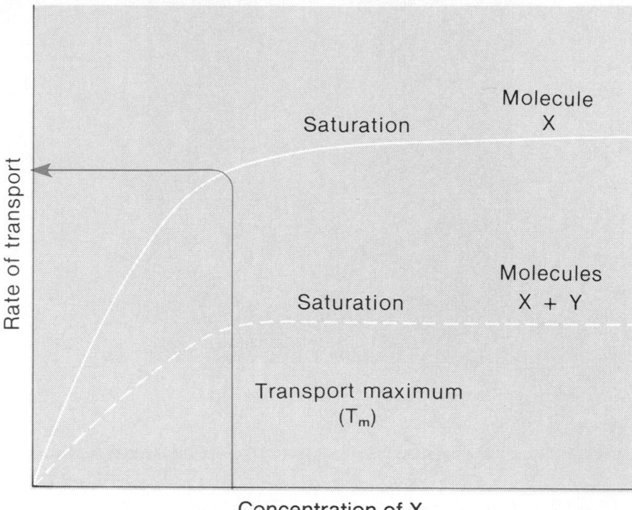

The kidneys transport a number of molecules from the blood filtrate (which will become urine) back into the blood. Glucose, for example, is normally completely reabsorbed so that urine normally is free of glucose. If the glucose concentration of the blood and filtrate is too high (a condition called *hyperglycemia*), however, the transport maximum will be exceeded. In this case, glucose will be found in the urine (a condition called *glycosuria [gli''ko-su're-ah]*). This may result from eating too many sweets or from the inadequate action of the hormone *insulin* (in the disease *diabetes mellitus [di''ah-bē'tez mel-li'tus]*).

Facilitated Diffusion

The transport of glucose from the blood across the cell membranes of tissue cells occurs by **facilitated** *(fah-sil'i-ta''tid)* **diffusion.** Facilitated diffusion, like simple diffusion, is powered by the thermal energy of the diffusing molecules and involves the net transport of substances through a cell membrane from the side of higher to the side of lower concentration. Active cellular metabolism is not required for either facilitated or simple diffusion.

Unlike simple diffusion of nonpolar molecules, water, and inorganic ions through a membrane, the diffusion of glucose through the cell membrane displays the properties of carrier-mediated transport: specificity, competition, and saturation. The diffusion of glucose through a cell membrane must therefore be mediated by protein carriers, even

Figure 5.14. A model of facilitated diffusion, where a molecule is transported across the cell membrane by a carrier protein.

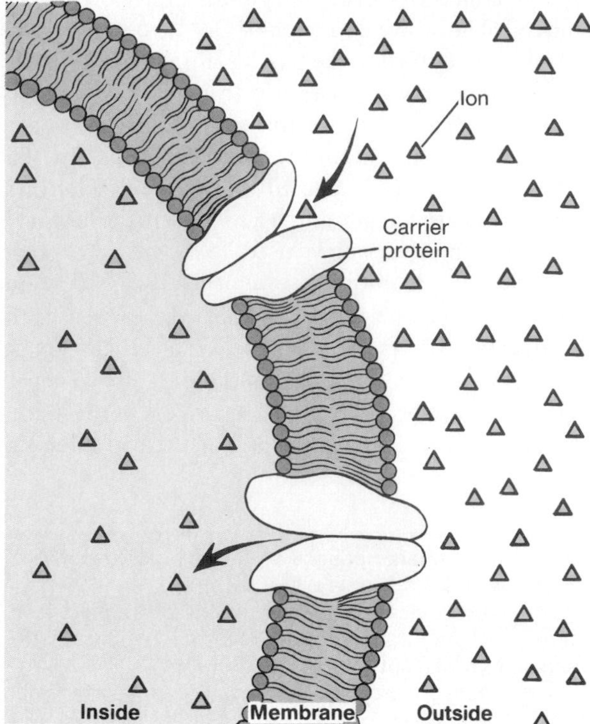

though these are too small to be directly observed. One conceptual model of the transport carriers is that they may each be composed of two protein subunits that interact with glucose in a specific way that creates a channel through the membrane (fig. 5.14), so that the glucose can move from the side of higher to the side of lower concentration.

The transport of glucose from the blood plasma into tissue cells occurs by facilitated diffusion. In this process, the membrane protein carriers help to transport glucose from the plasma and tissue fluid, where the glucose concentration is higher, into the cells, where the glucose concentration is lower. Since the carriers are seldom saturated, the rate of this transport depends directly on the plasma glucose concentration. When the plasma glucose concentration is abnormally low—a condition called *hypoglycemia*—the rate of transport of glucose into brain cells may be inadequate for the metabolic needs of the brain. Severe hypoglycemia, as may be produced in a diabetic person by an overdose of insulin, can thus result in the loss of consciousness and even death.

Active Transport

Some aspects of cell transport cannot be explained by simple or facilitated diffusion. The epithelial lining of the intestine and of the kidney tubules, for example, moves glucose from the side of lower to the side of higher concentration (from the lumen to the blood). Similarly, all

Figure 5.15. A model of active transport, showing the hingelike motion of the integral protein subunits.

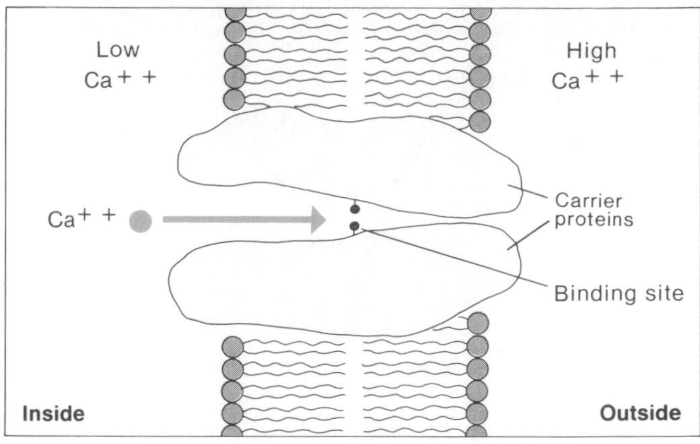

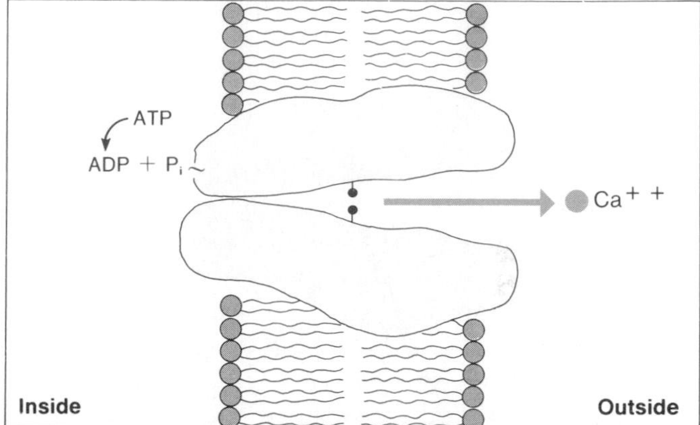

cells extrude Ca^{++} into the extracellular environment and, by this means, maintain an intracellular Ca^{++} concentration that is one thousand to ten thousand times lower than the extracellular Ca^{++} concentration.

The movement of molecules and ions against their concentration gradients, from lower to higher concentrations, requires the expenditure of cellular energy, which is obtained from ATP. This type of transport is thus termed **active transport.** If a cell is poisoned with cyanide (which inhibits oxidative phosphorylation), active transport is inhibited. This contrasts with passive transport, which can continue even when metabolic poisons kill the cell by preventing the formation of ATP.

Active transport, like facilitated diffusion, is carrier-mediated. These carriers appear to be integral proteins that span the thickness of the membrane. According to one theory of active transport, the following events may occur: (1) the molecule or ion to be transported bonds to a specific "recognition site" on one side of the protein carrier; (2) this bonding stimulates the breakdown of ATP, which in turn results in phosphorylation of the carrier protein; (3) as a result of phosphorylation, the carrier protein undergoes a conformational change, like the change in enzyme proteins as a result of allosteric effects; and (4) the carrier protein undergoes a hingelike motion, which releases the transported molecule or ion on the other side of the membrane. This model of active transport is illustrated in figure 5.15.

Figure 5.16. The Na^+/K^+ pump actively exchanges intracellular Na^+ for K^+. The carrier itself is an ATPase that breaks down ATP for energy. *Dotted lines* indicate the direction of passive transport (diffusion); *solid arrows* indicate the direction of active transport.

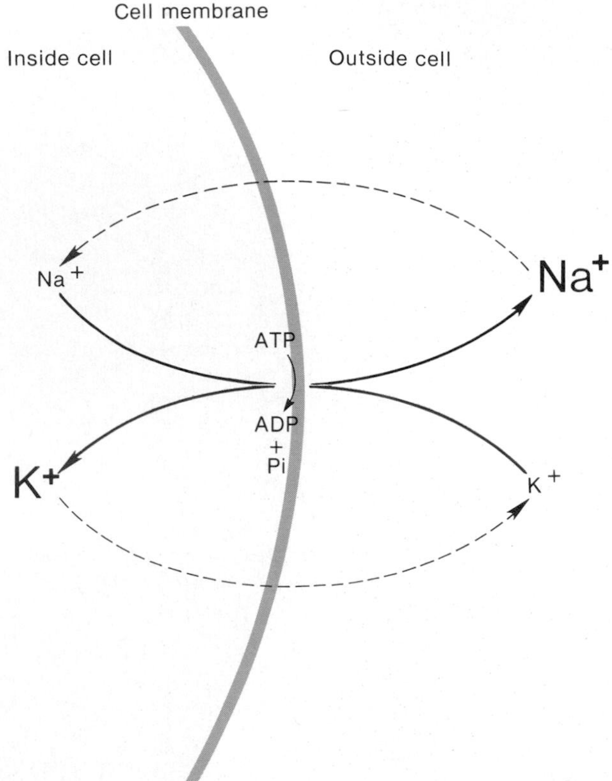

The Sodium-Potassium Pump. Active transport carriers are often referred to as "pumps." Although some of these carriers transport only one molecule or ion at a time, other carriers exchange one molecule or ion for another. The most important of the latter type of carriers is the **Na^+/K^+ pump.** This protein carrier, which is also an ATPase enzyme that converts ATP to ADP and P_i, actively extrudes three Na^+ ions from the cell as it transports two K^+ into the cell. This transport is energy dependent because Na^+ is more highly concentrated outside the cell and K^+ is more concentrated within the cell. Both ions, in other words, are moved against their concentration gradients (fig. 5.16).

All cells have numerous Na^+/K^+ pumps that are constantly active. This represents an enormous expenditure of energy used to maintain a steep gradient of Na^+ and K^+ across the cell membrane. This steep gradient serves three known functions: (1) the steep Na^+ gradient is used to provide energy for the "co-transport" of other molecules; (2) the activity of the Na^+/K^+ pumps can be adjusted (primarily by thyroid hormones) to regulate the resting calorie expenditure and basal metabolic rate of the body; and (3) the Na^+ and K^+ gradients across the cell membranes of nerve and muscle cells are used to produce electrical impulses.

Coupled Transport (Co-transport). In co-transport, the energy needed for the "uphill" movement of a molecule or ion is obtained from the "downhill" transport of Na^+ into the cell. The active extrusion of Ca^{++} from some cells, for example, is coupled to the passive diffusion of Na^+ into the cell. Cellular energy, obtained from ATP, is not used to move Ca^{++} directly out of the cell, but energy is constantly required to maintain the steep Na^+ gradient. The inward diffusion of Na^+ due to this concentration gradient, in turn, is used to power the active extrusion of Ca^{++}.

A similar mechanism is used to actively transport glucose across the membranes of epithelial cells that line the digestive tract and kidney tubules. In this case, the inward diffusion of Na^+ is coupled to the cellular uptake of glucose (as well as amino acids) against their concentration gradients (fig. 5.17). Active transport in the kidney tubules accounts for as much as 6% of the body's energy expenditure at rest.

1. List the three characteristics of facilitated diffusion that distinguish it from simple diffusion.
2. Draw a figure that illustrates two of the characteristics of carrier-mediated transport, and explain how this differs from simple diffusion.
3. Describe active transport, including co-transport in your description. Explain how active transport differs from facilitated diffusion.
4. Explain the functional significance of the Na^+/K^+pump.

The Membrane Potential

Large, negatively charged molecules are concentrated within the cell by their inability to pass through the cell membrane. This creates a polarity across the membrane, with the inside of the cell negatively charged in comparison to the outside of the cell. This polarity is maintained and increased by the activity of the Na^+/K^+ pumps. The magnitude of the difference in charge across the membrane is the membrane potential, measured in millivolts.

Objective 9. Describe how the permeability characteristics of the membrane result in the production of a potential difference across the membrane.

Objective 10. Explain how a potassium equilibrium potential can be theoretically produced.

Objective 11. Explain why the true membrane potential is close to, but slightly lower than, the potassium equilibrium potential.

Objective 12. Explain how the membrane potential is maintained, and in fact increased, by activity of Na^+/K^+ pumps.

Figure 5.17. Co-transport of glucose. Glucose accumulates inside the cell against its concentration gradient, using energy derived from the passive transport (diffusion) of Na^+ into the cell. Metabolic energy is indirectly required because the steep electrochemical gradient for Na^+ is maintained by the Na^+/K^+ pump. *Dotted arrows* indicate direction of diffusion; *solid arrows* show direction of active transport.

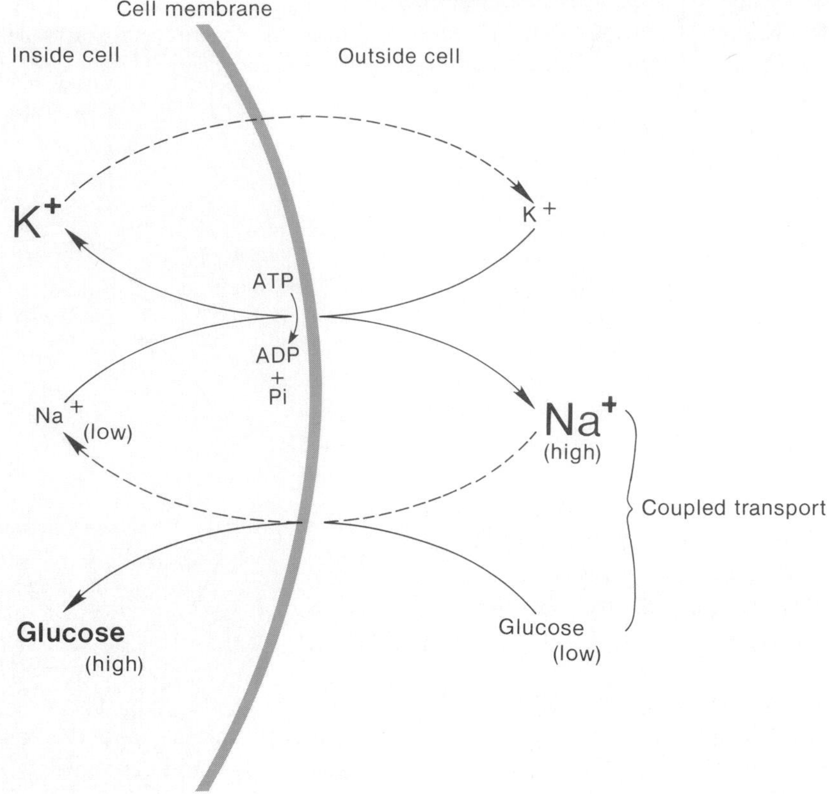

Cellular proteins and the phosphate groups of ATP and other organic molecules are negatively charged within the cell cytoplasm. These negative ions (anions *[an'i-onz]*) are "fixed" within the cell by the fact that they cannot penetrate the cell membrane. Since these negatively charged organic molecules cannot leave the cell, they attract positively charged inorganic ions (cations *[cat'i-onz]*) from the extracellular fluid that are small enough to diffuse through the membrane pores. The distribution of small, inorganic cations (mainly K^+, Na^+, and Ca^{++}) between the intracellular and extracellular compartments, in other words, is influenced by the negatively charged fixed ions within the cell.

Since the cell membrane is much more permeable to K^+ than to any other cation, K^+ accumulates within the cell more than the others as a result of its electrical attraction for the fixed anions (fig. 5.18). Instead of being evenly distributed between the intracellular and extracellular compartments, therefore, K^+ becomes more highly concentrated within the cell. In the human body the intracellular K^+ concentration is 155 mEq/L, compared to an extracellular concentration of 4 mEq/L (mEq = milliequivalents, which is the millimolar concentration multiplied by the valence of the ion—in this case, by one).

Figure 5.18. Proteins, organic phosphates, and other organic anions that cannot leave the cell create a fixed negative charge on the inside of the membrane. This attracts positively charged inorganic ions (cations), which therefore accumulate within the cell at a higher concentration than in the extracellular fluid. The amount of cations that accumulate within the cell is limited by the fact that a concentration gradient builds up, which favors the diffusion of the cations out of the cell.

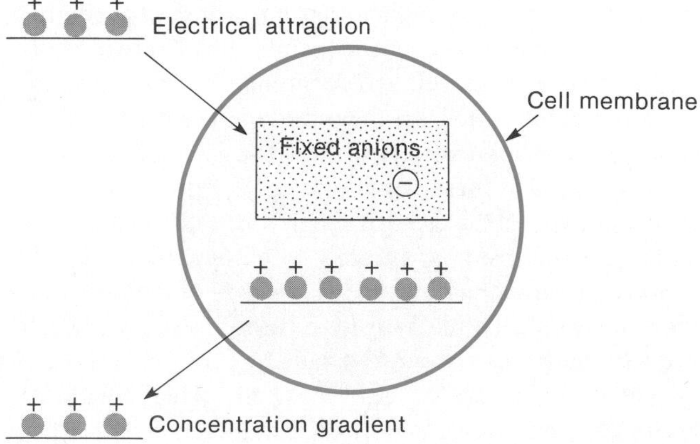

Even without the action of the Na^+/K^+ pumps, therefore, K^+ would be more highly concentrated within the cell than in the extracellular fluid. The magnitude of the intracellular K^+ concentration depends on the amount of fixed anions and on the K^+ concentrations that are normally found in the extracellular compartment (including plasma and tissue fluid). An increase in the extracellular K^+ concentration above normal, for example, would cause a corresponding increase in the intracellular K^+ concentration.

Equilibrium Potential

An equilibrium *(e''kwĭ-lib're-um)* potential is a theoretical voltage that would be produced across a cell membrane if only one ion were able to diffuse through the membrane. Since K^+ is the most permeable ion, we can construct a theoretical approximation to the true situation by considering what would happen if K^+ were the *only* ion able to cross the membrane. If this were the case, K^+ would diffuse until its concentration inside and outside of a cell became stable. An *equilibrium* would be established: if a certain amount of K^+ were to move inside the cell (by electrical attraction for the fixed anions), an identical amount of K^+ would diffuse out of the cell (down its concentration gradient).

At this equilibrium, the concentration of K^+ would be higher inside the cell than outside the cell; a concentration difference would exist across the cell membrane, which was stabilized by the attraction of K^+ to the fixed anions. At this point, we could ask the question: are the fixed anions neutralized; are the charges balanced? The answer to that question depends on how much K^+ gets into the cell, which in turn depends on the K^+ concentration in the extracellular fluid. At the K^+ concentrations that are, in fact, found in the body, the answer to the question is *no.* Not enough K^+ is present in the cell to neutralize the fixed anions (fig. 5.19).

At equilibrium, therefore, the inside of the cell membrane would have a higher concentration of negative charges than the outside of the membrane. There is a difference in charge, as well as a difference in concentration, across the membrane. The magnitude of the difference in charge, or **potential difference,** on the two sides of the membrane under these conditions is 98 millivolts (mV). This is shown with a negative sign (as −90 mV) to indicate that the inside of the cell is the negative pole. (If one were to write +90 mV, the magnitude of the potential difference would be unchanged, but the inside of the cell would be shown as the positive pole.)

The potential difference of −90 mV, which would be developed if K^+ were the only diffusible ion, is called the **K^+ equilibrium potential** (abbreviated E_K). This is a theoretical value, but is a good approximation to the truth because, in fact, K^+ *is* the ion to which the membrane is most permeable. The cell membrane, however, is slightly permeable to Na^+, and the concentration of Na^+ in the extracellular fluid (145 mEq/L) is much higher than it is in the cell (12 mEq/L). This concentration gradient, together with sodium's electrical attraction for the fixed anions, causes Na^+ to move into the cell at a slow rate permitted by the low permeability of the cell membrane to Na^+.

Figure 5.19. If K^+ was the only ion able to diffuse through the cell membrane, it would distribute itself between the intracellular and extracellular compartment until an equilibrium would be established. At equilibrium the K^+ concentration within the cell would be higher than outside the cell due to the attraction of K^+ for the fixed anions. Not enough K^+ would accumulate within the cell to neutralize these anions, however, so the inside of the cell would be 90 millivolts negative compared to the outside of the cell. This membrane voltage is the equilibrium potential (E_K) for potassium.

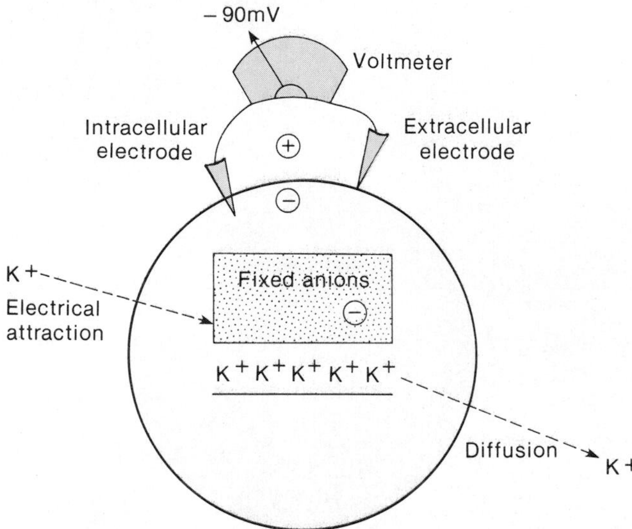

Resting Membrane Potential

If Na^+ were the only diffusible ion, it would diffuse into the cell until an equilibrium was established (analagous to the situation with K^+, described previously). In this case, so much Na^+ would move into the cell, because of its steep concentration gradient, that the inside of the cell would actually have a higher concentration of positive charges than the outside of the cell. At this equilibrium, the potential difference across the membrane (the E_{Na}) would be +66 mV. This situation actually occurs for a very short period of time during the production of a nerve impulse, when the membrane briefly becomes very permeable to Na^+ (the nerve impulse is discussed in chapter 14).

In a normal resting cell, however, the membrane permeability to Na^+ is so low that only a "trickle" of Na^+ can enter the cell. In a real cell, therefore, the potential difference across the membrane is close to the value predicted by the K^+ equilibrium potential, but it is slightly less negative than E_K as a result of the diffusion of some Na^+ into the cell (fig. 5.20). The true **resting membrane potential** is in the range of −85 mV to −65 mV in nerve and muscle cells.

Figure 5.20. Because some Na^+ leaks into the cell by diffusion, the actual resting membrane potential is less than the K^+ equilibrium potential. As a result, some K^+ diffuses out of the cell (*dotted lines*).

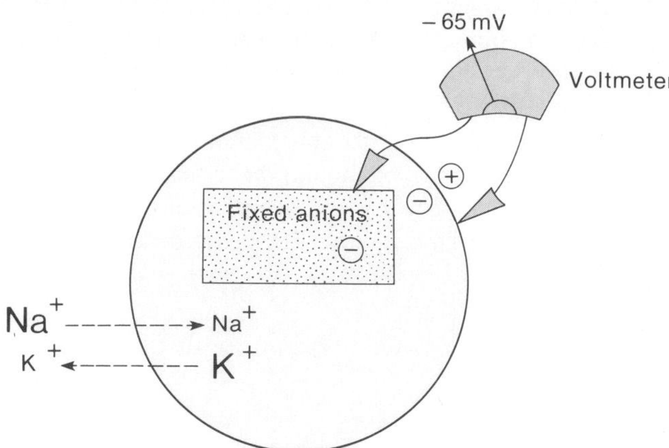

Figure 5.21. The concentrations of Na^+ and K^+ both inside and outside the cell do not change as a result of diffusion (*dotted arrows*) because of active transport (*solid arrows*) by the Na^+/K^+ pump. Since the pump transports three Na^+ for every two K^+, the pump itself helps to create a charge separation (a potential difference, or voltage) across the membrane.

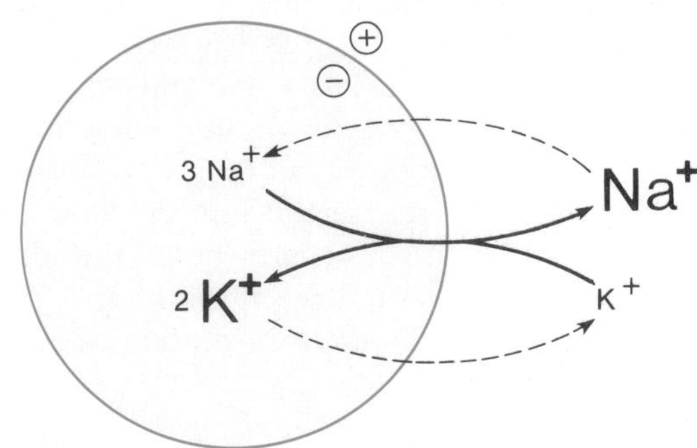

Table 5.1 Inherited defects of transport carriers in the kidney and intestine

Disease	Defect	Clinical significance
Cystinuria	Excessive urinary excretion of cystine, lysine, arginine, and ornithine	Calculi (stones) in urinary tract
Phosphaturia	Excessive urinary excretion of phosphate	Rickets: treated with large doses of vitamin D
Renal glycosuria	Kidney tubules have lower than normal T_m for glucose	None known
Glucose malabsorption	Dietary glucose not absorbed from intestine	Sometimes fatal; must use fructose as only dietary carbohydrate
Hartnup disease	Delayed intestinal absorption of tryptophan and related molecules	Cerebellum dysfunction; photosensitive dermatitis

The Na^+/K^+ Pump. Since the membrane potential is reduced below E_K as a result of Na^+ entry, some K^+ leaks out of the cell. The cell is *not* at equilibrium with respect to K^+ and Na^+ concentrations. Despite this, the concentrations of K^+ and Na^+ are maintained constant; this is due to the constant expenditure of energy in active transport by the Na^+/K^+ pump. The Na^+/K^+ pump acts to counter the leaks and thus maintains the membrane potential.

Actually, the Na^+/K^+ pump does more than simply work against the ion leaks, since it transports *three* Na^+ ions out of the cell for every *two* K^+ ions that it moves in, its action helps generate a potential difference across the membrane (fig. 5.21). As a result of all of these activities, a real cell has (1) a relatively constant intracellular concentration of Na^+ and K^+; and (2) a constant membrane potential (in the absence of stimulation) which, in nerves and muscles, is in the range of -65 mV to -85 mV.

1. Define the term *membrane potential,* and describe how it is measured.
2. Describe how an equilibrium potential is produced when potassium is the only diffusible cation.
3. Explain why the resting membrane potential is different from the potassium equilibrium potential.
4. Explain the role of the Na^+/K^+ pump in the generation and maintenance of the resting membrane potential.

Clinical Considerations

Inherited Defects in Membrane Carriers

Since membrane transport carriers are proteins that are coded by specific genes, inherited defects in these carriers can result when there is an alteration in the genetic code. Defective protein carriers in the cell membranes of epithelial cells that line the intestine may produce diseases that result from an inadequate absorption of ingested molecules. Defects in transport carriers within the epithelial cells of kidney tubules may result in the abnormal excretion of particular molecules in the urine (table 5.1).

Hyperkalemia and the Membrane Potential

Although changes in the extracellular concentration of many ions can affect the membrane potential, this potential is particularly sensitive to changes in plasma potassium. Since the maintenance of a particular membrane potential is critical for the generation of electrical events in nerves and muscles (including the heart), the body has a variety of mechanisms that serve to maintain plasma K^+ concentrations within very narrow limits. These mechanisms act primarily through the kidneys, which can excrete K^+ in the urine or reabsorb it into the blood. The excretion of K^+ is stimulated by hormones of the adrenal cortex (particularly by aldosterone); if the adrenal glands of an experimental animal are removed, the animal may die as a result of an accumulation of K^+ in the blood. An abnormal increase in the blood concentration of K^+ is called **hyperkalemia** *(hi''per-kah-le'me-ah).*

When hyperkalemia occurs, the diffusion gradient that favors the extrusion of K^+ from the cell is reduced. As a result, more K^+ can enter the cell and neutralize more of the fixed negative charges. This reduces the membrane potential (brings it closer to zero) and thus alters the function of many organs, particularly the heart. For these reasons, the blood electrolyte concentrations are monitored very carefully in patients with heart or kidney disease.

hyperkalemia: Gk. *hyper,* over; L. *kalium,* potash; Gk. *haima,* blood

Chapter Summary

I. Simple Diffusion and Osmosis
  A. Diffusion is the net movement of molecules or ions from regions of high to regions of low concentration.
  B. The rate of diffusion is dependent on the concentration gradient, the membrane permeability, and the surface area of the membrane.
  C. Simple diffusion is the type of passive transport in which small molecules and inorganic ions, such as Na^+ and K^+, move through membrane.
  D. Osmosis is the simple diffusion of solvent (water) through a membrane that is more permeable to the solvent than it is to the solute.

II. Carrier-Mediated Transport
  A. The passage of glucose, amino acids, and other substances through the cell membrane is mediated by carrier proteins in the cell membrane.
  B. The transport of molecules such as glucose from the side of higher to the side of lower concentration by means of membrane carriers is called facilitated diffusion.
  C. The active transport of molecules and ions across a membrane requires the expenditure of cellular energy (ATP).

III. The Membrane Potential
  A. The cytoplasm of the cell contains negatively charged organic ions (anions) that cannot leave the cell; they are "fixed anions."
  B. The slow rate of Na^+ entry is accompanied by a slow rate of K^+ exit from the cell.

Review Activities

Objective Questions

1. The movement of water across a cell membrane occurs by
   (a) active transport
   (b) facilitated diffusion
   (c) simple diffusion (osmosis)
   (d) all of the above
2. Which of the following statements about the facilitated diffusion of glucose is *true*?
   (a) There is a net movement from the region of low to the region of high concentration.
   (b) Protein carriers in the cell membrane are required for this transport.
   (c) This transport requires energy obtained from ATP.
   (d) This is an example of co-transport.
3. If a poison such as cyanide stops the production of ATP, which of the following transport processes would cease?
   (a) the movement of Na^+ out of a cell
   (b) osmosis
   (c) the movement of K^+ out of a cell
   (d) all of these
4. Red blood cells crenate in
   (a) a hypotonic solution
   (b) an isotonic solution
   (c) a hypertonic solution
5. Plasma has an osmolality of about 300 mOsm. Isotonic saline has an osmolality of
   (a) 150 mOsm
   (b) 300 mOsm
   (c) 600 mOsm
   (d) none of the above
6. A 0.5 m NaCl solution and a 1.0 m glucose solution
   (a) have the same osmolality
   (b) have the same osmotic pressure
   (c) are isotonic to each other
   (d) all of the above
7. The diffusible ion that is most important in the establishment of the membrane potential is
   (a) K^+
   (b) Na^+
   (c) Ca^{++}
   (d) Cl^-
8. An increase in blood osmolality
   (a) can occur as a result of dehydration
   (b) causes a decrease in blood osmotic pressure
   (c) is accompanied by a decrease in ADH secretion
   (d) all of the above
9. In hyperkalemia, the membrane potential
   (a) increases (becomes more negative)
   (b) decreases (becomes less negative)
   (c) is not changed
10. Which of the following statements about the Na^+/K^+ pump is *true*?
   (a) Na^+ is actively transported into the cell.
   (b) K^+ is actively transported out of the cell.
   (c) An equal number of Na^+ and K^+ ions are transported with each cycle of the pump.
   (d) The pumps are constantly active in all cells.

Essay Questions

1. Describe the conditions required to produce osmosis, and explain why osmosis occurs under these conditions.
2. Explain how simple diffusion can be distinguished from facilitated diffusion and how active transport can be distinguished from passive transport.
3. Compare the theoretical membrane potential that occurs at K^+ equilibrium with the true resting membrane potential. Explain the reasons for the differences between these.
4. Explain how the Na^+/K^+ pump contributes to the resting membrane potential.

6 HISTOLOGY

Outline

Concepts

Histology is an integral part of anatomy and physiology because it imparts an understanding of the structure and function of organs at the tissue level of study. Tissues are classified into four principal kinds on the basis of their cellular composition and histological appearance.

The various types of tissues are established during embryonic development, and as differentiation continues, organs form, each of which is composed of a specific arrangement of tissues.

Epithelia are classified according to the physical features of the tightly packed cells that comprise these tissues. Epithelia line all body surfaces, cavities, and lumina and are adapted for protection, absorption, and secretion.

Connective tissues are classified according to the characteristics of the matrix that binds the cells. Connective tissues provide structural and metabolic support for other tissues and organs of the body.

Muscle tissues are responsible for the movement of materials through the body, the movement of one part of the body with respect to another, and for locomotion. Fibers in the three kinds of muscle tissue are adapted to contract in response to stimuli.

Nervous tissue is composed of neurons, which respond to stimuli and conduct impulses to and from all body organs, and neuroglia, which functionally support and physically bind neurons.

Definition and Classification of Tissues

Histology is an integral part of anatomy and physiology because it imparts an understanding of the structure and function of organs at the tissue level of study. Tissues are classified into four principal kinds on the basis of their cellular composition and histological appearance.

Objective 1. Define tissue, and discuss the importance of histology.

Objective 2. Explain the functional relationship between cells and tissues.

Objective 3. Classify the tissues of the body into four major types, and give the distinguishing characteristics of each type.

Although cells are the structural and functional units of the body, the cells of a multicellular organism are so specialized that they do not function independently. *Tissues* are aggregations of similar cells that perform specific functions. The study of tissues is referred to as **histology** *(his-tol'o-je)*. The various types of tissues are established during embryonic development, and as differentiation continues, organs form, each composed of a specific arrangement of tissues. Many adult organs contain the original cells and tissues, which were not replaced through mitotic activity during further growth and development. Some functional changes may occur, however, as the tissues of an organ are acted upon by hormones or as their effectiveness diminishes with age.

Although histology is actually microscopic anatomy, it is an essential part of anatomy and physiology because it imparts an understanding of the structure and function of organs at the tissue level. Many diseases profoundly alter the tissues within an affected organ; therefore, by knowing the normal tissue structure, a medical person can recognize the abnormal. In medical schools a course in histology is usually followed by a course in *pathology,* which is primarily concerned with identifying diseased tissues.

Although histologists utilize many different techniques for preparing, staining, and sectioning tissues, basically only two kinds of microscopes are used to view the prepared tissues. *Light microscopy (mi-kros'ko-pe)* is used for the general observation of tissue structure (fig. 6.1), and *electron microscopy* permits observation of the fine

histology: Gk. *histos,* web (tissue); *logos,* study

pathology: Gk. *pathos,* suffering, disease; *logos,* study

Figure 6.1. The appearance of skin at various magnifications: (*a*) = 10×, (*b*) = 25×, (*c*) = 50× through a compound light microscope; and (*d*) a hair emerging from a follicle as seen through a scanning electron microscope (SEM) at 280×.

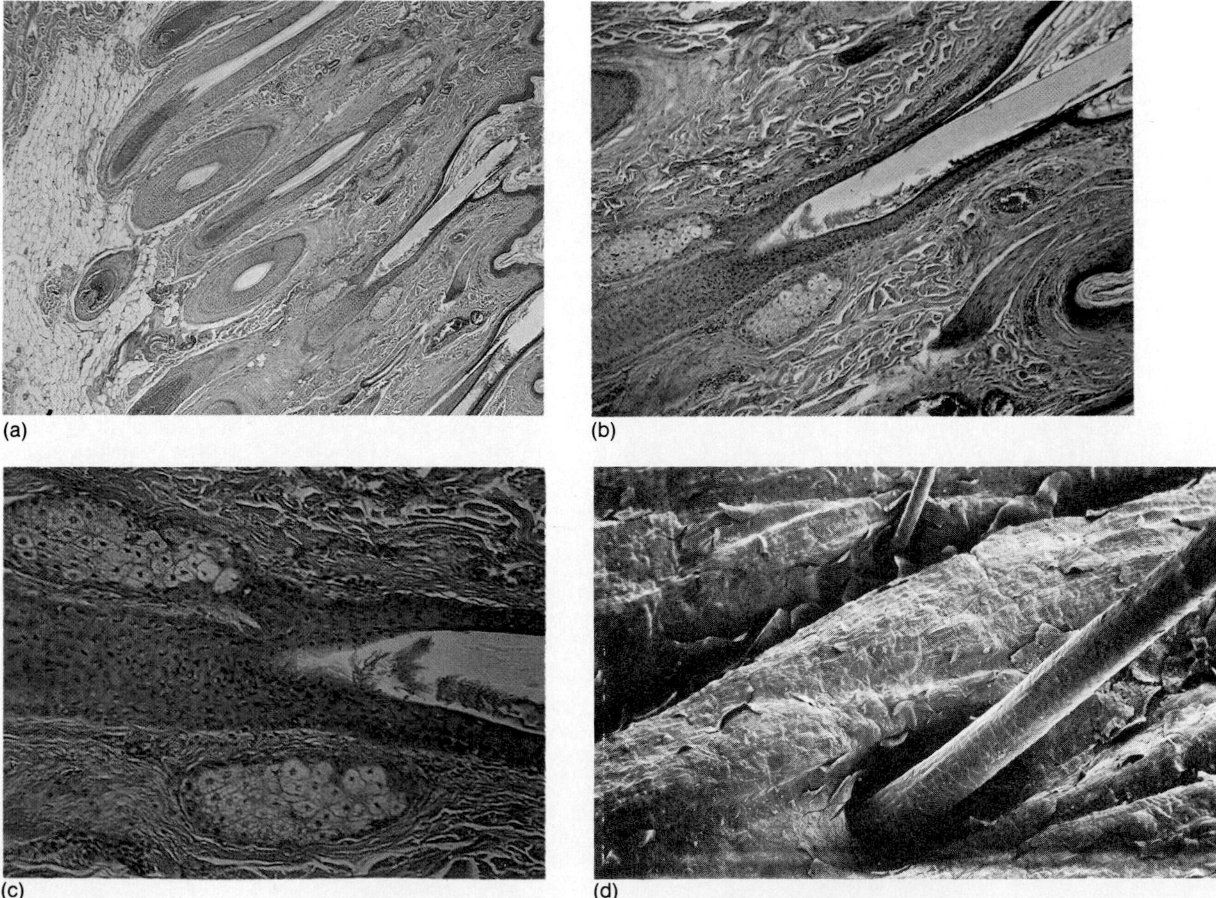

details of tissue and cellular structure. Most of the histological photomicrographs in this text will be at the light microscopic level to present an overview of tissue structure. A few electron micrographs from an electron microscope are used to depict the fine structural detail needed to understand a particular function.

Tissue cells are separated and bound together by a nonliving intercellular **matrix** *(ma'triks)* that the cells secrete. Matrix varies in composition from one tissue to another and may take the form of a liquid, semisolid, or solid. Blood tissue, for example, has a liquid matrix, permitting this tissue to flow through vessels, whereas bone cells are separated by a solid matrix, permitting this tissue to support the body.

The tissues of the body are classified into four principal types, determined by structure and function: (1) *epithelial (ep''i-the'le-al) tissues* cover body and organ surfaces, line body and lumen cavities, and form various glands; (2) *connective tissues* bind, support, and protect body parts; (3) *muscle tissues* contract to produce movement; and (4) *nervous tissues* initiate and transmit nerve impulses from one body part to another.

1. Define the term *tissue,* and explain why histology is important to the study of anatomy and physiology and medicine.
2. Cells are the functional units of the body. Explain how the matrix permits specific kinds of cells to be even more effective and functional as tissues.
3. List and give the distinguishing characteristics of the four principal types of tissues in the body, and explain their basic functions.

Development of Tissues

The various types of tissues are established during embryonic development, and as differentiation continues, organs form, each of which is composed of a specific arrangement of tissues.

Objective 4. Describe the location of the three primary germ layers during early embryonic development.

Objective 5. List the tissues and body organs derived from each of the three primary germ layers.

Human prenatal development is initiated with the fertilization of an ovulated ovum (egg) from a female by a sperm cell from a male. The chromosomes within the nucleus of a **zygote** *(zi'gōt)* (fertilized egg) contain all the genetic information necessary for the differentiation and development of all body structures.

Within thirty hours after fertilization, the zygote undergoes a mitotic division as it moves through the uterine tube toward the uterus (chapter 29). After several more cellular divisions, the embryonic mass consists of sixteen or more cells and is called a **morula** *(mor'u-lah)* (fig. 6.2). Three or four days after conception, the morula enters the uterine cavity, where it remains unattached for about three days. During this time, the center of the morula fills with fluid passing in from the uterine cavity. As the fluid-filled space develops inside the morula, two distinct groups of

matrix: L. *matris,* mother
zygote: Gk. *zygotos,* yolked
morula: Gk. *morus,* mulberry

Figure 6.2. The early stages of embryonic development. (*a*) the zygote immediately following conception. (*b*) the morula at about the third day as it enters the uterine cavity. (*c*) the early blastocyst at the time of implantation between the fifth and seventh day. (*d*) a cross section of an implanted blastocyst at two weeks. (*e*) a cross section of a blastocyst at three weeks showing the three primary germ layers, which constitute the embryonic disc.

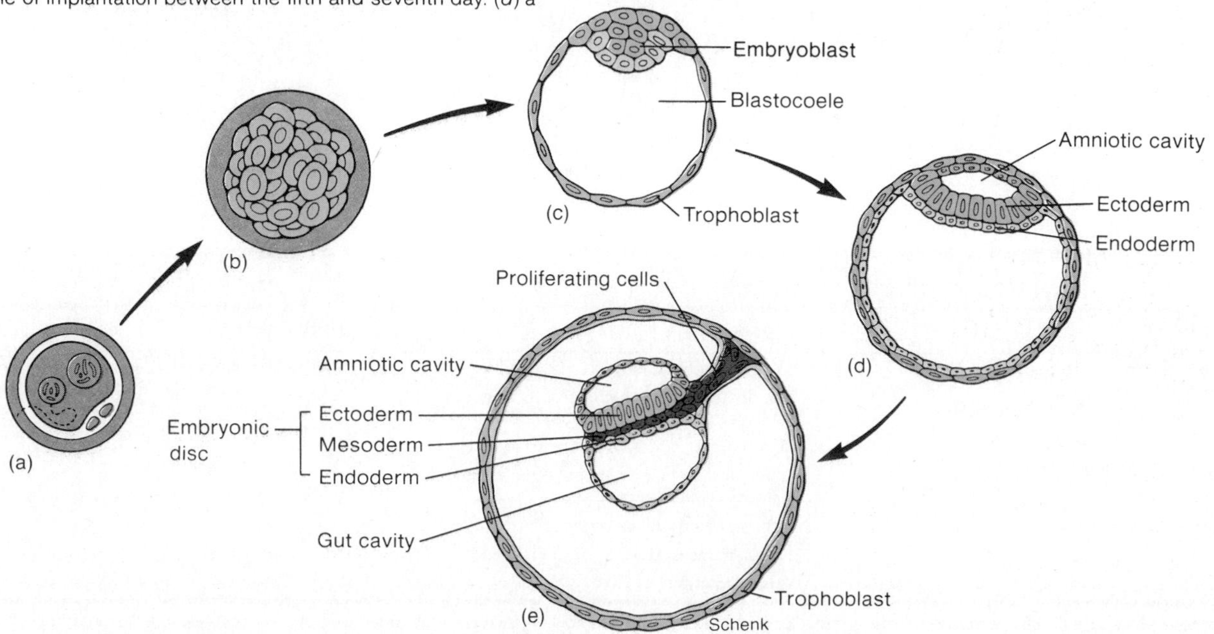

Figure 6.3. The body systems and the primary germ layers from which they develop.

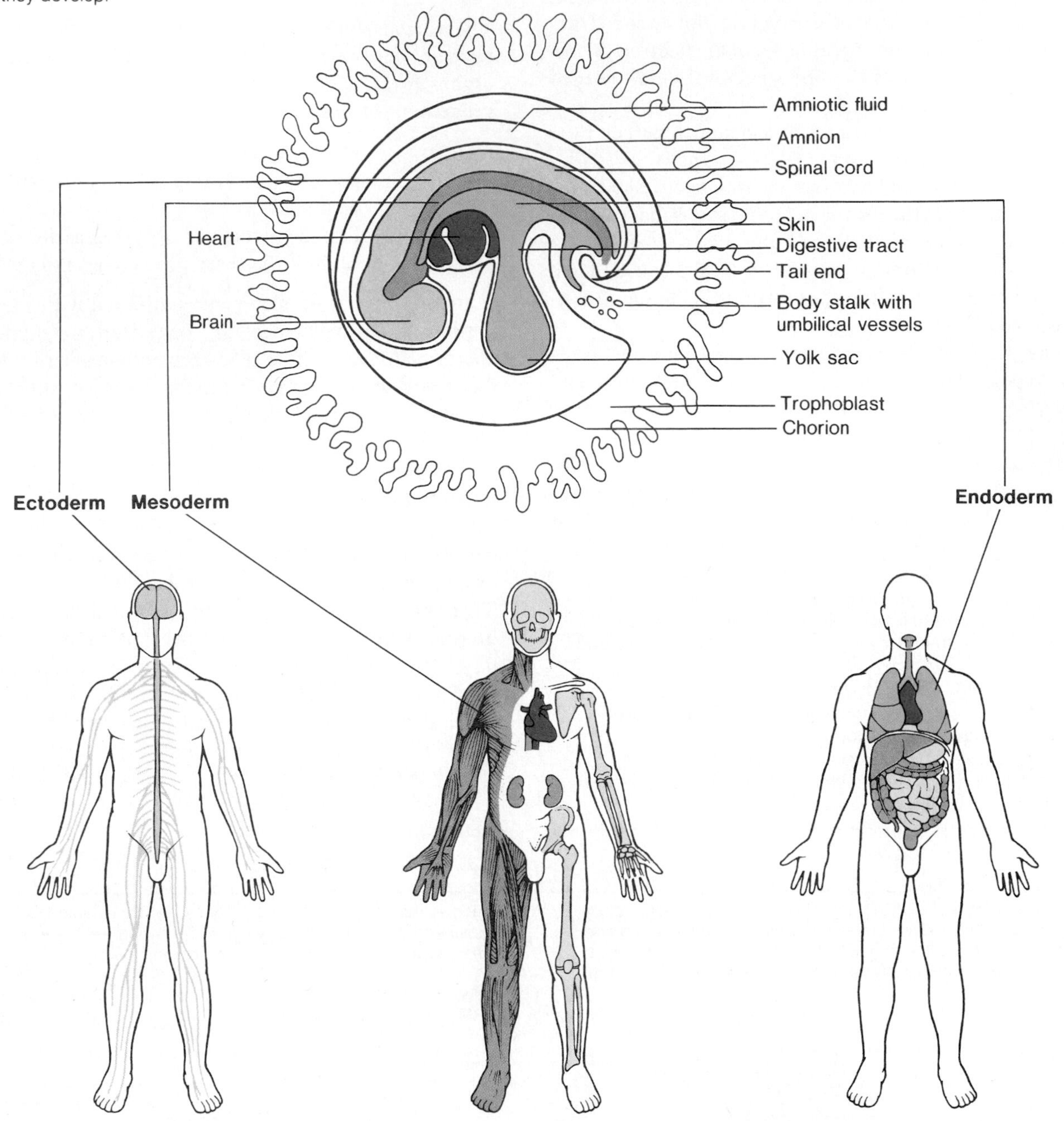

Table 6.1 Derivatives of the germ layers

Ectoderm	Mesoderm	Endoderm
Epidermis of skin and epidermal derivatives: hair, nails, glands of the skin; linings of oral, nasal, anal, and vaginal cavities Nervous tissue; sense organs Lens of eye; enamel of teeth Pituitary gland Adrenal medulla	Muscle: smooth, cardiac, and skeletal Connective tissue: embryonic, connective tissue proper, cartilage, bone, blood Dermis of skin; dentin of teeth Epithelium of blood vessels, lymphatic vessels, body cavities, joint cavities Internal reproductive organs Kidneys and ureters Adrenal cortex	Epithelium of pharynx, auditory canal, tonsils, thyroid, parathyroid, thymus, larynx, trachea, lungs, digestive tract, urinary bladder and urethra, and vagina Liver and pancreas

cells form. The single layer of cells, called *trophoblast cells,* forming the outer wall becomes the **trophoblast,** and the small, inner aggregation of cells becomes the **embryoblast,** or **inner cell mass.** After further development the trophoblast becomes a portion of the placenta, and the embryoblast becomes the embryo. With the establishment of these two groups of cells, the morula becomes known as a **blastocyst** *(blas'to-sist).* Implantation of the blastocyst begins between the fifth and seventh day (chapter 29).

As the blastocyst completes implantation during the second week of development, the embryoblast undergoes marked differentiation. A slitlike space called the **amniotic** *(am'ne-ot-ic)* **cavity** forms within the embryoblast adjacent to the trophoblast (see fig. 6.2). The embryoblast now consists of two layers: an upper **ectoderm,** which is closer to the amniotic cavity, and a lower **endoderm,** which borders the blastocyst cavity. A short time later, a third layer, called the **mesoderm,** forms between the endoderm and ectoderm. These three layers constitute the **primary germ layers.**

The primary germ layers are important because all the cells and tissues of the body are derived from them. Ectodermal cells form the nervous system; the outer layer of skin (epidermis), including hair, nails, and skin glands; and portions of the sensory organs. Mesodermal cells form the skeleton, muscles, blood, reproductive organs, dermis of the skin, and connective tissue. Endodermal cells produce the lining of the digestive tract, the digestive organs, the respiratory tract and lungs, and the urinary bladder and urethra.

Figure 6.3 illustrates the organs and body systems that derive from each of the three primary germ layers. Table 6.1 lists the derivatives of the primary germ layers.

1. What is a morula? How does it differ from a blastocyst?
2. At what developmental age (weeks following conception) are the germ layers present, and where are they located?
3. Referring to chapter 1, list the ten body systems, and indicate the principal germ layers from which the organs of each system derive.

Epithelial Tissue

Epithelia are classified according to the physical features of the tightly packed cells that comprise these tissues. Epithelia line all body surfaces, cavities, and lumina and are adapted for protection, absorption, and secretion.

trophoblast: Gk. *trophe,* nourishment; *blastos,* germ
embryoblast: Gk. *embryon,* to be full, swell; *blastos,* germ
ectoderm: Gk. *ecto,* outside; *derm,* skin
endoderm: Gk. *endo,* within; *derm,* skin
mesoderm: Gk. *meso,* middle; *derm,* skin

Objective 6. Differentiate between the kinds of epithelia.
Objective 7. Describe how epithelial cells are held together.
Objective 8. Define *gland,* and distinguish between the various types of glands in the body.

Characteristics of Epithelia

Epithelia are located throughout the body and form such structures as the outer layer of the skin, the inner lining of body cavities and lumina (hollow portions of body tubes), the covering of visceral organs, and the secretory portion of glands.

One side of epithelia is always exposed to a body cavity, lumen, or skin surface. Some epithelia are derived from ectoderm, such as the outer layer of the skin and integumentary glands; some from mesoderm, such as the inside lining of blood vessels; and others from endoderm, such as the inside lining of the digestive tract.

Epithelia may be one layer or several layers thick. The upper surface of epithelia may be exposed to gases, as in the case of epithelium in the integumentary and respiratory systems; to liquids, as in the circulatory and urinary systems; or to semisolids, as in the digestive system. The deep surface of epithelia is bound to underlying supportive tissue by a **basement membrane,** consisting of glycoprotein from the epithelial cells and a meshwork of collagen and reticular fibers from the underlying connective tissue. With few exceptions, epithelia are avascular (without blood vessels) and must be nourished by diffusion from underlying connective tissues. Cells comprising epithelia are tightly packed together, and there is little intercellular matrix between them.

The cells of epithelia are tightly bonded in one of three ways (fig. 6.4): (1) Some epithelial cells have a bonding matrix of *glycoprotein deposits.* Glycoprotein is a combination of polysaccharides and protein secreted by the cells to bind them to other like cells in a process similar to that which binds epithelia to the basement membrane. (2) Certain epithelial cells are bonded with *desmosomes.* Desmosomes are characterized by V-shaped **tonofilaments** *(ton''o-fil'ah-mentz),* reinforcing the cell membrane where glycoprotein is secreted to bond adjacent cells. (3) *Tight junctions* and *intermediate junctions* occur in the same type of epithelium. A tight junction exists where the cell membranes of adjacent cells form a serrated pattern. In an intermediate junction, the cell membranes of adjacent cells are bonded with glycoprotein.

Some of the functions of epithelial tissues are quite specific, but certain generalities can be made. Epithelia that cover or line surfaces provide *protection* from pathogens, physical injury, toxins, and desiccation. Epithelia

epithelium: Gk. *epi,* upon; *thelium,* to cover

Figure 6.4. Types of cellular bonding in epithelia. (*a*) glycoprotein deposits. (*b*) desmosomes. (*c*) tight and intermediate junctions.

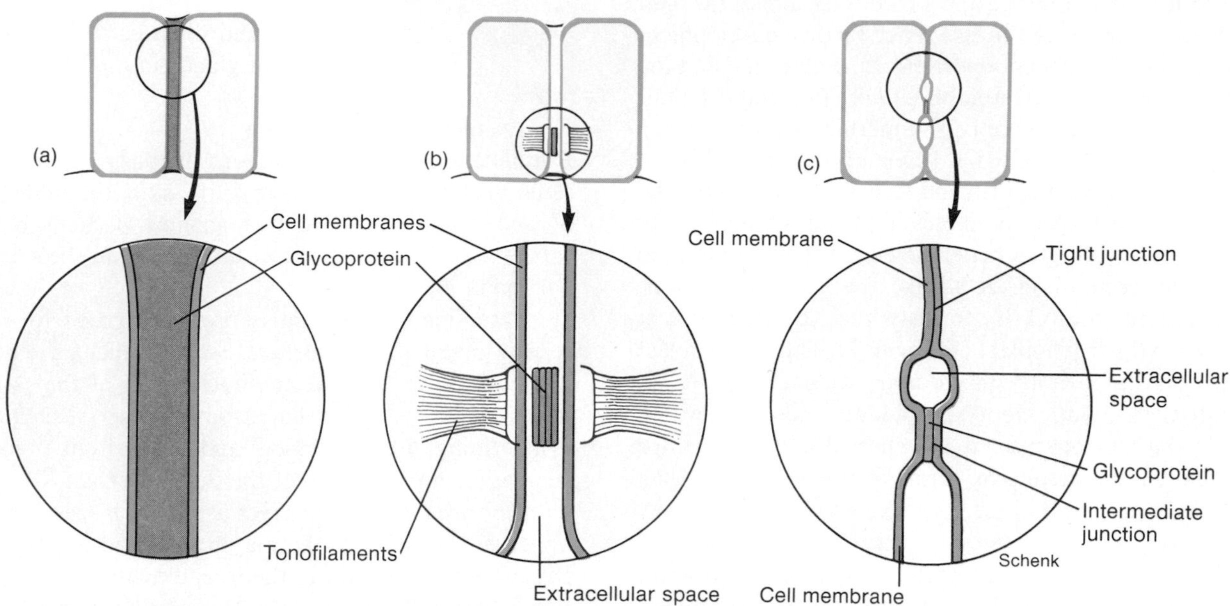

lining the lumen of the digestive tract function in *absorption* and *secretion.* Glandular epithelia elsewhere in the body are also secretory. The epithelium of the kidneys provides *filtration,* whereas the epithelium within the air sacs of the lungs allows *diffusion.* Highly specialized *neuroepithelium* in the taste buds and in the nasal region functions as *chemoreceptors (ke''mo-re-sep'torz).*

Many epithelial tissues are exposed and, therefore, subject to trauma and destruction. For this reason, epithelial tissues have remarkable regenerative abilities. The mitotic replacement of the outer layer of skin and the lining of the digestive tract, for example, is a continuous process.

Epithelial tissues are histologically classified by the number of layers of cells and the shape of the cells along the exposed surface. Epithelial tissues that are composed of a single layer of cells are called *simple,* and those that are layered are said to be *stratified. Squamous* cells are flattened; *cuboidal* cells are cube shaped; and *columnar* cells are taller than they are wide.

In describing and discussing the epithelial tissues, the specific types will be presented under three major headings: simple epithelia, stratified epithelia, and glandular epithelia. A summary of simple and stratified epithelia is presented in table 6.2.

squamous: L. *squamosus,* scaly

Simple Epithelia

Simple epithelial tissues are a single layer thick and are located where diffusion, filtration, and secretion occur. The cells that constitute simple epithelia range in size from thin, flattened cells to tall, columnar cells, depending on function. These cells may also exhibit surface specializations, such as cilia and microvilli, which facilitate specific surface functions.

Simple Squamous Epithelium. Simple squamous *(skwa'mus)* epithelium is composed of flattened, irregularly shaped cells that are tightly bound together in a mosaiclike pattern (fig. 6.5). Each cell contains an oval, centrally located nucleus. Simple squamous epithelium is adapted for diffusion and filtration and occurs in such places as the lining of alveoli within the lungs (where gaseous exchange occurs), portions of the kidney (where blood is filtered), the inside lining of the walls of blood vessels, the lining of body cavities, and in the covering of the viscera. The simple squamous epithelium lining the lumina of blood and lymphatic vessels is sometimes referred to as **endothelium** (fig. 6.5b). That which covers visceral organs and lines body cavities is called **mesothelium.**

endothelium: Gk. *endon,* within; *thelium,* to cover
mesothelium: Gk. *meso,* middle; *thelium,* to cover

Figure 6.5. Simple squamous epithelium. (*a*) as the name implies, this type of epithelium consists of a single layer of flattened cells and is highly adapted to diffusion, osmosis, and filtration. (*b*) where it lines the lumina of blood and lymphatic vessels, it is referred to as endothelium.

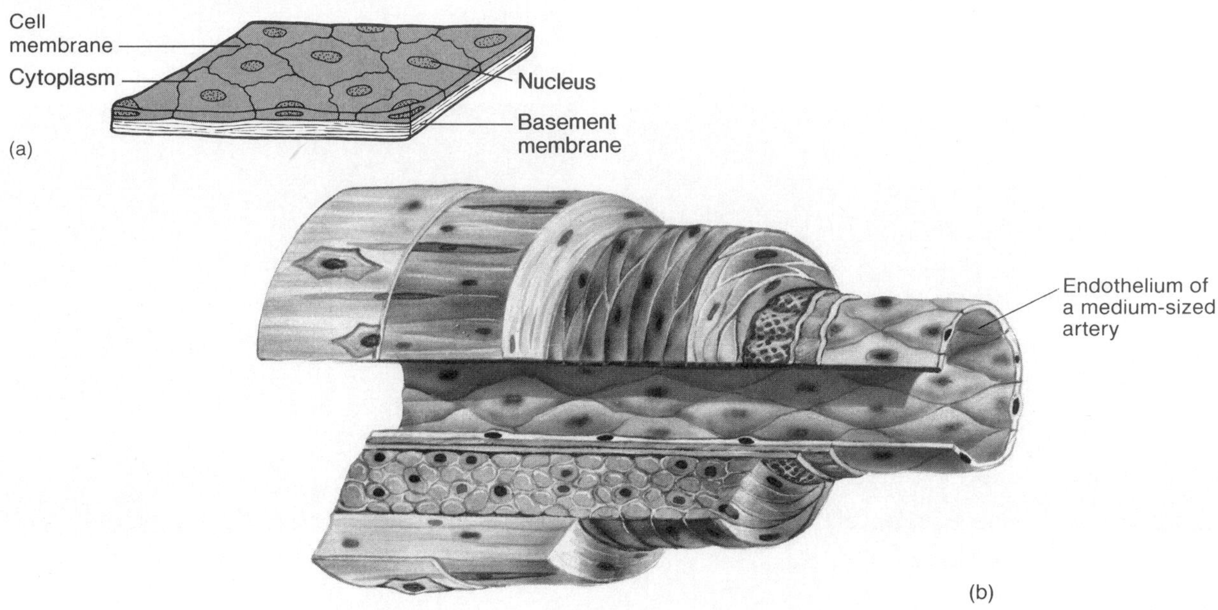

Table 6.2 Summary of epithelial tissues		
Type	**Structure and function**	**Location**
Simple epithelia	Single layer of cells; diffusion and filtration	Covering visceral organs, linings of lumina and body cavities
Simple squamous epithelium	Single layer of flattened, tightly bound cells; diffusion and filtration	Capillary walls, air sacs of lungs, covering visceral organs, linings of body cavities
Simple cuboidal epithelium	Single layer of cube-shaped cells; excretion, secretion, or absorption	Surface of ovaries; linings of kidney tubules, salivary ducts, and pancreatic ducts
Simple columnar epithelium	Single, nonciliated layer of tall, columnar-shaped cells; protection, secretion, and absorption	Lining of digestive tract
Simple ciliated columnar epithelium	Single, ciliated layer of columnar-shaped cells; transportive role through ciliary motion	Lining the lumen of the uterine tubes
Pseudostratified ciliated columnar epithelium	Single layer of ciliated, irregularly shaped cells, many goblet cells; protection, secretion, ciliary movement	Lining of respiratory passageways
Stratified epithelia	Two or more layers of cells; protection, strengthening, or distension	Epidermal layer of skin; linings of body openings, ducts, urinary bladder
Stratified squamous epithelium (keratinized)	Numerous layers, contains keratin, outer layers flattened and dead; protection	Epidermis of skin
Stratified squamous epithelium (nonkeratinized)	Numerous layers, lacks keratin, outer layers moistened and alive; protection and pliability	Linings of oral and nasal cavities, vagina, and anal canal
Stratified cuboidal epithelium	Usually two layers of cube-shaped cells; strengthen luminal walls	Larger ducts of sweat glands, salivary glands, and pancreas
Transitional epithelium	Numerous layers of rounded, nonkeratinized cells; distension	Luminal walls of ureters and urinary bladder

Figure 6.6. Simple cuboidal epithelium lines the lumina of ducts, as seen in the photomicrographs of kidney tubules. (*a*) a diagrammatic drawing of this type of epithelium that consists of a single layer of tightly packed, cube-shaped cells, each with a large, centrally located nucleus; (*b*) longitudinal view of a tubule; (*c*) cross-sectional view of several tubules.

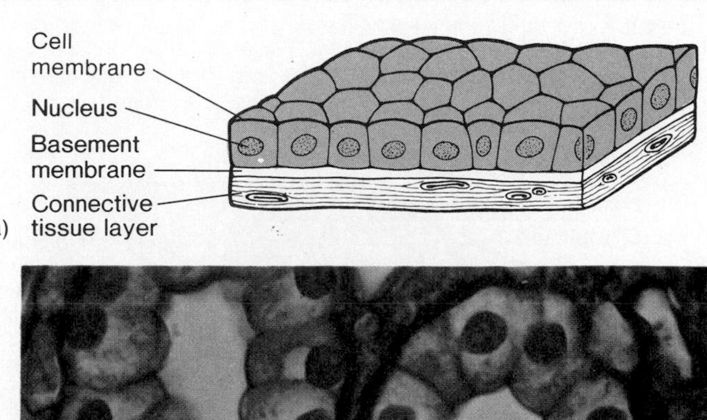

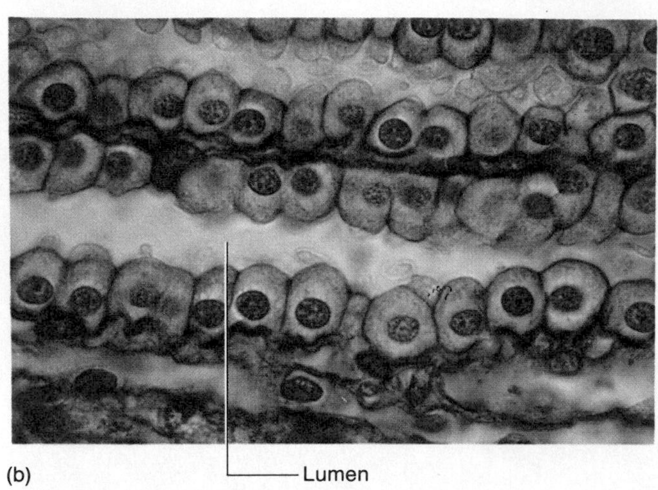

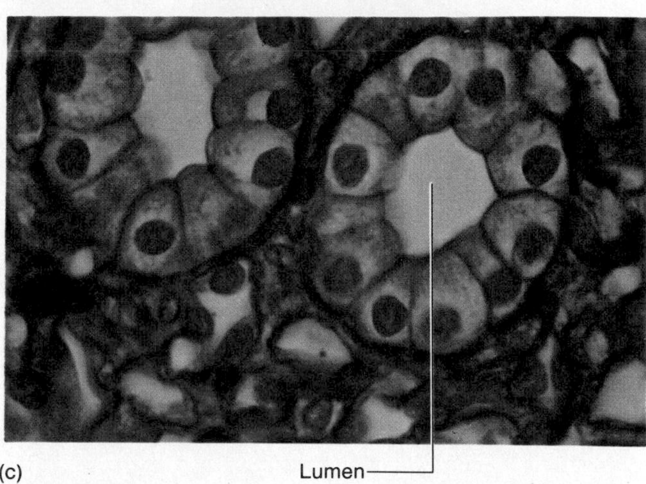

Figure 6.7. (*a*) simple columnar epithelium lines the lumen of the digestive tract. (*b*) a photomicrograph of this epithelium. (*c*) a diagrammatic drawing of the tall, columnar-shaped cells and the associated goblet cells.

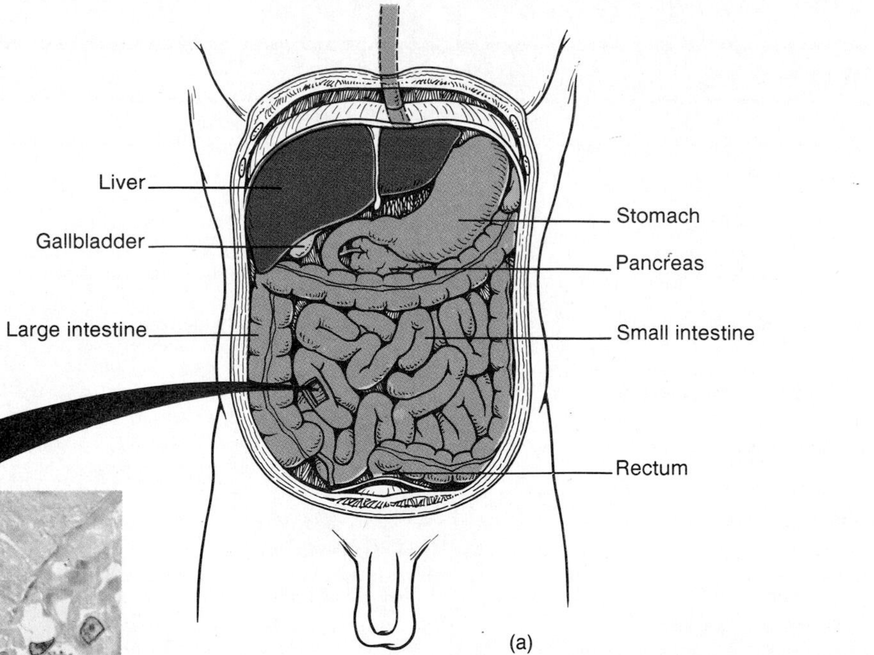

Liver
Gallbladder
Large intestine
Stomach
Pancreas
Small intestine
Rectum
(a)

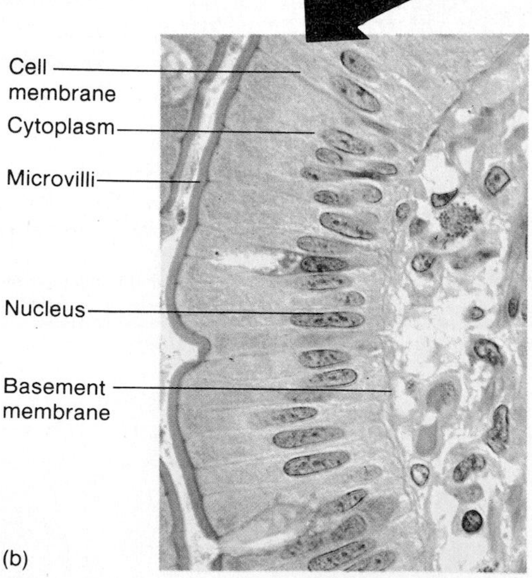

Cell membrane
Cytoplasm
Microvilli
Nucleus
Basement membrane
(b)

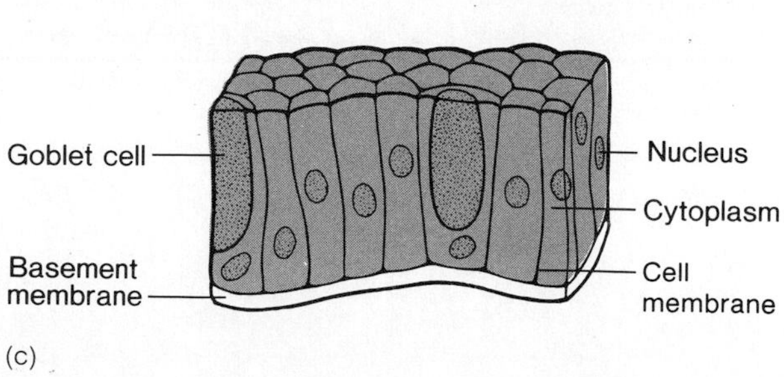

Goblet cell
Basement membrane
Nucleus
Cytoplasm
Cell membrane
(c)

Figure 6.8. (*a*) simple ciliated columnar epithelium occurs within the uterine tubes of the female reproductive system, where it moves the ovum toward the uterine cavity. (*b*) a photomicrograph showing the cilia and (*c*) a diagram.

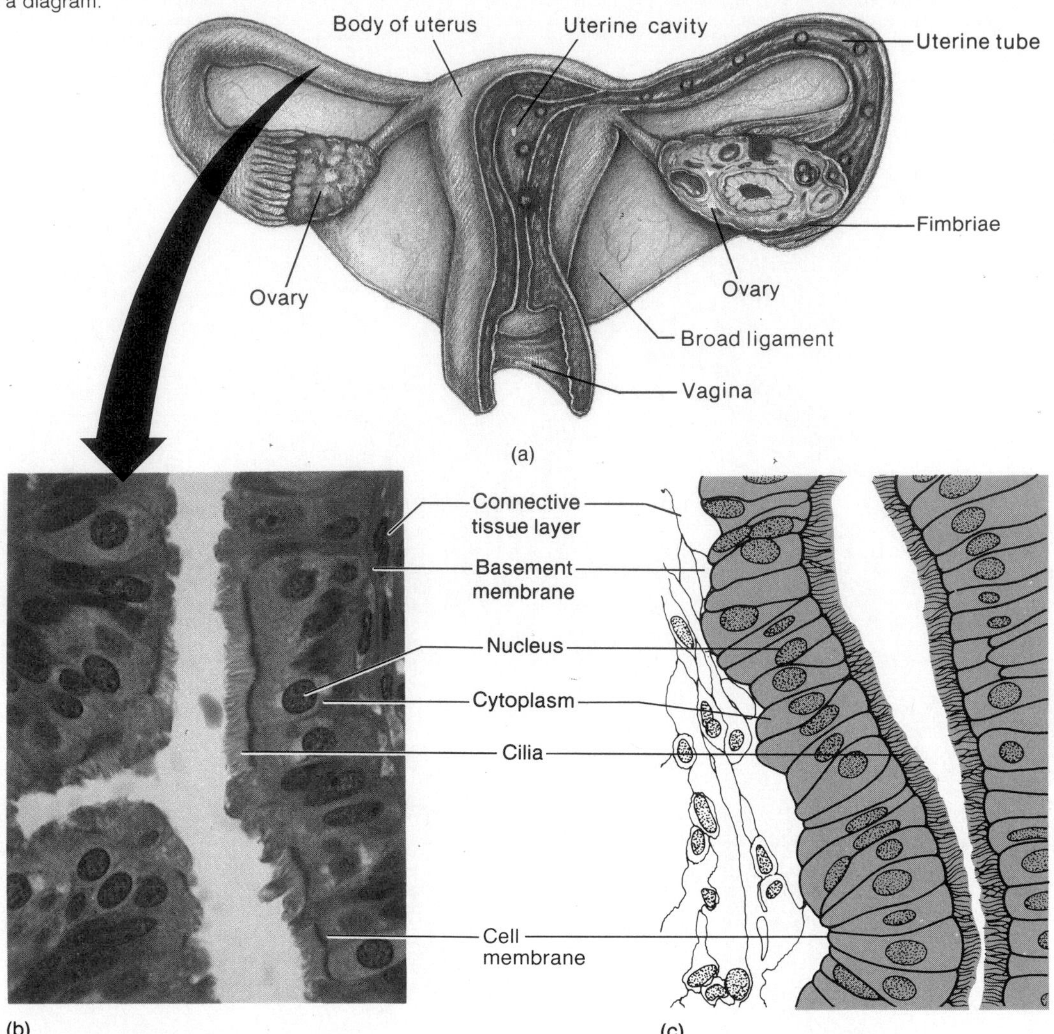

Simple Cuboidal Epithelium. Simple cuboidal epithelium is composed of a single layer of tightly fitted, hexagonal cells (fig. 6.6). This type of epithelium is found lining small ducts and tubules that may have excretory, secretory, or absorptive functions. It occurs on the surface of the ovaries, forms a portion of the tubules within the kidney, and lines the ducts of the salivary glands and pancreas.

Simple Columnar Epithelium. Simple columnar epithelium is composed of tall, columnar cells (fig. 6.7). The height of the cells varies, depending on the site and function of the tissue. Each cell contains a single nucleus usually located near the basement membrane. Specialized unicellular glands, called **goblet cells,** are dispersed throughout this tissue and secrete a lubricative and protective mucus along the free surfaces of the cells. This type of epithelium is found lining the lumen of the stomach and intestine. In the digestive system, simple columnar epithelium forms a highly absorptive surface and also secretes certain digestive substances. Within the stomach, simple columnar epithelium has a tremendous rate of mitotic activity—replacing itself every two or three days.

Simple Ciliated Columnar Epithelium. Simple ciliated columnar epithelium differs from the simple columnar type by the presence of cilia along the free surface (fig. 6.8). Cilia produce wavelike movements that transport materials through tubes or passageways. This type of epithelium occurs in the uterine tubes of the female, where the currents generated by the cilia propel the ovum toward the uterus.

Not only do cilia of the columnar epithelium move the ovum, but recent evidence indicates that sperm introduced during sexual intercourse may be moved along the return currents, or eddies, produced by ciliary movement. This gives fertilization a much greater chance to occur.

Figure 6.9. (*a*) pseudostratified ciliated columnar epithelium lines most of the respiratory tract. The action of the cilia and the mucus secreted by goblet cells trap foreign material and move it away from the alveoli of the lungs. (*b*) this type of tissue appears stratified, but (*c*) all the cells are in contact with the basement membrane.

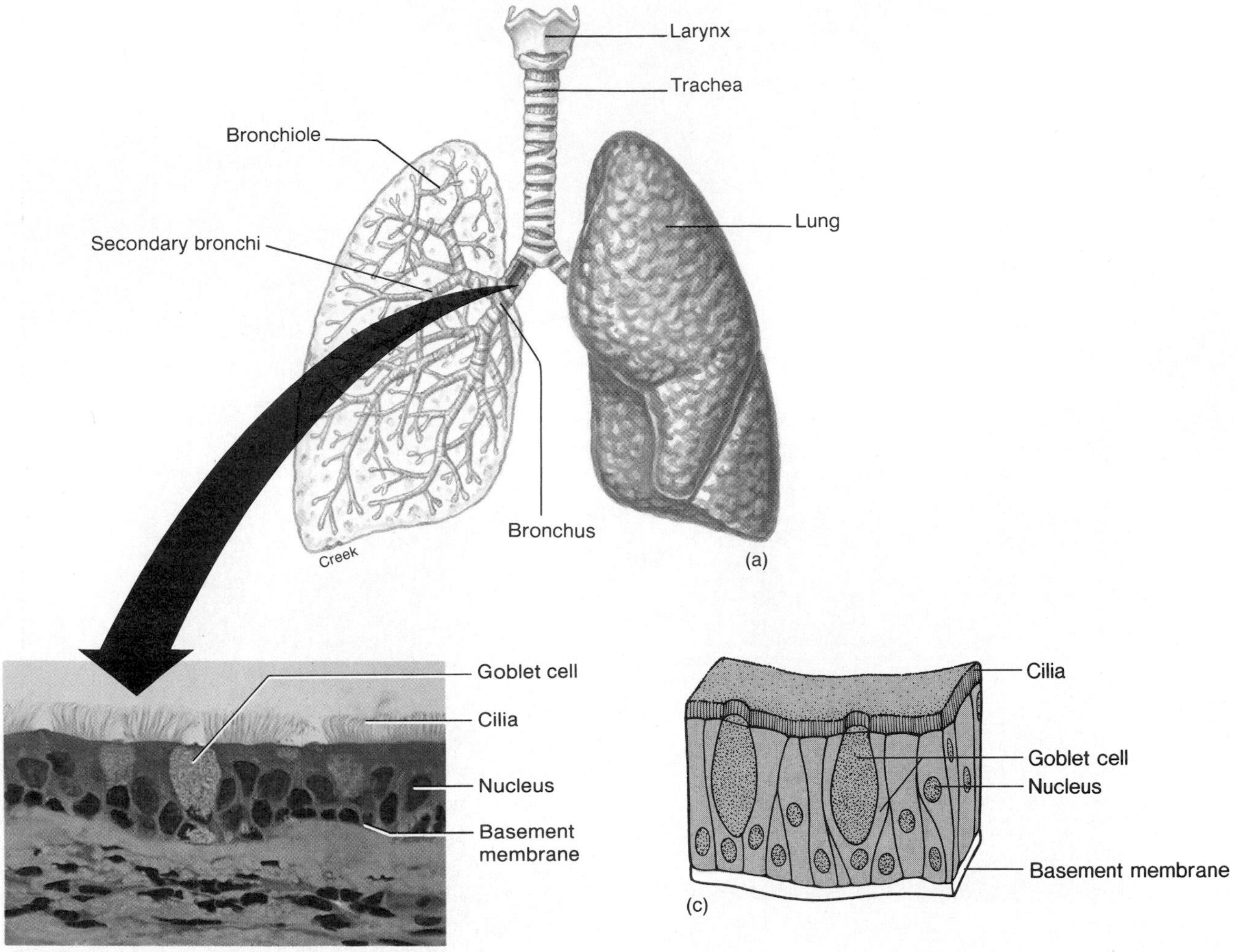

Pseudostratified Ciliated Columnar Epithelium. As the name implies, this type of epithelium appears stratified but is actually simple since each cell is in contact with the basement membrane, though not all cells are exposed to the surface (fig. 6.9). The epithelium has a stratified appearance because the nuclei of these cells are located at different levels. Numerous goblet cells and a ciliated, exposed surface are characteristic of this epithelium. The lumina of the trachea and the bronchial tubes are lined with this tissue; hence, it is frequently called respiratory epithelium. Its function is to remove foreign dust and bacteria entrapped in mucus from the lower respiratory system.

Coughing, sneezing, or simply "clearing the throat" are protective reflex mechanisms for clearing the respiratory passages of obstruction or of inhaled particles that have been trapped in the mucus along the ciliated lining. The material that is coughed up consists of the mucous-entrapped particles.

Stratified Epithelia

Stratified epithelia are tissues consisting of two or more layers of cells. In contrast to simple epithelia, stratified epithelia are poorly suited for absorption and secretion because of their thickness. Stratified epithelia have a primarily protective function that is enhanced by a characteristic rapid mitotic activity. Stratified epithelia are classified according to the shape of the surface layer of cells, since the layer in contact with the basement membrane is cuboidal or columnar in shape.

Figure 6.10. Stratified squamous epithelium is a multilayered, protective tissue that forms the outer layer of skin and the lining of body openings. In the moistened areas such as in the vagina (*a*), it is nonkeratinized, whereas in the epidermis of the skin it is keratinized. The lower layers are cube shaped as can be seen in (*b*) the photomicrograph (100×) and (*c*) the diagram, whereas the outer layers are flattened and scalelike.

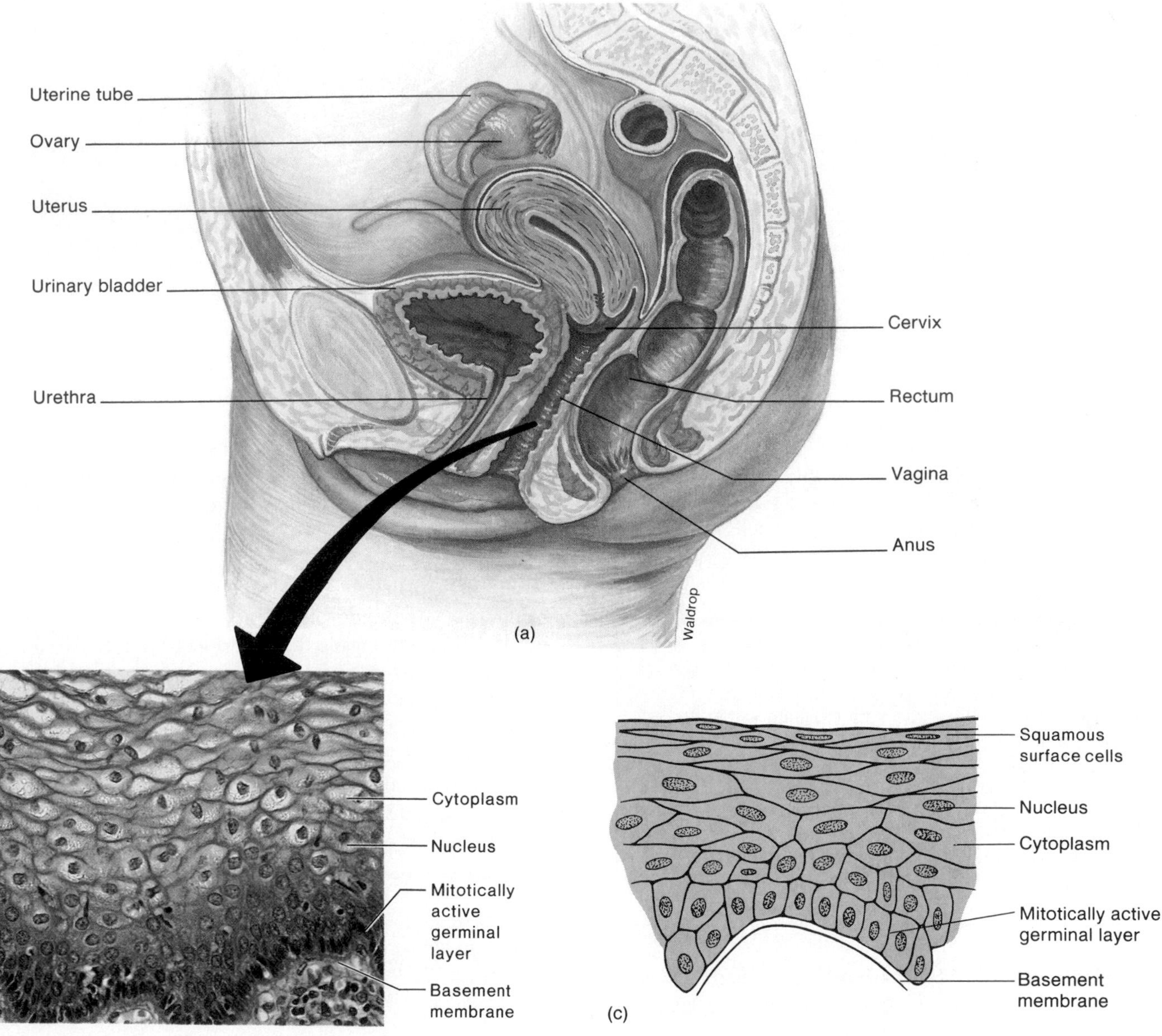

Stratified Squamous Epithelium. Stratified squamous epithelium is composed of a variable number of cell layers that tend to flatten near the surface (fig. 6.10). Only at the deepest layer, called the **stratum basale,** does mitosis occur. The mitotic rate approximates the rate at which cells are sloughed off at the surface. As the newly produced cells grow in size, they are pushed toward the surface, where they will replace the cells that are sloughed off. Movement away from the source of nutrition, provided by the supportive basement membrane, causes the cells to become progressively dehydrated and flattened.

There are two types of stratified squamous epithelia: *keratinized* and *nonkeratinized.* Stratified squamous epithelium that is keratinized forms the outer layer, or *epidermis,* of the skin (see chapter 8). **Keratin** *(ker'ah-tin)* is a protein that strengthens the tissue. This type of epithelium is especially durable and can generally withstand physical abrasion, desiccation, and bacterial invasion. The outer layers of the stratified squamous epithelium of the skin are dead but are kept moist by local glandular secretions.

keratin: Gk. *keras,* horn

Figure 6.11. Stratified cuboidal epithelium consists of two or more layers of cube-shaped cells surrounding a lumen. (*a*) a photomicrograph of this tissue and (*b*) a diagram.

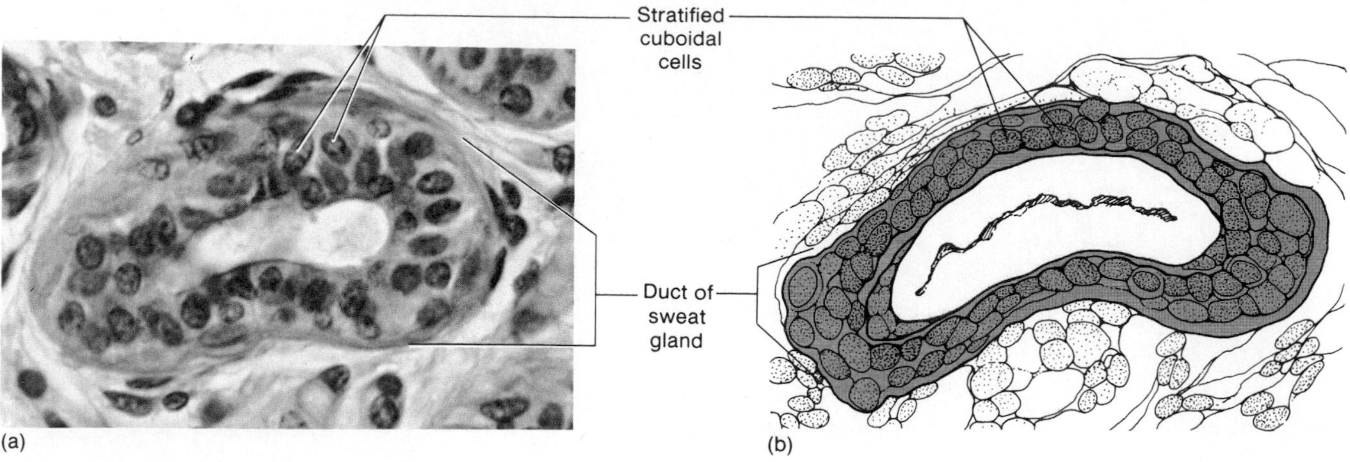

Nonkeratinized stratified squamous epithelium lines the oral cavity and pharynx, nasal cavity, vagina, and anal canal. This type of epithelium is well adapted to withstand moderate abrasion but not fluid loss. The cells on the free surface of this tissue remain alive and are always moistened.

Stratified squamous epithelium is the first line of defense against the entry of living organisms into the body. Stratification as well as rapid mitotic activity and keratinization within the epidermis of the skin are important protective features. An acidic pH along the surfaces of this tissue also helps prevent disease. The pH of the skin is between 4 and 6.8. The pH in the oral cavity ranges between 5.8 and 7.1, which tends to retard the growth of microorganisms. The pH of the anal region is about 6, and the pH along the surface of the vagina is 4 or lower.

Stratified Cuboidal Epithelium. Stratified cuboidal epithelium usually consists of only two or three layers of cuboidal cells forming the lining around a lumen (fig. 6.11). This type of epithelium is confined to the linings of the larger ducts of sweat glands, salivary glands, and the pancreas. The stratification of this tissue probably provides a more robust lining than would be afforded by simple epithelium.

Transitional Epithelium. Transitional epithelium is similar to nonkeratinized stratified squamous epithelium except that the surface cells are large and round rather than flat, and some may have two nuclei (fig. 6.12). Transitional epithelium is located only within the urinary system, particularly in the luminal surface of the urinary bladder and the walls of the ureters. This tissue is specialized to permit distension (stretching) of the urinary bladder and to withstand the toxicity of urine.

Glandular Epithelia

During the prenatal development of epithelial tissue, certain epithelial cells invade the underlying connective tissue, forming specialized secretory accumulations called *exocrine (ek'so-krin) glands.* Exocrine glands retain a connection to the surface in the form of a duct through which secretions flow. Exocrine glands within the integumentary system include sebaceous (oil) glands, sweat glands, and mammary glands. Within the digestive system, they include the salivary and pancreatic glands.

Exocrine glands are derived from epithelial tissue and secrete a substance through ducts to the surface of the skin or the lumen of a body cavity. Exocrine glands should not be confused with endocrine glands, which are ductless and secrete hormones into the blood.

Exocrine glands are classified according to the structure of the gland and its means of discharging the secretory product. Structurally, there are two types, unicellular and multicellular.

Unicellular Glands. Unicellular glands are single-celled glands interspersed with the various columnar epithelia. A mucus-secreting goblet cell is a good example of a unicellular gland. Goblet cells are found in the epithelial linings of the respiratory, digestive, urinary, and reproductive systems, where the mucus secretion lubricates and protects the surface linings (fig. 6.13).

exocrine: Gk. *exo,* outside; *krinein,* to separate

Figure 6.12. (*a*) transitional epithelium lines the lumina of the urinary bladder and the ureters. (*b, c*) the cells of this tissue are stratified in a unique way to permit distension (400×).

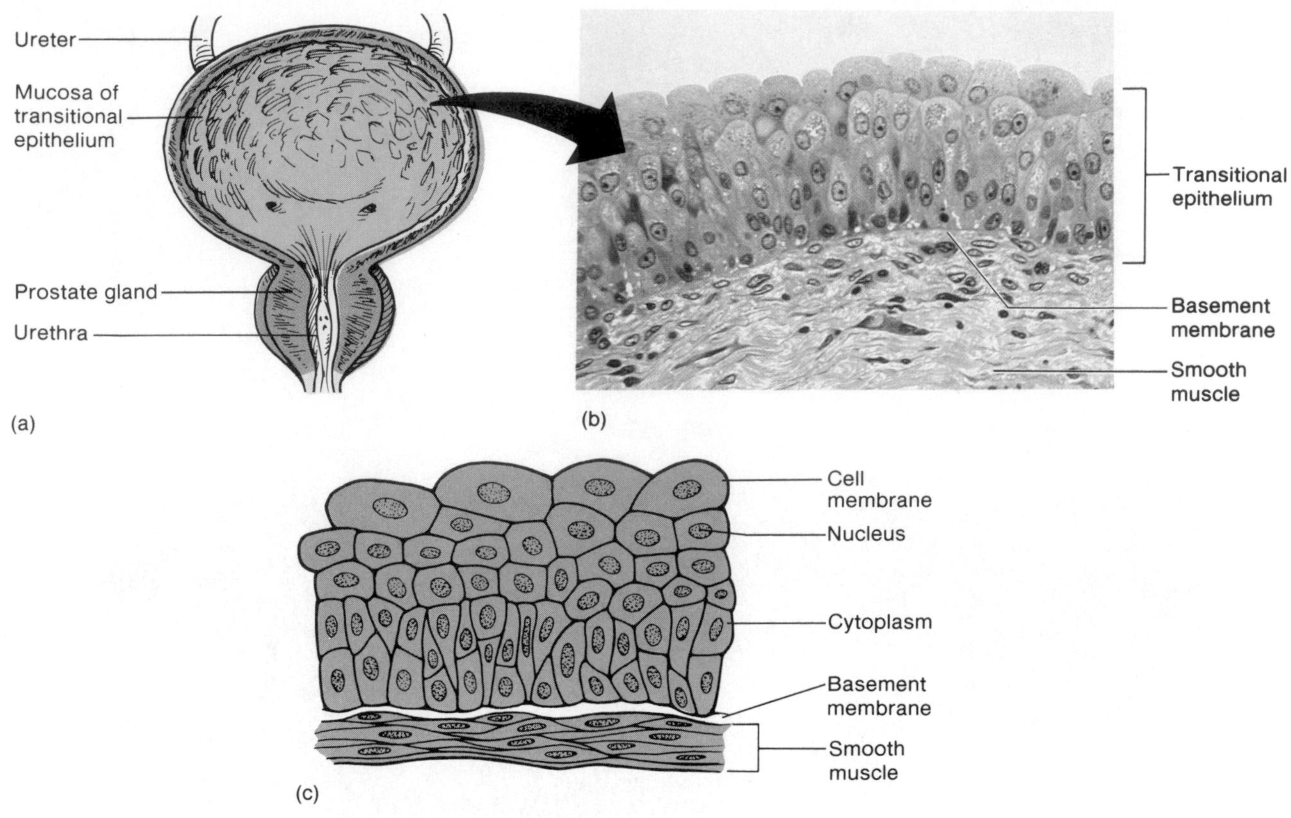

Figure 6.13. The goblet cell is a unicellular gland that secretes mucus to lubricate and protect surface linings. (*a*) a photomicrograph through the ileum of the small intestine shows the numerous goblet cells (*arrow*). (*b*) columnar epithelia contain mucus-secreting goblet cells.

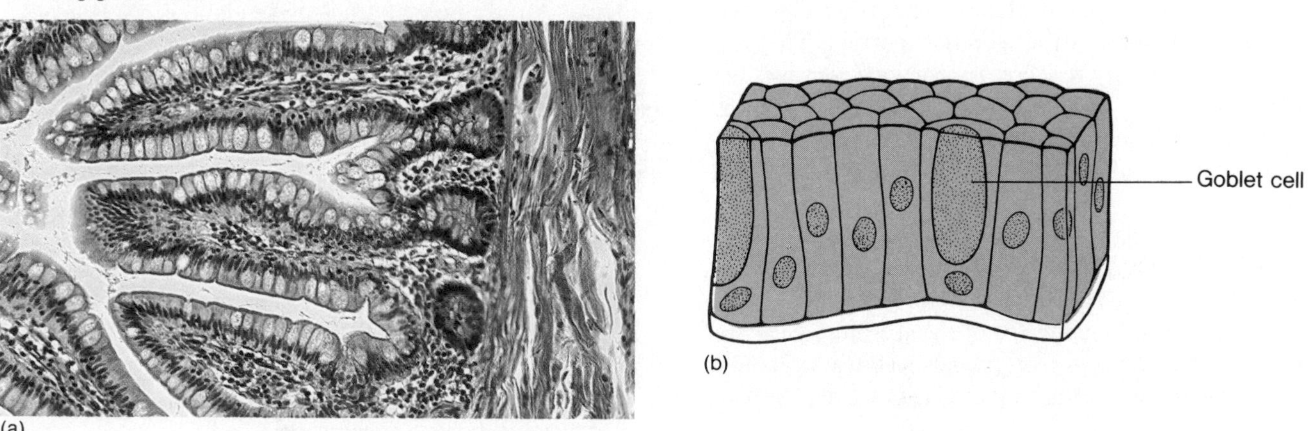

Figure 6.14. Structural classification of multicellular exocrine glands. The secretory portions of the simple glands either do not branch or have a few branches, whereas those of the compound type have multiple branches.

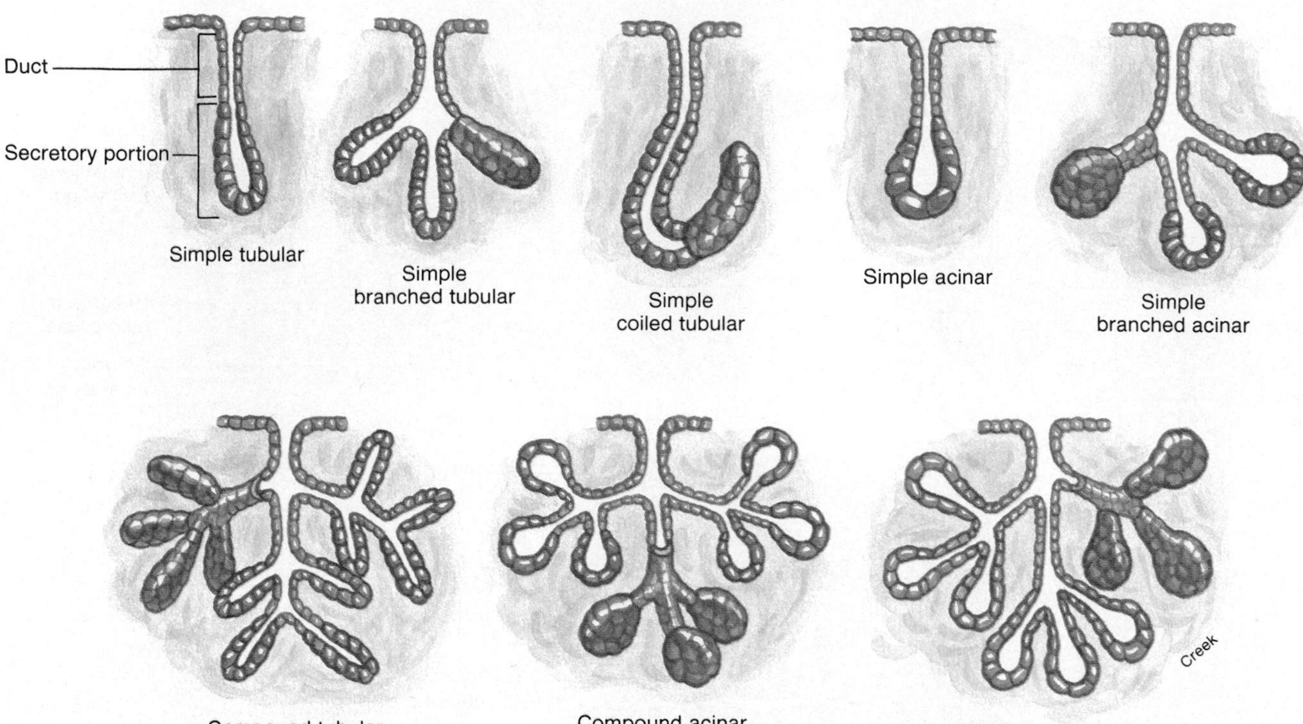

Multicellular Glands. Multicellular glands, as their name implies, are composed of numerous secretory cells as well as of cells forming the walls of the ducts. Multicellular glands are divided into *simple* and *compound* glands. The ducts of the simple gland do not branch or have a few branches, whereas those of the compound type do (fig. 6.14). Multicellular glands are further classified according to the shape of the secretory portion. They are identified as *tubular* if the secretory portion resembles the ductule portion and as *acinar* if the secretory portion is flasklike. Multicellular glands with a secretory portion that resembles both a tube and a flask are termed *tubuloacinar*.

Multicellular glands are also classified according to the means by which they discharge the secretory product (fig. 6.15). Glands that secrete their products by exocytosis through the cell membrane of the secretory cells are called **merocrine** *(mer'o-krīn)* **glands.** Salivary glands, pancreatic glands, and certain sweat glands are of this type. In **apocrine** *(ap'o-krīn)* **glands** the secretion accumulates on the surface of the secretory cell, and then a portion of the cell, along with the secretion, is pinched off to be discharged. Mammary glands and certain sweat glands are apocrine glands. In a **holocrine** *(hol'o-krīn)* **gland,** the entire secretory cell is discharged along with the secretory product. An example of a holocrine gland is a sebaceous, or oil-secreting, gland of the skin.

Table 6.3 summarizes the exocrine glands and their secretions.

1. List the functions of simple squamous epithelia.
2. What are the three types of columnarlike epithelia? What do they have in common? How are they different?
3. What are the two types of stratified squamous epithelia, and how do they differ?
4. Describe the three ways that epithelial cells are bonded together.
5. Distinguish between unicellular and multicellular glands. Explain how multicellular glands are classified according to their mechanism of secretion.
6. In what ways are mammary glands and certain sweat glands similar?

merocrine: Gk. *meros,* part; *krinein,* to separate

apocrine: Gk. *apo,* off; *krinein,* to separate
holocrine: Gk. *holos,* whole; *krinein,* to separate

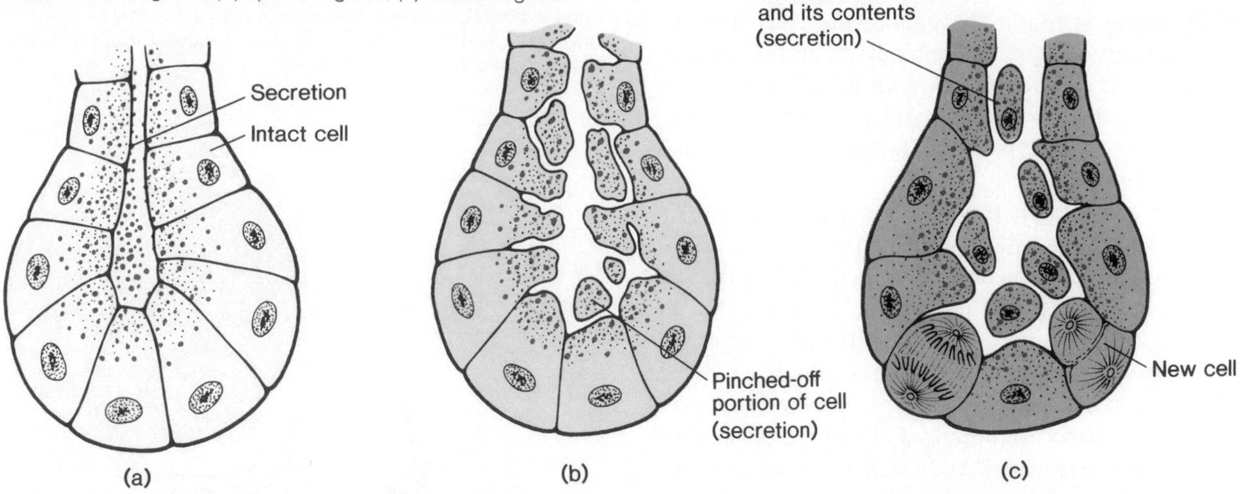

Figure 6.15. Secretory classification of multicellular exocrine glands: (*a*) merocrine gland; (*b*) apocrine gland; (*c*) holocrine gland.

Table 6.3 Summary of glandular epithelia

Type	Function	Example
Structural classification of exocrine glands		
I. Unicellular	Lubricate and protect	Goblet cells of digestive, respiratory, urinary, and reproductive systems
II. Multicellular	Protect, cool body, lubricate, aid in digestion, maintain body homeostasis	Sweat glands, digestive glands, mammary glands, sebaceous glands
A. Simple		
1. Tubular	Aid in digestion	Intestinal glands
2. Branched tubular	Protect, aid in digestion	Uterine glands, gastric glands
3. Coiled tubular	Regulate temperature	Certain sweat glands
4. Acinar	Additive to spermatozoa	Seminal vesicle of male reproductive system
5. Branched acinar	Skin conditioner	Sebaceous skin glands
B. Compound		
1. Tubular	Lubricate urethra of male, assist body digestion	Bulbourethral gland of male reproductive system, liver
2. Acinar	Nourishment to infant, aid in digestion	Mammary gland, salivary gland (sublingual and submandibular)
3. Tubuloacinar	Aid in digestion	Salivary gland (parotid), pancreas
Secretory classification of exocrine glands		
Type	**Description of secretion**	**Example**
Merocrine glands	Watery secretion for regulating temperature or enzymes that promote digestion	Salivary and pancreatic glands, certain sweat glands
Apocrine glands	Portion of secretory cell and secretion are discharged; provides nourishment to infant, assists in regulating temperature	Mammary glands, certain sweat glands
Holocrine glands	Entire secretory cell with enclosed secretion is discharged; skin conditioner	Sebaceous glands of the skin

CONNECTIVE TISSUE

Connective tissues are classified according to the characteristics of the matrix that binds the cells. Connective tissues provide structural and metabolic support for other tissues and organs of the body.

Objective 9. Describe the general characteristics, locations, and functions of connective tissue.

Objective 10. Explain the functional relationship between embryonic and adult connective tissue.

Objective 11. List the various ground substances, fiber types, and cells that constitute connective tissue, and explain their functions.

Characteristics and Classification of Connective Tissue

Connective tissue is found throughout the body and, as the name indicates, supports or binds other tissues and provides for the metabolic needs of all body organs. Certain types of connective tissue store nutritional substances, whereas other types manufacture protective and regulatory materials.

Although connective tissue varies tremendously in structure and function, all connective tissues have similarities. With the exception of cartilage, connective tissues are highly vascular and well nourished. They are able to replicate and, by so doing, are responsible for the repair of body organs. Unlike epithelial tissues, which are composed of tightly fitted cells, connective tissues contain considerably more matrix than cells. Connective tissues do not occur on free surfaces of body cavities or on the surface of the body as do epithelial tissues. Furthermore, connective tissues are embryonically derived from mesoderm, whereas epithelial tissues derive from ectoderm, mesoderm, and endoderm.

The classification of connective tissues is not exact, and several schemes have been devised. In general, however, they are named according to the kind and arrangement of the matrix. The following are the basic kinds of connective tissues.

A. Embryonic connective tissue
B. Connective tissue proper
   1. Loose (areolar)
   2. Dense fibrous
   3. Elastic
   4. Reticular
   5. Adipose
C. Cartilage
   1. Hyaline
   2. Fibrocartilage
   3. Elastic
D. Bone
E. Vascular (blood) tissue

Embryonic Connective Tissue

The embryonic period of development, which lasts six weeks (from the beginning of the third to the end of the eighth week), is characterized by a tremendous amount of tissue differentiation and organ formation. At the beginning of the embryonic period, all connective tissue appears the same and is referred to as **mesenchyme** *(mes'en-kīm)*. Mesenchyme is undifferentiated embryonic connective tissue, which is derived from mesoderm and consists of irregularly shaped cells lying in large amounts of a homogeneous, jellylike matrix (fig. 6.16). In certain areas of development, mesenchyme migrates to predisposed sites where it interacts with other tissues to form organs. Before the end of the embryonic period, once mesenchyme is in the appropriate position, it differentiates, and from it all other kinds of connective tissues are formed.

Figure 6.16. Mesenchyme is undifferentiated embryonic mesodermal connective tissue that can migrate and give rise to all other kinds of connective tissue. (*a*) it is found within an early developing embryo and (*b*) consists of irregularly shaped cells lying in a homogeneous, jellylike matrix.

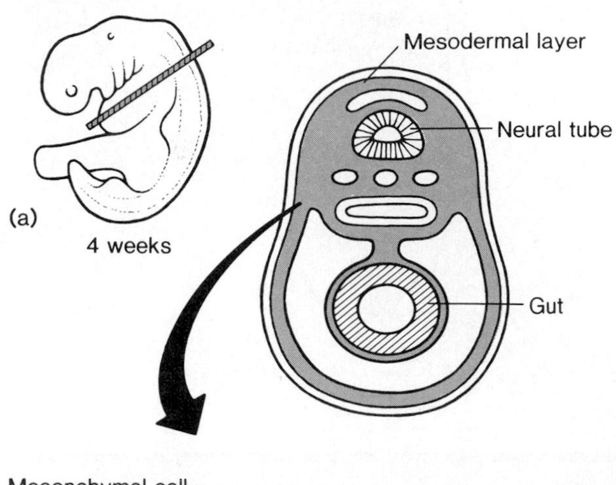

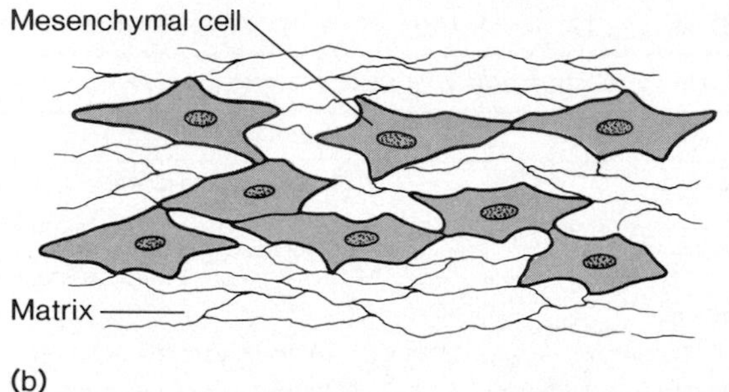

In summary, mesenchymal tissue is undifferentiated embryonic connective tissue, which can migrate and give rise to all other types of connective tissue. The causes of tissue differentiation are not fully understood.

Some mesenchymal-like tissue persists past the embryonic period in certain sites within the body. Good examples are the undifferentiated cells that surround blood vessels and form fibroblasts if the vessels are traumatized. Fibroblasts assist in healing wounds (chapter 8).

Another kind of prenatal connective tissue exists only in the fetus (the fetal period is from nine weeks to birth) and is called *mucous connective tissue,* or *Wharton's jelly.* It provides a turgid consistency to the umbilical cord.

Connective Tissue Proper

Connective tissue proper has a loose, flexible matrix, frequently called **ground substance.** The most common cell within connective tissue proper is called a **fibroblast** *(fi'bro-blast)*. Fibroblasts are large, star-shaped cells that produce collagenous, elastic, and reticular *(rĕ-tik'u-lar)* fibers. **Collagenous** *(kol-laj'ĕ-nus),* or **white fibers,** are composed of a protein called *collagen (kol'ah-jen);* they are flexible, yet they have tremendous strength. **Elastic,** or **yellow fibers,** are composed of a protein called *elastin,*

reticular: L. *rete,* net or netlike
collagen: Gk. *kolla,* glue
elastin: Gk. *elasticus,* to drive

Figure 6.17. Loose connective tissue consists of collagenous and elastic fibers as well as fibroblasts and mast cells. This type of connective tissue is an important packing and binding material. It surrounds muscle, nerves, and vessels and binds the skin to the underlying muscles. (*a*) a photomicrograph and (*b*) a diagram.

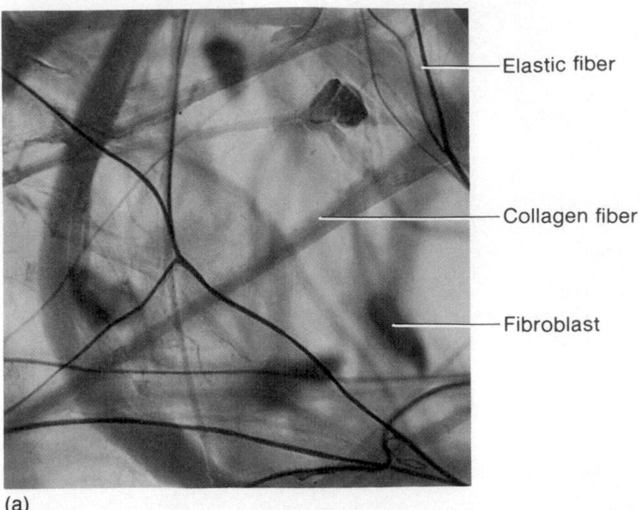

(a)

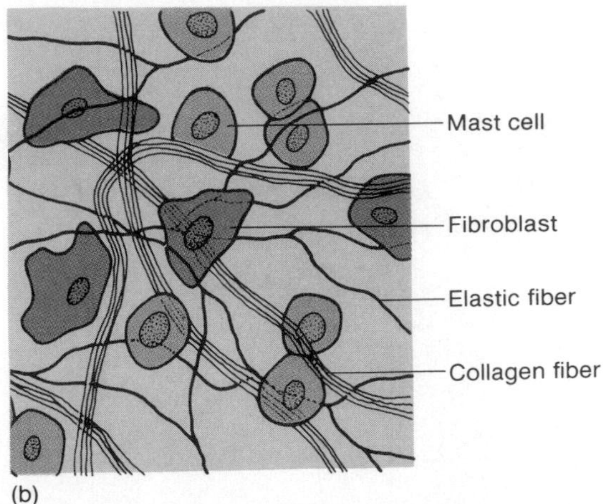

(b)

which gives elasticity to certain tissues. Collagenous and elastic fibers may be either sparse and irregularly arranged, as in loose connective tissue, or tightly packed, as in dense connective tissue. Tissues with loosely arranged fibers generally form packing material that cushions and protects various organs, whereas those that are tightly arranged form the binding and supportive connective tissues of the body. Dense connective tissue can also vary in the orientation of the fibers. If the fibers are interwoven to provide tensile strength in any direction, they are referred to as *irregularly arranged.* Dense, irregularly arranged connective tissue forms the periosteum, or covering, around bone, the perichondrium surrounding cartilage, and the fibrous capsules around certain organs (testes, parts of the liver, and lymph nodes). *Regularly arranged* fibers are parallel to the direction of force. Dense, regularly arranged connective tissues are found where strong, flexible bindings are required, such as in tendons, ligaments, and aponeuroses.

Tensile strength in tissues that contain elastic fibers is extremely important for several physical functions of the body. Consider, for example, that elastic fibers are found in the walls of arteries and in the walls of the lower respiratory passageways. As these walls are expanded by blood moving through vessels or by inspired air, the elastic fibers must first stretch and then recoil. This maintains the pressures of the fluid or air moving through the lumina, which provides adequate flow rates and, thus, adequate rates of diffusion through capillary and lung surfaces.

Reticular fibers reinforce by forming thin, short threads that branch and join to form a delicate lattice or reticulum. Reticular fibers are common in lymphatic glands, where they form a meshlike **stroma.**

Five basic types of connective tissue proper are generally recognized. These tissues are distinguished by the consistency of the ground substance and the type and arrangement of the reinforcement fibers.

Loose Connective (Areolar) Tissue. Loose connective tissue is distributed throughout the body as a binding and packing material. It binds the skin to the underlying muscles and is highly vascular, providing nutrients to the skin. Loose connective tissue surrounding muscle fibers and muscle groups is known as **fascia** *(fash'e-ah).* It also surrounds blood vessels and nerves, where it provides both protection and nourishment. Specialized cells called **mast cells** are dispersed throughout the loose connective tissue surrounding blood vessels. Mast cells produce *heparin (hep'ah-rin),* an anticoagulant that prevents blood from clotting within the vessels.

The cells of loose connective tissue are predominantly fibroblasts, with collagenous and elastic fibers dispersed throughout the ground substance (fig. 6.17). The irregular arrangement of this tissue provides flexibility and yet strength in any direction. It is this tissue layer, for example, that permits the skin to move when a part of the body is rubbed.

stroma: Gk., *stroma,* a couch or bed
fascia: L. *fascia,* band or girdle
heparin: Gk. *hepatos,* the liver

Figure 6.18. Dense fibrous connective tissue forms tendons (*a*). In this tissue type, the collagenous, or white, fibers, are arranged in bundles, and fibroblasts are positioned between the bundles. The structure of this flexible, strong tissue is seen in (*b*) a photomicrograph and (*c*) a diagram.

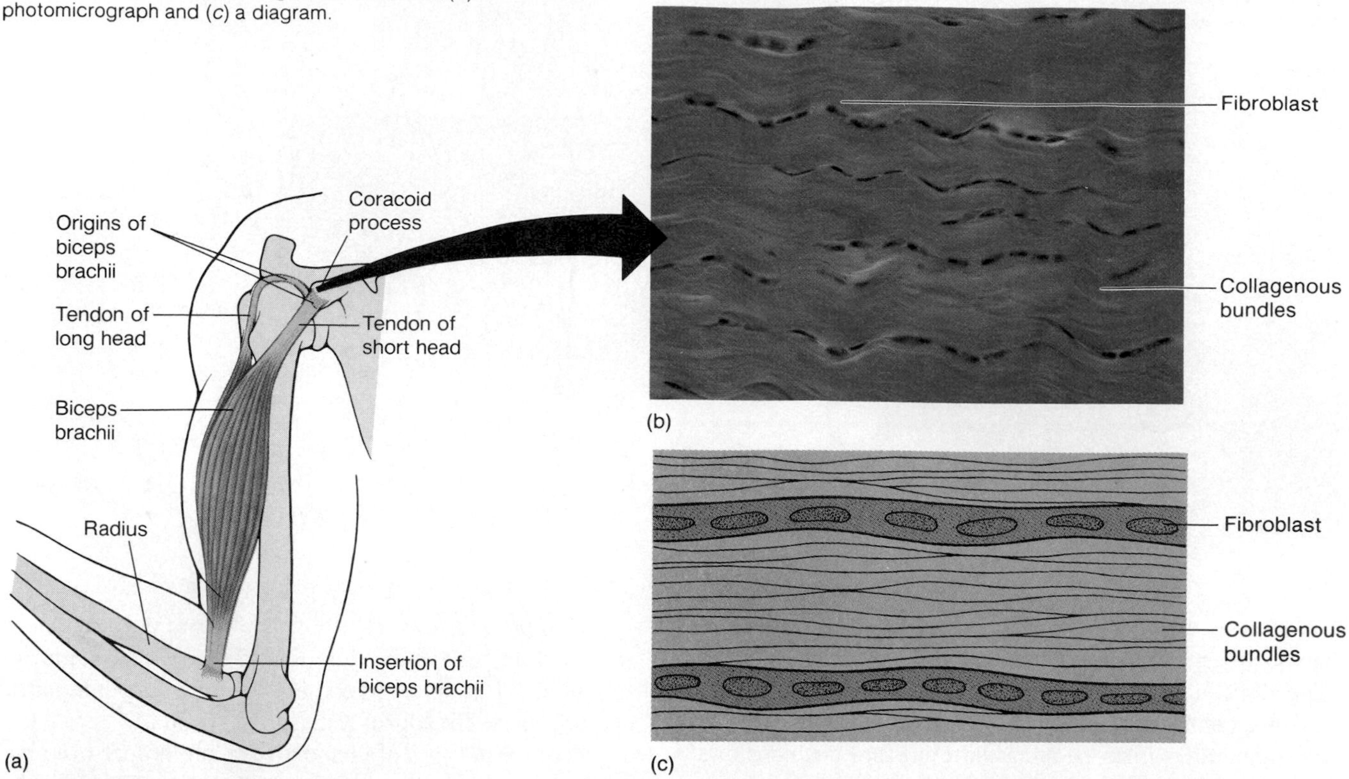

> Much of the fluid of the body is found within loose connective tissue and is called *tissue fluid.* Sometimes excessive tissue fluid accumulates, causing a swelled condition called *edema (ĕ-de'mah).* Edema is a symptom of a variety of dysfunctions or disease processes.

Dense Fibrous Connective Tissue. Dense fibrous connective tissue is characterized by large amounts of densely packed collagenous fibers. Because this tissue is silvery white in appearance, it is sometimes called dense, white fibrous connective tissue (fig. 6.18). The blood supply to dense fibrous connective tissue is poor, so if this tissue is injured it heals relatively slowly.

Dense fibrous connective tissue occurs where strong, flexible support is necessary. **Tendons,** which attach muscles to bones and transfer the forces of muscle contractions, and **ligaments,** which connect bone to bone across articulations, are composed of this type of tissue. The sclera of the eye and deep connective tissues of the skin are also made of dense fibrous connective tissue.

tendon: L. *tendere,* to stretch
ligament: L. *ligare,* bind

Elastic Connective Tissue. Elastic connective tissue has a predominance of elastic fibers that are irregularly arranged and yellowish in color (fig. 6.19). They can be stretched to one and a half times their original lengths and will snap back to their former size. Elastic connective tissue is found in the walls of large arteries, in portions of the larynx, and in the trachea and bronchial tubes of the lungs. It is also present between the arches of vertebrae, which make up the vertebral column.

Reticular Connective Tissue. Reticular connective tissue is characterized by a network of reticular fibers woven through a jellylike matrix (fig. 6.20). Certain specialized cells within reticular tissue are *phagocytic (fag"o-sit'ik)* and therefore ingest foreign materials. The liver, spleen, lymph nodes, and bone marrow contain reticular connective tissue.

Adipose Tissue. Adipose tissue is a specialized type of loose fibrous connective tissue that contains large quantities of **adipose** *(ad'i-pos)* **cells.** Adipose cells form from fibroblasts and are, for the most part, formed prenatally and during the first year of life. Adipose cells store droplets of fat within their cytoplasm, causing them to swell and forcing their nuclei to one side (fig. 6.21).

adipose: L. *adiposus,* fat

Figure 6.19. (*a*) the wall of an artery is composed of several tissue types. (*b*) the cellular structure. (*c*) elastic connective tissue is found within the inner coat. This type of connective tissue allows stretching as the blood flows through. The outer coat consists of loose fibrous connective tissue, and the middle coat contains smooth muscle.

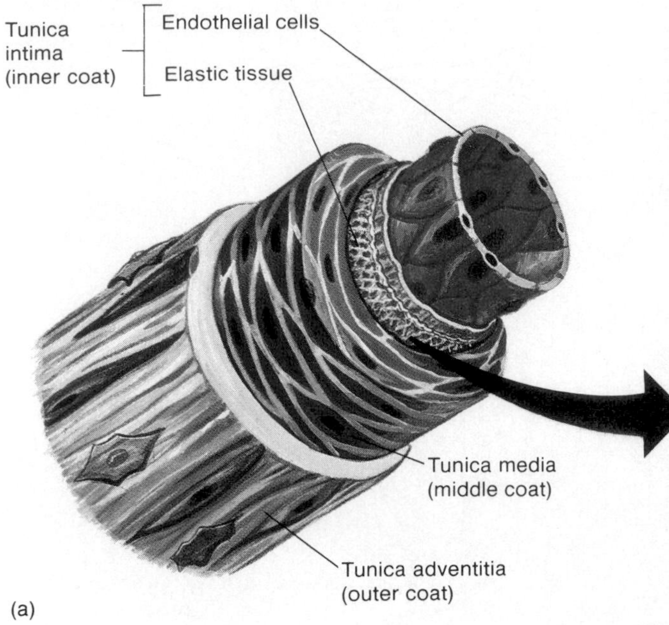

Tunica intima (inner coat)
Endothelial cells
Elastic tissue
Tunica media (middle coat)
Tunica adventitia (outer coat)
(a)

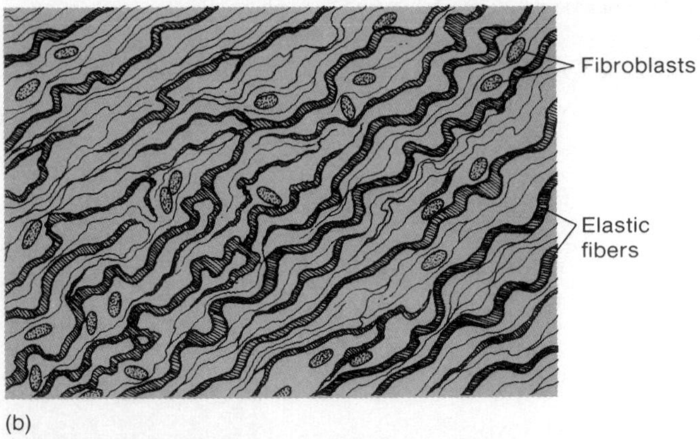

Fibroblasts
Elastic fibers
(b)

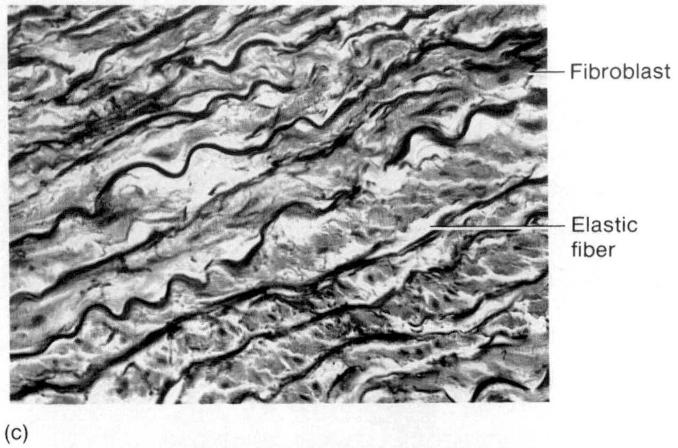

Fibroblast
Elastic fiber
(c)

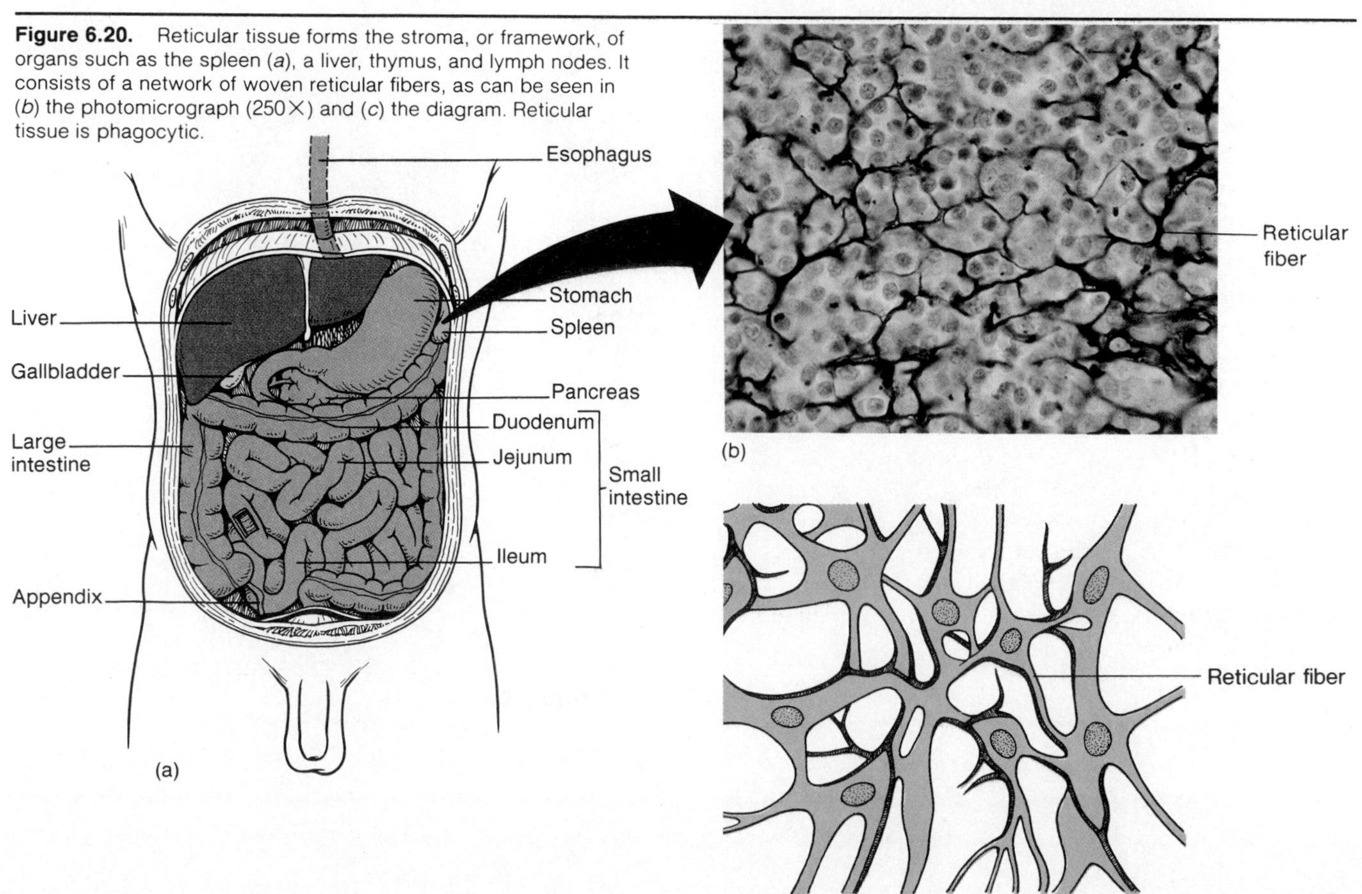

Figure 6.20. Reticular tissue forms the stroma, or framework, of organs such as the spleen (*a*), a liver, thymus, and lymph nodes. It consists of a network of woven reticular fibers, as can be seen in (*b*) the photomicrograph (250×) and (*c*) the diagram. Reticular tissue is phagocytic.

Figure 6.21. Adipose connective tissue is abundant in (*a*) the hypodermis of the skin and around various internal organs. It is protective and serves to store fat. The fat globules (*b, c*) are stored in the adipose cells (250×).

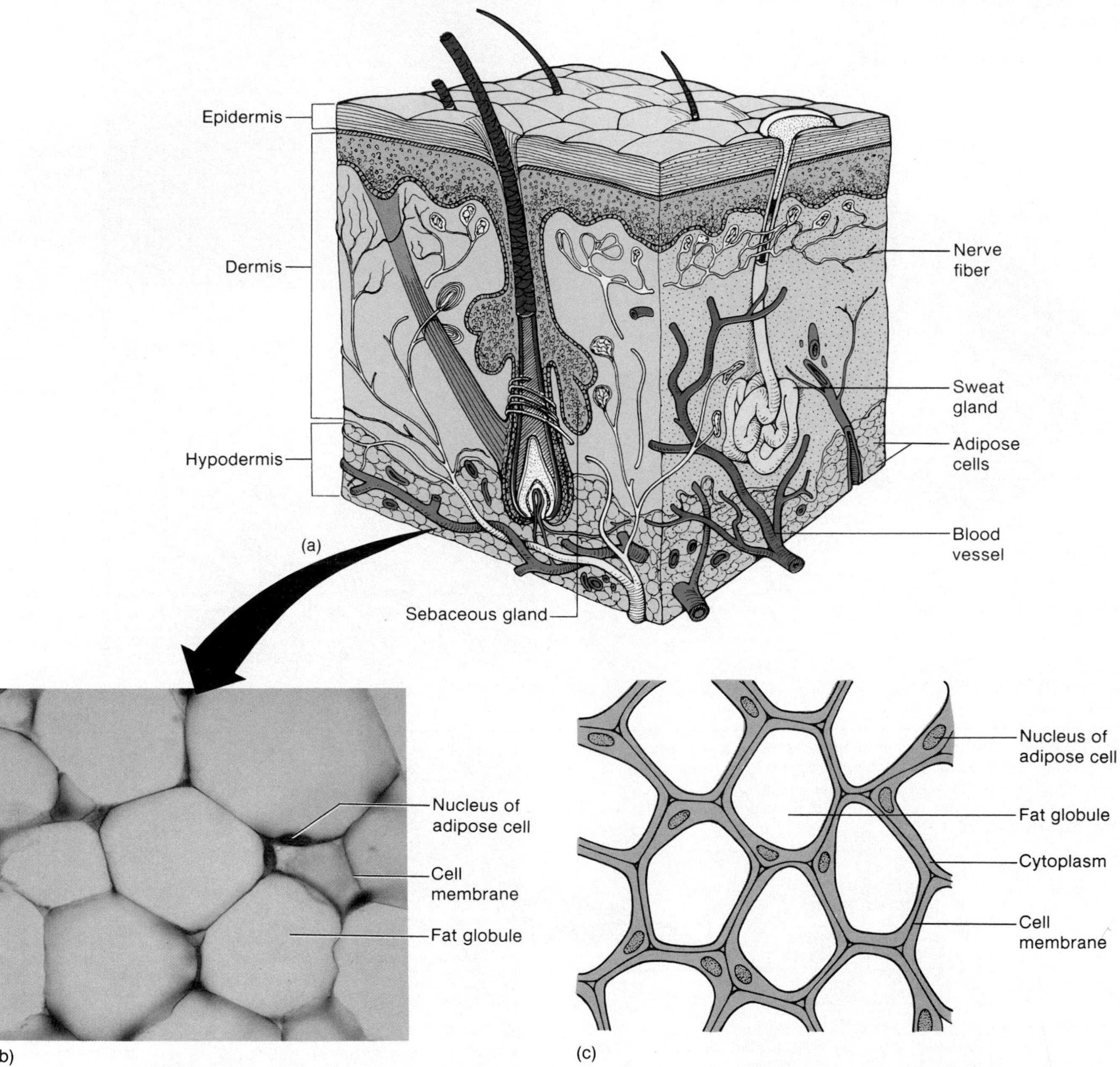

Table 6.4 Summary of connective tissue proper

Type	Structure and function	Location
Loose connective (areolar) tissue	Predominantly fibroblast cells with lesser amounts of collagen and elastin cells; binds organs, holds tissue fluids, diffusion	Surrounding nerves and vessels, between muscles, beneath the skin
Dense fibrous connective tissue	Densely packed collagen fibers; provides strong, flexible support	Tendons, ligaments, sclera of eye, deep skin layers
Elastic connective tissue	Predominantly irregularly arranged elastic fibers; supports, provides framework	Large arteries, lower respiratory tract, between vertebrae
Reticular connective tissue	Reticular fibers forming supportive network; stores, phagocytic	Lymph nodes, liver, spleen, thymus, bone marrow
Adipose connective tissue	Adipose cells; protects, stores fat, insulates	Hypodermis of skin, surface of heart, omentum, around kidneys, back of eyeball, surrounding joints

Feeding an infant an excessive amount during its first year, when adipose cells are forming, may cause a greater amount of adipose tissue to develop. A person with large amounts of adipose tissue is more susceptible to developing obesity later in life than a person with a lesser amount of adipose tissue. Dieting eliminates the fat stored within the tissue but not the tissue itself.

Adipose tissue is found throughout the body but is concentrated around the kidney, in the hypodermis of the skin, on the surface of the heart, surrounding joints, and in the breasts of sexually mature females. Fat functions not only as a food reserve but supports and protects various organs. Fat is a good insulator against cold because it is a poor conductor of heat.

Excessive fat can be both unsightly and unhealthy, placing a strain on the heart and perhaps causing early death. For these reasons, good exercise programs and sensible diets are extremely important. Adipose tissue can also retain environmental pollutants that are ingested or absorbed through the skin.

The surgical procedure of *suction lipectomy* may be used to remove small amounts of adipose tissue from certain localized body areas, such as the breasts, abdomen, buttocks, or thighs. Suction lipectomy is used for cosmetic purposes rather than as a treatment for obesity. Potential candidates should have good skin elasticity, be only about 15 to 20 pounds overweight, and be between thirty and forty years old.

Table 6.4 summarizes the characteristics, functions, and locations of connective tissue proper.

Cartilage

Cartilage tissue consists of cartilage cells, called **chondrocytes** *(kon'dro-sītz),* and a semisolid ground substance that imparts marked elastic properties to the tissue. Cartilage is a type of supportive and protective connective tissue commonly called "gristle." Cartilage is frequently associated with bone. It forms a precursor to one type of bone and persists at the articular surfaces on the bones of all movable joints.

The chondrocytes within cartilage may occur singly but are frequently clustered. Chondrocytes occupy spaces, called **lacunae** *(lah-ku'ne),* within the matrix. Most cartilage tissue is surrounded by a dense fibrous connective tissue called a **perichondrium** *(per"ĭ-kon'dre-um).* Cartilage at the articular surfaces of bones (articular cartilage) lacks perichondria. Because cartilage is avascular, it must receive nutrients through diffusion from the surrounding tissue. For this reason, cartilaginous tissue has a slow rate of mitotic activity and, if damaged, heals with difficulty.

There are three kinds of cartilage, each of which is distinguished by the types of fibers embedded within the matrix: hyaline cartilage, fibrocartilage, and elastic cartilage.

Hyaline Cartilage. Hyaline *(hi'ah-lın)* cartilage has a homogeneous, bluish-stained matrix in which the collagenous fibers are so fine that they can be observed only with an electron microscope. When viewed through a microscope, hyaline cartilage has a clear, glassy appearance (fig. 6.22).

Hyaline cartilage is the most abundant cartilage within the body. It covers the articular surfaces of bones, supports the tubular trachea and bronchi of the respiratory system, reinforces the nose, and forms the flexible bridge, called **costal cartilage,** between the ventral end of each of the first ten ribs and the sternum. Most of the bones of the body form first as hyaline cartilage and later become bone through ossification.

---

lacuna: L. *lacuna,* hole or pit
hyaline: Gk. *hyalos,* glass

Figure 6.22. Hyaline cartilage is the most abundant cartilage within the body. It occurs in places such as the larynx and trachea (*a*), articular ends of bones, ends of ribs, and embryonic skeleton. (*b*) the homogeneous matrix of hyaline cartilage can be seen in the photomicrograph (100×). Its structure is diagrammed in (*c*).

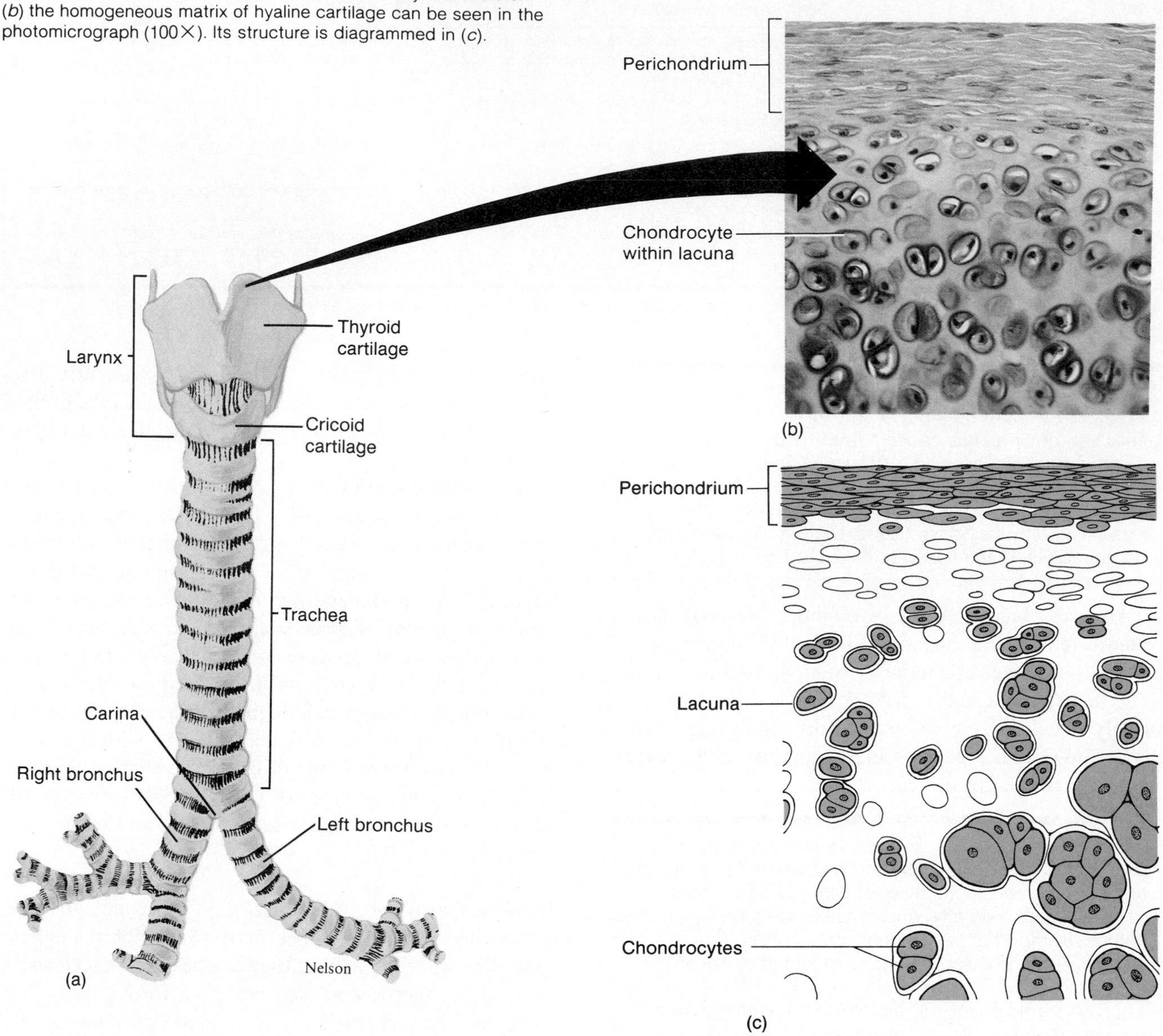

By the end of the day the intervertebral discs of the vertebral column are somewhat compacted. So a person is actually slightly shorter in the evening than in the morning following a recuperative rest. Aging brings a gradual compression of the fibrocartilaginous discs that is irreversible.

Fibrocartilage. Fibrocartilage has its matrix reinforced with numerous collagenous fibers (fig. 6.23). It is a durable tissue adapted to withstand tension and compression. It is found at the symphysis pubis, where the two pelvic bones articulate, and between the vertebrae as intervertebral discs. It also forms the cartilaginous wedges within the knee joint.

Elastic Cartilage. Elastic cartilage is similar to hyaline except for the presence of abundant elastic fibers, which make it very flexible while maintaining its strength (fig. 6.24). The numerous elastic fibers also give it a yellowish appearance. This tissue is found in the external ear, portions of the larynx, and the auditory canal.

Table 6.5 summarizes the structure, function, and location of cartilage within the body.

Figure 6.23. Fibrocartilage is located at the symphysis pubis, within the knee joint, and between the vertebrae as the intervertebral discs (*a*). It consists of chondrocytes within a matrix that is reinforced with collagenous fibers (*b, c*) to withstand stress and compression as it provides durable support (250×).

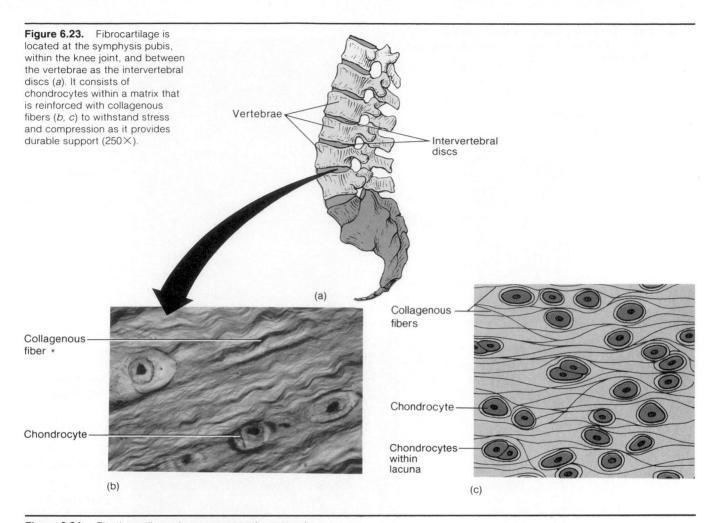

Figure 6.24. Elastic cartilage gives support to the external ear, auditory canal, and parts of the larynx. (*a*) the matrix of elastic tissue is laced with elastic fibers to provide flexibility while maintaining shape. (*b*) a photomicrograph of elastic cartilage (100×).

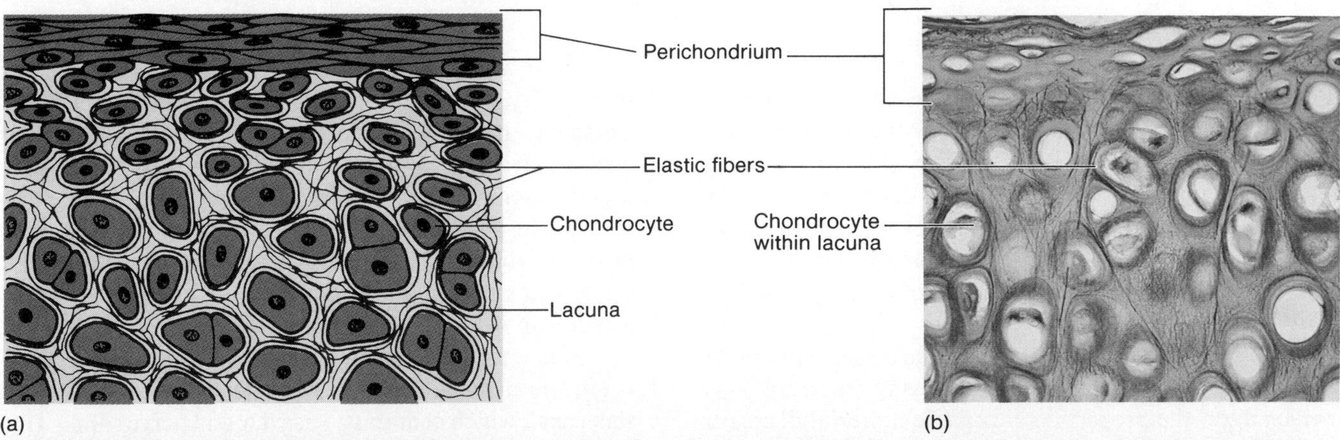

Table 6.5 Summary of cartilage tissue

Type	Structure and function	Location
Hyaline cartilage	Homogeneous matrix with extremely fine collagenous fibers; flexible support, protection, precursor to bone	Articular surfaces of bones, nose, walls of respiratory passages, fetal skeleton
Fibrocartilage	Abundant collagenous fibers within matrix; support, withstand compression	Symphysis pubis, intervertebral discs, knee joint
Elastic cartilage	Abundant elastic fibers within matrix; support, flexibility	Framework of external ear, auditory canal, portions of larynx

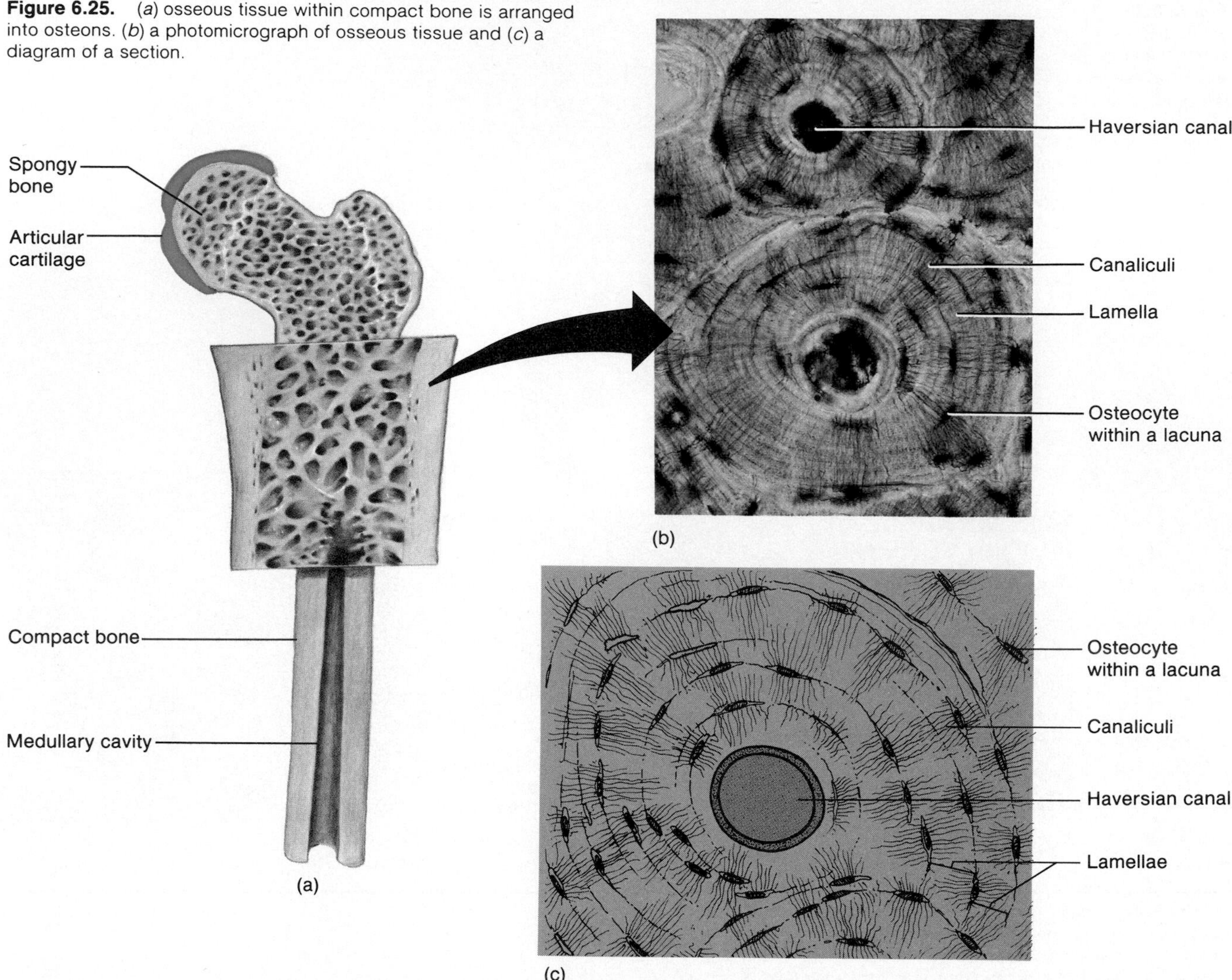

Figure 6.25. (*a*) osseous tissue within compact bone is arranged into osteons. (*b*) a photomicrograph of osseous tissue and (*c*) a diagram of a section.

Bone

Bone, or **osseous connective tissue,** is the most rigid of the connective tissues. Unlike cartilage, bone has a rich vascular supply and is the site of considerable metabolic activity. The hardness of bone is largely due to the inorganic calcium phosphate and calcium carbonate salts deposited within the intercellular matrix. Numerous collagenous fibers, also embedded within the matrix, give some flexibility to bone.

In healthy osseous tissue, a balance exists between the organic living and the inorganic nonliving materials. As a person ages, the proportion of organic material decreases and the bones become brittle. The bones of elderly persons not only fracture easily but do not heal readily.

When bone is placed in a weak acid, the calcium salts dissolve away and the bone becomes pliable. It retains its basic shape but can be easily bent and twisted. In calcium deficiency diseases, such as rickets, the osseous tissue becomes pliable and bends under the weight of the body.

There are two kinds of bone (osseous) tissue, based on porosity, and most bones have both types (fig. 6.25). **Compact,** or **dense, bone tissue** is the hard, outer layer, whereas **spongy,** or **cancellous, bone tissue** is the porous, highly vascular, inner portion. Compact bone tissue is covered by the periosteum, serves for attachment of muscles, provides protection, and gives durable strength to the bone. Spongy bone tissue makes the bone lighter and provides a space for bone marrow where blood cells are produced.

In compact bone tissue, the bone cells, called **osteocytes,** are arranged in concentric layers around a **haversian canal,** which contains a vascular and nerve supply (fig. 6.26). Each osteocyte occupies a space called a **lacuna.** Radiating from each lacuna are numerous minute canals, called **canaliculi,** that traverse the dense matrix of the bone tissue to adjacent lacunae. Nutrients diffuse through the canaliculi to reach each osteocyte. The inorganic matrix is deposited in concentric layers called **lamellae.**

haversian canal: from Clopton Havers, English anatomist, 1650–1702

Figure 6.26. A scanning electron micrograph of a Haversian canal of osseous tissue. The lacunae (*La*) provide a space for the osteocytes, which are connected to one another by canaliculi (*Ca*). Note the divisions between concentric lamellae (*arrows*). (From: *Tissues and Organs: A Text Atlas of Scanning Electron Microscopy* by R. G. Kessel and R. Kardon. W. H. Freeman and Company. © 1979.)

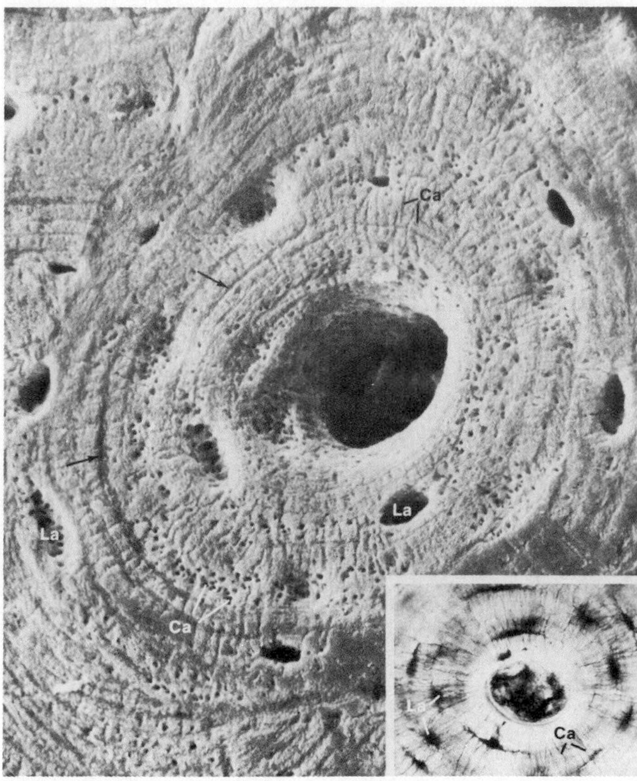

Figure 6.27. Vascular connective tissue, or blood, is composed of a fluid matrix, called plasma, and formed elements of which there are three types: erythrocytes (red blood cells), thrombocytes (platelets), and leukocytes (white blood cells).

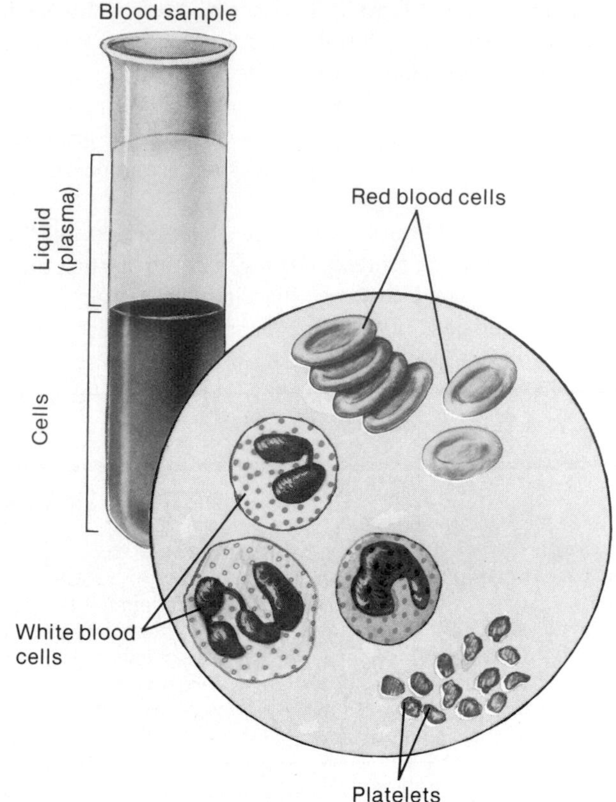

A haversian canal, with its surrounding osteocytes, lacunae, canaliculi, and concentric lamellae, constitutes a **haversian system,** or an **osteon** (fig. 6.25). Metabolic activity within osseous tissue occurs at the osteon level. Areas between the osteons contain **interstitial** *(in''ter-stish'al)* **lamellae.** These areas possess osteocytes within lacunae and associated canaliculi but are arranged irregularly. Transverse channels, called **Volkmann's canals** (not illustrated), penetrate compact bone and connect various osteons with blood vessels and nerves.

Electron microscopy is a relatively new technique for examining tissues. Many students of histology do not understand or appreciate the added dimension provided by the electron microscope. The structure of the haversian canal, for example, shown in figure 6.26, can be more accurately described using a scanning electron micrograph rather than a conventional photomicrograph. Electron microscopy is an extension of light microscopy and should be used to demonstrate the continuity of structural organization within tissues.

osteon: Gk. *osteon,* bone
Volkmann's canals: from Alfred W. Volkmann, German physiologist, 1800–1877

Vascular Connective Tissue

Blood is a highly specialized, viscous connective tissue. **Formed elements** (red blood cells, white blood cells, and platelets) are suspended in the liquid plasma matrix. Blood plays a vital role in maintaining internal body homeostasis. The three types of formed elements found within the blood are erythrocytes, leukocytes, and thrombocytes (fig. 6.27).

Erythrocytes. Erythrocytes *(ĕ-rith'ro-sītz),* or **red blood cells** (RBC), are minute, nonnucleated, biconcave discs that transport respiratory gases. The red color results from the pigment *hemoglobin.* In an infant, erythrocytes are produced in the spleen and bone marrow, but in a mature person, the bone marrow is the only production site. Erythrocytes live between 90 and 120 days and are disposed of in the liver.

erythrocyte: Gk. *erythros,* red; *kytos,* hollow (cell)
leukocyte: Gk. *leukos,* white; *kytos,* hollow (cell)
thrombocyte: Gk. *thrombos,* a clot; *kytos,* hollow (cell)
hemoglobin: Gk. *haima,* blood; *globus,* globe

Leukocytes. Leukocytes *(loo'ko-sītz)*, or **white blood cells** (WBC), are amoeboid in movement and slightly larger than erythrocytes. Leukocytes primarily defend the body against invasions by microorganisms. They are produced in bone marrow and lymphatic tissue and have a life span that ranges from 3 to 300 days. There are five kinds of leukocytes: **neutrophils, eosinophils** *(e''o-sin'o-filz)*, **basophils** *(ba'so-filz)*, **lymphocytes** *(lim'fo-sītz)*, and **monocytes.** The first three can be further classified as **granular leukocytes** and the remaining two as **nongranulocytes.**

Thrombocytes. Thrombocytes are the **platelets** of the blood; along with the protein *fibrinogen* of the plasma, they play a role in blood clotting. Platelets arise from large, multinucleated cells in the bone marrow, which are called **megakaryocytes** *(meg''ah-kar'e-o-sītz)*. As these large cells fragment, numerous platelets form. Platelets, like leukocytes, have amoeboid movement.

An injury to a portion of the body may stimulate tissue repair activity, usually involving connective tissue. A minor scrape or cut results in platelet and plasma activity of the exposed blood and the formation of a scab. The epidermis of the skin regenerates beneath the scab. A severe open wound brings about connective tissue granulation, in which collagenous fibers form from surrounding fibroblasts to strengthen the traumatized area. The healed area is known as a *scar.*

1. List the kinds of connective tissues, and describe the structure, function, and location of each.
2. Which of the previously discussed connective tissues function to protect body organs? Which type is phagocytic? Which types bind and support various structures? Which types are associated in some way with the skin?
3. What is the developmental significance of mesenchyme, and how does it functionally differ from adult connective tissue?
4. Define or describe ground substance, reticular fibers, fibroblasts, collagenous fibers, elastic fibers, and mast cells.

Muscle Tissue

Muscle tissues are responsible for the movement of materials through the body, the movement of one part of the body with respect to another, and for locomotion. Fibers in the three kinds of muscle tissue are adapted to contract in response to stimuli.

megakaryocyte: Gk. *megas,* great; *karyon,* nut or kernel; *kytos,* hollow (cell)

Objective 12. Describe the structure, location, and function of the three types of muscle tissue.

Muscle tissues are unique in possessing the property of *contractility.* The muscle cells, or *fibers,* are elongated in the direction of contraction. Movement is accomplished through the shortening of the fibers in response to a stimulus. Muscle tissue is derived from mesoderm. There are three types of muscle tissue in the body: smooth, cardiac, and skeletal (fig. 6.28).

Smooth Muscle. Smooth muscle tissue is common throughout the body, occurring in many of the systems. For example, in the wall of the alimentary canal it provides the motive power for mechanically digesting food and the peristaltic movements through the digestive tract. Smooth muscle is also found in the walls of arteries, the walls of respiratory passages, and in the urinary and reproductive ducts. The contraction of smooth muscle is under autonomic (involuntary) nervous control.

Smooth muscle fibers are long, spindle-shaped cells that contain a single nucleus and lack striations. These cells are usually grouped together in flattened sheets, forming the muscular portion of a wall around a lumen.

Cardiac Muscle. Cardiac muscle tissue makes up most of the wall of the heart. This tissue is characterized by bifurcating (branching) fibers, a centrally positioned nucleus, and transversely positioned **intercalated** *(in-ter'kah-lāt-ed)* **discs.** Intercalated discs help to hold adjacent cells together and transmit the force of contraction from cell to cell. Like skeletal muscle, cardiac muscle is striated, but unlike skeletal muscle it experiences rhythmical involuntary contractions. Cardiac muscle is further discussed in chapter 20.

Skeletal Muscle. Skeletal muscle tissue attaches to the skeleton and is responsible for voluntary body movements. Each fiber is elongated, multinucleated, and has distinct transverse striations. Fibers of this muscle tissue are grouped into parallel fasciculi (bundles), which can be seen without a microscope in fresh muscle. Both cardiac and skeletal muscle fibers are striated and are so specialized at contraction that they cannot replicate once tissue formation is completed shortly after birth. Skeletal muscle tissue is further discussed in chapter 12. Table 6.6 summarizes the three types of muscle tissue.

1. Identify the general characteristics of muscle tissue. What is meant by a voluntary or involuntary muscle?
2. Distinguish between smooth, cardiac, and skeletal muscle tissue on the bases of structure, location, and function.

Figure 6.28. The three types of muscle tissue: (*a*) smooth muscle fibers teased apart; (*b*) cardiac muscle; and (*c*) skeletal muscle.

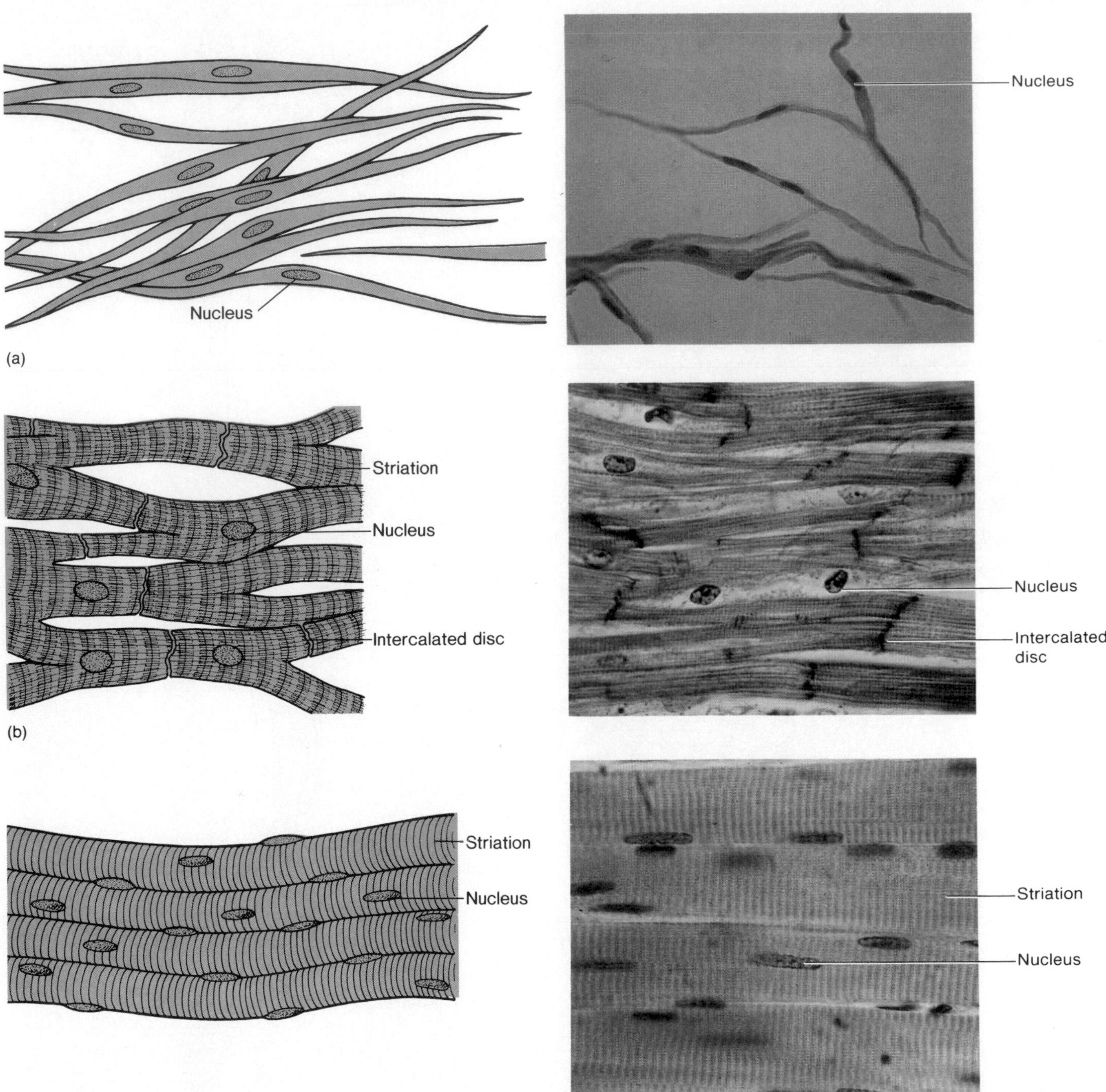

Table 6.6 Summary of muscle tissue

Type	Structure and function	Location	Type	Structure and function	Location
Smooth	Elongated, spindle-shaped fiber with single nucleus; involuntary movements of internal organs	Walls of hollow internal organs	Skeletal	Multinucleated, striated, cylindrical fiber that occurs in fasciculi; voluntary movement of skeletal parts	Associated with skeleton; spans joints of skeleton via tendons
Cardiac	Branched, striated fiber with single nucleus and intercalated discs; involuntary rhythmic contraction	Heart muscle			

Figure 6.29. The neuron. (*a*) a photomicrograph of nerve tissue. (*b*) a simplified diagram of a neuron and its principal parts. The arrows indicate the direction of the nerve impulse.

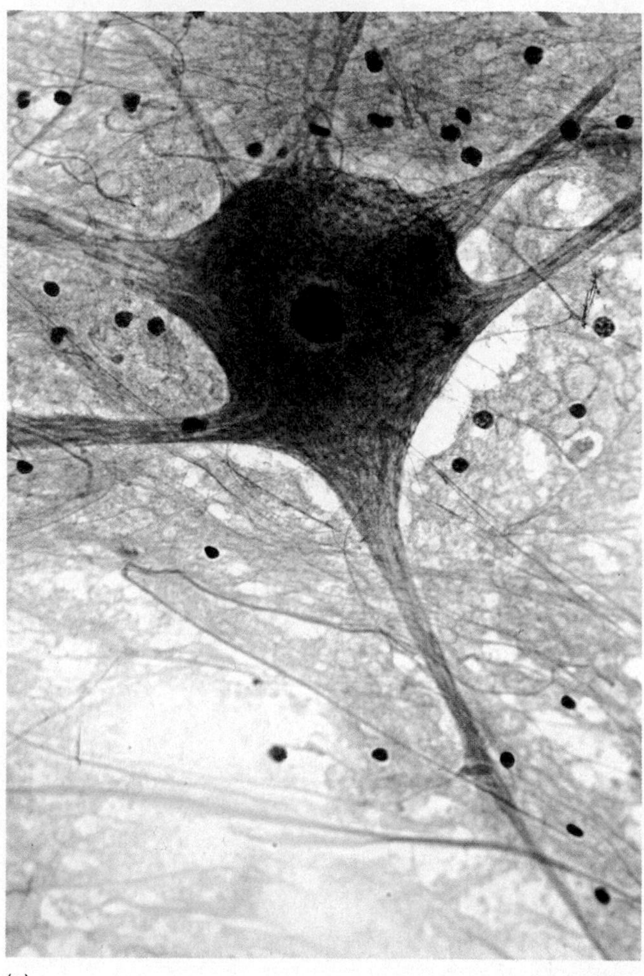

(a)

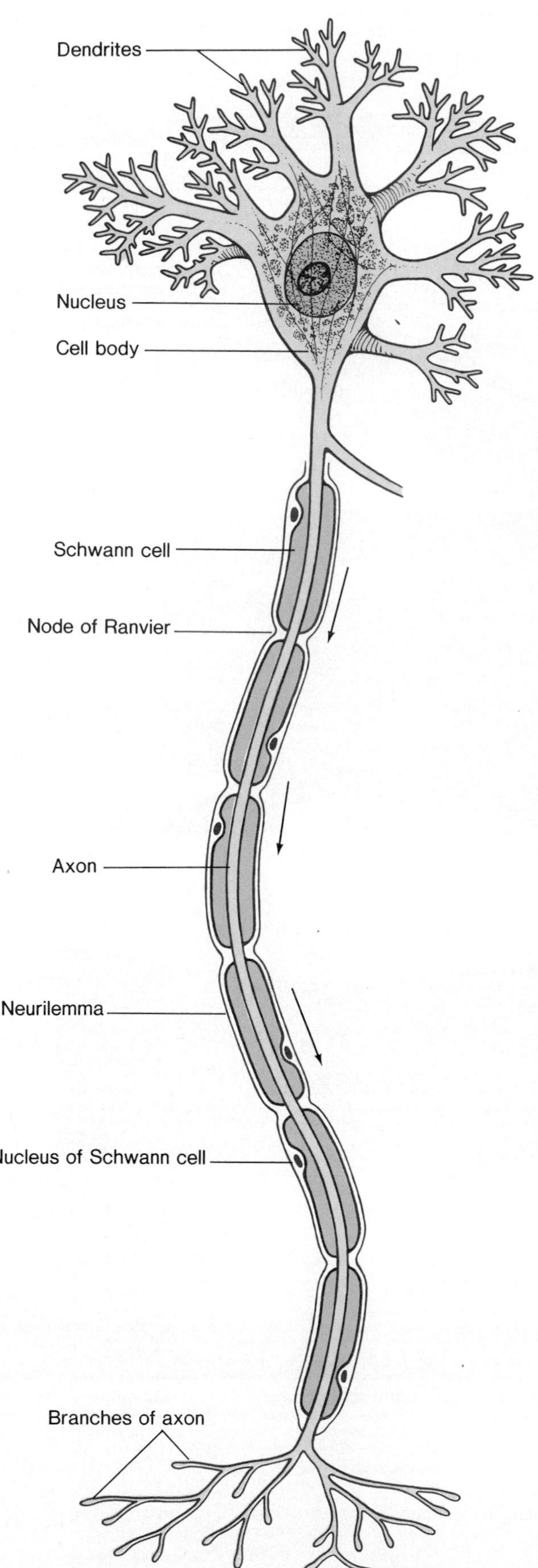

(b)

Figure 6.30. Various types of neuroglial cells that connect, support, and nourish neurons. (*a*) an astrocyte in contact with a capillary of a blood vessel serving the brain; (*b*) a microglia with processes extending to two nerve cell bodies; (*c*) oligodendrocytes near a nerve cell body.

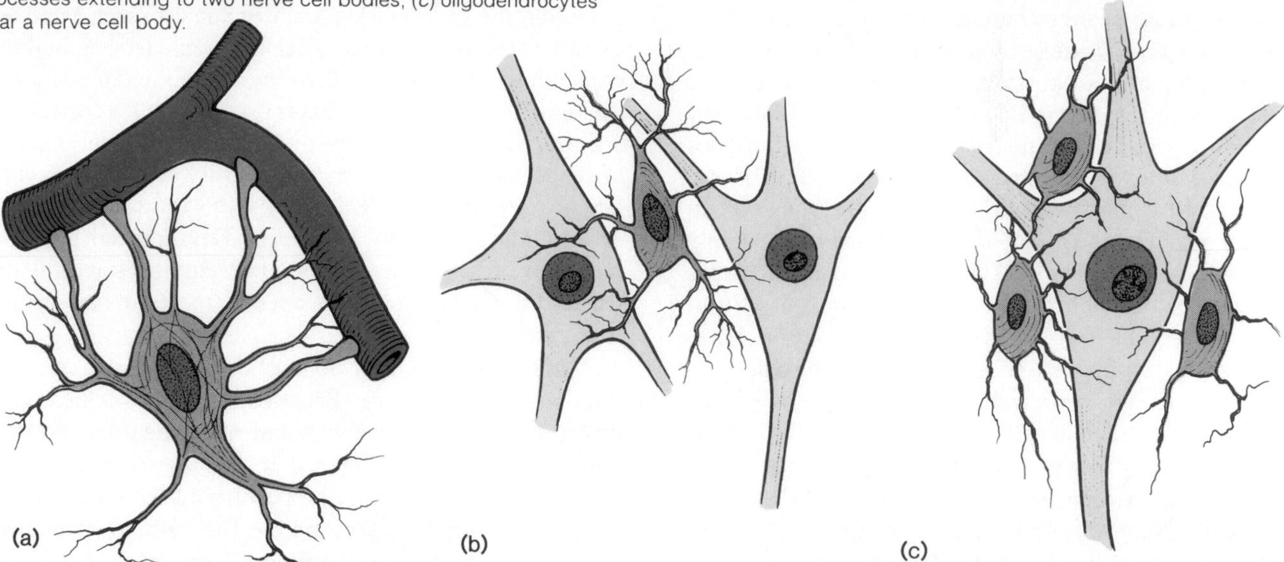

Nervous Tissue

Nervous tissue is composed of neurons, which respond to stimuli and conduct impulses to and from all body organs, and neuroglia, which functionally support and physically bind neurons.

Objective 13. Give the basic characteristics and functions of nervous tissue.

Objective 14. Distinguish between neurons and neuroglia.

Neurons. Although there are several kinds of neurons *(nu'rons)* in nervous tissue, they all have three principal components: (1) a cell body, or perikaryon; (2) dendrites; and (3) an axon (fig. 6.29). **Dendrites** function to receive a stimulus and conduct the impulse to the cell body. The **cell body,** or **perikaryon** *(per''ĭ-kar'e-on),* contains the nucleus and specialized organelles and microtubules. The **axon** is a cytoplasmic extension that conducts an impulse away from the cell body. The term **nerve fiber** refers to any process extending from the cell body of a neuron.

Neurons derive from ectoderm and are the basic structural and functional units of the nervous system. They are specialized to respond to physical and chemical stimuli, conduct nerve impulses, and perform other functions such as storing memory, thinking, and regulating the activity of other organs or glands. Of all the body's cells, neurons are probably the most specialized. As with muscle cells, the number of neurons is established shortly after birth, and thereafter they lack the ability to undergo mitosis, although under certain circumstances a severed portion can regenerate.

Neuroglia. In addition to neurons, nervous tissue is composed of neuroglia *(nu'rog'le-ah)* (fig. 6.30). Neuroglial cells are about five times as abundant as neurons and have limited mitotic abilities. They do not transmit impulses but support and bind neurons together. Certain neuroglial cells are phagocytic, and others assist in providing sustenance to the neurons.

1. Compare the structure, function, and location of neurons and neuroglia, which comprise nervous tissue.
2. List the structures of a neuron in the sequence that a nerve impulse would pass through the cell.

Clinical Considerations

As discussed in the beginning of this chapter, understanding histology is extremely important in determining organ and system structure and function. Histology has immense clinical importance as well. Many diseases are diagnosed through microscopic examination of tissue sections. Even in performing an autopsy, an examination of various tissues is vital for establishing the cause of death.

Several sciences are concerned with specific aspects of tissues. *Histopathology* is the study of diseased tissues. *Histochemistry* is concerned with the biochemical physiology of tissues as they function to maintain body homeostasis. *Histotechnology* studies the various ways tissues can be better stained and observed. In all of these disciplines, a thorough understanding of normal, or healthy, tissues is imperative for recognizing altered, or abnormal, tissues.

neuron: Gk. *neuron,* sinew or nerve
perikaryon: Gk. *peri,* around; *karyon,* nut or kernel

Changes in Tissue Composition

Most diseases alter tissue structure *locally* where the disease is prevalent, though some diseases, called *general conditions,* cause changes remote from the location of the disease. **Atrophy** (wasting of body tissue), for example, may be confined to a particular organ where the disease interferes with the metabolism of that organ; or it may involve an entire limb if nourishment or nerve impulses are decreased or prohibited. *Muscle atrophy,* for example, can be caused by a disease of the nervous system like polio or can be the result of a diminished blood supply to a muscle. *Senile atrophy* is the natural aging of tissues and organs within the body. *Disuse atrophy* is a local atrophy that results from the inactivity of a tissue or organ. Muscular dystrophy causes a disuse atrophy that decreases muscle size and strength due to the loss of sarcoplasm within the muscle.

Necrosis *(nĕ-kro'sis)* is cellular or tissue death within the living body. It can be recognized by changes in the dead tissues. Necrosis can be caused by a number of factors, such as severe injury, physical agents (trauma, heat, radiant energy, chemical poisons), or interference with the nutrition of tissues. When histologically examined, the necrotic tissue usually appears opaque, and a whitish or yellow color is assumed. **Gangrene** is a massive necrosis of tissue accompanied by an invasion of microorganisms that live on decaying tissues.

Somatic death is the death of the body as a whole. Following somatic death, tissues undergo irreversible changes such as **rigor mortis** (muscular rigidity), clotting of the blood, and cooling of the body. Postmortem changes occur under varying conditions at predictable rates of time, which help in estimating the approximate time of death.

Tissue Analysis

In diagnosing a disease, it is frequently important to examine tissues from a living person histologically. When this is necessary, a **biopsy,** the removal of a section of living tissue, is taken. There are several techniques for obtaining a biopsy. *Surgical removal* is usually done on large masses or tumors. *Curettage* involves cutting and scraping tissue, as may be done in examining for uterine cancer. In a *percutaneous needle biopsy,* a biopsy needle is inserted through a small skin incision, and tissue samples are aspirated. Both normal and diseased tissues are removed for purposes of comparison.

There are several steps to preparing tissues for examination. **Fixation** is fundamental for all histological preparation. It is the rapid killing, hardening, and preservation of tissue to maintain its existing structure. **Embedding** the tissue in a supporting medium such as paraffin wax usually follows fixation. The next step is **sectioning** the tissue into extremely thin slices, followed by **mounting** the specimen on a slide. Some tissues are fixed by rapid freezing and then sectioned while frozen, making embedding unnecessary. Frozen sections enable the pathologist to make a quick diagnosis during a surgical operation. These are done frequently, for example, in cases of suspected breast cancer. **Staining** is the next step. The hematoxylin and eosin (H & E) stains are routinely used on all tissue specimens. They give a differential blue and red color to the basic and acidic structures within the tissue. Other dyes may be needed to stain for specific structures.

Examination is first done with the unaided eye and then with a microscope. Practically all histological conditions can be diagnosed with low magnification (100×). Higher magnification is used to clarify specific details. Further examination may be performed with an electron microscope, which makes visible the cellular structures that are the morphological bases of metabolic processes. Histological observation is the foundation of subsequent diagnosis, prognosis, treatment, and reevaluation.

Tissue Transplantation

In the last two decades, medical science has made tremendous advancements in tissue transplants. Tissue transplants are necessary for replacing nonfunctional, damaged, or lost body parts. The most successful transplant is one where tissue is taken from one place on a person's body and moved to another place, such as a skin graft from the thigh to replace burned tissue of the hand. This type of transplant is termed an **autotransplant. Isotransplants** are transplants between genetically closely related persons. Identical twins have the best acceptance success in this type of transplant. **Homotransplants** (between individuals of the same species) and **heterotransplants** (between two different species) have low acceptance percentages because of a *tissue rejection reaction.* When this occurs, the recipient's immune mechanisms are triggered, and the donor's tissue is identified as foreign and is destroyed. The reaction can be minimized by "matching" recipient and donor tissue. Immunosuppressive drugs also may lessen the rejection rate. These drugs act by interfering with the recipient's immune mechanisms. Unfortunately, immunosuppressive drugs may lower the recipient's resistance to infections as well. New techniques involving blood transfusions from donor to recipient before transplant are proving successful. In any event, tissue transplants are an important aspect of medical research, and significant breakthroughs are on the horizon.

---

atrophy: Gk. *a,* without; *trophe,* nourishment
necrosis: Gk. *nekros,* corpse
gangrene: Gk. *gangraina,* gnaw or eat

Chapter Summary

I. Definition and Classification of Tissues
   A. Tissues are aggregations of similar cells that perform specific functions. The study of tissues is histology.
   B. Cells are separated and bound together by an intercellular matrix; its composition varies from solids to liquids.
   C. The four principal types of tissues are epithelial tissue, connective tissue, muscle tissue, and nervous tissue.

II. Development of Tissues
   A. Prenatal development following fertilization results in a blastocyst, which becomes implanted in the uterine wall between the fifth and seventh day.
   B. The three primary germ layers, from which all body tissues and organs eventually derive, form shortly after implantation.
   C. The various types of tissues are established during embryonic development, and as differentiation continues, organs form, each of which is composed of a specific arrangement of tissues.

III. Epithelial Tissue
   A. Epithelia derive from all three germ layers and may be one or several layers thick; the lower surface is supported by a basement membrane.
   B. Epithelial cells are bonded by glycoprotein deposits, desmosomes, or tight and intermediate junctions.
   C. Simple epithelia vary in shape and surface characteristics; they are located where diffusion, filtration, and secretion occur.
   D. Stratified epithelia consist of two or more layers of cells and are adaptive for protection.
   E. Transitional epithelium lines the urinary bladder and is adapted for distension.
   F. Glandular epithelia derive from developing epithelial tissue and function as secretory exocrine glands.

IV. Connective Tissue
   A. Connective tissues derive from mesenchymal mesoderm and, with the exception of cartilage, are highly vascular.
   B. Connective tissue proper contains fibroblasts, collagenous fibers, and elastic fibers within a flexible ground substance.
   C. Cartilage provides a flexible framework for many organs and consists of a semisolid matrix of chondrocytes and various fibers.
   D. Bone (osseous) tissue consists of osteocytes, collagenous fibers, and a durable matrix of mineral salts.
   E. Vascular (blood) tissue consists of formed cellular elements (erythrocytes, leukocytes, and thrombocytes) suspended in a fluid plasma matrix.

V. Muscle Tissue
   A. Muscle tissues (smooth, cardiac, and skeletal) are responsible for the movement of materials through the body, the movement of one part of the body with respect to another, and for locomotion.
   B. Fibers in the three kinds of muscle tissue are adapted to contract in response to stimuli.

VI. Nervous Tissue
   A. Neurons are the functional units of the nervous system; they respond to stimuli and conduct impulses to and from all body organs.
   B. Neuroglia support and bind neurons; some are phagocytic, and others provide sustenance to neurons.

Review Activities

Objective Questions

1. Which of the following is *not* a principal type of body tissue?
   (a) nervous
   (b) integumentary
   (c) connective
   (d) muscular
   (e) epithelial
2. The correct sequence of the prenatal stages of development is
   (a) ovum, zygote, blastocyst, morula
   (b) ovum, morula, blastocyst, zygote
   (c) ovum, zygote, morula, blastocyst
   (d) zygote, ovum, blastocyst, morula
3. Connective tissue, muscle, and the dermis of the skin derive from embryonic
   (a) mesoderm
   (b) endoderm
   (c) ectoderm
4. Which statement is *false* regarding epithelia?
   (a) They are derived from both ectoderm and endoderm.
   (b) They are strengthened by elastic and collagenous fibers.
   (c) One side is exposed to the lumen, cavity, or external environment.
   (d) They have very little extracellular matrix-binding cells.
5. A gastric ulcer of the stomach would involve
   (a) simple cuboidal epithelium
   (b) transitional epithelium
   (c) simple ciliated columnar epithelium
   (d) simple columnar epithelium
6. Which structural and secretory designation describes mammary glands?
   (a) acinar, apocrine
   (b) tubular, holocrine
   (c) tubular, merocrine
   (d) acinar, holocrine
7. Dense fibrous connective tissue is in
   (a) blood vessels
   (b) the spleen
   (c) tendons
   (d) the wall of the uterus
8. The phagocytic connective tissue found in the lymph nodes, liver, spleen, and bone marrow is
   (a) reticular
   (b) areolar
   (c) mesenchyme
   (d) elastic
9. Cartilage is slow in healing following an injury because
   (a) it is located in body areas that are under constant physical strain
   (b) it is avascular
   (c) its chondrocytes cannot reproduce
   (d) it has a semisolid matrix
10. Cardiac muscle tissue has
   (a) striations
   (b) intercalated discs
   (c) rhythmical involuntary contractions
   (d) all of the above

Essay Questions

1. Define *tissue*. What are the differences between cells, tissues, glands, and organs?
2. What physiological functions are epithelial tissues adapted to perform?
3. Identify the epithelial tissue found in each of the following structures or organs, and give the function of the tissue in each case.
   (a) in the alveoli of the lungs
   (b) lining the lumen of the digestive tract
   (c) in the outer layer of skin
   (d) in the urinary bladder
   (e) in the uterine tube
   (f) lining the lumina of the lower respiratory tract
4. Why are both keratinized and nonkeratinized epithelia found within the body?
5. Describe how epithelial glands are classified according to structural complexity and secretory function.
6. Identify the connective tissue found in each of the following structures or organs, and give the function of the tissue in each case.
   (a) on the surface of the heart and surrounding the kidneys
   (b) within the lumen of the aorta
   (c) forming the symphysis pubis
   (d) supporting the external ear
   (e) lymph node
   (f) Achilles tendon
7. Distinguish among the following: reticular fibers, collagen, elastin, fibroblasts, and mast cells.
8. What is the relationship between adipose cells and fat? Discuss the function of fat, and explain the potential danger of excessive fat.
9. Discuss the mitotic abilities of each of the four principal types of tissues.
10. Distinguish among the following terms: atrophy, necrosis, gangrene, and somatic death.

7

Homeostasis and Control Mechanisms

Outline

Concepts

The numerous regulatory mechanisms of the body can be understood in terms of a single, shared function: that of maintaining homeostasis, or a relative constancy of the internal environment. In order to share in this common function, the actions of effector organs are regulated by sensory information from the internal environment. The activity of effector organs is thus controlled by the very changes these organs help to produce—that is, by feedback control mechanisms.

The effectors of most negative feedback loops include the actions of nerves and hormones. In both neural and endocrine regulation, particular chemical regulators stimulate target cells by interacting with specific receptor proteins. The nervous and endocrine systems usually function together to maintain homeostasis.

Homeostasis and Feedback Control

The numerous regulatory mechanisms of the body can be understood in terms of a single, shared function: that of maintaining homeostasis, or a relative constancy of the internal environment. In order to share in this common function, the actions of effector organs are regulated by sensory information from the internal environment. The activity of effector organs is thus controlled by the very changes these organs help to produce—that is, by feedback control mechanisms.

Objective 1. Define homeostasis, and describe the importance of this concept in physiology and medicine.

Objective 2. Describe negative feedback and positive feedback loops.

Objective 3. Explain how feedback control mechanisms help to maintain homeostasis.

Over a century ago the French physiologist Claude Bernard observed that the *milieu intérieur* (internal environment) remains remarkably constant despite changing conditions in the external environment. In a book entitled *The Wisdom of the Body* (published in 1932), Walter Cannon coined the term **homeostasis** *(ho″me-o-sta′sis)* to describe this internal constancy. Cannon further suggested that mechanisms of physiological regulation exist for one purpose—the maintenance of internal constancy.

The concept of homeostasis has been of inestimable value in the study of anatomy and physiology because it allows diverse regulatory mechanisms to be understood in terms of their "why" as well as their "how." The concept of homeostasis also provides a major foundation for medical diagnostic procedures. When a particular measurement of the internal environment, such as blood measurements (table 7.1), deviates significantly from the normal range of values, it can be concluded that homeostasis is not maintained and the person is sick. A number of such measurements, combined with clinical observations, may allow the particular defective mechanism to be identified.

Negative Feedback Loops

To maintain internal constancy the body must have **sensors** that are able to detect deviations from a *set point.* The set point is analogous to the temperature set on a house thermostat. In a similar manner, there is a set point for body temperature (which is set upward when a fever is produced), blood glucose concentration, the tension on a tendon, and so on. When a sensor detects a deviation from a particular set point, it must relay this information to an **integrating center,** which usually receives information from many different sensors. The integrating center is often a particular region of the brain or spinal cord, but in some cases it can also be cells of endocrine glands. The relative strengths of different sensory inputs are weighed in the integrating center, and, in response, the integrating center either increases or decreases the activity of particular **effectors** (muscles and glands).

In response to sensory information about a deviation from a set point, therefore, effectors act to promote a reverse change in the internal environment. If the initial deviation was an increase in body temperature, for example, the effectors act to lower the temperature. If the initial deviation was a decrease in blood glucose, the effectors act to increase the blood glucose. Since the activity of the effectors is regulated by the effects they produce (by means of sensors acting on the integrating centers), and since this regulation is in a negative, or reverse, direction, this type of control system is known as a **negative feedback loop** (fig. 7.1).

It is important to realize that these negative feedback loops are continuous, ongoing processes. Thus, a particular nerve fiber that is part of an effector mechanism may always display some activity, and a particular hormone, which is part of another effector mechanism, may always be present in the blood. The nerve activity and hormone concentration may decrease in response to deviations of the internal environment in one direction (fig. 7.1), or they may increase in response to deviations in the opposite direction (fig. 7.2). Changes from the normal range in either direction are thus compensated by reverse changes in effector activity.

Table 7.1 Approximate normal ranges for measurements of some blood values

Measurement	Normal range	Measurement	Normal range
Arterial pH	7.35–7.43	Urea	12–35 mg/100 ml
Bicarbonate	21.3–28.5 mEq/L	Amino acids	3.3–5.1 mg/100 ml
Sodium	136–151 mEq/L	Protein	6.5–8.0 g/100 ml
Calcium	4.6–5.2 mEq/L	Total lipids	350–850 mg/100 ml
Oxygen content	17.2–22.0 ml/100 ml	Glucose	75–110 mg/100 ml

homeostasis: Gk. *homoios,* like; *stasis,* a standing

Figure 7.1. A rise in some factor of the internal environment (↑X) is detected by a sensor. Acting through an integrating center, this caused an effector to produce a change in the opposite direction (↓X). The initial deviation is thus reversed, completing a negative feedback loop (shown by dashed arrow and negative sign). The numbers indicate the sequence of changes.

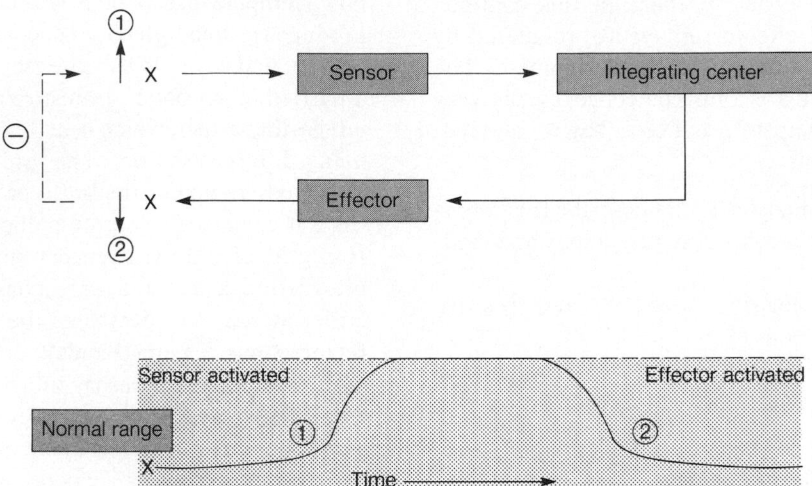

Figure 7.2. A negative feedback loop that compensates for a fall in some factor of the internal environment (↓X). Compare this figure with figure 7.1.

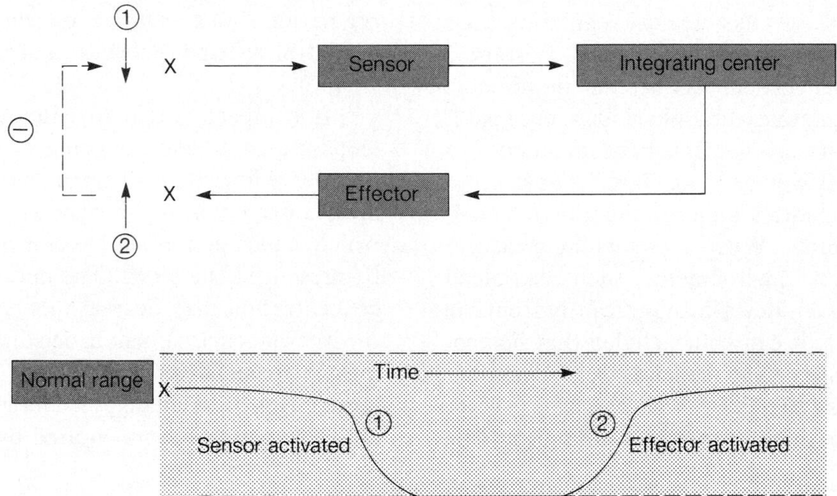

Homeostasis is best conceived as a state of **dynamic constancy,** rather than as a state of absolute constancy. The values of particular measurements of the internal environment fluctuate above and below the set point, which can be taken as the average value within the normal range of measurements (fig. 7.3). This state of dynamic constancy results from the greater or lesser activation of effectors in response to sensory feedback, and from the competing actions of antagonistic effectors.

Antagonistic Effectors. Most factors in the internal environment are controlled by several effectors, which often have antagonistic actions. Control by antagonistic effectors is sometimes described as "push-pull," where the increasing activity of one effector is accompanied by decreasing activity of an antagonistic effector. This affords a finer degree of control than could be achieved by simply switching one effector on and off. Normal body temperature, for example, is maintained at a set point of about 37°C by the antagonistic effects of sweating, shivering, and other mechanisms (fig. 7.4).

antagonistic: Gk. *anti,* against; *agonizomai,* to fight

Figure 7.3. Negative feedback loops (indicated by negative signs) maintain a state of dynamic constancy within the internal environment.

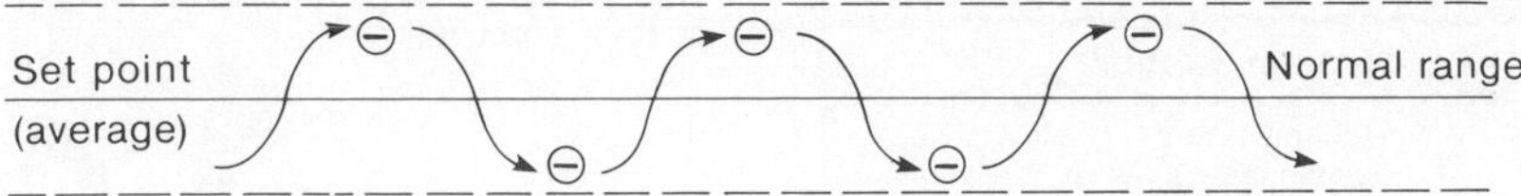

Figure 7.4. A simplified scheme by which body temperature is maintained within the normal range (with a set point of 37°C) by two antagonistic mechanisms—shivering and sweating. Shivering is induced when the body temperature falls too low and gradually subsides as the temperature rises. Sweating occurs when the body temperature is too high and diminishes as the temperature falls. Most aspects of the internal environment are regulated by the antagonistic actions of different effector mechanisms.

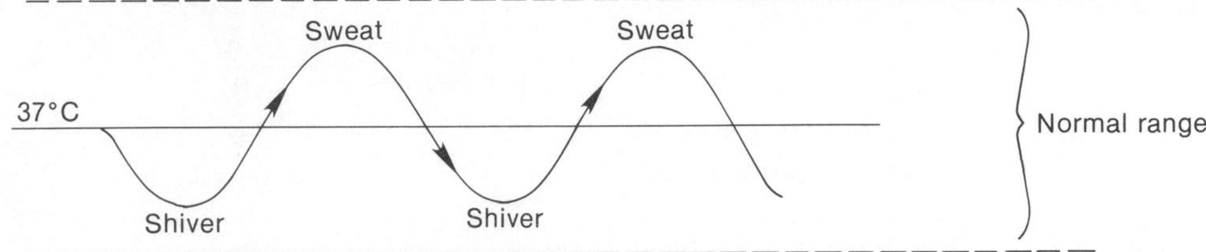

The blood concentrations of glucose, calcium, and other substances are regulated by negative feedback loops that involve hormones which promote opposite effects. While insulin, for example, lowers blood glucose, other hormones raise the blood glucose concentration. The heart rate, similarly, is controlled by nerve fibers that produce opposite effects: stimulation of one group of nerve fibers increases heart rate, and stimulation of another group slows the heart rate.

Positive Feedback

Constancy of the internal environment is maintained by effectors that act to compensate for the change that served as the stimulus for their activation; in short, by negative feedback loops. A thermostat, for example, maintains a constant temperature by increasing heat production when it is cold and decreasing heat production when it is warm. The opposite occurs during **positive feedback**—in this case, the action of effectors *amplifies* those changes that stimulated the effectors. A thermostat that works by positive feedback, for example, would increase heat production in response to a rise in temperature.

It is clear that homeostasis must ultimately be maintained by negative rather than by positive feedback mechanisms. The effectiveness of some negative feedback loops, however, is increased by positive feedback mechanisms that amplify the actions of a negative feedback response. Blood clotting, for example, occurs as a result of a sequential activation of clotting factors; the activation of one clotting factor results in activation of many in a positive feedback, avalanchelike manner. In this way, a single change is amplified to produce a blood clot. Formation of the clot, however, can prevent further loss of blood and thus represents the completion of a negative feedback loop.

1. Define homeostasis, and describe how this concept can be used to understand physiological control mechanisms, and explain its importance to medicine.
2. Describe the meaning of the term *negative feedback*, illustrate this concept by drawing a negative feedback loop, and explain how it contributes to homeostasis.
3. Explain how the actions of antagonistic effectors provide a greater control of the internal environment than could be afforded by a single negative feedback loop.
4. Define the term *positive feedback*, and explain how this control system can contribute to homeostasis.

Neural and Endocrine Regulation

The effectors of most negative feedback loops include the actions of nerves and hormones. In both neural and endocrine regulation, particular chemical regulators stimulate target cells by interacting with specific receptor proteins. The nervous and endocrine systems usually function together to maintain homeostasis.

Objective 4. Define the terms *endocrine gland, hormone,* and *neurotransmitter.*

Objective 5. Describe the nature and importance of receptor proteins.

Objective 6. Explain how the secretion of hormones is controlled by negative feedback inhibition.

Objective 7. Describe a negative feedback loop involving neural regulation, and identify the sensory, integrating center, and effector parts of this loop.

endocrine: Gk. *endon,* within; *krinein,* to separate

Table 7.2 Comparison of the basic functions of the endocrine and nervous systems

Endocrine system	Nervous system
Secretes hormones that are transported to target tissues via the blood	Transmits neurochemical impulses via nerve fibers
Causes changes in the metabolic activities in specific cells	Causes muscles to contract or glands to secrete
Exerts effects relatively slowly (seconds or even days)	Exerts effects relatively rapidly (milliseconds)
Has generally prolonged effects	Has generally brief effects

From Kent M. Van De Graaff, *Human Anatomy,* 2d ed. Copyright © 1988 Wm. C. Brown Publishers, Dubuque, Iowa. All Rights Reserved. Reprinted by permission.

Homeostasis is maintained by two general categories of regulatory mechanisms: (1) those that are **intrinsic,** or "built-in," to the organ; these are usually due to the effects of chemicals that act within the organs that produce them; and (2) those that are **extrinsic;** that is, regulation of an organ by the nervous and endocrine systems.

The endocrine system functions closely with the nervous system in regulating and integrating body processes and maintaining homeostasis (table 7.2). The nervous system controls the secretion of many endocrine glands, and some hormones in turn affect the function of the nervous system. Together, the nervous and endocrine systems regulate the activities of most of the other systems of body.

Regulation by the endocrine system is achieved by the secretion of chemical regulators called **hormones** into the blood. This type of secretion distinguishes endocrine glands from exocrine glands, which secrete their products into a duct leading to the outside of an epithelial membrane (fig. 7.5). Since hormones are secreted into the blood, the blood carries them to all organs in the body. Only specific organs can respond to a particular hormone, however; these are known as the *target organs* of that hormone.

Nerve fibers, or axons (see chapter 14) are said to *innervate* the organs that they regulate. When stimulated, these axons produce electrochemical nerve impulses that are conducted from the origin of the axon at the cell body to the end of the axon in the target organ innervated by the axon. These target organs can be muscles or glands, which may function as effectors in the maintenance of homeostasis.

Common Aspects of Neural and Endocrine Regulation

It may seem as if neural regulation is electrical and therefore distinct from the chemical nature of endocrine control systems. This idea is incorrect for two reasons. First, nerve fibers do not conduct electrons like copper wires; the nerve impulse is electrochemical, involving the diffusion of ions through the membrane of the axon. Second, most nerve fibers do not stimulate their target cells electrically. Instead, the nerve endings release chemicals known as **neurotransmitters** *(nu''ro-trans'mit-er).* Neurotransmitters differ from hormones in that they do not travel in the blood but instead diffuse only a very short distance from the nerve ending to the target cell (fig. 7.6). In other respects, however, the actions of neurotransmitters and hormones are very similar.

Figure 7.5. Comparison of (*a*) an endocrine gland with (*b*) an exocrine gland such as a sweat gland in the skin. An endocrine gland is a ductless gland that releases hormones into the blood, whereas exocrine glands secrete their products into ducts that lead to body surfaces.

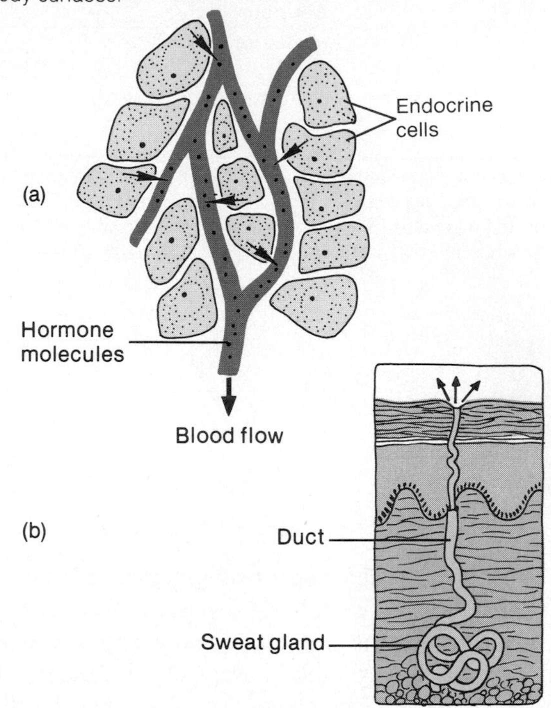

To respond to a particular neurotransmitter chemical or hormone, a target cell must have specific **receptor proteins** for that neurotransmitter or hormone. Cells that are not targets for the action of a particular hormone or neurotransmitter do not have these receptor proteins. The receptor proteins interact with the regulatory molecules in a very specific fashion, similar to the specificity seen in enzyme proteins (chapter 4) or carrier proteins (chapter 5). Receptor proteins are not enzymes, however; nor do they transport the regulator molecules into the target cell. Instead, receptor proteins bind the regulatory molecules with a high affinity and, through that binding, initiate specific changes in the target cell. These changes may involve changes in membrane permeability to specific ions, changes in enzyme activity, or changes in genetic expression.

hormone: Gk. *hormon,* to set in motion
exocrine: Gk. *exo,* outside; *krinein,* to separate

Figure 7.6. Nerve endings release chemical regulators called neurotransmitters, which diffuse across very short distances to the target cells. This differs from hormones, which generally travel long distances in the blood to reach their target cells.

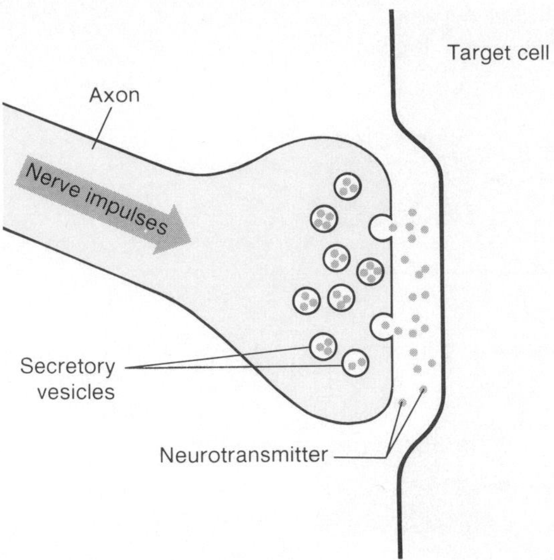

Figure 7.7. The regulation of hormone secretion by the hypothalamus, anterior pituitary, and target glands (thyroid, adrenal cortex, and gonads). Hormones secreted by the target glands exert negative feedback inhibition on the release of hormones from the hypothalamus and anterior pituitary.

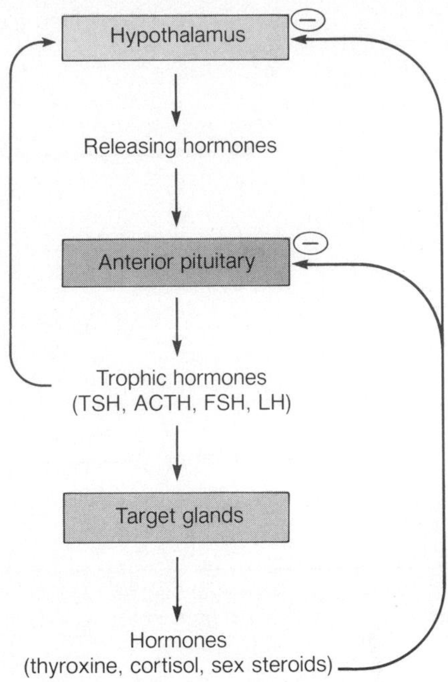

The receptor proteins for lipid-soluble hormones (steroid hormones and thyroid hormones) are located within the target cell. The receptors for other hormones and for neurotransmitters, however, are located in the cell membrane of the target cell. In these cases, the binding sites of the receptors face the extracellular fluid so that the regulator molecule can bind to the receptor without entering the target cell. This may cause changes in membrane permeability, or it may cause the target cell to produce some chemical within it that more directly exerts the regulatory effect of the hormone or neurotransmitter. These mechanisms are described in detail in chapter 14 (for neural regulation) and chapter 19 (for endocrine regulation).

Regardless of whether a particular chemical is acting as a neurotransmitter or as a hormone, in order for physiological regulation to be achieved there must be a mechanism to turn off the action of the regulator. Without an "off switch," physiological control is impossible. This process involves rapid removal or chemical inactivation of the regulatory molecules. This occurs largely at the site of the target cell and nerve ending, but it may also involve (in the case of hormones) metabolic inactivation by other organs, such as the liver, which remove the active hormone from the blood. As a result, continuous activation of effectors that are responsive to negative feedback control is required to maintain a given response, and homeostasis can be maintained.

Control of Hormone Secretion

Hormones are secreted in response to specific chemical stimuli. The presence of proteins in the stomach, for example, stimulates the secretion of gastrin, and a rise in plasma glucose concentration stimulates insulin secretion from the pancreatic islets (islets of Langerhans) in the pancreas. Hormones are also secreted in response to neurotransmitters released from nerve endings and to stimulation by other hormones.

The nervous system helps to regulate hormone secretion in a variety of ways. Autonomic nerves, for example, stimulate the secretion of epinephrine from the adrenal medulla and also participate in the regulation of insulin and glucagon secretion from the pancreatic islets in the pancreas. Hormonal secretion by the adrenal cortex, thyroid, and gonads is stimulated by hormones secreted by the anterior pituitary gland. Secretions of the anterior pituitary, in turn, are regulated by hormones released by neurons in the hypothalamus of the brain (chapters 16 and 18). This control is not one-way; many neural centers in the brain are affected by the actions of hormones (fig. 7.7).

islets of Langerhans: from Paul Langerhans, German anatomist, 1847–1888

Figure 7.8. The negative feedback control of insulin secretion by changes in the blood glucose concentration. Dashed arrows and negative signs indicate that negative feedback loops compensate the initial changes in blood glucose concentrations produced by eating or fasting.

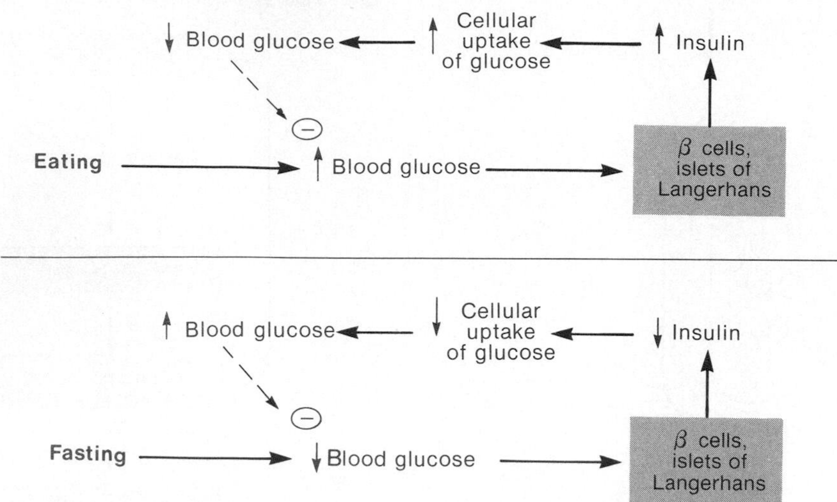

Figure 7.9. A schematic diagram of the flow of information from sensory neurons into the CNS and from motor neurons to effector organs.

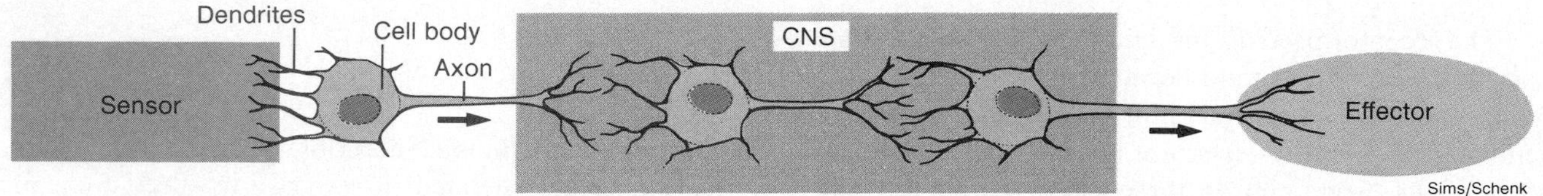

The secretion of hormones is controlled by inhibitory as well as by stimulatory influences. The effect of a given hormone's action can inhibit its own secretion. The secretion of insulin—which acts to lower the plasma glucose concentration—is stimulated by a rise in glucose concentration, for example, and is inhibited by a fall in blood glucose. The lowering of blood glucose levels by insulin thus has an inhibitory feedback effect on further insulin secretion. This closed loop control system, called **negative feedback inhibition** (see fig. 7.8), serves as a fine example of how negative feedback loops help to maintain homeostasis.

Neural Control of Homeostasis

The functional unit of the nervous system is the nerve cell, or neuron, which consists of dendrites, cell body, and axon (see chapter 14). As previously described, the axon innervates another cell; this other cell can be another neuron or an effector cell (muscle or gland). The functional connection between the axon and the other cell is called a **synapse** *(sin'aps)*. Neurotransmitters are released by the ending of the axon and diffuse across a tiny synaptic gap to reach the other (postsynaptic) cell (see fig. 7.6).

The brain and spinal cord are together known as the **central nervous system (CNS).** Axons and some entire neurons that are located outside the brain and spinal cord are part of the **peripheral nervous system (PNS).** A **nerve** is a collection of axons, packaged together in connective tissue, which is located in the PNS. A nerve might be thought of as a cable containing many wires (axons) that lead from or to the central junction box (CNS). Those axons that carry nerve impulses from sensory organs into the CNS are **sensory,** and those that carry impulses from the CNS to effector organs are **motor.** By means of these neural connections, effectors can respond appropriately to sensory information (fig. 7.9).

Electrical events in neurons are produced by the movements of ions—usually of Na^+ and K^+, but sometimes of Ca^{++} or Cl^-—across a limited region of the cell membrane of a neuron. As these charges diffuse down their concentration gradients, either entering or leaving the neuron, they produce electrical currents across the membrane. Such electrical events are conducted to the axon endings, where they stimulate the release of neurotransmitter chemicals at the synapse.

Figure 7.10. A schematic diagram showing how the homeostasis of blood pressure can be maintained by a negative feedback loop involving the nervous system.

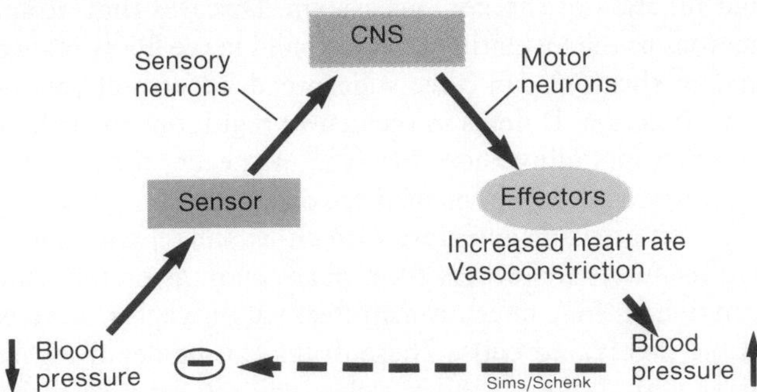

The neurotransmitters diffuse across a tiny gap to the membrane of the next cell, where receptor proteins for those transmitters are located. The bonding of the neurotransmitters to their receptor proteins in a synapse is analogous to the bonding of hormones to receptor proteins on the outer surface of target cell membranes. After the neurotransmitters have bonded to their receptor proteins, intermediate processes are stimulated that can result in the production of new electrical events. If the synapse is with an effector cell, the effector is activated, which can lead to muscle contraction or gland secretion.

Specificity of regulation is insured by the proper "wiring" (the location of synapses made by particular neurons) and by the fact that receptor proteins for neurotransmitters and hormones can combine with only specific molecules. Skeletal muscles, for example, can only be stimulated by one chemical—acetylcholine—which is released by specific motor neurons. Contractions of the heart, in contrast, are stimulated by epinephrine and norepinephrine. Both of these chemicals are released into the blood as hormones from the adrenal medulla gland, and norepinephrine is released as a neurotransmitter from the endings of some nerve fibers that innervate the heart. Through these mechanisms and the specific actions of sensors and effectors, the nervous system can mediate the negative feedback loops that help maintain homeostasis (fig. 7.10).

1. Distinguish between endocrine and exocrine glands, and describe the distinguishing characteristics of hormones.
2. Compare the properties and actions of hormones and neurotransmitters.
3. Describe the location and explain the functions of receptor proteins.
4. Explain how the hormonal secretions of the anterior pituitary are regulated by the brain and by negative feedback mechanisms
5. Explain how negative feedback inhibition controls the secretion of insulin.
6. Explain, in general terms, how the nervous system functions in the negative feedback control of blood pressure.

Clinical Considerations

Sampling the Internal Environment

Medical personnel obtain numerous measurements of conditions in the internal environment in order to compare these measurements to normal values. These normal values are obtained from large populations of people who are believed to be normal for the factors being tested. This circular definition of normal values can lead to some misinterpretation. (For example, an endurance-trained athlete in fine health may have a lower cardiac rate than the general population.) However, proper statistical analysis and frequent reevaluation of these measurements help to produce very useful reference values. When various measurements, obtained during physical examination and by laboratory tests of body tissues (usually of blood), differ from the established normal ranges, it is probable that homeostasis is not maintained. More important, the particular disease process that caused deviations from a homeostatic state can often be determined through such measurements of the internal environment.

Endocrine Regulation

Measurements of Hormone Levels in Blood. There are many endocrine disorders that are characterized by either abnormally high or low levels of hormone secretion. This may result from damage to the particular gland involved (as by a tumor) or to abnormal stimulation (or a lack of stimulation) of the gland. Since most hormones are normally present in the blood at extremely low concentrations (in the nanogram or picogram range per volume of blood sample), it has been difficult in the past to accurately measure the levels of many hormones. Today, however, even amounts as small as picograms can be accurately and easily measured using a technique known as **radioimmunoassay (RIA).** This technique, which is now a relatively routine clinical procedure, utilizes the specificity of antibodies and the sensitivity of radioactivity measurements to measure levels of hormones that would previously have been undetectable.

Clinical Uses of Hormone Receptors. Since receptor proteins bond with high specificity to particular hormones, the receptors can be used to measure extremely low concentrations of hormones in a technique analogous to radioimmunoassay, where antibodies are used. Some pregnancy tests, for example, utilize receptors for a hormone (human chorionic gonadotrophin, hCG) secreted by the developing embryo. In other applications the technique is reversed, and radioactively labeled hormones are used to detect the presence of receptor proteins in a tissue.

This technique may be used, for example, on a mammary tumor to determine if it is able to respond to estrogen (a female sex hormone); this information is useful in determining the proper course of treatment.

Hormone Receptors and Disease. There are some rare diseases that are known to be associated with hormone receptor proteins. In a disease called testicular feminization syndrome, for example, a male embryo lacks receptors for male sex hormones, even though it has normal and active testes. The baby, as a result, is born with female genitalia. Other diseases associated with hormone receptors include those in which antibodies are produced that bond to the particular receptors. These antibodies may mimic the normal hormone and cause stimulation of the target tissue (as in Graves' disease, in which the thyroid is stimulated), or the antibodies may prevent the normal action of a hormone on its receptors. The latter situation is seen in a rare form of diabetes mellitus, in which the action of insulin is blocked, and with some forms of allergy, in which the effectiveness of epinephrine is decreased.

Graves' disease: from Robert James Graves, Irish physician, 1796–1853

Neural Regulations

There is a tremendous number of health problems associated with congenital or acquired defects in the structure and function of the nervous system. Diseases that attack neurons or the insulating layers around nerve fibers (called myelin sheaths) can have widespread adverse effects on body function. Defects in the neural regulation of skeletal muscles, including those involved in speech, produce the most obvious symptoms of these disorders.

Certain drugs interfere with either the release of neurotransmitter molecules from nerve endings or the combination of these neurotransmitters with receptor proteins in the innervated cells. These drugs can be deadly when nerve stimulation of the muscles required for respiration is blocked. Some of these drugs, however, are put to beneficial uses by modern medicine. Curare, a deadly poison that South American Indians apply to darts, is commonly used in a controlled fashion to promote muscle relaxation in a hospital setting. Other drugs that act to enhance the effect of a neurotransmitter are used to strengthen neuromuscular transmission when this is therapeutically appropriate.

Chapter Summary

I. Homeostasis and Feedback Control
   A. Homeostasis refers to the dynamic constancy of the internal environment.
   B. Positive feedback loops serve to amplify changes and may be part of the action of an overall negative feedback mechanism.

II. Neural and Endocrine Regulation
   A. Neural and endocrine control share many similarities and complement each other.
   B. The secretion of hormones is controlled by the brain and by particular changes in the internal environment.
   C. Sensory neurons transmit impulses into the central nervous system, and motor neurons conduct impulses out of the central nervous system to an effector organ.

Review Activities

Objective Questions

1. Which of the following statements about homeostasis is *true*?
   (a) The internal environment is maintained absolutely constant.
   (b) Negative feedback mechanisms act to correct deviations from a normal state.
   (c) Homeostasis is maintained by switching regulator mechanisms on and off.
   (d) All of the above are true.
2. In a negative feedback loop the effector organ produces changes that are
   (a) similar in direction to that of the initial stimulus
   (b) opposite in direction to that of the initial stimulus
3. A hormone called *parathyroid hormone* acts to help raise the blood calcium concentration. According to the principles of negative feedback, an effective stimulus for parathyroid hormone secretion would be
   (a) a fall in blood calcium
   (b) a rise in blood calcium
4. The act of breathing raises the blood oxygen level, lowers the blood carbon dioxide concentration, and raises the blood pH. According to the principles of negative feedback, sensors that regulate breathing should respond to
   (a) a rise in blood oxygen
   (b) a rise in blood pH
   (c) a rise in blood carbon dioxide concentration
   (d) all of these
5. Which of the following statements about neural and endocrine regulation is *true*?
   (a) Hormonal stimulation generally lasts longer than neural stimulation.
   (b) Regulation by nerves is generally faster than regulation by hormones.
   (c) Neural and endocrine regulation are both achieved by the interaction of a chemical with a receptor protein.
   (d) All of the above are true.

Essay Questions

1. Explain the role of antagonistic negative feedback processes in the maintenance of homeostasis.
2. Compare and contrast, in a general way, regulation by the nervous and endocrine systems.
3. Explain, using examples, how the secretion of a hormone is controlled by the effects of that hormone's actions.
4. Describe the interaction of neural and endocrine regulation in the control of the anterior pituitary gland.

Reference Figures

The Human Organism

Figure 1.
Anterior view of the human torso with the superficial muscles exposed.

Figure 2.
The torso with the deep muscles exposed.

Figure 3.
The torso with the anterior abdominal wall removed to expose the abdominal viscera.

Figure 4.
The torso with the anterior thoracic wall removed to expose the thoracic viscera.

Figure 5.
The torso as viewed with the thoracic viscera sectioned in a coronal plane, and the abdominal viscera as viewed with most of the small intestine removed.

Figure 6.
The torso as viewed with the heart, liver, stomach, and portions of the small and large intestines removed.

Figure 7.
The torso with the anterior thoracic and abdominal walls removed, along with the viscera, to expose the posterior walls and body cavities.

Figure 1. Anterior view of the human torso with the superficial muscles exposed.

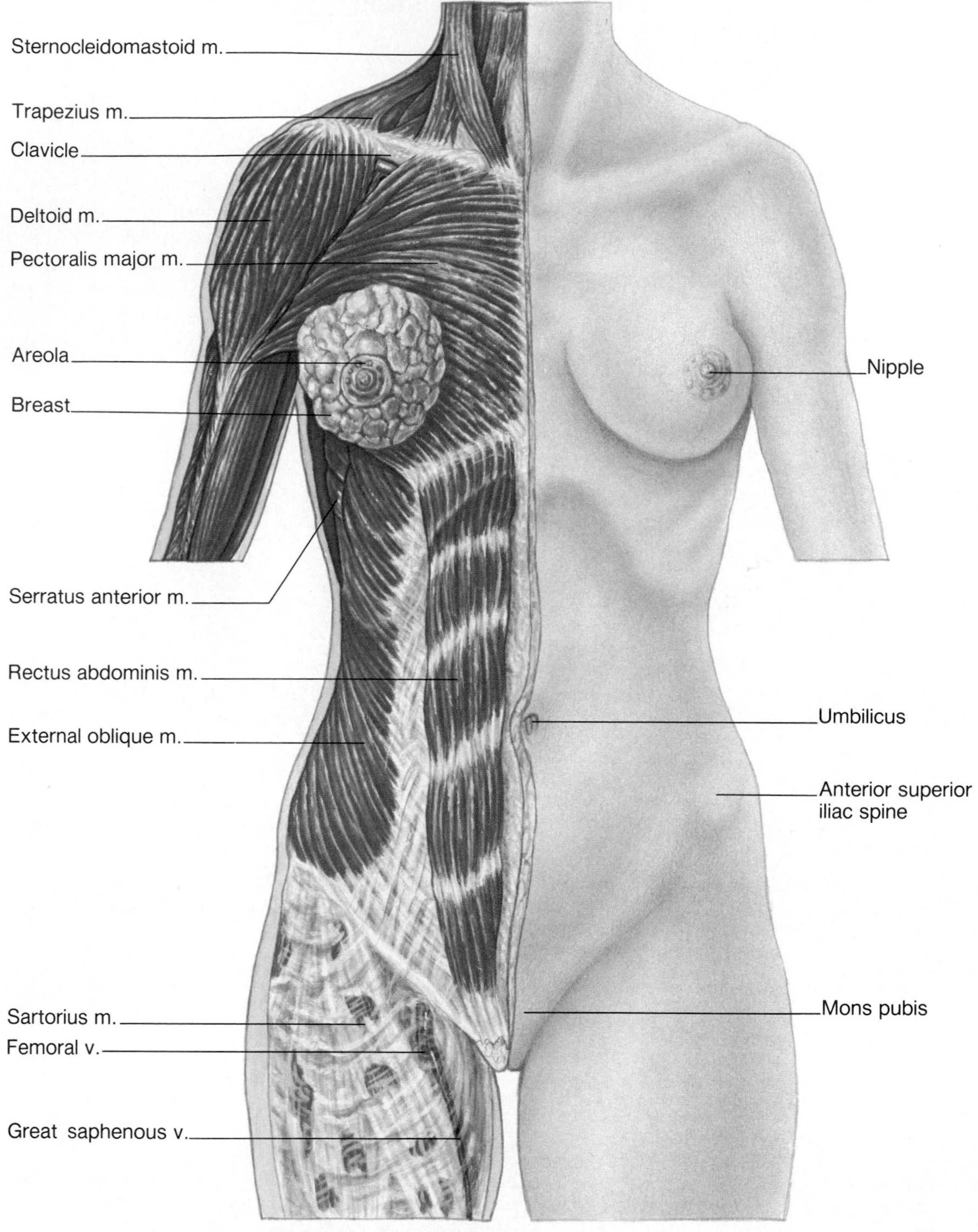

Figure 2. The torso with the deep muscles exposed.

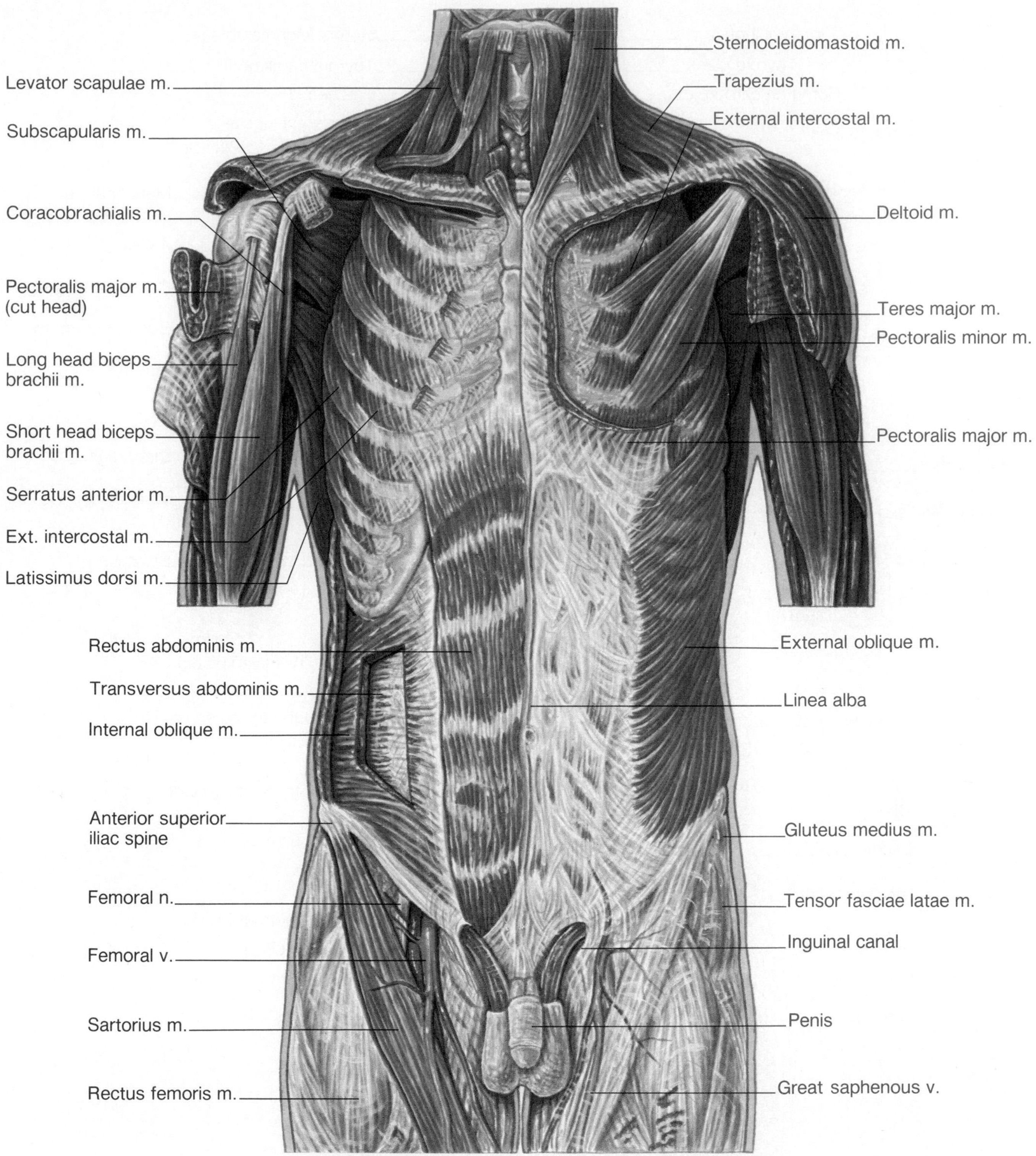
Sternocleidomastoid m.
Levator scapulae m.
Trapezius m.
External intercostal m.
Subscapularis m.
Coracobrachialis m.
Deltoid m.
Pectoralis major m.
(cut head)
Teres major m.
Pectoralis minor m.
Long head biceps
brachii m.
Short head biceps
brachii m.
Pectoralis major m.
Serratus anterior m.
Ext. intercostal m.
Latissimus dorsi m.
Rectus abdominis m.
External oblique m.
Transversus abdominis m.
Linea alba
Internal oblique m.
Anterior superior
iliac spine
Gluteus medius m.
Femoral n.
Tensor fasciae latae m.
Inguinal canal
Femoral v.
Sartorius m.
Penis
Rectus femoris m.
Great saphenous v.

Figure 3. The torso with the anterior abdominal wall removed to expose the abdominal viscera.

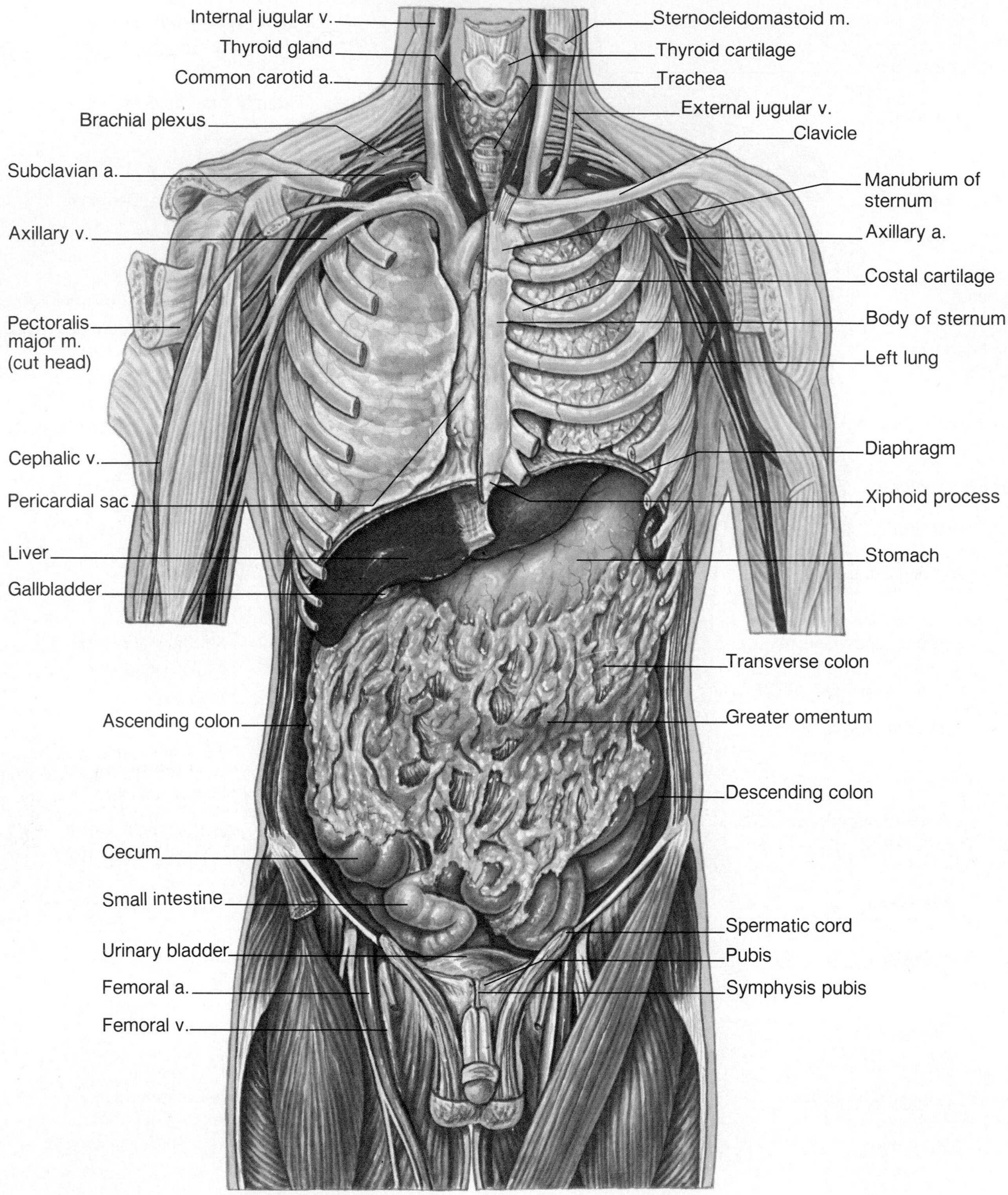

Figure 4. The torso with the anterior thoracic wall removed to expose the thoracic viscera.

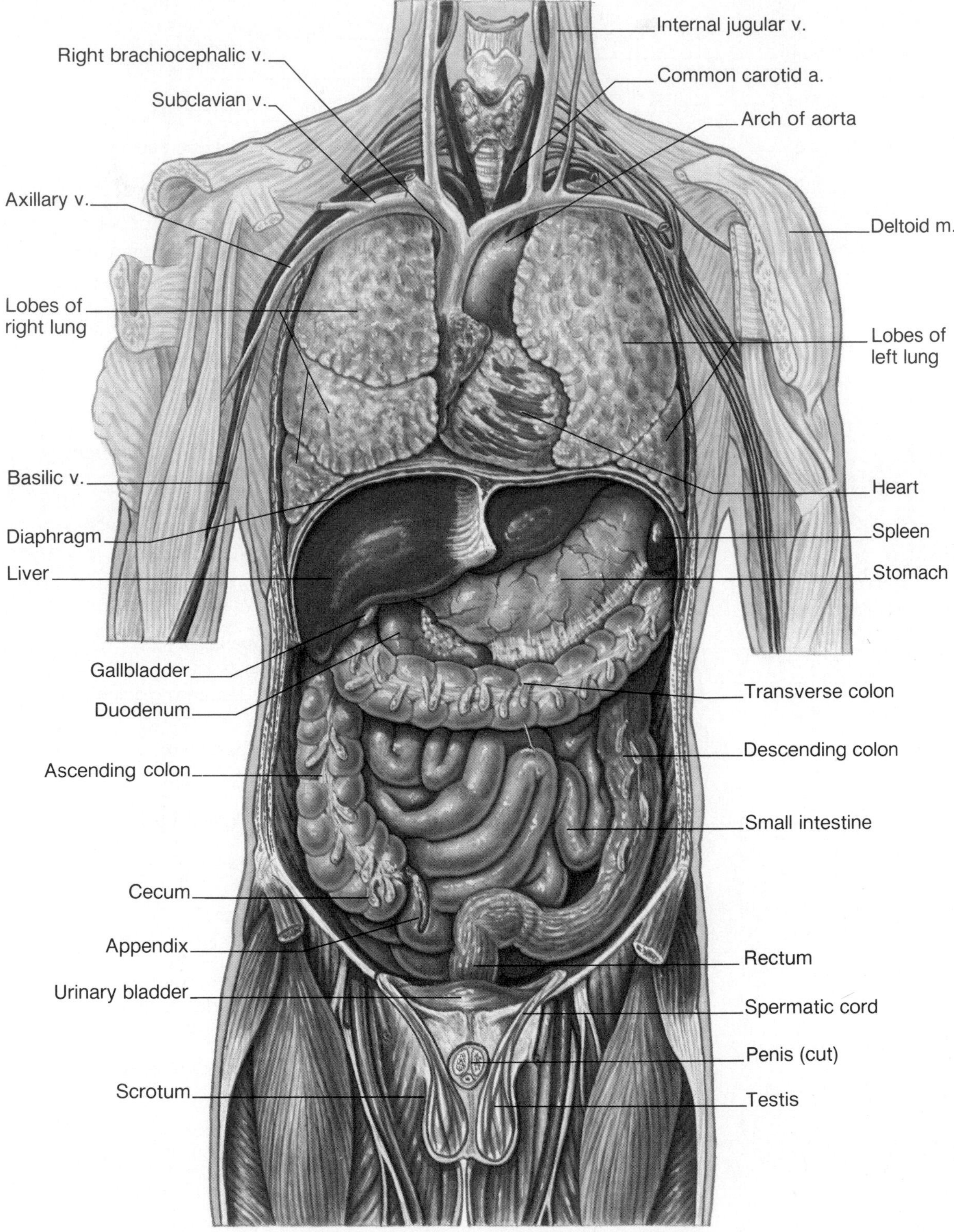

Figure 5. The torso as viewed with the thoracic viscera sectioned in a coronal plane, and the abdominal viscera as viewed with most of the small intestine removed.

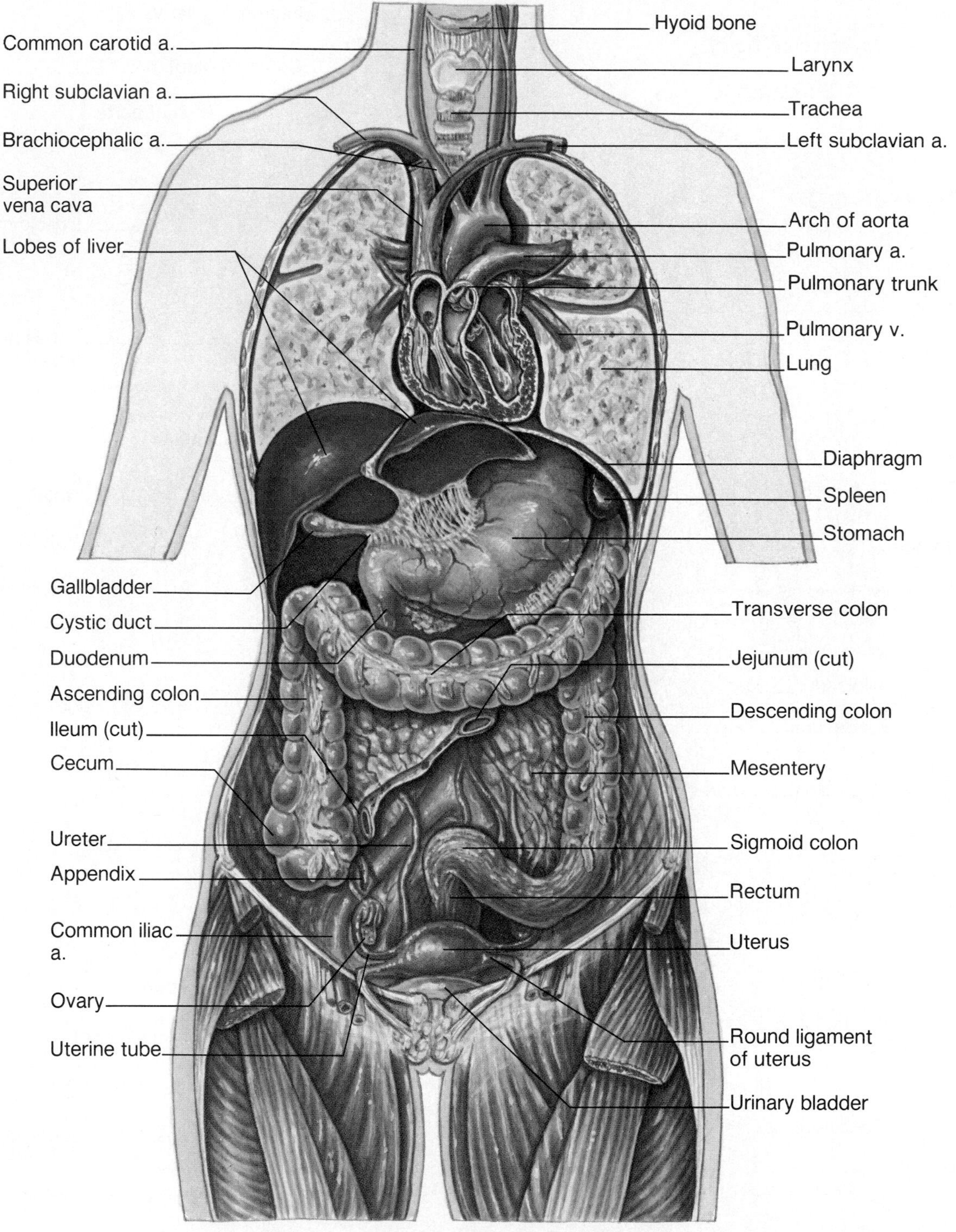

Figure 6. The torso as viewed with the heart, liver, stomach, and portions of the small and large intestines removed.

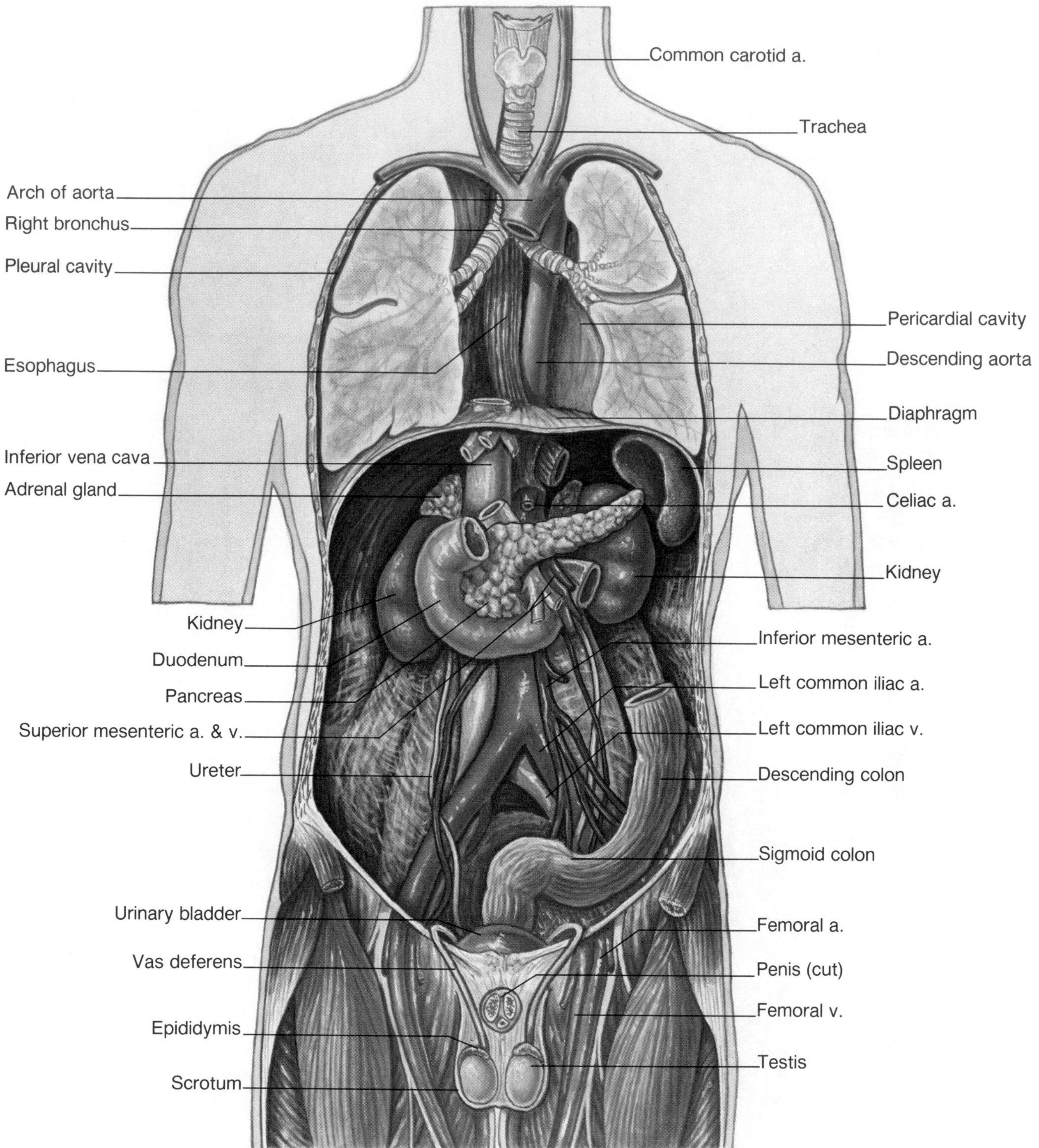

Figure 7. The torso with the anterior thoracic and abdominal walls removed, along with the viscera, to expose the posterior walls and body cavities.

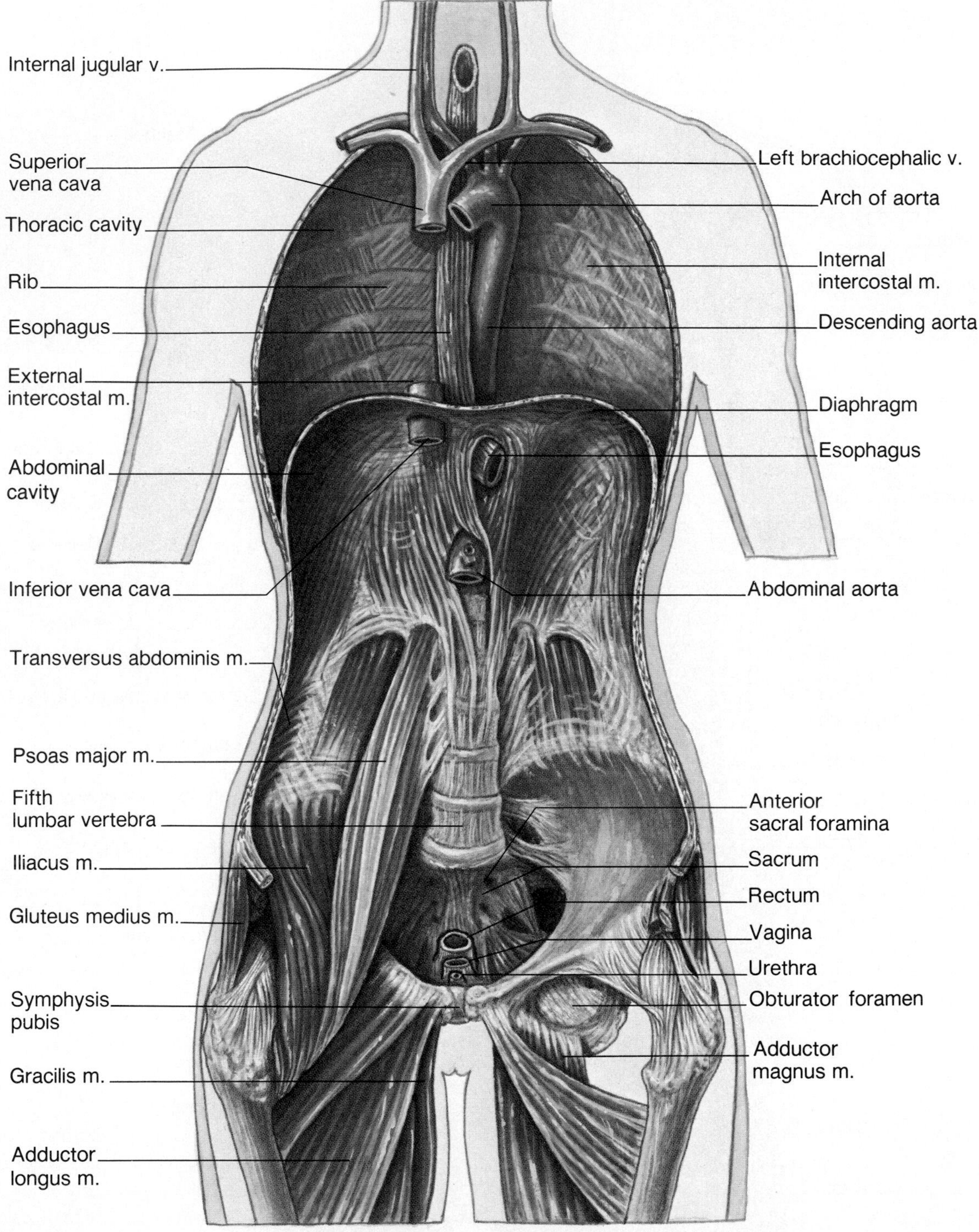

UNIT 2

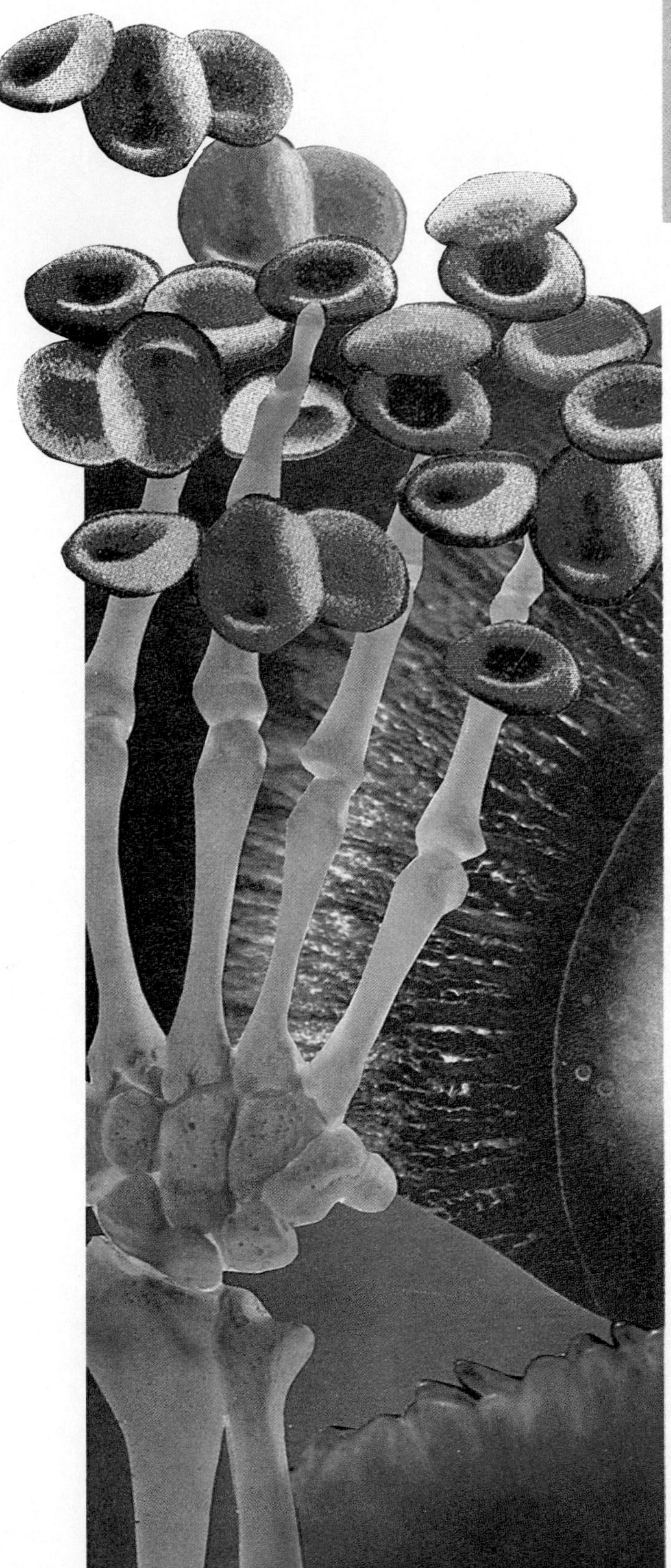

The chapters of unit 2 are concerned principally with those organ systems that provide support and movement. They describe how the integumentary and skeletal systems support and protect the body. The structural and functional relationships among the bones, articulations, and muscles are discussed, and the mechanisms of muscle contraction and the regulation of contraction are explained. The structures that can be seen or palpated on the body's surface are discussed. These chapters present the basic anatomical terminology of the integumentary, muscular, and skeletal systems and explain the roles of these systems in the maintenance of homeostasis.

Support and Movement of the Human Body

8 Integumentary System
The layered integument supports hair and glands, protects the body from pathogens and injury, and is important in maintaining homeostasis.

9 Skeletal System: The Axial Skeleton
The axial skeleton, consisting of the skull, vertebral column, and rib cage, forms the axis of the body and supports and protects the organs of the head, neck, and torso.

10 Skeletal System: The Appendicular Skeleton
The appendicular skeleton, consisting of the bones of the upper and lower extremities and the bony girdles, anchors the appendages to the axial skeleton and provides attachments for muscles.

11 Articulations
The articulations between the bones of the skeleton are classified as synarthroses (immovable joints), amphiarthroses (slightly movable), or diarthroses (freely movable).

12 Muscular System
The many skeletal muscles of the body are adapted to contract in response to electrochemical stimuli in order to carry out the functions of motion, heat production, posture, and bodily support.

13 Surface Anatomy and Body Topography
Surface anatomy is concerned with identifying body structures through observation and palpation; it has tremendous application in physical fitness and in medical diagnosis and treatment.

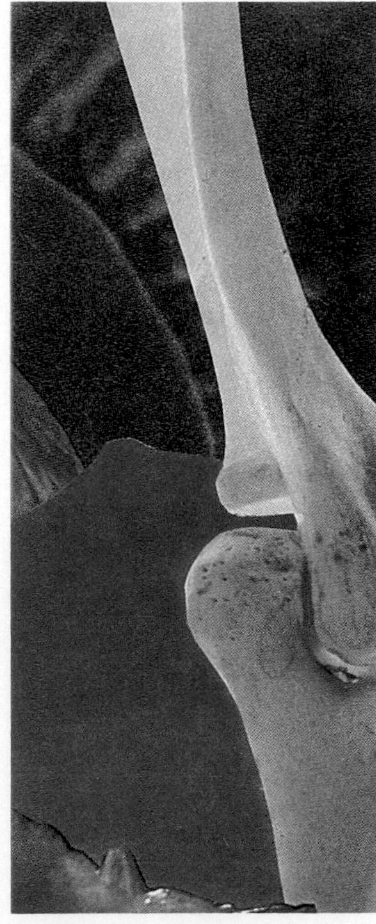

8 Integumentary System

Outline

Concepts

The integument is the largest organ of the body, and together with its epidermal structures (hair, glands, and nails), it constitutes the integumentary system. It has adaptive modifications in certain body areas that accommodate protective or metabolic functions. The integument is a dynamic interface between the continually changing external environment and the body's internal environment and aids in maintaining homeostasis.

All of the structures of the integumentary system are derived from ectodermal and mesodermal germ layers. Although integumentary formation is an ongoing process through prenatal development, the integument as a body covering is established early during the embryonic period.

The integument consists of three principal layers. The outer epidermis is stratified into four or five structural and functional layers, and the thick and deeper dermis consists of two layers. The hypodermis, a single layer, binds the skin to underlying tissues or organs.

The integument is a highly dynamic organ that not only protects the body from pathogens and external injury but is also extremely important for body homeostasis.

Hair, nails, and integumentary glands form from the epidermal layer and are therefore of ectodermal derivation. Hair and nails are structural features of the integument and have a limited functional role, whereas integumentary glands are extremely important in body defense and maintaining homeostasis.

The Integument as an Organ

The integument is the largest organ of the body, and together with its epidermal structures (hair, glands, and nails), it constitutes the integumentary system. It has adaptive modifications in certain body areas that accommodate protective or metabolic functions. The integument is a dynamic interface between the continually changing external environment and the body's internal environment and aids in maintaining homeostasis.

Objective 1. Explain why the integument is considered an organ and a component of the integumentary system.

Objective 2. Describe some common clinical conditions of the integument that result from nutritional deficiencies or body dysfunctions.

We are more aware of and concerned with our integumentary system than perhaps any other system of our body. One of our first actions in the morning is to examine ourselves in a mirror and assess how to make our appearance acceptable. Periodically, we examine our skin for wrinkles or gray hairs as signs of aging. We recognize other persons to a large degree by features of the skin.

Unfortunately, the appearance of the skin frequently determines first impressions and social acceptance. For example, social rejection during teenage years, imagined or real, can be directly associated with skin problems such as acne. One's image of oneself and consequent personality may be closely associated with physical appearance.

Even clothing styles are somewhat determined by how much skin we, or the designers, want to expose. But our skin is much more than a showpiece; it protects and regulates structures within the body.

The skin, or *integument (in-teg'u-ment),* and its associated structures (hair, glands, and nails) constitute the integumentary system. Included in this system are the millions of sensory receptors and the vascular network of the skin. This system accounts for approximately 7% of a person's body weight. The skin is a dynamic interface between the body and the external environment. It protects the body from the environment and at the same time allows communication with the environment.

The skin is considered an organ since it consists of several kinds of tissues that are structurally arranged to function together. It is the largest organ of the body, covering over 7,600 sq cm (3,000 sq in.) in the average adult. The skin is of variable thickness, averaging between 1.0 and 2.0 mm. It is thickest on the parts of the body exposed to wear and abrasion, such as the soles of the feet and palms of the hand, where it is about 6 mm. It is thinnest on the eyelids, external genitalia, and tympanum, where it is approximately 0.5 mm. Even the texture varies from the rough, callous skin covering the elbow and knuckle joints to the soft, sensitive areas of the eyelids, nipples, and genitalia.

The general appearance and condition of the skin are clinically important because they provide clues to certain body conditions or dysfunctions. Pale skin may indicate shock, whereas red, flushed, overwarm skin may indicate fever and infection. A rash may indicate allergies or local infections. Abnormal textures of the skin may be the result of glandular or nutritional problems (table 8.1). Even chewed fingernails may be a clue to emotional problems.

1. Explain why the skin is considered an organ and why the skin plus the integumentary derivatives is considered a system.
2. Which vitamins and minerals are important for healthy skin?
3. Describe the appearance of the skin that may accompany each of the following conditions: allergy; shock; infection; dry, stiff hair; hyperpigmentation; and general dermatitis.

integument: L. *integumentum,* a covering

Table 8.1 Variations of the skin and associated structures indicating nutritional deficiencies or body dysfunctions

Condition	Deficiency	Comments
General dermatitis	Zinc	Redness and itching
Scrotal or vulval dermatitis	Riboflavin	Inflammation in genital region
Hyperpigmentation	Vitamin B_{12}, folic acid, or starvation	Dark pigmentation occurs on backs of hands and feet
Dry, stiff, brittle hair	Protein, calories, and other nutrients	Usually in young children or infants
Follicular hyperkeratosis	Vitamin A, unsaturated fatty acids	Rough skin due to keratotic plugs from hair follicles
Pellagrous dermatitis	Niacin and tryptophan	Lesions develop on areas exposed to sun
Thickened skin at pressure points	Niacin	Noted at belt area at the hips
Spoon nails	Iron	Nails are thin and concave or spoon-shaped
Dry skin	Water or thyroid hormone	Dehydration, hypothyroidism, rough skin
Oily skin (acne)		Hyperactivity of sebaceous glands

Development of the Integumentary System

All of the structures of the integumentary system are derived from ectodermal and mesodermal germ layers. Although integumentary formation is an ongoing process through prenatal development, the integument as a body covering is established early during the embryonic period.

Objective 3. Describe the development of the ectodermal and mesodermal components of the integument.

The embryonic germ layers involved in forming the various body tissues were introduced in chapter 6 (see figs. 6.2, 6.3). Just as the germ layers contribute to tissue formation, they also contribute to the formation of body systems.

The skin is composed of three structural and functional layers (see figs. 8.5, 8.6). The **epidermis** *(ep″ĭ-der′mis)* is the thinnest, outermost layer and is derived from the embryonic *ectoderm.* The **dermis** *(der′mis)* underlies the epidermis, and the **hypodermis** (subcutaneous *[sub″ku-ta′ne-us]* layer) affixes the dermis of the skin to underlying tissues or organs. Both the dermis and hypodermis derive from embryonic *mesoderm.*

Development of the Layers of the Integument

By the fourth week following conception, the embryo is surrounded by a single-cell layer of ectoderm (fig. 8.1). Just beneath the ectoderm is a thicker layer of undifferentiated mesoderm called *mesenchyme (mez′en-kīm″).*

The ectodermal layer differentiates by six weeks into an outer flattened **periderm** and an inner, cuboidal **germinal (basal) layer** in contact with the mesenchyme. The periderm eventually sloughs off, forming the **vernix caseosa** *(ka″se-o′sah),* a cheeselike protective coat that covers the skin of the fetus. The germinal layer gives rise to the entire epidermis as well as all the glands, nails, hair, and hair follicles of the integument.

By eleven weeks, the mesenchymal cells below the germinal cells have differentiated into the distinct collagenous and elastic connective tissue fibers of the dermis. The tensile properties of these fibers cause a buckling of the epidermis and the formation of **dermal papillae** (fig. 8.1c). During the early fetal period (about ten weeks),

---

epidermis: Gk. *epi,* on; *derma,* skin
hypodermis: Gk. *hypo,* under; *derma,* skin
mesenchyme: Gk. *mesos,* middle; *enchyma,* infusion
periderm: Gk. *peri,* around; *derm,* skin
vernix caseosa: L. *vernix,* varnish; *caseus,* cheese

Figure 8.1. The development of the integument. (*a*) at four weeks only the surface ectoderm and mesenchyme of the mesoderm are apparent. (*b*) at seven weeks the surface ectoderm has differentiated into periderm and germinal layers. (*c*) by the eleventh week, the epidermis is stratified (layered) and the dermis is forming. (*d*) the integument at birth has distinct layers in both the epidermis and dermis. Note the protective vernix casseosa over the epidermis and the developing melanocytes within the deeper epidermal layers.

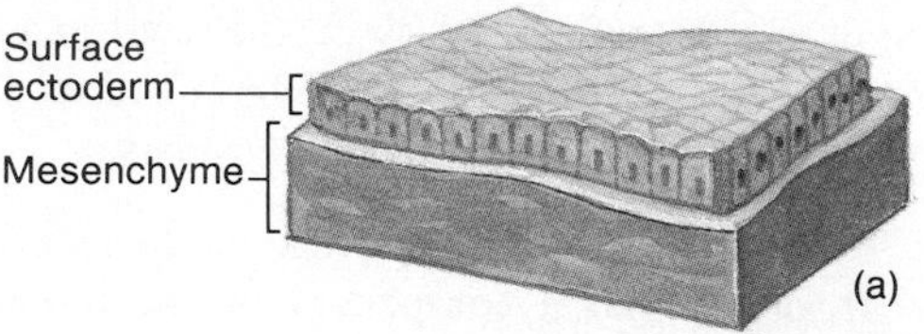

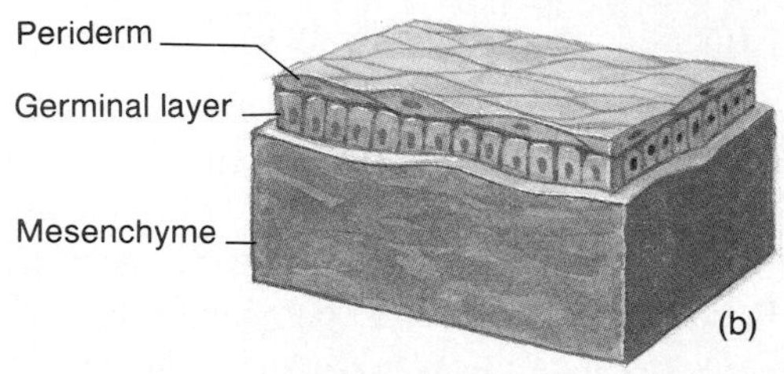

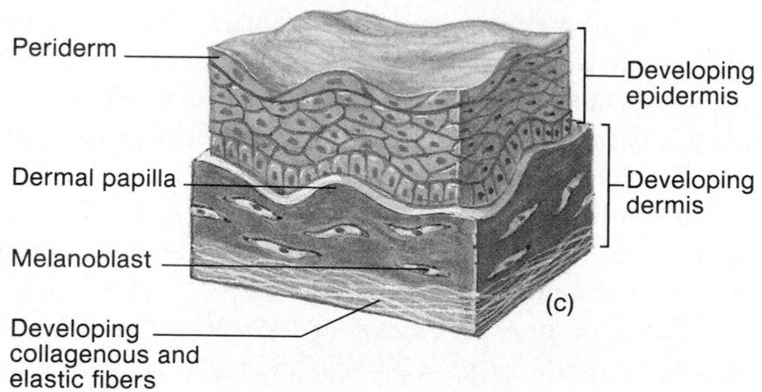

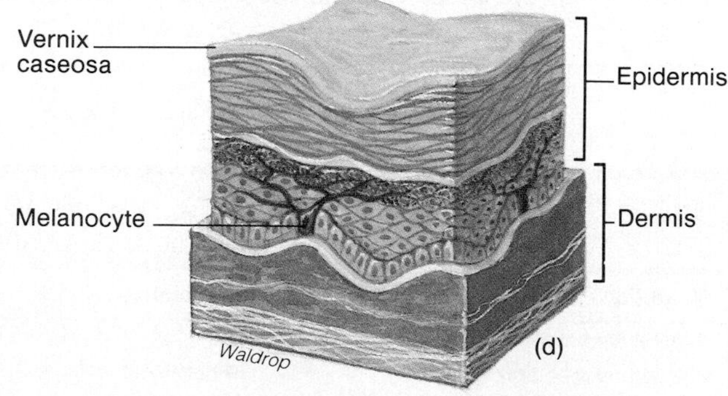

Figure 8.2. The development of hair and the integumentary glands. (*a*) twelve weeks. (*b*) fifteen weeks. (*c*) at birth.

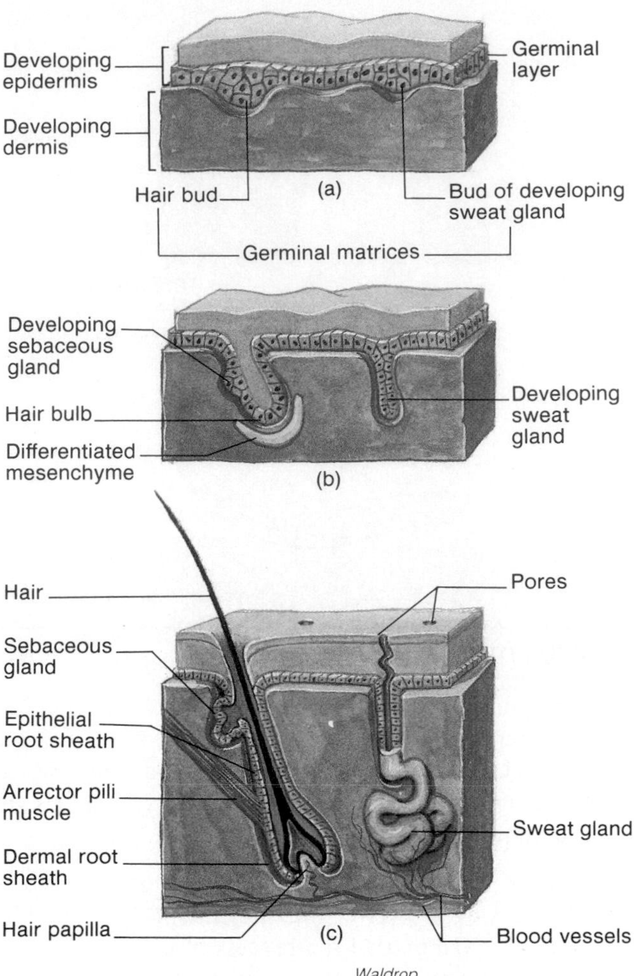

Figure 8.3. The development of mammary glands. (*a*) twelve weeks. (*b*) sixteen weeks. (*c*) about twenty-eight weeks.

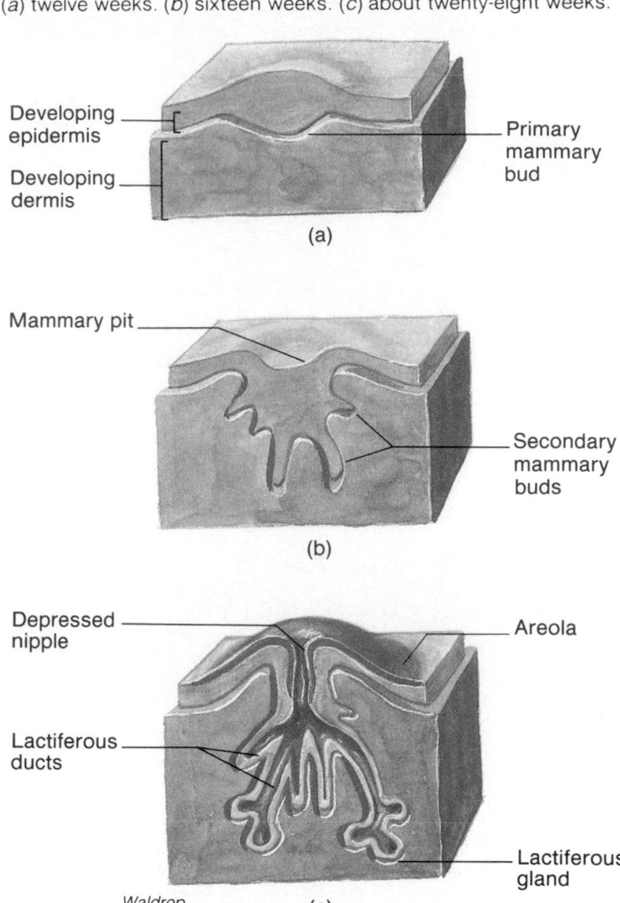

specialized *neural crest cells* called **melanoblasts** *(mel'ah-no-blastz'')* migrate into the developing dermis and differentiate into **melanocytes** *(mel'ah-no-sītz'')*. The melanocytes soon migrate to the germinal layer of the epidermis (fig. 8.1d), where they produce *melanin (mel'ah-nin)*, which colors the epidermis.

Development of the Associated Structures of the Integument

Hair, glands, and nails develop from the germinal layer of the epidermis and are therefore ectodermal in origin. Before hair can form, a hair follicle must be present. Each **hair follicle** *(fol'lĭ-k'l)* begins to develop at about twelve weeks as a mass of germinal cells, called a **hair bud,** proliferates into the underlying mesenchyme (fig. 8.2). As the hair bud becomes club-shaped, it is known as a **hair bulb.** The **hair follicle,** which physically supports and provides nourishment to the hair, is derived from specialized mesenchyme, called the **hair papilla** *(pah-pil'ah)*, which is localized around the hair bulb, and from the epithelial cells of the hair bulb, called the **germinal matrix.** The wall of the hair follicle, known as the **epithelial root sheath,** is supported by the specialized, mesenchymal **dermal root sheath.** Continuous mitotic activity in the epithelial cells of the hair bulb results in the growth of the hair.

Sebaceous *(sĕ-ba'shus)* **glands** and **sweat glands** are the two principal types of integumentary glands. Both develop from the germinal layer of the epidermis. Sebaceous glands develop as proliferations from the sides of the developing hair follicle (fig. 8.2). In a mature sebaceous gland, the oily secretions empty onto the shaft of a hair in a hair follicle. Sweat glands (fig. 8.2) become coiled as the secretory portion of the developing gland proliferates into the dermal mesenchyme.

Mammary glands are modified sweat glands, which develop, like sweat glands, as downward proliferations of the germinal epithelium *(ep''ĭ-the'le-um)* (fig. 8.3). The site of development is specific, however, along a **mammary ridge** (fig. 8.4). In humans, generally only one pair of pectoral (chest) mammary glands develops. As the mammary bud extends into the dermis, it branches (rather than coils as sweat glands) to form the secretory and ductal portions of the mammary glands (see fig. 8.3).

melanin: Gk. *melas*, black

Figure 8.4. The mammary ridge and accessory nipples. Mammary glands are positioned along a mammary ridge. Occasionally in humans, additional nipples (polythelia) develop elsewhere along the mammary ridge.

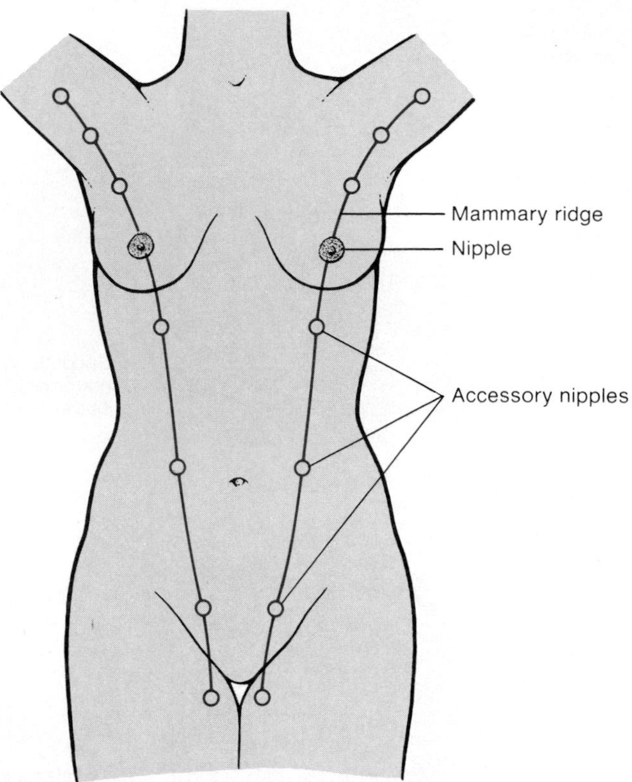

The mammary glands of newborns (males and females) are frequently swollen, and small amounts of secretion, called "witch's milk," may be discharged. Stimulation of the infant's mammary glands is caused by maternal hormones passing into the fetal circulation via the placenta.

Nails begin developing at about ten weeks along the dorsal aspect of each digit (fig. 8.5). The development of fingernails precedes that of toenails by about two weeks. The appearance of a thickened area of epidermis, called the **nail field,** is the first indication of nail development. The proximal and lateral borders of the nail field become thickened as the **nail folds.** Continued mitotic activity at the proximal nail fold causes a toughened **nail plate (nail)** to grow over the nail bed.

1. Discuss the role of the ectodermal germinal layer in the formation of the epidermis and associated structures of the skin, as well as the role of the mesodermal mesenchymal layer in the formation of the dermis and hypodermis of the skin.
2. Describe the sequence involved in establishing melanocytes in the germinal layer of the epidermis.

Figure 8.5. The development of nails. (*a*) 10 weeks. (*b*) about 14 weeks. (*c*) sagittal view of distal digit and nail at birth.

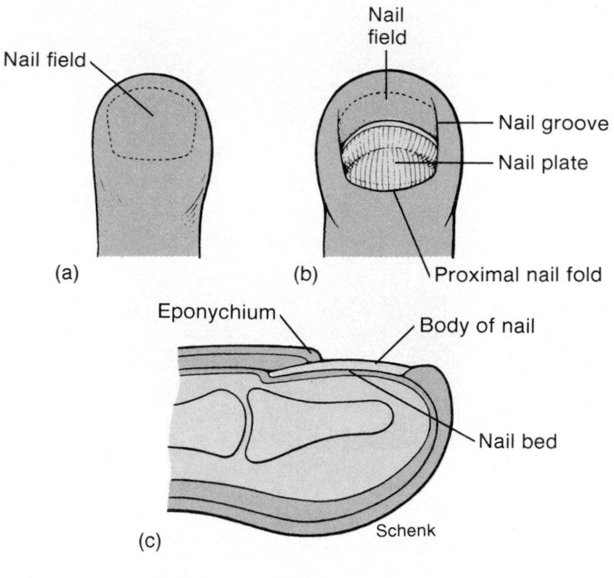

Layers of the Integument

The integument consists of three principal layers. The outer epidermis is stratified into four or five structural and functional layers, and the thick and deeper dermis consists of two layers. The hypodermis, a single layer, binds the skin to underlying tissues or organs.

Objective 4. Describe the histological characteristics of each layer of the integument.

Objective 5. Summarize the process of cellular replacement within the integument, and explain the transitional events that occur within each of the epidermal layers.

Epidermis

The **epidermis** is the superficial protective layer of the skin and is composed of stratified squamous epithelium. Over most of the body, this layer is between thirty and fifty cell layers thick, or 0.007–0.12 mm. All but the deepest layers of the epidermis are dead cells that contain a protein called **keratin** *(ker'ah-tin),* which toughens and waterproofs the skin. The epidermis is composed of either four or five layers, depending on its location within the body (figs. 8.6, 8.7). The epidermis of the palms and soles has five layers because these areas are exposed to greater friction. The epidermis of all other areas of the body has only four layers. The names and characteristics of the epidermal layer are as follows.

1. **Stratum basale.** The stratum basale is composed of a single layer of cuboidal cells in contact with the dermis. These cells are constantly dividing

stratum: L. *stratum,* something spread out
basale: Gk. *basis,* base

Figure 8.6. A diagram of the skin from the back of the hand.

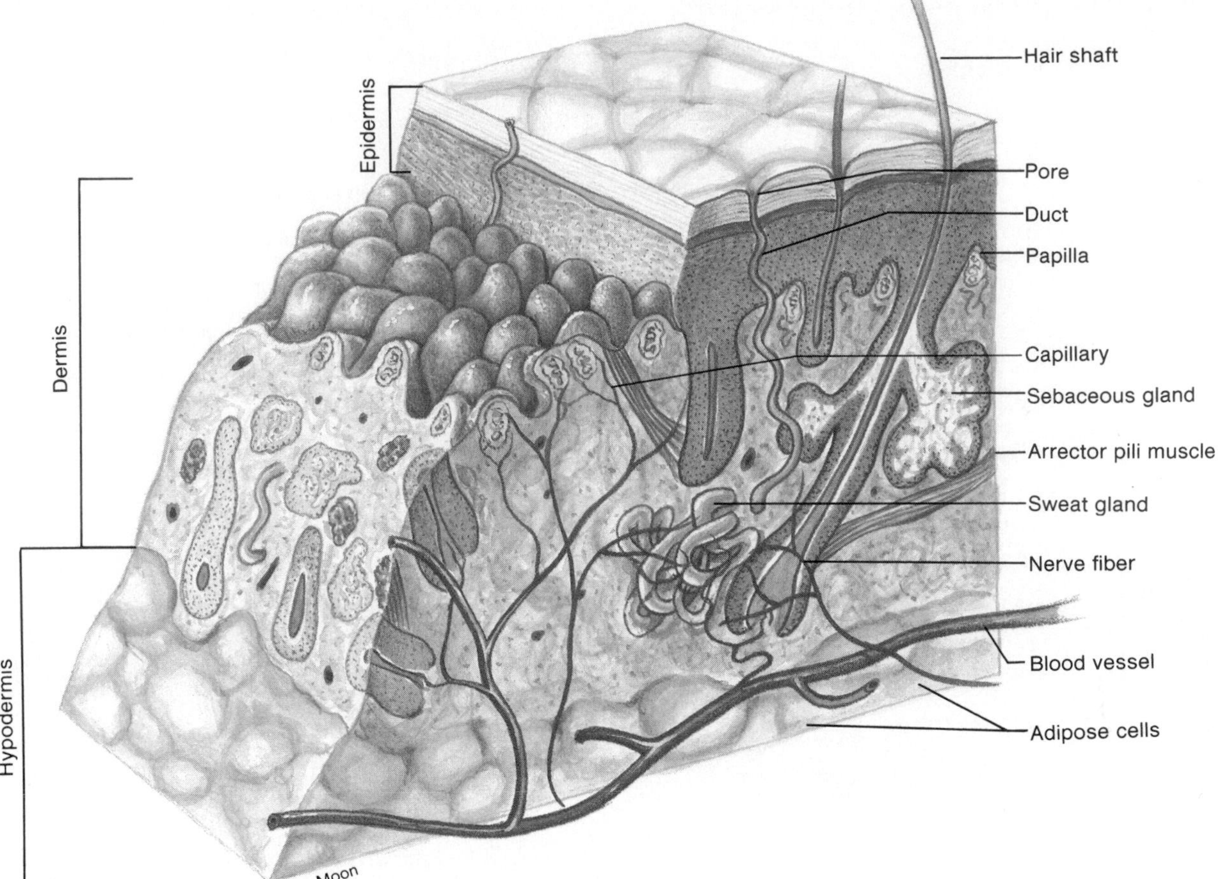

mitotically and moving outward to renew the epidermis. As the cells push away from the dermis and move away from the vascular nutrient and oxygen supply, their nuclei degenerate and the cells die. In approximately six to eight weeks, the cells are sloughed off from the top layer of the epidermis.

2. **Stratum spinosum** (stratum malpighii). The stratum spinosum contains several stratified layers of polygonal (many-sided) cells, which are tightly attached by spinelike projections. Since there is limited mitosis in this layer as well as in the stratum basale, these two layers are collectively referred to as the **stratum germinativum.**
3. **Stratum granulosum.** The cells of the stratum granulosum are flattened and contain dark-staining granules (the source of this layer's name). The process of *keratinization (ker''ah-tin''i-za'shun),* which is associated with cellular death, is initiated in this layer.
4. **Stratum lucidum.** The nuclei, organelles, and cell membranes are no longer visible in the cells of the stratum lucidum, and so histologically this layer appears clear. This layer exists only in the thickened skin of the soles and palms.
5. **Stratum corneum** *(kor'ne-um).* The stratum corneum is composed of twenty-five to thirty layers of flattened, scalelike cells, which are continuously shed as flakelike residues of cells. This surface layer is cornified and is the real protective layer of the skin. *Cornification* is brought on by keratinization, and the hardening, flattening process takes place as the cells migrate to the surface.

Friction at the surface of the skin stimulates additional mitotic activity of the stratum germinativum, resulting in the formation of a *callus* for additional protection. Table 8.2 describes the specific characteristics of each epidermal layer.

Tattooing colors the skin permanently because pigmented dyes are injected below the mitotic germinative layer into the dermis. Because of frequent nonsterile conditions, those who administer the tatoo must take care not to introduce infections along with the dye.

spinosum: L. *spina,* thorn
germinativum: L. *germinare,* sprout or growth
granulosum: L. *granum,* grain
lucidum: L. *lucidus,* light

corneum: L. *corneus,* horny

Figure 8.7. The layers of the epidermis. (*a*) an illustration showing the relative thickness of all of the epidermal layers. (*b*) a photomicrograph of the epidermis from the sole of the foot (160×).

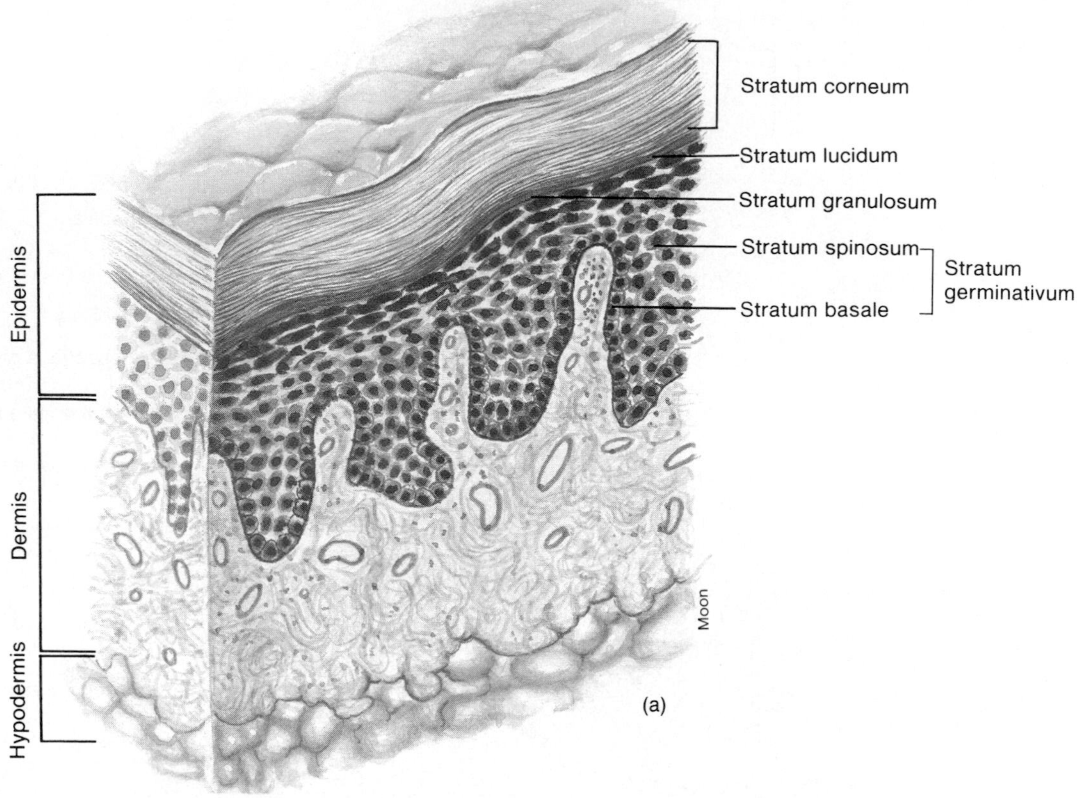

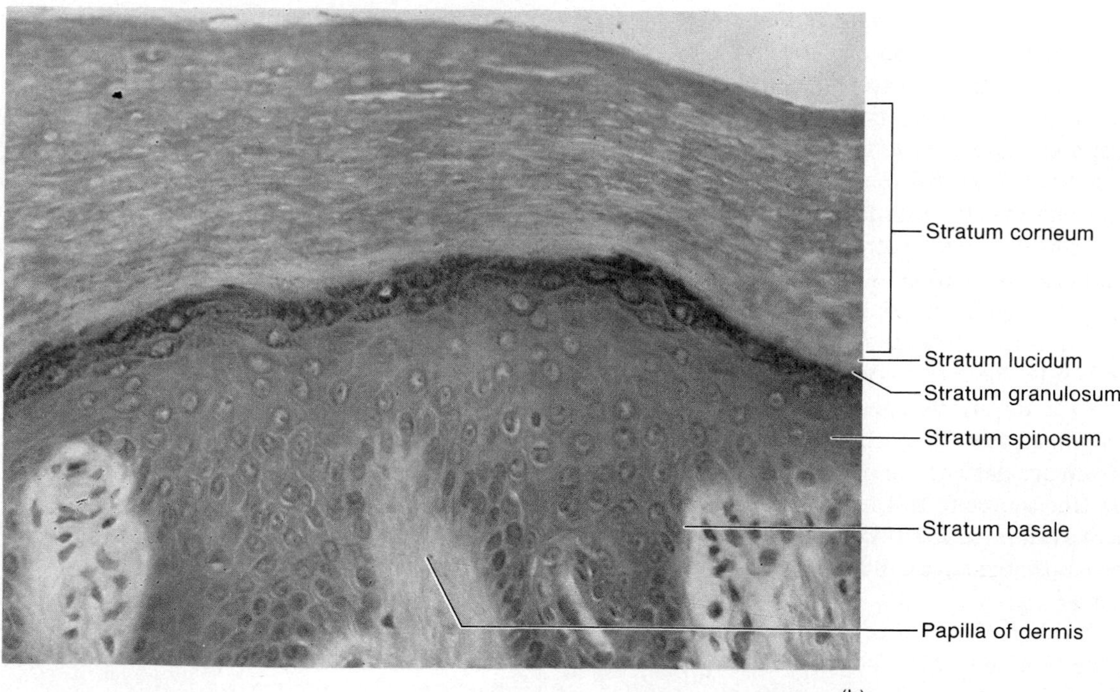

Table 8.2 Layers of the epidermis		
Layer	**Location**	**Characteristics**
Stratum basale	Deepest layer	Consists of a single layer of cuboidal cells that undergo mitosis; contains pigment-producing melanocytes
Stratum spinosum	Above the stratum basale	Composed of several layers of cells with centrally located, large, oval nuclei and spinelike processes; limited mitosis
Stratum granulosum	Above the stratum spinosum	Composed of one or more layers of granular cells that contain fibers of keratin and shriveled nuclei
Stratum lucidum	Above the stratum granulosum	A thin, clear layer found only in the epidermis of the palms and soles
Stratum corneum (horny layer)	Outermost layer	Consists of many layers of keratinized, dead cells that are flattened and nonnucleated; cornified

Figure 8.8. Melanocytes throughout the strata basale and spinosum (see arrow) produce melanin.

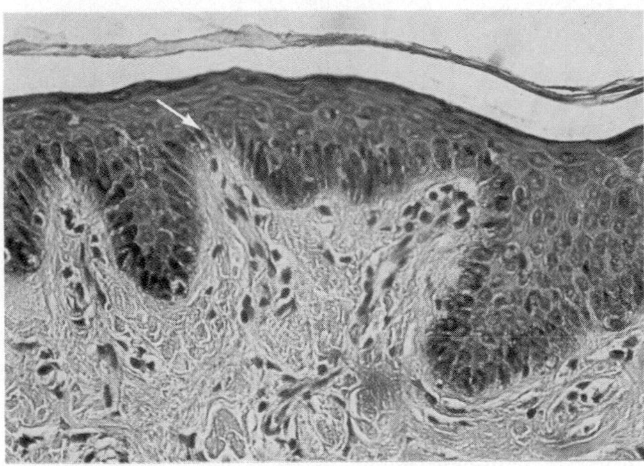

Coloration of the Skin. Normal skin color is caused by the expression of a combination of three pigments: **melanin** *(mel'ah-nin),* **carotene,** and **hemoglobin** *(he''mo-glo'bin).* Melanin is a brown-black pigment formed in the pigment cells, called **melanocytes.** Melanocytes are found throughout the strata basale and spinosum (fig. 8.8). All races have virtually the same number of melanocytes, but the amount of melanin produced and the degree of granular aggregation of the melanin determine whether an individual's skin color is black, brown, red, tan, or white. Melanin is a protective device that guards against the damaging effect of the ultraviolet rays in sunlight. A gradual exposure to the sunlight promotes the increased production of melanin within the melanocytes and hence tanning of the skin. The skin of a genetically determined albino has the normal complement of melanocytes in the epidermis but lacks the enzyme tyrosinase, which converts the amino acid tyrosine to melanin.

carotene: L. *carota,* carrot (referring to orange coloration)
hemoglobin: Gk. *haima,* blood; *globus,* globe

There are other genetical expressions of melanocytes that are more common than albinism. *Freckles,* for example, are caused by aggregated patches of melanin. A lack of melanocytes in localized areas of the skin causes distinct white spots in the condition called *vitiligo (vit-i-li'gō).*

Excessive exposure to the sunlight can cause *basal cell carcinoma,* or *malignant melanoma.* While in sunlight, one's skin absorbs two wavelengths of ultraviolet rays known as UV-A and UV-B. The DNA within the basal skin cells may be damaged as the sun's more dangerous UV-B rays penetrate the skin. Although it was once believed that UV-A rays were harmless, recent findings indicate that excessive exposure to UV-A rays may inhibit the DNA repair process that follows exposure to UV-B. Therefore, individuals who are exposed solely to UV-A rays in tanning salons are still in danger of basal cell carcinoma since they will later be exposed to UV-B rays of sunlight when they are out-of-doors.

Carotene is a yellowish pigment found in the epidermal cells and fatty parts of the dermis. Carotene is abundant in the skin of Oriental people and, together with melanin, accounts for their yellow-tan skin.

Hemoglobin is not a pigment of the skin but the oxygen-binding pigment found in red blood cells. Oxygenated blood flowing through the dermis gives the skin its pinkish tones.

Certain physical conditions or diseases cause symptomatic discoloration of the skin. Cyanosis is a bluish discoloration of the skin that appears in people with certain cardiovascular or respiratory diseases. People also become cyanotic during an interruption of breathing. In jaundice, the skin appears yellowish because of an excess of bile pigment in the bloodstream. Jaundice is usually symptomatic of liver dysfunction and sometimes of liver immaturity, as in a jaundiced newborn.

cyanosis: Gk. *kyanosis,* dark-blue color
jaundice: L. *galbus,* yellow

Figure 8.9. Basic dermatoglyphic patterns in the digits: (*a*) arch, (*b*) whorl, (*c*) loop, and (*d*) combination. Individual variations are commonly understood, but it is neither known why basic differences exist between men and women nor why the prints of mentally handicapped persons consistently deviate from normal patterns.

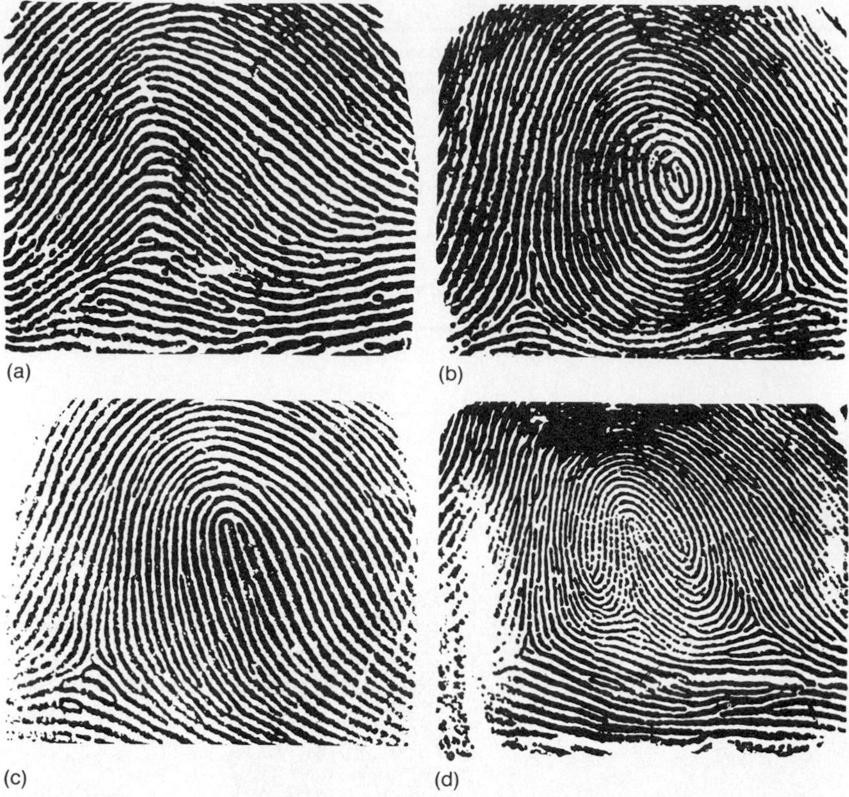

Surface Patterns. The exposed surface of the skin has recognizable patterns that are either congenital or acquired. Congenital patterns called fingerprints, or **friction ridges,** are present on the finger and toe pads as well as on the palms and soles. The designs formed by these lines have basic similarities but are not identical in any two individuals (fig. 8.9). They are formed by the pull of elastic fibers within the dermis and are well established prenatally. As the name implies, friction ridges function to prevent slippage when grasping objects. Because they are precise and easy to reproduce, fingerprints are customarily used for identifying individuals.

Acquired lines include the deep **flexion** *(flek'shun)* **creases** on the palms and the shallow **flexion lines** that can be seen on the knuckles and on the surface of other joints. Furrows on the forehead and face are acquired from continual contraction of facial muscles, such as from smiling or squinting in bright light or against the wind. Facial lines become more strongly delineated as a person ages.

> The science known as dermatoglyphics is concerned with the classification and identification of fingerprints. As seen in figure 8.9, there are four basic dermatoglyphic patterns. Every individual's prints are unique, including those of identical twins. Fingerprints, however, are not exclusive to humans. All other primates have fingerprints, and even dogs have a characteristic "nose print" that is used for identification in the military canine corps and in certain dog kennels.

Dermis

The **dermis,** or *corium (kor'e-um),* is deeper and thicker than the epidermis (see fig. 8.6). Vessels within the dermis nourish the living portion of the epidermis, and numerous collagenous, elastic, and reticular fibers give support to the skin. The fibers within the dermis radiate in definite directions and are responsible for producing lines of tension on the surface of the skin (fig. 8.10). Elastic fibers are more superficial and provide skin tone. There are considerably more elastic fibers in the dermis of a young person than in an elderly one. The decreasing amount of elastic fiber is apparently directly associated with aging. The dermis is highly vascular and glandular and contains many nerve endings and hair follicles.

Layers of the Dermis. The dermis is composed of two layers. The upper layer, called the **stratum papillarosum** (papillary layer), is in contact with the epidermis and accounts for about one-fifth of the entire dermis. Numerous projections, called papillae, extend from the upper portion of the dermis into the epidermis. Papillae form the base for the friction ridges on the fingers and toes.

The deeper and thicker layer of the dermis is called the **stratum reticularosum** (reticular layer). Fibers within this layer are more dense and regularly arranged to form a tough, flexible meshwork. It is quite distensible, as is evident in pregnant women or obese persons, but it can be

corium: L. *corium,* hide or leather
papilla: L. *papula,* swelling or pimple

Figure 8.10. Lines of tension within the skin. These lines are caused by the pull of elastic and collagen fibers within the dermis and are of special interest to surgeons. Incisions made parallel to the lines of tension heal more rapidly and create less scar tissue than transverse incisions.

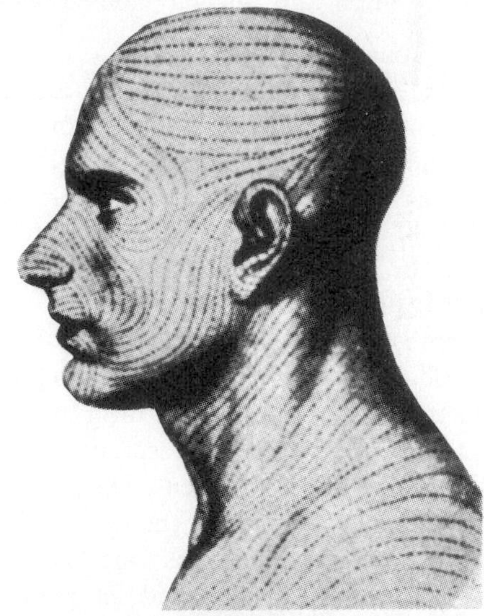

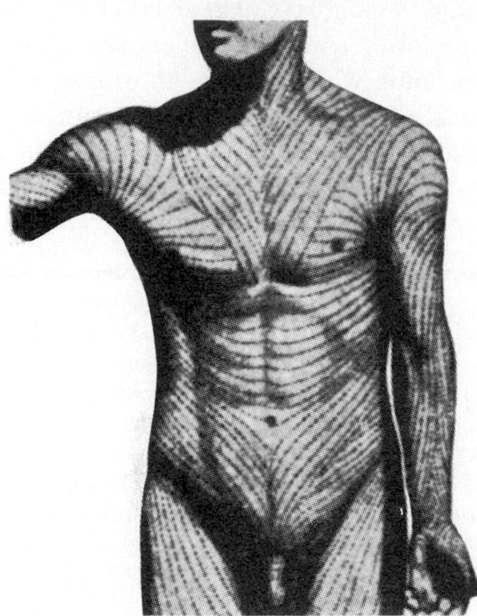

stretched too far, causing "tearing" of the dermis. The repair of a strained dermal area leaves a white streak called a stretch mark, or **linea albicans.** Lineae albicantes are frequently found on the buttocks, thighs, abdomen, and breasts (fig. 8.11).

It is the strong, resilient reticular layer of domestic mammals that is used in making leather and suede. In the tanning process, the hide of an animal is treated with various chemicals that cause the epidermis with its hair and the papillary layer of the dermis to separate from the underlying reticular layer. The reticular layer is then softened and treated with protective chemicals to make it usable.

Figure 8.11. Stretch marks (linea albicantes) on the abdomen of a pregnant woman. Stretch marks generally fade with time but frequently leave permanent integumentary markings.

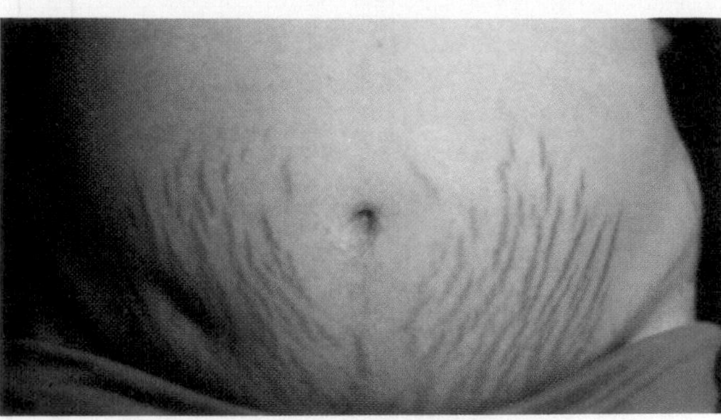

Innervation of the Skin. The dermis of the skin has extensive innervation. Specialized integumentary *effectors* consist of muscles or glands within the dermis that respond to efferent or motor impulses transmitted from the central nervous system to the skin by autonomic nerve fibers.

Several types of **afferent** *(af'er-ent),* or **sensory, receptors** respond to various tactile (touch), pressure, temperature, tickle, or pain sensations. Some are exposed nerve endings, some form a network around hair follicles, and some extend into the papillae of the dermis. Certain areas of the body, such as the palms, soles, lips, and external genitalia, have a greater concentration of sensory receptors and are therefore more sensitive to touch. Chapter 18 contains a detailed structural and functional account of the various sensory receptors.

Vascular Supply of the Skin. Blood vessels within the dermis supply nutrients to the mitotically active stratum germinativum of the epidermis and to the cellular structures of the dermis such as glands and hair follicles. Dermal blood vessels play an important role in regulating body temperature and even blood pressure. Autonomic vasoconstriction or vasodilation responses can either shunt the blood away from the superficial dermal arterioles or permit it to flow freely throughout dermal vessels. Fever or shock can be detected by the color and temperature of the skin. Blushing is the result of involuntary vasodilation of dermal blood vessels.

A healthy circulating blood flow in debilitated bed patients is important for the prevention of bedsores, or *decubitus ulcers.* When a person lies in one position for an extended period, the dermal blood flow is restricted where the body presses against the bed. As a consequence, cells die and open wounds develop (fig. 8.12). Changing the position of the patient frequently and periodically massaging the skin to stimulate blood flow are good preventive measures against decubitus ulcers.

decubitus: L. *decumbere,* lie down
ulcer: L. *ulcus,* sore

Figure 8.12. A decubitus ulcer on the medial surface of the ankle of the right leg. The most frequent sites for decubitus ulcers are in the skin overlying a bony projection, such as on the hip, ankle, heel, shoulder, or elbow.

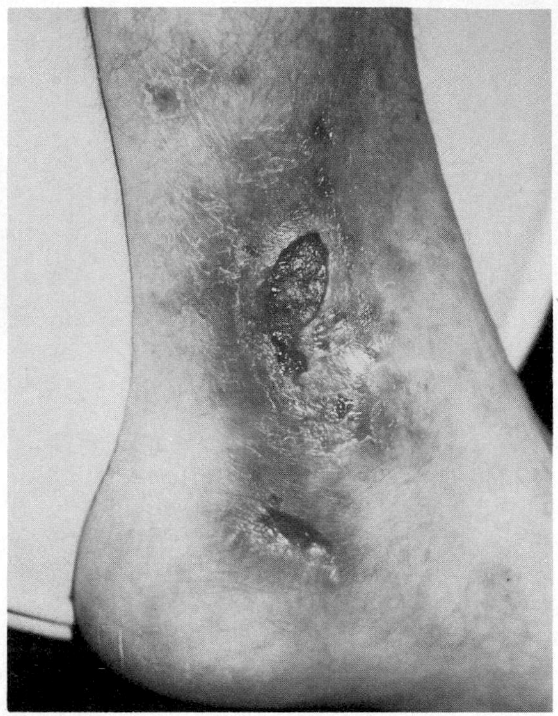

Hypodermis

The deepest layer of the skin is called the **hypodermis,** or subcutaneous layer. It binds the dermis to underlying organs. This layer is composed primarily of loose fibrous connective tissue and adipose cells interlaced with blood vessels (see fig. 8.6). Collagenous and elastic fibers reinforce the hypodermis—particularly on the palms and soles, where the skin is firmly attached to underlying structures. The amount of adipose in the hypodermis varies with the sex, age, region of the body, and nutritional state of the individual. Females generally have about an 8% thicker hypodermis than do males. This layer also functions to store lipids, insulate and cushion the body, and regulate temperature.

1. List the layers of the epidermis and dermis, and explain how they differ in structure and function.
2. Describe the sequence of cellular replacement within the epidermis and the processes of keratinization and cornification.
3. How do both the dermis and hypodermis function in thermoregulation?
4. What two basic types of innervation are found within the dermis?

Physiology of the Integument

The integument is a highly dynamic organ that not only protects the body from pathogens and external injury but is also extremely important for body homeostasis.

Objective 6. Discuss the role of the integument in the protection of the body from disease and external injury; the regulation of body fluids and temperature; absorption; synthesis; sensory reception; and communication.

Protection of the Body from Disease and External Injury

The integument is a physical barrier to most microorganisms, water, and excessive ultraviolet (UV) light. Oily secretions onto the surface of the skin form an acidic (pH 4.0–6.8) protective film, which waterproofs the body and retards the growth of most pathogens. The protein keratin in the epidermis also waterproofs the skin. Cornification of the outer layers of the epidermis toughens the mostly dead cells to withstand abrasion and the penetration of microorganisms. Upon exposure to UV light, the melanocytes in the lower epidermal layers are stimulated to synthesize melanin, which in turn absorbs and disperses sunlight. Surface friction causes the epidermis to thicken, as a protective response, by increasing mitosis in the cells of the stratum germinativum, which results in the formation of a protective *callus.*

Regardless of skin pigmentation, everyone is susceptible to skin cancer if his or her exposure to sunlight is sufficiently intense and continuous. There are an estimated eight hundred thousand new cases of skin cancer yearly in the United States, and approximately ninety-three hundred of these are diagnosed as the potentially life-threatening *melanoma* (cancer of melanocytes). Melanomas are usually termed malignant because they may spread rapidly. Sunscreens are advised for persons who must be in direct sunlight for long periods of time.

Regulation of Body Fluids and Temperature

Fluid Loss. Animals that are terrestrial (land-dwelling), such as reptiles, birds, and mammals, face a serious problem of desiccation (dehydration). The epidermis of the skin of terrestrial animals is adapted for continuous exposure to the air by being thickened, keratinized, and cornified. In addition, the outer layers are dead and scale-like, and a protein-polysaccharide basement membrane adheres the stratum basale to the dermis. Human skin is virtually waterproof, protecting the body from desiccation on dry land and even from water absorption when immersed in water.

Figure 8.13. A thermogram of the hand showing differential heat radiation. Hair and body fat are good insulators. Red and yellow indicate the warmest portions of the body, whereas blue, green, and white indicate the coolest.

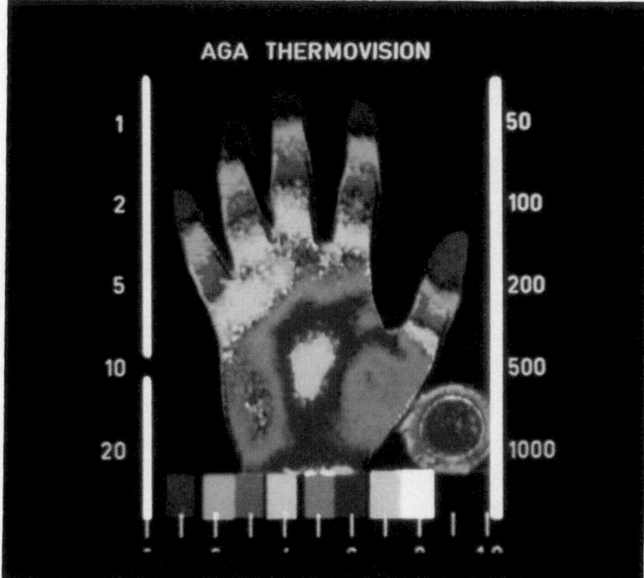

Temperature Regulation. The integument plays a crucial role in the regulation of body temperature. Body heat comes from cellular metabolism, particularly in muscle cells as they maintain tone or a degree of tension. A normal body temperature of 37°C is maintained by the antagonistic effects of sweating and shivering involving feedback mechanisms. Excess heat is actually lost from the body in three ways, all involving the skin: (1) through radiation from dilated blood vessels; (2) through excretion and the evaporation of sweat; and (3) through convection and the conduction of heat directly through the skin (fig. 8.13). Sweat excretion increases approximately 100–150 mL/day for each 1°C elevation in body temperature. Up to 10 L of sweat may be excreted to cool the body of a person doing hard physical work out-of-doors in the summertime.

A serious danger of continued exposure to heat and excessive water and salt loss is *heat exhaustion,* characterized by nausea, weakness, dizziness, headache, and a decreased blood pressure. *Heat stroke* is similar to heat exhaustion, with the exception that sweating is inhibited (for reasons that are not clear) and body temperature rises. Convulsions, brain damage, and death may follow.

Excessive heat loss triggers a shivering response in muscles, which increases cellular metabolism. Not only do skeletal muscles contract, but tiny smooth muscles attached to hair follicles contract involuntarily, causing goose bumps. The tiny muscle attached to a hair follicle is called an **arrector pili** *(ah-rek'tor pi'li)* muscle (see fig. 8.16). When these specialized smooth muscles are collectively stimulated to contract in the skin of mammals that have denser fur, the individual hairs are erected. This creates an insulating dead air space around the skin.

When the body's heat-producing mechanisms cannot keep pace with heat loss, *hypothermia* results. A lengthy exposure to temperatures below 20°C and dampness may lead to this condition. This is why it is so important that a hiker, for example, dress appropriately for the weather conditions, especially on cool, rainy spring or fall days. The initial symptoms of hypothermia are numbness, paleness, delirium, and uncontrolled shivering. If the core temperature falls below 32°C (90°F), the heart loses its ability to pump blood and will go into fibrillation (erratic contractions). If the victim is not warmed, extreme drowsiness, coma, and death follow.

Cutaneous Absorption

Because of the effective protective barriers of the integument already described, cutaneous (through the skin) absorption is limited. Some gases, such as oxygen and carbon dioxide, may pass through the skin and enter the bloodstream. Small amounts of UV light, necessary for synthesis of vitamin D, are absorbed readily. The skin is no barrier to steroid hormones, such as cortisol, and fat-soluble vitamins (A, D, E, and K). Of clinical consideration is the fact that certain toxins and pesticides enter the body through cutaneous absorption.

Synthesis

The integumentary system synthesizes melanin and keratin, which remain in the skin, and vitamin D, which is used elsewhere in the body. The integumentary cells contain a compound called *dehydrocholesterol (de-hi''dro-ko-les'tah-rol),* from which they synthesize vitamin D in the presence of UV light. Only small amounts of UV light are necessary for vitamin D synthesis, but these amounts are very important to a growing child (fig. 8.14). Synthesized vitamin D enters the blood and helps regulate the metabolism of calcium and phosphorus, which are important for development of strong and healthy bones. *Rickets* is a disease caused by vitamin D deficiency.

Sensory Reception

Highly specialized sensory receptors that respond to thermal (heat and cold), mechanical (pressure, touch, and vibration), and noxious (pain) stimuli are located abundantly throughout the dermis and hypodermis of the integument. These receptors, referred to as **cutaneous receptors,** are abundant in the skin in parts of the face, palms and fingers of the hands, soles of the feet, and the genitalia. They are less abundant along the back and on the back of the neck and are sparse in the skin over joints, especially the elbow. Generally speaking, the thinner the skin, the greater the sensitivity.

Figure 8.14. (*a*) A case of rickets in a child who lives in a village in Nepal, where the people reside in windowless huts. During the five- to six-month rainy season, the children are kept indoors. (*b*) An X ray of rickets in a ten-month-old child. Rickets develop from improper diets and also from lack of UV light in the sunlight necessary to synthesize vitamin D.

(a)

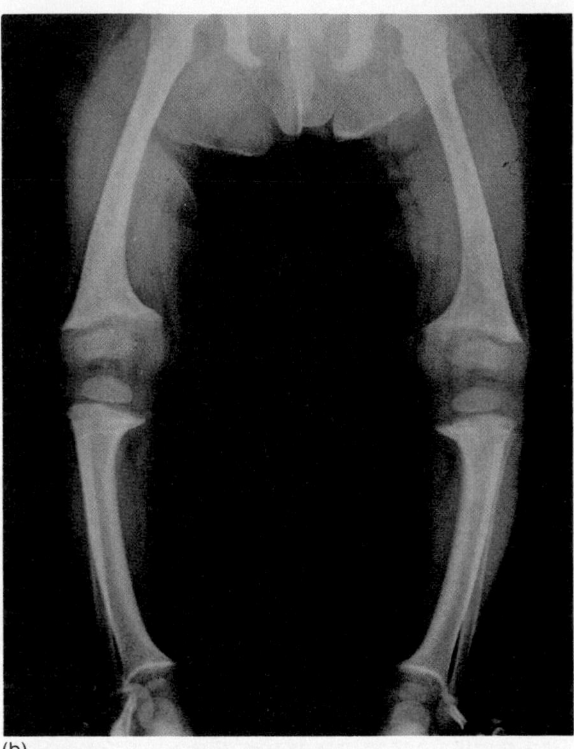

(b)

Communication

Humans are highly social animals, and the integument plays an important role in communication. Various emotions such as anger or embarrassment may be reflected in changes of skin color. Contracting specific facial muscles causes facial expressions, which convey an array of emotions, including love, surprise, happiness, sadness, or despair. Secretions from certain integumentary glands frequently elicit subconscious responses from those that detect the odors.

1. List five modifications of the integument that are structurally or functionally protective.
2. Explain how the integument participates in regulating body fluids and temperature.
3. What substances are synthesized in the integument?

Epidermal Derivatives

Hair, nails, and integumentary glands form from the epidermal layer and are therefore of ectodermal derivation. Hair and nails are structural features of the integument and have a limited functional role, whereas integumentary glands are extremely important in body defense and maintaining homeostasis.

Objective 7. Describe the structure of hair, and list the three principal types.

Objective 8. Discuss the structure and function of nails.

Objective 9. Compare the structure and function of the three principal kinds of integumentary glands.

Hair

Hair, or *pili (pi'li),* is characteristic of all mammals, but its distribution, function, density, and texture varies among different species of mammals. Humans are relatively hairless, with only the scalp, face, pubis, and axilla being densely haired. Men and women have about the same density of hair on their bodies, but it is generally more obvious on men due to male hormones (fig. 8.15). Certain structures and regions of the body are hairless, such as the palms of the hands, soles of the feet, lips, nipples, penis, and labia minora.

The primary function of hair is protection, even though its effectiveness is limited. Hair on the scalp and eyebrows protects against sunlight. The eyelashes and the hair in the nostrils protect against airborne particles. Hair on the scalp may also protect against mechanical injury. Some secondary functions of hair are to distinguish individuals and to serve as an ornamental sexual attractant.

Each hair consists of a diagonally positioned **shaft, root,** and **bulb** (fig. 8.16). The shaft is the visible but dead portion of the hair projecting above the surface of the skin.

pili: L. *pilus,* hair

Figure 8.15. The difference between male and female in the expression of hair on the body.

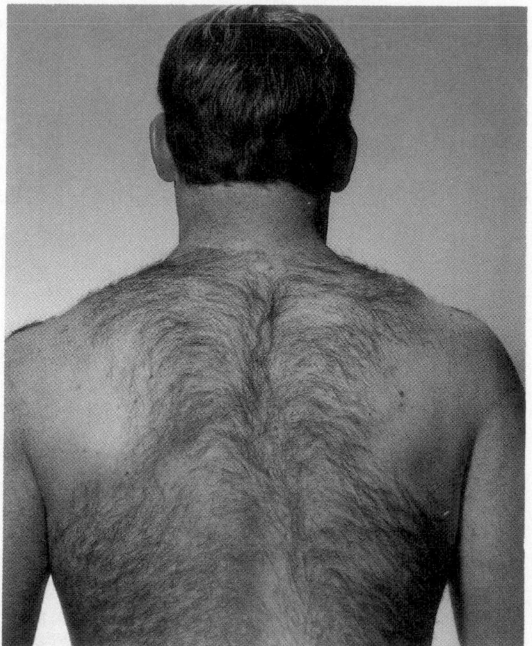

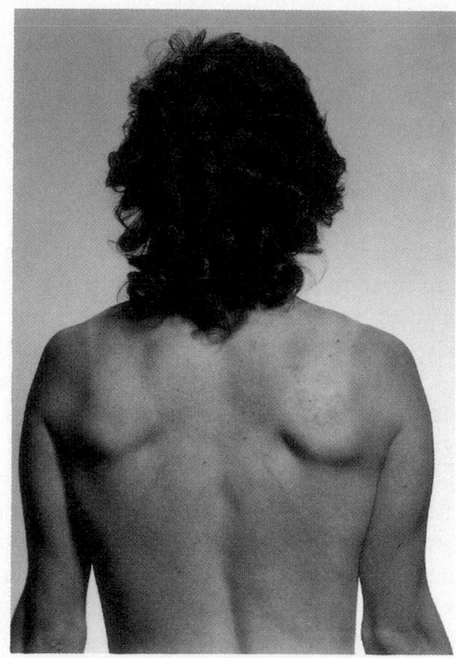

Figure 8.16. The structure of hair and the hair follicle. (*a*) a photomicrograph (63×) of the bulb and root of a hair within a hair follicle. (*b*) a scanning electron micrograph (280×) of a hair as it extends from a follicle. (*c*) a diagram of hair, a hair follicle, a sebaceous gland, and an arrector pili muscle.

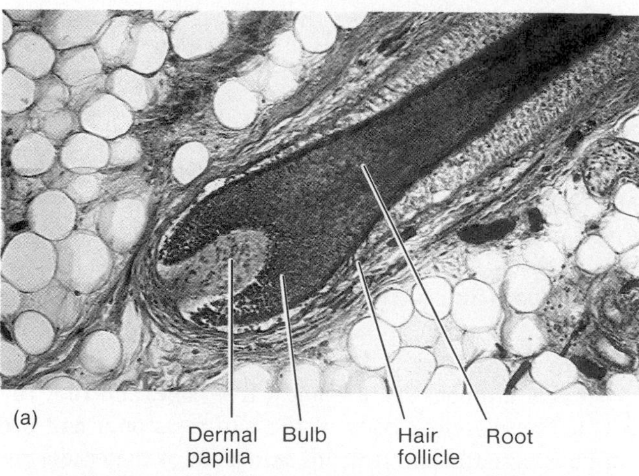

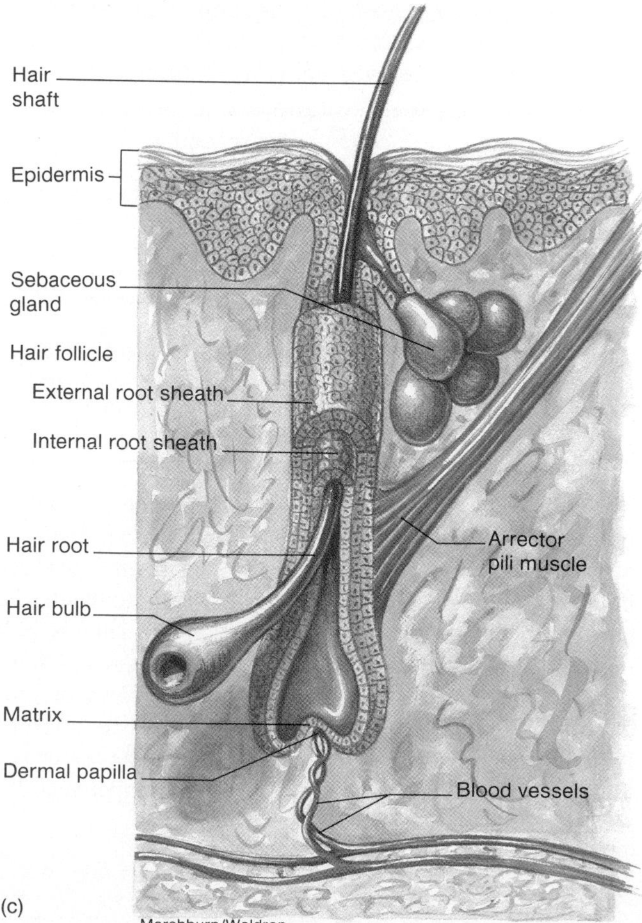

The bulb is the enlarged base of the root within the **hair follicle.** Each hair develops from stratum germinativum cells within the bulb of the hair, where nutrients are received from dermal blood vessels. As the cells divide, they are pushed away from the nutrient supply toward the surface, and cellular death and keratinization occur. In a healthy person, hair grows at the rate of approximately one millimeter every three days. As the hair becomes longer, however, it goes through a resting period during which it is anchored in its follicle.

The life span of a hair varies from three to four months for an eyelash to three to four years for a scalp hair. Each hair lost is replaced by a new hair that grows from the base of the follicle and pushes the old hair out. Between 10 and 100 hairs are lost each day through replacement. Baldness results when hair is lost and not replaced. This condition may be disease-related, but it is generally inherited and most frequently occurs in males because of genetic influences combined with the action of the male sex hormone *testosterone (tes-tos'tĕ-rōn).* No treatment is available for genetic baldness, other than grafting "plugs" of skin containing healthy follicles from hairy parts of the body to hairless regions.

Three layers can be observed in hair that is cut in cross section. An inner **medulla** *(mĕ-dul'ah)* is composed of loosely arranged cells separated by many air cells. The thick median layer, called the **cortex,** consists of hardened, tightly packed cells. A **cuticle** layer covers the cortex and forms the toughened outer portion of the hair. Cells of the cuticle have serrated edges that give a hair a scaly appearance when observed under a dissecting scope.

People exposed to heavy metals such as lead, mercury, arsenic, or cadmium will have concentrations in the hair ten times as great as that found in their blood or urine. Because of this, hair samples may be extremely important in certain diagnostic tests.

Even certain metabolic diseases or nutritional deficiencies may be detected in hair samples. For example, the hair of children with cystic fibrosis will be deficient of calcium and have excessive sodium. There is a deficiency of zinc in the hair of malnourished individuals.

Hair color is determined by the type and amount of pigment produced in the stratum germinativum at the base of the hair follicle. Varying amounts of melanin produce hair ranging in color from blond to brunette to black; the more abundant the melanin, the darker the hair. A pigment with an iron base (trichosiderin) causes red hair. Gray and white hair is caused by lack of pigment production and by air spaces within the layers of the shaft of the hair and generally accompanies aging. The texture of hair is determined by the cross-sectional shape; straight hair is round in cross section, wavy hair is oval, and kinky hair is flat.

A sebaceous gland and specialized smooth muscle, called **arrector pili,** are attached to the hair follicle (fig. 8.16). When these muscles involuntarily contract due to thermal or psychological stimuli, the hairs are pulled into a more vertical position called goose bumps.

Humans have three distinct kinds of hair.

1. *Lanugo.* Lanugo is a fine, silky, fetal hair that appears during the last trimester of development. Lanugo is usually not evident on the baby at birth unless the baby is born prematurely.
2. *Angora.* Angora hair grows continuously in length, as on the scalp of males and females, and on the face of males.
3. *Definitive.* Definitive hair grows to a certain length and then ceases to grow. It is the most common type of hair. Eyelashes, eyebrows, and pubic and axillary hair are examples.

Anthropologists have referred to humans as the "naked apes" because of our relative hairlessness. The clothing that we wear over the exposed surface areas of our body functions to insulate and protect us as hair does in other mammals. The nakedness of our skin does lead to some problems. Skin cancer occurs frequently in humans, particularly in regions of the skin exposed to the sun. Acne, another problem unique to humans, is partly related to the fact that hair is not present to dissipate the oily secretion from the sebaceous glands.

Nails

Nails are found on the distal dorsum of each of the fingers and toes. Both fingernails and toenails serve to protect the digits, and fingernails also aid in grasping and picking up small objects. Nails form from a hardened, transparent stratum corneum of the epidermis. The hardness of the nail is due to a dense, parallel arrangement of keratin fibrils between the cells.

Each nail consists of a **body, free edge,** and **root** (fig. 8.17). The platelike body of the nail rests on a **nail bed,** which is actually the stratum spinosum of the epidermis. The body and nail bed appear pinkish because of the underlying vascular tissue. The sides of the body of the nail are protected by a **nail fold,** and the furrow between the two is the **nail groove.** The free edge is the distal exposed border, which is attached to the undersurface by the **hyponychium** *(hi"po-ni'kē-um)* (quick). The root of the nail is attached proximally.

An **eponychium** (cuticle) covers above the root of the nail. The eponychium frequently splits, causing a hangnail. The growth areas of the nail are the **matrix** and the **lunule** *(loon'ūl),* which is the white half-moon-shaped area

medulla: L. marrow
cortex: L. bark
cuticle: L. *cuticula,* small skin

hyponychium: Gk. *hypo,* under; *onyx,* nail
lunule: L. *lunula,* small moon

Figure 8.17. A fingertip showing the associated structures of the nail. (*a*) a diagram; (*b*) a photomicrograph of nail of neonatal human (10×).

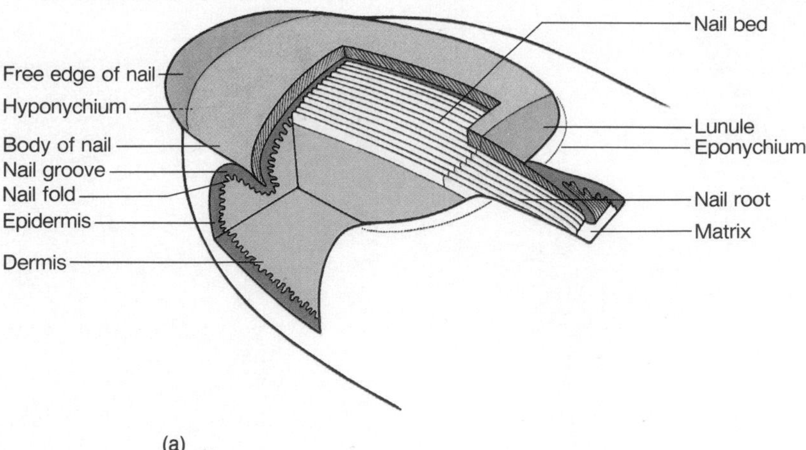

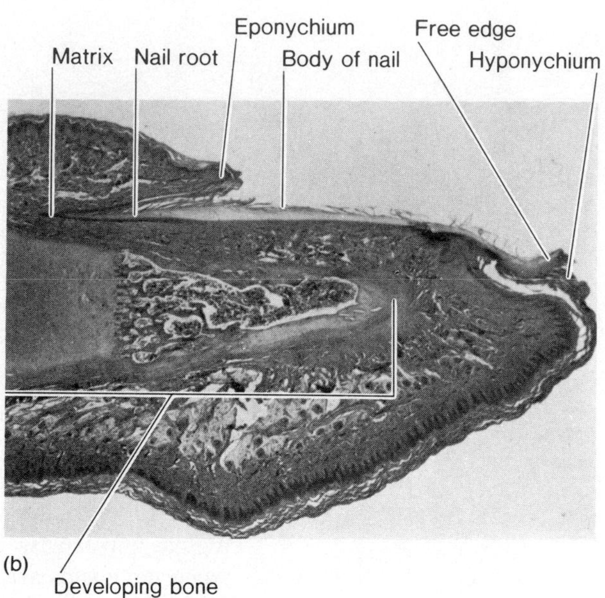

at the base of the nail. The nail grows by the transformation of the superficial cells of the matrix into nail cells. These harder, transparent cells are then pushed forward over the strata basale and spinosum of the nail bed. Fingernails grow at the rate of approximately one millimeter per week. The growth rate of toenails is somewhat slower.

The condition of nails can be an indication of the general health and personality of a person. Nails should appear pinkish, showing the rich vascular capillaries beneath the translucent nail. A yellowish hue may indicate certain glandular dysfunctions or nutritional deficiencies. Split nails may also be caused by nutritional deficiencies. A prominent bluish tint may indicate improper oxygenation of the blood. Spoon nails (concave body) may be the result of iron-deficiency anemia, and "clubbing" at the base of the nail may be caused by lung cancer. Dirty or ragged nails may indicate poor personal hygiene, and chewed nails may suggest emotional problems.

Glands

Although they originate in the epidermal layer, all of the glands of the skin are located in the dermis, where they receive physical support and nutritive sustenance. Glands of the skin are referred to as *exocrine* since they excrete substances through ducts. The glands of the skin are of three basic types: sebaceous, sudoriferous *(soo'dor-if'er-us)*, and ceruminous *(sĕ-roo'mĭ-nus)*.

Sebaceous. Sebaceous, or oil, glands are associated with a hair follicle since they develop from the follicular epithelium of the hair. They are simple, branched glands that are connected to hair follicles, where they secrete **sebum** onto the shaft of the hair (fig. 8.16). Sebum, which consists mainly of lipids, is dispersed along the shaft of the hair to the surface of the skin, where it lubricates and waterproofs the stratum corneum layer. Sebum also prevents the hair from splitting and becoming brittle. If the drainage pathway for sebaceous glands becomes blocked

sebum: L. *sebum*, tallow or grease

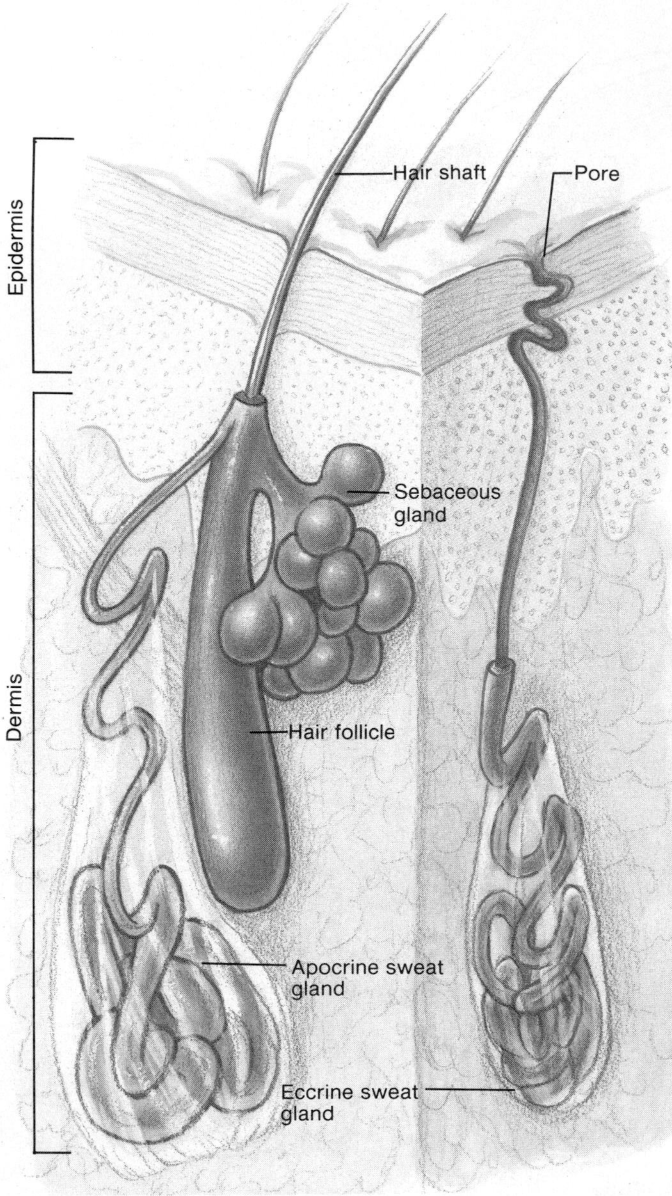

Figure 8.18. Types of skin glands.

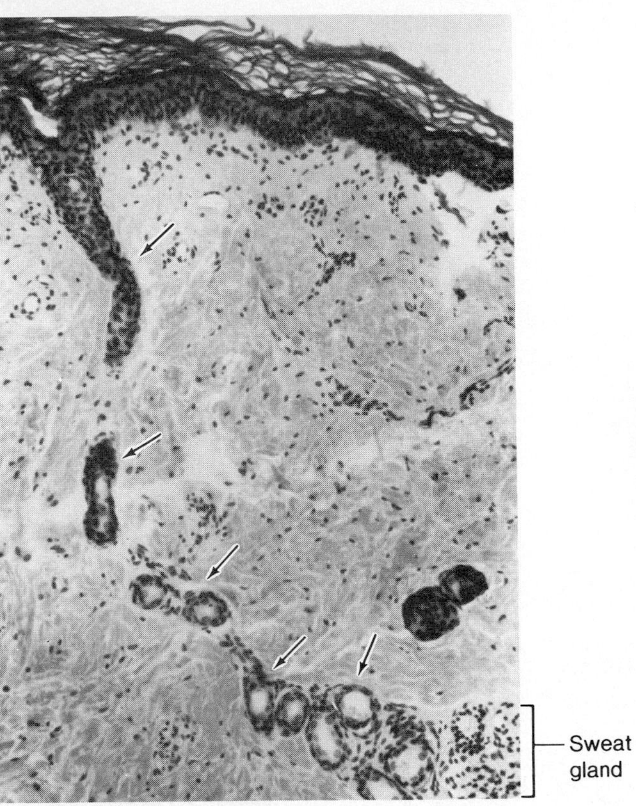

Figure 8.19. A photomicrograph of an eccrine sweat gland (100×). The coiled structure of the ductule portion of the gland accounts for its irregular appearance.

for some reason, the glands may become infected, resulting in *acne*. Sex hormones regulate the production and secretion of sebum, and hyperactivity of sebaceous glands can result in serious acne problems, particularly during teenage years.

Sudoriferous. Sudoriferous, or **sweat glands,** excrete perspiration, or sweat, onto the surface of the skin. Sweat glands are most numerous on the palms of the hands, soles of the feet, axillary and pubic regions, and the forehead. They are coiled and tubular shaped (fig. 8.18) and are of two types.

1. **Eccrine** *(ek'rin)* **sweat glands** are widely distributed over the body, especially on the forehead, back, palms, and soles. These glands are formed totally before birth and function in evaporative cooling in response to thermal or psychological stimuli (figs. 8.18, 8.19).
2. **Apocrine** *(ap'o-krin)* **sweat glands** are much larger, localized glands found in the axillary and pubic regions and are associated with hair follicles. Apocrine glands are not functional until puberty, and their odoriferous secretion is thought to act as a sexual attractant.

sudoriferous: L. *sudorifer,* sweat; *ferre,* to bear

Perspiration is composed of water, salts, urea, uric acid, and traces of other elements. Perspiration is therefore valuable not only for evaporative cooling but also for the excretion of certain wastes.

Mammary glands, found within the breasts, are specialized sudoriferous, or sweat, glands that secrete milk during lactation periods (see fig. 8.20). The breasts of the human female reach their greatest development during the childbearing years under the stimulus of pituitary and ovarian hormones.

Ceruminous Glands. These highly specialized glands are found only in the external auditory meatus (ear canal). They secrete **cerumen** *(sĕ-roo'men),* or earwax, which is a water and insect repellent and also keeps the tympanum (eardrum) from drying out. Excessive amounts of cerumen may interfere with hearing.

cerumen: L. *cera,* wax

Figure 8.20. Mammary gland within the breast of a human female.

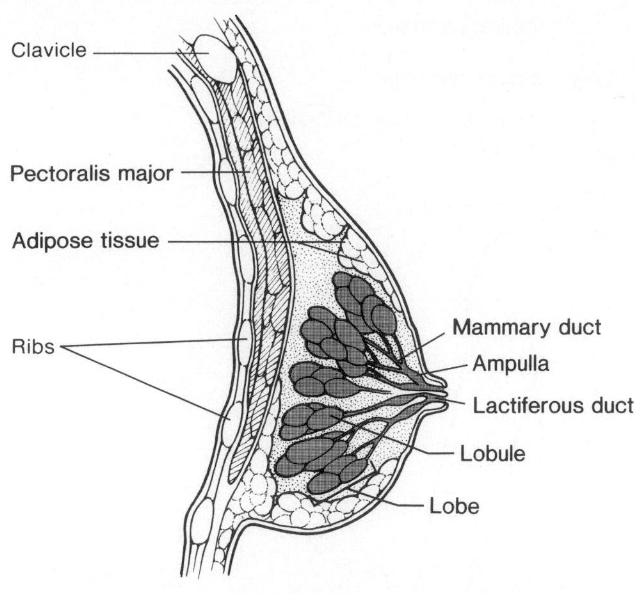

Good routine hygiene is very important for health and social reasons. Washing away the dried residue of perspiration and sebum eliminates dirt. Excessive bathing, however, can wash off the natural sebum and dry the skin, causing it to itch or crack. The commercial lotions used for dry skin are, for the most part, refined and perfumed lanolin, which is sebum from sheep.

1. Draw and label a hair. Indicate which portion is alive, and discuss what causes the cells in a hair to die. What are the three principal types of hair?
2. Describe the structure and function of nails, and explain how nails grow and harden.
3. List the types of integumentary glands, and discuss the structure and function of each.
4. Are skin glands mesodermal or ectodermal in derivation? Are they epidermal or dermal in functional position?

Clinical Considerations

The skin is a buffer against the external environment and is therefore subject to a variety of disease-causing microorganisms and physical assaults. A few of the many diseases and disorders of the integumentary system are briefly discussed here.

Inflammatory (Dermatitis) Conditions

Inflammatory skin disorders are caused by immunologic hypersensitivity or infectious agents. Some persons are allergic to certain foreign proteins and, because of this inherited predisposition, experience hypersensitive reactions, such as asthma, hay fever, hives, drug and food allergies, and eczema. These conditions are characterized by redness, itching, and swollen vascular lesions that become dry, scaly, and crusted. **Lesions** occur commonly with skin disorders and are defined as more or less circumscribed pathologic changes in the tissue. Figure 8.21 identifies the common lesions of the skin and their usual sites.

There are a number of *infectious diseases* of the skin, which is not surprising considering that we are highly social and communal animals. Effective prevention and treatment are now available for most of these diseases, but too frequently persons do not avail themselves of safeguards or treatments. Infectious diseases of the skin include childhood viral infections (measles and chicken pox), bacteria such as staphylococcus (impetigo), venereal diseases, leprosy, fungi (ringworm, athlete's foot, candida), and mites (scabies).

Neoplasms

Both benign and malignant neoplastic conditions or diseases are common in the skin. Pigmented moles (nevi), for example, are a type of benign neoplastic growth of melanocytes. Dermal cysts and benign viral infections are also common. Warts are virally caused abnormal growths of tissue that occur frequently on the hands and feet. A different type of wart, called a venereal wart, occurs in the anogenital region of sexual partners. Both types of warts are easy to treat by excision or various drugs. Aging spots on elderly persons, which appear as pigmented patches on the surface of the skin, are a benign growth of melanin pigmented germinativum cells. Usually no treatment is required, unless for cosmetic purposes.

Skin cancer (fig. 8.22) is the most common malignancy in the United States, but except for malignant melanomas, which arise from melanocytes within the epidermis, they are generally not life threatening. Excessive exposure to ultraviolet light from the sun is a known cause of skin cancer. The preferred treatment for this disease is complete surgical excision of the cancerous portion.

Burns

A burn is an epithelial injury caused by contact with thermal, radioactive, chemical, or electrical agents. Burns generally occur on the skin, but burns can involve the linings of the respiratory and digestive tracts. The extent and location of a burn is frequently less important than the degree to which it disrupts body homeostasis. Burns that have a **local effect** (local tissue destruction) are not as serious as those that have a **systemic effect.** Systemic effects directly or indirectly involve the entire body and are a threat to life. Possible systemic effects include body dehydration, shock, reduced circulation and urine production, and bacterial infections.

neoplasm: Gk. *neo,* new; *plasma,* something formed
benign: L. *benignus,* good-natured
malignant: L. *malignus,* acting from malice

Figure 8.21. Common lesions of the skin and their usual site of occurrence.

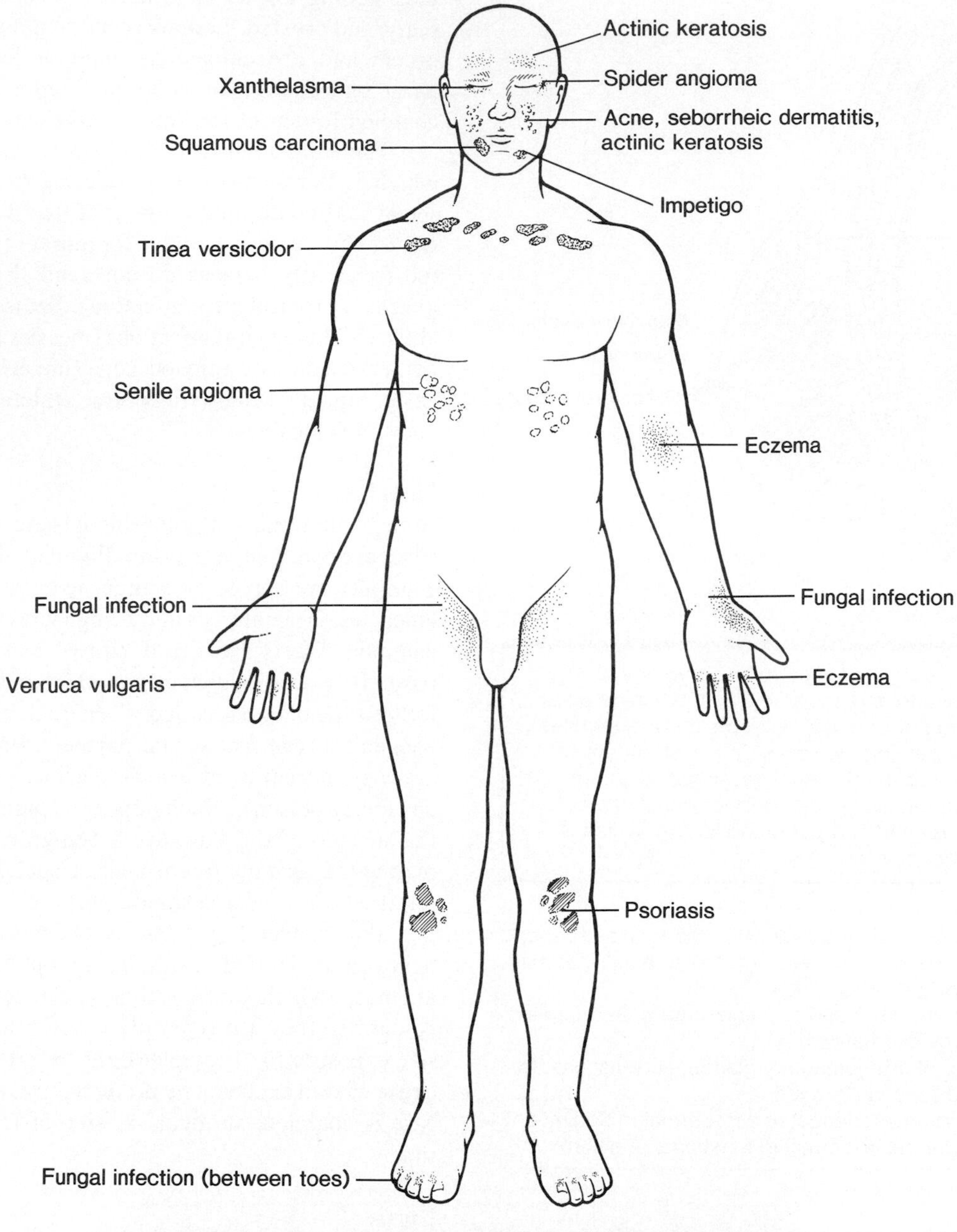

Actinic keratosis
Xanthelasma
Spider angioma
Acne, seborrheic dermatitis, actinic keratosis
Squamous carcinoma
Impetigo
Tinea versicolor
Senile angioma
Eczema
Fungal infection
Fungal infection
Verruca vulgaris
Eczema
Psoriasis
Fungal infection (between toes)

Figure 8.22. Skin cancer.

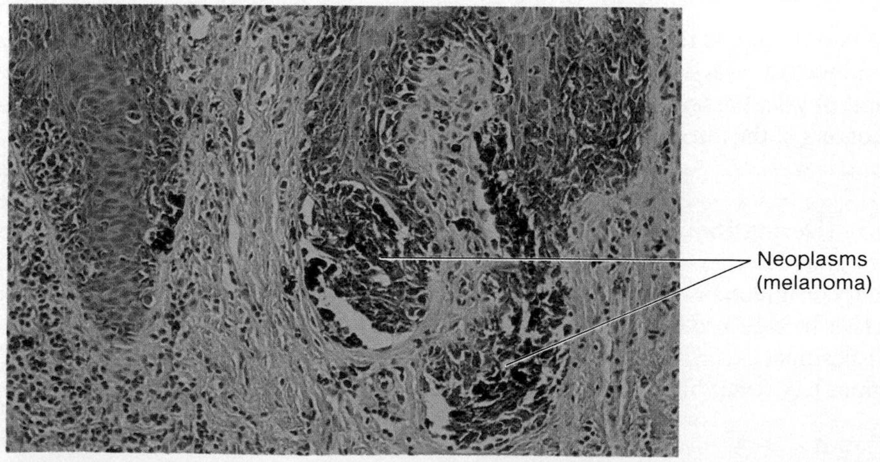

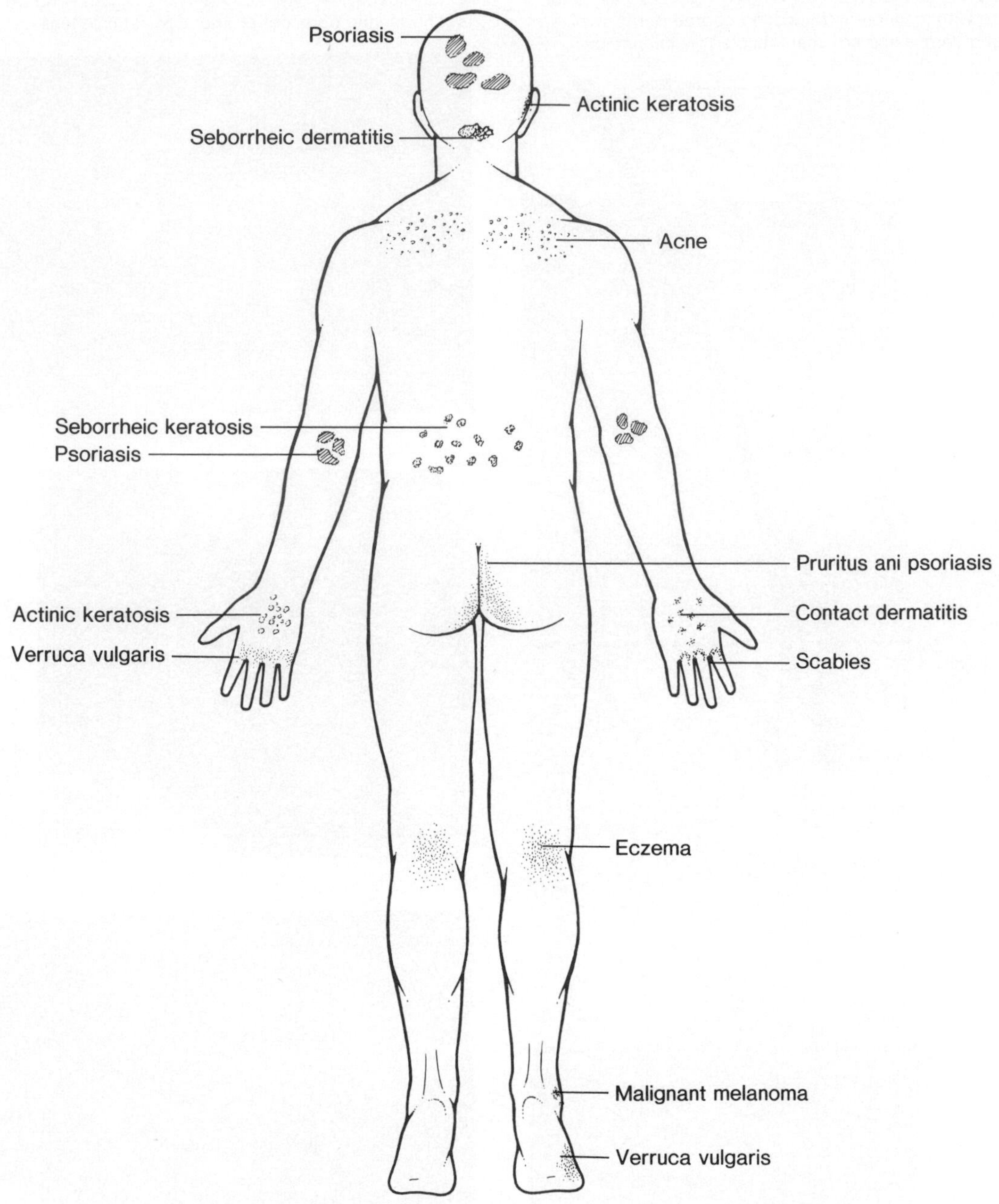

Burns are classified into three types, according to their severity: first-degree, second-degree, and third-degree (fig. 8.23). In **first-degree** burns, the epidermal layers of the skin are damaged and symptoms are restricted to local effects such as redness, pain, and edema (swelling). A shedding of the surface layers (desquamation) generally follows in a few days. A sunburn is an example. **Second-degree** burns involve both the epidermis and dermis. Blisters appear and recovery is usually complete, although slow. **Third-degree** burns destroy the entire thickness of the skin and frequently some of the underlying muscle. The skin appears charred and is insensitive to touch. As a result, ulcerating wounds develop and the body attempts to heal itself by forming scar tissue. Skin grafts are frequently used to assist recovery.

A procedure for estimating the extent of damaged integument suffered in burned patients is to apply the "*rule of nines*" (fig. 8.24), in which the surface area of the body is divided into regions accounting for about 9% of the area (or a multiple of 9%). Collectively, these areas account for 100% of the integumentary (body) surface. An estimation of the percentage of surface area damaged is important in treating with intravenous fluid, which replaces the fluids lost from tissue damage.

Figure 8.23. The classification of burns. (*a*) first-degree burns involve the epidermis and are characterized by redness, pain, and edema—such as with a sunburn; (*b*) second-degree burns involve the epidermis and dermis and are characterized by intense pain, redness, and blistering; (*c*) third-degree burns destroy the entire skin and frequently expose the underlying organs. The skin is charred and numb and does not protect against fluid loss.

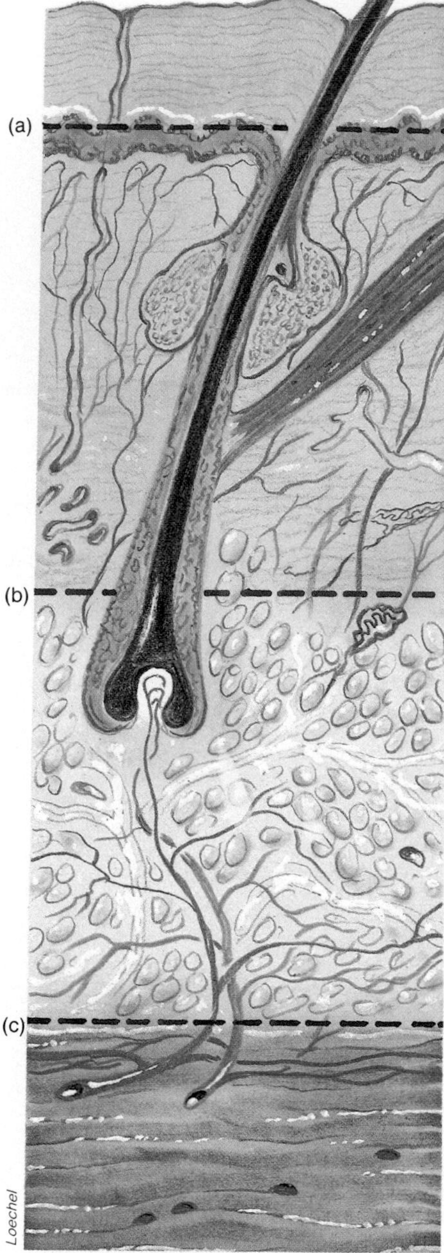

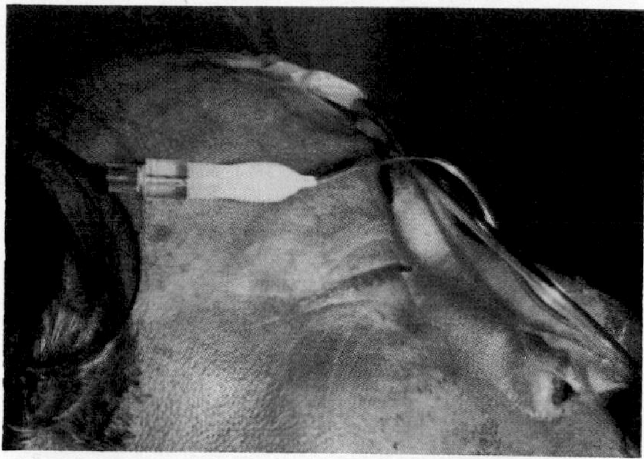

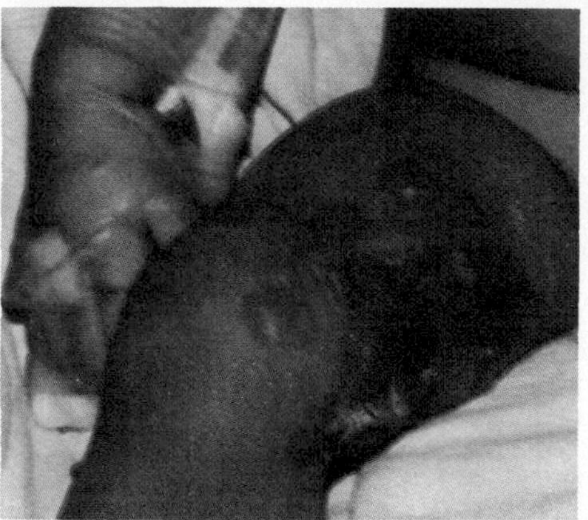

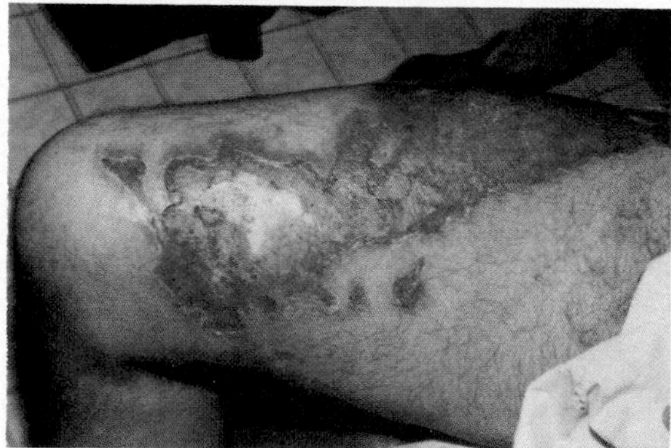

Frostbite

Frostbite is a local destruction of the skin resulting from freezing. Like burns, frostbite is classified by its degree of severity: first-degree, second-degree, and third-degree. In **first-degree** frostbite the skin will appear cyanotic (bluish) and swollen. Vesicle formation and hyperemia (swollen with blood) are symptoms of **second-degree** frostbite. As the affected area is warmed, there will be further swelling, and the skin will redden and blister. In **third-degree** frostbite, there will be severe edema, some bleeding, and numbness followed by intense throbbing pain and necrosis of the affected tissue. Gangrene will follow untreated third-degree frostbite.

Skin Grafts

If extensive areas of the stratum germinativum of the epidermis are destroyed in second-degree or third-degree burns or frostbite, new skin cannot grow back. In order for this type of wound to heal, a skin graft must be performed.

A **skin graft** is a segment of skin that has been excised from a *donor site* and transplanted to the *recipient site,* or *graft bed.* An *autotransplant* is the most successful graft. This type of transplant involves taking a thin sheet of healthy epidermis from a donor site of the burn or frostbite patient and moving it to the recipient site (fig. 8.25). A *heterotransplant* (between two different species) can be a temporary treatment to prevent infection and fluid loss.

Figure 8.24. The extent of burns, as estimated by the "rule of nines." (*a*) anterior. (*b*) posterior.

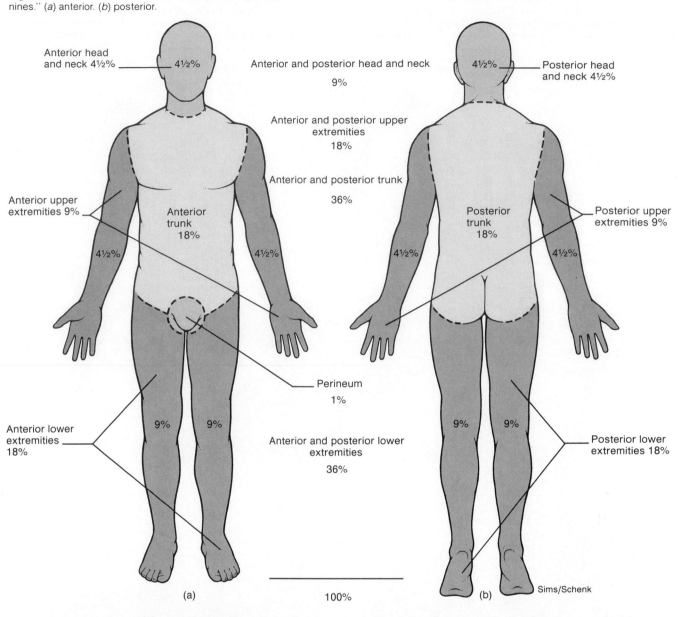

Anterior head and neck 4½%

4½%

Anterior and posterior head and neck

9%

4½%

Posterior head and neck 4½%

Anterior and posterior upper extremities

18%

Anterior and posterior trunk

36%

Anterior upper extremities 9%

Anterior trunk 18%

Posterior trunk 18%

Posterior upper extremities 9%

4½%

4½%

4½%

4½%

Perineum

1%

Anterior lower extremities 18%

9%

9%

Anterior and posterior lower extremities

36%

9%

9%

Posterior lower extremities 18%

100%

(a)

(b)

Sims/Schenk

Figure 8.25. Skin graft to the scalp. (*a*) traumatized tissue is excised and (*b*) healthy tissue from another body location is transplanted.

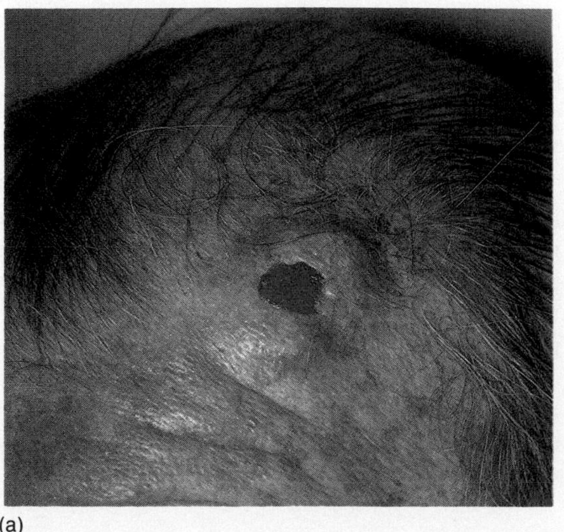

(a)

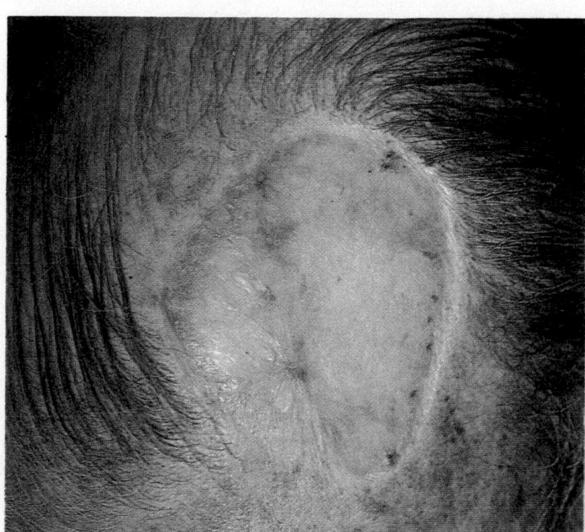

(b)

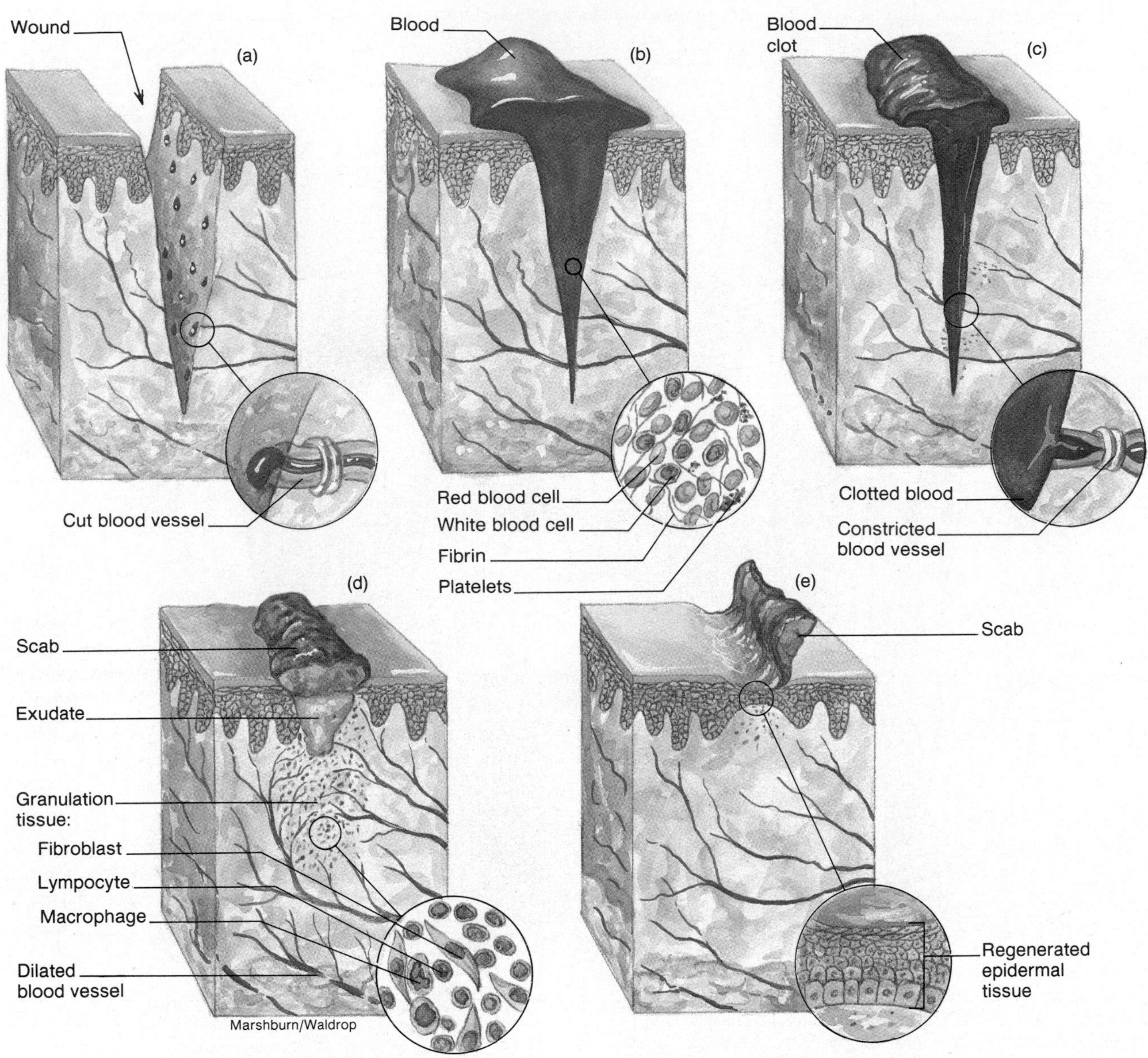

Figure 8.26. The process of wound healing. (*a*) a penetrating wound into the dermis ruptures blood vessels. (*b*) blood cells, fibrinogen, and fibrin flow out of the wound. (*c*) vessels constrict and a clot blocks the flow of blood. (*d*) a protective scab is formed from the clot, and granulation tissue forms within the site of the wound. (*e*) the scab sloughs off as the epidermal layers are regenerated.

Three types of grafts may be used to heal severe skin wounds. For relatively small wounds, a *pinch graft* is used, consisting of a conical piece of skin including the entire thickness of the epidermis and dermis. The epidermal area is broader than the tapered dermal area in this type of graft. A *full thickness graft,* used in extensive burns, consists of a broad sheet of skin that includes both epidermal and dermal layers. A *split thickness graft* is also used in extensive burns but includes only the upper portion of the dermis. A successful skin transplant is referred to as a *graft take.* Blood flow is usually well established by the third or fourth day in a graft take.

Recently, laboratory-cultured skin from the patient's healthy epidermal cells has become available. Progress is even being made to produce synthetic skin. Both of these advancements are desirable because they minimize the problem of tissue rejection from the person's own immunity system.

Wound Healing

The skin effectively protects against many abrasions, but if a wound does happen, a sequential chain of events promotes rapid healing. The process of wound healing depends on the extent and severity of the injury. Trauma to the epidermal layers stimulates an increased mitotic activity in the stratum germinativum, whereas injuries that extend to the dermis or subcutaneous layer elicit activity throughout the body as well as within the wound itself. General body responses include a temporary elevation of temperature and pulse rate.

Figure 8.27. Scars for body adornment on the face of this Buduma man from the islands of Lake Chad are created by instruments that make crescent-shaped incisions into the skin in beadlike patterns. Special ointments are applied to the cuts to retard healing and promote the forming of scars.

Figure 8.28. The albino individual in this photograph has melanocytes within his skin, but due to a mutant gene he lacks the ability to synthesize melanin.

In an open wound (fig. 8.26), blood vessels are broken and bleeding occurs. Through the action of **blood platelets** *(plat'letz)* and protein molecules, called **fibrinogen** *(fi-brin'o-jen),* a clot forms and soon blocks the flow of blood. A scab forms and covers and protects the damaged area. Mechanisms are activated to destroy bacteria, dispose of dead or injured cells, and isolate the injured area. These responses are collectively referred to as *inflammation* and are characterized by redness, heat, edema, and pain. Inflammation is a response that confines the injury and promotes healing.

The next step in healing is the differentiation of binding **fibroblasts** from connective tissue at the wound margins. Together with new branches from surrounding blood vessels, **granulation tissue** is formed. Phagocytic cells migrate into the wound and ingest dead cells and foreign debris. Eventually the damaged area is repaired, and the protective scab is sloughed off.

If the wound is severe enough, the granulation tissue may develop into **scar tissue** (fig. 8.27). Scar tissue differs from normal skin in that its collagen fibers are more dense and it has no stratified squamous epidermal layer. Scar tissue has fewer blood vessels and may lack hair, glands, and sensory receptors. The closer the edges of a wound, the less granulation tissue develops and the less obvious a scar. This is one reason for suturing a large break in the skin.

Important Clinical Terminology

acne An inflammatory condition of sebaceous glands. Acne is effected by gonadal hormones and is therefore more common during puberty and adolescence. Pimples and blackheads on the face, chest, and back are expressions of this condition.

albinism A congenital, genetic deficiency of the pigment of the skin, hair, and eyes due to a metabolic block in the synthesis of melanin (fig. 8.28).

alopecia Loss of hair, baldness. Baldness is usually due to genetic factors and cannot be treated. Baldness may signify anatomical maturity.

athlete's foot *(tinea pedis)* A skin fungus disease of the foot.

blister A collection of fluid between the epidermis and dermis, caused by excessive friction or a burn.

boil (furuncle) A localized bacterial infection originating in a hair follicle or skin gland.

callus A localized buildup of the stratum corneum layer of the epidermis due to excessive friction.

carbuncle A bacterial infection similar to a boil, except that it infects the subcutaneous tissues.

cold sore *(fever blister)* A lesion on the lip or oral mucous membrane, caused by type I herpes simplex virus (HSV), transmitted by oral or respiratory exposure.

comedo A plug of sebum and epithelial debris in the hair follicle and excretory duct of the sebaceous gland. Also called a blackhead or whitehead.

corn A type of callus that is localized on the foot, usually over toe joints.

dandruff Common dandruff is the continual shedding of epidermal cells of the scalp, which can be removed by normal washing and brushing of the hair. Abnormal dandruff may be caused by certain skin diseases such as seborrhea or psoriasis.

decubitus ulcer A bedsore. An exposed ulcer caused by a continual pressure that restricts dermal blood flow to a localized portion of the skin (see fig. 8.12).

dermabrasion A procedure for removing tattoos or acne scars by high speed sanding or scrubbing.

dermatitis An inflammation of the skin.

dermatology A specialty of medicine concerned with the study of the skin, its anatomy, physiology, histopathology, and the relationship of cutaneous lesions to systemic disease.

dermatome An instrument for cutting thin slices of skin for grafting or excising small lesions.
eczema *(eg'ze-mah)* A noncontagious inflammatory condition of the skin producing red, itching, vesicular lesions, which may be crusty or scaly.
erythema *(er''i-the'mah)* Redness of the skin caused by vasodilation from skin injury, infection, or inflammation.
furuncle A boil; a localized abscess resulting from an infected hair follicle.
gangrene Necrosis of tissue due to the obstruction of blood flow. It may be localized or extensive and may secondarily be infected with anaerobic microorganisms.
impetigo A contagious skin infection that results in lesions followed by scaly patches. It generally occurs on the face and is caused by staphylococci or streptococci.
keratosis Any abnormal growth and hardening of the stratum corneum layer of the skin.
melanoma A cancerous tumor originating from proliferating melanocytes within the epidermis of the skin.
nevus *(ne'vus)* A mole or birthmark; congenital pigmentation of a certain area of the skin.
papilloma A benign epithelial neoplasm, such as a wart or a corn.
papule A small inflamed elevation of the skin, such as a pimple.
pruritus Itching. It may be symptomatic of systemic disorders but is generally due to dry skin.
psoriasis *(so-ri'ah-sis)* An inherited inflammatory skin disease usually expressed as circular scaly patches of skin.
pustule A small, localized elevation of the skin containing pus.
seborrhea *(seb''o-re'ah)* A disease characterized by an excessive activity of the sebaceous glands and accompanied by oily skin and dandruff.
ulcer An open sore. A decubitus ulcer denotes a bedsore.
urticaria (hives) A skin eruption of reddish weals, usually with extreme itching. It may be caused by an allergic reaction, stress, or contact with some external or internal precipitating factor.
wart A horny projection of epidermal cells caused by a virus.

Chapter Summary

I. The Integument as an Organ
   A. The skin is the largest body organ, and together with the epidermal structures (hair, glands, and nails), it constitutes the integumentary system.
   B. The appearance of the skin is clinically important because it provides clues to certain body conditions or dysfunctions.

II. Development of the Integumentary System
   A. Epidermal layers, hair, nails, and cutaneous glands, including mammary glands, derive from ectoderm that specializes into a mitotically active germinal layer.
   B. Hair and sebaceous glands develop together during the formation of a hair follicle.
   C. Sweat glands develop independently from the germinal layer of the epidermis, and mammary glands are modified sweat glands that develop along a mammary ridge.
   D. The dermis and hypodermis derive from mesoderm. Specialized mesoderm, called mesenchyme, gives rise to specific dermal structures.

III. Layers of the Integument
   A. The outer epidermis consists of four or five layers, the dermis consists of two layers, and the hypodermis is a single layer.
   B. The stratified squamous epithelium of the epidermis is divisible into five structural and functional layers: stratum basale, stratum spinosum, stratum granulosum, stratum lucidum, and stratum corneum.
   C. The thick dermis of the skin is composed of fibrous connective tissue interlaced with elastic fibers. The two layers of the dermis are the upper stratum papillarosum and the deeper stratum reticularosum.
   D. The hypodermis, composed of adipose and fibrous connective tissue, binds the dermis to underlying organs.

IV. Physiology of the Integument
   A. Physical features of the skin provide protection of the body from disease and external injury; keratin and an acidic oily secretion on the surface protect the skin from water and microorganisms; cornification of the skin protects against abrasion; and melanin is a barrier to UV light.
   B. The skin regulates body fluids and temperatures: fluid loss is minimal due to keratinization and cornification; and temperature regulation is maintained by radiation, convection, and the antagonistic effects of sweating and shivering.
   C. The skin permits the absorption of UV light, respiratory gases, steroids, fat-soluble vitamins, and certain toxins and pesticides.
   D. The integument synthesizes melanin and keratin, which remain in the skin, and vitamin D, which is used elsewhere in the body.
   E. Cutaneous receptors throughout the dermis and hypodermis provide sensory reception in the skin. Cutaneous receptors respond to precise sensory stimuli and are more sensitive in thin skin.

V. Epidermal Derivatives
   A. Hair is characteristic of all mammals, but its distribution, function, density, and texture varies among species of different mammals.
   B. Hardened, keratinized nails are found on the distal dorsum of each digit, where they protect the digits; fingernails aid in grasping and picking up small objects.
   C. Integumentary glands are exocrine since they either secrete or excrete substances through ducts; sebaceous glands secrete sebum onto the shaft of the hair; sudoriferous (sweat) glands are eccrine and apocrine; mammary glands are specialized sudoriferous glands that secrete milk during lactation; and ceruminous glands secrete cerumen (earwax).

Review Activities

Objective Questions

1. Hair, nails, integumentary glands, and the epidermis of the skin are derived from embryonic
   (a) ectoderm
   (b) mesoderm
   (c) endoderm
   (d) mesenchyme
2. Spoon-shaped nails may result when a person has a dietary deficiency of
   (a) zinc
   (b) iron
   (c) niacin
   (d) vitamin B_{12}
3. The epidermal layer *not* present in the thin skin of the face is the stratum
   (a) granulosum
   (b) lucidum
   (c) spinosum
   (d) corneum
4. Which of the following does *not* contribute to skin color?
   (a) keratin
   (b) melanin
   (c) carotene
   (d) hemoglobin
5. Integumentary cells synthesize vitamin D in the presence of UV light and
   (a) keratin
   (b) cortisol
   (c) trichosiderin
   (d) dehydrocholesterol
6. Integumentary glands that empty their secretions into hair follicles are
   (a) sebaceous glands
   (b) sudoriferous glands
   (c) eccrine glands
   (d) ceruminous glands
7. Fetal hair that is present during the last trimester of development is referred to as
   (a) angora
   (b) definitive
   (c) lanugo
   (d) replacement
8. Which of these conditions is potentially life threatening?
   (a) acne
   (b) melanoma
   (c) eczema
   (d) seborrhea
9. The skin of a burn victim has been severely damaged through the epidermis and into the dermis; integumentary regeneration will be slow, with some scarring, but complete. What degree burn is it?
   (a) first
   (b) second
   (c) third
10. The technical name for a blackhead or whitehead is
    (a) carbuncle
    (b) melanoma
    (c) nevus
    (d) comedo

Essay Questions

1. Discuss the development of the skin and its associated hair, glands, and nails. What role do the ectoderm and mesoderm play in integumentary development?
2. What are some physical and chemical features of the skin that make it an effective protective organ?
3. Of what practical value is it for the outer layers of the epidermis and hair to be composed of dead cells?
4. Contrast the structure and function of sebaceous, sudoriferous, mammary, and ceruminous glands.
5. Discuss what is meant by an inflammatory lesion. What are some frequent causes of skin lesions?

9

Skeletal System: The Axial Skeleton

Outline

Concepts

The axial and appendicular components of the skeletal system of an adult human consist of 206 individual bones arranged into a strong, flexible body framework.

The bones of the skeleton perform the mechanical functions of support, protection, and body movement and the metabolic functions of hemopoiesis and mineral storage.

All bones derive from specialized mesenchymal mesoderm through endochondral ossification or through intramembranous ossification. Ossification begins by the fourth week of prenatal development and is not completed in certain bones until about age twenty-five or thirty.

Although bones have commonalities of histological structure, each bone has a characteristic pattern of ossification and growth, a characteristic shape, and diagnostic surface features that indicate its functional relationship to other bones, muscles, and to the body structure as a whole.

The human skull, consisting of eight cranial and fourteen facial bones, contains several cavities that house the brain and sensory organs. Each bone of the skull articulates with the adjacent bones and has diagnostic and functional processes or other surface features and foramina.

The supporting vertebral column consists of twenty-six vertebrae, separated by fibrocartilaginous intervertebral discs, which lend flexibility and absorb the stress of movement. Vertebrae enclose and protect the spinal cord, support and permit movement of the skull, articulate with the rib cage, and provide for the attachment of the trunk muscles.

The cone-shaped and flexible rib cage consists of the thoracic vertebrae, twelve paired ribs, costal cartilages, and the sternum. It encloses and protects the thoracic viscera and is directly involved in the mechanics of breathing.

Organization of the Skeletal System

The axial and appendicular components of the skeletal system of an adult human consist of 206 individual bones arranged into a strong, flexible body framework.

Objective 1. Describe the structural organization of the skeletal system, and list the bones of the axial and appendicular portions.

Osteology is the science concerned with the study of bones. Each bone is an organ that plays a part in the total functioning of the skeletal system. The skeletal system of an adult human is composed of approximately 206 bones. Actually, the number of bones differs from person to person, depending on age and genetic variations. At birth, the skeleton consists of approximately 270 bones. As further bone development (ossification) occurs during infancy, the number increases. During adolescence, however, the number of bones decreases as there is a gradual fusion of separate bones.

Some adults have extra bones within the joints (sutures) of the skull called **sutural** *(soo'cher-ahl),* or **Wormian, bones.** Additional bones may develop in tendons in response to stress as the tendons repeatedly move across a joint. Bones formed this way are called **sesamoid** *(ses'ah-moid)* **bones.** Sesamoid bones, like the sutural bones, vary in number. The patellae (kneecaps) are two sesamoid bones present in all people.

For the convenience of study, the skeleton is divided into axial and appendicular portions. Anterior and posterior views of the skeleton are shown in figure 9.1. Table 9.1 lists the divisions of the skeleton and the number of bones in each portion.

The **axial skeleton** consists of the bones that form the axis of the body and that support and protect the organs of the head, neck, and torso:

1. **Skull.** The skull consists of two sets of bones: the cranial bones that form the cranium, or braincase, and the facial bones that support the eyes, nose, and jaws.
2. **Auditory ossicles.** Three auditory ossicles are present in the middle ear chamber of each ear and serve to transmit sound impulses.
3. **Hyoid bone.** The hyoid *(hi'oid)* bone is located above the larynx and below the lower jaw. It supports the tongue and assists in swallowing.
4. **Vertebral column.** The vertebral column (backbone) consists of twenty-six individual vertebrae separated by cartilaginous intervertebral discs. In the pelvic region several vertebrae are fused to form the sacrum, which provides attachment for the pelvic girdle. A few terminal vertebrae are fused to form the coccyx, the so-called tailbone.
5. **Rib cage.** The rib cage, or thoracic cage, forms the bony and cartilaginous framework of the thorax. The rib cage articulates posteriorly with the thoracic vertebrae and includes the twelve pairs of **ribs,** the flattened **sternum,** and the **costal cartilages,** which connect the ribs to the sternum on the anterior side.

The **appendicular skeleton** is composed of the bones of the upper and lower extremities and the bony girdles, which anchor the appendages to the axial skeleton:

1. **Pectoral girdle.** The paired **scapulae** and **clavicles** constitute the pectoral girdle. It is not a complete girdle, having only an anterior attachment to the axial skeleton at the sternum via the clavicles. The primary function of the pectoral girdle is to provide attachment for the muscles that move the brachium and forearm.
2. **Upper extremities.** Each upper extremity consists of a proximal **humerus** within the brachium, an **ulna** and **radius** within the forearm, the **carpal** bones of the wrist, and the **metacarpal** and **phalangeal** *(fah-lan'je-al)* bones of the hand.
3. **Pelvic girdle.** The pelvic girdle is formed by two **ossa coxae** (hipbones) united anteriorly by the **symphysis** *(sim'fi-sis)* **pubis** and posteriorly by the sacrum of the vertebral column. The pelvic girdle supports the weight of the body through the vertebral column and protects the lower viscera within the pelvic cavity.
4. **Lower extremities.** Each lower extremity consists of a proximal **femur** within the thigh, a **tibia** and **fibula** within the lower leg, the **tarsal** bones of the ankle, and the **metatarsal** and **phalangeal** bones of the foot. In addition, the **patella** *(pah-tel'ah)* is located on the anterior surface of the knee joint between the thigh and lower leg regions.

---

1. List the bones of the body which you can palpate on yourself. Indicate which are of the axial skeleton and which are of the appendicular skeleton.
2. What are sesamoid bones, and where are they found?
3. Describe the locations and functions of the pectoral and pelvic girdles.

Functions of the Skeletal System

The bones of the skeleton perform the mechanical functions of support, protection, and body movement and the metabolic functions of hemopoiesis and mineral storage.

Objective 2. Discuss the principal functions of the skeletal system, and identify the body systems that these functions serve.

sesamoid: Gk. *sesamon,* like a sesame seed
ossicle: L. *ossiculum,* little bone

Figure 9.1. The human skeleton. (*a*) an anterior view. (*b*) a posterior view. The axial and appendicular portions are distinguished with color.

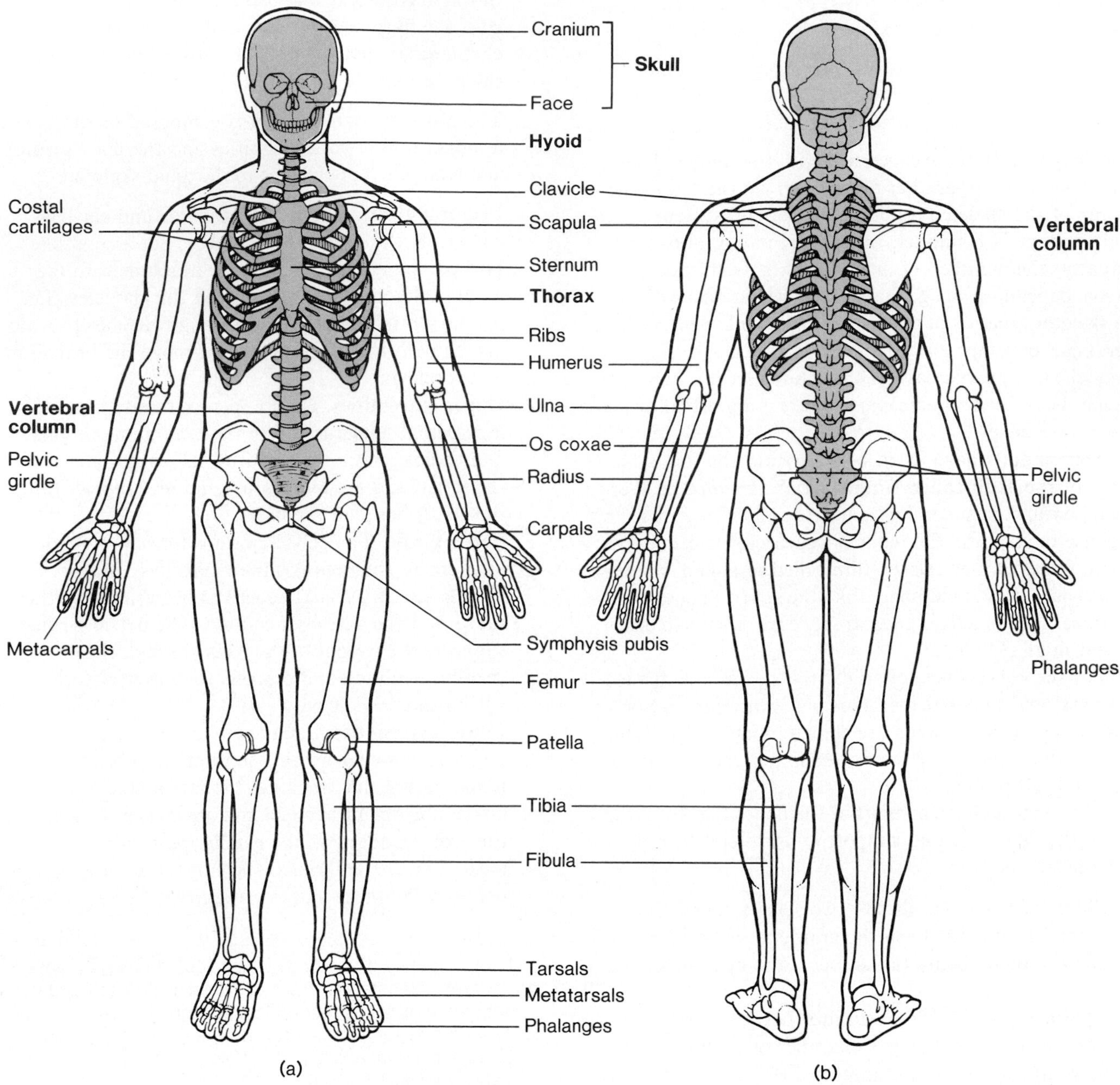

The strength of bone comes from its inorganic components, which resist decomposition even after death. Much of what we know of prehistoric animals, including humans, has been determined from preserved skeletal remains. Frequently when we think of bone, we think of a hard, dry structure. In fact, the term *skeleton* comes from a Greek word meaning "dried up." Living bone is not inert material, however; it is dynamic and adaptable in performing many body functions, including support, protection, body movement, hemopoiesis, and mineral storage.

1. **Support.** The skeleton forms a rigid framework to which are attached the softer tissues and organs of the body.
2. **Protection.** The skull and vertebral column enclose the central nervous system; the rib cage protects the heart, lungs, great vessels, liver, and spleen; and the pelvic cavity supports and protects the pelvic viscera. Even the site where red blood cells are produced is protected within the central, hollow portion of certain bones.

Table 9.1 Classification of the bones of the adult skeleton

I. Axial skeleton	
a. Skull	22 bones
8 cranial bones:	
frontal 1	
parietal 2	
occipital 1	
temporal 2	
sphenoid 1	
ethmoid 1	
13 facial bones:	
maxilla 2	
palatine 2	
zygomatic 2	
lacrimal 2	
nasal 2	
vomer 1	
inferior nasal concha 2	
1 mandible	
b. Middle ear ossicles	6 bones
malleus 2	
incus 2	
stapes 2	
c. Hyoid	1 bone
hyoid bone 1	
d. Vertebral column	26 bones
cervical vertebra 7	
thoracic vertebra 12	
lumbar vertebra 5	
sacrum 1 (5 fused bones)	
coccyx 1 (3–5 fused bones)	
e. Rib cage	25 bones
rib 24	
sternum 1	
II. Appendicular skeleton	
a. Pectoral girdle	4 bones
scapula 2	
clavicle 2	
b. Upper extremities	60 bones
humerus 2	
radius 2	
ulna 2	
carpal 16	
metacarpal 10	
phalanx 28	
c. Pelvic girdle	2 bones
os coxa 2 (each os coxa contains 3 fused bones)	
d. Lower extremities	60 bones
femur 2	
tibia 2	
fibula 2	
patella 2	
tarsal 14	
metatarsal 10	
phalanx 28	
Total	206 bones

3. **Body movement.** Bones serve as anchoring attachments for most skeletal muscles. In this capacity, the bones act as levers with the joints functioning as pivots when muscles contract to cause body movement.
4. **Hemopoiesis** *(he″mo-poi-e′sis)*. The red bone marrow of an adult produces white blood cells, red blood cells, and platelets. In an infant, the spleen and liver produce red blood cells, but as the bones mature, the bone marrow assumes the performance of this formidable task. It is estimated that an average of one million red blood cells are produced every second by the bone marrow to replace those that are worn out and destroyed by the liver.
5. **Mineral storage.** The inorganic matrix of bone is composed primarily of the minerals calcium and phosphorus. These minerals give bone its rigidity and account for approximately two-thirds of the weight of bone. About 95% of the calcium and 90% of the phosphorus within the body are deposited in the bones and teeth. Although the concentration of these organic salts within the blood is kept within narrow limits, both of these mineral salts are essential for other body functions. Calcium is necessary for muscle contraction, blood clotting, and the movement of ions and nutrients across cell membranes. Phosphorus is required for the activities of the nucleic acids, DNA and RNA, as well as for ATP utilization. If mineral salts are not present in the diet in sufficient amounts, they may be withdrawn from the bones until they are replenished through proper nutrition. In addition to calcium and phosphorus, lesser amounts of magnesium and sodium salts are stored in bone tissue.

Vitamin D assists in the absorption of calcium and phosphorus from the intestine into the blood. As the bones develop in children, it is immensely important that their diet contain an adequate amount of these two minerals and vitamin D. If the diet is deficient in these essentials, the blood level falls below that necessary for calcification and a condition known as *rickets* develops (see fig. 8.14). Rickets is characterized by soft bones that may result in bowlegs and malformation of the head, chest, and pelvic girdle.

In summary, the skeletal system is not an isolated body system. It functions with the muscle system since it stores the calcium needed for muscular contraction and provides an attachment for muscles as they span the movable joints. The skeletal system serves the circulatory system by producing blood cells in protected sites. Also, many of the vessels of the circulatory system are named according to the bones they parallel. The skeletal system supports and protects, to varying degrees, all of the systems of the body.

1. List the functions of the skeletal system.
2. Discuss two ways the skeletal system serves the circulatory system in the production of blood. What two ways does it serve the muscular system?

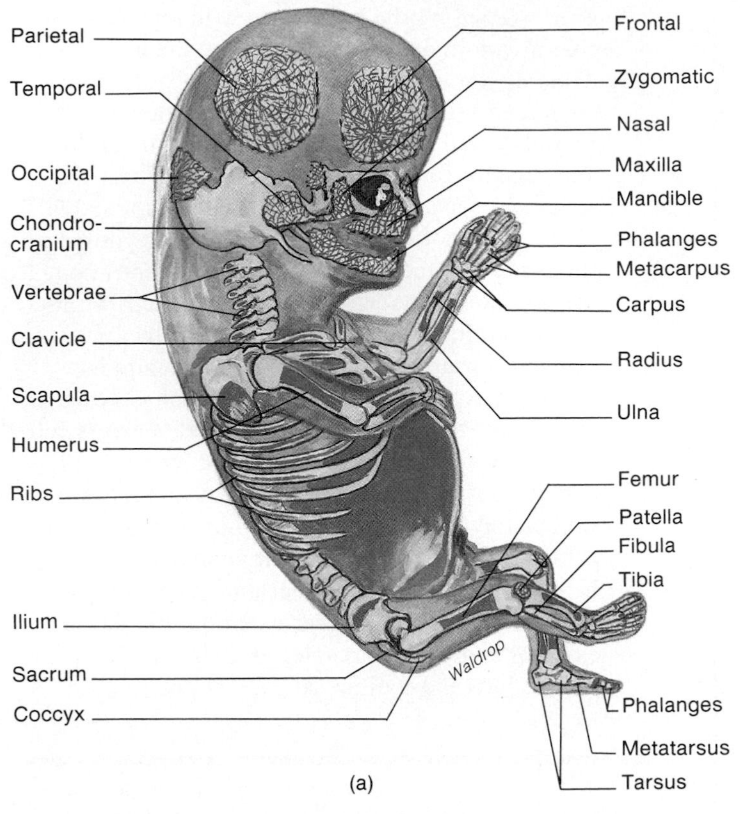

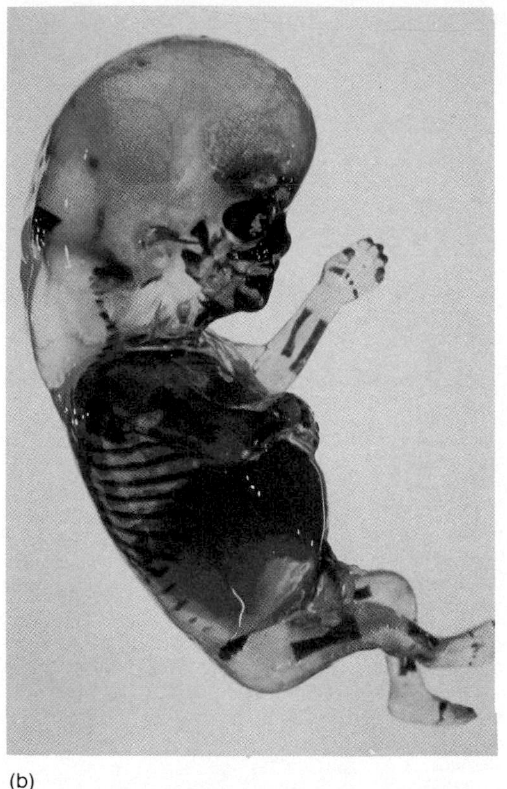

Figure 9.2. Ossification centers of the skeleton of a 10-week-old fetus. (*a*) the diagram depicts endochondral ossification in red and intramembranous ossification in a stippled pattern. The cartilaginous portions of the skeleton are shown in gray. (*b*) the photograph shows the ossification centers stained with a red indicator dye.

Development of the Axial Skeletal System

All bones derive from specialized mesenchymal mesoderm through endochondral ossification or through intramembranous ossification. Ossification begins by the fourth week of prenatal development and is not completed in certain bones until about age twenty-five or thirty.

Objective 3. Distinguish between endochondral ossification and intramembranous ossification.

Objective 4. Explain the processes of axial and appendicular skeletal development.

Objective 5. Describe the fontanels of a fetal skull, and explain their importance.

Bone formation, or *ossification,* begins about the fourth week of embryonic development, but ossification centers cannot be observed until about the eighth week (fig. 9.2). Bone tissue derives from specialized migratory cells known as *mesenchyme.* Some of the embryonic mesenchymal cells will transform into *chondroblasts (kon'dro-blastz)* and develop a cartilage matrix that is later replaced by bone in a process known as **endochondral** *(en''do-kon'dral)* **ossification.** Most of the skeleton is formed in this fashion—first it goes through a hyaline cartilage stage and then ossifies as bone.

A smaller number of mesenchymal cells develop directly into bone without first going through a cartilage stage. This type of bone formation process is referred to as **intramembranous** *(in''trah-mem'brah-nus)* **ossification.** Facial bones and certain bones of the cranium are formed this way. **Sesamoid bones** are specialized intramembranous bones that develop in tendons.

Development of the Vertebral Column, Rib Cage, and Sternum

The development of the vertebral column is a sequential process involving embryonic somites and the notochord. **Somites** are paired blocklike condensations of mesoderm that form on both sides of the neural tube toward the end of the third week of embryonic development (fig. 9.3). The **notochord** is a flexible rod of mesodermal tissue that extends the length of the back of an embryo (see fig. 1.8).

During the fourth week of development the somites differentiate into three separate components (fig. 9.4), which give rise to specific structures: the dorsolateral **dermatome** gives rise to the dermis of the skin; the medial **myotome** forms most of the skeletal muscle of the body; and the ventromedial **sclerotome** *(skle'rah-tōm)* forms the **vertebral column** and certain other bones of the body.

somite: Gk. *soma,* body
dermatome: Gk. *dermia,* skin; *tome,* a cutting
myotome: Gk. *myos,* muscle; *tome,* a cutting
sclerotome: Gk. *scleros,* hard; *tome,* a cutting

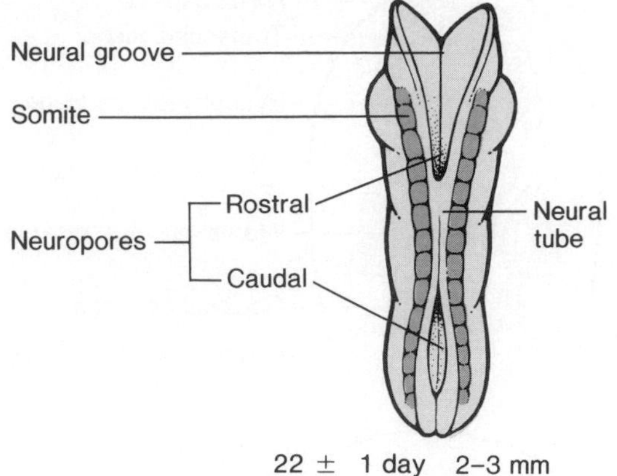

Figure 9.3. A dorsal view of a twenty-eight-day-old embryo showing the position of the somites lateral to the neural tube forming along the neural groove.

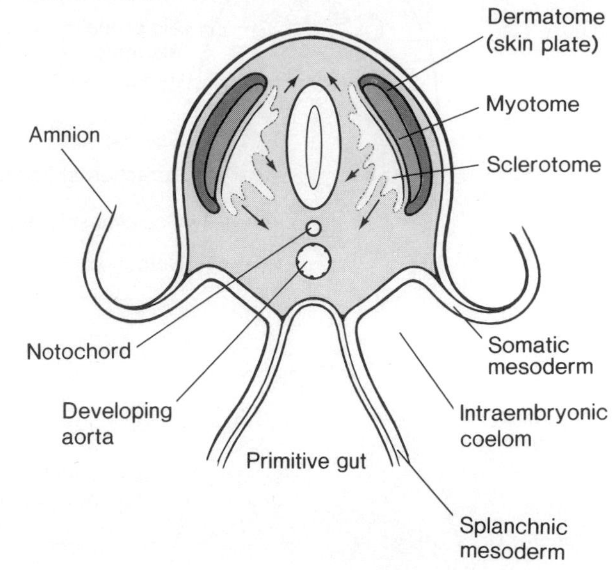

Figure 9.4. A transverse view through the dorsal aspect of a twenty-eight-day-old embryo showing the dermatome, myotome, and sclerotome of a somite. The arrows represent the direction of migration of the mesenchymal cells of the sclerotome, which give rise to the vertebral column.

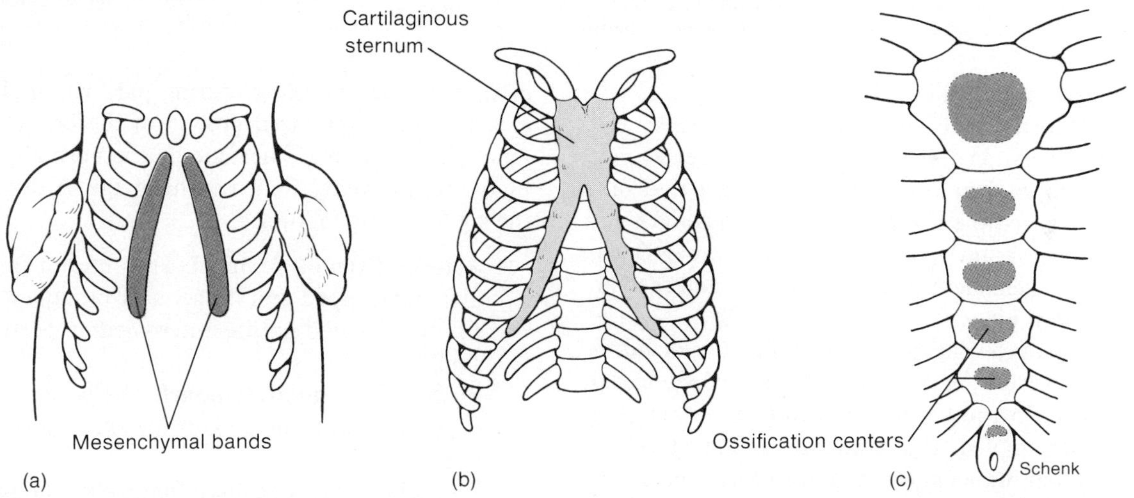

Figure 9.5. The endochondral development of the sternum. (*a*) an anterior view of the thorax of an embryo at thirty-five days showing the paired mesenchymal bands. (*b*) fusion of the mesenchymal bands and differentiation of mesenchymal cells into chondroblasts forms an unsegmented cartilaginous sternum. (*c*) the sternum becomes segmented as endochondral ossification progresses in specific ossification sites.

At about twenty-eight days of development, the sclerotome portion of the somites begins migrating in three main directions. (1) Mesenchymal cells migrate ventromedially to surround the notochord and form the bodies of the vertebrae (fig. 9.4). The notochord eventually degenerates as the vertebrae ossify. Only small portions of the notochord persist as gelatinous centers of the intervertebral discs called the **nucleus pulposus** *(pul-po'sus).* (2) Mesenchymal cells migrate dorsally to form the **vertebral,** or **neural, arches,** covering the neural tube. (3) A group of mesenchymal cells pass ventrolaterally as **costal processes** into the developing body wall to form the ribs and costal cartilage of the rib cage.

The development of the sternum is independent of the ribs and is first apparent at about thirty-five days as a pair of mesenchymal bands form lateral to the ventral midline in the thoracic region (fig. 9.5). Gradually the mesenchymal bands migrate together and fuse as the mesenchymal cells become cartilaginous. Segmentation of the sternum occurs secondarily as ossification centers develop along the cartilaginous sternum.

Figure 9.6. The embryological development of the skull. (*a–c*) superior views showing cartilaginous development and fusion in the formation of the chondrocranium. (*d*) lateral view at about twenty weeks as the viscerocranium and dermatocranium are developing. (*a*) a view of an embryo at six weeks, (*b*) at seven weeks, and (*c*) at twelve weeks.

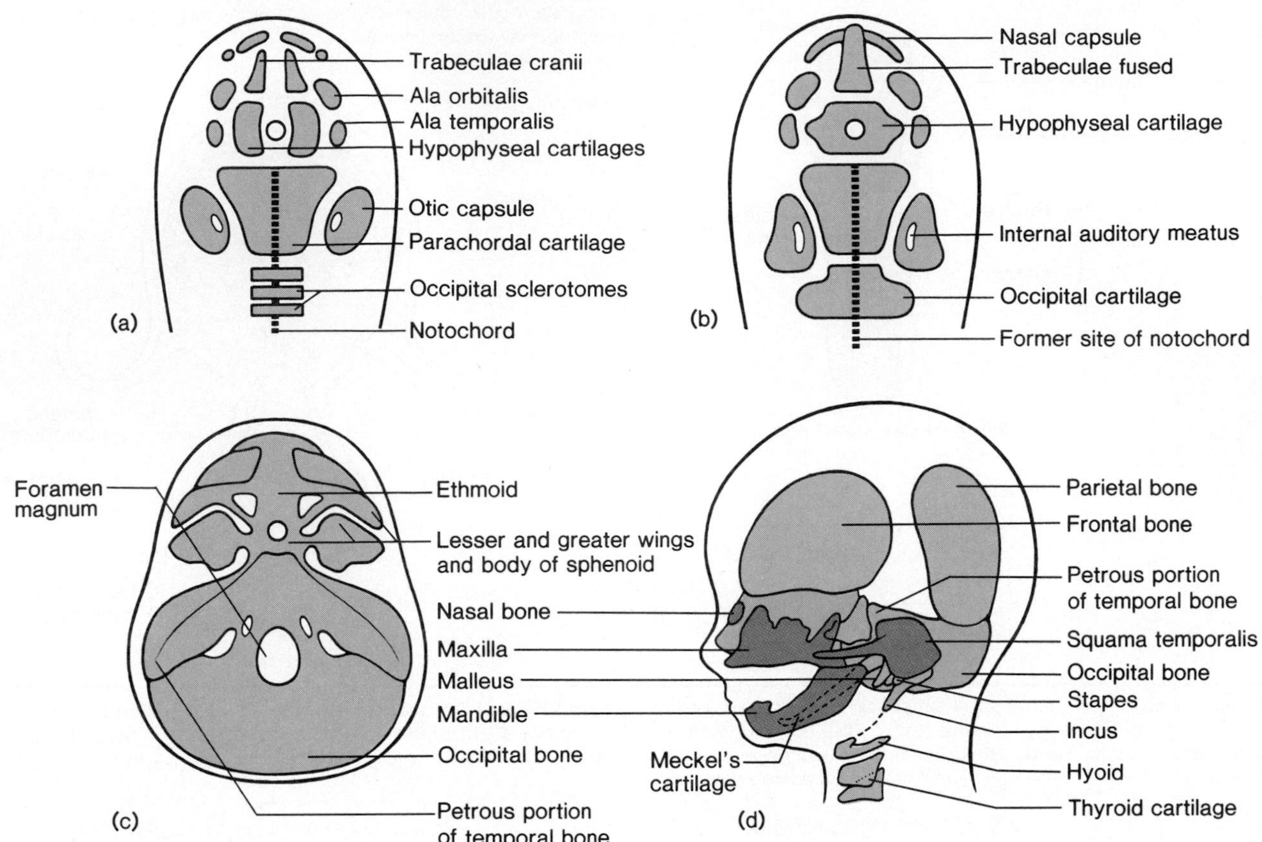

Development of the Skull

Two distinct portions of mesodermal germ cells contribute to the formation of the skull around the developing brain. The **neurocranium** forms a protective case that surrounds the brain, and the **viscerocranium** *(vis''er-o-kra'ne-um)* (splanchnocranium) portion forms the ear ossicles, hyoid bone, laryngeal and tracheal cartilages, and specific processes of the skull.

One portion of the neurocranium, called the **chondrocranium** *(kon'dro-kra''ne-um),* undergoes endochondral ossification to form the bones supporting the brain, and another portion called the **dermatocranium** develops through intramembranous ossification to form the bones covering the brain and facial region (fig. 9.6).

During fetal development and infancy, the bones of the neurocranium covering the brain are separated by fibrous sutures. There are also six large membranous areas of the skull that provide spaces between the developing bones (fig. 9.7). These areas are called **fontanels** *(fon''tah-nel)* ("soft spots") and permit the skull to undergo changes of shape, called *molding,* during parturition. The fontanels also allow for rapid growth of the brain during infancy. Ossification of the fontanels is normally complete by twenty to twenty-four months of age. A description of the six fontanels follows:

1. **Anterior (frontal) fontanel.** The anterior fontanel is diamond-shaped and is the most prominent of the six. It is located on the anteromedian portion of the skull.
2. **Posterior (occipital) fontanel.** The posterior fontanel is positioned at the back of the skull on the median line.
3. **Anterolateral (sphenoidal) fontanels.** The paired anterolateral fontanels are found on either side of the skull directly below the anterior fontanel.
4. **Posterolateral (mastoid) fontanels.** The paired posterolateral fontanels are located on the posterolateral sides of the skull.

A prominent **sagittal suture** extends the anteroposterior median length of the skull between the anterior and posterior fontanels. A coronal suture extends from the anterior fontanel to the anterolateral fontanel. A **lambdoidal**

neurocranium: Gk. *neuron,* nerve; *kranion,* skull
viscerocranium: L. *viscera,* soft parts; Gk. *kranion,* skull
fontanel: Fr. *fontaine,* little fountain

lambdoidal: Gk. *lambda,* letter (λ) in Greek alphabet

Figure 9.7. The fetal skull showing the six fontanels and the sutures. (*a*) a right lateral view; (*b*) a superior view.

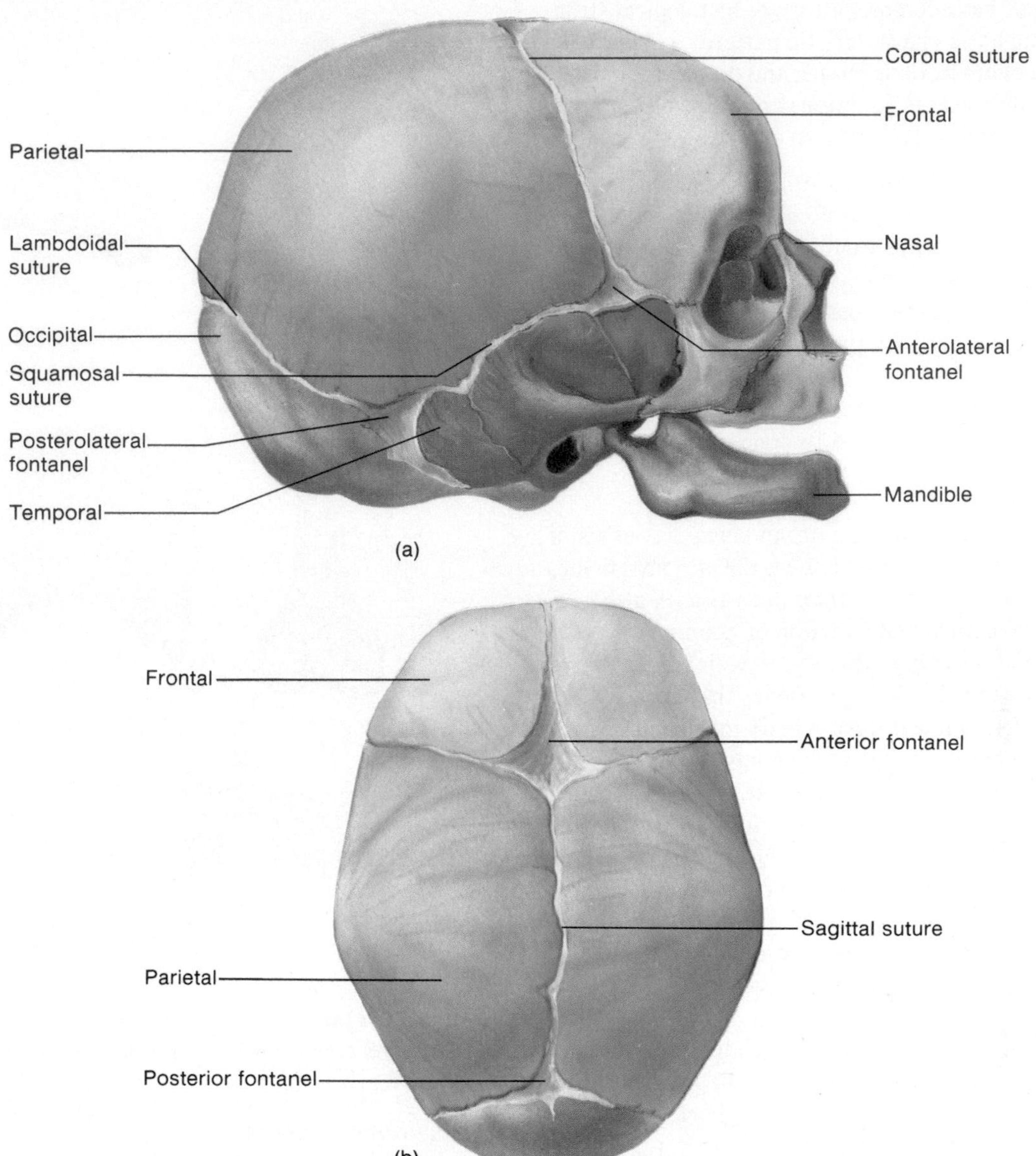

suture extends from the posterior fontanel to the posterolateral fontanel. A **squamosal suture** connects the posterolateral fontanel to the anterolateral fontanel.

During normal parturition, the *molding* of the fetal skull is such that the occipital bone is usually pressed under the two parietal bones. In addition, one parietal bone overlaps the other, with the depressed one against the promontory of the mother's sacrum. If a baby is born *breech* (buttocks first), molding does not occur and delivery is more difficult.

The viscerocranium is the portion of the skull that forms from the **branchial** *(brang'ke-al),* or **pharyngeal, arches,** which are apparent in the neck region of an embryo during the fourth week of development. In aquatic vertebrates (fish and amphibians), the branchial arches participate in the formation of gills. In humans and other mammals, these arches form the jaws and ear ossicles of the skull and the hyoid bone and larynx within the neck (fig. 9.6). The viscerocranium, like the neurocranium, develops through both endochondral and intramembranous ossification.

1. Discuss how endochondral ossification differs from intramembranous ossification.
2. What are somites, and how do they participate in the formation of the vertebral column?
3. Referring to the neurocranium, viscerocranium, chondrocranium, and dermatocranium, describe the formation of the skull, ear ossicles, and hyoid bone.
4. Describe the fontanels, and list their two functions.

branchial: L. *branchion,* gill

Bone Structure and Growth

Although bones have commonalities of histological structure, each bone has a characteristic pattern of ossification and growth, a characteristic shape, and diagnostic surface features that indicate its functional relationship to other bones, muscles, and to the body structure as a whole.

Objective 6. Classify bones according to their shapes, and give an example of each type.
Objective 7. Describe the various markings on the surfaces of bones.
Objective 8. Describe the gross features of a typical long bone, and list the functions of each structure.
Objective 9. Describe the process of endochondral ossification as related to bone growth.

Bone Structure

Each bone of the skeleton is an organ since it consists of several types of tissue. Osseous tissue is the principal tissue, but nervous, vascular, and cartilaginous tissues also contribute to the structure and function of bone.

The shape and surface features of each bone indicate its functional role in the skeleton. Bones that are long, for example, function as levers during body movement. Bones that support the body are massive and have large articular surfaces and processes for muscle attachment. Roughened areas on these bones may serve for the attachment of ligaments, tendons, or muscles. A flattened surface provides placement for a large muscle or may serve for protection. Grooves around an articular end of a bone are where a tendon or nerve passes, and openings through a bone permit the passage of nerves or vessels.

Shapes of Bones. The bones of the skeleton are classified into four principal types on the basis of shape rather than size. The four classes are long bones, short bones, flat bones, and irregular bones (fig. 9.8).

1. **Long bones.** Long bones are longer than they are wide and function as levers. Most of the bones of the upper and lower extremities are of this type (e.g., the humerus, radius, ulna, metacarpals, femur, tibia, fibula, metatarsals, and phalanges).
2. **Short bones.** Short bones are somewhat cube-shaped and are found in confined spaces where they transfer forces (e.g., the wrist and ankle).
3. **Flat bones.** Flat bones have a broad, dense surface for muscle attachment or protection of underlying organs (e.g., the cranium, ribs, and bones of the shoulder girdle).
4. **Irregular bones.** Irregular bones have varied shapes and have many surface markings for muscle attachment or articulation (e.g., the vertebrae and certain bones of the skull).

Figure 9.8. Shapes of bones. (*a*) a long bone; (*b*) a short bone; (*c*) a flat bone; (*d*) an irregular bone.

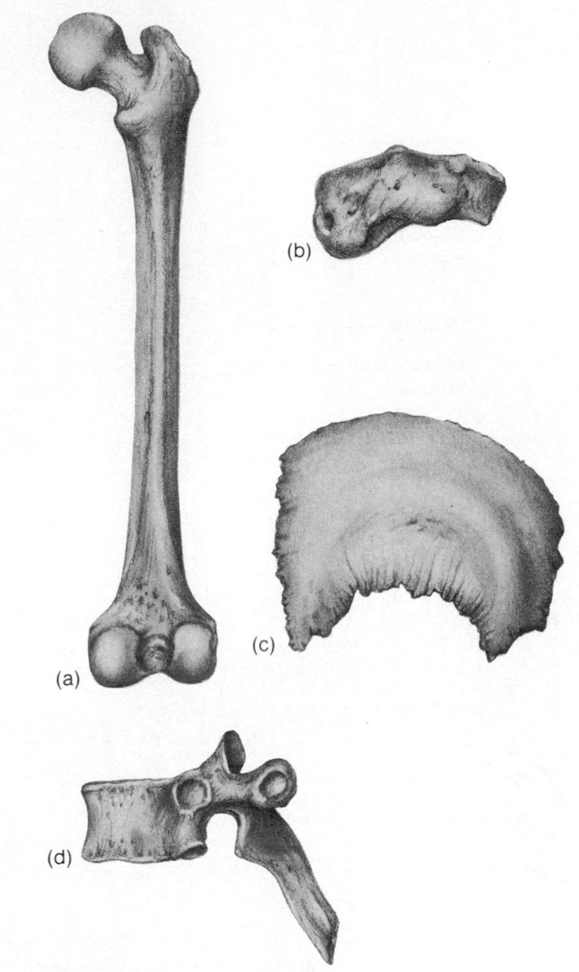

Surface Features of Bone. The following are the various descriptive terms (with examples) used to identify the surface features of bone:

Articulating surfaces

condyle A large, rounded, articulating knob (the occipital condyle of the occipital bone)

facet *(fas'et)* A flattened or shallow articulating surface (the costal facet of a thoracic vertebra)

head A prominent, rounded, articulating end of a bone (the head of the femur)

Nonarticulating prominences

crest A narrow, ridgelike projection (the iliac crest of the os coxa)

epicondyle A projection above a condyle (the medial epicondyle of the femur)

process Any marked, bony prominence (the mastoid process of the temporal bone)

spine A sharp, slender process (the spine of the scapula)

trochanter *(tro-kan'ter)* A massive process found only on the femur (the greater trochanter of the femur)

---

facet: Fr. *facette,* little face
trochanter: Gk. *trochanter,* runner

Figure 9.9. A section through the skull showing diploe. Diploe is a layer of cancellous bone sandwiched between two surface layers of compact bone.

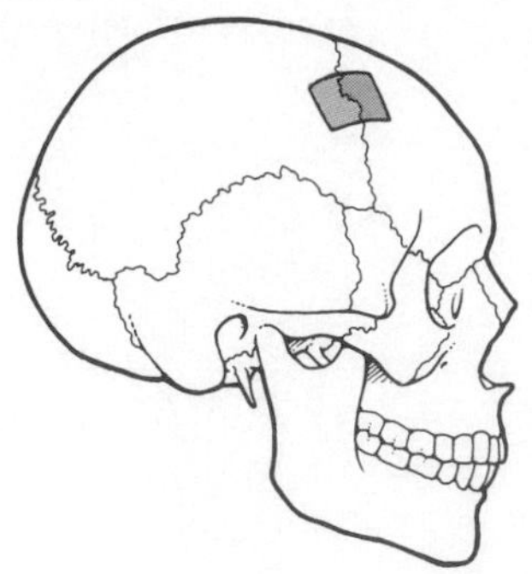

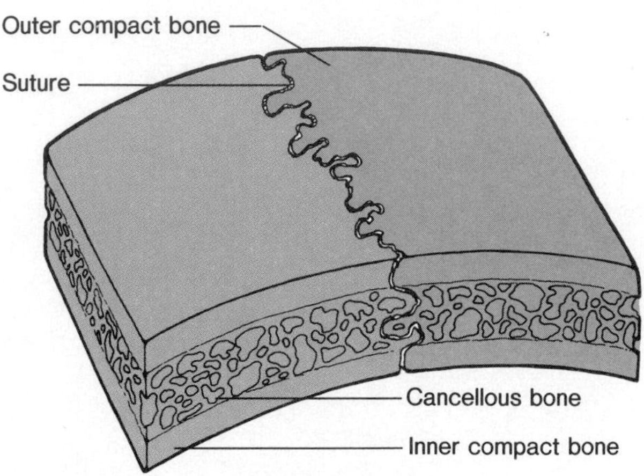

tubercle *(tu'ber-kl)* A small, rounded process (the greater tubercle of the humerus)

tuberosity A large, roughened process (the radial tuberosity of the radius)

Depressions and openings

alveolus *(al-ve'o-lus)* A deep pit or socket (the alveoli for teeth in the maxilla)

fissure A narrow, slitlike opening (the superior orbital fissure of the sphenoid bone)

foramen *(fo-ra'men)*, pl. **foramina;** A rounded opening through a bone (the foramen magnum of the occipital bone)

fossa A flattened or shallow surface (the mandibular fossa of the temporal bone)

fovea A small pit or depression (the fovea capitis of the femur)

meatus *(me-a'tus)*, or **canal;** A tubelike passageway through a bone (the external auditory meatus of the temporal bone)

sinus A cavity or hollow space in a bone (the frontal sinus of the frontal bone)

sulcus *(sul'kus)* A groove that accommodates a vessel, nerve, or tendon (the intertubercular sulcus of the humerus)

tuberosity: L. *tuberosus*, lump

Figure 9.10. A diagram of a long bone shown in longitudinal section.

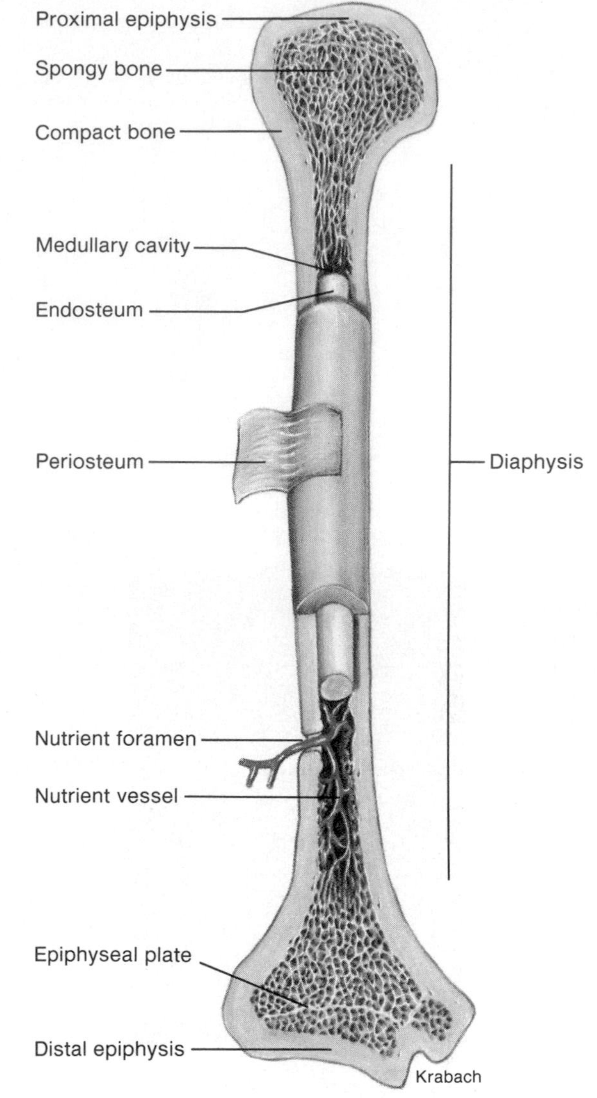

Gross Structure of Bone. The arrangement of **osteocytes,** or bone cells, within **Haversian systems,** which form **osseous tissue,** was described in chapter 6. Two types of osseous tissue are found in bone: a dense **compact** type and a **spongy,** or **cancellous,** type. In a flat intramembranous bone of the skull, the cancellous bone is sandwiched between compact bone and is called **diploe** *(dip'lo-e)* (fig. 9.9). In a long bone from an appendage, the bone shaft, or **diaphysis** *(di-af'ĭ-sis)*, consists of compact bone forming a cylinder that surrounds a central cavity called the **medullary cavity** (fig. 9.10). The medullary cavity is lined with a thin layer of connective tissue called the **endosteum** *(en-dos'te-um)* and contains **yellow bone marrow,** so named because of the large amounts of fat it contains. On each end of the diaphysis is an **epiphysis** *(ĕ-pif'ĭ-sis)*, consisting of cancellous bone surrounded by a layer of compact bone. **Red bone marrow** is found within

diploe: Gk. *diplous*, double
diaphysis: Gk. *dia*, throughout; *physis*, growth
epiphysis: Gk. *epi*, upon; *physis*, growth

Figure 9.11. The process of endochondral ossification, beginning with (*a*) the cartilaginous model as it occurs in a fetus. The bone develops (*b–e*) through intermediate stages to (*f*) adult bone. (*g*) the activity of ossification is shown in a photomicrograph from an epiphyseal region (63×).

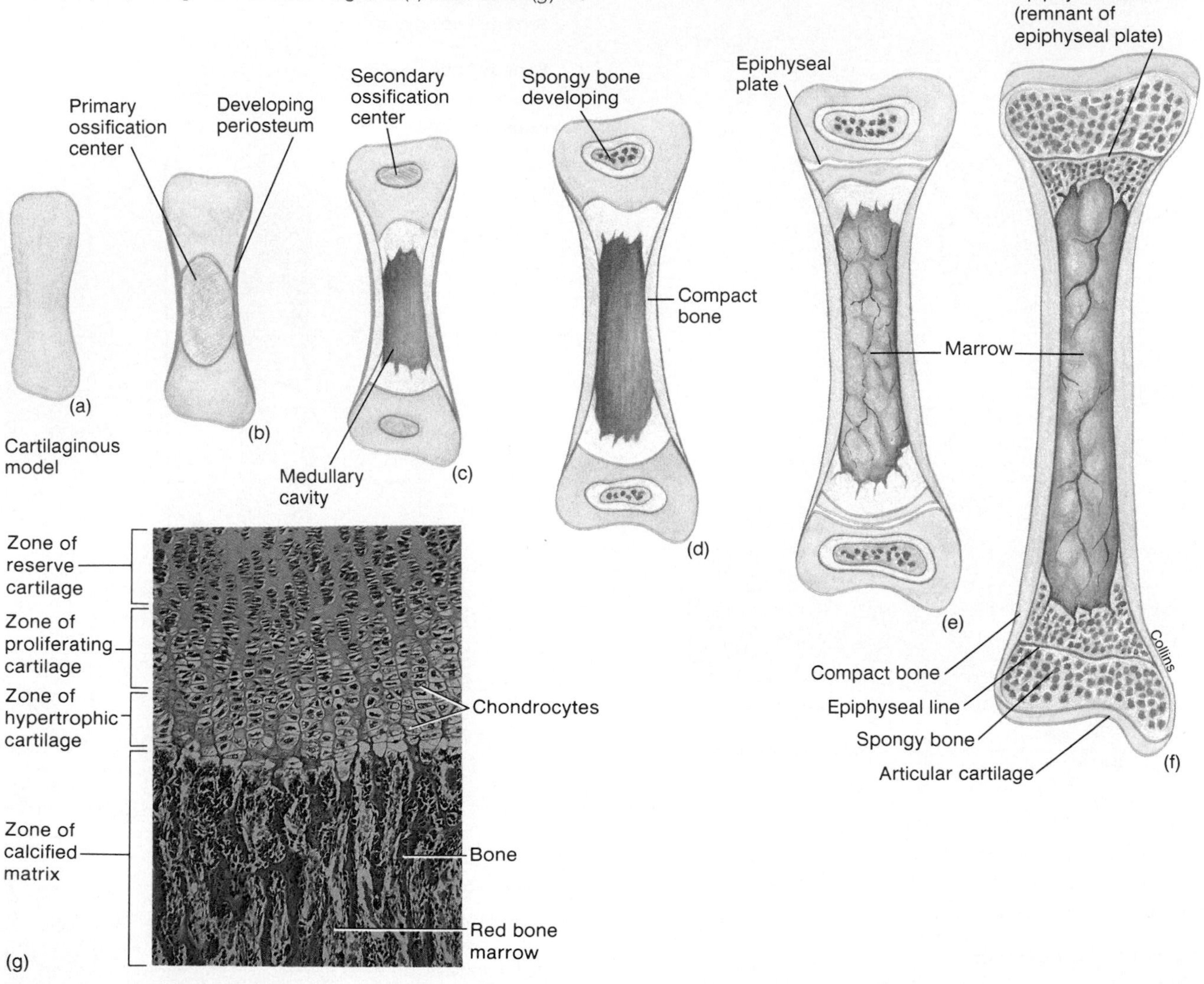

the porous chambers of spongy bone. In an adult, erythropoiesis, the production of red blood cells, occurs in the red bone marrow, especially that of the sternum, vertebrae, portions of the ossa coxae, and the proximal epiphyses of the femora and humeri. The red bone marrow is also responsible for the formation of certain white blood cells and platelets and for the phagocytosis of worn-out red blood cells. **Articular cartilage,** which is composed of thin hyaline cartilage, caps each epiphysis and facilitates joint movement. Along the diaphysis are **nutrient foramina,** small openings into the bone that allow passage of nutrient vessels into the bone for nourishment of the living tissue.

Between the diaphysis and epiphysis is an **epiphyseal** *(ep''ĭ-fiz'e-al)* **plate** of cartilage, a region of mitotic activity that is responsible for elongation of bone. As bone growth is completed, an **epiphyseal line** replaces the plates and ossification occurs between the epiphysis and the diaphysis. A **periosteum** *(per''e-os'te-um)* of dense, white fibrous tissue covers the surface of the bone. This highly vascular layer serves as a place for a tendon-muscle attachment and is responsible for diametric (width) bone growth.

Fracture of a long bone in a young person may be especially serious if it results in displacement of an epiphyseal plate. If such an injury is untreated, or treated improperly, longitudinal growth of the bone may be arrested or retarded, resulting in permanent shortening of the limb.

Bone Growth

The development of bone from embryonic to adult size depends on the orderly processes of mitotic divisions, growth, and the structural remodeling determined by genetics, hormonal secretions, and nutritional supply. In most bone development, a cartilaginous model is gradually replaced by osseous tissue during endochondral bone formation (fig. 9.11). As the cartilage model grows, the chondrocytes (cartilage cells) in the center of the shaft hypertrophy, and minerals are deposited within the matrix in a process called

periosteum: Gk. *peri,* around; *osteon,* bone

Figure 9.12. The presence of epiphyseal plates, as seen in an X ray of a child's hand, indicates that bones are still growing in length.

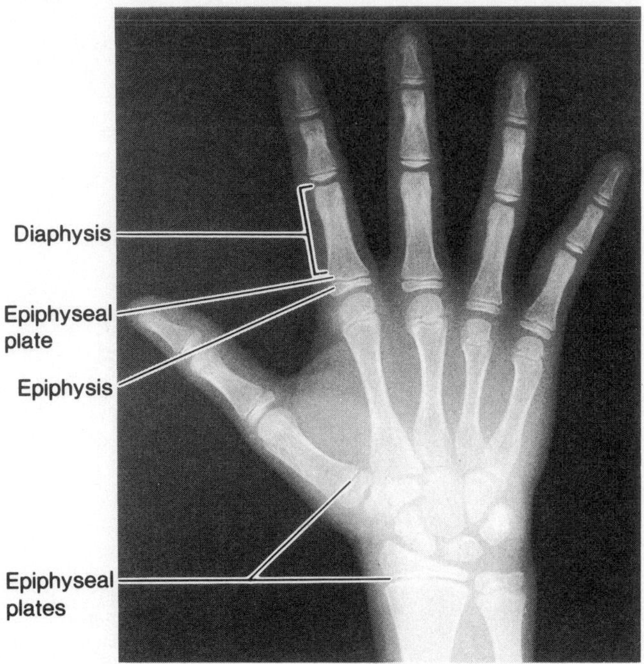

calcification. Calcification restricts the passage of nutrients to the chondrocytes, causing them to die. At the same time, some cells of the perichondrium (dense fibrous connective tissue surrounding cartilage) differentiate into primordial bone cells called **osteoblasts,** which secrete **osteoid,** the organic component of bone. As the perichondrium calcifies, it gives rise to a thin plate of compact bone, called the **periosteal bone collar,** that is surrounded by the periosteum.

A **periosteal bud,** consisting of osteoblasts and blood vessels, invades the disintegrating center of the cartilage model from the periosteum. Once in the center, the osteoblasts secrete osteoid, and a **primary ossification center** is established from which ossification expands into deteriorating cartilage. This process is repeated in both the proximal and distal epiphyses, forming **secondary ossification centers.**

Bone growth continues as long as the cartilage cells at the epiphyseal plate between the two ossification centers continue to divide (fig. 9.12). When mitosis of the chondrocytes gradually decreases, ossification finally occurs within the plate, thus prohibiting the further lengthwise growth of the bone. The time when epiphyseal plates ossify varies greatly from bone to bone, but generally it occurs between the ages of eighteen and twenty within the long bones of most persons (table 9.2).

Bone remodeling is a continual process throughout a person's life. The diagnostic processes on the surface of bones develop as stress is applied to the periosteum, resulting in the osteoblastic secretion of osteoid and the formation of new bone tissue. These processes may continue to change somewhat in persons who are athletically active even though they have stopped growing in height.

Specialized bone cells called **osteoclasts** can enzymatically cause bone resorption. As new bone layers are deposited on the outside surface of the bone, osteoclasts dissolve bone tissue adjacent to the medullary cavity. In this way, the size of the cavity keeps pace with the increased growth of the bone.

Normal bone growth and the maintenance of healthy bones is dependent on certain physiological and even physical factors. The physiological factors and their effect on bone growth and function are summarized in table 9.3.

osteoblast: Gk. *osteon,* bone; *blastos,* offspring or germ

osteoclast: Gk. *osteon,* bone; *klastos,* broken

Table 9.2 Average age of completion of bone ossification

Bone	Chronological age of fusion
Scapula	18–20
Clavicle	23–31
Bones of upper extremity (brachium, forearm, hand)	17–20
Os coxa	18–23
Bones of lower extremity (thigh, leg, foot)	18–22
Vertebra	25
Sacrum	23–25
Sternum (body)	23
Sternum (manubrium, xiphoid)	after 30

From Kent M. Van De Graaff, *Human Anatomy,* 2d ed.

Table 9.3 Factors affecting bone physiology

Substance	Effect
Growth hormone	Stimulates osteoblast activity and collagen synthesis
Thyroid hormones	Stimulate osteoblast activity, collagen synthesis, and formation of ossification centers
Sex hormones (especially androgens)	Stimulate osteoblast activity and bone growth
Adrenocorticoid hormones	Stimulate osteoclast activity
Vitamin A	Promotes chondrocyte function; synthesis of lysosomal enzymes for osteoclast activity
Vitamin C	Promotes synthesis of collagen

Radiologists can determine the ages of persons who are still growing by examining X-ray pictures of their bones. Ossification at the epiphyseal plates occurs at predictable periods of development and is therefore a reliable indicator of age. Large discrepancies between bone age and chronological age may indicate a genetic or endocrine dysfunction.

A normal hormonal balance maintained by feedback mechanisms and a continual source of specific essential minerals and vitamins are necessary for healthy bones. Consider, for example, that *hypothyroidism* results in decreased bone resorption and slowed ossification, and *hyperthyroidism* results in increased bone resorption and premature calcification of the epiphyseal plates. The decrease in sex hormones that accompanies aging causes an increase in bone resorption and loss of bone mass (*osteoporosis* [*os″te-o-po-ro′sis*]). Excessive glucocorticoids also results in osteoporosis, as evidenced in *Cushing's syndrome*. Although vitamin A is needed for chondrocyte and osteoclastic activities, an excess of vitamin A is more serious than a deficiency because it increases demineralization, making bones brittle and easily fractured.

The size, shape, and processes of bones depend on the physical stresses and pressures applied. Initial bone formation, bone growth, and the adaptation of mature bone to units of applied force is reflected in the **stress lines,** or structure of ossification within a bone (fig. 9.13). Stress lines provide maximum strength to the bone in the direction of greatest applied pressure. **Compression lines** are stress lines that are linear and parallel to the applied weight, as in weight-bearing bones like the tibia. **Tension lines** develop in bones subjected to tensile forces such as in the head and neck of the femur, which is positioned on a supportive angle lateral to the os coxa. The diagnostic processes on some bones, such as the greater trochanter of the femur, develop in response to forces of stress applied to the periosteum of the bone where the tendons of muscles attach. In other words, as a muscle is contracted, a pulling stress is applied to the periosteum of the bone where the tendon is attached. This mechanical tension causes an increase in osteoblastic activity and bone deposition, and a bony process is gradually formed.

Bone is highly dynamic and is continually being remodeled in response to mechanical stress or even the absence of stress. The effect of the absence of stress can best be seen in the bones of inactive patients confined to bed or the bones of persons who are paralyzed. X-ray examination shows the loss of bone mass or even osteoporosis. The absence of gravity that accompanies space flight may result in mineral loss from bones if an exercise program is not maintained.

The movement of teeth in orthodontics involves bone remodeling. The teeth sockets (alveoli) are reshaped through the activity of osteoclast and osteoblast cells as stress is applied through the application of braces. The use of traction in treating certain skeletal disorders has a similar effect.

Figure 9.13. A longitudinal section of the proximal end of the femur showing stress lines.

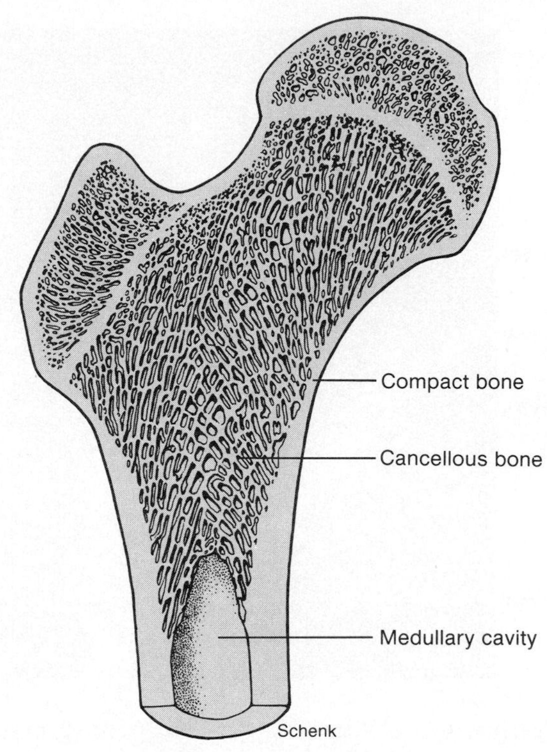

1. Using examples, discuss the function of each of the four kinds of bones as determined by shape.
2. Define each of the following surface markings on bones: condyle, head, facet, process, crest, epicondyle, fossa, alveolus, foramen, and sinus.
3. Diagram a sagittal view of a typical long bone, and label the following: diaphysis, medullary cavity, epiphyses, articular cartilages, nutrient foramen, periosteum, and epiphyseal plates. Explain the function of each of these structures.
4. Define osteocytes, osteoblasts, and osteoclasts, and explain the function of each as related to the processes of endochondral ossification and bone growth.

Skull

The human skull, consisting of eight cranial and fourteen facial bones, contains several cavities that house the brain and sensory organs. Each bone of the skull articulates the adjacent bones and has diagnostic and functional processes or other surface features and foramina.

Objective 10. Identify the cranial and facial bones of the skull, and describe their structural characteristics.

Objective 11. Describe the location of each of the bones of the skull, and name the articulations that affix them together.

Table 9.4 Summary of major foramina of the skull

Foramen	Location	Structures transmitted
Carotid canal	Petrous portion of temporal bone	Internal carotid artery and sympathetic nerves
Greater palatine	Palatine bone of hard palate	Greater palatine nerve and descending palatine vessels
Hypoglossal foramen/canal	Anterolateral edge of the occipital condyle	Hypoglossal nerve and branch of ascending pharyngeal artery
Incisive	Anterior region of hard palate, posterior to the incisor teeth	Branches of descending palatine vessels and nasopalatine nerve
Inferior orbital	Between maxilla and greater wing of sphenoid	Maxillary branch of trigeminal nerve, zygomatic nerve, and infraorbital vessels
Infraorbital	Inferior to orbit in maxilla	Infraorbital nerve and artery
Jugular	Between petrous portion of temporal and occipital, posterior to carotid canal	Internal jugular vein; vagus, glossopharyngeal, and accessory nerves
Lacerum	Between petrous portion of temporal and sphenoid	Branch of ascending pharyngeal artery
Lesser palatine	Posterior to greater palatine foramen in hard palate	Lesser palatine nerves
Magnum	Occipital bone	Union of medulla oblongata and spinal cord, meningeal membranes, accessory nerves; vertebral and spinal arteries
Mandibular	Medial surface of ramus of mandible	Inferior alveolar nerve and vessels
Mental	Below the second premolar on the lateral side of mandible	Mental nerve and vessels
Nasolacrimal canal	Lacrimal bone	Nasolacrimal (tear) duct
Olfactory	Cribriform plate of the ethmoid	Olfactory nerves
Optic	Back of orbit in lesser wing of sphenoid	Optic nerve and ophthalmic artery
Ovale	Greater wing of sphenoid	Mandibular branch of trigeminal nerve
Rotundum	Within body of sphenoid	Maxillary branch of trigeminal nerve
Spinosum	Posterior angle of sphenoid	Middle meningeal vessels
Stylomastoid	Between styloid and mastoid processes of temporal	Facial nerve and stylomastoid artery
Superior orbital fissure	Between greater and lesser wings of sphenoid	Four cranial nerves (oculomotor, trochlear, ophthalmic branch of trigeminal and abducens)
Supraorbital	Supraorbital ridge of orbit	Supraorbital nerve and artery
Zygomaticofacial	Anterior surface of zygomatic bone	Zygomaticofacial nerve and vessels

The skull consists of **cranial bones** and **facial bones.** The eight bones of the cranium join firmly with one another to enclose and protect the brain and associated sense organs. The fourteen facial bones form the framework for the facial region and support the teeth. A variation in shape and density of the facial bones is a major contributor to the individuality of each human face. The facial bones, with the exception of the bone within the lower jaw, are also firmly interlocked with one another and the cranial bones.

The skull has several cavities. The **cranial cavity** is the largest, with a capacity of about 1,300–1,350 cc. The **nasal cavity** is formed by both cranial and facial bones and is partitioned into two chambers, or **nasal fossae,** by a **nasal septum** of bone and cartilage. Four sets of **paranasal sinuses** are located within the bones surrounding the nasal area and communicate via ducts into the nasal cavity. **Middle** and **inner ear chambers** are positioned inferior to the cranial cavity and house the organs of hearing and balance. The two **orbits** for the eyeballs are formed by facial and cranial bones. The **oral,** or **buccal, cavity** (mouth), which is only partially formed by bone, is completely within the facial region.

The bones of the skull contain numerous foramina to accommodate nerves, vessels, and other structures. Table 9.4 summarizes the foramina of the skull. Figures 9.14 through 9.20 show various views of the skull. The diagrams of the bones of the skull are color-coded to facilitate learning their relative positions. X rays (roentgenograms) of the skull are shown in figure 9.21.

Although the hyoid bone and the three paired ear ossicles are not considered part of the skull, they are within the axial skeleton and will be described immediately following the discussion of the skull.

---

roentgenogram: from Wilhelm K. Roentgen, German physicist, 1845–1923

Figure 9.14. An anterior view of the skull.

Frontal bone
Parietal bone
Coronal suture
Supraorbital foramen
Supraorbital margin
Orbital portion of sphenoid
Ethmoid bone
Lacrimal bone
Perpendicular plate
Infraorbital foramen
Inferior nasal concha
Maxilla
Mental foramen
Optic foramen
Temporal bone
Zygomatic process of frontal
Superior orbital fissure
Nasal bone
Lacrimal foramen
Inferior orbital fissure
Zygomatic bone
Middle nasal concha of ethmoid
Vomer
Mandible

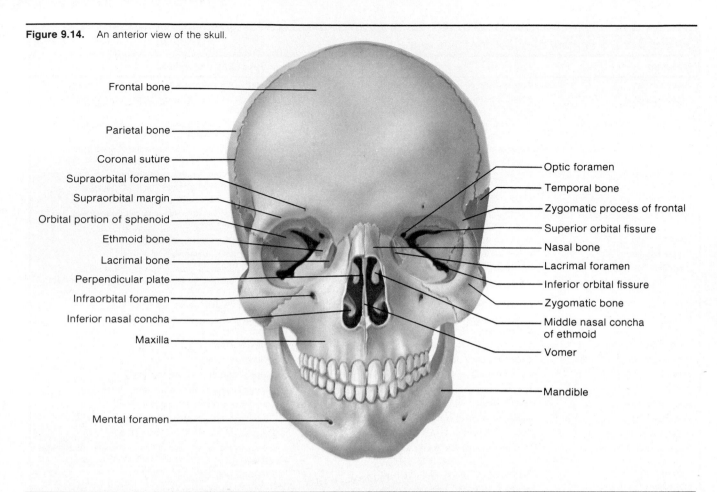

Figure 9.15. A lateral view of the skull.

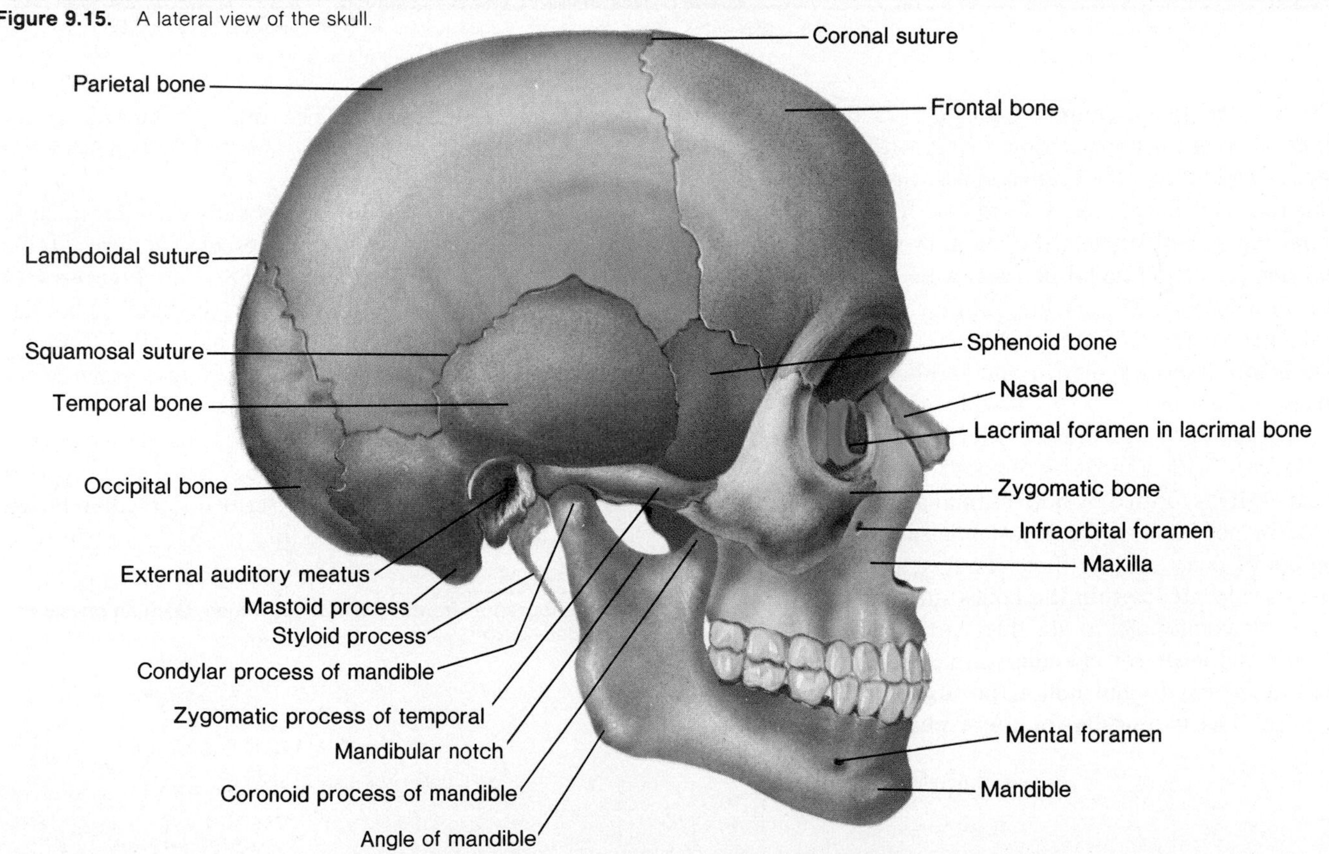

Figure 9.16. An inferior view of the skull.

Incisors
Canine
Premolars
Incisive fossa
Molars
Median palatine suture
Palatine process of maxilla
Zygomatic bone
Sphenoid bone
Palatine bone
Greater palatine foramen
Zygomatic arch
Pterygoid process
Foramen ovale
Vomer
Mandibular fossa
Foramen lacerum
Carotid canal
External auditory meatus
Jugular foramen
Styloid process
Stylomastoid foramen
Mastoid process
Foramen magnum
Occipital condyle
Mastoid foramen
Temporal bone
Lambdoidal suture
Condyloid canal
Superior nuchal line
Occipital
External occipital protuberance

Figure 9.17. A midsagittal view of the skull.

Coronal suture
Sphenoparietal suture
Greater wing of sphenoid bone
Arterial groove
Sphenosquamosal suture
Temporal bone
Frontal bone
Parietal foramen
Parietal bone
Lesser wing of sphenoid bone
Squamosal suture
Sphenofrontal suture
Lambdoidal suture
Orbital portion of frontal bone
Occipital bone
Frontal sinus
Internal auditory meatus
Crista galli
Groove of transverse sinus
Fronto-ethmoidal suture
Cribriform plate of ethmoid bone
External occipital protuberance
Nasal bone
Perpendicular plate of ethmoid bone (nasal septum)
Jugular foramen
Inferior nasal concha
Occipitomastoidal suture
Maxilla
Sella turcica
Anterior nasal spine
Hypoglossal foramen
Petro-occipital fissure
Incisive canal
Mastoid process
Foramen magnum
Styloid process
Palatine process of maxilla
Sphenoidal sinus
Pterygoid process
Posterior nasal spine
Mandible
Vomer

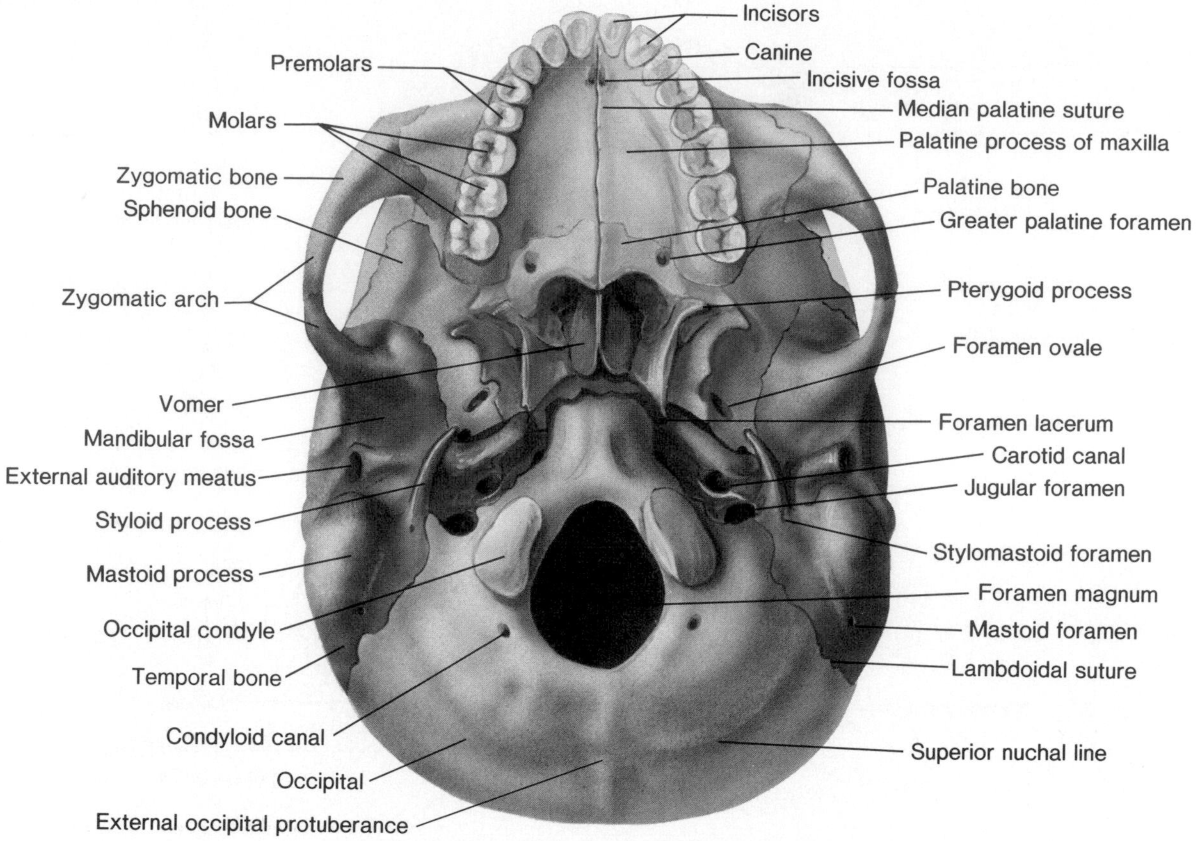

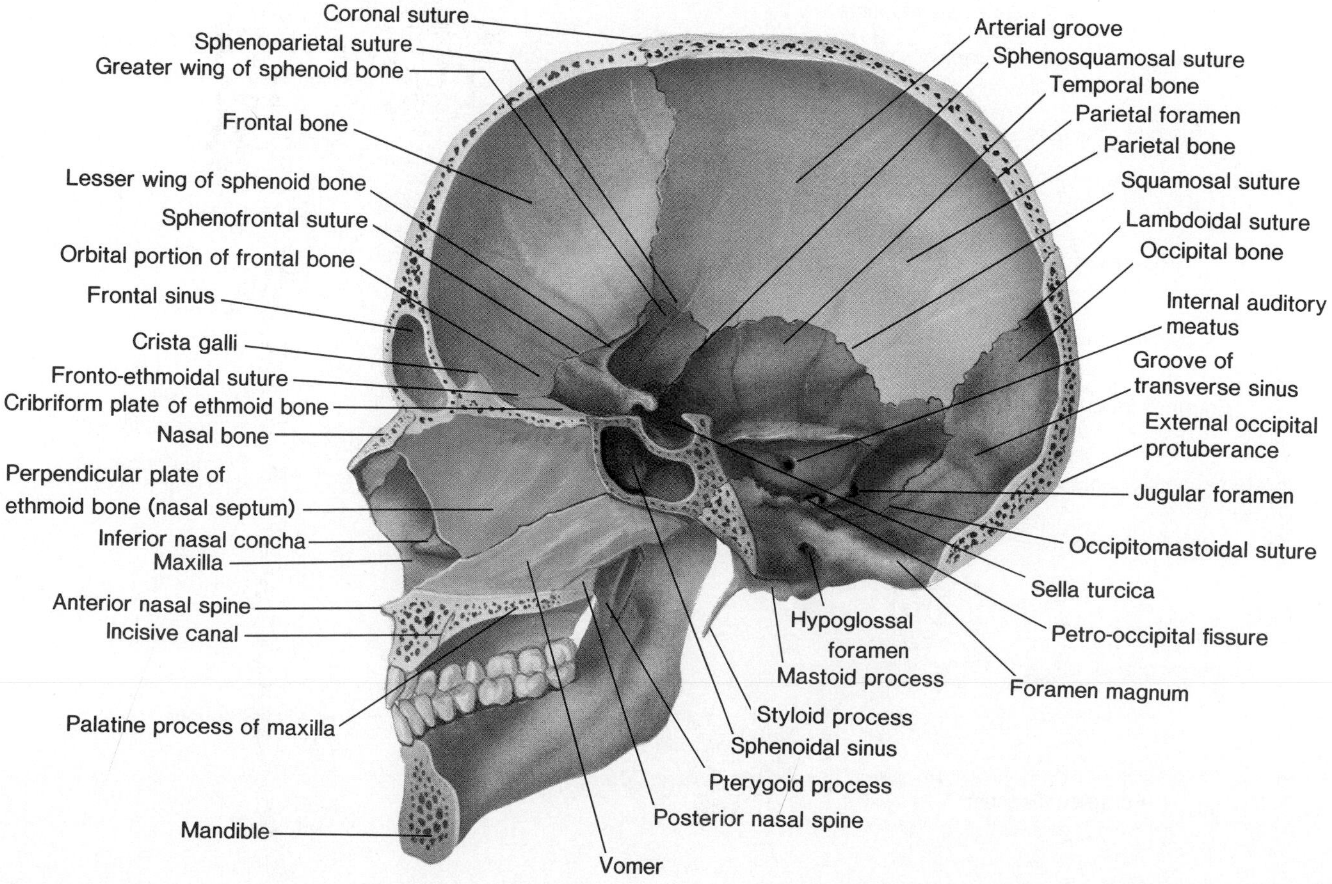

Figure 9.18. A posterior view of a frontal (coronal) section of the skull.

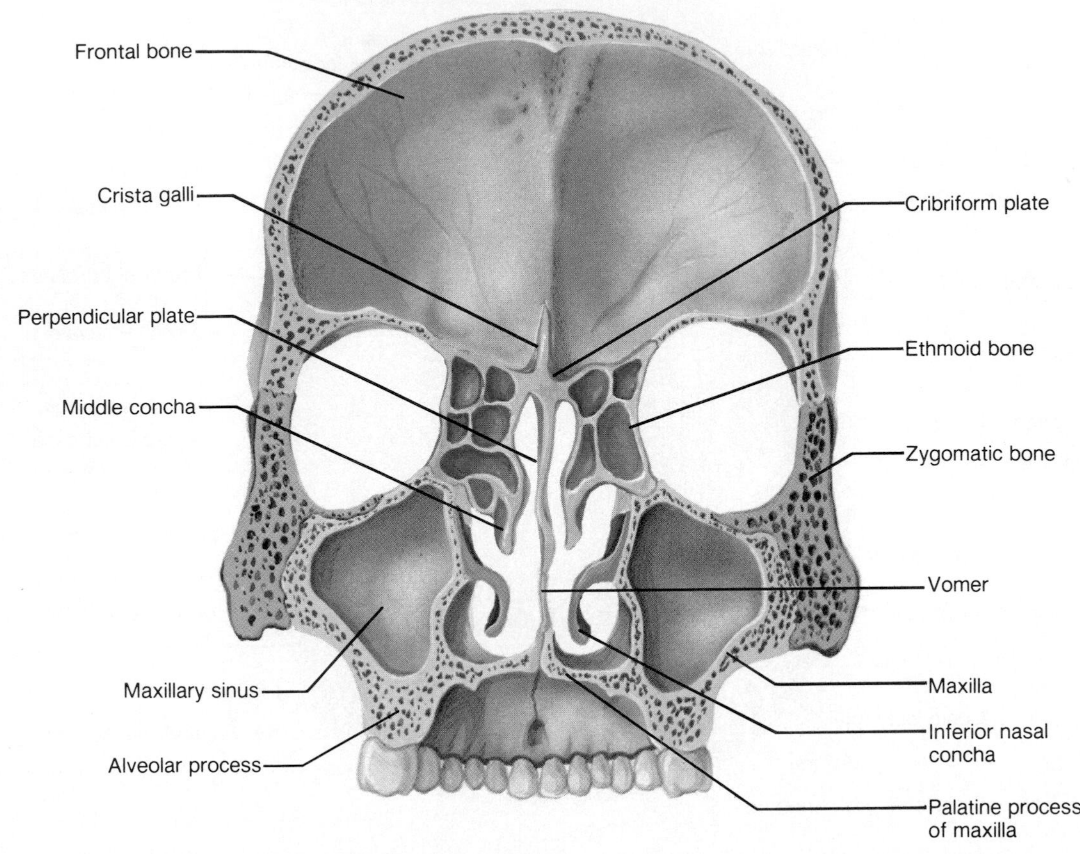

Figure 9.19. The floor of the cranial cavity.

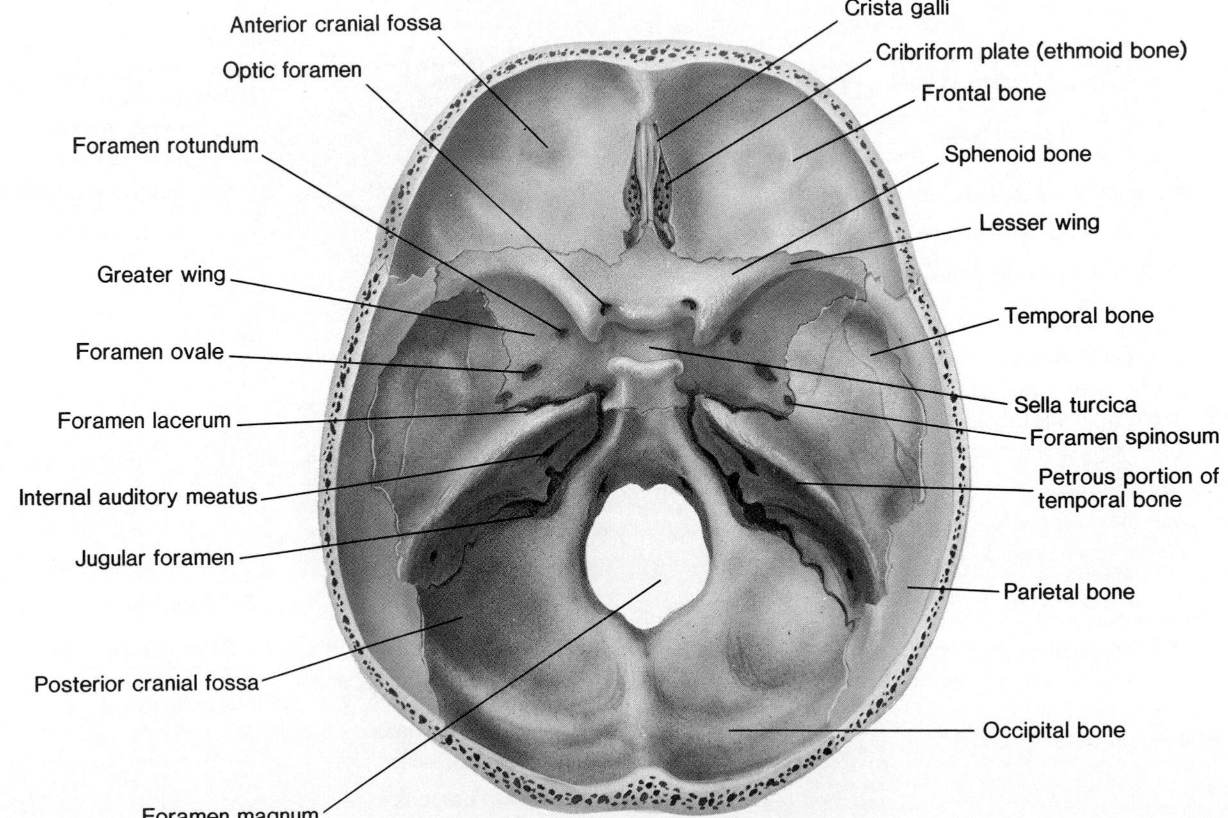

Figure 9.20. Bones of the orbit of the eye.

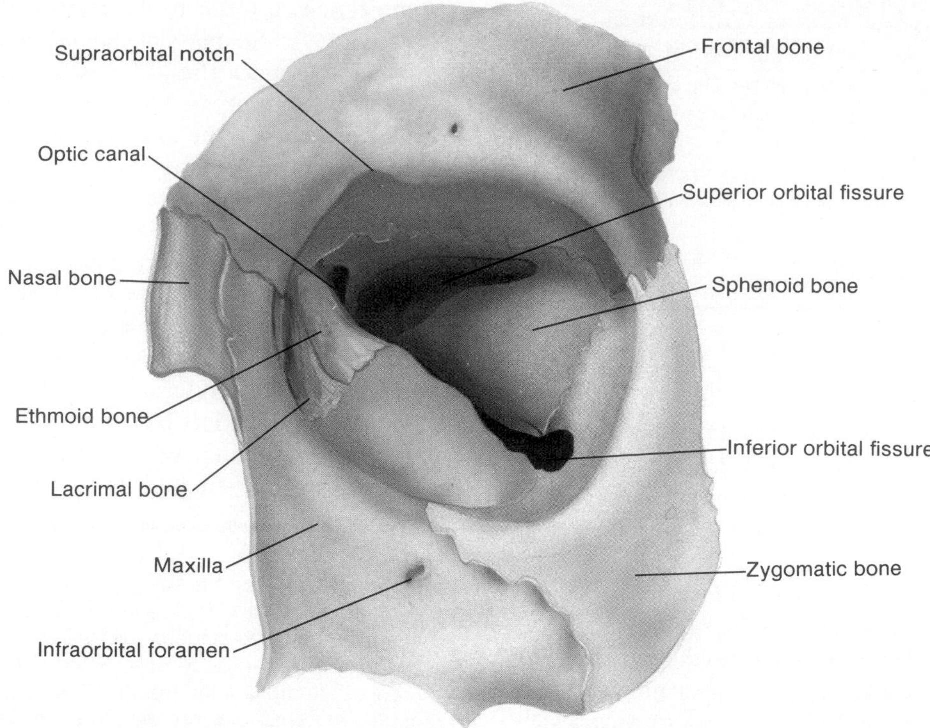

Figure 9.21. X rays of the skull showing paranasal sinuses. (*a*) an anteroposterior view; (*b*) a right lateral view.

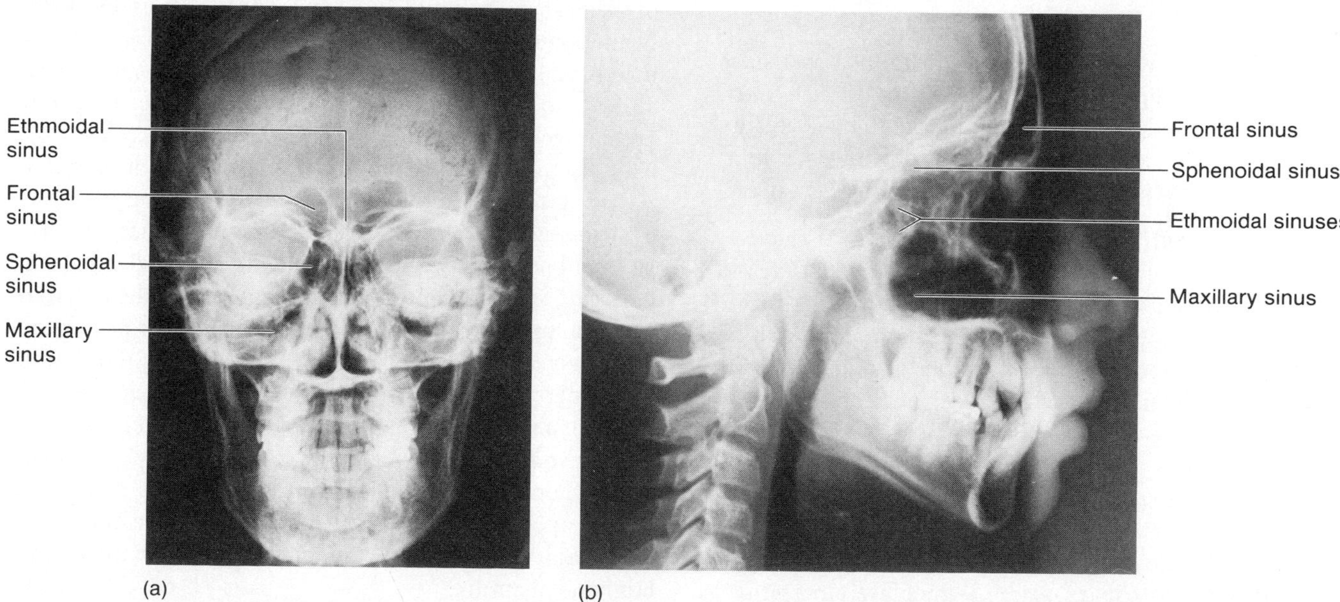

Cranium

The cranial bones enclose the brain and consist of one frontal, two parietals, two temporals, one occipital, one sphenoid, and one ethmoid.

Frontal. The frontal bone forms the anterior roof of the cranium, the forehead, the roof of the nasal cavity, and the superior arch of the **bony orbits,** which contain the eyeballs. Table 9.5 summarizes the bones of the orbit. The frontal bone develops in two halves that grow together and are usually completely fused by age five or six. Occasionally a complete suture persists between these two portions beyond age six and is referred to as a **metopic** *(me-top'ik)* **suture.** The **supraorbital margin** is a prominent bony ridge over the orbit. Openings along this ridge, called supraorbital foramina, allow passage of small nerves and vessels.

cranium: Gk. *kranion,* skull

metopic suture: Gk. *metopon,* forehead; L. *sutura,* sew

Table 9.5 Bones forming the orbit

Region of the orbit	Contributing bones
Roof (superior)	Frontal; lesser wing of sphenoid
Floor (inferior)	Maxilla; zygomatic
Lateral wall	Zygomatic
Posterior wall	Greater wing of sphenoid
Medial wall	Maxilla; lacrimal; ethmoid
Superior rim	Frontal
Lateral rim	Zygomatic
Medial rim	Maxilla

The frontal bone contains a **frontal sinus,** which is connected to the nasal cavity (fig 9.21). This sinus, along with the other paranasal sinuses, lessens the weight of the skull and acts as a sound chamber for voice resonance.

Parietal. The **coronal** *(ko-ro'nal)* **suture** separates the frontal bone from the parietals, and the **sagittal** *(saj'ĭ-tal)* **suture** along the dorsal midline separates the right and left parietal. The parietal bones form the upper sides and roof of the cranium (figs. 9.15, 9.17). The inner concave surface of the parietal bone, as well as the inner concave surfaces of other cranial bones, is marked by shallow impressions from convolutions of the brain and vessels serving the brain.

Temporal. The two temporal bones form the lower sides of the cranium (figs. 9.15, 9.16, 9.17, and 9.22). Each temporal is joined to its adjacent parietal bone by the **squamosal** *(skwa-mo'sal)* **suture.** Structurally, each temporal has four parts.

1. **Squamous portion.** The squamous portion is the flattened plate of bone at the sides of the skull. Projecting forward is a **zygomatic process** that forms the posterior portion of the **zygomatic arch.** On the inferior surface of the squamous portion is the **mandibular fossa,** which receives the articular condyle of the mandible. This articulation is referred to as the **temporomandibular** *(tem''po-ro-man-dib'u-lar)* **joint.**
2. **Tympanic portion.** The tympanic portion of the temporal bone contains the **external auditory meatus,** or ear canal, located immediately posterior to the mandibular fossa. A thin, pointed **styloid process** (figs. 9.16, 9.21) projects downward from the tympanic portion.
3. **Mastoid portion.** The **mastoid process,** a rounded projection posterior to the external auditory meatus, accounts for the mass of the mastoid portion. The **mastoid foramen** (fig. 9.16) is directly posterior to the mastoid process. The **stylomastoid foramen,** located between the mastoid and styloid processes, is the passage for part of the facial nerve.
4. **Petrous portion.** The petrous *(pet'rus)* portion is viewed inferiorly (fig. 9.19). The structures of the middle and inner ear are housed in this dense, bony portion. The **carotid** *(kah-rot'id)* **canal** and the **jugular foramen** border on the medial side of the petrous portion. The carotid canal allows blood into the brain via the internal carotid artery, and the jugular foramen lets blood drain from the brain via the internal jugular vein. Three cranial nerves also pass through the jugular foramen.

The mastoid process of the temporal bone can be easily palpated as a bony knob immediately behind the earlobe. This process contains a number of small sinuses that are clinically important because they can become infected in *mastoiditis.* A tubular communication from the mastoid sinuses to the middle ear chamber may permit prolonged ear infections to spread to this region.

Occipital. The occipital bone forms the back and much of the base of the skull. It is fastened to the parietal bones by the **lambdoidal suture.** The **foramen magnum** is the large hole in the occipital bone through which the spinal cord attaches to the brain. On each side of the foramen magnum are the **occipital condyles** *(ok-sip'ĭ-tal kon'dīlz)* (fig. 9.16), which articulate with the atlas of the vertebral column. At the anterolateral edge of the occipital condyle is the **hypoglossal foramen** (fig. 9.17), through which the hypoglossal nerve passes. A **condyloid** *(kon'dĭ-loid)* **canal** lies posterior to the occipital condyle (fig. 9.16). The **external occipital protuberance** is a prominent posterior projection on the occipital bone that can be felt as a definite bump just under the skin. The **superior nuchal** *(nu'kal)* **line** is a ridge of bone extending laterally from the occipital protuberance to the mastoid portion of the temporal bone. **Sutural,** or **wormian, bones** are small clusters of irregularly shaped bones that may be found between the joints of certain cranial bones but generally occur along the lambdoidal suture.

Sphenoid. The sphenoid *(sfe'noid)* bone forms the anterior base of the cranium and can be viewed laterally and inferiorly (figs. 9.15, 9.16). This bone resembles a bat with

zygomatic: Gk. *zygoma,* yolk
styloid: Gk. *stylos,* pillar

mastoid: Gk. *mastos,* breast
petrous: Gk. *petra,* rock
magnum: L. *magnum,* great
nuchal: Fr. *nuque,* nape of neck

Figure 9.22. A lateral view of the right temporal bone.

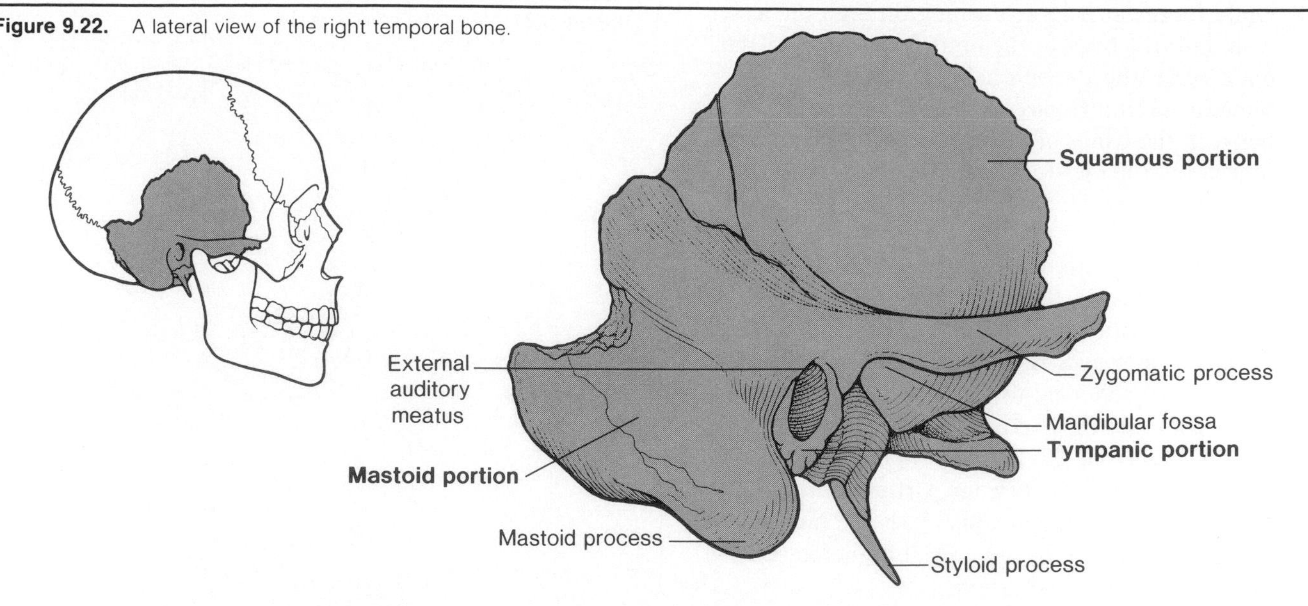

Figure 9.23. The sphenoid bone. (*a*) a superior view; (*b*) a posterior view.

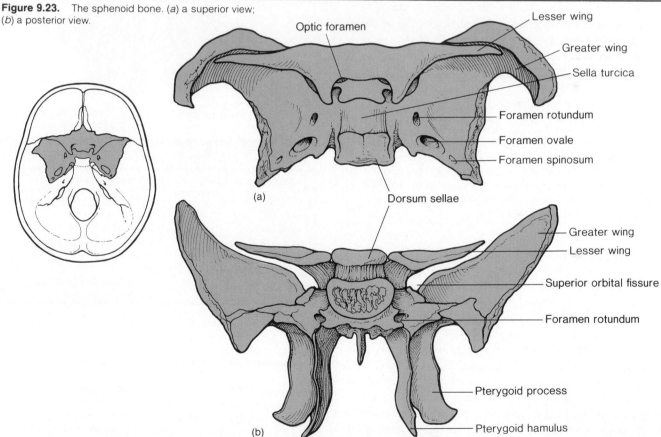

Optic foramen
Lesser wing
Greater wing
Sella turcica
Foramen rotundum
Foramen ovale
Foramen spinosum
(a)
Dorsum sellae
Greater wing
Lesser wing
Superior orbital fissure
Foramen rotundum
Pterygoid process
Pterygoid hamulus
(b)

outstretched wings (fig. 9.23). It consists of a **body** with laterally projecting **greater** and **lesser wings,** which form part of the bony orbit. The body is a wedgelike central portion that contains the **sphenoidal** *(sfe-noi'dal)* **sinuses** and a prominent depression called the **sella turcica** *(sel'ah tur'si-kah),* which supports the pituitary gland. The sella turcica (meaning "Turk's saddle") is seen on the floor of the cranium (fig. 9.19). A pair of **pterygoid** *(ter'i-goid)* **processes** project inferiorly from the sphenoid bone to help form the lateral walls of the nasal cavity. Several foramina (figs. 9.16, 9.19, 9.23) are associated with the sphenoid bone.

sphenoid: Gk. *sphenoeides,* wedgelike

1. **Optic foramen.** A large opening through the lesser wing into the back of the orbit for passage of the optic nerve and the ophthalmic artery.
2. **Superior orbital fissure.** A triangular opening between the wings of the sphenoid for passage of four cranial nerves (oculomotor, trochlear, ophthalmic branch of the trigeminal, and abducens).
3. **Foramen ovale.** An opening at the base of the pterygoid process, through which passes the mandibular branch of the trigeminal nerve.
4. **Foramen spinosum.** A small opening at the posterior angle of the sphenoid for passage of the middle meningeal vessels.
5. **Foramen lacerum** *(las'er-um)*. An opening between the sphenoid and the petrous portion of the temporal bone, through which pass the internal carotid artery and the meningeal branch of the ascending pharyngeal artery.
6. **Foramen rotundum.** An opening located just posterior to the superior orbital fissure at the junction of the anterior and medial portions of the sphenoid bone. The maxillary branch of the trigeminal nerve passes through this foramen.

Located on the inferior side of the cranium, the sphenoid bone would seemingly be well protected from trauma. Actually just the opposite is true, and in fact the sphenoid is the most frequently fractured bone of the cranium. It has several broad, thin platelike extensions that are perforated and weakened with numerous foramina. A blow to almost any portion of the skull causes the buoyed, fluid-filled brain to rebound against the vulnerable sphenoid bone, often causing it to fracture.

Ethmoid. The ethmoid bone is located in the anterior portion of the floor of the cranium between the orbits, where it forms the roof of the nasal cavity (figs. 9.18, 9.24, 9.25). An inferior projection of the ethmoid, called the **perpendicular plate,** contributes in part to the nasal septum that separates the nasal cavity into two chambers referred to as **nasal fossae.** A spine of the perpendicular plate, the **crista galli,** projects superiorly into the cranial cavity and serves as an attachment for the meninges covering the brain. On both lateral walls of the nasal cavity are two scroll-shaped plates of the ethmoid, called the **superior** and **middle nasal conchae.** At right angles to the perpendicular plate, within the floor of the cranium, is the **cribriform plate,** which has numerous perforations for the passage of olfactory nerves from the nasal cavity. Table 9.6 summarizes the bones of the nasal cavity.

conchae: L. *conchae,* shells
cribriform: L. *cribrum,* sieve; *forma,* like

Figure 9.24. An anterior view of the ethmoid bone.

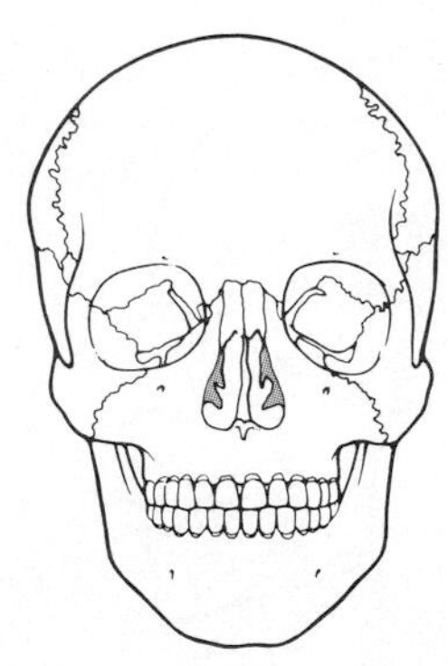

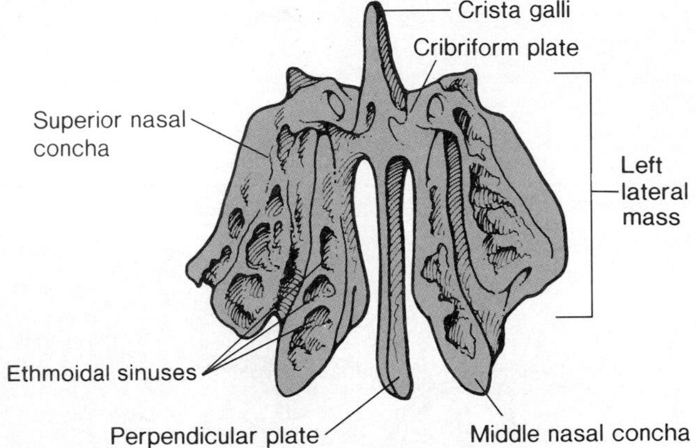

Table 9.6 Bones forming the nasal cavity

Region of nasal cavity	Contributing bones
Roof (superior)	Ethmoid (cribriform plate); frontal
Floor (inferior)	Maxilla; palatine
Lateral wall	Maxilla; palatine
Nasal septum (medial)	Ethmoid (perpendicular plate); vomer; nasal
Bridge	Nasal
Conchae	Ethmoid (superior and middle); inferior nasal conchae

The moist, warm vascular lining within the nasal cavity is susceptible to infections, particularly if a person is not in good health. Infections of the nasal cavity can spread to several surrounding areas. The paranasal sinuses connect to the nasal cavity and are especially prone to infection. The eyes may become reddened and swollen during a nasal infection because of the connection of the nasolacrimal duct, through which tears drain from the orbit to the nasal cavity. Organisms may spread via the auditory canal from the nasopharynx to the middle ear. With prolonged nasal infections, organisms may even ascend to the meninges covering the brain, along the sheaths of the olfactory nerves, and through the cribriform plate to produce *meningitis*.

Figure 9.25. Lateral wall of the nasal cavity.

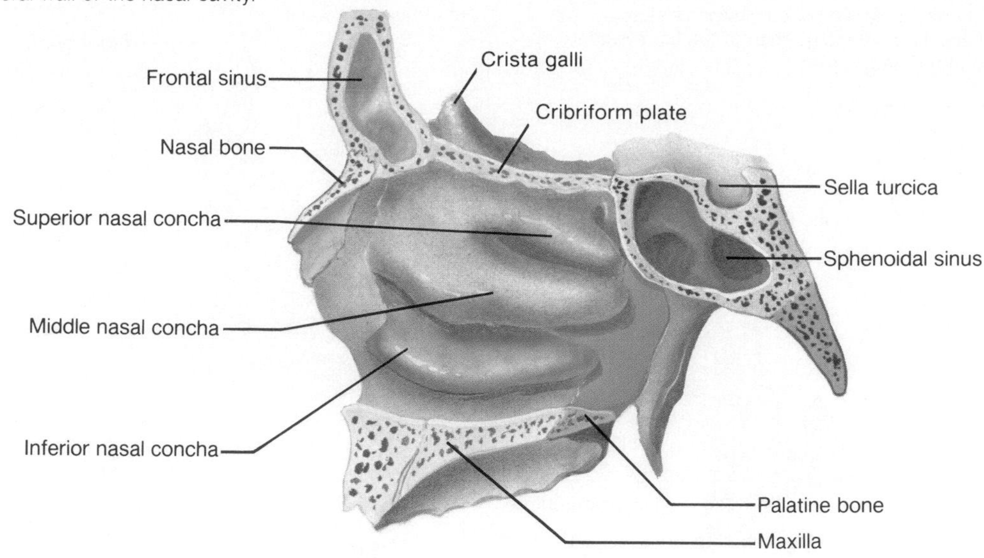

Figure 9.26. The right maxilla in (*a*) lateral view and (*b*) medial view.

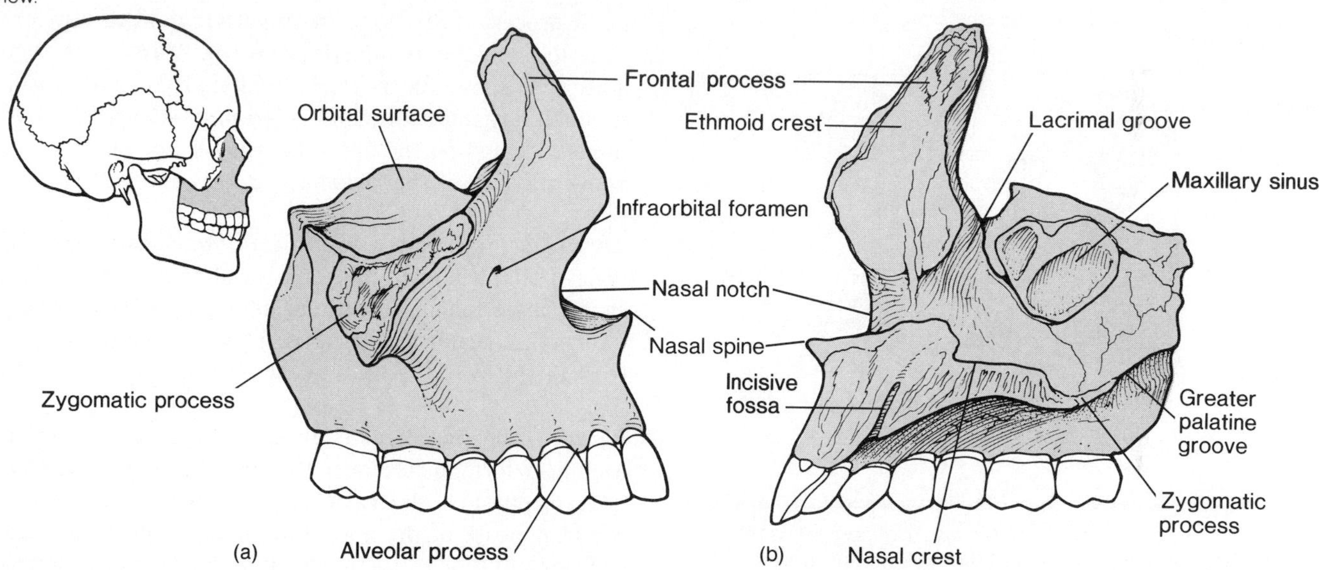

Facial Bones

The fourteen bones of the skull not in contact with the brain are called **facial bones.** These bones, together with certain cranial bones (frontal and portions of the ethmoid and temporals), provide the basic shape of the face. Facial bones also support the teeth and provide attachments for various muscles that move the jaw and cause facial expressions. All the facial bones are paired except the vomer and mandible. The articulated facial bones can be seen in figures 9.14 through 9.20.

Maxilla. The two maxillae unite at the midline to form the upper jaw, which supports the upper teeth. **Incisors** *(in-si'zorz),* **canines** *(ka'nīnz),* **premolars,** and **molars** are contained in sockets, or **alveoli,** within the **alveolar** *(al-ve'o-lar)* **process** of the maxilla (fig. 9.26). The palatine process, a horizontal plate of the maxilla, forms the greater portion of the **hard palate** *(pal'at),* or roof of the mouth. The **incisive fossa** (fig. 9.16) is located in the anterior region of the hard palate behind the incisor teeth. An **infraorbital foramen** is located under each orbit and serves as a passageway for the infraorbital nerve and artery to the nose (figs. 9.14, 9.15, 9.20, 9.26). A final opening within the maxilla is the **inferior orbital fissure.** It is located between the maxilla and the greater wing of the sphenoid (fig. 9.14) and is the opening for the maxillary branch of the trigeminal nerve and infraorbital vessels. The large **maxillary sinus** located within the maxilla is one of the four paranasal sinuses (figs. 9.18, 9.21).

incisor: L. *incidere,* to cut
canine: L. *canis,* dog
molar: L. *mola,* millstone

alveolus: L. *alveus,* little cavity

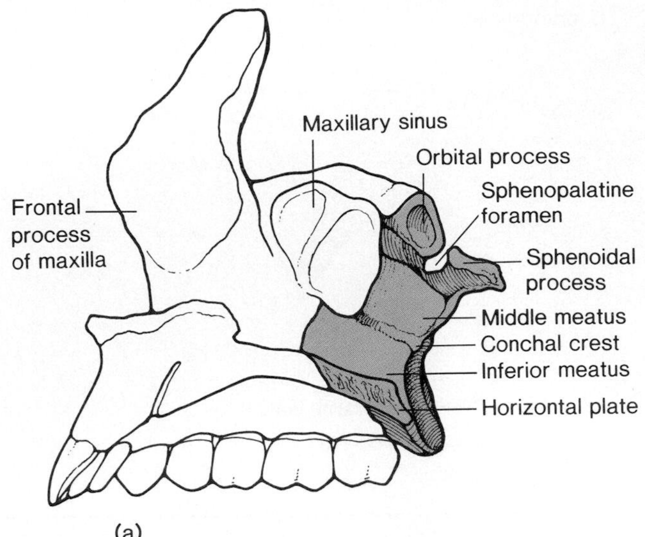

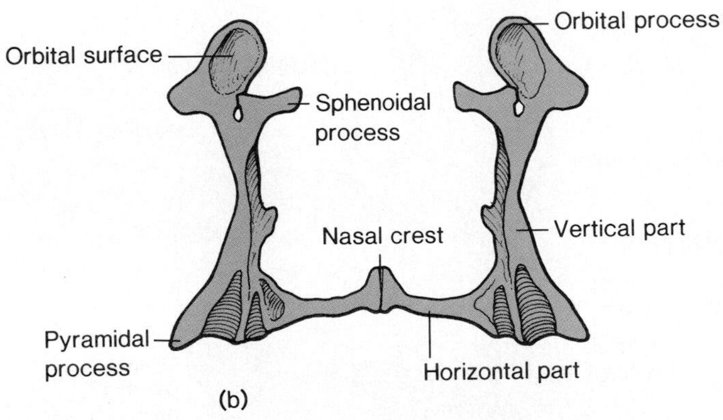

Figure 9.27. (*a*) the right palatine bone as it articulates anteriorly with the maxilla. (*b*) the two palatine bones viewed posteriorly. The two palatine bones form the posterior portion of the hard palate where they articulate at the nasal crest.

If the two palatine processes fail to join during early prenatal development (about twelve weeks), a *cleft palate* results. A cleft palate may be accompanied by a *cleft lip* (harelip) lateral to the midline. These conditions can be surgically and cosmetically treated with excellent success. A more immediate problem, however, is that a newborn with a cleft palate may have a difficult time swallowing while nursing because it is unable to create the necessary suction within the oral cavity.

Palatine. The L-shaped palatine bones form the posterior third of the hard palate, a portion of the orbits, and a part of the nasal cavity. The **horizontal plates** of the palatines contribute to the formation of the hard palate (fig. 9.27). On the hard palate of each palatine bone is a large **greater palatine foramen,** which permits the passage of the greater palatine nerve and descending palatine vessels (fig. 9.16). Two or more smaller **lesser palatine foramina** are positioned posterior to the greater palatine foramina. Branches of the lesser palatine nerve pass through these openings.

Figure 9.28. A lateral view of the bony and cartilaginous structure of the external nose.

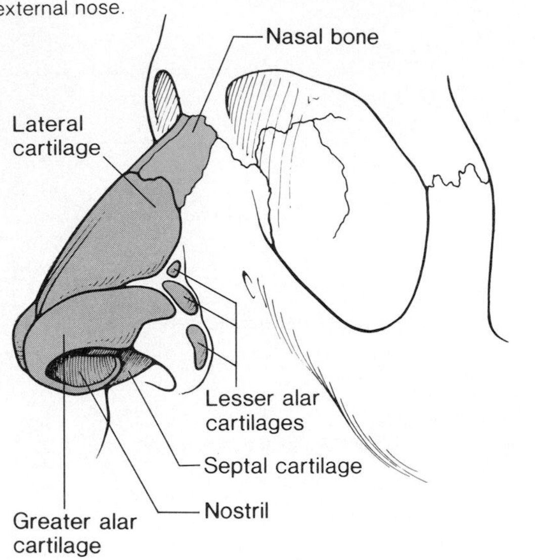

Zygomatic (Malar). The two zygomatic bones form the cheekbones of the face. A posteriorly extending **zygomatic process** of this bone unites with that of the temporal bone to form the **zygomatic arch** (fig. 9.16). The zygomatic bone also forms the lateral margin of the orbit. A small **zygomaticofacial** *(zi''go-mat''ĭ-ko-fa'shal)* **foramen,** located on the anterolateral surface of this bone, allows passage of the zygomatic nerves and vessels.

Lacrimal. The small lacrimals *(lak'rĭ-mal)* are thin bones that form the anterior part of the medial wall of each orbit. Each has a **lacrimal sulcus,** a groove that helps to form the **nasolacrimal canal.** This opening permits the tears of the eye to drain into the nasal cavity (fig. 9.15).

Nasal. The small, rectangular nasal bones join in the midline to form the bridge of the nose (fig. 9.28). The nasal bones support the **lateral cartilages,** which are a part of the framework of the nose. Fractures of the nasal bones or fragmentation of the associated cartilage are common facial injuries.

Inferior Nasal Concha. The two inferior nasal conchae are fragile, scroll-like bones that project horizontally and medially from the lateral walls of the nasal cavity (fig. 9.14). They extend into the nasal cavity just below the superior and middle nasal conchae, which are part of the ethmoid bone (fig. 9.24). The inferior nasal conchae are the largest of the three conchae, and, like the other two, they are covered with mucous membranes to warm, moisten, and cleanse inhaled air.

Vomer. The vomer *(vo'mer)* is a thin, plowshare-shaped bone that forms the lower part of the nasal septum (figs. 9.16, 9.29). The vomer, along with the perpendicular plate of the ethmoid, supports the septal cartilage to complete the nasal septum.

---

vomer: L. *vomer,* plowshare

Figure 9.29. The bones and cartilages that comprise the nasal septum.

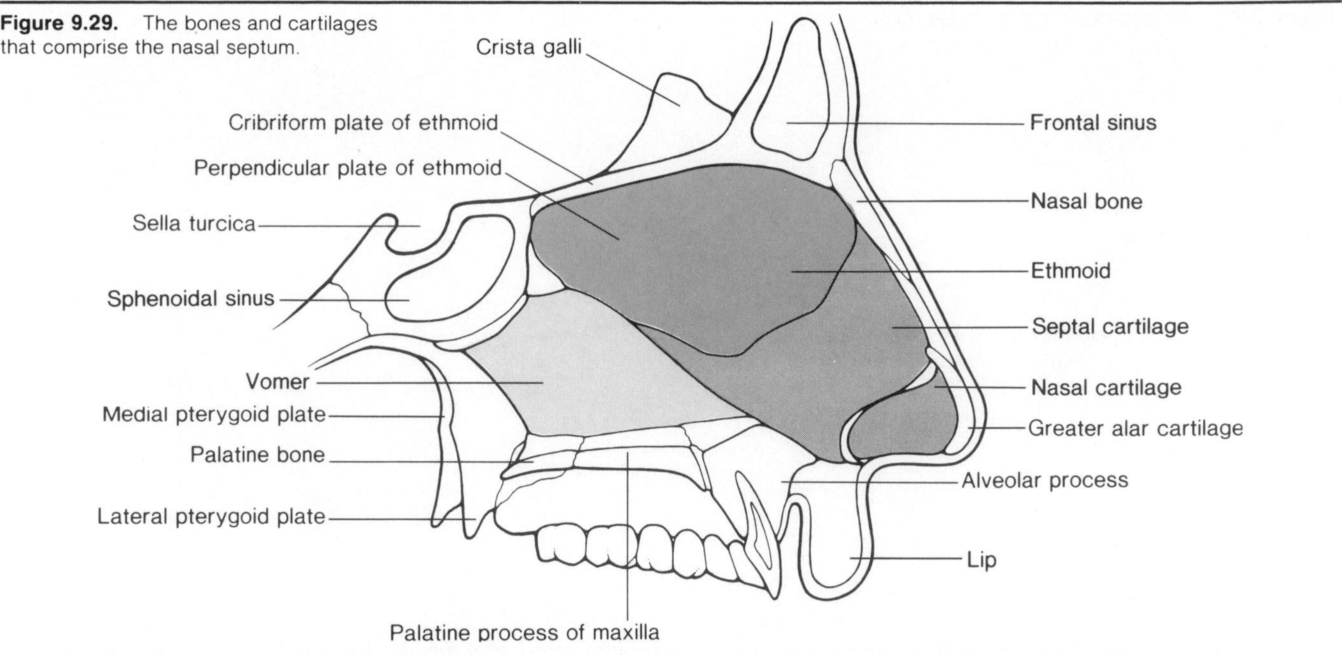

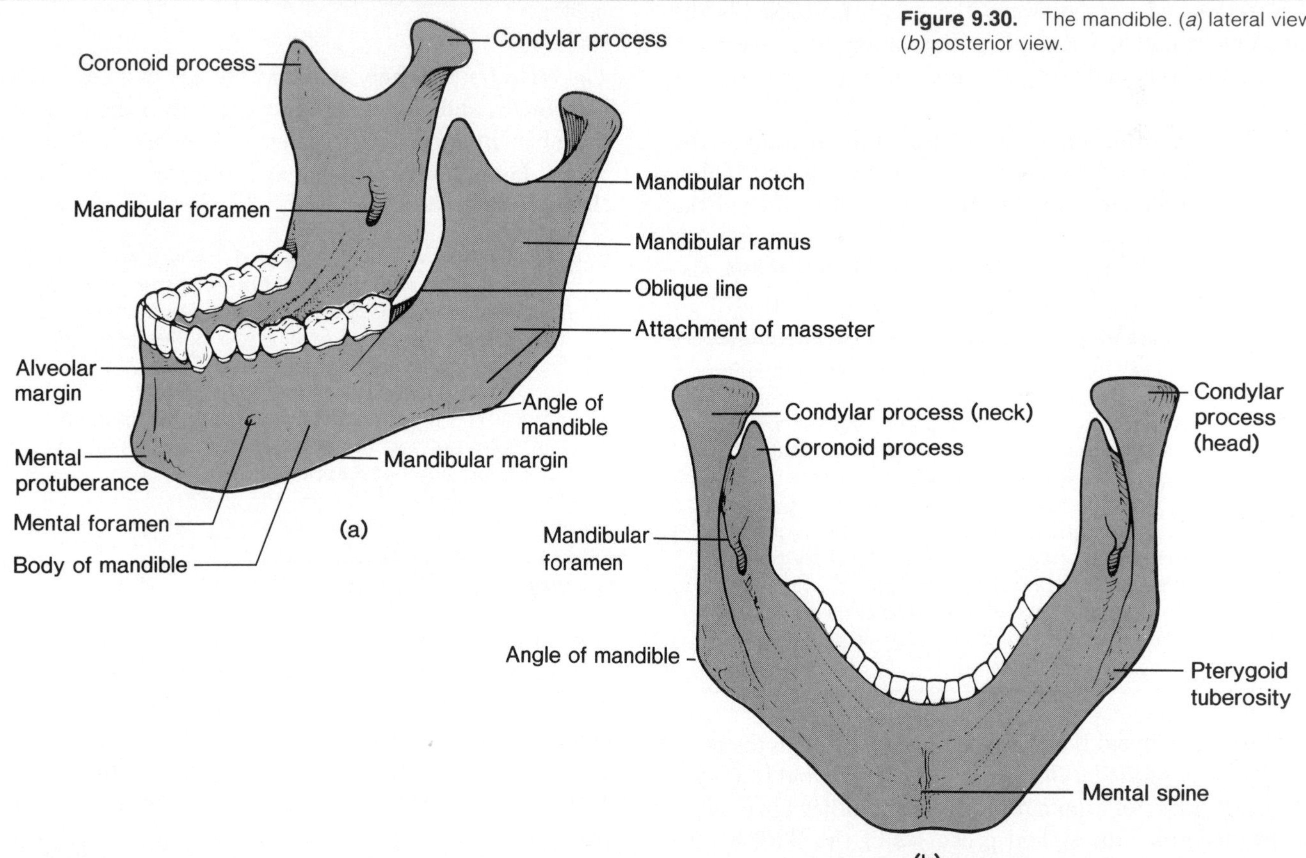

Figure 9.30. The mandible. (*a*) lateral view; (*b*) posterior view.

Mandible. The mandible, or lower jawbone, is attached to the skull by a temporomandibular articulation and is the only movable bone of the skull. Several muscles that close the jaw extend from the skull to the mandible and are discussed in chapter 12. The mandible of an adult supports sixteen teeth within alveoli, which occlude with those of the maxilla.

The horseshoe-shaped front and horizontal lateral sides of the mandible are referred to as the **body** (fig. 9.30). Extending vertically from the posterior portion of the body are two **rami** *(ra'mi;* singular, *ramus).* Each ramus has a knoblike **condyloid process,** which articulates with the

mandible: L. *mandere,* to chew

ramus: L. *ramus,* branch
condyloid: L. *condylus,* knucklelike

Figure 9.31. An anterior view of the hyoid bone.

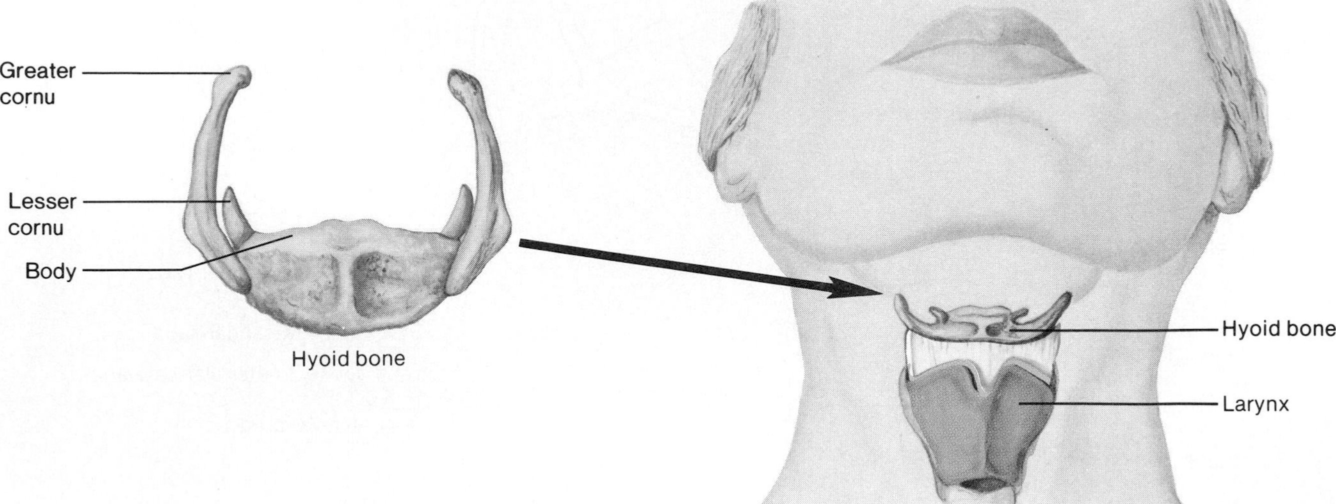

mandibular fossa of the temporal bone, and a pointed **coronoid process** for the attachment of the temporalis muscle. The depressed area between these two processes is the **mandibular notch.** The angle of the mandible is where the horizontal body and vertical ramus meet at the corner of the jaw.

Two sets of foramina are found on the mandible: the **mental foramen** on the lateral side below the first molar, and the **mandibular foramen** on the medial surface of the ramus. The mental nerve and vessels pass through the mental foramen, and the inferior alveolar nerve and vessels are transmitted through the mandibular foramen.

Dentists use bony landmarks of the facial region to locate the nerves that traverse the foramina so that anesthetics can be injected. For example, the trigeminal nerve is composed of three large branches, the lower two of which convey sensations from the teeth, gums, and jaws. The mandibular teeth can be desensitized by an injection near the mandibular foramen, called a *third division,* or *lower, nerve block.* An injection near the foramen rotundum of the skull, called a *second division nerve block,* desensitizes all the upper teeth on one side of the maxilla.

Hyoid. The hyoid is a U-shaped bone located in the neck just superior to the larynx (voice box). The hyoid is unique in that it does not attach directly to any other bone but is suspended from the styloid processes of the skull by the **stylohyoid** *(sti''lo-hi'oid)* **ligaments.** The hyoid has a **body,** two **lesser cornua** extending anteriorly, and two **greater cornua,** which project posteriorly to the stylohyoid ligaments (fig. 9.31). Several neck and tongue muscles attach to the hyoid bone. The hyoid may be palpated by placing a thumb and a finger on either side of the upper neck under the lateral portions of the mandible and firmly squeezing medially.

Ear Ossicles. Three small, paired ear ossicles, the **malleus, incus,** and **stapes,** are located within the middle ear chambers in the petrous portion of the temporal bones (fig. 9.32). These bones transfer and amplify sound impulses through the middle ear.

1. List which facial and cranial bones of the skull are paired and which are unpaired. Also, indicate at least two structural features for each of the bones of the skull.
2. Describe the location of each of the bones of the skull, and indicate the sutures that join them.
3. What is the function of the sella turcica, foramen magnum, petrous bone, crista galli, and nasal conchae?
4. Which facial bones support the teeth?

Vertebral Column

The supporting vertebral column consists of twenty-six vertebrae, separated by fibrocartilaginous intervertebral discs, which lend flexibility and absorb the stress of movement. Vertebrae enclose and protect the spinal cord, support and permit movement of the skull, articulate with the rib cage, and provide for the attachment of trunk muscles.

Objective 12. Identify the bones of the five regions of the vertebral column, and describe the characteristic curves of each region.

Objective 13. Describe the structure of a typical vertebra.

coronoid: Gk. *korone,* like a crow's beak
cornu: L. *cornu,* horn

malleus: L. *malleus,* hammer
incus: L. *incus,* anvil
stapes: L. *stapes,* stirrup

Figure 9.32. The three ear ossicles within the middle ear cavity.

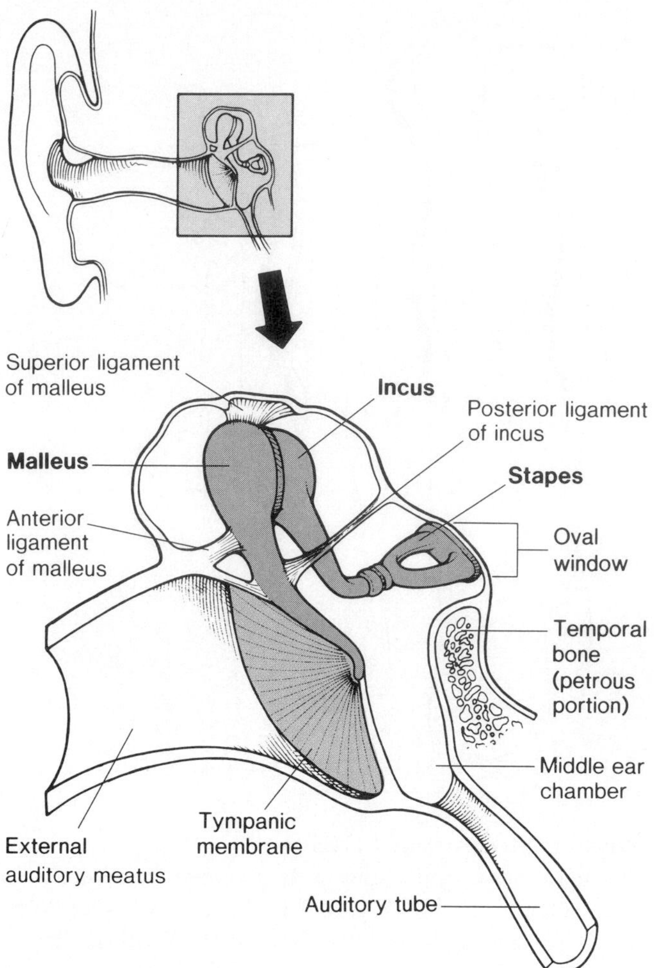

Figure 9.33. The vertebral column of an adult has four curves named according to the region in which they occur. The vertebrae are separated by intervertebral discs, which allow flexibility.

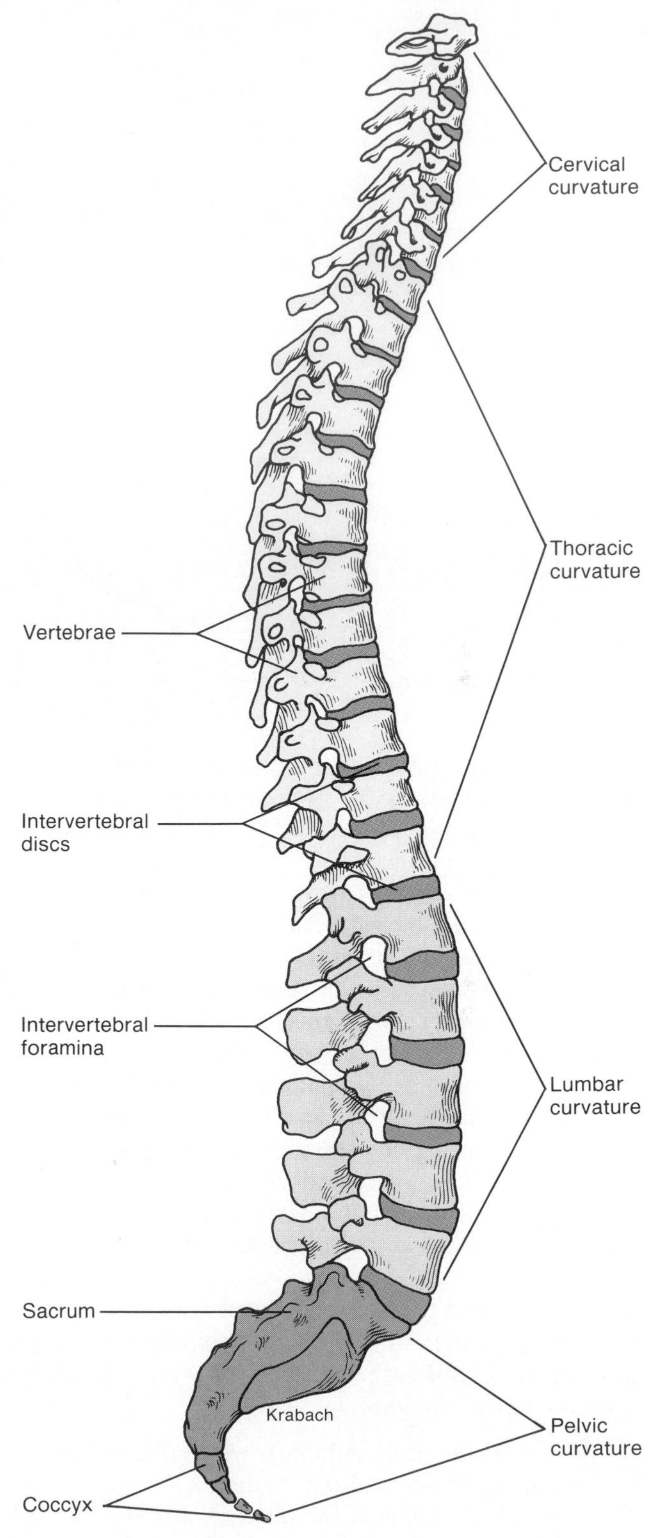

The vertebral column is composed of thirty-three individual vertebrae. There are seven **cervical,** twelve **thoracic** *(tho-ras'ik),* five **lumbar,** five fused **sacral,** and four fused **coccygeal** *(kok-sij'e-al)* vertebrae; thus, the vertebral column is composed of a total of twenty-six movable parts. Vertebrae are separated by fibrocartilaginous intervertebral discs and are secured to one another by interlocking processes and binding ligaments. This structural arrangement provides limited movements between vertebrae but extensive movements for the entire vertebral column. Between the vertebrae are openings called **intervertebral** *(in"ter-ver'te-bral)* **foramina** that permit passage of spinal nerves.

Four curvatures of the vertebral column of an adult can be identified and viewed from the side (fig. 9.33). The **cervical, thoracic,** and **lumbar curves** are identified by the type of vertebrae they include. The **pelvic curve** is formed by the shape of the sacrum and coccyx *(kok'siks).* The

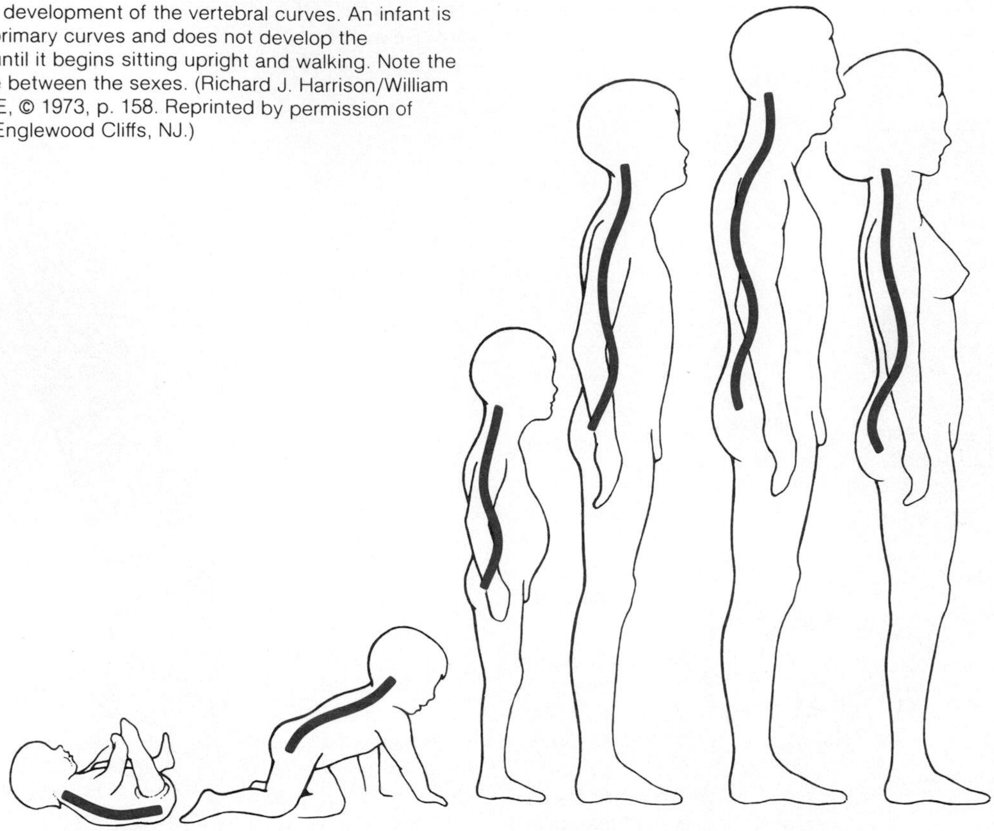

Figure 9.34. The development of the vertebral curves. An infant is born with the two primary curves and does not develop the secondary curves until it begins sitting upright and walking. Note the differences in curve between the sexes. (Richard J. Harrison/William Montagna, *Man*, 2/E, © 1973, p. 158. Reprinted by permission of Prentice-Hall, Inc., Englewood Cliffs, NJ.)

curves of the vertebral column play an important functional role in increasing the strength and maintaining the balance of the upper portion of the body and also make possible a bipedal stance.

The four vertebral curves are not present in a newborn baby. Instead, the vertebral column is somewhat anteriorly concave, and except for the cervical region, it remains this way even as an infant learns to crawl (fig. 9.34). The cervical curve begins to develop at about three months of age, as a baby begins holding up its head, and the curve becomes more pronounced as the baby learns to sit up. The lumbar curve develops as a child begins to walk. The thoracic and pelvic curves are called **primary curves** because they retain the anteriorly concave shape of the fetus. The cervical and lumbar curves are called **secondary curves** because they are modifications of the fetal shape.

The vertebral column is commonly called the "backbone" and together with the spinal cord of the nervous system constitutes the spinal column. The vertebral column has three basic functions:

1. to support the head and upper extremities while permitting freedom of movement;
2. to provide attachment for various muscles, ribs, and visceral structures; and
3. to protect the spinal cord and permit passage of the spinal nerves.

General Structure of Vertebrae

Vertebrae show similarities in their general structure from one region to another. Figure 9.35 illustrates a typical vertebra. A vertebra is usually composed of an anterior drum-shaped **body (centrum)** adapted to withstand compression. The body is in contact with **intervertebral discs** on each end. The **vertebral,** or **neural, arch** is affixed to the posterior surface of the body and is composed of two supporting **pedicles** *(ped'ĭ-k'l)* and two arched **laminae.** The hollow space formed by the vertebral arch and body is the **vertebral foramen,** through which the spinal cord passes. Between the pedicles of adjacent vertebrae are the **intervertebral foramina,** through which spinal nerves emerge as they branch off from the spinal cord.

Seven processes arise from the vertebral arch: the **spinous** *(spi'nus)* **process,** two **transverse processes,** two **superior articular processes,** and two **inferior articular processes** (fig. 9.36). The first two processes serve for muscle attachment, and the latter two pairs limit twisting of the vertebral column. The spinous process protrudes posteriorly and inferiorly from the vertebral arch. The transverse process extends laterally from each side of a vertebra at the point where the lamina and pedicle join. The superior articular processes of a vertebra have interlocking articulations with the inferior articular processes of the adjacent anterior vertebra.

pedicle: L. *pediculus,* small foot
lamina: L. *lamina,* thin layer

Figure 9.35. Cervical vertebrae. (*a*) an X ray of the cervical region. (*b*) a typical cervical vertebra. (*c*) the atlas and axis as they articulate.

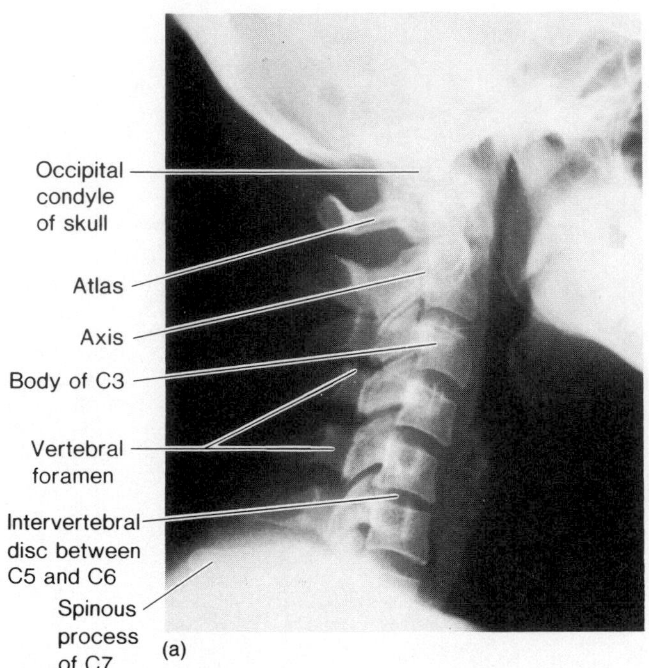

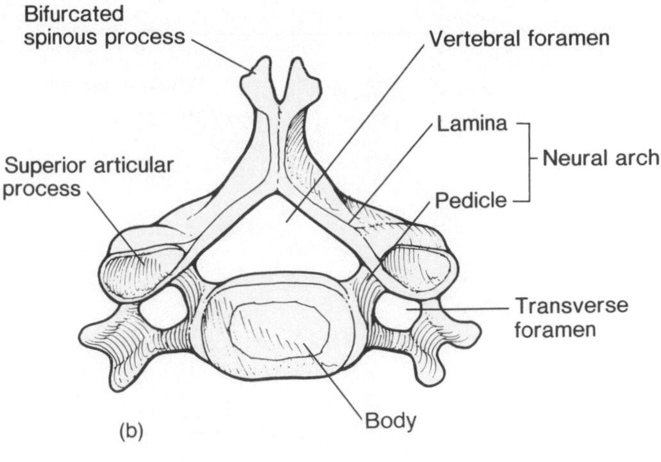

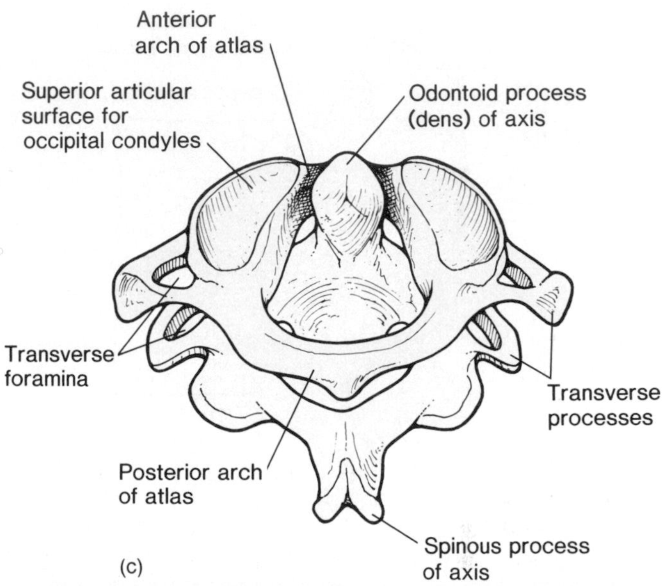

A *laminectomy* is the surgical removal of the spinous processes and their supporting vertebral laminae in a particular region of the vertebral column. A laminectomy may be performed to relieve pressure on the spinal cord caused by a blood clot, a tumor, or a herniated disc. It may also be performed on a cadaver to expose the spinal cord and its surrounding meninges.

Regional Characteristics of Vertebrae

Cervical Vertebrae. The seven cervical vertebrae form the flexible framework of the neck region and support the head. The osseous tissue of cervical vertebrae is more dense than that of the other vertebral regions, and, except for the coccygeal region, the cervical vertebrae are smallest. Cervical vertebrae are distinguished by the presence of a **transverse foramen** in the transverse process (fig. 9.35). The vertebral vessels pass through this opening as they transfer blood to and from the brain. The spinous processes of the third, fourth, and fifth cervical vertebrae are **bifid** *(bi'fid),* or notched, for the attachment of the strong **nuchal ligament,** which attaches to the back of the skull for added support.

The first cervical vertebra, the **atlas,** is adapted to articulate with the occipital condyles of the skull while supporting the head. The atlas has concave **superior articular surfaces** to articulate with the oval-shaped occipital condyles. This joint permits the nodding of the head in a "yes" movement. The atlas lacks a body. It has a short, bifurcated spinous process.

The second cervical vertebra is called the **axis** and is easily identified by the presence of a peglike projection called the **odontoid** *(o-don'toid)* **process,** or **dens.** This process extends superiorly to provide a pivot for rotation with the atlas. This joint permits rotation, or the turning of the head to the side, as in a "no" movement.

atlas: from Gk. mythology, Atlas—the Titan who supported the heavens
axis: L. *axis,* axle
odontoid: Gk. *odontos,* tooth

Figure 9.36. Thoracic vertebrae. (*a*) representative vertebrae in a right lateral view; (*b*) a superior view.

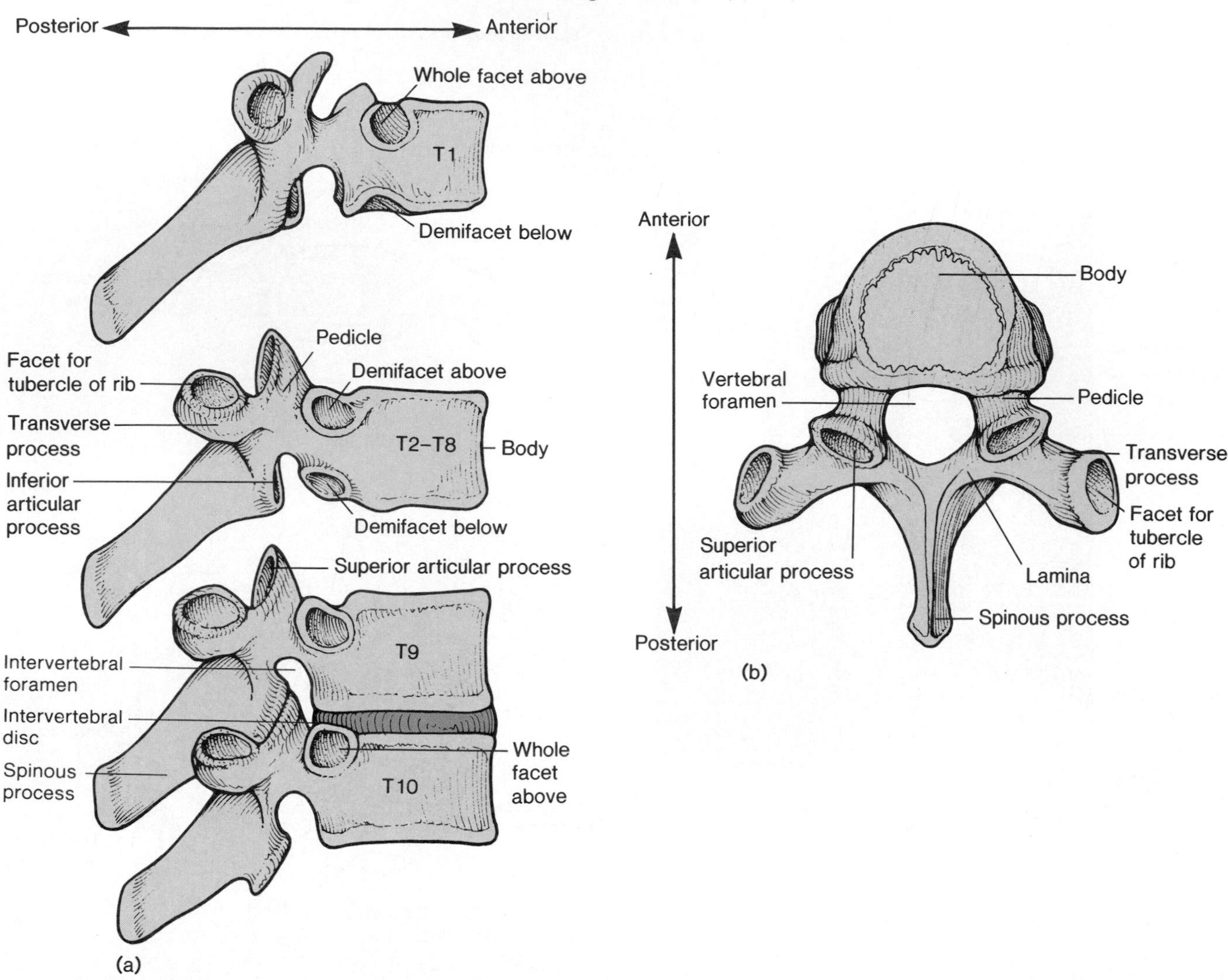

> *"Whiplash"* of the neck is a common, general term for injury to the cervical region. Muscle, bone, or ligament injury in this portion of the spinal column is relatively common in individuals involved in automobile wrecks and sports accidents. Joint dislocation without vertebral fracture occurs commonly between the fourth and fifth or fifth and sixth cervical vertebrae, where mobility is greatest. Bilateral dislocations are particularly dangerous because of the probability of spinal cord injury. Compression fractures of the bodies of the first three cervical vertebrae are common and follow abrupt forced flexion of the neck. Fractures of this type may be extremely painful because of pinched spinal nerves.

Thoracic Vertebrae. The thoracic vertebrae serve as attachments of the ribs to form the posterior anchor of the rib cage. Thoracic vertebrae are larger than cervical vertebrae and increase in size from superior (T1) to inferior (T12). Each thoracic vertebra has a long spinous process that slopes obliquely downward and **facets** or **demifacets** for articulation with the ribs (fig. 9.36). The first thoracic vertebra has a superior facet and an inferior demifacet on both sides. The second through eighth have two demifacets on each side so that each head of a rib articulates with two vertebrae. The ninth vertebra has a single demifacet on each side, and the tenth through the twelfth have facets. The first ten vertebrae also have facets on their transverse processes for articulation with the tubercles of the ribs.

Lumbar Vertebrae. The five lumbar vertebrae are easily identified by their heavy bodies and thick, blunt spinous processes (fig. 9.37) for attachment of powerful back muscles. They are the largest vertebrae of the vertebral column. Their articular processes are also distinctive in that the superior articular processes are directed medially instead of superiorly, and the inferior articular processes are directed laterally instead of inferiorly.

---

lumbar: L. *lumbus,* loin

Figure 9.37. Lumbar vertebrae. (*a*) an X ray; (*b*) a superior view; (*c*) a right lateral view.

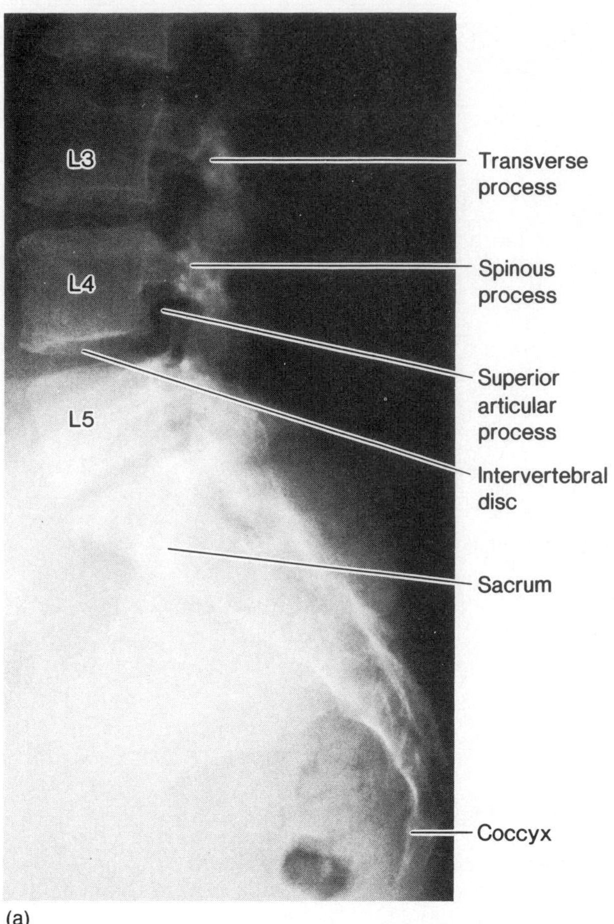

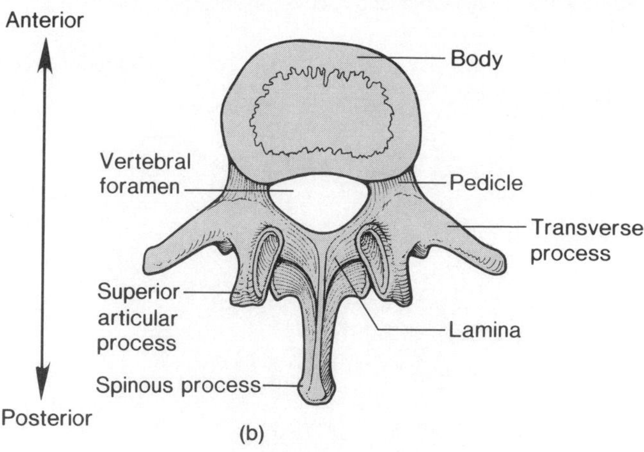

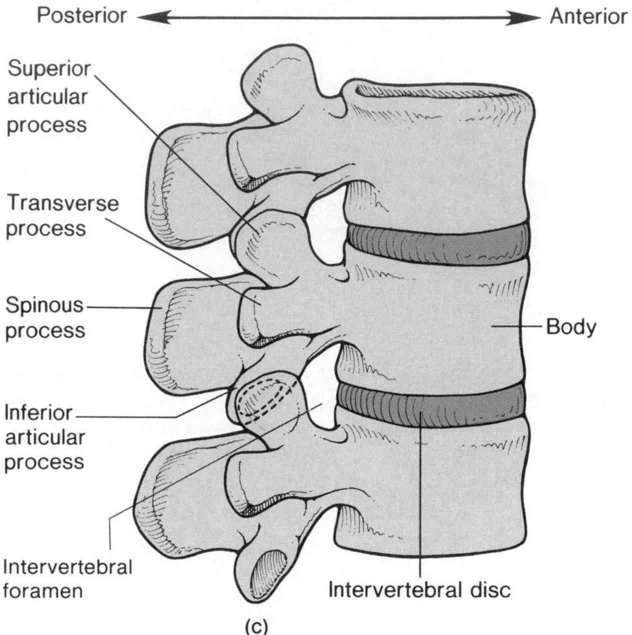

Sacrum. The wedge-shaped sacrum (fig. 9.38) consists of five sacral vertebrae, which become fused after age twenty-six. The sacrum is functionally adapted to provide a strong foundation for the pelvic girdle. The sacrum has an extensive **auricular surface** on each side for the formation of an immovable **sacroiliac** *(sa''kro-il'e-ak)* **joint** with the os coxa. A **median sacral crest** is formed along the dorsal surface by the fusion of the spinous processes. **Dorsal sacral foramina** on either side of the crest allow the passage of nerves from the spinal cord. The **sacral canal** is the tubular cavity within the sacrum that is continuous with the vertebral canal. Paired **superior articular processes,** which articulate with the fifth lumbar vertebra, arise from the roughened **sacral tuberosity** along the dorsal surface.

The smooth anterior surface of the sacrum forms the posterior surface of the pelvic cavity. It has four **transverse lines** denoting the fusion of the vertebral bodies. On either side of the transverse lines are the paired **pelvic foramina.** The superior border of the anterior surface of the sacrum, called the **sacral promontory** *(prom'on-to''re),* is an important obstetrical landmark for pelvic measurements.

Coccyx. The coccyx *(kok'siks)* is the so-called tailbone. It is composed of four or five fused coccygeal vertebrae, which form a triangular-shaped structure. The first vertebra of the fused coccyx has two long **coccygeal cornua,** which are attached by ligaments to the sacrum (fig. 9.38). Lateral to the cornua are the transverse processes.

The regions of the vertebral column are summarized in table 9.7.

sacrum: L. *sacris,* sacred

coccyx: Gk. *kokkyx,* like a cuckoo's beak

Figure 9.38. The sacrum and coccyx. (*a*) an anterior view; (*b*) a posterior view.

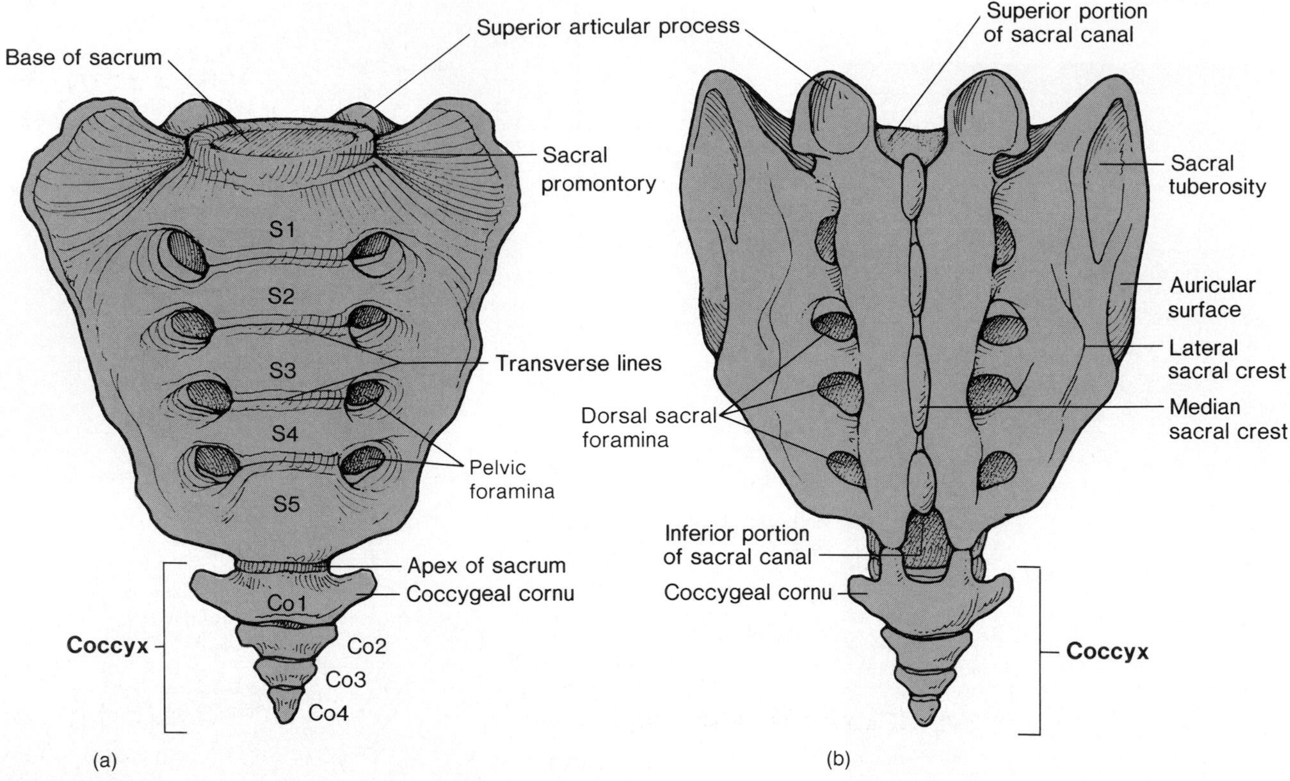

Table 9.7 Regions of the vertebral column

Region	Number of bones	Diagnostic features
Cervical	7	Transverse foramina; superior facets of atlas articulate with occipital condyle; odontoid process of axis; spinous processes of 3d through 5th vertebrae are bifid
Thoracic	12	Long spinous processes that slope obliquely inferiorly; facets and demifacets for articulation with ribs
Lumbar	5	Large bodies; prominent transverse processes; short, thick spinous processes
Sacrum	5 fused vertebrae	Extensive auricular surface; median sacral crest; dorsal sacral foramina; sacral promontory; sacral canal
Coccyx	4 fused vertebrae	Small, triangular; coccygeal cornua

When a person sits, the coccyx flexes anteriorly somewhat, acting as a shock absorber. An abrupt fall on the coccyx, however, may cause a painful subperiosteal bruising, fracture, or fracture-dislocation of the sacrococcygeal joint. An especially difficult childbirth can even injure the coccyx of the mother. Coccygeal trauma is painful and may require months to heal.

1. Which are the primary curves of the vertebral column, and which are the secondary curves? Describe the characteristic curves of each region.
2. What is the function of the transverse foramina of the cervical vertebrae?
3. Describe the diagnostic differences between a thoracic and a lumbar vertebra. Which structures are similar and could therefore be characteristic of a typical vertebra?

Rib Cage

The cone-shaped and flexible rib cage consists of the thoracic vertebrae, twelve paired ribs, costal cartilages, and the sternum. It encloses and protects the thoracic viscera and is directly involved in the mechanics of breathing.

Objective 14. Identify the parts of the rib cage, and distinguish between the various types of ribs.

The **sternum, ribs, costal cartilages,** and the previously described thoracic vertebrae form the rib cage, or **thoracic cage,** of the thorax (fig. 9.39). The rib cage is anteroposteriorly compressed and more narrow superiorly than inferiorly. It supports the pectoral girdle and upper extremities, protects and supports the thoracic and upper abdominal viscera, and plays a major role in breathing. Certain bones of the rib cage contain active sites for blood cell production in the red bone marrow.

sternum: Gk. *sternon,* chest

Figure 9.39. The rib cage.

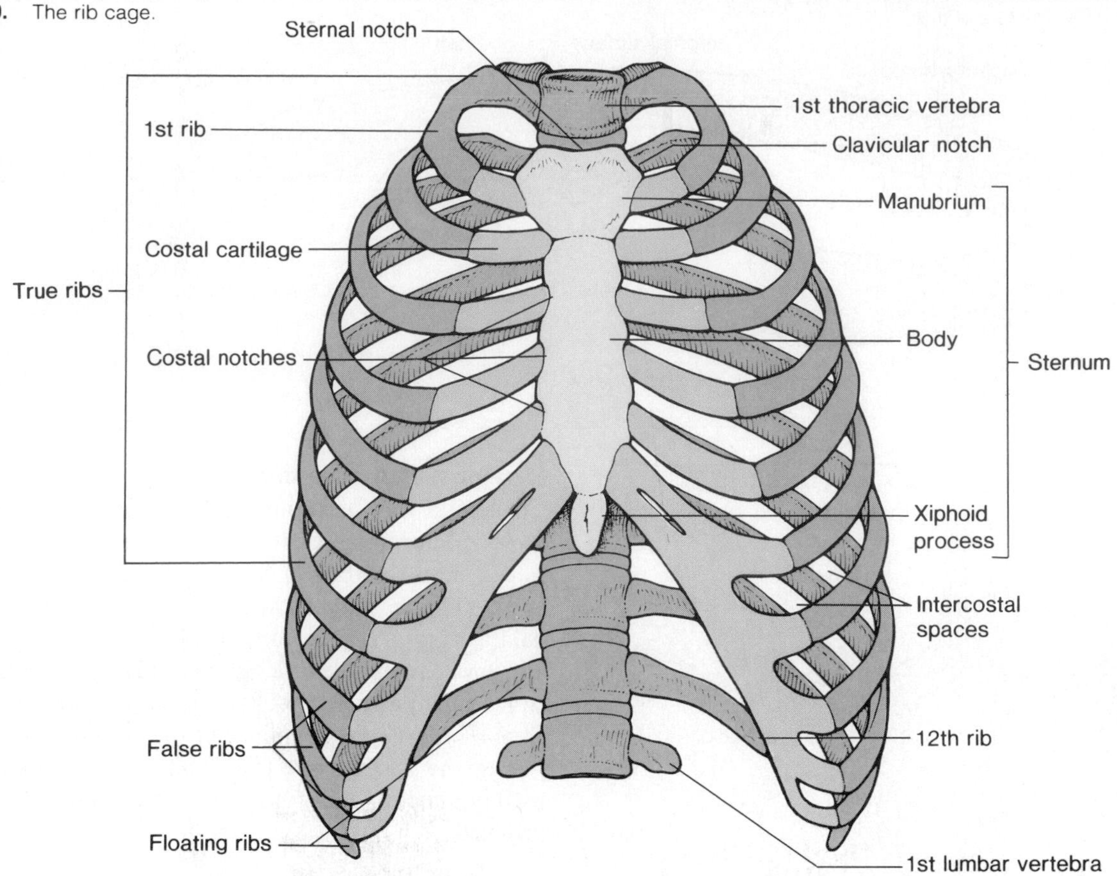

Sternum

The **sternum** (breastbone) is an elongated, flattened bony plate consisting of three separate bones: the upper **manubrium,** the central **body,** and the lower **xiphoid** *(zi'foid)* **process.** On the lateral sides of the sternum are **costal notches** where the costal cartilages attach. A **sternal notch** is formed at the superior end of the manubrium, and a **clavicular** *(klah-vik'u-lar)* **notch** for articulation with the clavicle is present on both sides of the sternal notch. The manubrium articulates with the costal cartilages of the first and second ribs. The body of the sternum attaches to the costal cartilages of the second through the tenth ribs. The xiphoid process does not attach to ribs but is an attachment for abdominal muscles. The costal cartilages of the eighth, ninth, and tenth ribs fuse to form the costal margin. A costal angle is formed where the two costal margins come together at the xiphoid process. The **sternal angle** (angle of Louis) may be palpated as a depression between the manubrium and body of the sternum at the level of the second rib. The costal angle, costal margins, and sternal angle are important surface landmarks of the chest and abdomen.

manubrium: L. *manubrium,* a handle
xiphoid: Gk. *xiphos,* sword
costal: L. *costa,* rib

Ribs

There are twelve pairs of **ribs,** each pair being attached posteriorly to a thoracic vertebra. Anteriorly, the first seven pairs are anchored to the sternum by individual costal cartilages and are called **true ribs.** Ribs 8, 9, and 10 are attached to the costal cartilage of rib 7 and are termed **false ribs.** The remaining two paired ribs do not attach to the sternum and are referred to as **floating ribs.** They are attached to the muscles of the body wall.

Although the structure of ribs varies, each of the first ten pairs has a **head** and a **tubercle** for articulation with a vertebra. The last two have a head but no tubercle. In addition, each of the twelve pairs has a **neck, angle,** and **shaft** (fig. 9.40). The head of a rib projects posteriorly and articulates with the body of a thoracic vertebra (fig. 9.41). The tubercle is a knoblike process just lateral to the head. It articulates with the facet of the transverse process. The neck is the constricted area between the head and the tubercle. The **shaft,** or **body,** is the main, curved part of the rib. Along the inner surface of the shaft is a depressed canal called the **costal groove,** which protects the costal vessels and nerve. Spaces between the ribs are called **intercostal spaces** and are occupied by the intercostal muscles.

Figure 9.40. The structure of a rib.

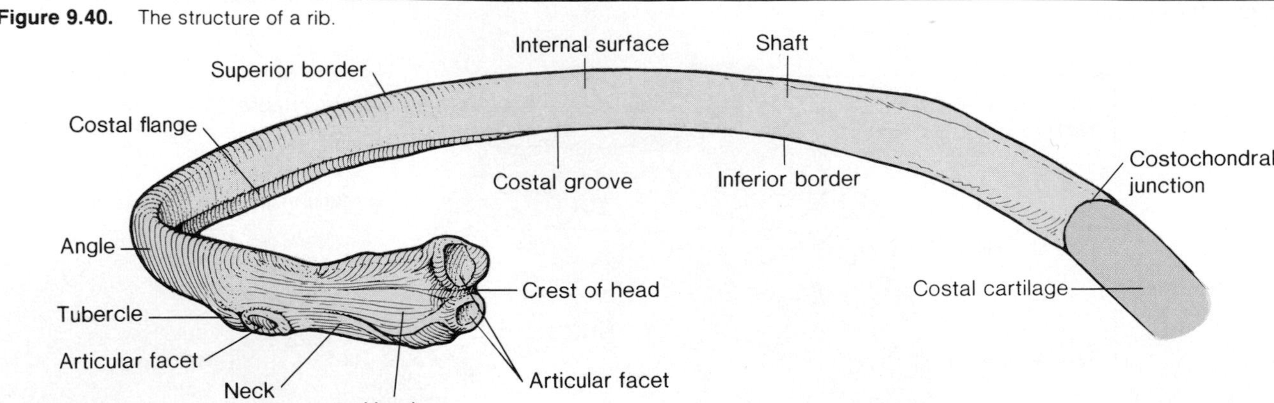

Figure 9.41. Articulation of a rib with a thoracic vertebra as seen in a superior view.

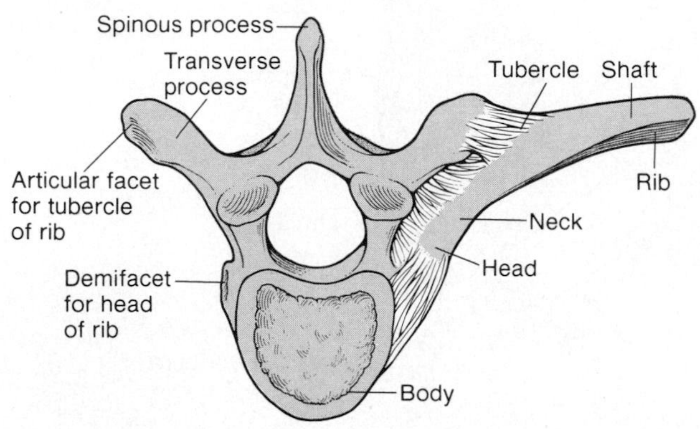

Fractures of the ribs are relatively common injuries and most frequently occur between ribs 3 and 10. The first two pairs of ribs are protected by the clavicles, and the last two pairs move freely and will give with an impact. Little can be done to assist the healing of broken ribs other than binding them tightly to restrict movement.

1. Describe the rib cage, and list its functions. What determines if a rib is designated as true, false, or floating?
2. Define the costal margin and the costal angle.

Clinical Considerations

Bone is a dynamic living tissue that is susceptible to hormonal or nutritional deficiency, diseases, and changes brought on by age. Since the development of bone is genetically governed, congenital conditions can occur. The hardness of bone gives it strength, yet it lacks the resiliency to avoid fracture if it undergoes excessive trauma. All of these aspects of bone provide some important and interesting clinical considerations.

Developmental Disorders

Congenital malformations account for several types of skeletal deformities. Certain bones may fail to form during osteogenesis, or they may form abnormally. **Cleft palate** and **cleft lip** are malformations of the palate and face. Cleft palates vary in severity and seem to have a mixed genetic and environmental cause. **Spina bifida** *(spi'nah bif'i-dah)* is a congenital defect of the vertebral column resulting from a failure of the laminae of the vertebrae to fuse, exposing the spinal cord (fig. 9.42). The lumbar area is mainly affected, and frequently only a single vertebra is involved.

Nutritional and Hormonal Disorders

Several bone disorders result from nutritional deficiencies or from excessive or deficient amounts of the hormones that regulate bone development and growth. Vitamin D has a tremendous influence on proper bone structure and function. When there is a deficiency of this vitamin, the body is unable to metabolize calcium and phosphorus. Vitamin D deficiency in children causes **rickets.** The bones of a child with rickets remain soft and are deformed from the weight of the body (see fig. 8.14).

A deficiency in vitamin D in the adult causes the bones to demineralize, or to give up stored calcium and phosphorus. This condition is called **osteomalacia** *(os''te-o-mah-la'she-ah)*. Osteomalacia is prevalent in women who have repeated pregnancies and poor diets and are seldom exposed to the sun.

The consequences of endocrine disorders are described in chapter 19. Since the impact of hormones on bone development is great, however, a few endocrine disorders will be briefly mentioned. Hypersecretion of the growth hormone from the pituitary gland leads to **gigantism** in young persons if it starts before ossification of their epiphyseal plates and to **acromegaly** *(ak''ro-meg'ah-le)* in adults. Acromegaly is characterized by hypertrophy of the bones of the face, hands, and feet. In contrast, in a child, a hyposecretion of the growth hormone can lead to **dwarfism.**

Osteoporosis is a bone disorder characterized by marked demineralization, which weakens bone. The causes of osteoporosis include aging, prolonged inactivity, malnutrition, and an unbalanced secretion of hormones. It is

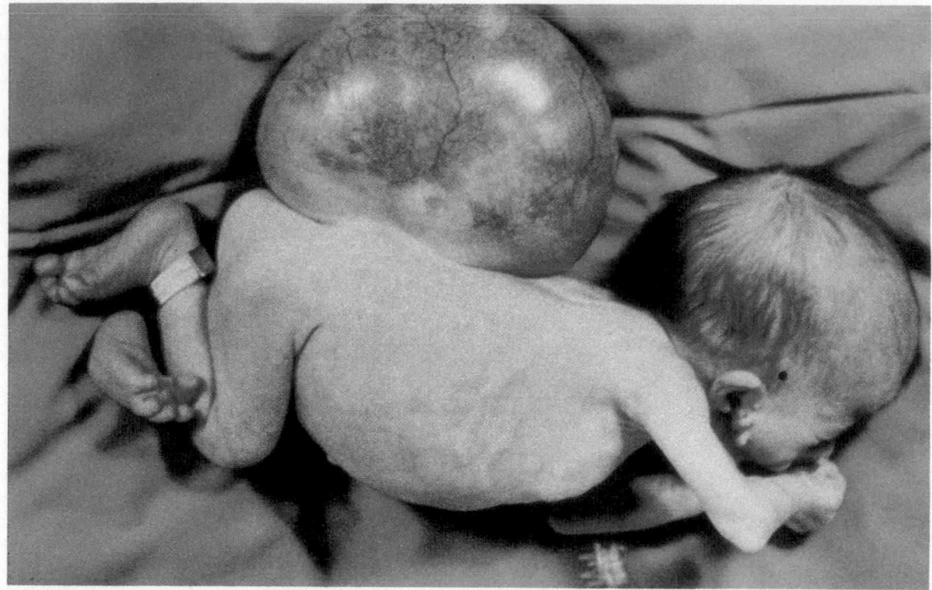

Figure 9.42. In spina bifida failure of the vertebral arches to fuse permits a herniation of the meninges that cover the spinal cord through the vertebral column, resulting in a condition called meningomyelocele.

most common in postmenopausal women. People with osteoporosis are prone to bone fracture, particularly at the pelvic girdle and vertebrae, because the bones become too brittle to support the body. Although there is no known cure for osteoporosis, it can be somewhat prevented in younger adults through proper diet, exercise, and good general health habits. Treatment of the disease in women through dietary calcium, exercise, and estrogens has limited positive results.

Paget's disease is a disease of disorganized metabolic processes within bone. The activity of osteoblasts and osteoclasts becomes irregular, producing areas with thickened osseous deposits and other areas where too much bone is removed. The etiology of the disease is unknown, but it is a relatively common affliction in persons over age fifty, and it occurs more frequently in males than in females.

Trauma and Injury

There are a variety of types of trauma to the skeletal system, ranging from injury to the bone itself to damage of the joints in the form of sprains or dislocations. Fractures and the healing of fractures will be discussed in the next chapter (chap. 10), and joint injuries will be discussed in chapter 11.

Neoplasms of Bone

Malignant bone tumors are three times more common than benign tumors. Pain is the usual symptom of either type of osseous neoplasm, although benign tumors may not have associated pain.

Two types of benign bone tumors are **osteoma,** which is the more frequent and often involves the skull, and **osteoid osteoma,** which is a painful neoplasm of the long bones, usually in children.

Figure 9.43. A bone scan.

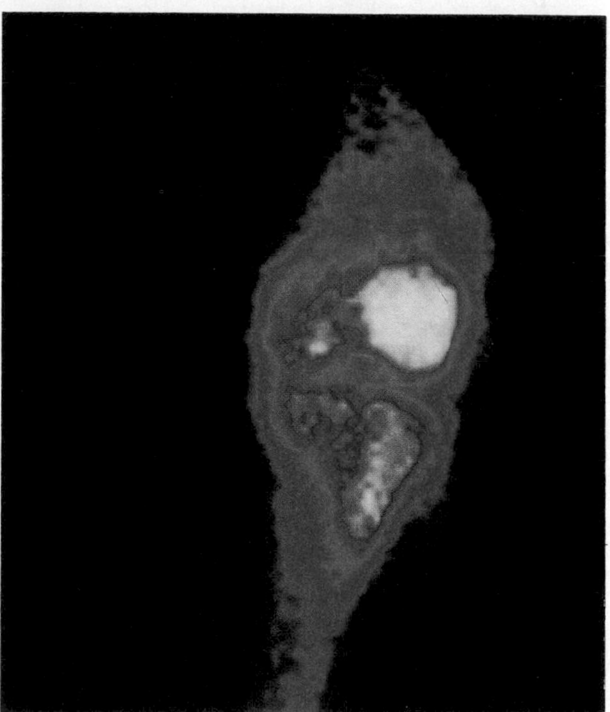

Osteogenic sarcoma is the most virulent type of bone cancer and frequently metastasizes through the blood to the lungs. This disease usually originates in the long bones and is accompanied by aching and persistent pain.

A **bone scan** (fig. 9.43) is a diagnostic procedure frequently done on a person who has had a malignancy elsewhere in the body that may have metastasized to the bone. The patient receiving a bone scan may be injected with a radioactive substance that accumulates more rapidly in malignant tissue than normal tissue. Entire body radiographs show malignant bone areas as intensely dark dots.

Important Clinical Terminology

achondroplasia *(ah-kon″dro-pla′ze-ah)* A genetic defect resulting in the retarded formation of cartilaginous bone during fetal development.

craniotomy Surgical cutting into the cranium to provide access to the brain.

epiphysiolysis *(ep″ĭ-fiz″e-ol′ĭ-sis)* A weakening and separation of the epiphysis from the diaphysis of a long bone.

laminectomy The surgical removal of the posterior arch of a vertebra, usually to repair a herniated intervertebral disc.

orthopedics A branch of medicine concerned with the diagnosis and treatment of trauma, diseases, and abnormalities involving the skeletal and muscular systems.

osteitis An inflammation of bone tissue.

osteoblastoma A benign tumor produced from bone-forming cells, most frequently in the vertebrae of young children.

osteochondritis An inflammation of bone and cartilage tissues.

osteomyelitis An inflammation of bone marrow caused by bacteria or fungi.

osteonecrosis The death of bone tissue, usually caused by obstructed arteries.

osteopathology The study of bone diseases.

osteosarcoma A malignant tumor of bone tissue.

osteotomy The surgical removal of a bone.

Chapter Summary

I. Organization of the Skeletal System
   A. The axial skeleton consists of the skull, auditory ossicles, hyoid bone, vertebral column, and rib cage.
   B. The appendicular skeleton consists of the pectoral girdle, upper extremities, pelvic girdle, and lower extremities.

II. Functions of the Skeletal System
   A. The mechanical functions of bones include the support and protection of softer body tissues and organs, and certain bones function as levers during body movement.
   B. The metabolic functions of bones include hemopoiesis and mineral storage.

III. Development of the Axial Skeletal System
   A. The development of the vertebral column, rib cage, and sternum involves the notochord, somites, and sclerotomes.
   B. The development of the skull involves the neurocranium, viscerocranium, and chondrocranium.

IV. Bone Structure and Growth
   A. Bone structure includes the shape and surface features of each bone as well as gross internal components.
   B. The surface features of bones can be broadly classified into articulating surfaces, nonarticulating prominences, and depressions and openings.
   C. A typical long bone has a diaphysis filled with marrow in the medullary cavity, epiphyses, epiphyseal plates for linear growth, and a covering of periosteum for diametric growth and the attachment of ligaments and tendons.
   D. Bone growth from embryonic to adult size is an orderly process determined by genetics, hormonal secretions, and nutritional supply.
   E. Bone remodeling is a continual process involving osteoclasts in bone resorption and osteoblasts in the formation of new osseous tissue.

V. Skull
   A. The cranium encloses and protects the brain and provides for the attachment of muscles.
      1. Sutures are immovable joints between cranial bones.
      2. The eight cranial bones include the frontal, parietals, temporals, occipital, sphenoid, and ethmoid.
   B. Facial bones form the basic shape of the face, support the teeth, and provide for the attachment of the facial muscles.
      1. The fourteen facial bones are the nasals, maxillae, zygomatics, mandible, lacrimals, palatines, inferior nasal conchae, and vomer.
      2. The hyoid bone is located in the neck between the mandible and the larynx.
      3. The three paired ear ossicles are located within the middle ear chambers of the petrous portion of the temporal bones.

VI. Vertebral Column
   A. The vertebral column consists of seven cervical, twelve thoracic, five lumbar, four or five fused sacral, and four fused coccygeal vertebrae.
   B. Cervical vertebrae have transverse foramina; thoracic vertebrae have facets and demifacets for articulation with ribs; lumbar vertebrae have heavy bodies; sacral vertebrae are triangularly fused and articulate with the pelvic girdle; the coccygeal vertebrae form a small triangular bone.

VII. Rib Cage
   A. The sternum consists of a manubrium, body, and xiphoid process.
   B. There are seven pairs of true ribs, three pairs of false ribs, and two pairs of floating ribs.

Review Activities

Objective Questions

1. A bone is considered to be a(n)
   (a) tissue
   (b) cell
   (c) organ
   (d) system
2. Which of the following statements is *false?*
   (a) Bones are important in the synthesis of vitamin D.
   (b) Bones and teeth contain about 99% of the body's calcium.
   (c) Red bone marrow is the primary site for hemopoiesis.
   (d) Most bones develop through endochondral ossification.
3. Bone tissue derives from specialized migratory mesodermal cells called
   (a) dermatomes
   (b) mesenchyme
   (c) myotomes
   (d) somites
4. Small portions of the embryonic notochord persist in the adult as the
   (a) primary curves
   (b) nucleus pulposus
   (c) pedicles
   (d) centrum
5. Which of the following is *not* a long bone? The
   (a) talus
   (b) proximal phalanx
   (c) metatarsal
   (d) fibula
6. Specialized bone cells that enzymatically reabsorb bone tissue are
   (a) osteoblasts
   (b) osteocytes
   (c) osteons
   (d) osteoclasts
7. The mandibular fossa is located in which structural portion of the temporal bone? The
   (a) squamous
   (b) tympanic
   (c) mastoid
   (d) petrous
8. The crista galli is a structural feature of which bone? The
   (a) sphenoid
   (b) ethmoid
   (c) palatine
   (d) temporal
9. Transverse foramina are characteristic of
   (a) lumbar vertebrae
   (b) sacral vertebrae
   (c) thoracic vertebrae
   (d) cervical vertebrae
10. The bone disorder common in aged persons, particularly if they have prolonged inactivity, malnutrition, or an unbalanced secretion of hormones, is
   (a) osteitis
   (b) osteonecrosis
   (c) osteoporosis
   (d) osteomalacia

Essay Questions

1. Explain the significance of each of the following structures involved in bone development: mesenchyme, somites, costal processes, neurocranium, viscerocranium, branchial arches, and limb buds.
2. List the bones of the skull that are paired. Which are unpaired? Identify the bones of the skull that can be palpated.
3. Diagram a typical long bone, and label the epiphyses, diaphysis, epiphyseal plates, medullary cavity, nutrient foramina, periosteum, and articular cartilages.
4. List the bones that form the cranial cavity, the orbit, and the nasal cavity. Describe the location of the paranasal sinuses, the mastoidal sinus, and the inner-ear cavity.
5. Describe how bones grow in length and in circumference. How are these processes similar, and how do they differ? Explain how X rays can be used to determine normal bone growth.
6. Explain the process of endochondral ossification of a long bone. Why is it important that a balance be maintained between osteoblast activity and osteoclast activity?
7. Describe the curvature of the vertebral column. What is meant by primary curves as compared to secondary curves?
8. List two or more characteristics by which vertebrae from each of the five regions of the vertebral column can be identified.
9. Identify the bones that form the rib cage. What functional role do the bones and the costal cartilages have in respiration?
10. Explain why a proper balance of vitamins, hormones, and minerals is essential in maintaining healthy osseous tissue. Give examples of diseases or skeletal conditions that may occur if there is an imbalance of any of these three essential substances.

10

Skeletal System: The Appendicular Skeleton

Outline

Concepts

The development of the appendicular skeleton begins at the end of the fourth week and progresses through the embryonic period as the limb buds follow the growth of the apical ectodermal ridges.

The structure of the pectoral girdle and upper extremities is adaptive for freedom of movement and extensive muscle attachment.

The structure of the pelvic girdle and lower extremities is adaptive for support and locomotion. There are extensive processes and surface features on certain bones of the pelvic girdle and lower extremities that accommodate massive muscles for posture and locomotion.

Development of the Appendicular Skeleton

The development of the appendicular skeleton begins at the end of the fourth week and progresses through the embryonic period as the limb buds follow the growth of the apical ectodermal ridges.

Objective 1. Explain the process of appendicular skeletal development.

The development of the upper and lower extremities is initiated toward the end of the fourth week as four small elevations called **limb buds** appear (fig. 10.1). The anterior pair are the arm buds, which precede the development of the posterior pair of leg buds by a few days. Each limb bud consists of a mass of undifferentiated mesoderm partially covered with a layer of ectoderm, called the **apical** *(ap'e-kal)* **ectodermal ridge,** which promotes bone and muscle development.

As the limb buds elongate, migrating mesenchymal tissues differentiate into specific cartilaginous bones. Soon primary ossification centers form in each bone, and the hyaline cartilage tissue is gradually replaced by bony tissue in the process of endochondral ossification.

Initially, the developing limbs are directed caudally, but later there is a lateral rotation in the upper extremity and a medial rotation in the lower extremity. As a result, the elbows are directed backward and the knees directed forward.

Digital rays that will form the hands and feet are apparent by the fifth week, and the individual digits separate by the end of the sixth week.

A large number of limb deformities occurred in children born between 1957 and 1962 as a result of mothers ingesting thalidomide during early pregnancy to relieve "morning sickness." It is estimated that 7,000 infants were malformed by thalidomide. The malformations ranged from *micromelia* (short limbs) to *amelia* (absence of limbs).

1. Define limb buds, apical ectodermal ridge, and digital rays.
2. During which weeks of pregnancy is an embryo most susceptible to deformities of the appendages from the maternal ingestion of teratogenic drugs?

Figure 10.1. The development of the appendicular skeleton. (*a*) limb buds are apparent in an embryo by 28 days, and (*b*) an ectodermal ridge is the precursor of the skeletal and muscular structures. (*c*) mesenchymal primordial cells are present at 33 days. (*d*) hyaline cartilaginous models of individual bones develop early in the sixth week. (*e*) later in the sixth week, the cartilaginous skeleton of the upper extremity is well formed.

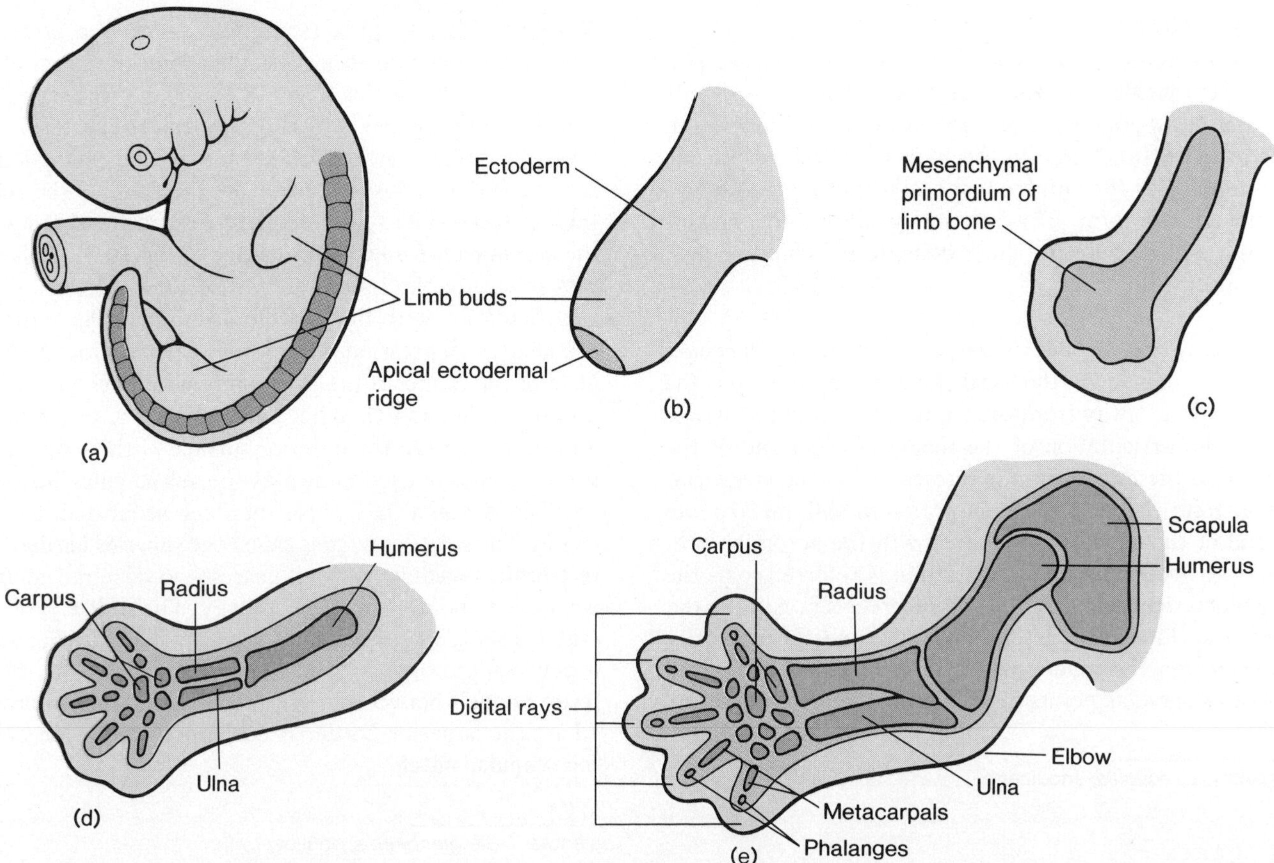

Figure 10.2. The right clavicle. (*a*) a superior view; (*b*) an inferior view.

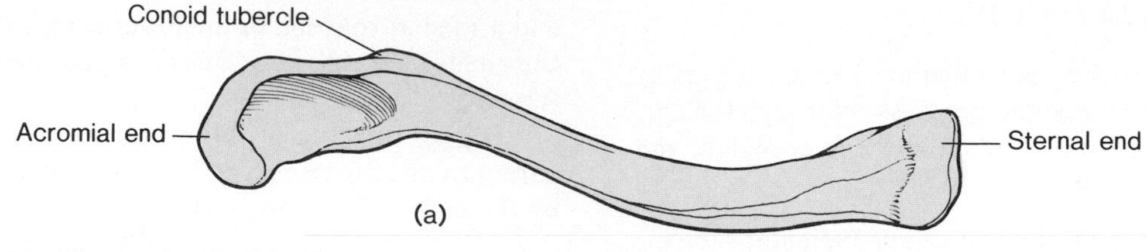

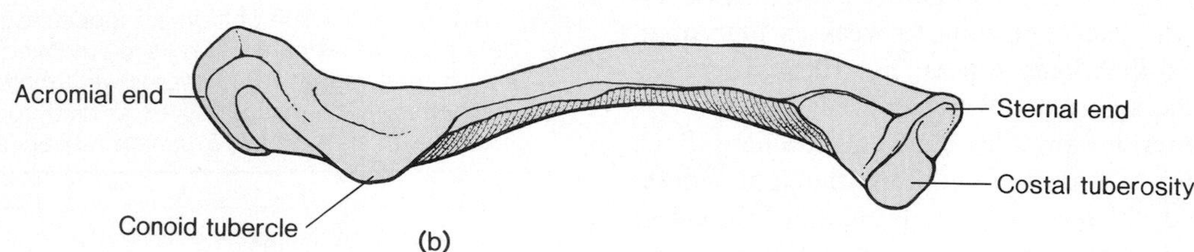

Pectoral Girdle and Upper Extremity

The structure of the pectoral girdle and upper extremities is adaptive for freedom of movement and extensive muscle attachment.

Objective 2. Describe the bones of the pectoral girdle and the positions of articulations.

Objective 3. Identify the bones of the upper extremity, and list the diagnostic features of each bone.

> The long, delicate clavicle is the most commonly fractured bone in the body. Blows to the shoulder or an attempt to break a fall with an outstretched hand cause the force to be displaced to the clavicle. The most vulnerable area for a fracture of this bone is through its center, immediately proximal to the conoid tubercle. Because the clavicle is subcutaneous and not covered with muscle, a fracture can easily be palpated.

Pectoral Girdle

The two scapulae and two clavicles make up the **pectoral** (shoulder) **girdle.** It is not a complete girdle, having only an anterior attachment to the axial skeleton at the sternum. The primary function of the pectoral girdle is to provide attachment for the numerous muscles that move the brachium and forearm. The pectoral girdle is not weight-bearing and is therefore more delicate in structure than the pelvic girdle.

Clavicle. The slender S-shaped clavicle (collarbone) binds the shoulder to the axial skeleton and positions the shoulder joint away from the trunk for freedom of movement. The articulation of the medial **sternal end** of the clavicle to the manubrium is referred to as the **sternoclavicular joint** (fig. 10.2). The lateral **acromial** *(ah-kro'me-al)* **end** of the clavicle articulates with the acromion process of the scapula. This articulation is referred to as the **acromioclavicular joint.** A **conoid tubercle** is present on the inferior surface of the lateral end, and a **costal tuberosity** is present on the inner surface of the medial end. Both processes serve as points of attachment for ligaments.

scapula: L. *scapula,* shoulder

Scapula. The scapula (shoulder blade) is a large triangular flat bone positioned on the posterior aspect of the rib cage, overlying ribs 2 to 7. The **spine** of the scapula is a prominent diagonal bony ridge seen on the posterior surface (fig. 10.3). Above the spine is the **supraspinous** *(su"pra-spi'nus)* **fossa,** and below the spine is the **infraspinous fossa.** The spine broadens toward the shoulder as the **acromion** *(ah-kro'me-on)* **process** (fig. 10.4). This process serves for the attachment of several muscles as well as articulation with the clavicle. Inferior to the acromion is a shallow depression, the **glenoid cavity,** into which the head of the humerus fits. The **coracoid process** is a thick, upward projection that lies superior and anterior to the glenoid cavity. On the anterior surface of the scapula is a slightly concave area known as the **subscapular fossa.**

The scapula has three borders separated by two angles. The superior edge is called the **superior border.** The **vertebral,** or **medial, border** is nearest the vertebral column, positioned about 5 cm (2 in.) away. The **axillary,** or **lateral, border** is directed toward the arm. The **superior angle** is between the superior and vertebral borders, and the **inferior angle** is between the vertebral and axillary borders. Along the superior border is a distinct depression called the **scapular notch.**

glenoid: Gk. *glenoeides,* shallow form

Figure 10.3. The right scapula. (*a*) an anterior view; (*b*) a posterior view.

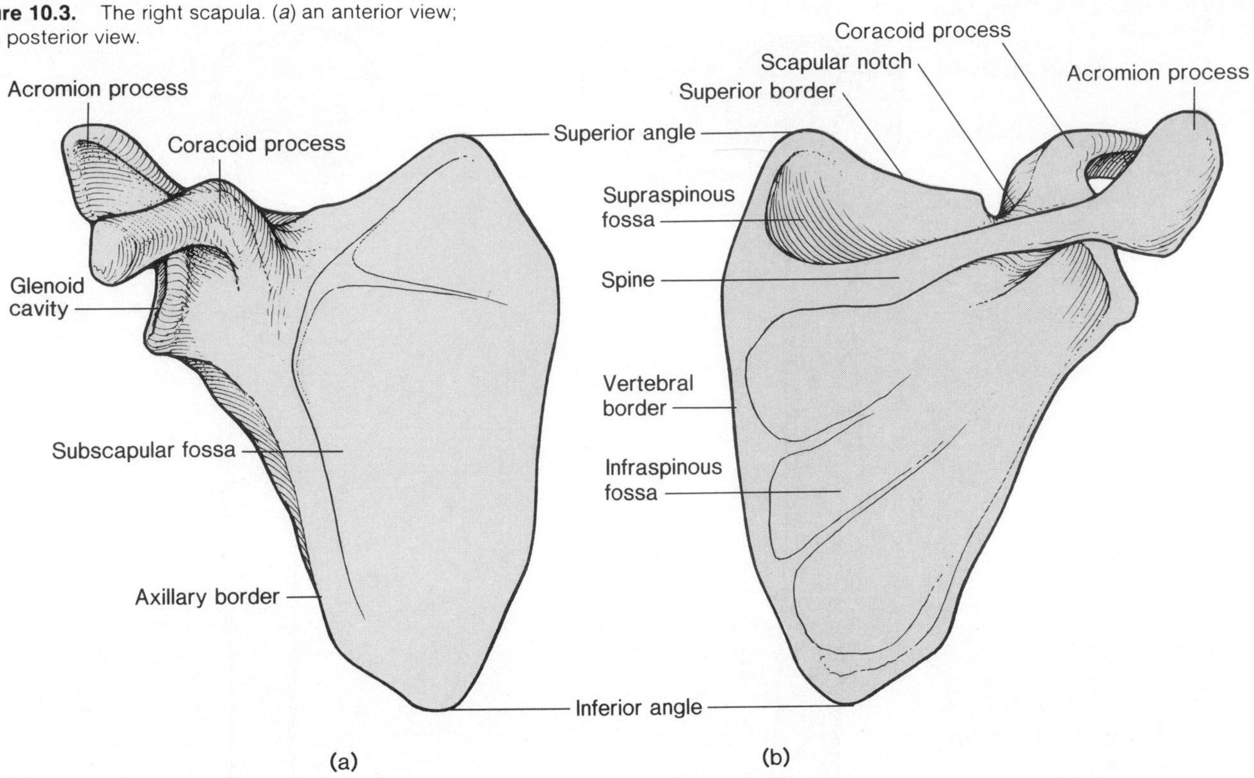

Acromion process
Coracoid process
Glenoid cavity
Subscapular fossa
Axillary border
Superior angle
Inferior angle
(a)

Coracoid process
Scapular notch
Superior border
Acromion process
Supraspinous fossa
Spine
Vertebral border
Infraspinous fossa
(b)

The anatomy of the scapula is important to know because some fifteen muscles attach to its processes and fossae. Clinically, the pectoral girdle is significant because the clavicle and acromion process of the scapula are frequently fractured in trying to break a fall. The acromion process is palpated when locating the proper site for an intramuscular injection of the arm.

Brachium

The **brachium** is the upper arm and contains a single bone, the humerus.

Humerus. The humerus (fig. 10.5) is the longest bone of the upper extremity. It consists of a proximal **head,** which articulates with the glenoid cavity of the scapula; a **shaft** (body); and a distal end, which is modified to receive the two bones of the forearm. Surrounding the margin of the head is a slightly indented groove denoting the **anatomical neck.** The region where the shaft begins to taper is referred to as the **surgical neck,** a frequent site of fractures. Lateral to the head is a large eminence, the **greater tubercle.** The **lesser tubercle** is slightly anterior to the greater and is separated from the greater by an **intertubercular (bicipital) groove,** through which passes the tendon from the biceps brachii muscle.

Figure 10.4. An X ray of the right shoulder shows the articulation of the clavicle, scapula, and humerus forming the shoulder joint.

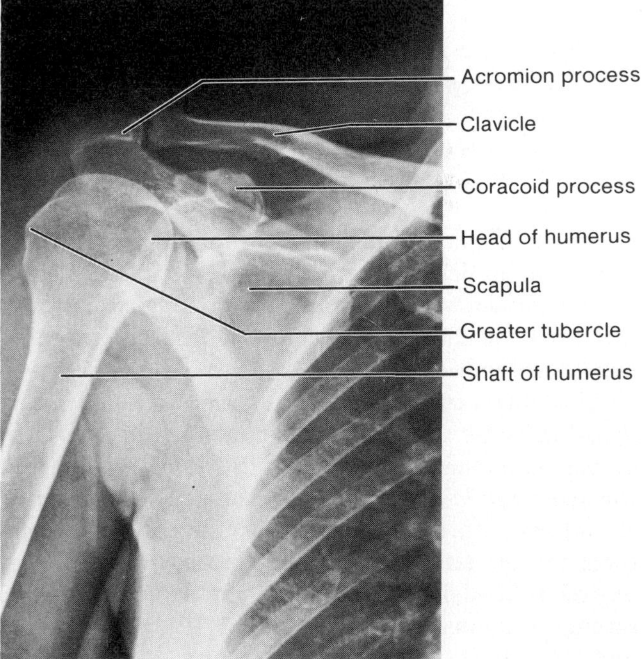

Figure 10.5. The right humerus. (*a*) an anterior view; (*b*) a posterior view

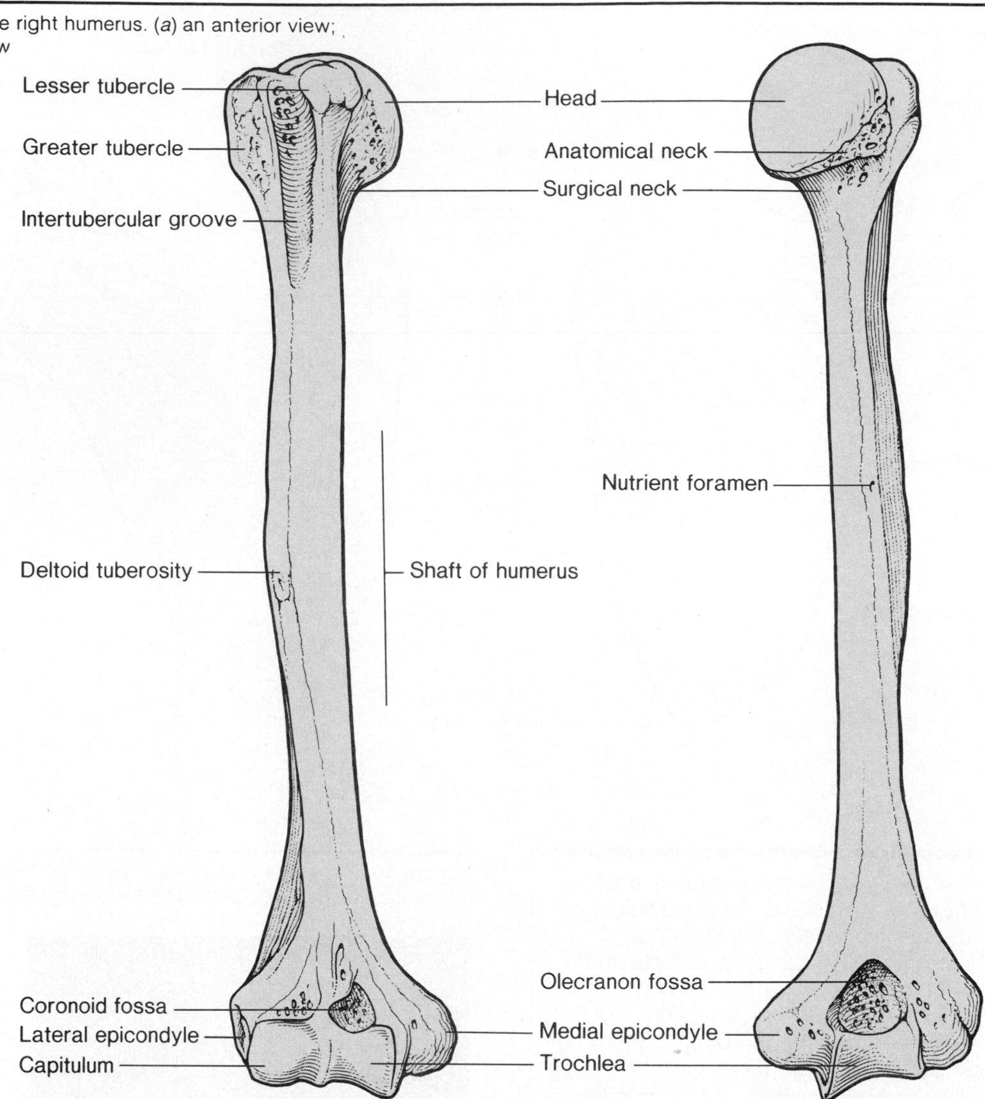

Along the lateral midregion of the shaft is a roughened area, the **deltoid tuberosity** for the attachment of the deltoid muscle, which elevates the arm horizontally to the side. Small openings in the bone along the shaft are called **nutrient foramina.**

The distal end of the humerus has two rounded condyloid articular surfaces. The **capitulum** *(kah-pit'u-lum)* is the lateral rounded condyle that receives the radius. The **trochlea** *(trok'le-ah)* is the pulleylike medial surface that articulates with the ulna. On either side above the condyles are the **lateral** and **medial epicondyles.** The large medial epicondyle protects the ulnar nerve that passes through the ulnar sulcus (see fig. 13.19). The **coronoid fossa** is a depression above the trochlea on the anterior surface. The **olecranon** *(o-lek'rah-non)* **fossa** is a depression on the distal posterior surface. Both fossae are adapted to receive parts of the ulna during movement of the forearm.

The medical term for tennis elbow is *lateral epicondylitis,* which means an inflammation of the tissues surrounding the lateral epicondyle of the humerus. Six muscles that control backward (extension) movement of the hand and fingers originate on the lateral epicondyle. Repeated strenuous contractions of these muscles, as in stroking with a tennis racket, may cause a strain on the periosteum and tendinous muscle attachments resulting in tenderness and pain around the epicondyle. Binding usually eases the pain, but only rest can eliminate the causative factor, and recovery generally follows.

Figure 10.6. An anterior view of the right radius and ulna. A transverse section shows the binding interosseous ligament.

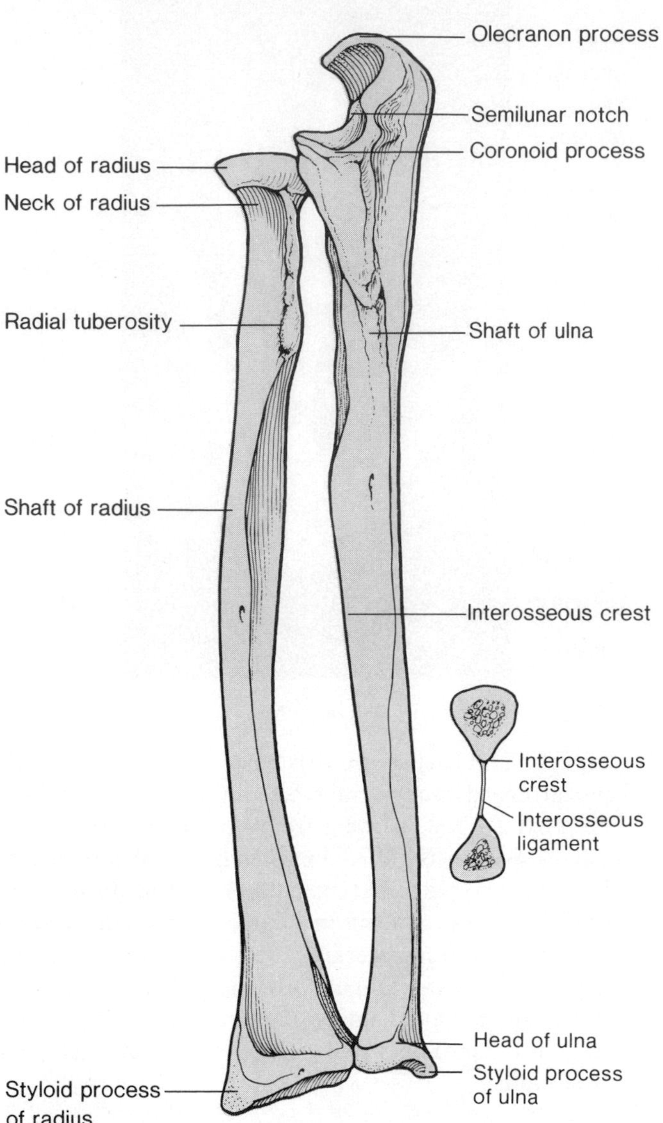

Figure 10.7. A posterior view of the right radius and ulna.

Olecranon process
Semilunar notch
Coronoid process
Head of radius
Neck
Radial tuberosity
Shaft of ulna
Shaft of radius
Interosseous crest
Styloid process of ulna
Styloid process of radius

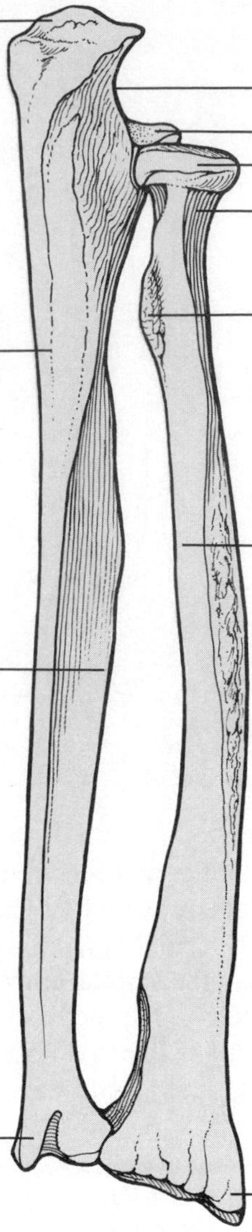

Forearm

The skeletal structures of the **forearm** are the ulna on the medial side and the radius on the lateral (thumb) side (figs. 10.6, 10.7). The ulna is longer, and it is more firmly connected to the humerus than the radius. The radius, however, contributes more significantly at the wrist joint than does the ulna.

Ulna. The proximal end of the ulna articulates with the humerus and radius. A distinct depression, the **semilunar (trochlear) notch,** articulates with the trochlea of the humerus. The **coronoid process** forms the anterior lip of the semilunar notch, and the **olecranon process** forms the posterior portion, or elbow. Lateral and inferior to the coronoid process is the **radial notch,** which accommodates the head of the radius.

Along the anterolateral surface of the shaft is a sharp ridge called the **interosseous crest.** The **interosseous ligament** extends from this crest to the radius to bind these two bones together (fig. 10.6).

The tapered distal end of the ulna has a knobbed portion, the **head,** and a posteromedial projection, the **styloid process.** The ulna articulates with the radius distally as well as proximally.

Radius. The radius consists of a **shaft** that has a small proximal end and a large distal end. A proximal disc-shaped **head** articulates with the capitulum of the humerus and the radial notch of the ulna. The prominent **radial tuberosity,** for attachment of the biceps muscle, is on the medial side of the shaft just below the head. The

Figure 10.8. A posterior view of the right wrist and hand. (*a*) a drawing; (*b*) a photograph.

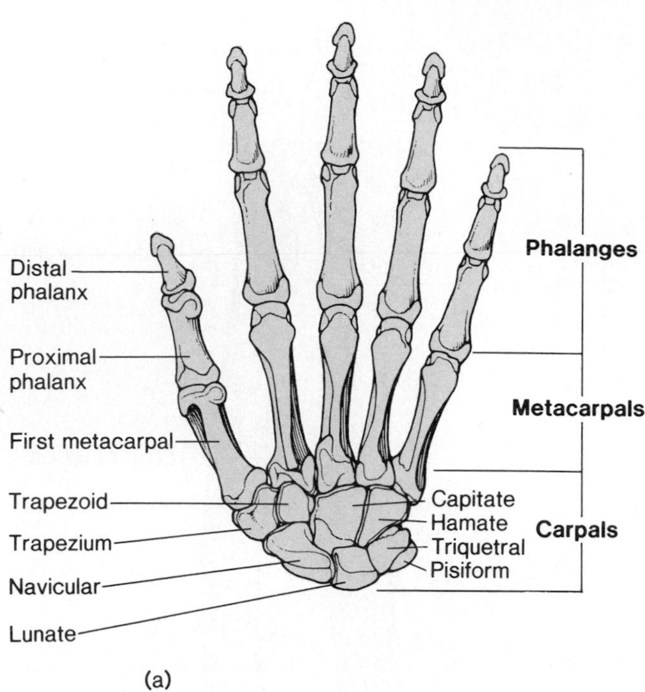

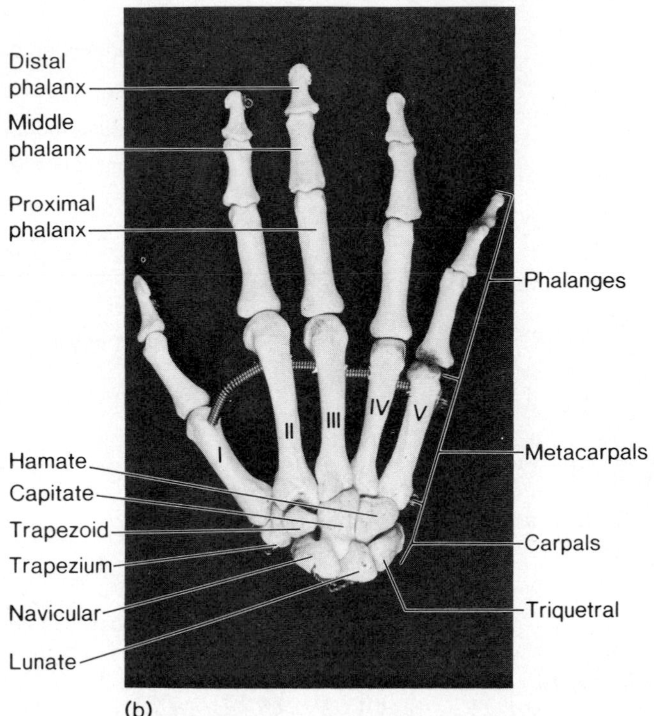

distal end of the radius has a double-faceted surface for articulation with the proximal carpal bones: a lateral **navicular facet** and a medial **lunate facet.** The distal end of the radius also has the **styloid process** on the lateral tip and the **ulnar notch** on the medial side for receiving the distal end of the ulna. The styloid processes on the ulna and radius provide lateral and medial support for articulation at the wrist.

When a person falls, the natural tendency is to extend the hand to break the fall. This reflexive movement frequently results in fractured bones. Common fractures of the radius include a fracture of the head as it is driven forcefully against the capitulum, a fracture of the neck, or a fracture of the distal end *(Colles' fracture)* caused by landing on an outstretched hand.

When falling, it is less traumatic to the body to withdraw the appendages, bend the knees, and let the entire body hit the surface. Athletes learn that this is the safe way to fall.

Wrist (Carpus) and Hand (Manus)

The **wrist** and **hand** contain twenty-seven bones arranged into the carpus, metacarpus, and phalanges (figs. 10.8, 10.9, 10.10).

Carpus. The carpus, or wrist, consists of eight carpal bones arranged into two transverse rows of four bones each. The proximal row, naming from the lateral (thumb) to medial side, consists of the **navicular** (scaphoid), **lunate, triquetral** *(tri-kwe'tral)* (triangular), and **pisiform** *(pi'sĭ-form)*. The pisiform forms in a tendon as a sesamoid bone. The distal row, from lateral to medial, consists of the **trapezium** (greater multangular), **trapezoid** (lesser multangular), **capitate,** and **hamate** *(ham'at)*. The navicular and lunate of the proximal row articulate with the distal end of the radius.

Metacarpus. The metacarpus, or palm of the hand, is composed of five metacarpal bones. Each metacarpal bone consists of a proximal **base,** a **shaft,** and a distal **head,** which is rounded for articulation with the base of each

carpus: Gk. *karpos,* wrist
navicular: L. *navicula,* small ship
lunate: L. *lunare,* crescent or moon-shaped
triquetral: L. *triquetrus,* three-cornered
pisiform: Gk. *pisos,* pea
trapezium: Gk. *trapesion,* small table
capitate: L. *capitatus,* head
hamate: L. *hamatus,* hook

Figure 10.9. An anterior view of the right wrist and hand.

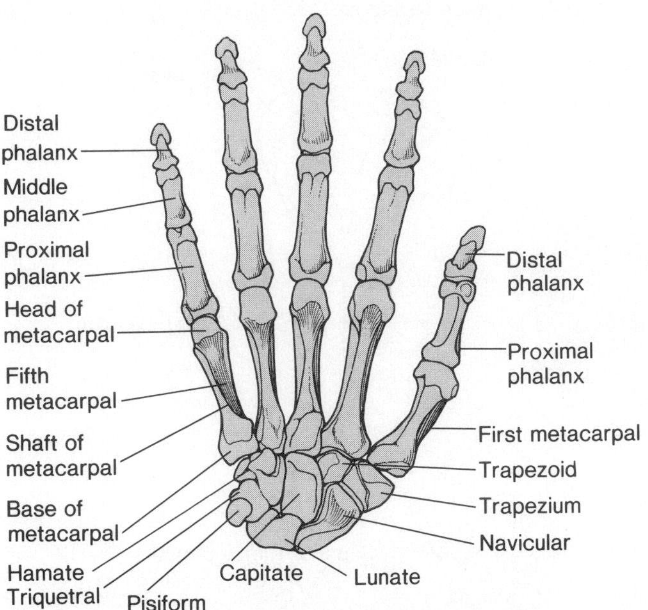

Figure 10.10. An X ray of the right wrist and hand shown in an anteroposterior projection. (Note the presence of a sesamoid bone at the thumb joint.)

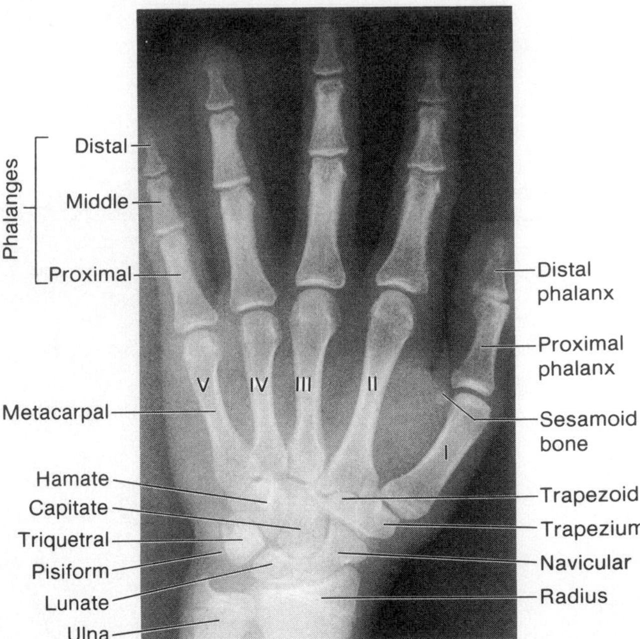

Table 10.1 Bones of the pectoral girdle and upper extremities

Name and number	Location	Diagnostic features
Clavicle (2)	Anterior base of neck, between sternum and scapula	S-shaped; sternal and acromial ends
Scapula (2)	Upper back forming part of the shoulder	Triangular-shaped; spine; acromion and corocoid processes
Humerus (2)	Brachium between scapula and elbow	Longest bone of upper extremity; greater and lesser tubercles; surgical neck; deltoid tuberosity; capitulum; trochlea; coronoid and olecranon fossae
Ulna (2)	Medial side of forearm	Semilunar notch; olecranon and styloid processes
Radius (2)	Lateral side of forearm	Head; radial tuberosity; styloid process
Carpal (16)	Wrist	Short bones arranged in two rows of four bones each
Metacarpal (10)	Palm of hand	Long bones arranged one in line with each digit
Phalanx (28)	Digits	Three in each finger, except two in thumb

proximal phalanx. The heads of the metacarpals are distally located and form the knuckles of a clenched fist. The metacarpal bones are numbered from one to five, the lateral, or thumb, side being one.

Phalanges. The fourteen phalanges are the skeletal elements of the fingers. A single finger bone is called a **phalanx** *(fa'lanks)*. The phalanges of the fingers are arranged into a proximal row, a middle row, and a distal row. The thumb has only a proximal and a distal phalanx.

phalanx: Gk. *phalanx,* finger bone or toe bone

A summary of the bones of the upper extremities is presented in table 10.1.

The hand is a marvel of structural organization that despite its complexity is able to sustain considerable abuse. Other than sprained fingers and dislocation, the most common osseous injury is a fracture to the navicular bone of the wrist (about 70% of carpal fractures occur here). When immobilizing the wrist joint with a plaster cast, the wrist is positioned in the plane of relaxed function. This is the position in which the hand is about to grasp an object between the thumb and index finger.

Figure 10.11. An anterior view of the pelvic girdle.

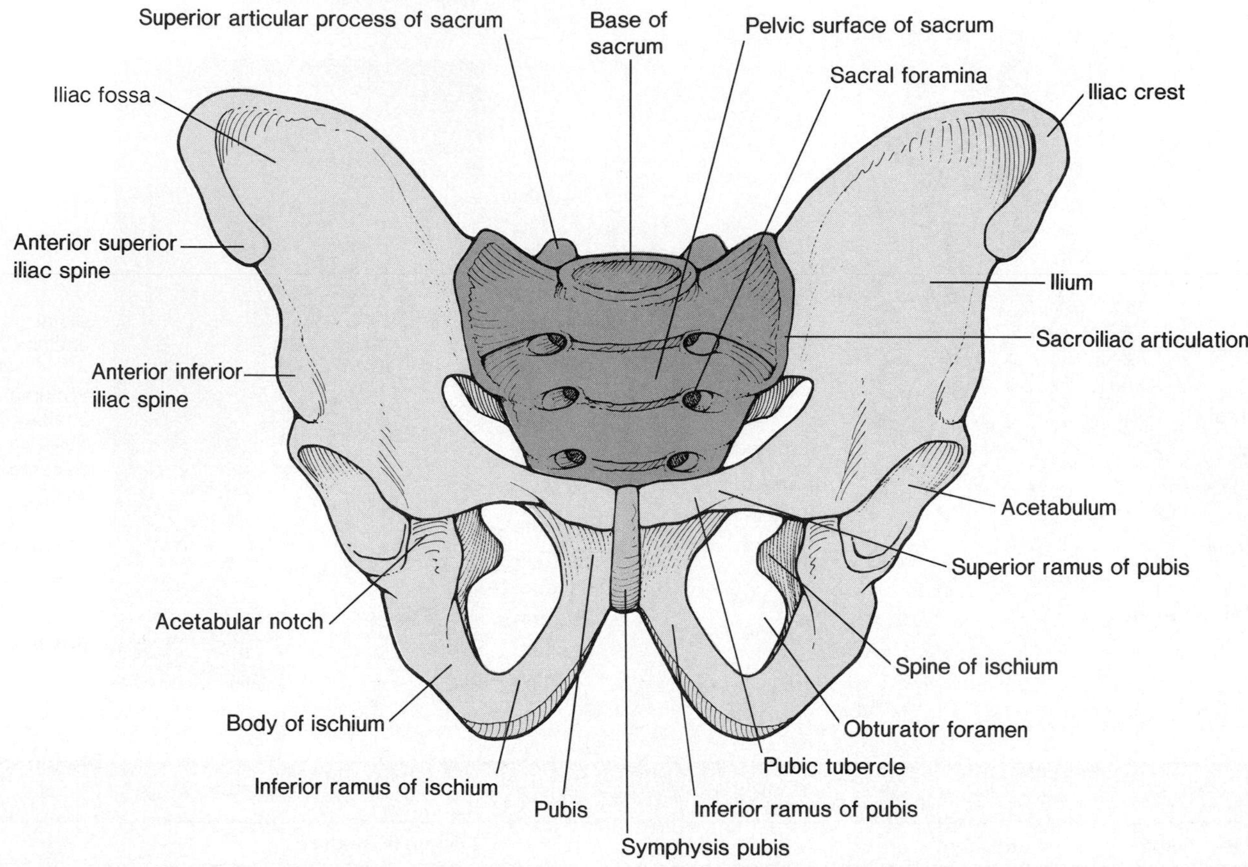

1. Describe the structure of the pectoral girdle. Why is the pectoral girdle considered an incomplete girdle?
2. Name the fossae found on the scapula.
3. Describe each of the long bones of the upper extremity.
4. Where are the styloid processes of the wrist area? What are their functions?
5. Name the bones in the proximal row of the carpus. Which of these bones articulate with the radius?

Pelvic Girdle and Lower Extremity

The structure of the pelvic girdle and lower extremities is adaptive for support and locomotion. There are extensive processes and surface features on certain bones of the pelvic girdle and lower extremities that accommodate massive muscles for posture and locomotion.

Objective 4. Describe the structure of the pelvic girdle, and list its functions.

Objective 5. Describe the structural differences between the male and female pelvis.

Objective 6. Identify the bones of the lower extremity, and list the diagnostic features of each bone.

Objective 7. Describe the structural features and functions of the arches of the foot.

Pelvic Girdle

The **pelvic girdle,** or **pelvis,** is formed by two **ossa coxae** (hipbones) united anteriorly by the **symphysis pubis** (fig. 10.11). It is attached posteriorly to the sacrum of the vertebral column. The pelvic girdle and its associated ligaments support the weight of the body from the vertebral column. The pelvic girdle also supports and protects the lower viscera, including the urinary bladder, the reproductive organs, and in a pregnant woman, the developing fetus.

Clinically, the basinlike pelvis is frequently divided into a **greater,** or **false, pelvis** and a **lesser,** or **true, pelvis** (see fig. 10.14). These two components are divided by the **pelvic brim,** a curved bony rim passing inferiorly from the sacral promontory to the upper margin of the symphysis pubis. The greater pelvis is the expanded portion of the pelvis superior to the pelvic brim. The pelvic brim not only

coxae: L. *coxae,* hips

Figure 10.12. An X ray of the pelvic girdle and the articulating femora.

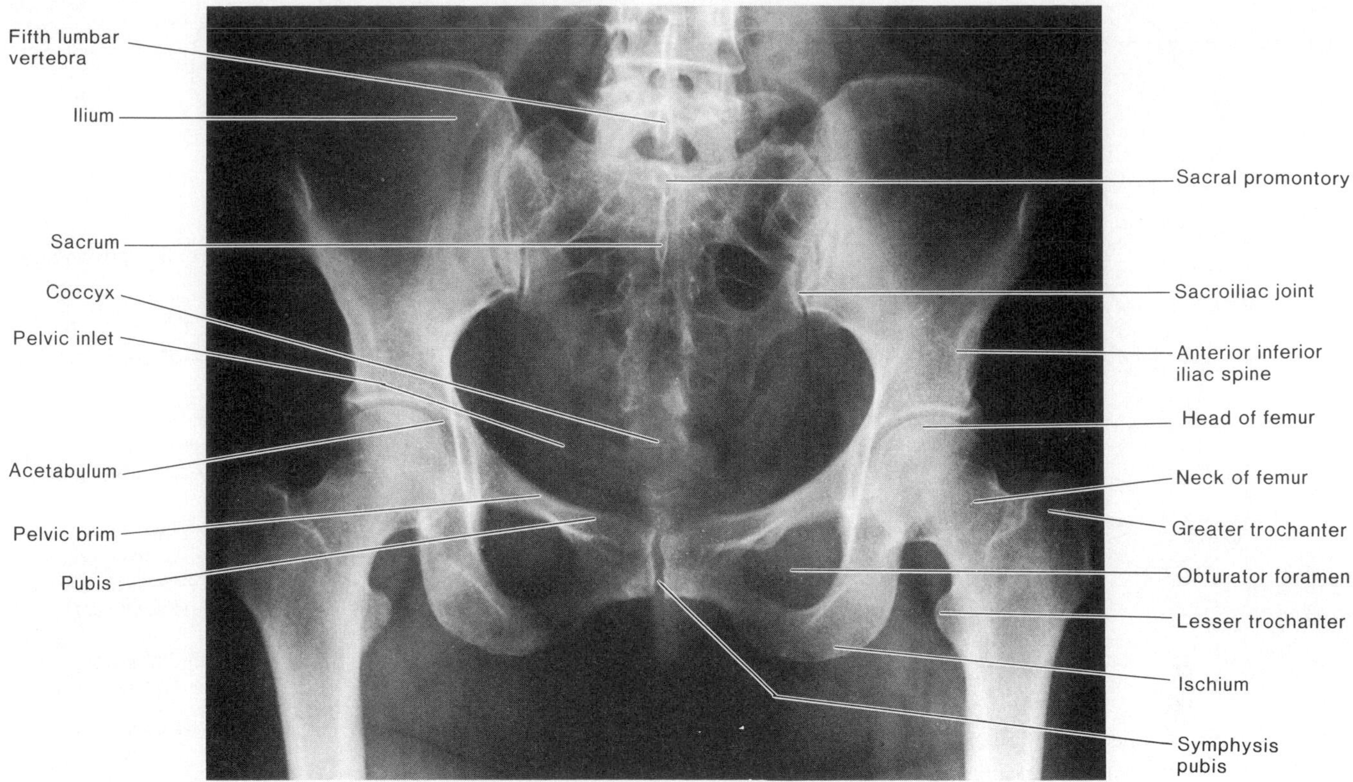

divides the two portions but surrounds the **pelvic inlet** of the lesser pelvis. The lower circumference of the lesser pelvis bounds the pelvic outlet. During parturition, a child must pass through its mother's lesser pelvis for a natural delivery. *Pelvimetry* measures the dimension of the lesser pelvis to determine if a cesarean delivery might be necessary. Diameters may be determined by vaginal palpation or by X-ray measurements (fig. 10.12).

Each os coxa actually consists of three separate bones: the **ilium,** the **ischium** *(is'ke-um),* and the **pubis** (fig. 10.13). These bones are fused together in the adult. On the lateral surface of the os coxa where the three bones ossify together is a large circular depression, the **acetabulum** *(as"ĕ-tab'u-lum),* which receives the head of the femur. Although in the adult both ossa coxae are single bones, the three components are considered separately for descriptive purposes.

Ilium. The ilium is the largest and uppermost of the three pelvic bones. The ilium presents a crest and four angles, or spines, which serve for muscle attachment and are important surface landmarks. The **iliac crest** forms the prominence of the hip. This crest terminates anteriorly as the **anterior superior iliac spine.** Just below this spine is the **anterior inferior iliac spine.** The posterior termination of the iliac crest is the **posterior superior iliac spine,** and just below this is the **posterior inferior iliac spine.**

Below the posterior inferior iliac spine is the **greater sciatic** *(si-at'ik)* **notch.** On the medial surface of the ilium is the roughened **auricular surface** that articulates with the sacrum. The **iliac fossa** is the smooth, concave surface on the anterior portion of the ilium. The iliacus muscle

ilium: L. *ilia,* loin
ischium: Gk. *ischion,* hip joint
pubis: L. *pubis,* genital area
acetabulum: L. *acetabulum,* vinegar cup

Figure 10.13. The lateral aspect of the right os coxa.

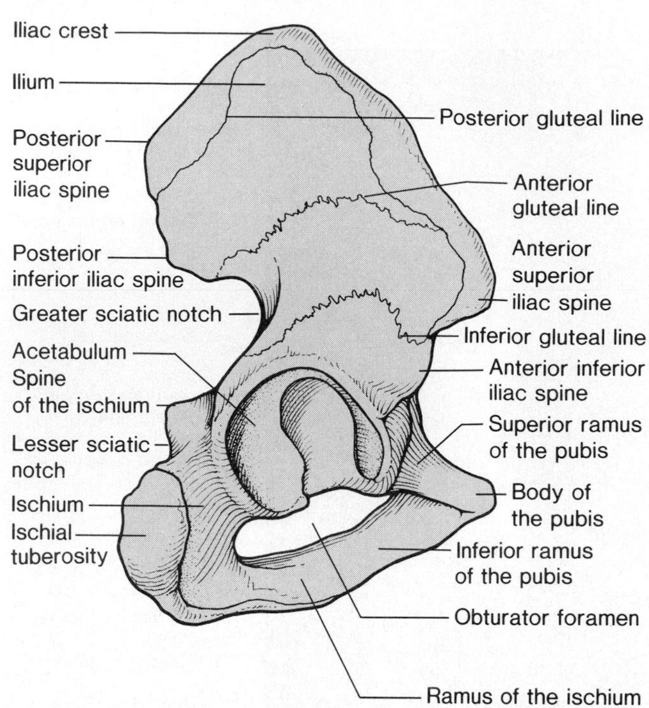

Figure 10.14. A comparison of the (*a*) male and (*b*) female pelvic girdle.

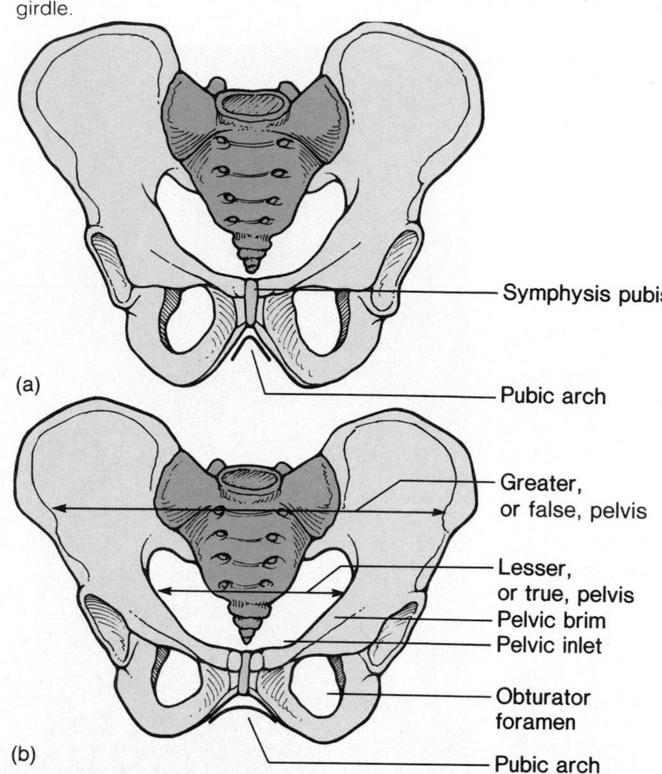

originates from this fossa. The **iliac tuberosity,** for the attachment of the sacroiliac ligament, is positioned posterior to the iliac fossa. Three roughened ridges are present on the **gluteal surface** of the posterior aspect of the ilium. These ridges serve to attach the gluteal muscles and are the **inferior, anterior,** and **posterior gluteal lines.**

Ischium. The ischium is the posterior-inferior component of the os coxa. This bone has several significant features. The **spine of the ischium** is the projection immediately posterior to the greater sciatic notch of the ilium and ischium. Inferior to this spine is the **lesser sciatic notch** of the ischium. The **ischial tuberosity** is the bony projection that supports the weight of the body in the sitting position. A deep **acetabular notch** is present on the inferior portion of the acetabulum. The large **obturator foramen** is formed by the **ramus** of the ischium together with the pubis. The obturator foramen is covered by the obturator membrane, to which several muscles attach.

Pubis. The pubis is the anterior component of the os coxa. This bone consists of a **superior ramus** and an **inferior ramus** that supports the **body** of the pubis. The body contributes to the formation of the symphysis pubis—the joint between the two ossa coxae.

Table 10.2 Sexual differences of the pelvic girdle

Characteristics	Male pelvis	Female pelvis
General structure	More massive; prominent processes	More delicate; processes not so prominent
Pelvic inlet	Heart shaped	Round or oval
Pelvic outlet	Narrower	Wider
Anterior superior iliac spines	Less wide apart	More wide apart
Obturator foramen	Oval	Triangular
Acetabulum	Faces laterally	Faces more anteriorly
Symphysis pubis	Deeper, longer	Shallower, shorter
Pubic arch	Acute (less than 90°)	Obtuse (greater than 90°)

Sex Differences in the Pelvis. There are structural differences between the pelvis of an adult male and that of an adult female (fig. 10.14, and table 10.2) that reflect the female's role in pregnancy and childbirth.

In addition to the osseous differences listed in table 10.2, the symphysis pubis and sacroiliac joints stretch during pregnancy and parturition.

Figure 10.15. The right femur. (*a*) an anterior view; (*b*) a posterior view

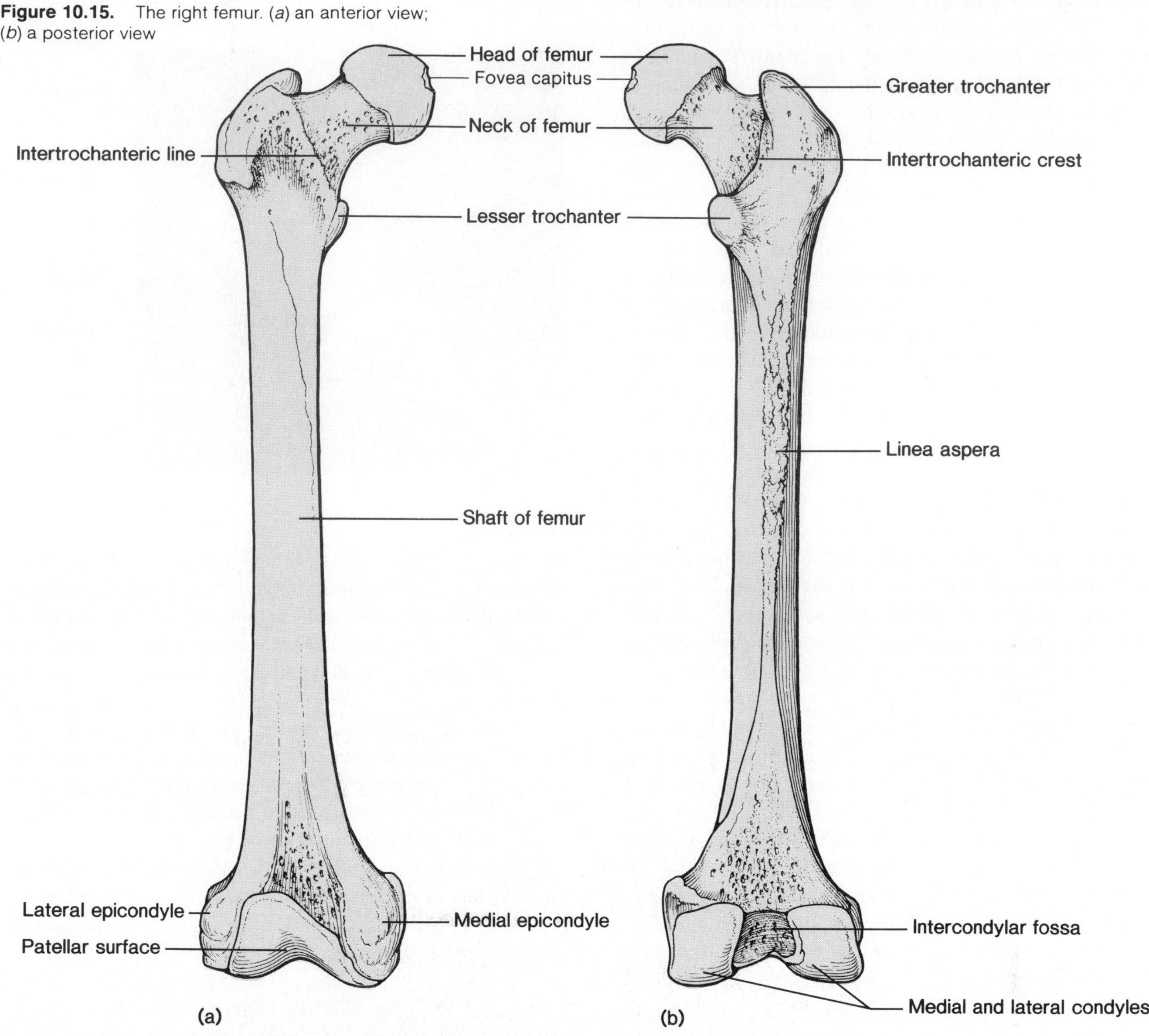

The structure of the human pelvis, in its attachment to the vertebral column, permits an upright posture and locomotion on two legs (bipedal) rather than on four legs like other mammals. Although these structures are well adapted for *bipedal locomotion,* an upright posture may cause problems. The sacroiliac joint may weaken with age, causing lower back pains. The weight of the viscera may weaken the walls of the lower abdominal area and cause hernias. Some of the problems of childbirth are related to the structure of the mother's pelvis. The hip joint tends to deteriorate with age, and many older persons suffer from fractured hips.

Thigh

The patella will be considered along with the femur in the discussion of the thigh, even though the femur is the only bone of the thigh.

Femur. The femur (thighbone) is the longest, heaviest, and strongest bone in the body (fig. 10.15). The proximal rounded **head** of the femur articulates with the acetabulum of the os coxa. A roughened, shallow pit, called the **fovea capitis,** is in the lower center of the head of the femur.

femur: L. *femur,* thigh

The fovea capitis provides the point of attachment for the ligamentum teres, which helps to support the head of the femur against the acetabulum. The constricted region supporting the head is called the **neck** and is a common site for fractures in aged persons.

The **shaft** of the femur has a slight medial bow so that it converges with the femur of the opposite thigh and brings the knee joints more in line with the body's plane of gravity. The degree of convergence is even greater in the female because of the wide pelvis. The shaft has several important structures for muscle attachment. On the proximolateral side of the shaft is the **greater trochanter,** and on the medial side is the **lesser trochanter.** On the anterior side between the trochanters is the **intertrochanteric line.** On the posterior side between the trochanters is the **intertrochanteric crest.** The **linea aspera** is a roughened vertical ridge on the posterior surface of the shaft.

The distal end of the femur is expanded for articulation with the tibia. The **medial** and **lateral condyles** *(kon'dīlz)* are the articular processes for this joint. The depression between the condyles on the posterior aspect is called the **intercondylar fossa.** The **patellar surface** is between the condyles on the anterior side. Above the condyles on the lateral and medial sides are the **epicondyles,** which serve for ligament and tendon attachment.

Patella. The patella, or kneecap, is a sesamoid bone positioned on the anterior side of the knee joint (figs. 10.16, 10.17). It develops in response to stress in the tendon of the quadriceps femoris muscle. The patella is a triangular bone with a broad **base** and an inferiorly pointed **apex. Articular facets** on the posterior surface of this bone articulate with the medial and lateral condyles of the femur.

The functions of the patella are to protect the knee joint and to strengthen the quadriceps tendon. It also increases the leverage of the quadriceps femoris muscle as it straightens (extends) the leg.

The patella can be fractured by a direct blow. It usually does not fragment, however, because it is confined within the tendon. Dislocations of the patella may result from injury or may be congenital due to underdevelopment of the lateral condyle of the femur.

Lower Leg

The tibia and fibula are the skeletal elements of the lower leg. The tibia is the larger and more medial of the two bones. Figure 10.17 illustrates the skeletal structure of the lower leg.

linea aspera: L. *linea,* line; *asperare,* rough

Figure 10.16. A lateral X ray of the right knee region.

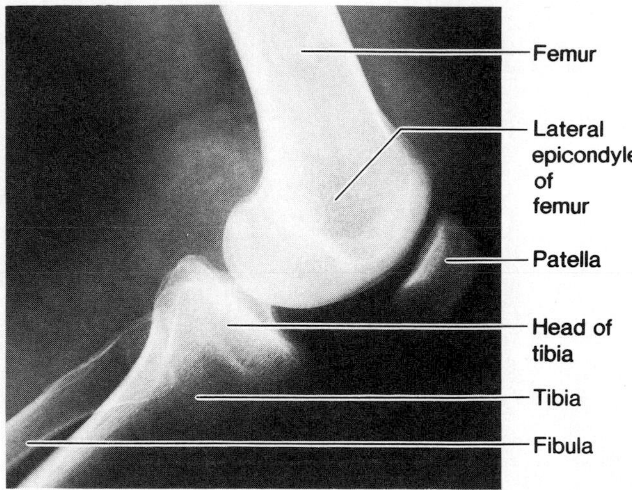

Tibia. The tibia (shinbone) articulates proximally with the femur at the knee joint to bear the weight of the body. On the distal end, the tibia articulates with the talus of the ankle. Two slightly concave surfaces on the proximal end of the tibia, the **medial** and **lateral condyles,** articulate with the condyles of the femur. Between the condyles is a slight upward projection called the **intercondylar eminence.** The **tibial tuberosity,** for attachment of the patellar ligament, is located on the proximal-anterior portion of the shaft. There is a sharp ridge along the anterior surface of the shaft called the **anterior crest.**

The **medial malleolus** *(mal-le'o-lus)* is a prominent medial knob of bone located on the distal end of the tibia. A **fibular notch,** for articulation with the fibula, is located on the distal-lateral end.

Fibula. The fibula is a long, narrow bone that is more important for muscle attachment than for support. The **head** of the fibula articulates with the lateral-proximal end of the tibia. The distal end has a prominent knob called the **lateral malleolus.**

The lateral and medial malleoli are positioned on either side of the talus and help stabilize the ankle joint. Both processes can be seen as prominent surface features and are easily palpated. Fractures to either or both malleoli are common in skiers. These fractures, clinically referred to as *Pott's fractures,* result from a shearing force occurring at a vulnerable spot on the lower leg.

tibia: L. *tibia,* shinbone, pipe, flute
malleolus: L. *malleolus,* small hammer
fibula: L. *fibula,* clasp or brooch

Figure 10.17. The right tibia, fibula, and patella. (*a*) an anterior view; (*b*) a posterior view.

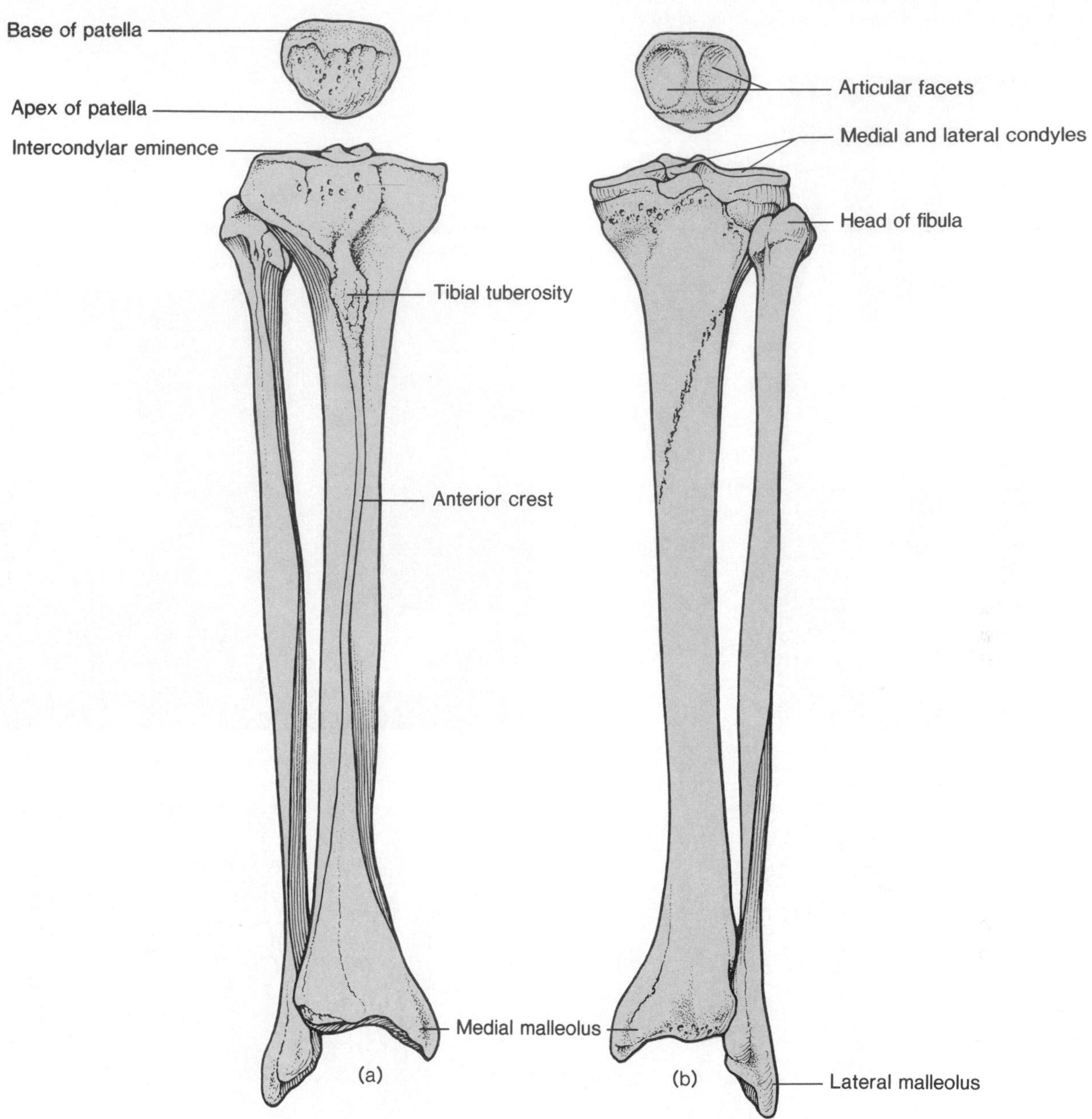

Ankle and Foot (Pes)

The **ankle** and **foot** contain twenty-six bones, arranged into the tarsus, metatarsus, and phalanges (figs. 10.18, 10.19). The bones of the ankle and foot are basically similar to those of the wrist and hand. They do, however, have distinct structural differences in order to provide weight support and leverage during walking.

Tarsus. There are seven tarsal bones. The **talus** is the tarsal bone that articulates with the tibia to form the ankle joint. The **calcaneus** *(kal-ka'ne-us)* is the largest of the tarsal bones and forms the heel of the foot. It has a large posterior extension, called the **tuberosity of the calcaneus,** for the attachment of the calf muscles. Anterior to the talus is the block-shaped **navicular** *(nah-vik'u-lar)* bone.

tarsus: G. *tarsos,* flat of the foot
talus: L. *talus,* ankle
calcaneus: L. *calcis,* heel

Figure 10.18. (*a*) a superior view of the right foot. (*b*) a medial X ray of the right foot and ankle. (*c*) a superior view of the bones of the right ankle and foot. (*d*) an inferior view of the bones of the right ankle and foot. (Note the presence of a sesamoid bone in (*b*) at the base of the big toe.)

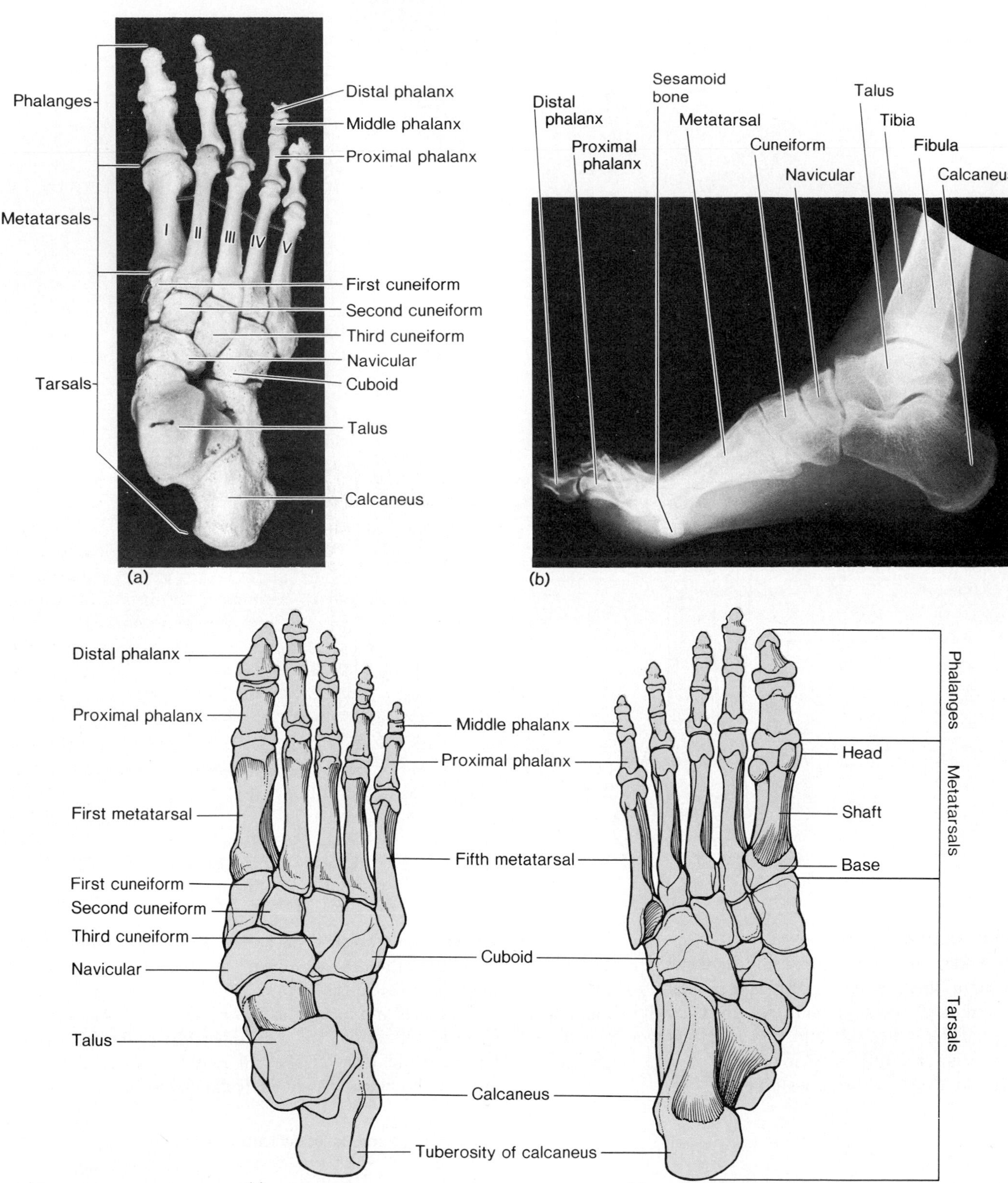

The remaining four tarsal bones form a distal series that articulate with the metatarsals. They are, from the medial to lateral side, the **first, second,** and **third cuneiforms** *(ku-ne'ĭ-formz)* and the **cuboid.**

Metatarsus. The metatarsus, or sole of the foot, is composed of five metatarsal bones. The metatarsals are numbered from one to five, with the medial, or big toe, side being one. The first metatarsal is larger than the others because of its weight-bearing function.

Figure 10.19. The arches of the foot. (*a*) a medial view of the left foot showing both arches. (*b*) a transverse view through the bases of the metatarsals showing a portion of the transverse arch.

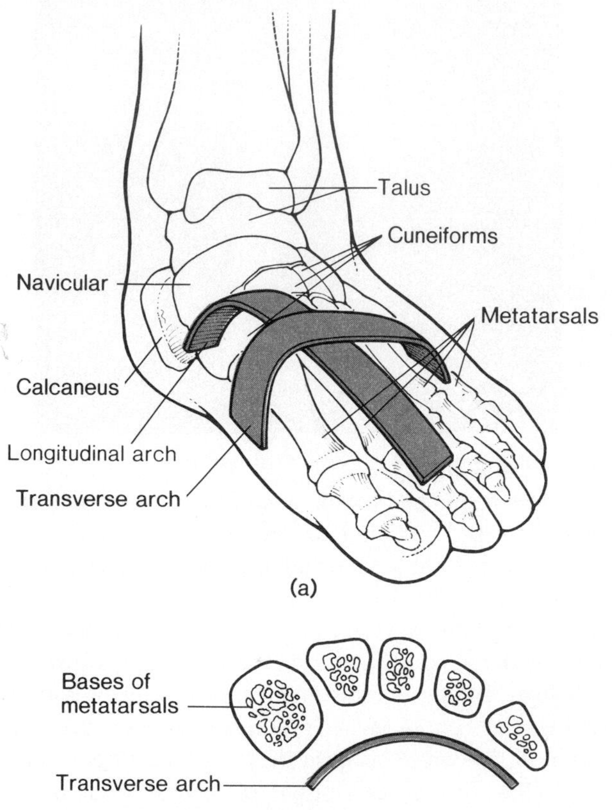

The metatarsals are long bones, each having a **base, shaft,** and **head.** The proximal bases of the first, second, and third metatarsals articulate proximally with the cuneiforms. The heads of the metatarsals articulate distally with the proximal phalanges. The proximal joints are called **tarsometatarsal joints,** and the distal joints are called **metatarsophalangeal joints.** The ball of the foot is formed by the heads of the first two metatarsal bones.

Phalanges. The fourteen phalanges are the skeletal elements of the toes. As with the fingers of the hand, the phalanges of the toes are arranged into a proximal row, a middle row, and a distal row. The great toe has only a proximal and a distal phalanx.

Arches of the Foot. The foot has two arches that support the weight of the body and provide leverage when walking. These arches are formed by the structure and arrangement of the bones held in place by ligaments and tendons. The arches are not rigid; they yield when weight is placed on the foot and spring back as the weight is lifted (fig. 10.19).

The **longitudinal arch** is divided into medial and lateral portions. The medial, or inner, portion is the larger of the two. It is supported by the calcaneus proximally and by the heads of the first three metatarsals distally. The wedge, or "keystone," of this portion of the longitudinal arch is the talus. The shallower lateral portion consists of the calcaneus, cuboid, and fourth and fifth metatarsals. The cuboid is the "keystone" of this portion.

The **transverse arch** extends across the width of the foot and is formed by the distal portion of the calcaneus, the navicular, the cuboid, and the proximal portions of all five metatarsals.

A weakening of the ligaments and tendons of the foot decreases the height of the longitudinal arch in a condition called *pes planus,* or flatfoot.

Table 10.3 contains a summary of the bones of the lower extremities.

Table 10.3 Bones of the pelvic girdle and the lower extremities

Name and number	Location	Diagnostic features
Os coxa (2)	Hip, part of the pelvic girdle; composed of 3 fused bones	Iliac crest, acetabulum, anterior superior iliac spine, ischial tuberosity, obturator foramen
Femur (2)	Bone of the thigh between hip and knee	Head, fovea capitis, neck, greater and lesser trochanters, lateral and medial condyles
Patella (2)	Anterior surface of knee	Triangular sesamoid bone
Tibia (2)	Medial side of leg between knee and ankle	Medial and lateral condyles, tibial crest, medial malleolus
Fibula (2)	Lateral side of leg between knee and ankle	Head, lateral malleolus
Tarsal (14)	Ankle	Large talus and calcaneus to receive the weight of leg; five other wedge-shaped bones to help form arches of foot
Metatarsal (10)	Sole of foot	Long bones, one in line with each digit
Phalanx (28)	Digits	Three in each toe, two in big toe

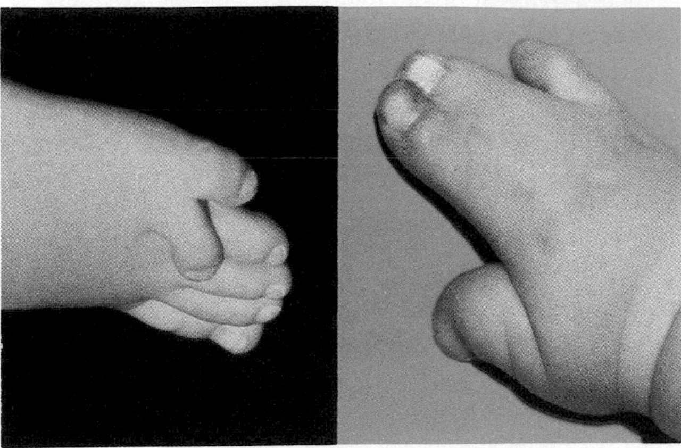

Figure 10.20. Polydactyly is having extra digits. It is the most common congenital deformity of the foot, although it also occurs in the hand. Syndactyly is when two or more digits are webbed together. It is a common congenital deformity of the hand, although it also occurs in the foot. Both conditions can be surgically corrected.

1. Describe the structure of the pelvic girdle that reflects its weight-bearing role.
2. How can female and male pelves be distinguished? What is the clinical importance of the lesser pelvis in females?
3. Describe the structure of each of the long bones of the lower extremity and the position of each of the tarsal bones.
4. Which bones of the foot participate in the formation of the arches of the foot? What are the functions of the arches?

Clinical Considerations

Developmental Disorders

Minor defects of the extremities are relatively common malformations. Extra digits, a condition called **polydactyly** *(pol″e-dak′tĭ-le)* (fig. 10.20), is the most common limb deformity. Usually an extra digit is incompletely formed and does not function. **Syndactyly,** or webbed digits, is likewise a relatively common limb malformation. Polydactyly is inherited as a dominant trait, whereas syndactyly is a recessive trait.

Talipes *(tal′ĭ-pēz),* or clubfoot (fig. 10.21), is a congenital malformation in which the sole of the foot is twisted medially. It is not certain if abnormal positioning or restricted movement *in utero* causes this condition, but both genetics and environmental conditions are involved in most cases.

Figure 10.21. Talipes, or clubfoot, is a congenital malformation of a foot or both feet. The condition can be effectively treated surgically if done at a young age.

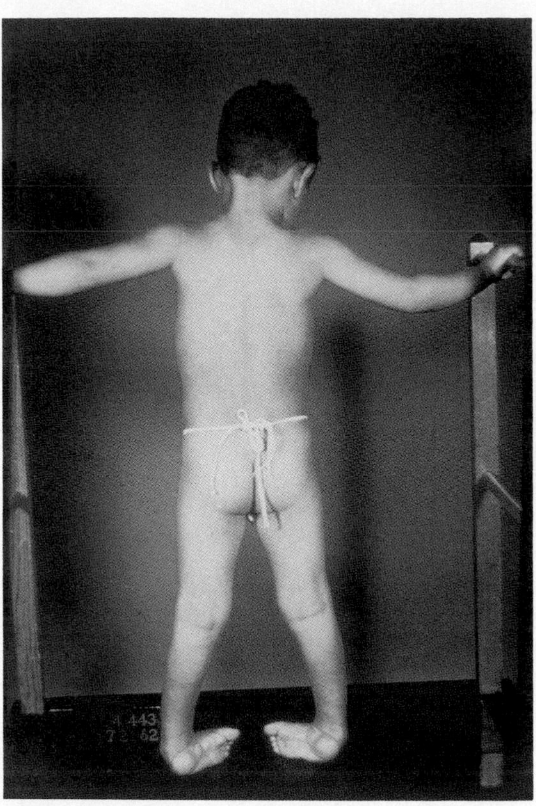

Trauma and Injury

The most common type of bone injury is a **fracture.** A fracture is the cracking or breaking of a bone. In *radiology,* X rays are used to diagnose the position and extent of a fracture. Fractures may be classified in several ways, and the type and severity of the fracture varies with the age and the general health of the body. **Spontaneous,** or **pathologic, fractures,** for example, result from diseases that weaken the bones. Most fractures, however, are called **traumatic fractures** because they are caused by injuries. The following are descriptions of several kinds of traumatic fractures (fig. 10.22).

1. **Simple,** or **closed.** The fractured bone does not break through the skin.
2. **Compound,** or **open.** The fractured bone is exposed to the outside through an opening in the skin.
3. **Partial (fissured).** The bone is incompletely broken.
4. **Complete.** The fracture has separated the bone into two portions.
5. **Capillary.** A hairlike crack occurs within the bone.
6. **Comminuted** *(kom′ĭ-nut″ed).* The bone is splintered into small fragments.

Figure 10.22. Examples of types of traumatic fractures.

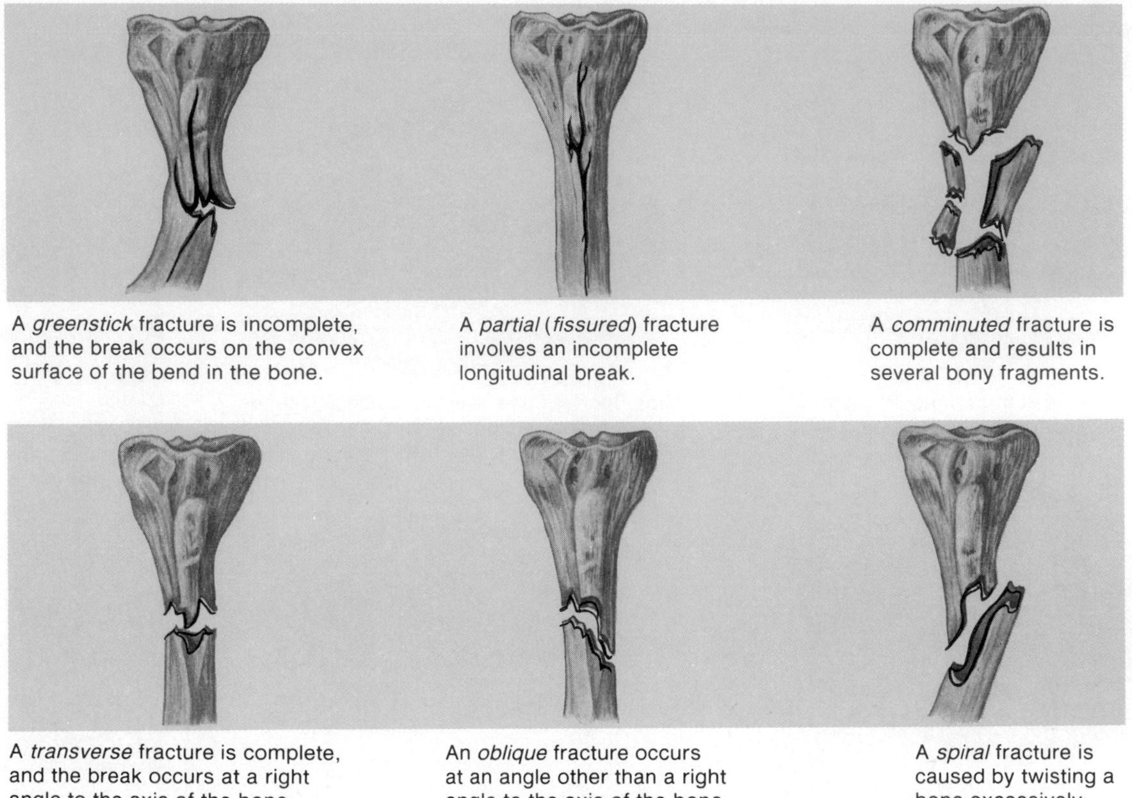

A *greenstick* fracture is incomplete, and the break occurs on the convex surface of the bend in the bone.

A *partial* (*fissured*) fracture involves an incomplete longitudinal break.

A *comminuted* fracture is complete and results in several bony fragments.

A *transverse* fracture is complete, and the break occurs at a right angle to the axis of the bone.

An *oblique* fracture occurs at an angle other than a right angle to the axis of the bone.

A *spiral* fracture is caused by twisting a bone excessively.

7. **Spiral.** The bone is twisted as it is broken.
8. **Greenstick.** In this incomplete break one side of the bone is broken, and the other side is bowed.
9. **Impacted.** One broken end of a bone is driven into the other.
10. **Transverse.** A fracture occurs across the bone at right angles to the shaft.
11. **Oblique.** A fracture occurs across the bone at an oblique angle to the axis of the bone.
12. **Colles'.** The styloid process of the radius is fractured.
13. **Pott's.** The lateral malleolus of the fibula is fractured.
14. **Avulsion.** In this fracture a structure, a finger for example, is torn off.
15. **Depressed.** The broken portion of the bone is driven inward, as in certain skull fractures.
16. **Displaced.** In this fracture the bone fragments are not in anatomical alignment.
17. **Nondisplaced.** In this fracture the bone fragments are in anatomical alignment.

When a bone fractures, medical treatment involves realigning the broken ends and then immobilizing them until new bone tissue is formed and the fracture is healed. The site and severity of the fracture and the age of the patient will determine the type of immobilization. The methods of immobilization include tape, splints, casts, straps, wires, and steel pins. Even with these various methods of treatment, certain fractures heal poorly. Ongoing research into these problems is producing promising results. It has been found, for example, that applying weak electrical currents to fractured bones promotes healing and reduces the time of immobilization by half. Furthermore, new glasslike material called "bioglass" has been used to bond broken bones. This may be especially useful in elderly patients because their bones are slow in healing.

Figure 10.23. *(a–d)* stages of the repair of a fracture. *(e)* an X ray of a healing fracture.

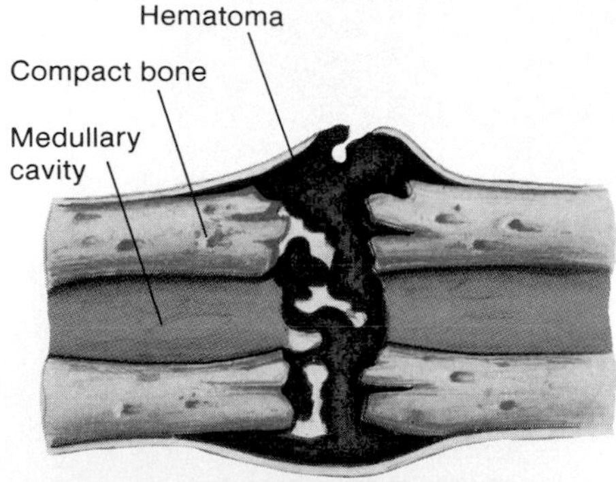

(a) Blood escapes from ruptured blood vessels and forms a hematoma.

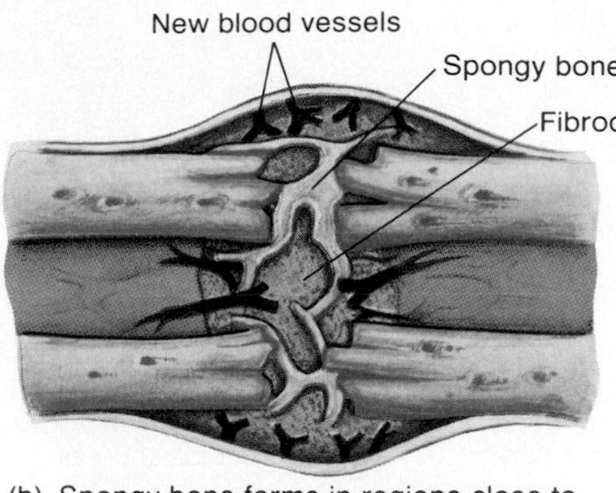

(b) Spongy bone forms in regions close to developing blood vessels, and fibrocartilage forms in more distant regions.

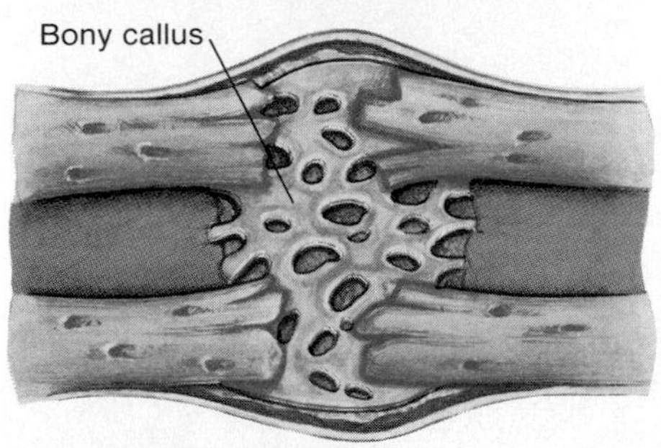

(c) Fibrocartilage is replaced by a bony callus.

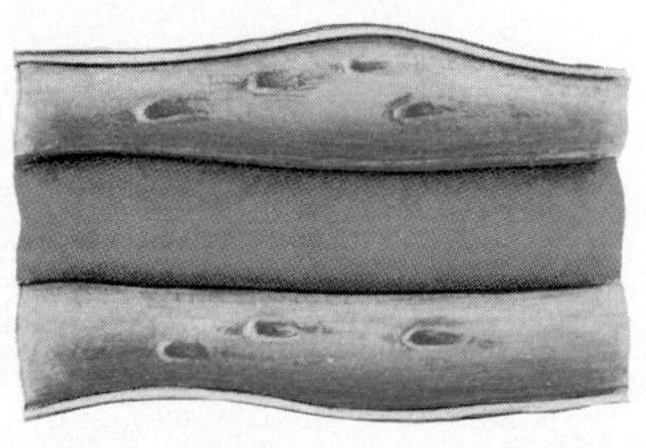

(d) Osteoclasts remove excess bony tissue, making new bone structure much like the original.

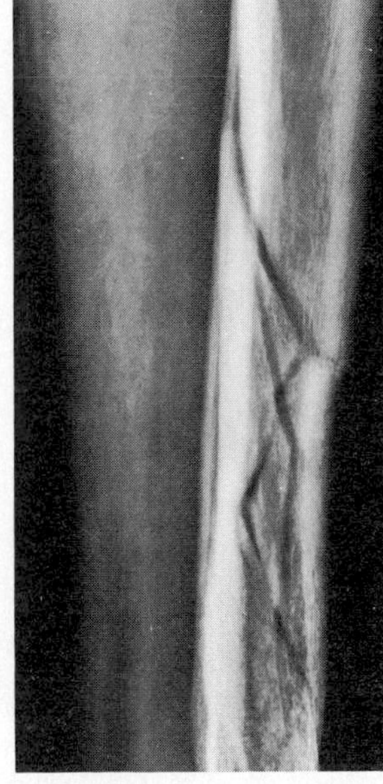

(e)

Physicians can realign and immobilize a fracture, but the ultimate repair of the bone occurs naturally within the bone itself. Several steps occur in the repair of a fracture (fig. 10.23).

1. When a bone is fractured, the surrounding periosteum is usually torn, and blood vessels in both tissues are ruptured. A blood clot called a **fracture hematoma** *(hem-ah-to'mah)* soon forms throughout the damaged area. A disrupted blood supply to osteocytes and periosteal cells at the fracture site causes localized cellular death. This is followed by swelling and inflammation.
2. The traumatized area is "cleaned up" by the activity of phagocytic cells within the blood and osteoclasts that resorb bone fragments. As the debris is removed, fibrocartilage fills the gap within the fragmented bone, and a cartilaginous mass called a **callus** is formed. The callus becomes the precursor to the formation of bone in much the same way that hyaline cartilage is the precursor to developing bone.
3. The remodeling of the callus is the final step in the healing process. The cartilaginous callus is broken down, a new vascular supply is established, and compact bone develops around the periphery of the fracture. A healed fracture line is frequently undetectable by X ray, except that the bone in this area is usually slightly thicker.

Chapter Summary

I. Development of the Appendicular Skeleton
   A. The development of the appendicular skeleton begins at the end of the fourth week and is completed by the end of the sixth week.
   B. Development progresses as the limb buds follow the growth of apical ectodermal ridges.

II. Pectoral Girdle and Upper Extremity
   A. The pectoral girdle is composed of two scapulae and two clavicles. The clavicles attach the pectoral girdle to the axial skeleton at the sternum.
   B. The brachium contains the humerus, which extends from the scapula to the elbow.
   C. The forearm contains the medial ulna and the lateral radius.
   D. The wrist and hand contain twenty-seven bones arranged into the carpus, metacarpus, and phalanges.

III. Pelvic Girdle and Lower Extremity
   A. The pelvic girdle is formed by two ossa coxae united anteriorly by the symphysis pubis.
      1. The pelvis is divided into a greater pelvis, which helps to support the pelvic viscera, and a lesser pelvis, which forms the walls of the birth canal.
      2. Each os coxa consists of an ilium, ischium, and pubis.
   B. The thigh contains the femur, which extends from the hip to the knee, where it articulates with the tibia and the patella.
   C. The lower leg contains the medial tibia and the lateral fibula.
   D. The ankle and foot contain twenty-six bones arranged into the tarsus, metatarsus, and phalanges.

Review Activities

Objective Questions

1. The cartilaginous bones of the appendicular skeleton derive from
   (a) mesoderm
   (b) ectoderm
   (c) endoderm
   (d) chondroderm
2. The clavicle articulates with the
   (a) scapula and the humerus
   (b) humerus and the manubrium
   (c) manubrium and the scapula
   (d) manubrium, the scapula, and the humerus
3. Which of the following bones has a conoid tubercle?
   (a) scapula
   (b) humerus
   (c) radius
   (d) clavicle
   (e) ulna
4. The "elbow" of the ulna is formed by the
   (a) lateral epicondyle
   (b) olecranon process
   (c) coronoid process
   (d) styloid process
   (e) medial epicondyle
5. Which of the following statements concerning the carpus is *false*?
   (a) There are eight carpal bones arranged into two transverse rows of four bones each.
   (b) All of the carpal bones are considered sesamoid bones.
   (c) The navicular and the lunate articulate with the radius.
   (d) The trapezium, trapezoid, capitate, and hamate articulate with the metacarpal bones.
6. Pelvimetry is a measurement of the
   (a) os coxa
   (b) symphysis pubis
   (c) pelvic brim
   (d) lesser pelvis
7. Which of the following is *not* a structural feature of the os coxa?
   (a) obturator foramen
   (b) acetabulum
   (c) auricular surface
   (d) greater sciatic notch
   (e) linea aspera
8. A fracture across the intertrochanteric line would involve the
   (a) ilium
   (b) femur
   (c) tibia
   (d) fibula
   (e) patella
9. As compared to the male pelvis, the female pelvis
   (a) is more massive
   (b) is more narrow at the pelvic outlet
   (c) is tilted backward
   (d) has a more shallow symphysis pubis
10. Clubfoot is a congenital malformation that is medically referred to as
   (a) talipes
   (b) syndactyly
   (c) pes planus
   (d) polydactyly

Essay Questions

1. Explain the significance of the limb buds, apical ectodermal ridges, and digital rays in limb development. When does limb development begin, and when is it completed?
2. Compare the pectoral and pelvic girdles in structure, articulation to the axial skeleton, and function.
3. Explain why the clavicle is more frequently fractured than the scapula.
4. List the processes of the bones of the upper and lower extremities that can be palpated. Why are these bony landmarks important to know?
5. There are some basic similarities and specific differences between the bones of the hands and those of the feet. Contrast and compare these appendages, taking into account the functional role of each.
6. Define *bipedal locomotion,* and discuss the adaptations of the pelvic girdle and lower extremities that permit this type of movement.
7. What are the structural differences between male and female pelves?
8. What is meant by a congenital skeletal malformation? Give two examples of such abnormalities that occur within the appendicular skeleton.
9. What are the differences between spontaneous and traumatic fractures? List some kinds of traumatic fractures.
10. How does a fractured bone repair itself? Why is it important that the fracture be immobilized?

11 ARTICULATIONS

Outline

Concepts

The articulations between the bones of the skeleton are classified into three types according to structure or into three types according to function, or the degree of movement permitted.

Diarthroses form by the third month as the epiphyses of adjacent endochondral bones develop distinct configurations and as the muscles that move the joints undergo contractions.

Articulating bones in synarthroses are tightly connected by either fibrous tissue or cartilage. Synarthrotic joints are rigid and immovable and are of two types: sutures and synchondroses.

Amphiarthroses allow limited motion in response to twisting, compression, or stress. The two types of amphiarthroses are symphyses and syndesmoses.

Diarthroses are freely movable joints enclosed by joint capsules that contain synovial fluid. Types of diarthroses include gliding, hinge, pivot, condyloid, saddle, and ball-and-socket.

Movements at diarthroses are produced by the contraction of skeletal muscles spanning the joints and attaching to or near the bones forming the articulations. In these actions, the bones act as levers, the muscles provide the force, and the joints are the fulcra, or pivots.

Of the numerous joints in the body, some have special structural features that enable them to perform particular functions. Furthermore, these joints are somewhat vulnerable to trauma and are therefore clinically important.

Classification of Joints

The articulations between the bones of the skeleton are classified into three types according to structure or into three types according to function, or the degree of movement permitted.

Objective 1. Define the terms *arthrology* and *kinesiology.*

Objective 2. Distinguish between a structural and a functional classification of joints.

One of the functions of the skeletal system is to permit body movement. It is not the rigid bones that allow movement but the **articulations** *(ar-tik"u-la'shunz),* or **joints,** between the bones. The structure of a joint determines the range of movement it permits. Not all joints are flexible, however, and as one part of the body moves, other joints remain rigid to stabilize the body and maintain balance. The coordinated activity of all of the joints permits the sinuous, elegant movements of a gymnast or ballet dancer, as well as the mundane activities of walking, eating, writing, and speaking.

Arthrology is the science concerned with the study of joints. Generally speaking, an arthrologist is interested in the structure, classification, and function of joints as well as any dysfunction that may develop. In the more practical and dynamic science of **kinesiology** *(ki-ne"se-ol'o-je),* a kinesiologist studies the functional relationship, or biomechanics, of the skeleton, joints, muscles, and innervation as they work together to produce coordinated movement.

To study the joints, one should adopt a kinetic approach and be able to demonstrate the various movements permitted at each of the movable joints and understand the adaptive advantage as well as the limitations of each type of movement.

The joints of the body may be classified according to structure or function. Structural classification is based on the presence or absence of a joint cavity and the kind of supportive connective tissue surrounding the joint. In classification by structure, there are three types of joints:

1. **Fibrous** *(fi'brus).* A fibrous joint lacks a joint cavity, and fibrous connective tissue connects articulating bones.
2. **Cartilaginous** *(kar"tĭ-laj'ĭ-nus).* A cartilaginous joint lacks a joint cavity, and cartilage binds articulating bones.
3. **Synovial** *(sĭ-no've-al).* A synovial joint has a joint cavity, and ligaments help to support the articulating bones.

arthrology: Gk. *arthron,* joint; *logos,* study
kinesiology: Gk. *kinesis,* movement; *logos,* study

The functional classification of joints is based on the degree of movement permitted within the joint. Using this type of classification, there are three kinds of articulations:

1. **Synarthroses.** Immovable joints.
2. **Amphiarthroses.** Slightly movable joints.
3. **Diarthroses.** Freely movable joints.

This book uses the functional classification scheme in discussing the various kinds of joints of the body. But continual reference will be made to the structural classification as the specific anatomy of a particular joint is presented.

1. Explain the statement that kinesiology is applied arthrology.
2. According to both structural and functional classifications, how would a joint be classified that is tightly bound with hyaline cartilage so that only slight movement is possible?
3. List the three types of functional joints, and speculate as to which provides the greatest support and which is the most vulnerable to trauma.

Development of Freely Movable Joints

Diarthroses form by the third month as the epiphyses of adjacent endochondral bones develop distinct configurations and as the muscles that move the joints undergo contractions.

Objective 3. Describe the development of diarthroses, and discuss the importance of fetal movement to the normal development of freely movable joints.

The sites of developing diarthroses *(di"ar-thro'sēz)* (freely movable joints) are discernible at six weeks as mesenchyme becomes concentrated in the areas where precartilage cells differentiate (fig. 11.1). The future joints at this stage appear as intervals of less concentrated mesenchymal cells. As cartilage cells develop within a forming bone, a thin flattened sheet of cells forms around the cartilaginous model to become the perichondrium. These same cells are continuous across the gap between the adjacent developing bone. Surrounding the gap, the flattened mesenchymal cells differentiate to become the **joint capsule.**

During the early part of the third month of development, the mesenchymal cells still remaining within the joint capsule begin migrating toward the epiphyses of the adjacent developing bones. The cleft eventually enlarges to become the **joint cavity.** Thin pads of hyaline cartilage develop on the surfaces of the epiphyses that contact the joint cavity. These pads become the **articular cartilages** of

Figure 11.1. Development of diarthroses. (*a*) at six weeks, different densities of mesenchyme denote where the bones and joints will form. (*b*) at nine weeks, a basic diarthrotic model is present. At twelve weeks, the diarthroses are formed and have either (*c*) a free joint cavity (e.g., interphalangeal joint); (*d*) a cavity containing menisci (e.g., knee joint); or (*e*) a cavity with a complete articular disc (e.g., sternoclavicular joint).

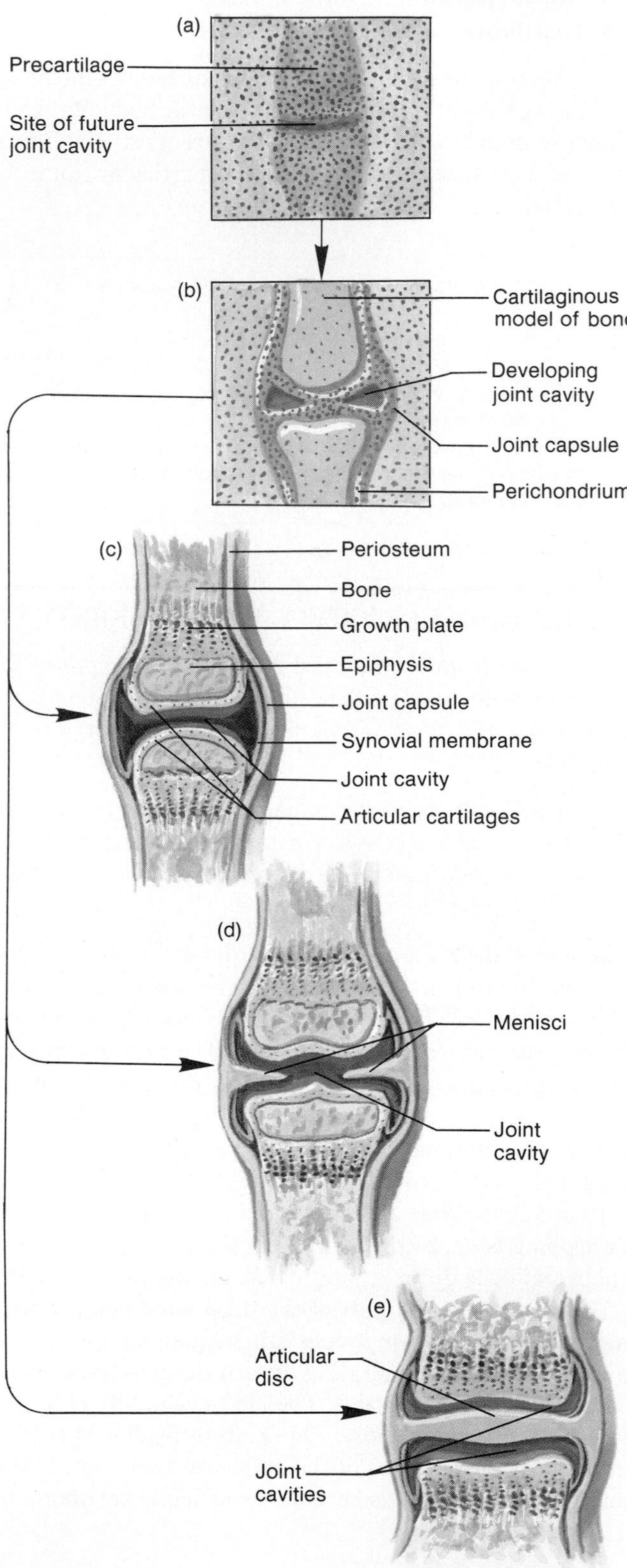

the functional joint. As the joint continues to develop, a highly vascular **synovial membrane** forms on the inside of the joint capsule and begins secreting a watery *synovial fluid* into the joint cavity.

In certain developing diarthroses, the mesenchymal cells do not migrate away from the center of the joint cavity but, rather, give rise to cartilaginous wedges, called **menisci,** as in the knee joint (fig. 11.1), or to complete cartilaginous pads, called **articular discs,** as in the sternoclavicular joint.

The formation of most diarthroses is completed by the end of the third month. Shortly after this time, there are fetal muscle contractions, known as *quickening,* that cause movement at these joints. Joint movement enhances the nutrition of the articular cartilage and prevents the fusion of connective tissues within the joint.

1. Describe the development of a *diarthrosis,* and give the approximate prenatal date of each event.
2. Describe the variation of cartilaginous separations between the articular cartilages of diarthroses.
3. Explain the importance of quickening to the health of a newly formed *diarthrosis.*

Synarthroses

Articulating bones in synarthroses are tightly connected by either fibrous tissue or cartilage. Synarthrotic joints are rigid and immovable and are of two types: sutures and synchondroses.

Objective 4. Describe the structure of a suture, and indicate where sutures are located.

Objective 5. Describe the structure of a synchondrosis; indicate where synchondroses are located and how they change as a person ages.

Sutures

Sutures *(su'churz),* one type of synarthroses *(sin'ar-thro'sēz),* are found only within the skull and are characterized by a thin layer of dense fibrous connective tissue that binds the articulating bones (fig. 11.2). Sutures form at about eighteen months of age and replace the pliable fontanels of an infant's skull (see fig. 9.7).

There are several types of sutures, which can be distinguished on the basis of the appearance of the articulating margin of bone. A **serrate suture** is characterized by interlocking articulations. This is the most common type of suture, an example being the sagittal suture between the two parietal bones. In a **lap (squamous) suture,** the margin of one bone overlaps that of the articulating bone.

suture: L. *sutura,* sew

Figure 11.2. A section across the skull showing a suture.

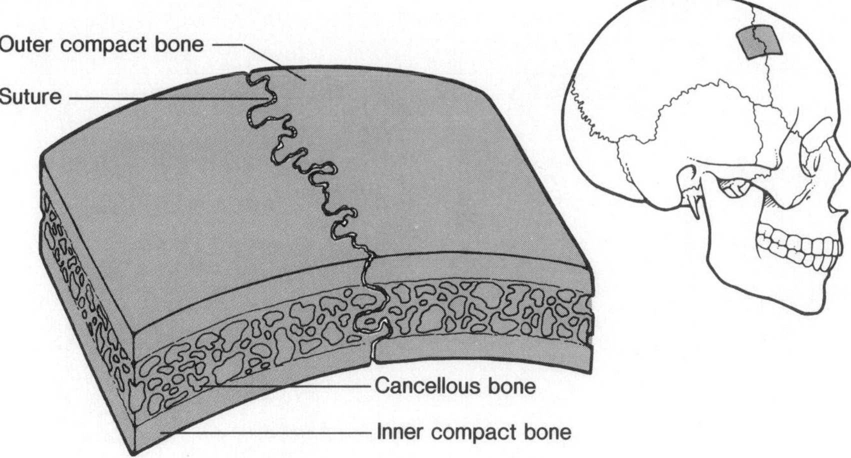

The squamous suture formed between the temporal and parietal bones is an example. In a **plane (butt) suture,** the margins of the articulating bones are fairly smooth. An example is the maxillary suture, where the two maxillary bones articulate to form the hard palate. A **synostosis** *(sin''os-to'sis),* a unique type of suture, is present during growth of the skull but becomes totally ossified in the adult. The union between the right and left portions of the frontal bone is an example of synostosis.

Fractures of the skull are fairly common in an adult but much less so in a child. The skull of a child is resilient to blows because of the nature of the bone and the layer of fibrous connective tissue within the sutures. The skull of an adult is much like an eggshell in its lack of resilience and will frequently splinter on impact.

Synchondroses

Synchondroses *(sin''kon-dro'sēs)* are synarthrotic joints that have hyaline cartilage between the bone segments. These joints are typically temporary, forming the growth lines (epiphyseal plates) between the diaphyses and epiphyses in the long bones of children. When growth is complete, the synchondrotic joints ossify. A totally ossified synchondrosis can also be referred to as a synostosis. A synchondrosis can be clearly seen in the X ray of a long bone of a child in figure 11.3.

1. Distinguish between the three principal kinds of sutures, and give an example and the location of each. What does each have in common? How do synchondroses change with age?
2. Explain why synchondroses are found on bones that have epiphyses and a diaphysis.

Figure 11.3. An X ray of the left humerus of a ten-year-old child, showing a synchondrotic joint. In a long bone, this type of joint occurs at both the proximal and distal epiphyseal plates. The mitotic activity at a synchondrotic joint is responsible for bone growth in length.

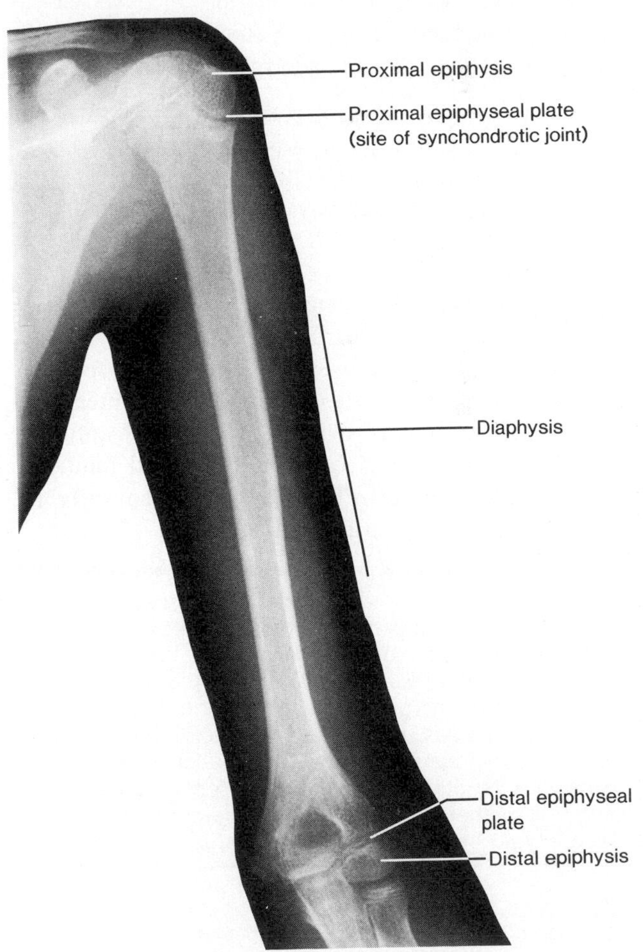

synchondrosis: Gk. *syn,* together; *chondros,* cartilage
synostosis: Gk. *syn,* together; *osteon,* bone

Figure 11.4. Amphiarthrotic joints. (*a*) the symphysis pubis and (*b*) the intervertebral joints between vertebral bodies.

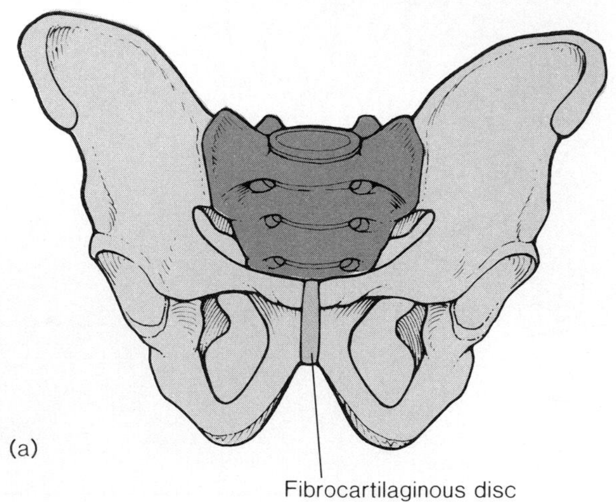

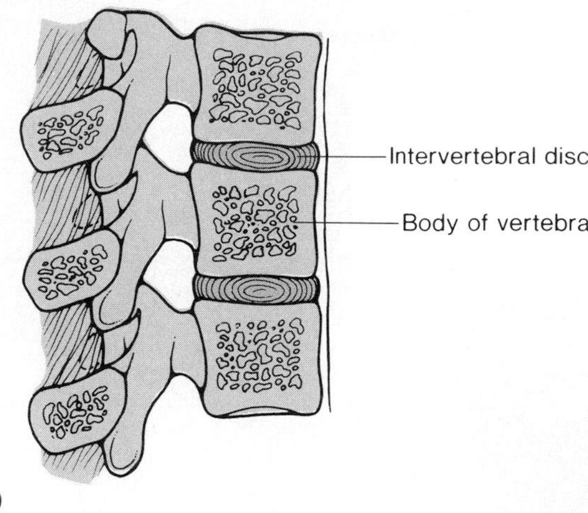

AMPHIARTHROSES

Amphiarthroses allow limited motion in response to twisting, compression, or stress. The two types of amphiarthroses are symphyses and syndesmoses.

Objective 6. Describe the structure of a symphysis, and indicate where symphyses occur.

Objective 7. Describe the structure of a syndesmosis, and indicate where syndesmoses occur.

Symphyses

The adjoining bones of a **symphysis** *(sim'fi-sis)* joint are separated by a pad of fibrocartilage. This pad cushions the joint and allows limited movement. The symphysis pubis and the intervertebral joints (fig. 11.4) at the intervertebral discs are examples of symphyses. Although only limited motion is possible at each intervertebral joint, the combined movement of all the joints of the vertebral column results in extensive spinal action.

Hormonal action in a pregnant woman causes the symphysis pubis, sacroiliac, and sacrococcygeal joints to soften and become more flexible. This relaxation of the pelvic joints increases the potential capacity of the pelvic cavity and reduces trauma during childbirth by increasing the diameter of the pelvic outlet. These joints will strengthen postpartum, but following the birth of a mother's first child, the diameter of her pelvis will generally be slightly wider than before.

symphysis: Gk. *symphysis,* growing together

Figure 11.5. The distal articulation of the ulna and radius forms a syndesmotic joint. Interosseous ligaments tightly bind these bones and permit only slight movement between them.

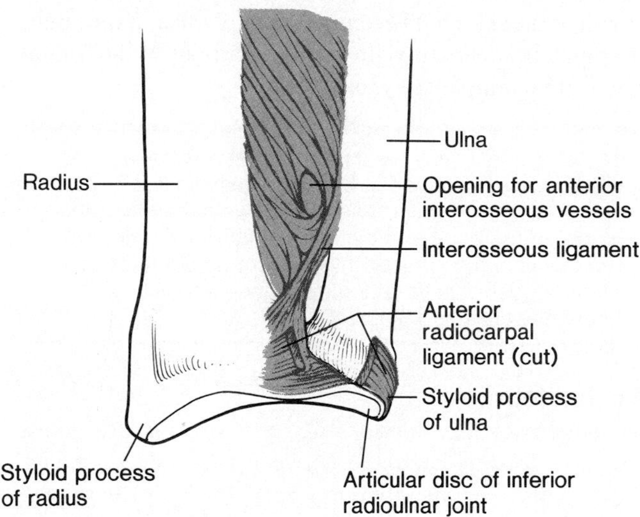

Syndesmoses

In **syndesmoses** adjacent bones are held together by collagenous fibers or interosseous ligaments. A syndesmosis *(sin''des-mo'sis)* is characteristic of the distal ends of the tibia-fibula and the radius-ulna (fig. 11.5).

1. Discuss the function of the pad of fibrocartilage in a symphysis joint, and give two examples of symphyses.
2. What structural feature is characteristic to all syndesmoses? Give two examples of syndesmoses.

syndesmosis: Gk. *syndesmos,* binding together

Diarthroses

Diarthroses are freely movable joints enclosed by joint capsules that contain synovial fluid. Types of diarthroses include gliding, hinge, pivot, condyloid, saddle, and ball-and-socket.

Objective 8. Describe the structure of a diarthrotic joint, and indicate where diarthroses occur.

Objective 9. List and discuss the various kinds of diarthroses, where they occur, and the movements they permit.

The most obvious type of articulation in the body is the freely movable **diarthrosis.** The dual function of diarthrotic joints is to provide a wide range of precise, smooth movements and yet maintain stability, strength, and, in certain aspects, rigidity in the body.

Diarthroses are the most complex and varied of the three major types of joints. A diarthrotic joint's range of movement is limited by three factors: (1) the structure of the bones participating in the articulation (certain processes on bones, for example, the oleocranon process of the ulna, actually limit the range of motion and "lock" the articulation to prevent overextension of the joint); (2) the strength and tautness of the associated ligaments, tendons, and joint capsule; (3) the size, arrangement, and action of the muscles that span the joint. There is tremendous individual variation in joint motility, most of which is related to body conditioning. "Double-jointed" is a misnomer because such a joint is not double, although it does permit extreme maneuverability.

Structure of a Diarthrotic Joint

In structural terms, diarthroses are synovial joints because they are enclosed by a fibroelastic joint capsule, which is filled with lubricating **synovial fluid,** or **synovium** (fig. 11.6). Synovial fluid is secreted by a thin **synovial membrane** that lines the inside of the capsule. Synovial fluid is similar to interstitial fluid (fluid surrounding cells of a tissue), and has a high concentration of hyaluronic acid, a lubricating substance. The bones that articulate in a diarthrosis are capped with a smooth **articular cartilage** *(ar-tik'u-lar kar'tĭ-lij).* The avascular articular cartilage is only about 2 mm thick and depends upon the alternating compression and decompression during joint activity for the exchange of nutrients and waste products with the synovial fluid. Ligaments help to bind a diarthrosis and may be located within the joint cavity or on the outside of the capsule. Tough, fibrous, cartilaginous pads called **menisci**—singular, *meniscus (mĕ-nis'kus)*—are located within the capsule of certain diarthroses (for example, the knee joint) and serve to cushion as well as to guide the articulating bones.

meniscus: Gk. *meniskos,* small moon

Figure 11.6. A diarthrotic joint is represented by this diagrammatic lateral view of the knee joint.

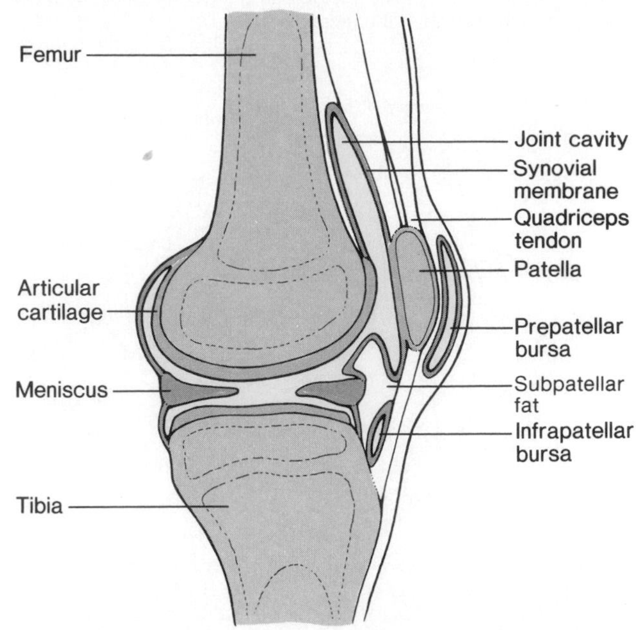

Articulating bones of diarthroses do not come in contact with one another. Articular cartilage caps the articular surface of each bone, and synovial fluid circulates through the joint during movement. Both of these joint structures minimize friction as well as cushion the articulating bones. Trauma or disease may render either of these two joint structures nonfunctional, and the two articulating bones will come in contact. Bony deposits will then form, and a type of arthritis will develop within the joint.

Located near certain diarthrotic joints are flattened, pouchlike sacs called **bursae**—singular, *bursa (ber'sah)*—which are filled with synovial fluid (fig. 11.6). These closed sacs are commonly located between muscles or within an area where a tendon passes over a bone. The functions of bursae are to cushion certain muscles and to facilitate the movement of tendons or muscles over bony or ligamentous surfaces.

A **tendon sheath** is a modified bursa that surrounds and lubricates the tendons of certain muscles, particularly those that cross the wrist and ankle joints (fig. 11.7). The tendons that are surrounded by sheaths are strong and cordlike and are usually associated with important flexor muscles.

bursa: Gk. *byrsa,* bag or purse

Figure 11.7. A tendon sheath facilitates the gliding of a tendon as it traverses a fibrous or bony tunnel. Tendon sheaths are closed sacs, one layer of the synovial membrane lining the tunnel, the other folding over the surface of the tendon.

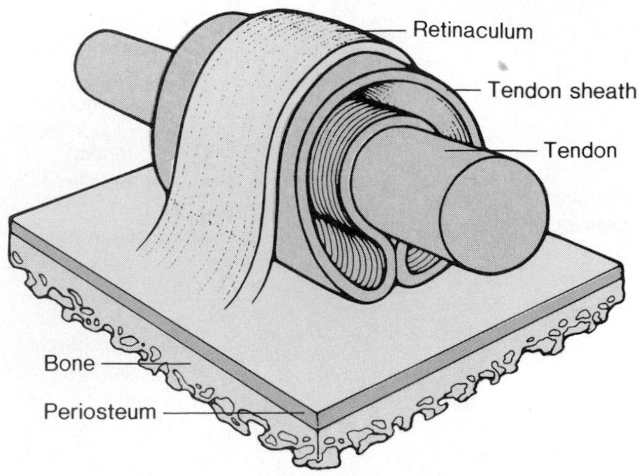

Kinds of Diarthroses

Diarthroses are classified into six main types according to their structure and the kinds of motion they permit. The six types are gliding, hinge, pivot, condyloid, saddle, and ball-and-socket.

Gliding. Gliding joints allow only side-to-side and back-and-forth movements with some slight rotational movement. This is the simplest type of joint movement. The articulating surfaces can be nearly flat, or one may be slightly concave and the other slightly convex. The intercarpal and intertarsal joints, as well as the sternoclavicular joint and the joint between the articular processes of adjacent vertebrae, are examples of gliding joints (fig. 11.8).

Hinge. The structure of a hinge joint permits bending in only one plane, much like the hinge of a door. In this type of articulation the surface of one bone is always concave, the other convex (fig. 11.9). Hinge joints are the most common type of diarthroses and include such specific joints as the knee, the elbow (humeroulnar articulation), and the joints between the phalanges.

Figure 11.8. The locations of both gliding and condyloid joints. (*a*) gliding joints are located between the proximal and distal carpals and between the individual carpals. (*b*) condyloid joints are located between the metacarpal bones and the phalanges. Note the diagrammatic representation of these joints showing the directions of possible movements.

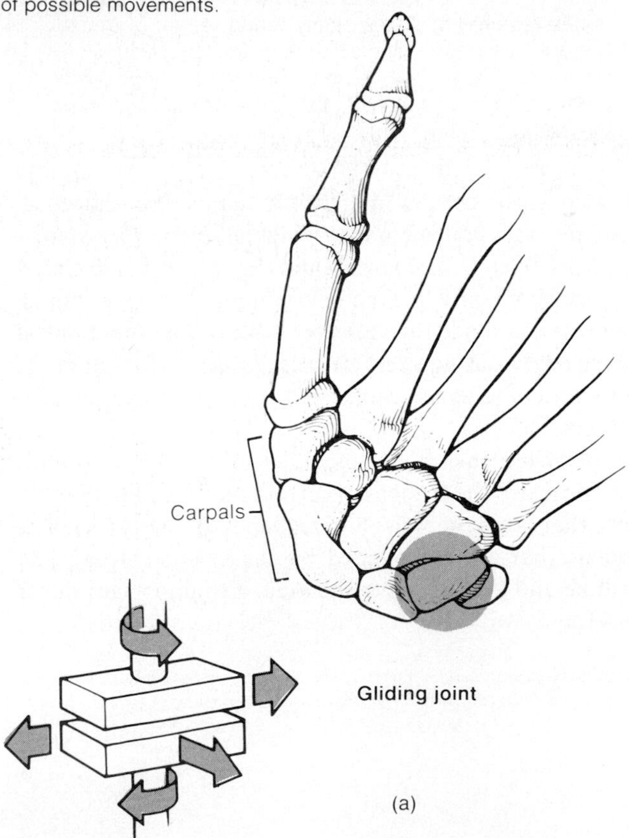

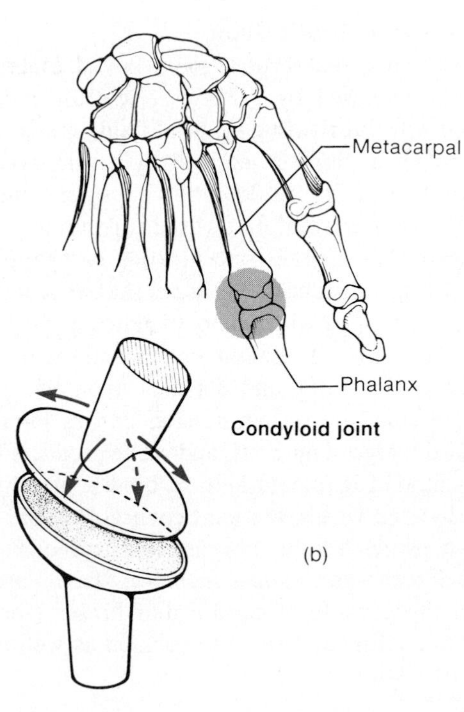

Pivot. The movement in a pivot joint is limited to rotation about a central axis. In this type of articulation the articular surface on one bone is conical or rounded and fits into a depression on another bone. Examples are the proximal articulation of the radius and ulna for rotation of the forearm, as in turning a doorknob, and the articulation between the atlas and axis that makes rotational movement of the head possible (fig. 11.10).

Condyloid (Ellipsoid). A condyloid *(kon'dĭ-loid)* articulation is structured so that an oval, convex articular surface of one bone fits into an elliptical, concave depression on another bone. This permits angular movement in two directions (biaxial), as in an up-and-down and side-to-side motion. A condyloid joint does not permit rotational movement. The radiocarpal joint of the wrist is an example (see fig. 11.8).

Saddle. Each articular process of a saddle-shaped joint has a concave surface in one direction and a convex surface in another. This unique articulation is a modified condyloid joint that allows a wide range of movement and is associated only with the thumb. Specifically, the saddle joint is found at the articulation of the trapezium of the carpus with the first metacarpal bone (fig. 11.11).

Figure 11.10. The atlas articulating with the axis forms a pivot joint that permits a rotational movement in one axis. Note the diagrammatic representation showing the direction of possible movement. Refer to figure 11.9 and determine which articulating bones of the elbow region form a pivot joint.

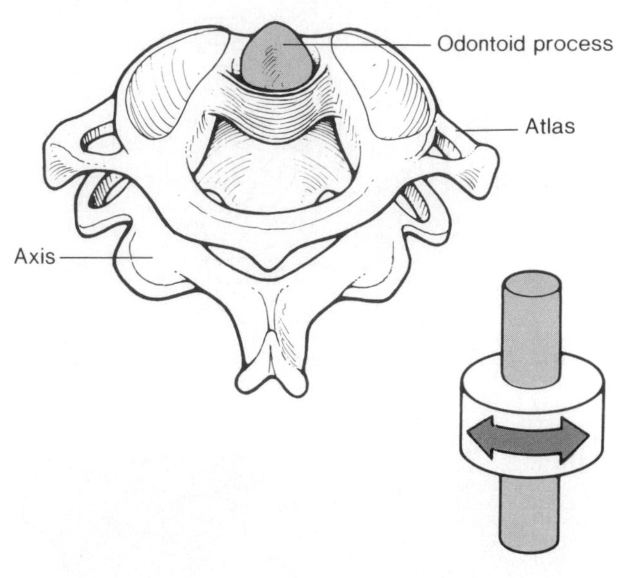

Figure 11.9. A hinge joint permits only a bending movement (flexion and extension). The hinge joint of the elbow involves the distal end of the humerus articulating with the proximal end of the ulna. Note the diagrammatic representation of this joint showing the direction of possible movement.

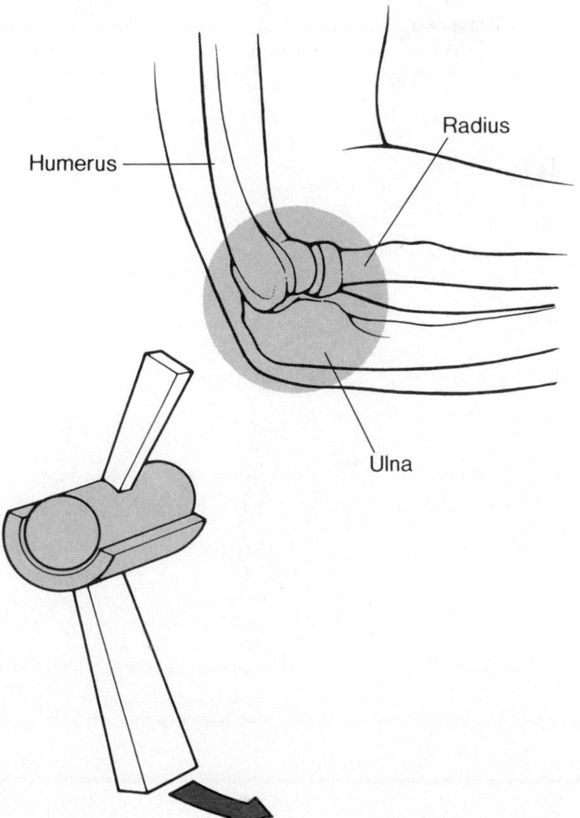

Figure 11.11. A saddle joint is formed as the trapezium articulates with the base of the first metacarpal bone. Note the diagrammatic representation of this joint.

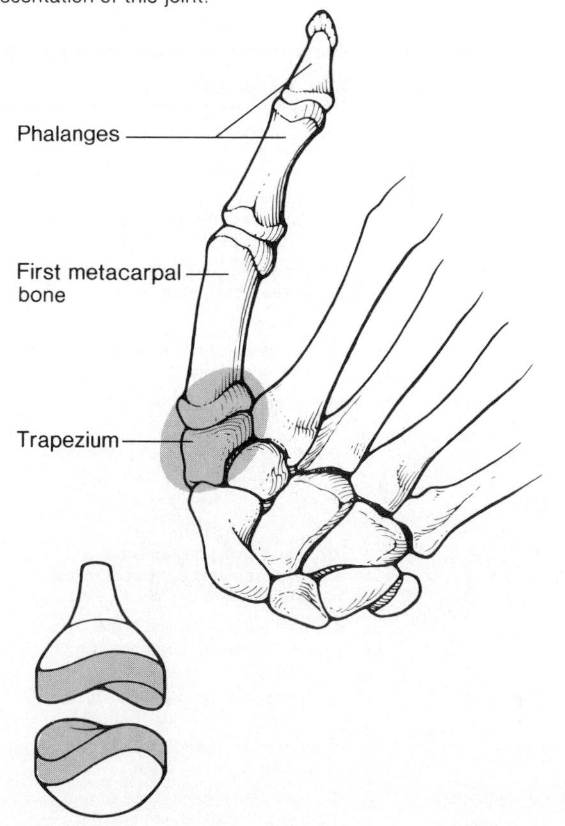

Figure 11.12. A ball-and-socket articulation illustrated by the hip joint. Note the diagrammatic representation showing the directions of possible movement.

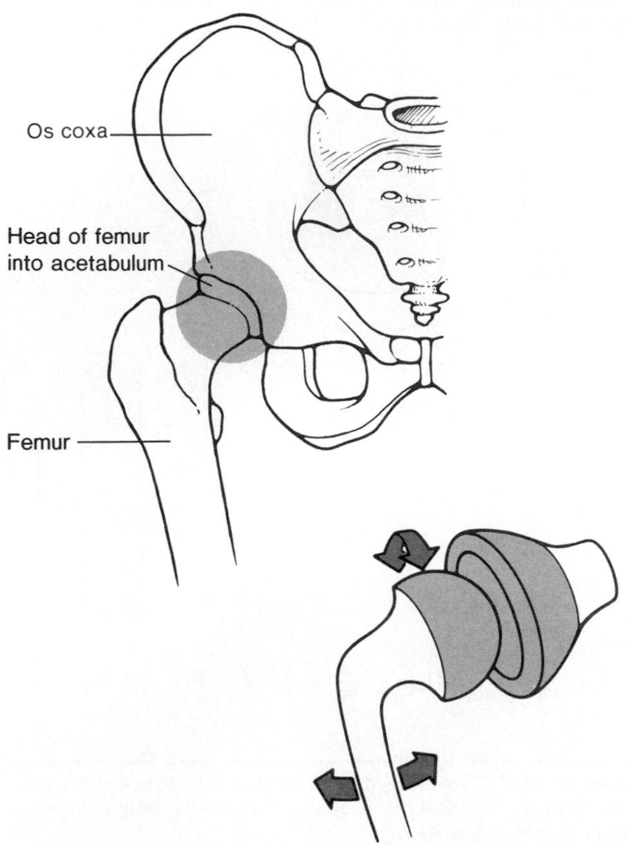

Ball-and-Socket. Ball-and-socket joints are formed by the articulation of a rounded convex surface with a cuplike cavity (fig. 11.12). This type of articulation provides the greatest range of movement of all of the diarthrotic joints. Examples are the hip and shoulder joints.

A summary of the various types of joints within the body is presented in table 11.1.

Synovial fluid within the diarthrotic joints serves to lubricate the articular surfaces and provide nourishment to the articular cartilage. Trauma to the joint causes the excessive production of synovial fluid in an attempt to cushion and immobilize the joint. This leads to swelling and discomfort to the joint. The most frequent type of joint injury is a *sprain,* in which the supporting ligaments or joint capsule are damaged to varying degrees.

1. List the structures of a diarthrosis, and explain the function of each.
2. What three factors limit the range of movement within diarthroses?
3. Give an example of each of the six kinds of diarthroses, and describe the range of movement possible at each.

Table 11.1 Types of articulations

Type	Structure	Movements	Example
Synarthroses			
Suture	Frequently serrated edges of articulating bones; separated by thin layer of fibrous tissue	None	Sutures between bones of the skull
Synchondroses	Mitotically active hyaline cartilage between bony segments	None	Epiphyseal plates within long bones
Amphiarthroses			
Symphyses	Articulating bones separated by pad of fibrocartilage	Slightly movable	Intervertebral joints; symphysis pubis and sacroiliac joint
Syndesmoses	Articulating bones bound by interosseous ligament	Slightly movable	Joints between tibia-fibula and radius-ulna
Diarthroses	Joint capsule containing synovial membrane and synovial fluid	Freely movable	
Gliding	Flattened or slightly curved articulating surfaces	Sliding	Intercarpal and intertarsal joints
Hinge	Concave surface of one bone articulates with convex surface of another	Bending motion in one plane	Knee; elbow; joints of phalanges
Pivot	Conical surface of one bone articulates with a depression of another	Rotation about a central axis	Atlantoaxial joint; proximal radioulnar joint
Condyloid	Oval condyle of one bone articulates with elliptical cavity of another	Biaxial movement	Radiocarpal joint
Saddle	Concave and convex surface on each articulating bone	Wide range of movements	Carpometacarpal joint of the thumb
Ball-and-Socket	Rounded convex surface of one bone articulates with cuplike socket of another	Movement in all planes and rotation	Shoulder and hip joints

Movements at Diarthroses

Movements at diarthroses are produced by the contraction of skeletal muscles spanning the joints and attaching to or near the bones forming the articulations. In these actions, the bones act as levers, the muscles provide the force, and the joints are the fulcra, or pivots.

Objective 10. List and discuss the various kinds of movements possible within diarthrotic joints.

Objective 11. Describe the components of a lever, and explain the role of diarthrotic joints in lever systems.

Objective 12. Compare the structure of first-, second-, and third-class levers.

As previously mentioned, the range of movement is determined by the structure of the individual joint and the arrangement of the associated muscle and bone. The movement at a hinge joint, for example, occurs in only one plane, whereas the structure of a ball-and-socket joint permits movement around many axes. Movements within joints are broadly classified as **angular** and **circular.** Within each of these categories are specific types of movements, and certain special movements may involve several of the specific types.

Angular

Angular movements increase or decrease the joint angle produced by the articulating bones. The four types of angular movements are flexion, extension, abduction, and adduction.

Flexion. Flexion *(flek'shun)* is a movement that decreases the joint angle on an anterior-posterior plane (fig. 11.13). Examples of flexion are the bending of the elbow or knee. Flexion of the elbow is a forward movement, whereas flexion of the knee is a backward movement. Flexion in most joints is simple to understand, such as flexion of the head as it bends forward or the flexing of a digit, but flexion of the ankle and shoulder joints needs further explanation. In the ankle joint, flexion occurs as the dorsum of the foot is elevated. This movement is frequently called **dorsiflexion** (fig. 11.14). Pressing the foot downward (as in rising on the toes) is **plantar flexion.** The shoulder joint is flexed when the arm is brought forward, thus decreasing the joint angle.

Extension. In extension *(ek-sten'shun),* which is the reverse of flexion, the joint angle is increased (fig. 11.13). Extension returns a body part to the anatomical position. In an extended joint, the angle between the articulating bones is 180°. The exception to this is the ankle joint, in which a 90° angle exists between the foot and the lower leg in the anatomical position. Examples of extension are the straightening of the elbow or knee joints from flexion

Figure 11.13. Examples of flexion and extension. Contraction of the posterior thigh muscles with the femur held rigid results in flexion at the knee joint as the lower leg is moved posteriorly. Contraction of the anterior thigh muscles with the entire leg held rigid results in flexion at the hip joint as the leg is moved anteriorly. Contraction of the posterior thigh muscles with the entire leg held rigid results in extension at the hip joint as the leg is moved posteriorly.

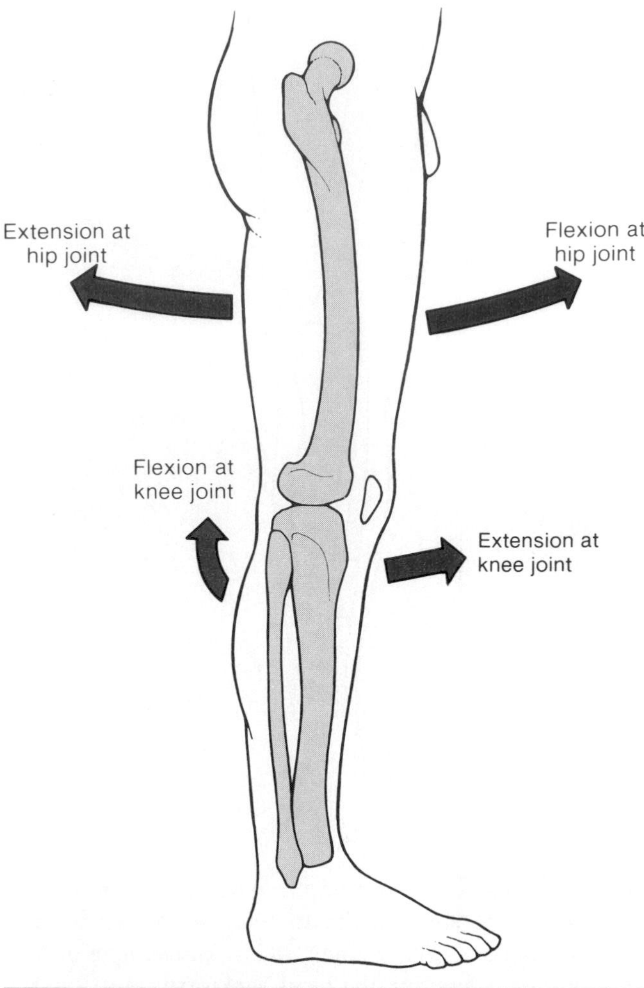

Figure 11.14. Dorsiflexion and plantar flexion of the foot at the ankle joint.

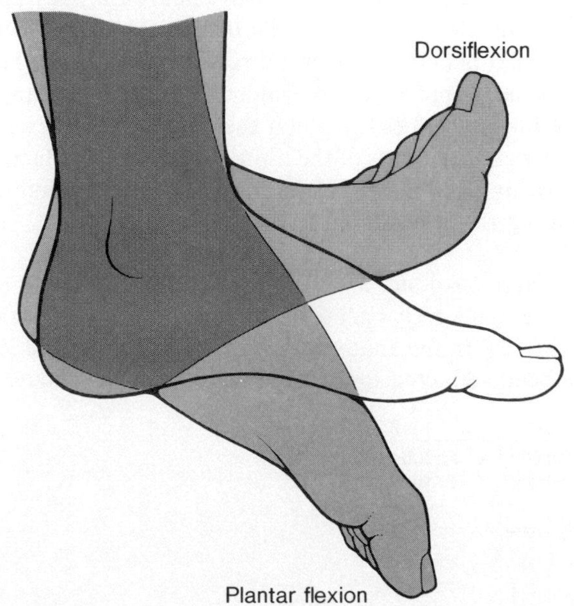

flexion: L. *flectere,* to bend

extension: L. *ex,* out, away from; *tendere,* stretch

Figure 11.15. Abduction and adduction of the lower extremities in reference to the main axis of the body.

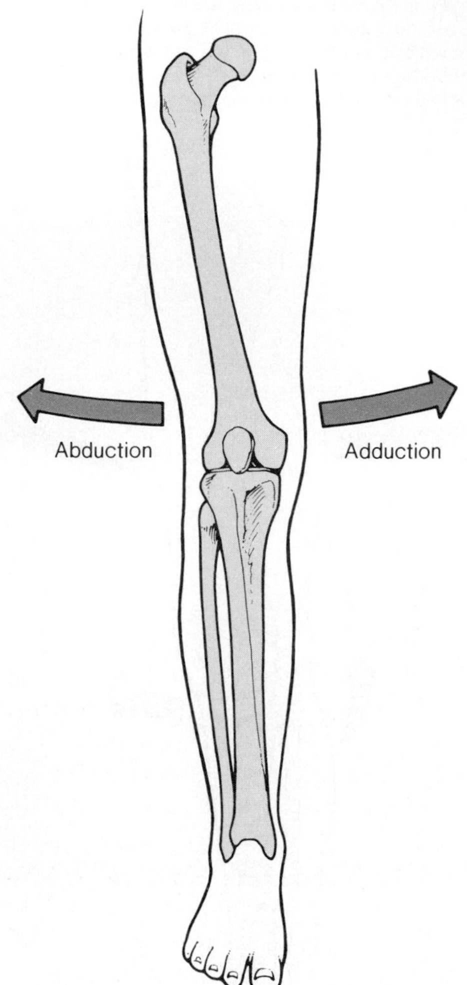

Figure 11.16. Supination (*a*) and pronation (*b*) of the hand. Note the relative position of the ulna and radius in both positions. Pronation requires medial rotation of the forearm relative to the anatomical position.

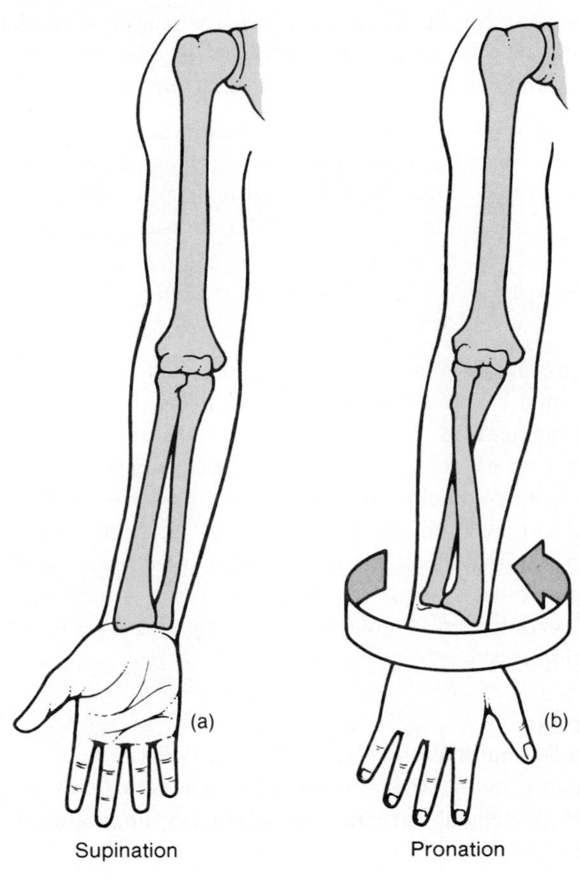

positions. **Hyperextension** occurs when a portion of the body is extended beyond the anatomical position so that the joint angle is greater than 180°. An example of hyperextension is bending the head backward.

Abduction. Abduction is the movement of a body part away from the main axis of the body, or away from the midsagittal plane, in a lateral direction (fig. 11.15). This term usually applies to the arm or leg but can also apply to the fingers or toes, in which case the line of reference is the longitudinal axis of the limb. Examples of abduction are moving the arms sideward and away from the body or spreading the fingers apart.

Adduction. Adduction, the opposite of abduction, is the movement of a body part toward the main axis of the body (fig. 11.15). In the anatomical position, the arms and legs have been adducted toward the midplane of the body.

abduction: L. *abducere*, lead away
adduction: L. *adductus*, bring to

Circular

Joints that permit **circular movement** are composed of a bone with a rounded or oval surface articulating with a corresponding cup or depression on another bone. The two basic types of circular movements are rotation and circumduction.

Rotation. Rotation is the movement of a bone around its own axis (see fig. 11.10). There is no lateral displacement during this movement. Examples are turning the head from side to side in a "no" motion and moving the forearm from a palm-up position to a palm-down position.

Supination *(su''pĭ-na'shun)* is a specialized rotation of the forearm that results in the palm of the hand being turned forward (anteriorly). In the supine position, the ulna and radius of the forearm are parallel and in the anatomical position. **Pronation** *(pro-na'shun)* is the opposite of supination (fig. 11.16). It is a rotational movement of the forearm that results in the palm of the hand being directed backward (posteriorly).

rotation: L. *rotare*, a wheel

Figure 11.17. Circumduction of the shoulder joint.

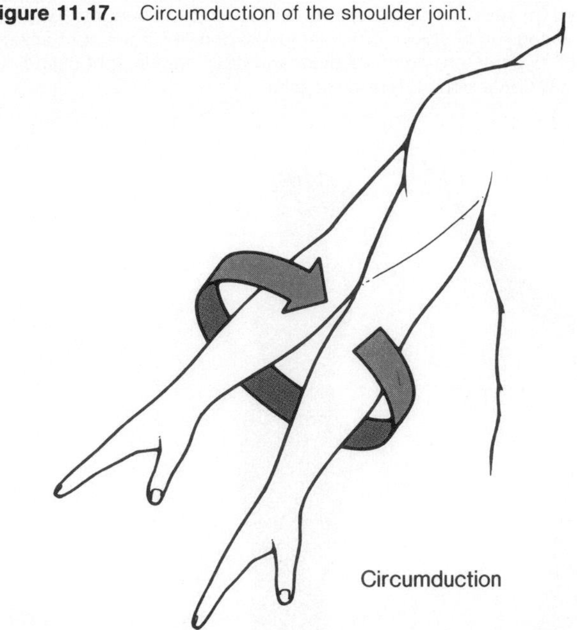

Figure 11.18. Inversion and eversion of the foot on the intertarsal joints.

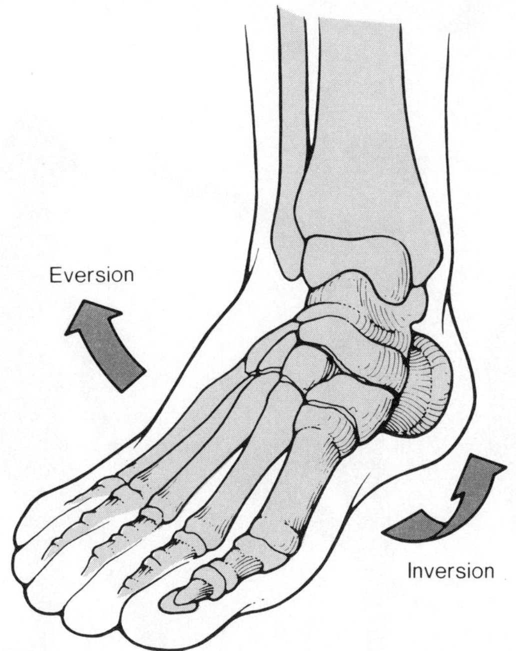

Circumduction. Circumduction is the circular, conelike movement of a body segment. The distal extremity forms the circular movement, and the proximal attachment forms the pivot (fig. 11.17). This type of motion is possible at the shoulder, wrist, trunk, hip, and ankle joints.

Special Movements

Because the terms used to describe generalized movements around axes do not apply to the structure of certain joints, other terms must be used to describe the motion of such joints.

Figure 11.19. Protraction and retraction of the lower jaw.

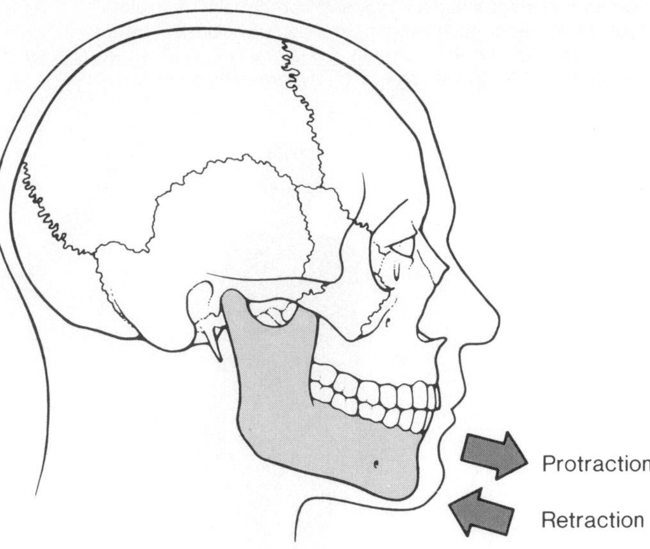

Inversion. Inversion is the movement of the sole of the foot inward or medially (fig. 11.18). The pivot axes are at the ankle and intertarsal joints.

Eversion. Eversion is the opposite of inversion and is the movement of the sole of the foot outward or laterally (fig. 11.18). Both inversion and eversion are clinical terms usually used to describe developmental abnormalities.

Protraction. Protraction is the movement of part of the body forward on a plane parallel to the ground. Examples are thrusting out the lower jaw (fig. 11.19) or movement of the pectoral girdle forward.

Retraction. Retraction is the pulling back of a protracted part of the body on a plane parallel to the ground (fig. 11.19). Retraction of the mandible brings the lower jaw back in alignment with the upper jaw so that the teeth occlude.

Elevation. Elevation is a movement that results in a portion of the body being lifted upward. Examples of elevation include elevating the mandible to close the mouth or lifting the shoulders to shrug.

Depression. Depression is the movement opposite of elevation. Both the mandible and shoulders are depressed when moved downward.

Figures 11.20, 11.21, and 11.22 present a visual summary of many of the movements permitted at diarthrotic joints.

Figure 11.20 A photographic summary of joint movements. (*a*) adduction of shoulder, hip, and carpophalangeal joints; (*b*) abduction of shoulder, hip, and carpophalangeal joints; (*c*) rotation of vertebral column; (*d*) lateral flexion of vertebral column; (*e*) flexion of vertebral column; (*f*) hyperextension of vertebral column; (*g*) flexion of shoulder, hip, and knee joints of right side of body; extension of elbow and wrist joints; dorsiflexion of right ankle joint; (*h*) hyperextension of shoulder and hip joints on right side of body; plantar flexion of right ankle joint.

Figure 11.21. A visual summary of some angular movements at diarthroses. (*a*) flexion, extension, and hyperextension in the cervical region; (*b*) flexion and extension at the knee joint, and dorsiflexion and plantar flexion at the ankle joint; (*c*) flexion, extension, and hyperextension at the wrist joint; (*d*) adduction and abduction of the arm and fingers; (*e*) posterior view of abduction and adduction of the hand at the wrist joint. Note that the range of abduction at the wrist joint is less than the range of adduction due to the length of the styloid process of the radius.

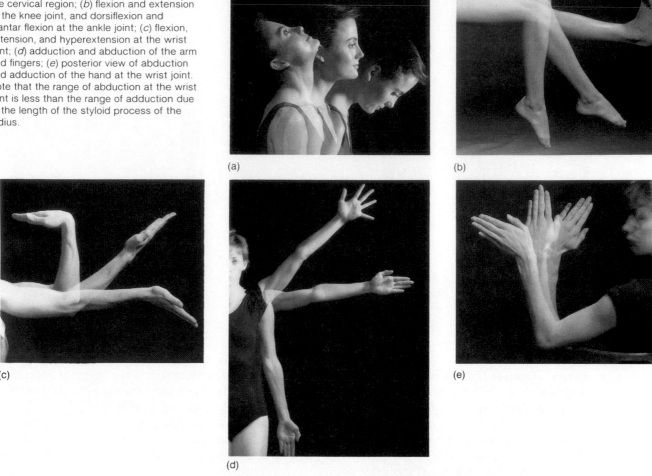

(a) 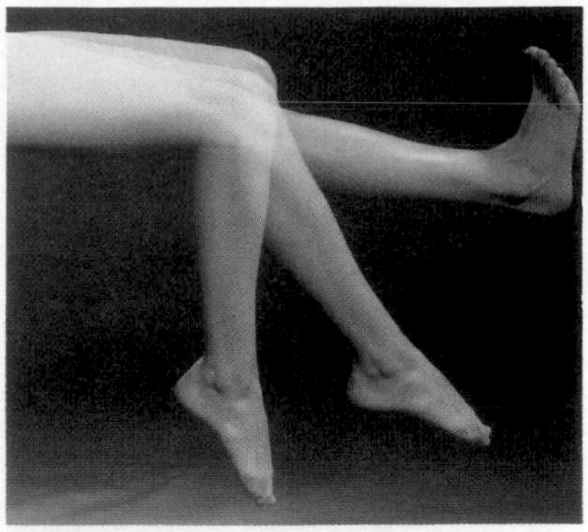 (b)

(c)  (d) 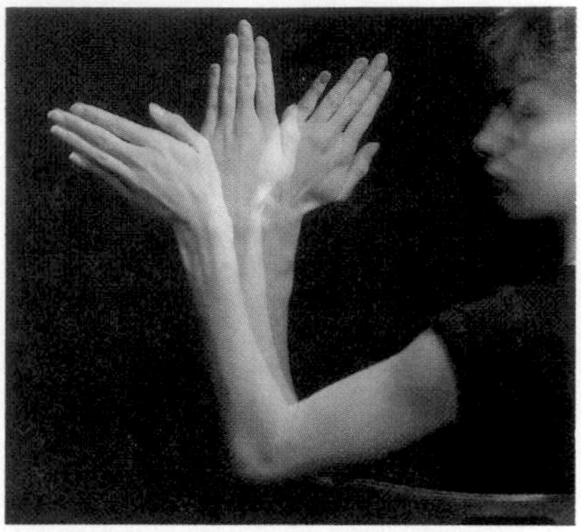 (e)

Figure 11.22. A visual summary of some rotational movements at diarthroses. (*a*) rotation of the head at the cervical vertebrae—especially at the atlantoaxial joint; (*b*) rotation of the forearm (antebrachium) at the proximal radioulnar joint.

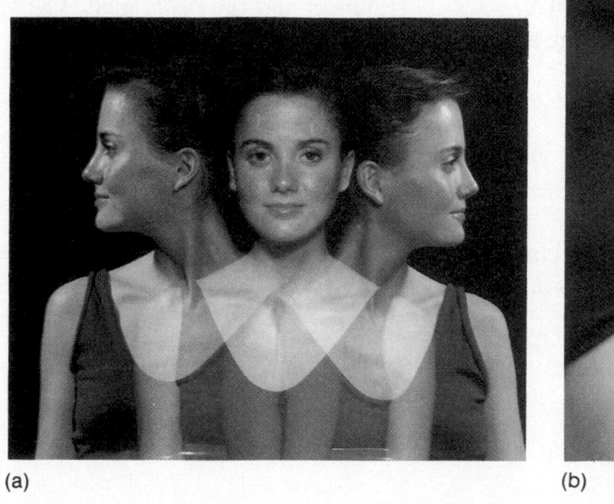

(a)

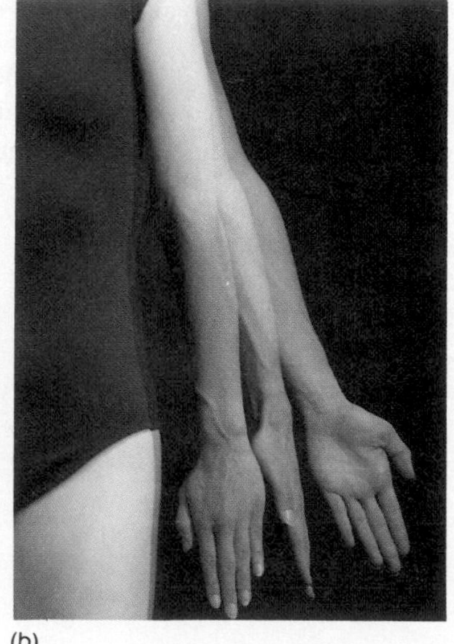

(b)

Figure 11.23. The three classes of levers. (*a*) in a first-class lever, the fulcrum (F) is positioned between the resistance (R) and the effort (E). (*b*) in a second-class lever, the resistance is between the fulcrum and the effort. (*c*) in a third-class lever, the effort is between the fulcrum and the resistance.

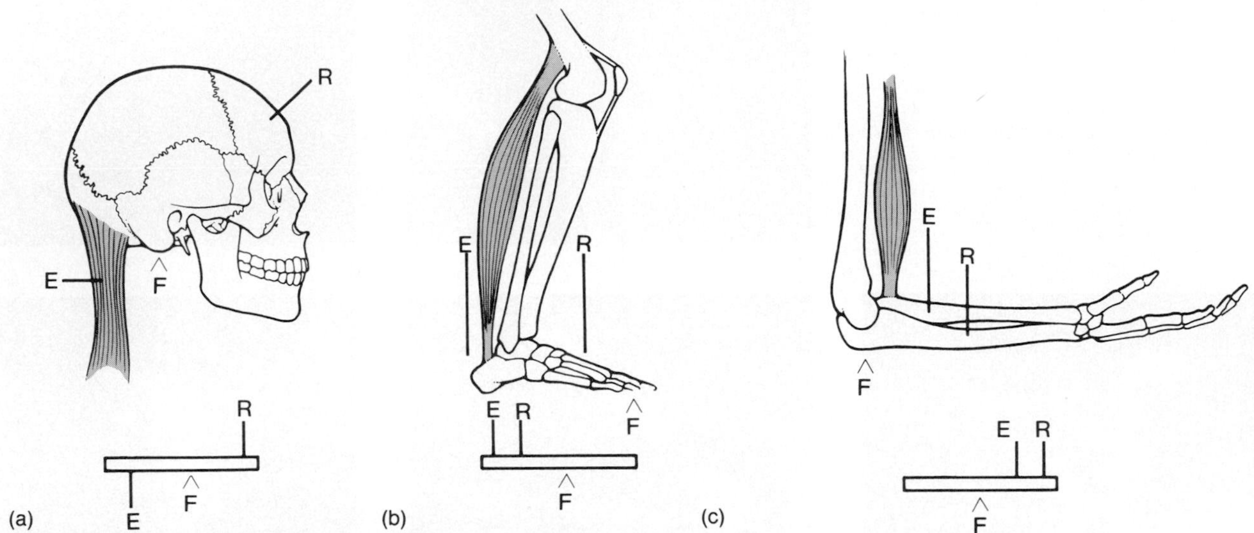

Biomechanics of Body Movement

A lever is any rigid structure that turns about a fulcrum when force is applied. Because bones are rigid structures that can be moved at diarthrotic joints in response to applied forces, they fit the criteria of levers. There are four basic components to a lever: (*a*) a rigid bar or other such structure; (*b*) a pivot or fulcrum; (*c*) an object or resistance that is moved; and (*d*) a force that is applied to one portion of the rigid structure.

Levers are generally associated with machines but can equally apply to other mechanical structures, such as the human body. Diarthrotic joints are always the fulcra (*F*), the muscles provide the force, or effort (*E*), and the bones are the rigid structures that move the resisting object (*R*).

There are three kinds of levers, determined by the arrangement of their parts (fig. 11.23).

1. In a **first-class lever,** the fulcrum is positioned between the effort and the resistance. The parts in a first-class lever are much like those of a seesaw—a sequence of resistance-pivot-effort. Scissors and hemostats are mechanical examples of the first-class levers. In the body, the head at the atlantooccipital *(at-lan''to-ok-sip'ĭ-tal)* joint is a first-class lever. The weight of the skull and facial portion of the head is the resistance, and the posterior neck muscles that contract to maintain the balance of the head on the joint are the effort.
2. In a **second-class lever,** the resistance is positioned between the fulcrum and the effort. The sequence of arrangement is pivot-resistance-effort, as in a wheelbarrow or the action of a crowbar when one end is placed under a rock and the other end lifted. Contracting the calf muscles (*E*) to elevate the body (*R*) on the toes, with the ball of the foot acting as the fulcrum, is another example.
3. In a **third-class lever,** the effort lies between the fulcrum and the resistance. The sequence of the parts is pivot-effort-resistance. An example is using a pair of forceps to grasp an object. The third-class lever is the most common type within the body. The flexion of the forearm at the elbow is an example. The effort occurs as the biceps muscle is contracted to move the resistance of the forearm with the elbow joint forming the fulcrum.

Figure 11.24. The position of a joint (fulcrum) relative to the length of a long bone (lever arm) and the point of attachment of a muscle (force) determines the mechanical advantage when movement occurs. (*a*) the elbow joint and extensor muscles of a human; (*b*) the elbow joint and extensor muscles of an armadillo.

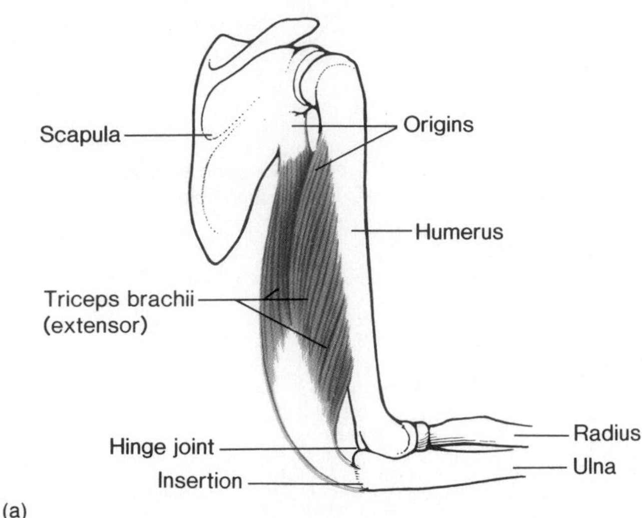

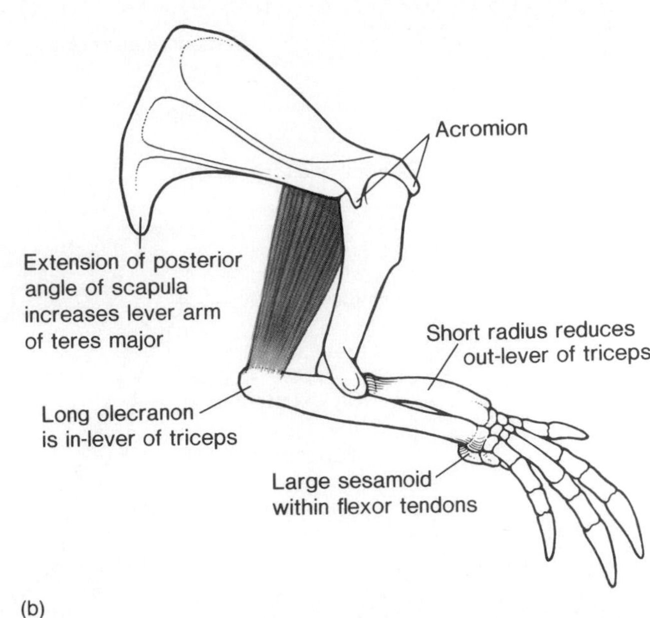

Each skeletal-muscular interaction at a diarthrotic joint forms some kind of lever system. The specific kind of lever is not always easy to identify. Certain joints are adapted for power at the expense of speed, whereas others are clearly adapted for speed (fig. 11.24). The specific attachment of muscles that span a joint plays an extremely important role in determining the mechanical advantage. The position of the insertion of a muscle relative to the joint is an important factor in the biomechanics of the contraction. An insertion close to the joint (fulcrum), for example, will produce a faster and greater range of movement than an insertion far away from the joint. An attachment far away from the joint takes advantage of the lever arm of the bone and increases power at the sacrifice of speed and range of movement.

1. Describe the structure of a joint that permits rotational movement.
2. What types of joints are involved in lever systems within the body?
3. Which is the most common type of lever in the body?
4. Considering the type of diarthrosis at the hip joint and the location of the gluteal muscles of the buttock, explain why this type of lever system is adapted for rapid, wide-ranging movements.

Specific Joints of the Body

Of the numerous joints in the body, some have special structural features that enable them to perform particular functions. Furthermore, these joints are somewhat vulnerable to trauma and are therefore clinically important.

Objective 13. Describe the structure, function, and possible clinical importance of the following joints: temporomandibular, sternoclavicular, humeroscapular, elbow, metacarpophalangeal, interphalangeal, coxal, tibiofemoral, and talocrural.

Temporomandibular Joint

The **temporomandibular** *(tem''po-ro-man-dib'u-lar)* **joint** is the only diarthrotic joint in the skull and is a unique combination of a hinge joint and a gliding joint (fig. 11.25). It is formed by the mandibular condyle and the mandibular fossa and the articular tubercle of the temporal bone. An **articular disc** separates the joint cavity into superior and inferior compartments.

Three ligaments support and reinforce the temporomandibular joint. The **temporomandibular ligament** is positioned on the lateral side of the joint capsule and is

Figure 11.25. The temporomandibular joint. (*a*) a lateral view; (*b*) a medial view; (*c*) a parasagittal view.

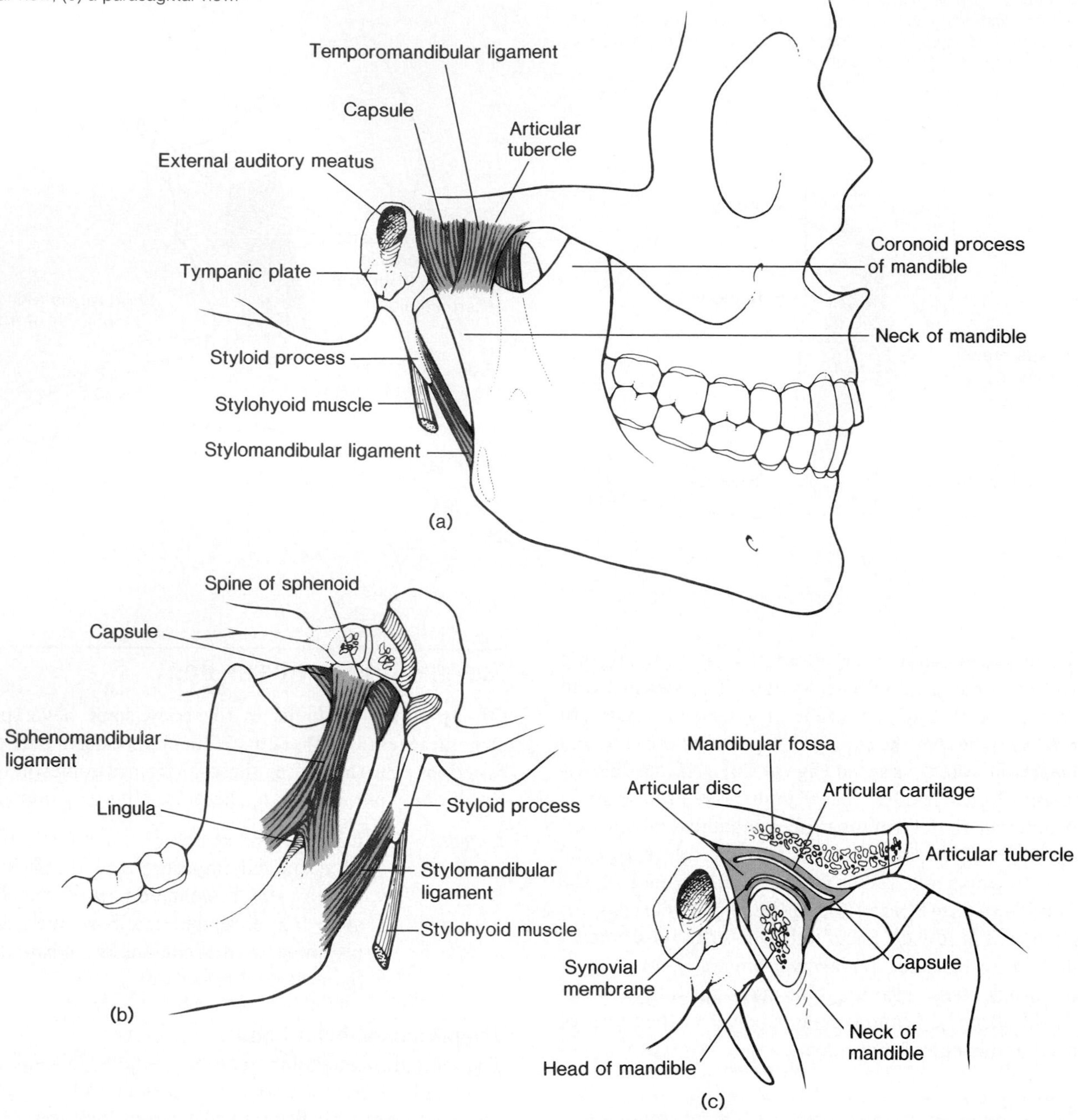

covered by the parotid salivary gland. This ligament prevents the head of the mandible from being displaced posteriorly and fracturing the tympanic plate when the chin suffers a severe blow. The **stylomandibular ligament** is not directly associated with the joint but extends inferiorly and anteriorly from the styloid process to the posterior border of the ramus of the mandible. A **sphenomandibular** *(sfe''no-man-dib'u-lar)* **ligament** crosses on the medial side of the joint from the spine of the sphenoid bone to the ramus of the mandible.

The movements of the temporomandibular joint include depression and elevation of the mouth as a hinge joint, protraction and retraction of the jaw as a gliding joint, and lateral rotatory movements. The lateral motion is made possible by the articular disc.

The temporomandibular joint can be easily palpated by applying firm pressure to the area in front of your auricle and opening and closing your mouth. This joint is most vulnerable to dislocation when the mandible is completely depressed, as in yawning. Relocating the jaw is usually a simple task, however, and is accomplished by pressing downward on the molar teeth while pushing the jaw backward.

Temporomandibular joint (TMJ) syndrome is a recently recognized ailment that may afflict an estimated 75 million Americans. The apparent cause of TMJ syndrome is a malalignment of one or both temporomandibular joints. The symptoms of the condition vary from moderate and intermittent pain to intense and continuous pain of the head, neck, shoulders, or back. Some vertigo (disorientation of coordination) and tinnitus (ringing in ear) may be experienced.

Figure 11.26. The sternoclavicular joint and associated ligaments. (*a*) an anterior view showing a frontal section; (*b*) a posterior view.

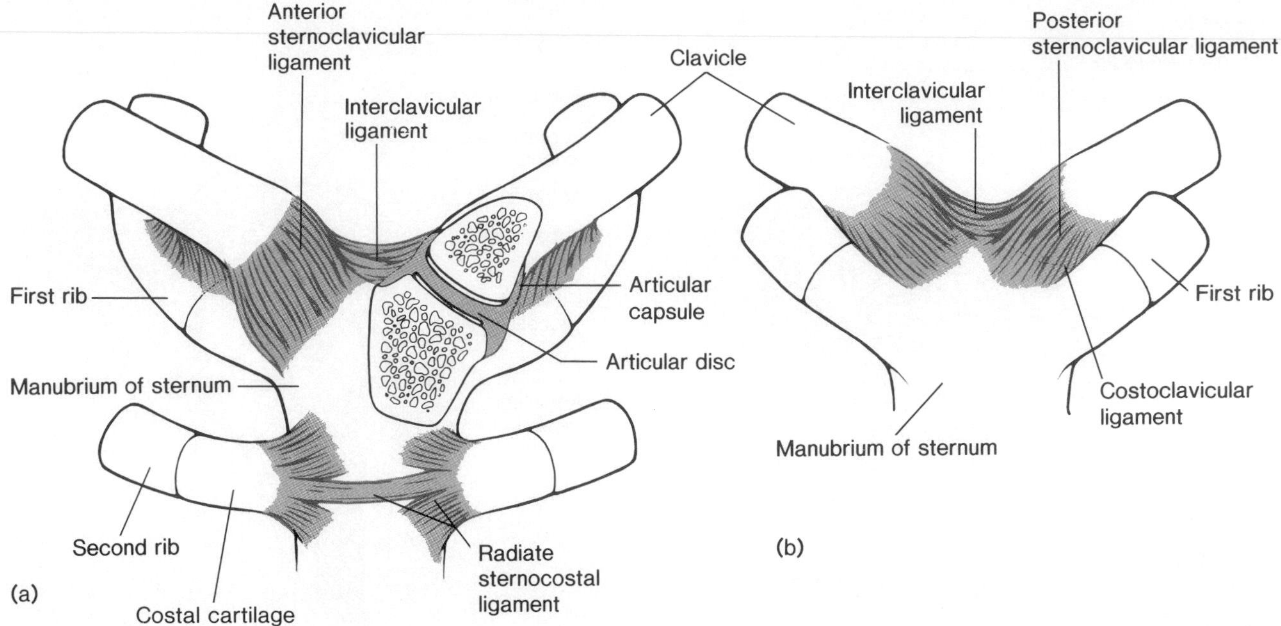

Sternoclavicular Joint

The **sternoclavicular** *(ster''no-klah-vik'u-lar)* **joint** is formed where the clavicle articulates with the manubrium of the sternum (fig. 11.26). Although a gliding joint, the sternoclavicular joint has a good range of movement because of the presence of an articular disc within the joint capsule.

Four ligaments support and give flexibility to the sternoclavicular joint. An **anterior sternoclavicular ligament** covers the anterior surface of the joint, and a **posterior sternoclavicular ligament** covers the posterior surface. Both ligaments extend from the sternal end of the clavicle to the manubrium. An **interclavicular ligament** extends between the sternal ends of both clavicles, binding them together. The **costoclavicular ligament** extends from the costal cartilage of the first rib to the costal tuberosity of the clavicle.

Of all the joints associated with the rib cage, the sternoclavicular joint is most frequently dislocated. Excessive force along the long axis of the clavicle may displace the clavicle forward and inferiorly. Injury to the costal cartilages is painful and is caused most frequently by a forceful, direct blow onto the costal cartilages.

Humeroscapular (Shoulder) Joint

The **shoulder joint** is formed by the articulation of the head of the humerus with the glenoid fossa of the scapula (fig. 11.27). It is a ball-and-socket joint and the most freely movable joint in the body. A circular band of fibrocartilage called the **glenoid** *(gle'noid)* **labrum** passes around the rim of the shoulder joint to deepen the concavity of the glenoid fossa (fig. 11.27). The shoulder joint is protected from above by an arch formed by the coracoid and acromion processes of the scapula and by the clavicle.

The shoulder joint is vulnerable to dislocations from sudden jerks of the arm, especially in children before strong shoulder muscles have developed. Because of the weakness of this joint in children, parents should be careful not to force a child to follow by yanking on the arm. Dislocation of the shoulder is extremely painful and may cause permanent damage or perhaps muscle atrophy due to disuse.

Although three ligaments surround and support the shoulder joint, most of the stability of this joint depends on the powerful muscles and tendons that cross over it. Thus it is an extremely mobile joint, in which stability has

labrum: L. *labrum,* lip

Figure 11.27. The humeroscapular joint. (*a*) an anterior view; (*b*) a coronally sectioned anterior view; (*c*) a posterior view; (*d*) a lateral view with the humerus removed.

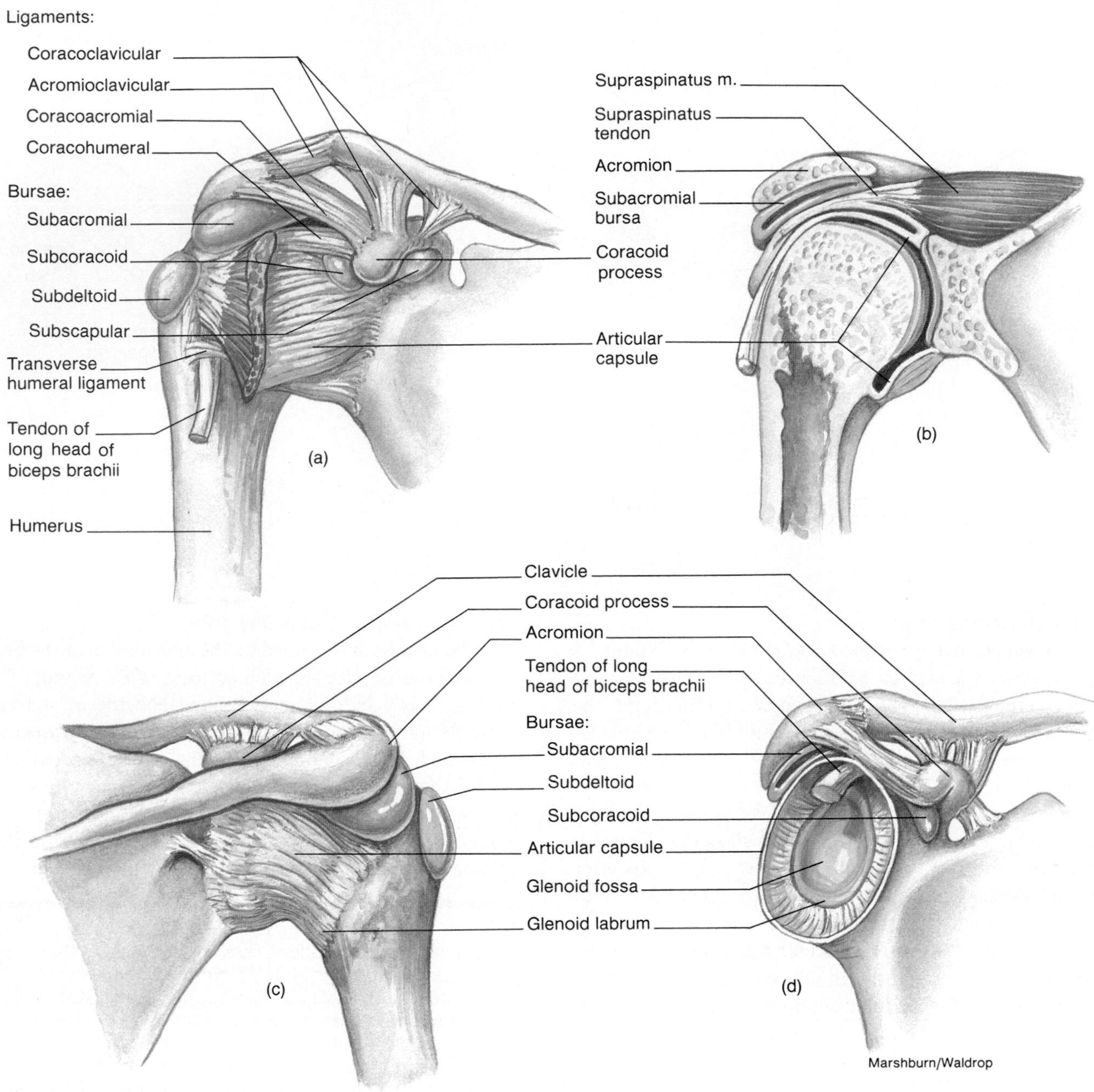

been sacrificed for mobility. The **coracohumeral** *(kor''ah-ko-hu'mer-al)* **ligament** extends from the coracoid process of the scapula to the greater tubercle of the humerus. The joint capsule is reinforced with three ligamentous bands called the **glenohumeral ligaments.** The final ligament of the shoulder joint is the **transverse humeral ligament,** a thin band that extends from the greater tubercle to the lesser tubercle of the humerus.

The stability of the shoulder joint is provided principally by the tendons of the subscapularis, supraspinatus, infraspinatus, and teres minor muscles together as they form the *musculotendinous (rotator) cuff.* The cuff is fused to the underlying capsule except in its inferior aspect. Because of the lack of inferior stability, most dislocations (subluxations) occur in this direction. The shoulder is most vulnerable to trauma when the joint is fully abducted and a sudden force from the superior direction is applied to the appendage. Degenerative changes in the musculotendinous cuff produce an inflamed, painful condition known as *pericapsulitis.*

Figure 11.28. The right elbow region. (*a*) an anterior view; (*b*) a posterior view; (*c*) a midsagittal section; (*d*) a lateral view; (*e*) a medial view.

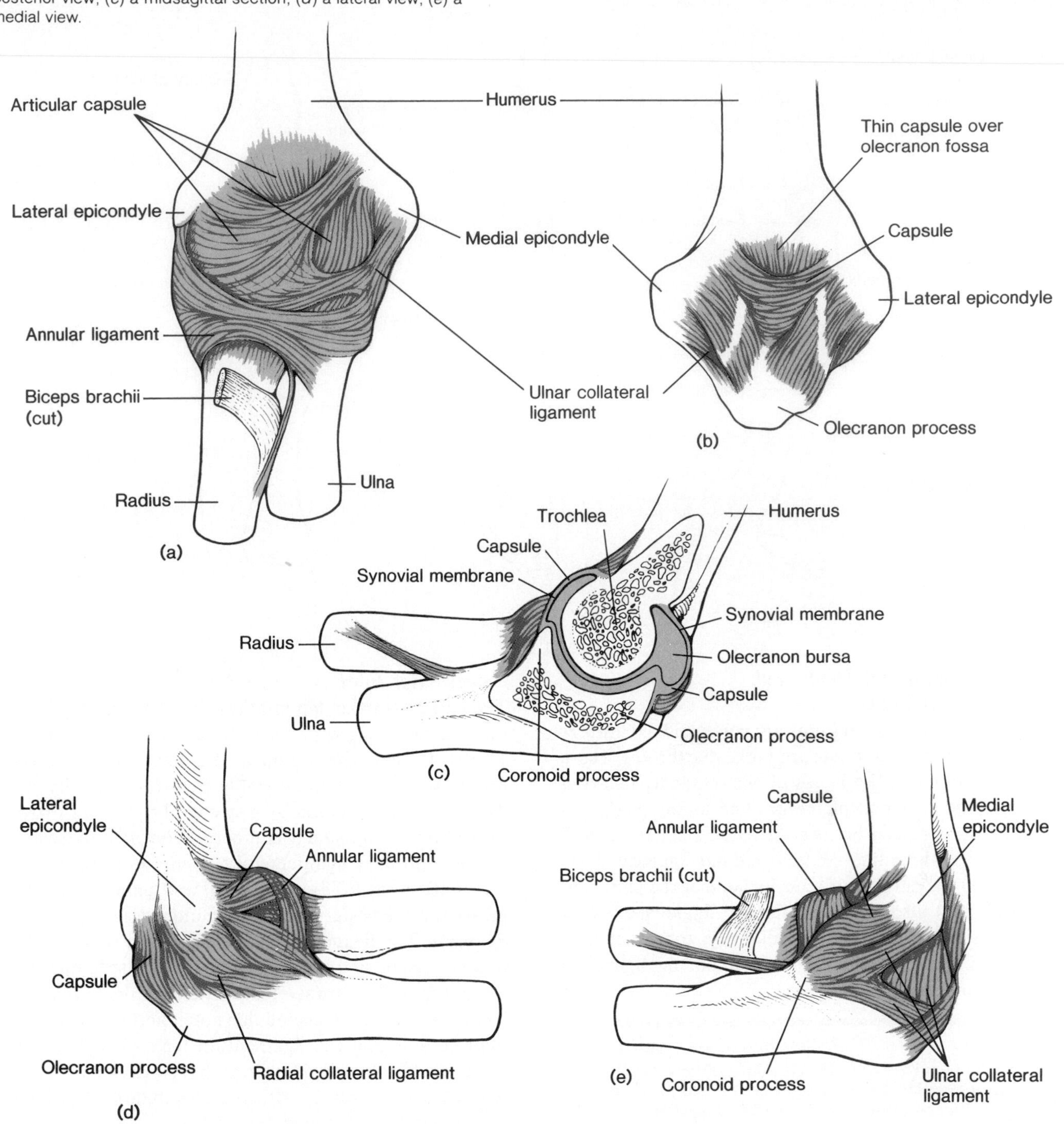

Two major and two minor bursae are associated with the shoulder joint. The larger bursae are the **subdeltoid bursa** between the deltoid muscle and the joint capsule and the **subacromial bursa** between the acromion and joint capsule. The **subcoracoid bursa** lies between the coracoid process and the joint capsule and is frequently considered an extension of the subacromial bursa. A small **subscapular bursa** is located between the tendon of the subscapularis muscle and the joint capsule.

Elbow Joint

There are three sets of articulations in the elbow region, two of which constitute the **elbow joint** (fig. 11.28). The elbow joint is a hinge joint, formed by the trochlea of the humerus articulating with the trochlear notch of the ulna and the capitulum of the humerus articulating with the head of the radius. Although there are two sets of articulations at the elbow joint, there is only one joint capsule and a large **olecranon bursa** to lubricate this area. A **radial (lateral) collateral ligament** reinforces the elbow joint on the lateral side, and an **ulnar (medial) collateral ligament** strengthens the medial side.

Figure 11.29. Metacarpophalangeal and interphalangeal joints. (*a*) a lateral view; (*b*) an anterior (palmar) view; (*c*) a posterior view.

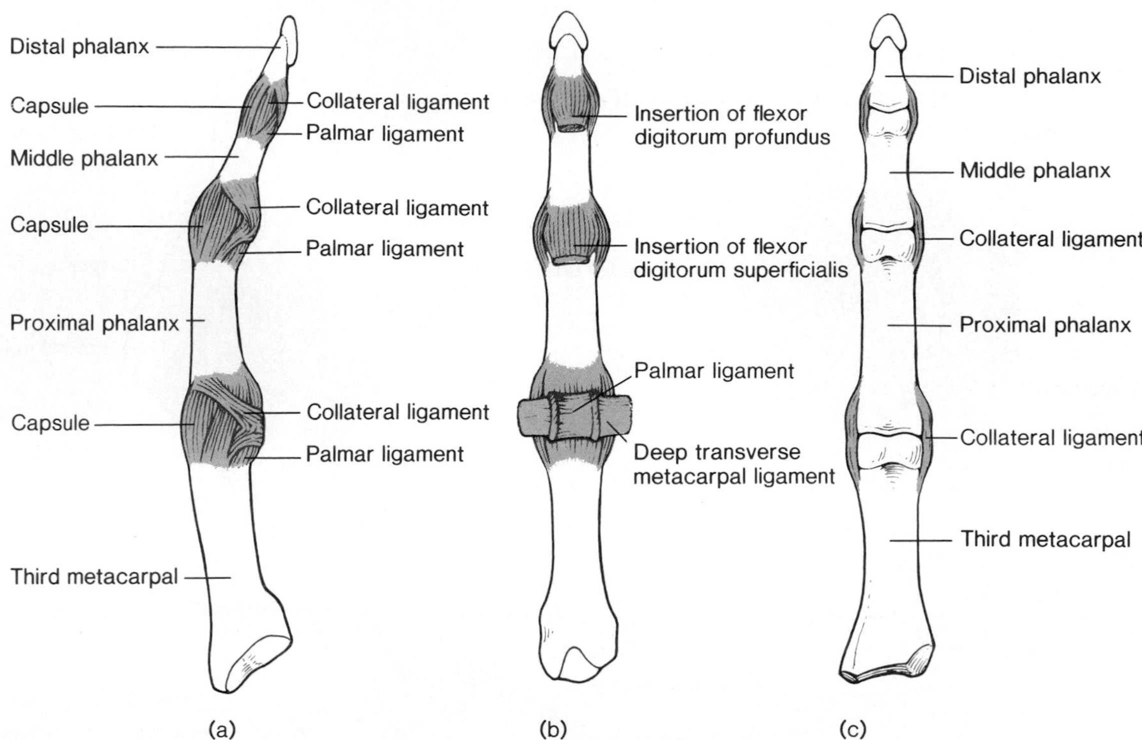

Metacarpophalangeal Joints and Interphalangeal Joints

The **metacarpophalangeal joints** are condyloid diarthroses and the **interphalangeal joints** are hinge diarthroses. These joints are formed as the heads of the metacarpals articulate with the proximal phalanges and as the phalanges articulate with one another (fig. 11.29). Each joint in both joint types has three ligaments. A **palmar ligament** spans each joint on the palmar, or anterior, side of the joint capsule. Each joint also has two **collateral ligaments,** one on the lateral side and one on the medial side, to further reinforce the joint capsule. There are no supporting ligaments on the posterior side.

Athletes frequently "jam a finger." It occurs when a ball forcefully strikes a distal phalanx as the fingers are extended, causing a sharp flexion at the joint between the middle and distal phalanges. No ligaments support the joint on the posterior side, but there is a tendon from the digital extensor muscles of the forearm. It is this tendon that is damaged when the finger is jammed. Treatment involves splinting the finger for a period of time. If splinting is not effective, however, surgery is necessary or the person will suffer a permanent crook in the finger.

Coxal (Hip) Joint

The ball-and-socket **hip joint** is formed by the head of the femur articulating with the acetabulum of the os coxa (fig. 11.30). It bears the weight of the body and is therefore much stronger and more stable than the shoulder joint. The hip joint is secured by a strong fibrous joint capsule, several ligaments, and a number of powerful muscles.

The primary ligaments of the hip joint are the anterior **iliofemoral** *(il''e-o-fem'or-al)* and **pubofemoral ligaments** and the posterior **ischiofemoral** *(is''ke-o-fem'or-al)* **ligament.** The **ligamentum teres** is located within the articular capsule and attaches the head of the femur to the acetabulum. The **transverse acetabular** *(as''ĕ-tab'u-lar)* **ligament** crosses the acetabular notch and connects to the joint capsule and the ligamentum teres. The **acetabular labrum,** a fibrocartilaginous rim that rings the head of the femur as it articulates with the acetabulum, is attached to the margin of the acetabulum.

Osteoarthritis is a degenerative disease of the articular cartilage of diarthrotic joints accompanied by the formation of bony spurs in the joint cavities. Immobility results if the hip joint is severely afflicted with this disease. Fortunately, the entire hip joint can be replaced in a procedure called *hip arthroplasty.* During this surgery, the acetabulum is replaced by a low-friction polyethylene socket, fit into the os coxa using bone cement. The head of the femur is replaced by a stainless steel, ball-shaped prosthesis (see fig. 11.35).

Figure 11.30. The right coxal (hip) joint. (*a*) an anterior view; (*b*) a posterior view; (*c*) a frontal section.

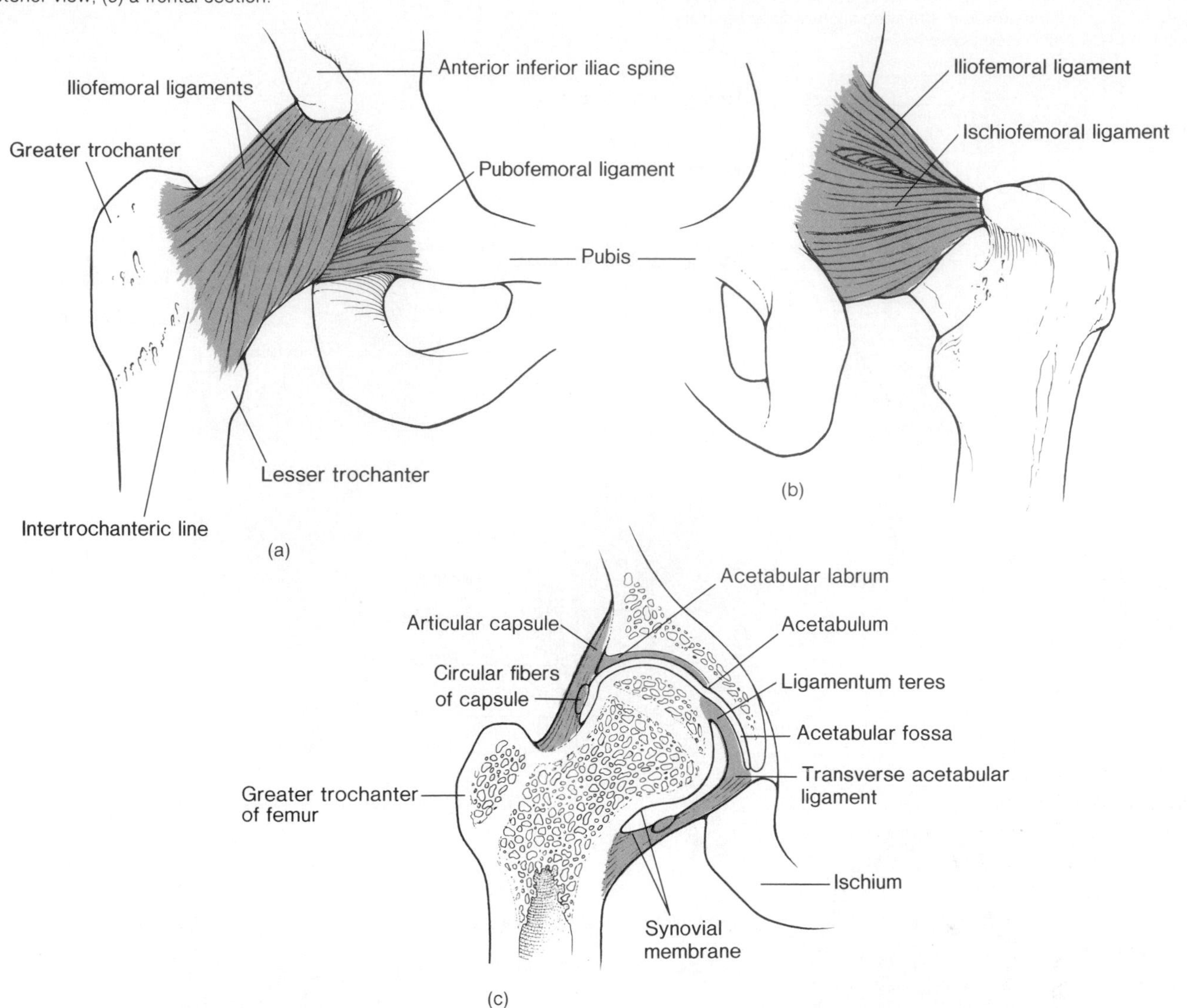

Tibiofemoral (Knee) Joint

The **knee joint** is the largest, most complex, and probably the most vulnerable joint in the body. The knee joint is formed as the femur and the tibia articulate. It is a complex hinge diarthrosis which, in addition to flexion and extension, permits limited rolling and gliding movements. On the anterior side, the knee joint is stabilized and protected by the patella and the **patellar** *(pah-tel'ar)* **ligament,** which forms a gliding **patellofemoral joint.**

Because of the complexity of the knee joint, only the relative positions of the ligaments, menisci, and bursae will be presented. Although the detailed attachments and functions will not be discussed, the locations of these structures can be seen in figure 11.31.

In addition to the patella and the patellar ligament on the anterior surface, the tendinous insertion of the quadriceps femoris muscle forms two supportive bands called the **lateral** and **medial patellar retinacula** *(ret''ĭ-nak'u-lah).* Four bursae are associated with the anterior aspect of the knee: the **superficial infrapatellar** *(in''frah-pah-tel'ar)* **bursa,** the **suprapatellar bursa,** the **prepatellar bursa,** and the **deep infrapatellar bursa.**

The posterior aspect of the knee is referred to as the **popliteal region.** The broad **oblique popliteal ligament** and the **arcuate** *(ar'cu-āt)* **popliteal ligament** are superficial in position, whereas the **anterior** and **posterior cruciate** *(kroo'she-āt)* **ligaments** are deep within the joint. The **popliteal bursa** and the **semimembranosus bursa** are the two bursae associated with the back of the knee.

Strong **collateral ligaments** support both the medial and lateral sides of the knee joint. Two fibrocartilaginous discs called the **lateral** and **medial menisci** are located

cruciate: L. *crucis,* cross

Figure 11.31. The right tibiofemoral (knee) joint. (*a*) an anterior view; (*b*) a superficial posterior view; (*c*) a lateral view showing the bursae; (*d*) an anterior view with the knee slightly flexed and the patella removed; (*e*) a deep posterior view.

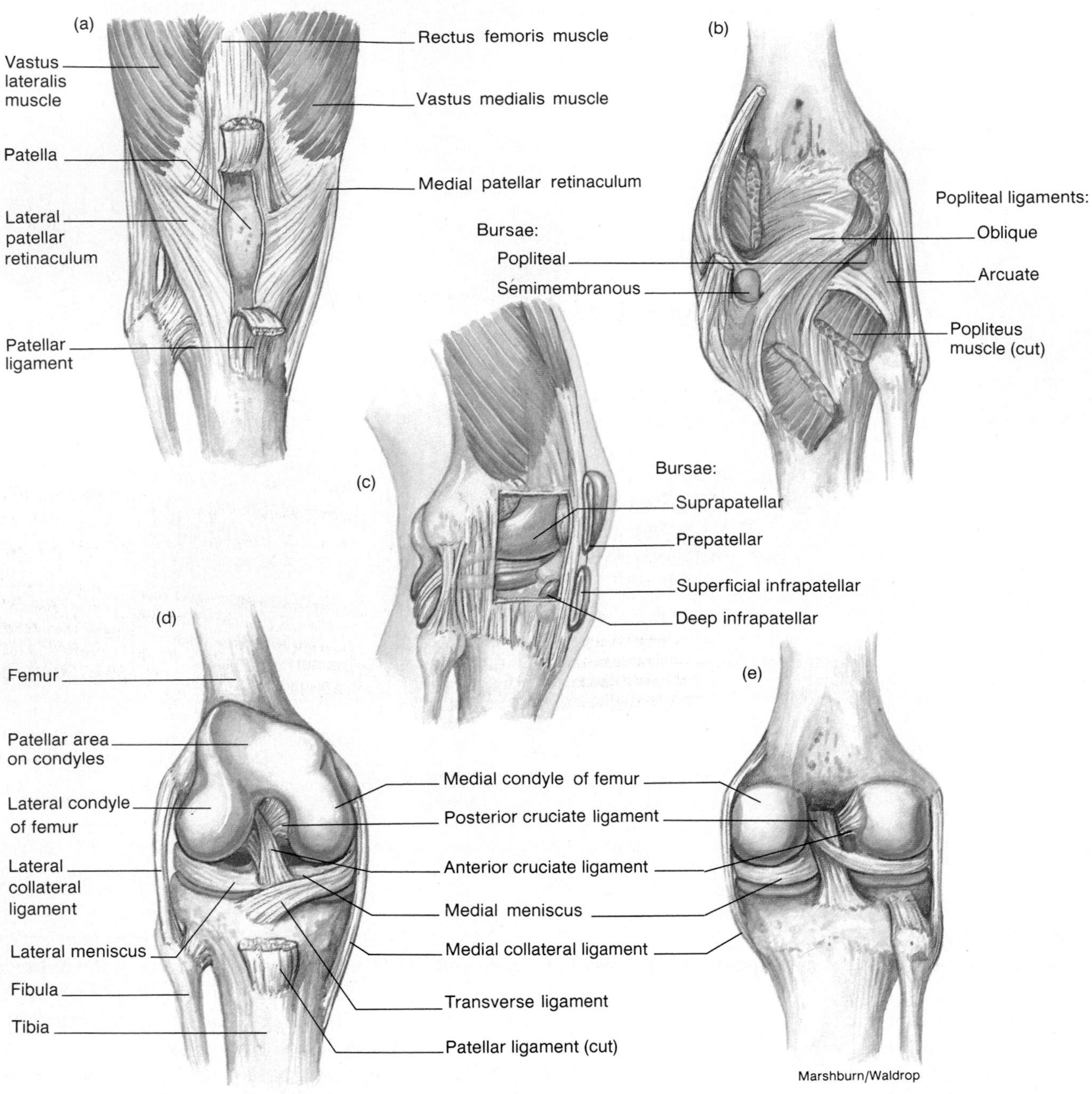

Figure 11.32. The right talocrural (ankle) joint. (*a*) an anterior view; (*b*) a posterior view.

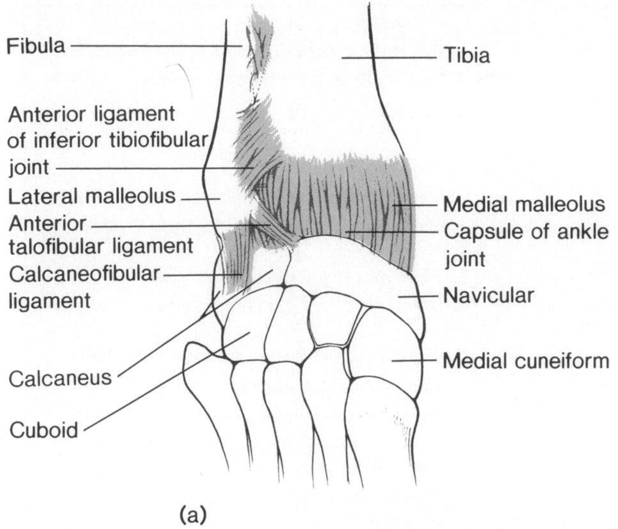

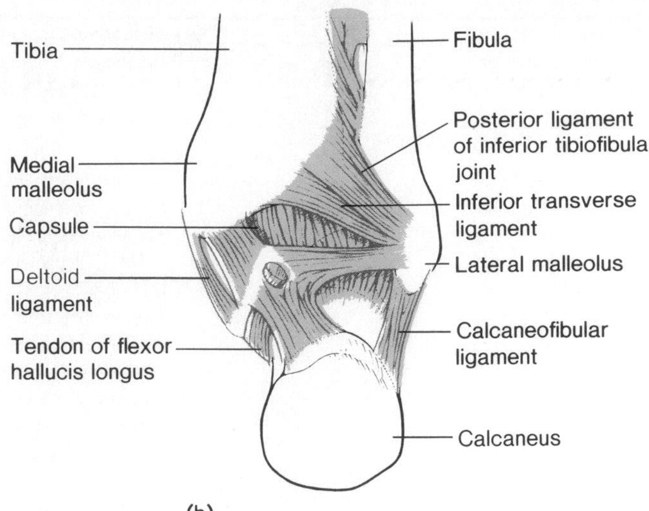

within the knee joint interposed between the distal femoral and proximal tibial condyles. The two menisci are connected by a **transverse ligament.** Several other bursae are associated with the knee joint. In addition to the four on the anterior side and the two on the posterior side, there are seven bursae on the lateral and medial sides, making a total of thirteen.

> During normal walking, running, and supporting of the body, the knee joint functions superbly. It can tolerate even moderate stress without tissue damage. However, the knee lacks bony support to withstand sudden forceful stresses, such as commonly occur among professional athletes. Knee injuries frequently require surgery and heal with difficulty due to the avascularity of the cartilaginous tissue. Because this joint is potentially so vulnerable, it is important to realize its limitations by understanding its anatomy.

Talocrural (Ankle) Joint

There are actually two articulations within the **ankle joint,** both of which are hinge diarthroses. One articulation is formed as the distal end of the tibia and its medial malleolus articulate with the talus, and the other is formed as the lateral malleolus of the fibula articulates with the talus (fig. 11.32).

One joint capsule surrounds the articulations of the three bones, and four ligaments support the ankle joint on the outside of the capsule. The strong **deltoid ligament** is associated with the tibia, whereas the **anterior talofibular** *(ta''lo-fib'u-lar)* **ligament, posterior talofibular ligament,** and **calcaneofibular** *(kal-ka''ne-o-fib'u-lar)* **ligament** are associated with the fibula.

The malleoli form a cap over the upper surface of the talus that prohibits side-to-side movement at the ankle joint. Unlike the condyloid joint at the wrist, the movements of the ankle are limited to flexion and extension. Dorsiflexion of the ankle is checked primarily by the Achilles tendon, whereas plantar flexion, or ankle extension, is checked by the tension of the extensor tendons on the front of the joint and the anterior portion of the joint capsule.

> *Ankle sprains* are a common type of locomotor injury. They vary extensively in seriousness but generally occur in certain locations. The most common cause of ankle sprain is excessive inversion of the foot, resulting in partial tearing of the anterior talofibular ligament and the calcaneofibular ligament. Less commonly, the deltoid ligament is injured by excessive eversion of the foot. Torn ligaments are extremely painful and are accompanied by immediate local swelling. Reducing the swelling and immobilizing the joint are about the only treatments for moderate sprains. Extreme sprains may require surgery and casting of the joint to facilitate healing.

Table 11.2 presents a summary of the principal joints of the body and their movement.

Table 11.2 Principal articulations

Joint	Type	Movement
Most skull joints	Synarthrosis (suture)	Immovable
Temporomandibular	Diarthrosis (hinge; gliding)	Elevation, depression; protraction, retraction
Atlantooccipital	Diarthrosis (condyloid)	Flexion, extension, circumduction
Atlantoaxial	Diarthrosis (pivot)	Rotation
Intervertebral		
bodies of vertebrae	Amphiarthrosis (symphysis)	Slight movement
articular processes	Diarthrosis (gliding)	Flexion, extension, slight rotation
Sacroiliac	Amphiarthrosis (synchondrosis); diarthrosis (gliding)	Slight gliding movement; may fuse in adults
Costovertebral	Diarthrosis (gliding)	Slight movement during breathing
Sternocostal	Amphiarthrosis (synchondrosis); diarthrosis (gliding)	Slight movement during breathing
Sternoclavicular	Diarthrosis (gliding)	Slight movement when shrugging shoulders
Sternal	Amphiarthrosis (symphysis)	Slight movement during breathing
Acromioclavicular	Diarthrosis (gliding)	Protraction, retraction; elevation, depression
Humeroscapular (shoulder)	Diarthrosis (ball-and-socket)	Flexion, extension; adduction, abduction; rotation, circumduction
Elbow	Diarthrosis (hinge)	Flexion, extension
Proximal radioulnar	Diarthrosis (pivot)	Rotation
Distal radioulnar	Amphiarthrosis (syndesmosis)	Slight movement
Radiocarpal (wrist)	Diarthrosis (condyloid)	Flexion, extension; adduction, abduction; circumduction
Intercarpal	Diarthrosis (gliding)	Slight movement
Carpometacarpal		
fingers	Diarthrosis (condyloid)	Flexion, extension; adduction, abduction
thumb	Diarthrosis (saddle)	Flexion, extension; adduction, abduction
Metacarpophalangeal and interphalangeal	Diarthrosis (hinge)	Flexion, extension
Symphysis pubis	Amphiarthrosis (symphysis)	Slight movement
Coxal (hip)	Diarthrosis (ball-and-socket)	Flexion, extension; adduction, abduction; rotation; circumduction
Tibiofemoral (knee)	Diarthrosis (hinge)	Flexion, extension; slight rotation when flexed
Proximal tibiofibular	Diarthrosis (gliding)	Slight movement
Distal tibiofibular	Amphiarthrosis (syndesmosis)	Slight movement
Talocrural (ankle)	Diarthrosis (hinge)	Dorsiflexion, plantar flexion; slight circumduction
Intertarsal	Diarthrosis (gliding)	Inversion, eversion
Tarsometatarsal	Diarthrosis (gliding)	Flexion, extension; adduction, abduction

1. Which diarthrotic joints have menisci?
2. What are the two types of joints found in the shoulder region? Why is the shoulder joint so vulnerable?
3. Which joints are reinforced with muscles that span the joint?
4. Describe the structure of the knee joint, and indicate which structures protect and reinforce its anterior surface.

Clinical Considerations

A diarthrotic joint is a remarkable biologic system that acts as a self-lubricating bearing surface, able to move with almost frictionless precision under tremendous loads and impacts. Under normal circumstances and in most people, the many joints of the body perform without problems throughout life. Joints are not indestructible, however, and are subject to various forms of trauma and disease. Although not all of the diseases of joints are fully understood, medical science has made remarkable progress in the treatment of arthrological problems.

Figure 11.33. Scoliosis is a lateral curvature of the spine, usually in the thoracic region. It may be congenital, acquired, or disease related.

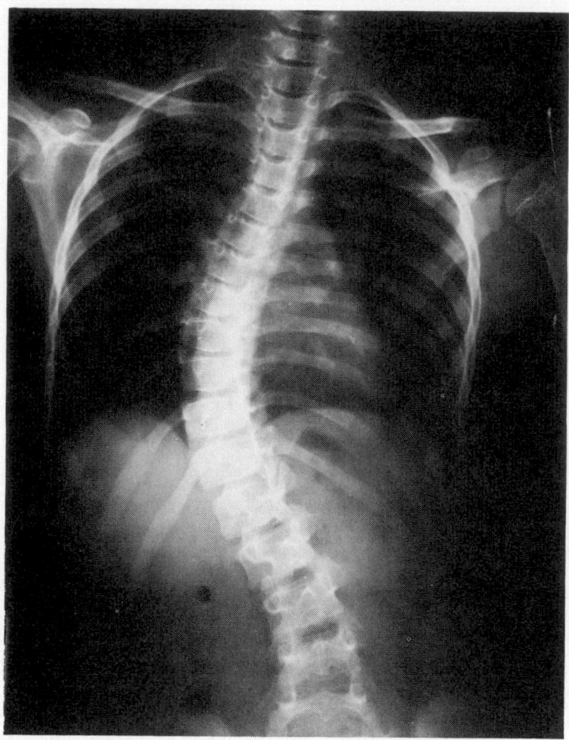

Trauma to Joints

Joints are well adapted to withstand compression and tension forces. Torsion or sudden impact to the side of a joint, however, can be devastating. These types of injuries frequently occur in athletes.

In a **strained joint,** unusual or excessive exertion stretches the tendons spanning a joint or the surrounding muscles but causes no serious damage. Strains frequently result from not "warming up," or activating joints and muscles, prior to strenuous use. A **sprain** is a tearing of the ligaments or tendons that surround a joint. There are various grades of sprains, and the severity will determine the treatment. Severe sprains of the knee joint are frequently accompanied by damage to articular cartilages and menisci, which generally requires surgery. Sprains are usually accompanied by **synovitis** *(sin''o-vi'tis),* an inflammation of the joint capsule.

Luxation, or **joint dislocation,** is derangement of the articulating bones that compose the joint. Joint dislocation is more serious than a sprain and is usually accompanied by sprains. The shoulder and knee joints are the most vulnerable to dislocation. Self-healing of a dislocated joint, like the knee joint, may be incomplete, leaving the person with a "trick knee" that may unexpectedly give way.

Subluxation is partial dislocation of a joint. Subluxation of the hip joint is a common type of birth defect that can be treated by bracing or casting the hip joints to promote suitable bone development.

Bursitis *(ber-si'tis)* is an inflammation of the bursa associated with a joint. Because of its close proximity to the joint, bursitis may affect the joint capsule as well. Bursitis may be caused by excessive stress on the bursa from overexertion, or it may be a local or systemic inflammatory process. As the bursa swells, the surrounding muscles become sore and stiff. **Tendonitis** involves the tendon, in or out of a tendon sheath, and usually has the same causes as bursitis.

The flexible vertebral column is a marvel of mechanical engineering. Not only do the individual vertebrae articulate one with another, but together they form the portion of the axial skeleton with which the head, ribs, and pelvic girdle articulate. The vertebral column also encloses the spinal cord and provides exits for thirty-one pairs of spinal nerves. With all the articulations within the vertebral column and the physical abuse it receives, it is no wonder that back ailments are second only to headaches as the most common physical complaint. Our way of life causes many of the problems associated with the vertebral column. Improper shoes, athletic exertion, sudden stops in vehicles, or improper lifting can all cause the back to go awry. Body weight, age, and general body condition influence a person's susceptibility to back problems.

The most common cause of back pain is *strained muscles,* generally the result of overexertion. The second most frequent back ailment is a *herniated disc.* The dislodged nucleus pulposus of a disc may push against a spinal nerve and cause excruciating pain. The third most frequent back problem is a *dislocated articular facet* between two vertebrae, caused by sudden torsion of the vertebral column. The treatment of back ailments varies from bed rest to spinal manipulation to extensive surgery.

Curvature disorders are another problem of the vertebral column. **Kyphosis** *(ki-fo'sis)* (hunchback) is an exaggeration of the thoracic curve. **Lordosis** (swayback) is an abnormal anterior convexity of the lumbar curve. **Scoliosis** (crookedness) is an abnormal lateral curvature of the vertebral column (fig. 11.33), which may be caused by one leg being longer than the other and uneven muscular development on the two sides of the vertebral column.

luxation: L. *luxus,* out of place

kyphosis: Gk. *kyphos,* hunched
lordosis: Gk. *lordos,* curving forward
scoliosis: Gk. *skoliosis,* crookedness

Figure 11.34. Arthroscopy. In this technique, a needlelike viewing arthroscope is threaded into the joint capsule through a tiny incision. The arthroscope has a fiberoptic light source, which illuminates the interior of a joint and the position of the surgical instruments that may be inserted through other small incisions.

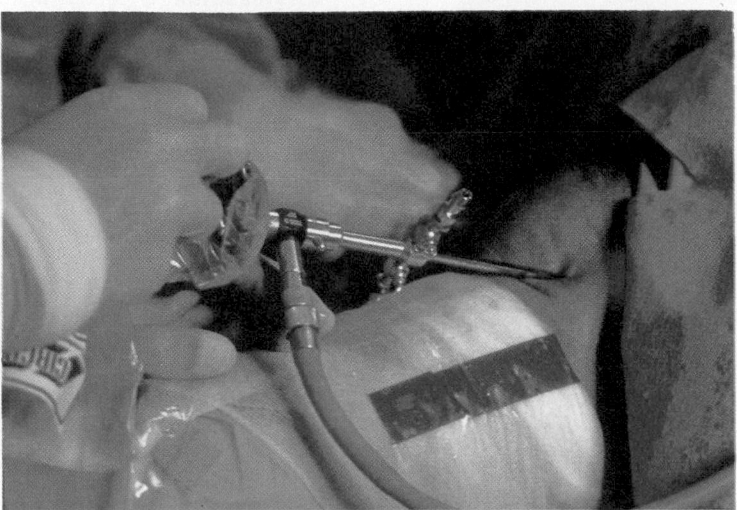

Diseases of Joints

Arthritis is a generalized term for over fifty different joint diseases, all of which have the symptoms of edema, inflammation, and pain. The causes of arthritis are unknown, but certain types follow joint trauma or bacterial infections. There is evidence that some types of arthritis are the result of hormonal or metabolic disorders. The most common forms are rheumatoid arthritis, osteoarthritis, and gouty arthritis.

In **rheumatoid** *(roo'mah-toid)* **arthritis,** the synovial membrane thickens and becomes tender and synovial fluid accumulates. This change is usually followed by an invasion of fibrous tissue and deterioration of the articular cartilage. When the cartilage is destroyed, the exposed osseous tissue is joined by the fibrous tissue and instigates ossification of the joint. It is the joint ossification that causes the crippling effect of rheumatoid arthritis. This disease affects females slightly more than males and occurs most commonly between the ages of thirty and fifty.

Osteoarthritis is a degenerative joint disease that results from aging and irritation of the joints. Although osteoarthritis is far more common than rheumatoid arthritis, it is usually less damaging. Osteoarthritis is a slow, progressive disease in which the articular cartilages gradually soften and disintegrate. The affected joints seldom swell, and the synovial membrane is rarely damaged. As the articular cartilage deteriorates, ossified spurs are deposited on the exposed bone, causing pain and restricting the movement of articulating bones. Osteoarthritis most frequently affects the knee, hip, and intervertebral joints.

Gouty arthritis results from a metabolic disorder in which an abnormal amount of uric acid is retained in the blood and sodium urate crystals are deposited in the joints. The salt crystals irritate the articular cartilage and synovial membrane, causing swelling, tissue deterioration, and pain. If gout is not treated, the affected joint fuses. Males have a greater incidence of gout than females, and apparently the disease is genetically determined.

Treatment of Joint Disorders

Arthroscopy *(ar-thros'ko-pe)* is widely used in diagnosing and, to a limited extent, treating joint disorders. Arthroscopic inspection involves making a local incision through the skin and into the joint capsule through which the tubelike arthroscopic instrument is threaded (fig. 11.34). In arthroscopy of the knee, the articular cartilage, synovial membrane, menisci, and cruciate ligaments can be observed. Samples can be extracted, and pictures taken for further evaluation.

Remarkable advancements have been made in the last fifteen years in **joint prostheses** *(pros-the'sēz)* (fig. 11.35). Joint prostheses, or artificial articulations, do not take the place of normal, healthy joints, but they are a valuable option for chronically disabled arthritis patients.

---

rheumatoid: Gk. *rheuma,* a flowing

gout: L. *gutta,* a drop (thought to be caused by "drops of viscous humors")

prosthesis: Gk. *pros,* in addition to; *thesis,* a setting down

Figure 11.35. Two examples of joint prostheses. (*a,b*) the hip joint; (*c,d*) the knee joint.

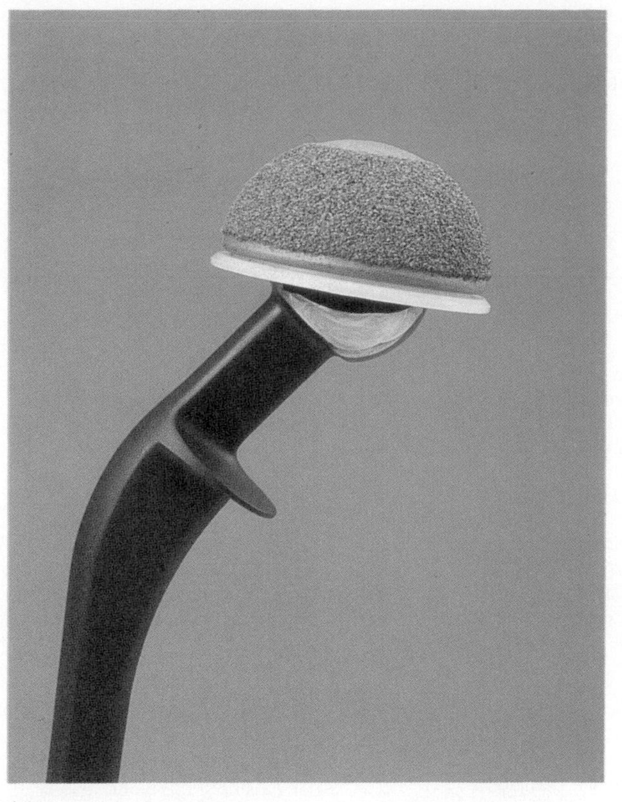

(a)

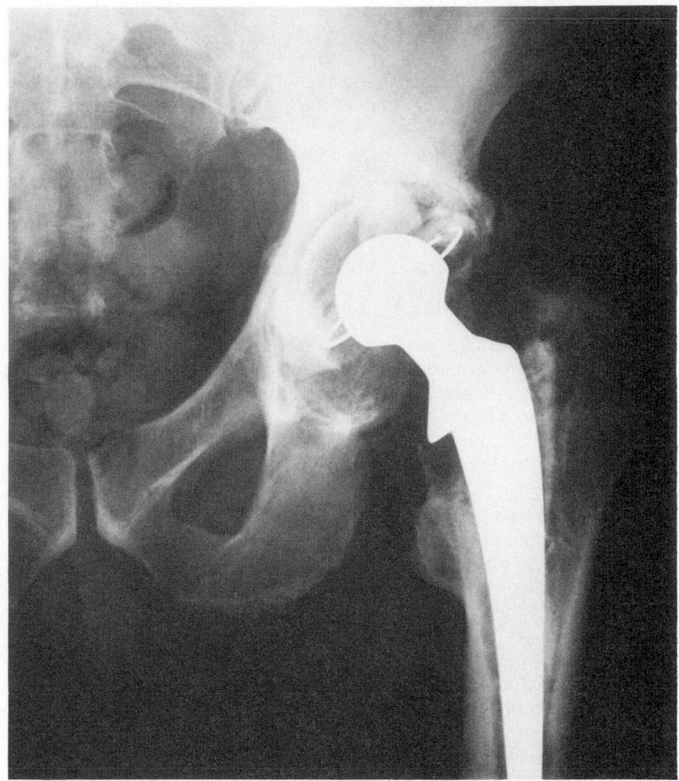

(b)

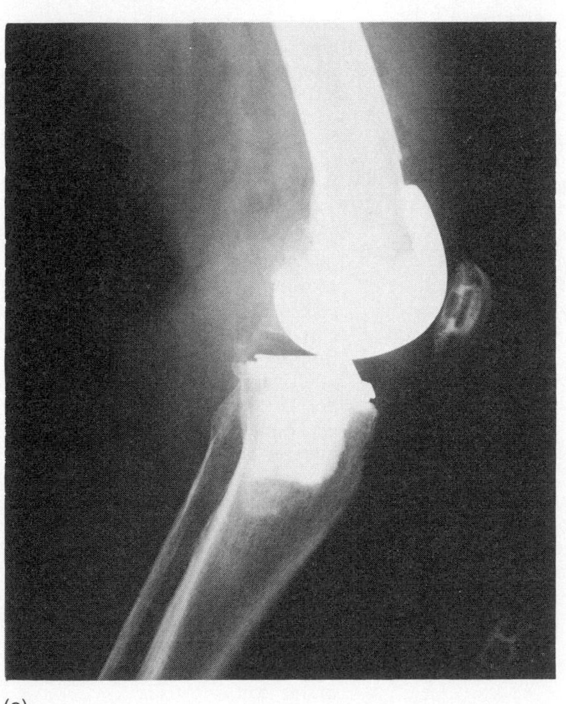

(c)

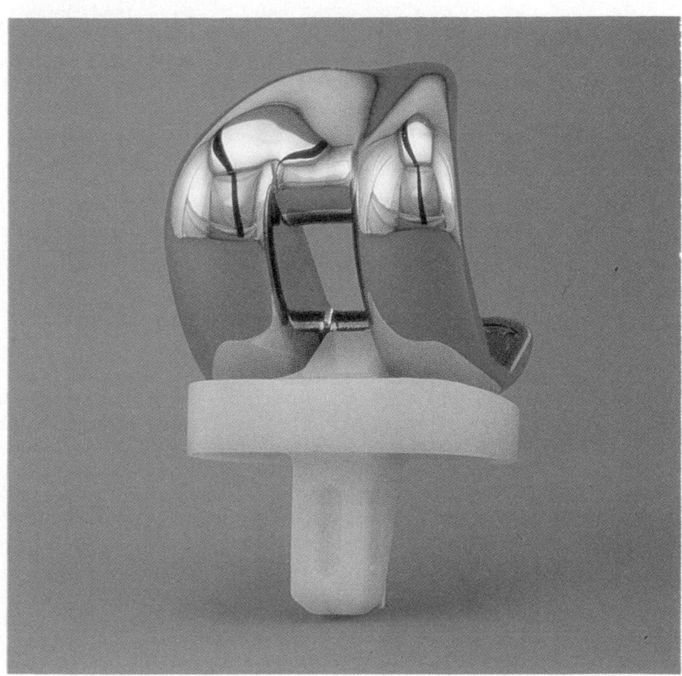

(d)

Important Clinical Terminology

ankylosis Stiffening of a joint resulting in severe or complete loss of movement.
arthralgia (also *arthrodynia*) Severe pain within a joint.
arthrolith A gouty deposit in a joint.
arthrometry The measurement of the range of movement in a joint.
arthroncus Swelling of a joint due to trauma or disease.
arthropathy Any disease affecting a joint.
arthroplasty The surgical repair of a joint.
arthrosis A joint or an articulation; also, a degenerative condition of a joint.
arthrosteitis An inflammation of the body structure of a joint.
chondritis An inflammation of the articular cartilage of a joint.
coxarthrosis The hip joint; also, a degenerative condition of the hip joint.
hemarthrosis An accumulation of blood in a joint cavity.
rheumatology The medical speciality concerned with the diagnosis and treatment of arthritis.
spondylitis An inflammation of one or several vertebrae.
synovitis The inflammation of the synovial membrane lining the inside of a joint capsule.

Chapter Summary

I. Classification of Joints
  A. Joints are formed as adjacent bones articulate together. Arthrology is the science concerned with the study of joints, and kinesiology is the study of movements involving certain joints.
  B. Joints can be classified according to structure or function.
    1. The structural classification divides joints into fibrous, cartilaginous, and synovial types.
    2. The functional classification divides joints into synarthroses, amphiarthroses, and diarthroses.

II. Development of Freely Movable Joints
  A. The development of each movable joint is initiated during the sixth week as the perichondrium from adjacent developing bones spans the future joint cavity and gives rise to the joint capsule.
  B. Diarthroses form by the third month as the epiphyses of adjacent endochondral bones develop distinct configurations and as the muscles that move the joints undergo contractions.
  C. Articular cartilages form on the epiphyses, and the synovial membrane secretes synovial fluid into the joint cavity.

III. Synarthroses
  A. Articulating bones in immovable joints are connected by either fibrous tissue or cartilage.
  B. Sutures are found only in the skull and are classified as serrate, lap, plane, or synostosis.
  C. Synchondroses are temporary joints formed in the growth lines (epiphyseal plates) between the diaphyses and epiphyses in the long bones of children.

IV. Amphiarthroses
  A. Amphiarthroses allow limited motion in response to twisting, compression, or stress.
  B. The articulating bones of amphiarthroses are connected by ligaments or fibrocartilaginous pads.
  C. The symphysis pubis and the intervertebral disc joints are examples of symphyses; the distal articulations of the tibia-fibula and the ulna-radius are examples of syndesmoses.

V. Diarthroses
  A. Diarthroses are freely movable joints enclosed by joint capsules that contain synovial fluid. They also contain joint cavities, synovial membranes, and articular cartilages.
  B. The range of movement of a diarthrosis is determined by the structure of the articulating bones, the ligaments and tendons, and the muscles that act on the joint.

VI. Movements at Diarthroses
  A. Movements at diarthroses are produced by the contraction of the skeletal muscles spanning the joints and attaching to or near the bones forming the articulations. In these actions, the bones act as levers, the muscles provide the force, and the joints are the fulcra, or pivots.
  B. Angular movements increase or decrease the joint angle produced by the articulating bones.
    1. Flexion decreases the joint angle on an anterior-posterior plane; extension increases the same joint angle.
    2. Abduction is the movement of a body part away from the main axis of the body; adduction is the movement of a body part toward the main axis of the body.
  C. Circular movements can occur only in joints that are composed of a bone with a rounded surface articulating with a corresponding depression on another bone.
    1. Rotation is the movement of a bone around its own axis.
    2. Circumduction is a conelike movement of a body segment.
  D. Special joint movements include inversion and eversion, protraction and retraction, and elevation and depression.
  E. Diarthroses can be classified as first-, second-, or third-class levers.

VII. Specific Joints of the Body
  A. The temporomandibular joint, a combined hinge and gliding joint, is of clinical importance because of temporomandibular joint (TMJ) syndrome.
  B. The sternoclavicular joint, a gliding joint, is frequently dislocated in injuries to the rib cage.
  C. The humeroscapular (shoulder) joint, a ball-and-socket joint, is vulnerable to dislocations from sudden jerks of the arm, especially in children before strong shoulder muscles have developed.
  D. There are two diarthroses at the elbow as the distal end of the humerus articulates with the proximal ends of the ulna and radius. Strain on the elbow joint is common during certain sports.
  E. Both metacarpophalangeal and interphalangeal joints are condyloid and hinge diarthroses, respectively, which may be traumatized when an athlete "jams a finger."
  F. The ball-and-socket coxal (hip) joint is especially subject to osteoarthritis in elderly people.
  G. The hinged tibiofemoral (knee) joint is the largest, most complex, and most vulnerable joint in the body.
  H. There are two hinged articulations within the talocrural (ankle) joint. Ankle sprains are common injuries of this joint.

Review Activities

Objective Questions

1. Which statement regarding joints is *false*?
   (a) Joints are the locations where two or more bones articulate.
   (b) The structural classification of joints includes fibrous, membranous, and cartilaginous types.
   (c) Arthrology is the study of joints; kinesiology is the study of the biomechanics of joint movement.
2. Synchondroses are a type of
   (a) synarthrosis
   (b) diarthrosis
   (c) amphiarthrosis
   (d) syndesmosis
3. A synostosis is a unique type of
   (a) synchondrosis
   (b) syndesmosis
   (c) symphysis
   (d) suture
4. Which of the following joint type-function word pairs is *incorrect*?
   (a) synchondrosis—growth at the epiphyseal plate
   (b) symphysis—movement at the intervertebral joint
   (c) suture—strength and stability in the skull
   (d) syndesmosis—movement of the jaw
5. Structurally, diarthroses are
   (a) synovial joints
   (b) fibrous joints
   (c) membranous joints
   (d) cartilaginous joints
6. Which of the following is *not* characteristic of all diarthroses?
   (a) articular cartilage
   (b) synovial fluid
   (c) joint capsule
   (d) meniscus
7. The atlantoaxial and the proximal radioulnar diarthrotic joints are specifically classified as
   (a) hinge
   (b) gliding
   (c) pivotal
   (d) condyloid
8. Which of the following joints can be readily and comfortably hyperextended?
   (a) interphalangeal
   (b) coxal
   (c) tibiofemoral
   (d) sternocostal
9. Which of the following is most vulnerable to luxation? The
   (a) elbow
   (b) humeroscapular
   (c) coxal
   (d) tibiofemoral
10. A thickening and tenderness of the synovial membrane and the accumulation of synovial fluid are signs of the development of
   (a) arthroscopitis
   (b) gouty arthritis
   (c) osteoarthritis
   (d) rheumatoid arthritis

Essay Questions

1. What is meant by a structural classification of joints, as compared to a functional classification?
2. Why is the anatomical position so important in explaining the movements that are possible at joints?
3. What are the structural components of a diarthrotic joint that determine or limit the range of movement at that joint?
4. What are the advantages of a hinge joint over a ball-and-socket type? If ball-and-socket joints allow a greater range of movement, why are not all the diarthrotic joints of this type?
5. What is synovial fluid? Where is it produced, and what are its functions?
6. What is a bursa? What is the difference between a bursa and a tendon sheath? What are the functions of each?
7. Identify four types of diarthrotic joints found in the wrist and hand region, and state the types of movement permitted by each.
8. Discuss the articulations of the pectoral and pelvic regions to the axial skeleton with regard to range of movement, ligamentous attachments, and potential clinical problems.
9. What is meant by a sprained ankle? How does a sprain differ from a strain or a luxation?
10. What occurs within the joint capsule in rheumatoid arthritis? How does rheumatoid arthritis differ from osteoarthritis?

12 Muscular System

Outline

Concepts

The muscles of the skeletal system are adapted to contract in order to carry out the functions of motion, heat production, and posture and body support.

Skeletal muscles begin development at four weeks after conception from blocks of mesoderm, called myotomes, in the trunk area, and from loosely organized mesenchyme, in the head and appendage areas. Each muscle fiber is derived from several embryonic myoblast cells.

Skeletal muscle tissue and its binding connective tissue are arranged in a highly organized pattern so that the forces of the contracting muscle fibers are united and directed onto the structure being moved.

Muscle fiber contraction in response to a motor impulse results from a sliding movement within the myofibrils in which the length of the sarcomeres is reduced.

The recruitment of motor units results in graded muscle contractions. Contraction is produced by sliding of filaments within the muscle fibers. Electrical excitation of muscle fibers is coupled to contraction through the effects of calcium ions.

Aerobic cellular respiration is required for the production of ATP needed for the cross-bridge activity of the various types of skeletal muscle fibers.

Skeletal muscles are named on the basis of shape, location, attachment, orientation of fibers, relative position, or function.

Muscles of the axial skeleton include those of facial expression, mastication, eye movement, tongue movement, neck movement, respiration, the abdominal wall, the pelvic outlet, and movement of the vertebral column.

The muscles of the appendicular skeleton include those of the pectoral girdle, arm, forearm, wrist, hand, and fingers, and those of the pelvic girdle, thigh, lower leg, ankle, foot, and toes.

Organization and General Functions

The muscles of the skeletal system are adapted to contract in order to carry out the functions of motion, heat production, and posture and body support.

Objective 1. Define the term *myology,* and describe the three principal functions of muscles.

Objective 2. Explain how muscles are described according to their anatomical location and cooperative function.

Myology is the study of muscles. Myology by itself does not have much meaning, however, except as it relates to the skeletal system and joints in the performance of body movements and to the nervous system in the performance of motor control. Muscles are dynamic, and an understanding of them requires an applied approach.

Functions of Muscles

Muscle cells (fibers) are adapted to contract when stimulated by electrochemical impulses. The stimulation of several fibers is not enough to cause a noticeable effect within a muscle, but these types of isolated fiber contractions are important and occur continuously within a muscle. When many fibers of a skeletal muscle are activated, the muscle contracts and causes body movement. Muscles perform three general functions: (1) motion, (2) heat production, and (3) posture and body support.

1. **Motion.** The most obvious type of motion performed by the skeletal muscles is to move the body and/or its appendages, as in walking, running, writing, chewing, and swallowing. The contraction of skeletal muscle is equally important in breathing and in moving body fluids. The stimulation of functional groups of fibers within the muscles maintains muscle tonus and is important in the movement of venous blood and lymphatic fluid. The eye and even the ear ossicles have associated skeletal muscles responsible for various movements.

   The primary impetus for blood flow is the contraction of cardiac muscles within the heart. All of the involuntary body systems (urinary, digestive, respiratory, circulatory, etc.) contain smooth muscles for the involuntary movement of materials through the body. The act of speaking involves several different muscle groups that control air volume, tense the vocal cords, move the tongue, and purse the lips. This is a highly coordinated effort of several muscle groups controlled by several different nerve centers.
2. **Heat production.** Body temperature is remarkably consistent. Metabolism within the cells releases heat as an end product. Since muscles constitute approximately 40 percent of the body weight and are in a continuous state of fiber activity, they are very important in the production of heat. The rate of heat production increases immensely as a person exercises strenuously.
3. **Posture and body support.** The skeletal system gives form and stability to the body, but skeletal muscles maintain posture and support around the flexible joints. Certain muscles are active postural muscles whose primary function is to work in opposition to gravity. Some postural muscles are working even when you think you are relaxed. As you are sitting, for example, the weight of your head is balanced at the atlantooccipital joint through the efforts of the muscles located at the back of the neck. If you start to get sleepy, your head will suddenly nod forward as the postural muscles relax and the weight (resistance) overcomes the effort.

There are three types of muscle tissue in the body: smooth, cardiac, and skeletal. Although these three types differ in structure and function, all muscle tissue possesses some basic characteristic properties:

1. **Irritability.** A muscle tissue receives and responds to a stimulus from a nerve.
2. **Contractility.** A muscle tissue responds to a stimulus by contracting, or shortening, its length.
3. **Extensibility.** Once a stimulus has subsided and the fibers within muscle tissue are shortened and relaxed, they may be stretched back to their original length by the contraction of an opposing muscle. The fibers are then stretched in readiness for another contraction.
4. **Elasticity.** Muscle tissue has a tendency to return to its initial length after being stretched.

A histological description of each of the three muscle types was presented in chapter 6 and should be reviewed at this time. Cardiac muscle is further discussed with the heart in chapter 20. Smooth muscle functions with various involuntary organs throughout the body and is discussed, when appropriate, with these organs. The remaining information presented in this chapter pertains only to skeletal muscle and the skeletal muscular system of the body.

Skeletal muscle constitutes a body system by itself and accounts for approximately 40% of body weight. Over six hundred individual muscles comprise the skeletal muscular system. The principal superficial muscles are shown in figures 12.1 and 12.2. Muscles are usually described in groups according to their anatomical location and cooperative function. The *muscles of the axial skeleton* have

myology: Gk. *myos,* muscle; *logos,* study of
muscle: L. *muris,* mouse (from the appearance of certain muscles)

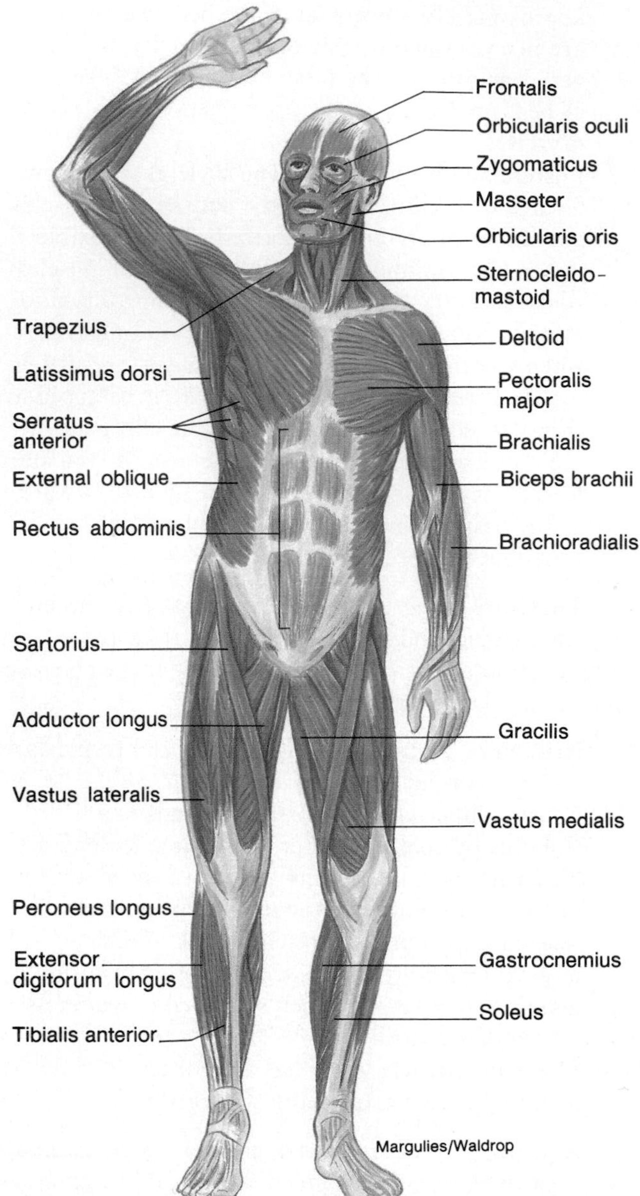

Figure 12.1. An anterior view of some of the superficial skeletal muscles.

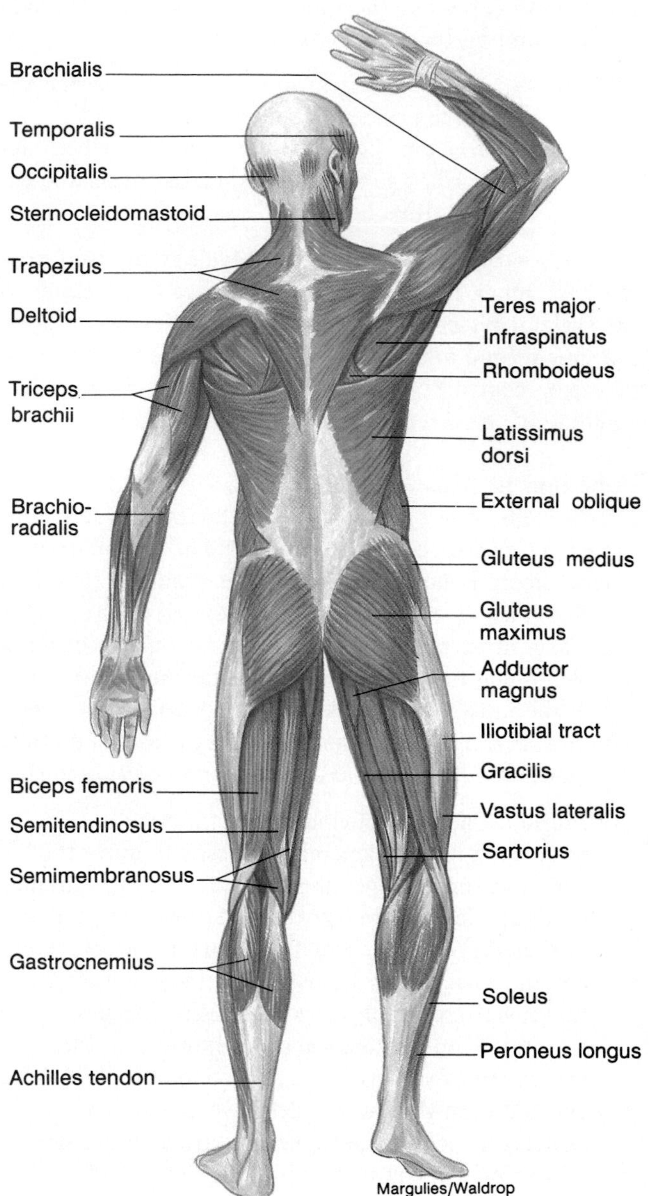

Figure 12.2. A posterior view of some of the superficial skeletal muscles.

their attachments to the bones of the axial skeleton and include facial muscles, neck muscles, and anterior and posterior trunk muscles. The *muscles of the appendicular skeleton* include those that act on the pectoral and pelvic girdles and those that move the segments of the appendages.

1. Discuss how the functions of muscles aid in maintaining body homeostasis.
2. What is meant by a postural muscle?
3. Distinguish between the axial and the appendicular muscles.

Development of Skeletal Muscles

Skeletal muscles begin development at four weeks after conception from blocks of mesoderm, called myotomes, in the trunk area, and from loosely organized mesenchyme, in the head and appendage areas. Each muscle fiber is derived from several embryonic myoblast cells.

Objective 3. Describe the formation of a muscle fiber from myoblast cells.

Objective 4. Describe the development of skeletal muscles from myotomes.

Figure 12.3. The development of skeletal muscle fibers. (*a*) at four weeks, mesodermal cells differentiate into myoblasts that aggregate into a developing syncytial tube. (*b*) at five weeks, the syncytial myotube is formed as individual plasma membranes are broken down and longitudinal myofilaments appear. Myotubes grow in length by incorporating additional myoblasts; each adds an additional nucleus. (*c*) muscle fibers are distinct at nine weeks, but the nuclei are still centrally located, and growth in length continues through the addition of myoblasts. (*d*) at five months, thin (actin) and thick (myosin) myofilaments are present, and moderate growth in length still continues. (*e*) by birth, the striated myofilaments have aggregated into bundles, the fiber has thickened, and the nuclei have shifted to the periphery. Myoblast activity ceases, and all the muscle fibers a person will have are formed.

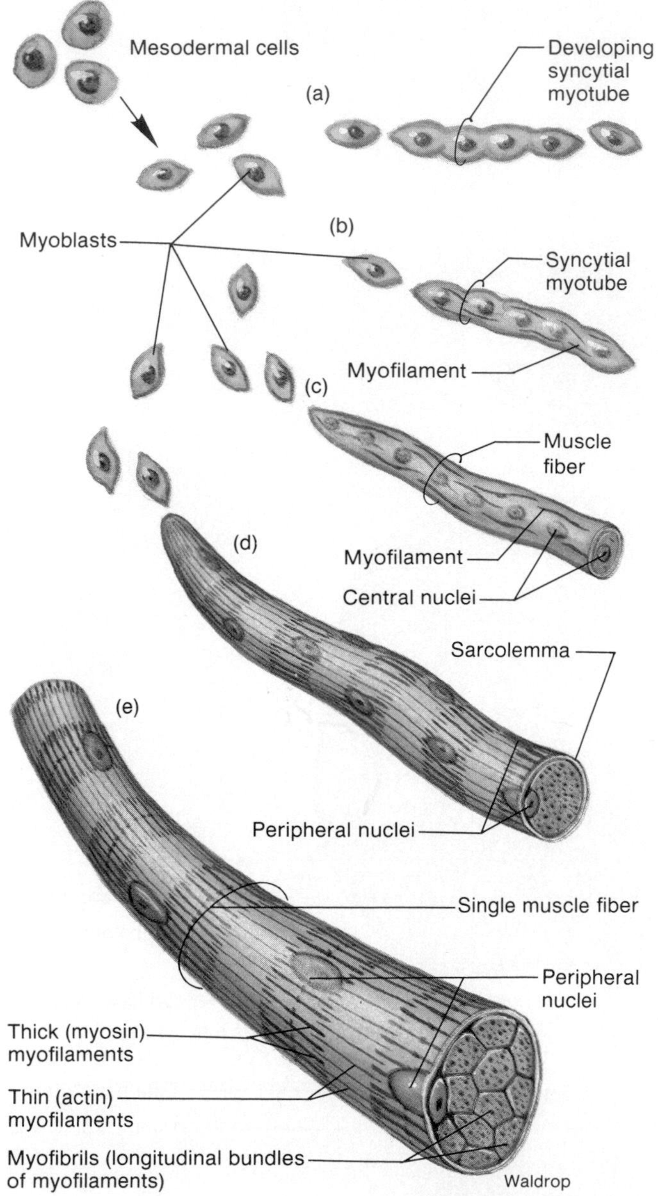

The formation of skeletal muscle tissue begins during the fourth week of embryonic development as specialized mesodermal cells, called **myoblasts** *(mi'o-blasts'')*, begin rapid mitotic division (fig. 12.3). The proliferation of new cells continues while the myoblast cells fuse together into **syncytial** *(sin-sish'al)* **myotubes.** A syncytium is a multinucleated protoplasmic mass formed by the union of originally separate cells. At nine weeks, primitive myofilaments course through the myotubes and the nuclei of the contributing myoblasts are centrally located. Growth in length continues through the addition of myoblasts.

It is not certain when skeletal muscle is sufficiently developed to sustain contractions, but by the seventeenth week the fetal movements known as **quickening** are strong enough to be recognized by the mother. At this time, the individual muscle fibers have thickened, the nuclei have moved peripherally, and the filaments (myofilaments) have differentiated into alternating thin *(actin)* and thick *(myosin)* bands. Growth in length still continues through addition of myoblasts.

Shortly before a baby is born, the formation of myoblast cells ceases, and all of the muscle cells of a person have been determined. Differences in strength, endurance, and coordination are somewhat genetically determined but are primarily the result of individual body conditioning. Muscle coordination is an ongoing process of achieving a fine neural control of muscle fibers. A mastery of muscle movement is comparatively slow in humans. It is several months before a newborn has the coordination to crawl, and about a year before it achieves bipedal posture and locomotion. By contrast, most mammals can walk and run within a few hours after they are born. Refined, precise movements of the hands for delicate work, such as playing a violin or performing surgery, may be learned only after many hours of disciplined practice, and some persons may not be able to develop these movements at all.

The process of muscle fiber development occurs within specialized mesodermal masses, called **myotomes** *(mi'o-tōms)*, in the embryonic trunk area and from loosely organized masses of mesoderm in the head and appendage areas. At six weeks, the trunk of an embryo is segmented into distinct myotomes (fig. 12.4) that are associated dorsally with specific sclerotomes. Sclerotomes give rise to vertebrae. As will be explained in chapter 16, spinal nerves arise from the spinal cord and exit between vertebrae to innervate (serve with motor and sensory nerve branches) developing muscles in the adjacent myotomes. As myotomes develop, they elongate anteriorly toward the midline of the body, or distally into the developing limbs. The muscles of the entire muscular system are differentiated and in correct position by the eighth week (fig. 12.4). The orientation of the developing muscles is preceded and influenced by cartilaginous models of bones.

1. Describe the process by which muscle fibers become multinucleated.
2. Distinguish between the terms *myoblasts, myotubes,* and *muscle fibers.*
3. Briefly describe the embryonic events that occur between weeks six and eight in the formation of specific muscles.

myoblast: Gk. *myos,* muscle; *blastos,* germ

syncytial: Gk. *syn,* with; *cyto,* cell

Figure 12.4. The development of skeletal muscles. (*a*) the distribution of embryonic myotomes at six weeks. Segmental myotomes give rise to the muscles of the trunk area and girdles. Loosely organized masses of mesoderm form the muscles of the head and extremities. (*b*) the arrangement of skeletal muscles at eight weeks. The development of muscles is influenced by the preceding cartilaginous models of bones. The innervation of muscles corresponds to the development of spinal nerves and dermatome arrangement.

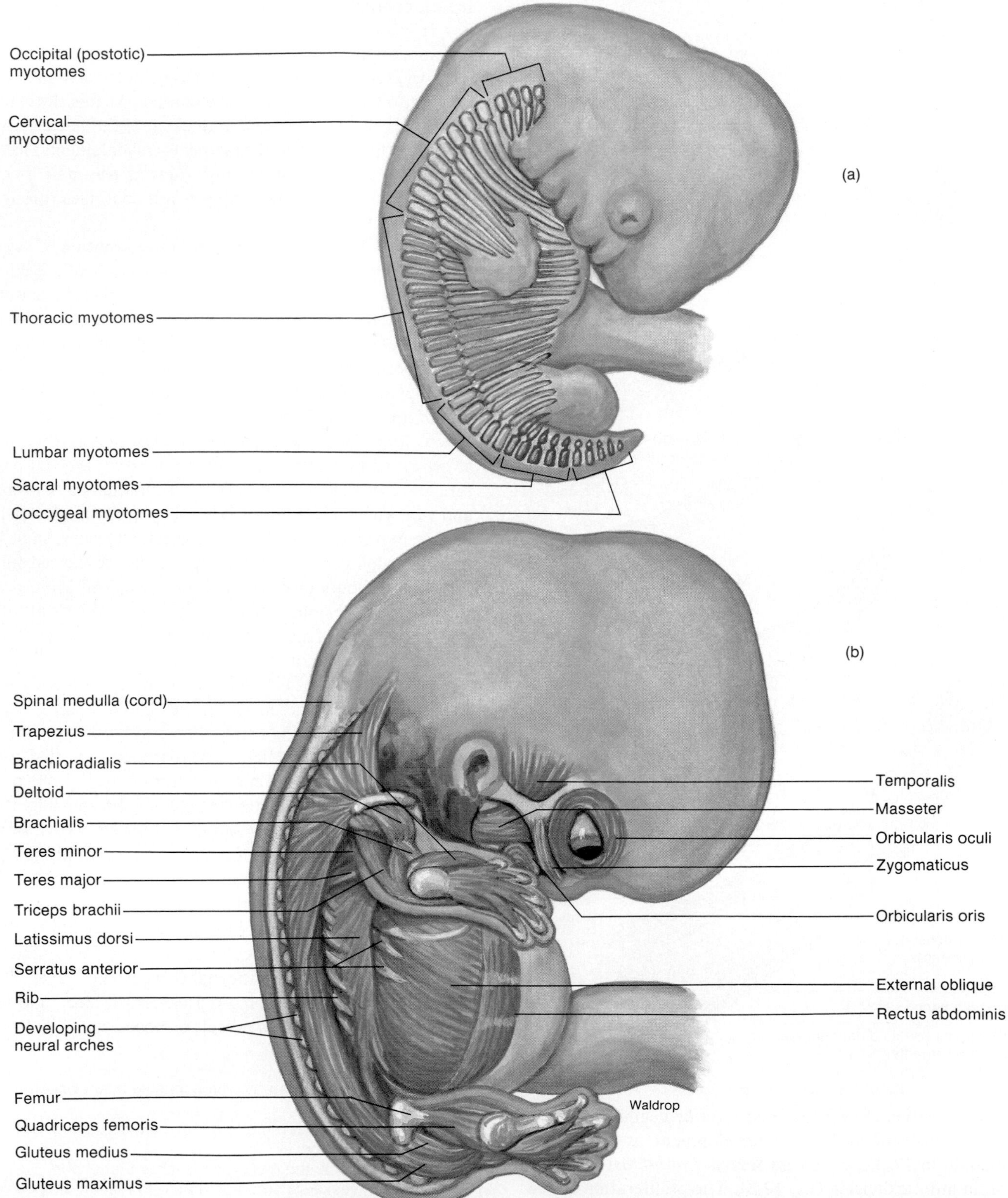

Structure of Skeletal Muscles

Skeletal muscle tissue and its binding connective tissue are arranged in a highly organized pattern so that the forces of the contracting muscle fibers are united and directed onto the structure being moved.

Objective 5. Distinguish between the various binding connective tissues associated with skeletal muscles.

Objective 6. Explain what is meant by synergistic and antagonistic muscle groups.

Objective 7. Describe the various types of muscle fiber architecture, and discuss the biomechanical advantage of each.

Attachments

Skeletal muscles are usually long and narrow, span a joint, and are attached to a bone at either end by a tendon (fig. 12.5). As the muscle contracts, one of the bones moves relative to the other at the joint. The more fixed, or stationary, attachment is designated as the **origin** of a muscle, whereas the movable end is its **insertion.** In muscles associated with the girdles and appendages, the origin is generally the proximal attachment and the insertion the distal attachment. The fleshy, thickened portion of a muscle is referred to as its **belly,** or **gaster.** Generally the belly of a muscle is located on the bone proximal to the bone that is to be moved. The joint is spanned by a **tendon** from the muscle. A tendon is a structure, composed of dense fibrous connective tissue, that connects a muscle to the periosteum of a bone. It functions to transfer the force of contraction from the muscle across the joint and onto the bone that is to be moved. Flattened, sheetlike tendons are called **aponeuroses** *(ap''o-nu-ro'sēz)* and occur where attachment is over a broad line. In certain places, especially in the wrist and ankle, the tendons are not only enclosed by protective **tendon sheaths,** which lubricate the tendons with synovial fluid, but the entire group of tendons is covered by a thin but strong band of connective tissue called a **retinaculum** *(ret''ĭ-nak'u-lum)* (see fig. 12.47).

Associated Connective Tissue

Contracting muscle fibers would not be effective if they worked as isolated units. Each fiber is bound to adjacent fibers to form bundles, and the bundles in turn are bound to other bundles. With this arrangement, the contraction in one area of a muscle works in conjunction with contracting fibers elsewhere in the muscle. The binding substances within muscles are the associated connective tissues.

aponeurosis: Gk. *aponeurosis,* change into a tendon
retinaculum: L. *retinere,* to hold back (retain)

Figure 12.5. The skeletomuscular relationship. The more proximal and fixed point of muscle attachment is the origin, whereas the distal, maneuverable point of attachment is the insertion. The contraction of muscle fibers causes one bone to move relative to another around a joint.

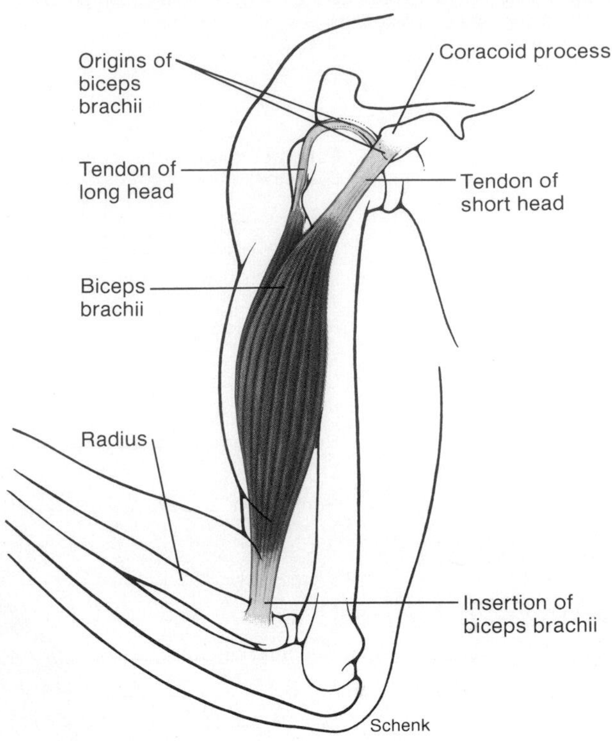

Connective tissue is structurally arranged within muscle to protect, strengthen, and bind muscle fibers into bundles and bind the bundles together (fig. 12.6). The individual fibers of skeletal muscles are surrounded by fine sheaths of connective tissue called **endomysium** *(en''do-mis'e-um).* The endomysium binds adjacent fibers and supports capillaries and nerve endings serving the muscle. Another connective tissue, called **perimysium,** binds groups of fibers together into bundles called **fasciculi** *(fah-sik'u-li).* The perimysium supports blood vessels and nerve fibers serving the various fasciculi. The entire muscle is covered by the **epimysium,** which in turn is continuous with a tendon.

A fibrous connective tissue, called **fascia** *(fash'e-ah),* is continuous with the epimysium and binds separate muscles together. Fascia may be divided into *superficial fascia* and *deep fascia.* Superficial fascia is a packinglike tissue that secures the hypodermis of the skin to the underlying muscles. It varies in thickness throughout the body. For

endomysium: Gk. *endon,* within; *myos,* muscle
perimysium: Gk. *peri,* around; *myos,* muscle
fasciculus: L. *fascis,* bundle
epimysium: Gk. *epi,* upon; *myos,* muscle
fascia: L. *fascia,* a band or girdle

Figure 12.6. The relationship between muscle tissue and connective tissue. (*a*) the fascia and tendon attaches a muscle to the periosteum of a bone. (*b*) the epimysium surrounds the entire muscle, and the perimysium separates and binds the fasciculi (muscle bundles). (*c*) the endomysium surrounds and binds individual muscle fibers. (*d*) an individual muscle fiber is composed of myofibrils consisting of filaments of actin and myosin.

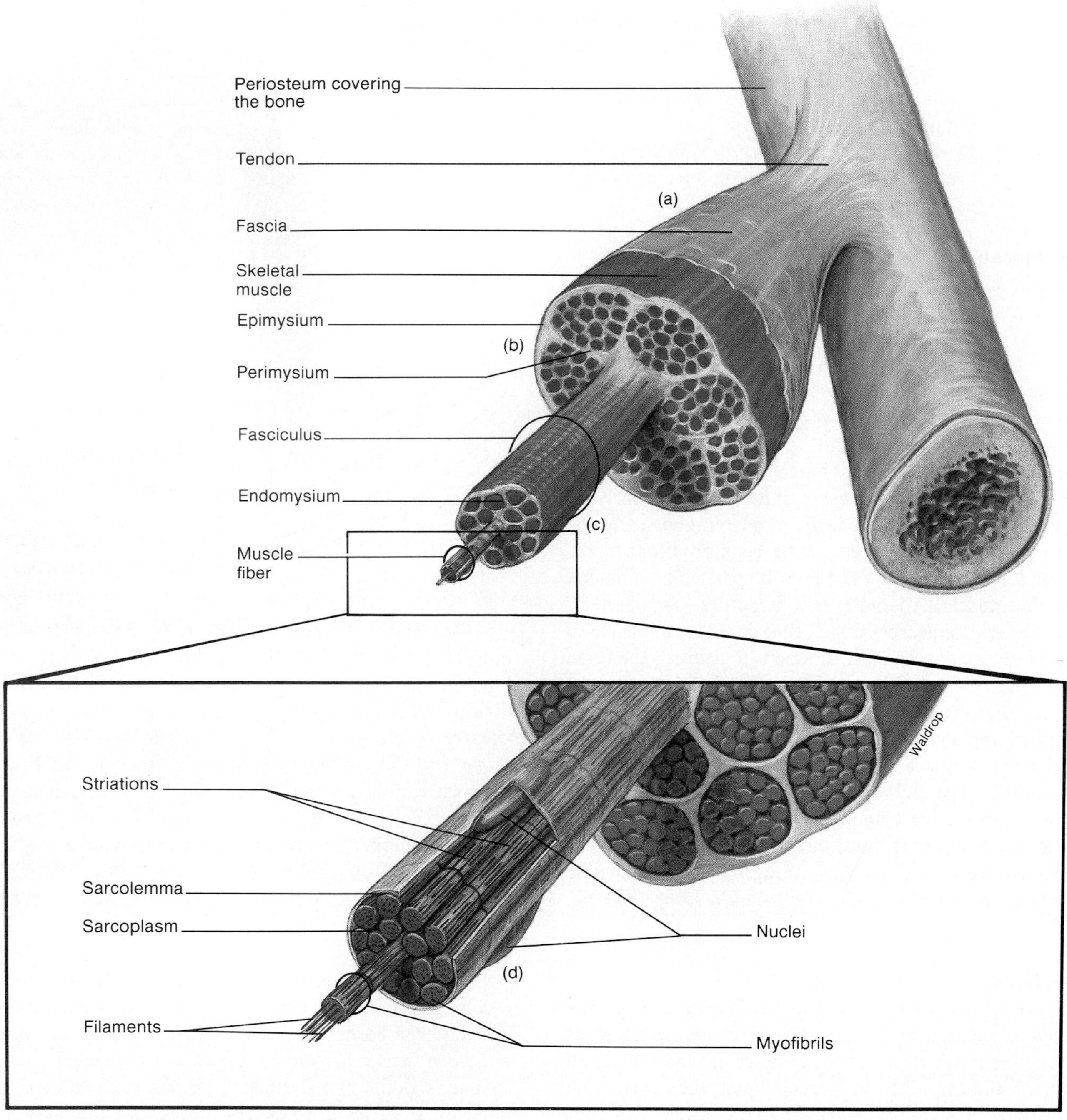

Figure 12.7. Examples of synergistic and antagonistic muscles. The two heads of the biceps are synergistic to each other, as are the three heads of the triceps. The biceps is antagonistic to the triceps, and the triceps is antagonistic to the biceps. When one antagonistic group contracts, the other one must relax, or movement does not occur.

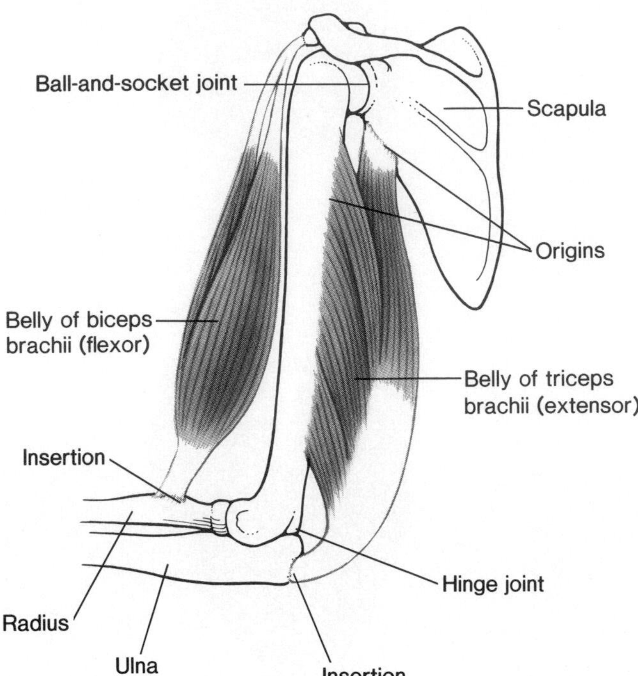

example, superficial fascia over the buttocks and anterior abdominal wall is thick and laced with adipose tissue. The superficial fascia under the skin of the dorsum of the hand and facial region is thin. Deep fascia lacks adipose tissue, occurs between individual muscle, and also surrounds adjacent muscles to bind them into functional groups.

Muscle Groups

Just as individual muscle fibers seldom contract independently, muscles generally do not contract separately but work as functional groups. Muscles that contract together and are coordinated in accomplishing a particular movement are said to be *synergistic* (fig. 12.7). *Antagonistic* muscles perform opposite functions and are generally located on the opposite sides of the limb. The two heads of the biceps brachii muscle along with the brachialis muscle, for example, contract together to *flex* the elbow joint. The triceps brachii muscle, the antagonist to the biceps brachii and brachialis muscles, *extends* the elbow as it is contracted.

synergistic: Gk. *synergein,* cooperate
antagonistic: Gk. *antagonistes,* struggle against

Seldom does a single muscle cause a movement at a joint. A division of labor is achieved by having several synergistic muscles rather than one massive muscle. One muscle may be an important postural muscle, for example, whereas another may be adapted for rapid, powerful contraction. When total output of the synergistic muscles is required, the contraction of several smaller muscles provides more strength than the contraction of one large one.

Muscle Architecture

Skeletal muscles may be classified on the basis of fiber arrangement (fig. 12.8). Certain advantages are inherent in each of the types of muscles. The following are the major types of fiber arrangement.

1. **Parallel,** or **fusiform.** Parallel muscles have relatively long excursions (contract over a great distance) and good endurance but are not especially strong. They are long, straplike muscles such as the sartorius and rectus abdominis muscles.
2. **Convergent.** Convergent-fibered muscles are so named because the fibers converge at the insertion point to maximize contraction. They are fan-shaped muscles such as the deltoid and pectoralis major muscles.
3. **Pennate.** Pennate-fibered muscles have many fibers per unit area and hence are strong. They provide dexterity. They have short excursions but generally tire quickly. There are three kinds of pennate-fibered muscles: *unipennate, bipennate,* and *multipennate.* Examples of each of the pennate types can be found in the forearm.
4. **Circular.** Circular-fibered muscles surround a body opening, or **orifice,** and act as a sphincter when contracted. Examples are the orbicularis oculi around the eye and orbicularis oris surrounding the mouth.

Muscle-fiber architecture can be observed on a cadaver or other dissection specimen. If you have the opportunity to learn the muscles of the body from a cadaver, observe the fiber architecture of specific muscles, and try to determine the advantages afforded to each muscle by its location and action.

pennate: L. *pennatus,* feather
orifice: L. *orificium,* mouth; *facere,* to make

Figure 12.8. Skeletal muscle architecture.

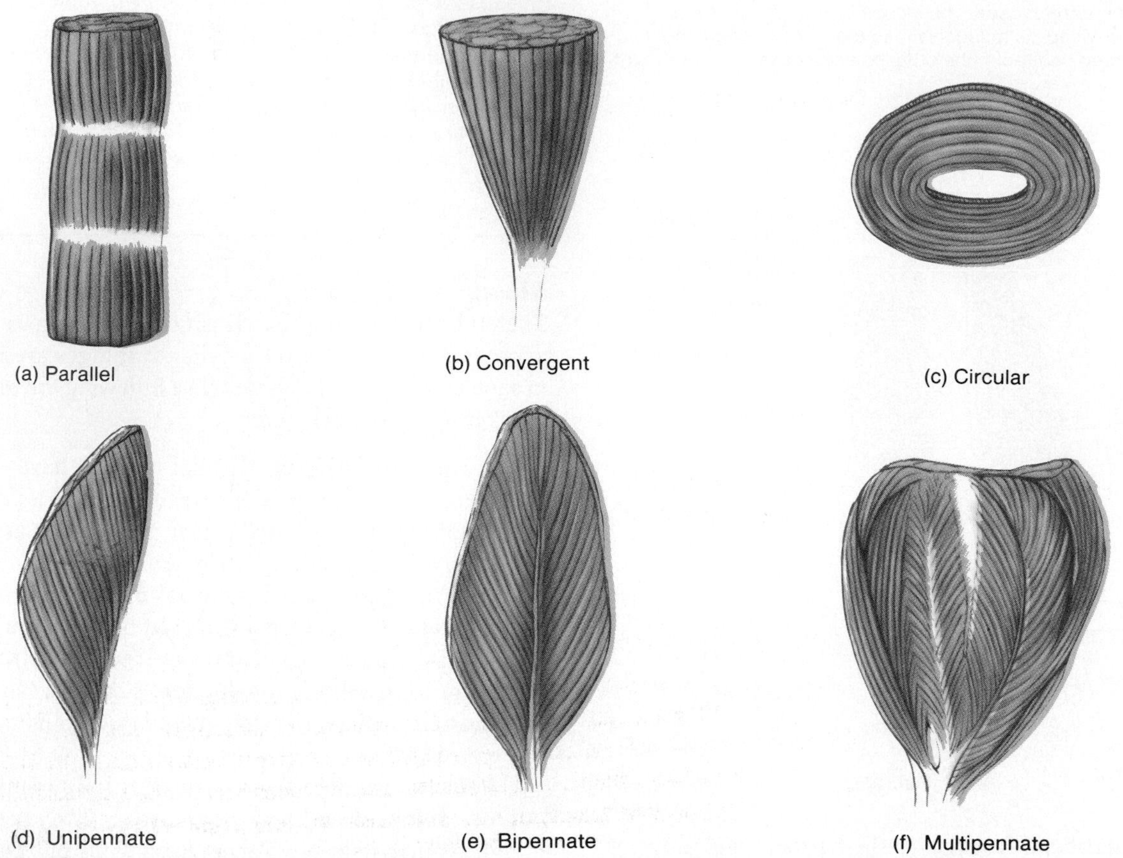

Blood and Nerve Supply to Skeletal Muscle

Muscle cells have a high rate of metabolic activity and therefore require extensive vascularity to receive nutrients and oxygen and to eliminate waste products. Smaller muscles generally have a single artery supplying blood and perhaps two veins returning blood (fig. 12.9). Large muscles may have several arteries and veins. The microscopic capillary exchange between arteries and veins occurs throughout the endomysium that surrounds individual fibers.

A skeletal muscle fiber cannot contract unless it is stimulated by a nerve. This means that there must be extensive innervation to a muscle to ensure the connection of each muscle fiber to a nerve cell. Actually there are two nerve pathways associated with each muscle. A **motor (efferent) neuron** is a nerve cell that conducts nerve impulses to the muscle fiber to stimulate it to contract. A **sensory (afferent) neuron** conducts nerve impulses away from the muscle fiber to the central nervous system, conveying information about the muscle's state of contraction. Muscle fibers will atrophy if they are not periodically stimulated to contract. The names of both the nerves and vessels usually refer to the muscle they serve.

The feeling of stiffness the day after strenuous exercise is due to trauma to muscle fibers and the accumulation of lactic acid within the muscle. Lactic acid is a product of anaerobic respiration (chapter 4) and is removed by the venous flow from the muscle. If a person is in good physical condition, the vascularity can readily remove the metabolic wastes, and stiffness is not as apt to occur. Being in good physical condition not only improves vascularity but enlarges muscle fibers and allows them to work more efficiently over a longer duration.

1. Contrast the following terms: *endomysium* and *epimysium; fascia* and *tendon; aponeurosis* and *retinaculum.*
2. Discuss the biomechanical advantage of having synergistic muscles. Give some examples of synergistic muscles, and state which muscles are antagonistic.
3. Which type of architecture provides dexterity and strength?

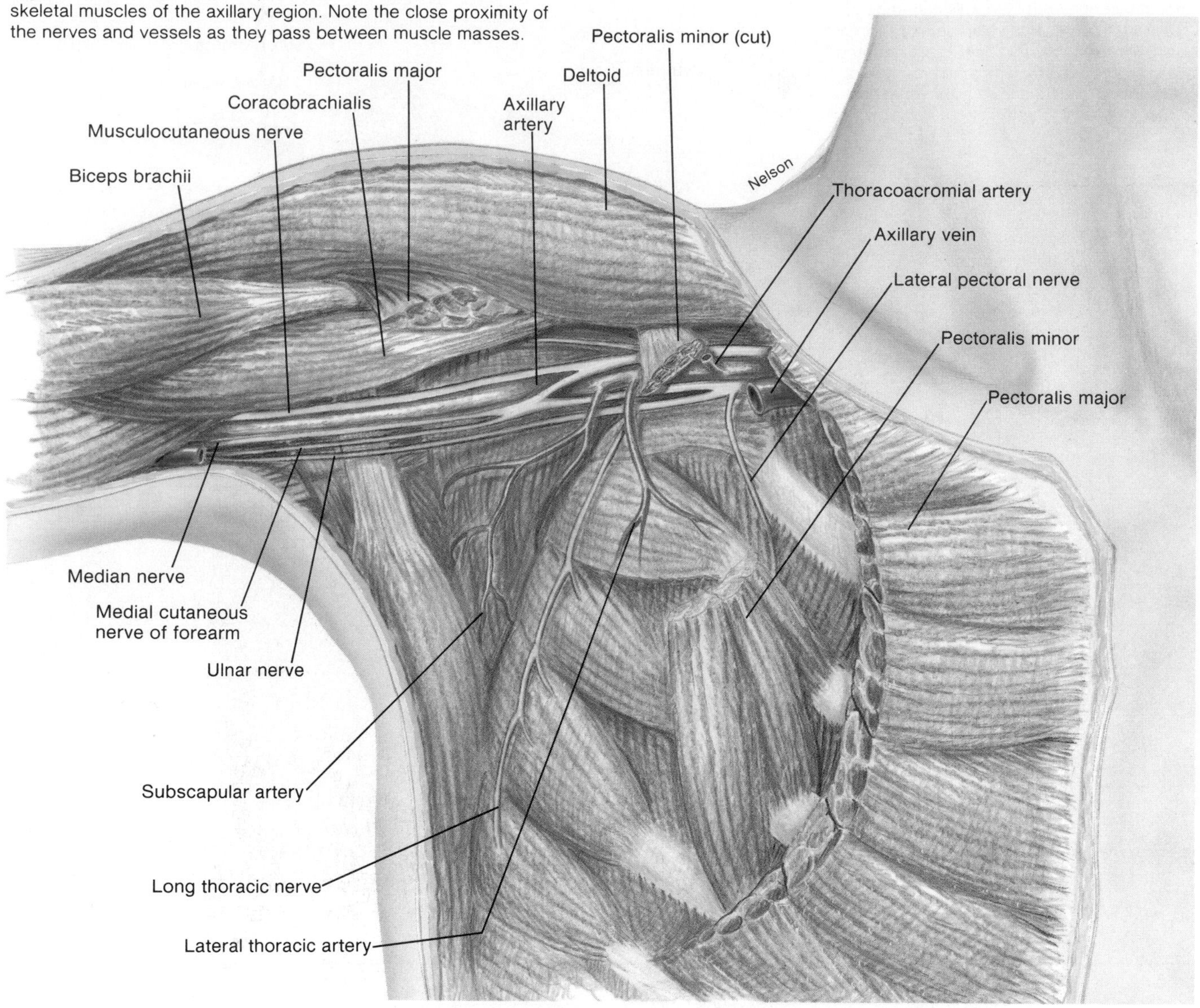

Figure 12.9. The relationship of blood vessels and nerves to skeletal muscles of the axillary region. Note the close proximity of the nerves and vessels as they pass between muscle masses.

Skeletal Muscle Fibers and Mode of Contraction

In response to stimulation by a motor neuron, muscle fibers contract as the myofibrils within them attempt to shorten. Shortening in the myofibrils is produced by shortening of their sarcomeres, which results from sliding of the myofilaments.

Objective 8. Identify the major components of a muscle fiber, and discuss the function of each part.

Objective 9. Discuss the sliding filament theory of muscle contraction.

Objective 10. Describe the length-tension relationship of muscle contraction, and explain how the relationship is produced.

Objective 11. Contrast isotonic and isometric contractions.

Skeletal Muscle Fibers

A skeletal muscle fiber is an elongated, cylindrical cell that may reach a length of 30 cm (12 in.) and have a diameter of 10 to 100 μm. Despite their unusual shape, muscle fibers have the same organelles that are present in other cells: mitochondria, intracellular membranes, glycogen granules, and others. Unlike other cells in the body, however, skeletal muscle fibers are multinucleated and striated (fig. 12.10). Each fiber is surrounded by a cell membrane, called a **sarcolemma,** and the cytoplasm within the cell is called **sarcoplasm.** A network of membranous channels,

sarcolemma: Gk. *sarkos,* flesh; *lemma,* to peel, husk

Figure 12.10. (*a*) a skeletal muscle fiber is composed of numerous threadlike strands of myofibrils that contain the filaments of actin and myosin. A skeletal muscle fiber is striated and multinucleated. (*b*) an electron micrograph of skeletal muscle fibers showing the striations and the peripheral location of the nuclei.

Nuclei
Sarcolemma
Sarcoplasm
Filaments
Myofibrils
Striations
(a)

Nuclei
Muscle fiber
(b)

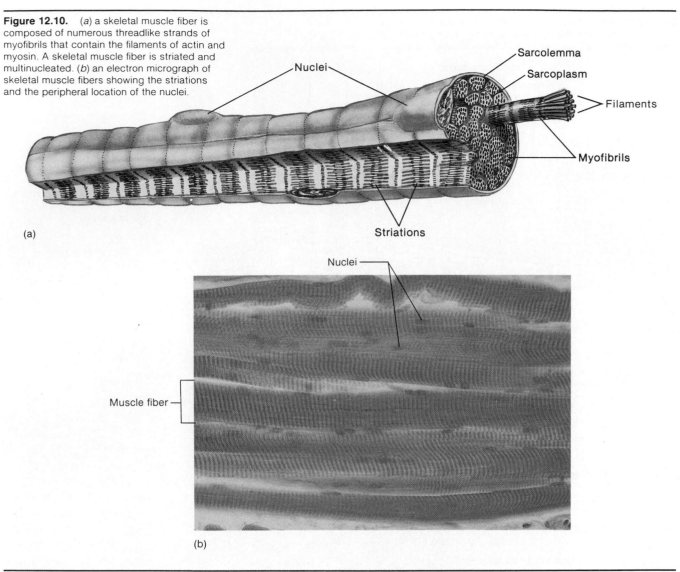

Figure 12.11. The structural relationship of the myofibrils of a muscle fiber to the sarcolemma, transverse tubules, and sarcoplasmic reticulum. Note the position of the mitochondria.

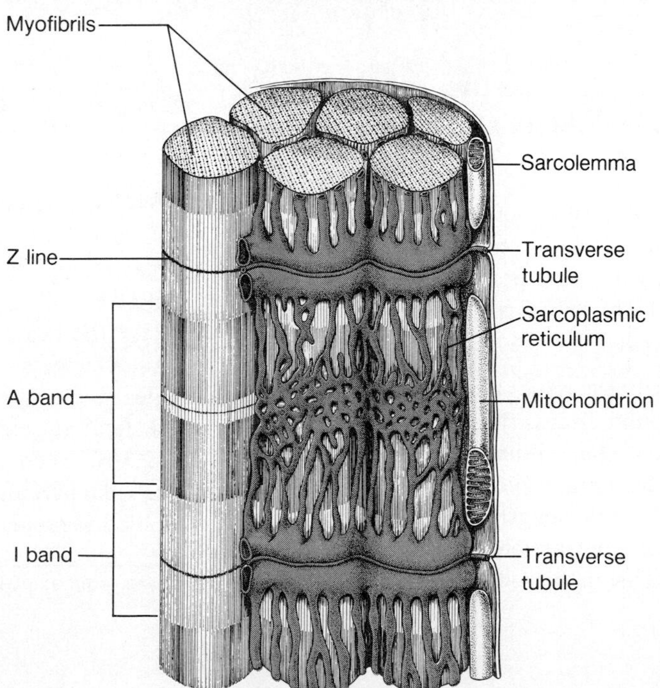

Figure 12.12. (*a*) a skeletal muscle fiber contains many threadlike structures called myofibrils, each arranged (*b*) into compartments called sarcomeres. (*c*) the characteristic striations of a sarcomere are due to the arrangement of thin and thick filaments composed of actin and myosin, respectively.

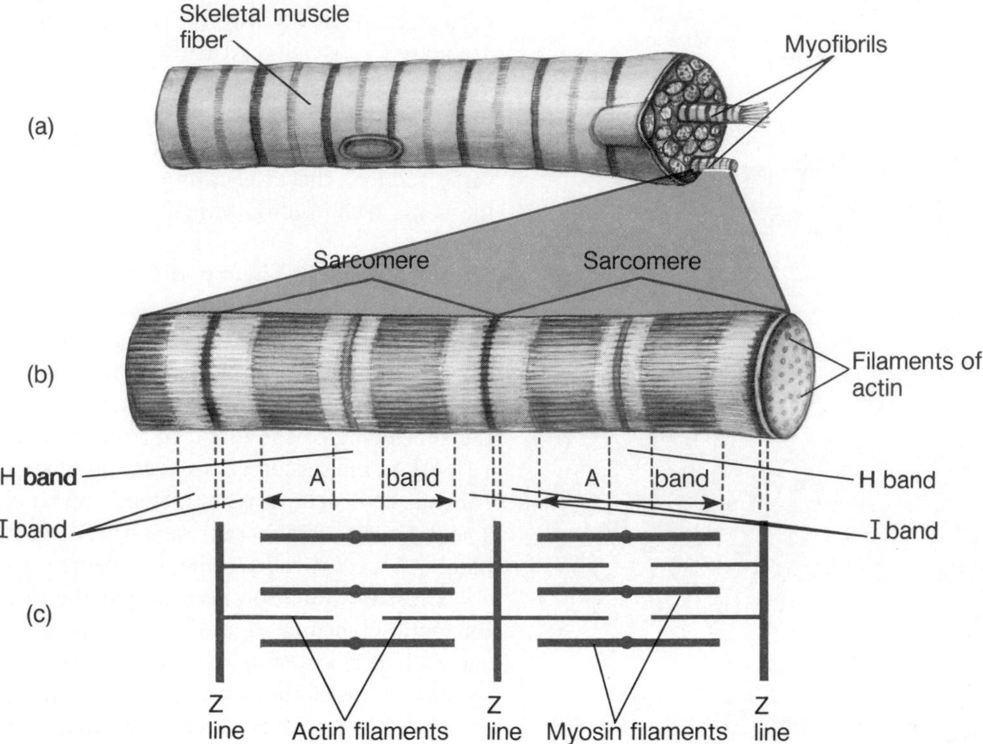

called the **sarcoplasmic reticulum,** extends throughout the sarcoplasm (fig. 12.11). A system of **transverse tubules** (T-tubules) is formed by narrow invaginations of the sarcolemma. These tubules run perpendicular to the sarcoplasmic reticulum and are open to the extracellular fluid. Embedded in the sarcoplasm and extending the entire length of the fiber are many parallel, threadlike structures called **myofibrils** *(mi''o-fi'brils)* (fig. 12.12). Each myofibril is composed of even smaller protein strands called **myofilaments.** *Thin filaments* are about 6nm in diameter and are composed of the protein **actin** *(ak'tin). Thick filaments* are about 16nm in diameter and are composed mostly of the protein **myosin** *(mi'o-sin).*

The characteristic dark and light striations of skeletal muscle myofibrils are due to the arrangement of these filaments. The dark crossbands, produced by overlapping myosin filaments, are able to polarize visible light (are therefore anisotropic) and have been named *A bands.* The light crossbands, composed of thin actin filaments, do not polarize light (are therefore isotropic) and are called *I bands.*

The I bands within a myofibril extend from the edge of one stack of thick myosin filaments to the edge of the next stack of thick filaments. They are light bands because they contain only thin actin filaments. Each thin filament, however, extends partway into the A bands on each side (between the thick filaments of each stack). Since thick and thin filaments overlap at the edges of each A band, the edges are darker in appearance than the central region of the A band. These central lighter regions of the A bands are called the *H bands* (for *helle,* a German word for bright). The central H bands thus contain only myosin that is not overlapped with thin filaments.

At high magnification, thin dark lines can be seen in the middle of the I bands. These are labeled *Z lines* (for *Zwischenscheibe,* a German word meaning "between disc"). The arrangement of thick and thin filaments between a pair of Z lines forms a repeating pattern that serves as the basic subunit of skeletal muscle contraction. These subunits, from Z line to Z line, are known as **sarcomeres** *(sar'ko-mērs).* A longitudinal section of a myofibril thus presents a side view of successive sarcomeres (fig. 12.13a,b).

---

actin: L. *actus,* motion, doing
myosin: L. *myosin,* within muscle
anisotropic: Gk. *anisos,* uneven; *tropos,* a turning
isotropic: Gk. *isos,* equal; *tropos,* a turning

Figure 12.13. Electron micrographs of myofibrils of a muscle fiber. (*a*) at low power (1,600×), a single muscle fiber containing numerous myofibrils. (*b*) at high power (53,000×), myofibrils in longitudinal section. Notice the sarcomeres and overlapping thick and thin filaments. (*c*) the hexagonal arrangement of thick and thin filaments as seen in cross section (arrows point to cross-bridges; SR=sarcoplasmic reticulum). ([*c*] From *Tissues and Organs: A Text Atlas of Scanning Electron Microscopy* by R. G. Kessel and R. Kardon. W. H. Freeman and Company. © 1979.)

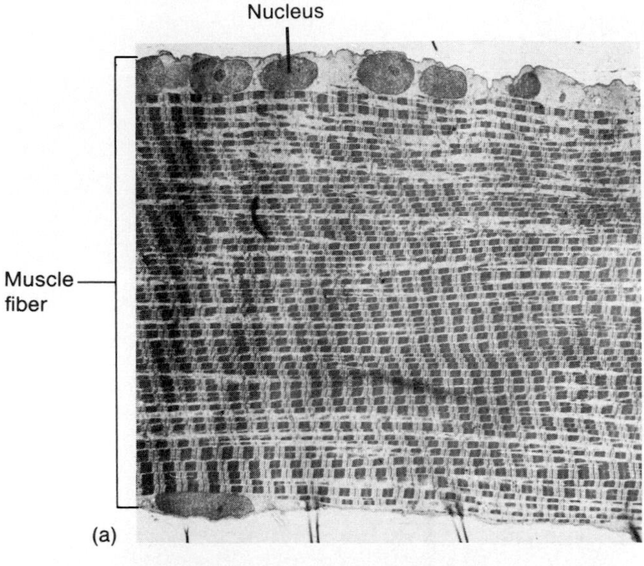

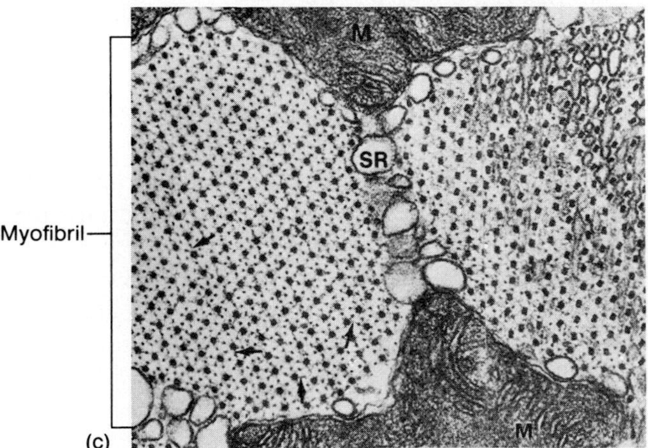

Kessel and Kardon

This side view is, in a sense, misleading; there are numerous sarcomeres within each myofibril that are out of the plane of the section (and out of the picture). A better appreciation of the three-dimensional structure of a myofibril can be obtained by viewing the myofibril in transverse section. In this view, it can be seen that the Z lines are actually disc shaped and that the thin filaments that penetrate these Z discs surround the thick filaments in a hexagonal arrangement (see fig. 12.13c). If one concentrates on a single row of dark thick filaments in this transverse section, the alternating pattern of thick and thin filaments seen in longitudinal section becomes apparent.

Sliding Filament Theory of Contraction

When a muscle is stimulated to contract, it decreases in length as a result of the shortening of its individual fibers. Shortening of the muscle fibers, in turn, is produced by shortening of their myofibrils, which occurs as a result of the shortening of the distance from Z line to Z line (fig. 12.14). As the sarcomeres shorten in length, however, the A bands do *not* shorten but instead appear closer together. The I bands—which represent the distance between A bands of successive sarcomeres—decrease in length.

Close examination reveals that the thick myosin and thin actin filaments remain the same length during muscle contraction. Shortening of the sarcomeres is produced, not by shortening of the filaments, but rather by the *sliding* of thin filaments over thick filaments. In the process of contraction, the thin filaments on either side of the A bands extend deeper and deeper toward the center, producing increasing amounts of overlap with the thick filaments. The central H bands thus get shorter and shorter during contraction.

Cross-Bridges. Sliding of the filaments is produced by the action of numerous **cross-bridges** that extend out from the myosin toward the actin. These cross-bridges are part of the myosin proteins that extend from the axis of the thick filaments to form "arms" that terminate in globular "heads" (fig. 12.15). The orientation of cross-bridges on one side of a sarcomere is opposite to that on the other side, so that when they attach to actin on each side of the sarcomere they can pull the actin from each side toward the center.

Relaxed muscles are easily stretched (although this is opposed *in vivo* by the stretch reflex), demonstrating that the myosin cross-bridges are not attached to actin when the muscle is at rest. Each globular head of a cross-bridge contains an ATP-bonding site and an actin-binding site (fig. 12.16). The globular heads function as **myosin ATPase** enzymes, splitting ATP into ADP and P_i. These products remain attached to their binding sites as the cross-bridges, which were activated by the splitting of ATP, bind with actin.

Figure 12.14. The sliding filament model of contraction. As the filaments slide, the Z lines are brought closer together. The A bands remain the same length during contraction, but the I and H bands get progressively narrower and may eventually become obliterated.

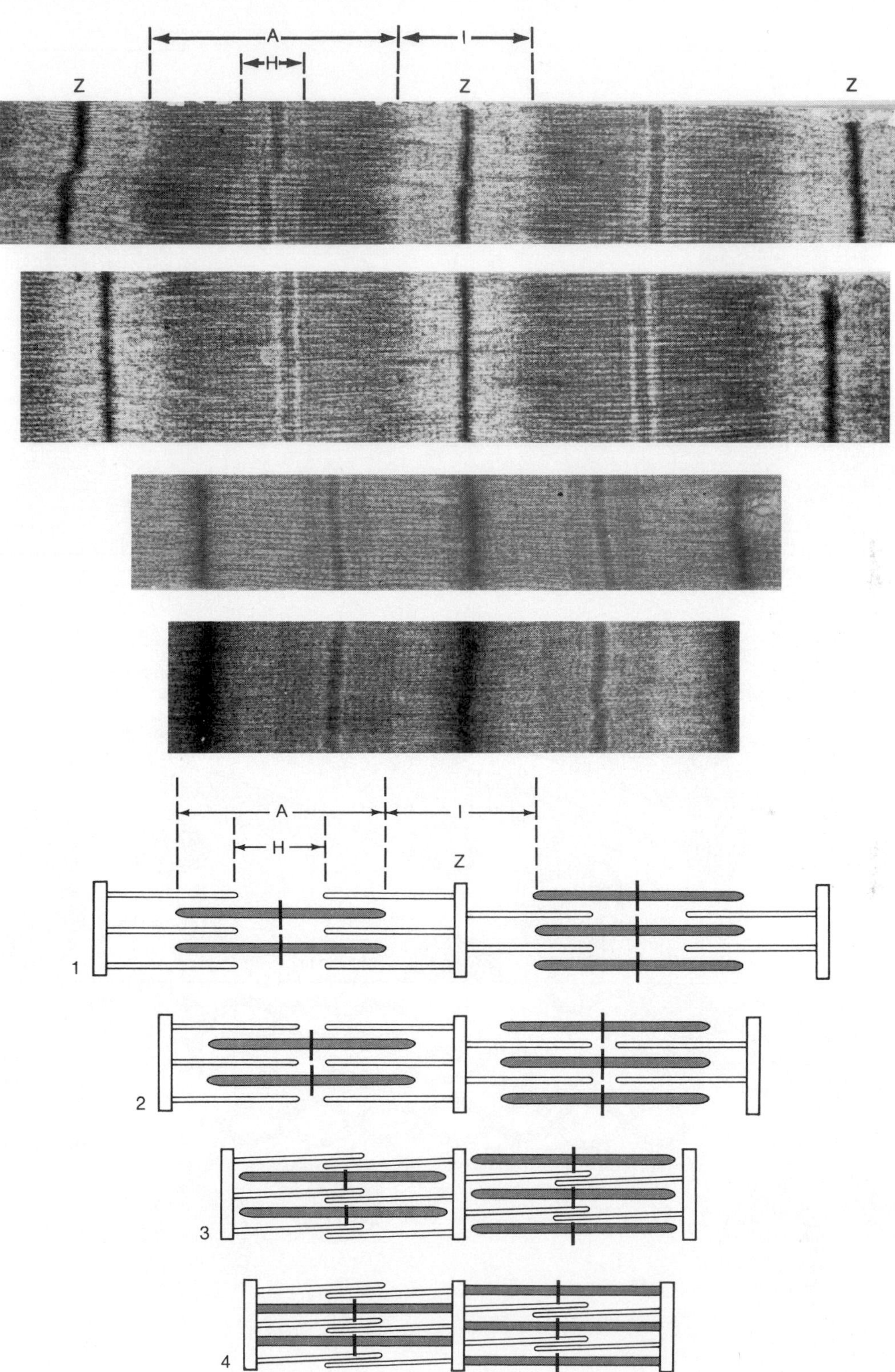

Figure 12.15. Myosin cross-bridges are oriented in opposite directions on either side of a sarcomere.

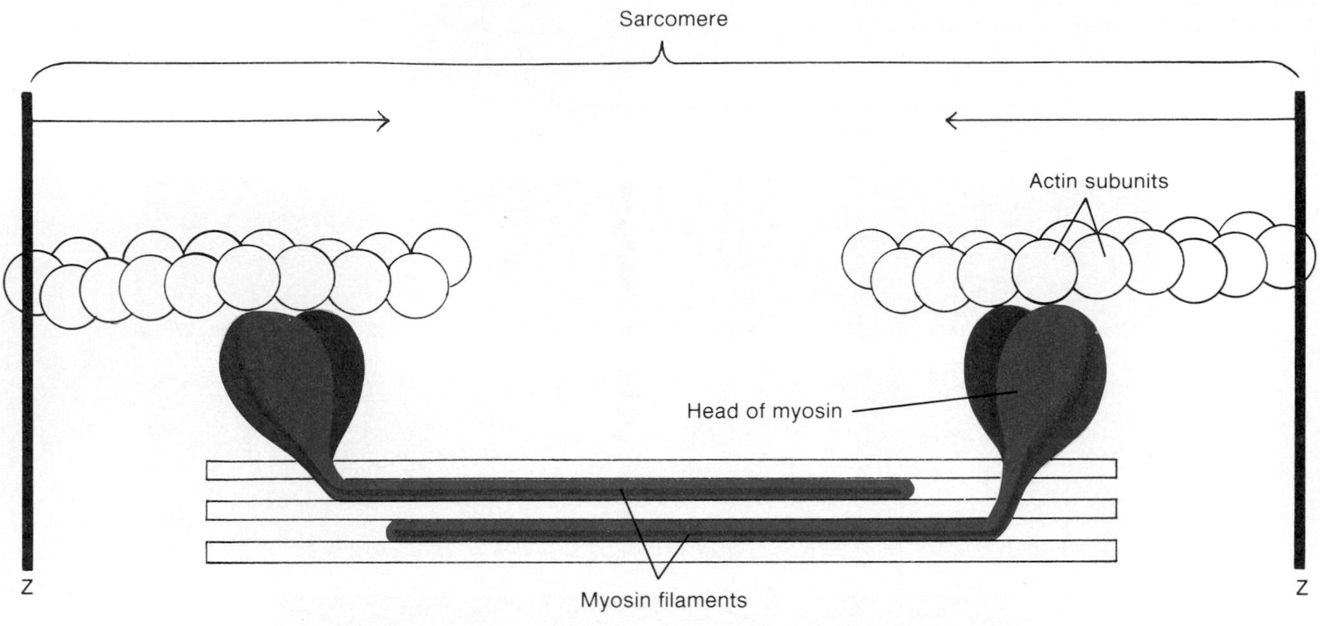

Figure 12.16. The structure of myosin, showing its binding sites for ATP and for actin.

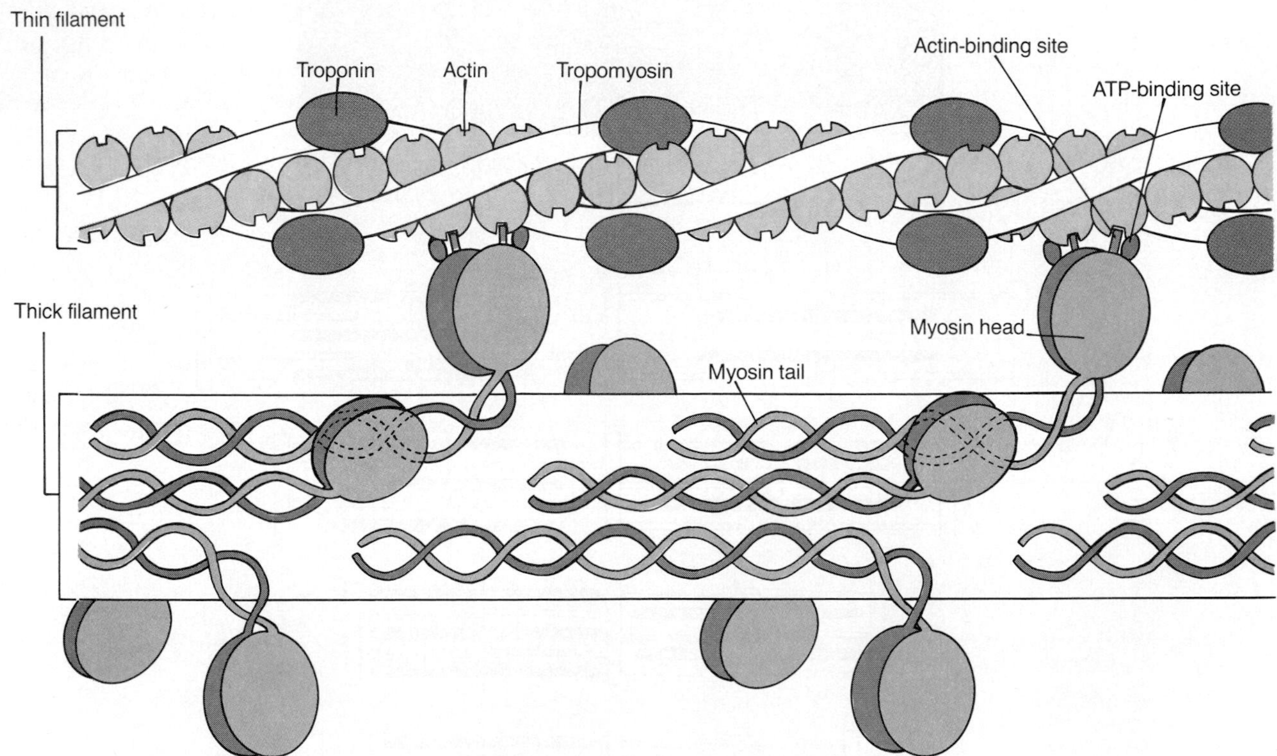

Table 12.1	Summary of the sliding filament theory of contraction
1.	A myofiber, together with all its myofibrils, shortens by movement of the insertion toward the origin of the muscle.
2.	Shortening of the myofibrils is caused by shortening of the sarcomeres—the distance between Z lines (or discs) is reduced.
3.	Shortening of the sarcomeres is accomplished by sliding of the myofilaments—each filament remains the same length during contraction.
4.	Sliding of the filaments is produced by asynchronous power strokes of myosin cross-bridges, which pull the thin filaments (actin) over the thick filaments (myosin).
5.	The A bands remain the same length during contraction, but are pulled toward the origin of the muscle.
6.	Adjacent A bands are pulled closer together as the I bands between them shorten.
7.	The H bands shorten during contraction as the thin filaments from each end of the sarcomeres are pulled toward the middle.

When the cross-bridges bond to actin, they undergo a conformation change. This has two effects: (1) ADP and P_i are released; and (2) the cross-bridges change their orientation, resulting in a *power stroke,* which pulls the thin filaments toward the center of the A bands. At the end of the power stroke each cross-bridge bonds to a fresh ATP molecule. This causes the cross-bridge to break its bond with actin and resume its resting orientation. The myosin ATPase will then split ATP and activate the cross-bridges as in the previous cycle. Note that the splitting of ATP is required before a cross-bridge can attach to actin and undergo a power stroke and that the attachment of a new ATP is needed for the cross-bridge to release from actin at the end of a power stroke.

A single contraction cycle and power stroke of all the cross-bridges in a muscle would shorten the muscle by only about 1% of its resting length. Since muscles can shorten up to 60% of their resting lengths, it is obvious that the contraction cycles must be repeated many times. In order for this to occur, the cross-bridges must detach from the actin at the end of a power stroke, reassume their resting orientation, and then reattach to the actin and repeat the cycle.

The detachment of a cross-bridge from actin at the end of a power stroke requires the attachment of a "fresh" ATP to the cross-bridge. The importance of this process is illustrated by the muscular contracture called *rigor mortis* that is observed in a dead person and occurs due to lack of ATP when the muscle dies. This results in the formation of "rigor complexes" between myosin and actin that cannot detach.

In *rigor mortis,* all of the cross-bridges are attached to actin at the same time. During normal contraction, however, only about 50% of the cross-bridges are attached at any given time. The power strokes are thus not in synchrony, as the strokes of a competitive rowing team are. Rather, they are like the actions of a team engaged in a tug-of-war, in which the pulling action of the members is asynchronous. Some cross-bridges are engaged in power strokes at all times during the contraction.

Table 12.1 summarizes the events of the sliding filament theory of muscle contraction.

Length-Tension Relationship

The strength of a muscle's contraction is affected by a number of factors. These include the number of muscle fibers within the muscle that are stimulated to contract, the thickness of each muscle fiber (thicker fibers have more myofibrils and thus can exert more power), the muscle architecture, and the initial length of the muscle fibers when they are at rest.

There is an "ideal" resting length of muscle fibers. When the resting length is greater than this ideal, the overlap between actin and myosin is so little that few cross-bridges can attach. When the muscle is stretched to the point that there is no overlap of actin with myosin, no cross-bridges can attach to the thin filaments and the muscle cannot contract. When the muscle is shortened to about 60% of its resting length, the Z lines abut against the thick filaments so that further contraction cannot occur.

The strength of a muscle's contraction can be measured by the force required to prevent it from shortening. Under these conditions, the strength of contraction, or *tension,* can be measured when the muscle length at rest is varied (fig. 12.17). Maximum tension is produced when the muscle is at its normal resting length *in vivo* (within the body).

Isotonic and Isometric Contractions

In order for muscle fibers to shorten when they contract, they must generate a force that is greater than the opposing forces that act to prevent movement of the muscle's insertion. Flexion of the forearm at the elbow joint, for example, occurs against the force of gravity and the weight of the object being lifted. The tension produced by the contraction of each muscle fiber separately is insufficient to overcome the opposing force, but the combined contractions of large numbers of muscle fibers may be sufficient to overcome the opposing force and flex the elbow joint as the muscle fibers shorten in length.

The contraction that results in muscle shortening is called *isotonic contraction,* so called because the force of contraction remains relatively constant throughout the

isotonic: Gk. *isos,* equal; *tonos,* tension

Figure 12.17. The length-tension relationship in skeletal muscles. Maximum relative tension (1.0) is achieved when the muscle is 100%–120% of its resting length (sarcomere lengths from 2.0 to 2.25 μm). Increases or decreases in muscle (and sarcomere) lengths result in rapid decreases in tension.

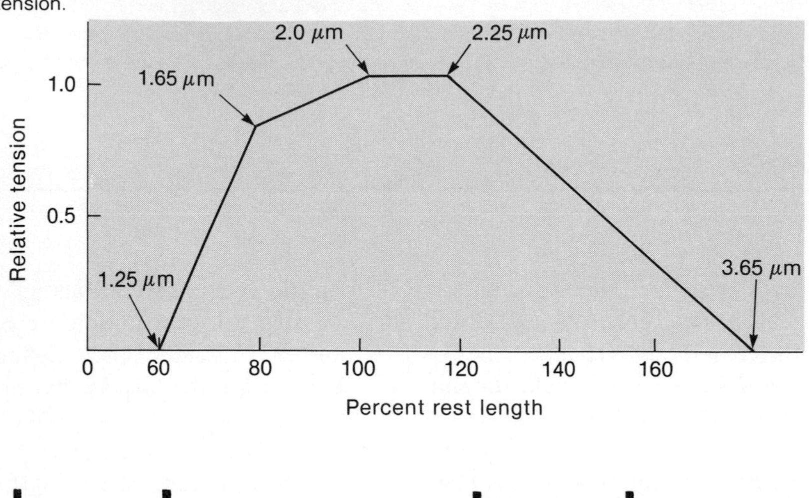

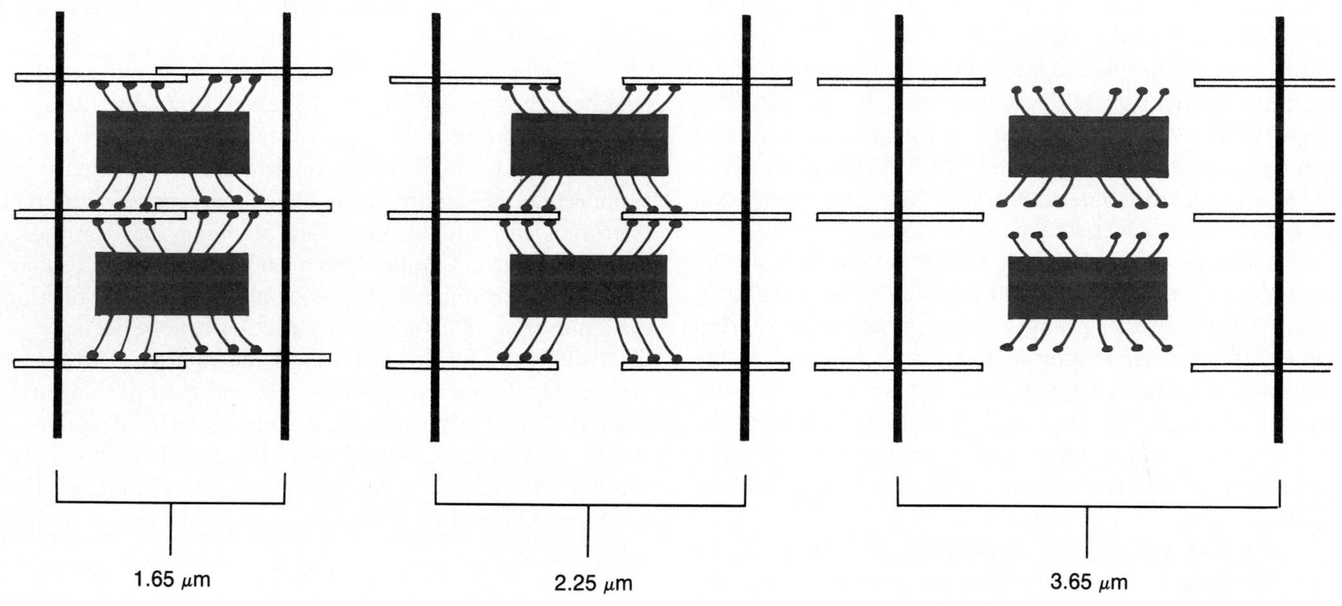

Figure 12.18. Types of muscle contraction. (*a*) in an isometric contraction, the muscle develops tension but does not shorten. (*b*) in an isotonic contraction, the muscle shortens while under a constant load.

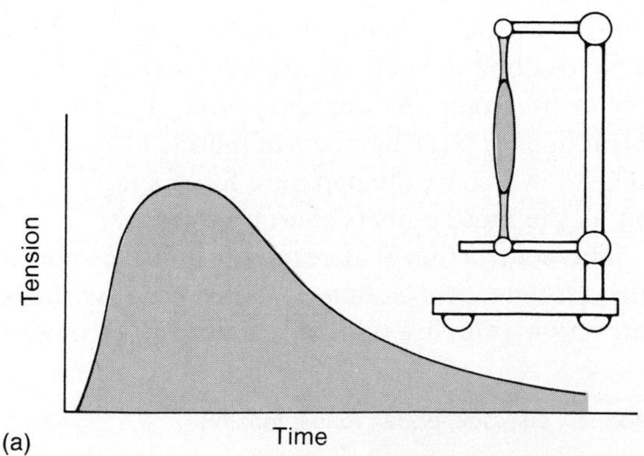

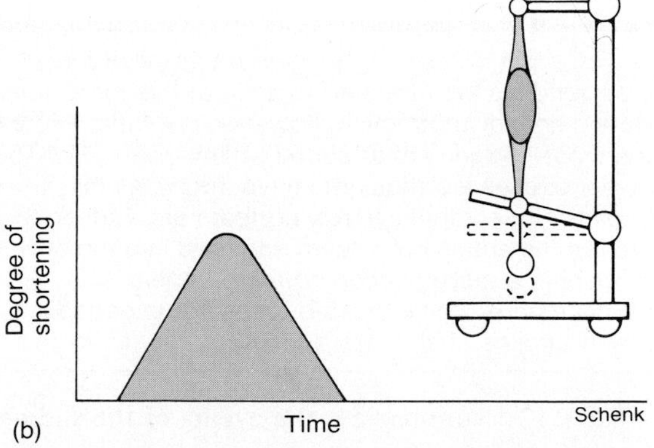

shortening process (fig. 12.18). In an *isometric contraction,* the length of a muscle remains constant because the antagonist force equals the force in the muscle being contracted. An isometric contraction occurs when a person supports an object in a fixed position. An isometric contraction is converted to an isotonic contraction when increased force is generated within the muscle overcoming the resistance and resulting in muscle movement.

---

1. Draw three successive sarcomeres in a myofibril of a resting muscle fiber. Label the myofibril, sarcomeres, A bands, I bands, H bands, and Z lines.
2. Why do the A bands appear darker than the I bands?
3. Draw three successive sarcomeres in a myofibril of a contracted fiber. Indicate which bands get shorter during contraction, and explain how this occurs.
4. Explain why the maximum relative tension in a muscle is achieved when it is 100%–120% of its resting length.
5. Describe how the antagonistic muscles in the brachium can be exercised through both isotonic and isometric contractions.

Stimulation, Regulation, and the Mechanics of Contraction

The recruitment of motor units results in graded muscle contractions. Contraction is produced by sliding of filaments within the muscle fibers. Electrical excitation of muscle fibers is coupled to contraction through the effects of calcium ions.

Objective 12. Define a motor unit, and discuss the role of motor units in muscular contraction.

Objective 13. Discuss the events that occur during excitation-contraction coupling.

Objective 14. Define the all-or-none principle of muscular contraction.

Objective 15. Describe the various types of contractions performed by skeletal muscle.

Neuromuscular Junction

A nerve serving a muscle is composed of both motor and sensory neurons. As the motor portion of the nerve penetrates a muscle, it splays into a number of branching neuron processes called **axons.** The terminal ends of axons contact the sarcolemma of muscle fibers by means of **motor end plates** (fig. 12.19). Each axon may split into enough branches to serve dozens of muscle fibers. The area consisting of the motor end plate and the sarcolemma of a muscle fiber is known as the **neuromuscular (myoneural) junction.**

Acetylcholine is a *neurotransmitter chemical* stored in **synaptic vesicles** at the terminal ends of the axons. A nerve impulse reaching the terminal end of an axon causes the release of acetylcholine into the *neuromuscular cleft* of the neuromuscular junction. As this chemical mediator contacts the sarcolemma, it initiates physiological activity within the muscle fiber, resulting in contraction.

Motor Unit

A **motor unit** consists of a single motor neuron and the aggregation of muscle fibers innervated by the motor neuron (fig. 12.19b). When a nerve impulse travels through a motor unit, all of the fibers served by it contract simultaneously to their maximum. Most muscles have an innervation ratio of one motor neuron per 100–150 muscle fibers. Muscles that are capable of precise, dexterous movements, such as an eye muscle, may have an innervation ratio of 1:10. Massive muscles that are responsible for gross body movements, like those of the thigh, may have an innervation ratio exceeding 1:500.

All of the motor units controlling a particular muscle, however, are not the same size. Neurons that innervate smaller numbers of muscle fibers have smaller cell bodies and axon diameters than neurons that have larger innervation ratios. The smaller neurons also are activated by lower levels of motor stimulus. The small motor units, as a result, are the ones that are used most often. The larger motor units are only activated when very forceful contractions are required.

Skeletal muscles are voluntary in that they can be consciously contracted. The magnitude of the task determines the number of motor units that are activated. Performing a light task, such as lifting a book, requires few motor units, whereas lifting a table requires many. Muscles with pennate architecture have many motor units, are strong and dexterous, but generally fatigue more readily than muscles with fewer motor units. Being mentally "psyched up" to accomplish an athletic feat involves voluntary activation of more motor units within the muscles. Seldom does a person utilize all the motor units within a muscle, but the secretion of *epinephrine (ep''ĭ-nef'rin)* from the adrenal gland does promote an increase in the force that can be produced when a given number of motor units is activated.

Regulation of Contraction

When the cross-bridges attach to actin, the myosin ATPase enzymes are activated, resulting in power strokes and muscle contraction. In order for muscle relaxation to occur, the attachment of the myosin cross-bridges to actin must be prevented. The regulation of cross-bridge attachment to actin is a function of two proteins found associated with actin in the thin filaments.

---

isometric: Gk. *isos,* equal; *metron,* measure
axon: Gk. *axon,* axis

Figure 12.19. A motor end plate at the neuromuscular junction. (*a*) a neuromuscular junction is the site where the nerve fiber and muscle fiber meet. The motor end plate is the specialized portion of the sarcolemma of a muscle fiber surrounding the terminal end of the axon. Note the slight gap between the membrane of the axon and that of the muscle fiber. (*b*) a photomicrograph of muscle fibers and motor end plates. A motor neuron and the muscle fibers it innervates constitute a motor unit.

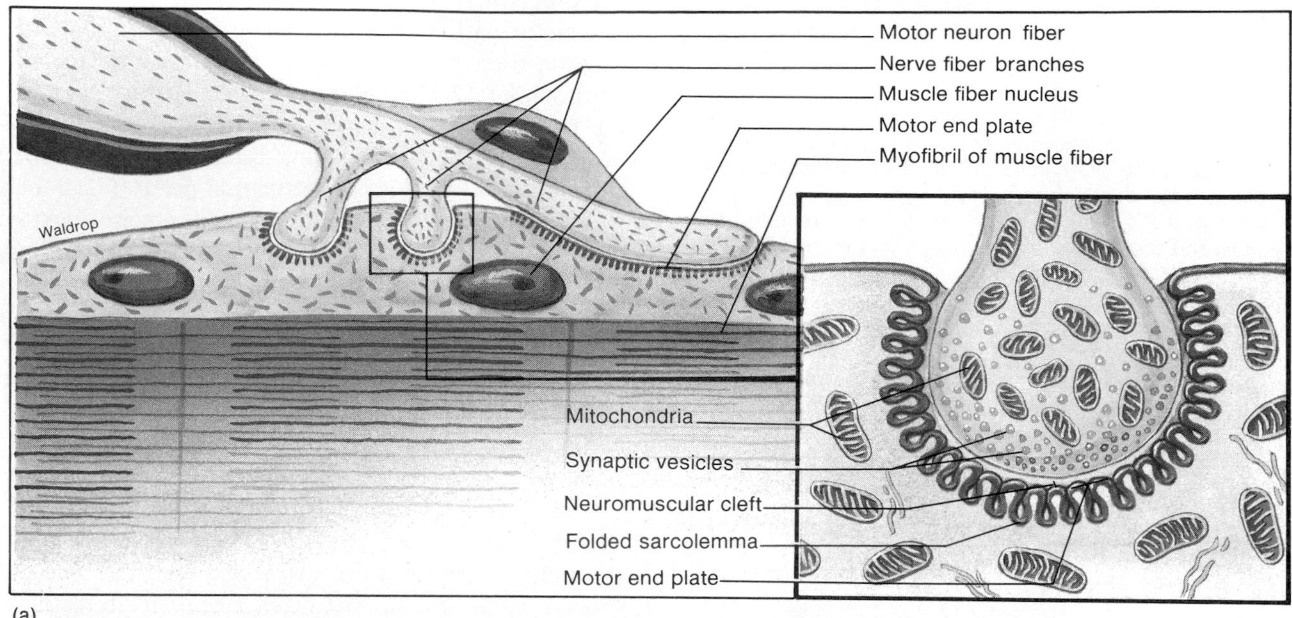

(a)

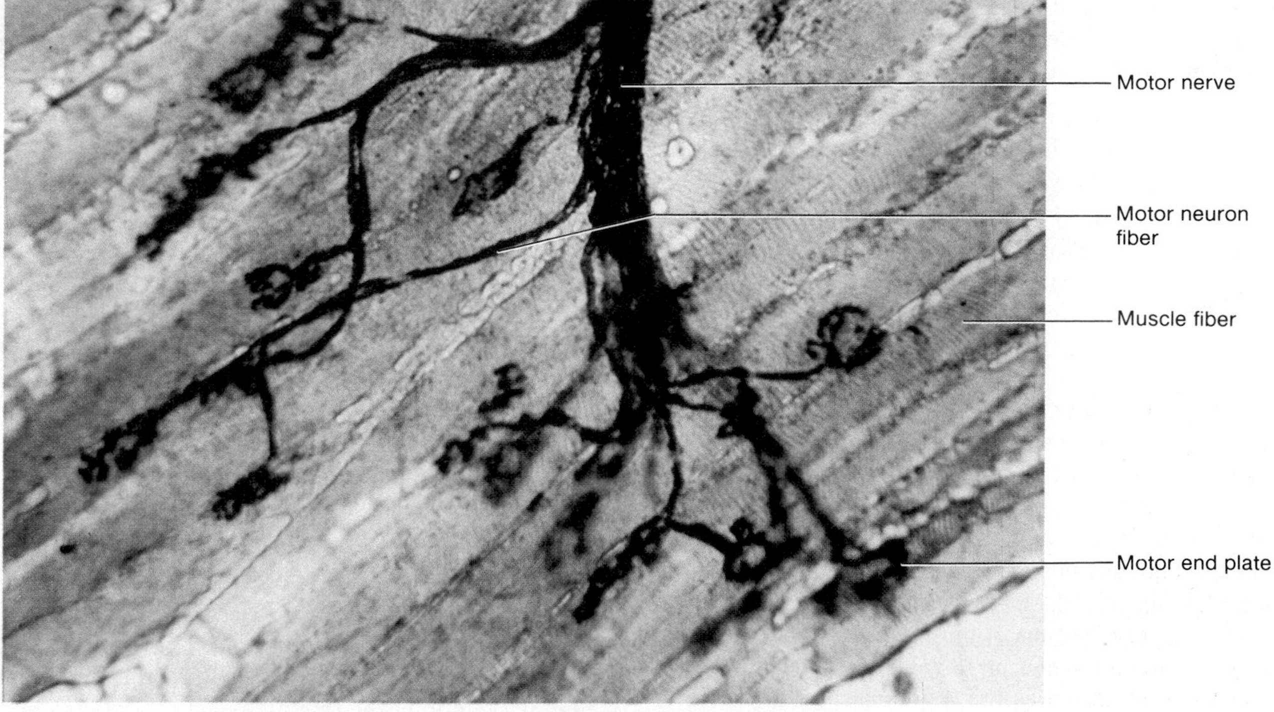

(b)

Figure 12.20. The relationship of troponin and tropomyosin to actin in the thin filaments. The tropomyosin is attached to actin, whereas the troponin complex of three subunits is attached to tropomyosin (not directly to actin).

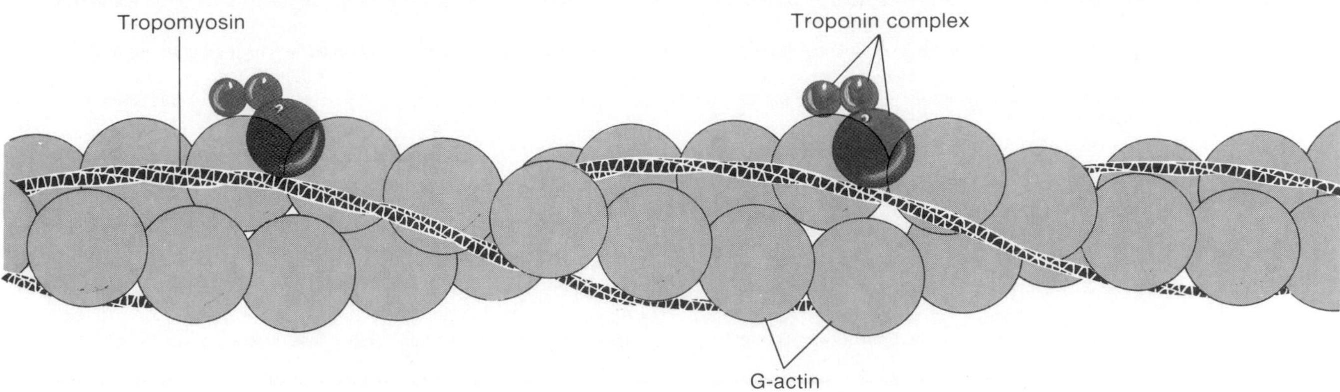

The actin filament—or *F-actin*—is a polymer formed of three hundred to four hundred globular subunits (*G-actin*) arranged in a double row and twisted to form a helix (fig. 12.20). A different type of protein, known as **tropomyosin** *(tro″po-mi′o-sin),* lies within the groove between the double row of G-actin. There are forty to sixty tropomyosin molecules per thin filament, with each tropomyosin spanning a distance of approximately seven actin subunits.

Within the thin filaments, attached to tropomyosin rather than directly to actin, is a third type of protein, called **troponin** *(tro′po-nin).* Troponin and tropomyosin work together to regulate the attachment of the cross-bridges to actin and thus serve as a switch for muscle contraction and relaxation. In a relaxed muscle, tropomyosin is in a position that physically blocks attachment of the cross-bridges to actin. In order for the myosin cross-bridges to attach to actin, tropomyosin must be moved.

Role of Ca^{++} in Muscle Contraction. In a relaxed muscle, when tropomyosin blocks the attachment of the cross-bridges to actin, the concentration of Ca^{++} in the sarcoplasm (cytoplasm of muscle cells) is very low. When the muscle cell is stimulated to contract, the concentration of Ca^{++} in the sarcoplasm quickly rises above 10^{-6} molar. Some of this Ca^{++} attaches to a subunit of troponin, causing a conformational change that moves troponin and its attached tropomyosin out of the way so that the cross-bridges can now attach to actin (fig. 12.21).

The position of the troponin-tropomyosin complexes in the thin filaments is thus adjustable. When Ca^{++} is not attached to troponin, tropomyosin is in a position that inhibits the attachment of the cross-bridges to actin; muscle contraction is thus prevented. When Ca^{++} attaches to troponin, the troponin-tropomyosin complexes shift position,

Figure 12.21. The attachment of Ca^{++} to troponin causes movement of the troponin-tropomyosin complex, which exposes binding sites on the actin. The myosin cross-bridges can then attach to actin and undergo a power stroke.

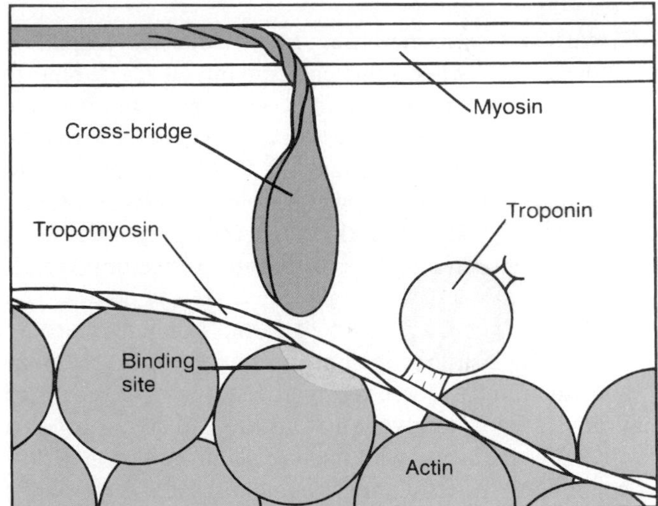

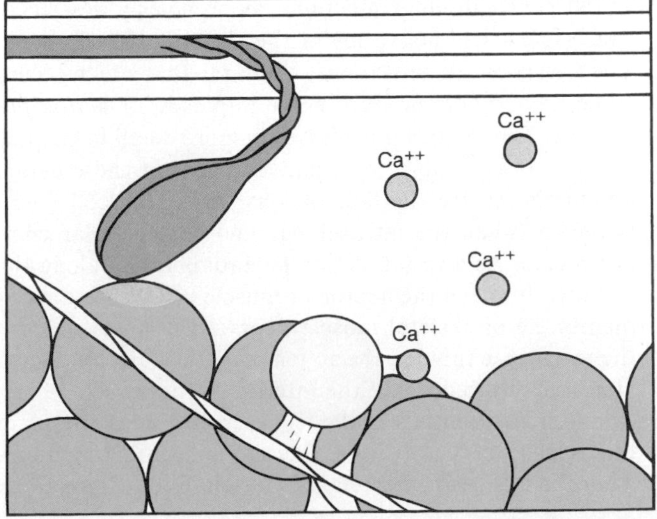

Table 12.2	Summary of events that occur during excitation-contraction coupling
1.	Action potentials in a somatic motor nerve cause the release of acetylcholine neurotransmitter at the neuromuscular junction (one myoneural junction per myofiber).
2.	Acetylcholine, through its interaction with receptors in the muscle cell membrane (sarcolemma), produces action potentials that are regenerated across the sarcolemma.
3.	The membranes of the transverse tubules (T-tubules) are continuous with the sarcolemma and conduct action potentials deep into the muscle fiber.
4.	Action potentials in the T-tubules, by a mechanism that is poorly understood, stimulate the release of Ca^{++} from the terminal cisternae of the sarcoplasmic reticulum.
5.	Ca^{++} released into the sarcoplasm attaches to troponin, causing a change in its structure.
6.	The shape change in troponin causes its attached tropomyosin to shift position in the actin filament, thus exposing bonding sites for the myosin cross-bridges.
7.	Myosin cross-bridges, previously activated by the hydrolysis of ATP, attach to actin.
8.	Hydrolysis of ATP provides energy for the power stroke by which cross-bridges pull the thin filaments over the thick filaments.
9.	Attachment of fresh ATP allows the cross-bridges to detach from actin and repeat the contraction cycle as long as Ca^{++} remains attached to troponin.
10.	When action potentials stop being produced, the sarcoplasmic reticulum actively accumulates Ca^{++} and tropomyosin moves again to its inhibitory position.

the cross-bridges attach to actin, and muscle contraction occurs. Cross-bridges can again attach to actin, produce a power stroke, detach from actin, and can continue these contraction cycles as long as Ca^{++} is attached to troponin.

Excitation-Contraction Coupling. Muscle contraction is "turned on" when sufficient amounts of Ca^{++} bind to troponin. This occurs when the Ca^{++} concentration of the sarcoplasm rises above 10^{-6} molar. In order for muscle relaxation to occur, therefore, the Ca^{++} concentration of the sarcoplasm must be lowered below this level. Muscle relaxation is produced by the active transport of Ca^{++} out of the sarcoplasm into the sarcoplasmic reticulum (see fig. 12.11).

Most of the Ca^{++} is stored within expanded portions of the sarcoplasmic reticulum known as **terminal cisternae** (not illustrated). These terminal cisternae are separated by only a very narrow gap from the transverse tubules, which are continuous with the sarcolemma and which thus open to the extracellular environment through pores in the cell surface.

The release of acetylcholine from the axon terminals at the neuromuscular junctions, as previously described, causes electrical activation of the skeletal muscle fibers. The nature of this activation is similar to that which occurs when nerve fibers produce nerve impulses, or *action potentials.* Though this is described in more detail in chapter 14, it is important here to understand that these action potentials involve the flow of ions across the cell membrane, between the intracellular and extracellular compartments. As a result, action potentials travel along the cell membrane of the neuron or muscle cell. Since the cell membrane of skeletal muscle fibers invaginates to form the transverse tubules, the action potentials are conducted along these tubules into the interior of the muscle fiber.

Action potentials in the transverse tubules stimulate the release of Ca^{++} from the sarcoplasmic reticulum, where it is stored in a relaxed muscle fiber. Since these two structures do not actually touch, it is believed that a chemical mediator (possibly inositol triphosphate, described in chapter 19) is secreted by the tubules. This causes Ca^{++} to diffuse out of the sarcoplasmic reticulum and into the sarcomeres, where it binds to troponin and causes the displacement of tropomyosin. The cross-bridges can now bond to actin and cause contraction. As long as action potentials continue to be produced—which is as long as the neural innervation to the muscle continues to be active—Ca^{++} will remain attached to troponin, and cross-bridges will be able to undergo contraction cycles.

When the somatic motor nerve to a muscle is active, therefore, Ca^{++} is attached to troponin and muscle contraction occurs. When the nerve activity ceases, Ca^{++} is actively transported into the sarcoplasmic reticulum. ATP is required for three steps in this process: (1) ATP provides the energy for the power strokes of the cross-bridges; (2) ATP must attach to the cross-bridges at the end of the power stroke, so that they can detach from actin; and (3) ATP is needed for active transport of Ca^{++} into the sarcoplasmic reticulum so that a muscle can relax when the action potentials stop.

Although ATP is required for both contraction and relaxation of a muscle, the quantity of ATP in a muscle fiber does not serve to regulate its activity. A living muscle fiber always has sufficient ATP for its needs. Instead, neural regulation of muscle contraction and relaxation is mediated by calcium ions. Electrical excitation and sliding of the filaments (contraction) are "coupled" through adjustments of the sarcoplasmic Ca^{++} concentration. The events that occur in this *excitation-contraction coupling* are summarized in table 12.2.

Twitch, Summation, and Tetanus

The contractile behavior of skeletal muscles is more easily studied *in vitro* (outside the body) than *in vivo.* In these studies a muscle such as the gastrocnemius (calf muscle) of a frog is usually mounted so that one end is fixed and the other is movable. In the classic laboratory studies, the

Figure 12.22. The setup for observing the contractile behavior of the isolated gastrocnemius muscle of a frog.

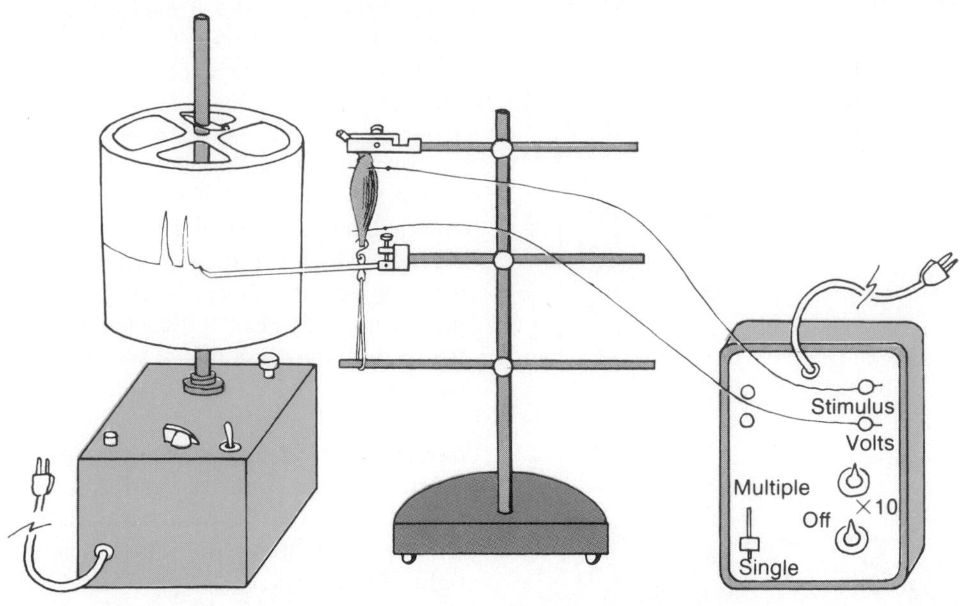

Figure 12.23. The physiograph Mark III recorder. This is one of many types of electronic recording devices that receive electrical signals produced by transducers and record these signals on moving paper.

movable end of the muscle directly produces deflections of a pen, which writes on a rotating drum recorder (fig. 12.22). When more modern equipment is used, the mechanical force of muscle contraction is transduced into an electric current, which can be amplified and displayed as pen deflections in a multi-channel recorder (a physiograph or polygraph—fig. 12.23). In this way, the contractile behavior of the whole muscle in response to experimentally administered electric shocks can be studied.

When the muscle is stimulated with a single electric shock of a sufficient voltage, it quickly contracts and relaxes. This response is called a **twitch.** Increasing the stimulus voltage increases the strength of the twitch up to a maximum. The strength of a muscle contraction can thus be *graded,* or varied—an obvious requirement for the proper control of skeletal movements. If a second electric shock is delivered immediately after the first, it will produce a second twitch that may partially "ride piggyback" on the first. This response is called **summation.**

If the stimulator is set to deliver an increasing frequency of electric shocks automatically, the relaxation time between successive twitches will get shorter and shorter, as the strength of contraction increases in amplitude. This

Figure 12.24. (*a*) a recording of summation of muscle twitches using a physiograph. (*b*) an illustration of twitch summation, tetanus, and fatigue in an isolated muscle.

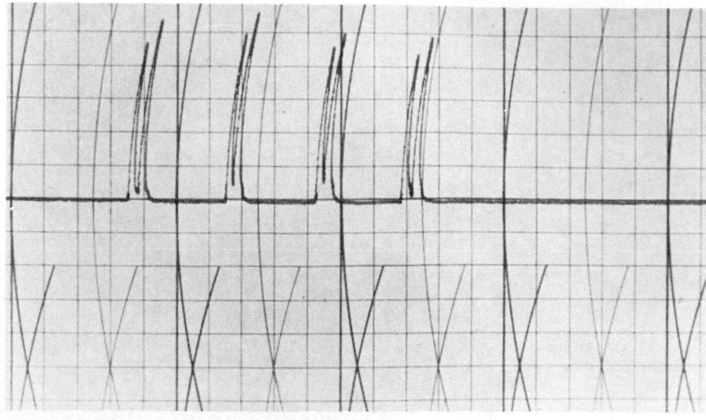

(a)

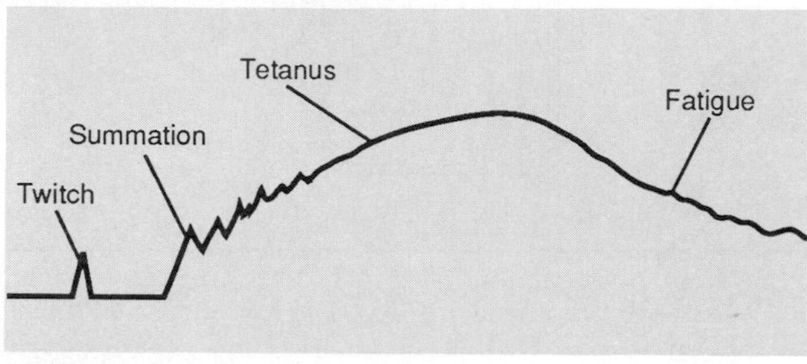

(b)

is **incomplete tetanus.** Finally, at a particular "fusion frequency" of stimulation, there is no visible relaxation between successive twitches (fig. 12.24). Contraction is smooth and sustained, very much like normal muscle contraction *in vivo.* This smooth, sustained contraction is called **complete tetanus.** The term *tetanus* should not be confused with the disease of the same name, which is accompanied by a painful state of muscle contracture, or *tetany.*

Research laboratories, using an experimental setup analogous to the one previously described, can study the contractile behavior of *an isolated muscle fiber.* The isolated fiber produces a twitch of submaximal strength in response to a single action potential. Since the fiber recovers electrically before it has finished its twitch, it can be stimulated a number of times to produce a summation of twitches. Continued stimulation of the isolated fiber can then result in tetanus, analogous to the behavior of the whole muscle previously described.

Stimulation of the whole muscle *in vitro* with an electric stimulator or stimulation of muscle *in vivo* by motor axons, however, results in the production of many action potentials in the stimulated muscle fibers. Within the whole muscle, therefore, muscle fibers are normally stimulated maximally and thus produce *all-or-none* contractions. In this case, gradations in the strength of skeletal muscle contractions result from summation of the contractions of different *numbers* of muscle fibers and not from variations in the strength of the contractions of individual muscle fibers. Stronger muscle contractions, in other words, are produced by the stimulation of greater numbers of muscle fibers. *In vivo,* this occurs when greater numbers of motor units are recruited.

Even when muscles are at rest, they respond to continued stimuli from the nervous system. This results in muscle **tonus** *(to'nus),* or a state of partial, sustained contraction, which keeps the muscles slightly taut in readiness for activity. Tonus is also extremely important in generating body heat. If muscle fibers are not periodically stimulated, they lose tonus, become flaccid, atrophy, and soon die.

Series Elastic Component. In order for a muscle to shorten when it contracts, and thus to move its insertion toward its origin, the noncontractile parts of the muscle and the connective tissue of its tendons must first be pulled tight. These structures, particularly the tendons, have elasticity—they resist distension, and when the distending force is released, they tend to spring back to their resting lengths. Since the tendons are in series with the force of muscle contraction, they provide the **series elastic component** of muscle contraction.

When the gastrocnemius muscle was stimulated with a single electric shock in the previously described experiment, the amplitude of the twitch was reduced because some of the force of contraction was used to stretch the series elastic component. Delivery of a second shock quickly after the first thus produced a greater degree of muscle shortening than the first shock, culminating at the fusion frequency of stimulation with complete tetanus, in which the strength of contraction was much greater than that of individual twitches.

Some of the energy used to stretch the series elastic component during muscle contraction is released when the muscle relaxes. This elastic recoil helps the muscles to return to their resting length, and is of particular importance for the muscles involved in breathing. As described in chapter 23, inspiration is produced by muscle contraction, and expiration is produced by the elastic recoil of the thoracic structures that were stretched during inspiration.

1. Explain why motor units are considered the basic functional units of muscle contraction.
2. Draw a figure of the molecular structure of the thick and thin filaments, showing the cross-bridges, G-actin, troponin, and tropomyosin. Describe the roles of Ca^{++}, troponin, and tropomyosin in muscle contraction.
3. List, in order of occurrence, the events that occur during the contraction cycles of the cross-bridges. Draw a flowchart (using arrows) of the events that occur between release of ACh at the neuromuscular junction and the attachment of Ca^{++} to troponin.
4. Draw a myogram of a twitch and one of tetanus. Explain why a sustained contraction would eventually fatigue.

Figure 12.25. The production and utilization of phosphocreatine in muscles.

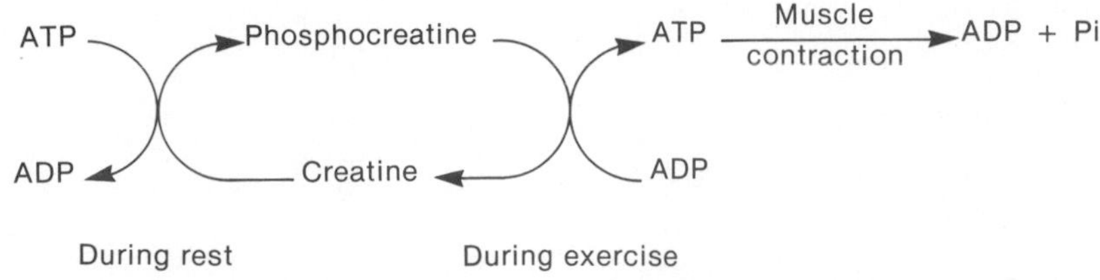

Figure 12.26. Creatine kinase in plasma is separated from other proteins and identified by the rate of its movement in an electric field. The height of each peak indicates concentration, and the position of each peak indicates the isoenzyme form. As can be seen in the *lower figure*, the plasma concentration of creatine kinase is elevated in both heart disease (myocardial infarction) and skeletal muscle disease, but the two types of disease may be distinguished in this test by the appearance of a different isoenzyme in myocardial infarction than in muscle disease.

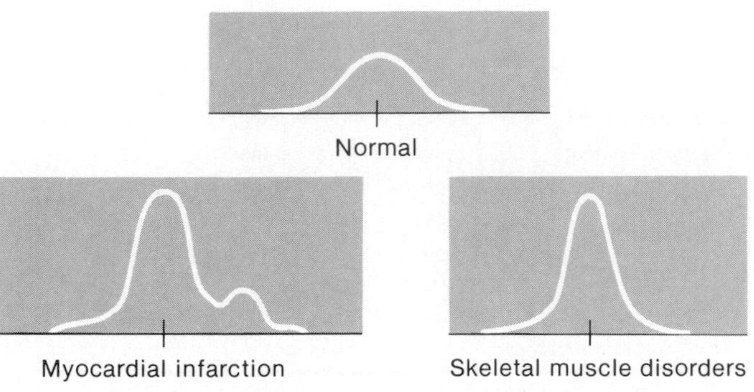

Figure 12.27. A cross section of skeletal muscle stained to indicate the activity of myosin ATPase. Different types of fibers can be distinguished by the intensity of the stain.

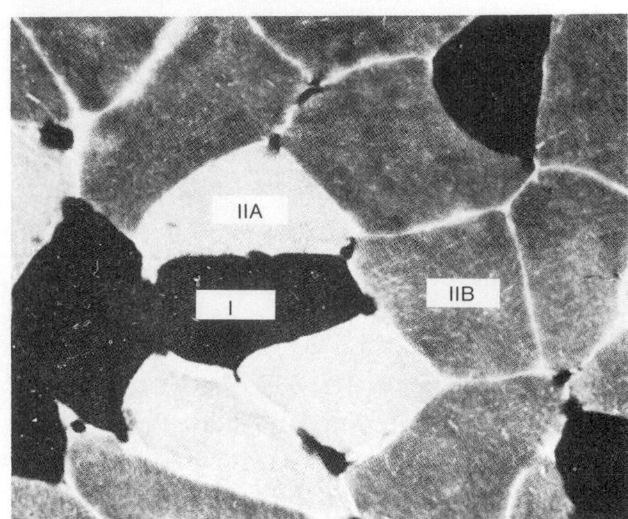

Energy Usage by Skeletal Muscle

Aerobic cellular respiration is required for the production of ATP needed for the cross-bridge activity of the various types of skeletal muscle fibers.

Objective 16. Discuss the relationship between ATP and phosphocreatine in muscle physiology.

Objective 17. Describe the differences between slow- and fast-twitch muscle fibers.

Objective 18. Define muscle *fatigue,* and explain its causes.

Objective 19. Describe the changes which occur in muscle tissue as a result of endurance training.

Skeletal muscles at rest obtain most of their energy from the aerobic respiration of fatty acids. During exercise, muscle glycogen and blood glucose are also used as energy sources. Energy thus obtained by cell respiration is used to make ATP, which serves as the immediate source of energy for movement of the cross-bridges and muscle contraction.

During sustained muscle contraction, the utilization of ATP may occur at a faster rate than the rate of ATP production through cell respiration. At these times the rapid renewal of ATP is accomplished by the combination of ADP with phosphate, derived from another "high-energy phosphate" compound called **phosphocreatine** *(fos''fo-kre'ah-tin),* or **creatine phosphate.**

The phosphocreatine concentration within muscle cells is greater than three times the concentration of ATP and represents a ready reserve of high-energy phosphate that can be donated directly to ADP. During times of rest, the depleted reserve of phosphocreatine can be restored by the reverse reaction—the phosphorylation of creatine with phosphate derived from ATP (fig. 12.25).

The enzyme that transfers phosphate between creatine and ATP is called **creatine kinase** *(ki'nās).* Skeletal muscle and heart muscle have two different forms of this enzyme (they have different isoenzymes).

The skeletal muscle isoenzyme is found to be elevated in the blood of people with *muscular dystrophy* (degenerative disease of skeletal muscles). The plasma concentration of the isoenzyme characteristic of heart muscle is elevated as a result of *myocardial infarction,* and measurements of this enzyme are thus used as a means of diagnosing this condition (fig. 12.26).

Slow- and Fast-Twitch Fibers

Skeletal muscle fibers can be grouped on the basis of their contraction speed (time required to reach maximum tension) into **slow-twitch,** or **type I fibers,** and **fast-twitch,** or **type II fibers.** These differences are associated with different myosin ATPase isoenzymes, designated "slow" and "fast," by which the two fiber types can be distinguished when they are appropriately stained (fig. 12.27). The extraocular muscles that position the eyes, for example, have

Figure 12.28. A comparison of the rates with which maximum tension is developed in three muscles. These include the relatively fast-twitch extraocular (*a*) and gastrocnemius (*b*) muscles and the slow-twitch soleus muscle (*c*).

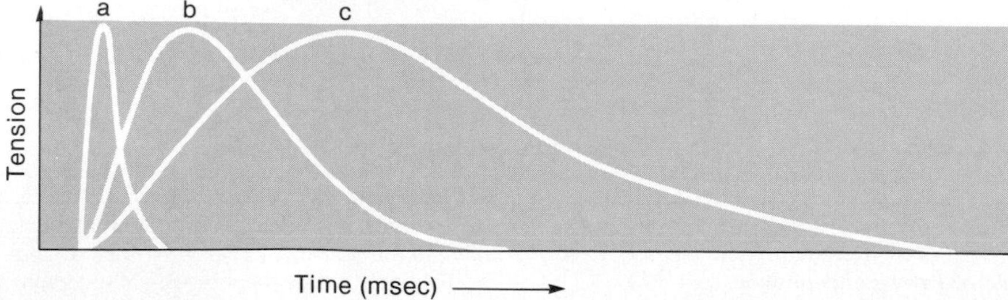

Figure 12.29. A cross section of skeletal muscle stained to show the activity of glycolytic enzymes in different fiber types.

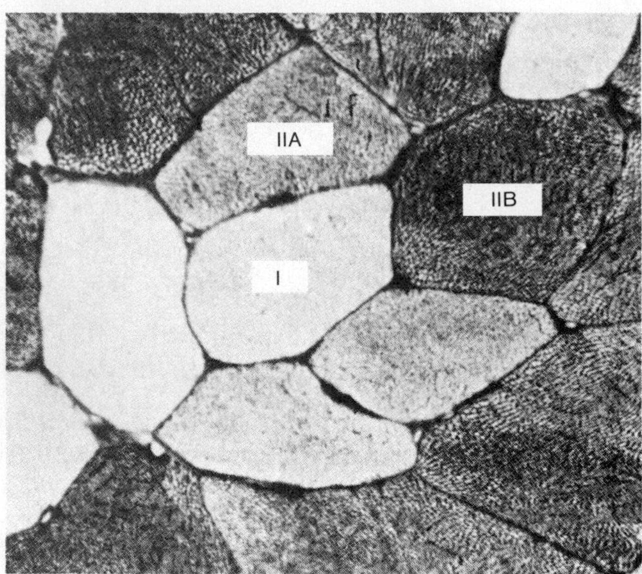

Table 12.3 Characteristics of red, intermediate, and white muscle fibers

	Red (Type I)	Intermediate (Type IIA)	White (Type IIB)
Diameter	Small	Intermediate	Large
Z-line thickness	Wide	Intermediate	Narrow
Glycogen content	Low	Intermediate	High
Resistance to fatigue	High	Intermediate	Low
Capillaries	Many	Many	Few
Myoglobin content	High	High	Low
Respiration type	Aerobic	Aerobic	Anaerobic
Twitch rate	Slow	Fast	Fast
Myosin ATPase content	Low	High	High

a high proportion of fast-twitch fibers and reach maximum tension in about 7.3 msec (milliseconds—thousandths of a second); the soleus muscle in the leg, in contrast, has a high proportion of slow-twitch fibers and requires about 100 msec to reach maximum tension (fig. 12.28).

Muscles like the soleus are *postural muscles* that must be able to sustain a contraction for a long period of time without fatigue. This is aided by other characteristics of slow-twitch (type I) fibers that endow them with a high capacity for aerobic respiration. Slow-twitch fibers are served by large numbers of capillaries, have numerous mitochondria and aerobic respiratory enzymes, and have a high concentration of *myoglobin* pigment. Myoglobin is a red pigment—hence the alternate name of *red fibers* for these muscle cells—that is related to the hemoglobin pigment of blood and serves to improve the delivery of oxygen to the slow-twitch fibers.

The thicker, fast-twitch (type II) fibers have a lower capillary supply, fewer mitochondria, and less myoglobin; hence, these fibers are also called *white fibers.* Fast-twitch fibers are adapted to respire anaerobically by a large store of glycogen and a high concentration of glycolytic enzymes, which enable these fibers to be distinguished when appropriately stained (fig. 12.29). In addition to the type I (slow-twitch) and type II (fast-twitch) fibers, human muscles may also have an intermediate form of fibers, which are fast-twitch but also have a high aerobic capability. These are sometimes called type IIA, to distinguish them from the anaerobically adapted fast-twitch fibers (which are then labeled IIB). A comparison of these fiber types is summarized in table 12.3

Interestingly, the conduction rate of motor neurons that innervate fast-twitch fibers is faster (80–90 meters per second) than the conduction rate to slow-twitch fibers (60–70 meters per second). The fiber type indeed seems to be determined by the motor neuron. When the motor neurons to different fiber types are switched in experimental animals, the previously fast-twitch fibers become slow and the slow-twitch fibers become fast. As expected from these observations, all of the muscle fibers innervated by the same motor neuron (that are part of the same motor unit) are of the same type.

Table 12.4	Summary of the effects of endurance training (long-distance running, swimming, bicycling, etc.) on skeletal muscles

1. Improved ability to obtain ATP from oxidative phosphorylation
2. Increased size and number of mitochondria
3. Less lactic acid produced per given amount of exercise
4. Increased myoglobin content
5. Increased intramuscular triglyceride content
6. Increased lipoprotein lipase (enzyme needed to utilize lipids from blood)
7. Increased proportion of energy derived from fat, less from carbohydrates
8. Lower rate of glycogen depletion during exercise
9. Improved efficiency in extracting oxygen from blood

From Stuart Ira Fox, *Human Physiology,* 2d ed.

A muscle such as the gastrocnemius (calf muscle) contains both fast- and slow-twitch fibers, although fast-twitch fibers predominate. A given somatic motor axon, however, only innervates muscle fibers of one type, and the motor units composed of slow-twitch fibers tend to be smaller (have fewer fibers) than the motor units of fast-twitch fibers. Since motor units are recruited from smaller to larger when increasing effort is required, as previously described, the smaller motor units with slow-twitch fibers would be used most often in routine activities. Larger motor units with fast-twitch fibers, which can exert a great deal of force but which respire anaerobically and thus fatigue quickly, would be used relatively infrequently and for only short periods of time.

Muscle Fatigue

Muscle fatigue may be defined as the inability to maintain a particular muscle tension when the contraction is sustained or to reproduce a particular tension during rhythmic contraction over time. Fatigue during a sustained maximal contraction, when all the motor units are used and the rate of neural firing is maximal, appears to be due to an accumulation of extracellular K^+. This reduces the membrane potential of muscle fibers and interferes with their ability to produce action potentials. Fatigue under these circumstances lasts only a short time, and maximal tension can again be produced after less than a minute's rest.

Fatigue during moderate exercise involving rhythmical contractions occurs as the slow-twitch fibers deplete their reserve glycogen and fast-twitch fibers are increasingly recruited. Fast-twitch fibers obtain their energy through anaerobic respiration, converting glucose to lactic acid, and this results in a rise in intracellular H^+ and a fall in pH. The decrease in muscle pH, in turn, inhibits the activity of key glycolytic enzymes so that the rate of ATP production is reduced. Since ATP is needed for all active transport, the decrease in ATP is believed to result in a loss of intracellular Ca^{++} to the extracellular environment, so that excitation-contraction coupling is hindered in the muscle cells. The ability of the muscle to maintain a particular tension is thus decreased, first due to loss of the contribution of smaller, slow-twitch motor units and then due to the inability of the fast-twitch muscle fibers to contract as lactic acid accumulates.

Note that the decrease in ATP within the muscle cell causes it to fatigue due to interference with excitation-contraction coupling, not due to direct interference with the cross-bridge cycle. This is supported by the observation that even at maximal exhaustion, the amount of ATP in muscle fibers is only reduced by about 25% (although the amount of phosphocreatine may be completely depleted). Actually, the interference with excitation-contraction coupling can be thought of as a protective mechanism because if the ATP is significantly depleted, the muscle enters a state analogous to rigor mortis. This does not occur in a living muscle, even during maximal exercise, because ATP can be quickly reformed, using the phosphate group from creatine phosphate, and if ATP does become depleted by a certain amount, the muscle becomes fatigued due to interference with excitation-contraction coupling.

Adaptations to Exercise

When exercise is performed at low levels of effort, such that the body does not need to consume more oxygen than 50% of its maximum oxygen consumption rate, the energy for muscle contraction is obtained almost entirely from aerobic cell respiration. Anaerobic cell respiration, with its consequent production of lactic acid, contributes to the energy requirements as the exercise level rises and more than 60% of the maximal oxygen uptake is required. The maximal oxygen uptake is about ten times higher than the resting oxygen uptake rate in average adults, but it can reach up to twenty times higher than the resting rate in top endurance-trained athletes. These athletes can produce less lactic acid at a given level of exercise than the average person, and they are therefore less subject to fatigue than the average person.

Since the fiber types are determined by their innervations, endurance-training cannot change fast-twitch (type II) fibers to slow-twitch (type I) fibers. All fiber types, however, adapt to endurance training by an increase in myoglobin and aerobic respiratory enzymes, so that the maximal oxygen uptake can be increased by up to 20% through this training. In addition to changes in aerobic capacity, fibers show an increase in their content of triglycerides, which serves as an alternate energy source helping to spare their stores of glycogen. A summary of the changes that occur as a result of endurance training is presented in table 12.4.

Endurance training does not increase the size of muscles. Muscle enlargement is produced only by frequent bouts of muscle contraction against a high resistance—as in weight lifting. As a result of this latter type of training, muscle fibers become thicker, and the muscle therefore grows by hypertrophy (increase in cell size, rather than number). This happens as the myofibrils within a muscle fiber get thicker, due to the addition of new sarcomeres and myofibrils. After a myofibril attains a certain thickness, it may split into two myofibrils, which may then each become thicker due to the addition of sarcomeres. Muscle hypertrophy, in summary, is due to an increase in the number and size of myofibrils within the muscle fibers.

1. Draw a figure illustrating the relationship between ATP and creatine phosphate, and explain the physiological significance of this relationship.
2. Describe the characteristics of slow- and fast-twitch fibers (including intermediate fibers). Explain how the fiber types are determined and the functions of different fiber types.
3. Explain the different causes of muscle fatigue, and relate the causes of fatigue to the fiber types.
4. Describe the effects of endurance training and strength training on the fiber characteristics of the muscles.

Naming of Muscles

Skeletal muscles are named on the basis of shape, location, attachment, orientation of fibers, relative position, or function.

Objective 20. Describe, using examples, the various ways in which muscles are named.

One of the tasks of a myologist is to learn the names of the more than six hundred skeletal muscles within the body. This task is somewhat simplified by the fact that most of the muscles are paired; that is, the right side is the mirror image of the left. Further simplifying the task are the descriptive names of muscles. Only the principal muscles of the body are described in the remaining pages of this chapter.

As you learn the muscles of the body, understand how they are named and be able to identify them on yourself. Learn them as functional groups so that you will better understand their actions. If you can identify a muscle on yourself, you will be able to contract the muscle and describe its action. Learning anatomy this way will be easier, more meaningful, and will improve retention.

The following are some ways in which the names of muscles have been logically derived, with examples of each given.

1. **Shape:** rhomboideus (like a rhomboid); trapezius (like a trapezoid); or denoting the number of heads of origin: triceps (three heads), biceps (two heads)
2. **Location:** pectoralis (in the chest, or pectus); intercostal (between ribs); brachii (upper arm)
3. **Attachment:** many facial muscles (zygomaticus, temporalis, nasalis); sternocleidomastoid (sternum, clavicle, and mastoid process of the skull)
4. **Size:** maximus (larger, largest); minimus (smaller, smallest); longus (long); brevis (short)
5. **Orientation of fibers:** rectus (straight); transverse (across); oblique (in an oblique direction)
6. **Relative position:** lateral, medial, abdominal, internal, and external
7. **Function:** adductor, flexor, extensor, pronator, and levator (lifter)

1. Refer to chapter 1, and review the location of the following body regions: cervical, pectoral, abdominal, gluteal, perineal, brachial, antebrachial, inguinal, thigh, and popliteal.
2. Refer to chapter 11, and review the movements permitted at diarthrotic joints.

Muscles of the Axial Skeleton

Muscles of the axial skeleton include those of facial expression, mastication, eye movement, tongue movement, neck movement, respiration, the abdominal wall, the pelvic outlet, and movement of the vertebral column.

Objective 21. Locate the major muscles of the axial skeleton; identify synergistic and antagonistic muscles, and describe the action of each.

Muscles of Facial Expression

Humans have a well-developed facial musculature that provides complex facial expressions as a means of social communication. Indeed, many messages can be conveyed by a person without speaking a word. The following muscles produce facial movements (figs. 12.30, 12.31).

Epicranius. The broad, extensive epicranius is divisible into two muscles: the **frontalis** *(frun-ta'lis),* over the frontal bone, or forehead; and the **occipitalis** *(ok-sip''ĭ-ta'lis),* covering the occipital region of the skull. These muscles are united by an aponeurosis covering the skull, called the *galea aponeurotica.* The frontalis originates on the galea aponeurotica, inserts on the soft tissue of the eyebrow, and functions to elevate the eyebrow and wrinkle the forehead. The occipitalis originates on the occipital bone and mastoid process, inserts on the galea aponeurotica, and functions to retract the scalp. The sustained contraction of this muscle frequently causes headaches.

Figure 12.30. Superficial facial muscles involved in facial expression.

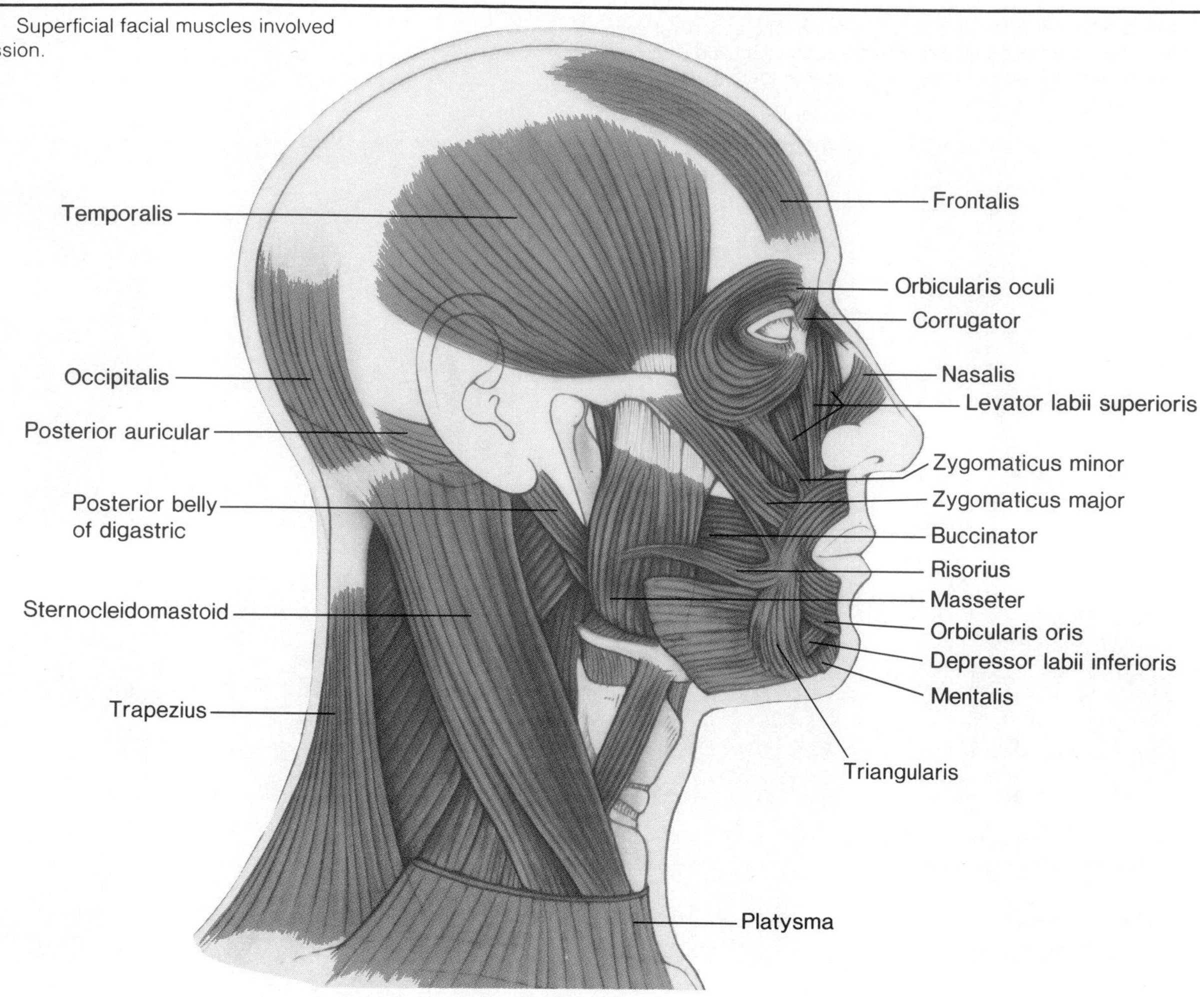

Corrugator. The corrugator muscle originates on the fleshy tissue above the eyebrow and inserts over the root of the nose. It causes a wrinkling between the eyes, or "frown lines."

Orbicularis Oculi. The orbicularis oculi muscle is a sphincter muscle that encircles the eye. It originates on the bones of the orbit (frontal and maxillary) and inserts on the tissue of the eyelids (palpebrae). The orbicularis oculi permits blinking, winking, and squinting. The contraction of this muscle also compresses the *lacrimal (tear) gland,* keeping the eyeball continuously moistened.

Nasalis. The nasalis extends over the bridge of the nose and acts to compress the nostrils.

Orbicularis Oris. The orbicularis oris is a sphincter muscle encircling the lips of the mouth. Its origin is the surrounding tissue, and it inserts on the mucosa of the lips. The orbicularis oris closes the mouth, forms words, and puckers or purses the mouth, as in kissing.

Levator Labii Superioris. The levator labii superioris is a broad, flat muscle originating in three heads from the upper maxilla and a portion of the zygomatic bone. It inserts on the superior margin of the orbicularis oris muscle so that when contracted it elevates the upper lip.

Zygomaticus. The zygomaticus is so named because it originates on the zygomatic bone. It inserts on the superior corner of the orbicularis oris muscle. When contracted, the zygomaticus elevates the corner of the mouth, as in smiling and laughing.

Risorius. The risorius muscle originates on fascia on the side of the face and inserts on the orbicularis oris muscle at the angle of the mouth. It draws the angle of the mouth laterally.

Triangularis (depressor anguli oris). The triangular muscle below either corner of the mouth is the triangularis. It originates on the mandible and inserts on the orbicularis oris muscle just below where the risorius inserts. The triangularis depresses the corner of the mouth, as in frowning.

Depressor Labii Inferioris. The depressor labii inferioris is medial to the triangularis. It originates on the mandible lateral to the midline and inserts on the orbicularis oris muscle and skin of the lower lip. It depresses the bottom lip.

Figure 12.31. Muscles of facial expression. The actions of each of the muscles of facial expression can be easily identified as a person contracts them. In each of these photographs, identify the muscles that are being contracted.

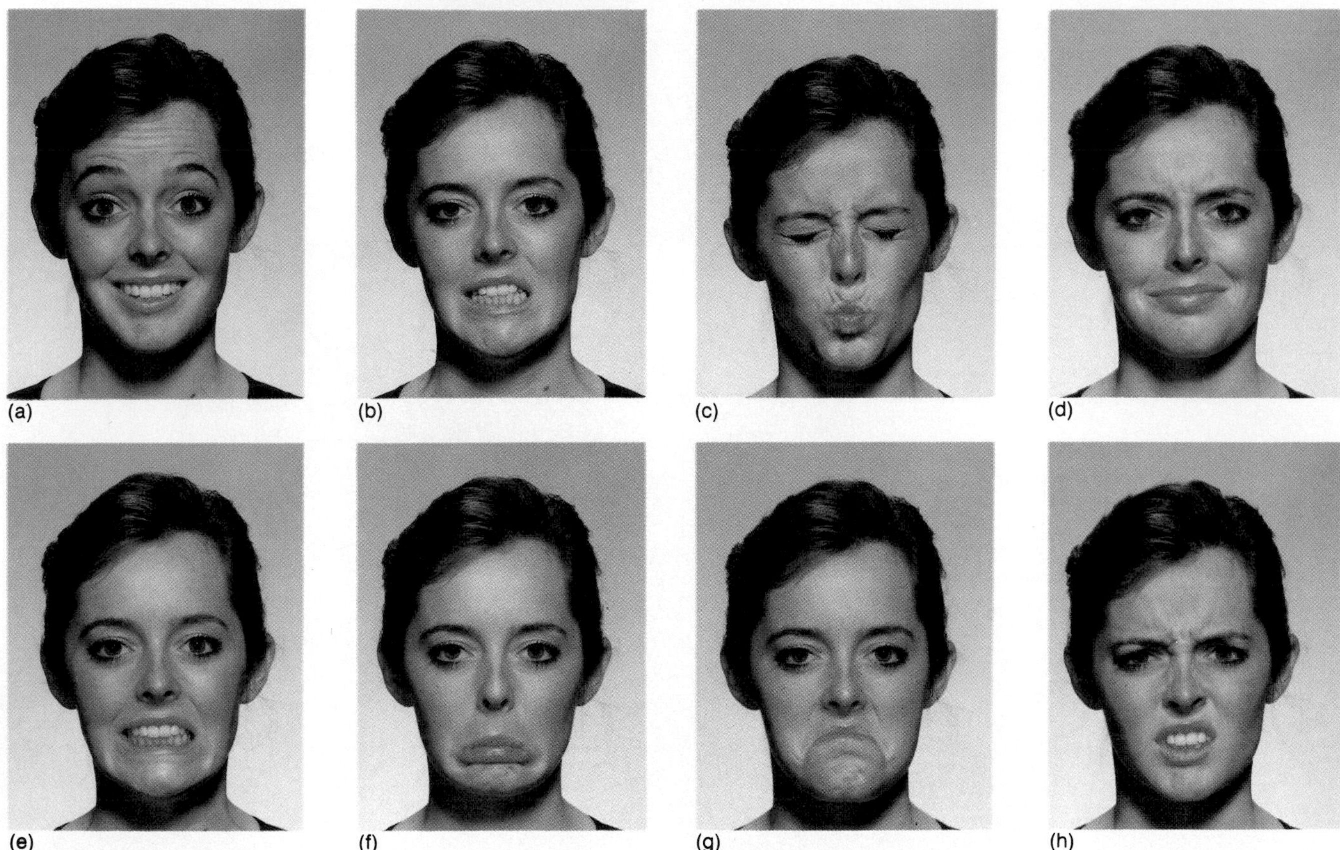

(a) (b) (c) (d) (e) (f) (g) (h)

Mentalis. The mentalis originates on the chin and inserts on the orbicularis oris muscle. The mentalis elevates the lower lip and protrudes it, as in pouting.

Platysma. The platysma is a thin, superficial, sheetlike muscle extending from its origin on the fascia of the pectoralis and deltoid muscles to the lower border of the mandible. The platysma draws the corners of the mouth backward, widening the mouth, as in an expression of horror. It may also assist in opening the mouth.

Buccinator. The buccinator *(buk'sĭ-na''tor)* is a deep cheek muscle that originates on the maxilla and mandible and crosses horizontally to insert on the orbicularis oris muscle. It compresses the cheek, and so aids the manipulation of food.

The location and action of the facial muscles are summarized in table 12.5. All the muscles of facial expression are innervated by the facial nerve (VII) and receive their blood supply from branches of the external carotid artery.

Muscles of Mastication

Four pairs of muscles are responsible for the biting and grinding movements of the lower jaw during mastication (figs. 12.30, 12.32). Each extends from the skull to the mandible, is innervated by the mandibular division of the trigeminal nerve (V), and is served blood from branches of the external carotid artery.

Temporalis. The temporalis is the largest of the four pairs. It is a convergent-fibered, fan-shaped muscle that originates on the temporal fossa, passes medially to the zygomatic arch, and inserts on the coronoid process of the mandible. The temporalis closes the mouth by elevating the mandible.

Masseter. The masseter *(mas-se'ter)* is positioned at the angle of the jaw. It originates on the zygomatic arch and inserts on the lateral surface of the ramus of the mandible. It acts synergistically with the temporalis in closing the mouth by elevating the mandible.

Medial Pterygoid. Both pterygoid muscles are positioned on the medial (internal) side of the mandible (fig. 12.32). The medial pterygoid originates on the pterygoid process of the sphenoid bone and inserts on the ramus of the mandible. It elevates the mandible and moves it laterally.

Table 12.5 Muscles of facial expression

Muscle	Derivation	Origin	Insertion	Action	Innervation
Epicranius	*epi* = above; *crani* = skull	Galea aponeurotica and occipital bone	Skin of eyebrow and galea aponeurotica	Wrinkles forehead and moves scalp	Facial
Frontalis	Referring to frontal bone	Galea aponeurotica	Skin of eyebrow	Wrinkles forehead and elevates eyebrow	Facial
Occipitalis	Referring to occipital bone	Occipital bone and mastoid process	Galea aponeurotica	Moves scalp backward	Facial
Corrugator	*ruga* = a wrinkle	Fascia above eyebrow	Root of nose	Draws eyebrows toward midline	Facial
Orbicularis oculi	*orbis* = orbit, circular; *oculus* = eye	Bones of medial orbit	Tissue of eyelid	Closes eye	Facial
Nasalis	*nasus* = nose	Maxilla and nasal cartilage	Aponeurosis of nose	Compresses nostrils	Facial
Orbicularis oris	*orb* = circular; *or* = mouth	Fascia surrounding lips	Mucosa of lips	Closes and purses lips	Facial
Levator labii superioris	*levator* = lifter; *labii* = lip; *superior* = upper	Upper maxilla and zygomatic bone	Orbicularis oris and skin above lips	Elevates upper lip	Facial
Zygomaticus	Referring to zygomatic bone	Zygomatic bone	Superior corner of orbicularis oris	Elevates corner of mouth	Facial
Risorius	*risor* = laughter	Fascia of cheek	Orbicularis oris at corner of mouth	Draws angle of mouth laterally	Facial
Triangularis	Referring to triangular shape	Mandible	Inferior corner of orbicularis oris	Depresses corner of mouth	Facial
Depressor labii inferioris	*depressor* = depress; *labii* = lip; *inferior* = lower	Mandible	Orbicularis oris and skin of lower lip	Depresses lower lip	Facial
Mentalis	*mentum* = chin	Mandible (chin)	Orbicularis oris	Elevates and protrudes lower lip	Facial
Platysma	*platy* = broad	Fascia of neck and chest	Inferior border of mandible	Depresses lower lip	Facial
Buccinator	*bucca* = cheek	Maxilla and mandible	Orbicularis oris	Compresses cheek	Facial

Figure 12.32. Muscles of mastication. (*a*) a superficial view; (*b*) a deep view.

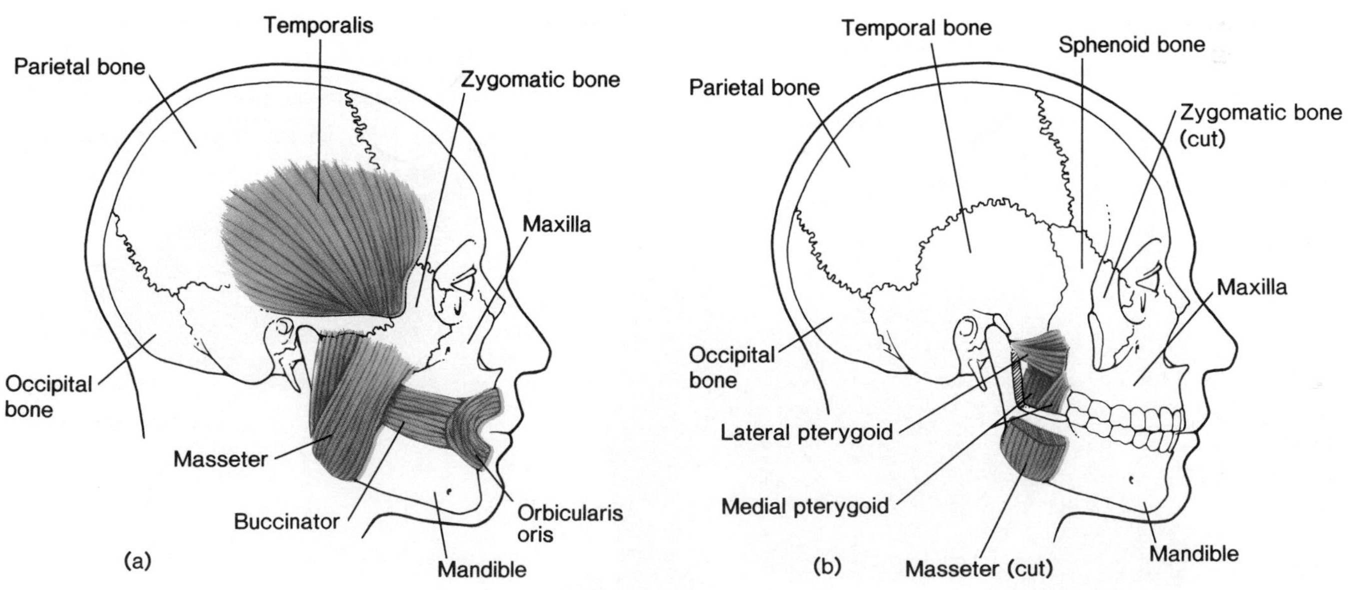

Lateral Pterygoid. The lateral pterygoid muscle also originates on the pterygoid process of the sphenoid. It inserts near the condyle of the mandible and acts to protract the mandible and provide limited lateral movement of the mandible.

A summary of the muscles of mastication is presented in table 12.6.

> Tetanus is a bacterial disease caused by anaerobic *Clostridium tetani* being introduced into the body, usually from a puncture wound. The bacteria produce a neurotoxin that is carried to the spinal cord by sensory nerves. The motor impulses relayed back cause certain muscles to contract continuously (tetany). The muscles that move the mandible are affected first, which has given the disease the common name of *lockjaw*.

Ocular Muscles

The movements of the eyeball are controlled by six extrinsic eye muscles called the extrinsic ocular muscles (fig. 12.33). Five of these muscles arise from the margin of the optic foramen at the back of the orbital cavity and insert on the outer layer (sclera) of the eyeball. Four **rectus muscles** maneuver the eyeball in the direction indicated by their names (**superior, inferior, lateral,** and **medial**), and two **oblique muscles** (**superior** and **inferior**) rotate the eyeball on its axis. The medial rectus on one side contracts with the medial rectus of the opposite eye when focusing on close objects. When looking to the side, the lateral rectus of one eyeball works with the medial rectus of the opposite eyeball to keep both eyes focused together. The superior oblique muscle passes through a pulleylike cartilaginous loop, the *trochlea,* before attaching to the eyeball. The ocular muscles are innervated by three cranial nerves (table 12.7).

Table 12.6 Muscles of mastication

Muscle	Derivation	Origin	Insertion	Action	Innervation
Temporalis	*templus* = time, temple	Temporal fossa	Coronoid process of mandible	Elevates mandible	Trigeminal
Masseter	*maseter* = chew	Zygomatic arch	Lateral ramus of mandible	Elevates mandible	Trigeminal
Medial pterygoid	*medial* = toward midline; *pteryg* = winglike (referring to process of sphenoid bone)	Sphenoid bone	Medial ramus of mandible	Elevates mandible and moves mandible laterally	Trigeminal
Lateral pterygoid	*latus* = side	Sphenoid bone	Anterior side of mandibular condyle	Protracts mandible	Trigeminal

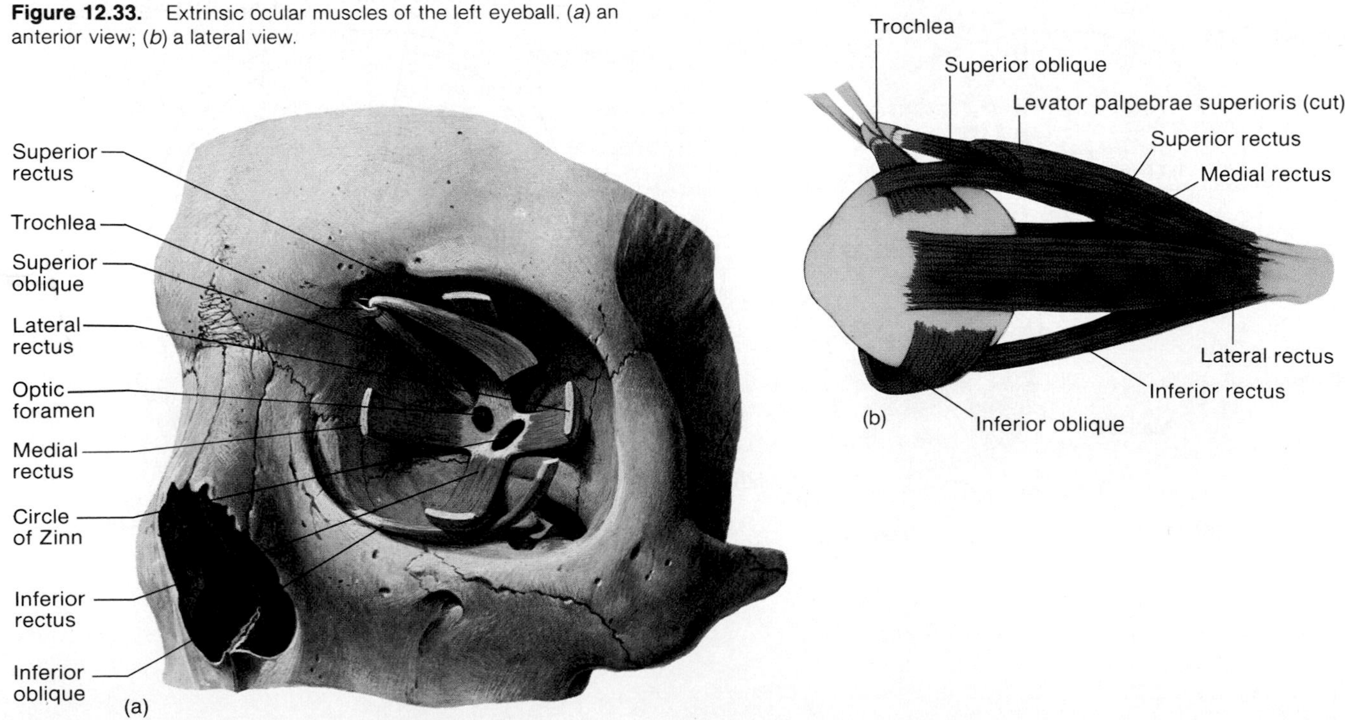

Figure 12.33. Extrinsic ocular muscles of the left eyeball. (*a*) an anterior view; (*b*) a lateral view.

Another muscle, the **levator palpebrae** *(le-va'tor pal'pĕ-bre)* **superioris,** is located in the ocular region but is not attached to the eyeball. It extends into the upper eyelid and raises the eyelid when contracted.

Table 12.7 Ocular muscles

Muscle	Cranial nerve innervation	Movement of eyeball
Lateral rectus	Abducens	Lateral
Medial rectus	Oculomotor	Medial
Superior rectus	Oculomotor	Superior and medial
Inferior rectus	Oculomotor	Inferior and medial
Inferior oblique	Oculomotor	Superior and lateral
Superior oblique	Trochlear	Inferior and lateral

Muscles That Move the Tongue

The tongue is a highly specialized muscular organ that functions in speaking, manipulating food during mastication, cleansing the teeth, and swallowing. Two groups of muscles are responsible for tongue movement: intrinsic and extrinsic. The *intrinsic muscles* are confined within the tongue and are responsible for its mobility and changes of shape. This group includes numerous unnamed bundles. *Extrinsic tongue muscles* are those that originate on structures away from the tongue and insert onto it to cause gross tongue movement (see fig. 12.34 and table 12.8). The four paired extrinsic muscles are the genioglossus, styloglossus, hyoglossus, and stylohyoid.

Table 12.8 Extrinsic tongue muscles

Muscle	Derivation	Origin	Insertion	Action	Innervation
Genioglossus	*geneion* = chin; *glossus* = tongue	Mental spine of mandible	Undersurface of tongue	Depresses and protracts tongue	Hypoglossal
Styloglossus	*stylo* = referring to styloid process of skull	Styloid process of temporal bone	Lateral side and undersurface of tongue	Elevates and retracts tongue	Hypoglossal
Hyoglossus	*hyo* = referring to hyoid bone	Body of hyoid	Side of tongue	Depresses sides of tongue	Hypoglossal
Stylohyoid		Styloid process of temporal bone	Body of hyoid	Elevates and retracts hyoid bone and tongue	Facial

Figure 12.34. Extrinsic muscles of the tongue.

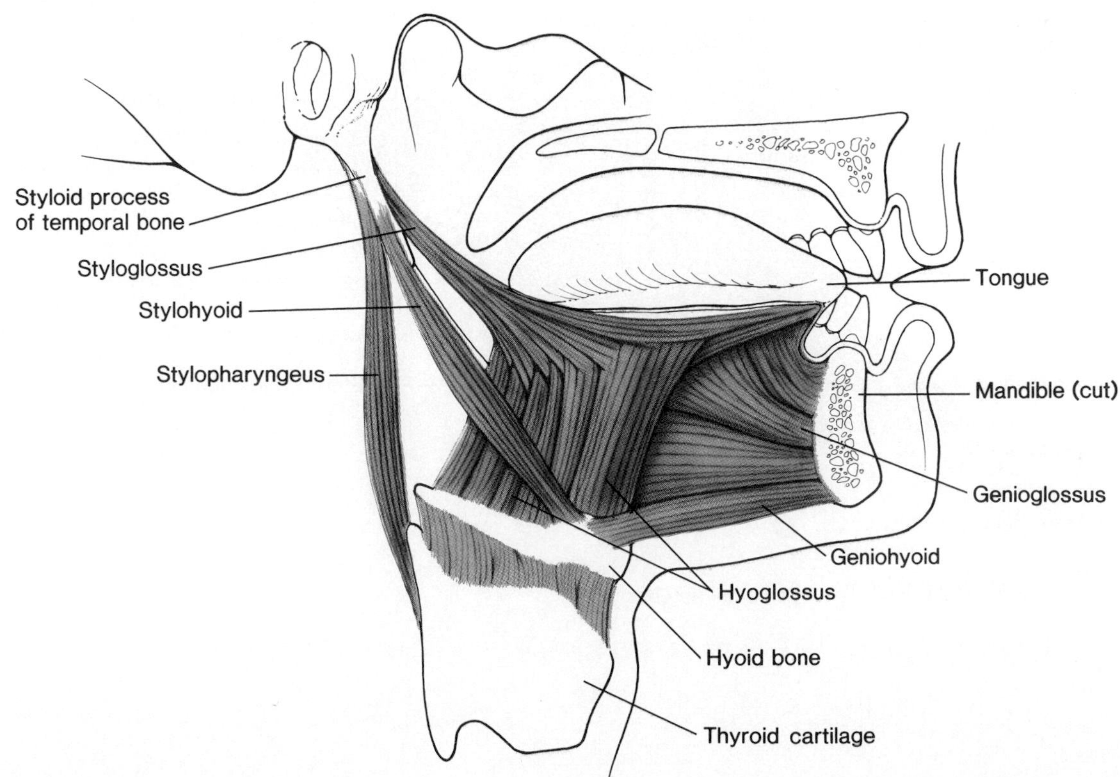

Figure 12.35. Deep muscles in the back of the neck.

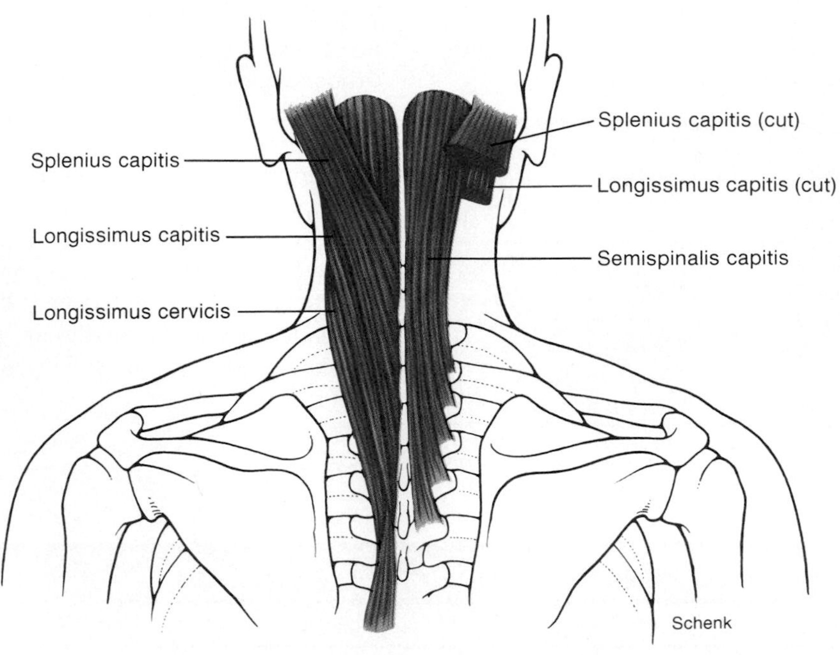

Figure 12.36. Muscles of the anterior and lateral neck regions.

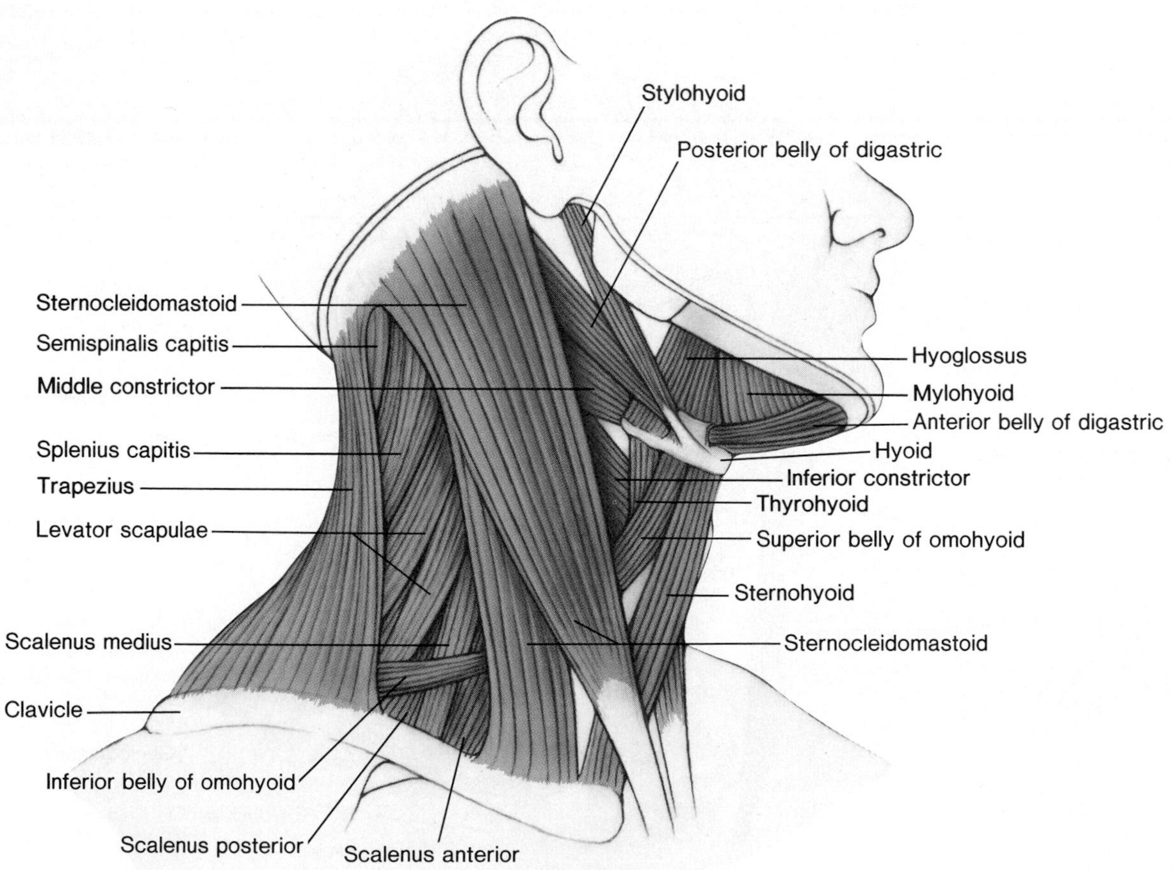

Table 12.9 Muscles of the neck

Muscle	Derivation	Origin	Insertion	Action	Innervation
Sternocleidomastoid	*sternum* = referring to sternal bone; *cleido* = clavicle; *mastoid* = mastoid process of skull	Sternum; clavicle	Mastoid process of temporal bone	Turns head to side; flexes neck and head	Accessory
Digastric	*di* = two; *gaster* = belly	Inferior border of mandible; mastoid groove	Hyoid bone	Opens mouth; elevates hyoid	Trigeminal (ant. belly); facial (post. belly)
Mylohyoid	*mylos* = akin to; *hyo*=hyoid bone	Inferior border of mandible	Body of hyoid and median raphe	Elevates hyoid bone and floor of mouth	Trigeminal
Stylohyoid	*stylo* = referring to styloid process	Styloid process of temporal bone	Body of hyoid	Elevates and retracts tongue	Facial
Hyoglossus	*hyo* = hyoid bone; *glossus*=tongue	Body of hyoid bone	Side of tongue	Depresses side of tongue	Hypoglossal
Sternohyoid		Manubrium	Body of hyoid	Depresses hyoid	Hypoglossal
Sternothyroid	*thyro* = shield shaped	Manubrium	Thyroid cartilage	Depresses thyroid cartilage	Hypoglossal
Thyrohyoid		Thyroid cartilage	Great cornu of hyoid	Depresses hyoid; elevates thyroid	Hypoglossal
Omohyoid	*omo* = shoulder	Superior border of scapula	Clavicle; body of hyoid	Depresses hyoid	Hypoglossal

Genioglossus. The thick genioglossus *(je''ne-o-glos'us)* muscle arises from the mandible on the inner side of the chin and fans out to insert upon the underside of the tongue. When the posterior portion of the genioglossus is contracted, the tongue is depressed and thrust forward. If both genioglossus muscles are contracted together along their entire lengths, the dorsal surface of the tongue becomes transversely concave. This muscle is extremely important in infants for sucking; the tongue is positioned around the nipple with a concave groove channeled toward the pharynx.

Styloglossus. The straplike styloglossus muscle originates from the styloid process of the temporal bone and inserts on the side and undersurface of the tongue. It is an antagonist to the genioglossus in that it retracts and elevates the tongue.

Hyoglossus. The short hyoglossus muscle originates from the hyoid bone and inserts on the side of the tongue. This muscle depresses the tongue, especially along the sides.

These three extrinsic tongue muscles (genioglossus, styloglossus, hyoglossus) are innervated with motor fibers by the hypoglossal nerve (XII).

Stylohyoid. The stylohyoid muscle does not attach directly to the tongue but has an indirect effect on the tongue as it elevates the hyoid bone where it inserts. The stylohyoid originates from the styloid process of the temporal bone and is innervated by the facial nerve (VII).

Muscles of the Neck

Muscles of the neck either support and move the head or are associated with structures within the neck region, such as the hyoid bone and larynx. Only the more obvious neck muscles will be considered in this chapter.

Several of the muscles in this section and the sections to follow can be observed on yourself. Refer to chapter 13 to determine which muscles form important surface landmarks. These muscles are illustrated in figures 12.35 and 12.36 and are summarized in table 12.9.

Posterior Muscles. The posterior muscles include the sternocleidomastoid (originates anteriorly), trapezius, splenius capitis, semispinalis capitis, and longissimus capitis.

As the name implies, the **sternocleidomastoid** *(ster''no-kli''do-mas'toid)* muscle originates on the sternum and clavicle and inserts on the mastoid process of the skull (fig. 12.36). When contracted on one side, it turns the head sideways in a direction opposite the side on which the muscle is located. If both sternocleidomastoid muscles are contracted, the head is pulled forward and down. The sternocleidomastoid is covered by the platysma, which has already been described (see fig. 12.30), and is innervated by the accessory nerve (XI).

Although a portion of the **trapezius** muscle extends over the posterior neck region, it is primarily a superficial muscle of the back and will be described later. The trapezius is served by the accessory nerve and C3 and C4 spinal nerves.

The **splenius capitis** is a broad muscle positioned deep to the trapezius. It originates on the ligamentum nuchae and the spinous processes of the seventh cervical and first three thoracic vertebrae. It inserts on the back of the skull below the superior nuchal line and on the mastoid process of the temporal bone. When the splenius capitis contracts on one side, the head rotates and extends to one side. Contracted together, these muscles extend the head at the neck. Further contraction causes hyperextension of the neck and head.

The broad, sheetlike **semispinalis capitis** muscle extends upward from the seventh cervical and first six thoracic vertebrae to insert on the occipital bone. When the two semispinalis capitis muscles contract together, they function to extend the head at the neck along with the splenius capitis muscle. If one of the muscles acts alone, the head is rotated to the side.

The narrow, straplike **longissimus capitis** muscle ascends from processes of the lower four cervical and upper five thoracic vertebrae and inserts on the mastoid process of the temporal bone. This muscle extends the head at the neck, bends it to the one side, or rotates it slightly.

Suprahyoid Muscles. The group of suprahyoid muscles located above the hyoid bone includes the digastric, mylohyoid, stylohyoid, and hyoglossus (fig. 12.36).

The **digastric** is a two-bellied muscle of double origin that inserts on the hyoid bone. The anterior origin is on the mandible at the point of the chin, and the posterior origin is near the mastoid process of the skull. The digastric can open the mouth or elevate the hyoid.

The **mylohyoid** forms the floor of the mouth. It originates on the inferior border of the mandible and inserts on the median raphe and body of the hyoid. As this muscle contracts, the floor of the mouth is elevated. It aids swallowing by forcing the food toward the back of the mouth.

The slender **stylohyoid** muscle extends from the styloid process of the skull to the hyoid bone, which it elevates as it contracts. The secondary effect of this muscle on tongue movement has already been described.

Infrahyoid Muscles. Infrahyoid muscles are paired, thin, straplike muscles located below the hyoid bone. They are individually named on the basis of their origin and insertion and include the sternohyoid, sternothyroid, thyrohyoid, and omohyoid (fig. 12.36).

The **sternohyoid** muscle originates on the manubrium of the sternum and inserts on the hyoid. It depresses the hyoid bone as it contracts.

The **sternothyroid** (not illustrated) muscle also originates on the manubrium but inserts on the thyroid cartilage of the larynx. When this muscle contracts, the larynx is pulled downward.

The short **thyrohyoid** muscle extends from the thyroid cartilage to the hyoid bone. Its actions are to elevate the larynx and lower the hyoid bone.

The long, thin **omohyoid** muscle originates on the superior border of the scapula and inserts on the clavicle bone and on the hyoid bone. It acts to depress the hyoid bone.

The coordinated movements of the hyoid bone and the larynx are impressive. The hyoid bone does not articulate with any other bone, yet it has eight paired muscles attached to it. Two paired muscles involve tongue movement, one paired muscle lowers the jaw, one muscle elevates the floor of the mouth, and four paired muscles depress the hyoid or elevate the thyroid cartilage of the larynx.

Muscles of Respiration

The muscles of respiration are skeletal muscles that continually and rhythmically contract, usually involuntarily. Breathing, or *pulmonary ventilation,* is divided into two phases: inspiration (inhalation) and expiration (exhalation).

During normal, relaxed inspiration, the important muscles are the **diaphragm** *(di'ah-fram)* and the **external intercostal** muscles (fig. 12.37). A contraction of the dome-shaped diaphragm downward causes a vertical increase in thoracic dimension. A simultaneous contraction of the external intercostals produces an increase in the lateral dimension of the thorax. The external intercostal muscles arise from the lower portion of the first eleven ribs, and the fibers extend downward and forward to insert upon the superior borders of the ribs below. The external intercostal muscles are innervated by the intercostal nerves, and the diaphragm receives its stimuli through the phrenic nerves.

Expiration is primarily a passive process, occurring as the muscles of inspiration are relaxed and the rib cage recoils to its original position. During forced expiration, the **internal intercostal** muscles contract, causing the rib cage to be depressed. The internal intercostals lie under the external intercostals, and their fibers are directed downward and backward. They also have intercostal nerve innervation. The abdominal muscles may also contract during forced expiration, which increases pressure within the abdominal cavity and forces the diaphragm superiorly, squeezing additional air out of the lungs.

The muscles of respiration are summarized in table 12.10.

Muscles of the Abdominal Wall

The anterolateral abdominal wall is composed of four pairs of flat, sheetlike muscles: the **external oblique, internal oblique, transversus abdominis,** and **rectus abdominis** (figs. 12.37, 12.38, and see fig. 12.43). These muscles support and protect the organs of the abdominal cavity and aid in breathing. When they contract, the pressure in the abdominal cavity increases, which can aid in defecation and in stabilizing the spine during heavy lifting.

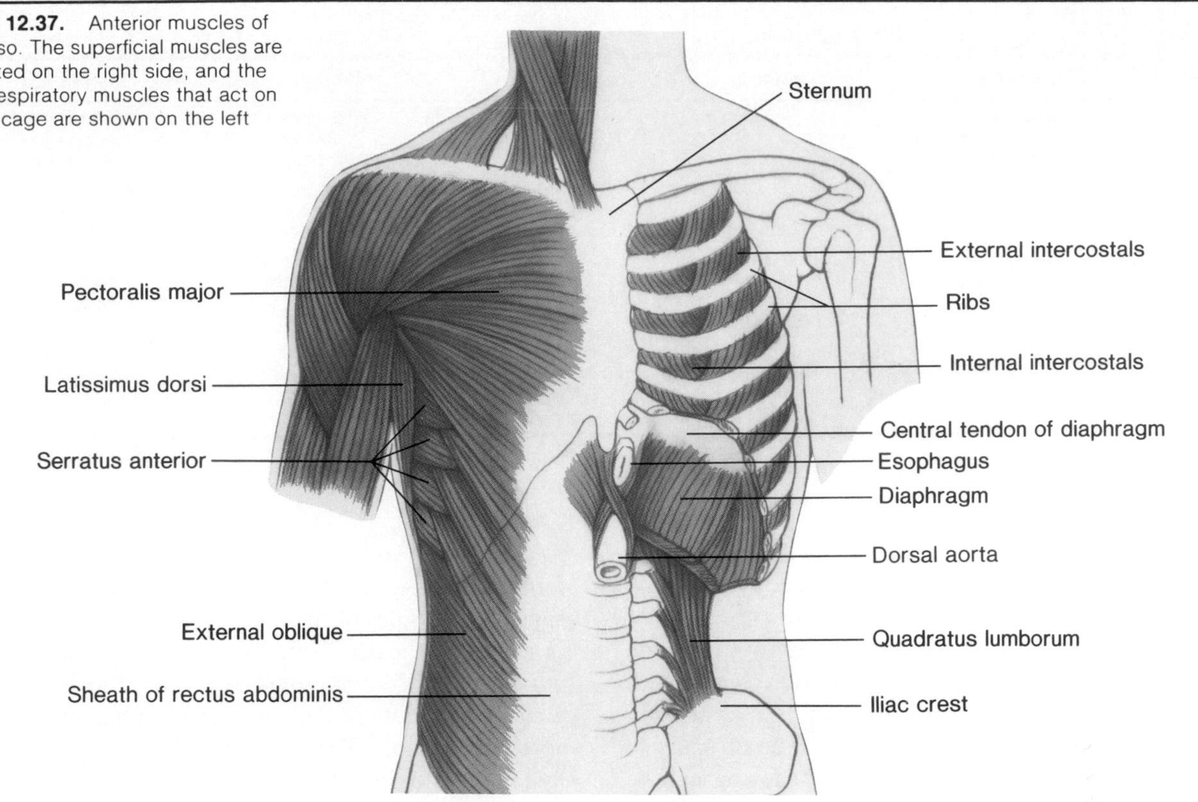

Figure 12.37. Anterior muscles of the torso. The superficial muscles are illustrated on the right side, and the deep respiratory muscles that act on the rib cage are shown on the left side.

Figure 12.38. Muscles of the anterior abdominal wall as seen in transverse section.

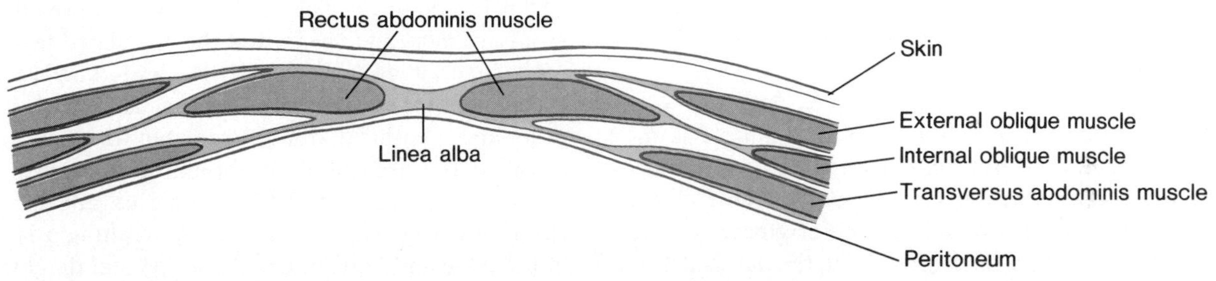

Table 12.10 Muscles of respiration

Muscle	Derivation	Origin	Insertion	Action	Innervation
Diaphragm	*dia* = across; *phragm* = wall	Xiphoid process, costal cartilages of last six ribs, and lumbar vertebrae	Central tendon	Pulls central tendon inferiorly and increases vertical dimension of thorax	Phrenic
External intercostals	*inter* = between; *costa* = rib	Inferior border of a rib	Superior border of a rib below	Elevates and draws ribs together	Intercostal
Internal intercostals		Superior border of a rib	Inferior border of a rib above	Draws ribs together and depresses thorax	Intercostal

Table 12.11 Muscles of the abdominal wall

Muscle	Derivation	Origin	Insertion	Action	Innervation
External oblique	*external* = closer to surface; *oblique* = diagonal	Lower eight ribs	Iliac crest, linea alba	Compresses abdomen; lateral rotation	Intercostals, iliohypogastric, and ilioinguinal
Internal oblique	*internal* = deep	Iliac crest, inguinal ligament, and lumbodorsal fascia	Linea alba, costal cartilage of last three or four ribs	Compresses abdomen; lateral rotation	Intercostals, iliohypogastric, and ilioinguinal
Transversus abdominis	*transverse* = horizontal; *abdomino* = belly	Iliac crest, inguinal ligament, lumbar fascia, and costal cartilage of last six ribs	Xiphoid process, linea alba, and pubis	Compresses abdomen	Intercostals, iliohypogastric, and ilioinguinal
Rectus abdominis	*rectus* = straplike; straight	Pubic crest and symphysis pubis	Costal cartilage of fifth to seventh ribs and xiphoid process	Flexes vertebral column	Intercostals

External Oblique. The external oblique muscle is the strongest and most superficial of the three layered muscles of the lateral abdominal wall. Its fibers are directed inferiorly and medially. The external oblique originates on the inferior surfaces of the lower eight ribs in contact with the serratus anterior muscle. Its insertion is a broad, thin aponeurosis that unites the external oblique of one side with its counterpart on the opposite side. The aponeurosis forms the *linea alba* ("white line"), a prominent band of connective tissue on the midline of the abdomen, and the *inguinal ligament* (see fig. 12.51), which extends from the anterior superior spine of the ilium to the pubic tubercle. This muscle has extensive innervation from branches of intercostal nerves eight through twelve and from the iliohypogastric and ilioinguinal nerves.

Internal Oblique. The internal oblique muscle is located between the external oblique and the transversus abdominis. Its fibers run at right angles to those of the external oblique. The internal oblique originates on the inguinal ligament, the iliac crest, and the lumbodorsal fascia of the back. Its primary insertion is on the linea alba and the pubic crest. An anterior portion of this muscle inserts upon the costal cartilages of the lower three ribs. The internal oblique is innervated by branches of the same nerves that serve the external oblique.

Transversus Abdominis. The deepest of the abdominal muscles is the transversus abdominis. It has an extensive origin on the inguinal ligament, the iliac crest, the lumbodorsal fascia, and the last six ribs. The fibers of the transversus abdominis muscle pass directly medially around the abdominal wall to insert upon the linea alba and the pubic crest. The transversus abdominis is innervated by branches of intercostal nerves seven through twelve and by iliohypogastric and ilioinguinal nerves.

Rectus Abdominis. The long, straplike rectus abdominis muscle (see fig. 12.43) extends from the pubis to the rib cage on both sides of the linea alba. It is entirely enclosed in a fibrous sheath formed from the aponeuroses of the other three abdominal muscles. Contraction of the rectus abdominis results in flexion of the lumbar portion of the vertebral column. The rectus abdominis muscle is served by branches of intercostal nerves seven through twelve.

Refer to table 12.11 for a summary of the muscles of the abdominal wall.

Muscles of the Pelvic Outlet

The pelvic outlet, or floor of the pelvis, is formed by two groups of muscular sheets and the associated fascia. The more superior group is called the **pelvic diaphragm** and consists of the levator ani and coccygeus muscles. The anterior and superficial group, associated with the genitalia, is called the **urogenital diaphragm** and consists of the transversus perinei, the bulbospongiosus, and the ischiocavernosus (fig. 12.39). The pelvic diaphragm is similar in the male and female, but the urogenital diaphragm of each sex differs markedly. The muscles of the pelvic outlet are innervated by the pudendal nerve or perineal branches of the pudendal nerve.

Levator Ani. The levator ani muscle arises from the pubic bone and the spine of the ischium and inserts at the midline on the coccyx and on the anterior fibers of the same muscle of the opposite side. The two levator ani muscles together form a thin sheet of muscle that helps to support the pelvic viscera and constrict the lower part of the rectum, pulling it forward and aiding defecation.

Coccygeus. The fan-shaped coccygeus muscle arises from the spine of the ischium and inserts on the sacrum and the coccyx (not shown in the illustration). The coccygeus muscle aids the levator ani in its functions. Either or both of these muscles are occasionally stretched and torn during parturition.

Figure 12.39. Muscles of the pelvic outlet. (*a*) male; (*b*) female.

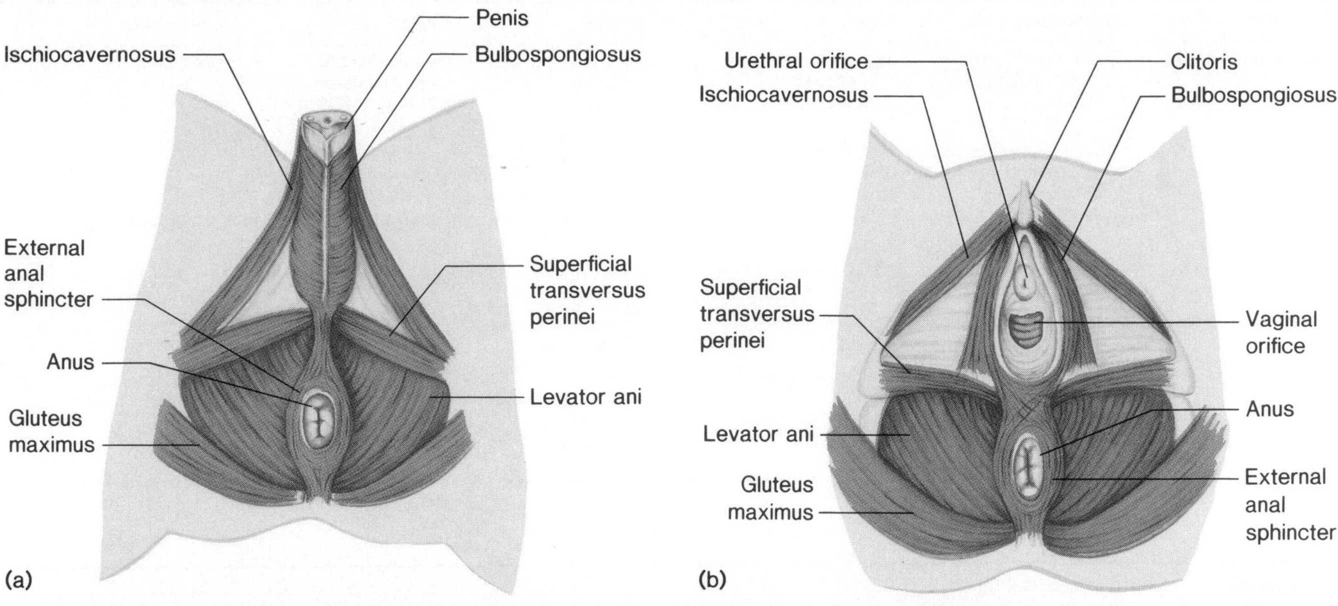

Table 12.12 Muscles of the pelvic outlet

Muscle	Derivation	Origin	Insertion	Action	Innervation
Levator ani	*anus* = ring	Spine of ischium and pubic bone	Coccyx	Supports pelvic viscera; aids in defecation	Pudendal plexus
Coccygeus	*coccyx* = like a cuckoo's bill	Ischial spine	Sacrum and coccyx	Supports pelvic viscera; aids in defecation	Perineal branch of pudendal nerve
Transversus perinei	*perineum* = referring to area between genitals and anus	Ischial tuberosity	Central tendon	Supports pelvic viscera	Pudendal plexus
Bulbospongiosus	*bulbus* = little bulb; *spongia* = spongelike	Central tendon	Males: base of penis; females: root of clitoris	Constricts urethral canal; constricts vagina	Perineal branch of pudendal nerve
Ischiocavernosus	*ischio* = ischium; *caverna* = hollow spaces	Ischial tuberosity	Males: pubic arch and crus of the penis Females: pubic arch and clitoris	Aids erection of penis or clitoris	Perineal branch of pudendal nerve

Transversus Perinei. The transversus perinei are small straplike muscles that arise from the ischial tuberosity, pass the genitalia, and insert on the central tendon of the midline. They assist other muscles in supporting the pelvic viscera.

Bulbospongiosus (Bulbocavernosus). In males, the bulbospongiosus muscle of one side unites with that of the opposite side to form a muscular constriction surrounding the base of the penis. When contracted, the two muscles constrict the urethral canal and assist in emptying the urethra. In females, these muscles are separated by the vaginal orifice, which they constrict as they contract.

Ischiocavernosus. The ischiocavernosus muscle originates from the tuberosity of the ischium and inserts onto the pubic arch and crus of the penis in the male and the pubic arch and clitoris of the female. This muscle aids the erection of the penis in the male and the clitoris in the female.

External Anal Sphincter. Although the external anal sphincter is not generally considered part of the pelvic outlet, it is in this region and is a skeletal muscle. This funnel-shaped constrictor muscle surrounds the anal canal and functions to keep the anal canal closed.

The muscles of the pelvic outlet are summarized in table 12.12.

Table 12.13 Muscles of the vertebral column

Muscle	Derivation	Origin	Insertion	Action	Innervation
Quadratus lumborum	*quad* = four; *lumb* = referring to lumbar region	Iliac crest and lower three lumbar vertebrae	Twelfth rib and upper four lumbar vertebrae	Extends lumbar region; lateral flexion of vertebral column	T12, L1–L4
Erector spinae	Consists of three groups of muscles: iliocostalis, longissimus, and spinalis. The iliocostalis and longissimus are further subdivided into three groups on the basis of location along the vertebral column.				
Iliocostalis lumborum	*ilium* = flank	Crest of ilium	Lower six ribs	Extends lumbar region	Dorsal rami of lumbar nerves
Iliocostalis thoracis	*thorax* = chest	Lower six ribs	Upper six ribs	Extends thoracic region	Dorsal rami of thoracic nerves
Iliocostalis cervicis	*cervix* = neck	Angles of third to sixth rib	Transverse processes of fourth to sixth cervical vertebrae	Extends cervical region	Dorsal rami of cervical nerves
Longissimus thoracis		Transverse processes of lumbar vertebrae	Transverse processes of all the thoracic vertebrae and lower nine ribs	Extends thoracic region	Dorsal rami of spinal nerves

Figure 12.40. Muscles of the vertebral column.

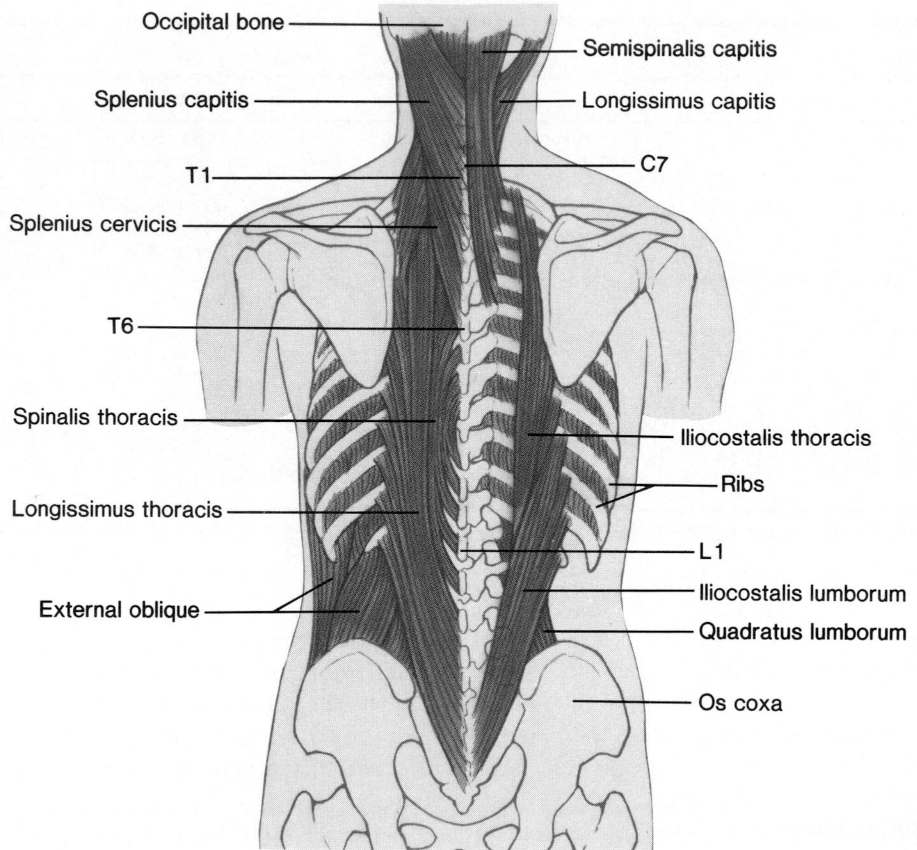

Muscle	Derivation	Origin	Insertion	Action	Innervation
Longissimus cervicis		Transverse processes of upper four or five thoracic vertebrae	Transverse processes of second to sixth cervical vertebrae	Extends cervical region and lateral flexion	Dorsal rami of spinal nerves
Longissimus capitis		Transverse processes of upper five thoracic vertebrae and articular processes of lower three cervical vertebrae	Posterior margin of cranium and mastoid processes	Extends head	Dorsal rami of middle and lower cervicals
Spinalis thoracis		Spinous processes of upper lumbar and lower thoracic vertebrae	Spinous processes of upper thoracic vertebrae	Extends vertebral column	Dorsal rami of spinal nerves

Muscles of the Vertebral Column

The muscles that move the vertebral column are strong and complex because they have to provide support and movement in resistance to the effect of gravity.

The vertebral column can be flexed, extended, abducted, adducted, and rotated. The muscle that flexes the vertebral column, the rectus abdominis, has already been described as a paired, straplike muscle of the anterior abdominal wall. The extensor muscles located on the posterior side of the vertebral column have to be stronger than the flexors because extension (such as lifting an object) is in opposition to gravity. The extensor muscles consist of a superficial group and a deep group. Only some of the muscles of the vertebral column will be described.

Quadratus Lumborum. The quadratus lumborum muscle is shown in figures 12.40 and 12.51. It originates on the iliac crest and the lower three lumbar vertebrae. It inserts on the transverse processes of the first four lumbar vertebrae and the inferior margin of the twelfth rib. When the right and left quadratus lumborum contract together, the vertebral column in the lumbar region extends. Separate contraction causes lateral flexion of the spine.

Erector Spinae (sacrospinalis). The erector spinae is a massive muscle group that extends from the sacrum to the skull. It actually consists of three groups of muscles: **iliocostalis, longissimus,** and **spinalis** (fig. 12.40). Each of these groups, in turn, consists of overlapping slips of muscle. The iliocostalis is the most lateral group, the longissimus is intermediate in position, and the spinalis, in medial position, is positioned in contact with the spinous processes of the vertebrae.

The erector spinae muscles are innervated by dorsal rami of spinal nerves. A summary of these muscles is presented in table 12.13.

The erector spinae muscles are frequently strained through improper lifting. A heavy object should not be lifted with the vertebral column flexed; instead, the thighs and knees should be flexed so that the pelvic and leg muscles can aid in the task.

The erector spinae muscles are also frequently strained in women during pregnancy. Pregnant women will try to counterbalance the effect of the protruding abdomen by hyperextending the vertebral column. This causes an exaggerated lumbar curvature, strained muscles, and a peculiar gait.

1. Identify the facial muscles that do the following: (a) wrinkle the forehead; (b) purse the lips; (c) protrude the lower lip; (d) produce frown lines; (e) smile; (f) frown; (g) wink; and (h) elevate the upper lip to show the teeth.
2. Describe the actions of the extrinsic muscles that move the tongue.
3. Which muscles of the neck either originate from or insert on the hyoid bone?
4. Describe the actions of the muscles of inspiration. Which muscles participate in forced expiration?
5. Which muscles of the pelvic outlet support the floor of the pelvic cavity, and which are associated with the genitalia?
6. What is the extent and the compartmentalization of the erector spinae muscle?

Muscles of the Appendicular Skeleton

The muscles of the appendicular skeleton include those of the pectoral girdle, arm, forearm, wrist, hand, and fingers, and those of the pelvic girdle, thigh, lower leg, ankle, foot, and toes.

Objective 22. Locate the major muscles of the appendicular skeleton; identify synergistic and antagonistic muscles, and describe the action of each.

Muscles That Act on the Pectoral Girdle

The shoulder is attached to the axial skeleton only at the sternoclavicular joint, so strong, straplike muscles are necessary. Furthermore, muscles that move the brachium originate on the scapula, and during brachial movement the scapula has to be held fixed. The muscles that act on the pectoral girdle originate on the axial skeleton and can be divided into anterior and posterior groups.

Anterior. The anterior group of muscles that act on the pectoral girdle includes the serratus anterior, pectoralis minor, and subclavius (fig. 12.41).

The **serratus** *(ser-ra'tus)* **anterior** muscle originates on the lateral surface of the first eight or nine ribs and inserts on the anterior portion of the vertebral border of the scapula. This muscle pulls the scapula forward and downward, holding it against the rib cage. The serratus anterior muscle is innervated by the long thoracic nerve.

The **pectoralis minor** muscle is deep to the pectoralis major on the anterior thoracic wall. It is a flat, triangular muscle originating on the anterior surface of the third, fourth, and fifth ribs and inserting on the coracoid process of the scapula. The pectoralis minor has several actions; the major one is to depress the shoulder while drawing the scapula laterally. It is innervated by the medial pectoral nerve.

The **subclavius** is a thin, straplike muscle that originates on the anterior surface of the first rib and inserts on the subclavian groove of the clavicle. This muscle depresses the clavicle and hence draws the shoulder downward.

Posterior. The posterior group of muscles that act on the pectoral girdle includes the trapezius, levator scapulae, and rhomboideus (fig. 12.42).

Treatment of advanced stages of *breast cancer* requires the surgical removal of both pectoralis major and pectoralis minor muscles in a procedure called a *radical mastectomy.* Postoperative physical therapy is primarily geared toward strengthening the synergistic muscles of this area. As the muscles that act on the brachium are learned, determine which are synergists with the pectoralis major.

Figure 12.41. Deep anterior thoracic muscles that move the pectoral girdle.

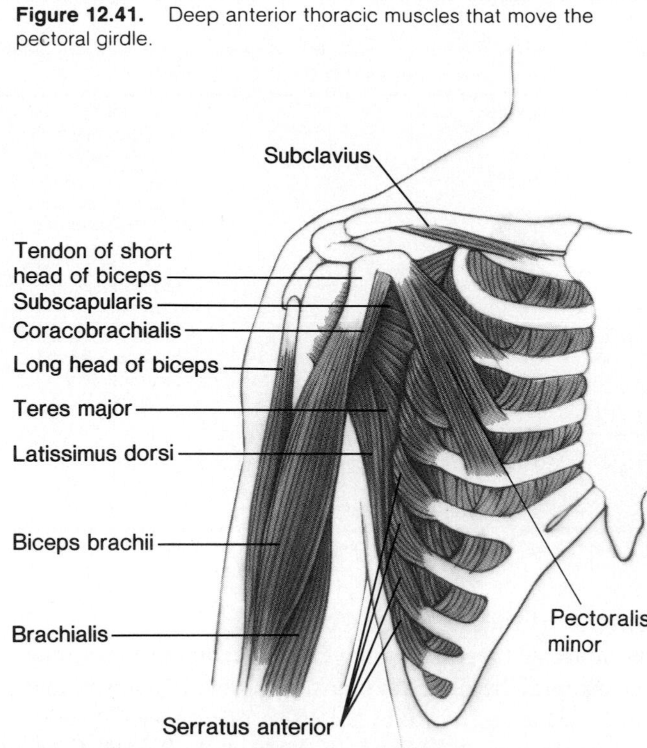

The **trapezius** is a large, superficial, triangular muscle of the neck and upper back. It arises on the occipital bone, the spinous process of the seventh cervical, and all of the thoracic vertebrae. It inserts on the clavicle, the acromion process of the scapula, and the spine of the scapula. The trapezius can adduct the scapula, elevate the scapula (as in shrugging the shoulder), and hyperextend the head. The trapezius is innervated by the accessory nerve (XI) and by branches of C3 and C4 spinal nerves.

The long, thin, straplike **levator scapulae** muscle originates on the first to fourth cervical vertebrae and attaches to the upper vertebral border of the scapula. When contracted, it elevates the scapula. This muscle is served by the dorsal scapular nerve.

The **major** and **minor rhomboids** are two separate but continuous muscles. They both extend from the spinous processes of the lower cervical and uppermost thoracic vertebrae to the vertebral border of the scapula. The rhomboideus major muscle is larger and positioned inferior to the minor muscle. Working together, they elevate and adduct the scapula. They are both served by the dorsal scapular nerve.

Table 12.14 summarizes the muscles that act on the pectoral girdle.

Figure 12.42. Posterior muscles of the neck, thorax, and abdomen. (Superficial muscles removed from right side.)

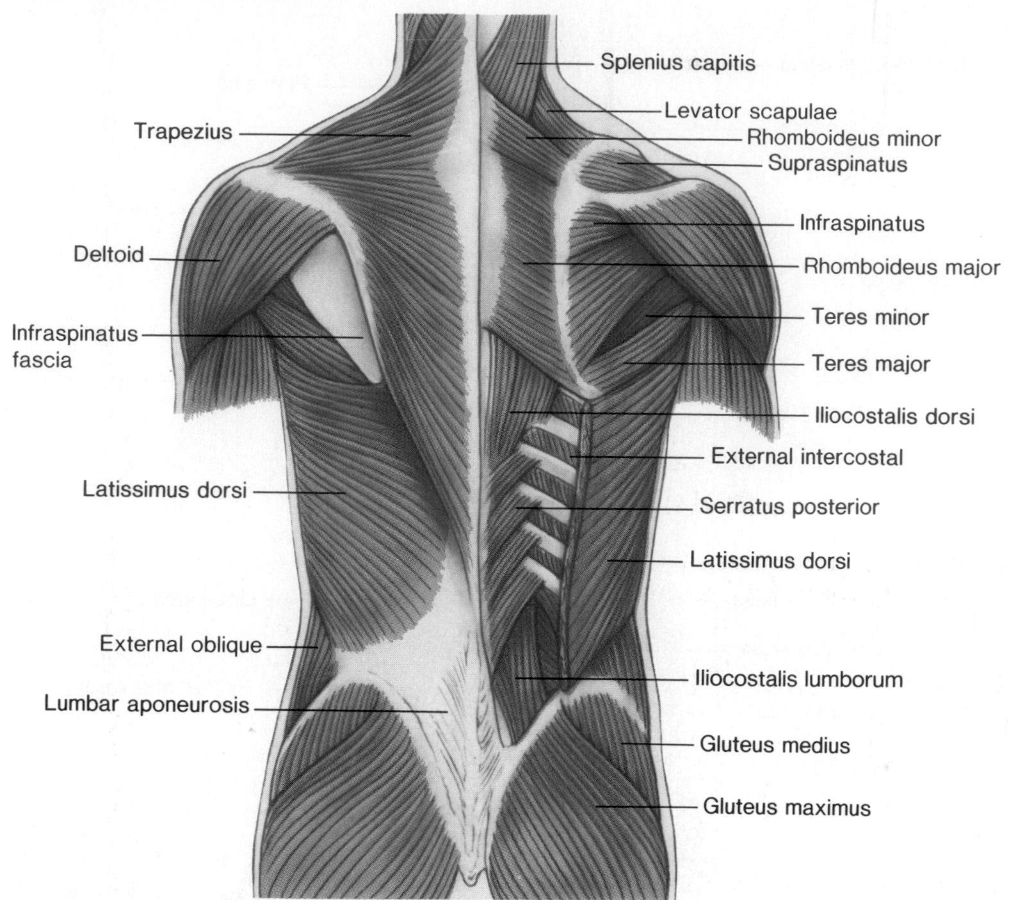

Table 12.14 Muscles that act on the pectoral girdle

Muscle	Derivation	Origin	Insertion	Action	Innervation
Serratus anterior	*serratus* = serrated; *anterior* = front	Upper eight or nine ribs	Anterior vertebral border of scapula	Pulls scapula forward and downward	Long thoracic
Pectoralis minor	*pectus* = chest; *minor* = lesser	Sternal ends of third, fourth, and fifth ribs	Coracoid process of scapula	Pulls scapula forward and downward	Medial pectoral
Subclavius	*sub* = below, under	First rib	Subclavian groove of clavicle	Draws clavicle downward	C5, C6
Trapezius	*trapezoeides* = trapezoid shaped	Occipital bone and spines of seventh cervical and all thoracic vertebrae	Clavicle, spine of scapula, and acromion process	Elevates scapula, draws head back, adducts scapula, braces shoulder	Accessory, C3, C4
Levator scapulae	*levator* = elevator; *scapulae* = scapula	First to fourth cervical vertebrae	Vertebral border of scapula	Elevates scapula	Dorsal scapular
Rhomboideus major	*rhomboides* = rhomboid shaped	Spines of second to fifth thoracic vertebrae	Vertebral border of scapula	Elevates and adducts scapula	Dorsal scapular
Rhomboideus minor		Seventh cervical and first thoracic vertebrae	Vertebral border of scapula	Elevates and adducts scapula	Dorsal scapular

Figure 12.43. Anterior muscles of the neck, thorax, and abdomen. (Superficial muscles removed from the right side.)

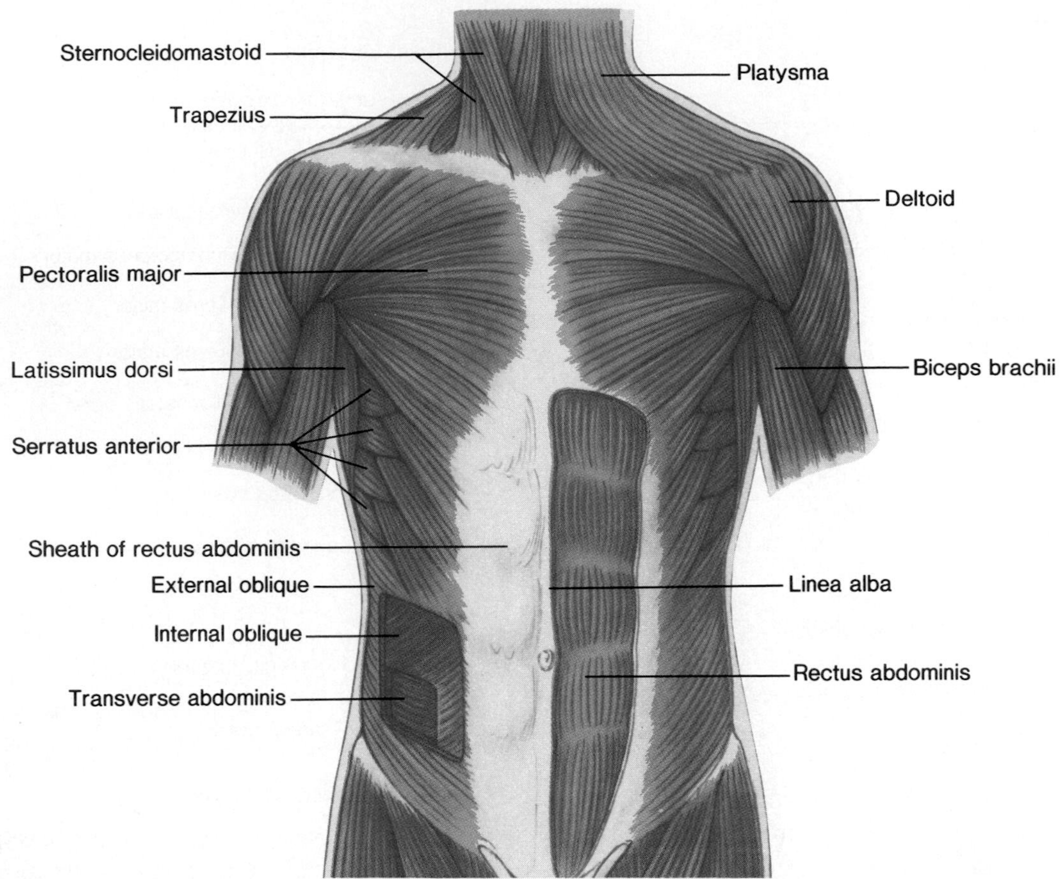

Muscles That Move the Humerus

Of the nine muscles that span the shoulder joint to insert on the humerus, only two of them, the pectoralis major and latissimus dorsi, do not originate on the scapula. These two are designated as axial muscles, whereas the remaining seven are scapular muscles. Figures 12.42 and 12.43 show the muscles of this region, and figure 12.44 shows the attachments of all the muscles that either originate or insert on the scapula.

In terms of their development, the pectoralis major and the latissimus dorsi muscles are not axial muscles at all. They develop in the forelimb and extend to the trunk secondarily. They are considered axial muscles only because their origins are on the axial skeleton.

Axial Muscles. The axial muscles include the pectoralis major and latissimus dorsi.

The **pectoralis major** is a large chest muscle that originates on the clavicle, sternum, costal cartilages, and rectus sheath (fig. 12.43). It inserts on the crest of the greater tubercle of the humerus. Its primary functions are to adduct, flex, and rotate the brachium medially. The pectoralis major is innervated by the medial and lateral anterior pectoral nerves.

The large, flat, triangular **latissimus dorsi** muscle covers the lower half of the thoracic region of the back (fig. 12.42). It arises from the broad lumbar aponeurosis that is attached to the spines of the lower thoracic, lumbar, and sacral vertebrae, and from the iliac crest and lower four ribs. The insertion of this muscle is on the intertubercular groove of the humerus. The latissimus dorsi is frequently called the "swimmer's muscle" because it powerfully adducts the arm, drawing it downward and backward while it rotates medially. It is innervated by the thoracodorsal nerve.

A latissimus dorsi muscle, conditioned with pulsated electrical impulses, will in time come to resemble cardiac muscle tissue in that it is nonfatigable and uses oxygen at a steady rate. Following conditioning, the muscle may be used in an autotransplant to repair a surgically removed portion of a patient's diseased heart. The procedure involves detaching the latissimus dorsi from its vertebral origin, leaving the blood supply and innervation intact, and slipping it into the pericardial cavity, where it is wrapped around the heart like a towel. A pacemaker is required to provide the continuous, rhythmic contractions.

Figure 12.44. A posterior view of the scapula and humerus, showing the areas of attachment of the associated muscles.

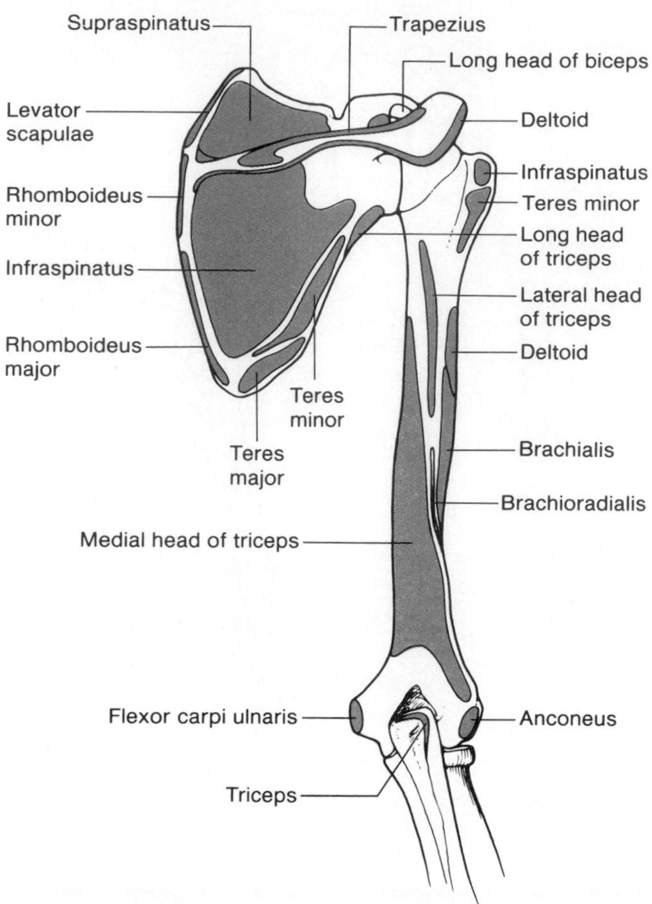

Scapular Muscles. The nonaxial, scapular muscles include the deltoid, supraspinatus, infraspinatus, teres major, teres minor, subscapularis, and coracobrachialis.

The **deltoid** is a thick, powerful muscle that caps the shoulder joint. It originates on the clavicle, acromion, and spine of the scapula and inserts on the deltoid tuberosity of the humerus (figs. 12.43, 12.44). It is innervated by the axillary nerve, and its function is to abduct, extend, and flex the brachium. The deltoid muscle is a common site for intramuscular injections.

The **supraspinatus** arises from the supraspinous fossa of the scapula, passes over the shoulder joint, and inserts on the greater tubercle of the humerus (see figs. 12.42 and 12.44). In addition to stabilizing the shoulder joint, it laterally rotates the humerus and is synergistic with the deltoid in abducting the brachium. The supraspinatus is innervated by the suprascapular nerve.

The **infraspinatus** muscle originates on the infraspinous fossa of the scapula and inserts on the posterior side of the greater tubercle of the humerus (see figs. 12.42, 12.44). Its function is to rotate the arm laterally, and its innervation is the suprascapular nerve.

The **teres major** arises from the inferior angle and lateral border of the scapula and attaches to the crest of the lesser tubercle immediately medial where the latissimus dorsi inserts (figs. 12.42, 12.44). The action of the teres major is similar to that of the latissimus dorsi, adducting and medially rotating the humerus. The teres major is innervated by the lower subscapular nerve.

The **teres minor** arises from the axillary border of the scapula and inserts on the posterior portion of the greater tubercle of the humerus (figs. 12.42, 12.44). This muscle works with the infraspinatus in laterally rotating the brachium. It is served by the axillary nerve.

The **subscapularis** is a deep muscle of the shoulder, located between the serratus anterior and the anterior surface of the scapula (see fig. 12.41). It originates on the subscapular fossa and inserts on the lesser tubercle of the humerus. The subscapularis is a strong stabilizer of the shoulder and also aids in medially rotating the brachium. It is innervated by the subscapular nerve.

The mass of the **coracobrachialis** is seen on the upper medial side of the brachium (see fig. 12.41). It has a tendinous origin on the coracoid process of the scapula and inserts on the medial surface of the humerus, about two-thirds of the way down the brachium. It is innervated by the musculocutaneous nerve and functions to flex and adduct the shoulder joint.

A summary of the muscles that act on the humerus is presented in table 12.15. The sites of attachments of the muscles of this region are shown in figure 12.44.

Muscles That Act on the Forearm

The powerful muscles of the brachium are responsible for movement of the forearm at the elbow and at the radioulnar joint. These muscles are the biceps brachii, brachialis, brachioradialis, and triceps brachii (fig. 12.45). A transverse section through the brachium in figure 12.46 provides a different perspective of the brachial region.

Biceps Brachii. The powerful biceps brachii muscle, positioned on the anterior surface of the humerus, is the most familiar muscle of the arm; yet it has no attachments on the humerus. The biceps has a dual origin: a medial tendinous head, the **short head,** arises from the coracoid process of the scapula, and the **long head** originates on the superior tuberosity of the glenoid fossa and passes through the shoulder joint and descends in the intertubercular groove on the humerus. Both heads of the biceps insert on the radial tuberosity and are served by the musculocutaneous nerve. When contracted, the biceps brachii flexes the forearm and also assists supination of the forearm and hand.

Brachialis. The brachialis is located on the distal anterior half of the humerus, deep to the biceps brachii. It originates directly from the periosteum of the humerus and inserts on the coronoid process of the ulna. The brachialis works synergistically with the biceps in flexing the forearm. It is innervated by the musculocutaneous, radial, and median nerves.

Figure 12.45. The right shoulder and brachium. (*a*) an anterior view; (*b*) a posterior view.

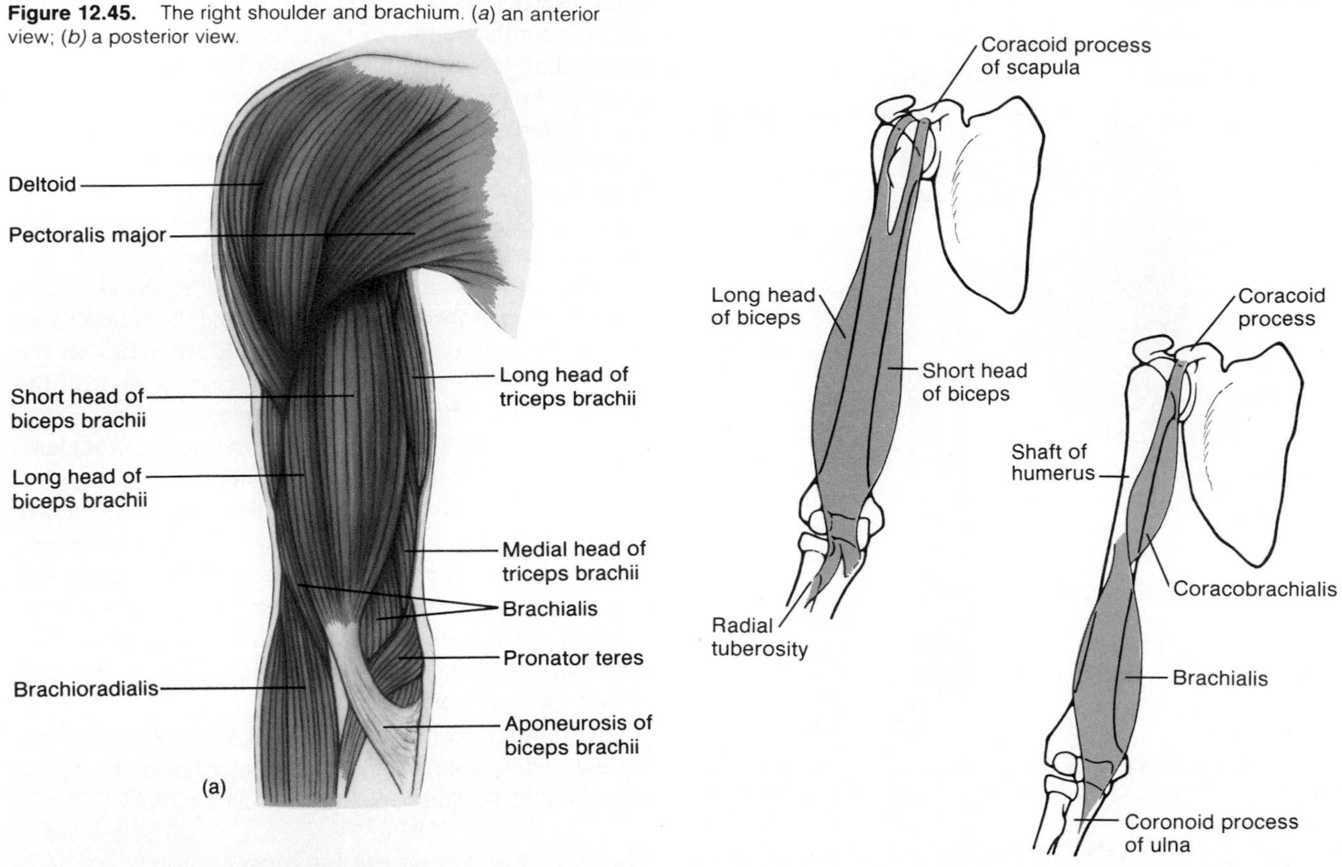

Table 12.15 Muscles that move the humerus

Muscle	Derivation	Origin	Insertion	Action	Innervation
Pectoralis major	*pectus* = chest; *major* = greater	Clavicle, sternum, costal cartilages of second to sixth rib; rectus sheath	Crest of greater tubercle of humerus	Flexes, adducts, and rotates arm medially	Medial and lateral pectoral
Latissimus dorsi	*latissimus* = widest; *dorsum* = back	Spines of sacral, lumbar, and lower thoracic vertebrae; iliac crest and lower four ribs	Intertubercular groove of humerus	Extends, adducts, and rotates humerus medially; retracts shoulder	Thoracodorsal
Deltoid	*delta* = triangular	Clavicle, acromion process; spine of scapula	Deltoid tuberosity of humerus	Abducts arm; extends or flexes humerus	Axillary
Supraspinatus	*supra* = above; *spina*=spine of scapula	Fossa—superior to spine of scapula	Greater tubercle of humerus	Abducts and laterally rotates humerus	Suprascapular
Infraspinatus	*infra* = below	Fossa—inferior to spine of scapula	Greater tubercle of humerus	Rotates arm laterally	Suprascapular
Teres major		Inferior angle and lateral border of scapula	Crest of lesser tubercle of humerus	Extends humerus, or adducts and rotates arm medially	Lower subscapular
Teres minor	*teres* = rounded; *minor* = lesser	Axillary border of scapula	Greater tubercle and groove of humerus	Rotates arm laterally	Axillary
Subscapularis	*sub* = below, under; *scapula* = scapula	Subscapular fossa	Lesser tubercle of humerus	Rotates arm medially	Subscapular
Coracobrachialis	*coraco* = referring to coracoid process; *brachium* = arm	Coracoid process of scapula	Shaft of humerus	Flexes and adducts shoulder joint	Musculocutaneous

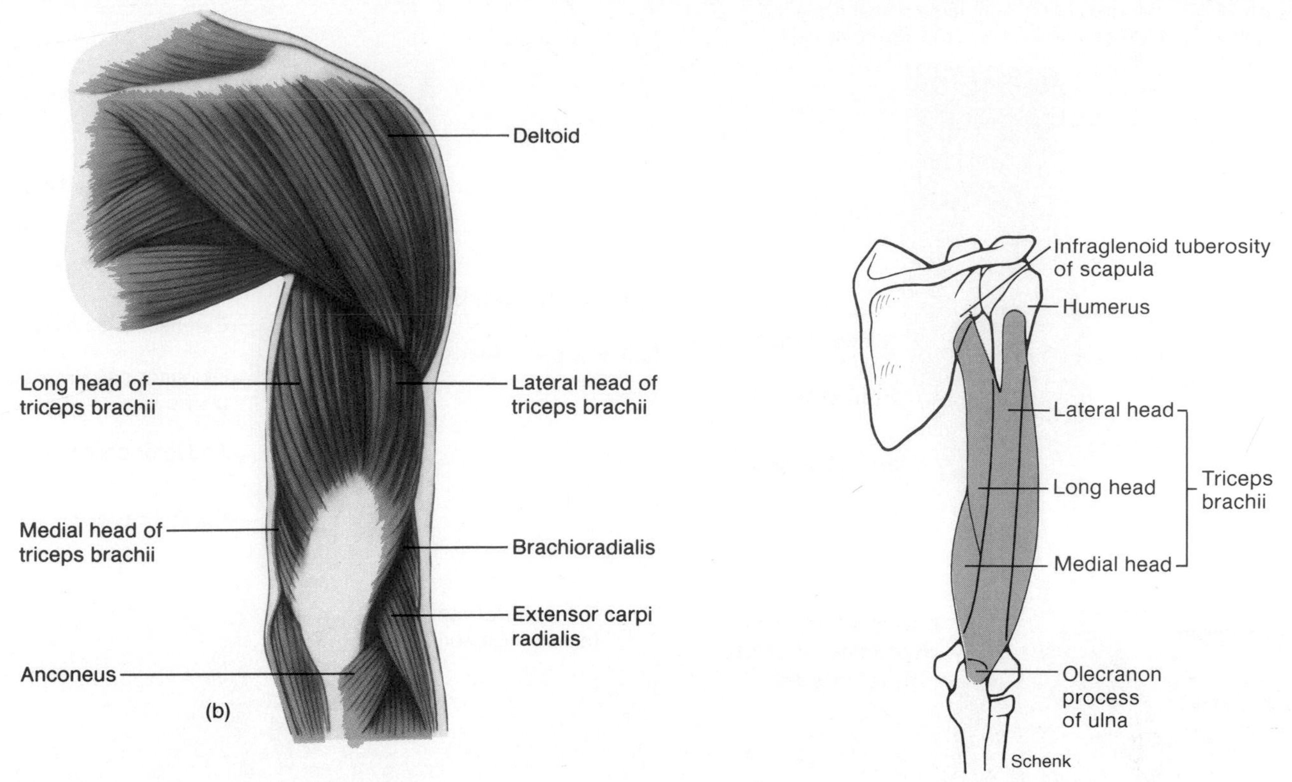

Figure 12.46. The axillary region and a transverse section through the brachium.

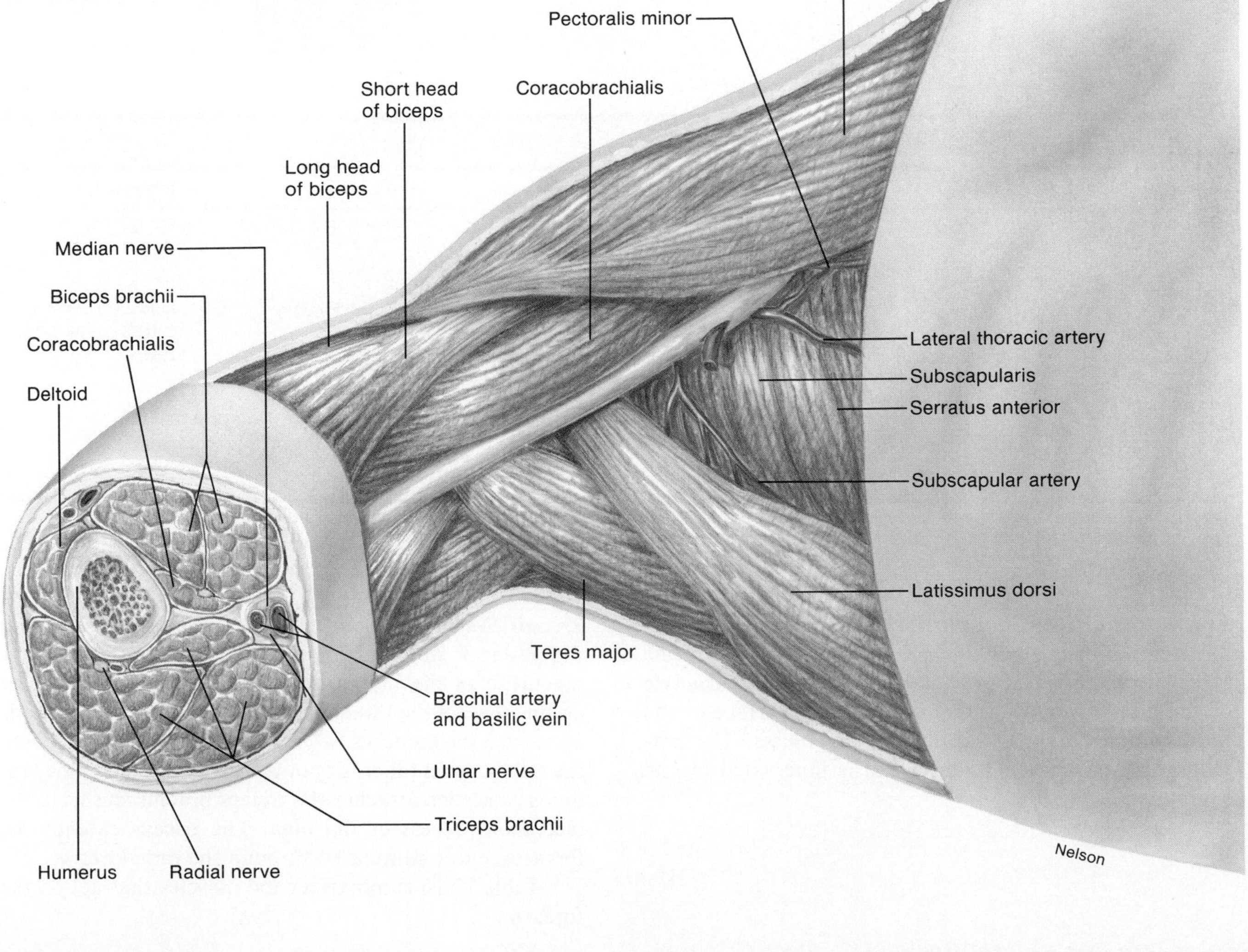

Figure 12.47. Muscles of the right forearm that move the wrist, hand, and digits. (*a*) an anterior view; (*b*) a posterior view.

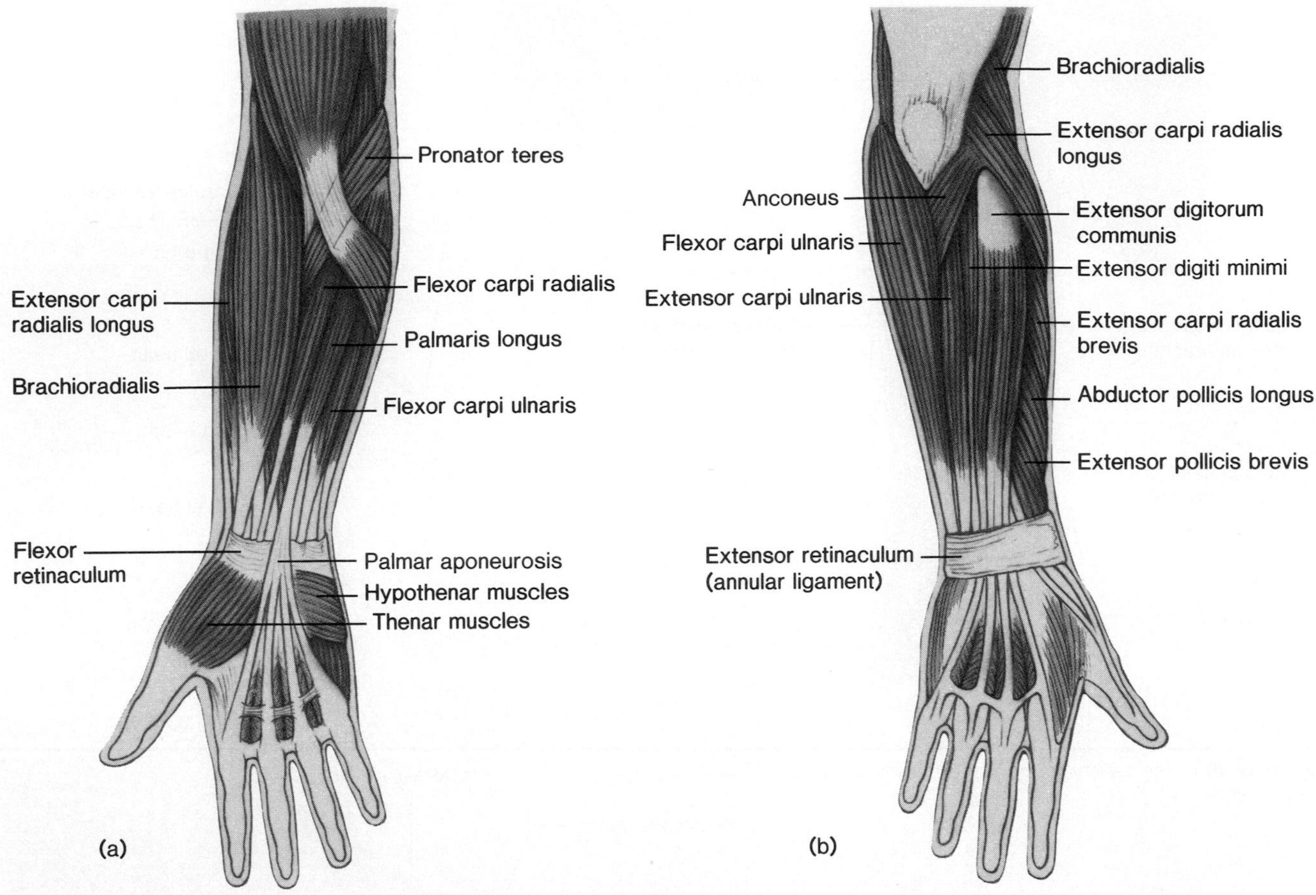

Table 12.16 Muscles that act on the forearm

Muscle	Derivation	Origin	Insertion	Action	Innervation
Biceps brachii	*biceps* = two heads; *brachium* = upper arm	Coracoid process; tuberosity above glenoid cavity of scapula	Radial tuberosity	Flexes and supinates forearm and hand	Musculocutaneous
Brachialis		Anterior shaft of humerus	Coronoid process of ulna	Flexes forearm	Musculocutaneous, median, and radial
Brachioradialis	*radialis* = pertaining to radius	Supracondylar ridge of humerus	Proximal to styloid process of radius	Flexes forearm	Radial
Triceps brachii	*triceps* = three heads	Tuberosity below glenoid cavity; lateral and medial surfaces of humerus	Olecranon process of ulna	Extends forearm	Radial

Brachioradialis. The brachioradialis is the prominent muscle positioned along the lateral (radial) surface of the forearm. It originates proximal to the lateral epicondyle of the humerus and inserts on the lateral surface of the radius immediately above the styloid process. The brachioradialis flexes the forearm and is innervated by the radial nerve.

Triceps Brachii. The triceps brachii muscle, located on the posterior surface of the brachium, is the muscle antagonistic to the biceps brachii. It has three heads, or origins. Two of the three, the **lateral head** and **medial head,** arise from the humerus, whereas the **long head** arises from the infraglenoid tuberosity of the scapula. A common tendinous insertion attaches the triceps brachii muscle to the olecranon process of the ulna. The triceps extends the forearm and is stimulated through the radial nerve.

Table 12.16 summarizes the muscles that act on the forearm.

Figure 12.48. Deep muscles of the forearm. (*a*) rotators; (*b*) flexors; (*c*) extensors.

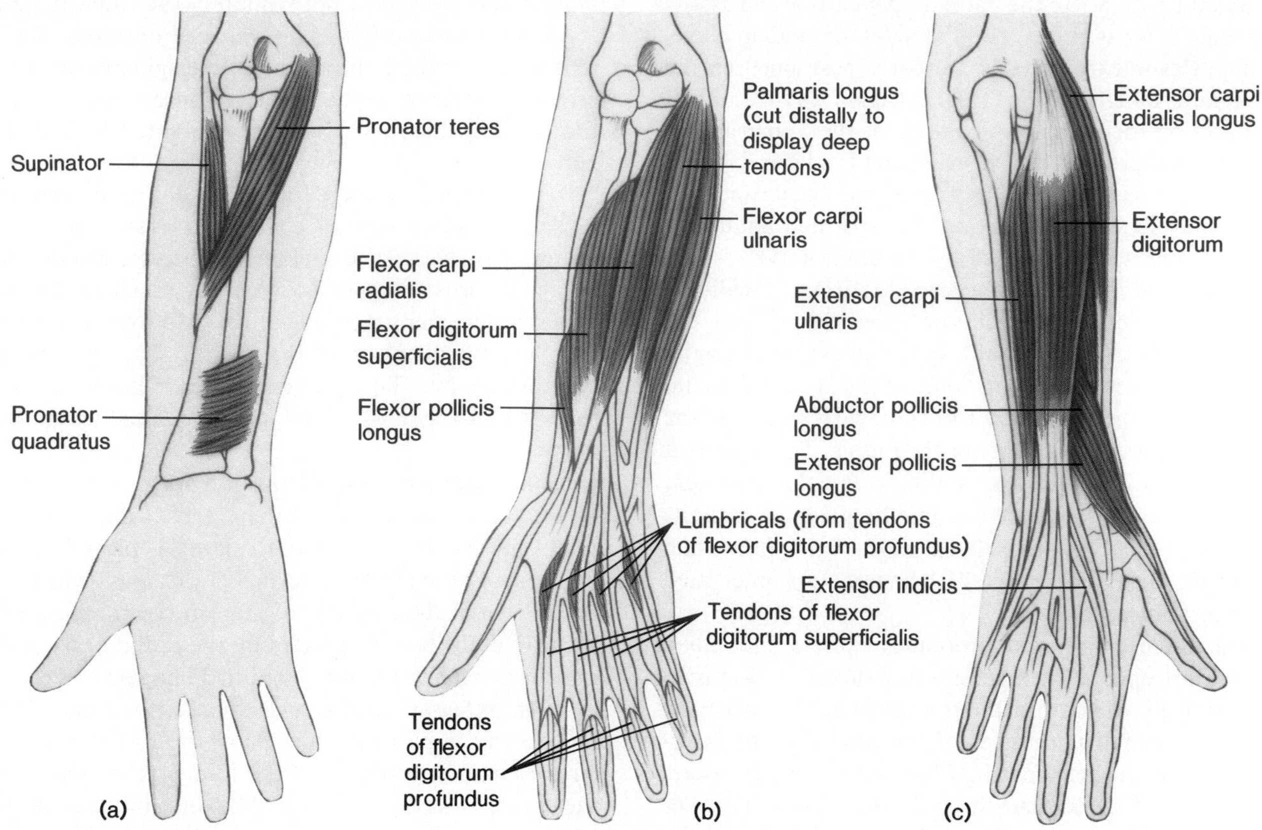

Muscles of the Forearm That Move the Wrist, Hand, and Fingers

The muscles that cause wrist, hand, and most finger movements are positioned along the forearm (figs. 12.47, 12.48). Several of these muscles act on two joints, the elbow and wrist. Others act on the joints of the wrist, hand, and digits. Still others produce rotational movement at the radioulnar joint. The precise actions of these muscles are complex, so only the basic movements will be described here. Most of these muscles perform four primary actions on the hand and digits: supination, pronation, flexion, and extension. Other actions of the hand include adduction and abduction.

Supination of the Hand. The **supinator** muscle is positioned around the upper posterior portion of the radius. It originates from the lateral epicondyle of the humerus and posterior ulna, and inserts on the lateral edge of the radial tuberosity and upper posterior surface of the radius. The supinator muscle works synergistically with the biceps brachii to supinate the hand. The supinator is innervated by the radial nerve (fig. 12.48).

Pronation of the Hand. Two muscles are responsible for pronating the hand, the pronator teres and pronator quadratus.

The **pronator teres** muscle is located on the upper medial side of the forearm (fig. 12.48). It arises on the medial epicondyle of the humerus and inserts on the upper lateral shaft of the radius. It acts to pronate the hand as it is stimulated by the median nerve.

The deep, anteriorly positioned **pronator quadratus** muscle extends between the ulna and radius on the distal fourth of the forearm (fig. 12.48). It originates on the ulna and inserts on the distal lateral portion of the radius. This muscle works synergistically with the pronator teres to rotate the palm of the hand posteriorly and position the thumb medially. It is served by the anterior interosseous branch of the median nerve.

Flexion of the Wrist, Hand, and Fingers. Six of the muscles that flex the wrist, hand, and fingers will be described (fig. 12.47). They will be discussed from lateral to medial and from superficial to deep. The brachioradialis, already described, is an obvious reference muscle for locating the muscles of the forearm that flex the wrist, hand, and fingers.

The **flexor carpi radialis** muscle arises from the medial epicondyle of the humerus, extends diagonally across the anterior surface of the forearm, and inserts onto the base of the second and third metacarpal bones (fig. 12.48). Three joints spanned by this muscle permit flexion and abduction of the hand. The flexor carpi radialis muscle itself flexes and abducts the hand. It is innervated by the median nerve.

The narrow **palmaris longus** muscle arises from the medial epicondyle of the humerus and has a long, slender tendon that attaches to the palmar aponeurosis and assists in flexing the wrist as it is stimulated by the median nerve.

The **flexor carpi ulnaris** muscle is positioned on the medial anterior side of the forearm (figs. 12.47, 12.48). It originates on the medial epicondyle of the humerus and the olecranon process, and it inserts on the lateral carpal bones and the base of the fifth metacarpal. The flexor carpi ulnaris muscle assists in flexing the wrist and adducting the hand, and it is innervated by the ulnar nerve.

The broad **flexor digitorum superficialis (sublimis)** muscle lies directly beneath the three flexors just described (see figs. 12.47, 12.48). It has an extensive origin, involving all three of the long bones of the arm: the medial epicondyle of the humerus, the coronoid process of the ulna, and the anterior border of the radius. The tendon at the distal end of this muscle is united across the wrist joint but then splits to attach to the middle phalanx of digits two through five. This muscle flexes the wrist, hand, and the second, third, fourth, and fifth fingers. It is innervated by the median nerve.

The **flexor digitorum profundus** muscle (not illustrated) lies deep to the flexor digitorum superficialis muscle just described. It is a bipennate muscle arising from the anterior, proximal two-thirds of the ulna and the interosseous ligament between the ulna and radius. It inserts on the bases of the second through fifth distal phalanges (fig. 12.48). This muscle flexes the wrist, hand, and the distal phalanges of all digits except the first (thumb), and it is innervated by both the median and ulnar nerves.

The **flexor pollicis longus** is a deep, lateral muscle originating on the anterior surface of the radius, its adjacent interosseous ligament, and the coronoid process of the ulna (fig. 12.48). Its tendinous insertion is on the base of the distal phalanx of the thumb. The flexor pollicis longus muscle flexes the thumb, assisting the grasping mechanism of the hand. It is innervated by the anterior interosseous branch of the median nerve.

> The tendons of the muscles that flex the hand can be seen on the wrist as a fist is made. These tendons are securely positioned by the *flexor retinaculum* (fig. 12.47), which crosses the wrist area transversely.

Extension of the Hand. The muscles that extend the hand are located on the posterior side of the forearm. Most of the primary extensors of the hand can be seen superficially in fig. 12.47 and will be discussed from lateral to medial.

The long, tapered **extensor carpi radialis longus** muscle is located posterior to the brachioradialis. It arises from the lateral supracondylar ridge of the humerus and inserts on the base of the second metacarpal. The extensor carpi radialis longus muscle extends and abducts the hand. It is innervated by the radial nerve.

The **extensor carpi radialis brevis** muscle is positioned immediately medial to the extensor carpi radialis longus muscle and performs approximately the same functions. Its origin and insertion, however, are different. The extensor carpi radialis brevis muscle originates on the lateral epicondyle of the humerus and inserts on the base of the third metacarpal. It too is innervated by the radial nerve.

The bipennate-fibered **extensor digitorum communis** muscle is positioned in the center of the forearm along the posterior surface. It originates on the lateral epicondyle of the humerus. Its tendon of insertion divides at the wrist, beneath the extensor retinaculum, into four tendons that attach to the distal tip of the medial phalanges of digits two through five. The function of this muscle is to extend the wrist and phalanges, and it is innervated by the radial nerve.

The **extensor digiti minimi** is a long, narrow muscle located on the ulnar side of the extensor digitorum communis muscle. It arises from the lateral epicondyle of the humerus. The tendinous insertion of this muscle fuses with the tendon of the extensor digitorum communis going to the fifth digit. It is innervated by the radial nerve and assists in extending the wrist and little finger.

The **extensor carpi ulnaris** muscle is the most medial muscle on the posterior surface of the forearm. It arises on the lateral epicondyle of the humerus and portions of the ulna. It inserts on the base of the fifth metacarpal bone, where it functions to extend and adduct the hand. It too is served by the radial nerve.

The **extensor pollicis longus** muscle extends from the mid-ulnar region, across the lower two-thirds of the forearm, and inserts onto the base of the distal phalanx of the thumb (see fig. 12.48). It extends the thumb and abducts the hand and is innervated by the radial nerve.

The **extensor pollicis brevis** muscle arises from the lower midportion of the radius and inserts on the base of the proximal phalanx of the thumb (fig. 12.47). The action of this muscle is similar to that of the extensor pollicis longus, and it is served by the same nerve.

As its name implies, the **abductor pollicis longus** muscle abducts the thumb and hand. It originates on the interosseous ligament, between the ulna and radius, and inserts on the base of the first metacarpal bone. The radial nerve innervates this muscle as well.

The muscles that act on the wrist, hand, and digits are summarized in table 12.17.

> Notice that your hand is partially contracted even when relaxed. The muscles that extend your hand are not as strong as the muscles that flex it. This is why persons who receive strong electrical shocks through the arms will tightly flex their hands and hold on. All the muscles of the arm are stimulated to contract, but the flexors, being stronger, cause the hands to close tightly.

Table 12.17 Muscles of the forearm that act on the wrist, hand, and digits

Muscle	Derivation	Origin	Insertion	Action	Innervation
Supinator	*supin* = bend back; palms up	Lateral epicondyle of humerus and crest of ulna	Lateral surface of radius	Supinates forearm	Radial
Pronator teres	*pron* = bend forward; palms down; *teres* = long, round	Medial epicondyle of humerus	Lateral surface of radius	Pronates forearm	Median
Pronator quadratus	*quad* = four (square)	Distal fourth of ulna	Distal fourth of radius	Pronates hand	Median
Flexor carpi radialis	*flexor* = decrease angle; *carpus* = wrist; *radial* = radius bone	Medial epicondyle of humerus	Base of second and third metacarpals	Flexes and abducts hand	Median
Palmaris longus	*palma* = flat of hand; *longus* = long	Medial epicondyle of humerus	Palmar aponeurosis	Flexes wrist	Median
Flexor carpi ulnaris		Medial epicondyle and olecranon process	Carpal and metacarpal bones	Flexes and adducts wrist	Ulnar
Flexor digitorum superficialis	*digitis* = finger or toe; *superficialis* = superficial	Medial epicondyle, coronoid process, and anterior border of radius	Middle phalanges of digits	Flexes wrist, hand, and digits	Median
Flexor digitorum profundus	*profundus* = deep	Proximal two-thirds of ulna and interosseous membrane	Distal phalanges	Flexes wrist, hand, and digits	Median and ulnar
Flexor pollicis longus	*pollex* = thumb	Shaft of radius, interosseous membrane, and coronoid process of ulna	Distal phalanx of thumb	Flexes thumb	Median
Extensor carpi radialis longus	*extensor* = increase joint angle	Lateral supracondylar ridge of humerus	Second metacarpal	Extends and abducts hand	Radial
Extensor carpi radialis brevis	*brevis* = short	Lateral epicondyle of humerus	Third metacarpal	Extends and abducts hand	Radial
Extensor digitorum communis	*communis* = common	Lateral epicondyle of humerus	Posterior surfaces of phalanges II-V	Extends wrist and phalanges	Radial
Extensor digiti minimi	*minimus* = small	Lateral epicondyle of humerus	Extensor aponeurosis of fifth digit	Extends fifth digit and wrist	Radial
Extensor carpi ulnaris		Lateral epicondyle of humerus and olecranon process	Base of fifth metacarpal	Extends and adducts wrist	Radial
Extensor pollicis longus		Middle shaft of ulna, lateral side	Base of distal phalanx of thumb	Extends thumb; abducts hand	Radial
Extensor pollicis brevis		Distal shaft of radius and interosseous membrane	Base of first phalanx of thumb	Extends thumb; abducts hand	Radial
Abductor pollicis longus		Distal radius and ulna and interosseous membrane	Base of first metacarpal bone	Abducts thumb and hand	Radial

Muscles of the Hand

The hand is a marvelously complex structure adapted to permit an array of intricate movements. Flexion and extension movements of the hand and phalanges are accomplished by the muscles of the forearm just described. Precise finger movements that require coordinating abduction and adduction with flexion and extension are the function of the small intrinsic muscles of the hand. These muscles and associated structures of the hand are depicted in figure 12.49. The position and actions of the muscles of the hand are listed in table 12.18.

The muscles of the hand are divided into **thenar** *(the'nar),* **hypothenar** *(hi-poth'ĕ-nar),* and **intermediate** groups. The **thenar eminence** is the fleshy base of the thumb and is formed by three muscles: the **abductor pollicis brevis,** the **flexor pollicis brevis,** and the **opponens pollicis.** All three muscles are innervated by the median nerve. The most important of the thenar muscles is the opponens pollicis, which opposes the thumb to the palm of the hand.

Figure 12.49. Intrinsic muscles of the hand and associated structures. (*a*) an anterior view; (*b*) a cross section.

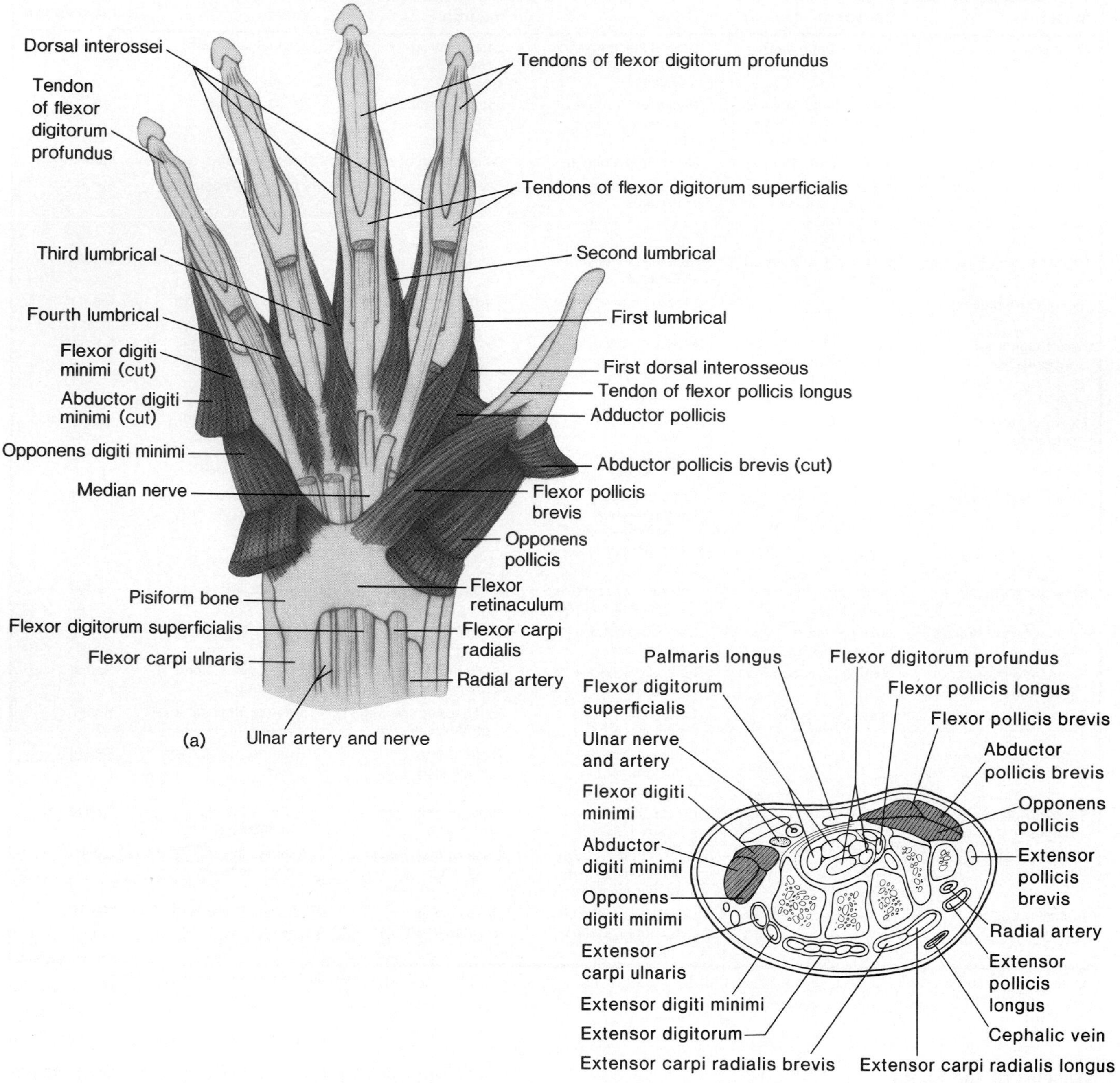

The *hypothenar eminence* is the elongated, fleshy bulge at the base of the little finger. It also is formed by three muscles: the **abductor digiti minimi,** the **flexor digiti minimi,** and the **opponens digiti minimi.** All of these muscles are supplied by branches of the ulnar nerve.

Muscles of the intermediate group are positioned between the metacarpal bones in the region of the palm. This group includes the **adductor pollicis,** the **lumbricales** *(lum'brĭ-kalz),* and the **palmar** and **dorsal interossei.** All the muscles of the intermediate group, except two lumbricales, are innervated by branches of the ulnar nerve. The first and second lumbricales receive branches from the median nerve.

Muscles That Move the Thigh

The muscles that move the thigh originate from the pelvic girdle and insert on various places on the femur. These muscles stabilize a highly movable hip joint and provide

Table 12.18 Intrinsic muscles of the hand

Muscle	Derivation	Origin	Insertion	Action	Innervation
Thenar Muscles					
Abductor pollicis brevis	*pollex* = thumb; *brevis* = short	Flexor retinaculum, scaphoid, and trapezium	Proximal phalanx of thumb	Abducts thumb	Median
Flexor pollicis brevis		Flexor retinaculum and trapezium	Proximal phalanx of thumb	Flexes thumb	Median
Opponens pollicis	*opponens* = against	Trapezium and flexor retinaculum	First metacarpal	Opposes thumb	Median
Intermediate Muscles					
Adductor pollicis (oblique and transverse heads)		Oblique head, capitate; transverse head, second and third metacarpals	Proximal phalanx of thumb	Adducts thumb	Ulnar
Lumbricales (4)	*lumbricalis* = referring to digits	Tendons of flexor digitorum profundus	Extensor expansions of digits 2–5	Flexes digits at metacarpophalangeal joints; extends digits at interphalangeal joints	Median and ulnar
Palmar interossei (3)	*palma* = palm	Medial side of second metacarpal; lateral sides of fourth and fifth metacarpals	Proximal phalanges of index, ring, and little fingers and extensor digitorum communis	Adducts fingers toward middle finger at metacarpophalangeal joints	Ulnar
Dorsal interossei (4)		Adjacent sides of metacarpals	Proximal phalanges of index and middle fingers (lateral sides) plus proximal phalanges of middle and ring fingers (medial sides) and extensor digitorum communis	Abducts fingers away from middle finger at metacarpophalangeal joints	Ulnar
Hypothenar Muscles					
Abductor digiti minimi	*minimus* = very small	Pisiform and tendon of flexor carpi ulnaris	Proximal phalanx of digit 5	Abducts digit 5	Ulnar
Flexor digiti minimi		Flexor retinaculum and hook of hamate	Proximal phalanx of digit 5	Flexes digit 5	Ulnar
Opponens digiti minimi		Flexor retinaculum and hook of hamate	Fifth metacarpal	Opposes digit 5	Ulnar

support for the body during bipedal stance and locomotion. Found in this region are the most massive muscles of the body as well as some extremely small muscles. The muscles that move the thigh are divided into anterior, posterior, and medial groups.

Anterior Muscles. The anterior muscles that move the thigh are the iliacus and psoas major (figs. 12.50, 12.51).

The triangular **iliacus** muscle arises from the iliac fossa and inserts on the lesser trochanter of the femur. It is innervated by the femoral nerve.

The long, thick **psoas major** muscle originates on the bodies and transverse processes of the lumbar vertebrae and inserts, along with the iliacus, on the lesser trochanter. The psoas major muscle is innervated by the spinal nerves L2 and L3. The psoas major muscle and the iliacus work synergistically in flexing and rotating the thigh and flexing the vertebral column. They are frequently referred to as a single muscle, the **iliopsoas** *(il''e-o-so'as).*

Posterior and Lateral (Buttock) Muscles. The posterior muscles that move the thigh include the gluteus maximus, gluteus medius, gluteus minimus, and tensor fasciae latae.

The large **gluteus maximus** muscle forms much of the prominence of the buttock (fig. 12.52). It is a powerful extensor muscle of the thigh and is very important for bipedal stance and locomotion. The gluteus maximus originates on the ilium, sacrum, coccyx, and aponeurosis of the lumbar region. It inserts on the gluteal tuberosity of the femur and the *iliotibial tract,* a tendinous band extending down the thigh (see fig. 12.54). The gluteus maximus is served by the inferior gluteal nerve. The mass of this muscle is of clinical significance as a site for intramuscular injections.

The **gluteus medius** muscle is located immediately deep to the gluteus maximus (fig. 12.52). It originates on the lateral surface of the ilium and inserts on the greater trochanter of the femur. The gluteus medius abducts and medially rotates the thigh. It is innervated by the superior gluteal nerve.

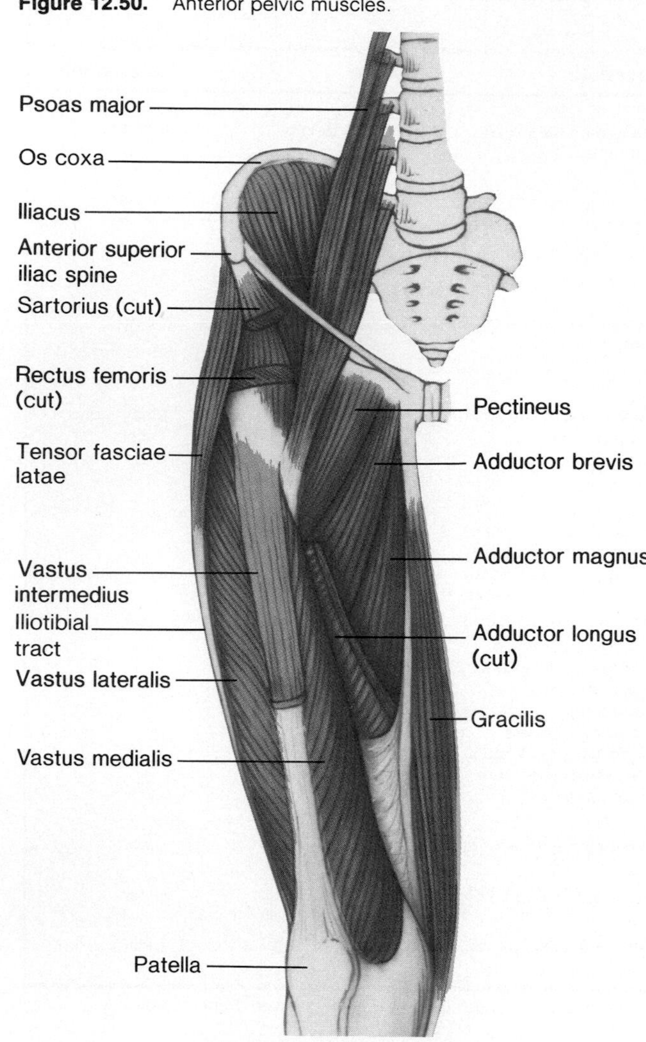

Figure 12.50. Anterior pelvic muscles.

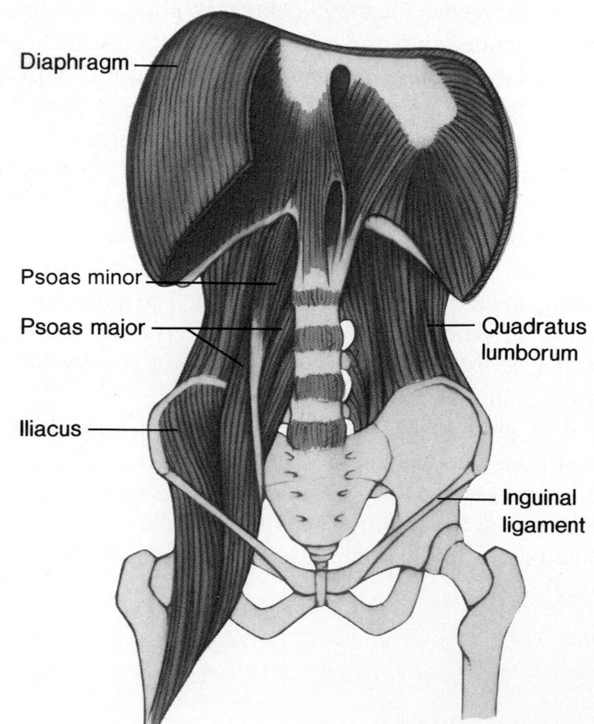

Figure 12.51. Anterior pelvic muscles that move the hip.

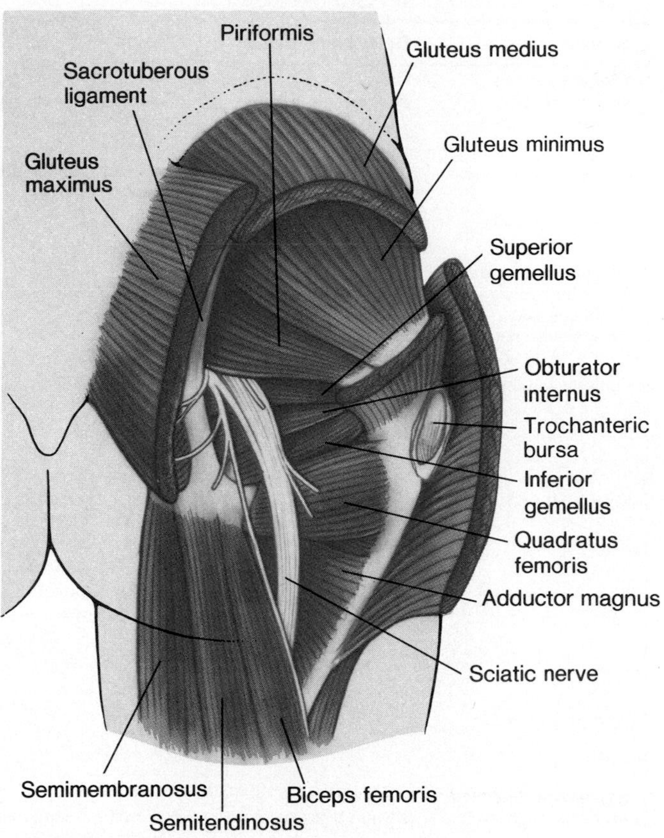

Figure 12.52. Deep gluteal muscles.

The **gluteus minimus** is the smallest and deepest of the gluteal muscles (fig. 12.52). It also arises from the lateral surface of the ilium and inserts on the lateral surface of the greater trochanter, where it acts synergistically with the gluteus medius to abduct and medially rotate the thigh. It too is innervated by the superior gluteal nerve.

The quadrangular **tensor fasciae latae** muscle is positioned superficially on the lateral surface of the hip (see figs. 12.50, 12.54). It originates on the iliac crest and inserts on a broad lateral fascia of the thigh called the *fascia lata.* The fascia lata is continuous with the iliotibial tract. The tensor fasciae latae is innervated by the superior gluteal nerve and, when stimulated, causes abduction of the thigh.

A deep group of six lateral rotators of the thigh is positioned directly over the posterior aspect of the hip joint. These muscles are not discussed but are identified in figures 12.52 and 12.53 from superior to inferior as the **piriformis, superior gemellus, obturator internus, inferior gemellus, obturator externus,** and **quadratus femoris.**

The anterior and posterior group of muscles that move the thigh are summarized in table 12.19.

Medial, or Adductor, Muscles. The medial muscles that move the thigh include the gracilis, pectineus, adductor longus, adductor brevis, and adductor magnus (figs. 12.53, 12.54).

Table 12.19 Anterior and posterior muscles that move the thigh

Muscle	Derivation	Origin	Insertion	Action	Innervation
Iliacus	*iliacus* = pertaining to ilium (hip)	Iliac fossa	Lesser trochanter of femur	Flexes and rotates thigh laterally; flexes vertebral column	Femoral
Psoas major	*psoa* = muscle of loin	Transverse processes of all lumbar vertebrae	Lesser trochanter with iliacus	Flexes and rotates thigh laterally; flexes vertebral column	L2, L3
Gluteus maximus	*gloutos* = rump; *maximus* = greatest	Iliac crest, sacrum, coccyx, and aponeurosis of the lumbar region	Gluteal tuberosity and iliotibial tract	Extends and rotates thigh laterally	Inferior gluteal
Gluteus medius	*medius* = middle	Lateral surface of ilium	Greater trochanter	Abducts and rotates thigh medially	Superior gluteal
Gluteus minimus	*minimus* = smallest	Lateral surface of lower half of ilium	Greater trochanter	Abducts and rotates thigh medially	Superior gluteal
Tensor fasciae latae	*tensor* = tightener; *fascia* = band; *latus* = wide	Anterior border of ilium and iliac crest	Iliotibial tract	Abducts thigh	Superior gluteal

Figure 12.53. Adductor muscles of the right thigh.

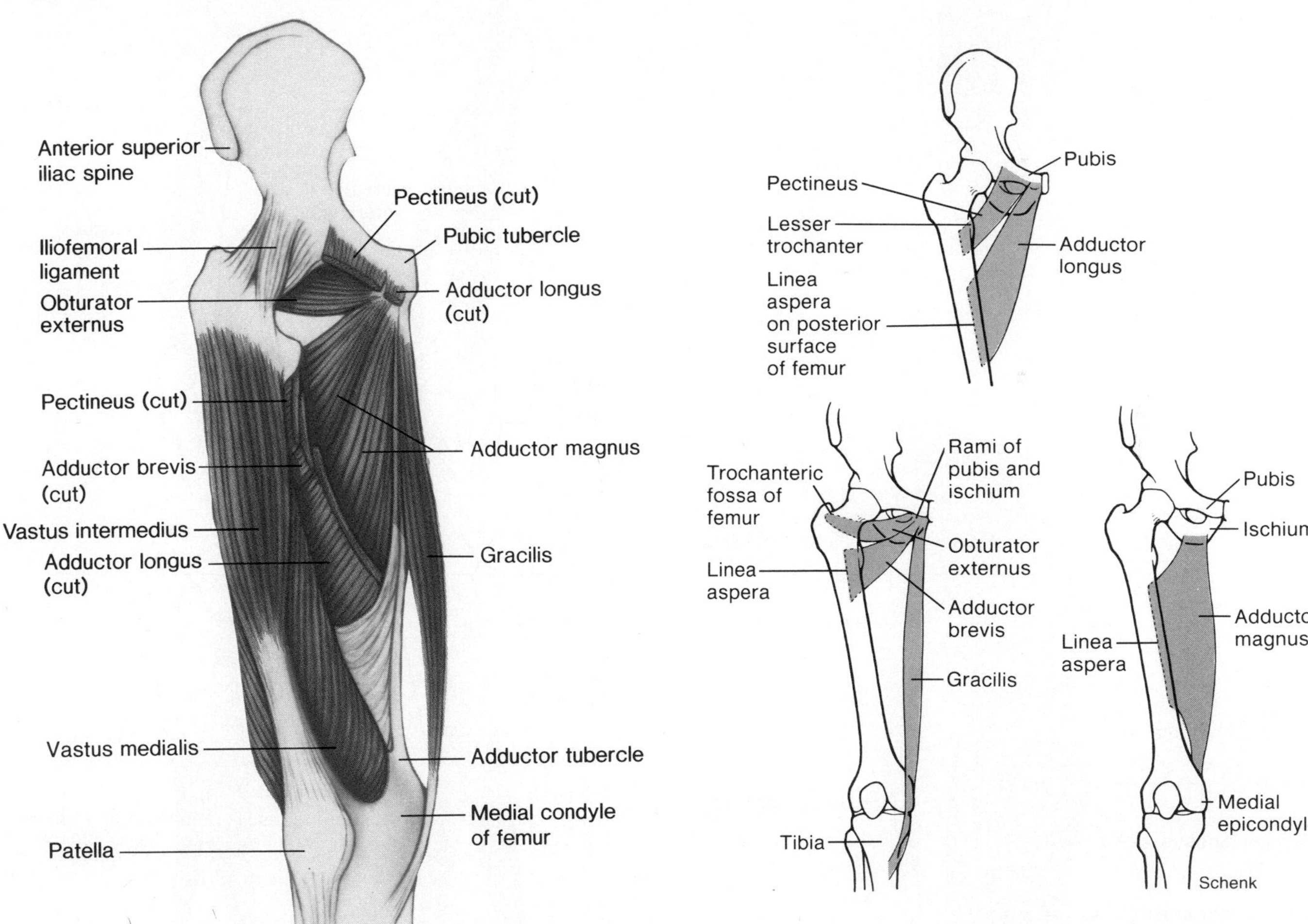

The long, thin **gracilis** *(gras'il-is)* muscle is the most superficial of the medial thigh muscles. It arises from the lower ramus of the pubis immediately below the symphysis pubis and inserts on the proximal medial surface of the tibia along with the sartorius. The gracilis is a two-joint muscle and can adduct the thigh or flex the leg. It is innervated by the obturator nerve.

The **pectineus** is the uppermost of the medial muscles that move the thigh. It is a flat, quadrangular muscle that originates on the pubis and inserts just below the lesser trochanter of the femur. The pectineus flexes and adducts the thigh and is served by the femoral nerve.

Figure 12.54. Muscles of the right thigh. (*a*) an anterior view; (*b*) a posterior view.

Gluteus medius
Sartorius
Tensor fasciae latae
Rectus femoris
Vastus lateralis
Patellar tendon
Inguinal ligament
Iliopsoas
Pectineus
Adductor longus
Gracilis
Vastus medialis
(a)

Anterior inferior iliac spine
Greater trochanter
Intertrochanteric line
Rectus femoris
Vastus lateralis
Patella
Tibia
Patellar ligament

Anterior superior iliac spine
Intertrochanteric line
Vastus intermedius
Vastus medialis
Sartorius
Patella
Tibia
Patella

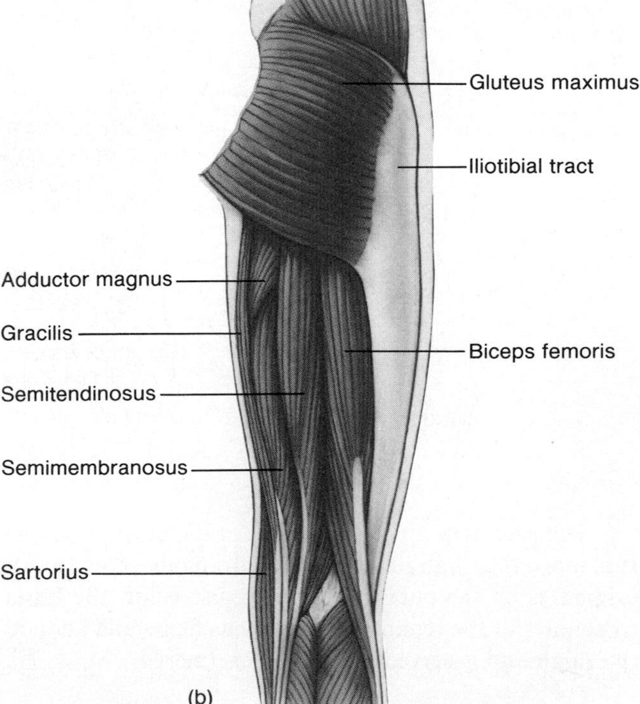

Ischial tuberosity
Semi-tendinosus
Long head of biceps femoris
Ischial tuberosity
Linea aspera
Semi-membranosus
Short head of biceps femoris
Head of fibula
Fibula
Schenk

Table 12.20	Medial muscles that move the thigh				
Muscle	**Derivation**	**Origin**	**Insertion**	**Action**	**Innervation**
Gracilis	*gracilis* = slender	Inferior edge of symphysis pubis	Proximal medial surface of tibia	Adducts thigh; flexes and rotates leg at knee	Obturator
Pectineus	*pecten* = comb	Pectineal line of pubis	Distal to lesser trochanter of femur	Adducts and flexes thigh	Femoral
Adductor longus	*ad* = toward; *longus* = long	Pubis—below pubic crest	Linea aspera of femur	Adducts, flexes, and laterally rotates thigh	Obturator
Adductor brevis	*brevis* = short	Inferior ramus of pubis	Linea aspera of femur	Adducts, flexes, and laterally rotates thigh	Obturator
Adductor magnus	*magnus* = large	Inferior ramus of ischium and pubis	Linea aspera and medial epicondyle of femur	Adducts, flexes, and laterally rotates thigh	Obturator

The **adductor longus** muscle is immediately lateral to the gracilis on the upper third of the thigh. It originates near the symphysis pubis and inserts on the linea aspera of the femur. The adductor longus and the following two adductors are synergistic in adducting, flexing, and laterally rotating the thigh. They are all innervated by the obturator nerve.

The thick, triangular **adductor brevis** muscle is located behind the adductor longus. It arises from the inferior ramus of the pubis and inserts on the linea aspera of the femur.

As the name implies, the **adductor magnus** is a large, thick muscle, somewhat triangular in shape. It has an extensive origin along the inferior ramus of the ischium and pubis. It has a dual insertion along the linea aspera and the medial epicondyle of the femur.

Table 12.20 summarizes the muscles that adduct the thigh.

Muscles of the Thigh That Move the Lower Leg

The muscles that move the lower leg originate on the pelvic girdle or thigh and are surrounded and compartmentalized by tough fascial sheets, which are a continuation of the fascia lata and iliotibial tract. They are divided according to function and position into two groups: anterior extensors and posterior flexors.

Anterior, or Extensor, Muscles. The anterior muscles that move the lower leg are the sartorius and quadriceps femoris (fig. 12.54).

The long, straplike **sartorius** muscle obliquely crosses the anterior aspect of the thigh. It arises from the anterior superior spine of the ilium and extends medially to insert on the proximal medial surface of the tibia. The sartorius can act on both the hip and knee joints to flex and rotate the thigh laterally and also to assist in flexing the leg and rotating it laterally. The sartorius muscle is the longest muscle of the body and is frequently called the "tailor's muscle" because it enables one to assume a cross-legged sitting position. It is innervated by the femoral nerve.

The **quadriceps femoris** is actually a composite of four distinct muscles that have separate origins but a common insertion on the patella via the *patellar tendon.* The patellar tendon is continuous over the patella and becomes the *patellar ligament* as it attaches to the head of the tibia (fig. 12.54a). These muscles function synergistically to extend the leg, as in kicking a football. The four muscles of the quadriceps femoris muscle are the rectus femoris, vastus lateralis, vastus medialis, and vastus intermedius. All four are innervated by the femoral nerve.

The **rectus femoris** occupies a superficial position and is the only one of the four quadriceps that functions in both the hip and knee joints. It has a dual origin on the anterior inferior spine of the ilium and the lip of the acetabulum.

The laterally positioned **vastus lateralis** is the largest muscle of the quadriceps femoris. It arises from the greater trochanter and linea aspera of the femur.

The **vastus medialis** occupies a medial position along the thigh. It originates on the medial surface of the linea aspera of the femur.

The **vastus intermedius** (fig. 12.54a) lies deep to the rectus femoris and originates on the anterior surface of the shaft of the femur.

The anterior thigh muscles that move the lower leg are summarized in table 12.21.

Posterior, or Flexor, Muscles. There are three posterior thigh muscles antagonistic to the quadriceps femoris muscles in flexing the knee. These muscles are known as the **hamstrings** (fig. 12.54), and each is innervated by the tibial nerve that branches from the sciatic nerve.

The **biceps femoris** muscle occupies the posterior lateral aspect of the thigh. It has a superficial long head and a deep short head. The long head arises from the ischial tuberosity, and the short head arises from the linea aspera. The biceps femoris inserts on the head of the fibula and the lateral condyle of the tibia. The action of this muscle is a bit complicated because it works over both the hip and knee joints. It flexes the leg at the knee as well as extends and laterally rotates the thigh at the hip.

The superficial **semitendinosus** muscle is fusiform and is located on the posterior medial aspect of the thigh. It originates on the ischial tuberosity from a common tendon with the biceps femoris. The long and tapering tendon inserts onto the proximal medial portion of the tibia. The

Table 12.21 Anterior thigh muscles that move the lower leg

Muscle	Derivation	Origin	Insertion	Action	Innervation
Sartorius	*sartor* = tailor	Spine of ilium	Medial surface of tibia	Flexes leg and thigh, abducts thigh, rotates thigh laterally, and rotates leg medially	Femoral
Quadriceps femoris	*quad* = four; *cephalic* = head; *femoris* = femur		Patella by common tendon, which continues as patellar tendon to tibial tuberosity	Extends leg at knee	Femoral
Rectus femoris	*rectus* = straight fibers	Spine of ilium and lip of acetabulum			
Vastus lateralis	*vastus* = large; *lateralis* = lateral	Greater trochanter and linea aspera of femur			
Vastus medialis	*medialis* = medial	Medial surface and linea aspera of femur			
Vastus intermedius	*intermedius* = middle	Anterior and lateral surfaces of femur			

Table 12.22 Posterior thigh muscles that move the lower leg

Muscle	Derivation	Origin	Insertion	Action	Innervation
Biceps femoris	*bi* = two; *cephal* = head; *femoris* = femur	Long head—ischial tuberosity; short head—linea aspera of femur	Head of fibula and lateral condyle of tibia	Flexes leg at knee; extends and laterally rotates thigh at hip	Tibial
Semitendinosus	*semi* = half; *tendo* = tendon	Ischial tuberosity	Proximal portion of medial surface of body of tibia	Flexes leg at knee; extends and medially rotates thigh at hip	Tibial
Semimembranosus	*membran* = membrane	Ischial tuberosity	Medial condyle of tibia	Flexes leg at knee; extends and medially rotates thigh at hip	Tibial

semitendinosus muscle also works over two joints, providing extension and medial rotation of the thigh at the hip and flexion of the leg at the knee.

The **semimembranosus** lies deep to the semitendinosus on the posterior medial aspect of the thigh. It is a flat muscle that arises from the ischial tuberosity and inserts on the medial condyle of the tibia. The action of the semimembranosus is similar to that of the semitendinosus.

The posterior thigh muscles that move the lower leg are summarized in table 12.22. The relationship of the muscles of the thigh is illustrated in figure 12.55.

Hamstring injuries are a common occurrence in some sports. The injury usually occurs when sudden lateral or medial stress to the knee joint tears the muscles or tendons. Because of its structure and the stress applied to it in competition, the knee joint is highly susceptible to injury. To reduce the number of knee injuries, the rules in contact sports should be altered or additional support and protection should be provided for this vulnerable joint.

Muscles of the Lower Leg That Move the Ankle, Foot, and Toes

The muscles of the leg, the **crural muscles,** are responsible for the movements of the foot. There are three groups of crural muscles: anterior, lateral, and posterior. The anteromedial aspect of the leg along the shaft of the tibia lacks muscle attachment. Muscles in the anterior compartment are innervated by the deep peroneal nerve, those in the lateral compartment by the superficial peroneal nerve, and those in the posterior compartment by the tibial nerve.

Anterior Crural Muscles. The anterior crural muscles include the tibialis anterior, extensor digitorum longus, extensor hallucis longus, and peroneus tertius (figs. 12.56, 12.57).

The large, superficial **tibialis anterior** muscle can be easily palpated on the anterior lateral portion of the tibia (fig. 12.56). It arises from the lateral condyle and lateral surface of the proximal half of the tibia. It has a long tendon of insertion that traverses to the medial plantar surface of the foot and attaches to the first cuneiform and

Figure 12.55. A transverse section of the right thigh as seen from above. Note the position of the vessels and nerves.

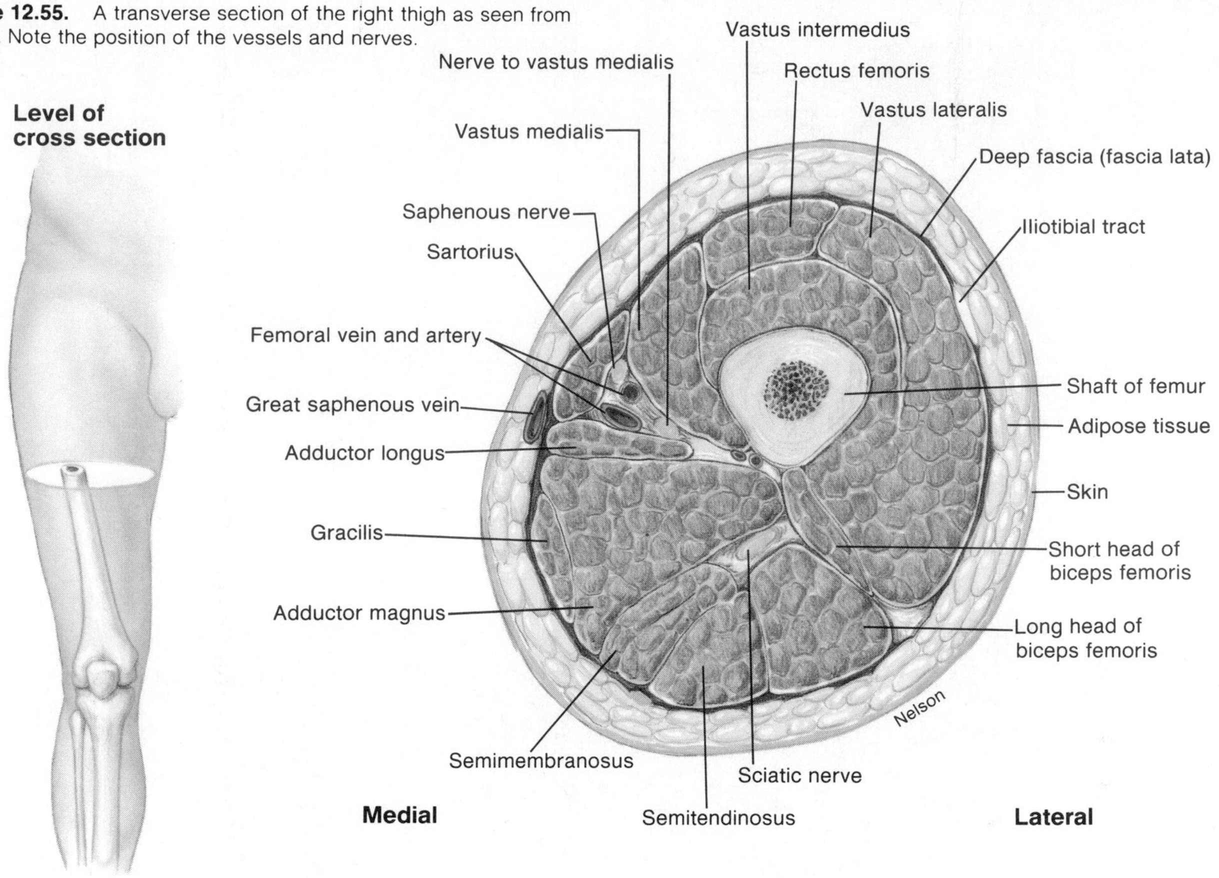

Figure 12.56. Anterior crural muscles.

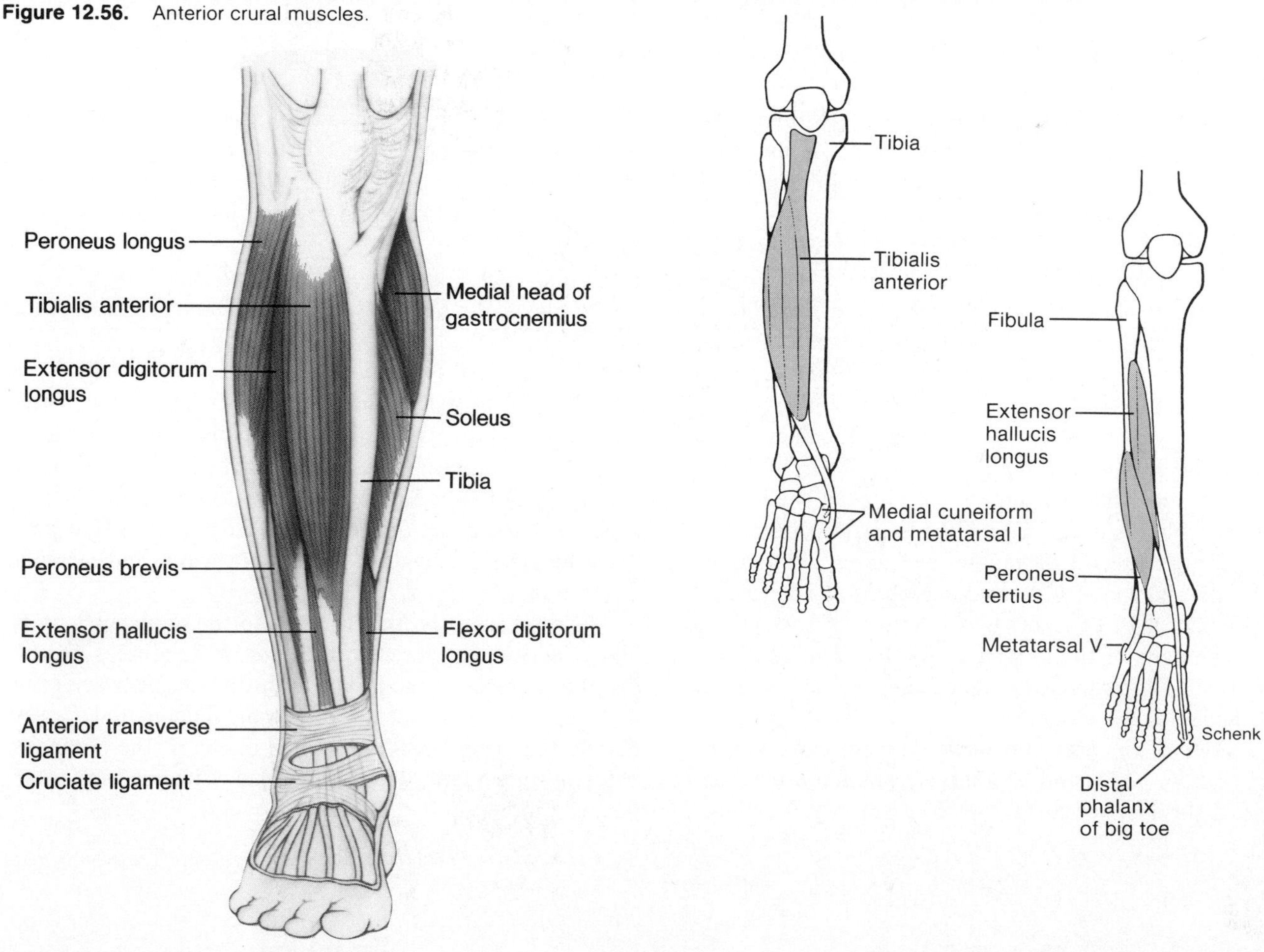

Figure 12.57. Lateral crural muscles.

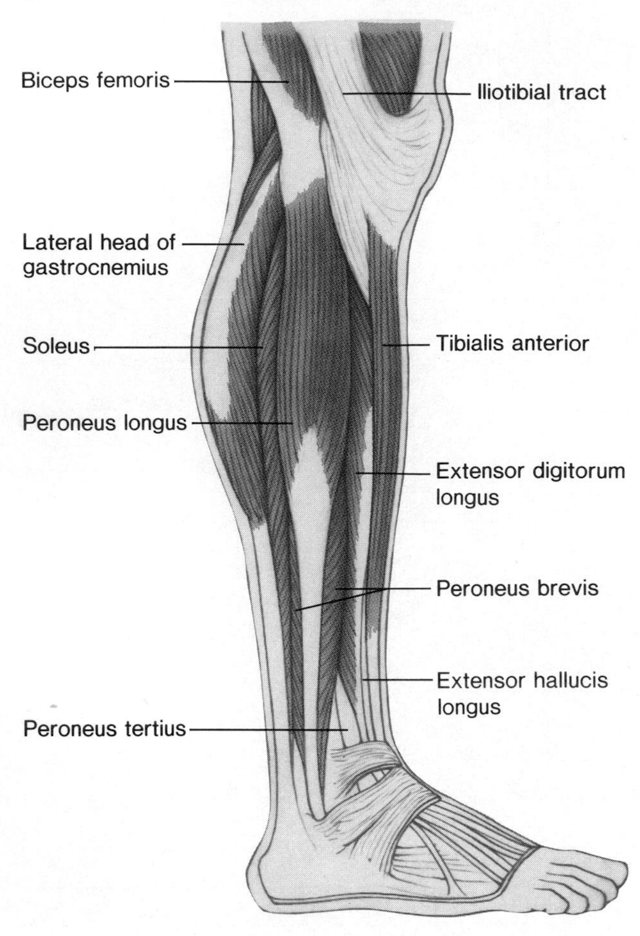

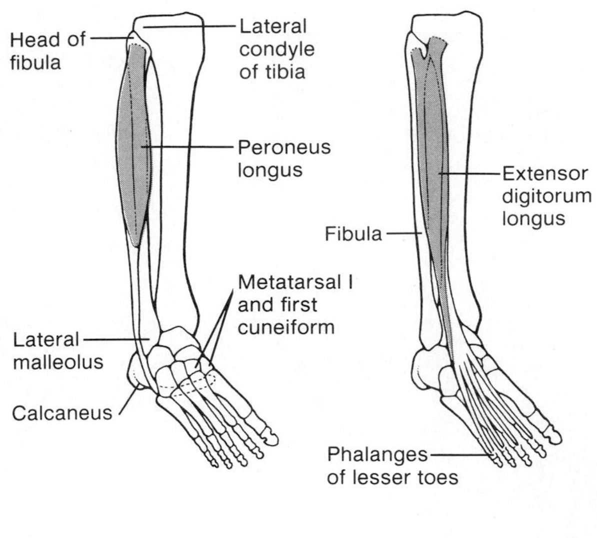

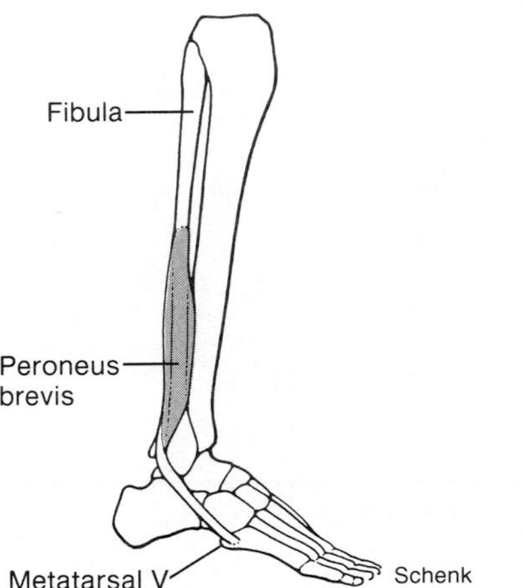

base of the first metatarsal (see fig. 12.62). In this position, the tendon of the tibialis anterior aids in supporting the longitudinal arch of the foot. The tibialis anterior dorsiflexes the ankle and inverts the foot.

The **extensor digitorum longus** muscle lies lateral to the tibialis anterior (fig. 12.56). It originates on the lateral condyle of the tibia and the anterior portion of the fibula. Its tendon of insertion branches into four parts that attach to the dorsal surface of the distal phalanges of the second, third, fourth, and fifth toes (see fig. 12.62). The extensor digitorum longus is responsible for the dorsiflexion of the ankle and the extension of the toes.

The **extensor hallucis longus** muscle is positioned deep between the tibialis anterior and the extensor digitorum longus (see fig. 12.62). It originates on the middle portion of the fibula and the interosseous membrane. The tendinous insertion of this muscle is on the dorsal surface of the distal phalanx of the great toe. As the name implies, the extensor hallucis longus extends the big toe and aids in dorsiflexion of the foot.

A portion of the **peroneus tertius** muscle can be seen in figures 12.56, 12.57, and 12.62. It is a small muscle that arises from the anterior surface of the distal one-fourth of the fibula and interosseous membrane. Its tendinous insertion is on the dorsal surface of the base of the fifth metatarsal. The peroneus tertius works over the ankle joint, assisting in dorsiflexion and eversion of the foot.

Lateral Crural Muscles. The lateral crural muscles are the peroneus longus and peroneus brevis (figs. 12.56, 12.57).

The long, flat **peroneus longus** muscle originates from the proximal two-thirds of the fibula and a portion of the lateral condyle of the tibia. On the distal end, the long tendon of this muscle passes behind the lateral malleolus and under the arch of the foot to insert on the first metatarsal and first cuneiform (see fig. 12.61c and d). The peroneus longus is responsible for plantar flexion and eversion of the foot.

The **peroneus brevis** lies deep to the peroneus longus and is positioned closer to the foot. It originates on the distal two-thirds of the fibula. It inserts on the base of the fifth metatarsal by means of a long tendon that passes behind the lateral malleolus. The action of the peroneus brevis is synergistic with the peroneus longus in plantar flexion and eversion of the foot.

Figure 12.58. Posterior crural muscles and popliteal region.

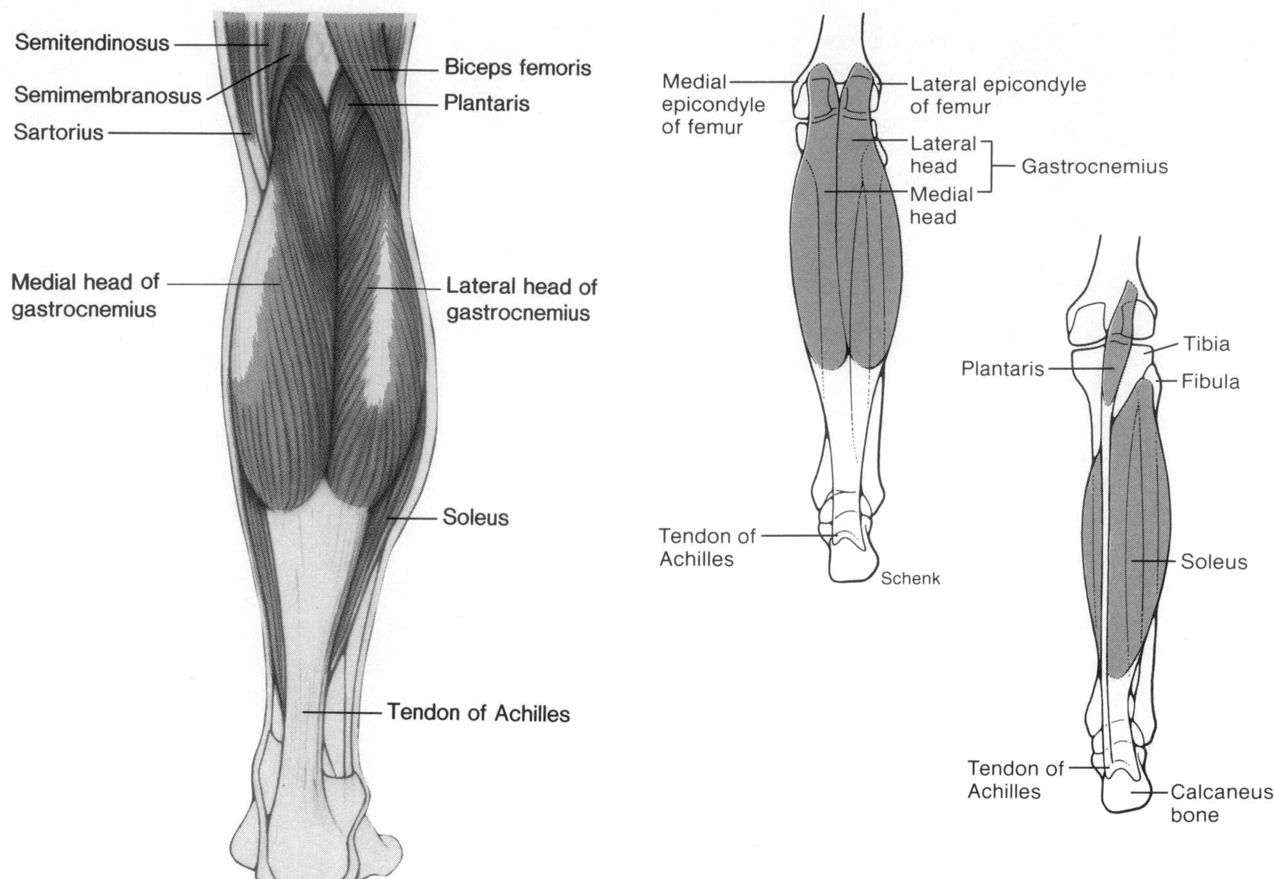

Posterior Crural Muscles. The seven posterior crural muscles can be grouped into a superficial and a deep group. The superficial group is composed of the gastrocnemius, soleus, and plantaris muscles (fig. 12.58). The four deep posterior crural muscles are the popliteus, flexor hallucis longus, flexor digitorum longus, and tibialis posterior (fig. 12.59).

The **gastrocnemius** *(gas''trok-ne'me-us)* is a large superficial muscle that forms the major portion of the calf of the leg. It consists of two distinct heads, which arise from the posterior surfaces of the medial and lateral condyles of the femur. This muscle and the deeper soleus muscle insert onto the calcaneus bone via the common *tendon of Achilles (tendo calcaneus).* The gastrocnemius acts over two joints to cause flexion of the knee and plantar flexion of the foot.

The **soleus** lies deep to the gastrocnemius. These two muscles are frequently referred to as a single muscle, the **triceps surae.** The soleus originates on the proximal heads of the fibula and tibia. It has a common insertion with the gastrocnemius but acts on only one joint, the ankle, in plantar flexing the foot.

The small **plantaris** muscle arises just above the origin of the lateral head of the gastrocnemius on the lateral condyle of the femur. It has a very long, slender tendon of insertion onto the calcaneus bone. The tendon of this muscle is frequently mistaken for a nerve by those dissecting it for the first time. The plantaris is a weak muscle with limited ability to flex the leg and plantar flex the foot.

The thin, triangular **popliteus** muscle is situated deep to the heads of the gastrocnemius, where it forms part of the floor of the popliteal fossa. The *popliteal fossa* is the depression on the back side of the knee joint (fig. 12.60). The popliteus arises from the lateral condyle of the femur and inserts on the posterior surface of the proximal portion of the tibia (fig. 12.59). The popliteus muscle is a medial rotator of the tibia on the femur.

The **flexor hallucis longus** muscle lies deep to the soleus on the posterolateral side of the leg. It is a bipennate muscle that arises from the distal two-thirds of the fibula. Its long tendon of attachment passes posterior to the medial malleolus and along the lateral sole of the foot to insert upon the hallux (the big toe). The flexor hallucis longus flexes the big toe and assists in plantar flexing and inverting the foot.

Figure 12.59. Deep posterior crural muscles.

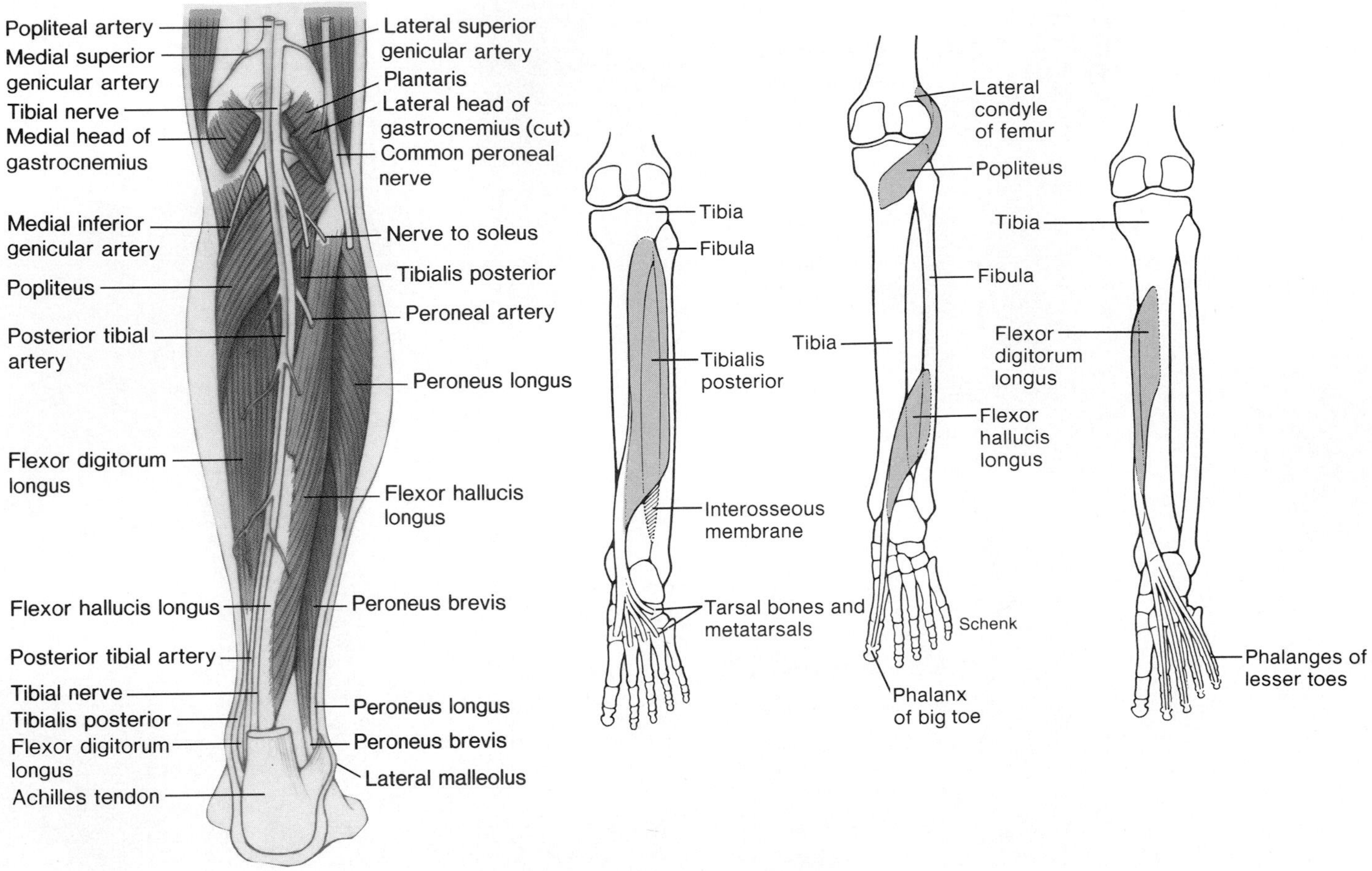

Popliteal artery
Medial superior genicular artery
Tibial nerve
Medial head of gastrocnemius
Medial inferior genicular artery
Popliteus
Posterior tibial artery
Flexor digitorum longus
Flexor hallucis longus
Posterior tibial artery
Tibial nerve
Tibialis posterior
Flexor digitorum longus
Achilles tendon
Lateral superior genicular artery
Plantaris
Lateral head of gastrocnemius (cut)
Common peroneal nerve
Nerve to soleus
Tibialis posterior
Peroneal artery
Peroneus longus
Flexor hallucis longus
Peroneus brevis
Peroneus longus
Peroneus brevis
Lateral malleolus
Tibia
Fibula
Tibialis posterior
Interosseous membrane
Tarsal bones and metatarsals
Lateral condyle of femur
Popliteus
Fibula
Tibia
Flexor hallucis longus
Schenk
Phalanx of big toe
Tibia
Flexor digitorum longus
Phalanges of lesser toes

Figure 12.60. Muscles that surround the popliteal fossa.

Semitendinosus
Gracilis
Semimembranosus
Sartorius
Great saphenous vein
Saphenous nerve
Small saphenous vein
Medial head of gastrocnemius
Biceps femoris
Common peroneal nerve
Tibial nerve
Popliteal artery
Popliteal vein
Sural nerve
Sural communicating branch
Lateral head of gastrocnemius
Lateral sural cutaneous nerve

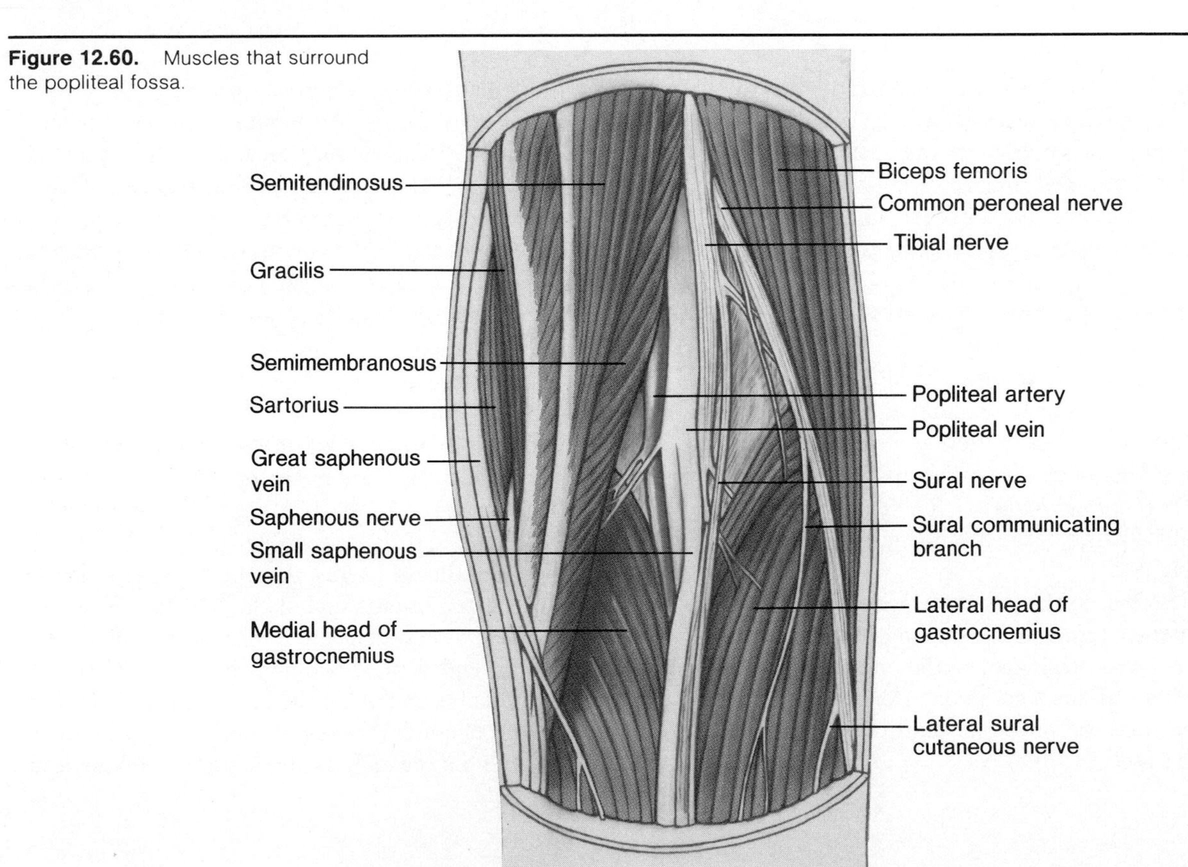

Figure 12.61. The four musculotendinous layers of the plantar aspect of the foot. (*a*) superficial layer; (*b*) second layer; (*c*) third layer; (*d*) deep layer.

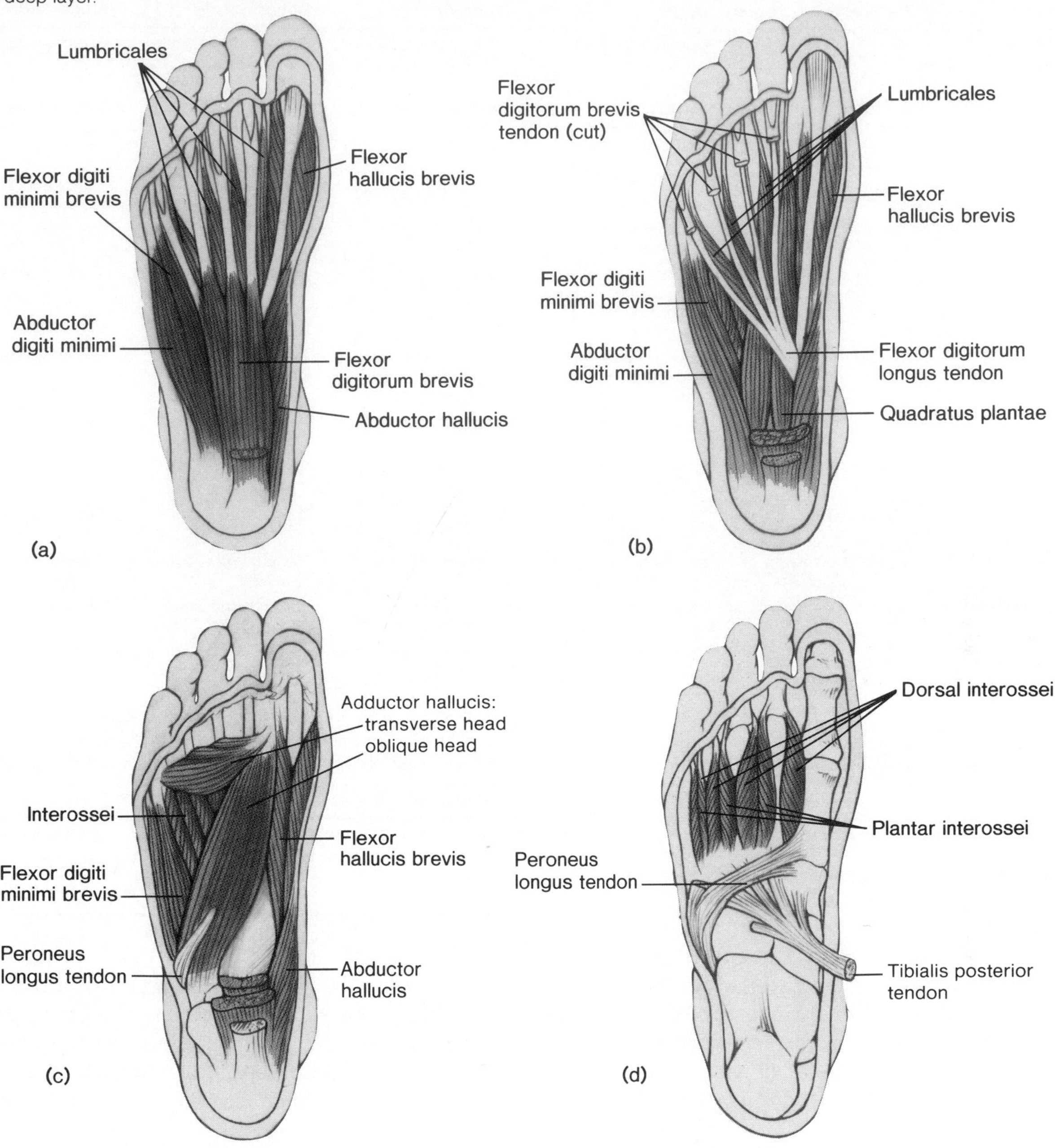

The **flexor digitorum longus** muscle also lies deep to the soleus and parallels the flexor hallucis longus on the medial side of the leg. It originates on the posterior surface of the shaft of the tibia. Its distal tendon passes behind the medial malleolus and continues along the plantar surface of the foot, where it branches into four tendinous slips. These four slips attach to the bases of the terminal phalanges of the second, third, fourth, and fifth toes (fig. 12.61b). The flexor digitorum longus works over several joints, flexing four of the digits and assisting in plantar flexing and inverting the foot.

The **tibialis posterior** muscle is located deep to the soleus, directly on the interosseous membrane on the posterior surface of the leg. It originates on the interosseous membrane and the adjacent proximal surfaces of the tibia and fibula. The distal tendon of the tibialis posterior passes behind the medial malleolus and inserts on the plantar surfaces of the navicular, the cuneiforms, the cuboid, and the second, third, and fourth metatarsals (fig. 12.61d). The tibialis posterior plantar flexes and inverts the foot and gives support to the arches of the foot.

The crural muscles are summarized in table 12.23.

Table 12.23 Muscles of the lower leg that move the ankle, foot, and toes

Muscle	Derivation	Origin	Insertion	Action	Innervation
Tibialis anterior	*tibialis* = tibia; *ante* = in front	Lateral condyle and body of tibia	First metatarsal and first cuneiform	Dorsiflexion of ankle and inversion of foot	Deep peroneal
Extensor digitorum longus	*extensor* = increase joint angle; *digit* = finger or toe; *longus* = long	Lateral condyle of tibia and anterior surface of fibula	Extensor expansions of digits II-V	Extends digits II-V and dorsiflexes foot	Deep peroneal
Extensor hallucis longus	*hallux* = great toe	Anterior surface of fibula and interosseous membrane	Distal phalanx of digit I	Extends big toe and assists dorsiflexion of foot	Deep peroneal
Peroneus tertius	*perone* = fibula; *tertius* = third	Anterior surface of fibula and interosseous membrane	Dorsal surface of fifth metatarsal	Dorsiflexes and everts foot	Deep peroneal
Peroneus longus	*longus* = long	Lateral condyle of tibia and head and shaft of fibula	First cuneiform and metatarsal I	Plantar flexes and everts foot	Superficial peroneal
Peroneus brevis	*brevis* = short	Lower aspect of fibula	Metatarsal V	Plantar flexes and everts foot	Superficial peroneal
Gastrocnemius	*gaster* = belly; *kneme* = leg	Lateral and medial condyle of femur	Posterior surface of calcaneus	Plantar flexes foot; flexes knee	Tibial
Soleus	*soleus* = sole of foot	Posterior aspect of fibula and tibia	Calcaneus	Plantar flexes foot	Tibial
Plantaris	*planta* = plantar surface of foot	Lateral supracondylar ridge of femur	Calcaneus	Plantar flexes foot	Tibial
Popliteus	*poples* = ham of the knee	Lateral condyle of femur	Upper posterior aspect of tibia	Flexes and medially rotates leg	Tibial
Flexor hallucis longus	*flexor* = decrease angle	Posterior aspect of fibula	Distal phalanx of big toe	Flexes distal phalanx of big toe	Tibial
Flexor digitorum longus		Posterior surface of tibia	Distal phalanges of digits II-V	Flexes distal phalanges of digits II-V	Tibial
Tibialis posterior	*posterior* = back	Tibia and fibula and interosseous membrane	Navicular, cuneiforms, cuboid, and metatarsals II-IV	Plantar flexes and inverts foot; supports arches	Tibial

Muscles of the Foot

With the exception of one additional intrinsic muscle, the **extensor digitorum brevis,** the muscles of the foot are similar in number and nomenclature to those of the hand. The functions of the muscles of the foot are different, however, because the foot is adapted to provide support while bearing body weight, whereas the hand is adapted for grasping.

The muscles of the foot are topographically arranged into four layers (fig. 12.61) that are difficult to dissociate even in dissection. The muscles function either to move the toes or to support the arches of the foot through their contraction. Because of their complexity, the muscles of the foot will be presented only in illustrations. (See figs. 12.61 and 12.62.)

1. List all the muscles that either originate from or insert on the scapula.
2. On the basis of function, list the muscles of the upper extremity into the categories of flexors, extensors, abductors, adductors, and rotators. (Each muscle may fit into two or more categories.)
3. Which muscles of the lower extremity span two joints and therefore have two different actions?

Clinical Considerations

Compared to the other systems of the body, the muscular system has few clinical afflictions. If properly exercised, the muscles of the body adequately serve a person for a lifetime. Muscles are capable of doing incredible amounts of work and through conditioning can become even stronger.

There are, however, several clinical considerations of muscles, including the diagnosis of muscle conditions, physiological conditions in muscles, and other diseases of muscles.

Diagnosis of Muscle Condition

The clinical symptoms of muscle diseases are usually weakness, loss of muscle mass (atrophy), and pain. The diagnosis and extent of the disease depends on the severity of the symptoms.

Several standard procedures are used to assess skeletal muscle dysfunction and disease. The most obvious is a clinical examination of the patient. Following this, it may be necessary to test the muscle function using **electromyography (EMG)** to measure conduction rates and motor unit activity within a muscle. Laboratory tests may be

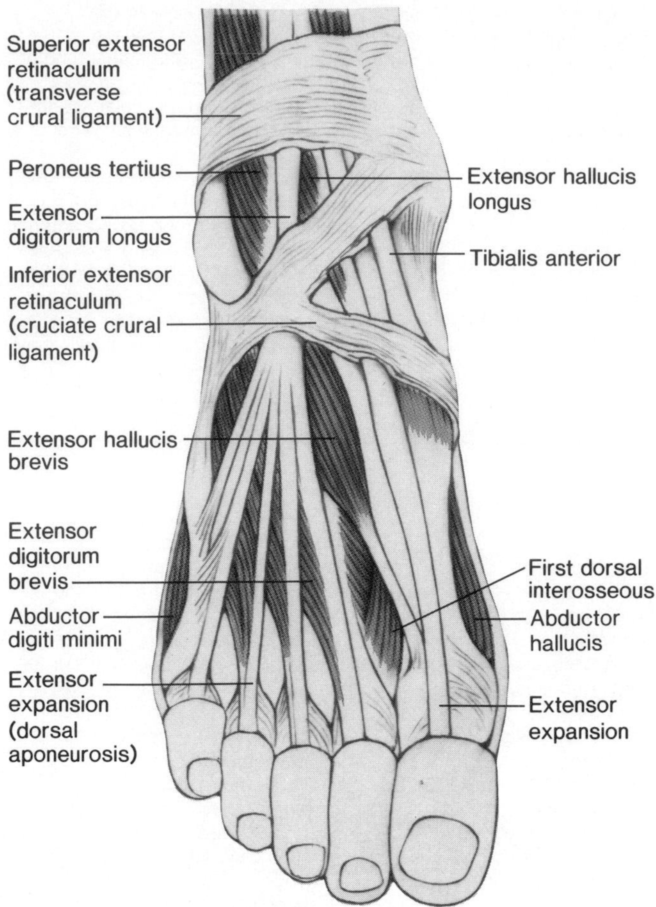

Figure 12.62. An anterior view of the dorsum of the foot.

conducted, such as serum enzyme assays that reflect muscle destruction or muscle biopsy examinations under the microscope. Muscle biopsy is perhaps the definitive source of diagnostic information. Progressive atrophy, polymyositis, and metabolic diseases of muscles can be determined through a biopsy.

Functional Conditions in Muscles

Muscles depend on systematic, periodic contraction to maintain optimal health. Obviously overuse or disease will cause a change in muscle tissue. The immediate effect of overexertion of muscle tissue is the accumulation of lactic acid, resulting in fatigue and soreness. Excessive contraction of a muscle can also damage the fibers or associated connective tissue, resulting in a **strained muscle.**

When skeletal muscles are not contracted, either because the motor nerve supply is blocked or because the limb is immobilized (as when a broken bone is in a cast), the muscle fibers **atrophy,** or diminish in size. Atrophy is reversible if exercise is resumed, as after a healed fracture, but tissue death is inevitable if the nerves cannot be stimulated.

The fibers in healthy muscle tissue increase in size, or **hypertrophy,** if a muscle is systematically exercised. This increase in muscle size and strength is due not to an increase in the number of muscle cells but to the production of more myofibrils accompanied by a strengthening of the associated connective tissue.

A **cramp** within a muscle is an involuntary, painful, and prolonged contraction. Cramps can occur while muscles are in use or at rest. The precise cause of cramps is unknown, but evidence indicates that cramps may be related to conditions within the muscle (e.g., calcium or oxygen deficiencies) or to stimulation of the motor neurons.

A condition called **rigor mortis** (rigidity of death) affects skeletal muscle tissue several hours after death as depletion of ATP within the fibers causes a state of muscle contracture and stiffness of the joint. In a few days, however, as the muscle proteins decompose, the rigidity of the corpse disappears.

Diseases of Muscles

Fibromyositis is an inflammation of both skeletal muscular tissue and the associated connective tissue. Its causes are not fully understood. Fibromyositis frequently occurs in the extensor muscles of the lumbar region of the vertebral column where extensive aponeuroses exist. Fibromyositis of this region is called **lumbago, or rheumatism.**

Muscular dystrophy is a genetic disease characterized by a gradual atrophy and weakening of muscle tissue. There are several kinds of muscular dystrophy, none of whose etiology is completely understood. The most frequent type affects children and is sex-linked to the male child. As muscular dystrophy progresses, the muscle fibers atrophy and are replaced by adipose tissue. Most children who have muscular dystrophy die before the age of twenty.

The disease **myasthenia gravis** is characterized by extreme muscle weakness and low endurance. There is a defective transmission of impulses at the neuromuscular junction. Myasthenia gravis is believed to be an autoimmune disease, and it typically affects women between the ages of twenty and forty.

Poliomyelitis (polio) is actually a viral disease of the nervous system that causes a paralysis of muscles. The viruses are usually localized in the ventral (anterior) horn of the spinal cord, where they affect the motor nerve impulses to skeletal muscles.

Neoplasms of muscle are rare, but when they do occur they are usually malignant. **Rhabdomyosarcoma** *(rab''do-mi''o-sar-ko'mah)* is a malignant tumor of skeletal muscle. It can arise in any skeletal muscle, and most often afflicts young children and elderly persons.

lumbago: L. *lumbus,* loin
myasthenia: Gk. *myos,* muscle; *astheneia,* weakness
poliomyelitis: Gk. *polios,* gray; *myolos,* marrow
rhabdomyosarcoma: Gk. *rhabdos,* rod; *myos,* muscle; *oma,* a growth

Important Clinical Terminology

convulsion An involuntary, spasmodic contraction of skeletal muscle.

fibrillation *(fi-bri-la'shun)* A series of rapid, uncoordinated, and spontaneous contractions of individual motor units within a muscle.

hernia The rupture or protrusion of a portion of the underlying viscera through muscle tissue. The most common hernias occur in the normally weak places of the abdominal wall. There are four basic types of hernias:
1. **femoral**—viscera passing through the femoral ring
2. **hiatal**—the superior portion of the stomach protruding through the esophageal opening of the diaphragm
3. **inguinal**—viscera protruding through the inguinal canal
4. **umbilical**—a hernia occurring at the navel

intramuscular injection A hypodermic injection at certain heavily muscled areas to avoid damaging nerves. The most common site is the buttock.

myalgia Pain within a muscle resulting from any muscular disorder or disease.

myokymia *(mi''o-kim'e-ah)* Continual quivering of a muscle.

myoma A tumor of muscle tissue.

myopathy Any muscular disease.

myotomy *(mi-ot'o-me)* Surgical cutting of muscle tissue.

myotonia A prolonged muscular spasm.

paralysis The loss of nervous control of a muscle.

shin splints Tenderness and pain on the anterior surface of the lower leg, caused by straining the tibialis anterior or extensor digitorium longus muscles.

torticollis (wryneck) A persistent contraction of a sternocleidomastoid muscle, drawing the head to one side and distorting the face. Torticollis may be acquired or congenital.

Chapter Summary

I. Organization and General Functions
- A. The contraction of skeletal muscle fibers results in body motion, heat production, and the maintenance of posture and body support.
- B. The four basic properties characteristic of all muscle tissue are irritability, contractility, extensibility, and elasticity.
- C. Axial muscles include facial muscles, neck muscles, and trunk muscles; appendicular muscles include those that act on the girdles and those that move the segments of the appendages.

II. Development of Skeletal Muscles
- A. Skeletal muscles begin development at four weeks after conception from blocks of mesoderm, called myotomes, in the trunk area and from loosely organized mesenchyme in the head area.
- B. The union of myoblasts forms a syncytial myotube, which becomes a skeletal muscle fiber as contractile filaments develop.

III. Structure of Skeletal Muscles
- A. The origin of a muscle is the more stationary attachment. The insertion is the more movable attachment.
- B. Individual muscle fibers are covered by endomysium. Fasciculi are covered by perimysium. The entire muscle is covered by epimysium.
- C. Synergistic muscles contract together. Antagonistic muscles perform in opposition to a particular group of muscles.
- D. Muscles may be classified according to fiber arrangement as parallel, convergent, pennate, or circular.
- E. Each muscle has motor and sensory innervation.

IV. Skeletal Muscle Fibers and Mode of Contraction
- A. Each skeletal muscle fiber is a multinucleated, striated cell, composed of myofibrils and enclosed by a sarcolemma.
  1. Myofibrils are characterized by alternating A and I bands; each I band is bisected by a Z line, and the portion between two Z lines is the sarcomere.
  2. Extending through the sarcoplasm is a network of membranous channels called the sarcoplasmic reticulum and a system of transverse tubules (T-tubules).
- B. The activity of the cross-bridges causes sliding of the filaments.
  1. The actin on each side of the A bands is pulled toward the center.
  2. The H bands thus appear to be shorter as more actin overlaps the myosin.
  3. The I bands also appear to be shorter as adjacent A bands are pulled closer together.
  4. The A bands stay the same length because the filaments (both thick and thin) do not shorten during muscle contraction.
- C. Maximum tension in a muscle is produced when the muscle is at its normal resting length *in vivo.*
- D. When a muscle exerts tension without shortening, the contraction is termed isometric; when shortening does occur, the contraction is isotonic.

V. Stimulation, Regulation, and the Mechanics of Contraction
- A. The neuromuscular junction is the area consisting of the motor end plate and the sarcolemma of a muscle fiber. In response to a nerve impulse, the synaptic vesicles of the terminal end of an axon secretes a neurotransmitter, which diffuses across the neuromuscular cleft of the neuromuscular junction and stimulates the muscle fiber.
- B. The motor neuron and the muscle fibers innervated by the motor neuron is called a motor unit.
  1. When a muscle is composed of many motor units (such as in the hand), there is a fine control of muscle contraction.
  2. The large muscles of the leg have relatively few motor units, which are correspondingly large.
  3. Graded contractions are produced by the asynchronous stimulation of different motor units.
- C. Calcium ions serve to couple electrical excitation of the muscle fiber with contraction.
  1. When a muscle is at rest, the Ca^{++} concentration of the sarcoplasm is very low, and cross-bridges are prevented from attaching to actin by complexes of troponin and tropomyosin in the thin filaments.
  2. Action potentials are conducted by transverse tubules into the muscle fiber, where they stimulate the release of Ca^{++} from the sarcoplasmic reticulum.

3. The binding of Ca^{++} to troponin causes movement of tropomyosin, which allows cross-bridges to attach and muscle contraction to occur.
4. When action potentials cease, Ca^{++} is removed from the sarcoplasm and stored in the sarcoplasmic reticulum.

D. The contraction of separate muscle fibers is all-or-none.
1. The contraction of separate muscle fibers is called a twitch.
2. A whole muscle also produces a twitch in response to a single electrical pulse *in vitro*.
3. A summation of fiber twitches can occur so rapidly that the muscle produces a smooth, sustained contraction known as tetanus.

VI. Energy Usage by Skeletal Muscle

A. ATP is produced from the combination of ADP with phosphate derived from phosphocreatine.
1. The phosphocreatine represents a ready reserve of high-energy phosphate during sustained muscle contractions.
2. The phosphocreatine is produced at rest from creatine, and phosphate is derived from ATP.

B. There are two types of muscle fibers.
1. Slow-twitch, thin fibers are adapted for aerobic respiration and are resistant to fatigue.
2. Fast-twitch, thicker fibers are adapted for anaerobic respiration.

C. Physical training affects the enzyme content and characteristics of the muscle fibers.
1. Strength training increases the enzymes of anaerobic respiration and makes the fast and intermediate fibers larger and stronger.
2. Endurance training increases the capacity for aerobic respiration of the muscle fibers.

VII. Naming of Muscles

A. Skeletal muscles are named on the basis of shape, location, attachment, orientation of fibers, relative position, and function.

B. Most of the muscles are paired; that is, the right side of the body is an image of the left.

VIII. Muscles of the Axial Skeleton

The muscles of the axial skeleton include those of facial expression, mastication, eye movement, tongue movement, neck movement, respiration, the abdominal wall, the pelvic outlet, and movement of the vertebral column.

IX. Muscles of the Appendicular Skeleton

The muscles of the appendicular skeleton include those of the pectoral girdle, humerus, forearm, wrist, hand, and fingers, and those of the pelvic girdle, thigh, lower leg, ankle, foot, and toes.

Review Activities

Objective Questions

1. Which of the following is *not* used as a means of naming muscles?
   (a) location
   (b) action
   (c) shape
   (d) attachment
   (e) strength of contraction
2. A graded muscle contraction is produced by variations in
   (a) the strength of the fiber's contraction
   (b) the number of fibers that are contracting
   (c) both of the above
   (d) neither of the above
3. Sustained muscle contraction (tetanus) is produced by
   (a) a sustained contraction of muscle fibers
   (b) asynchronous twitches of the muscle fibers
   (c) both of the above
   (d) neither of the above
4. A flexor of the shoulder joint is the
   (a) pectoralis major
   (b) supraspinatus
   (c) teres major
   (d) trapezium
   (e) latissimus dorsi
5. Which of the following muscles have motor units with the lowest innervation ratio?
   (a) brachial muscles
   (b) muscles of the forearm
   (c) thigh muscles
   (d) abdominal muscles
6. Neurotransmitters are stored in synaptic vesicles within
   (a) motor end plates
   (b) motor units
   (c) myofibrils
   (d) terminal ends of axons
7. When a skeletal muscle shortens during contraction, which of the following statements is *false?*
   (a) The A bands shorten.
   (b) The H bands shorten.
   (c) The I bands shorten.
   (d) The sarcomeres shorten.
8. Electrical excitation of a muscle fiber most directly causes
   (a) movement of tropomyosin
   (b) attachment of the cross-bridges to actin
   (c) the release of Ca^{++} from the sarcoplasmic reticulum
   (d) the splitting of ATP
9. The energy for muscle contraction is most directly obtained from
   (a) phosphocreatine
   (b) ATP
   (c) anaerobic respiration
   (d) aerobic respiration
10. Which of the following muscles does *not* have either an origin or insertion upon the humerus? The
   (a) teres minor
   (b) biceps brachii
   (c) supraspinatus
   (d) brachialis
   (e) pectoralis major

Essay Questions

1. Describe how muscle fibers are formed, and explain why the fibers are multinucleated.
2. Discuss the structure of a muscle fiber and the sliding filament theory of contraction.
3. What is a motor unit, and what is its role in muscle contraction?
4. Explain how electrical stimulation of a muscle fiber causes contraction and how relaxation is produced when the electrical stimulation has stopped.
5. Discuss the position of flexor and extensor muscles relative to the shoulder, elbow, and wrist joints.
6. Describe exercises that strengthen the following muscles: (a) the pectoralis major; (b) the deltoid; (c) the triceps; (d) the pronator teres; (e) the rhomboideus major; (f) the trapezius; (g) the serratus anterior; and (h) the latissimus dorsi.
7. Give three examples of synergistic muscle groups within the lower extremity, and identify the antagonistic muscle group for each.

13 SURFACE ANATOMY AND BODY TOPOGRAPHY

Outline

Concepts

Surface anatomy is a branch of gross anatomy concerned with identifying body structures through observation and palpation. The study of surface anatomy has tremendous application to physical fitness and medical diagnosis and treatment.

The surface anatomy of a neonatal infant differs from that of an adult because it represents an early stage of human life. Certain aspects of the surface anatomy of a neonate are of clinical importance in ascertaining the degree of physical development, general health, and possible congenital abnormalities.

The surface anatomy of the head is of concern because the head contains vital sense organs and provides openings into the respiratory and digestive systems. Of further concern are aesthetic appearances and the many medical problems that involve this part of the body, including trauma, diseases, and dysfunctions.

The flexible neck contains major organs, and several structures (the spinal cord, vessels, trachea, and esophagus) that are essential for body sustenance pass through the neck.

The locations of vital visceral organs in the cavities of the torso make the surface anatomy of this body region especially important.

The surface features of the pelvic region are important primarily to identify reproductive organs and clinical problems of these organs.

The surface anatomy of the shoulder and upper extremity is of clinical importance because of frequent trauma to these body regions. Vessels of the upper extremity are also used as pressure points and for intravenous injections or blood withdrawal.

The massive bones and muscles of the buttock and lower extremity are weight bearers and locomotors. Many of the surface features of these regions are important in relation to locomotion or locomotor dysfunctions.

INTRODUCTION TO SURFACE ANATOMY

Surface anatomy is a branch of gross anatomy concerned with identifying body structures through observation and palpation. The study of surface anatomy has tremendous application to physical fitness and medical diagnosis and treatment.

Objective 1. Describe the value of surface anatomy in learning internal anatomical structures.

Objective 2. Discuss how surface anatomy is important in diagnosing and treating various diseases or conditions of the body.

Objective 3. Describe subcutaneous differences in adult males and females that may have a bearing on their surface anatomy.

It is amazing how much anatomical information can be learned by studying the surface anatomy of one's own body. Surface anatomy is the study of the structure and markings of the surface of the body that can be identified through observation or palpation *(pal-pa'shun).* Surface features can be readily identified through visual *observation,* and anatomical features beneath the skin can be located by *palpation* (feeling with firm pressure). Knowledge of surface anatomy is clinically important in locating precise sites for *percussion* (tapping sharply to detect resonating vibrations) and *auscultation* (listening to sounds emitted from organs).

With the exception of certain cranial bones, the bones of the entire skeleton can be palpated. Once the position, shape, and processes of these bones are identified, these skeletal features can serve as landmarks for other anatomical structures. One can observe the location of many skeletal muscles and their tendinous attachments as they are contracted and made to bulge. The location and range of movement of each of the joints of the body can be determined as the articulating bones are moved by muscle contractions. On some persons, one can locate the positions of superficial veins and trace their courses. Even the location and function of the valves within the veins can be demonstrated on the surface of the skin (fig. 13.1). The locations of some of the arteries can be seen as they pulsate beneath the skin. The locations of these arterial pressure points are an important clinical aspect of surface anatomy. Other structures can be identified from the surface, including certain nerves, lymph nodes, glands, and other internal organs.

Figure 13.1. A demonstration of the presence and function of valves within the veins of the forearm, conducted by the great English anatomist William Harvey. (After William Harvey, *On the Motion of the Heart and Blood in Animals,* 1628.) Surface anatomy was extremely important to understanding the concept of a closed circulatory system (e.g., blood contained within vessels).

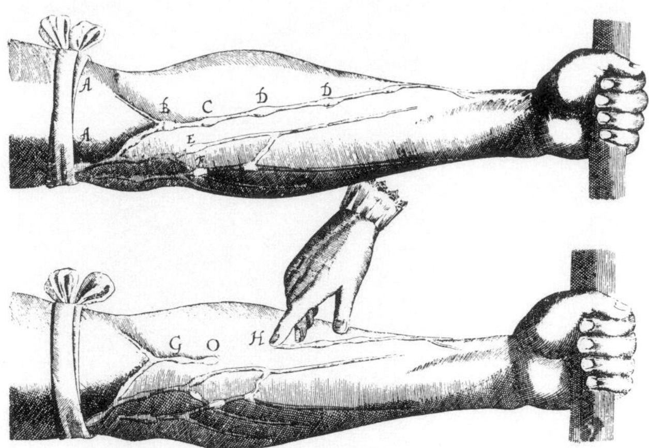

Surface anatomy is an essential part of the study of anatomy. Knowing the location of muscles and muscle groups as observed in surface anatomy can be extremely important in physical fitness and body conditioning. In many medical and paramedical professions the surface anatomy of a patient is of immeasurable value in diagnosis and treatment. Knowing where to record a pulse, insert needles and tubes, listen to the functioning of internal organs, take X rays, and perform physical therapy requires a knowledge of surface body landmarks.

The effectiveness of observation and palpation in studying a person's surface anatomy is dependent on the amount of subcutaneous adipose tissue present (fig. 13.2). In examining an obese person, it may be extremely difficult to observe or palpate certain internal structures that are readily discernible in a thin person. The hypodermis and subcutaneous tissue in a woman is normally thicker than in a man (fig. 13.3). This tends to smooth the surface contour of a woman and obscure the internal features, such as muscles, veins, and bony prominences, that are apparent in men.

This chapter will be of great value in reviewing the bones of the skeleton, articulations, and muscles already studied. Refer back to this chapter to learn the anatomy of the remaining body systems. Reviewing this way will reinforce an external perspective of where the various organs and structures are located.

Figure 13.2. Subcutaneous adipose tissue. (*a*) an anterior view of the left brachial region. (*b*) a posterior view of the left lower leg.

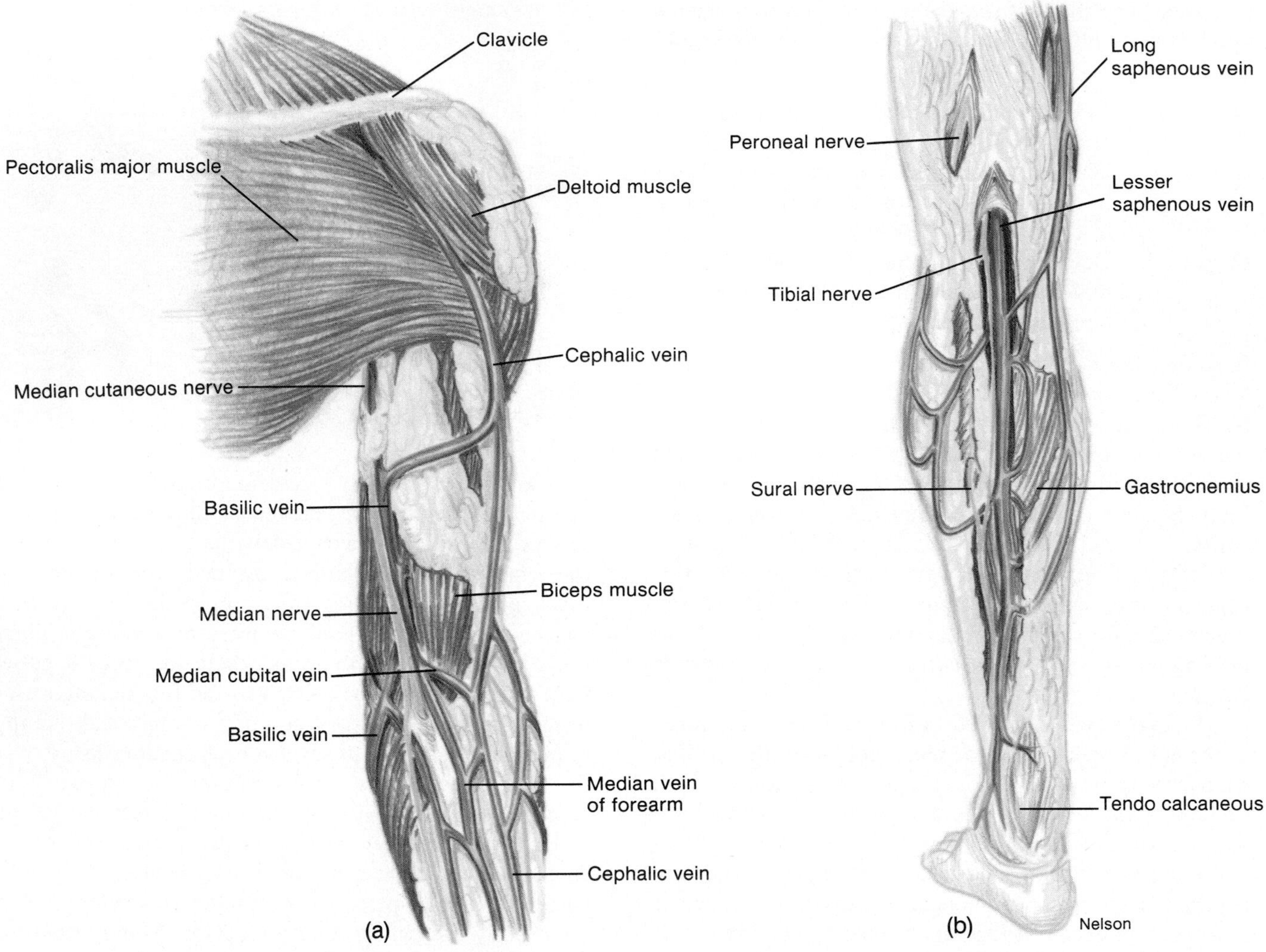

Make anatomy relevant and applicable by identifying your surface anatomy. Use yourself as a model from which to learn and review, and anatomy as a science will take on a new meaning. As you learn about a bone or a process on a bone, palpate that part of your body. Contract the muscles you are studying so that you better understand their locations, attachments, and actions. Do this and you will become better acquainted with your body, and anatomy will become more enjoyable and easier to learn. Your body is one "crib sheet" that can be taken with you to exams.

1. Define the terms *observation* and *palpation,* and explain the value of surface anatomy in learning the location of internal structures.
2. Discuss the medical importance of surface anatomy.
3. Describe the subcutaneous difference seen in the surface anatomy of men and women.

Figure 13.3. Principal areas of adipose deposition of a female. (*a*) an anterior view; (*b*) a lateral view. The outline of a male is superposed in the lateral view. There is significantly more adipose tissue interlaced in the fascia covering the muscles, vessels, and nerves in a female than in a male. The hypodermis layer of the skin is also approximately 8 percent thicker in a female than in a male.

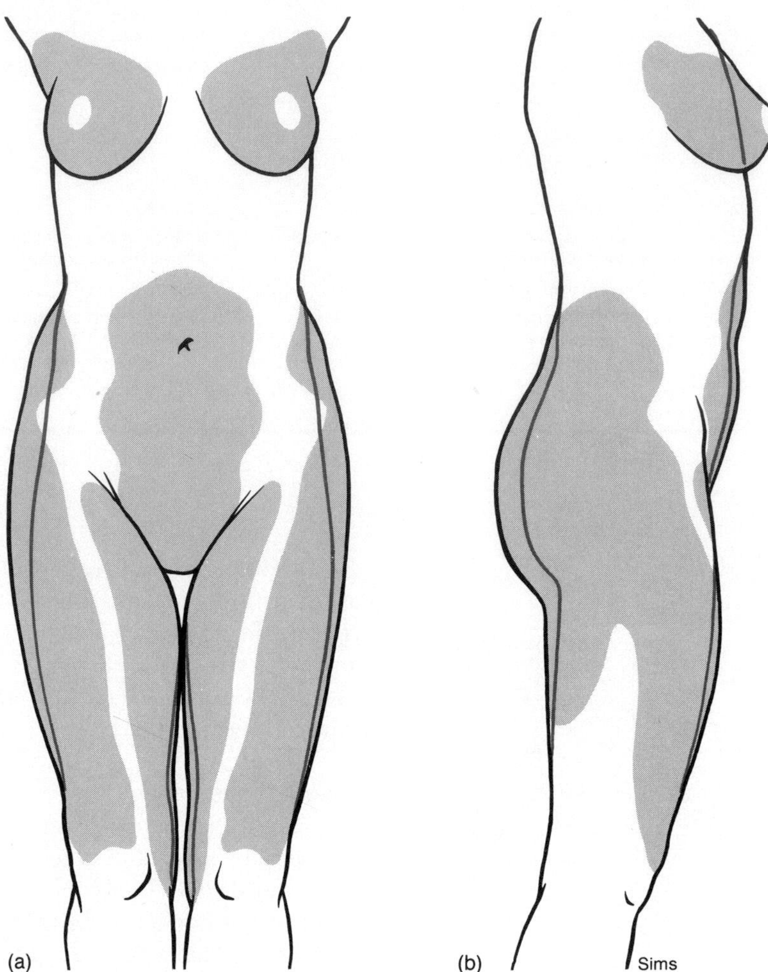

Surface Anatomy of the Newborn

The surface anatomy of a neonatal infant differs from that of an adult because it represents an early stage of human life. Certain aspects of the surface anatomy of a neonate are of clinical importance in ascertaining the degree of physical development, general health, and possible congenital abnormalities.

Objective 4. Describe the surface anatomy of a normal, full-term neonate.

Objective 5. List some internal anatomical structures that can be palpated in a neonate.

The birth of a baby is a dramatic culmination of a nine-month gestation, during which the miraculous development of the fetus prepares it for extrauterine life. The normal, full-term neonate is physiologically prepared for life but is totally dependent on parental care. The physical assessment of the neonate is extremely important to insure its survival. Much of the assessment is performed through inspection and palpation of its surface anatomy. The surface anatomy of a neonate is obviously different from that of an adult because of the transitional stage of development from fetus to infant.

Table 13.1 Surface anatomy of the neonate

Body structure	Normal conditions	Common variations
General posture	Flexion of vertebral column and extremities	Extended legs and neck; abducted and rotated thighs (breech birth)
Skin	Red or pink skin with vernix caseosa and lanugo; edematous face, extremities, and genitalia	Neonatal jaundice; integumentary blisters; Mongolian spots
Skull	Fontanels large, flat, and firm but soft to touch	Molded skull, bulging fontanels; cephalhematoma
Eyes	Lids edematous; color—gray, dark blue, or brown; absence of tears; corneal, pupillary, and blink reflexes	Conjunctivitis, subconjunctival hemorrhage
Ears	Auricle flexible, cartilage present; top of auricle positioned on horizontal line with outer canthus of eye	Auricle flat against head
Neck	Short and thick surrounded by neck folds	Torticollis
Chest	Equal anterioposterior and lateral dimensions; xiphoid process evident; breast enlargement	Funnel or pigeon chest; additional nipples (polythelia); secretions from breast (witch's milk)
Abdomen	Cylindric in shape; liver and kidneys palpable	Umbilical hernia
Genitalia	(♂ and ♀) Edematous and darkly pigmented; (♂) testes palpable in scrotum; periodic erection of penis	(♀) Blood-tinged discharge (pseudomenstruation); hymenal tag; (♂) testes palpable in inguinal canal; inability to retract prepuce; inguinal hernia
Extremities	Symmetry of extremities; 10 fingers and toes; soles flat with moderate to deep creases	Partial syndactyly; asymmetric length of toes

Figure 13.4. The flexion position of a neonate (*a* and *b*) is an indication of a healthy gestation and a normal delivery.

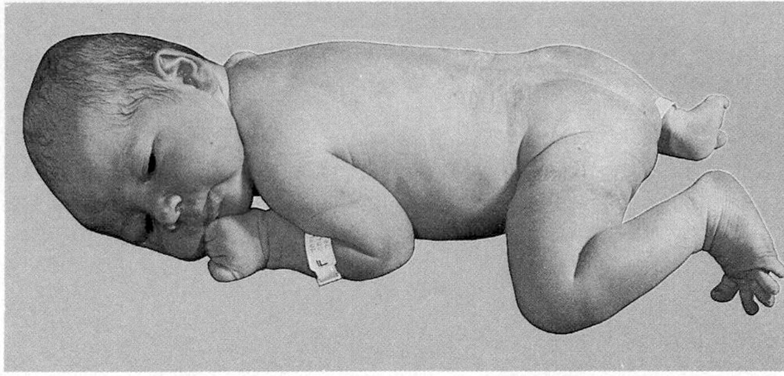

(a)

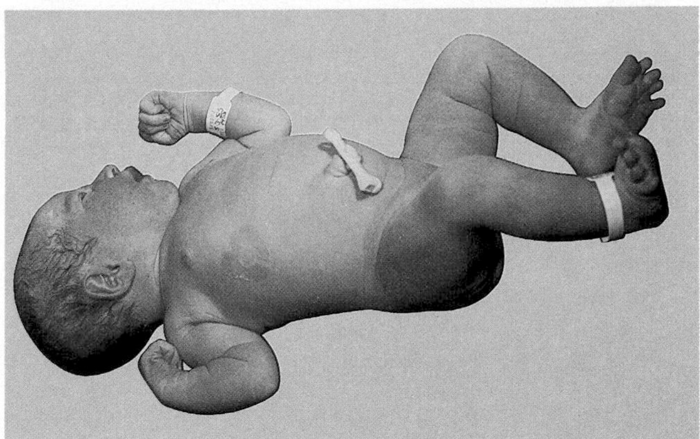

(b)

Although the surface anatomy of a neonate is discussed at this point in the book, prenatal development is discussed in chapter 29, and body growth with its accompanying physiological changes is discussed in chapter 30. A summary of the surface anatomy of the neonate is presented in table 13.1.

General Appearance

As a result of *in utero* position, the posture of the full-term neonate is one of flexion (fig. 13.4). The neonate born vertex (head first) keeps the neck and vertebral column flexed, with the chin resting on the upper chest. The hands are flexed (clenched), as are the arms, and held toward the chest. The legs are flexed at the knees, and the hips are flexed bringing the thighs toward the abdomen. The feet are dorsiflexed.

The skin is the one organ of the neonate that is completely visible and is therefore a source of considerable information concerning the state of development and the clinical condition of the newborn. At birth, the skin is covered with a grayish, cheeselike substance called **vernix caseosa.** If it is not washed away during bathing, it will dry and disappear within a couple of days. Fine, silklike hair, called **lanugo** *(lah-nu'go),* may be present on the forehead, cheeks, shoulders, and back. Distended sebaceous glands, called **milia** *(mil'e-ah),* may appear as tiny white papules on the nose, cheeks, and chin. Skin color depends on genetic background, although certain areas such as the genitalia, areolae, and linea alba may appear more darkly

neonate: Gk. *neos,* new; L. *natalis,* birth
vernix caseosa: L. *vernix,* varnish; *caseus,* cheese
milia: L. *miliarius,* relating to millet

Figure 13.5. Sole creases at different ages of gestation as seen from footprints of two premature babies (*a* and *b*) and a full-term baby (*c*). (*a*) at twenty-six weeks, only an anterior transverse crease is present; (*b*) by thirty-three weeks, creases develop along the medial instep; and (*c*) the entire sole is creased by thirty-eight weeks.

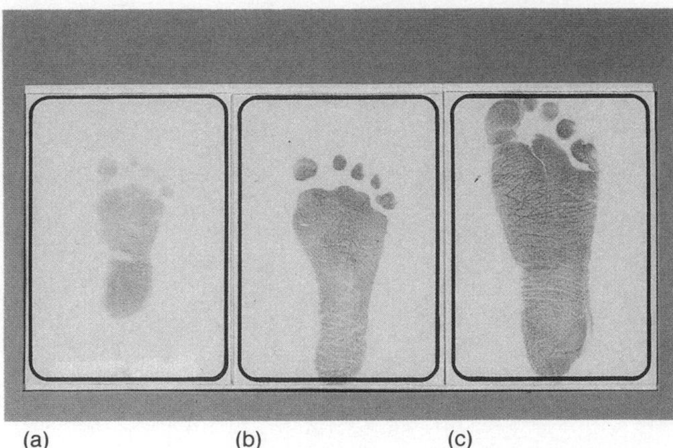

(a) (b) (c)

pigmented because of a response to the maternal and placental hormones that enter the fetal circulation. **Mongolian spots** occur in about 90% of newborn Negroes, Orientals, and American Indians. These blue-gray pigmented areas vary in size and are usually located in the lumbosacral region. Mongolian spots generally fade within the first year or two.

Abnormal skin color is clinically important in the physical assessment of the neonate. *Cyanosis* (bluish discoloration) is usually due to a pulmonary disease such as atelectasis or pneumonia or to congenital heart disease. Although *jaundice* (yellowish discoloration) is common in infants and is usually of no concern, it may indicate liver or bone marrow problems. *Pallor* (paleness) may indicate anemia, edema, or shock.

The appearance of the nails and nail beds is especially valuable in determining body dysfunctions, certain genetical conditions, and even normal gestations. Cyanosis, pallor, and capillary pulsations are best observed at the nails. Yellow nails are found in postmature neonates.

Local edema (swelling) is common, particularly in the skin of the face, legs, hands, feet, and genitalia. Creases on the palms of the hands and soles of the feet should be prominent; the absence of creases accompanies prematurity (fig. 13.5). The nose is usually flattened after birth, and because of immense vascularity, bruises here or on other areas of the face are common. The auricle of the ear is flexible, with the top edge positioned on a horizontal line with the outer canthus (corner) of the eye.

The neck of a neonate is short and thick and is surrounded by neck folds. The chest is rounded in cross section, and the abdomen is cylindrical. The abdomen may bulge in the upper right quadrant due to the large liver.

If the newborn is thin, peristaltic intestinal waves may be observed. At birth, the **umbilical** *(um-bil'ĭ-kal)* **cord** appears bluish white and moist. After clamping, it begins to dry and appears yellowish brown. It progressively shrivels and becomes greenish black prior to falling off by the second week.

The genitalia of both sexes may appear darkly pigmented due to maternal hormonal influences. In a female neonate, a **hymenal** *(hi'men-al)* **tag** is frequently present and is visible from the posterior opening of the vagina. It is composed of tissue from the hymen and labia minora. The hymenal tag usually disappears by the end of the first month.

Palpable Structures

The six **fontanels** can be lightly palpated as the "soft spots" on the infant's head. The liver is palpable 2–3 cm (1 in.) below the right costal margin. A physician doing a physical examination of a neonate will palpate both kidneys soon after delivery, before the intestines fill with air. The suprapubic area is also palpated for an abnormal distended urinary bladder. The newborn should void urine within the first twenty-four hours after birth.

The testes of a male child should always be palpated in the scrotum. If the neonate is small or premature, the testes may be palpable in the inguinal canals. An examination for inguinal hernias is facilitated by the infant's crying, which creates abdominal pressure.

1. Describe the appearance of each of the following in a normal neonate: skin, head, thorax, abdomen, genitalia, and extremities. What is meant by the normal flexion position of a neonate?
2. Which internal body organs are palpable in a neonate?

Head

The surface anatomy of the head is of concern because the head contains vital sense organs and provides openings into the respiratory and digestive systems. Of further concern are aesthetic appearances and the many medical problems that involve this part of the body, including trauma, diseases, and dysfunctions.

Objective 6. Identify by observation or palpation various surface features of the cranial and facial regions.

The **head,** or **caput,** contains the brain and sense organs—the eyes, ears, and nose. It also provides openings into the respiratory and digestive systems. The head is structurally and developmentally divided into the cranium and the face.

Figure 13.6. The surface anatomy of the facial region: (*a*) an anterior view; (*b*) a lateral view.

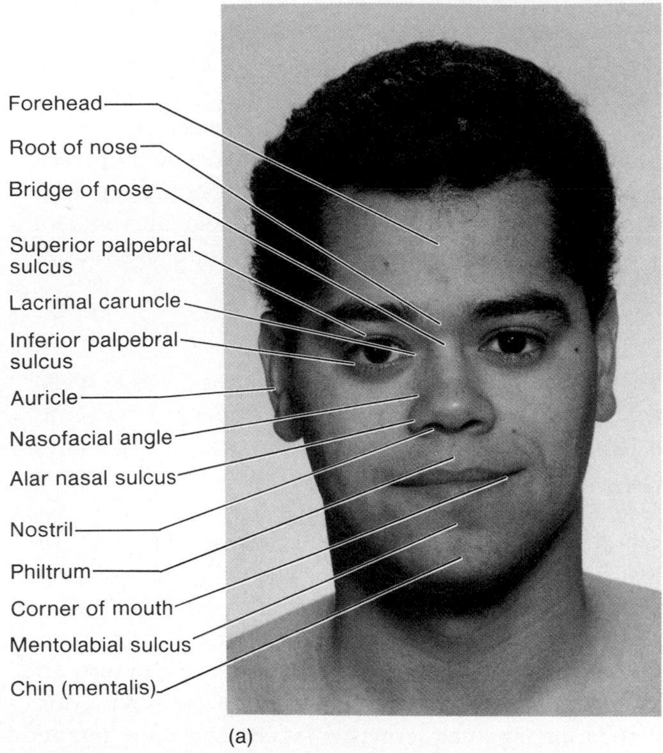

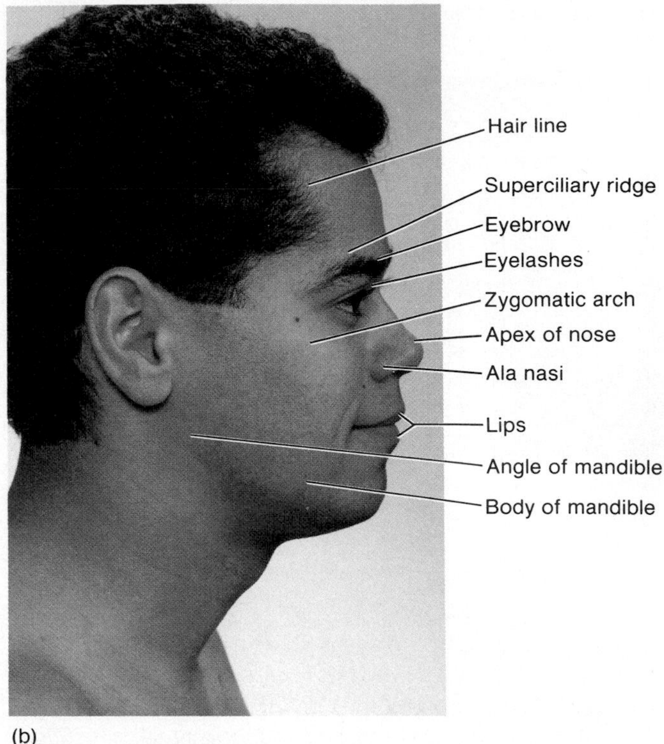

Cranium

The **cranium** *(kra'ne-um),* also known as the **braincase,** is covered by the *scalp.* The scalp is attached anteriorly, at the level of the **eyebrows,** to the **supraorbital ridges.** The scalp continues posteriorly over the area commonly called the forehead and across the crown (vertex) of the top of the head to the **superior nuchal** *(nu'kal)* **line,** a ridge on the back of the skull. Both the supraorbital ridge above the **orbit,** or socket of the eye, and the superior nuchal line at the back of the skull can be easily palpated. Laterally, the scalp covers the **temporal region** and terminates at the fleshy portion of the ear called the **auricle,** or **pinna.** The temporal region is the attachment for the **temporalis muscle,** which can be palpated when repeatedly clenching the jaw. This region is clinically important because it is a point of entrance to the cranial cavity in many surgical procedures.

Only a portion of the scalp is covered with hair, and the variable **hairline** is genetically determined. The scalp is clinically important because of the dense connective tissue layer that supports nerves and vessels beneath the skin. When the scalp is cut, the wound is held together by the connective tissue, but at the same time the vessels are held open, resulting in profuse bleeding.

cranium: Gk. *kranion,* skull
nuchal: L. *nucha,* nape of neck

Face

The **face,** or **facies** *(fa'she-ēz)* (fig. 13.6), is composed of four regions: the **ocular region,** which includes the eye and associated structures; the **auricular region,** which includes the ear; the **nasal region,** which includes the external and internal structures of the nose; and the **oral region,** which includes the mouth and associated structures.

> The two most commonly fractured bones of the face are the nasal and mandible. Trauma to either of these bones generally results in simple fracture, which is not usually serious. If the nasal septum or cribriform plate is fractured, however, careful treatment is required. If the cribriform plate is severely fractured, there may be a tear in the meninges, causing a sudden loss of cerebrospinal fluid and death.

The skin of the face is relatively thin and contains many sensory receptors, particularly in the oral region. Certain facial regions also have numerous **sweat glands** and **sebaceous** *(se-ba'shus)* **glands,** which secrete *sebum,* or oil. Acne is a serious facial dermatological problem for many teenagers. Facial hair appears over most of the facial region in males after they go through puberty; unwanted facial hair may occur sparsely on some females and can become a social problem.

Figure 13.7. The surface anatomy of the ocular region.

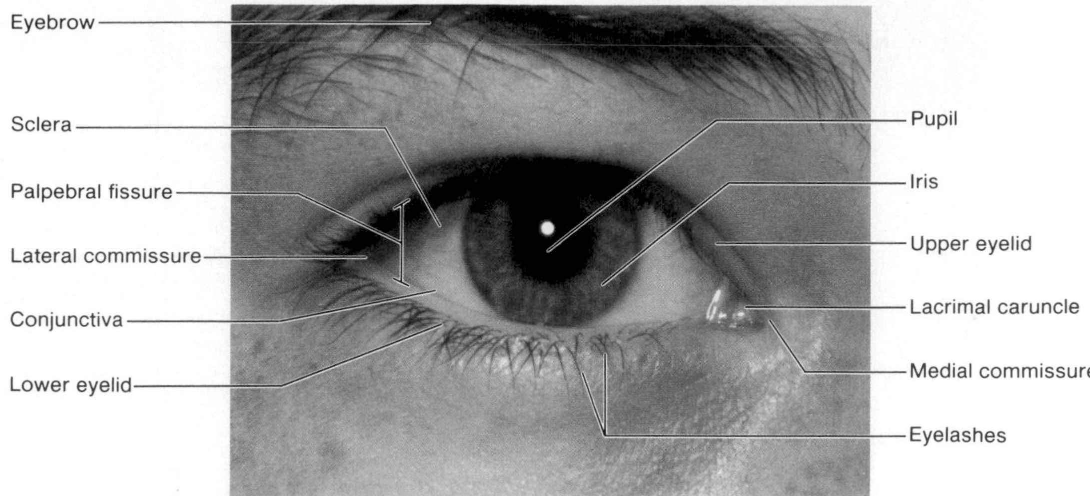

Table 13.2 Surface anatomy of the ocular region

Structure	Comments	Structure	Comments
Eyebrow	Ridge of hair that superiorly arches the eye. It protects the eye against sunlight and is important in facial expression.	Iris	Circular, colored, muscular portion of the eyeball that surrounds the pupil. It reflexly regulates the amount of incoming light.
Eyelids	Movable folds of skin and muscle that cover the eyeball anteriorly. They assist in lubricating the anterior surface of the eyeball and reflexly close to protect the eyeball.	Pupil	Opening in the center of the iris through which light enters the eyeball
		Palpebral fissure	Space between the eyelids when they are open
Eyelashes	Row of hairs on the margin of eyelid. They prevent airborne substances from contacting eyeball.	Subtarsal sulcus	Groove beneath the eyelid that parallels the margin of the lid. It traps small foreign particles that contact the conjunctiva.
Conjunctiva	Thin mucous membrane that covers the anterior surface of the eyeball and lines the undersurface of the eyelids. It aids in reducing friction during blinking.	Medial commissure	Medial corner of the eye where the upper and lower eyelids come together
		Lateral commissure	Lateral corner of the eye where the upper and lower eyelids come together
Sclera	Outer fibrous layer of eyeball; the "white" of the eye that gives form to the eyeball	Lacrimal caruncle	Fleshy, pinkish elevation at the medial commissure. It contains sebaceous and sweat glands.
Cornea	Transparent anterior portion of eyeball. It is slightly convex to refract incoming light waves.		

The muscles of facial expression are important for the way they affect surface features. Various emotions are conveyed when these muscles are contracted. These muscles originate on the different facial bones and insert into the dermis (second major layer) of the skin. Excessive contraction of these muscles may cause permanent crease lines in the skin.

Because the organs of the facial region are so complex and specialized, there are professional fields of speciality associated with the various regions. *Optometry* and *ophthalmology* are concerned with the structure and function of the eye. *Dentistry* is entirely devoted to the health and functional and cosmetic problems of the oral region, particularly the teeth. An *otorhinolaryngologist* is an ear, nose, and throat specialist.

The *ocular region* includes the eyeball and associated structures. Most of the surface features of the ocular region protect the eye. **Eyebrows** protect against potentially damaging sunlight and mechanical blows; **eyelids** reflexly close to protect against objects or visual stimuli; **eyelashes** prevent airborne objects from contacting the eyeball; and the **lacrimal** *(lak'rĭ-mal)* **secretions** (tears) wash away chemical or foreign materials and prevent the surface of the eyeball from drying. Figure 13.7 depicts many of the surface features of the ocular region, and table 13.2 describes these structures.

Figure 13.8. The surface anatomy of the auricular region.

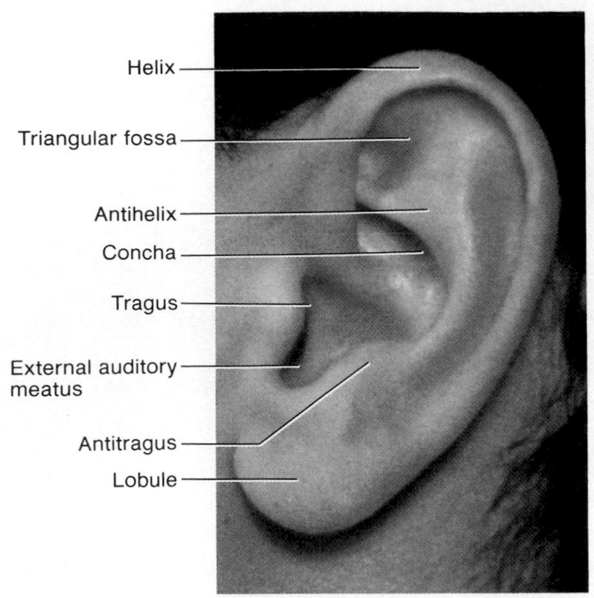

Table 13.3	Surface anatomy of the auricular region
Structure	**Comments**
Auricle (pinna)	Expanded, fleshy portion of the ear projecting from the side of the head that funnels sound waves into the external auditory meatus
Helix	Outer rim of auricle. It gives form and shape to the pinna.
Lobule	Inferior, fleshy portion of auricle
Tragus	Posterior cartilaginous projection of auricle. It partially covers and protects the external auditory meatus.
Antitragus	Small, cartilaginous anterior projection opposite the tragus
Antihelix	Semicircular ridge anterior to the greater portion of the helix
Concha	Depressed hollow of auricle. It funnels sound waves.
External auditory meatus	Canal extending inward to the tympanic membrane. It is slightly S-shaped and contains glands that secrete earwax for protection.
Triangular fossa	Triangular depression in superior portion of antihelix

The *auricular region* includes the visible surface structures as well as internal organs that function in hearing and in maintaining equilibrium. The fleshy **auricle,** or **pinna** *(pin'nah),* and the tubular opening into the middle ear, called the **external auditory meatus** *(me-a'tus),* are the only observable surface features of the auricular region. The rim of the auricle is called the **helix,** and the inferior portion is referred to as the **lobule.** The lobule is composed primarily of connective and fatty tissue and therefore can be easily pierced. For this reason, it is sometimes used when obtaining blood for a blood count. The **tragus** *(tra'gus)* is a small, posteriorly directed projection partially covering and protecting the external auditory meatus. Further protection is provided by the many fine hairs that surround the opening into this canal. The head of the mandible can be palpated at the opening of the external auditory meatus by placing the little finger in the opening and then vigorously moving the jaw. The **mastoid process** can be palpated as a bony knob on the skull immediately posterior to the lobule of the auricle. This process serves as the point of attachment for the muscle that turns the head. Refer to figure 13.8 and table 13.3 for an illustration and discussion of other surface features of the auricular region.

meatus: L. *meatus,* path
helix: Gk. *helix,* to turn or roll
tragus: Gk. *tragos,* goat (because of the tuft of hairs on it)

The inspection of some of the internal structures of the ear is part of a routine physical examination and is performed using an otoscope. Earwax may accumulate in the canal, but this is a protective substance. It waterproofs the eardrum and because of its bitter taste is thought to be an insect repellent.

A few structural features of the *nasal region* are apparent from its surface anatomy (fig. 13.9 and table 13.4). The principal function of the nose is associated with the respiratory system, and the need for a permanent body opening to permit gaseous ventilation accounts for its surface features. The **root** (nasion) is located at about the level of the eyebrows where the nose begins to protrude from the forehead. The firm, narrow portion between the eyes is the **bridge** of the nose and is formed by the union of the nasal bones. The nose below this level has a pliable cartilaginous framework that maintains an opening. The tip of the nose is called the **apex,** and the region between the bridge and the apex is the **dorsum nasi.** As the lateral portion of the nose blends with the face, it is called the **nasofacial angle.** The **external nares,** or **nostrils,** are the paired openings into the nose. The **nasal septum** forms a partition between the nostrils, and the **ala** *(a'lah)* forms the flared outer margin of each nostril.

ala: L. *ala,* winglike

Figure 13.9. The surface anatomy of the nasal and oral regions: (*a*) the nose and lips; (*b*) the teeth.

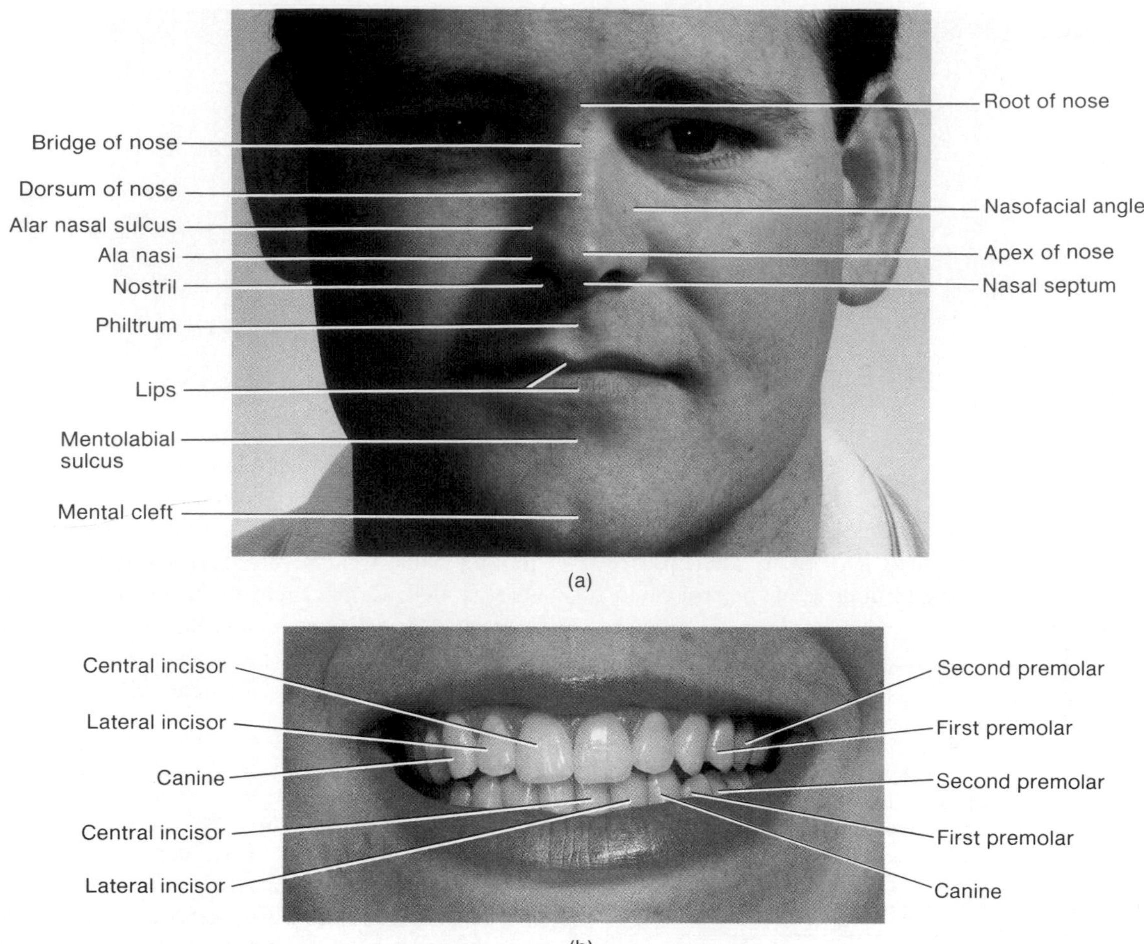

Table 13.4 Surface anatomy of the nasal and oral regions

Structure	Comments	Structure	Comments
Root (nasion)	Superior attachment of nose to the cranium	External nares	External openings into nasal cavity
Bridge	Anterior bony framework of nose formed by union of nasal bones	Nasal septum	Partition between external nares
Dorsum nasi	Anterior ridge of nose	Ala	Laterally expanded border of external nare
Nasofacial angle	Anteriolateral boundary of nose as it blends with tissues of the face	Philtrum	Vertical depression in medial portion of upper lip
Apex	Terminal tip of nose	Lips	Upper and lower anterior borders of the mouth
		Chin (mental)	Anterior portion of lower jaw

Figure 13.10. Surface structures of the oral cavity (*a*) with the mouth open, and (*b*) with the mouth open and the tongue elevated.

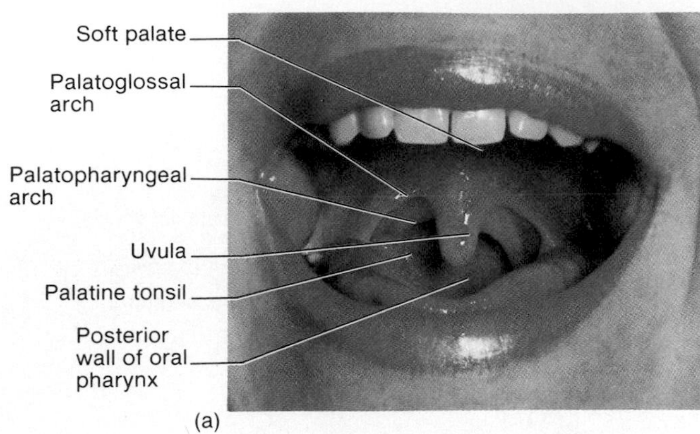

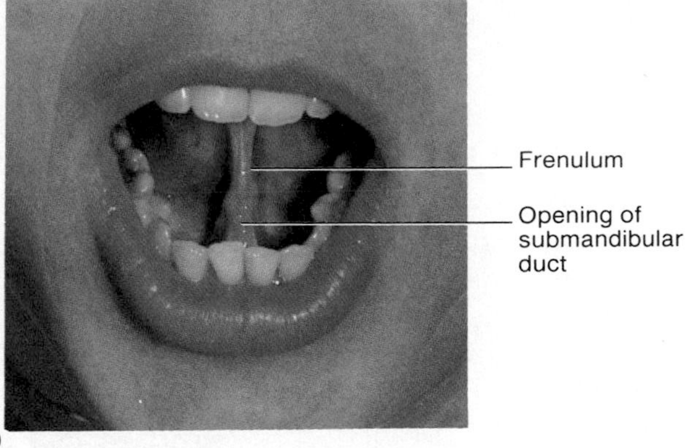

Structures of the *oral region* that are important in surface anatomy include the fleshy upper and lower **labia** *(la'be-ah),* or lips, and the structures of the **oral cavity** that can be observed when the mouth is open. The lips are shown in figure 13.9, and the structures of the oral cavity are seen in figure 13.10.

The color of the lips and other mucous membranes of the oral cavity are diagnostic of certain body dysfunctions. The lips may appear pale in patients with severe anemia and bluish or cyanotic in persons lacking in oxygen. Pernicious anemia may cause a lemon yellow tint to the lips, as does jaundice. In *Addison's disease,* the normally pinkish mucous membranes of the cheeks have brownish areas of pigmentation.

1. What are the boundaries of the cranial region, and why is this region clinically important?
2. Why do scalp wounds bleed so freely? How might this relate to infections?
3. What are the subdivisions of the facial region?

NECK

The flexible neck contains major organs, and several structures (the spinal cord, vessels, trachea, and esophagus) that are essential for body sustenance pass through the neck.

Objective 7. Discuss the functions of the neck.
Objective 8. List by name and location the triangles of the neck and the structures contained within these triangles.

The **neck,** or **collum,** is a complex region of the body that connects the head to the thorax. The spinal cord, digestive and respiratory tracts, and major vessels traverse this highly flexible area. Several organs and glands are located here as well. Remarkable musculature in the neck produces an array of movements. Because of this complexity, the neck is a clinically important area. The surface features provide landmarks for determining the location of internal structures.

Surface Features

The neck is divided into four regions: (1) an *anterior region,* called the **cervix,** which contains the digestive and respiratory tracts, the **larynx** *(lar'inks)* (voice box), vessels to and from the head, nerves, and the **thyroid** and **parathyroid glands;** (2) right and (3) left *lateral regions,* each composed of major neck muscles and **cervical** *(ser'vĭ-kal)* **lymph nodes;** and (4) a *posterior region,* referred to as the **nucha** *(nu'kah),* which includes the spinal cord, cervical vertebrae, and associated structures.

The most prominent structure of the cervix of the neck is the **thyroid cartilage** (Adam's apple) of the larynx (fig. 13.11). The thyroid cartilage is associated with the vocal cords and is larger in sexually mature males than in females, which accounts for the male's deeper voice. The **hyoid bone** can be palpated just above the larynx. Both of these structures are elevated during swallowing, which is one of the actions that directs food and fluid into the esophagus. Note this action on yourself by gently cupping your fingers on the larynx, then swallowing. Below the thyroid cartilage is the **cricoid** *(kri'koid)* **cartilage,** followed by the **trachea** *(tra'ke-ah)* (windpipe), both of which can be palpated. The trachea is clinically important in case

Addison's disease: from Thomas Addison, English physician, 1793–1860

cervix: L. *cervix,* neck
larynx: Gk. *larynx,* upper windpipe
hyoid: Gk. *hyoeides,* U-shaped

Figure 13.11. An anterolateral view of the neck.

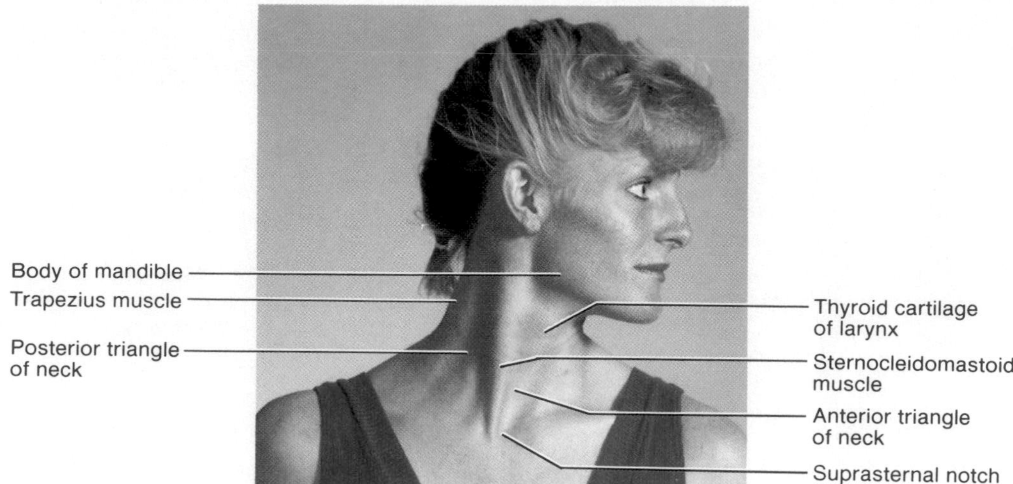

a respiratory tube has to be inserted during a *tracheotomy.* The thyroid gland can be palpated on either side just below the level of the larynx. Pulsations of the **common carotid artery** can be observed and palpated on either side of the neck just lateral and a bit superior to the level of the larynx.

The arteries of the head and neck are rarely damaged because of their elasticity. In a severe lateral blow to the head, however, the internal carotid artery may rupture, resulting in the perception of a roaring sound as blood rushes into the cavernous sinuses of the temporal bone. Containment of carotid hemorrhage within the sinuses may actually be lifesaving.

The **suprasternal notch** is the depression in the midline of the cervix just superior to the sternum. The two **clavicles** are obvious in all persons because they are subcutaneous (just under the skin).

The **sternocleidomastoid** and **trapezius muscles** are the prominent structures of each lateral region (fig. 13.11). The sternocleidomastoid muscle can be palpated its entire length when the head is turned to the side. The tendon of this muscle is especially prominent to the side of the suprasternal notch. The trapezius muscle can be felt if the shoulders are shrugged. An inflammation of the trapezius muscle causes a "stiff neck." If a person is angry or if a shirt collar is too tight, the **external jugular vein** can be seen as it courses obliquely across the sternocleidomastoid muscle. **Cervical lymph nodes** of the lateral neck region may become swollen and painful from infectious diseases of the oral or pharyngeal regions.

Most of the structures of the nucha are too deep to be of importance in surface anatomy. The spines of the lower cervical vertebrae (especially C7), however, can be observed and palpated when the neck is flexed. In this same position, a firm ridge is raised, called the **ligamentum nuchae** (not shown), that extends superiorly from C7 to the external occipital protuberance of the skull. Clinically the nucha is tremendously important because of the debilitating damage to it from whiplash injury or a broken neck.

Triangles of the Neck

The triangles of the neck, created by the arrangement of specific muscles and bones, are clinically important because of the specific structures included in each. The structures of the neck that are important in surface anatomy have already been identified, however, so only a summary of the two major and six minor triangles will be depicted in figure 13.12 and presented in table 13.5. The sternocleidomastoid muscle obliquely transects the neck, dividing it into an **anterior triangle** and a **posterior triangle.** The apex of the anterior triangle is directed inferiorly. The median line of the neck forms the anterior boundary of the anterior triangle, the inferior border of the mandible forms its superior boundary, and the sternocleidomastoid muscle forms its posterior boundary. The posterior triangle is formed by the sternocleidomastoid muscle anteriorly and the trapezius muscle posteriorly, and the clavicle forms its base inferiorly.

Figure 13.12. Triangles of the neck. (*a*) the two large triangular divisions; (*b*) the six lesser triangular subdivisions; (*c*) the detailed muscular anatomy of the neck.

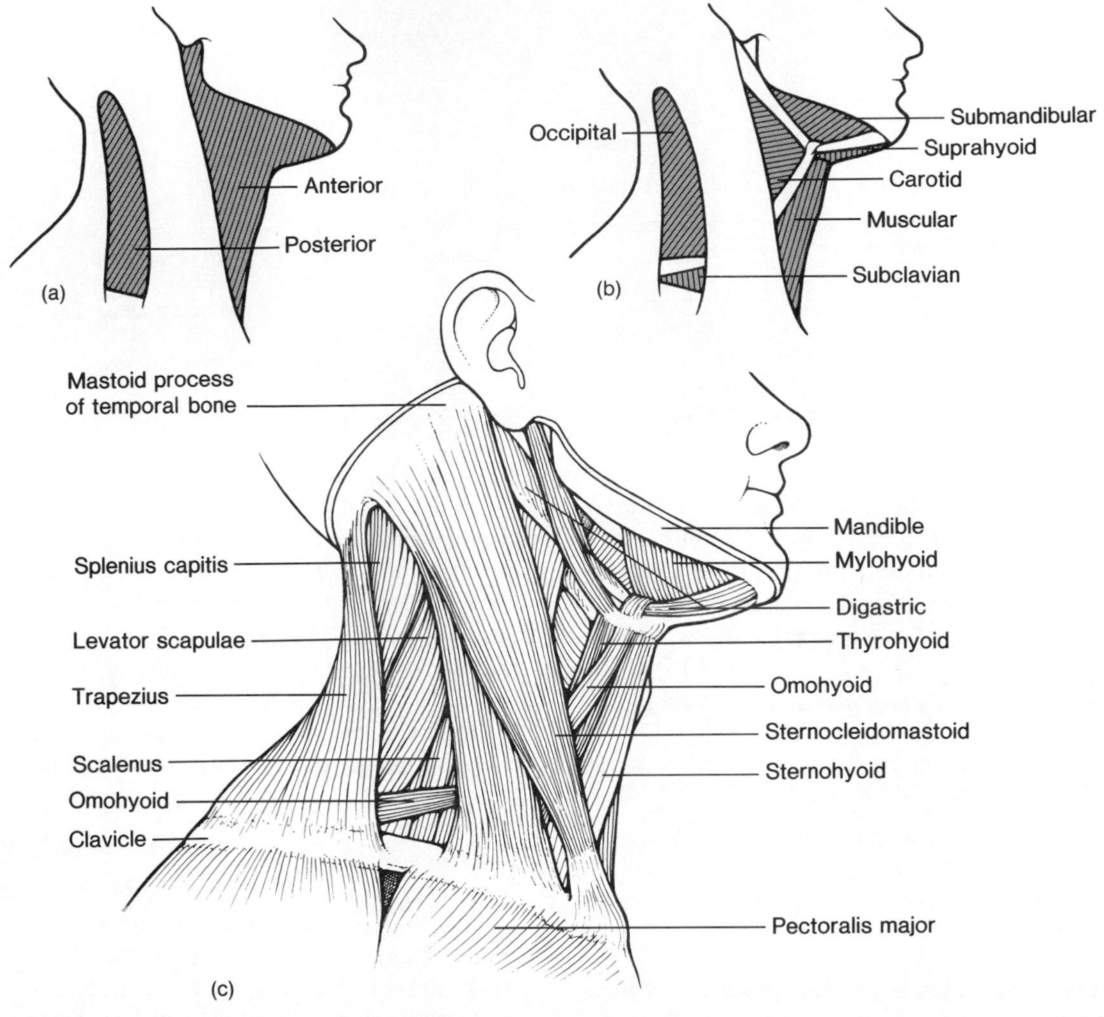

Table 13.5 Location and contents of the triangles of the neck

Triangle	Boundaries	Contents
Anterior	Sternocleidomastoid muscle, median line of neck, inferior border of mandible	Four lesser triangles, salivary glands, larynx, trachea, thyroid glands, various vessels and nerves
Carotid	Sternocleidomastoid, posterior digastric, and omohyoid muscles	Carotid arteries, internal jugular vein, vagus nerve
Submandibular	Digastric muscle (both heads), inferior border of mandible	Salivary glands
Suprahyoid	Digastric muscle, hyoid bone (This is the only unpaired triangle of the neck.)	Muscles of the floor of the mouth, salivary glands and ducts
Muscular	Sternocleidomastoid and omohyoid muscles, midline of neck	Larynx, trachea, thyroid gland, carotid sheath
Posterior	Sternocleidomastoid and trapezius muscles, clavicle	Nerves and vessels
Occipital	Sternocleidomastoid, trapezius, and omohyoid muscles	Cervical plexus, accessory nerve
Subclavian	Sternocleidomastoid and omohyoid muscles, clavicle	Brachial plexus, subclavian artery

Three structures traversing the neck are extremely important and potentially vulnerable. These structures are the common carotid artery, which carries blood to the head, the internal jugular vein, which drains blood from the head, and the vagus nerve, which conducts nerve impulses to visceral organs. These structures are protected in the neck by their deep position near the sternocleidomastoid muscle and by being enclosed in a tough connective tissue sheath called the *carotid sheath.*

1. List four functions of the neck, and state which body systems are located, in part, within the neck.
2. What are the structural regions of the neck? What structures are included in each region?
3. Using your knowledge of the triangles of the neck, describe where to palpate to feel a pulse, the trachea, cervical lymph nodes, and the thyroid gland.

Torso

The locations of vital visceral organs in the cavities of the torso make the surface anatomy of this body region especially important.

Objective 9. Identify by observation or palpation various surface features of the torso.

Objective 10. List the auscultation sites of the thorax and abdomen.

The **torso** *(tor'so),* or **trunk,** is divided into the **back** (dorsum), **thorax** *(tho'raks)* (chest), **abdomen** *(ab'do-men)* (venter), and **pelvis.** A region called the **perineum** *(per"i-ne'um)* forms the floor of the pelvis and includes the external genitalia. The surface anatomy of these regions is particularly important in determining the location and condition of the visceral organs. The usability of some of these surface features varies depending on age, sex, and body weight.

Back

No matter how obese a person is, a **median furrow** can be seen on the back along with some of the **vertebral spines** (fig. 13.13). The entire series of vertebral spines can be observed if the vertebral column is flexed. This position is important in determining vertebral-column defects (see clinical comments in chapters 11 and 15). The back of the **scapula** presents other important surface landmarks. The base of the spine of the scapula is level with the third thoracic vertebra, and the inferior angle of the scapula is even with the seventh thoracic vertebra. Several muscles of the scapula can be observed on a lean, muscular person and are identified in figure 13.13. Many of the ribs and muscles that attach to the ribs can be seen in a lateral view (fig. 13.14).

There are two pairs of clinically important triangles on the back. The **triangle of auscultation** (fig. 13.13) is bound by the trapezius muscle, the latissimus dorsi muscle, and the medial (vertebral) border of the scapula. Because there is a space between the superficial back muscles in this area, heart and respiratory sounds are not muffled by the muscles when a stethoscope is placed here. The **lumbar triangle** (not illustrated) is bound medially by the latissimus dorsi muscle, laterally by the external oblique muscle, and inferiorly by the iliac crest.

Thorax

The leading causes of death in the United States are associated with disease or dysfunction of the thoracic organs. With the exception of the breasts and surrounding lymph nodes, the organs of the thorax are within the rib cage. Because of the location of the thoracic visceral organs and their clinical importance, the surface anatomy of the thorax is extremely important.

The flexible rib cage presents several bony landmarks that can be observed or palpated (fig. 13.15). The paired clavicles and the suprasternal notch have already been identified as being important surface features of the neck. These structures are also important in the thoracic region as reference points for counting the ribs. Many of the ribs can be seen on a thin person. All of the ribs except the first, and at times the twelfth, can be palpated. The sternum is composed of three separate bones (manubrium, body, xiphoid process), each of which can be palpated. The **sternal angle** (angle of Louis) is felt as a depression between the manubrium and body of the sternum. The sternal angle is important because it is located at the level of the second rib. The articulation between the body of the sternum and the xiphoid process, called the **xiphisternal** *(zif"ĭ-ster'nal)* **joint,** is positioned over the lower border of the heart and the diaphragm. The **costal margin** of the rib cage is the lower oblique boundary and can be easily identified when a person inhales and holds his breath.

The costal cartilages in elderly people may undergo some ossification, reducing the flexibility of the rib cage. Furthermore, with ossification the ribs become radiopaque and may cause some confusion when medical personnel examine a chest X ray.

Figure 13.13. The surface anatomy of the back. (*a*) flexion of the upper extremities; (*b*) abduction of the shoulders and adduction of the scapulae. (Note the three distinct heads of the deltoid muscle: posterior, middle, and anterior.)

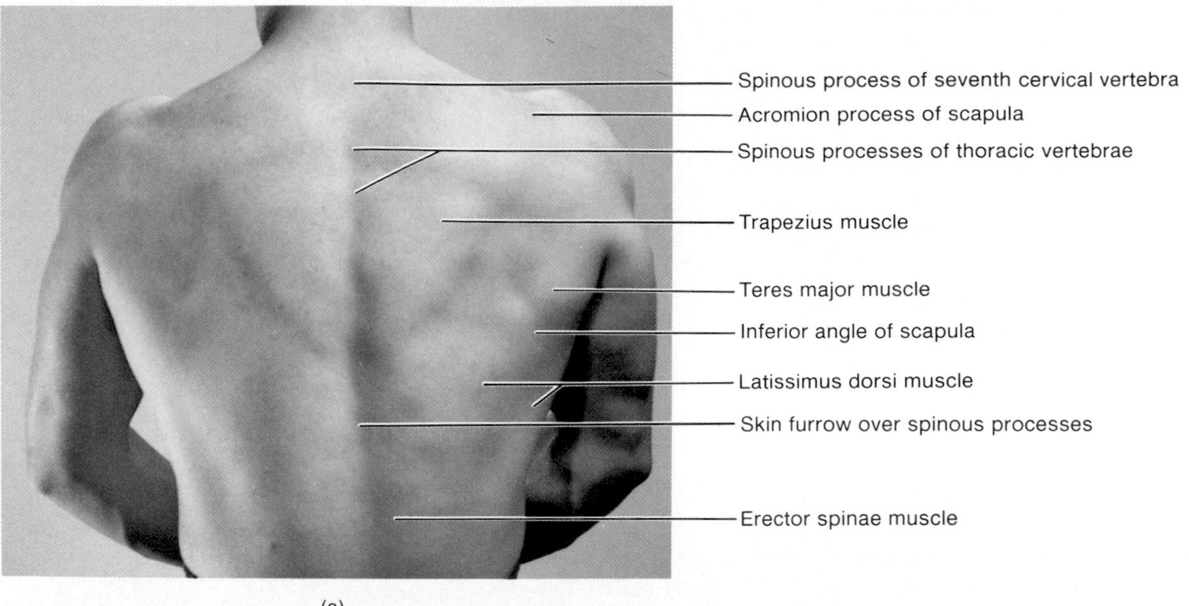

(a)

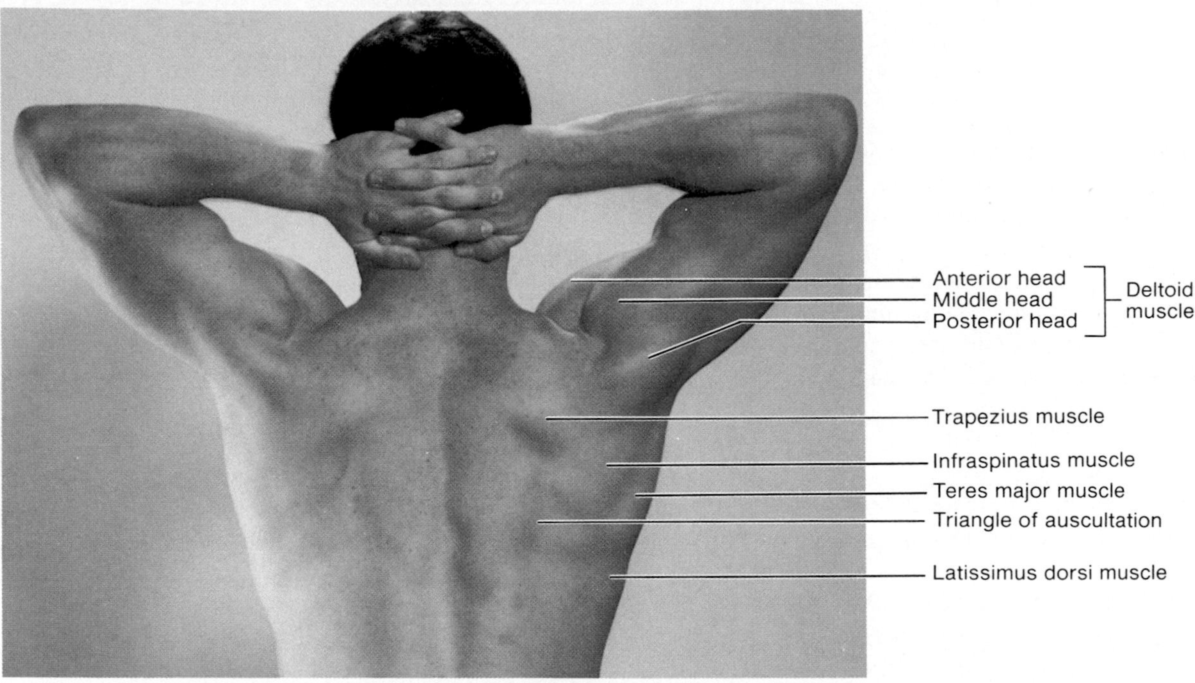

(b)

Figure 13.14. An anterolateral view of the torso and axilla.

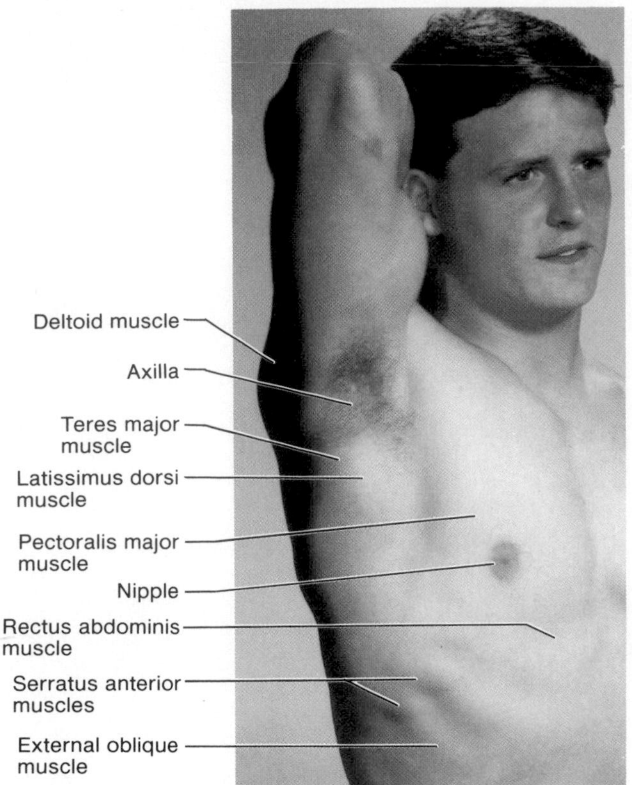

The nipples in the male are located at the fourth intercostal spaces (the area between the fourth and fifth ribs) and about 10 cm (4 in.) from the midline. They vary in position in sexually mature women according to age, size, and the pendulousness of the breasts. The position of the left nipple in males is an important landmark for knowing where to listen to various heart sounds and for determining if the heart is enlarged. For diagnostic purposes, an imaginary line, called the **midclavicular line,** can be extended vertically from the middle of the clavicle through the nipple. Several superficial chest muscles can be observed or palpated and are therefore important surface features. These muscles and the structures described above are depicted in figure 13.15.

In addition to helping one know where to listen with a stethoscope to heart sounds, surface features of the thorax are important for auscultations of the lungs, X rays, tissue biopsies, sternal taps for bone marrow studies, or thoracic surgery. Although the anatomical features of the rib cage are quite consistent, slight deformities and asymmetries do occur. These generally cause no disability and require no treatment. Most of the abnormalities are congenital and include conditions such as a projecting sternum ("pigeon breast") or a receding sternum ("funnel chest").

Figure 13.15. The surface anatomy of the anterior thoracic region of the male.

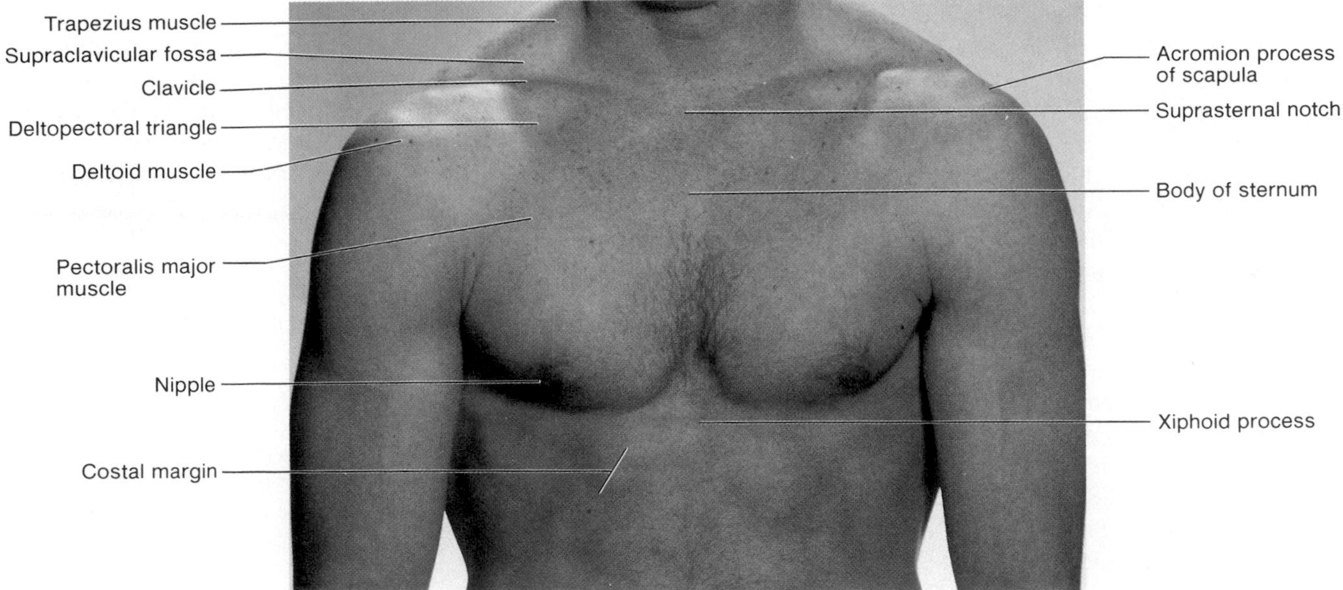

Figure 13.16. The surface anatomy of the anterior abdominal region.

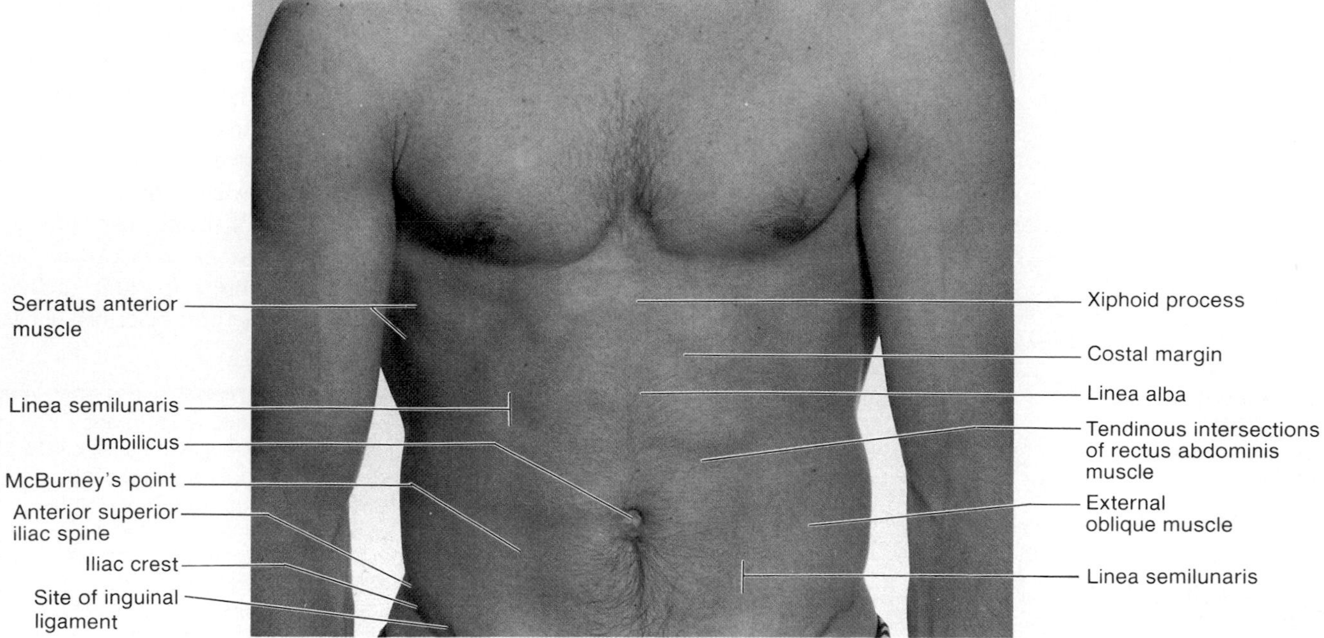

Abdomen

The **abdomen** is the portion of the body between the diaphragm and the pelvis. The abdomen does not have a bony framework as does the thorax, so the surface anatomical features are not as well defined. Bony landmarks of both the thorax and pelvis are used when referring to abdominal structures (fig. 13.16). The right costal margin of the rib cage is located over the liver on the right side, and the left costal margin is positioned over the stomach on the left. The xiphoid process is important because from this point a tendinous, midventral raphe, called the **linea alba,** extends the length of the abdomen to attach to the **symphysis** *(sim'fĭ-sis)* **pubis.** The symphysis pubis can be palpated at the anterior union of the two halves of the pelvic girdle. The **navel,** or **umbilicus,** is the former site of attachment of the fetal umbilical cord and is located along the linea alba. The linea alba separates the paired, straplike **rectus abdominis** muscles, which can be seen when a person flexes the abdomen (such as when doing sit-ups).

Clinically, the linea alba is a favored site for abdominal surgery because an incision made along this line severs no muscles and few vessels and nerves. The linea alba also heals readily (it has been said that only a zipper would provide a more convenient entry to the abdominal cavity).

The lateral margin of the rectus abdominis muscle can be observed on some persons, and the surface line produced is called the **linea semilunaris.** The **external oblique** muscle is the superficial layer of the muscular abdominal wall. The **iliac crest** is subcutaneous and can be palpated along its entire length. The highest point of the crest lies opposite the body of the fourth lumbar vertebra, an important level in spinal anesthesia. Another important landmark is **McBurney's point,** located one-third of the distance from the right anterior-superior iliac crest to the navel (see fig. 13.16). This point overlies the appendix of the intestinal tract.

The abdominal region is frequently divided into nine regions or four quadrants in order to describe the location of internal organs and to clinically identify the sites of various pains or conditions. These regions have been adequately described in chapter 1 (see figs. 1.15, 1.16).

> Although the position of the umbilicus is relatively consistent in all persons, its shape and health is not. Embryological remains of the umbilicus may cause clinical problems such as an opening to the outside, called a *fistula,* or herniation of some of the abdominal contents. Acquired umbilical hernias may develop in children who have a weak abdominal wall in this area, or umbilical hernias may develop in pregnant women because of the extra pressure exerted at this time. A depressed umbilicus on an obese person is difficult to keep clean, and so various types of infections may occur there.

1. Which structures of the torso can be readily observed? Which can be palpated?
2. Describe the location of the common auscultation sites of the torso.
3. Define linea alba, costal margin, linea semilunaris, and McBurney's point.

linea alba: L. *linea,* line; *alba,* white
navel: O. E. *nafela,* umbilicus
umbilicus: L. *umbilicus,* navel

McBurney's point: from Charles McBurney, U.S. surgeon, 1845–1914

Pelvis and Perineum

The surface features of the pelvic region are important primarily to identify reproductive organs and clinical problems of these organs.

Objective 11. Describe the location of the perineum, and list the organs of the pelvic and perineal regions.

The important bony structures of the **pelvis** *(pel''vis)* include the crest of the ilium and symphysis pubis, located anteriorly, and the ischium and os coccyx, which are palpable posteriorly. An **inguinal ligament** extends from the crest of the ilium to the symphysis pubis and is clinically important because hernias occur along it. Although the inguinal ligament cannot be seen, an oblique groove overlying the ligament is an apparent surface feature.

The **perineum** *(per''i-ne'um)* (see figs. 1.17 and 12.39) is the region that contains the external sex organs and the anal opening. The surface features of this region are further discussed in chapters 27 and 28. The surface anatomy of the perineum of a female becomes particularly important during parturition (childbirth).

1. Define the term *perineum.* What structures are located within the perineum?
2. List three body systems that have openings within the pelvic region.

Shoulder and Upper Extremity

The surface anatomy of the shoulder and upper extremity is of clinical importance because of frequent trauma to these body regions. Vessels of the upper extremity are also used as pressure points and for intravenous injections or blood withdrawal.

Objective 12. Identify by observation or palpation various surface features of the shoulder and upper extremity.

Objective 13. Discuss the clinical importance of the axilla, cubital fossa, and wrist.

Shoulder

The scapula, clavicle, and proximal portion of the humerus form the **shoulder,** and portions of each of these bones are important surface landmarks in this region. Posteriorly, the spine of the scapula and acromion process are subcutaneous and easily located. The **angle of acromion** *(ah-kro'me-on)* is frequently used by clinicians to make measurements down the brachium (fig. 13.17).

The acromion process and the clavicle, as well as several large shoulder muscles, can be seen anteriorly (fig. 13.18). The rounded curve of the shoulder is formed by

inguinal: L. *inguinalis,* groin
acromion: Gk. *akros,* extreme, tip; *omion,* small shoulder

Figure 13.17. A posterior view of the shoulder showing the spine of the scapula, acromion, and angle of acromion. The angle of acromion is used as a reference for locating certain brachial structures such as the surgical neck of the humerus or the axillary nerve, which is positioned against the humerus, 5 cm below the angle of acromion.

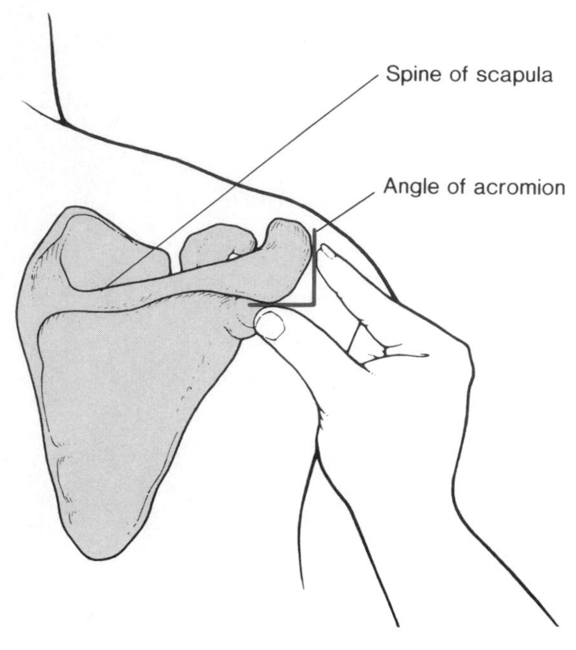

Figure 13.18. An anterior view of the shoulder region.

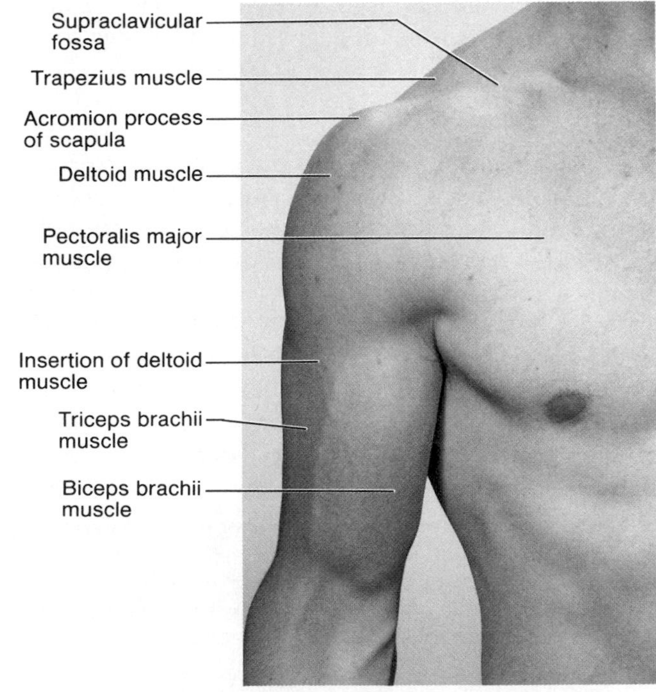

Figure 13.19. An anterior view of the upper extremity.

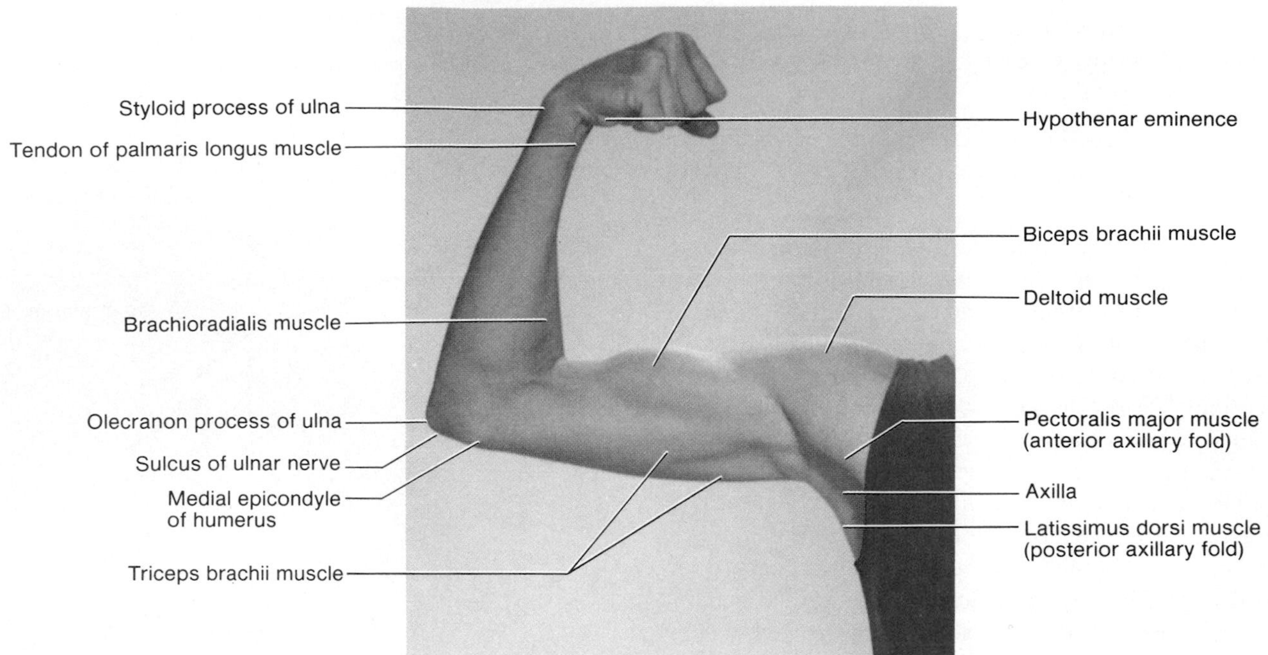

the deltoid muscle covering the greater tuberosity of the humerus. The deltoid muscle is frequently a site for intramuscular injections. The large pectoralis major muscle is prominent as it crosses the shoulder joint and attaches to the humerus. A small depression, called the **deltopectoral triangle** (see fig. 13.15), is situated below the outer third of the clavicle and is bounded on either side by the deltoid and pectoralis major muscles.

Axilla

The **axilla** is commonly called the armpit. This depressed region of the shoulder supports axillary hair in sexually mature individuals. The axilla is clinically important because of the subcutaneous position of vessels, nerves, and lymph nodes here. Two muscles form the anterior and posterior borders of this region (fig. 13.19). The **anterior axillary** *(ak'sĭ-lar''e)* **fold** is formed by the pectoralis major muscle, and the **posterior axillary fold** consists primarily of the latissimus dorsi muscle extending from the lumbar vertebrae to the humerus. Axillary lymph nodes are palpable in some persons.

In sexually mature females, the axillary tail of the mammary gland, which is positioned on the pectoralis major muscle, extends partially into the axilla. In doing a *breast self-examination* (see chapter 28), a woman should palpate the axillary area as well as the entire breast because the lymphatic drainage pathway is toward the axilla.

Brachium

Several muscles are clearly visible in the **brachium** (figs. 13.19, 13.20). The belly of the biceps brachii muscle becomes prominent when the elbow is flexed with the palm upward. While the arm is in this position, the deltoid muscle can be traced as it inserts upon the humerus. The triceps brachii muscle forms the bulk of the posterior surface of the brachium. A groove forms on the medial side of the brachium between the biceps and triceps muscles where pulsations of the brachial artery may be felt as it carries blood toward the forearm. This region is clinically important because it is where arterial blood pressure is taken with a sphygmomanometer. It is also the place to apply pressure in case of severe arterial hemorrhage in the forearm or hand.

Three bony prominences can be located in the region of the elbow (fig. 13.21). The medial and lateral epicondyles are processes on the humerus, whereas the olecranon is a proximal process of the ulna. When the elbow is extended, these prominences lie on the same straight plane; when the elbow is flexed, these points form a triangle. The ulnar nerve can be palpated in the groove behind the medial epicondyle. This area is commonly known as the "funny bone" or "crazy bone."

The **cubital** *(ku'bĭ-tal)* **fossa** is the triangular depression on the anterior surface of the elbow region where the median cubital vein links the cephalic and basilic veins. These veins are subcutaneous and become more conspicuous if a proximal compression is applied. For this reason, these veins, particularly the median cubital, are an important location for the removal of venous blood for analyses and transfusions or for intravenous therapy (fig. 13.22).

Figure 13.20. A lateral view of the upper extremity.

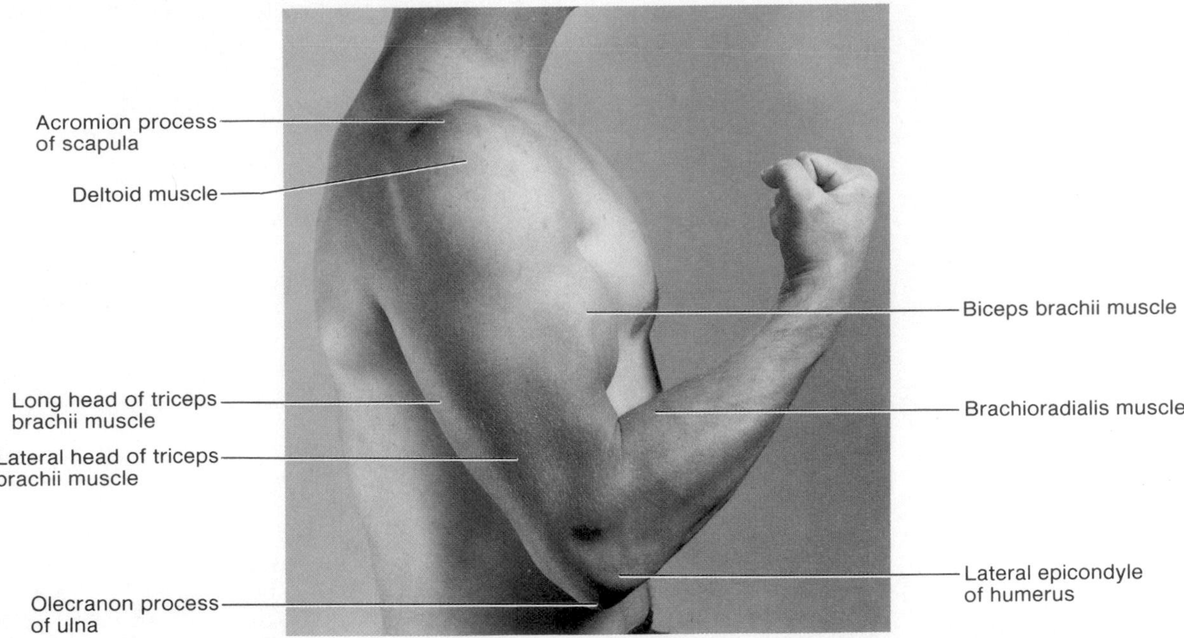

Figure 13.21. A posterior view of the elbow (joint slightly extended).

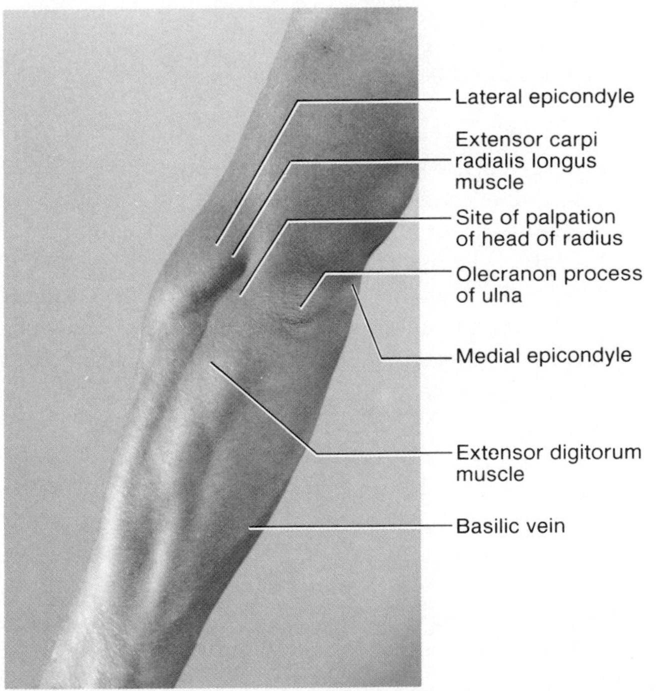

Figure 13.22. An anterior view of the forearm and hand.

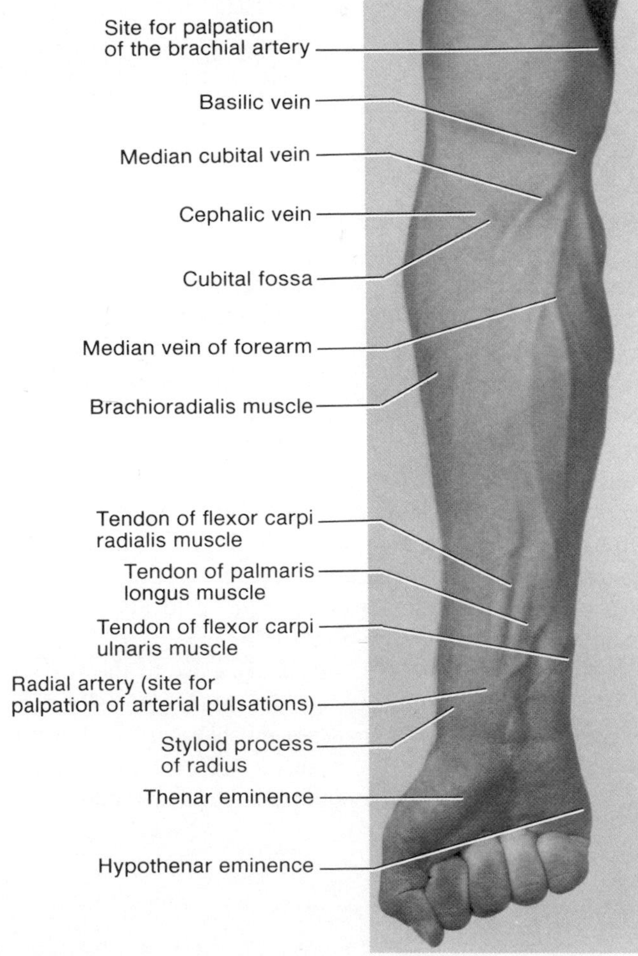

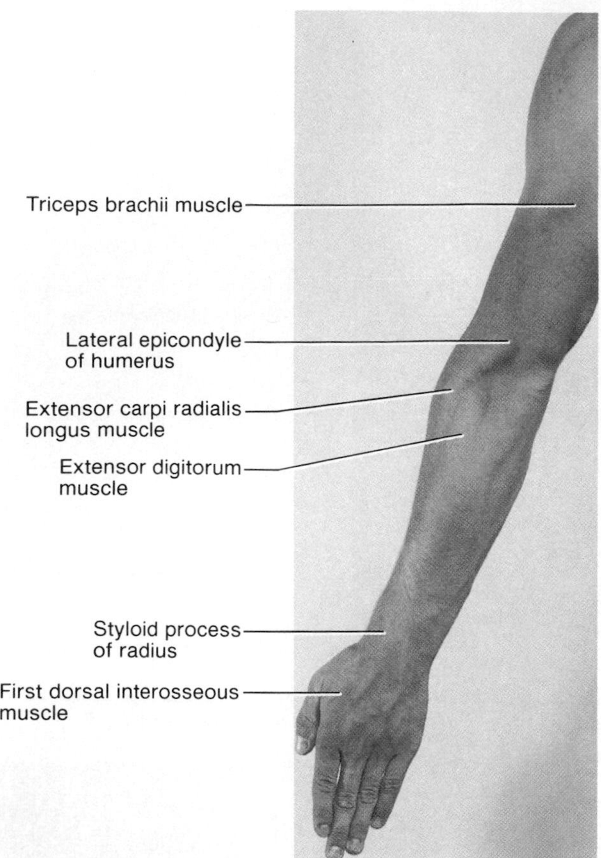

Figure 13.23. A posterior view of the forearm and hand.

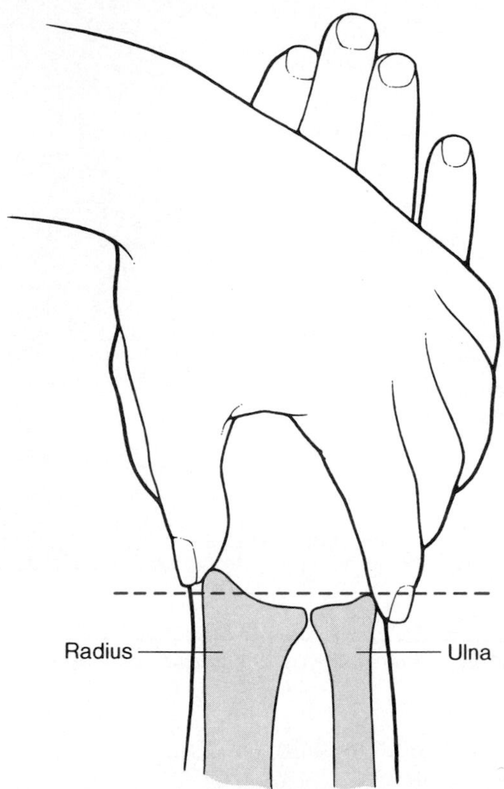

Figure 13.24. Palpation of the styloid processes of the ulna and radius. The radial styloid is always about 1 cm lower than that of the ulna, an important fact for a physician to know when treating a fractured bone of the forearm.

Forearm

Contained within the **forearm** are two parallel bones (the ulna and radius) and the muscles that control the movements of the hand. The forearm tapers distally toward the wrist, where the muscles give way to tendinous cords that attach to various bones of the hand. Several muscles of the forearm can be identified as surface features and are depicted in figures 13.22 and 13.23.

Because of the frequency of fractures involving the forearm, bony landmarks are clinically important when setting broken bones. The ulna can be palpated its entire length from the olecranon to the distal styloid process. The distal one-half of the radius is palpable as the forearm is rotated, and its styloid process can be located using a technique illustrated in figure 13.24.

Nerves, tendons, and vessels are close to the surface at the wrist, making cuts to this area potentially dangerous. Tendons from four flexor muscles can be observed as surface features if the anterior forearm muscles are strongly contracted while making a fist. The tendons that can be observed along this surface, from lateral to medial, are from the following muscles: flexor carpi radialis, palmaris longus, flexor digitorum superficialis, and flexor carpi ulnaris. The median nerve going to the hand is located under the tendon of the palmaris longus muscle (see fig. 13.22), and the ulnar nerve is lateral to the tendon of the flexor carpi ulnaris. The radial artery lies along the surface of the radius immediately lateral to the tendon of the flexor carpi radialis. This is the artery commonly used when taking a pulse. By careful palpation, pulsations can also be detected in the ulnar artery lateral to the tendon of the flexor carpi ulnaris.

Two tendons that attach to the thumb can be seen on the posterior surface of the wrist as the thumb is extended backward. The tendon of the extensor pollicis brevis is positioned anterolaterally along the thumb, and the tendon of the extensor pollicis longus lies posteromedially (fig. 13.25). The depression created between these two tendons as they are pulled taut is referred to as the "anatomical snuffbox." Pulsations of the radial artery can be detected in this depression.

Figure 13.25. The anatomical "snuffbox."

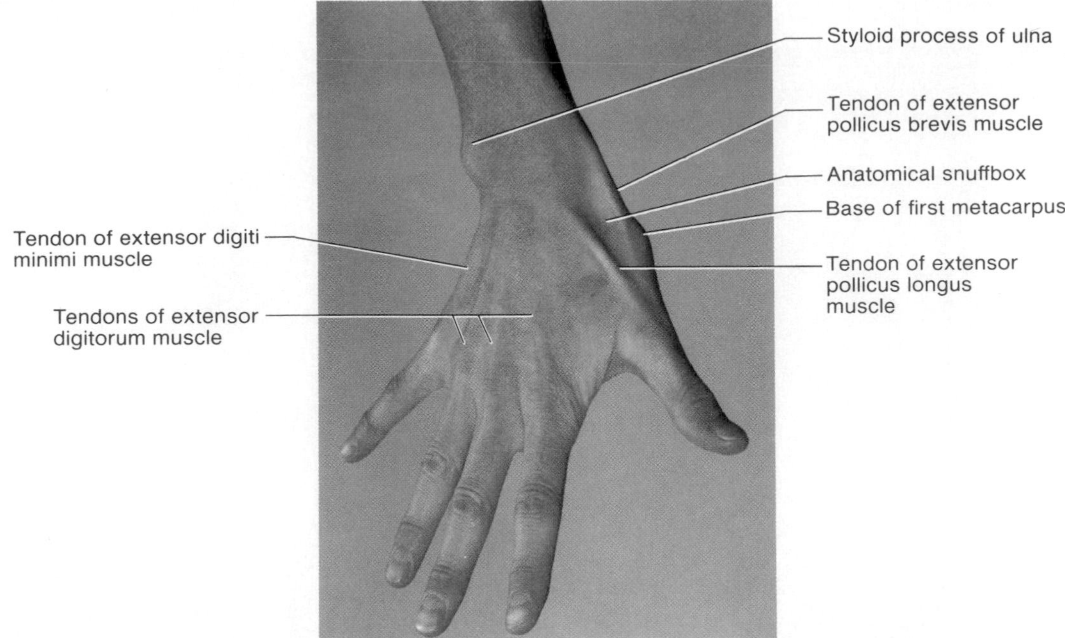

The median nerve, which serves the thumb, is the nerve in the forearm most commonly injured by stab wounds or the penetration of glass. If this nerve is severed, the muscles of the thumb are paralyzed and waste away, resulting in an inability to oppose the thumb in grasping.

Hand

Much of the surface anatomy of the **hand,** such as flexion creases, fingerprints, and fingernails, includes features of the skin that are discussed in chapter 8. Other surface features are the extensor tendons from the extensor digitorum muscle, which can be seen going to each of the fingers on the back side of the hand as the hand is extended (fig. 13.26). The "knuckles" of the hand are the distal ends of the second through the fifth metacarpal bones. Each of the joints of the fingers and the individual phalanges can be palpated. The **thenar** *(the'nar)* **eminence** is the thickened, muscular portion of the hand forming the base of the thumb (fig. 13.27).

Figure 13.26. A posterior view of the hand.

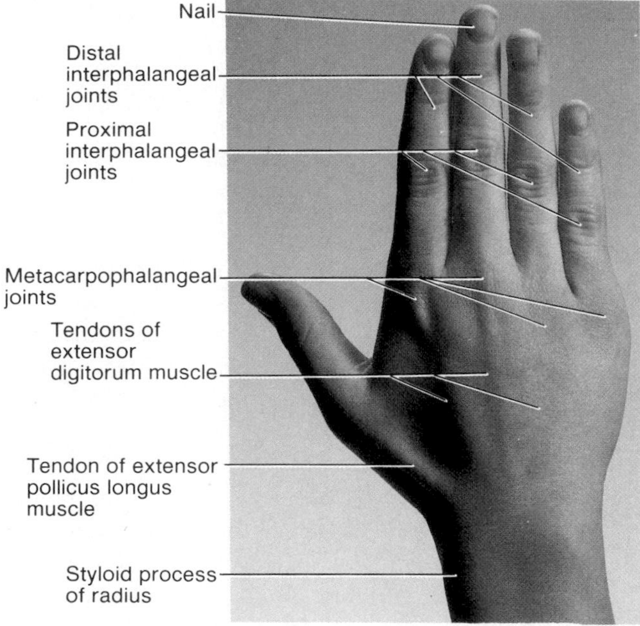

1. List the clinically important structures that can be observed or palpated in the shoulder and upper extremity.
2. Describe the locations of the axilla, brachium, cubital fossa, and wrist.
3. Bumping the ulnar nerve causes a "tingling" sensation along the medial portion of the forearm to the little finger of the hand. What does this tell you about its distribution?
4. Which of the two bones of the forearm is more stationary as the arm is rotated?

thenar: Gk. *thenar,* palm

Figure 13.27. An anterior view of the wrist and hand: (*a*) with the hand open and (*b*) with a clenched fist.

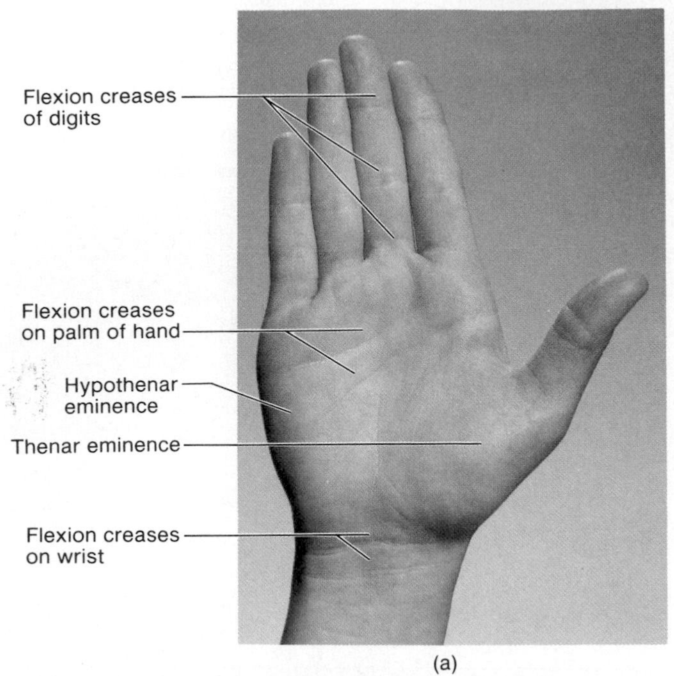

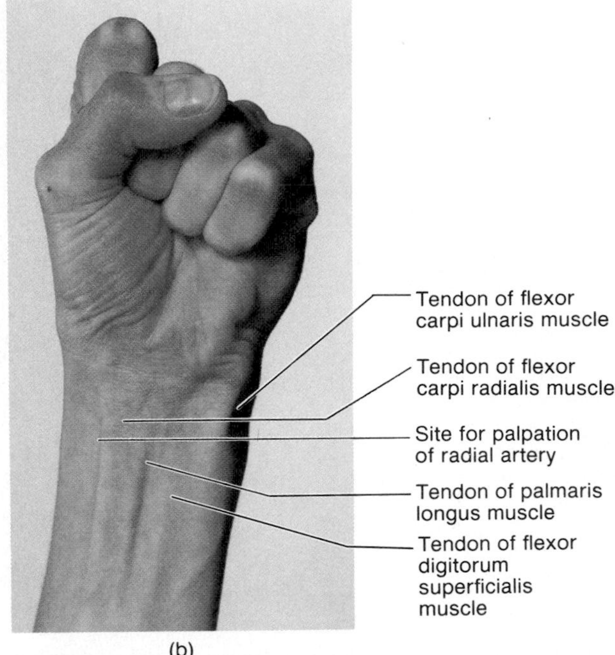

Buttock and Lower Extremity

The massive bones and muscles of the buttock and lower extremity are weight bearers and locomotors. Many of the surface features of these regions are important in relation to locomotion or locomotor dysfunction.

Objective 14. Identify by observation or palpation various surface features of the buttock and lower extremity.

Objective 15. Discuss the clinical importance of the buttock, femoral triangle, popliteal space, ankle, and arches of the foot.

Buttock

The superior borders of the **buttocks** *(but'oks),* or **gluteal** *(gloo'te-al)* **region,** are formed by the iliac crests (fig. 13.28). Each crest can be palpated medially to the level of the second sacral vertebra. From this point, the **natal cleft** extends vertically to separate the buttocks into two prominences, each formed by pads of fat as well as by the massive gluteal muscles. An ischial tuberosity can be palpated in the lower portion of each buttock. In a person who is seated, the ischial tuberosities support the weight of the body; but when the person is standing, these processes are covered by the gluteal muscles. The sciatic nerve, which is the major nerve to the lower extremity, lies under the gluteus maximus muscle. The inferior border of the gluteus maximus muscle forms the **fold of the buttock.**

buttock: O.E. *buttuc,* end or rump

Because of the thickness of the gluteal muscles and the rich blood supply, the buttock is a preferred site for intramuscular injections. Care must be taken, however, not to inject into the sciatic nerve. For this reason, the surface landmark of the iliac crest is important. The injection is usually administered 5–7 cm (2–3 in.) below the iliac crest in what is known as the upper lateral quadrant of the buttock.

Thigh

The femur is the only bone of the **thigh,** but there are three groups of thigh muscles. The anterior group of muscles, referred to as the **quadriceps,** extend the leg when they are contracted (fig. 13.29). The medial muscles are the **adductors,** and when contracted they draw the thigh medially. The **hamstrings** are positioned on the posterior aspect of the thigh (see fig. 13.28) and serve to extend the hip as well as flex the leg when they are contracted. The tendinous attachments of the hamstrings can be palpated along the posterior aspect of the knee joint when the leg is flexed. The hamstrings or their attachments are often injured in athletic events.

The **femoral** *(fem'or-al)* **triangle** is an extremely important part of the surface anatomy of the thigh. It can be seen as a depression inferior to the location of the inguinal *(ing'gwĭ-nal)* ligament on the anterior surface in the upper part of the thigh (see fig. 20.36). The major vessels of the leg as well as the femoral nerve traverse through this region. Hernias are frequent in this area.

Figure 13.28. The buttocks and the posterior aspect of the thigh. (Note the relation of the angle of the elbow joint to the pelvic region, which is characteristic of females.)

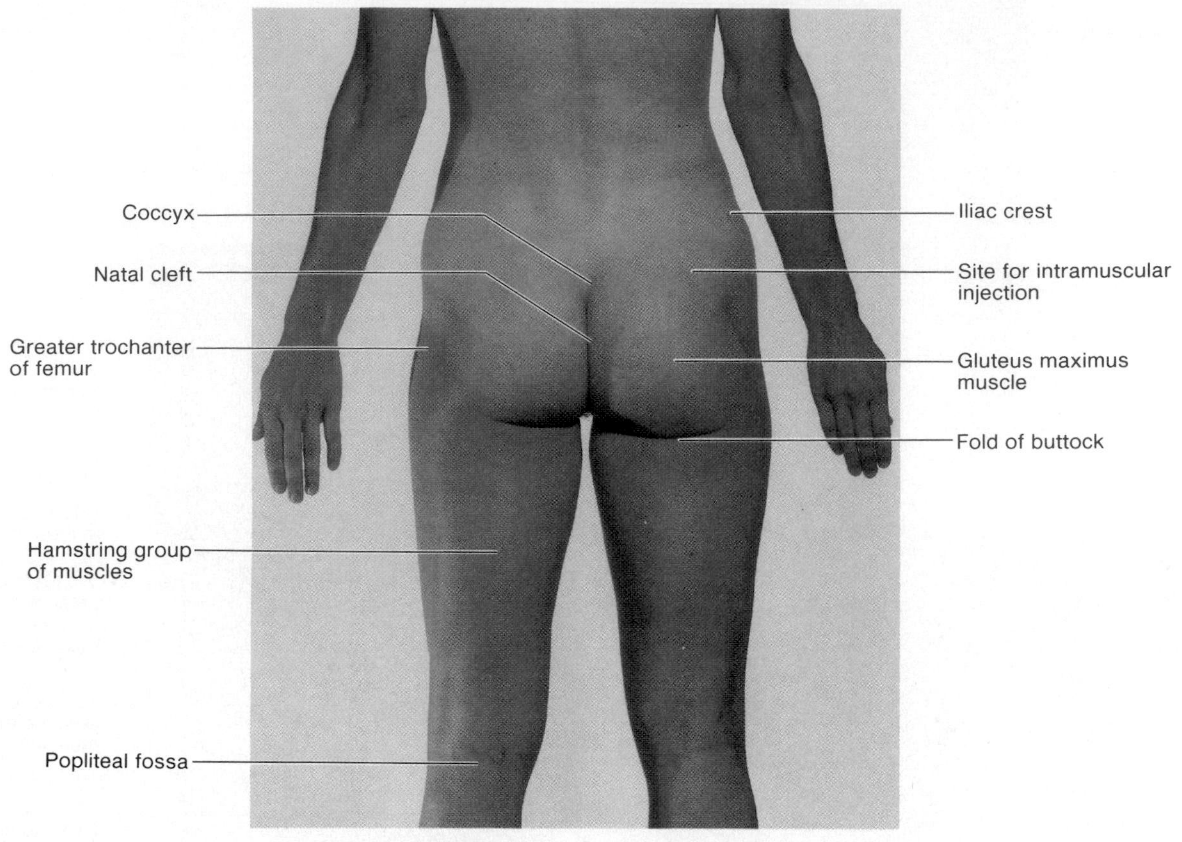

Figure 13.29. An anterior view of the thigh and knee.

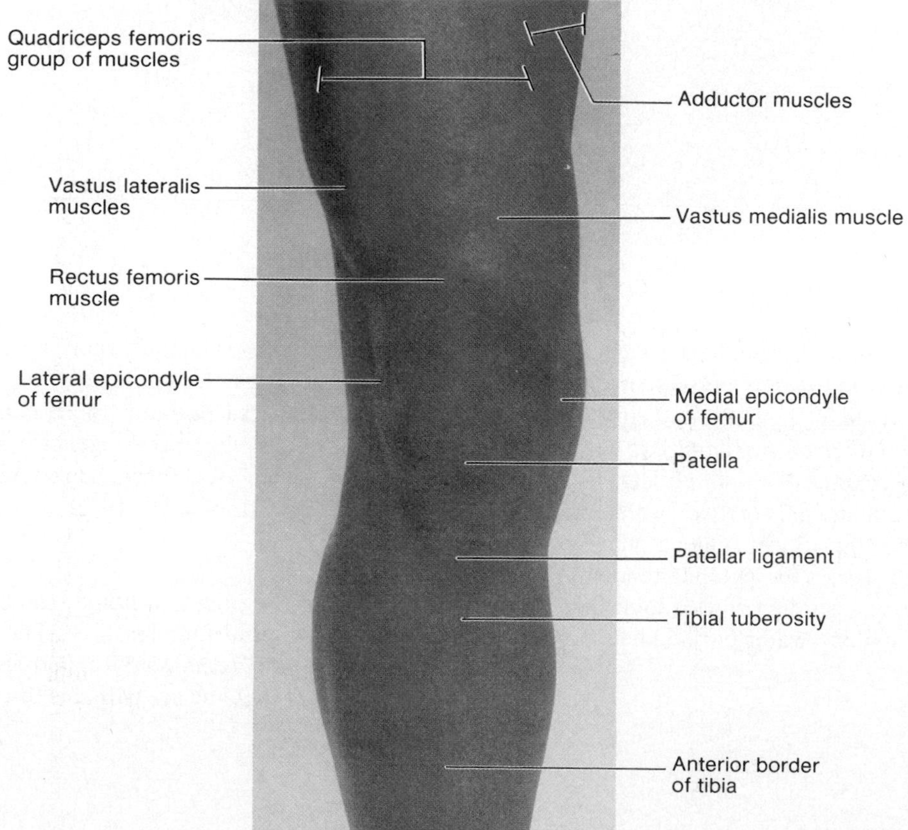

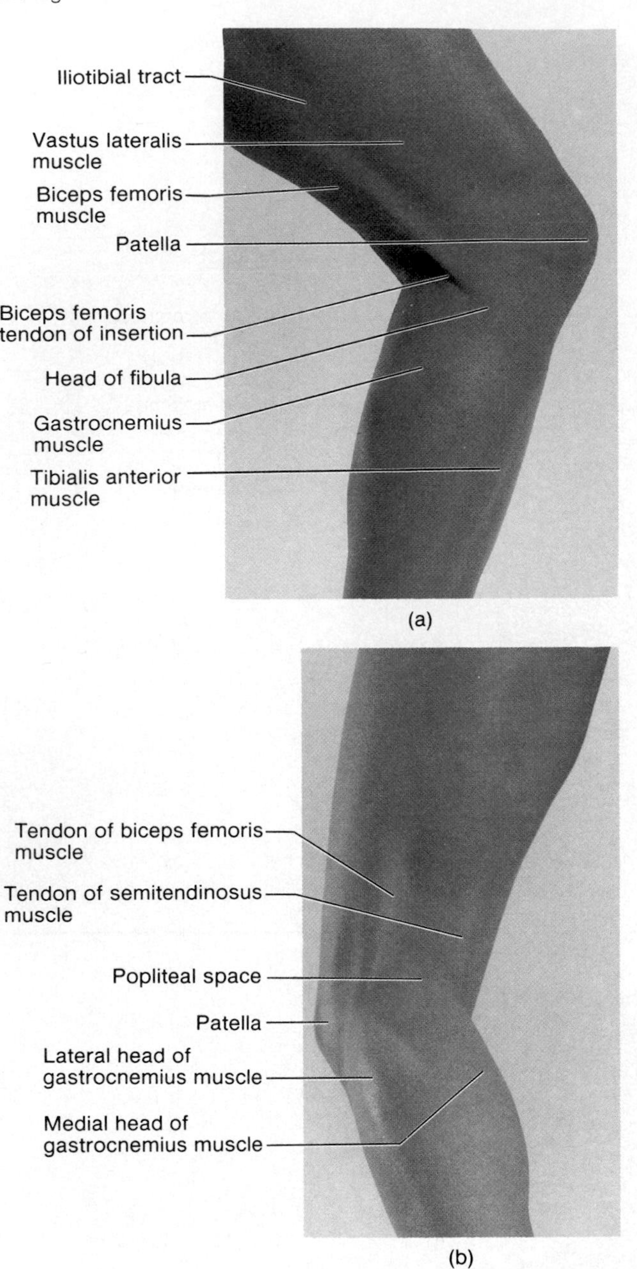

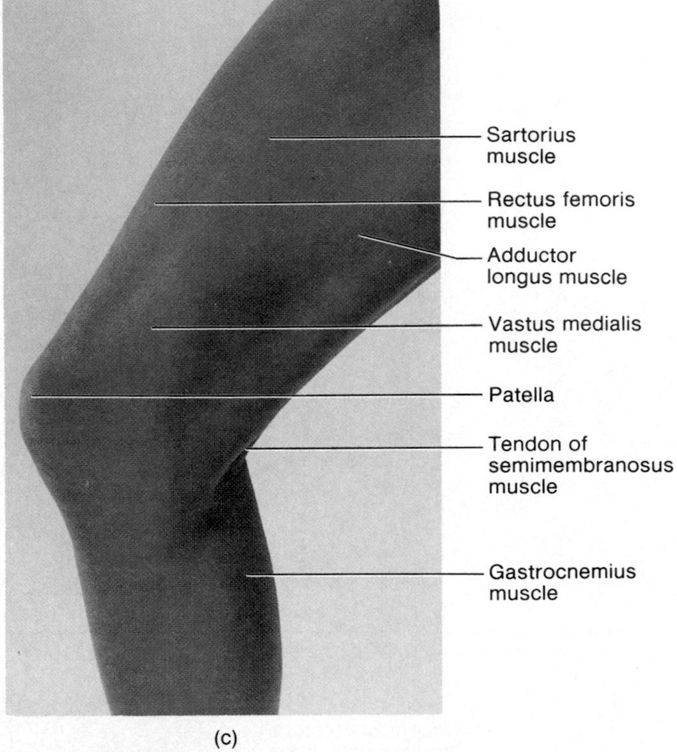

Figure 13.30. (*a*) lateral, (*b*) posterior, and (*c*) medial surfaces of the leg.

More important, the femoral triangle is an arterial pressure point where it is vital to apply pressure in the case of uncontrolled hemorrhage of the lower extremity.

The greater trochanter of the femur can be palpated on the lateral, upper surface of the thigh (fig. 13.28). At the knee, the lateral and medial condyles of the femur and tibia can be identified (fig. 13.29). The patella (kneecap) can be easily located within the patellar tendon anterior to the knee joint. Stress or injury to this joint may cause swelling, commonly called "water on the knee."

The depression on the posterior aspect of the knee joint is referred to as the **popliteal** *(pop''li-te'al)* **space** (fig. 13.30). This area becomes more clinically important in elderly persons who suffer degenerative conditions. Aneurysms of the popliteal artery are common, as are popliteal abscesses due to infected lymph nodes.

Leg

Portions of the tibia and fibula, the bones of the **leg,** can be observed as surface features. The medial surface and anterior border (commonly called the shin) of the tibia are subcutaneous and are palpable throughout their length.

Figure 13.31. The leg and foot: (*a*) a lateral view; (*b*) a medial view; (*c*) an anterior view; (*d*) a posterior view.

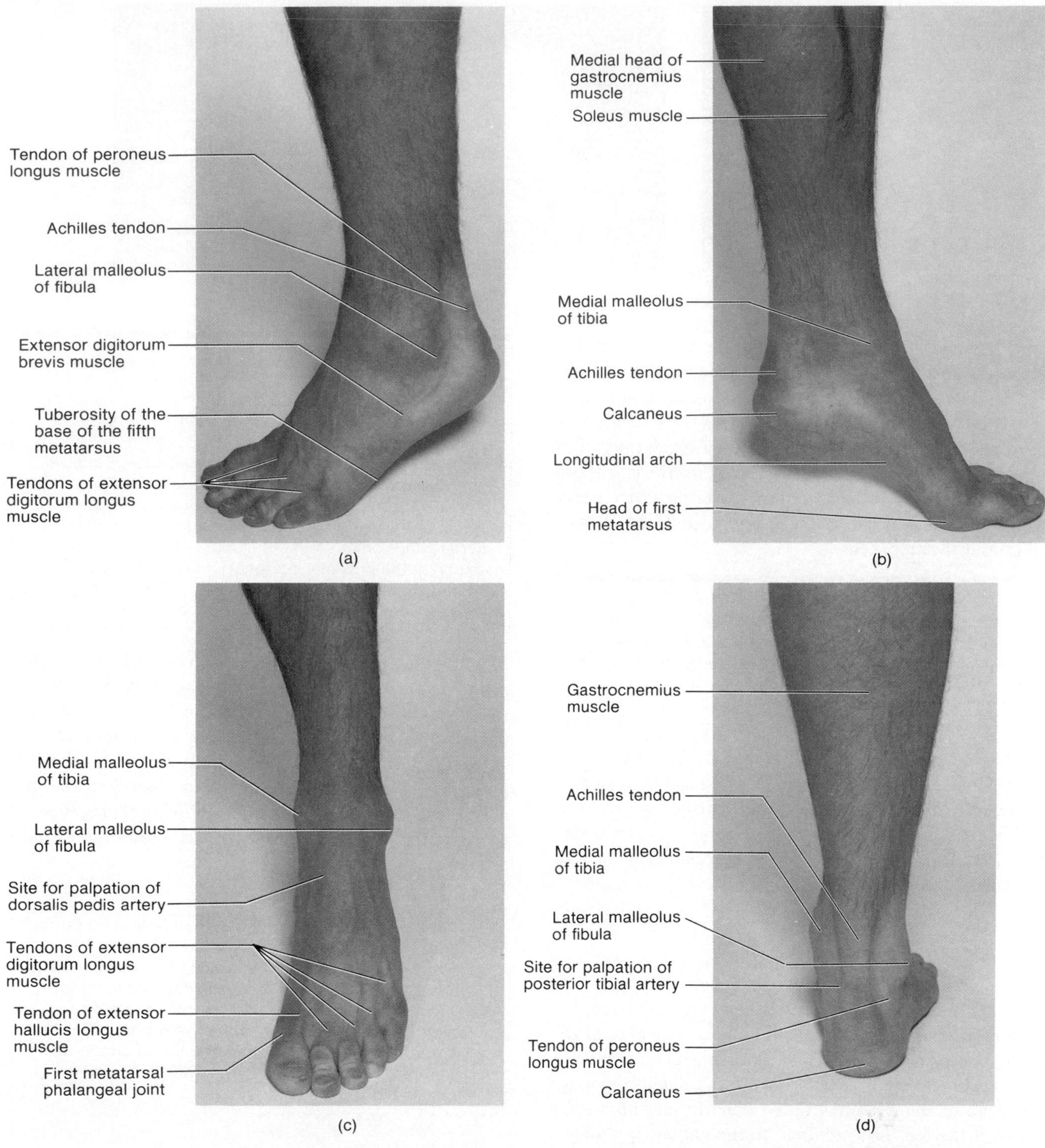

At the ankle, the medial malleolus of the tibia and the lateral malleolus of the fibula are easy to observe as prominent eminences (fig. 13.31). Of clinical importance in setting fractures of the leg is knowing that the top of the medial malleolus lies about 1.3 cm (0.6 in.) proximal to the level of the tip of the lateral malleolus.

Leg injuries are common among athletes. *Shin splints,* probably the result of a stress fracture or periosteum damage of the tibia, is a common condition in runners. A fracture of one or both malleoli is caused by a severe twisting of the ankle region. Skiing fractures are generally caused by strong torsion forces on the shafts of the tibia or fibula.

Figure 13.32. Common clinical conditions of the foot and toes: (*a*) ingrown toenail, (*b*) hammer toe (second digit), and (*c*) corn.

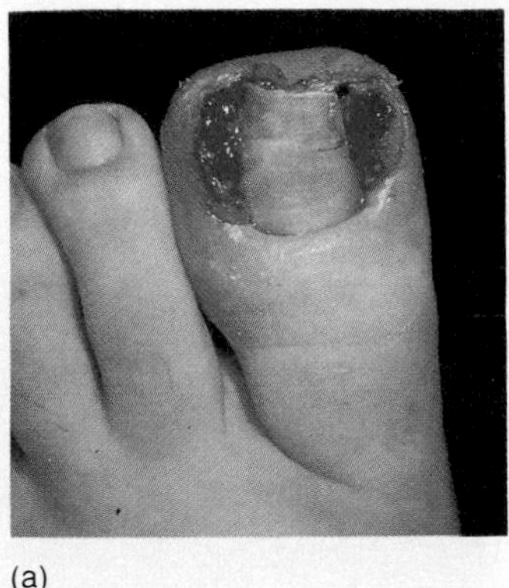
(a)

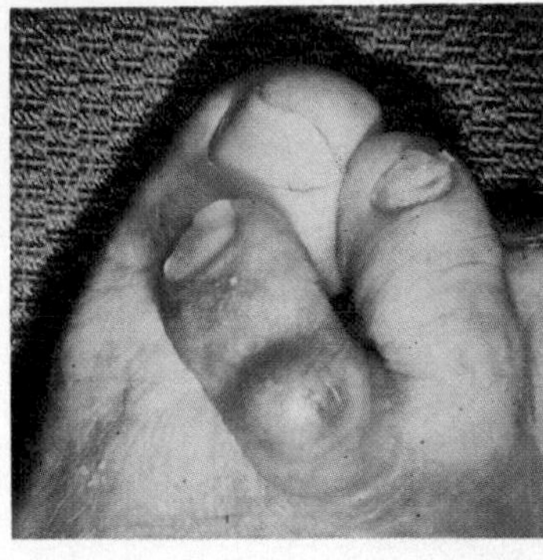
(b)

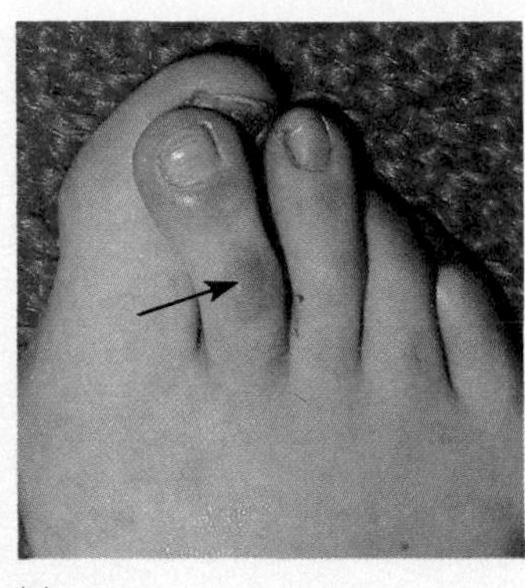
(c)

The heel is not part of the leg but the posterior portion of the calcaneus bone. It needs to be mentioned with the leg, however, because of its functional relationship to it. The **tendo calcaneus** (Achilles tendon) is the strong, cordlike tendon that attaches to the calcaneus from the calf of the leg. The muscle forming the "belly" of the calf is the gastrocnemius. Pulsations from the posterior tibial artery can be detected by palpating between the medial malleolus and the calcaneus.

Because arterial occlusive disease is common in elderly people, palpation of the posterior tibial artery is clinically important in general physical assessment. This can be accomplished by gently palpating between the medial malleolus and the gastrocnemius tendon.

The superficial veins of the leg can be observed on many persons. The great saphenous vein can be seen subcutaneously along the medial aspect of the leg. The less conspicuous, small saphenous vein drains the lateral surface of the leg. If these veins become excessively enlarged, they are called *varicose veins*.

Foot

The feet are adapted to support the weight of the body, to maintain balance, and to function mechanically during locomotion. The structural features and surface anatomy of the **foot** are indicative of these functions. The **arch of the foot** is located on the medial portion of the plantar surface (fig. 13.31) and provides a spring effect when locomoting. The head of the first metatarsal bone forms the medial "ball" of the foot just proximal to the hallux (big toe).

The feet and toes are adapted to endure tremendous compression forces during locomotion. Appropriate shoes help to minimize trauma to the feet and toes, but still there is an array of common clinical conditions (fig. 13.32) that may impede walking or running. An *ingrown toenail* occurs as the sharp edge of a toenail embeds into and injures the skin fold, resulting in inflammation and infection. *Hammer toe* is a condition resulting from a forceful hyperextension at the metatarsophalangeal joint with flexion at the proximal interphalangeal joint. A *corn* is a thickening of skin (callus) resulting from recurrent pressure on the skin over a bony prominence.

The fifth metatarsal bone forms much of the lateral border of the plantar surface of the foot. The tendons of the extensor digitorum longus muscle can be seen along the dorsal surface of the foot, especially if the toes are elevated. Pulsations of the dorsal pedis artery can be palpated on the dorsal surface of the foot between the first and second metatarsal bones. The individual phalanges of the toes, the joints between these bones, and the toenails are obvious surface landmarks.

1. What surface features form the boundaries of a buttock?
2. List the clinically important structures that can be observed or palpated in the buttock and lower extremity.
3. Describe the anatomical location where each of the following could be observed or palpated: the distal tendinous attachments of the hamstring muscles; the greater trochanter; the greater and lesser saphenous veins; the femoral, posterior tibial, and dorsal pedis arteries; and the medial malleolus.

Chapter Summary

I. Introduction to Surface Anatomy
   A. Surface anatomy is concerned with identifying body structures through observation and palpation. Surface anatomy has tremendous application to physical fitness and to medical diagnosis and treatment.
   B. Most of the bones of the skeleton are palpable and provide landmarks for other anatomical structures.
   C. The effectiveness of observation and palpation in studying a person's surface anatomy is dependent on the thickness of the hypodermis, which is due to the amount of subcutaneous adipose tissue present.

II. Surface Anatomy of the Newborn
   A. Certain aspects of the surface anatomy of a neonate are of clinical importance in ascertaining the degree of physical development, general health, and possible congenital abnormalities.
   B. The posture of a full-term, normal neonate is one of flexion.
   C. Portions of the skin and subcutaneous tissues are edematous. Vernix caseosa covers

the body of a neonate, and lanugo may be present on the head, neck, and back.

D. The fontanels, liver, and kidneys, and the testes of a male, should be palpable.

III. Head

A. Surface features of the cranium include the forehead, the crown, the temporalis muscles, and the hair and hairline.

B. The face is composed of the ocular region that surrounds the eye, the auricular region of the ear, the nasal region serving the respiratory system, and the oral region of the mouth.

IV. Neck

A. Major organs are located within the flexible neck, and structures that are essential for body sustenance pass through the neck to the torso.

B. The neck consists of an anterior cervix, right and left lateral regions, and a posterior nucha.

C. Two major and six minor triangles, which contain specific structures, are located on both sides of the neck.

V. Torso

A. Vital visceral organs in the torso make the surface anatomy of this region especially important.

B. The median furrow is observable and the vertebral spines and scapulae are palpable on the back.

C. Palpable structures of the thorax include the sternum, the ribs, and the costal margins.

D. The important surface anatomy of the abdomen includes the linea alba, umbilicus, costal margins, iliac crests, and pubis.

VI. Pelvis and Perineum

A. The crest of the ilium, symphysis pubis, and inguinal ligament are important pelvic landmarks.

B. The perineum is the region that contains the external genitalia and the anal opening.

VII. Shoulder and Upper Extremity

A. The surface anatomy of the shoulder and upper extremity is important because of frequent trauma to these regions. Vessels of the upper extremity are also used as pressure points and for intravenous injections or blood withdrawal.

B. The scapula, clavicle, and humerus are palpable in the shoulder.

C. The axilla is clinically important because of the vessels, nerves, and lymph nodes located there.

D. The brachial artery is an important pressure point in the brachium, and the median cubital vein is important for the removal of blood or for intravenous therapy.

E. The ulna, radius, and their processes are palpable landmarks of the forearm.

F. The knuckles, fingernails, and tendons for the extensor muscles of the forearm can be observed on the posterior aspect of the hand.

G. Flexion creases and the thenar eminence are important surface features on the anterior surface of the hand.

VIII. Buttock and Lower Extremity

A. The massive bones and muscles in the buttock and lower extremity serve as weight bearers and locomotors. Many of the surface features of these regions are important in relation to locomotion or locomotor dysfunction.

B. The prominences of the buttocks are formed by the gluteal muscles and are separated by the natal cleft.

C. The thigh has three muscle groups: quadriceps, adductors, and hamstrings. The femoral triangle and popliteal spaces are clinically important surface landmarks.

D. The structures of the leg include the tibia and fibula, the muscles of the calf, and the saphenous veins.

E. The surface anatomy of the foot includes structures adapted to support the weight of the body, maintain balance, and function during locomotion.

Review Activities

Objective Questions

1. Eyebrows are located on the
   (a) palpebral fissure
   (b) subtarsal sulcus
   (c) scalp
   (d) supraorbital ridges
   (e) c and d
2. Which of the following structures is *not* part of the auricle (pinna) of the ear? The
   (a) tragus
   (b) ala
   (c) lobule
   (d) helix
3. Which of the following clinical-structural word pairs is *incorrectly* matched?
   (a) cleft lip—philtrum
   (b) broken nose—nasion
   (c) pierced ears—lobule
   (d) "black eye"—concha
4. The conjunctiva
   (a) covers the entire eyeball
   (b) is a thick nonmucous membrane
   (c) secretes tears
   (d) none of the above
5. Which of the following could *not* be palpated within the cervix of the neck?
   (a) larynx
   (b) hyoid bone
   (c) trachea
   (d) cervical vertebrae
6. Palpation of a pulse to the head is best accomplished within the
   (a) carotid triangle
   (b) occipital triangle
   (c) suprahyoid triangle
   (d) submandibular triangle
   (e) muscular triangle
7. Which nerve traverses behind the medial epicondyle of the humerus?
   (a) ulnar
   (b) median
   (c) radial
   (d) brachial
   (e) cephalic
8. Which of the following surface features could *not* be observed on obese people?
   (a) suprasternal notch
   (b) scapular muscles
   (c) clavicles
   (d) vertebral spines
   (e) natal cleft
9. Which pairs of muscles form the anterior and posterior borders of the axilla?
   (a) deltoid—pectoralis minor
   (b) biceps brachii—triceps brachii
   (c) latissimus dorsi—pectoralis major
   (d) triceps brachii—pectoralis major
   (e) latissimus dorsi—deltoid
10. Varicose veins occur when which of the following becomes excessively enlarged?
    (a) saphenous veins
    (b) tibial veins
    (c) external iliac vein
    (d) popliteal vein
    (e) all of the above

Essay Questions

1. List four surface features of the cranium, and explain the relationship of the scalp to the cranium.
2. Identify the four regions of the face, and list at least two surface features of each region.
3. Which surface features can be observed on the torso on any person, regardless of how obese he or she is?
4. Identify the two major triangles of the neck, and list the associated six minor triangles. Discuss the importance of knowing the contents and boundaries of these triangles.
5. Describe the locations of the arteries that can be palpated as they pulsate in the following regions: neck, brachium, forearm, thigh, and ankle. Which of these are considered clinical pressure points?
6. Describe the locations and clinical importance of the following regions: cubital fossa, femoral triangle, axilla, perineum, and popliteal space.

Reference Figures
Regional Anatomy

Figure 8. Anterior view of the muscles of the head.

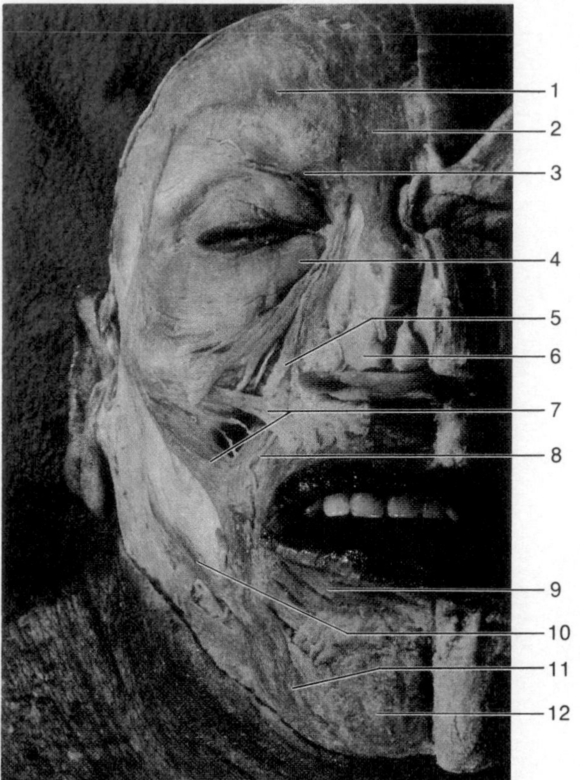

1 Frontalis m.
2 Supratrochlear a.
3 Corrugator m.
4 Orbicularis oculi m.
5 Levator labii superioris m.
6 Alar cartilage
7 Zygomaticus m.
8 Facial a.
9 Orbicularis oris m.
10 Risorius m.
11 Triangularis m.
12 Mentalis m.

Figure 9. Lateral view of the deep muscles of the head.

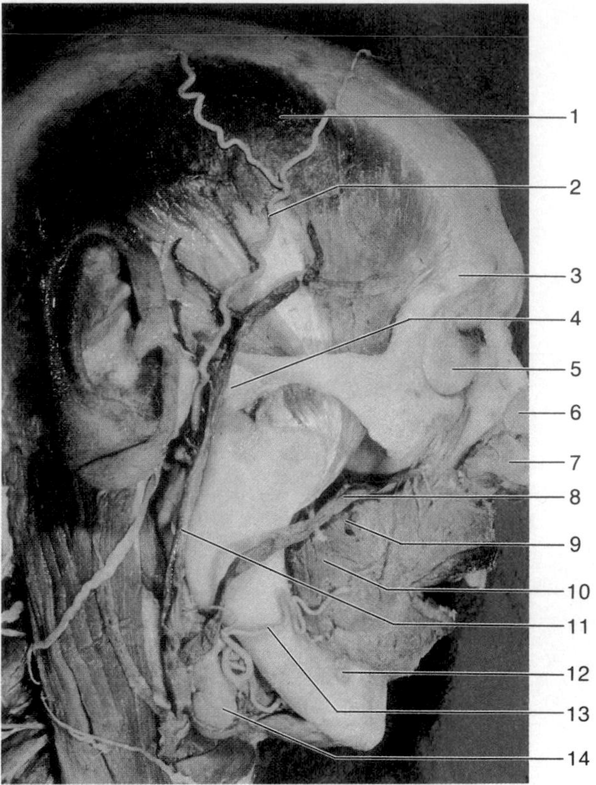

1 Temporalis m.
2 Superficial temporal a.
3 Supraorbital ridge
4 Temporomandibular joint
5 Orbital fat
6 Greater alar cartilage
7 Lateral alar cartilage
8 Facial v.
9 Parotid duct
10 Buccinator m.
11 Retromandibular v.
12 Mental a., v., n.
13 Facial a.
14 Submandibular gland

Figure 10. Anterior view of the right neck region.

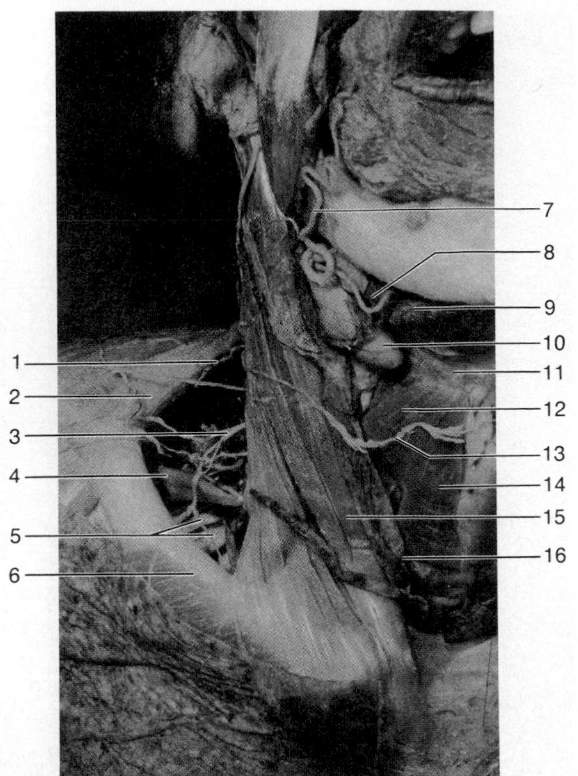

1 Accessory n.
2 Trapezius m.
3 Supraclavicular n.
4 Omohyoid m.
5 Brachial plexus
6 Clavicle
7 Facial a.
8 Mylohyoid m.
9 Digastric m.
10 Submandibular gland
11 Hyoid bone
12 Omohyoid m.
13 Transverse cervical n.
14 Sternohyoid m.
15 Sternocleidomastoid m.
16 External jugular v.

Figure 11. Posterior view of the deep muscles of the neck.

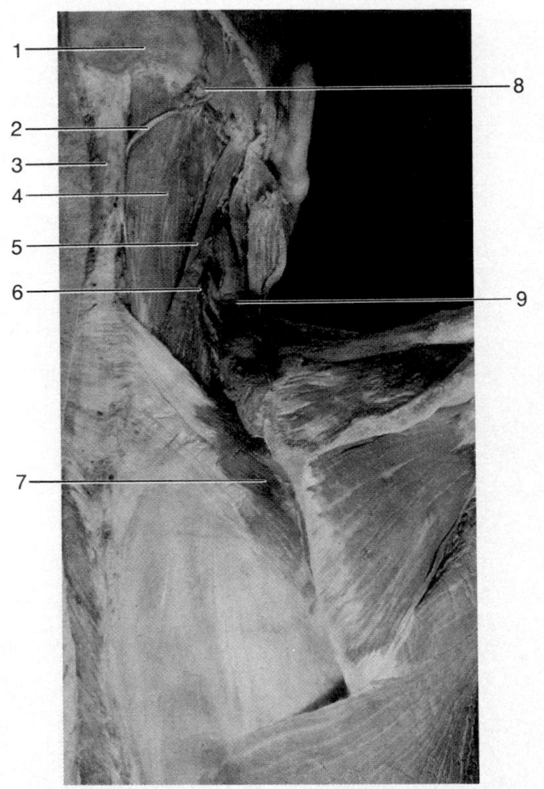

1 Occipital bone
2 Greater occipital n.
3 Ligamentum nuchae
4 Semispinalis capitis m.
5 Longissimus capitis m.
6 Longissimus cervicis m.
7 Serratus posterior m.
8 Occipital a.
9 Levator scapulae m.

Figure 12. Anterior view of the right thorax, shoulder, and brachium.

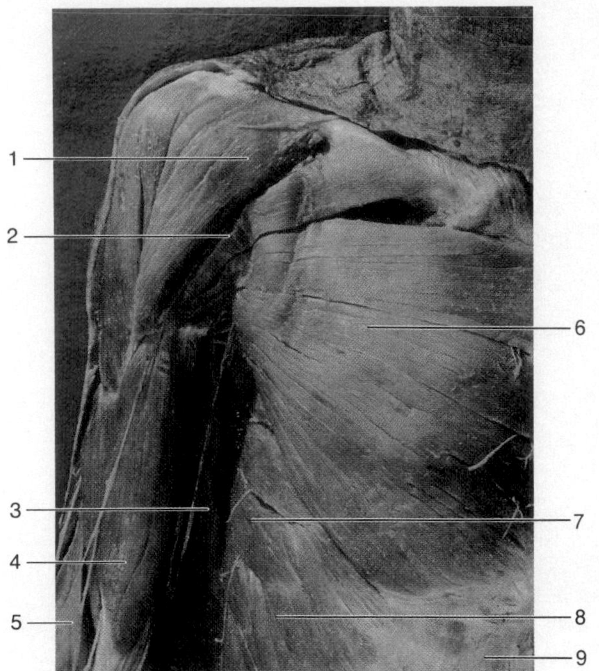

1 Deltoid m.
2 Cephalic v.
3 Latissimus dorsi m.
4 Biceps brachii m.
5 Brachioradialis m.
6 Pectoralis major m.
7 Serratus anterior m.
8 External oblique m.
9 Rectus sheath

Figure 13. Anterior view of the deep muscles of the right thorax, shoulder, and brachium.

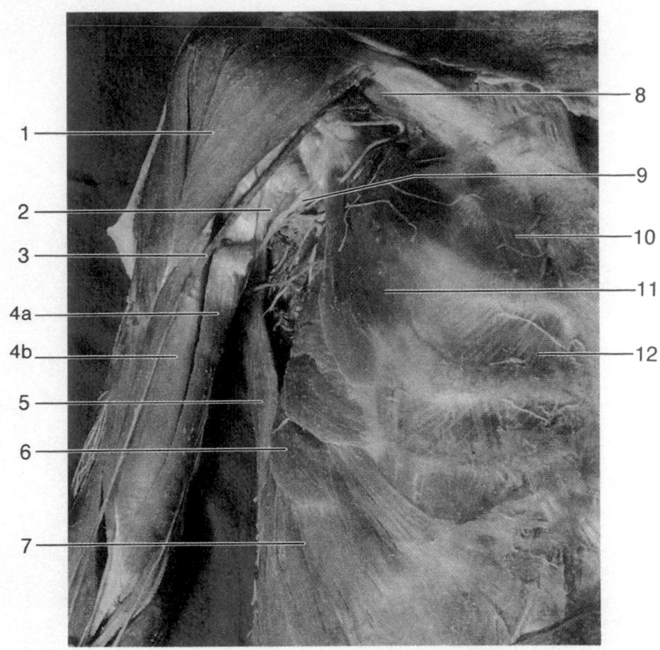

1 Deltoid m.
2 Coracobrachialis m.
3 Cephalic v.
4 Biceps brachii m:
  4a Short head
  4b Long head
5 Latissimus dorsi m.
6 Serratus anterior m.
7 External oblique m.
8 Subclavius m.
9 Brachial plexus
10 Internal intercostal m.
11 Pectoralis minor m.
12 External intercostal m.

Figure 14. Viscera of the thorax.

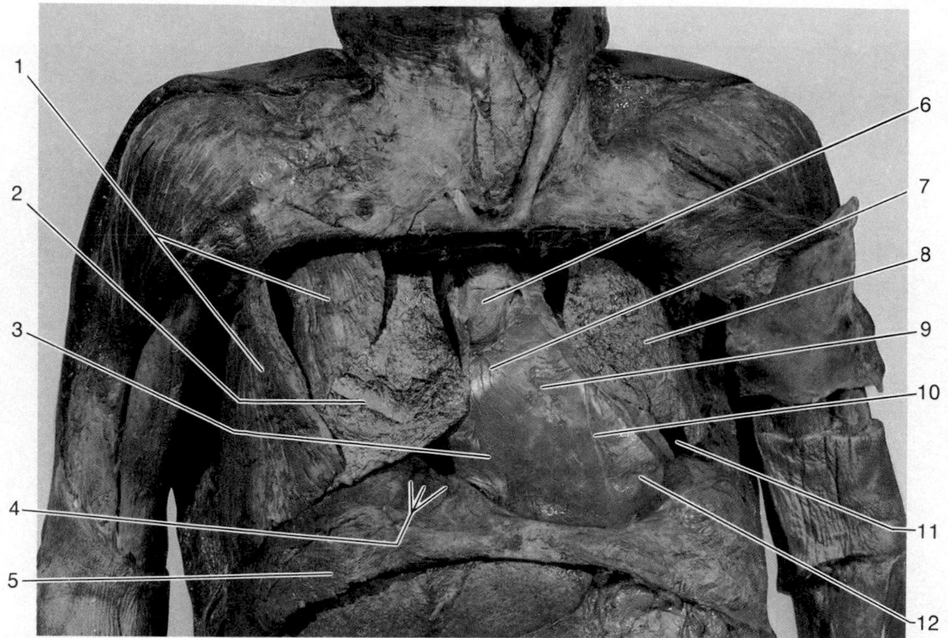

1 Pectoralis major and minor mm.
2 Right lung
3 Pericardium (cut)
4 Right phrenic arteries and veins
5 Diaphragm
6 Pulmonary trunk
7 Anterior cardiac arteries and veins
8 Left lung
9 Heart
10 Left coronary artery and cardiac vein (branches)
11 Cardiac notch
12 Apex of heart

Figure 15. Posterior view of the right thorax and neck.

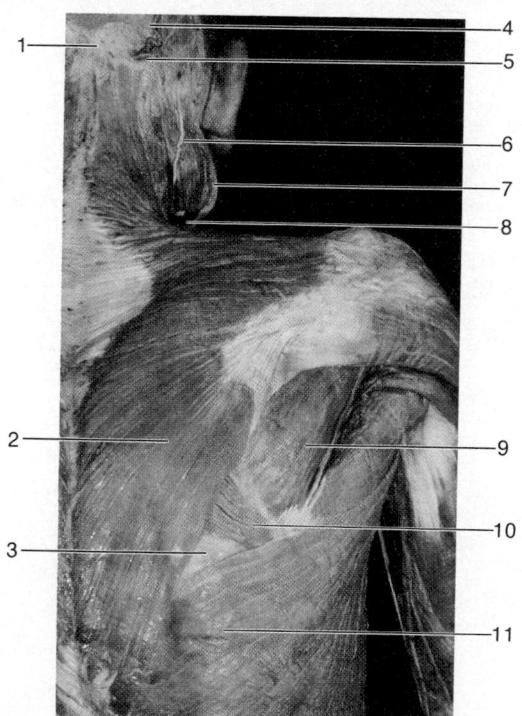

1 External occipital protuberance
2 Trapezius m.
3 Triangle of auscultation
4 Occipital a.
5 Greater occipital n.
6 Lesser occipital n.
7 Sternocleidomastoid m.
8 Greater auricular n.
9 Infraspinatus m.
10 Rhomboideus major m.
11 Latissimus dorsi m.

Figure 16. Posterior view of the deep muscles of the right thorax and neck.

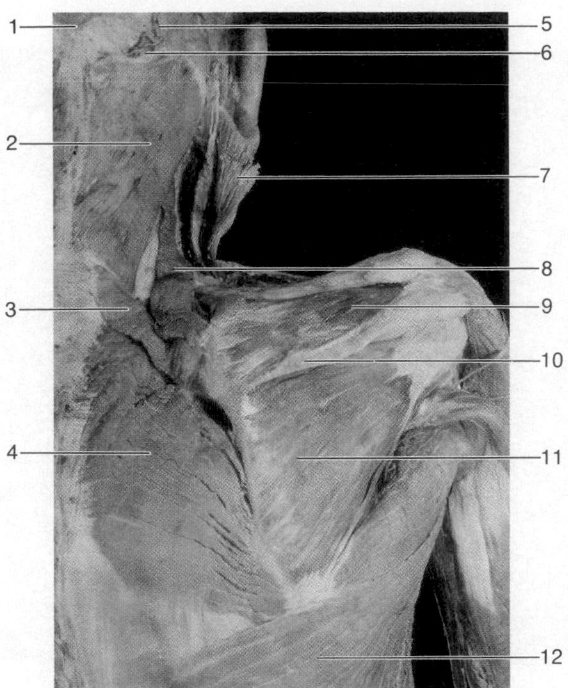

1 External occipital protuberance
2 Splenius capitis m.
3 Rhomboideus minor m.
4 Rhomboideus major m.
5 Occipital a.
6 Greater occipital n.
7 Sternocleidomastoid m. (cut)
8 Levator scapulae m.
9 Supraspinatus m.
10 Spine of scapula
11 Infraspinatus m.
12 Latissimus dorsi m.

Figure 17. Anterior view of the muscles of the abdominal wall.

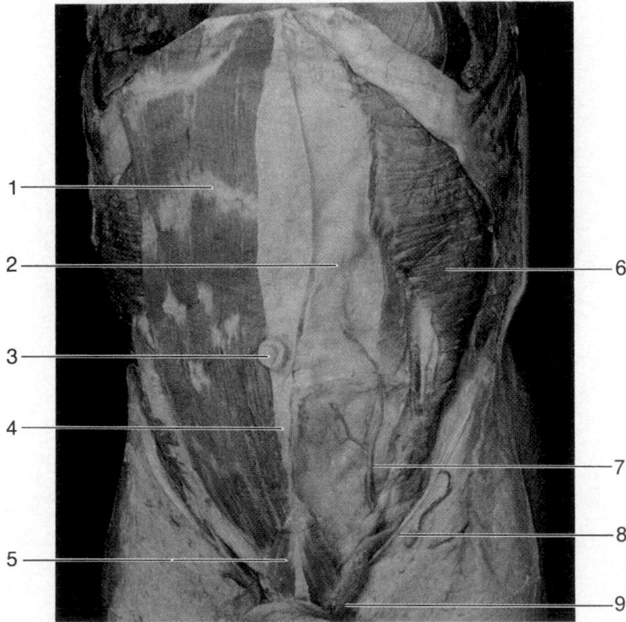

1 Rectus abdominis m.
2 Rectus sheath
3 Umbilicus
4 Linea alba
5 Pyramidalis m.
6 Transverse abdominis m.
7 Inferior epigastric a.
8 Inguinal ligament
9 Spermatic cord

Figure 18. Viscera of the abdomen.

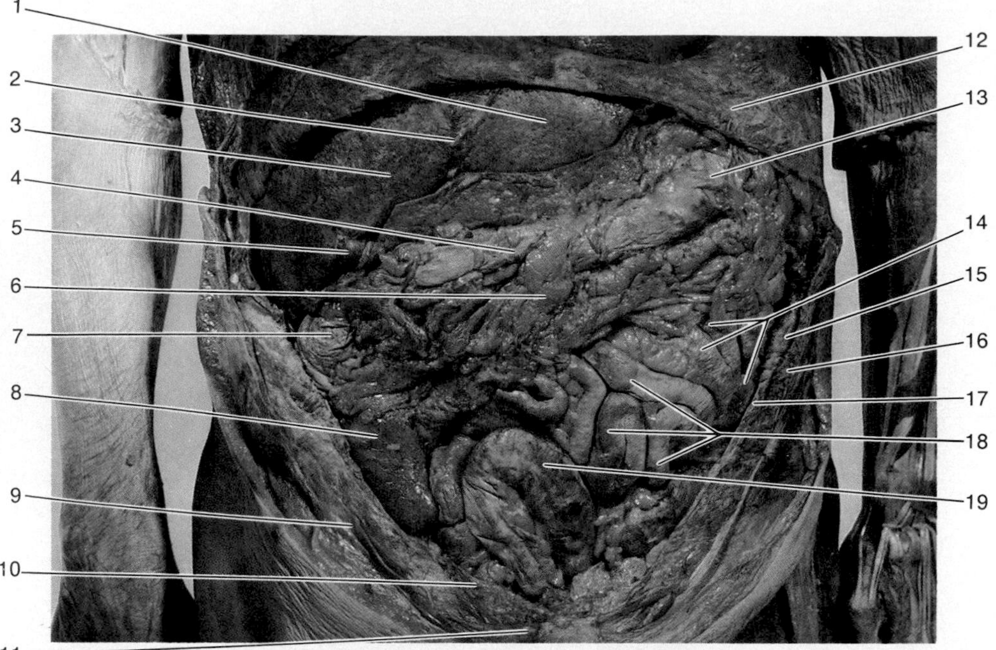

1 Left lobe of liver
2 Falciform ligament
3 Right lobe of liver
4 Transverse colon
5 Gallbladder
6 Greater omentum
7 Hepatic flexure of colon
8 Fat globule on greater omentum
9 Aponeurosis of internal oblique m.
10 Rectus abdominis m. (cut)
11 Rectus sheath (cut)
12 Diaphragm
13 Splenic flexure of colon
14 Jejunum
15 Transversus abdominis m. (cut)
16 Internal and external oblique mm. (cut)
17 Parietal peritoneum (cut)
18 Ileum
19 Anterior sigmoid colon (normal variation)

Figure 19. Posterior view of the right shoulder and brachium.

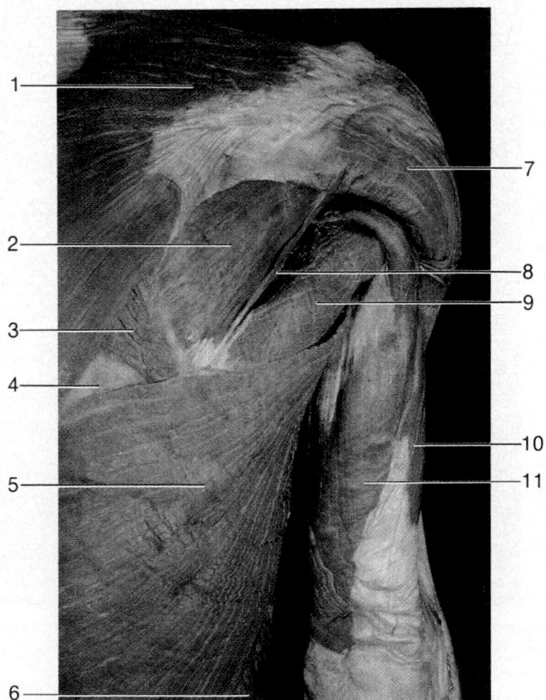

1 Trapezius m.
2 Infraspinatus m.
3 Rhomboideus major m.
4 Triangle of auscultation
5 Latissimus dorsi m.
6 External oblique m.
7 Deltoid m.
8 Teres minor m.
9 Teres major m.
10 Lateral head of triceps brachii m.
11 Long head of triceps brachii m.

Figure 20. Posterior view of the deep muscles of the right shoulder and brachium.

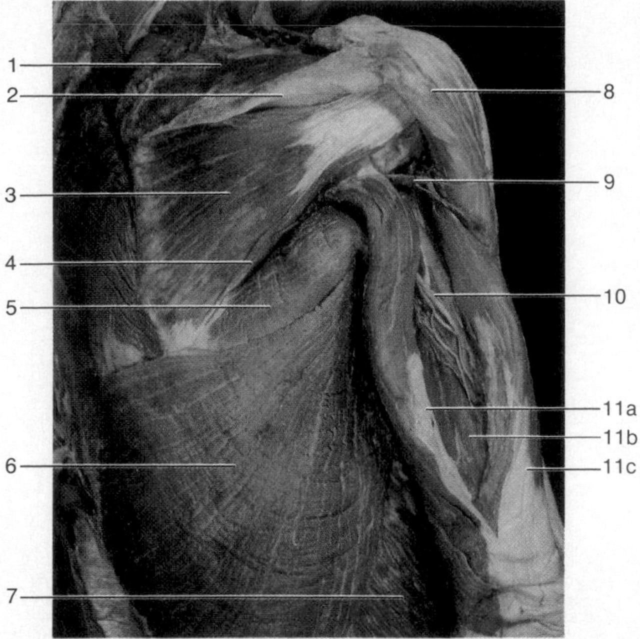

1 Supraspinatus m.
2 Spine of scapula
3 Infraspinatus m.
4 Teres minor m.
5 Teres major m.
6 Latissimus dorsi m.
7 External oblique m.
8 Deltoid m.
9 Axillary n.
10 Radial n.
11 Triceps brachii m.:
11a long head
11b medial head
11c lateral head

Figure 21. Axillary region.

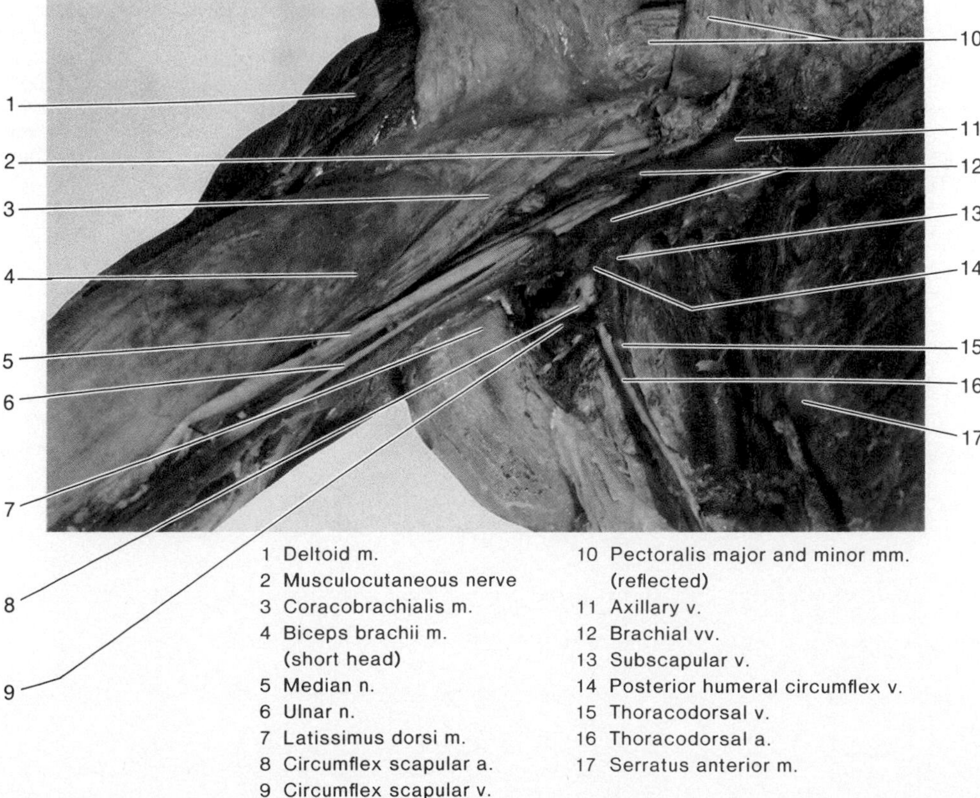

1 Deltoid m.
2 Musculocutaneous nerve
3 Coracobrachialis m.
4 Biceps brachii m. (short head)
5 Median n.
6 Ulnar n.
7 Latissimus dorsi m.
8 Circumflex scapular a.
9 Circumflex scapular v.
10 Pectoralis major and minor mm. (reflected)
11 Axillary v.
12 Brachial vv.
13 Subscapular v.
14 Posterior humeral circumflex v.
15 Thoracodorsal v.
16 Thoracodorsal a.
17 Serratus anterior m.

Figure 22. Posterior view of the left forearm and hand.

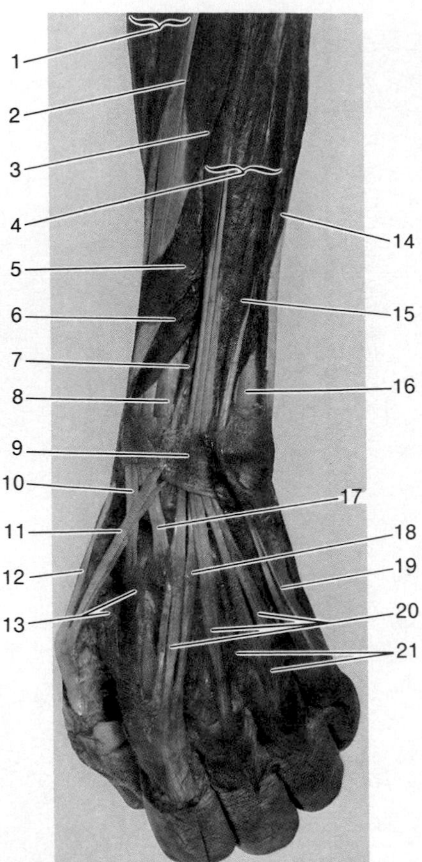

1 Brachioradialis m.
2 Extensor carpi radialis longus tendon
3 Extensor carpi radialis brevis m.
4 Extensor digitorum m.
5 Abductor pollicis longus m.
6 Extensor pollicis brevis m.
7 Extensor pollicis longus m.
8 Radius
9 Extensor retinaculum
10 Extensor carpi radialis longus tendon
11 Extensor pollicis longus tendon
12 Extensor pollicis brevis tendon
13 First dorsal interosseous m.
14 Extensor carpi ulnaris m.
15 Extensor digiti minimi m.
16 Ulna
17 Extensor carpi radialis brevis tendon
18 Extensor indicis tendon
19 Extensor digiti minimi tendon
20 Extensor digitorum tendons
21 Intertendinous connections

Figure 23. Anterior view of the left forearm and hand.

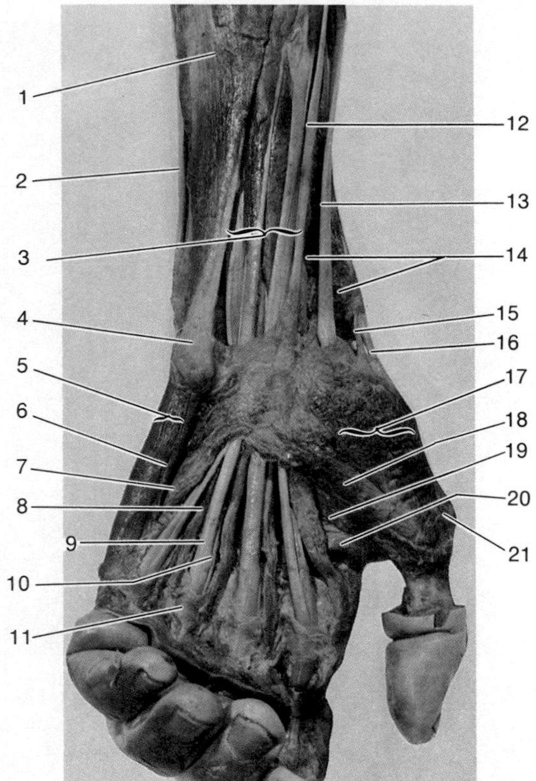

1 Flexor carpi ulnaris m.
2 Extensor carpi ulnaris m.
3 Flexor digitorum superficialis m.
4 Pisiform bone
5 Abductor digiti minimi m.
6 Flexor digiti minimi m.
7 Opponens digiti minimi m.
8 Lumbrical m.
9 Flexor digitorum superficialis tendon
10 Flexor digitorum profundus tendon
11 Fibrous digital sheath
12 Palmaris longus tendon
13 Flexor carpi radialis tendon
14 Pronator quadratus m.
15 Extensor pollicis brevis tendon
16 Extensor pollicis longus tendon
17 Abductor pollicis brevis m.
18 Flexor pollicis brevis m.
19 Adductor pollicis m. (oblique head)
20 Adductor pollicis m. (transverse head)
21 Opponens pollicis m.

Figure 24. Posterior view of the right abdominal and gluteal regions.

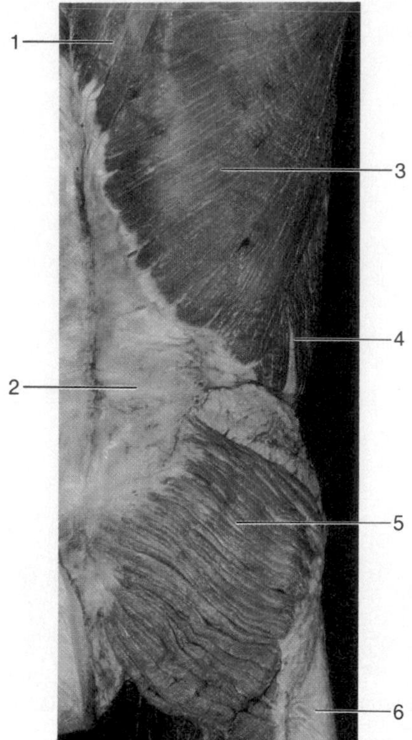

1 Trapezius m.
2 Lumbar aponeurosis
3 Latissimus dorsi m.
4 External oblique m.
5 Gluteus maximus m.
6 Fascia lata

Figure 25. Posterior view of the deep muscles of the right abdominal and gluteal regions.

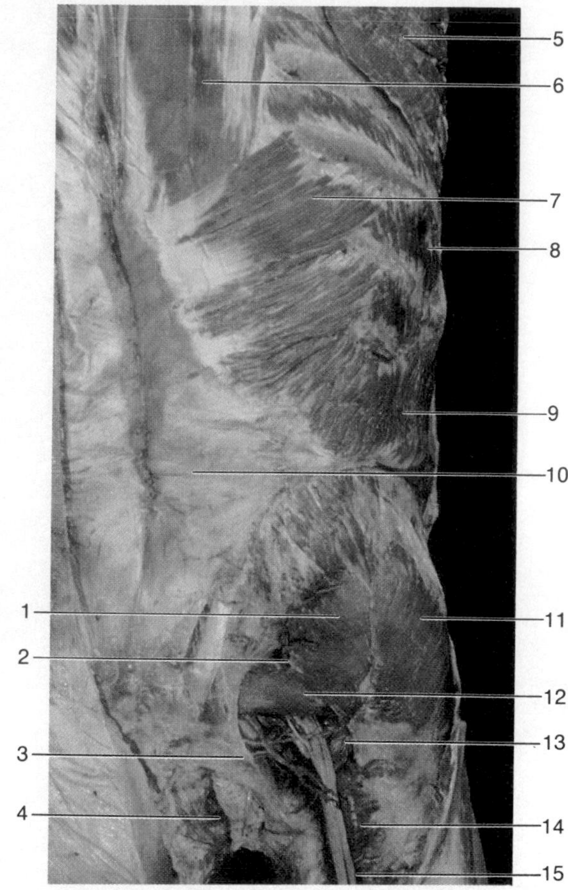

1 Superior gluteal vessels
2 Inferior gluteal vessels
3 Sacrotuberous ligament
4 Levator ani m.
5 Serratus anterior m.
6 Erector spinae m.
7 Serratus posterior m.
8 External intercostal m.
9 Internal oblique m.
10 Lumbar aponeurosis
11 Gluteus medius m.
12 Piriformis m.
13 Obturator internus m.
14 Quadratus femoris m.
15 Sciatic n.

Figure 26. Anterior view of the left thigh.

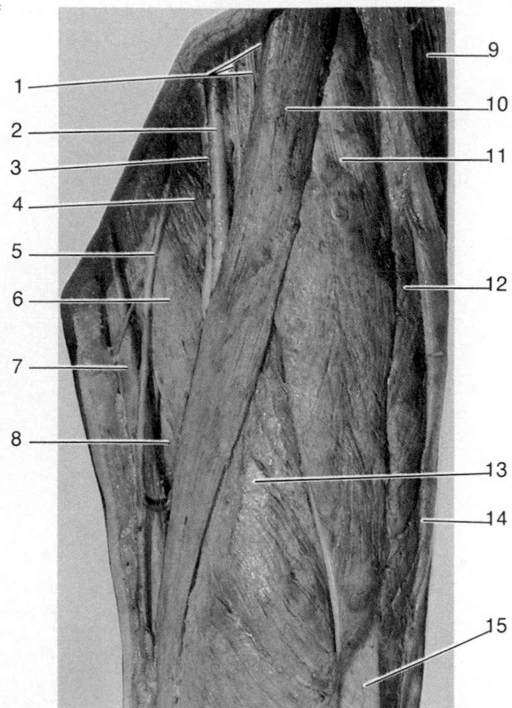

1 Branches of femoral n.
2 Femoral a.
3 Femoral v.
4 Pectineus m.
5 Great saphenous v.
6 Adductor longus m.
7 Gracillis m.
8 Adductor magnus m.
9 Tensor fasciae latae m.
10 Sartorius m.
11 Rectus femoris m.
12 Vastus lateralis m.
13 Vastus medialis m.
14 Iliotibial tract
15 Rectus femoris tendon

Figure 27. Medial view of the right knee and surrounding musculature.

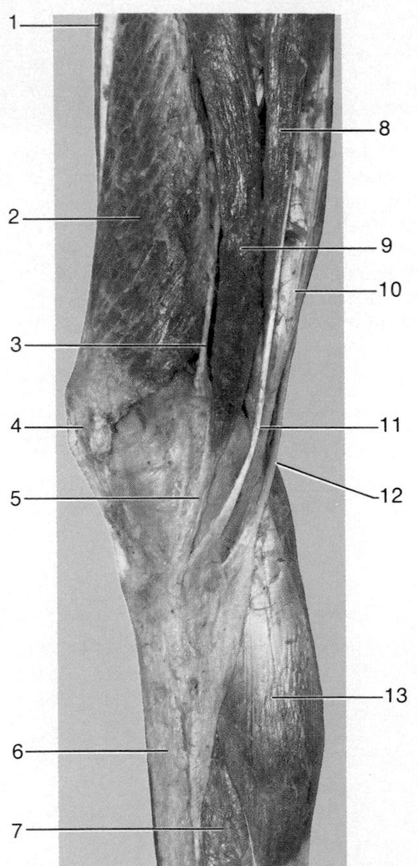

1 Rectus femoris m.
2 Vastus medialis m.
3 Adductor magnus tendon
4 Patella
5 Sartorius tendon
6 Tibia
7 Soleus m.
8 Gracilis m.
9 Sartorius m.
10 Semimembranosus m.
11 Gracilis tendon
12 Semitendinosus tendon
13 Gastrocnemius m.

Figure 28. Postero-lateral view of the right knee musculature and popliteal space.

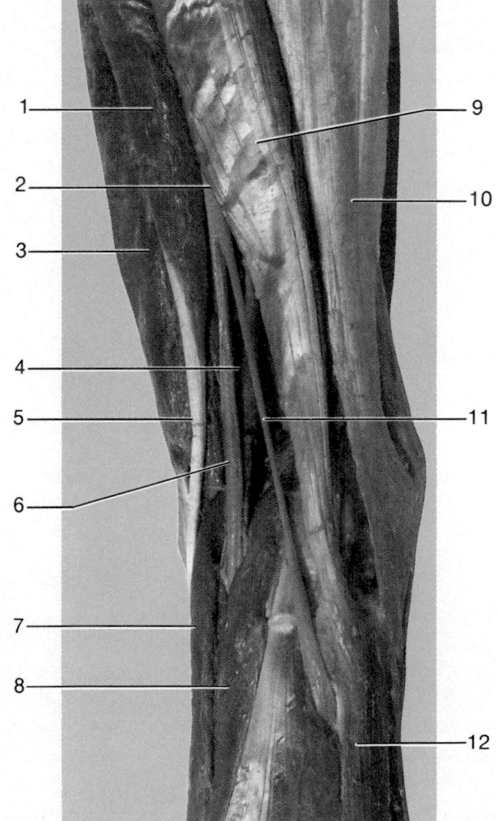

1 Semitendinosus m.
2 Sciatic n.
3 Semimembranosus m.
4 Popliteal a.
5 Semitendinosus tendon
6 Tibial n.
7 Medial head of gastrocnemius m.
8 Lateral head of gastrocnemius m.
9 Biceps femoris m. (long head)
10 Fascia lata
11 Common peroneal n.
12 Peroneus longus m.

Figure 29. Anterior view of the right ankle and foot.

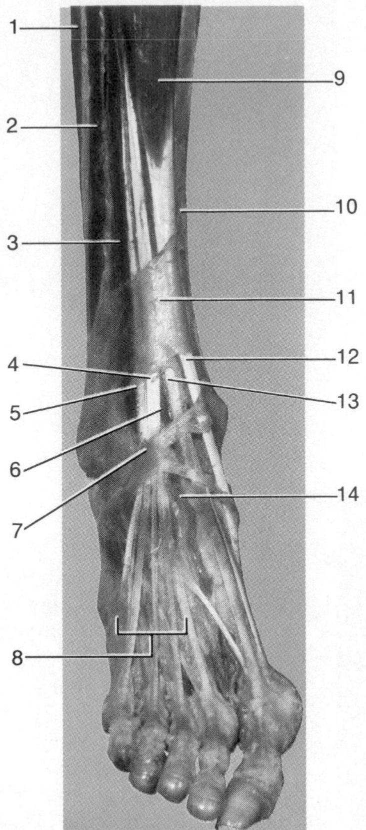

1 Peroneus longus m.
2 Peroneus brevis m.
3 Peroneus tertius m.
4 Extensor digitorum longus tendon
5 Peroneus tertius tendon
6 Anterior tibial a.
7 Inferior extensor retinaculum
8 Tendons of extensor digitorum longus
9 Tibialis anterior m.
10 Tibia
11 Superior extensor retinaculum
12 Tibialis anterior tendon
13 Extensor hallicus longus tendon
14 Dorsalis pedis a.

Figure 30. Medial view of the right ankle and foot.

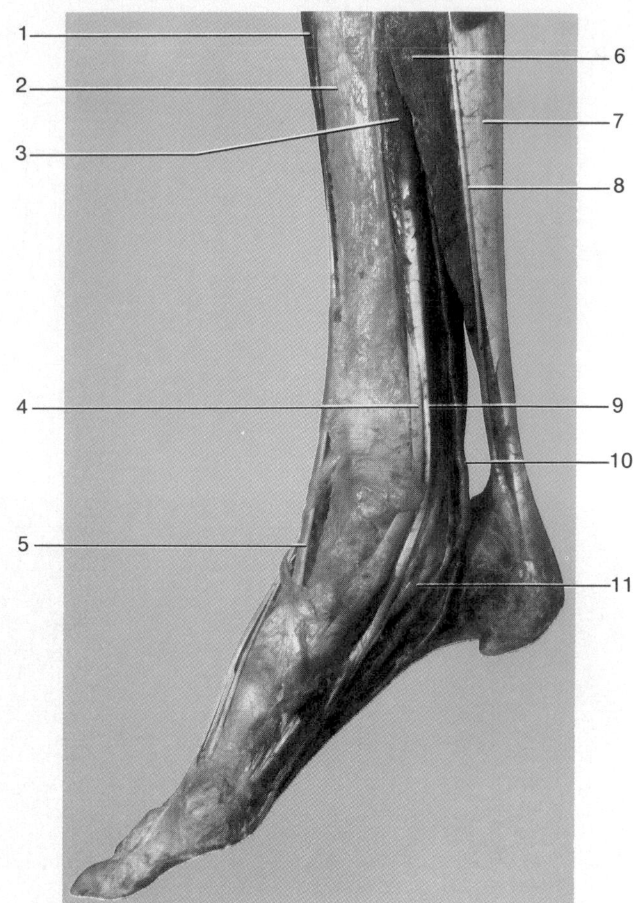

1 Tibialis anterior m.
2 Tibia
3 Flexor digitorum longus m.
4 Tibialis posterior tendon
5 Tibialis anterior tendon
6 Soleus m.
7 Gastrocnemius tendon
8 Plantaris tendon
9 Flexor digitorum longus tendon
10 Posterior tibial a.
11 Flexor hallicus longus tendon

UNIT 3

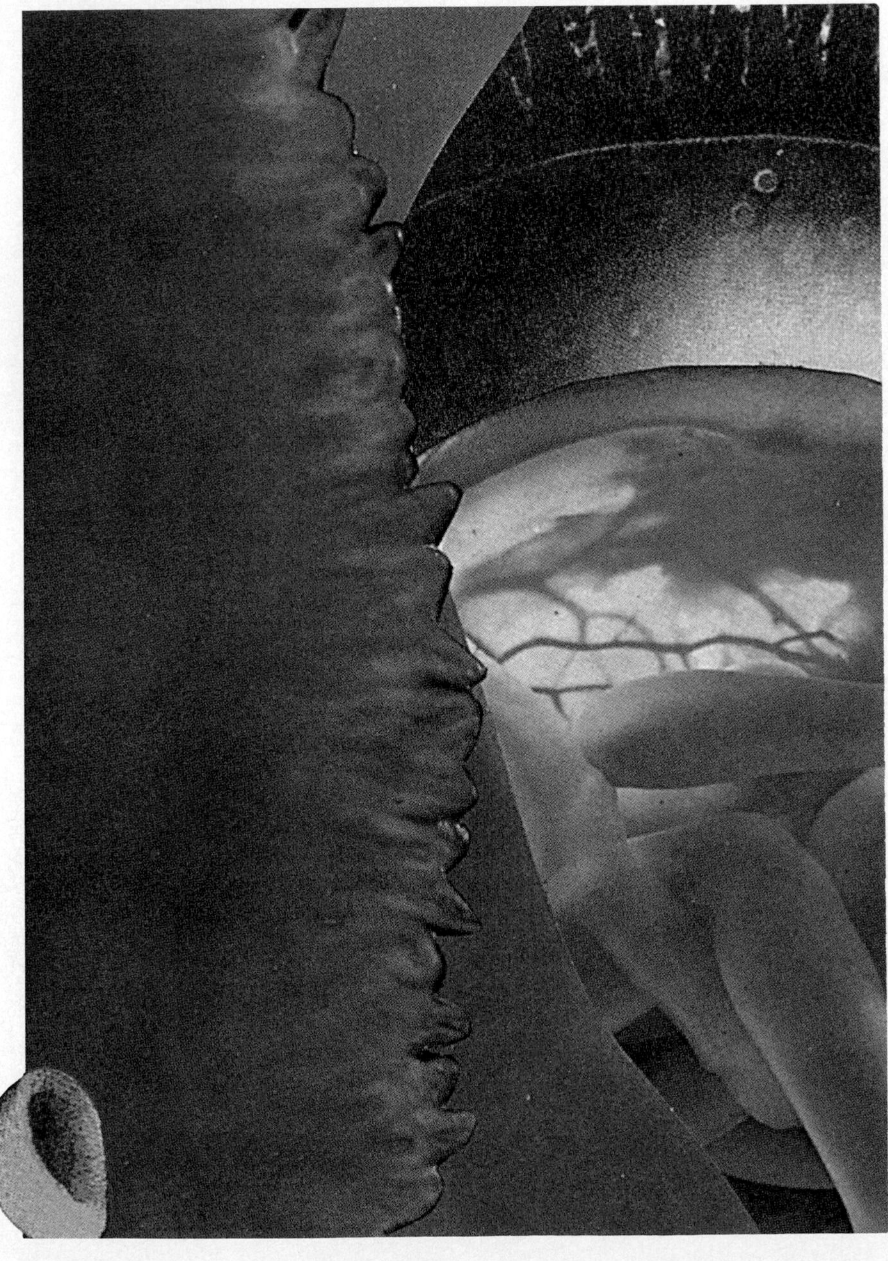

The chapters of unit 3 are concerned with the regulatory systems of the body. These specifically include the nervous system, with all of its divisions, and the endocrine system. The nervous and endocrine systems interact with each other, and the manner in which they regulate the other systems of the body is similar. The anatomical structures of the nervous system and endocrine system are described, and the function of neural and endocrine tissue is explained. These chapters emphasize the critical importance of neural and endocrine regulation in the maintenance of homeostasis. The central themes of negative feedback regulation and neurotransmitter and hormone functions provide the basis for later study of the anatomy and physiology of other systems.

Integration and Control Systems of the Human Body

14 Functional Organization of the Nervous System
Neurons and neuroglial cells comprise the nervous system, which is specialized to transmit nerve impulses along neurons and to activate precise chemical responses at synapses.

15 Central Nervous System
The brain and spinal cord comprise the central nervous system (CNS), which functions in the body's orientation and coordination, assimilation of experiences, and programming of instinctual behavior.

16 Peripheral Nervous System
The peripheral nervous system (PNS) consists of nerves that convey impulses to and from the central nervous system. Reflex arcs are specific pathways, involving neurons of both the PNS and the CNS, that respond rapidly to potentially threatening stimuli.

17 Autonomic Nervous System
The autonomic nervous system (ANS), composed of sympathetic and parasympathetic divisions, helps to maintain homeostasis through adrenergic and cholinergic effects.

18 Sensory Organs
The sensory organs—the eyes, ears, olfactory mucosa, taste buds, and somatic receptors—are specialized extensions of the nervous system that respond to specific stimuli and conduct nerve impulses to the CNS.

19 Endocrine System
The endocrine system secretes regulatory molecules, called hormones, into the blood where they are carried to target tissues, bond to specific receptor proteins, and elicit specific physiological responses.

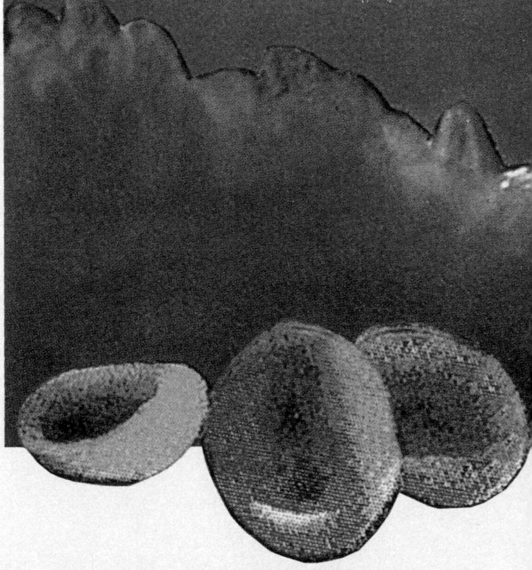

14 Functional Organization of the Nervous System

Outline

Concepts

The central nervous system and the peripheral nervous system are structural components of the nervous system, whereas the autonomic nervous system is a functional component. Together they coordinate body activities, permit the assimilation of experiences, and program instinctual behavior.

The embryonic neural tube forms the brain and spinal cord, and the cranial and spinal nerves are formed predominately from the embryonic neural crest.

Neurons have many forms, but all contain a cell body, dendrites for the reception of stimuli, and an axon for the conduction of nerve impulses. Neuroglia support the neurons, both structurally and functionally.

The permeability of the cell membrane of axons to specific ions is regulated by membrane gates. Increases in the permeability to these ions at specific locations in the membrane allow diffusion through the membrane, either into or out of the cell, producing changes in the membrane potential and the nerve impulse.

Synaptic transmission is either electrical or chemical. The predominant type are chemical synapses, in which neurotransmitters are released from endings of nerve fibers in response to stimulation by action potentials.

Synapses that use acetylcholine (ACh) as a neurotransmitter combine with receptor proteins in the postsynaptic membrane and electrically activate the postsynaptic cell by causing the chemically regulated Na^+ and K^+ gates to open. This causes the production of small, graded depolarizations, which in turn can stimulate the opening of voltage-regulated gates and the production of action potentials.

A variety of chemicals found in the CNS are believed to function as neurotransmitters. These include ACh, the catecholamines (norepinephrine and dopamine), certain amino acids, and a number of polypeptides. Particular neural pathways in the brain utilize the use of specific neurotransmitters, and the actions of some of these neurotransmitters have been modified clinically in the treatment of diseases.

A number of EPSPs must summate in order to stimulate the postsynaptic cell to produce action potentials. This stimulatory effect can be reduced or inhibited by hyperpolarization (IPSPs) in a process known as postsynaptic inhibition.

ORGANIZATION AND FUNCTIONS OF THE NERVOUS SYSTEM

The central nervous system and the peripheral nervous system are structural components of the nervous system, whereas the autonomic nervous system is a functional component. Together they coordinate body activities, permit the assimilation of experiences, and program instinctual behavior.

Objective 1. Describe the divisions of the nervous system.
Objective 2. Define *neurology;* define *neuron.*
Objective 3. List the functions of the nervous system.

The immensely complex brain and its myriad of connecting pathways constitute the nervous system. The nervous system, along with the endocrine system, regulates the functions of other body systems. The brain, however, does much more than that, and its potential is perhaps greatly underestimated. It is incomprehensible that one's personality, thoughts, and aspirations result from the functioning of a body organ. Plato referred to the brain as "the divinest part of us." The thought processes of this organ have devised the technology for launching rockets into space, curing diseases, and splitting atoms. But with all of these achievements, the brain still remains amazingly ignorant of its own workings.

Neurology, or the study of the nervous system, has been referred to as "the last frontier of functional anatomy." Some basic questions concerning the functioning of the nervous system remain unanswered: How do nerve cells store and retrieve memory? What are the roles of the many chemical compounds within the brain? What causes mental illness or senility? Scientists are still developing the technology and skills necessary to understand the functional complexity of the nervous system. The next few decades should be very interesting as scientists come to better understand the nervous system.

Organization of the Nervous System

The nervous system is divided into the **central nervous system (CNS),** which includes the **brain** and **spinal cord,** and the **peripheral** *(pe-rif'er-al)* **nervous system (PNS),** which includes the **cranial nerves,** arising from the brain, and the **spinal nerves,** arising from the spinal cord (fig. 14.1 and table 14.1).

The **autonomic nervous system (ANS)** is a functional subdivision of the entire nervous system. The controlling centers of the ANS are within the brain and are considered part of the CNS, and the nerve portions of the ANS are subdivided into the **sympathetic** and **parasympathetic** divisions.

neurology: Gk. *neuron,* nerve; L. *logos,* study of

Figure 14.1. The nervous system consists of the central nervous system (brain and spinal cord) and the peripheral nervous system (cranial nerves and spinal nerves). The nerves within the extremities are also part of the peripheral nervous system.

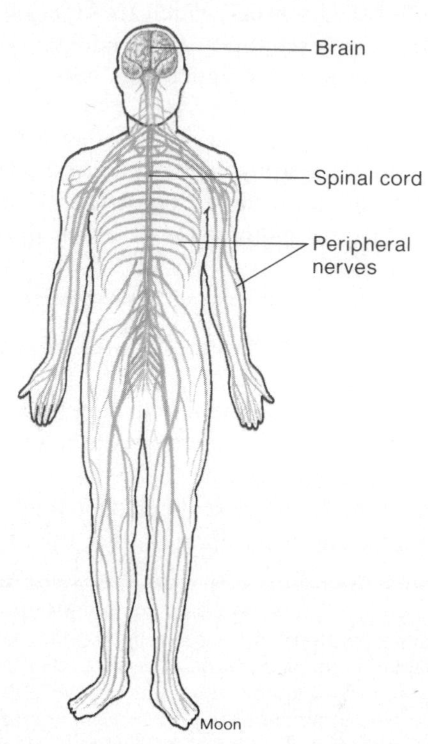

Table 14.1 Anatomical terms used in describing the nervous system

Term	Definition
Central nervous system (CNS)	Brain and spinal cord
Peripheral nervous system (PNS)	Nerves and ganglia
Sensory nerve fiber	Neuron that transmits impulses from a sensory receptor into the CNS (an afferent fiber)
Motor nerve fiber	Neuron that transmits impulses from the CNS to an effector organ, e.g., muscle (efferent)
Nerve	Cablelike collection of nerve fibers; may be "mixed" (contain both sensory and motor fibers)
Plexus	Network of interlaced nerves
Somatic motor nerve	Nerve that stimulates contraction of skeletal muscles
Autonomic motor nerve	Nerve that stimulates contraction (or inhibits contraction) of smooth muscle and cardiac muscles, and stimulates secretion of glands
Ganglion	Collection of neuron cell bodies located outside the CNS
Nucleus	Grouping of neuron cell bodies within the CNS
Tract	Collections of nerve fibers that interconnect regions of the CNS

Functions of the Nervous System

The nervous system is specialized for perceiving and responding to events in our internal and external environments. An awareness of one's environment is made possible by nerve cells called **neurons,** which are highly specialized in the properties of excitability and conductivity. The nervous system is present throughout the body, and it functions with the endocrine system (see chap. 19) to coordinate closely the activities of the other body systems. In addition to integrating body activities, the nervous system has the ability to store experiences *(memory)* and to establish patterns of response based on prior experiences *(learning).*

The functions of the nervous system are the

1. monitoring of conditions in the internal and external environments;
2. coordination and control of body activities;
3. assimilation of experiences requisite to memory, learning, and intelligence;
4. programming of instinctual behavior (apparently more important in vertebrates other than humans).

An instinct may also be called a fixed action pattern and is typically genetically specified, little modified by the environment, and is triggered only by a specific stimulus. Some of the basic instincts in humans include survival, feeding, drinking, voiding, and specific vocalization. To some ethologists—those who study animal behavior—following puberty, reproduction becomes an instinctive behavior.

1. State the meaning of ANS, PNS, and CNS. Which of these is a functional component of the nervous system and is further subdivided into sympathetic and parasympathetic divisions?
2. Explain why neurology is considered a dynamic science.

Development of the Nervous System

The embryonic neural tube forms the brain and spinal cord, and the cranial and spinal nerves are formed predominately from the embryonic neural crest.

Objective 4. List the five developmental regions of the brain, and discuss the embryonic formation of these regions.

Objective 5. Describe the development of the spinal cord.

Objective 6. Define *dermatome,* and describe the clinical significance of the pattern of dermatome innervation in the body.

The first indication of nervous tissue development occurs about seventeen days following conception, when a thickening appears along the entire dorsal length of the embryo. This thickening, called the **neural plate** (fig. 14.2), differentiates and eventually gives rise to all of the neurons and to most of the **neuroglial** *(nu-rog'le-al)* **cells,** which are the supporting cells of the nervous system.

As development progresses, the midline of the neural plate invaginates to become the **neural groove.** At the same time, there is a proliferation of cells along the lateral margins of the neural plate, which become the thickened **neural folds.** The neural groove continues to deepen as the neural folds elevate. By the twentieth day, the neural folds meet and fuse at the midline, and the neural groove becomes a **neural tube.** The neural tube, for a short time, is open both cranially and caudally. These openings, called **neuropores** (fig. 14.2), close during the fourth week. Once formed, the neural tube separates from the surface ectoderm and eventually develops into the CNS (brain and spinal cord).

The **neural crest** forms from the neural folds as they fuse longitudinally along the dorsal midline. The neural crest is positioned between the surface ectoderm and the neural tube. Most of the PNS (cranial and spinal nerves) forms from the neural crest. Some neural crest cells break away from the main tissue mass and migrate to other locations, where they differentiate into motor nerve cells of the sympathetic nervous system or into Schwann cells, which are a type of neuroglial cell important in the PNS.

Development of the Brain

The brain begins its embryonic development as the cephalic end of the neural tube starts to grow rapidly and to differentiate (fig. 14.3). By the middle of the fourth week, three distinct swellings are evident: the **prosencephalon** *(pros''en-sef'ah-lon)* (forebrain), the **mesencephalon** (midbrain), and the **rhombencephalon** (hindbrain). Further development during the fifth week results in the formation of five specific regions: the **telencephalon** and the **diencephalon** *(di''en-sef'ah-lon)* derive from the forebrain; the mesencephalon remains unchanged; and the **metencephalon** and **myelencephalon** form from the hindbrain.

The rapid growth of the brain during the fourth and fifth weeks causes marked bendings or flexures to appear (fig. 14.4). The most prominent of these is the **midbrain flexure** along the ventral aspect of the mesencephalon. The **cervical flexure** occurs at the junction of the myelencephalon and the spinal cord. The less prominent **pontine** *(pon'tēn)* **flexure** occurs between the midbrain and the cervical flexure in the roof of the hindbrain.

Figure 14.2. The early development of the nervous system from embryonic ectoderm. (*a*) a dorsal view of an eighteen-day-old embryo showing the formation of the neural plate and the position of a transverse cut indicated in (a^1). (*b*) a dorsal view of a twenty-two-day-old embryo showing cranial and caudal neuropores and the positions of three transverse cuts indicated in (b^1–b^3). Note the amount of fusion of the neural tube at the various levels of the twenty-two-day-old embryo. Note also the relationship of the notochord to the neural tube.

Neural plate
Notochordal plate
(a)
Neural crest
Neural groove
Neural plate
Surface ectoderm
Mesoderm
(a_1)
Wall of yolk sac
Notochordal plate
Neural crest
Neural groove
(b_1)
Notochord
Cranial neuropore
Neural crest
Neural canal
Neural tube
(b_2)
Notochord
Caudal neuropore
Neural crest
Neural groove
Neural groove
Neural crest
Notochord
(b)
(b_3)
Wall of yolk sac
Waldrop

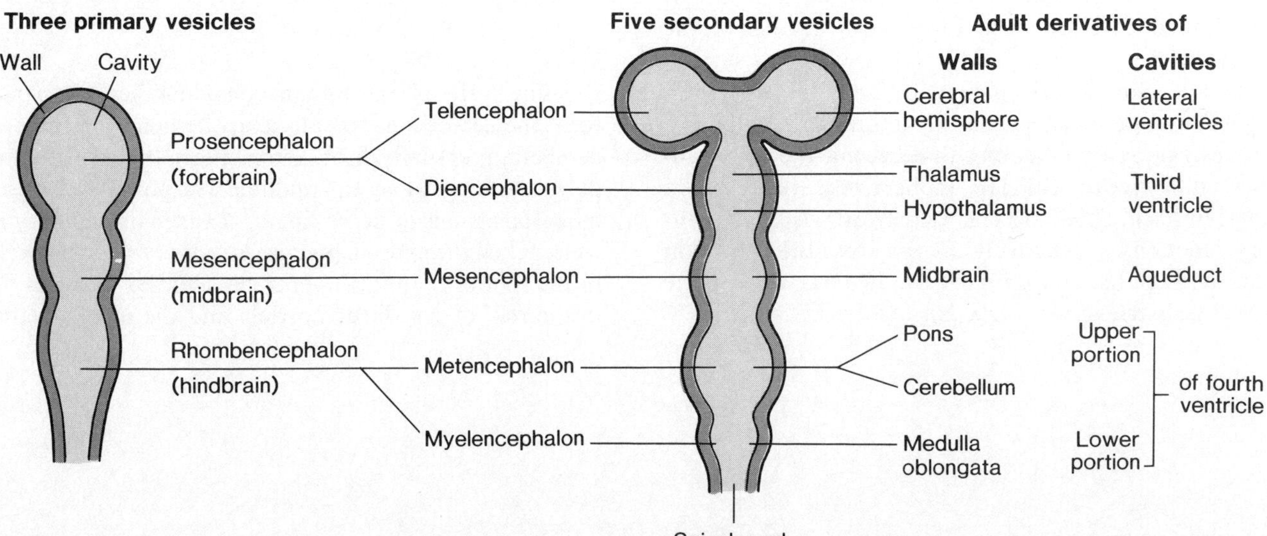

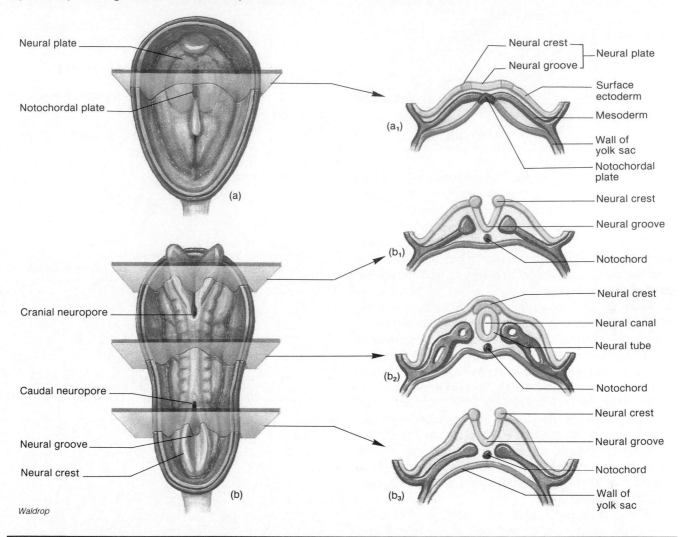

Figure 14.3. The developmental sequence of the brain. During the fourth week, the three principal regions of the brain are formed. During the fifth week, a five-regioned brain develops, and specific structures begin to form.

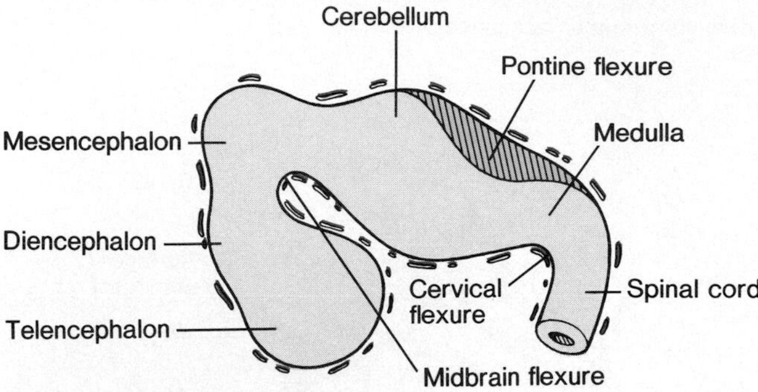

Figure 14.4. Pronounced flexures form in the brain by the end of the fifth week.

During the developmental transformation of the brain, hollow chambers known as **ventricles** form within each region. The **first** and **second ventricles** (lateral ventricles) form in the forebrain, a narrow **third ventricle** forms in the forebrain and midbrain, and the cavity of the hindbrain becomes the **fourth ventricle.** The ventricles of the brain are continuous with each other and with the **central canal** of the spinal cord.

Once the five principal regions of the brain are developed, rapid differentiation occurs in each of the regions, and shortly after that, all of the major structures of the brain are recognizable. The greatest amount of growth occurs within the telencephalon as its dorsal portions rapidly expand to form the two **hemispheres** of the **cerebrum** (fig. 14.5). The swelling of the cerebrum causes it to cover the diencephalon, the midbrain, and a portion of the hindbrain. By the thirty-fifth week, the lobes of the cerebrum are formed, and the surface of the cerebrum is distinctly convoluted. The outer, convoluted portion of the cerebrum consists of gray matter, and the inner portion consists of white matter.

Within the diencephalon three swellings form in the walls of the third ventricle and become the **epithalamus** *(ep″ĭ-thal′ah-mus)*, the **thalamus,** and the **hypothalamus** *(hi″po-thal′ah-mus)*. The **pineal** *(pin′e-al)* **gland** also forms in the diencephalon as a midline diverticulum develops from the roof of the third ventricle. A portion of the **pituitary gland** develops from the ventral portion of the diencephalon.

Few developmental changes occur within the mesencephalon. The most obvious structures to develop are the four aggregates of neurons that become the paired **superior** and **inferior colliculi** (collectively, the corpora quadrigemina), which are concerned with visual and auditory functions, respectively. Fibers from the cerebrum extend through the midbrain to form two tracts called the **cerebral peduncles** *(pe-dung′k′lz)*.

Figure 14.5. The sequential development of the cerebrum within the telencephalon. (*a*) thirteen weeks, (*b*) twenty-six weeks, (*c*) thirty-five weeks, (*d*) newborn. Note the gradual formation of the cerebral lobes, including the internally positioned insula, and the progressive convolutions of the cerebral cortex.

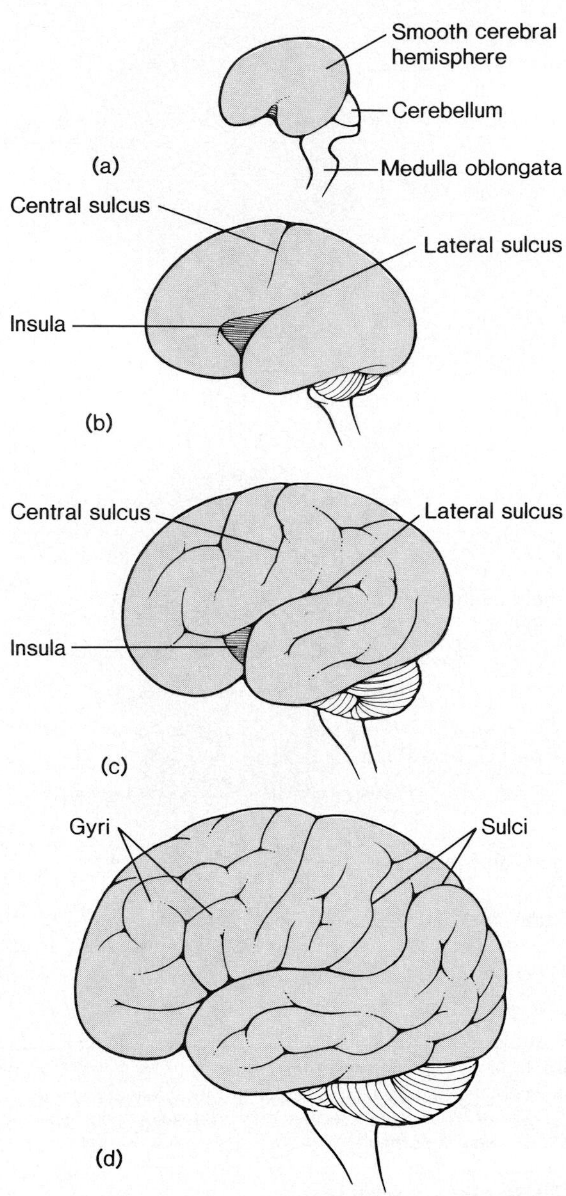

The walls of the metencephalon greatly expand to form the **cerebellum** dorsally and the **pons** ventrally. The cerebellum actually forms from the joining of a pair of dorsal swellings along the midline. The pons develops from numerous bands of nerve fibers. A vascular capillary network called the **choroid plexus** *(ko′roid plek′sus)* develops in the roof of the fourth ventricle. Similar plexuses form in the roof of the third ventricle and the medial walls of

Figure 14.6. The development of the spinal cord. (*a*) a dorsal view of an embryo at twenty-three days and the position of a transverse cut indicated in (*b*). (*c*) the formation of the alar and basal plates is evident in a transverse section through the spinal cord at six weeks. (*d*) the central canal has reduced in size, and functional dorsal and ventral horns have formed at nine weeks.

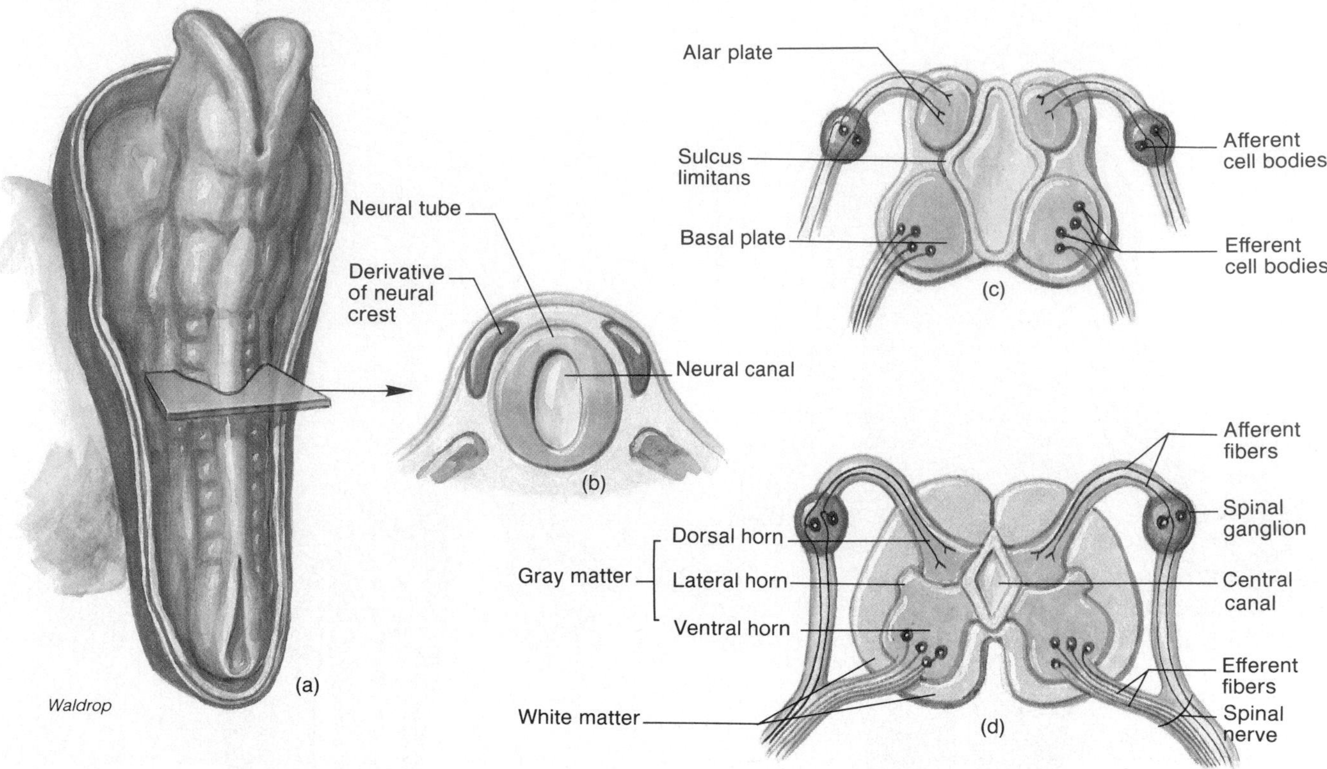

the lateral ventricles. These plexuses secrete **cerebrospinal** *(ser''ĕ-bro-spi'nal)* **fluid,** which circulates throughout the cavities of the CNS and within spaces outside the brain and spinal cord.

The caudal portion of the myelencephalon is continuous with and resembles the spinal cord. The entire adult myelencephalon is known as the **medulla oblongata** and forms from an aggregate of nerve fibers. The ventral **pyramids** of the myelencephalon form from the axon tracts that extend from the developing cerebral cortex.

Development of the Spinal Cord

The spinal cord, like the brain, develops as the neural tube undergoes differentiation and specialization. Throughout the developmental process, the hollow central canal persists while the specialized white and gray matter forms (fig. 14.6). Changes in the neural tube become apparent during the sixth week as the lateral walls thicken to form a groove, called the **sulcus limitans,** along each lateral wall of the central canal. A pair of **alar** *(a'lar)* **plates** forms dorsal to the sulcus limitans, and a pair of **basal plates** forms ventrally. By the ninth week, the alar plates have specialized to become the **dorsal horns,** containing fibers of the afferent cell bodies, and the basal plates have specialized to form the **ventral** and **lateral horns,** containing efferent cell bodies. Afferent fibers of spinal nerves conduct impulses toward the spinal cord, whereas efferent fibers conduct impulses away from the spinal cord.

Mitotic activity within nervous tissue is completed during prenatal development. Thus, a person is born with all the neurons he or she is capable of producing. However, nervous tissue continues to grow and specialize after a person is born, particularly in the first several years of postnatal life.

Figure 14.7. The pattern of dermatomes and the peripheral distribution of spinal nerves. (*a*) an anterior view; (*b*) a posterior view.

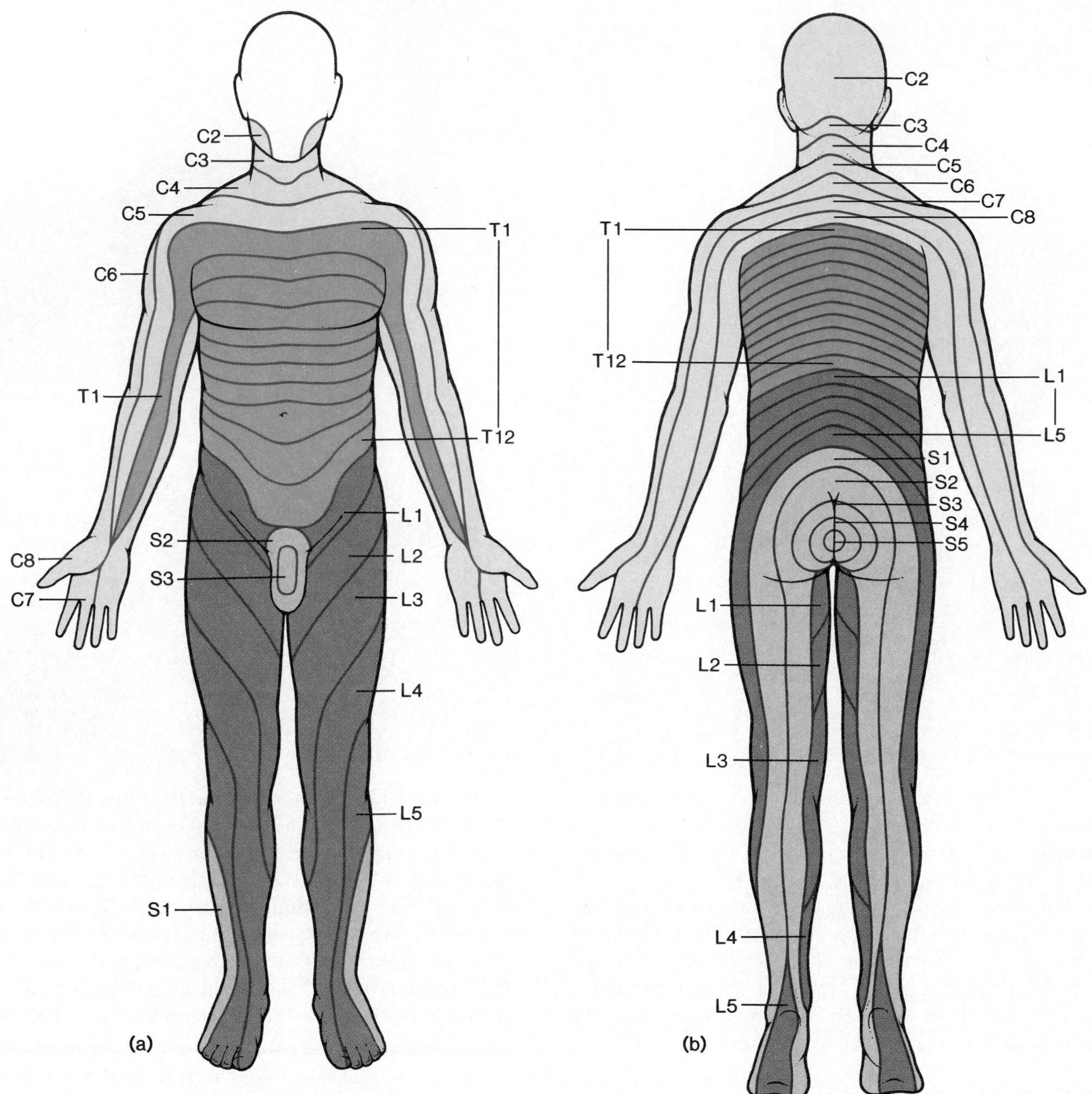

Development of the Peripheral Nervous System

Development of the peripheral nervous system produces the pattern of dermatomes within the body (fig. 14.7). A **dermatome** *(der'mah-tōm)* is an area of the skin innervated by all the cutaneous neurons of a certain spinal or cranial nerve. Most of the scalp and face is innervated by sensory fibers from the trigeminal nerve. With the exception of the first cervical nerve (Cl), all of the spinal nerves are associated with specific dermatomes. Dermatomes are consecutive in the neck and torso regions. In the appendages, however, adjacent dermatome innervations overlap. The apparently uneven dermatome arrangement in the appendages is due to the uneven rate of nerve growth into the limb buds. Actually the limbs are segmented, and dermatomes overlap only slightly.

dermatome: Gk. *derma,* skin; *tomia,* a cutting

The pattern of dermatome innervation is of clinical importance when a physician desires to anesthetize a particular portion of the body. Because adjacent dermatomes overlap in the appendages, at least three spinal nerves must be blocked to produce complete anesthesia in these regions. Abnormally functioning dermatomes provide clues about injury to the spinal cord or specific spinal nerves. If a dermatome is stimulated but no sensation is perceived, the physician can infer that the injury involves the innervation to that dermatome.

1. List the five regions of the brain that develop during the fifth week, and discuss their formation.
2. Describe the locations of the ventricles within the brain.
3. Describe the embryonic origin of afferent and efferent cell bodies of spinal nerves.
4. Discuss the derivatives of the alar plates and the basal plates in the formation of the spinal cord.
5. Describe the dermatomes, and explain their clinical significance.

Neurons and Neuroglia

Neurons have many forms, but all contain a cell body, dendrites for the reception of stimuli, and an axon for the conduction of nerve impulses. Neuroglia cells support the neurons, both structurally and functionally.

Objective 7. Describe the microscopic structure of a neuron.

Objective 8. Explain the different classification of neurons.

Objective 9. List the different types of neuroglial cells and describe their functions.

Objective 10. Describe the structure of a sheath of Schwann and a myelin sheath, and explain how they are formed.

Objective 11. Explain why the gray matter and the white matter of the CNS have their characteristic colors, and distinguish between the myelination of axons of the PNS and those of the CNS.

The highly specialized and complex nervous system is composed of only two principal categories of cells, neurons and neuroglia. **Neurons** are the basic structural and functional units of the nervous system. They are specialized to respond to physical and chemical stimuli, conduct impulses, and release specific chemical regulators. Through these activities, neurons perform functions such as storing memory, thinking, and regulating other organs and glands. Neurons cannot divide mitotically, although some neurons can regenerate a severed portion or sprout small new branches under some conditions.

Neuroglia, or **glial** *(gli'al)* **cells,** are supportive cells in the nervous system that aid the function of neurons. Glial cells are about five times more abundant than neurons and have limited mitotic abilities.

Neurons

Although neurons vary considerably in size and shape, they generally have three principal components: (1) a cell body; (2) dendrites; and (3) an axon (fig. 14.8).

The **cell body,** or **perikaryon** *(per''i-kar'e-on),* is the enlarged portion of the neuron that contains the nucleus and nucleolus surrounded by cytoplasm. Besides containing organelles typically found in cells, the cytoplasm of neurons is characterized by the presence of **Nissl bodies** and filamentous strands of protein called **neurofibrils** *(nu''ro-fi'bril).* Nissl bodies are specialized layers of granular (rough) endoplasmic reticulum that synthesize fibrils and minute microtubules, which appear to be involved in transporting material within the cell. The cell bodies within the CNS are frequently clustered into regions called **nuclei** (not to be confused with the nucleus of a cell). Cell bodies in the PNS generally occur in clusters called **ganglia** (see table 14.1).

Figure 14.8. A photomicrograph of neurons from the anterior column of gray matter of the spinal cord (120X).

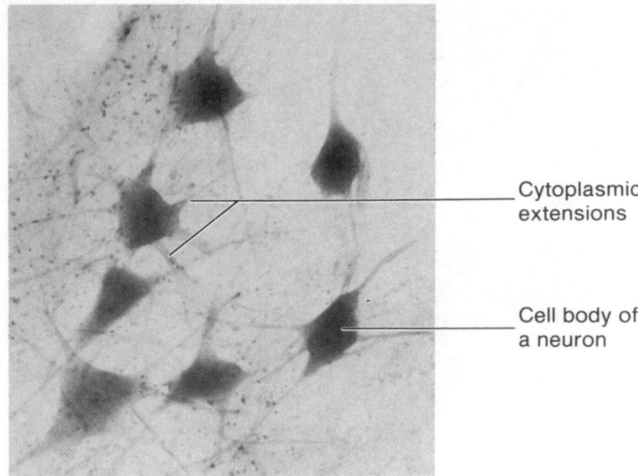

Dendrites are branched processes that extend from the cytoplasm of the cell body. Dendrites function to receive a stimulus and conduct impulses to the cell body. Some dendrites are covered with minute **dendritic spines,** which greatly increase their surface area and provide contact points for other neurons. The volume occupied by dendrites is referred to as the **dendritic zone** of a neuron.

The **axon,** or **nerve fiber,** is the second type of cytoplasmic extension from the cell body. An axon is a relatively long, cylindrical process that conducts impulses away from the cell body. The conical tapering region of the axon, where it originates from the cell body, is referred to as the **axon hillock.** Axons vary in length from a few millimeters in the CNS to over a meter between the distal portions of the extremities and the spinal cord. Side branches called **axon collaterals** extend a short distance from the axon. The cytoplasm of an axon contains many mitochondria, microtubules, and neurofibrils.

Classification of Neurons and Nerves

Neurons may be classified according to structure or function. The functional classification is based on the direction of conducted impulses. Sensory impulses originate in sensory receptors and are conducted by **sensory,** or **afferent, neurons** to the CNS. Motor impulses originate in the CNS and are conducted by **motor,** or **efferent, neurons** to a muscle or gland (figs. 14.9, 14.10). **Association neurons,**

perikaryon: Gk. *peri,* around; *karyon,* nucleus
Nissl body: from Franz Nissl, German neuroanatomist, 1860–1919

ganglion: Gk. *ganglion,* swelling
dendrite: Gk. *dendron,* tree branch
axon: Gk. *axon,* axis

Figure 14.9. The structure of two kinds of neurons. (*a*) a motor neuron; (*b*) a sensory neuron.

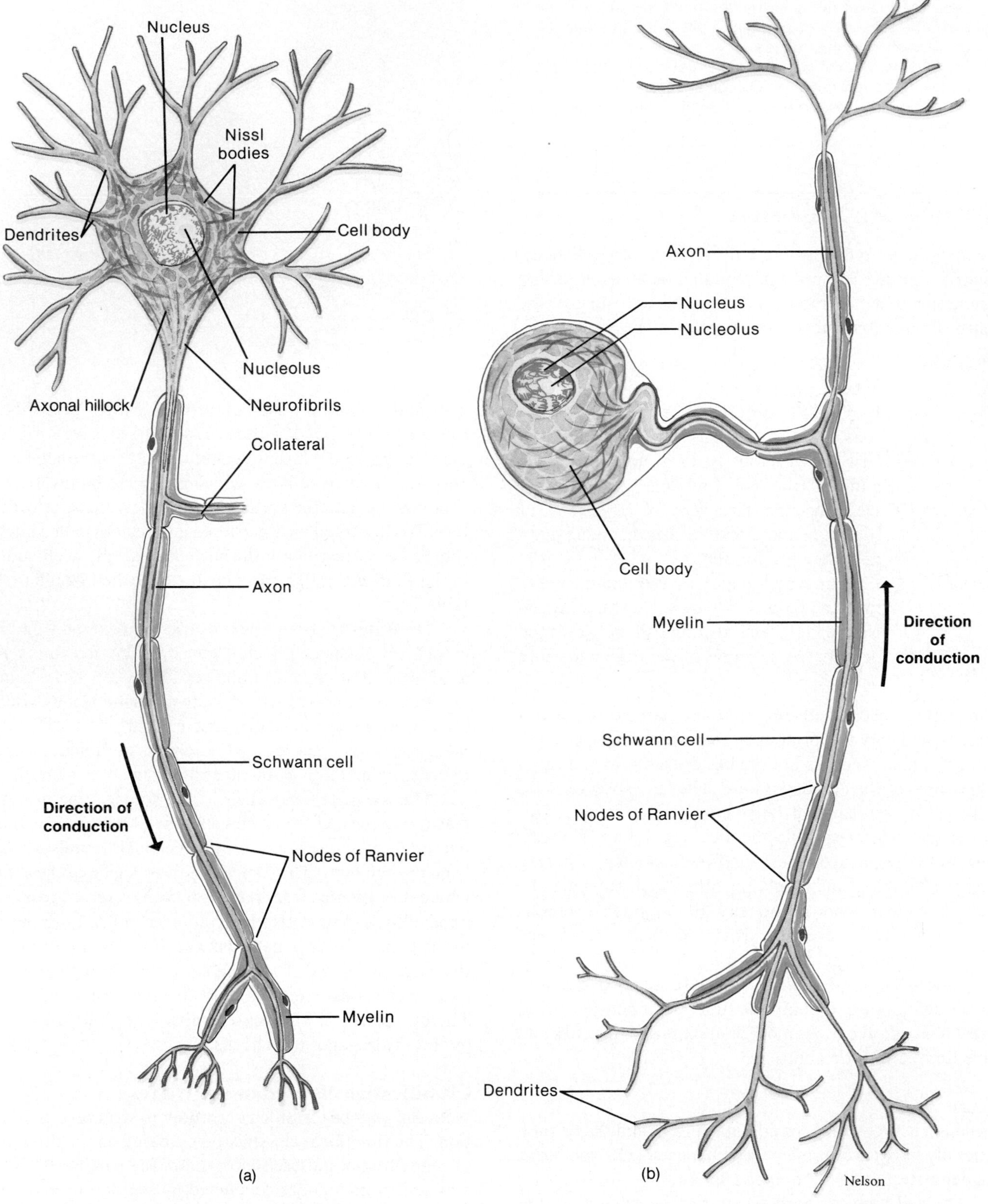

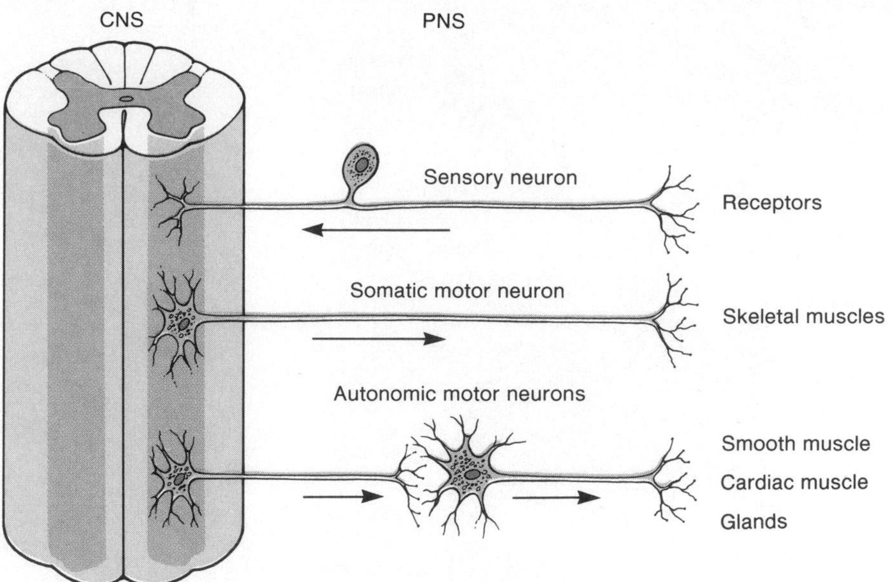

Figure 14.10. The relationship between sensory and motor fibers of the peripheral nervous system (PNS) and the central nervous system (CNS).

or **interneurons,** are located between sensory and motor neurons and are found within the spinal cord and brain.

The structural classification of neurons is based upon the number of processes that extend from the cell body of the neuron (fig. 14.11). The spindle-shaped **bipolar neuron** has a process at either end; this type occurs in the retina of the eye. **Multipolar neurons** are the most common type and are characterized by several dendrites and one axon extending from the cell body. Motor neurons are a good example of this type. A **pseudounipolar neuron** has a single process that divides into two. Sensory neurons are pseudounipolar and have their cell bodies located in the dorsal root ganglia of spinal and cranial nerves.

A **nerve** is a collection of nerve fibers outside the CNS. Fibers within a nerve are held together and strengthened by loose fibrous connective tissue (fig. 14.12). Each individual nerve fiber is enclosed in a connective tissue sheath called the **endoneurium** *(en''do-nu're-um).* A group of fibers, called a **fasciculus** *(fah-sik'u-lus),* is surrounded by a connective tissue sheath called a **perineurium.** The entire nerve is surrounded and supported by connective tissue called the **epineurium,** which contains tiny blood vessels and often adipose cells. Perhaps less than a quarter of the bulk of a nerve consists of nerve fibers: more than half is associated connective tissue, and approximately a quarter is the myelin that surrounds the nerve fibers.

Most nerves are composed of both motor and sensory fibers and thus are called **mixed nerves.** Some of the cranial nerves, however, are composed either of sensory neurons only (**sensory nerves**) or of motor neurons only (**motor nerves**). Sensory nerves serve the special senses, such as taste, smell, sight, and hearing. Motor nerves conduct impulses to muscles, causing them to contract, or to glands, causing them to secrete.

fasciculus: L. diminutive of *fascis,* bundle

Figure 14.11. Three different types of neurons.

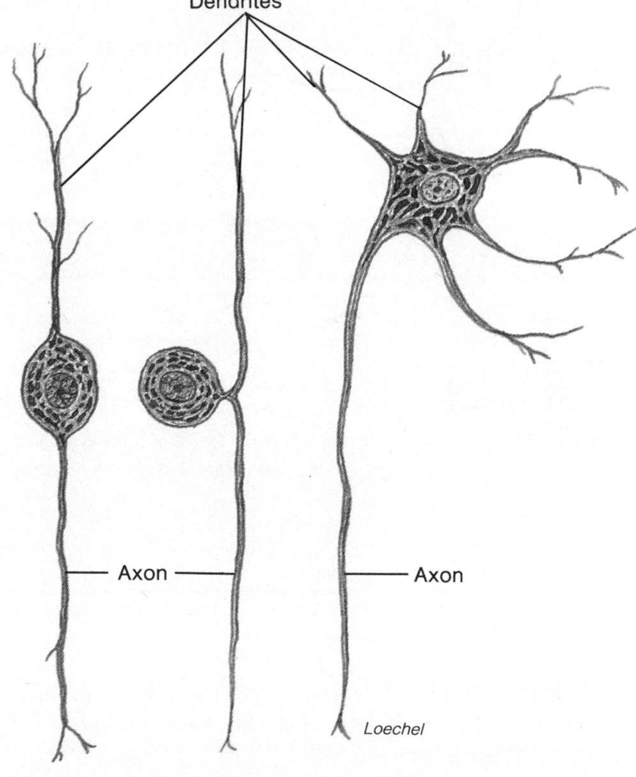

Neurons and their fibers within nerves may be classified according to the area of innervation into the following scheme (fig. 14.13):

1. **Somatic afferent.** Sensory receptors within the skin, muscles, and joints receive stimuli and convey nerve impulses through somatic afferent fibers to the CNS for interpretation.

Figure 14.12. The structure of a nerve.

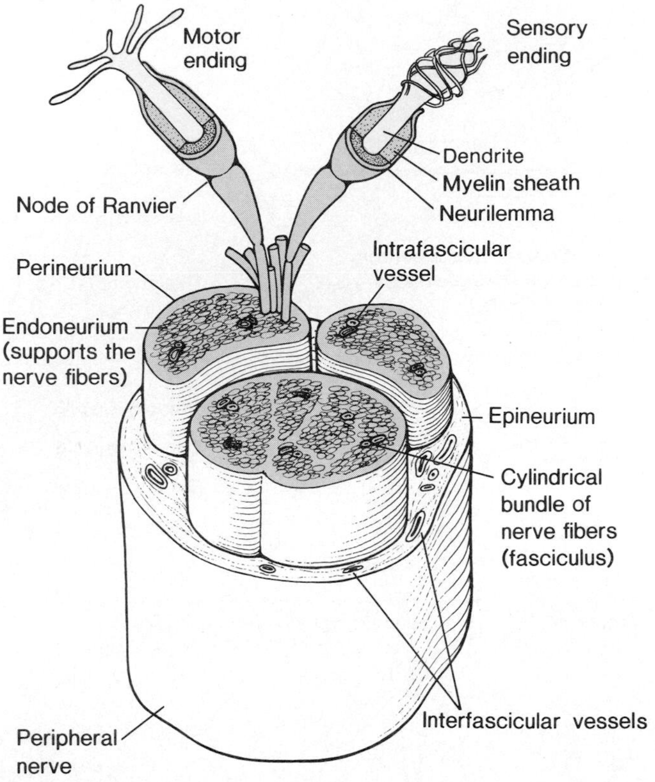

Figure 14.13. The classification of nerve fibers by origin and function.

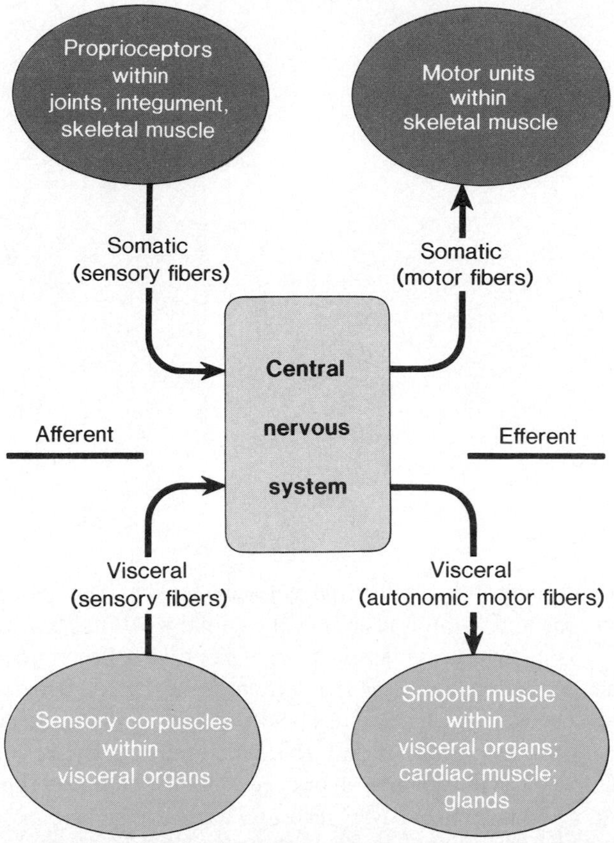

Table 14.2 Structure and function of neuroglia

Type	Structure	Function
Astrocytes	Stellate with numerous processes	Form structural support between capillaries and neurons within the CNS; blood-brain barrier
Oligodendrocytes	Similar to astrocytes but with shorter and fewer processes	Form myelin in CNS; guide development of neurons within CNS
Microglia	Minute cells with few short processes	Phagocytize pathogens and cellular debris within CNS
Ependyma	Columnar cells that may have ciliated free surfaces	Line ventricles and central canal within CNS where cerebrospinal fluid is circulated by ciliary motion
Satellite cells	Small, flattened cells	Support ganglia within PNS
Schwann cells	Flattened cells arranged in series around axons or dendrites	Form myelin within the PNS

2. **Somatic efferent.** Impulses from the CNS through somatic efferent fibers cause the contraction of skeletal muscles.
3. **Visceral afferent.** Visceral afferent fibers convey impulses from visceral organs and blood vessels to the CNS for interpretation.
4. **Visceral efferent.** Visceral efferent fibers, also called **autonomic fibers,** are part of the autonomic nervous system. They originate in the CNS and innervate cardiac muscle, glands, and smooth muscle within the visceral organs. The autonomic system is described in detail in chapter 17.

Neuroglia

Unlike other organs that are "packaged" in connective tissue derived from mesoderm (the middle layer of embryonic tissue), the supporting neuroglial cells of the nervous system are derived from the same embryonic tissue layer (ectoderm) that produces neurons. There are six categories of neuroglia: (1) **Schwann cells,** which form myelin sheaths around axons of the PNS; (2) **oligodendrocytes** *(ol″i-go-den′dro-sīts),* which form myelin

Schwann cell: from Theodor Schwann, German histologist, 1810–1882

oligodendrocyte: Gk. *oligos,* few; L. *dens,* tooth; Gk. *kytos,* hollow (cell)

sheaths around axons of the CNS; (3) **microglia** *(mi-krog'le-ah)*, which are phagocytic cells that migrate through the CNS and remove foreign and degenerated material; (4) **astrocytes,** which may help regulate the passage of molecules from the blood to the brain; (5) **ependyma** *(e-pen'di-mah)*, which line the ventricles of the brain and the central canal of the spinal cord; and (6) **satellite cells,** which support neuron cell bodies within the ganglia of the PNS (table 14.2).

Sheath of Schwann and Myelin Sheath. Some axons in the CNS and PNS are surrounded by a myelin sheath and are known as *myelinated* axons. Other axons do not have a myelin sheath and are *unmyelinated.*

Unmyelinated axons in the PNS are surrounded by a living sheath of Schwann cells, known as the **sheath of Schwann.** The outer surface of this layer of Schwann cells is encased in a glycoprotein *basement membrane,* often called the *neurilemma (nu''ril-lem'mah),* which is analogous to the basement membrane that underlies epithelial membranes. The unmyelinated axons of the CNS, in contrast, lack a sheath of Schwann (because Schwann cells are found only in the PNS) and also lack a continuous basement membrane.

Axons that are less than two micrometers in diameter are usually unmyelinated. Larger axons are generally surrounded by a **myelin** *(mi'e-lin)* **sheath** (fig. 14.14), which is composed of successive wrappings of the cell membrane of Schwann cells or oligodendrocytes. In the process of myelin formation, Schwann cells (in the PNS) wrap around the axon so that their cytoplasm becomes squeezed toward the outermost region (fig. 14.15), like toothpaste rolled from the bottom of a tube. Each Schwann cell wraps only about 1 mm of axon, leaving gaps of exposed axon between the adjacent Schwann cells. These gaps in the myelin sheath are known as the **nodes of Ranvier** *(rahn-ve-a')*. The successive wrappings of Schwann cell membrane provide insulation around the axon, leaving only the nodes of Ranvier exposed to produce nerve impulses.

The Schwann cells remain alive as their cytoplasm is squeezed to the outside of the myelin sheath. As a result, myelinated axons of the PNS, like their unmyelinated counterparts, are surrounded by a living sheath of Schwann and a continuous basement membrane (fig. 14.16).

The myelin sheaths of the CNS are formed by oligodendrocytes. Unlike a Schwann cell, which forms a myelin sheath around only one axon, each oligodendrocyte has extensions that form myelin sheaths around several axons. Myelinated axons of the CNS, as a result, are not surrounded by a continuous basement membrane. The

Figure 14.14. A myelinated neuron. (*a*) a myelin sheath is formed by Schwann cells around the axons of many peripheral neurons. (*b*) the myelin sheath is composed of wrappings of Schwann cell membrane. The Schwann cell nucleus and most of the cytoplasm, together with an outermost basement membrane, is located to the outside of the myelin sheath.

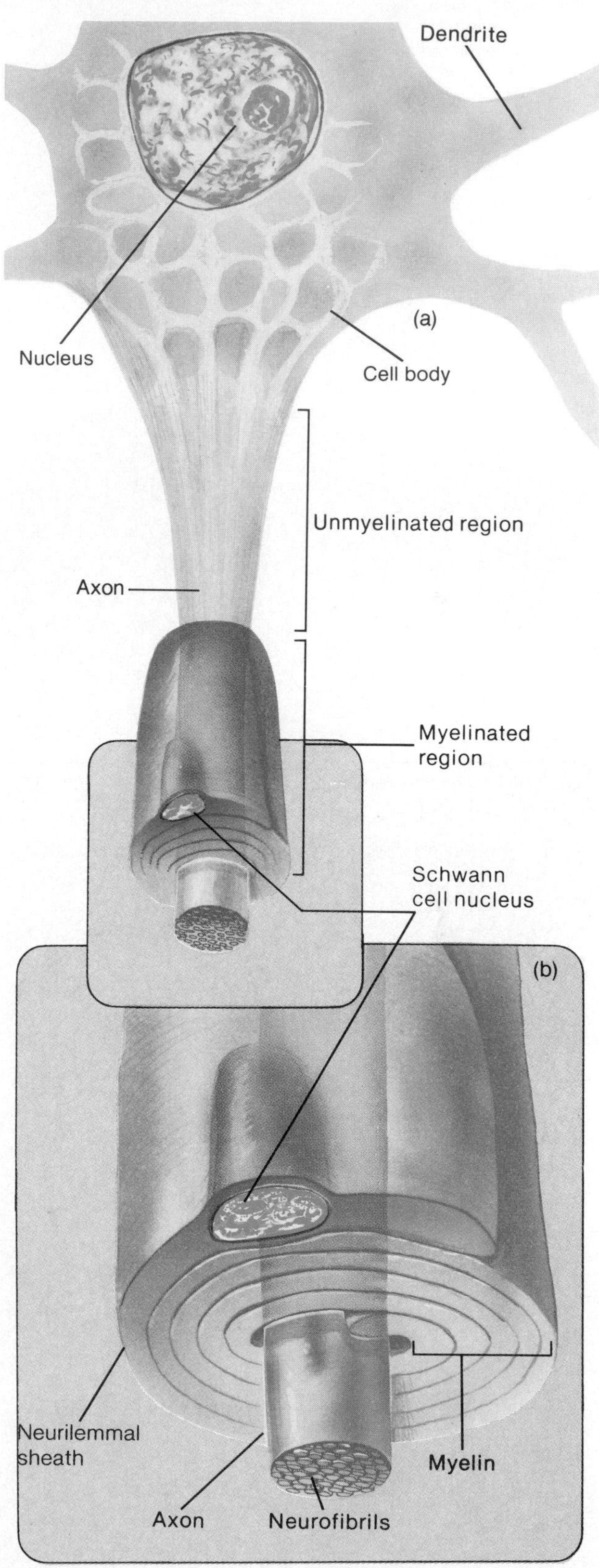

microglia: Gk. *mikros,* small; *glia,* glue
ependyma: Gk. *ependyma,* upper garment
satellite: L. *satelles,* attendant
myelin: Gk. *myelos,* marrow
nodes of Ranvier: from Louis A. Ranvier, French pathologist, 1835–1922

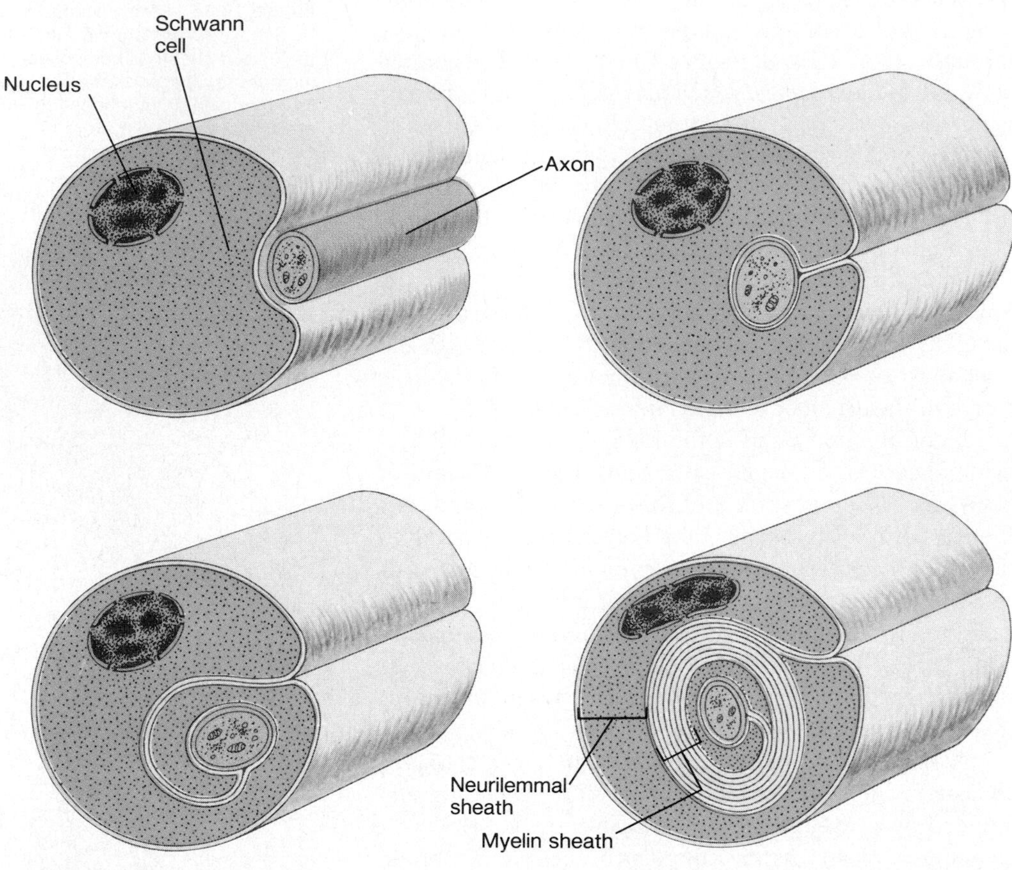

Figure 14.15. The formation of a myelin sheath in a peripheral axon. The myelin sheath is formed by successive wrappings of the Schwann cell membranes, leaving most of the Schwann cell cytoplasm outside the myelin. The neurilemmal sheath of Schwann cells is thus located outside the myelin sheath.

Figure 14.16. An electron micrograph of unmyelinated (*left*) and myelinated axons (*right*).

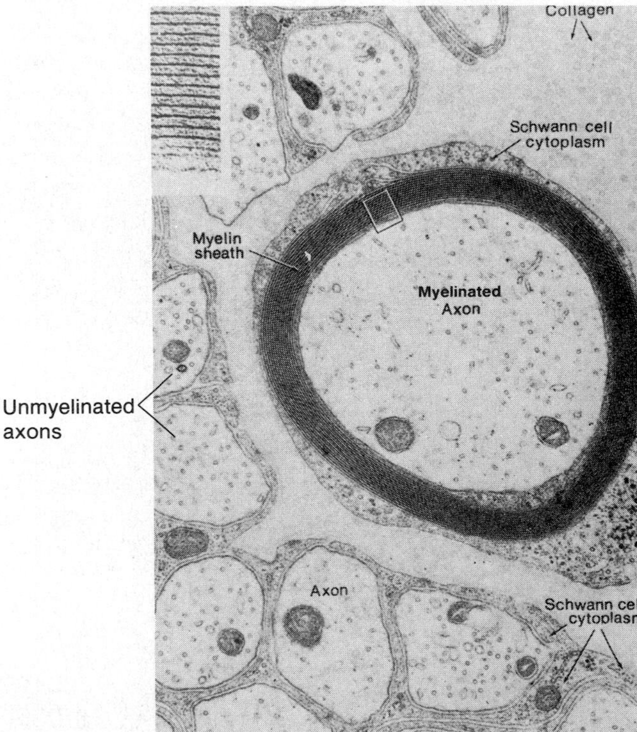

myelin sheaths around axons of the CNS give this tissue a white color; areas of the CNS that contain a high concentration of axons thus form the *white matter.* The *gray matter* of the CNS is composed of high concentrations of cell bodies and dendrites, which lack myelin sheaths.

Regeneration of a Cut Axon. When an axon in a peripheral nerve is cut, the distal portion of the axon that was severed from the cell body degenerates and is phagocytosed by Schwann cells. The Schwann cells, surrounded by the basement membrane, then form a *regeneration tube* (fig. 14.17), as the part of the axon that is connected to the cell body begins to grow and exhibit amoeboid movement. The Schwann cells of the regeneration tube are believed to secrete chemicals that attract the growing axon tip, and the regeneration tube helps to guide the regenerating axon to its proper destination. Even a severed major nerve may be surgically reconnected and the function of the nerve largely reestablished if the surgery is performed before tissue death.

Injury in the CNS stimulates growth of axon collaterals, but central axons have a much more limited ability to regenerate than peripheral axons. This is believed to be primarily due to the absence of a continuous basement membrane, so that a regeneration tube cannot be formed.

Figure 14.17. The process of neuron regeneration. (*a*) if a neuron is severed through a myelinated axon, the proximal portion may survive, but (*b*) the distal portion degenerates through phagocytosis. The myelin sheath provides a pathway for (*c* and *d*) the regeneration of an axon, and (*e*) innervation is restored.

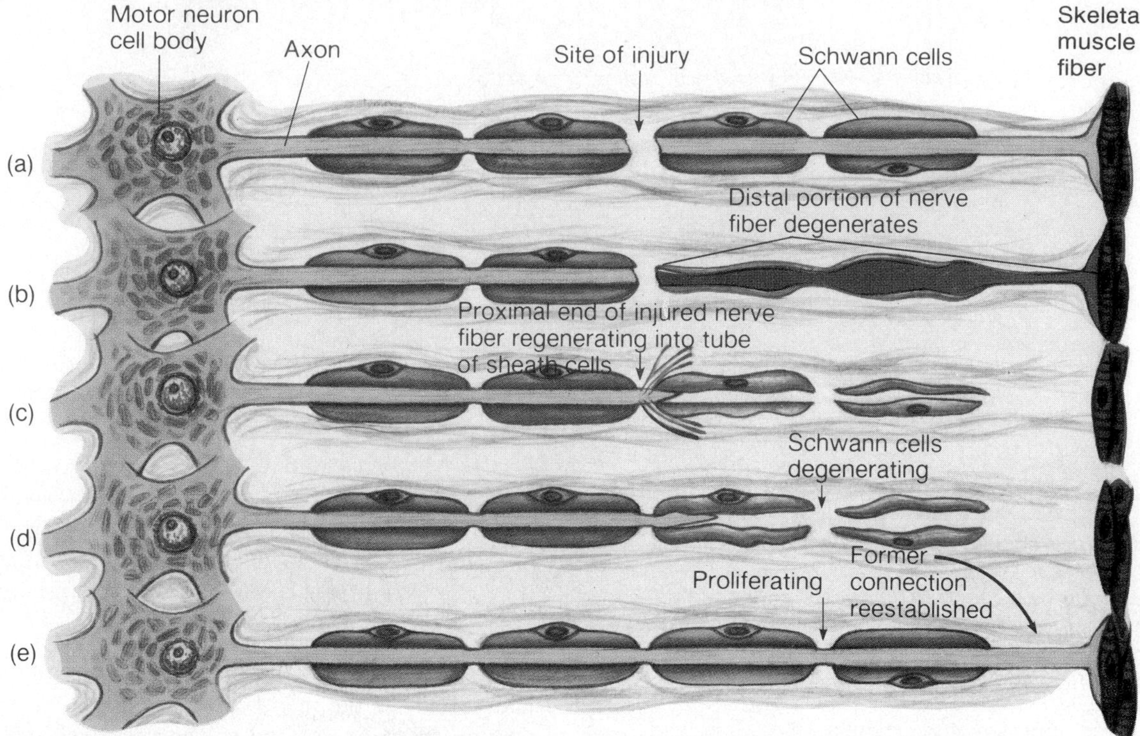

> The lack of a continuous basement membrane around the axons of neurons in the CNS is believed to contribute to the fact that central neurons do not regenerate as readily as peripheral neurons. Experiments *in vitro* suggest that central axons can regenerate if they are provided with the appropriate environment. In a developing fetal brain, chemicals that include *nerve growth factor* promote axon growth, and such chemicals have been shown in some experimental animals to promote neuron regeneration in adult brains. Grafts of fetal brain tissue into adult brains have similar effects. In another approach to the problem, researchers have shown that providing a basement membrane derived from a human amnion (the membrane surrounding a fetus) can promote axon regeneration in the adult brains of experimental animals. With continued experimentation, it may someday be possible to apply these findings clinically in the regeneration of brain tissue that has been damaged by disease or injury.

Astrocytes and the Blood-Brain Barrier. Astrocytes are large, stellate cells with numerous cytoplasmic processes that radiate outward. They are the most abundant of the neuroglial cells in the CNS, constituting up to 90% of the nervous tissue in some areas of the brain.

Capillaries in the brain, unlike those of most other organs, do not have pores between adjacent endothelial cells (the cells that compose the walls of capillaries). Unlike other organs, therefore, the brain cannot obtain molecules from the blood plasma by a nonspecific filtering process. Instead, molecules within brain capillaries must be moved through the endothelial cells by active transport, endocytosis, and exocytosis. This imposes a very selective **blood-brain barrier.** Astrocytes may contribute to this blood-brain barrier because the brain capillaries are surrounded by extensions of astrocytes known as *perivascular feet* (fig. 14.18). Before molecules in the blood can enter neurons in the CNS, they may have to pass through both the endothelial cells of the capillaries and the astrocytes.

> The blood-brain barrier presents difficulties in the chemotherapy of brain diseases because drugs that could enter other organs may not be able to enter the brain. In the treatment of *Parkinson's disease,* for example, patients who need a chemical called dopamine in the brain must be given a precursor molecule called levodopa (L-dopa). This is because dopamine cannot cross the blood-brain barrier, whereas L-dopa can enter the neurons and be changed to dopamine in the brain.

astrocyte: Gk. *aster,* star; *kytos,* hollow (cell)

Figure 14.18. A photomicrograph showing the perivascular feet of astrocytes (a type of neuroglial cell), which cover most of the surface area of brain capillaries.

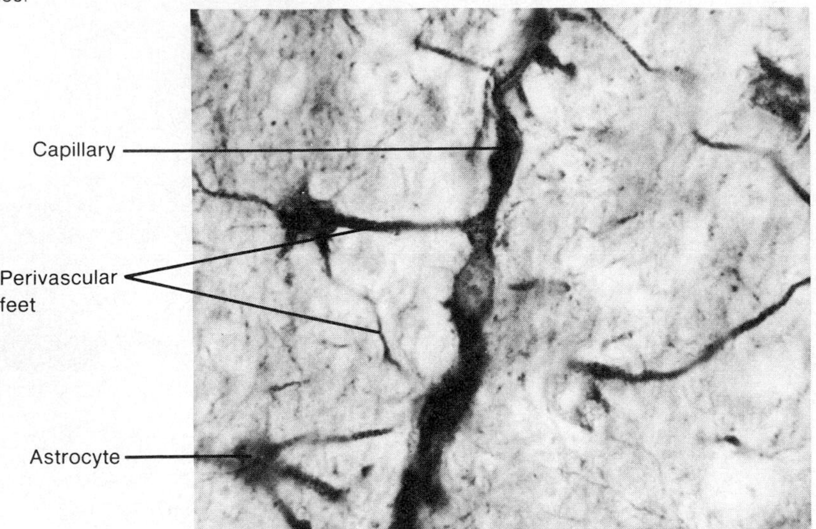

1. Draw a neuron, label its parts, and describe the functions of these parts.
2. Distinguish between sensory neurons, motor neurons, and interneurons. Describe the classifications of these neurons in terms of their number of dendrites.
3. Describe the location and function of microglia and ependyma.
4. Describe the structure of the sheath of Schwann, and explain how a myelin sheath in the PNS is formed.
5. Describe how myelin sheaths in the CNS are formed, how they contribute to the color differences seen in the CNS, and how these sheaths differ from myelin sheaths in the PNS.

Electrical Activity of Axons

The permeability of the cell membrane of axons to specific ions is regulated by membrane gates. Increases in the permeability to these ions at specific locations in the membrane allow diffusion through the membrane, either into or out of the cell, producing changes in the membrane potential, and the nerve impulse.

Objective 12. Describe how voltage-regulated gates for Na^+ and K^+ function when an axon is at rest and when it is depolarized.

Objective 13. Explain how diffusion of Na^+ and K^+ through the axon membrane can result in the action potential.

Objective 14. Describe the all-or-none nature of an action potential, and explain the nature of refractory periods.

Objective 15. Explain how the strength of a stimulus is coded by the conduction of action potentials in an axon.

Objective 16. Describe how action potentials are conducted in an unmyelinated and in a myelinated axon, and explain the differences described.

All cells in the body maintain a potential difference (voltage) across the membrane, or *membrane potential,* in which the inside of the cell is negatively charged in comparison to the outside of the cell (for example, −65 mV). As explained in chapter 4, this potential difference is largely the result of the permeability properties of the cell membrane, which traps large, negatively charged organic molecules within the cell and which permits only limited diffusion of positively charged inorganic ions. Although all cells have a membrane potential, however, only a few types of cells have been shown to alter their membrane potential in response to stimulation. Such alterations in membrane potential are achieved by varying the membrane permeability to specific ions in response to stimulation. A central aspect of the physiology of neurons and muscle cells is their ability to produce and conduct these changes in membrane potential.

An increase in membrane permeability to a specific ion results in the diffusion of that ion down its concentration gradient, either into or out of the cell. These *ion currents* occur only across limited patches of membrane (located fractions of a millimeter apart), where specific ion channels are located. Changes in the potential difference across the membrane at these points can be measured by the voltage developed between two electrodes—one placed inside the cell, the other placed outside the cell membrane at the region being recorded. The voltage between these two *recording electrodes* can be visualized by connecting these electrodes to an oscilloscope (fig. 14.19).

In an oscilloscope, electrons from a cathode ray "gun" are sprayed across a fluorescent screen, producing a line of light. Changes in the potential difference between the

Figure 14.19. The difference in potential (in millivolts) between an intracellular and extracellular recording electrode is displayed on an oscilloscope screen. The resting membrane potential (*rmp*) of the axon may be reduced (depolarization) or increased (hyperpolarization).

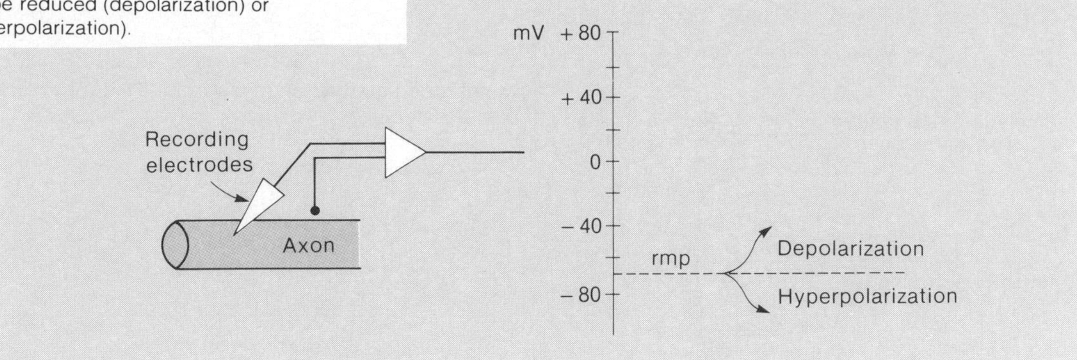

two recording electrodes produce a deflection of this line. The oscilloscope can be calibrated in such a way that an upward deflection of this line indicates that the inside of the membrane has become less negative (or more positive) compared to the outside of the membrane. A downward deflection of the line, conversely, indicates that the inside of the cell has become more negative.

If both recording electrodes were placed outside of the cell, the potential difference between the two would be zero (there is no charge separation). When one of the two electrodes penetrates the cell membrane, the oscilloscope shows that the intracellular electrode is electrically negative with respect to the extracellular electrode. If appropriate stimulation causes positive charges to flow into the cell, the line would deflect upward; this change is called **depolarization** *(de-po″lar-i-za′shun),* or hypopolarization, because the potential difference between the two recording electrodes is reduced. If the inside of the membrane becomes more negative as a result of stimulation, the line on the oscilloscope would deflect downward; this is called **hyperpolarization.**

Ion "Gating" in Nerve Fibers

As discussed in chapter 4, the intracellular potassium concentration is much higher than the concentration of potassium in the extracellular fluid. The sodium concentration, conversely, is higher in extracellular fluid than it is within the cell. These concentration differences are a result of (1) electrical attraction by negatively charged proteins and organic phosphates that are fixed within the cell; (2) the greater permeability of the resting cell membrane to K^+ than to Na^+; and (3) active transport by the Na^+/K^+ pump. Because of the uneven distribution of ions, the inside of a neuron is about 65 millivolts negative (−65 mV) compared to the extracellular fluid (the exact value of this membrane potential is different in different cells).

The permeability of the membrane to Na^+, K^+, and other ions can be regulated by parts of the ion channels through the membrane called **gates.** Gates are believed to be composed of polypeptide chains that can open or close a membrane channel according to specific conditions. When the gates of specific ion channels are closed, the membrane is not very permeable to that ion, and when the gates are opened, the permeability to that ion can be greatly increased.

It is believed that there are two types of channels for K^+; one type lacks gates and is always open, whereas the other type has gates that are closed in the resting cell. Channels for Na^+, in contrast, always have gates, and these gates are closed in the resting cell. The resting cell is thus more permeable to K^+ than to Na^+ (some Na^+ does leak into the cell, as described in chapter 4; this leakage may occur in a nonspecific manner through open K^+ channels).

Whether the gates for the Na^+ and K^+ channels are open or closed depends on the membrane potential. The gated channels are closed at the resting membrane potential of −65 mV, but they open when the membrane is depolarized to a certain threshold level. Since the opening and closing of these gates is regulated by the membrane voltage, the gates are said to be **voltage regulated.**

Depolarization of a small region of an axon can be experimentally induced by a pair of stimulating electrodes. If a pair of recording electrodes is placed in the same region (one electrode within the axon and one outside), an upward deflection of the oscilloscope line will be observed as a result of this depolarization. If a certain level of depolarization (from −65 mV to −55 mV, for example) is achieved by this artificial stimulation, a sudden and very rapid change in the membrane potential will be observed. This is because *depolarization to a threshold level causes the voltage-regulated* Na^+ *gates to open.* Now the permeability properties of the membrane are changed, and Na^+ diffuses down its concentration gradient into the cell.

A fraction of a second after the Na^+ gates open, they close again. At this time, the depolarization stimulus causes the K^+ gates to open. This makes the membrane more permeable to K^+ than it is at rest, and K^+ diffuses down its concentration gradient out of the cell. The K^+ gates will then close, and the permeability properties of the membrane will return to what they were at rest.

Figure 14.20. Depolarization of an axon has two effects: (1) Na^+ gates open and Na^+ diffuses into the cell, and (2) after a brief period, K^+ gates open and K^+ diffuses out of the cell. An inward diffusion of Na^+ causes further depolarization—this causes further opening of Na^+ gates in a positive feedback ⊕ fashion. The opening of K^+ gates and outward diffusion of K^+ make the inside of the cell more negative and thus have a negative feedback effect ⊖ on the initial depolarization.

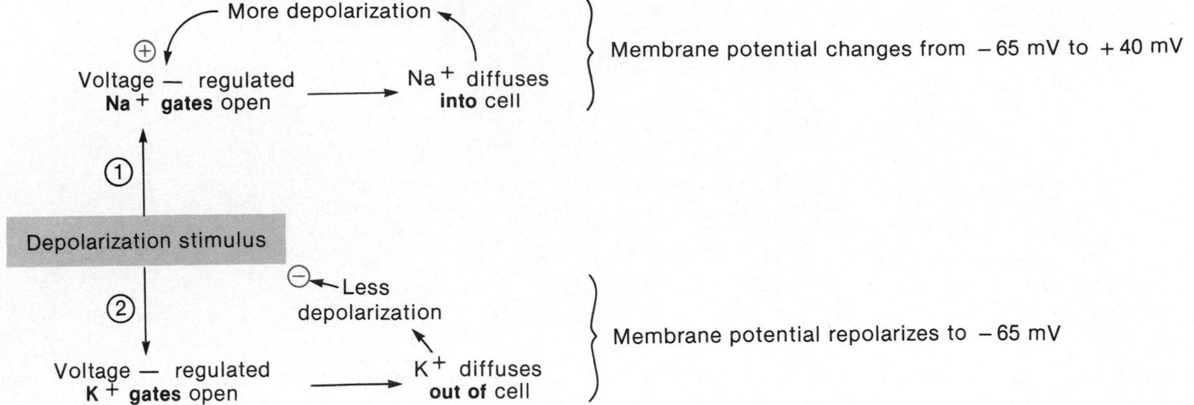

Figure 14.21. An action potential (*upper figure*) is produced by an increase in sodium conductance followed, with a short time delay, by an increase in potassium conductance (*lower figure*). This drives the membrane potential first toward the sodium equilibrium potential and then toward the potassium equilibrium potential.

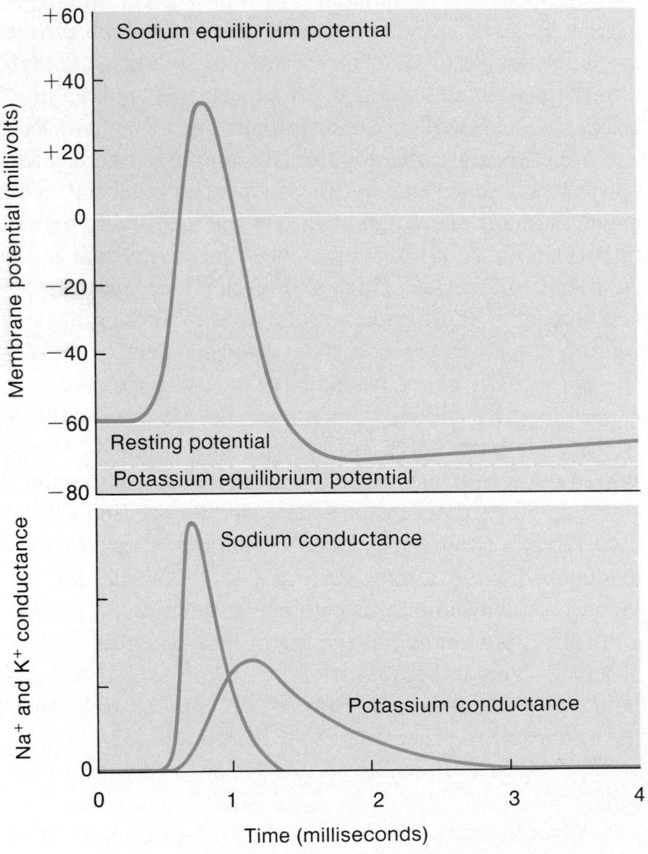

Action Potentials

When the axon membrane has been depolarized to a threshold level—by stimulating electrodes, in the previous example—the Na^+ gates open and the membrane becomes permeable to Na^+. This permits Na^+ to enter the axon by diffusion, which further depolarizes the membrane (makes the inside less negative, or more positive). Since the Na^+ gates of the axon are voltage regulated, this further depolarization makes the membrane even more permeable to Na^+, so that even more Na^+ can enter the cell and open even more voltage-regulated Na^+ gates. A *positive feedback loop* (fig. 14.20) is thus created, which causes the rate of Na^+ entry and depolarization to accelerate in an explosive fashion.

After a slight time delay, depolarization of the axon membrane also causes the opening of voltage-regulated K^+ gates, and the diffusion of K^+ out of the cell. Since K^+ is positively charged, the diffusion of K^+ out of the cell makes the inside of the cell less positive, or more negative, and acts to restore the original resting membrane potential. This process is called **repolarization** and represents the completion of a *negative feedback loop* (fig. 14.20).

Figure 14.21 (bottom) illustrates the conductance of Na^+ and K^+ through the axon membrane in response to a depolarization stimulus. Notice that the explosive increase in Na^+ conductance causes rapid depolarization to 0 mV and then *overshoot* of the membrane potential so that the inside of the membrane actually becomes positively charged (almost +40 mV) compared to the outside (fig. 14.21, top). The Na^+ conductance then rapidly decreases as the conductance of K^+ increases, resulting in repolarization to the resting membrane potential. These changes in Na^+ and K^+ conductances and the resulting changes in the membrane potential that they produce constitute an event called the **action potential,** or **nerve impulse.**

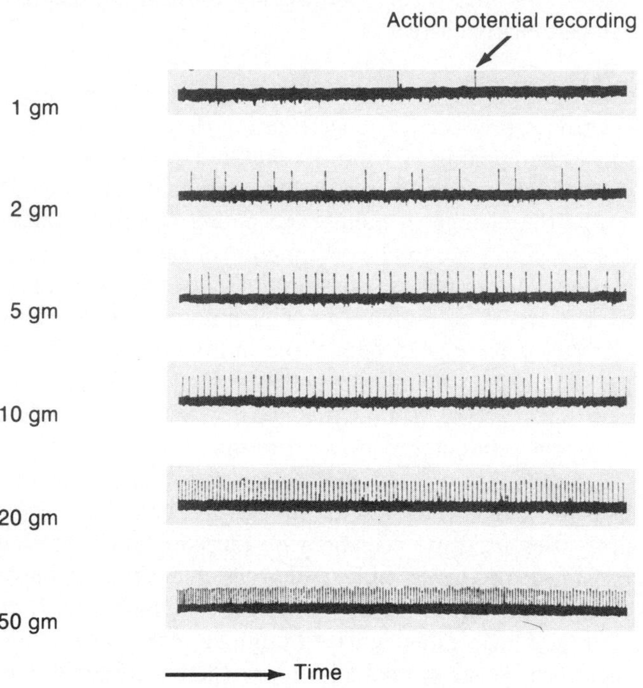

Figure 14.22. Recordings form a single sensory fiber of a sciatic nerve of a frog stimulated by varying degrees of stretch of the gastrocnemius muscle. Note that increasing amounts of stretch (indicated by increasing weights attached to the muscle) result in an increased frequency of action potentials.

Once an action potential is completed, the Na^+/K^+ pumps will, by active transport, extrude the extra Na^+ that has entered the axon and recover the K^+ that has diffused out of the axon. This occurs very quickly because the events described occur across only a very small area of membrane, and so only a relatively small amount of Na^+ and K^+ actually diffuse through the membrane during the production of an action potential. The total concentrations of Na^+ and K^+ in the axon and in the extracellular fluid are not significantly changed during an action potential. Even during the overshoot phase, for example, the concentration of Na^+ remains higher outside the axon; repolarization thus requires the outward diffusion of K^+, which has a concentration gradient in a direction opposite to the Na^+ gradient.

Notice that active transport processes are not directly involved in the production of an action potential; both depolarization and repolarization are produced by the diffusion of ions down their concentration gradients. A neuron poisoned with cyanide, so that it cannot produce ATP, can still produce action potentials for a period of time. After a while, however, the lack of ATP for active transport by the Na^+/K^+ pumps will result in a decline in the ability of the axon to produce action potentials. This shows that the Na^+/K^+ pumps are not directly involved but are instead required to maintain the concentration gradients needed for the diffusion of Na^+ and K^+ during action potentials.

All-or-None Law. Once a region of axon membrane has been depolarized to a threshold value, the positive feedback effect of depolarization on Na^+ permeability and of Na^+ permeability on depolarization causes the membrane potential to shoot toward about +40 mV. It does not normally become more positive because of the fact that the Na^+ gates quickly close and the K^+ gates open. The length of time that the Na^+ and K^+ gates stay open is independent of the strength of the depolarization stimulus.

The amplitude of action potentials is therefore **all-or-none.** When depolarization is below a threshold value, the voltage-regulated gates are closed; when depolarization reaches threshold, a maximum potential change (the action potential) is produced. Since the change from −65 mV to +40 mV and back to −65 mV lasts only about three msec (milliseconds), the image of an action potential on an oscilloscope screen looks like a spike. Action potentials are therefore sometimes called *spike potentials.*

Both the rising (depolarization) and falling (repolarization) phases of an action potential are, in summary, produced by gating the *diffusion* of ions down their opposing concentration gradients. Since the gates are open for a fixed period of time, the duration of each action potential is about the same. Likewise, since the concentration gradient for Na^+ is relatively constant, the amplitude of each action potential is about the same in all axons at all times (from −65 mV to +40 mV, or about 100 mV in total amplitude).

Coding for Stimulus Intensity. If one depolarization stimulus is greater than another, the greater stimulus strength is not coded by a greater amplitude of action potentials. The code for stimulus strength in the nervous system is not *amplitude modulated* **(AM).** When a greater stimulus strength is applied to a neuron, identical action potentials are produced more frequently (more are produced per minute). Therefore, the code for stimulus strength in the nervous system is *frequency modulated* **(FM).** This is shown in figure 14.22.

When an entire collection of axons (in a nerve) is stimulated, different axons will be stimulated at different stimulus intensities. A low-intensity stimulus will only activate those few fibers with low thresholds, whereas high-intensity stimuli can activate fibers with higher thresholds. As the intensity of stimulation increases, more and more fibers will become activated. This process, called **recruitment,** represents another mechanism by which the nervous system can code for stimulus strength.

Refractory Periods. If a stimulus of a given intensity is maintained at one point of an axon and depolarizes it to threshold, action potentials will be produced at that point at a given frequency (number per minute). As the stimulus strength is increased, the frequency of action potentials produced at that point will increase accordingly. As

Figure 14.23. The absolute and relative refractory periods. While a segment of axon is producing an action potential, the membrane is absolutely or relatively resistant (refractory) to further stimulation.

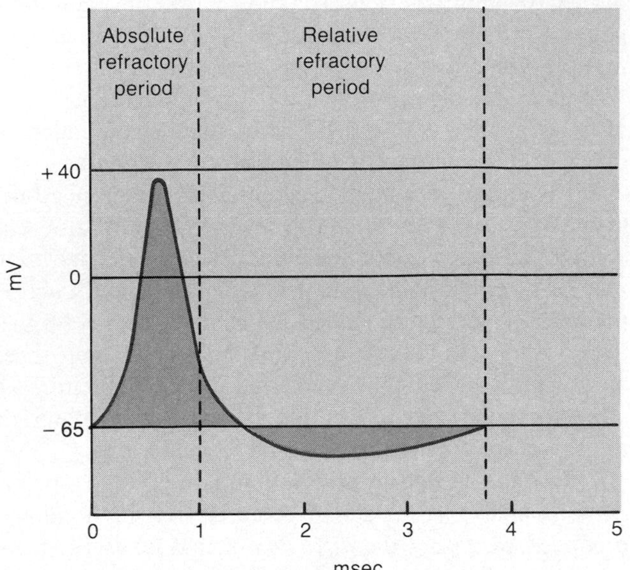

action potentials are produced with increasing frequency, the time between successive action potentials will decrease—but only up to a minimum time interval. The interval between successive action potentials will never become so short that one action potential is produced before the preceding one has finished.

During the time that a patch of axon membrane is producing an action potential, it is incapable of responding—or *refractory (re-frak'to-re)*—to further stimulation. If a second stimulus is applied, for example, while the Na^+ gates are open in response to a first stimulus, the second stimulus cannot have any effect (the gates are already open). During the time that the Na^+ gates are open, therefore, the membrane is in an **absolute refractory period** and cannot respond to any subsequent stimulus. If a second stimulus is applied while the K^+ gates are open (and the membrane is in the process of repolarizing), the membrane is in a **relative refractory period.** During this time only a very strong stimulus can depolarize the membrane and produce a second action potential (fig. 14.23).

As a result of the fact that the cell membrane is refractory during the time it is producing an action potential, each action potential remains a separate, all-or-none event. In this way, as a continuously applied stimulus increases in intensity, its strength can be coded strictly by the frequency of the action potential it produces at each point of the axon membrane.

One might think that as an axon produces a large number of action potentials, the relative concentrations of Na^+ and K^+ would be changed in the extracellular and intracellular compartments. This is not true. In a typical mammalian axon that is one micrometer in diameter, for example, only one intracellular K^+ ion in three thousand would be exchanged for an Na^+. Since a typical neuron has about one million Na^+/K^+ pumps, able to transport nearly 200 million ions per second, these small changes can be quickly corrected.

Cable Properties of Neurons. If a pair of stimulating electrodes produces a depolarization that is too weak to cause the opening of voltage-regulated Na^+ gates—that is, if the depolarization is below threshold (about −55 mV)—the change in membrane potential will be *localized* to within one to two millimeters of the point of stimulation. For example, if the stimulus causes depolarization from −65 mV to −60 mV at one point, and the recording electrodes are placed only three millimeters away from the stimulus, the membrane potential recorded will remain at −65 mV (the resting potential). The axon is thus a very poor conductor, compared to metal wires.

The ability of a neuron to transmit charges through its cytoplasm is known as its *cable properties.* These cable properties are quite poor because there is a high internal resistance to the spread of charges and because many charges leak out of the axon through its membrane. If an axon had to conduct only through its cable properties, therefore, no axon could be more than a millimeter in length. The fact that some axons are a meter or more in length suggests that the conduction of nerve impulses does not rely on the cable properties of the axon.

Conduction of Nerve Impulses

When stimulating electrodes artificially depolarize one point of an axon membrane to a threshold level, voltage-regulated gates open and an action potential is produced at that small region of axon membrane containing those gates. For about the first millisecond of the action potential, when the membrane voltage changes from −65 mV to +40 mV, a current of Na^+ enters the cell by diffusion due to the opening of the Na^+ gates. Each action potential thus "injects" positive charges (Na^+ ions) into the axon.

These positively charged Na^+ ions are conducted, by the cable properties of the axon, to an adjacent region that still has a membrane potential of −65 mV. Within the limits of the cable properties of the axon (one to two millimeters), this helps to depolarize the adjacent region of axon membrane. When this adjacent region of membrane reaches a threshold level of depolarization, it too produces an action potential as its voltage-regulated gates open.

Each action potential thus acts as a stimulus for the production of another action potential at the next region of membrane that contains voltage-regulated gates. In the previous description of action potentials, the stimulus for their production was artificial—depolarization produced by a pair of stimulating electrodes. Now it can be seen

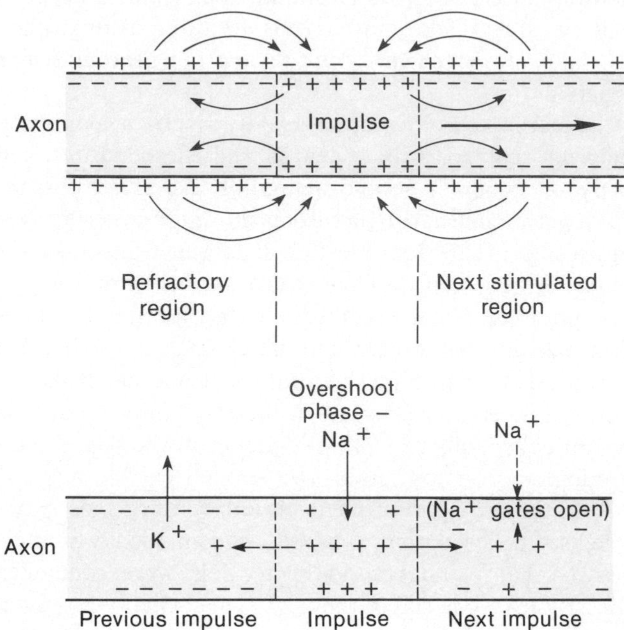

Figure 14.24. The conduction of a nerve impulse (action potential) in an unmyelinated nerve fiber (axon). Each action potential "injects" positive charges that spread to adjacent regions. The region that has previously produced an action potential is refractory. The previously unstimulated region is partially depolarized. As a result, its voltage-regulated Na^+ gates open, repeating the process.

that an action potential at one point along an axon is produced by depolarization that results from the production of a preceding action potential. This line of reasoning explains how all action potentials along an axon are produced after the first action potential is generated.

Conduction in an Unmyelinated Nerve Fiber. In an unmyelinated axon, every patch of membrane that contains Na^+ and K^+ gates can produce an action potential. Action potentials are thus produced at locations only a fraction of a micrometer apart all along the length of the axon.

The cablelike spread of depolarization induced by Na^+ influx during one action potential helps to depolarize the adjacent regions of membrane. This process is also aided by movements of ions on the outer surface of the axon membrane (fig. 14.24). This process would depolarize the adjacent membranes on each side of the region producing an action potential, but the area that had previously produced an action potential cannot produce a new one at this time because it is still in its refractory period. In this way, action potentials are passed in one direction only along the axon.

Notice that action potentials are not really conducted, although it is convenient to use that word. Each action potential is a separate, complete event (flow of Na^+ and K^+ across the membrane to produce an all-or-none change in potential at a limited region of membrane) that is repeated, or *regenerated,* along the axon's length. The

Figure 14.25. The conduction of the nerve impulse in a myelinated nerve fiber. Since the myelin sheath prevents inward Na^+ current, action potentials can only be produced at the interruptions in the myelin sheath, or nodes of Ranvier. This "leaping" of the action potential from node to node is known as saltatory conduction.

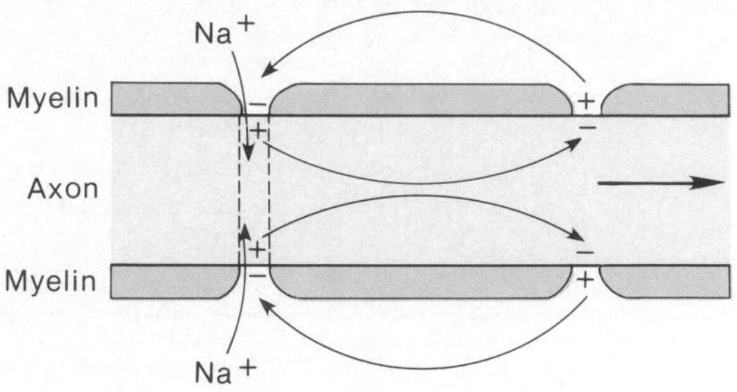

action potential produced at the end of the axon is thus a completely new event that was produced in response to depolarization from the previous action potential. The last action potential has the same amplitude as the first. Action potentials are thus said to be **conducted without decrement** (without decreasing in amplitude).

The spread of depolarization by the cable properties of an axon is fast compared to the time involved in producing an action potential. Since action potentials are produced at every fraction of a micrometer in an unmyelinated axon, the conduction rate is relatively slow. This conduction rate is somewhat faster if the unmyelinated axon is thicker, because the ability of fibers to conduct charges by cable properties improves with increasing diameter. The conduction rate is substantially faster if the axon is myelinated.

Conduction in a Myelinated Axon. The myelin sheath provides insulation for the axon, preventing movements of Na^+ and K^+ through the membrane. If the myelin sheath were continuous, therefore, action potentials could not be produced. Fortunately, there are interruptions in the myelin known as the *nodes of Ranvier.*

Since the cable properties of axons can only conduct depolarizations a very short distance (1 mm–2 mm), it follows that the nodes of Ranvier must be very close together (usually they are about 1 mm apart). Studies have shown that Na^+ channels are highly concentrated at the nodes (estimated at 10,000 per square micrometer) and almost absent in the regions of axon membrane between the nodes. Action potentials, therefore, occur only at the nodes of Ranvier (fig. 14.25) and seem to "leap" from node to node; this is called **saltatory conduction** *(sal'tah-to"re).*

Since the cablelike spread of depolarization between the nodes is very fast and fewer action potentials need to be produced per given length of axon, saltatory conduction allows a *faster rate of conduction* than is possible in an unmyelinated fiber. Conduction rates in the human

saltatory: L. *saltatio,* to leap

Table 14.3 Examples of conduction velocities and functions of mammalian nerves of different diameters

Diameter (μm)	Conduction velocity (m/sec)	Examples of functions served
12–22	70–120	Somatic motor fibers
5–13	30–70	Sensory: muscle position pressure
3–8	15–40	Sensory: touch, pressure
1–5	12–30	Sensory: pain, temperature
1–3	3–15	Autonomic fibers to ganglia
0.3–1.3	0.7–2.2	Autonomic fibers to muscles

nervous system vary from 1.0 m/sec—in thin, unmyelinated fibers that mediate slow, visceral responses—to greater than 100 m/sec (225 miles per hour)—in thick myelinated fibers involved in quick stretch reflexes in skeletal muscles (table 14.3).

1. Define the meaning of the terms *depolarization* and *repolarization,* and illustrate these processes with a figure.
2. Describe how the permeability of the axon membrane to Na^+ and K^+ is regulated and how changes in permeability to these ions affects the membrane potential.
3. Describe how gating of Na^+ and K^+ in the axon membrane results in the production of an action potential.
4. Define the all-or-none law of action potentials, and describe the effect of increased stimulus strength on action potential production. Explain how the refractory periods affect the maximum frequency of action potential production.
5. Describe how action potentials are conducted by unmyelinated nerve fibers, and explain why saltatory conduction in myelinated fibers is more rapid.

Synaptic Transmission

Synaptic transmission is either electrical or chemical. The predominant type are chemical synapses, in which neurotransmitters are released from the endings of nerve fibers in response to stimulation by action potentials.

Objective 17. Describe the different types of synapses.
Objective 18. Describe the structure and function of gap junctions.
Objective 19. Describe the structure of presynaptic nerve endings of chemical synapses, and explain how neurotransmitters are released.

A **synapse** *(sin'aps)* is the functional connection between a neuron and a second cell. In the CNS this other cell is also a neuron. In the PNS the other cell may be a neuron or an *effector cell* within either a muscle or a gland. Although the physiology of neuron-neuron synapses and neuron-muscle synapses is similar, the synapses between neurons and skeletal muscle cells are often distinguished by the names **myoneural** *(mi''o-nu'ral)* or **neuromuscular junctions.**

Neuron-neuron synapses usually involve a connection between the axon of one neuron and the dendrites, cell body, or axon of a second neuron. These are called, respectively, **axodendritic, axosomatic,** and **axoaxonic synapses** (fig. 14.26). In almost all synapses, transmission is in one direction only—from the axon of the first (or presynaptic) cell to the second (or postsynaptic) cell. **Dendrodendritic synapses** do not fit this classic pattern; in these synapses, two dendrites from different neurons make reciprocal innervations—some of these synapses conduct in one direction, whereas others conduct in the opposite direction.

In the early part of the twentieth century, most physiologists believed that synaptic transmission was *electrical*—that is, that action potentials were conducted directly from one cell to the next. This belief was reasonable in view of the facts that nerve endings appeared to touch the postsynaptic cells and that the delay in synaptic conduction was extremely short (about 0.5 msec). Improved histological techniques, however, revealed tiny gaps in the synapses, and experiments demonstrated that the actions of autonomic nerves could be duplicated by certain chemicals. This led to the suspicion that synaptic transmission might be *chemical*—that the presynaptic nerve endings might release chemical **neurotransmitters** that stimulated action potentials in the postsynaptic cells.

In 1921, a physiologist named Otto Loewi published the results of an experiment suggesting that synaptic transmission was chemical, at least at the junction between a branch of the parasympathetic vagus nerve and the heart. He had isolated the heart of a frog and—while stimulating the branch of the vagus that innervates the heart—perfused the heart with an isotonic salt solution. Stimulation of this nerve slowed the heart rate, as expected. More important, application of this salt solution to the heart of a second frog caused the second heart to slow its rate of beat.

Loewi concluded that the nerve endings of the vagus must have released a chemical—which he called *vagusstoff (va'gus-tof)*—that inhibited the heart rate. This chemical was subsequently identified as **acetylcholine** *(as''ĕ-til-ko'len),* or **ACh.** In the decades following Loewi's discovery, many other examples of chemical synapses were discovered, and the theory of electrical synaptic transmission fell into disrepute. More recent evidence,

Otto Loewi: German pharmacologist (1873–1961)

Figure 14.26. Different types of synapses: (*a*) axodendritic, (*b*) axoaxonic, (*c*) dendrodendritic, and (*d*) axosomatic.

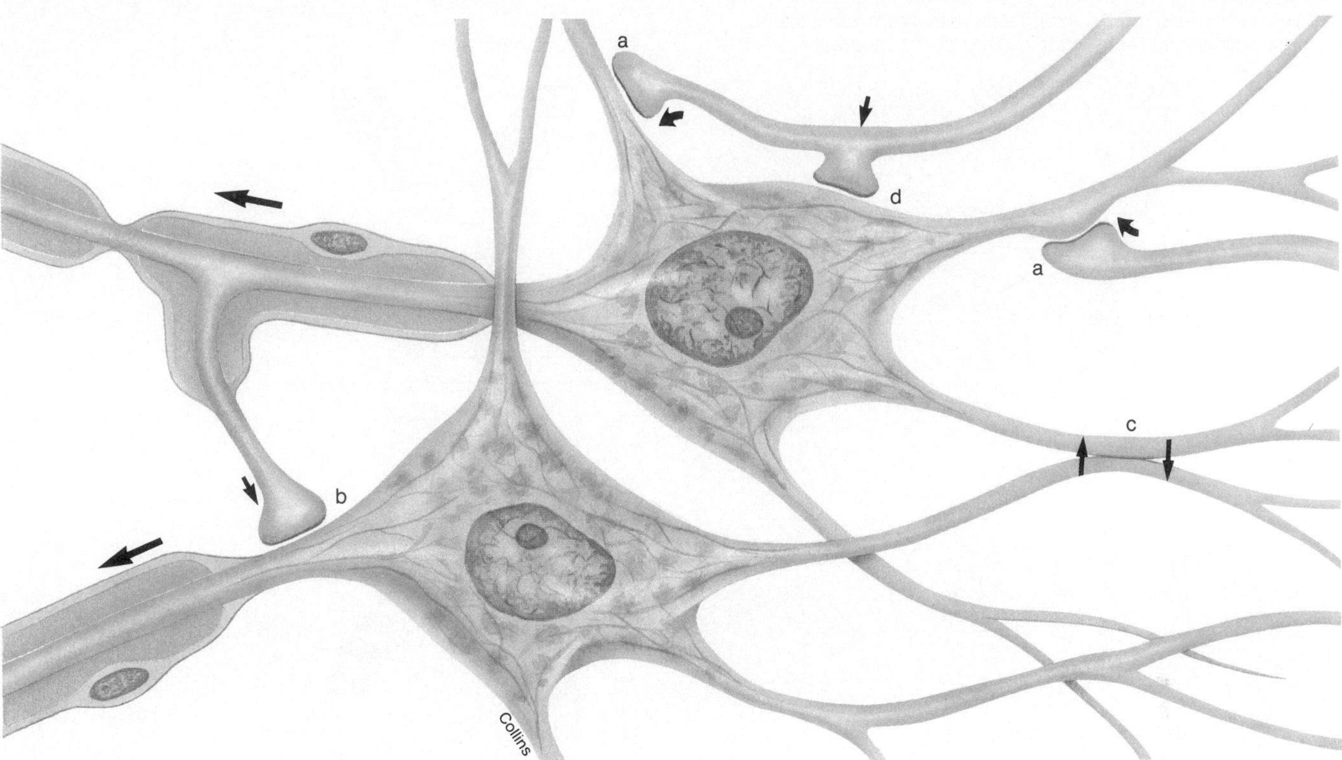

ironically, has shown that electrical synapses do exist (though they are the exception) in the nervous system, within smooth muscles, and between cardiac cells in the heart.

Electrical Synapses: Gap Junctions

In order for two cells to be electrically coupled they must be approximately equal in size and must be joined by areas of contact with low electrical resistance. In this way, impulses can be regenerated from one cell to the next without interruption.

Adjacent cells that are electrically coupled are joined together by **gap junctions.** In gap junctions, the membranes of the two cells are separated by only two nanometers (1 nanometer equals 10^{-9} meter). When gap junctions are observed in the electron microscope from a surface view, hexagonal arrays of particles are seen that are believed to be channels through which ions and molecules may pass from one cell to the next (fig. 14.27).

Gap junctions are present in smooth and cardiac muscle, where they allow excitation and rhythmic contraction of large masses of muscle cells. Gap junctions have also been observed in various regions of the brain. Although their functional significance here is unknown, it has been speculated that they may allow a two-way transmission of impulses (in contrast to chemical synapses, which are always one-way). Gap junctions have also been observed between neuroglial cells, which do not produce electrical impulses; perhaps these act as channels for the passage of informational molecules between cells. It is interesting in this regard that many embryonic tissues have gap junctions that disappear as the tissue becomes more specialized.

Chemical Synapses

Transmission across the majority of synapses in the nervous system is one-way and occurs through the release of chemical neurotransmitters from presynaptic axon endings. These presynaptic endings, which are called **terminal boutons** *(boo-tanz')* because of their swollen appearance, are separated from the postsynaptic cell by a **synaptic cleft** so narrow that it can only be seen clearly with an electron microscope (fig. 14.28).

Neurotransmitter molecules within the presynaptic neuron endings are contained within many small, membrane-enclosed **synaptic vesicles.** In order for the neurotransmitter within these vesicles to be released into the synaptic cleft, the vesicle membrane must fuse with the axon membrane in the process of *exocytosis (eks''o-si-to'sis)*. The neurotransmitter is released in *quanta*—that

bouton: Fr. *bouton,* button
exocytosis: Gk. *exo,* outside; *kytos,* hollow (cell); *osis,* a condition

Figure 14.27. Gap junctions are shown in electron micrographs, (*a*) and (*b*). (*a*) the photograph shows that cell membranes of two cells are fused together in the gap junction; (*b*) a surface view of a gap junction is seen. (*c*) the information presented in these and other electron micrographs is interpreted by the illustration.

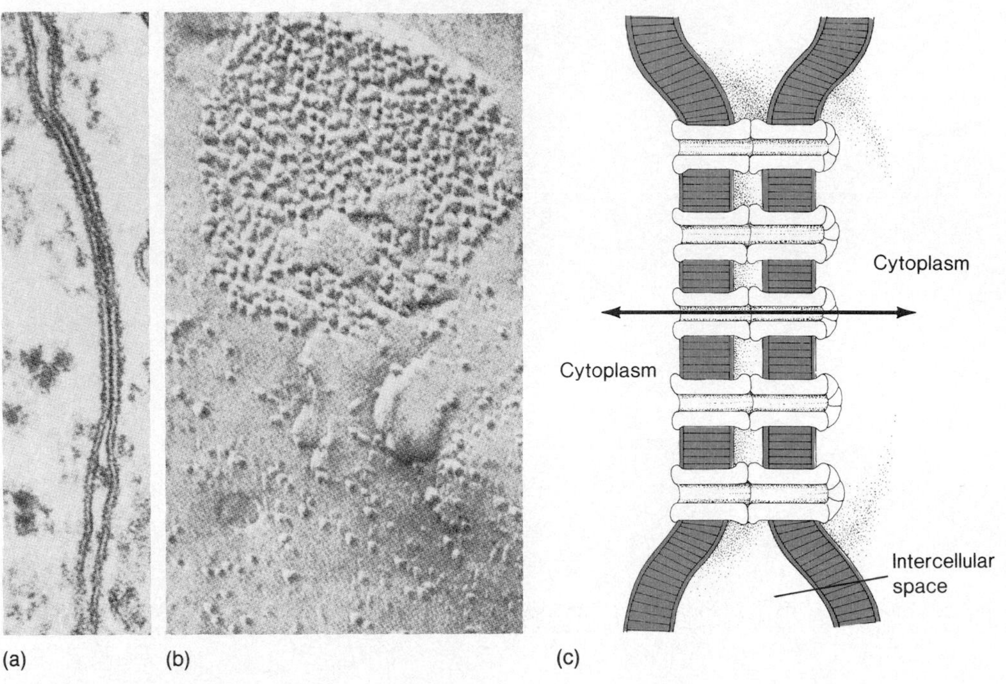

Figure 14.28. An electron micrograph of a chemical synapse, showing synaptic vesicles at the end of an axon.

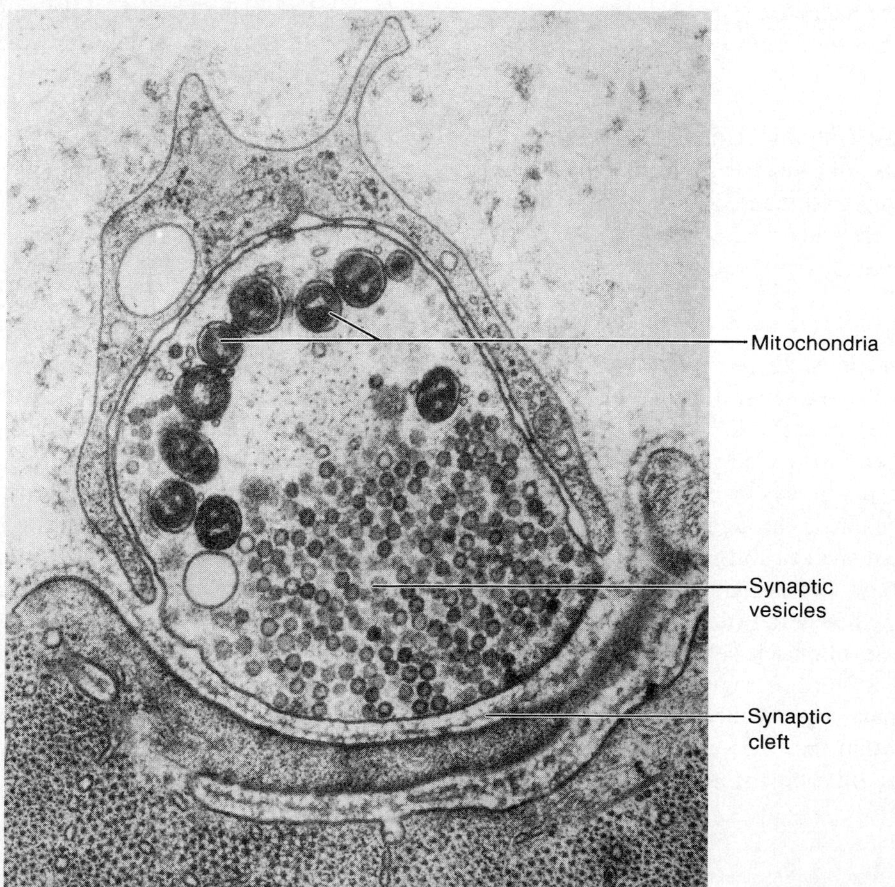

is, in multiples of the amount contained in one vesicle. The number of vesicles that undergo exocytosis is directly related to the frequency of action potentials produced at the presynaptic axon ending.

The release of neurotransmitters following electrical excitation of the presynaptic endings accounts for most of the 0.5 msec delay in synaptic transmission. During this time interval there is a sudden, transient inflow of Ca^{++} into the presynaptic endings. This inflow of Ca^{++} is apparently due to opening of Ca^{++} gates in response to electrical excitation and is required for the release of neurotransmitters. The Ca^{++} is believed to activate previously inactive enzymes within the axon terminals. This process may produce a change in the ratio of phospholipids in the axon membrane, which may change the fluidity of the membrane and thus allow exocytosis to occur. Recent evidence has suggested, alternatively, that the Ca^{++} influx may stimulate the production of specific transporters of ACh in the axon membrane. By either mechanism, Ca^{++} serves to couple electrical excitation of the axon ending to the release of neurotransmitter.

1. Define a synapse, and describe the structures that form synapses.
2. Describe the structure, locations, and functions of gap junctions.
3. Explain how neurotransmitter chemicals are stored and released by presynaptic neuron endings.

ACETYLCHOLINE

Synapses that use acetylcholine (ACh) as a neurotransmitter combine with receptor proteins in the postsynaptic membrane and electrically activate the postsynaptic cell by causing the chemically regulated Na^+ and K^+ gates to open. This causes the production of small, graded depolarizations, which in turn can stimulate the opening of voltage-regulated gates and the production of action potentials.

Objective 20. Describe how ACh is released from axon endings, and describe the fate of ACh after being released.

Objective 21. Explain how ACh stimulates the production of end plate potentials, or EPSPs, in the postsynaptic cell.

Objective 22. Compare the characteristics of EPSPs and action potentials.

Objective 23. Explain how EPSPs cause action potentials to be produced.

Acetylcholine (ACh) is used as a neurotransmitter by some neurons in the CNS, by somatic motor neurons at the neuromuscular junction, and by parasympathetic nerve endings. The effects of this chemical are excitatory in the first two synapses and either excitatory or inhibitory in the third. The excitatory effects of ACh will be discussed in this section; its inhibitory effects are considered in a later section on synaptic integration.

Acetylcholine at the Neuromuscular Junction

The neuromuscular junction—the synapse between a somatic motor neuron and a skeletal muscle fiber—is the most accessible synapse to study and thus the best understood. The presynaptic neuron endings have synaptic vesicles, which each contain about ten thousand molecules of ACh. Once these molecules are released by exocytosis, they quickly diffuse across the narrow synaptic cleft to the membrane of the postsynaptic cell. Here they chemically bond to **receptor proteins** that are built into the postsynaptic membrane. These receptor proteins combine with ACh in a specific manner, analogous to the specific interaction between transport proteins and their substrates.

Acetylcholine is not, however, transported into the postsynaptic cell. Instead, the interaction between ACh and its receptor proteins causes changes in membrane structure that result in the opening of gates that regulate the membrane's permeability to Na^+ and K^+. These are **chemically regulated gates** and contrast with the voltage-regulated gates previously discussed in the axon, which open in response to depolarization.

In contrast to voltage-regulated gates, where the outward flow of K^+ occurs after the inward flow of Na^+, chemically regulated gates allow the simultaneous diffusion of Na^+ and K^+ (fig. 14.29). The depolarizing effect of Na^+ diffusion predominates because the electrochemical gradient for Na^+ is greater than for K^+. The outflow of K^+, however, does prevent the "overshoot" characteristic of action potentials—the membrane potential can reach 0 mV but cannot reverse polarity (as occurs in action potentials). Because of the characteristics of chemically regulated gates, neurotransmitters do not directly produce action potentials. They can only produce depolarization, which may stimulate the opening of voltage-regulated gates and production of action potentials a short distance away from the site of the synapse.

Acetylcholinesterase. The bond between ACh and its receptor protein exists for only a brief instant. The ACh-receptor complex quickly dissociates but can be quickly reformed as long as free ACh is in the vicinity. In order for activity in the postsynaptic cell to be controlled (in this case, for control of skeletal muscle contraction), free ACh must be inactivated very soon after it is released. The inactivation of ACh is achieved by means of an enzyme called **acetylcholinesterase** *(as″ĕ-til-ko″lin-es′ter-ās),* or **AChE,** which is on the postsynaptic membrane or immediately outside the membrane, with its active site facing the synaptic cleft (fig. 14.30).

Figure 14.29. (*a* and *b*) the binding of acetylcholine to receptor proteins causes the opening of chemically regulated gates in the postsynaptic membrane. (*c*) this results in the increased diffusion of Na^+ and K^+ through the membrane.

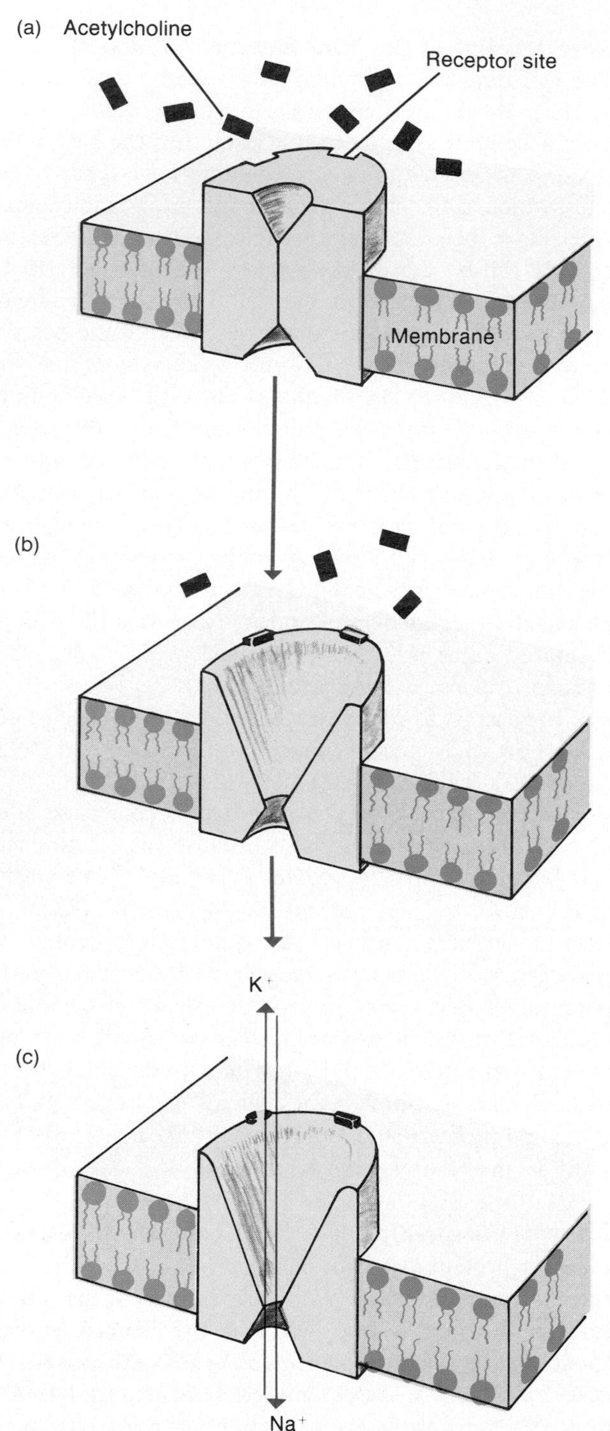

Nerve gas exerts its odious effects by inhibiting AChE in skeletal muscles. Since ACh is not degraded, it can continue to combine with receptor proteins and can continue to stimulate the postsynaptic cell, leading to spastic paralysis. Clinically, cholinesterase inhibitors (such as *neostigmine [ne''o-stig'min]*) are used to enhance the effects of ACh on muscle contraction when neuromuscular transmission is weak, as in the disease *myasthenia gravis.*

End Plate Potential. The interaction of ACh with its receptors in the muscle cell membrane opens chemically regulated gates and depolarizes that region of membrane. The area of the muscle cell membrane innervated by somatic motor nerve endings and affected in this way by ACh is called a **motor end plate.** Depolarizations of this membrane in response to ACh are known as **end plate potentials.** The depolarizations of the end plate potentials do not overshoot 0 mV, for reasons previously discussed, and differ from action potentials in a number of other respects.

Unlike action potentials, end plate potentials have *no threshold;* a single quantum of ACh (released from a single synaptic vesicle) produces a tiny depolarization of the end plate. When more quanta of ACh are released, the depolarization of the motor end plate is correspondingly greater. End plate potentials are therefore *graded* in magnitude, unlike all-or-none action potentials. Since end plate potentials can be graded, they are capable of *summation.* This is quite different from action potentials, which are prevented from summating by their all-or-none nature and by the presence of refractory periods.

When the end plate becomes sufficiently depolarized, the depolarization serves as a stimulus for the opening of voltage-regulated gates in adjacent areas of the skeletal muscle cell membrane. The opening of voltage-regulated gates causes action potentials to be produced. These action potentials in turn serve as stimuli for the production of other action potentials in the next area of membrane. Skeletal muscle fibers conduct action potentials in the same manner as unmyelinated axons. Electrical excitation of a muscle fiber stimulates muscle contraction.

If any stage in the process of neuromuscular transmission is blocked, muscle weakness—sometimes leading to paralysis and death—may result. The drug *curare (koo-rah're),* for example, competes with ACh for attachment to the receptor proteins and thus reduces the size of the end plate potentials. This drug was first used on poison darts by South American Indians because it produced flaccid paralysis of their victims. Clinically, curare is used as a muscle relaxant during anesthesia and to prevent muscle damage during electroconvulsive therapy.

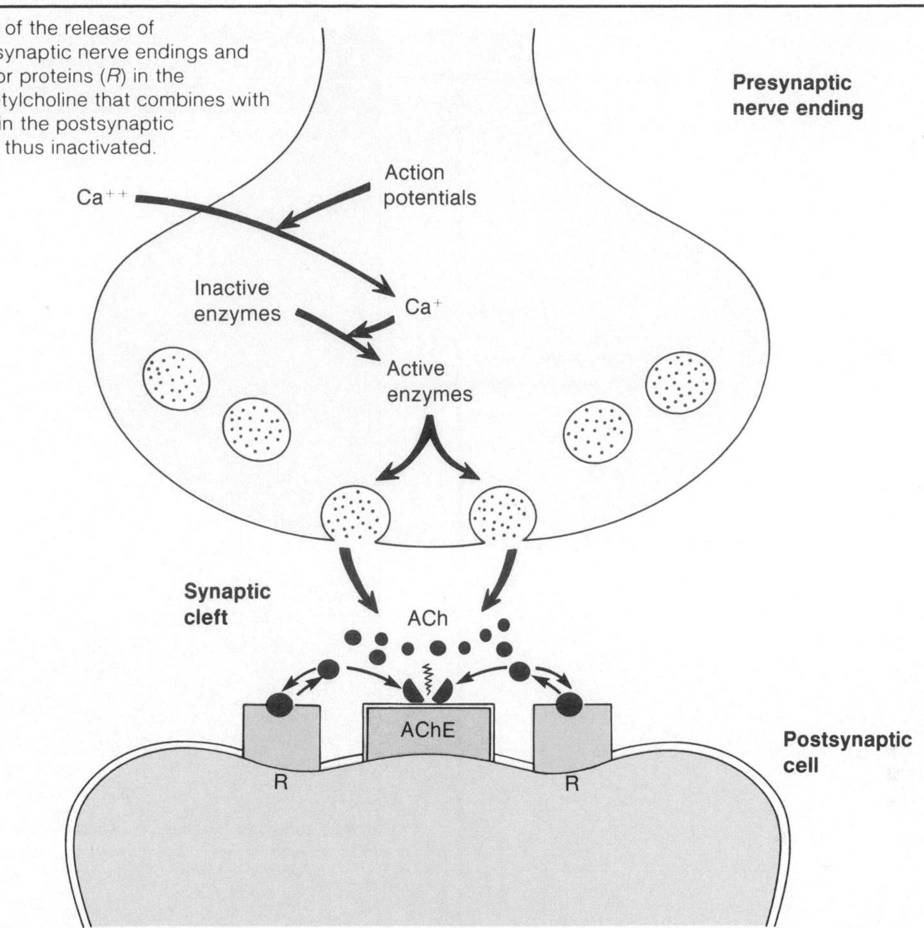

Figure 14.30. Mechanisms of the release of acetylcholine (*ACh*) from presynaptic nerve endings and the binding of ACh to receptor proteins (*R*) in the postsynaptic membrane. Acetylcholine that combines with acetylcholinesterase (*AChE*) in the postsynaptic membrane is hydrolyzed and thus inactivated.

Table 14.4 Derivation and effect of various drugs that affect neural control of skeletal muscles

Drug	Origin	Effects
Botulinus toxin	Produced by *Clostridium botulinum* (bacteria)	Inhibits release of acetylcholine (ACh)
Curare	Resin from a South American tree	Prevents interaction of ACh with the postsynaptic receptor protein
α-bungarotoxin	Venom of *Bungarus* snakes	Binds to ACh receptor proteins
Saxitoxin	Red tide *(Gonyaulax)*	Blocks voltage-regulated Na^+ channels
Tetrodotoxin	Pufferfish	Blocks voltage-regulated Na^+ channels
Nerve gas	Artificial	Inhibits acetylcholinesterase in postsynaptic cell
Prostigmine	Nigerian bean	Inhibits acetylcholinesterase in postsynaptic cell
Strychnine	Seeds of an Asian tree	Prevents IPSPs in spinal cord that inhibit contraction of antagonistic muscles

Muscle weakness in the disease *myasthenia (mi''as-the'ne-ah) gravis* is due to the fact that ACh receptors are blocked by antibodies secreted by the immune system of the affected person. Paralysis in people who eat shellfish poisoned with saxitoxin, produced by unicellular organisms that cause the red tides, results from inhibition of the chemically regulated gates. The effects of these and other poisons on neuromuscular transmission are summarized in table 14.4.

Acetylcholine in Neuron-Neuron Synapses

Within the nervous system, the axon terminals of one neuron typically synapse with the dendrites or cell body of another. The dendrites and cell body thus serve as the receptive area of the neuron, and it is in these regions that receptor proteins for neurotransmitters and chemically regulated gates are located. The first voltage-regulated gates are located at the beginning of the axon, at the axon hillock. It is here that action potentials are first produced (fig. 14.31).

myasthenia: Gk. *myos,* muscle; *asthenia,* weakness

Figure 14.31. A diagram illustrating the functional specialization of different regions in a "typical" neuron.

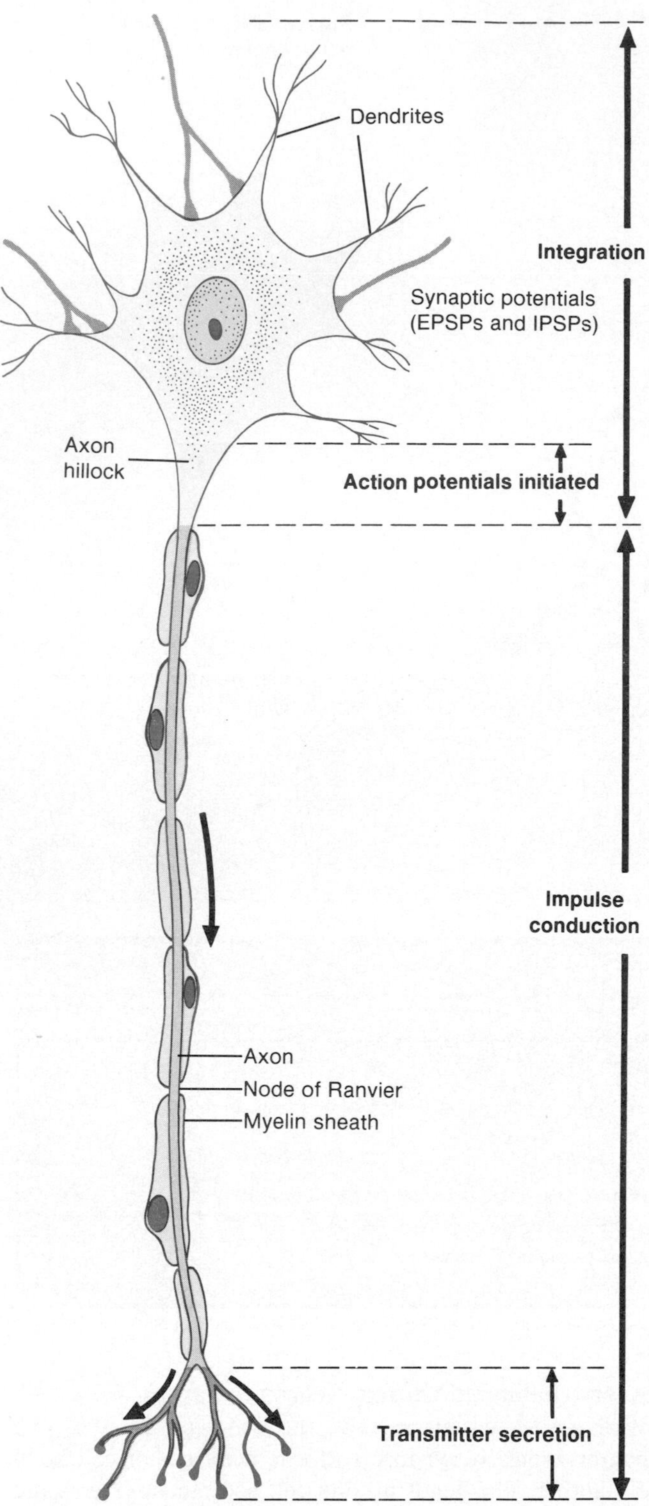

Table 14.5 Comparison of action potentials with synaptic potentials at the neuromuscular junction

Characteristic	Action potential	Excitatory postsynaptic potential
Stimulus for opening of ionic gates	Depolarization	Acetylcholine (ACh)
Initial effect of stimulus	Na^+ gates open	Na^+ and K^+ gates open
Production of repolarization	Opening of K^+ gates	Loss of intracellular positive charges with time and distance
Conduction distance	Not conducted—regenerated over length of axon	1–2mm; a localized potential
Positive feedback between depolarization and opening of Na^+ gates	Present	Absent
Maximum depolarization	+40 mV	Close to zero
Summation	No summation—is all-or-none	Summation of EPSPs, producing graded depolarizations
Refractory period	Present	Absent
Effect of drugs	Inhibited by tetrodotoxin, not by curare	Inhibited by curare, not by tetrodotoxin

Release of ACh at these synapses results, through opening of chemically regulated gates, in the production of graded depolarizations in the dendrites and cell body. These graded depolarizations are known as **excitatory postsynaptic potentials (EPSPs)** and are analagous to the end plate potentials of skeletal muscle fibers previously discussed. A comparison of EPSPs and action potentials is provided in table 14.5.

Voltage-regulated gates are absent in most dendrites. Although some may be present in the cell body, they are normally too few in number to produce an action potential. The *initial segment* of the axon, which is the unmyelinated region of the axon around the axon hillock, on the other hand, has a high density of voltage-regulated gates and a low threshold for the production of action potentials. Since the dendrites and cell body cannot produce a depolarization that regenerates itself due to voltage-regulated gates, depolarization in these regions must spread by cable properties to the initial segment of the axon. If the depolarization is at or above threshold by the time it reaches the initial segment of the axon, the EPSP will stimulate the production of action potentials, which

Figure 14.32. The graded nature of excitatory postsynaptic potentials is shown, in which stimuli of increasing strength produce increasing amounts of depolarization. When a threshold level of depolarization is produced, action potentials are generated in the axon.

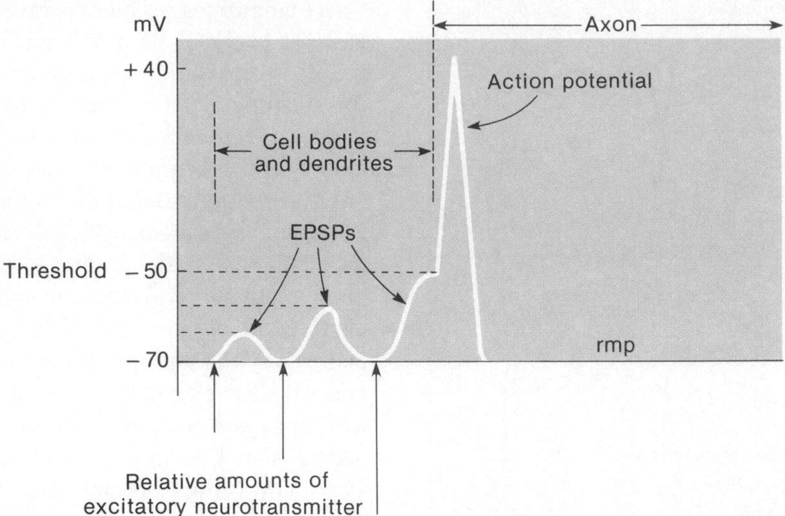

can regenerate themselves along the axon. If, however, the EPSP is below threshold at the initial segment, no action potentials will be produced in the postsynaptic cell (fig. 14.32).

A skeletal muscle fiber receives synaptic input from only one neuron. A neuron, in contrast, may receive synaptic inputs from as many as one thousand other neurons. So, whereas the synaptic input from a single axon must be sufficient to excite a muscle cell, the excitation of a neuron requires the summation of many EPSPs before a threshold depolarization is produced at the initial segment of the axon. Summation, which can only occur between graded synaptic potentials (action potentials cannot summate), allows the postsynaptic cell to "weigh" different combinations of synaptic inputs and respond appropriately.

Alzheimer's disease is the most common cause of senile dementia, which often begins in middle age and produces progressive mental deterioration. The cause of Alzheimer's disease is not known, but there is evidence that it is associated with a loss of cholinergic neurons (those that use acetylcholine as a neurotransmitter), which terminate in the hippocampus and cerebral cortex of the brain (areas concerned with memory storage). Since acetylcholine is produced from acetyl coenzyme A and choline, attempts have been made to increase ACh in the brain by having patients ingest large amounts of lecithin (which contains choline). Thus far such nutritional treatments and possible drug treatments to increase ACh by inhibiting acetylcholinesterase have met with only limited success.

1. Describe the events that occur between the release of ACh and the production of end plate potentials in a skeletal muscle fiber.
2. Describe the function of acetylcholinesterase and its physiological significance.
3. Compare the properties of EPSPs and action potentials, and describe where these events occur in a postsynaptic neuron.
4. Explain how EPSPs produce action potentials in the postsynaptic neuron.

Neurotransmitters of the Central Nervous System

A variety of chemicals found in the CNS are believed to function as neurotransmitters. These include ACh, the catecholamines (norepinephrine and dopamine), certain amino acids, and a number of polypeptides. Particular neural pathways in the brain utilize specific neurotransmitters, and the actions of some of these neurotransmitters have been modified clinically in the treatment of diseases.

Objective 24. Identify the catecholamines, and describe how they are inactivated at the synapse.

Objective 25. Explain how cyclic AMP functions as a second messenger in the action of catecholamines.

Objective 26. Describe the significance of dopamine, GABA, and endorphins as neurotransmitters.

Table 14.6 Examples of chemicals that are either proven or putative neurotransmitters

Category	Chemicals
Amines	Acetylcholine
	Histamine
	Serotonin
Catecholamines	Dopamine
	Epinephrine
	Norepinephrine
Amino acids	Aspartic acid
	GABA (gamma-aminobutyric acid)
	Glutamic acid
	Glycine
	Glucagon
	Insulin
	Somatostatin
	Substance P
Polypeptides	ACTH (adrenocorticotrophic hormone)
	Angiotensin II
	Endorphins
	LHRH (luteinizing hormone-releasing hormone)
	TRH (thyrotrophin-releasing hormone)
	Vasopressin (antidiuretic hormone)

Objective 27. Explain how glycine produces hyperpolarization of the postsynaptic membrane.
Objective 28. Explain the meaning of the term *synaptic plasticity.*

In addition to acetylcholine, a great variety of other molecules serve as neurotransmitters in the CNS. **Catecholamines** *(kat''e-kol'ah-men)* are a group of regulatory molecules, derived from the amino acid tyrosine, that include *dopamine, norepinephrine,* and *epinephrine.* Dopamine and norepinephrine function as neurotransmitters; epinephrine and, to a lesser degree, norepinephrine additionally serve as hormones secreted by the adrenal medulla gland.

The catecholamines, together with a related molecule called *serotonin,* are included in a larger category of molecules called *monoamines.* There is good evidence that serotonin and particular amino acids and polypeptides function as important neurotransmitters in the CNS. It is extremely difficult, however, to prove conclusively that a particular molecule functions as a neurotransmitter at a central synapse. Because of such lack of conclusive proof of their transmitter function, dopamine, serotonin, histamine, glycine, glutamic acid, aspartic acid, and many polypeptides are often referred to as **putative neurotransmitters** (table 14.6). These molecules, together with the well-established neurotransmitters (acetylcholine, norepinephrine, and GABA), provide the diversity needed for the complex functions of the central nervous system.

Catecholamines as Neurotransmitters

Like ACh, catecholamine neurotransmitters are released by exocytosis from presynaptic vesicles and diffuse across the synaptic cleft to interact with specific receptor proteins in the membrane of the postsynaptic cell. The stimulatory effects of these catecholamines, like those of ACh, are quickly inhibited. The inhibition of ACh stimulation is accomplished by acetylcholinesterase; the inhibition of catecholamine action is due to (1) *reuptake* of catecholamines into the presynaptic neuron endings; (2) enzymatic degradation of catecholamines in the presynaptic neuron endings by *monoamine (mon''o-am'en) oxidase (MAO);* and (3) their enzymatic degradation in the postsynaptic neuron by *catecholamine-O-methyltransferase (COMT).* This is shown in figure 14.33. Drugs that inhibit MAO and COMT thus promote the effects of catecholamine action.

The interaction of catecholamines with their receptors does not directly open chemically regulated Na^+ and K^+ gates. Instead, these neurotransmitters activate an enzyme in the postsynaptic cell membrane known as *adenylate cyclase (ad'ĕ-nil si'klās).* This enzyme converts ATP to **cyclic AMP (cAMP)** and pyrophosphate (two inorganic phosphates) within the postsynaptic cell cytoplasm. Cyclic AMP in turn activates other enzymes that open the chemically regulated ionic gates in the postsynaptic membrane. Cyclic AMP is thus a **second messenger** in the action of these neurotransmitters (fig. 14.34).

In addition to promoting the opening of ionic gates and the production of electrical impulses, cyclic AMP may also promote more long-term modifications of the postsynaptic neuron. These long-term modifications may involve cAMP-mediated changes in genetic expression.

Dopamine and Norepinephrine. Neurons that use **dopamine** as a neurotransmitter and postsynaptic neurons with dopamine receptor proteins in their membranes can be identified in postmortem brain tissue. More recently, the position of dopamine receptor proteins has been observed in the living brain using the technique of *positron emission tomography (PET).* These investigations have been spurred by the great clinical interest in the effects of **dopaminergic neurons** (those that use dopamine as a neurotransmitter).

Dopaminergic neurons are highly concentrated in the *substantia nigra* (literally, the "dark substance," so-called because it contains melanin pigment) of the brain. Many neurons in the substantia nigra send fibers to the *basal ganglia,* which are large masses of cell bodies deep in the cerebrum that are involved in the coordination of skeletal

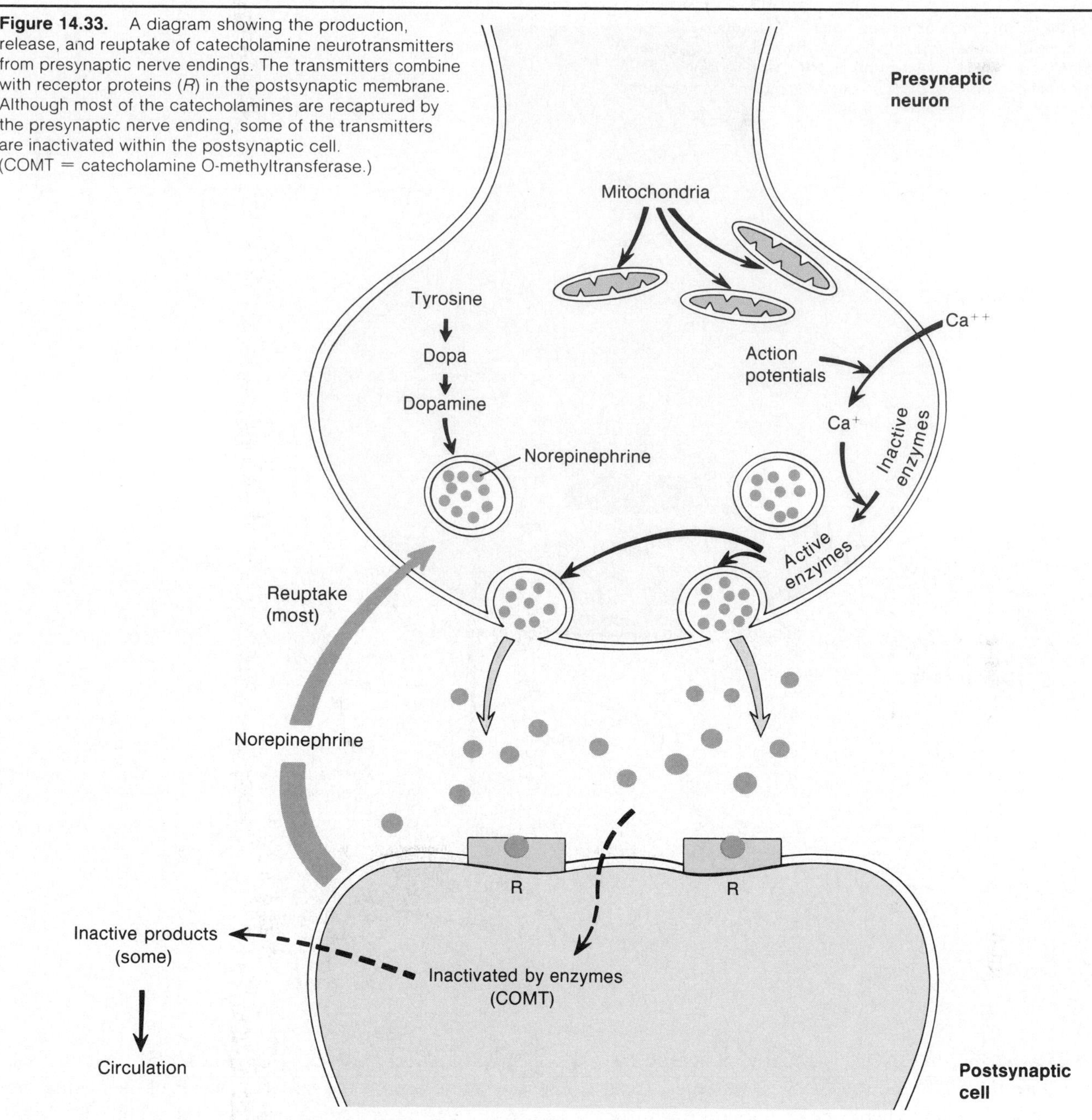

Figure 14.33. A diagram showing the production, release, and reuptake of catecholamine neurotransmitters from presynaptic nerve endings. The transmitters combine with receptor proteins (*R*) in the postsynaptic membrane. Although most of the catecholamines are recaptured by the presynaptic nerve ending, some of the transmitters are inactivated within the postsynaptic cell. (COMT = catecholamine O-methyltransferase.)

movements. There is much evidence that *Parkinson's disease,* or *paralysis agitans,* is caused by degeneration of the dopaminergic neurons in the substantia nigra. Parkinson's disease, a major cause of neurological disability in people over sixty years of age, is associated with such symptoms as muscle tremors and rigidity, difficulty in the initiation of movements and in speech, and other severe problems. These patients are treated with L-dopa to increase the production of dopamine in the brain, as described previously in this chapter, and with anticholinergic drugs to decrease the production of acetylcholine (ACh is believed to antagonize the actions of dopaminergic neurons in the basal ganglia).

Parkinson's disease: from James Parkinson, English physician, 1755–1824

A side effect of L-dopa treatment in some patients with Parkinson's disease is the appearance of symptoms characteristic of *schizophrenia.* This effect is not surprising, in view of the fact that the drugs used to treat schizophrenic patients (chlorpromazine and related compounds) act as specific antagonists of dopamine receptors. As might be predicted from these observations, schizophrenic patients treated with these drugs often develop symptoms of Parkinson's disease. It seems reasonable to suppose, from this evidence, that schizophrenia may be caused, at least in part, by overactivity of the dopaminergic pathways. This hypothesis is strengthened by additional evidence from PET scans that the brains of schizophrenic patients appear to have an abnormally high content of dopamine receptors.

Figure 14.34. Catecholamine neurotransmitters induce the intracellular production of cyclic AMP (cAMP) in the postsynaptic cell. The cAMP, in turn, stimulates the opening of ionic gates (short-term effects) and genetic expression (long-term effects) through the activation of previously inactive enzymes.

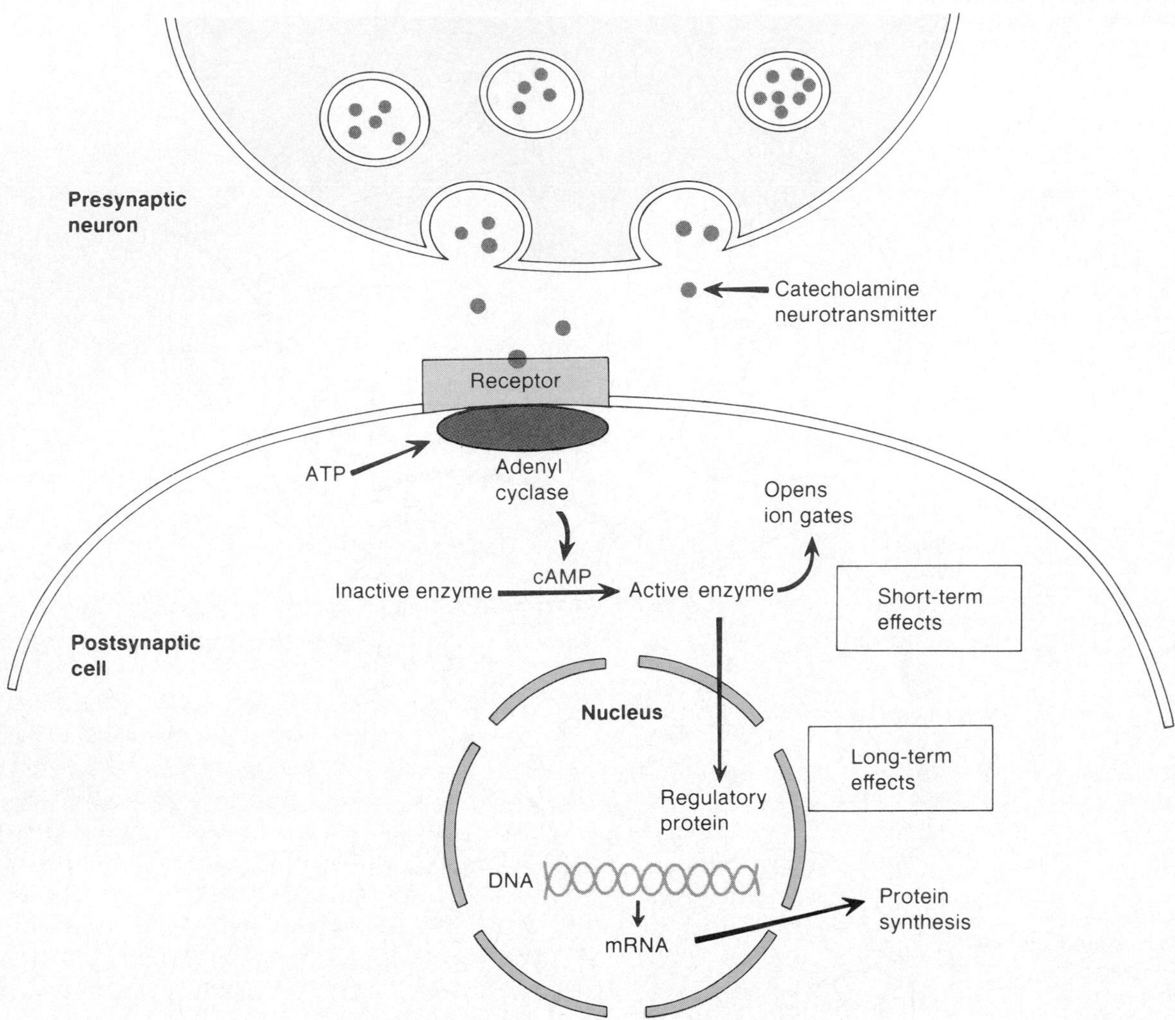

Norepinephrine, like ACh, is used as a neurotransmitter in both the PNS and the CNS. Sympathetic neurons of the PNS use norepinephrine as a neurotransmitter at their synapse with smooth muscles, cardiac muscle, and glands. Some neurons in the CNS also appear to use norepinephrine as a neurotransmitter; these neurons seem to be involved in general behavioral arousal. This would help to explain the effects of such drugs as *amphetamines* and *cocaine,* which specifically stimulate pathways that use norepinephrine (and dopamine) as a neurotransmitter.

Amino Acids as Neurotransmitters

The amino acids **glutamic acid** and **aspartic acid** function as excitatory neurotransmitters in some neurons of the CNS. The amino acid **glycine,** in contrast, is inhibitory; instead of depolarizing the postsynaptic membrane and producing an EPSP, it hyperpolarizes the postsynaptic membrane. This hyperpolarization is produced by the opening of *chloride gates,* which allows Cl^- to enter the postsynaptic neuron and make the inside of the membrane even more negative than it is at rest (changing the membrane potential from -65 mV to, for example, -85 mV). This hyperpolarization, known as an **inhibitory postsynaptic potential (IPSP),** will be discussed in more detail in the next section.

The inhibitory effects of glycine are very important in the spinal cord, where they help in the control of skeletal movements. Flexion of an arm, for example, involves stimulation of the flexor muscles. The motor neurons that innervate these flexor muscles are stimulated in the spinal cord; the motor neurons that innervate the antagonistic extensor muscles are inhibited by IPSPs produced by glycine released from other neurons. The importance of the inhibitory actions of glycine is revealed by the deadly effects of the poison *strychnine,* which causes spastic paralysis by specifically blocking the glycine receptor proteins. Animals poisoned with strychnine die from asphyxiation due to their inability to contract the diaphragm muscle.

The neurotransmitter **GABA (gamma-aminobutyric acid)** is a derivative of the amino acid glutamic acid. GABA is the most prevalent neurotransmitter in the brain; in fact, as many as one-third of all the neurons in the brain use GABA as a neurotransmitter. Like glycine, GABA is inhibitory—it opens the chloride gates and hyperpolarizes the postsynaptic membrane. Also like glycine, some neurons that use GABA are involved in motor control; a deficiency in these neurons produces the uncontrolled movements seen in people with *Huntington's chorea.*

In addition to its involvement in motor control, GABA also appears to function as a neurotransmitter involved in mood and emotion. Drugs given to treat anxiety—including the *benzodiazepines* (e.g., Valium)—act by binding to part of the GABA receptors and thus increasing the effectiveness of GABA in the brain. Drugs that antagonize the actions of GABA, conversely, can produce extreme feelings of anxiety.

Polypeptide Neurotransmitters

Many polypeptides of various sizes are found in the brain and are believed to function as neurotransmitters. Interestingly, some of the polypeptides that function as hormones secreted by the intestine and other endocrine glands are also produced in the brain and may function there as neurotransmitters (table 14.6).

Synaptic Plasticity. Although some of the polypeptides released from neurons may function as neurotransmitters in the traditional sense—by stimulating the opening of ionic gates and causing changes in the membrane potential—others may have more subtle and poorly understood effects. The name *neuromodulators* has been proposed to identify compounds with such alternative effects. An exciting recent discovery is that some neurons in both the PNS and CNS produce both a classical neurotransmitter (ACh or a catecholamine) and a polypeptide neurotransmitter. Additionally, there is evidence that these neurons may secrete one or the other of their transmitters, depending upon the environmental conditions.

Recent evidence suggests that synapses are more changeable than was previously believed. This has been called **synaptic plasticity,** and it exists at both the molecular and cellular levels. At the molecular level, it has been shown that some neurons may secrete either a classical neurotransmitter or a putative neurotransmitter, as previously discussed. At the cellular level, there is evidence that axonal sprouting over short distances produces a turnover of synapses even in the mature CNS. This breakdown and reforming of synapses may occur within a time span of only a few hours. The physiological significance of these interesting discoveries is not yet fully understood.

Endorphins. The ability of opium and its analogues—that is, the opioids—to relieve pain (promote analgesia) has been known for centuries. Morphine, for example, has long been used for this purpose. The discovery in 1973 of opioid receptor proteins in the brain suggested that the effects of these drugs might be due to the stimulation of specific neuron pathways. This implied that opioids—like LSD, mescaline, and other mind-altering drugs—might resemble neurotransmitters produced by the brain.

The analgesic effects of morphine are blocked in a specific manner by a drug called *naloxone.* In the same year that opioid receptor proteins were discovered, it was found that naloxone also blocked the analgesic effect of electrical brain stimulation. Subsequently, evidence suggested that the analgesic effects of hypnosis and acupuncture could also be blocked by naloxone. These experiments indicated that the brain might be producing its own morphinelike analgesic compounds.

These compounds have been identified as a family of chemicals called **endorphins** (for "endogenously produced morphinelike compounds") produced by the brain and pituitary gland. The endorphins include a group of five-amino-acid peptides called **enkephalins,** *(en-kef'ah-lin)* which may function as neurotransmitters, and a thirty-one-amino-acid polypeptide, produced by the pituitary gland, called *β-endorphin.*

Endorphins have been shown to block the transmission of pain. Current evidence for this includes results obtained both from neurophysiological studies—in which endorphins blocked the release of substance P (the chemical transmitter believed to be released by nerve fibers that mediate painful sensations)—and from behavioral studies. The pain threshold of pregnant rats, for example, has been found to decrease when they are treated with naloxone.

Endorphins—like opium and morphine—may also provide pleasant sensations and thus mediate reward or positive reinforcement pathways. Overeating in genetically obese mice, for example, appears to be blocked by naloxone. It has been found that blood levels of β-endorphin are increased in exercise. Some people have suggested that the "jogger's high" may thus be due to endorphins. Although evidence for this is poor, it does appear that endorphins may promote some type of psychic reward system as well as analgesia.

1. Describe three ways that catecholamine neurotransmitters are inactivated at the synapse.
2. Explain how catecholamines stimulate the production of cyclic AMP, and why cAMP is called a second messenger in the action of these neurotransmitters.
3. Describe the known significance of dopaminergic neurons, and explain how their function relates to the etiology of Parkinson's disease and schizophrenia.
4. Describe the nature and explain the significance of inhibitory neurotransmitters in the CNS.
5. Explain the meaning of the term *synaptic plasticity* at the molecular and cellular level.
6. Describe the nature and explain the proposed functions of the endorphins.

Synaptic Integration

A number of EPSPs must summate in order to stimulate the postsynaptic cell to produce action potentials. This stimulatory effect can be reduced or inhibited by hyperpolarization (IPSPs) in a process known as postsynaptic inhibition.

Objective 29. Describe the processes of spatial summation and temporal summation of EPSPs.
Objective 30. Explain the process and significance of post-tetanic potentiation.
Objective 31. Explain how EPSPs and IPSPs interact and how postsynaptic inhibition occurs.
Objective 32. Explain the mechanism of presynaptic inhibition.

Since voltage-regulated Na^+ and K^+ gates are absent in the dendrites and cell bodies of multipolar neurons, changes in membrane potential induced by neurotransmitters in these areas do not have the all-or-none characteristics of action potentials. Synaptic potentials are graded and can add together, or summate. **Spatial summation** occurs because many presynaptic nerve fibers converge on a single postsynaptic neuron (fig. 14.35). In spatial summation, synaptic depolarizations (EPSPs) produced at different synapses may summate in the postsynaptic dendrites and cell body. In **temporal summation,** the successive activity of presynaptic axon terminals, causing successive waves of transmitter release, may result in the summation of EPSPs in the postsynaptic neuron. The summation of EPSPs helps to insure that the depolarization that reaches the axon hillock will be sufficient to generate new action potentials in the postsynaptic nerve fiber (fig. 14.36).

When a presynaptic neuron is stimulated continuously, even for as short a time as a few seconds, its ability to excite a postsynaptic neuron is enhanced—or potentiated—when this neuron pathway is subsequently stimulated. This potentiation of synaptic transmission may last for hours or even weeks following the tetanic (continuous) stimulation of the presynaptic neuron. This phenomenon is called **post-tetanic** *(tĕ-tan'ik)* **potentiation.**

Post-tetanic potentiation is due to the increased release of neurotransmitter as a result of previous tetanic stimulation of the presynaptic neuron. Experiments suggest that there is increased Ca^{++} concentration in the axon terminals following tetanic stimulation. Since Ca^{++} mediates exocytosis of the synaptic vesicles in response to action potentials, this observation may explain the improved synaptic efficiency that results. Post-tetanic potentiation may favor transmission along frequently used neural pathways and thus may represent a mechanism of neural "learning."

Figure 14.35. A diagram illustrating the convergence of large numbers of presynaptic fibers on the cell body of a spinal motor neuron.

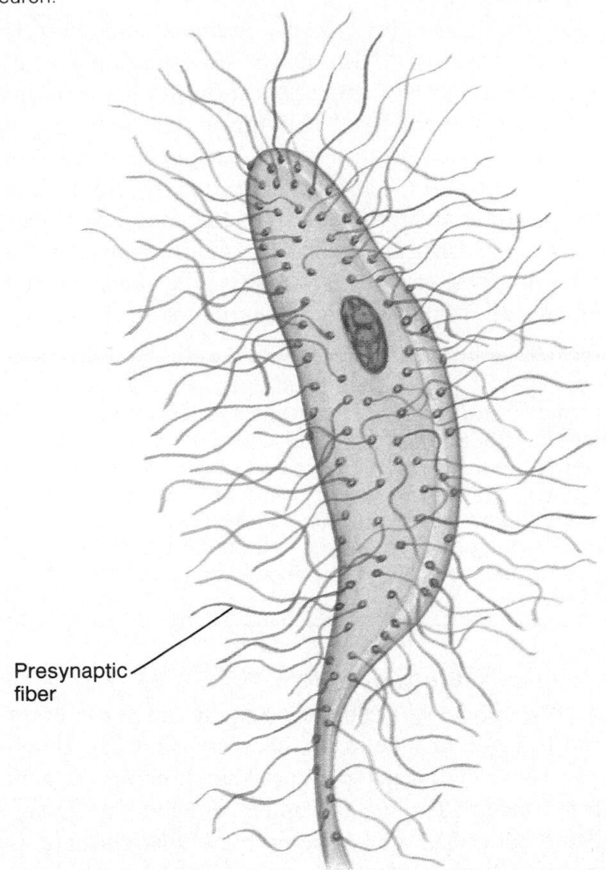

Synaptic Inhibition

Although most neurotransmitters depolarize the postsynaptic membrane (produce EPSPs), some transmitters have the opposite effect. These inhibitory neurotransmitters cause *hyperpolarization* of the postsynaptic membrane: they make the inside of the membrane more negative than it is at rest. Since hyperpolarization (from -65 mV to, for example, -85 mV) takes the membrane potential farther away from the threshold depolarization required to stimulate action potentials, such hyperpolarization inhibits the activity of the postsynaptic neuron. Hyperpolarizations produced by neurotransmitters are therefore called **inhibitory postsynaptic potentials (IPSPs).** The inhibition produced in this way is called **postsynaptic inhibition.**

Excitatory and inhibitory inputs (EPSPs and IPSPs) to a postsynaptic neuron can summate in an algebraic fashion (fig. 14.37). The effects of IPSPs in this way reduce, or may even abolish, the ability of EPSPs to generate action potentials in the postsynaptic cell. Considering that the nervous system contains approximately 10^{12} neurons and that a given neuron may receive as many as 1,000 presynaptic inputs, one can see that the possibilities for synaptic integration in this way are awesome.

Figure 14.36. Excitatory postsynaptic potentials (EPSPs) can summate over distance (spatial summation) and time (temporal summation). When summation results in a threshold level of depolarization at the axon hillock, voltage-regulated Na^+ gates are opened and an action potential is produced.

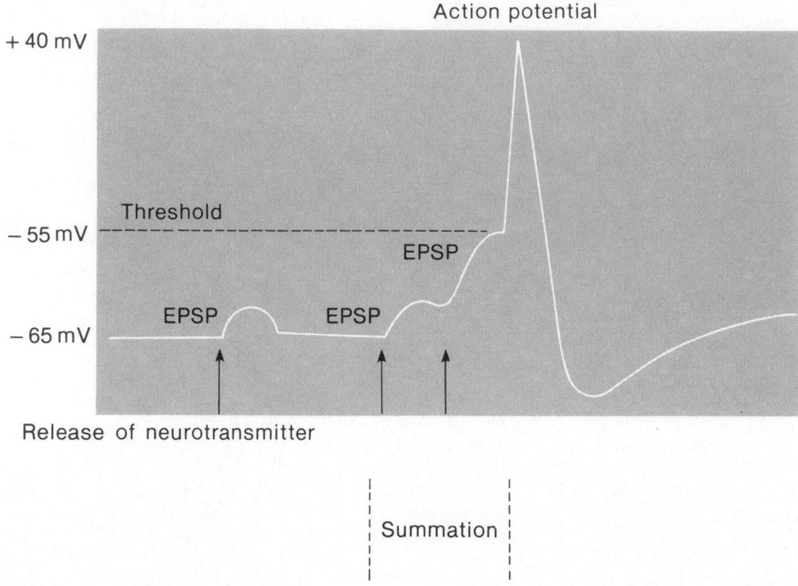

Figure 14.37. An inhibitory postsynaptic potential (IPSP) makes the inside of the postsynaptic membrane more negative than the resting potential—it hyperpolarizes the membrane. Subsequent or simultaneous excitatory postsynaptic potentials (EPSPs), which are depolarizations, must thus be stronger to reach the threshold required to generate action potentials at the axon hillock.

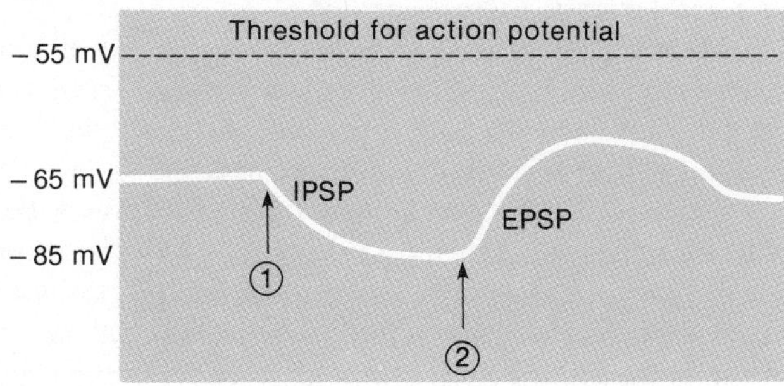

In addition to serving an integrative function in the CNS, IPSPs also have more specialized roles. Although ACh produces depolarization at the neuromuscular junction and stimulates skeletal muscle contraction, for example, the same transmitter causes hyperpolarization in the heart. This is due to the fact that the combination of ACh with its receptor protein in the myocardial cells causes opening of only K^+ gates, and the outward diffusion of K^+ causes hyperpolarization. The baseline membrane potential is thus lowered, so that a longer time is required to reach threshold. The parasympathetic fibers that innervate the heart in this way cause a slowing of the heart rate.

Glycine and GABA, as previously discussed, hyperpolarize the postsynaptic membrane by promoting the inward diffusion of Cl^-. This action produces postsynaptic inhibition within the central nervous system.

In **presynaptic inhibition** (fig. 14.38), the release of neurotransmitter from the axon of one neuron is reduced by a second neuron, which makes an excitatory axoaxonic synapse with the first. Depolarization of the first neuron's axon produced by this axoaxonic synapse puts it in a partial refractory period. This reduces the frequency of action potentials in the first neuron's axon, thus reducing the amount of neurotransmitter it is stimulated to release.

1. Define spatial summation, temporal summation, and post-tetanic potentiation. Explain their functional importance.
2. Explain how postsynaptic inhibition is produced and how IPSPs and EPSPs can interact.
3. Explain the mechanism of presynaptic inhibition.

Clinical Considerations

The clinical aspects of the nervous system are extensive and usually complex. Numerous diseases and developmental problems directly involve the nervous system, and the nervous system is indirectly involved with most of the diseases that afflict the body because of the perception of pain. Pain receptors are free nerve endings that elicit pain in disease or trauma; such pain is important in diagnosing specific diseases or dysfunctions.

Figure 14.38. A diagram illustrating postsynaptic and presynaptic inhibition.

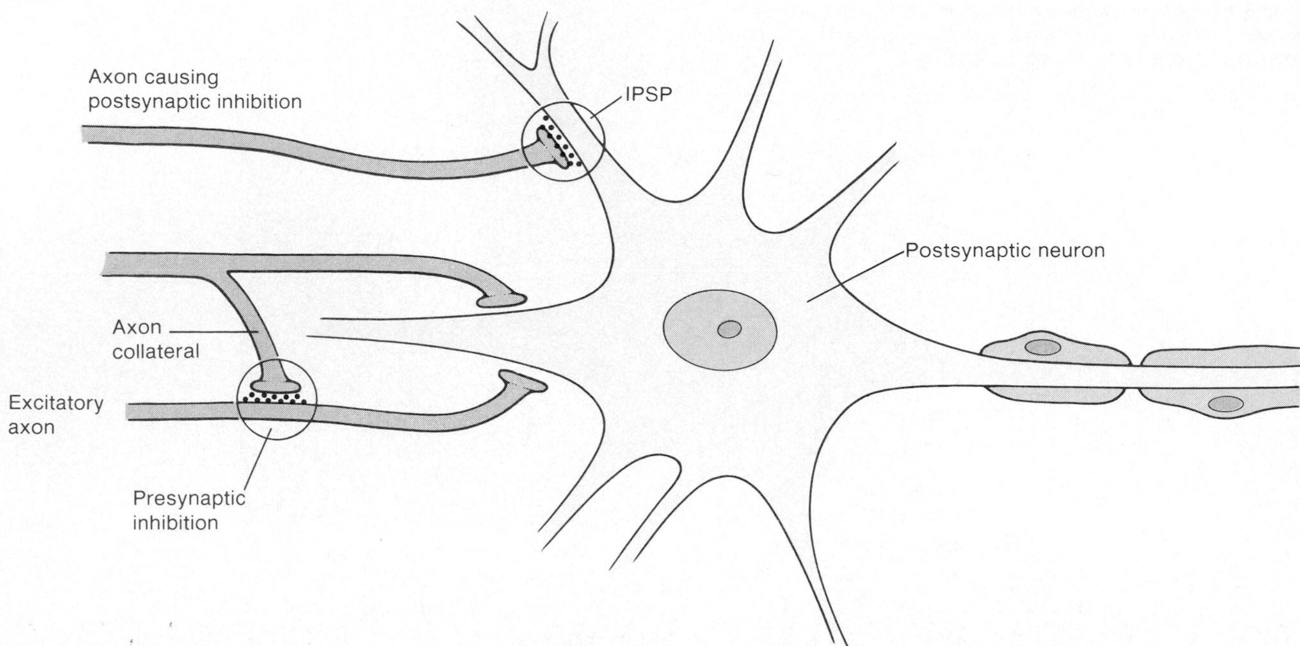

Only some of the diseases involving developmental problems of the nervous system and clinical aspects of nerve conduction and synaptic transmission will be considered in this section. Other clinical aspects of neurology will be discussed in later chapters.

Developmental Problems

Congenital malformations of the CNS are common and frequently involve overlying bone, muscle, and connective tissue. The more severe abnormalities make life impossible, and the less severe malformations frequently result in functional disability.

Most congenital malformations of the nervous system occur during the sensitive embryonic period. Neurological malformations are generally caused by genetic abnormalities but may result from environmental factors such as anoxia, infectious agents, drugs, and ionizing radiation.

Spina bifida is a defective fusion of the vertebral elements and may or may not involve the spinal cord. **Spina bifida occulta** is the most common and least serious type of spina bifida. This defect usually involves few vertebrae, is not externally apparent except for perhaps a pigmented spot with a tuft of hair, and usually does not cause neurological disturbances. **Spina bifida cystica,** a severe type of spina bifida, is a saclike protrusion of skin and underlying meninges, which may contain portions of the spinal cord and nerve roots. Spina bifida cystica is most common in the lower thoracic, lumbar, and sacral regions. The position and extent of the defect determines the degree of neurological impairment.

Anencephalia *(an''en-sĕ-fa'le-ah)* is a markedly defective development of the brain and the surrounding cranial bones. Anencephalia occurs in one per thousand births and makes sustained extrauterine life impossible. This congenital defect apparently results from the failure of the neural folds at the cranial portion of the neural plate to fuse and form the prosencephalon.

Microcephaly *(mi''kro-sef'ah-le)* is an uncommon condition in which brain development is not completed. If enough neurological tissue is present, the infant will survive but will be severely mentally retarded.

Defective skull development frequently causes **cranial encephalocele** *(en-sef'ah-lo-se'le).* This condition occurs approximately once per two thousand births. Cranial encephalocele is characterized by a herniated portion of the brain and meninges through a defect in the skull, usually in the occipital region. Occasionally the herniation contains fluid within the meninges but not the brain tissue. In this case, it is referred to as a **cranial meningocele** *(mĕ-ning'go-sēl).*

Hydrocephalus is the abnormal accumulation of cerebrovascular fluid in the ventricles and subarachnoid or subdural space. Hydrocephalus may be caused by the excessive production or blocked flow of cerebrospinal fluid. It may also be associated with other congenital problems such as spina bifida cystica or encephalocele. Hydrocephalus frequently causes cranial bones to thin and the cerebral cortex to atrophy.

Many congenital disorders cause an impairment of intelligence, known as **mental retardation.** Chromosomal abnormalities, maternal and fetal infections such as syphilis and German measles, and excessive irradiation of the fetus are common causes of mental retardation.

Figure 14.39. The presence of scleroses, or scars, can be seen as white spots on this MR scan of the brain of a person who has multiple sclerosis (MS).

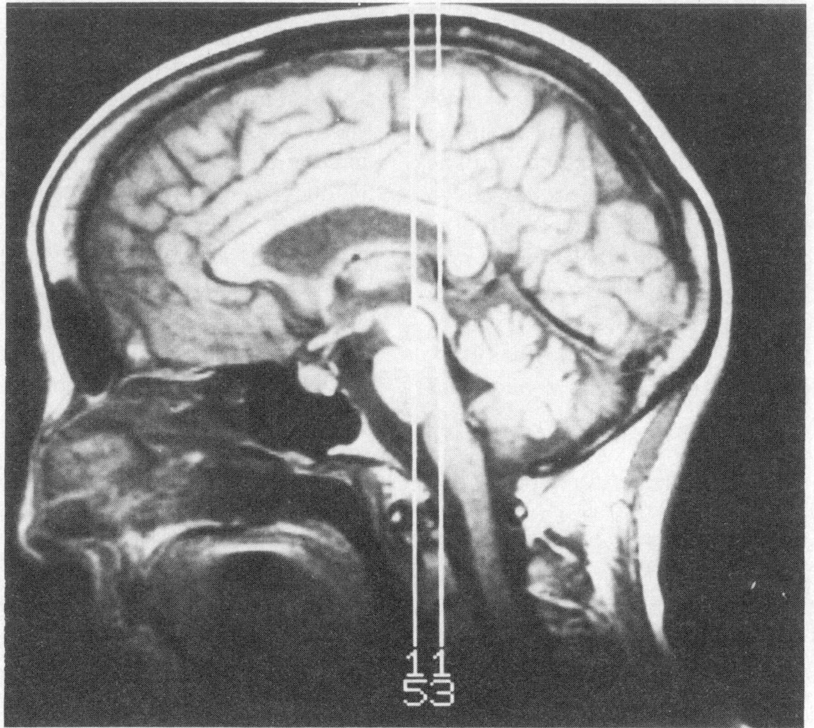

Diseases of the Myelin Sheath

Multiple sclerosis (MS) is a relatively common neurological disease in persons between the ages of twenty and forty. MS is a chronic, degenerating, remitting, and relapsing disease that progressively destroys the myelin sheaths of neurons in multiple areas of the CNS (fig. 14.39). Initially, lesions form on the myelin sheaths and soon develop into hardened *scleroses,* or scars (hence the name). Destruction of the myelin sheaths prohibits the normal conduction of impulses, resulting in a progressive loss of functions. Because myelin degeneration is widely distributed, MS has a greater variety of symptoms than any other neurological disease. This characteristic, coupled with remissions, frequently causes misdiagnosis of this disease.

During the early stages of MS, many patients are considered neurotic because of the wide variety and temporary nature of their symptoms. As the disease progresses, the symptoms may include double vision (diplopia), spots in the visual field, blindness, tremor, numbness of appendages, and locomotor difficulty. Eventually the patient is bedridden, and death may occur anytime from seven to thirty years after the first symptoms appear.

In **Tay-Sachs disease** the myelin sheaths are destroyed by the excessive accumulation of one of the lipid components of the myelin. This results from an enzyme defect due to the inheritance of genes that are carried by the parents in a recessive state. Tay-Sachs disease, which is inherited primarily in Jews of Eastern European descent, appears when the infant is under one year of age. It causes blindness, loss of mental and motor ability, and ultimately death by the age of three. Potential parents can tell if they are carriers for this condition by having a special blood test for the defective enzyme.

Blood-Brain Barrier

Substances that are more lipid soluble can leave the blood and enter the brain much more readily than can more polar, water-soluble molecules. Injection of a hypertonic solution (of glucose, for example) has been shown to reduce the blood-brain barrier and thus increase the ease with which molecules can enter the brain. Then, chemotherapeutic drugs that otherwise could not as effectively enter the brain can be administered (for example, drugs that treat disorders such as brain tumors).

Diseases Involving Neurotransmitters

Parkinson's Disease. As previously described, Parkinson's disease results from the degeneration of particular dopaminergic neurons in the brain. Traditional treatment of this disease with L-dopa and anticholinesterase drugs, unfortunately, has met with only very limited success. A

multiple sclerosis: L. *multiplus,* many parts; Gk. *skleros,* hardened

Tay-Sachs disease: from Warren Tay, English physician, 1843–1927, and Bernard Sachs, U.S. neurologist, 1858–1944

recent surgical treatment, however, has evoked widespread interest and excitement. In this procedure, surgeons remove the right adrenal gland of a patient, excise a portion of the adrenal medulla (the inner part of the gland, which is derived from embryonic neural tissue), and implant it on top of the caudate nucleus in the right hemisphere of the brain. The adrenal medulla normally produces L-dopa, which it then converts into the hormone epinephrine and lesser amounts of norepinephrine. The surgeons originally expected the adrenal medulla cells to develop into dopaminergic neurons. To their surprise, the patient's improvement was far greater than could be predicted from this mechanism. Some scientists now hypothesize that the implant may secrete L-dopa, and/or a neuron growth-promoting chemical, into the cerebrospinal fluid. These exciting possibilities and their clinical applications are currently under investigation.

Alzheimer's Disease. This disease, as previously discussed, is associated with a selective loss of specific cholinergic neurons in the brain and with a deficiency in the enzyme responsible for producing acetylcholine from acetyl coenzyme A and choline. Treatment for Alzheimer's has been reported to be fairly good with different drugs that inhibit the activity of acetylcholinesterase; by inhibiting this enzyme, the breakdown of ACh in synapses is reduced so that the action of ACh is improved.

Autopsies of people who have died of Alzheimer's disease reveal "neuritic plaques," which are composed of degenerating axons and deposits of amyloid protein. Similar plaques are seen in the brains of people with Down's syndrome, a genetic disease caused by an extra chromosome number 21. Recently, scientists have discovered that the gene that codes for this amyloid protein in people with Alzheimer's disease is located in chromosome number 21, suggesting that this disease may be caused by genetic defects located on this chromosome.

Alzheimer's disease: from Alois Alzheimer, German neurologist, 1864–1915

Chapter Summary

I. Organization and Functions of the Nervous System
   A. The central nervous system (CNS), consisting of the brain and spinal cord, is covered with meninges and bathed in cerebrospinal fluid and contains gray and white matter.
   B. The functions of the nervous system are orientation, coordination, assimilation, and instinctual behavior.

II. Development of the Nervous System
   A. The prosencephalon, mesencephalon, and rhombencephalon develop from the embryonic neural tube.
      1. The telencephalon and diencephalon develop from the forebrain (prosencephalon).
      2. The metencephalon and myelencephalon develop from the hindbrain (rhombencephalon).
   B. The spinal cord develops from the neural tube as afferent and efferent neurons are formed that conduct into and out of the spinal cord, respectively.
   C. The development of the peripheral nervous system from neural crest tissue produces a pattern of dermatomes in the skin.

III. Neurons and Neuroglia
   A. Every neuron contains a cell body, dendrites, and an axon.
      1. Collections of cell bodies in the CNS are called nuclei, and in the PNS they are called ganglia.
      2. On the basis of the number of processes extending from the cell body, neurons can be classified as pseudounipolar, bipolar, or multipolar.
      3. Neurons in the PNS that conduct impulses into the CNS are sensory; those that conduct impulses out of the CNS are motor.
      4. Motor neurons that innervate skeletal muscles are somatic; those that innervate the heart, smooth muscles, and glands are autonomic.
   B. Neuroglial cells include Schwann cells and satellite cells in the PNS and, in the CNS, include oligodendrocytes, microglia, astrocytes, and ependyma.
      1. In the PNS, Schwann cells surround axons to form a sheath of Schwann, which provides a continuous basement membrane around the axon. In addition, many axons have a myelin sheath, which is formed by successive wrappings of the Schwann cell membrane. In a myelinated axon, the myelin is located to the inside of the sheath of Schwann. Gaps in the myelin sheath are known as nodes of Ranvier.
      2. In the CNS, myelin sheaths are formed by oligodendrocytes. One oligodendrocyte forms myelin around several axons, so there is no continuous basement membrane as is provided by the sheath of Schwann in the PNS. Collections of myelinated axons in the CNS produce white matter; the gray matter consists of cell bodies and dendrites, which are not myelinated.
      3. The sheath of Schwann allows damaged peripheral axons to regenerate; the absence of a sheath of Schwann in the CNS hinders regeneration of central axons.
      4. Astrocytes send processes that surround capillaries in the brain; these capillaries form a continuous barrier that provides a blood-brain barrier.

IV. Electrical Activity of Axons
   A. The permeability of the axon membrane to Na^+ and K^+ is regulated by gates.
      1. At the resting membrane potential of -65 mV, the membrane is relatively impermeable to Na^+ and only slightly permeable to K^+.
      2. These gates are voltage regulated; the Na^+ and K^+ gates open in response to the stimulus of depolarization.
      3. When the membrane is depolarized to a threshold level, the Na^+ gates open first, followed quickly by the K^+ gates.
   B. The opening of voltage-regulated gates produces an action potential.
      1. The opening of Na^+ gates in response to depolarization allows Na^+ to diffuse into the axon, thus further depolarizing the membrane in a positive feedback fashion.
      2. The inward diffusion of Na^+ causes a reversal of the membrane potential from -65 mV to $+40$ mV.
      3. The opening of K^+ gates and outward diffusion of K^+ causes the reestablishment of the resting membrane potential; this is called repolarization.
      4. Action potentials are all-or-none events.
      5. The refractory periods of an axon membrane prevent action potentials from running together.

6. Stronger stimuli produce action potentials at greater frequency.

C. One action potential serves as the depolarization stimulus for production of the next action potential in the axon.
   1. In unmyelinated axons, action potentials are produced fractions of a micrometer apart.
   2. In myelinated axons, action potentials are produced only at the nodes of Ranvier; this saltatory conduction is faster than conduction in an unmyelinated nerve fiber.

V. Synaptic Transmission
   A. Gap junctions are electrical synapses and are found in cardiac muscle, smooth muscle, and some synapses in the CNS.
   B. In chemical synapses, neurotransmitters are packaged in synaptic vesicles and released by exocytosis into the synaptic cleft.

VI. Acetylcholine
   A. The combination of ACh with its receptor protein in the postsynaptic membrane causes the chemically regulated gates to open and produces depolarizations called end plate potentials.
      1. ACh is inactivated by acetylcholinesterase.
      2. Depolarizations called end plate potentials are produced by ACh in skeletal muscle cells.
      3. End plate potentials are graded depolarizations that stimulate the production of action potentials in skeletal muscle fibers.
   B. Depolarization produced by neurotransmitters in neurons is called an excitatory postsynaptic potential (EPSP).
      1. EPSPs are graded and capable of summation, and they decrease in amplitude as they are conducted.
      2. EPSPs produced at synapses in the dendrites or cell body travel to the axon hillock, stimulate opening of voltage-regulated gates, and generate action potentials in the axon.

VII. Neurotransmitters of the Central Nervous System
   A. Catecholamines include dopamine, norepinephrine, and epinephrine.
      1. These neurotransmitters are inactivated after being released, primarily by reuptake into the presynaptic nerve endings.
      2. Catecholamines activate adenylate cyclase in the postsynaptic cell, which catalyzes the formation of cyclic AMP.
      3. Cyclic AMP, formed in the postsynaptic cell, produces the effects of the neurotransmitter.
   B. In addition to the classical neurotransmitters, neurons produce a large number of other chemicals that are believed to have a neurotransmitter function.
      1. These chemicals include glycine and GABA, which have inhibitory effects.
      2. The putative neurotransmitters include the endorphins and many other polypeptides.

VIII. Synaptic Integration
   A. Spatial and temporal summation of EPSPs allows a sufficient depolarization to be produced to cause the stimulation of action potentials in the postsynaptic neuron.
   B. Neurotransmitters that cause hyperpolarization of the postsynaptic membrane produce inhibitory postsynaptic potentials (IPSPs).
      1. IPSPs and EPSPs from different synaptic inputs can summate.
      2. The production of IPSPs is called postsynaptic inhibition.
      3. Presynaptic inhibition occurs in an axoaxonic synapse and reduces the amount of neurotransmitter released by the inhibited neuron.

Review Activities

Objective Questions

Match the following structures of the brain to the region in which they are located:

1. Cerebellum
2. Cerebral cortex
3. Medulla oblongata

(a) telencephalon
(b) diencephalon
(c) mesencephalon
(d) metencephalon
(e) myelencephalon

4. The neuroglial cells that form myelin sheaths in the peripheral nervous system are
   (a) oligodendrocytes
   (b) satellite cells
   (c) Schwann cells
   (d) astrocytes
   (e) microglia
5. A collection of neuron cell bodies located outside the CNS is called a
   (a) tract
   (b) nerve
   (c) nucleus
   (d) ganglion
6. Which of the following neurons is pseudounipolar?
   (a) sensory neurons
   (b) somatic motor neurons
   (c) neurons in the retina
   (d) autonomic motor neurons
7. Depolarization of an axon is produced by the
   (a) inward diffusion of Na^+
   (b) active extrusion of K^+
   (c) outward diffusion of K^+
   (d) inward active transport of Na^+
8. Repolarization of an axon during an action potential is produced by the
   (a) inward diffusion of Na^+
   (b) active extrusion of K^+
   (c) outward diffusion of K^+
   (d) inward active transport of Na^+
9. As the strength of a depolarizing stimulus to an axon is increased,
   (a) the amplitude of action potentials increases
   (b) the duration of action potentials increases
   (c) the speed with which action potentials are conducted increases
   (d) the frequency with which action potentials are produced increases
10. The conduction of action potentials in a myelinated nerve fiber is
   (a) saltatory
   (b) without decrement
   (c) faster than in an unmyelinated fiber
   (d) all of the above
11. Which of the following is *not* a characteristic of synaptic potentials?
   (a) They are all-or-none in amplitude.
   (b) They decrease in amplitude with distance.
   (c) They are produced in dendrites and cell bodies.
   (d) They are graded in amplitude.
   (e) They are produced by chemically regulated gates.
12. Which of the following is *not* a characteristic of action potentials?
   (a) They are produced by voltage-regulated gates.
   (b) They are conducted without decrement.
   (c) Na^+ and K^+ gates open at the same time.
   (d) The membrane potential reverses polarity during depolarization.
13. A drug that inactivates acetylcholinesterase
   (a) inhibits the release of ACh from presynaptic endings
   (b) inhibits the attachment of ACh to its receptor protein
   (c) increases the ability of ACh to stimulate muscle contraction
   (d) all of the above
14. Postsynaptic inhibition is produced by
   (a) depolarization of the postsynaptic membrane
   (b) hyperpolarization of the postsynaptic membrane
   (c) axoaxonic synapses
   (d) post-tetanic potentiation
15. Hyperpolarization of the postsynaptic membrane in response to glycine or GABA is produced by the opening of
   (a) Na^+ gates
   (b) K^+ gates
   (c) Ca^{++} gates
   (d) Cl^- gates

Essay Questions

1. Compare the characteristics of action potentials with those of synaptic potentials.
2. Explain how voltage-regulated gates produce an all-or-none action potential.
3. Explain how action potentials are regenerated along an axon.
4. Explain why conduction in a myelinated axon is faster than in an unmyelinated axon.
5. Trace the course of events between the production of an EPSP and the generation of action potentials at the axon hillock. Explain the effect of spatial and temporal summation on this process.
6. Explain how an IPSP is produced and how IPSPs can inhibit activity of the postsynaptic neuron.

15

Central Nervous System

Outline

Concepts

The central nervous system is covered with meninges, is bathed in cerebrospinal fluid, and contains gray and white matter. The tremendous metabolic rate of the brain requires a continuous flow of blood amounting to approximately 20% of the total resting cardiac output.

The cerebrum, consisting of five paired lobes within two convoluted hemispheres, is concerned with higher brain functions, such as the perception of sensory impulses, the instigation of voluntary movement, the storage of memory, thought processes, and reasoning ability. The cerebrum is also concerned with instinctual and limbic (emotional) functions.

The diencephalon is a major autonomic region of the brain that consists of vital structures such as the thalamus, hypothalamus, epithalamus, and pituitary gland.

The mesencephalon contains the corpora quadrigemina, concerned with visual and auditory reflexes, and the cerebral peduncles, composed of fiber tracts. It also contains specialized nuclei that help to control posture and movement.

The metencephalon contains the pons, which relays impulses, and the cerebellum, which coordinates skeletal muscle contractions.

The medulla oblongata, contained within the myelencephalon, connects to the spinal cord and contains nuclei for the cranial nerves and vital autonomic functions.

The CNS is covered by protective meninges, consisting of a dura mater, an arachnoid membrane, and a pia mater.

The ventricles, central canal, and subarachnoid space contain cerebrospinal fluid, formed by the active transport of substances from blood plasma in the choroid plexuses.

The spinal cord consists of centrally located gray matter involved in reflexes, and peripherally located ascending and descending tracts of white matter, which conduct impulses to and from the brain.

Characteristics of the Central Nervous System

The central nervous system is covered with meninges, is bathed in cerebrospinal fluid, and contains gray and white matter. The tremendous metabolic rate of the brain requires a continuous flow of blood amounting to approximately 20% of the total resting cardiac output.

Objective 1. Describe the general characteristics of the brain and spinal cord.
Objective 2. Discuss the basic metabolic demands of the brain.

The central nervous system (CNS) consists of the brain and spinal cord. The entire delicate CNS is protected by a bony encasement—the cranium surrounding the brain (fig. 15.1) and the vertebral column surrounding the spinal cord. The **meninges** *(mĕ-nin'jēz)* form a protective membrane between the bone and the soft tissue of the CNS. The CNS is bathed in **cerebrospinal** *(ser''ĕ-bro-spi'nal)* **fluid,** which circulates within the hollow **ventricles** of the brain, the **central canal** of the spinal cord (fig. 15.2), and the **subarachnoid** *(sub''ah-rak'noid)* **space** surrounding the entire CNS.

Figure 15.1. A magnetic resonance image (MR image) of the brain in sagittal section.

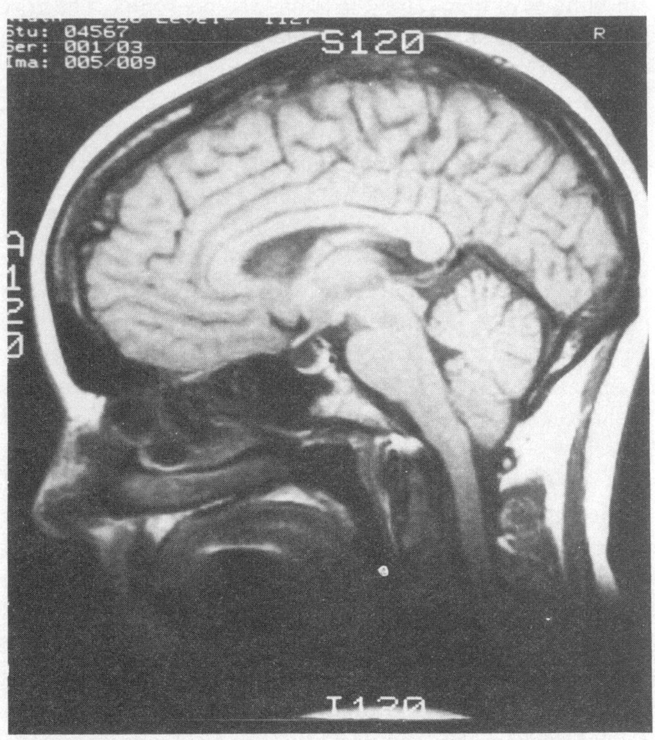

Figure 15.2. The CNS consists of the brain and the spinal cord, both of which are covered with meninges and bathed in cerebrospinal fluid.

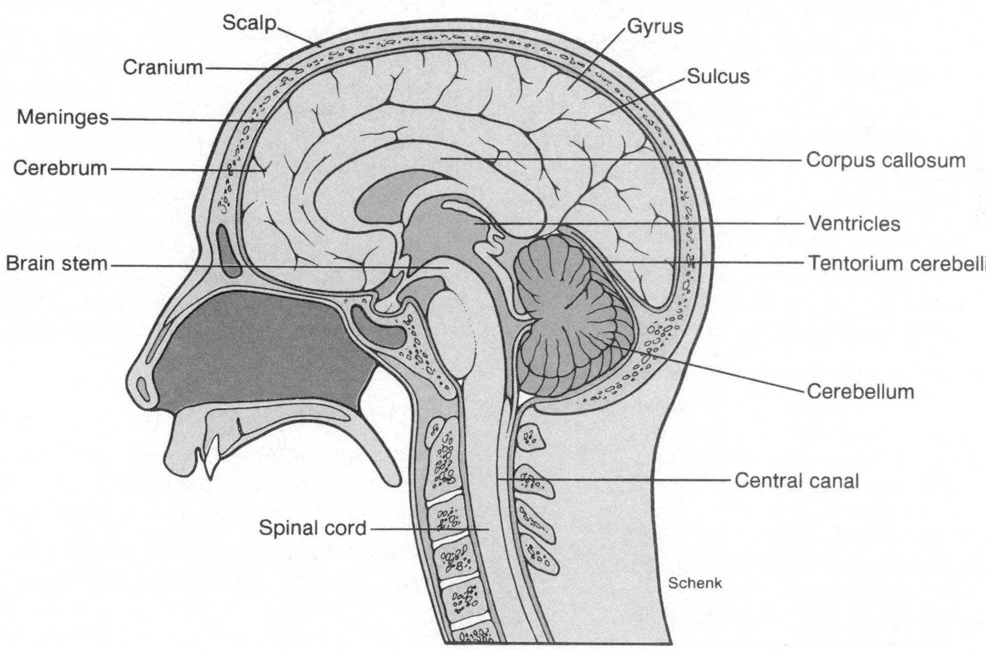

The CNS is composed of gray and white matter. **Gray matter** consists of either nerve cell bodies and dendrites or bundles of unmyelinated axons and neuroglia. The gray matter of the brain exists as the outer convoluted **cortex layer** of the cerebrum and cerebellum. There are also specialized gray matter clusters of nerve cells, called **nuclei,** deep within the white matter. **White matter** forms the tracts within the CNS and consists of aggregations of myelinated axons and associated neuroglia.

The brain of an adult weighs nearly 1.5 kg (3–3.5 lbs) and is composed of an estimated 100 billion (10^{11}) neurons. Neurons communicate with one another by means of innumerable synapses between the axons and dendrites within the brain. Neurotransmission within the brain is regulated by many different neurotransmitter chemicals (chapter 14) that are found in specific brain regions and tracts.

The brain has a tremendous metabolic rate and must have a continual supply of oxygen and nutrients. The brain accounts for only 2% of a person's body weight, and yet it receives approximately 20% of the total resting cardiac output. This amounts to a flow of about 750 ml of blood per minute. The volume remains relatively constant even during physical or mental activity. This continuous flow is so crucial that a failure of cerebral circulation for as short as ten seconds causes unconsciousness. The brain is composed of perhaps the most sensitive tissue of the body. Not only must it have continuous oxygen, but because of its high metabolic rate, it also must have a continuous nutrient supply and rapid removal of wastes. The brain is also very sensitive to toxins and drugs. The cerebrospinal fluid aids the metabolic needs of the brain through the distribution of nutrients and the removal of wastes. Cerebrospinal fluid also maintains a protective homeostatic environment within the brain. The blood-brain barrier (chapter 14) and the secretory activities of neural tissue also help to maintain homeostasis. The brain has an extensive vascular supply through the paired internal carotid and vertebral arteries that unite at the arterial circle (circle of Willis) (see chapter 20 and fig. 20.30).

The brain of a newborn is especially sensitive to oxygen deprivation or to excessive oxygen. If complications arise during childbirth and the oxygen supply from the mother's blood to the baby is interrupted while it is still in the birth canal, the infant may be stillborn or suffer brain damage that can result in cerebral palsy, epilepsy, paralysis, or mental retardation. Excessive oxygen administered to a newborn may cause blindness.

Figure 15.3. Positron emission tomographic (PET) brain scan of an unmedicated schizophrenic patient. Red areas indicate high glucose use (uptake of 18-F-deoxyglucose). The scan shows highest glucose uptake in the posterior region, where the brain's visual center is located.

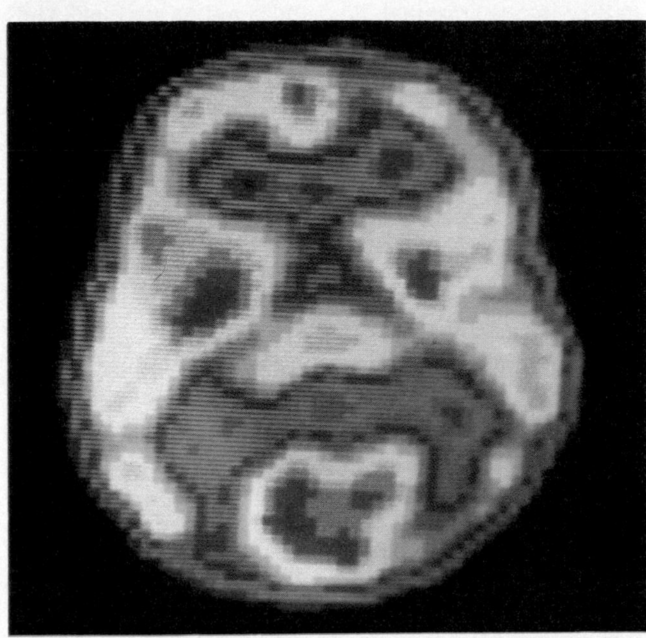

There are measurable increases in regional blood flow within the brain and in glucose and oxygen metabolism that accompany mental functions, including perception and emotion. These metabolic changes can be assessed through the use of *positron emission tomography* (*PET*). The technique of a PET scan (fig. 15.3) is based on injecting radioactive tracer molecules labeled with carbon-11, fluorine-18, and oxygen-15 into the bloodstream and photographing the gamma rays that are emitted from the patient's brain through the skull. PET scans are of value in studying neurotransmitters and neuroreceptors as well as the substrate metabolism of the brain. They are also of value in the diagnosis and treatment of several neurological disorders, including schizophrenia and Parkinson's disease.

The development of the five basic regions of the brain—telencephalon, diencephalon, mesencephalon, metencephalon, and myelencephalon—was discussed in chapter 14. Distinct functional structures are formed from these regions (table 15.1), which will be discussed in greater detail in the following sections.

1. What characteristics do the brain and spinal cord have in common? Describe the general features of each.
2. Using specific examples, describe the metabolic requirements of the brain.

Table 15.1 Derivation and functions of the major brain structures

	Region	Structure	Function
Forebrain	Telencephalon	Cerebrum	Controls most sensory and motor activities; reasoning, memory, intelligence, etc.; instinctual and limbic functions
	Diencephalon	Thalamus	Relay center: all impulses (except olfactory) going into cerebrum synapse here; some sensory interpretation; initial autonomic response to pain
		Hypothalamus	Regulation of renal water flow, body temperature, hunger, heartbeat, etc.; control of secretory activity in anterior pituitary gland; instinctual and limbic functions
		Pituitary gland	Regulation of other endocrine glands
Midbrain	Mesencephalon	Superior colliculus	Visual reflexes
		Inferior colliculus	Auditory reflexes
		Cerebral peduncles	Coordinating reflexes; contain many motor fibers
Hindbrain	Metencephalon	Cerebellum	Balance and motor coordination
		Pons	Relay center; contains nuclei (pontine nuclei)
	Myelencephalon	Medulla oblongata	Relay center; contains many nuclei; visceral autonomic center (e.g., respiration, heart rate, vasoconstriction)

From Kent M. Van De Graaff, *Human Anatomy,* 2d ed.

Cerebrum

The cerebrum, consisting of five paired lobes within two convoluted hemispheres, is concerned with higher brain functions, such as the perception of sensory impulses, the instigation of voluntary movement, the storage of memory, thought processes, and reasoning ability. The cerebrum is also concerned with instinctual and limbic (emotional) functions.

Objective 3. Describe the structure of the cerebrum, and list the functions of the cerebral lobes.

Objective 4. Define the term *electroencephalogram,* and discuss its clinical importance.

Objective 5. Describe the fiber tracts within the cerebrum.

Objective 6. Describe the locations of the language centers and discuss the function of each.

Objective 7. Compare short-term and long-term memory.

Structure of the Cerebrum

The **cerebrum** *(ser'ĕ-brum),* located in the region of the telencephalon, is the largest and most obvious portion of the brain. It accounts for about 80% of the mass of the brain and is responsible for the higher mental functions, including memory and reason. The cerebrum consists of the **right** and **left hemispheres,** which are incompletely separated by a **longitudinal fissure** (fig. 15.4). Portions of the two hemispheres are connected internally by the **corpus callosum** *(kah-lo'sum),* a large tract of white matter (see fig. 15.2). A portion of the meninges, called the **falx** *(falks)* **cerebri,** extends into the longitudinal fissure. Each cerebral hemisphere contains a central cavity called the **lateral ventricle** (fig. 15.5), which is lined with ependymal cells and filled with cerebrospinal fluid.

cerebrum: L. *cerebrum,* brain

The two cerebral hemispheres carry out different functions. In most people, the left hemisphere controls analytical and verbal skills such as reading, writing, and mathematics (see fig. 18.43). The right hemisphere is the source of spatial and artistic kinds of intelligence. The corpus callosum unifies attention and awareness between the two hemispheres and permits a sharing of learning and memory. Severing the corpus callosum is a radical treatment to control severe epileptic seizures. Although this surgery is successful, it results in the cerebral hemispheres functioning as separate structures, each with its own memories and thoughts, competing for control. A more recent and effective technique of controlling epileptic seizures is a precise laser treatment of the corpus callosum.

The cerebrum consists of two layers. The surface layer, referred to as the **cerebral cortex,** is composed of gray matter that is 2–4 mm (0.08–0.16 in.) thick (fig. 15.5). Beneath the cerebral cortex is the thick **white matter** of the cerebrum, which constitutes the second layer. The cerebral cortex is characterized by numerous folds and grooves called **convolutions** *(kon-vo-lu'shuns).* Convolutions form during early fetal development, when brain size increases rapidly and the cortex enlarges out of proportion to the underlying white matter. The elevated folds of the convolutions are the **gyri** (singular, *gyrus*), and the depressed grooves are the **sulci** *(sul'si)* (singular, *sulcus*). The convolutions greatly increase the area of the gray matter, which is comprised of nerve cell bodies.

gyrus: Gk. *gyros,* circle
sulcus: L. *sulcus,* a furrow or ditch

Figure 15.4. The cerebrum. (*a*) a lateral view; (*b*) a superior view.

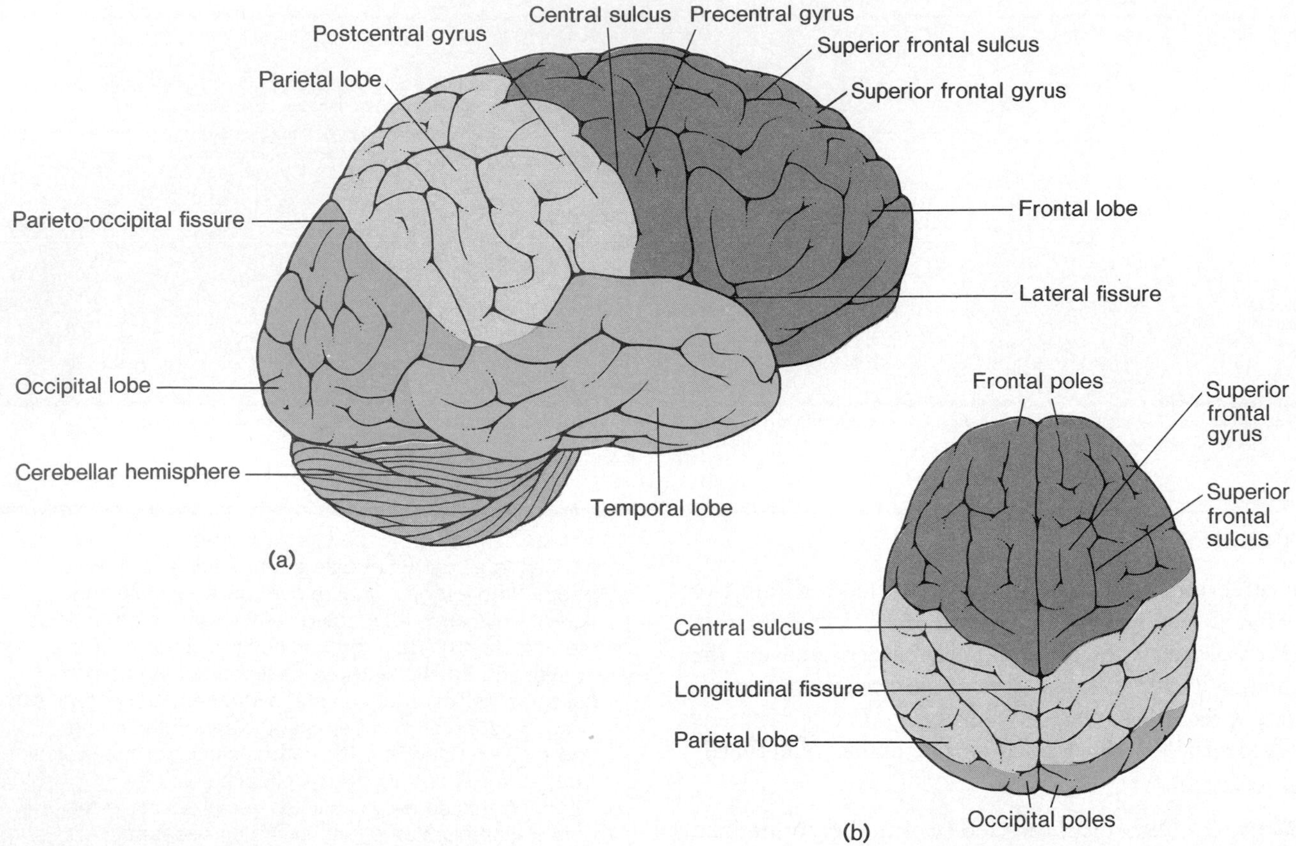

Recent studies indicate that with increased learning, there is an increase in the number of synapses between neurons within the cerebrum. Although the number of neurons is established during prenatal development, the number of synapses is variable depending upon the learning process. The number of cytoplasmic extensions from the cell body of a neuron determines the extent of nerve impulse conduction and the associations which can be made to cerebral areas already containing stored information.

Lobes of the Cerebrum

Each cerebral hemisphere is subdivided by deep sulci, or fissures, into the five lobes, four of which appear on the surface of the cerebrum and are named according to the overlying cranial bones (fig. 15.6). The reasons for the separate cerebral lobes, as well as two cerebral hemispheres, have to do with specificity of function (table 15.2).

Frontal Lobe. The frontal lobe forms the anterior portion of each cerebral hemisphere (fig. 15.6). A prominent deep furrow, called the **central sulcus (fissure of Rolando),** separates the frontal lobe from the parietal lobe. The central sulcus extends at right angles from the longitudinal fissure to the lateral sulcus. The **lateral sulcus (fissure of Sylvius)** extends laterally from the inferior surface of the cerebrum to separate the frontal and temporal lobes. The **precentral gyrus** (see figs. 15.4, 15.6), an important motor area, is positioned immediately in front of the central sulcus. The frontal lobe's functions include initiating voluntary motor impulses for the movement of skeletal muscles, analyzing sensory experiences, and providing responses relating to personality. The frontal lobes also involve responses related to memory, emotions, reasoning, judgment, planning, and verbal communication.

fissure of Rolando: from Luigi Rolando, Italian anatomist, 1773–1831
fissure of Sylvius: from Franciscus Sylvius de la Boe, Dutch anatomist, 1614–72

Figure 15.5. Sections through the cerebrum and diencephalon. (*a*) a coronal section. (*b*) a transverse section.

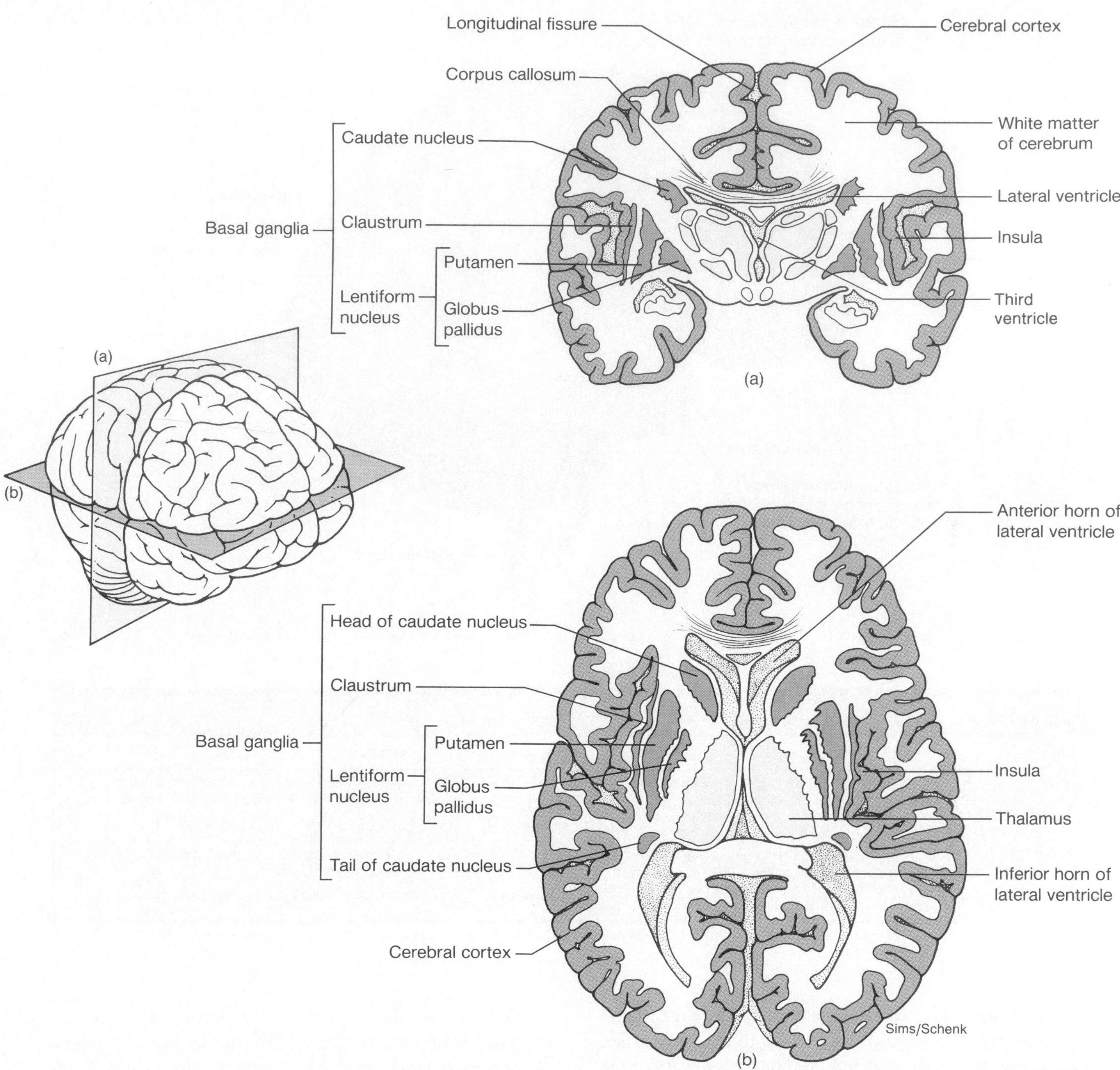

Figure 15.6. The lobes of the left cerebral hemisphere showing the principal motor and sensory areas of the cerebral cortex. Dotted lines show the division of the lobes.

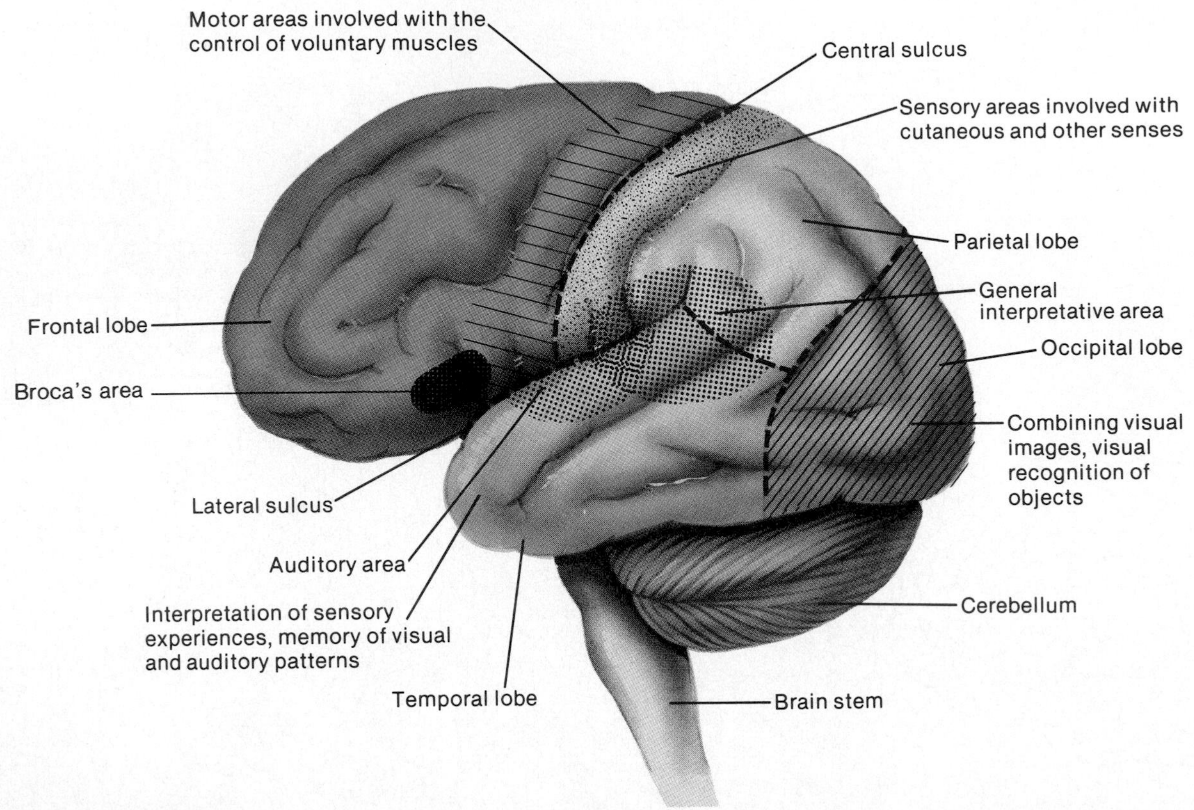

Table 15.2 Functions of the cerebral lobes

Lobe	Functions	Lobe	Functions
Frontal	Voluntary motor control of skeletal muscles; personality; higher intellectual processes (e.g., concentration, planning, and decision making); verbal communication	Temporal	Interpretation of auditory sensations; storage (memory) of auditory and visual experiences
		Occipital	Integrates movements in focusing the eye; correlating visual images with previous visual experiences and other sensory stimuli; conscious perception of vision
Parietal	Somatesthetic interpretation (e.g., cutaneous and muscular sensations); understanding speech and formulating words to express thoughts and emotions; interpretation of textures and shapes	Insula	Memory; integrates other cerebral activities

Parietal Lobe. The parietal lobe is posterior to the frontal lobe. Except for the central sulcus along its anterior border, the parietal lobe lacks distinct boundaries. An important sensory area called the **postcentral gyrus** (see figs. 15.4, 15.6) is positioned immediately behind the central sulcus. The postcentral gyrus is designated as a somatesthetic area because it responds to stimuli from cutaneous and muscular receptors throughout the body.

The size of the portions of the precentral gyrus responsible for motor movement and the size of the portions of the postcentral gyrus that respond to sensory stimuli do not correspond to the size of the part of the body being served but to the number of motor units activated or the density of receptors (fig. 15.7). For example, because the hand has many motor units and sensory receptors, larger portions of the precentral and postcentral gyri serve it than serve the thorax, which is much larger in size.

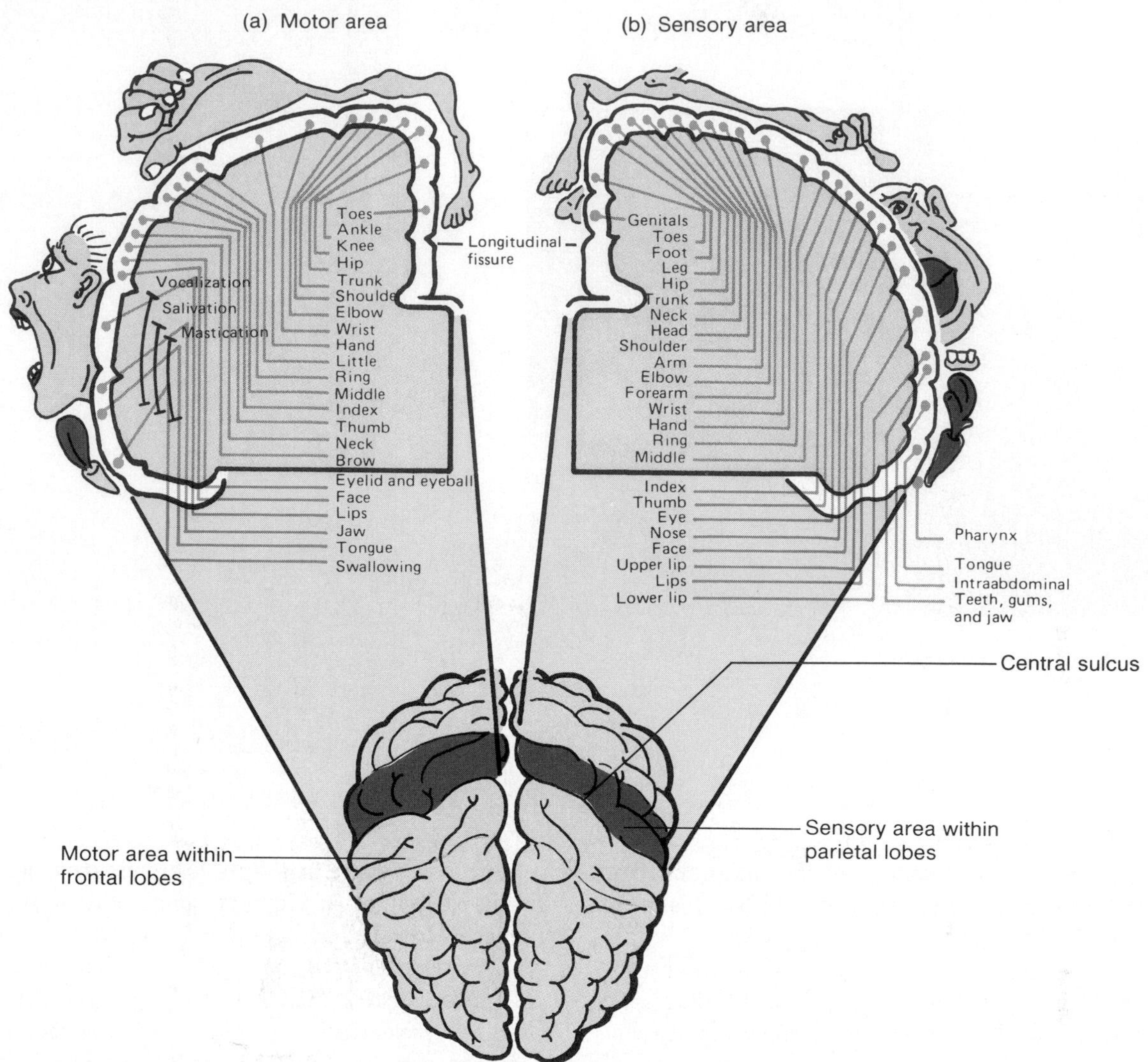

Figure 15.7. Motor and sensory areas of the cerebral cortex. (*a*) motor areas that control skeletal muscles. (*b*) sensory areas that receive somatesthetic sensations. Note that larger areas of the cortex are devoted to the hands and face than to the rest of the body.

In addition to responding to somatesthetic stimuli, the parietal lobe functions in the understanding of speech and in verbal articulation of thoughts and emotions. The parietal lobe also interprets the textures and shapes of objects as they are handled.

Temporal Lobe. The temporal lobe is located below the parietal lobe and the posterior portion of the frontal lobe. It is separated from both by the lateral sulcus (see fig. 15.6). The temporal lobe contains auditory centers that receive sensory fibers from the cochlea of the ear. This lobe also interprets some sensory experiences and stores memories of both auditory and visual experiences.

Occipital Lobe. The occipital lobe forms the posterior portion of the cerebrum and has no distinct separation from the temporal and parietal lobes (see fig. 15.6). The occipital lobe is superior to the cerebellum and is separated from it by a shell-like infolding of the meningeal layer called the **tentorium cerebelli** (see fig. 15.2). The principal functions of the occipital lobe concern vision. The occipital lobe integrates eye movements by directing and focusing the eye. It is also responsible for visual association, correlating visual images with previous visual experiences and other sensory stimuli.

Figure 15.8. Brain waves. (*a*) a technician using an electroencephalograph to take the EEG of a teenaged boy. (*b*) types of EEG waves.

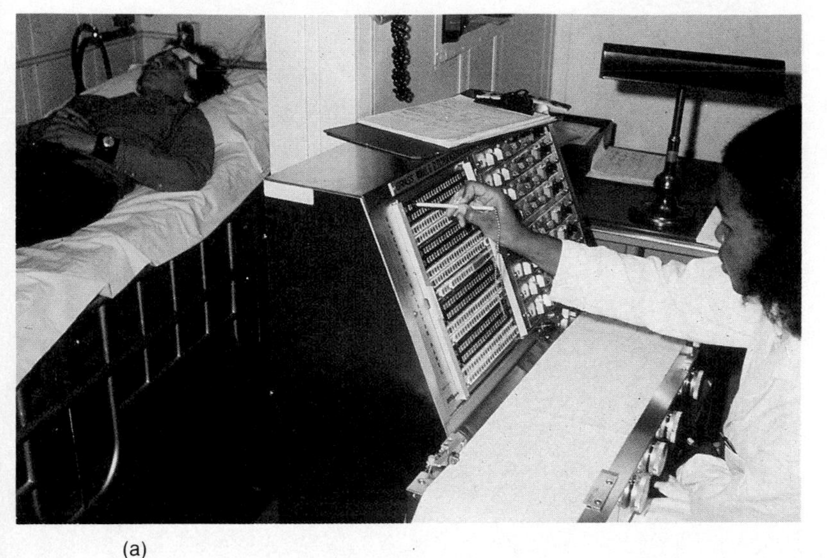

(a)

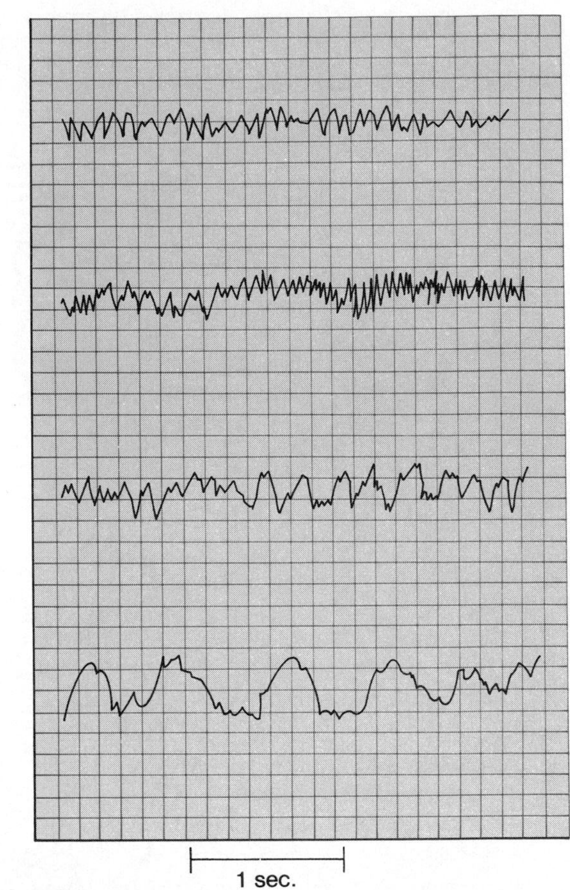

(b)

Insula. The insula is a deep lobe of the cerebrum that cannot be viewed on the surface (see fig. 15.5). It is deep to the lateral sulcus and is covered by portions of the frontal, parietal, and temporal lobes. Little is known of the function of the insula, though it primarily integrates other cerebral activities and may have some function in memory.

> Because of its size and position, portions of the cerebrum frequently suffer brain trauma. A concussion to the brain may cause a temporary or permanent impairment of cerebral functions; a stroke usually affects cerebral function. Much of what is known about cerebral function comes from observing body dysfunctions when specific regions of the cerebrum are traumatized.

Brain Waves. Neurons within the cerebral cortex continuously generate electrical activity, which can be recorded by electrodes attached to precise locations on the scalp, producing an **electroencephalogram (EEG).** An EEG pattern, commonly called *brain waves,* is the collective expression of millions of action potentials from neurons.

Brain waves are first emitted from a developing brain during early fetal development and continue throughout a person's life. The cessation of brain-wave patterns may be a decisive factor in the legal determination of death.

Certain distinct EEG patterns signify healthy mental functions. Deviations from these patterns are of clinical significance in diagnosing trauma, mental depression, hematomas, and various diseases such as tumors, infections, and epilepsy. Normally, there are four kinds of EEG patterns (fig. 15.8).

1. **Alpha waves** are best recorded from the parietal and occipital regions while a person is awake and relaxed but with the eyes closed. These waves are rhythmic oscillations of about 10–12 cycles/second. The alpha rhythm of a child younger than eight years old occurs at a slightly lower frequency of 4–7 cycles/second.
2. **Beta waves** are strongest from the frontal lobes, especially the area near the precentral gyrus. These waves are sensory evoked and respond to visual and mental activity. Because they respond to stimuli from receptors and are superimposed on the continuous activity patterns of the alpha waves, they constitute *evoked activity.* The frequency of beta waves is 13–25 cycles/second.

---

insula: L. *insula,* island

Figure 15.9. Types of fiber tracts within the white matter associated with the cerebrum. (*a*) association fibers of a given hemisphere. (*b*) commissural fibers connecting the hemispheres and projection fibers connecting the hemispheres with other structures of the CNS. Note the decussation (crossing over) of projection fibers within the medulla oblongata.

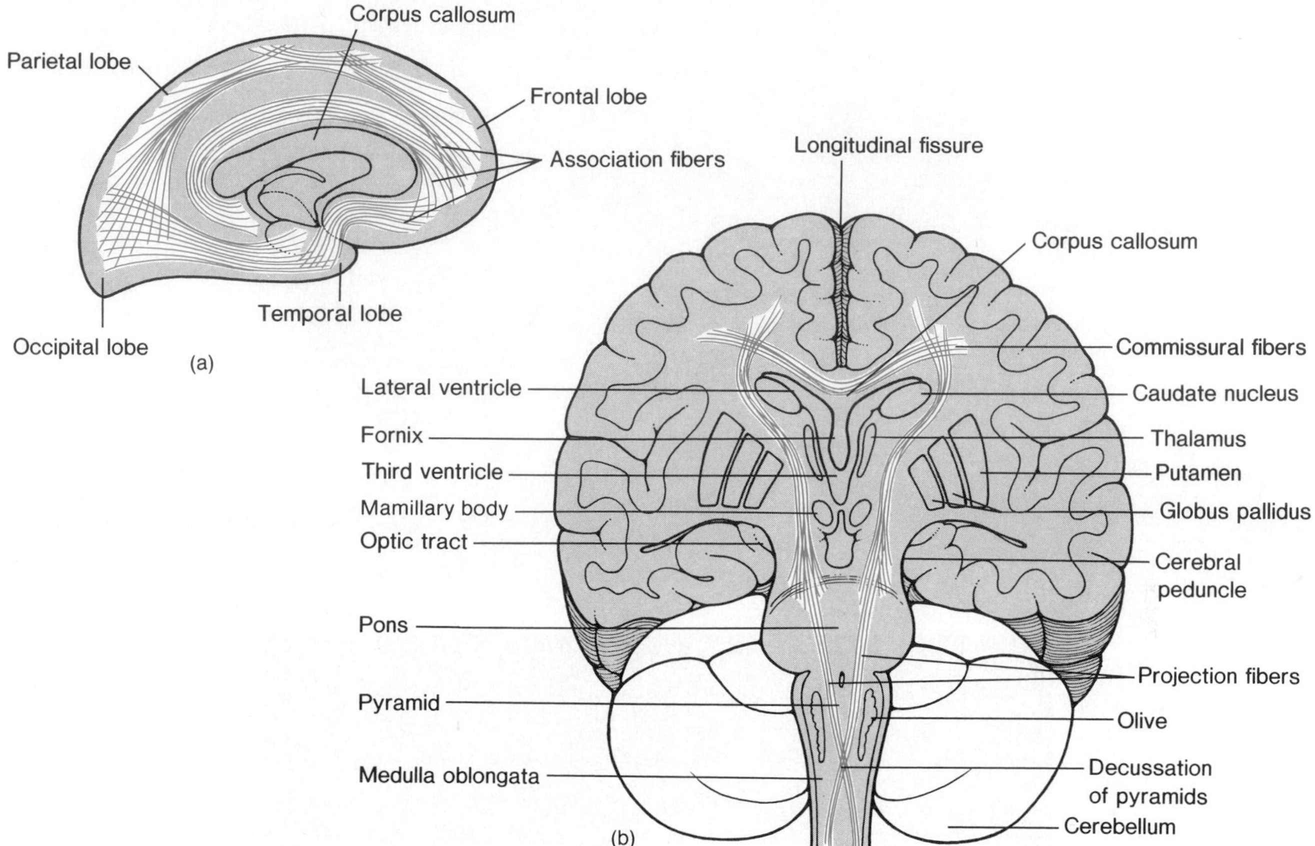

3. **Theta waves** are emitted from the temporal and occipital lobes. They have a frequency of 5–8 cycles/second and are common in newborn infants. The recording of theta waves in adults generally indicates severe emotional stress and can be a forewarning of a nervous breakdown.
4. **Delta waves** seem to be emitted in a general pattern from the cerebral cortex. These waves have a frequency of 1–5 cycles/second and are common during sleep and in an awake infant. The presence of delta waves in an awake adult indicates brain damage.

White Matter of the Cerebrum

The thick **white matter** of the cerebrum is deep to the cortex (see fig. 15.5) and consists of dendrites, myelinated axons, and associated neuroglia. These fibers form the billions of connections within the brain by which information in the form of electrical impulses is transmitted to the appropriate places. There are three types of fiber tracts within the white matter, which are named according to location and the direction in which they conduct impulses (fig. 15.9).

1. **Association fibers** are confined to a given cerebral hemisphere and conduct impulses between neurons within that hemisphere.
2. **Commissural** *(kom-mis'u-ral)* **fibers** connect the neurons and gyri of one hemisphere with those of the other. The **corpus callosum** and **anterior commissure** *(kom'ĭ-shŭr)* (fig. 15.10) are composed of commissural fibers.
3. **Projection fibers** form the ascending and descending tracts that transmit impulses from the cerebrum to other parts of the brain and spinal cord and from the spinal cord and other parts of the brain to the cerebrum.

Figure 15.10. A midsagittal section through the brain. (a) a diagram. (b) a photograph.

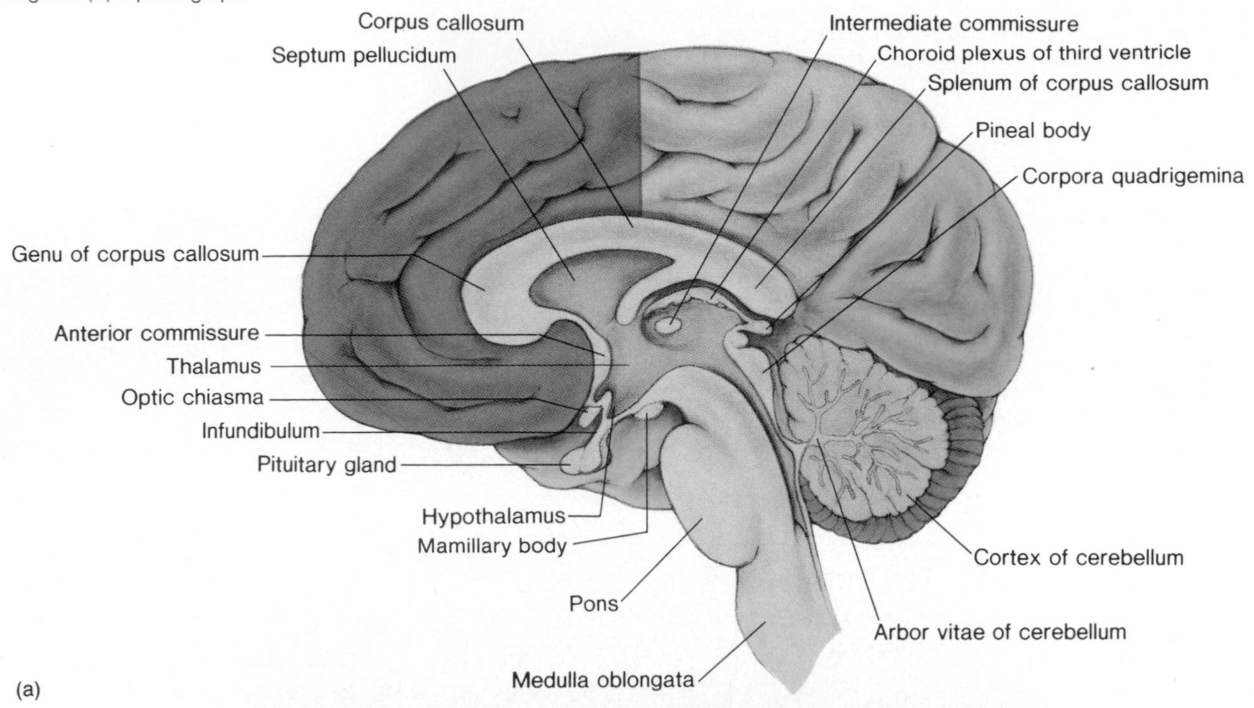

(a)

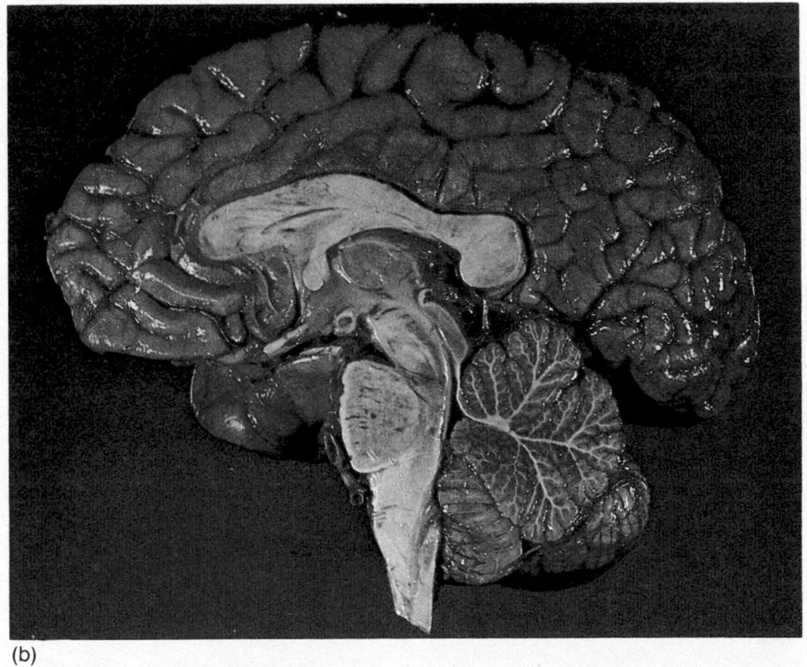

(b)

Basal Ganglia

The **basal ganglia** are specialized paired masses of gray matter located deep within the white matter of the cerebrum (see fig. 15.5). The most prominent of the basal ganglia is the **corpus striatum,** so named because of its striped appearance. The corpus striatum is composed of several masses of nuclei. The **caudate nucleus** is the upper mass. A thick band of white matter lies between the caudate nucleus and the next lower two masses, collectively called the **lentiform nucleus.** The lentiform nucleus consists of a lateral portion, called the **putamen** *(pu-ta'men),* and a medial portion, called the **globus pallidus** (fig. 15.5). The

corpus striatum: L. *corpus,* body; *striare,* striped

lentiform: L. *lentis,* elongated
putamen: L. *putare,* to cut, prune
globus pallidus: L. *globus,* sphere; *pallidus,* pale

Figure 15.11. Brain areas involved in the control of speech. Arrows indicate the direction of communication between these areas.

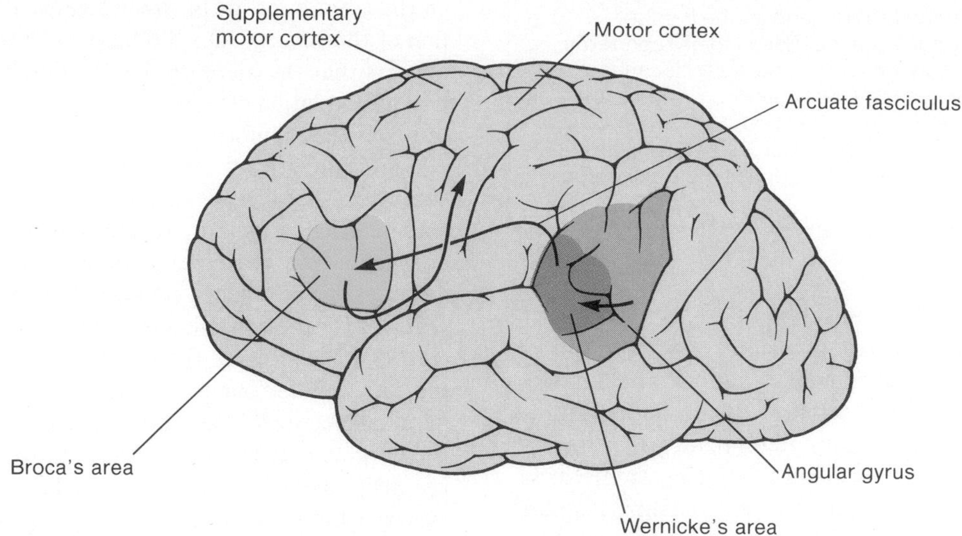

claustrum is another portion of the basal ganglia. It is a thin layer of gray matter just deep to the cerebral cortex of the insula.

The basal ganglia are associated with other structures of the brain, particularly within the mesencephalon. The caudate nucleus and putamen of the basal ganglia control unconscious contractions of certain skeletal muscles, such as those of the upper extremities involved in involuntary arm movements during walking. The globus pallidus regulates the muscle tone necessary for specific, intentional body movements. Neural diseases or physical trauma to the basal ganglia generally cause a variety of motor movement dysfunctions, including rigidity, tremor, and rapid and aimless movements.

Language

Knowledge of the brain regions involved in language has been gained primarily by the study of *aphasias*—speech and language disorders caused by damage to specific language areas of the brain. These areas (fig. 15.11) are generally located in the cerebral cortex of the left hemisphere in both right-handed and left-handed people.

Broca's area, or the **motor speech area,** is located in the left inferior gyrus of the frontal lobe. Neural activity in Broca's area causes selective stimulation of motor impulses in motor centers elsewhere in the frontal lobe, which in turn causes coordinated skeletal muscle movement in the pharynx and larynx. At the same time, motor impulses are sent to the respiratory muscles to regulate air movement across the vocal cords. The combined muscular stimulation translates thought patterns into speech.

Wernicke's area is located in the superior gyrus of the temporal lobe and is directly connected to Broca's area by a fiber tract called the **arcuate fasciculus.** People with *Wernicke's aphasia* produce speech that has been described as a "word salad." The words used may be real words that are chaotically mixed together, or they may be made-up words. Language comprehension has been destroyed in people with Wernicke's aphasia; they cannot understand either spoken or written language.

It appears that the concept of words to be spoken originates in Wernicke's area and is then communicated to Broca's area through the arcuate fasciculus. Damage to the arcuate fasciculus produces *conduction aphasia,* which is fluent but nonsensical speech as in Wernicke's aphasia, even though both Broca's and Wernicke's areas are intact.

The **angular gyrus,** located at the junction of the parietal, temporal, and occipital lobes, is believed to be a center for the integration of auditory, visual, and somatesthetic information. Damage to the angular gyrus produces aphasias, which suggests that this area projects to Wernicke's area. Some patients with damage to the left angular gyrus can speak and understand spoken language but cannot read or write. Other patients can write a sentence but cannot read it, presumably due to damage to the projections from the occipital lobe (involved in vision) to the angular gyrus.

aphasia: L. *a,* without; Gk. *phasis,* speech
Broca's area: from Pierre P. Broca, French neurologist, 1824–80

Wernicke's area: from Karl Wernicke, German neurologist, 1848–1905

Recovery of language ability, by transfer to the right hemisphere after damage to the left hemisphere, is very good in children but decreases after adolescence. Recovery is reported to be faster in left-handed people, possibly because language ability is more evenly divided between the two hemispheres in left-handed people. Some recovery usually occurs after damage to Broca's area, but damage to Wernicke's area produces more severe and permanent aphasias.

Memory

Clinical studies of *amnesia* suggest that several different brain regions are involved in memory storage and retrieval. Amnesia has been found to result from damage to the temporal lobe of the cerebral cortex, hippocampus, head of the caudate (in Huntington's disease), or the dorsomedial thalamus (in alcoholics suffering from Korsakoff's syndrome with thiamine deficiency). Clinical studies also suggest that there are two major categories of memory: **short-term memory** and **long-term memory.** People with head trauma, for example, and patients with suicidal depression who are treated by *electroconvulsive shock* (*ECS*) therapy, may lose their memory of recent events but retain their older memories.

The **hippocampus** (see fig. 17.12) appears to be required for short-term memory and for the consolidation of that memory into a long-term form. Surgical removal of the left hippocampus impairs the consolidation of short-term verbal memories, and removal of the right hippocampus impairs the consolidation of nonverbal memories. Surgical removal of both the right and left hippocampus was performed in one patient, designated "H.M.," in an effort to treat his epilepsy. After the surgery he was unable to consolidate any short-term memory. He could repeat a phone number and carry out a normal conversation; he could not remember the phone number if momentarily distracted, however, and if the person to whom he was talking left the room and came back a few minutes later, H.M. would have no recollection of seeing that person or of having had a conversation with that person before. Although his memory of events that occurred before the operation was intact, all subsequent events in his life seemed as if they were happening for the first time.

The cerebral cortex is thought to store factual information, with verbal memories lateralized to the left hemisphere and visuospatial information in the right hemisphere. The neurosurgeon Wilder Penfield has electrically stimulated various regions in the brain of awake patients, often evoking visual or auditory memories that were extremely vivid. Electrical stimulation of specific points in the temporal lobe evoked specific memories that were so detailed the patient felt that he was reliving the experience. Surgical removal of these regions did not, however, abolish the memory. The amount of memory destroyed by ablation of brain tissue appears to depend more on the amount of brain tissue removed than on the location of the surgery. On the basis of these observations, it appears that the memory may be diffusely located in the brain; stimulation of the correct location of the cortex then retrieves the memory.

Since long-term memory is not abolished by electroconvulsive shock, it seems reasonable to conclude that the consolidation of memory depends on relatively permanent changes in the chemical structure of neurons and their synapses. Experiments suggest that protein synthesis is required for the consolidation of the "memory trace." According to one theory, these proteins may be secreted into the extracellular environment, where they influence synaptic connections. According to another theory, new receptor proteins in the membrane of the postsynaptic neuron are made available as a result of high-frequency stimulation of the presynaptic neuron. This would help to account for the increased sensitivity of postsynaptic neurons to neurotransmitter, as seen in post-tetanic potentiation (discussed in chapter 14). Much more research is obviously needed in this exciting area of physiology before memory can be fully explained at a cellular and molecular level.

1. Diagram a lateral view of the cerebrum, and label the four superficial lobes and the fissures that separate them.
2. List the functions of each of the paired cerebral lobes.
3. What is a brain-wave pattern? How are these patterns monitored clinically?
4. Describe the arrangement of the fiber tracts within the cerebrum.
5. Define *basal ganglia,* and list their functions.
6. Describe the aphasias produced by damage to Broca's and Wernicke's areas, by damage to the arcuate fasciculus, and by damage to the angular gyrus. Explain how these areas may interact in the production of speech.
7. Explain the difference between short-term and long-term memory, and describe the possible roles of different brain regions in memory.

Diencephalon

The diencephalon is a major autonomic region of the brain that consists of vital structures such as the thalamus, hypothalamus, epithalamus, and pituitary gland.

Objective 8. List the autonomic functions of the thalamus and the hypothalamus.

Objective 9. Describe the location and structure of the pituitary gland.

The **diencephalon** *(di''en-sef'ah-lon)* is the second subdivision of the forebrain and is almost completely surrounded by the cerebral hemispheres of the telencephalon.

The third ventricle (see fig. 15.22) forms a midplane cavity within the diencephalon. The most important structures of the diencephalon are the thalamus, hypothalamus, epithalamus, and pituitary gland.

Thalamus

The **thalamus** *(thal'ah-mus)* is a large ovoid mass of gray matter, constituting nearly four-fifths of the diencephalon. It is actually a paired organ, with each portion positioned immediately below the lateral ventricle of its respective cerebral hemisphere (see figs. 15.5, 15.10). The principal function of the thalamus is to act as a relay center for all sensory impulses, except smell, to the cerebral cortex. Specialized masses of nuclei relay the incoming impulses to precise locations within the cerebral lobes for interpretation.

The thalamus also performs some sensory interpretation. The cerebral cortex discriminates pain and other tactile stimuli, but the thalamus responds to general sensory stimuli and provides crude awareness. The thalamus probably plays a role in the initial autonomic response of the body to intense pain and is, therefore, partially responsible for the physiological shock that frequently follows serious trauma.

Hypothalamus

The **hypothalamus** *(hi''po-thal'ah-mus)* is a small portion of the diencephalon located below the thalamus, where it forms the floor and part of the lateral walls of the third ventricle (fig. 15.10). The hypothalamus consists of several masses of nuclei interconnected with other parts of the nervous system. Despite its small size, the hypothalamus performs numerous vital functions, most of which relate directly or indirectly to the regulation of visceral activities. It also performs emotional and instinctual functions.

The hypothalamus has been described as an autonomic nervous center because of its role in accelerating or decreasing certain body functions. Another major function of the hypothalamus is to regulate the release of hormones from the pituitary gland. The principal autonomic and limbic functions of the hypothalamus are described below.

1. **Cardiovascular regulation.** Although the heart has an innate pattern of contraction, impulses from the hypothalamus cause autonomic acceleration or deceleration of the heart. Impulses from the posterior hypothalamus produce a rise in arterial blood pressure and an increase of the heart rate. Impulses from the anterior portion have an opposite effect. The impulses from these regions do not travel directly to the heart but pass first to the cardiovascular centers of the medulla oblongata.
2. **Body-temperature regulation.** Specialized nuclei within the anterior portion of the hypothalamus are sensitive to changes in body temperature. If the arterial blood flowing through this portion of the hypothalamus is above normal temperature, the hypothalamus initiates impulses that cause heat loss through sweating and vasodilation of cutaneous vessels of the skin. A blood temperature below normal causes the hypothalamus to relay impulses that result in heat production and retention through the initiation of shivering, the contraction of cutaneous blood vessels, and the cessation of sweating.
3. **Regulation of water and electrolyte balance.** Specialized *osmoreceptors* in the hypothalamus continuously monitor the osmotic concentration of the blood. An increased osmotic concentration due to lack of water causes antidiuretic hormone (ADH) to be produced by the hypothalamus and released from the posterior pituitary gland. At the same time, a *thirst center* within the hypothalamus produces feelings of thirst.
4. **Regulation of hunger and control of gastrointestinal activity.** The *feeding center* is a specialized portion of the lateral hypothalamus that monitors the blood glucose, fatty-acid, and amino-acid levels. Low levels of these substances in the blood are partially responsible for a sensation of hunger elicited from the hypothalamus. When sufficient amounts of food have been ingested, the *satiety (sah-ti'e-te) center* in the midportion of the hypothalamus inhibits the feeding center. The hypothalamus also receives sensory impulses from the abdominal viscera and regulates glandular secretions and the peristaltic movements of the digestive tract.
5. **Regulation of sleeping and wakefulness.** The hypothalamus has both a *sleep center* and a *wakefulness center* that function with other parts of the brain to determine the level of conscious alertness.
6. **Sexual response.** Specialized *sexual center* nuclei within the dorsal portion of the hypothalamus respond to sexual stimulation of the tactile receptors within the genital organs. The experience of orgasm involves neural activity within the sexual center of the hypothalamus.
7. **Emotions.** Found within the hypothalamus are a number of nuclei associated with specific emotional responses such as anger, fear, pain, and pleasure.
8. **Control of endocrine functions.** The hypothalamus produces neurosecretory chemicals that stimulate the anterior pituitary to release various hormones, which in turn regulate other endocrine glands (chapter 19). The hypothalamus also produces the two hormones secreted by the posterior pituitary gland.

---

thalamus: L. *thalamus,* inner room

Figure 15.12. The pituitary gland is positioned within the sella turcica of the sphenoid bone and is attached to the brain by the infundibulum.

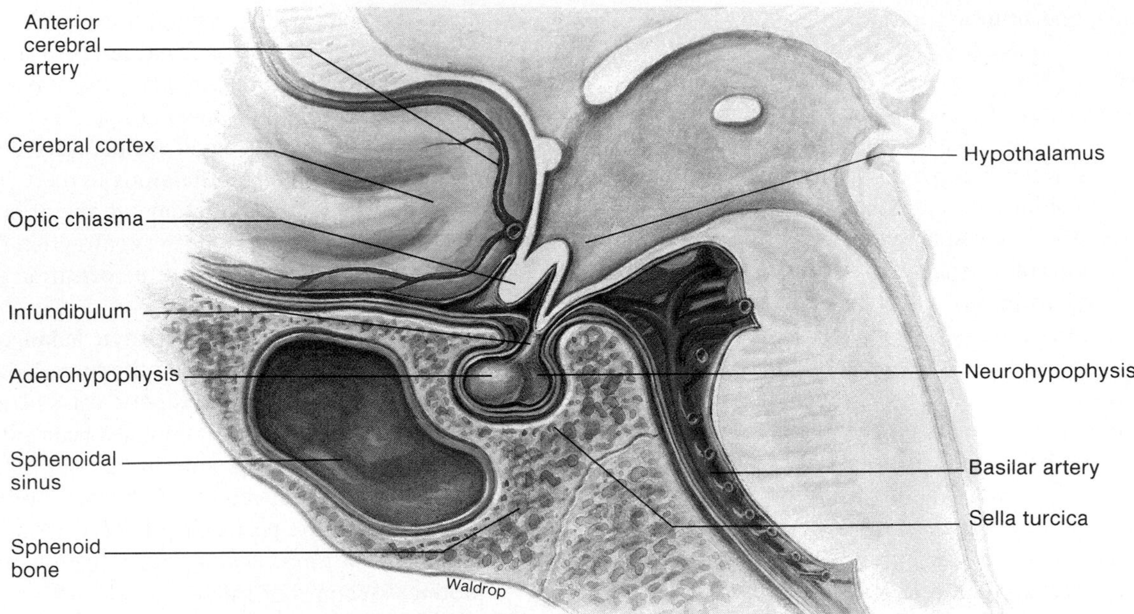

Obviously the hypothalamus is one of the most vital structures of the body. Dysfunction of the hypothalamus may seriously affect the autonomic, somatic, or psychic body functions. Unsurprisingly, the hypothalamus is implicated as a principal factor in *psychosomatic illness*. Insomnia, peptic ulcers, palpitation of the heart, diarrhea, and constipation are a few symptoms of psychosomatic problems.

Epithalamus

The **epithalamus** is the dorsal portion of the diencephalon that includes a thin roof over the third ventricle (see fig. 15.22). The inside lining of the roof consists of a vascular **choroid plexus** where cerebrospinal fluid is produced (see fig. 15.10). A small cone-shaped mass, called the **pineal gland,** or body (**epiphysis),** which has a neuroendocrine function, extends outward from the posterior end of the epithalamus (see fig. 15.10). The posterior commissure, located ventral to the pineal gland, is a tract of commissural fibers that connects the superior colliculi (see fig. 15.15).

Pituitary Gland

The **pituitary** *(pĭ-tu'ĭ-tār''e)* **gland,** or **hypophysis** *(hi-pof'ĭ-sis),* is positioned on the inferior aspect of the diencephalon and is attached to the hypothalamus by a stalk-like structure called the **infundibulum** *(in''fun-dib'u-lum)* (see fig. 15.10). The pituitary is a rounded, pea-shaped gland measuring about 1.3 cm (0.5 in.) in diameter. It is covered by the dura mater and is supported by the sella turcica of the sphenoid bone (fig. 15.12). The arterial circle (circle of Willis) (see fig. 20.30) surrounds the highly vascular pituitary gland, providing it with a rich blood exchange. The pituitary, which has an endocrine function, is structurally and functionally divided into an anterior portion, called the **adenohypophysis** *(ad''ĕ-no-hi-pof'ĭ-sis),* and a posterior portion, called the **neurohypophysis.**

pineal: L. *pinea,* pine cone
pituitary: L. *pituita,* phlegm (this gland was originally thought to secrete mucus into nasal cavity)
infundibulum: L. *infundibulum,* funnel

1. Discuss what is meant by the statement that the thalamus is the pain center of the brain.
2. List the body systems and the functions over which the hypothalamus has some control.
3. Describe the location of the pituitary gland relative to the rest of the diencephalon and the rest of the brain. What are the two portions of the pituitary gland, and how are they supported?

Mesencephalon

The mesencephalon contains the corpora quadrigemina, concerned with visual and auditory reflexes, and the cerebral peduncles, composed of fiber tracts. It also contains specialized nuclei that help to control posture and movement.

Objective 10. List the structures of the mesencephalon, and explain their functions.

The **mesencephalon** *(mes''en-sef'ah-lon),* or midbrain, is a short section of the brain stem between the diencephalon and the pons (see fig. 15.15). Within the midbrain is the

Figure 15.13. Nuclei within the pons and medulla oblongata that constitute the respiratory center.

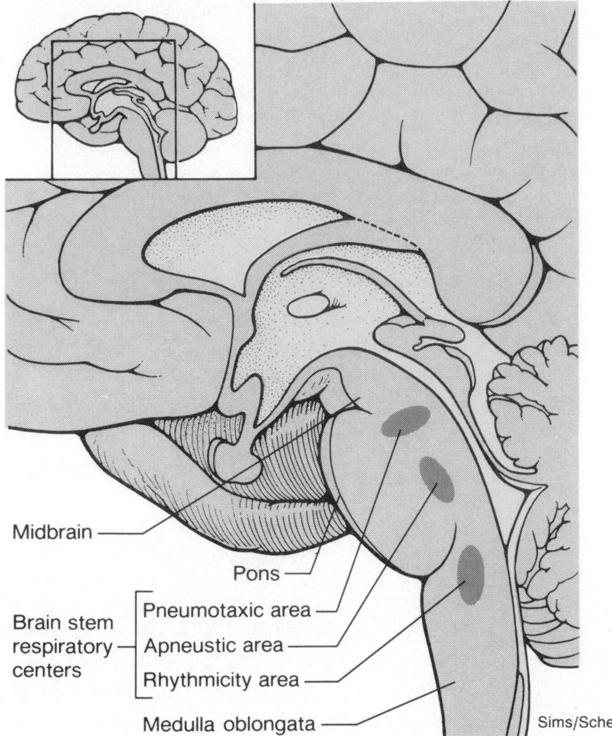

cerebral aqueduct (aqueduct of Sylvius) (see fig. 15.22), which connects the third and fourth ventricles. The midbrain also contains the corpora quadrigemina (fig. 15.10), cerebral peduncles (see figs. 15.9, 15.15), red nucleus, and substantia nigra.

The **corpora quadrigemina** *(kwod"rĭ-jem'ĭ-nah)* are the four rounded elevations on the dorsal portion of the midbrain. The two upper eminences, the **superior colliculi,** are concerned with visual reflexes. The two posterior eminences, the **inferior colliculi,** are responsible for auditory reflexes. The **cerebral peduncles** *(pe-dung'k'l)* are a pair of cylindrical structures composed of ascending and descending projection fiber tracts that support and connect the cerebrum to the other regions of the brain.

The **red nucleus** is deep within the midbrain between the cerebral peduncle and the cerebral aqueduct. The red nucleus is gray matter that connects the cerebral hemispheres and the cerebellum and functions in reflexes concerned with motor coordination and maintaining posture. Another nucleus called the **substantia nigra** is ventral to the red nucleus. The substantia nigra is thought to inhibit involuntary movements.

aqueduct of Sylvius: from Jacobus Sylvius, French anatomist, 1478–1555
corpora quadrigemina: L. *corpus,* body; *quadri,* four; *geminus,* twin
colliculus: L. *colliculus,* small mound

1. Discuss what possible symptoms might be apparent as a result of a tumor in the midbrain.
2. Which structures of the midbrain function with the cerebellum in controlling posture and movement?

Metencephalon

The metencephalon contains the pons, which relays impulses, and the cerebellum, which coordinates skeletal muscle contractions.

Objective 11. Describe the location and structure of the pons and cerebellum, and list their functions.

The **metencephalon** *(met"en-sef'ah-lon)* is the most superior portion of the hindbrain. Two vital structures of the metencephalon are the pons and cerebellum. The cerebral aqueduct of the mesencephalon enlarges to become the **fourth ventricle** (see fig. 15.22) within the metencephalon and myelencephalon.

Pons

The **pons** can be observed as a rounded bulge on the underside of the brain, between the midbrain and the medulla oblongata (fig. 15.13). The pons consists of white fiber tracts that course in two principal directions. The surface fibers extend transversely to connect with the cerebellum through the middle cerebellar peduncles. The deeper longitudinal fibers are part of the motor and sensory tracts that connect the medulla with the tracts of the midbrain.

Scattered throughout the pons are several nuclei associated with specific cranial nerves. The cranial nerves that have nuclei within the pons include the trigeminal (V), which transmits impulses for chewing and sensory sensations from the head; the abducens (VI), which controls certain movements of the eyeball; the facial (VII), which transmits impulses for facial movements and sensory sensations from the taste buds; and the vestibular branches of the vestibulocochlear (VIII), which maintain equilibrium.

Other nuclei of the pons function with nuclei of the medulla oblongata to regulate the rate and depth of breathing. The two respiratory centers of the pons are called the **apneustic** and the **pneumotaxic areas** (fig. 15.13).

Cerebellum

The **cerebellum** *(ser"ĕ-bel'um)* is the second largest structure of the brain. It is located in the metencephalon and occupies the inferior and posterior aspect of the cranial cavity. The cerebellum is separated from the overlying cerebrum by a **transverse fissure** (see fig. 15.2). A portion

pons: L. *pons,* bridge
cerebellum: L. *cerebellum,* diminutive of *cerebrum,* brain

Figure 15.14. The structure of the cerebellum. (*a*) a superior view; (*b*) an inferior view.

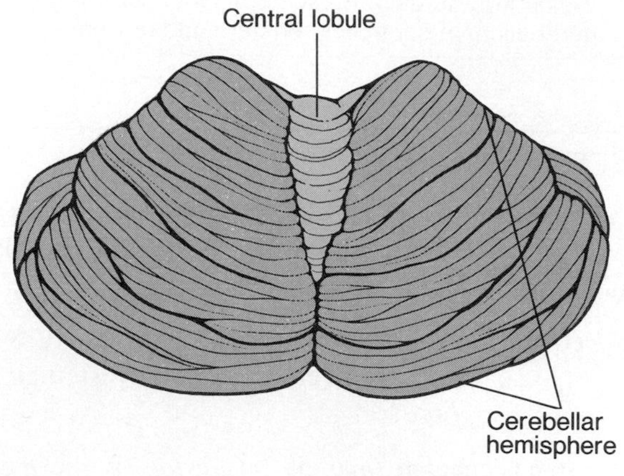

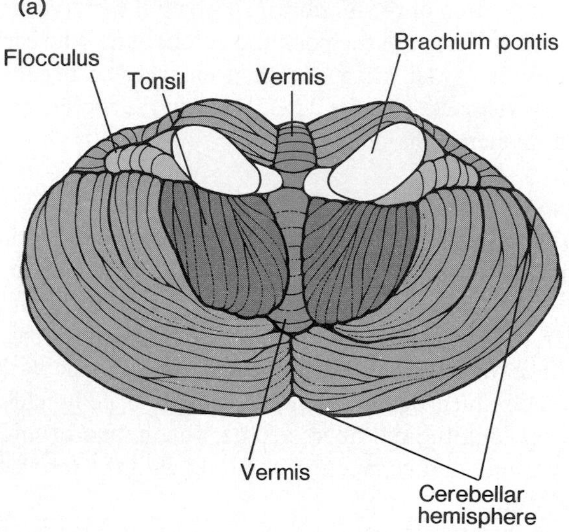

of the meninges called the **tentorium cerebelli** extends into the transverse fissure. The cerebellum consists of two **hemispheres** and a central constricted area called the **vermis** (fig. 15.14). The **falx cerebelli** is the portion of the meninges that partially extends between the hemispheres (see fig. 15.18).

Like the cerebrum, the cerebellum has a thin, outer layer of gray matter, called the **cerebellar cortex,** and a thick, deeper layer of white matter. The cerebellum is convoluted into a series of slender, parallel **gyri.** The tracts of white matter within the cerebellum have a distinctive branching pattern called the **arbor vitae,** which can be seen in a sagittal view (see fig. 15.10).

vermis: L. *vermis,* worm
arbor vitae: L. *arbor,* tree; *vitae,* life

Figure 15.15. The cerebellar peduncles can be seen when the cerebellum has been removed from its attachment to the brain stem.

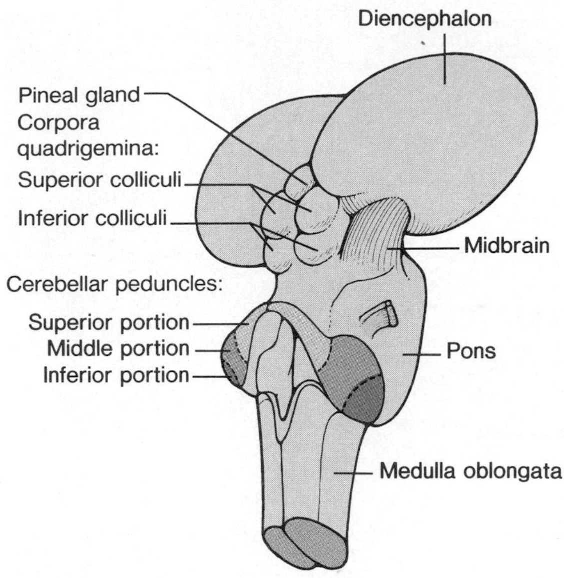

Three paired bundles of nerve fibers, called **cerebellar peduncles,** support the cerebellum and provide it with tracts for communicating with the rest of the brain (fig. 15.15). The cerebellar peduncles are as follows.

1. **Superior cerebellar peduncles** connect the cerebellum with the midbrain. The fibers within these peduncles originate primarily from specialized **dentate nuclei** within the cerebellum and pass through the red nucleus to the thalamus and then to the motor areas of the cerebral cortex. Impulses through the fibers of these peduncles provide feedback to the cerebrum.
2. **Middle cerebellar peduncles** convey impulses of voluntary movement from the cerebrum through the pons and to the cerebellum.
3. **Inferior cerebellar peduncles** connect the cerebellum with the medulla oblongata and the spinal cord. They contain both incoming vestibular and proprioceptive fibers and outgoing motor fibers.

The principal function of the cerebellum is coordinating skeletal muscle contractions. The cerebellum does this by recruiting precise motor units within the muscles voluntarily, subconsciously, or reflexively. Impulses for voluntary muscular movement originate in the cerebral

peduncle: L. *peduncle,* diminutive of *pes,* foot

Figure 15.16. A sagittal section of the medulla oblongata and pons showing the cranial nerve nuclei of gray matter.

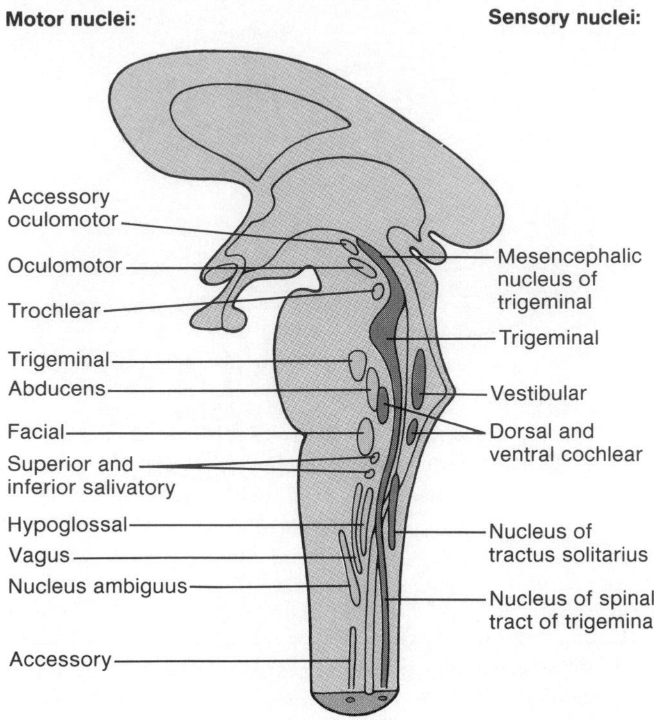

cortex and are coordinated by the cerebellum. The cerebellum constantly initiates impulses to selective motor units for maintaining posture and muscle tone. The cerebellum also adjusts to incoming impulses from **proprioceptors** *(pro''pre-o-sep'tor)* within muscles, tendons, joints, and special sense organs.

Trauma or diseases of the cerebellum, such as *cerebral palsy* or a *stroke,* frequently cause an impairment of skeletal muscle function. Movements become jerky and uncoordinated in a condition known as *ataxia.* There is also a loss of equilibrium, resulting in a disturbance of gait. *Alcohol intoxication* causes similar uncoordinated body movements.

1. Describe the locations and relative sizes of the pons and the cerebellum.
2. Which cranial nerves have nuclei located within the pons?
3. Define the terms *tentorium cerebelli, vermis, arbor vitae,* and *cerebellar peduncles.*
4. List the functions of the pons and the cerebellum.

Myelencephalon

The medulla oblongata, contained within the myelencephalon, connects to the spinal cord and contains nuclei for the cranial nerves and vital autonomic functions.

Objective 12. Describe the location and structure of the medulla oblongata, and list its functions.

Objective 13. Define the term *reticular formation,* and explain its function.

Medulla Oblongata

The **medulla oblongata** *(mĕ-dul'ah ob''long-gah'tah),* or simply **medulla,** is a bulbous structure about 3 cm (1 in.) long that is continuous with the pons anteriorly and the spinal cord posteriorly at the level of the foramen magnum (see figs. 15.9, 15.10). Externally, the medulla resembles the spinal cord, except for the two triangular, elevated structures, called **pyramids,** on the ventral side and an oval enlargement, called the **olive** (see fig. 15.9), on each lateral surface. The **fourth ventricle,** the space within the medulla, is continuous posteriorly with the central canal of the spinal cord and anteriorly with the cerebral aqueduct (see fig. 15.22).

The medulla is composed of vital nuclei and white matter that form all of the descending and ascending tracts communicating between the spinal cord and various parts of the brain. Most of the fibers within these tracts cross over to the opposite side through the pyramidal region of the medulla, permitting one side of the brain to receive information from and send information to the opposite side of the body (see fig. 15.9).

The gray matter of the medulla consists of several important nuclei for the cranial nerves, sensory relay, and for autonomic functions (fig. 15.16). The **nucleus ambiguus** and the **hypoglossal nucleus** are the centers from which arise the vestibulocochlear (VIII), glossopharyngeal (IX), accessory (XI), and hypoglossal (XII) nerves. The vagus nerves (X) arise from **vagus nuclei,** one on each lateral side of the medulla adjacent to the fourth ventricle. The **nucleus gracilis** and the **nucleus cuneatus** relay sensory information to the thalamus, and then the impulses are relayed to the cerebral cortex via the thalamic nuclei (not illustrated). The **inferior olivary nuclei** and the **accessory olivary nuclei** of the olive mediate impulses passing from the forebrain and midbrain through the inferior cerebellar peduncles to the cerebellum.

medulla: L. *medulla,* marrow

Three other nuclei within the medulla function as autonomic centers for controlling vital visceral functions.

1. **Cardiac center.** Both *inhibitory* and *accelerator fibers* arise from nuclei of the cardiac center. Inhibitory impulses constantly travel through the vagus nerves to slow the heartbeat. Accelerator impulses travel through the spinal cord and eventually innervate the heart through fibers within spinal nerves T1–T5.
2. **Vasomotor center.** Nuclei of the vasomotor center send impulses via the spinal cord and spinal nerves to the smooth muscles of arteriole walls, causing them to constrict or dilate, thus regulating blood pressure and blood flow.
3. **Respiratory center.** The respiratory center of the medulla controls the rate and depth of breathing and functions with nuclei of the pons (see fig. 15.13) to produce rhythmic breathing.

Other nuclei of the medulla function as centers for nonvital respiratory movements such as sneezing, coughing, swallowing, and vomiting. Some of these activities, such as swallowing, may be initiated voluntarily, but once they progress to a certain point they become involuntary and cannot be stopped.

Reticular Formation

The **reticular formation** is a complex network of nuclei and nerve fibers within the brain stem that functions as the *reticular activating system* (*RAS*), which arouses the cerebrum. Portions of the reticular formation are located in the spinal cord, pons, midbrain, and parts of the hypothalamus and thalamus (fig. 15.17). The reticular formation contains ascending and descending fibers from most of the structures within the brain.

Nuclei within the reticular formation generate a continuous flow of impulses unless they are inhibited by other parts of the brain. The principal functions of the RAS are to keep the cerebrum in a state of alert consciousness and to monitor selectively the afferent impulses perceived by the cerebrum. The RAS also helps the cerebellum activate selected motor units to maintain muscle tonus and produce smooth, coordinated contractions of skeletal muscles.

The RAS is sensitive to changes in and trauma to the brain. The sleep response is thought to occur because of a decrease in activity within the RAS, perhaps due to the secretion of specific neurotransmitters. A blow to the head or certain drugs and diseases may damage the RAS, causing unconsciousness. A *coma* is a state of unconsciousness and inactivity of the RAS that even the most powerful external stimuli cannot disturb.

Figure 15.17. The reticular activating system. The arrows indicate the direction of impulses along nerve pathways that connect with the RAS.

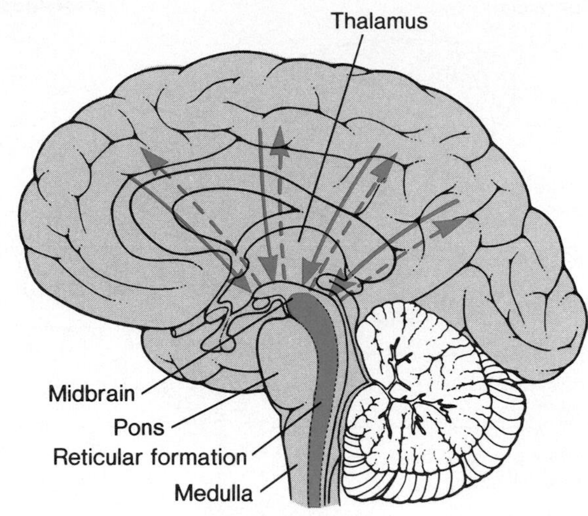

1. When a nurse or physician prepares a case study, the location, structure, function, and possible dysfunction of an afflicted structure or organ is reported. Prepare a brief case study of a patient who has suffered severe trauma to the medulla oblongata from a blow to the back of the skull.
2. Which cranial nerves arise from the medulla oblongata?
3. Describe the location of the reticular formation. What are two functions of the RAS?

Meninges of the Central Nervous System

The CNS is covered by protective meninges, consisting of a dura mater, an arachnoid membrane, and a pia mater.

Objective 14. Describe the position of the meninges as they protect the CNS.

The central nervous system is protected by three connective tissue membranous coverings called the meninges. See figures 15.18, 15.19, 15.20, 15.21. Individually, from the outside in, they are known as the dura mater, the arachnoid membrane, and the pia mater.

meninges: L. plural form of *meninx,* membrane

Figure 15.18. The meninges of the brain.

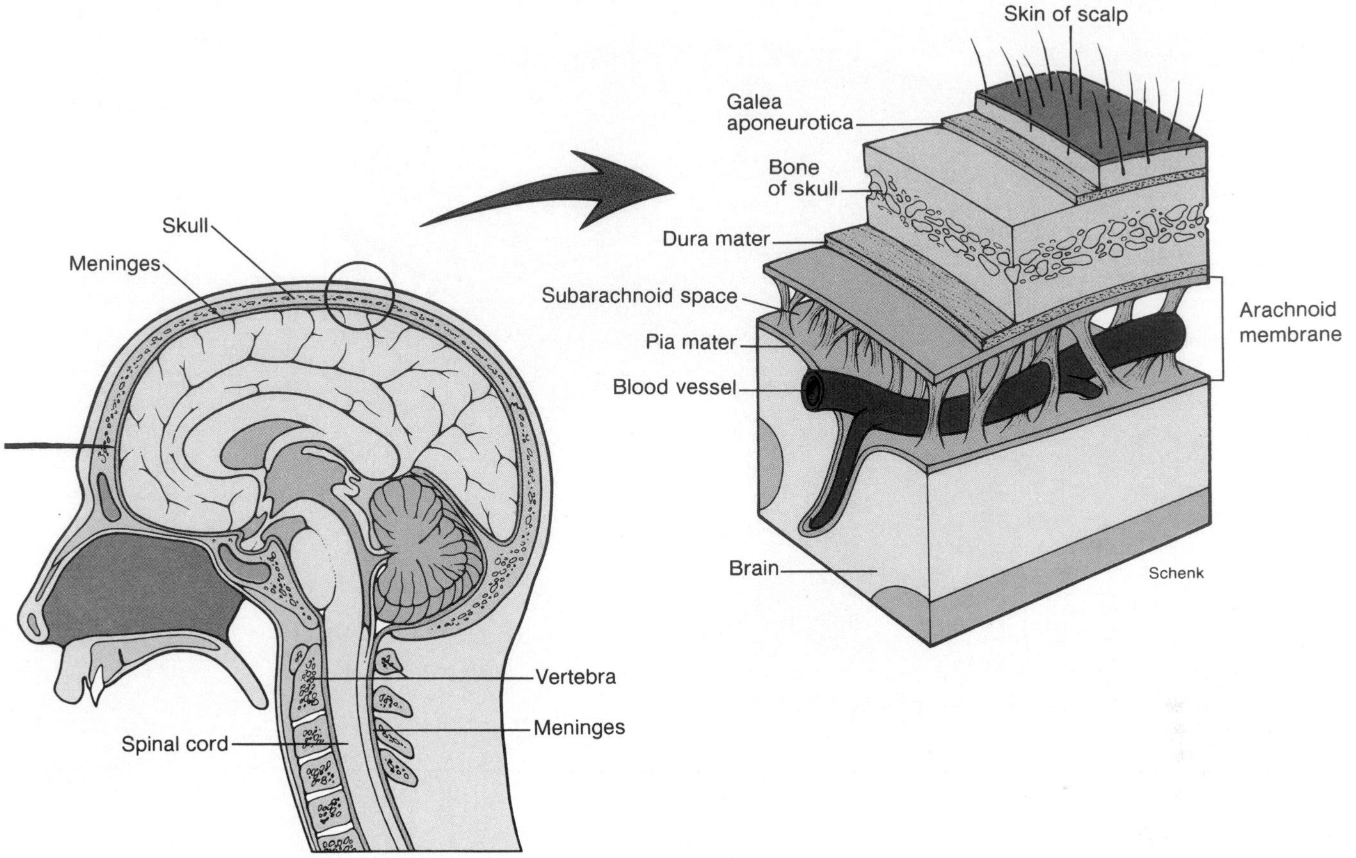

Dura Mater

The **dura mater** is in contact with the bone and is composed primarily of tough, white fibrous connective tissue. The **cranial dura mater** is a double-layered structure. The thicker, outer **periosteal** *(per''e-os'te-al)* **layer** adheres lightly to the cranium where it is the periosteum (fig. 15.19). The thinner, inner **meningeal layer** follows the general contour of the brain. The **spinal dura mater** is not double layered but is similar to the meningeal layer of the cranial dura mater (fig. 15.20).

The two layers of the cranial dura mater are fused and cover most of the brain. In certain regions, however, the layers are separated, enclosing **dural sinuses** (see fig. 15.19), which collect venous blood and drain it to the internal jugular veins of the neck.

In four locations, the meningeal layer of the cranial dura forms distinct septa to partition major structures on the surface of the brain and anchor the brain to the inside of the cranial case. These septa have been previously identified in this chapter and are reviewed in table 15.3.

dura mater: L. *dura,* hard; *mater,* mother

The spinal dura mater forms a tough, tubular **dural sheath** that continues into the vertebral canal and surrounds the spinal cord. There is no connection between the dural sheath and the vertebrae forming the vertebral canal, but instead there is a potential cavity called the **epidural space** (see fig. 15.20). The epidural space is highly vascular and contains areolar and adipose connective tissue, which form a protective pad around the spinal cord.

Arachnoid Membrane

The **arachnoid membrane** is the middle of the three meninges. This delicate, netlike membrane spreads over the CNS but generally does not extend into the sulci or fissures of the brain. The **subarachnoid space,** located between the arachnoid membrane and the deepest meninx, the pia mater, contains cerebrospinal fluid. The subarachnoid space is maintained by delicate, weblike strands that connect the arachnoid membrane and pia mater (see fig. 15.19).

arachnoid: L. *arachnoides,* like a cobweb

Figure 15.19. The dural sinuses.

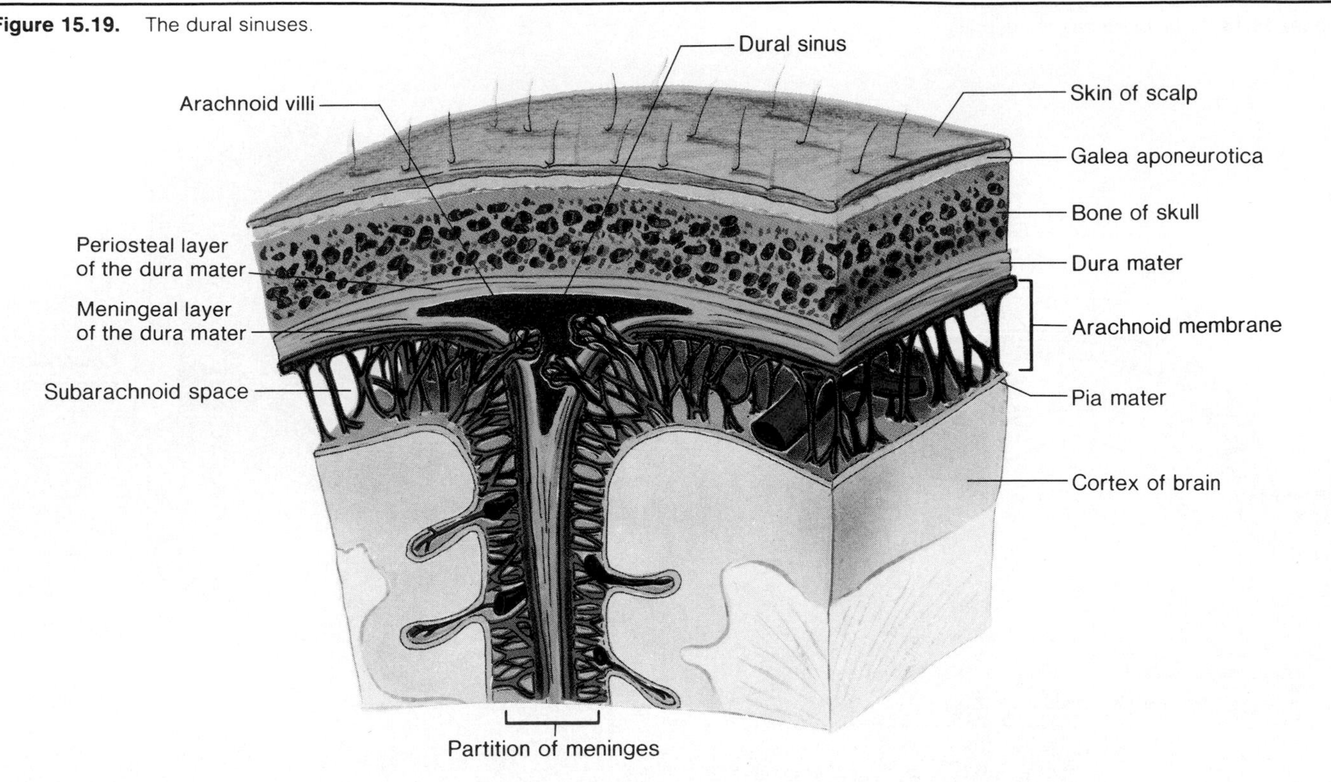

Figure 15.20. A transverse section showing the relationship of the spinal cord and meninges to the protective fat in the epidural space.

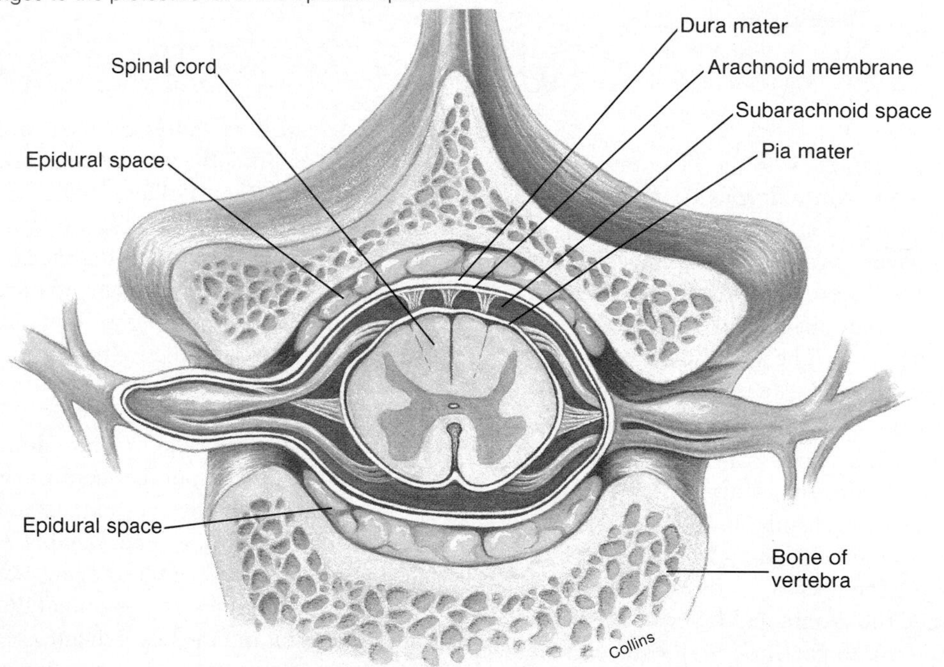

Figure 15.21. The spinal cord and the meninges.

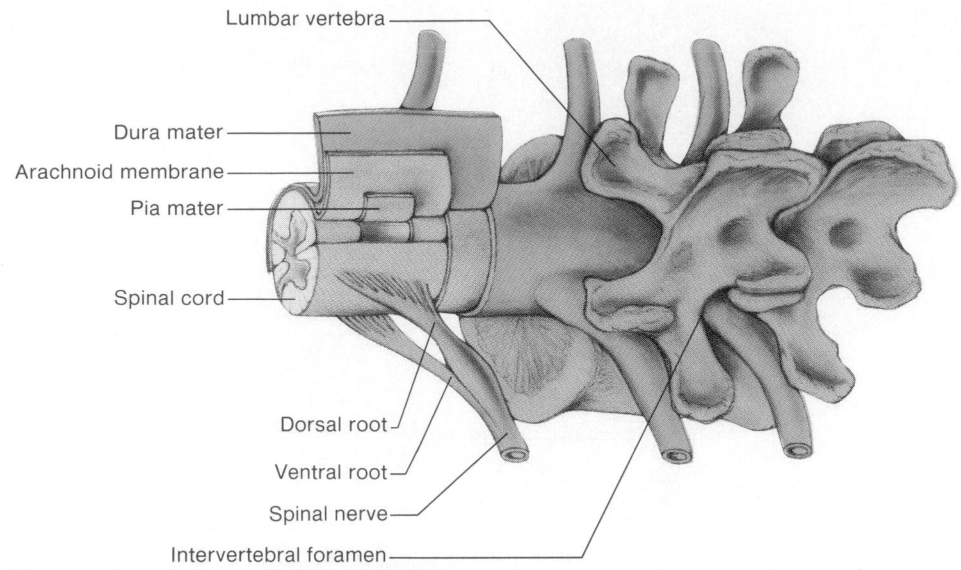

Table 15.3 Septa of the cranial dura mater

Septa	Location
Falx cerebri	Extends downward into the longitudinal fissure to partition the right and left cerebral hemispheres; anchored anteriorly to the crista galli of the ethmoid bone and posteriorly to the tentorium
Tentorium cerebelli	Separates the occipital lobes of the cerebrum from the cerebellum; anchored to the tentorium, petrous bones, and occipital bone
Falx cerebelli	Partitions the right and left cerebellar hemispheres; anchored to occipital crest
Diaphragma sellae	Forms the roof of the sella turcica

Pia Mater

The thin **pia mater** is attached to the surfaces of the CNS and follows the irregular contours of the brain and spinal cord. The pia mater is composed of modified loose fibrous connective tissue. It is highly vascular and functions to support the vessels that nourish the underlying cells of the brain and spinal cord. The pia mater is specialized over the roofs of the ventricles, where along with the arachnoid membrane, it contributes to the formation of the choroid plexuses. Lateral extensions of the pia mater along the spinal cord form the **ligamentum denticulatum,** which attaches the cord to the dura mater.

pia mater: L. *pia,* soft or tender; *mater,* mother

Meningitis is an inflammation of the meninges, usually caused by certain bacteria or viruses. The arachnoid membrane and the pia mater are the two meninges most frequently affected. Meningitis is accompanied by high fever and severe headache. Complications may cause sensory impairment, paralysis, or mental retardation. Untreated meningitis generally results in coma and death.

1. Contrast cranial and spinal dura mater.
2. Explain how the meninges support and protect the CNS.
3. Describe the location of the dural sinuses and the epidural space.

Ventricles and Cerebrospinal Fluid

The ventricles, central canal, and subarachnoid space contain cerebrospinal fluid, formed by the active transport of substances from blood plasma in the choroid plexuses.

Objective 15. Discuss the formation, function, and flow of cerebrospinal fluid.

Cerebrospinal fluid *(ser″ĕ-bro-spi′nal floo′id), CSF,* is a clear, lymphlike fluid that forms a protective cushion around and within the CNS. The fluid also buoys the brain. CSF circulates through the various **ventricles** of the brain, the **central canal** of the spinal cord, and the **subarachnoid space** around the entire CNS. The cerebrospinal fluid returns to the circulatory system by draining through the walls of the **arachnoid villi,** which are venous capillaries.

Figure 15.22. The ventricles of the brain. (*a*) an anterior view; (*b*) a lateral view.

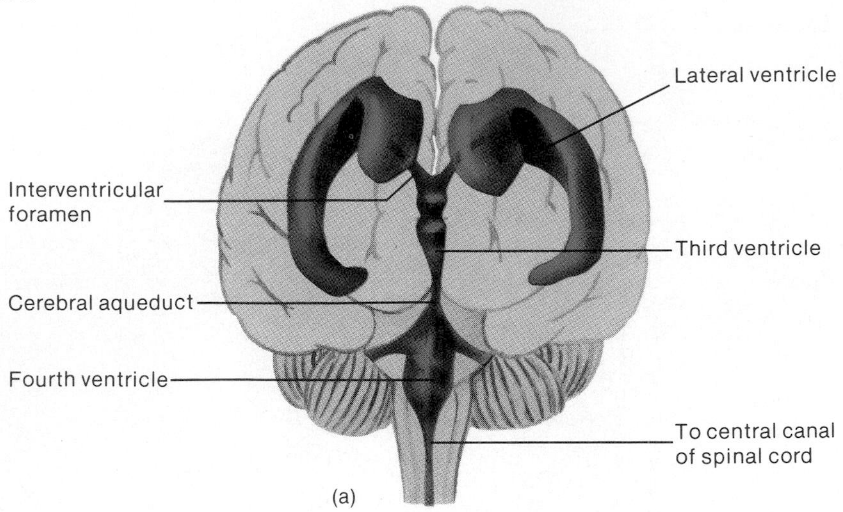

Lateral ventricle
Interventricular foramen
Third ventricle
Cerebral aqueduct
Fourth ventricle
To central canal of spinal cord
(a)

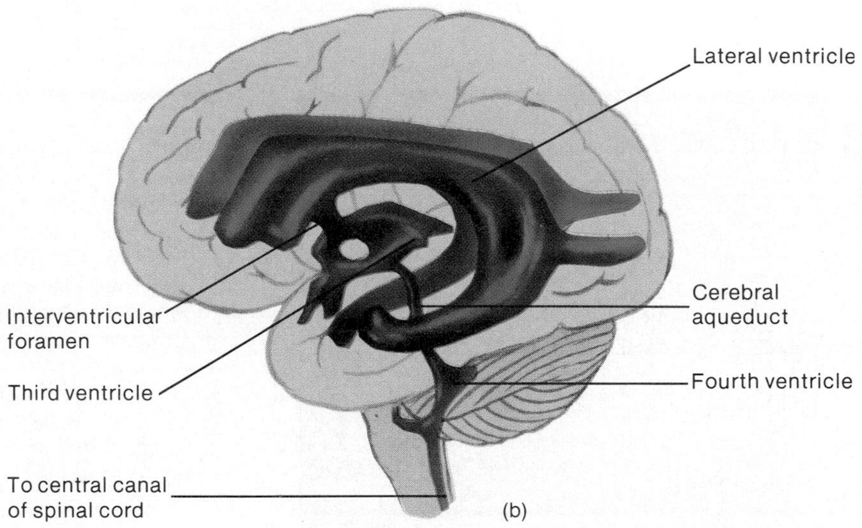

Lateral ventricle
Cerebral aqueduct
Interventricular foramen
Third ventricle
Fourth ventricle
To central canal of spinal cord
(b)

Figure 15.23. CT scans of the brain showing (*a*) a normal configuration of the ventricles and (*b*) an abnormal configuration of the ventricles due to hydrocephalism.

Ventricles

(a)

Ventricles

(b)

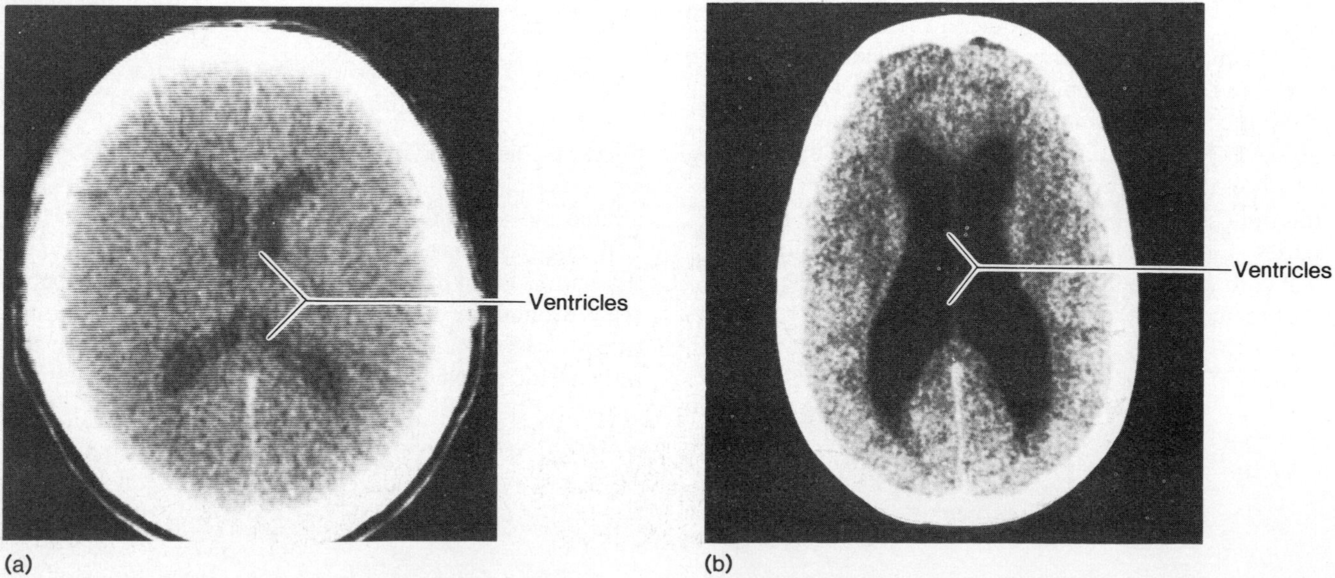

Figure 15.24. The flow of cerebrospinal fluid. Cerebrospinal fluid is secreted by choroid plexuses in the ventricular walls. The fluid circulates through the ventricles and central canal, enters the subarachnoid space, and is reabsorbed into the blood of the dural sinuses through the arachnoid villi.

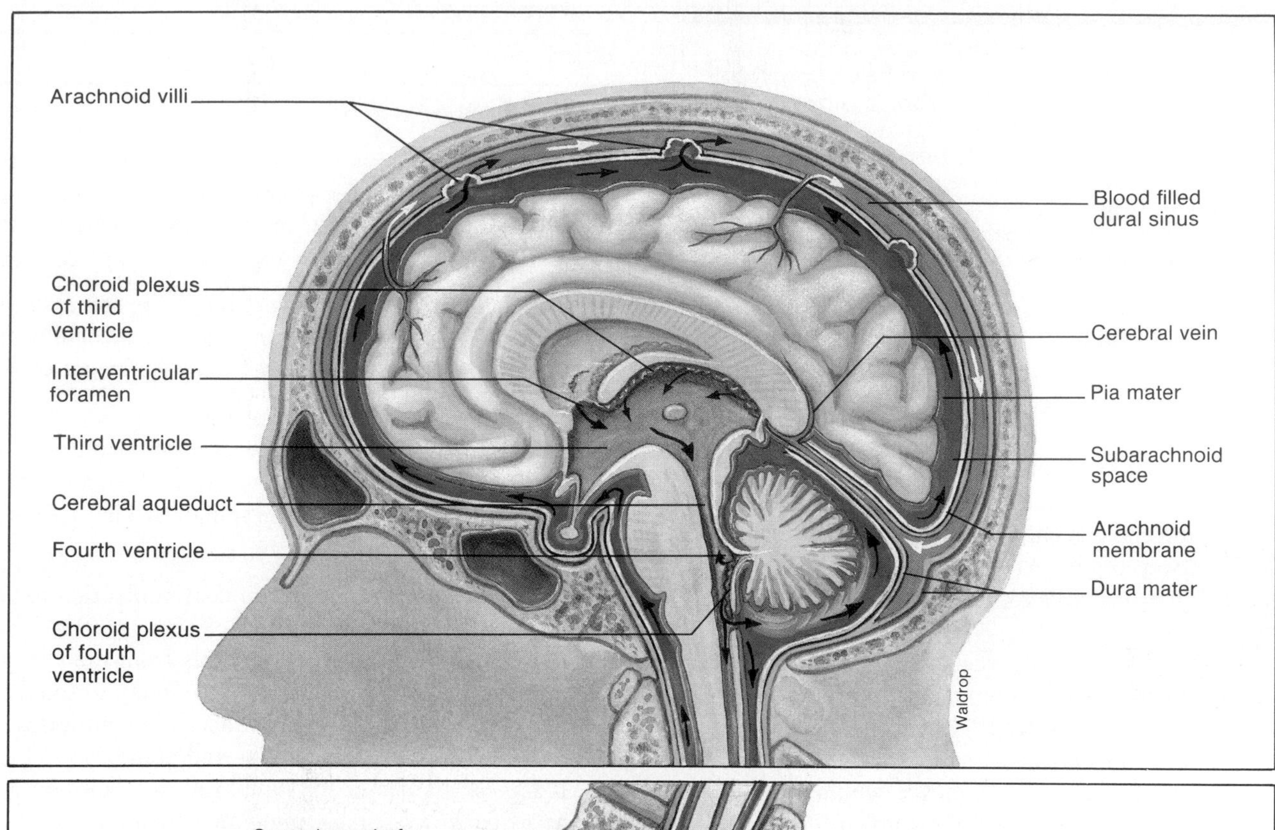

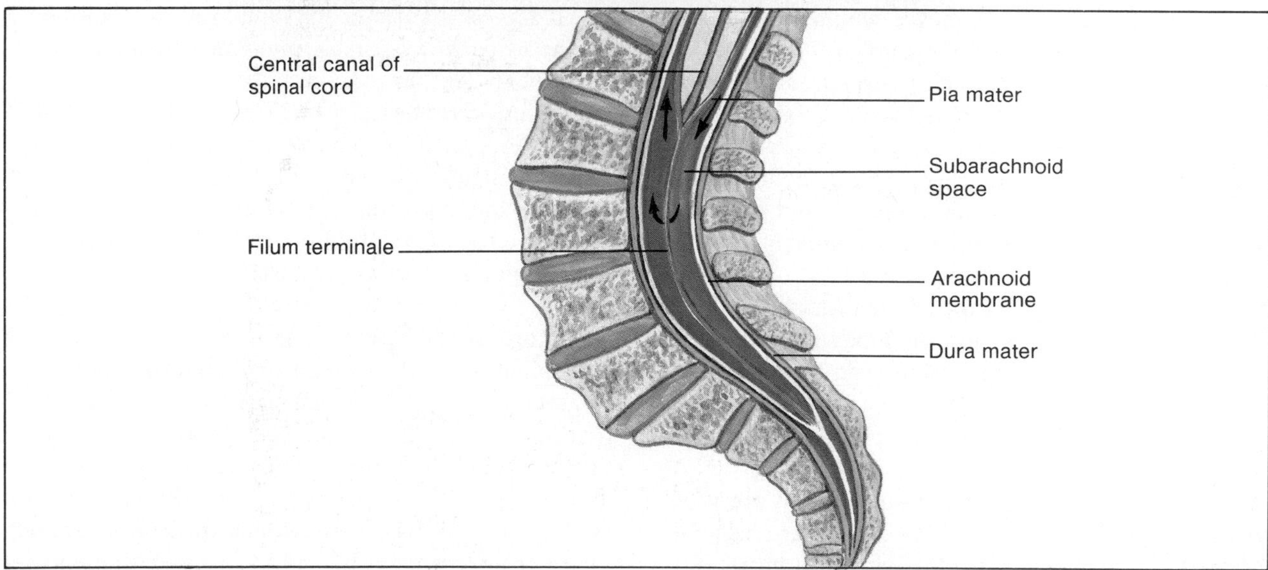

Ventricles of the Brain

The ventricles of the brain are connected to one another and to the central canal of the spinal cord (figs. 15.22, 15.23). Each of the two **lateral ventricles** (first and second ventricles) is located in each hemisphere of the cerebrum, inferior to the corpus callosum. The **third ventricle** is located in the diencephalon between the thalami. Each lateral ventricle is connected to the third ventricle by a narrow, oval opening called the **interventricular foramen** (foramen of Monro). The **fourth ventricle** is located in the brain stem, within the pons, cerebellum, and medulla oblongata. The **cerebral aqueduct** (aqueduct of Sylvius) passes through the midbrain to link the third and fourth ventricles. The fourth ventricle also communicates posteriorly with the central canal. Cerebrospinal fluid exits from the fourth ventricle into the subarachnoid space (fig. 15.24) through three foramina: the **foramen of Magendie,**

foramen of Monro: from Alexander Monro, Jr., Scottish anatomist, 1733–1817
foramen of Magendie: from Francois Magendie, French physiologist, 1783–1855

a medial opening, and two lateral **foramina of Luschka** (not illustrated). Cerebrospinal fluid returns to the venous blood through the arachnoid villi.

Internal hydrocephalus is a condition in which cerebrospinal fluid builds up within the ventricles of the brain (fig. 15.23b). It is more common in infants whose cranial sutures have not yet strengthened or ossified. If the pressure is excessive, the condition may have to be treated surgically.

External hydrocephalus, an accumulation of fluid within the subarachnoid space, usually results from an obstruction of drainage at the arachnoid villi.

Cerebrospinal Fluid

CSF buoys the CNS and protects it from mechanical injury. CSF has a specific gravity of 1.007, which is a density close to that of brain tissue. The brain weight is about 1,500 grams, but suspended in CSF its buoyed weight is about 50 grams. This means that the brain has a near neutral buoyancy and can therefore function effectively as a relatively heavy organ. At a true neutral buoyancy, an object does not float or sink but is suspended in its fluid environment.

In addition to buoying the CNS, CSF reduces the damaging effect of an impact to the head by spreading the force over a larger area. CSF also helps to remove metabolic wastes from nervous tissue. Since the CNS lacks lymphatic circulation, the CSF moves cellular wastes into the venous return at its places of drainage.

The clear, watery CSF is continuously produced from materials within the blood by masses of specialized capillaries called **choroid plexuses** *(ko'roid plek'sus-es)* and, to a lesser extent, by secretions of the ependymal cells. The ciliated ependymal cells cover the choroid plexuses as well as line the central canal and presumably aid the movement of the CSF.

CSF is formed mainly by the active transport and ultrafiltration of substances within the blood plasma. CSF has more sodium, chloride, magnesium, and hydrogen ions than blood plasma and less calcium, potassium, and glucose. In addition, CSF contains some proteins, urea, and white blood cells.

As much as 800 ml of cerebrospinal fluid are produced each day, although only 140–200 ml are bathing the CNS at any moment. A person lying in a horizontal position has a slow but continuous circulation of cerebrospinal fluid, with a fluid pressure of about 10 mm Hg.

The homeostatic consistency of the CSF composition is critical, and a chemical unbalance may have marked effects on CNS functions. An increase in amino acid glycine concentration, for example, produces hypothermia and hypotension as temperature and blood pressure regulatory mechanisms are disrupted. A slight change in pH may affect the respiratory rate and depth.

foramen of Luschka: from Hubert Luschka, German anatomist, 1820–75

1. Describe the location of the ventricles within the brain.
2. What are the functions of cerebrospinal fluid?
3. Where is cerebrospinal fluid produced, and where does it drain?

Spinal Cord

The spinal cord consists of centrally located gray matter involved in reflexes, and peripherally located ascending and descending tracts of white matter, which conduct impulses to and from the brain.

Objective 16. Describe the structure of the spinal cord.
Objective 17. Describe the arrangement of ascending and descending tracts within the spinal cord.

The **spinal cord** is the portion of the CNS that extends through the neural canal of the vertebral column (fig. 15.25). It is continuous with the brain through the foramen magnum of the skull. The spinal cord has two principal functions:

1. It provides a means of neural communication to and from the brain through tracts of white matter. **Ascending tracts** conduct impulses from the peripheral sensory receptors of the body to the brain. **Descending tracts** conduct motor impulses from the brain to the muscles and glands.
2. It serves as a center for spinal reflexes. Specific nerve pathways enable some movements to be reflexive rather than initiated voluntarily by the brain. Movements of this type are not confined to skeletal muscles; reflexive movements of cardiac and smooth muscles control heart rate, breathing rate, blood pressure, and digestive activities. Spinal nerve pathways are also involved in swallowing, coughing, sneezing, and vomiting.

Structure of the Spinal Cord

The spinal cord extends inferiorly from the position of the foramen magnum of the occipital bone to the level of the first lumbar vertebra (L1). The spinal cord is somewhat flattened dorsoventrally, making it oval in cross section. Two prominent enlargements can be seen in a dorsal view (see fig. 15.25). The **cervical enlargement** is located between the third cervical and the second thoracic vertebrae. Nerves emerging from this region serve the upper extremities. The **lumbar enlargement** lies between the ninth and twelfth thoracic vertebrae. Nerves from the lumbar enlargement supply the lower extremities.

The embryonic spinal cord develops more slowly than the associated vertebral column; thus, in the adult, the cord does not extend beyond L1. The tapering, terminal portion of the spinal cord is called the **conus medullaris.** The **filum terminale,** a fibrous strand composed mostly of pia mater, extends inferiorly from the conus medullaris at the level of L1 to the coccyx (see fig. 15.24). Nerve roots also

filum terminalis: L. *filum,* filament; *terminus,* end

Figure 15.25. The spinal cord and plexuses.

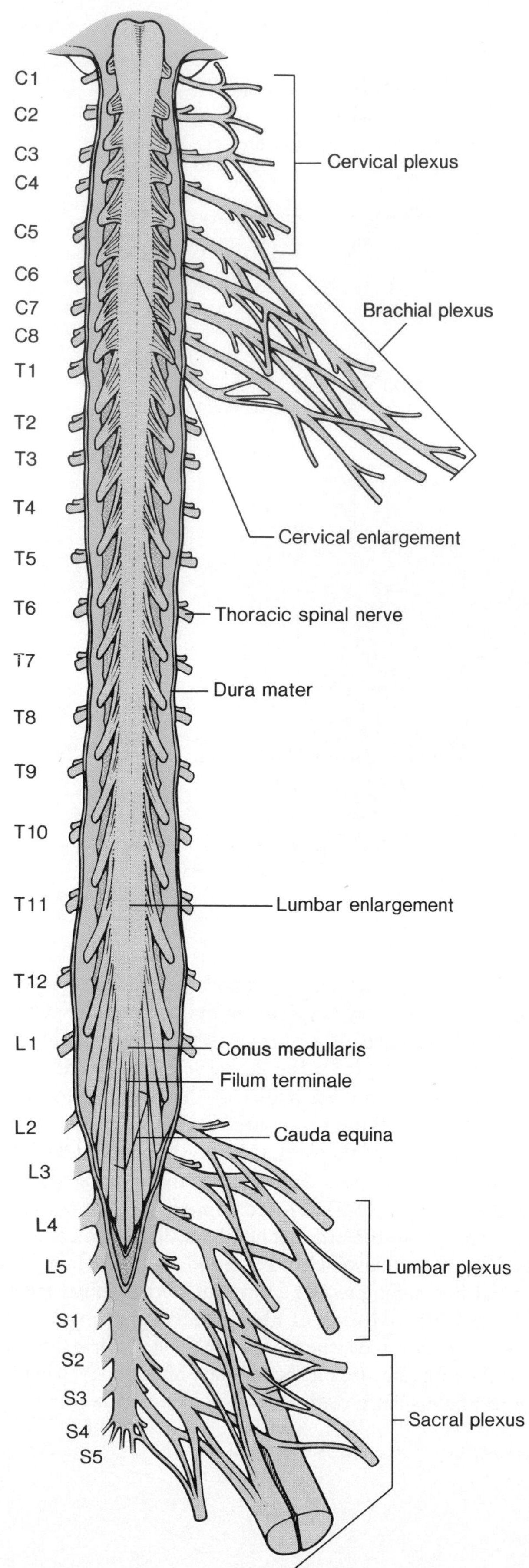

radiate inferiorly from the conus medullaris through the vertebral canal. These nerves are collectively referred to as the **cauda equina** because they resemble a horse's tail.

The spinal cord develops as thirty-one segments, each of which gives rise to a pair of **spinal nerves** that emerge from the cord through the intervertebral foramina. Two grooves, an **anterior median fissure** and a **posterior median sulcus,** extend the length of the spinal cord and partially divide the cord into right and left portions. The spinal cord, like the brain, is protected by three distinct meninges and is cushioned by cerebrospinal fluid. The pia mater contains an extensive vascular network.

The **gray matter** of the spinal cord is centrally located and surrounded by white matter. It is composed of nerve cell bodies, neuroglia, and unmyelinated internuncial (association) neurons. The **white matter** consists of bundles, or tracts, of myelinated fibers of sensory and motor neurons.

The relative size and shape of the gray and white matter varies throughout the spinal cord. The amount of white matter increases toward the brain as the nerve tracts become thicker. More gray matter exists in the cervical and lumbar enlargements where innervations from the upper and lower extremities respectively make connections.

The core of gray matter roughly resembles the letter H (fig. 15.26). Projections of the gray matter within the spinal cord are called horns and are named according to the direction in which they project. The paired **posterior (dorsal) horns** extend posteriorly, and the paired **anterior (ventral) horns** project anteriorly. A pair of short **lateral horns** extend to the sides and are located between the other two pairs. Lateral horns are prominent only in the thoracic and upper lumbar regions. The transverse bar of gray matter that connects the paired horns across the center of the spinal cord is called the **gray commissure.** Contained within the gray commissure is the **central canal,** which is continuous with the ventricles of the brain and filled with cerebrospinal fluid.

Spinal Cord Tracts

Impulses are conducted through the ascending and descending tracts of the spinal cord within the columns of white matter. The spinal cord has six columns of white matter called **funiculi** *(fu-nik'u-li),* which are named according to their relative position which form the cord. The two **anterior (ventral) funiculi** are located between the two anterior horns of gray matter to either side of the anterior median fissure (fig. 15.26). The two **posterior (dorsal) funiculi** are located between the two posterior horns of gray matter to either side of the posterior median sulcus. Two **lateral funiculi** are located between the anterior and posterior horns of gray matter.

---

cauda equina: L. *cauda,* tail; *equus,* horse
commissure: L. *commissura,* a joining
funiculus: L. diminutive of *funis,* cord, rope

Figure 15.26. The (*a*) spinal cord in a transverse section. (*b*) a photograph of the spinal cord and nerve.

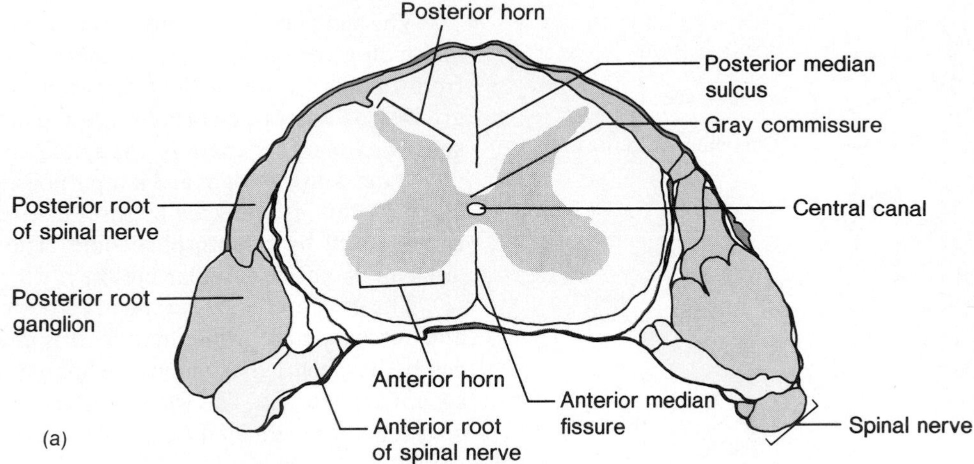

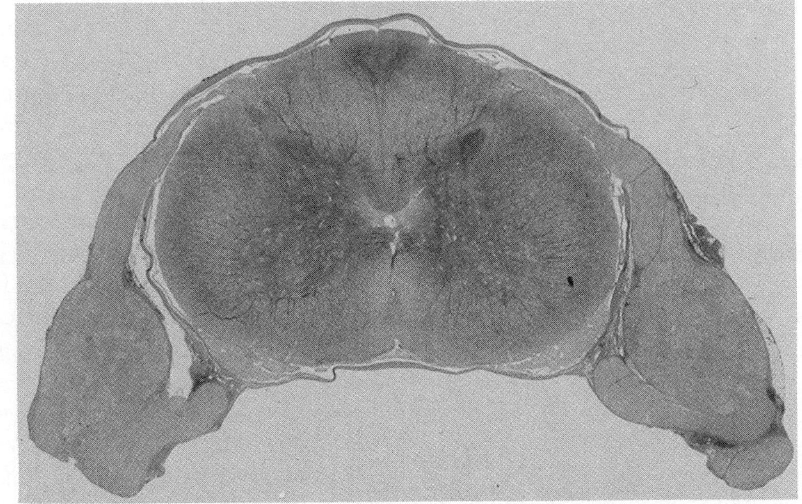

Each funiculus consists of both ascending and descending tracts. The nerve fibers within the tracts are generally myelinated and have specific sites of origin and termination. In fact, the names of the various tracts reflect their origin and termination. The fibers of the tracts either remain on the same side of the brain and spinal cord or cross over within the medulla or the spinal cord. The crossing over of nerve tracts is referred to as *decussation (de''kus-sa'shun)*. Figure 15.27 illustrates a descending tract that decussates within the medulla, and figure 15.28 illustrates an ascending tract that decussates within the medulla.

The principal ascending and descending tracts within the funiculi are presented with their functions in table 15.4 and are illustrated in figure 15.29.

decussation: L. *decussare*, to form an X intersection

Descending tracts are grouped according to place of origin as either corticospinal or extrapyramidal. **Corticospinal (pyramidal) tracts** descend directly, without synaptic interruption, from the cerebral cortex to the lower motor neurons. The cell bodies of the neurons that contribute fibers to these tracts are located primarily in the precentral gyrus of the frontal lobe. Most (about 85%) of the corticospinal fibers decussate in the pyramids of the medulla (see fig. 15.9). The remaining 15% do not cross from one side to the other. The fibers that cross comprise the **lateral corticospinal tracts,** and the remaining uncrossed fibers comprise the **anterior corticospinal tracts.** Because of the crossing of fibers from higher motor neurons in the pyramids, the right hemisphere primarily controls the musculature on the left side of the body, whereas the left hemisphere controls the right musculature.

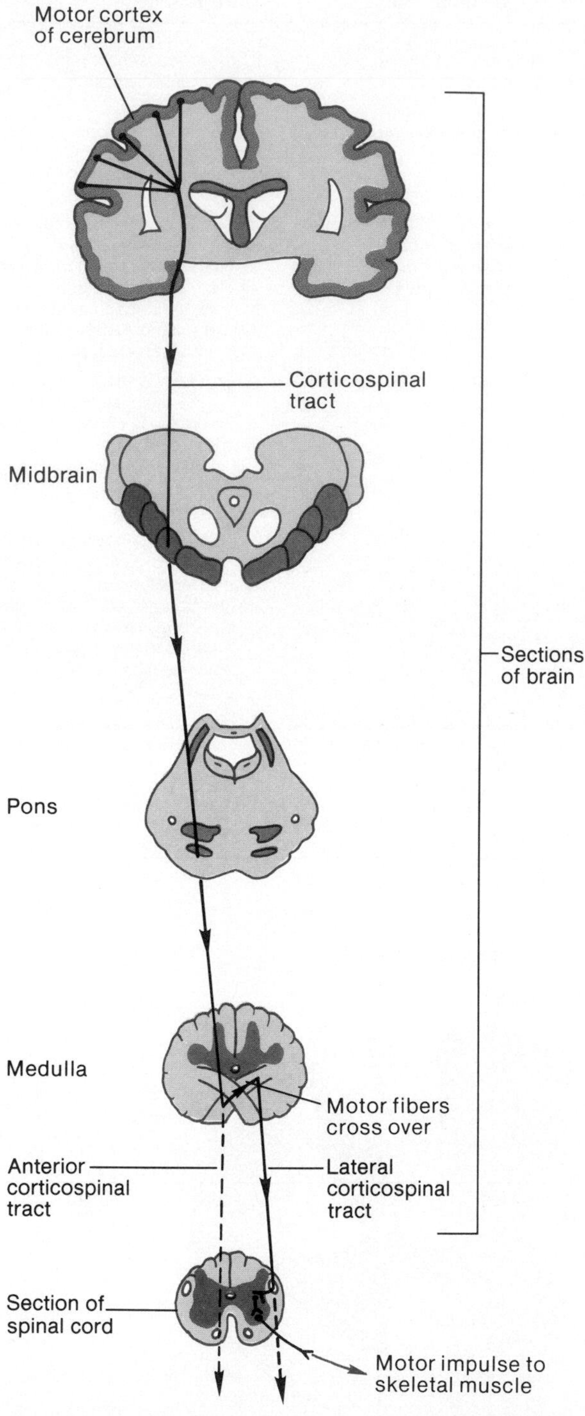

Figure 15.27. Descending tracts composed of motor fibers that cross over within the medulla oblongata.

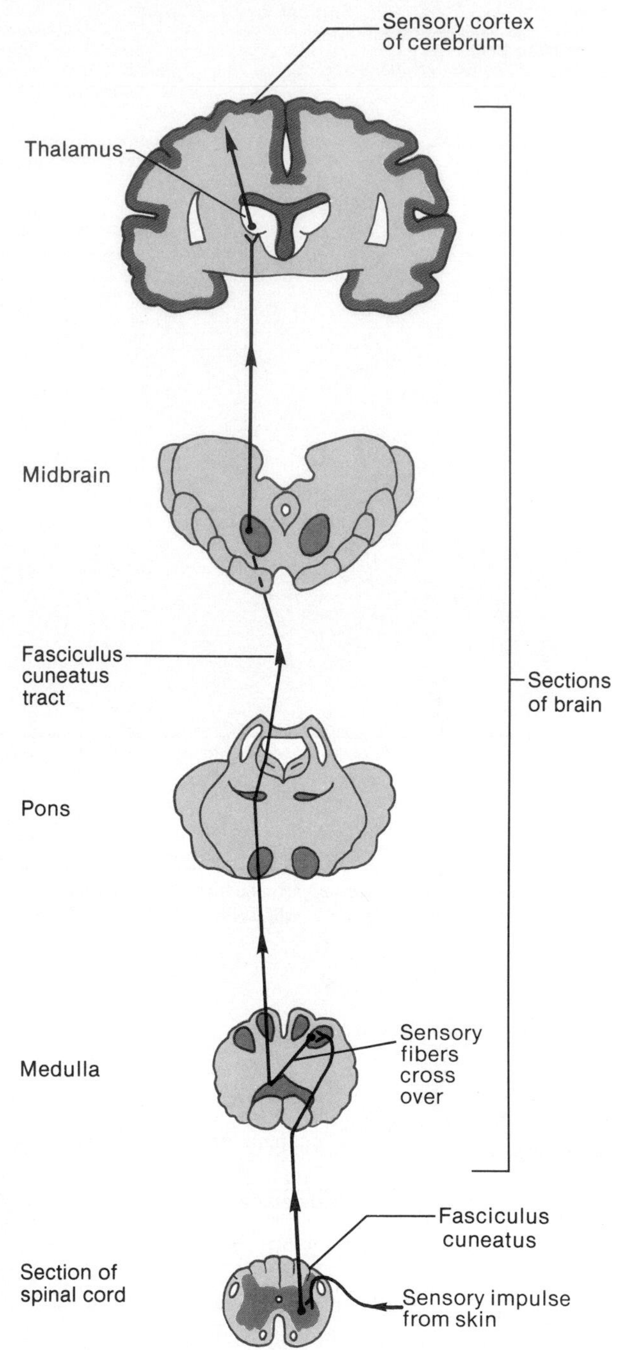

Figure 15.28. Ascending tracts composed of sensory fibers that cross over within the medulla.

The corticospinal tracts are particularly important in voluntary movements that require correlation between the motor cortex and sensory input. Speech, for example, is impaired when the corticospinal tracts are damaged in the thoracic region of the spinal cord, whereas involuntary breathing continues. Damage to the pyramidal motor system can be detected clinically by a positive *Babinski's reflex,* in which stimulation of the sole of the foot causes extension (upward movement) of the toes.

The remaining descending tracts are **extrapyramidal tracts,** which originate in the midbrain and brain stem regions (fig. 15.30). Electrical stimulation of the cerebral cortex, the cerebellum, and the basal ganglia indirectly evokes movements because of synaptic connections within extrapyramidal tracts.

The **reticulospinal** *(rĕ-tik''u-lo-spi'nal)* **tracts** are the major descending pathways of the extrapyramidal system. These tracts originate in the reticular formation of the

Table 15.4 Principal ascending and descending tracts of spinal cord

Tract	Funiculus	Origin	Termination	Function
Ascending tracts				
Anterior spinothalamic	Anterior	Posterior horn on one side of cord but crosses to opposite side	Thalamus, then cerebral cortex	Conducts sensory impulses for crude touch and pressure
Lateral spinothalamic	Lateral	Posterior horn on one side of cord but crosses to opposite side	Thalamus, then cerebral cortex	Conducts pain and temperature impulses that are interpreted within cerebral cortex
Fasciculus gracilis and fasciculus cuneatus	Posterior	Peripheral afferent neurons; does not cross over	Nucleus gracilis and nucleus cuneatus of medulla; eventually thalamus, then cerebral cortex	Conducts sensory impulses from skin, muscles, tendons, and joints, which are interpreted as sensations of fine touch, precise pressures, and body movements
Posterior spinocerebellar	Lateral	Posterior horn; does not cross over	Cerebellum	Conducts sensory impulses from one side of body to same side of cerebellum for subconscious proprioception necessary for coordinated muscular contractions
Anterior spinocerebellar	Lateral	Posterior horn; some fibers cross, others do not	Cerebellum	Conducts sensory impulses from both sides of body to cerebellum for subconscious proprioception necessary for coordinated muscular contractions

Figure 15.29. A transverse section showing the principal ascending and descending tracts within the spinal cord.

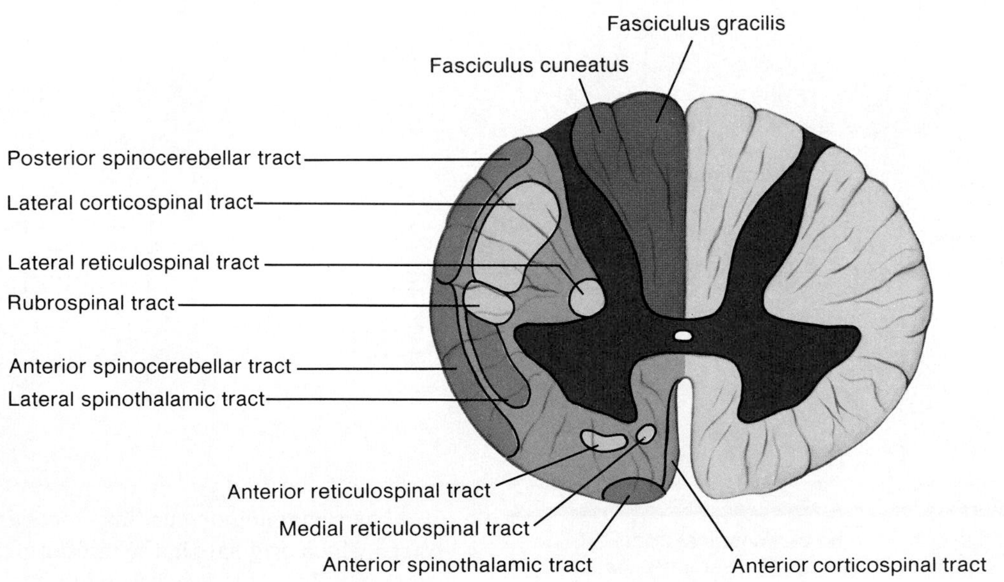

Tract	Funiculus	Origin	Termination	Function
Descending tracts				
Anterior corticospinal	Anterior	Cerebral cortex on one side of brain; crosses to opposite side of cord	Anterior horn	Conducts motor impulses from cerebrum to spinal nerves and outward to cells of anterior horns for coordinated, precise voluntary movements of skeletal muscle
Lateral corticospinal	Lateral	Cerebral cortex on one side of brain; crosses in base of medulla to opposite side of cord	Anterior horn	Conducts motor impulses from cerebrum to spinal nerves and outward to cells of anterior horns for coordinated, precise voluntary movements
Tectospinal	Anterior	Midbrain; crosses to opposite side of cord	Anterior horn	Conducts motor impulses to cells of anterior horns and eventually to muscles that move the head in response to visual, auditory, or cutaneous stimuli
Rubrospinal	Lateral	Midbrain (red nucleus); crosses to opposite side of cord	Anterior horn	Conducts motor impulses concerned with muscle tone and posture
Vestibulospinal	Anterior	Medulla; does not cross over	Anterior horn	Conducts motor impulses that regulate body tone and posture (equilibrium) in response to movements of head
Anterior and medial reticulospinal	Anterior	Reticular formation of brain stem; does not cross over	Anterior horn	Conducts motor impulses that control muscle tone and sweat gland activity
Lateral reticulospinal	Lateral	Reticular formation of brain stem; does not cross over	Anterior horn	Conducts motor impulses that control muscle tone and sweat gland activity

Figure 15.30. Areas of the brain containing neurons involved in the control of skeletal muscles (higher motor neurons). The thalamus is a relay center between the motor cortex and other brain areas.

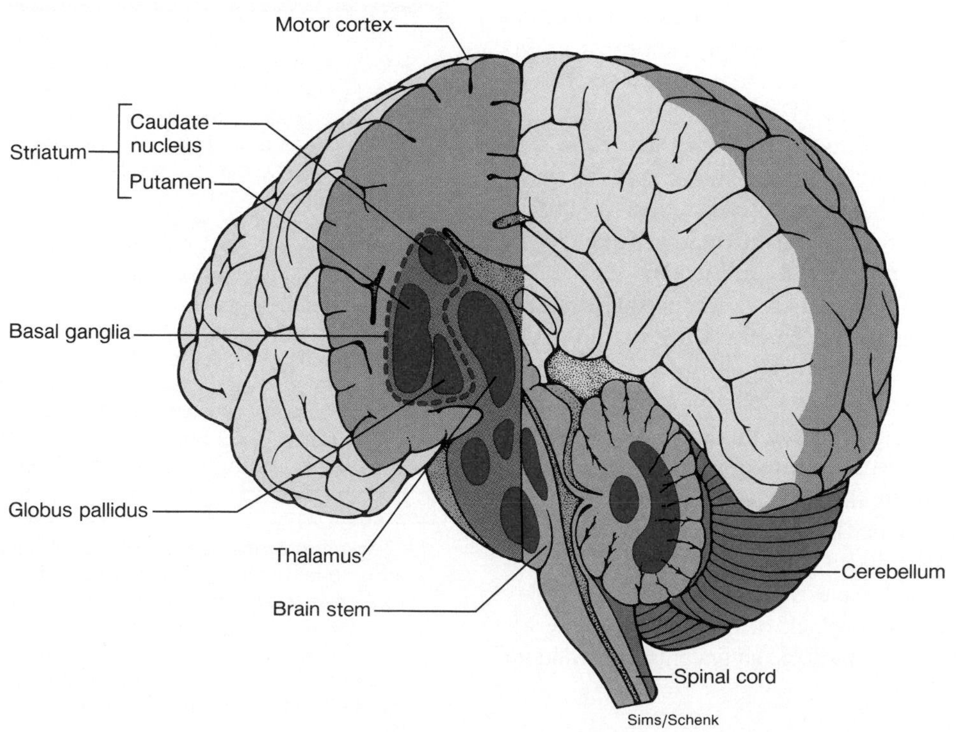

Figure 15.31. Pathways involved in the higher motor neuron control of skeletal muscles.

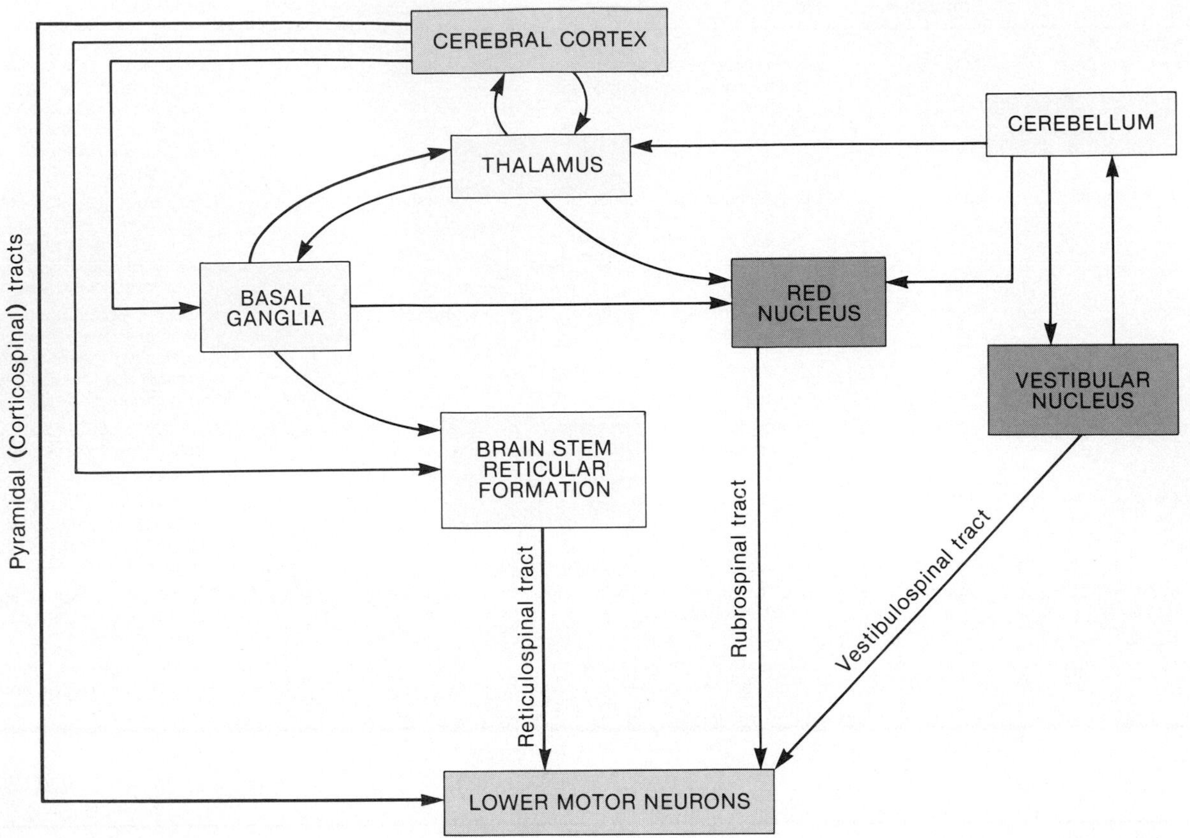

brain stem. Neurostimulation of the reticular formation by the cerebrum or cerebellum either facilitates or inhibits (depending on the area stimulated) the activity of lower motor neurons (fig. 15.31).

There are no descending tracts from the cerebellum. The cerebellum can only influence motor activity indirectly, through the vestibular nuclei, red nucleus, and basal ganglia. These structures, in turn, affect lower motor neurons via the **vestibulospinal tracts, rubrospinal tracts,** and reticulospinal tracts. Damage to the cerebellum interferes with the coordination of movements with spatial judgment. Underreaching or overreaching for an object may occur, followed by *intention tremor,* in which the limb moves back and forth in a pendulumlike motion.

The basal ganglia, acting through synapses in the reticular formation particularly, appear normally to exert an inhibitory influence on the activity of lower motor neurons. Damage to the basal ganglia thus results in increased muscle tone. People with such damage display *akinesia (ah''ki-ne'ze-ah)* (lack of desire to use the affected limb) and *chorea*—sudden and uncontrolled random movements.

akinesia: Gk. *a,* without; *kinesis,* movement
chorea: Fr. *choros,* a dance

Paralysis agitans, better known as *Parkinson's disease,* is a disorder of the basal ganglia involving the degeneration of fibers from the substantia nigra. These fibers, which use dopamine as a neurotransmitter, are required to antagonize the effects of other fibers that use acetylcholine (ACh) as a transmitter. The relative deficiency of dopamine compared to ACh is believed to produce the symptoms of Parkinson's disease, including *resting tremor.* This "shaking" of the limbs tends to disappear during voluntary movements and then reappear when the limb is again at rest.

Parkinson's disease is treated with drugs that block the effects of ACh and by the administration of L-dopa, which can be converted to dopamine in the brain (dopamine cannot be given directly because it does not cross the blood-brain barrier).

1. Diagram a cross section of the spinal cord, and label the structures of the gray matter and the white matter. Describe the location of the spinal cord.
2. List the structures and function of the corticospinal tracts. Make a similar list for the extrapyramidal tracts.
3. Explain why damage to the right side of the brain primarily affects motor activities on the left side of the body.

Parkinson's disease: from James Parkinson, English physician, 1755–1824

Clinical Considerations

Neurological Assessment and Drugs

Neurological assessment has become exceedingly sophisticated and accurate in the past few years. In most physical examinations, only the basic aspects such as reflexes and sensory functions are assessed. But if the physician suspects abnormalities involving the nervous system, further neurological tests are done, employing the following techniques.

A **lumbar puncture** is performed by inserting a fine needle between the third and fourth lumbar vertebrae and withdrawing a sample of cerebrospinal fluid from the subarachnoid space (fig. 15.32). A **cisternal puncture** is similar to a lumbar puncture except that the cerebrospinal fluid is withdrawn from the cisterna magna near the foramen magnum of the skull. The pressure of the cerebrospinal fluid, which is normally about 10 mm of mercury, is measured with a *manometer.* Samples of cerebrospinal fluid may also be examined for abnormal constituents. Also, excessive fluids, accumulated as a result of disease or trauma, may be drained.

The condition of the arteries of the brain can be determined through a **cerebral angiogram.** In this technique, a radiopaque substance is injected into the common carotid arteries and allowed to disperse through the cerebral vessels. Aneurysms and vascular constrictions or displacements by tumors may then be revealed on X rays.

The development of the **CT scanner,** or **computerized axial tomographic scanner,** revolutionized the diagnosis of brain disorders. More recently, the use of the **MRI,** or **magnetic resonance imaging,** has immensely improved neurological diagnostic procedures. The MRI projects a sharply detailed image of a patient's brain or spinal cord onto a television screen. The versatile MRI allows quick and accurate diagnoses of tumors, aneurysms, blood clots, and hemorrhage. MRI may also be used to detect certain types of birth defects, brain damage, scar tissue, and old or recent strokes.

Another machine that also has greater potential than the CT scanner is the **DSR,** or **dynamic spatial reconstructor.** Like the CT scanner, the DSR is computerized to transform X-ray pictures into composite video images. The advantage of the DSR over the CT is the three-dimensional view it presents and the speed with which the image is portrayed. DSR can produce 75,000 cross-sectional images in five seconds, whereas CT can produce only one. At that speed, body functions as well as structures may be studied. For example, blood flow through blood vessels of the brain can be observed. This type of data is important in detecting early symptoms of a stroke or other disorders.

Certain disorders of the brain may be diagnosed more simply by examining brain-wave patterns using an **electroencephalogram.** Sensitive electrodes placed on the scalp record particular EEG patterns being emitted from evoked cerebral activity. EEG recordings are used to monitor epileptic patients to predict seizures and determine proper drug therapy and also to monitor comatose patients.

Figure 15.32. (*a*) a lumbar puncture is performed by inserting a needle between the third and fourth lumbar vertebrae (L3–L4) and (*b*) withdrawing cerebrospinal fluid from the subarachnoid space.

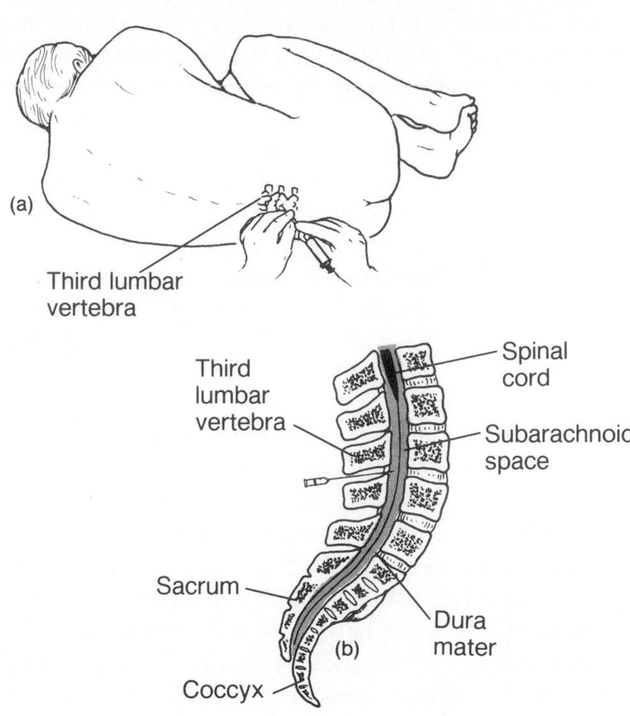

The fact that the nervous system is extremely sensitive to various drugs is both fortunate for and potentially disastrous to a person. *Drug abuse* is a major clinical concern because of the addictive and devastating effect that certain drugs have on the nervous system. Much has been written on drug abuse, and it is beyond the scope of this text to elaborate on the effects of drugs. A positive aspect of drugs is their administration in medicine to temporarily interrupt the passage or perception of sensory impulses. Injecting an anesthetic drug near a nerve, as in dentistry, desensitizes a specific area and causes a *nerve block.* Nerve blocks of a limited extent occur if an appendage is cooled or if a nerve is compressed for a period of time. Before the discovery of pharmacological drugs, physicians would frequently cool an affected appendage with ice or snow before performing surgery. **General anesthetics** affect the brain and render a person unconscious. A **local anesthetic** causes a nerve block by desensitizing a specific area.

Injuries

Although the brain and spinal cord seem to be well protected within a bony encasement, they are sensitive organs highly susceptible to injury.

Certain symptomatic terms are used when determining possible trauma within the CNS. **Headaches** are the most common ailment of the CNS. Most headaches are due to dilated blood vessels within the meninges of the brain. Headaches are generally asymptomatic of brain disorders, associated rather with physiological stress, eyestrain, or fatigue. Persistent and intense headaches may indicate a more serious problem such as a brain tumor. A **migraine** is a specific type of headache commonly preceded or accompanied by visual impairments and gastrointestinal unrest. It is not known why only 5%–10% of the population periodically suffer from migraines or why they are more common in women. Fatigue, allergy, or emotional stress tends to trigger migraines.

Fainting is a brief loss of consciousness that may result from a rapid pooling of blood in the lower extremities. It may occur when a person rapidly arises from a reclined position, receives a blow to the head, or experiences an intense psychologic stimulus, such as viewing a cadaver for the first time. Fainting is of more concern when it is symptomatic of a particular disease.

A **concussion** is a sudden movement of the brain caused by a violent blow to the head, which may or may not fracture bones of the skull. A concussion usually results in a brief period of unconsciousness followed by mild **delirium,** in which the patient is in a state of confusion. **Amnesia** is a more intense disorientation in which the patient suffers various degrees of memory loss.

A person who survives a severe head injury may be **comatose** for a short or an extended period of time. A coma is a state of unconsciousness from which the patient cannot be aroused by even the most intense external stimuli. The area of the brain most likely to cause a coma from trauma is the reticular activating system. Although a head injury is the most common cause of coma, certain diseases, drugs, poisons, infections, and some chemical imbalances (e.g., diabetes) may also be responsible.

The flexibility of the vertebral column is essential for body movements, but because of this flexibility the spinal cord and spinal nerves are somewhat vulnerable to trauma. Falls or severe blows to the back are a common cause of injury. A skeletal injury, such as a fracture, dislocation, or compression of the vertebrae, usually traumatizes nervous tissue as well. Other frequent causes of trauma to the spinal cord include gunshot wounds, stabbings, herniated discs, and birth injuries. The consequences of the trauma depend upon the severity and location of the injury and the medical treatment the patient receives. If nerve fibers of the spinal cord are severed, motor or sensory functions will be permanently lost.

Paralysis is a permanent loss of motor control, usually resulting from disease or a lesion of the spinal cord or specific nerves. Paralysis of both lower extremities is called **paraplegia.** Paralysis of both the upper and lower extremity on the same side is called **hemiplegia,** and paralysis of all four extremities is **quadriplegia.** Paralysis may be flaccid or spastic. **Flaccid** *(flak'sid)* **paralysis** generally results from a lesion of the anterior horn cells and is characterized by noncontractible muscles that atrophy. **Spastic paralysis** results from lesions of the corticospinal tracts of the spinal cord and is characterized by hypertonicity of the skeletal muscles.

amnesia: L. *amnesia,* forgetfulness
comatose: Gk. *koma,* deep sleep
paralysis: Gk. *paralysis,* loosening
paraplegia: Gk. *para,* beside; *plessein,* to strike

Figure 15.33. Whiplash varies in severity from muscle and ligament strains to dislocation of the vertebrae and compression of the spinal cord. Injuries such as this may cause permanent loss of some or all of the spinal cord functions.

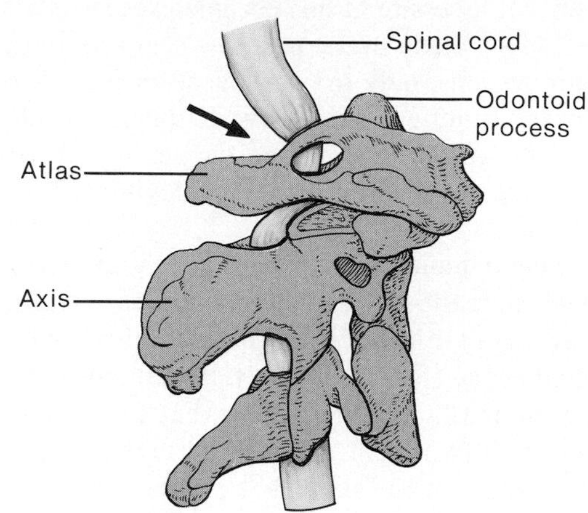

Whiplash is a sudden hyperextension and flexion of the cervical vertebrae (fig. 15.33) such as may occur during a rear-end automobile collision. Recovery of a minor whiplash (muscle and ligament strains) is generally complete but requires a long time. Severe whiplash (spinal cord compression) may cause permanent paralysis to the structures below the level of injury.

Diseases and Infections

Mental illness is a major clinical consideration of the nervous system and is perhaps the least understood. The two principal categories of mental disorders are neurosis and psychosis. In *neurosis,* a maladjustment to certain aspects of life interferes with normal functioning, but contact with reality is maintained. An irrational fear (*phobia*) is an example of neurosis. Neurosis frequently causes intense anxiety or abnormal distress that brings about increased sympathetic stimulation. *Psychosis,* a more serious mental condition, is typified by a withdrawal from reality and is usually socially unacceptable. The more common forms of psychosis include *schizophrenia,* in which a person withdraws into a world of fantasy; *paranoia,* in which a person has systematized delusions often of a persecutory nature; and *manic-depressive psychosis,* in which a person's moods swing widely from intense elation to deepest despair.

Epilepsy is a relatively common brain disorder with a strong hereditary basis, but it is also caused by head injuries, tumors, and childhood infectious diseases or can be idiopathic (without demonstrable cause). A person with epilepsy may periodically suffer from an *epileptic seizure,* which has various symptoms depending on the type of epilepsy.

epilepsy: Gk. *epi,* upon; *lepsis,* seize

The most common kinds of epilepsy are petit mal, psychomotor epilepsy, and grand mal. **Petit mal** *(pĕ-te' mahl')* occurs almost exclusively in children between the ages of three and twelve. A child experiencing a petit mal seizure loses contact with reality for 5–30 seconds but does not lose consciousness or display convulsions. There may, however, be slight uncontrollable facial gestures or eye movements, and the child stares as if in a daydream. During a petit mal seizure, the thalamus and hypothalamus produce an extremely slow EEG pattern of 3 waves per second. Children with petit mal usually outgrow the condition by age nine or ten and generally require no medication.

Psychomotor epilepsy is often confused with mental illness because of the symptoms characteristic of the seizure. During such a seizure, EEG activity accelerates in the temporal lobes, causing a person to become disoriented and lose contact with reality. Occasionally during a seizure, specific cerebral motor areas cause a person to smack his lips or clap his hands involuntarily. If motor areas in the brain are not stimulated, a person having a psychomotor epileptic seizure may wander aimlessly until the seizure subsides.

Grand mal is a more serious form of epilepsy characterized by periodic convulsive seizures that generally render a person unconscious. Grand mal epileptic seizures are accompanied by rapid EEG patterns of 25–30 waves per second. This sudden increase of EEG patterns from the normal of about 10 waves per second may cause an extensive stimulation of motor units and, therefore, uncontrollable skeletal-muscle activity. During a grand mal seizure, a person loses consciousness, convulses, and may lose urinary and bowel control. The unconsciousness and convulsions usually last a few minutes, after which the muscles relax and the person awakes but remains disoriented for a short time.

Epilepsy almost never affects intelligence and can be effectively treated with drugs in about 85% of the patients.

Cerebral Palsy. Cerebral palsy is a condition of motor disorders characterized by paresis (partial paralysis) and lack of muscular contraction. It is caused by damage to the motor areas of the brain during prenatal development, birth, or infancy. During neural development within an embryo, radiation or bacterial toxins (such as from German measles), transferred through the placenta of a pregnant mother, may cause cerebral palsy. Oxygen deprivation due to complications at birth and hydrocephalus in a newborn may also cause cerebral palsy. The three areas of the brain most severely affected by this disease are the cerebral cortex, the basal ganglia, and the cerebellum. The type of cerebral palsy is determined by which region of the brain is affected.

Some degree of mental retardation occurs in 60%–70% of cerebral palsy victims. Partial blindness, deafness, and speech problems frequently accompany this disease. Cerebral palsy is nonprogressive (i.e., these impairments do not worsen as a person ages), but neither are there organic improvements.

Neoplasms of the CNS. Neoplasms of the CNS are either **intracranial tumors,** which affect cells within or associated with the brain, or they are **intravertebral (intraspinal) tumors,** which affect cells within or near the spinal cord. **Primary neoplasms** develop within the CNS. Approximately one-half are benign, but they may become lethal because of the pressure they apply upon vital centers as they grow. Patients with **secondary,** or **metastatic, neoplasms** within the brain have a poor prognosis because the cancer has already established itself in another body organ, frequently the liver, lung, or breast, and has only secondarily spread to the brain. The symptoms of a brain tumor include headache, convulsions, pain, paralysis, or a change in behavior.

Neoplasms of the CNS are classified according to the tissues in which the cancer occurs. Tumors arising in neuroglia cells are called **gliomas** *(gli-o'mahz)* and account for about one-half of all primary neoplasms within the brain. Gliomas are frequently spread throughout cerebral tissue, develop rapidly, and usually cause death within a year after diagnosis. **Astrocytoma** *(as''tro-si-to'mah),* **oligodendroglioma** *(ol''ĭ-go-den''dro-gli-o'mah),* and **ependymoma** *(ĕ-pen''dĭ-mo'mah)* are common types of gliomas.

Meningiomas arise from meningeal coverings of the brain and account for about 15% of primary intracranial tumors. Meningiomas are usually benign if they can be treated readily.

Intravertebral tumors are classified as **extramedullary** when they develop on the outside of the spinal cord and as **intramedullary** when they develop within the substance of the spinal cord. Extramedullary neoplasms may cause pain and numbness in body structures away from the tumor as the growing tumor compresses the spinal cord. An intramedullary neoplasm causes a gradual loss of sensory, temperature, and motor function below the spinal-segmental level of the affliction.

The methods of detecting and treating cancers within the CNS have greatly improved in the last few years. Early detection and competent treatment have lessened the likelihood of death from this disease and reduced the probability of physical impairment.

Dyslexia. Dyslexia is a defect in the language center within the brain. In dyslexia, otherwise intelligent people reverse the order of letters in syllables, of syllables in words, and of words in sentences. The sentence: "The man saw a red dog," for example might be read by the dyslexic as "A red god was the man." Dyslexia is believed to result from the failure of one cerebral hemisphere to respond to written language, perhaps due to structural defects. Dyslexia can usually be overcome by intense remedial instruction in reading and writing.

petit mal: L. *pitinnus,* small child; *malus,* bad

dyslexia: Gk. *dys,* bad; *lexis,* speech

Meningitis. The nervous system is vulnerable to a variety of organisms and viruses that may cause abscesses or infections. Meningitis is an infection of the meninges, which may be confined to the spinal cord, in which case it is referred to as **spinal meningitis,** or it may involve the brain and associated meninges, in which case it is known as **encephalitis** or **encephalomyelitis** *(en-sef''ah-lo-mi''ĕ-li'tis),* respectively. The microorganisms that most commonly cause meningitis are meningococci, streptococci, pneumococci, and tubercle bacilli. Viral meningitis is more serious than bacterial meningitis. Nearly 20% of viral encephalitides are fatal. The organisms that cause meningitis probably enter the body through respiratory passageways.

Poliomyelitis. Poliomyelitis, or infantile paralysis, is primarily a childhood disease caused by a virus that destroys nerve cell bodies within the anterior horn of the spinal cord, especially those within the cervical and lumbar enlargements. This degenerative disease is characterized by fever, severe headache, stiffness and pain, and the loss of certain somatic reflexes. Muscle paralysis follows within several weeks, and eventually the muscles atrophy. Death results if the virus invades the vasomotor and respiratory nuclei within the medulla oblongata or anterior horn cells controlling respiratory muscles. Poliomyelitis has been effectively controlled with immunization.

Syphilis. Syphilis is a venereal disease that, if untreated, progressively destroys body organs. When syphilis causes organ degeneration, it is said to be in the *tertiary stage* (ten to twenty years after the primary infection). The organs of the nervous system are frequently infected, causing a condition called **neurosyphilis.** Neurosyphilis is classified according to the tissue involved, and the symptoms vary correspondingly. If the meninges are infected, the condition is termed **chronic meningitis. Tabes dorsalis** is a form of neurosyphilis in which there is a progressive degeneration of the posterior funiculi of the spinal cord and dorsal roots of spinal nerves. Motor control is gradually lost, and patients eventually become bedridden, unable even to feed themselves.

Degenerative Disorders of the Nervous System

Degenerative diseases of the CNS are characterized by a progressive, symmetrical deterioration of vital structures of the brain or spinal cord. The etiologies of these diseases are poorly understood, but it is thought that most of them are genetic.

Cerebrovascular Accident (CVA). Cerebrovascular accident is the most common disease of the nervous system. It is the third highest cause of death in the United States and perhaps the major cause of disability. The term **stroke** is frequently used synonymously with CVA, but actually a stroke refers to the sudden and dramatic appearance of a neurological defect. Cerebral thrombosis, in which a thrombus, or clot, forms in an artery of the brain, is the most common cause of CVA. Other causes of CVA include intracerebral hemorrhages, aneurysms, atherosclerosis, and arteriosclerosis of the cerebral arteries.

Patients who recover from CVA frequently suffer partial paralysis and mental disorders, such as loss of language skills. The dysfunction depends upon the severity of the CVA and the regions of the brain that were injured. Patients surviving a CVA can often be rehabilitated, but approximately two-thirds die within three years of the initial damage.

Syringomyelia. Syringomyelia *(sĭ-ring''go-mi-e'le-ah)* is a relatively uncommon condition characterized by the appearance of cystlike cavities, called *syringes,* within the gray matter of the spinal cord. These syringes progressively destroy the cord from the inside out. Syringomyelia is a chronic, slow progressing disease of unknown cause. As the spinal cord deteriorates, the patient experiences muscular weakness and atrophy and sensory loss, particularly the senses of pain and temperature.

Chapter Summary

I. Characteristics of the Central Nervous System
   A. The central nervous system (CNS), consisting of the brain and spinal cord, is covered with meninges, is bathed in cerebrospinal fluid, and contains gray and white matter.
   B. The tremendous metabolic rate of the 1.5 kg brain requires a continuous flow of blood amounting to approximately 20% of the total cardiac output.

II. Cerebrum
   A. The cerebrum, consisting of two convoluted hemispheres, is concerned with higher brain functions, such as the perception of sensory impulses, the instigation of voluntary movement, the storage of memory, thought processes, and reasoning ability.
   B. The cerebral cortex of the two cerebral hemispheres is convoluted with gyri and sulci.
   C. Each cerebral hemisphere contains frontal, parietal, temporal, occipital, and insula lobes.
   D. Brain waves generated by the cerebral cortex are recorded as an electroencephalogram and may provide valuable diagnostic information.
   E. The white matter of the cerebrum consists of association, commissural, and projection fibers.
   F. The basal ganglia are specialized masses of gray matter located within the white matter of the cerebrum.
   G. Broca's area, Wernicke's area, the arcuate fasciculus, and the angular gyrus are the language areas of the brain and are generally located in the cerebral cortex of the left hemisphere.
   H. The consolidation of memory requires protein synthesis and probably involves changes in the chemical structure and function of synapses.

III. Diencephalon
   A. The diencephalon is a major autonomic region of the brain.
   B. The thalamus is an ovoid mass of gray matter that functions as a relay center for sensory impulses and responds to pain.
   C. The hypothalamus is an aggregation of specialized nuclei that regulate many visceral activities. It also performs emotional and instinctual functions.

D. The epithalamus contains the pineal gland and the vascular choroid plexus over the roof of the third ventricle.

IV. Mesencephalon

A. The mesencephalon contains the corpora quadrigemina, the cerebral peduncles, and specialized nuclei that help to control posture and movement.

B. The superior colliculi of the corpora quadrigemina are concerned with visual reflexes, and the inferior colliculi are concerned with auditory reflexes.

C. The red nucleus and the substantia nigra are concerned with motor activities.

V. Metencephalon

A. The pons consists of fiber tracts connecting the cerebellum and medulla to other structures of the brain. The pons also contains nuclei for certain cranial nerves and the regulation of respiration.

B. The cerebellum consists of two hemispheres connected by the vermis and supported by three paired peduncles.

1. It is composed of white matter surrounded by a thin convoluted cortex of gray matter.
2. The cerebellum is concerned with coordinated contractions of skeletal muscle.

VI. Myelencephalon

A. The medulla oblongata is composed of the ascending and descending tracts of the spinal cord and contains nuclei for several autonomic functions.

B. The reticular formation functions as the reticular activating system in arousing the cerebrum.

VII. Meninges of the Central Nervous System

A. The cranial dura mater consists of an outer periosteal layer and an inner meningeal layer. The spinal dura mater is a single layer that is surrounded by the vascular epidural space.

B. The arachnoid membrane is a netlike meninx surrounding the subarachnoid space, which contains cerebrospinal fluid.

C. The thin pia mater adheres to the contour of the CNS.

VIII. Ventricles and Cerebrospinal Fluid

A. The lateral (first and second), third, and fourth ventricles are interconnected chambers within the brain that are continuous with the central canal of the spinal cord.

B. These chambers are filled with cerebrospinal fluid, which also flows throughout the subarachnoid space.

C. Cerebrospinal fluid is continuously secreted by the choroid plexuses and is absorbed into the blood at the arachnoid villi.

IX. Spinal Cord

A. The spinal cord is composed of thirty-one segments, each of which gives rise to a pair of spinal nerves.

1. It is characterized by a cervical enlargement, a lumbar enlargement, and two longitudinal grooves, which partially divide it into right and left halves.
2. The conus medullaris is the terminal portion of the spinal cord, and the cauda equina are nerve roots which radiate inferiorly from that point.

B. Ascending and descending spinal cord tracts are referred to as funiculi.

1. Descending tracts are grouped as either corticospinal (pyramidal) or extrapyramidal.
2. Many of the fibers in the funiculi decussate (cross over) in the spinal cord or in the medulla oblongata.

Review Activities

Objective Questions

1. Which of the following is *not* a lobe of the cerebrum?
   (a) parietal
   (b) sphenoid
   (c) temporal
   (d) insula
   (e) occipital
2. The principal connection between the cerebral hemispheres is the
   (a) corpus callosum
   (b) pons
   (c) intermediate mass
   (d) vermis
   (e) precentral gyrus
3. The structure of the brain that is most directly involved in the autonomic response to pain is the
   (a) pons
   (b) hypothalamus
   (c) medulla oblongata
   (d) thalamus
4. Which statement is *false* concerning the basal ganglia?
   (a) They are located within the cerebrum.
   (b) They regulate the basal metabolic rate.
   (c) They consist of the caudate nucleus, lentiform nucleus, putamen, and globus pallidus.
   (d) They indirectly exert an inhibitory influence on lower motor neurons.
5. The corpora quadrigemina, red nucleus, and substantia nigra are structures of the
   (a) diencephalon
   (b) metencephalon
   (c) mesencephalon
   (d) myelencephalon
6. The fourth ventricle is contained within the
   (a) cerebrum
   (b) cerebellum
   (c) midbrain
   (d) metencephalon
7. The right cerebral cortex controls voluntary movements on the left side of the body because
   (a) most persons are right-handed
   (b) the right hemisphere dominates
   (c) many of the fibers in the funiculi decussate in the medulla oblongata
   (d) of distinct cerebral specialization of hemispheres
8. A patient experiencing a fluctuating body temperature, lack of hunger and thirst, and psychosomatic disorders may have a malfunctioning
   (a) hypothalamus
   (b) midbrain
   (c) cerebellum
   (d) medulla oblongata
9. Spinal cord tracts that descend from the cerebral cortex to the lower motor neurons without synaptic interruption are called
   (a) reticulospinal tracts
   (b) corticospinal tracts
   (c) rubrospinal tracts
   (d) vestibulospinal tracts
10. The disease characterized by the destruction of the myelin sheaths of neurons in the CNS and the formation of plaques is
    (a) syringomyelia
    (b) neurosyphilis
    (c) poliomyelitis
    (d) multiple sclerosis

Essay Questions

1. List the types of brain waves recorded on an electroencephalogram, and explain the diagnostic value of each.
2. List the functions of the hypothalamus. Why is the hypothalamus considered a major part of the autonomic nervous system?
3. What structures are within the midbrain? List the nuclei located in the midbrain, and give the function of each.
4. Describe the location and structure of the medulla oblongata. List the nuclei found within this structure. What are the functions of the medulla oblongata?
5. What is cerebrospinal fluid? Where is it produced, and what is its pathway of circulation?
6. Define the following abbreviations: EEG, ANS, CSF, PNS, RAS, CT scan, MS, DSR, and CVA.
7. Describe the various techniques available for conducting a neurological assessment.
8. List and define the various psychological terms used to describe mental illness.
9. What is epilepsy? What causes it, and how is it controlled?
10. What do meningitis, poliomyelitis, and neurosyphilis have in common, and how do these conditions differ?

16

Peripheral Nervous System

Outline

Concepts

The peripheral nervous system consists of nerves that convey impulses to and from the central nervous system.

Twelve pairs of cranial nerves emerge from the inferior surface of the brain and pass through the foramina of the skull to innervate structures in the head, neck, and visceral organs of the torso.

Each of the thirty-one pairs of spinal nerves is formed by the union of a dorsal and a ventral spinal root, which emerges from the spinal cord through an intervertebral foramen to innervate a body dermatome.

Except in the thoracic nerves T2–T12, the ventral rami of the spinal nerves combine and then split again as networks of nerves referred to as plexuses. There are four plexuses of spinal nerves: the cervical, the brachial, the lumbar, and the sacral. Nerves emerge from the plexuses and are named according to the structures they innervate or the general course they take.

The conduction pathway of a reflex arc consists of a receptor, a sensory neuron, a motor neuron and its innervation in the PNS, and an association neuron in the CNS. The reflex arc provides the mechanism for a rapid, automatic response to a potentially threatening stimulus.

Introduction to the Peripheral Nervous System

The peripheral nervous system consists of nerves that convey impulses to and from the central nervous system.

Objective 1. Define *peripheral nervous system,* and distinguish between sensory, motor, and mixed nerves.

The **peripheral nervous system (PNS)** is that portion of the nervous system outside the central nervous system. The PNS functions to convey impulses to and from the brain or spinal cord. Sensory receptors within the sensory organs, neurons, nerves, ganglia, and plexuses are all part of the PNS, which serves virtually every part of the body (fig. 16.1). The sensory receptors are discussed in chapter 18.

The nerves of the PNS are classified as cranial nerves or spinal nerves, depending on whether they arise from the brain or the spinal cord. The terms *sensory nerve, motor nerve,* and *mixed nerve* relate to the direction in which the nerve impulses are being conducted. **Sensory nerves** consist of sensory (afferent) fibers that convey impulses toward the CNS. **Motor nerves** consist of motor (efferent) fibers that convey impulses away from the CNS. **Mixed nerves** are composed of both sensory and motor fibers and therefore convey impulses in both directions. Reflexes that are

Figure 16.1. The peripheral nervous system consists of cranial nerves, spinal nerves, plexuses, ganglia (not shown), and peripheral nerves.

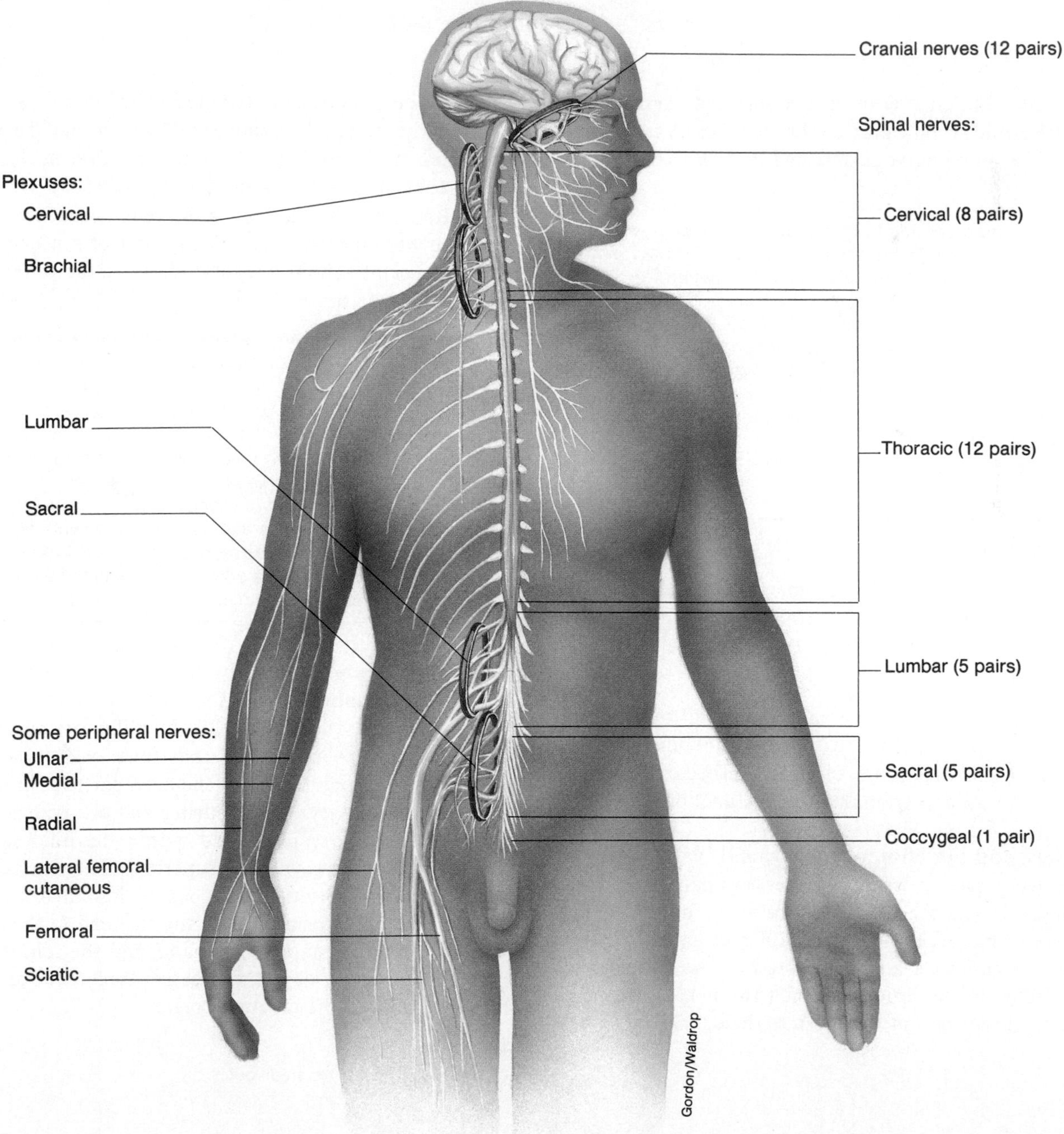

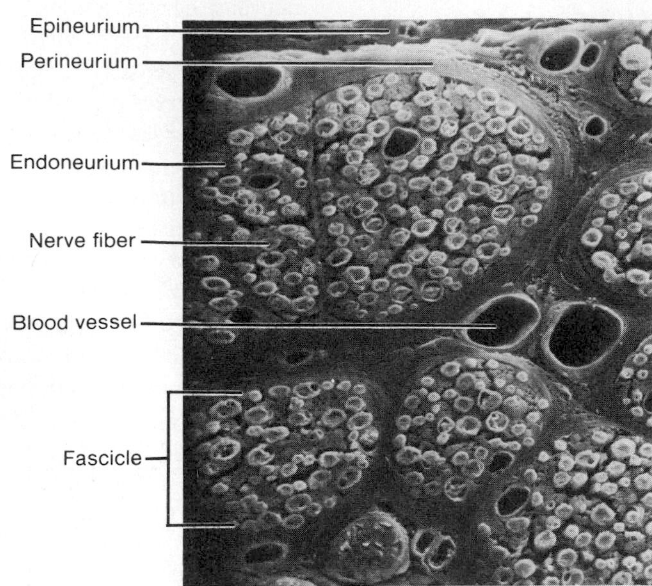

Figure 16.2. Scanning electron micrograph of a spinal nerve seen in transverse section (about 1000X). (*Tissues and Organs: A Text-Atlas of Scanning Electron Microscopy*, by R. G. Kessel and R. Kardon. © 1979 W. H. Freeman and Company.)

considered in this chapter involve sensory and motor nerves of the PNS and specific portions of the CNS. A transverse section of a spinal nerve is depicted in figure 16.2.

1. The tongue responds to tastes and pain and moves to manipulate food. Make a quick sketch of the brain and the tongue to depict the relationship between the CNS and the PNS. Indicate diagrammatically with lines and arrows the sensory and motor innervation of the tongue. Define *mixed nerve.* What kinds of sensory stimulation arise from the tongue? What type of response is caused by motor stimulation to the tongue?
2. List the structures of the nervous system that are considered a part of the PNS.

Cranial Nerves

Twelve pairs of cranial nerves emerge from the inferior surface of the brain and pass through the foramina of the skull to innervate structures in the head, neck, and visceral organs of the torso.

Objective 2. List the twelve pairs of cranial nerves, and describe the location and function of each.

Objective 3. Describe the clinical methods for determining cranial nerve dysfunction.

Structure and Function of the Cranial Nerves

Of the twelve pairs of cranial nerves, two pairs arise from the forebrain and ten pairs arise from the midbrain and brain stem (fig. 16.3). The cranial nerves are designated by Roman numerals and with names. The Roman numerals refer to the order in which the nerves are positioned from the front of the brain to the back. The names indicate the structures innervated or the principal functions of the nerves. A summary of the cranial nerves is presented in table 16.1, and the locations of the nuclei from which they arise are illustrated in figure 15.16.

Although most cranial nerves are mixed, some are associated with special senses and consist of sensory fibers only. The cell bodies of sensory fibers are located in ganglia outside the brain.

Generations of anatomy students have used a mnemonic device to help them remember the order in which the cranial nerves emerge from the brain: "On old Olympus' towering top, a Finn and German viewed a hop." The initial letter of each word in this jingle corresponds to the initial letter of each pair of cranial nerves. A problem with this classic verse is that the eighth cranial nerve represented by *and* in the jungle, which used to be referred to as auditory, is currently recognized as the vestibulocochlear cranial nerve.

I Olfactory Nerve. Actually, numerous olfactory nerves relay sensory impulses of smell from the mucous membranes of the nasal cavity (fig. 16.4). Olfactory nerves are composed of bipolar neurons that function as *chemoreceptors,* responding to volatile chemical particles breathed into the nasal cavity. The dendrites and cell bodies of olfactory neurons are positioned within the mucosa, primarily that which covers the superior nasal conchae and adjacent nasal septum. The axons of these neurons pass through the cribriform plate of the ethmoid to the **olfactory bulb** where synapses are made, and the sensory impulses are passed through the **olfactory tract** to the primary olfactory area in the cerebral cortex.

olfactory: L. *olfacere,* smell out

Figure 16.3. Cranial nerves.

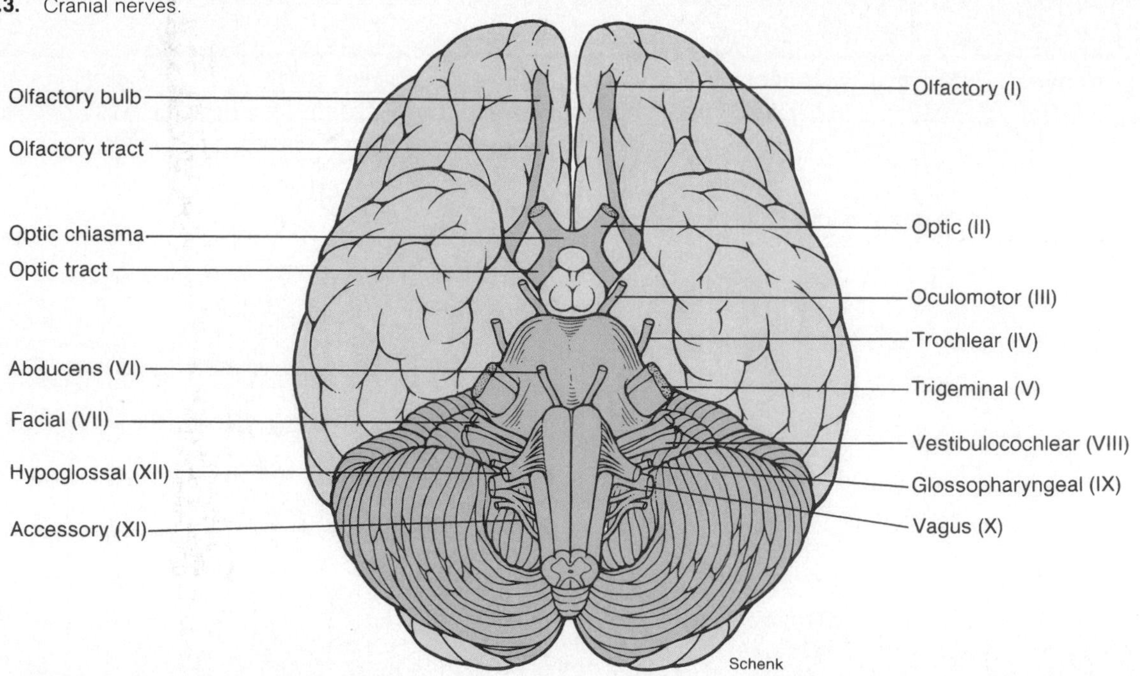

Figure 16.4. The olfactory nerve.

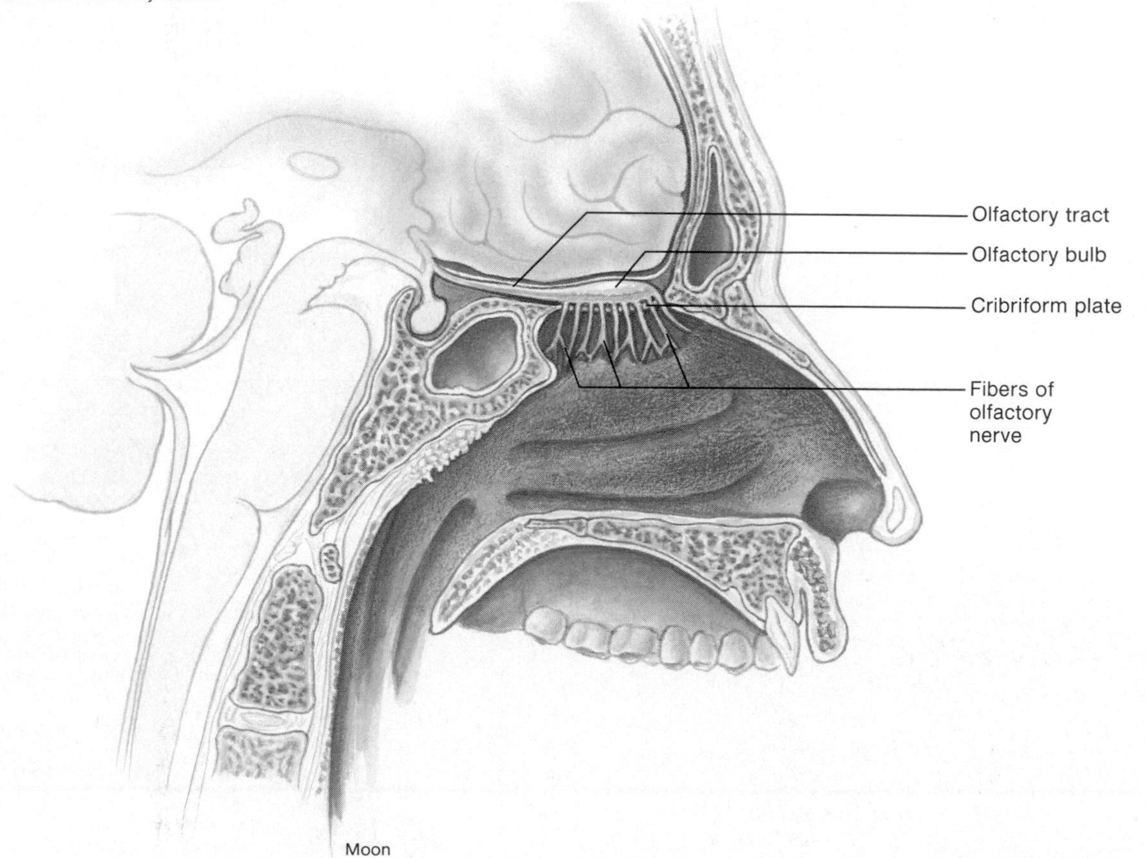

Table 16.1 Summary of cranial nerves

Number and name	Foramen transmitting	Composition	Location of cell bodies	Function
I Olfactory	Foramina in ethmoidal cribriform plate	Sensory	Bipolar cells in nasal mucosa	Olfaction
II Optic	Optic canal	Sensory	Ganglion cells of retina	Vision
III Oculomotor	Superior orbital fissure	Motor	Oculomotor nucleus	Motor impulses to levator palpebrae superioris and extrinsic eye muscles except superior oblique and lateral rectus; innervation to muscles that regulate amount of light entering eye and that focus the lens
		Sensory: proprioception		Proprioception from muscles innervated with motor fibers
IV Trochlear	Superior orbital fissure	Motor	Trochlear nucleus	Motor impulses to superior oblique muscle of eyeball
		Sensory: proprioception		Proprioception from superior oblique muscle of eyeball
V Trigeminal				
Ophthalmic division	Superior orbital fissure	Sensory	Semilunar ganglion	Sensory impulses from cornea, skin of nose, forehead, and scalp
Maxillary division	Foramen rotundum	Sensory	Semilunar ganglion	Sensory impulses from nasal mucosa, upper teeth and gums, palate, upper lip, and skin of cheek
Mandibular division	Foramen ovale	Sensory	Semilunar ganglion	Sensory impulses from temporal region, tongue, lower teeth and gum, and skin of chin and lower jaw
		Sensory: proprioception		Proprioception from muscles of mastication
		Motor	Motor trigeminal nucleus	Motor impulses to muscles of mastication and muscle that tenses tympanum
VI Abducens	Superior orbital fissure	Motor	Abducens nucleus	Motor impulses to lateral rectus muscle of eyeball
		Sensory: proprioception		Proprioception from lateral rectus muscle of eyeball
VII Facial	Stylomastoid foramen	Motor	Motor facial nucleus	Motor impulses to muscles of facial expression and muscle that tenses the stapes
		Motor: parasympathetic	Superior salivatory nucleus	Secretion of tears from lacrimal gland and salivation from sublingual and submandibular salivary glands
		Sensory	Geniculate ganglion	Sensory impulses from taste buds on anterior two-thirds of tongue; nasal and palatal sensation
		Sensory: proprioception		Proprioception from muscles of facial expression

Number and name	Foramen transmitting	Composition	Location of cell bodies	Function
VIII Vestibulocochlear	Internal acoustic meatus	Sensory	Vestibular ganglion	Sensory impulses associated with equilibrium
			Spiral ganglion	Sensory impulses associated with hearing
IX Glossopharyngeal	Jugular foramen	Motor	Nucleus ambiguus	Motor impulses to muscles of pharynx used in swallowing
		Sensory: proprioception	Petrosal ganglion	Proprioception from muscles of pharynx
		Sensory	Petrosal ganglion	Sensory impulses from taste buds on posterior one-third of tongue; pharynx, middle-ear cavity, carotid sinus
		Parasympathetic	Inferior salivatory nucleus	Salivation from parotid salivary gland
X Vagus	Jugular foramen	Motor	Nucleus ambiguus	Contraction of muscles of pharynx (swallowing) and larynx (phonation)
		Sensory: proprioception		Proprioception from visceral muscles
		Sensory	Nodose ganglion	Sensory impulses from taste buds on rear of tongue; sensations from auricle of ear; general visceral sensations
		Motor: parasympathetic	Dorsal motor nucleus	Regulate visceral motility
XI Accessory	Jugular foramen	Motor	Nucleus ambiguus	Laryngeal movement; soft palate
			Spinal accessory nucleus	Motor impulses to trapezius and sternocleidomastoid muscles for movement of head, neck, and shoulders
		Sensory: proprioception		Proprioception from muscles that move head, neck, and shoulders
XII Hypoglossal	Hypoglossal canal	Motor	Hypoglossal nucleus	Motor impulses to intrinsic and extrinsic muscles of tongue and infrahyoid muscles
		Sensory: proprioception		Proprioception from muscles of tongue

II Optic Nerve. The optic nerve, another sensory nerve, conducts impulses from the *photoreceptors* (rods and cones) in the retina of the eye. Each optic nerve is composed of an estimated 1.25 million nerve fibers that converge at the back of the eyeball and enter the cranial cavity through the optic foramen. The two optic nerves unite on the floor of the diencephalon to form the **optic chiasma** *(ki-as'mah)* (fig. 16.5). Nerve fibers that arise from the medial half of each retina cross at the chiasma to the opposite side of the brain, whereas fibers arising from the lateral half remain on the same side of the brain. The optic nerve fibers pass posteriorly from the chiasma in the **optic tracts.** The optic tracts lead to the thalami, where a majority of the fibers terminate within certain thalamic nuclei. A few of the ganglion-cell axons that reach the thalamic nuclei have collaterals that convey impulses to the superior colliculi. Synapses within the thalamic nuclei, however, permit impulses to pass through neurons to the **visual cortex** within the occipital lobes. Other synapses permit impulses to reach the nuclei for the oculomotor, trochlear, and abducens nerves, which regulate intrinsic (internal) and extrinsic (from orbit to eyeball) eye muscles. The visual pathway into the eyeball functions reflexively to produce motor responses to light stimuli. If an optic nerve is damaged, the eyeball served by that nerve is blinded.

Figure 16.5. The optic nerve.

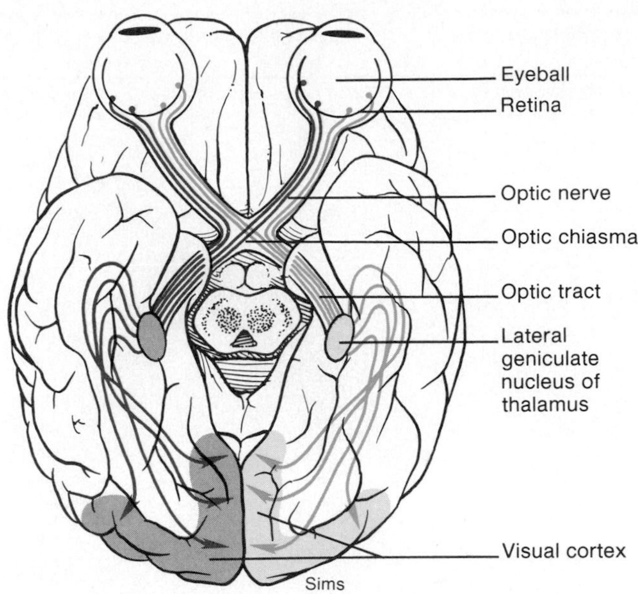

III Oculomotor. The oculomotor nerve produces certain extrinsic and intrinsic movements of the eyeball. The oculomotor is primarily a motor nerve that arises from nuclei within the midbrain. The oculomotor divides into superior and inferior branches as it passes through the superior orbital fissure in the orbit (fig. 16.6). The superior branch innervates the **superior rectus** muscle, which moves the eyeball superiorly, and the **levator palpebrae** *(le-va'tor pal'pĕ-bre)* **superioris** muscle, which raises the upper eyelid. The inferior branch innervates the **medial rectus, inferior rectus,** and **inferior oblique** eye muscles for medial, inferior, and superior and lateral movement of the eyeball, respectively. In addition, fibers from the inferior branch of the oculomotor enter the eyeball to supply motor innervation to the intrinsic smooth muscles of the iris for pupil dilation and to the muscles within the ciliary body for lens accommodation.

A few sensory fibers of the oculomotor originate from proprioceptors within the intrinsic muscles of the eyeball. These fibers convey impulses that affect the position and activity of the muscles they serve. A person whose oculomotor nerve is damaged may have a drooping upper eyelid or a dilated pupil or be unable to move the eyeball in the directions permitted by the four extrinsic muscles innervated by this nerve.

optic: L. *optica,* see
chiasma: Gk. *chiasma,* an X-shaped arrangement

IV Trochlear. The trochlear is a very small mixed nerve that emerges from a nucleus within the midbrain and passes from the cranium through the superior orbital fissure of the orbit. The trochlear innervates the **superior oblique** muscle of the eyeball with both motor and sensory fibers (fig. 16.6). Motor impulses to the superior oblique cause the eyeball to rotate downward and away from the midline. Sensory impulses originate in proprioceptors of the superior oblique muscle and provide information about its position and activity. Damage to the trochlear nerve impairs movement in the direction permitted by the superior oblique eye muscle.

V Trigeminal. The large trigeminal nerve is a mixed nerve with motor functions originating from the nuclei within the pons and sensory functions terminating in nuclei within the midbrain, pons, and medulla. Two roots of the trigeminal are apparent as they emerge from the ventrolateral side of the pons (see fig. 16.3). The larger, **sensory root** immediately enlarges into a swelling called the **semilunar (gasserian) ganglion,** located in a bony depression on the inner surface of the petrous portion of the temporal bone. Three large branches arise from the semilunar ganglion (fig. 16.7): the **ophthalmic branch** enters the orbit through the superior orbital fissure; the **maxillary branch** extends through the foramen rotundum; and the **mandibular branch** passes through the foramen ovale. The smaller,

trochlear: Gk. *trochos,* a wheel
trigeminal: L. *trigeminus,* three born together
gasserian ganglion: from Johann L. Gasser, Viennese anatomist, 18th century
ophthalmic: L. *ophthalmia,* region of the eye

Figure 16.6. The oculomotor, trochlear, and abducens cranial nerves.

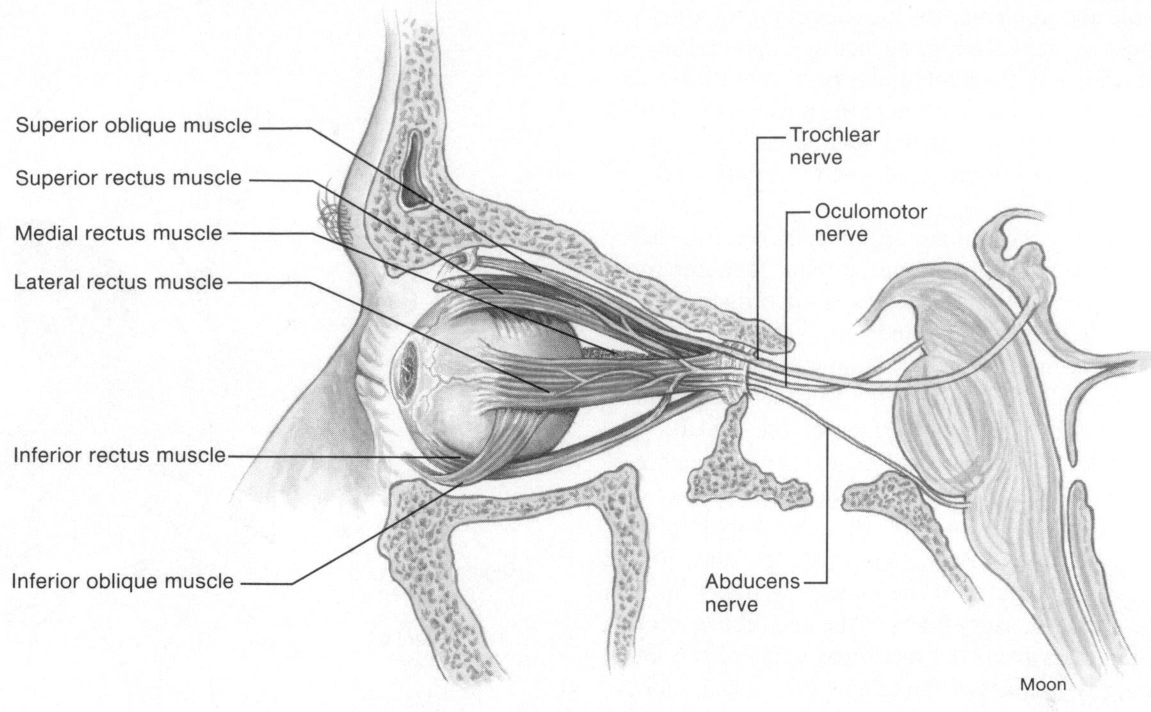

Figure 16.7. The trigeminal nerve and its branches.

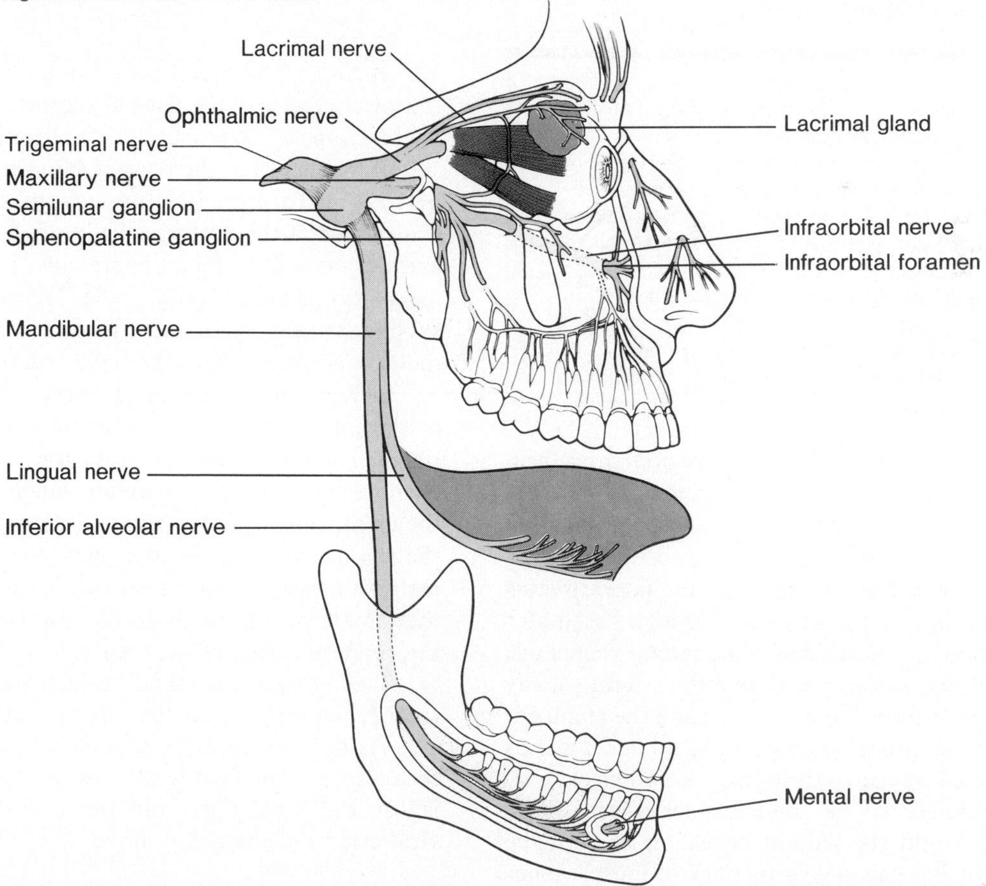

motor root consists of motor fibers of the trigeminal nerve, which accompany the mandibular branch through the foramen ovale and innervate the muscles of mastication and certain muscles in the floor of the mouth. Impulses through the motor branch of the mandibular portion of the trigeminal nerve stimulate contraction of the muscles involved in chewing, including the medial and lateral pterygoids, masseter, temporalis, mylohyoid, and the anterior belly of the digastric muscle.

Although the trigeminal is a mixed nerve, its sensory functions are much more extensive than its motor functions. The three sensory branches of the trigeminal respond to touch, temperature, and pain sensations from the face. More specifically, the ophthalmic branch consists of sensory fibers from the anterior half of the scalp, skin of the forehead, upper eyelid, surface of the eyeball, lacrimal (tear) gland, side of the nose, and upper mucosa of the nasal cavity. The maxillary branch is composed of sensory fibers from the lower eyelid, lateral and inferior mucosa of the nasal cavity, palate and portions of the pharynx, teeth and gums of the upper jaw, upper lip, and skin of the cheek. Sensory fibers of the mandibular branch transmit impulses from the teeth and gums of the lower jaw, anterior two-thirds of the tongue (not taste), mucosa of the mouth, auricle of the ear, and lower part of the face. Trauma to the trigeminal nerve causes a lack of sensation from specific facial structures, whereas damage to the mandibular branch impairs chewing.

The trigeminal nerve is the principal nerve relating to the practice of dentistry. Before teeth are filled or extracted, anesthetic is injected into the appropriate nerve to block sensation. A *maxillary nerve block,* or *second-division nerve block,* performed by injecting near the sphenopalatine (pterygopalatine) ganglion (see fig. 16.7), desensitizes the teeth in the upper jaw. A *mandibular,* or *third-division, nerve block* desensitizes the lower teeth. This is performed by injecting anesthetic into the inferior alveolar branch of the mandibular nerve where it enters the mandible through the mandibular foramen.

VI Abducens. The small abducens nerve originates from a nucleus within the pons and emerges from the lower portion of the pons and the anterior border of the medulla oblongata. It is a mixed nerve that traverses the superior orbital fissure of the orbit to innervate the **lateral rectus** eye muscle (see fig. 16.6). Impulses through the motor fibers of the abducens nerve cause the lateral rectus eye muscle to contract and the eyeball to move laterally away from the midline. Sensory impulses through the abducens nerve originate in proprioceptors in the lateral rectus muscle and are conveyed to the pons, where muscle contraction is mediated. If the abducens nerve were damaged, not only would the patient be unable to turn the eyeball laterally, but because of the lack of muscle tonus to the lateral rectus, the eyeball would be pulled medially.

Figure 16.8. Facial nerve and its branches.

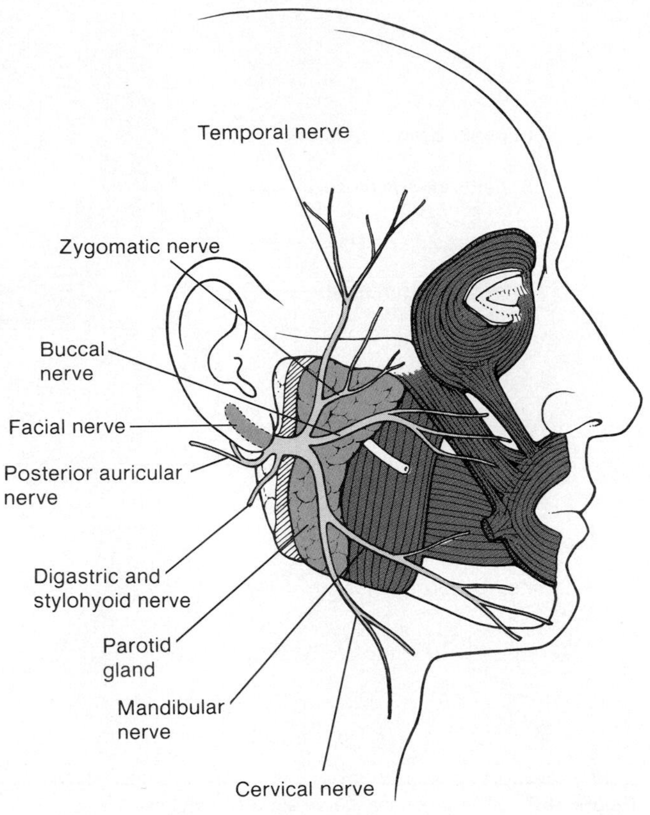

VII Facial. The facial nerve arises from nuclei within the lower portion of the pons, traverses the petrous portion of the temporal bone (see fig. 16.9), and emerges on the side of the face near the parotid salivary gland. The facial nerve is mixed. Impulses through the motor fibers cause contraction of the posterior belly of the digastric muscle and the muscles of facial expression, including the scalp and platysma muscles (fig. 16.8). The submandibular and sublingual salivary glands also receive some autonomic motor innervation from the facial nerve.

Sensory fibers of the facial nerve arise from taste buds on the anterior two-thirds of the tongue. Taste buds function as *chemoreceptors* because they respond to specific chemical stimuli. The **geniculate ganglion** *(gang'gle-on)* is the enlargement of the facial nerve prior to the entry of the sensory portion into the pons. Sensory sensations of taste are conveyed into nuclei within the medulla, through the thalamus, and finally to the gustatory (taste) area of the cerebral cortex of the insula.

Trauma to the facial nerve destroys the ability to contract facial muscles on the affected side of the face and distorts taste perception, particularly of sweets. The affected side of the face tends to sag because muscle tonus is lost. *Bell's palsy* is a functional disorder (probably of viral origin) of the facial nerve.

Bell's palsy: from Sir Charles Bell, Scottish physician, 1774–1842

Figure 16.9. The vestibulocochlear nerve.

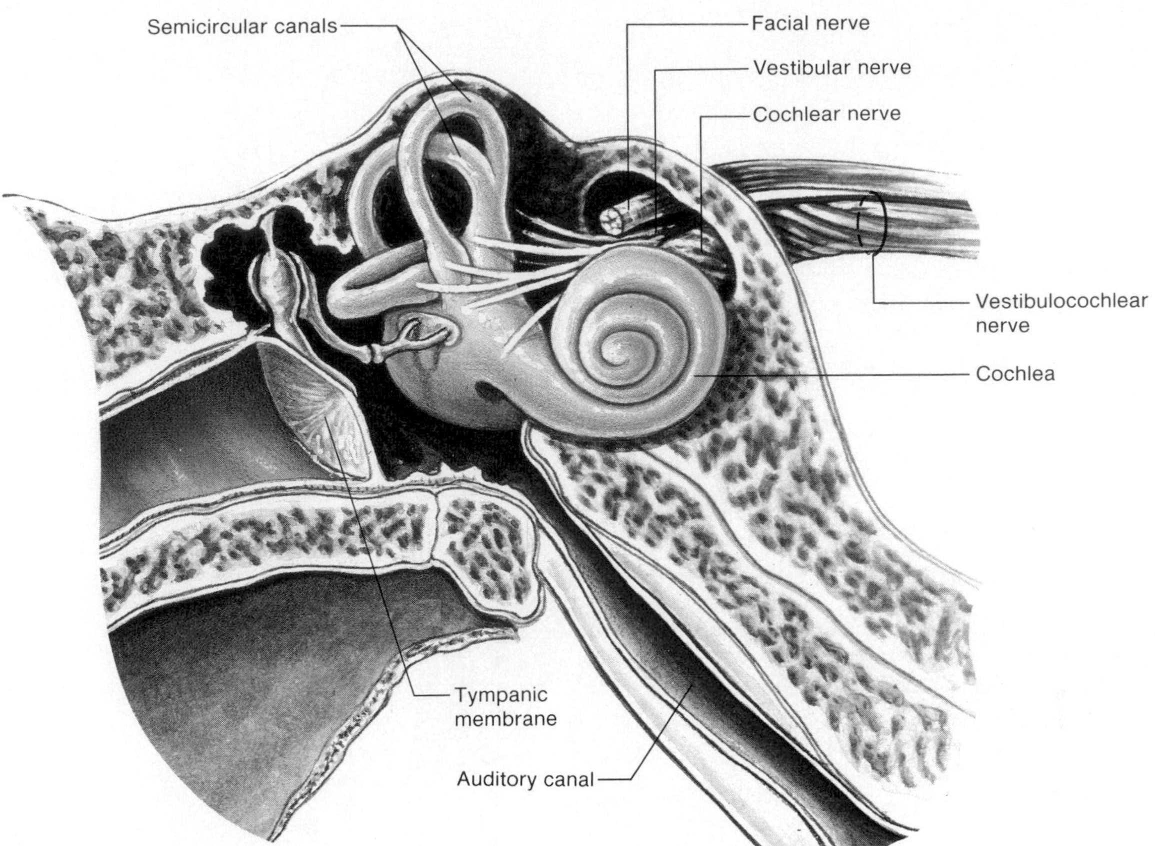

VIII Vestibulocochlear (Auditory). The vestibulocochlear nerve is also referred to as the **auditory, acoustic,** or **statoacoustic** nerve. It is purely sensory and is composed of two branches that arise within the inner ear (fig. 16.9). The **vestibular branch** arises from the **vestibular organs** of equilibrium and balance. Bipolar neurons from the vestibular organs (saccule, utricle, and semicircular canals) extend to the **vestibular ganglion,** where cell bodies are contained. From there, fibers convey impulses to the **vestibular nuclei** *(nu'kle-i)* within the pons and medulla oblongata. Fibers from there extend to the thalamus and the cerebellum.

The **cochlear branch** arises from the **organ of Corti** within the cochlea and is associated with hearing. The cochlear branch is composed of bipolar neurons that convey impulses through the **spiral ganglion** to the **cochlear nuclei** within the medulla oblongata. From there, fibers extend to the thalamus and synapse with neurons that convey the impulses to the auditory areas of the cerebral cortex.

Injury to the cochlear portion of the vestibulocochlear nerve results in perception deafness, whereas damage to the vestibular portion causes dizziness and an inability to maintain balance.

IX Glossopharyngeal. The glossopharyngeal nerve is a mixed nerve that innervates part of the tongue and pharynx (fig. 16.10). The motor fibers of this nerve originate in a nucleus within the medulla oblongata and pass through the jugular foramen. The motor fibers innervate the muscles of the pharynx and the parotid salivary gland to stimulate the swallowing reflex and the secretion of saliva.

The sensory fibers of the glossopharyngeal nerve arise from the pharyngeal region, the parotid salivary gland, the middle ear cavity, and the taste buds on the posterior one-third of the tongue. These taste buds, like those innervated by the facial nerve, are *chemoreceptors.* Some sensory fibers also arise from sensory receptors within the carotid sinus of the neck and help regulate blood pressure. Impulses from the glossopharyngeal nerve travel through the medulla and into the thalamus, where they synapse with fibers that convey the impulses to the gustatory area of the cerebral cortex.

Damage to the glossopharyngeal nerve results in the loss of perception of bitter and sour taste from taste buds on the posterior portion of the tongue. If the motor portion of this nerve is damaged, swallowing becomes difficult.

---

vestibulocochlear: L. *vestibulum,* chamber; *cochlea,* snail shell
organ of Corti: from Alfonso Corti, Italian anatomist, 1822–88

glossopharyngeal: L. *glossa,* tongue; Gk. *pharynx,* throat

Figure 16.10. The glossopharyngeal nerve.

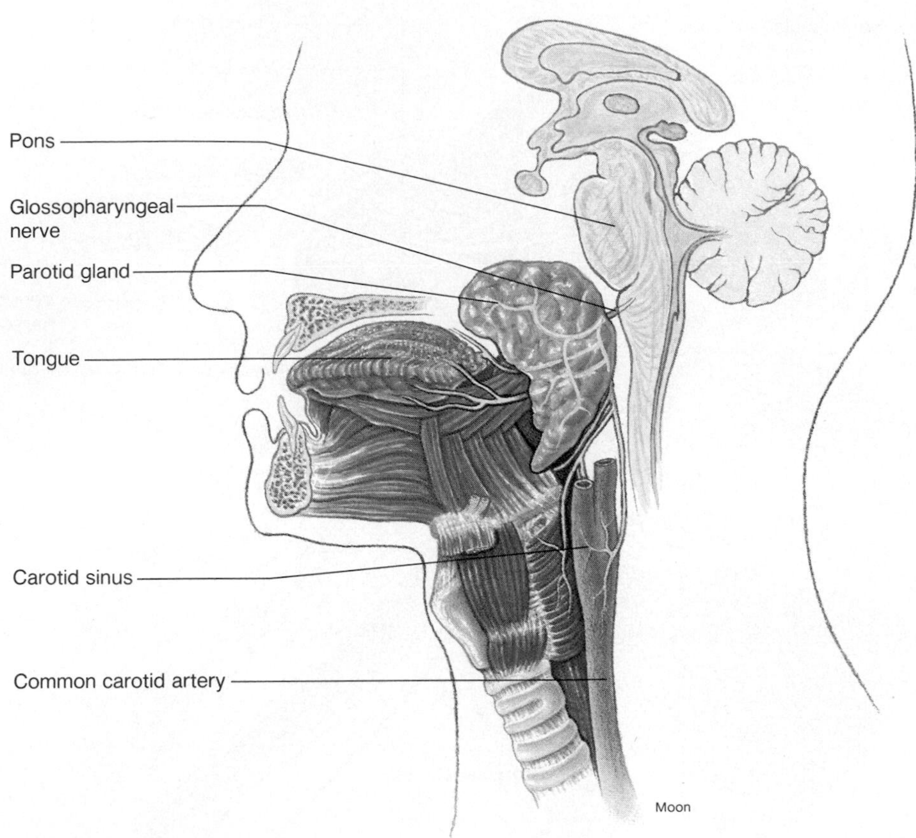

X Vagus. The vagus nerve has motor and sensory fibers that innervate visceral organs of the thoracic and abdominal cavities (fig. 16.11). The vagus nerve passes through the jugular foramen. The motor portion of the vagus arises from the **nucleus ambiguus** and **dorsal motor nucleus** of the vagus within the medulla. Through various branches, it innervates the muscles of the pharynx, larynx, respiratory tract, lungs, heart, esophagus, and abdominal viscera with the exception of the lower portion of the large intestine. One motor branch of the vagus nerve, the **recurrent laryngeal nerve,** innervates the larynx, enabling one to speak.

Sensory fibers of the vagus convey impulses from essentially the same organs served by motor fibers. Impulses through the sensory fibers relay specific sensations such as hunger pangs, distension, intestinal discomfort, or laryngeal movements. Sensory fibers also arise from proprioceptors in the muscles innervated by the motor fibers of this nerve.

If both vagus nerves are damaged, death ensues rapidly because vital autonomic functions stop, but the injury of one nerve causes vocal impairment, difficulty in swallowing, or other visceral disturbances.

XI Accessory. The accessory nerve is principally a motor nerve, but it does contain some sensory fibers from proprioceptors within the muscles it innervates. The accessory nerve is unique in that it arises from both the brain and the spinal cord (fig. 16.12). The **cranial motor component,** also called the **bulbar (medullary) portion,** arises from nuclei within the medulla (ambiguus and spinal accessory), passes through the jugular foramen with the vagus nerve, and innervates the skeletal muscles of the soft palate, pharynx, and larynx, which contract reflexively during swallowing. The **spinal motor component** arises from the first five segments of the cervical portion of the spinal cord, passes cranially through the foramen magnum to join with the bulbar portion, and passes through the jugular foramen. The spinal motor component of the accessory nerve innervates the sternocleidomastoid and the trapezius muscles that move the head, neck, and shoulders. Damage to an accessory nerve makes it difficult to move the head or shrug the shoulders.

vagus: L. *vagus,* wandering

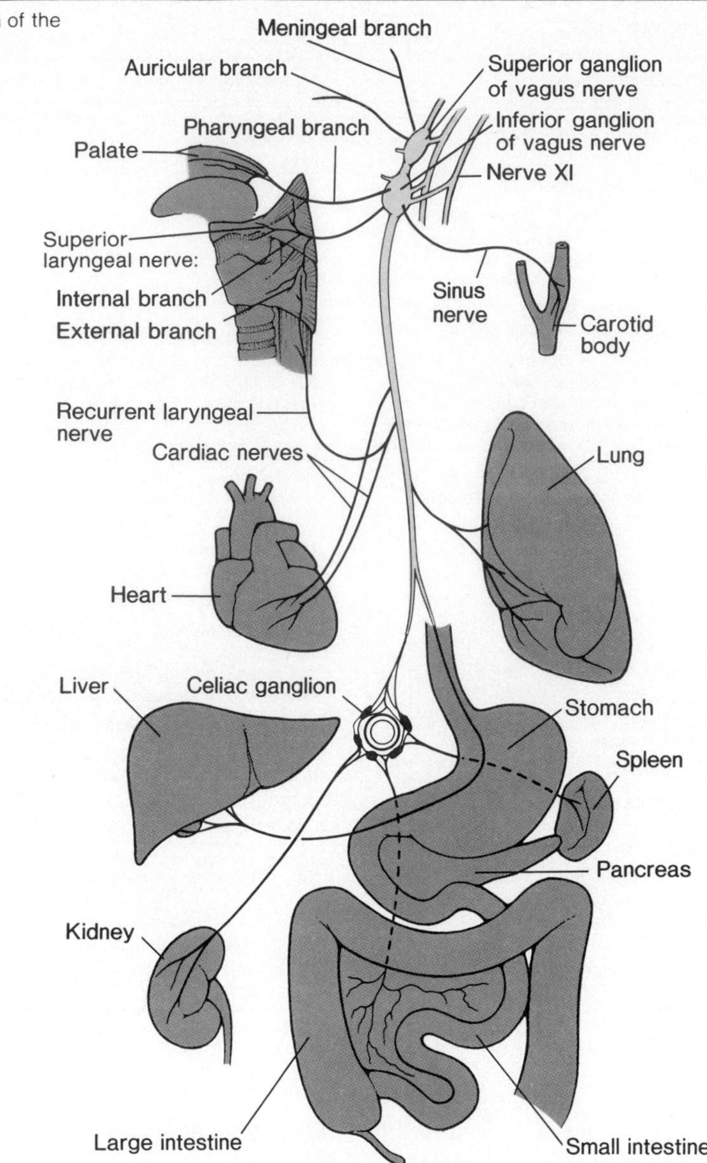

Figure 16.11. The distribution of the vagus nerve.

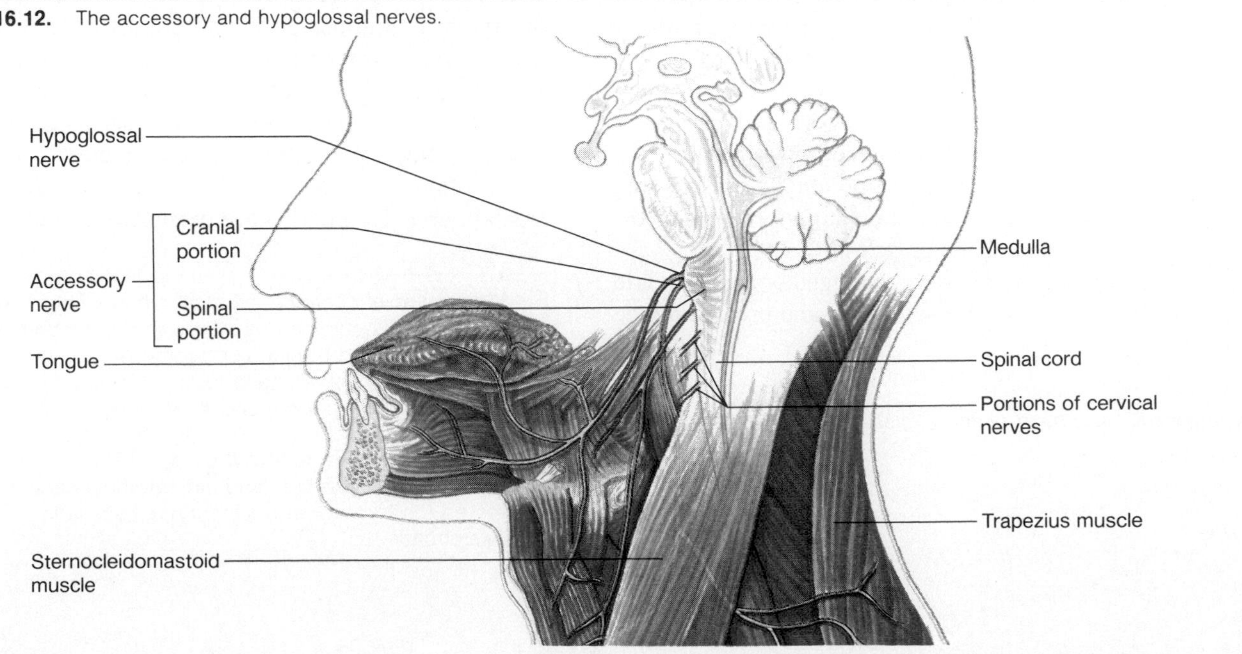

Figure 16.12. The accessory and hypoglossal nerves.

Table 16.2 Methods of determining cranial-nerve dysfunction

Nerve	Techniques of examination	Comments
Olfactory	With eyes closed, patient differentiates odors (tobacco, coffee, soap, etc.).	Nasal passages must be patent and tested separately by occluding the opposite side.
Optic	Examine the optic fundi with ophthalmoscope; test visual acuity by reading eye charts.	Visual acuity must be determined with lenses on, if patient wears them.
Oculomotor	Have patient follow examiner's finger movement with eyes—especially cross-eyed movement; observe pupillary change by shining light into each eye separately.	Examiner should note rate of pupillary change and coordinated constriction of pupils. Light in one eye should cause a similar pupillary change in other eye but to a lesser degree.
Trochlear	Have patient follow examiner's finger movement with eyes—especially lateral and downward movement.	
Trigeminal	Motor portion: examiner palpates temporalis and masseter muscles as patient clenches teeth; patient is asked to open mouth against resistance applied by examiner.	Muscles of both sides of the jaw should show equal contractile strength.
	Sensory portion: tactile and pain receptors are tested by lightly touching entire face with cotton and then with pin stimulus.	Patient's eyes should be closed and innervation areas for all three trigeminal branches should be tested.
Abducens	Have patient follow examiner's finger movement—especially lateral movement.	Motor functioning of cranial nerves III, IV, and VI may be tested simultaneously through selective movements of eyeball.
Facial	Motor portion: have patient raise eyebrows, frown, tightly constrict eyelids, smile, puff out cheeks, and whistle.	Examiner should note lack of tonus expressed by sagging regions of face.
	Sensory portion: examiner should place sugar on each side of tip of tongue.	Not reliable test for specific facial-nerve dysfunction because of tendency to stimulate taste buds on both sides of tip of tongue.
Vestibulocochlear	Vestibular portion: patient asked to walk a straight line.	Not usually tested unless patient complains of dizziness or balance problems.
	Cochlear portion: test with tuning fork.	Note ability to discriminate sounds.
Glossopharyngeal and vagus	Motor: examiner should note disturbances in swallowing, talking, and movement of soft palate; stimulate back of throat and note gag reflex.	Visceral innervation of vagus cannot be examined except for innervation to larynx, which is also served by glossopharyngeal.
Accessory	Patient is asked to shrug shoulders against resistance of examiner's hand and to rotate head against resistance.	Sides should show uniformity of strength.
Hypoglossal	Patient is requested to protrude tongue; tongue thrust may be resisted with tongue blade.	Tongue should protrude straight; deviation to side indicates ipsilateral-nerve dysfunction; asymmetry, atrophy, or lack of strength should be noted.

XII Hypoglossal. The hypoglossal nerve is a mixed nerve. The motor fibers arise from the hypoglossal nucleus within the medulla oblongata and pass through the hypoglossal canal of the skull to innervate both the extrinsic and intrinsic muscles of the tongue (see fig. 16.12). Motor impulses along these fibers account for the coordinated contraction of the tongue muscles necessary for such activities as food manipulation, swallowing, and speech.

The sensory portion of the hypoglossal nerve arises from proprioceptors within the same tongue muscles and conveys impulses to the medulla oblongata regarding the position and function of the muscles.

If a hypoglossal nerve were damaged, a person would have difficulty in speaking, swallowing, and protruding the tongue.

hypoglossal: Gk. *hypo,* under; L. *glossa,* tongue

Neurological Assessment of Cranial Nerves

Head injuries and brain concussions are common occurrences in automobile accidents. The cranial nerves would seem to be well protected on the inferior side of the brain. But the brain, immersed in and filled with cerebrospinal fluid, is like a water-sodden log; a blow to the top of the head can cause a serious rebound of the brain from the floor of the cranium. Routine neurological examinations involve testing for cranial-nerve dysfunction.

Commonly used clinical methods for determining cranial-nerve dysfunction are presented in table 16.2.

1. Which cranial nerves consist of sensory fibers only?
2. Which cranial nerves pass through the superior orbital fissure? through the jugular foramen?
3. Which cranial nerves have to do with tasting, chewing and manipulating food, and swallowing?
4. Which cranial nerves have to do with the structure, function, or movement of the eyeball?
5. List the cranial nerves, and indicate how each would be tested (both motor and sensory) for possible dysfunction.

Figure 16.13. The distribution of the spinal nerves.

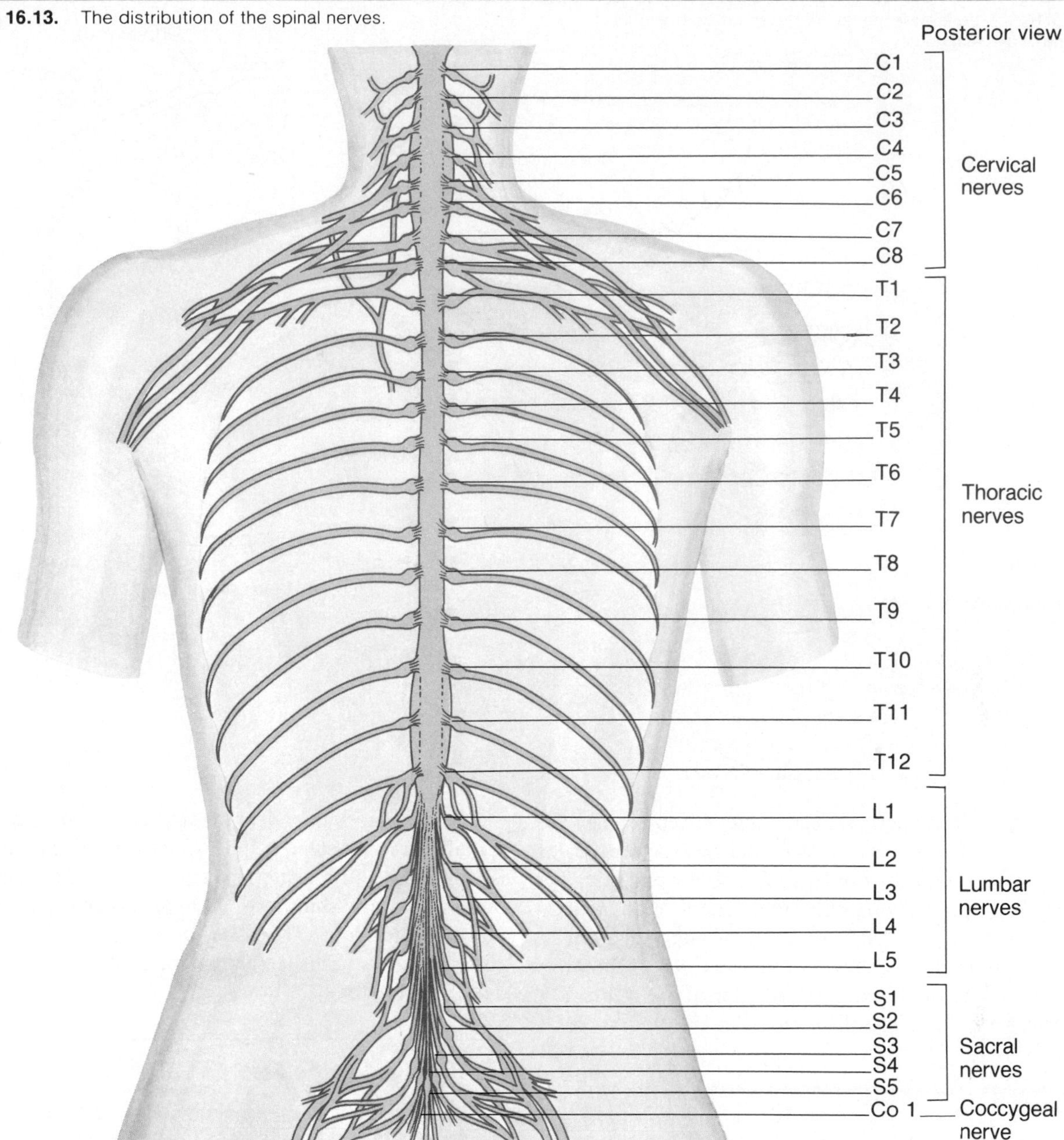

Spinal Nerves

Each of the thirty-one pairs of spinal nerves is formed by the union of a dorsal and a ventral spinal root, which emerges from the spinal cord through an intervertebral foramen to innervate a body dermatome.

Objective 4. Discuss how the spinal nerves are grouped.
Objective 5. Describe the general distribution of a spinal nerve.

The thirty-one pairs of **spinal nerves** (fig. 16.13) are grouped as follows: eight cervical, twelve thoracic, five lumbar, five sacral, and one coccygeal. With the exception of the first cervical nerve, the spinal nerves leave the spinal cord and vertebral canal through intervertebral foramina. The first pair of cervical nerves emerges between the occipital bone of the skull and the atlas vertebra. The second through the seventh pairs of cervical nerves emerge above the vertebrae for which they are named, whereas the eighth pair of cervical nerves passes between the seventh cervical and first thoracic vertebrae. The remaining pairs of spinal nerves emerge below the vertebrae for which they are named.

A spinal nerve is a mixed nerve attached to the spinal cord by a **dorsal (posterior) root** composed of sensory fibers and a **ventral (anterior) root** composed of motor fibers (fig. 16.14). The dorsal root contains an enlargement called the

Figure 16.14. A transverse section of the spinal cord and the distribution of a spinal nerve.

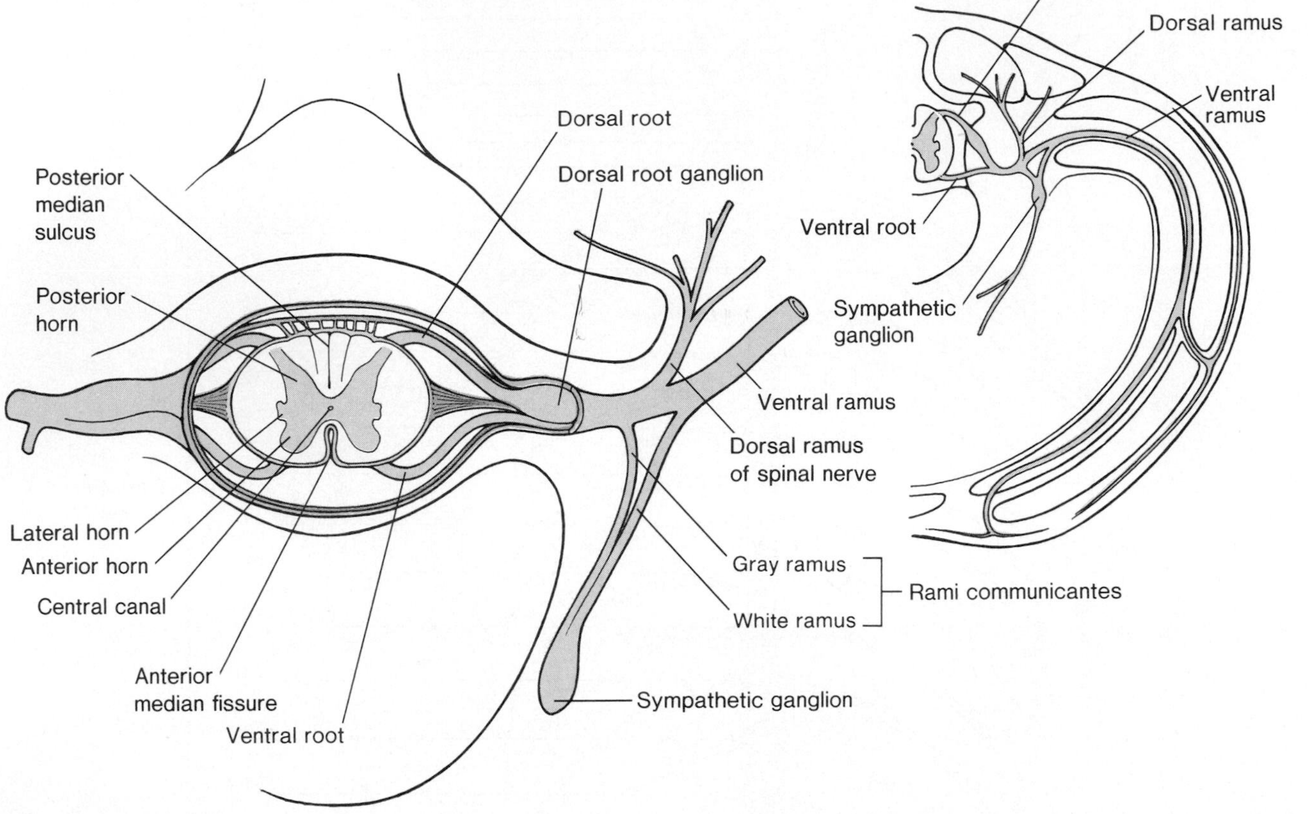

dorsal-root ganglion, where the cell bodies of sensory neurons are located. The axons of sensory neurons convey sensory impulses through the dorsal root and into the spinal cord where synapses occur with dendrites of other neurons. The ventral root consists of axons of motor neurons that convey motor impulses away from the CNS. A spinal nerve is formed as the fibers from the dorsal and ventral roots converge and emerge through an intervertebral foramen.

The disease *herpes zoster,* also known as *shingles,* is a viral infection of the dorsal-root ganglia. Herpes zoster causes painful, often unilateral, clusters of fluid-filled vesicles in the skin along the paths of the affected peripheral sensory neurons. This disease requires no special treatment, as the lesions gradually heal, and is usually not serious except in elderly debilitated patients, who may die from exhaustion.

A spinal nerve divides into several branches immediately after it emerges through the intervertebral foramen. The small **meningeal branch** reenters the vertebral canal to innervate the meninges, vertebrae, and vertebral ligaments. A larger branch, called the **dorsal ramus,** innervates the muscles, joints, and skin of the back along the vertebral column (fig. 16.14). A **ventral ramus** of a spinal nerve innervates the muscles and skin on the lateral and ventral (anterior) side of the torso. Combinations of ventral rami innervate the limbs.

The **rami communicantes** are two branches off the spinal nerve that connect to a **sympathetic ganglion,** which is part of the autonomic nervous system. The rami communicantes are composed of a **gray ramus,** containing unmyelinated fibers, and a **white ramus,** containing myelinated fibers. This arrangement is described in more detail in chapter 17.

1. List the number of nerves in each of the five regions of the vertebral column.
2. What are the four principal branches, or rami, from a spinal nerve, and what does each innervate?

Nerve Plexuses

Except in the thoracic nerves T2–T12, the ventral rami of the spinal nerves combine and then split again as networks of nerves referred to as plexuses. There are four plexuses of spinal nerves: the cervical, the brachial, the lumbar, and the sacral. Nerves emerge from the plexuses and are named according to the structures they innervate or the general course they take.

Objective 6. List the spinal nerve composition of each of the plexuses arising from the spinal cord.

Objective 7. List the principal nerves that emerge from the plexuses, and describe their general innervation.

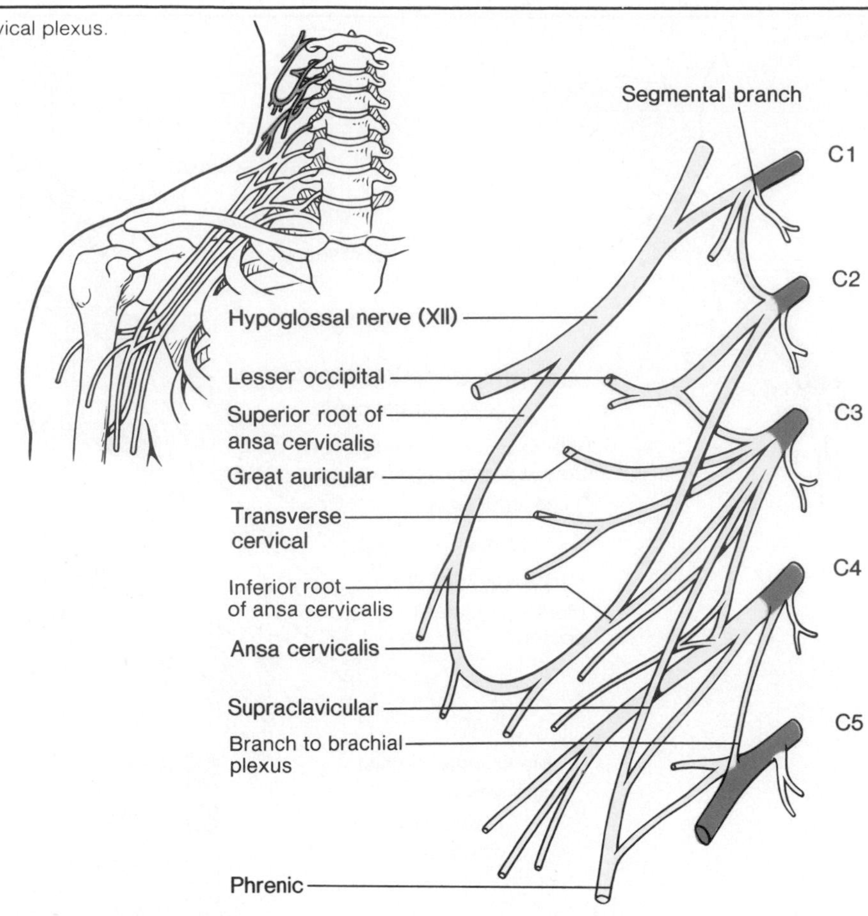

Figure 16.15. The cervical plexus.

Cervical Plexus. The cervical plexus *(plek'sus)* is positioned deep on the side of the neck, lateral to the first four cervical vertebrae (fig. 16.15). It is formed by the ventral rami of the first four cervical nerves (C1–C4) and a portion of C5. Branches from the cervical plexus innervate the skin and muscles of the neck and portions of the head and shoulders. Branches of the cervical plexus also combine with the accessory and hypoglossal cranial nerves to supply dual innervation to some specific neck and pharyngeal muscles. Fibers from the third, fourth, and fifth cervical nerves unite to become the **phrenic** *(fren'ik)* **nerve,** which innervates the diaphragm. Motor impulses through the paired phrenic nerves cause the diaphragm to contract, inspiring air into the lungs.

The nerves of the cervical plexus are summarized in table 16.3.

Brachial Plexus. The brachial plexus is positioned to the side of the last four cervical and first thoracic vertebrae. It is formed by the ventral rami of C5 through T1 and some fibers from C4 and T2. From its emergence, the brachial plexus extends downward and laterally, passes over the first rib behind the clavicle, and enters the axilla. Each

Table 16.3 Branches of cervical plexus

Nerve	Spinal component	Innervation
Superficial Cutaneous Branches		
Lesser occipital	C2,C3	Skin of scalp above and behind ear
Greater auricular	C2,C3	Skin in front of, above, and below ear
Transverse cervical	C2,C3	Skin of anterior aspect of neck
Supraclavicular	C3,C4	Skin of upper portion of chest and shoulder
Deep Motor Branches		
Ansa cervicalis		
Superior root	C1,C2	Geniohyoid, thyrohyoid, and infrahyoid muscles of neck
Inferior root	C3,C4	Omohyoid, sternohyoid, and sternothyroid muscles of neck
Phrenic	C3,C4,C5	Diaphragm
Segmental branches	C1–C5	Deep muscles of neck (levator scapulae ventralis, trapezius, scalenus, and sternocleidomastoid)

Figure 16.16. The brachial plexus.

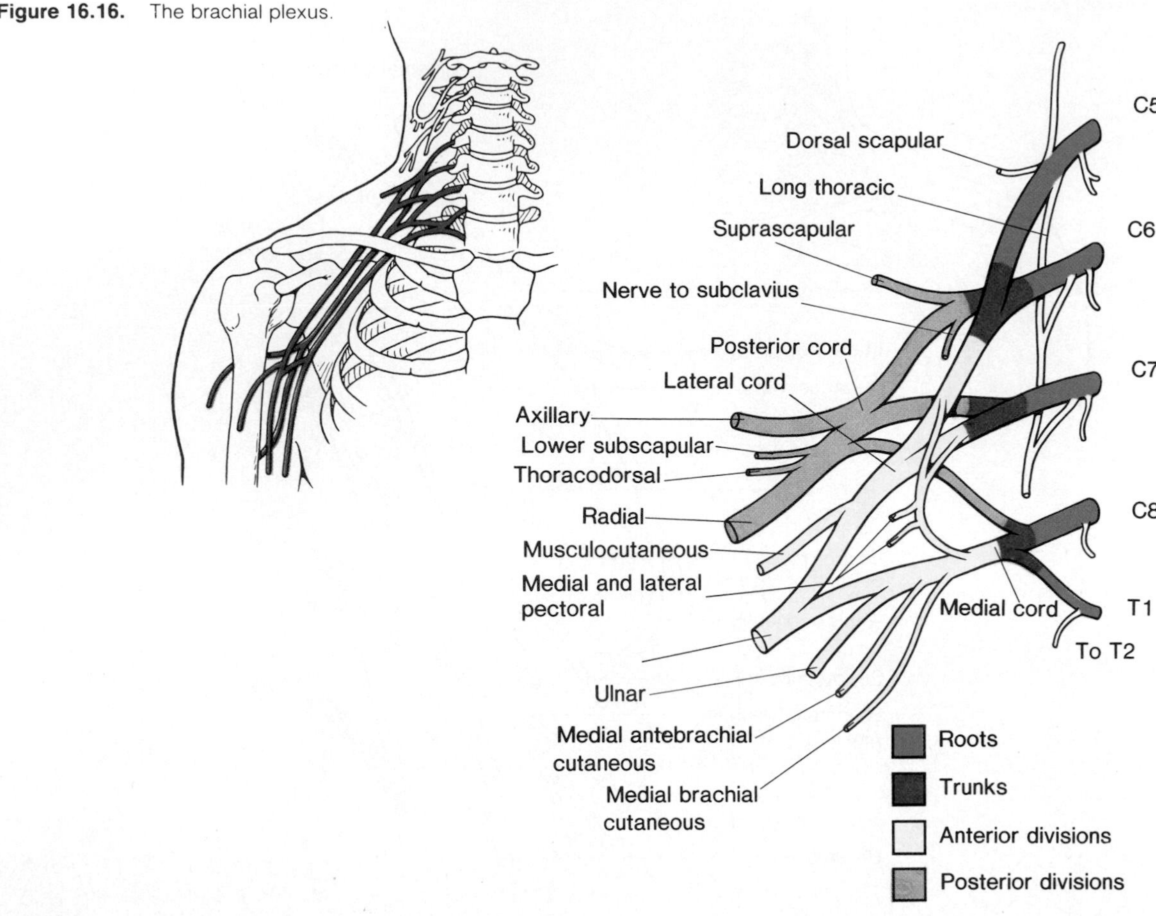

brachial plexus innervates the entire upper extremity of one side plus a number of shoulder and neck muscles.

Structurally, the brachial plexus is divided into *roots, trunks, divisions,* and *cords* (fig. 16.16). The roots of the brachial plexus are simply continuations of the ventral rami of the cervical nerves. The ventral rami of C5 and C6 converge to become the **upper trunk,** the C7 ramus becomes the **middle trunk,** and the ventral rami of C8 and T1 converge to become the **lower trunk.** Each of the three trunks immediately divides into an **anterior division** and a **posterior division.** The divisions then converge to form three cords. The **posterior cord** is formed by the convergence of the posterior divisions of the upper, middle, and lower trunks and hence contains fibers from C5 through C8. The **medial cord** is a continuation of the anterior division of the lower trunk and primarily contains fibers from C8 and T1. The **lateral cord** is formed by the convergence of the anterior division of the upper and middle trunk and consists of fibers from C5 through C7.

In summary, the brachial plexus is composed of nerve fibers from spinal nerves C5 through T1 and a few fibers from C4 and T2. Roots are continuations of the ventral rami. The roots converge to form trunks, and the trunks branch into divisions. The divisions in turn form cords, and the nerves of the upper extremity arise from the cords.

Five major nerves and several smaller ones arise from the three cords of the brachial plexus. The principal nerves of the brachial plexus are summarized in table 16.4.

Trauma to the brachial plexus is not uncommon, especially if the clavicle, upper ribs, or lower cervical vertebrae are seriously fractured. Occasionally, the brachial plexus of a newborn is severely strained during a difficult delivery when the baby is pulled through the birth canal. In such cases, the arm of the injured side is paralyzed and withers as the muscles atrophy in relation to the extent of the injury.

Lumbar Plexus. The lumbar plexus is positioned to the side of the first four lumbar vertebrae. It is formed by the ventral rami of spinal nerves L1–L4 and some fibers from T12 (fig. 16.17). The nerves that arise from the lumbar plexus innervate structures of the lower abdomen and anterior and medial portions of the lower extremity. The lumbar plexus is not as complex as the brachial plexus, having only roots and divisions rather than the roots, trunks, divisions, and cords as in the brachial plexus.

Structurally, the **posterior division** of the lumbar plexus passes obliquely outward, deep to the psoas major muscle, whereas the **anterior division** is superficial to the quadratus lumborum muscle. From these two divisions arise the large **femoral nerve,** which innervates the anterior muscles of the thigh, and the **obturator nerve,** which innervates the adductor muscles of the leg. Several smaller nerves also arise from the lumbar plexus and are summarized in table 16.5.

Figure 16.17. The lumbar plexus.

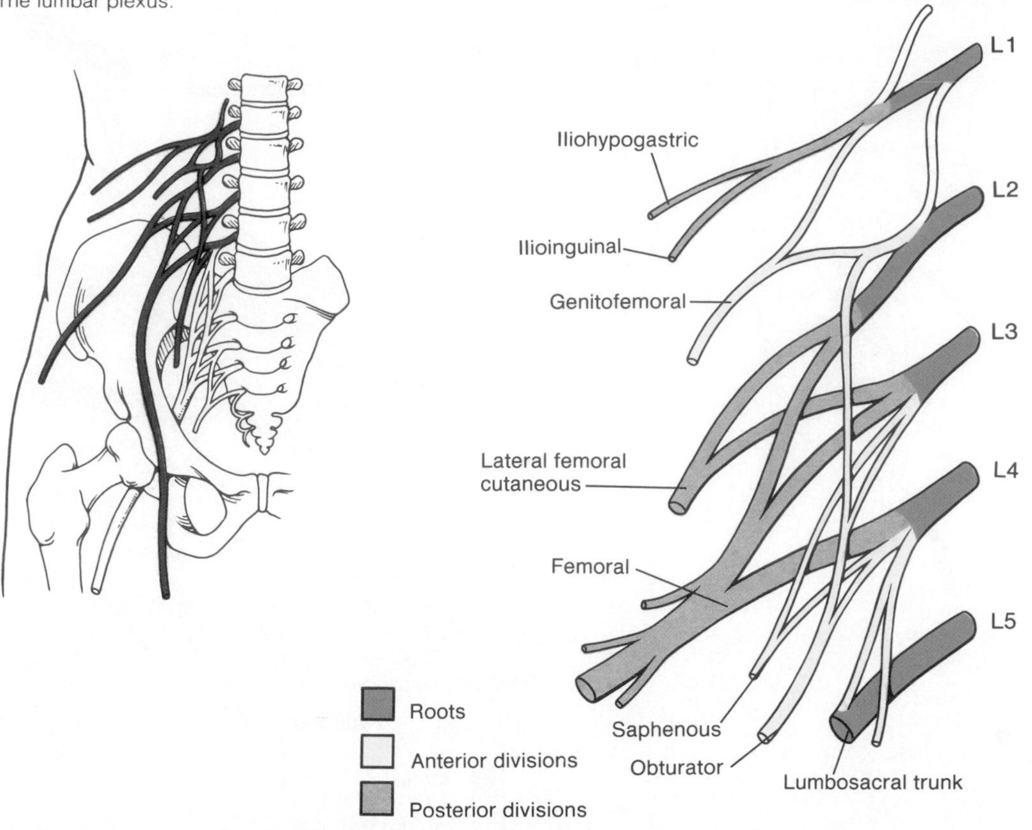

Table 16.4 Branches of brachial plexus

Nerve	Cord and spinal components	Innervation
Axillary	Posterior cord (C5,C6)	Skin of shoulder; shoulder joint, deltoid and teres minor muscles
Radial	Posterior cord (C5–C8,T1)	Skin of posterior lateral surface of arm, forearm, and hand; extensor muscles of posterior upper arm and forearm (triceps brachii, supinator, anconeus, brachioradialis, extensor carpi radialis brevis, extensor carpi radialis longus, extensor carpi ulnaris)
Musculocutaneous	Lateral cord (C5,C6,C7)	Skin of lateral surface of forearm; muscles of anterior upper arm (coracobrachialis, biceps brachii, brachialis)
Ulnar	Medial cord (C8,T1)	Skin of medial third of hand; flexor muscles of anterior forearm (flexor carpi ulnaris, flexor digitorum), medial palm, and intrinsic flexor muscles of hand (profundus, third and fourth lumbricales)
Median	Medial cord (C6,C7,C8,T1)	Skin of lateral two-thirds of hand; flexor muscles of anterior forearm, lateral palm, and first and second lumbricales

Table 16.5 Branches of lumbar plexus

Nerve	Spinal components	Innervation
Iliohypogastric	T12–L1	Skin of lower abdomen and buttock; muscles of anterolateral abdominal wall (external oblique, internal oblique, transversus abdominis)
Ilioinguinal	L1	Skin of upper median thigh, scrotum and root of penis in male, and labia majora in female; muscles of anterolateral abdominal wall with iliohypogastric nerve
Genitofemoral	L1,L2	Skin of middle anterior surface of thigh, scrotum in male, and labia majora in female; cremaster muscle in male
Lateral femoral cutaneous	L2,L3	Skin of anterior, lateral, and posterior aspects of thigh
Femoral	L2–L4	Skin of anterior and medial aspect of thigh and medial aspect of leg and foot; anterior muscles of thigh (iliacus, psoas major, pectineus, rectus femoris, sartorius) and extensor muscles of leg (rectus femoris, vastus lateralis, vastus medialis, vastus intermedius)
Obturator	L2–L4	Skin of medial aspect of thigh; adductor muscles of leg (external obturator, pectineus, adductor longus, adductor brevis, adductor magnus, gracilis)
Saphenous	L2–L4	Skin of medial aspect of leg

Figure 16.18. The sacral plexus.

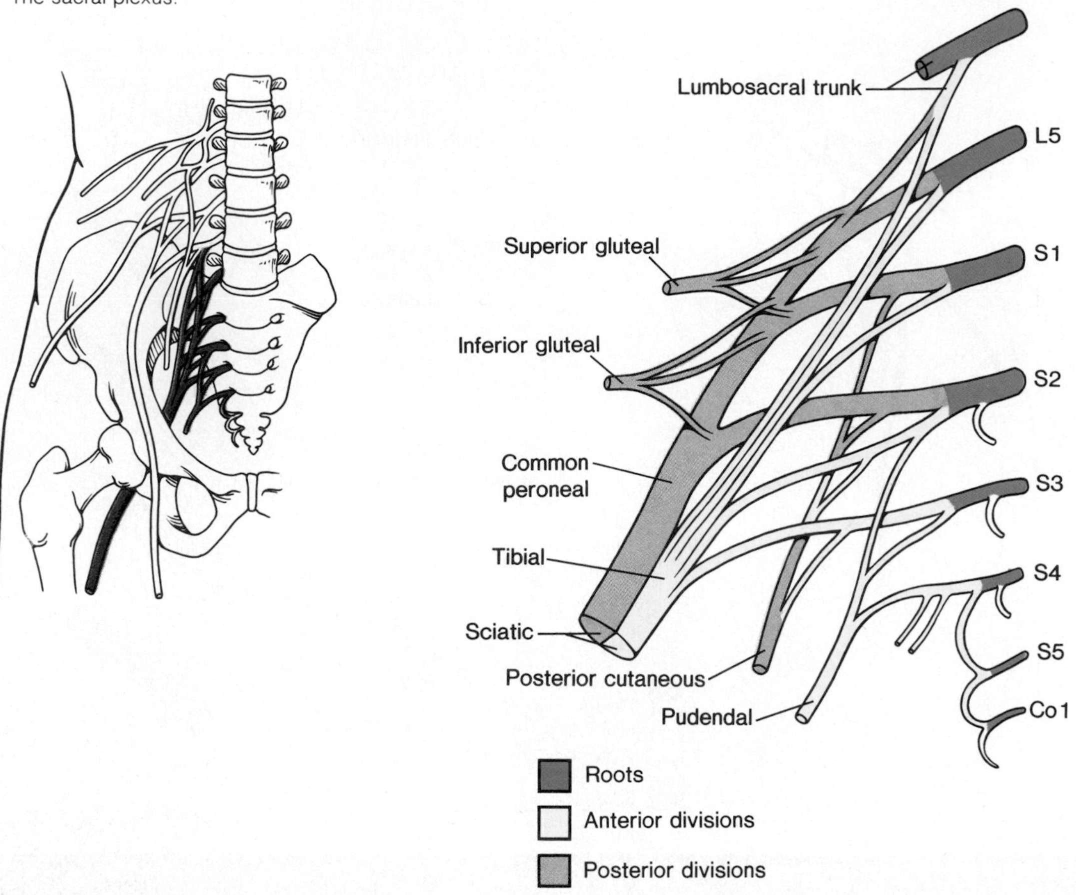

A herniated intervertebral disc (see fig. 16.19) in the lumbar region may cause a condition called *sciatica (si-at'i-kah)*. This condition is characterized by a sharp pain in the gluteal region and by pains that extend down the posterior side of the thigh, caused by the disc compressing the spinal roots that form the sciatic nerve.

Sacral Plexus. The sacral plexus is positioned immediately caudal to the lumbar plexus. It is formed by the ventral rami of spinal nerves L4, L5, and S1–S4 (fig. 16.18). The nerves arising from the lumbar plexus innervate the lower back, pelvis, perineum, posterior surface of the thigh and leg, and the dorsal and ventral surfaces of the foot. Like the lumbar plexus, the sacral plexus consists of **roots** and **anterior** and **posterior divisions** from which nerves arise. Because some of the nerves of the sacral plexus also contain fibers from the nerves of the lumbar plexus through the **lumbosacral trunk,** these two plexuses are frequently described collectively as the **lumbosacral plexus.**

The **sciatic** *(si-at'ik)* **nerve** is the largest nerve arising from the sacral plexus and, in fact, is the largest nerve in the body. The sciatic nerve passes from the pelvis through the greater sciatic notch of the os coxa and extends down the posterior aspect of the thigh. The sciatic nerve is actually composed of two nerves, the common peroneal and tibial nerves, wrapped in a common sciatic sheath.

A summary of the nerves of the sacral plexus and their distribution is presented in table 16.6.

sacral: L. *sacris,* sacred

sciatic: L. *sciaticus,* hip joint
peroneal: Gk. *perone,* pin, fibula

Table 16.6 Branches of sacral plexus

Nerve	Spinal components	Innervation
Superior gluteal	L4,L5,S1	Abductor muscles of thigh (gluteus minimus, gluteus medius, tensor fasciae latae)
Inferior gluteal	L5–S2	Extensor muscle of thigh (gluteus maximus)
Nerve to piriformis	S1,S2	Abductor and rotator of thigh (piriformis)
Nerve to quadratus femoris	L4,L5,S1	Rotators of thigh (gemellus inferior, quadratus femoris)
Nerve to internal obturator	L5–S2	Rotators of thigh (gemellus superior, internal obturator)
Perforating cutaneous	S2,S3	Skin over lower medial surface of buttock
Posterior cutaneous	S1–S3	Skin over lower lateral surface of buttock, anal region, upper posterior surface of thigh, upper aspect of calf, scrotum in male, and labia majora in female
Sciatic	L4–S3	Composed of two nerves (tibial and common peroneal) within sciatic sheath; splits into two portions at popliteal fossa; branches from sciatic in thigh region to hamstring muscles (biceps femoris, semitendinosus, semimembranosus) and adductor magnus
Tibial (sural, medial and lateral plantar)	L4–S3	Skin of posterior surface of leg and sole of foot; muscle innervation includes gastrocnemius, soleus, flexor digitorum longus, flexor hallucis longus, tibialis posterior, popliteus, and intrinsic muscles of the foot
Common peroneal (superficial and deep peroneal)	L4–S2	Skin of anterior surface of the leg and dorsum of foot; muscle innervation includes peroneus tertius, peroneus brevis, peroneus longus, tibialis anterior, extensor hallucis longus, extensor digitorum longus, extensor digitorum brevis
Pudendal	S2–S4	Skin of penis and scrotum in male and skin of clitoris, labia majora, labia minora, and lower vagina in female; muscles of perineum

The sciatic nerve in the buttock lies deep to the gluteus maximus muscle, midway between the greater trochanter and the ischial tuberosity. Because of its position, the sciatic nerve is of tremendous clinical importance. A posterior dislocation of the hip joint will generally injure the sciatic nerve. A herniated disc (fig. 16.19), or pressure from the uterus during pregnancy may damage the roots of the nerves forming the sciatic nerve. An improperly administered injection into the buttock may injure the sciatic nerve itself.

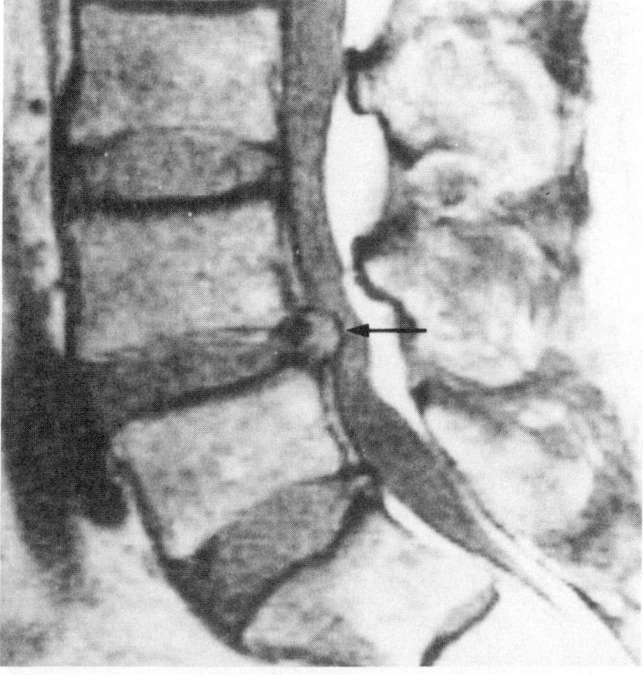

Figure 16.19. An MR scan of a herniated disc in the lumbar region.

1. Define *nerve plexus.* What are the four spinal nerve plexuses, and which spinal nerves contribute to each?
2. Which of the spinal nerves are not involved in a plexus?
3. Distinguish between a dorsal ramus and a ventral ramus. Which is involved in the formation of plexuses?
4. Construct a table that lists the plexus of origin and the general region of innervation for the following nerves: pudendal, phrenic, femoral, ulnar, median, sciatic, saphenous, axillary, radial, ansa cervicalis, tibial, and common peroneal.

Figure 16.20. The reflex arc.

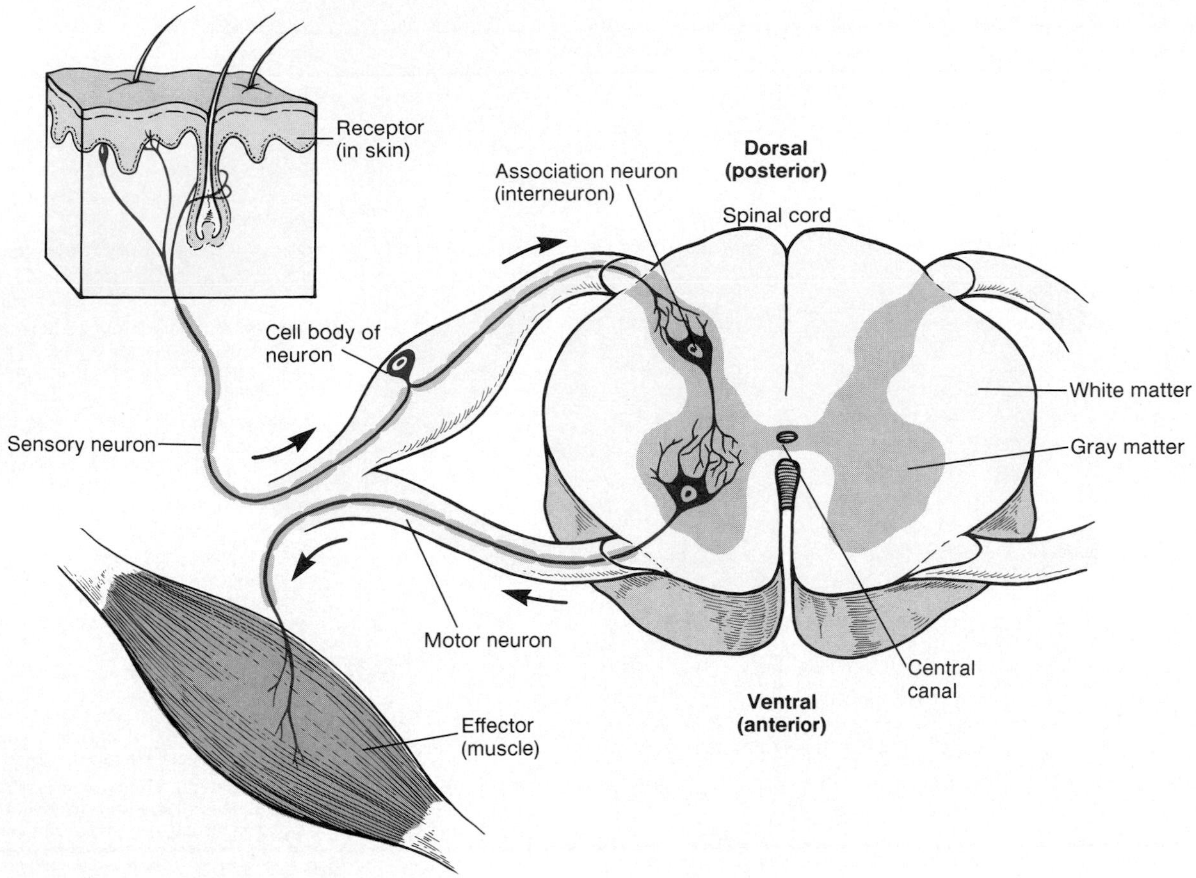

Reflex Arcs and Reflexes

The conduction pathway of a reflex arc consists of a receptor, a sensory neuron, a motor neuron and its innervation in the PNS, and an association neuron in the CNS. The reflex arc provides the mechanism for a rapid, automatic response to a potentially threatening stimulus.

Objective 8. Define *reflex arc,* and list its five components.

Objective 9. Distinguish between the various kinds of reflexes.

Specific **nerve pathways** provide routes by which impulses travel through the nervous system. Frequently, a nerve pathway begins with impulses being conducted to the CNS through sensory receptors and sensory neurons of the PNS. Once within the CNS, impulses may immediately travel back through motor portions of the PNS to activate specific skeletal muscles, glands, or smooth muscles, and also be sent simultaneously to other parts of the CNS through ascending tracts within the spinal cord.

Components of the Reflex Arc

The simplest type of nerve pathway is a **reflex arc** (fig. 16.20). A reflex arc implies an automatic, unconscious, protective response to a situation in an attempt to maintain body homeostasis. A reflex arc leads by a short route from sensory to motor neurons and includes only two or three neurons. The five components of a reflex arc are the receptor, sensory neuron, center, motor neuron, and effector. The **receptor** includes the dendrite of a sensory neuron and the place where the electrical impulse is initiated. The sensory neuron relays the impulse through the posterior root to the CNS. The **center** is within the CNS and usually involves an association neuron (interneuron). It is here that the arc is made and other impulses are sent through synapses to other parts of the body. The **motor neuron** conducts the impulse to an effector organ (generally a skeletal muscle) that responds. The result of an impulse through a reflex arc is called a reflex action or, simply, a *reflex*.

Figure 16.21. The knee-jerk reflex is an ipsilateral reflex. The receptor and effector organs are on the same side of the spinal cord. The knee-jerk reflex is also a monosynaptic reflex because it involves only two neurons and one synapse.

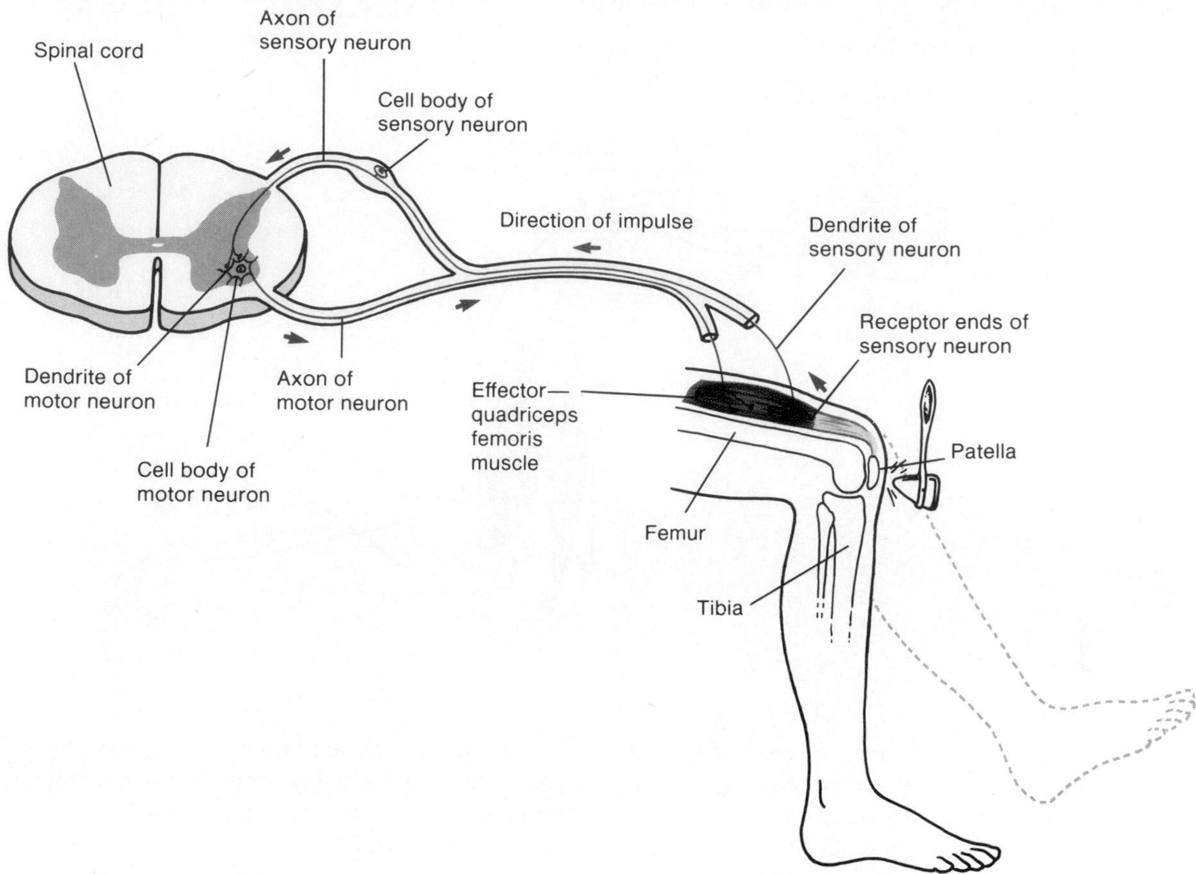

Kinds of Reflexes

Visceral Reflexes. Reflexes that cause smooth or cardiac muscle to contract or glands to secrete are **visceral (autonomic) reflexes.** Visceral reflexes help control the body's many involuntary processes such as heart rate, respiratory rate, blood pressure, and digestion. Swallowing, sneezing, coughing, and vomiting may also be reflexive, although they involve skeletal muscles that function involuntarily.

Somatic Reflexes. Somatic reflexes are those that result in the contraction of skeletal muscles. There are three principal kinds of somatic reflexes, named according to the response they produce.

The **stretch reflex** involves only two neurons and one synapse in the pathway and is therefore called a *monosynaptic reflex arc.* Slight stretching of neuromuscular spindle receptors (described in chapter 18) within a muscle initiates an impulse along an afferent neuron to the spinal cord. A synapse with a motor neuron occurs in the anterior gray horn and a motor unit is activated, causing specific muscle fibers to contract. Since the receptor and effector organs of the stretch reflex involve structures on the same side of the spinal cord, the reflex arc is an *ipsilateral reflex arc.* The knee-jerk reflex is an ipsilateral reflex (fig. 16.21), as are all monosynaptic reflex arcs.

A **flexor reflex,** or **withdrawal reflex,** consists of a *polysynaptic reflex arc* (fig. 16.22). Flexor reflexes involve association neurons in addition to the sensory and motor neurons. A flexor reflex is initiated as a person encounters a painful stimulus such as a hot or sharp object.

Figure 16.22. The flexor, or withdrawal, reflex is a polysynaptic reflex because it involves association neurons (interneurons) in addition to sensory and motor neurons.

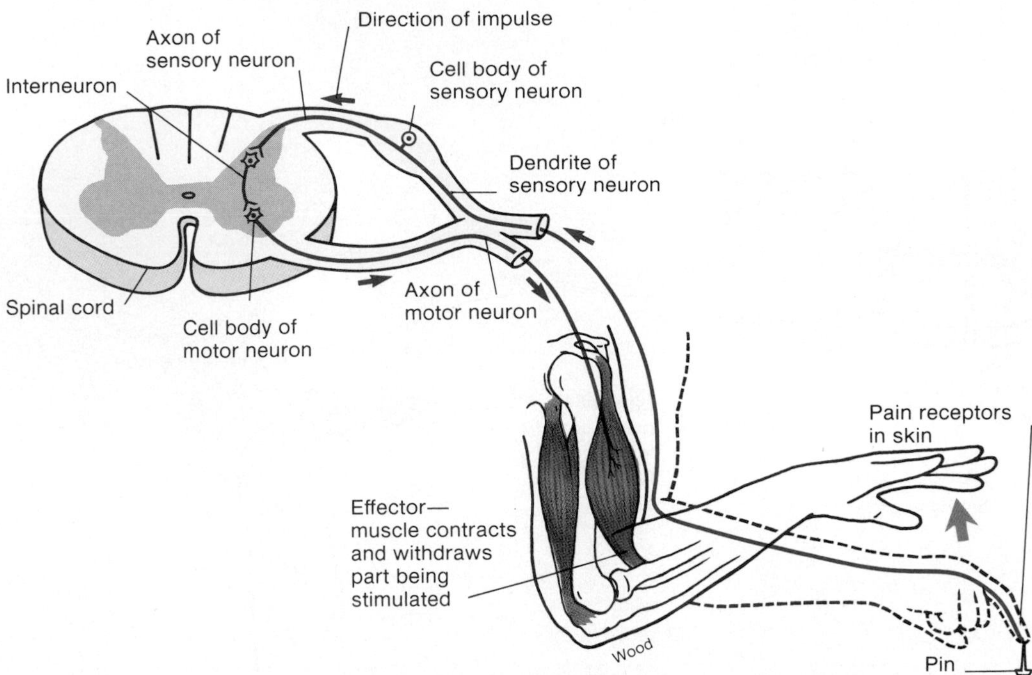

As a receptor organ is stimulated, afferent neurons transmit the impulse to the spinal cord where association neurons are activated. There the impulses are directed through motor neurons to flexor muscles that contract in response. Simultaneously, antagonistic muscles are inhibited (relaxed) so that the traumatized extremity can be quickly withdrawn from the harmful source of stimulation.

Several additional reflexes may be activated while a flexor reflex is in progress. In an *intersegmental reflex arc,* motor units from several segments of the spinal cord are activated by impulses coming in from the receptor organ. An intersegmental reflex arc produces stimulation of more than one effector organ. Frequently, afferent impulses from a receptor organ cross over through the spinal cord to activate effector organs within the opposite (*contralateral*) limb. This type of reflex is called a **crossed extensor reflex** (fig. 16.23) and is important for maintaining body balance while a flexor reflex is in progress. The reflexive inhibition of certain muscles to contract, called *reciprocal inhibition,* also helps maintain balance while either flexor or crossed extensor reflexes are in progress.

Certain reflexes are important for physiological functions, while others are important for avoiding injury. Some common reflexes are presented in table 16.7 and illustrated in figure 16.24.

Part of a routine physical examination involves testing a person's reflexes. The condition of the nervous system, particularly the functioning of the synapses, may be determined by examining reflexes. In case of injury to some portion of the nervous system, testing certain reflexes may indicate the location and extent of the injury. Also, an anesthesiologist may try to initiate a reflex to ascertain the effect of an anesthetic.

1. How are reflexes important in maintaining body homeostasis?
2. List the five components of a reflex arc.
3. Define or describe the following: visceral reflex, somatic reflex, stretch reflex, flexor reflex, crossed extensor reflex, ipsilateral reflex, and contralateral reflex.

anesthesia: Gk. *an,* without; *aisthesis,* sensation

Figure 16.23. A crossed extensor reflex causes a reciprocal inhibition of muscles of the opposite appendage. This type of reflex inhibition is important in maintaining balance.

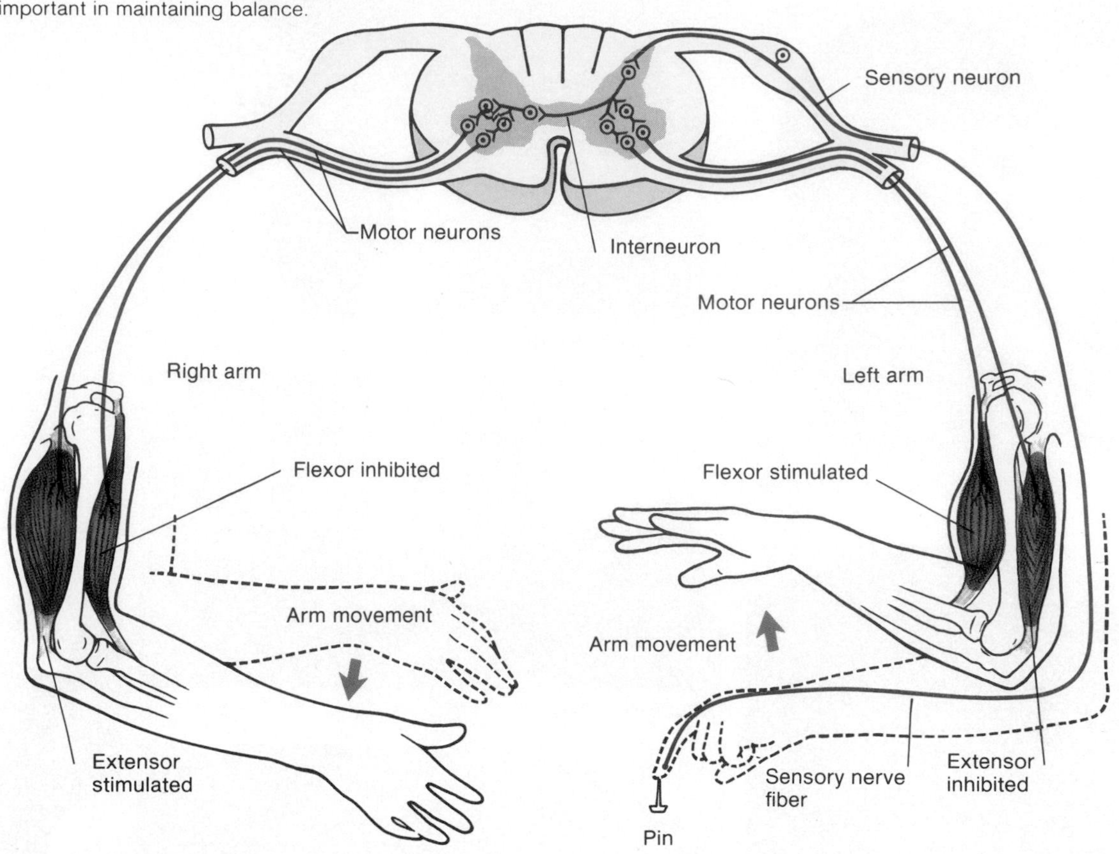

Table 16.7 Some clinically important reflexes

Reflex	Spinal segment	Site of receptor stimulation	Effector action
Biceps reflex	C5, C6	Biceps tendon near attachment on radial tuberosity	Contracts biceps brachii to flex elbow
Triceps reflex	C7, C8	Triceps tendon near attachment on olecranon process	Contracts triceps brachii to extend elbow
Supinator or brachioradialis reflex	C5, C6	Radial attachment of supinator and brachioradialis muscles	Supinates forearm and hand
Knee reflex	L2, L3, L4	Patellar tendon just below patella	Contracts quadriceps to extend the knee
Ankle reflex	S1, S2	Achilles tendon near attachment on calcaneus	Plantar flexes ankle
Plantar reflex	L4, L5, S1, S2	Lateral aspect of sole from heel to ball of foot	Plantar flexes foot and flexes toes
Babinski reflex*	L4, L5, S1, S2	Lateral aspect of sole from heel to ball of foot	Dorsiflexes great toe and fans other toes
Abdominal reflexes	T8, T9, T10 above umbilicus and T10, T11, T12 below umbilicus	Sides of abdomen above and below level of umbilicus	Contracts abdominal muscles and deviates umbilicus toward stimulus
Cremasteric reflex	L1, L2	Stroking upper inside of thigh in males	Contracts cremasteric muscle and elevates testis on same side of stimulation

*If the Babinski reflex rather than the plantar reflex occurs as the sole of the foot is stimulated, it may indicate damage to the corticospinal tract within the spinal cord. However, the Babinski reflex is present in infants up to 12 months of age because of the immaturity of their corticospinal tracts.

Babinski reflex: from Joseph F. Babinski, French neurologist, 1857–1932

Figure 16.24. Some reflexes of clinical importance.

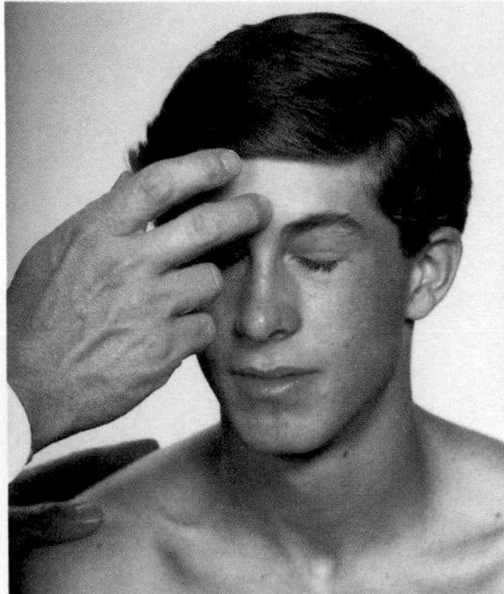

Glabellar reflex

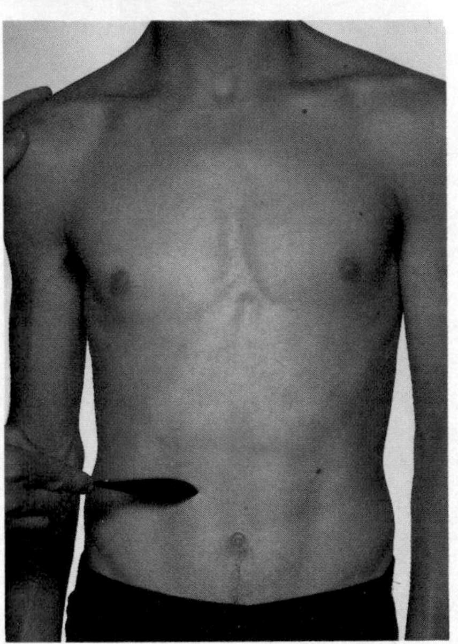

Abdominal reflex

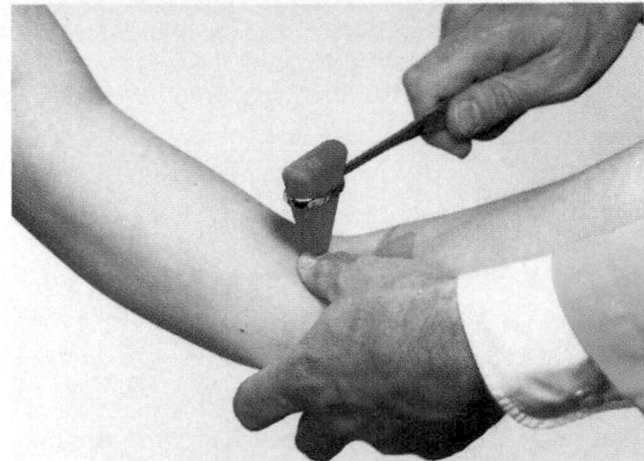

Biceps reflex

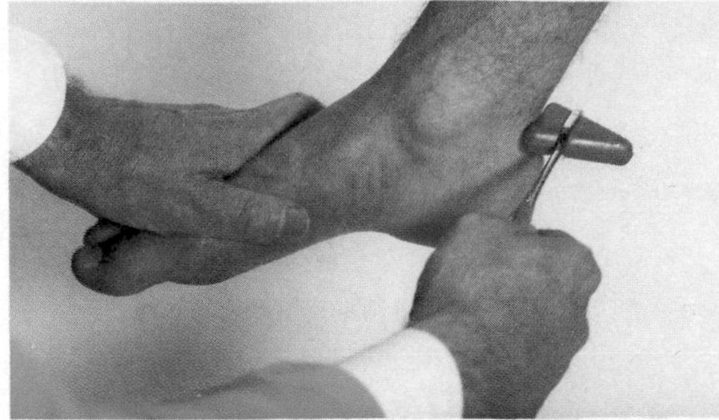

Ankle (Achilles) reflex

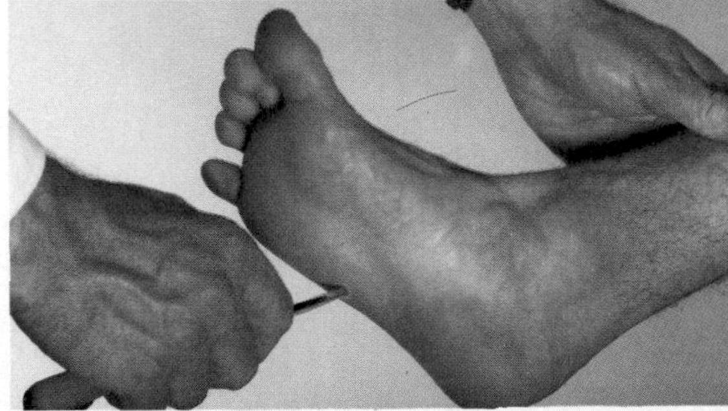

Babinski reflex

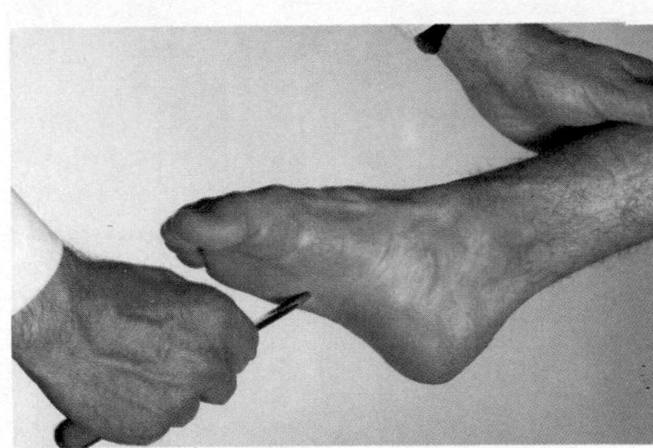

Plantar reflex

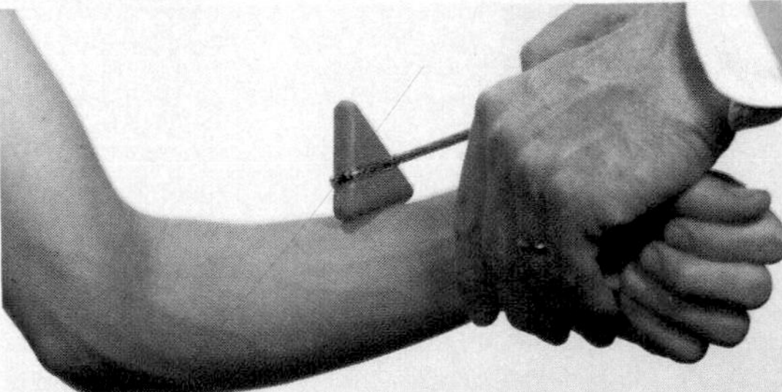

Supinator (brachioradialis) reflex

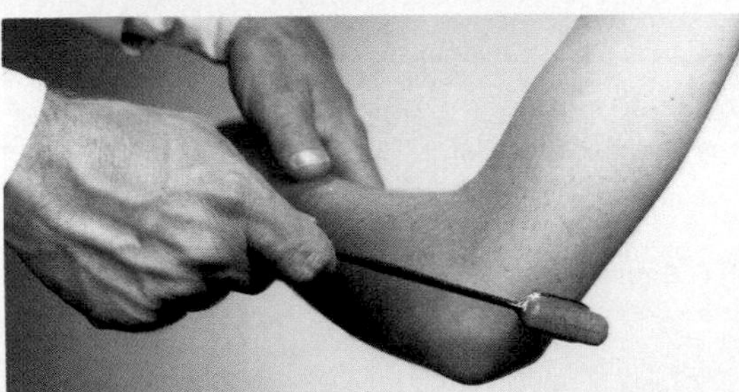

Triceps reflex

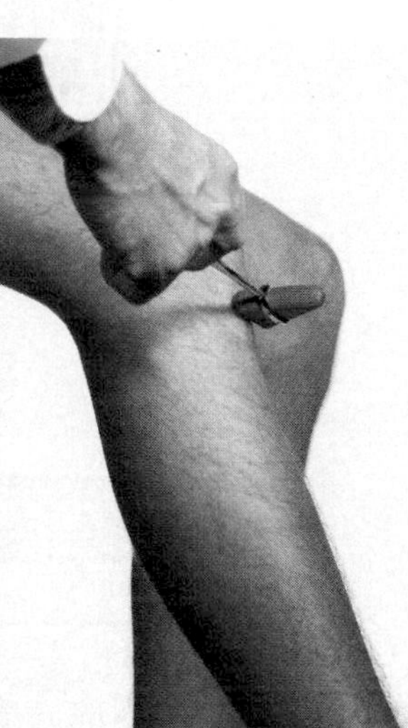

Knee (patellar) reflex

Chapter Summary

I. Introduction to the Peripheral Nervous System
  A. The peripheral nervous system consists of nerves that convey impulses to and from the central nervous system.
  B. The cranial nerves arise from the brain, and the spinal nerves arise from the spinal cord.
  C. Sensory (afferent) nerves convey impulses toward the CNS, motor (efferent) nerves convey impulses away from the CNS, and mixed nerves are composed of both sensory and motor fibers.

II. Cranial Nerves
  A. Twelve pairs of cranial nerves emerge from the inferior surface of the brain and pass through foramina of the skull to innervate structures in the head, neck, and visceral organs of the torso.
  B. The names of the cranial nerves indicate their primary function or the general distribution of their fibers.
  C. Most of the cranial nerves are mixed, some consist of sensory fibers only, and others are primarily motor.
  D. Some of the cranial nerve fibers are somatic, and others are visceral.
  E. A test for cranial-nerve dysfunction is clinically important in a neurological examination.

III. Spinal Nerves
  A. Each of the thirty-one pairs of spinal nerves is formed by the union of a dorsal and ventral spinal root, which emerges from the spinal cord through an intervertebral foramen to innervate a body dermatome.
  B. The spinal nerves are grouped according to the levels of the spinal column from which they arise, and they are numbered in sequence.
  C. Each spinal nerve is a mixed nerve consisting of a dorsal root of sensory fibers and a ventral root of motor fibers.
  D. Just beyond its intervertebral foramen, each spinal nerve divides into several branches.

IV. Nerve Plexuses
  A. Except in the thoracic nerves T2–T12, the ventral rami of the spinal nerves combine and then split again as networks of nerves referred to as plexuses. There are four plexuses of spinal nerves: the cervical, the brachial, the lumbar, and the sacral. Nerves emerge from the plexuses and are named according to the structures they innervate or the general course they take.
  B. The cervical plexus is formed by the ventral rami of C1–C4 and by a portion of C5.
  C. The brachial plexus is formed by the ventral rami of C5–T1 and by some fibers from C4 and T2.
    1. The brachial plexus is divided into roots, trunks, divisions, and cords.
    2. The axillary, radial, musculocutaneous, ulnar, and median are the principal nerves arising from the brachial plexus.
  D. The lumbar plexus is formed by the ventral rami of L1–L4 and by some fibers from T12.
    1. The lumbar plexus is divided into roots and divisions.
    2. The femoral and obturator are the principal nerves arising from the lumbar plexus.
  E. The sacral plexus is formed by the ventral rami of L4, L5, and S1–S4.
    1. The sacral plexus is divided into roots and divisions.
    2. The sciatic nerve, composed of the common peroneal and tibial nerves, arises from the sacral plexus.

V. Reflex Arcs and Reflexes
  A. The conduction pathway of a reflex arc consists of a receptor, a sensory neuron, a motor neuron and its innervation in the PNS, and an association neuron in the CNS. The reflex arc provides mechanisms for a rapid, automatic response to a potentially threatening stimulus.
  B. A reflex arc is the simplest type of nerve pathway.
  C. Visceral reflexes cause smooth or cardiac muscle to contract or glands to secrete.
  D. Somatic reflexes cause skeletal muscles to contract.
    1. The stretch reflex is a monosynaptic reflex arc.
    2. The flexor reflex is a polysynaptic reflex arc.

Review Activities

Objective Questions

1. Which of the following is a *false* statement concerning the peripheral nervous system?
   (a) It consists of cranial and spinal nerves only.
   (b) It contains components of the autonomic nervous system.
   (c) Sensory receptors, neurons, nerves, ganglia, and plexuses are all part of the PNS.
2. An inability to look cross-eyed would most likely indicate a problem with which cranial nerve? The
   (a) optic
   (b) oculomotor
   (c) abducens
   (d) facial
3. Which cranial nerve innervates the muscle that raises the upper eyelid? The
   (a) trochlear
   (b) oculomotor
   (c) abducens
   (d) facial
4. The inability to walk a straight line may indicate damage to which cranial nerve? The
   (a) trigeminal
   (b) facial
   (c) vestibulocochlear
   (d) vagus
5. Which cranial nerve passes through the stylomastoid foramen? The
   (a) facial
   (b) glossopharyngeal
   (c) vagus
   (d) hypoglossal
6. Which of the following cranial nerves does not contain parasympathetic fibers? The
   (a) oculomotor
   (b) accessory
   (c) vagus
   (d) facial
7. Which of the following is not a spinal nerve plexus? The
   (a) cervical
   (b) brachial
   (c) thoracic
   (d) lumbar
   (e) sacral
8. Roots, trunks, divisions, and cords are characteristic of the
   (a) sacral plexus
   (b) thoracic plexus
   (c) lumbar plexus
   (d) brachial plexus
9. Which of the following nerve-plexus associations is incorrect?
   (a) median—sacral
   (b) phrenic—cervical
   (c) axillary—brachial
   (d) femoral—lumbar
10. Extending the leg when the patellar tendon is tapped is an example of a(n)
   (a) visceral reflex
   (b) flexor reflex
   (c) ipsilateral reflex
   (d) crossed extensor reflex

Essay Questions

1. Explain the structural and functional relationship between the central nervous system, the autonomic nervous system, and the peripheral nervous system.
2. List the cranial nerves, and describe the major function(s) of each. How is each cranial nerve tested for dysfunction?
3. Describe the structure of a spinal nerve.
4. List the roots of each of the spinal plexuses, and describe where each is located and the nerves which originate from them.
5. Distinguish between monosynaptic, polysynaptic, ipsilateral, stretch, and flexor reflexes.

17 Autonomic Nervous System

Outline

Concepts

The action of effectors—muscles and glands—is controlled to a large extent by motor neuron impulses. Skeletal muscles, which are the voluntary effectors, are regulated by somatic motor impulses. The involuntary effectors, including smooth muscle, cardiac muscle, and glands, are regulated by autonomic motor impulses through the autonomic nervous system.

The sympathetic and parasympathetic divisions of the autonomic system consist of preganglionic neurons that originate in the CNS and postganglionic neurons that originate outside of the CNS in ganglia. The specific origin of the preganglionic fibers and the location of the ganglia, however, are different in the two subdivisions of the autonomic nervous system.

The actions of the autonomic nervous system, together with the actions of hormones, help to maintain a state of dynamic constancy in the internal environment. The sympathetic division activates the body to "fight or flight" through adrenergic effects; the parasympathetic division often has antagonistic actions through cholinergic effects. Homeostasis thus depends, in large part, on the complementary and often antagonistic effects of sympathetic and parasympathetic innervation.

Visceral functions are mainly regulated by autonomic reflexes. In most autonomic reflexes, sensory input is directed to brain centers, which in turn regulate the activity of descending pathways to preganglionic autonomic neurons. The neural centers that directly control the activity of autonomic nerves are influenced by higher brain areas as well as by sensory input.

The Autonomic Nervous System and Its Visceral Effectors

The action of effectors—muscles and glands—is controlled to a large extent by motor neuron impulses. Skeletal muscles, which are the voluntary effectors, are regulated by somatic motor impulses. The involuntary effectors, including smooth muscle, cardiac muscle, and glands, are regulated by autonomic motor impulses through the autonomic nervous system.

Objective 1. Define the terms *preganglionic neuron* and *postganglionic neuron,* and explain the difference between the efferent pathways of the somatic motor and autonomic motor systems.

Objective 2. Describe how the neural regulation of visceral effector organs can differ from the neural regulation of skeletal muscles.

Objective 3. Compare the structure and regulation of single-unit smooth muscle with that of multiunit smooth muscle.

Organization of the Autonomic Nervous System

The function of the autonomic portion of the nervous system is to maintain homeostasis within the body by increasing or decreasing the activity of various organs in response to changing physiological conditions. Although the autonomic nervous system (ANS) is composed of portions of both the central nervous system and peripheral nervous system, it functions independently and without the conscious control of the person.

Autonomic motor nerves innervate organs whose functions are not usually under voluntary control. The effectors that respond to autonomic regulation include **cardiac muscle** (the heart), **smooth** (visceral) **muscles,** and **glands.** These effectors are part of the organs of the *viscera* and of blood vessels, and specialized structures within other organs. The involuntary effects of autonomic innervation contrast with the voluntary control of skeletal muscles by way of somatic motor innervation.

Unlike somatic motor neurons, which conduct impulses along a single axon from the spinal cord to the neuromuscular junction, the autonomic motor pathway involves two neurons in the efferent flow of impulses (table 17.1).

The first of these neurons has its cell body in the gray matter of the brain or spinal cord. The axon of this neuron does not directly innervate the effector organ but instead synapses with a second neuron within an *autonomic ganglion* (a ganglion is a collection of nerve cell bodies outside the CNS). The first neuron is thus called a **preganglionic,** or **presynaptic, neuron.** The second neuron in this pathway, called a **postganglionic,** or **postsynaptic, neuron,** has an axon that leaves the autonomic ganglion and synapses with the cells of an effector organ (fig. 17.1).

Preganglionic autonomic fibers originate in the midbrain and hindbrain and in the upper thoracic to the fourth sacral levels of the spinal cord. Autonomic ganglia are located in the head, neck, and abdomen; chains of autonomic ganglia also parallel the right and left sides of the spinal cord. The origin of the preganglionic fibers and the

autonomic: Gk. *auto,* self; *nomos,* law
viscus, pl. viscera: L. *viscera,* internal organs
ganglion: Gk. *ganglion,* a swelling or knot

Table 17.1 Comparisons of the somatic motor system with the autonomic motor system

Feature	Somatic motor	Autonomic motor
Effector organs	Skeletal muscles	Cardiac muscle, smooth muscle, and glands
Presence of ganglia	No ganglia outside CNS	Cell bodies of postganglionic autonomic fibers located in paravertebral, prevertebral (collateral), and terminal ganglia
Number of neurons from CNS to effector	One	Two
Type of neuromuscular junction	Specialized motor end plate	No specialization of postsynaptic membrane; all areas of smooth muscle cells contain receptor proteins for neurotransmitters
Effect of nerve impulse on muscle	Excitatory only	Either excitatory or inhibitory
Type of nerve fibers	Fast-conducting, thick (9–13 μm), and myelinated	Slow conducting; preganglionic fibers, lightly myelinated but thin (3 μm); postganglionic fibers, unmyelinated and very thin (about 1.0 μm)
Effect of denervation	Flaccid paralysis and atrophy	Much muscle tone and function remains; target cells show denervation hypersensitivity

Figure 17.1. A comparison of a somatic motor reflex with an autonomic motor reflex. Although, for the sake of clarity, each is shown on different sides of the spinal cord, both visceral and somatic sensory neurons are found bilaterally in the dorsal roots of spinal nerves, and somatic and autonomic motor fibers are found bilaterally in the ventral roots.

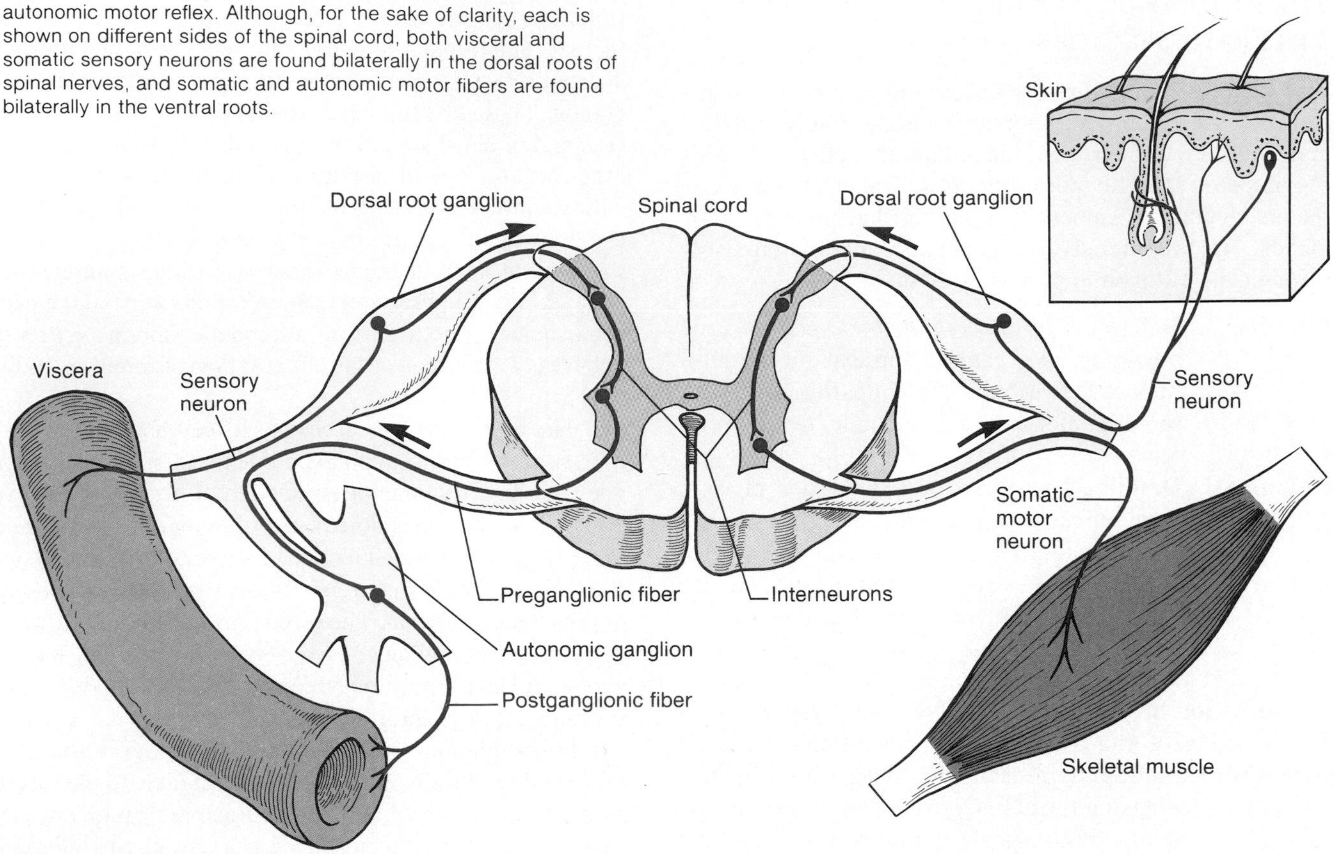

location of the autonomic ganglia help to differentiate the **sympathetic** and **parasympathetic** subdivisions of the autonomic system, which will be discussed in later sections of this chapter.

Visceral Effectors

Unlike skeletal muscles, which enter a state of flaccid paralysis when their motor nerves are severed, the involuntary effectors are somewhat independent of their innervation. Smooth muscles maintain a resting tone (tension) in the absence of nerve stimulation. Damage to an autonomic nerve, in fact, makes its target muscle more sensitive than normal to stimulating agents. This phenomenon is called **denervation** *(de''ner-va'shun)* **hypersensitivity.**

In addition to their intrinsic ("built-in") muscle tone, many smooth muscles and the cardiac muscle take their autonomy a step further. These muscles can contract rhythmically, even in the absence of nerve stimulation, in response to electrical waves of depolarization initiated by the muscles themselves. Autonomic nerves also maintain a resting "tone," in the sense that they maintain a baseline firing rate that can be either increased or decreased. Changes in tonic neural activity produce changes in the intrinsic activity of the effector organ. A decrease in the excitatory input to the heart, for example, will slow its rate of beat.

Unlike somatic motor neurons—where release of the neurotransmitter (ACh) always stimulates the effector organ (skeletal muscle)—some autonomic nerves release transmitters that inhibit the activity of their effectors. Increases or decreases in the inhibitory effect of parasympathetic nerve fibers to the heart, for example, results in decreases or increases, respectively, in the heart rate.

Cardiac Muscle. Like skeletal muscle cells, cardiac muscle cells, or *myocardial cells,* are striated; they contain actin and myosin filaments arranged in the form of sarcomeres, and they contract by means of the sliding filament mechanism. The long, fibrous skeletal muscle cells, however, are structurally and functionally separated from each other, whereas the myocardial cells are short, branched, and interconnected. Adjacent myocardial cells are joined by electrical synapses, or **gap junctions.** Gap junctions have an affinity for stain that gives them the appearance of **intercalated** *(in-ter'kah-lat-ed)* **discs** in a light microscope (see chapter 6).

Figure 17.2. An electron micrograph of the thick and thin filaments of smooth muscle. A longitudinal section of a complete long myosin filament is shown between the arrows.

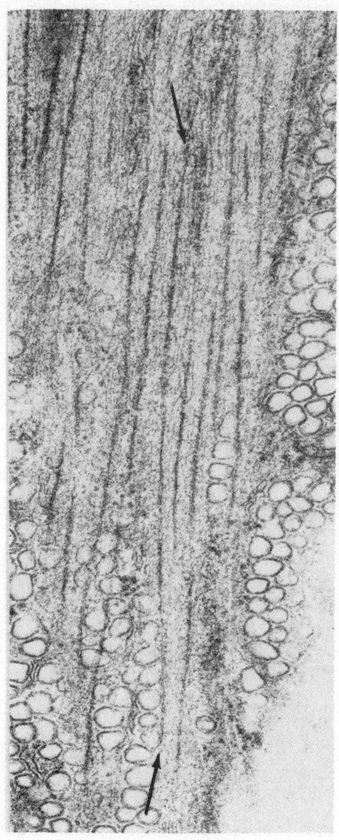

Electrical impulses that originate at any point in the mass of myocardial cells, called the **myocardium** *(mi''o-kar'de-um),* can spread to all cells in the mass that are joined by gap junctions. Because all cells in a myocardium are electrically joined, the myocardium behaves as a single functional unit, or a *functional syncytium (sin-sish'e-um).* Unlike skeletal muscles, which can produce graded contractions with a strength that depends on the number of cells stimulated, the heart contracts with an *all-or-none contraction.*

Unlike skeletal muscles, which require external stimulation through somatic motor nerves before they can produce action potentials and contract, cardiac muscle is able to produce action potentials automatically. Cardiac action potentials normally originate in a specialized group of cells called the *sinoatrial node (pacemaker).* However, the rate of this spontaneous depolarization and, thus, the rate of the heartbeat is regulated by autonomic innervation.

Smooth Muscles. Smooth (visceral) muscles are arranged in circular layers around the walls of blood vessels, bronchioles (small air passages in the lungs), and in the sphincter muscles of the digestive tract. Both circular and longitudinal smooth muscle layers are found in the tubular digestive tract, the ureters (which transport urine), and the ductus deferentia (which transport sperm), and the uterine tubes (which transport ova). The alternate contraction of circular and longitudinal smooth muscle layers produces **peristaltic waves,** which propel the contents of these tubes in one direction.

Smooth muscle cells do not contain sarcomeres (which produce striations in skeletal and cardiac muscle). Smooth muscle cells do, however, contain a great amount of actin and some myosin, which produces a ratio of thin-to-thick filaments of about 16:1 (in striated muscles the ratio is 2:1). The myosin filaments in smooth muscle cells are quite long (fig. 17.2).

The long length of myosin filaments and the fact that they are not organized into sarcomeres may be advantageous for the function of smooth muscles. Smooth muscles must be able to exert tension even when greatly stretched—in the urinary bladder, for example, the smooth muscle cells may be stretched up to two and a half times their resting length. Skeletal muscles, in contrast, lose their ability to contract when the sarcomeres are stretched to the point where actin and myosin no longer overlap.

As in skeletal muscles, the contraction of smooth muscles is triggered by a sharp rise in the Ca^{++} concentration within the cytoplasm of the muscle cells. Unlike skeletal muscles, the sarcoplasmic reticulum of smooth muscles is poorly developed, and Ca^{++} uptake primarily occurs across the cell membrane from the extracellular environment. The events that follow are also somewhat different in smooth muscles. In smooth muscle cells, Ca^{++} combines with a protein in the cytoplasm called **calmodulin.** The calmodulin-Ca^{++} complex thus formed activates an enzyme called *myosin light chain kinase,* which catalyzes the phosphorylation of myosin ATPase. Unlike the situation in skeletal muscles, the phosphorylation of myosin ATPase is required for its activation in smooth muscle cells. These events in smooth muscle cells substitute for the binding of Ca^{++} to troponin, which occurs in skeletal muscle cells.

Despite their differences, it is currently believed that both smooth muscles and skeletal muscles contract by means of a *sliding filament* mechanism. The sliding filament mechanism in smooth muscle, however, is not as well understood as it is in skeletal muscle. Some comparisons of skeletal muscle, cardiac muscle, and smooth muscle are given in table 17.2.

Table 17.2 Some comparisons of skeletal, cardiac, and smooth muscles

Skeletal muscle	Cardiac muscle	Smooth muscle
Striated; actin and myosin arranged in sarcomeres	Striated; actin and myosin arranged in sarcomeres	Not striated; more actin than myosin; actin inserts into dense bodies and cell membrane
Well-developed sarcoplasmic reticulum and transverse tubules	Moderately developed sarcoplasmic reticulum and transverse tubules	Sarcoplasmic reticulum poorly developed; no transverse tubules
Contains troponin in the thin filaments	Contains troponin in the thin filaments	Contains a Ca^{++} binding protein; may be located in thick filaments
Ca^{++} released into cytoplasm from sarcoplasmic reticulum	Ca^{++} enters cytoplasm from sarcoplasmic reticulum and extracellular fluid	Ca^{++} enters cytoplasm from extracellular fluid, sarcoplasmic reticulum, and perhaps mitochondria
Cannot contract without nerve stimulation; denervation results in muscle atrophy	Can contract without nerve stimulation; action potentials originate in pacemaker cells of heart	Maintains tone in absence of nerve stimulation; visceral smooth muscle produces pacemaker potentials; denervation results in hypersensitivity to stimulation
Muscle fibers stimulated independently; no gap junctions	Gap junctions present as intercalated discs	Gap junctions present in most smooth muscles

Table 17.3 Some comparisons between single-unit and multi-unit smooth muscles

	Single-unit muscle	Multi-unit muscle
Location	Gastrointestinal tract; uterus; ureter; small arteries (arterioles)	Arrector pili muscles of hair follicles; ciliary muscle (attached to lens); iris; vas deferens; large arteries
Origin of electrical activity	Spontaneous activity by pacemakers (myogenic)	Not spontaneously active; potentials are neurogenic
Type of potentials	Action potentials	Graded depolarizations
Response to stretch	Responds to stretch by contraction; not dependent on nerve stimulation	No inherent response to stretch
Presence of gap junctions	Numerous gap junctions join all cells together electrically	Few (if any) gap junctions
Type of contraction	Slow, sustained contractions	Slow, sustained contractions

Single-Unit and Multi-Unit Smooth Muscles. Smooth muscles are often grouped into two functional categories: **single-unit** and **multi-unit.** Single-unit smooth muscles have numerous gap junctions (electrical synapses) between adjacent cells that weld them together electrically; they thus behave as a single unit. Multi-unit smooth muscles have few, if any, gap junctions; the individual cells must thus be stimulated separately by impulses through nerve endings. This is similar to the numerous motor units required for the control of skeletal muscles.

Single-unit smooth muscles display *pacemaker* activity in which certain cells stimulate others in the mass. Single-unit smooth muscles also display intrinsic, or *myogenic (mi''o-jen'ik)* electrical activity and contraction in response to stretch. For example, the stretch induced by an increase in the volume of a small artery or a section of the gastrointestinal tract can stimulate myogenic contraction. Such contraction does not require stimulation by autonomic nerves. Contraction of multi-unit smooth muscles, in contrast, requires nerve stimulation. Comparisons of single-unit and multi-unit smooth muscles are given in table 17.3.

myogenic: Gk. *mys,* muscle; *genesis,* origin

Autonomic Innervation of Smooth Muscles. There are significant differences between the neural control of skeletal muscles and that of smooth muscles. A skeletal muscle fiber has only one junction with a somatic nerve fiber, and the receptors for the neurotransmitter are located only at the membrane of the neuromuscular junction. In contrast, the entire surface of smooth muscle cells contains transmitter receptor proteins. Neurotransmitter molecules are released along a stretch of an autonomic nerve fiber that is located some distance from the smooth muscle cells. The regions of the autonomic fiber that release transmitters appear as bulges, or *varicosities,* and the neurotransmitters released from these varicosities stimulate a number of smooth muscle cells. Since there are a number of varicosities along a stretch of an autonomic nerve ending, they form synapses "in passing"—or *synapses en passant*—with the smooth muscle cells.

Glands. Exocrine glands are derived from cells of epithelial membranes that cover and line the body surfaces. The secretions of these cells are expressed to the outside of the epithelial membranes (and hence to the outside of the body) through *ducts.* This is in contrast to endocrine glands, which lack ducts and which therefore secrete into the bloodstream at the capillary level.

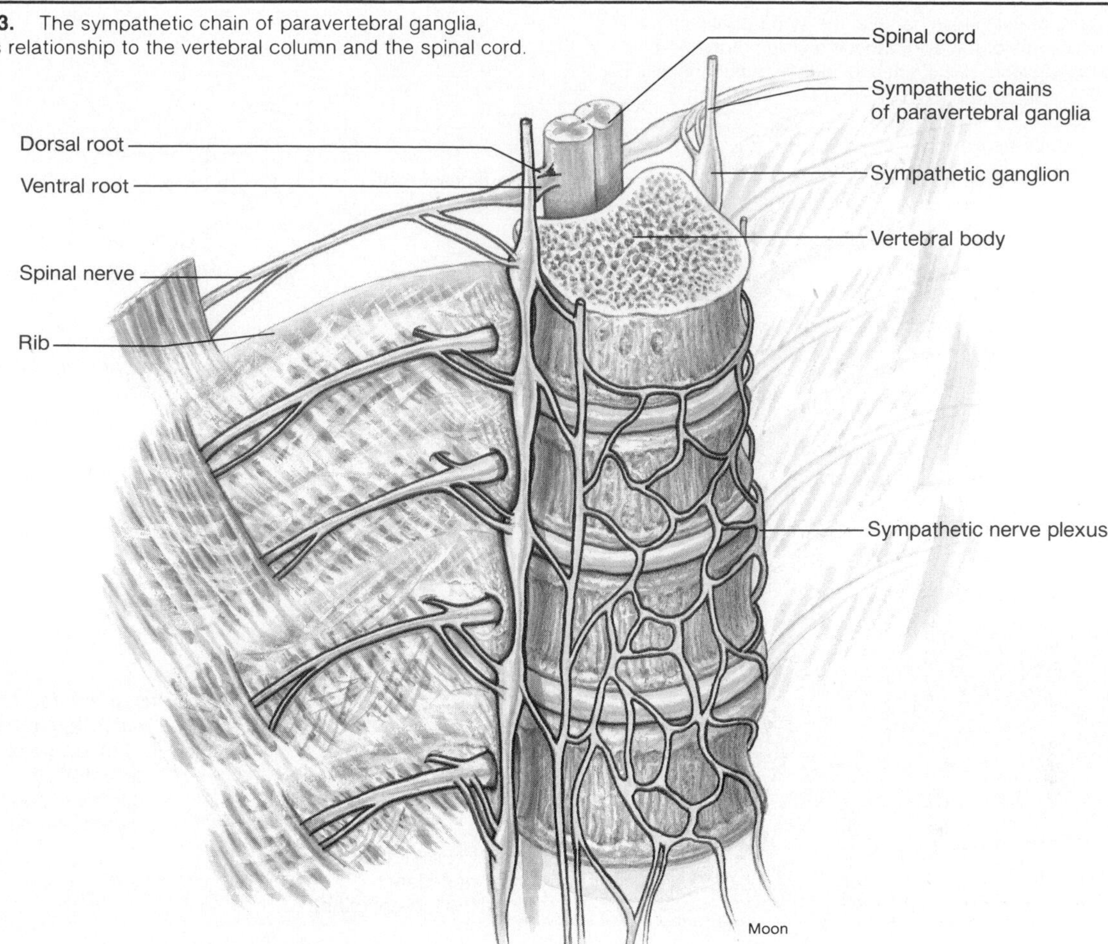

Figure 17.3. The sympathetic chain of paravertebral ganglia, showing its relationship to the vertebral column and the spinal cord.

1. Describe how the regulation of the contraction of cardiac and smooth muscle cells differs from that of skeletal muscle cells and how these muscles are affected by the experimental removal of their innervation.
2. Define the terms *preganglionic* and *postganglionic neurons* in the autonomic system, and use a diagram to illustrate the difference in efferent outflow between somatic and autonomic nerves.
3. Compare the structure of single-unit and multi-unit smooth muscles, and explain how these two types differ in the way they are regulated by autonomic nerves.
4. Explain how synapses between autonomic nerves and smooth muscles differ from the neuromuscular junctions of somatic motor nerves with skeletal muscle fibers.

Structure of the Autonomic System

The sympathetic and parasympathetic divisions of the autonomic system consist of preganglionic neurons that originate in the CNS and postganglionic neurons that originate outside of the CNS in ganglia. The specific origin of the preganglionic fibers and the location of the ganglia, however, are different in the two subdivisions of the autonomic system.

Objective 4. Describe the origin of preganglionic sympathetic neurons and the location of sympathetic ganglia.

Objective 5. Explain the relationship between the sympathetic system and the adrenal medulla.

Objective 6. Describe the origin of the preganglionic parasympathetic neurons and the location of the parasympathetic ganglia.

Objective 7. Describe the distribution of the vagus nerve, and explain its significance within the parasympathetic division.

Sympathetic (Thoracolumbar) Division

The **sympathetic division** is also called the thoracolumbar division of the autonomic system because its preganglionic fibers exit the spinal cord from the first thoracic (T1) to the second lumbar (L2) levels. Most sympathetic nerve fibers, however, separate from the somatic motor fibers and synapse with postganglionic neurons within a double chain of sympathetic ganglia, or **paravertebral ganglia,** located on either side of the spinal cord (fig. 17.3).

Figure 17.4. Sympathetic chain ganglia, the sympathetic chain, and rami communicantes of the sympathetic division of the ANS. (Solid lines = preganglionic fibers; dashed lines = postganglionic fibers.)

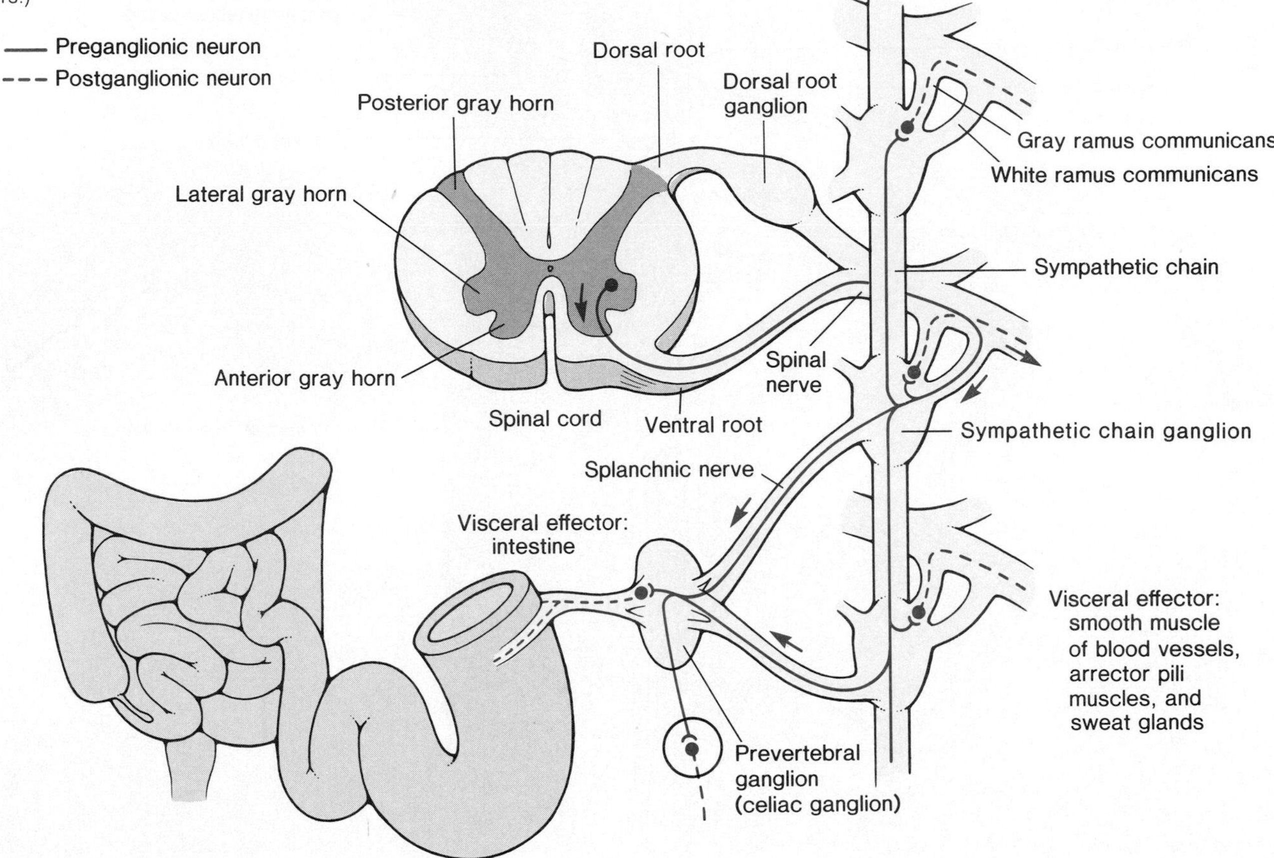

Since the preganglionic sympathetic fibers are myelinated and thus appear white, they are called **white rami communicantes.** Some of these preganglionic sympathetic fibers synapse with postganglionic neurons located at their same level in the chain of sympathetic ganglia. Other preganglionic fibers travel up or down within the sympathetic chain before synapsing with postganglionic neurons. Since the fibers of the postganglionic sympathetic neurons are unmyelinated and thus appear gray, they form the **gray rami communicantes.** Postganglionic axons in the gray rami communicantes extend directly back to the ventral roots of the spinal nerves and travel distally within the spinal nerves to innervate their effector organs (fig. 17.4).

Within the chain of paravertebral ganglia, *divergence* is seen when preganglionic fibers branch to synapse with many postganglionic neurons located at different levels in the chain. *Convergence* is also seen when a given postganglionic neuron receives synaptic input from a large number of preganglionic fibers. The divergence of impulses from the spinal cord to the ganglia and the convergence of impulses within the ganglia usually results in the **mass activation** of almost all of the postganglionic fibers. This explains why the sympathetic system is usually activated as a unit and affects all of its effector organs at the same time.

Many preganglionic fibers that exit the spinal cord in the upper thoracic level travel through the sympathetic chain into the neck, where they synapse in cervical sympathetic ganglia (fig. 17.5). Postganglionic fibers from here innervate the smooth muscles and glands of the head and neck.

Collateral Ganglia. Many preganglionic fibers that exit the spinal cord below the level of the diaphragm pass through the sympathetic chain of ganglia without synapsing. Beyond the sympathetic chain, these preganglionic fibers form **splanchnic** *(splank'nik)* **nerves** (fig. 17.4). Preganglionic fibers in the splanchnic nerves synapse in **collateral ganglia,** which are also called **prevertebral ganglia.** These include the **celiac** *(se'le-ak),* **superior mesenteric** *(mes''en-ter'ik),* and **inferior mesenteric ganglia** (figs. 17.6, 17.7).

ramus: L. *ramus,* a branch

splanchnic: Gk. *splanchno-,* relating to viscera

Figure 17.5. The cervical sympathetic ganglia.

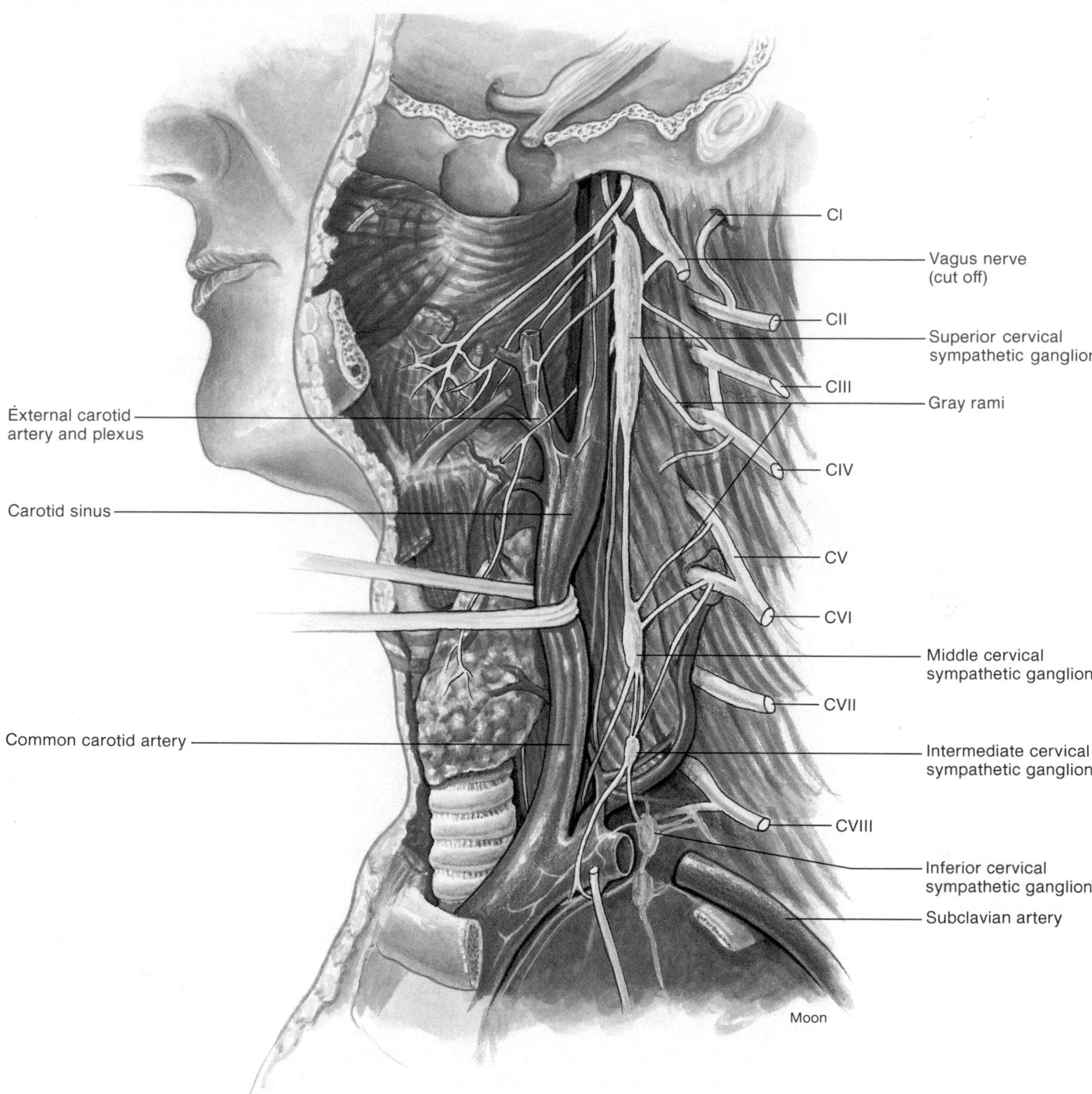

The *greater splanchnic nerve* arises from preganglionic sympathetic fibers T4–T9 and synapses in the celiac ganglion. These fibers contribute to the **celiac (solar) plexus.** Postganglionic fibers from the celiac ganglion innervate the stomach, spleen, liver, small intestine, and kidneys. The **lesser splanchnic nerve** terminates in the superior mesenteric ganglion. Postganglionic fibers from here innervate the small intestine and colon. The **lumbar splanchnic nerve** synapses in the inferior mesenteric ganglion, and the post ganglionic fibers innervate the distal colon and rectum, urinary bladder, and genital organs.

Adrenal Glands. The paired adrenal glands are located above each kidney. Each adrenal is composed of two parts: an outer **cortex** and an inner **medulla.** These two parts are really two functionally different glands with different embryonic origins, different hormones, and different regulatory mechanisms. The adrenal cortex secretes steroid hormones; the adrenal medulla secretes the hormone **epinephrine** (adrenalin) and, to a lesser degree, **norepinephrine,** when it is stimulated by the sympathetic system.

adrenal: L. *ad,* to; *renes,* kidney
cortex: L. *cortex,* bark
medulla: L. *medulla,* marrow

Figure 17.6. The collateral sympathetic ganglia: the celiac and the superior and inferior mesenteric ganglia.

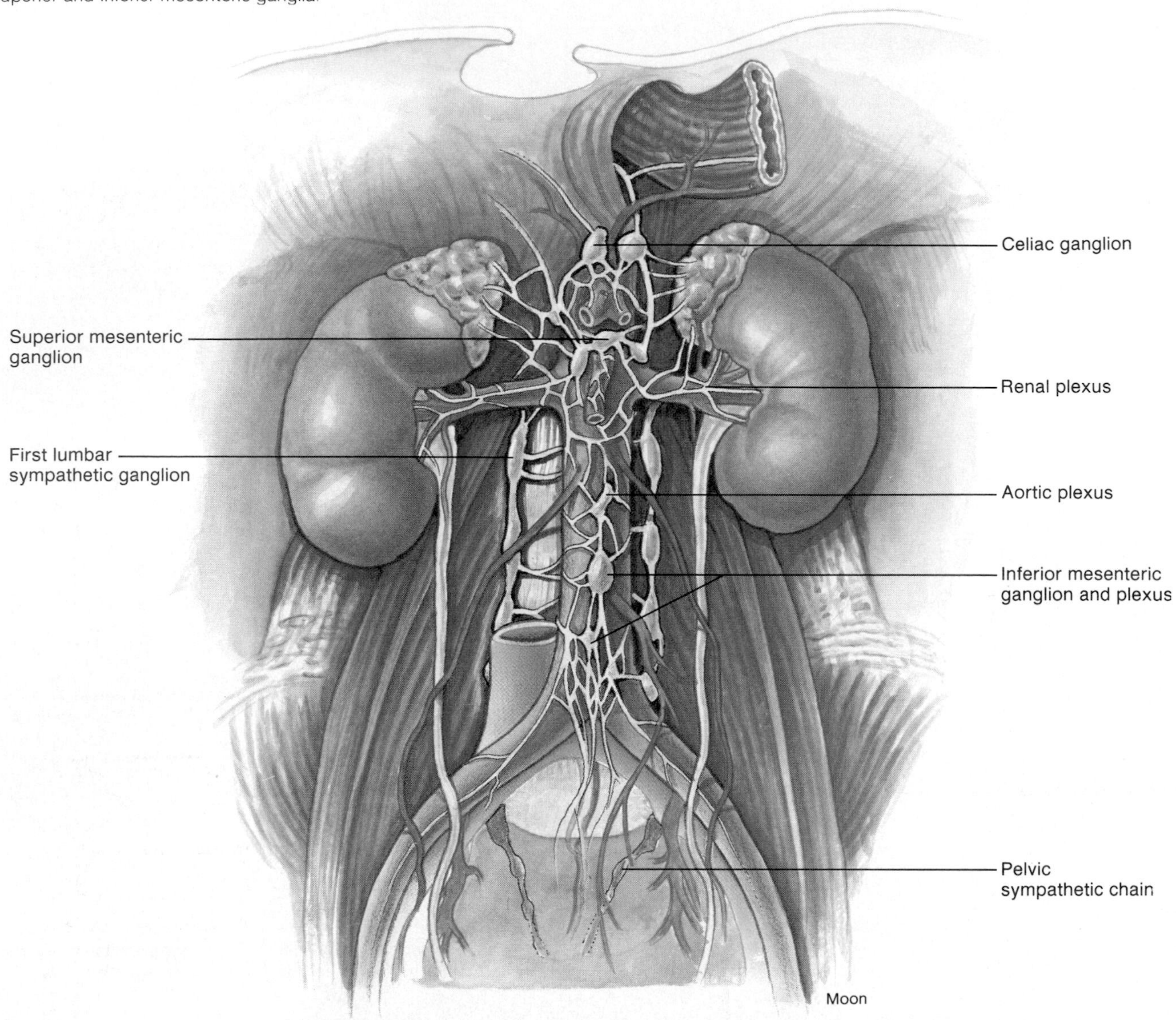

The adrenal medulla is a modified sympathetic ganglion whose cells are derived from postganglionic sympathetic neurons. The cells of the adrenal medulla are innervated by preganglionic sympathetic fibers originating in the thoracic level of the spinal cord; they secrete epinephrine into the blood in response to this neural stimulation. The effects of epinephrine are complementary to those of the neurotransmitter norepinephrine, which is released from postganglionic sympathetic nerve endings. For this reason and because the adrenal medulla is stimulated as part of the mass activation of the sympathetic system, the two are often grouped together as the **sympathoadrenal system.**

Parasympathetic (Craniosacral) Division

The **parasympathetic division** is also known as the *craniosacral division* of the autonomic system. This is because its preganglionic fibers originate in the brain (specifically, the midbrain and the medulla oblongata of the brain stem) and in the second through fourth sacral levels of the spinal cord. These preganglionic parasympathetic fibers synapse in ganglia that are located next to—or actually within—the organs innervated. These parasympathetic ganglia, which are called **terminal ganglia,** supply the postganglionic fibers that synapse with the effector cells. Tables 17.4 and 17.5 show the comparative structures of the sympathetic and parasympathetic divisions. It should be noted that unlike sympathetic fibers, most parasympathetic fibers do not travel within spinal nerves. Cutaneous effectors (blood vessels, sweat glands, and arrector pili muscles) and blood vessels in skeletal muscles thus receive sympathetic but not parasympathetic innervation.

Figure 17.7. The autonomic nervous system. The sympathetic division is shown in brown, the parasympathetic in blue. The solid lines indicate preganglionic fibers, and the dotted lines indicate postganglionic fibers.

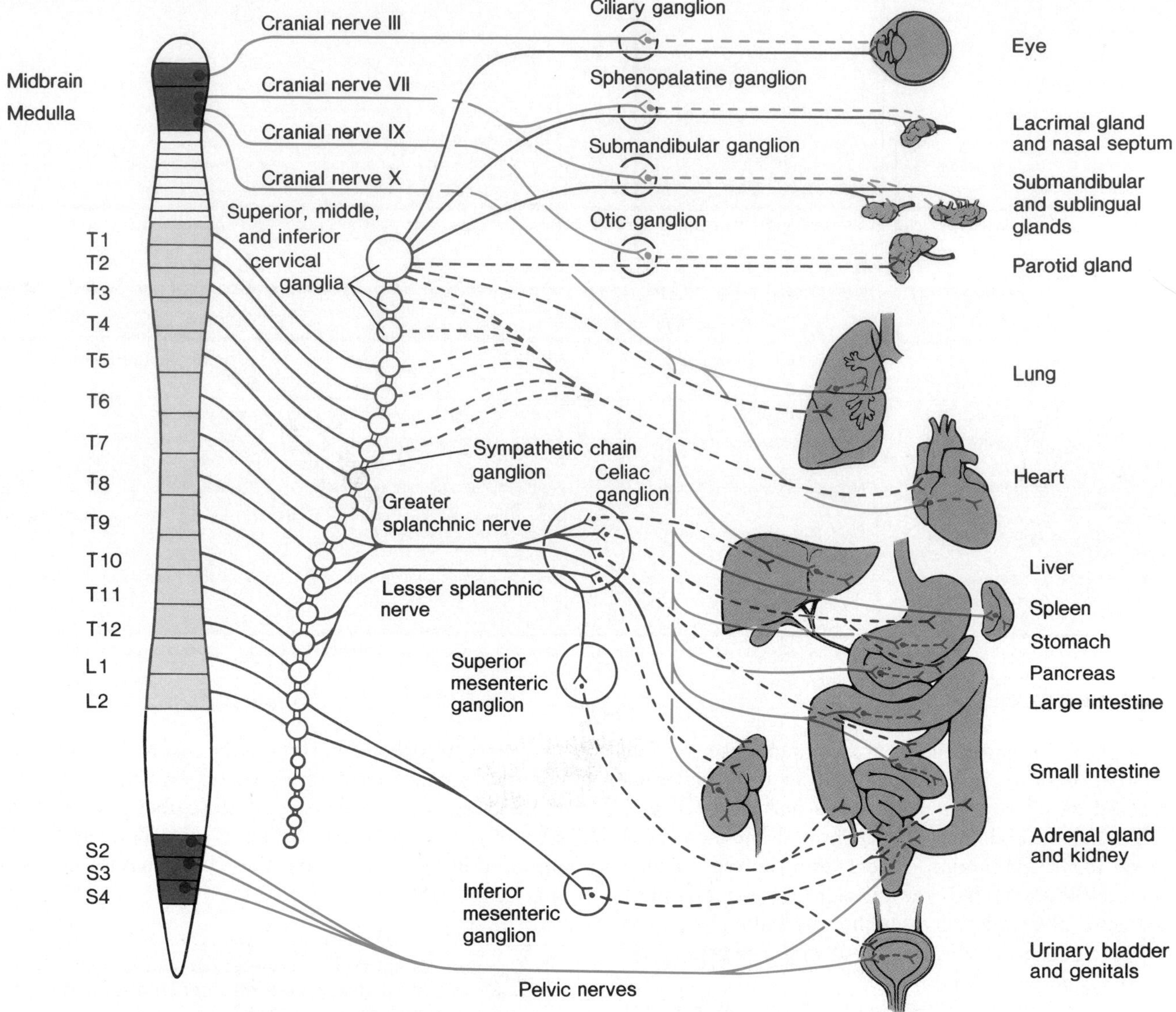

Four of the twelve pairs of cranial nerves contain preganglionic parasympathetic fibers. These are the oculomotor (III), facial (VII), glossopharyngeal (IX), and vagus (X) nerves. Parasympathetic fibers within the first three of these cranial nerves synapse in ganglia located in the head; fibers in the vagus nerve synapse in terminal ganglia located in many regions of the body.

The oculomotor nerve contains somatic motor and parasympathetic fibers that originate in the oculomotor nuclei of the midbrain. These parasympathetic fibers synapse in the **ciliary ganglion,** whose postganglionic fibers innervate the ciliary muscle and constrictor fibers in the iris of the eye. Preganglionic fibers that originate in the medulla oblongata travel in the facial nerve to the **pterygopalatine** *(ter″ĭ-go-pal′ah-tēn)* **ganglion,** also called the sphenopalatine ganglion, which sends postganglionic fibers to the nasal mucosa, pharynx, palate, and lacrimal glands. Another group of fibers in the facial nerve terminate in the **submandibular ganglion,** which sends postganglionic fibers to the submandibular and sublingual salivary glands. Preganglionic fibers of the glossopharyngeal nerve synapse in the **otic** *(o′tik)* **ganglion,** which sends postganglionic fibers to innervate the parotid salivary gland.

Other nuclei in the medulla oblongata contribute preganglionic fibers to the very long vagus nerves, which provide the most extensive parasympathetic innervation in the body. As the paired vagus nerves pass through the thorax, they contribute to the *pulmonary plexuses* within the mediastinum. Branches of the pulmonary plexuses accompany blood vessels and bronchi into the lungs. Below the pulmonary plexuses, the vagus nerves merge to form the *esophageal (ĕ-sof″ah-je′al) plexuses.*

vagus: L. *vagus,* wandering

Table 17.4 The sympathetic (thoracolumbar) system

Parts of body innervated	Spinal origin of preganglionic fibers	Origin of postganglionic fibers
Eye	C-8 and T-1	Cervical ganglia
Head and neck	T-1 to T-4	Cervical ganglia
Heart and lungs	T-1 to T-5	Upper thoracic (paravertebral) ganglia
Upper extremity	T-2 to T-9	Lower cervical and upper thoracic (paravertebral) ganglia
Upper abdominal viscera	T-4 to T-9	Celiac and superior mesenteric (collateral) ganglia
Adrenal	T-10 and T-11	Adrenal medulla
Urinary and reproductive systems	T-12 to L-2	Celiac and inferior mesenteric (collateral) ganglia
Lower extremities	T-9 to L-2	Lumbar and upper sacral (paravertebral) ganglia

Table 17.5 The parasympathetic (craniosacral) system

Effector organs	Origin of preganglionic fibers	Nerve	Origin of postganglionic fibers
Eye (ciliary and iris muscles)	Midbrain (cranial)	Oculomotor (third cranial) nerve	Ciliary ganglion
Lacrimal, mucus, and salivary glands in head	Medulla oblongata (cranial)	Facial (seventh cranial) nerve	Sphenopalatine and submandibular ganglia
Parotid (salivary) gland	Medulla oblongata (cranial)	Glossopharyngeal (ninth cranial) nerve	Otic ganglion
Heart, lungs, gastrointestinal tract, liver, pancreas	Medulla oblongata (cranial)	Vagus (tenth cranial) nerve	Terminal ganglia in or near organ
Lower half of large intestine, rectum, urinary bladder, and reproductive organs	S-2 to S-4 (sacral)	Through pelvic spinal nerves	Terminal ganglia near organs

At the lower end of the esophagus, vagal fibers collect to form an **anterior** and **posterior vagal** *(va'gal)* **trunk,** each composed of fibers from both vagus nerves. The vagal trunks enter the abdominal cavity through the esophageal opening in the diaphragm. Fibers from the vagal trunks innervate the stomach on the anterior and posterior sides. Branches of the vagus nerves within the abdominal cavity also contribute to the *celiac plexus* and *plexuses of the abdominal aorta.*

The preganglionic fibers in the vagus synapse with postganglionic neurons that are actually located *within* the innervated organs. These preganglionic fibers are thus quite long, and they provide parasympathetic innervation to the heart, lungs, esophagus, stomach, pancreas, liver, small intestine, and upper half of the large intestine. Postganglionic parasympathetic fibers arise from terminal ganglia within these organs and synapse with effector cells.

Preganglionic fibers from the sacral levels of the spinal cord provide parasympathetic innervation to the lower half of the large intestine, rectum, and to the urinary and reproductive systems. These fibers, like those of the vagus, synapse with terminal ganglia located within the effector organs.

Parasympathetic nerves to the visceral organs thus consist of preganglionic fibers, whereas sympathetic nerves to these organs contain postganglionic fibers. A composite view of the sympathetic and parasympathetic systems is provided in figure 17.7, and these comparisons are summarized in table 17.6.

1. Compare the origin of preganglionic sympathetic and parasympathetic fibers and the location of sympathetic and parasympathetic ganglia.
2. Using a simple line drawing, illustrate the sympathetic pathway from the spinal cord to the heart. Label the preganglionic neuron, postganglionic neuron, and the ganglion.
3. Use a simple line diagram to show the parasympathetic innervation of the heart. Label the preganglionic and postganglionic neurons, the name of the nerve involved, and the terminal ganglion.
4. Describe the distribution of the vagus nerve and its physiological significance.
5. Define the terms *white ramus communicantes* and *gray ramus communicantes,* and explain why blood vessels in the skin and skeletal muscles receive sympathetic but not parasympathetic innervation.
6. Describe the structure of the adrenal gland, and explain its relationship to the sympathetic nervous system.

Table 17.6 Some comparisons between the structure of the sympathetic and parasympathetic systems

	Sympathetic	Parasympathetic
Origin of preganglionic outflow	Thoracolumbar levels of spinal cord	Midbrain, hindbrain, and sacral levels of spinal cord
Location of ganglia	Chain of paravertebral ganglia and prevertebral (collateral) ganglia	Terminal ganglia in or near effector organs
Distribution of postganglionic fibers	Throughout the body	Mainly limited to the head and the viscera of the chest, abdomen, and pelvis
Divergence of impulses from pre- to postganglionic fibers	Great divergence (one preganglionic may activate twenty postganglionic fibers)	Little divergence (one preganglionic only activates a few postganglionic fibers)
Mass discharge of system as a whole	Yes	Not normally

Functions of the Autonomic Nervous System

The actions of the autonomic nervous system, together with the actions of hormones, help to maintain a state of dynamic constancy in the internal environment. The sympathetic division activates the body to "fight or flight" through adrenergic effects; the parasympathetic division often has antagonistic actions through cholinergic effects. Homeostasis thus depends, in large part, on the complementary and often antagonistic effects of sympathetic and parasympathetic innervation.

Objective 8. List the neurotransmitters of the preganglionic and postganglionic neurons of the sympathetic and parasympathetic systems.

Objective 9. Describe the actions of the sympathoadrenal system, and explain the significance of alpha-adrenergic, beta-adrenergic, and cholinergic stimulation.

Objective 10. Describe the effects of acetylcholine released by postganglionic parasympathetic nerve fibers.

Objective 11. Explain the antagonistic, complementary, and cooperative effects of sympathetic and parasympathetic innervation.

The sympathetic and parasympathetic divisions of the autonomic system (fig. 17.7) affect the visceral organs in different ways. Mass activation of the *sympathetic division* prepares the body for intense physical activity in

Table 17.7 Effects of autonomic nerve stimulation on various visceral effector organs

Effector organ	Sympathetic effect	Parasympathetic effect
Eye		
Iris (radial muscle)	Dilates pupil	—
Iris (sphincter muscle)	—	Constricts pupil
Ciliary muscle	Relaxes (for far vision)	Contracts (for near vision)
Glands		
Lacrimal (tear)	—	Stimulates secretion
Sweat	Stimulates secretion	—
Salivary	Decreases secretion; saliva becomes thick	Increases secretion; saliva becomes thin
Stomach	—	Stimulates secretion
Intestine	—	Stimulates secretion
Adrenal medulla	Stimulates secretion of hormones	—
Heart		
Rate	Increases	Decreases
Conduction	Increases rate	Decreases rate
Strength	Increases	—
Blood vessels	Mostly constricts; affects all organs	Dilates in a few organs (e.g., penis)
Lungs		
Bronchioles (tubes)	Dilates	Constricts
Mucous glands	Inhibits secretion	Stimulates secretion
Gastrointestinal tract		
Motility	Inhibits movement	Stimulates movement
Sphincters	Stimulates closing	Inhibits closing
Liver	Stimulates hydrolysis of glycogen	—
Adipose (fat) cells	Stimulates hydrolysis of fat	—
Pancreas	Inhibits exocrine secretions	Stimulates exocrine secretions
Spleen	Stimulates contraction	—
Urinary bladder	Helps set muscle tone	Stimulates contraction
Piloerector muscles	Stimulates erection of hair and "goosebumps"	—
Uterus	If pregnant: contraction if not pregnant: relaxation	—
Penis	Erection; ejaculation	Erection (due to vasodilation)

emergencies; the heart rate increases, blood glucose rises, and blood is diverted to the skeletal muscles (away from the visceral organs and skin). These and other effects are listed in table 17.7. The "theme" of the sympathetic system has been aptly summarized in a phrase: *fight or flight.*

The effects of *parasympathetic nerve* stimulation are in many ways opposite to the effects of sympathetic stimulation. The parasympathetic system, however, is not normally activated as a whole. Stimulation of separate

Figure 17.8. Neurotransmitters of the autonomic motor system. ACh = acetylcholine; NE = norepinephrine; E = epinephrine. Those nerves that release ACh are called cholinergic; those nerves that release NE are called adrenergic. The adrenal medulla secretes both epinephrine (85%) and norepinephrine (15%) as hormones into the blood.

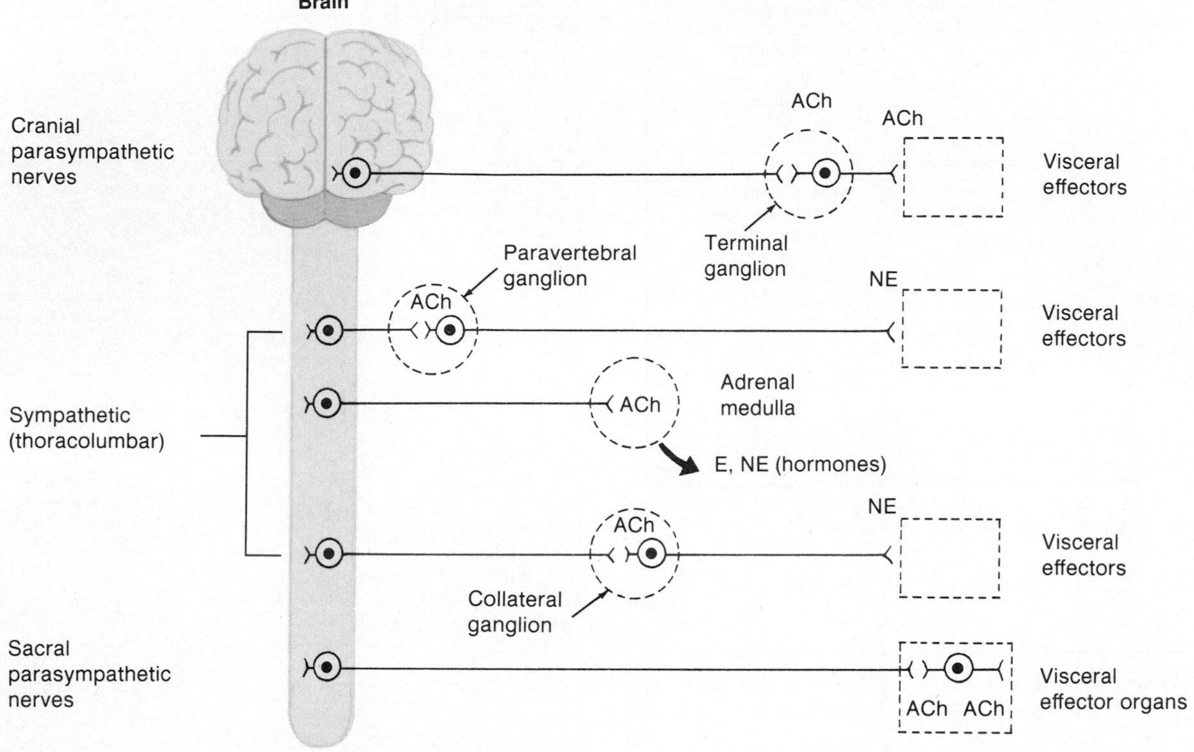

Figure 17.9. The structure of the catecholamines norepinephrine and epinephrine.

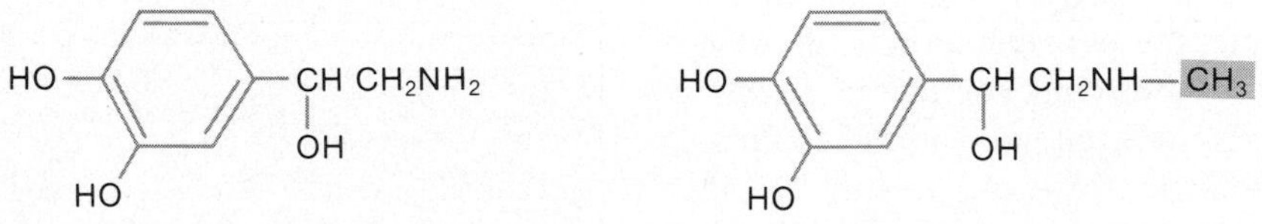

HO, HO — CH CH_2NH_2 — OH

Norepinephrine

HO, HO — CH CH_2NH—CH_3 — OH

Epinephrine

parasympathetic nerves can result in slowing of the heart, dilation of visceral blood vessels, and an increased activity of the digestive tract (table 17.7). The different responses of visceral organs to sympathetic and parasympathetic nerve activity is due to the characteristic way that each organ responds to the different neurotransmitters released by the postganglionic fibers of these two divisions of the autonomic system.

Neurotransmitters of the Autonomic System

Acetylcholine *(as''ĕ-til-kō'lēn) (ACh)* is the neurotransmitter of all preganglionic fibers (both sympathetic and parasympathetic). Acetylcholine is also the transmitter released by all parasympathetic postganglionic fibers at their synapses with effector cells (fig. 17.8). Transmission at the autonomic ganglia and at synapses of postganglionic nerve fibers is thus said to be **cholinergic** *(ko''lin-er'jik).*

The neurotransmitter released by most postganglionic sympathetic nerve fibers is **norepinephrine** *(noradrenalin).* Transmission at these synapses is thus said to be **adrenergic** *(ad''ren-er'jik).* There are a few exceptions to this rule: some sympathetic fibers that innervate blood vessels in skeletal muscles, as well as sympathetic fibers to sweat glands, release ACh (are cholinergic).

In view of the fact that the cells of the adrenal medulla are derived from postganglionic sympathetic neurons, it is not surprising that the hormones they secrete (normally about 85% epinephrine and 15% norepinephrine) are similar to the transmitter of postganglionic sympathetic neurons. Epinephrine differs from norepinephrine only in the presence of an additional methyl (CH_3) group, as shown in figure 17.9. Epinephrine, norepinephrine, and

cholinergic: Gk. *chole,* bile; *ergon,* work

Table 17.8 Adrenergic and cholinergic effects of sympathetic and parasympathetic nerves

	Effect of			
	Sympathetic		Parasympathetic	
Organ	Action	Receptor	Action	Receptor
Eye				
Iris				
Radial muscle	Contracts	α	. . .	. . .
Ciliary muscle	. . .	. . .	Contracts	M
Heart				
Sinoatrial node	Accelerates	β_1	Decelerates	M
Contractility	Increases	β_1	Decreases (atria)	M
Vascular smooth muscle				
Skin, splanchnic vessels	Contracts	α	. . .	. . .
Skeletal muscle vessels	Relaxes	β_2	. . .	. . .
	Relaxes	M*	. . .	. . .
Bronchiolar smooth muscle	Relaxes	β_2	Contracts	M
Gastrointestinal tract				
Smooth muscle				
Muscular tunic	Relaxes	β_2	Contracts	M
Sphincters	Contracts	α	Relaxes	M
Secretion	. . .	. . .	Increases	M
Myenteric plexus	Inhibits	α	. . .	. . .
Genitourinary smooth muscle				
Detrusor muscle	Relaxes	β_2	Contracts	M
Urethral sphincter	Contracts	α	Relaxes	M
Uterus, pregnant	Relaxes	β_2	. . .	. . .
	Contracts	α	. . .	. . .
Penis, seminal vesicles	Ejaculation	α	Erection	M
Skin				
Arrector pili muscle	Contracts	α	. . .	. . .
Sweat glands				
Thermoregulatory	Increases	M	. . .	. . .
Apocrine (stress)	Increases	α	. . .	. . .

*Vascular smooth muscle in skeletal muscle has sympathetic cholinergic dilator fibers.
Adrenergic receptors are indicated as alpha (α) or beta (β); cholinergic receptors are indicated as muscarinic (M).
Reproduced and modified, with permission, from Katzung, B. G.: *Basic and Clinical Pharmacology,* 3rd edition, copyright Appleton & Lange, 1987.

dopamine (a transmitter within the CNS) are collectively termed **catecholamines** *(kat''ĕ-kol-am'in)* (see chapter 14). The catecholamines, together with serotonin (another transmitter within the CNS), are often referred to as *monoamines,* or *biogenic amines.*

Responses to Adrenergic Stimulation

Adrenergic stimulation—by epinephrine in the blood and by norepinephrine released from sympathetic nerve endings—has both excitatory and inhibitory effects. The heart, dilatory muscles of the iris, and the smooth muscles of many blood vessels are stimulated to contract. The smooth muscles of the bronchioles and of some blood vessels, however, are inhibited from contracting; adrenergic chemicals, therefore, cause these structures to dilate.

Since excitatory and inhibitory effects can be produced in different tissues by the same chemical, the responses clearly depend on the biochemistry of the tissue cells rather than on the intrinsic properties of the chemical. Included in the biochemical differences among the target tissues for catecholamines are differences in the *membrane receptor proteins* for these chemical agents. There are two major classes of these receptor proteins, designated **alpha** (α) and **beta** (β) **receptors.**

Further experiments have revealed that there are two subtypes of each category of adrenergic receptor. These are designated by subscripts: α_1 and α_2; β_1 and β_2. Scientists have developed compounds that selectively bind more to one or the other type of adrenergic receptor and, by this means, either promote or inhibit the normal action produced when epinephrine binds to the receptor. By using these selective compounds, it has been possible to determine the adrenergic receptor subtype in each organ and to correlate that receptor with the physiological effect of autonomic neurotransmitters in the organs (table 17.8). Both types of beta receptors appear to produce these effects by stimulating the production of cyclic AMP (discussed in chapter 14) within the target cells. The activation of the α_2 receptors has the opposite effect—cyclic AMP production is blocked. The response to α_1 receptors may be mediated by a rise in the cytoplasmic concentration of Ca^{++}, which serves as a "second messenger" in the target cell instead of cyclic AMP. Each of these changes, it should be remembered, ultimately results in the characteristic response of the tissue to the autonomic neurotransmitter.

Review of table 17.8 reveals certain generalities about the actions of adrenergic receptors. The stimulation of alpha-adrenergic receptors consistently causes constriction of vascular smooth muscles. One can thus state that the vasoconstrictor effect of sympathetic nerves always results from the activation of alpha-adrenergic receptors. The vasodilation effect of sympathetic nerves—which is of relatively less importance—may be produced by the activation of beta-adrenergic receptors or by means of cholinergic receptors (for vessels in skeletal muscles). The effects of beta-adrenergic activation are more complex; these receptors stimulate the relaxation of smooth muscles (in the digestive tract, bronchioles, and uterus, for example), but stimulate contraction of cardiac muscle and promote an increase in cardiac rate.

The use of drugs that selectively stimulate or block (act as *agonists* or *antagonists*) adrenergic receptors is of great clinical benefit. People with hypertension, for example, may receive a beta-blocking drug, such as *propranolol,* which reduces the cardiac rate. People with asthma may be treated with epinephrine, which stimulates bronchodilation (a β_2 effect), but since epinephrine also stimulates the heart (a β_1 effect), more selective β_2 agonist drugs are available for the treatment of asthma. Compounds that stimulate α_1 adrenergic receptors, such as *phenylephrine,* are often part of medications to relieve a stuffy nose (by causing vasoconstriction in the nasal mucosa).

Responses to Cholinergic Stimulation

Somatic motor neurons, all preganglionic autonomic neurons, and all postganglionic parasympathetic neurons are cholinergic—they release acetylcholine as a neurotransmitter. The cholinergic effects of somatic motor neurons and preganglionic autonomic neurons are always excitatory. The cholinergic effects of postganglionic parasympathetic fibers are usually excitatory, but there are notable exceptions—the parasympathetic fibers innervating the heart, for example, cause slowing of the heart rate. It is useful to remember that the effects of parasympathetic stimulation are, in general, opposite to the effects of sympathetic stimulation.

There are two known subtypes of cholinergic receptors. The drug *muscarine (mus'kar-in),* derived from some poisonous mushrooms, stimulates the cholinergic receptor proteins in the target organs of postganglionic parasympathetic nerve fibers (such as in the heart, eye, and digestive system). Muscarine does not stimulate ACh receptor proteins in autonomic ganglia or at the neuromuscular junction of skeletal muscle fibers. The ACh receptors stimulated by muscarine are therefore called **muscarinic receptors,** and the effects produced by parasympathetic nerves in their target organs are called *muscarinic effects* (indicated in table 17.8).

The drug *nicotine,* derived from the tobacco plant, specifically stimulates cholinergic transmission of preganglionic fibers at the autonomic ganglia and activation of the neuromuscular junction of skeletal muscles. These ACh receptors are thus called **nicotinic receptors** to distinguish them from the muscarinic receptors. The drug *curare* specifically blocks nicotinic receptors but has little effect on muscarinic receptors.

The muscarinic effects of ACh are specifically inhibited by the drug *atropine,* derived from the deadly nightshade plant *(Atropa belladonna).* Indeed, extracts of this plant were used by women during the middle ages to dilate their pupils (atropine inhibits parasympathetic stimulation of the iris). This was done to enhance their beauty (belladonna—beautiful woman). Atropine is used clinically today to dilate pupils during eye examinations, to dry mucous membranes of the respiratory tract prior to general anesthesia, and to inhibit spasmodic contractions of the lower digestive tract.

Organs with Dual Innervation

Most organs receive a dual innervation—they are innervated by both sympathetic and parasympathetic fibers. When this occurs, the effects of these two divisions may be antagonistic, complementary, or cooperative.

Antagonistic Effects. The effects of sympathetic and parasympathetic innervation of the pacemaker region of the heart is the best example of the antagonism of these two systems. In this case sympathetic and parasympathetic fibers innervate the same cells. Adrenergic stimulation from sympathetic fibers increases the heart rate, and cholinergic stimulation from parasympathetic fibers inhibits the pacemaker cells and, thus, decreases the heart rate. Antagonism is also seen in the digestive tract, where sympathetic nerves inhibit and parasympathetic nerves stimulate intestinal movements and secretions.

The effects of sympathetic and parasympathetic stimulation on the diameter of the pupil of the eye is analogous to the reciprocal innervation of flexor and extensor skeletal muscles by somatic motor neurons. This is because the iris contains antagonistic muscle layers. Contraction of the radial muscles, which is stimulated by sympathetic nerves, causes dilation; contraction of the circular muscles, which are innervated by parasympathetic nerve endings, causes constriction of the pupils (fig. 17.10).

Complementary Effects. The effects of sympathetic and parasympathetic stimulation on salivary gland secretion are complementary. The secretion of watery saliva is stimulated through parasympathetic nerves, which also stimulate the secretion of other exocrine glands in the digestive tract. Impulses through sympathetic nerves stimulate the constriction of blood vessels throughout the digestive tract. The resultant decrease in blood flow to the salivary glands causes the production of a thicker, more viscous saliva.

Figure 17.10. Reciprocal innervation of the iris muscles by the sympathetic and parasympathetic systems. Stimulation of sympathetic nerves produces contraction of the dilator (radial) muscles, which enlarges the pupil. Stimulation of the parasympathetic nerves produces contraction of the constrictor (circular) muscle layer, which makes the pupil smaller.

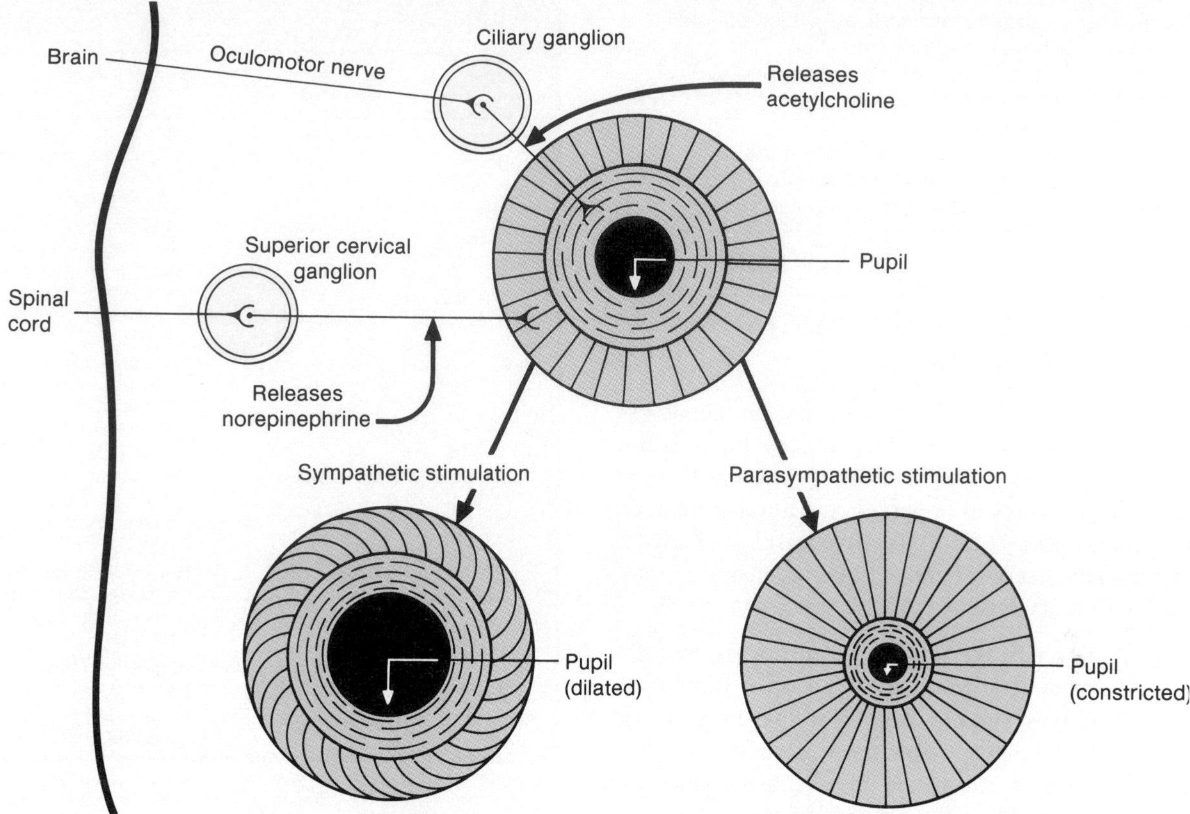

Cooperative Effects. The effects of sympathetic and parasympathetic stimulation on the urinary and reproductive systems are cooperative. Erection of the penis, for example, is due to vasodilation resulting from parasympathetic nerve stimulation; ejaculation is due to stimulation through sympathetic nerves. Although the contraction of the urinary bladder is myogenic (independent of nerve stimulation), it is promoted in part by the action of parasympathetic nerves. This *micturition (mik"tu-rish'un),* or urination, urge and reflex is also enhanced by sympathetic nerve activity, which increases the tone of the bladder muscles. Emotional states that are accompanied by high sympathetic nerve activity may thus result in reflex urination at bladder volumes that are normally too low to trigger this reflex.

Organs without Dual Innervation

Although most organs are innervated by both sympathetic and parasympathetic nerves, some—including the adrenal medulla, arrector pili muscles, sweat glands, and most blood vessels—receive only sympathetic innervation. In these cases regulation is achieved by increases or decreases in the "tone" (firing rate) of the sympathetic fibers. Constriction of blood vessels, for example, is produced by increased sympathetic activity, which stimulates alpha-adrenergic receptors, and vasodilation results from decreased sympathetic nerve stimulation.

The sympathoadrenal system is required for *nonshivering thermogenesis:* animals deprived of their sympathetic system and adrenals cannot tolerate cold stress. The sympathetic system itself is required for proper thermoregulatory responses to heat. In a hot room, for example, decreased sympathetic stimulation produces dilation of the blood vessels in the surface of the skin, which increases cutaneous blood flow and provides better heat radiation. During exercise, in contrast, there is increased sympathetic activity, which causes constriction of the blood vessels in the skin of the limbs and stimulation of sweat glands in the trunk.

The eccrine sweat glands in the trunk secrete a watery fluid in response to sympathetic stimulation. Evaporation of this dilute sweat helps to cool the body. The eccrine sweat glands also secrete a chemical called **bradykinin** *(brad"e-ki'nin)* in response to sympathetic stimulation. Bradykinin stimulates dilation of the surface blood vessels near the sweat glands, helping to radiate heat. At the conclusion of exercise, sympathetic stimulation is reduced and blood flow to the surface of the limbs is increased, which aids in the elimination of metabolic heat. Notice that all of these thermoregulatory responses are achieved without the direct involvement of the parasympathetic system.

micturition: L. *micturire,* to urinate

1. Define the meaning of the terms *adrenergic* and *cholinergic,* and use these terms to describe the neurotransmitters of different autonomic nerve fibers.
2. List the effects of sympathoadrenal stimulation on different effector organs, and indicate which effects are due to alpha or beta receptor stimulation.
3. Describe the effects of the drug *atropine,* and explain these effects in terms of the actions of the parasympathetic system.
4. Explain how the sympathetic and parasympathetic systems can have antagonistic, cooperative, and complementary effects, and give examples.

Control of the Autonomic System by Higher Brain Centers

Visceral functions are mainly regulated by autonomic reflexes. In most autonomic reflexes, sensory input is directed to brain centers, which in turn regulate the activity of descending pathways to preganglionic autonomic neurons. The neural centers that directly control the activity of autonomic nerves are influenced by higher brain areas as well as by sensory input.

Objective 12. Describe the area of the brain that most directly controls the activity of autonomic nerves, and describe the higher brain areas that influence autonomic activity.

Objective 13. Explain, in terms of the structures involved, how the activity of the autonomic system and the activity of the endocrine system can be coordinated.

Objective 14. Explain, in terms of the structures involved, how autonomic functions can be affected by emotions.

The **medulla oblongata** of the brain stem is the area that most directly controls the activity of the autonomic system. Almost all autonomic responses can be elicited by experimental stimulation of the medulla, which contains centers for the control of the cardiovascular, pulmonary, urinary, reproductive, and digestive systems. Much of the sensory input to these centers travels in the afferent fibers of the vagus nerve, which is a mixed nerve containing both sensory and motor fibers. These reflexes are listed in table 17.9.

Table 17.9 Some reflexes stimulated by sensory input from afferent fibers in the vagus that are transmitted to centers in the medulla oblongata

Organs	Type of receptors	Reflex effects
Lungs	Stretch receptors	Inhibits further inhalation; stimulates an increase in cardiac rate and vasodilation
	Type J receptors	Stimulated by pulmonary congestion—produces feelings of breathlessness, and causes a reflex fall in cardiac rate and blood pressure
Aorta	Chemoreceptors	Stimulated by rise in CO_2 and fall in O_2—produces increased rate of breathing, fall in heart rate, and vasoconstriction
	Baroreceptors	Stimulated by increased blood pressure—produces a reflex decrease in heart rate
Heart	Atrial stretch receptors	Inhibits antidiuretic hormone secretion, thus increasing the volume of urine excreted
	Stretch receptors in ventricles	Produces a reflex decrease in heart rate and vasodilation
Gastrointestinal tract	Stretch receptors	Feelings of satiety, discomfort, and pain

From Stuart Ira Fox, *Human Physiology,* 2d ed.

Hypothalamus

The hypothalamus (fig. 17.11) is an extremely important brain region located just above (superior to) the pituitary gland, which in turn is located above the posterior portion of the roof of the mouth. By means of efferent fibers to the brain stem and posterior pituitary and by means of hormones that regulate the anterior pituitary, the hypothalamus serves to orchestrate somatic, autonomic, and endocrine responses during various behavioral states.

Experimental stimulation of different areas of the hypothalamus can evoke the autonomic responses characteristic of aggression, sexual behavior, eating, or satiety. Chronic stimulation of the lateral hypothalamus, for example, can make an animal eat and become obese, whereas stimulation of the medial hypothalamus inhibits eating. Other areas contain osmoreceptors that stimulate thirst and the secretion of antidiuretic hormone (ADH) from the posterior pituitary.

The hypothalamus is also where the body's "thermostat" is located. Experimental cooling of the preoptic-anterior hypothalamus causes shivering (a somatic response) and nonshivering thermogenesis (sympathetic responses). Experimental heating of this hypothalamic area results in hyperventilation (stimulated by somatic motor nerves), vasodilation, salivation, and sweat gland secretion (stimulated by autonomic nerves).

The coordination of sympathetic and parasympathetic reflexes by the medulla oblongata is thus integrated with the control of somatic and endocrine responses by the hypothalamus. The activities of the hypothalamus are in turn influenced by higher brain centers.

Figure 17.11. (*a*) a photograph of a sagittal section of the brain; (*b*) an illustration of the area of the hypothalamus.

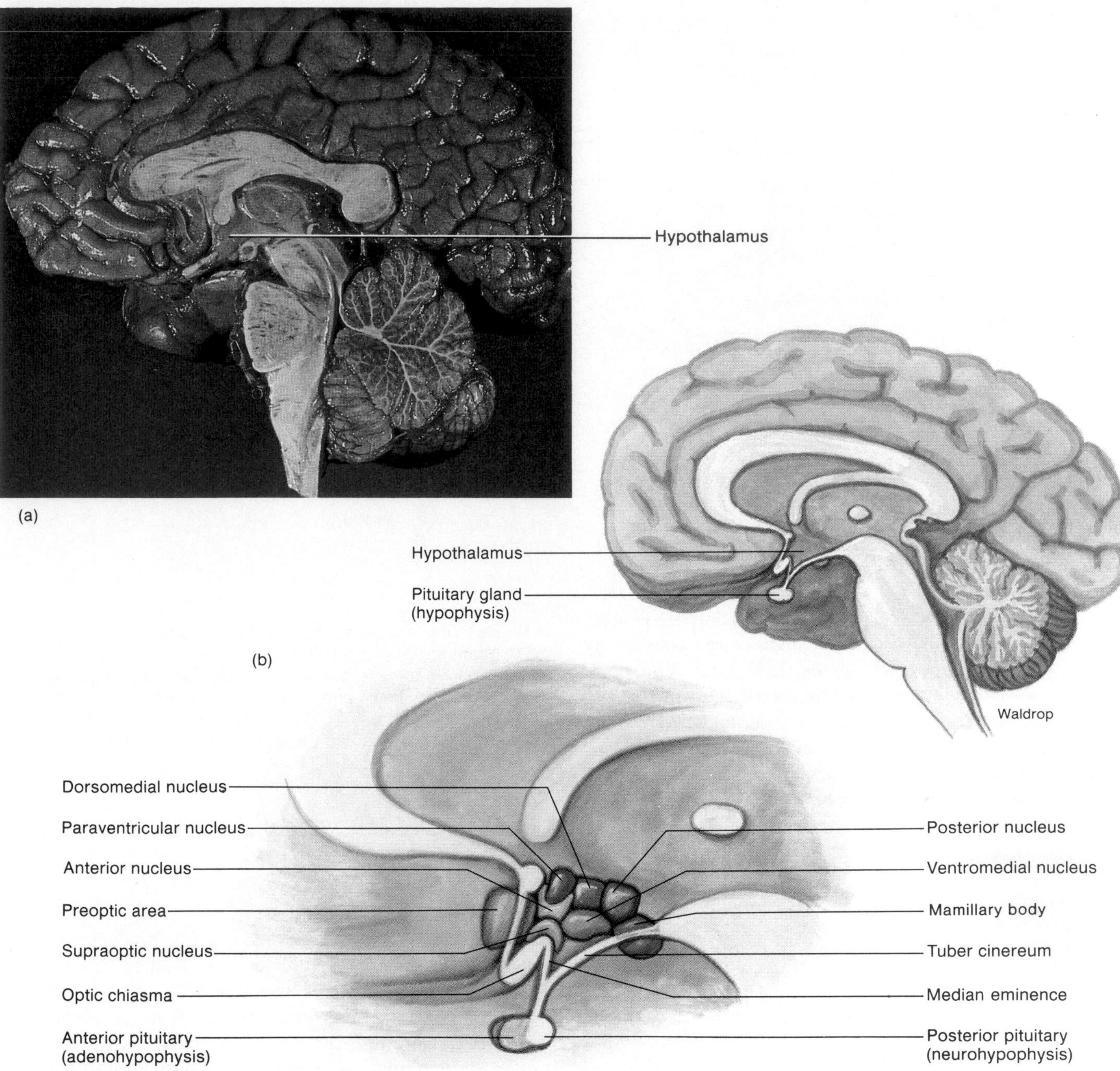

Limbic System, Cerebellum, and Cerebrum

The **limbic system** is a group of fiber tracts and nuclei that form a ring (a limbus) around the brain stem. It includes the cingulate gyrus of the cerebral cortex, the hypothalamus, the fornix (a fiber tract), the hippocampus, and the amygdaloid nucleus (fig. 17.12). These structures, which were derived early in the course of vertebrate evolution, were once called the *rhinencephalon (ri''nen-sef'ah-lon),* or "smell brain," because of their importance in the central processing of olfactory information.

In higher vertebrates, these structures are now recognized as centers involved in basic emotional drives—such as anger, fear, sex, and hunger—and in short-term memory. Complex circuits between the hypothalamus and other parts of the limbic system (illustrated in figure 17.12) contribute visceral responses to emotions.

limbic: L. *limbus,* edge or border

Figure 17.12. The limbic system and the pathways that interconnect the structures of the limbic system. (Note: the left temporal lobe of the cerebral cortex has been removed.)

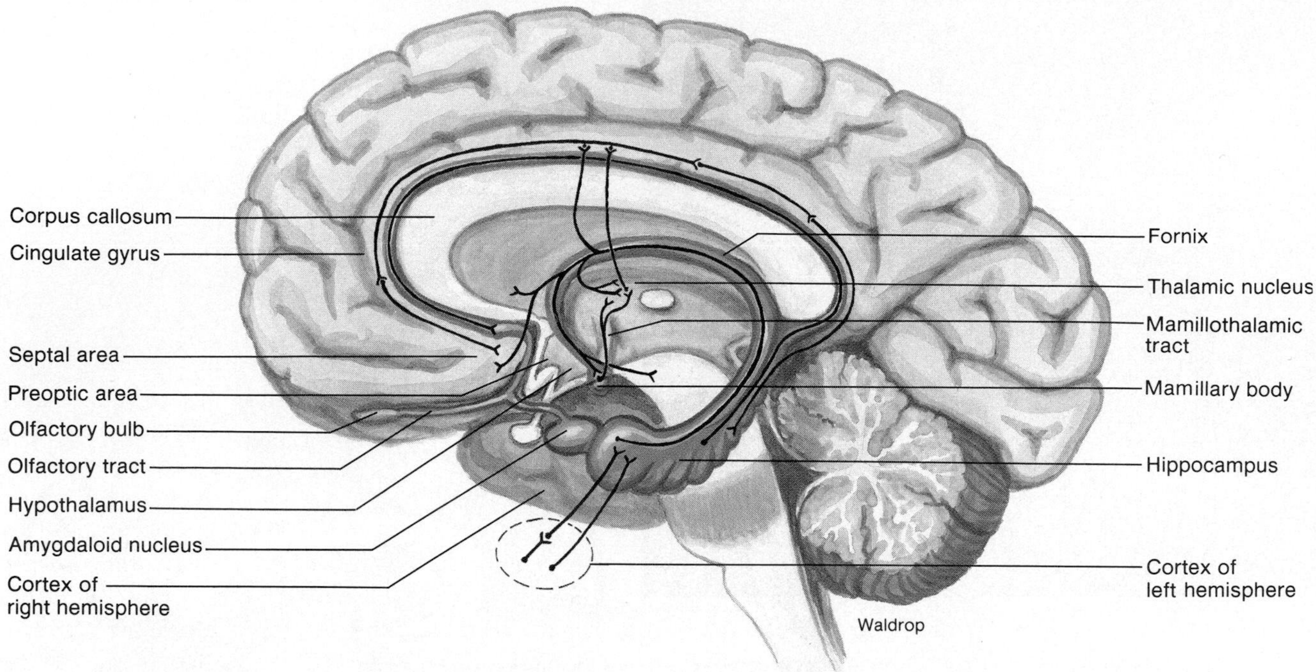

The autonomic correlates of motion sickness—nausea, sweating, and cardiovascular changes—are abolished by cutting the efferent tracts of the cerebellum. This demonstrates that impulses from the cerebellum to the medulla influence activity of the autonomic system. Experimental and clinical observations have also demonstrated that the frontal and temporal lobes of the cerebral cortex influence lower brain areas as part of their involvement in emotion and personality.

One of the most dramatic examples of the role of higher brain areas in personality and emotion is the famous crowbar accident, which occurred in 1848. A twenty-five-year-old railroad foreman, Phineas P. Gage, was tamping gunpowder into a hole in a rock when it exploded. The rod—three feet, seven inches long and one and one-fourth inches thick—was driven through his left eye, passed through his brain, and emerged in the back of his skull.

After a few minutes of convulsions, Gage got up, rode a horse three-quarters of a mile into town, and walked up a long flight of stairs to see a doctor. He recovered well, with no noticeable sensory or motor deficits. His associates, however, noted striking personality changes. Before the accident Gage was a responsible, capable, and financially prudent man. Afterwards, he appeared to lose social inhibitions, engaged in gross profanity (which he did not do before the accident), and seemed to be tossed about by chance whims. He was eventually fired from his job, and his previous friends remarked that he was "no longer Gage."

1. Describe the role of the medulla oblongata in the regulation of the autonomic system.
2. Describe the role of the hypothalamus in the regulation of the autonomic and endocrine systems.
3. Explain, in terms of the neural pathways involved, how emotional states can be associated with particular autonomic responses (such as blushing).

Clinical Considerations

Autonomic Dysreflexia

Autonomic dysreflexia, a serious condition producing rapid elevations in blood pressure that can lead to stroke (cerebrovascular accident), occurs in 85% of people with quadriplegia and others with spinal cord lesions above the sixth thoracic level. Lesions to the spinal cord first produce the symptoms of spinal shock, characterized by the loss of both skeletal muscle and autonomic reflexes. After a period of time both types of reflexes return in an exaggerated state; the skeletal muscles may become spastic, due to absence of higher inhibitory influences, and the visceral organs experience denervation hypersensitivity. Patients in this state have difficulty emptying their urinary bladders and must often be catheterized.

Noxious stimuli, such as overdistension of the urinary bladder, can result in reflex activation of the sympathetic nerves below the spinal cord lesion. This produces goose bumps, cold skin, and vasoconstriction in the regions served by the spinal cord below the level of the lesion. The rise in blood pressure resulting from this vasoconstriction activates pressure receptors that transmit impulses along sensory nerve fibers to the medulla. In response to this sensory input, the medulla directs a reflex slowing of the heart and vasodilation. Since descending impulses are blocked by the spinal lesion, however, the skin is warm and moist (due to vasodilation and sweat gland secretion) above the lesion, but cold below the level of spinal cord damage.

Pharmacology of the Autonomic System

Epinephrine, norepinephrine, acetylcholine, and some chemicals that are not normally found in the body can either enhance or block the physiological effects of the autonomic system. Drugs that enhance a particular effect are called **agonists,** and those that block an effect are called **antagonists.** Drugs that affect the functions of the autonomic system can be further divided into those that enhance or block adrenergic or cholinergic effects.

Adrenergic Drugs. The use of β-blocking drugs, such as propranolol, to slow the heart, and α_1 agonist drugs to cause vasoconstriction in mucous membranes, has been previously discussed. Use of these drugs is easily understood, since stimulation of β-receptors in the heart increases its rate of beat, and stimulation of α_1 receptors in blood vessels causes vasoconstriction.

Although it may at first seem paradoxical, some people with hypertension are treated with an α_2 agonist drug—*clonidine.* Stimulation of the α_2 receptors, it may be recalled, results in the inhibition of cyclic AMP production and thus has just the opposite effect to that of stimulation of beta-adrenergic receptors. It is currently believed that presynaptic axon terminals in the brain contain α_2 receptors and that stimulation of these receptors inhibits the release of neurotransmitter from the nerve terminals; this may represent a negative feedback control of the amount of neurotransmitter released. It is known that clonidine, by stimulating α_2 receptors in the CNS, inhibits the central activation of the sympathetic nervous system and thus serves to decrease the heart rate and lower the blood pressure of hypertensive patients.

Cholinergic Drugs. Acetylcholine is used as a neurotransmitter by somatic motor neurons, preganglionic autonomic fibers, all postganglionic parasympathetic nerve fibers, and some sympathetic nerve fibers. Drugs that are similar in structure and action to acetylcholine (methacholine and bethanecholine) thus can promote the effects of these nerves. Drugs that inhibit the action of acetylcholinesterase (an enzyme that degrades ACh)—such as physostigmine and neostigmine—enhance the action of ACh. Because of the effects of these drugs on the neuromuscular junction, they are used in the treatment of *myasthenia gravis* and other muscular disorders. Some of these drugs are also used as *parasympathomimetics (par''ah-sim''pah-tho-mi-met'ik)*—drugs that duplicate the action of parasympathetic nerves—in the treatment of *glaucoma.* Atropine and related drugs block the action of ACh released by postganglionic parasympathetic neurons and thus serve to block the actions of the parasympathetic system.

Chapter Summary

I. The Autonomic Nervous System and Its Visceral Effectors
   A. Preganglionic autonomic neurons originate from the brain or spinal cord; postganglionic neurons originate from ganglia located outside the CNS.
   B. Smooth muscle, cardiac muscle, and glands receive autonomic innervation.
      1. The involuntary effectors are somewhat independent of their innervation and become hypersensitive when their innervation is removed.
      2. Myocardial cells are interconnected by electrical synapses, or gap junctions, to form a functional syncytium with independent pacemaker activity.
      3. Single-unit smooth muscles have gap junctions and pacemaker activity; multi-unit smooth muscles lack gap junctions and pacemaker activity.

II. Structure of the Autonomic System
   A. Preganglionic neurons of the sympathetic division originate in the spinal cord between the thoracic and lumbar levels.
      1. Many of these fibers synapse with postganglionic neurons whose cell bodies are located in a double chain of sympathetic (paravertebral) ganglia outside the spinal cord.
      2. Some preganglionic fibers synapse in collateral ganglia; these are the celiac, superior mesenteric, and the inferior mesenteric ganglia.
      3. Some preganglionic fibers innervate the adrenal medulla, which secretes epinephrine (and some norepinephrine) into the blood in response to this stimulation.
   B. Preganglionic parasympathetic fibers originate in the brain and in the sacral levels of the spinal cord.
      1. Preganglionic parasympathetic fibers contribute to cranial nerves III, VII, IX, and X.
      2. Preganglionic fibers of the vagus (X) nerve are very long and synapse in terminal ganglia located next to or within the innervated organ; short postganglionic fibers then innervate the effector cells.
      3. The vagus provides parasympathetic innervation to the heart, lungs, esophagus, stomach, liver, small intestine, and upper half of the large intestine.
      4. Parasympathetic outflow from the sacral levels of the spinal cord innervates terminal ganglia in the lower half of the large intestine, the rectum, and the urinary and reproductive systems.

III. Functions of the Autonomic Nervous System
   A. The sympathetic division of the autonomic system activates the body to "fight or flight" through adrenergic effects; the parasympathetic division often has antagonistic actions through cholinergic effects.

B. All preganglionic autonomic nerve fibers are cholinergic (use ACh as a neurotransmitter).
   1. All postganglionic parasympathetic fibers are cholinergic.
   2. Most postganglionic sympathetic fibers are adrenergic (use norepinephrine as a neurotransmitter).
   3. Sympathetic fibers that innervate sweat glands and those that innervate blood vessels in skeletal muscles are cholinergic.

C. Adrenergic effects include stimulation of the heart, vasoconstriction in the viscera and skin, bronchodilation, and glycogenolysis in the liver.
   1. There are two main groups of adrenergic receptor proteins: alpha and beta receptors.
   2. Different organs have only alpha or beta receptors, and some organs (such as the heart) have both receptors.
   3. There are two subtypes of alpha receptors (α_1 and α_2) and two subtypes of beta receptors (β_1 and β_2), which can be selectively stimulated or blocked by therapeutic drugs.

D. Cholinergic effects of parasympathetic nerves are promoted by the drug muscarine and inhibited by atropine.

E. In organs with dual innervation, the actions of the sympathetic and parasympathetic divisions can be antagonistic, complementary, or cooperative.
   1. The effects are antagonistic in the heart and pupils.
   2. The actions are complementary in the regulation of salivary gland secretion and are cooperative in the regulation of the reproductive and urinary systems.

F. In organs without dual innervation (such as most blood vessels), regulation is achieved by variations in sympathetic nerve activity.

IV. Control of the Autonomic System by Higher Brain Centers

A. The medulla oblongata of the brain stem is the area that most directly controls the activity of the autonomic system.
   1. The medulla oblongata is in turn influenced by sensory input and by input from the hypothalamus.
   2. The hypothalamus orchestrates somatic, autonomic, and endocrine responses during various behavioral states.

B. The activity of the hypothalamus is influenced by input from the limbic system, cerebellum, and cerebrum; these interconnections provide an autonomic component to changes in body position, emotion, and various expressions of personality.

Review Activities

Objective Questions

1. Which of the following statements about the superior mesenteric ganglion is *true*?
   (a) It is a parasympathetic ganglion.
   (b) It is a paravertebral sympathetic ganglion.
   (c) It is located in the head.
   (d) It contains postganglionic sympathetic neurons.
2. The pterygopalatine, ciliary, submandibular, and otic ganglia are
   (a) collateral sympathetic ganglia
   (b) cervical sympathetic ganglia
   (c) parasympathetic ganglia that receive fibers from the vagus
   (d) parasympathetic ganglia that receive fibers from the third, seventh, and ninth cranial nerves
3. Which of the following is *not* a result of parasympathetic nerve simulation?
   (a) increased movement of the digestive tract
   (b) increased mucus secretion
   (c) constriction of the pupils
   (d) constriction of visceral blood vessels
4. Which of the following statements about cardiac muscle is *true*?
   (a) It is striated.
   (b) It contracts automatically.
   (c) Its cells are joined by electrical synapses.
   (d) All of the above are true.
5. Which of the following types of smooth muscle is most like cardiac muscle in its electrical behavior?
   (a) single-unit smooth muscle
   (b) multi-unit smooth muscle
6. When a visceral organ is denervated,
   (a) it ceases to function
   (b) it becomes less sensitive to subsequent stimulation by neurotransmitters
   (c) it becomes hypersensitive to subsequent stimulation
7. The pancreas is both an exocrine and endocrine gland. If its duct is ligated (tied),
   (a) its endocrine secretion would be blocked
   (b) its exocrine secretion would be blocked
   (c) both its endocrine and exocrine secretions would be blocked
8. Parasympathetic ganglia are located
   (a) in a chain parallel to the spinal cord
   (b) in the dorsal roots of spinal nerves
   (c) next to or within the organs innervated
   (d) in the brain
9. The neurotransmitter of preganglionic sympathetic fibers is
   (a) norepinephrine
   (b) epinephrine
   (c) acetylcholine
   (d) dopamine
10. Which of the following results from stimulation of alpha-adrenergic receptors?
    (a) constriction of blood vessels
    (b) dilation of bronchioles
    (c) increased heart rate
    (d) sweat gland secretion
11. Which of the following fibers release norepinephrine?
    (a) preganglionic parasympathetic fibers
    (b) postganglionic parasympathetic fibers
    (c) postganglionic sympathetic fibers in the heart
    (d) postganglionic parasympathetic fibers in sweat glands
    (e) all of the above
12. The actions of sympathetic and parasympathetic fibers are cooperative in the
    (a) heart
    (b) reproductive system
    (c) digestive system
    (d) eyes
13. Propranolol is a "beta-blocker." It would therefore cause
    (a) vasodilation
    (b) slowing of the heart rate
    (c) increased blood pressure
    (d) secretion of saliva
14. Atropine blocks parasympathetic nerve effects. It would therefore cause
    (a) dilation of the pupils
    (b) decreased mucus secretion
    (c) decreased movements of the digestive tract
    (d) increased heart rate
    (e) all of the above
15. The area of the brain that is most directly involved in the reflex control of the autonomic system is the
    (a) hypothalamus
    (b) cerebral cortex
    (c) medulla oblongata
    (d) cerebellum

Essay Questions

1. Compare the sympathetic and parasympathetic systems in terms of the location of their ganglia and the distribution of their nerves.
2. Explain the anatomical and physiological relationship between the sympathetic nervous system and the adrenal glands.
3. Compare the effects of adrenergic and cholinergic stimulation on the cardiovascular and digestive systems.
4. Explain how effectors that receive only sympathetic innervation are regulated by the autonomic system.
5. Distinguish between the different types of adrenergic receptors, and explain how their differences are clinically exploited.

18

Sensory Organs

Outline

Overview of Sensory Perception
Classification of the Senses
Physiology of Sensory Receptors
Law of Specific Nerve Energies
Sensory Neurons and Receptor Potentials
Somatic Senses
Tactile and Pressure Receptors
Thermoreceptors
Pain Receptors
Proprioceptors
Receptive Fields and Sensory Acuity
Neural Pathways for Somatic Sensation
Olfactory Sense
Gustatory Sense
Visual Sense
Development of the Eye
Associated Structures of the Eye
Structure of the Eyeball
Physiology of Vision
Effect of Light on the Rods
Electrical Activity of Retinal Cells
Cones and Color Vision
Visual Acuity and Sensitivity
Neural Pathways of Vision, Eye Movements, and Processing Visual Information
Senses of Hearing and Balance
Development of the Ear
Structure of the Ear
Sound Waves and Neural Pathways for Hearing
Mechanics of Equilibrium
Clinical Considerations
Diagnosis of Sensory Organs
Developmental Problems of the Eyes and Ears
Functional Impairments of the Eye
Infections and Diseases of the Eye
Infections, Diseases, and Functional Impairments of the Ear
Chapter Summary
Review Activities

Concepts

Sensory organs are highly specialized extensions of the nervous system in that they contain sensory neurons adapted to respond to specific stimuli and conduct nerve impulses to the brain for interpretation.

The senses are classified as general or special according to the simplicity or complexity of the receptors and neural pathways. They are also classified as somatic or visceral according to the location of the receptors. A functional classification groups sensory receptors as phasic or tonic or according to the type of stimulus energy they transduce.

A sensory receptor responds to stimulus energy and initiates nerve impulses in a sensory neuron. Stimulation of a particular sensory receptor produces a particular modality of sensation.

Somatic senses include cutaneous receptors and proprioceptors. The perception of somatic sensations is determined by the density of the receptors in the stimulated receptive field and the intensity of the sensation. Accommodation, afterimage, and projection are also factors of somatic perception.

Olfactory receptors are the dendritic endings of the olfactory (first cranial) nerve that respond to chemical stimuli and transmit the sensation of olfaction directly to the olfactory portion of the cerebral cortex.

Taste receptors are specialized epithelial cells clustered together into taste buds that respond to chemical stimuli and transmit the sense of taste through the glossopharyngeal (ninth cranial) nerve or the facial (seventh cranial) nerve to the cortex of the parietal cerebral lobe for interpretation.

Rods and cones are the photoreceptors within the eyeball that are sensitive to light energy and are stimulated to transmit nerve impulses through the optic nerve and optic tract to the visual cortex of the occipital lobes, where the interpretation of vision occurs. The sensory components of the eye are formed by twenty weeks, and the accessory structures are formed by thirty-two weeks.

Structures of the outer, middle, and inner ear are involved in the sense of hearing. The inner ear also contains structures that provide a sense of balance, or equilibrium. The development of the ear begins during the fourth week and is completed by the thirty-second week.

Overview of Sensory Perception

Sensory organs are highly specialized extensions of the nervous system in that they contain sensory neurons adapted to respond to specific stimuli and conduct nerve impulses to the brain for interpretation.

Objective 1. Explain the requirements for perceiving a sensation.
Objective 2. Discuss the selectivity of sensory receptors for specific stimuli.

The sense organs are actually extensions of the nervous system that allow humans to perceive their internal and external environments. These sense organs have been described as "windows for the brain" because it is through them that awareness of the environment is possible. A stimulus must first be received before the sensation can be interpreted and the necessary body adjustments dictated by the central nervous system can be made. The sense organs permit daily experiences of pleasure and are vital to the daily survival of humans by allowing them to hear sounds, see dangers, taste undesirable foods, and perceive pain, hunger pangs, or intestinal aches.

A **sensation** *(sen-sa'shun)* is the arrival of a sensory impulse to the brain. The interpretation of a sensation is referred to as **perception** *(per-sep'shun)*. In other words, one feels, sees, hears, tastes, and smells with the brain. In order to perceive a sensation, the following four conditions are necessary.

1. A **stimulus** sufficient to initiate a response in the nervous system must be present.
2. A **receptor** must convert the stimulus to a nerve impulse. A receptor is a specialized, peripheral dendritic ending of an afferent nerve fiber or the specialized receptor cell associated with it.
3. The **conduction of the nerve impulse** must occur from the receptor to the brain along a nervous pathway.
4. The **interpretation of the impulse** in the form of a perception must occur within a specific portion of the brain.

Only impulses reaching the cerebral cortex of the brain are consciously interpreted as sensations. If impulses terminate in the spinal cord or brain stem, they initiate a reflexive motor activity response rather than a conscious sensation. Impulses reaching the higher brain centers travel through nerve fibers composing *sensory*, or *ascending, tracts.* Clusters of neuron cell bodies, called **nuclei,** are synaptic sites along sensory tracts within the CNS. The nuclei that sensory impulses pass through before reaching the cerebral cortex are located in the spinal cord, medulla oblongata, pons, and thalamus.

Through the use of scientific instruments, it is known that the senses act as energy filters that allow perception of only a narrow range of energy. Vision, for example, is limited to light waves in the visible spectrum. Other types of waves of the same type of energy as visible light, such as X rays, radio waves, and ultraviolet and infrared light, normally cannot excite the sensory receptors in the eyes. Because there is no such thing as cold—only varying degrees of heat—the perception of cold is entirely a function of the nervous system. The perception of cold has obvious survival value. Although filtered and distorted by the limitations of sensory functions, perceptions allow humans effective interactions with the environment.

1. Distinguish between sensation and perception.
2. List the four conditions necessary for perception, and identify which of the four must always involve consciousness in order for perception to occur.
3. Using examples, explain the statement that each of the senses acts as a filter, allowing the perception of only a narrow range of energy.

Classification of the Senses

The senses are classified as general or special according to the simplicity or complexity of the receptors and neural pathways. They are also classified as somatic or visceral according to the location of the receptors. A functional classification groups sensory receptors as phasic or tonic or according to the type of stimulus energy they transduce.

Objective 3. Compare and contrast somatic, visceral, and special senses.
Objective 4. Define and give examples of *exteroceptors, visceroceptors,* and *proprioceptors.*
Objective 5. Distinguish between phasic and tonic receptors.

Structurally, the sensory receptor can be the dendrites of sensory neurons, which are either free (such as those in the skin that mediate pain and temperature) or are encapsuled within non-neural structures, such as lamellated corpuscles, or pressure receptors in the skin (see fig. 18.5). Other receptors, including taste buds, photoreceptors in the eyes, and hair cells in the inner ears derive from epithelial cells that synapse with sensory dendrites.

The senses of the body can be classified as general or special according to the simplicity or complexity of the receptors and neural pathways involved. **General senses** are widespread through the body and are simple in structure. Examples are touch, pressure, cold, heat, and pain. **Special senses** are localized in complex receptor organs and have extensive neural pathways. Among the special senses are taste, smell, sight, hearing, and balance.

The senses can also be classified as somatic or visceral according to the location of the receptors. **Somatic senses** are those in which the receptors are localized within the body wall. These include the cutaneous receptors and those within muscles, tendons, and joints. **Visceral senses** are those in which the receptors are located within visceral organs. These classes of senses intersect in certain organs. Hearing, for example, is a special somatic sense, whereas pain from the digestive tract is a general visceral sense.

Senses are also classified according to the location of the receptors and the types of stimuli to which they respond. There are three basic kinds of receptors: exteroceptors, visceroceptors, and proprioceptors.

Exteroceptors. Exteroceptors *(eks''ter-o-sep'tor)* are located near the surface of the body where they respond to stimuli from the external environment. They include the following.

1. Cones and rods in the retina of the eye—*photoreceptors (fo''to-re-sep'tor).*
2. Hair cells in the organ of Corti within the inner ear—*pressure receptors.*
3. Olfactory receptors in the nasal epithelium of the nasal cavity—*chemoreceptors (ke''mo-re-sep'tor).*
4. Taste receptors on the tongue—*chemoreceptors.*
5. Skin receptors within the dermis—*tactile receptors* for touch; *pressure receptors,* or *mechanoreceptors (mek''ah-no-re-sep'tor),* for pressure; *thermoreceptors* for temperature; and *nocioreceptors* (pain receptors).

Pain receptors are stimulated by chemicals released from damaged tissue cells and thus are a type of chemoreceptor. Although there are specific pain receptors, nearly all types of receptors when intensely stimulated transmit impulses that are perceived as pain. Pain receptors exist throughout the body, but only those located within the skin are classified as exteroceptors.

Visceroceptors (Enteroceptors). As the name implies, visceroceptors *(vis''er-o-sep'tor)* produce sensations arising from the viscera, such as internal pain, hunger, thirst, fatigue, or nausea. Specialized visceroceptors located within the circulatory system are sensitive to changes in blood pressure (baroreceptors).

Proprioceptors. Proprioceptors *(pro''pre-o-sep'tor)* relay information about body position, equilibrium, and movement. They are located in the inner ear, joints, tendons, and muscles.

The three types of receptors may be further classified on the basis of sensory adaptation (accommodation). Some

somatic: Gk. *somatikos,* body
visceral: L. *viscera,* body organs
proprioceptor: L. *proprius,* one's own; *ceptus,* taken

Figure 18.1. Tonic receptors (*a*) continue to fire at a relatively constant rate as long as the stimulus is maintained. These produce slowly adapting sensations. Phasic receptors (*b*) respond with a burst of action potentials when the stimulus is first applied but quickly reduce their rate of firing when the stimulus is maintained. This produces rapidly adapting sensations.

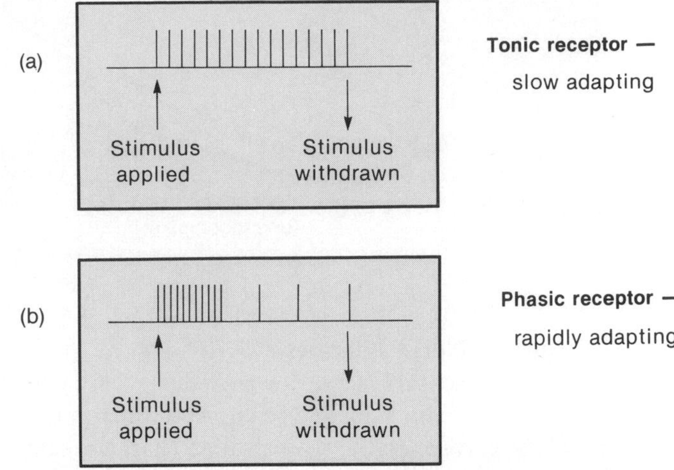

receptors respond with a burst of activity when a stimulus is first applied but quickly decrease their firing rate as they adapt to the stimulus even though the stimulus is maintained. Receptors with this response pattern are called **phasic receptors.** Receptors that produce a relatively constant rate of firing as long as the stimulus is maintained are known as **tonic receptors** (fig. 18.1).

Phasic receptors alert us to changes in sensory stimuli and are in part responsible for the fact that we can cease paying attention to constant stimuli. This ability is called *sensory adaptation.* Odor and touch, for example, adapt rapidly; bath water feels hotter when we first enter it. Sensations of pain, in contrast, adapt little if at all.

1. Using examples, explain how sensory receptors can be classified according to complexity, location, structure, and the type of stimuli to which they respond.
2. Describe sensory adaptation in olfactory and pain receptors. Using a line drawing, relate sensory adaptation to the responses of phasic and tonic receptors.

Physiology of Sensory Receptors

A sensory receptor responds to stimulus energy and initiates nerve impulses in a sensory neuron. Stimulation of a particular sensory receptor produces a particular modality of sensation.

Objective 6. Define the law of specific nerve energies.
Objective 7. Explain how action potentials are generated in sensory neurons in response to sensory stimuli.

Figure 18.2. The conduction of sensory impulses by a pseudounipolar neuron into the gray matter of the spinal cord.

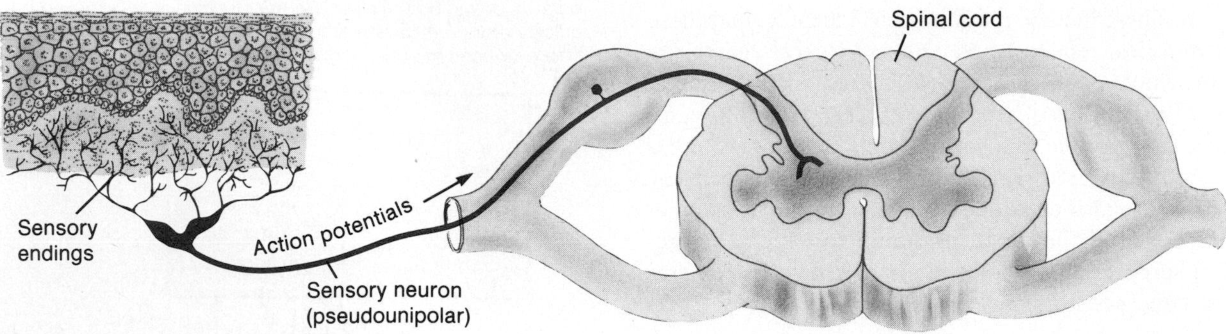

Law of Specific Nerve Energies

Stimulation of a sensory nerve fiber produces only one sensation—touch, cold, pain, and so on. According to the **law of specific nerve energies,** the sensation characteristic of each type of sensory neuron is that produced by its normal, or *adequate stimulus.* The adequate stimulus for the photoreceptors of the eye, for example, is light. If these receptors are stimulated by some other means—such as by pressure produced by a punch to the eye—a flash of light (the adequate stimulus) may be perceived.

Paradoxical cold provides another example of the law of specific nerve energies. First, a receptor for cold is located by touching the tip of a cold metal rod to the skin. Sensation gradually disappears as the rod warms to body temperature. Applying the tip of a rod heated to 45°C to the same spot, however, causes the sensation of cold to reappear. This paradoxical cold is produced because the heat slightly damages receptor endings, and by this means produces an "injury current" that stimulates the receptor.

Regardless of how a sensory neuron is stimulated, therefore, only one sensory modality will be perceived. This specificity is due to the synaptic pathways within the brain that are activated by the sensory neuron. The ability of receptors to function as sensory filters and be stimulated normally by only one type of stimulus (the adequate stimulus) usually allows the brain to perceive the stimulus accurately.

Sensory Neurons and Receptor Potentials

Sensory neurons are *pseudounipolar* (unipolar). Neurons of this type have a *peripheral branch,* comparable to a dendrite, and a *central branch,* comparable to the axon (fig. 18.2). Sensory neurons thus form a continuous pathway from the sensory endings, past the dorsal root ganglia (which contain the cell bodies of sensory neurons), to the gray matter of the spinal cord. Excluding the sensory endings, the entire length of a sensory neuron contains voltage-regulated gates that produce action potentials.

Figure 18.3. Sensory stimuli result in the production of local, graded potential changes known as the receptor, or generator potential (number 1 through 4). If the receptor potential reaches a threshold value of depolarization, it generates action potentials (number 5) in the sensory neuron.

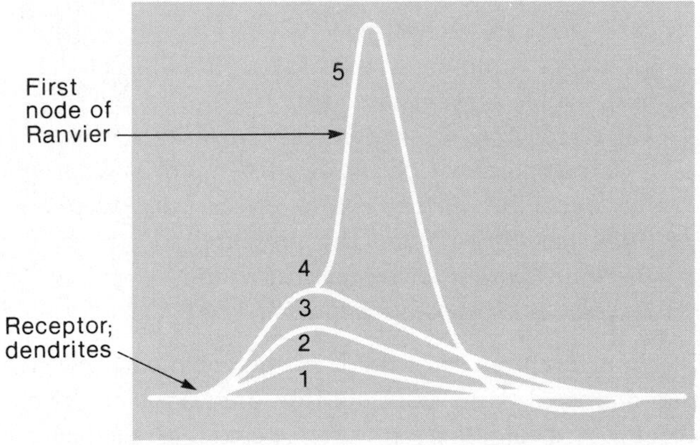

The electrical behavior of the sensory endings is similar to that of the dendrites of other neurons. Upon stimulation, dendrites exhibit a local, graded depolarization that increases in proportion to the stimulus intensity. This potential is analogous to an excitatory postsynaptic potential (EPSP). In sensory endings, however, it is known as a **receptor potential,** or a **generator potential,** because it stimulates opening of voltage-regulated gates and thus serves to generate action potentials (fig. 18.3) in response to sensory stimuli.

The lamellated (Pacinian) corpuscle, a cutaneous receptor for pressure (see fig. 18.5), can serve as an example. When a light touch is applied to the receptor, a small depolarization (the generator potential) is produced. Increasing the pressure on the lamellated corpuscle increases the magnitude of the generator potential until it reaches the threshold required to produce an action potential (fig. 18.4). The lamellated corpuscle, however,

corpuscle: L. *corpusculum,* diminutive of corpus, body

Figure 18.4. The response of a tonic receptor to stimuli. As the strength of stimulation is increased, the generator potential increases (*a, b, c*). The amplitude of the generator potential and the length of time it remains above threshold determine the frequency and duration of action potentials (*b, c*).

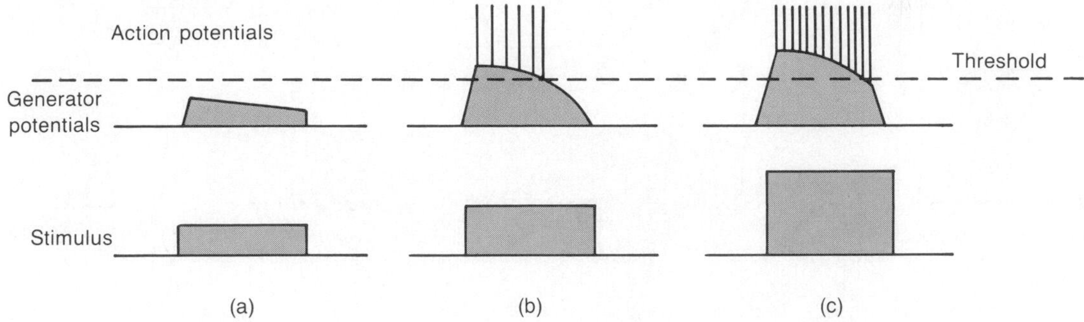

is a phasic receptor; if the pressure is maintained, the size of the generator potential produced quickly diminishes. It is interesting to note that this phasic response is a result of the onionlike covering around the dendritic nerve ending; if these layers are peeled off and the nerve ending is stimulated directly, it responds in a tonic fashion.

When a tonic receptor is stimulated, the generator potential it produces is proportional to the intensity of the stimulus. After a threshold depolarization is produced, increases in the amplitude of the generator potential result in increases in the *frequency* with which action potentials are produced (fig. 18.4). In this way, the frequency of action potentials that are conducted into the central nervous system serves to code for the strength of the stimulus. This frequency code is needed since the amplitude of action potentials is constant (all-or-none). Acting through changes in action potential frequency, tonic receptors thus provide information about the relative intensity of a stimulus.

1. Our perceptions are products of our brains; they are incompletely and inconstantly related to physical reality. Explain this statement, using examples of vision and perceptions of cold.
2. Describe how the magnitude of a sensory stimulus is transduced into a receptor potential and how the magnitude of the receptor potential is coded in the sensory nerve process.

Somatic Senses

Somatic senses include cutaneous receptors and proprioceptors. The perception of somatic sensations is determined by the density of the receptors in the stimulated receptive field and the intensity of the sensation. Accommodation, afterimage, and projection are also factors of somatic perception.

Objective 8. Describe the structure, function, and location of the various tactile receptors and the neural pathway for somatic sensation.
Objective 9. Explain the purpose of pain, the receptors that respond to pain, and the impulse pathway to the brain.
Objective 10. Define and give examples of *referred pain* and *phantom pain.*
Objective 11. Describe the physiological factors involved in the perception of somatic sensations.

The **somatic senses,** or **somasthetic senses,** include the receptors that are located in the skin and cause cutaneous sensations such as touch, tickling, pressure, cold, heat, and pain. The somatic senses also include proprioceptors that are located in the inner ear, joints, tendons, and muscles, and relay information about body position, equilibrium, and movement.

Tactile and Pressure Receptors

Both tactile receptors and pressure receptors are sensitive to mechanical forces that distort or displace the tissue in which they are located. **Tactile receptors** respond to *fine,* or *light,* touch and are located primarily in the dermis and hypodermis of the skin. **Pressure receptors** respond to *deep pressure* and are commonly found in the hypodermis of the skin and in the tendons and ligaments of joints. There are three kinds of tactile receptors and one kind of pressure receptor (fig. 18.5).

Meissner's Corpuscles. A Meissner's corpuscle *(kor' pus'l)* is an oval receptor composed of a mass of dendritic endings from two or three nerve fibers enclosed by connective tissue sheaths. These corpuscles are numerous in the hairless portions of the body, such as the eyelids, lips,

Meissner's corpuscle: from George Meissner, German histologist, 1829–1905

Figure 18.5. A diagramatic section of the skin showing the general location and magnified structure of cutaneous receptors.

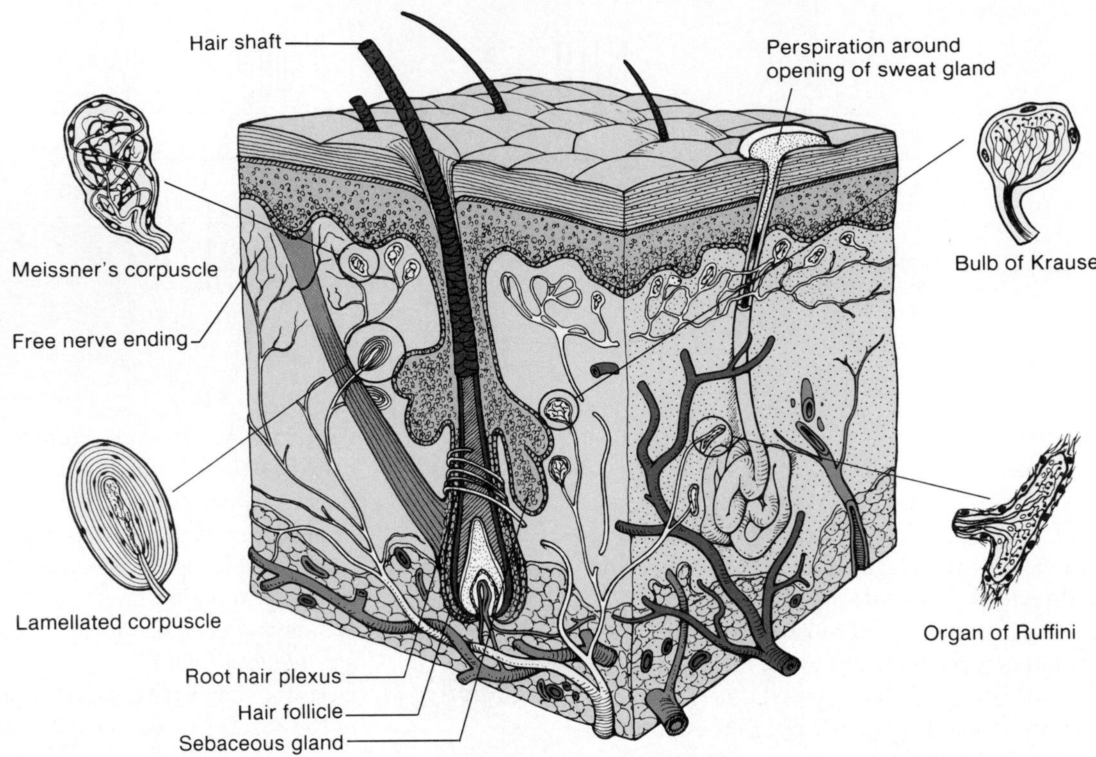

tip of the tongue, fingertips, palms of the hands, nipples, and external genital organs. Meissner's corpuscles lie within the papillary layer of the dermis, where they are especially sensitive to the motion of objects that barely contact the skin. Sensations of fine or light touch are perceived as these receptors are stimulated. They also function when a person touches an object to determine its texture.

Free Nerve Endings. Free nerve endings are the least modified and the most superficial of the tactile receptors. These receptors extend into the lower layers of the epidermis, where they end as knobs between the epithelial cells. Free nerve endings are most important as pain receptors, although they also respond to objects that are in continuous contact with the skin, such as clothing.

Hair Root Plexuses. Hair root plexuses are a specialized type of free nerve ending. They are coiled around hair follicles, where they respond to movement of the hair.

Lamellated (Pacinian) Corpuscles. Lamellated corpuscles are large, onion-shaped receptors composed of the dendritic endings of several sensory nerve fibers enclosed by connective tissue layers. They are commonly found within the synovial membranes of diarthrotic joints, in the perimysium of muscle tissue, and in certain visceral organs. Lamellated corpuscles are also abundant in the skin of the palms and fingers of the hand, soles of the feet, external genital organs, and breasts.

Lamellated corpuscles respond to heavy pressures, generally those that are constantly applied. They can also detect vibrations in tissues and organs.

Thermoreceptors

Thermoreceptors are widely distributed throughout the dermis of the skin but are especially abundant in the lips and the mucous membranes of the mouth and anal regions. There are two kinds of thermoreceptors—one that responds to heat and the other to cold (fig. 18.5).

pacinian corpuscle: from Filippo Pacini, Italian anatomist, 1812–1883

Table 18.1 Cutaneous receptors			
Type	**Location**	**Function**	**Sensation**
Meissner's corpuscles (mechanoreceptors)	Papillae of dermis; numerous in hairless portions of body (eyelids, fingertips, lips, nipples, external genitalia)	Detect light motion against surface of skin	Fine touch; texture
Free nerve endings (pressure receptors; pain receptors)	Lower layers of epidermis	Detect changes in pressure; detect tissue damage	Touch, pressure; pain
Hair root plexuses (tactile receptors)	Around hair follicles	Detect movement of hair	Touch
Lamellated (Pacinian) corpuscles (mechanoreceptors)	Hypodermis; synovial membranes; perimysium; certain visceral organs	Detect changes in pressure	Deep pressure; vibrations
Organs of Ruffini (thermoreceptors)	Lower layers of dermis	Detect changes in temperature	Heat
Bulbs of Krause (thermoreceptors)	Dermis	Detect changes in temperature	Cold

Organs of Ruffini. The organs of Ruffini are heat receptors located deep within the dermis. Heat receptors are elongated, oval structures that are most sensitive to temperatures from 25°C (77°F). Temperatures above 45°C (113°F) elicit impulses through the organs of Ruffini that are perceived as painful, burning sensations.

Bulbs of Krause. The bulbs of Krause are receptors for the sensation of cold. They are more abundant than heat receptors and are closer to the surface of the skin. The bulbs of Krause are most sensitive to temperatures below 20°C (68°F). Temperatures below 10°C elicit responses through the bulbs of Krause that are perceived as painful, freezing sensations.

Pain Receptors

The principal receptors for pain are the **free nerve endings.** Several million free nerve endings are distributed throughout the skin and internal tissues. Pain receptors are sparse in certain visceral organs and absent within the nervous tissue of the brain. Although the free nerve endings are specialized to respond to tissue damage, all of the cutaneous receptors will relay impulses that are interpreted as pain if stimulated excessively. The cutaneous receptors are summarized in table 18.1.

The protective value of pain receptors is obvious. Unlike other cutaneous receptors, free nerve endings have little accommodation, so impulses are relayed continuously to the CNS as long as the irritating stimulus is present. Although pain receptors can be activated by all types of stimuli, they are particularly sensitive to chemical stimulation. Muscle spasms, muscle fatigue, or an inadequate supply of blood to an organ may also cause pain.

organs of Ruffini: from Angelo Ruffini, Italian anatomist, 1864–1929

bulbs of Krause: from Wilhelm J. F. Krause, German anatomist, 1833–1910

Sensory impulses for pain are conducted to the spinal cord through afferent nerve fibers. Impulses are then conducted to the thalamus along the *lateral spinothalamic tract* of the spinal cord and from there to the somatesthetic area of the cerebral cortex. Although an awareness of pain occurs in the thalamus, the type and intensity of pain is interpreted in the specialized areas of the cerebral cortex.

The sensation of pain can be clinically classified as **somatic pain** and **visceral pain.** Stimulation of the cutaneous pain receptors results in the perception of superficial somatic pain. Deep somatic pain comes from stimulation of receptors in skeletal muscles, joints, and tendons.

Stimulation of the receptors within the viscera causes the perception of visceral pain. Through precise neural pathways, the brain is able to perceive the area of stimulation and project the pain sensation back to that area. The sensation of pain from certain visceral organs, however, may not be perceived as arising from those organs but from other somatic locations. This phenomenon is known as **referred pain** (fig. 18.6). The sensation of referred pain is consistent from one person to another and is clinically important in diagnosing organ dysfunctions. The pain of a heart attack, for example, may be perceived subcutaneously over the heart and down the medial side of the left arm. Ulcers of the stomach may cause pain that is perceived as coming from the upper central (epigastric) region of the torso. Pain from problems of the liver or gallbladder may be perceived as localized visceral pain or perceived as referred pain arising from the right neck and shoulder regions.

Referred pain is not totally understood but seems to be related to the development of the tracts within the spinal cord. It is thought that there are some *common nerve pathways* used by sensory impulses coming from both the cutaneous areas and from visceral organs (fig. 18.7). Consequently, impulses along these pathways may be incorrectly interpreted as arising cutaneously rather than from a visceral organ.

Figure 18.6. Sites of referred pain are perceived cutaneously but actually originate from specific visceral organs.

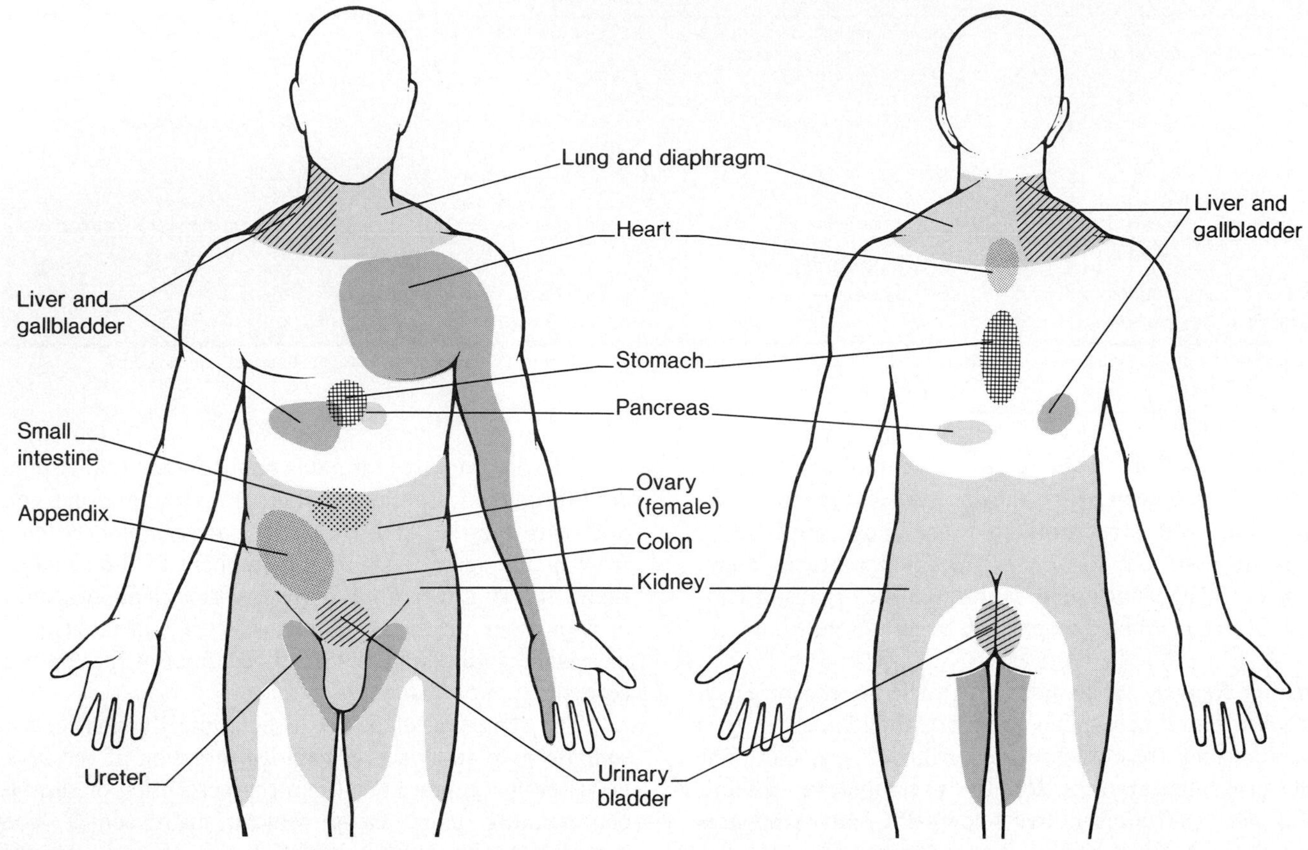

The perception of pain has survival value because it alerts the body to an injury, disease, or organ dysfunction. *Acute pain* is sudden, usually short-term, and can generally be endured and attributed to a known cause. *Chronic pain,* however, is long-term and tends to weaken a person as it interferes with the ability to function effectively. Certain diseases, such as arthritis, are characterized by chronic pain. In these patients, relief of pain is of paramount concern. Treatment of chronic pain often requires the use of moderate pain-reducing drugs (analgesics) or intense narcotic drugs. Treatment in severely tormented chronic pain patients may include severing sensory nerves or implanting stimulating electrodes in appropriate nerve tracts.

Phantom pain is frequently experienced by an amputee who continues to feel pain from the body part that was amputated. After amputation, the severed sensory neurons heal and function in the remaining portion of the appendage. Although it is not known why impulses that are interpreted as pain are sent periodically through these neurons, the sensations evoked in the brain are interpreted as arising from their normal source, the amputated region, thus resulting in phantom pain.

Proprioceptors

Proprioceptors monitor internal conditions, frequently those under voluntary control. They are located within skeletal muscle tissue, tendons, and the synovial membranes and connective tissue surrounding joints. Proprioceptors are sensitive to changes in stretch and tension. Some of the sensory impulses from proprioceptors reach the level of consciousness as the **kinesthetic sense,** by which the position of the body parts is perceived. Other proprioceptor information is not consciously interpreted and is used to adjust the intensity and timing of muscle contractions to provide coordinated movements. With the kinesthetic sense, we can determine the position and movement of our limbs without visual sensations, such as when we dress or walk in the dark. The kinesthetic sense, along with hearing, becomes keenly developed in a blind person.

High-speed transmission is a vital characteristic of the kinesthetic sense, since rapid feedback to various body parts is essential for quick, smooth, coordinated body movements.

There are three types of proprioceptors: joint kinesthetic receptors, neuromuscular spindles, and neurotendinous receptors.

Figure 18.7. An explanation of referred pain. Pain originating from the myocardium of the heart may be perceived as coming from the skin of the left arm because sensory impulses from these two organs are conducted through common nerve pathways to the brain.

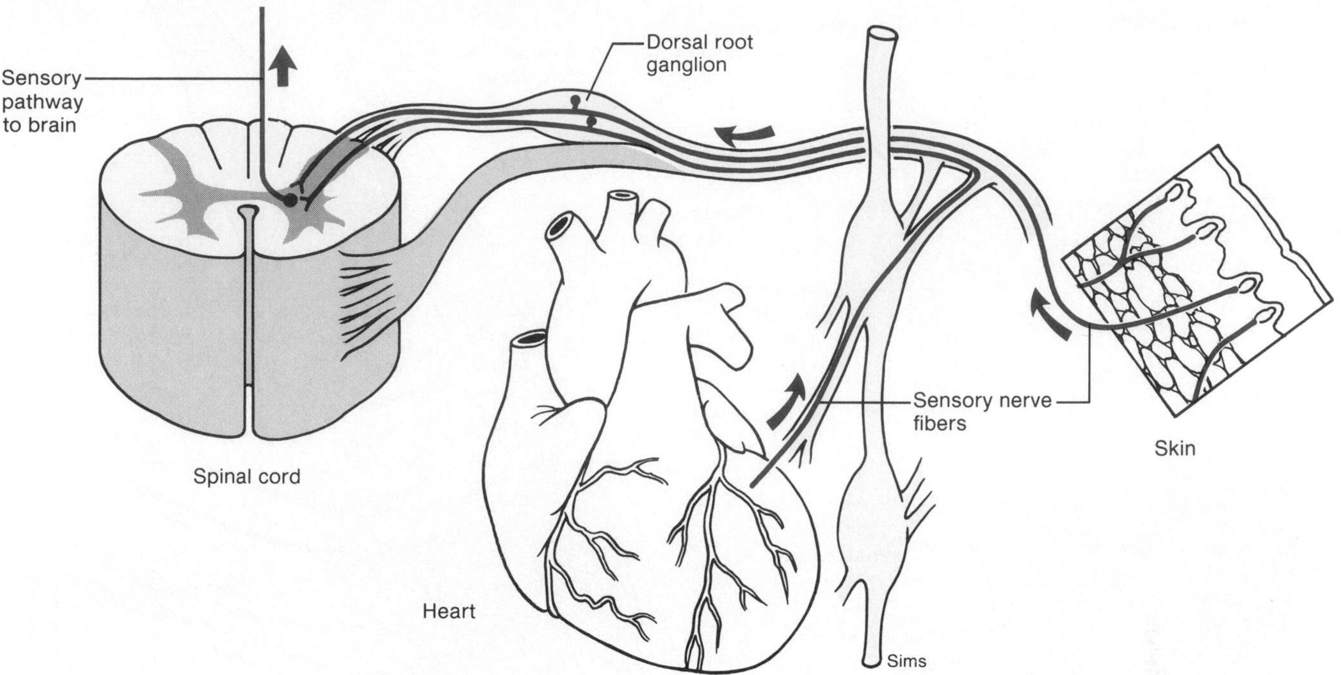

Joint kinesthetic receptors are located in the connective tissue capsule in diarthrotic joints, where they are stimulated by changes in position caused by movement at the joints. The other proprioceptors are more complex in their actions and will be described separately.

Neuromuscular Spindles. Each neuromuscular spindle, or spindle apparatus, contains several thin muscle cells packaged within a connective tissue sheath. Like the stronger and more numerous "ordinary" muscle fibers the spindles insert into tendons on each end of the muscle. Spindles are therefore said to be in parallel with the ordinary fibers.

There are two types of spindle fibers. One type, the *nuclear bag fibers,* has nuclei arranged in a loose aggregate in the central regions of the fibers. The other type of fibers has nuclei arranged in rows and are called *nuclear chain fibers.* There are likewise two types of sensory nerve endings that serve these spindle fibers. **Primary,** or **annulospiral, endings** wrap around the central regions of the nuclear bag and chain fibers (fig. 18.8), and **secondary,** or **flower-spray, endings** are located near the ends of the nuclear chain fibers.

Since the spindles are arranged in parallel with the muscle fibers, stretching a muscle causes its spindles to stretch. This stimulates both the primary and secondary sensory endings. The primary endings, however, are most stimulated at the onset of stretch, whereas the secondary endings respond in a more tonic fashion as stretch is maintained. Sudden, rapid stretching of a muscle activates both types of sensory endings and is thus a more powerful stimulus for the muscle spindles than slower, more gradual stretching. Since the activation of the sensory endings in muscle spindles produces a reflex contraction (the stretch reflex, fig. 16.21), the force of this reflex contraction is greater in response to rapid stretch than to gradual stretch.

Neurotendinous Receptors. The neurotendinous receptors (Golgi tendon organs) (fig. 18.8) continuously monitor tension in the tendons that is produced by muscle contraction or passive stretching of a muscle. Sensory neurons from these receptors synapse with interneurons in the spinal cord; these interneurons, in turn, have *inhibitory synapses* (via IPSPs and postsynaptic inhibition) with motor neurons that innervate the muscle. This helps prevent excessive muscle contractions or excessive passive muscle stretching. Indeed, if a muscle is stretched extensively, it will actually relax as a result of the inhibitory effects of the neurotendinous receptors.

Figure 18.8. (*a*) the relationship of a muscle spindle and neurotendinous receptor (Golgi tendon organ) to a skeletal muscle. (*b*) enlarged view of a muscle spindle.

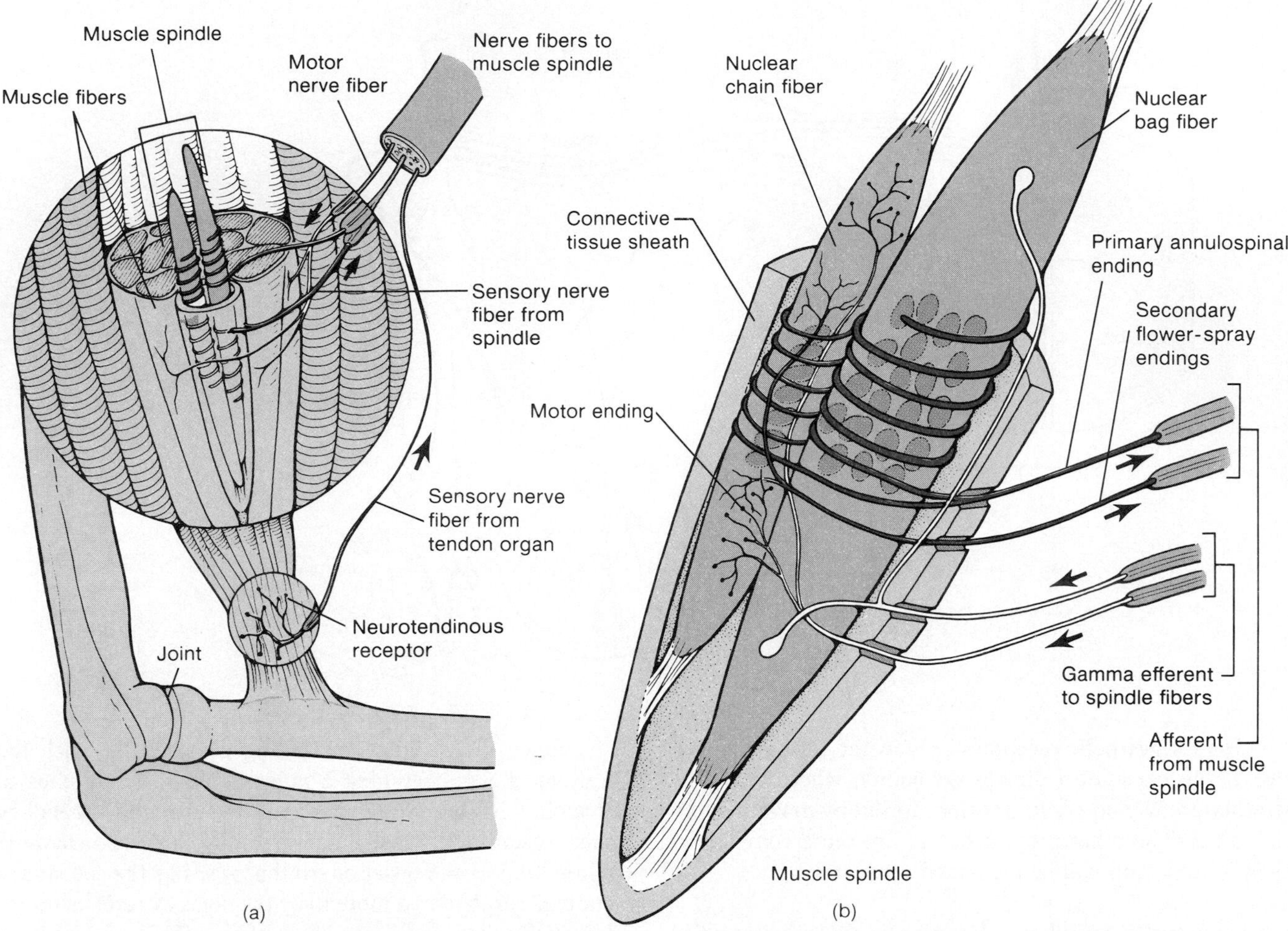

Rapid stretching of skeletal muscles produces very forceful muscle contractions as a result of the activation of primary and secondary endings in the muscle spindles and the monosynaptic stretch reflex. This can result in painful muscle spasms, as may occur, for example, when muscles are forcefully pulled in the process of setting broken bones. Painful muscle spasms may be avoided in physical exercise by stretching slowly and thereby stimulating mainly the secondary endings in the muscle spindles. A slower rate of stretch also provides time for the inhibitory neurotendinous receptor reflex to occur and promote muscle relaxation.

Receptive Fields and Sensory Acuity

The *receptive field* of a neuron serving cutaneous sensation is the area of the skin whose stimulation results in changes in the firing rate of the neuron. The area of each receptive field in the skin varies inversely with the *density of the receptors* in the region. Cutaneous receptors are abundant in parts of the face, palms and fingers of the hands, soles of the feet, and the genitalia. They are less abundant in the skin of the torso, particularly along the back and on the back of the neck, and are sparse in the skin over joints, especially the elbow. Generally speaking, the thinner the skin, the greater the sensitivity. The *intensity of a sensation* depends not only on the density of the receptors but also on the number of receptors stimulated in a given amount of time.

The high tactile acuity of the fingertips is exploited in the reading of Braille. *Braille symbols* consist of dots that are raised 1 mm from the page and separated from each other by 2.5 mm. Experienced Braille readers can scan words at about the same speed that a sighted person can read aloud—a rate of about 100 words per minute.

When a blunt object touches the skin, a number of receptive fields may be stimulated. Those receptive fields in the center areas where the touch is strongest will be stimulated more than in neighboring fields where the touch is lighter. A sensation of light touch surrounding a center

Braille: from Louis Braille, French teacher of the blind, 1809–1852

Figure 18.9. When an object touches the skin (*a*), receptors in the center of the touched skin are stimulated more than neighboring receptors (*b*). As a result of lateral inhibition within the central nervous system (*c*), input from these neighboring sensory neurons is reduced. Sensation, as a result, is more sharply localized to the area of skin that was stimulated the most (*d*).

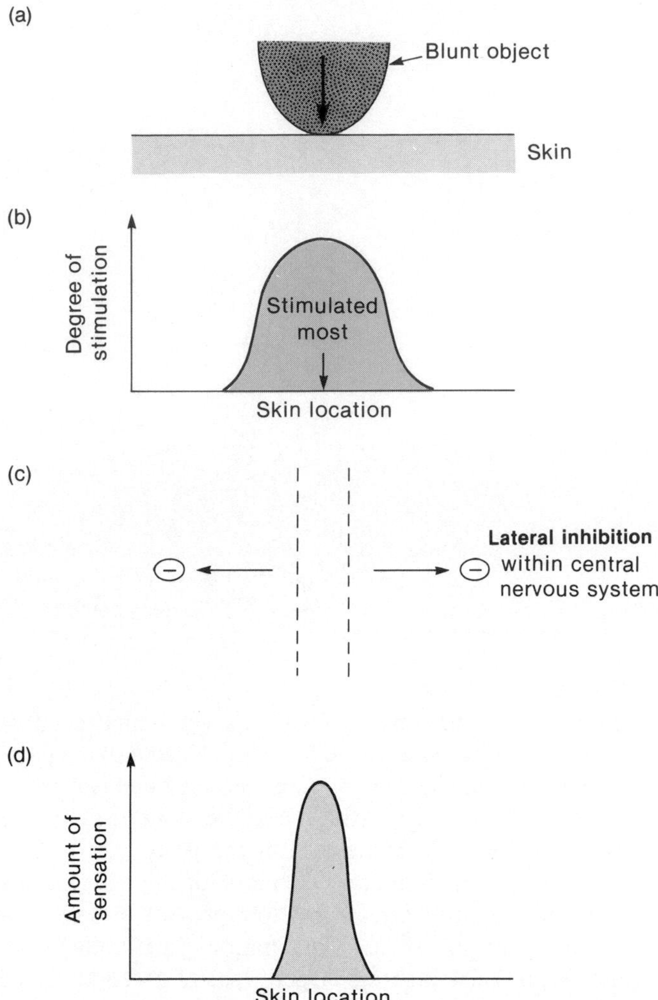

of stronger touch is, however, not perceived. Instead, only a single touch is felt, which is somewhat sharper than the actual shape of the blunt object. This sharpening of sensation is due to a process called *lateral inhibition* (fig. 18.9).

Lateral inhibition and the sharpening of sensation that results from it occur within the central nervous system. Those sensory neurons whose receptive fields are stimulated most strongly inhibit—by means of interneurons that pass "laterally" within the CNS—sensory neurons that serve neighboring receptive fields. Lateral inhibition similarly plays a prominent role in pitch discrimination.

When the intensity of a given stimulus does not change once the sensation has been perceived in the CNS, *accommodation* may occur. This means that the sensation is not perceived continuously at the same level even though the stimulus is continuous. Accommodation occurs, for example, when a person gets into a tub of hot water and perceives a burning sensation. As the sensory neurons accommodate, the stimulus becomes one of comfortable warmth, even though the temperature of the water has not changed.

Afterimage is the reverse of accommodation, such as when a sensation continues to be perceived even though the stimulus is no longer present. The sensation of a hat on the head after it has been removed is an example of afterimage.

As the brain perceives a sensation, it interprets or identifies the site from which the stimulus is being applied. This process is called *projection* because the brain projects the sensation back to the apparent source of the sensation. Because of projection, one attributes the function of sight to the eyes, hearing to the ears, feeling to the skin, and so forth, when actually these perceptions are functions of the brain.

One method of determining sensitivity of the skin is the *two-point touch threshold test* (fig. 18.10). In this test, the points of two pins are applied to the skin. The distance between the points of stimulation is then consistently narrowed until the minimum distance is found at which two distinct sites of stimulation can be distinguished. For example, the receptors in the tip of the tongue can discriminate two points of stimulation at a minimum distance of 1.4 mm (0.06 in.). If the test is performed on the back of the neck, the two points are perceived until a minimum distance of 36.2 mm (1.43 in.). At a distance less than this, a single point of stimulation is perceived. The two-point touch threshold is thus an indication of *tactile acuity,* or the sharpness of touch perception. Table 18.2 indicates the two-point threshold for different regions of the body.

Neural Pathways for Somatic Sensation

The conduction pathways for the somatic senses are shown in figure 18.11. Proprioception and pressure are carried by large, myelinated nerve fibers that ascend in the dorsal columns of the spinal cord on the ipsilateral (same) side. These fibers do not synapse until they reach the medulla oblongata of the brain stem; fibers that carry these sensations from the feet are thus incredibly long. After synapsing in the medulla with other, second-order sensory neurons, information in the latter neurons crosses over to the contralateral (opposite) side as it ascends via a fiber tract, called the **medial lemniscus,** to the thalamus. Third-order sensory neurons in the thalamus that receive this input in turn project to the **postcentral gyrus.**

acuity: L. *acuo,* sharpen

Figure 18.10. The two-point touch threshold test. If each point touches the receptive fields of different sensory neurons, two separate points of touch will be felt. If both caliper points touch the receptive field of one sensory neuron, only one point of touch will be felt.

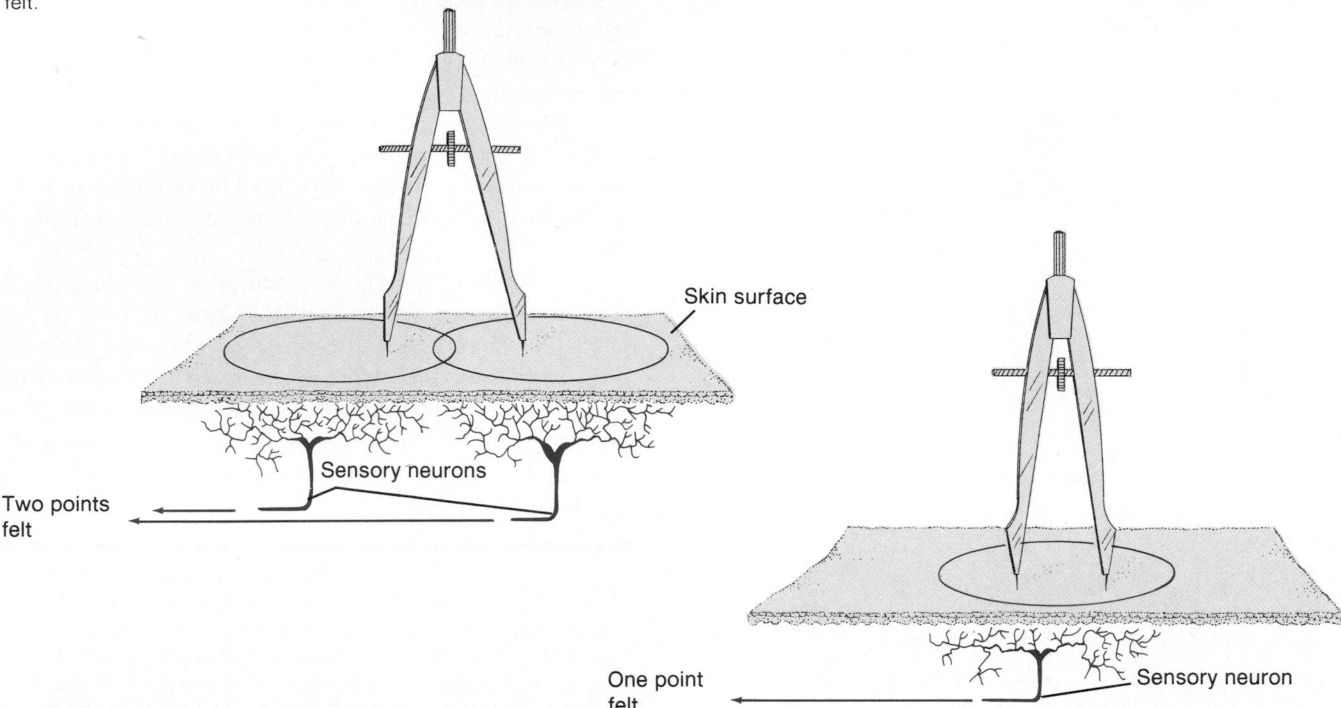

Table 18.2 The two-point touch threshold for different regions of the body

Body region	Two-point touch threshold (mm)
Big toe	10
Sole of foot	22
Calf	48
Thigh	46
Back	42
Abdomen	36
Upper arm	47
Forehead	18
Palm of hand	13
Thumb	3
First finger	2

From Dan R. Kenshalo, *The Skin Senses,* Copyright © 1968. Courtesy of Charles C Thomas, Publisher, Springfield, Illinois.

Sensations of hot, cold, and pain are carried by thin, unmyelinated sensory neurons into the spinal cord. These synapse within the spinal cord with second-order interneurons, which cross over to the contralateral side and ascend to the brain in the **lateral spinothalamic tract.** Fibers that mediate touch and pressure ascend in the **ventral spinothalamic tract.** Fibers of both spinothalamic tracts synapse in the thalamus with third-order neurons, which in turn project to the postcentral gyrus of the parietal cerebral lobe. Note that in all cases, somatic information is carried to the postcentral gyrus in third-order neurons. Also, because of crossing over, somatic information from each side of the body is projected to the postcentral gyrus of the contralateral cerebral hemisphere.

All somatic information from the same area of the body projects to the same area of the postcentral gyrus, so that a *sensory homunculus* "map" of the body can be drawn on the postcentral gyrus to represent sensory projection points (fig. 18.12). This map is greatly distorted, however, because it shows larger areas of cerebral cortex devoted to sensation in the face and hands than in other areas of the body. This disproportionately larger area of the cortex devoted to the face and hands reflects the fact that there is a higher density of sensory receptors in these regions.

1. List the types of tactile receptors, and state where they are located. What portion of the brain interprets tactile sensations?
2. Discuss the importance of pain. List the receptors that respond to pain and the structures of the brain that are particularly important in the perception of pain sensation.
3. Using examples, distinguish between referred pain and phantom pain. Discuss why it is important for a physician to know the referred pain sites.
4. Describe the meaning of sensory acuity, and explain how acuity is affected by the density of receptive fields and by the process of lateral inhibition.
5. Using a flow diagram, describe the neural pathways leading from cutaneous pain and pressure receptors to the postcentral gyrus. Indicate where crossing over occurs.

Figure 18.11. Pathways that lead from the cutaneous receptors and proprioreceptors into the postcentral gyrus in the cerebral cortex.

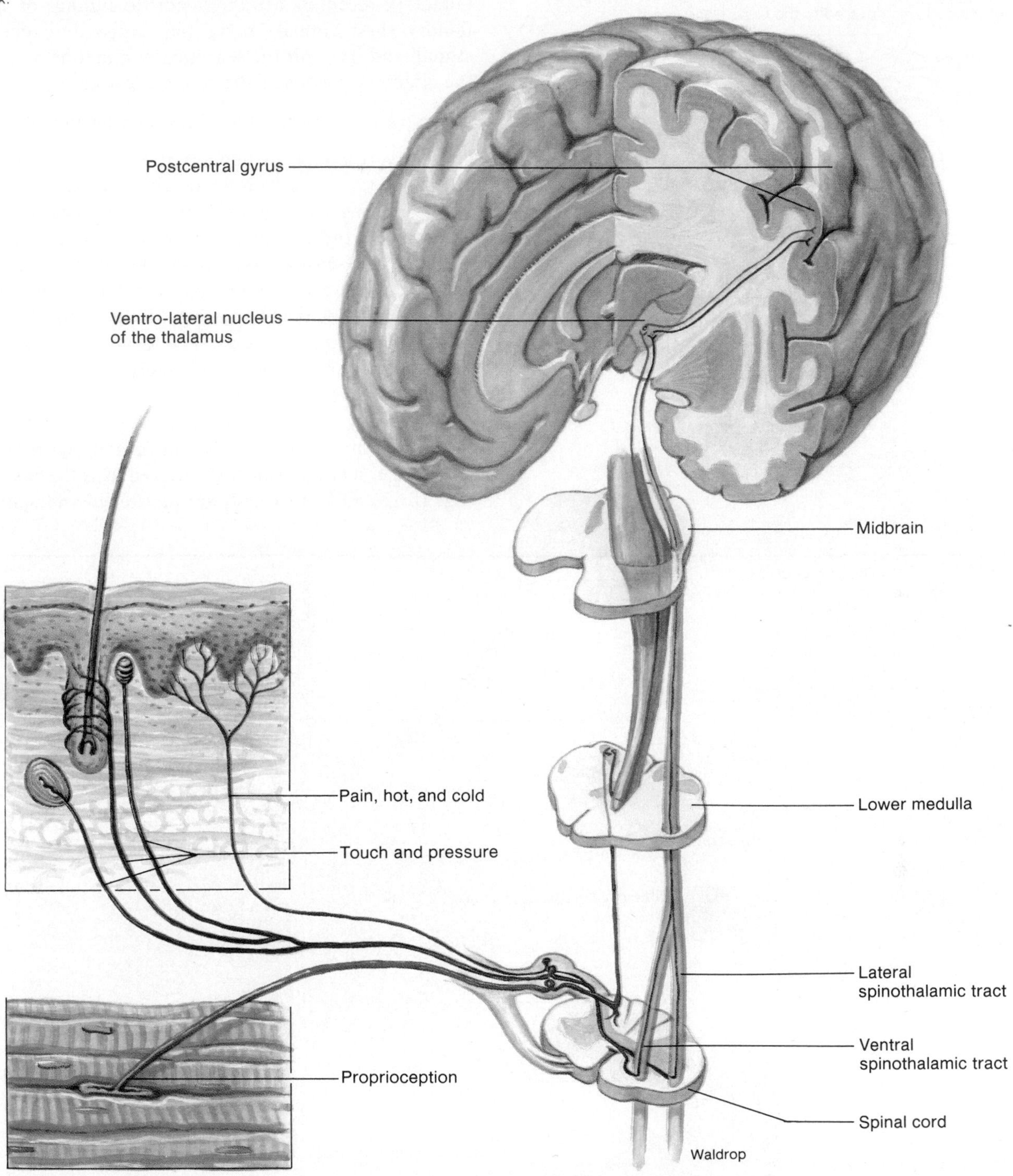

Figure 18.12. A sensory homunculus map depicts the areas of the postcentral gyrus of the cerebral cortex devoted to sensation in different parts of the body. Notice that a disproportionately large area of the cortex is devoted to the fingers and lips.

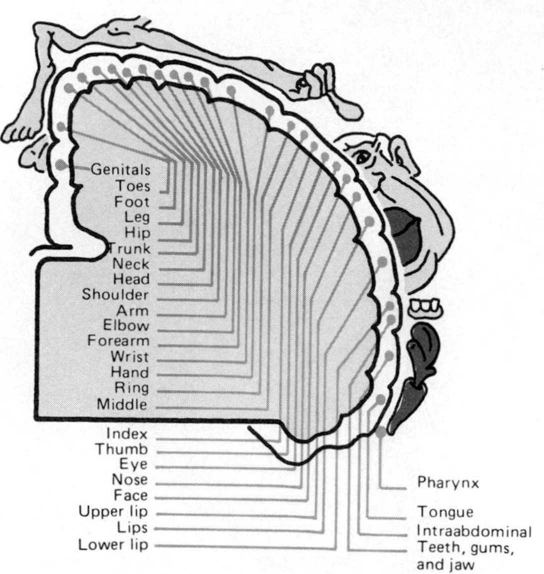

Olfactory Sense

Olfactory receptors are the dendritic endings of the olfactory (first cranial) nerve that respond to chemical stimuli and transmit the sensation of olfaction directly to the olfactory portion of the cerebral cortex.

Objective 12. Describe the afferent pathway of olfaction.

Olfactory reception in humans is not highly developed compared to that of certain other vertebrates. Because humans do not rely on smell for communicating or for finding food, the olfactory sense is probably the least important of the senses; it is more important in detecting the presence of an odor rather than its intensity. Accommodation occurs relatively rapidly with this sense. Olfaction *(ol-fak'shun)* functions closely with gustation (taste) in that the receptors of both are chemoreceptors and require dissolved substances for stimuli.

Olfactory receptor cells are located in the nasal epithelium within the roof of the nasal cavity on both sides of the nasal septum in an area referred to as the **nasal cleft** (fig. 18.13). Olfactory cells are moistened and stabilized

Figure 18.13. The olfactory receptive area. (*a*) a frontal section through the nose showing the relationship of the nasal cavity and the nasal septum to the olfactory sensory structures within the nasal cleft. (*b*) a sagittal section of the forebrain and the nasal cavity showing the location of olfactory receptors.

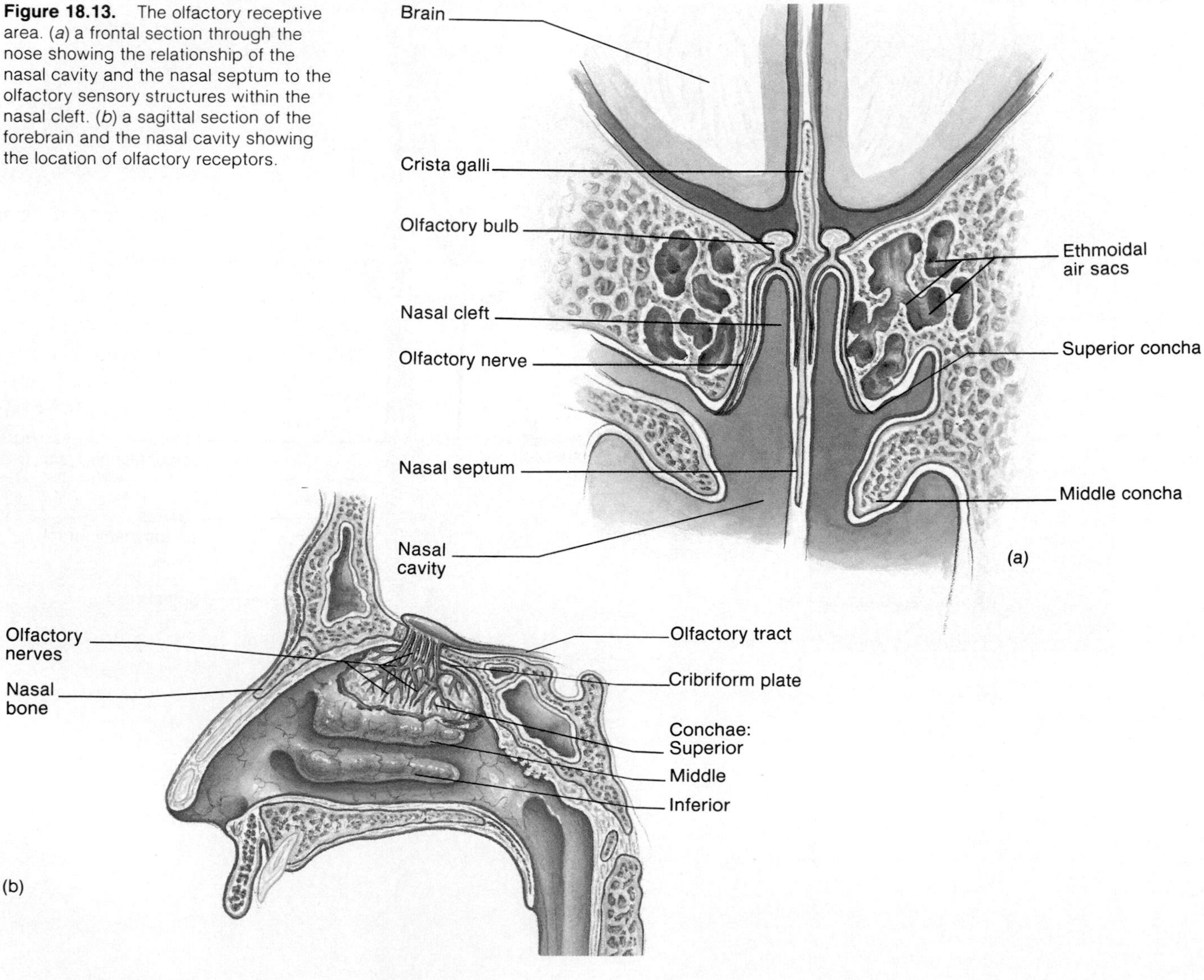

by the surrounding glandular goblet cells and the supporting columnar epithelial cells. The olfactory cells are actually bipolar neurons whose cell bodies lie between the supporting columnar cells. The free end of each olfactory cell contains several dendritic endings called **olfactory hairs,** which are the sensitive portion of the receptor cell.

The afferent pathway for olfaction consists of several neural segments. The unmyelinated axons of the olfactory cells unite to form the **olfactory nerves,** which traverse the foramina of the cribriform plate and terminate in the paired masses of gray and white matter called the **olfactory bulbs.** The olfactory bulbs lie on both sides of the crista galli of the ethmoid bone beneath the frontal lobes of the cerebrum. The neurons of the olfactory nerves synapse within the olfactory bulb with dendrites of the **olfactory tract.** Sensory impulses are conveyed along the olfactory tract and into the olfactory portion of the cerebral cortex, where they are interpreted as odor and cause the sensation of smell.

Unlike taste, which is divisible into only four modalities, thousands of distinct odors can be distinguished by people who are trained in this capacity (as in the perfume industry). The molecular basis of olfaction is not understood, but it is known that a single odorant molecule is sufficient to excite an olfactory receptor.

Only about 2% of inhaled air comes in contact with the olfactory receptors, which are positioned above the main stream of air flow. Olfactory sensitivity can be increased by forceful sniffing, which draws the air into contact with the receptors.

Certain chemicals activate the trigeminal (fifth) as well as the olfactory (first) cranial nerves and cause reactions. Pepper, for example, may cause sneezing; onions cause the eyes to water; and smelling salts (ammonium salts) initiate respiratory reflexes and are used to revive unconscious persons.

Gustatory Sense

Taste receptors are specialized epithelial cells clustered together into taste buds that respond to chemical stimuli and transmit the sense of taste through the glossopharyngeal (ninth cranial) nerve or the facial (seventh cranial) nerve to the cortex of the parietal cerebral lobe for interpretation.

Objective 13. List the three types of papillae, and explain how they function in the perception of taste.

Objective 14. Identify the cranial nerves and the afferent pathways of gustation.

The receptors of taste are located in the **taste buds.** Taste buds are specialized sensory organs that are most numerous on the dorsum of the tongue but are also present on the soft palate and on the walls of the oropharynx. The cylindrical taste bud is composed of many sensory **gustatory cells** that are encapsuled by **supporting cells** (fig. 18.14). Each gustatory cell contains a dendritic ending called a **gustatory hair** that projects to the surface through an opening in the taste bud called the **taste pore.** The gustatory hairs are the sensitive portion of the receptor cells. Saliva provides a moistened environment necessary for a chemical stimulus to activate the gustatory cells.

Taste buds are elevated by surrounding connective tissue and epithelium to form *papillae (pah-pil'ah)* (fig. 18.14). Three types of papillae can be identified.

1. **Circumvallate** *(ser''kum-val'āt)* **papillae.** The largest but least numerous are the circumvallate papillae, which are arranged in an inverted V-shape pattern on the back of the tongue.
2. **Fungiform papillae.** Knoblike fungiform papillae are present on the tip and sides of the tongue.
3. **Filiform papillae.** Short, thickened, threadlike filiform papillae are located on the anterior two-thirds of the tongue.

There are only four basic modalities of taste, known as *gustation (gus-ta'shun),* which are sensed most acutely on particular regions of the tongue (fig. 18.15). These are *sweet* (tip of tongue), *sour* (sides of tongue), *bitter* (back of tongue), and *salty* (over most of the tongue but concentrated on the sides).

Sour taste is produced by hydrogen ions (H^+); all acids therefore taste sour. Most organic molecules, particularly sugars, taste sweet to varying degrees. Only pure table salt ($NaCl$) has a pure salty taste. Other salts, such as KCl (commonly used in place of $NaCl$ by people with hypertension), taste salty but have bitter overtones. Bitter taste is evoked by quinine and seemingly unrelated molecules.

The afferent pathway of taste receptors to the brain involves mainly two cranial nerves (fig. 18.16). Taste buds on the posterior one-third of the tongue have a sensory pathway through the glossopharyngeal nerve, whereas the anterior two-thirds of the tongue is served by the chorda tympani branch of the facial nerve.

Taste sensations passing through the nerves just mentioned are conveyed through the medulla and thalamus to the parietal lobe of the cerebral cortex, where they are interpreted.

gustatory: L. *gustare,* to taste
papilla: L. *papilla,* nipple

Figure 18.14. Papillae of the tongue and associated taste buds. (*a*) numerous taste buds are positioned within each papilla. (*b*) each gustatory cell and its associated gustatory hair is encapsuled by supporting cells. (*c*) a photomicrograph of some taste buds.

Papillae

(a)

Taste buds

Sensory nerve fiber

Connective tissue

Gustatory cell

Supporting cell

Taste pore

Gustatory hair

Epithelium of tongue

(b)

(c)

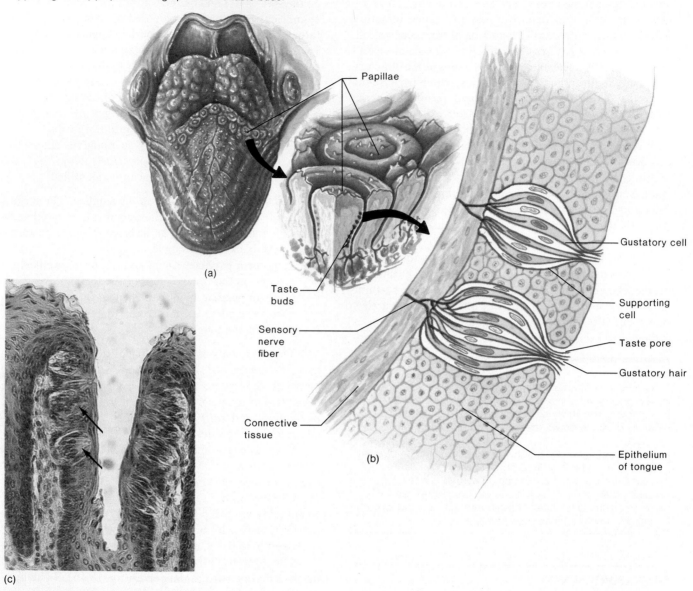

Figure 18.15. Patterns of taste receptor distribution on the dorsum of the tongue. (*a*) sweet receptors; (*b*) sour receptors; (*c*) salt receptors; (*d*) bitter receptors.

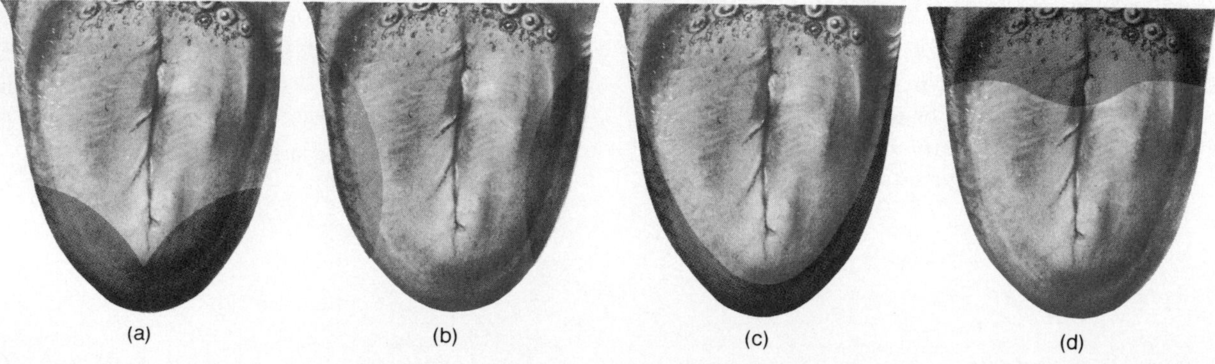

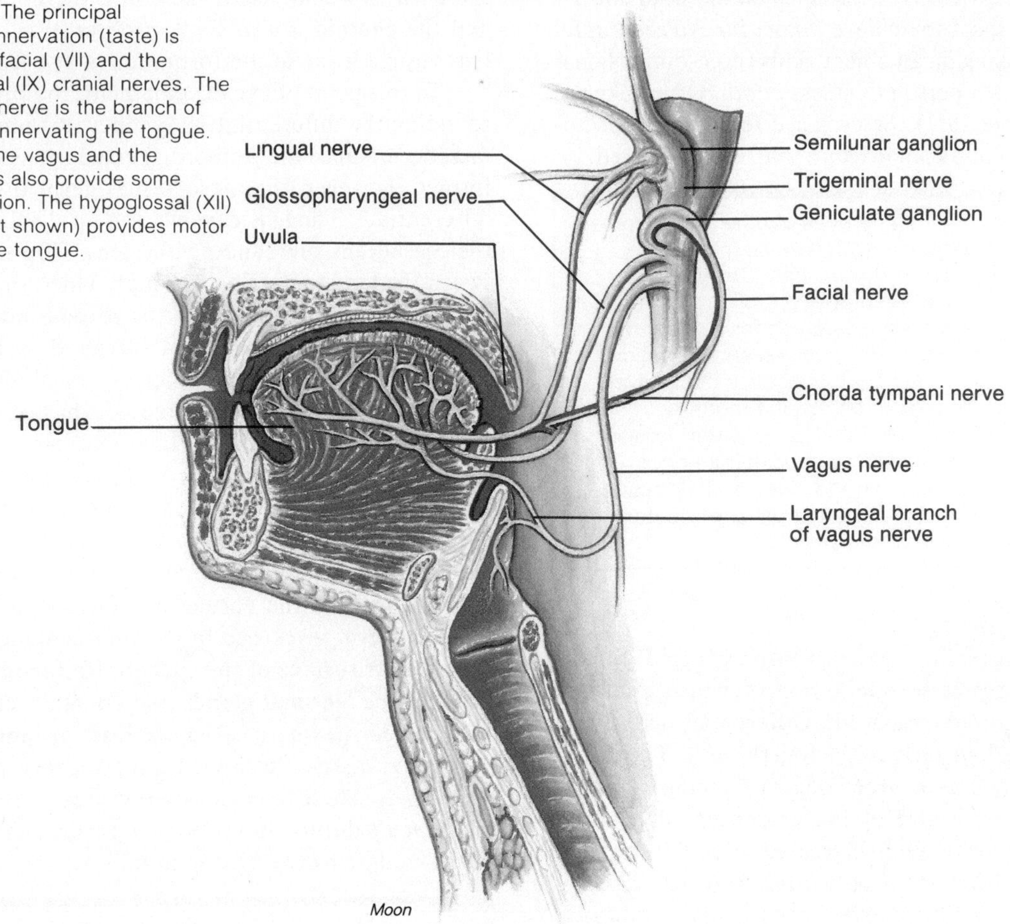

Figure 18.16. The principal viscerosensory innervation (taste) is provided by the facial (VII) and the glossopharyngeal (IX) cranial nerves. The chorda tympani nerve is the branch of the facial nerve innervating the tongue. Branches from the vagus and the trigeminal nerves also provide some sensory innervation. The hypoglossal (XII) cranial nerve (not shown) provides motor innervation to the tongue.

Because taste and smell are both chemoreceptors, they complement each other. We often confuse a substance's smell with its taste; and if we have a head cold or hold our nose while eating, food seems to lose its flavor.

1. Distinguish between papillae, taste buds, and gustatory cells. Discuss the function of each as related to taste.
2. List the three types of papillae.
3. Which cranial nerves have sensory innervation associated with taste? What are the afferent pathways to the brain where the perception of taste occurs?

Visual Sense

Rods and cones are the photoreceptors within the eyeball that are sensitive to light energy and are stimulated to transmit nerve impulses through the optic nerve and optic tract to the visual cortex of the occipital lobes, where the interpretation of vision occurs. The sensory components of the eye are formed by twenty-weeks, and the accessory structures are formed by thirty-two weeks.

Objective 15. Describe the development of the eye in the human embryo.
Objective 16. Describe the associated structures of the eye and the structure of the eyeball.
Objective 17. Trace the path of light waves through the eye, and explain how they are focused on distant and near objects.
Objective 18. Describe the neural pathway of a visual impulse, and discuss the neural processing of visual information.

The eyes are organs that refract (bend) and focus the incoming light waves onto the sensitive photoreceptors at the back of each eye. Nerve impulses from the stimulated photoreceptors are conveyed through visual pathways within the brain to the occipital lobes of the cerebrum where the sense of vision is perceived. The specialized photoreceptor cells can respond to an incredible one billion different stimuli each second. Further, these cells are sensitive to about 10 million gradations of light intensity and 7 million different shades of color.

The eyes are anteriorly positioned on the skull and set just far enough apart to achieve *binocular (stereoscopic) vision* when focusing on an object. This three-dimensional perspective allows a person to assess depth. Often likened to a camera (table 18.3) the eyes are responsible for approximately 80% of all knowledge that is assimilated.

The eyes of other vertebrates are basically similar to yours. Certain species, however, have adaptive modifications. Consider, for example, the extremely keen eyesight of a hawk, which soars high in the sky searching for food, or the eyesight of a nocturnal animal such as an owl, which feeds only at night. Note how the location of the eyes on the head corresponds to behavior. Predatory species, like cats, have eyes that are directed forward, allowing depth perception. Prey species, like deer, have eyes positioned high on the sides of their heads, allowing panoramic vision to detect distant threatening movements, even while grazing.

Development of the Eye

The development of the eye is a complex, rapid process that involves the precise interaction of neuroectoderm, surface ectoderm, and mesoderm. Differentiation of these germ layers is evident early in the fourth week. The initial differentiation involves neuroectoderm forming a lateral diverticulum on each side of the prosencephalon (forebrain). As the diverticulum increases in size, the distal portion dilates to become the **optic** *(op'tik)* **vesicle,** and the proximal portion constricts to become the **optic stalk** (fig. 18.17). Once the optic vesicle is formed, the overlying surface ectoderm thickens and invaginates. The thickened portion is the **lens placode** *(plak'ōd),* and the invagination is the **lens pit.**

During the fifth week, the lens placode is depressed and eventually cut off from the surface ectoderm, causing the formation of the **lens vesicle.** Simultaneously with the formation of the lens vesicle, the optic vesicle invaginates and differentiates into the two-layered **optic cup.** A groove called the **optic fissure** appears along the inferior surface of the optic cup and is continuous with a depression along the optic stalk. The **hyaloid artery** and **hyaloid vein** traverse the optic fissure longitudinally to serve the developing eyeball. The walls of the optic fissure eventually close so that the hyaloid vessels are within the tissue of the optic stalk. They become the **central vessels of the retina** *(ret'ĭ-nah)* of the mature eye. The optic stalk eventually becomes the **optic nerve,** composed of sensory axons from the retina.

By the end of the sixth week and the early part of the seventh week, the optic cup has differentiated into two sheets of epithelial tissue that become the sensory and pigmented layers of the **retina.** Both of these layers also line the entire vascular coat, including the **ciliary body, iris,** and the **choroid** *(ko'roid).* A proliferation of cells in the lens vesicle leads to the formation of the **lens.**

To this point of eye development, the tissue mass surrounding the differentiating ectoderm is undifferentiated mesoderm. Once the primordial ectodermal structures are formed, however, the mesodermal germ tissue begins to differentiate. The **lens capsule** (see fig. 18.25) forms from the mesoderm surrounding the lens. The mesoderm between the lens and retina, through which the hyaloid vessels pass, is transformed into the **vitreous humor** (see figs. 18.25 and 18.37). Mesoderm surrounding the optic cup differentiates into two distinct layers of the developing eyeball. The inner layer of mesoderm becomes the vascular **choroid,** and the outer layer becomes the toughened **sclera** *(skle'rah)* posteriorly and the transparent **cornea** anteriorly. The anterior chamber and the posterior chamber are the fluid-filled spaces that form between the cornea and the iris and between the iris and the lens, respectively. Once the cornea has formed, additional surface ectoderm gives rise to the thin **conjunctiva** covering the anterior surface of the eyeball. Epithelium of the **eyelids** and the **lacrimal glands** and **ducts** develop from surface ectoderm, whereas the **extrinsic eye muscles** and all connective tissue associated with the eye develop from mesoderm. These accessory structures of the eye gradually develop during the embryonic period and into the fetal period as late as the fifth month.

The developing eye is extremely sensitive to and may be impaired by teratogenic agents, particularly *rubella (German measles).* If a pregnant woman contracts rubella, there is a 90% chance the embryo or fetus will contract it also. An embryo afflicted with rubella has more than a 30% chance of being aborted, stillborn, or congenitally deformed. Rubella interferes with the mitotic process and thus causes underdeveloped organs. An embryo with rubella may suffer from a number of physical deformities, *cataracts* and *glaucoma* being common deformities of the eye.

Associated Structures of the Eye

Associated structures of the eye either protect the eyeball or provide eye movement. Protective structures include the bony orbit, eyebrow, facial muscles, eyelids, eyelashes, conjunctiva, and the lacrimal apparatus that produces tears. Eyeball movements are made possible by the actions of the extrinsic ocular eye muscles that arise from the orbit and insert on the outer layer of the eyeball.

Orbit. Each eyeball is positioned in a bony depression in the skull called the orbit (see fig. 9.14 and table 9.5). Seven bones of the skull (frontal, lacrimal, ethmoid, zygomatic, maxilla, sphenoid, and palatine) form the walls of the orbit that support and protect the eye.

optic: L. *optica,* see
hyaloid: Gk. *hyalos,* glass; *eiodos,* form

Figure 18.17. The development of the eye. (*a*) an anterior view of the developing head of a twenty-two-day-old embryo and the formation of the optic vesicle from the neuroectoderm of the prosencephalon (forebrain). (*b*) the development of the optic cup. The lens vesicle is formed (*c*) as the ectodermal lens placode invaginates during the fourth week. The hyaloid vessels become enclosed (c_1) and (e_1) within the optic nerve as there is fusion of the optic fissure. (*d*) the basic shape of the eyeball and the position of its internal structures are established during the fifth week. The successive development of the eye is shown at (*e*) six weeks and at (*f*) twenty weeks, respectively. (*g*) the eye of the newborn.

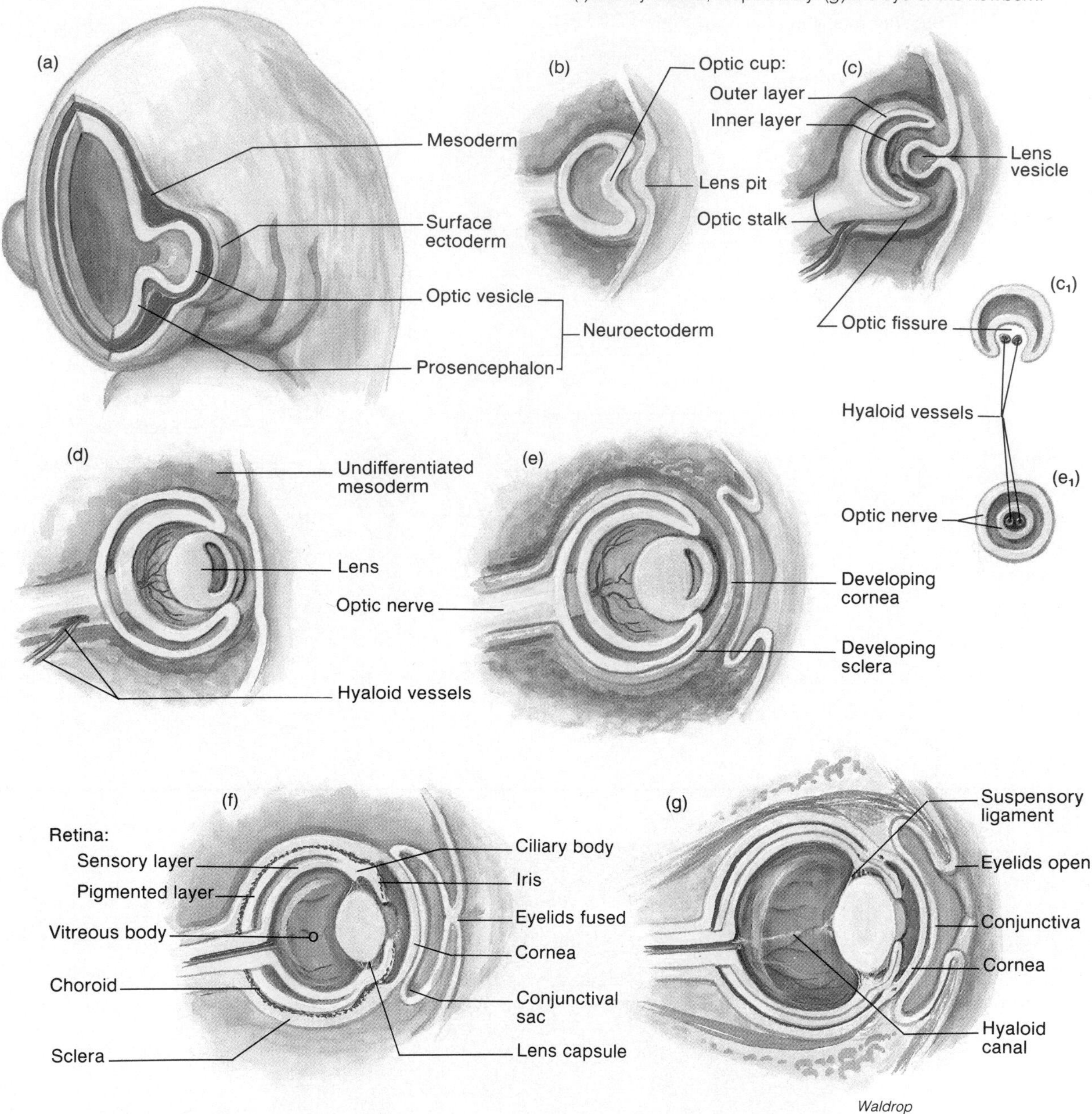

Eyebrow. Eyebrows consist of short, thick hair positioned transversely above both eyes along the superior orbital ridges of the skull (see figs. 13.6 and 18.18). Located here they effectively shade the eyes from the sun and prevent perspiration or falling particles from getting into the eyes. Underneath the skin of each eyebrow is the orbital portion of the orbicularis oculi muscle and a portion of the corrugator muscle (see fig. 12.30). Contraction of either of these muscles causes the eyebrow to move, often reflexively, to protect the eye.

Eyelids and Eyelashes. **Eyelids** (**palpebrae**) develop as reinforced folds of skin with attached skeletal muscle so that they are movable. In addition to the orbicularis oculi muscle attached to the skin that surrounds the front of the eye, the **levator palpebrae** *(le-va'tor pal'pĕ-bre)* **superioris** muscle attaches along the upper eyelid and provides it with greater movability than the lower eyelid. Contraction of

palpebra: L. *palpebra,* eyelid (related to *palpare,* to pat gently)

Levator palpebrae superioris m.
Superior rectus m.
Retina
Anterior chamber
Suspensory ligament of lens
Posterior chamber
Lens
Iris
Ciliary body
Orbicularis oculi m.
Eyebrow
Cornea
Palpebral conjunctiva
Superior tarsal plate
Bulbar conjunctiva
Eyelashes
Conjunctival sac
Orbicularis oculi m.

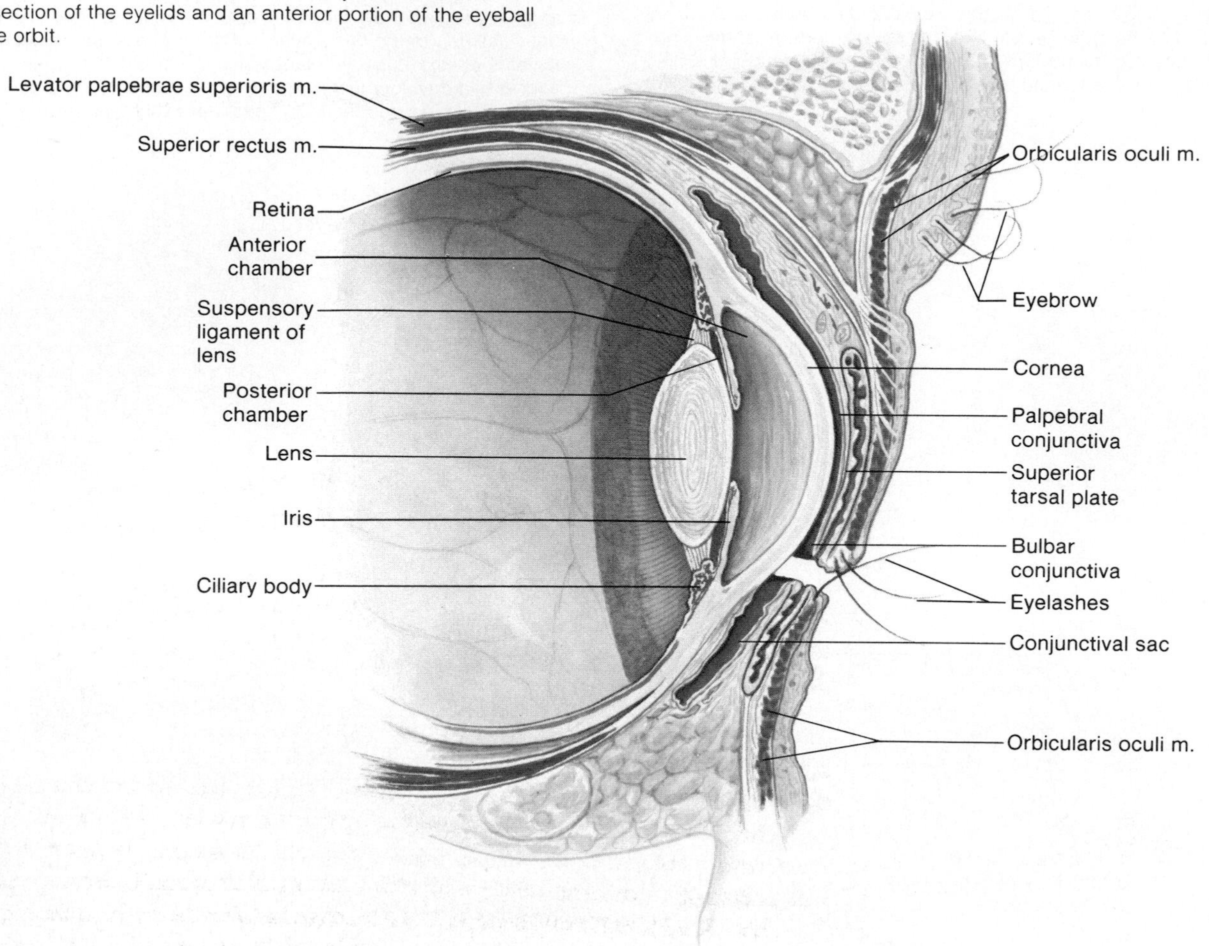

Figure 18.18. Accessory structures of the eyeball as seen in a sagittal section of the eyelids and an anterior portion of the eyeball within the orbit.

Eyebrow
Sclera
Palpebral fissure
Lateral commissure
Conjunctiva
Lower eyelid
Pupil
Iris
Upper eyelid
Lacrimal caruncle
Medial commissure
Eyelashes

Figure 18.19. The surface anatomy of the eye.

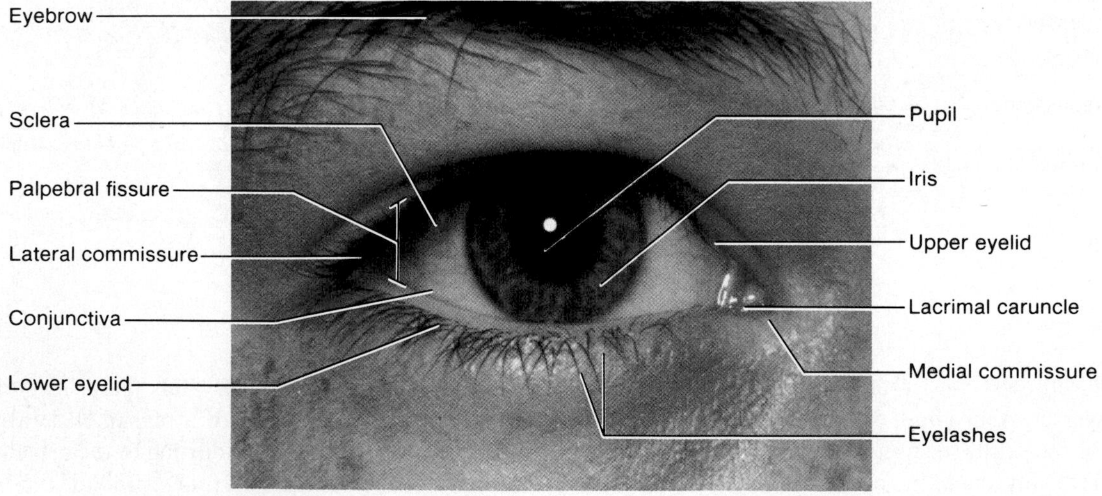

Figure 18.20. The lacrimal apparatus consists of the lacrimal gland and the ducts that drain lacrimal fluid into the nasal cavity. The lacrimal gland produces lacrimal fluid, which moistens and cleanses the conjunctiva that lines the underside of the eyelids and covers the anterior surface of the eyeball.

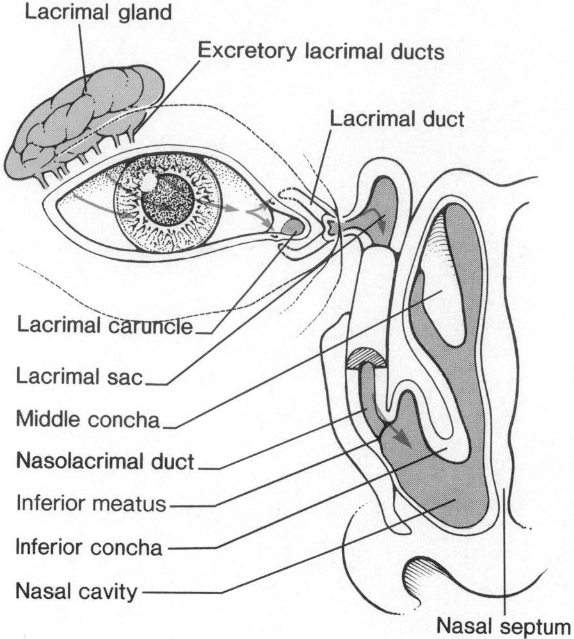

the orbicularis oculi muscle closes the eyelids over the eye, and contraction of the levator palpebrae superioris muscle elevates the upper eyelid to expose the eye. The eyelids protect the eyeball from desiccation by reflexively blinking about every seven seconds and moving fluid across the anterior surface of the eyeball. To avoid a blurred image, the eyelid will blink when the eyeball moves to a new position of fixation.

The **palpebral fissure** (fig. 18.19) is the interval between the upper and lower eyelids. The **commissures** (**canthi**) of the eye are the medial and lateral angles where the eyelids come together. The **medial commissure** is broader than the **lateral commissure** and is characterized by a small, reddish, fleshy elevation called the **lacrimal caruncle** *(kar'ung-kl)* (fig. 18.20), which contains sebaceous and sudoriferous glands. The shape of the palpebral fissure is elliptical when the eyes are open.

Each eyelid supports a row of numerous **eyelashes,** which protect the eye from airborne particles. The shaft of each eyelash is surrounded by a root hair plexus that provides the hair with the sensitivity necessary to elicit a reflexive closure of the lids. Eyelashes of the upper lid are long and turn upward, whereas those of the lower lid are short and turn downward.

commissure: L. *commissura,* a joining
caruncle: L. *caruncula,* diminutive of caro, flesh

In addition to the layers of the skin and the underlying connective tissue and orbicularis oculi muscle fibers, each eyelid contains a tarsal plate, tarsal glands, and conjunctiva. The **tarsal plates,** composed of dense fibrous connective tissue, are important in maintaining the shape of the eyelids (fig. 18.18). Specialized sebaceous glands called **tarsal meibomian** *(mi-bo'me-an)* **glands** are embedded within the tarsal plates along the exposed inner surfaces of the eyelids. The ducts of the tarsal glands open onto the edges of the eyelids, and their oily secretions help keep the eyelids from adhering to each other. A *chalazion (kah-la'ze-on)* is a tumor or cyst on the eyelid that results from an infection of the tarsal glands. Modified sweat glands called **ciliary glands** are also located within the eyelids along with additional sebaceous glands at the bases of the hair follicles of the eyelashes. An infection of these sebaceous glands is referred to as a *sty.*

Conjunctiva. The conjunctiva is a thin mucus-secreting epithelial membrane that lines the interior surface of each eyelid and exposed anterior surface of the eyeball (fig. 18.18). It consists of stratified squamous epithelium that varies in thickness in different regions. The **palpebral conjunctiva** is thick and adheres to the tarsal plates of the eyelids. As the conjunctiva reflects onto the anterior surface of the eyeball, it is known as the **bulbar conjunctiva.** This portion is transparent and especially thin where it covers the cornea. Because the conjunctiva is continuous and reflects from the eyelids to the anterior surface of the eyeball, a space called the **conjunctival sac** exists when the eyelids are closed. The conjunctival sac protects the eyeball by preventing objects (including a contact lens) from passing beyond the confines of the sac. The conjunctiva can repair itself rapidly if it is scratched.

Lacrimal Apparatus. The lacrimal apparatus consists of the lacrimal gland, which secretes the *lacrimal fluid* (tears), and a series of ducts that drain the secretion into the nasal cavity (fig. 18.20). The **lacrimal gland,** which is about the size and shape of an almond, is located in the superolateral portion of the orbit. It is a compound tubuloacinar gland that secretes lacrimal fluid through several excretory lacrimal ducts into the conjunctival sac of the upper eyelid. With each blink of the eyelids, lacrimal fluid passes medially and downward and drains into two small openings, called **puncta lacrimalia,** on both sides of the lacrimal caruncle. From here the lacrimal fluid drains through the lacrimal duct into the lacrimal sac and continues through the nasolacrimal duct to the inferior meatus of the nasal cavity.

tarsal: Gk. *tarsos,* flat basket
meibomian glands: from Heinrich Meibom, German anatomist, 1638–1700
chalazion: Gk. *chalazion,* hail; a small tubercle

Figure 18.21. The extrinsic ocular muscles of the right eyeball.

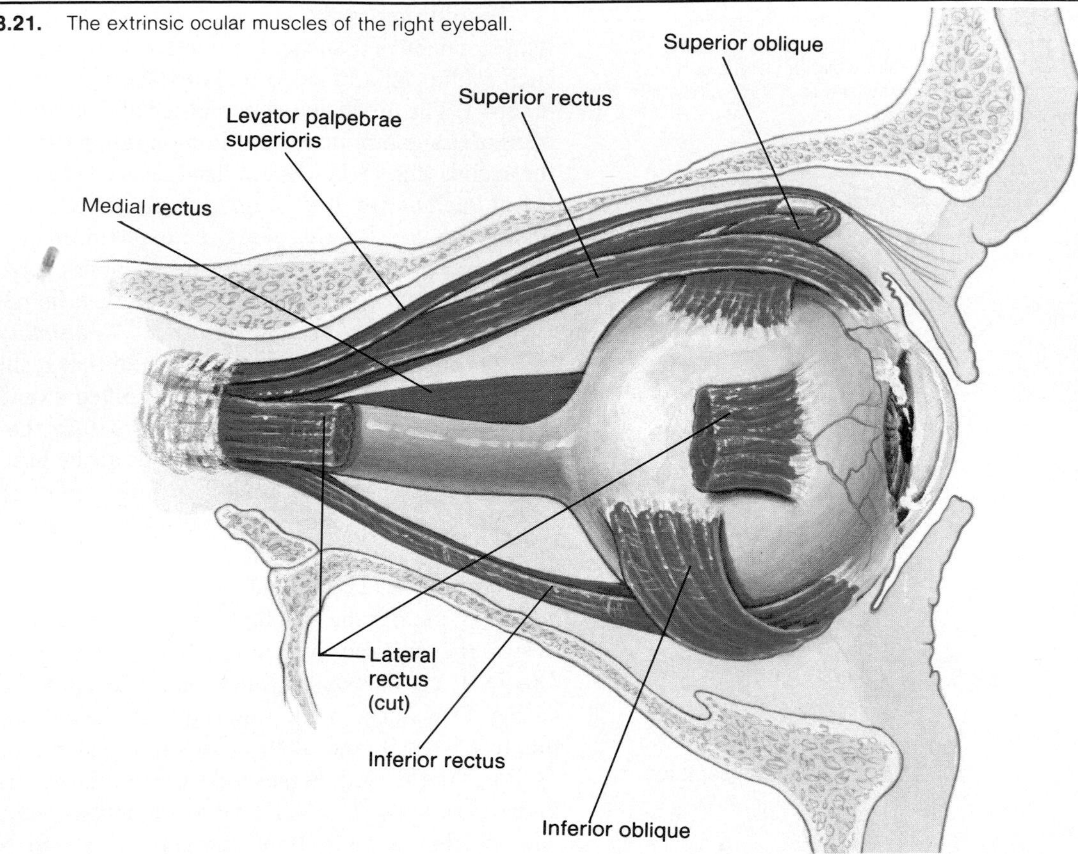

Lacrimal fluid is an aqueous, mucus secretion that contains a bactericidal enzyme called *lysozyme*. Lacrimal fluid not only moistens and lubricates the conjunctival sac but also reduces the chance of eye infections. Normally about one milliliter of lacrimal fluid is produced each day by the lacrimal gland of each eye. If irritating substances, such as particles of sand or chemicals from onions, make contact with the conjunctiva, the lacrimal glands are stimulated to oversecrete. The extra lacrimal fluid protects the eye by diluting and washing away the irritating substance.

Humans are the only animals known to weep in response to emotional stress. While crying, the volume of lacrimal secretion is so great that the tears may spill over the edges of the eyelids and the nasal cavity fill with fluid. The crying response is an effective means of communicating one's emotions and results from stimulation of the lacrimal glands by parasympathetic motor neurons.

Extrinsic Eye Muscles. The movements of the eyeball are controlled by six extrinsic eye muscles called the **extrinsic ocular muscles** (figs. 18.21, 18.22). Each extrinsic ocular muscle originates from the bony orbit and inserts by a tendinous attachment to the tough outer tunic of the eyeball. Four **recti muscles** maneuver the eyeball in the direction indicated by their names (**superior, inferior, lateral,** and **medial**), and two **oblique muscles** (**superior** and **inferior**) rotate the eyeball on its axis. One of the extrinsic ocular muscles, the superior oblique, passes through a pulleylike cartilaginous loop, called the **trochlea** *(trok'le-ah),* before attaching to the eyeball. Although stimulation of each muscle causes a precise movement of the eyeball, most of the movements involve the combined contraction of usually two muscles.

The motor units of the intrinsic ocular muscles are the smallest in the body. This means that a single motor neuron serves about ten muscle fibers, resulting in precision movements. The eyes move in synchrony by contracting synergistic muscles while relaxing antagonistic muscles.

The extrinsic ocular muscles are innervated by three cranial nerves (table 18.3). Innervation of the other skeletal and smooth muscles that serve the eye is also indicated in table 18.3.

A physical examination may include an eye movement test. As the patient's eyes follow the movement of a physician's finger, the physician can assess weaknesses in specific muscles or dysfunctions of specific cranial nerves. The patient experiencing *double vision (diplopia)* when moving his eyes may be suffering from muscle weakness. Looking laterally tests the abducens cranial nerve; looking inferiorly and laterally tests the trochlear cranial nerve; and looking cross-eyed tests the oculomotor and trochlear cranial nerves of both eyes.

trochlea: Gk. *trochos,* a wheel

Figure 18.22. The positions of the eyes as the ocular muscles are contracted. (*a*) right eye, inferior oblique muscle; left eye, superior and medial recti muscles. (*b*) both eyes, superior recti and inferior oblique muscles. (*c*) right eye, superior and medial recti muscles; left eye, inferior oblique muscle. (*d*) right eye, lateral rectus muscle; left eye, medial rectus muscle. (*e*) primary position with the eyes fixed on a distant fixation point. (*f*) right eye, medial rectus muscle; left eye, lateral rectus muscle. (*g*) right eye, superior oblique muscle; left eye; inferior and medial recti muscles. (*h*) both eyes, inferior recti and superior oblique muscles. (*i*) right eye, inferior and medial recti muscles; left eye, superior oblique muscle.

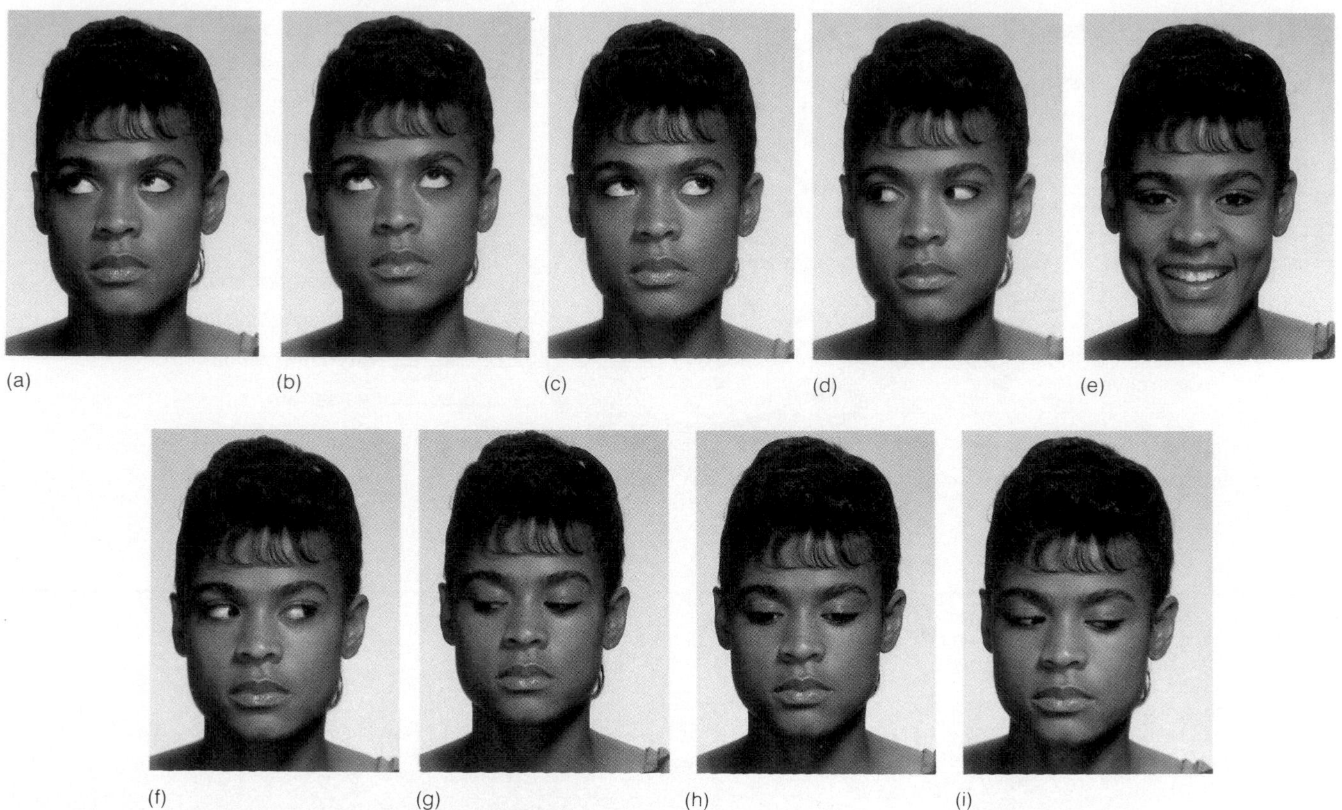

Table 18.3 Comparisons between eye structures and analogous structures in a camera

Eye structure	Analogous camera structure
Cornea and lens	Lens system
Iris and pupil	Variable aperture system
Eyelid	Lens cap
Sclera	Camera frame
Pigment epithelium and choroid	Black interior of camera
Retina*	Film

*Since neural processing begins in the retina, this eye structure may also be considered analogous (in part) to the photographer.

From Stuart Ira Fox, *Human Physiology.* Copyright © 1984 Wm. C. Brown Publishers, Dubuque, Iowa. All Rights Reserved. Reprinted by permission.

Structure of the Eyeball

The eyeball of an adult is essentially spherical and is approximately 25 mm (1 in.) in diameter. About four-fifths of the eyeball is positioned within the orbit of the skull. The eyeball consists of three basic layers: the fibrous tunic, the vascular tunic, and the internal tunic, or retina (figs. 18.23, 18.24).

Fibrous Tunic. The fibrous tunic is the outer layer of the eyeball. It is divided into two regions: the posterior five-sixths is the opaque sclera, and the anterior one-sixth is the transparent cornea.

The toughened **sclera** is the "white of the eye." It is composed of tightly bound elastic and collagenous fibers, which give shape to the eyeball and protect its inner structures. The sclera is avascular but does contain sensory receptors for pain. The large **optic nerve** exits through the sclera at the posterior portion of the eyeball.

The **cornea** is transparent and convex to permit the passage and cause the refraction (bending) of incoming light waves. The transparency of the cornea is due to tightly packed, avascular, dense connective tissue. Also, the relatively few cells that are present in the cornea are arranged in unusually regular patterns. The circumferential edge of the cornea is continuous structurally with the sclera. The outer surface of the cornea is covered with a thin, nonkeratinized, stratified squamous epithelial layer called the **bulbar (corneal) epithelium,** which is actually a continuation of the conjunctiva of the sclera (fig. 18.18).

sclera: Gk. *skleros,* hard
cornea: L. *cornu,* horn

Figure 18.23. The internal anatomy of the eyeball.

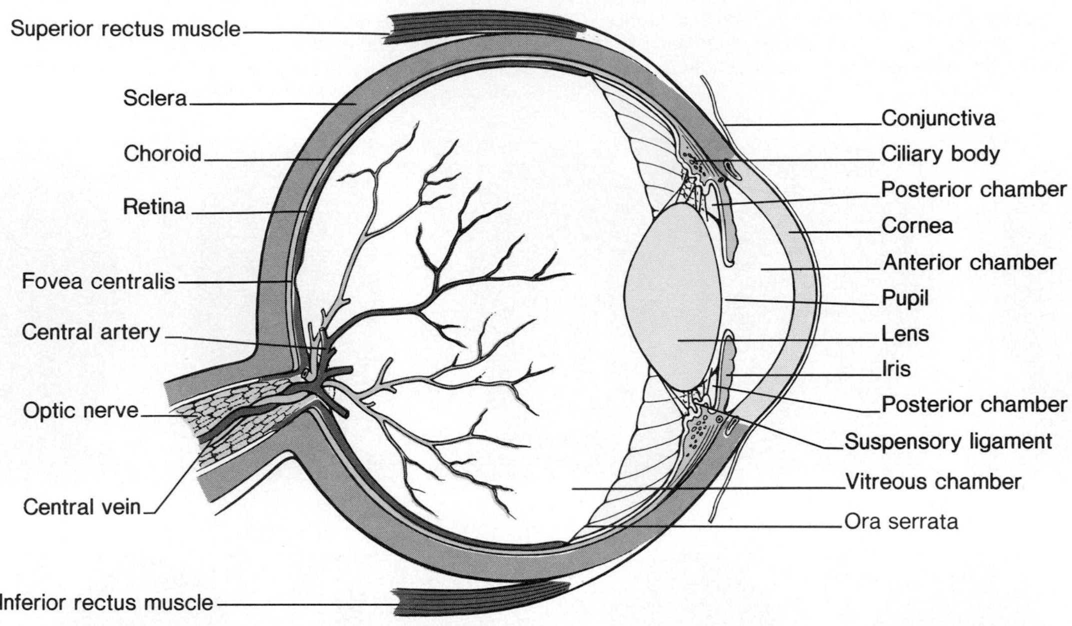

Figure 18.24. Sagittal sections of the eyeball. (*a*) an anterior portion showing the cornea and the lens, and (*b*) a posterior portion showing the retina, optic disc, and optic nerve.

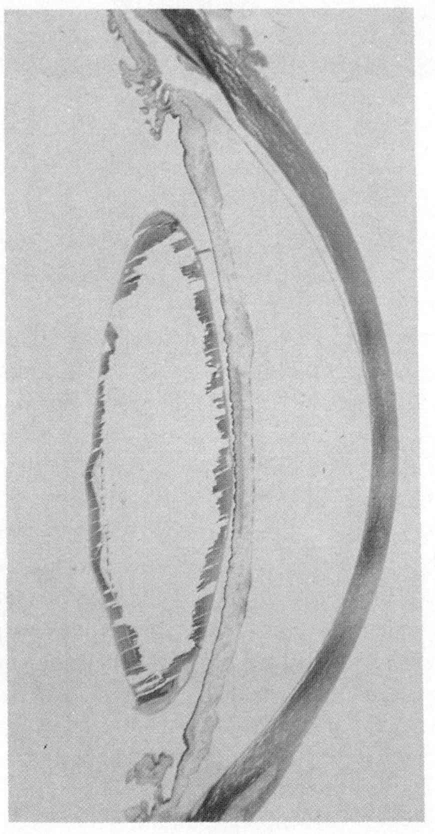

(a)

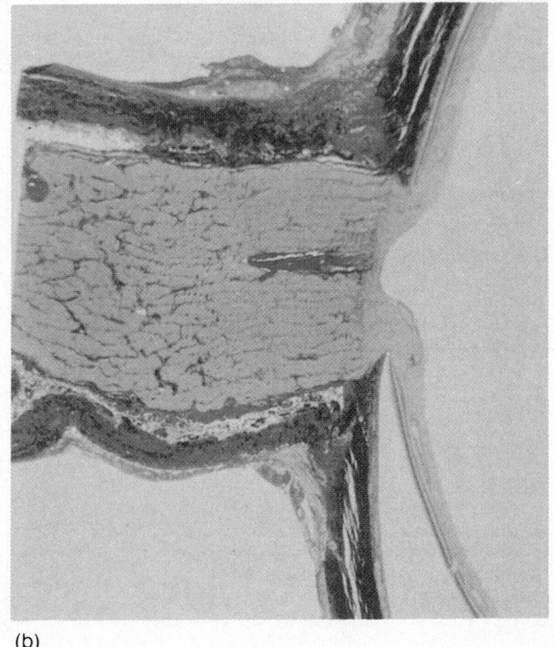

(b)

Figure 18.25. The detailed structure of the anterior portion of the eyeball.

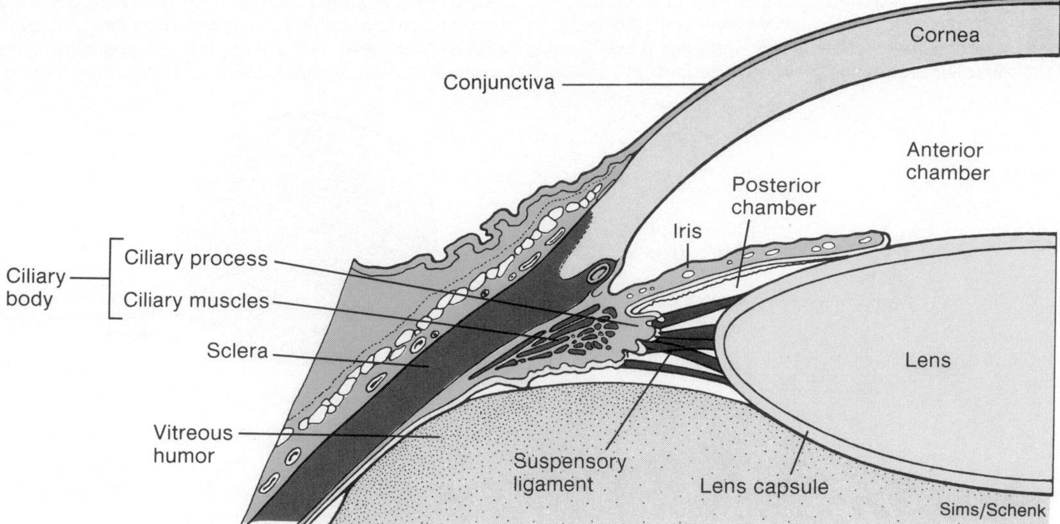

A defective cornea can be replaced with a donor cornea in a surgical procedure called a *corneal transplant (keratoplasty).* A defective cornea is one that does not transmit or refract light effectively due to its shape, scars, or disease. During a corneal transplant, the defective cornea is excised and replaced with a transplanted cornea that is sewn into place with nylon sutures. It is considered to be the most successful type of homotransplant (between individuals of the same species).

Vascular Tunic. The vascular tunic, or **uvea** *(u've-ah)* of the eyeball consists of the choroid, the ciliary body, and the iris (fig. 18.23).

The **choroid** is a thin, highly vascular layer that lines most of the internal surface of the sclera. The choroid contains numerous pigment-producing melanocytes, which give it a dark brownish color to prevent light waves from being reflected out of the eyeball. There is an opening in the choroid at the back of the eyeball where the optic nerve is located.

The **ciliary body** is the thickened, anterior portion of the vascular tunic that forms an internal muscular ring toward the front of the eyeball (figs. 18.25, 18.26). Three distinct planes of smooth muscle fibers called **ciliary muscles** are found within the ciliary body.

Numerous extensions of the ciliary body called **ciliary processes** attach to the **zonular fibers,** which in turn attach to the lens. Collectively, the zonular fibers constitute the **suspensory ligament.** The transparent **lens** consists of tight layers of protein fibers arranged like the layers of an onion. A thin, clear **lens capsule** encloses the lens and provides attachment for the suspensory ligament (see fig. 18.25).

Figure 18.26. The ciliary body, suspensory ligament, and the lens viewed from posterior to anterior.

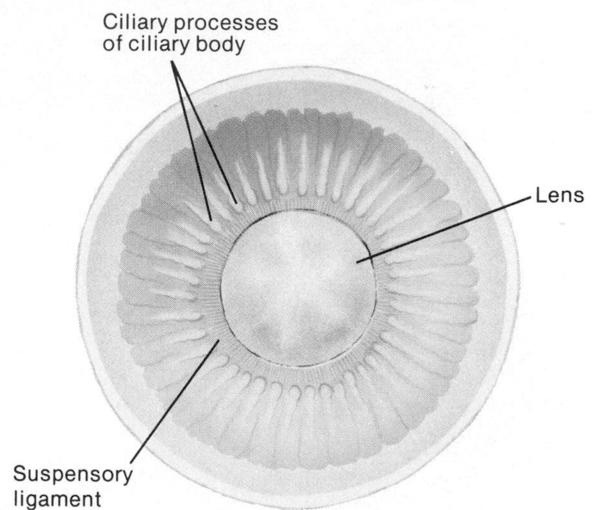

The shape of the lens determines how much the light waves that pass through will be refracted. Constant tension of the suspensory ligament, when the ciliary muscles are relaxed, flattens the lens somewhat (fig. 18.27). Contraction of the ciliary muscles relaxes the suspensory ligament and makes the lens more spherical. The constant tension within the lens capsule results in the surface of the lens becoming more convex when the suspensory ligament is not taut. A flattened lens facilitates viewing a distant object, whereas a rounded lens permits viewing a close object.

uvea: L. *uva,* grape
choroid: Gk. *chorion,* membrane
zonular fibers: L. *zona,* a girdle

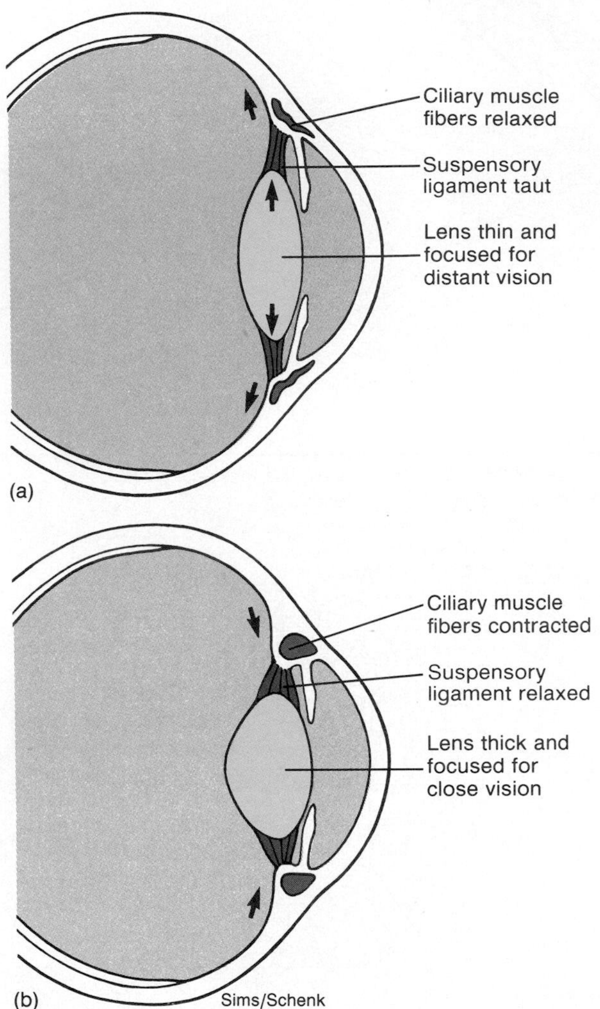

Figure 18.27. Changes in the shape of the lens during accommodation. (*a*) the lens is flattened for distant vision when the ciliary muscle fibers are relaxed and the suspensory ligament is taut. (*b*) the lens is more spherical for close-up vision when the ciliary muscle fibers are contracted and the suspensory ligament is relaxed.

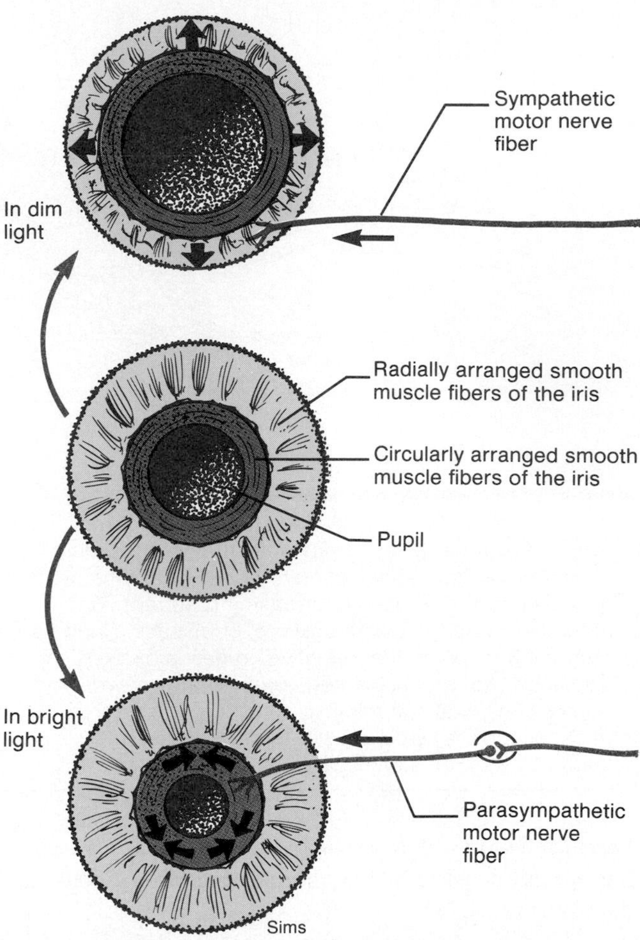

Figure 18.28. Dilation and constriction of the pupil. In dim light, the radially arranged smooth muscle fibers are stimulated to contract by sympathetic stimulation, dilating the pupil. In bright light, the circularly arranged smooth muscle fibers are stimulated to contract by parasympathetic stimulation, constricting the pupil.

The **iris** is the anterior portion of the vascular tunic and is continuous with the choroid. The iris is viewed from the outside as the colored portion of the eyeball (figs. 18.23, 18.25). It consists of smooth muscle fibers arranged in a circular and a radial pattern. The contraction of the smooth muscle fibers regulates the diameter of the **pupil** (table 18.5 and fig. 18.28), an opening in the center of the iris. Contraction of the circularly arranged fibers of the iris, stimulated by bright light, constricts the pupil and diminishes the amount of light entering the eyeball. Contraction of the radially arranged fibers, in response to dim light, dilates the pupil and permits more light to enter.

iris: Gk. *irid,* rainbow

Retina. The retina covers the choroid as the innermost layer of the eyeball. It consists of a thin, outer **pigmented layer** in contact with the choroid, and a thick, inner **nervous layer,** or visual portion. The retina is principally in the posterior portion of the eyeball (fig. 18.23). The visual layer of the retina terminates in a jagged margin near the ciliary body called the **ora serrata.** The pigmented layer extends anteriorly over the back of the ciliary body and iris.

The pigmented layer and nervous layer of the retina are not attached to each other except surrounding the optic nerve and at the ora serrata. Because of this loose connection, the two layers may become separated as a *detached retina.* Such a separation can be corrected by fusing the layers with a laser.

ora serrata: L. *ora,* margin; *serra,* saw

Figure 18.29. The layers of the retina. The retina is inverted, so that light must pass through various layers of nerve cells before reaching the photoreceptors (rods and cones). (*a*) a schematic view diagram; (*b*) a light micrograph.

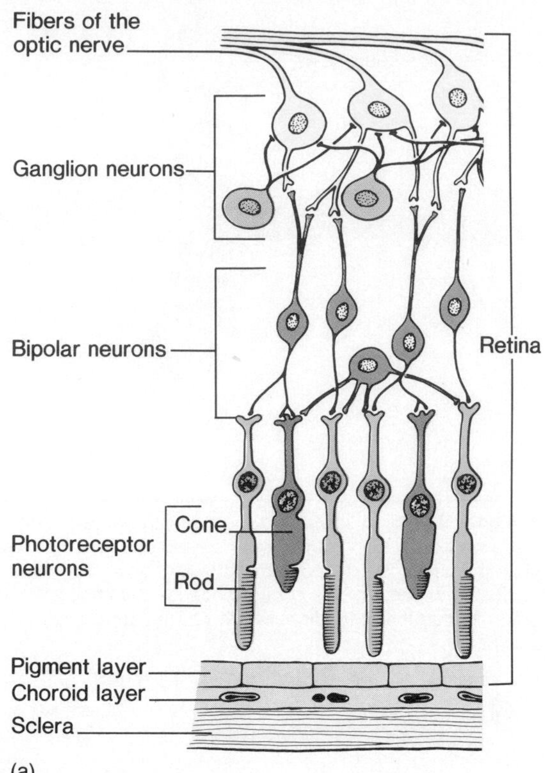

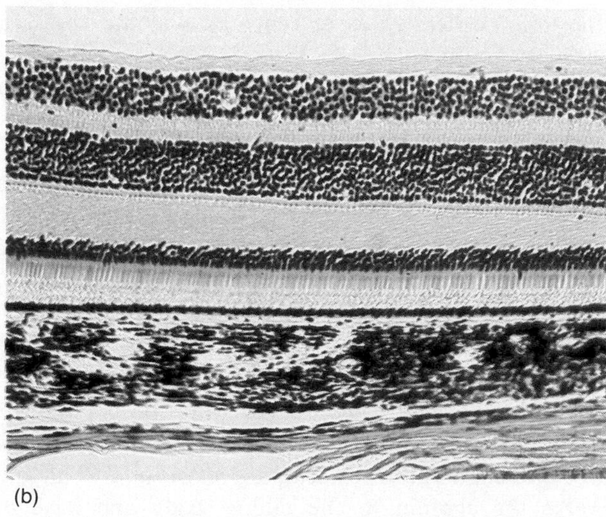

Figure 18.30. Photoreceptor cells of the eyeball.

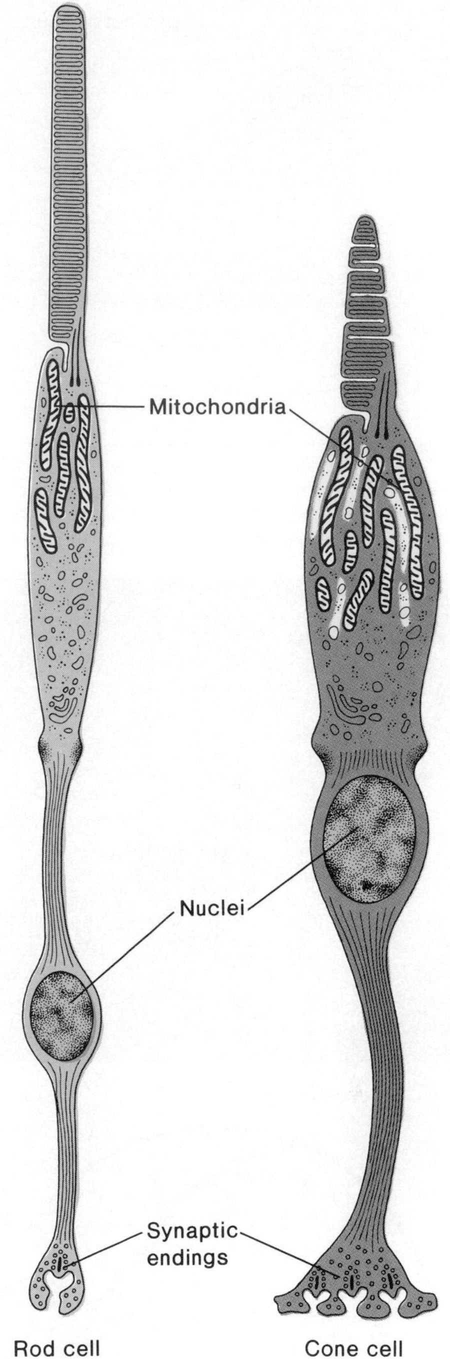

The nervous layer of the retina is composed of three principal layers of neurons. Listing them in the order in which they conduct impulses, they are rod and cone cells, bipolar neurons, and ganglion neurons (fig. 18.29). In terms of the passage of light, however, the order is reversed. Light must first pass through the layer of ganglion cells and then bipolar cells before reaching the photoreceptors.

Rods and **cones** are photoreceptors. Rods number over 100 million per eye and are more slender and elongated than cones (fig. 18.30). Rods are positioned on the peripheral parts of the retina, where they respond to dim light for black-and-white vision. They also respond to form and movement but provide poor visual acuity. Cones number about 7 million per eye. They provide daylight color vision and are responsible for visual acuity. The photoreceptors synapse with **bipolar neurons,** which in turn synapse with the **ganglion neurons.** The axons of ganglion neurons leave the eye as the optic nerve.

Figure 18.31. A diagram of the retina sectioned through the fovea centralis.

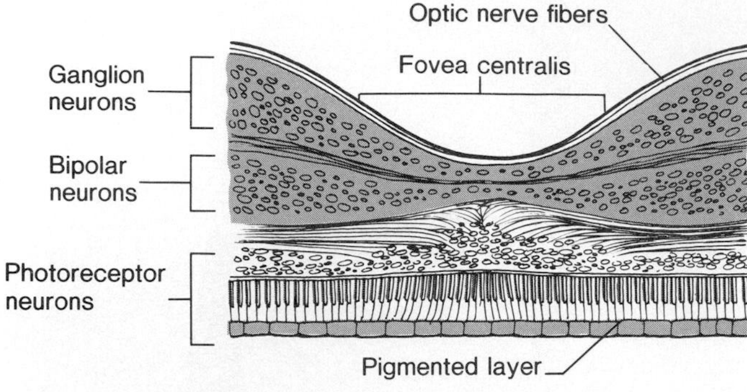

Figure 18.32. A view of the retina (*a*) as seen with an *ophthalmoscope;* (*b*) a diagrammatic view. Optic nerve fibers leave the eyeball at the optic disc to form the optic nerve. Note the blood vessels that can be seen entering the eyeball at the optic disc.

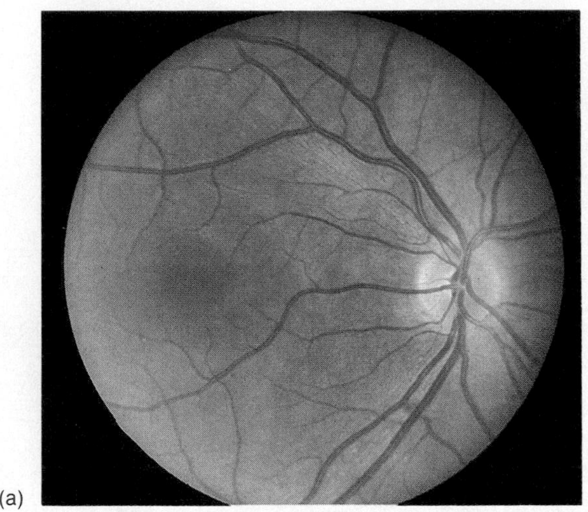

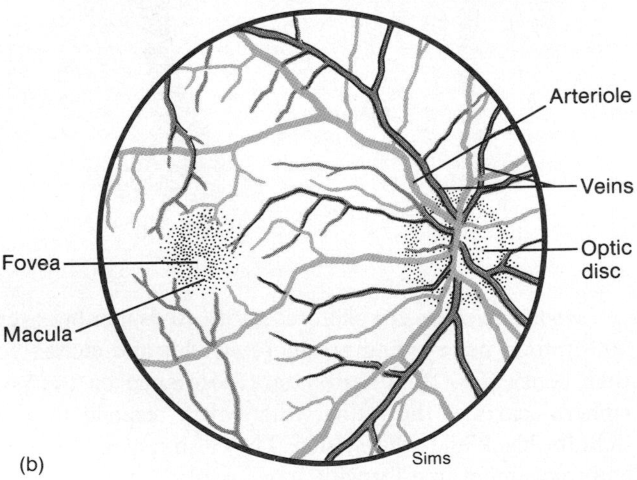

Figure 18.33. The optic disc is a small area of the retina where the fibers of the ganglion neurons emerge forming the optic nerve. The optic disc is frequently called the "blind spot" because it is devoid of rods and cones.

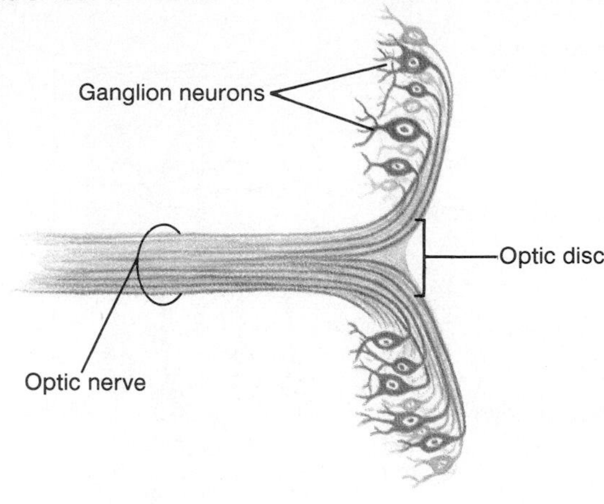

Figure 18.34. The blind spot. Hold the drawing about *twenty inches* from your face, with your left eye closed and your right eye focused on the circle. Slowly move the drawing closer to your face until the cross disappears. This occurs because the image of the cross is focused on the optic disc, where photoreceptors are absent.

Cones are concentrated in a depression near the center of the retina called the **fovea centralis,** which is the area of keenest vision (figs. 18.23, 18.31). Surrounding the fovea centralis is the yellowish **macula lutea,** which also has an abundance of cones (fig. 18.32). There are no photoreceptors at the point where the optic nerve is attached. This area is a "blind spot" and is referred to as the **optic disc** (fig. 18.33). Normally a person is unaware of the blind spot because the eyes continually move about and because an object is viewed from a different angle with each eye. The blind spot can easily be demonstrated as described in figure 18.34.

Blood Supply to the Eyeball. Both the choroid and the retina are richly supplied with blood. Two **ciliary arteries** pierce the sclera at the posterior aspect of the eyeball and traverse the choroid to the ciliary body and base of the iris. Although the ciliary arteries enter the eyeball independently, they anastomose (connect) extensively throughout the choroid.

The **central artery (central retinal artery)** branches from the ophthalmic artery and enters the eyeball in contact with the optic nerve. As the central artery passes through the optic disc, it divides into superior and inferior branches, each of which then divides into temporal and nasal branches to serve the inner layers of the retina (see

fovea: L. *fovea,* small pit
macula lutea: L. *macula,* spot; *luteus,* yellow

Figure 18.35. Aqueous humor maintains the intraocular pressure within the anterior and posterior chambers. It is secreted into the posterior chamber, flows through the pupil into the anterior chamber, and drains from the eyeball through the venous sinus (canal of Schlemm).

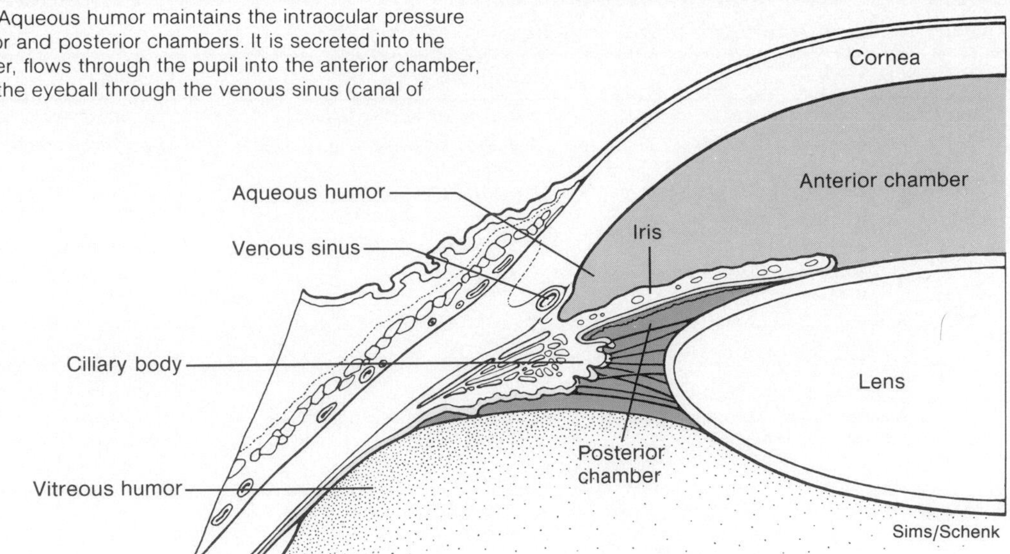

fig. 18.23). The **central vein** drains blood from the eyeball through the optic disc. The branches of the central artery can be observed within the eyeball through an ophthalmoscope (fig. 18.32).

> An internal examination of the eyeball with an ophthalmoscope is frequently part of a routine physical examination. Capillary vessels, which can be seen within the eyeball, can reveal certain diseases or body dysfunctions. Diseases such as arteriosclerosis, diabetes, cataracts, or glaucoma can be detected.

Chambers of the Eyeball. The interior of the eyeball is separated by the lens into two main cavities (see fig. 18.23). The cavity anterior to the lens is subdivided by the iris into an **anterior chamber** and a **posterior chamber.** The anterior chamber is located in front of the iris and behind the cornea. The posterior chamber is located between the iris and the suspensory ligament and the lens. The anterior and posterior chambers connect through the pupil and are filled with a watery fluid called **aqueous humor.** The constant production of aqueous humor maintains an *intraocular pressure* of about 24 mm of mercury within the anterior and posterior chambers. Aqueous humor also provides nutrients and oxygen to the avascular lens and cornea. An estimated 5–6 ml of aqueous humor is secreted each day from the vascular epithelium of the ciliary body (fig. 18.35). From its site of secretion within the posterior chamber, the aqueous humor passes through the pupil into the anterior chamber. From here, it drains from the eyeball through the **venous sinus** (**canal of Schlemm**) into the bloodstream. The venous sinus is located at the junction of the cornea and iris.

Between the lens and retina is the large **vitreous chamber,** which is filled with a transparent jellylike **vitreous humor.** Vitreous humor also contributes to the intraocular pressure to maintain the shape of the eyeball and to hold the retina against the choroid. Unlike aqueous humor, vitreous humor is not continuously produced but formed prenatally.

The location and functions of the structures within the eyeball are summarized in table 18.4.

> Puncture wounds to the eyeball are especially dangerous and frequently cause blindness. Protective equipment such as goggles, shields, and shatterproof lenses should be used in hazardous occupations and certain sports. If the eye is punctured, the principal thing to remember is to *not remove the object* if it is still impaling the eyeball. Removal may allow the fluids to drain from the eyeball and cause the loss of intraocular pressure. This is particularly serious if the nonreplaceable vitreous humor drains from the eyeball.

Physiology of Vision

The focusing of light waves and stimulation of photoreceptors of the retina require five basic processes:

1. The transmission of light waves through transparent media of the eyeball.
2. The refraction of light waves through media of different densities.
3. Accommodation of the lens to focus the light waves.
4. Constriction of the pupil by the iris to regulate the amount of light entering the vitreous chamber.
5. Convergence of the eyeballs so that visual acuity is maintained.

---

canal of Schlemm: from Friedrich S. Schlemm, German anatomist, 1795–1858

vitreous: L. *vitreus,* glassy

Table 18.4 Location and functions of the structures of the eyeball

Tunic and structure	Location	Composition	Function
Fibrous tunic	Outer layer of eyeball	Avascular connective tissue	Provides shape of eyeball
Sclera	Posterior, outer layer; "white of the eye"	Tightly bound elastic and collagen fibers	Supports and protects eyeball
Cornea	Anterior surface of eyeball	Tightly packed dense connective tissue—transparent and convex	Transmits and refracts light
Vascular tunic (uvea)	Middle layer of eyeball	Highly vascular pigmented tissue	Supplies blood; prevents reflection
Choroid	Middle layer in posterior portion of eyeball	Vascular layer	Supplies blood to eyeball
Ciliary body	Anterior portion of vascular tunic	Smooth muscle fibers and glandular epithelium	Supports the lens through suspensory ligament and determines its shape; secretes aqueous humor
Iris	Anterior portion of vascular tunic continuous with ciliary body	Pigment cells and smooth muscle fibers	Regulates the diameter of the pupil and hence the amount of light entering the vitreous chamber
Retina	Inner layer in posterior portion of eyeball	Photoreceptor neurons (rods and cones), bipolar neurons, and ganglion neurons	Photoreception; transmits impulses
Lens (not part of any tunic)	Between posterior and vitreous chambers; supported by suspensory ligament of ciliary body	Tightly arranged protein fibers; transparent	Refracts light and focuses onto fovea centralis

Visual impairment may exist if one or more of these processes does not function properly (see clinical considerations).

Transmission of Light Waves. Light waves entering the eyeball pass through four transparent media before they stimulate the photoreceptors. In sequence, the media through which light waves pass are the cornea, aqueous humor, lens, and vitreous humor. The cornea and lens are solid media composed of tightly packed, avascular protein fibers. An additional thin, transparent membranous continuation of the conjunctiva covers the outer surface of the cornea. The aqueous humor is a low viscosity fluid, and the vitreous humor is jellylike in consistency.

Refraction of Light Waves. Refraction is the bending of light waves. Refraction occurs as light waves pass at an oblique angle from a medium of one optical density to a medium of another optical density (fig. 18.36). The convex cornea is the principal refractive medium; the fluids within the various chambers produce minimal refraction. The lens is particularly important for refining and altering refraction. Of the refractive media, only the lens can be altered in shape to achieve precise refraction.

The refraction of light waves is so extensive that the visual image is formed upside down on the retina (fig. 18.37). Nerve impulses of the image in this position are relayed to the visual cortex of the occipital lobe, where the inverted image is interpreted as right side up.

Accommodation of the Lens. Accommodation is the automatic adjustment of the curvature of the lens by contraction of ciliary muscles to bring light waves into sharp focus on the retina. The lens of the eyeball is biconvex. When an object is viewed closer than about twenty feet, the lens must make an adjustment, or accommodation, if a clear focus is to appear on the retina. Contraction of the smooth muscle fibers of the ciliary body causes the suspensory ligament to relax and the lens to become thicker (see fig. 18.27). A thicker, more convex lens causes a greater convergence of the refracted light waves (see fig. 18.36) necessary for viewing close objects.

Constriction of the Pupil. Constriction of the pupil through parasympathetic stimulation that causes contraction of the circular muscle fibers of the iris (see fig. 18.28) is important for two reasons. One is the reduction in the amount of light that enters the vitreous chamber. A reflexive pupil constriction protects the retina from sudden or intense bright light. Another more important function is that a reduced pupil diameter prevents light waves from entering the vitreous chamber through the periphery of the lens. Light waves refracted from the periphery would not be brought into focus on the retina and would cause blurred vision. Autonomic constriction of the pupil and accommodation of the lens occur simultaneously.

Convergence of the Eyeballs. Convergence of the eyeballs means that the eyes rotate more medially when viewing closer objects. In fact, focusing on an object close to the tip of the nose causes a person to appear cross-eyed. The eyeballs must converge when viewing close objects because only then can the light waves focus on the same portions in both retinas.

Figure 18.36. The shape of a lens determines the refraction of the light waves that pass through. (*a*) a lens with a convex surface causes light waves to converge to a focal point. (*b*) a lens with a concave surface causes light waves to diverge.

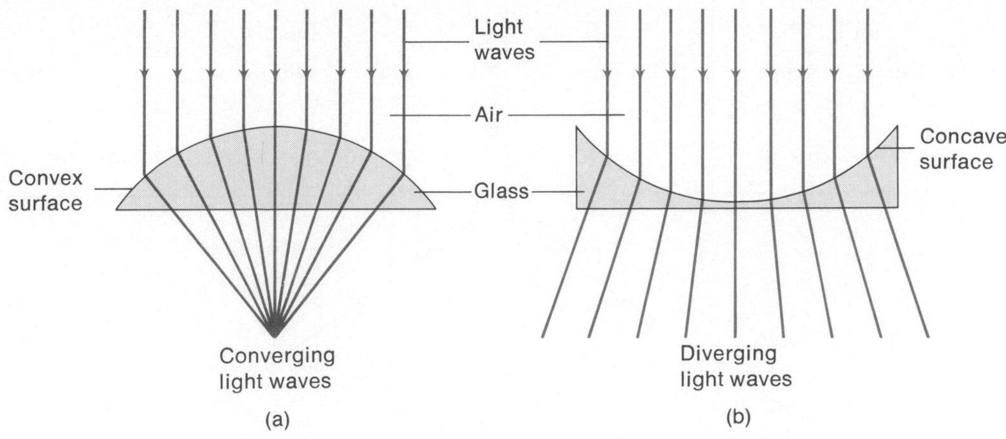

Figure 18.37. The refraction of light waves within the eyeball causes the image of an object to be inverted on the retina.

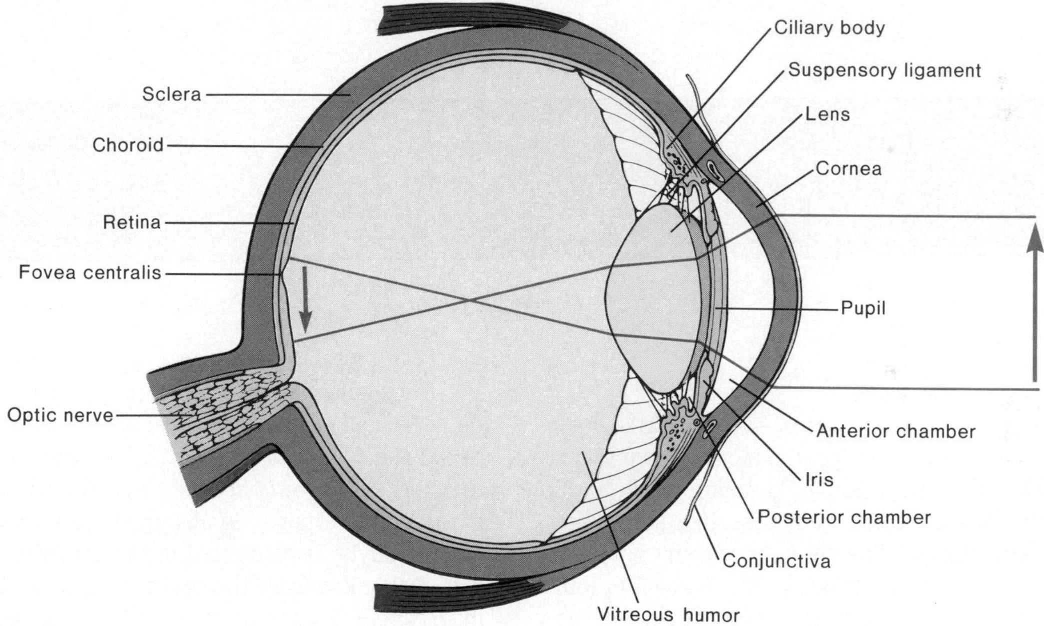

Amblyopia exanopsia, commonly called "lazy eye," is a condition of ocular muscle weakness causing a deviation of one eye so that there is not a concurrent convergence of both eyeballs. Because of disconjugate fixation, two images are received by the optic cortex—one of which is suppressed to avoid *diplopia (double vision)* or images of unequal clarity. A person who has amblyopia will experience dimness of vision and partial loss of sight. Amblyopia is frequently tested for in young children because if untreated before age six, there is little that can be done to strengthen the afflicted muscle.

amblyopia: Gk. *amblys,* dull; *ops,* vision

Visual Spectrum. The eyes transduce energy in the *electromagnetic spectrum* (fig. 18.38) into nerve impulses. Only a limited part of this spectrum can excite the photoreceptors. Electromagnetic energy with wavelengths between 400 and 700 nanometers (nm) comprise *visible light.* Light of longer wavelengths, which are in the infrared regions of the spectrum, do not have sufficient energy to excite photoreceptors but are perceived as heat. Ultraviolet light, which has shorter wavelengths and more energy than visible light, is filtered out by the yellow color of the eyes' lenses. Certain insects, such as honeybees, and people who have had their lens removed, can see light in the ultraviolet range.

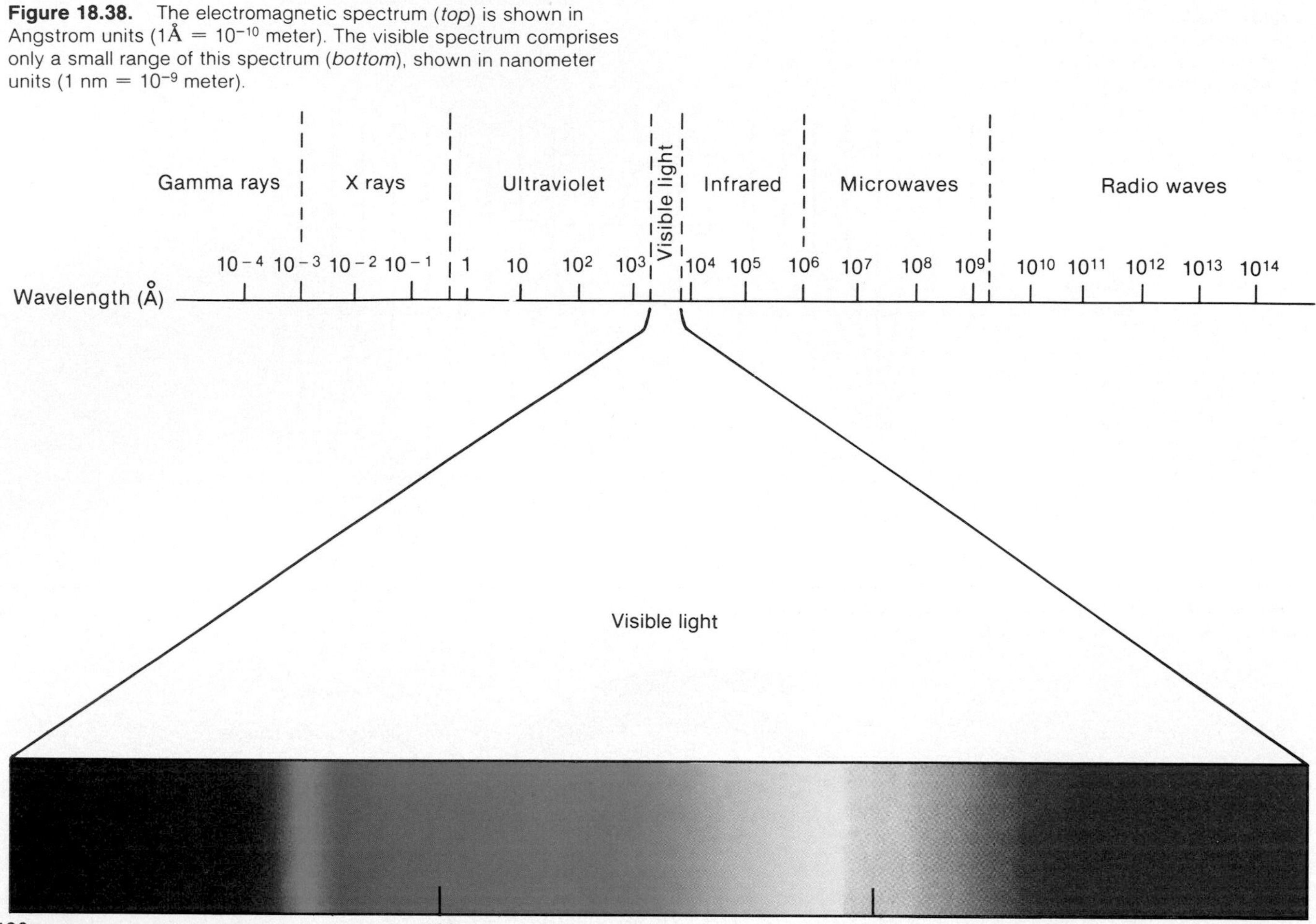

Figure 18.38. The electromagnetic spectrum (*top*) is shown in Angstrom units (1Å = 10^{-10} meter). The visible spectrum comprises only a small range of this spectrum (*bottom*), shown in nanometer units (1 nm = 10^{-9} meter).

Effect of Light on the Rods

The photoreceptors—rods and cones—are activated when light produces a chemical change in molecules of pigment contained within the membranous lamellae of the outer segments of the receptor cells. Rods contain a purple pigment known as **rhodopsin.** The pigment appears purple (a combination of red and blue), because it transmits light in the red and blue regions of the spectrum, while absorbing light energy in the green region. The wavelength of light that is absorbed best—the *absorption maximum*—is about 500 nm (a green-colored light).

Cars and other objects that are green in color are seen more easily at night (when rods are used for vision) than are red objects. This is because red light is not well absorbed by rhodopsin, and only absorbed light can produce the photochemical reaction that results in vision. In response to absorbed light, rhodopsin dissociates into its two components: a pigment called **retinaldehyde,** derived from vitamin A, and a protein called **opsin.** This reaction is known as the *bleaching reaction.*

rhodopsin: Gk. *rhodon,* rose; *ops,* eye
opsin: Gk. *ops,* eye

Retinaldehyde can exist in two possible configurations (shapes)—one known as the all-*trans* form and one called the 11-*cis* form (fig. 18.39). The all-*trans* form is the most stable, but only the 11-*cis* form is found attached to opsin. In response to absorbed light energy, the 11-*cis* retinaldehyde is converted to the all-*trans* form, causing it to dissociate from the opsin. This dissociation reaction in response to light initiates changes in the ionic permeability of the rod cell membrane and ultimately results in the production of nerve impulses in the ganglion cells. As a result of these effects, rods provide black-and-white vision under conditions of low light intensity (as described in a later section).

Dark Adaptation. The bleaching reaction that occurs in the light results in a lowered amount of rhodopsin in the rods and lowered amounts of visual pigments in the cones. When a light-adapted person first enters a darkened room, therefore, sensitivity to light is low and vision is poor. A gradual increase in photoreceptor sensitivity, known as *dark adaptation,* then occurs, reaching maximal sensitivity at about twenty minutes. The increased sensitivity

Figure 18.39. The photopigment rhodopsin consists of the protein opsin combined with 11-*cis* retinaldehyde (*a*). In response to light the retinaldehyde is converted to a different form, called all-*trans*, and dissociates from the opsin (*b*). This photochemical reaction induces changes in ionic permeability that ultimately results in stimulation of ganglion cells in the retina.

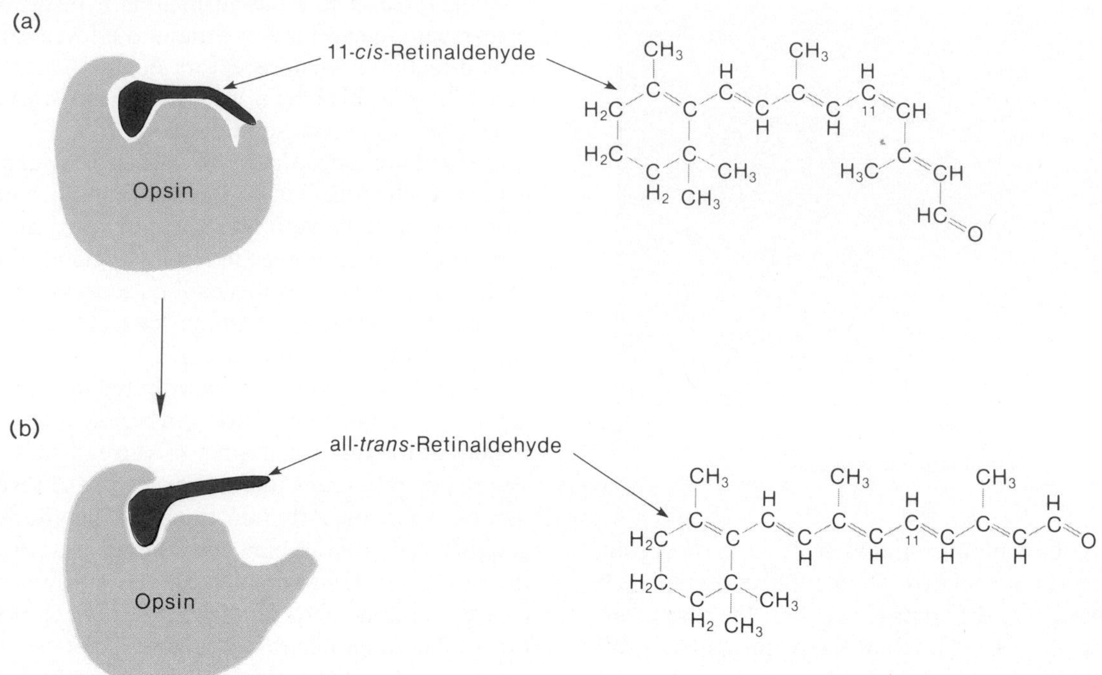

to low light intensity is due partly to increased amounts of visual pigments produced in the dark. Increased pigments in the cones produce a slight dark adaptation in the first five minutes. Increased rhodopsin in the rods produces a much greater increase in sensitivity to low light levels and is partly responsible for the adaptation that occurs after about five minutes in the dark. In addition to the increased concentration of rhodopsin, other more subtle (and less well-understood) changes occur in the rods that ultimately result in a 100,000-fold increase in light sensitivity in dark-adapted as compared to light-adapted eyes.

Electrical Activity of Retinal Cells

The only neurons in the retina that produce all-or-none action potentials are ganglion cells and amacrine cells. The photoreceptors, bipolar cells, and horizontal cells (see fig. 18.29) instead produce only graded depolarizations or hyperpolarizations, analogous to EPSPs and IPSPs.

In the dark, the photoreceptors have a resting membrane potential that is less negative (closer to zero) than that of most other neurons. This is caused by a constant current of Na^+ into the cell, called a *dark current,* through special Na^+ channels. Light causes these Na^+ channels to become blocked; as a result, the photoreceptors become less depolarized than they are in the dark. Light, therefore, causes the photoreceptors to become *hyperpolarized* in comparison to their membrane potential in the dark.

Photoreceptors in the dark release a neurotransmitter chemical at a constant rate at their synapses with bipolar cells. Hyperpolarization of the photoreceptors in response to light is similar to the hyperpolarization of other neurons that occurs during postsynaptic inhibition; the release of neurotransmitter by the photoreceptors is inhibited. The bipolar cells, as a result, may be less stimulated (if the neurotransmitter was excitatory) or less inhibited (if the neurotransmitter was inhibitory). There are both excitatory and inhibitory synapses between photoreceptors and bipolar cells, but it is not known if different transmitter chemicals are involved in this effect or if different bipolar cells respond in opposite ways to the same neurotransmitter. In either case, the bipolar cells may, through the mechanisms described, be either stimulated (by less inhibitory transmitter) or inhibited (by less excitatory transmitter) as a result of the hyperpolarization of photoreceptors in response to light.

Cones and Color Vision

Cones are less sensitive to light than rods are but provide color vision and greater visual acuity, as described in the next section. During the day, therefore, the high light intensity bleaches out the rods, and color vision with high acuity is provided by the cones. According to the **trichromatic theory** of color vision, our perception of a multitude of colors is due to stimulation of only three types of cones.

Figure 18.40. There are three types of cones. Each type of cone contains retinaldehyde combined with a different type of protein, producing a pigment that absorbs light maximally at a different wavelength. Color vision, according to the trichromatic theory, is produced by activity of these blue cones, green cones, and red cones.

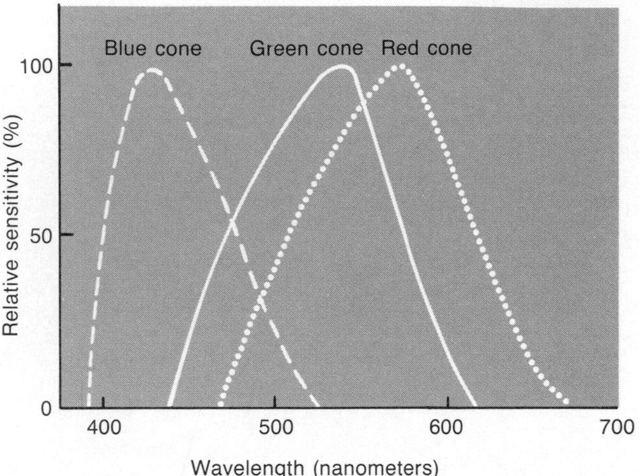

Each type of cone contains retinaldehyde, as in rhodopsin, but this molecule is associated with a different protein than opsin. The protein is different for each of the three cone pigments, and as a result, each of the pigments has a different color. The three colors are blue, green, and red, which correspond to the region of the visible spectrum in which each cone pigment absorbs light maximally (fig. 18.40). Our perception of any given color is produced by the relative degree to which each cone is stimulated by any given wavelength of visible light.

Suppose a person has become dark-adapted in a photographic darkroom over a period of twenty minutes or longer, but needs an increase in light to examine some prints. Since rods do not absorb red light but red cones do, a red light in a photographic darkroom allows vision (because of the red cones), but does not cause bleaching of the rods. When the light is turned off, therefore, the rods will still be dark-adapted and the person will still be able to see.

Color blindness which affects about 5% of American men (women are far less frequently affected because this is a sex-linked trait), is due to a congenital lack of one or more types of cones. People with normal color vision are *trichromats;* people with only two types of cones are *dichromats.* They may be missing red cones (have *protanopia*), or green cones (have *deuteranopia*), or blue cones (have *tritanopia*). Such a person would have difficulty distinguishing red from green. People who are *monochromats* (which is very rare), have only one cone system and can see only black, white, and shades of gray.

Visual Acuity and Sensitivity

While reading or similarly focusing visual attention on objects in daylight, each eye is oriented so that the image falls within the fovea centralis. The fovea is a pinhead-sized pit within a yellow area of the retina called the *macula lutea.* The pit is formed as a result of the displacement of neural layers around the fovea, so that light falls directly on photoreceptors in this region—whereas light falling on other areas must pass through several layers of neurons, as previously described.

There are more than 100 million rods and 7 million cones in each retina, but only about 1.2 million nerve fibers enter the optic nerve of each eye. This gives an overall convergence of photoreceptors on ganglion cells of about 105:1. This number is misleading, however, because the degree of convergence is much lower for cones than for rods, and it is 1:1 in the fovea.

The photoreceptors are distributed in such a way that the fovea contains only cones, whereas more peripheral regions of the retina contain a mixture of rods and cones. Approximately four thousand cones in the fovea provide input to approximately four thousand ganglion cells; each ganglion cell in this region, therefore, has a private line to the visual field. Each ganglion cell thus receives input from an area of retina corresponding to the diameter of one cone (about 2 μm). Peripheral to the fovea, however, many rods synapse with a single bipolar cell, and many bipolar cells synapse with a single ganglion cell. A single ganglion cell outside the fovea thus may receive input from large numbers of rods, corresponding to an area of about 1 mm^2 on the retina.

Since each cone in the fovea has a private line to a ganglion cell, and since each ganglion cell receives input from only a tiny region of the retina, visual acuity is greatest and sensitivity to low light is poorest when light falls on the fovea. In dim light only the rods are activated, and vision is best out of the corners of the eye when the image falls away from the fovea. Under these conditions, the convergence of many rods on a single bipolar cell and the convergence of many bipolar cells on a single ganglion cell increase sensitivity to dim light at the expense of visual acuity. Night vision is therefore less distinct than day vision.

Neural Pathways of Vision, Eye Movements, and Processing Visual Information

The photoreceptor neurons, rods and cones, are the functional units of sight in that they respond to light and produce nerve impulses through bipolar neurons to ganglion neurons (see fig. 18.29). The optic nerve consists of axons of aggregated ganglion neurons that emerge through the posterior aspect of the eyeball. The two optic nerves (one from each eyeball) converge at the **optic chiasma** *(ki-as'mah)* (fig. 18.41). The fibers partially cross at the optic chiasma so that all the fibers arising from the medial (nasal) half of each retina cross to the opposite side. The

Figure 18.41. Visual fields of the eyes and neural pathways for vision. An overlapping of the visual field of each eye provides binocular vision, which is the ability to perceive depth.

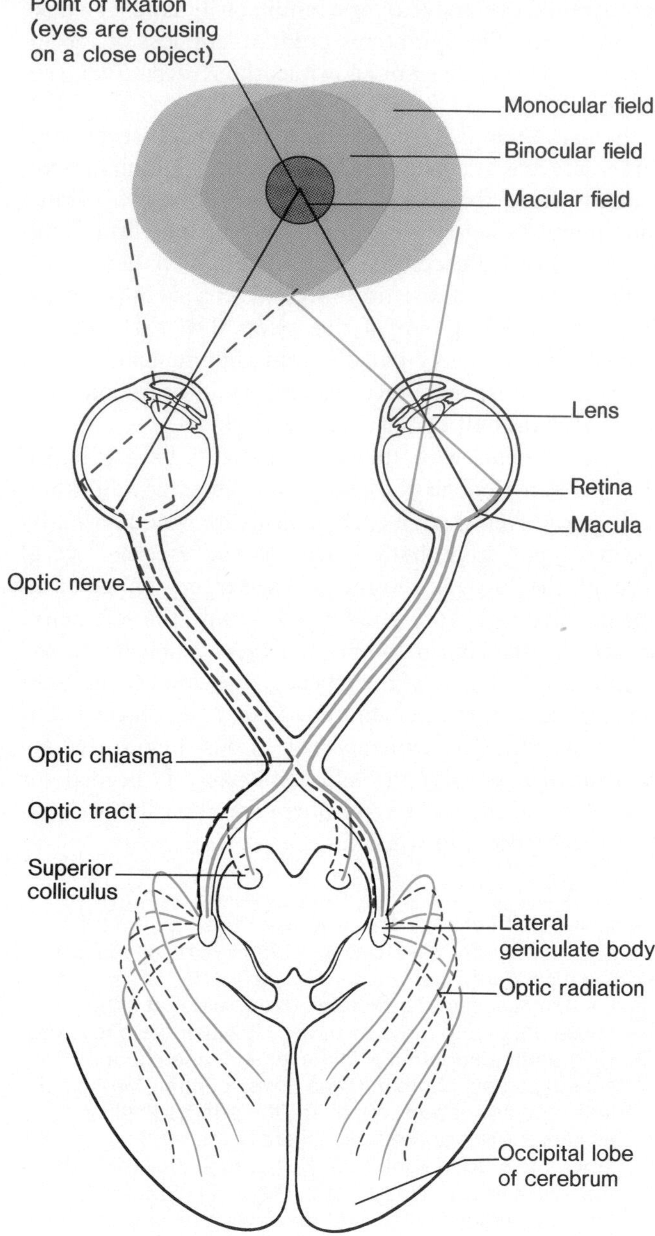

Figure 18.42. The lateral geniculate body. Each lateral geniculate consists of six layers (numbered *1* through *6* in this figure). Each of these layers receives input from only one eye, with right and left eyes alternating. An arrow through these six layers (see figure) of the left lateral geniculate, for example, encounters corresponding projections from a part of the right visual field in right and left eyes, alternatively, as it passes from the outer to the inner layers.

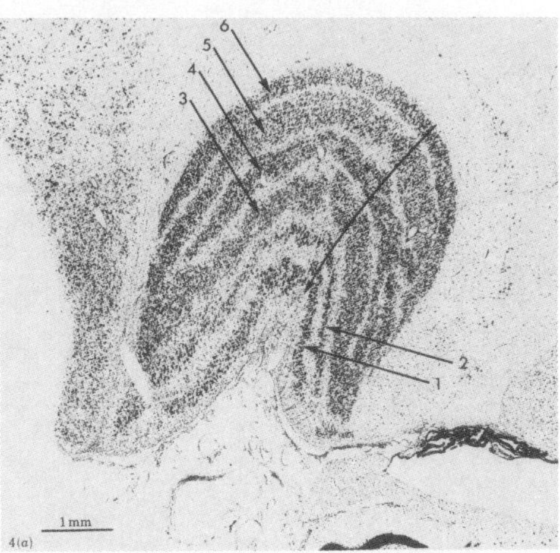

fibers of the optic nerve that arise from the lateral (temporal) half of the retina do not cross. The **optic tract** is a continuation of optic nerve fibers from the optic chiasma and is composed of fibers arising from the retinas of both eyeballs.

As the optic tracts enter the brain, some of the fibers in the tracts terminate in the **superior colliculi** *(ko-lik'u-li)*. These fibers and the motor pathways they activate constitute the **tectal system,** which is responsible for body-eye coordination.

Approximately 70%–80% of the fibers in the optic tract pass to the **lateral geniculate** *(jĕ-nik'u-lat)* **body** of the thalamus (fig. 18.42). Here the fibers synapse with neurons whose axons constitute a pathway called the **optic radiation.** The optic radiation transmits impulses to the **striate cortex** area of the occipital cerebral lobe. This entire arrangement of visual fibers is known as the **geniculostriate system,** which is responsible for perception of the visual field.

The nerve fibers that cross at the optic chiasma arise from the retinas in the medial portions of the eyeballs. The photoreceptors of these fibers are stimulated by light entering the eyeball from the periphery. If the optic chiasma were severed longitudinally, peripheral vision would be lost and leave only "tunnel vision." If an optic tract were severed, both eyes would be partially blind, the lateral field of vision lost for one eye and the medial field of vision lost for the other.

The Superior Colliculi and Eye Movements. Neural pathways from the superior colliculi to motor neurons in the spinal cord help mediate the "startle" response to the sight of an unexpected intruder. Other nerve fibers from the superior colliculi stimulate the **extrinsic eye muscles** (table 18.5), which are the skeletal muscles that move the eyes.

Table 18.5 Muscles of the eye

Extrinsic muscles (skeletal)		
Superior rectus	Oculomotor nerve (III)	Rotates eye upward and toward midline
Inferior rectus	Oculomotor nerve (III)	Rotates eye downward and toward midline
Medial rectus	Oculomotor nerve (III)	Rotates eye toward midline
Lateral rectus	Abducens nerve (VI)	Rotates eye away from midline
Superior oblique	Trochlear nerve (IV)	Rotates eye downward and away from midline
Inferior oblique	Oculomotor nerve (III)	Rotates eye upward and away from midline
Intrinsic muscles (smooth)		
Ciliary muscles	Oculomotor nerve (III) parasympathetic fibers	Causes suspensory ligament to relax
Iris, circular muscles	Oculomotor nerve (III) parasympathetic fibers	Causes size of pupil to decrease
Iris, radial muscles	Sympathetic fibers	Causes size of pupil to increase

There are two types of eye movements coordinated by the superior colliculi. *Smooth pursuit movements* track moving objects and keep the image focused on the fovea centralis. *Saccadic (sah-kad'ik) eye movements* are short (lasting 20 to 50 msec), jerky movements that occur while the eyes appear to be still. These saccadic movements are believed to be important in maintaining visual acuity.

The tectal system is also involved in the control of the **intrinsic eye muscles**—the iris and the muscles of the ciliary body. Shining a light into one eye stimulates the *pupillary reflex,* in which both pupils constrict. This is caused by activation of parasympathetic neurons by fibers from the superior colliculi. Postganglionic neurons in the ciliary ganglia behind the eyes, in turn, stimulate constrictor fibers in the iris. Contraction of the ciliary body during *accommodation* also involves stimulation of the superior colliculi.

Processing of Visual Information. In order for visual information to have meaning it must be associated with past experience and integrated with information from other senses. Some of this higher processing occurs in the inferior temporal lobes of the cortex. Experimental removal of these areas from monkeys impairs their ability to remember visual tasks that they previously learned and hinders their ability to associate visual images with the significance of the object. Monkeys with their inferior temporal lobes removed, for example, will handle a snake without fear. The symptoms produced by loss of the inferior temporal lobes are known as the *Klüver-Bucy syndrome.*

In an attempt to reduce the symptoms of severe epilepsy, surgeons at one time cut the corpus callosum in some patients. This fiber tract, as previously described, transmits impulses between the right and left cerebral hemispheres. The right cerebral hemisphere of patients with such *split brains,* therefore, receives sensory information from only the left half of the external world. The left hemisphere, similarly cut off from communication with the right hemisphere, receives sensory information from only the right half of the external world.

Experiments with split-brain patients have revealed that the functions of the two hemispheres are different. This is true even though each hemisphere would normally receive input from both halves of the external world through the corpus callosum. If the sensory image of an object, such as a key, is delivered to only the left hemisphere (by showing it to only the right visual field), the object can be named. If the object is presented to the right cerebral cortex, the person knows what the object is but cannot name it. Experiments such as this suggest that (in right-handed people) the left hemisphere is needed for language and the right hemisphere is responsible for pattern recognition (fig. 18.43).

1. List four structures of the eye that develop from the neuroectoderm. Which layers of the eye develop from the mesoderm?
2. List the associated structures of the eye that either cause the eye to move or protect it within the bony orbit.
3. Diagram the structure of the eye, and label the following: sclera, cornea, choroid, retina, fovea centralis, iris, pupil, lens, and ciliary body. What are the principal cells or tissues in each of the three layers of the eye?
4. Trace the path of light through the three chambers of the eye, and explain the mechanism of light refraction and how the eye is focused for viewing distant and near objects.
5. List the different layers of the retina, and describe the path of light and of nerve activity through these layers. Continue tracing the path of a visual impulse to the cerebral cortex, and list in order the structures traversed.
6. Describe the photochemical reaction that leads to the activation of rods and the role of vitamin A in vision.
7. Distinguish between rods and cones in terms of function, visual acuity, and sensitivity. Why is dark adaptation not lost when a red light is used in a photographic dark room?

Figure 18.43. Different functions of the right and left cerebral hemispheres, as revealed by experiments with people who have had the tract connecting the two hemispheres (the corpus callosum) surgically split.

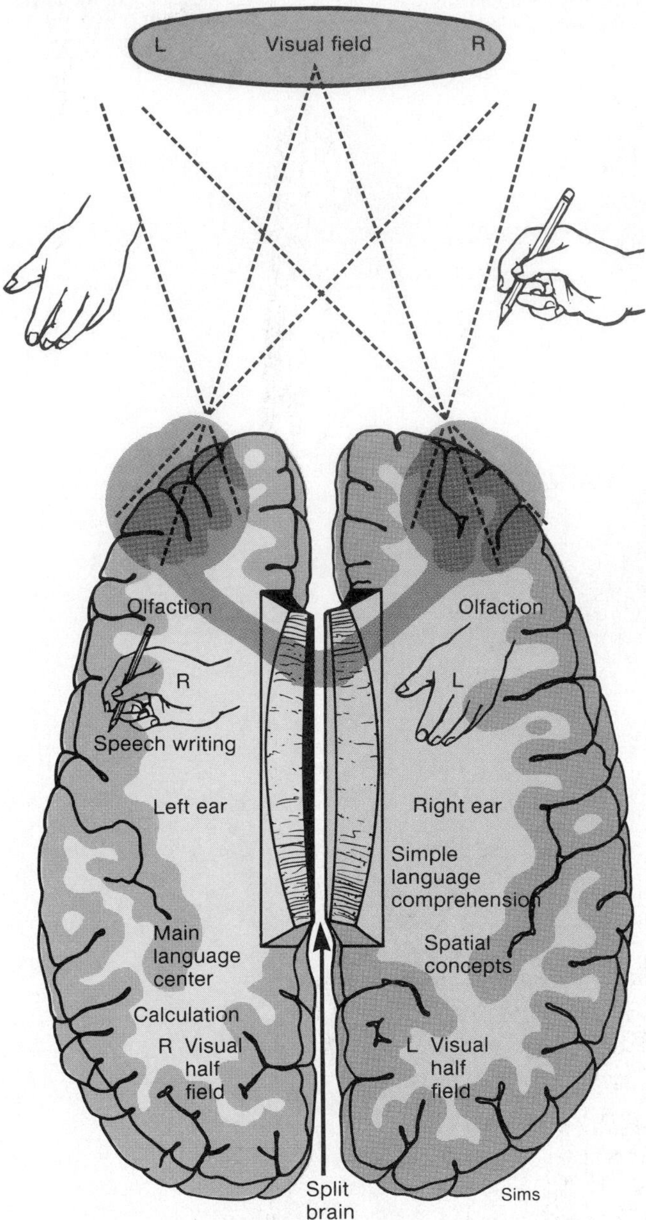

Senses of Hearing and Balance

Structures of the outer, middle, and inner ear are involved in the sense of hearing. The inner ear also contains structures that provide a sense of balance, or equilibrium. The development of the ear begins during the fourth week and is completed by the thirty-second week.

Objective 19. Describe the development of the ear.

Objective 20. Describe the structures of the ear that relate to hearing, and list their locations and functions.

Objective 21. Trace the path of sound waves through the ear, and describe how they are transmitted and converted to nerve impulses.

Objective 22. Explain the mechanisms by which equilibrium is maintained.

Development of the Ear

The ear begins to develop at the same time as the eye, early during the fourth week. All three embryonic germ layers—ectoderm, mesoderm, and endoderm—are involved in the formation of the ear. Both types of ectoderm (neuroectoderm and surface ectoderm) play a role.

The ear of an adult is structurally and functionally divided into an **external ear,** a **middle ear,** and an **inner ear,** each of which has a separate embryonic origin. The inner ear does not develop from deep embryonic tissue as one might expect but rather begins to form early in the fourth week when a plate of surface ectoderm called the **otic** *(o'tik)* **placode** appears lateral to the developing hindbrain (fig. 18.44). The otic placode soon invaginates and forms an **otic pit.** Toward the end of the fourth week, the outer edges of the invaginated otic pit come together and fuse to form an **otocyst (otic vesicle).** The otocyst soon pinches off and separates from the surface ectoderm. The otocyst further differentiates to form a dorsal **utricular portion** and a ventral **saccular portion.** Three separate diverticula extend outward from the utricular portion and develop into the **semicircular canals,** which later function in balance and equilibrium. A tubular diverticulum, called the **cochlear duct,** extends in a coiled fashion from the saccular portion and forms the membranous portion of the **cochlea** of the ear (fig. 18.44). The **organ of Corti,** which is the functional portion of the cochlea, differentiates from cells along the wall of the cochlear duct (fig. 18.45). The sensory nerves that innervate the inner ear are derived from neuroectoderm from the developing brain.

The differentiating otocyst is surrounded by mesodermal tissue that soon forms a cartilaginous **otic capsule** (fig. 18.46). As the otocyst and surrounding otic capsule grow in size, vacuoles containing the fluid **perilymph** form within the otic capsule. The vacuoles soon enlarge and coalesce to form the **perilymphatic space,** which divides into the **scala tympani** and the **scala vestibuli.** Eventually, the cartilaginous otic capsule ossifies to form the **bony (osseous) labyrinth** of the inner ear. The middle ear chamber is referred to as the **tympanic cavity** and derives from the first pharyngeal pouch (see fig. 18.46). The **ossicles,** which amplify incoming sound waves, derive from the first and second pharyngeal arch cartilages. As the tympanic cavity

otic: Gk. *otikos,* ear

Figure 18.44. Development of the inner ear. (*a*) a lateral view of a twenty-two-day-old embryo showing the position of a transverse cut through the otic placode. (a_1) the otic placode of surface ectoderm begins to invaginate at twenty-two days. (a_2) by twenty-four days, a distinct otic pit is formed and the neural ectoderm is positioned to give rise to the brain. (*b*) a lateral view of a twenty-eight-day-old embryo showing the position of a transverse cut through the otocyst. (b_1) by twenty-eight days, the otic pit has become a distinct otocyst. (b_2) the otocyst is in position in the thirty-day-old embryo where it gives rise to the structures of the inner ear. (*c-e*) lateral views of the differentiating otocyst into the cochlea and semicircular canals from the fifth to the eighth week.

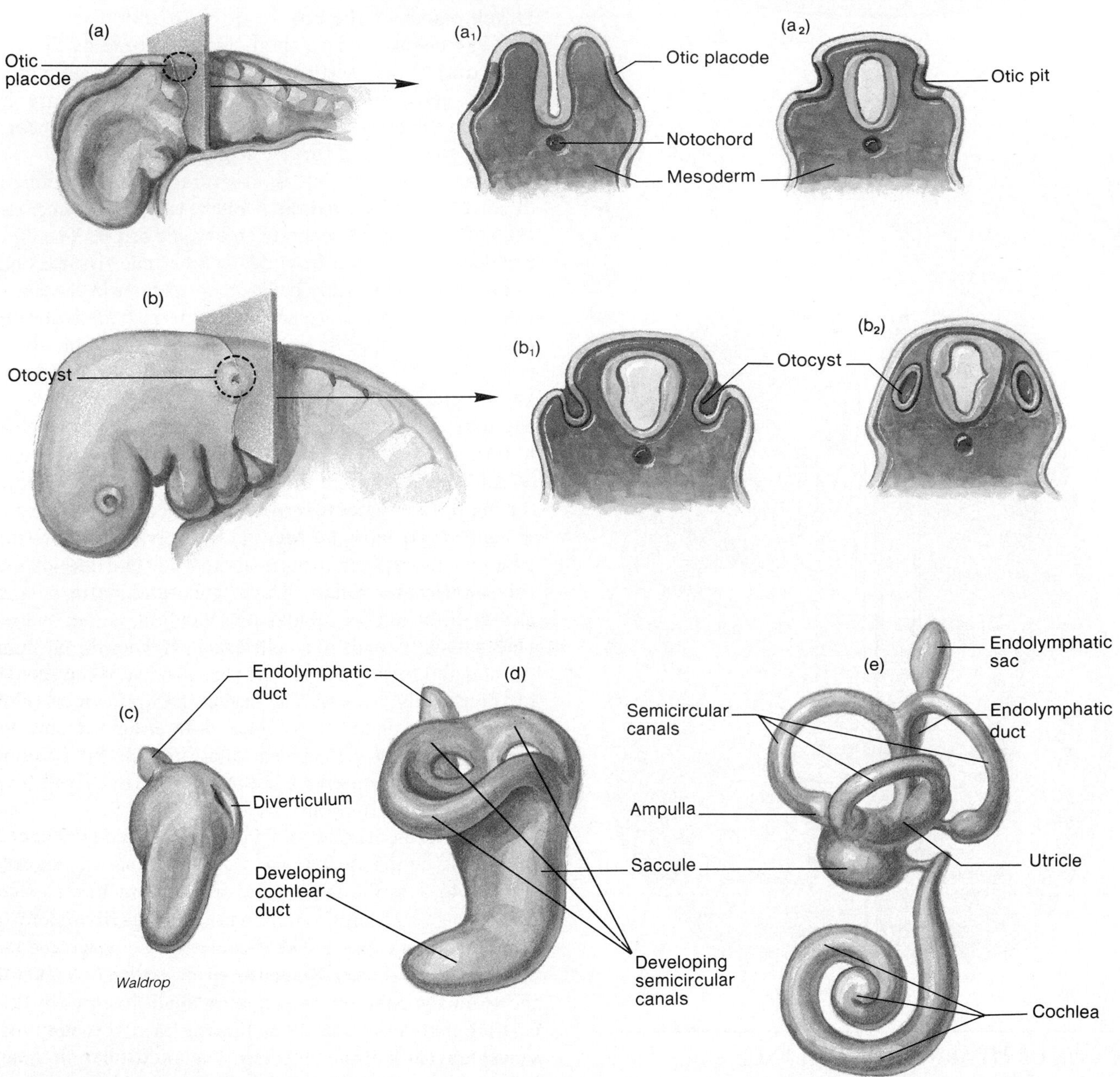

enlarges, it surrounds and encloses the developing ossicles (fig. 18.46). The connection of the tympanic cavity to the pharynx gradually elongates to develop into the **auditory (eustachian) tube,** which remains patent throughout life and is important in maintaining an equilibrium of air pressure between the pharyngeal and tympanic cavities.

The external ear includes the fleshy **auricle** attached to the side of the head and the tubular **external auditory meatus** that extends into the temporal bone of the skull. The external auditory meatus forms from the surface ectoderm that covers the dorsal end of the first branchial groove (see fig. 18.46). A solid epithelial plate called the

Figure 18.45. The formation of the cochlea and the organ of Corti from the otic capsule. (*a–d*) successive stages of development of the perilymphatic space and the organ of Corti from the eighth to the twentieth week.

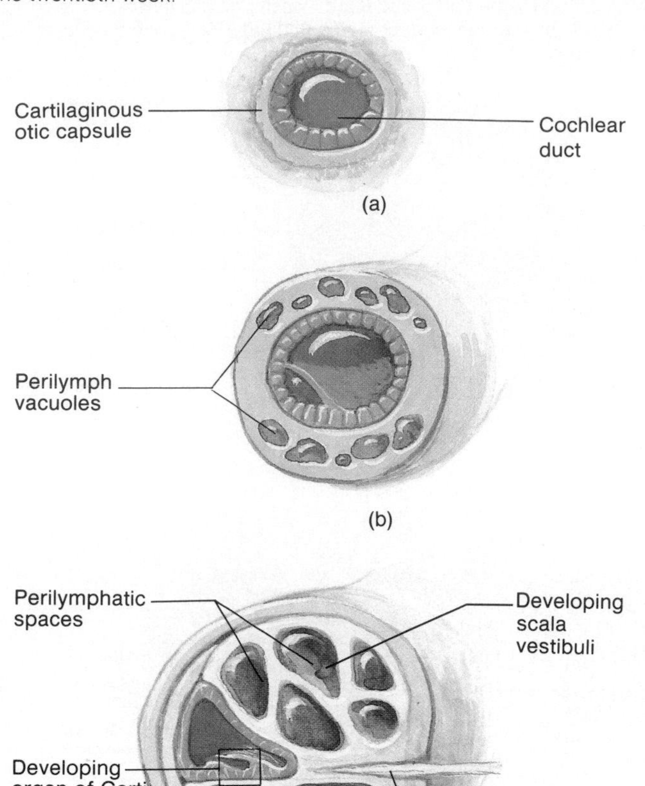

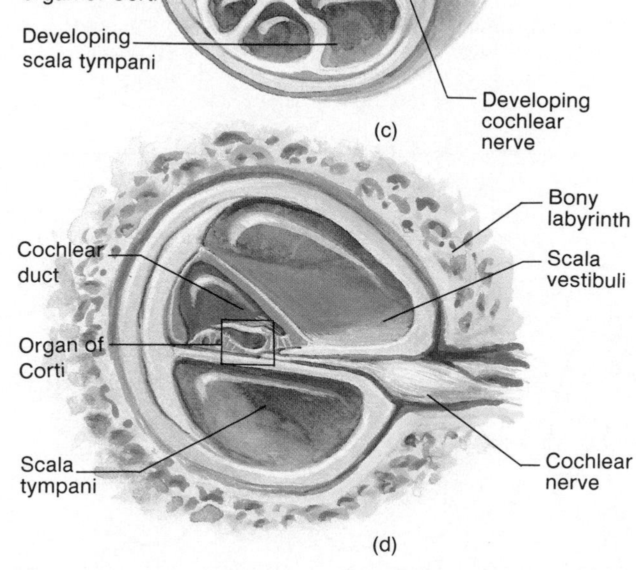

meatal plug soon develops at the bottom of the funnel-shaped branchial groove. The meatal plug is involved in the formation of the inner wall of the external auditory meatus and contributes to the **tympanic membrane** (eardrum). The tympanic membrane has a dual origin from surface ectoderm and the endoderm lining the first pharyngeal pouch (see fig. 18.46).

The auricle develops from six separate swellings of surface ectoderm, called **auricular hillocks,** that form around the first branchial groove (fig. 18.47). As the auricular hillocks enlarge, they coalesce and are supported by mesodermal mesenchyme that differentiates into fibrous tissue. The auricles migrate superiorly as the face of the embryo and fetus forms, so that by thirty-two weeks the auricles have ascended to the dorsoventral level of the eyes.

Structure of the Ear

The ear is the organ of hearing and equilibrium. It contains receptors that respond to movements of the head and receptors that convert sound waves into nerve impulses. Impulses from both receptor types are transmitted through the vestibulocochlear (VIII) cranial nerve to the brain for interpretation. The ear consists of three principal regions: the external ear, the middle ear, and the inner ear (fig. 18.48).

External Ear. The external ear consists of the auricle, or pinna, and the external auditory meatus. The **auricle** *(aw'rĕ-kl)* is the visible fleshy appendage attached to the side of the head. It consists of a cartilaginous framework of elastic connective tissue covered with skin. The rim of the auricle is called the **helix,** and the inferior fleshy portion is the **lobe** (fig. 18.49). The lobe is the only portion of the auricle that is not supported with cartilage. The auricle has a ligamentous attachment to the skull and poorly developed auricular muscles inserting anteriorly, superiorly, and posteriorly within it. The blood supply to the auricle is from the posterior auricular and occipital arteries, which branch from the external carotid and superficial temporal arteries, respectively. The structure of the auricle directs sound waves into the external auditory meatus.

The **external auditory meatus** is a slightly S-shaped canal about 2.5 cm (1 in.) in length, extending slightly upward from the auricle to the tympanic membrane (fig. 18.48). The skin that lines the meatus contains fine hairs and sebaceous glands near the entrance. Specialized wax-secreting glands, called **ceruminous** *(sĕ-roo'mĭ-nus)* **glands,** are located in the skin, deep within the meatus. *Cerumen* (earwax) secreted from ceruminous glands keeps the tympanic membrane soft and waterproof. Cerumen and the hairs also help to prevent small foreign objects from reaching the tympanic membrane. The bitter cerumen is probably an insect repellent as well.

The **tympanic** *(tim-pan'ik)* **membrane** (eardrum) is a thin, double-layered, epithelial partition between the external auditory meatus and the middle ear. It is approximately 1 cm in diameter and is composed of an outer concave layer of stratified squamous epithelium and an

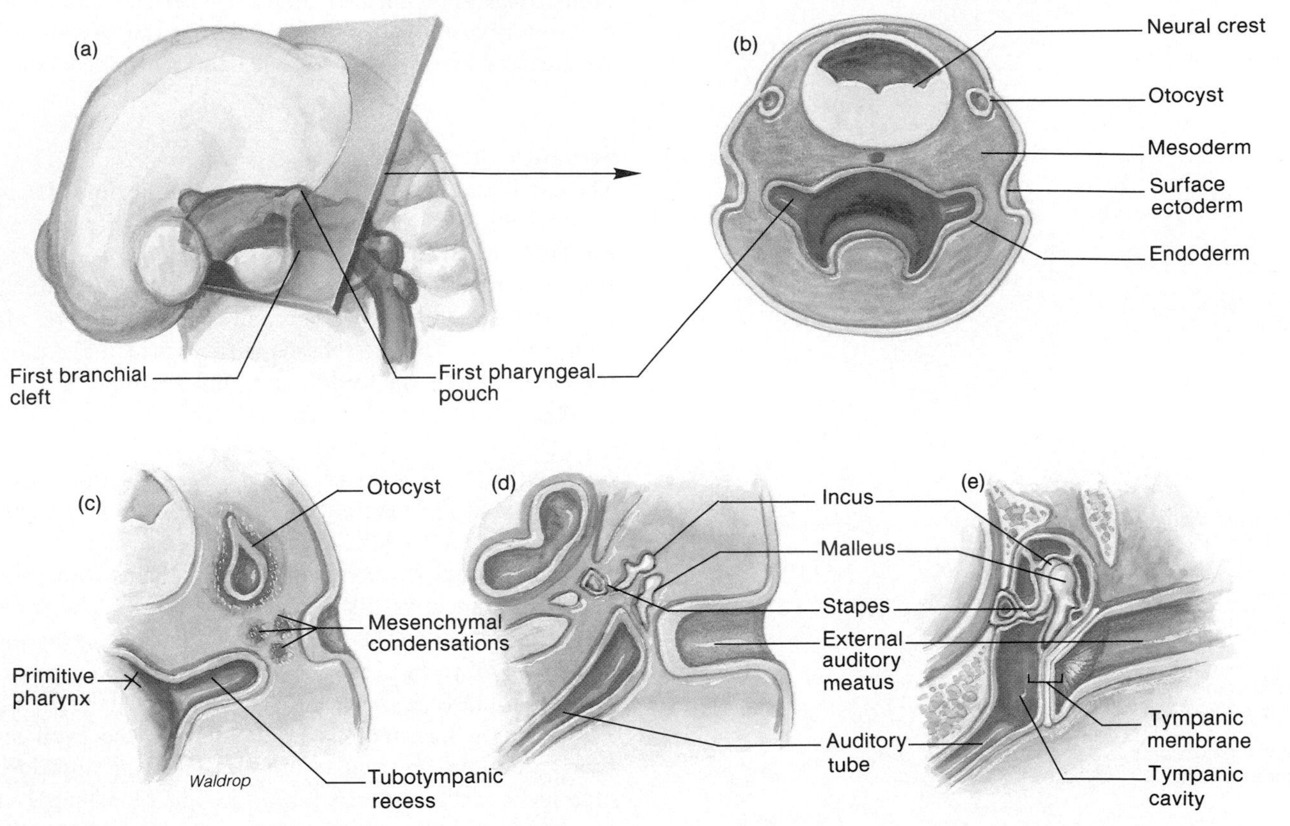

Figure 18.46. Development of the outer and middle ear regions and the ear ossicles. (*a*) a lateral view of a four-week-old embryo showing the position of the cut depicted in the sequential development (*b-e*). (*b*) the embryo at four weeks illustrating the invagination of the surface ectoderm and the evagination of the endoderm at the level of the first pharyngeal pouch. (*c*) during the fifth week, mesenchymal condensations are apparent which will derive the ear ossicles. (*d*) further invagination and evagination at six weeks correctly positions the structures of the outer and middle ear regions. (*e*) by the end of the eighth week, the ear ossicles, tympanic membrane, auditory tube, and external auditory meatus are formed.

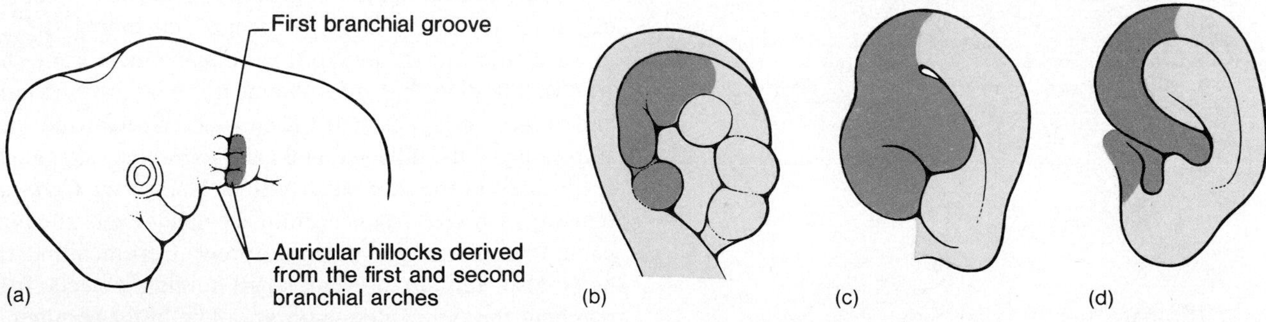

Figure 18.47. The progressive development of the auricle of the ear: (*a*) five weeks; (*b*) six weeks; (*c*) eight weeks; (*d*) thirty-two weeks. The ears develop from structures of the neck region but gradually migrate to the side of the head.

inner convex layer of low columnar epithelium. The tympanic membrane is extremely sensitive to pain and is innervated by the auriculotemporal nerve (a branch of the mandibular portion of the trigeminal [V] cranial nerve) and the auricular nerve (a branch of the vagus [X] cranial nerve).

Inspecting the tympanic membrane with an otoscope during a physical examination provides significant information about the condition of the middle ear. A diagram of the normal tympanic membrane is presented in figure 18.50. The color, curvature, presence of lesions, or position of the malleus of the middle ear are features of particular importance. If ruptured, the tympanic membrane can generally regenerate and readily heal itself.

otoscope: Gk. *otikos,* ear; *skopein,* to examine

Figure 18.48. The ear. Note the outer, middle, and inner regions of the ear.

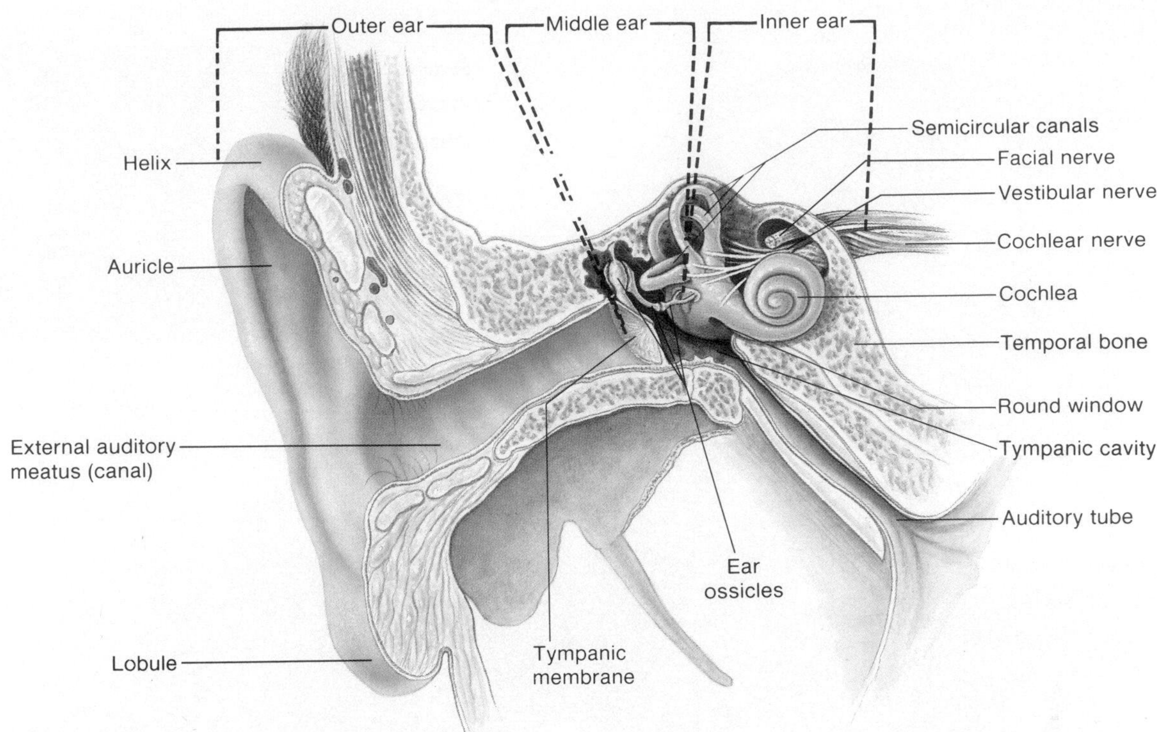

Figure 18.49. The surface anatomy of the auricle of the ear.

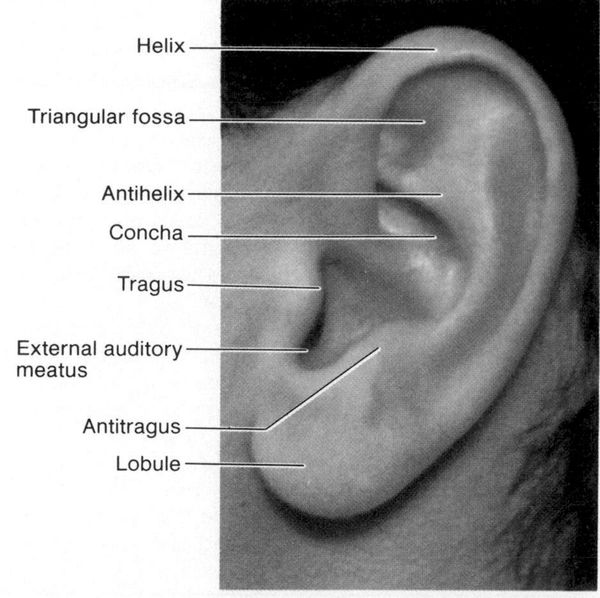

Figure 18.50. The right tympanic membrane as seen with an *auriscope* (a form of otoscope). A small depression, the *umbo,* occurs at the tip of the handle of the malleus. When the tympanic membrane is illuminated with an auriscope, the concavity of the umbo produces a "cone of light" that radiates anteriorly and inferiorly.

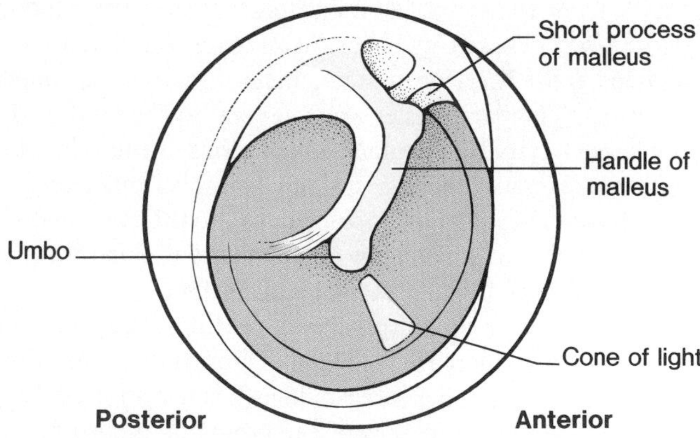

Middle Ear. The laterally compressed middle ear is an air-filled chamber called the **tympanic cavity** in the petrous portion of the temporal bone (see figs. 18.48, 18.51). The tympanic membrane separates the middle ear from the external auditory meatus of the outer ear. A bony partition containing the **oval window (fenestra vestibuli)** and the **round window (fenestra cochlea)** separates the middle ear from the inner ear.

There are two openings into the tympanic cavity. The **epitympanic recess** in the posterior wall connects the tympanic cavity to the **mastoidal air cells** within the mastoid process of the temporal bone. The **auditory** *(aw'di-to"re),* or **eustachian, tube** connects the tympanic cavity anteriorly with the nasopharynx and equalizes air pressure on both sides of the tympanic membrane.

Figure 18.51. The ear ossicles and associated structures within the (*a*) tympanic cavity; (*b*) the stapedius muscle arises from a bony protrusion called the pyramid.

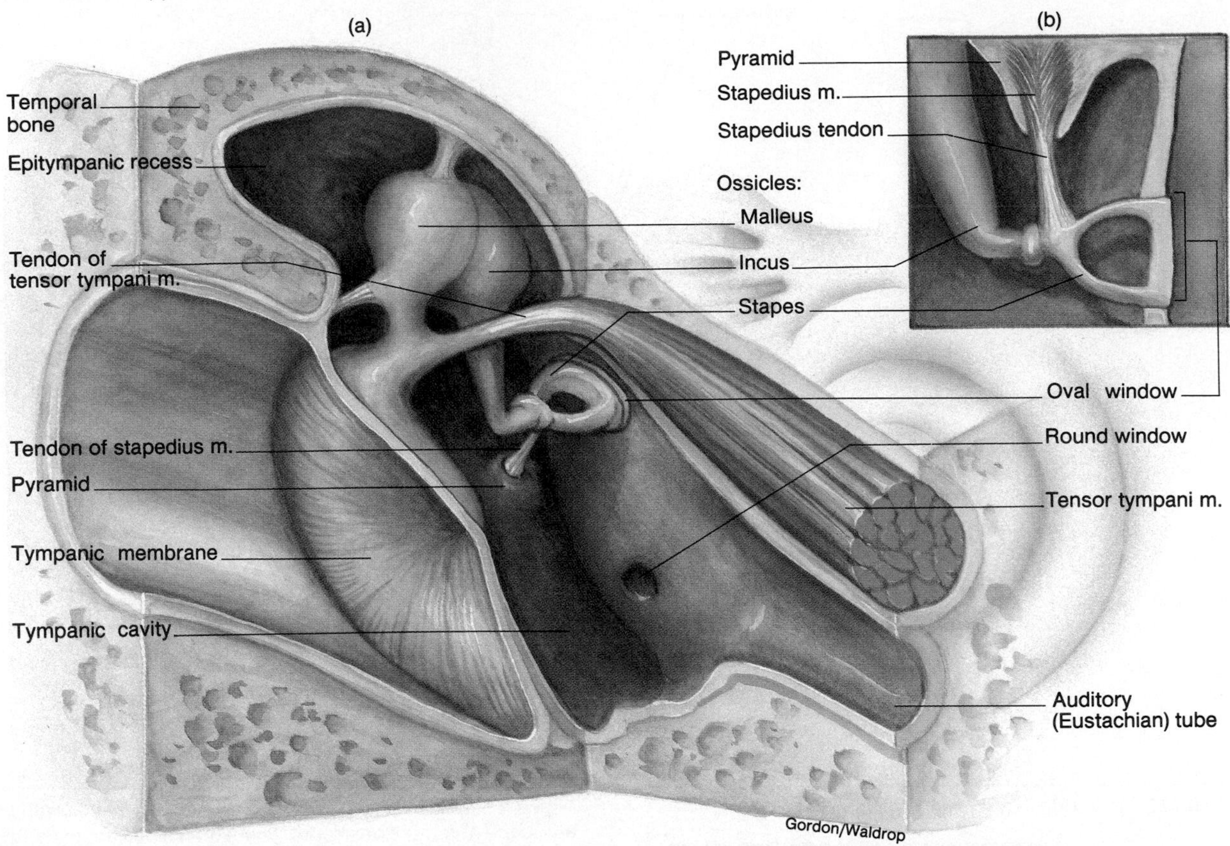

A series of three **ear ossicles** crosses the tympanic cavity from the tympanic membrane to the oval window (fig. 18.51). These tiny bones, from outer to inner, are the **malleus** *(mal'e-us)* (hammer), **incus** (anvil), and **stapes** (stirrup). The ear ossicles are attached to the wall of the tympanic cavity by ligaments. Vibrations of the tympanic membrane cause the ear ossicles to move and transmit sound waves across the tympanic cavity to the oval window. Vibration of the oval window moves a fluid within the inner ear and stimulates the receptors of hearing.

As the ear ossicles transmit vibrations, they act as a lever system to increase the force of vibration. In addition, the force of vibration is intensified as it is transmitted from the relatively large surface of the tympanic membrane to the smaller surface area of the oval window. The combined effect increases the force of the vibrations by about twenty times.

Two small skeletal muscles, the **tensor tympani** and the **stapedius** *(sta-pe'de-us)* (fig. 18.51), attach to the malleus and stapes, respectively, and contract reflexively to protect the inner ear against loud noises. When contracted, the tensor tympani pulls the malleus inward, and the stapedius pulls the stapes outward. This combined action reduces the force of vibration of the ossicles.

The mucous membranes that line the tympanic cavity, the mastoidal air cells, and the auditory tube are continuous with those of the nasopharynx. For this reason, infections of the nose or throat may spread to the tympanic cavity and cause a middle ear infection or even to the mastoidal air cells and cause *mastoiditis.* Forcefully blowing the nose advances the spread of the infection.

An equalization of air pressure on both sides of the tympanic membrane is of functional importance to hearing. When atmospheric pressure is reduced, as occurs at higher altitudes, the tympanic membrane bulges outward in response to the greater air pressure within the tympanic cavity. The bulging is painful and may impair hearing by reducing flexibility. The auditory tube, which is collapsed most of the time in adults, opens during swallowing or yawning and allows the air pressures to equalize.

Inner Ear. The inner ear, or **labyrinth,** contains the functional organs for hearing and equilibrium. The labyrinth consists of two parts: an outer **bony labyrinth** *(lab'ĭ-rinth)* and a **membranous labyrinth** contained within the bony labyrinth (fig. 18.52). The space between the bony labyrinth and the membranous labyrinth is filled with a fluid, called *perilymph,* secreted by cells lining the bony canals. Within the tubular chambers of the membranous labyrinth is another fluid called *endolymph.* These two

Figure 18.52. The labyrinths of the inner ear. The membranous labyrinth is contained within the bony labyrinth.

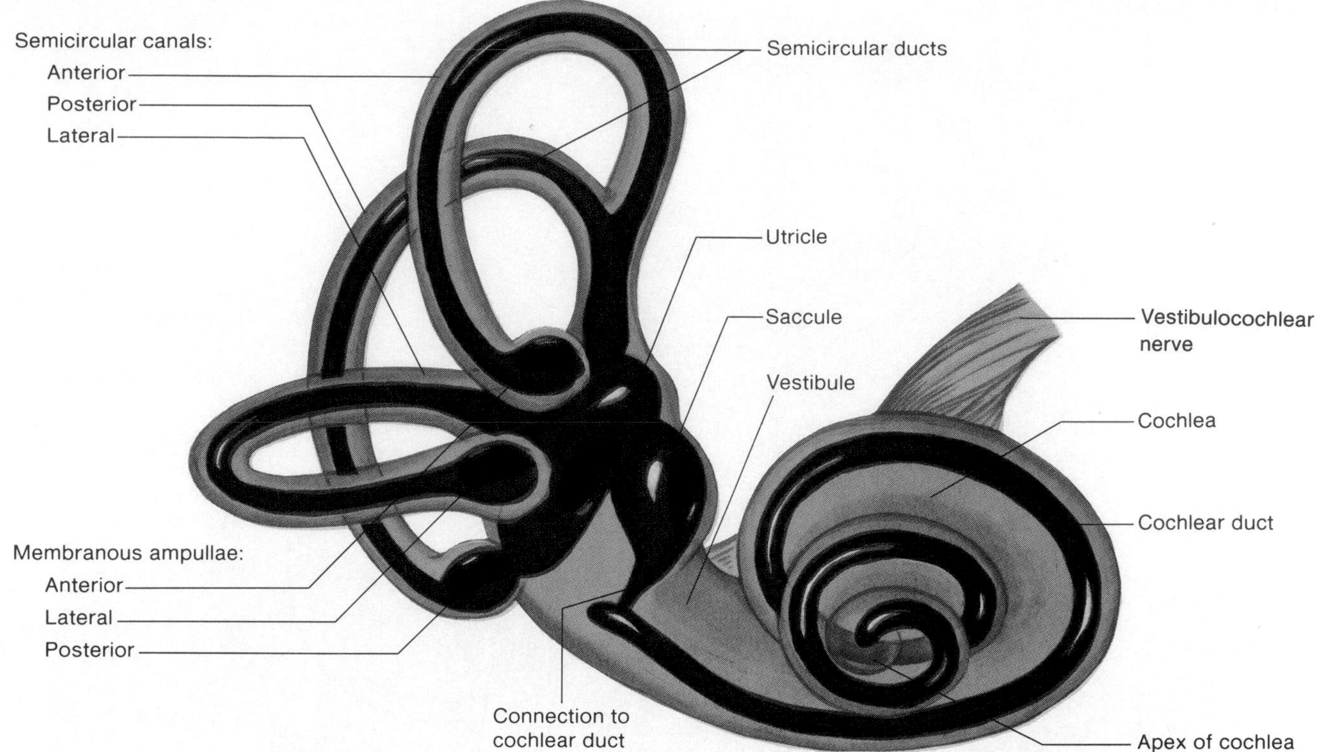

fluids provide a liquid-conducting medium for the vibrations involved in hearing and the maintenance of equilibrium.

The bony labyrinth is structurally and functionally divided into three areas: vestibule, semicircular canals, and cochlea.

Vestibule. The vestibule is the central portion of the bony labyrinth that contains the oval window, into which the stapes fits, and the round window on the opposite end (fig. 18.52).

The membranous labyrinth within the vestibule consists of two connected sacs called the **utricle** *(u'tre-k'l)* and the **saccule** *(sak'ūl.)* The utricle is larger than the saccule and lies in the upper back portion of the vestibule. Both the utricle and saccule contain receptors that are sensitive to gravity and linear movement of the head.

Semicircular Canals. The three bony semicircular canals of each ear are at right angles to each other and are positioned posterior to the vestibule. The thinner **semicircular ducts** form the membranous labyrinth within the semicircular canals (fig. 18.52). Each of the three semicircular ducts has a **membranous ampulla** *(am-pul'lah)* at one end and connects with the upper back part of the utricle. Receptors within the semicircular ducts are sensitive to angular acceleration and deceleration of the head, as in rotational movement.

Cochlea. The snail-shaped cochlea *(kok'le-ah)* is coiled two and a half times around a central axis of bone (fig. 18.53). There are three chambers in the cochlea (fig. 18.54). The upper chamber, the **scala** *(ska'lah)* **vestibuli,** begins at the oval window and is continuous with the vestibule. The lower chamber, the **scala tympani,** terminates at the round window. The **cochlear duct** is the triangular middle chamber. The roof of the cochlear duct is called the **vestibular membrane,** and the floor is called the **basilar** *(bas'ĭ-lar)* **membrane.** The scala vestibuli and the scala tympani contain perilymph and are completely separated, except at the narrow apex of the cochlea, called the **helicotrema** *(hel"ĭ-ko-tre'mah),* where they are continuous (see fig. 18.56). The cochlear duct is filled with endolymph and ends at the helicotrema.

The **organ of Corti** is contained within the cochlear duct of the cochlea. The sound receptors that transform mechanical vibrations into nerve impulses are located along the basilar membrane of this structure, making it the functional unit of hearing. The epithelium of the organ of Corti consists of supporting cells and hair cells (figs. 18.54, 18.55). The bases of the hair cells are anchored in the basilar membrane, and their tips are embedded in the tectorial membrane that forms a gelatinous canopy over them.

---

cochlea: L. *cochlea,* snail shell
scala: Gk. *scala,* staircase
helicotrema: Gk. *helix,* a spiral; *trema,* a hole
organ of Corti: from Alfonso Corti, Italian anatomist, 1822–1888

Figure 18.53. A cross section of the cochlea showing its three turns and its three compartments—scala vestibuli, cochlear duct (scala media), and scala tympani.

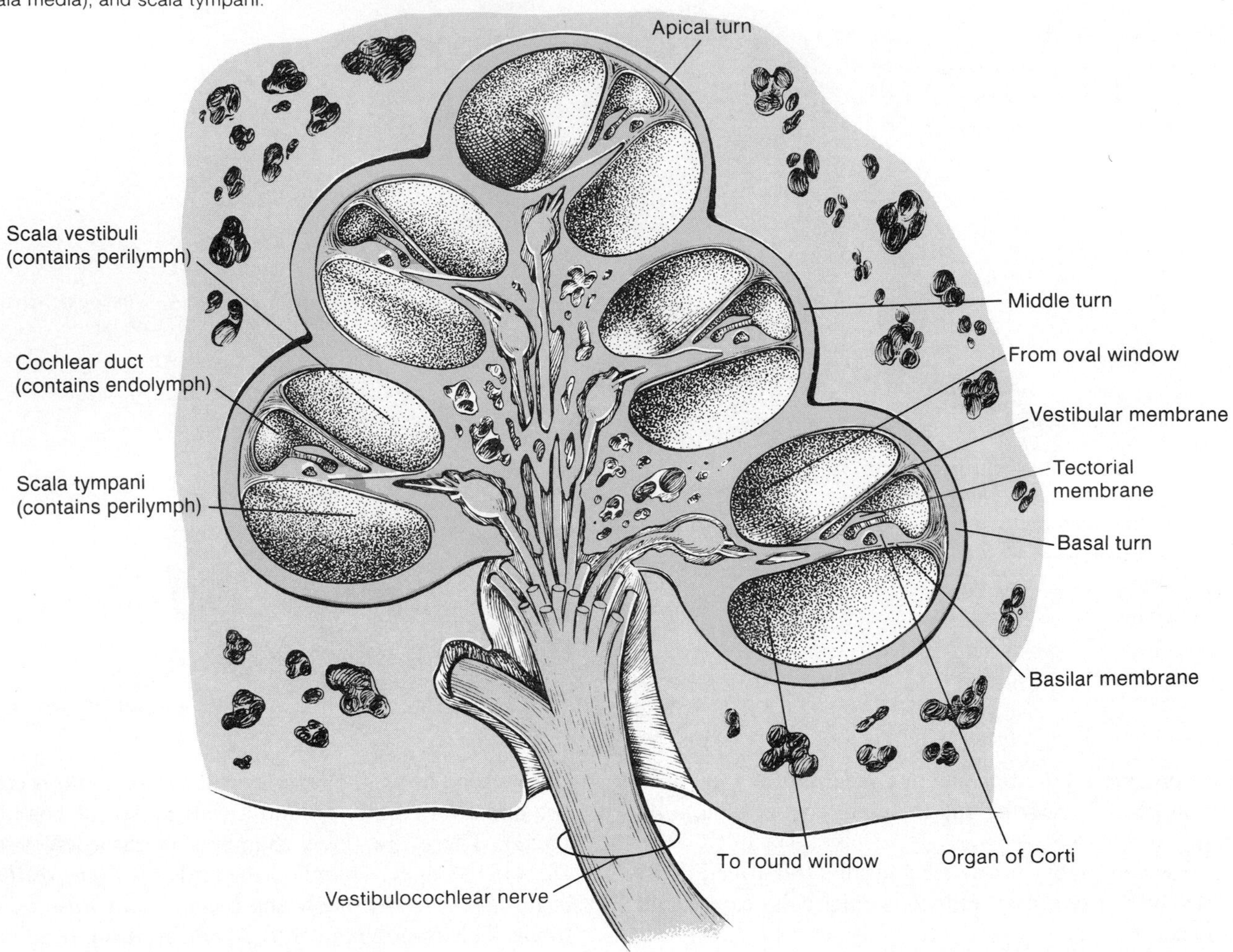

Sound Waves and Neural Pathways for Hearing

Sound Waves. Sound waves travel in all directions from their source, like ripples in a pond after a stone is dropped. These waves are characterized by their frequency and their intensity. The **frequency,** or distances between crests of the sound waves, is measured in *hertz* (*Hz*), which is the modern designation for *cycles per second* (*cps*). The *pitch* of a sound is directly related to its frequency—the higher the frequency of a sound, the higher its pitch.

The **intensity,** or loudness of a sound, is directly related to the amplitude of the sound waves. This is measured in units known as *decibels* (*db*). A sound that is barely audible—at the threshold of hearing—has an intensity of zero decibels. Every 10 decibels indicates a tenfold increase in sound intensity: a sound is ten times higher than threshold at 10 db, 100 times higher at 20 db, a million times higher at 60 db and ten billion times higher at 100 db.

The ear of a trained, young individual can hear sound over a frequency range of 20,000–30,000 Hz, yet can distinguish between two pitches that have only a 0.3% difference in frequency. The human ear can detect differences in sound intensities of only 0.1 to 0.5 db, whereas the range of audible intensities covers twelve orders of magnitude (10^{12}), from the barely audible to the limits of painful loudness.

Sound waves funneled through the external auditory meatus produce extremely small vibrations of the tympanic membrane. Movements of the tympanum during speech (average intensity of 60 db) are estimated to be equal to the diameter of a molecule of hydrogen.

Sound waves passing through the solid medium of the ear ossicles are amplified by about twenty times as they reach the footplate of the stapes against the oval window. As the oval window is displaced, pressure waves pass

Figure 18.54. The organ of Corti (*a*) is located within the cochlear duct of the cochlea (*b*).

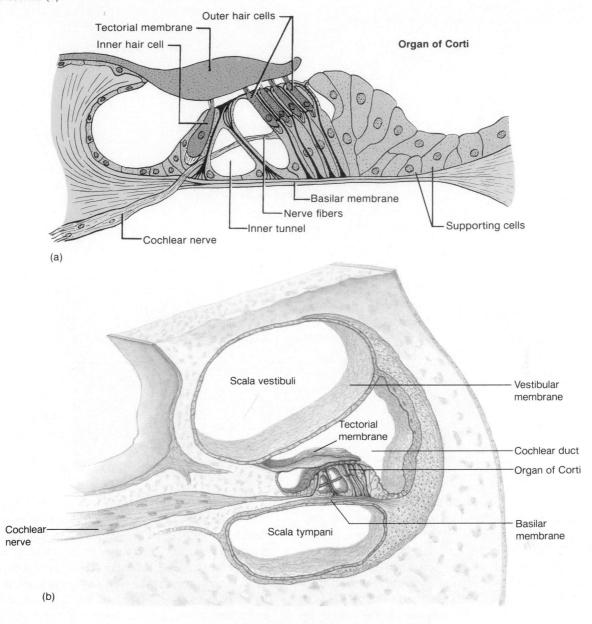

Outer hair cells
Tectorial membrane
Inner hair cell
Organ of Corti
Basilar membrane
Nerve fibers
Inner tunnel
Supporting cells
Cochlear nerve
(a)

Scala vestibuli
Vestibular membrane
Tectorial membrane
Cochlear duct
Organ of Corti
Basilar membrane
Cochlear nerve
Scala tympani
(b)

Figure 18.55. A scanning electron micrograph of the hair cells of the organ of Corti.

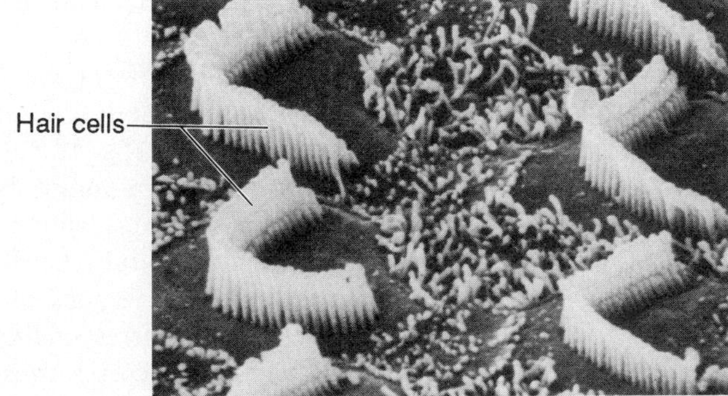

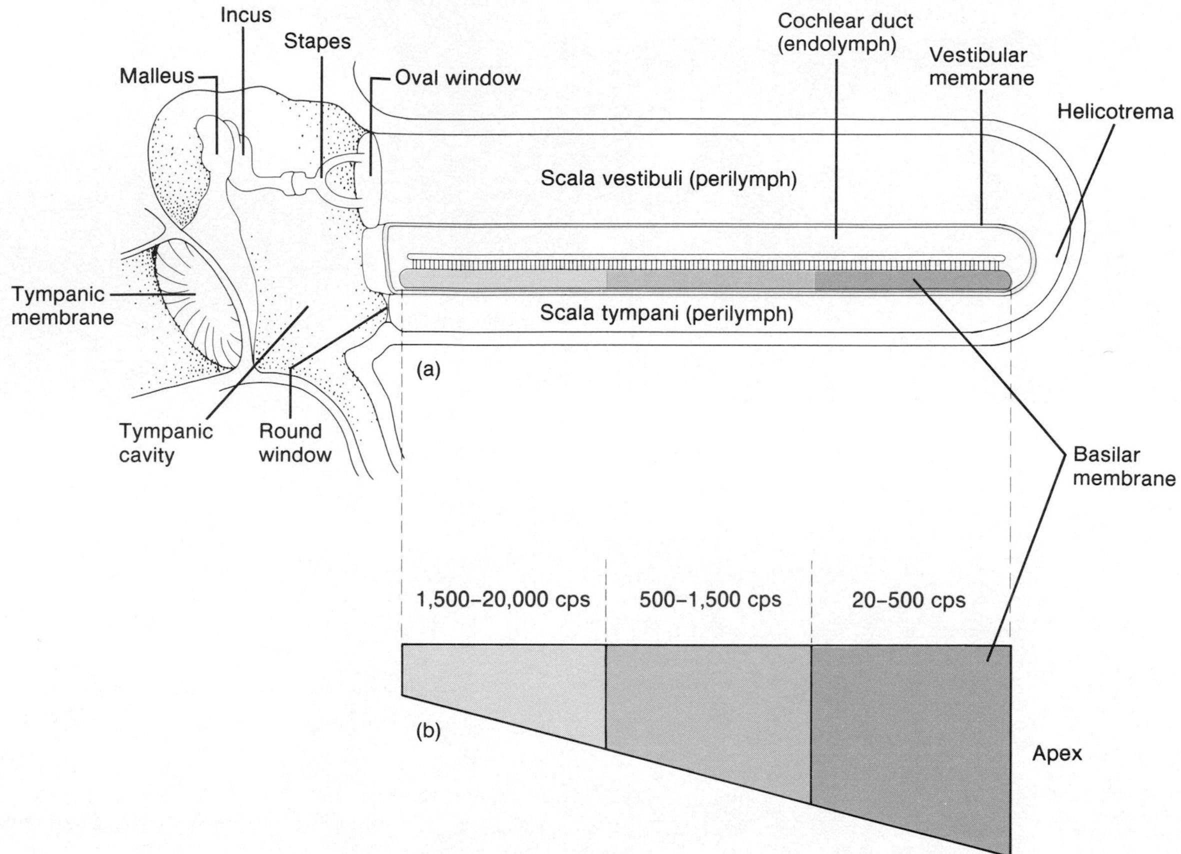

Figure 18.56. (*a*) the cochlea is illustrated as a straightened structure to show that the perilymph in the scala vestibuli and scala tympani is continuous at the helicotrema. These two chambers are separated by the blind-ending cochlear duct (scala media). The basilar membrane of the cochlear duct is narrower at the base and wider at the apex. (*b*) sounds of different frequencies (pitches) produce maximum displacement of the basilar membrane in these regions.

through the fluid medium of the scala vestibuli (fig. 18.56), which pass around the helicotrema to the scala tympani. Movements of perilymph within the scala tympani, in turn, displace the round window into the tympanic cavity.

When the sound frequency (pitch) is sufficiently low, there is adequate time for the pressure waves of perilymph within the upper scala vestibuli to travel through the helicotrema to the scala tympani. As the sound frequency increases, however, pressure waves of perilymph within the scala vestibuli do not have time to travel all the way to the apex of the cochlea. Instead, they are transmitted through the *vestibular membrane,* which separates the scala vestibuli from the cochlear duct, and through the *basilar membrane,* which separates the cochlear duct from the scala tympani, to the perilymph of the scala tympani. The distance that these pressure waves travel, therefore, decreases as the sound frequency increases.

Sounds of low pitch (with frequencies below about 50 Hz) cause movements of the entire length of the basilar membrane—from the base to the apex. Higher sound frequencies result in maximum displacement of the basilar membrane closer to its base, as illustrated in figure 18.57.

Displacement of the basilar membrane and hair cells by movements of perilymph causes the microvilli that are embedded in the tectorial membrane to bend. This stimulation excites the sensory cells, which causes the release of an unknown neurotransmitter that excites sensory endings of the cochlear nerve.

Neural Pathways for Hearing. Cochlear sensory neurons in the eighth cranial nerve synapse with neurons in the medulla (fig. 18.58), which project to the inferior colliculi of the midbrain. Neurons in this area in turn project to the thalamus, which sends axons to the auditory cortex of the temporal lobe. By means of this pathway, neurons in different regions of the basilar membrane stimulate neurons in corresponding areas of the auditory cortex; each area of this cortex thus represents a different part of the basilar membrane and a different pitch (fig. 18.59).

Figure 18.57. Sounds of low frequency cause pressure waves of perilymph to pass through the helicotrema. Sounds of higher frequency cause pressure waves to "short cut" through the cochlear duct. This sets up traveling waves in the basilar membrane.

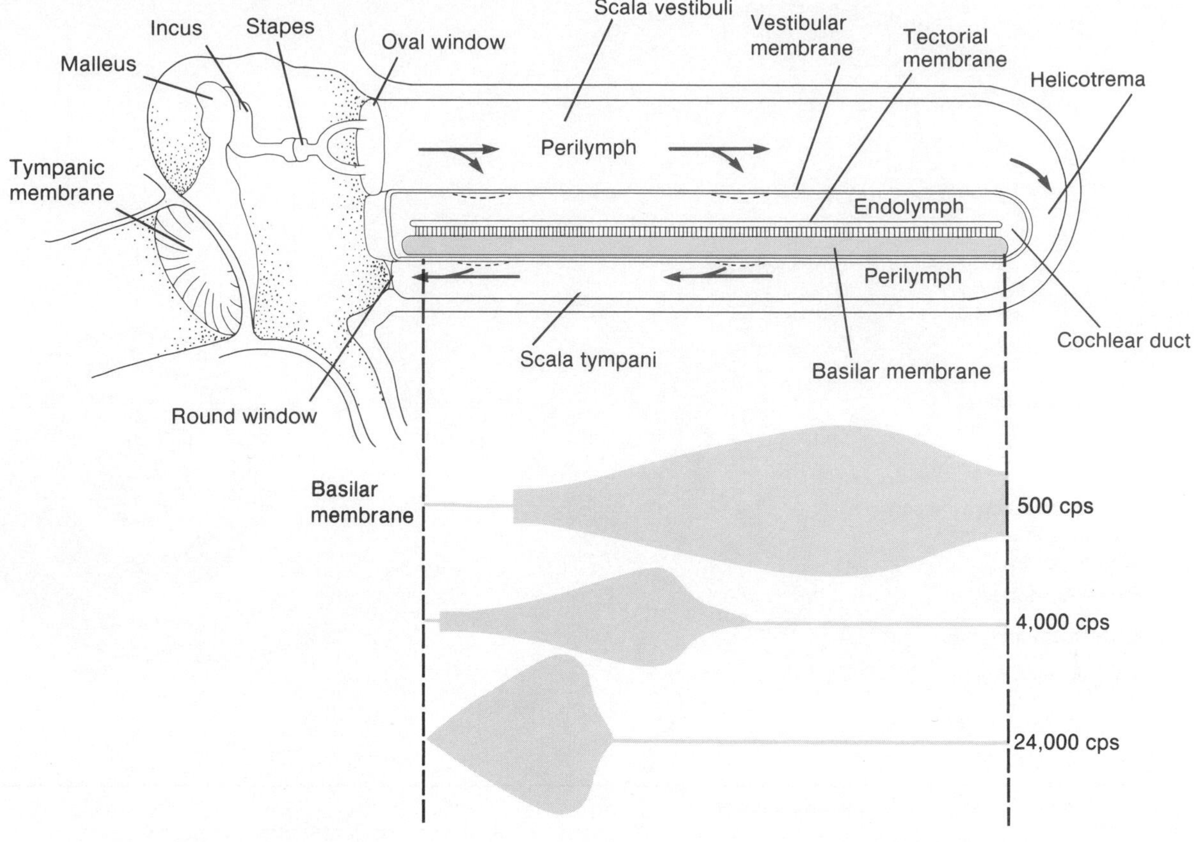

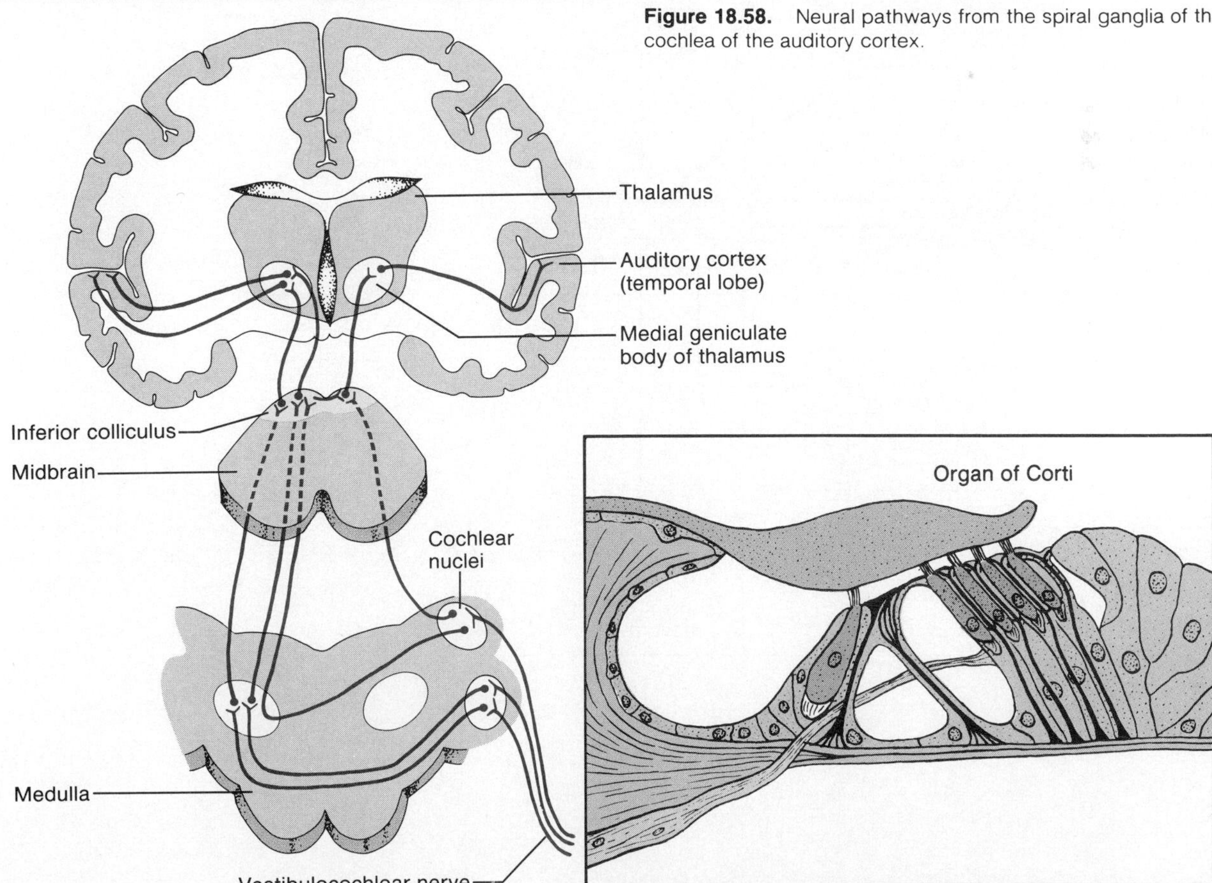

Figure 18.58. Neural pathways from the spiral ganglia of the cochlea of the auditory cortex.

Figure 18.59. Sounds of different frequencies (pitches) excite different sensory neurons in the cochlea; these in turn send their input to different regions of the auditory cortex.

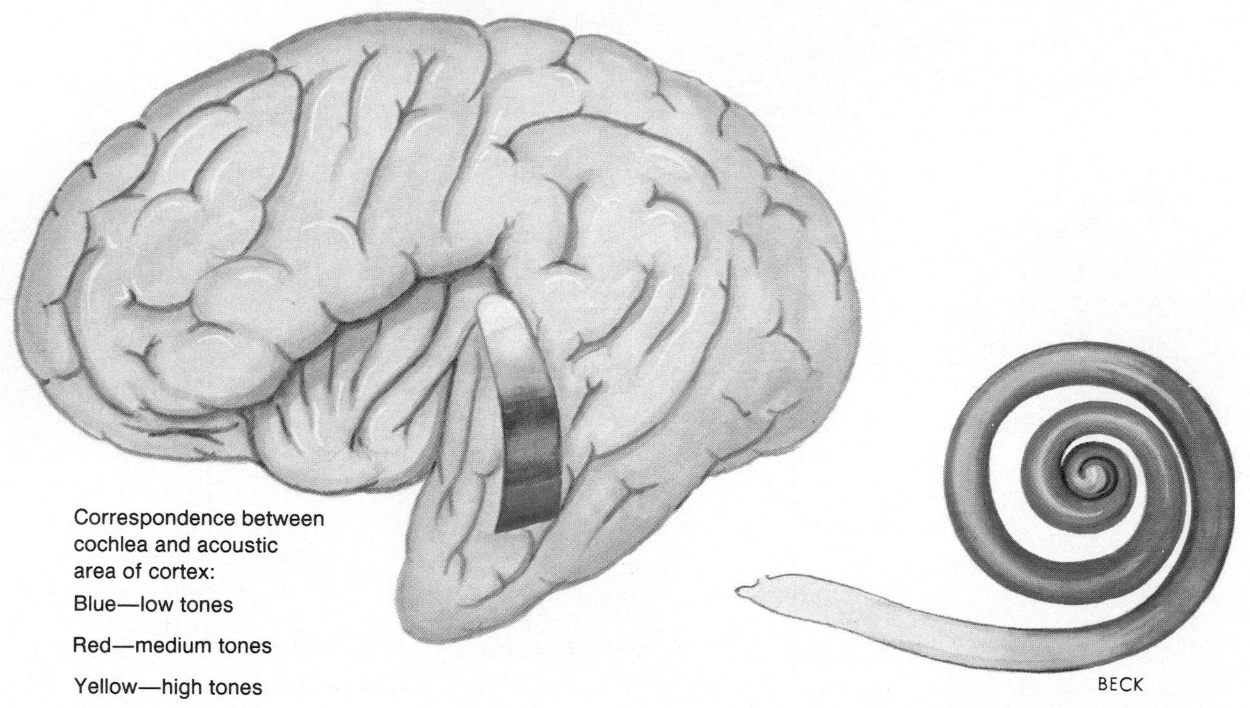

Correspondence between cochlea and acoustic area of cortex:
Blue—low tones
Red—medium tones
Yellow—high tones
BECK

Figure 18.60. Neural processing involved in maintenance of equilibrium and balance.

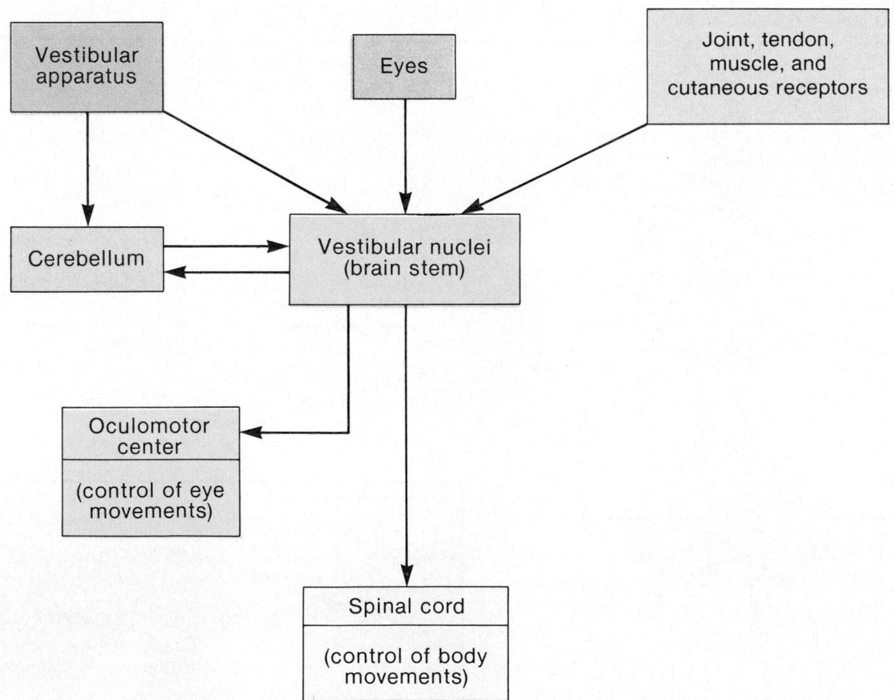

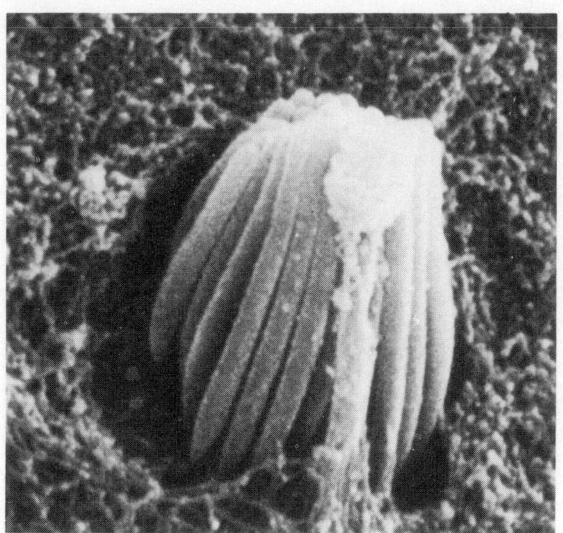

Figure 18.61. A scanning electron photomicrograph of hairs and kinocilium within the vestibular apparatus.

Mechanics of Equilibrium

Maintaining equilibrium is a complex process that depends on constant input from the vestibular organs of both inner ears. Although the vestibular organs are the principal source of sensory information for equilibrium, visual input from the eyes, tactile receptors within the skin, and proprioceptors of tendons, muscles, and joints also provide sensory input that is vital for maintaining equilibrium (fig. 18.60).

The vestibular organs provide the CNS with two kinds of receptor information. One kind is provided by receptors within the saccule and utricle, which are sensitive to gravity and to linear acceleration and deceleration of the head, such as in riding in a car. The other is provided by receptors within the semicircular canals, which are sensitive to angular acceleration and deceleration, such as turning the head, spinning, or tumbling.

The receptor hair cells of the vestibular organs contain twenty to fifty microvilli and one cilium, called a **kinocilium** *(ki''no-sil'e-um)* (fig. 18.61). When the hair cells are displaced in the direction of the kinocilium, the cell membrane is depressed and becomes depolarized (fig. 18.62). When the hair cells are displaced in the opposite direction, the membrane becomes hyperpolarized.

Figure 18.62. (*a*) sensory hair cells in the vestibular apparatus contain hairs (microvilli) and one kinocilium. (*b*) when hair cells are bent in the direction of the kinocilium, the cell membrane is depressed (see arrow) and the sensory neuron innervating the hair cell is stimulated. (*c*) when the hairs are bent in a direction opposite to the kinocilium, the sensory neuron is inhibited.

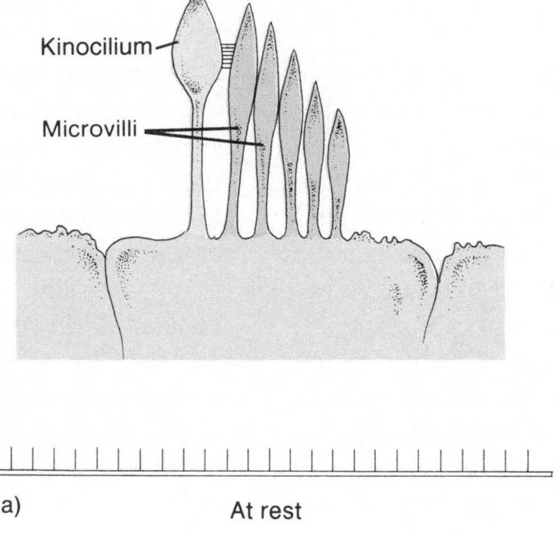

(a) At rest

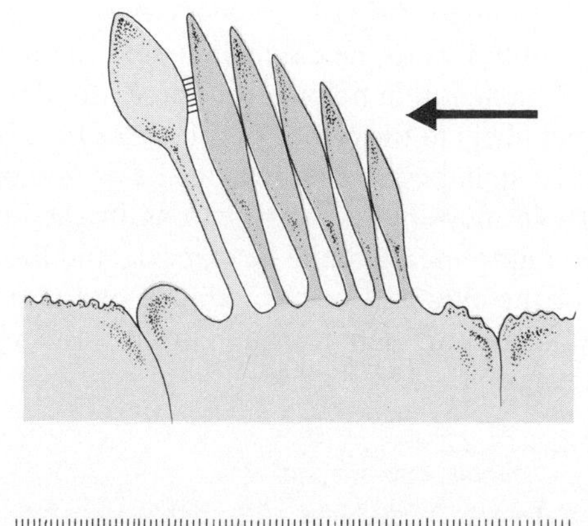

(b) Stimulated

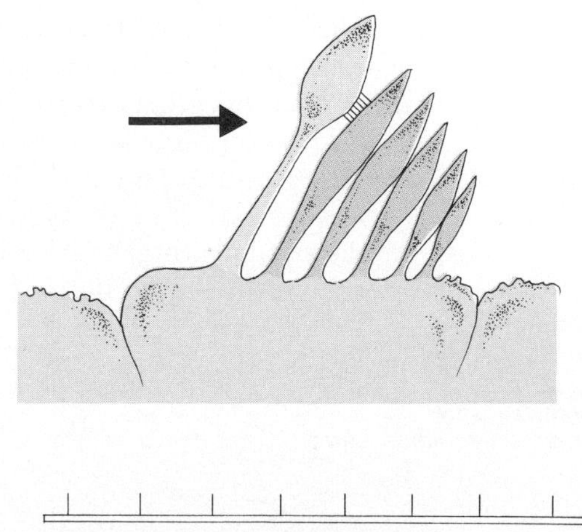

(c) Inhibited

Figure 18.63. The macula of the utricle. (*a*) when the head is in an upright position, the weight of the otoliths applies direct pressure to the sensitive cytoplasmic extensions of the hair cells. (*b*) as the head is tilted forward, the extensions of the hair cells bend in response to gravitational force and cause the hair cells to be stimulated.

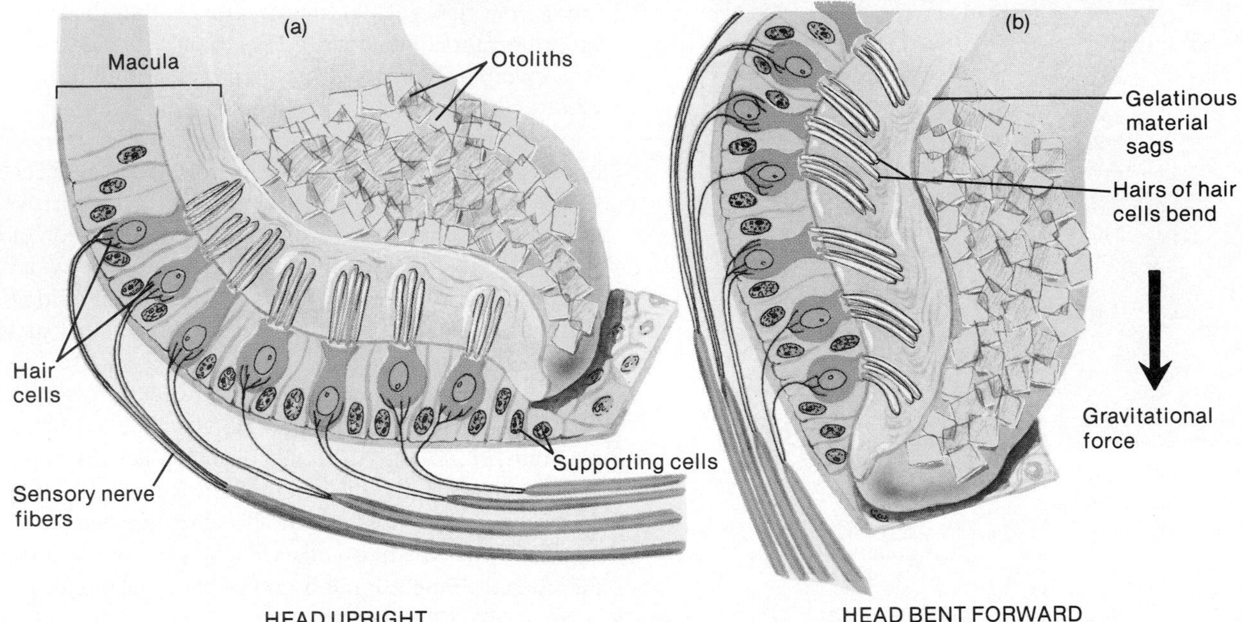

Saccule and Utricle. Receptor hair cells of the saccule and utricle are located in a small, thickened area of the walls of these organs called the **macula** (fig. 18.63). Cytoplasmic extensions of the hair cells project into a gelatinous material that supports microscopic crystals of calcium carbonate, called **otoliths.** The otoliths increase the weight of the gelatinous mass, which results in a higher *inertia* (resistance to change in movement).

When a person is upright, the hairs of the utricle are oriented vertically, whereas those of the saccule are oriented horizontally into the otolith membrane. During forward acceleration, the otolith membrane lags behind the hair cells, so the hairs are pushed backward. This is similar to the backward thrust of the body when a car accelerates rapidly forward. The inertia of the otolith membrane similarly causes the hairs of the utricle to be pushed upward when a person jumps from a raised platform. These effects, and the opposite ones that occur when a person accelerates backward or upward, allow us to maintain our equilibrium with respect to gravity during linear acceleration.

Sensory impulses from the vestibular organs are conveyed to the brain by way of the vestibular portion of the vestibulocochlear nerve.

macula: L. *macula,* spot
otoliths: Gk. *otos,* ear; *lithos,* stone

Semicircular Canals. Receptors of the semicircular canals are contained within the ampulla at the base of each semicircular duct. An elevated area of the ampulla called the **crista ampullaris** *(am″pu-lar′is)* contains numerous hair cells and supporting cells (fig. 18.64). Like the saccule and utricle, the hair cells have cytoplasmic extensions that project into a dome-shaped gelatinous mass called the **cupula** *(ku′pu-lah).* When the hair cells within the cupula are bent by rapid displacement of the fluid within the semicircular ducts, as in spinning around, sensory impulses travel to the brain by way of the vestibular portion of the vestibulocochlear nerve.

Neural Pathways. Stimulation of the hair cells in the vestibular apparatus activates the sensory neurons of the *vestibulocochlear* (*eighth cranial*) *nerve.* These fibers transmit impulses to the cerebellum and to the vestibular nuclei of the medulla oblongata. The vestibular nuclei, in turn, send fibers to the oculomotor center of the brain stem and to the spinal cord. Neurons in the oculomotor center control eye movements, and neurons in the spinal cord stimulate movements of the head, neck, and limbs. Movements of the eyes and body produced by these pathways serve to maintain balance and "track" the visual field during rotation.

cupula: L. *cupula,* cup-shaped

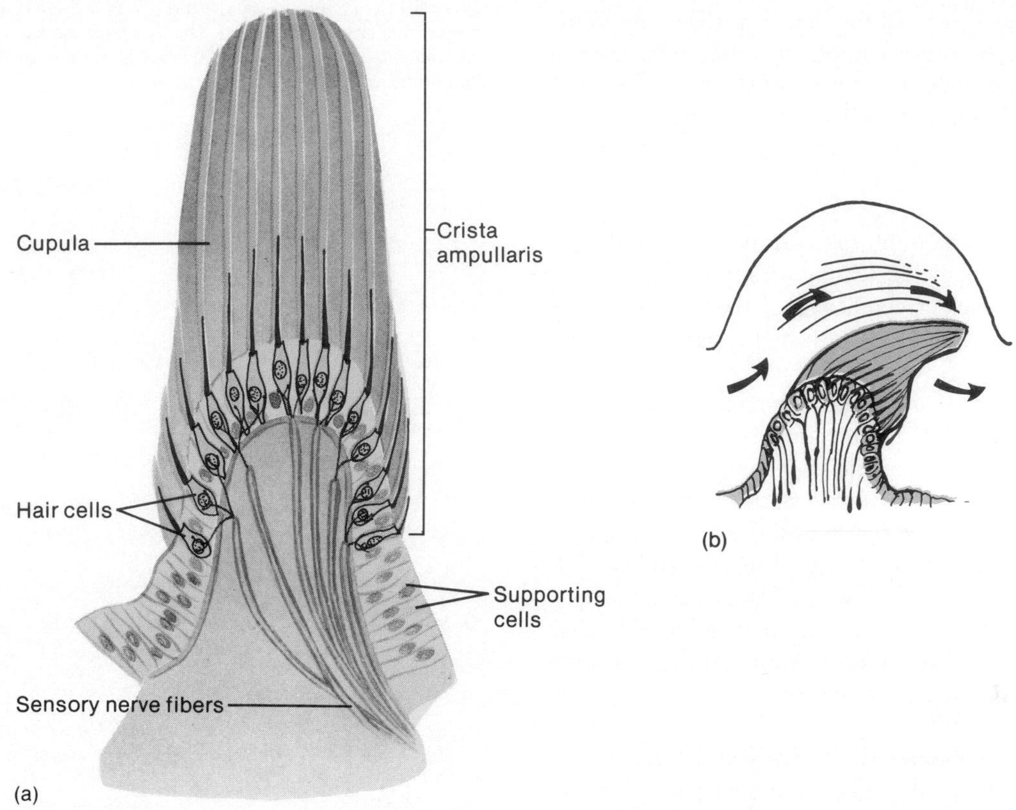

Figure 18.64. (*a*) a crista ampullaris within an ampulla. (*b*) movement of the endolymph during rotation causes the cupula to displace, thus stimulating the hair cells.

1. List the structures of the external ear, middle ear, and inner ear, and explain the function of each as related to hearing.
2. Use a flowchart to describe how sound waves in air within the external auditory meatus are transduced into movements of the basilar membrane.
3. Explain how movements of the basilar membrane of the organ of Corti can code for different sound frequencies (pitches).
4. Explain how the vestibular organs maintain a sense of balance and equilibrium.

Clinical Considerations

Numerous disorders and diseases afflict the sensory organs. Some of these occur during the sensitive period of prenatal development and have a congenital effect. Other sensory impairments, some of which are avoidable, can occur at any time of life. Still others are the result of changes during the natural aging process. The loss of a sense is frequently a traumatic adjustment. Fortunately, however, when a sensory function is impaired or lost, the other senses seem to become more keen to lessen the extent of the handicap. A blind person, for example, compensates somewhat for the loss of sight by developing a remarkable hearing ability.

Many books have been written on diseases and dysfunctions of the sensory organs, and entire specialities within medicine are devoted to treating the disorders of the specific sensory organs. It is beyond the scope of this book to attempt a comprehensive discussion of the numerous aspects of the sensory organs. Some general comments will be made, however, on the diagnosis of sensory disorders and on developmental problems that can affect the eyes or ears. In addition, the more common diseases and dysfunctions of these organs will be discussed.

Diagnosis of Sensory Organs

Eye. There are two distinct professional specialities concerned with the structure and function of the eye. *Optometry* is the paramedical profession concerned with assessing vision and treating visual problems. An *optometrist* is not a physician and does not treat eye diseases but prescribes lenses or visual training. *Ophthalmology (of″thal-mol′o-je)* is the speciality of medicine concerned with diagnosing and treating eye diseases.

Although the eyeball is an extremely complex organ, it is quite accessible to examination. Various techniques and instruments are used during an eye examination, but the following are used most frequently: (1) a *cycloplegic drug,* which is instilled into the eyes to dilate the pupils and temporarily inactivate the ciliary muscles; (2) a *Snellen's chart,* which is used to determine the visual acuity of a person standing 20 feet from the chart (a reading of 20/20 is considered normal for the test); (3) an *ophthalmoscope,* which contains a light, mirrors, and lenses to illuminate and magnify the interior of the eyeball so that the structures within may be examined; and (4) a *tonometer,* used to measure ocular tension, which is important in detecting glaucoma.

Ear. *Otorhinolaryngology (o''to-ri''no-lar''in-gol'o-je)* is the speciality of medicine dealing with the diagnosis and treatment of diseases or conditions of the ear, nose, and throat. *Audiology* is the study of hearing, particularly assessing the ear and its functioning.

There are three common instruments or techniques used to examine the ears to determine auditory function: (1) an *otoscope* is an instrument used to examine the tympanic membrane of the ear; abnormalities of this membrane are informative when diagnosing specific auditory problems, including middle ear infections; (2) *tuning fork tests* are useful in determining hearing acuity and especially for discriminating the various kinds of hearing loss; (3) *audiometry* is a functional examination for hearing sensitivity and speech discrimination.

Developmental Problems of the Eyes and Ears

Although there are many congenital abnormalities of the eyes and ears, most of them are rare. The sensitive period of development of these organs is from about twenty-four to forty-five days after conception. Most congenital disorders of the eyes and ears are caused by genetic factors or intrauterine infections such as *rubella virus.*

Eye. Most **congenital cataracts** are hereditary, but they may also be caused by maternal rubella infection during the critical fourth to sixth week of eye development. In this condition, the lens is opaque and frequently appears grayish white.

Cyclopia is a rare condition in which the eyes are partially fused into a median eye enclosed by a single orbit. Other severe malformations, which are incompatible with life, are generally expressed with this condition.

Ear. **Congenital deafness** is generally caused by an autosomal recessive gene but may also be caused by a maternal rubella infection. The actual functional impairment is generally either a defective set of ossicles or improper development of the neurosensory structures of the inner ear.

Although the shape of the auricle varies widely, **auricular abnormalities** are not uncommon, especially in infants with chromosomal syndromes causing mental deficiencies. In addition, the external auditory meatus frequently does not develop in these children, producing a condition called **atresia** *(ah-tre'ze-ah)* of the external auditory meatus.

Figure 18.65. In a normal eye, parallel rays of light are brought to focus on the retina by refraction in the cornea and lens. If the eye is too long, as in myopia, the focus is in front of the retina. This can be corrected by a concave lens. If the eye is too short, as in hyperopia, the focus is behind the retina. This is corrected by a convex lens. In astigmatism, light refraction is uneven due to an abnormal shape of the cornea or lens.

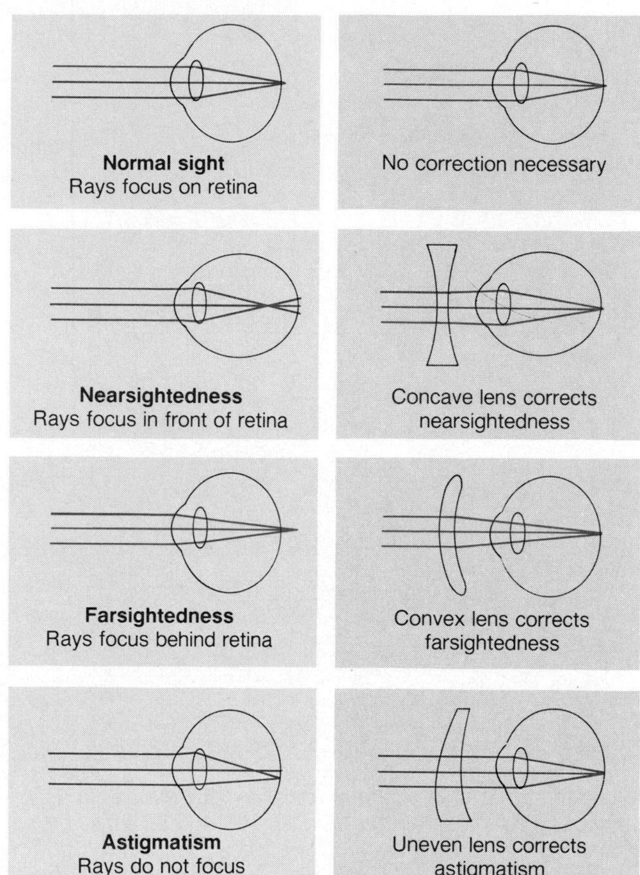

Functional Impairments of the Eye

Few people have perfect vision. Slight variations in the shape of the eyeball or curvature of the cornea or lens cause an imperfect focal point of light waves onto the retina. Most variations are slight, however, and the error of refraction goes unnoticed. Severe deviations that are not corrected may cause blurred vision, fatigue, chronic headaches, and depression.

There are four principal clinical considerations caused by defects in the refractory structures or general shapes of the eyeball. **Myopia** (nearsightedness) is an elongation of the eyeball that causes light waves to focus at a point in the vitreous humor in front of the retina (fig. 18.65). Only light waves from close objects can be focused clearly

myopia: Gk. *myein,* to shut; *ops,* eye

on the retina; distant objects appear blurred. **Hyperopia** (farsightedness) is a condition in which the eyeball is too short, which causes light waves to have a focal point behind the retina. Although visual accommodation aids a hyperopic person, it generally does not help enough for the person to clearly view very close or distant objects. **Presbyopia** is a condition in which the lens tends to lose its elasticity and ability to accommodate. It is relatively common in elderly people. In order to read print on a page, a person with presbyopia must hold the page farther from the eyes. **Astigmatism** is a condition in which an irregular curvature of the cornea or lens of the eye distorts the refraction of light waves. If a person with astigmatism views a circle, the image will not appear clear in all 360 degrees; the parts of the circle that appear blurred can thus be used to map the astigmatism.

Various glass or plastic lenses can generally aid persons with the above visual impairments. Myopia may be corrected with a biconcave lens; hyperopia with a biconvex lens; and presbyopia with bifocals, or a combination of two lenses adjusted for near and distant vision. Correction for astigmatism requires a careful assessment of the irregularities and a prescription of a specially ground corrective lens.

The condition of myopia may also be treated by a surgical procedure called **radial keratotomy.** In this technique, eight to sixteen microscopic slashes, like the spokes of a wheel, are made in the cornea from the center to the edge. The ocular pressure inside the eyeball bulges the weakened cornea and flattens its center, changing the focal length of the eyeball. Side effects may include vision fluctuations, sensitivity to glare, and incorrect corneal alteration.

Cataracts. A cataract is a clouding of the lens that leads to a gradual blurring of vision and the eventual loss of sight. A cataract is not a growth within or upon the eye but a chemical change in the protein of the lens caused by injury, poisons, infections, or age degeneration.

Cataracts are the leading cause of blindness. A cataract can be removed surgically, however, and vision restored by implanting a tiny intraocular lens that either clips to the iris or is secured into the vacant lens capsule. Special contact lenses or thick lenses for glasses are other options.

Detachment of the Retina. Retinal detachment is a separation of the nervous or visual layer of the retina from the underlying pigment epithelium. It generally begins as a minute tear in the retina that gradually extends as vitreous fluid accumulates between the layers. Retinal detachment may result from hemorrhage, a tumor, degeneration, or trauma from a violent blow to the eye. A detached retina may be repaired by using laser beams, cryoprobes, or intense heat to destroy the tissue beneath the tear and rejoin the layers.

Strabismus. Strabismus is a condition in which both eyes do not focus upon the same axis of vision. This prevents stereoscopic vision, and the persons afflicted will have varied visual impairments. Strabismus is usually caused by a weakened extrinsic eye muscle.

Strabismus is assessed while the patient attempts to look straight ahead. If the afflicted eye is turned toward the nose, the condition is called convergent strabismus (**esotropia**). If the eye is turned outward, it is called divergent strabismus (**exotropia**). Disuse of the afflicted eye causes a visual impairment called **amblyopia.** Visual input from the normal eye and the eye with strabismus results in **diplopia,** or double vision. A normal, healthy person who has excessively consumed alcohol may experience diplopia.

Infections and Diseases of the Eye

Infections. Infections and inflammation can occur in any of the associated structures of the eye and in structures within or on the eyeball itself. The causes of infections are usually microorganisms, mechanical irritation, or sensitivity to particular substances.

Conjunctivitis (inflammation of the conjunctiva) may result from sunburn or an infection by organisms such as staphylococci, viruses, or streptococci. Bacterial conjunctivitis is commonly called "pinkeye."

Keratitis (inflammation of the cornea) may develop secondarily from conjunctivitis or be caused by diseases such as tuberculosis, syphilis, mumps, or measles. Keratitis is painful and may cause blindness if untreated.

A **chalazion** *(kah-la'ze-on)* is a cyst caused by an infection and a subsequent blockage in the ducts of the sebaceous glands along the free edge of the eyelids.

Styes (**hordeola**) are relatively common, but mild, infections of the follicle of an eyelash or the sebaceous gland of the follicle. Styes may spread readily from one eyelash to another if untreated. Poor hygiene and the excessive use of cosmetics may contribute to development of styes.

Diseases. **Trachoma** *(trah-ko'mah)* is a highly contagious bacterial disease of the conjunctiva and cornea. Although rare in the United States, it is estimated that over 500 million people are afflicted by this disease. Trachoma may be treated readily with sulfonamides and some antibiotics, but if untreated it will spread progressively until it covers the cornea. At this stage, vision is lost and the eye undergoes degenerative changes.

hyperopia: Gk. *hyper,* over; *ops,* eye
presbyopia: Gk. *presbys,* old man; *ops,* eye
astigmatism: Gk. *a,* without; *stigma,* point
cataract: Gk. *katarrhegnynai,* to break down

Glaucoma is the second leading cause of blindness and is particularly common in underdeveloped countries. Although it can afflict persons of any age, 95% of the cases involve persons over the age of forty. Glaucoma is an abnormal increase in the intraocular pressure of the eyeball. Aqueous humor does not drain through the venous sinus as quickly as it is produced. Glaucoma can be caused by several things, including malnutrition. Glaucoma causes compression of the blood vessels in the eyeball and compression of the optic nerve. Retinal cells die and the optic nerve may atrophy, producing blindness.

Infections, Diseases, and Functional Impairments of the Ear

Disorders of the ear are common and may affect both hearing and the vestibular functions. The ear is afflicted by numerous diseases as well as environmental factors—some of which can be prevented.

Infections and Diseases. **External otitis** is a general term for infections of the external ear. The causes of external otitis range from dermatitis to fungal and bacterial infections.

Acute purulent otitis media is a middle ear infection. Pathogens of this disease usually enter through the auditory tube, most often following a cold or tonsillitis. Children frequently have middle ear infections because of their susceptibility to infections and their short and straight auditory tubes. As a middle ear infection progresses to the inflammatory stage, the auditory tube is closed and drainage prohibited. An intense earache is a common symptom of a middle ear infection. The pressure from the inflammation may eventually rupture the tympanic membrane to permit drainage.

Repeated middle ear infections, particularly in children, usually necessitate an incision of the tympanic membrane, called a **myringotomy** *(mir''in-got'o-me),* and the implantation of a tiny tube within the tympanic membrane (fig. 18.66) to assist the patency of the auditory tube. The tubes, which are eventually sloughed out of the ear, permit the infection to heal and help prohibit further infections by keeping the auditory tube open.

Perforation of the tympanic membrane may occur as the result of infections or trauma. Sudden, intense noise might rupture the membrane. Spontaneous perforation of the membrane usually heals rapidly, but scar tissue may form and lessen the sensitivity to sound vibrations.

Otosclerosis is a progressive deterioration of the normal bone of the bony labyrinth and its replacement with vascular spongy bone (fig. 18.67). This frequently causes hearing loss as the ear ossicles are immobilized. Surgical scraping of the bone growth and replacing the stapes with a prosthesis is the most frequent treatment of otosclerosis.

glaucoma: Gk. *glaukos,* gray

Figure 18.66. An implanted ventilation tube in the tympanic membrane following a *myringotomy.*

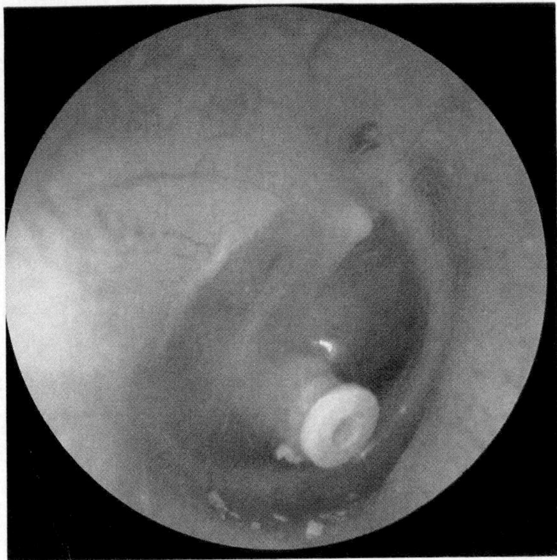

Meniere's disease afflicts the inner ear and may cause hearing loss as well as equilibrium disturbance. The causes of Meniere's disease vary and are not completely understood, but they are thought to be related to a dysfunction of the autonomic nervous system that causes a vasoconstriction within the inner ear. The disease is characterized by recurrent periods of **vertigo** (dizziness and a sensation of rotation), **tinnitus** *(tĭ-ni'tus)* (ringing in the ear), and progressive deafness in the affected ear. Meniere's disease is chronic and affects both sexes equally. It is more common in middle-aged and elderly persons.

Nystagmus and Vertigo. When a person first begins to spin, the inertia of the endolymph within the semicircular canals causes the cupula to bend in the opposite direction. As the spin continues, however, the inertia of the endolymph is overcome and the cupula straightens. At this time the endolymph and the cupula are moving in the same direction and at the same speed. If movement is suddenly stopped, the greater inertia of the endolymph causes it to continue moving in the previous direction of spin and to bend the cupula in that direction.

Bending of the cupula after movement has stopped affects muscular control of the eyes and body. The eyes slowly drift in the direction of the previous spin and are then rapidly jerked back to the midline position, producing involuntary oscillations. These movements are called post-rotatory **vestibular nystagmus** *(nis-tag'mus).*

Meniere's disease: from Prosper Meniere, French physician, 1799–1862
vertigo: L. *vertigo,* dizziness
tinnitus: L. *tinnitus,* ring or tingle

Figure 18.67. The development of otosclerosis in the middle ear immobilizes the stapes, leading to conduction deafness.

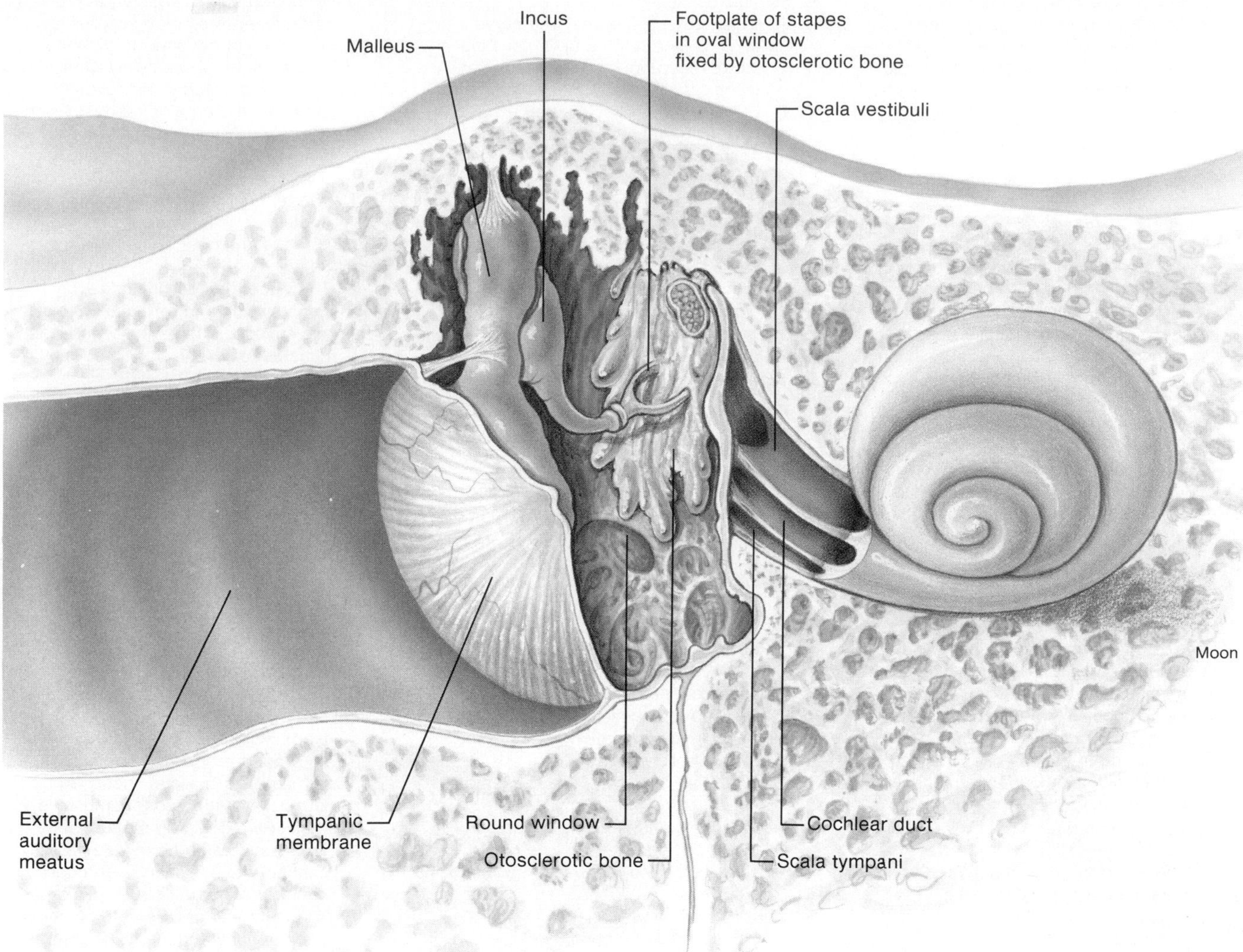

People experiencing this may feel that they, or the room, are spinning. The loss of equilibrium that results is called **vertigo.** If the vertigo is sufficiently severe or the person particularly susceptible, the autonomic system may become involved. This can produce dizziness, pallor, sweating, and nausea.

Auditory Impairment. Loss of hearing results from disease, trauma, or developmental problems involving any portion of the auditory apparatus, cochlear nerve and auditory pathway, or areas of auditory perception within the brain. Hearing impairment varies from slight disablement, which may or may not worsen, to total deafness. Some types of hearing impairment, including deafness, can be mitigated through hearing aids or surgery.

There are two principal types of auditory impairments. **Conduction deafness** is caused by an interference with the sound waves through the external or middle ear. Conduction problems include impacted cerumen (wax) ruptured tympanic membrane, a severe middle ear infection, or adhesions of one or more ear ossicles *(otosclerosis).* Conductive deafness can be successfully treated.

Perceptive (sensorineural) deafness results from disorders that affect the inner ear, the cochlear nerve or nerve pathway, or auditory centers within the brain. Perceptive impairment ranges in severity from the inability to hear certain frequencies to deafness. Such deafness may be caused by a number of factors, including diseases, trauma, and genetic or developmental problems. Hearing impairments of this type, which are common in elderly persons, are related to age. The ability to perceive high-frequency sounds is generally affected first. Hearing aids may help patients with perceptive deafness. This type of deafness is permanent, however, because it involves sensory structures that cannot heal themselves through regeneration or replication.

Chapter Summary

I. Overview of Sensory Perception
   A. Sensory organs are specialized extensions of the nervous system that respond to specific stimuli and conduct nerve impulses.
   B. A stimulus to a receptor that conducts an impulse to the brain is necessary for perception.
   C. Senses act as energy filters that permit perception of only a narrow range of energy.

II. Classification of the Senses
   A. The senses are classified according to receptors or on the basis of the stimulus energy they transduce.
   B. General senses are widespread throughout the body and are simple in structure. Special senses are localized in complex receptor organs and have extensive neural pathways.
   C. Somatic senses are localized within the body wall, and visceral senses are located within the visceral organs.
   D. Phasic receptors respond quickly to a stimulus but then adapt and decrease firing rate. Tonic receptors produce a constant rate of firing.

III. Physiology of Sensory Receptors
   A. A sensory receptor responds to stimulus energy and initiates the nerve impulse that produces a particular modality of sensation.
   B. According to the law of specific nerve energies, the sensation characteristic of each type of sensory neuron is that produced by its adequate stimulus.
   C. The entire length of a sensory neuron contains voltage-regulated gates that produce action potentials.

IV. Somatic Senses
   A. Meissner's corpuscles, free nerve endings, and hair root plexuses are tactile receptors, responding to light touch.
   B. Lamellated corpuscles are pressure receptors in the dermis.
   C. Thermoreceptors include the organs of Ruffini (heat) and bulbs of Krause (cold).
   D. Free nerve endings are the principal pain receptors, serving to protect the body from injury.
   E. Joint kinesthetic receptors, neuromuscular spindles, and neurotendinous receptors are proprioceptors that are sensitive to changes in stretch and tension.
   F. The acuity of sensation depends on receptor density and lateral inhibition.

V. Olfactory Sense
   A. Olfactory receptors of the olfactory nerve respond to chemical stimuli and transmit the sensation of olfaction to the cerebral cortex.
   B. Olfaction functions closely with gustation in that the receptors of both are chemoreceptors, requiring dissolved substances for stimuli.

VI. Gustatory Sense
   A. Taste receptors in taste buds respond to chemical stimuli and transmit the sensation of gustation to the cerebral cortex.
   B. Gustatory cells compose the taste buds located in circumvallate, fungiform, and filiform papillae.
   C. The four modalities of taste sensation are sweet, salty, sour, and bitter.

VII. Visual Sense
   A. The sensory components of the eye are formed by twenty weeks, and the accessory structures are formed by thirty-two weeks.
   B. Accessory structures of the eye (the orbit, eyebrow, eyelids, eyelashes, conjunctiva, lacrimal gland, and ocular muscles) either protect the eyeball or provide eye movement.
   C. The eyeball consists of the fibrous tunic, which is divided into the sclera and cornea; the vascular tunic, which consists of the choroid, the ciliary body, and the iris; and the internal tunic, or retina, which consists of an outer pigmented layer and an inner nervous layer.
   D. Rods and cones, which are the photoreceptors in the nervous layer of the retina, respond to dim and colored light, respectively. Cones are concentrated in the fovea centralis, the area of keenest vision.
   E. Rods and cones contain specific pigments that provide sensitivity to different light waves.
   F. The anterior and posterior chambers contain aqueous humor, and the vitreous chamber contains vitreous humor.
   G. The visual process includes the transmission and refraction of light waves, accommodation of the lens, constriction of the pupil, and convergence of the eyes.
   H. Neural pathways from the retina to the superior colliculus help regulate eye and body movements. Most fibers from the retina project to the lateral geniculate body and from there to the striate cortex.

VIII. Senses of Hearing and Balance
   A. The development of the ear begins during the fourth week and is completed by the thirty-second week.
   B. The external ear consists of the auricle and the external auditory meatus.
   C. The middle ear, bounded by the tympanic membrane and the oval and round windows, contains the ear ossicles (malleus, incus, and stapes) and the auditory muscles (tensor tympani and stapedius).
   D. The vestibule of the middle ear connects to the pharynx through the auditory tube.
   E. The inner ear, or labyrinth, contains the functional organs for hearing and equilibrium: the organ of Corti in the cochlea, and the semicircular canals, saccule, and utricle.

Review Activities

Objective Questions

1. Which of the following must occur for the perception of a sensation to take place?
   (a) stimulus present
   (b) nerve impulse conduction
   (c) receptor activated
   (d) all of the above
2. The cutaneous receptor sensitized for deep pressure detection is the
   (a) hair root plexus
   (b) Pacinian corpuscle
   (c) organ of Ruffini
   (d) free nerve ending
3. Cutaneous receptive fields are smallest in the
   (a) back
   (b) fingertips
   (c) thighs
   (d) arms
4. The sensation of visceral pain perceived as arising from another somatic location is known as
   (a) related pain
   (b) phantom pain
   (c) referred pain
   (d) parietal pain
5. When a person with normal vision views an object from a distance of at least 20 feet;
   (a) the ciliary muscles are relaxed
   (b) the suspensory ligament is taut
   (c) the lens is flat, having the least convex shape
   (d) all of the above
6. Which of the following is an avascular ocular tissue?
   (a) sclera
   (b) cornea
   (c) ciliary body
   (d) iris
7. The stimulation of hair cells in the semicircular canals results from the movement of
   (a) endolymph
   (b) perilymph
   (c) the otolith membrane
8. The middle ear is separated from the inner ear by the
   (a) round window
   (b) tympanic membrane
   (c) oval window
   (d) both a and c
9. Glasses with concave lenses help to correct
   (a) presbyopia
   (b) myopia
   (c) hyperopia
   (d) astigmatism

Essay Questions

1. Outline the major events in the development of the eye and ear. When are congenital deformities most likely to occur?
2. Name the six senses of the body. Differentiate between the special and somatic senses. What are the similarities between these two senses?
3. List the functions of proprioceptors, and differentiate between the three types. What role do proprioceptors play in the kinesthetic sense?
4. Diagram the structure of the eyeball, and label the sclera, cornea, choroid, macula lutea, ciliary body, suspensory ligaments, lens, iris, pupil, retina, optic disc, and fovea centralis.
5. Outline and explain the process of focusing light waves onto the fovea centralis.
6. Trace a sound wave through the structures of the ear, and explain how the mechanism of hearing functions.

19 Endocrine System

Outline

Concepts

Hormones are regulatory molecules secreted into the blood by endocrine glands. The effect of a specific hormone on target cells is relatively precise but is affected by the concentration of the hormone and by the concentration of other hormones in the blood.

Although each hormone has its own characteristic effects on specific target cells, hormones that are in the same chemical category have similar mechanisms of action. These similarities involve the location of cellular receptor proteins and the events in the target cells after hormone-receptor interaction has occurred.

Prostaglandins are autocrine regulatory molecules derived from arachidonic acid, and they exert their effects within the same tissues in which they are produced. The actions of prostaglandins supplement and complement the regulatory effects of the endocrine and nervous systems.

The neurohypophysis secretes hormones that are produced by the hypothalamus, whereas the adenohypophysis secretes its own hormones in response to regulation from hypothalamic hormones. The secretions of the pituitary gland are thus controlled by the hypothalamus, as well as by negative feedback influences from the target glands.

The adrenal medulla secretes catecholamine hormones, which complement the action of the sympathetic nervous system. The adrenal cortex secretes corticosteroids, which participate in the regulation of mineral balance, energy balance, and reproductive function.

The thyroid gland secretes thyroxine and triiodothyronine, which participate in the regulation of metabolism and are critically important for proper growth and development. The thyroid also secretes calcitonin, which may antagonize the action of parathyroid hormone in the regulation of calcium and phosphate balance.

The pancreatic islets secrete two hormones, insulin and glucagon, which are critically involved in the regulation of metabolic balance in the body. Additionally, many other organs secrete hormones that help regulate digestion, metabolism, growth, immune function, and reproduction.

HORMONES: ACTIONS AND INTERACTIONS

Hormones are regulatory molecules secreted into the blood by endocrine glands. The effect of a specific hormone on target cells is relatively precise but is affected by the concentration of the hormone and by the concentration of other hormones in the blood.

Objective 1. Define the terms *endocrine gland* and *hormone,* and identify the different chemical classes to which hormones belong.

Objective 2. Define the terms *prohormone* and *prehormone,* and give examples of each.

Objective 3. Describe how the concentration of a given polypeptide hormone can affect the response of its target tissue to subsequent exposures to the same hormone.

Objective 4. Explain the different ways that one hormone can affect the response of a target tissue to a different hormone.

Endocrine *(en'do-krin)* **glands** (fig. 19.1*a*), which lack the ducts present in exocrine glands, secrete biologically active chemicals called **hormones** into the blood (table 19.1). Hormones may be steroids, derivatives of amino acids, polypeptides, or glycoproteins. Whatever their chemical nature, hormones combine with specific receptor proteins in their target cells and modify the production or action of specific cellular enzymes and other proteins.

Hormones affect the metabolism of their target organs and, by this means, help regulate (1) total body metabolism, (2) growth, and (3) reproduction. The effects of hormones on body metabolism and growth are discussed in chapter 26; the regulation of reproductive functions by hormones is included in chapters 27 and 28.

Organs such as the adrenal, thyroid, parathyroid, pituitary, and pineal glands appear to be exclusively endocrine in function. Other organs of the endocrine system secrete hormones while they serve additional functions. The stomach and small intestine, for example, secrete hormones that help regulate their digestive functions. The pancreas is both an exocrine and an endocrine gland—pancreatic juice is an exocrine secretion that enters the pancreatic duct, whereas the hormones insulin and glucagon are secreted into the blood by groups of cells known as the pancreatic islets (islets of Langerhans) (fig. 19.1*b*). Likewise, the skin, thymus, liver, kidneys, and brain secrete hormones and perform other specialized functions.

endocrine: Gk. *endon,* within; *krinein,* to separate
exocrine: Gk. *exo,* outside; *krinein,* to separate
hormone: Gk. *hormon,* to set in motion

Figure 19.1. (*a*) the anatomy of some of the endocrine glands. (*b*) the pancreatic islets (islets of Langerhans) within the pancreas.

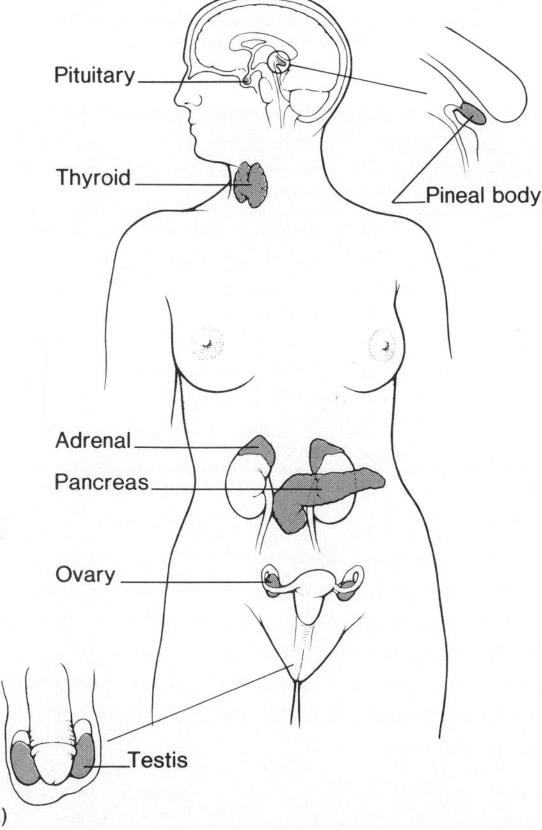

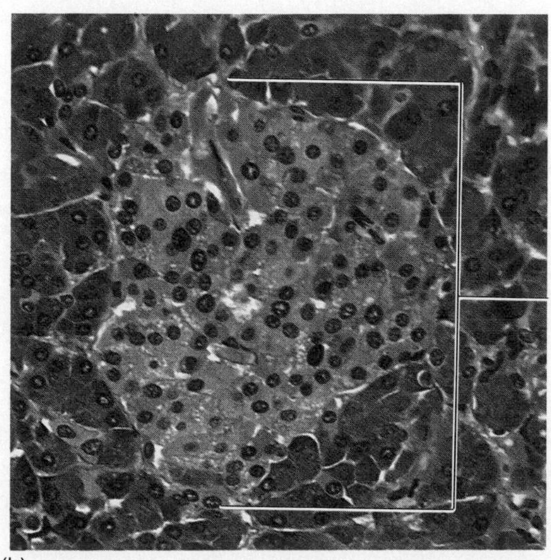

Table 19.1 A partial list of the endocrine glands

Endocrine gland	Major hormones	Primary target organs	Primary effects
Adrenal cortex	Cortisol Aldosterone	Liver, muscles Kidneys	Glucose metabolism; Na^+ retention, K^+ excretion
Adrenal medulla	Epinephrine	Heart, bronchioles, blood vessels	Adrenergic stimulation
Hypothalamus	Releasing and inhibiting hormones	Anterior pituitary	Regulates secretion of anterior pituitary hormones
Intestine	Secretin and cholecystokinin	Stomach, liver, and pancreas	Inhibits gastric motility; stimulates bile and pancreatic juice secretion
Islets of Langerhans (pancreas)	Insulin Glucagon	Many organs Liver and adipose tissue	Insulin promotes cellular uptake of glucose and formation of glycogen and fat; glucagon stimulates hydrolysis of glycogen and fat
Ovaries	Estradiol-17β and progesterone	Female genital tract and mammary glands	Maintains structure of genital tract; promotes secondary sexual characteristics
Parathyroids	Parathyroid hormone	Bone, intestine, and kidneys	Increases Ca^{++} concentration in blood
Pineal	Melatonin	Hypothalamus and anterior pituitary	Affects secretion of gonadotrophic hormones
Pituitary, anterior	Trophic hormones	Endocrine glands and other organs	Stimulates growth and development of target organs; stimulates secretion of other hormones
Pituitary, posterior	Antidiuretic hormone Oxytocin	Kidneys, blood vessels Uterus, mammary glands	Antidiuretic hormone promotes water retention and vasoconstriction; oxytocin stimulates contraction of uterus and mammary secretory units
Stomach	Gastrin	Stomach	Stimulates acid secretion
Testes	Testosterone	Prostate, seminal vesicles, other organs	Stimulates secondary sexual development
Thymus	Thymosin	Lymph nodes	Stimulates white blood cell production
Thyroid	Thyroxine (T_4) and triiodothyronine (T_3)	Most organs	Growth and development; stimulates basal rate of cell respiration (basal metabolic rate or BMR)

Table 19.2 Some examples of polypeptide and glycoprotein hormones

Hormone	Structure	Gland	Primary effects
Antidiuretic hormone	8 amino acids	Posterior pituitary	Water retention and vasoconstriction
Oxytocin	8 amino acids	Posterior pituitary	Uterine and mammary contraction
Insulin	21 and 30 amino acids (double chain)	β cells in pancreatic islets	Cellular glucose uptake, lipogenesis, and glycogenesis
Glucagon	29 amino acids	α cells in pancreatic islets	Hydrolysis of stored glycogen and fat
ACTH	39 amino acids	Anterior pituitary	Stimulation of adrenal cortex
Parathyroid hormone	84 amino acids	Parathyroid	Increases blood Ca^{++} concentration
FSH, LH, TSH	Glycoproteins	Anterior pituitary	Stimulates growth, development, and secretion of target glands

Chemical Classification of Hormones

Hormones secreted by different endocrine glands are diverse in chemical structure. All hormones, however, can be grouped into three general chemical categories: (1) **amines,** such as catecholamines (epinephrine and norepinephrine) and other derivatives of amino acids; (2) **polypeptides** and **glycoproteins,** including shorter chain polypeptides such as antidiuretic hormone and insulin and large glycoproteins such as thyroid-stimulating hormone (table 19.2); and (3) **steroids,** such as cortisol and testosterone.

Steroid hormones, which are derived from cholesterol (fig. 19.2), are lipids and thus are not water-soluble. The gonads—testes and ovaries—secrete *sex steroids;* the adrenal cortex secretes *corticosteroids* such as cortisol, aldosterone, and others.

Figure 19.2. Simplified biosynthetic pathways for steroid hormones. Notice that progesterone (a hormone secreted by the ovaries) is a common precursor in the formation of all other steroid hormones and that testosterone (the major androgen secreted by the testes) is a precursor in the formation of estradiol–17*β*, the major estrogen secreted by the ovaries.

Cholesterol

Pregnenolone

Progesterone

Secreted by ovaries

Cortisol (hydrocortisone)

Secreted by adrenal cortex

Androstenedione

Testosterone

Secreted by testes

Estradiol - 17β

Secreted by ovaries

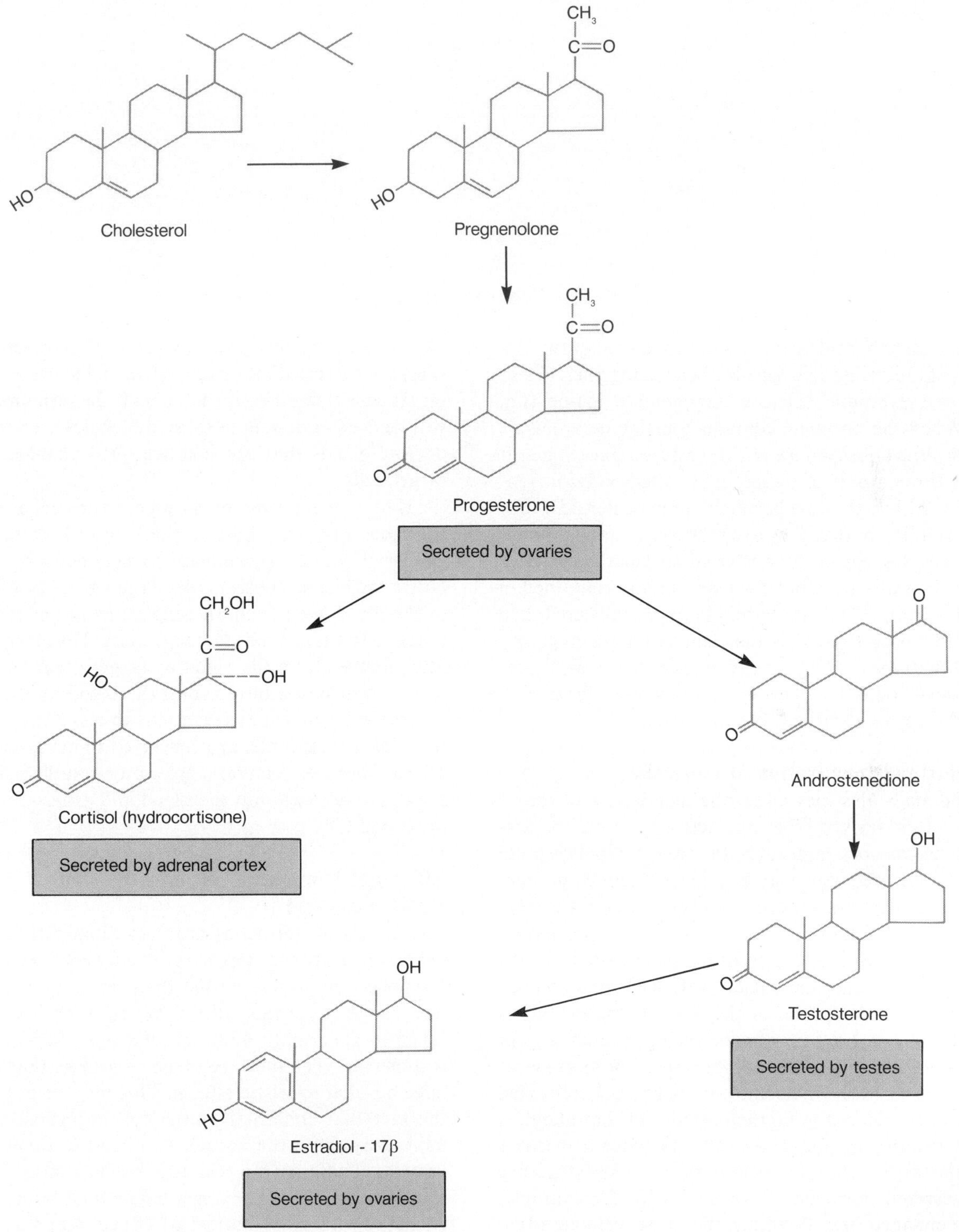

Figure 19.3. The thyroid hormones: thyroxine (T_4) and triiodothyronine (T_3) are secreted in a ratio of 9:1.

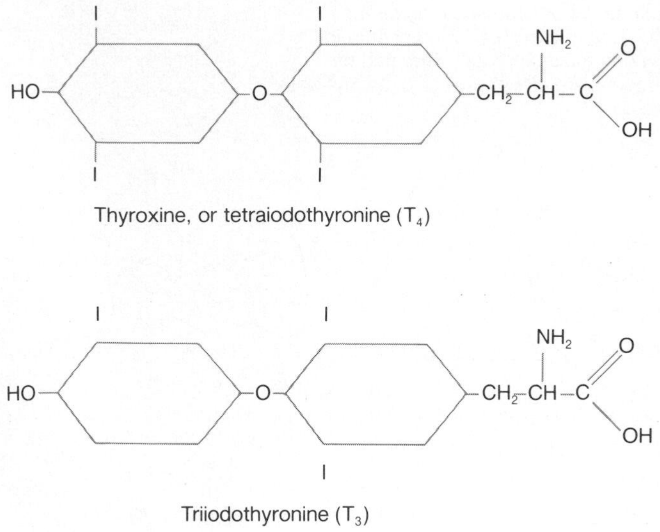

The major thyroid hormones are composed of two derivatives of the amino acid tyrosine bonded together. These hormones are unique because they contain iodine (fig. 19.3). When the hormone contains four iodine atoms, it is called *tetraiodothyronine* (T_4), or *thyroxine.* When it contains three atoms of iodine, it is called *triiodothyronine* (T_3). Although these hormones are not steroids, they are like steroids in that they are relatively small and nonpolar molecules. Steroid and thyroid hormones are active when taken orally (as a pill); sex steroids are contained in the contraceptive pill, and thyroid hormone pills are taken by people whose thyroid is deficient (who are hypothyroid). Other types of hormones cannot be taken orally because they would be digested into inactive fragments before they were absorbed into the blood.

Activation and Inactivation of Hormones

Hormone molecules that affect the metabolism of target cells are often derived from less active "parent," or *precursor (pre' kur-sor),* molecules. In the case of polypeptide hormones, the precursor may be a longer chained **prohormone** that is cut and spliced together to make the hormone. Insulin, for example, is produced from *proinsulin* within the endocrine beta cells of the pancreatic islets. In some cases the prohormone itself is derived from an even larger precursor molecule; in the case of insulin, this is called *pre-proinsulin.* Many endocrinologists use the term *prehormone* to indicate such precursors of prohormones.

In some cases the molecule secreted by the endocrine gland, and considered to be the hormone of that gland, is actually inactive in the target cells. In order to become active, the target cells must modify the chemical structure of the secreted hormone. Thyroxine (T_4), for example, must be changed into T_3 within the target cells to affect the metabolism of the target cells. Similarly, testosterone (secreted by the testes) and vitamin D_3 (secreted by the skin) are converted into more active molecules within their target cells (table 19.3). In this text, the term **prehormone** will be used to designate those molecules secreted by endocrine glands that are inactive until changed by their target cells.

The concentration of hormones (or prehormones) in the blood primarily reflects the rate of secretion by the endocrine glands—hormones do not generally accumulate in the blood. This is because the *half-life* of hormones in the blood is quite short, ranging from under two minutes to a few hours for most hormones. Hormones are rapidly removed from the blood by target organs and by the liver. Hormones removed from the blood by the liver are converted by enzymatic reactions to less active products. Steroids, for example, are converted to more polar derivatives. These less active, more water-soluble polar derivatives are released into the blood and are excreted in the urine and bile.

Effects of Hormone Concentrations on Tissue Response

The effects of hormones are very concentration dependent. Normal tissue responses are only produced when the hormones are present within their normal, or *physiological,* range of concentrations. When some hormones are taken in abnormally high, or *pharmacological,* concentrations (as when they are taken as drugs), they may produce noncharacteristic effects. This may be partly due to the fact that abnormally high concentrations of a hormone may cause the hormone to bond to tissue receptor proteins of different but related hormones. Also, since some steroid hormones can be converted by their target cells into products with other biological effects (such as the conversion of androgens into estrogens), the administration of large amounts of one steroid can result in the production of a significant amount of other steroids with different effects.

Table 19.3 Conversion of some prehormones into biologically active derivatives			
Endocrine gland	**Prehormone**	**Active products**	**Comments**
Skin	Vitamin D_3	1,25-dihydroxyvitamin D_3	Hydroxylation reactions occur in the liver and kidneys.
Testes	Testosterone	Dihydrotestosterone (DHT)	DHT and other 5α-reduced androgens are formed in most androgen-dependent tissue.
		Estradiol-17β (E_2)	E_2 is formed in the brain from testosterone, where it is believed to affect both endocrine function and behavior; small amounts of E_2 are also produced in the testes.
Thyroid	Thyroxine (T_4)	Triiodothyronine (T_3)	Conversion of T_4 to T_3 occurs in almost all tissues.

Pharmacological doses of hormones, particularly of steroids, can thus have widespread and often damaging "side effects." People with inflammatory diseases who are treated with high doses of cortisone over long periods of time, for example, may develop characteristic changes in bone and soft tissue structure. Contraceptive pills, which contain sex steroids, have a number of potential side effects that could not have been predicted at the time the pill was first introduced.

Priming Effects. Variations in hormone concentration within the normal, physiological range can affect the responsiveness of target cells. This is due in part to the effects of polypeptide and glycoprotein hormones on the number of their receptor proteins in target cells. Small amounts of gonadotropin-releasing hormone (GnRH), secreted by the hypothalamus, for example, increase the sensitivity of anterior pituitary cells to further GnRH stimulation. This is a priming effect; subsequent stimulation by GnRH thus causes a greater response from the anterior pituitary.

Desensitization and Downregulation. Prolonged exposure to high concentrations of polypeptide hormones has been found to *desensitize* the target cells. Subsequent exposure to the same concentration of the same hormone thus produces less of a target tissue response. This desensitization may be partially due to the fact that high concentrations of these hormones cause a decrease in the number of receptor proteins in their target cells, a phenomenon called *downregulation.* Such desensitization and downregulation of receptors has been shown to occur, for example, in adipose cells exposed to high concentrations of insulin and in testicular cells exposed to high concentrations of luteinizing hormone (LH).

Hormone Interactions

A given target tissue is usually responsive to a number of different hormones, which may antagonize each other or work together to produce effects that are additive or complementary. The responsiveness of a target tissue to a particular hormone is thus affected not only by the concentration of that hormone, but also by the effects of other hormones on that tissue. Terms used to describe hormone interactions include *synergistic, permissive,* and *antagonistic.*

Synergistic and Permissive Effects. When two or more hormones work together to produce a particular result, their effects are said to be **synergistic** *(sin''er-jis'tik).* These effects may be additive or complementary. The action of epinephrine and norepinephrine on the heart is a good example of an additive effect. Each of these hormones separately produces an increase in cardiac rate; if the same concentrations of these hormones are given together, the stimulation of cardiac rate is increased. The synergistic action of FSH and testosterone is an example of a complementary effect; each hormone separately stimulates a different stage of spermatogenesis during puberty, so that both hormones together are needed at that time to complete sperm development. Likewise, the ability of mammary glands to produce and secrete milk requires the synergistic action of many hormones—estrogen, cortisol, prolactin, oxytocin, and others.

A hormone is said to have a **permissive** effect on the action of a second hormone when it enhances the responsiveness of a target organ to the second hormone or when it increases the activity of the second hormone. Prior exposure of the uterus to estrogen, for example, induces the formation of receptor proteins for progesterone, which increases the response of the uterus when it is subsequently exposed to progesterone. Estrogen thus has a permissive effect on the responsiveness of the uterus to progesterone. Prior exposure of the gonads to FSH, as another example, stimulates the production of LH receptors, which enhances the responsiveness of the testes and ovaries to LH. Glucocorticoids (a class of corticosteroids including cortisol) exert permissive effects on the actions of catecholamines (epinephrine and norepinephrine). When there is an absence of these permissive effects due to abnormally low glucocorticoids, the inadequate action of catecholamines may result in low blood pressure.

Vitamin D_3 is a prehormone that must first be converted by enzymes in the kidneys and liver into the active hormone 1,25-dihydroxyvitamin D_3, which helps to raise blood calcium levels. Parathyroid hormone (PTH) has a permissive effect on the actions of vitamin D_3 because it stimulates the production of these hydroxylating enzymes in the kidneys and liver. By this means, an increased secretion of PTH results in an increase in the ability of vitamin D_3 to stimulate the intestinal absorption of calcium.

Table 19.4 Functional categories of hormones, based on the location of their receptor proteins and the mechanisms of their action

Types of hormones	Secreted by	Location of receptors	Effects of hormone–receptor interaction
Catecholamines, polypeptides, glycoproteins	All glands except adrenal cortex, gonads, and thyroid	Outer surface of cell membrane	Stimulates production of intracellular "second messenger," which activates previously inactive enzymes
Steroids	Adrenal cortex, testes, ovaries	Cytoplasm of target cells	Stimulates translocation of hormone–receptor complex to nucleus and activation of specific genes
Thyroxine (T_4)	Thyroid	Nucleus of target cells	After conversion to triiodothyronine (T_3), activates specific genes

Antagonistic Effects. In some situations the actions of one hormone antagonize the effects of another. Lactation during pregnancy, for example, is prevented because the high concentration of estrogen in the blood inhibits the secretion and action of prolactin. Another example of antagonism is the action of insulin and glucagon (two hormones from the pancreatic islets) on adipose tissue; the formation of fat is promoted by insulin, whereas glucagon promotes fat breakdown.

1. Define the terms *prohormone* and *prehormone,* and give examples of each.
2. Distinguish between the physiological effect and the pharmacological effect of a hormone.
3. Explain how the response of the body to a given hormone can be affected by the concentration of that hormone in the blood.
4. Explain how the effects of one hormone can influence the response of a target organ to other hormones.

Mechanisms of Hormone Action

Although each hormone has its own characteristic effects on specific target cells, hormones that are in the same chemical category have similar mechanisms of action. These similarities involve the location of cellular receptor proteins and the events in the target cells after hormone-receptor interaction has occurred.

Objective 5. Describe the differences in cellular location of the receptor proteins for different hormones.

Objective 6. Explain how steroid hormones and thyroxine exert their effects on target cells.

Objective 7. Describe what is meant by a "second messenger" in hormone action, and explain the sequence of events that occurs after a hormone bonds to a receptor when cyclic AMP functions as a second messenger.

Objective 8. Explain how calcium may function as a second messenger in the action of some hormones.

Hormones are delivered by the blood to every cell in the body, but only the **target cells** are able to respond to each hormone. In order to respond to a hormone, a target cell must have specific receptor proteins for that hormone. Receptor protein–hormone interaction is highly specific, much like the interaction of an enzyme with its substrate. Receptor proteins are not enzymes, however, and since they cannot be detected by the techniques used to assay enzymes, other methods must be used. These methods are based on the observation that hormones bond to receptors with a *high affinity* (high bond strength, as compared to the nonspecific binding of hormones) and with a *low capacity.* The latter characteristic refers to the fact that there are only a limited number of receptors per target cell (usually a few thousand), so that it is possible to saturate receptors with hormones. Notice that the characteristics of specific bonding and saturation of receptor proteins are similar to the characteristics of enzyme and transport carrier proteins discussed in previous chapters.

The location of a hormone's receptor proteins in its target cells depends on the chemical nature of the hormone. Based on the location of the receptor proteins, hormones can be grouped into three categories: (1) receptor proteins within the nucleus of target cells—*thyroid hormones;* (2) receptor proteins within the cytoplasm of target cells—*steroid hormones;* and (3) receptor proteins in the outer surface of the target cell membrane—*amine, polypeptide, and glycoprotein hormones.* This information is summarized in table 19.4.

Mechanisms of Steroid and Thyroid Hormone Action

Steroid and thyroid hormones are similar in size and in the fact that they are nonpolar and thus are not very water-soluble. Unlike other hormones, therefore, steroids and thyroid hormones (primarily thyroxine) do not travel dissolved in the aqueous portion of the plasma but instead are transported to their target cells attached to plasma carrier proteins. These hormones then dissociate from the carrier proteins in the blood and easily pass through the lipid component of the target cell's membrane.

Figure 19.4. The mechanism of the action of a steroid hormone (*H*) on the target cells.

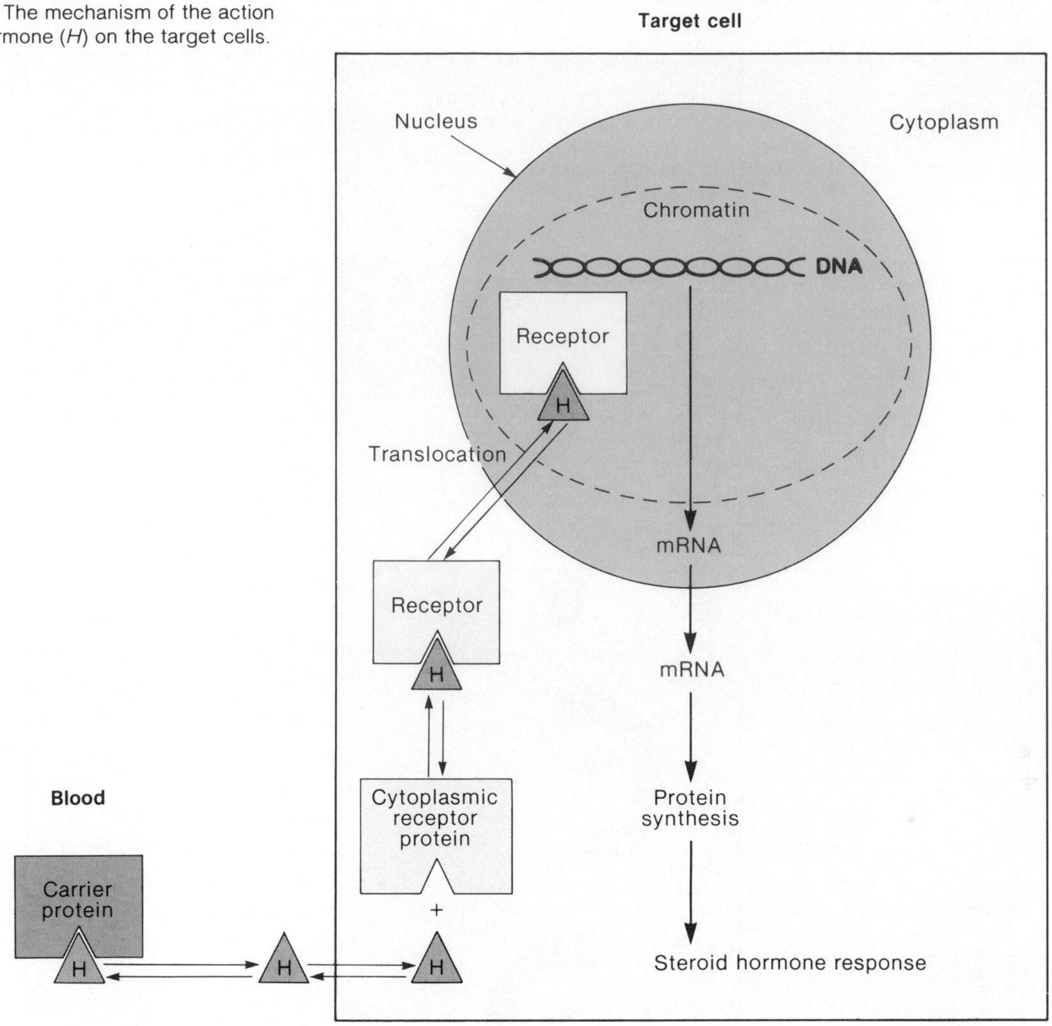

Steroid Hormones. Once through the cell membrane, steroid hormones attach to *cytoplasmic receptor proteins* in the target cells. The steroid hormone–receptor protein complex then *translocates* to the nucleus and attaches by means of the receptor proteins to the chromatin. The sites of attachment, termed *acceptor sites,* are specific for the target tissue. This specificity is believed to be determined by acidic (nonhistone) proteins in the chromatin. According to one theory, part of the receptor bonds to an acidic protein, while a different part of the receptor bonds to DNA.

The attachment of the receptor protein–steroid complex to the acceptor site "turns on genes." Specific genes become activated by this process and produce nuclear RNA, which is then processed into messenger RNA (mRNA). This new mRNA enters ribosomes and codes for the production of new proteins. Since some of these newly synthesized proteins may be enzymes, the metabolism of the target cell is changed in a specific manner (fig. 19.4). Steroid hormones, in short, affect their target cells by stimulating genetic transcription (RNA synthesis) followed by genetic translation (protein synthesis).

Thyroxine. The major hormone secreted by the thyroid gland is thyroxine, or **tetraiodothyronine** *(tet''rah-i''o-do-thi'ro-nēn)* (**T_4**). Like steroid hormones, thyroxine travels in the blood attached to carrier proteins (primarily attached to *thyroxine-binding globulin,* or *TBG*). The thyroid also secretes a small amount of **triiodothyronine** *(tri''i-o''do-tri'ro-nēn),* or **T_3**. The carrier proteins have a higher affinity for T_4 than for T_3, however, and as a result, the amount of unbound (or "free") T_3 is about ten times greater than the amount of free T_4 in the plasma.

Approximately 99.96% of the thyroxine in the blood is attached to carrier proteins in the plasma; the rest is free. Only the free thyroxine and T_3 can enter target cells; the protein-bound thyroxine serves as a "reservoir" of this hormone in the blood (this is why it takes a couple of weeks after surgical removal of the thyroid for the symptoms of hypothyroidism to develop). Once the free thyroxine passes into the target cell cytoplasm, it is enzymatically converted into T_3. Thyroxine, therefore, is not the chemical form of the hormone that is active in the target cells.

Figure 19.5. The mechanism of the action of T_3 (triiodothyronine) on the target cells.

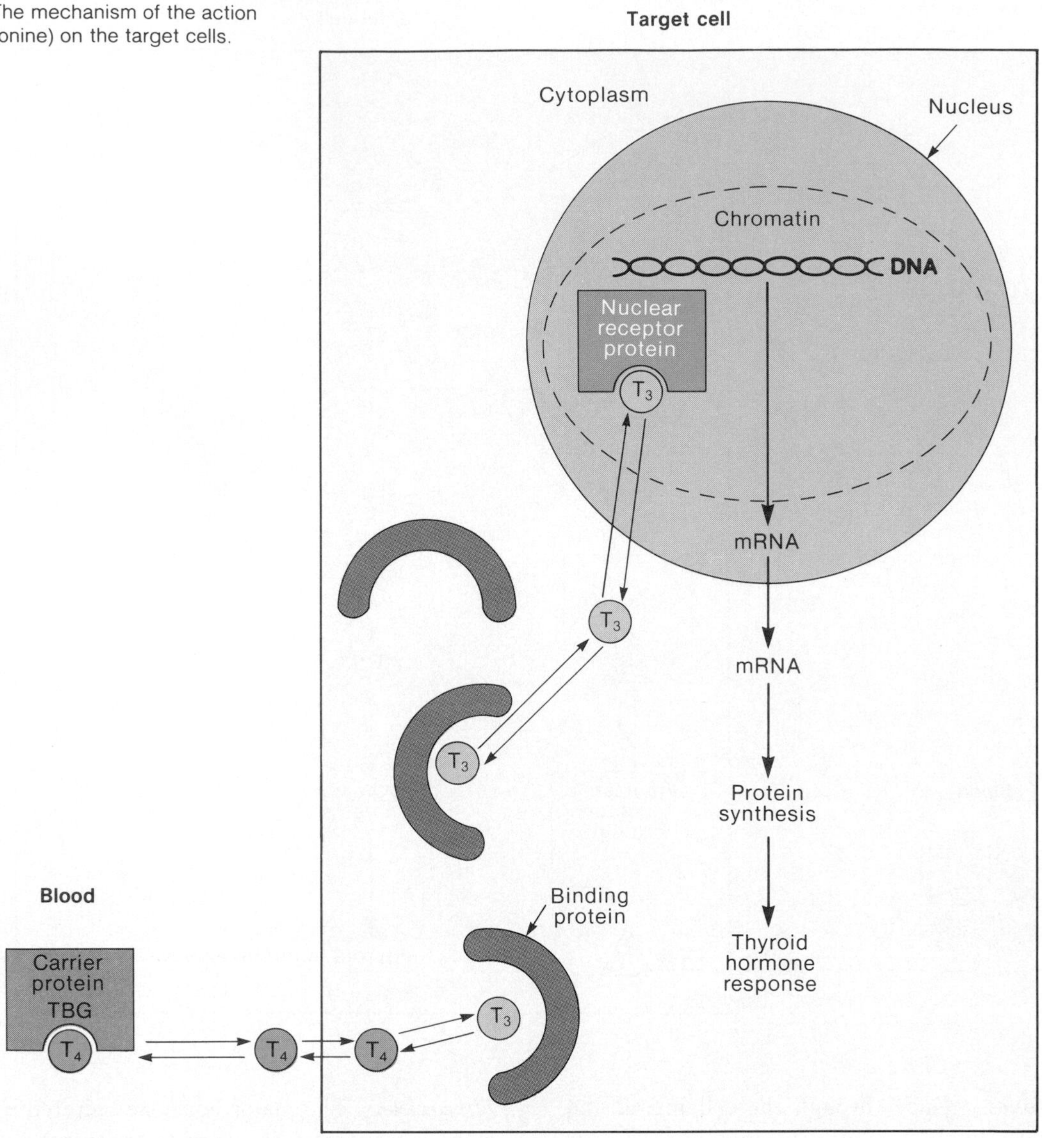

Inactive T_3 receptor proteins are already in the nucleus attached to chromatin. These receptors are inactive until T_3 enters the nucleus from the cytoplasm. The attachment of T_3 to the chromatin-bound receptor proteins activates genes and results in the production of new mRNA and new proteins. This sequence of events is summarized in figure 19.5.

Mechanisms of Amine and Polypeptide Hormone Action: Second Messengers

Amine, polypeptide, and glycoprotein hormones cannot pass through the lipid barrier of the target cell membrane. Although some of these hormones may enter the cell by pinocytosis, most of the effects of these hormones are believed to result from interaction of these hormones with receptor proteins in the outer surface of the target cell membrane. Since these hormones do not have to enter the target cells to exert their effects, other molecules must mediate the actions of these hormones within the target cells. If you think of hormones as "messengers" from the endocrine glands, the intracellular mediators of the hormone's action can be called **second messengers.**

Cyclic AMP as a Second Messenger. Cyclic adenosine monophosphate (abbreviated cAMP) was the first "second messenger" to be discovered and is the best understood. The hormonal effects of epinephrine (adrenalin) and norepinephrine are due to the actions of cAMP, and the effects of norepinephrine released as a neurotransmitter by sympathetic nerve endings are also due to cAMP production. It was later discovered that the effects of many (but not all—see table 19.5) polypeptide and glycoprotein hormones are also mediated by cAMP.

Figure 19.6. Cyclic AMP (cAMP) as a second messenger in the action of catecholamine, polypeptide, and glycoprotein hormones.

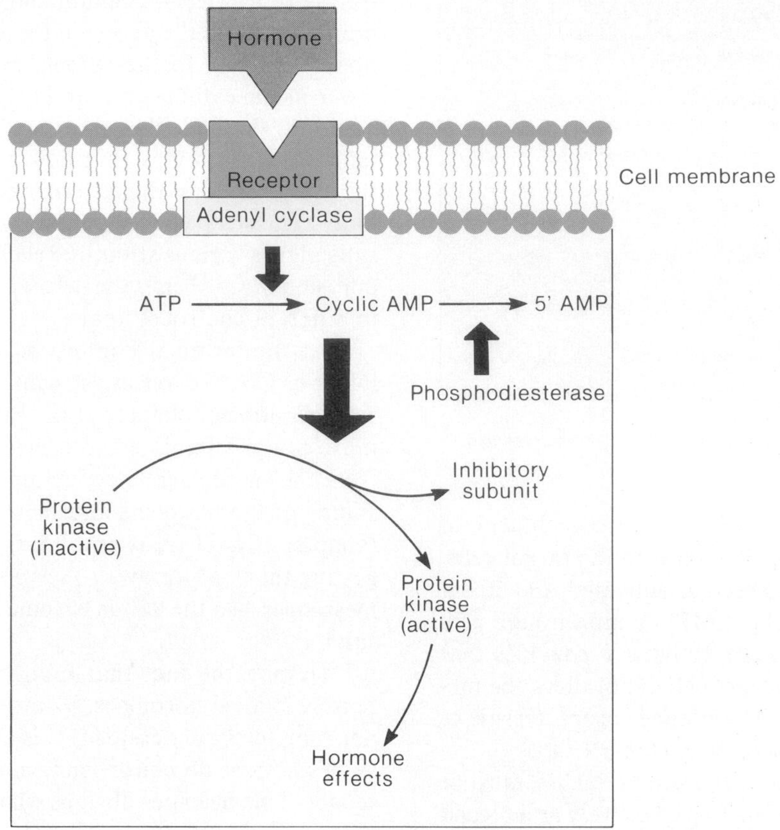

The bonding of these hormones to their membrane receptor proteins activates an enzyme called **adenylate cyclase** *(ah-den'ĭ-lāt si'klās)*. This enzyme is built into the cell membrane, and when activated, it catalyzes the following reaction:

$$ATP \rightarrow cAMP + PP_i.$$

Adenosine triphosphate (ATP) is thus converted into cAMP plus two inorganic phosphates (*pyrophosphate,* abbreviated PP_i). As a result of the interaction of the hormone with its receptor and the activation of adenylate cyclase, therefore, the intracellular concentration of cAMP is increased. Cyclic AMP activates a previously inactive enzyme in the cytoplasm called **protein kinase.** The inactive form of this enzyme consists of two subunits: a catalytic subunit and an inhibitory subunit. The enzyme is produced in an inactive form and becomes active only when cAMP attaches to the inhibitory subunit and causes it to dissociate from the catalytic subunit, which then becomes active (fig. 19.6). The hormone, in summary—acting through an increase in cAMP production—causes an increase in protein kinase enzyme activity within its target cells.

Table 19.5 Hormones that activate adenylate cyclase and use cAMP as a second messenger and hormones that use other second messengers

Hormones that use cAMP as a second messenger	Hormones that use other second messengers
Adrenocorticotropic hormone (ACTH)	Catecholamines (α-adrenergic)
Calcitonin	Growth hormone (GH)
Epinephrine (β-adrenergic)	Insulin
Follicle-stimulating hormone (FSH)	Oxytocin
Glucagon	Prolactin
Luteinizing hormone (LH)	Somatomedin
Parathyroid hormone	Somatostatin
Thyrotropin-releasing hormone (TRH)	
Thyroid-stimulating hormone (TSH)	
Antidiuretic hormone	

Table 19.6	Sequence of events that occurs with cyclic AMP as a second messenger

1. The hormones combine with their receptors on the outer surface of target cell membranes.
2. Hormone-receptor interaction stimulates activation of adenylate cyclase on the cytoplasmic side of the membranes.
3. Activated adenyl cyclase catalyzes the conversion of ATP to cyclic AMP (cAMP) within the cytoplasm.
4. Cyclic AMP activates protein kinase enzymes that were already present in the cytoplasm in an inactive state.
5. Activated cAMP-dependent protein kinase transfers phosphate groups to (phosphorylates) other enzymes in the cytoplasm.
6. The activity of specific enzymes is either increased or inhibited by phosphorylation.
7. Altered enzyme activity mediates the target cell's response to the hormone.

Active protein kinase catalyzes the attachment of phosphate groups to different proteins in the target cells. This causes some enzymes to become activated, and others to become inactivated. Cyclic AMP, acting through protein kinase, thus modulates the activity of enzymes that are already present in the target cell. This alters the metabolism of the target tissue in a manner characteristic of the actions of that specific hormone (table 19.6).

Like all biologically active molecules, cAMP must be rapidly inactivated for it to function effectively as a second messenger in hormone action. This function is served by an enzyme called **phosphodiesterase** *(fos″fo-di-es′ter-as)* within the target cells, which hydrolyzes cAMP into inactive fragments. Through the action of phosphodiesterase, the stimulatory effect of a hormone that uses cAMP as a second messenger depends upon the continuous generation of new cAMP molecules, and thus depends upon the level of secretion of the hormone.

Drugs that inhibit the activity of phosphodiesterase thus prevent the breakdown of cAMP and result in increased concentrations of cAMP within the target cells. The drug *theophylline (the′o-fil′in),* for example, is used clinically to raise cAMP levels within bronchiolar smooth muscle. This duplicates the effect of epinephrine on the bronchioles (producing dilation) in people who suffer from asthma.

In addition to cyclic AMP, other cyclic nucleotides may have second messenger functions. **Cyclic guanosine monophosphate (cGMP),** in particular, has been shown to antagonize the action of cAMP is some cases. The control of cell division and the cell cycle, for example, appears to be related to the ratio of cAMP to cGMP.

Ca^{++} as Second Messenger. The concentration of Ca^{++} in the cytoplasm is kept very low by the action of active transport carriers—calcium pumps—in the cell membrane. Through the action of these pumps, the concentration of calcium in the cytoplasm is 5,000–10,000 times lower in the cytoplasm than in the extracellular fluid. In addition, the endoplasmic reticulum of many cells contains calcium pumps that actively transport Ca^{++} from the cytoplasm into the cisternae of the endoplasmic reticulum. The steep concentration gradient for Ca^{++} that results allows various stimuli to evoke a rapid, though brief, diffusion of Ca^{++} into the cell, which can serve as a signal in different control systems.

At the terminal boutons of axons, for example, the influx of Ca^{++} serves as the signal for the release of neurotransmitters (chapter 14). Similarly, when skeletal muscles are stimulated to contract, the release of Ca^{++} from the sarcoplasmic reticulum serves as the signal for sliding of the filaments and shortening of the sarcomeres (chapter 12). There is now a large body of evidence suggesting that Ca^{++} may also serve as the signal—or second messenger—in the action of some hormones, including insulin.

Despite the fact that insulin is one of the most intensely studied hormones, its mechanism of action is still not completely understood. It is known, however, that insulin's actions do not depend on cAMP as a second messenger. This becomes obvious when one considers the fact that insulin's actions on adipose and liver cells are antagonistic to those of epinephrine, which does use cAMP as a second messenger.

When insulin combines with its receptor protein on the surface of a target cell, there is a rapid and transient inflow of Ca^{++}. This Ca^{++} binds to a cytoplasmic protein called **calmodulin** *(kal″mod-u′lin).* Once Ca^{++} binds to calmodulin, the now-active calmodulin in turn activates specific protein kinase enzymes, which modify the actions of other enzymes in the cell. Activation of specific enzymes by insulin then causes at least some of the effects of insulin in the target cell.

The activation of calmodulin-dependent protein kinase by hormones that use Ca^{++} as a second messenger is analogous to the activation of cAMP-dependent protein kinase by hormones that use cAMP as a second messenger. This is illustrated in figure 19.7, in which the actions of two classes of hormones on liver cells is compared. In one case, epinephrine and glucagon stimulate glycogen breakdown using cAMP as a second messenger; in the other case, vasopressin and angiotension II stimulate glycogen breakdown using Ca^{++} as a second messenger. The effects of cAMP and Ca^{++} in different tissues are summarized in table 19.7.

Figure 19.7. Hormones can stimulate the hydrolysis of liver glycogen by way of two second messenger systems. In this instance, the action of epinephrine and glucagon via the cAMP system is physiologically more significant than the action of vasopressin and angiotensin II via the Ca^{++} calmodulin system.

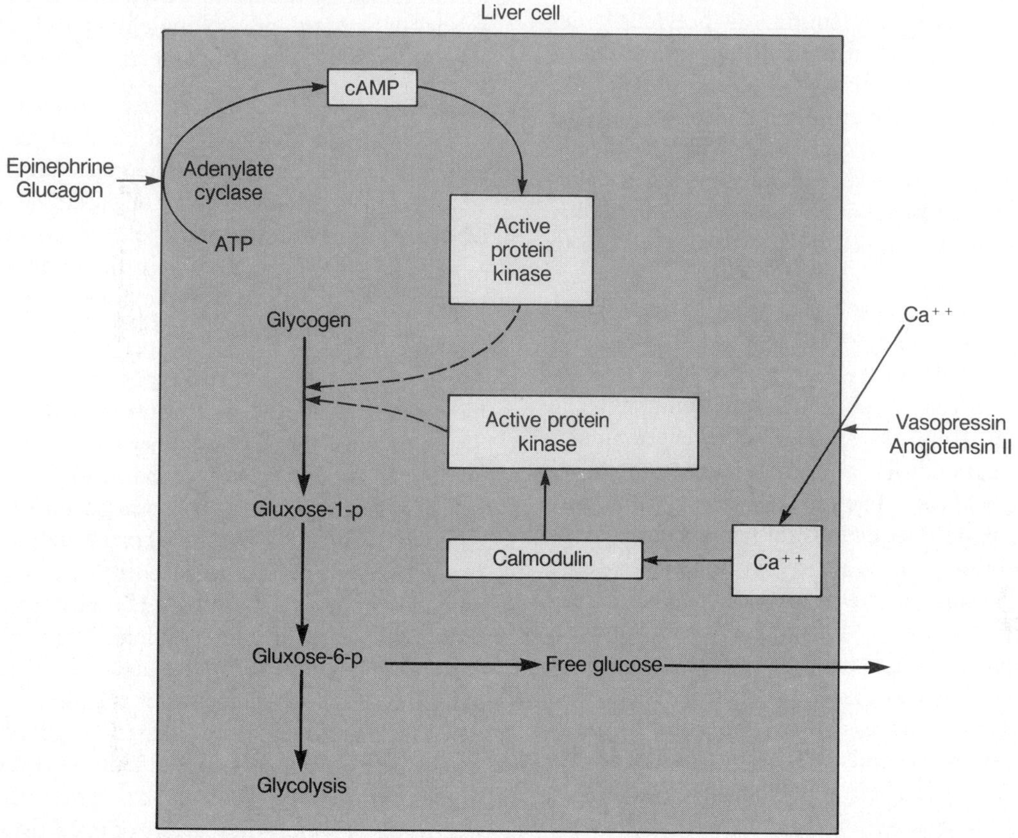

Table 19.7 Comparison of the effects of cyclic AMP (cAMP) and Ca^{++} as second messengers in different tissues

Second messenger	Mechanism	Effects: nerves	Effects: muscles	Effects: hormones
cAMP	Activation of cAMP-dependent protein kinase	Opens ion channels in postsynaptic membrane	Phosphorylation of myosin in skeletal muscle in response to depolarization (significance controversial)	Activation or inactivation of target cell enzymes in response to many polypeptide and glycoprotein hormones
Ca^{++}	Activation of Ca^{++}-dependent regulatory protein (calmodulin)	Mediates release of neurotransmitter in response to depolarization	Phosphorylation of myosin activates actomyosin ATPase in smooth muscles and stimulates contraction	May mediate the effects of insulin on target cells; may increase cAMP by stimulation of adenylate cyclase or decrease cAMP by stimulation of phosphodiesterase

Interactions among Second Messengers. The study of endocrine physiology has come to focus increasingly on events that occur at the molecular level. This study is producing a more complex—but because of this, a more satisfying—picture of endocrine regulation. Since a given hormone can have numerous effects on a particular target cell, and since a given target cell responds to numerous hormones that may use different second messengers, a more complex picture of molecular events is required for adequate explanation of endocrine regulation. This brief section on interactions among second messengers is intended to provide an overview of some of these more advanced concepts.

Evidence has accumulated in recent years to support the presence of another class of second messengers that interacts with the calcium messenger system. Again, insulin can serve as an example. When insulin combines with its receptor protein, two events occur: (1) Ca^{++}enters the cell and combines with calmodulin, as previously described; and (2) an enzyme in the cell membrane called *phospholipase C* becomes activated. The phospholipase C enzyme converts a class of membrane phospholipids (the phosphoinositides) into two additional second messengers. These phospholipid-derived second messengers are **inositol** *(in-o'si-tol)* **triphosphate** and **diacylglycerol** *(di'as''il-glis'er-ol)* (the latter is composed of glycerol attached to two fatty acids).

Inositol triphosphate is released from the cell membrane and travels into the cytoplasm, where it stimulates the endoplasmic reticulum to release its store of Ca^{++}. This Ca^{++} derived from the endoplasmic reticulum, like the Ca^{++} that comes from the extracellular environment, combines with calmodulin and activates calmodulin-dependent protein kinase. The diacylglycerol, which was produced at the same time as inositol triphosphate, stays in the cell membrane, where its action is believed to be required for the sustained effects of the hormone.

The interaction of the Ca^{++} messenger system and the membrane phospholipid-derived second messengers is well illustrated in the action of insulin. Similarly, the Ca^{++}-calmodulin system has been shown to interact in many ways with the cAMP second messenger system. Ca^{++}-activated calmodulin, for example, has been shown to stimulate adenylate cyclase and thus raise cAMP levels in some cases, and to stimulate phosphodiesterase (thus lowering cAMP levels) in other cases. Conversely, cAMP has been shown to modify the transport of Ca^{++} across the cell membrane and endoplasmic reticulum in some tissues. These and other interactions of second messengers probably form the basis for the myriad effects of a given hormone on a target cell, and for the complex interactions that a number of hormones can have—synergistic, permissive, and antagonistic—on their target tissues.

1. Using diagrams, describe how steroid hormones and thyroxine exert their effects on their target cells.
2. Using a diagram, describe how cyclic AMP is produced within a target cell in response to hormone stimulation. List three hormones that use cAMP as a second messenger.
3. Explain how cAMP functions as a second messenger in the action of some hormones.
4. Explain how Ca^{++} can have regulatory functions in different tissues and why it may also be considered to be a second messenger in the action of some hormones.
5. Explain how the membrane-derived phospholipid second messengers interact with the Ca^{++} messenger system.

PROSTAGLANDINS

Prostaglandins are autocrine regulatory molecules derived from arachidonic acid, and they exert their effects within the same tissues in which they are produced. The actions of prostaglandins supplement and complement the regulatory effects of the endocrine and nervous systems.

Objective 9. Define the term *paracrine* or *autocrine,* and contrast this with endocrine regulation.

Objective 10. Describe the chemical nature and biosynthetic pathway of prostaglandins.

Objective 11. Describe some of the effects of different prostaglandins in the body.

Objective 12. Explain how the nonsteroidal, anti-inflammatory drugs work.

In chapters 14 and 19, two types of regulatory molecules have been considered—neurotransmitters and hormones. These two classes of regulatory molecules cannot simply be defined by differences in chemical structure, since the same molecule (such as norepinephrine) may be in both categories, but they must rather be defined by function. Neurotransmitters are released by axons, travel across a narrow synaptic cleft, and affect a postsynaptic cell. Hormones are secreted into the blood by an endocrine gland and, through transport in the blood, come to influence the activities of one or more target organs.

There is yet another class of regulatory molecules, which are distinguished by the fact that they are produced in many different organs and are generally *active within the organ in which they are produced.* Molecules of this type are called *autocrine,* or *paracrine,* regulators. The major group of such molecules is the **prostaglandins** *(pros''tah-glan'din).* Since prostaglandins do not generally function as hormones, they are not properly considered a part of the endocrine system. Nevertheless, as a result of their regulatory function and their close association with hormonal control mechanisms, prostaglandins will be considered in this chapter as a corollary of endocrine regulation.

A prostaglandin is a twenty-carbon-long fatty acid that contains a five-membered carbon ring. This molecule is derived from the precursor molecule *arachidonic acid,* which can be released from phospholipids in the cell membrane under hormonal or other stimulation. Arachidonic acid can then enter one of two possible metabolic pathways. In one case, the arachidonic acid may be converted by the enzyme *cyclo-oxygenase* into a prostaglandin, which can then be changed by other enzymes into other prostaglandins. In the other case, arachidonic acid may be converted by the enzyme *lipoxygenase* into **leukotrienes** *(loo''ko-tri-ēnz),* which are compounds that are closely related to the prostaglandins (fig. 19.8).

Figure 19.8. Formation and actions of leukotrienes and prostaglandins (PG = prostaglandin; TX = thromboxane).

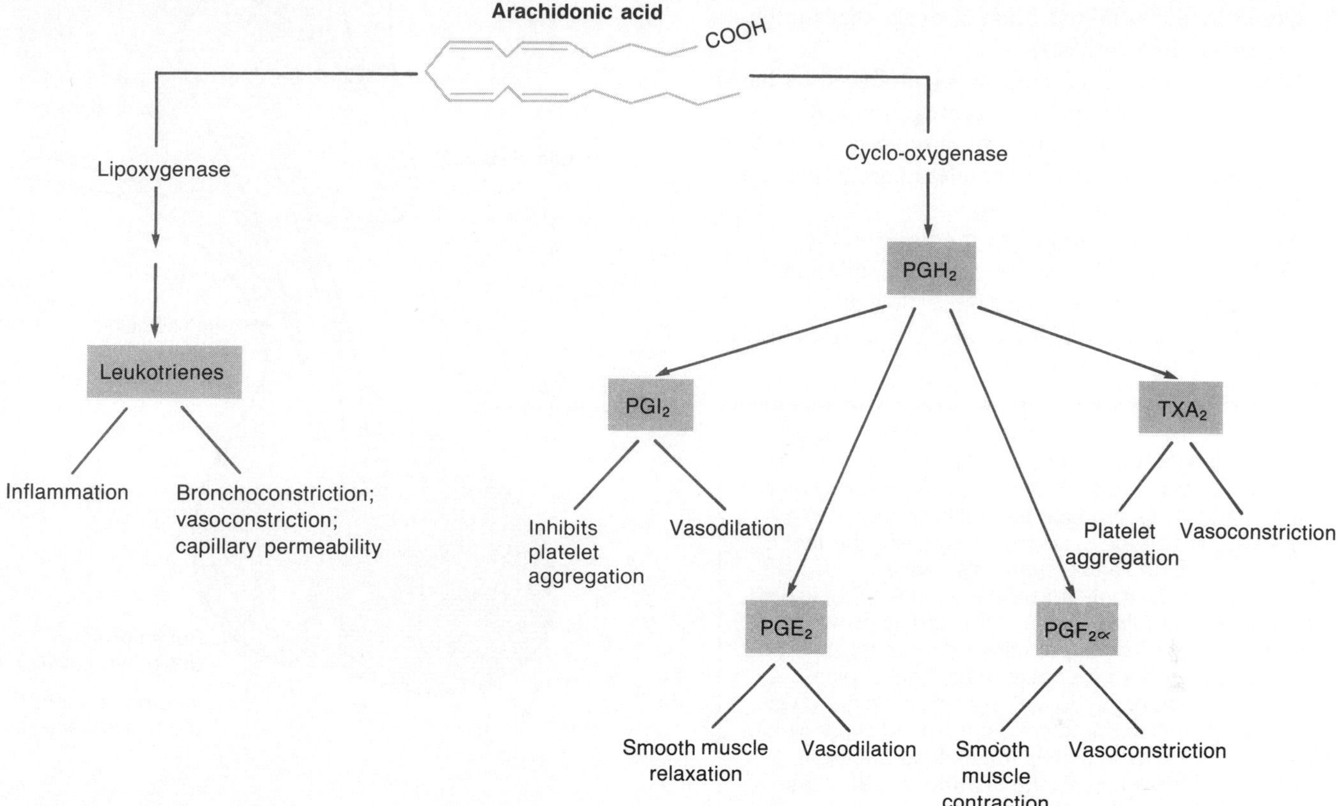

Prostaglandins are produced in almost every organ and have been implicated in a wide variety of regulatory functions. The study of prostaglandin function can be confusing, in part because a given prostaglandin may have opposite effects in different organs. Prostaglandins of the E series (PGE), for example, cause relaxation of smooth muscle in the urinary bladder, bronchioles, intestine, and uterus, but the same molecules cause contraction of vascular smooth muscle. A different prostaglandin, designated $PGF_{2\alpha}$, has exactly the opposite effects.

The antagonistic effects of prostaglandins on blood clotting make good physiological sense. Blood platelets, which are required for blood clotting, produce *thromboxane* A_2. This prostaglandin promotes clotting by stimulating platelet aggregation and vasoconstriction. The endothelial cells of blood vessels, in contrast, produce a different prostaglandin, known as PGI_2, or *prostacyclin,* which has the opposite effects—it inhibits platelet aggregation and causes vasodilation. These antagonistic effects help to promote clotting but to insure that clots do not normally form on the walls of intact blood vessels.

The following are some of the regulatory functions proposed for prostaglandins in different organs and systems:

1. **Inflammation.** Prostaglandins promote many aspects of the inflammatory process, including the development of pain and fever. Drugs that inhibit prostaglandin synthesis help to alleviate these symptoms.
2. **Reproductive system.** Prostaglandins may play a role in ovulation and corpus luteum function in the ovaries and in contraction of the uterus. Excessive prostaglandin production may be involved in premature labor, endometriosis, premenstrual tension, and other gynecological disorders.
3. **Gastrointestinal tract.** The stomach and intestines produce prostaglandins, which are believed to inhibit gastric secretions and influence intestinal motility and fluid absorption. Since prostaglandins inhibit gastric secretion, drugs that suppress prostaglandin production may make a patient more susceptible to peptic ulcers.
4. **Respiratory system.** Some prostaglandins cause constriction, whereas others cause dilation of blood vessels in the lungs and of bronchiolar smooth muscle. The leukotrienes are potent bronchoconstrictors, and these compounds, together with some prostaglandins, may cause respiratory distress and contribute to bronchoconstriction in asthma.
5. **Blood vessels.** Some prostaglandins are vasoconstrictors, and others are vasodilators. In a fetus, PGE_2 is believed to promote dilation of the ductus arteriosus, a short vessel that connects the pulmonary trunk with the aorta. After birth the

ductus arteriosus normally closes as a result of a rise in blood oxygen when the baby breathes. If the ductus remains patent (open), however, it can be closed by the administration of drugs that inhibit prostaglandin synthesis.

6. **Blood clotting.** Thromboxane A_2, produced by blood platelets, promotes platelet aggregation and vasoconstriction. Prostacyclin, produced by vascular endothelial cells, inhibits platelet aggregation and promotes vasodilation.
7. **Kidneys.** Prostaglandins are produced in the medulla of the kidneys and cause vasodilation, resulting in increased renal blood flow and increased excretion of water and electrolytes in the urine.

Aspirin is the most widely used member of a class of drugs known as *nonsteroidal anti-inflammatory drugs.* Aspirin and other drugs of this class produce their effects because they specifically inhibit the cyclo-oxygenase enzyme that is needed for prostaglandin synthesis. Since prostaglandins promote inflammation, therefore, aspirin helps to reduce inflammation, pain, and fever. Acting through the inhibition of prostaglandin synthesis, which is needed for proper blood clotting, aspirin significantly increases the clotting time for eight days after administration (platelets are replaced every eight days). For this reason aspirin may be used to treat venous embolisms and arterial thromboses and has been used in patients with myocardial ischemia in the hope that it would prevent the formation of blood clots in the coronary vessels.

1. Describe the chemical nature of prostaglandins, and explain why these molecules are not considered to be hormones.
2. Describe the metabolic pathway by which prostaglandins are produced, and describe the different types of functions performed by different prostaglandins.
3. Explain how aspirin works, and describe the effects of aspirin on different body processes.

Pituitary Gland

The neurohypophysis secretes hormones that are produced by the hypothalamus, whereas the adenohypophysis secretes its own hormones in response to regulation from hypothalamic hormones. The secretions of the pituitary gland are thus controlled by the hypothalamus, as well as by negative feedback influences from the target glands.

Objective 13. Describe the structure and embryonic development of the pituitary gland.

Objective 14. List the hormones secreted by the adenohypophysis and neurohypophysis.

Figure 19.9. The structure of the pituitary gland as seen in the sagittal view.

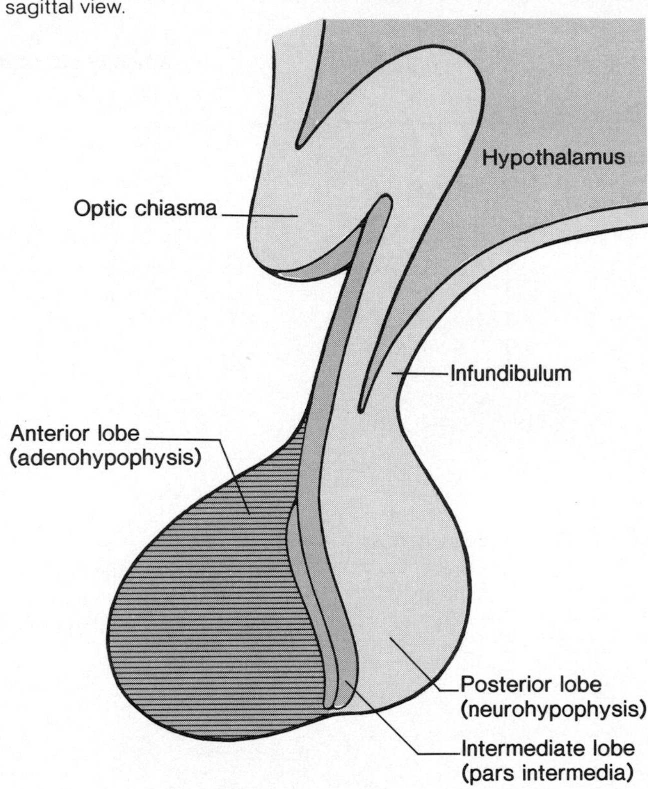

Objective 15. Describe, in a general way, the actions of anterior pituitary hormones, and explain why the term "master gland" has been used in the past to describe the adenohypophysis.

Objective 16. Describe how the secretions of anterior and posterior pituitary hormones are controlled by the hypothalamus.

Objective 17. Explain how the secretion of anterior pituitary hormones is regulated by negative feedback.

Structure of the Pituitary Gland

The **pituitary gland,** or **hypophysis** *(hi-pof'i-sis),* is located on the inferior aspect of the brain in the region of the diencephalon, and is attached to the hypothalamus by a stalk-like structure called the *infundibulum* (fig. 19.9). The pituitary is a rounded, pea-shaped gland measuring about 1.3 cm (0.5 in.) in diameter. It is covered by the dura mater and is supported by the sella turcica of the sphenoid bone. The arterial circle surrounds the highly vascular pituitary gland, providing it with a rich blood supply.

pituitary: L. *pituita,* phlegm (this gland was originally thought to secrete mucus into the nasal cavity)

Figure 19.10. The development of the pituitary gland. (*a*) the head end of an embryo at four weeks showing the position of a midsagittal cut seen in the developmental sequence (*b–e*). The pituitary gland arises from a specific portion of the neuroectoderm, called the neurohypophyseal bud, which evaginates downward during the fourth and fifth weeks, respectively, in (*b*) and (*c*) and a specific portion of the oral ectoderm, called the hypophyseal (Rathke's) pouch, which evaginated upward from a specific portion of the primitive oral cavity. By eight weeks (*d*), the hypophyseal pouch is no longer connected to the pharyngeal roof of the oral cavity. During the fetal stage (*e*), the development of the pituitary gland is completed.

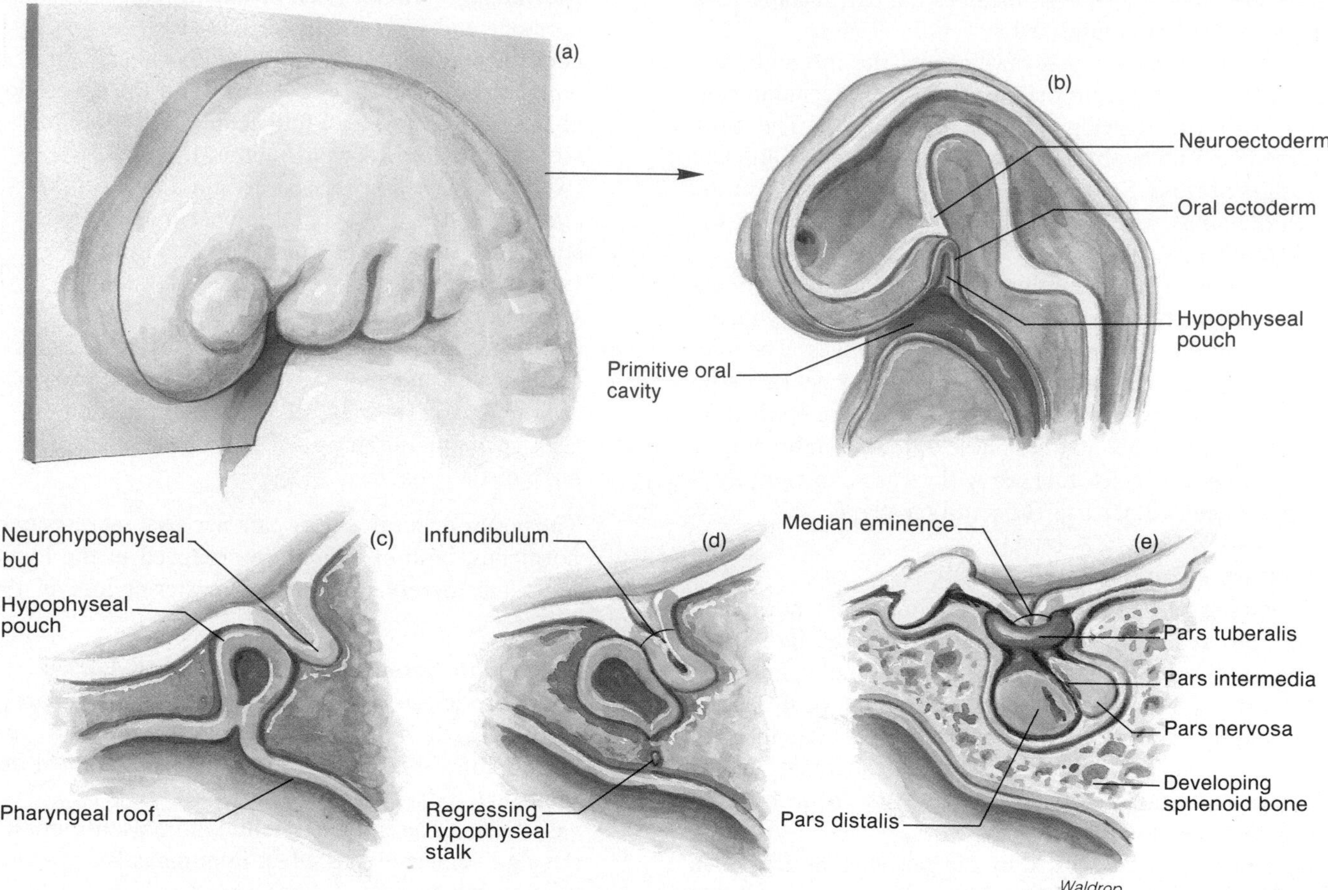

The pituitary gland is structurally and functionally divided into an anterior lobe, or **adenohypophysis** *(ad"ĕ-no-hi-pof'ĭ-sis)*, and a posterior lobe called the **neurohypophysis.** These two parts have different embryonic origins. The adenohypophysis is derived from epithelial tissue, whereas the neurohypophysis is formed as a downgrowth of the brain. The adenohypophysis consists of three parts: (1) the **pars distalis** is the bulbar portion; (2) the **pars tuberalis** is the thin extension in contact with the infundibulum; and (3) the **pars intermedia** is between the anterior and posterior parts of the pituitary. These parts are illustrated in figure 19.9.

The neurohypophysis is the neural part of the pituitary gland. It consists of the bulbar **pars nervosa,** which is in contact with the pars intermedia and the pars distalis of the adenohypophysis, and the **infundibulum** *(in"fun-dib'u-lum)*, which is the connecting stalk to the hypothalamus. Nerve fibers extend through the infundibulum along with minute neuroglia-like cells, called **pituicites** *(pĭ-tu'ĭ-sīt)*.

adenohypophysis: Gk. *adeno,* gland; *hypo,* under; *physis,* a growing
infundibulum: L. *infundibulum,* a funnel

The pituitary is the structure of the brain perhaps most subject to neoplasms. A tumor of the pituitary is generally easily detected as it begins to grow and interfere with hormonal activity. If a tumor of the pituitary grows superiorly, it may exert sufficient pressure on the optic chiasma to cause bitemporal hemianopia, which is blindness in the temporal field of vision of both eyes, or "tunnel vision." Surgical removal of a neoplasm of the pituitary gland (hypophysectomy) may be performed transcranially through the frontal bone or through the nasal cavity and sphenoidal air sinus.

Development of the Pituitary Gland

The adenohypophysis begins to develop during the third week as a diverticulum *(di"ver-tik'u-lum),* a pouchlike extension, called the **hypophyseal (Rathke's) pouch** (fig. 19.10). It arises from the roof of the primitive oral cavity and grows toward the brain. At the same time, another

Rathke's pouch: from Martin H. Rathke, German anatomist, 1793–1860

diverticulum called the infundibulum forms from the diencephalon on the inferior aspect of the brain. As the two diverticula come in contact, the hypophyseal pouch loses its connection with the oral cavity, and the primordial tissue of the adenohypophysis is formed. The fully developed adenohypophysis includes the pars distalis, pars tuberalis, and pars intermedia.

The neurohypophysis develops as the infundibulum extends inferiorly from the diencephalon to come in contact with the developing adenohypophysis. The fully formed neurohypophysis consists of the infundibulum and the pars nervosa. Specialized nerve fibers that connect the hypothalamus with the pars nervosa develop within the infundibulum.

Notice that the neurohypophysis is essentially an extension of the brain, and indeed the entire pituitary gland, like the brain, is surrounded by the meninges. The adenohypophysis, in contrast, is derived from nonneural tissue (the same embryonic tissue that will form the epithelium over the roof of the mouth). These different embryologies have important consequences with regard to the physiology of the two parts of the pituitary gland.

Pituitary Hormones

The pituitary gland secretes eight important hormones. The first six in the following list are secreted by the anterior pituitary (the pars distalis of the adenohypophysis). The last two hormones in the list are produced in the hypothalamus, transported through axons in the infundibulum to the posterior pituitary, and secreted by the posterior pituitary (the pars nervosa of the neurohypophysis) into the blood.

The hormones secreted by the pars distalis are called **trophic** *(trof'ik)* **hormones.** The term *trophic* means "food." This term is used because high amounts of the anterior pituitary hormones make their target glands hypertrophy, while low amounts cause the target glands to atrophy. When names are applied to the hormones of the pars distalis, therefore, the "trophic" term—conventionally shortened to *tropic*—is incorporated into these names and as the suffix *-tropin.* The hormones of the pars distalis are:

1. **Growth hormone (GH, or somatotropin** [*so-mah-to-tro'pin*]**).** This hormone stimulates growth in all body organs. It also promotes the movement of amino acids into tissue cells and the incorporation of these amino acids into tissue proteins.
2. **Thyroid-stimulating hormone (TSH, or thyrotropin).** This hormone stimulates the thyroid gland to produce and secrete thyroxine (tetraiodothyronine, or T_4).
3. **Adrenocorticotropic hormone (ACTH, or corticotropin).** This hormone stimulates the adrenal cortex to secrete the glucocorticoids, such as hydrocortisone (cortisol).
4. **Follicle-stimulating hormone (FSH, or folliculotropin).** This hormone stimulates the growth and secretion of ovarian follicles in the ovaries of females and the production of sperm in the testes of males.
5. **Luteinizing hormone (LH, or luteotropin).** This hormone and FSH are collectively called **gonadotropic** *(gon''ah-do-tro'pik)* **hormones.** In females, LH stimulates ovulation and the conversion of the ovulated ovarian follicle into an endocrine structure called a corpus luteum. In males, LH (which is sometimes also called interstitial *[in''ter-stish'al]* cell-stimulating hormone, or ICSH) stimulates the secretion of male sex hormones (mainly testosterone) from the interstitial cells of Leydig in the testes.
6. **Prolactin.** This hormone is secreted by both males and females, but its function is only well understood in females, where it stimulates the production of milk in the mammary glands of women after the birth of their babies.

The posterior pituitary, or pars nervosa, secretes only two hormones, both of which are produced in the hypothalamus and merely stored in the posterior lobe of the pituitary.

1. **Antidiuretic hormone (ADH, or vasopressin).** Antidiuretic *(an''ti-di-u-ret'ik)* hormone stimulates the kidneys to retain water so that less water is excreted in the urine and more water is retained in the blood. This hormone also causes vasoconstriction in experimental animals, although the significance of this effect in humans is controversial.
2. **Oxytocin.** This hormone has no known function in males, but in females it is known to stimulate contractions of the uterus during labor and contractions of the mammary gland alveoli and ducts, which result in the milk-ejection reflex during lactation.

Oxytocin, like ADH, is produced by neurons in the hypothalamus and travels down axons in the infundibulum to the pars nervosa, where it is stored and secreted in response to nerve impulses from the hypothalamus.

> Injections of oxytocin may be given to a woman during labor if she is having difficulties in parturition. Increased amounts of oxytocin assist uterine contractions and generally speed up delivery. Oxytocin administration after parturition causes the uterus to regress in size and squeezes the blood vessels, thus minimizing the danger of hemorrhage.

Figure 19.11. The posterior pituitary, or neurohypophysis, stores and secretes hormones (vasopressin and oxytocin) produced in neuron cell bodies within the supraoptic and paraventricular nuclei of the hypothalamus. These hormones are transported to the posterior pituitary by nerve fibers of the hypothalamo-hypophyseal tract.

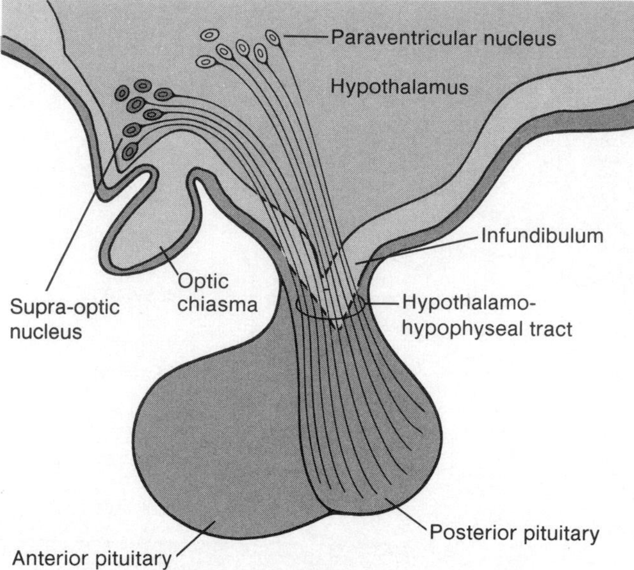

Hypothalamic Control of the Neurohypophysis

The posterior pituitary (pars nervosa of the neurohypophysis) secretes two hormones: antidiuretic hormone (ADH) and oxytocin. These two hormones, however, are actually produced in neuron cell bodies of the *supraoptic (su''prah-op'tik) nuclei* and *paraventricular nuclei* of the hypothalamus. These nuclei within the hypothalamus are thus endocrine glands; the hormones they produce are transported along axons of the **hypothalamo-hypophyseal** *(hi''po-fiz'e-al)* **tract** (fig. 19.11) to the posterior pituitary, which stores and later secretes these hormones. The posterior pituitary is thus more a storage organ than a true gland.

The secretion of ADH and oxytocin from the posterior pituitary is controlled by **neuroendocrine** *(nu''ro-en'do-krin)* **reflexes.** In nursing mothers, for example, the stimulus of sucking acts via sensory nerve impulses to the hypothalamus to stimulate the reflex secretion of oxytocin. The secretion of ADH is stimulated by osmoreceptor neurons in the hypothalamus in response to a rise in blood osmotic pressure; its secretion is inhibited by sensory impulses from stretch receptors in the left atrium of the heart in response to a rise in blood volume.

Hypothalamic Control of the Adenohypophysis

At one time the anterior pituitary was called the "master gland" because it secretes hormones that regulate some other endocrine glands (see fig. 19.12 and table 19.8). Adrenocorticotropic hormone (ACTH), thyroid-stimulating hormone (TSH), and the gonadotropic hormones (FSH and LH) stimulate the adrenal cortex, thyroid, and gonads, respectively, to secrete their hormones. The anterior pituitary hormones also have a "trophic" effect on their target glands, in that the structure and health of the target glands depend on adequate stimulation by anterior pituitary hormones. The anterior pituitary, however, is not really the master gland, because secretion of its hormones is controlled by hormones secreted by the hypothalamus. Both hypothalamic hormones and anterior pituitary hormones are in turn regulated by feedback effects of the hormones of the target glands. Control, therefore, is not vested in any one gland, and thus the entire concept of a "master gland" is inappropriate.

Releasing and Inhibiting Hormones. Since axons do not enter the anterior pituitary, hypothalamic control of the anterior pituitary is achieved through hormonal rather than neural regulation. Neurons in the hypothalamus produce releasing and inhibiting hormones, which are transported to axon endings in the basal portion of the hypothalamus. This region, known as the *median eminence,* contains blood capillaries that are drained by venules in the stalk of the pituitary.

The venules that drain the median eminence deliver blood to a second capillary bed in the anterior pituitary. Since this second capillary bed receives venous rather than arterial blood, the vascular link between the median eminence and the anterior pituitary comprises a portal system. (This is analogous to the hepatic portal system that delivers venous blood from the intestine to the liver.) The vascular link between the hypothalamus and the anterior pituitary is thus called the **hypothalamo-hypophyseal portal system.**

Neurons of the hypothalamus secrete polypeptide hormones into this portal system that regulate the secretions of the anterior pituitary (table 19.9). Thyrotropin-releasing hormone (**TRH**) stimulates the secretion of TSH, and corticotropin *(kor''ti-ko-tro'pin)*-releasing hormone (**CRH**) stimulates the secretion of ACTH from the anterior pituitary. A single releasing hormone, gonadotropin-releasing hormone, or **GnRH,** appears to stimulate the secretion of both gonadotropic hormones (FSH and LH) from the anterior pituitary. The secretion of prolactin and of growth hormone from the anterior pituitary is regulated by hypothalamic inhibitory hormones, known as **PIH** (prolactin-inhibiting hormone) and **somatostatin,** respectively. Recently, a specific hypothalamic releasing hormone that stimulates growth hormone secretion has also been identified as a polypeptide consisting of forty-four amino acids. Experiments suggest that a releasing hormone for prolactin may also exist, but no such specific releasing hormone has yet been discovered (although there is evidence that TRH may stimulate the secretion of prolactin as well as of TSH).

Figure 19.12. The hormones secreted by the pars distalis of the pituitary and the target organs for those hormones. (Note: MSH is secreted by the pars intermedia.)

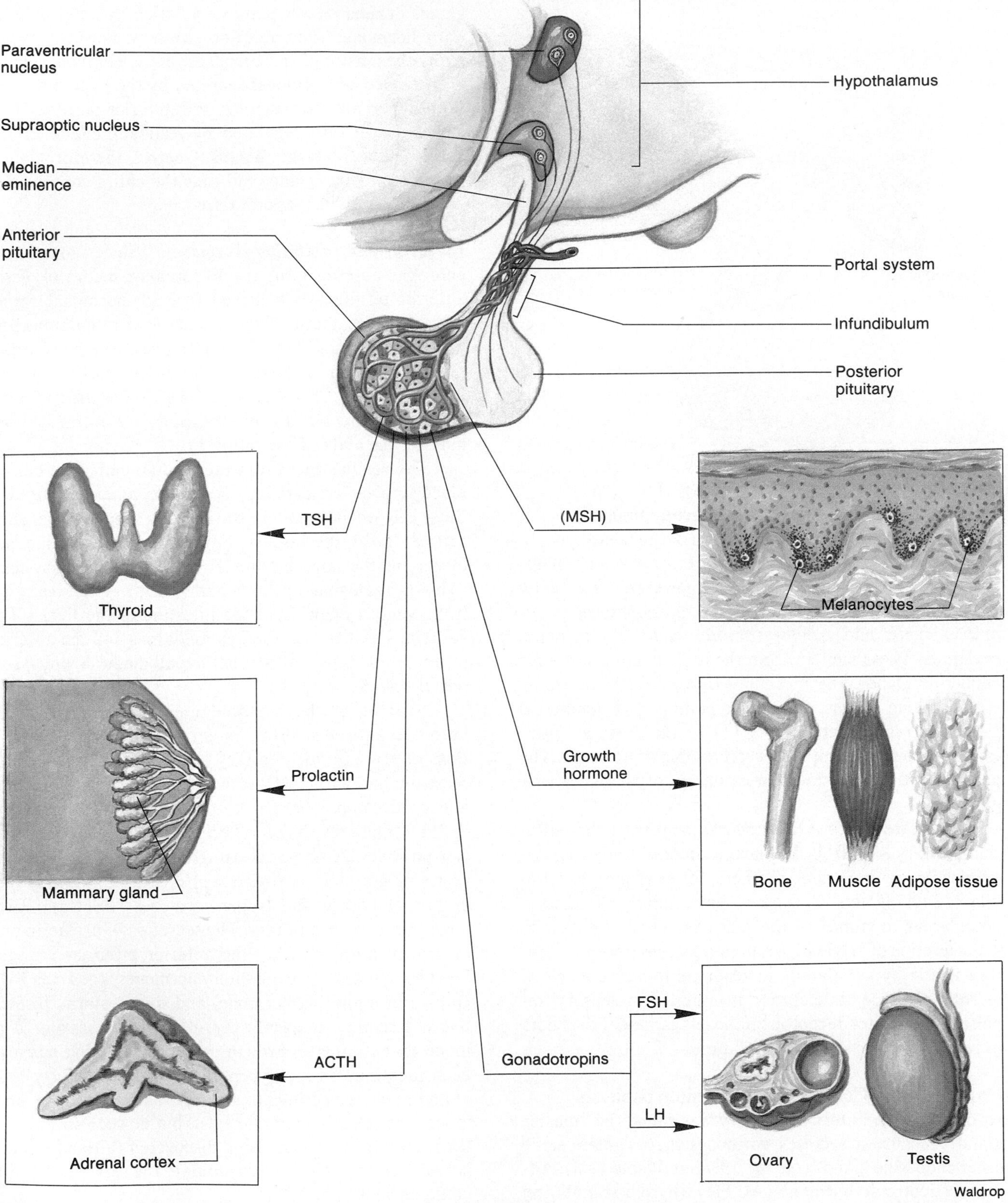

Table 19.8 Hormones secreted by the anterior pituitary

Hormone	Target tissue	Stimulated by hormone	Regulation of secretion
ACTH (adrenocorticotropic hormone)	Adrenal cortex	Secretion of glucocorticoids	Stimulated by CRH (corticotropin-releasing hormone); inhibited by glucocorticoids
TSH (thyroid-stimulating hormone)	Thyroid gland	Secretion of thyroid hormones	Stimulated by TRH (thyrotropin-releasing hormone); inhibited by thyroid hormones
GH (growth hormone)	Most tissue	Protein synthesis and growth; lipolysis and increased blood glucose	Inhibited by somatostatin; stimulated by growth hormone-releasing hormone
FSH (follicle-stimulating hormone) and LH (luteinizing hormone)	Gonads	Gamete production and sex steroid hormone secretion	Stimulated by GnRH (gonadotropin-releasing hormone); inhibited by sex steroids
Prolactin	Mammary glands and other sex accessory organs	Milk production Controversial actions in other organs	Inhibited by PIH (prolactin-inhibiting hormone)
LH (luteinizing hormone)	Gonads	Sex hormone secretion: ovulation and corpus luteum formation	Stimulated by GnRH

Table 19.9 Some hypothalamic hormones involved in the control of the anterior pituitary

Hypothalamic hormone	Structure	Effect on anterior pituitary	Action of anterior pituitary hormone
Corticotropin-releasing hormone (CRH)	41 amino acids	Stimulates secretion of adrenocorticotropic hormone (ACTH)	Stimulates secretions of adrenal cortex
Gonadotropin-releasing hormone (GnRH)	10 amino acids	Stimulates secretion of follicle-stimulating hormone (FSH) and luteinizing hormone (LH)	Stimulates gonads to produce gametes (sperm and ova) and secrete sex steroids
Prolactin-inhibiting hormone (PIH)	Controversial (may be dopamine)	Inhibits prolactin secretion	Stimulates production of milk in mammary glands
Somatostatin	14 amino acids	Inhibits secretion of growth hormone	Stimulates anabolism and growth in many organs
Thyrotropin-releasing hormone (TRH)	3 amino acids	Stimulates secretion of thyroid-stimulating hormone (TSH)	Stimulates secretion of thyroid gland
Growth hormone-releasing hormone (GRH)	44 amino acids	Stimulates growth hormone secretion	Stimulates anabolism and growth in many organs

Feedback Control of the Adenohypophysis

In view of its secretion of releasing and inhibiting hormones, the hypothalamus might be considered the "master gland." The chain of command, however, is not linear; the hypothalamus and anterior pituitary are controlled by the effects of their own actions. In the endocrine system, to use an analogy, the general takes orders from the private. The hypothalamus and anterior pituitary are not master glands because their secretions are controlled by the target glands they regulate.

Anterior pituitary secretion of ACTH, TSH, and the gonadotropins (FSH and LH) is controlled by **negative feedback inhibition** from the target gland hormones. Secretion of ACTH is inhibited by a rise in corticosteroid secretion, for example, and TSH is inhibited by a rise in the secretion of thyroxine from the thyroid. These negative feedback relationships are easily demonstrated by removal of the target glands. Castration (surgical removal of the gonads), for example, produces a rise in the secretion of FSH and LH. In a similar manner, removal of the adrenals or the thyroid would result in an abnormal increase in ACTH or TSH secretion from the anterior pituitary.

These effects demonstrate that under normal conditions the target glands exert an inhibitory effect on the anterior pituitary. This inhibitory effect can occur at two levels: (1) the target gland hormones could act on the hypothalamus and inhibit the secretion of releasing hormones, and (2) the target gland hormones could act on the anterior pituitary and inhibit its response to the releasing hormones. Thyroxine, for example, appears to inhibit the response of the anterior pituitary to TRH and thus acts to reduce TSH secretion (fig. 19.13). Sex steroids, in contrast, reduce the secretion of gonadotropins by inhibiting both GnRH secretion and the ability of the anterior pituitary to respond to stimulation by GnRH (see fig. 19.14).

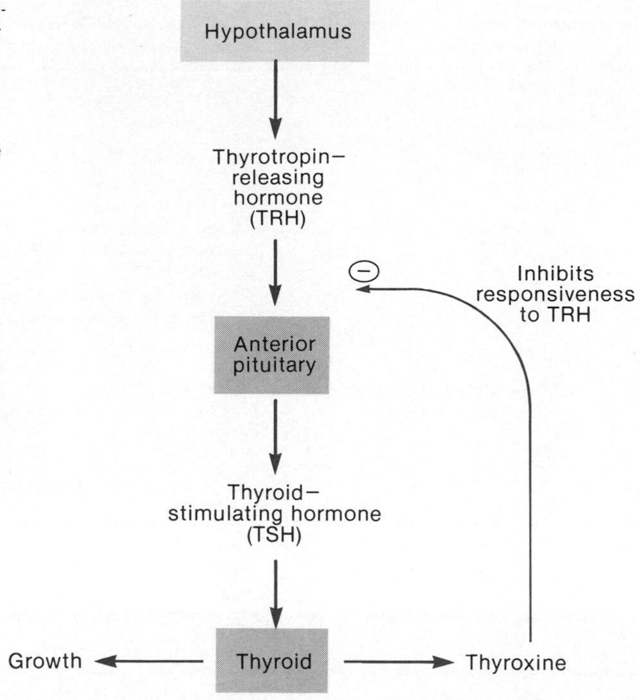

Figure 19.13. The secretion of thyroxine from the thyroid is stimulated by the thyroid-stimulating hormone (TSH) from the anterior pituitary. The secretion of TSH is stimulated by the thyrotropin-releasing hormone (TRH) secreted from the hypothalamus into the hypothalamo-hypophyseal portal system. This stimulation is balanced by the negative feedback inhibition of thyroxine, which decreases the responsiveness of the anterior pituitary to stimulation by TRH.

Recent evidence suggests that there may be retrograde transport of blood from the anterior pituitary to the hypothalamus. This may permit a *short feedback loop* where a particular trophic hormone inhibits the secretion of its releasing hormone from the hypothalamus. A high secretion of TSH, for example, may inhibit further secretion of TRH by this means.

In addition to negative feedback control of the anterior pituitary, there is one example where a hormone from a target organ actually stimulates the secretion of an anterior pituitary hormone. This occurs toward the middle of the menstrual cycle when the rising secretion of estradiol from the ovaries stimulates the anterior pituitary to secrete a "surge" of LH, which results in ovulation. This case is commonly referred to as a *positive feedback* effect to distinguish it from the more usual negative feedback inhibition of target gland hormones on anterior pituitary secretion. Interestingly, higher levels of estradiol at a later stage of the menstrual cycle exert the opposite effect—negative feedback inhibition—on LH secretion.

Higher Brain Function and Pituitary Secretion

The feedback effect of estradiol on the secretion of gonadotropic hormones is believed to be exerted at the level of the pituitary gland and hypothalamus. Since the hypothalamus receives neural input from "higher brain centers," however, it is not surprising that the pituitary-gonad axis can be affected by emotions, so that intense emotions may alter the timing of ovulation or menstruation. The influences of higher brain centers on the pituitary-gonad axis also help to explain the "dormitory effect," in which researchers have noted a tendency for the menstrual cycles to synchronize in girls who room together.

The effect of stress on the pituitary-adrenal axis is another good example of the influence of higher brain centers on pituitary function. Stressors, as described later in this chapter, produce an increase in CRH secretion from the hypothalamus, which in turn results in elevated ACTH and corticosteroid secretion. In addition, the influence of higher brain centers produces *circadian (ser''kah-de'an)* ("about a day") ***rhythms*** in the secretion of many anterior pituitary hormones. The secretion of growth hormone, for example, is highest during sleep and decreases during wakefulness, although its secretion is also stimulated, following a meal, by the absorption of particular amino acids.

1. Describe the embryonic origins of the adenohypophysis and neurohypophysis, and list the parts of each. Indicate which of these parts are also called the "anterior pituitary" and "posterior pituitary."
2. List the hormones secreted by the posterior pituitary. Describe the site of origin of these hormones and the mechanisms by which their secretions are regulated.
3. List the hormones secreted by the anterior pituitary, and describe how the hypothalamus controls the secretion of each hormone.
4. Draw a negative feedback loop showing the control of ACTH secretion. Explain how this system would be affected by (a) an injection of ACTH; (b) surgical removal of the pituitary; (c) an injection of corticosteroids; and (d) surgical removal of the adrenal glands.

Figure 19.14. Negative feedback control of gonadotropin secretion. The possible short negative feedback loop involving the inhibition of GnRH secretion by the retrograde transport of gonadotropins is shown by a dotted arrow.

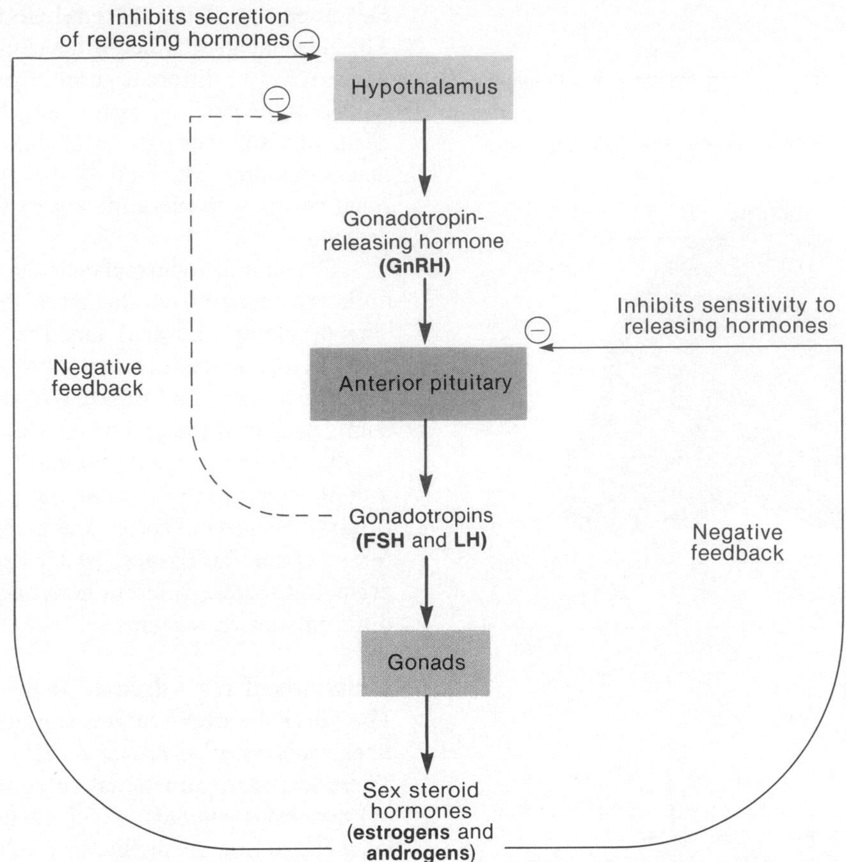

Adrenal Glands

The adrenal medulla secretes catecholamine hormones, which complement the action of the sympathetic nervous system. The adrenal cortex secretes corticosteroids, which participate in the regulation of mineral balance, energy balance, and reproductive function.

Objective 18. Describe the structure and embryonic development of the adrenal cortex and adrenal medulla.

Objective 19. Describe the action of epinephrine and norepinephrine, and explain how the secretion of these hormones is regulated.

Objective 20. Describe the different categories of corticosteroids, identify the parts of the adrenal cortex that secretes each, and explain how the secretion of these hormones is regulated.

Objective 21. Explain the relationship between stress and the secretions of the adrenal cortex.

Structure of the Adrenal Glands

The **adrenal** *(ah-dre'nal)* **glands** (also called **suprarenal glands**) are paired organs that cap the superior borders of the kidneys (fig. 19.15). The adrenal glands, along with the kidneys, are embedded against the muscles of the back in a protective pad of fat.

Each adrenal gland is about 50 mm (2 in.) in length, 30 mm (1.1 in.) in width, and 10 mm (0.4 in.) in depth, and is generally pyramidal in appearance. Each adrenal gland consists of an outer cortex and inner medulla (fig. 19.15), which function as separate glands.

The **adrenal cortex** makes up the bulk of the gland and is histologically subdivided into three zones: an outer **zona glomerulosa** *(glo-mer''u-lo'sah),* an intermediate **zona fasciculata** *(fah-sik''u-la'tah),* and an inner **zona reticularis.** The **adrenal medulla** is composed of tightly packed clusters of **chromaffin** *(kro-maf'in)* **cells,** which are arranged around blood vessels. Each cluster of chromaffin cells receives direct autonomic innervation.

adrenal: L. *ad,* to; *renes,* kidney

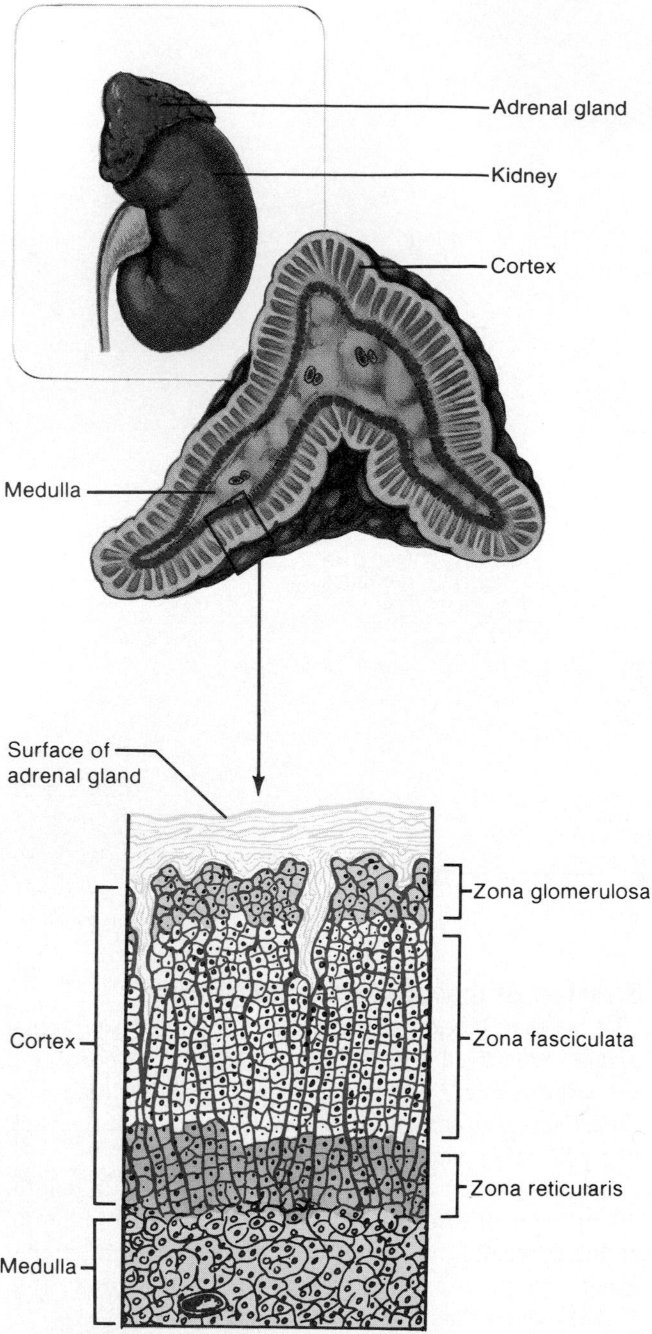

Figure 19.15. The structure of the adrenal gland showing the three zones of the adrenal cortex.

Like other endocrine glands, the adrenal glands are highly vascular. Three separate suprarenal arteries supply blood to each adrenal gland. One arises from the inferior phrenic artery, another from the aorta, and a third is a branch of the renal artery. The venous drainage goes through a suprarenal vein into the inferior vena cava for the right adrenal gland and through the suprarenal vein into the left renal vein for the left adrenal gland.

The adrenal glands are innervated by preganglionic fibers of the splanchnic nerves and by fibers of the celiac and associated sympathetic plexuses.

Development of the Adrenal Glands

The adrenal glands begin development during the fifth week from two different germ layers. Each adrenal gland has an outer part, or cortex, which develops from mesoderm, and an inner part, or medulla, which develops from neuroectoderm. The mesodermal ridge that forms the adrenal cortex is in the same region from which the gonads develop.

The neuroectodermal cells that form the adrenal medulla are derived from the neural crest of the neural tube. The developing adrenal medulla is gradually encapsulated by the adrenal cortex, a process that continues into the fetal stage. The formation of the adrenal gland is not completed until the end of the third year of age.

Notice that the adrenal gland, like the pituitary, has a dual origin; part is neural and part is not. Like the pituitary, the adrenal cortex and medulla are really two different endocrine tissues, which are located in the same organ but secrete different hormones and are regulated by different control systems.

Functions of the Adrenal Cortex

The adrenal cortex secretes steroid hormones called **corticosteroids** *(kor''tĭ-ko-ste'roids),* or **corticoids,** for short. There are three functional categories of corticosteroids: (1) **mineralocorticoids,** which regulate Na^+ and K^+ balance (by acting on the kidneys [chapter 26]); (2) **glucocorticoids,** which regulate the metabolism of glucose and other organic molecules; and (3) **sex steroids,** which are weak androgens (and lesser amounts of estrogens) that supplement the sex steroids secreted by the gonads. These hormones are secreted by the different zones of the adrenal cortex.

Aldosterone is the most potent mineralocorticoid. The mineralocorticoids are produced in the zona glomerulosa (fig. 19.16). The predominant glucocorticoid in humans is cortisol (hydrocortisone), which is secreted by the zona fasciculata and perhaps also by the zona reticularis. The secretion of aldosterone by the zona glomerulosa is stimulated by angiotensin II and by blood K^+. The secretion of cortisol by the zona fasciculata is stimulated by ACTH (fig. 19.17).

Stress and the Adrenal Gland. In 1936, Hans Selye discovered that injections of cattle ovaries into rats (1) stimulated growth of the adrenal cortex; (2) caused atrophy of the lymphoid tissue of the spleen, lymph nodes, and thymus; and (3) produced bleeding peptic ulcers. At

Figure 19.16. Simplified pathways for the synthesis of steroid hormones in the adrenal cortex. The adrenal cortex produces steroids that regulate Na^+ and K^+ balance (mineralocorticoids), steroids that regulate glucose balance (glucocorticoids), and small amounts of sex steroid hormones.

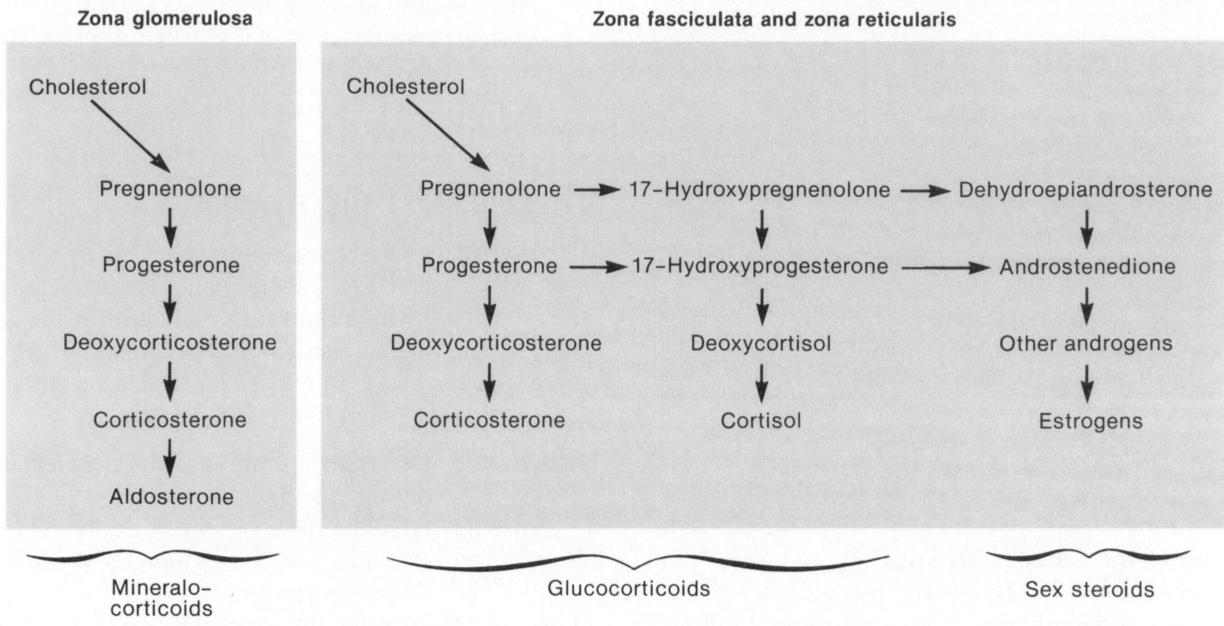

Figure 19.17. The activation of the pituitary-adrenal axis by nonspecific stress.

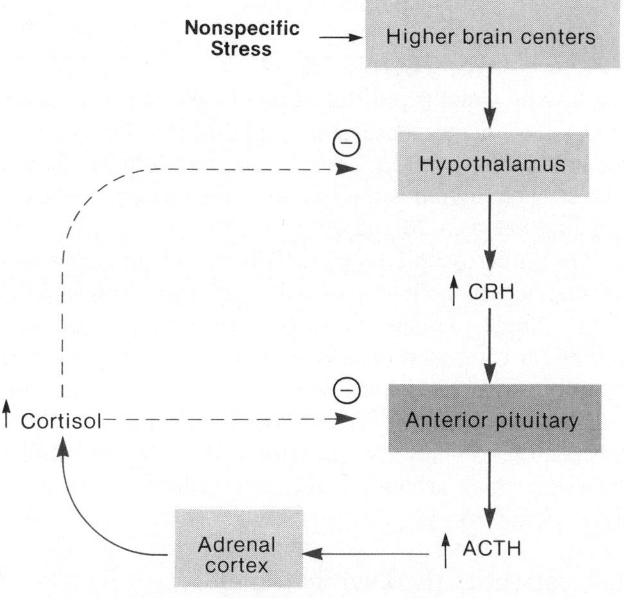

first he thought that these ovarian extracts contained a specific hormone that caused these effects. He later discovered that injections of a variety of substances, including foreign chemicals such as formaldehyde, could produce the same effects. Indeed, the same pattern of effects occurred when he placed rats in cold environments or when he dropped them into water and made them swim until they were exhausted.

The specific pattern of effects produced by these procedures suggested that these effects were the result of something that the procedures shared in common. Selye reasoned that all of the procedures were *stressful* and that the pattern of changes he observed represented a specific response to any stressful agent. He later discovered that all forms of stress produce these effects because all stressors stimulate the pituitary-adrenal axis. Under stressful conditions, there is increased secretion of ACTH and thus increased secretion of corticosteroids from the adrenal cortex.

Since stress is so very difficult to define, many people prefer to define stress operationally as any stimulus that activates the pituitary-adrenal axis. Using this criterion, it has been found that pleasant changes in one's life—such as marriage, a recent promotion, and so on—can be as stressful as unpleasant changes. On this basis, Selye has stated that there is "a nonspecific response of the body to readjust itself following any demand made upon it."

Selye termed this nonspecific response the **general adaptation syndrome (GAS).** Stress, in other words, produces GAS. There are three stages in the response to stress: (1) the *alarm reaction,* when the adrenal glands are activated; (2) the *stage of resistance,* in which readjustment occurs; and (3) if the readjustment is not complete, the *stage of exhaustion* may follow, leading to sickness and possibly death.

Table 19.10 Comparison of the hormones from the adrenal medulla

Epinephrine	Norepinephrine
Elevates blood pressure because of increased cardiac output and peripheral vasoconstriction	Elevates blood pressure because of generalized vasoconstriction
Accelerates respiratory rate and dilates respiratory passageways	Similar effect but to a lesser degree
Increases efficiency of muscular contraction	Similar effect but to a lesser degree
Increases rate of glycogen breakdown into glucose, so level of blood glucose rises	Similar effect but to a lesser degree
Increases rate of fatty acid released from fat, so level of blood fatty acids rises	Similar effect but to a lesser degree
Increases release of ACTH and TSH from the adenohypophysis of the pituitary gland	No effect

Functions of the Adrenal Medulla

The chromaffin cells of the adrenal medulla secrete **epinephrine** and **norepinephrine** in an approximate ratio of 4:1, respectively. These hormones are classified as amines (more specifically, as catecholamines), and are derived from the amino acid tyrosine.

The effects of these hormones are similar to those caused by stimulation of the sympathetic nervous system, except that the hormonal effect lasts about ten times longer. The hormones from the adrenal medulla increase cardiac output and heart rate, dilate coronary blood vessels, increase mental alertness, increase the respiratory rate, and elevate metabolic rate. A comparison of the effects of epinephrine and norepinephrine are presented in table 19.10.

The adrenal medulla is innervated by sympathetic nerve fibers. The impulses are initiated from the hypothalamus via the spinal cord when the sympathetic nervous system is stimulated. Many stressors, therefore, activate the adrenal medulla as well as the adrenal cortex. Activation of the adrenal medulla together with the sympathetic nervous system prepares the body for greater physical performance—the *fight-or-flight* response (discussed in chapter 17).

Excessive stimulation of the adrenal medulla can result in depletion of the body's energy reserves, and high levels of corticosteroid secretion from the adrenal cortex can significantly impair the immune system. It is reasonable to expect, therefore, that prolonged stress can result in increased susceptibility to disease. Indeed, many studies show that prolonged stress results in an increased incidence of cancer and other diseases.

1. List the categories of corticosteroids and the zones of the adrenal cortex that secrete these hormones.
2. List the hormones of the adrenal medulla, and describe their effects.
3. Explain how the secretions of the adrenal cortex and adrenal medulla are regulated.
4. Describe how stress affects the secretions of the adrenal cortex and medulla, and explain how adrenal hormones can result in increased susceptibility to disease.

Thyroid and Parathyroids

The thyroid gland secretes thyroxine and triiodothyronine, which participate in the regulation of metabolism and are critically important for proper growth and development. The thyroid also secretes calcitonin, which may antagonize the action of parathyroid hormone in the regulation of calcium and phosphate balance.

Objective 22. Describe the embryonic development of the thyroid gland.

Objective 23. Describe the structure of the thyroid follicles and how these follicles produce the thyroid hormones.

Objective 24. Describe the metabolic effects of thyroxine and triiodothyronine, and explain how the secretion of these hormones is regulated.

Objective 25. Describe the structure and location of the parathyroid glands, and list the actions of parathyroid hormone.

Structure of the Thyroid Gland

The **thyroid gland** is positioned just below the larynx (fig. 19.18). This gland consists of two lobes that lie on either side of the trachea and are connected anteriorly by a broad **isthmus.** The thyroid is the largest of the endocrine glands, weighing between 20 and 25 g.

On a microscopic level, the thyroid gland consists of many spherical hollow sacs called **thyroid follicles** (fig. 19.19). These follicles are lined with a simple cuboidal epithelium composed of **principal cells,** which synthesize the principal thyroid hormones. The interior of the follicles contains **colloid** *(kol'oid),* which is a protein-rich fluid. Between the follicles are epithelial cells called **parafollicular cells,** which produce a hormone called calcitonin (or thyrocalcitonin).

Development of the Thyroid Gland

The thyroid gland is derived from endoderm and begins its development during the third week as a thickening in the floor of the primitive pharynx. The thickening soon out-pouches downward as the **thyroid diverticulum** (fig. 19.20). A narrow **thyroglossal duct** connects the descending primordial thyroid tissue to the pharynx. As the descent continues, the tongue starts to develop, and the opening into the thyroglossal duct, called the **foramen cecum** *(se'kum),* pierces through the base of the tongue.

Figure 19.18. The thyroid gland. (*a*) its relationship to the larynx and trachea. (*b*) a scan of the thyroid gland twenty-four hours after the intake of radioactive iodine.

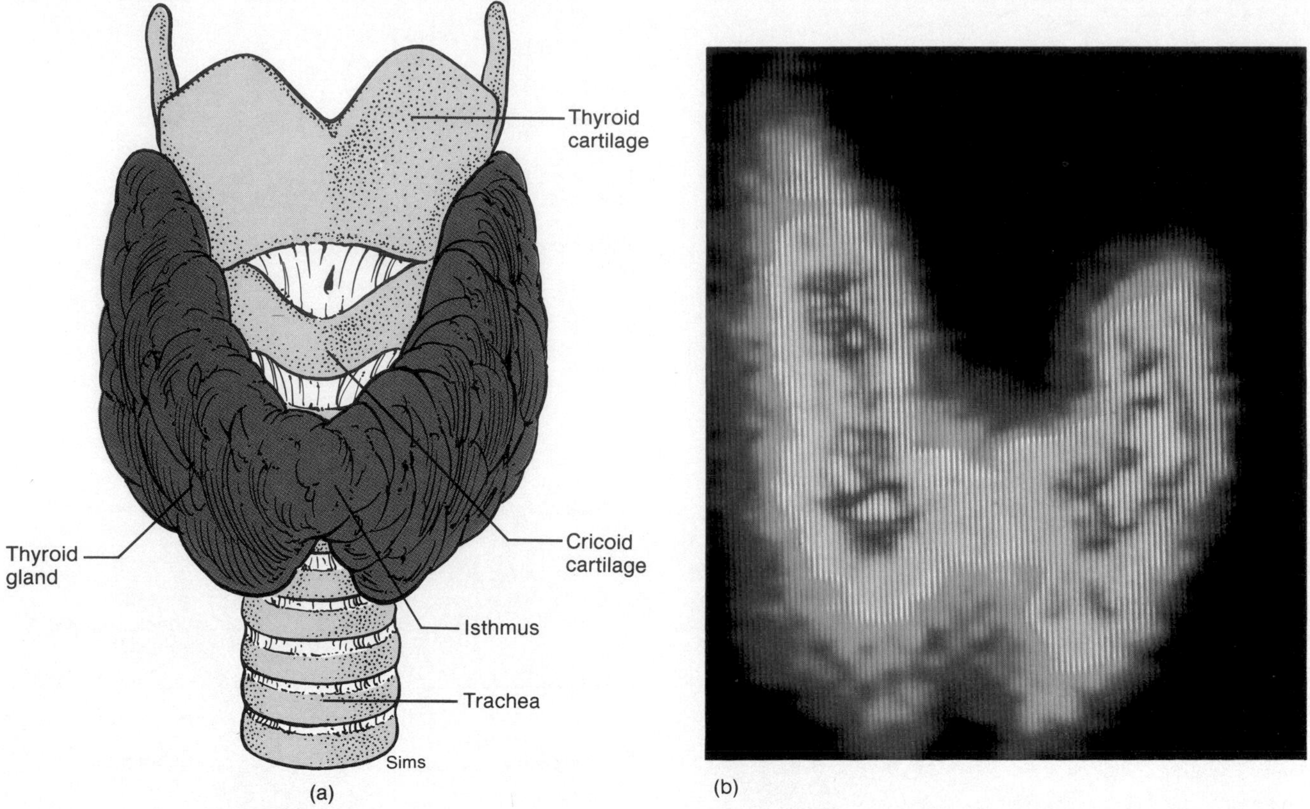

Figure 19.19. A photomicrograph (magnification ×250) of a thyroid gland, showing numerous thyroid follicles. Each follicle consists of follicular cells surrounding the fluid known as colloid, which contains thyroglobulin.

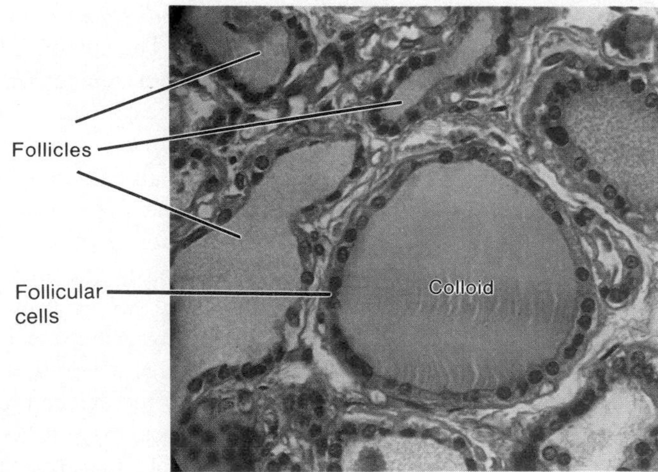

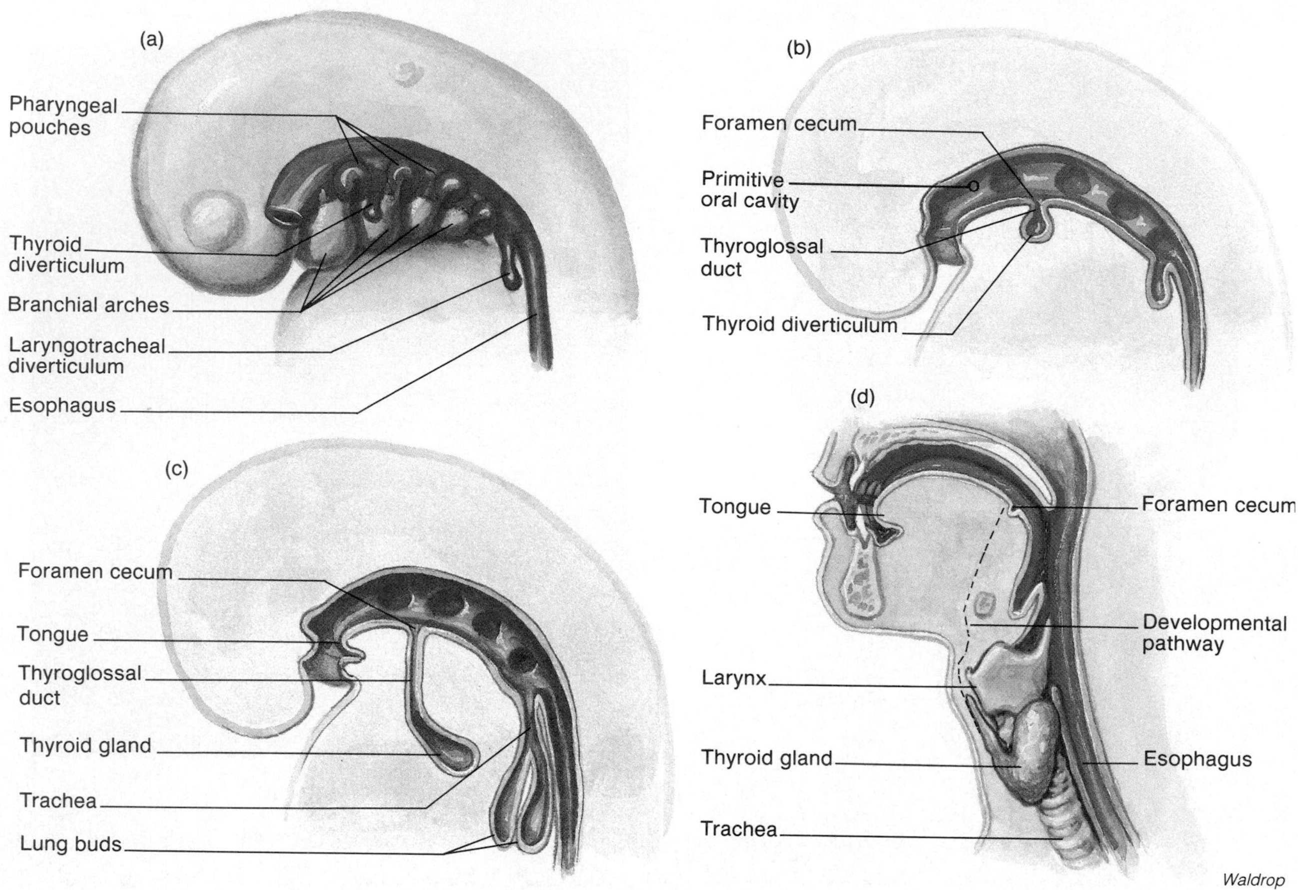

Figure 19.20. The embryonic development of the thyroid gland. (*a*) at four weeks, a thyroid diverticulum begins to form in the floor of the pharynx at the level of the second brachial arch. (*b,c*) the thyroid diverticulum extends downward during the fifth and sixth weeks. (*d*) a sagittal section through the head and neck of an adult shows the path of development and final position of the thyroid gland.

By the seventh week, the thyroid gland occupies a position immediately inferior to the larynx and around the front and lateral sides of the trachea. Once in position, the thyroglossal duct disappears, and the foramen cecum regresses in size to a vestigial pit that persists throughout life.

Production and Action of Thyroid Hormones

The thyroid follicles actively accumulate iodide (I^-) from the blood and secrete it into the colloid. Once the iodide is in the colloid it is oxidized to iodine and attached to specific amino acids (tyrosines) within the polypeptide chain of a protein called **thyroglobulin** *(thi-ro-glob'u-lin)*. The attachment of one iodine to tyrosine produces *monoiodotyrosine (mon''o-i-o''do-ti'ro-sēn)* (**MIT**); the attachment of two iodines produces *diiodotyrosine* (**DIT**).

Within the colloid, enzymes modify the structure of MIT and DIT and couple them together (fig. 19.21). When two DIT molecules that are appropriately modified are coupled together, a molecule of **tetraiodothyronine** *(tet''rah-i''o-di-thi'ro-nēn)*, T_4, or **thyroxine** *(thi-rok'sin)*, is produced. The combination of one MIT with one DIT forms **triiodothyronine** *(tri''i-o''do-thi'ro-nēn)*, T_3. Note that within the colloid, T_4 and T_3 are still attached to thyroglobulin. Upon stimulation by TSH, the principal cells take up a small volume of colloid by pinocytosis, hydrolyze the T_3 and T_4 from the thyroglobulin, and secrete the T_3 and T_4 into the blood.

As previously described, most of the thyroxine in blood is attached to carrier proteins. Only the very small percentage of thyroxine that is free in the plasma can enter the target cells, where it is converted to triiodothyronine and attached to nuclear receptor proteins. Through activation of genes, thyroid hormones stimulate protein synthesis, promote maturation of the nervous system, and increase the rate of energy utilization by the body (described in chapter 25).

Figure 19.21. Stages in the formation and secretion of thyroid hormones. Iodide is actively accumulated by the follicular cells. In the colloid it is converted into iodine and attached to tyrosine amino acids within the thyroglobulin protein. Pinocytosis of iodinated thyroglobulin, coupling of MIT and DIT, and the release of thyroid hormones are stimulated by TSH from the anterior pituitary.

Follicle cell

Colloid

I^- Iodide in plasma

Active transport

I^-

I^-

Peroxidase

H_2O_2

I_2

Iodine

+

Thyroglobulin

Triiodothyronine (T_3)

MIT × DIT

DIT × DIT

Tetraiodothyronine (T_4)

Monoiodotyrosine (MIT)

Diiodotyrosine (DIT)

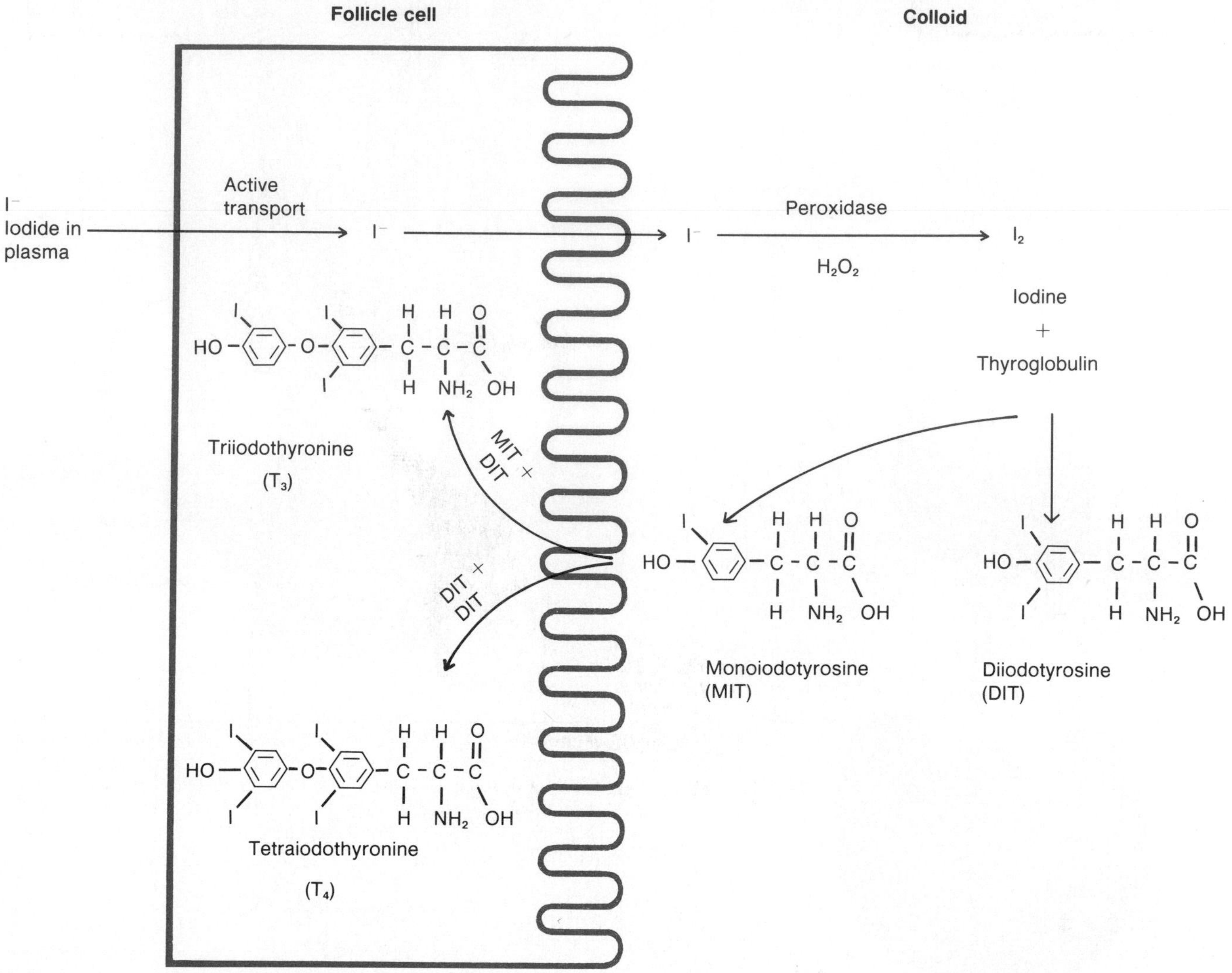

> Thyroid-stimulating hormone (TSH) from the anterior pituitary stimulates the thyroid to secrete thyroxine and exerts a trophic effect on the thyroid gland. This trophic effect is dramatically revealed in people who develop an *iodine-deficiency (endemic) goiter*. In the absence of sufficient dietary iodine the thyroid cannot produce adequate amounts of T_4 and T_3. The resulting lack of negative feedback inhibition causes abnormally high levels of TSH secretion, which in turn stimulate the abnormal growth of the thyroid (a goiter). These events are summarized in figure 19.22.

The parafollicular cells of the thyroid secrete a hormone called **calcitonin** *(kal''sĭ-to'nin)* (also called **thyrocalcitonin**). Under certain conditions, calcitonin promotes a decrease in blood calcium levels and thus antagonizes the effects of parathyroid hormone and vitamin D_3.

Parathyroid Glands

The small, flattened **parathyroid glands** are embedded in the posterior surfaces of the lateral lobes of the thyroid gland (see fig. 19.23). There are usually four parathyroid glands: a **superior** and an **inferior** pair. Each parathyroid gland is a small yellowish brown body measuring 3–8 mm (0.1–0.3 in.) in length, 2–5 mm (.07–0.2 in.) in width, and about 1.5 mm (0.05 in.) in depth.

On a microscopic level, the parathyroids are composed of two types of epithelial cells. The cells that synthesize parathyroid hormone are called **chief cells** and are scattered among **oxyphil** *(ok'se-fil)* **cells.** Oxyphil cells support the chief cells and are believed to produce reserve quantities of parathyroid hormone.

Figure 19.22. The mechanism of goiter formation in iodine deficiency. Low negative feedback inhibition results in excessive TSH secretion, which stimulates abnormal growth of the thyroid.

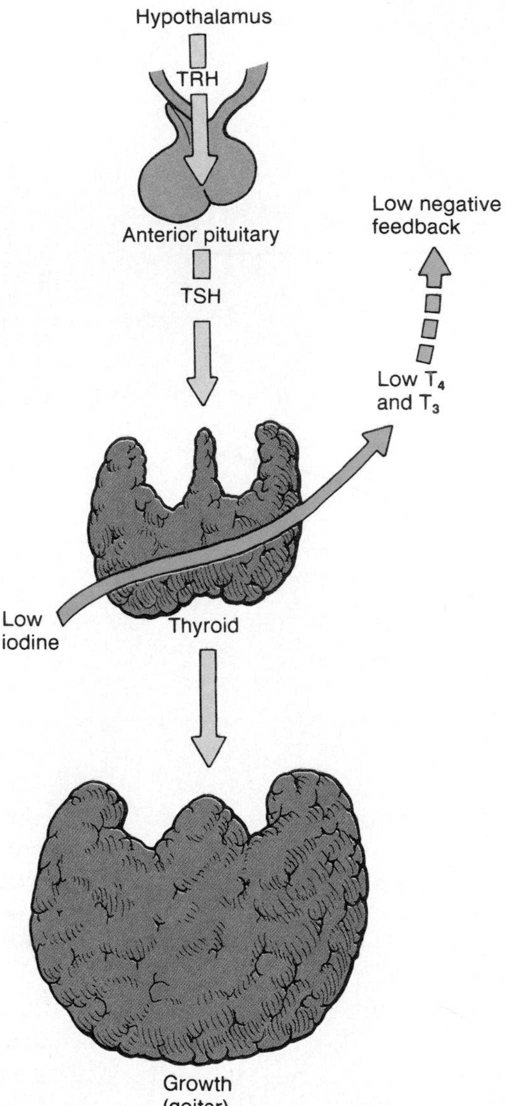

Figure 19.23. A posterior view of the parathyroid glands.

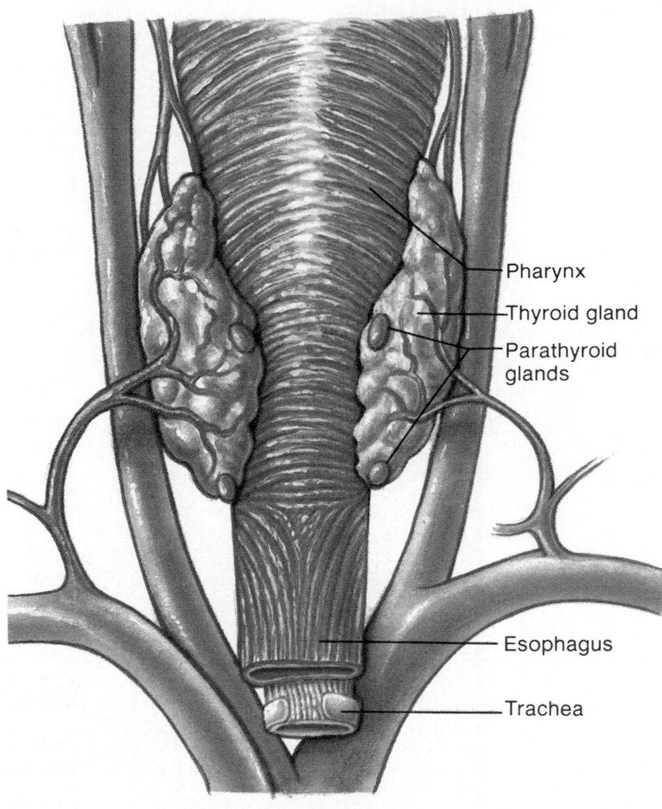

Figure 19.24. Actions of parathyroid hormone. An increased level of parathyroid hormone causes the bones to release calcium, the kidneys to conserve calcium loss through the urine, and the absorption of calcium through the intestinal wall. Negative feedback of increased calcium levels in the blood inhibits the secretion of this hormone.

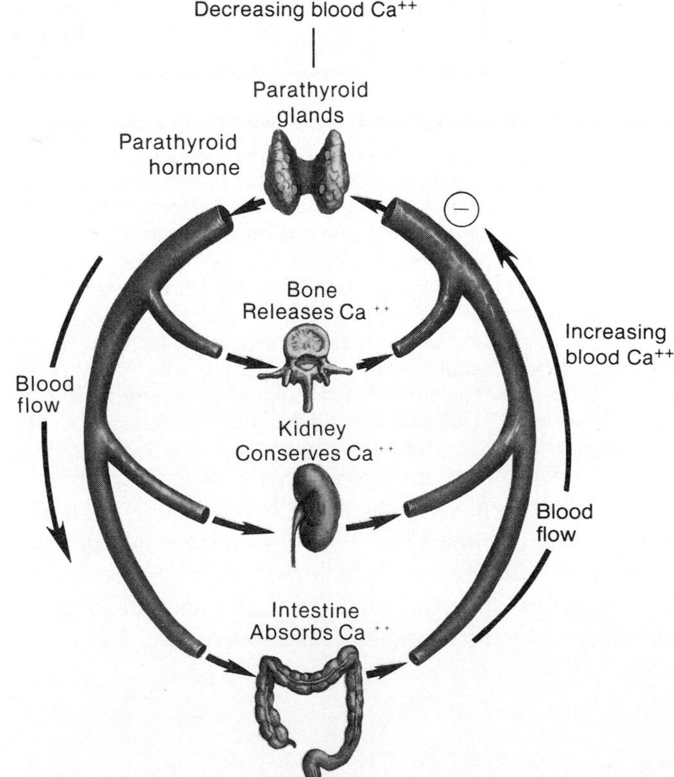

The parathyroid glands secrete one hormone called **parathyroid hormone (PTH).** This hormone promotes a rise in blood calcium levels by acting on the bones, kidneys, and intestine (fig. 19.24). Regulation of calcium balance is described in more detail in chapter 25.

1. Discuss the development of the thyroid gland and describe its structure once it is formed.
2. List the effects of thyroid hormones.
3. Describe how thyroid hormones are produced and how their secretion is regulated, and explain the consequences of an inadequate dietary intake of iodine.
4. Describe the structure of the parathyroid glands, and explain the actions of parathyroid hormone.

Figure 19.25. The pancreatic islets.

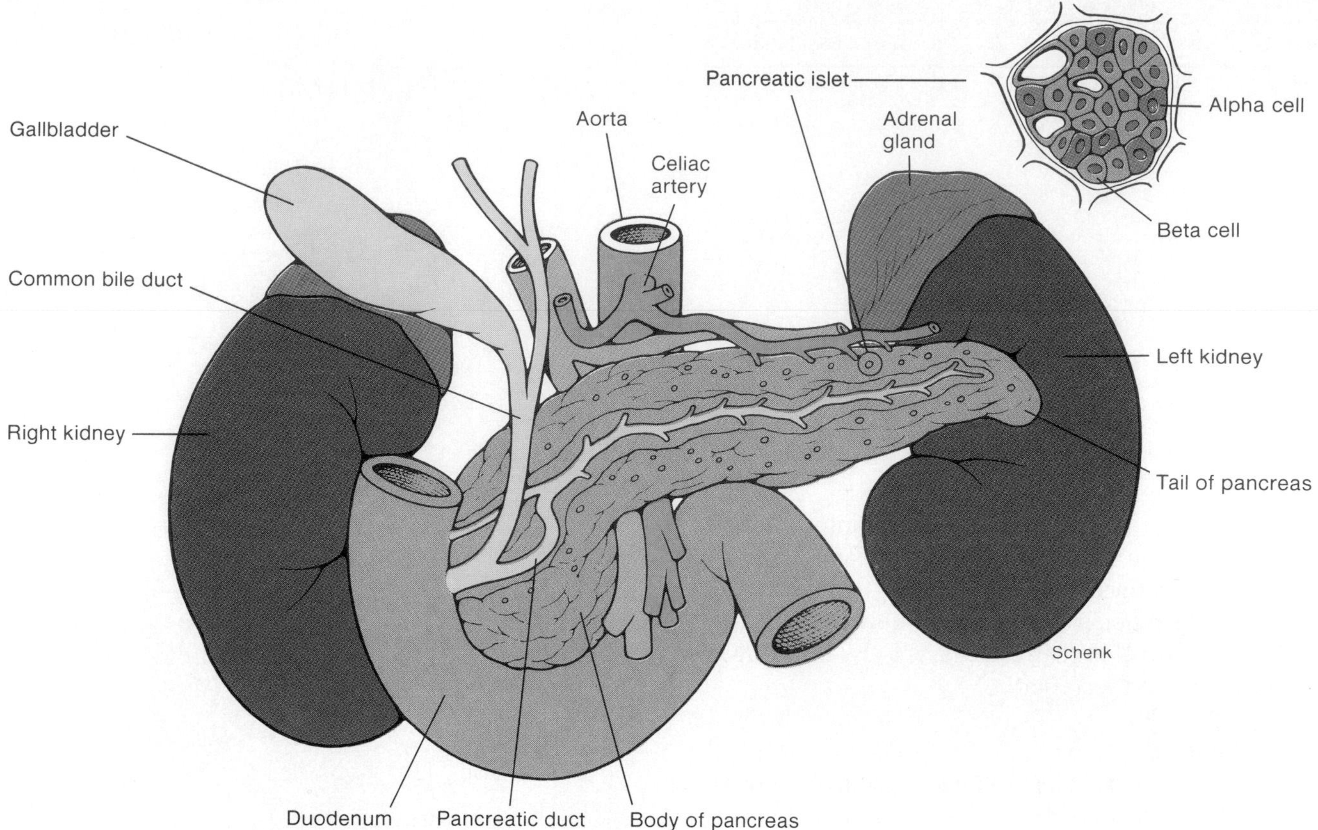

Pancreas and Other Endocrine Glands

The pancreatic islets secrete two hormones, insulin and glucagon, which are critically involved in the regulation of metabolic balance in the body. Additionally, many other organs secrete hormones that help regulate digestion, metabolism, growth, immune function, and reproduction.

Objective 26. Describe the structure of the endocrine portion of the pancreas and the origin of insulin and glucagon.

Objective 27. Describe the actions of insulin and glucagon.

Objective 28. Describe the structure and anatomic location of the pineal and thymus glands, and indicate their endocrine functions.

Pancreatic Islets (Islets of Langerhans)

The pancreas is both an endocrine and an exocrine gland. The gross structure of this gland and its exocrine functions in digestion are described in chapter 25. The endocrine portion of the pancreas consists of scattered clusters of cells called the **pancreatic islets.** These endocrine structures are most common in the body and tail of the pancreas (fig. 19.25).

Histologically, the most conspicuous cells in the islets are the *alpha* and *beta* cells. The alpha cells secrete the hormone **glucagon** *(gloo'kah-gon),* and the beta cells secrete **insulin.** The actions of these hormones are discussed in detail in chapter 26 and will only be briefly summarized here.

Alpha cells secrete glucagon in response to a fall in the blood glucose concentrations. Glucagon stimulates the liver to convert glycogen to glucose (**glycogenolysis** [*gli''ko-jĕ-nol'ĭ-sis*]), which causes the blood glucose level to rise. This effect represents the completion of a negative feedback loop. Glucagon also stimulates the hydrolysis of stored fat (**lipolysis**) and the consequent release of free fatty acids into the blood. This effect helps to provide energy substrates for the body during fasting, when blood glucose levels decrease. Glucagon, together with other hormones, also stimulates the conversion of fatty acids to ketone bodies, which can be secreted by the liver into the blood and used by many organs as an energy source.

Beta cells secrete insulin in response to a rise in the blood glucose concentrations. Insulin promotes the entry of glucose into tissue cells, and the conversion of this glucose into energy storage molecules of glycogen and fat.

islets of Langerhans: from Paul Langerhans, German anatomist, 1847–1888

insulin: L. *insula,* island

Table 19.11 Hormones of the pancreas

Hormone	Action	Source of regulation
Glucagon	Stimulates the liver to convert glycogen into glucose, causing the blood glucose level to rise	Blood glucose level through negative feedback in pancreas
Insulin	Promotes movement of glucose through plasma membranes; stimulates liver to convert glucose into glycogen; promotes the transport of amino acids into cells; assists in synthesis of proteins and fats	Blood glucose level through negative feedback in pancreas

Insulin also aids the entry of amino acids into cells and the production of cellular protein. The actions of insulin and glucagon are thus antagonistic (table 19.11). After a meal, insulin secretion is increased and glucagon secretion is decreased; fasting, in contrast, causes a rise in glucagon and a fall in insulin secretion.

Pineal Gland

The small, cone-shaped **pineal gland,** or **epiphysis cerebri,** is located in the roof of the third ventricle near the corpora quadrigemina, where it is encapsulated by the meninges covering the brain. The pineal gland of a child weighs about 0.2 g and is 5–8 mm (0.2–0.3 in.) long and 9 mm wide. The gland begins to regress in size at about age seven and in the adult appears as a thickened strand of fibrous tissue. Histologically, the pineal gland consists of specialized parenchymal and neuroglial cells. Although the pineal gland lacks direct nervous connections to the rest of the brain, it is highly innervated by the sympathetic nervous system from the superior cervical ganglion.

The principal hormone of the pineal is **melatonin** *(mel"ah-to'nin).* The secretion of melatonin is inhibited by light and is therefore maximal at night. It has long been suspected that this hormone inhibits the pituitary-gonad axis. Melatonin secretion is highest in children of ages one to five and decreases thereafter, reaching its lowest levels at the end of puberty, where concentrations are 75% lower than during early childhood. The secretion of melatonin is, therefore, believed by some researchers to play a role in the onset of puberty, but this possibility is highly controversial.

pineal: L. *pinea,* pine cone

Figure 19.26. The thymus is a bilobed organ within the mediastinum of the thorax.

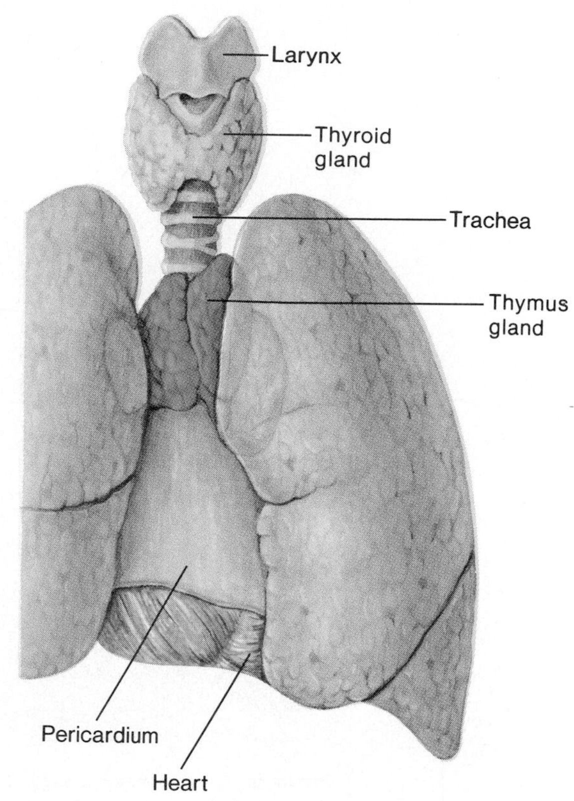

Thymus

The **thymus** *(thi'mus)* is a bilobed organ positioned in the upper mediastinum, in front of the aorta and behind the manubrium of the sternum (fig. 19.26). Although the size of the thymus varies considerably from person to person, it is relatively large in newborns and children and sharply regresses in size after puberty. Besides decreasing in size, the thymus of adults becomes infiltrated with strands of fibrous and fatty connective tissue.

The thymus serves as the site of production of **T cells (thymus-dependent cells),** which are the lymphocytes involved in cell-mediated immunity (chapter 22). In addition to providing T cells, the thymus secretes a number of hormones that are believed to stimulate T cells after they leave the thymus.

Gastrointestinal Tract

The stomach and small intestine secrete a number of hormones that act on the gastrointestinal tract itself and on the pancreas and gallbladder (discussed in chapter 24). The effects of these hormones act together with regulation by the autonomic nervous system to coordinate the activities of different regions of the digestive tract and the secretions of pancreatic juice and bile.

Leydig cells: from Franz von Leydig, German anatomist, 1821–1908

Gonads and Placenta

The **gonads (testes** and **ovaries**) secrete *sex steroids.* These include male sex hormones, or *androgens (an'dro-jen),* and female sex hormones—*estrogens* and *progestogens.* The principal hormones in each of these categories are *testosterone (tes-tos'tĕ-rōn), estradiol-17β,* and *progesterone,* respectively.

The testes consist of two compartments: **seminiferous tubules,** which produce sperm, and **interstitial tissue** between the convolutions of the tubules. Within the interstitial tissue are **Leydig cells,** which secrete testosterone and lesser amounts of weaker androgens (as well as small amounts of estradiol-17β). Testosterone is needed for the development and maintenance of the male genitalia (penis and scrotum) and male sex accessory organs (prostate, seminal vesicles, epididymides, and ductus deferens), as well as for the development of male secondary sexual characteristics (chapter 27).

During the first half of the menstrual cycle, estrogen is secreted by many small structures within the ovary called **ovarian follicles.** These follicles contain the egg cell, or *ovum,* and *granulosa cells* that secrete estrogen. By about mid-cycle, one of these follicles grows very large and, in the process of ovulation, extrudes its ovum from the ovary. The empty follicle, under the influence of luteinizing hormone (LH) from the anterior pituitary, then becomes a new endocrine structure called a *corpus luteum.* The corpus luteum secretes progesterone as well as estradiol-17β.

The **placenta** is the organ responsible for nutrient and waste exchange between the fetus and mother. The placenta is also an endocrine gland; it secretes large amounts of estrogens and progesterone, as well as a number of polypeptide and protein hormones that are similar to some hormones secreted by the anterior pituitary gland. These latter hormones include **human chorionic gonadotropin (hCG),** which is similar to LH, and **somatomammotropin,** which is similar in action to growth hormone and prolactin. Detailed aspects of femal reproduction physiology are discussed in chapter 28.

1. Describe the structure of the endocrine pancreas, and indicate the sites of origin of insulin and glucagon.
2. Describe how insulin and glucagon secretion is affected by eating and by fasting, and explain the actions of these two hormones.
3. Describe the location of the pineal gland and the possible function of melatonin.
4. Describe the location and function of the thymus gland.
5. Explain how the hormones secreted by the gonads and placenta are categorized, and describe which hormones of these categories are secreted by each gland.

Clinical Considerations

Disorders of the Pituitary Gland

Panhypopituitarism. A reduction in the activity of the pituitary gland is called **hypopituitarism** *(hi''po-pĭ-tu'ĭ-tah-rizm'')* and can result from intracranial hemorrhage, a blood clot, prolonged steroid treatments, or a tumor. Total pituitary impairment, termed **panhypopituitarism,** brings about a progressive and general loss of hormonal activity. For example, the gonads stop functioning and the person suffers from amenorrhea (lack of menstruation) or aspermia (no sperm production) and loss of pubic and axillary hair. The thyroid and adrenals also eventually stop functioning. People with this condition, and those who have had their pituitary surgically removed (a procedure called **hypophysectomy**) receive thyroxine, cortisone, growth hormone, and gonadal hormones throughout life to maintain normal body function.

Abnormal Growth Hormone Secretion. Inadequate growth hormone secretion during childhood causes **pituitary dwarfism.** Hyposecretion of growth hormone in an adult produces a rare condition called **pituitary cachexia** *(kah-kek'se-ah)* **(Simmonds' disease).** One of the symptoms of this disease is premature aging caused by tissue atrophy. Oversecretion of growth hormone during childhood, in contrast, causes **gigantism.** Excessive growth hormone secretion in an adult does not cause further growth in length because the epiphyseal discs have ossified. Hypersecretion of growth hormone in an adult causes **acromegaly** *(ak''ro-meg'ah-le),* in which the person's appearance gradually changes as a result of thickening of bones and growth of soft tissues, particularly in the face, hands, and feet.

Inadequate ADH Secretion. A dysfunction of the neurohypophysis results in a deficiency in ADH secretion, causing a condition called **diabetes insipidus.** Symptoms of this disease include polyuria (excessive urination), polydipsia (drinking large amounts of water), and severe ionic imbalances. Diabetes insipidus is treated by injections of ADH.

Disorders of the Adrenal Glands

Tumors of the Adrenal Medulla. Tumors of the chromaffin cells of the adrenal medulla are referred to as **pheochromocytomas** *(fe-o-kro''mo-si-to'mah).* These tumors cause hypersecretion of epinephrine and norepinephrine, which produce an effect similar to continuous sympathetic nervous stimulation. The symptoms of this condition are

Simmonds' disease: from Morris Simmonds, German physician, 1855–1925
acromegaly: Gk. *akron,* extremity; *megas,* large
diabetes: Gk. *diabetes,* to pass through a siphon
pheochromocytomas: Gk. *phaios,* dusky; *chroma,* color; *oma,* tumor

hypertension, elevated metabolism, hyperglycemia and sugar in the urine, nervousness, digestive problems, and sweating. It does not take long for the body to become totally fatigued under these conditions, making the patient susceptible to other diseases.

Addison's Disease. This disease is caused by inadequate secretion of both glucocorticoids and mineralocorticoids, which results in hypoglycemia, sodium and potassium imbalance, dehydration, hypotension, rapid weight loss, and generalized weakness. A person with this condition who is not treated with corticosteroids will die within a few days because of the severe electrolyte imbalance and dehydration. Another symptom of this disease is darkening of the skin. This is caused by high secretion of ACTH and possibly MSH (because MSH is derived from the same parent molecule as ACTH), which is a result of lack of negative feedback inhibition of the pituitary by corticosteroids.

Cushing's Syndrome. Hypersecretion of corticosteroids results in Cushing's syndrome. This is generally caused by a tumor of the adrenal cortex or by oversecretion of ACTH from the adenohypophysis. Cushing's syndrome is characterized by changes in carbohydrate and protein metabolism, hyperglycemia, hypertension, and muscular weakness. Metabolic problems give the body a puffy appearance and can cause structural changes characterized as "buffalo hump" and "moon face." Similar effects are also seen when people with chronic inflammatory diseases receive prolonged treatment with corticosteroids, which are given to reduce inflammation and inhibit the immune response.

Adrenogenital Syndrome. Usually associated with Cushing's syndrome, this condition is caused by hypersecretion of adrenal sex hormones, particularly the androgens. Adrenogenital syndrome in young children causes premature puberty and enlarged genitals, especially the penis in a male and the clitoris in a female. An increase in body hair and a deeper voice are other characteristics. This condition in a mature woman can cause growth of a beard.

Disorders of the Thyroid and Parathyroids

Hypothyroidism. The infantile form of hypothyroidism is known as **cretinism** *(kre'tin-izm)* (fig. 19.27). An affected child usually appears normal at birth because thyroxine is received from the mother through the placenta.

Figure 19.27. Cretinism is a disease of infancy caused by an underactive thyroid gland.

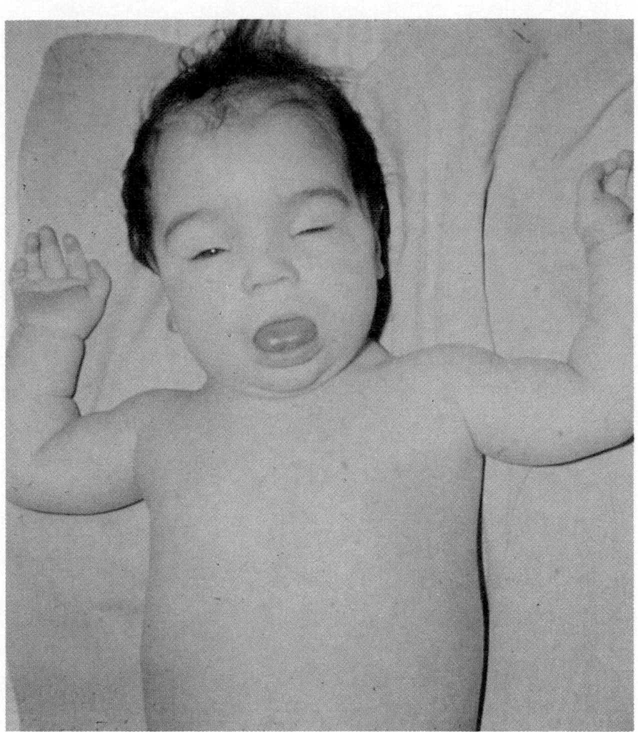

The clinical symptoms of cretinism are stunted growth, thickened facial features, abnormal bone development, mental retardation, low body temperature, and general lethargy. If cretinism is diagnosed early, it can be successfully treated by administering thyroxine.

Myxedema. Hypothyroidism in an adult causes myxedema *(mik"sĕ-de'mah)*. This disorder affects body fluids, causing edema and increasing blood volume, hence increasing blood pressure. A person with myxedema has a low metabolic rate, lethargy, and a tendency to gain weight. This condition is treated with thyroxine or with triiodothyronine, which are taken orally (as pills).

Endemic Goiter. A goiter is an abnormal growth of the thyroid gland. When this is a result of inadequate dietary intake of iodine, the condition is called endemic goiter (fig. 19.28). In this case, growth of the thyroid is due to excessive TSH secretion, which results from low levels of thyroxine secretion. Endemic goiter is thus associated with hypothyroidism.

Addison's disease: from Thomas Addison, English physician, 1793–1860
Cushing's syndrome: from Harvey Cushing, U.S. physician, 1869–1939

myxedema: Gk. *myxa,* mucus; *oidema,* swelling

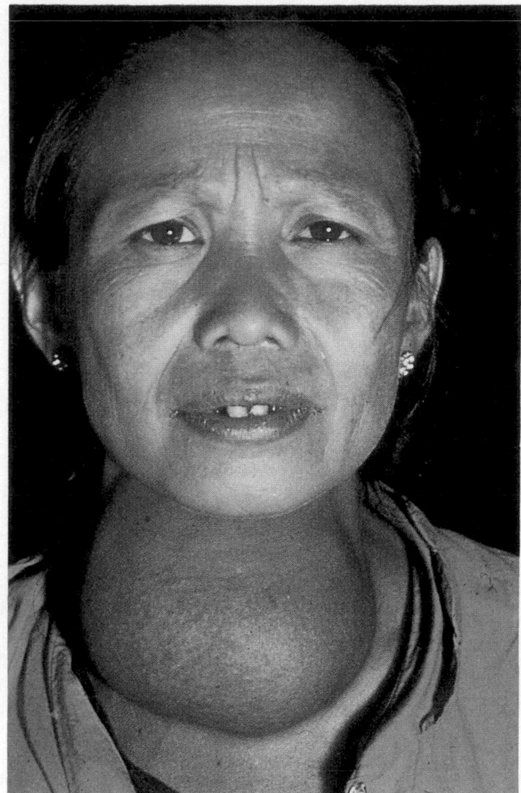

Figure 19.28. A simple or endemic goiter is caused by insufficient iodine in the diet.

Graves' Disease. Graves' disease, also called **toxic goiter,** involves growth of the thyroid associated with hypersecretion of thyroxine. This hyperthyroidism is produced by antibodies that act like TSH and stimulate the thyroid; it is an autoimmune disease. As a consequence of high levels of thyroxine secretion, the metabolic rate and heart rate increase, there is loss of weight, and the autonomic nervous system induces excessive sweating. In about half of the cases, **exophthalmos** *(ek''sof-thal'mos)* (bulging of the eyes) also develops (fig. 19.29) because of edema in the tissues of the eye sockets and swelling of the extrinsic eye muscles.

Disorders of the Parathyroid Glands. Surgical removal of the parathyroids sometimes unintentionally occurs when the thyroid is removed because of a tumor or the presence of Graves' disease. The resulting fall in parathyroid hormone (PTH) causes a decrease in plasma calcium levels, which can lead to severe muscle tetany. Hyperparathyroidism is usually caused by a tumor that secretes excessive amounts of PTH. This stimulates demineralization of bone, which makes the bones soft and raises the blood levels of calcium and phosphate. As a result of these changes, bones are subject to deformity and fracture, and stones composed of calcium phosphate are likely to develop in the urinary tract.

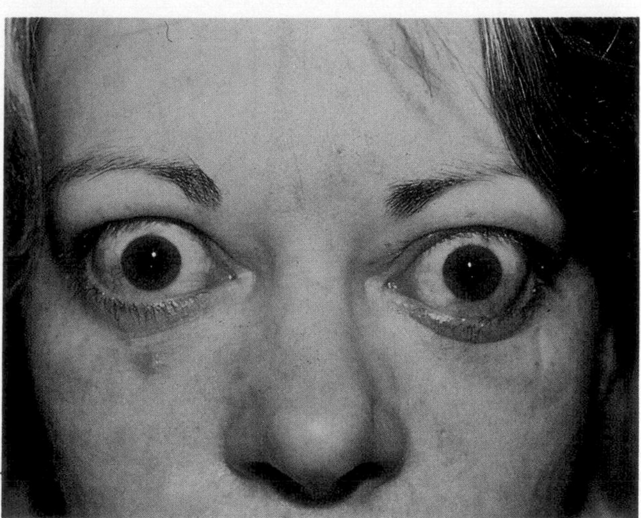

Figure 19.29. Hyperthyroidism is characterized by an increased metabolic rate, weight loss, muscular weakness, and nervousness. The eyes may also protrude.

Disorders of the Pancreatic Islets

Diabetes Mellitus. Diabetes mellitus is characterized by fasting hyperglycemia and the presence of glucose in the urine. There are two forms of this disease. **Type I,** or insulin-dependent diabetes mellitus, is caused by destruction of the beta cells and the resulting lack of insulin secretion. **Type II,** or non-insulin-dependent diabetes mellitus (which is the more common form), is caused by decreased tissue sensitivity to the effects of insulin, so that larger amounts of insulin are required to produce a normal effect. Both types of diabetes mellitus are also associated with abnormally high levels of glucagon secretion.

Reactive Hypoglycemia. People who have a genetic predisposition for type II diabetes mellitus often first develop reactive hypoglycemia. In this condition, the rise in blood glucose that follows the ingestion of carbohydrates stimulates excessive secretion of insulin, which in turn causes the blood glucose levels to fall below the normal range. This can result in weakness, changes in personality, and mental disorientation.

exophthalmos: Gk. *ex,* out; *opthalmos,* eyeball
Graves' disease: from Robert James Graves, Irish physician, 1796–1853

Chapter Summary

I. Hormones: Actions and Interactions
  A. Precursors of active hormones may be called either prohormones or prehormones.
    1. Prohormones are relatively inactive precursor molecules made in the endocrine cells.
    2. Prehormones are the normal secretions of an endocrine gland that must be converted to other derivatives by target cells to be active.
  B. The effects of a hormone in the body depend on its concentration.
    1. Abnormally high amounts of a hormone can result in effects that do not occur under normal conditions.
    2. Target tissues can become desensitized by high hormone concentrations.
  C. Hormones can interact in permissive, synergistic, or antagonistic ways.

II. Mechanisms of Hormone Action
  A. Steroid and thyroid hormones enter their target cells.
    1. Thyroid hormones attach to chromatin-bound receptors located in the nucleus.
    2. Steroid hormones bond to cytoplasmic receptor proteins and translocate to the nucleus.
    3. Attachment of the hormone-receptor protein complex to the chromatin activates genes and thereby stimulates RNA and protein synthesis.
  B. Amine, polypeptide, and glycoprotein hormones bond to receptor proteins on the outer surface of the target cell membrane.
    1. In many cases, this leads to the intracellular production of cyclic AMP, which serves as a second messenger in the action of these hormones.
    2. In other cases, cyclic GMP, or Ca^{++} may serve as a second messenger in the action of the hormone.
    3. When Ca^{++} is used as a second messenger, hormonal interaction with its membrane receptor causes a rapid diffusion of Ca^{++} into the cytoplasm.
    4. Ca^{++} combines with a cytoplasmic protein called calmodulin, and this in turn activates or inactivates specific enzymes.
    5. Inositol triphosphate and diacyglyceride are second messengers that interact with the Ca^{++}-calmodulin system.

III. Prostaglandins
  A. Prostaglandins are special, twenty-carbon-long fatty acids produced by many different organs that usually have regulatory functions within the organ in which they are produced.

IV. Pituitary Gland
  A. The pituitary secretes eight hormones.
    1. The anterior pituitary secretes growth hormone, thyroid stimulating hormone, adrenocorticotropic hormone, follicle-stimulating hormone, luteinizing hormone, and prolactin.
    2. The posterior pituitary secretes antidiuretic hormone (also called vasopressin) and oxytocin.
  B. Secretions of the posterior pituitary are controlled by the hypothalamo-hypophyseal tract.
  C. Secretions of the anterior pituitary are controlled by hypothalamic hormones that stimulate or inhibit secretions of the anterior pituitary.
    1. Hypothalamic hormones include TRH, CRH, GnRH, PIH, somatostatin, and a growth hormone–releasing hormone.
    2. These hormones are carried to the anterior pituitary by the hypothalamo-hypophyseal portal system.
  D. Secretions of the anterior pituitary are also regulated by the feedback (usually negative feedback) of hormones from the target glands.
  E. Higher brain centers, acting through the hypothalamus, can influence pituitary secretion.

V. Adrenal Glands
  A. The adrenal cortex secretes mineralocorticoids (mainly aldosterone), and sex steroids (primarily weak androgens).
    1. The glucocorticoids help regulate energy balance; they also can inhibit inflammation and suppress immune function.
    2. The pituitary-adrenal axis is stimulated by stress as part of the general adaptation syndrome.
  B. The adrenal medulla secretes epinephrine and lesser amounts of norepinephrine, which complement the action of the sympathetic nervous system.

VI. Thyroid and Parathyroids
  A. The thyroid follicles secrete tetraiodothyronine (T_4, or thyroxine) and lesser amounts of triiodothyronine (T_3).
    1. These hormones are formed within the colloid of the thyroid follicles.
    2. The parafollicular cells of the thyroid secrete the hormone calcitonin, which may act to lower blood calcium levels.
  B. The parathyroids are small structures embedded within the thyroid gland; the parathyroids secrete a hormone that promotes a rise in blood calcium levels.

VII. Pancreas and Other Endocrine Glands
  A. Beta cells in the islets secrete insulin; alpha cells secrete glucagon.
    1. Insulin lowers blood glucose and stimulates the production of glycogen, fat, and protein.
    2. Glucagon raises blood glucose by stimulating the breakdown of liver glycogen; glucagon also promotes lipolysis and the formation of ketone bodies.
    3. The secretion of insulin is stimulated by a rise in blood glucose following meals; the secretion of glucagon is stimulated by a fall in blood glucose during periods of fasting.
  B. The pineal gland, located on the roof of the third ventricle of the brain, secretes melatonin; this hormone may play a role in regulating reproductive function.
  C. The thymus is located in front of the aorta in the mediastinum; it is the site of the production of T cell lymphocytes and secretes a number of hormones that may help regulate the immune system.
  D. The gastrointestinal tract secretes a number of hormones that help regulate functions of the digestive system.
  E. The gonads secrete sex steroid hormones.
    1. Leydig cells in the interstitial tissue of the testes secrete testosterone and other androgens.
    2. Granulosa cells of the ovarian follicles secrete estrogen.
    3. The corpus luteum of the ovaries secretes progesterone as well as estrogen.
  F. The placenta secretes estrogen, progesterone, and a variety of polypeptide hormones that have actions similar to some anterior pituitary hormones.

Review Activities

Objective Questions

Match the gland to its embryonic origin.

1. Adenohypophysis
2. Neurohypophysis
3. Adrenal medulla
4. Pancreas
5. Thyroid gland

(a) endoderm of pharynx
(b) diverticulum from brain
(c) endoderm of foregut
(d) neural crest ectoderm
(e) hypophyseal pouch

6. Hypothalamic releasing hormones
   (a) are secreted into capillaries in the median eminence
   (b) are transported by portal veins to the anterior pituitary
   (c) stimulate the secretion of specific hormones from the anterior pituitary
   (d) all of the above
7. The hormone primarily responsible for setting the basal metabolic rate and for promoting the maturation of the brain is
   (a) cortisol
   (b) ACTH
   (c) TSH
   (d) thyroxine
8. Which of the following statements about the adrenal cortex is *true?*
   (a) It is not innervated by nerve fibers.
   (b) It secretes some androgens.
   (c) The zona granulosa secretes aldosterone.
   (d) The zona fasciculata is stimulated by ACTH.
   (e) All of the above are true.
9. The hormone insulin
   (a) is secreted by alpha cells in the pancreatic islets
   (b) is secreted in response to a rise in blood glucose
   (c) stimulates the production of glycogen and fat
   (d) both *a* and *b*
   (e) both *b* and c

Match the hormone with the primary agent that stimulates its secretion.

10. Epinephrine
11. Thyroxine
12. Corticosteroids
13. ACTH

(a) TSH
(b) ACTH
(c) growth hormone
(d) sympathetic nerves
(e) CRH

14. Steroid hormones are secreted by
   (a) the adrenal cortex
   (b) the gonads
   (c) the thyroid
   (d) both *a* and *b*
   (e) both *b* and *c*
15. The secretion of which of the following hormones would be *increased* in a person with endemic goiter?
   (a) TSH
   (b) thyroxine
   (c) triiodothyronine
   (d) all of the above
16. Which of the following hormones use cAMP as a second messenger?
   (a) testosterone
   (b) corticol
   (c) insulin
   (d) epinephrine
17. Which of the following terms best describes the type of interaction between the effects of insulin and glucagon?
   (a) synergistic
   (b) permissive
   (c) antagonistic
   (d) cooperative

Essay Questions

1. Explain how the regulation of the neurohypophysis and adrenal medulla are related to their embryonic origins.
2. Compare steroid and polypeptide hormones in terms of their mechanism of action in target organs.
3. Explain the significance of the term *trophic* in regard to the actions of anterior pituitary hormones.
4. Suppose a drug blocks the conversion of T_4 to T_3. Explain what the effects of this drug would be on (*a*) TSH secretion, (*b*) thyroxine secretion, and (*c*) the size of the thyroid gland.
5. Explain why the phrase "master gland" is sometimes used to describe the anterior pituitary, and why this term is misleading.
6. Suppose a person's immune system made antibodies against insulin receptor proteins. Describe the possible effect of this condition on carbohydrate and fat metabolism.

UNIT 4

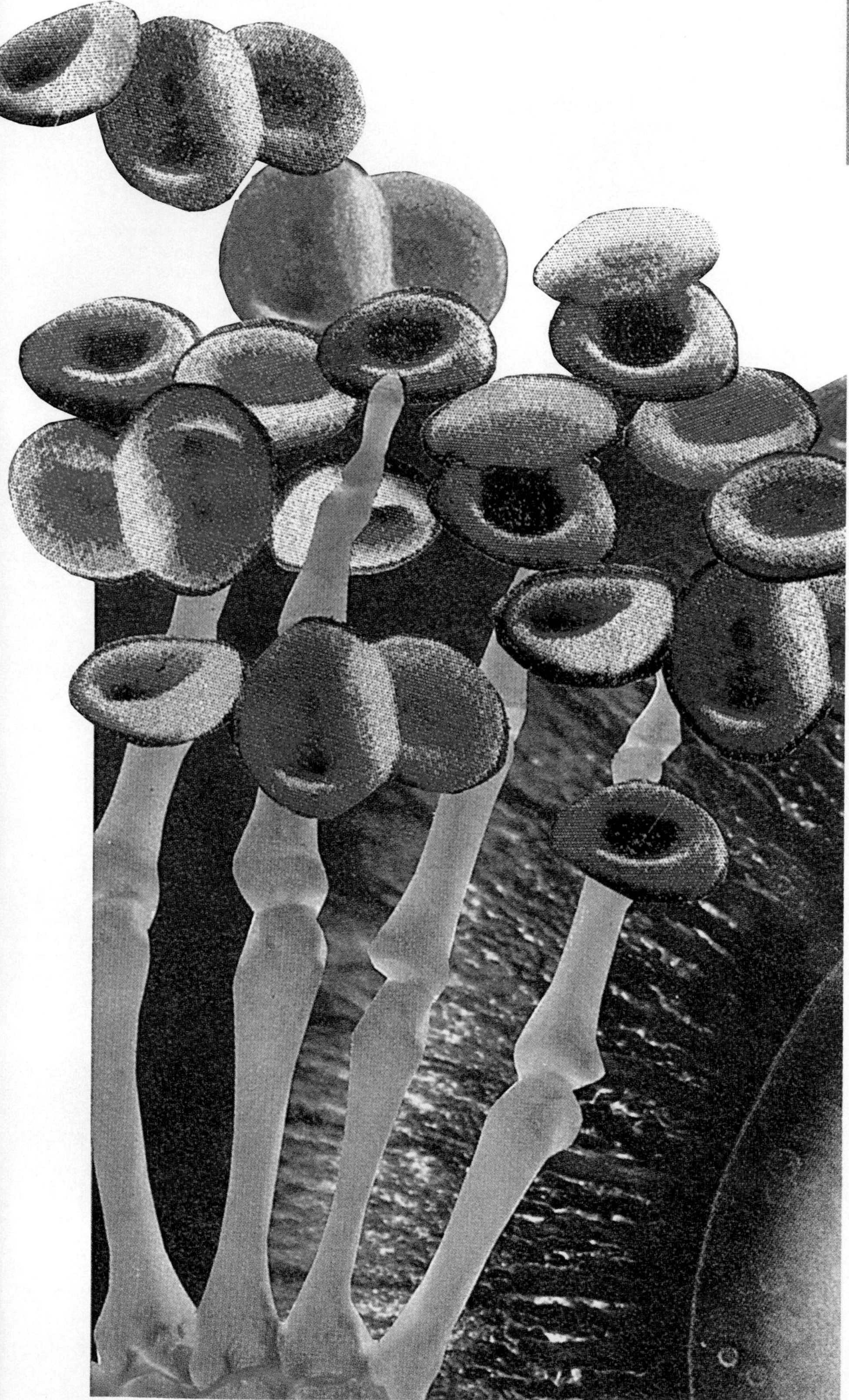

Unit 4 is concerned with the anatomy and physiology of those body systems whose principal function is to maintain homeostasis. Blood transports oxygen, nutrients, and regulatory molecules to the tissue cells and removes waste products. The cardiovascular system, as regulated by the nervous and endocrine system, serves to pump and deliver blood to the tissue capillaries. One function of the lungs, kidneys, and liver is to eliminate waste substances from the blood. The digestive system delivers nutrients to the blood, and the lungs deliver oxygen and remove carbon dioxide from the blood. Homeostasis is further advanced by the actions of the immune system, which acts to protect the body from many diseases.

Regulation and Maintenance of the Human Body

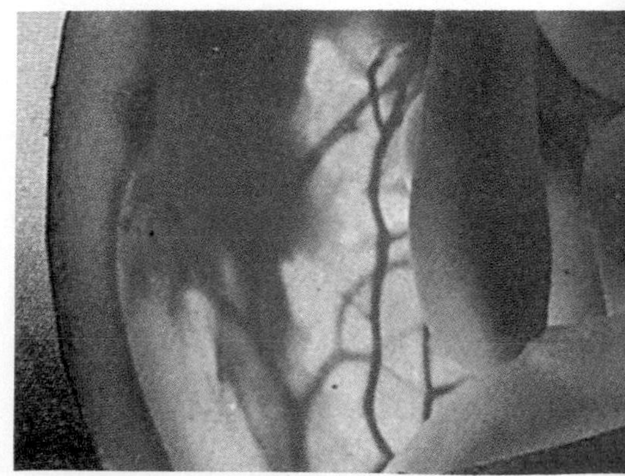

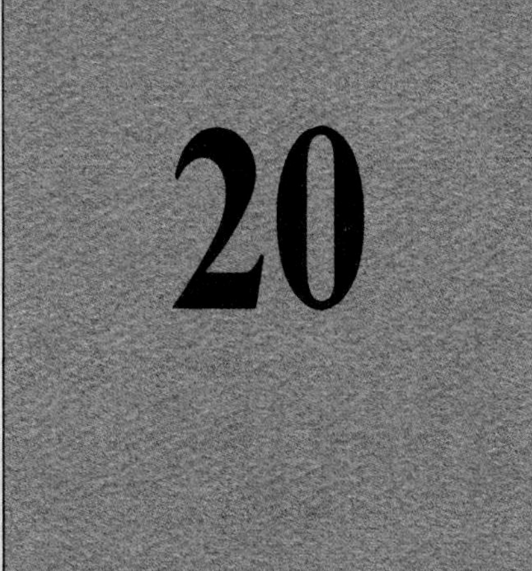

20 Circulatory System

Outline

Concepts

The circulatory system provides functions essential to the life of a complex, multicellular organism.

During a seven-day period, cardiogenic mesoderm migrates to become the heart tube, which differentiates into an embryonic heart, pumping blood by day 25.

The structure of the heart and the action of its valves allow it to pump blood low in oxygen to the lungs and to pump oxygen-rich blood to the body.

The cardiac cycle results from electrical activity, which can be monitored with an electrocardiogram. The changes in intraventricular pressure produced during the cardiac cycle can be followed through the sounds created by the closing of the heart valves.

The structure of arteries and veins allows them to transport blood from the heart to the capillaries and back to the heart. The structure of capillaries permits the exchange of plasma fluid and dissolved molecules between the blood and surrounding tissues.

The aorta ascends from the left ventricle, arches to the left, and descends through the torso. Branches of the aorta carry oxygenated blood to all the cells of the body.

After systemic blood has passed through the tissues and become depleted of oxygen, it returns through veins of progressively larger diameters to the right atrium of the heart.

The fetal circulation is adapted to the fact that the fetal lungs are nonfunctional and that oxygen and nutrients are obtained from the placenta.

Blood consists of formed elements that are suspended and carried in plasma. The constituents of blood function in gas transport, transport of regulatory molecules and nutrients, immunity, and blood-clotting mechanisms.

Trauma to a blood vessel initiates a sequence of events resulting in the formation of a clot, followed by the healing of the blood vessel and the dissolution of the clot.

Functions and Major Components of the Circulatory System

The circulatory system provides functions essential to the life of a complex, multicellular organism.

Objective 1. Describe the functions of the circulatory system.

Objective 2. Describe the major components of the circulatory system.

A unicellular organism can provide for its own maintenance and continuity by performing the wide variety of functions needed for life. By contrast, the complex human body is composed of trillions of specialized cells that demonstrate a division of labor. Cells of a multicellular organism depend on one another for the very basics of their existence. The majority of the cells of the body are firmly implanted in tissues and are incapable of procuring food and oxygen or even moving away from their own wastes. Therefore, a highly specialized and effective means of transporting materials within the body is needed.

The blood serves this transportation function. An estimated 60,000 miles of vessels throughout the body of an adult ensure that continued sustenance reaches each of the trillions of living cells. The blood can also, however, serve to transport disease-causing viruses, bacteria, and their toxins. To ensure against this, the circulatory system has protective mechanisms: the white blood cells and lymphatic system. In order to perform its various functions, the circulatory system works together with the respiratory, urinary, digestive, endocrine, and integumentary systems in maintaining homeostasis.

Functions of the Circulatory System

The functions of the circulatory system can be divided into three areas: transportation, regulation, and protection.

1. **Transportation.** All of the substances essential for cellular metabolism are transported by the circulatory system. These substances can be categorized as follows:
   a. Respiratory. Red blood cells, called **erythrocytes** *(ĕ-rith'ro-sīt),* transport oxygen to the tissue cells. In the lungs, oxygen from the inhaled air attaches to hemoglobin molecules within the erythrocytes and is transported to the cells for aerobic respiration. Carbon dioxide produced by cell respiration is carried by the blood to the lungs for elimination in the exhaled air.
   b. Nutritive. The digestive system is responsible for the mechanical and chemical breakdown of food so that it can be absorbed through the intestinal wall and into the blood vessels of the circulatory system. The blood then carries these absorbed products of digestion through the liver and to the cells of the body (fig. 20.1).
   c. Excretory. Metabolic wastes, excessive water and ions, as well as other molecules in plasma (the fluid portion of blood), are filtered through the capillaries of the kidneys and excreted in urine.
2. **Regulation.** The blood carries hormones and other regulatory molecules from their site of origin to distant target tissues.
3. **Protection.** The circulatory system protects against injury and foreign microbes or toxins introduced into the body. The clotting mechanism protects against blood loss when vessels are damaged, and **leukocytes** *(lu'ko-sīt)* (white blood cells) provide immunity to many disease-causing agents.

Major Components of the Circulatory System

The circulatory system is frequently divided into the **cardiovascular system,** which consists of the heart and blood vessels, and the **lymphatic system,** which consists of lymph vessels and lymph nodes. The lymphatic system and its immunological functions is described in chapter 22.

The **heart** is a four-chambered, double pump. Its pumping action creates the pressure needed to push blood in vessels to the lungs and body cells. At rest, the heart of an adult pumps about 5.0 L/min of blood. It takes about one minute for blood to be circulated to the most distal extremity and back to the heart.

Blood vessels form a tubular network that permits blood to flow from the heart to all the living cells of the body and then back to the heart. The circulatory system is said to be "continuous" because the blood cells remain within vessels throughout the journey (some of the fluid does leave the vessels, however, becoming tissue fluid). *Arteries* carry blood away from the heart, while *veins* return blood to the heart. Arteries and veins are continuous with each other through smaller blood vessels.

Arteries branch extensively to form a "tree" of progressively smaller vessels. Those that are microscopic in diameter are called *arterioles.* Conversely, microscopic-sized veins, called *venules (ven'ūl),* deliver blood into ever larger vessels that empty into the large veins. Blood passes from the arterial to the venous system in *capillaries (kap'ĭ-lar"e),* which are the thinnest and most numerous blood vessels. All exchanges of fluid, nutrients, and wastes between the blood and tissues occur across the walls of capillaries.

Fluid derived from plasma that passes out of capillary walls into the surrounding tissues is called *tissue fluid,* or *interstitial (in"ter-stish'al) fluid.* Some of this fluid returns directly to capillaries, and some enters into **lymph vessels** located in the connective tissues around the blood vessels. Fluid in lymph vessels is called *lymph (limf).* This fluid is returned to the venous blood at particular sites. **Lymph nodes,** positioned along the way, cleanse the lymph prior to its return to the venous blood.

Figure 20.1. A schematic diagram of the circulatory system.

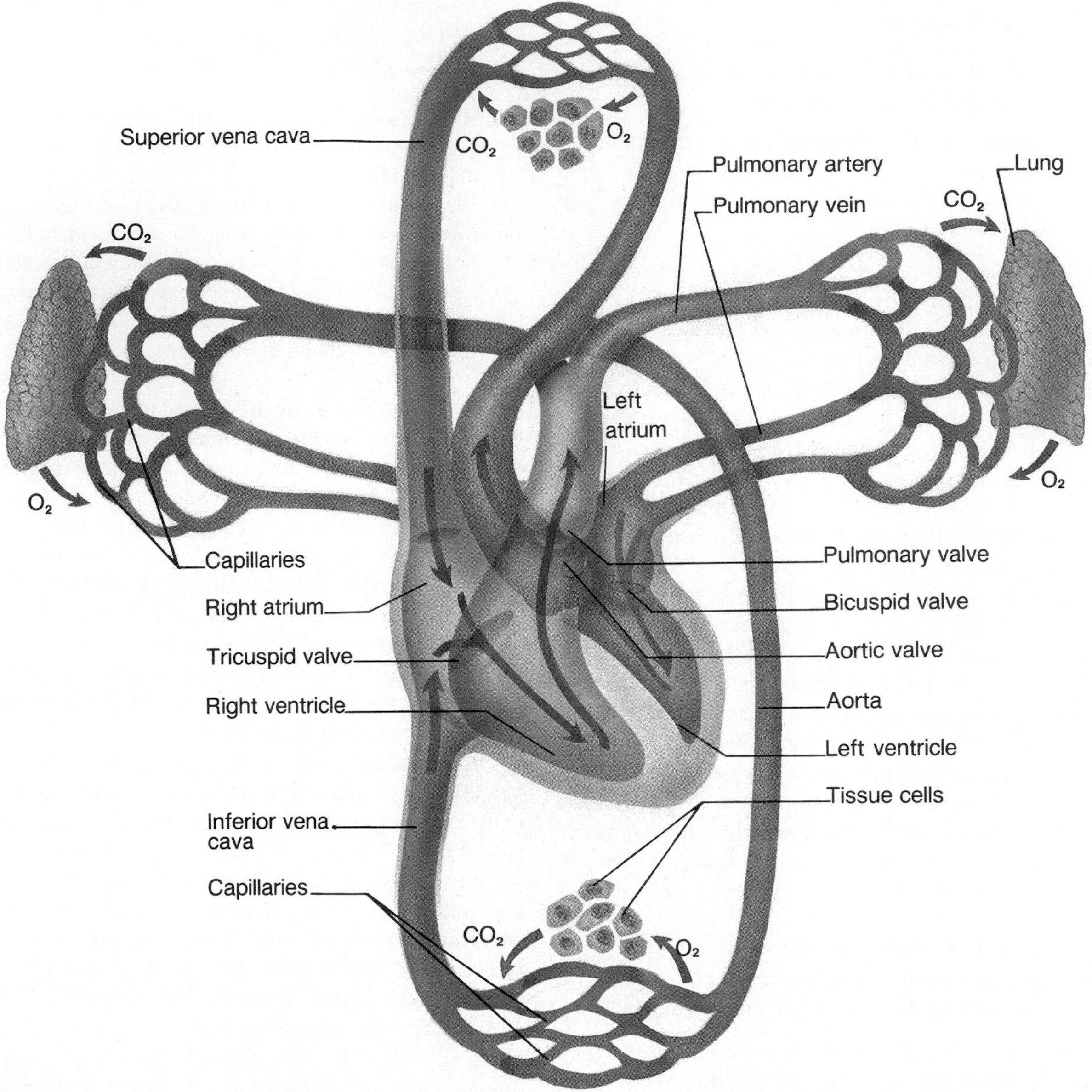

Endothermic (warm-blooded) animals, including humans, need an efficient circulatory system to transport oxygen-rich blood rapidly to all parts of the body. Of the vertebrates, only birds and mammals with their consistently warm body temperatures are considered endothermic, and only birds, mammals, and a few reptiles (crocodiles and alligators) have a four-chambered heart. This is far more efficient than the three-chambered hearts of other reptiles and amphibians or the two-chambered hearts of fishes.

1. Name the components of the circulatory system that function in oxygen transport, in the transport of nutrients from the digestive system, and in protection.
2. Define and describe the function of arteries, veins, and capillaries.
3. Define the terms *interstitial fluid* and *lymph*. Describe their relationships to blood plasma.

Development of the Heart

During a seven-day period, cardiogenic mesoderm migrates to become the heart tube, which differentiates into an embryonic heart, pumping blood by day 25.

Objective 3. Describe the events in the development of the embryonic heart from the eighteenth through the twenty-fifth day.

The cardiovascular system is one of the first systems to form in the embryo, delivering nutrients to the mitotically active cells and disposing of waste products through its association with the maternal blood vessels in the placenta. Blood is formed and is circulated through the vessels by the pumping action of the heart on about the twenty-fifth day after conception.

Figure 20.2. The early development of the heart from embryonic mesoderm. (*a*) a dorsal view of an embryo at day 20 showing the position of a transverse cut depicted in (a_1) (*b*), and (*c*). (a_1) at day 20, the endocardial heart tubes are formed from the heart cords. (*b*) by day 21, the medial migration of the endocardial heart tubes brings them together within the forming pericardial cavity. (*c*) a single endocardial heart tube is completed at day 22 and is suspended in the pericardial cavity by the dorsal mesocardium.

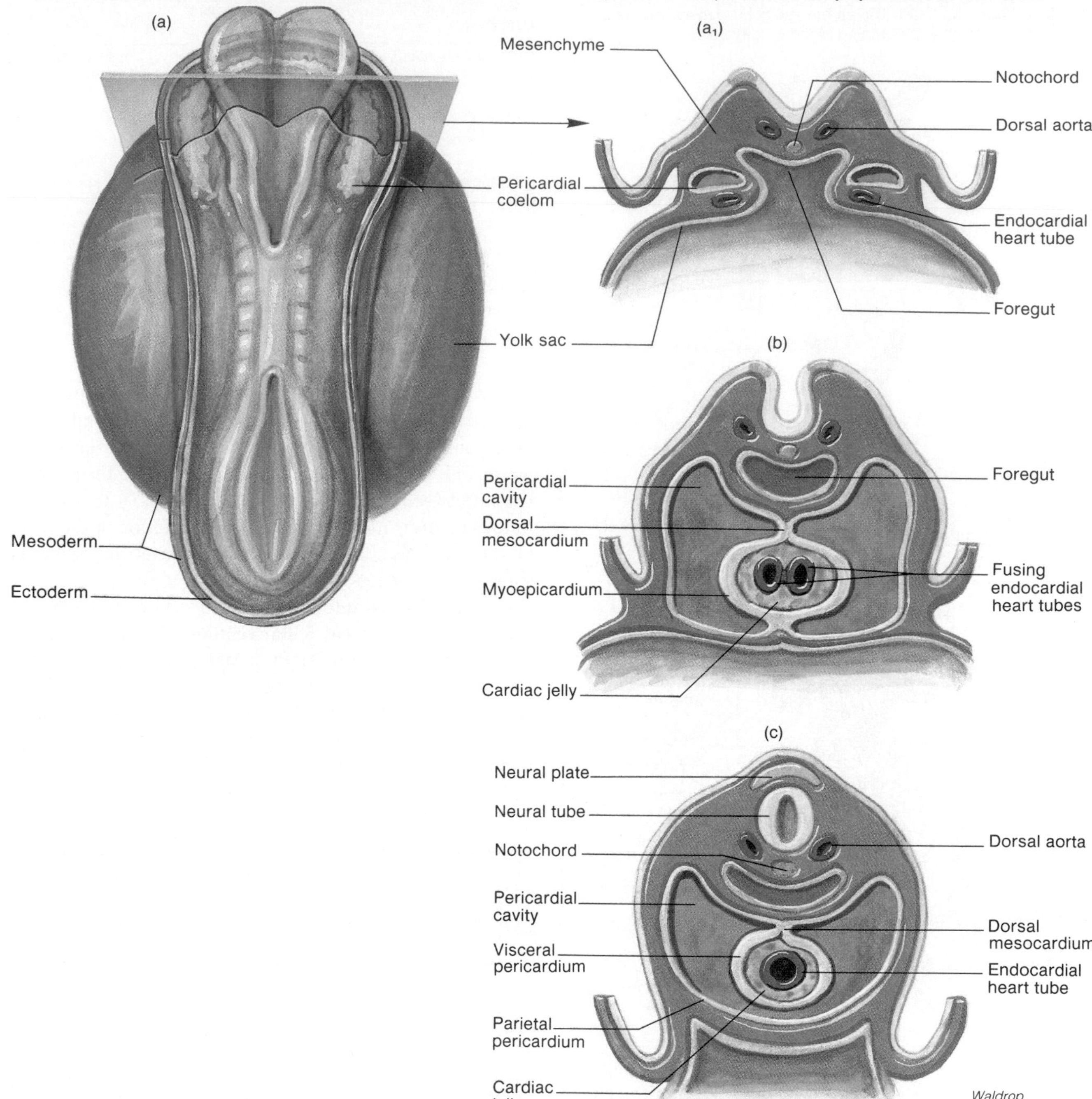

Throughout pregnancy, the fetus is dependent on the mother's circulatory system for nutrient, gaseous, and waste exchange. Therefore, some unique structures of the fetal circulatory system develop to accommodate the placental arrangement. This section of the chapter is concerned with the development of the fetal heart. The circulation of blood through the fetal cardiovascular system is discussed later in the chapter after some basic concepts of the circulatory system are introduced.

The remarkable development of the heart requires only six to seven days. Heart development is first apparent at the eighteenth or nineteenth day in the *cardiogenic (kar''de-o-jen'ik) area* of the mesoderm layer (fig. 20.2). A small paired mass of specialized cells called *heart cords* form here. Shortly after, a hollow center develops in each heart cord, and each structure is then referred to as a *heart tube.* The heart tubes begin to migrate toward each other

cardiogenic: Gk. *kardia,* heart; *genesis,* be born (origin)

Figure 20.3. Formation of the heart chambers. (*a*) the heart tubes fuse during days 21 and 22. (*b*) the developmental chambers are formed during day 23. (*c*) differential growth causes folding between the chambers during day 24, and vessels are developed to transport blood to and from the heart. The embryonic heart generally begins rhythmic contractions and pumping blood by day 25.

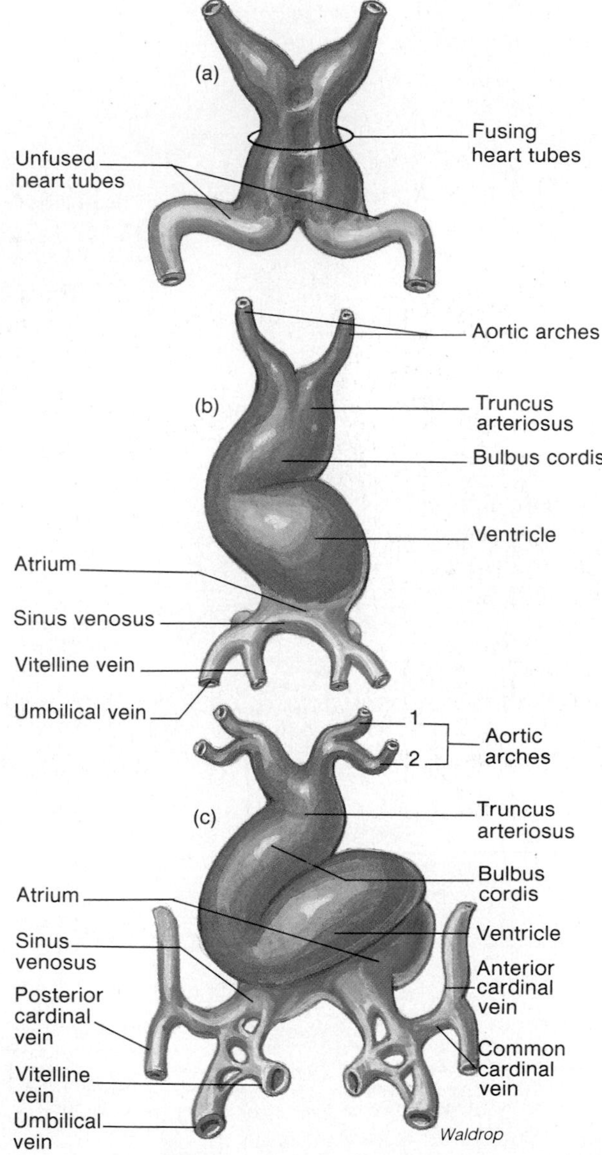

during the twenty-first day and soon fuse to form a single median *endocardial heart tube.* During this time the endocardial heart tube undergoes dilations and constrictions so that when fusion is completed during the fourth week, five distinct regions of the heart can be identified. These are the **truncus arteriosus, bulbus cordis, ventricle, atrium,** and **sinus venosus** (fig. 20.3).

After the fusion of the heart tubes and the formation of distinct dilations, the heart begins to pump blood. Partitioning of the heart chambers begins during the middle of the fourth week and is completed by the end of the fifth week. It is during this crucial time that many congenital heart problems develop.

Table 20.1 Structural changes in the development of the heart

Primitive cardiac dilation	Developmental fate
Truncus arteriosus	Aorticopulmonary septum, which forms the partition between the ascending aorta and the pulmonary trunk
Bulbus cordis	Incorporated into the walls of the ventricles
Sinus venosus	Differentiates to form the coronary sinus and a portion of the wall of the right atrium
Ventricle	Development of the interventricular septum to form the right and left ventricles
Atrium	Development of the septum secundum to partially partition the right and left atria, and formation of the foramen ovale with the valve of the foramen ovale

Major changes occur in each of the five primitive dilations of the developing heart during the week-and-a-half embryonic period beginning in the middle of the fourth week. The truncus arteriosus differentiates to form a partition between the aorta and the pulmonary trunk. The bulbus cordis is incorporated in the formation of the walls of the ventricles. The sinus venosus forms the **coronary sinus** and a portion of the wall of the right atrium. The ventricle is divided into the right and left chambers by the growth of the **interventricular septum.** The atrium is partially partitioned into right and left chambers by the **septum secundum.** An opening between the two atria called the **foramen ovale** persists throughout fetal development. This opening is covered by a flexible valve, which permits blood to pass from the right to the left side of the heart.

The development of the heart is summarized in table 20.1.

Congenital defects of the cardiac septa are a relatively common form of birth abnormality. Approximately 0.7% of live births and 2.7% of stillbirths show cardiac abnormalities. Ventricular septal defects are the most common of the cardiac defects. An infant with a congenital cardiac defect may suffer from inadequate oxygenation of blood and is termed a "blue baby."

1. When does the heart begin development, and when does it begin to pump blood?
2. List the five regions of the embryonic heart tube, and indicate which regions of the heart they help form.

Structure of the Heart

The structure of the heart and the action of its valves allow it to pump blood low in oxygen to the lungs and to pump oxygen-rich blood to the body.

Objective 4. Describe the location of the heart in relation to other organs of the thoracic cavity and their associated serous membranes.

Objective 5. Describe the structure and functions of the three layers of the heart wall.

Objective 6. Describe the nature, location, and significance of the fibrous skeleton of the heart.

Objective 7. Identify the chambers, valves, and associated vessels of the heart.

Objective 8. Trace the flow of blood through the heart, and distinguish between the pulmonary and systemic circulations.

The heart is a hollow, four-chambered muscular organ, which is roughly the size of a clenched fist and averages 255 grams in adult females and 310 grams in adult males. It is estimated that the heart contracts some 42 million times and ejects 700,000 gallons of blood during a year.

The heart is located within the thoracic cavity between the lungs in the mediastinum (fig. 20.4). About two-thirds of the heart is located left of the midline, with its *apex,* or cone-shaped end, pointing downward in contact with the diaphragm. The *base* of the heart is the broad superior end where the large vessels attach.

The heart is enclosed and protected by a loose-fitting, serous sac with dense fibrous connective tissue called the **parietal pericardium** *(per''ĭ-kar'de-um),* or **pericardial sac** (fig. 20.5). The parietal pericardium separates the heart

pericardium: Gk. *peri,* around; *kardia,* heart

Figure 20.4. An X ray of the heart situated within the mediastinum between the lungs.

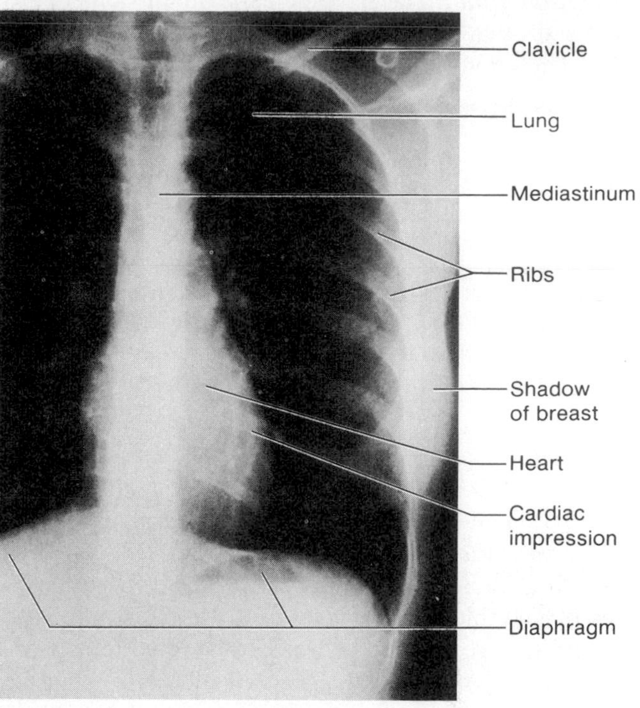

Figure 20.5. The position of the heart and associated serous membranes within the thoracic cavity.

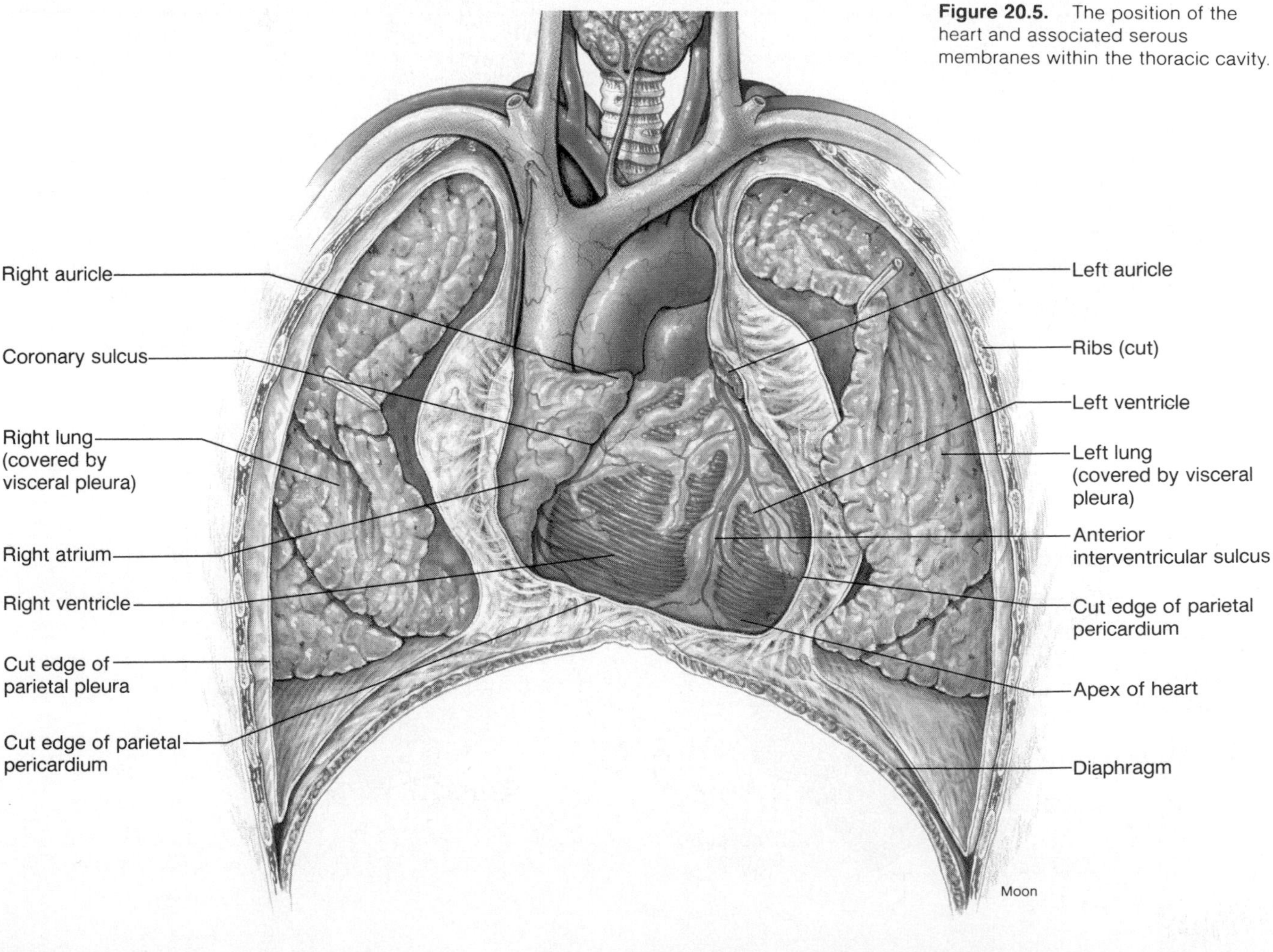

from the other thoracic organs and forms the wall of the *pericardial cavity,* which contains a watery, lubricating *pericardial fluid.* The parietal pericardium is actually composed of an outer *fibrous layer* and an inner *serous layer.* The serous layer secretes the pericardial fluid.

Pericarditis is an inflammation of the parietal pericardium associated with increases in the secretion of fluid into the pericardial cavity. Because the tough, fibrous portion of the parietal pericardium is inelastic, an increase in fluid pressure impairs the movement of blood into and out of the chambers of the heart. Some of the pericardial fluid may be withdrawn for analysis by injecting a needle to the left of the xiphoid process to pierce the parietal pericardium.

Heart Wall

The wall of the heart is composed of three distinct layers (fig. 20.6). The outer layer is the **epicardium,** also called the *visceral pericardium.* The space between this layer and the parietal pericardium is the pericardial cavity just described. The thick middle layer of the heart is called the **myocardium.** It is composed of cardiac muscle and arranged in such a way that the contraction of the muscle bundles results in squeezing or wringing of the heart chambers (fig. 20.7). The thickness of the myocardium varies, depending on the force needed to eject blood from the particular chambers. The thickest portion of the myocardium, therefore, surrounds the left ventricle. The atrial walls are relatively thin. The inner layer, called the **endocardium,** is continuous with the endothelium of blood vessels. The endocardium also forms part of the valves of the heart. Inflammation of the endocardium is called *endocarditis.*

Table 20.2 summarizes the characteristics of the three layers of the heart wall.

Chambers and Valves

The heart is a four-chambered, double pump, consisting of upper right and left **atria** (singular, **atrium**) and lower right and left **ventricles;** the atria contract and empty simultaneously into the ventricles (fig. 20.8), which also contract in unison. The atria are separated by the thin muscular *interatrial septum,* while the ventricles are separated by the thick muscular *interventricular septum.* **Atrioventricular** *(a''tre-o-ven-trik'u-lar)* **valves** are located between the atria and ventricles, and **semilunar valves** are present at the bases of the two large vessels leaving the heart. These valves maintain the flow of blood in one direction.

---

atrium: L. *atrium,* chamber
ventricle: L. *ventriculus,* diminutive of *venter,* belly

Figure 20.6. The structure of the heart wall and the pericardium. A section of the heart has been cut out and rotated to show the layers of the heart and the pericardium.

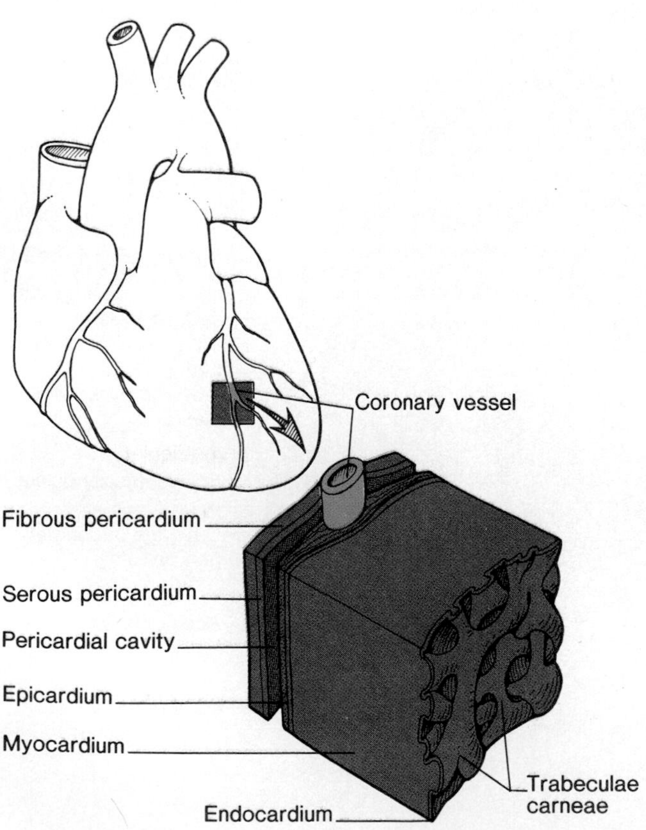

Figure 20.7. The structural arrangement of the cardiac muscles forming the myocardium of the heart. The fiber arrangement around both ventricles (*a*) and (*b*) is a whorled pattern.

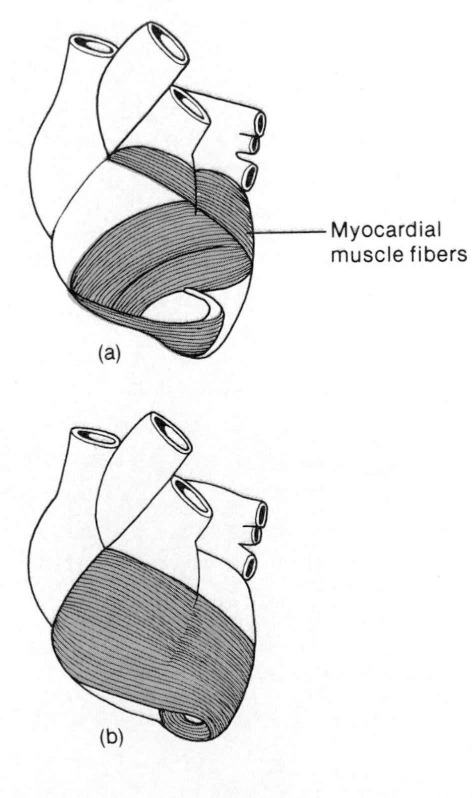

Table 20.2 Layers of the heart wall		
Layer	**Characteristics**	**Function**
Epicardium (visceral pericardium)	Serous membrane of connective tissue covered with epithelium and including blood capillaries, lymph capillaries, and nerve fibers	Lubricative outer covering
Myocardium	Cardiac muscle tissue separated by connective tissues and including blood capillaries, lymph capillaries, and nerve fibers	Muscular contractions that eject blood from the heart chambers
Endocardium	Serous membrane of epithelium and connective tissues, including elastic and collagenous fibers, blood vessels, and specialized muscle fibers	Protective inner lining of the chambers and valves

From Kent M. Van De Graaff, *Human Anatomy,* 2d ed.

Each atrium has a pouchlike extension, or appendage, that is visible when the external surface of the heart is viewed. These appendages of the atria are the right and left **auricles.** Grooved depressions on the surface of the heart indicate the partitions between the chambers and also contain **coronary vessels** that supply blood to the wall of the heart. The most prominent groove is the *atrioventricular* (*coronary*) *sulcus* that encircles the heart and marks the division between the atria and ventricles. The partition between right and left ventricles is denoted by two (anterior and posterior) *interventricular sulci.*

The following discussion follows the sequence in which blood flows through the structures of the atria, ventricles, and valves.

Right Atrium. The right atrium *(a'tre-um)* receives systemic venous blood from the **superior vena cava,** which drains the upper portion of the body, and from the **inferior vena cava,** which drains the lower portion (fig. 20.8). There is an additional venous opening into the right atrium from the *coronary sinus,* which returns venous blood from the myocardium of the heart itself.

auricle: L. *auricula,* a little ear
vena cava: L. *vena,* vein; *cava,* empty

Figure 20.8. The structure of the heart. (*a*) an anterior view; (*b*) a posterior view; (*c*) an internal view.

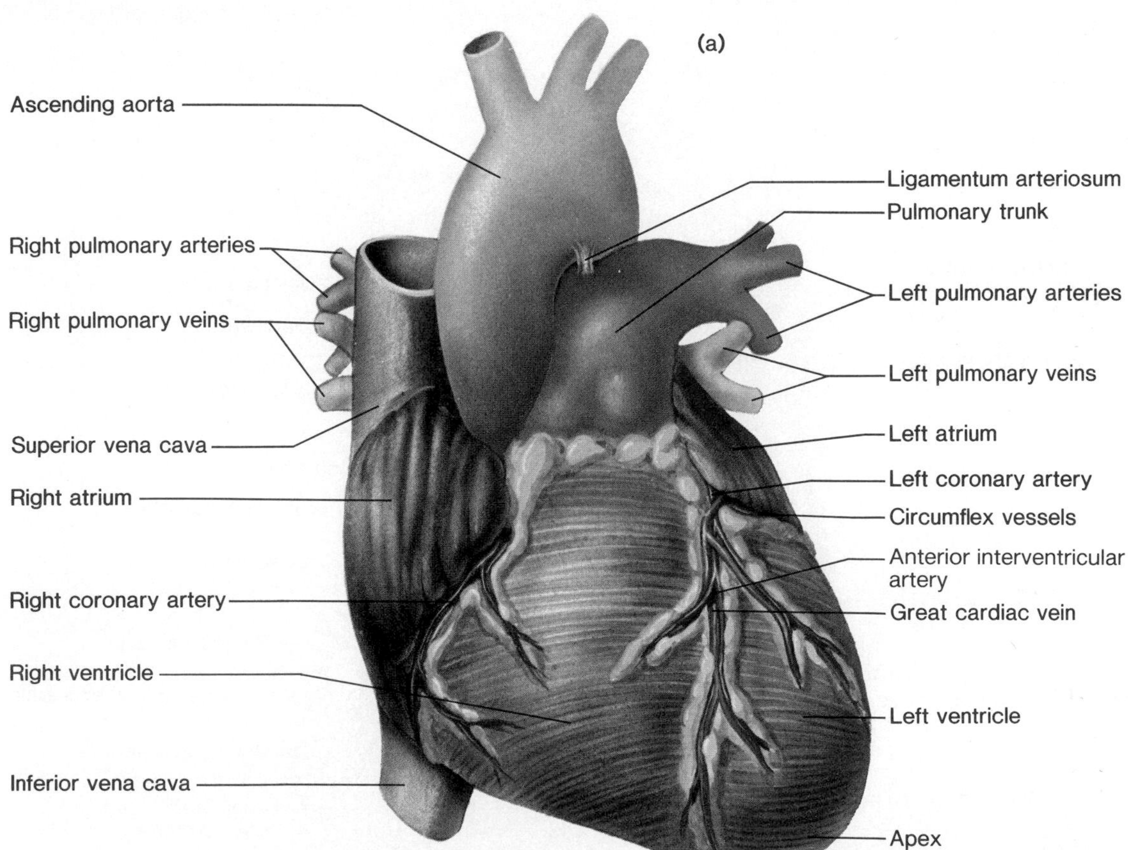

Figure 20.8. *continued*

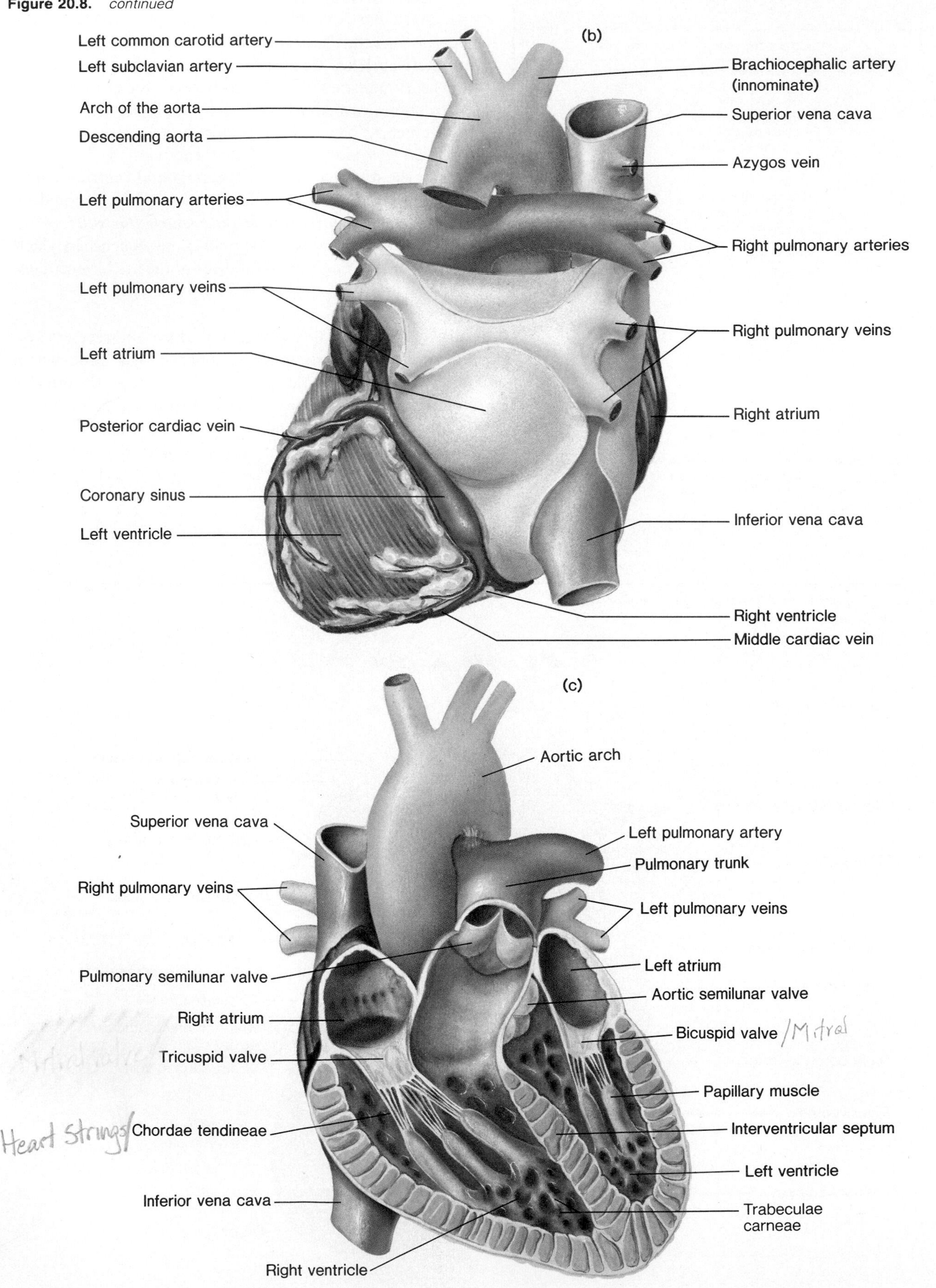

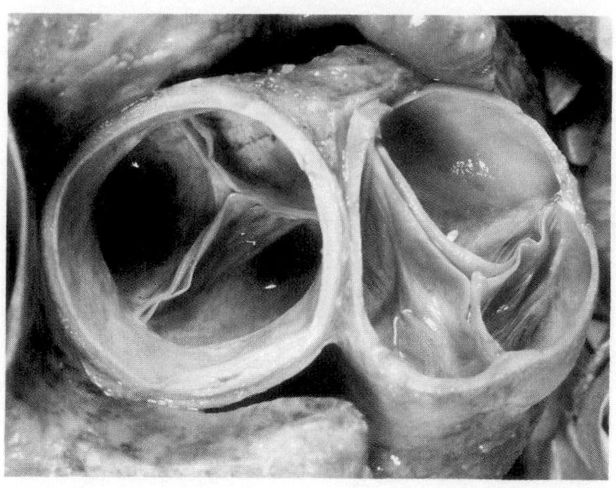

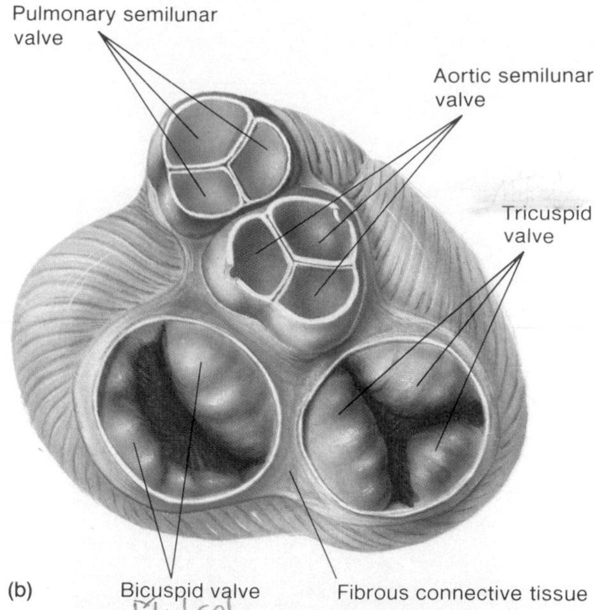

(a) (b)

Figure 20.9. A superior view of the heart valves. (*a*) a photograph of the semilunar valves. (*b*) a diagram of the relative position of the valves and the supporting fibrous connective tissue.

Right Ventricle. Blood from the right atrium passes through the **tricuspid valve** to fill the right ventricle *(ven'tri-k'l)*. The tricuspid valve is an atrioventricular (AV) valve that gets its name from its three valve leaflets, or cusps. Each **cusp** is held in position by strong tendinous cords called **chordae tendineae,** which are secured to the ventricular wall by cone-shaped **papillary muscles.** These structures prevent the valves from everting, like an umbrella in a strong wind, when the ventricles contract.

Ventricular contraction causes the tricuspid valve to close and the blood to leave the right ventricle through the **pulmonary trunk** and to enter the lungs through the right and left **pulmonary arteries.** The **pulmonary semilunar valve** lies at the base of the pulmonary trunk as it leaves the right ventricle.

Left Atrium. After gas exchange has occurred within the capillaries of the lungs, oxygenated blood is transported to the left atrium through four **pulmonary veins,** two from each lung.

Left Ventricle. The left ventricle receives blood from the left atrium. These two chambers are separated by the **bicuspid,** or **mitral** *(mi'tral)*, **valve.** When the left ventricle is relaxed, the valve is open and allows blood to flow from the atrium to the ventricle; when the left ventricle contracts, the valve closes. Closing of the valve during ventricular contraction prevents backflow of blood into the atrium.

The walls of the left ventricle are thicker than those of the right ventricle. The endocardium of both chambers has distinct ridges called **trabeculae** *(trah-bek'u-le)* **carneae** (see figs. 20.6, 20.8c). Oxygenated blood leaves the left ventricle through the **ascending aorta** *(a-or'tah)*. The **aortic semilunar valve,** located at the base of the ascending aorta, closes as a result of the pressure of the blood when the left ventricle relaxes and thus prevents backflow of blood into the relaxed ventricle.

The appearance of the valves is shown in figure 20.9, and their actions are summarized in table 20.3.

Table 20.3 Valves of the heart

Valve	Location	Comments
Tricuspid valve	Between right atrium and right ventricle, surrounding atrioventricular orifice	Composed of three cusps that prevent a backflow of blood from right ventricle into right atrium during ventricular contraction
Pulmonary semilunar valve	Entrance to pulmonary trunk	Composed of three half-moon-shaped flaps that prevent a backflow of blood from pulmonary trunk into right ventricle during ventricular relaxation
Bicuspid (mitral) valve	Between left atrium and left ventricle, surrounding atrioventricular orifice	Composed of two cusps that prevent a backflow of blood from left ventricle to left atrium during ventricular contraction
Aortic semilunar valve	Entrance to ascending aorta	Composed of three half-moon-shaped flaps that prevent a backflow of blood from aorta into left ventricle during ventricular relaxation

chordae tendineae: L. *chorda*, string; *tendere*, to stretch
semilunar: L. *semi*, half; *luna*, moon
bicuspid: L. *bi*, two; *cuspis*, tooth point or spike
mitral: L. *mitra*, like a bishop's mitre
trabeculae carneae: L. *trabecula*, small beams; *carneus*, flesh

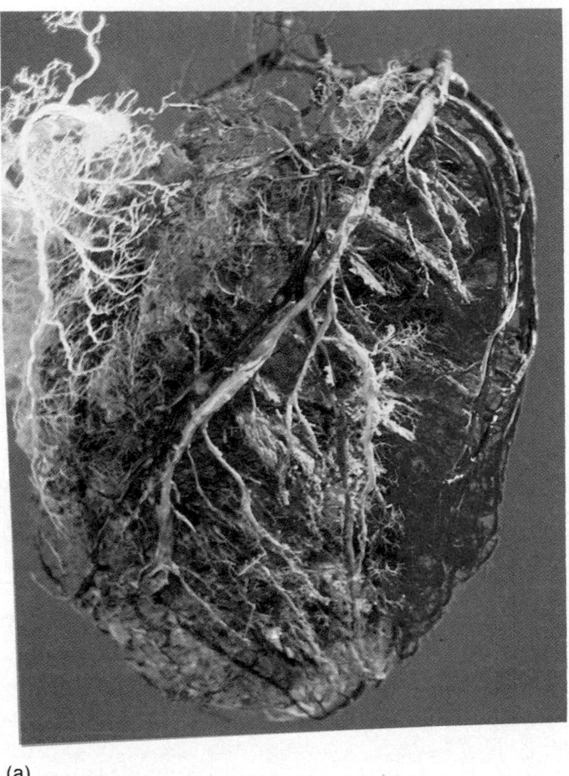

(a)

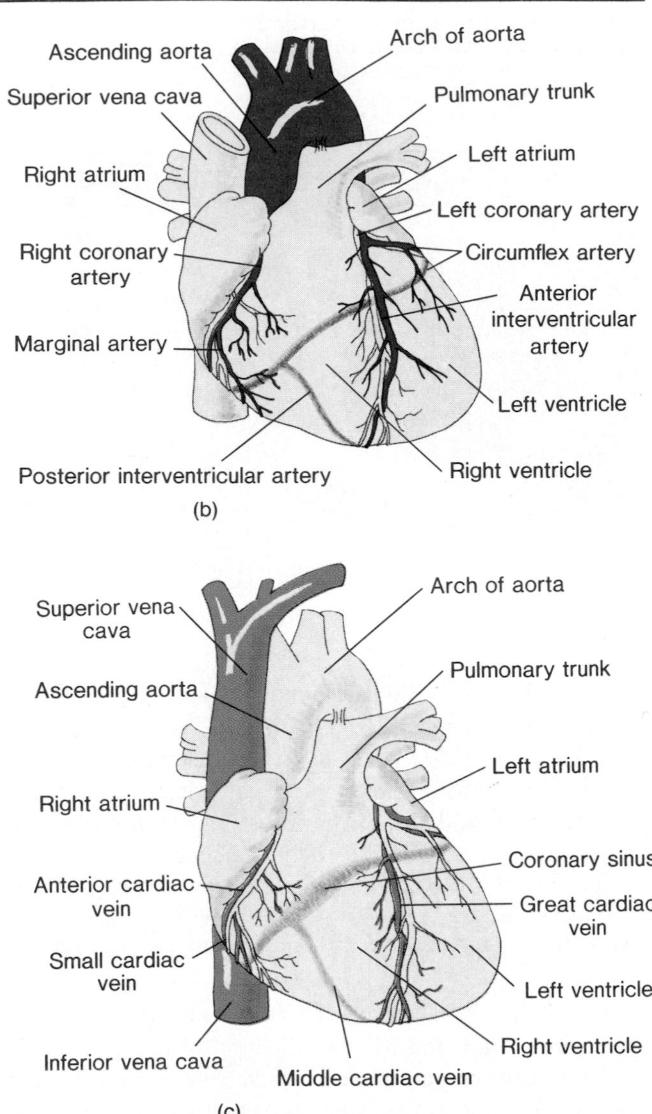

Figure 20.10. Coronary circulation. (*a*) a plastic cast of the coronary arteries and their major branches. (*b*) an anterior view of the arterial supply to the heart. (*c*) an anterior view of the venous drainage.

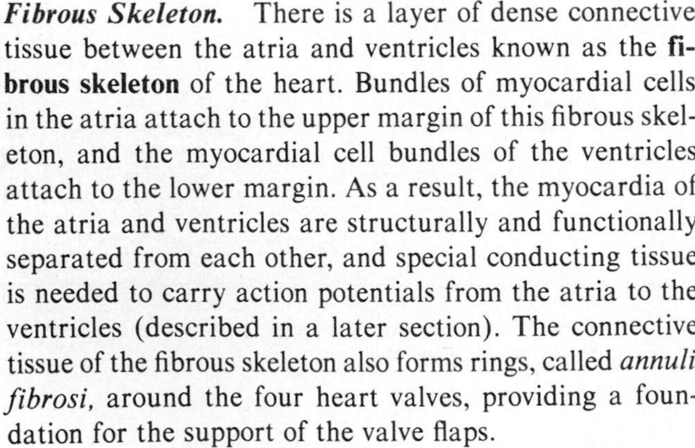

Fibrous Skeleton. There is a layer of dense connective tissue between the atria and ventricles known as the **fibrous skeleton** of the heart. Bundles of myocardial cells in the atria attach to the upper margin of this fibrous skeleton, and the myocardial cell bundles of the ventricles attach to the lower margin. As a result, the myocardia of the atria and ventricles are structurally and functionally separated from each other, and special conducting tissue is needed to carry action potentials from the atria to the ventricles (described in a later section). The connective tissue of the fibrous skeleton also forms rings, called *annuli fibrosi,* around the four heart valves, providing a foundation for the support of the valve flaps.

Circulatory Routes

Figure 20.1 illustrates the circulatory routes of the blood. There are two principal divisions of the circulatory blood flow: the pulmonary and systemic circulations.

The **pulmonary circulation** consists of blood vessels that transport blood from the right ventricle to the lungs for gas exchange and then to the left atrium of the heart. It consists of the pulmonary trunk with its semilunar valve, the pulmonary arteries that transport blood to the lungs, the pulmonary capillaries within each lung, and four pulmonary veins that transport oxygenated blood back to the heart.

The **systemic** *(sis-tem'ik)* **circulation** is composed of all the remaining vessels of the body that are not part of the pulmonary circulation. This includes the aorta with its semilunar valve, all of the branches of the aorta, all capillaries other than those in the lungs, and all veins other than the pulmonary veins. The right atrium receives all the venous return of oxygen-depleted blood from the systemic veins.

Coronary Circulation. The myocardium is supplied with blood by the right and left **coronary arteries** (fig. 20.10). These two vessels arise from the aorta immediately beyond the aortic semilunar valve. The coronaries encircle the heart within the atrioventricular sulcus, the depression between the atria and ventricles. Two branches arise from both the right and left coronaries to serve the atrial and ventricular walls. The left coronary artery gives rise to the **anterior interventricular artery,** which courses within the anterior interventricular sulcus to serve both ventricles, and the **circumflex artery,** which supplies oxygenated blood to the walls of the left atrium and left ventricle. The right coronary artery gives rise to the **posterior interventricular artery,** which courses through the posterior interventricular sulcus to serve the two ventricles, and a **marginal artery,** which serves the walls of the right atrium and right ventricle. The main trunks of the right and left coronaries anastomose (join together) on the posterior surface of the heart.

From the capillaries in the myocardium, the blood enters the **coronary veins.** The course of these vessels parallels that of the coronary arteries. The coronary veins, however, have thinner walls and are more superficial than the arteries. The two principal coronary veins are the **great cardiac vein,** which returns blood from the anterior aspect of the heart, and the **middle cardiac vein,** which drains the posterior aspect of the heart. These coronary veins converge into a large venous channel on the posterior surface of the heart called the **coronary sinus** (fig. 20.10). The coronary venous blood then enters the heart through an opening into the right atrium.

Heart attacks are the most common cause of death in the United States. The most common type of heart attack involves an occlusion of a coronary artery, which reduces the delivery of oxygen to the myocardium. There are several things you can do to minimize the risk of heart attack: (1) reduce excess weight; (2) avoid hypertension through proper diet, stress reduction, or medication if necessary; (3) avoid smoking; (4) eat foods low in cholesterol and saturated fat; and (5) exercise regularly.

1. List the three layers of the heart wall, and describe the structures associated with each layer.
2. List the valves that control blood flow through the heart, and describe their locations and functions.
3. Trace the flow of a drop of blood from the lungs, through the heart, and back to the lungs.
4. Explain the structure and functional significance of the fibrous skeleton of the heart.

circumflex: L. *circum,* around; *flectere,* to bend

Cardiac Cycle, Heart Sounds, and Electrocardiogram

The cardiac cycle results from electrical activity, which can be monitored with an electrocardiogram. The changes in intraventricular pressure produced during the cardiac cycle can be followed through the sounds created by the closing of the heart valves.

Objective 9. Describe the cycle of contraction and relaxation of the atria and ventricles and how these events are related in time.

Objective 10. Describe how pressure changes in the heart affect the AV valves and semilunar valves and how these events produce the heart sounds.

Objective 11. Describe the electrical activity of the SA node, and explain why the SA node functions as the normal pacemaker.

Objective 12. Identify the conducting tissue of the heart, and describe the pathway of electrical conduction.

Objective 13. Explain the correlation between the first and second heart sounds and the waves of the ECG.

The cardiac cycle refers to the repeating pattern of contraction and relaxation of the heart. The phase of contraction is called **systole** *(sis'to-le),* and the phase of relaxation is called **diastole** *(di-as'to-le).* When these terms are used alone, they refer to contraction and relaxation of the ventricles. It should be noted, however, that the atria also contract and relax. There is an atrial systole and diastole. Atrial contraction occurs when the ventricles are relaxed in diastole; when the ventricles contract during systole, the atria are relaxed.

The heart thus has a two-step pumping action. The right and left atria contract almost simultaneously, followed about 0.1–0.2 seconds later by contraction of the right and left ventricles. During the time when both the atria and ventricles are relaxed, the venous return of blood fills the atria and (because the AV valves are open) fills the ventricles. It has been estimated that the ventricles are about 80 percent filled with blood even before the atria contract. Contraction of the atria adds the final one-fifth to the *end-diastolic volume* of blood in the ventricles.

Contraction of the ventricles in systole ejects about two-thirds of the blood in the ventricles, leaving one-third of the initial amount as the *end-systolic volume.* The ventricles then fill with blood during the next cycle. At an average **cardiac rate** of seventy-five beats per minute, each cycle lasts 0.8 second; 0.5 second is spent in diastole, and systole takes 0.3 second (fig. 20.11).

systole: Gk. *systole,* contraction
diastole: Gk. *diastole,* prolonged or expansion

Figure 20.11. The cardiac cycle of ventricular systole and diastole. Contraction of the atria occurs in the last 0.1 second of ventricular diastole. Relaxation of the atria occurs during ventricular systole. The durations of systole and diastole given are accurate for a cardiac rate of seventy-five beats per minute.

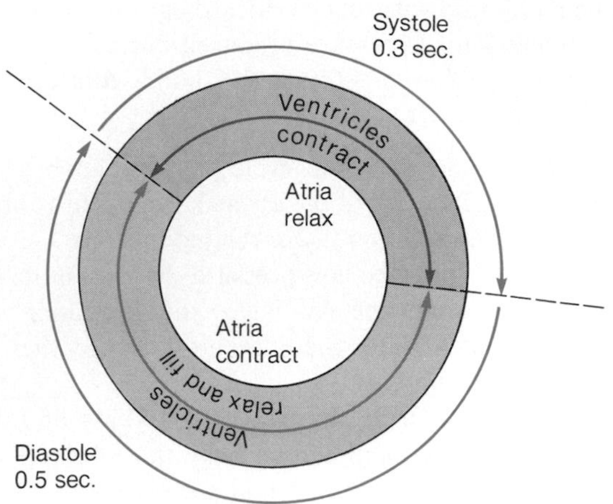

Figure 20.12. The relationship between the heart sounds and the left intraventricular pressure and volume. Closing of the AV valves occurs during the early part of contraction, when the intraventricular pressure rises prior to ejection of blood. Closing of the semilunar valves occurs at the beginning of ventricular relaxation, just prior to filling. The first and second sounds thus appear during the stages of isovolumetric contraction and isovolumetric relaxation (*iso*=same).

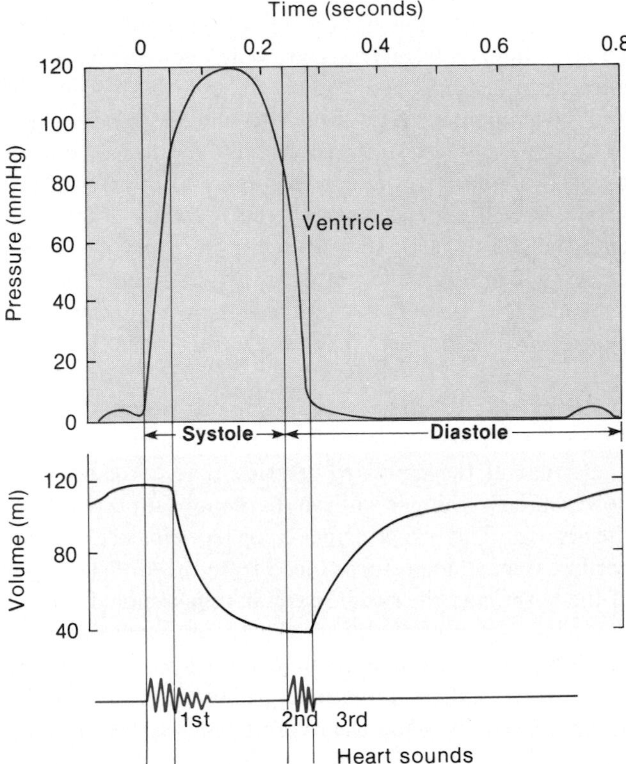

Pressure Changes during the Cardiac Cycle

When the heart is in diastole, pressure in the systemic arteries averages about 80 mm Hg (millimeters of mercury). The following events in the cardiac cycle then occur.

1. *Isovolumetric contraction* of the ventricles occurs when the inflow valves (atrioventricular) and outflow valves (semilunar) are closed.
2. When the pressure in the left ventricle becomes greater than the pressure in the aorta, the phase of *ejection* begins as the semilunar valves open. The pressure in the left ventricle and aorta rises to about 120 mm Hg (fig. 20.12) as the ventricular volume decreases.
3. As the pressure in the left ventricle falls below the pressure in the aorta, the back pressure closes the semilunar valves. The pressure in the aorta falls to 80 mm Hg, while pressure in the left ventricle falls to 0 mm Hg.
4. During *isovolumetric relaxation* the AV and semilunar valves are closed. This phase lasts until the pressure in the ventricles falls below the pressure in the atria.
5. When the pressure in the ventricles falls below the pressure in the atria, a phase of *rapid filling* of the ventricles occurs.
6. *Atrial contraction* (*systole*) empties the final amount of blood into the ventricles immediately prior to the next phase of isovolumetric contraction of the ventricles.

Similar events occur in the right ventricle and pulmonary circulation, but the pressures are lower. The maximum pressure produced at systole of the right ventricle is 25 mm Hg, which falls to a low of 8 mm Hg at diastole.

Heart Sounds

Closing of the AV and semilunar valves, and the resulting turbulence and vibrations that are produced, cause sounds that can be heard at the surface of the chest with a stethoscope. These sounds are often described phonetically as *lub-dub*. The "lub," or **first sound,** is produced by closing of the AV valves. The "dub," or **second sound,** is produced by closing of the semilunar valves. The first sound is thus heard when the ventricles contract at systole, and the second sound is heard when the ventricles relax at the beginning of diastole.

Heart sounds are of clinical importance because they provide information about the condition of the heart valves and other heart problems. Abnormal sounds are referred to as *heart murmurs* and are caused by valvular leakage or turbulence of the blood as it passes through the heart.

The valves of the heart are positioned directly deep to the sternum, which tends to obscure and dissipate valvular sounds. For this reason, a physician will listen with a stethoscope for the heart sounds at locations designated as **valvular auscultatory areas,** which are named according

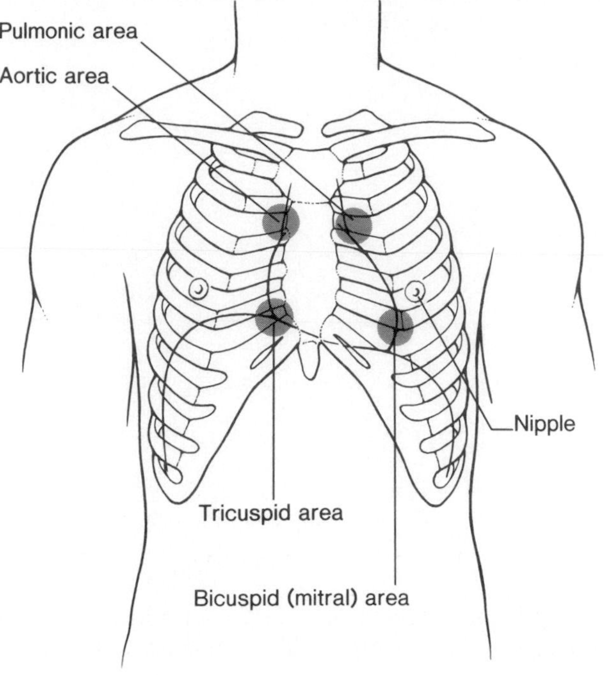

Figure 20.13. The valvular auscultatory areas are the routine stethoscope positions for listening to the heart sounds.

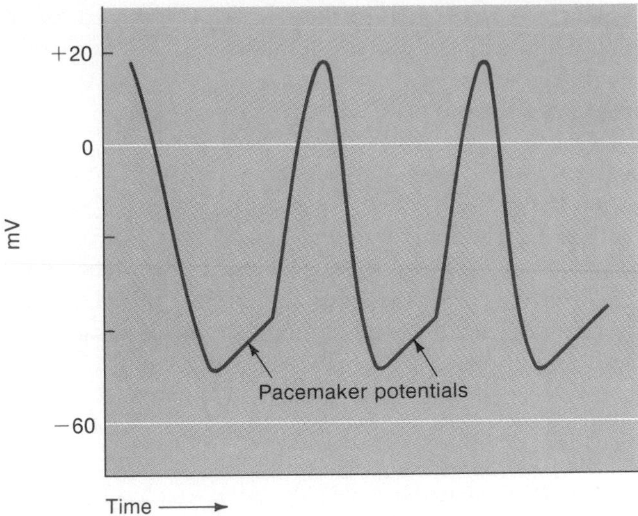

Figure 20.14. Pacemaker potentials and action potentials in the SA node.

to the valve that can be detected (fig. 20.13). The **aortic area** is at the right second intercostal space near the sternum. The **pulmonic area** is directly across from the aortic area at the left second intercostal space near the sternum. The **tricuspid** and **bicuspid (mitral) areas** are both at the fifth intercostal space, with the bicuspid area further out. Surface landmarks are extremely important in identifying auscultatory areas.

Electrical Activity of the Heart

As demonstrated experimentally using isolated myocardial cells in tissue culture, most myocardial cells have the potential for spontaneous electrical activity; they have an intrinsic rhythm. In the normal heart, however, spontaneous electrical activity is limited to a small region (about 2 × 5 × 15 mm) in the right atrium near the opening of the superior vena cava. This region is the **sinoatrial** *(si''no-a'tre-al),* or **SA, node** and serves as the normal *pacemaker* of the heart.

Cells of the SA node are never in an electrically resting state; as soon as they have repolarized from a previous action potential, they begin to depolarize. This depolarization, resulting from spontaneous changes in membrane permeability to cations, is called a *pacemaker potential.* When the pacemaker potential produces a depolarization to about −55 mV, an action potential is stimulated. As in action potentials produced by nerve and muscle fibers, the membrane potential first reverses polarity (to about +20 mV) due to the inflow of Na^+ and then moves toward about −60 mV due to the outflow of K^+ and the action of the Na^+/K^+ pumps. Permeability to K^+ gradually diminishes as the action potential ends, but instead of stabilizing at a resting potential, the membrane gradually depolarizes again as a new pacemaker potential is produced by the inward leakage of Na^+. This pattern is repeated for the lifetime of a normal heart (fig. 20.14).

Some other regions of the heart, including the area around the SA node and the atrioventricular bundle, can potentially produce pacemaker potentials. The rate of spontaneous depolarization of these cells, however, is slower than that of the SA node. The potential pacemaker cells are therefore stimulated by action potentials from the SA node before they can depolarize spontaneously. If action potentials from the SA node are prevented from reaching these areas, they will generate their own pacemaker potentials and serve as sites for the origin of action potentials; they will function as pacemakers. A pacemaker other than the SA node is called an *ectopic (ek-top'ik) pacemaker* (also called an *ectopic focus*). From this discussion, it is clear that the rhythm set by such an ectopic pacemaker is usually slower than that normally set by the SA node.

Myocardial cells stimulated by action potentials from the SA node produce and conduct their own action potentials. The majority of myocardial cells have resting potentials of about −90 mV. When stimulated by action potentials from a pacemaker region, these cells depolarize to +20 mV as a result of the inward diffusion of Na^+. Following the rapid reversal of the membrane polarity, the voltage quickly declines to about −10 to −20 mV. Unlike the action potentials of other cells, however, this level of depolarization is maintained for 200–300 msec (milliseconds) before repolarization (fig. 20.15). This *plateau phase* results from a slow inward diffusion of Ca^{++}, which balances a slow outward diffusion of cations. Rapid repolarization at the end of the plateau phase is achieved, as in other cells, by the opening of K^+ gates and the rapid outward diffusion of K^+ that results.

Conducting Tissues of the Heart. Action potentials that originate in the SA node spread to adjacent myocardial cells of the right and left atria through the *gap junctions* (chapter 14) between these cells. Since the myocardium of the atria is separated from that of the ventricles by the fibrous skeleton of the heart, however, the impulse cannot be conducted directly from the atria to the ventricles. Specialized conducting tissue, composed of modified myocardial cells, is thus required.

Once the impulse spreads through the atria, it passes to the **atrioventricular node (AV node),** which is located on the inferior portion of the interatrial septum (fig. 20.16). From here the impulse continues through the **atrioventricular bundle,** or **bundle of His,** beginning at the top of the interventricular septum. This conducting tissue pierces the fibrous skeleton of the heart and continues to descend along the interventricular septum. The atrioventricular bundle divides into right and left bundle branches, which are continuous with the **Purkinje** *(pur-kin'je)* **fibers** within the ventricular walls. Stimulation of these fibers causes the ventricles to contract simultaneously and eject blood into the pulmonary and systemic circulations.

Conduction of the Impulse. Action potentials spread very quickly—at a rate of 0.8 to 1.0 meter (m) per second—across the myocardial cells of both atria. The conduction rate then slows considerably as the impulse passes into the AV node. Slow conduction of impulses (0.03 to 0.05 m per second) through the AV node accounts for over half of the time delay between excitation of the atria and ventricles. This slow conduction allows time for the atria to finish their contraction before the impulse is conducted into the ventricles. After the impulses spread through the AV node, the conduction rate increases greatly in the atrioventricular bundle and reaches very high velocities (5 m per second) in the Purkinje fibers. As a result of this rapid conduction of impulses, ventricular contraction begins about 0.1–0.2 seconds after contraction of the atria.

Unlike skeletal muscles, the heart cannot sustain a contraction. This is because the atria and ventricles behave as if each were composed of only one muscle cell; the entire myocardium of each is electrically stimulated as a single unit and contracts as a unit. This contraction, which corresponds to the long myocardial action potential and lasts almost 300 msec, is analogous to a twitch produced by a single skeletal muscle fiber (which lasts only 20–100 msec in comparison). The heart normally cannot be stimulated again until after it has relaxed from its previous contraction, because myocardial cells have *long refractory periods* (fig. 20.17), which correspond to the long duration of their action potentials. The rhythmic pumping action of the heart is thus insured.

Figure 20.15. An action potential in a myocardial cell from the ventricles. The plateau phase of the action potential is maintained by a slow inward diffusion of Ca^{++}. The cardiac action potential, as a result, has a duration that is about one hundred times longer than the "spike potential" of an axon.

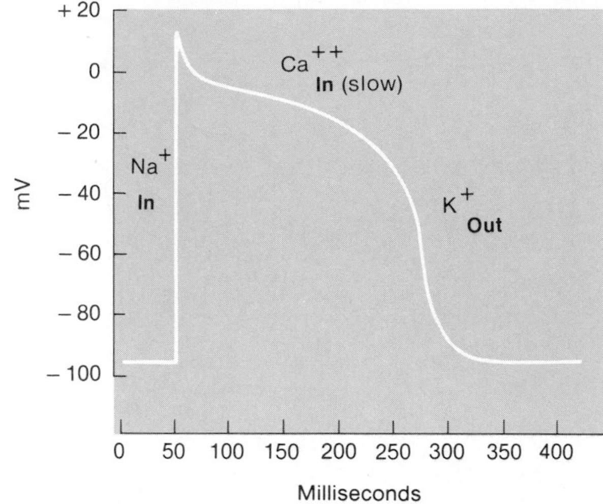

Abnormal patterns of electrical conduction in the heart can produce abnormalities of the cardiac cycle and seriously compromise the function of the heart. These *arrhythmias* may be treated with a variety of drugs that inhibit specific aspects of the cardiac action potentials and, in this way, inhibit the production or conduction of impulses along abnormal pathways. Drugs used to treat arrhythmias may (1) block the fast Na^+ channel (quinidine, procainamide, lidocaine); (2) block the slow Ca^{++} channel (verapamil); or (3) block β-adrenergic receptors (propranolol), since the rate of impulse production and conduction is stimulated by catecholamines.

The Electrocardiogram

A pair of surface electrodes placed directly on the heart will record a repeating pattern of potential changes. As action potentials spread from the atria to the ventricles, the voltage measured between these two electrodes will vary in a way that provides a "picture" of the electrical activity of the heart. The nature of this picture can be

bundle of His: from Wilhelm His, Jr., Swiss physician, 1863–1934
Purkinje fibers: from Johannes E. von Purkinje, Bohemian anatomist, 1787–1869

Figure 20.16. The conduction system of the heart.

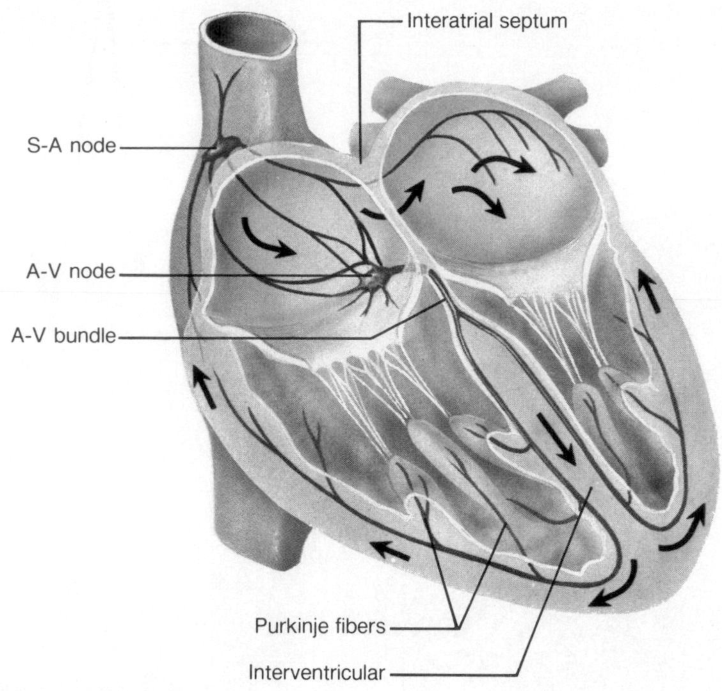

Figure 20.17. The time course for the myocardial action potential (*A*) is compared with the duration of contraction (*B*). Notice that the long action potential results in a correspondingly long absolute refractory period (*ARP*) and relative refractory period (*RRP*). These refractory periods last almost as long as the contraction, so that the myocardial cells cannot be stimulated a second time until they have finished their contraction from the first stimulus.

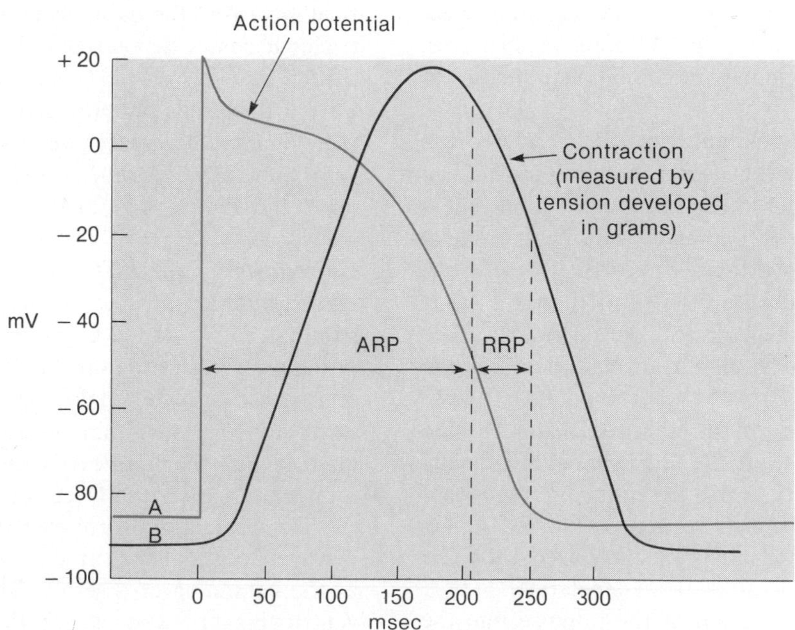

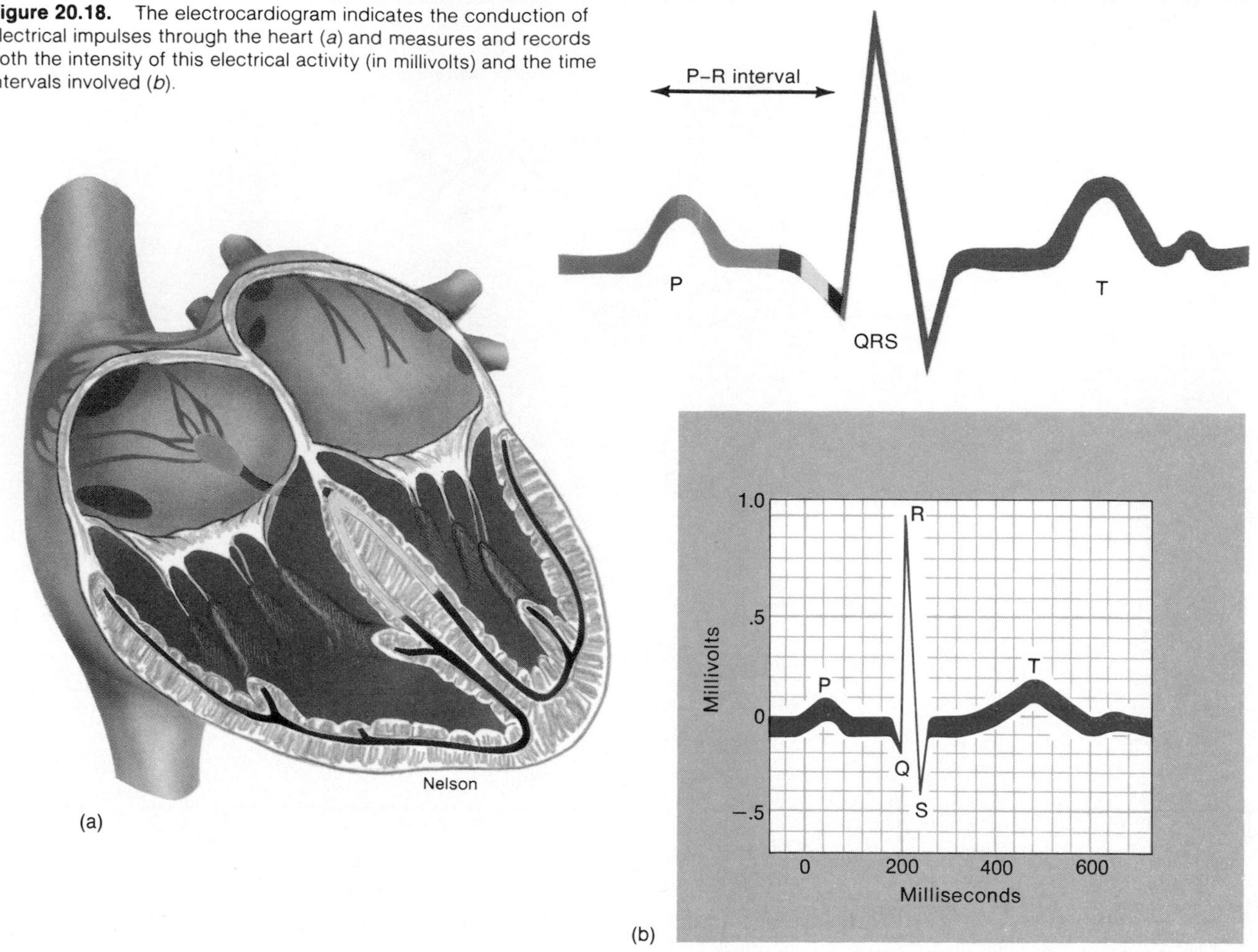

Figure 20.18. The electrocardiogram indicates the conduction of electrical impulses through the heart (*a*) and measures and records both the intensity of this electrical activity (in millivolts) and the time intervals involved (*b*).

varied by changing the position of the recording electrodes; different positions provide different perspectives, enabling an observer to gain a more complete picture of the electrical events.

The body is a good conductor of electricity because tissue fluids contain a high concentration of ions that can move (creating a current) in response to potential differences. Potential differences generated by the heart are thus conducted to the body surface, where they can be recorded by electrodes placed on the skin (these electrode positions are discussed under Clinical Considerations). The recording thus obtained is called an **electrocardiogram (ECG** or **EKG)** (fig. 20.18).

Each cardiac cycle produces three distinct ECG waves, designated P, QRS, and T. The spread of depolarization through the atria causes a potential difference that is indicated by an upward deflection of the ECG line. When the entire mass of the atria is depolarized, the ECG returns to baseline; the spread of atrial depolarization thus creates the **P wave.** Conduction of the impulse into the ventricles similarly creates a potential difference that results in a sharp upward deflection of the ECG line, which then returns to the baseline as the entire mass of the ventricles becomes depolarized. The spread of the depolarization into the ventricles is thus represented by the **QRS wave.** During this time the atria repolarize, but this event is hidden by the greater depolarization occurring in the ventricles. Finally, repolarization of the ventricles produces the **T wave** (fig. 20.19).

Correlation of the ECG with Heart Sounds. Depolarization of the ventricles, as indicated by the QRS wave, stimulates contraction by promoting the uptake of Ca^{++} into the regions of the sarcomeres. The QRS wave is thus seen to occur at the beginning systole. The rise in intraventricular pressure that results causes the AV valves to close, so that the first heart sound (S_1, or lub) is produced immediately after the QRS wave (fig. 20.20).

Repolarization of the ventricles, as indicated by the T wave, occurs at the same time that the ventricles relax at the beginning of diastole. The resulting fall in intraventricular pressure causes the aortic and pulmonary semilunar valves to close, so that the second heart sound (S_2, or dub) is produced shortly after the T wave in an electrocardiogram begins.

Figure 20.19. The conduction of electrical impulses in the heart, as indicated by the electrocardiogram (ECG). The direction of the arrows in (*e*) indicates that depolarization of the ventricles occurs from the inside (endocardium) out (to the epicardium), whereas the arrows in (*g*) indicate that repolarization of the ventricles occurs in the opposite direction.

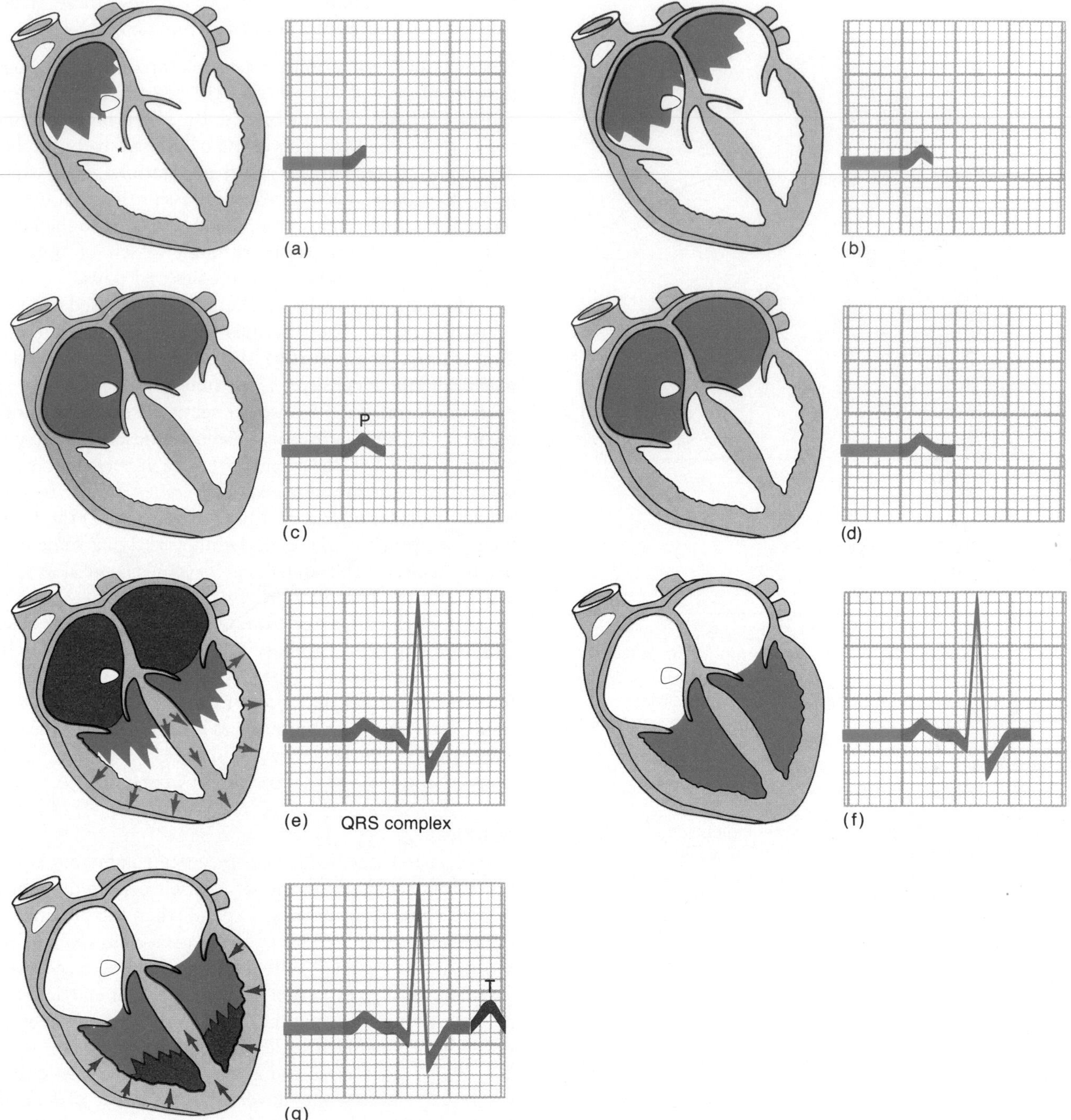

Figure 20.20. The relationship between changes in left intraventricular pressure and the electrocardiogram during the cardiac cycle. The QRS wave (representing depolarization of the ventricles) occurs at the beginning of systole, while the T wave (representing repolarization of the ventricles) occurs at the beginning of diastole.

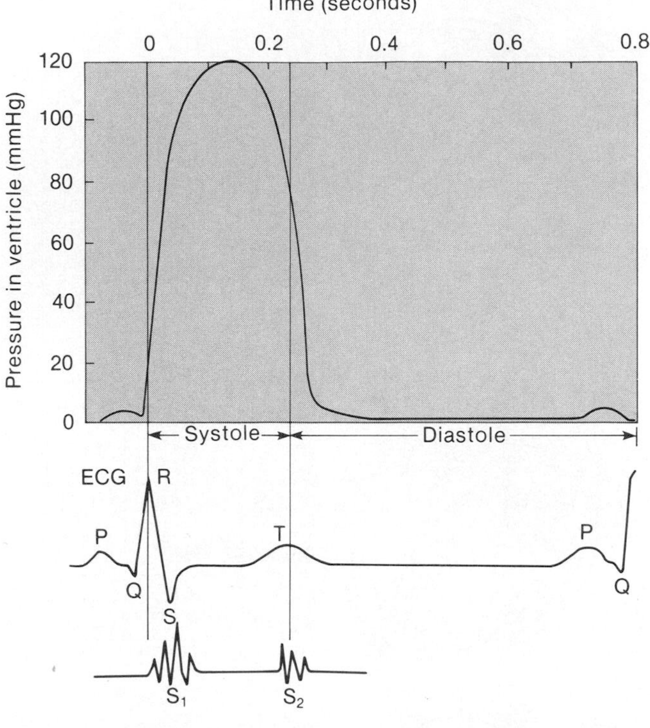

1. Describe the electrical activity of the cells of the SA node, and explain how the SA node functions as the normal pacemaker.
2. Using a line diagram, illustrate a myocardial action potential and the time course for myocardial contraction. Describe how the relationship between these two events prevents the heart from sustaining a contraction and how it normally prevents circus rhythms of electrical activity.
3. Draw an ECG and label the waves. Indicate the electrical events in the heart that produce these waves.
4. Draw a figure illustrating the relationship between ECG waves and the heart sounds. Explain this relationship.
5. Using a flowchart (arrows), describe the pathway of electrical conduction of the heart, starting with the SA node. Explain how damage to the AV node affects this conduction pathway and the ECG.

Blood Vessels

The structure of arteries and veins allows them to transport blood from the heart to the capillaries and back to the heart. The structure of capillaries permits the exchange of plasma fluid and dissolved molecules between the blood and surrounding tissues.

Objective 14. Describe the structures of muscular and elastic arteries, and explain how these structures aid their functions.

Objective 15. Describe the structure of veins, and explain how this structure aids their function.

Objective 16. Describe the structure of different types of capillaries, and explain how these structures relate to their function.

Blood vessels form a tubular network throughout the body that permits blood to flow from the heart to all the living cells of the body and then back to the heart. Blood leaving the heart passes through vessels of progressively smaller diameters referred to as **arteries, arterioles,** and **capillaries.** Capillaries are microscopic vessels that join the arterial flow to the venous flow. Blood returning to the heart from the capillaries passes through vessels of progressively larger diameters called **venules** and **veins.**

The walls of arteries and veins are composed of three coats, or "tunics." The outermost layer is the **tunica** *(tu'ni-kah)* **externa** (or **adventitia**), the middle layer is the **tunica media,** and the inner layer is the **tunica intima.** The tunica externa is composed of loose connective tissue. The tunica media is composed primarily of smooth muscle. The tunica intima consists of three parts: (1) an innermost simple squamous epithelium, the *endothelium (en''do-the'le-um),* which lines the lumina of all blood vessels; (2) the basement membrane of the endothelium overlying some connective tissue fibers; and (3) a layer of elastic fibers, or elastin, forming an *internal elastic lamina.*

Although both arteries and veins have the same basic structure (fig. 20.21), there are some important differences between the two types. Arteries have more muscle for their diameters than do comparably sized veins. As a result, arteries appear more round in cross section, whereas veins are usually partially collapsed. In addition, many veins have valves, which are absent in arteries.

Arteries

The aorta and other large arteries contain numerous layers of elastin fibers between smooth muscle cells in the tunica media. These large arteries expand when the pressure of the blood rises as a result of the heart's contraction; they recoil, like a stretched rubber band, when the blood pressure falls during relaxation of the heart. This elastic recoil helps to produce a smoother, less pulsatile flow of blood through the smaller arteries and arterioles.

The small arteries and arterioles are less elastic than the larger arteries and have a thicker layer of smooth muscle for their diameters. Unlike the larger *elastic arteries,* therefore, the smaller *muscular arteries* retain almost the same diameter as the pressure of the blood rises

tunica: L. *tunica,* covering or coat
lumen: L. *lumen,* opening

Figure 20.21. The structure of a medium-sized artery and vein showing the relative thickness and composition of the tunicas.

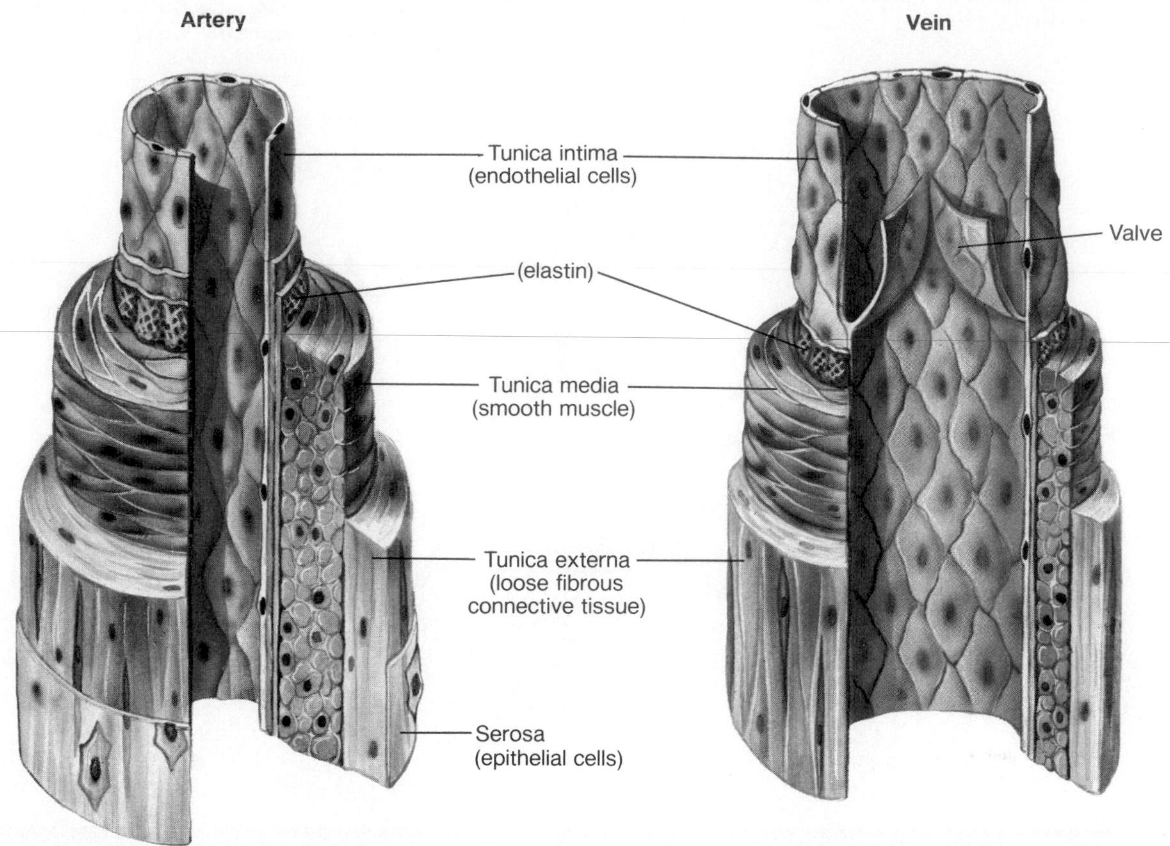

and falls during the heart's pumping activity. Since arterioles and small muscular arteries have narrow lumina, they provide the greatest resistance to blood flow through the arterial system.

Small muscular arteries that are 100 μm or less in diameter branch to form smaller arterioles (20–30 μm in diameter). In some tissues, blood from the arterioles can enter the venules directly through **arteriovenous anastomoses** *(ah-nas″to-mo′sis)*. In most cases, however, blood from arterioles passes into capillaries (fig. 20.22). Capillaries are the narrowest of blood vessels (7–10 μm in diameter), and serve as the "functional units" of the circulatory system in which exchanges of gases and nutrients between the blood and the tissues occur.

Capillaries

The arterial system branches extensively (table 20.4) to deliver blood to over a billion capillaries in the body. The extensiveness of these branchings is indicated by the fact that all tissue cells are located within a distance of only 60–80 μm of a capillary and by the fact that capillaries provide a total surface area of 1,000 square miles for exchanges between blood and tissue fluid.

Despite their large number, capillaries contain only about 250 ml of blood at any time, out of a total blood volume of about 5,000 ml (most is contained in the venous system). The amount of blood flowing through a particular capillary bed is determined in part by the action of the **precapillary sphincter muscles.** These muscles allow only 5%–10% of the capillary beds in skeletal muscles, for example, to be open at rest. Blood flow to an organ is regulated by the action of these precapillary sphincters and by the degree of resistance to blood flow (due to constriction or dilation) provided by the small arteries and arterioles in the organ.

Unlike the vessels of the arterial and venous systems, the walls of capillaries are composed of only one cell layer—a simple squamous epithelium, or endothelium (fig. 20.23). The absence of smooth muscle and connective tissue layers permits a more rapid rate of transport of materials between the blood and the tissues.

anastomosis: Gk. *anastomoo,* to furnish with a mouth (coming together)

Figure 20.22. The microcirculation. Metarterioles (arteriovenous anastomoses) provide a path of least resistance between arterioles and venules. Precapillary sphincter muscles regulate the flow of blood through the capillaries.

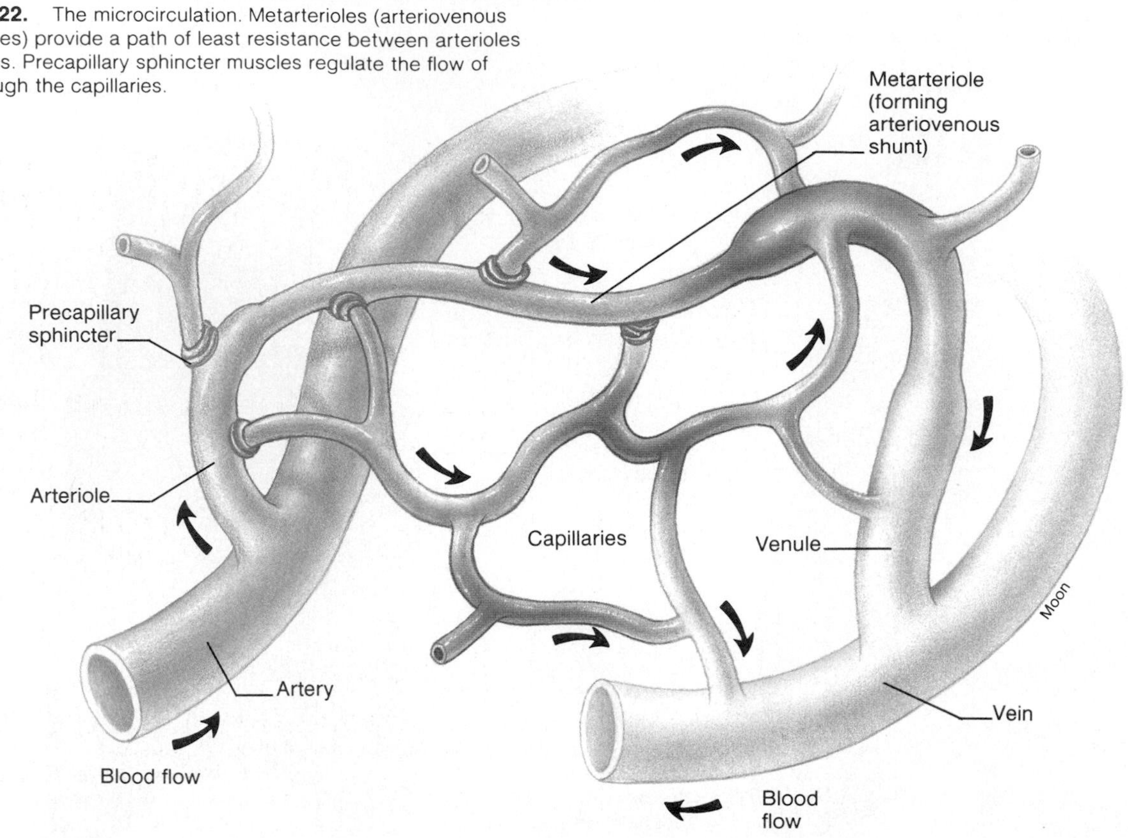

Table 20.4	Characteristics of the vascular supply to the mesenteries in a dog (pattern is similar in a human)				
Kind of vessel	**Diameter (mm)**	**Number**	**Total cross-sectional area (cm^2)**	**Length (cm)**	**Total volume (cm^3)**
Aorta	10	1	0.8	40	30
Large arteries	3	40	3.0	20	60
Main artery branches	1	600	5.0	10	50
Terminal branches	0.6	1,800	5.0	1	25
Arterioles	0.02	40,000,000	125	0.2	25
Capillaries	0.008	1,200,000,000	600	0.1	60
Venules	0.03	80,000,000	570	0.2	110
Terminal veins	1.5	1,800	30	1	30
Main venous branches	2.4	600	27	10	270
Large veins	6.0	40	11	20	220
Vena cava	12.5	1	1.2	40	50
					930

Types of Capillaries. Different organs have different types of capillaries, which are distinguished by significant differences in structure. In terms of their endothelial lining, these capillary types include those that are continuous, those that are discontinuous, and those that are fenestrated *(fen'es-trāt''ed)*.

Continuous capillaries are those in which adjacent endothelial cells are closely joined together. These are found in muscles, lungs, adipose tissue, and in the central nervous system. Continuous capillaries in the CNS lack intercellular channels; this fact contributes to the blood-brain barrier. Continuous capillaries in other organs have narrow intercellular channels (about 40–45 Å in width) that allow the passage of molecules other than protein between the capillary blood and tissue fluid (fig. 20.24).

Examination of endothelial cells with an electron microscope has revealed the presence of pinocytotic vesicles, which suggests that the intracellular transport of material

Figure 20.23. An electron micrograph of a capillary in the heart. Notice the thin intercellular channel (middle left) and the fact that the capillary wall is only one cell thick. Arrows show some of the many pinocytic vesicles.

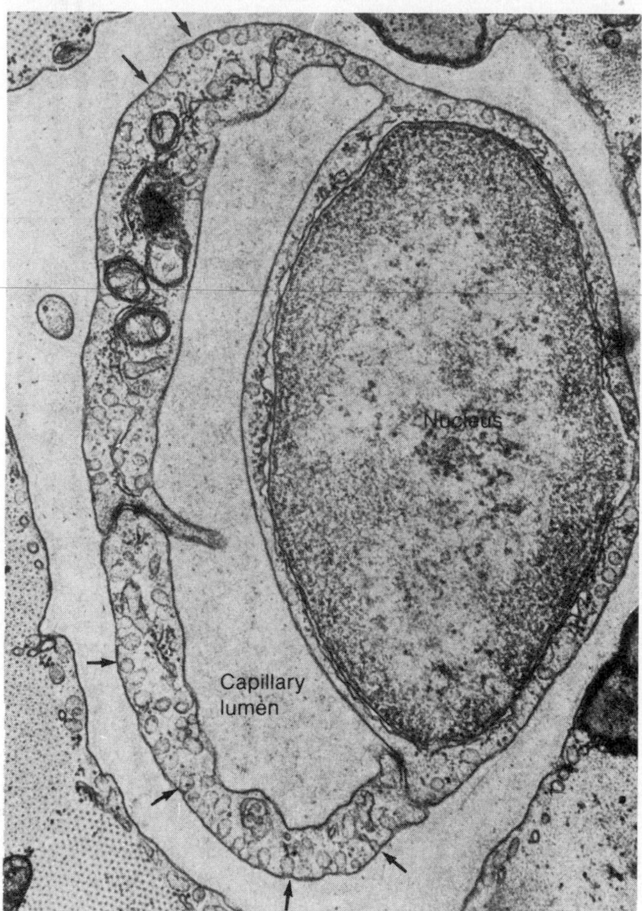

may occur across the capillary walls. This type of transport appears to be the only mechanism of capillary exchange available within the central nervous system and may account, in part, for the selective nature of the blood-brain barrier.

The kidneys, endocrine glands, and intestines have *fenestrated capillaries,* characterized by wide intercellular pores (800–1,000 Å) that are covered by a layer of mucoprotein, which may serve as a diaphragm. In the bone marrow, liver, and spleen, the distance between endothelial cells is so great that these *discontinuous capillaries* appear as *sinusoids* (little cavities) in the organ.

Veins

The average pressure in the veins is only 2 mm Hg (millimeters of mercury), compared to a much higher average arterial pressure of about 100 mm Hg. These pressures represent the hydrostatic pressure that the blood exerts on the walls of the vessels, and the numbers indicate the differences from atmospheric pressure.

The low venous pressure is insufficient to return blood to the heart, particularly from the lower limbs. Veins, however, pass between skeletal muscle groups that produce a massaging action as they contract (fig. 20.25). As the veins are squeezed by contracting skeletal muscles, a

Figure 20.24. Diagrams of continuous, fenestrated, and discontinuous capillaries as they appear in the electron microscope. This classification is derived from the continuity of the endothelial layer. (Dark circles in the cytoplasm indicate pinocytotic vesicles.)

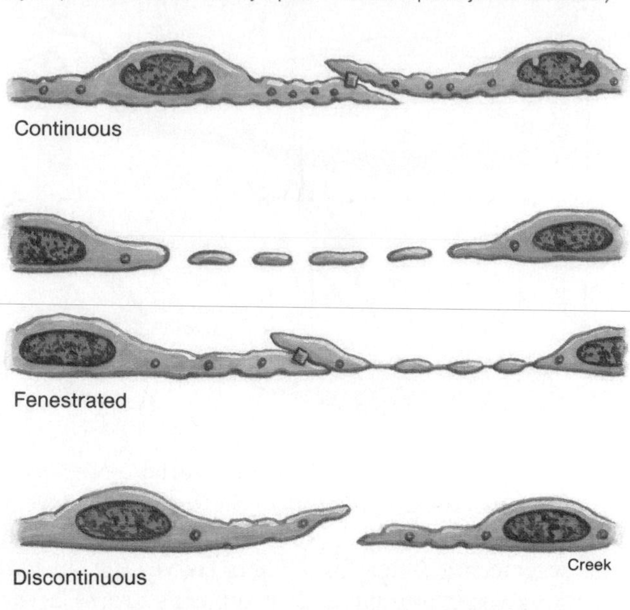

Figure 20.25. The action of the one-way venous valves. Contraction of skeletal muscles helps to pump blood toward the heart, but is prevented from pushing blood away from the heart by closure of the venous valves.

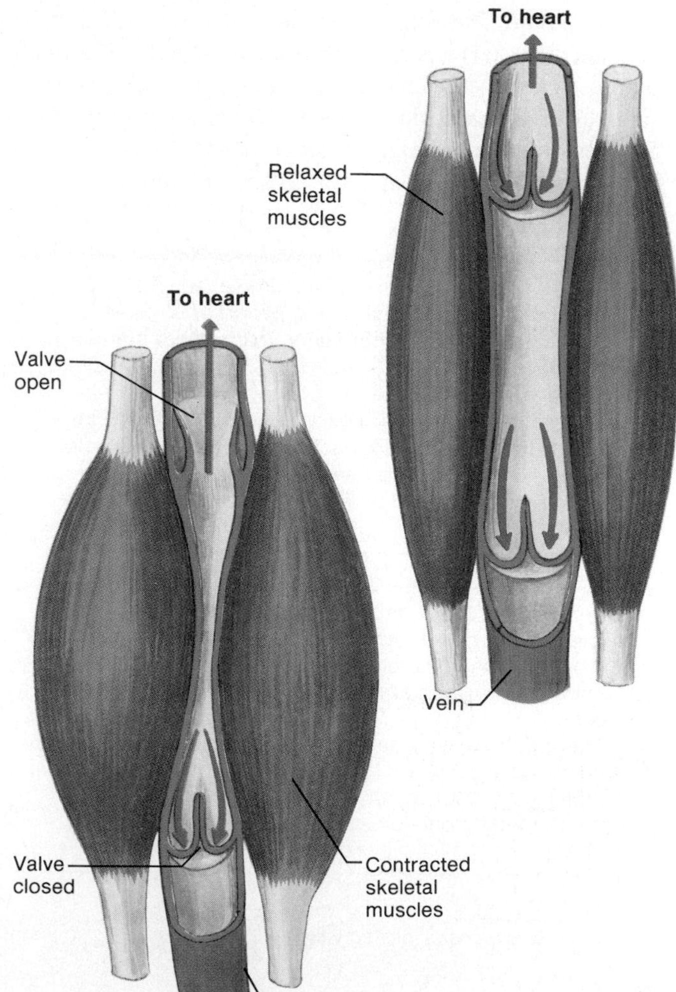

Figure 20.26. A classical demonstration by William Harvey of the existence of venous valves that prevent the flow of blood away from the heart.

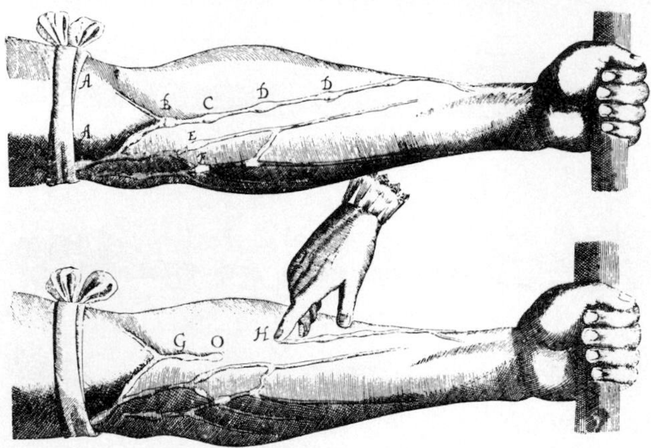

one-way flow of blood to the heart is insured by the presence of **venous valves.** The ability of these valves to prevent the flow of blood away from the heart was demonstrated in the seventeenth century by William Harvey (fig. 20.26). After applying a tourniquet to a subject's arm, Harvey found that he could push the blood in a bulging vein toward the heart but not in the reverse direction.

The effect of the massaging action of skeletal muscles on venous blood flow is often described as the **skeletal muscle pump.** The rate of venous return to the heart is dependent, in large part, on the action of skeletal muscle pumps. When these pumps are less active, as when a person stands still or is bedridden, blood accumulates in the veins and causes them to bulge. When a person is more active, blood returns to the heart at a faster rate and less is left in the venous system.

The accumulation of blood in the veins of the legs over a long period of time, as may occur in people with occupations that require standing still all day, can cause the veins to stretch to the point where the venous valves are no longer efficient. This can produce *varicose (var'i-kōs) veins.* During walking the movements of the foot activate the soleus muscle pump. This effect can be produced in bedridden people by upward and downward manipulations of the feet.

1. Describe the basic structural pattern of arteries and veins. Describe how arteries and veins differ in structure and how these differences contribute to the resistance function of arteries and the capacitance function of veins.
2. Describe the functional significance of the "skeletal muscle pump," and explain the action of venous valves.
3. Explain the functions of capillaries, and describe the structural differences between capillaries in different organs.

varicose: L. *varicosus,* dilated vein

Principal Arteries of the Body

The aorta ascends from the left ventricle, arches to the left, and descends through the torso. Branches of the aorta carry oxygenated blood to all the cells of the body.

Objective 17. In the form of a flowchart, list the arterial branches of the ascending aorta and aortic arch.
Objective 18. Describe the arterial supply to the brain.
Objective 19. Describe the arterial pathways that supply the upper extremities.
Objective 20. Describe the major arteries to the thorax, abdomen, and lower extremities.

Contraction of the left ventricle forces oxygenated blood into the arteries of the systemic circulation. The principal arteries of the body are shown in figure 20.27. They will be described by region and identified in order from largest to smallest, or as the blood flows through the system. The major systemic artery is the **aorta,** from which all the primary arteries arise.

Arch of the Aorta

The systemic vessel that ascends from the left ventricle of the heart is called the **ascending aorta.** The right and left **coronary arteries** serving the myocardium of the heart with blood are the only branches that arise off the ascending aorta. The aorta arches to the left and posteriorly over the pulmonary arteries as the **aortic arch** (fig. 20.28). Three vessels arise from the aortic arch: the **brachiocephalic** *(brak''e-o-sĕ-fal'ik)* **(innominate),** the **left common carotid** *(kah-rot'id),* and the **left subclavian.**

The brachiocephalic is the first vessel to branch from the aortic arch and, as its name suggests, supplies blood to the tissue of the arm and head on the right side of the body. It is a short vessel rising superiorly through the mediastinum to a point near the junction of the sternum and the right clavicle. There it bifurcates into the **right common carotid,** which extends to the right side of the neck and head, and the **right subclavian,** which carries blood to the right upper extremity. Note that the brachiocephalic artery is unpaired; the left common carotid and subclavian arteries, unlike the right, arise directly from the aorta.

The remaining two branches from the arch of the aorta are the left common carotid and the left subclavian arteries. The left common carotid transports blood to the left side of the neck and head, and the left subclavian supplies the left upper extremity.

carotid: Gk. *karotikos,* stupefying (a state that can be induced by finger pressure in a region of the carotid called the carotid sinus)

Figure 20.27. Principal arteries of the body (a. = artery).

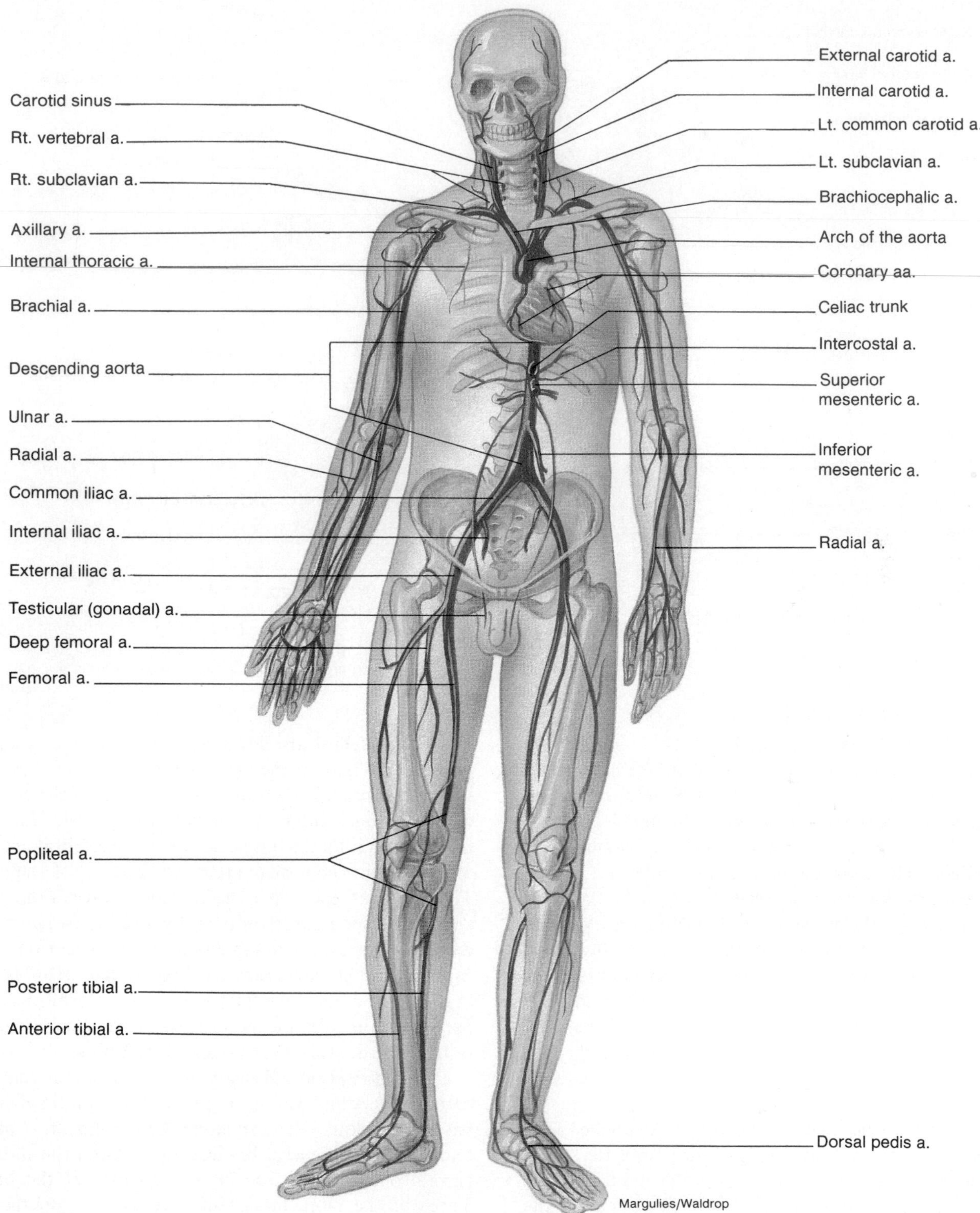
Carotid sinus
Rt. vertebral a.
Rt. subclavian a.
Axillary a.
Internal thoracic a.
Brachial a.
Descending aorta
Ulnar a.
Radial a.
Common iliac a.
Internal iliac a.
External iliac a.
Testicular (gonadal) a.
Deep femoral a.
Femoral a.
Popliteal a.
Posterior tibial a.
Anterior tibial a.
External carotid a.
Internal carotid a.
Lt. common carotid a.
Lt. subclavian a.
Brachiocephalic a.
Arch of the aorta
Coronary aa.
Celiac trunk
Intercostal a.
Superior mesenteric a.
Inferior mesenteric a.
Radial a.
Dorsal pedis a.
Margulies/Waldrop

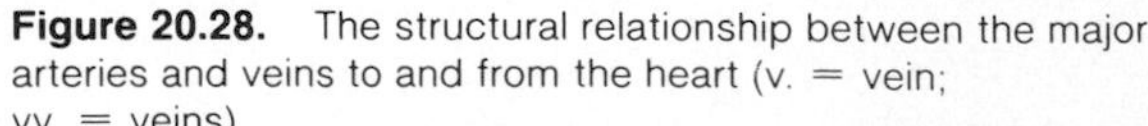

Figure 20.28. The structural relationship between the major arteries and veins to and from the heart (v. = vein; vv. = veins).

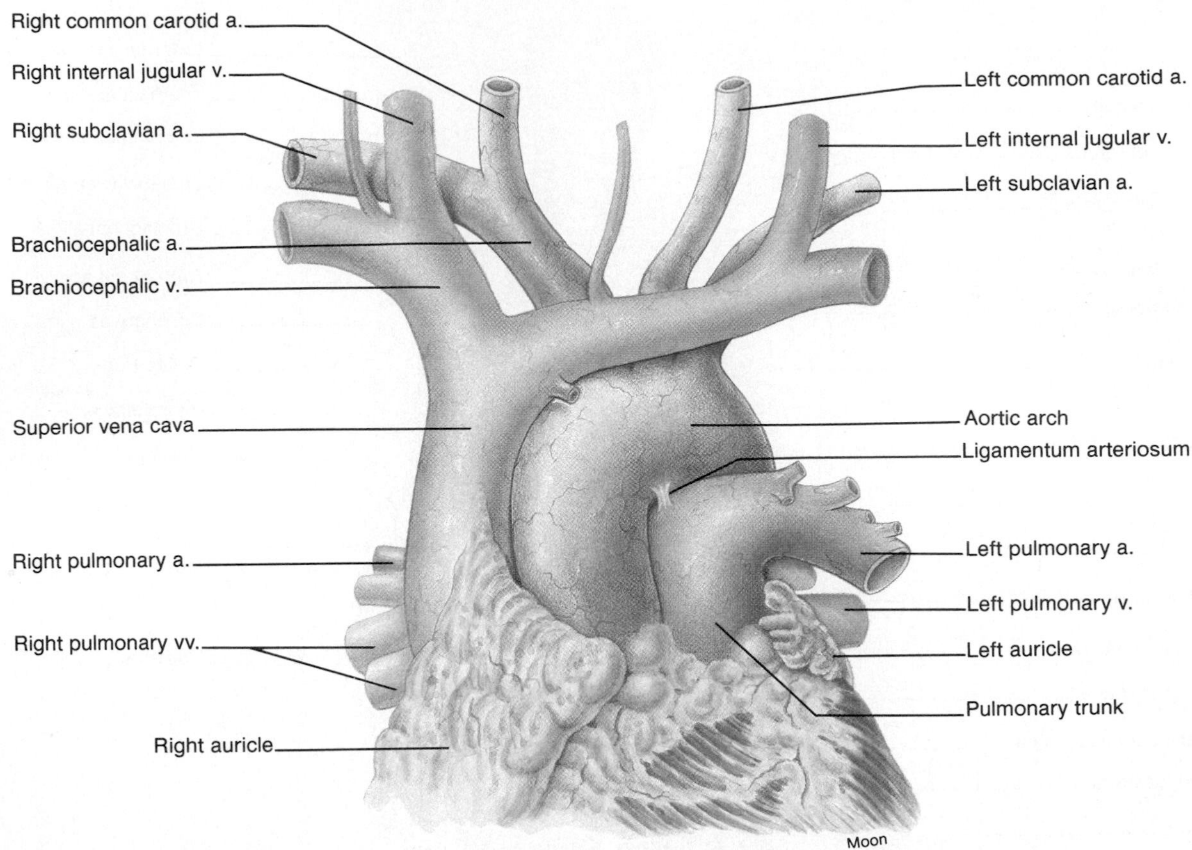

Arteries of the Neck and Head

The common carotids course upward in the neck along either side of the trachea (fig. 20.29). Several small vessels arise from the common carotid to supply blood to the larynx, thyroid, anterior neck muscles, and lymph glands of the neck. The common carotid bifurcates into the **internal** and **external carotid arteries** slightly below the angle of the mandible. By pressing gently in this area you can detect your pulse. At the base of the internal carotid, near the bifurcation, is a slight dilation called the **carotid sinus.** The carotid sinus contains *baroreceptors,* which monitor blood pressure, and *chemoreceptors (ke''mo-re-sep'tor)* within the **carotid body,** which respond to chemical changes in the blood.

Blood Supply to the Brain. The brain is supplied with arterial blood that arrives through four vessels that eventually unite on the inferior aspect of the brain surrounding the pituitary gland (fig. 20.30). The four vessels are the paired internal carotid arteries and the paired vertebral arteries. The value of having four separate vessels that anastomose (come together) at one location is that if one becomes occluded, the three alternate routes may insure an adequate blood supply to the brain.

The **vertebral arteries** arise from the subclavian arteries at the base of the neck (see fig. 20.29). They pass superiorly through the transverse foramina of the cervical vertebrae and enter the skull through the foramen magnum. Within the braincase, the two vertebral arteries unite to form the **basilar artery** at the level of the pons. The basilar ascends along the inferior surface of the brain stem and inner ear; it terminates by forming two **posterior cerebral arteries,** which supply the posterior portion of the cerebrum. The **posterior communicating arteries** are branches that arise from the posterior cerebral arteries and participate in forming the **arterial circle** around the pituitary gland, also called the **circle of Willis.**

The **internal carotid artery** branches from the common carotid artery and ascends in the neck until it reaches the base of the skull, where it enters the carotid canal of the temporal bone. Several branches arise from the internal carotid once it is on the inferior surface of the brain. Three of the more important ones are the **ophthalmic** *(of'thal'mik)* **artery,** which supplies the eye and associated structures, and the **anterior** and **middle cerebral arteries,** which provide blood to the cerebrum. The internal carotids are connected to the posterior cerebral arteries at the arterial circle.

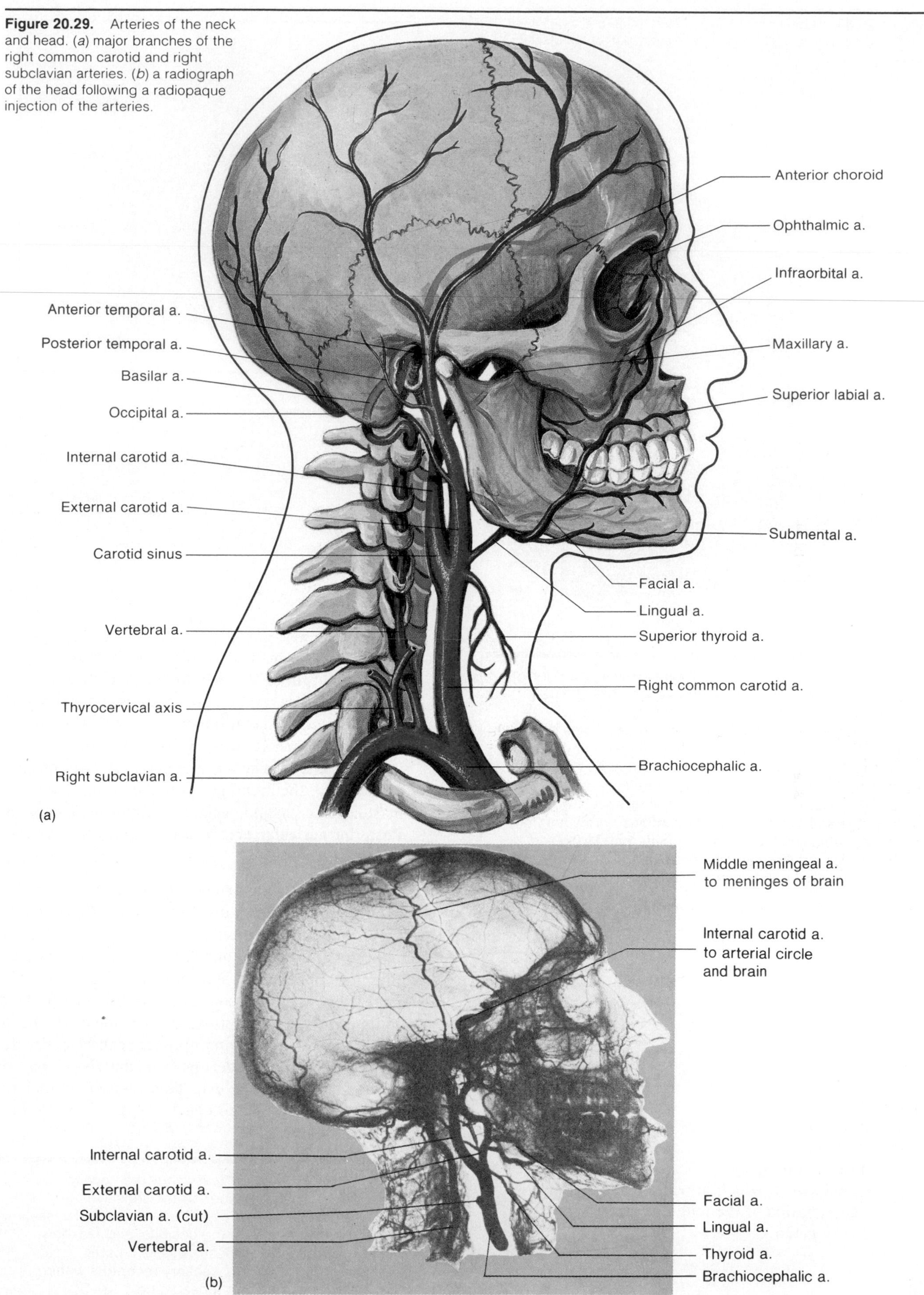

Figure 20.29. Arteries of the neck and head. (*a*) major branches of the right common carotid and right subclavian arteries. (*b*) a radiograph of the head following a radiopaque injection of the arteries.

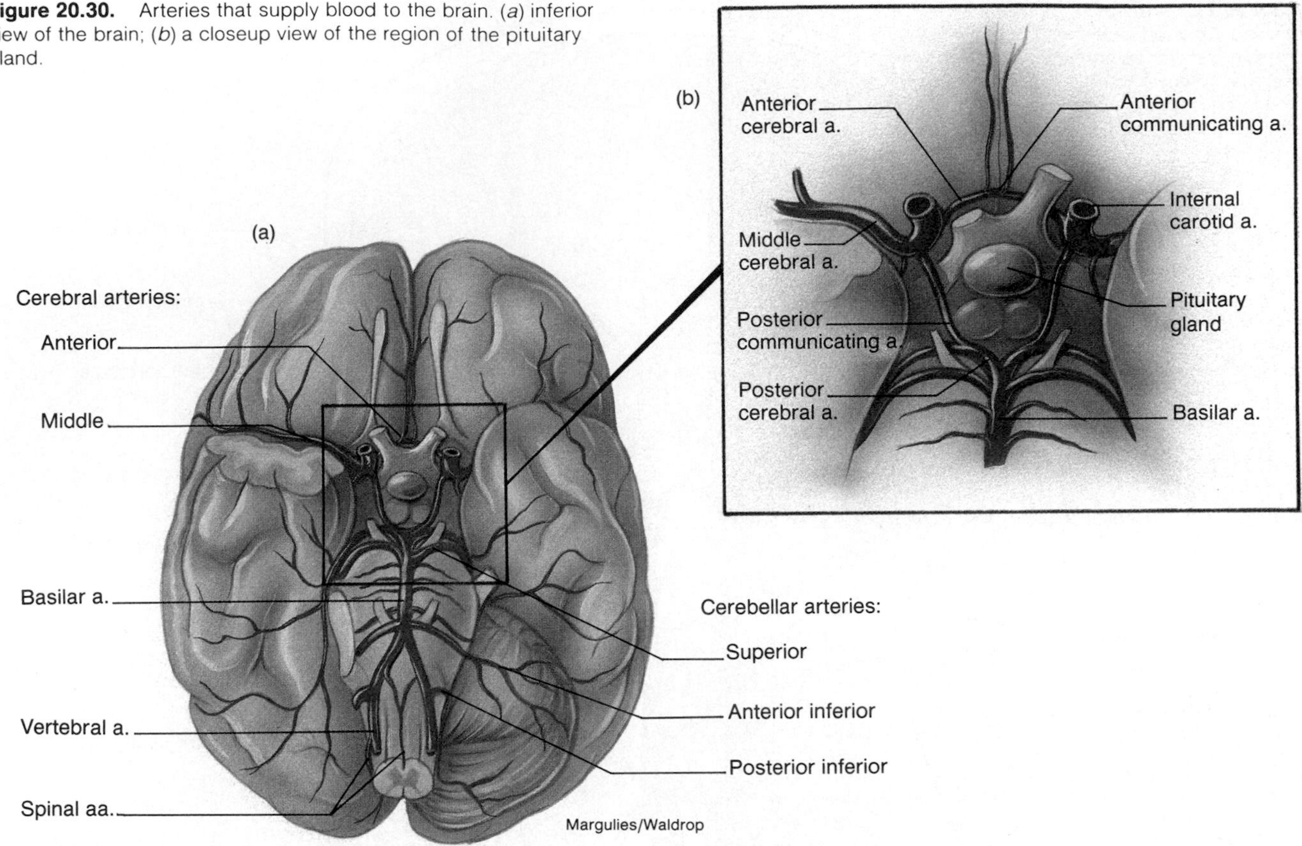

Figure 20.30. Arteries that supply blood to the brain. (*a*) inferior view of the brain; (*b*) a closeup view of the region of the pituitary gland.

Capillaries within the pituitary gland receive both arterial and venous blood. The venous blood arrives from venules immediately superior to the pituitary, which drain capillaries in the hypothalamus of the brain. This arrangement of two capillary beds in series—where the second capillary bed receives venous blood from the first—is called a *portal system.* The venous blood that travels from the hypothalamus to the pituitary contains hormones from the hypothalamus that help regulate pituitary gland hormone secretion, as described in chapter 19.

External Carotid Artery. The external carotid artery gives off several branches as it extends upward along the side of the neck and head (see fig. 20.29). The names of these branches are determined by the areas or structures that they serve. The principal vessels that arise from the external carotid are the following.

1. **Superior thyroid artery,** which serves the muscles of the hyoid region, the larynx and vocal cords, and the thyroid gland.
2. **Ascending pharyngeal artery** (not seen in illustration), which serves the pharyngeal area and various lymph nodes.
3. **Lingual artery,** which provides extensive vascularization of the tongue and sublingual salivary gland.
4. **Facial artery,** which traverses a notch on the inferior margin of the mandible to serve the pharyngeal area, palate, chin, lips, and nasal region. The facial artery is an important vessel to apply pressure to when controlling bleeding from the face.
5. **Occipital artery,** which serves the posterior portion of the scalp, the meninges over the brain, the mastoid process, and certain posterior neck muscles.
6. **Posterior auricular artery,** which serves the ear and scalp over the ear.

The external carotid artery terminates at a level near the mandibular condyle by dividing into **maxillary** and **superficial temporal arteries.** The maxillary artery gives off branches to the teeth and gums, the muscles of mastication, nasal cavity, eyelids, and meninges. The superficial temporal artery supplies blood to the parotid salivary gland and to the superficial structures on the side of the head. Pulsations through the temporal artery can be easily detected by placing your fingertips immediately in front of the ear at the level of the eye. This vessel is frequently used by anesthesiologists to check a patient's pulse rate during surgery.

Headaches are usually caused by vascular pressure on the sensitive meninges covering the brain. The two principal vessels serving the meninges are the occipital and maxillary arteries. Vasodilation of these vessels creates excessive pressure on the sensory receptors within the meninges, resulting in a headache.

Figure 20.31. Arteries of the upper extremity. (*a*) a radiograph of the forearm and hand following a radiopaque injection of the arteries. (*b*) an anterior view of major arteries.

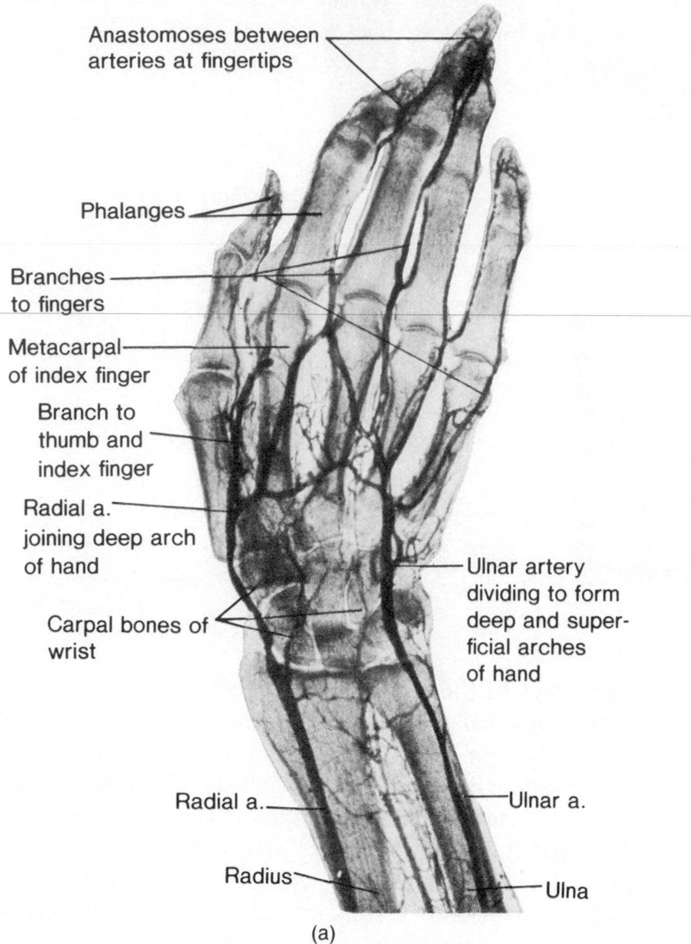

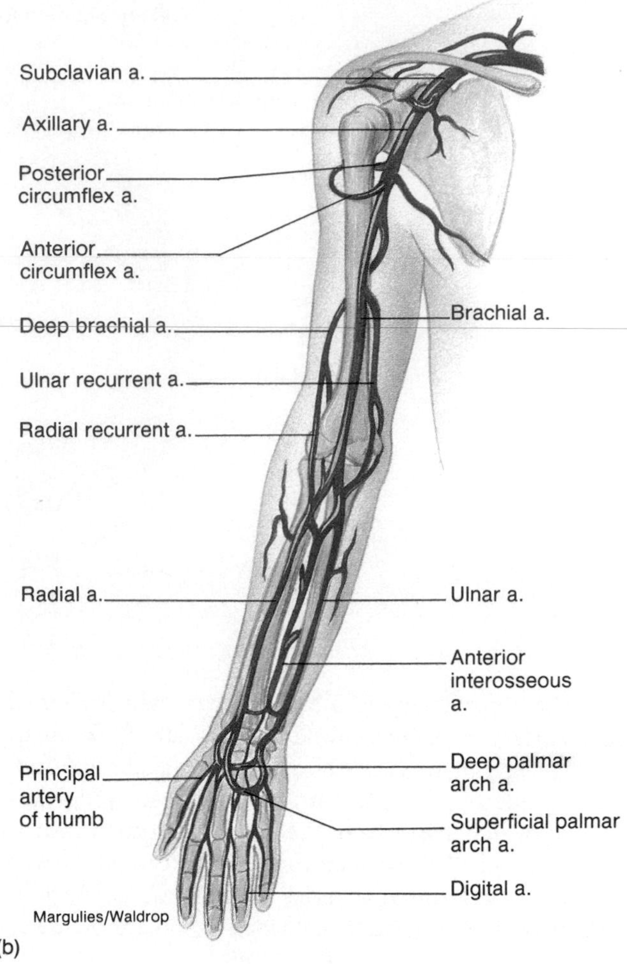

Arteries of the Upper Extremity

The right subclavian artery branches off the brachiocephalic, and the left subclavian artery arises directly from the arch of the aorta (see fig. 20.27). The **subclavian** *(sub-cla've-an)* **artery** passes laterally deep to the clavicle, carrying blood toward the arm (fig. 20.31). The pulsations of the subclavian artery can be detected by pressing firmly on the tissue just above the medial portion of the clavicle. From each subclavian artery arises a **vertebral artery** that carries blood to the brain (already described), a short **thyrocervical trunk** that serves the thyroid gland, trachea, and larynx, and an **internal thoracic (mammary) artery** that descends into the thorax to serve the thoracic wall, thymus, and pericardium. A branch of the internal thoracic artery supplies blood to the muscles and tissues (mammary glands) of the anterior thorax.

The subclavian artery becomes the **axillary** *(ak'sĭ-lar''e)* **artery** as it passes into the axillary region. The axillary artery is that portion of the major artery of the upper extremity between the outer border of the first rib and the lower border of the teres major muscle. Several small branches arise from the axillary artery and supply blood to the tissues of the upper thorax and shoulder region.

The **brachial** *(bra'ke-al)* **artery** is the continuation of the axillary artery through the brachial region. The brachial artery courses on the medial side of the humerus, where it is a major pressure point and the most common site for determining blood pressure. A **deep brachial artery** branches from the brachial artery and curves posteriorly near the radial nerve to supply the triceps muscle. Two additional branches from the brachial, the **anterior** and **posterior circumflex arteries,** form a continuous ring of vessels around the proximal and distal portions of the arm, respectively.

Just proximal to the cubital fossa, the brachial bifurcates into the **radial** and **ulnar arteries,** which supply blood to the forearm and a portion of the hand and digits. The radial artery courses down the lateral, or radial, side of the arm, where it sends numerous small branches to the muscles of the forearm. The **radial recurrent artery** serves the region of the elbow and is the first and largest off of the radial artery. The radial artery is important as a site to record the pulse near the wrist.

Figure 20.32. Arteries that serve the thoracic wall.

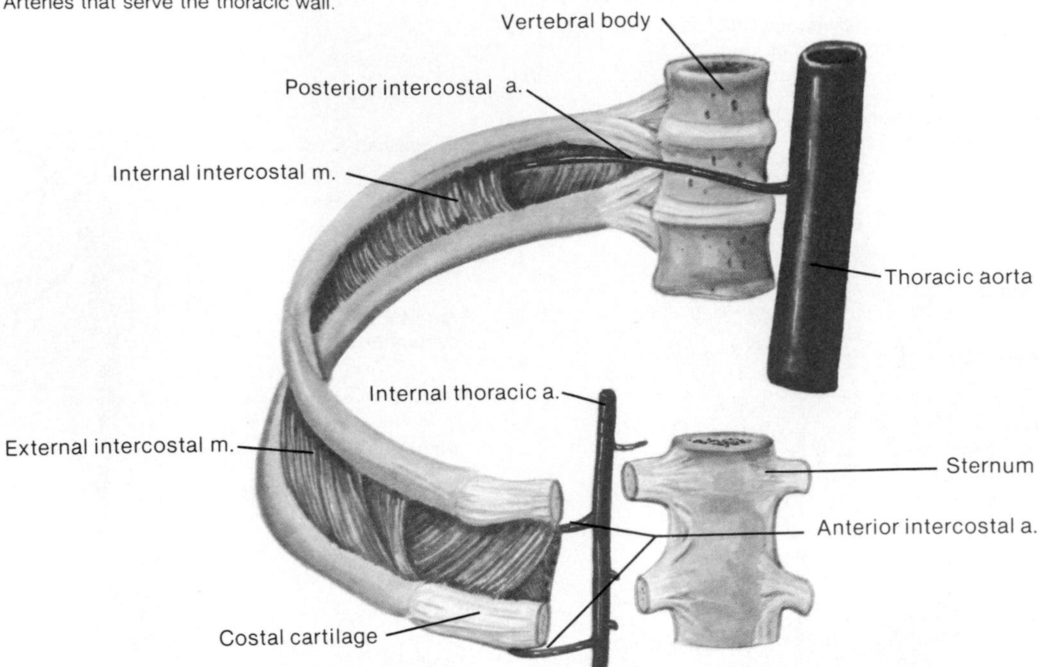

The ulnar artery is a bit larger in caliber than the radial artery. It extends down the ulnar side of the forearm and gives off many small branches to the muscles on the medial side of the forearm. It too has an initial large branch, which arises from the proximal portion near the elbow and is called the **ulnar recurrent artery.** At the wrist, the ulnar and radial arteries anastomose to form the **palmar arch** in the hand, from which **digital arteries** extend into the fingers.

Branches of the Thoracic Aorta

The **thoracic aorta** is a continuation of the arch of the aorta as it descends through the thoracic cavity to the diaphragm. This large vessel gives off branches to the organs and muscles of the thoracic region. These branches include **pericardial arteries,** going to the pericardium of the heart; **bronchial arteries** for systemic circulation to the lungs; **esophageal** *(ĕ-sof″ah-je′al)* **arteries,** going to the esophagus as it passes through the mediastinum; segmental **intercostal arteries,** serving the intercostal muscles and structures of the wall of the thorax (fig. 20.32); and **phrenic** *(fren′ik)* **arteries,** supplying blood to the diaphragm.

These vessels are summarized according to their location and function in table 20.5.

Branches of the Abdominal Aorta

The **abdominal aorta** is the segment of the aorta between the diaphragm and the level of the fourth lumbar vertebra, where it divides into the right and left **common iliac** *(il′e-ak)* **arteries.** The first branch of the abdominal aorta is called the **celiac** *(se′le-ak)* **artery.** It is a short, thick trunk that arises anteriorly just below the diaphragm. The celiac divides immediately into three arteries: the **splenic** *(splen′ik),* going to the spleen, the **left gastric,** going to the stomach, and the **hepatic** *(hĕ-pat′ik),* going to the liver (fig. 20.33).

The **superior mesenteric artery** is another unpaired vessel and arises anteriorly from the abdominal aorta just below the celiac. The superior mesenteric supplies blood to the small intestine (except for a portion of the duodenum), the cecum, the appendix, the ascending colon, and the transverse colon.

The next major vessels to arise from the abdominal aorta are the paired **renal arteries,** which carry blood to the kidneys. Smaller **suprarenal arteries** are located just above the renal arteries and serve the suprarenal (adrenal) glands. The **testicular** (internal spermatic) **arteries** in the male and the **ovarian arteries** in the female are small, paired vessels that arise from the abdominal aorta just below the renals and serve the gonads.

The **inferior mesenteric artery** is the last major branch of the abdominal aorta. It is an unpaired anterior vessel that arises just before the iliac bifurcation. The inferior mesenteric supplies blood to the descending colon and rectum.

Figure 20.33. An anterior view of the abdominal aorta and its principal branches. (*a*) the abdominal viscera have been removed; (*b*) the abdominal viscera are intact.

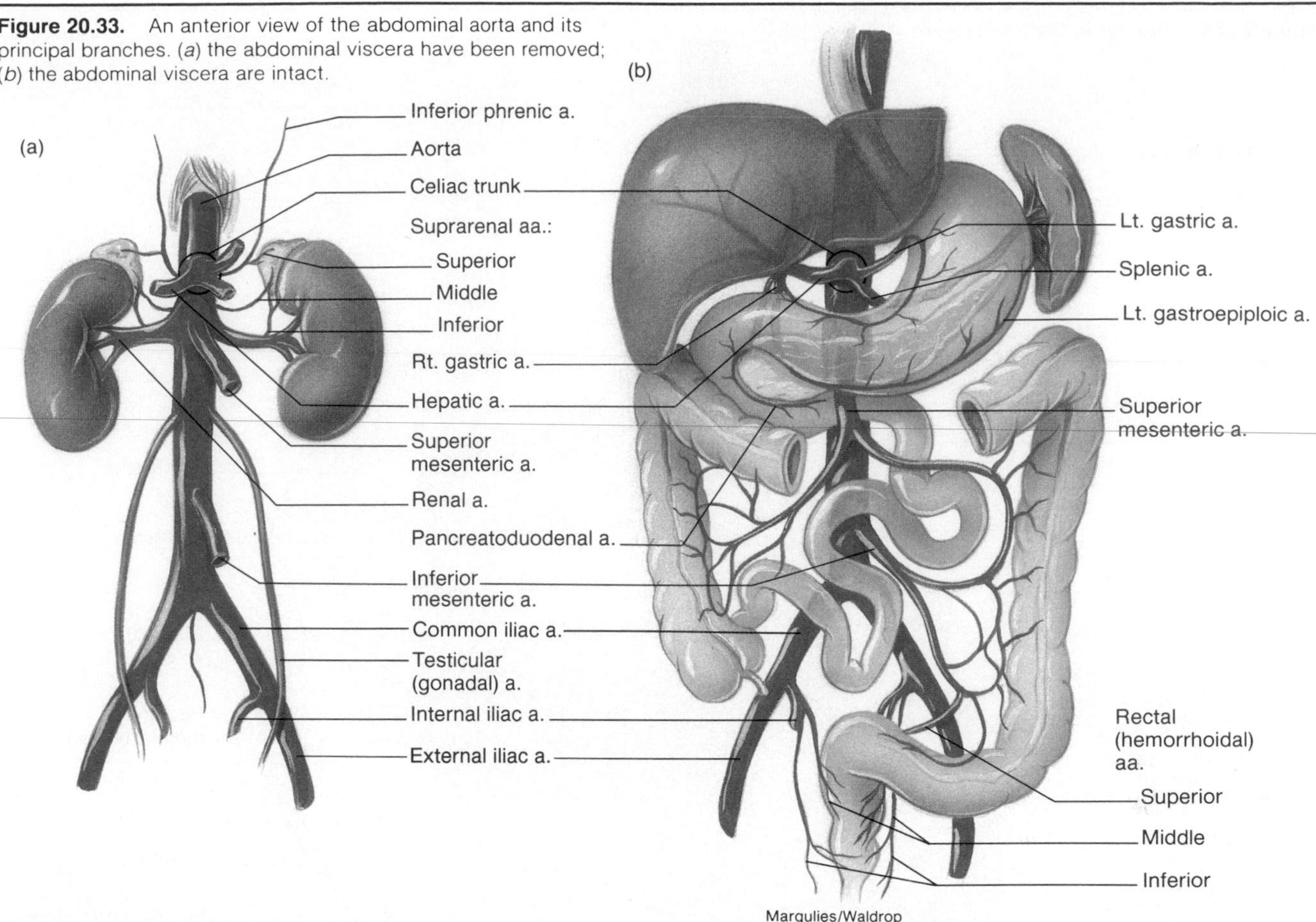

Table 20.5 Segments and branches of the aorta

Segment of aorta	Arterial branch	General region or organ served	Segment of aorta	Arterial branch	General region or organ served
Ascending aorta	Right and left coronary	Heart	Abdominal aorta	Celiac	
				Common hepatic	Liver, upper pancreas, duodenum
Arch of aorta	Brachiocephalic			Left gastric	Stomach, esophagus
	Right common carotid	Right side of head and neck		Splenic	Spleen, pancreas, stomach
	Right subclavian	Right shoulder and right upper extremity		Superior mesenteric	Small intestine, pancreas, cecum, ascending and transverse colons
	Left common carotid	Left side of head and neck			
	Left subclavian	Left shoulder and left upper extremity		Suprarenals	Suprarenal (adrenal) glands
Thoracic aorta	Pericardials	Pericardium of heart		Renals	Kidneys
	Intercostals	Intercostal and thoracic muscles, pleurae		Gonadals	
				Testiculars	Testes
				Ovarians	Ovaries
	Superior phrenics	Diaphragm		Inferior mesenteric	Transverse, descending, and sigmoid colons; rectum
	Bronchials	Bronchi of lungs			
	Esophageals	Esophagus			
	Inferior phrenics	Inferior surface of diaphragm		Common iliacs	
				External iliacs	Lower extremities
				Internal iliacs (hypogastrics)	Genital organs, gluteal muscles

Figure 20.34. Arteries of the pelvic region.

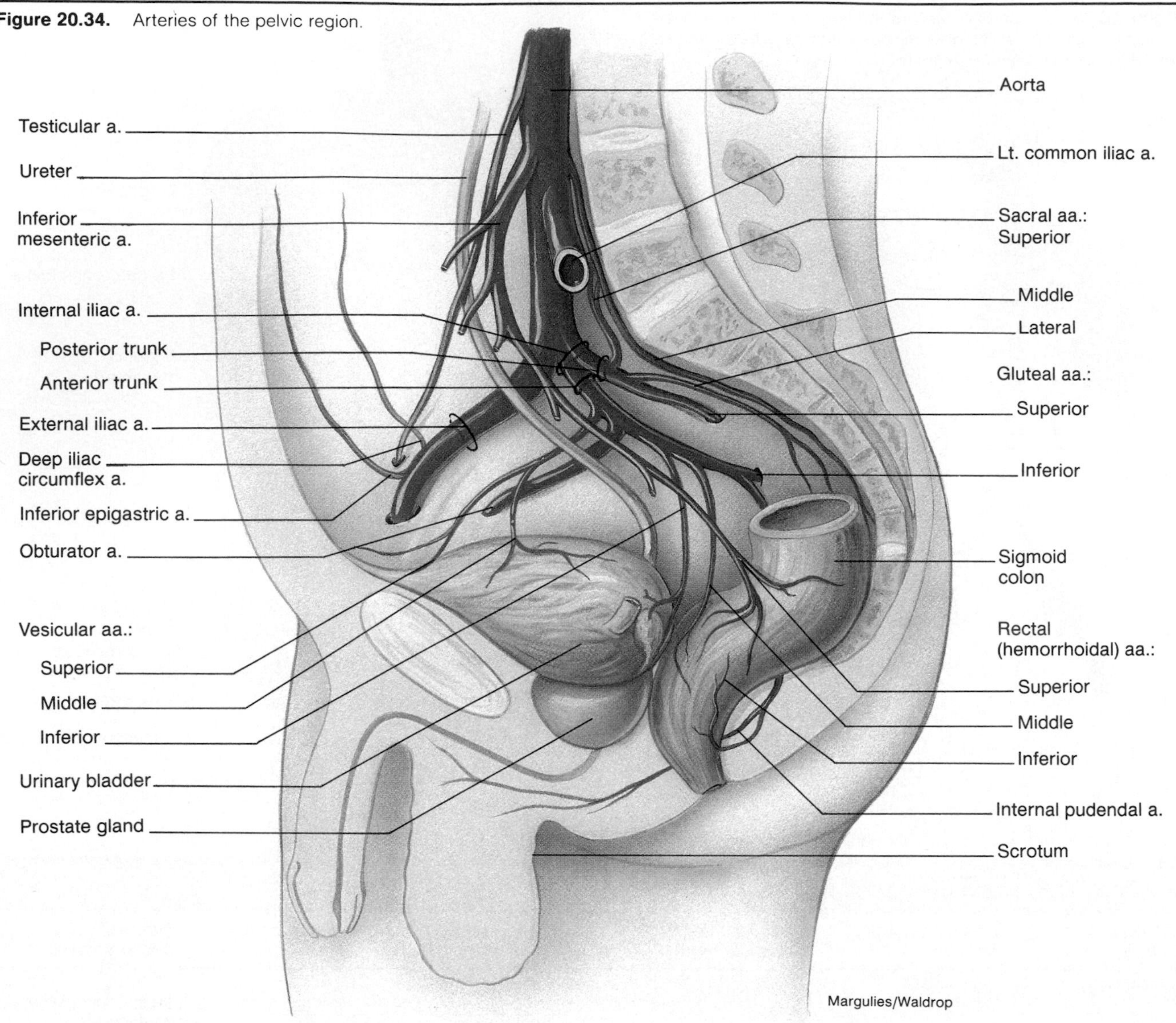

Several **lumbar arteries** branch posteriorly from the abdominal aorta throughout its length and serve the muscles and the spinal cord in the lumbar region. In addition, an unpaired **middle sacral artery** (fig. 20.34) arises from the posterior terminal portion of the abdominal aorta to supply the sacrum and coccyx.

Arteries of the Pelvis and Lower Extremities

The aorta terminates in the posterior pelvic area as it bifurcates into the right and left common iliac arteries. These vessels pass downward approximately 5 cm on their respective sides and terminate by dividing into the **internal** and **external iliac arteries.**

The internal iliac has extensive branches to supply arterial blood to the gluteal muscles and the organs of the pelvic region (fig. 20.34). The wall of the pelvis is served by the **iliolumbar** and **lateral sacral arteries.** The internal visceral organs of the pelvis are served by the **middle rectal** and the **superior, middle,** and **inferior vesicular arteries** to the urinary bladder. In addition, in females **uterine** and **vaginal** *(vaj'ĭ-nal)* **arteries** branch from the internal iliacs to serve the reproductive organs. The muscles of the buttock are served by the **superior** and **inferior gluteal arteries.** Some of the upper medial thigh muscles are supplied with blood from the **obturator artery.** The **internal pudendal artery** of the internal iliac serves the external genitalia of male and female and is an important vessel in sexual activity. Erection of the penis in the male and corresponding vascular engorgement of the genitalia of the female are vascular phenomena controlled by the autonomic nervous system.

The external iliac artery passes out of the pelvic cavity beneath the inguinal ligament (fig. 20.34) and becomes the **femoral artery.** Two branches arise from the external iliac, however, before it passes beneath the inguinal ligament. An **inferior epigastric artery** branches immediately from the external iliac and passes superiorly to supply

Figure 20.35. Arteries of the right lower extremity. (*a*) an anterior view; (*b*) a posterior view.

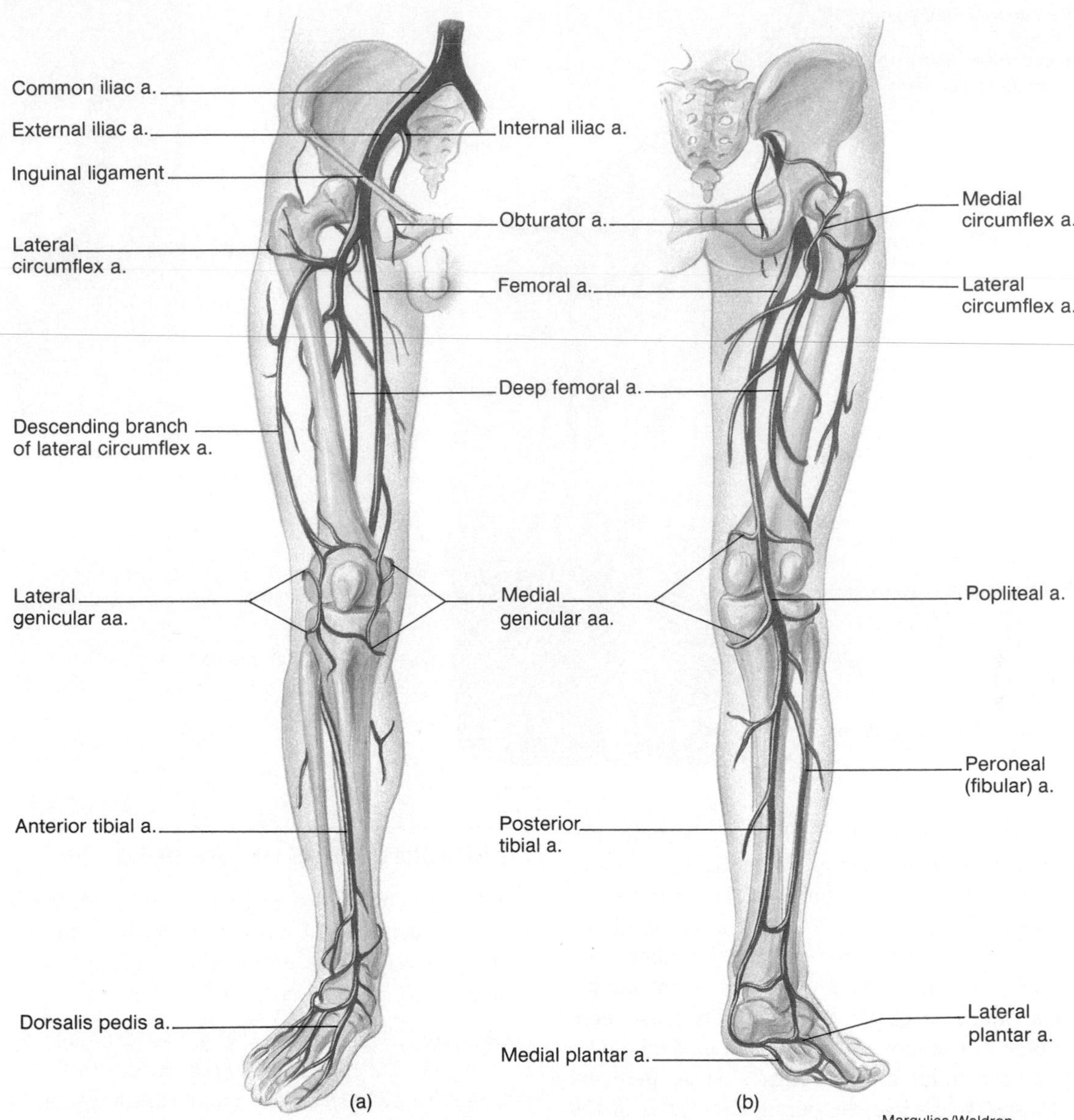

the skin and muscles of the abdominal wall. The **deep circumflex iliac artery** is a small branch that extends laterally to supply the muscles attached to the iliac fossa.

The femoral artery passes through an area called the **femoral triangle** on the upper medial portion of the thigh (figs. 20.35, 20.36). At this point, the femoral artery is close to the surface and not only can be palpated but is an important pressure point. Several vessels arise from the femoral to serve the thigh region. The largest of these, the **deep femoral artery,** passes posteriorly to serve the hamstring muscles. The **lateral** and **medial femoral circumflex arteries** encircle the proximal end of the femur and serve muscles in this region. The femoral artery becomes the **popliteal** *(pop-lit'e-al)* **artery** as it passes across the posterior aspect of the knee.

Hemorrhage can be a serious problem in many accidents. Therefore, one should know the pressure points where the arterial blood flow can be curtailed to prevent a victim from bleeding to death. The pressure points for the appendages are the brachial artery on the medial side of the arm and the femoral artery in the groin. Firmly applied pressure to these regions greatly diminishes the flow of blood to traumatized areas below. A tourniquet may have to be applied if bleeding is severe enough to endanger life.

The popliteal artery supplies small branches to the knee joint and then divides into an **anterior tibial artery** and **posterior tibial artery** (fig. 20.35). These vessels traverse the anterior and posterior aspects of the leg, respectively, providing blood to the muscles of these regions and to the foot.

Figure 20.36. The femoral triangle.

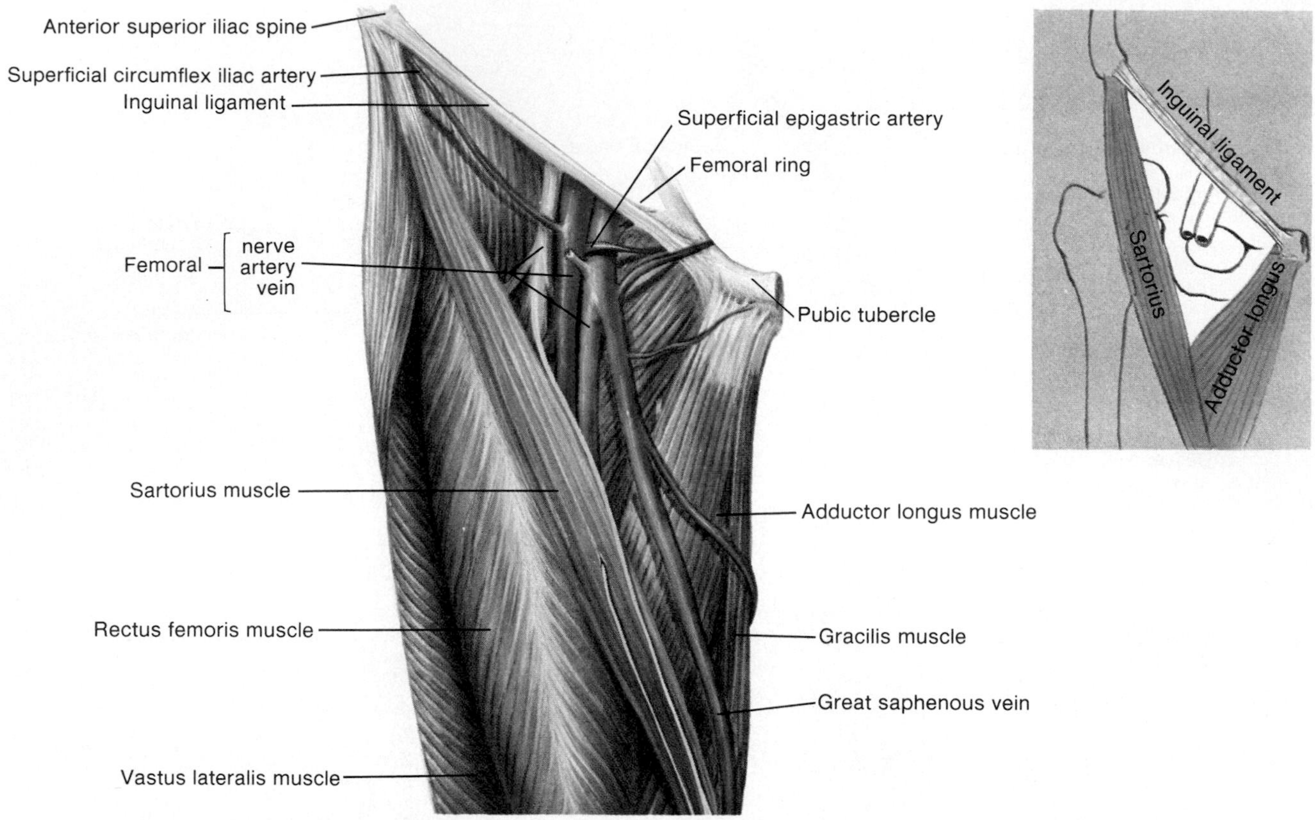

Anterior superior iliac spine
Superficial circumflex iliac artery
Inguinal ligament
Superficial epigastric artery
Femoral ring
Femoral — nerve, artery, vein
Pubic tubercle
Sartorius muscle
Adductor longus muscle
Rectus femoris muscle
Gracilis muscle
Great saphenous vein
Vastus lateralis muscle
Inguinal ligament
Sartorius
Adductor longus

At the ankle, the anterior tibial artery becomes the **dorsalis pedis artery,** which serves the ankle and dorsum of the foot, after which it contributes to the formation of the **plantar arch** of the foot. Clinically, palpation of the dorsalis pedis artery can provide information about circulation to the foot; but more important, it can provide information about the circulation in general because these pulses are taken at the most distal portion of the body.

The posterior tibial artery sends a large **peroneal** *(per''o-ne'al)* **artery** to serve the peroneal muscles of the leg. At the ankle, the posterior tibial bifurcates into the **lateral** and **medial plantar arteries,** which supply the muscles and structures on the sole of the foot. The lateral plantar artery anastomoses with the dorsal pedis artery to form the plantar arch, similar to the arterial arrangement in the hand. **Digital arteries** arise from the plantar arch to supply the toes with blood.

1. Describe the blood supply to the brain. Where is the arterial circle located, and how is it formed?
2. Describe the clinical significance of the brachial and radial arteries.
3. Describe the arterial pathway from the subclavian artery to the digital arteries.
4. List the arteries that supply blood to the lower abdominal wall, the external genitalia, the hamstring muscles, the knee joint, and the dorsum of the foot.

Principal Veins of the Body

After systemic blood has passed through the tissues and become depleted of oxygen, it returns through veins of progressively larger diameters to the right atrium of the heart.

Objective 21. Describe the venous drainage of the head, neck, and upper extremities.

Objective 22. Describe venous drainage of the thorax, lower extremities, and abdominal region.

Objective 23. Describe the vessels involved in the hepatic portal system.

In the venous portion of the systemic circulation, blood flows from smaller vessels into larger ones, so that a vein receives smaller tributaries instead of giving off branches as an artery does. The veins from all parts of the body converge into two major vessels that empty into the right atrium: the **superior** and **inferior vena cavae.** Veins are more numerous than arteries and are both superficial and deep. Superficial veins can generally be seen just beneath the skin and are clinically important in drawing blood and giving injections. Deep veins are close to the principal arteries and are usually similarly named. As with arteries, veins are named according to the region in which they are found or the organ that they serve (when a vein serves an organ, it drains blood away from it).

The principal systemic veins of the body are illustrated in figure 20.37.

Figure 20.37. Principal veins of the body. Superficial veins are depicted in the left extremities and deep veins in the right extremities.

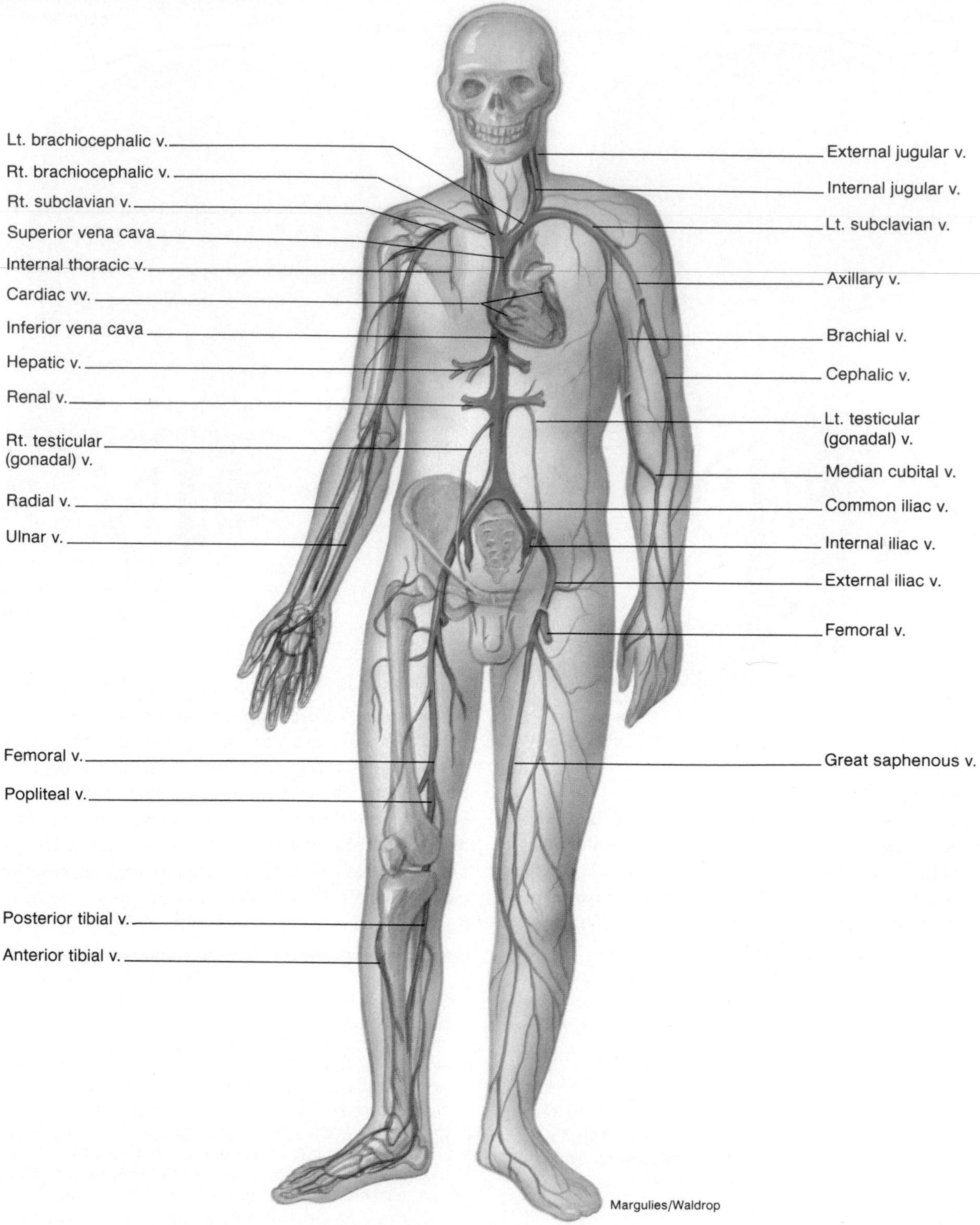

Figure 20.38. Veins that drain the head and neck.

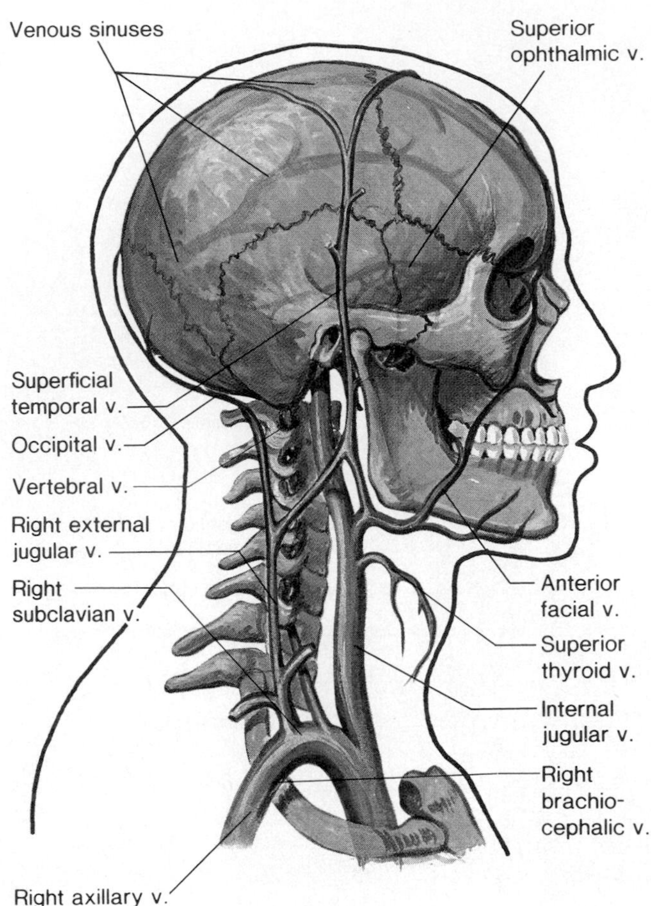

Figure 20.39. Cranial venous sinuses and the internal jugular vein.

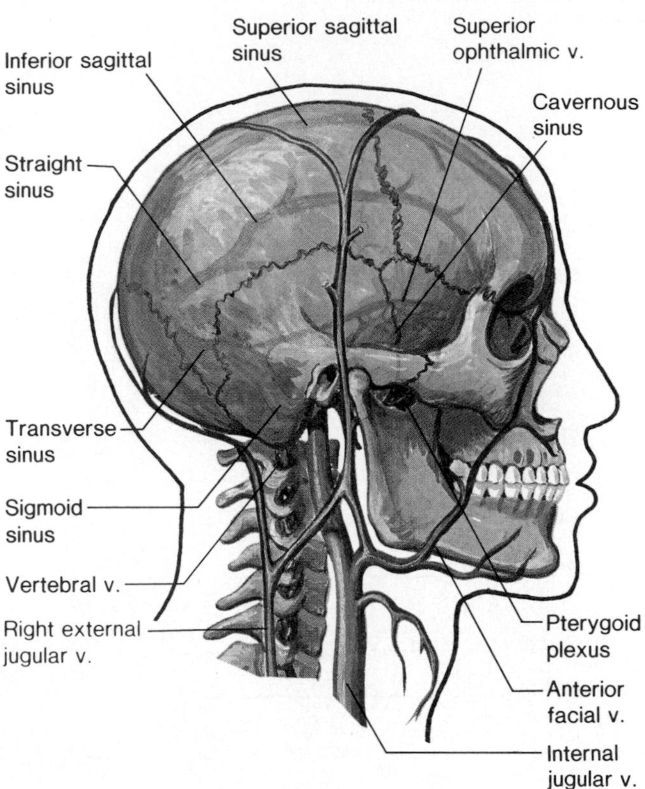

Veins Draining the Head and Neck

Blood from the scalp, portions of the face, and the superficial neck regions is drained by the **external jugular veins** (fig. 20.38). These vessels descend on either side of the neck, superficial to the sternocleidomastoid muscle and deep to the platysma. They drain into the right and left **subclavian veins** located just behind the clavicles.

The paired **internal jugular veins** drain blood from the brain, meninges, and deep regions of the face and neck. The internal jugular veins are larger and deeper than the external jugular veins. They arise from numerous cranial **venous sinuses,** which are a series of both paired and unpaired channels positioned between the two layers of dura mater. The venous sinuses, in turn, receive venous blood from the **cerebral,** the **cerebellar,** the **ophthalmic,** and the **meningeal** *(mĕ-nin'je-al)* **veins.** Figure 20.39 illustrates the principal cranial venous sinuses and the formation of the internal jugular vein.

The internal jugular vein passes inferiorly down the neck adjacent to the common carotid artery and the vagus nerve. All three are surrounded by the protective *carotid sheath* and are positioned beneath the sternocleidomastoid muscle. The internal jugular empties into the subclavian vein, and the union of these two vessels forms the large **brachiocephalic (innominate) vein** on each side. The two brachiocephalic veins then merge to form the **superior vena cava,** which drains into the right atrium of the heart (see fig. 20.37).

The external jugular vein is the vessel that is seen on the side of the neck when a person is angry or wearing a tight collar. You can voluntarily distend this vein by performing the *Valsalva maneuver.* To do this, take a deep breath and hold it while you forcibly contract your abdominal muscles as in a forced exhalation. This procedure is automatically performed when lifting a heavy object or defecating. The increased thoracic pressure that results compresses the vena cavae and interferes with the return of blood to the right atrium. The Valsalva maneuver, as a result, can be dangerous if performed by people with cardiovascular disease.

jugular: L. *jugulum,* throat or neck

Valsalva maneuver: from Antonio Valsalva, Italian anatomist, 1666–1723

Figure 20.40. An anterior view of the veins that drain the upper right extremity. (*a*) superficial veins; (*b*) deep veins.

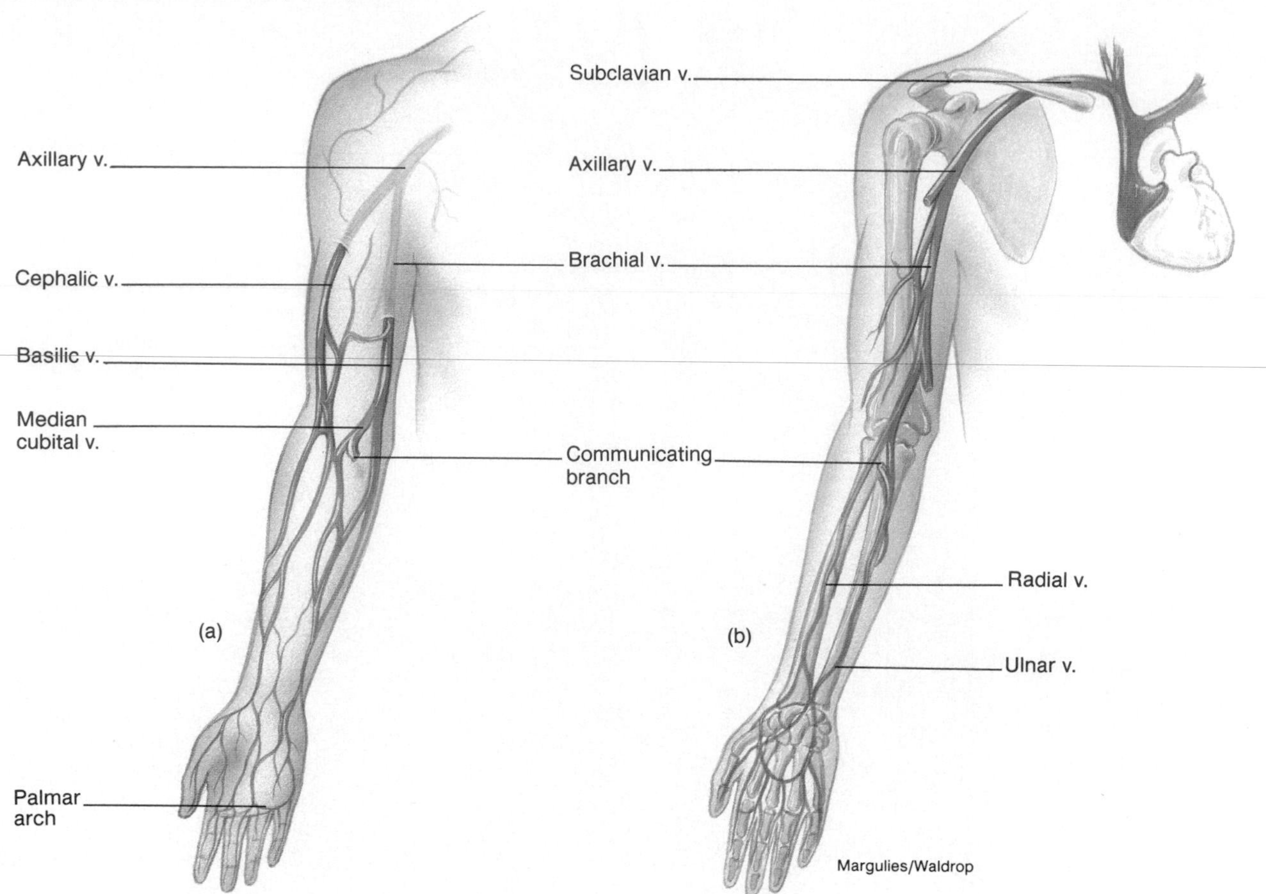

Veins of the Upper Extremity

The upper extremity has both superficial and deep venous drainage (fig. 20.40). The superficial veins are highly variable and form an extensive network just below the skin. The deep veins accompany the arteries of the same region and are given similar names. The deep veins of the upper extremity will be described first. Both the **radial vein** on the lateral side of the forearm and the **ulnar vein** on the medial side drain blood from the **deep** and **superficial volar arches** of the hand. The radial and ulnar veins join in the cubital fossa to form the **brachial vein,** which continues up the medial side of the brachium.

The main superficial vessels of the upper extremity are the **basilic vein** and the **cephalic vein.** The basilic vein passes on the ulnar side of the forearm and the medial side of the arm. Near the head of the humerus, the basilic vein merges with the brachial vein and forms the **axillary vein.**

The cephalic vein drains the superficial portion of the hand and forearm on the radial side and then continues up the lateral side of the arm. In the shoulder region, the cephalic vein pierces the fascia and joins the axillary vein. The axillary vein then passes the first rib to form the subclavian vein, which unites with the external jugular to form the brachiocephalic vein of that side.

Superficially, in the cubital fossa of the elbow, the **median cubital vein** ascends from the cephalic vein on the lateral side to connect with the basilic vein on the medial side. The median cubital vein is a frequent site for venipuncture in order to remove a sample of blood or add fluids to the blood.

Veins of the Thorax

The **superior vena cava,** formed by the union of the two brachiocephalic veins, empties venous blood from the head, neck, and upper extremities directly into the right atrium of the heart. These large vessels lack the valves that are characteristic of most other veins in the body.

In addition to receiving blood from the brachiocephalic veins, the superior vena cava collects blood from the azygos system of veins arising from the posterior thoracic wall (fig. 20.41). The **azygos** *(az'ĭ-gos)* **vein** extends superiorly along the dorsal abdominal and thoracic walls on the right side of the vertebral column. The azygos vein ascends through the mediastinum to join the superior vena cava at the level of the fourth thoracic vertebra. Tributaries of the azygos vein include the **ascending lumbar veins** that drain the lumbar and sacral regions, **intercostal veins**

azygos: Gk. *a,* without; *zygon,* yoke

Figure 20.41. Veins of the thoracic region. The lungs and heart have been removed.

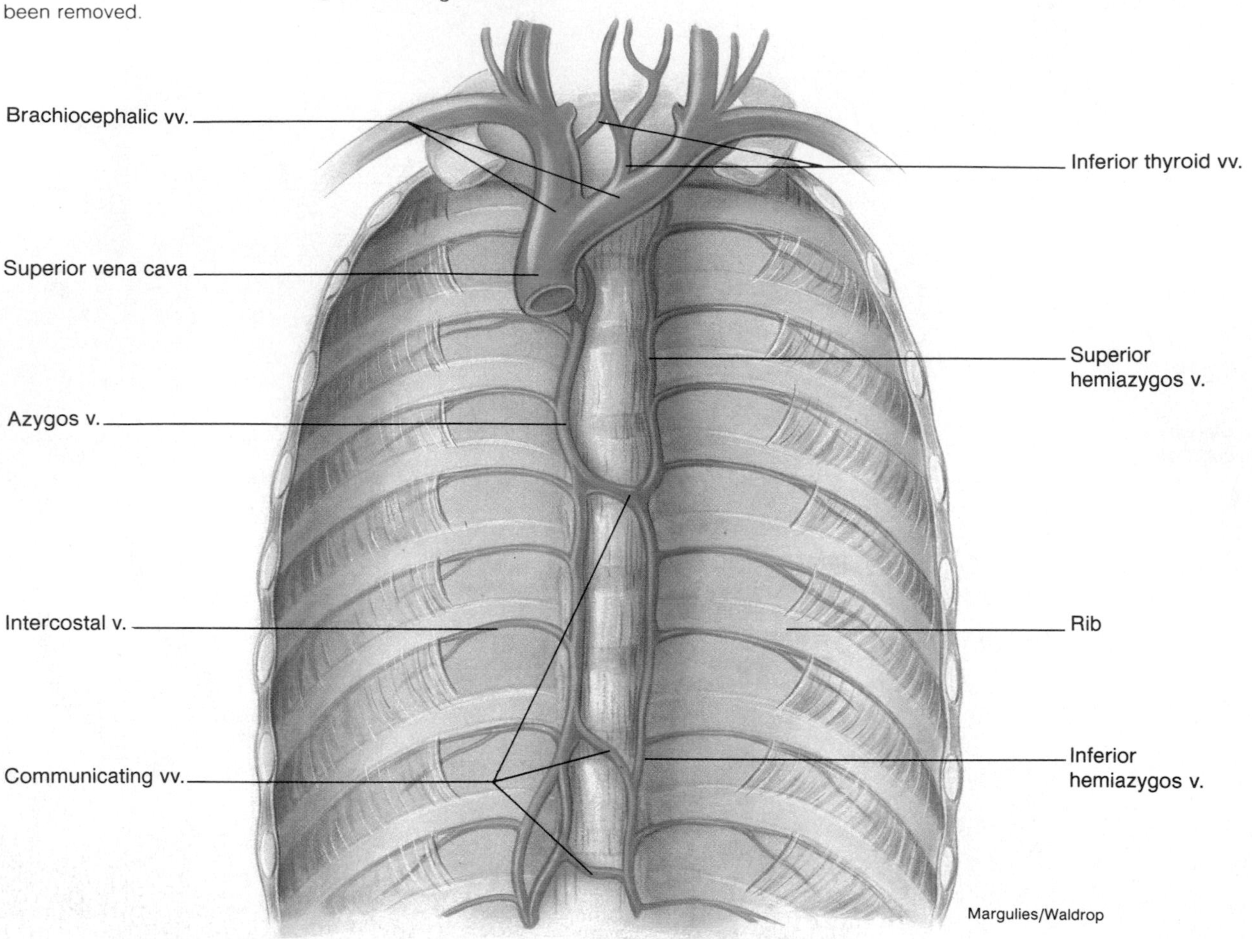

draining from the intercostal muscles, and the **accessory hemiazygos** and **hemiazygos veins,** which form the major tributaries to the left of the vertebral column.

Veins of the Lower Extremity

The lower extremities, like the upper extremities, have both a deep and a superficial group of veins (fig. 20.42). The deep veins accompany corresponding arteries and have more valves than do the superficial veins. The deep veins will be described first.

The **posterior** and **anterior tibial veins** originate in the foot and course upward behind and in front of the tibia to the back of the knee, where they merge to form the **popliteal vein.** The popliteal vein receives blood from the knee region. Just above the knee, this vessel becomes the **femoral vein.** The femoral vein in turn continues up the thigh and receives blood from the **deep femoral vein** near the groin. Just above this, the femoral vein receives blood from the **great saphenous** *(sah-fe'nus)* **vein** and then becomes the **external iliac vein** as it passes under the inguinal ligament. The external iliac curves upward to the level of the sacroiliac joint. There it merges with the **internal iliac** (hypogastric) **vein** at the pelvic and genital regions to form the **common iliac vein.** At the level of the fifth lumbar vertebra, the right and left common iliacs unite to form the large **inferior vena cava** (fig. 20.42).

The superficial veins of the lower extremity are the **small** and **great saphenous veins.** The small saphenous vein arises from the lateral side of the foot, courses posteriorly along the surface of the calf of the leg, and descends deep to enter the popliteal vein behind the knee. The great saphenous vein is the longest vessel in the body. It originates at the arch of the foot and ascends superiorly along the medial aspect of the leg and thigh before draining into the femoral vein.

Veins of the Abdominal Region

The **inferior vena cava** parallels the abdominal aorta on the right side as it ascends through the abdominal cavity to penetrate the diaphragm and enter the right atrium (see fig. 20.37). It is the largest vessel in diameter in the body and is formed by the union of the two common iliac veins draining the lower extremities. As the inferior vena cava ascends through the abdominal cavity, it receives tributaries from veins that correspond in name and position to arteries previously described.

saphenous: L. *saphena,* the hidden one

Figure 20.42. Veins of the lower extremity. (*a*) superficial veins, medial and posterior aspects. (*b*) deep veins, medial view.

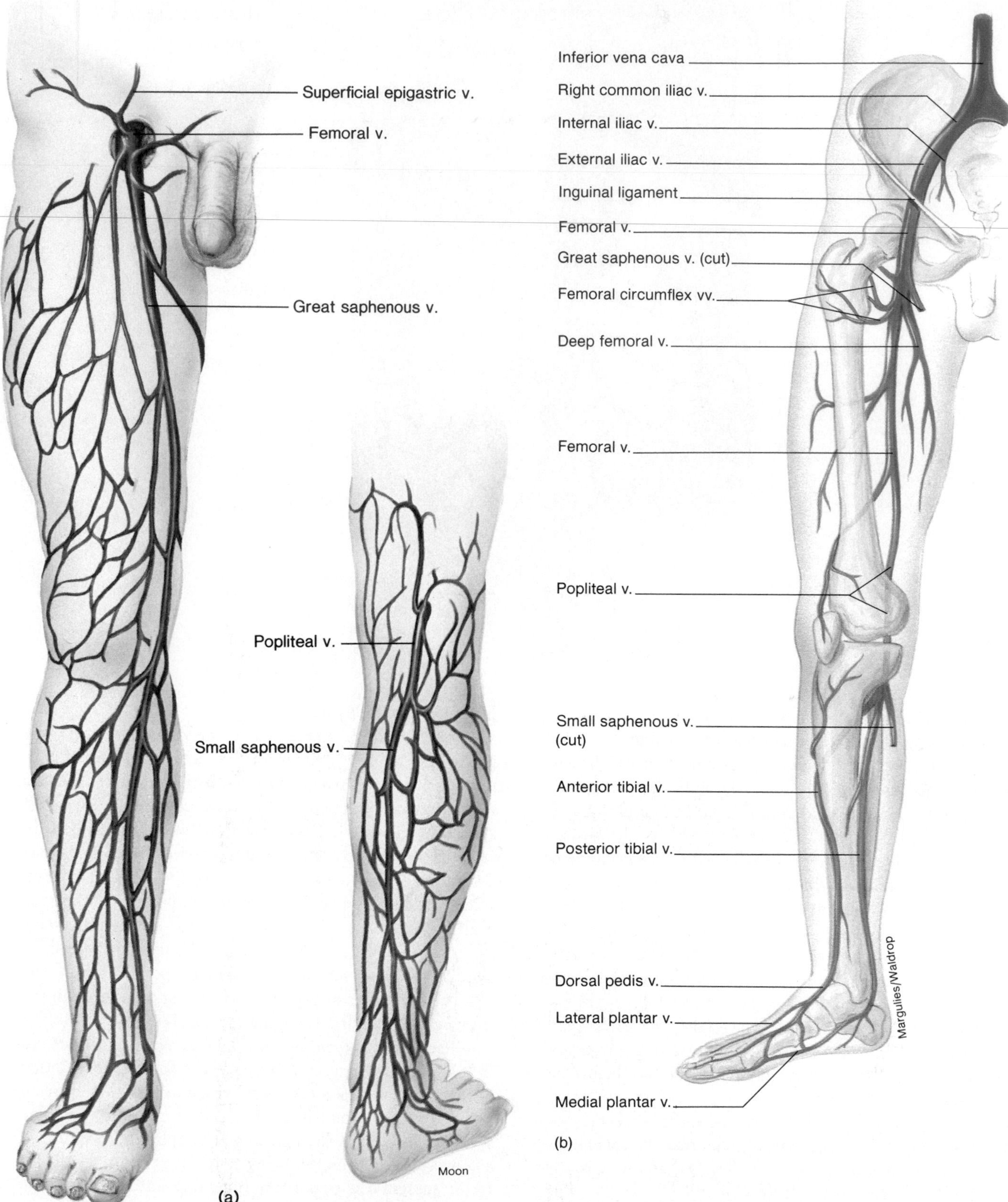

Figure 20.43. The hepatic portal system.

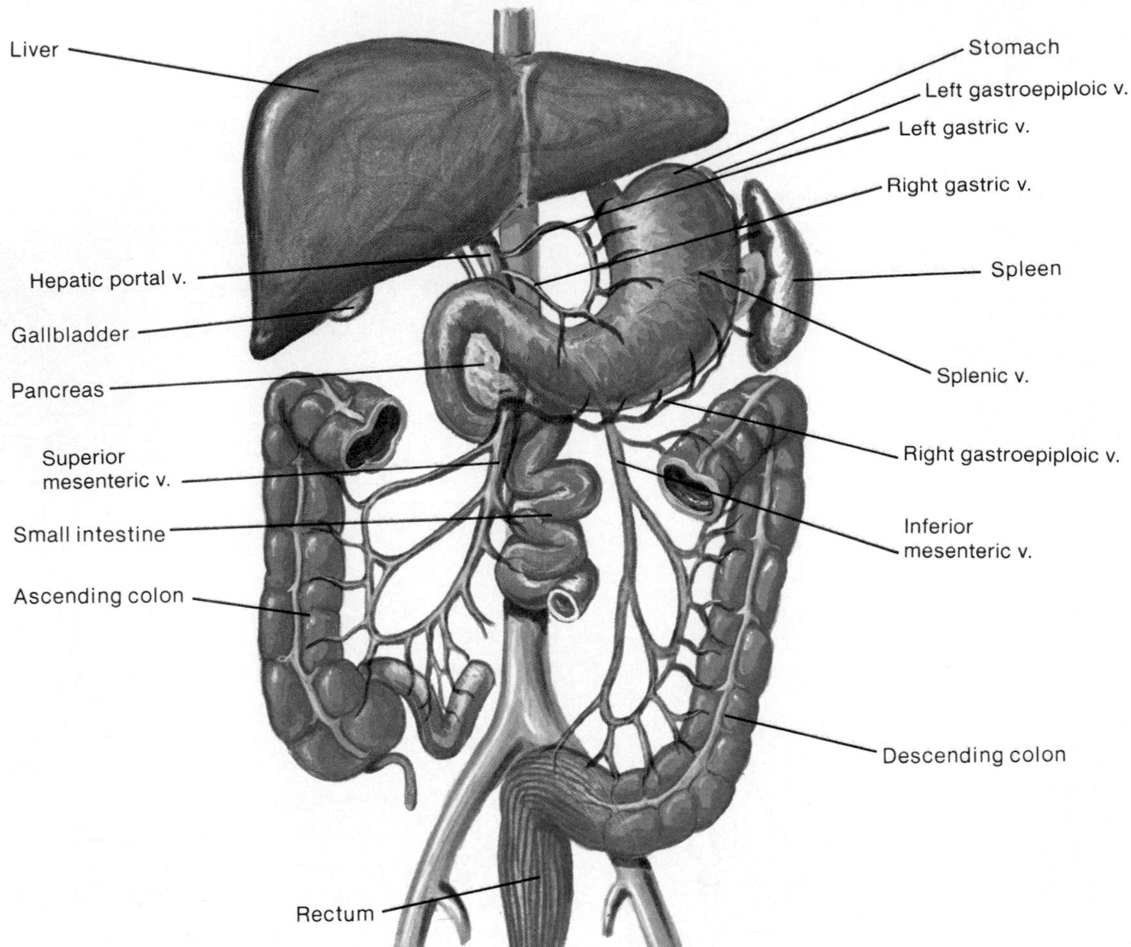

Four paired **lumbar veins** (not shown) drain the posterior abdominal wall, the vertebral column, and the spinal cord. In the male, the **testicular (internal spermatic) veins** receive blood from the testes. The right testicular vein empties into the left renal vein. The drainage pathway is similar in the female, but the vessels are called the **ovarian veins** and receive blood from the ovaries. The **renal veins** drain blood from the kidneys and ureters into the inferior vena cava. Small **suprarenal veins** drain the suprarenal (adrenal) glands. The right suprarenal vein empties directly into the inferior vena cava, and the left suprarenal connects to the left renal vein. The **inferior phrenic veins** receive blood from the inferior side of the diaphragm and drain into the inferior vena cava. Right and left **hepatic veins** originate from the capillary sinusoids of the liver and empty into the inferior vena cava immediately below the diaphragm.

Note that the inferior vena cava does not receive blood directly from the digestive tract, pancreas, or spleen. Instead, the venous outflow from these organs first passes through capillaries in the liver.

Hepatic Portal System

A *portal system* is one in which the veins that drain one group of capillaries deliver blood to another group of capillaries, which in turn are drained by more usual systemic veins that carry blood to the vena cavae and the right atrium of the heart. There are thus two capillary beds in series. The hepatic portal system is composed of veins that drain blood from capillaries in the intestines, pancreas, spleen, stomach, and gallbladder into capillaries in the liver (called *sinusoids*) and of the right and left **hepatic veins** that drain the liver and empty into the inferior vena cava (fig. 20.43). As a result of the hepatic portal system, the absorbed products of digestion must first pass through the liver before entering the general circulation.

The **hepatic portal vein** is the large vessel that receives blood from the digestive organs. It is formed by a union of the **superior mesenteric vein,** which drains nutrient-rich blood from the small intestine, and the **splenic vein.** The splenic vein drains the spleen but is enlarged because of a convergence of the following four tributaries: (1) the **inferior mesenteric vein** from the large intestine, (2) the

pancreatic vein from the pancreas, (3) the **left gastroepiploic vein** from the stomach, and (4) the **right gastroepiploic vein,** also from the stomach.

Three additional veins empty into the portal vein. The **right** and **left gastric veins** drain the lesser curvature of the stomach, and the **cystic vein** drains blood from the gallbladder.

One of the functions of the liver is to detoxify harmful substances, such as alcohol, that are absorbed into the blood from the small intestine. However, excessive quantities of alcohol cannot be processed during a single pass through the liver, and so a person becomes intoxicated. Eventually, the liver is able to process the alcohol as the circulating blood is repeatedly exposed to the liver sinusoids via the hepatic artery. Alcoholics may eventually suffer from *cirrhosis* of the liver as the normal liver tissue is destroyed.

In summary, it is important to note that the sinusoids of the liver receive blood from two sources. The hepatic artery supplies oxygen-rich blood to the liver, whereas the hepatic portal vein transports nutrient-rich blood from the intestine for processing. These two blood sources become mixed in the liver sinusoids. Liver cells exposed to this blood obtain nourishment from it and are uniquely qualified (because of their anatomical position and enzymatic ability) to modify the chemical nature of the venous blood that enters the general circulation from the digestive tract.

1. Using a flowchart, list the venous drainage from the head and neck to the superior vena cava. Indicate which vein may bulge in the side of the neck when a person performs the Valsalva maneuver and which vein is commonly used as a site for venipuncture.
2. Describe the positions, sources, and drainages of the small and great saphenous veins.
3. Describe the hepatic portal system, and explain the functional importance of this system.

Fetal Circulation

The fetal circulation is adapted to the fact that the fetal lungs are nonfunctional and that oxygen and nutrients are obtained from the placenta.

Objective 24. Describe the fetal circulation to and from the placenta.

Objective 25. Describe the structure and function of the foramen ovale, ductus venosus, and ductus arteriosus.

The circulation of blood through a fetus is by necessity different from that of a newborn infant (fig. 20.44). Respiration, the procurement of nutrients, and the elimination of metabolic wastes occur through the maternal blood instead of through the organs of the fetus. The capillary exchange between the maternal and fetal circulation occurs within the **placenta.** This remarkable structure includes part of the uterus of the mother during pregnancy and is discharged following delivery as the afterbirth.

The **umbilical cord** is the connection between the placenta and the fetal umbilicus. It includes one **umbilical vein** and two **umbilical arteries** surrounded by a gelatinous substance. Oxygenated and nutrient-rich blood flows through the umbilical vein toward the inferior surface of the liver. At this point, the umbilical vein bifurcates into one branch that joins with the portal vein, while the other branch, called the **ductus venosus,** enters the inferior vena cava. Thus, oxygenated blood is mixed with venous blood returning from the lower extremities of the fetus before it enters the heart. The umbilical vein is the only vessel of the fetus that carries fully oxygenated blood.

The inferior vena cava empties into the right atrium of the fetal heart. From here, most of the blood is directed through the **foramen ovale,** an opening between the two atria. Here it mixes with a small quantity of blood returning through the pulmonary circulation. The blood then passes into the left ventricle, from which it is pumped into the aorta and through the body of the fetus. Some blood entering the right atrium passes into the right ventricle and out of the heart via the pulmonary trunk. Since the lungs of the fetus are not functional, only a small portion of blood continues through the pulmonary circulation (the resistance to blood flow is very high in the collapsed fetal lungs). Most of the blood in the pulmonary trunk passes through the **ductus arteriosus** into the aortic arch, where it mixes with blood coming from the left ventricle. Blood is returned to the placenta by the two **umbilical arteries** that arise from the internal iliac arteries.

Notice that in the fetus oxygen-rich blood is transported by the inferior vena cava to the heart and via the foramen ovale and ductus arteriosus to the systemic circulation.

Important changes occur in the cardiovascular system at birth. The foramen ovale, ductus arteriosus, ductus venosus, and the umbilical vessels are no longer necessary. The foramen ovale abruptly closes with the first breath of air because the reduced pressure in the right side of the heart causes a flap to cover the opening. This reduced pressure is due to the fact that (1) when the lungs fill with air, the vascular resistance to blood flow in the pulmonary circulation falls far below that of the systemic circulation; and (2) the pressure in the inferior vena cava and right atrium falls as a result of the loss of the placental circulation.

gastroepiploic: Gk. *gastros,* stomach; *epiplein,* to float on (referring to greater omentum)

Figure 20.44. Fetal circulation. Arrows indicate the direction of blood flow.

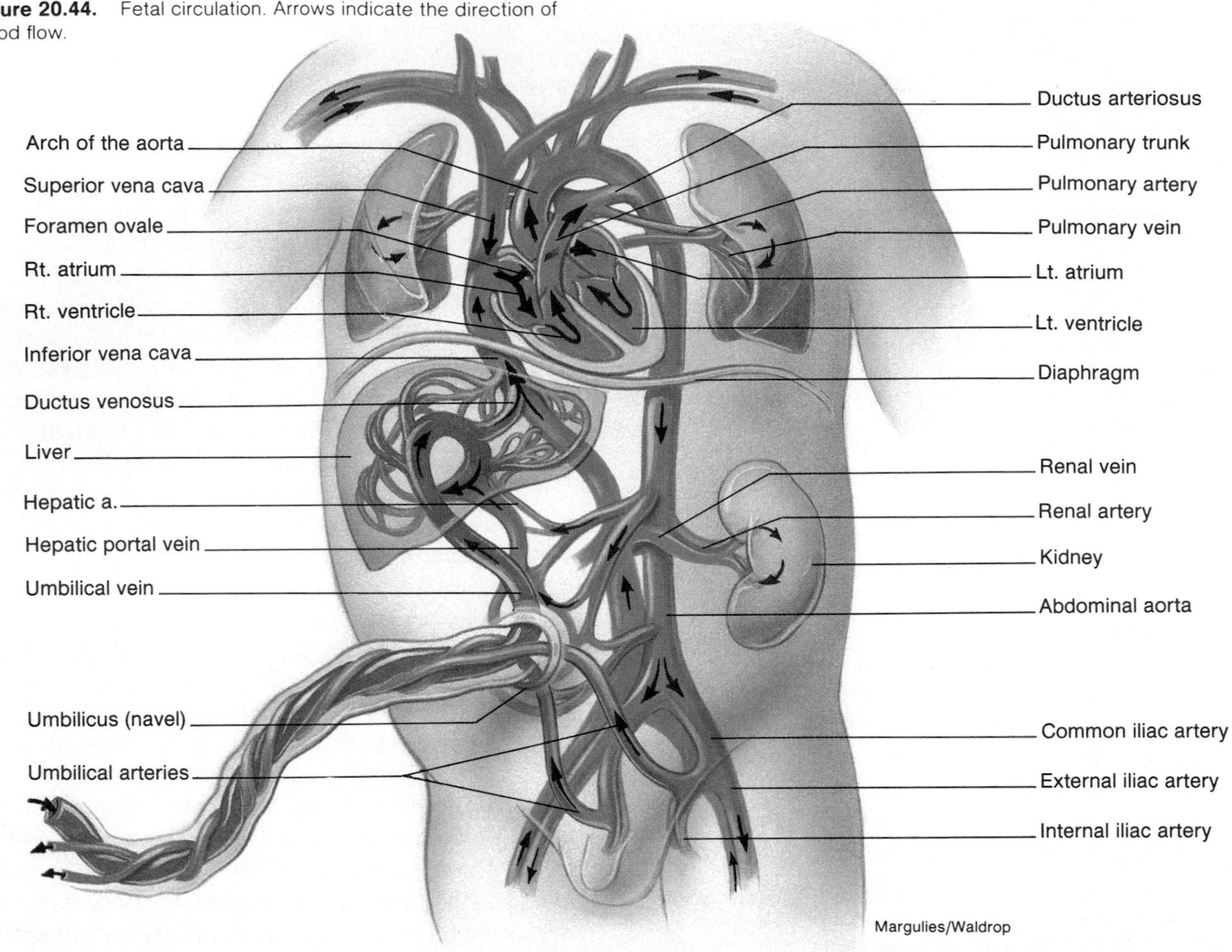

Table 20.6 Cardiovascular structures of the fetus and changes in the neonatal infant

Structure	Location	Function	Fate in neonatal infant
Umbilical vein	Connects the placenta to the liver; forms a major portion of umbilical cord	Transports nutrient-rich, oxygenated blood from the placenta to the fetus	Forms the round ligament of the liver
Ductus venosus	Venous shunt within the liver to connect with the inferior vena cava	Transports oxygenated blood directly into the inferior vena cava	Forms the ligamentum venosum, a fibrous cord in the liver
Foramen ovale	Opening between the right and left atria	A shunt to bypass the pulmonary circulation	Closes at birth and becomes the fossa ovalis, a depression in the interatrial septum
Ductus arteriosus	Between pulmonary trunk and arch of the aorta	A shunt to bypass the pulmonary circulation	Closes shortly after birth, atrophies, and becomes the ligamentum arteriosum
Umbilical arteries	Arise from internal iliac arteries and associated with umbilical cord	Transports blood from the fetus to the placenta	Atrophies to become the lateral umbilical ligaments

The constriction of the ductus arteriosus occurs gradually over a period of about six weeks after birth as the vascular muscle cells constrict in response to the higher oxygen concentration in the postnatal blood. The remaining structure of the ductus gradually atrophies and becomes nonfunctional as a blood vessel. The fate of the unique fetal cardiovascular structures is summarized in table 20.6.

1. Describe the path of blood from the fetal heart, through the placenta, and back to the fetal heart.
2. Trace the pathway of oxygenated blood through the fetal circulation.
3. Describe the foramen ovale and ductus arteriosus, and explain why blood flows the way it does through these structures in the fetal circulation.

Figure 20.45. Blood cells become packed at the bottom of the test tube when whole blood is centrifuged, leaving the fluid plasma at the top of the tube. Red blood cells are the most abundant of the blood cells—white blood cells and platelets form only a thin, light-colored "buffy coat" at the interface between the packed red blood cells and the plasma.

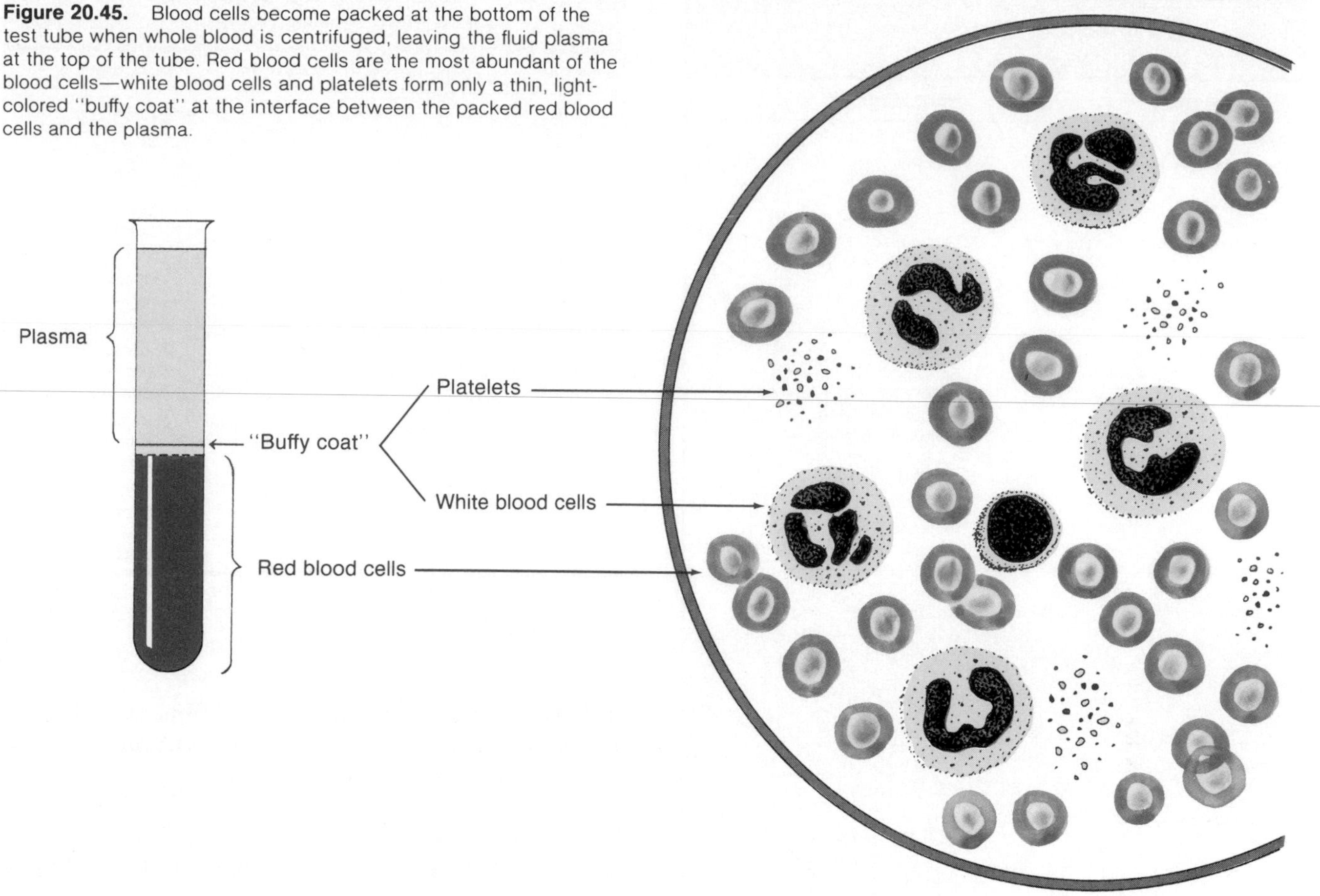

Plasma

"Buffy coat"

Platelets

White blood cells

Red blood cells

Composition of the Blood

Blood consists of formed elements that are suspended and carried in plasma. The constituents of blood function in gas transport, transport of regulatory molecules and nutrients, immunity, and blood-clotting mechanisms.

Objective 26. List the different types of substances present in plasma.
Objective 27. Describe the origin and function of the different categories of plasma proteins.
Objective 28. List the different types of formed elements of blood, and describe their appearance.
Objective 29. Describe the origin of erythrocytes, leukocytes, and thrombocytes.

The average-sized adult has about 5 liters of blood, which constitute about 8% of the total body weight. Blood leaving the heart is referred to as *arterial blood.* Arterial blood, with the exception of that going to the lungs, is bright red in color due to a high concentration of oxyhemoglobin (the combination of oxygen with hemoglobin) in the erythrocytes. *Venous blood* is blood returning to the heart and, except for the venous blood from the lungs, has a darker color (due to increased amounts of hemoglobin that are no longer combined with oxygen). Blood has a viscosity that ranges between 4.5 and 5.5. This means that it flows thicker than water, which has a viscosity of 1. Blood has a pH range of 7.35 to 7.45 and a temperature within the thorax of the body of about 38°C (100.4°F).

Blood is composed of a cellular portion, called **formed elements,** and a fluid portion, called **plasma.** When a blood sample is centrifuged, the heavier formed elements are packed into the bottom of the tube, leaving plasma at the top (fig. 20.45). The formed elements constitute approximately 45% of the total blood volume (the *hematocrit*), and the plasma accounts for the remaining 55%.

Plasma

Plasma is a straw-colored liquid consisting of water and dissolved solutes. Sodium ion is the major solute of plasma in terms of its concentration (Na^+ contributes most to the total osmolality of plasma). In addition to Na^+, plasma contains many other salts and ions, as well as organic molecules such as metabolites, hormones, enzymes, antibodies, and other proteins. The values of some of these constituents of plasma are shown in table 20.7.

hematocrit: Gk. *haima, blood; krino,* to separate
plasma: Gk. *plasma,* to form or mold

Table 20.7 Representative normal plasma values

Measurement	Normal range
Blood volume	80–85 ml/kg body weight
Blood osmolality	280–296 mOsm
Blood pH	7.35–7.45
Enzymes	
Creatine phosphokinase (CPK)	Female: 10–79 U/L Male: 17–148 U/L
Lactic dehydrogenase (LDH)	45–90 U/L
Phosphatase (acid)	Female: 0.01–0.56 Sigma U/ml Male: 0.13–0.63 Sigma U/ml
Hematology values	
Hematocrit	Female: 37%–48% Male: 45%–52%
Hemoglobin	Female: 12–16 g/100 ml Male: 13–18 g/100 ml
Red blood cell count	4.2–5.9 million/mm^3
White blood cell count	4,300–10,880/mm^3
Hormones	
Testosterone	Male: 300–1,100 ng/100 ml Female: 25–90 ng/100 ml
Adrenocorticotropic hormone (ACTH)	15–70 pg/ml
Growth hormone	Children: over 10 ng/ml Adult male: below 5 ng/ml
Insulin	6–26 μU/ml (fasting)
Ions	
Bicarbonate	24–30 mmol/l
Calcium	2.1–2.6 mmol/l
Chloride	100–106 mmol/l
Potassium	3.5–5.0 mmol/l
Sodium	135–145 mmol/l
Organic molecules (other)	
Cholesterol	120–220 mg/100 ml
Glucose	70–110 mg/100 ml (fasting)
Lactic acid	0.6–1.8 mmol/l
Protein (total)	6.0–8.4 g/100 ml
Triglyceride	40–150 mg/100 ml
Urea nitrogen	8–25 mg/100 ml
Uric acid	3–7 mg/100 ml

Excerpted from material originally appearing in *The New England Journal of Medicine*, "Case Records of the Massachusetts General Hospital," Vol. 302, No. 1, pp. 37–48, January 3, 1980 and "Case Records of the Massachusetts General Hospital," Vol. 314, No. 1, pp. 39–49.

Plasma Proteins. Plasma proteins constitute 7%–9% of the plasma. The three types of proteins are albumins, globulins, and fibrinogen. **Albumins** *(al-bu'min)* account for most (60%–80%) of the plasma proteins and are the smallest in size. They are produced by the liver and serve to provide the osmotic pressure needed to draw water from the surrounding tissue fluid into the capillaries. This action is needed to maintain blood volume and pressure. **Globulins** *(glob'u-lin)* are divided into three subtypes: **alpha globulins, beta globulins,** and **gamma globulins.** The alpha and beta globulins are produced by the liver and function to transport lipids and fat-soluble vitamins in the blood. Gamma globulins are antibodies produced by lymphocytes (one of the formed elements found in blood and lymphoid tissues) and function in immunity. **Fibrinogen** *(fi-brin'o-jen)* accounts for only about 4% of the protein content of plasma. Fibrinogens, which are synthesized in the liver (table 20.8), are converted to insoluble threads of **fibrin** when blood clots. The fluid from clotted blood, which is called **serum,** is therefore identical to plasma except for the absence of fibrinogen.

albumin: L. *albumen*, white
globulin: L. *globulus*, small globe
fibrinogen: L. *fibra*, fibrous

Table 20.8 Plasma proteins

Protein	Percentage of total	Origin	Function
Albumin	60	Liver	Gives viscosity to blood and assists in maintaining blood osmotic pressure
Globulin	36		
Alpha globulins		Liver	Transport lipids and fat-soluble vitamins
Beta globulins		Liver	Transport lipids and fat-soluble vitamins
Gamma globulins		Lymphoid tissue	Constitute antibodies of immunity
Fibrinogen	4	Liver	Assist thrombocytes in the formation of clots

From Kent M. Van De Graaff, *Human Anatomy*, 2d ed.

The Formed Elements of Blood

The formed elements of blood include two types of blood cells: **erythrocytes** *(ĕ-rith'ro-sīt),* or red blood cells, and **leukocytes** *(loo'ko-sīt),* or white blood cells. Erythrocytes are by far the most numerous of these two types: a cubic millimeter of blood contains 5.1 million to 5.8 million erythrocytes in males and 4.3 million to 5.2 million erythrocytes in females. The same volume of blood, in contrast, contains only 5,000 to 9,000 leukocytes.

Erythrocytes. Erythrocytes are flattened, biconcave discs, about 7 μm in diameter and 2.2 μm thick. Their unique shape relates to their function of transporting oxygen and provides an increased surface area through which gas can diffuse (fig. 20.46). Erythrocytes lack a nucleus and mitochondria (they get energy from anaerobic respiration). Because of these deficiencies, erythrocytes have a circulating life span of only about 120 days before they are destroyed by phagocytic cells in the liver, spleen, and bone marrow.

Each erythrocyte contains approximately 280 million *hemoglobin* molecules, which give blood its red color. Each hemoglobin molecule consists of a protein, called globin, and an iron-containing pigment, called heme. The iron group of heme is able to combine with oxygen in the lungs and release oxygen in the tissues.

serum: L. *serum*, liquid
erythrocytes: Gk. *erythros*, red; *kytos*, hollow (cell)
hemoglobin: Gk. *haima*, blood; *globus*, globe

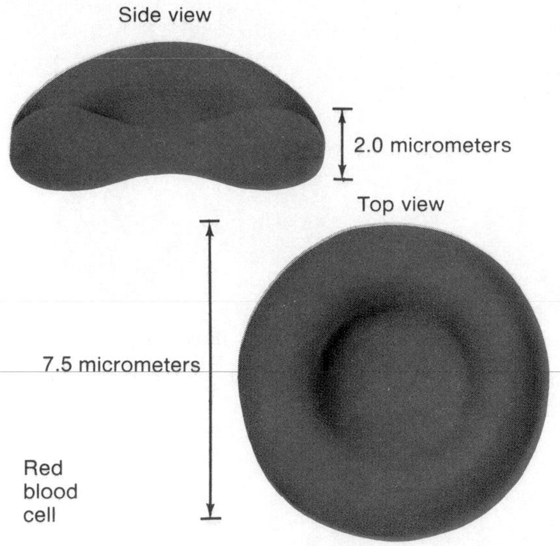

Figure 20.46. The structure of an erythrocyte.

Anemia is present when there is an abnormally low hemoglobin concentration and/or red blood cell count. The most common cause of this condition is a deficiency in iron *(iron-deficiency anemia),* which is an essential component of the hemoglobin molecule. In *pernicious anemia,* there is inadequate availability of vitamin B_{12}, which is needed for red blood cell production. This results from atrophy of the glandular mucosa of the stomach, which normally secretes a substance, called *intrinsic factor,* that is needed for absorption of vitamin B_{12} obtained in the diet. *Aplastic anemia* is anemia due to destruction of the bone marrow, which may be caused by chemicals (including benzene and arsenic) and by X rays.

Leukocytes. Leukocytes, or white blood cells, differ from erythrocytes in several ways. Leukocytes contain nuclei and mitochondria and can move in an amoeboid fashion (erythrocytes are not able to move independently). Because of their amoeboid ability, leukocytes can squeeze through pores in capillary walls and get to a site of infection, whereas erythrocytes usually remain confined within blood vessels. The movement of leukocytes through capillary walls is called *diapedesis.*

Leukocytes, which are almost invisible under the microscope unless they are stained, are classified according to their stained appearance. Those leukocytes that have granules in their cytoplasm are called **granular leukocytes,** and those that do not are **agranular** (or nongranular) **leukocytes.** The granular leukocytes are also identified by their oddly shaped nuclei, which in some cases are contorted into lobes separated by thin strands. The granular leukocytes are therefore known as **polymorphonuclear** *(pol''e-mor''fo-nu'kle-ar)* **leukocytes** (abbreviated PMN).

The stain used to identify leukocytes is usually a mixture of a pink-to-red stain called eosin and a blue-to-purple stain called a "basic stain." Granular leukocytes with pink-staining granules are therefore called **eosinophils** *(e''o-sin'o-fil),* and those with blue-staining granules are called **basophils.** Those with granules that have little affinity for either stain are **neutrophils.** Neutrophils are the most abundant type of leukocyte, comprising 50%–70% of the leukocytes in the blood.

There are two types of agranular leukocytes: lymphocytes and monocytes. **Lymphocytes** are usually the second most numerous type of leukocyte; they are small cells with round nuclei and little cytoplasm. **Monocytes,** in contrast, are the largest of the leukocytes and generally have kidney- or horseshoe-shaped nuclei. In addition to these two cell types, there are smaller numbers of large lymphocytes, or **plasma cells,** that may be difficult to distinguish from monocytes. Plasma cells produce and secrete large amounts of antibodies.

Thrombocytes. Thrombocytes, or **platelets** *(plāt'let),* are the smallest of the formed elements and are actually fragments of large cells called **megakaryocytes** *(meg''ah-kar'e-o-sīt),* found in bone marrow. (This is why the term *formed elements* is used rather than *blood cells* to describe erythrocytes, leukocytes, and thrombocytes.) The fragments that enter the circulation as platelets lack nuclei but, like leukocytes, are capable of amoeboid movement. The platelet count per cubic millimeter of blood is 130,000 to 360,000. Platelets survive about five to nine days and then are destroyed by the spleen and liver.

Platelets play an important role in blood clotting. They constitute the major portion of the mass of the clot, and phospholipids in their cell membranes serve to activate the clotting factors in plasma that result in threads of fibrin, which reinforce the platelet plug. Platelets that attach together in a blood clot also release a chemical called *serotonin (ser''o-to'nin),* which stimulates constriction of blood vessels and thus reduces the flow of blood to the injured area.

The appearance of the formed elements of the blood is shown in figure 20.47, and a summary of the characteristics of these formed elements is presented in table 20.9.

Blood cell counts are an important source of information in determining the health of a person. An abnormal increase in erythrocytes, for example, is termed *polycythemia (pol''e-si-the'me-ah)* and is indicative of several dysfunctions. An abnormally low red blood cell count is *anemia.* An elevated leukocyte count, called *leukocytosis,* is often associated with localized infection. A large number of immature leukocytes within a blood sample is diagnostic of the disease *leukemia.*

leukocytes: Gk. *leukos,* white; *kytos,* hollow (cell)
diapedesis: Gk. *dia,* through; *pedester,* on foot

Figure 20.47. Types of formed elements in blood.

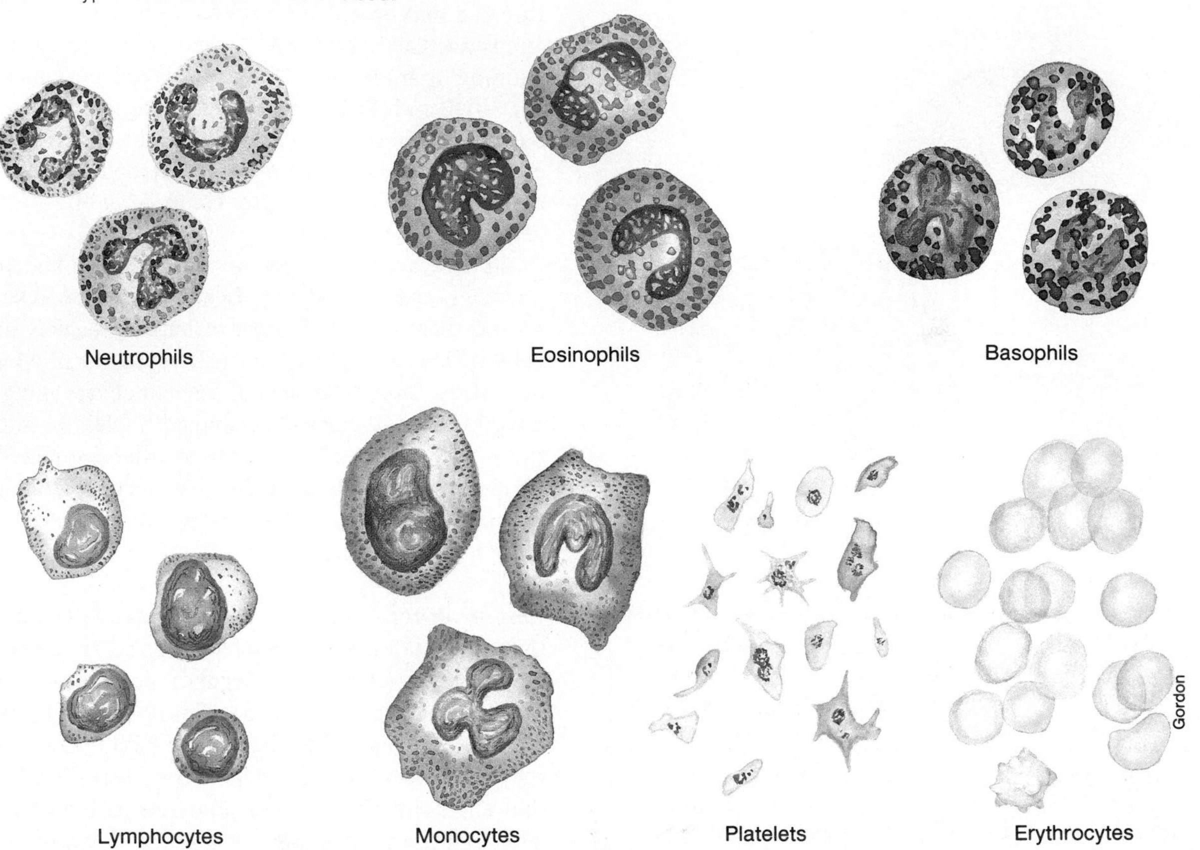

Table 20.9 Formed elements of the blood

Component	Description	Number present	Function
Erythrocyte (red blood cell)	Biconcave disc without nucleus; contains hemoglobin; survives 100–120 days	4,000,000 to 6,000,000/mm^3	Transports oxygen and carbon dioxide
Leukocytes (white blood cells)		5,000 to 10,000/mm^3	Aid in defense against infections by microorganisms
Granulocytes	About twice the size of red blood cells; cytoplasmic granules present; survive 12 hours to 3 days		
1. Neutrophil	Nucleus with 2–5 lobes; cytoplasmic granules stain slightly pink	50%–70% of white cells present	Phagocytic
2. Eosinophil	Nucleus bilobed; cytoplasmic granules stain red in eosin stain	1%–3% of white cells present	Helps to detoxify foreign substances; secretes enzymes that break down clots
3. Basophil	Nucleus lobed; cytoplasmic granules stain blue in hematoxylin stain	Less than 1% of white cells present	Releases anticoagulant heparin
Agranulocytes	Cytoplasmic granules absent; survive 100–300 days		
1. Monocyte	2–3 times larger than red blood cell; nuclear shape varies from round to lobed	3%–9% of white cells present	Phagocytic
2. Lymphocyte	Only slightly larger than red blood cell; nucleus nearly fills cell	25%–33% of white cells present	Provides specific immune response (including antibodies)
Thrombocyte (platelet)	Cytoplasmic fragment; survives 5–9 days	130,000 to 360,000/mm^3	Clotting

Hemopoiesis

Blood cells are constantly formed through a process called *hemopoiesis* (fig. 20.48). The term **erythropoiesis** refers to the formation of erythrocytes, and **leukopoiesis** to the formation of leukocytes. These processes occur in two classes of tissues. **Myeloid tissue** is the red bone marrow of the humeri, femora, ribs, sternum, bodies of vertebrae, and portions of the skull. **Lymphoid tissue** includes the lymph nodes, tonsils, spleen, and thymus. The bone marrow produces all of the different types of blood cells, while one type—lymphocytes—is also produced in the lymphoid tissue (discussed in chapter 22).

During embryonic development hemopoiesis first occurs in the wall of the yolk sac; then the liver becomes the major site for blood cell production. Toward the end of fetal development, and after birth, the bone marrow and lymphoid organs assume their roles as hemopoietic centers, and the liver and spleen serve as sites for blood cell destruction.

Erythropoiesis is an extremely active process. It is estimated that about 2.5 million erythrocytes are produced every second in order to replace the number that are continuously destroyed by the spleen and liver. During the destruction of erythrocytes, iron is salvaged and returned to the red bone marrow, where it is used again in the formation of erythrocytes. The life span of an erythrocyte is approximately 120 days. Agranular leukocytes remain functional for 100–300 days under normal body conditions. Granular leukocytes, in contrast, have an extremely short life span of 12 hours to three days.

Hemopoiesis begins the same way in both myeloid and lymphoid tissues (fig. 20.48). Undifferentiated mesenchymal-like cells develop into stem cells called **hemocytoblasts.** These stem cells are able to divide rapidly; some of the daughter cells become new stem cells (so the stem cell population is never depleted), whereas other daughter cells become specialized along different paths of blood cell formation. Hemocytoblasts, for example, may develop into **proerythroblasts** *(pro''ē-rith'ro-blast)* that form erythrocytes, **myeloblasts** that form granular leukocytes (neutrophils, eosinophils and basophils), **lymphoblasts** that form lymphocytes, **monoblasts** that form monocytes, or **megakaryoblasts** *(meg''ah-kar'e-o-blast)* that form platelets (thrombocytes).

Red Blood Cell Antigens and Blood Typing

All cells of the body contain, on their surfaces, certain molecules that can be recognized as foreign by the immune system of another individual. These molecules are known as *antigens*. As part of the immune response, particular lymphocytes secrete a class of proteins, called *antibodies*, which bond in a specific fashion to antigens. The specificity of antibodies for antigens is analogous to the specificity of enzymes for substrates, or of receptor proteins for regulatory molecules. A complete description of antibodies and antigens is provided in chapter 22.

The distinguishing antigens on other cells are far more varied than the antigens on red blood cells. Red blood cell antigens, however, are of extreme clinical importance because their types must be matched between donors and recipients for blood transfusions. There are several groups of red blood cell antigens, but the major group is known as the ABO system. In terms of the antigens present on the red blood cell surface, a person may be *type A* (with only A antigens), *type B* (with only B antigens), *type AB* (with both A and B antigens), or *type O* (with neither A nor B antigens). It should again be noted that the blood type denotes the class of antigens (chemically, a type of glycolipid) found on the red blood cell surface.

ABO System. Each person inherits two genes (one from each parent) that control the production of the ABO antigens. The gene for A or B antigens is dominant to the gene for O, since the latter simply means the absence of A or B. A person who is type A, therefore, may have inherited the A gene from each parent (may have the genotype AA) or the A gene from one parent and the O gene from the other parent (and have the genotype AO). Likewise, a person who is type B may have the genotype BB or BO. It follows that a type O person inherited the O gene from each parent (has genotype OO), whereas a type AB person inherited the A gene from one parent and the B gene from the other (there is no dominance-recessive relationship between A and B).

The immune system is tolerant of its own red blood cell antigens. A person who is type A, for example, does not produce anti-A antibodies. Surprisingly, however, people with type A blood do make antibodies against the B antigen, and conversely, people with type B blood make antibodies against the A antigen. This is believed to result from the fact that antibodies made in response to some common bacteria can cross-react with the A or B antigens. A person who is type A, therefore, acquires antibodies that can react with B antigens by exposure to these bacteria but does not develop antibodies that can react with A antigens, because this is prevented by tolerance mechanisms.

People who are type AB develop tolerance to both of these antigens and thus do not produce either anti-A or anti-B antibodies. Those who are type O, in contrast, do not develop tolerance to either antigen and, therefore, have both anti-A and anti-B antibodies in their plasma (table 20.10).

---

hemopoiesis: Gk. *haima,* blood; *poiesis,* production
erythropoiesis: Gk. *erythros,* red; *poiesis,* production
leukopoiesis: Gk. *leukos,* white; *poiesis,* production
megakaryoblasts: Gk. *megas,* great; *karyon,* nut; *blastos,* germ

Figure 20.48. The processes of hemopoiesis. Formed elements begin as hemocytoblasts and differentiate into the various kinds of blood cells depending on the needs of the body.

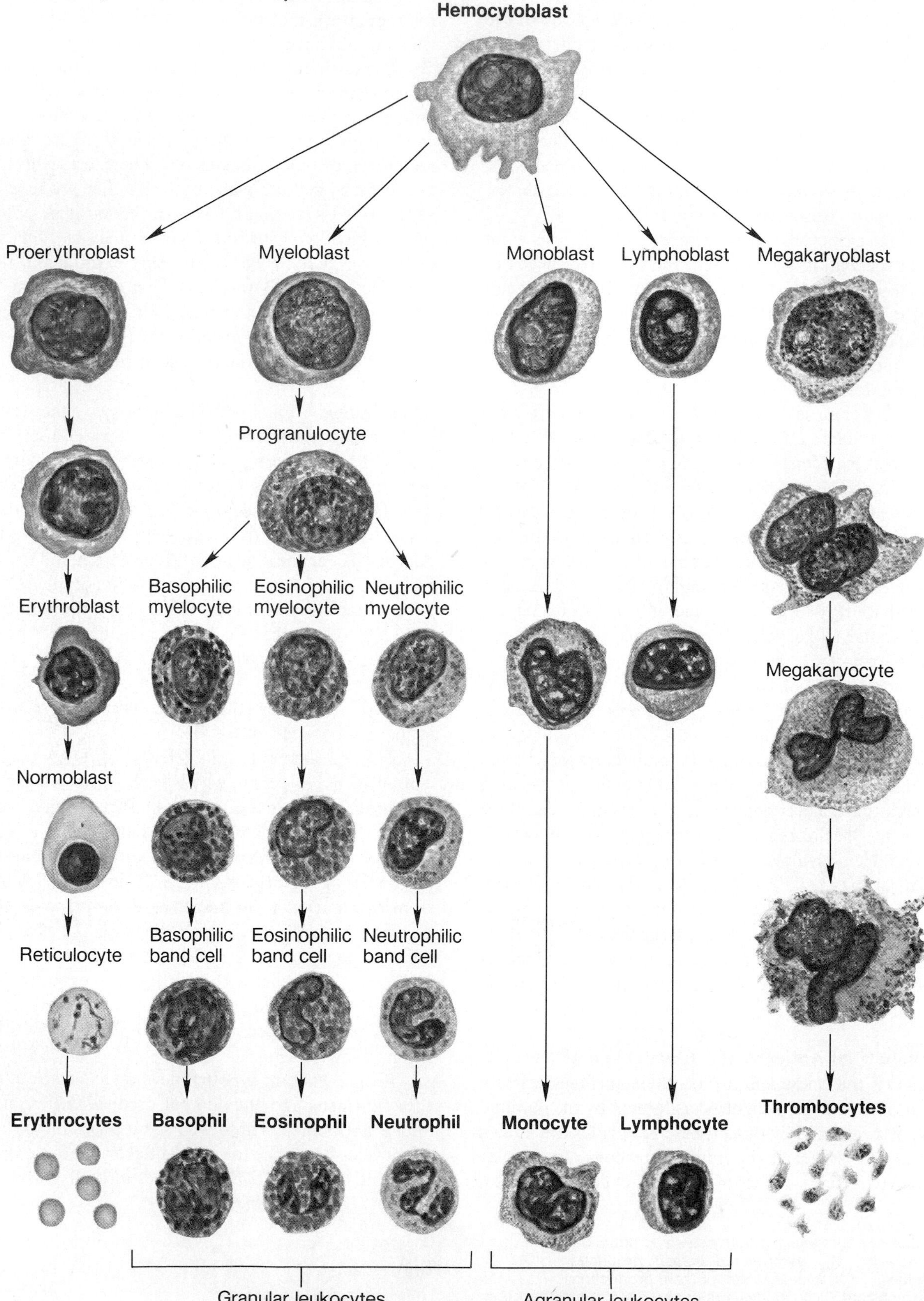

Figure 20.49. The agglutination (clumping) of red blood cells occurs when cells with A-type antigens are mixed with anti-A antibodies and when cells with B-type antigens are mixed with anti-B antibodies. No agglutination would occur with type O blood (not shown).

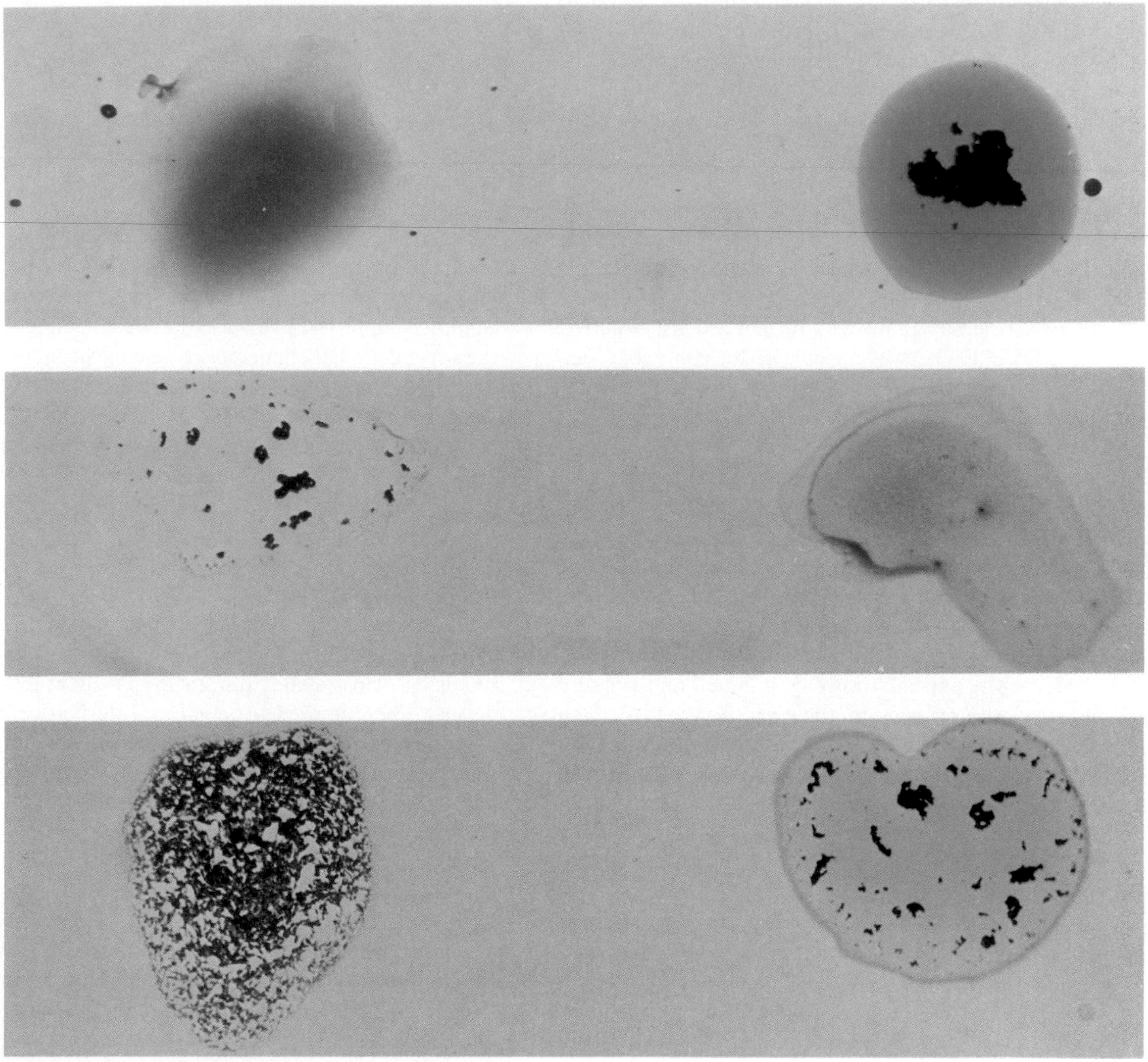

Table 20.10	The ABO system of antigens on red blood cells
Antigen on RBCs	**Antibody in plasma**
A	Anti-B
B	Anti-A
O	Anti-A and anti-B
AB	Neither anti-A nor anti-B

From Stuart Ira Fox, *Human Physiology,* 2d ed. Copyright © 1987 Wm. C. Brown Publishers, Dubuque, Iowa. All Rights Reserved. Reprinted by permission.

Transfusion Reactions. Before transfusions are performed, a *major crossmatch* is made by mixing serum from the recipient with blood cells from the donor. If the types do not match—if the donor is type A, for example, and the recipient is type B—the recipient's antibodies attach to the donor's red blood cells and form bridges that cause the cells to clump together, or **agglutinate** (fig. 20.49). Because of this agglutination *(ah-gloo''ti-na'shun)* reaction, the A and B antigens are sometimes called *agglutinogens (ag''loo-tin'o-jen),* and the antibodies against them are called *agglutinins (ah-gloo'ti-nin).* Transfusion errors that result in such agglutination in the blood can produce a blockage of small blood vessels and organ damage.

In emergencies, type O blood has been given to people who are type A, B, AB, or O. Since type O red blood cells lack A or B antigens, the recipient's antibodies cannot cause agglutination of the donor red blood cells. Type O is, therefore, a *universal donor,* but only as long as the volume of plasma donated is small, because plasma from a type O person would agglutinate type A, type B, and type AB red blood cells. Likewise, type AB people are *universal recipients* because they lack anti-A and anti-B antibodies and thus cannot agglutinate donor red blood cells. (Donor plasma could agglutinate recipient red blood cells if the transfusion volume were too large.) Because of the dangers involved, the universal donor and recipient concept in blood transfusions is strongly discouraged.

Rh Factor. Another important group of antigens found in most red blood cells is the *Rh factor* (Rh stands for Rhesus monkey, in which these antigens were first discovered). People who have these antigens are said to be **Rh positive,** whereas those who do not are **Rh negative.** There are fewer Rh negative people because this condition is recessive to Rh positive. The Rh factor is of particular significance when Rh negative mothers give birth to Rh positive babies.

Since the fetal and maternal blood are normally kept separate across the placenta, the Rh negative mother is not usually exposed to the Rh antigen of the fetus during the pregnancy. At the time of birth, however, a variable degree of exposure may occur, and the mother's immune system may become sensitized and produce antibodies against the Rh antigen. This does not always occur, however, because the exposure may be minimal and because Rh negative women vary in their sensitivity to the Rh factor. If the woman does produce antibodies against the Rh factor, these antibodies can cross the placenta in subsequent pregnancies and cause hemolysis of the Rh positive red blood cells of the fetus. The baby is therefore born anemic, with a condition called *erythroblastosis (e-rith"ro-blas-to'sis) fetalis.*

Erythroblastosis fetalis can be prevented by injecting the Rh negative mother with *antibodies against the Rh factor* (one trade name for this is RhoGAM) within seventy-two hours after the birth of each Rh positive baby. This is a type of passive immunization in which the injected antibodies inactivate the Rh antigens and thus prevent the mother from becoming actively immunized to them.

1. List the different classes of plasma proteins, and describe their functions.
2. Describe erythropoiesis in terms of its location, rate, and the specific steps that lead to the production of mature erythrocytes.
3. Describe the different types of leukocytes, and explain where and how these different types are produced.
4. Explain how platelets differ from other formed elements of blood, and describe how they are produced.
5. Explain the meaning of the term "blood type," and explain why matching of donor and recipient blood is required for blood transfusions.

Blood Clotting

Trauma to a blood vessel initiates a sequence of events resulting in the formation of a clot followed by the healing of the blood vessel and the dissolution of the clot.

Objective 30. Describe the mechanisms that lead to the formation of a platelet plug and the role of platelets in the activation of clotting factors.
Objective 31. Describe the intrinsic and extrinsic clotting pathways.
Objective 32. Explain how blood clots are dissolved.

When a blood vessel is injured, a number of physiological mechanisms are activated that promote **hemostasis** *(he-mos'tah-sis),* or the cessation of bleeding. Breakage of the endothelial lining of a vessel exposes collagen proteins from the subendothelial connective tissue to the blood. This initiates three separate, but overlapping, hemostatic mechanisms: (1) vasoconstriction; (2) the formation of a platelet plug; and (3) the production of a web of fibrin proteins around the platelet plug. Fibrin is produced by the successive activation of a number of clotting factors, most of which are produced by the liver.

Functions of Platelets

In the absence of vessel damage, platelets are repelled from each other and from the endothelial lining of vessels. The repulsion of platelets from an intact endothelium is believed to be due to *prostacyclin,* a derivative of prostaglandins, produced within the endothelium. Mechanisms that prevent platelets from sticking to the blood vessels and to each other are obviously needed to prevent inappropriate blood clotting.

Damage to the endothelium of vessels exposes subendothelial tissue to the blood. Platelets are able to stick to exposed collagen proteins that have become coated with a protein (*von Willebrand factor*) secreted by endothelial cells. Platelets contain secretory granules; when platelets stick to collagen, they *degranulate* as the secretory granules release their products. These products include *ADP* (adenosine diphosphate), *serotonin,* and a prostaglandin called *thromboxane* A_2. This event is known as the **platelet release reaction.**

Serotonin and thromboxane A_2 stimulate vasoconstriction, which helps to decrease blood flow to the injured vessel. Phospholipids that are exposed on the platelet membrane participate in the activation of clotting factors.

The release of ADP and thromboxane A_2 from platelets that are stuck to exposed collagen makes other platelets in the vicinity "sticky," so that they adhere to those stuck to the collagen. The second layer of platelets, in turn,

hemostasis: Gk. *haima,* blood; *stasis,* a standing

Figure 20.50. A scanning electron micrograph showing threads of fibrin.

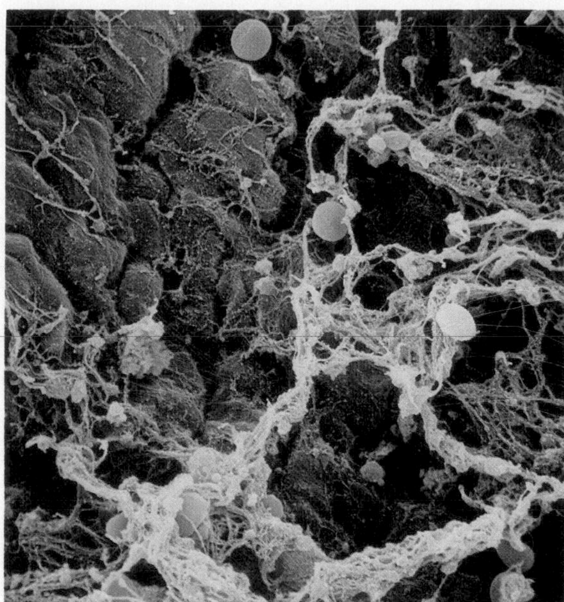

Figure 20.51. The sequence of events leading to platelet aggregation and the formation of a blood clot.

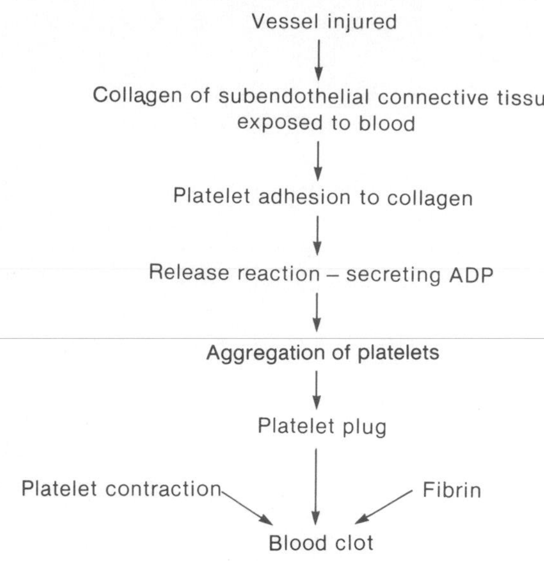

undergoes a platelet release reaction, and the ADP and thromboxane A_2 that are secreted cause additional platelets to aggregate at the site of injury. This produces a **platelet plug** in the damaged vessel, which is strengthened by the activation of plasma-clotting factors.

In order to undergo a release reaction, the production of prostaglandins by the platelets is required. *Aspirin* inhibits the conversion of arachidonic acid (a cyclic fatty acid) into prostaglandins and thus inhibits the release reaction and consequent formation of a platelet plug. The ingestion of excessive amounts of aspirin can thus significantly prolong bleeding time, which is why blood donors and women in the last trimester of pregnancy are advised to avoid aspirin.

Clotting Factors: Formation of Fibrin

The platelet plug is strengthened by a meshwork of insoluble protein fibers known as **fibrin** (fig. 20.50). Blood clots therefore contain platelets, fibrin, and usually trapped red blood cells that give the clot a red color (clots formed in arteries generally lack red blood cells and are gray in color). Finally, contraction of the platelets in the process of *clot retraction* forms a more compact and effective plug (fig. 20.51). Fluid squeezed from the clot as it retracts is called **serum,** which is plasma without fibrinogen (the soluble precursor of fibrin).

There are two pathways that result in the conversion of fibrinogen into fibrin. Blood left in a test tube will clot without the addition of any external chemicals; the pathway that produces this clot is thus called the **intrinsic pathway.** The intrinsic pathway also produces clots in damaged blood vessels when collagen is exposed to plasma. Damaged tissues, however, release a chemical that initiates a "shortcut" to the formation of fibrin. Since this chemical is not part of blood, the shorter pathway is called the **extrinsic pathway.**

Intrinsic Pathway. The intrinsic pathway is initiated by the exposure of plasma to negatively charged surfaces, such as that provided by collagen or glass. This activates a plasma protein called **factor XII** (table 20.11), which is a protein-digesting enzyme (protease). Active factor XII in turn activates another plasma protein—**factor XI**—by cleaving part of its inactive precursor. Activated factor XI, in turn, activates **factor IX.**

The next steps in the sequence require the presence of phospholipids, which are provided by platelets, and Ca^{++}. The combination of these with active factor IX and **factor VIII** forms a complex that activates **factor X.** A complex is then formed between active factor X, **factor V,** platelet phospholipids, and Ca^{++} that converts **prothrombin** (inactive **factor II**) to **thrombin.**

Thrombin is a protease that converts the soluble protein **fibrinogen** (**factor I**) into fibrin monomers. These monomers are joined together to form insoluble fibrin polymer by the action of **factor XIII.** The intrinsic clotting sequence is shown on the right side of figure 20.52.

Figure 20.52. The extrinsic and intrinsic clotting pathways that lead to the formation of insoluble fibrin polymers.

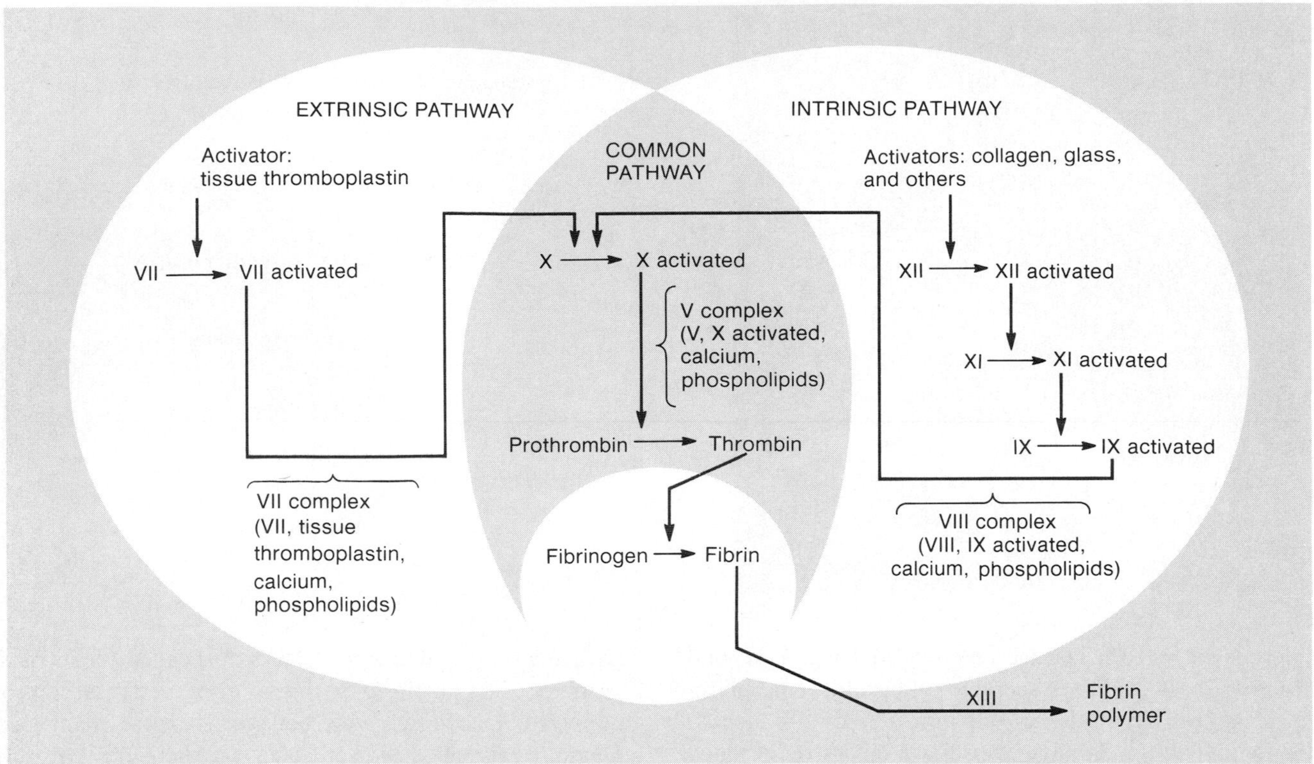

Table 20.11 The plasma-clotting factors

Factor	Name	Function	Pathway
I	Fibrinogen	converted to fibrin	common
II	Prothrombin	enzyme	common
III	Tissue thromboplastin	cofactor	extrinsic
IV	Calcium ions (Ca^{++})	cofactor	intrinsic, extrinsic, and common
V	Proaccelerin	cofactor	common
VII	Proconvertin	enzyme	extrinsic
VIII	Antihemophilic factor	cofactor	intrinsic
IX	Plasma thromboplastin component; Christmas factor	enzyme	intrinsic
X	Stuart-Prower factor	enzyme	common
XI	Plasma thromboplastin antecedent	enzyme	intrinsic
XII	Hageman factor	enzyme	intrinsic
XIII	Fibrin stabilizing factor	enzyme	common

Extrinsic Pathway. The formation of fibrin can occur more rapidly as a result of the release of **tissue thromboplastin** *(throm''bo-plas'tin),* or **factor III,** from damaged tissue cells. Tissue thromboplastin activates and combines with **factor VII;** factor VII, together with phospholipids and Ca^{++}, forms a complex (the VII complex) that activates factor X. The extrinsic clotting sequence is shown on the left side of figure 20.52.

The extrinsic and intrinsic pathways overlap at this point (see fig. 20.52) into a common pathway. The V complex, formed by active factor X, factor V, phospholipids, and Ca^{++} converts prothrombin to thrombin, which in turn converts fibrinogen to fibrin.

Dissolution of Clots

As the damaged blood vessel wall is repaired, activated factor XII promotes the conversion of another inactive molecule in plasma, **prekallikrein,** to the active form called **kallikrein** *(kal''ĭ-kre'in).* Kallikrein, in turn, catalyzes the conversion of **plasminogen** *(plaz-min'o-jen)* into the active molecule called **plasmin** (or, **fibrinolysin**). Plasmin is an enzyme that digests fibrin into "split products," thus promoting dissolution of the clot (fig. 20.53). Since a clotting factor (factor XII) initiates the pathway leading to dissolution of the clot, the action of plasmin represents the completion of a delayed negative feedback loop.

Figure 20.53. Events that produce dissolution of the blood clot and vasodilation.

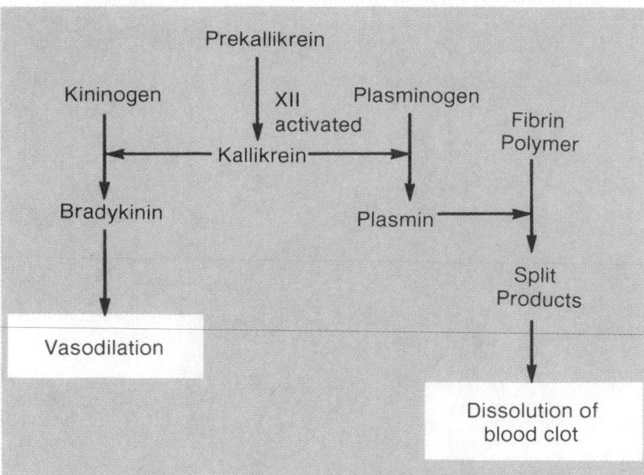

Tissue cells produce a number of compounds that activate plasminogen (converting it to plasmin), and thus aid in the dissolution of blood clots. An exciting recent development in genetic engineering technology is the commercial availability of one of these endogenous compounds, called *tissue plasminogen activator.* This compound may be used in the future to eliminate blood clots that obstruct the flow of blood in coronary vessels. Currently, an exogenous compound (produced by bacteria) which activates plasminogen, called *streptokinase,* is used for this purpose. Streptokinase is either injected into the general circulation or into a specific coronary vessel that has become occluded by a thrombus (blood clot).

Kallikrein, in addition to its function in the formation of plasmin, also stimulates the formation of a molecule called **bradykinin** *(brad"e-ki'nin)* from its precursor (**kininogen**). Bradykinin stimulates vasodilation, which helps to counter the vasoconstrictor effects of prostaglandins and serotonin released by platelets during the formation of a clot.

Anticoagulants. Clotting of blood in test tubes can be prevented by the addition of *citrate* or *EDTA,* which chelates (binds to) calcium. By this means Ca^{++} levels in the blood that can participate in the clotting sequence are lowered, and clotting is inhibited. A mucoprotein called **heparin** can also be added to the tube to prevent clotting. Heparin activates a plasma protein called antithrombin III, which combines with and inactivates thrombin. Heparin is also given intravenously during certain medical procedures to prevent clotting. Patients may also be given **coumarins** *(koo'mah-rin)* as anticoagulants. The coumarins prevent blood clotting by competing with vitamin K, which is needed for the formation of factors II, VII, IX, and X by the liver. In contrast to the immediate effects of heparin, therefore, coumarin must be given to a patient for several days to be effective.

bradykinin: Gk. *bradys,* slow; *kinesis,* movement

Table 20.12 Some acquired and inherited defects in the clotting mechanism

Category	Cause of disorder	Comments
Acquired clotting disorders	Vitamin K deficiency	Inadequate formation of prothrombin and other clotting factors in the liver
	Aspirin	Inhibits prostaglandin production, resulting in defective platelet release reaction
Anticoagulants	Coumarin	Competes with the action of vitamin K
	Heparin	Inhibits activity of thrombin
	Citrate	Combines with Ca^{++} and thus inhibits the activity of many clotting factors
Inherited clotting disorders	Hemophilia A (defective factor VIII_{AHF})	Recessive trait carried on X chromosome; results in delayed formation of fibrin
	Von Willebrand's disease (defective factor VIII_{VWF})	Dominant trait carried on autosomal chromosome; impaired ability of platelets to adhere to collagen in subendothelial connective tissue
	Hemophilia B (defective factor IX), also called Christmas disease	Recessive trait carried on X chromosome; results in delayed formation of fibrin

A number of hereditary diseases involve the clotting system. Examples of hereditary clotting disorders include two different genetic defects in factor VIII. A defect in one subunit of factor VIII prevents this factor from participating in the intrinsic clotting pathway. This genetic disease, called *hemophilia A,* is an X-linked recessive trait that is prevalent in the royal families of Europe. A defect in another subunit of factor VIII results in *von Willebrand's disease.* In this disease, rapidly circulating platelets are unable to stick to collagen, and a platelet plug cannot be formed. Some acquired and inherited defects in the clotting system are summarized in table 20.12.

1. Describe how a platelet plug is formed when a vessel is cut.
2. List the steps shared in common by the intrinsic and extrinsic clotting pathways.
3. Explain the meaning of the terms *intrinsic* and *extrinsic* in terms of the clotting pathways, and describe how these pathways differ from each other.
4. Describe the steps that lead to the formation of plasmin, and explain how this might be regarded as a negative feedback mechanism.

von Willebrand's disease: from E. A. von Willebrand, Finnish physician, 1870–1949

Figure 20.54. The placement of the bipolar limb leads and the exploratory electrode for the unipolar chest leads in an electrocardiogram (ECG).

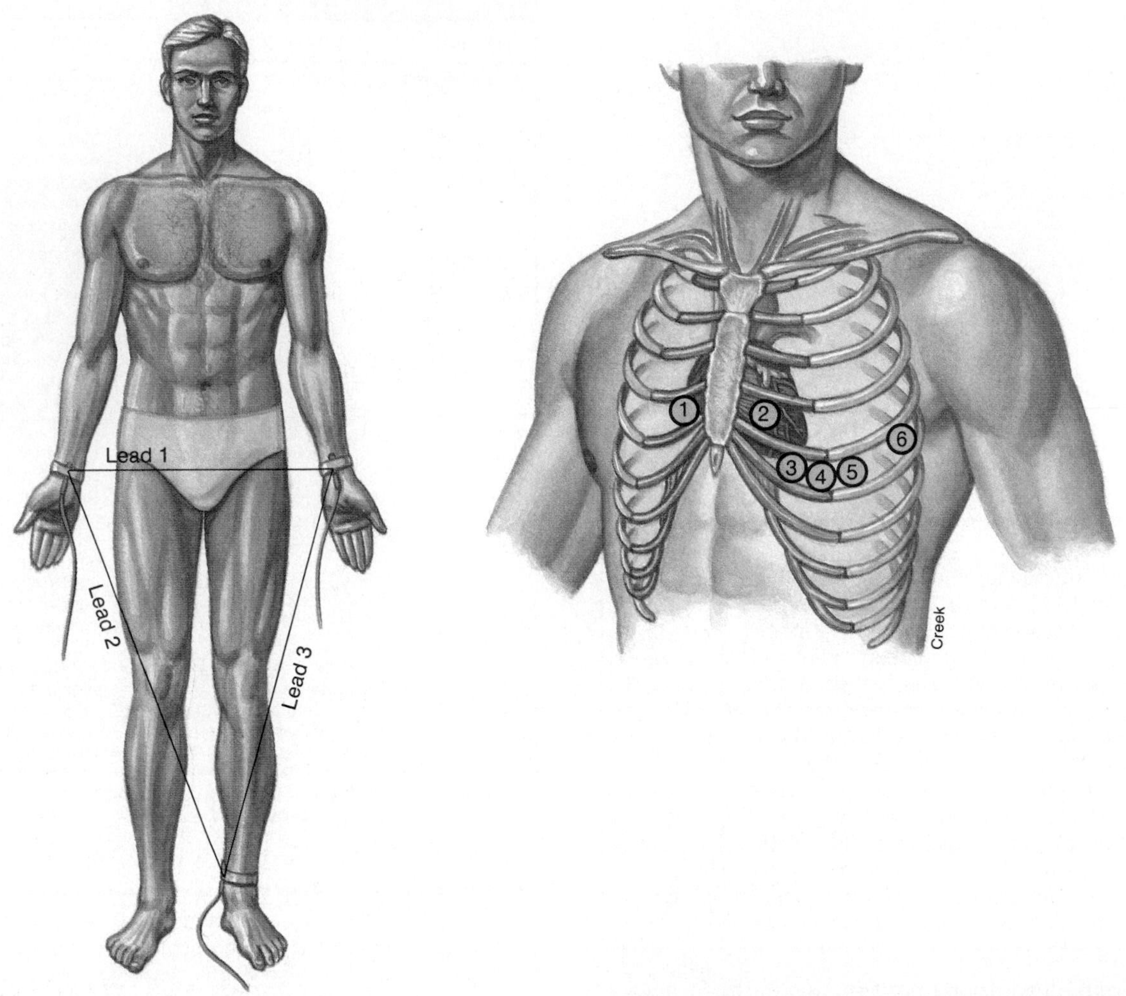

Clinical Considerations

ECG Leads

There are two types of ECG recording electrodes, or "leads." The *bipolar limb leads* record the voltage between electrodes placed on the wrists and legs. These bipolar leads include lead I (right arm to left arm), lead II (right arm to left leg), and lead III (left arm to left leg). In the *unipolar leads,* voltage is recorded between a single "exploratory electrode" placed on the body and an electrode that is built into the electrocardiograph and maintained at zero potential (ground).

The unipolar limb leads are placed on the right arm, left arm, and left leg, and are abbreviated AVR, AVL, and AVF, respectively. The unipolar chest leads are labeled one through six, starting from the midline position (fig. 20.54). There are thus a total of twelve standard ECG leads that "view" the changing pattern of the heart's electrical activity from different perspectives (table 20.13). This is important because certain abnormalities are best seen with particular leads and may not be visible at all with other leads.

Table 20.13 The electrocardiograph (ECG) leads

Name of lead	Placement of electrodes
Bipolar limb leads	
I	Right arm and left arm
II	Right arm and left leg
III	Left arm and left leg
Unipolar limb leads	
AVR	Right arm
AVL	Left arm
AVF	Left leg
Unipolar chest leads	
V_1	4th intercostal space right of sternum
V_2	4th intercostal space left of sternum
V_3	5th intercostal space to the left of the sternum
V_4	5th intercostal space in line with the middle of the clavicle
V_5	5th intercostal space to the left of V_4
V_6	5th intercostal space in line with the middle of the axilla

From Stuart Ira Fox, *Human Physiology,* 2d ed. Copyright © 1987 Wm. C. Brown Publishers, Dubuque, Iowa. All Rights Reserved. Reprinted by permission.

Figure 20.55. In (*a*) the heartbeat is paced by the normal pacemaker—the S-A node (hence the name sinus rhythm). This can be abnormally slow (bradycardia—46 beats per minute in this example) or fast (tachycardia—136 beats per minute in this example). Compare the pattern of tachycardia in (*a*) with the tachycardia in (*b*), which is produced by an ectopic pacemaker in the ventricles. This dangerous condition can quickly lead to ventricular fibrillation, also shown in (*b*).

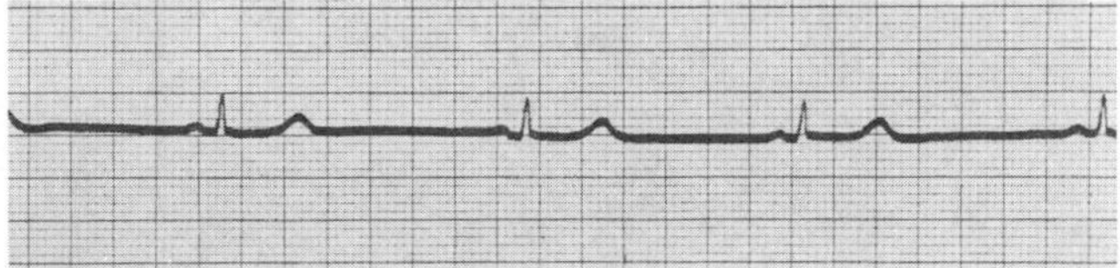

Sinus Bradycardia

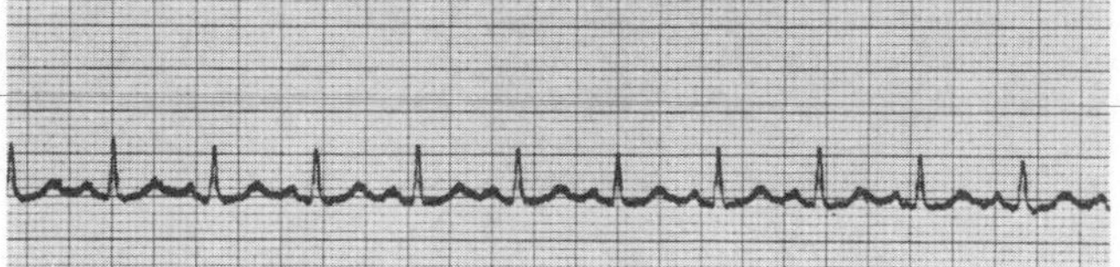

Sinus Tachycardia

(a)

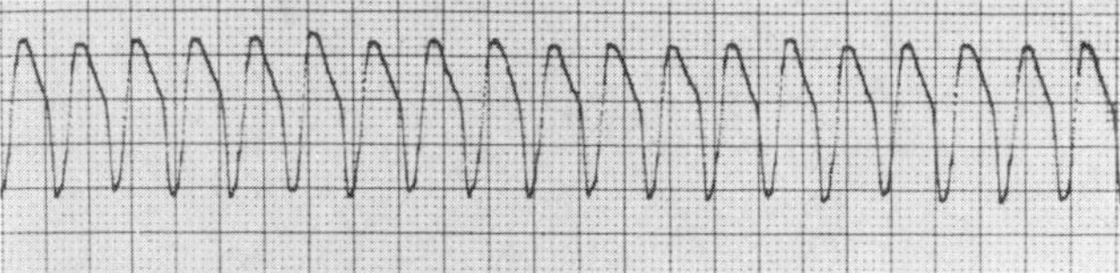

Ventricular tachycardia

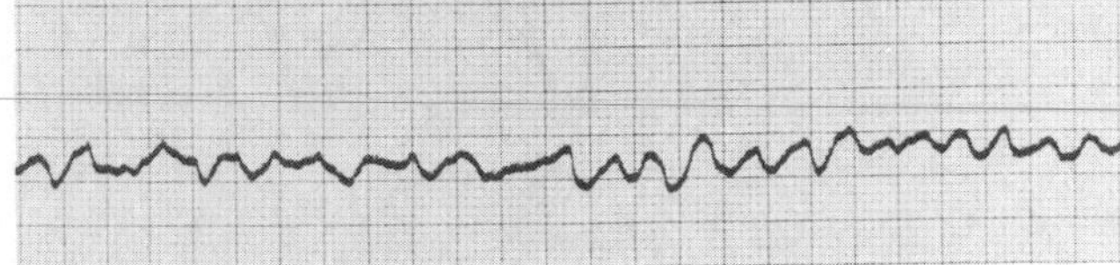

Ventricular fibrillation

(b)

Arrhythmias Detected by the Electrocardiograph

Arrhythmias, or abnormal heart rhythms, can be detected and described by the abnormal ECG patterns they produce. Although the proper clinical interpretation of electrocardiograms requires more knowledge than is presented in this chapter, some knowledge of abnormal rhythms is interesting in itself and is useful in understanding normal physiology.

Since a heartbeat occurs whenever a normal QRS complex is seen and since the ECG chart paper moves at a known speed, so that its x-axis indicates time, the cardiac rate (beats per minute) can easily be obtained from the ECG recording. A cardiac rate slower than 60 beats per minute indicates **bradycardia** *(brad″e-kar′de-ah);* a rate faster than 100 beats per minute is described as **tachycardia.**

Both bradycardia and tachycardia can occur normally. Endurance-trained athletes, for example, commonly have a slower heart rate than the general population. This *athlete's bradycardia* occurs as a result of higher levels of parasympathetic inhibition of the SA node and is a beneficial adaptation. Activation of the sympathetic system during exercise or emergencies causes a normal tachycardia to occur.

Abnormal tachycardia occurs when a person is at rest. This may result from abnormally fast pacing by the atria, due to drugs or to the development of abnormally fast *ectopic pacemakers*—cells located outside the SA node that assume a pacemaker function. This abnormal atrial tachycardia thus differs from normal "sinus" (SA node) tachycardia. *Ventricular tachycardia* results when abnormally fast ectopic pacemakers in the ventricles cause them to beat rapidly and independently of the atria (fig. 20.55). This is very dangerous because it can quickly degenerate into a lethal condition known as ventricular fibrillation.

Ventricular Fibrillation. Fibrillation is caused by a continuous recycling of electrical waves through the myocardium. This recycling is normally prevented by the fact that the myocardium enters a refractory period simultaneously at all regions. If some cells emerge from their refractory periods before others, however, electrical waves can be continuously regenerated and conducted. The recycling of electrical waves along continuously changing pathways produces uncoordinated contraction and an impotent pumping action. These effects can be produced by damage to the myocardium.

Fibrillation can sometimes be stopped by a strong electric shock delivered to the chest, a procedure called **electrical defibrillation.** The electric shock depolarizes all the myocardial cells at the same time, causing them to enter a refractory state. The conduction of random, recirculating impulses thus stops, and—within a short time—the SA node can begin to stimulate contraction in a normal fashion. This does not correct the initial problem that caused the abnormal electrical patterns, but it does keep the person alive long enough to take other corrective measures.

Structural Heart Disorders

Congenital heart problems result from abnormalities in embryonic development and may be attributed to heredity, nutritional problems of the pregnant mother, or viral infections such as rubella. Congenital heart diseases occur in approximately 3 of every 100 births and account for about 50% of early childhood deaths.

Heart murmurs are both congenital and acquired. Nearly 10% of the population have heart murmurs, but most are not clinically significant. In general, three basic conditions cause murmurs: (1) *valvular insufficiency,* in which the cusps of the valves do not form a tight seal;

Figure 20.56. Abnormal patterns of blood flow due to septal defects. Left-to-right shunting of blood is shown *(circled areas)* because the left pump is at a higher pressure than the right pump. Under certain conditions, however, the pressure in the right atrium may exceed that of the left, causing right-to-left shunting of blood through a septal defect in the atria.

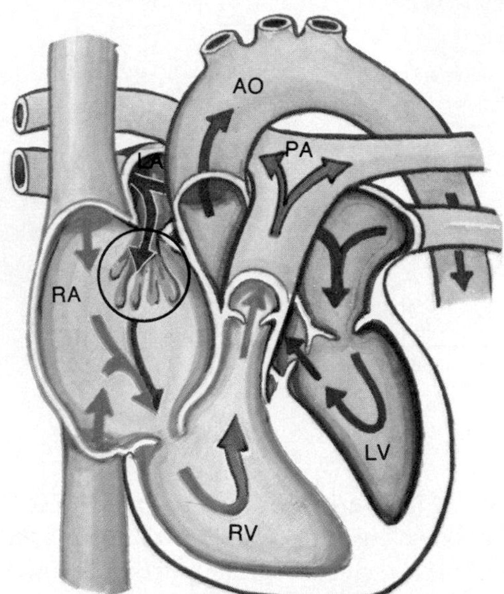

Septal defect
in atria

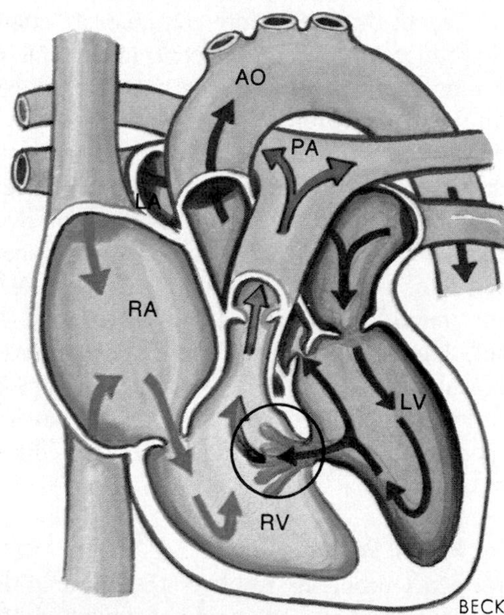

Septal defect
in ventricles

Figure 20.57. The flow of blood through a patent (open) ductus arteriosus.

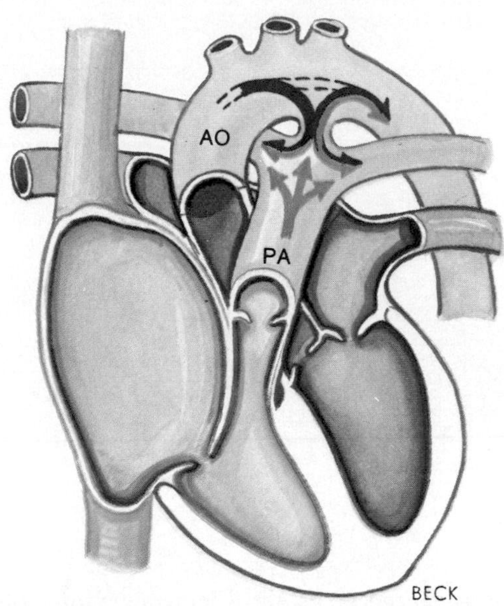

(2) *stenosis,* in which the walls surrounding a valve are roughened or constricted; and (3) a *functional murmur,* which is frequent in children and is caused by turbulence of the blood moving through the heart during heavy exercise. Functional murmurs are not pathological and are considered normal.

A **septal defect** (fig. 20.56) is the most common type of congenital heart problem. An atrial septal defect, or **patent foramen ovale,** is a failure of the fetal foramen ovale to close after birth. A **ventricular septal defect** is caused by an abnormal development of the interventricular septum. **Pulmonary stenosis** is a narrowing of the opening into the pulmonary trunk from the right ventricle. It may lead to a pulmonary embolism and is usually recognized by extreme lung congestion. A **patent ductus arteriosus** (fig. 20.57) is a failure of the ductus arteriosus to close after birth, allowing a backflow of blood into the pulmonary circulation from the arch of the aorta.

The **tetralogy of Fallot** is a combination of four defects within a newborn and immediately causes a cyanotic condition, leading to the newborn being termed a "blue baby." The four characteristics of this anomaly are (1) a ventricular septal defect, (2) an overriding aorta, (3) pulmonary stenosis, and (4) right ventricular hypertrophy (fig. 20.58). Pulmonary stenosis obstructs blood flow to the lungs and causes hypertrophy of the right ventricle. Open-heart surgery is required to correct this condition, and the overall mortality is about 5%.

Acquired heart diseases include those that result from arterial damage and those that result from bacterial infection. *Bacterial endocarditis* is a disease of the lining of the heart, especially of the valve cusps. It is caused by bacteria that enter the bloodstream. The most common cause of valve defects results from infection by the same organisms that cause rheumatic fever.

stenosis: Gk. *stenosis,* a narrowing
tetralogy of Fallot: from Etienne-Louis A. Fallot, French physician, 1850–1922

Figure 20.58. The tetralogy of Fallot. The four defects of this anomaly are (*1*) a ventricular septal defect; (*2*) an overriding aorta; (*3*) pulmonary stenosis; and (*4*) right ventricular hypertrophy.

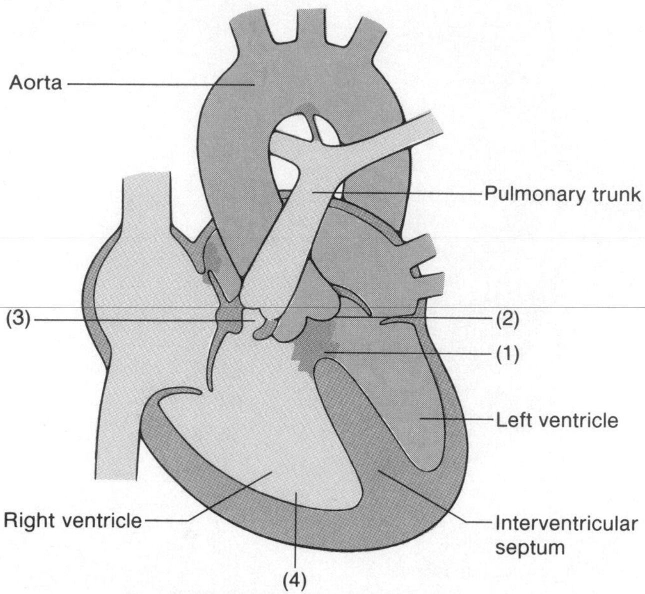

Atherosclerosis

Atherosclerosis is the most common form of arteriosclerosis (hardening of the arteries) and, through its contribution to heart disease and stroke, is responsible for about 50% of the deaths in the United States. In atherosclerosis, localized *plaques,* or **atheromas,** protrude into the lumen of the artery and thus reduce blood flow. The atheromas additionally serve as sites for *thrombus* (a blood clot attached to the wall of a blood vessel) formation, which can further occlude the blood supply to an organ (fig. 20.59).

It is currently believed that atheromas begin as smooth muscle cells that migrate from the tunica media to the intima and proliferate by cell division. In later stages, these cells accumulate cholesterol and other lipids, giving them a "foamy cell" appearance. In fully developed plaques, cholesterol and other lipids accumulate outside the smooth muscle cells, and these accumulations can become calcified. Damage to the endothelium that covers the atheromas results in the exposure of subendothelial connective tissue to the blood. This may contribute to further growth of the atheroma and to the formation of blood clots.

Risk factors in the development of atherosclerosis include advanced age, smoking, hypertension, and high blood cholesterol concentrations. It is currently believed that an intact and properly functioning endothelium protects against atherosclerosis. If the endothelium in a particular region of an artery is removed, or if it is damaged in some way, growth factors, or *mitogens* (chemicals that stimulate mitosis), may stimulate proliferation of the smooth muscle cells and the growth of an atheroma. These growth factors may be derived from blood platelets, endothelial cells, and/or monocytes.

Cholesterol and Plasma Lipoproteins. There is good evidence that high blood cholesterol is associated with an increased risk of atherosclerosis. This high blood cholesterol can be produced by a diet rich in cholesterol and saturated fat, or it may be the result of an inherited condition known as *familial hypercholesteremia.* This condition is inherited as a single dominant gene; individuals who inherit two of these genes have extremely high cholesterol concentrations (regardless of diet) and usually suffer heart attacks during childhood.

Cholesterol is carried to the arteries by plasma proteins called **low-density lipoproteins (LDL).** These particles, produced by the liver, consist of a core of cholesterol surrounded by a layer of phospholipids (to make the particle water soluble) and a protein. Cells in various organs contain receptors for the protein in LDL; when LDL attaches to its receptors, the cell engulfs the LDL by receptor-mediated endocytosis (described in chapter 5) and utilizes the cholesterol for different purposes. Most of the LDL in the blood is removed in this way by the liver.

Once LDL has passed through the endothelium of an artery, it may stimulate monocytes to enter the area and engulf the cholesterol (thereby becoming "foam cells"). The monocytes may then be stimulated by LDL to secrete a growth factor that either begins or contributes to the development of an atheroma. A high blood concentration of LDL favors these events. Recent evidence shows that people whose diet is high in cholesterol and saturated fat and people with familial hypercholesteremia have a high blood LDL concentration because their tissues (principally the liver) have a low number of LDL receptors. With fewer LDL receptors the liver is less able to remove the LDL from the blood, the blood LDL concentration is raised, and the risk of atherosclerosis is greatly increased.

Excessive cholesterol may be released from cells and travel in the blood as **high-density lipoproteins (HDL),** which are removed by the liver. Since the cholesterol in HDL does not travel to the blood vessels, it does not contribute to atherosclerosis. Indeed, a high proportion of cholesterol in HDL as compared to LDL is beneficial, since it indicates that cholesterol may be traveling away from the blood vessels to the liver. The concentration of HDL-cholesterol appears to be higher and the risk of atherosclerosis lower in people who exercise regularly. The HDL-cholesterol concentration, for example, is higher in marathon runners than in joggers and is higher in joggers than in inactive men. Women in general have higher HDL-cholesterol concentrations and a lower risk of atherosclerosis than men.

Ischemic Heart Disease. A tissue is said to be **ischemic** *(is-kem'ik)* when it receives an inadequate supply of oxygen because of an inadequate blood flow. The most common cause of myocardial ischemia is atherosclerosis of the coronary arteries. The adequacy of blood flow is relative—it depends on the metabolic requirements of the

Figure 20.59. (*a*) the lumen (opening) of a human coronary artery is partially occluded by an atheroma and (*b*) almost completely occluded by a thrombus. (*c*) close-up view of the cleared left anterior descending coronary artery containing calcified atherosclerotic plaques. The heart is of an eighty-five-year-old female. (*d*) the structure of an atheroma is diagramed.

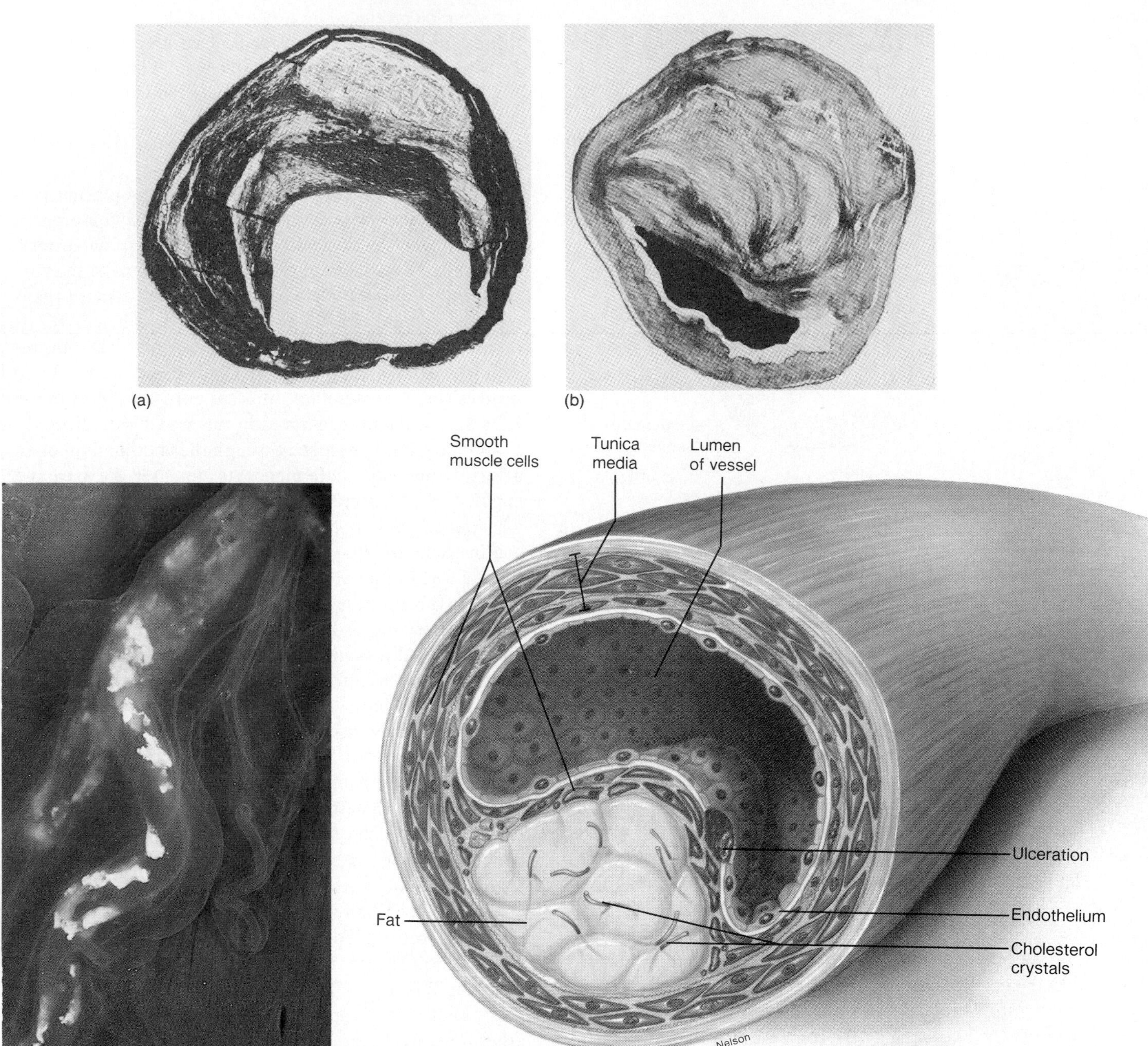

tissue for oxygen. An obstruction in a coronary artery, for example, may allow sufficient blood flow at rest but may produce ischemia when the heart is stressed by exercise or emotional conditions.

Myocardial ischemia is associated with increased concentrations of blood lactic acid produced by anaerobic respiration of the ischemic tissue. This condition often causes substernal pain, which may also be referred to the left shoulder and arm, as well as other areas. This referred pain is called **angina pectoris.** People with angina frequently take nitroglycerin or related drugs that help relieve the ischemia and pain. These drugs are effective because they stimulate vasodilation, which improves circulation to the heart and decreases the work that the heart must perform to eject blood into the arteries.

Myocardial cells are adapted to respire aerobically and cannot respire anaerobically for more than a few minutes. If ischemia and anaerobic respiration continue for

Figure 20.60. Depression of the S-T segment of the electrocardiogram as a result of myocardial ischemia.

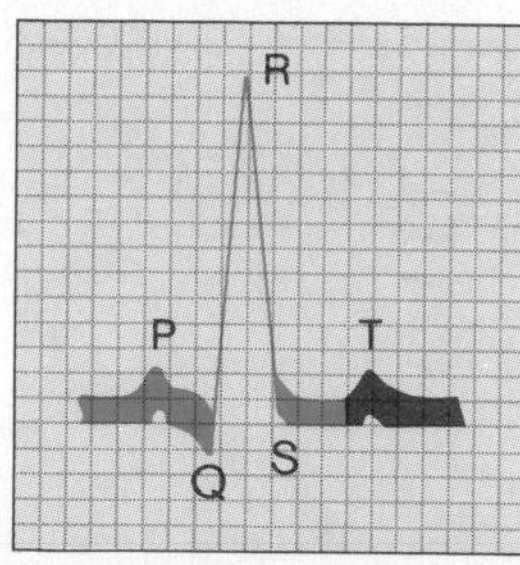

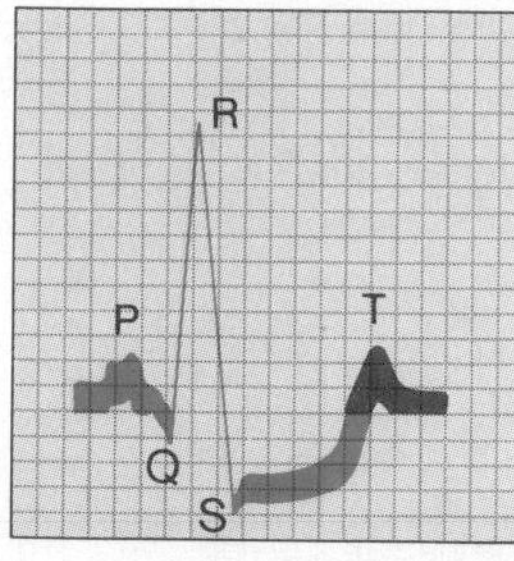

Table 20.14 Changes in the enzyme activity in plasma following a myocardial infarction

Serum enzyme	Earliest increase (hr)	Maximum concentration (hr)	Return to normal (days)	Amplitude of increase (× normal)
Creatine kinase	3–6	24–36	3	7
Malate dehydrogenase	4–6	24–48	5	4
AST	6–8	24–48	4–6	5
Lactate dehydrogenase	10–12	48–72	11	3
α-Hydroxybutyrate dehydrogenase	10–12	48–72	13	3–4
Aldolase	6–8	24–48	4	4
ALT	Usually normal, unless there are other complications			
Isocitrate dehydrogenase	Usually normal			

more than a few minutes, *necrosis* (cellular death) may occur in the areas most deprived of oxygen. A sudden, irreversible injury of this kind is called a **myocardial infarction,** or **MI.**

Myocardial ischemia may be detected by changes in the S-T segment of the electrocardiogram (fig. 20.60). The diagnosis of myocardial infarction is aided by measurement of the concentration of enzymes in the blood that are released by the infarcted tissue. Plasma concentrations of *creatine phosphokinase* (*CPK*), for example, increase within three to six hours after the onset of symptoms and return to normal after three days. Plasma levels of *lactate dehydrogenase* (*LDH*) reach a peak within forty-eight to seventy-two hours after the onset of symptoms and remain elevated for about eleven days (table 20.14).

Other Vascular Disorders

An **aneurism** *(an'u-rizm)* is an expansion or bulging of the heart, aorta, or any other artery. A **coarctation** is a constriction in a segment of a vessel, usually the aorta, and is frequently caused by a remnant of the ductus arteriosus tightening around the vessel. **Varicose veins** are weakened veins that become stretched and swollen. Varicose veins are most common in the legs because the force of gravity tends to weaken the valves and overload the veins. Varicose veins can also occur in the rectum, in which case they are called **hemorrhoids.** *Vein stripping* is the surgical removal of superficial weakened veins. **Phlebitis** *(flĕ-bi'tis)* is inflammation of a vein. It may develop as a result of trauma or an aftermath of surgery. Frequently, however, it appears for no apparent reason. Phlebitis interferes with normal venous circulation.

Chapter Summary

I. Functions and Major Components of the Circulatory System
   A. The blood transports oxygen and nutrients to the tissue cells and removes waste products from the tissues; the blood also serves a regulatory function through its transport of hormones.
      1. Oxygen is carried by red blood cells, or erythrocytes.
      2. The white blood cells, or leukocytes, serve to protect the body from disease.
   B. The circulatory system consists of the cardiovascular system (heart, blood vessels, and blood) and the lymphatic system.

II. Development of the Heart
   A. The heart develops from a heart tube that becomes subdivided into regions called the truncus arteriosus, bulbus cordis, ventricle, atrium, and sinus venosus.
   B. The ventricle is divided into right and left chambers by the growth of an interventricular septum.
   C. A septum also forms in the atrium, but in the fetus it contains an opening called the foramen ovale, which allows communication between the right and left atria.

III. Structure of the Heart
   A. The right and left sides of the heart pump blood through the pulmonary and systemic circulations.
      1. The right ventricle pumps blood to the lungs; this blood then returns to the left atrium.

   2. The left ventricle pumps blood into the aorta and systemic arteries; this blood then returns to the right atrium.
- B. The heart contains two pairs of one-way valves.
   1. The atrioventricular valves allow blood to go from the atria to the ventricles, but not in the reverse direction.
   2. The semilunar valves allow blood to leave the ventricles and enter the pulmonary and systemic circulations, but these valves prevent blood from returning from the arteries to the ventricles.
- C. The pulmonary circulation involves the pulmonary arteries and pulmonary veins; all other arteries and veins in the body are part of the systemic circulation.

IV. Cardiac Cycle, Heart Sounds, and Electrocardiogram
- A. The heart is a two-step pump; first the atria contract, and then the ventricles contract.
   1. During diastole, first the atria and then the ventricles fill with blood.
   2. The ventricles are about 80% filled before the atria contract and add the final 20% to the end-diastolic volume.
- B. When the ventricles contract at systole, the pressure within them first rises sufficiently to close the AV valves and then rises sufficiently to open the semilunar valves.
   1. Blood is ejected from the ventricles until the pressure within them falls below the pressure in the arteries; at this point the semilunar valves close and the ventricles begin relaxation.
   2. When the pressure in the ventricles falls below the pressure in the atria, a phase of rapid filling of the ventricles occurs, followed by the final filling caused by contraction of the atria.
- C. Closing of the AV valves produces the first heart sound, or "lub", at systole; closing of the semilunar valves produces the second heart sound, or "dub", at diastole. Abnormal valves can cause abnormal sounds called murmurs.
- D. The electrical impulse begins in the sinoatrial node and spreads through both atria by electrical conduction from one myocardial cell to another.
   1. The impulse then excites the atrioventricular node, from which it is conducted by the atrioventricular bundle into the ventricles.
   2. The Purkinje fibers transmit the impulse into the ventricular muscle and cause it to contract.
- E. The regular pattern of conduction in the heart produces a changing pattern of potential differences between two points on the body surface.
   1. Measurements of the voltage between two points on the surface of the body caused by the electrical activity of the heart is called an electrocardiogram (ECG).
   2. The P wave is caused by depolarization of the atria; the QRS wave is caused by depolarization of the ventricles; the T wave is produced by repolarization of the ventricles.

V. Blood Vessels
- A. Arteries contain three layers, or tunics: the intima, media, and externa.
   1. The tunica intima consists of a layer of endothelium, which is separated from the tunica media by a band of elastin fibers.
   2. The tunica media consists of smooth muscle.
- B. Capillaries are the narrowest but the most numerous of the blood vessels.
   1. Capillary walls consist of only a single layer of endothelial cells, which provides for the exchange of molecules between the blood and the surrounding tissues.
   2. The flow of blood from arterioles to capillaries is regulated by precapillary sphincter muscles.
   3. The capillary wall may be continuous, fenestrated, or discontinuous.
- C. Veins have the same three tunics as arteries, but they generally have a thinner muscular layer than comparably sized arteries.
   1. Veins are more distensible than arteries and can expand to hold a larger quantity of blood.
   2. Many veins have venous valves that permit a one-way flow of blood to the heart.

VI. Principal Arteries of the Body
- A. Three arteries arise from the aortic arch: the brachiocephalic, the left common carotid, and the left subclavian; the brachiocephalic divides into the right common carotid and the right subclavian.
- B. The head and neck receive an arterial supply from branches of the internal and external carotid arteries and the vertebral arteries.
- C. The upper extremity is served by the subclavian artery and its derivatives.
- D. The abdominal aorta produces the following branches: the celiac, superior mesenteric, renal, suprarenal, testicular (or ovarian), and inferior mesenteric arteries; the abdominal aorta divides into the right and left iliac arteries.
- E. The common iliac arteries divide into the internal and external iliac arteries, which supply blood to the pelvis and lower extremities.

VII. Principal Veins of the Body
- A. Blood from the head and neck is drained by the external and internal jugular veins; blood from the brain is drained by the internal jugular vein.
- B. The upper extremity is drained by superficial and deep veins.
- C. In the thorax, the superior vena cava is formed by the union of the two brachiocephalic veins and also receives blood from the azygos vein.
- D. The lower extremity is drained by both superficial and deep veins; at the level of the fifth lumbar vertebra, the right and left iliac veins unite to form the inferior vena cava.
- E. Blood from capillaries in the digestive tract is drained via the hepatic portal vein to the liver.

VIII. Fetal Circulation
- A. Fully oxygenated blood is carried only in the umbilical vein, which drains the placenta; this blood is carried via the ductus venosus to the inferior vena cava of the fetus.
- B. Partially oxygenated blood is shunted from the right to the left atrium via the foramen ovale and from the pulmonary trunk to the aorta via the ductus arteriosus.

IX. Composition of the Blood
- A. Plasma is the fluid part of the blood, containing dissolved ions and various organic molecules.
   1. Hormones are found in the plasma portion of the blood.
   2. Plasma proteins are divided into albumins and alpha, beta, and gamma globulins.
- B. The formed elements of the blood include erythrocytes, leukocytes, and platelets.
   1. Erythrocytes, or red blood cells, contain hemoglobin and transport oxygen.
   2. Leukocytes may be granular (also called polymorphonuclear) or agranular; they function in immunity.
   3. Thrombocytes, or platelets, are required for blood clotting.
- C. Hemopoiesis occurs in the bone marrow and lymphoid tissue; all blood cells derive from a single type of stem cell.
- D. The ABO and Rh blood groups refer to antigens present on the surface of red blood cells.
   1. Except for people who are type AB, the plasma contains antibodies against blood types other than the person's own.
   2. Mixing one person's red blood cells with plasma from a person with a different blood type causes the red blood cells to agglutinate.

X. Blood Clotting
   A. When a blood vessel is damaged, platelets adhere to the exposed subendothelial collagen proteins.
   B. In the formation of a blood clot, a soluble protein called fibrinogen is converted into insoluble threads of fibrin.
      1. This reaction is catalyzed by the enzyme thrombin.
      2. Thrombin is derived from its inactive precursor, called prothrombin, by either an intrinsic or an extrinsic pathway.
      3. The clotting sequence requires Ca^{++} and phospholipids to be present in the platelet cell membranes.
   C. Dissolution of the clot eventually occurs by the digestive action of plasmin, which cleaves fibrin into split products.

Review Activities

Objective Questions

1. All arteries in the body contain oxygen-rich blood with the exception of the
   (a) aorta
   (b) pulmonary arteries
   (c) renal arteries
   (d) coronary arteries
2. Most blood from the coronary circulation directly enters the
   (a) inferior vena cava
   (b) superior vena cava
   (c) right atrium
   (d) left atrium
3. The second heart sound immediately follows the occurrence of the
   (a) P wave
   (b) QRS wave
   (c) T wave
   (d) U wave
4. Which of the following arteries does *not* arise from the aortic arch? The
   (a) brachiocephalic
   (b) coronary
   (c) left common carotid
   (d) left subclavian
5. Which of the following arteries do *not* supply blood to the brain? The
   (a) external carotid
   (b) internal carotid
   (c) vertebral
   (d) basilar
6. The maxillary and superficial temporal arteries are branches from the
   (a) external carotid artery
   (b) internal carotid artery
   (c) vertebral artery
   (d) facial artery
7. Which of the following statements is *false*?
   (a) Most of the total blood volume is contained in veins.
   (b) Capillaries have a greater total surface area than any other type of vessel.
   (c) Exchanges between blood and tissue fluid occur across the walls of venules.
   (d) Small arteries and arterioles present great resistance to blood flow.
8. The "lub," or first heart sound, is produced by closing of
   (a) the aortic semilunar valve
   (b) the pulmonary semilunar valve
   (c) the tricuspid valve
   (d) the bicuspid valve
   (e) both AV valves
9. The first heart sound is produced at the
   (a) beginning of systole
   (b) end of systole
   (c) beginning of diastole
   (d) end of diastole
10. Changes in the cardiac rate primarily reflect changes in the duration of
    (a) systole
    (b) diastole
11. The QRS wave of an ECG is produced by
    (a) depolarization of the atria
    (b) repolarization of the atria
    (c) depolarization of the ventricles
    (d) repolarization of the ventricles
12. The cells that normally have the fastest rate of spontaneous diastolic depolarization are located in the
    (a) SA node
    (b) AV node
    (c) atrioventricular bundle
    (d) Purkinje fibers
13. Which of the following statements is *true?*
    (a) The heart can produce a graded contraction.
    (b) The heart can produce a sustained contraction.
    (c) The action potentials produced at each cardiac cycle normally travel around the heart in circus rhythms.
    (d) All of the myocardial cells in the ventricles are normally in a refractory period at the same time.
14. The activation of factor X is
    (a) part of the intrinsic pathway only
    (b) part of the extrinsic pathway only
    (c) part of both the intrinsic and extrinsic pathways
    (d) not part of either the intrinsic or extrinsic pathways
15. Platelets
    (a) form a plug by sticking to each other
    (b) release chemicals that stimulate vasoconstriction
    (c) provide phospholipids needed for the intrinsic pathway
    (d) all of the above

Essay Questions

1. Explain why the beat of the heart is automatic and why the SA node functions as the normal pacemaker.
2. Compare the duration of the heart's contraction with the myocardial action potential and refractory period. Explain the significance of these relationships.
3. Describe the pressure changes that occur during the cardiac cycle, and relate these changes to the occurrence of the heart sounds.
4. Describe the causes of the P, QRS, and T waves of an ECG, and indicate when each of these waves occurs in the cardiac cycle. Explain why the first heart sound occurs immediately after the QRS wave and why the second sound occurs at the time of the T wave.
5. Can a defective valve be detected by an ECG? Can a partially damaged AV node be detected by auscultation with a stethoscope? Explain.
6. Describe the functions of the foramen ovale and ductus arteriosus in a fetus, and explain why these can be dangerous if they remain patent after birth.
7. Trace the flow of blood from the left ventricle to the upper teeth.
8. Trace the flow of blood from the intestine, to the heart, and back to the intestine.
9. Define the term *portal system,* and explain the functional significance of this system in the liver and pituitary gland.
10. Explain how a cut in the skin initiates both the instrinsic and extrinsic clotting pathways. Which pathway finishes first? Why?

21 Cardiac Output, Blood Flow, and Blood Pressure

Outline

Concepts

The cardiac output is a measure of the pumping ability of the heart and is affected by mechanisms that regulate the cardiac rate and stroke volume.

Fluid in the extracellular environment of the body is distributed between the plasma and the tissue fluid compartments. Blood volume is affected by the function of the kidneys. Through their actions on the kidneys, ADH and aldosterone help regulate the blood volume.

The rate of blood flow is dependent in part on vascular resistance. Resistance is regulated by vasoconstriction and vasodilation, produced by extrinsic and intrinsic mechanisms.

Blood flow to the heart and skeletal muscles is regulated by both intrinsic and extrinsic mechanisms that increase the rate of flow when the metabolic requirements of these tissues increase during exercise.

Control mechanisms help to maintain relatively constant rates of cerebral blood flow. Blood flow to the skin is varied according to the needs of the body.

The blood pressure rise from diastolic to systolic levels provides the driving force for blood flow. Blood pressure is regulated by a variety of control mechanisms and is usually measured indirectly by the auscultatory method.

Cardiac Output

The cardiac output is a measure of the pumping ability of the heart and is affected by mechanisms that regulate the cardiac rate and stroke volume.

Objective 1. Describe how the cardiac output is affected by cardiac rate and stroke volume.
Objective 2. Describe the effects of autonomic stimulation of the heart.
Objective 3. Explain the Frank-Starling law of the heart.
Objective 4. Explain how the venous return of blood to the heart is regulated.

The cardiac output is equal to the milliliters (ml) of blood pumped per minute by each ventricle. The average resting **cardiac rate** in an adult is 70 beats per minute; the average **stroke volume** (ml of blood pumped per beat by each ventricle) is 70–80 ml per beat. The product of these two variables gives an average **cardiac output** of 4900–5600 ml per minute:

cardiac output (ml/min) = stroke volume (ml/beat) × cardiac rate (beats/min).

The **total blood volume** is also equal to about 5–6 liters. This means that each ventricle pumps the equivalent of the total blood volume each minute under resting conditions. Put another way, it takes about a minute for a drop of blood to complete the pulmonary and systemic circuits. An increase in cardiac output, as occurs during exercise, must thus be accompanied by an increased rate of blood flow through the circulation. This is accomplished, in part, by factors that regulate the cardiac rate and stroke volume.

Regulation of Cardiac Rate

In the complete absence of neural influences, the heart will continue to beat according to the rhythm set by the SA node. This automatic rhythm is produced by the spontaneous decay of the resting membrane potential to a threshold depolarization, at which point voltage-regulated membrane gates are opened, action potentials are produced, and contraction occurs.

Normally, however, sympathetic and parasympathetic (vagus) nerve fibers to the heart are continuously active. Norepinephrine released primarily by sympathetic nerve endings and epinephrine secreted by the adrenal medulla stimulate an increase in the spontaneous rate of firing of the SA node. Acetylcholine released from parasympathetic endings hyperpolarizes the SA node and thus decreases the rate of its spontaneous firing (fig. 21.1). The actual pace set by the SA node at any time depends on the net effect of these antagonistic influences.

Autonomic innervation of the SA node represents the major means by which cardiac rate is regulated. Autonomic nerves do, however, affect cardiac rate by other mechanisms to a lesser degree. Sympathetic endings in the

Figure 21.1. The effects of sympathetic and parasympathetic activity on the pacemaker function of the SA node. Sympathetic stimulation causes action potentials to be generated more rapidly, and parasympathetic inhibition causes action potentials to be generated more slowly, than would occur in the absence of autonomic influences (*shaded areas*).

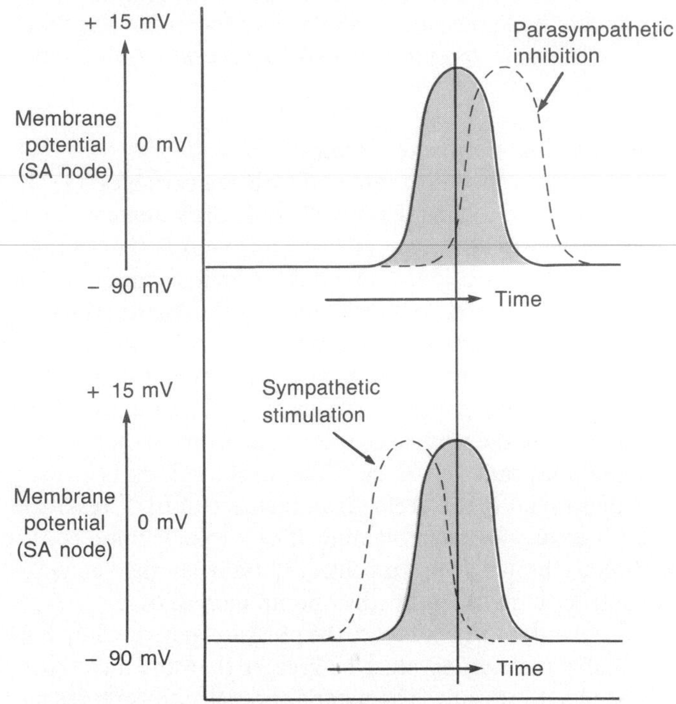

Table 21.1 Effects of autonomic nerve activity on the heart

Region affected	Sympathetic nerve effects	Parasympathetic nerve effects
SA node	Increased rate of diastolic depolarization; increased cardiac rate	Decreased rate of diastolic depolarization; decreased cardiac rate
AV node	Increased conduction rate	Decreased conduction rate
Atrial muscle	Increased strength of contraction	Decreased strength of contraction
Ventricular muscle	Increased strength of contraction	No significant effect

musculature of the atria and ventricles increase the strength of contraction and decrease slightly the time spent in systole when the cardiac rate is high (table 21.1).

In exercise the cardiac rate increases first as a result of decreased vagus nerve inhibition and then due to increased sympathetic nerve stimulation. The resting bradycardia of endurance-trained athletes is due largely to increased vagus nerve activity. These changes are coordinated by *cardiac control centers* in the *medulla oblongata* of the brain stem. There may be separate cardioaccelerator and cardioinhibitory centers (this is currently controversial) in the medulla that control autonomic innervation to the heart. These centers, in turn, are affected by higher brain centers and by sensory feedback from pressure receptors, called *baroreceptors,* in a reflex fashion.

Regulation of Stroke Volume

The stroke volume is regulated by three variables: (1) the **end-diastolic volume (EDV),** which is the volume of blood in the ventricles at the end of diastole; (2) the average, or mean, aortic blood pressure, which is measured as the **mean arterial pressure;** and (3) the **contractility,** or strength, **of ventricular contraction.**

The end-diastolic volume is the amount of blood in the ventricles just prior to contraction. This is a work load imposed on the ventricles prior to contraction and is thus sometimes called a *preload.* The stroke volume is directly proportional to the preload; an increase in EDV results in an increase in stroke volume. The stroke volume is also directly proportional to contractility; when the ventricles contract more strongly, they pump more blood.

In order to eject blood, the pressure generated by ventricular contraction must be greater than the mean arterial blood pressure. The arterial pressure thus represents an impedence to the ejection of blood from the ventricles, or an *afterload* imposed on the ventricles after contraction has begun. The stroke volume is inversely proportional to the afterload; the greater the mean arterial pressure (other factors being equal), the lower the stroke volume.

The proportion of the EDV that is ejected against a given afterload depends upon the strength of ventricular contraction. Normally, contraction strength is sufficient to eject 70–80 ml of blood out of a total end-diastolic volume of 110–130 ml. The *ejection fraction* is thus about two-thirds. Amazingly, this fraction remains relatively constant as the EDV increases. This implies that the strength of ventricular contraction must increase as the EDV increases.

Intrinsic Control of Contraction Strength. Two physiologists, Frank and Starling, demonstrated that normally the strength of ventricular contraction varies directly with the end-diastolic volume. Even in experiments where the heart is removed from the body (and is thus not subject to neural or hormonal regulation) and where the heart is filled with blood flowing from a reservoir, an increase in EDV within the physiological range results in increased contraction strength and, therefore, in increased stroke volume. This relationship between EDV, contraction strength, and stroke volume is thus an **intrinsic,** or built-in, **property of heart muscle.**

Frank-Starling Law of the Heart. The intrinsic control of contraction strength and stroke volume is due to variations in the degree to which the myocardium is stretched by the EDV. As the EDV rises within the physiological range, the myocardium is increasingly stretched and, as a result, contracts more forcibly. This mechanism that allows the heart to adjust intrinsically to changing demands is known as the **Frank-Starling law of the heart.**

Stretch can also increase the contraction strength of skeletal muscles. The resting length of skeletal muscles, however, is close to ideal, so that significant stretching decreases contraction strength. This is not true of the heart. Prior to filling with blood during diastole, the sarcomere lengths of myocardial cells are only about 1.5 μm. At this length the actin filaments from each side overlap in the middle of the sarcomeres, and the cells can only contract weakly (fig. 21.2).

As the ventricles fill with blood, the myocardium stretches so that the actin filaments overlap with myosin only at the edges of the A bands (fig. 21.2). This allows more force to be developed during contraction. Since this more advantageous overlapping of actin and myosin is produced by stretching of the ventricles, and since the degree of stretching is controlled by the degree of filling (the end-diastolic volume), the strength of contraction is intrinsically adjusted so that more blood will be ejected as the EDV rises.

Extrinsic Control of Contractility. The *contractility* is the contraction strength at any given fiber length. At any given degree of stretch, the strength of ventricular contraction depends on the activity of the sympathoadrenal system. Norepinephrine from sympathetic endings and epinephrine from the adrenal medulla produce an increase in contraction strength. This is often called a **positive inotropic** *(in''o-trop'ik)* **effect.**

The cardiac output is thus affected in two ways by the activity of the sympathoadrenal system: the positive inotropic effect on contractility, and the **positive chronotropic** *(kron''o-trop'ik)* **effect** that increases cardiac rate (fig. 21.3). Stimulation through the parasympathetic nerve to the heart has a **negative chronotropic effect** (decreases cardiac rate), but does not directly affect the contraction strength of the ventricles.

Starling: Ernest Henry Starling, English physiologist, 1866–1927
chronotropic: Gk. *chronos,* time; *trope,* turn, change
inotropic: Gk. *inos,* fiber; *trope,* turn, change

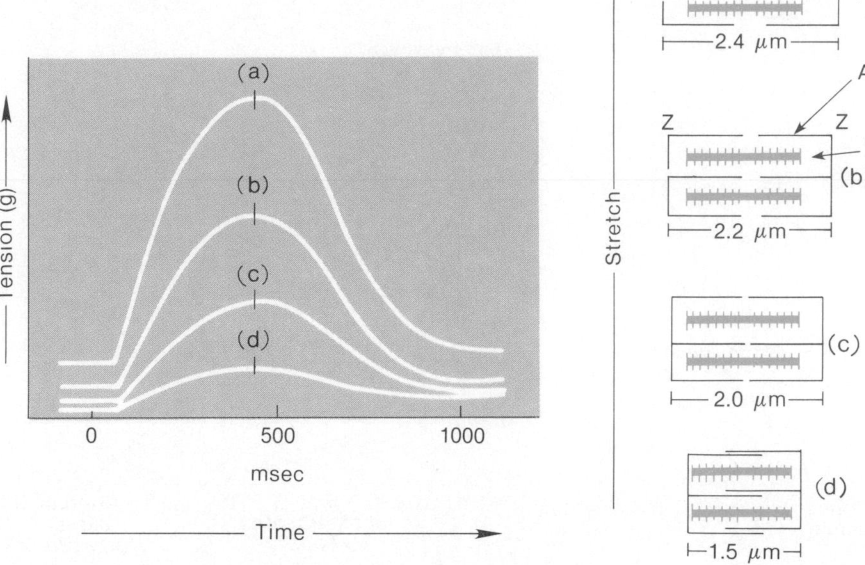

Figure 21.2. The Frank-Starling mechanism (law of the heart). When the heart muscle is subject to increased amounts of stretch, it contracts more strongly. As a result of the increased contraction strength (shown as tension), the time required to reach maximum contraction is the same regardless of the degree of stretch.

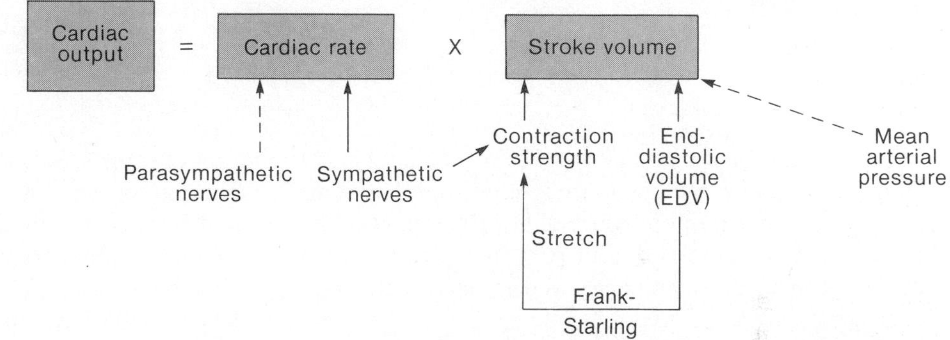

Figure 21.3. The regulation of cardiac output. Factors that stimulate cardiac output are shown as solid arrows; factors that inhibit cardiac output are shown as dotted arrows.

Venous Return

The end-diastolic volume—and thus the stroke volume and cardiac output—is controlled by factors that affect the *venous return* of blood to the heart. The rate at which the atria are filled with venous blood depends on the total blood volume and the venous pressure, which serves as the driving force for the return of blood to the heart.

Veins have thinner, less muscular walls than do arteries and thus have a higher *compliance.* This means that a given amount of pressure will cause more distension (expansion) in veins than in arteries, so that the veins can hold more blood. Approximately two-thirds of the total blood volume is located in the veins (fig. 21.4). Veins are therefore called *capacitance vessels,* by analogy with capacitors in electronics, which can accumulate electrical charges. Muscular arteries and arterioles expand less under pressure (are less compliant) and thus are called *resistance vessels.*

Figure 21.4. The distribution of blood within the circulatory system at rest. (From *Circulation* by Bjorn Folkow and Eric Neil. Copyright © 1971 by Oxford University Press, Inc. Reprinted by permission.)

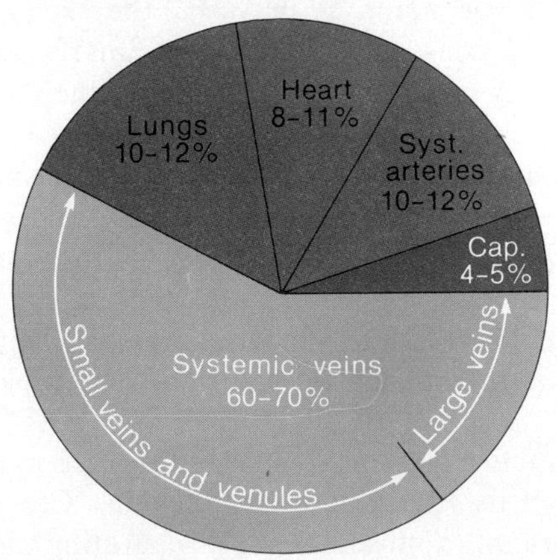

Figure 21.5. Variables that affect venous return and thus end-diastolic volume. Direct relationships are indicated by solid arrows; inverse relationships are shown with dashed arrows.

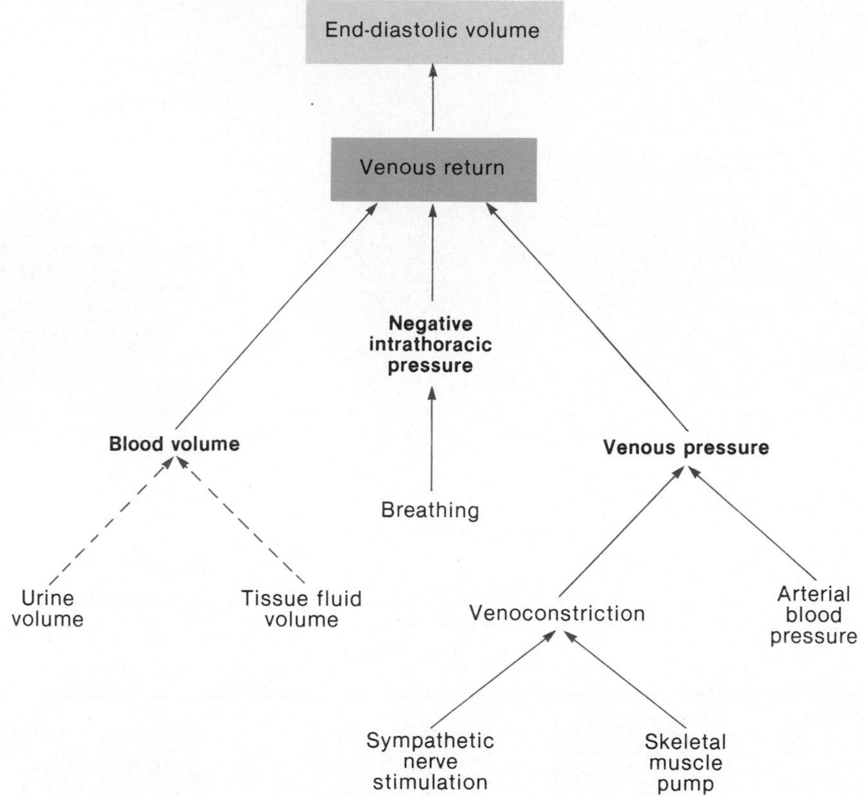

Although veins contain almost 70% of the total blood volume, the mean venous pressure is only 2 mm Hg, compared to a mean arterial pressure of 90–100 mm Hg. The lower venous pressure is due in part to a pressure drop between arteries and capillaries and in part to the high venous compliance.

The venous pressure is highest in the venules (10 mm Hg) and lowest at the junction of the vena cavae with the right atrium (0 mm Hg). In addition to this pressure difference, the venous return to the heart is aided by (1) sympathetic nerve activity; (2) the skeletal muscle pump, which constricts veins during muscle contraction; and (3) the pressure difference between the thoracic and abdominal cavities.

Sympathetic stimulation of the veins causes some contraction of the smooth muscle walls. This does not cause much constriction of the lumina of veins, in contrast to the effects on arteries; it does, however, make the veins less compliant. This reduced venous compliance helps to raise venous pressure and aid the return of venous blood to the heart.

Contraction of skeletal muscles functions as a "pump" by virtue of its squeezing action on veins. Contraction of the diaphragm (a muscular sheet separating the thoracic and abdominal cavities) during inhalation also improves venous return. This results from the fact that as the diaphragm contracts, it lowers to increase the thoracic volume and decrease the abdominal volume. This creates a partial vacuum in the thoracic cavity and a higher pressure in the abdominal cavity. The pressure difference thus produced favors blood flow from abdominal to thoracic veins (fig. 21.5).

1. Describe how the stroke volume is intrinsically regulated by the end-diastolic volume, and explain how this regulation is accomplished.
2. Describe the effects of autonomic nerve stimulation on the cardiac rate and stroke volume.
3. Define the terms *preload* and *afterload,* and explain how these affect the cardiac output.
4. List the factors that affect venous return, and draw a flowchart to show how an increased venous return can result in an increased cardiac output.

Blood Volume

Fluid in the extracellular environment of the body is distributed between the plasma and the tissue fluid compartments. Blood volume is affected by the function of the kidneys. Through their actions on the kidneys, ADH and aldosterone help regulate the blood volume.

Figure 21.6. The daily intake and excretion of body water and its distribution between different intracellular and extracellular compartments. (From *Circulation* by Bjorn Folkow and Eric Neil. Copyright © 1971 by Oxford University Press, Inc. Reprinted by permission.)

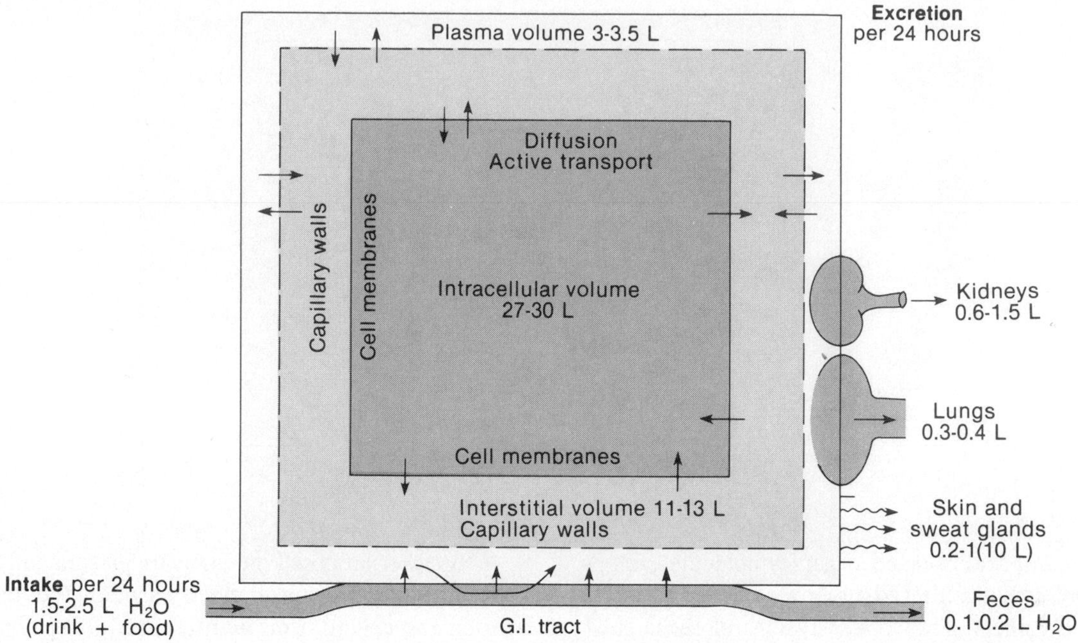

Objective 5. Describe how tissue fluid is formed from and returns to the blood capillaries.

Objective 6. Describe the regulation of antidiuretic hormone (ADH) secretion, and the effects of ADH on the blood volume.

Objective 7. Describe the regulation of aldosterone secretion and the effects of aldosterone on blood volume and pressure.

Blood volume represents one part, or "compartment," of the total body water. Approximately two-thirds of the total body water is located within cells—in the **intracellular compartment.** The remaining one-third is in the **extracellular compartment.** This extracellular fluid is normally distributed so that about 80% is in the tissues—as tissue or **interstitial** *(in''ter-stish'al)* **fluid**—and 20% is in the blood plasma (fig. 21.6).

The distribution of water between these compartments is determined by a balance between opposing forces. Blood pressure, for example, promotes formation of tissue fluid at the expense of plasma, while osmotic forces draw water from tissues into the vascular system. The total volume of intracellular and extracellular fluid is normally maintained constant by a balance between water loss and water gain. Mechanisms that affect drinking, urine volume, and the distribution of water between plasma and tissue fluid thus help regulate blood volume, and by this means help regulate cardiac output and blood flow.

Exchange of Fluid between Capillaries and Tissues

The distribution of extracellular fluid between the plasma and interstitial compartments is in a state of dynamic equilibrium. Tissue fluid is not normally a "stagnant pond" but is rather a continuously circulating medium, formed from and returning to the vascular system. In this way, the tissue cells receive a continuously fresh supply of glucose and other plasma solutes that are filtered through tiny endothelial channels in the capillary walls.

Filtration results from the blood pressure within the capillaries. This hydrostatic pressure, which is exerted against the inner capillary wall, is equal to about 30 mm Hg at the arteriolar end of systemic capillaries and decreases to 10–15 mm Hg at the venular end of the capillaries. This produces an average capillary pressure (near the middle) of about 17 mm Hg.

The **net filtration pressure** is equal to the hydrostatic pressure of the blood in the capillaries minus the hydrostatic pressure of tissue fluid outside the capillaries. If, as an extreme example, these two values were equal, there would be no filtration. The normal value of tissue hydrostatic pressure is currently controversial. Most researchers believe that this value is actually a *negative pressure,* meaning that it is lower than the pressure of the atmosphere. If the tissue fluid hydrostatic pressure is -6.5 mm Hg, as believed by some authorities, the average net filtration pressure would be $17 - (-6.5) = 23.5$ mm Hg.

Figure 21.7. Tissue or interstitial fluid is formed by filtration as a result of blood pressures at the arteriolar ends of capillaries and returned to venular ends of capillaries by the colloid osmotic pressure of plasma proteins. Solid arrows indicate the net direction of flow.

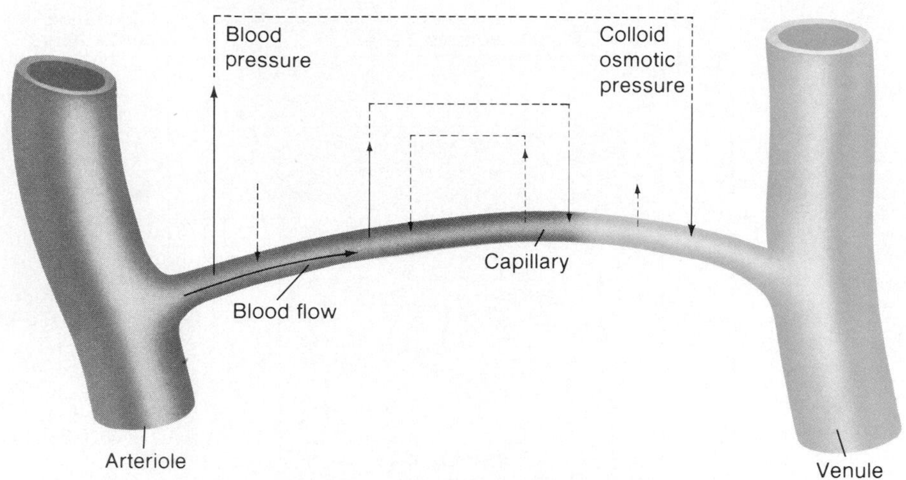

Glucose, comparably sized organic molecules, inorganic salts, and ions are filtered along with water through the capillary channels. The concentrations of these substances in tissue fluid are thus the same as in plasma. The protein concentration of tissue fluid (2 g/100 ml), however, is less than the protein concentration of plasma (6–8 g/100 ml). This difference is due to the fact that filtration of proteins is restricted by the capillary pores. The osmotic pressure exerted by plasma proteins—called the **colloid osmotic pressure** of the plasma—is, therefore, greater than the colloid osmotic pressure of tissue fluid.

The colloid osmotic pressure of plasma averages 28 mm Hg, and that of tissue fluid averages 5 mm Hg. The difference between these two values (23 mm Hg)—called the *oncotic pressure*—represents the net force favoring osmosis of water into the capillaries. If these values are correct, this pressure is lower than the filtration pressure at the arteriolar ends of capillaries and greater than the filtration pressure at the venular ends of capillaries. Fluid and solutes would thus be filtered out of capillaries at the arteriolar ends, and fluid would be returned by osmosis to the venular ends of capillaries (fig. 21.7).

This "classic" view of capillary dynamics has recently been challenged by some investigators who believe that when capillaries are open, the net filtration force exceeds the force for the osmotic return of water throughout the length of the capillary. They believe that the opposite is true in closed capillaries (capillaries can be opened or closed by the action of precapillary sphincter muscles). Net filtration, in summary, would occur in open capillaries, whereas net absorption of water would occur in closed capillaries.

By either proposed mechanism, plasma and tissue fluid are continuously interchanged through the action of filtration and colloid osmotic pressure forces. The return of fluid to the vascular system by osmosis, however, does not exactly equal the amount filtered. According to some estimates, approximately 85% of the capillary filtrate is directly returned to the capillaries by osmosis; the remaining 15% (amounting to at least 2 L per day) is returned to the vascular system by way of **lymphatic vessels** (fig. 21.8). Although the lymphatic system is covered in detail in chapter 22, a brief description will be presented here in the context of extracellular fluid regulation.

Lymphatic System and Edema

Excessive accumulation of tissue fluid and filtered proteins is normally prevented by drainage into highly permeable, blind-ending lymphatic capillaries located in the connective tissues. Tissue fluid that enters these lymphatic capillaries is called *lymph (limf),* and it is drained into *lymphatic vessels.* These vessels, like veins, have one-way valves, and skeletal muscle contraction serves as a pump by squeezing lymphatic vessels as well as veins. In addition to the pressure generated by the skeletal muscle pump, lymphatic vessels display automatic, rhythmic contraction that helps to move the lymph.

Lymph is eventually transported to the *right lymphatic duct* and the *thoracic duct,* which drain into large veins. As lymph passes through this system it percolates through *lymph nodes,* which contain antibody-producing lymphocytes. The functions of lymph nodes and lymphocytes will be described in detail in chapter 22.

Figure 21.8. The lymphatic system returns excess tissue fluid to the venous system.

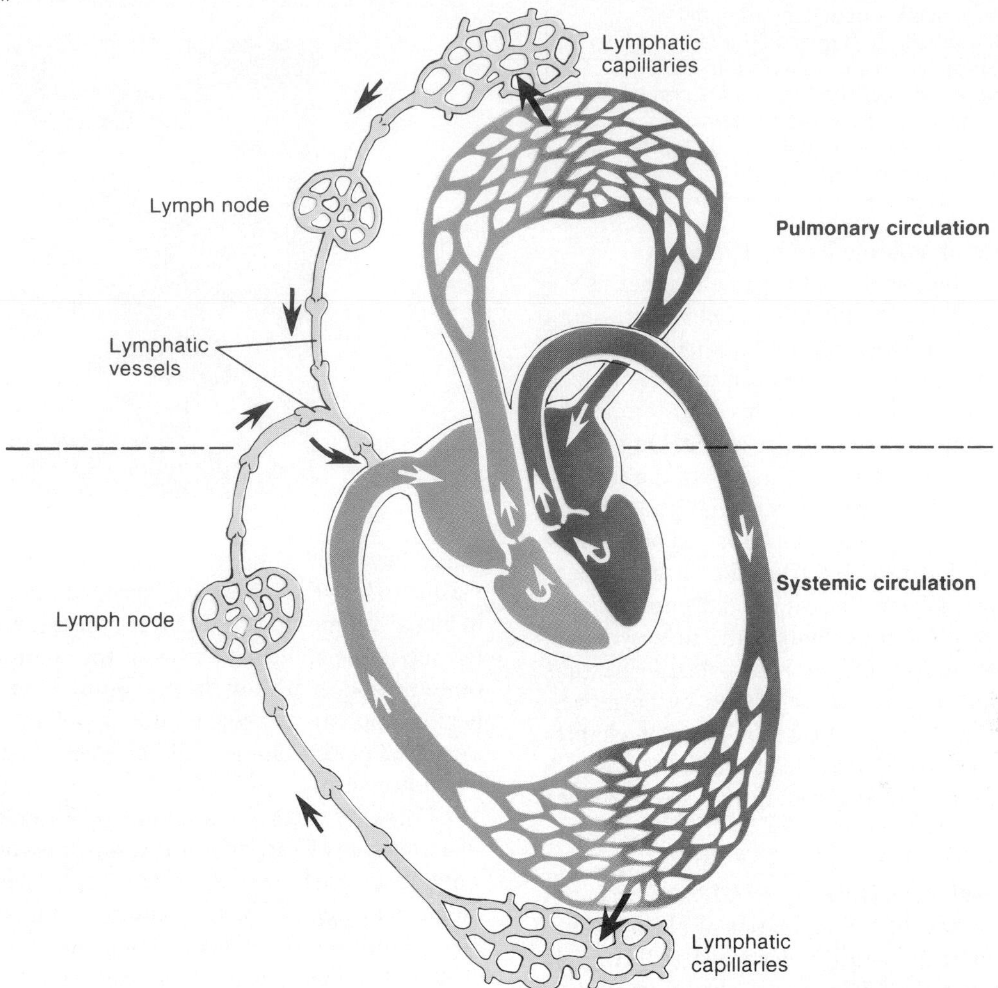

Causes of Edema. Excessive accumulation of tissue fluid is known as **edema** *(ĕ-de'mah)*. This condition is normally prevented by a proper balance between capillary filtration and osmotic uptake of water and by proper lymphatic drainage. Edema may thus result from (1) *high blood pressure,* which increases capillary pressure and causes excessive filtration; (2) *leakage of plasma proteins into tissue fluid,* which causes reduced osmotic flow of water into the capillaries (this occurs during inflammation and allergic reactions as a result of increased capillary permeability); (3) *excessive production of particular glycoproteins (mucin) in the interstitial spaces caused by hypothyroidism*—this produces a characteristic type of edema called *myxedema;* (4) *decreased plasma protein concentration* as a result of liver disease (the liver makes most of the plasma proteins), or as a result of kidney disease in which proteins are excreted in the urine; and (5) *obstruction of the lymphatic drainage* (table 21.2).

Table 21.2 The causes of edema

Variables that cause edema	Effects
Increased blood pressure	Increases filtration pressure so that more tissue fluid is formed at the arteriolar ends of capillaries
Increased tissue protein concentration	Decreases osmosis of water into the venular ends of capillaries. Usually a localized tissue edema due to leakage of plasma proteins through capillaries during inflammation and allergic reactions. Myxedema due to hyperthyroidism is also in this category.
Decreased plasma protein concentration	Decreases osmosis of water into the venular ends of capillaries. May be caused by liver disease (which can be associated with insufficient plasma protein production), kidney disease (due to leakage of plasma protein into urine), or protein malnutrition
Obstruction of lymphatic vessels	Infections by filaria roundworms (nematodes) transmitted by a certain species of mosquito block lymphatic drainage, causing edema and tremendous swelling of the affected areas (a condition called elephantiasis)

In the tropical disease *filariasis (fil''ah-ri'ah-sis),* mosquitoes transmit a nematode worm parasite to humans. The larvae of these worms invade lymphatic vessels and block lymphatic drainage. The edema that results can be so severe that the tissues swell to produce an elephantlike appearance—a condition aptly termed *elephantiasis (el''ĕ-fan-ti'ah-sis)* (fig. 21.9).

Regulation of Blood Volume by the Kidneys

The formation of urine begins in the same manner as the formation of tissue fluid—by filtration of plasma through capillary channels. The kidneys produce about 180 L per day of blood filtrate, but since there is only 5.5 L of blood in the body, it is clear that most of this filtrate must be returned to the vascular system and recycled. Only about 1–2 L per day of urine is excreted; 98%–99% of the amount filtered is *reabsorbed* back into the vascular system.

The volume of urine excreted can be varied by changes in the reabsorption of filtrate. If 99% of the filtrate is reabsorbed, for example, 1% must be excreted. Decreasing the reabsorption by only 1% thus doubles the urine volume. The percent of the kidney filtrate reabsorbed—and thus the urine and blood volume—is regulated by hormones. Although kidney function is discussed in detail in chapter 26, some information about the role of the kidneys is presented here for proper understanding of cardiovascular physiology.

Regulation by Antidiuretic Hormone (ADH). One of the major hormones involved in the regulation of blood volume is **antidiuretic** *(an''tĭ-di''u-ret'ik)* **hormone (ADH),** also known as *vasopressin.* This hormone is produced by neurons in the hypothalamus, transported by axons into the neurohypophysis (posterior pituitary), and released from this storage gland in response to hypothalamic stimulation. The secretion of ADH from the posterior pituitary occurs when neurons called **osmoreceptors** in the hypothalamus detect an increase in plasma osmolality (osmotic pressure).

An increase in plasma osmolality occurs when the plasma becomes more concentrated. This can be produced either by *dehydration* or by *excessive salt intake.* Stimulation of osmoreceptors produces sensations of thirst, leading to increased water intake, and increased ADH secretion from the posterior pituitary. Through mechanisms that will be discussed in conjunction with kidney physiology in chapter 24, ADH stimulates water reabsorption from the filtrate. A smaller volume of urine is thus excreted as a result of the action of ADH (fig. 21.10).

A person who is dehydrated or who eats excessive amounts of salt thus drinks more and urinates less. This raises the blood volume and, in the process, dilutes the plasma to lower its previously elevated osmolality. The rise in blood volume that results from these mechanisms is extremely important in stabilizing the condition of a dehydrated person with low blood volume and pressure. In a person who eats excessive amounts of salt, however, the increased blood volume could produce hypertension (high blood pressure).

Figure 21.9. Parasitic larvae that block lymphatic drainage produce tissue edema, resulting in elephantiasis.

Drinking excessive amounts of water without excessive amounts of salt does not result in hypertension. The water does enter the blood from the intestine and momentarily raises the blood volume; at the same time, however, it dilutes the blood, which decreases the plasma osmolality and thus inhibits ADH secretion. With less ADH there is less reabsorption of filtrate in the kidneys—a larger volume of urine is excreted. Water is therefore a *diuretic (di''u-ret'ik)*—a substance that promotes urine formation—because it inhibits the secretion of ADH.

Since Na^+ and Cl^- are easily filtered in the kidneys, a mechanism must exist to promote the reabsorption and retention of salt when the dietary salt intake is too low. Throughout most of human history, salt was in short supply and was thus highly valued. Moorish merchants in the sixth century traded an ounce of salt for an ounce of gold; salt cakes were used as money in Abyssinia; part of a Roman soldier's pay was given in salt—it is from this practice that the word salary (*sal* = salt) is derived. The phrase "worth his salt" is derived from the fact that Greeks and Romans sometimes purchased slaves with salt.

diuretic: Gk. *dia,* completely; *ouresis,* urination

Figure 21.10. The role of antidiuretic hormone (ADH) in the regulation of blood osmolality and volume. Both dehydration and salt ingestion stimulate drinking and increased water retention. In dehydration, this provides a negative feedback mechanism (shown by dashed arrows) that helps to counter the loss of blood volume. As a result of salt ingestion, this same mechanism produces a rise in blood volume. Numbers indicate the sequence of cause-and-effect steps.

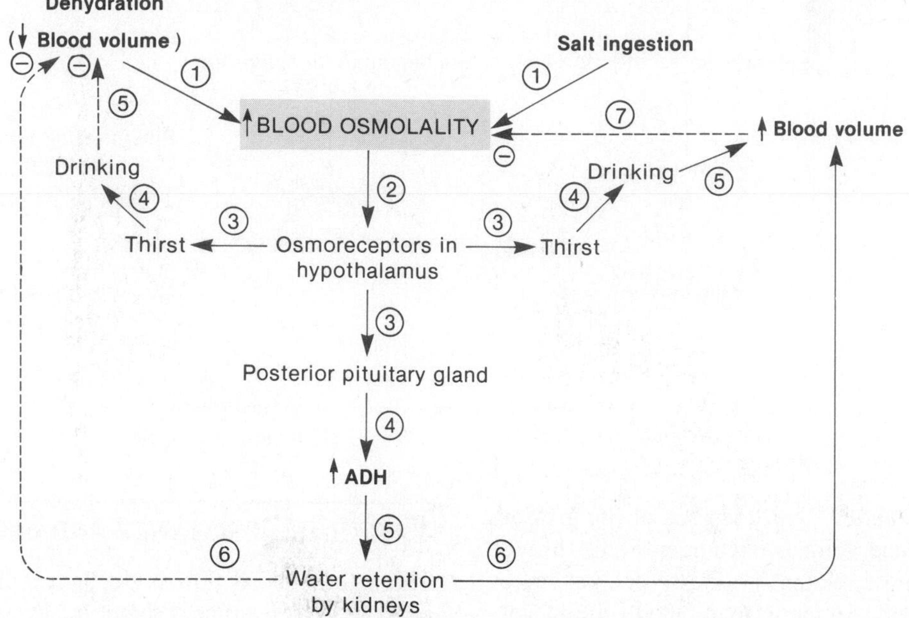

Regulation by Aldosterone. **Aldosterone** *(al-dos'ter-ōn),* a steroid hormone secreted by the adrenal cortex, stimulates the retention of salt by the kidneys. Aldosterone is thus a "salt-retaining hormone." Retention of salt indirectly promotes retention of water (in part, by the action of ADH, as previously discussed). The action of aldosterone thus causes an increase in blood volume, but—unlike the effects of ADH—does not cause a change in plasma osmolality. The secretion of aldosterone is stimulated during salt deprivation, when the blood volume and pressure are reduced. Because the adrenal cortex is not directly stimulated by these conditions, however, an intermediate mechanism is required.

Renin-Angiotensin System. When the blood flow and pressure are reduced in the renal artery (as they would be in the low blood volume state of salt deprivation), a group of cells in the kidneys called the **juxtaglomerular** *(juks"tah-glo-mer'u-lar)* **apparatus** secretes the enzyme **renin** into the blood. This enzyme cleaves a ten-amino-acid polypeptide called *angiotensin I* from a plasma protein called *angiotensinogen.* As angiotensin I passes through different organs, a *converting enzyme* removes two amino acids. This leaves an eight-amino-acid polypeptide called **angiotensin** *(an"je-o-ten'sin)* **II** (fig. 21.11). Conditions of salt deprivation, low blood volume, and low blood pressure, in summary, cause increased production of angiotensin II in the blood.

Angiotensin II has numerous effects that cause a rise in blood pressure. This rise in pressure is due both to constriction of muscular arteries and arterioles (vasoconstriction) and to increases in blood volume. Vasoconstriction is produced directly by the effects of angiotensin II on blood vessels and indirectly by the augmentation of sympathetic nerve activity by angiotensin II (causing increased release of norepinephrine in the blood vessels).

Angiotensin II promotes a rise in blood volume by means of two mechanisms: (1) thirst centers in the hypothalamus are stimulted by angiotensin II, so that more water is ingested; and (2) secretion of aldosterone by the adrenal cortex is stimulated by angiotensin II, so that more salt and water are retained by the kidneys. The relationship between angiotensin II and aldosterone is sometimes described as the *renin-angiotensin-aldosterone system.*

This mechanism can also work in the opposite direction: high salt intake, leading to high blood volume and pressure, normally inhibits renin secretion. With less angiotensin II formation and less aldosterone secretion, less salt is retained by the kidneys and more is excreted in the urine. Unfortunately, many people with chronically high blood pressure may have normal or even elevated levels of renin secretion. In these cases, the intake of salt must be lowered to match the impaired ability to excrete salt in the urine.

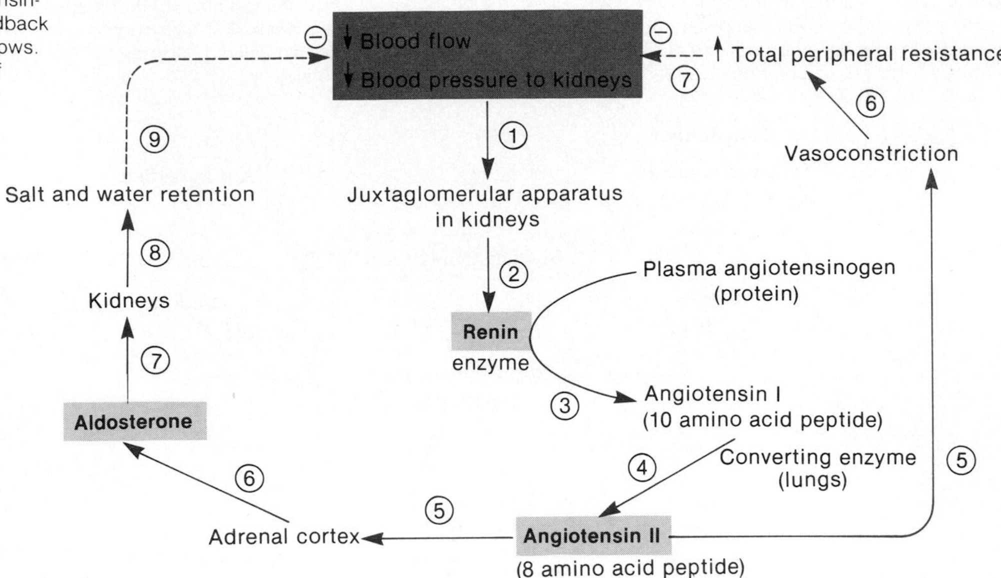

Figure 21.11. The renin-angiotensin-aldosterone system. Negative feedback effects are shown with dashed arrows. Numbers indicate the sequence of cause-and-effect steps.

Atrial Natriuretic Factor. As described in the previous section, a fall in blood volume is compensated, through activation of the renin-angiotensin-aldosterone system, by renal retention of fluid. An increase in blood volume, conversely, is compensated by renal excretion of a larger volume of urine. Experiments suggest that the increase in water excretion under conditions of high blood volume is at least partly due to an increase in the excretion of Na^+ in the urine, or *natriuresis.*

Natriuresis may be produced by a decline in aldosterone secretion, but evidence suggests that the action of a separate hormone causes this condition. This **natriuretic** *(na″tre-u-ret′ik)* **hormone** would thus be antagonistic to aldosterone and would promote Na^+ and water excretion in response to a rise in blood volume. The atria of the heart have recently been shown to produce a hormone with these properties, which is now identified as *atrial natriuretic factor.* By promoting salt and water excretion in the urine (through mechanisms discussed in chapter 26), atrial natriuretic hormone can act to lower the blood pressure in a manner analogous to the action of diuretic drugs taken by people with hypertension.

1. Describe the composition of tissue fluid. Using a flow diagram, explain how tissue fluid is formed and how it is returned to the vascular system.
2. Define the term *edema,* and describe four different mechanisms that can produce this condition.
3. Describe the effects of dehydration on blood and urine volumes, and explain the cause-and-effect mechanism involved.
4. Describe how salt deprivation causes increased salt and water retention by the kidneys. Explain how a high-salt diet can be compensated by increased urinary excretion of salt.

natriuresis: L. *natrium,* sodium; Gk. *ouresis,* urination

Vascular Resistance and Blood Flow

The rate of blood flow is dependent in part on vascular resistance. Resistance is regulated by vasoconstriction and vasodilation, produced by extrinsic and intrinsic mechanisms.

Objective 8. Describe how blood flow is affected by blood pressure and vascular resistance.
Objective 9. Describe how vascular resistance is affected by the autonomic system and by other extrinsic agents.
Objective 10. Describe the intrinsic regulation of vascular resistance.

The amount of blood that the heart pumps per minute is equal to the rate of venous return and thus is equal to the rate of blood flow through the entire circulation. The cardiac output of 5–6 L per minute is distributed unequally to the different organs. At rest, blood flow is about 2,500 ml per minute through the liver, kidneys, and gastrointestinal tract, 1,200 ml per minute through the skeletal muscles, 750 ml per minute through the brain, and 250 ml per minute through the coronary arteries of the heart. The balance of the cardiac output (500–1,100 ml per minute) is distributed to the other organs (table 21.3).

Physical Laws Describing Blood Flow

The flow of blood through the vascular system, like the flow of any fluid through a tube, depends in part on the difference in pressure at the two ends of the tube. If the pressure at both ends of the tube is the same, there will be no flow. If the pressure at one end is greater than at the other, blood will flow from the region of higher to the region of lower pressure. The rate of blood flow therefore

Table 21.3 Estimated distribution of the cardiac output at rest

Organs	Blood flow	
	Milliliters per minute	Percent total
Gastrointestinal tract and liver	1,400	24
Kidneys	1,100	19
Brain	750	13
Heart	250	4
Skeletal muscles	1,200	21
Skin	500	9
Other organs	600	10
Total organs	5,800	100

From O. L. Wade, and J. M. Bishop, *Cardiac Output and Regional Blood Flow.* Copyright © 1962 Blackwell Scientific Publications, Limited, London, England. Reprinted by permission.

Figure 21.12. The flow of blood in the systemic circulation is ultimately dependent on the pressure difference (△ *P*) between the origin of the flow (mean pressure of about 100 mm Hg in the aorta) and the end of the circuit—zero mm Hg in the vena cava where it joins the right atrium (*RA*). (*LA* = left atrium, *RV* = right ventricle, *LV* = left ventricle.)

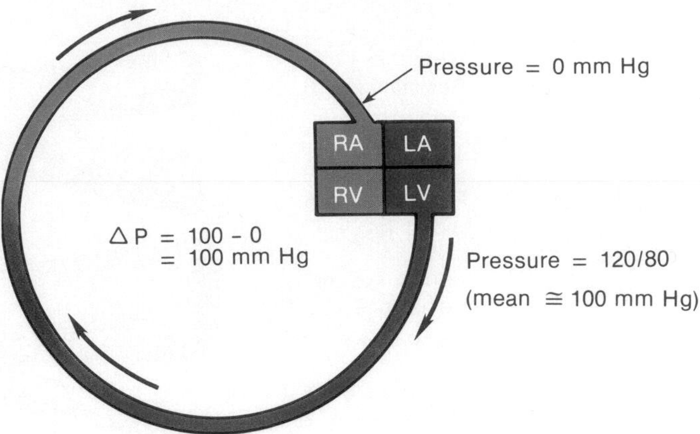

is proportional to the pressure difference ($P_1 - P_2$) between the two ends of the tube. The term **pressure difference** is abbreviated ΔP, in which the Greek letter Δ (*delta*) means "change in" (fig. 21.12).

If the systemic circulation is pictured as a single tube leading from and back to the heart (fig. 21.12), blood flow through this system would occur as a result of the pressure difference between the beginning of the tube (the aorta) and the end of the tube (the junction of the venae cavae with the right atrium). The average, or mean, arterial pressure is about 100 mm Hg; the pressure at the right atrium is 0 mm Hg. The "pressure head," or driving force (ΔP), is therefore about 100 − 0 = 100 mm Hg.

Blood flow is directly proportional to the pressure difference between the two ends of the tube (ΔP) but is *inversely proportional* to the frictional resistance to blood flow through the vessels. Inverse proportionality is expressed by showing one of the factors in the denominator of a fraction, since a fraction decreases when the denominator increases:

$$\text{Blood flow} \propto \frac{\Delta P}{\text{Resistance}}.$$

The **resistance** to blood flow through a vessel is directly proportional to the length of the vessel and to the viscosity of the blood (the "thickness," or ability of molecules to "slip over" each other). Vascular resistance is inversely proportional to the fourth power of the radius of the vessel:

$$\text{Resistance} \propto \frac{L\eta}{r^4}$$

where: L = length of vessel
η = viscosity of blood
r = radius of vessel.

If one vessel has half the radius of another and if all other factors are the same, the smaller vessel would have sixteen times (2^4) the resistance of the larger vessel. Blood flow through the larger vessel, as a result, would be sixteen times greater than in the smaller vessel (fig. 21.13).

When physical constants are added to this relationship, the rate of blood flow can be calculated according to **Poiseuille's** *(pwah-zuh'yez)* **law:**

$$\text{Blood flow} = \frac{\Delta P\ r^4\ (\pi)}{\eta\ L\ (8)}.$$

Since vessel length and blood viscosity do not vary significantly in normal physiology, the major factors affecting blood flow through an organ are the mean arterial pressure and the vascular resistance. At a given mean arterial pressure, blood can be diverted from one organ to another by variations in the degree of vasoconstriction and vasodilation in the arterial supplies of the organs. Since arterioles are the smallest arteries and can become narrower by vasoconstriction, they provide the greatest resistance to blood flow. Blood flow to an organ can be increased by dilation of its arterioles and can be decreased by constriction of its arterioles.

Total Peripheral Resistance. The sum of all the vascular resistances within the systemic circulation is called the *total peripheral resistance.* The arterial supplies to the organs, which contribute to this total, are generally in parallel rather than in series with each other. That is, arterial blood passes through only one set of resistance vessels (arterioles) before returning to the heart (fig. 21.14).

Poiseuille's law: from Jean Poiseuille, French physiologist, 1799–1869

Figure 21.13. Relationships between blood flow, vessel radius, and resistance. (*a*) the resistance and blood flow is equally divided between two branches of a vessel. (*b*) a doubling of the radius of one branch and halving of the radius of the other produces a sixteenfold increase in blood flow in the former and a sixteenfold decrease of blood flow in the latter.

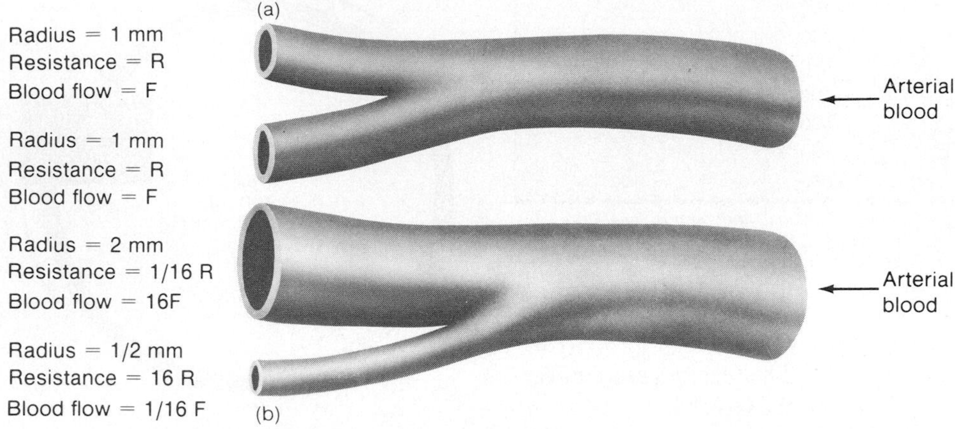

Figure 21.14. A diagram of the systemic and pulmonary circulations. Notice that with few exceptions (such as the blood flow in the renal circulation), the flow of arterial blood is in parallel rather than in series (arterial blood does not usually flow from one organ to another).

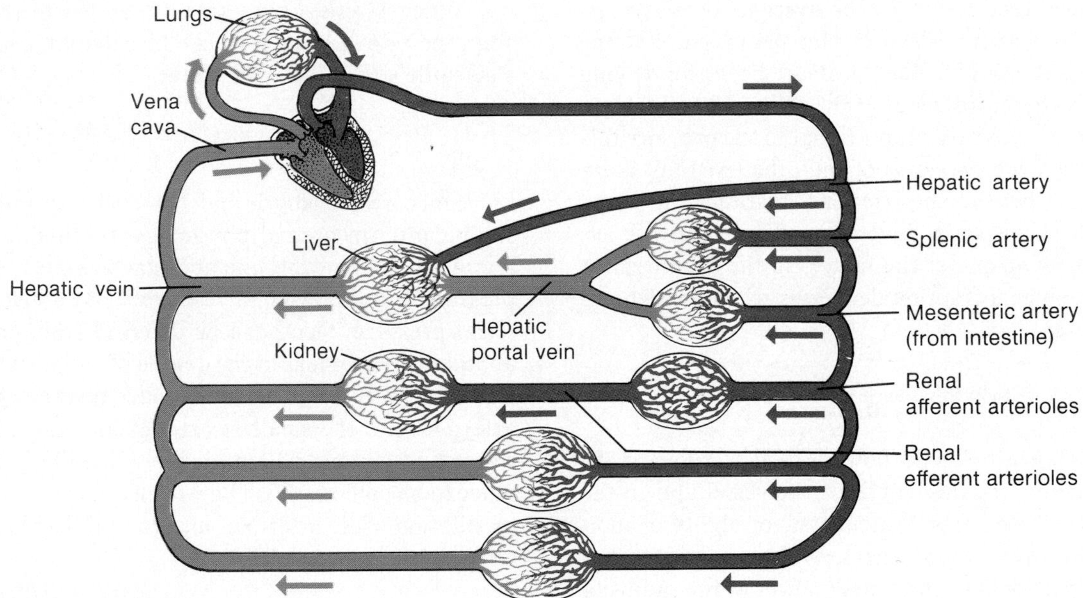

Table 21.4 Extrinsic control of vascular resistance and blood flow

Extrinsic agent	Effect	Comments
Sympathetic nerves		
Alpha-adrenergic	Vasoconstriction	It occurs throughout the body. This is the dominant effect of sympathetic nerve stimulation on the vascular system.
Beta-adrenergic	Vasodilation	There is some activity in arterioles in skeletal muscles and in coronary vessels, but effects are masked by dominant alpha-receptor-mediated constriction.
Cholinergic	Vasodilation	Effects are localized to arterioles in skeletal muscles and are only activated during defense (fight or flight) reaction.
Parasympathetic nerves	Vasodilation	Effects are primarily restricted to the gastrointestinal tract, external genitals, and salivary glands and have little effect on total peripheral resistance.
Angiotensin II	Vasoconstriction	A powerful vasoconstrictor produced as a result of secretion of renin from the kidneys, it may function to help maintain adequate filtration pressure in kidneys when systemic blood flow and pressure is reduced.
ADH (vasopressin)	Vasoconstriction	Although the effects of this hormone on vascular resistance and blood pressure in anesthetized animals are well documented, the importance of these effects in conscious humans is controversial.
Histamine	Vasodilation	It promotes localized vasodilation during inflammation and allergic reactions.
Bradykinins	Vasodilation	Bradykinins are polypeptides secreted by sweat glands that promote local vasodilation.
Prostaglandins	Vasodilation or vasoconstriction	Prostaglandins are cyclic fatty acids that can be produced by most tissues, including blood vessel walls. Prostaglandin I_2 is a vasodilator, whereas thromboxane A_2 is a vasoconstrictor. The physiological significance of these effects is presently controversial.

Since one organ is not "downstream" from another in terms of its arterial supply, changes in resistance within one organ directly affect blood flow in only that organ.

Vasodilation in a large organ might, however, significantly decrease the total peripheral resistance and, by this means, might decrease the mean arterial pressure. If this is not prevented by compensations, the driving force for blood flow through all organs might be reduced. This is normally prevented by an increase in the cardiac output and by vasoconstriction in other areas. During exercise of large muscles, for example, the arterioles in the exercising muscles are dilated, the cardiac output is increased, and there is constriction of arterioles in the viscera and skin.

Extrinsic Regulation of Blood Flow

Extrinsic regulation refers to control by the autonomic nervous system and the endocrine system. Angiotensin II, for example, directly stimulates vascular smooth muscle to produce generalized vasoconstriction. Antidiuretic hormone (ADH) also has a vasoconstrictor effect at high concentrations; this is why it is also called **vasopressin.**

Regulation by Sympathetic Nerves. Stimulation of the sympathoadrenal system produces an increase in the cardiac output (as previously discussed) and an increase in total peripheral resistance. The latter effect is due to alpha-adrenergic stimulation (chapter 17) of vascular smooth muscle by norepinephrine and, to a lesser degree, by epinephrine. This produces vasoconstriction of arterioles in the viscera and skin.

Even when a person is calm, the sympathoadrenal system is active to a certain degree and helps set the "tone" of vascular smooth muscles. In this case, **adrenergic** *(ad''ren-er'jik)* **sympathetic fibers** (those that release norepinephrine) cause a basal level of vasoconstriction in skeletal muscles and throughout the body. During the "fight-or-flight" reaction the activity of adrenergic fibers increases, so that vasoconstriction is produced in the digestive tract, kidneys, and skin.

Arterioles in skeletal muscles receive **cholinergic** *(ko''lin-er'jik)* **sympathetic fibers,** which release acetylcholine as a neurotransmitter. During the "fight-or-flight" reaction, the activity of cholinergic fibers increases, causing vasodilation. Thus, while the total peripheral resistance and blood pressure are increased by adrenergic fibers that stimulate vasoconstriction in the viscera and skin, vasodilation in skeletal muscles is produced by cholinergic fibers. Blood flow is thus diverted to skeletal muscles during emergency conditions, which may produce an "extra edge" for the skeletal muscles' response to the emergency.

Parasympathetic Control of Blood Flow. Parasympathetic endings in arterioles are always cholinergic and always promote vasodilation. Parasympathetic innervation of blood vessels, however, is limited to the digestive tract, external genitals, and salivary glands. Because of this limited distribution, the parasympathetic system is less important than the sympathetic system in the control of total peripheral resistance. The extrinsic control of blood flow is summarized in table 21.4.

Intrinsic Control of Blood Flow

Extrinsic control mechanisms affect resistance and flow in many regions of the body. In contrast to these more generalized effects, intrinsic ("built-in") mechanisms within individual organs provide a more localized regulation of vascular resistance and blood flow. Intrinsic mechanisms are: (1) **myogenic** and (2) **metabolic.** Some organs, the brain and kidneys in particular, utilize these intrinsic mechanisms to maintain relatively constant flow rates despite wide fluctuations in blood pressure. This ability is termed **autoregulation.**

Myogenic Control Mechanisms. If the arterial blood pressure and flow through an organ is inadequate—if the organ is inadequately *perfused with blood*—the metabolism of the organ cannot be maintained beyond a limited period of time. Excessively high blood pressure can also be dangerous, particularly in the brain, because this may cause fine blood vessels to burst (cerebrovascular accident—CVA, or stroke).

Changes in systemic arterial pressure are compensated for in the brain and some other organs by the appropriate responses of vascular smooth muscle. A decrease in arterial pressure causes cerebral vessels to dilate, so that adequate rates of blood flow can be maintained despite the decreased pressure. High blood pressure, in contrast, causes cerebral vessels to constrict so that finer vessels downstream are protected from the elevated pressure. These responses are myogenic; they are direct responses by the vascular smooth muscle to changes in pressure.

Metabolic Control Mechanisms. Local vasodilation within an organ can occur as a result of the chemical environment created by the organ's metabolism. The localized chemical conditions that promote vasodilation include (1) **decreased oxygen concentrations** that result from increased metabolic rate; (2) **increased carbon dioxide concentrations;** (3) **decreased tissue pH** (due to CO_2, lactic acid, and other metabolic products); and (4) the **release of adenosine or K^+** from the tissue cells.

The vasodilation that occurs in response to tissue metabolism can be demonstrated by constricting the blood supply to an area for a short time and then removing the constriction. The constriction allows metabolic products to accumulate by preventing venous drainage of the area. When the constriction is removed and blood flow resumes, the metabolic products that have accumulated cause vasodilation. The tissue thus appears red. This response is called **reactive hyperemia** *(hi''per-e'me-ah)*. A similar increase in blood flow occurs in skeletal muscles and other organs as a result of increased metabolism. This is called **active hyperemia.** Intrinsic control mechanisms are summarized in table 21.5.

Table 21.5 The intrinsic control of vascular resistance and blood flow

Category	Agent (↑ = increase; ↓ = decrease)	Mechanisms	Comments
Myogenic	↑ Blood pressure	Stretching of the arterial wall as the blood pressure rises directly stimulates increased smooth muscle tone (vasoconstriction).	It helps to maintain relatively constant rates of blood flow and pressure within an organ despite changes in systematic arterial pressure (autoregulation).
Metabolic	↓ Oxygen ↑ Carbon dioxide ↓ pH ↑ Adenosine ↑ K	Local changes in gas and metabolic concentrations act directly on vascular smooth muscle walls to produce vasodilation in the systemic circulation. The importance of different agents varies in different organs.	It aids in autoregulation of blood flow and also helps to shunt increased amounts of blood to organs with higher metabolic rates (active hyperemia).

1. Describe the relationship between blood flow rate, arterial blood pressure, and vascular resistance.
2. Describe the relationship between vascular resistance and the radius of a vessel. Explain how blood flow can be diverted from one organ to another.
3. Explain how vascular resistance and blood flow is regulated by (1) sympathetic adrenergic fibers, (2) sympathetic cholinergic fibers, and (3) parasympathetic fibers.
4. Define *autoregulation,* and explain how this is accomplished through myogenic and metabolic mechanisms.

Blood Flow to the Heart and Skeletal Muscles

Blood flow to the heart and skeletal muscles is regulated by both extrinsic and intrinsic mechanisms that increase the rate of flow when the metabolic requirements of these tissues increase during exercise.

Objective 11. Describe how coronary blood flow is regulated, and explain how it is affected by the heart's contraction and relaxation.

Objective 12. Describe how skeletal muscle blood flow is regulated, and explain how this flow is increased during exercise.

Objective 13. Explain the mechanisms that contribute to increased cardiac output during exercise.

Table 21.6 Some vascular and metabolic comparisons between heart and skeletal muscle

	Cardiac muscle	Skeletal muscle
Number of capillaries	4×	1×
Mean blood flow	10–20×	1×
Myolemma (sarcolemma)	Thin, low resistance	Thicker, higher resistance
Sarcomere length	1×	1.7×
Glycogen concentration	Maintained	Depleted by fasting, diabetes
Glycolytic enzyme systems	1×	2×, strongly developed
Creatine phosphate concentration	1×	6×
Anaerobic energy production	Beating: 2 min Arrested: 30–90 min	Up to 40% total energy
Lactic acid production	Terminal mechanism	Frequent when incurring O_2 debt
Ability to incur O_2 debt	1×	4×
Increased O_2 requirement met primarily by	Increased flow	Increased extraction
Oxygen consumption	3×	1×
Oxygen extraction at rest	Near maximal	Significant reserve
Increased O_2 consumption with increased work	2×	30×
Myoglobin	Present	Present in red skeletal muscle
Mitochondria	Abundant, giant	Fewer, smaller
Krebs cycle enzymes	2–3×	1×
Cytochrome-C	6×	1×
Myosin ATPase activity	1×	3×

From Norman Brachfeld, "The Physiology of Muscular Exercise," in *Primary Cardiology*, page 112, June 1979. Copyright © 1979 P. W. Communications, Secaucus, NJ. Reprinted by permission.

Survival requires that the heart and brain receive adequate rates of blood flow at all times. The ability of skeletal muscles to respond quickly in emergencies and to maintain continued high levels of activity may also be critically important for survival. During such times, high rates of blood flow to the skeletal muscles must be maintained without compromising blood flow to the heart and brain. This is accomplished by mechanisms that increase the cardiac output and that divert a higher proportion of the cardiac output to the heart, skeletal muscles, and brain, and away from the viscera and skin.

Aerobic Requirements of the Heart

The coronary arteries supply an enormous number of capillaries, which are packed within the myocardium at a density of about 2,500–4,000 per cubic millimeter of tissue. Fast-twitch skeletal muscles, in contrast, have a capillary density of 300–400 per cubic millimeter of tissue. Each myocardial cell, as a consequence, is within 10 μm of a capillary (compared to an average distance in other organs of 60–80 μm). The exchange of gases by diffusion between myocardial cells and capillary blood thus occurs very quickly.

Contraction of the myocardium squeezes the coronary arteries. Unlike blood flow in all other organs, flow in the coronary vessels thus decreases in systole and increases during diastole. The myocardium, however, contains large amounts of **myoglobin** *(mi''o-glo'bin),* a pigment related to hemoglobin, which stores oxygen during diastole and releases its oxygen during systole. In this way, the myocardial cells can receive a continuous supply of oxygen even though coronary blood flow is temporarily reduced during systole.

In addition to containing large amounts of myoglobin, heart muscle contains numerous mitochondria and aerobic respiratory enzymes. This indicates that—even more than slow-twitch skeletal muscles—the heart is extremely specialized for aerobic respiration (table 21.6). The normal heart always respires aerobically, even during heavy exercise when the metabolic demand for oxygen can rise to five times resting levels. This increased oxygen requirement is met by a corresponding increase in coronary blood flow, from about 80 ml at rest to about 400 ml per minute per 100 g tissue during heavy exercise.

Regulation of Coronary Blood Flow

Sympathetic nerve fibers, through stimulation of alpha-adrenergic receptors in the coronary arterioles, produce a relatively high vascular resistance in the coronary circulation at rest. Vasodilation of coronary vessels may be produced in part by sympathetic nerve activation of beta-adrenergic receptors. Most of the vasodilation that occurs during exercise, however, is due to intrinsic metabolic control mechanisms. As the metabolism of the myocardium increases, local accumulation of carbon dioxide and adenosine *(ah-den'o-sēn)* acts directly on the vascular smooth muscles to cause relaxation and vasodilation.

Under abnormal conditions, the blood flow to the myocardium may be inadequate, resulting in *myocardial ischemia.* This can result from blockage by atheromas and/or blood clots or from muscular spasm of a coronary artery (fig. 21.15). Occlusion of a coronary artery can be visualized by a technique called *selective coronary arteriography.* In this procedure, a catheter (plastic tube) is inserted into a brachial or femoral artery all the way to the opening of the coronary arteries in the aorta, and radiographic contrast material is injected. The picture thus obtained is called an *angiogram (an'je-o-gram'').*

If the occlusion is sufficiently great, a coronary bypass operation may be performed. In this procedure a length of blood vessel, usually taken from the saphenous vein in the leg, is sutured to the aorta and to the coronary artery at a location beyond the site of the occlusion (fig. 21.16).

Figure 21.15. An arteriogram of the left coronary artery in a patient (*a*) when the ECG was normal and (*b*) when the ECG showed evidence of myocardial ischemia. Notice that a coronary artery spasm—see arrow in (*b*)—appears to accompany the ischemia.

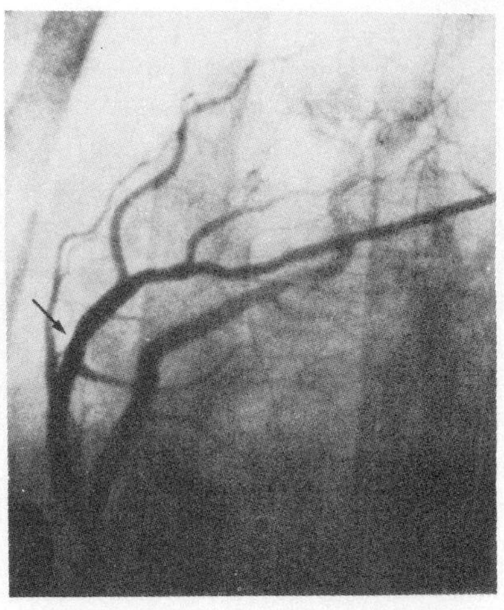

(a)

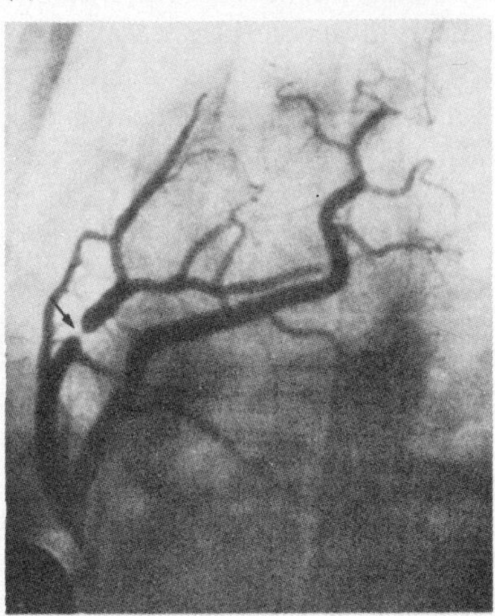

(b)

Figure 21.16. A diagram of coronary artery bypass surgery.

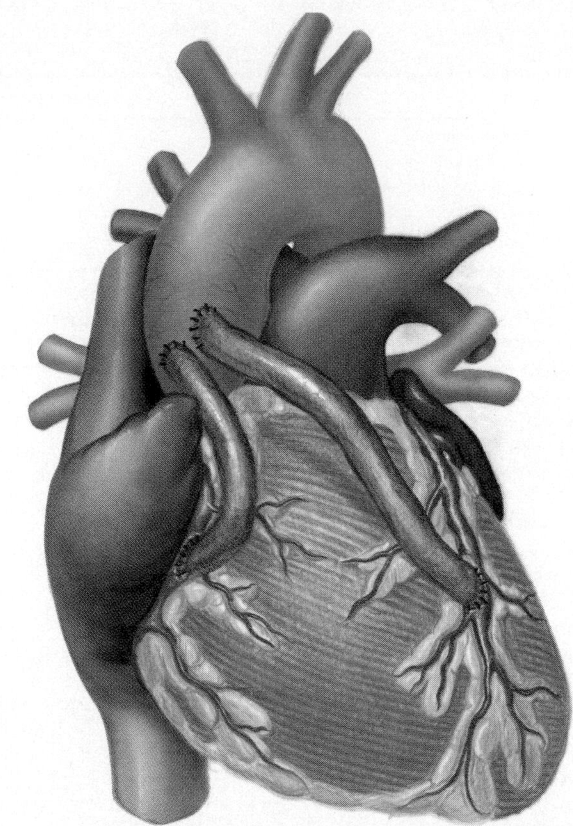

Regulation of Blood Flow through Skeletal Muscles

The arterioles in skeletal muscles, like those of the coronary circulation, have a high vascular resistance at rest as a result of alpha-adrenergic sympathetic stimulation. This produces a relatively low rate of blood flow per tissue weight (4–6 ml per minute per 100 g tissue), but because muscles have such a large mass, this still accounts for 20–25% of the total blood flow in the body at rest.

In addition to adrenergic fibers (those that release norepinephrine), there are also sympathetic cholinergic fibers in skeletal muscles. The release of acetylcholine (ACh) from these fibers stimulates vasodilation as part of the "fight-or-flight" response to any stressful state, including that existing just prior to exercise (table 21.7). The increased blood flow that results from this vasodilation may provide an extra edge that improves skeletal muscle performance once exercise begins.

As exercise progresses, the vasodilation and increased skeletal muscle blood flow that occur are almost entirely due to intrinsic metabolic control. The high metabolic rate of skeletal muscles during exercise causes local changes such as increased carbon dioxide concentrations, decreased pH (due to carbonic acid and lactic acid), decreased oxygen, increased extracellular K^+, and the secretion of adenosine. Like the intrinsic control of the coronary circulation, these changes cause vasodilation of arterioles in skeletal muscles. This decreases the vascular resistance and increases the rate of blood flow. This effect is combined with the recruitment of capillaries by the opening of precapillary sphincter muscles (only 5–10% of the skeletal muscle capillaries are open at rest). As a result of these changes, skeletal muscles can receive as much as 85% of the total blood flow in the body during maximal exercise.

While the vascular resistance in skeletal muscles decreases during exercise, the resistance to flow through visceral organs and skin increases. This increased resistance occurs because of vasoconstriction stimulated by adrenergic sympathetic fibers and results in decreased rates of blood flow through these organs. During exercise, therefore, the blood flow to skeletal muscles increases because of increased total blood flow (cardiac output) and because blood is diverted away from the viscera and skin. Blood flow to the heart also increases during exercise, whereas blood flow to the brain does not appear to change significantly (fig. 21.17).

Although the skeletal muscles consume large amounts of oxygen during exercise, the oxygen concentration of arterial blood is usually not reduced. This is because the rate

Figure 21.17. The distribution of blood flow (cardiac output) during rest and heavy work. At rest, the cardiac output is 5 L per minute (*bottom of figure*); during heavy work the cardiac output increases to 25 L per minute (*top of figure*). During rest, for example, the brain receives 15% of 5 L per minute (= 750 ml/min.), whereas during exercise it receives 3% to 4% of 25 L per minute (0.03 × 25 = 750 ml/min.). Flow to the skeletal muscles increases more than twentyfold, because the total cardiac output increases (from 5 L/min. to 25 L/min.) and because the percent of the total received by the muscles increases from 15% to 80%.

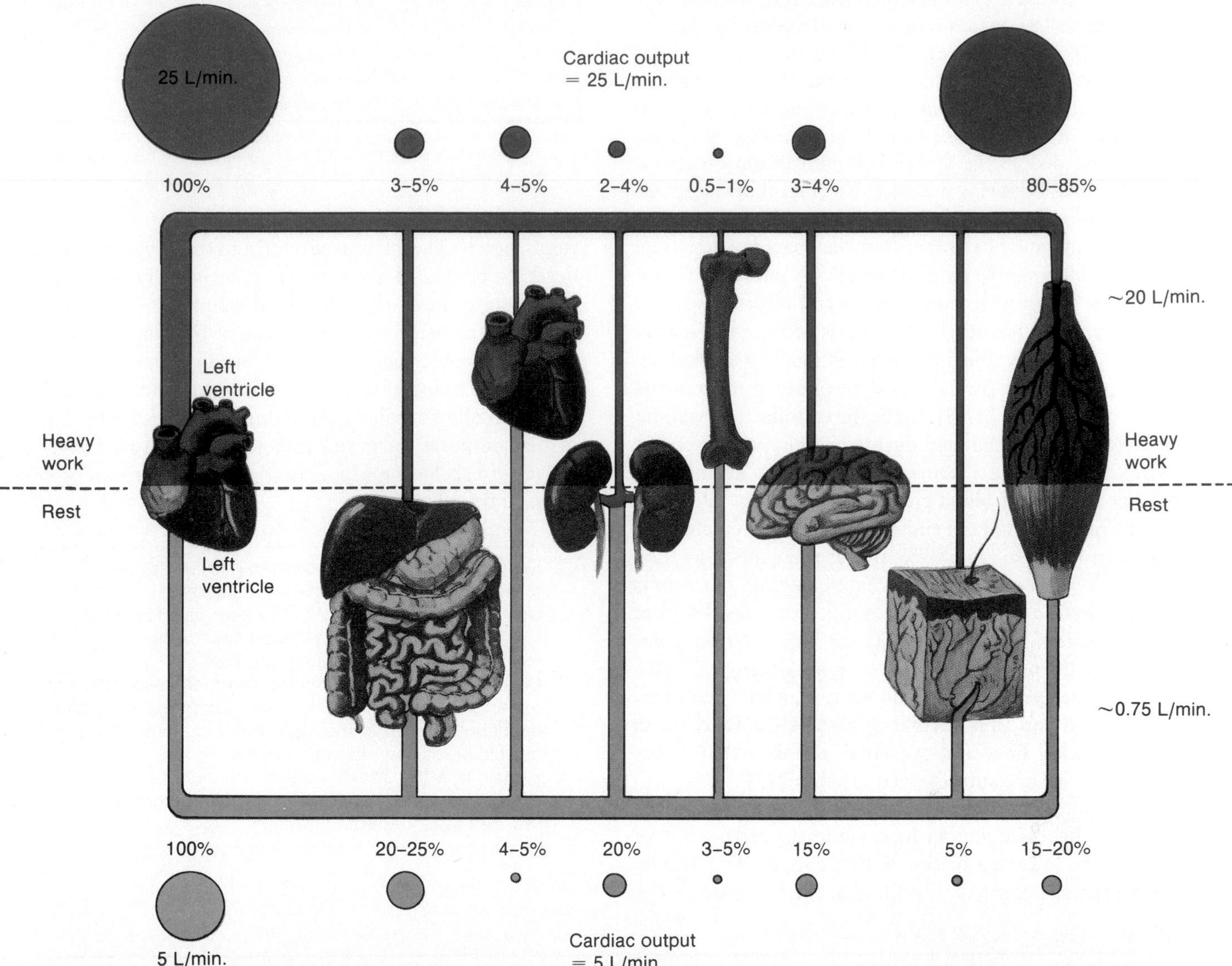

Table 21.7 Changes in skeletal muscle blood flow under conditions of rest and exercise

Condition	Blood flow (ml/min)	Mechanism
Rest	1,000	High adrenergic sympathetic stimulation of vascular alpha-receptors causing vasoconstriction
Beginning exercise	Increased	Dilation of arterioles in skeletal muscles due to cholinergic sympathetic nerve activity
Heavy exercise	20,000	Fall in alpha-adrenergic activity Increased sympathetic cholinergic activity Increased metabolic rate of exercising muscles producing intrinsic vasodilation

and depth of breathing is also increased during exercise and because the pulmonary blood flow increases to match the increased systemic blood flow. The oxygen concentration of venous blood, however, can be decreased to one-half or one-third the resting levels by exercising muscles, which indicates that the muscles are extracting a much higher proportion of the oxygen delivered to them in the arterial blood. The ability of exercising muscles to extract oxygen from arterial blood is improved by endurance training, which results in increased capillary density and aerobic enzymes in the trained muscles.

Changes in Cardiac Output during Exercise

During exercise the cardiac output can increase fivefold, from about 5 L per minute to about 25 L per minute. This is primarily due to an increase in cardiac rate. The cardiac rate, however, can only increase up to a maximum value (table 21.8), which is determined mainly by the person's age. In very well-trained athletes the stroke volume can also increase significantly, allowing these individuals to achieve a cardiac output during strenuous exercise as much as six or seven times greater than their resting values.

In most people the stroke volume can only increase from 10% to 35% during exercise. The fact that the stroke volume can increase at all during exercise may at first be surprising, in view of the fact that the heart has less time to fill with blood between beats when it is pumping faster. Despite the faster beat, however, the end-diastolic volume during exercise is not decreased. This is because the venous return is aided by the improved action of the skeletal muscle pumps and by increased respiratory movements during exercise (fig. 21.18). Since the end-diastolic volume is not significantly changed during exercise, any increase in stroke volume that occurs must be due to an increase in the proportion of blood ejected per stroke.

The proportion of the end-diastolic volume ejected per stroke can increase from 67% at rest to as much as 90% during heavy exercise. This increased *ejection fraction* is produced by the increased contractility that results from sympathoadrenal stimulation. There may also be a decrease in total peripheral resistance as a result of vasodilation in the exercising skeletal muscles, which decreases the afterload and thus further augments the increase in stroke volume. The cardiovascular changes that occur during exercise are summarized in table 21.9.

Endurance training often results in a lowering of the resting cardiac rate and an increase in the resting stroke volume. The lowering of the resting cardiac rate results from a greater degree of inhibition of the SA node by the vagus nerve. The increased resting stroke volume is believed to be due to an increase in blood volume; indeed, studies have shown that the blood volume can increase by about 500 ml after only eight days of training. These adaptations enable the trained athlete to produce a larger proportionate increase in cardiac output and achieve a higher absolute cardiac output during exercise. This large cardiac output is the major factor in the improved oxygen delivery to skeletal muscles that occurs as a result of endurance training.

Table 21.8 Relationship between age and maximum cardiac rate

Age	Maximum cardiac rate
20–29	190 beats/min
30–39	160 beats/min
40–49	150 beats/min
50–59	140 beats/min
60 and above	130 beats/min

1. Describe blood flow and oxygen delivery to the myocardium during systole and diastole.
2. Describe how blood flow to the heart is affected by exercise, and explain how blood flow to the heart is regulated at rest and during exercise.
3. Describe the mechanisms that produce vasodilation of the arterioles in skeletal muscles during exercise, and describe three other mechanisms that increase skeletal muscle blood flow during exercise.
4. Explain how the stroke volume can increase during exercise despite the fact that the filling times are reduced at high cardiac rates.

Figure 21.18. Cardiovascular adaptations to exercise.

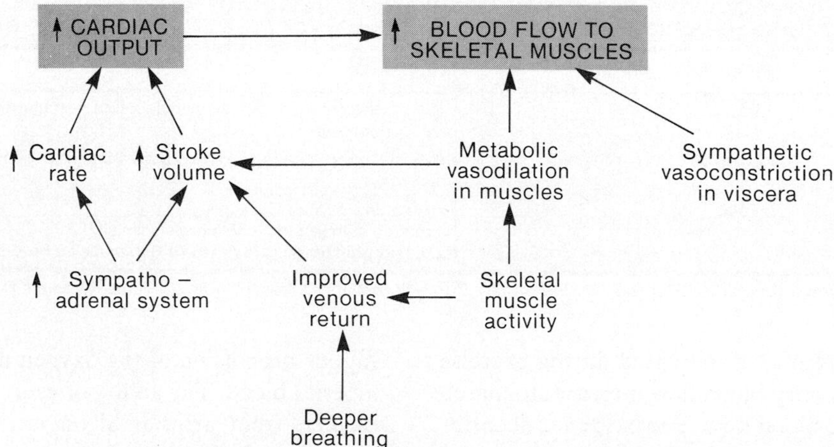

Table 21.9 Cardiovascular changes that occur during moderate exercise

Variable	Change	Mechanisms
Cardiac output	Increased	Cardiac rate and stroke volume increased
Cardiac rate	Increased	Increased sympathetic nerve activity; decreased activity of the vagus nerve
Stroke volume	Increased	Increased myocardial contractility due to stimulation by sympathoadrenal system; decreased total peripheral resistance
Total peripheral resistance	Decreased	Vasodilation of arterioles in skeletal muscles (and in skin when thermoregulatory adjustments are needed)
Arterial blood pressure	Increased	Increased systolic and pulse pressure due primarily to increased cardiac output; diastolic pressure rises less due to decreased total peripheral resistance
End-diastolic volume	Unchanged	Decreased filling time at high cardiac rates is compensated by increased venous pressure, increased activity of the skeletal muscle pump, and decreased intrathoracic pressure aiding the venous return
Blood flow to heart and muscles	Increased	Increased muscle metabolism produces intrinsic vasodilation; aided by increased cardiac output and increased vascular resistance in visceral organs
Blood flow to visceral organs	Decreased	Vasoconstriction in digestive tract, liver, and kidneys due to sympathetic nerve stimulation
Blood flow to skin	Increased	Metabolic heat produced by exercising muscles produces reflex (involving hypothalamus) that reduces sympathetic constriction of arteriovenous shunts and arterioles
Blood flow to brain	Unchanged	Autoregulation of cerebral vessels, which maintains constant cerebral blood flow despite increased arterial blood pressure

Blood Flow to the Brain and Skin

Control mechanisms help to maintain relatively constant rates of cerebral blood flow. Blood flow to the skin is varied according to the needs of the body.

Objective 14. Describe the mechanisms that control cerebral blood flow. Explain the significance of the autoregulation of cerebral blood flow.

Objective 15. Describe the cutaneous circulation, and explain how it is affected by ambient temperature, exercise, and emotional states.

The examination of cerebral *(ser'ĕ-bral)* and cutaneous *(ku-ta'ne-us)* blood flow is a study in contrasts. Cerebral blood flow is regulated primarily by intrinsic mechanisms; cutaneous blood flow is regulated by extrinsic mechanisms. Cerebral blood flow is relatively constant; cutaneous blood flow exhibits more variation than blood flow in any other organ. The brain is the organ that can least tolerate and the skin is the organ that can most tolerate low rates of blood flow.

Cerebral Circulation

When the brain is deprived of oxygen for a few seconds, the person will lose consciousness; irreversible brain injury may occur after a few minutes. For these reasons, the cerebral blood flow is remarkably constant at about 750 ml per minute. This amounts to about 15% of the total cardiac output at rest.

Unlike the coronary and skeletal muscle blood flow, cerebral blood flow is not normally influenced by sympathetic nerve activity. Only when the mean arterial pressure rises to about 200 mm Hg do sympathetic nerves cause a significant degree of vasoconstriction in the cerebral circulation. This vasoconstriction helps to protect small, thin-walled arterioles from bursting under the pressure and thus helps to prevent cerebrovascular accident (stroke).

In the normal range of arterial pressures, cerebral blood flow is regulated almost exclusively by intrinsic mechanisms. These mechanisms help insure a constant rate of blood flow despite changes in systemic arterial pressure—a process called *autoregulation.* The autoregulation of cerebral blood flow is achieved by both myogenic and metabolic mechanisms.

Myogenic regulation occurs when there is variation in systemic arterial pressure. The cerebral arteries automatically dilate when the blood pressure falls and constrict when the pressure rises. This helps to maintain a constant flow rate during the normal pressure variations that occur during rest, exercise, and emotional states.

The cerebral vessels are also sensitive to the carbon dioxide concentration of arterial blood. When the carbon dioxide concentration rises, as a result of inadequate ventilation (hypoventilation), the cerebral arterioles dilate. This is believed to be due to decreases in the pH of cerebrospinal fluid rather than to a direct effect of CO_2 on the cerebral vessels. Conversely, when the arterial CO_2 falls below normal during hyperventilation, the cerebral vessels constrict. The resulting decrease in cerebral blood flow is responsible for the dizziness that occurs during hyperventilation.

Figure 21.19. Computerized picture of blood flow distribution in the brain after injecting the carotid artery with a radioactive isotope. In (*a*), on the left, the subject followed a moving object with his eyes. High activity is seen over the occipital lobe of the brain. In (*a*), on the right, the subject listened to spoken words. Notice that the high activity is seen over the temporal lobe (the auditory cortex). In (*b*), on the left, the subject moved his fingers on the side of the body opposite to the cerebral hemisphere being studied. In (*b*), on the right, the subject counted to twenty. High activity is shown over the mouth area of the motor cortex, the supplementary motor area, and the auditory cortex.

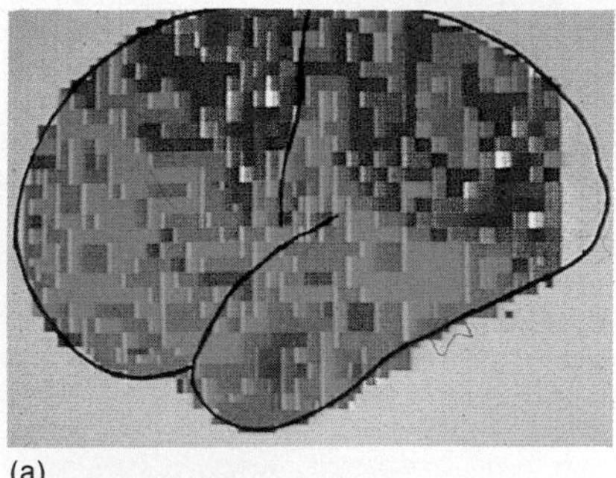

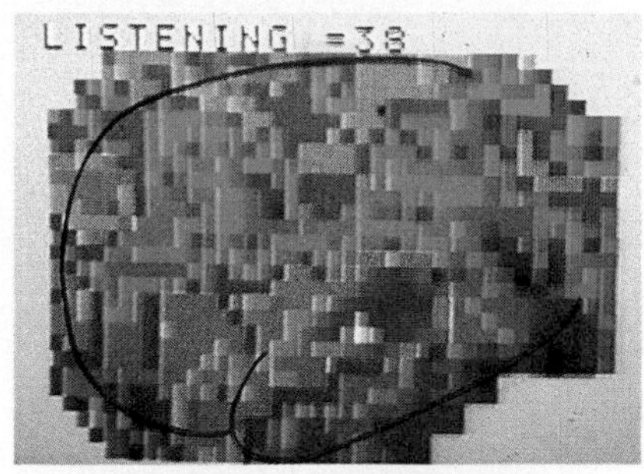

(a)

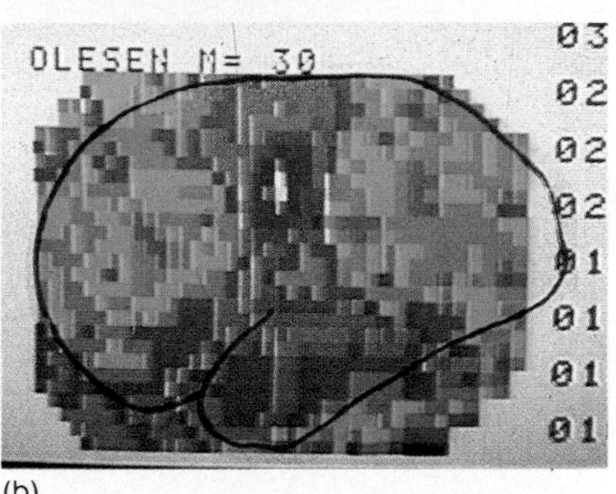

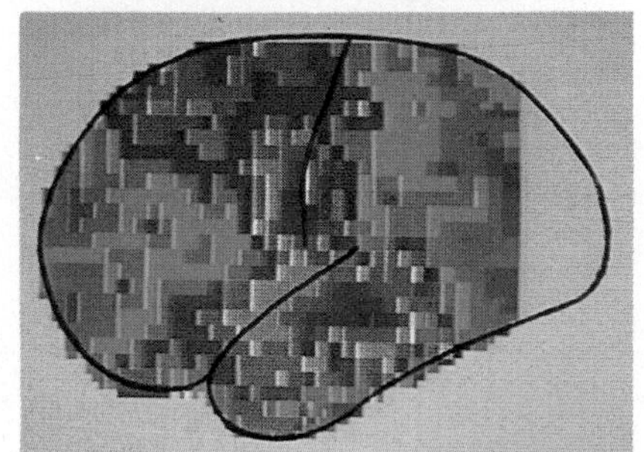

(b)

The cerebral arterioles are exquisitely sensitive to local changes in metabolic activity, so that those brain regions with the highest metabolic activity get the most blood. Indeed, areas of the brain that control specific processes have been mapped by the changing patterns of blood flow that result when these areas are activated. Vision and hearing, for example, increase blood flow to the appropriate sensory areas of the cerebral cortex, whereas motor activities such as movements of the eyes, arms, and organs of speech result in different patterns of blood flow (fig. 21.19).

The exact mechanism whereby increases in neural activity in a particular area of the brain elicit local vasodilation is not completely understood. There is evidence, however, that local cerebral vasodilation may be caused by K^+, which is released from active neurons during repolarization. It has been proposed that astrocytes may take up this extruded K^+ near the active neurons and then release the K^+ through their perivascular feet (chapter 14) surrounding arterioles, thereby causing the arterioles to dilate.

Cutaneous Blood Flow

The skin is the outer covering of the body and as such serves as the first line of defense against invasion by disease-causing organisms. The skin, as the interface between the internal and external environments, also serves to help maintain a constant deep-body temperature despite changes in the ambient (external) temperature, a process called *thermoregulation.* The small thickness and large size of the skin (1.0–1.5 mm thick; 1.7–1.8 square meters in surface area) make it an effective radiator of heat when the body temperature is greater than the ambient temperature. The transfer of heat from the body to the external environment is aided by the flow of warm blood through capillary loops near the surface of the skin.

Blood flow through the skin is adjusted to maintain deep-body temperature at about 37°C (98.6°F). These adjustments are made by variations in the degree of constriction or dilation of ordinary arterioles and of unique **arteriovenous anastomoses** *(ah-nas''to-mo'sēs)* (fig. 21.20). These latter vessels, found predominantly in the fingertips, palms of the hands, toes, soles of the feet, ears, nose, and lips, shunt (divert) blood directly from arterioles to deep venules, thus bypassing superficial capillary loops.

anastomosis: Gk. *anastomosis,* opening or outlet.

Figure 21.20. Circulation in the skin showing arteriovenous anastomoses, which function as shunts allowing blood to be diverted directly from the arteriole to the venule, thus bypassing superficial capillary loops.

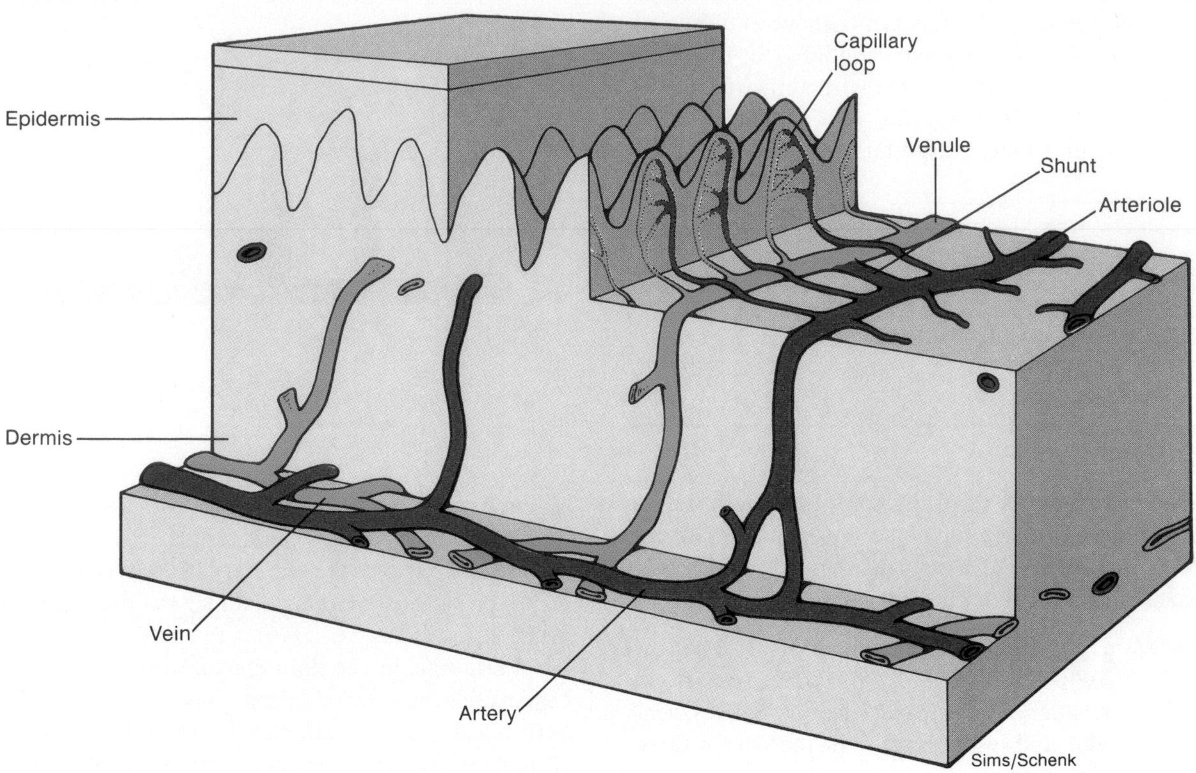

Both the ordinary arterioles and the arteriovenous anastomoses are innervated by sympathetic nerve fibers. When the ambient temperature is cold, sympathetic nerves stimulate cutaneous vasoconstriction; cutaneous blood flow is thus decreased, so that less heat will be lost from the body. Since the arteriovenous anastomoses also constrict, the skin may appear rosy as a result of the fact that blood is diverted to the superficial capillary loops. In spite of this rosy appearance, however, the total cutaneous blood flow and rate of heat loss is lower than under usual conditions.

Skin can tolerate an extremely low blood flow rate in cold weather because its metabolic rate decreases when the ambient temperature decreases. In cold weather, therefore, the skin requires less blood. In very cold weather, however, blood flow to the skin can be so severely restricted that the tissue does die: this is *frostbite.* Blood flow to the skin can vary from less than 20 ml per minute at maximal vasoconstriction to as much as 3–4 L per minute at maximal vasodilation.

As the temperature warms, cutaneous arterioles in the hands and feet dilate as a result of decreased sympathetic nerve activity. Continued warming causes dilation of arterioles in other areas of the skin. If the resulting increase in cutaneous blood flow is not sufficient to cool the body, secretion of the sweat glands may be stimulated. Sweat helps to cool the body as it evaporates from the surface of the skin. Also, the sweat glands secrete **bradykinin** *(brad''e-ki'nin),* a polypeptide that stimulates vasodilation. This increases blood flow to the skin and to the sweat glands, so that larger volumes of more dilute sweat are produced.

Under the usual conditions of ambient temperature, the cutaneous vascular resistance is high and the blood flow is low when a person is not exercising. In the pre-exercise state of "fight or flight," sympathetic nerve activity reduces cutaneous blood flow still further. During exercise, however, the need to maintain a deep-body temperature takes precedence over the need to maintain an adequate systemic blood pressure. As the body temperature rises during exercise, vasodilation in cutaneous vessels occurs together with vasodilation in the exercising muscles. This can produce an even greater lowering of total peripheral resistance. If exercise is performed in hot and humid weather and if restrictive clothing is worn that increases skin temperature and cutaneous vasodilation, a dangerously low blood pressure may be produced after exercise has ceased and the cardiac output has declined. People have lost consciousness and even died as a result.

Changes in cutaneous blood flow occur as a result of changes in sympathetic nerve activity. Since the activity of the sympathetic nervous system is controlled by the brain, emotional states, acting through control centers in the medulla oblongata, can affect sympathetic activity and cutaneous blood flow. During fear reactions, for example, vasoconstriction in the skin together with activation of the sweat glands can produce a pallor and a "cold sweat." Other emotions may cause vasodilation and blushing.

1. Define the term *autoregulation*, and describe how this is accomplished in the cerebral circulation.
2. Describe how hyperventilation can cause dizziness.
3. Explain how cutaneous blood flow is adjusted to maintain a constant deep-body temperature.

Blood Pressure

The blood pressure rise from diastolic to systolic levels provides the driving force for blood flow. Blood pressure is regulated by a variety of control mechanisms and is usually measured indirectly by the auscultatory method.

Objective 16. Describe how the blood pressure changes as blood passes from large arteries to smaller arteries, arterioles, capillaries, venules, and veins. Explain the reasons for these changes.

Objective 17. Describe the factors that directly affect the arterial blood pressure.

Objective 18. Describe the baroreceptor reflex, and explain its physiological significance.

Objective 19. Describe the auscultatory method of blood pressure measurement, and explain how it works.

Resistance to flow in the arterial system is greatest in the arterioles because these vessels have the smallest diameters. Although the total blood flow through a system of arterioles must be equal to the flow in the larger vessel that gave rise to those arterioles, the narrow diameter of each arteriole reduces the flow rate in each according to Poiseuille's law. Blood flow rate and pressure is thus reduced in the capillaries, which are located downstream of the high resistance imposed by the arterioles. The blood pressure upstream of the arterioles—in the medium and large arteries—is correspondingly increased (fig. 21.21).

The blood pressure and flow rate within the capillaries is further reduced by the fact that their total cross-sectional area is much greater, due to their large number, than the cross-sectional areas of arteries and arterioles (fig. 21.22). Thus, although each capillary is much narrower than each arteriole, the capillary beds served by arterioles do not provide as great a resistance to blood flow as do the arterioles.

Figure 21.21. A constriction increases blood pressure upstream (analogous to the arterial pressure) and decreases pressure downstream (analogous to capillary and venous pressure).

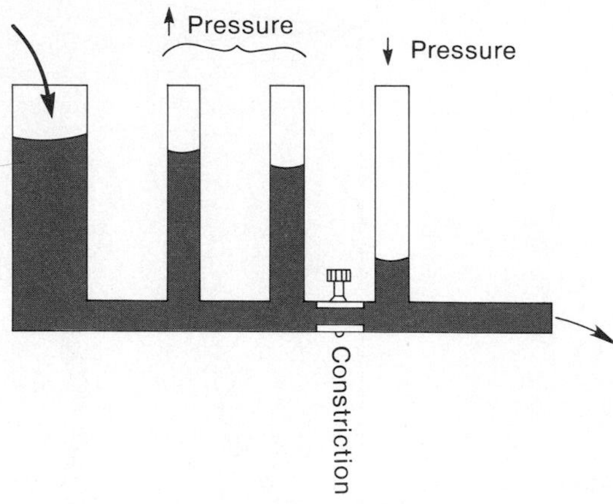

Variations in the diameter of arterioles due to vasoconstriction and vasodilation thus simultaneously affect both blood flow through capillaries and the *arterial* (upstream) *blood pressure.* An increase in total peripheral resistance due to vasoconstriction of arterioles can raise arterial blood pressure. Blood pressure can also be raised by an increase in the cardiac output. This may be due to elevations in cardiac rate or stroke volume, which in turn is affected by other factors. The three most important variables affecting blood pressure are the **cardiac rate, blood volume,** and the **total peripheral resistance.** An increase in any of these three, if not compensated by a decrease in another variable, will result in an increased blood pressure.

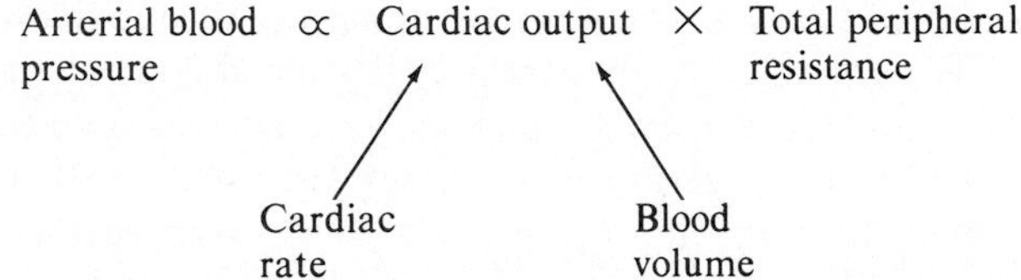

Blood pressure can thus be regulated by the kidneys, which control blood volume, and by the sympathoadrenal system. Increased activity of the sympathoadrenal system can raise blood pressure by stimulating vasoconstriction of arterioles (thus raising total peripheral resistance) and by promoting an increased cardiac output. Sympathetic stimulation can also affect blood volume indirectly, by stimulating constriction of renal blood vessels and thus reducing urine output.

Figure 21.22. As blood passes from the aorta to the smaller arteries, arterioles, and capillaries, the cross-sectional data increases as the pressure decreases.

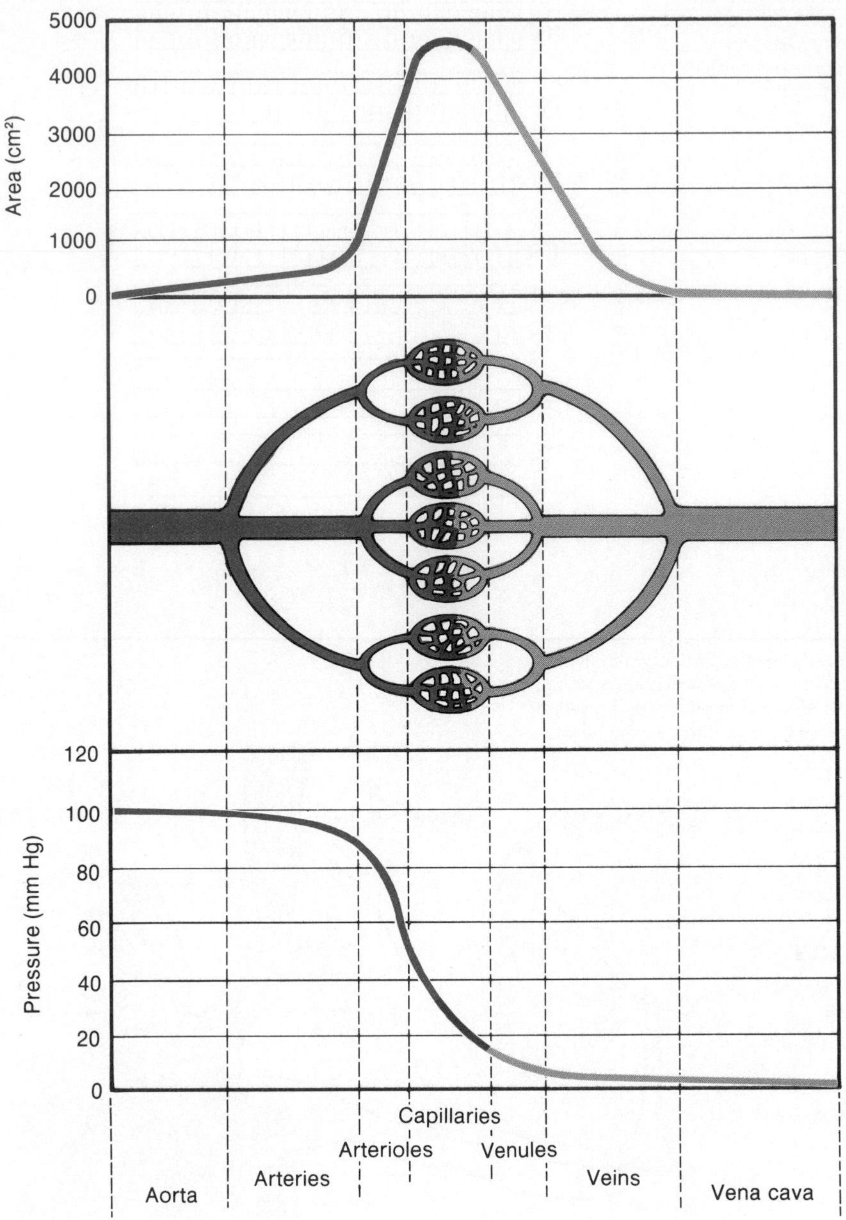

Regulation of Blood Pressure

In order for blood pressure to be maintained within limits, specialized receptors for pressure are needed. These **baroreceptors** *(bar''o-re-sep'tor)* are stretch receptors located in the **aortic arch** and in the **carotid sinuses.** An increase in pressure causes the walls of these arterial regions to stretch and stimulate the activity of sensory nerve endings (fig. 21.23). A fall in pressure below the normal range, in contrast, causes a decrease in the frequency of action potentials produced by these sensory nerve fibers.

Sensory nerve activity from the baroreceptors ascends via the vagus and glossopharyngeal nerves to the medulla oblongata, which directs the autonomic system to respond appropriately. **Vasomotor control centers** in the medulla control vasoconstriction/vasodilation and hence help regulate total peripheral resistance. **Cardiac control centers** (separate cardiostimulatory and cardioinhibitory regions may exist) in the medulla regulate the cardiac rate (fig. 21.24).

baroreceptor: Gk. *baros,* pressure; L. *receiver,* to receive

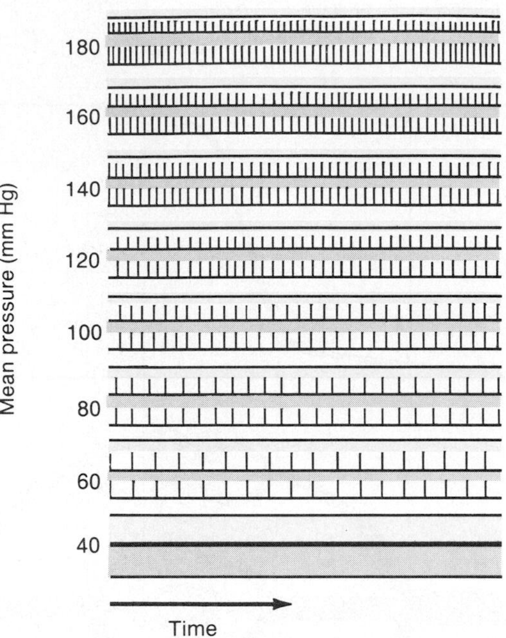

Figure 21.23. Action potential frequency in sensory nerve fibers from baroreceptors in the carotid sinus and aortic arch. As the blood pressure increases, the baroreceptors become increasingly stretched. This results in an increase in the frequency of action potentials that are transmitted to the cardiac and vasomotor control centers in the medulla oblongata.

Figure 21.24. The baroreceptor reflex. Sensory stimuli from baroreceptors in the carotid sinus and the aortic arch, acting via control centers in the medulla oblongata, affect the activity of sympathetic and parasympathetic nerve fibers in the heart.

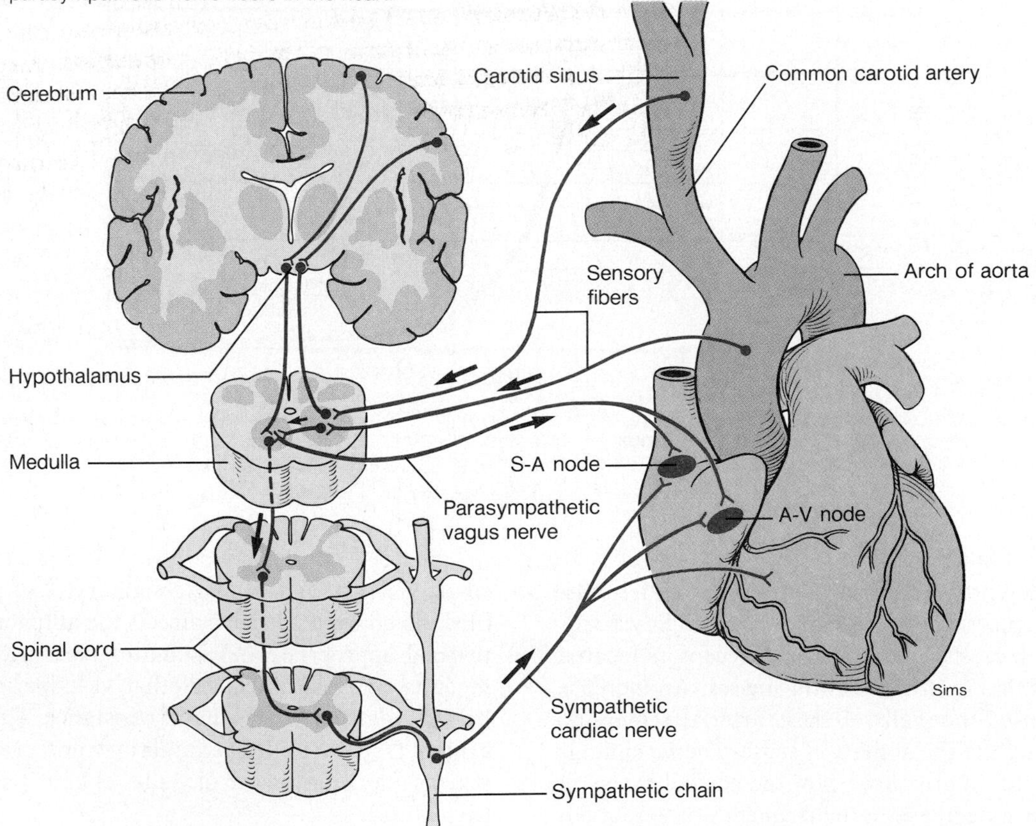

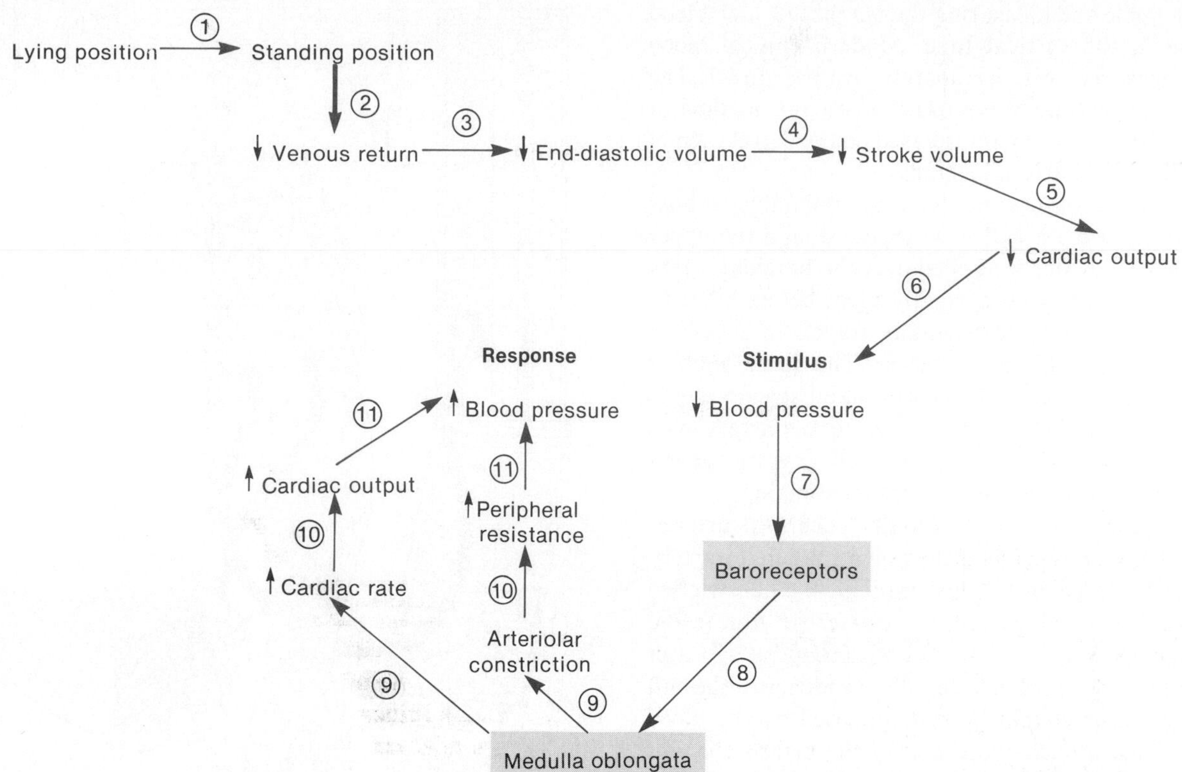

Figure 21.25. Compensations for the upright posture induced by the baroreceptor reflex. Numbers indicate the sequence of cause-and-effect steps.

Baroreceptor Reflex. When a person goes from a lying to a standing position, there is a shift of 500–700 ml of blood from the veins of the thoracic cavity to veins in the lower extremities, which expand to contain the extra volume of blood. This pooling of blood reduces the venous return and cardiac output. The resulting fall in blood pressure is almost immediately compensated by the baroreceptor reflex. A decrease in baroreceptor sensory information to the medulla oblongata inhibits parasympathetic activity and promotes sympathetic nerve activity, resulting in increased cardiac rate and vasoconstriction. These responses help maintain an adequate blood pressure upon standing (fig. 21.25).

An effective baroreceptor reflex helps to maintain adequate blood flow to the brain upon standing. If the baroreceptor sensitivity is reduced, perhaps by *atherosclerosis,* an uncompensated fall in pressure may occur when a person stands up too rapidly. This condition—called *postural,* or *orthostatic, hypotension* (low blood pressure)—can make a person feel dizzy or even faint because of inadequate perfusion of the brain.

The baroreceptor reflex can also mediate the opposite response. When the blood pressure rises above an individual's normal range, the baroreceptor reflex causes a slowing of the cardiac rate and vasodilation. Manual massage of the carotid sinus, a procedure sometimes employed by physicians to reduce tachycardia and lower blood pressure, also evokes this reflex. Such carotid massage should be used cautiously, however, because the intense vagus-nerve-induced slowing of the cardiac rate could cause loss of consciousness (as occurs in emotional fainting).

Other Reflexes Controlling Blood Pressure. The reflex control of ADH secretion by osmoreceptors in the hypothalamus and the control of angiotensin II production and aldosterone secretion by the juxtaglomerular apparatus of the kidneys have been previously discussed. Antidiuretic hormone and aldosterone increase blood volume, whereas angiotensin II stimulates vasoconstriction.

Another reflex that may be important to blood pressure regulation is initiated by stretch receptors located in the left atrium of the heart. These receptors are activated by increased venous return to the heart and produce (1) reflex tachycardia, as a result of increased sympathetic nerve activity; and (2) inhibition of ADH secretion, resulting in larger volumes of urine excretion and a lowering of blood volume.

Measurement of Blood Pressure

Stephen Hales accomplished the first documented measurement of blood pressure by inserting a cannula into the artery of a horse and measuring the height to which blood would rise in the vertical tube. Modern clinical blood pressure measurements, fortunately, are less direct. The indirect, or **auscultatory** *(aws-kul'tah-to''re),* **method** of blood pressure measurement is based on the correlation of blood pressure and arterial sounds.

In the auscultatory method, an inflatable rubber bladder within a cloth cuff is wrapped around the upper arm, and a stethoscope is applied over the brachial artery (fig. 21.26). The artery is normally silent before inflation of the cuff because blood normally travels in a smooth *laminar flow* through the arteries. The term laminar means layered—blood in the central axial stream moves the fastest, and blood flowing closer to the artery wall moves more slowly. There is little transverse movement between these layers that would produce mixing.

The laminar flow that normally occurs in arteries produces little vibration and is thus silent. When the artery is pinched, however, blood flow through the constriction becomes turbulent, which causes the artery to vibrate and produce sounds (much like the sounds produced by water through a kink in a garden hose). The tendency of the cuff pressure to constrict the artery is opposed by the blood pressure. Thus, in order to constrict the artery, the cuff pressure must be greater than the diastolic blood pressure. If the cuff pressure is also greater than the systolic blood pressure, the artery would be pinched off and silent. *Turbulent flow* and sounds produced by vibrations of the artery as a result of this flow, therefore, only occurs when the cuff pressure is greater than the diastolic and less than the systolic blood pressure.

Suppose that a person has a systolic pressure of 120 mm Hg and a diastolic pressure of 80 mm Hg (the average normal values). When the cuff pressure is between 80 and 120 mm Hg, the artery will be closed during diastole and opened during systole. As the artery begins to open with every systole, turbulent flow of blood through the constriction will create vibrations that are heard as the **sounds of Korotkoff** *(ko-rot'kof),* as shown in figure 21.27. These are usually "tapping" sounds because the artery becomes constricted and blood flow is stopped with every diastole. It should be understood that the sounds of Korotkoff are *not* "lub-dub" sounds produced by closing of the heart valves (those sounds can only be heard on the chest, not on the brachial artery).

Initially, the cuff is usually inflated to produce a pressure greater than the systolic pressure so that the artery is pinched off and silent. The pressure in the cuff is read from an attached meter called a *sphygmomanometer (sfig''mo-mah-nom'ĕ-ter).* A valve is then turned to allow the release of air from the cuff, causing a gradual decrease in cuff pressure. When the cuff pressure is equal to the systolic pressure, the *first sound of Korotkoff* is heard as blood passes in a turbulent flow through the constricted opening of the artery.

Stephen Hales: English physiologist, 1677–1761
auscultatory: L., *auscultare,* to listen to
Nicolai S. Korotkoff: Russian physician, 1874–1920

Figure 21.26. The use of a pressure cuff and sphygmomanometer to measure blood pressure.

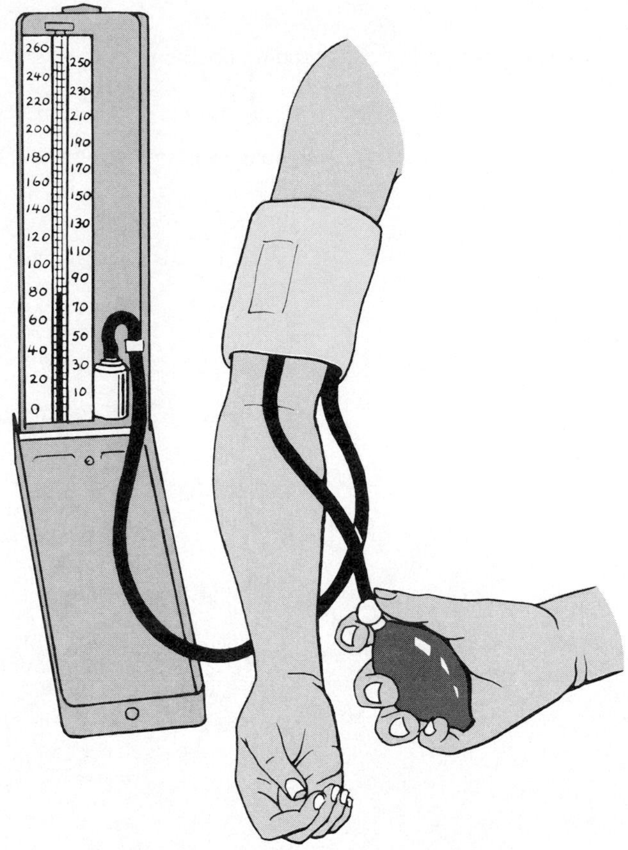

Korotkoff sounds will continue to be heard at every systole as long as the cuff pressure remains greater than the diastolic pressure. When the cuff pressure becomes equal to or less than the diastolic pressure, the sounds disappear because the artery remains open and laminar flow is resumed (fig. 21.28). The *last sound of Korotkoff* thus occurs when the cuff pressure is equal to the diastolic pressure.

Different phases in the measurement of blood pressure are identified on the basis of the quality of the Korotkoff sounds (fig. 21.29). In some people, the Korotkoff sounds do not disappear even when the cuff pressure is reduced to zero (zero pressure means that it is equal to

sphygmomanometer: Gr., *sphygmos,* pulse; *manos,* thin; *metro,* measure

Figure 21.27. Korotkoff sounds are produced by the turbulent flow of blood through the partially constricted brachial artery. This occurs when the cuff pressure is greater than the diastolic pressure but less than the systolic pressure.

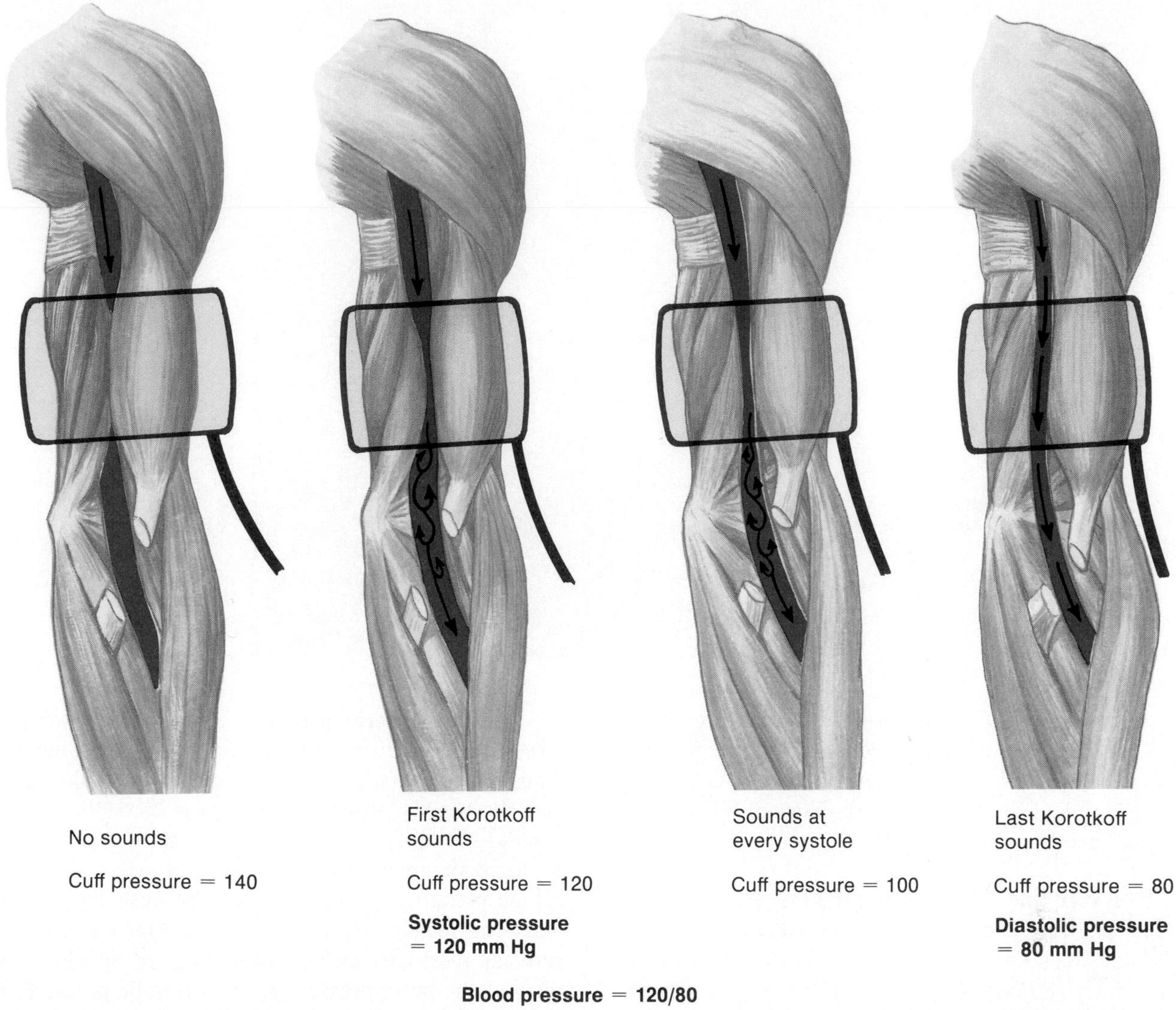

No sounds

Cuff pressure = 140

First Korotkoff sounds

Cuff pressure = 120

Systolic pressure = 120 mm Hg

Sounds at every systole

Cuff pressure = 100

Last Korotkoff sounds

Cuff pressure = 80

Diastolic pressure = 80 mm Hg

Blood pressure = 120/80

Figure 21.28. The indirect, or auscultatory, method of blood-pressure measurement. Korotkoff sounds, produced by turbulent blood flow through a constricted artery, occur whenever the cuff pressure is less than the systolic blood pressure and greater than the diastolic blood pressure. As a result, the first Korotkoff sound is heard when the cuff pressure is equal to the systolic blood pressure, and the last sound is heard when the cuff pressure and diastolic blood pressures are equal.

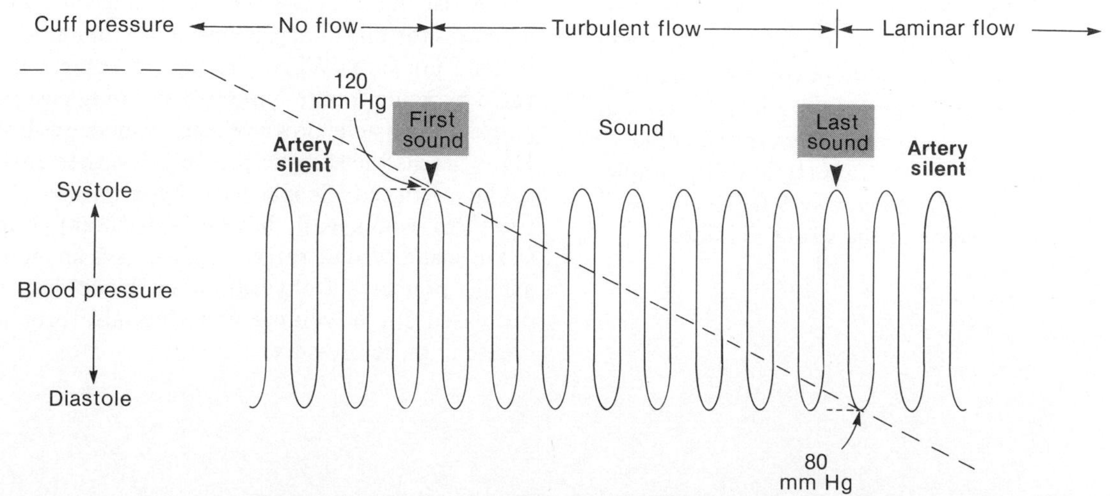

Figure 21.29. The five phases of blood-pressure measurement.

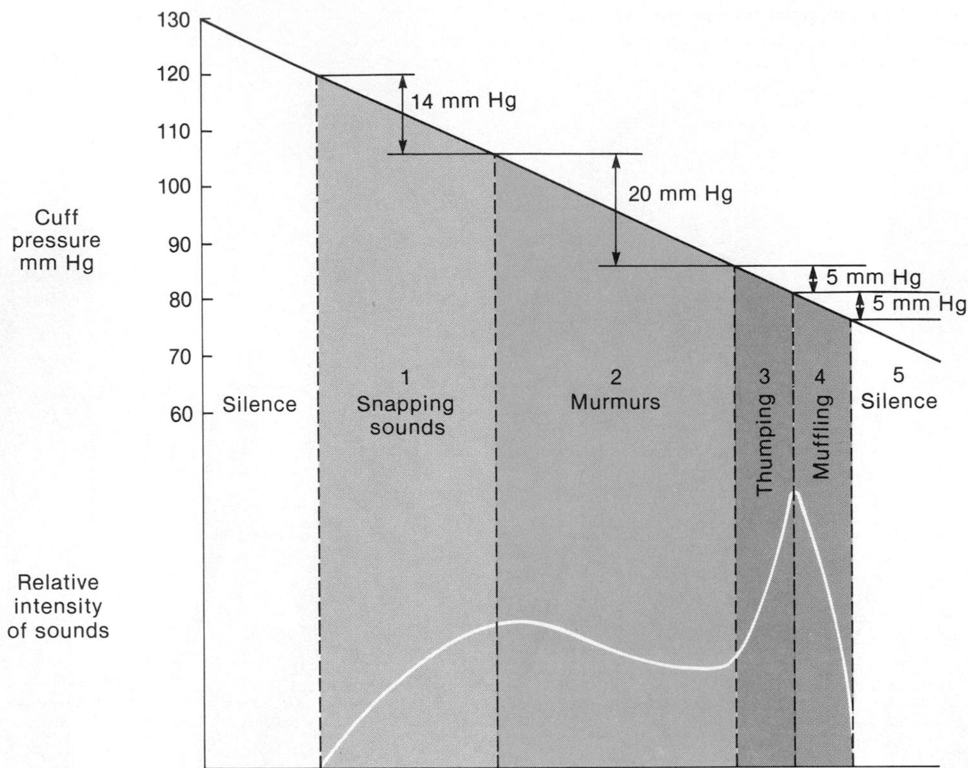

atmospheric pressure). In these cases—and often routinely—the onset of muffling of the sounds (phase 4 in fig. 21.29) is used rather than the onset of silence (phase 5) as an indication of diastolic pressure. Normal blood pressure values are shown in table 21.10.

Pulse Pressure and Mean Arterial Pressure

The **pulse pressure** is equal to the difference between the systolic and diastolic pressures. If a person has a blood pressure of 120/80 (systolic/diastolic), therefore, the pulse pressure would be 40 mm Hg.

Pulse pressure = Systolic pressure − Diastolic pressure

At diastole in this example the aortic pressure equals 80 mm Hg. When the left ventricle contracts, the intraventricular pressure rises above 80 mm Hg and ejection begins. As a result, the amount of blood in the aorta increases by the amount ejected from the left ventricle (the stroke volume). Due to the increase in volume there is an increase in blood pressure. The pressure in the brachial artery, where blood pressure measurements are commonly taken, therefore increases to 120 mm Hg in this example. The rise in pressure from diastolic to systolic levels (pulse pressure) is thus a reflection of the stroke volume.

The **mean arterial pressure** represents the average arterial pressure during the cardiac cycle. This value is significant because it is the difference between this pressure and the venous pressure that drives blood through the capillary beds of organs. The mean arterial pressure is not a simple arithmetic average (the sum of systolic and diastolic pressures divided by two) because the period of diastole is longer than that of systole. Mean arterial pressure can most correctly be approximated by adding one-third of the pulse pressure to the diastolic pressure. If a person has a blood pressure of 120/80, for example, the mean arterial pressure will be 80 + ⅓(40) = 93 mm Hg.

Mean arterial pressure = Diastolic pressure + ⅓ × Pulse pressure

A rise in total peripheral resistance and cardiac rate increases the diastolic pressure more than it increases the systolic pressure. When the baroreceptor reflex is activated by going from a lying to a standing position, for example, the diastolic pressure usually increases by 5–10 mm Hg, whereas the systolic pressure is either unchanged or slightly reduced. People with hypertension (high blood pressure), who usually have elevated total peripheral resistance and cardiac rates, likewise have an increase in diastolic pressure. Dehydration or blood loss results in decreased stroke volume and thus also produces a decrease in pulse pressure.

Table 21.10 Normal arterial blood pressure at different ages*

Age	Systolic Men	Systolic Women	Diastolic Men	Diastolic Women	Age	Systolic Men	Systolic Women	Diastolic Men	Diastolic Women
1 day	70†				16 years	118	116	73	72
3 days	72†				17 years	121	116	74	72
9 days	73†				18 years	120	116	74	72
3 weeks	77†				19 years	122	115	75	71
3 months	86†				20–24 years	123	116	76	72
6–12 months	89	93	60	62	25–29 years	125	117	78	74
1 year	96	95	66	65	30–34 years	126	120	79	75
2 years	99	92	64	60	35–39 years	127	124	80	78
3 years	100	100	67	64	40–44 years	129	127	81	80
4 years	99	99	65	66	45–49 years	130	131	82	82
5 years	92	92	62	62	50–54 years	135	137	83	84
6 years	94	94	64	64	55–59 years	138	139	84	84
7 years	97	97	65	66	60–64 years	142	144	85	85
8 years	100	100	67	68	65–69 years	143	154	83	85
9 years	101	101	68	69	70–74 years	145	159	82	85
10 years	103	103	69	70	75–79 years	146	158	81	84
11 years	104	104	70	71	80–84 years	145	157	82	83
12 years	106	106	71	72	85–89 years	145	154	79	82
13 years	108	108	72	73	90–94 years	145	150	78	79
14 years	110	110	73	74	95–106 years	145	149	78	81
15 years	112	112	75	76					

*Mean arterial blood pressure; derived from various sources.

†Value for both male and female.

Table 21.11 Effects of different variables on blood pressure measurements

Variable	Diastolic pressure	Systolic pressure	Pulse pressure
↑Cardiac rate	Increases (more)	Increases (less)	Decreases
↑Peripheral resistance	Increases (more)	Increases (less)	Decreases
↑Blood volume*	Increases (less)	Increases (more)	Increases
↑Stroke volume	Increases (less)	Increases (more)	Increases

*Conversely, a fall in blood volume (as in dehydration) produces a decrease in pulse pressure.

An increase in cardiac output, in contrast, raises the systolic pressure more than it raises the diastolic pressure (though both pressures do rise). This occurs during exercise, for example, when the blood pressure may rise to values as high as 200/100 (yielding a pulse pressure of 100 mm Hg). The effects of different variables on blood pressure are summarized in table 21.11.

1. Describe the relationship between blood pressure and the total cross-sectional area of arteries, arterioles, and capillaries. Describe how arterioles influence blood flow through capillaries and arterial blood pressure.
2. Describe how the baroreceptor reflex helps to compensate for a fall in blood pressure. Explain how a person who is severely dehydrated would have a rapid pulse.
3. Describe how the sounds of Korotkoff are produced and how these sounds are used to measure blood pressure.
4. Define pulse pressure and explain its physiological significance.

Clinical Considerations

Hypertension

Approximately 20% of all adults in the United States have *hypertension*—blood pressure in excess of the normal range for the person's age and sex. Hypertension that is a result of ("secondary to") known disease processes is logically called **secondary hypertension.** This comprises only about 10% of the hypertensive population. Hypertension that is the result of complex and poorly understood processes is not-so-logically called **primary,** or **essential, hypertension.**

Diseases of the kidneys and arteriosclerosis of the renal arteries can cause secondary hypertension because of high blood volume. More commonly, the reduction of renal

Table 21.12 Some possible causes of secondary hypertension

System involved	Examples	Mechanisms
Kidneys	Kidney disease Renal artery disease	Decreased urine formation Secretion of vasoactive chemicals
Endocrine	Excess catecholamines (tumor of adrenal medulla) Excess aldosterone (Conn's disease)	Increased cardiac output and total peripheral resistance Excess salt and water retention by the kidneys
Nervous	Increased intracranial pressure Damage to vasomotor center	Activation of sympathoadrenal system Activation of sympathoadrenal system
Cardiovascular	Complete heart block; patent ductus arteriosus Arteriosclerosis of aorta; coarctation of aorta	Increased stroke volume Decreased distensibility of aorta

blood flow can raise blood pressure by stimulating the secretion of vasoactive chemicals from the kidneys. Experiments in which the renal artery is pinched, for example, produce hypertension that is associated (at least initially) with elevated renin secretion. These and other causes of secondary hypertension are summarized in table 21.12.

Essential Hypertension. The vast majority of people with hypertension have essential hypertension. An increased total peripheral resistance is a universal characteristic of this condition. Cardiac rate and the cardiac output are elevated in many, but not all, of these cases.

The secretion of renin, which is correlated with angiotensin II production and aldosterone secretion, is likewise variable. Although some people with essential hypertension have low renin secretion, most have either normal or elevated levels of renin secretion. Renin secretion in the normal range is inappropriate for people with hypertension, since high blood pressure should inhibit renin secretion and, through a lowering of aldosterone, result in greater excretion of salt and water. Inappropriately high levels of renin secretion could thus contribute to hypertension by promoting (via stimulation of aldosterone secretion) salt and water retention and high blood volume.

Sustained high stress (acting via the sympathetic nervous system) and high salt intake appear to act synergistically in the development of hypertension. There is some evidence that Na^+ enhances the vascular response to sympathetic stimulation. Further, sympathetic nerves can cause constriction of the renal blood vessels and thus decrease the excretion of salt and water. In one study, hypertensive students given an IQ test had decreased renal blood flow, increased renin secretion, decreased excretion of salt and water, and a greater increase in blood pressure than did students with normal blood pressure. There also appears to be a genetic component to these responses; students who were normotensive but who had a family history of hypertension had responses that were intermediate between hypertensive students and students who were normotensive with no family history of hypertension.

As an adaptive response to prolonged high blood pressure, the arterial wall becomes thickened. This response can lead to arteriosclerosis and results in an even greater increase in total peripheral resistance, thus raising blood pressure still more in a positive feedback fashion.

The interactions between salt intake, sympathetic nerve activity, cardiovascular responses to sympathetic nerve activity, kidney function, and genetics make it difficult to sort out the cause-and-effect sequence that leads to essential hypertension. Many people have suggested that there is no single cause-and-effect sequence but rather a web of causes and effects (see fig. 21.30). This is currently controversial.

Dangers of Hypertension. If other factors remain constant, blood flow increases as arterial blood pressure increases. People with hypertension thus have adequate perfusion of their organs with blood until the hypertension causes vascular damage. Hypertension, as a result, is usually without symptoms until a dangerous amount of vascular damage is produced.

Hypertension is dangerous because (1) high arterial pressure increases the afterload, making it more difficult for the ventricles to eject blood; this increases the amount of work that the heart must perform and may result in pathological changes in heart structure and function, leading to congestive heart failure; (2) high pressure may damage cerebral blood vessels, leading to cerebrovascular accident (stroke); and (3) it contributes to the development of atherosclerosis, which can itself lead to heart disease and stroke as previously described.

Treatment of Hypertension. Hypertension is usually treated by restricting salt intake and by taking drugs that act in a variety of ways. Most commonly, these drugs are *diuretics* that increase urine volume, thus decreasing blood volume and pressure. Sympathetic-blocking drugs are also often used; drugs that block beta-adrenergic receptors (such as propranolol) decrease cardiac rate. Various vasodilators (see table 21.13) may also be used to decrease total peripheral resistance.

Figure 21.30. The sequence of events that has been proposed as a possible cause of essential hypertension.

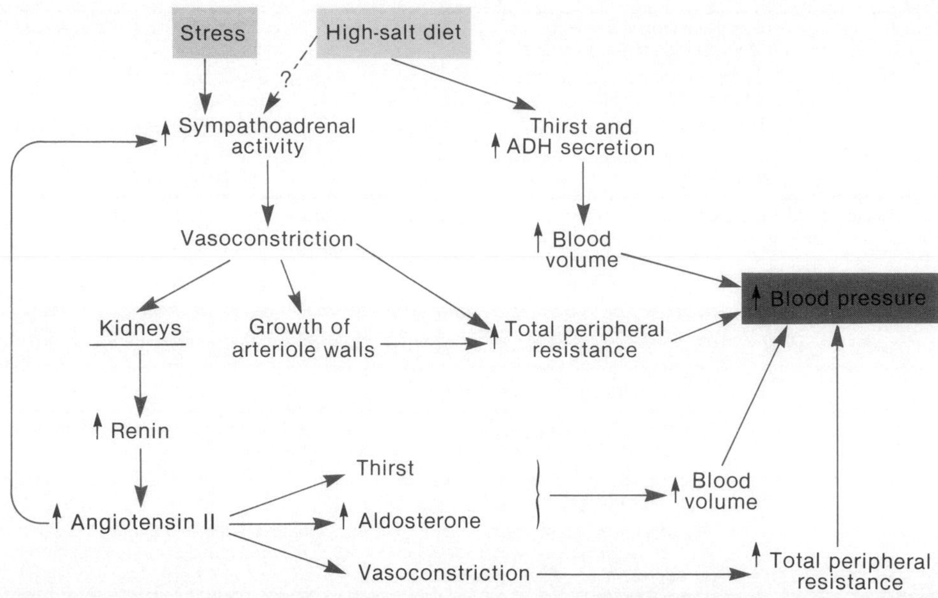

Table 21.13 Examples and mechanisms of action of some antihypertensive drugs

Category of drugs	Examples	Mechanisms
Extracellular fluid volume depletors	Thiazide diuretics	Increase volume of urine excreted, thus lowering blood volume
Sympathoadrenal system inhibitors	Clonidine; alpha-methyldopa Guanethidine; reserpine Propranolol	Acts on brain to decrease sympathoadrenal stimulation Depletes norepinephrine from sympathetic nerve endings Blocks beta-adrenergic receptors, decreasing cardiac output and/or renin secretion
	Phentolamine	Blocks alpha-adrenergic receptors, decreasing sympathetic vasoconstriction
Vasodilators	Hydralazine; sodium nitroprusside	Causes vasodilation by acting directly on vascular smooth muscle
Calcium channel blockers	Verapamil	Inhibits diffusion of Ca^{++} into vascular smooth muscle cells, causing vasodilation and reduced peripheral resistance
Angiotension II inhibitors	Captopril	Inhibits the conversion of angiotensin I into angiotensin II

Circulatory Shock

Circulatory shock occurs when there is inadequate blood flow and/or oxygen utilization by the tissues. Some of the signs of shock (table 21.14) are a result of inadequate tissue perfusion; other signs of shock are produced by cardiovascular responses that help to compensate for the poor tissue perfusion (table 21.15). When these compensations are effective, they (together with emergency medical care) are able to reestablish adequate tissue perfusion. In some cases, however, and for reasons that are not clearly understood, the shock may progress to an irreversible stage and death may result.

Hypovolemic Shock. The term **hypovolemic** *(hi″po-vo-le′mik)* **shock** refers to circulatory shock due to low blood volume, as might be caused by hemorrhage (bleeding), dehydration, or burns. This is accompanied by decreased blood pressure and decreased cardiac output. In response to these changes, the sympathoadrenal system is activated as a result of the baroreceptor reflex. As a result, tachycardia is produced and vasoconstriction occurs in the skin, digestive tract, kidneys, and muscles (table 21.15). Decreased blood flow through the kidneys stimulates renin secretion and activation of the renin-angiotensin-aldosterone system. A person in hypovolemic shock thus has low blood pressure, rapid pulse, cold, clammy skin, and little urine excretion.

Other Causes of Circulatory Shock. A rapid fall in blood pressure occurs in **anaphylactic** *(an″ah-fĭ-lak′tik)* **shock** as a result of severe allergic reaction (usually to bee stings or penicillin). This results from the widespread release of histamine, which causes vasodilation and thus decreases total peripheral resistance. A rapid fall in blood pressure

Table 21.14 Signs of shock

	Early sign	Late sign
Blood pressure	Decreased pulse pressure Increased diastolic pressure	Decreased systolic pressure
Urine	Decreased Na^+ concentration Increased osmolality	Decreased volume
Blood pH	Increased pH (alkalosis) due to hyperventilation	Decreased pH (acidosis) due to "metabolic" acids
Effects of poor tissue perfusion	Slight restlessness; occasionally warm, dry skin	Cold, clammy skin "Cloudy" senses

Table 21.15 Cardiovascular reflexes that help to compensate for circulatory shock

Organ(s)	Compensatory mechanisms
Heart	Sympathoadrenal stimulation produces increased cardiac rate and increased stroke volume, due to "positive inotropic effect" on myocardial contractility
Digestive tract and skin	Decreased blood flow due to vasoconstriction as a result of sympathetic nerve stimulation (alpha-adrenergic effect)
Kidneys	Decreased urine production as a result of sympathetic-nerve-induced constriction of renal arterioles; increased salt and water retention due to increased aldosterone and antidiuretic hormone (ADH) secretion

also occurs in **neurogenic shock,** in which sympathetic tone is decreased, usually because of upper spinal cord damage or spinal anesthesia. **Cardiogenic shock** results from cardiac failure, as defined by a cardiac output that is inadequate to maintain tissue perfusion.

Congestive Heart Failure

Cardiac failure occurs when the cardiac output is insufficient to maintain the blood flow required by the body. This may be due to heart disease—resulting from myocardial infarction or congenital defects—or to hypertension, which increases the afterload of the heart. The most common causes of left pump failure are myocardial infarction, aortic valve stenosis, and incompetence of the aortic and bicuspid (mitral) valves. Failure of the right pump is usually caused by prior failure of the left pump.

Heart failure can also result from disturbance in the electrolyte concentrations of the blood. Excessive plasma K^+ concentration decreases the resting membrane potential of myocardial cells; low blood Ca^{++} reduces excitation-contraction coupling. High blood K^+ and low blood Ca^{++} can thus cause the heart to stop in diastole. Conversely, low blood K^+ and high blood Ca^{++} can arrest the heart in systole.

The term *congestive* is often used in describing heart failure because of the increased venous volume and pressure that results. Failure of the left pump, for example, raises the left atrial pressure and produces pulmonary congestion and edema, which causes shortness of breath and fatigue. Failure of the right pump results in increased right atrial pressure, which produces congestion and edema in the systemic circulation.

Figure 21.31. The relationship between end-diastolic volume and ventricular pressure (a measure of contraction strength). At a given end-diastolic volume the ventricular pressure is decreased during heart failure and increased by sympathetic stimulation. Heart failure may be compensated by sympathetic stimulation.

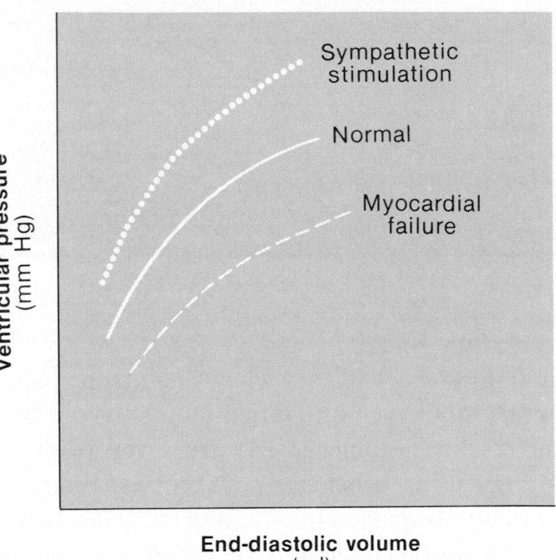

The compensatory responses that occur during congestive heart failure are similar to those that occur during hypovolemic shock. Activation of the sympathoadrenal system stimulates cardiac rate, contractility of the ventricles (fig. 21.31), and constriction of arterioles. As in hypovolemic shock, renin secretion is increased and urine output is reduced.

As a result of these compensations, chronically low cardiac output is associated with elevated blood volume and dilation and hypertrophy of the ventricles. These changes can themselves be dangerous. Elevated blood volume places a work overload on the heart, and the enlarged ventricles have a higher metabolic requirement for oxygen. These problems are often treated with drugs that increase myocardial contractility (such as digitalis), drugs that are vasodilators (such as nitroglycerin), and diuretic drugs that lower blood volume by increasing the volume of urine excreted.

Chapter Summary

I. Cardiac Output
  A. Cardiac rate is increased by sympathoadrenal stimulation and decreased by the effects of parasympathetic fibers that innervate the SA node.
  B. Stroke volume is regulated both extrinsically and intrinsically.
    1. The Frank-Starling law of the heart describes the way the end-diastolic volume, through various degrees of myocardial stretching, influences the contraction strength of the myocardium and thus the stroke volume.
    2. The end-diastolic volume is called the preload; the total peripheral resistance, through its effect on arterial blood pressure, provides an afterload that acts to reduce the stroke volume.
    3. At a given end-diastolic volume, the amount of blood ejected depends on contractility; strength of contraction is increased by sympathoadrenal stimulation.
  C. The venous return of blood to the heart is dependent largely on the total blood volume and mechanisms that improve the flow of blood in veins.
    1. The total blood volume is regulated by the kidneys.
    2. The venous flow of blood to the heart is aided by the action of skeletal muscle pumps and the effects of breathing.

II. Blood Volume
  A. Tissue fluid is formed from and returns to the blood.
  B. Extrinsic regulation of vascular resistance is provided mainly by the sympathetic nervous system, which stimulates vasoconstriction of arterioles in the viscera and skin.
  C. Intrinsic control of vascular resistance allows organs to autoregulate their own blood flow rates.
    1. Myogenic regulation occurs when vessels constrict or dilate as a direct response to a rise or fall in blood pressure.
    2. Metabolic regulation occurs when vessels constrict or dilate as a direct response to a rise or fall in blood pressure.

III. Vascular Resistance and Blood Flow
  A. Poiseuille's law describes the fact that blood flow is directly related to the pressure difference between the two ends of a vessel and is inversely related to the resistance to blood flow through the vessel.
  B. Extrinsic regulation of vascular resistance is provided mainly by the sympathetic nervous system, which stimulates vasoconstriction of arterioles in the viscera and skin.
  C. Intrinsic control of vascular resistance allows organs to autoregulate their own blood flow rates.
    1. Myogenic regulation occurs when vessels constrict or dilate as a direct response to a rise or fall in blood pressure.
    2. Metabolic regulation occurs when vessels dilate in response to the local chemical environment within the organ.

IV. Blood Flow to the Heart and Skeletal Muscles
  A. The heart normally always respires aerobically because of its high capillary supply, myoglobin content, and enzyme content.
  B. During exercise, when the heart's metabolism increases, intrinsic metabolic mechanisms stimulate vasodilation of the coronary vessels and thus increase coronary blood flow.
  C. Just prior to exercise and at the start of exercise, blood flow through skeletal muscles increases due to vasodilation caused by cholinergic sympathetic nerve fibers; during exercise intrinsic metabolic vasodilation occurs.
  D. Since cardiac output can increase by a factor of five or more during exercise, the heart and skeletal muscles receive an increased proportion of a higher total blood flow.
    1. The cardiac rate increases due to lower activity of the vagus nerve and higher activity of the sympathetic nerve.
    2. The venous return is faster because of higher activity of the skeletal muscle pumps and increased breathing.
    3. Increased contractility of the heart, combined with a decrease in total peripheral resistance, can result in a higher stroke volume.

V. Blood Flow to the Brain and Skin
  A. Cerebral blood flow is regulated both myogenically and metabolically.
    1. Cerebral vessels automatically constrict if the systemic blood pressure rises too high.
    2. Metabolic products cause local vessels to dilate and supply more active areas with more blood.
  B. The skin has unique arteriovenous anastomoses, which can shunt the blood away from surface capillary loops.
    1. Sympathetic nerve fibers cause constriction of cutaneous arterioles.
    2. As a thermoregulatory response, there is increased cutaneous blood flow and increased flow through surface capillary loops.

VI. Blood Pressure
  A. Baroreceptors in the aortic arch and carotid sinuses affect, via the sympathetic nervous system, the cardiac rate and the total peripheral resistance.
    1. The baroreceptor reflex causes pressure to be maintained when an upright posture is assumed and can cause a lowered pressure when the carotid sinuses are massaged.
    2. Other mechanisms that affect blood volume help to regulate blood pressure.
  B. Blood pressure is commonly measured indirectly by auscultation of the brachial artery when a pressure cuff is inflated and deflated.
    1. The first sound of Korotkoff, caused by turbulent flow of blood through a constriction in the artery, occurs when the cuff pressure equals the systolic pressure.
    2. The last sound of Korotkoff is heard when the cuff pressure equals the diastolic blood pressure.
  C. The pulse pressure, which is the difference between systolic and diastolic pressures, is a reflection of the stroke volume.

Review Activities

Objective Questions

1. According to the Frank-Starling law, the strength of ventricular contraction is
   (a) directly proportional to the end-diastolic volume
   (b) inversely proportional to the end-diastolic volume
   (c) independent of the end-diastolic volume
2. In the absence of compensations, the stroke volume will decrease when
   (a) blood volume increases
   (b) venous return increases
   (c) contractility increases
   (d) arterial blood pressure increases
3. Which of the following statements about tissue fluid is *false?*
   (a) It contains the same glucose and salt concentration as plasma.
   (b) It contains a lower protein concentration than plasma.
   (c) Its colloid osmotic pressure is greater than that of plasma.
   (d) Its hydrostatic pressure is less than that of plasma.
4. Edema may be caused by
   (a) high blood pressure
   (b) decreased plasma protein concentration
   (c) leakage of plasma protein into tissue fluid
   (d) blockage of lymphatic vessels
   (e) all of the above
5. Both ADH and aldosterone act to
   (a) increase urine volume
   (b) increase blood volume
   (c) increase total peripheral resistance
   (d) all of the above
6. The greatest resistance to blood flow occurs in the
   (a) large arteries
   (b) medium-sized arteries
   (c) arterioles
   (d) capillaries
7. If a vessel were to dilate to twice its previous radius and if pressure remained constant, blood flow through this vessel would
   (a) increase by a factor of 16
   (b) increase by a factor of 4
   (c) increase by a factor of 2
   (d) decrease by a factor of 2
8. The sounds of Korotkoff are produced by
   (a) closing of the semilunar valves
   (b) closing of the AV valves
   (c) the turbulent flow of blood through an artery
   (d) elastic recoil of the aorta
9. Vasodilation in the heart and skeletal muscles during exercise is primarily due to the effects of
   (a) alpha-adrenergic stimulation
   (b) beta-adrenergic stimulation
   (c) cholinergic stimulation
   (d) products released by the exercising muscle cells
10. Blood flow in the coronary circulation is
    (a) increased during systole
    (b) increased during diastole
    (c) constant throughout the cardiac cycle
11. Blood flow in the cerebral circulation
    (a) varies with systemic arterial pressure
    (b) is regulated primarily by the sympathetic system
    (c) is maintained constant within physiological limits
    (d) is increased during exercise
12. Which of the following organs is able to tolerate the greatest restriction in blood flow? The
    (a) brain
    (b) heart
    (c) skeletal muscles
    (d) skin
13. Arteriovenous shunts in the skin
    (a) divert blood to superficial capillary loops
    (b) are closed when the ambient temperature is very cold
    (c) are closed when the deep-body temperature rises much above 37°C.
    (d) all of the above

Essay Questions

1. Define the terms *contractility, preload,* and *afterload,* and explain how these factors affect the cardiac output.
2. Explain, using the Frank-Starling law, how the stroke volume is affected by (a) bradycardia and (b) a "missed beat."
3. Which part of the cardiovascular system contains the most blood? Which part provides the greatest resistance to blood flow? Which part provides the greatest surface area? Explain.
4. Explain how the kidneys regulate blood volume.
5. A person who is dehydrated drinks more and urinates less. Explain the mechanisms involved.
6. Explain, using Poiseuille's law, how arterial blood flow can be diverted from one organ system to another.
7. Describe the mechanisms that increase the cardiac output during exercise and that increase the rate of blood flow to the heart and skeletal muscles.
8. Explain how an anxious person may have cold, clammy skin and how the skin becomes hot and flushed on a hot, humid day.

Lymphatic System and Immunity

Outline

Concepts

The lymphatic system, consisting of lymphatic vessels and lymph nodes, returns tissue fluid to the venous system and helps protect the body from diseases.

The immune system includes all of the structures and processes that provide a defense against potential pathogens. These defenses can be grouped into nonspecific and specific categories.

B lymphocytes secrete protein antibodies that can bond in a specific fashion with antigens. The formation of antigen-antibody complexes activates other elements of the immune system and results in more effective defense against pathogens.

T cells assist all aspects of the immune system, including cell-mediated destruction by cytotoxic T cells and supporting roles by helper and suppressor T cells. T cells are only activated by antigens presented to them by macrophages; the activated T cells in turn produce lymphokines, which activate other cells of the immune system.

Resistance to pathogens can be gained by both active and passive immunizations. The change from a primary to a secondary response in active immunity is best explained by the clonal selection theory.

Tumor cells usually reveal antigens that activate an immune reaction that destroys the tumor. When cancers develop, this immunological surveillance system—primarily the function of T cells and natural killer cells—has failed to prevent the growth and metastasis of the tumor.

Lymphatic System

The lymphatic system, consisting of lymphatic vessels and lymph nodes, returns tissue fluid to the venous system and helps protect the body from diseases.

Objective 1. Describe the pattern of lymph flow from the lymphatic capillaries to the venous system.
Objective 2. Describe the structure and location of the lymph nodes, and list two lymphatic organs.

The lymphatic system has three basic functions: (1) it transports interstitial (tissue) fluid, which was initially formed as a blood filtrate, back to the bloodstream; (2) it serves as the route by which absorbed fat from the intestine is transported to the blood; and (3) its cells—called **lymphocytes** *(lim'fo-sīt)* (produced in lymph nodes, lymphatic organs, and bone marrow)—help provide immunological defenses against disease-causing agents.

Lymph and Lymph Capillaries

The smallest vessels of the lymphatic *(lim-fat'ik)* system are the **lymph capillaries** (fig. 22.1). Lymph capillaries are microscopic, closed-ended tubes that form vast networks in the intercellular spaces within most tissues. Within the villi of the small intestine, for example, lymph capillaries, called *lacteals (lak'te-al),* transport the products of fat absorption away from the digestive tract.

Because the walls of lymph capillaries are composed of endothelial cells with porous junctions, interstitial fluid, proteins, microorganisms, and absorbed fat (in the intestine) can easily enter. Once fluid enters the lymphatic capillaries, it is referred to as **lymph.**

Lymph Ducts

From merging lymph capillaries, the lymph is carried into larger **lymph ducts.** The walls of lymph ducts are similar to those of veins in that they have the same three layers and contain valves to prevent the backflow of lymph. Interconnecting lymph ducts eventually empty into one of the two principal vessels: the **thoracic duct** and the **right lymphatic duct** (fig. 22.2).

The larger thoracic duct drains lymph from the lower extremities, abdomen, left thoracic region, left upper extremity, and left side of the head and neck. The main trunk of this vessel ascends along the spinal column and drains

lacteal: L. *lacteus,* milk
lymph: L. *lympha,* clear water

Figure 22.1. A schematic diagram showing the structural relationship of a capillary bed and a lymph capillary.

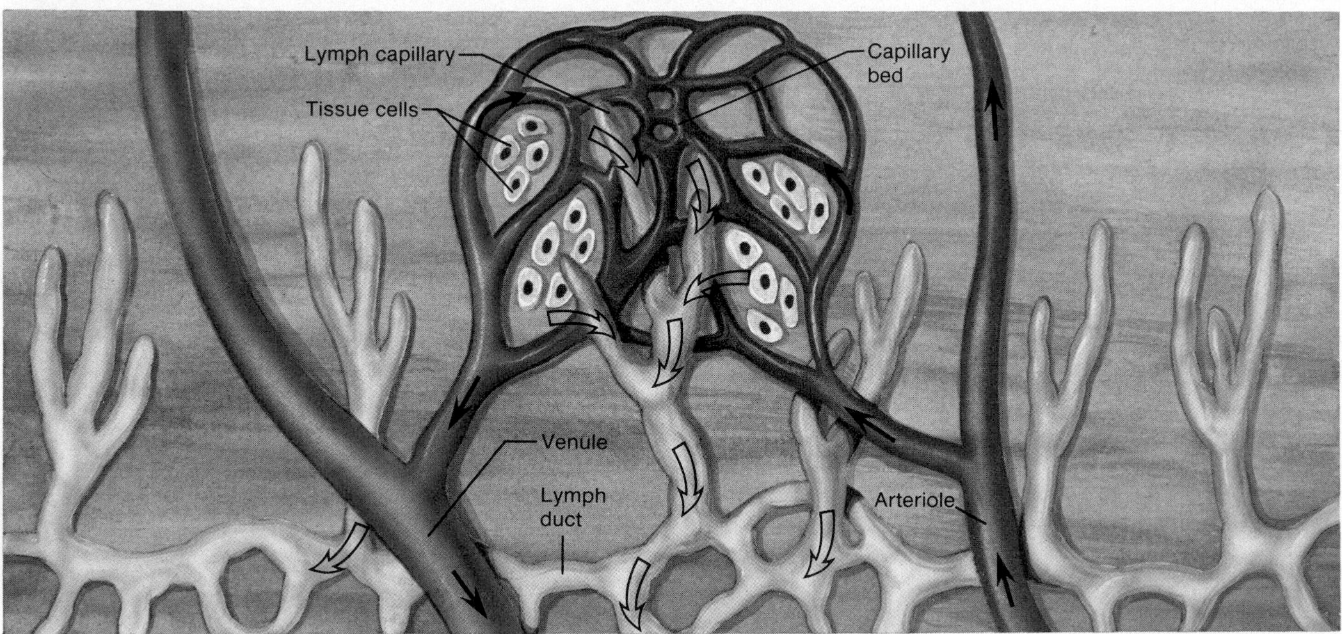

into the left subclavian vein. In the abdominal area is a saclike enlargement of the thoracic duct called the **cisterna chyli** *(sis-ter'nah ki'le)*. The shorter right lymphatic duct drains lymph vessels from the right upper extremity, right thoracic region, and right side of the head and neck. The right lymphatic duct empties into the right subclavian vein near the right internal jugular vein.

The pressure that pushes lymph through the lymph ducts comes from the massaging actions produced by skeletal muscle contractions, intestinal movements, and other body movements. The many valves keep lymph moving in one direction.

Lymph Nodes

Lymph filters through the reticular tissue of **lymph nodes** (fig. 22.3), which contain phagocytic cells that help purify the fluid. Lymph nodes are small, oval bodies enclosed within fibrous connective tissue *capsules*. Specialized connective tissue bands called *trabeculae* divide the node. *Afferent lymphatic vessels* carry lymph into the node, where it is circulated through sinuses in the *cortical tissue*. Lymph leaves the node through the *efferent lymphatic vessel*, which emerges from the *hilus*, the depression on the concave side. **Germinal centers** within the node are sites of lymphocyte production, and thus important in the development of an immune response, as discussed later in this chapter.

Lymph nodes usually occur in clusters in specific regions of the body (fig. 22.4). Some of the principal groups of lymph nodes are the **popliteal** and **inguinal nodes** of the lower extremity; the **lumbar nodes** of the pelvic region; the **cubital** and **axillary nodes** of the upper extremity; the **thoracic nodes** of the chest; and the **cervical nodes** of the neck. The submucosa of the small intestine contains numerous scattered lymphocytes and lymphatic nodules and larger aggregations of lymphatic tissue called **Peyer's patches.**

Migrating cancer cells (metastases) are especially dangerous if they enter the lymphatic system, which can disperse them widely. On entering the lymph nodes, the cancer cells can multiply and establish secondary tumors in organs that are away from the site of the primary tumor.

cisterna chyli: L. *cisterna*, box; Gk. *chylos*, juice

Peyer's patches: from Johann K. Peyer, Swiss anatomist, 1653–1712

Figure 22.2. Lymphatic vessels. (*a*) a magnified view of the upper right quadrant showing the lymph drainage of the right breast. (*b*) a major lymph drainage of body.

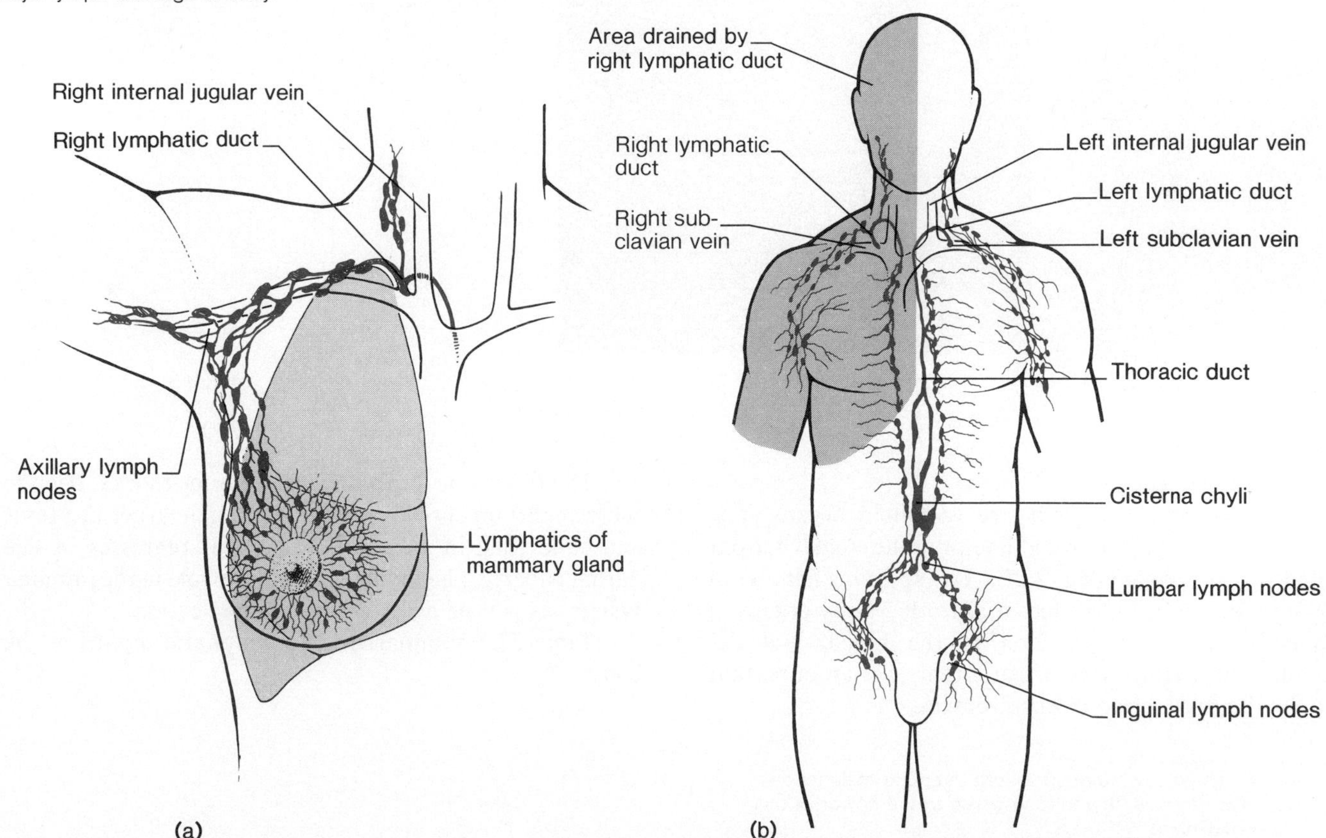

Figure 22.3. The structure of a lymph node. (*a*) a schematic diagram of a sectioned lymph node and associated vessels. (*b*) a photomicrograph of lymphatic tissue. (*c*) a photograph of a lymph node positioned near a blood vessel.

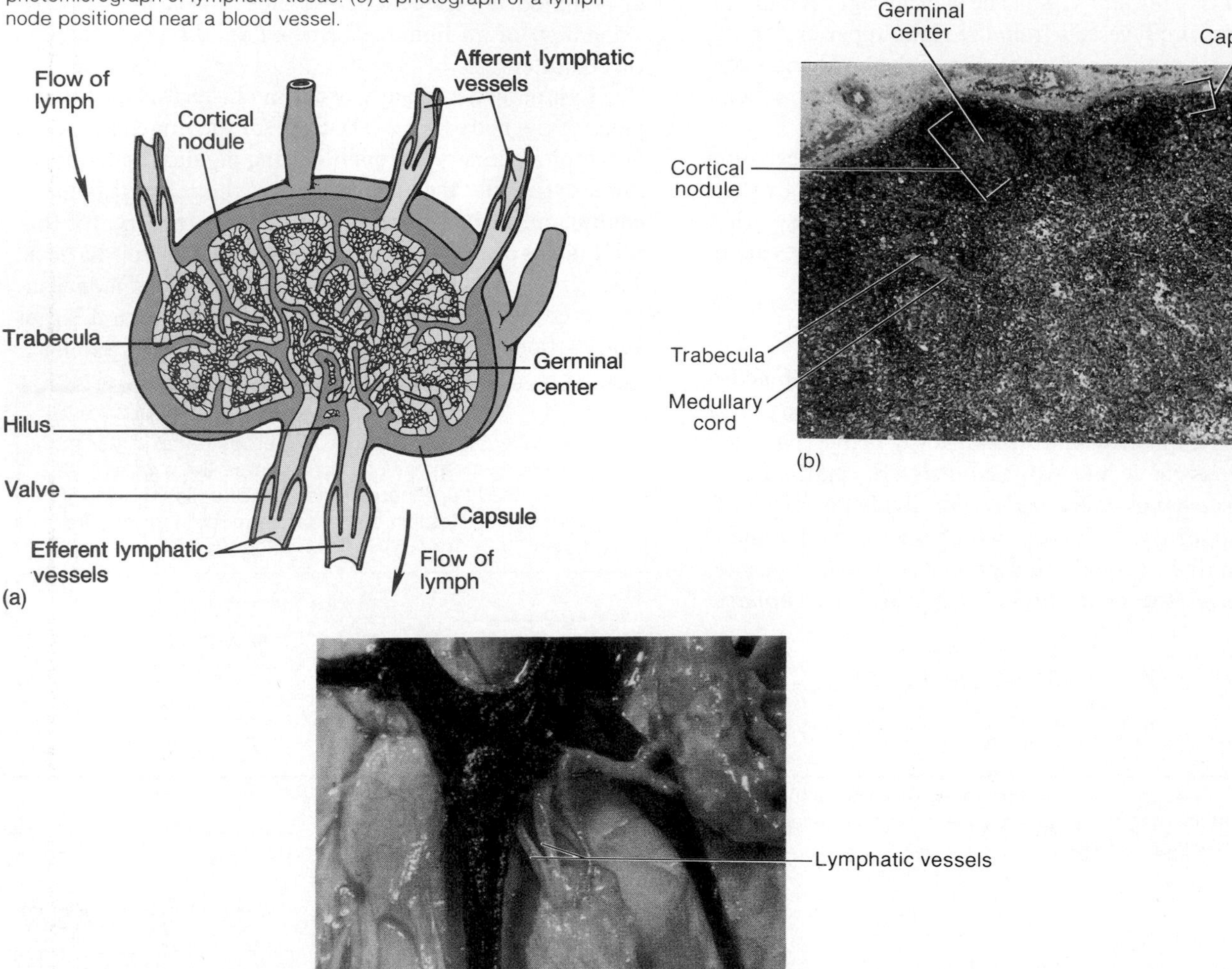

Lymphoid Organs

The **spleen** and the **thymus** are lymphoid organs. The spleen is located posterior and lateral to the stomach, from which it is suspended (fig. 22.5). The spleen is not a vital organ in an adult, but it does assist other body organs in producing lymphocytes, filtering the blood, and destroying old erythrocytes. In an infant, it is an important site for the production of erythrocytes.

The thymus is located in the anterior thorax, deep to the manubrium of the sternum. It is much larger in a fetus and child than in an adult, because it regresses in size during puberty. The thymus plays a key role in the immune system, as will be described in a later section.

Table 22.1 summarizes the lymphatic organs of the body.

spleen: L. *splen,* low spirits (thought to cause melancholy)
thymus: Gk. *thymos,* thyme (compared to the flower of this plant by Galen)

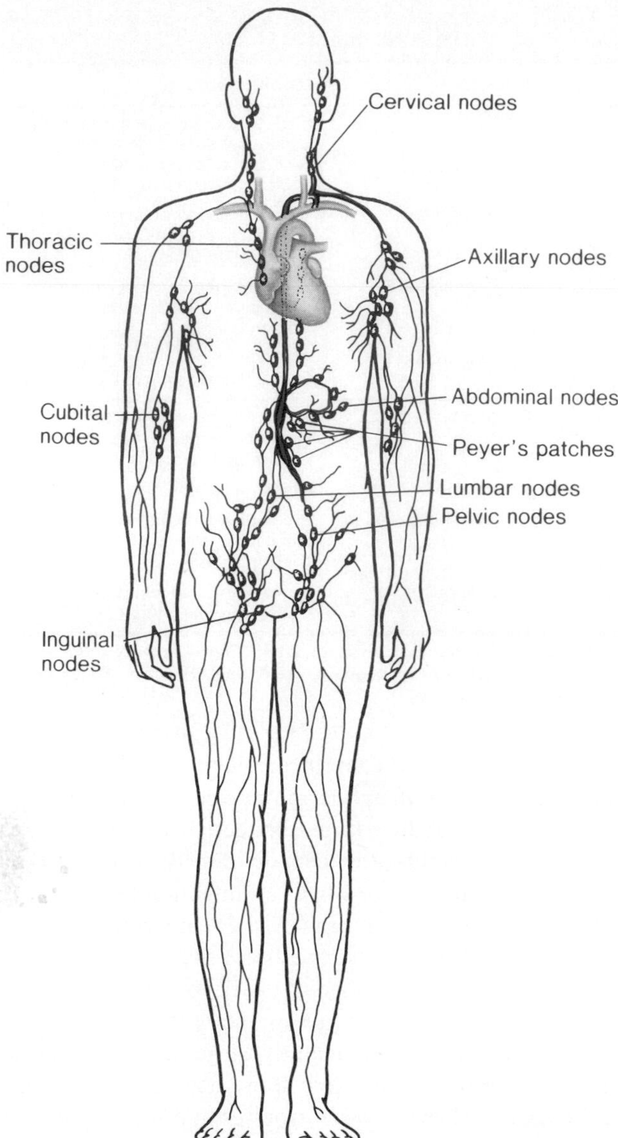

Figure 22.4. Major locations of lymph nodes.

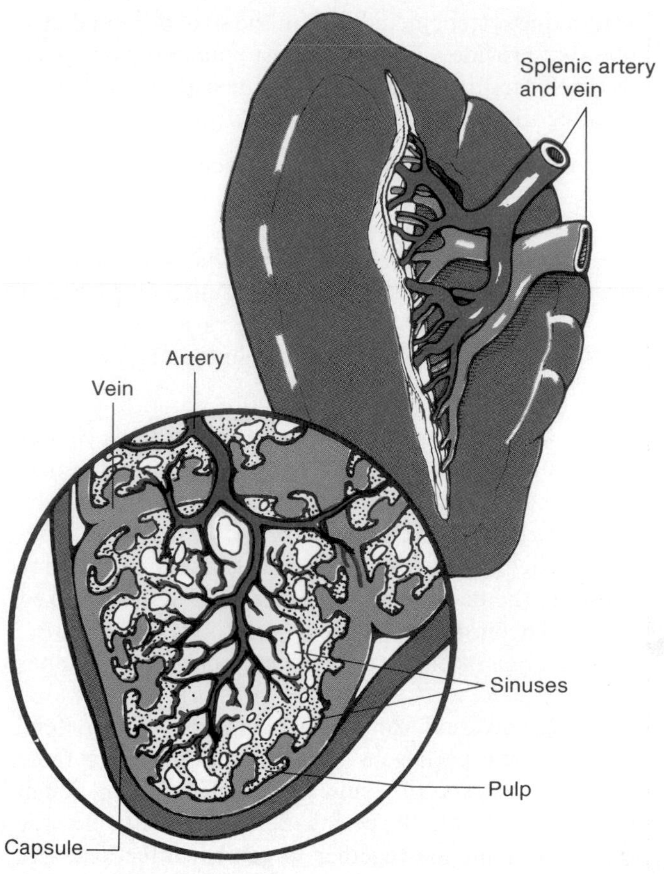

Figure 22.5. The structure of the spleen.

The tonsils, of which there are three pairs, are actually lymphatic organs of the pharyngeal region. The function of the tonsils is to combat infection of the ear, nose, and throat regions. Because of the persistent infections that some children suffer, however, the tonsils may become so overrun with infections that they actually become the source of infections. A *tonsillectomy* may then have to be performed. This operation is not as common as in the past because of the availability of powerful antibiotics and because the functional value of the tonsils is more greatly appreciated.

1. Describe the relationship between lymph and blood in terms of its origin, composition, and fate.
2. List the two major lymph vessels of the body, and describe their relationships to the vascular system.
3. Explain the functions of lymph nodes and lymphoid organs.

Table 22.1 Lymphatic organs

Organ	Location	Function
Lymph nodes	In clusters or chains along the paths of larger lymphatic vessels	Lymphocyte production; house T lymphocytes and B lymphocytes that are responsible for immunity; phagocytes filter foreign particles and cellular debris from lymph
Thymus	Within the mediastinum behind the manubrium	Important site of immunity in a child; houses lymphocytes; changes undifferentiated lymphocytes into T lymphocytes
Spleen	In upper left portion of abdominal cavity beneath the diaphragm and suspended from the stomach	Serves as blood reservoir; phagocytes filter foreign particles, cellular debris, and worn erythrocytes from the blood; houses lymphocytes

Defense Mechanisms

The immune system includes all of the structures and processes that provide a defense against potential pathogens. These defenses can be grouped into nonspecific and specific categories.

Objective 3. Describe some of the mechanisms involved in nonspecific immunity.
Objective 4. Explain what an antigen is, and define the terms *hapten* and *antigenic determinant site.*
Objective 5. Describe the origin and functions of B and T lymphocytes.
Objective 6. Distinguish between humoral and cell-mediated immunity.

Nonspecific defense mechanisms are inherited as part of the structure of each organism. Epithelial membranes that cover the body surfaces, for example, restrict infection by most pathogens. The strong acidity of gastric juice (pH 1–2) also helps to kill many microorganisms before they can invade the body. These external defenses are backed by internal defenses, such as phagocytosis *(fag''o-si-to'sis),* which function in both a specific and nonspecific manner (table 22.2).

Each individual can acquire the ability to defend against specific pathogens by a prior exposure to those pathogens. This specific immune response is a function of lymphocytes. Internal specific and nonspecific defense mechanisms function together to combat infection, with lymphocytes interacting in a coordinated effort with phagocytic cells.

Table 22.2 Structures and defense mechanisms of nonspecific immunity

	Structure	Mechanisms
External	Skin	Physical barrier to penetration by pathogens; secretions have lysozyme (enzyme that destroys bacteria); acidic pH (6.5)
	Digestive tract	High acidity of stomach (pH 1.5) Protection by normal bacterial population of colon
	Respiratory tract	Secretion of mucus; movement of mucus by cilia; alveolar macrophages
	Genitourinary tract	Acidity of urine (pH 6.0–6.5) Vaginal lactic acid (pH 4.0)
Internal	Phagocytic cells	Ingest and destroy bacteria, cellular debris, denatured proteins, and toxins
	Interferons	Inhibit replication of viruses
	Complement proteins	Promote destruction of bacteria and other effects of inflammation
	Endogenous pyrogen	Secreted by leukocytes and other cells; produces fever

Nonspecific Immunity

Invading pathogens that have crossed epithelial barriers enter connective tissues. These invaders—or chemicals called *toxins* secreted from them—may enter blood or lymphatic capillaries and be carried to other areas of the body. The invasion and spread of infection is fought in two stages: (1) nonspecific immunological defenses are employed; if these are sufficiently effective the pathogens may be destroyed without progression to the next step; (2) lymphocytes may be recruited and their specific actions used to reinforce the previously nonspecific immune defenses.

Phagocytosis. There are two major groups of phagocytic cells: (1) neutrophils; and (2) the cells of the *mononuclear phagocyte system.* This latter category includes *monocytes* in the blood, *macrophages* (derived from monocytes) in the connective tissues, and *organ-specific phagocytes* in the liver, spleen, lymph nodes, lungs, and brain (table 22.3).

pathogen: Gk. *pathema,* suffering; *gen,* to produce
toxin: Gk. *toxikon,* poison
phagocytosis: Gk. *phagein,* to eat; *kytos,* hollow (cell)

Connective tissues contain a resident population of all leukocyte types. Neutrophils and monocytes in particular can be highly mobile within connective tissues as they scavenge for invaders and cellular debris. These leukocytes are recruited to the site of an infection by a process known as **chemotaxis** *(ke''mo-tak'sis)*—movement toward chemical attractants. Neutrophils are the first to arrive at the site of an infection; monocytes arrive later and can be transformed into macrophages as the battle progresses.

If the infection is sufficiently large, new phagocytic cells from the blood may join those already in the connective tissue. These new neutrophils and monocytes are able to squeeze through the tiny gaps between adjacent endothelial cells in the capillary wall and enter the connective tissues. This process, called **diapedesis** *(di''ah-pē'de'sis),* is illustrated in figure 22.6.

Phagocytic cells engulf particles in a manner similar to the way an amoeba eats. The particle becomes surrounded by cytoplasmic extensions called pseudopods, which ultimately fuse together. The particle thus becomes surrounded by a membrane derived from the plasma membrane (fig. 22.7) and contained within an organelle analogous to a food vacuole in an amoeba. This vacuole then fuses with lysosomes (organelles that contain digestive enzymes), so that the ingested particle and the digestive enzymes remain separated from the cytoplasm by a continuous membrane. Often, however, lysosomal enzymes are released before the food vacuole has completely formed. When this occurs, free lysosomal enzymes may be released into the infected area and contribute to inflammation.

chemotaxis: Gk. *chemeia,* alchemy; *taxis,* orderly arrangement
diapedesis: Gk. *dia,* through; *pedesis,* a leaping
pseudopod: Gk. *pseudes,* false; *pous,* foot

Table 22.3 Phagocytic cells and their location

Phagocyte	Location
Neutrophils	Blood and all tissues
Monocytes	Blood and all tissues
Tissue macrophages (histiocytes)	All tissues (including spleen, lymph nodes, bone marrow)
Kupffer cells	Liver
Alveolar macrophages	Lungs
Microglia	Central nervous system

Figure 22.6. Diapedesis. White blood cells squeeze through openings between capillary endothelial cells to enter underlying connective tissues.

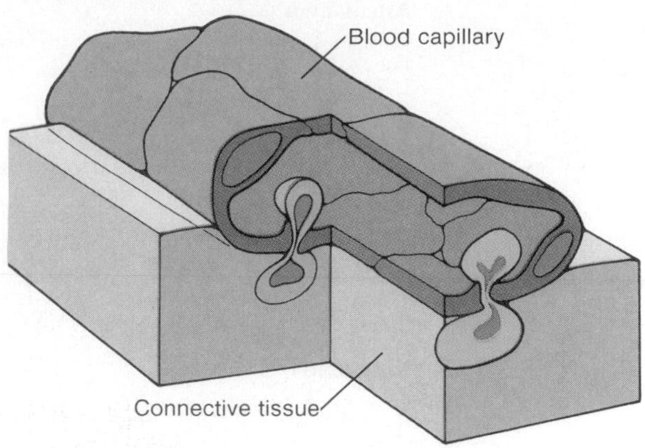

Figure 22.7. Phagocytosis by a neutrophil or macrophage. A phagocytic cell extends its pseudopods around the object to be engulfed (such as a bacterium). Dots represent lysosomal enzymes (*L* = lysosomes). If the pseudopods fuse to form a complete food vacuole (*1*), lysosomal enzymes are restricted to the organelle formed by the lysosome and food vacuole. If the lysosome fuses with the vacuole before fusion of the pseudopods is complete (*2*), lysosomal enzymes are released into the infected area of tissue.

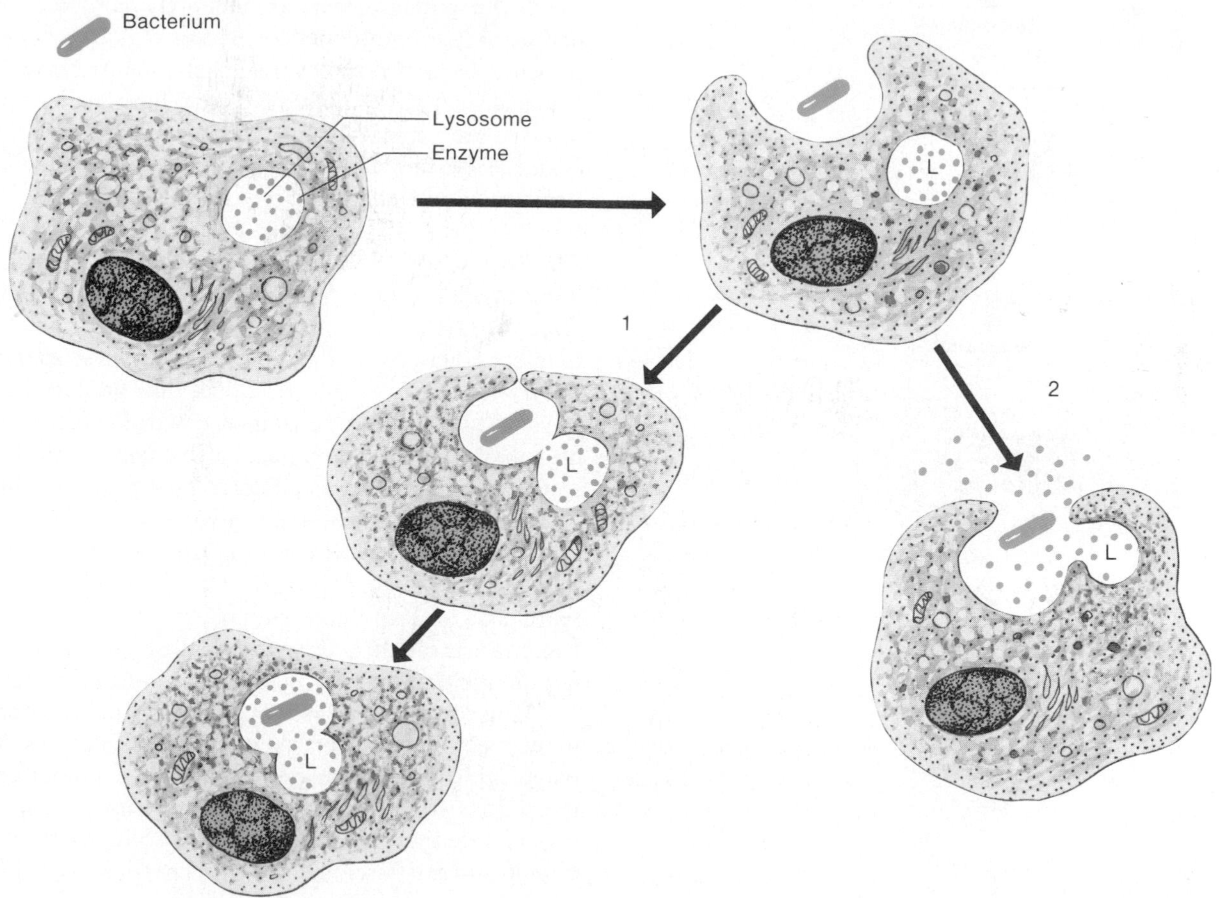

Figure 22.8. The sequence of events that can occur when human cells are infected with virus particles.

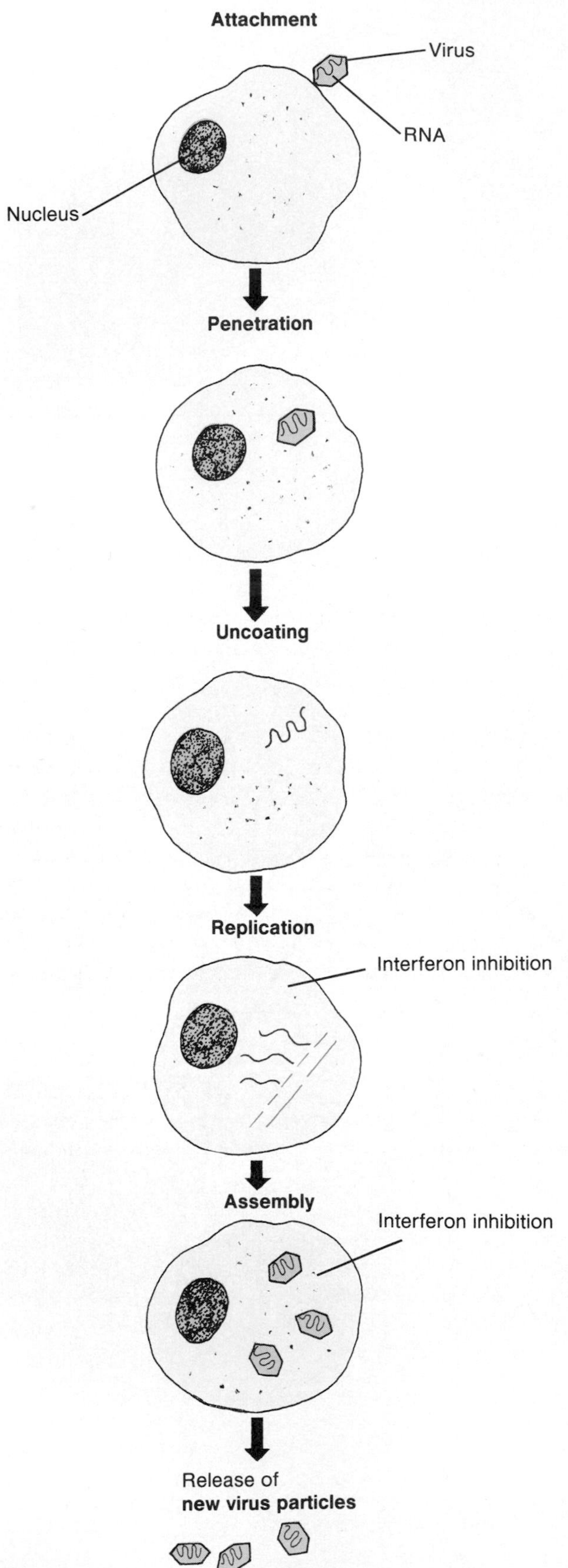

The *Kupffer cells* in the liver, together with phagocytic cells in the spleen and lymph nodes, are **fixed phagocytes.** This term refers to the fact that these cells are immobile ("fixed") in the channels within these organs. As blood flows through the liver and spleen and as lymph percolates through the lymph nodes, foreign chemicals and debris are removed by phagocytosis and chemically inactivated within the phagocytic cells. Invading pathogens are very effectively removed in this manner, so that blood is usually sterile after a few passes through the liver and spleen.

Fever. Fever may be a component of the nonspecific defense system. Body temperature is regulated by the hypothalamus, which contains a thermoregulatory control center (a "thermostat") that coordinates skeletal muscle shivering and the activity of the sympathoadrenal system to maintain body temperature at about 37°C. This thermostat is reset upward in response to a chemical called **endogenous pyrogen** *(en-doj'ĕ-nus pi'ro-jen),* secreted by leukocytes. Endogenous pyrogen secretion can be stimulated by a chemical called *endotoxin,* which is released by certain bacteria.

Although high fevers are definitely dangerous, many believe that a mild to moderate fever may be a beneficial response that aids recovery from bacterial infections. There is some evidence to support this view, but the mechanisms involved are not clearly understood. One theory is that elevated body temperature may interfere with the nutritional requirements of some bacteria.

Interferons. In 1957, researchers demonstrated that cells infected with a virus produced polypeptides that "interfered with" the ability of a second, unrelated strain of virus to infect other cells in the same culture. These **interferons** *(in''ter-fer'on)* as they were called, thus produced a nonspecific, short-acting resistance to viral infection. This discovery produced a great deal of excitement, but further research in this area was hindered by the fact that human interferons could be obtained only in very small quantities and that animal interferons had little effect in humans. In 1980, however, technological breakthroughs allowed researchers to introduce human interferon genes into bacteria—through a technique called *genetic recombination*—so that bacteria could act as interferon factories.

Leukocytes, fibroblasts, and probably many other cells make their own characteristic types of interferons. These polypeptides act as messengers that protect other cells in the vicinity from viral infection. The viruses are still able to penetrate these other cells, but the ability of the viruses to replicate and assemble new virus particles (fig. 22.8) is inhibited.

Kupffer cell: from Karl W. Kupffer, German anatomist, 1829–1902
interferon: L. *inter,* between; *ferio,* to strike

Lymphocytes called *T lymphocytes* release interferons in response to viral infections and perhaps as part of their immunological surveillance against cancer. Interferons may destroy cancer cells directly and indirectly by the activation of T lymphocytes and *natural killer cells* (described in a later section). Some of the proposed effects of interferons are summarized in table 22.4.

Recent clinical trials have suggested that interferons may be effective in the treatment of such viral infections as viral hepatitis and herpes-induced "cold sores." Current evidence suggests that interferons may not prove to be a "magic bullet" against cancer but may, when combined with other methods, significantly augment the treatment of some forms of cancer. The emphasis is on treatment rather than prevention here because the protective effects of interferons are of short duration.

Specific Immunity

In 1890, von Behring demonstrated that a guinea pig that had been previously injected with a sublethal dose of diphtheria toxin could survive subsequent injections of otherwise lethal doses of that toxin. Further, von Behring showed that this immunity could be transferred to a second, nonexposed animal by injections of serum from the immunized guinea pig. He concluded that the immunized animal had chemicals in its serum—which he called **antibodies**—that were responsible for the immunity. He also showed that these antibodies conferred immunity only to subsequent diphtheria infections; the antibodies were *specific* in their actions. It was later learned that antibodies are *proteins* produced by a particular type of lymphocyte.

Antigens. Antigens are molecules that stimulate antibody production and combine with specific antibodies. Most antigens are large molecules (such as proteins) with a molecular weight greater than about 10,000, and they are foreign to the blood and other body fluids (although there are exceptions to both descriptions). The ability of a molecule to function as an antigen depends not only on its size but also on the complexity of its structure. Proteins are therefore more antigenic than polysaccharides, which have a simpler structure. Plastics used in artificial implants are composed of large molecules but are not very antigenic because of their simple, repeating structures.

A large, complex foreign molecule can have a number of different **antigenic** *(an-tĭ-jen'ik)* **determinant sites,** which are areas of the molecule that stimulate production of and combine with different antibodies. Most naturally occurring antigens have many antigenic determinant sites and stimulate the production of different antibodies with specificities for these sites.

von Behring: Emil Adolph von Behring, German bacteriologist, 1854–1917

Table 22.4 Some of the proposed effects of interferons

Stimulation	Inhibition
Macrophage phagocytosis	Cell division
Activity of cytotoxic ("killer") T cells	Tumor growth
Activity of natural killer cells	Maturation of adipose cells
Production of antibodies	Maturation of erythrocytes

From Stuart Ira Fox, *Human Physiology*, 2d ed. Copyright © 1987 Wm. C. Brown Publishers, Dubuque, Iowa. All Rights Reserved. Reprinted by permission.

Haptens. Many small organic molecules are not antigenic by themselves but can become antigens if they bond to proteins (and thus become antigenic determinant sites on the proteins). This was discovered by Karl Landsteiner, the same man who discovered the ABO blood groups. By bonding these small molecules—which Landsteiner called **haptens** *(hap'ten)*—to proteins in the laboratory, new antigens could be created for research or diagnostic purposes. The bonding of foreign haptens to a person's own proteins can also occur in the body; by this means, derivatives of penicillin, for example, that would otherwise be harmless can produce fatal allergic reactions in susceptible people.

Immunoassays. When the antigen or antibody is attached to the surface of a cell or to particles of latex rubber (in commercial diagnostic tests), the antigen-antibody reaction becomes visible because the particles *agglutinate* (clump) as a result of antigen-antibody bonding (fig. 22.9). These agglutinated particles can be used to assay a variety of antigens, and tests that utilize this procedure are called *immunoassays (im"u-no-as'sa)*. Blood typing (described in chapter 20) and modern pregnancy tests are examples of such immunoassays.

Lymphocytes

Leukocytes, erythrocytes, and blood platelets are all ultimately derived from ("stem from") unspecialized cells in the bone marrow. These *stem cells* produce the specialized blood cells, and they replace themselves by cell division so that the stem cell population is not exhausted. Lymphocytes produced in this manner seed the thymus, spleen, and lymph nodes, producing self-replacing lymphocyte colonies in these organs.

The lymphocytes that are seeded in the thymus become **T lymphocytes.** These cells have surface characteristics and an immunological function that is different from other lymphocytes. The thymus, in turn, seeds other organs; about 65% to 85% of the lymphocytes in blood and most of the lymphocytes in lymph nodes are T lymphocytes. T lymphocytes therefore come from or had an ancestor that came from the thymus gland.

Karl Landsteiner: Austrian-born pathologist and immunologist in America, 1868–1943

Figure 22.9. Immunoassay using the agglutination technique. Antibodies against a particular antigen are adsorbed to red blood cells or latex particles. When these are mixed with a solution that contains the appropriate antigen, the formation of the antigen-antibody complexes produces clumping (agglutination) that can be seen with the unaided eye.

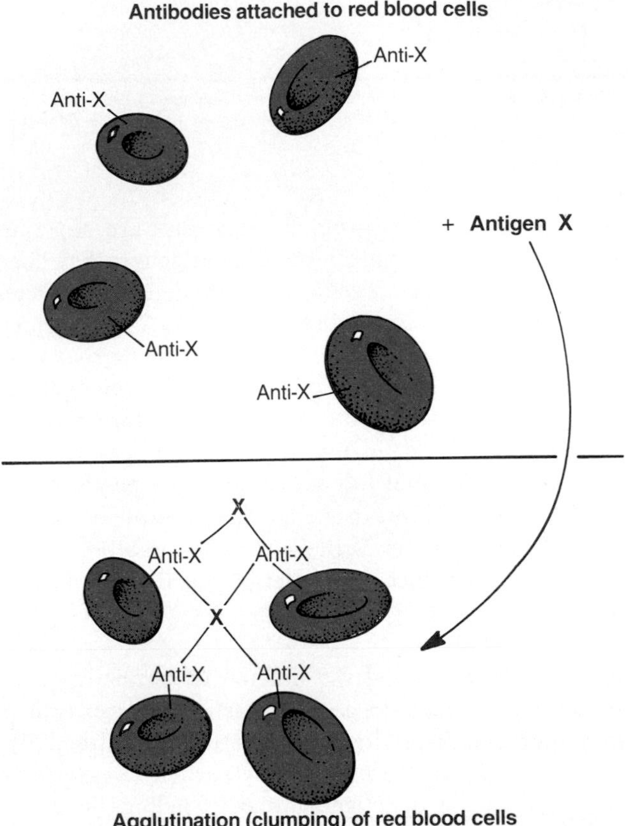

Most of the lymphocytes that are not T lymphocytes are called **B lymphocytes.** The letter *B* is derived from immunological research performed in chickens. Chickens have an organ called the *bursa of Fabricius* at the hind end of their digestive tracts, which is the equivalent of the thymus located at the anterior end. The thymus processes T lymphocytes; the bursa, in a chicken, processes B lymphocytes. Since mammals do not have a bursa, the *B* is often translated as the "bursa equivalent" for humans and other mammals. It is currently believed that the B lymphocytes in mammals are processed in the bone marrow, which conveniently also begins with the letter *B*.

Both B and T lymphocytes function in specific immunity. The B lymphocytes combat bacterial and some viral infections by secreting antibodies into the blood and lymph. They are therefore said to provide **humoral immunity.** T lymphocytes attack host cells that have become infected with viruses or fungi, transplanted human cells, and cancerous cells. The T lymphocytes do not secrete antibodies; they must have actual physical contact with the victim cell in order to destroy it. T lymphocytes are therefore said to provide **cell-mediated immunity** (table 22.5).

1. List the phagocytic cells in blood and lymph, and indicate which organs contain fixed phagocytes.
2. Describe the actions of interferons.
3. List four properties that antigens usually have. Explain why proteins are more antigenic than polysaccharides.
4. Describe the meaning of the term *hapten,* and give an example.
5. Distinguish between B and T lymphocytes in terms of their origins and immune functions.

bursa of Fabricius: from Hieronymus Fabricius, Italian anatomist, 1533–1619

Table 22.5 Comparison of B and T lymphocytes

Characteristic	B lymphocyte	T lymphocyte
Site where processed	Bursa equivalent (bone marrow)	Thymus
Type of immunity	Humoral (secretes antibodies)	Cell-mediated
Subpopulations	Memory cells and plasma cells	Cytotoxic (killer) T cells, helper cells, suppressor cells
Presence of surface antibodies	Yes—IgM or IgD	Not detectable
Receptors for antigens	Present—are surface antibodies	Present—nature unknown
Life span	Short	Long
Tissue distribution	High in spleen, low in blood	High in blood and lymph
Percent of blood lymphocytes	10% to 15%	75% to 80%
Transformed by antigens to	Plasma cells	Small lymphocytes
Secretory product	Antibodies	Lymphokines
Immunity to viral infections	Enteroviruses, poliomyelitis	Most others
Immunity to bacterial infections	Streptococcus, staphylococcus, many others	Tuberculosis, leprosy
Immunity to fungal infections	None known	Many
Immunity to parasitic infections	Trypanosomiasis, maybe to malaria	Most others

Functions of B Lymphocytes

B lymphocytes secrete protein antibodies that can bond in a specific fashion with antigens. The formation of antigen-antibody complexes activates other elements of the immune system and results in more effective defense against pathogens.

Objective 7. Describe the structure and origin of antibodies.

Objective 8. Explain how antibodies promote the destruction of invading pathogens.

Objective 9. Describe the complement system and its function.

Objective 10. Describe the events that occur in a local inflammation.

Exposure of a B lymphocyte to the appropriate antigens results in cell growth followed by many cell divisions. Some of the progeny become **memory cells,** which are indistinguishable from the original cell; others are transformed into **plasma cells** (fig. 22.10). Plasma cells are protein factories that produce about two thousand antibody proteins per second in their brief (five- to seven-day) life span.

The antibodies that are produced by plasma cells when B lymphocytes are exposed to a particular antigen react specifically with that antigen. Such antigens may be isolated molecules, or they may be molecules at the surface of an invading foreign cell. The specific bonding of antibodies to antigens serves to identify the enemy and to activate defense mechanisms that lead to the invader's destruction.

Figure 22.10. B lymphocytes have antibodies on their surface that function as receptors for specific antigens. The interaction of antigens and antibodies on the surface stimulates cell division and the maturation of the B cell progeny into memory cells and plasma cells. Plasma cells produce and secrete large amounts of the antibody (note the extensive rough endoplasmic reticulum in these cells).

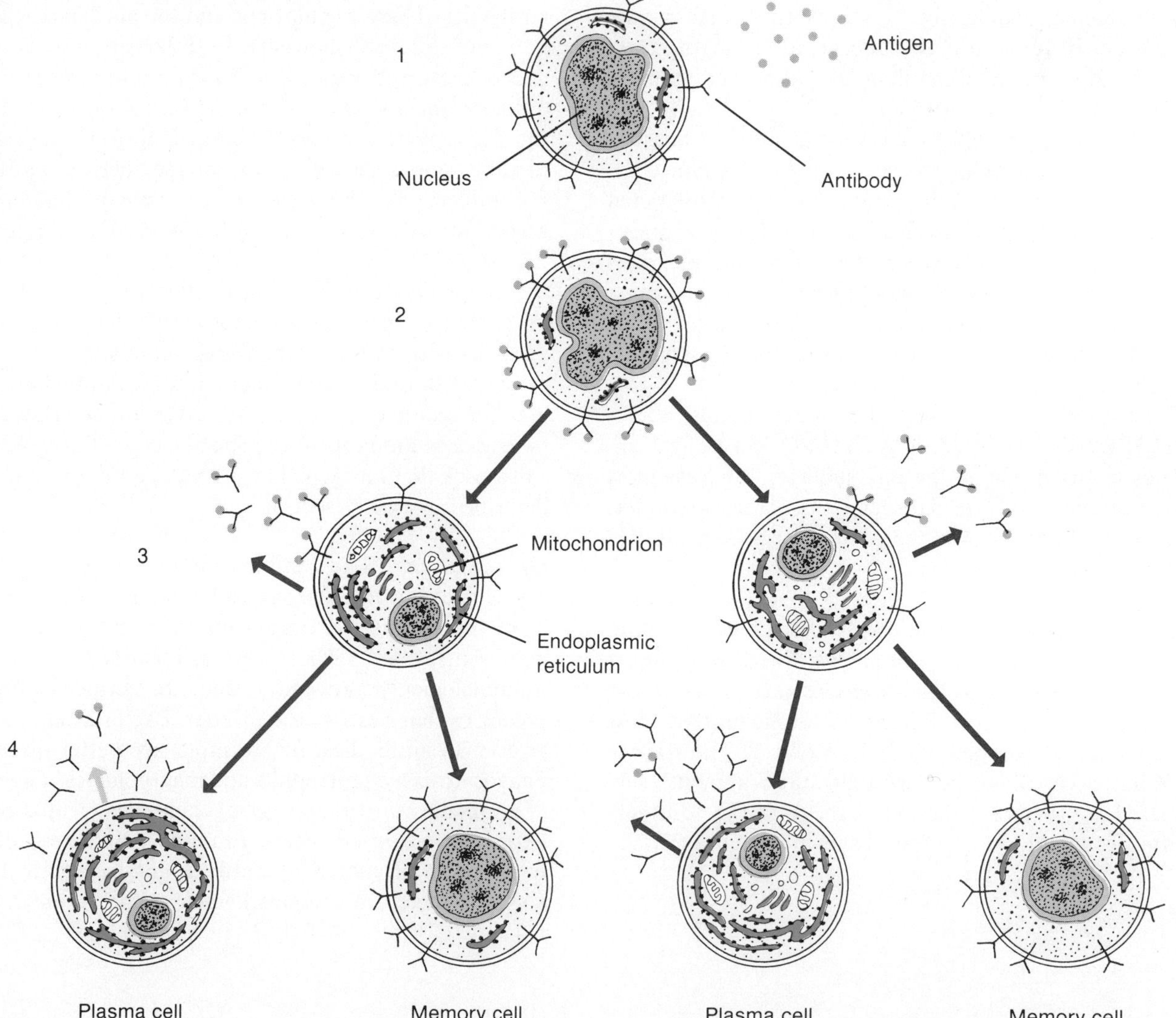

Figure 22.11. The separation of serum protein by electrophoresis. *A* = albumin, alpha-1 (α_1), alpha-2 (α_2), beta (β), and gamma (γ) globulin.

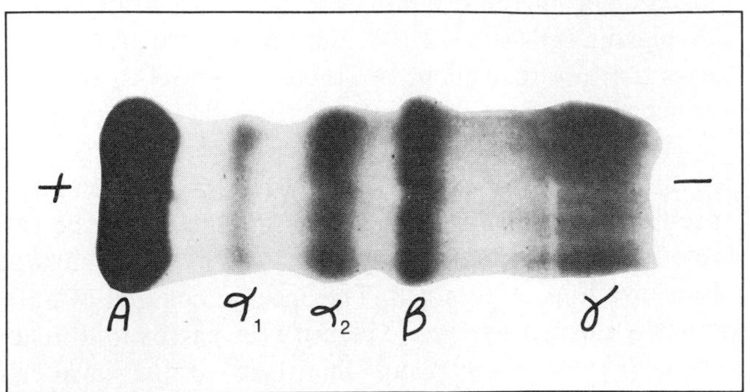

Table 22.6 The immunoglobulins

Immunoglobulin	Examples of functions
IgG	Main form of antibodies in circulation: production increased after immunization
IgA	Main antibody type in external secretions, such as saliva and mother's milk
IgE	Responsible for allergic symptoms in immediate hypersensitivity reactions
IgM	Function as antigen receptors on lymphocyte surface prior to immunization; secreted during primary response
IgD	Function as antigen receptors on lymphocyte surface prior to immunization; other functions unknown

Antibodies

Antibody proteins are also known as **immunoglobulins** *(im''u-no-glob'u-lin).* These are found in the gamma globulin class of plasma proteins, as identified by a technique called *electrophoresis (e-lek''tro-fo-re'sis),* in which classes of plasma proteins are separated by their movement in an electric field (fig. 22.11). With this technique, five distinct bands of proteins appear: albumin, alpha-1-globulin, alpha-2-globulin, beta globulin, and gamma globulin.

The gamma globulin band is wide and diffuse because it represents a heterogenous class of molecules. Since antibodies are specific in their actions, it could be predicted that different types of antibodies have different structures. An antibody against smallpox, for example, does not confer immunity to poliomyelitis and, therefore, must have a slightly different structure than an antibody against polio. Despite these differences, antibodies are structurally related and form only a few subclasses.

There are five subclasses of immunoglobulins (abbreviated Ig): *IgG, IgA, IgM, IgD,* and *IgE.* Most of the antibodies in serum are in the IgG subclass, whereas most of the antibodies in external secretions (saliva and milk) are IgA (table 22.6). Antibodies in the IgE subclass are involved in allergic reactions.

Antibody Structure. All antibody molecules consist of four interconnected polypeptide chains. Two longer, higher molecular weight chains (the *H chains*) are joined to two shorter, lighter *L chains.* Research has shown that these four chains are arranged in the form of a Y. The stalk of the Y has been called the "crystallizable fragment" (abbreviated F_c), whereas the top of the Y is the "antigen-binding fragment" (F_{ab}). This is shown in figure 22.12.

The amino acid sequences of some antibodies have been determined using antibodies derived from people with multiple myelomas. These lymphocyte tumors are derived from the division of a single B lymphocyte, forming a population of genetically identical cells that secrete identical antibodies. These populations and the antibodies they secrete are different, however, in different patients. Analyses of these antibodies have shown that the F_c regions of different antibodies are the same (are *constant*), whereas the F_{ab} regions are *variable.* Variability of the antigen-binding regions is required for the specificity of antibodies for antigens. The F_{ab} region of an antibody thus provides a specific site for bonding with a particular antigen (fig. 22.13).

B lymphocytes have antibodies on their cell membrane that serve as **receptors** for antigens. Combination of antigens with these antibody-receptors stimulates the B cell to divide and produce more of these antibodies, which are then secreted. Exposure to a given antigen thus results in increased amounts of the specific type of antibody that can attack that antigen. This provides active immunity, as described in a later section.

Opsonization. The combination of antibodies with antigens does not itself produce destruction of the antigens or of the pathogenic organisms that contain these antigens. Antibodies, rather, serve to identify the targets for immunological attack and to activate nonspecific immune processes that destroy the invader. Bacteria that are buttered with antibodies, for example, are better targets for phagocytosis by neutrophils and macrophages. The ability of antibodies to stimulate phagocytosis is termed **opsonization** *(op''so-ni-za'shun).* Immune destruction of bacteria is also promoted by antibody-induced activation of a system of serum proteins known as *complement.*

Figure 22.12. Antibodies are composed of four polypeptide chains—two are heavy (*H*) and two are light (*L*). (*a*) a computer-generated model of antibody structure. (*b*) a simplified diagram showing the constant and variable regions. The variable regions are abbreviated V, and the constant regions are abbreviated C. Antigens combine with the variable regions. Each antibody molecule is divided into an F_{ab} (antigen-binding) fragment and an F_c (crystallizable) fragment.

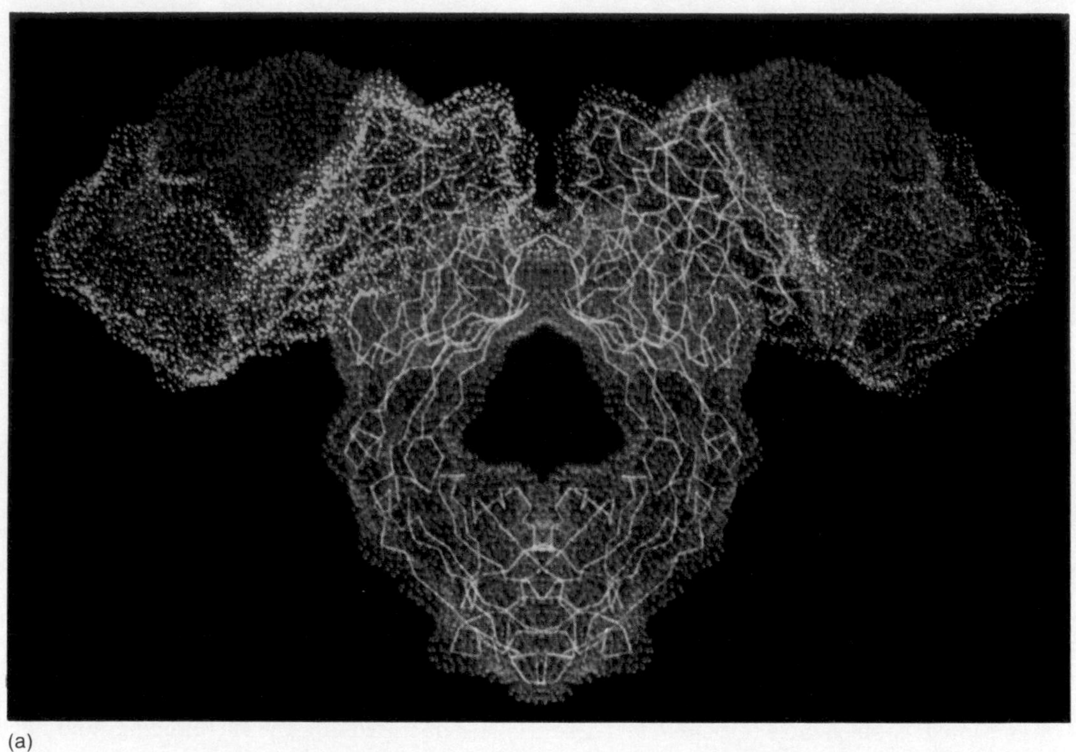

(a)

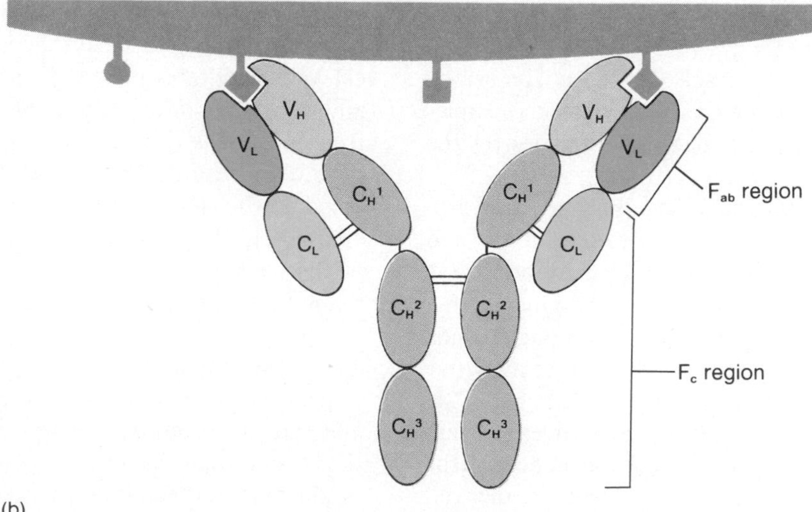

(b)

Figure 22.13. Structure of the F_{ab} portion of an antibody molecule and the antigen with which it combines as determined by X-ray diffraction. The heavy and light chains of the antibody are shown in blue and yellow, respectively, and the antigen is shown in green. Note the complementary shape at the region where the two join together (*a* to *b*).

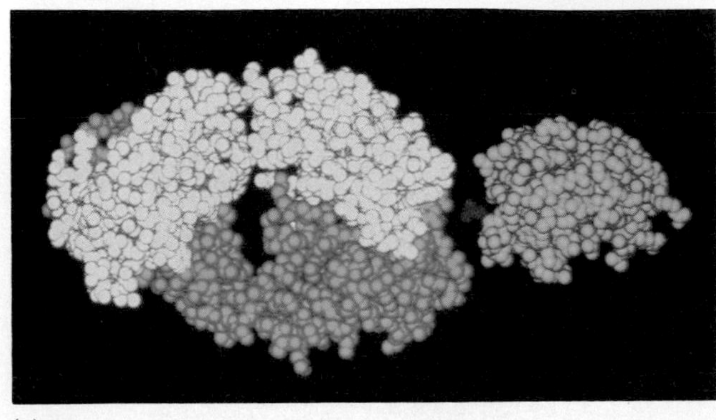

(a)

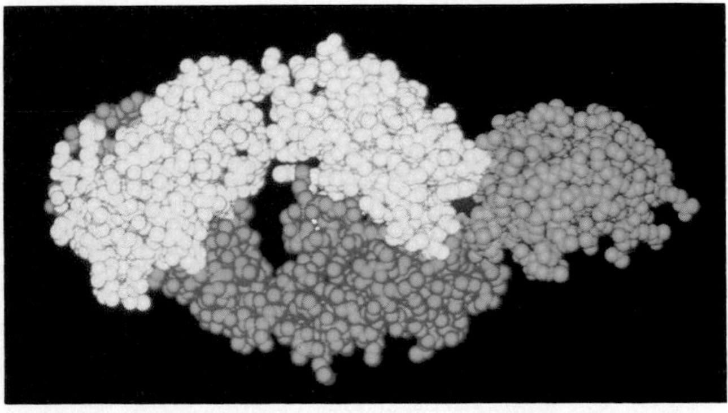

(b)

The Complement System

It was learned in the early part of the twentieth century that rabbit antibodies to sheep red blood cell antigens could not lyse (destroy) these cells unless certain protein components of serum were present. These proteins, called **complement,** are a nonspecific defense system that is activated by the bonding of antibodies to antigens and by this means is directed against specific invaders that have been identified by antibodies.

There are eleven complement proteins, designated C1 (which has three protein components) through C9. These proteins are present in an inactive state within plasma and other body fluids and become activated by the attachment of antibodies to antigens. In terms of their functions, the complement proteins can be subdivided into three components: (1) recognition (C1); (2) activation (C4, C2, and C3, in that order); and (3) attack (C5–C9). The attack phase consists of **complement fixation,** in which complement proteins attach to the cell membrane and destroy the victim cell.

Antibodies of the IgG and IgM subclasses attach to antigens on the invading cell's membrane, bond to C1, and by this means activate its enzyme activity. Activated C1 catalyzes the hydrolysis of C4 into two fragments (fig. 22.14), designated C4a and C4b. The C4b fragment bonds to the cell membrane (is "fixed") and becomes an active enzyme that splits C2 into two fragments, C2a and C2b. The C2a becomes attached to C4b and cleaves C3 into C3a and C3b. Fragment C3b becomes attached to the growing complex of complement proteins on the cell membrane. The C3b converts C5 to C5a and C5b. The C5b, and eventually C6 through C9, become fixed to the cell membrane.

Complement proteins C5 through C9 create large pores in the membrane (fig. 22.15). These pores allow the osmotic influx of water, so that the victim cell swells and bursts. Notice that the complement proteins, not the antibodies directly, kill the cell; antibodies only serve as activators of this process. Other molecules can also activate the complement system in an alternate nonspecific pathway that bypasses the early phases of the specific pathway described here.

Complement fragments that are not fixed but instead are liberated into the surrounding fluid have a number of effects. These effects include (1) *chemotaxis*—the liberated complement fragments attract phagocytic cells to the site of complement activation; (2) *opsonization*—phagocytic cells have receptors for C3b, so that this fragment may form bridges between the phagocyte and the victim cell that facilitates phagocytosis; and (3) fragments C3a and C5a *stimulate release of* **histamine** *(his'tah-mēn)* from mast cells (a connective tissue cell type) and basophils. As a result of histamine release, there is increased blood flow to the infected area due to vasodilation and increased capillary permeability. The latter effect can result in the leakage of plasma proteins into the surrounding tissue fluid, producing local edema.

Local Inflammation

Aspects of the nonspecific and specific immune responses and their interactions are well illustrated by the events that occur when bacteria enter a break in the skin and produce a local inflammation (table 22.7). The inflammatory reaction is initiated by the nonspecific mechanisms of phagocytosis and complement activation. Activated complement further increases this nonspecific response by attracting new phagocytes to the area and by increasing the activity of these phagocytic cells.

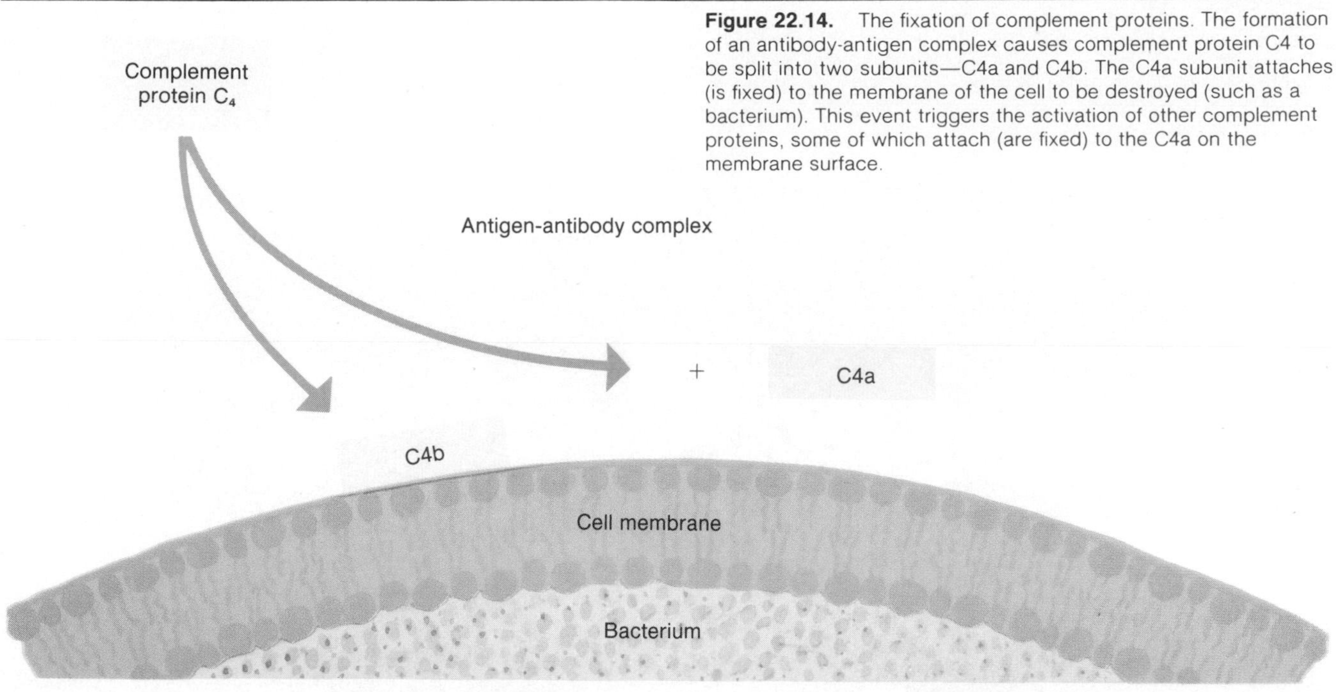

Figure 22.14. The fixation of complement proteins. The formation of an antibody-antigen complex causes complement protein C4 to be split into two subunits—C4a and C4b. The C4a subunit attaches (is fixed) to the membrane of the cell to be destroyed (such as a bacterium). This event triggers the activation of other complement proteins, some of which attach (are fixed) to the C4a on the membrane surface.

Figure 22.15. Complement proteins C5 through C9 (illustrated as a doughnut-shaped ring) puncture the membrane of the cell to which they are attached (fixed). This aids destruction of the cell.

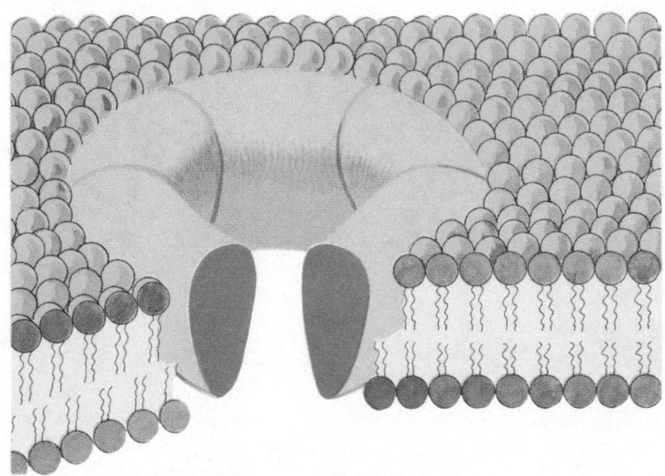

Table 22.7 Summary of events that occur in a local inflammation when a break in the skin permits entry of bacteria

Category	Events
Nonspecific immunity	Bacteria enter through break in anatomic barrier of skin. Resident phagocytic cells—neutrophils and macrophages—engulf bacteria. Nonspecific activation of complement proteins occurs.
Specific immunity	B cells are stimulated to produce specific antibodies. Phagocytosis is enhanced by antibodies attached to bacterial surface antigens. Specific activation of complement proteins occurs, which stimulates phagocytosis, chemotaxis of new phagocytes to the infected area, and secretion of histamine from tissue mast cells. Diapedesis allows new phagocytic leukocytes (neutrophils and monocytes) to invade the infected area. Vasodilation and increased capillary permeability (as a result of histamine secretion) produce redness and edema.

Figure 22.16. The entry of bacteria through a cut in the skin produces a local inflammatory reaction. In this reaction, antigens on the bacterial surface are coated with antibodies and ingested by phagocytic cells. Symptoms of inflammation are produced by the release of lysosomal enzymes and by the secretion of histamine and other chemicals from tissue mast cells.

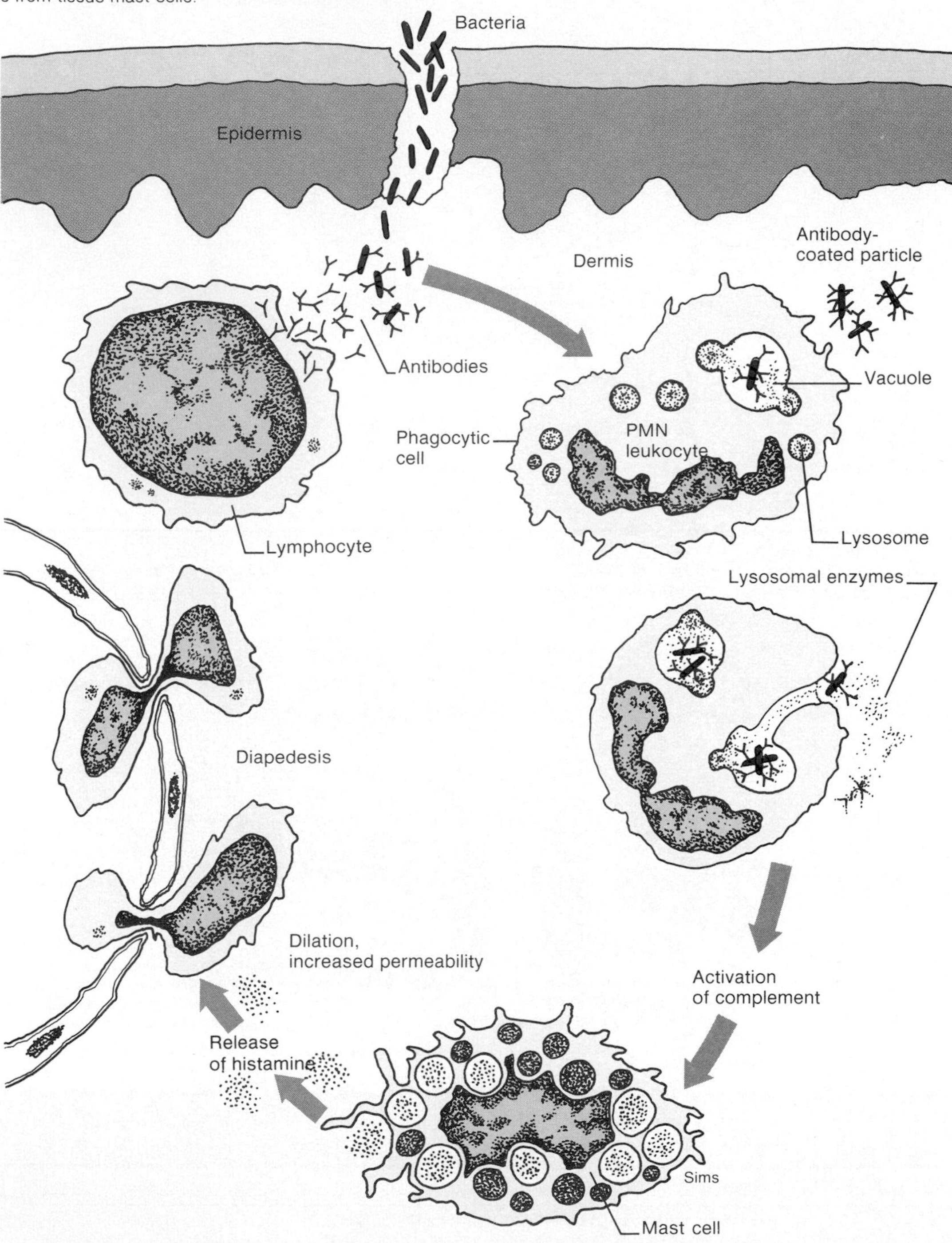

After some time, B lymphocytes are stimulated to produce antibodies against specific antigens that are part of the invading bacteria. Attachment of these antibodies to antigens in the bacteria greatly increases the previously nonspecific response. This occurs because of greater activation of complement, which directly destroys the bacteria and which—together with the antibodies themselves—promotes the phagocytic activity of neutrophils, macrophages, and monocytes (fig. 22.16).

As inflammation progresses, the release of lysosomal enzymes from macrophages causes the destruction of leukocytes and other tissue cells. These effects, together with those produced by histamine and other chemicals released from mast cells, produce the characteristic symptoms of a local inflammation: *redness* and *warmth* (due to vasodilation); *swelling* (edema); and *pain.* If the infection continues, the release of endogenous pyrogen from leukocytes and macrophages may produce a *fever.*

---

1. Illustrate the structure of an antibody molecule. Label the constant and variable regions, the F_c and F_{ab} parts, and the heavy and light chains.
2. Define opsonization, and name two types of molecules that promote this process.
3. Describe complement fixation, and explain the roles of complement fragments that do not become fixed.
4. Explain how nonspecific and specific immune mechanisms cooperate during a local inflammation.

Functions of T Lymphocytes

T cells assist all aspects of the immune system, including cell mediated destruction by cytotoxic T cells and supporting roles by helper and suppressor T cells. T cells are only activated by antigens presented to them by macrophages; the activated T cells in turn produce lymphokines, which activate other cells of the immune system.

Objective 11. Describe the functions of the thymus.
Objective 12. Describe the source and functions of lymphokines.
Objective 13. Explain the importance of histocompatibility antigens in the function of T cells.
Objective 14. Explain how interaction of helper T cells with macrophages can lead to stimulation of both cell-mediated and humoral immunity.
Objective 15. Explain how suppressor T cells may function to help regulate the immune response to foreign and self-antigens.

The thymus processes lymphocytes in such a way that their functions become quite distinct from those of B cells; lymphocytes that reside in the thymus, or that came from the thymus, or that are derived from cells that came from the thymus are all T lymphocytes that can be distinguished from B cells by special techniques. The T lymphocytes provide specific immune protection without secreting antibodies. This is accomplished in different ways by the three subpopulations of T lymphocytes, which are designated as cytotoxic (killer), helper, and suppressor cells.

Thymus Gland

The **thymus gland** extends from below the thyroid in the neck into the thoracic cavity. Lymphocytes from the fetal liver and spleen in prenatal life, and from the bone marrow postnatally, seed the thymus and become transformed into T cells. These lymphocytes in turn enter the blood and seed lymph nodes and other organs, where they divide to produce new T cells when stimulated by antigens.

Small T lymphocytes that have not yet been stimulated by antigens have very long life spans—months or perhaps years. Still, new T cells must be continuously produced to provide efficient cell-mediated immunity. Since the thymus atrophies after puberty, this organ may not be able to provide new T cells in later life. Colonies of T cells in the lymph nodes and other organs are apparently able to produce new T cells under the stimulation of various **thymus hormones.**

Two hormones that are believed to be secreted by the thymus—*thymopoietin (thi''mo-poi'ĕ-tin) I* and *thymopoietin II*—may promote the transformation of lymphocytes into T cells. Another thymus hormone, called *thymosin (thi'mo-sin),* may promote the maturation of T lymphocytes.

There is some experimental evidence supporting the notion that the administration of these thymus hormones may be able to restore cell-mediated immunity in some cases where T cell function has declined. This decline occurs in some congenital and acquired diseases as well as naturally in the course of aging in conjunction with an increased susceptibility to viral infections and cancer. Clinical manipulation of T cell function—through the administration of thymus hormones, interferon, interleukin-2, and other techniques—is an exciting possibility that may result from further experimental research.

Cytotoxic, Helper, and Suppressor T Lymphocytes

The **cytotoxic** *(si''to-tok'sik),* or **killer, T lymphocytes** destroy specific victim cells that are identified by specific antigens on their surface. In order to effect this *cell-mediated* destruction, the T lymphocytes must be in actual contact with their victim cells (in contrast to B cells, which kill at a distance by secreting antibodies). The mechanisms by which the cytotoxic lymphocytes kill their victim cells are presently not well understood. There is evidence, however, that once the cytotoxic T cells make contact with their target cell, they release polypeptides within the tiny space between the two cells, and that these polypeptides act—much like complement proteins do—to open pores in the victim cell membrane.

The cytotoxic lymphocytes defend against viral and fungal infections and are also responsible for transplant rejection reactions and for immunological surveillance against cancer. Although most bacterial infections are fought by B lymphocytes, some are the targets of cell-mediated attack by cytotoxic T lymphocytes. This is the case with the *tubercle bacilli* that cause tuberculosis. Injections of some of these bacteria under the skin produce inflammation after a latent period of forty-eight to seventy-two hours. This *delayed hypersensitivity* reaction is cell mediated rather than humoral, as shown by the fact that it can be induced in an unexposed guinea pig by an infusion of lymphocytes, but not of serum, from an exposed animal.

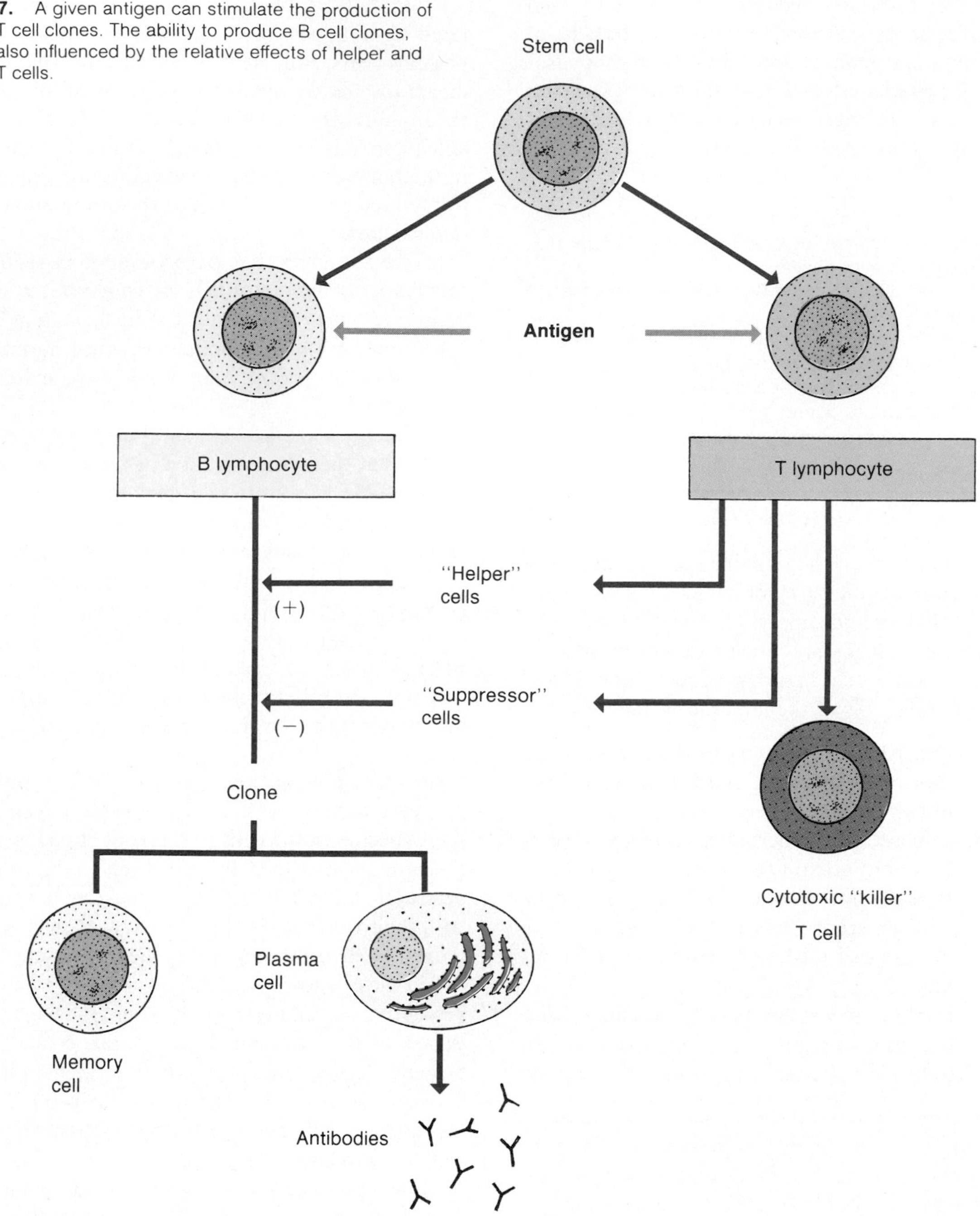

Figure 22.17. A given antigen can stimulate the production of both B and T cell clones. The ability to produce B cell clones, however, is also influenced by the relative effects of helper and suppressor T cells.

The **helper T lymphocytes** and **suppressor T lymphocytes** indirectly participate in the specific immune response by regulating the responses of the B cells (fig. 22.17) and the cytotoxic T cells. The activity of B cells and cytotoxic T cells is increased by helper T lymphocytes and decreased by suppressor T lymphocytes. The amount of antibodies secreted in response to antigens is thus affected by the relative numbers of helper to suppressor T cells that develop in response to a given antigen.

As a result of advances in recombinant DNA technology (genetic engineering) that allow the production of monoclonal antibodies (discussed later), it is now possible for clinical laboratories to distinguish between the different subcategories of lymphocytes by means of antigen "markers" on their surfaces. Counting the lymphocytes in each of these subcategories provides far more information about diseases and their causes than was previously available.

Acquired Immune Deficiency Syndrome (AIDS). *Acquired immune deficiency syndrome (AIDS)* is a disease that has increased overall adult male and female mortality in the United States by 0.7 and 0.07%, respectively. People at high risk for AIDS include homosexual and bisexual men, intravenous drug users, hemophiliacs who received blood prior to 1985 (before plasma and clotting factor solutions were tested for AIDS), and women who have had sexual relations with men at high risk. Most children with AIDS were infected *in utero* by mothers who were intravenous drug users. In countries such as Haiti and central Africa, heterosexual contact is the primary route of AIDS transmission.

Through the use of laboratory tests, it has been learned that AIDS is associated with an abnormally low helper T lymphocyte count. This results in decreased immunological function and thus an increased incidence of opportunistic infections. *Pneumocystis carinii pneumonia* and *Kaposi's sarcoma* are two previously rare conditions that account for a high percentage of deaths in AIDS victims. It has been shown that AIDS is caused by a specific strain of virus that infects human lymphocytes. This virus is now commonly referred to as **HIV** (human immunodeficiency virus).

Laboratory tests of AIDS include counting the relative number of helper and suppressor T cells. Helper cells have a particular antigen marker on their cell surface known as the T4 antigen, which can be distinguished from a different antigen (T8) on suppressor and cytotoxic cells. A normal person has a T4/T8 ratio of greater than 1.0; a person with AIDS has a ratio reduced to 0.5 or less, indicating a great reduction in helper cells.

The HIV virus is a *retrovirus;* it injects RNA instead of DNA into the host cell. This viral RNA is then transcribed into DNA by means of a virally produced enzyme known as *reverse transcriptase.* The viral DNA can then be incorporated into the host DNA and can direct the synthesis of new viral particles. At present, the only effective treatment (not a cure) available for AIDS is a drug called *AZT* (azidothymidine), which is an analogue of the nucleotide base thymidine. This drug binds to reverse transcriptase and blocks the transcription of RNA into DNA. Other drugs that can potentially block viral replication, including interferon, are currently being tested in an intensive immunological research effort directed to finding other treatments and a possible cure for AIDS.

Lymphokines. The T lymphocytes secrete a family of low-molecular weight polypeptides called **lymphokines** *(lim'fo-kins),* which are believed to mediate the actions of these cells. The lymphokines include (1) *interferon;* (2) *lymphotoxin,* which may be released from cytotoxic T cells and cause destruction of the victim cells; (3) *macrophage-activating factor,* which promotes phagocytic activity; (4) chemicals that act by chemotaxis to attract leukocytes to the infected area; (5) *macrophage-migration-inhibiting factor,* which prevents phagocytic cells from leaving the area; and (6) *interleukin-2,* which is secreted by helper cells and stimulates cell division and proliferation of cytotoxic T cells, suppressor T cells, B cells, and natural killer cells.

Genetic engineering techniques have been used to generate recombinant bacteria that can produce large amounts of interleukin-2 for research and clinical applications. In reports first published in 1985, interleukin-2 alone, and interleukin-2 activated T lymphocytes (called "lymphokine activated killer cells," or LAK cells) obtained from the patient's own blood, were used to treat cancer patients in whom other forms of treatment had failed. LAK cells plus interleukin-2 treatment resulted in complete remission in a few patients, and partial remissions in some others. Although much more research is required, this immunotherapy approach to the treatment of cancer represents an exciting beginning to a potentially new mode of cancer treatment.

T Cell Receptor Proteins. Unlike B cells, T cells do not make antibodies and thus do not have antibodies on their surface to serve as receptors for antigens. The T cells do, however, have specific receptors for antigens on their membrane surface, and these T cell receptors have recently been identified as molecules closely related to immunoglobulins. The T cell receptors differ from the antibody receptors on B cells in another, and very important, characteristic: they cannot bond to free antigens. In order for a T lymphocyte to respond to a foreign antigen, the antigen must be presented to the T lymphocyte on the membrane of an *antigen-presenting cell.* The chief antigen-presenting cells are macrophages, which present the foreign antigen together with other surface antigens, called *histocompatibility antigens,* to the T lymphocytes. Some knowledge of the histocompatibility antigens is thus required before T cell-macrophage interactions and T cell functions can be understood.

Histocompatibility Antigens. Tissue that is transplanted from one person to another contains antigens that are foreign to the host. This is because all tissue cells, with the exception of mature red blood cells, are genetically marked with a characteristic combination of **histocompatibility antigens** on the membrane surface. The greater the difference in these antigens between the donor and the recipient in a transplant, the greater will be the chance of transplant rejection. Prior to organ transplantation, therefore, the "tissue type" of the recipient is matched to that of potential donors. Since the person's white blood cells are used for this purpose, an alternate name for histocompatibility antigens in humans is **human leukocyte antigens,** abbreviated **HLA.**

Figure 22.18. There are four human histocompatibility antigens (or human leukocyte antigens—HLA). Each of these antigens is coded by a gene located on chromosome number 6.

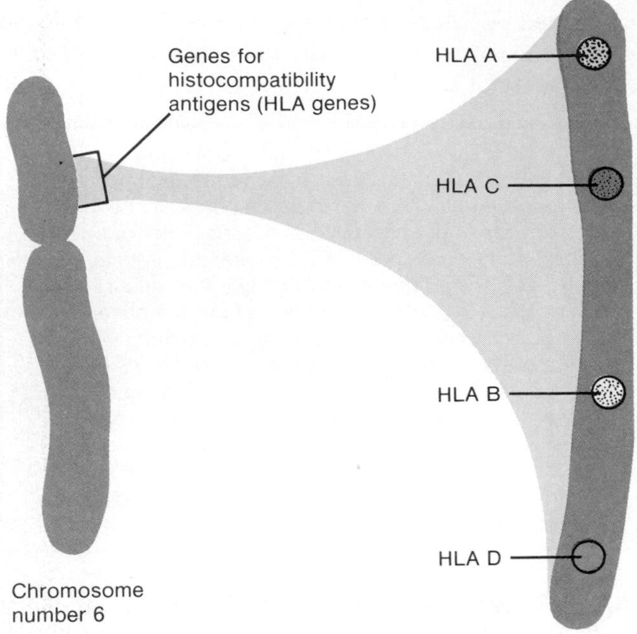

Table 22.8 Association between particular human histocompatibility antigens and diseases.

Disease	HLA antigen	Frequency in patients (%)	Frequency in controls (%)
Ankylosing spondylitis	B27	90	7
	B13	18	4
Psoriasis	B17	29	8
	B16	15	5
Graves' disease	B8	47	21
Coeliac disease	B8	78	24
Dermatitis herpetiformis	B8	62	27
Myasthenia gravis	B8	52	24
SLE (systemic lupus erythematosus)	B15	33	8
Multiple sclerosis	A3	36	25
	B7	36	25
Acute lymphatic leukemia	A2	63	37
	B35	25	16
Hodgkin's disease	A1	39	32
	B8	26	22
Chronic hepatitis	B8	68	18
Ragweed hay fever	B7	50	19

The histocompatibility antigens (HLA) are proteins that are coded by a group of genes, called the *major histocompatibility complex* (*MHC*), located on chromosome number 6 (fig. 22.18). These four genes are labeled A, B, C, and D. Each of these genes can code for only one protein in a given individual, but this protein can be different in different people. Two people, for example, may both have antigen A3, but one might have antigen B17 while the other has antigen B21. The closer two people are related, the more similar their histocompatibility antigens will be.

Clinical interest has been generated by the observation that certain diseases are much more common in people who have particular histocompatibility antigens (table 22.8). *Ankylosing spondylitis* (a type of rheumatoid arthritis), for example, is much more common in people who have antigen B27, and *psoriasis* (a skin disorder) is three times more common in people with antigen B17 than in the general population. Some other diseases that have a high correlation with particular histocompatibility antigens include *Hodgkin's disease* (a cancer of the lymph nodes), *myasthenia gravis*, *Graves' disease*, and *Type I*, or *insulin-dependent*, *diabetes*.

Interactions between Macrophages and Lymphocytes

The major histocompatibility complex of genes produces two classes of HLA antigens, designated *class-1* and *class-2*. The class-1 HLA antigens are made by all cells in the body except red blood cells, (which have their own unique set of antigens, used for blood typing and described in chapter 20). Class-2 HLA antigens are produced only by macrophages and B lymphocytes.

When a foreign particle, such as a virus, infects the body, it is taken into macrophages *(mak''ro-faj'es)* by phagocytosis and partially digested. One phase of macrophage-T cell interaction then occurs: the macrophage is stimulated to secrete a chemical called *interleukin-1*, which stimulates cell division and proliferation of T cells. Within the macrophage, the partially digested virus provides foreign antigens that are moved to the surface of the macrophage cell membrane, where they bond to class-2 HLA antigens. This combination of foreign and class-2 HLA antigens is needed for recognition by the receptor proteins on the surface of the helper T lymphocytes and is thus required for activation of helper T cells (fig. 22.19). It should be remembered that T cells are "blind" to free antigens; they can only respond to antigens presented to them by particular cells, primarily macrophages, in the manner described.

The helper T lymphocytes that are activated in this way proliferate and secrete the lymphokines, including interleukin-2, which greatly enhances other aspects of the immune reponse. For example, interleukin-2 stimulates

Figure 22.19. (*a*) an electron micrograph showing contact between a macrophage (*left*) and a lymphocyte (*right*). As illustrated in (*b*), such contact between a macrophage and a T cell requires that the helper T cell interact with both the foreign antigen and the HLA class-2 antigen on the surface of the macrophage.

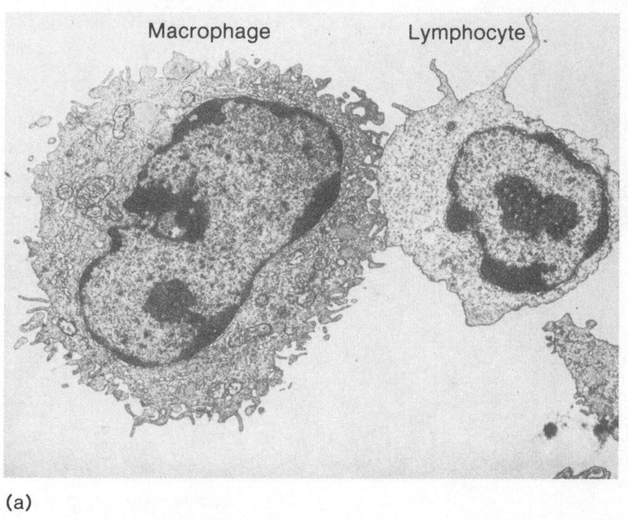

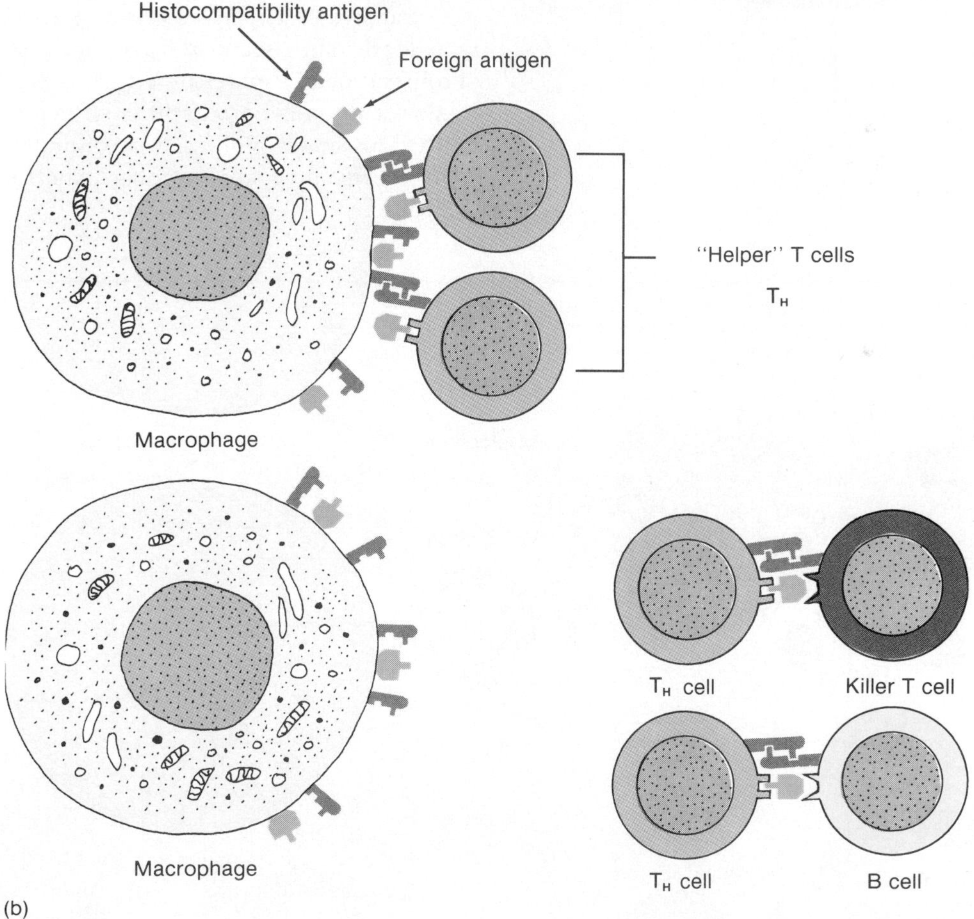

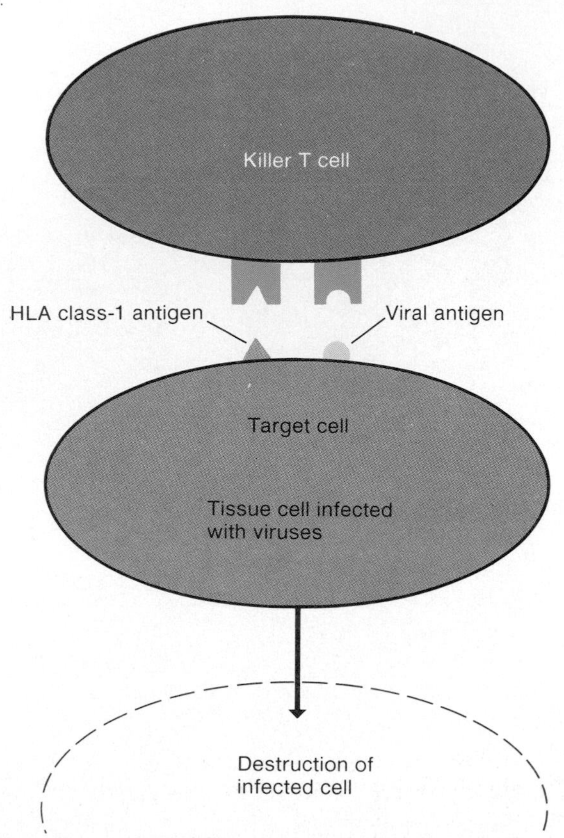

Figure 22.20. In order for a killer T cell to destroy a tissue cell infected with viruses, the T cell must interact with both the foreign antigen and the class-1 HLA antigen on the surface of the infected cell.

macrophages to become more phagocytic, and to secrete a chemical called *tumor necrosis factor,* which is particularly effective in killing cancer cells. Interleukin-2 also stimulates proliferation of cytotoxic T cells and B cells.

The cytotoxic T lymphocytes are only able to destroy infected cells if those infected cells display the foreign antigen together with class-1 HLA antigens (fig. 22.20). Interaction of the receptor proteins on the surface of specific cytotoxic T cells with the foreign antigen, in conjunction with the effects of interleukin-2 secreted by helper T cells, stimulates proliferation of the cytotoxic T cells. In this way, interaction of helper T cells with macrophages aids the cytotoxic T cell's response to a specific antigen (fig. 22.21).

Activation of helper T cells is also needed for the full B cell response to a foreign antigen. In this case, the membrane receptors on activated helper T cells combine with the class-2 HLA antigens and with foreign antigens attached to receptor immunoglobin molecules on the B cell surface (fig. 22.22). This interaction of helper T cells with B cells is believed to stimulate proliferation of the B cells, their conversion to plasma cells, and the secretion of antibodies directed against the foreign antigens.

Suppressor T lymphocytes may also be stimulated by the interleukin-2 secreted by the activated helper T cells. Proliferation of suppressor T cells is believed to occur at a slower rate than proliferation of cytotoxic T cells or B cells. This slower proliferation of suppressor T cells may help provide a negative feedback control of the immune response.

Figure 22.21. Interaction between macrophages, helper T lymphocytes, cytotoxic T lymphocytes, and infected cells in the immunological defense against viral infections.

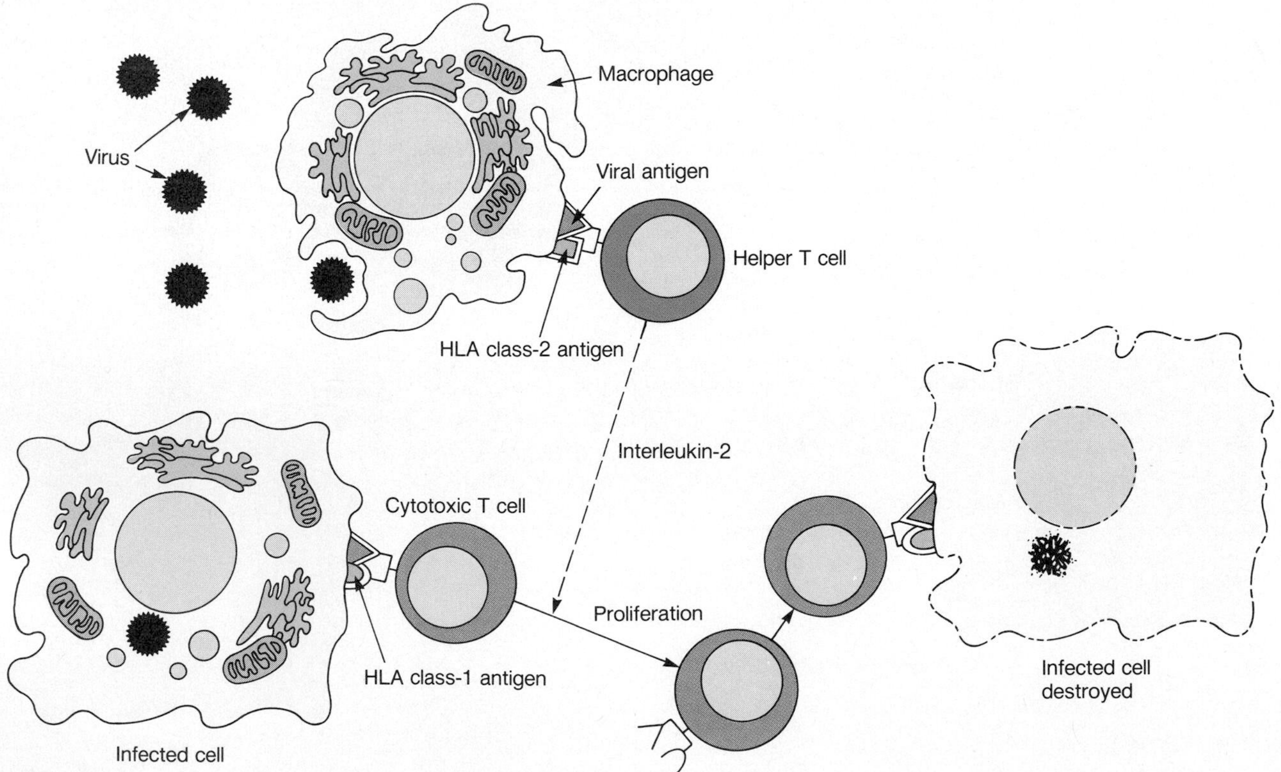

Figure 22.22. Schematic diagram of the events that are believed to occur in the interactions of macrophages, helper T lymphocytes, and B lymphocytes.

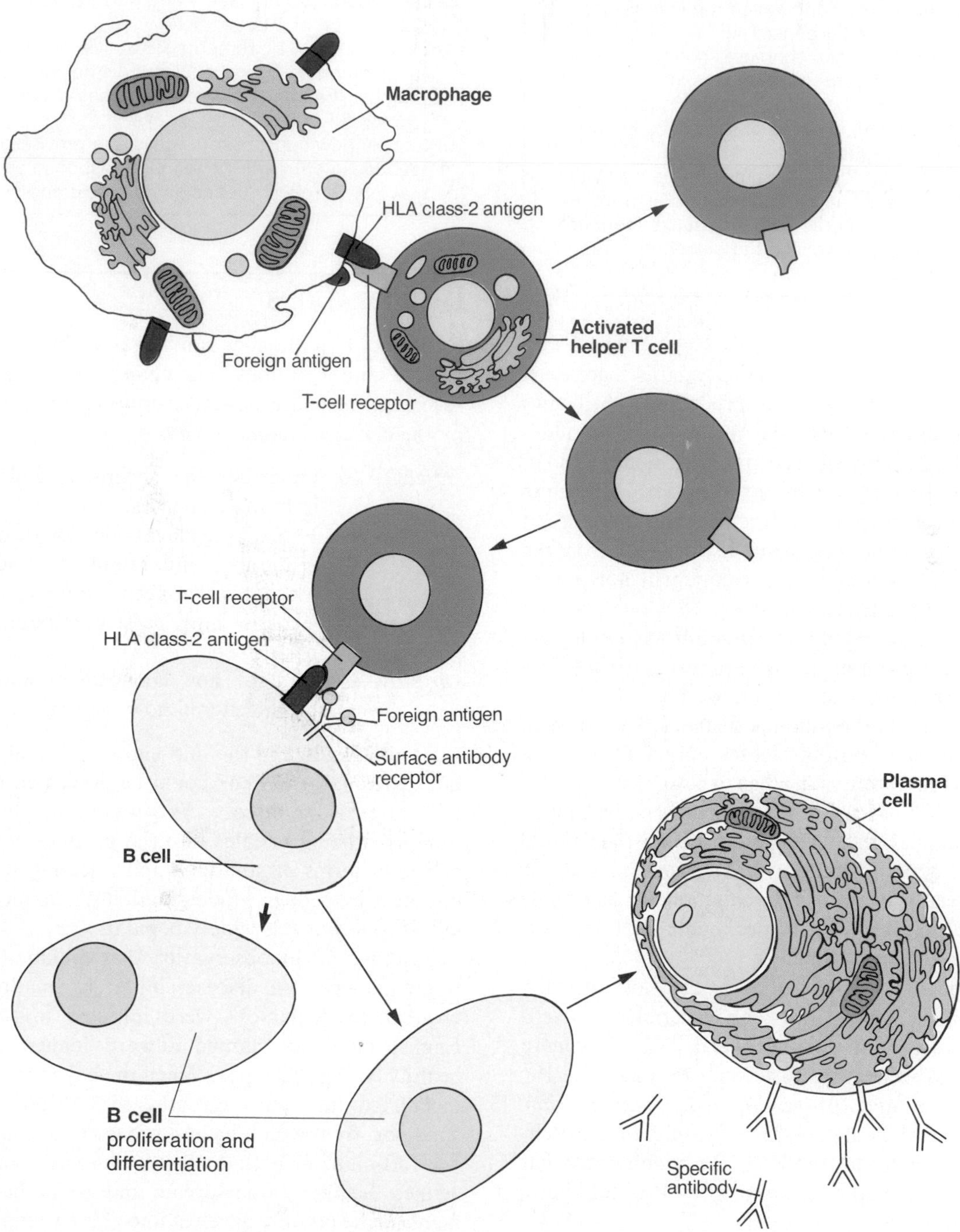

Glucocorticoids (such as hydrocortisone), secreted by the adrenal cortex, can act to suppress the activity of the immune system and inflammatory reactions. This is why *cortisone* and its analogues are used clinically in the treatment of inflammatory disorders and to inhibit the immune rejection of transplanted organs. The immunosuppression by glucocorticoid hormones may result from the fact that these hormones inhibit the secretion of interleukin-1 and the lymphokines. It therefore makes sense, in terms of the principles of negative feedback regulation, that secretion of interleukin-1 has been found to stimulate the pituitary to secrete ACTH, which in turn stimulates the adrenal cortex to secrete glucocorticoids. A negative feedback loop is thus completed involving the glucocorticoids and the lymphokines.

Tolerance

The ability to produce antibodies against **non-self-antigens** while tolerating (not producing antibodies against) **self-antigens** occurs early in life when immunological competence is established (the first month or so of postnatal life). If a fetal mouse of one strain receives transplanted antigens from a different strain, therefore, it will not recognize tissue transplanted later in life from the other strain as foreign and, as a result, will not immunologically reject the transplant.

The ability of an individual's immune system to recognize and tolerate self-antigens requires continuous exposure of the immune system to those antigens. If this exposure begins when the immune system is weak—such as in fetal and early postnatal life—tolerance is more complete and long lasting than when exposure occurs later in life. Some self-antigens, however, are normally hidden from the blood, such as thyroglobulin within the thyroid gland and lens protein in the eye. An exposure to these self-antigens results in antibody production just as if these proteins were foreign. Antibodies made against self-antigens are called **autoantibodies.**

The reasons why autoantibodies are not normally produced—that is, the reasons why "tolerance" to self-antigens occurs—are not well understood. Two of the more easily explained mechanisms to account for tolerance are: (1) *clone deletion;* and (2) *immunological suppression.* According to the clone deletion theory, tolerance to self-antigens is achieved by destruction of the lymphocytes that inherit the ability to make autoantibodies. It is not known how this might occur.

According to the immunological suppression theory, the lymphocytes that make autoantibodies are present throughout life but are normally inhibited from attacking self-antigens. This is presumably due to the effects of suppressor T lymphocytes. An alteration in the ratio of suppressor to helper T lymphocytes in later life, therefore, might result in the production of autoantibodies.

1. Describe the role of the thymus in cell-mediated immunity.
2. Define the term *lymphokines,* identify their origin, and list the different functions of these molecules.
3. Define the term *histocompatibility antigens,* and explain the importance of class-1 and class-2 HLA antigens in the function of T cells.
4. Describe the requirements for activation of helper T cells by macrophages, and explain how helper T cells promote the immunological defenses provided by cytotoxic T cells and by B cells.
5. Explain how suppressor T cells can provide a negative feedback control of the immune response and how they may play a role in mechanisms of tolerance.

Immunizations

Resistance to pathogens can be gained by both active and passive immunizations. The change from a primary to a secondary response in active immunity is best explained by the clonal selection theory.

Objective 16. Describe the primary and secondary immune responses.
Objective 17. Explain the clonal selection theory of active immunity, and relate this theory to the mechanism of active immunizations.
Objective 18. Explain how passive immunizations are performed.
Objective 19. Explain how monoclonal antibodies are produced and how they might be used.

It was first known in the mid-eighteenth century that the fatal effects of smallpox could be prevented by inducing mild cases of the disease. This was accomplished at that time by rubbing needles into the pustules of people who had mild forms of smallpox and injecting these needles into healthy people. Understandably, this method of immunization was not widely popular.

Acting on the observation that milkmaids who contracted cowpox—a disease similar to smallpox but less *virulent* (pathogenic)—were immune to smallpox, an English physician, named Edward Jenner, inoculated a healthy boy with cowpox. When the boy recovered, Jenner inoculated him with an otherwise deadly amount of smallpox, from which he also proved to be immune. (This was fortunate for both the boy—who was an orphan—and Jenner; Jenner's fame spread, and as the boy grew into manhood he proudly gave testimonials on Jenner's behalf.) This experiment, performed in 1796, began the first widespread immunization program.

A similar, but more sophisticated, demonstration of the effectiveness of immunizations was performed by Louis Pasteur almost a century later. Pasteur isolated the bacteria that cause anthrax and heated them until their ability

Edward Jenner: English physician, 1749–1823
Louis Pasteur: French chemist and bacteriologist, 1822–1895

Figure 22.23. Active immunity to a pathogen can be gained by exposure to the fully virulent form or by inoculation with a pathogen whose virulence (ability to cause disease) has been attenuated (reduced) but whose antigens are the same as in the virulent form.

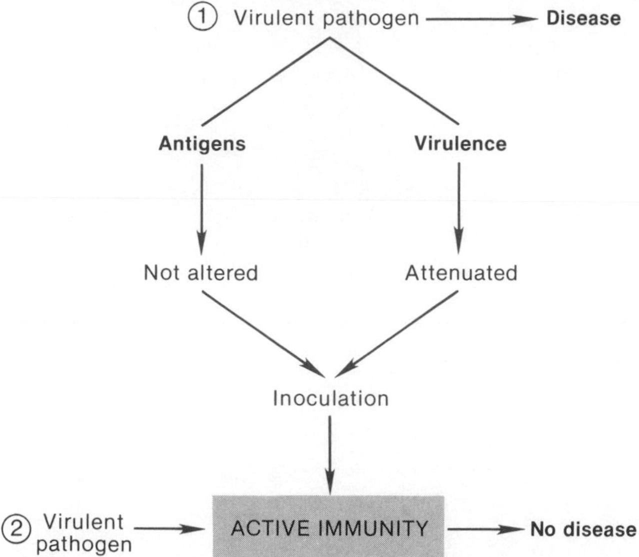

Figure 22.24. A comparison of antibody production in the primary response (upon first exposure to an antigen) to antibody production in the secondary response (upon subsequent exposure to the antigen). The greater secondary response is believed to be due to the development of lymphocyte clones produced during the primary response.

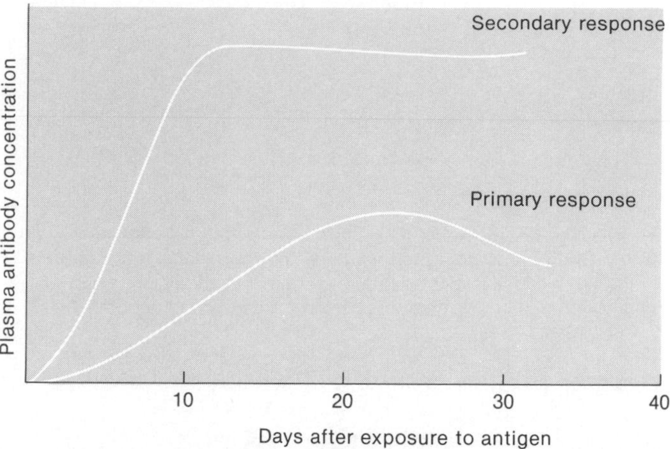

to cause disease was greatly reduced (their virulence was *attenuated*). He then injected these attenuated bacteria into twenty-five sheep, leaving another twenty-five unimmunized. Several weeks later, before a gathering of scientists, he injected all fifty sheep with the completely active anthrax bacteria. All twenty-five of the unimmunized sheep died—all twenty-five of the immunized sheep survived (fig. 22.23).

Active Immunity and the Clonal Selection Theory

When a person is exposed to a particular pathogen for the first time, there is a latent period of five to ten days before measurable amounts of specific antibodies appear in the blood. This sluggish **primary response** may not be sufficient to protect the individual against the disease caused by the pathogen. Antibody concentrations in the blood during this primary response reach a plateau in a few days and decline after a few weeks.

A subsequent exposure of the same individual to the same antigen results in a **secondary response** (fig. 22.24). Compared to the primary response, antibody production during the secondary response is much more rapid. Maximum antibody concentrations in the blood are reached in less than two hours and are maintained for a longer time than in the primary response. This rapid rise in antibody production is usually sufficient to prevent the disease.

Before stimulation by a particular antigen, the B cells that make the appropriate antibodies produce mainly IgM antibodies. These IgM antibodies therefore account for a high proportion of the antibodies made during the primary response. In contrast, most antibodies made during the secondary response are in the IgG subclass. This is shown in figure 22.25.

Clonal Selection Theory. The immunization procedures of Jenner and Pasteur were effective because the people who were inoculated produced a secondary rather than a primary response when exposed to the virulent pathogens. This protection is not simply due to accumulations of antibodies in the blood, because secondary responses occur even after antibodies produced by the primary response have disappeared. Immunizations, therefore, seem to produce a type of "learning" in which the ability of the immune system to combat a particular pathogen is improved by prior exposure.

The mechanisms by which secondary responses are produced are not completely understood; the **clonal selection theory,** however, appears to account for most of the evidence. According to this theory, B lymphocytes *inherit* the ability to produce particular antibodies (and T lymphocytes inherit the ability to respond to particular antigens). This inherited specificity is reflected also in the membrane receptor on the lymphocyte surface. Some lymphocytes, therefore, can respond to smallpox, for example, and produce antibodies against it even if the person has never been previously exposed to this disease.

Exposure to smallpox antigens stimulates these specific lymphocytes to divide many times until a large population of genetically identical cells—a clone—is produced. Some of these cells become plasma cells that secrete antibodies for the primary response; others become memory cells that can be stimulated to secrete antibodies during the secondary response (fig. 22.26).

Figure 22.25. The greater antibody production in the secondary response compared to the primary response is due mainly to increased amounts of IgG antibodies. Notice that the amount of IgM antibodies is about the same in both primary and secondary responses.

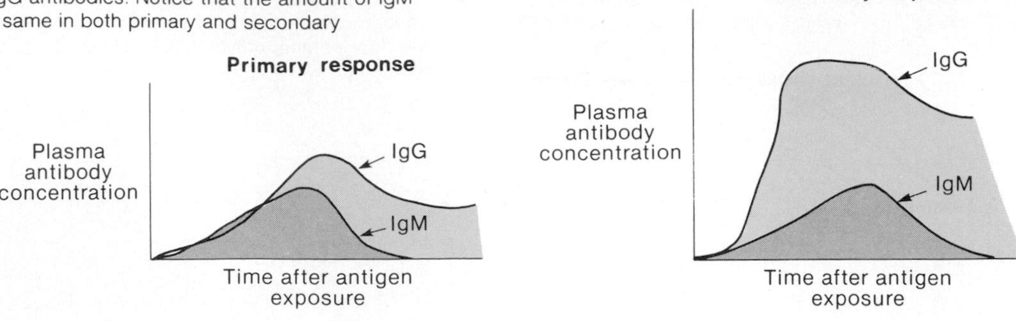

Figure 22.26. According to the clonal selection theory, exposure to an antigen stimulates the production of lymphocyte clones and the maturation of some members of B cell clones into antibody-secreting plasma cells.

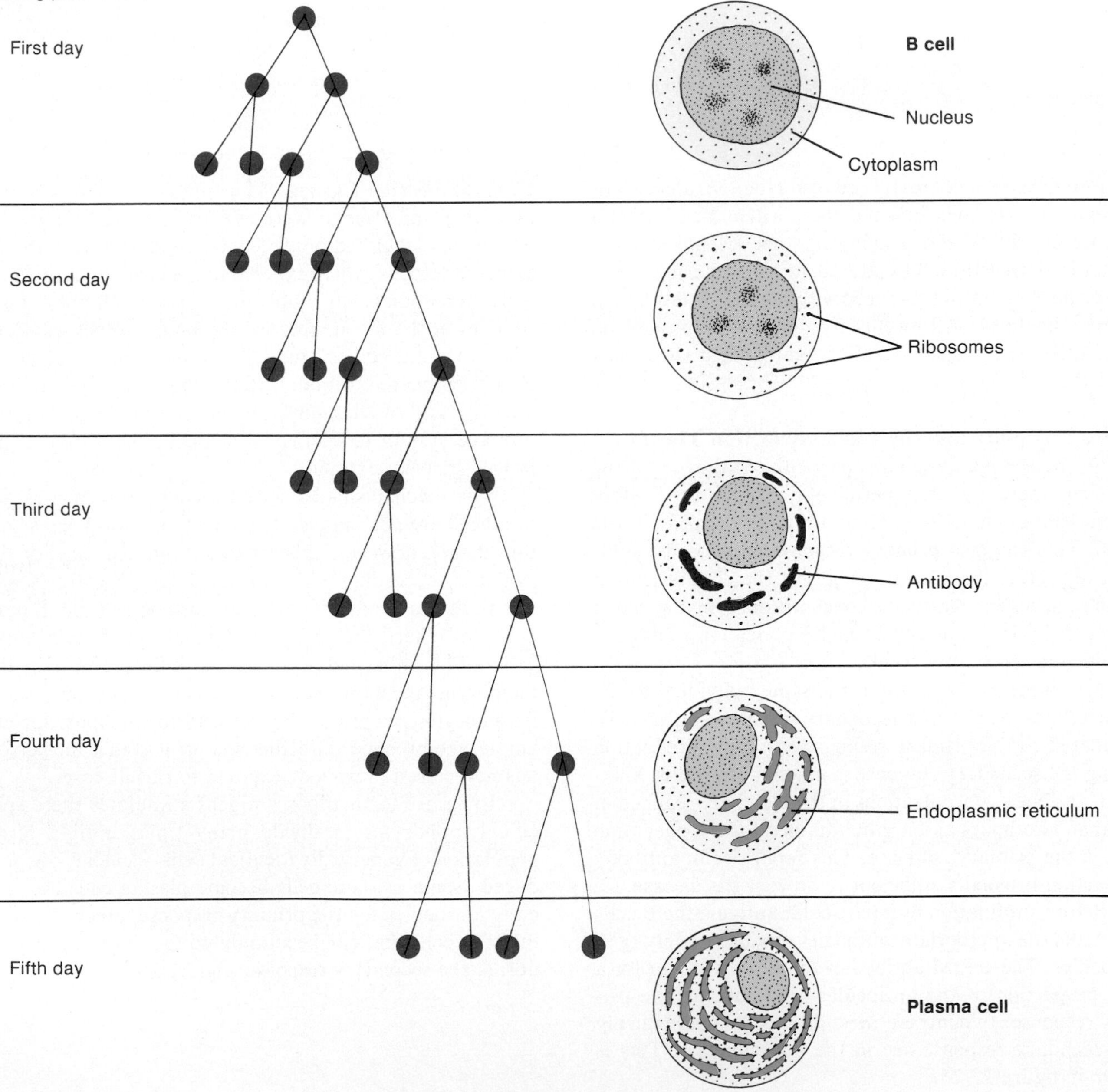

First day
B cell
Nucleus
Cytoplasm
Second day
Ribosomes
Third day
Antibody
Fourth day
Endoplasmic reticulum
Fifth day
Plasma cell

Table 22.9 Summary of the clonal selection theory (with regard to B cells)

Process	Results
Lymphocytes inherit the ability to produce specific antibodies.	Prior to antigen exposure, lymphocytes are already present in the body that can make the appropriate antibodies.
Antigens interact with antibody receptors on the lymphocyte surface.	Antigen-antibody interaction stimulates cell division and the development of lymphocyte clones containing memory cells and plasma cells that secrete antibodies.
Subsequent exposure to the specific antigens produce a more efficient response.	Exposure of lymphocyte clones to specific antigens results in greater and more rapid production of specific antibodies.

Notice that according to the clonal selection theory (table 22.9), antigens do not induce lymphocytes to make the appropriate antibodies. Rather, antigens select lymphocytes (through interaction with surface receptors) that are already able to make antibodies against that antigen. This is analogous to evolution by natural selection. An environmental agent (in this case, antigens) acts on the genetic diversity already present in a population of organisms (lymphocytes) to cause increasing numbers of the individuals that are selected.

Active Immunity. The development of a secondary response provides **active immunity** against the specific pathogens. The development of active immunity requires prior exposure to the specific antigens, during which time a primary response is produced and the person may get sick. Some parents, for example, deliberately expose their children to others who have measles, chicken pox, and mumps so that their children will be immune to these diseases in later life.

Clinical immunization programs induce primary responses by inoculating people with pathogens whose virulence has been attenuated or destroyed (such as Pasteur's heat-inactivated anthrax bacteria) or by using closely related strains of microorganisms that are antigenically similar but less pathogenic (such as Jenner's cowpox inoculations). These procedures cause the development of lymphocyte clones that can combat the virulent pathogens by producing secondary responses.

The first successful polio vaccine (the Salk vaccine) was composed of viruses that had been inactivated by treatment with formaldehyde. These "killed" viruses were injected into the body, in contrast to the currently used oral (Sabin) vaccine. The oral vaccine contains "living" viruses that have attenuated virulence. These viruses invade the epithelial lining of the intestine and multiply but do not invade nerve tissue. The immune system can, therefore, become sensitized to polio antigens and produce a secondary response if polio viruses that attack the nervous system are later encountered.

Salk vaccine: from Jonas Salk, American immunologist, 1914
Sabin vaccine: from Albert B. Sabin, American virologist, 1906

Passive Immunity

The ability to mount an immune response—called **immunological competence**—does not develop until about a month after birth. The fetus, therefore, cannot immunologically reject its mother. The immune system of the mother is fully competent but does not usually respond to fetal antigens, for reasons that are not completely understood. Some IgG antibodies from the mother do cross the placenta and enter the fetal circulation, however, and these serve to confer **passive immunity** to the fetus.

The fetus and the newborn baby are, therefore, immune to the same antigens as the mother. Since the baby did not itself produce the lymphocyte clones needed to form these antibodies, such passive immunity disappears when the infant is about one month old. If the baby is breast-fed, it can receive additional antibodies of the IgA subclass in its mother's first milk (the *colostrum*).

Passive immunizations are used clinically to protect people who have been exposed to extremely virulent infections or toxins, such as snake venom, tetanus, and others. In these cases the affected person is injected with **antiserum** (serum containing antibodies), also called **antitoxin** *(an''tĭ-tok'sin),* from an animal that has been previously exposed to the pathogen. The animal develops the lymphocyte clones and active immunity and thus has a high concentration of antibodies in its blood. Since the person who is injected with these antibodies does not develop active immunity, he or she must again be injected with antitoxin upon subsequent exposures. A comparison of active and passive immunity is shown in table 22.10.

Monoclonal Antibodies

In addition to their use in passive immunity, antibodies are also commercially prepared for use in research and clinical laboratory tests. In the past, antibodies were obtained by first chemically purifying a specific antigen and then injecting this antigen into animals. Since an antigen typically has many different antigenic determinant sites, however, the antibodies obtained by this method were polyclonal; they had different specificities. This decreased their sensitivity to a particular antigenic site and resulted in some degree of crossover with closely related antigen molecules.

Table 22.10 Comparison of active and passive immunity

Characteristic	Active immunity	Passive immunity
Injection of person with	Antigens	Antibodies
Source of antibodies	The person inoculated	An animal that is inoculated with the antigen; or the mother
Method	Injection with killed or attenuated pathogens or their toxins	Natural—transfer of antibodies across the placenta; artificial—injection with antibodies
Time to develop resistance	5 to 14 days	Immediately after injection
Duration of resistance	Long (perhaps years)	Short (days to weeks)
When used	Before exposure to pathogen	Before or after exposure to pathogen

In the preparation of monoclonal antibodies, an animal is injected with an antigen and then subsequently killed. B lymphocytes are then obtained from its spleen and placed in thousands of different *in vitro* incubation vessels. These cells soon die, however, unless they are hybridized with cancerous multiple myeloma cells. Cell fusion is promoted by a chemical, polyethylene glycol. The fusion of a B lymphocyte with a cancerous cell produces a hybrid that undergoes cell division and produces a clone, called a **hybridoma** *(hi''brĭ-do'mah)*. Each hybridoma secretes large amounts of identical, **monoclonal** *(mon''o-klōn'al)* **antibodies.**

The availability of large quantities of pure monoclonal antibodies has resulted in the development of much more sensitive clinical laboratory tests (of pregnancy, for example). These pure antibodies have also been used to pick one molecule (the specific antigen interferon, for example) out of a solution of many molecules and thus isolate and concentrate it. In the future, monoclonal antibodies against specific tumor antigens may aid the diagnosis of cancer. Even more exciting, cytotoxic drugs that can kill normal as well as cancerous cells might be aimed directly at a tumor by combining these drugs with monoclonal antibodies against specific tumor antigens.

1. Describe two methods used to induce active immunity.
2. Explain the characteristics of the primary and secondary immune responses, and draw graphs to illustrate your discussion.
3. Explain the clonal selection theory and how this theory accounts for the secondary response.
4. Describe passive immunity, and give natural and clinical examples of this type of immunization.

Tumor Immunology

Tumor cells usually reveal antigens that may activate an immune reaction that destroys the tumor. When cancers develop, this immunological surveillance system—primarily the function of T cells and natural killer cells—has failed to prevent the growth and metastasis of the tumor.

Objective 20. Describe the significance of the fact that cancer cells are dedifferentiated.

Objective 21. Identify the cells and mechanisms involved in the immunological surveillance against cancer.

Objective 22. Explain how stress and aging might result in increased susceptibility to cancer.

Oncology (the study of tumors) has revealed that tumor biology is similar to and interrelated with the functions of the immune system. Most tumors appear to be clones of single cells that have become transformed. This is similar to the development of lymphocyte clones in response to specific antibodies. Lymphocyte clones, however, are under complex inhibitory control systems—such as those exerted by suppressor T lymphocytes and negative feedback by antibodies. The division of tumor cells, in contrast, is not effectively controlled by normal inhibitory mechanisms. Tumor cells are also relatively unspecialized—they *dedifferentiate,* which means that they become similar to the less-specialized cells of an embryo.

Tumors are described as *benign* when they are relatively slow growing and limited to a specific location (warts, for example). Benign tumors do not undergo *metastasis (mĕ-tas'tah-sis),* a term that refers to the dispersion of tumor cells and the resultant seeding of new tumors in different locations. *Malignant* tumors grow more rapidly and do metastasize. The term **cancer,** as it is usually used, refers to malignant, life-threatening tumors.

As tumors dedifferentiate, they reveal surface antigens that can stimulate the immune destruction of the tumor cells. Consistent with the concept of dedifferentiation, some of these antigens are proteins produced in embryonic or fetal life that are not normally produced postnatally. Since they are absent at the time immunological competence is established, they are treated as foreign and fit subjects for immunological attack when they are produced by cancerous cells. The release of two such antigens into the blood has provided the basis for a laboratory diagnosis of some cancers. *Carcinoembryonic*

benign: L. *benignus,* kind
metastasis: Gk. *metastasis,* a removing
cancer: L. *cancer,* a crab

Figure 22.27. A killer T cell (*a*) contacts a cancer cell (the larger cell), in a manner that requires specific interaction with antigens on the cancer cell. The killer T cell releases lymphokines, including toxins that cause the death of the cancer cell (*b*). (Scanning electron micrographs © Andrejs Liepens.)

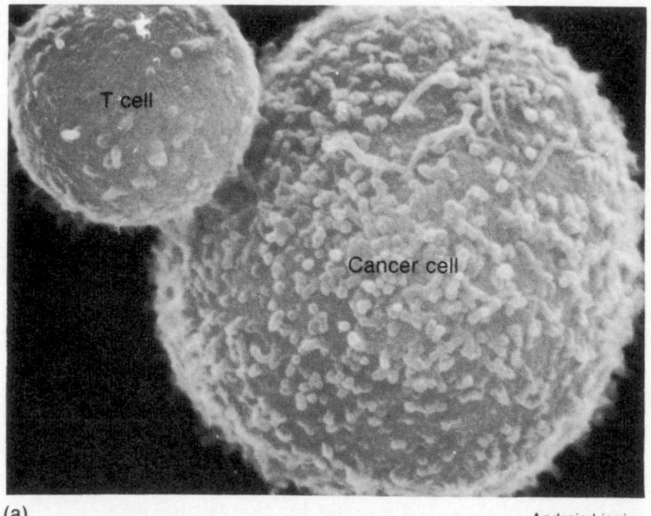

(a) Andrejs Liepins

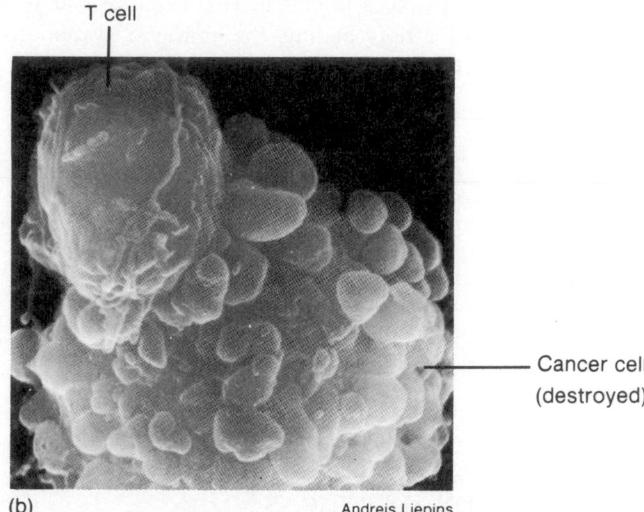

(b) Andrejs Liepins

(kar''sin-o-em''bre-on'ik) antigen tests are useful in the diagnosis of colon cancer, for example, and tests for *alpha-fetoprotein* (normally produced only by the fetal liver) help in the diagnosis of liver cancer.

Tumor antigens activate the immune system, initiating attack primarily by killer T lymphocytes (fig. 22.27) and natural killer cells (described below). The concept of **immunological surveillance** against cancer was introduced in the early 1970s to describe the proposed role of the immune system in fighting cancer. According to this concept, tumor cells originate frequently in the body but are normally recognized and destroyed by the immune system before they can cause cancer. There is evidence that immunological surveillance does prevent some types of cancer; this helps to explain why, for example, AIDS victims (with a depressed immune system) have a high incidence of Kaposi's sarcoma. It is not clear, however, why all types of cancers do not appear with high frequency in AIDs patients and others whose immune system is suppressed. For these reasons, the generality of the immunological surveillance concept is currently controversial.

Natural Killer Cells

Researchers observed that a strain of hairless mice, which genetically lacks a thymus and T lymphocytes, does not suffer from a particularly high incidence of tumor production. This surprising observation led to the discovery of **natural killer (NK) cells,** which are lymphocytes that are related to, but distinct from, T lymphocytes. Unlike killer T cells, NK cells destroy tumors in a nonspecific fashion and do not require prior exposure for sensitization to the tumor antigens. The NK cells thus provide a first line of cell-mediated defense, which is subsequently backed up by a specific response mediated by killer T cells. These two cell types interact, however; the activity of NK cells is stimulated by interferon released as one of the lymphokines from T lymphocytes.

Effects of Aging and Stress

Susceptibility to cancer varies greatly. The Epstein-Barr virus that causes Burkitt's lymphoma in a few individuals in Africa, for example, can also be found in healthy people throughout the world. In most cases the virus is harmless; in some cases mononucleosis (involving a limited proliferation of white blood cells) is produced. Only rarely does this virus cause the uncontrolled proliferation of leukocytes occurring in Burkitt's lymphoma. The reasons for these different responses to the Epstein-Barr virus and indeed for the different susceptibilities of people to other forms of cancer are not well understood.

It is known that cancer risk increases with age. According to one theory, this is due to the fact that aging lymphocytes accumulate genetic errors over the years that decrease their effectiveness. The secretion of thymus hormones also decreases with age in parallel with a decrease in cell-mediated immune competence. Both of these changes and perhaps others not yet discovered could increase susceptibility to cancer.

Numerous experiments have demonstrated that tumors grow faster in experimental animals subject to stress than in unstressed control animals. This is generally believed to result from the fact that stressed animals, including humans, have increased secretion of corticosteroid hormones, which act to suppress the immune system (this is why cortisone is given to people who receive organ transplants and to people with chronic inflammatory diseases). Some recent experiments, however, suggest that the stress-induced suppression of the immune system may also be due to other factors that do not involve the adrenal cortex. Future advances in cancer therapy may incorporate methods of strengthening the immune system together with methods that directly destroy tumors.

1. Define the term *immunological tolerance,* and explain how this might be produced.
2. Describe the nature, functions, and clinical applications of the histocompatibility antigens.
3. Describe the red blood cell antigens, and explain how the agglutination reactions are produced.
4. Describe how erythroblastosis fetalis is produced, and explain the technique used to prevent this condition.
5. Explain how T lymphocytes and natural killer cells recognize cancer cells.

Clinical Considerations

The ability of the normal immune system to tolerate self-antigens while identifying and attacking foreign antigens provides a specific defense against invading pathogens. In every individual, however, this system of defense against invaders at times commits domestic offenses. This can result in diseases that range in severity from the sniffles to sudden death.

Diseases caused by the immune system can be grouped into three interrelated categories: (1) *autoimmune diseases;* (2) *immune complex diseases;* and (3) *allergy,* or *hypersensitivity.* It is important to remember that these diseases are not caused by foreign pathogens but by abnormal responses of the immune system.

Autoimmunity

Autoimmune diseases are those produced by failure of the immune system to recognize and tolerate self-antigens. This failure results in the production of autoantibodies that can cause inflammation and organ damage. Such autoimmune destruction may occur as a result of the following mechanisms.

1. **An antigen that does not normally circulate in the blood may become exposed to the immune system.** Thyroglobulin protein that is normally trapped within the thyroid follicles, for example, can stimulate the production of autoantibodies that cause the destruction of the thyroid (fig. 22.28); this occurs in *Hashimoto's thyroiditis.* Similarly, autoantibodies developed against lens protein in a damaged eye may cause the destruction of a healthy eye (in *sympathetic ophthalmia*).
2. **A self-antigen, which is otherwise tolerated, may be altered by combining with a foreign hapten.** The disease *thrombocytopenia* (low platelet count), for example, can be caused by the autoimmune destruction of thrombocytes (platelets). This occurs when drugs such as aspirin, sulfonamide, antihistamines, digoxin, and others combine with platelet proteins to produce new antigens. The symptoms of this disease usually stop when the person stops taking these drugs.
3. **A self-antigen may become damaged and, as a result, may expose new antigenic sites.** This may occur in *rheumatoid arthritis.* An initial inflammation of the joints (perhaps through viral infection) may result in the release of digestive enzymes from lysosomes in phagocytic cells. When these enzymes digest part of IgG antibodies, they expose new antigenic sites. This stimulates the

Figure 22.28. Autoimmune thyroiditis in a rabbit, induced experimentally by injection with thyroglobulin. Compare the picture of a normal thyroid (*left*) with that of the diseased thyroid (*right*). The grainy appearance of the diseased thyroid is due to the infiltration of large numbers of lymphocytes and macrophages.

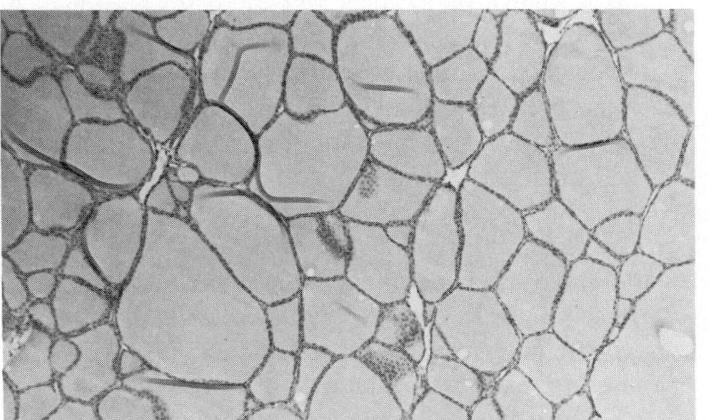

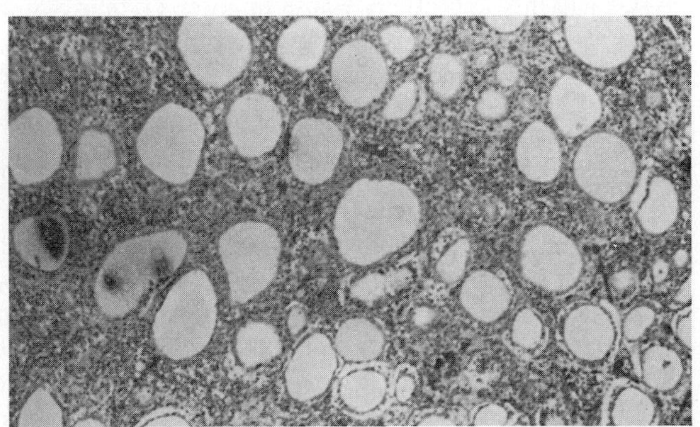

production of IgM antibodies directed against the altered IgG proteins (fig. 22.29). The IgM antibodies that a person develops against his or her own IgG antibodies are called *rheumatoid factor* and can be used to diagnose this condition.

4. **Antibodies produced against foreign antigens may cross-react with self-antigens.** Autoimmune diseases of this sort can occur, for example, as a result of *streptococcus* bacterial infections. Antibodies produced in response to antigens in this bacterium may cross-react with self-antigens in the heart and kidneys. The inflammation induced by such autoantibodies can produce heart damage (including the valve defects characteristic of *rheumatic fever*) and damage to the glomerular capillaries in the kidneys (*glomerulonephritis [glo-mer''u-lo-nĕ-fri'tis]*).
5. **Self-antigens, such as receptor proteins, may be presented to the helper T lymphocytes together with class-2 HLA antigens.** Normally, only macrophages and B lymphocytes produce class-2 HLA antigens, which are associated with foreign antigens and are recognized by helper T cells. Perhaps as a result of viral infection, however, cells that do not normally produce class-2 HLA antigens may start to do so and, in this way, present a self-antigen to the helper T cells. In *Graves' disease,* for example, the thyroid cells produce class-2 HLA antigens, and the immune system produces autoantibodies against the TSH receptor proteins in the thyroid cells. These autoantibodies, called *TSAb* for "thyroid-stimulating antibody," interact with the TSH receptors and overstimulate the thyroid gland. Similarly, in type I *diabetes mellitus,* the beta cells abnormally produce class-2 HLA antigens, resulting in autoimmune destruction of the insulin-producing cells.
6. **Genes that code for non-self-antibodies may mutate to produce autoantibodies.** This would not cause particular diseases but would lead to the increased frequency of autoimmune diseases in general. An increased frequency of autoimmune diseases does in fact occur with age, as predicted by this mutation theory. Table 22.11 provides some examples of autoimmune diseases.

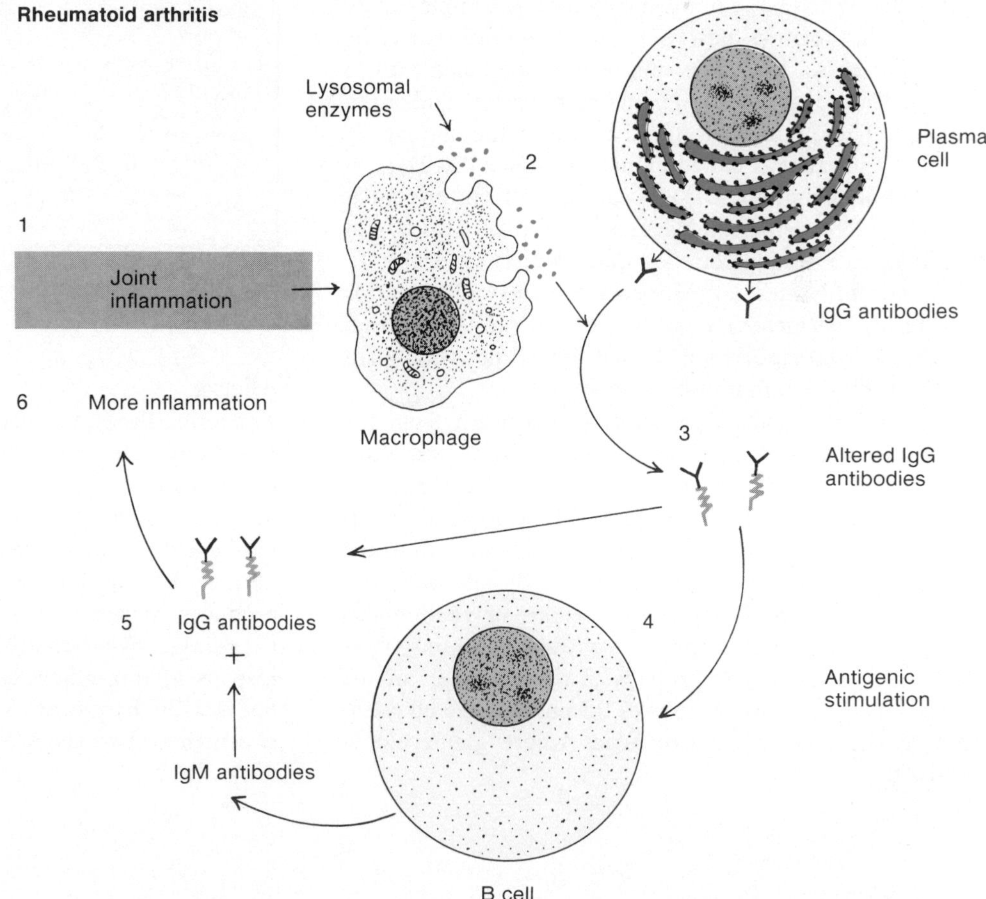

Figure 22.29. One proposed mechanism for the development of rheumatoid arthritis. An initial joint inflammation (*1*) results in the release of lysosomal enzymes (*2*) that cause damage to IgG antibodies (*3*). These damaged IgG antibodies act as antigens (*4*) to stimulate B lymphocytes, resulting in the production of IgM antibodies antigens. This, in turn, produces further inflammation.

Table 22.11 Examples of autoimmune diseases

Disease	Antigen	Ig and/or T cell response
Postvaccinal and postinfectious encephalomyelitis	Myelin, cross-reactive	T cell
Aspermatogenesis	Sperm	T cell
Sympathetic ophthalmia	Uvea	T cell
Hashimoto's disease	Thyroglobulin	IgG and T cell
Graves' disease	Receptor proteins for TSH	Long-acting thyroid stimulator (LATS)
Autoimmune hemolytic disease	I, Rh, and others on surface of RBCs	IgM and IgG
Thrombocytopenic purpura	Hapten-platelet or hapten-adsorbed antigen complex	IgG
Myasthenia gravis	Myosin	IgG
Rheumatic fever	Streptococcal cross-reactive with heart	IgG and IgM
Glomerulonephritis	Streptococcal cross-reactive with kidney	IgG and IgM
Rheumatoid arthritis	IgG	IgM to Fc(γ)
Systemic lupus erythematosus	DNA, nucleoprotein, RNA, etc.	IgG

Immune Complex Diseases

The term *immune complexes* refers to combinations of antibodies with antigens that are free, rather than attached to bacterial or other cells. Formation of such complexes activates complement proteins and promotes inflammation. This inflammation normally is self-limiting because the immune complexes are removed by phagocytic cells. When large numbers of immune complexes are continuously formed, however, the inflammation may be prolonged. Also, the dispersion of immune complexes to other sites can lead to widespread inflammations and organ damage. The damage produced by the inflammatory response to antigens is called **immune complex disease.**

Immune complex diseases can result from infections by bacteria, parasites, and viruses. In viral hepatitis B, for example, an immune complex that consists of viral antigens and antibodies can cause widespread inflammation of arteries *(periarteritis [per''e-ar''tĕ-ri'tis]).* Note that the arterial damage is not caused by the hepatitis virus itself but by the inflammatory process.

Immune complex diseases can also result from the formation of complexes between self-antigens and autoantibodies. This is the case in rheumatoid arthritis, in which the inflammation is produced by complexes of altered IgG antibodies (the antigens in this case) and IgM antibodies. Another immune complex disease that has an autoimmune basis is *systemic lupus erythematosus (SLE).* People with SLE produce antibodies against their own DNA and nuclear proteins. This can result in the formation of immune complexes throughout the body, including the glomerular capillaries where glomerulonephritis may be produced.

Table 22.12 Allergy: comparison between immediate and delayed hypersensitivity reaction

Characteristic	Immediate reaction	Delayed reaction
Time for onset of symptoms	Within several minutes	Within one to three days
Lymphocytes involved	B cells	T cells
Immune effector	IgE antibodies	Cell-mediated immunity
Allergies most commonly produced	Hay fever, asthma, and most other allergic conditions	Contact dermatitis (such as to poison ivy and poison oak)
Therapy	Antihistamines and adrenergic drugs	Corticosteroids (such as cortisone)

Allergy

The term **allergy,** usually used synonymously with *hypersensitivity,* refers to particular types of abnormal immune responses to antigens, which are called *allergens (al'er-jen)* in these cases. There are two major forms of allergy: (1) **immediate hypersensitivity,** which is due to an abnormal B lymphocyte response to an allergen that produces symptoms within seconds or minutes; and (2) **delayed hypersensitivity,** which is an abnormal T cell response that produces symptoms within about forty-eight hours after exposure to an allergen. A comparison between these two types of hypersensitivity is provided in table 22.12.

Table 22.13 Some of the chemicals released by mast cells and leukocytes that are responsible for immediate hypersensitivity

Chemical	Derivation	Action
Histamine	From histidine	Contracts smooth muscles in bronchioles; dilates blood vessels; increases capillary permeability
Serotonin	From tryptophane	Contracts smooth muscles
Prostaglandins and leukotrienes	Arachidonic acid	Prolonged contraction of smooth muscle; increased capillary permeability
Eosinophil chemotactic factor	Polypeptides	Attracts eosinophils
Bradykinin and related compounds	Polypeptides	Slow smooth muscle contraction

From Stuart Ira Fox, *Human Physiology,* 2d ed.

Figure 22.30. Allergy (immediate hypersensitivity) is produced when antibodies of the IgE subclass attach to tissue mast cells. The combination of these antibodies with allergens (antigens that provoke an allergic reaction) cause the mast cell to secrete histamine and other chemicals that produce the symptoms of allergy.

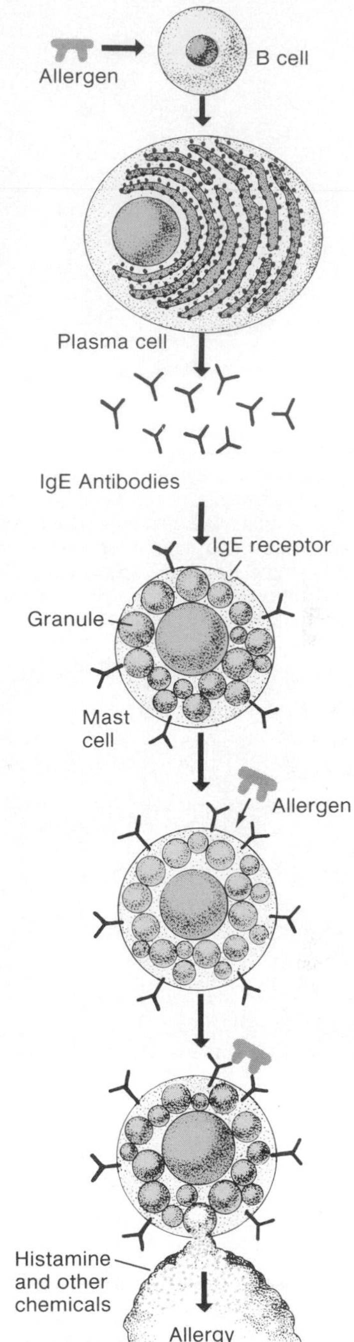

Immediate Hypersensitivity. Immediate hypersensitivity can produce such symptoms as allergic rhinitis (chronic runny or stuffy nose), conjunctivitis (red eyes), allergic asthma, atopic dermatitis (urticaria, or hives), and others. These symptoms result from the production of antibodies of the IgE subclass, instead of the normal IgG antibodies.

Unlike IgG antibodies, IgE antibodies do not circulate in the blood but instead attach to tissue mast cells (which have membrane receptors for these antibodies). When the person is again exposed to the same allergen, the allergen bonds to the antibodies attached to the mast cells. This stimulates the mast cells to secrete various chemicals, including **histamine** (fig. 22.30). During this process, leukocytes may also secrete **prostaglandins** and related molecules called **leukotrienes.** These chemicals (table 22.13) produce the symptoms of the allergic reactions.

The itching, sneezing, tearing, and runny nose of persons suffering from hay fever are produced largely by histamine and can be treated effectively by antihistamine drugs. Food allergies, causing diarrhea and colic, are mediated primarily by prostaglandins and can be treated with aspirin (these are the only allergies that respond positively to aspirin). Asthma, produced by smooth muscle constriction in the bronchioles in the lungs, is due to the release of leukotrienes. Since there are no antileukotriene drugs presently available, asthma is treated with epinephrine-like compounds (which cause bronchodilation) and corticosteroids.

Immediate hypersensitivity to a particular antigen is commonly tested by injecting various antigens under the skin (fig. 22.31). Within a short time a *flare-and-wheal reaction* is produced if the person is allergic to that antigen. This reaction is due to the release of histamine and other chemical mediators: the flare is due to vasodilation, and the wheal results from local edema.

Allergens that provoke immediate hypersensitivity include various foods, bee stings, and pollen grains. The most common allergy of this type is seasonal hay fever,

Figure 22.31. Skin test for allergy. If an allergen is injected into the skin of a sensitive individual, a typical flare-and-wheal response occurs within several minutes.

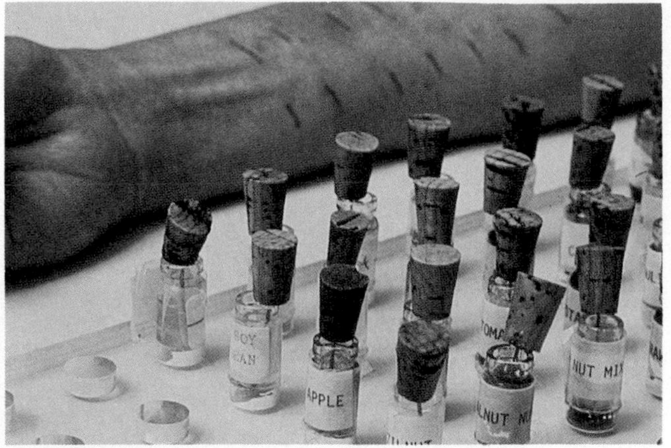

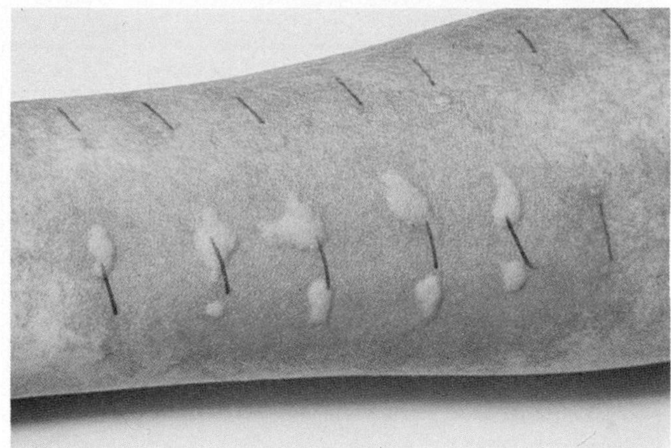

Figure 22.32. (*a*) a scanning electron micrograph of ragweed (*Ambrosia*), which is responsible for hay fever. (*b*) a scanning electron micrograph of the house dust mite (*Dermatophagoides farinae*), which lives in dust and is often responsible for yearlong allergic rhinitis and asthma. ([*a*] From: *Tissues and Organs: A Text Atlas of Scanning Electron Microscopy* by R. G. Kessel and R. Kardon. W. H. Freeman and Company. © 1979.)

(a)

(b)

which may be provoked by ragweed *(Ambrosia)* pollen grains. People with chronic allergic rhinitis due to an allergy to "dust" or "feathers" are actually allergic to a tiny mite (fig. 22.32) that lives in dust and eats the scales of skin that are constantly shed from the body.

Delayed Hypersensitivity. As the name implies, a longer time is required for the development of symptoms in delayed hypersensitivity (hours to days) than in immediate hypersensitivity. This may be due to the fact that immediate hypersensitivity is mediated by antibodies, whereas delayed hypersensitivity is a cell-mediated T lymphocyte response. Since the symptoms are caused by the secretion of lymphokines, rather than by the secretion of histamine, treatment with antihistamines provides little benefit. At present, corticosteroids are the only drugs that can effectively treat delayed hypersensitivity.

One of the best-known examples of delayed hypersensitivity is **contact dermatitis,** caused by poison ivy, poison oak, and poison sumac. The skin test for tuberculosis—the tine test and the Mantoux test—also rely on delayed hypersensitivity reactions. If a person has been exposed to the tubercle bacillus and has, as a result, developed T cell clones, skin reactions appear within a few days after the tubercle antigens are rubbed into the skin with small needles (tine test) or are injected under the skin (Mantoux test).

Chapter Summary

I. Lymphatic System
   A. Lymph capillaries drain excess interstitial fluid and are highly permeable to proteins and particles such as microorganisms.
   B. Lymph ducts receive lymph from the lymph capillaries and eventually empty into the thoracic duct and right lymphatic duct, which in turn empty into the left and right subclavian veins, respectively.
   C. Lymph nodes contain germinal centers of lymphocyte production.
   D. The spleen and thymus are lymphoid organs that contain germinal centers and serve other functions.

II. Defense Mechanisms
   A. Nonspecific defense mechanisms include barriers to penetration of the body and internal defenses.
      1. Phagocytic cells engulf invading pathogens.
      2. Interferons are polypeptides, secreted by cells infected with viruses, that help protect other cells from viral infections.
   B. Specific immune responses are directed against antigens.
      1. Antigens are molecules, or parts of molecules, that are usually large, complex, and foreign.
      2. A given molecule can have a number of antigenic determinant sites that stimulate the production of different antibodies.
   C. Specific immunity is a function of lymphocytes.
      1. B lymphocytes secrete antibodies and provide humoral immunity.
      2. T lymphocytes provide cell-mediated immunity.

III. Functions of B Lymphocytes
   A. There are five subclasses of antibodies, or immunoglobulins.
      1. Each type of antibody has two variable regions that combine with specific antigens.
      2. The combination of antibodies with antigens promotes phagocytosis, a process called opsonization.
   B. Antigen-antibody complexes activate a system of proteins called the complement system.
      1. This results in complement fixation, where complement proteins attach to a cell membrane and promote the destruction of the cell.
      2. Free complement proteins promote opsonization and chemotaxis and stimulate the release of histamine from tissue mast cells.
   C. Specific and nonspecific immune mechanisms cooperate in the development of a local inflammation.

IV. Functions of T Lymphocytes
   A. The thymus processes T lymphocytes and secretes hormones that are believed to be required for the proper function of the immune response of T lymphocytes throughout the body.
   B. There are three subcategories of T lymphocytes.
      1. Cytotoxic, or killer, T lymphocytes kill victim cells by a mechanism that does not involve antibodies but that does require close contact between the cytotoxic T cell and the victim cell.
      2. Cytotoxic T lymphocytes are responsible for transplant rejection and for the immunological defense against fungal and viral infections, as well as for the defense against some bacterial infections.
      3. Helper and suppressor T lymphocytes stimulate or suppress, respectively, the function of B lymphocytes and cytotoxic T lymphocytes.
      4. The T lymphocytes secrete a family of compounds called lymphokines, which promote the action of lymphocytes and macrophages; one of the more important lymphokines is interleukin-2, which stimulates proliferation of lymphocytes.
      5. Receptor proteins on the cell membrane of T lymphocytes must bond to both a foreign antigen and a histocompatibility antigen in order for the T cell to become activated.
      6 Histocompatibility antigens, or HLA antigens, are a family of molecules on the membranes of cells that are present in different combinations in different individuals.
   C. Macrophages partially digest a foreign body, such as a virus, and present the antigens to the lymphocytes on the surface of the macrophage in combination with class-2 HLA antigens.
      1. Helper T lymphocytes require such interaction with macrophages in order to be activated by a foreign antigen; when activated in this way, the helper T cells secrete interleukin-2.
      2. Interleukin-2 stimulates proliferation of cytotoxic T lymphocytes that are specific for the foreign antigen.
      3. The interleukin-2 secreted by helper T cells also stimulates proliferation of B lymphocytes and so promotes the secretion of antibodies in response to the foreign antigen.
   D. Tolerance to self-antigens may be due to the destruction of lymphocytes that can recognize the self-antigens, or it may be due to suppression of the immune response by the action of specific suppressor T lymphocytes.

V. Immunizations
   A. A primary response is produced when a person is first exposed to a pathogen; a subsequent exposure results in a secondary response.
      1. During the primary response IgM antibodies are produced slowly, and the person is likely to get sick.
      2. During the secondary response IgG antibodies are produced quickly, and the person has resistance to the pathogen.
      3. In active immunizations, the person is exposed to pathogens of attenuated virulence that have the same antigenicity as the virulent pathogen.
      4. The secondary response is believed to be due to the development of lymphocyte clones after the first exposure as a result of the antigen-stimulated proliferation of appropriate lymphocytes.
   B. Passive immunity is provided by antibodies made by a different organism.
      1. This occurs naturally from mother to fetus.
      2. Injections of antiserum provide passive immunity to some pathogenic organisms and toxins.
   C. Monoclonal antibodies are made by hybridomas, which are formed artificially by the fusion of B lymphocytes with multiple myeloma cells.

VI. Tumor Immunology
   A. Immunological surveillance against cancer is provided mainly by cytotoxic T lymphocytes and natural killer cells.
      1. Cancerous cells dedifferentiate and may produce fetal antigens; these or other antigens may be presented to lymphocytes in association with abnormally produced class-2 HLA antigens.
      2. Natural killer cells are nonspecific; T lymphocytes are directed against specific antigens on the cancer cell surface.
      3. Immunological surveillance against cancer is weakened by stress.

Review Activities

Objective Questions

1. Which of the following offers a nonspecific defense against viral infection?
   (a) antibodies
   (b) leukotrienes
   (c) interferon
   (d) histamine

Match the cell type with its secretion.

2. Killer T cells	(a) antibodies
3. Mast cells	(b) lymphokines
4. Plasma cells	(c) lysosomal enzymes
5. Macrophages	(d) histamine

6. Which of the following statements about the F_{ab} portion of antibodies is true?
   (a) It bonds to antigens.
   (b) Its amino acid sequences are variable.
   (c) It consists of both H and L chains.
   (d) All of the above are true.
7. Which of the following statements about complement proteins C3a and C5a is *false*?
   (a) They are released during the complement fixation process.
   (b) They stimulate chemotaxis of phagocytic cells.
   (c) They promote the activity of phagocytic cells.
   (d) They produce pores in the victim cell membrane.
8. Mast cell secretion during an immediated hypersensitivity reaction is stimulated when antigens combine with
   (a) IgG antibodies
   (b) IgE antibodies
   (c) IgM antibodies
   (d) IgA antibodies
9. During a secondary response
   (a) antibodies are made quickly and in great amounts
   (b) antibody production lasts longer than in a primary response
   (c) antibodies of the IgG class are produced
   (d) lymphocyte clones are believed to develop
   (e) all of the above
10. Which of the following cells aids the activation of lymphocytes by antigens?
    (a) macrophages
    (b) neutrophils
    (c) mast cells
    (d) natural killer cells
11. Which of the following statements about T lymphocytes is *false*?
    (a) Some T cells promote the activity of B cells.
    (b) Some T cells suppress the activity of B cells.
    (c) Some T cells secrete interferon.
    (d) Some T cells produce antibodies.
12. Delayed hypersensitivity is mediated by
    (a) T cells
    (b) B cells
    (c) plasma cells
    (d) natural killer cells

Match the cell type with its secretion.

13. Interleukin-1	(a) B lymphocytes
14. Interleukin-2	(b) helper T cells
15. Tumor necrosis factor	(c) suppressor T cells
16. Interferon	(d) macrophages

Essay Questions

1. Explain how antibodies help destroy invading bacterial cells.
2. Explain how killer T lymphocytes help destroy cells infected with viruses.
3. Explain the possible roles of helper and suppressor T lymphocytes in
   (a) defense against infections and
   (b) tolerance to self-antigens.
4. Describe the clonal selection theory, and use this theory to explain how active immunity is produced.
5. Explain the physiological and clinical significance of histocompatibility and red blood cell antigens.

23 RESPIRATORY SYSTEM

Outline

Concepts

The respiratory system can be divided into upper and lower divisions on the basis of function and embryological development.

Air is conducted through the oral and nasal cavities to the pharynx and then through the larynx to the trachea and bronchial tree. These structures deliver warmed and humidified air to the respiratory division in the lungs.

Alveoli are the functional units of the lungs where gas exchange occurs. Right and left lungs are separately contained in pleural membranes.

The movement of air into and out of the lungs occurs as a result of pressure differences induced by changes in lung volumes. Ventilation is thus influenced by the physical properties of the lungs, including their compliance, elasticity, and surface tension.

Normal, quiet inspiration results from muscle contraction, expiration from muscle relaxation, and elastic recoil. These actions can be forced by contractions of the accessory respiratory muscles. The amount of air inspired and expired can be measured by different procedures and used as tests of pulmonary function.

The lungs function to bring the blood into gaseous equilibrium with the alveolar air. The efficiency of this process is measured by the P_{O_2} and P_{CO_2} of arterial blood.

The rhythm of breathing is controlled by centers in the brain stem. These centers are influenced by higher brain areas and are regulated by sensory information that makes breathing responsive to the changing respiratory needs of the body.

Hemoglobin loads with oxygen in the lungs and unloads oxygen in the tissue capillaries. The extent of this unloading is influenced by a variety of factors that act to insure an adequate delivery of oxygen to the tissues while maintaining an adequate reserve of oxygen in the blood.

Carbonic acid and bicarbonate are the major transport forms of carbon dioxide and the major contributors to the acid-base balance of the blood. Since the concentration of these compounds is partly controlled by ventilation, pulmonary function is an important regulator of the acid-base balance in the body.

Changes in ventilation and oxygen delivery occur during exercise and during acclimatization to a high altitude. These changes help to compensate for the increased metabolic rate during exercise and for the decreased arterial P_{O_2} at a high altitude.

Functions and Development of the Respiratory System

The respiratory system can be divided into upper and lower divisions on the basis of function and embryological development.

Objective 1. Describe the functions included in the term *respiration.*
Objective 2. Identify the organs of the respiratory system, and describe their locations.
Objective 3. Compare the development of the conducting and respiratory divisions of the lungs.

The term *respiration* includes three separate but related functions: (1) **ventilation** (breathing); (2) **gas exchange** between the air and blood in the lungs and between the blood and tissues; and (3) **oxygen utilization** by the tissues in the energy-liberating reactions of cell respiration. The exchange of gases (oxygen and carbon dioxide) between the air and blood is called *external respiration.* Gas exchange between the blood and tissues is known as *internal respiration.*

It is estimated that a relaxed adult breathes about nine to twenty times a minute, ventilating about 5 to 6 liters of air during this period. Strenuous exercise increases the demand for oxygen and increases the respiratory rate fifteenfold to twentyfold, so that about 100 liters of air are breathed each minute. If breathing stops, a person will lose consciousness after four or five minutes, may suffer brain damage after seven to eight minutes, and will die after ten minutes. A knowledge of the structure and function of the respiratory system, therefore, is of the greatest clinical importance.

Ventilation is the mechanical process that moves air into and out of the lungs. Since air brought to the lungs has a higher oxygen concentration than the blood, oxygen diffuses from air to blood. Carbon dioxide, similarly, moves from the blood to the air within the lungs by diffusion. As a result of this gas exchange, the inspired air contains more oxygen and less carbon dioxide than the expired air. More important, blood leaving the lungs (in the pulmonary veins) contains a higher oxygen and a lower carbon dioxide concentration than the blood delivered to the lungs in the pulmonary arteries. This results from the fact that the lungs function to bring the blood into gaseous equilibrium with the air.

Gas exchange between the air and blood occurs entirely by diffusion through the millions of minute air sacs within the lungs known as *alveoli.* This diffusion occurs very rapidly because there is a high surface area within the lungs and a very short diffusion distance between blood and air.

respiration: L. *re,* back; *spirare,* to breathe
alveolus: L. diminutive of *alveus,* cavity

Figure 23.1. The basic anatomy of the respiratory system.

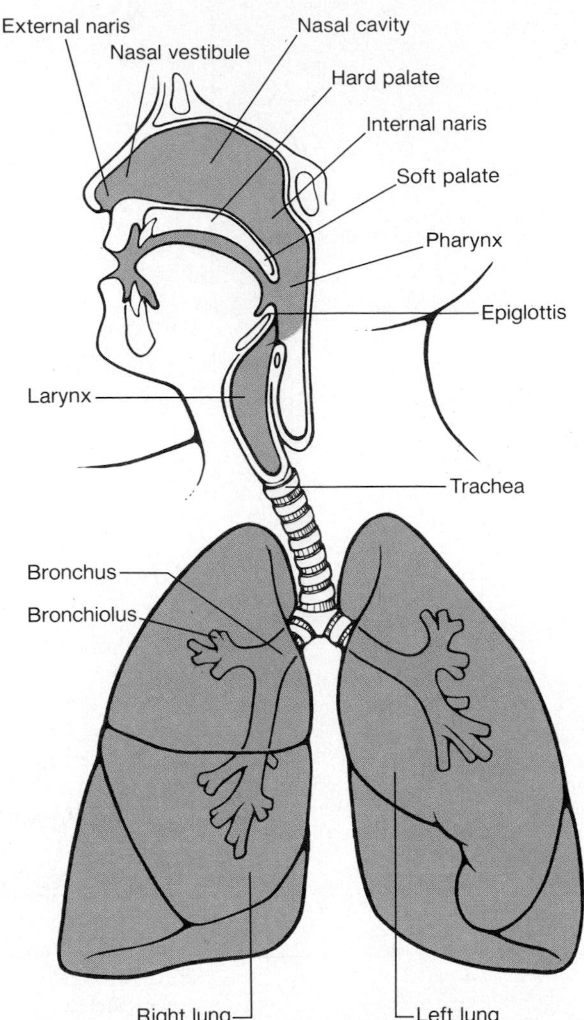

The major passages and structures of the respiratory system (fig. 23.1) are the **nasal cavity, pharynx, larynx,** and **trachea,** and the **bronchi, bronchioles,** and **alveoli** within the **lungs.** The respiratory system is frequently divided into the *conducting division* and the *respiratory division.* The conducting division includes all of the cavities and structures that transport gases to the respiratory division. The respiratory division includes the respiratory bronchioles and the alveoli.

Embryological Development of the Respiratory System

The development of the respiratory system is initiated early in embryonic development and involves both ectoderm and endoderm. Although the structures of the conducting division develop simultaneously with those of the respiratory division, it is best to discuss the development separately because of the different germ layers involved.

Figure 23.2. The development of the upper respiratory system. (*a*) an anterior view of the developing head of an embryo at four weeks, showing the position of a transverse cut depicted in (a_1, a_2, and a_3). (a_2) development at five weeks and (a_3) at five and one-half weeks. (*b*) an anterior view of the developing head of an embryo at six weeks, showing the position of a sagittal cut depicted in (b_1, b_2, b_3, and b_4) at fourteen weeks.

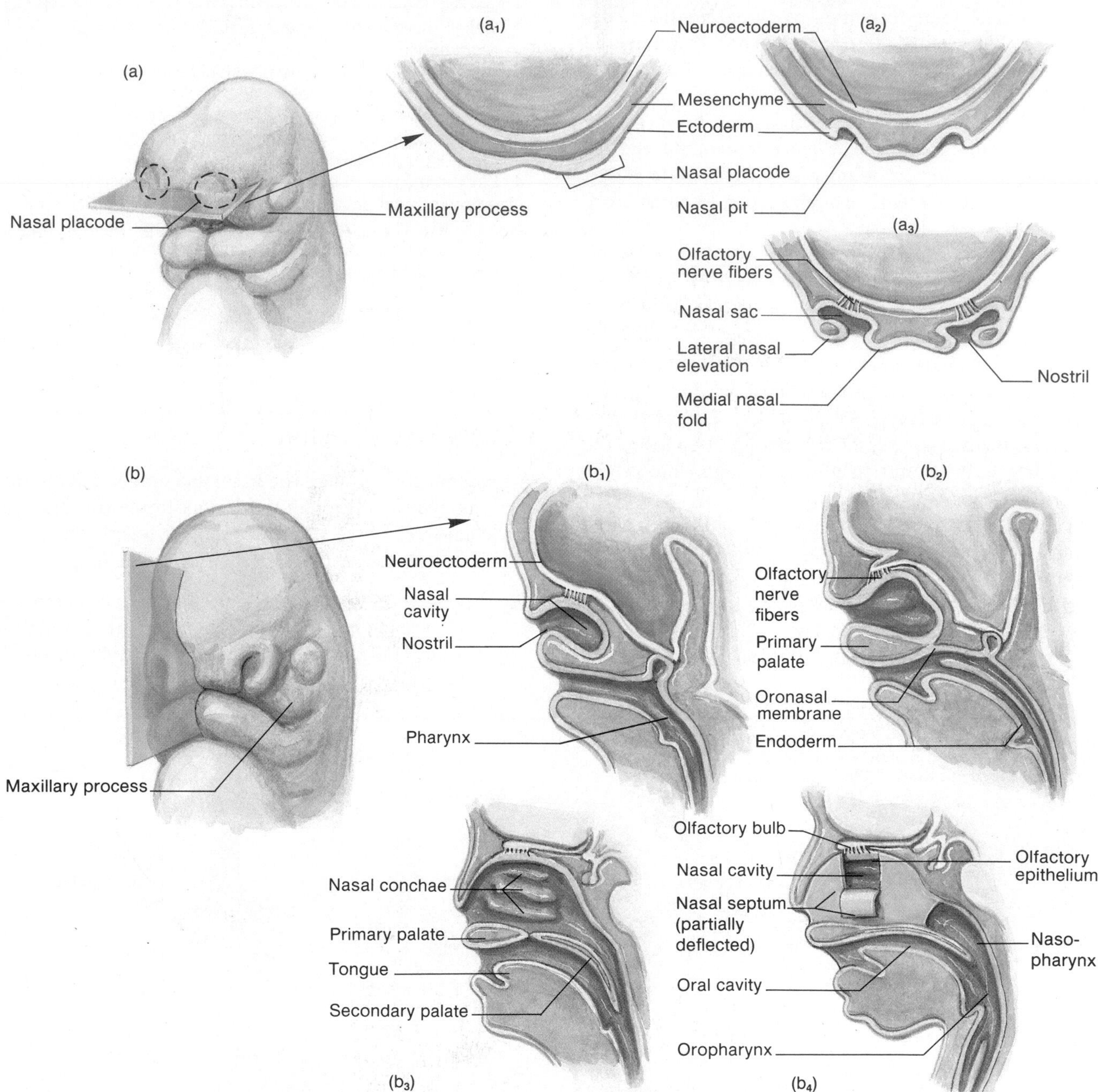

Development of the Conducting Division. *Cephalization (sef"al-i-za'shun)* means that the cephalic, or head, end of an organism structurally and functionally differentiates from the rest of the body. In humans, cephalization is apparent early in development. One of the initial events is the formation of the nasal cavity at three and one-half to four weeks of embryonic life. A region of thickened ectoderm called the **olfactory** (nasal) **placode** *(plak'ōd)* appears on the front and inferior part of the head (fig. 23.2). The placode invaginates to form the **olfactory pit,** which extends posteriorly to connect with the **foregut.** The foregut, derived of endoderm, later develops into the pharynx.

The mouth, or oral cavity, develops at the same time as the nasal cavity, and for a short time there is a thin **oronasal** *(or"o-na'zal)* **membrane** separating the two cavities. This membrane ruptures during the seventh week,

cephalization: Gk. *kephale,* head

and a single, large **oronasal cavity** forms. Shortly after, tissue plates of mesoderm begin to grow horizontally across the cavity. At approximately the same time, a vertical plate develops inferiorly from the roof of the nasal cavity. These plates complete their formation by three months of development. The vertical plate forms the nasal septum, and the horizontal plates form the hard palate.

A *cleft palate* forms when the horizontal plates fail to meet in the midline. This condition can be surgically and cosmetically treated. The more immediate and serious problem facing an infant with a cleft palate is that it may be unable to create enough suction to nurse properly.

Development of the Respiratory Division. The respiratory system begins as a diverticulum *(di''ver-tik'u-lum),* or outpouching, from the ventral surface of endoderm along the lower pharyngeal region (fig. 23.3). This diverticulum forms during the fourth week of development and is referred to as the **laryngotracheal** *(lah-ring''go-tra'ke-al)* **bud.** As the bud grows, the proximal portion forms the trachea and the distal portion bifurcates (splits) into a right and left bronchus.

The buds continue to elongate and split until all the tubular network within the lower respiratory tract is formed (fig. 23.3). As the terminal portion forms air sacs, called **alveoli,** at about eight weeks of development, the supporting lung tissue begins to form. The complete structure of the lungs, however, is not fully developed until about twenty-six weeks of age, so premature infants born prior to this time require special artificial respiratory equipment to live.

1. Define the terms *external respiration* and *internal respiration.*
2. Differentiate between the physical requirements and the functions of the respiratory system.
3. List in order the major passages and structures through which inspired air would pass from the external nares (nostrils) to the alveoli of the lungs.
4. Explain the role of ectoderm and endoderm in the embryological development of the respiratory pathway.
5. At what age might a cleft palate develop? At what age are the lungs sufficiently developed to sustain independent life?
6. Explain the difference between an invagination and a diverticulum in relation to the development of the respiratory system.

Conducting Division

Air is conducted through the nasal and oral cavities to the pharynx and then through the larynx to the trachea and bronchial tree. These structures deliver warmed and humidified air to the respiratory division in the lungs.

Objective 4. List the types of epithelial tissue present in each region of the respiratory tract, and discuss the significance of these differences.

Figure 23.3. The development of the lower respiratory system. (*a*) through (*h*) a sequence of diverticula development: (*a*) through (*d*) occur by the fourth week; (*e*) through (*h*) occur by the eighth week.

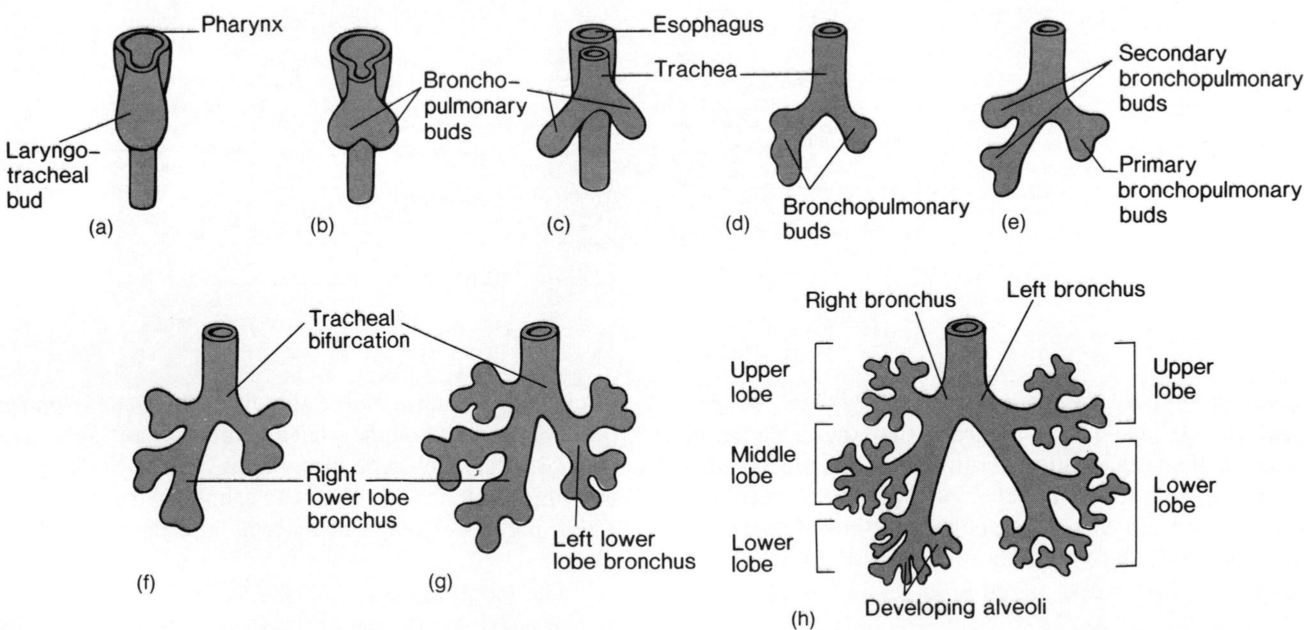

Objective 5. Identify the boundaries of the nasal cavity, and discuss the relationship of the paranasal sinuses to the rest of the respiratory system.

Objective 6. Describe the three regions of the pharynx, and identify the structures located in each.

Objective 7. Discuss the role of the laryngeal region in digestion and respiration.

Objective 8. Identify the anatomical features of the larynx associated with sound production and respiration.

The conducting passages serve to transport air to the respiratory structures of the lungs. The passageways are lined with various types of epithelia to prepare the air properly for utilization. The majority of the conducting passages are held permanently open by muscle or a bony or cartilaginous framework.

Nose

The **nose** includes an external portion that juts out from the face and an internal **nasal cavity** for the passage of air. The external portion of the nose is covered with skin and supported by paired **nasal bones,** forming the bridge, and pliable cartilage, forming the distal portions (figs. 23.4, 23.5). The **septal cartilage** forms the anterior portion of the **nasal septum,** and the paired **lateral cartilages** and **alar cartilages** form the framework around the **nostrils.**

nose: O.E. *nosu,* nose

Figure 23.4. The supporting framework of the external nose.

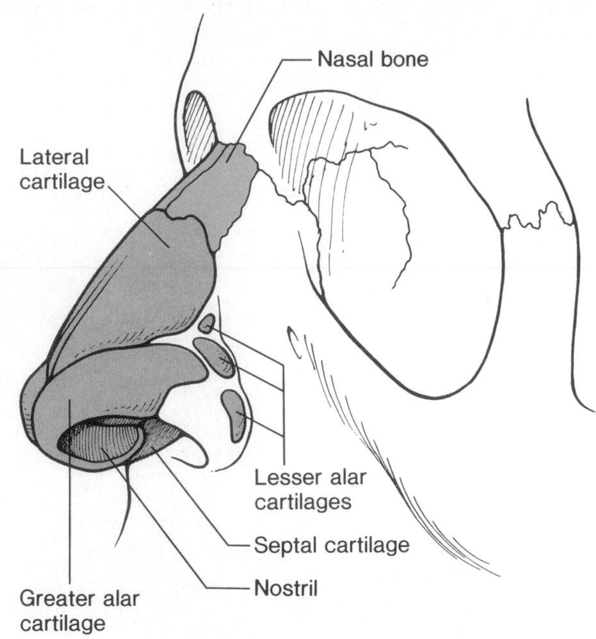

Figure 23.5. A sagittal section through the nose, showing the parts of the nasal septum.

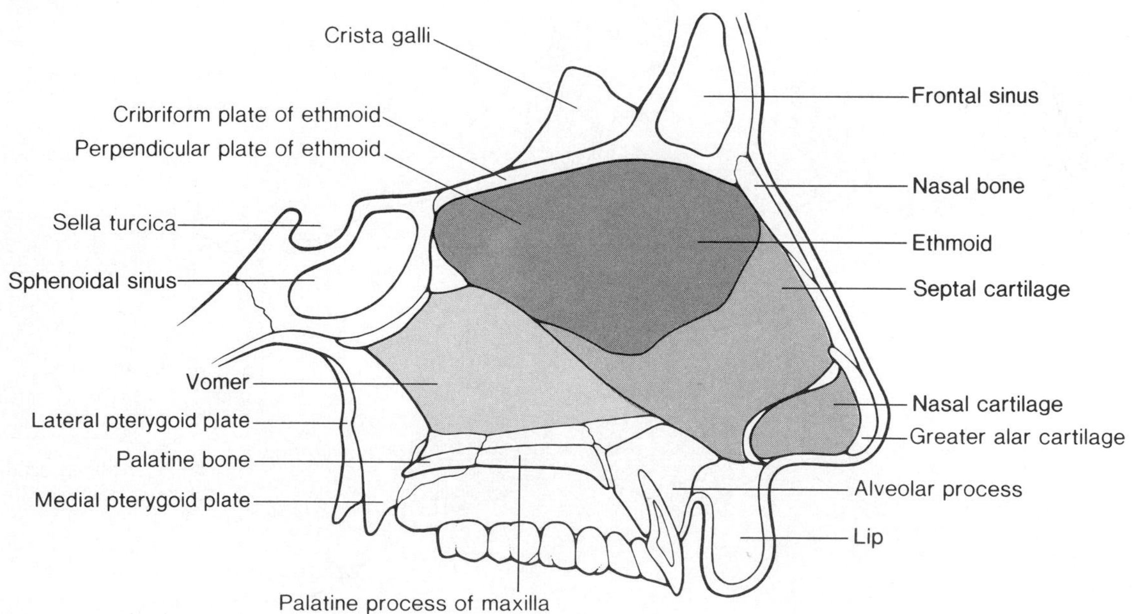

Figure 23.6. The lateral wall of the nasal cavity. There are several openings into the nasal cavity, including the openings of the various paranasal sinuses and the nasolacrimal ducts that drain from the eyes.

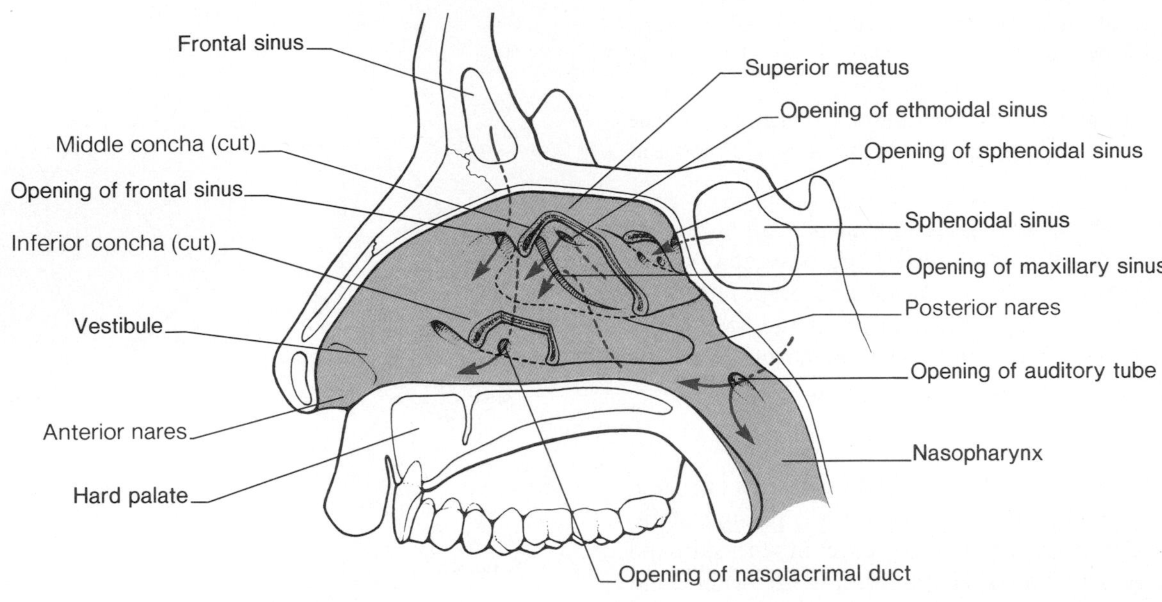

Figure 23.7. The various types of epithelial tissues present throughout the respiratory system.

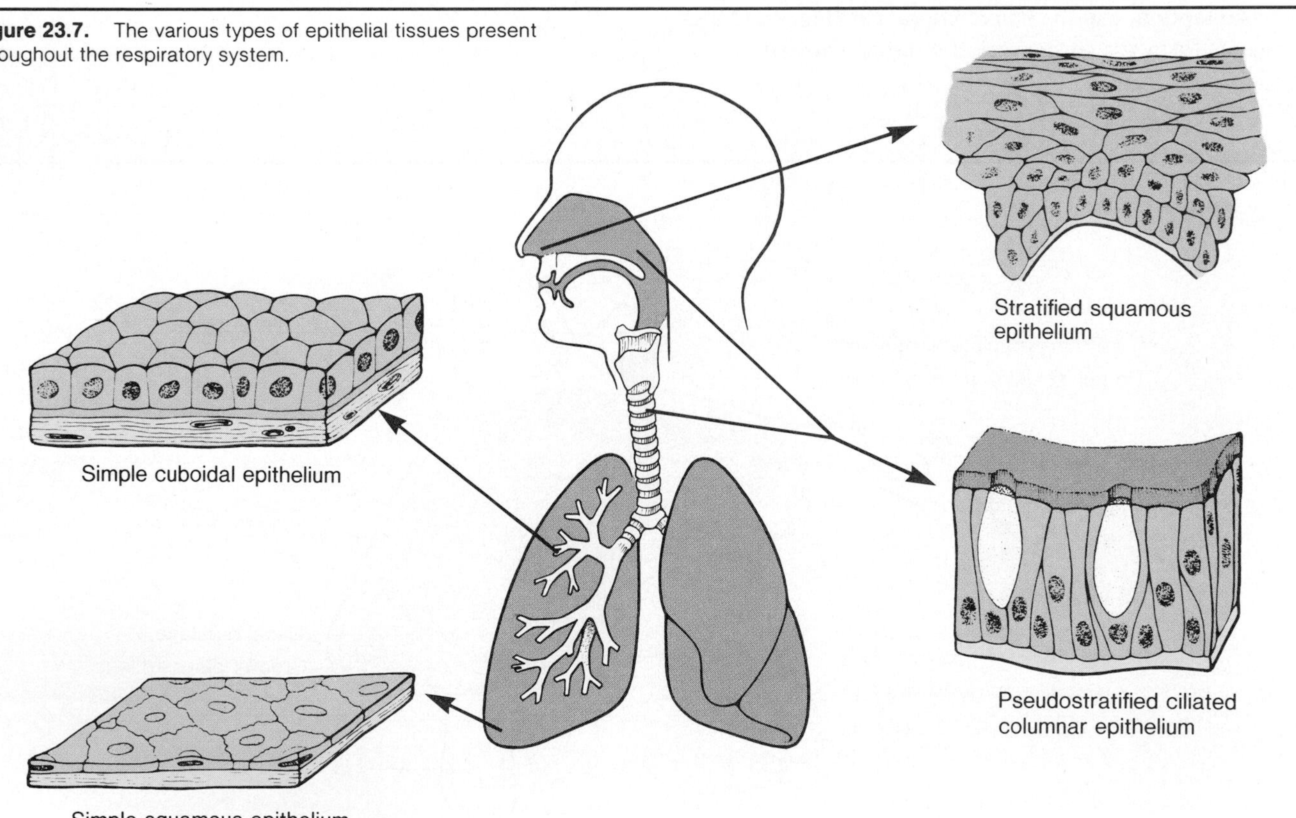

Figure 23.8. A scanning electron micrograph of a bronchial wall showing cilia. In the trachea and bronchi, there are about three hundred cilia per cell that move in a coordinated fashion, moving mucus and trapped particles toward the pharynx, where it can either be swallowed or expectorated.

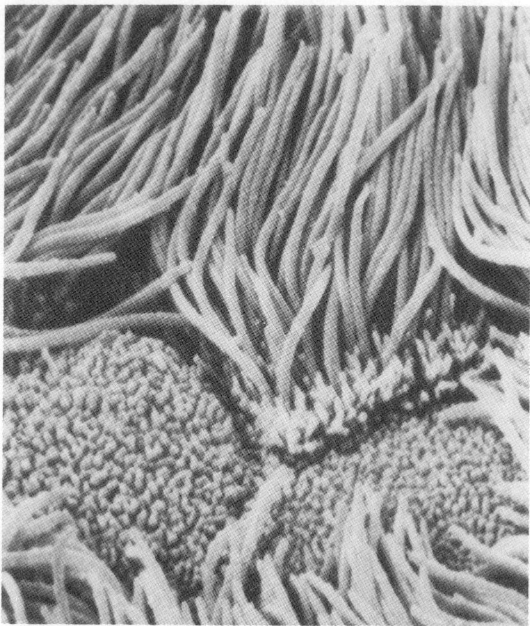

The perpendicular plate of the **ethmoid** bone and the **vomer** bone, together with the septal cartilage, constitute the supporting framework of the **nasal septum,** which divides the nasal cavity into two lateral halves, each of which is referred to as a **nasal fossa.** The vestibule is the anterior expanded portion of the nasal fossa (fig. 23.6). Each nasal fossa opens anteriorly through the **nostril,** or **anterior nares** *(na'rēz),* and communicates posteriorly with the **nasopharynx** through the **choanae** *(ko-a'ne),* or **posterior nares.** The roof of the nasal cavity is formed anteriorly by the frontal bone and paired nasal bones, medially by the cribriform plate of the ethmoid, and posteriorly by the sphenoid bone (see fig. 23.5). The palatine and maxillary bones form the floor of the cavity. On the lateral walls are the **superior, middle,** and **inferior conchae** *(kong'ke)* (turbinate bones). Air passages between the conchae are referred to as **meatuses** (fig. 23.6). The anterior openings of the nasal cavity are lined with stratified squamous epithelium, whereas the conchae are lined with pseudostratified ciliated columnar epithelium (figs. 23.7, 23.8). Mucus-secreting goblet cells are present in great abundance throughout both regions.

nostril: O.E. *nosu,* nose; *thyrel,* hole
choanae: Gk. *choane,* funnel
meatus: L. *meatus,* path

Three functions are associated with the nasal cavity and contents:

1. The nasal epithelium covering the conchae serves to warm, moisten, and cleanse the air. The nasal epithelium is highly vascular and covers an extensive surface area. This is important for warming the air but unfortunately also makes humans susceptible to nosebleeds. Nasal hairs called **vibrissae** *(vi-bris'e),* which often extend from the nostrils, filter macroparticles that might otherwise be inhaled. Fine particles such as dust, pollen, or smoke are trapped along the moist mucous membrane lining the nasal cavity.
2. Olfactory epithelium in the upper medial portion of the nasal cavity is concerned with the sense of smell.
3. The nasal cavity is associated with voice phonetics by functioning as a resonating chamber.

There are several drainage openings into the nasal cavity (see fig. 23.6). The paranasal ducts drain mucus from the paranasal (the frontal, ethmoidal, sphenoidal, and maxillary) sinuses, and the nasolacrimal ducts drain tears from the eyes. An excessive secretion of tears causes the nose to run as the tears drain into the nasal cavity. The auditory tube from the middle ear enters the upper respiratory tract posterior to the nasal cavity in the nasopharynx. With all these accessory connections, it is no wonder that infections can spread so easily from one chamber to another throughout the facial area. To avoid causing damage or spreading infections to other areas, one must be careful not to blow the nose too forcefully.

Paranasal Sinuses

Paired air spaces in certain bones of the skull are called **paranasal sinuses.** These sinuses are named according to the bones in which they are found: thus there are the **maxillary, frontal, sphenoidal,** and **ethmoidal sinuses** (fig. 23.9). Each sinus communicates via drainage ducts within the nasal cavity on its own side (see fig. 23.6). Paranasal sinuses may help to warm and moisten the inspired air. These sinuses are responsible for some sound resonance, but most important, they function to decrease the weight of the skull while giving structural strength.

You can observe your own paranasal sinuses. Face a mirror in a darkened room and shine a small flashlight into your face. The frontal sinuses will be illuminated by directing the light just below the eyebrow. The maxillary sinuses are illuminated by shining the light into the oral cavity and closing your mouth around the flashlight.

vibrissa: L. *vibrare,* to vibrate
sinus: L. *sinus,* bend or curve

Figure 23.9. Paranasal sinuses.

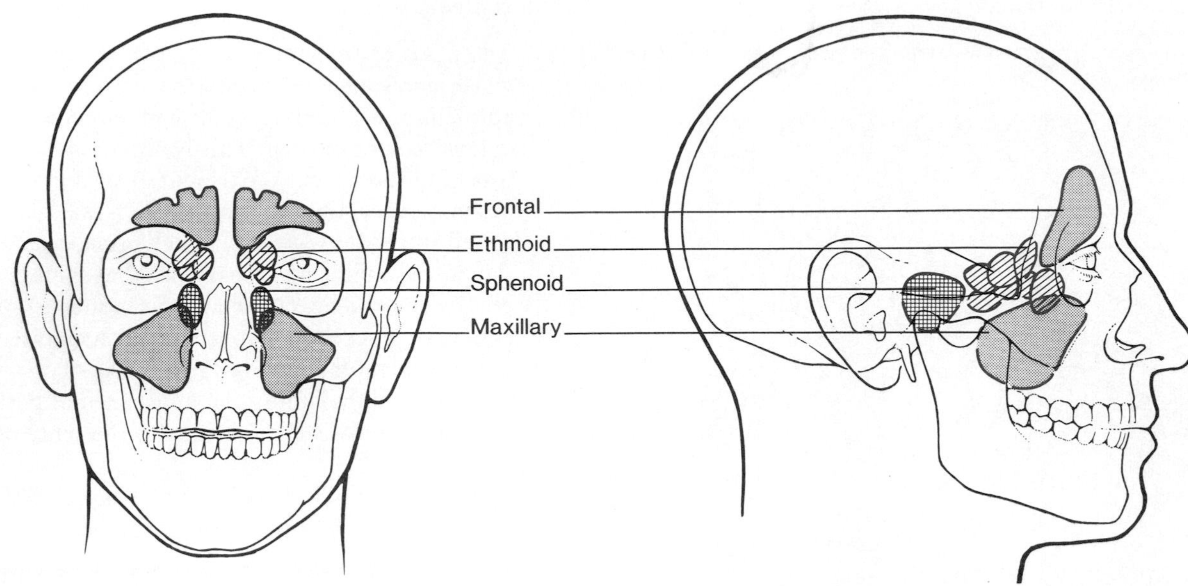

Pharynx

The **pharynx** *(far'inks)* is a funnel-shaped passageway, approximately 13 cm (5 in.) in length, that connects the nasal and oral cavities to the larynx at the base of the skull. The supporting walls of the pharynx are composed of skeletal muscle, and the lumen is lined with a mucous membrane. There are several paired lymphoid organs, called **tonsils,** in the pharynx (fig. 23.10). The pharynx, commonly referred to as the throat or gullet, has both respiratory and digestive functions. It also provides a resonating chamber for certain speech sounds. The pharynx is divided on the basis of location and function into three regions: nasal, oral, and laryngeal.

The **nasopharynx** *(na"zo-far'inks)* has only a respiratory function. It is the uppermost portion of the pharynx, directly posterior to the nasal cavity and above the soft palate. A pendulous **uvula** hangs from the middle, lower border of the soft palate. The paired **auditory,** or **eustachian, tubes** connect the nasopharynx with the middle ear cavities. **Pharyngeal tonsils,** or **adenoids,** are situated in the posterior wall of this cavity. During the act of swallowing, the soft palate and uvula are elevated to block the nasal cavity and prevent food from entering. Occasionally a person may suddenly exhale air (as with a laugh) while in the process of swallowing fluid. If this occurs before the uvula effectively blocks the nasopharynx, fluid will be discharged through the nasal cavity.

The **oropharynx** *(o"ro-far'inks)* is the middle portion of the pharynx between the soft palate and the level of the hyoid bone. The base of the tongue forms the anterior wall of the oropharynx. Paired **palatine tonsils** are located along the posterior lateral wall of the oropharynx, and the **lingual tonsils** are found on the base of the tongue. This portion of the pharynx has both a respiratory and a digestive function.

The **laryngopharynx** *(lah-ring"go-far'inks)* is the lowermost portion of the pharynx. It extends posteriorly from the level of the hyoid bone to the larynx and opens into the esophagus and larynx. It is at the lower laryngopharynx that the respiratory and digestive systems become distinct. Food is directed into the esophagus, whereas air is moved anteriorly into the larynx.

pharynx: L. *pharynx,* throat
tonsil: L. *toles,* goiter or swelling
uvula: L. *uvula,* small grape
adenoid: Gk. *adenoeides,* glandlike

During a physical examination, a physician commonly depresses the patient's tongue and examines the condition of the palatine tonsils. Tonsils are pharyngeal lymphoid organs and tend to become swollen and inflamed after persistent infections. Tonsils have to be surgically removed when they become so overrun with pathogens after repeated infections that they themselves become the source of the infection. The removal of the palatine tonsils is called a *tonsillectomy,* whereas the removal of the pharyngeal tonsils is called an *adenoidectomy.*

Larynx

The **larynx** *(lar'inks),* or voice box, forms the entrance into the lower respiratory system as it connects the laryngopharynx with the trachea. It is positioned in the anterior midline of the neck at the level of the fourth through sixth cervical vertebrae. The larynx has two functions. Its primary function is to prevent food or fluid from entering the trachea and lungs during swallowing and to permit passage of air while breathing. A secondary function is to produce sounds.

In structure, the larynx is shaped like a triangular box (fig. 23.11). It is composed of a framework involving nine cartilages: three are large, single structures and six are

larynx: Gk. *larynx,* upper windpipe

Figure 23.10. A sagittal section of the head, showing the structures of the upper respiratory tract.

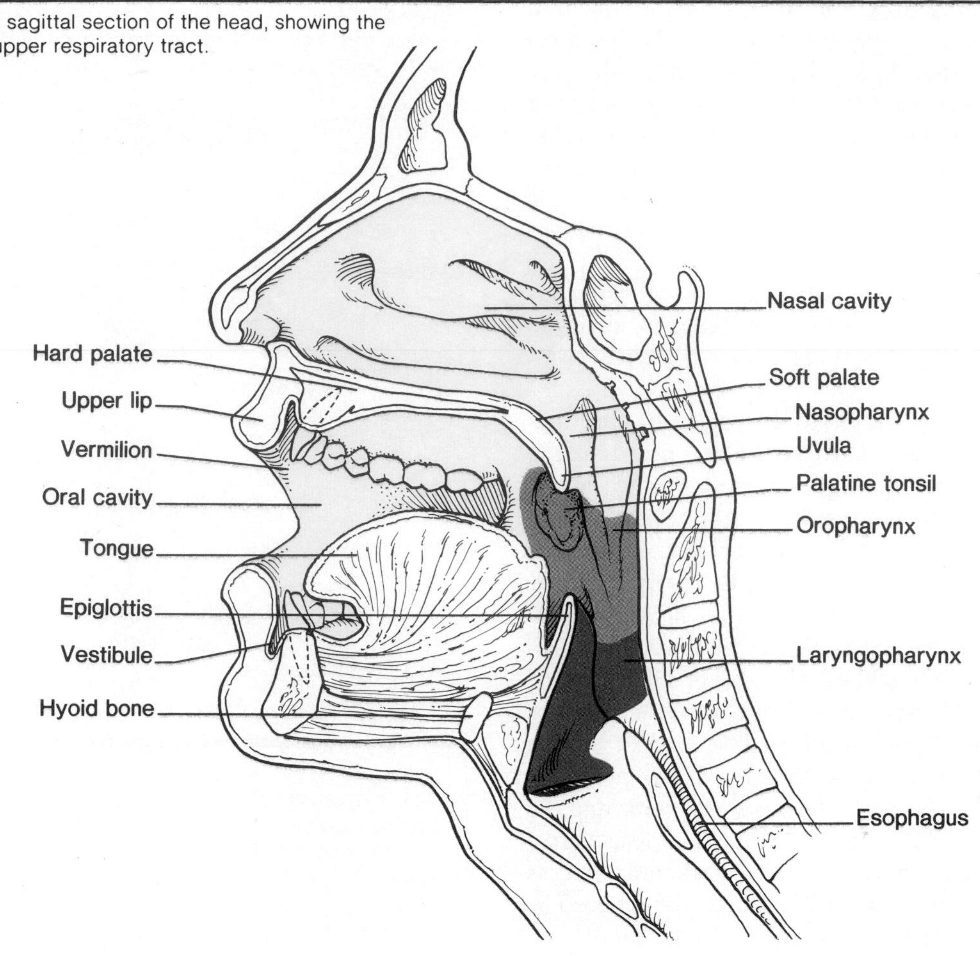

Figure 23.11. The structure of the larynx: (*a*) an anterior view; (*b*) a lateral view; (*c*) a sagittal view.

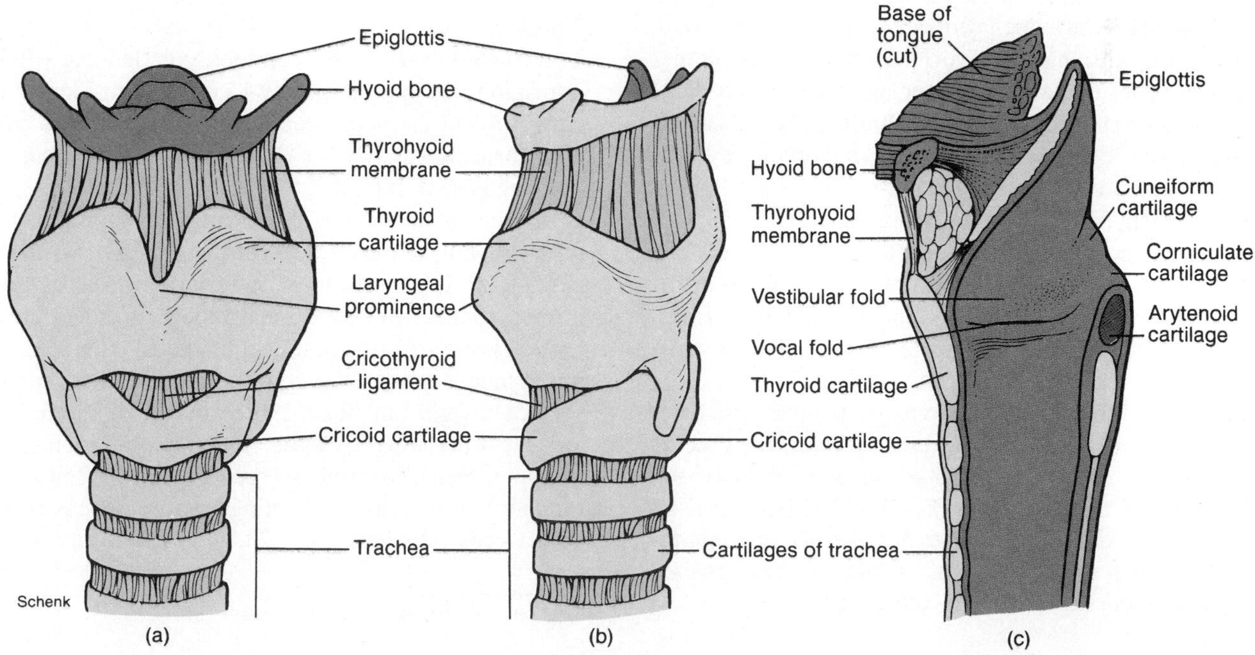

smaller, paired structures. The largest of the unpaired cartilages is the anterior **thyroid cartilage.** The **laryngeal prominence** of the thyroid cartilage, commonly called the "Adam's apple," forms an anterior vertical ridge along the larynx that can be palpated on the midline of the neck. The thyroid cartilage is typically larger and more prominent in males than females because of the effect of male sex hormones on the development of the larynx during puberty.

The spoon-shaped **epiglottis** has a cartilaginous framework. It is behind the root of the tongue and aids in closing the **glottis,** or laryngeal opening, during swallowing.

The lower end of the larynx is formed by the ring-shaped **cricoid** *(kri'koid)* **cartilage.** This third unpaired cartilage connects the thyroid cartilage above and the trachea below. The paired **arytenoid** *(ar"ĕ-te'noid)* **cartilages,** located above the cricoid and behind the thyroid, furnish the attachment of the **vocal cords.** The other paired **cuneiform cartilages** and **corniculate** *(kor-nik'u-lat)* **cartilages** are small accessory cartilages that are closely associated with the arytenoid cartilages (fig. 23.12).

Two pairs of strong connective tissue bands are stretched across the upper opening of the larynx from the thyroid cartilage anteriorly to the paired arytenoid cartilages posteriorly. These are the **true vocal cords** and the **false vocal cords** (figs. 23.12, 23.13). The false vocal cords, or **ventricular folds,** support the true vocal cords and, as their name implies, are not used in sound production. The true vocal cords vibrate in the production of sound.

The **laryngeal muscles** are extremely important in closing the glottis during swallowing and in speech. There are two groups of laryngeal muscles: **extrinsic muscles,** responsible for elevating the larynx during swallowing, and **intrinsic muscles** that when contracted, change the length, position, and tension of the vocal cords. Various pitches are produced as air passes over the altered vocal cords. If the vocal cords are taut, vibration is more rapid and causes a higher pitch. Less tension on the cords produces lower sounds. Mature males generally have thicker and longer vocal cords than females; therefore, the vocal cords of males vibrate more slowly and produce lower pitches. The loudness of vocal sound is determined by the force of the air passed over the vocal cords and the amount of vibration. The vocal cords do not vibrate when a person is whispering.

Sounds originate in the larynx, but other structures are necessary to convert sound into recognizable speech. Vowel sounds, for example, are produced by constriction of the walls of the pharynx. The pharynx, paranasal sinuses, and oral and nasal cavities act as resonating chambers. The final enunciation of words is accomplished through movements of the lips and tongue.

thyroid: Gk. *thyreos,* shieldlike
cricoid: Gk. *krikos,* ring; *eidos,* form
arytenoid: Gk. *arytaina,* ladle- or cup-shaped
cuneiform: L. *cuneus,* wedge-shaped
corniculate: L. *corniculum,* diminutive of *cornu,* horn

Figure 23.12. A superior view of the vocal cords as seen with a laryngoscope: (*a*) vocal cords are taut; (*b*) vocal cords are relaxed and the glottis is opened.

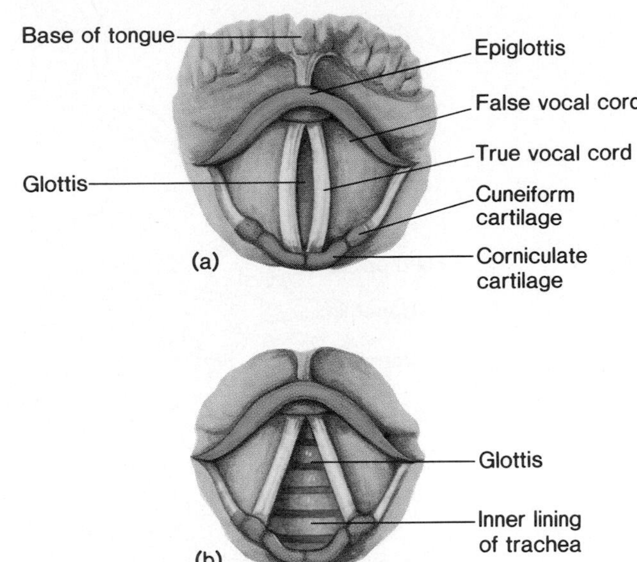

Swallowing is a complex action involving several neurological pathways and various groups of muscles. One of the swallowing responses is the elevation of the larynx to close the glottis against the epiglottis. This movement can be noted by cupping the fingers lightly over the larynx and then swallowing. Food may become lodged within the glottis if it is not closed as it should be. In this case, the *abdominal thrust (Heimlich) maneuver* can be used to prevent suffocation.

Trachea

The **trachea** *(tra'ke-ah),* commonly called the windpipe, is a rigid tube, approximately 12 cm (4½ in.) long and 2.5 cm (1 in.) in diameter, connecting the larynx to the primary bronchi (fig. 23.14). It is positioned anterior to the esophagus as it extends into the thoracic cavity. A series of sixteen to twenty C-shaped rings of hyaline cartilage form the walls of the trachea (figs. 23.15, 23.16). The open part of the C is positioned posteriorly and is covered by fibrous connective tissue and smooth muscle. This soft tissue allows the adjacent esophagus to expand as swallowed food is transported to the stomach. These cartilages provide a rigid but flexible tube in which the lumen, or airway, is permanently open. The mucosa (surface lining of the lumen) consists of pseudostratified ciliated columnar epithelium containing many mucus-secreting **goblet cells** (see figs. 23.7, 23.8). Inhaled dust particles

trachea: L. *trachia,* rough air vessel

Figure 23.13. A photograph in superior view of the larynx, showing the true and false vocal cords and the glottis.

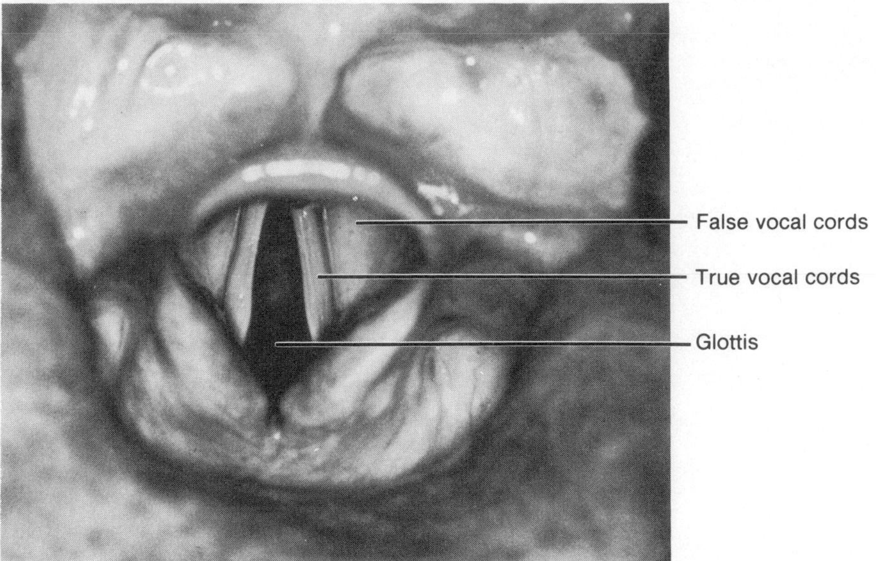

Figure 23.14. An anterior view of the larynx, trachea, and bronchi.

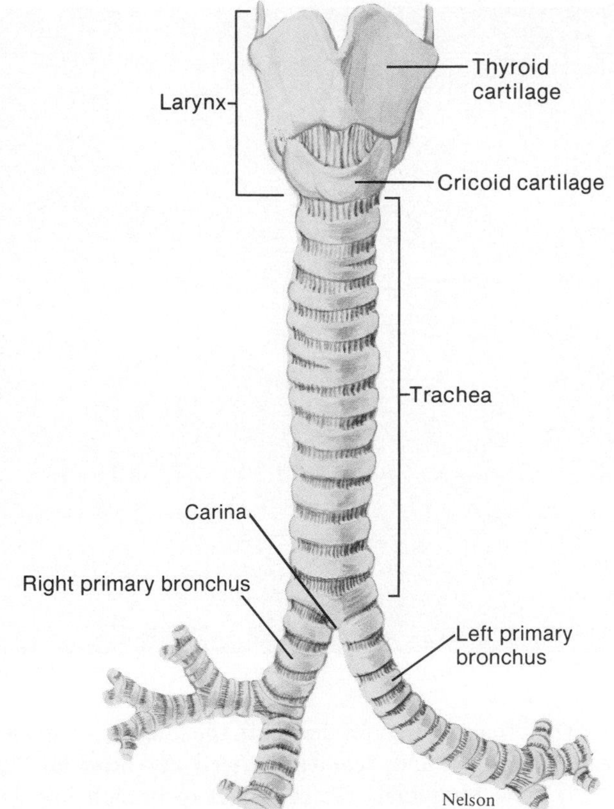

stick to the mucus, and the cilia sweep it upward to the pharynx, where it is removed through a cough reflex. The tracheal bifurcation forming the right and left **bronchi** is reinforced by the **carina** *(kah-ri'nah)*, a keel-like cartilage plate (fig. 23.14).

Figure 23.15. A transverse section of the trachea. The connective tissue portion faces posteriorly in contact with the esophagus.

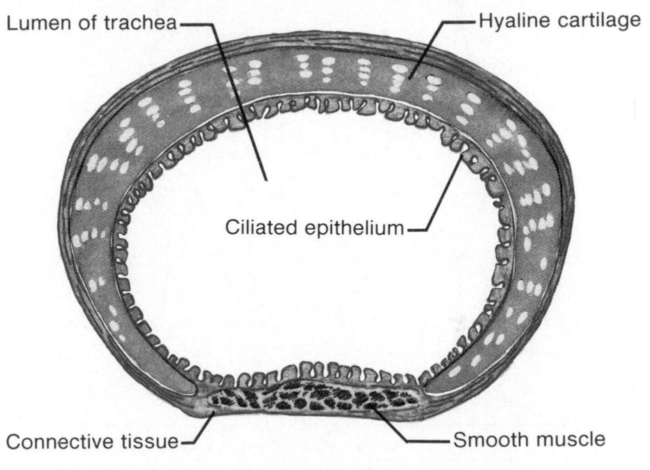

If the trachea becomes occluded through inflammation, excessive secretion, trauma, or aspiration of a foreign object, it may be necessary to create an emergency opening into this tube so that ventilation can still occur. A *tracheotomy* is the process of surgically opening the trachea, and a *tracheostomy* is the procedure of inserting a tube into the trachea to permit breathing and to keep the passageway open (fig. 23.17). A tracheotomy should be performed only by a competent physician, however, as there is great risk of cutting a recurrent laryngeal nerve or carotid artery, which is also in this area.

bronchus: L. *bronchus*, windpipe
carina: L. *carina*, keel

Figure 23.16. Histology of the trachea. (*a*) photomicrograph showing the relationship of the trachea to the esophagus (3X); (*b*) photomicrograph of tracheal cartilage (63X).

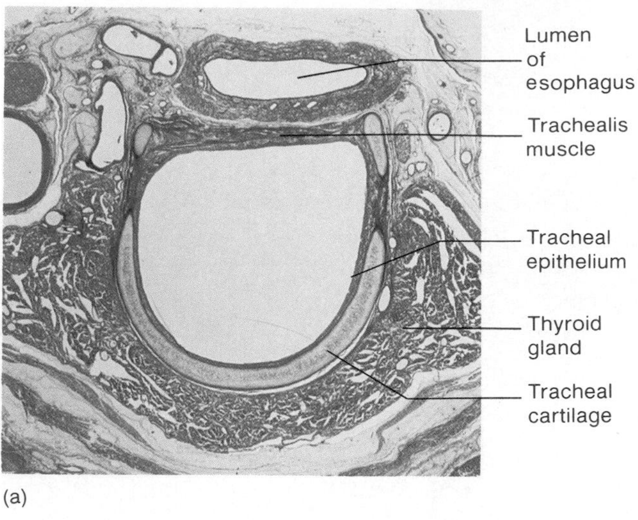

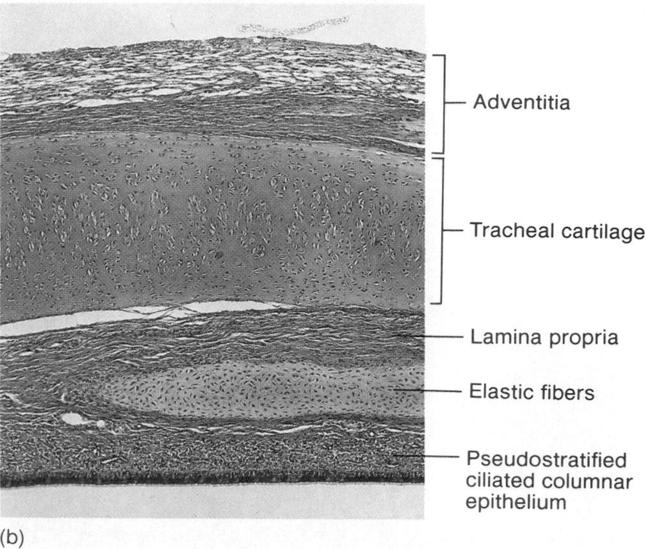

Figure 23.17. The site for a surgical tracheostomy.

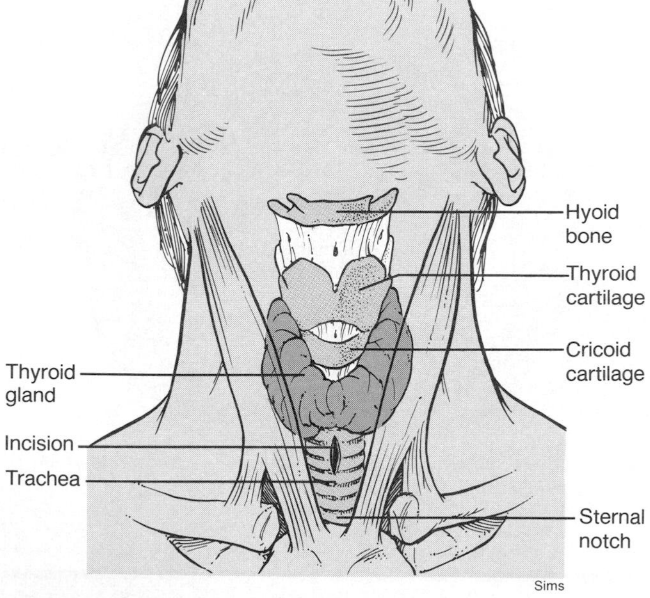

Figure 23.18. A photograph of a plastic cast of the conducting airways from the trachea to the terminal bronchioles.

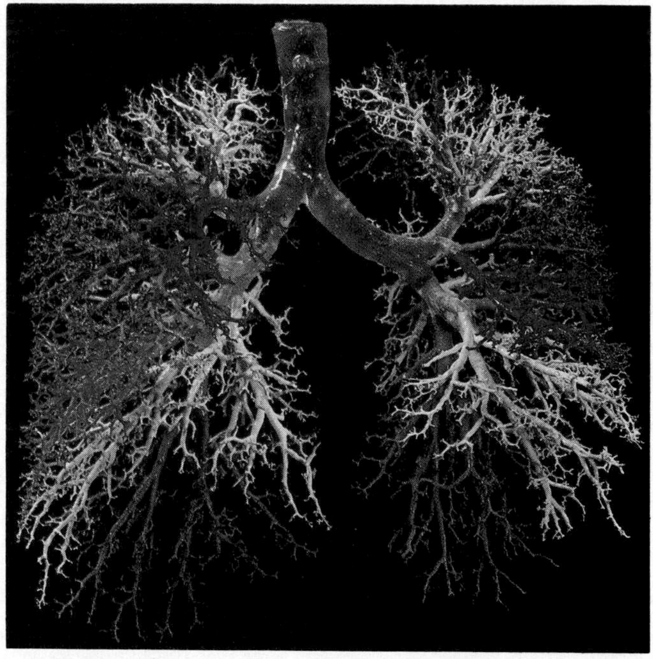

Bronchial Tree

The **bronchial tree** is so named because it is composed of a series of respiratory tubes that branch into progressively narrower tubes as they extend into the lung (fig. 23.18). The trachea bifurcates into a right and left **primary bronchus** *(brong'kus)* at the level of the sternal angle behind the manubrium. Each bronchus has hyaline cartilage rings surrounding its lumen to keep it open as it extends into the lung. Because of the more vertical position of the right bronchus, foreign particles are more likely to lodge here than in the left bronchus.

The bronchus divides deeper in the lungs to form **secondary bronchi** and **segmental (tertiary) bronchi** (fig. 23.19). The bronchial tree continues to branch into yet smaller tubules called **bronchioles** *(brong'ke-ōl)* (fig. 23.20). There is little cartilage in the bronchioles, which contain thick smooth muscle that can constrict or dilate these airways. Bronchioles provide the greatest resistance to air flow in the conducting passages and thus are analogous to the function of arterioles in the vascular system.

Figure 23.19. An anterior view of the lower respiratory system, showing the bronchial tree, alveoli, and lungs.

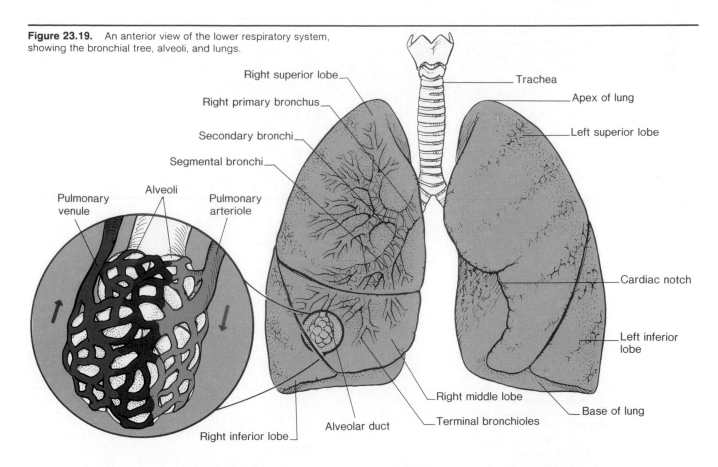

A simple cuboidal epithelium lines the bronchioles rather than a pseudostratified columnar epithelium as in the bronchi (see fig. 23.7). Numerous **terminal bronchioles** mark the end of the air-conducting pathway to the alveoli.

Asthma is a neurological or allergic condition that involves the bronchi. During an asthmatic attack, there is a spasm of the smooth muscles in the lower bronchioles. Because of an absence of cartilage at this level, the air passageways constrict.

A fluoroscopic examination of the bronchi with a radiopaque medium is termed *bronchography.* This technique enables the physician to visualize the bronchial tree by X-ray film. A bronchogram of the lungs is shown in figure 23.20, and figure 23.21 demonstrates the change in the bronchial tree during respiration.

1. List in order the types of epithelia through which inspired air would pass in going through the nasal cavity to the alveoli of the lungs. What is the function of each of these epithelia?
2. What are the functions of the nasal cavity?
3. Identify the structures that make up the nasal septum.
4. Describe the location of the nasopharynx, and list the structures found in this cavity.
5. List the paired and unpaired cartilages of the larynx, and describe the functions of the larynx.
6. Describe the structure of the conducting airways from the trachea to the terminal bronchioles.

Figure 23.20. An anteroposterior bronchogram of the lungs.

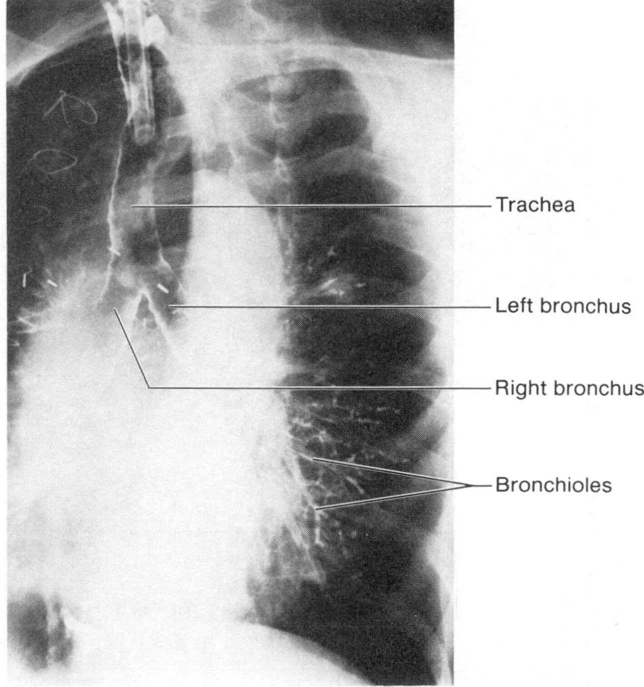

Figure 23.21. A diagram of the change in the bronchial tree during respiration. During inspiration, the bronchi lengthen and increase in diameter, whereas during expiration they return to their original length and diameter.

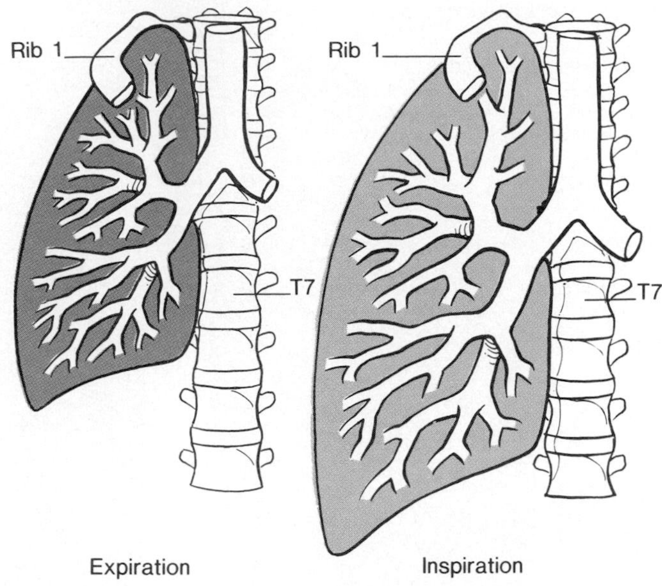

Figure 23.22. The respiratory division of the respiratory system. The respiratory tubes end in minute alveoli, each of which is surrounded by an extensive capillary network.

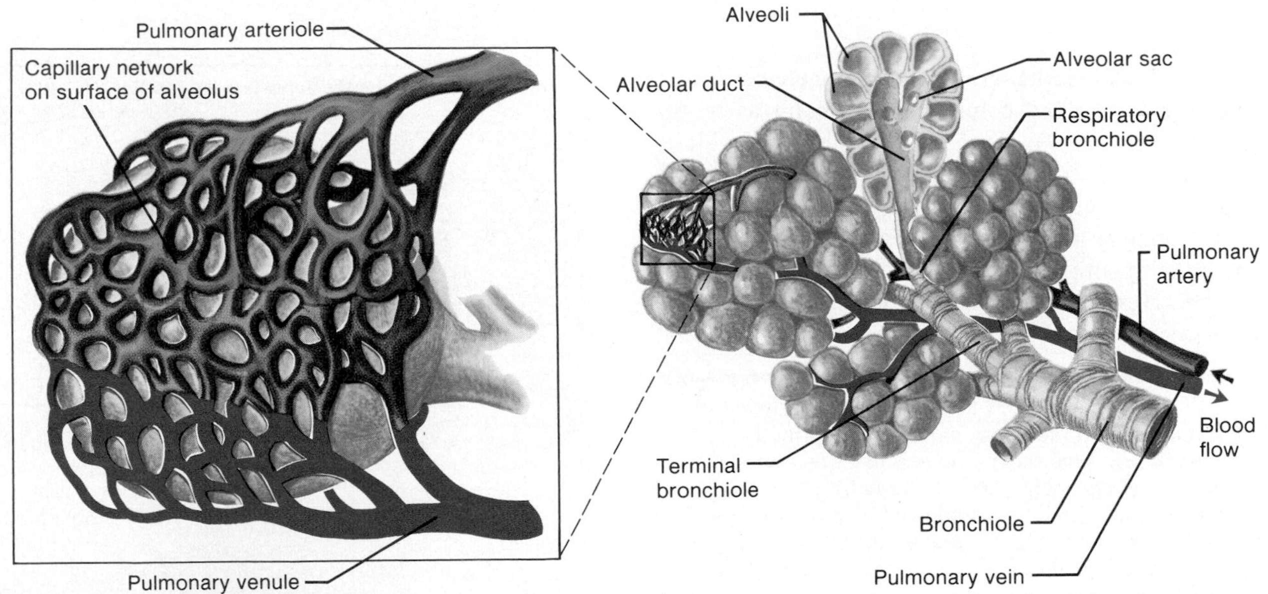

Alveoli, Lungs, and Pleura

Alveoli are the functional units of the lungs where gas exchange occurs. Right and left lungs are separately contained in pleural membranes.

Objective 9. Describe the structure and function of the alveoli.

Objective 10. Describe the surface anatomy of the lungs in relation to the thorax.

Objective 11. Discuss the structural arrangement of the thoracic serous membranes, and explain their functions.

Alveoli

Air from the terminal bronchioles enters the **alveolar ducts,** which contain individual **alveoli** as outpouchings along their length and which, at their ends, open into clusters of alveoli called alveolar sacs (fig. 23.22). The latter three structures comprise the *respiratory division* of the lungs.

Figure 23.23. A diagram showing the relationship between lung alveoli and pulmonary capillaries.

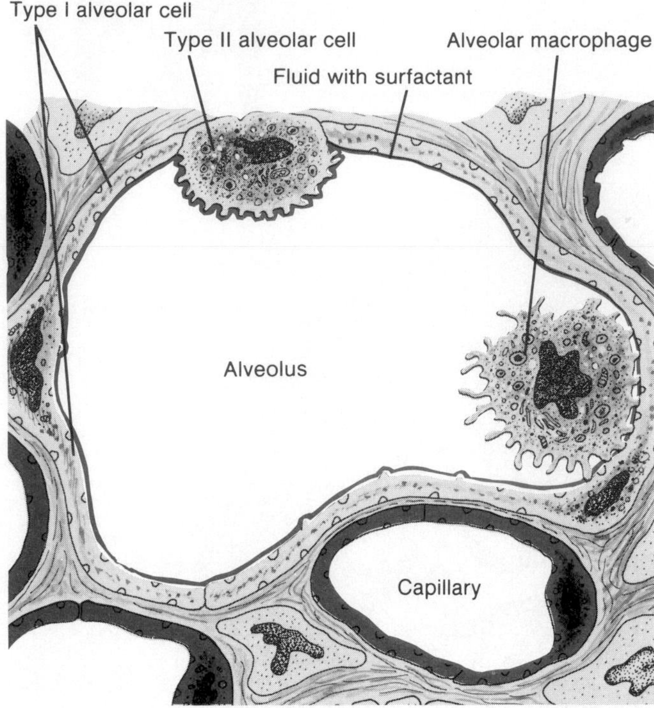

Figure 23.24. A scanning electron micrograph of lung tissue showing alveolar sacs and alveoli.

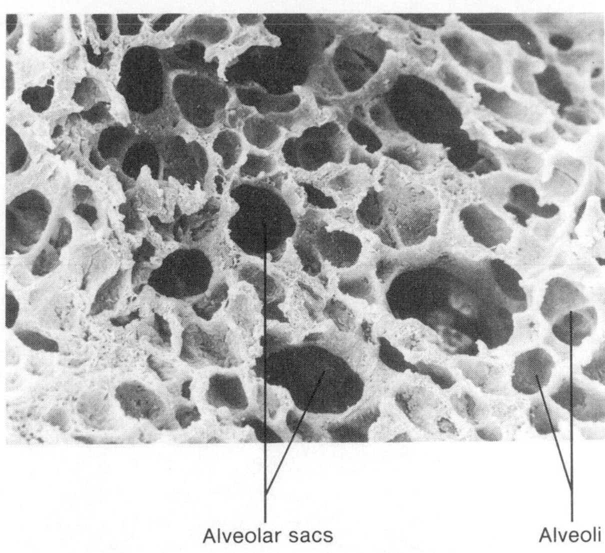

Figure 23.25. Scanning electron micrographs of plastic casts of alveolar capillaries. In the upper photograph, only the capillaries are visible. In the lower photograph, plastic also entered the alveoli so that the relationship between alveoli and capillaries can be seen. (From: *Tissues and Organs: A Text Atlas of Scanning Electron Microscopy* by R. G. Kessel and R. Kardon. W. H. Freeman and Company. © 1979.)

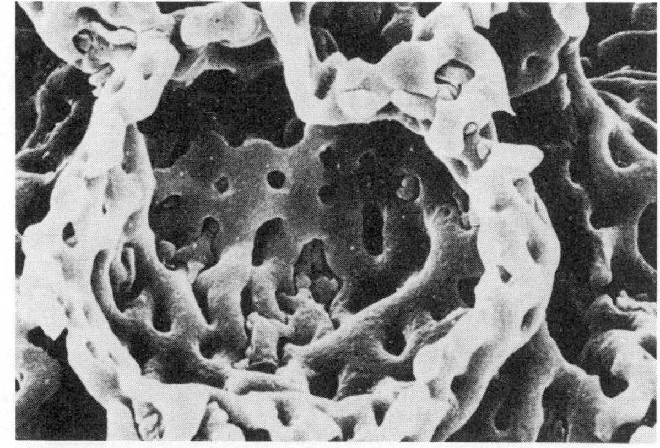

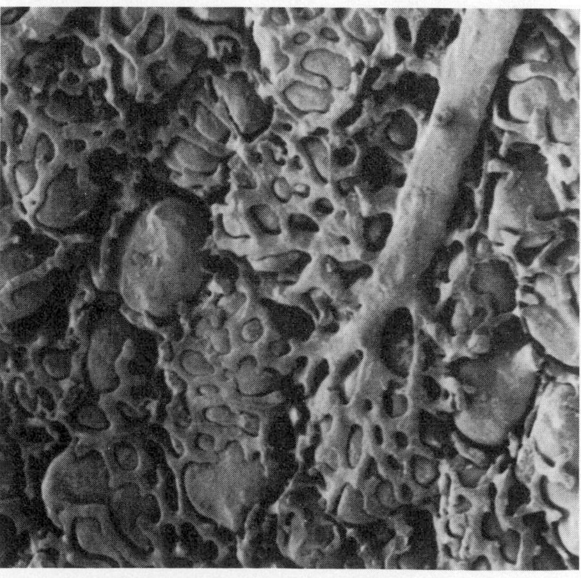

Gas exchange occurs across the walls of the tiny (0.25–0.50 mm in diameter) alveoli, which are therefore the functional units of the respiratory system. The enormous number of these structures (about 350 million per lung) provides a very high surface area (60–80 square meters, or 760 square feet) for the diffusion of gases. The diffusion rate is further increased by the fact that each alveolus *(al-ve'o-lus)* is only one cell layer thick, so that the total "air-blood barrier" is only one alveolar cell and one blood capillary cell across, or about 2 μm. This is an average distance because the type II alveolar cells are thicker than the type I alveolar cells (fig. 23.23). Type I alveolar cells permit diffusion, and type II alveolar cells secrete a lipoprotein substance called *surfactant* that lines the walls of the alveoli.

Alveoli are polyhedral in shape and are usually clustered together, like the units of a honeycomb, in groups called alveolar sacs (fig. 23.24). These clusters usually occur at the ends of alveolar ducts. Individual alveoli also occur as separate outpouchings along the length of the alveolar ducts. Although the distance between each alveolar duct and its terminal alveoli is only about 0.5 mm, these units together comprise most of the mass of the lungs.

The enormous surface area of alveoli and the short diffusion distance between alveolar air and the capillary blood quickly bring the blood into gaseous equilibrium with the alveolar air. This function is further aided by the fact that each alveolus is surrounded by so many capillaries that they form an almost continuous sheet of blood around the alveoli (fig. 23.25).

Figure 23.26. A cross section of the thoracic cavity, showing the mediastinum and pleural membranes.

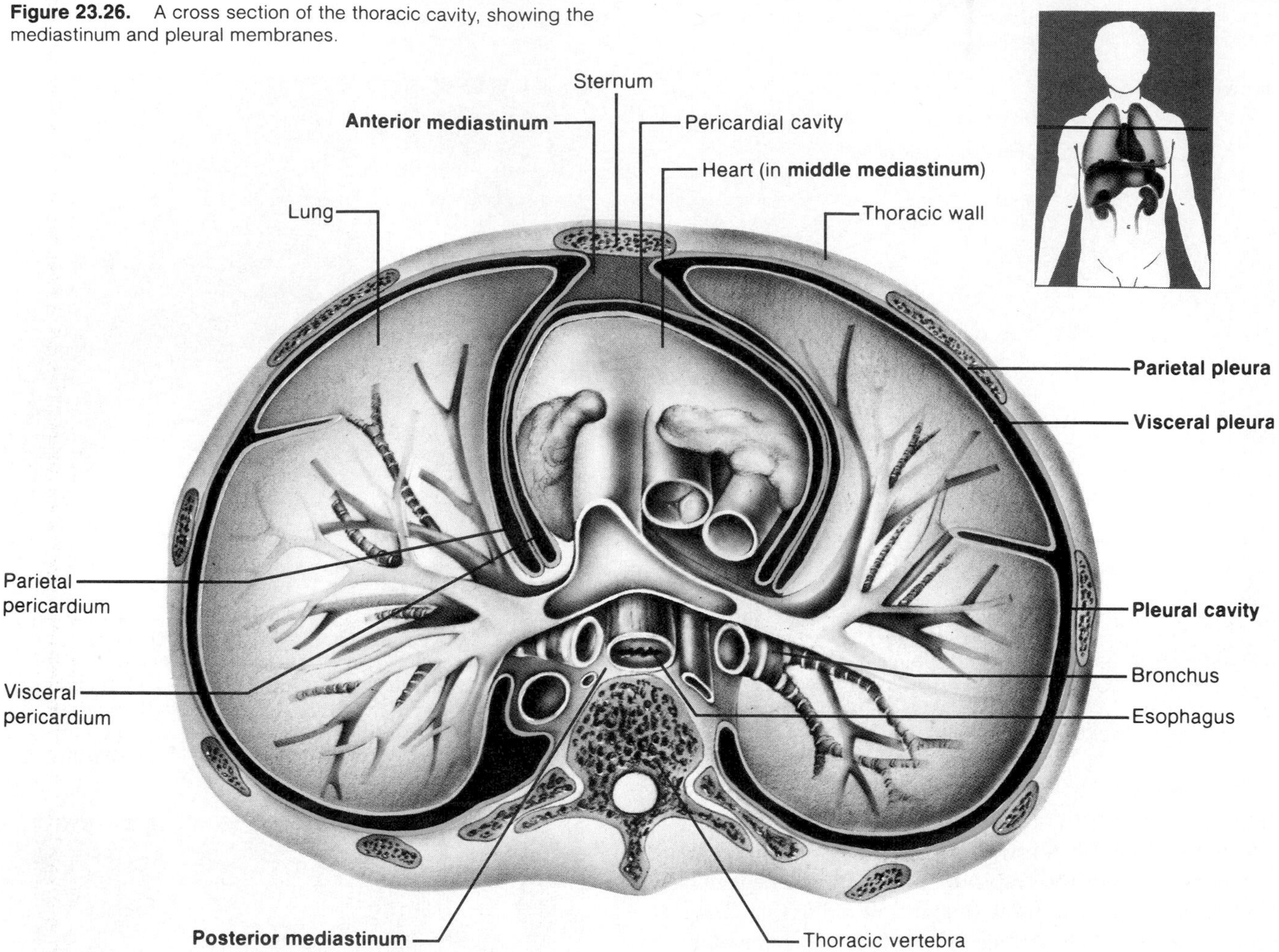

Lungs

The **lungs** are large, spongy, paired organs within the thoracic cavity. They lie against the rib cage anteriorly and posteriorly and extend from the diaphragm to a point just above the clavicles. The lungs are separated from one another by the heart and other structures of the **mediastinum** *(me″de-as-ti′num),* as shown in figure 23.26. The mediastinum is the area between the lungs. All structures of the respiratory system beyond the bronchi, including the bronchial tree and alveoli, are contained in the lungs.

There are some basic similarities and distinct differences between the right and left lung. Each lung presents four borders or surfaces that match the contour of the thoracic cavity. The **mediastinal (medial) surface** of each lung is slightly concave and contains a vertical slit, the **hilum** *(hi′lum),* through which pulmonary vessels, nerves, and bronchi traverse. The inferior border of the lung, called the **base,** is concave as it fits over the convex dome of the diaphragm. Anteriorly, the portion of the lungs that extends above the level of the clavicle is called the **apex,** or **cupola.** Finally, the broad, rounded surface in contact with the membranes covering the ribs is called the **costal surface.**

The left lung is somewhat smaller than the right and has a **cardiac notch** on its medial surface to accommodate the heart. The left lung is subdivided into a **superior lobe** and an **inferior lobe** by a single fissure. The right lung is subdivided by two fissures into three lobes: **superior, middle,** and **inferior** (see fig. 23.19). Each lobe of the lung is divided into many small lobules, which in turn contain the alveoli. Lobular divisions of the lungs comprise specific bronchial segments. The right lung contains ten bronchial segments, and the left lung contains eight (fig. 23.27).

The lungs of a newborn are pink in color but may become discolored in an adult as a result of smoking or air pollution. Smoking not only discolors the lungs, it may also cause deterioration of the alveoli. *Emphysema (em″fi-se′mah)* and *lung cancer* are diseases that have been linked to heavy smoking. If a person moves to a less polluted environment or gives up smoking, the lungs will become more pinkish and healthy.

mediastinum: L. *mediastinus,* intermediate
hilum: L. *hilum,* a trifle (having little significance)

cupola: L. *cupula,* diminutive of *cupa,* dome, tub

Figure 23.27. Lobes, lobules, and bronchopulmonary segments of the lungs.

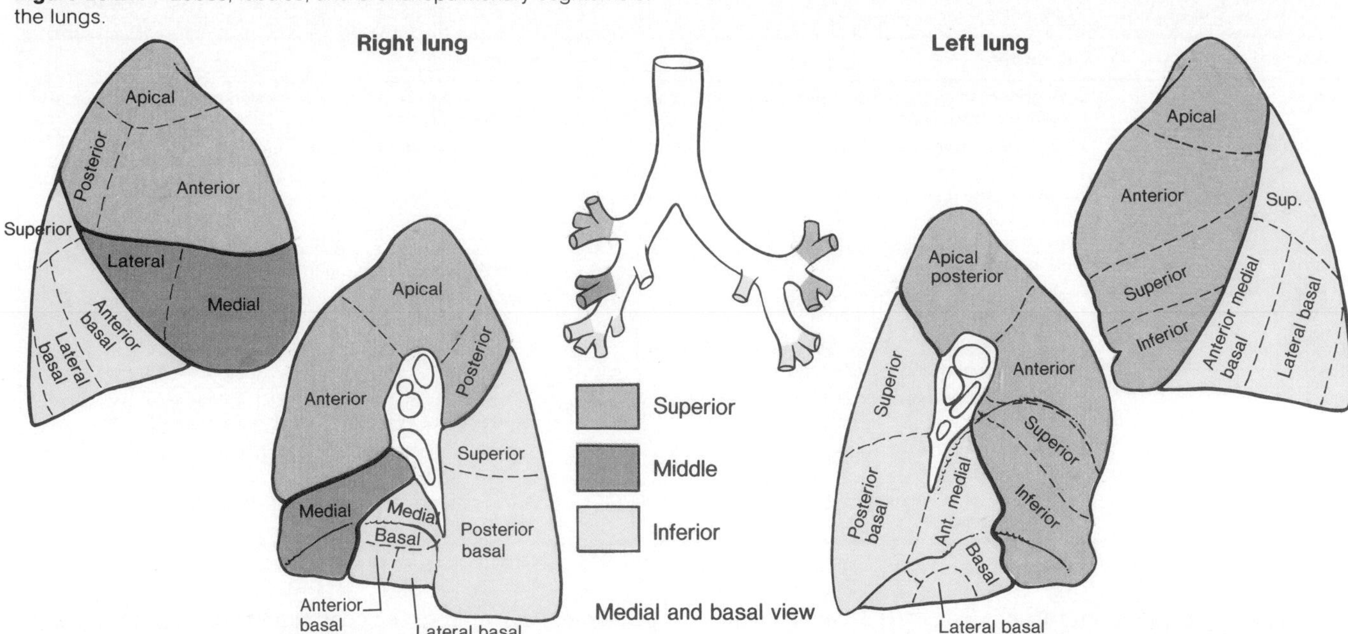

Pleurae

Pleurae are serous membranes surrounding the lungs (figs. 23.26, 23.28). The pleura of each lung is composed of both a visceral and parietal portion. The **visceral pleura** *(ploo'rah)* adheres to the outer surface of the lung and extends into each of the interlobar fissures. The **parietal pleura** lines the thoracic walls and the thoracic surface of the diaphragm. A continuation of the parietal pleura around the heart and between the lungs forms the boundary of the mediastinum. Between the visceral and parietal pleurae is a moistened space called the **pleural cavity.** An inferiorly extending reflection of the pleural layers around the roots of each lung is called the **pulmonary ligament.** The pulmonary ligaments help to support the lungs.

The arrangement of the pleurae serve three functions. First, a small amount of fluid within the pleural cavities acts as a lubricant to allow the lungs to slide along the chest wall. Second, the pressure in the pleural cavity is lower than the pressure in the lungs, which is needed for ventilation. The third function of the pleurae is the effective separation of the thoracic organs. Although the serous membrane is composed of simple squamous epithelium and fibrous connective tissue, it provides tight compartments for each major thoracic organ.

There are four distinct compartments in the thoracic cavity. A pleural cavity surrounds each lung, the heart is situated in the pericardial cavity, and the esophagus, thoracic duct, major vessels, various nerves, and portions of the respiratory tract are located in the mediastinum. This *compartmentalization* has a protective value because infections are usually confined to one compartment and damage to one organ will not usually involve another.

pleura: Gk. *pleura,* side or rib
pulmonary: Gk. *pleumon,* lung

Figure 23.28. The position of the lungs and associated pleurae.

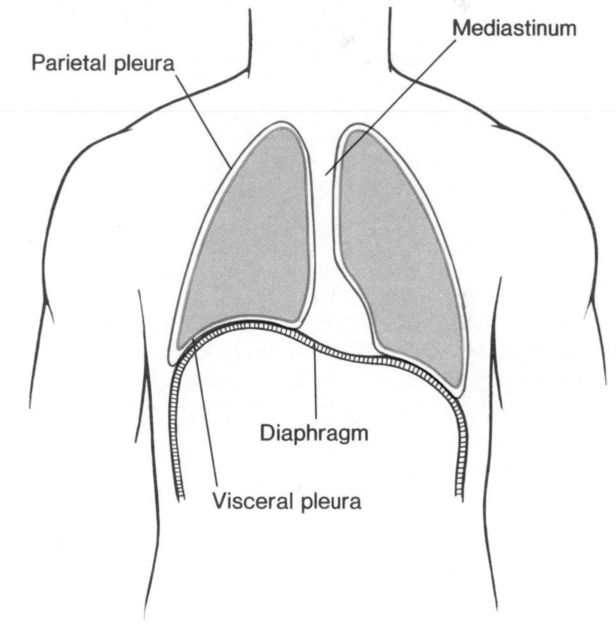

Pleurisy, for example, which is an inflamed pleura, is generally confined to one side. A penetrating injury to one side, like a knife wound, might cause one lung to collapse but not the other.

The moistened serous membranes of the visceral and parietal pleurae are normally against each other, so that the lungs are stuck to the thoracic wall. The pleural cavity (intrapleural space) between the two moistened membranes contains only a thin layer of fluid secreted by the pleural membranes. The pleural cavity in a healthy, living

Table 23.1 Major structures of the respiratory system

Structure	Description	Function
Nose	Jutting external portion that is part of the face, plus internal nasal cavity	Warms, moistens, and filters inhaled air as it is conducted to the pharynx
Paranasal sinuses	Air spaces in certain facial bones	Produce mucus, provide sound resonance, lighten the skull
Pharynx	Chamber connecting oral and nasal cavities to the larynx	Passageway for air into the larynx and for food into the esophagus
Larynx	Voice box; short passageway that connects the pharynx to the trachea	Passageway for air; produces sound; prevents foreign materials from entering the trachea
Trachea	Flexible, tubular connection between the larynx and bronchial tree	Passageway for air; pseudostratified ciliated columnar epithelium cleanses the air
Bronchial tree	Bronchi and branching bronchioles in the lung; tubular connection between the trachea and alveoli	Passageway for air; continued cleansing of air
Alveoli	Microscopic, membranous air sacs within the lungs	Functional units of respiration: site of gaseous exchange between the respiratory and circulatory systems
Lungs	Major organs of the respiratory system; located in pleural cavities in the thorax	Contain bronchial trees, alveoli, and associated pulmonary vessels
Pleurae	Serous membranes covering the lungs and lining the thoracic cavity	Compartmentalize, protect, and lubricate the lungs

organism is thus potential rather than real; it can become real only in abnormal situations when air enters the intrapleural space. Since the lungs normally remain against the thoracic wall, they get larger and smaller together with the thoracic cavity during respiratory movements.

The functions of the major organs of the respiratory system, along with a description of each, is presented in table 23.1.

1. Describe the structure of the respiratory division of the lungs, and explain how this structure aids in gas exchange.
2. Compare the structure of the right and left lung.
3. Describe the structure and location of the mediastinum, and list the organs it contains.
4. Describe the location of the visceral pleura, parietal pleura, and pleural cavity. Explain the functional significance of the pleural membranes.

Physical Aspects of Ventilation

The movement of air into and out of the lungs occurs as a result of pressure differences induced by changes in lung volumes. Ventilation is thus influenced by the physical properties of the lungs, including their compliance, elasticity, and surface tension.

Objective 12. Define intrapulmonary and intrapleural pressures, and describe how they change during breathing.

Objective 13. Describe Boyle's law, and explain how it relates to lung function.

Objective 14. Define the terms *compliance, elasticity,* and *surface tension,* and describe how they affect lung function.

Movement of air from the conducting zone to the terminal bronchioles occurs as a result of the pressure difference between the two ends of the airways. Air flow through bronchioles, like blood flow through blood vessels, is directly proportional to the pressure difference and inversely proportional to the frictional resistance to flow. The pressure differences in the pulmonary system are induced by changes in lung volumes. Ventilation is thus influenced by the physical properties of the lungs, including their compliance, elasticity, and surface tension. The wet, serous membranes of the visceral and parietal pleurae are normally against each other, so that the lungs are stuck to the chest wall in the same manner that two wet pieces of glass stick to each other. The *intrapleural space* between the two wet membranes contains only a thin layer of fluid secreted by the pleural membranes.

Intrapulmonary and Intrapleural Pressures

Air enters the lungs during inspiration because the atmospheric pressure is greater than the **intrapulmonary,** or **intra-alveolar pressure.** Since the atmospheric pressure does not usually change, the intrapulmonary pressure must fall below atmospheric pressure to cause inspiration. A pressure below that of the atmosphere is called a *subatmospheric pressure,* or *negative pressure.* During quiet inspiration, for example, the intrapulmonary pressure may become 3 mm Hg less than the pressure of the atmosphere. This subatmospheric pressure is commonly shown with a negative sign, as −3 mm Hg. Expiration, conversely, occurs when the intrapulmonary pressure is greater than the atmospheric pressure. During quiet expiration, for example, the intrapulmonary pressure may rise to at least +3 mm Hg over the atmospheric pressure.

The lack of air in the intrapleural space produces a subatmospheric **intrapleural pressure,** which is lower than the intrapulmonary pressure (table 23.2). There is thus a

Table 23.2 Intrapulmonary and intrapleural pressures in normal, quiet breathing, and the transpulmonary pressure (intrapulmonary minus intrapleural pressure) acting to expand the lungs

	Inspiration	Expiration
Intrapulmonary pressure (mm Hg)	−3 (to zero)	+3 (to zero)
Intrapleural pressure (mm Hg)	−6	−3
Transpulmonary pressure (mm Hg)	+3	+6

Note: Pressures indicate mm Hg below or above atmospheric pressure. Intrapleural pressure is normally always negative (subatmospheric).

Figure 23.29. The volume of air inspired as a function of the transpulmonary pressure. The slope of each line (change in volume per unit change in pressure) in the linear regions is called the *compliance* and is a measure of the distensibility of the lungs.

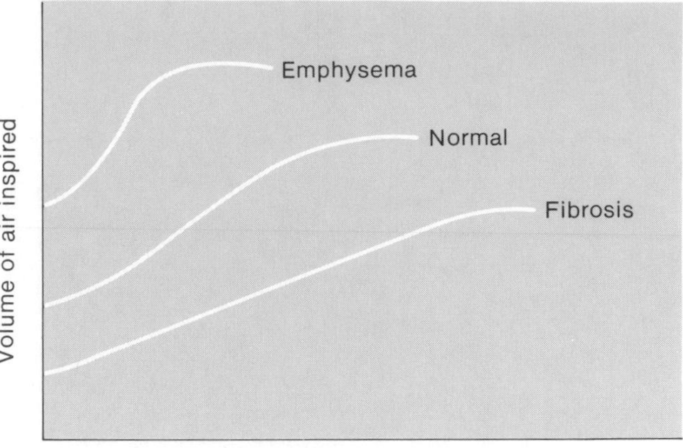

pressure difference across the wall of the lung—called the **transpulmonary pressure**—which is the difference between the intrapulmonary pressure and the intrapleural pressure. Since the pressure within the lungs (intrapulmonary pressure) is greater than that outside the lungs (intrapleural pressure), the difference in pressure (transpulmonary pressure) acts to expand the lungs as the thoracic volume expands during inspiration.

Boyle's Law. Changes in intrapulmonary pressure occur as a result of changes in lung volume. This follows from **Boyle's law,** which states that *the pressure of a gas is inversely proportional to its volume.* An increase in lung volume during inspiration decreases intrapulmonary pressure to subatmospheric levels; air therefore goes in. A decrease in lung volume raises the intrapulmonary pressure above that of the atmosphere, thus pushing air out. These changes in lung volume occur as a consequence of changes in thoracic volume produced by muscle contraction.

Physical Properties of the Lungs

In order for inspiration to occur the lungs must be able to expand when stretched; they must have high *compliance.* In order for expiration to occur, the lungs must get smaller when this stretching force is released; they must have *elasticity.* The tendency to get smaller is also aided by *surface tension* forces within the alveoli.

Compliance. The lungs are very distensible—they are, in fact, about one hundred times more distensible than a toy balloon. Another term for distensibility is **compliance** *(kom-pli'ans),* which is the change in lung volume per change in transpulmonary pressure. A given transpulmonary pressure, in other words, will cause greater or lesser expansion, depending on the compliance of the lungs.

The compliance of the lungs is reduced whenever there is more resistance to distension. If the lungs were filled with concrete (as an extreme example), a given transpulmonary pressure would produce no increase in lung volume and no air would enter; the compliance would be zero. The infiltration of lung tissue with connective tissue proteins, a condition called *pulmonary fibrosis,* similarly decreases lung compliance (fig. 23.29). In *emphysema,* in which alveolar tissue is destroyed, the lungs are less resistant to distension and have a greater compliance (if no fibrosis is present).

Elasticity. The term **elasticity** refers to the tendency of a structure to return to its initial size after being distended. The lungs are very elastic, due to a high content of elastin proteins, and resist distension. Since the lungs are normally stuck to the chest wall, they are always in a state of elastic tension. This tension increases during inspiration, when the lungs are stretched, and is reduced by elastic recoil during expiration. The elasticity of the lungs and of other thoracic structures thus aids in pushing the air out during expiration.

The elastic nature of lung tissue is revealed when air enters the intrapleural space (as a result of an open chest wound, for example). This condition is called a *pneumothorax (nu"mo-tho'raks),* which is shown in figure 23.30. As air enters the intrapleural space, the intrapleural pressure rises until it is equal to the atmospheric pressure. When the intrapleural pressure is the same as the intrapulmonary pressure, the lung can no longer expand. Not only does the lung not expand during inspiration, it actually collapses away from the chest wall as a result of elastic recoil. Fortunately, a pneumothorax usually occurs only in one lung because each lung is in a separate pleural compartment.

Boyle's law: from Robert Boyle, British physicist, 1627–1691

pneumothorax: Gk. *pneumon,* spirit (breath); L. *thorax,* thorax, chest

Figure 23.30. A pneumothorax of the right lung. The right side of the thorax appears uniformly dark because it is filled with air; the space between the ribs is also greater than on the left due to release from the elastic tension of the lungs. The left lung appears denser (less dark) because of shunting of blood from the right to the left lung.

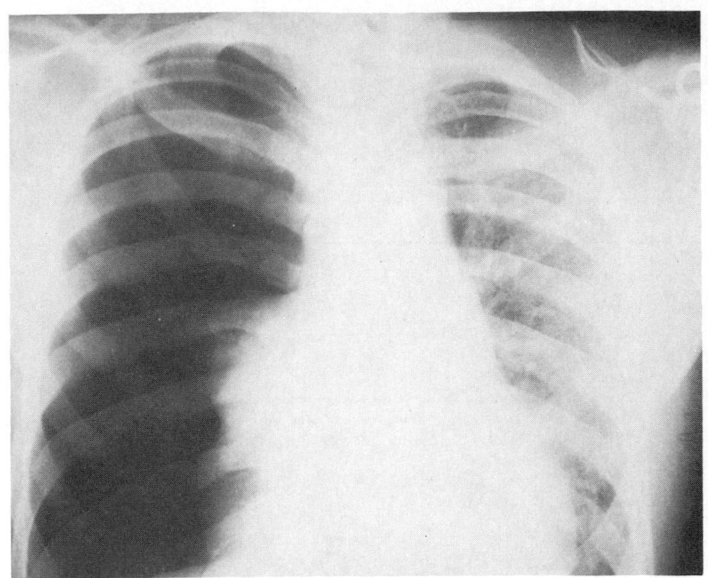

Figure 23.31. Water molecules at the surface have a greater attraction for other water molecules than for air. The surface molecules are thus attracted to each other and pulled tightly together by the attractive forces of water underneath. This produces surface tension.

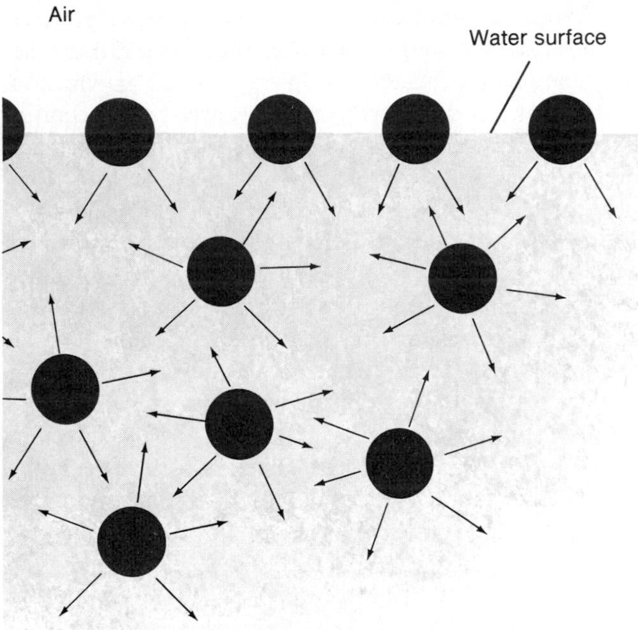

Figure 23.32. According to the Law of LaPlace, the pressure created by surface tension should be greater in the smaller alveolus (*right*) than in the larger alveolus (*left*). This implies that (without surfactant) smaller alveoli would collapse and empty their air into larger alveoli.

Law of LaPlace

$$P = \frac{2 \times T}{r}$$

Bronchioles

Alveoli

r = 1

r = 2

$$P = \frac{2 \times T}{2} \qquad P = \frac{2 \times T}{1}$$

$$P = T \qquad P = 2T$$

Surface Tension. The forces that act to resist distension include elastic resistance and **surface tension** forces exerted by fluid in the alveoli. Although the alveoli are relatively dry, they do contain a very thin film of fluid, much like soap bubbles. Surface tension is created by the fact that water molecules at the surface are attracted more to other water molecules than to air. As a result, the surface water molecules are pulled tightly together by attractive forces from underneath (fig. 23.31).

The surface tension of an alveolus produces a force that is directed inward and, as a result, creates pressure within the alveolus. As described by the **Law of LaPlace,** the pressure thus created is directly proportional to the surface tension and *inversely proportional to the radius* of the alveolus. According to this law, the pressure in a

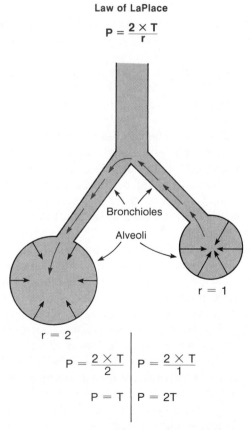

Figure 23.33. The production of pulmonary surfactant by type II alveolar cells. Surfactant appears to be composed of a derivative of lecithin combined with protein.

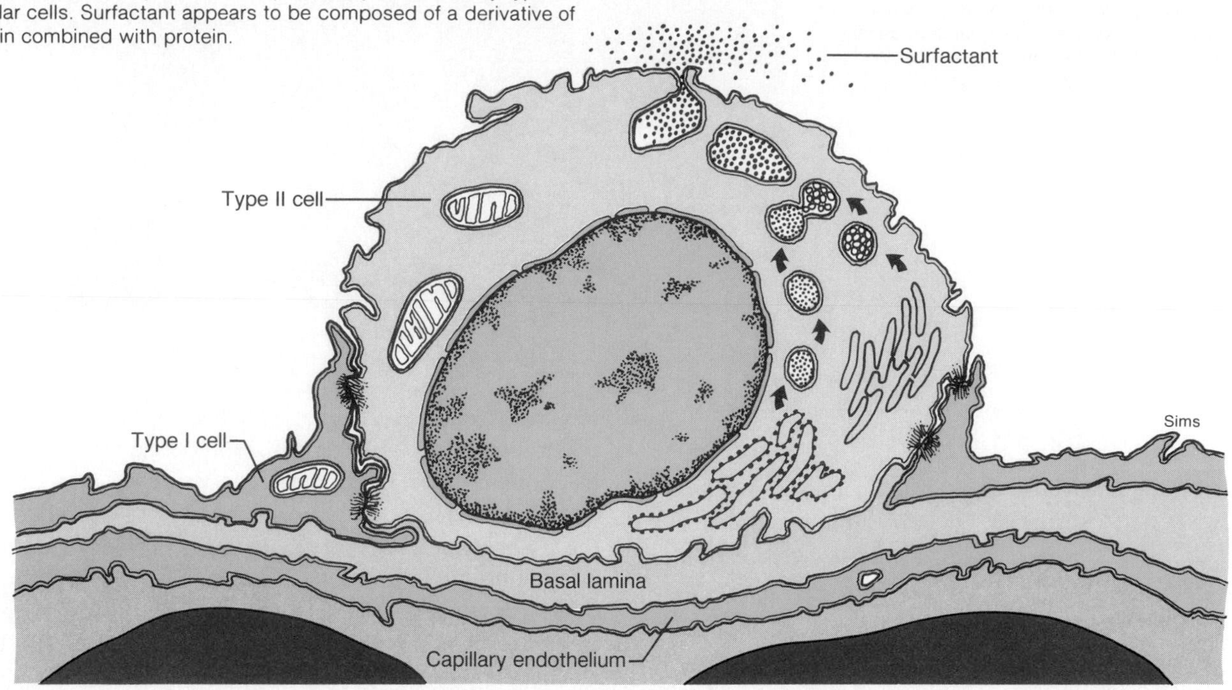

smaller alveolus would be greater than in a larger alveolus if the surface tension is the same in both. The greater pressure of the smaller alveolus would then cause it to empty its air into the larger one (fig. 23.32). This does not normally occur because as an alveolus decreases in size, its surface tension as well as its radius is reduced. The cause of the reduced surface tension is described in the next section.

Surfactant and the Respiratory Distress Syndrome

Alveolar fluid contains a phospholipid, known as dipalmitoyl lecithin, which is probably attached to a protein and functions to lower surface tension. This compound is called **lung surfactant** *(sur-fak'tant),* which is a contraction of the term *surface active agent.* Because of the presence of surfactant, the surface tension in the alveoli is lower at any given lung volume than would be predicted if surfactant were absent. Further, the ability of surfactant to lower surface tension improves as the alveoli get smaller during expiration. Surfactant thus helps to prevent the alveoli from collapsing as a result of surface tension. Even after a forceful expiration, the alveoli remain open and a *residual volume* of air remains in the lungs. Since the alveoli do not collapse, less surface tension has to be overcome to inflate them at the next inspiration.

Surfactant is produced by type II alveolar cells (fig. 23.33) in late fetal life. Since surfactant does not start to be produced until about the eighth month, premature babies are sometimes born with lungs that lack sufficient surfactant, and their alveoli are collapsed as a result. This condition is called *respiratory distress syndrome.* It is also called *hyaline (hi'ah-līn) membrane disease* because the high surface tension causes plasma fluid to leak into the alveoli, producing a glistening "membrane" appearance (and pulmonary edema). This condition does not occur in all premature babies; the rate of lung development depends on hormonal conditions (thyroxine and hydrocortisone primarily) and on genetic factors.

Even under normal conditions, the first breath of life is a difficult one because the newborn must overcome large surface tension forces in order to inflate its partially collapsed alveoli. The transpulmonary pressure required for the first breath is fifteen to twenty times that required for subsequent breaths, and an infant with *respiratory distress syndrome* must duplicate this effort with every breath. Fortunately, many babies with this condition can be saved by mechanical ventilators that keep them alive long enough for their lungs to mature and manufacture sufficient surfactant. In addition, surfactant derived from cow lungs or from the amniotic fluid of the baby can be delivered into the baby's lungs through an endotracheal tube.

1. Describe how the intrapulmonary and intrapleural pressures change during inspiration and expiration, and explain the reasons for these changes in terms of Boyle's law.
2. Define the terms *compliance* and *elasticity,* and explain how these lung properties affect inspiration and expiration.
3. Describe lung surfactant, and explain why the alveoli would collapse in the absence of surfactant.

Figure 23.34. A change in lung volume, as shown by radiographs, during expiration (*a*) and inspiration (*b*). The increase in lung volume during full inspiration is shown by comparison with the lung volume in full expiration (*dashed lines*).

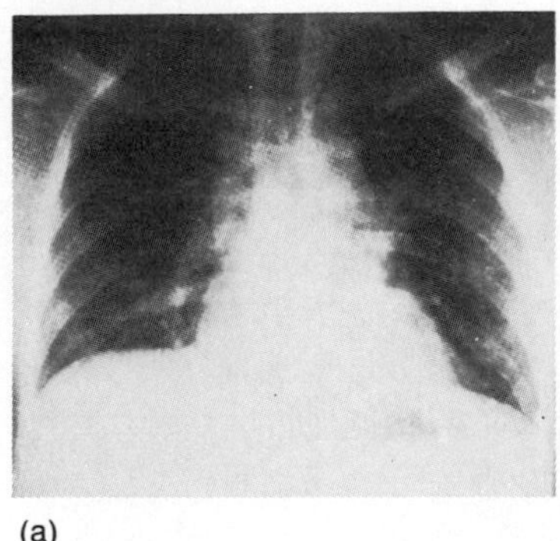

(a)

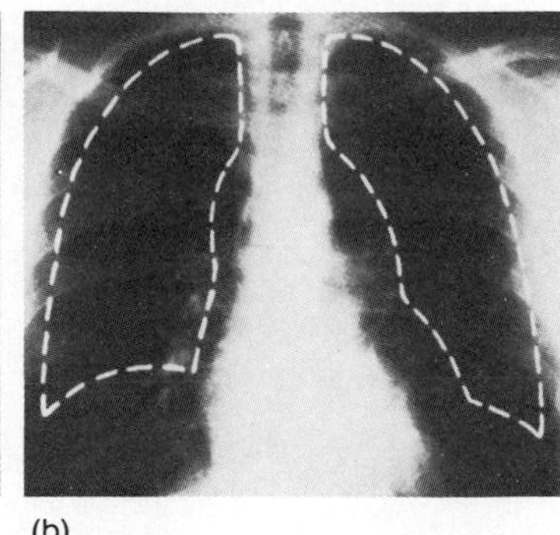

(b)

Mechanics of Breathing

Normal, quiet inspiration results from muscle contraction, expiration from muscle relaxation and elastic recoil. These actions can be forced by contractions of the accessory respiratory muscles. The amount of air inspired and expired can be measured by different procedures and used as tests of pulmonary function.

Objective 15. Identify and describe the actions of the muscles involved in both quiet and forced inspiration.

Objective 16. Describe how quiet expiration is produced, and identify and describe the actions of the muscles involved in forced expiration.

Objective 17. Define the concepts of restrictive and obstructive pulmonary disease, and give examples of each type.

Objective 18. Identify different lung volumes and capacities, and describe how pulmonary function tests help diagnose lung disorders.

Breathing, or **pulmonary ventilation,** refers to the movement of air into and out of the respiratory system. This movement of air occurs as a result of differences between the atmospheric and the intrapulmonary pressures; air goes in when the intrapulmonary pressure is subatmospheric, and air goes out when the intrapulmonary pressure rises above that of the atmosphere. As previously discussed, these pressure changes within the lungs occur as a consequence of changes in thoracic volume.

The thorax must be semirigid yet flexible. It must be sufficiently rigid so that it can protect vital organs and provide attachments for many short, powerful muscles. It must be flexible to function as a bellows during the ventilation cycle. The rigidity and the surfaces for muscle attachment are provided by the bony composition of the rib cage. The rib cage is pliable, however, because the ribs are separate from one another and because most ribs (the upper ten of the twelve pairs) are attached to the sternum by resilient costal cartilages. The vertebral attachments likewise provide considerable mobility. The structure of the rib cage and associated cartilages provide continuous elastic tension, so that when stretched by muscle contraction during inspiration, the rib cage can return passively to its resting dimensions when the muscles relax. This elastic recoil is greatly aided by the elasticity of the lungs.

Pulmonary ventilation consists of two phases, inspiration and expiration. Inspiration (inhalation) and expiration (exhalation) are accomplished by alternately increasing and decreasing the volumes of the thorax and lungs (fig. 23.34).

Inspiration and Expiration

The thoracic cavity increases in size during inspiration in three directions: anteroposteriorly, laterally, and vertically. This is accomplished by contractions of the **diaphragm** primarily, and also of the **external intercostal muscles** (fig. 23.35). A contraction of the dome-shaped diaphragm downward increases the thoracic volume vertically. A simultaneous contraction of the external intercostals increases the lateral and anteroposterior dimensions of the thorax.

Other thoracic muscles become involved in forced (deep) inspiration. The most important of these is the scalenus, followed by the pectoralis minor, and in extreme cases the sternocleidomastoid. Contraction of these muscles elevates the ribs in an anteroposterior direction, while at the same time the upper rib cage is stabilized so that the intercostals become more effective.

Quiet expiration is a passive process. After becoming stretched by contractions of the diaphragm and thoracic muscles, the thorax and lungs recoil as a result of their elastic tension when the respiratory muscles relax. The decrease in lung volume raises the pressure within the alveoli above the atmospheric pressure and pushes the air out.

diaphragm: Gk. *dia,* across; *phragma,* fence

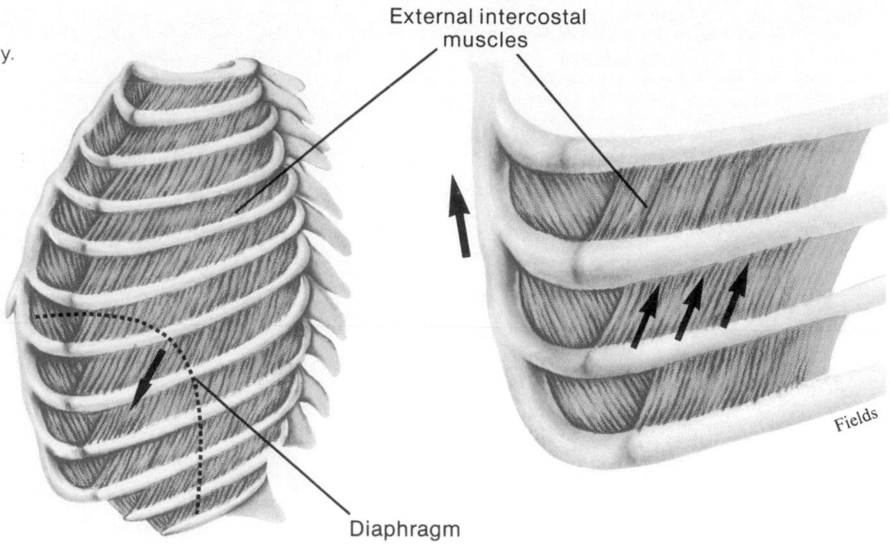

Figure 23.35. The position of the principal muscles of inspiration. Contraction of these muscles enlarges the size of the thoracic cavity.

Figure 23.36. Mechanics of pulmonary ventilation. (*a*) lungs and pleural cavities prior to inspiration. (*b*) during inspiration, the diaphragm is contracted and the rib cage elevated; intrathoracic and intrapulmonic pressures are reduced and air inflates the lungs. (*c*) inspiration is completed as intrapulmonic pressure is equal to atmospheric pressure. (*d*) as the muscles of inspiration are relaxed, the rib cage recoils and the intrathoracic and intrapulmonic pressures are raised after which the intrapulmonic pressure equals the atmospheric pressure (*a*).

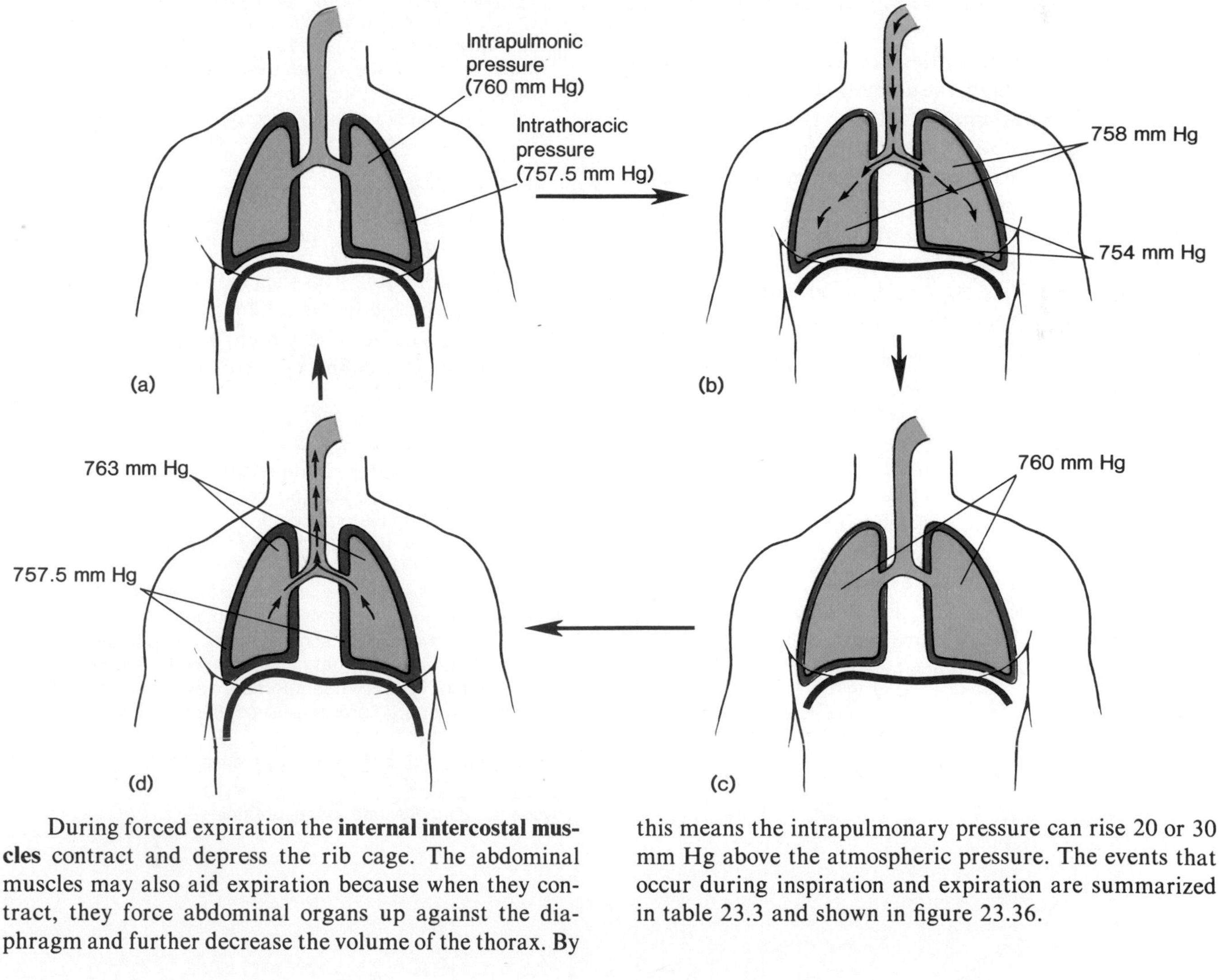

During forced expiration the **internal intercostal muscles** contract and depress the rib cage. The abdominal muscles may also aid expiration because when they contract, they force abdominal organs up against the diaphragm and further decrease the volume of the thorax. By this means the intrapulmonary pressure can rise 20 or 30 mm Hg above the atmospheric pressure. The events that occur during inspiration and expiration are summarized in table 23.3 and shown in figure 23.36.

Table 23.3 Definitions of terms used to describe lung volumes and capacities

Term	Definition	Term	Definition
Lung volumes	The four nonoverlapping components of the total lung capacity	Lung capacities	Measurements that are the sum of two or more lung volumes
Tidal volume	The volume of gas inspired or expired in an unforced respiratory cycle	Total lung capacity	The total amount of gas in the lungs at the end of a maximum inspiration
Inspiratory reserve volume	The maximum volume of gas that can be inspired from the end of a tidal inspiration	Vital capacity	The maximum amount of gas that can be expired after a maximum inspiration
Expiratory reserve volume	The maximum volume of gas that can be expired from the end of a tidal expiration	Inspiratory capacity	The maximum amount of gas that can be inspired at the end of a tidal expiration
Residual volume	The volume of gas remaining in the lungs after a maximum expiration	Functional residual capacity	The amount of gas remaining in the lungs at the end of a tidal expiration

Pulmonary Function Tests

Pulmonary function may be assessed clinically by means of a technique known as *spirometry (spi-rom'ĕ-tre).* In this procedure, a subject breathes in a closed system in which air is trapped within a light plastic bell floating in water. The bell moves up when the subject exhales and down when the subject inhales. The movements of the bell cause corresponding movements of a pen, which traces a record of the breathing on a rotating drum recorder (fig. 23.37).

The appearance of a spirogram taken during quiet, resting breathing is shown in figure 23.38, and the various lung volumes and capacities are defined in table 23.3. During quiet breathing, the amount of air expired in each breath (the **tidal volume**) is equal to about 500 ml. This is about 12% of the **vital capacity** (the maximum amount of air that can be expired after a maximum inspiration). Multiplying the tidal volume at rest times the number of breaths per minute yields a **total minute volume** of about 6 L per minute. During exercise, the tidal volume can increase to as much as 50% of the vital capacity, and the number of breaths per minute likewise increases to produce a total minute volume as high as 100–200 L per minute.

Spirometry is useful in the diagnosis of lung diseases. In purely *restrictive disorders,* such as pulmonary fibrosis, the vital capacity is reduced below normal. In disorders that are only obstructive, such as asthma, the vital capacity is normal because lung tissue is not damaged. Bronchoconstriction and the increased airway resistance that results, however, make expiration more difficult and causes it to take longer. *Obstructive disorders* are thus diagnosed by tests that measure the rate of expiration. One such test is the **forced expiratory volume (FEV),** in which the percent of the vital capacity that can be exhaled in the first second ($FEV_{1.0}$), second ($FEV_{2.0}$), and third ($FEV_{3.0}$) seconds is measured (fig. 23.39). An $FEV_{1.0}$ that is significantly less than 75% to 85% suggests the presence of obstructive pulmonary disease.

spirometry: L. *spiro,* to breathe; Gk. *metron,* measure

Bronchoconstriction often occurs in response to the inhalation of noxious agents in the air, such as smoke or smog. The $FEV_{1.0}$ has, therefore, been used by researchers to determine the effects of various components of smog and of passive cigarette smoke inhalation on pulmonary function. These studies have shown that it is unhealthy to exercise on very smoggy days and that inhalation of smoke from other people's cigarettes in a closed environment can measurably affect pulmonary function.

People with pulmonary disorders frequently complain of *dyspnea (disp'ne-ah),* which is a subjective feeling of "shortness of breath." Dyspnea may occur even when ventilation is normal, however, and may not occur even when total minute volume is very high, as in exercise. Some of the terms used to define ventilation are defined in table 23.4.

1. Describe the actions of the diaphragm and external intercostal muscles during inspiration, and explain how quiet expiration is produced.
2. Describe how forced inspiration and forced expiration are produced.
3. Define the terms *tidal volume* and *vital capacity.* Describe how the total minute volume is calculated, and explain how this value changes during exercise.
4. Describe how the vital capacity and the forced expiratory volume measurements are affected by asthma and pulmonary fibrosis, and explain why these measurements are affected.

dyspnea: Gk. *dys,* bad; *pnoe,* breathing

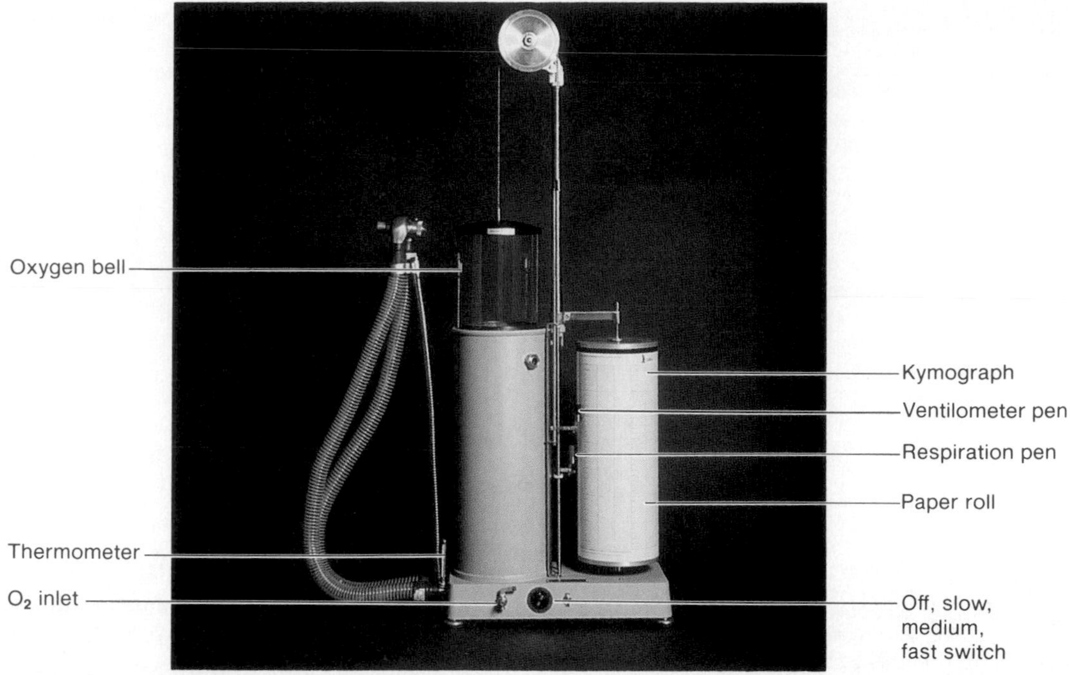

Figure 23.37. A spirometer (Collins 9L respirometer) used to measure lung volumes and capacities.

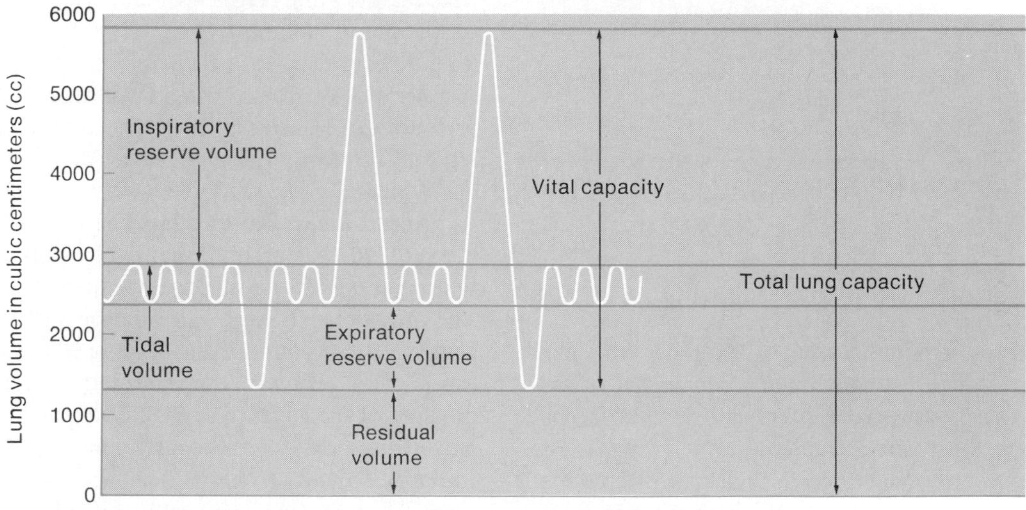

Figure 23.38. A spirogram showing respiratory air volumes.

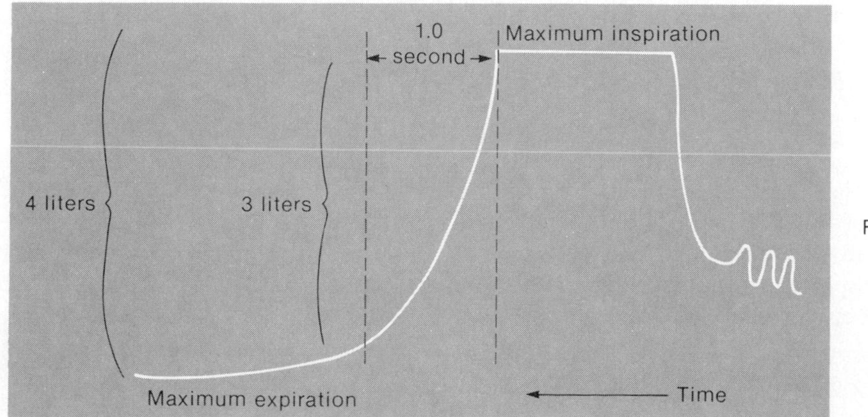

$$FEV_{1.0} = \frac{3L}{4L} \times 100\% = 75\%$$

Figure 23.39. An illustration of the one-second forced expiratory volume ($FEV_{1.0}$) spirometry test for detecting obstructive pulmonary disorders.

Table 23.4 Definitions of some terms used to describe ventilation

Term	Definition
Air spaces	Alveolar ducts, alveolar sacs, and alveoli
Airways	Structures that conduct air from the mouth and nose to the respiratory bronchioles
Alveolar ventilation	Removal and replacement of gas in alveoli; equal to the tidal volume minus the volume of dead space
Anatomical dead space	Volume of the conducting airways to the zone where gas exchange occurs
Apnea	Cessation of breathing
Dyspnea	Unpleasant subjective feeling of difficult or labored breathing
Eupnea	Normal, comfortable breathing at rest
Hyperventilation	An alveolar ventilation that is excessive in relation to metabolic rate; results in abnormally low alveolar CO_2
Hypoventilation	An alveolar ventilation that is low in relation to metabolic rate; results in abnormally high alveolar CO_2
Physiological dead space	Combination of anatomical dead space and underventilated alveoli that do not contribute normally to blood-gas exchange
Pneumothorax	Presence of gas in the pleural space (the space between the visceral and parietal pleural membranes) causing lung collapse
Torr	Synonymous with millimeters of mercury (760 mm Hg = 760 torr)

From Stuart Ira Fox, *Human Physiology,* 2d ed. Copyright © 1987 Wm. C. Brown Publishers, Dubuque, Iowa. All Rights Reserved. Reprinted by permission.

Gas Exchange in the Lungs

The lungs function to bring the blood into gaseous equilibrium with the alveolar air. The efficiency of this process is measured by the P_{O_2} and P_{CO_2} of arterial blood.

Objective 19. Describe how the partial pressures in a gas mixture are calculated, and explain how these values are affected by water vapor pressure and altitude.

Objective 20. Describe how the P_{O_2} and P_{CO_2} of blood is measured, and list the factors that affect these measurements.

Objective 21. Explain the significance of the blood P_{O_2} measurement in terms of the oxygen content of blood and the delivery of oxygen to the tissues.

Objective 22. Explain how oxygen toxicity, nitrogen narcosis, and decompression sickness are produced.

The atmosphere is an ocean of gas that exerts pressure on all objects within it. The amount of this pressure can be measured with a glass U-tube filled with fluid. One end of the U-tube is exposed to the atmosphere, while the other side is continuous with a sealed vacuum tube. Since the atmosphere presses on the exposed side, but not on the side connected to the vacuum tube, atmospheric pressure pushes fluid in the U-tube up on the vacuum side to a height determined by the atmospheric pressure and the density of the fluid. Water, for example, would be pushed up to a height of 33.9 feet (10,332 mm) at sea level, whereas mercury (Hg)—which is more dense—is raised to a height of 760 mm. As a matter of convenience, therefore, devices used to measure atmospheric pressure (barometers) use mercury rather than water. The atmospheric pressure at sea level is thus said to be equal to *760 mm Hg* (or *760 torr*), which is also described as a pressure of *one atmosphere* (fig. 23.40).

Figure 23.40. Atmospheric pressure at sea level can push a column of mercury to a height of 760 millimeters. This is also described as 760 torr, or one atmospheric pressure.

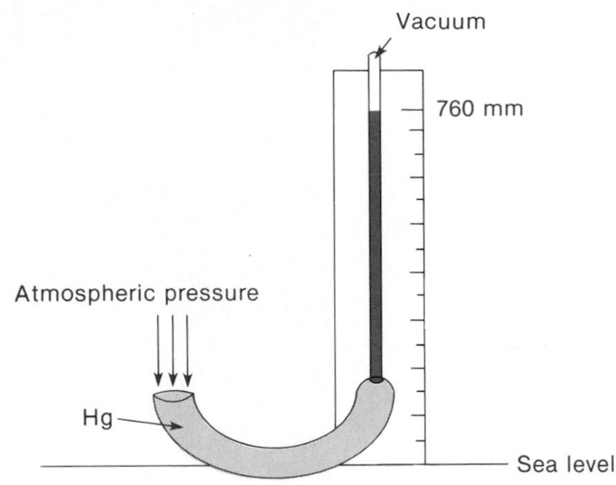

According to **Dalton's law,** the total pressure of a gas mixture (such as air) is equal to the sum of the pressures that each gas in the mixture would exert independently. The pressure that each gas would exert independently is equal to the product of the total pressure times the fractional composition of the gas in the mixture. The total pressure of the gas mixture is thus equal to the sum of the partial pressures of the constituent gases. Since oxygen comprises about 21 percent of the atmosphere, for example, its partial pressure (abbreviated P_{O_2}) is 21% of 760, or about 159 mm Hg. Since nitrogen comprises about 79% of the atmosphere, its partial pressure is equal to $0.79 \times 760 = 601$ mm Hg. These two gases thus contribute about 99% of the total pressure of 760 mm Hg:

$$P_{\text{dry atmosphere}} = P_{N_2} + P_{O_2} + P_{CO_2} = 760 \text{ mm Hg}$$

Calculation of P_{O_2}

As one goes to a higher altitude, the total atmospheric pressure and the partial pressure of the constituent gases decrease (table 23.5). At Denver, for example (5,000 feet above sea level), the atmospheric pressure is decreased to 619 mm Hg, and the P_{O_2} is therefore reduced to $619 \times 0.21 = 130$ mm Hg. At the peak of Mount Everest (at 29,000 feet), the P_{O_2} is only 42 mm Hg. As one

Dalton's law: from John Dalton, English chemist, 1766–1844

Table 23.5 The effect of altitude on P_{O_2}

Changes in P_{O_2} at various altitudes

Altitude (feet above sea level)	Atmospheric pressure (mm Hg)	P_{O_2} in air (mm Hg)	P_{O_2} in alveoli (mm Hg)	P_{O_2} in arterial blood (mm Hg)
0	760	159	105	100
2,000	707	148	97	92
4,000	656	137	90	85
6,000	609	127	84	79
8,000	564	118	79	74
10,000	523	109	74	69
20,000	349	73	40	35
30,000	226	47	21	19

Table 23.6 Some of the physical laws that are important in ventilation and gas exchange

Physical law	Description
Boyle's law	The volume of a gas is inversely proportional to its pressure when the temperature and mass are constant (PV = constant).
Dalton's law	The total pressure of a gas mixture is equal to the sum of the partial pressures of its constituent gases. The partial pressure of each gas is the pressure it would exert if it alone occupied the total volume of the mixture.
Graham's law	The rate of diffusion of a gas through a liquid is directly proportional to its solubility and inversely proportional to its density (or gram molecular weight).
Henry's law	The weight of a gas dissolved in a liquid at a given temperature is proportional to the partial pressure of the gas.

descends below sea level, as in ocean diving, the total pressure increases by one atmosphere for every 33 feet. At 33 feet therefore, the pressure equals 2 × 760 = 1,520 mm Hg. At 66 feet, the pressure equals three atmospheres.

Inspired air contains variable amounts of moisture. By the time the air has passed into the respiratory zone of the lungs, however, it is normally saturated with water vapor (has a relative humidity of 100%). The capacity of air to contain water vapor depends on its temperature; since the temperature of the respiratory zone is constant at 37°C, its water vapor pressure is also constant (47 mm Hg).

Water vapor, like the other constituent gases, contributes a partial pressure to the total atmospheric pressure. Since the total atmospheric pressure is constant (depending only on the height of the air mass), the water vapor "dilutes" the contribution of other gases to the total pressure. This follows from **Avogadro's law,** which states

Avogadro's law: from Amedeo Avogadro, Italian physicist, 1776–1856

Figure 23.41. Partial pressures of gases in the inspired air and the alveolar air.

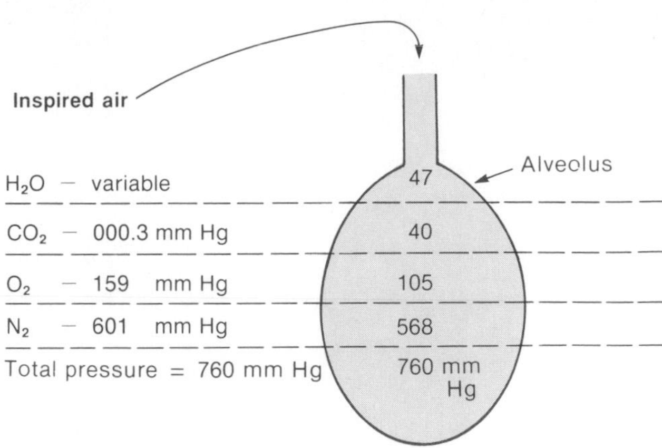

that all gases at the same temperature and pressure contain the same number of molecules in a given volume.

$$P_{\text{wet atmosphere}} = P_{N_2} + P_{O_2} + P_{CO_2} + P_{H_2O}$$

When the effect of water vapor pressure is considered, the partial pressure of oxygen in the inspired air is decreased at sea level to

$$P_{O_2} \text{ (sea level)} = 0.21 \times (760 - 47) = 150 \text{ mm Hg.}$$

As a result of gas exchange in the alveoli, the P_{O_2} of *alveolar air* is further reduced to about 105 mm Hg. A comparison of the partial pressures of the inspired air with the partial pressures of alveolar air is shown in figure 23.41.

Partial Pressures of Gases in Blood

When a fluid and a gas, such as blood and alveolar air, are at equilibrium, the amount of gas dissolved in the fluid reaches a maximum value. This value depends, according to **Henry's law** (table 23.6), on (1) the solubility of the gas in the fluid, which is a physical constant; (2) the temperature of the fluid—more gas can be dissolved in cold water than warm water; and (3) the partial pressure of the gas. Since the temperature of the blood does not vary significantly, the amount of dissolved gas depends directly on its partial pressure. When water—or plasma—is brought into equilibrium with air at a P_{O_2} of 100 mm Hg, for example, the fluid will contain 0.3 ml O_2 per 100 ml of fluid at 37°C. If the P_{O_2} of the gas is reduced by half, the amount of dissolved oxygen would also be reduced by half.

Blood-Gas Measurements. Measurement of the oxygen content of blood (in ml of O_2 per 100 ml blood) is a laborious procedure. Fortunately, an **oxygen electrode** that produces an electric current in proportion to the amount of *dissolved oxygen* has been developed. If this electrode is placed in a fluid while oxygen is artificially bubbled into

Henry's law: from William Henry, English chemist, 1775–1837

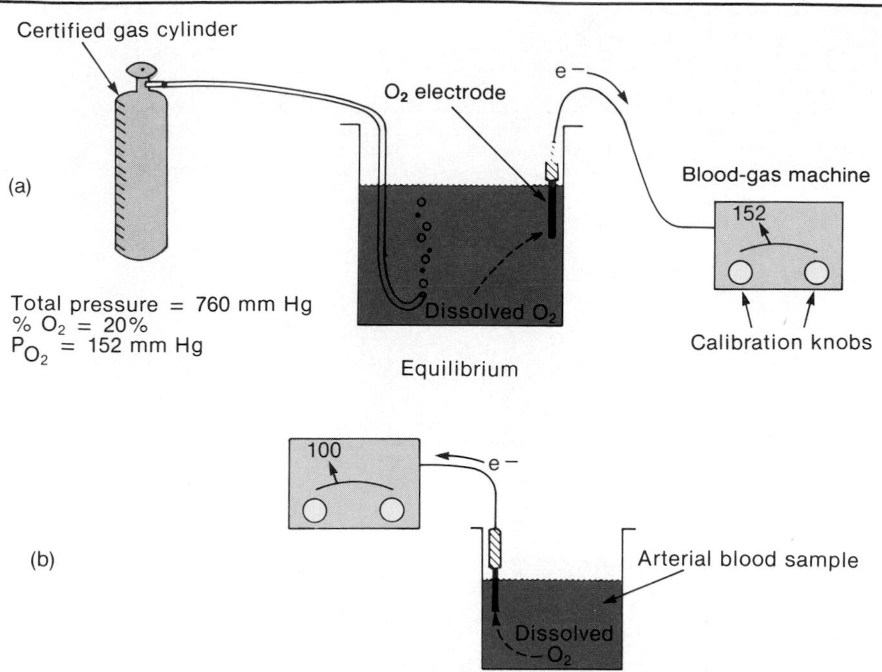

Figure 23.42. Blood-gas measurements using a P_{O_2} electrode. (*a*) the electrical current generated by the oxygen electrode is calibrated so that the needle of the blood-gas machine points to the P_{O_2} of the gas with which the fluid is in equilibrium. (*b*) once standardized in this way, the electrode can be inserted in a fluid, such as blood, and the P_{O_2} of this solution can be measured.

it, the current produced by the oxygen electrode will increase up to a maximum value. At this maximum value the fluid is *saturated* with oxygen—that is, all of the oxygen that can be dissolved at that temperature and P_{O_2} is dissolved. At a constant temperature, the amount dissolved and, thus, the electric current depend only on the P_{O_2} of the gas.

As a matter of convenience, it can now be said that the *fluid has the same P_{O_2} as the gas.* If it is known that the gas has a P_{O_2} of 150 mm Hg, for example, the deflection of the needle produced on a scale by the oxygen electrode can be calibrated at 150 mm Hg (fig. 23.42). The actual amount of dissolved oxygen under these circumstances is not known or very important (it can be looked up in solubility tables, if desired), since this is simply a linear function of the P_{O_2}. A lower P_{O_2} indicates that less oxygen is dissolved, and a higher P_{O_2} indicates that more oxygen is dissolved.

If the oxygen electrode is next inserted into an unknown sample of blood, the P_{O_2} of that sample can be read directly from the previously calibrated scale. Suppose, as illustrated in figure 23.42, the blood sample has a P_{O_2} of 100 mm Hg. Since alveolar air has a P_{O_2} of about 105 mm Hg, this reading indicates that the blood is almost in complete equilibrium with the alveolar air.

The oxygen electrode responds only to the oxygen dissolved in water or plasma; it cannot respond to the oxygen that is hidden in red blood cells. Most of the oxygen in blood, however, is located in the red blood cells attached to hemoglobin. The oxygen content of whole blood thus depends on both its P_{O_2} and its red blood cell and hemoglobin content. At a P_{O_2} of about 100 mm Hg, whole blood normally contains 20 ml O_2 per 100 ml blood; of this amount, only 0.3 ml O_2 is dissolved in the plasma and 19.7 ml O_2 is within the red blood cells. Since only the 0.3 ml O_2 per 100 ml blood affects the P_{O_2} measurement, this measurement would be unchanged if the red blood cells were removed from the sample.

Significance of Blood P_{O_2} and P_{CO_2} Measurements

Since blood P_{O_2} measurements are not directly affected by the oxygen in red blood cells, the P_{O_2} does not provide a measurement of the oxygen content of whole blood. It does, however, provide a good index of *lung function.* If the inspired air has a normal P_{O_2} but the arterial P_{O_2} is below normal, for example, gas exchange in the lungs must be impaired. Measurements of arterial P_{O_2} thus provide valuable information in the treatment of people with pulmonary diseases, during surgery (when breathing may be depressed by anesthesia), and in the care of premature babies with respiratory distress syndrome.

When the lungs are functioning properly, the P_{O_2} of systemic arterial blood is only 5 mm Hg less than the P_{O_2} of alveolar air. Hyperventilation, therefore, cannot significantly increase the blood P_{O_2}. At a normal P_{O_2} of about 100 mm Hg, hemoglobin is almost completely loaded with oxygen. An increase in blood P_{O_2}—produced, for example, by breathing 100% oxygen from a gas tank—thus cannot significantly increase the amount of oxygen contained in the blood. This can, however, significantly increase the amount of oxygen dissolved in the plasma (because the amount dissolved is directly determined by the P_{O_2}). If the P_{O_2} doubles, the amount of oxygen dissolved in the plasma also doubles, but the total oxygen content of whole blood increases only slightly since most of the oxygen by far is not in plasma but in the red blood cells.

Since the oxygen carried by red blood cells must first dissolve in plasma before it can diffuse to the tissue cells, however, a doubling of the blood P_{O_2} means that the *rate of oxygen delivery* to the tissues would double under these conditions. For this reason, breathing from a tank of 100%

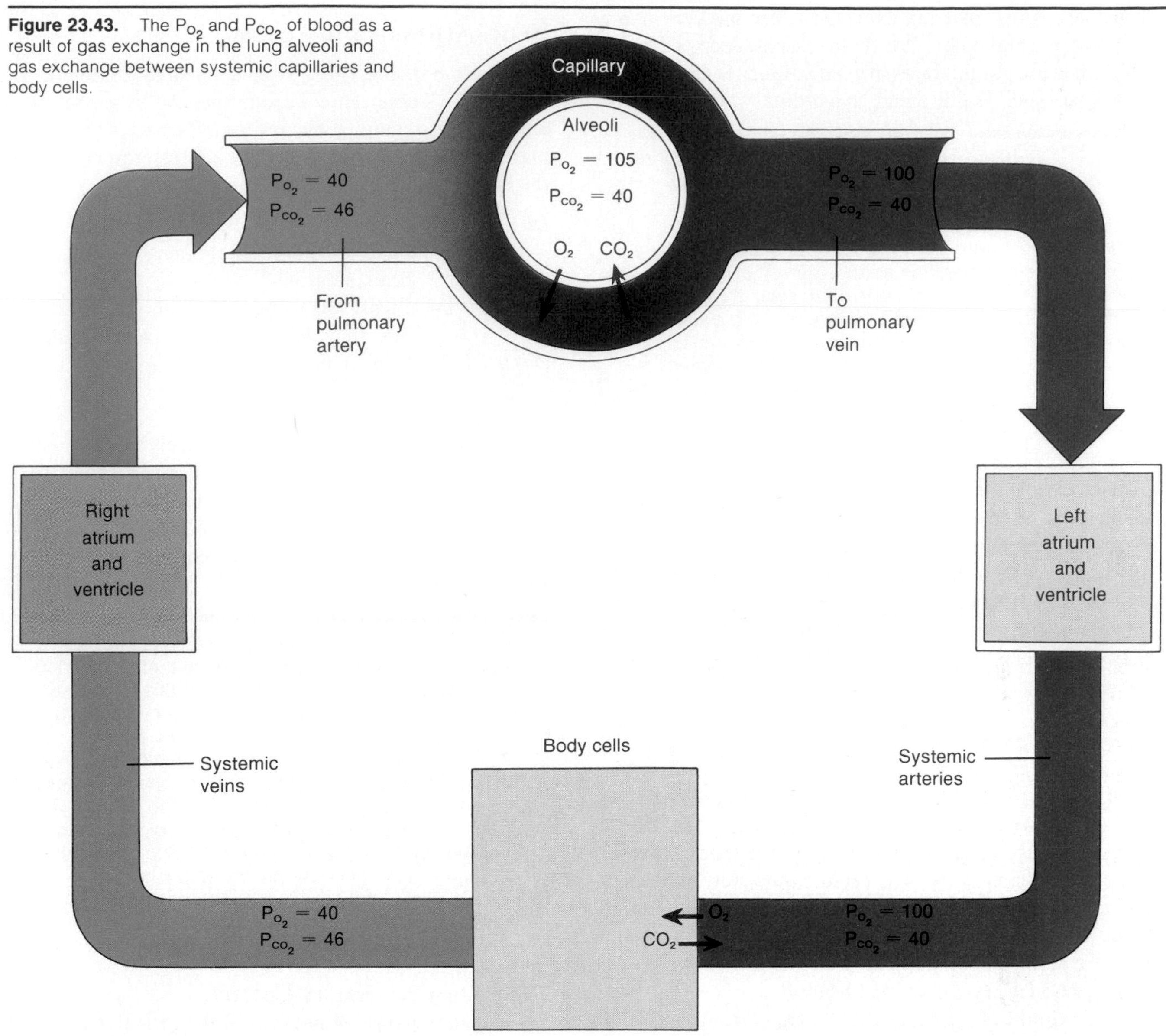

Figure 23.43. The P_{O_2} and P_{CO_2} of blood as a result of gas exchange in the lung alveoli and gas exchange between systemic capillaries and body cells.

oxygen (with a P_{O_2} of 760 mm Hg) would significantly increase oxygen delivery to the tissues, though it would have little effect on the total oxygen content of blood.

An electrode that produces a current in response to dissolved carbon dioxide has also been developed, so that the P_{CO_2} of blood can be measured together with its P_{O_2}. Blood in the systemic veins, which is delivered to the lungs by the pulmonary arteries, usually has a P_{O_2} of 40 mm Hg and a P_{CO_2} of 46 mm Hg. After gas exchange in the alveoli of the lungs, blood in the pulmonary veins and systemic arteries has a P_{O_2} of about 100 mm Hg and a P_{CO_2} of 40 mm Hg (fig. 23.43). The values in arterial blood are relatively constant and clinically significant, because they reflect lung function. Blood-gas measurements of venous blood are not performed clinically because these values are far more variable; venous P_{O_2} is much lower and P_{CO_2} much higher after exercise, for example, than at rest, whereas arterial values are not significantly affected by usual changes in physical activity.

Disorders Caused by High Partial Pressure of Gases

The total atmospheric pressure increases by one atmosphere (760 mm Hg) for every 10 m (33 feet) below sea level. If one dives 10 m below sea level, therefore, the partial pressures and amounts of dissolved gases in the plasma are twice that at sea level. At 66 feet they are three times, and at 100 feet they are four times, the values at sea level. The increased amounts of nitrogen and oxygen dissolved in the blood plasma under these conditions can have serious effects on the body.

Oxygen Toxicity. Although breathing 100% oxygen at one or two atmospheres pressure can be safely tolerated for a few hours, higher partial oxygen pressures can be very dangerous. *Oxygen toxicity* develops rapidly when the P_{O_2} rises above about 2.5 atmospheres. This is apparently caused by the oxidation of enzymes and other destructive changes that can damage the nervous system and lead to coma and death. For these reasons, deep-sea divers commonly use gas mixtures in which oxygen is diluted with inert gases such as nitrogen (as in ordinary air) or helium.

Hyperbaric oxygen—oxygen at greater than one atmosphere pressure—is often used to treat conditions such as carbon monoxide poisoning, circulatory shock, and gas gangrene. Before the dangers of oxygen toxicity were realized these hyperbaric oxygen treatments sometimes resulted in tragedy. Particularly tragic were the cases of *retrolental fibroplasia (re''tro-len'tal fi''bro-pla'se-ah),* in which damage to the retina and blindness resulted from the hyperbaric oxygen treatment of premature babies with hyaline membrane disease.

Nitrogen Narcosis. Although at sea level nitrogen is physiologically inert, larger amounts of dissolved nitrogen under hyperbaric conditions have deleterious effects. Since it takes time for the nitrogen to dissolve, these effects usually don't appear until the person has remained submerged over an hour. *Nitrogen narcosis* resembles alcohol intoxication; depending on the depth of the dive, the diver may experience "rapture of the deep" or may become so drowsy that he or she is totally incapacitated.

Decompression Sickness. The amount of nitrogen dissolved in the plasma decreases as the diver ascends to sea level, due to a progressive decrease in the P_{N_2}. If the diver surfaces slowly, a large amount of nitrogen can diffuse through the alveoli and be eliminated in the expired breath. If decompression occurs too rapidly, however, bubbles of nitrogen gas (N_2) can form in the blood and block small blood channels, producing muscle and joint pain as well as more serious damage. These effects are known as *decompression sickness,* or "the bends."

The cabins of airplanes that fly long distances at high altitudes (30,000 to 40,000 feet) are pressurized so that the passengers and crew are not exposed to the very low atmospheric pressures of these altitudes. If a cabin were to become rapidly depressurized at high altitude, much less nitrogen could remain dissolved at the greatly lowered pressure. People in this situation, like the divers that ascend too rapidly, would thus experience decompression sickness.

1. Describe how the P_{O_2} of air is calculated and how this value is affected by altitude, diving, and water vapor pressure.
2. Describe how blood P_{O_2} measurements are taken, and explain the physiological and clinical significance of these measurements.
3. Explain how the arterial P_{O_2} and the oxygen content of whole blood is affected by (1) hyperventilation; (2) breathing from a tank containing 100% oxygen; (3) anemia (low red blood cell count and hemoglobin concentration); and (4) high altitude.
4. Explain what happens to N_2 in blood when a person dives to a great depth and then ascends too rapidly.

Regulation of Breathing

The rhythm of breathing is controlled by centers in the brain stem. These centers are influenced by higher brain areas and are regulated by sensory information that makes breathing responsive to the changing respiratory needs of the body.

Objective 23. Describe the functions of the pneumotaxic, apneustic, and rhythmicity centers in the brain stem.

Objective 24. Identify the chemoreceptors, and describe how they help to regulate ventilation.

Objective 25. Explain how blood P_{O_2}, P_{CO_2}, and pH affect ventilation.

Inspiration and expiration are produced by the contractions and relaxations of skeletal muscles in response to activity in somatic motor neurons. The activity of these motor neurons, in turn, is controlled by neurons in the respiratory control centers of the brain stem and by neurons in the cerebral cortex.

The automatic control of breathing is regulated by nerve fibers that descend in the lateral and ventral white matter of the spinal cord from the medulla oblongata. The voluntary control of breathing is a function of the cerebral cortex and involves nerve fibers that descend in the corticospinal tracts. The separation of the voluntary and involuntary pathways is dramatically illustrated in the condition called *Ondine's curse,* in which neurological damage abolishes the automatic but not the voluntary control of breathing. People with this condition must consciously force themselves to breathe and be put on artificial respirators when they sleep.

Brain Stem Respiratory Centers

A loose aggregation of neurons in the reticular formation of the *medulla oblongata* forms the **rhythmicity** *(rith-mis'i-te)* **center** that controls automatic breathing. The rhythmicity center consists of interacting pools of neurons that fire either during inspiration (*I neurons*) or expiration (*E neurons*). The I neurons project to and stimulate spinal motoneurons that innervate the respiratory muscles. Expiration is a passive process that occurs when the I neurons are inhibited by the activity of the E neurons. The activity of I and E neurons varies in a reciprocal way, so that a rhythmic pattern of breathing is produced. The cycle of inspiration and expiration is thus intrinsic to the neural activity of the medulla. The rhythmicity center in the medulla is divided into a dorsal group of neurons, which regulates the activity of the phrenic nerves to the diaphragm, and a ventral group, which controls the motor neurons to the intercostal muscles.

The activity of the medullary rhythmicity center is influenced by centers in the *pons.* As a result of research in which the brain stem is destroyed at different levels, two respiratory control centers have been identified in the pons. One area—the **apneustic** *(ap-nu'stik)* **center**—

appears to promote inspiration by stimulating the I neurons in the medulla. The other pontine area—called the **pneumotaxic** *(nu''mo-tak'sik)* **center**—seems to antagonize the apneustic center and inhibit inspiration (fig. 23.44). The apneustic center is believed to provide a tonic, or constant, stimulus for inspiration which is cyclically inhibited by the activity of the pneumotaxic center.

The automatic control of breathing is also influenced by input from receptors sensitive to the chemical composition of the blood. There are two groups of *chemoreceptors (ke''mo-re-cep'tors)* that respond to changes in blood P_{CO_2}, pH, and P_{O_2}. These are the **central chemoreceptors** in the medulla oblongata and the **peripheral chemoreceptors.** The peripheral chemoreceptors include a couple of *aortic bodies,* located in the aortic arch, and a pair of *carotid bodies,* located in the carotid artery on each side near the division into the external and internal carotids (fig. 23.45).

The peripheral chemoreceptors control breathing indirectly via sensory nerve fibers to the medulla. The aortic bodies send sensory information to the medulla in the vagus (tenth) cranial nerve; the carotid bodies stimulate sensory fibers in the glossopharyngeal (ninth) cranial nerve. The neural and sensory control of ventilation is summarized in figure 23.46.

Figure 23.44. Approximate locations of the brain stem respiratory centers.

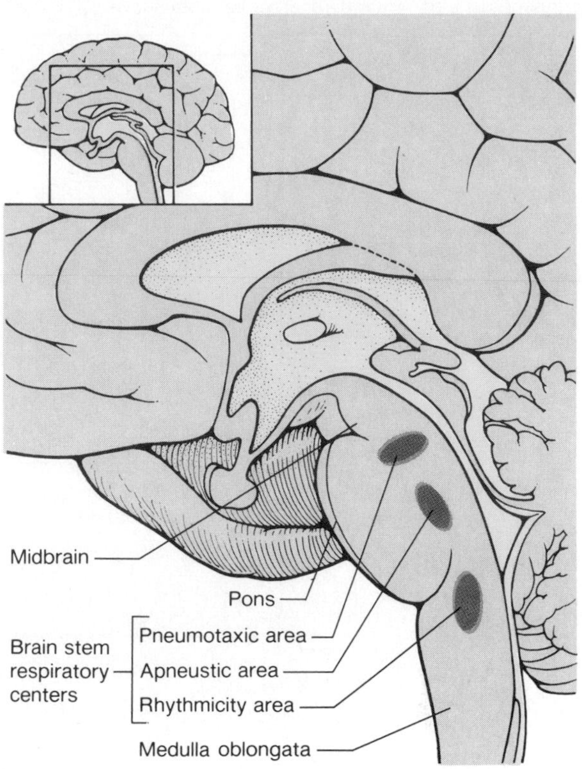

Figure 23.45. The peripheral chemoreceptors (aortic and carotid bodies) regulate the brain stem respiratory centers by means of sensory nerve stimulation.

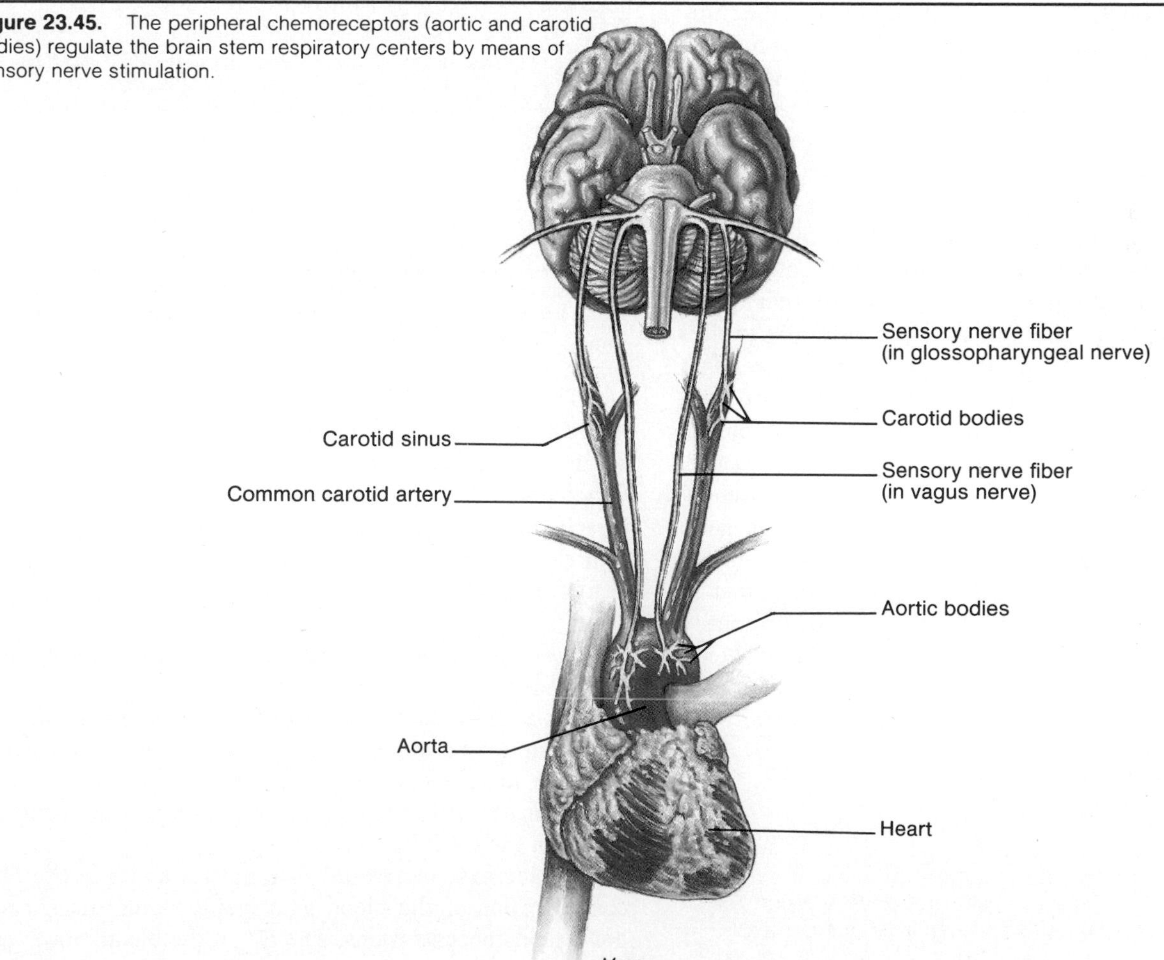

Figure 23.46. A schematic illustration of the control of ventilation by the central nervous system. The feedback effects of pulmonary stretch receptors and "irritant" receptors are not shown.

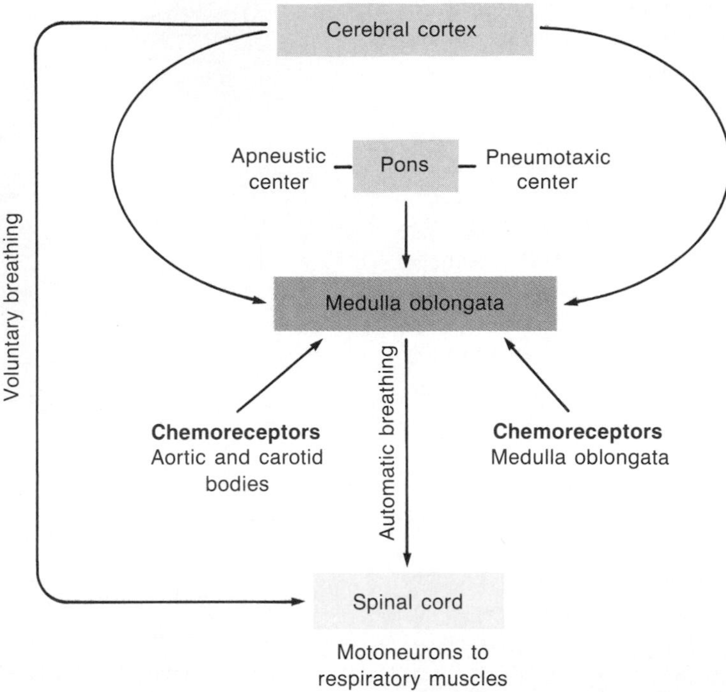

Effects of Blood P_{CO_2} and pH on Ventilation

Chemoreceptor input to the brain stem modifies the rate and depth of breathing so that, under normal conditions, arterial P_{CO_2}, pH, and P_{O_2} remain relatively constant. If hypoventilation (inadequate ventilation) occurs, P_{CO_2} quickly rises and pH falls. The pH falls because carbon dioxide can combine with water to form carbonic acid. P_{O_2} also falls quickly, but this reflects the O_2 content of plasma only. The total oxygen content of the blood decreases much more slowly because there is a large "reservoir" of oxygen attached to hemoglobin. During hyperventilation, conversely, blood P_{CO_2} quickly falls and pH rises due to the excessive elimination of carbonic acid. The oxygen content of blood, on the other hand, is not significantly increased by hyperventilation (hemoglobin in arterial blood is 97% saturated with oxygen during normal ventilation).

The blood P_{CO_2} and pH are, therefore, more immediately affected by changes in ventilation than is the oxygen content. Indeed, changes in P_{CO_2} provide a sensitive index of ventilation, as shown in table 23.7. In view of these facts, it is not surprising that changes in P_{CO_2} provide the most potent stimulus for the reflex control of ventilation. Ventilation, in other words, is adjusted to maintain a constant P_{CO_2}; proper oxygenation of the blood occurs naturally as a side product of this reflex control.

Table 23.7 The effect of ventilation, as measured by total minute volume (breathing rate × tidal volume), on the P_{CO_2} of arterial blood

Total minute volume	Arterial P_{CO_2}	Type of ventilation
2 L/min	80 mm Hg	Hypoventilation
4–5 L/min	40 mm Hg	Normal ventilation
8 L/min	20–25 mm Hg	Hyperventilation

The rate and depth of ventilation are normally adjusted to maintain an arterial P_{CO_2} of 40 mm Hg. Hypoventilation causes a rise in P_{CO_2}—a condition called *hypercapnia (hi''per-kap'ne-ah).* Hyperventilation, conversely, results in *hypocapnia.* Chemoreceptor regulation of breathing in response to changes in P_{CO_2} is illustrated in figure 23.47.

Chemoreceptors in the Medulla. The chemoreceptors most sensitive to changes in the arterial P_{CO_2} are located in the ventral area of the medulla, near the exit of the ninth and tenth cranial nerves. These chemoreceptor neurons are anatomically separate from, but synaptically communicate with, the neurons of the respiratory control center in the medulla.

An increase in arterial P_{CO_2} causes a rise in the H^+ concentration of the blood as a result of increased carbonic acid concentrations. The H^+ in the blood, however,

Figure 23.47. Negative feedback control of ventilation through changes in blood P_{CO_2} and pH.

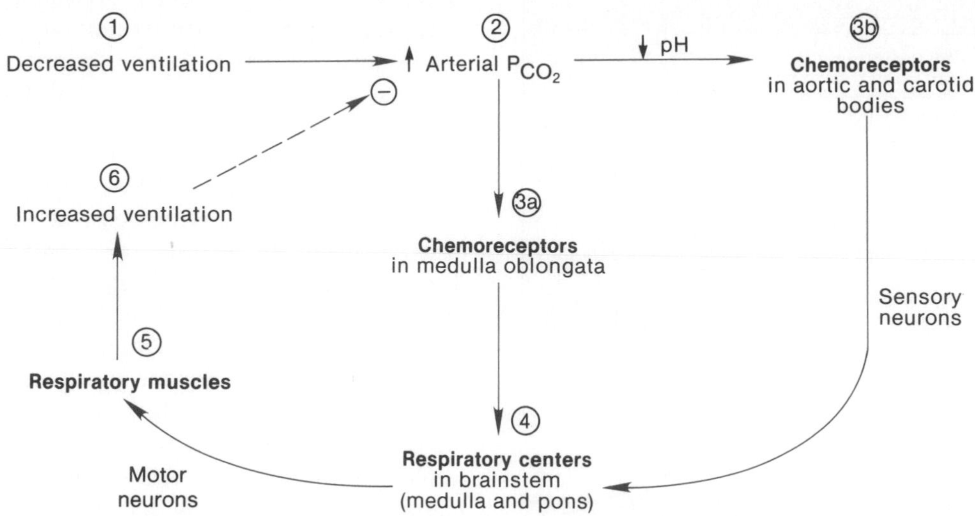

cannot cross the blood-brain barrier and, therefore, cannot influence the medullary chemoreceptors. Carbon dioxide in the arterial blood can cross the blood-brain barrier and, through the formation of carbonic acid, can lower the pH of cerebrospinal fluid. This fall in cerebrospinal fluid pH directly stimulates the medullary chemoreceptors when there is a rise in arterial P_{CO_2}.

The chemoreceptors in the medulla are ultimately responsible for 70%–80% of the increased ventilation that occurs in response to a sustained rise in arterial P_{CO_2}. This response, however, takes several minutes. The immediate increase in ventilation that occurs when P_{CO_2} rises is produced by stimulation of the peripheral chemoreceptors.

Peripheral Chemoreceptors. The aortic and carotid bodies are stimulated directly by a rise in the H^+ concentration (fall in pH) of arterial blood. This is normally produced by an increase in P_{CO_2} and carbonic acid. The retention of CO_2 during hypoventilation thus stimulates the medullary chemoreceptors through a lowering of cerebrospinal fluid pH and stimulates peripheral chemoreceptors through a lowering of blood pH.

People who *hyperventilate* during psychological stress are sometimes told to breathe into a paper bag so that they re-breathe their expired air that is enriched in CO_2. This procedure helps to raise their blood P_{CO_2} back up to the normal range. This is needed because *hypocapnia* causes cerebral vasconstriction. In addition to producing dizziness, the cerebral ischemia that results can lead to acidotic conditions in the brain that, through stimulation of the medullary chemoreceptors, can cause further hyperventilation. Breathing into a paper bag can thus relieve the hypocapnia and stop the hyperventilation.

Effects of Blood P_{O_2} on Ventilation

Under normal conditions, blood P_{O_2} affects breathing only indirectly, by influencing the chemoreceptor sensitivity to changes in P_{CO_2}. Chemoreceptor sensitivity to P_{CO_2} is augmented by a low P_{O_2} (so ventilation is increased at a high altitude, for example) and is decreased by a high P_{O_2}. If the blood P_{O_2} is raised by breathing 100% oxygen, therefore, the breath can be held longer because the response to increased P_{CO_2} is blunted.

When the blood P_{CO_2} is held constant by experimental techniques, the P_{O_2} of arterial blood must fall from 100 mm Hg to below 50 mm Hg before ventilation is significantly stimulated. This stimulation is apparently due to a direct effect of P_{O_2} on the carotid bodies. Since this degree of *hypoxemia (hi''pok-se'me-ah),* or low blood oxygen (table 23.8), does not normally occur even in breathholding, P_{O_2} does not normally exert this direct effect on breathing.

In emphysema, when there is a chronic retention of carbon dioxide, the chemoreceptors eventually lose their ability to respond to increases in the arterial P_{CO_2}. The abnormally high P_{CO_2}, however, enhances the sensitivity of the carotid bodies to a fall in P_{O_2}. For people with emphysema, breathing may thus be stimulated by a *hypoxic drive* rather than by increases in blood P_{CO_2}.

The effect of changes in the blood P_{CO_2}, pH, and P_{O_2} on chemoreceptors and the regulation of ventilation are summarized in table 23.9.

Table 23.8 Definitions of some terms used to describe blood oxygen and carbon dioxide levels

Term	Definition
Hypoxemia	The oxygen content or P_{O_2} is lower than normal in arterial blood.
Hypoxia	The oxygen content or P_{O_2} is lower than normal in the lungs, blood, or tissues. This is a more general term than hypoxemia. Tissues can be hypoxic, for example, even though there is no hypoxemia (as when the blood flow is occluded).
Hypercapnia, or hypercarbia	The P_{CO_2} of systemic arteries is higher than 40 mm Hg. Usually this occurs when the ventilation is inadequate for a given metabolic rate (hypoventilation). Antonyms are *hypocapnia* and *hypocarbia* (usually produced by hyperventilation).

Table 23.9 The sensitivity of chemoreceptors to changes in blood gases and pH

Stimulus	Chemoreceptor(s)	Comments
↑ P_{CO_2}	Medulla oblongata	Medullary chemoreceptors are sensitive to the pH of cerebrospinal fluid (CSF). Diffusion of CO_2 from the blood into the CSF lowers the pH of CSF by forming carbonic acid.
↓ pH	Aortic and carotid bodies	Peripheral chemoreceptors are stimulated by decreased blood pH independent of the effect of blood CO_2. Chemoreceptors in the medulla are not affected by changes in blood pH because H^+ cannot cross the blood-brain barrier.
↓ P_{O_2}	Carotid bodies	Low blood P_{O_2} (hypoxemia) augments the chemoreceptor response to blood P_{CO_2}, and can stimulate ventilation directly when the P_{O_2} falls below 50 mm Hg.

Pulmonary Stretch and Irritant Reflexes

The lungs contain various types of receptors that influence the brain stem respiratory control centers via sensory fibers in the vagus (table 23.10). Irritant receptors in the lungs, for example, stimulate reflex constriction of the bronchioles in response to smoke and smog. Similarly, sneezing, sniffing, and coughing may be stimulated by irritant receptors in the nose, larynx, and trachea.

The **Hering-Breuer reflex** is stimulated by pulmonary stretch receptors. The activation of these receptors during inspiration inhibits the respiratory control centers, making further inspiration increasingly difficult. This helps to prevent undue distension of the lungs and may contribute to the smoothness of the ventilation cycles. A similar inhibitory reflex may occur during expiration. The Hering-Breuer reflex appears to be important in the control of normal ventilation in the newborn. Pulmonary stretch receptors in adults, however, are probably not active at normal resting tidal volumes (500 ml per breath) but may contribute to respiratory control at high tidal volumes, as during exercise.

1. Describe the effects of voluntary hyperventilation and breath-holding on arterial P_{CO_2}, pH, and oxygen content. Indicate the relative degree of changes in these values.
2. Using a flow chart to show a negative feedback loop, explain the relationship between ventilation and arterial P_{CO_2}.
3. Explain the effect of increased arterial P_{CO_2} on (1) chemoreceptors in the medulla oblongata and (2) chemoreceptors in the aortic and carotid bodies.
4. Explain the role of arterial P_{CO_2} in the regulation of breathing. Explain why ventilation increases when a person goes to a high altitude.

Hering-Breuer reflex: from Heinrich E. Hering, German physiologist, 1866–1948, and Josef Breuer, German physician, 1842–1925

Table 23.10 Pulmonary receptors and reflexes served by sensory nerve fibers in the vagus (tenth cranial nerve)

Stimulus	Receptors	Comments
Stretch of lungs at inspiration	Stretch receptors	The Hering-Breuer reflex. This reflex is believed to be important in the control of breathing in the newborn but not in the adult at normal tidal volumes.
Stretch of lungs at expiration	Stretch receptors	Like the Hering-Breuer reflex, this is not believed to be significant in adults at normal tidal volumes.
Pulmonary congestion	Type J (juxta-capillary) receptors	This reflex may produce feelings of dyspnea at high altitudes and during severe exercise.
Irritation	"Irritant receptors"	This reflex causes reflex constriction of bronchioles in response to irritation from smoke, smog, and other noxious agents.

Hemoglobin and Oxygen Transport

Hemoglobin loads with oxygen in the lungs and unloads oxygen in the tissue capillaries. The extent of this unloading is influenced by a variety of factors that act to insure an adequate delivery of oxygen to the tissues while maintaining an adequate reserve of oxygen in the blood.

Objective 26. Distinguish between the different forms of hemoglobin.

Objective 27. Describe the loading and unloading reactions, and explain the effects of P_{O_2} and hemoglobin affinity for oxygen on these reactions.

Objective 28. Describe the effects of changes in pH, temperature, and 2,3-DPG on the unloading reaction, and explain the significance of these effects.

If the lungs are functioning properly, blood leaving in the pulmonary veins and traveling in systemic arteries has a P_{O_2} of about 100 mm Hg, indicating a plasma oxygen concentration of about 0.3 ml O_2 per 100 ml blood. The total oxygen content of the blood, however, is not known; this depends not only on the P_{O_2} but also on the hemoglobin concentration. If the P_{O_2} and hemoglobin concentration are normal, arterial blood contains about 20 ml of O_2 per 100 ml of blood (fig. 23.48).

Hemoglobin

Most of the oxygen in the blood is contained within the red blood cells, where it is chemically bonded to **hemoglobin.** Each hemoglobin molecule consists of (1) a protein globin part composed of four polypeptide chains; and (2) four nitrogen-containing, disc-shaped organic pigment molecules, called *hemes* (fig. 23.49).

hemoglobin: Gk. *haima,* blood; L. *globus,* globe

The protein part of hemoglobin is composed of two identical *alpha chains,* which are each 141 amino acids long, and two identical *beta chains,* which are each 146 amino acids long. Each of the four polypeptide chains is combined with one heme group. In the center of each heme group is a single atom of iron, which can combine with one molecule of oxygen (O_2). One hemoglobin molecule can thus combine with four molecules of oxygen; since there are about 280 million hemoglobin molecules per red blood cell, each red blood cell can carry over a billion molecules of oxygen.

Normal heme contains iron in the reduced form (Fe^{++}, or ferrous iron). In this form, the iron can share electrons and bond to oxygen to form **oxyhemoglobin.** When oxyhemoglobin dissociates to release oxygen to the tissues, the heme iron is still in the reduced (Fe^{++}) form and the hemoglobin is called **deoxyhemoglobin,** or **reduced hemoglobin.** The term *oxyhemoglobin* is thus not equivalent to *oxidized* hemoglobin; hemoglobin does not lose an electron (and become oxidized) when it combines with oxygen. Oxidized hemoglobin, or **methemoglobin** *(met-he'mo-glo'bin),* has iron in the oxidized (Fe^{+++}, or ferric) state. Methemoglobin thus lacks the electron it needs to form a bond with oxygen and cannot participate in oxygen transport. Blood normally contains only a small amount of methemoglobin, but certain drugs can increase this amount.

Carboxyhemoglobin is an abnormal form of hemoglobin in which the reduced heme is combined with *carbon monoxide* instead of oxygen. Since the bond with carbon monoxide is about 210 times stronger than the bond with oxygen, carbon monoxide tends to displace oxygen in hemoglobin and remains attached to hemoglobin as the blood passes through systemic capillaries. The transport of oxygen to the tissues is thus reduced.

Figure 23.48. Plasma and whole blood that are brought into equilibrium with the same gas mixture have the same P_{O_2} and thus the same amount of dissolved oxygen molecules (shown as black dots). The oxygen content of whole blood, however, is much higher than that of plasma because of the binding of oxygen to hemoglobin.

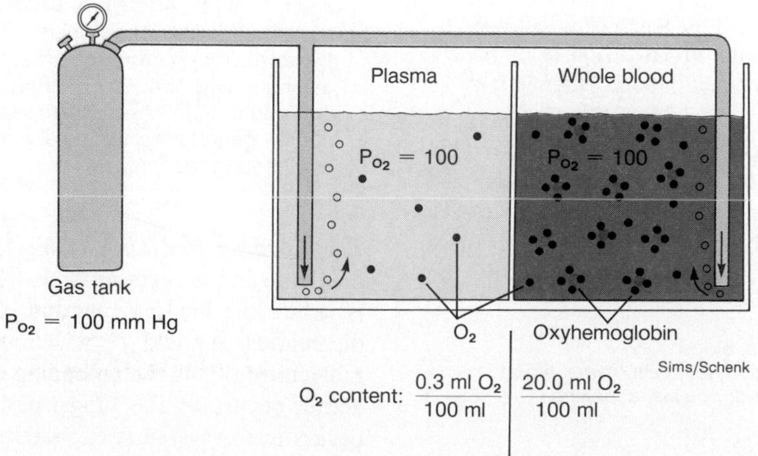

Figure 23.49. (*a*) an illustration of the three-dimensional structure of hemoglobin, in which the two alpha and two beta polypeptide chains are shown; the four heme groups are represented as flat structures with iron (*dark spheres*) in the centers. (*b*) the chemical structure of heme.

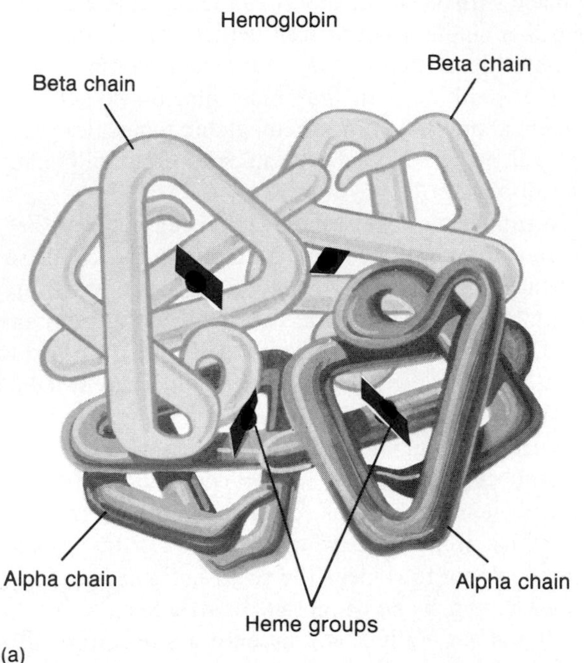

Heme

(b)

According to federal standards, the percent carboxyhemoglobin in the blood of active nonsmokers should not be higher than 1.5%. Studies have shown, however, that these values can range as high as 3% or more in nonsmokers and as high as 10% or more in smokers in some cities. Although these high levels may not cause immediate problems in healthy people, long-term effects on health might occur. People with respiratory or cardiovascular diseases would be particularly vulnerable to the negative effects of carboxyhemoglobin on oxygen transport.

Hemoglobin Concentration. The *oxygen-carrying capacity* of whole blood is determined by its concentration of normal hemoglobin. If the hemoglobin concentration is below normal—a condition called **anemia**—the oxygen concentration of the blood is reduced below normal. Conversely, when the hemoglobin concentration is increased above the normal range—as occurs in **polycythemia** *(pol″e-si-the′me-ah)* (high red blood cell count)—the oxygen-carrying capacity of blood is increased accordingly. This can occur as an adaptation to life at a high altitude.

The production of hemoglobin and red blood cells in bone marrow is controlled by a hormone called **erythropoietin** *(ĕ-rith″ro-poi′ĕ-tin),* produced primarily by the kidneys. This hormone stimulates cell division and differentiation of the erythrocyte stem cells in bone marrow (chapter 20). The production of erythropoietin—and thus the production of red blood cells—is stimulated when the delivery of oxygen to the kidneys and other organs is lower than normal. Red blood cell production is also promoted by androgens, which explains why the hemoglobin concentration in men averages 1–2 g per 100 ml higher than in women.

The kidney is an endocrine gland; it secretes erythropoietin, a glycoprotein hormone containing 166 amino acids. Considering this fact and the fact that erythropoietin is needed to stimulate red blood cell production, it is not surprising that patients with kidney disease usually also suffer from *anemia.* It has recently been shown that this anemia can be effectively treated by giving these patients human erythropoietin. This exciting therapeutic breakthrough was made possible by the commercial production of erythropoietin, using cultured mammalian cells which had incorporated the human gene for erythropoietin through genetic engineering techniques.

The Loading and Unloading Reactions. Reduced hemoglobin and oxygen combine to form oxyhemoglobin; this is called the **loading reaction.** Oxyhemoglobin, in turn, dissociates to yield deoxyhemoglobin and free oxygen molecules; this is the **unloading reaction.** The loading reaction occurs in the lungs, and the unloading reaction occurs in the systemic capillaries.

anemia: Gk. *a,* negative; *haima,* blood
polycythemia: Gk. *polys,* much; *kytos,* cell; *haima,* blood
erythropoietin: Gk. *erythros,* red; *poiesis,* a making

Figure 23.50. The percent of oxyhemoglobin saturation and the blood oxygen content are shown at different values of P_{O_2}. Notice that there is about a 25% decrease in percent oxyhemoglobin as the blood passes through the tissue from arteries to veins, resulting in the unloading of approximately 5 ml O_2 per 100 ml to the tissues.

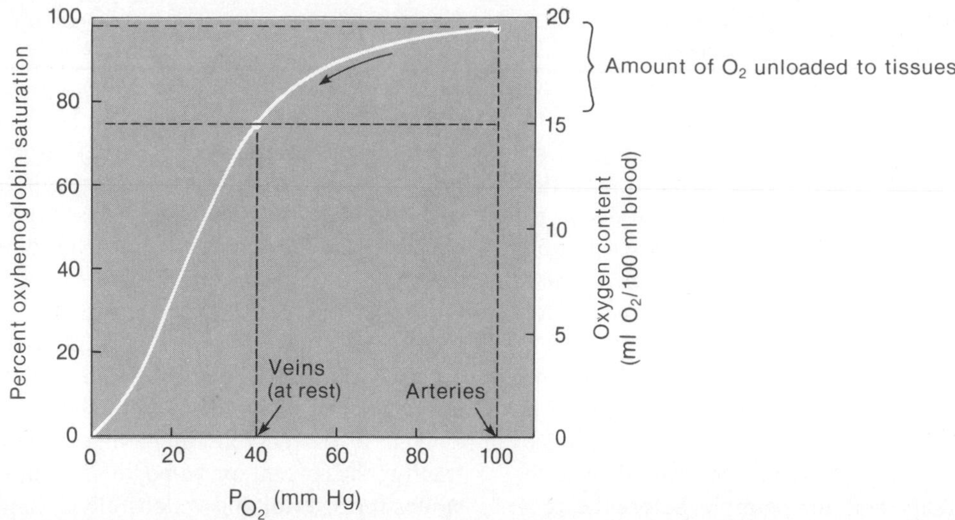

Table 23.11 The relationship between percent oxyhemoglobin saturation and P_{O_2} (at pH = 7.40 and temperature = 37°C)

P_{O_2}(mm Hg)	100	80	61	45	40	36	30	26	23	21	19
Percent oxyhemoglobin	97	95	90	80	75	70	60	50	40	35	30
	Arterial blood				Venous blood						

Loading and unloading can thus be shown as a reversible reaction:

$$\text{Deoxyhemoglobin} + O_2 \underset{\text{(tissues)}}{\overset{\text{(lungs)}}{\rightleftharpoons}} \text{Oxyhemoglobin.}$$

The extent that the reaction will go in each direction depends on two factors: (1) the P_{O_2} of the environment and (2) the *affinity,* or bond strength, between hemoglobin and oxygen. High P_{O_2} drives the loading reaction; at the high P_{O_2} of the pulmonary capillaries almost all the deoxyhemoglobin molecules combine with oxygen. Low P_{O_2} in the systemic capillaries drives the reaction in the opposite direction to promote unloading. The extent of this unloading depends on how low the P_{O_2} values are.

The affinity (bond strength) between hemoglobin and oxygen also influences the loading and unloading reactions. A very strong bond would favor loading but inhibit unloading; a weak bond would hinder loading but improve unloading. The bond strength between hemoglobin and oxygen is normally strong enough so that 97% of the hemoglobin leaving the lungs is in the form of oxyhemoglobin, yet the bond is sufficiently weak so that adequate amounts of oxygen are unloaded to sustain aerobic respiration in the tissues. Under normal resting conditions only about 22% of the oxygen is unloaded; this satisfies the tissue needs for oxygen yet maintains an oxygen reserve in the blood for emergency conditions.

The Oxyhemoglobin Dissociation Curve

Blood in the systemic arteries, at a P_{O_2} of 100 mm Hg, has a *percent oxyhemoglobin saturation* of 97% (which means that 97% of the hemoglobin is in the form of oxyhemoglobin). This blood is delivered to the systemic capillaries, where oxygen diffuses into the tissue cells and is consumed in aerobic respiration. Blood leaving in the systemic veins is thus reduced in oxygen; it has a P_{O_2} of about 40 mm Hg and a percent saturation of about 75% (table 23.11). In other words, blood entering the tissues contains 20 ml O_2 per 100 ml blood, and blood leaving the tissues contains 15.5 ml O_2 per 100 ml blood (fig. 23.50). Thus, 22%, or 4.5 ml of O_2 out of 20 ml O_2 per 100 ml blood, is unloaded to the tissues.

A graphic illustration of the percent oxyhemoglobin saturation at different values of P_{O_2} is called an **oxyhemoglobin dissociation curve** (fig. 23.50). The values in this graph are obtained by subjecting samples of blood *in vitro* (outside the body) to different partial oxygen pressures. The percent oxyhemoglobin saturation obtained by this procedure, however, can be used to predict what the percent unloading would be *in vivo* (within the body) with a given difference in arterial and venous P_{O_2} values.

Figure 23.50 shows the difference between the arterial and venous P_{O_2} and the percent oxyhemoglobin saturation at rest. The relatively large amount of oxyhemoglobin remaining in the venous blood at rest functions

Table 23.12 Effect of pH on hemoglobin affinity for oxygen and oxyhemoglobin "unloading"

pH	Affinity	Arterial O_2 content per 100 ml	Venous O_2 content per 100 ml	O_2 unloaded to tissues per 100 ml
7.40	Normal	19.8 ml O_2	14.8 ml O_2	5.0 ml O_2
7.60	Increased	20.0 ml O_2	17.0 ml O_2	3.0 ml O_2
7.20	Decreased	19.2 ml O_2	12.6 ml O_2	6.6 ml O_2

as an oxygen reservoir. If a person stops breathing, there is a sufficient reserve of oxygen in the blood to keep the brain and heart alive for approximately four to five minutes in the absence of cardiopulmonary resuscitation (CPR) techniques. This reserve supply of oxygen can also be tapped when the tissue's requirements for oxygen are raised.

The oxyhemoglobin dissociation curve is S-shaped, or *sigmoidal.* The fact that it is relatively flat at high P_{O_2} values indicates that changes in P_{O_2} within this range have little effect on the loading reaction. One would have to ascend as high as 10,000 feet, for example, before the oxyhemoglobin saturation or arterial blood would decrease from 97% to 93%. At more common elevations the percent oxyhemoglobin saturation would not be significantly different from the 97% value at sea level.

At the steep part of the sigmoidal curve, however, small changes in P_{O_2} values produce large differences in percent saturation. A decrease in *venous* P_{O_2} from 40 mm Hg to 30 mm Hg, as might occur during mild exercise, corresponds to a change in percent saturation from 75% to 58%. Since the *arterial* percent saturation is usually still 97% during mild exercise, this change in venous percent saturation indicates that more oxygen has been unloaded to the tissues. The difference between the arterial and venous percent saturations indicates the percent unloading: in these examples, 97% minus 75% = 22% unloading at rest, and 97% minus 58% = 39% unloading during exercise. During heavier exercise, the venous P_{O_2} can drop to 20 mm Hg or less, indicating a percent unloading in excess of 70%.

Effect of pH and Temperature on Oxygen Transport

In addition to changes in P_{O_2}, the loading and unloading reactions are influenced by changes in the bond strength, or affinity, of hemoglobin for oxygen. The affinity is decreased when the pH is lowered and increased when the pH is raised; this is called the **Bohr effect.** When the affinity of hemoglobin for oxygen is reduced, there is less loading of the blood with oxygen in the lungs but greater unloading of oxygen in the tissues. A weakening of the bond between hemoglobin and oxygen, however, has less effect at the higher P_{O_2} values in the lungs than at the lower P_{O_2} values of the tissues. A lowering of pH thus results in slightly less oxygen loading in the lungs and significantly more oxygen unloading in the tissues. The net effect is that the tissues receive more oxygen when the blood pH is lowered (table 23.12). Since the pH can be decreased by carbon dioxide (through the formation of carbonic acid), the Bohr effect helps to provide a little more oxygen to the tissues when their carbon dioxide output (and metabolism) is increased.

When the percent oxyhemoglobin saturation at different pH values is graphed as a function of P_{O_2}, the dissociation curve is shown to be shifted to the right by a lowering of pH and shifted to the left by a rise in pH (fig. 23.51). If the percent unloading is calculated by subtracting the percent oxyhemoglobin saturation at given P_{O_2} values for arterial and venous blood, it will be clear that a *shift to the right* of the curve indicates a greater oxygen unloading, whereas a *shift to the left* indicates less unloading but slightly more oxygen loading in the lungs.

When oxyhemoglobin dissociation curves are constructed at constant pH values but at different temperatures, it can be seen that the affinity of hemoglobin for oxygen is decreased by a rise in temperature. An increase in temperature weakens the bond between hemoglobin and oxygen and thus has the same effect as a fall in pH; the oxyhemoglobin dissociation curve is shifted to the right (fig. 23.52). At higher temperatures, therefore, more oxygen is unloaded to the tissues than would be the case if the bond strength were constant. This effect can significantly increase the delivery of oxygen to muscles that are warmed during exercise.

Effect of 2,3-DPG on Oxygen Unloading

Mature red blood cells lack both nuclei and mitochondria. Without mitochondria they cannot respire aerobically and thus cannot use the oxygen they carry. Red blood cells, therefore, obtain energy through the anaerobic respiration of glucose. At a certain point in the glycolytic pathway there is a "side reaction" in red blood cells that results in a unique product—**2,3-diphosphoglyceric acid (2,3-DPG).**

The enzyme that produces 2,3-DPG is inhibited by oxyhemoglobin. When the oxyhemoglobin concentration is decreased, therefore, the production of 2,3-DPG is increased. This increase in 2,3-DPG production can occur when the total hemoglobin concentration is low (in anemia) or when the P_{O_2} is low (at a high altitude, for example). 2,3-DPG combines with deoxyhemoglobin and makes it more stable. At the P_{O_2} values in the tissue capillaries, therefore, a higher proportion of the oxyhemoglobin will be converted to deoxyhemoglobin by unloading its oxygen. Increased concentrations of 2,3-DPG in red blood cells thus increase oxygen unloading (table 23.13) and shift the oxyhemoglobin dissociation curve to the right.

Figure 23.51. A decrease in blood pH (an increase in H^+ concentration) decreases the affinity of hemoglobin for oxygen at each P_{O_2} value, resulting in a "shift to the right" of the oxyhemoglobin dissociation curve. A curve that is shifted to the right has a lower percent oxyhemoglobin saturation at each P_{O_2}, but the effect is more marked at lower P_{O_2} values. This is called the *Bohr effect.*

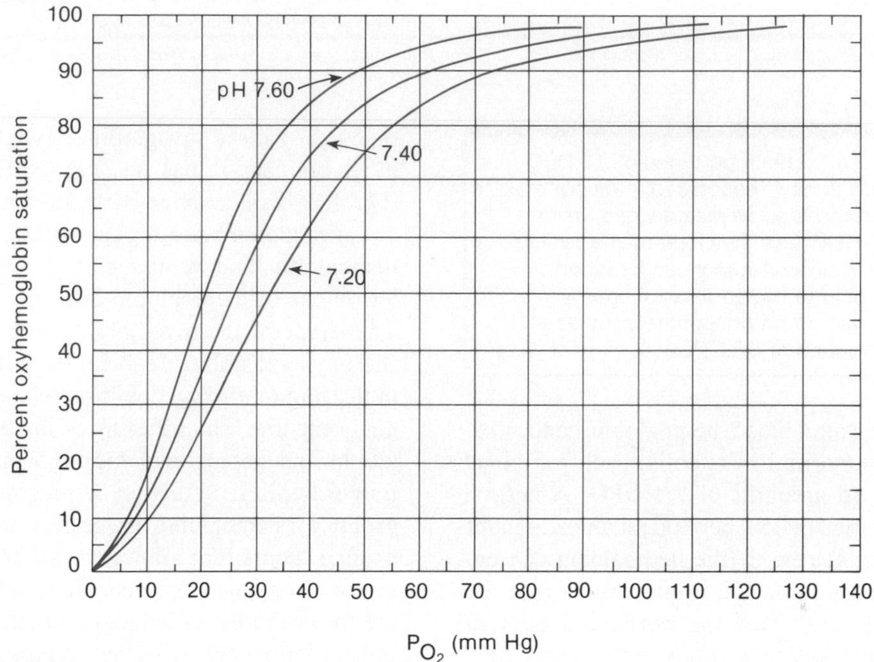

Figure 23.52. The oxyhemoglobin dissociation curve is shifted to the right as the temperature increases, indicating a lowered affinity of hemoglobin for oxygen at each P_{O_2}. This effect, like the Bohr effect (see figure 23.51), is more marked at lower P_{O_2} values.

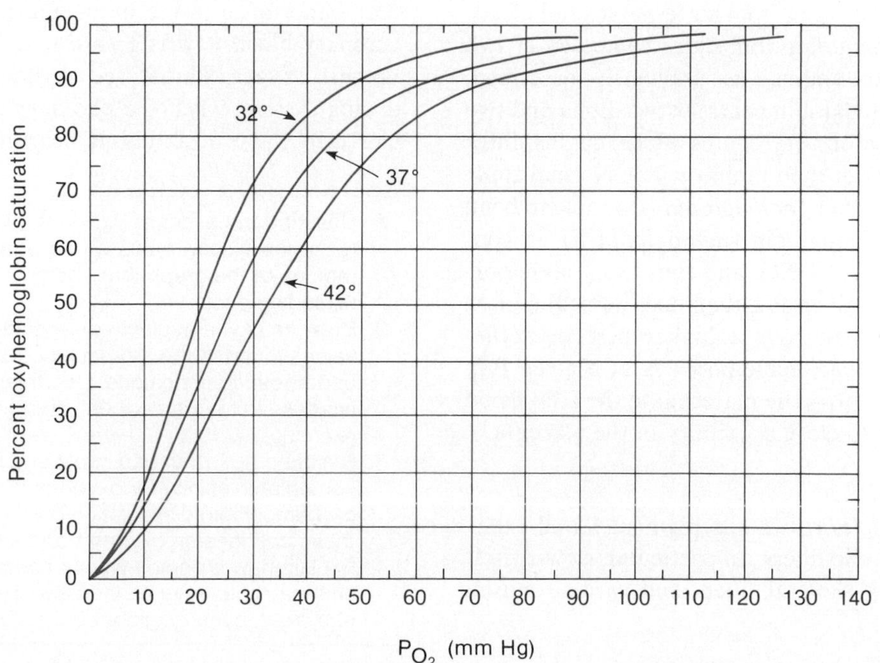

Table 23.13 Factors that affect the affinity of hemoglobin for oxygen and the position of the oxyhemoglobin dissociation curve

Factor	Affinity	Position of curve	Comments
↓ pH	Decreased	Shift to the right	Called the Bohr effect; increases oxygen delivery during hypercapnia
↑ Temperature	Decreased	Shift to the right	Increases oxygen unloading during exercise and fever
↑ 2,3-DPG	Decreased	Shift to the right	Increases oxygen unloading when there is a decrease in total hemoglobin or total oxygen content; an adaptation to anemia and high-altitude living

The importance of 2,3-DPG within red blood cells is now recognized in *blood banking.* Old, stored red blood cells can lose their ability to produce 2,3-DPG as they lose their ability to metabolize glucose. Modern techniques for blood storage, therefore, include the addition of energy substrates for respiration and phosphate sources needed for the production of 2,3-DPG.

Anemia. When the total blood hemoglobin concentration is reduced below normal in anemia, each red blood cell produces increased amounts of 2,3-DPG. A normal hemoglobin concentration of 15 g per 100 ml unloads about 4.5 ml O_2 per 100 ml at rest. If the hemoglobin concentration were reduced by half, you might expect that the tissues would receive only half the normal amount of oxygen (2.25 ml O_2 per 100 ml). It has been shown, however, that as much as 3.3 ml O_2 per 100 ml are unloaded to the tissues under these conditions. This occurs as a result of a rise in 2,3-DPG production that produces a decrease in hemoglobin affinity for oxygen.

Hemoglobin F. The effects of 2,3-DPG are also important in the transfer of oxygen from maternal to fetal blood. The mother has hemoglobin molecules composed of two alpha and two beta chains, as previously described, whereas the fetal hemoglobin contains two alpha and two *delta chains* in place of beta chains (delta chains differ from beta chains in 37 of their amino acids). Normal adult hemoglobin in the mother (*hemoglobin A*) is able to bond to 2,3-DPG. Fetal hemoglobin (*hemoglobin F*), in contrast, cannot bond to 2,3-DPG, and thus has a higher affinity for oxygen at a given P_{O_2} than does hemoglobin A. Since hemoglobin F can have a higher percent oxyhemoglobin saturation than hemoglobin A at a given P_{O_2}, oxygen is transferred from the maternal to the fetal blood as these two come into close proximity in the placenta.

Muscle Myoglobin

Myoglobin *(mi''o-glo'bin)* is a red pigment found exclusively in striated muscle fibers. In particular, slow-twitch, aerobically respiring skeletal fibers and cardiac muscle fibers are rich in myoglobin. Myoglobin is similar to hemoglobin, but it has one rather than four hemes and, therefore, can combine with only one molecule of oxygen.

Myoglobin has a higher affinity for oxygen than does hemoglobin, and its dissociation curve is therefore to the left of the oxyhemoglobin dissociation curve (fig. 23.53). The shape of the myoglobin curve is also different from the oxyhemoglobin dissociation curve; it is rectangular, indicating that oxygen will be released only when the P_{O_2} gets very low. The differences in behavior between myoglobin and hemoglobin highlight the importance of the four-subunit structure of hemoglobin. When one heme group of hemoglobin combines with oxygen, a shape change occurs that allows the other heme groups to combine with oxygen more easily. The same interaction beween the four subunits of hemoglobin allows the hemoglobin to unload its oxygen at higher values of P_{O_2} than can myoglobin.

Since the P_{O_2} in mitochondria is very low (because oxygen is incorporated into water here), myoglobin may act as a "middleman" in the transfer of oxygen from blood to the mitochondria within muscle cells. Myoglobin may also have an oxygen storage function, which is of particular importance in the heart. During diastole, when the coronary blood flow is greatest, the myoglobin can load up with oxygen. This stored oxygen can then be released during systole, when the coronary arteries are squeezed closed by the contracting myocardium.

1. Describe the effect of P_{O_2} on the loading and unloading reactions. Illustrate this with a graph, and label the regions of the graph that show when loading and unloading occur.
2. Draw an oxyhemoglobin dissociation curve, and label the values found in the arterial blood and venous blood under resting conditions. Use this graph to show an increase in the percent unloading that can occur during exercise.
3. Describe how changes in pH and temperature affect the hemoglobin affinity for oxygen and the position of the oxyhemoglobin dissociation curve. Explain the effect of these changes on oxygen transport.
4. Explain how a person who is anemic or a person at high altitude could have an increase in the percent unloading of oxygen by hemoglobin.

Figure 23.53. A comparison of the dissociation curves for hemoglobin and for myoglobin. At the P_{O_2} of venous blood, the myoglobin retains almost all of its oxygen, indicating a higher affinity than hemoglobin for oxygen. The myoglobin does, however, release its oxygen at the very low P_{O_2} values found inside the mitochondria.

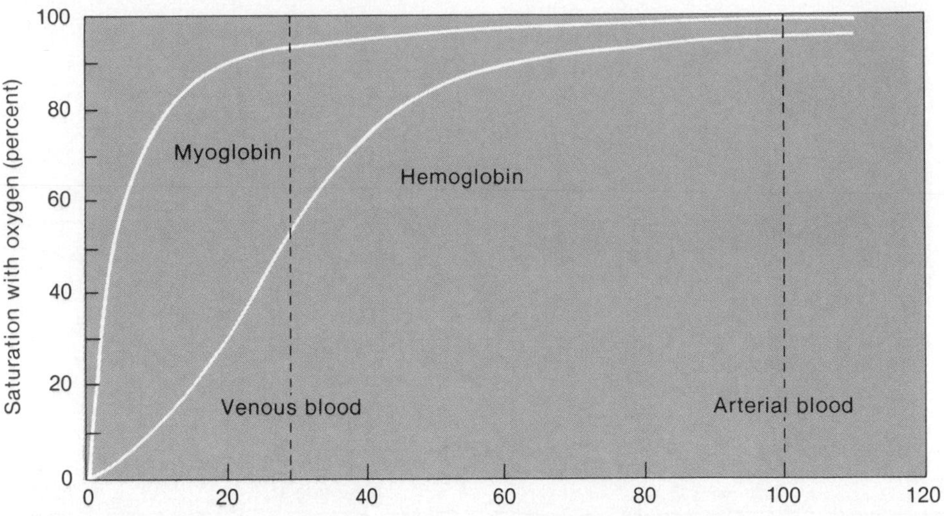

Carbon Dioxide Transport and Acid-Base Balance

Carbonic acid and bicarbonate are the major transport forms of carbon dioxide and the major contributors to the acid-base balance of the blood. Since the concentration of these compounds is partly controlled by ventilation, pulmonary function is an important regulator of the acid-base balance in the body.

Objective 29. Describe how carbon dioxide is transported in the blood, and describe the chloride shift.
Objective 30. Define the terms respiratory acidosis, respiratory alkalosis, metabolic acidosis, and metabolic alkalosis.
Objective 31. Explain how ventilation helps to regulate the acid-base balance of the blood.

Carbon dioxide is carried by the blood in three forms: (1) as *dissolved CO_2*—since carbon dioxide is about twenty-one times more soluble than oxygen in water, a substantial portion (about one-tenth) of the total blood CO_2 is dissolved in plasma; (2) as *carbaminohemoglobin (kar-bam''i-no-he''mo-glo'bin)*—about one-fifth of the total blood CO_2 is carried attached to an amino acid in hemoglobin (carbaminohemoglobin should not be confused with carboxyhemoglobin, which is a combination of hemoglobin with carbon monoxide); and (3) as *bicarbonate,* which account for about 70% of the CO_2 carried by the blood.

Carbon dioxide is able to combine with water to form carbonic acid. This reaction occurs spontaneously in the plasma at a slow rate but occurs much more rapidly within the red blood cells due to the catalytic action of the enzyme **carbonic anhydrase.** Since this enzyme is confined to the red blood cells, most of the carbonic acid is produced there rather than in the plasma. The formation of carbonic acid from CO_2 and water is favored by the high P_{CO_2} found in tissue capillaries (this is an example of the *law of mass action,* described in chapter 4).

$$CO_2 + H_2O \xrightarrow[\text{high } P_{CO_2}]{\text{carbonic anhydrase}} H_2CO_3$$

The Chloride Shift

As a result of catalysis by carbonic anhydrase within the red blood cells, large amounts of carbonic acid are produced as blood passes through systemic capillaries. The buildup of carbonic acid concentrations within the red blood cells favors the dissociation of these molecules into *H^+* (protons, which contribute to the acidity of a solution) and *HCO_3^-* (bicarbonate):

$$H_2CO_3 \longrightarrow H^+ + HCO_3^-.$$

The H^+ released by the dissociation of carbonic acid are largely buffered by their combination with deoxyhemoglobin within the red blood cells. Although the unbuffered H^+ are free to diffuse out of the red blood cells, more bicarbonate diffuses outward into the plasma than does H^+. As a result of the "trapping" of H^+ within the red blood cells by their attachment to hemoglobin and the outward

Figure 23.54. An illustration of carbon dioxide transport by the blood and the "chloride shift." Carbon dioxide is transported in two forms: as dissolved CO_2 gas, attached to hemoglobin as carbaminohemoglobin, and as carbonic acid and bicarbonate.

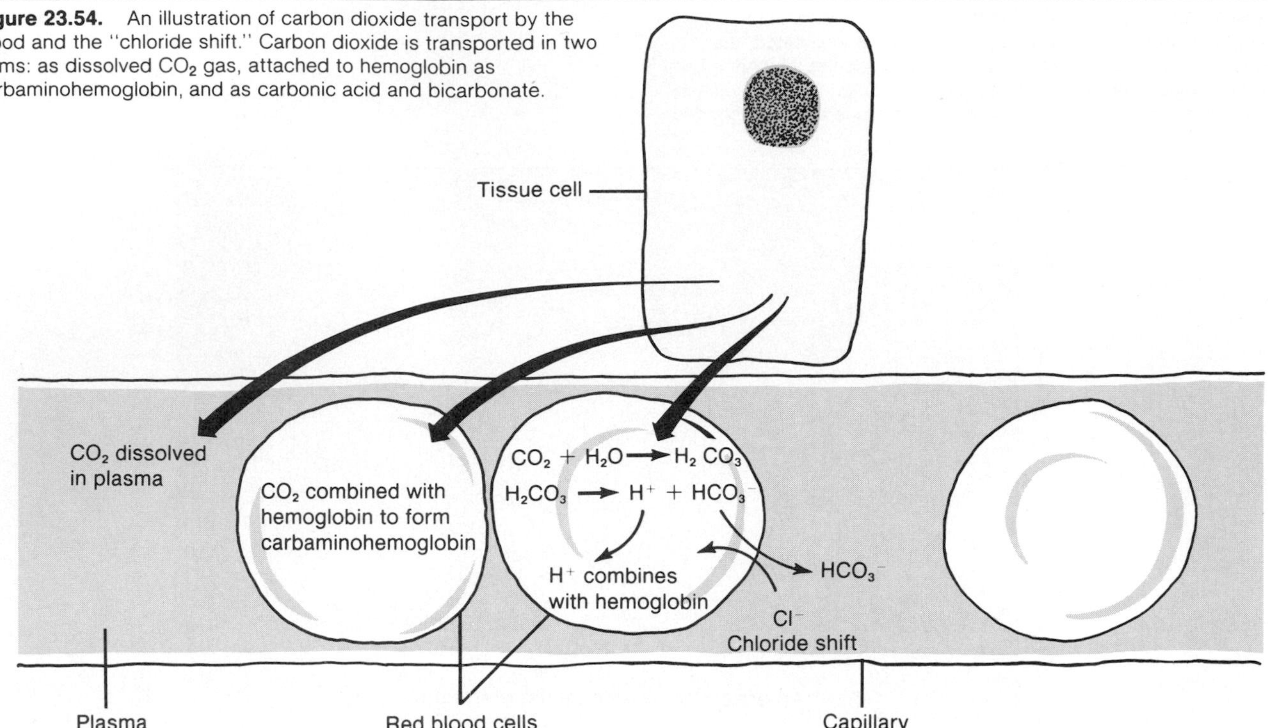

diffusion of bicarbonate, the inside of the red blood cell gains a net positive charge. This attracts chloride ions (Cl^-), which move into the red blood cells as HCO_3^- moves out. This exchange of anions as blood moves through the tissue capillaries is called the *chloride shift* (fig. 23.54).

When blood reaches the pulmonary capillaries, deoxyhemoglobin is converted to oxyhemoglobin. Since oxyhemoglobin has a lower affinity for H^+ than does deoxyhemoglobin, H^+ are released within the red blood cells. This attracts HCO_3^- from the plasma, which combines with H^+ to form carbonic acid:

$$H^+ + HCO_3^- \longrightarrow H_2CO_3.$$

Under conditions of low P_{CO_2}, as occurs in the pulmonary capillaries, carbonic anhydrase catalyzes the conversion of carbonic acid to carbon dioxide and water:

$$H_2CO_3 \xrightarrow[\text{low } P_{CO_2}]{\text{carbonic anhydrase}} CO_2 + H_2O.$$

In summary, the carbon dioxide produced by the tissue cells is converted within the systemic capillaries, mostly through the action of carbonic anhydrase in the red blood cells, to carbonic acid. With the buildup of carbonic acid concentrations in the RBCs, the carbonic acid dissociates into bicarbonate and H^+, which results in the chloride shift. A *reverse chloride shift* operates in the pulmonary capillaries to convert carbonic acid and bicarbonate to CO_2 gas, which is eliminated in the expired breath (fig. 23.55). The P_{CO_2}, carbonic acid, H^+, and bicarbonate concentrations in the systemic arteries are thus maintained relatively constant by normal ventilation.

Ventilation and Acid-Base Balance

Normal systemic arterial blood has a pH of 7.35 to 7.45. In other words, arterial blood has a H^+ concentration of about $10^{-7.4}$ molar. Some of these H^+ are derived from carbonic acid, and some are derived from nonvolatile *metabolic acids* (fatty acids, ketone bodies, lactic acid, and others) that cannot be eliminated in the expired breath.

Under normal conditions the H^+ released by metabolic acids do not affect blood pH because these H^+ combine with HCO_3^- and are thereby removed from solution. *Bicarbonate is the major buffer in the plasma* and acts to maintain a blood pH of 7.4 despite the constant production of nonvolatile metabolic acids by the tissues. In this buffering process, some of the HCO_3^- released from the red blood cells during the chloride shift is converted into H_2CO_3 in the plasma. Normally, however, there is still a buffer reserve of free bicarbonate that can help protect against unusually large additions of metabolic acids to the blood. These processes are illustrated in figure 23.56.

Normal plasma, therefore, contains free bicarbonate, carbonic acid, and H^+ concentrations indicated by a pH of 7.4. If the H^+ concentration of the blood should fall, the carbonic acid produced by the buffering reaction can dissociate and serve as a source of additional H^+. If the H^+ concentration should rise, bicarbonate can remove this excess H^+ from solution. Carbonic acid and bicarbonate are thus said to function as a *buffer pair.*

Acidosis and Alkalosis. A fall in blood pH below 7.35 is called **acidosis** *(as''ĭ-do'sis),* because the pH is to the acid-side of normal. Acidosis does not mean acidic (pH less than 7); a blood pH of 7.2, for example, represents

Figure 23.55. Carbon dioxide is released from the blood as it travels through the pulmonary capillaries. During this time a "reverse chloride shift" occurs and carbonic acid is transformed into CO_2 and H_2O.

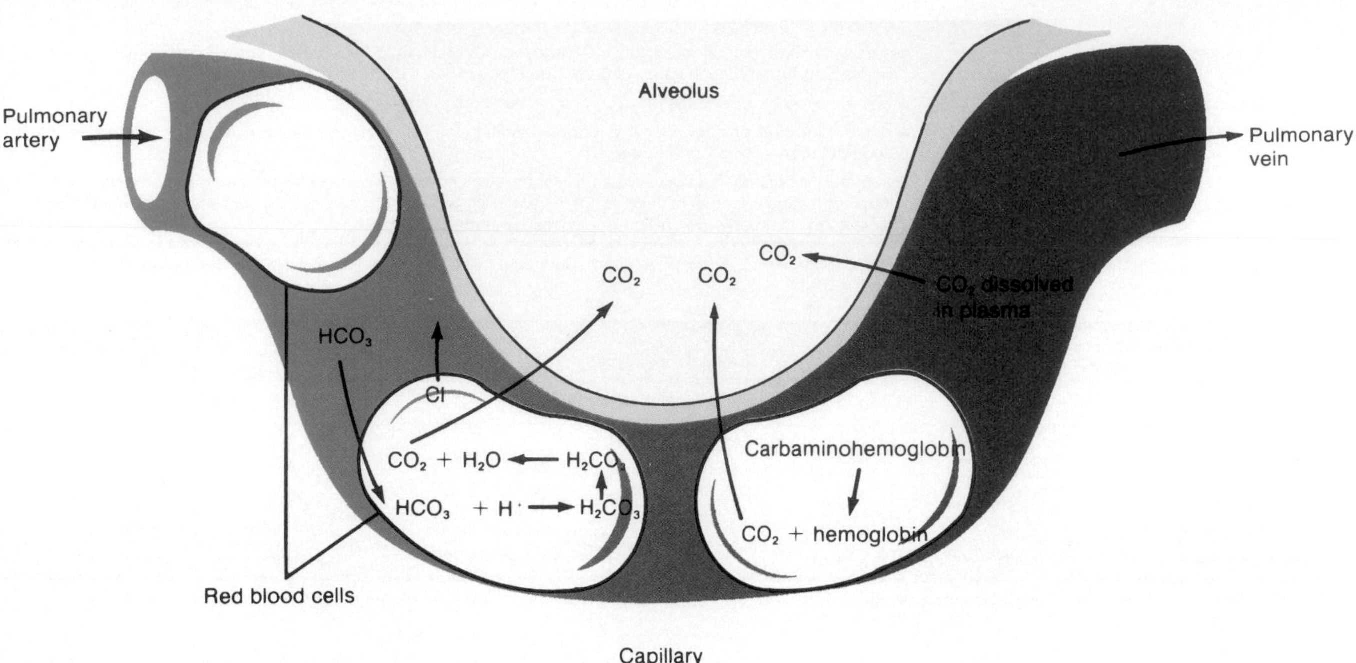

Figure 23.56. Bicarbonate released into the plasma from red blood cells functions to buffer H^+ produced by the ionization of metabolic acids (lactic acid, fatty acids, ketone bodies, and others).

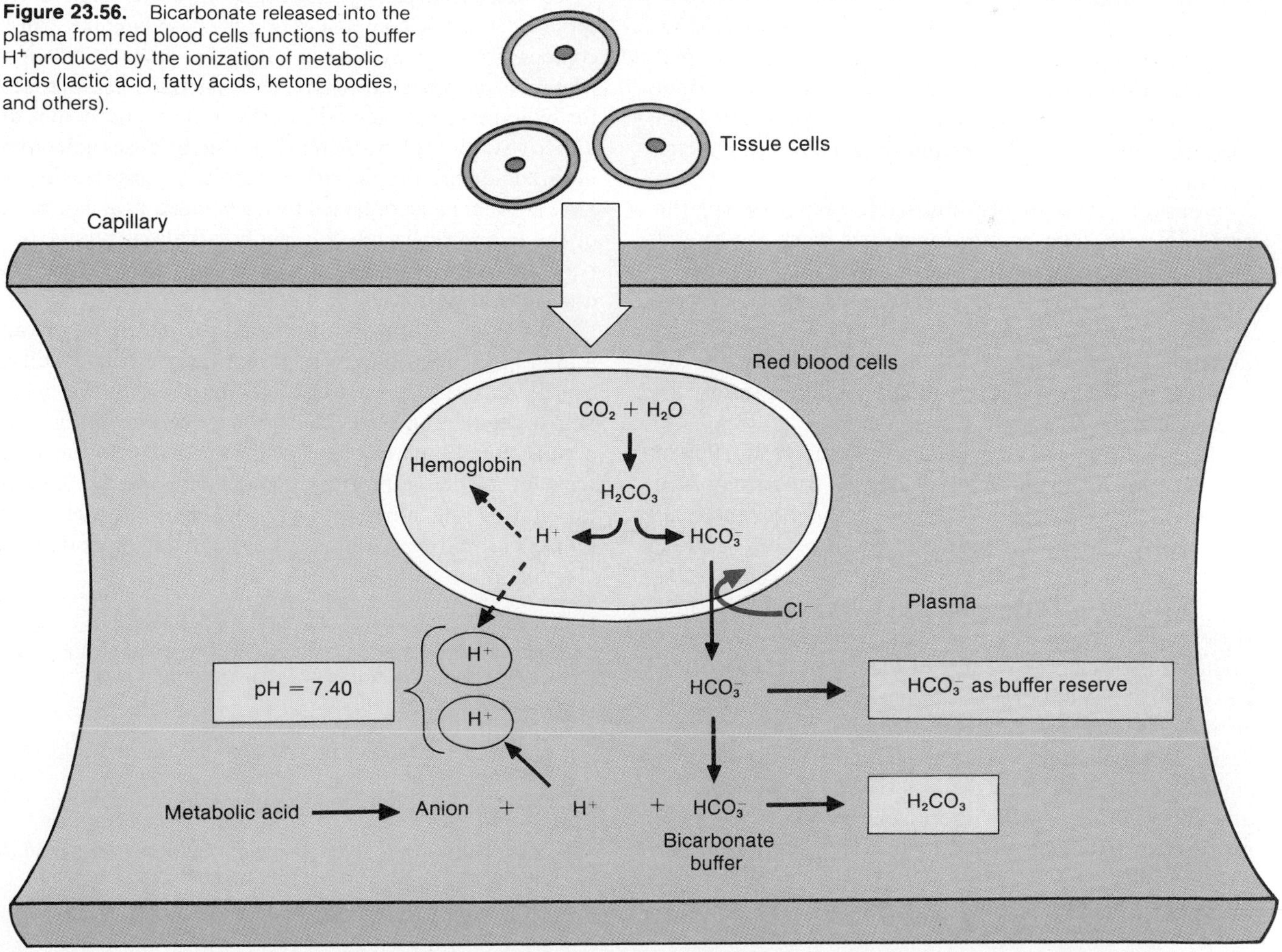

Tissue cells
Capillary
Red blood cells
CO_2 + H_2O
H_2CO_3
Hemoglobin
H^+
HCO_3^-
Cl^-
Plasma
H^+
H^+
pH = 7.40
HCO_3^-
HCO_3^- as buffer reserve
Metabolic acid → Anion + H^+ + HCO_3^- → H_2CO_3
Bicarbonate buffer

Table 23.14 Definition of terms used to describe acid-base balance

Terms	Definitions
Acidosis, respiratory	Increased carbon dioxide retention (due to hypoventilation), which can result in the accumulation of carbonic acid and thus a fall in blood pH below normal
Acidosis, metabolic	Increased production of "nonvolatile" acids such as lactic acid, fatty acids, and ketone bodies, or loss of blood bicarbonate (such as by diarrhea) resulting in a fall in blood pH below normal
Alkalosis, respiratory	A rise in blood pH due to loss of CO_2 and carbonic acid (through hyperventilation)
Alkalosis, metabolic	A rise in blood pH produced by loss of nonvolatile acids (as in excessive vomiting) or by excessive accumulation of bicarbonate base
Compensated acidosis or alkalosis	Metabolic acidosis or alkalosis is partially compensated by opposite changes in blood carbonic acid levels (through changes in ventilation). Respiratory acidosis or alkalosis is partially compensated by increased retention or excretion of bicarbonate in the urine.

Table 23.15 The effect of lung function on blood acid-base balance

Condition	pH	P_{CO_2}	Ventilation	Cause or compensation
Normal	7.35–7.45	39–41 mm Hg	Normal	Not applicable
Respiratory acidosis	Low	High	Hypoventilation	Cause of the acidosis
Respiratory alkalosis	High	Low	Hyperventilation	Cause of the alkalosis
Metabolic acidosis	Low	Low	Hyperventilation	Compensation for acidosis
Metabolic alkalosis	High	High	Hypoventilation	Compensation for alkalosis

serious acidosis. Similarly, a rise in blood pH above 7.45 is known as **alkalosis** *(al"kah-lo'sis).* There are two components to acid-base balance: *respiratory* and *metabolic.* Respiratory acidosis and alkalosis are due to abnormal concentrations of carbonic acid, as a result of abnormal ventilation; metabolic acidosis and alkalosis result from abnormal amounts of H^+ derived from nonvolatile metabolic acids (table 23.14).

Ventilation is normally adjusted to keep pace with the metabolic rate. *Hypoventilation* occurs when the rate of CO_2 production exceeds the rate at which it is "blown off" in ventilation. Under these conditions, carbonic acid production is excessively high compared to ventilation, and *respiratory acidosis* occurs. In *hyperventilation,* the depletion of carbonic acid raises the pH, and **respiratory alkalosis** occurs.

Metabolic acidosis can occur when the production of nonvolatile acids is abnormally increased. In uncontrolled diabetes mellitus, for example, ketone bodies (derived from fatty acids) may accumulate and produce *ketoacidosis (ke"to-ah"sĭ-do'sis).* In order for metabolic acidosis to occur, however, the buffer reserve of bicarbonate must first be depleted (this is why metabolic acidosis can also be produced by the excessive loss of bicarbonate, as in diarrhea). Until the buffer reserve is depleted, the pH remains normal—ketosis can occur, for example, without ketoacidosis. **Metabolic alkalosis,** less common than metabolic acidosis, can result when there is a loss of acidic gastric juice by vomiting or when there is an excessive intake of bicarbonate (from stomach antacids or from an intravenous solution).

Compensations for Acidosis and Alkalosis. A change in blood pH, produced by alterations in either the respiratory or metabolic component of acid-base balance, can be compensated for by a change in the other component. Metabolic acidosis, for example, is partially compensated for by hyperventilation (the aortic and carotid bodies are directly stimulated by blood H^+), which causes a decrease in carbonic acid. People with metabolic acidosis would thus have a low pH accompanied by a low blood P_{CO_2} as a result of the hyperventilation. Metabolic alkalosis, similarly, is partially compensated for by the retention of carbonic acid due to hypoventilation (table 23.15).

A person with respiratory acidosis has a low pH and a high blood P_{CO_2} due to hypoventilation. This condition can be partially compensated for by the kidneys, which help to regulate the blood bicarbonate concentration. Two organs thus regulate blood acid-base balance: the lungs (regulating the respiratory component) and the kidneys (regulating the metabolic component). The role of the kidneys in acid-base balance is discussed in more detail in chapter 24.

1. List the ways that carbon dioxide is carried by the blood. Using equations, show how carbonic acid and bicarbonate are formed.
2. Describe the events that occur in the chloride shift in the systemic capillaries, and describe the reverse chloride shift that occurs in the pulmonary capillaries.
3. Describe the functions of bicarbonate and carbonic acid in blood.
4. Describe how hyperventilation and hypoventilation affect the blood pH, and explain the mechanisms involved.

Effect of Exercise and High Altitude on Respiratory Function

Changes in ventilation and oxygen delivery occur during exercise and during acclimatization to a high altitude. These changes help to compensate for the increased metabolic rate during exercise and for the decreased arterial P_{O_2} at a high altitude.

Objective 32. Describe the effect of exercise on blood P_{O_2}, P_{CO_2}, and pH, and describe the theories that have been proposed to explain the increase in ventilation that occurs during exercise.

Objective 33. Explain the effect of endurance training on the anaerobic threshold.

Objective 34. Explain the respiratory adaptations that occur during acclimatization to life at a high altitude.

Ventilation during Exercise

Immediately upon exercise, the rate and depth of breathing increase to produce a total minute volume that is many times the resting value. This increased ventilation, particularly in well-trained athletes, is exquisitely matched to the simultaneous increase in oxygen consumption and carbon dioxide production by the exercising muscles. The arterial blood P_{O_2}, P_{CO_2}, and pH, thus, remain surprisingly constant during exercise (fig. 23.57).

It is tempting to suppose that ventilation increases during exercise as a result of the increased CO_2 production by the exercising muscles. Ventilation increases together with increased CO_2 production, however, so that blood measurements of P_{CO_2} during exercise are not significantly higher than at rest. The mechanisms responsible for the increased ventilation during exercise must therefore be more complex.

Two groups of mechanisms—*neurogenic* and *humoral*—have been proposed to explain the increased ventilation that occurs during exercise. Possible neurogenic mechanisms include the following: (1) sensory nerve activity from the exercising limbs may stimulate the respiratory muscles, either directly or via the brain stem respiratory centers; and/or (2) input from the cerebral cortex may stimulate the brain stem centers to modify ventilation. These neurogenic theories help to explain the immediate increase in ventilation that occurs at the beginning of exercise.

Rapid and deep ventilation continues after exercise has stopped, suggesting that humoral (chemical) factors in the blood may also stimulate ventilation during exercise. Since the P_{O_2}, P_{CO_2}, and pH of the blood samples from exercising subjects are within the resting range, these humoral theories propose that (1) the P_{CO_2} and pH in the region of the chemoreceptors may be different from these values "downstream" where blood samples are taken; and/or (2) there may be cyclic variations in these values that stimulate the chemoreceptors but cannot be detected by blood samples. The evidence suggests that both neurogenic and humoral mechanisms are involved in the **hyperpnea** *(hi"perp-ne'ah)*, or increased ventilation, of exercise.

Figure 23.57. The effect of moderate and heavy exercise on arterial blood gases and pH. Notice that there are no consistent and significant changes in these measurements during the first several minutes of moderate and heavy exercise and that only the P_{CO_2} changes (actually decreases) during more prolonged exercise.

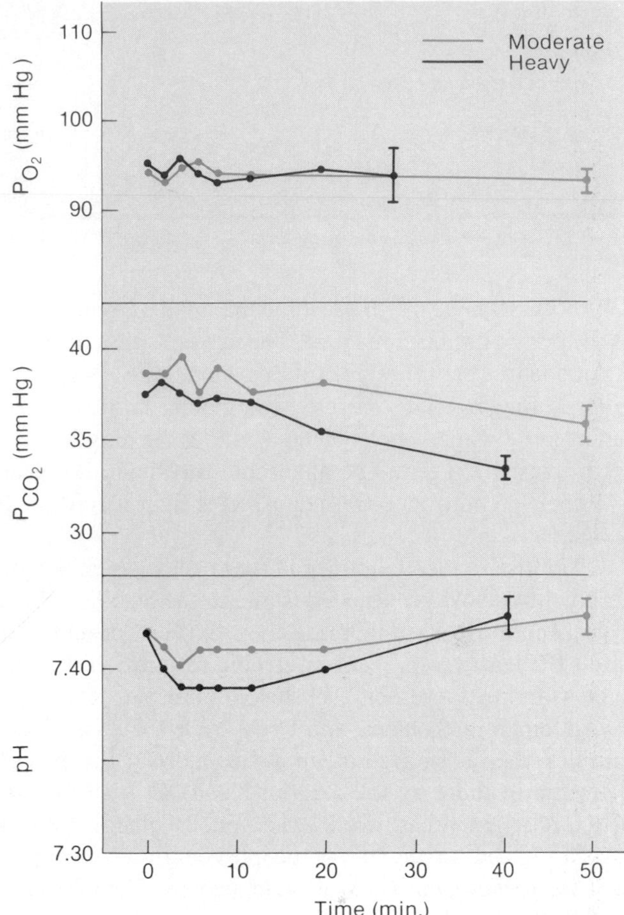

Anaerobic Threshold and Endurance Training. The ability of the cardiopulmonary system to deliver adequate amounts of oxygen to the exercising muscles at the beginning of exercise may be insufficient, as a result of the time lag required to make proper cardiovascular adjustments. During this time, therefore, the muscles respire anaerobically and a "stitch in the side"—probably due to hypoxia of the diaphragm—may develop. After the cardiovascular adjustments have been made, a person may experience a "second wind" when the muscles receive sufficient oxygen for their needs.

Continued heavy exercise can cause a person to reach the **anaerobic threshold,** which is the maximum rate of oxygen consumption that can be attained before blood lactic acid levels rise as a result of anaerobic respiration.

Table 23.16 Changes in respiratory function during exercise

Variable	Change	Comments
Ventilation	Increased	This is not hyperventilation because ventilation is matched to increased metabolic rate. Mechanisms responsible for increased ventilation are not well understood.
Blood gases	No change	Blood-gas measurements during light, moderate, and heavy exercise show little change because ventilation is increased to match increased muscle oxygen consumption and carbon dioxide production.
Oxygen delivery to muscles	Increased	Although the total oxygen content and P_{O_2} do not increase during exercise, there is an increased rate of blood flow to the exercising muscles.
Oxygen extraction by muscles	Increased	Increased oxygen consumption lowers the tissue P_{O_2} and lowers the affinity of hemoglobin for oxygen (due to the effect of increased temperature). More oxygen, as a result, is unloaded so that venous blood contains a lower oxyhemoglobin saturation than at rest. This effect is enhanced by endurance training.

This occurs when 50%–60% of the maximal oxygen uptake of the person has been reached. The rise in lactic acid levels is not due to a malfunction of the cardiopulmonary system; indeed, the arterial oxygen hemoglobin saturation remains at 97% and venous blood draining the muscles contains unused oxygen. The anaerobic threshold, however, is higher in endurance-trained athletes than it is in other people.

The rise in blood lactic acid that occurs when the anaerobic threshold is exceeded is due to the inability of the exercising muscles to increase their oxygen consumption rate sufficiently to prevent anaerobic respiration. Endurance training increases the skeletal muscle content of myoglobin, mitochondria, and Krebs cycle enzymes. These muscles, therefore, are able to utilize more of the oxygen delivered to them by the arterial blood. At a given level of exercise, in other words, the venous blood draining muscles in endurance-trained people contains a lower percent oxyhemoglobin than in other people. The effects of exercise and endurance training on respiratory function are summarized in table 23.16.

Acclimatization to High Altitude

When a person who is from a region near sea level moves to a significantly higher altitude, several adjustments in the respiratory system must be made to compensate for the decreased atmospheric pressure and P_{O_2} at the higher elevation. These adjustments include changes in ventilation, in the hemoglobin affinity for oxygen, and in the total hemoglobin concentration.

Changes in Ventilation. A decrease in arterial P_{O_2} makes the chemoreceptors more sensitive to increases in P_{CO_2}; a decrease in P_{O_2} below 50 mm Hg directly stimulates the carotid bodies to increase ventilation. These changes cause *hyperventilation,* which becomes stabilized after a few days at about 2–3 L per minute more than the total minute volume at sea level. As a result of this hyperventilation, the arterial P_{CO_2} decreases from 40 mm Hg (its value at sea level) to about 29 mm Hg.

Hyperventilation cannot, of course, increase blood P_{O_2} above that of the inspired air. The P_{O_2} of arterial blood is therefore low at high altitudes. In the Peruvian Andes, for example, the normal arterial P_{O_2} is reduced from 100 mm Hg (at sea level) to 45 mm Hg. The loading of hemoglobin with oxygen is therefore incomplete, producing an oxyhemoglobin saturation that is decreased from 97% (at sea level) to 81%.

Hemoglobin Affinity for Oxygen. Normal arterial blood at sea level only unloads about 22% of its oxygen to the tissues at rest; the percent saturation is reduced from 97% in arterial blood to 75% in venous blood. As a partial compensation for the decrease in oxygen content at high altitude, the affinity of hemoglobin for oxygen is reduced, so that a higher proportion of oxygen is unloaded. This occurs because the low oxyhemoglobin content of red blood cells stimulates the production of 2,3-DPG, which in turn decreases the hemoglobin affinity for oxygen.

At very high altitudes, however, the story becomes more complex. At the summit of Mount Everest, in one study, the very low arterial P_{O_2} (28 mm Hg) stimulated intense hyperventilation, so that the arterial P_{CO_2} was decreased to 7.5 mm Hg. The resultant respiratory alkalosis (arterial pH greater than 7.7) caused a shift to the left of the oxyhemoglobin dissociation curve despite the antagonistic effects of increased 2, 3-DPG concentrations. It was suggested that the increased affinity of hemoglobin for oxygen caused by the respiratory alkalosis may have been beneficial at such a high altitude because it would have increased the loading of hemoglobin with oxygen in the lungs.

Increased Hemoglobin and Red Blood Cell Production. In response to tissue hypoxia, the kidneys secrete the hormone erythropoietin, which stimulates the bone marrow to increase its production of hemoglobin and red blood cells. In the Peruvian Andes, for example, people have a total hemoglobin concentration that is increased from 15 g per 100 ml (at sea level) to 19.8 g per 100 ml. Although the percent oxyhemoglobin saturation is still lower than

Table 23.17 Changes in respiratory function during acclimatization to high altitude

Variable	Change	Comments
Partial oxygen pressure	Decreased	Due to decreased total atmospheric pressure
Percent oxyhemoglobin saturation	Decreased	Due to lower P_{O_2} in pulmonary capillaries
Ventilation	Increased	Due at first to low arterial P_{O_2}; later to changes in blood P_{CO_2} as chemoreceptor sensitivity increases
Total hemoglobin	Increased	Due to stimulation by erythropoietin; raises oxygen capacity of blood to partially or completely compensate for the reduced partial pressure
Oxyhemoglobin affinity	Decreased	Due to increased DPG within the red blood cells; results in a higher percent unloading of oxygen to the tissues, which may partially or completely compensate for the reduced arterial oxyhemoglobin saturation

at sea level, the total oxygen content of the blood is actually greater—22.4 ml O_2 per 100 ml compared to a sea level value of about 20 ml O_2 per 100 ml. These compensations of the respiratory system to high altitude are summarized in table 23.17

1. Describe the effect of exercise on the blood values of P_{O_2}, P_{CO_2}, and pH, and explain how ventilation might be increased during exercise.
2. Explain why endurance-trained athletes have a higher anaerobic threshold than other people.
3. Describe the changes that occur in the respiratory system during acclimatization to life at a high altitude.

Clinical Considerations

The respiratory system is particularly vulnerable to inflammatory and infectious diseases simply because many pathogens are airborne, humans are highly social, and the warm, moistened environment along the respiratory tract is very susceptible. Injury and trauma are likewise frequent problems. Protruding noses may become broken, the large, spongy lungs are easily penetrated by broken ribs, and portions of the respiratory tract are liable to become occluded because they also have a digestive function.

Developmental Problems

Birth defects, inherited disorders, and premature births commonly cause problems in the respiratory system of infants. A **cleft palate** is a developmental deformity of the hard palate of the mouth in which an opening persists between the oral and nasal cavities, making it difficult, if not impossible, for an infant to nurse. A cleft palate may be hereditary or a complication of some disease (e.g., German measles) contracted by the mother during pregnancy. A **cleft lip** is a genetically based developmental disorder in which the two sides of the upper lip fail to fuse. Cleft palates and cleft lips can be treated very effectively with cosmetic surgery.

As mentioned earlier in this chapter, **hyaline membrane disease** is a fairly common respiratory disorder that accounts for about one-third of neonatal deaths. This condition results from the deficient production of surfactant. **Cystic fibrosis** *(sis'tik fi-bro'sis)* is a genetic disorder that affects the respiratory system, as well as other systems of the body, and accounts for approximately one childhood death in twenty. The effect of this disease on the respiratory system is usually a persistent inflammation and infection of the respiratory tract.

There are a number of hemoglobin diseases that are produced by inherited defects in the protein part of hemoglobin. **Sickle-cell anemia**—a disease that affects 8%–11% of the Black population of the United States—is caused by an abnormal form of hemoglobin called *hemoglobin S*. This abnormal hemoglobin becomes gel-like when the P_{O_2} of the blood decreases; this causes a characteristic sickling of the red blood cells and organ damage as a result of ischemia and the release of hemoglobin from the damaged erythrocytes. A related hemoglobin disorder is **thalassemia** *(thal''ah-se'me-ah)*, found predominantly among people of Mediterranean ancestry.

Trauma or Injury

Humans are especially susceptible to **epistaxis** *(ep''ĭ-stak'sis)* (nosebleeds) because the nose is located where it can be easily bumped and because the nasal mucosa has extensive vascularity for warming the inspired air. Certain conditions or diseases such as high blood pressure or leukemia may cause epistaxis.

When air enters the pleural cavity surrounding either lung and causes the lung to collapse, it is referred to as a **pneumothorax.** A pneumothorax can result from an external injury, such as a stabbing, bullet wound, or penetrating fractured rib, or it can occur internally. A severely diseased lung, as in emphysema, can create a pneumothorax as the wall of the lung deteriorates and permits air to enter the pleural cavity.

epistaxis: Gk. *epi*, upon; *stazein*, spill in drops
pneumothorax: Gk. *pneumon*, spirit (breath); L. *thorax*, thorax, chest

Choking on a foreign object such as aspirated food is a common serious trauma to the respiratory system. More than eight Americans choke to death each day on food lodged in their trachea. A simple process termed the **abdominal thrust (Heimlich) maneuver** can save the life of a person who is choking. The abdominal thrust maneuver is performed as follows:

If the victim is standing or sitting:

1. Stand behind the victim or the victim's chair, and wrap your arms around his or her waist.
2. Grasp your fist with your other hand, and place the fist against the victim's abdomen, slightly above the navel and below the rib cage.
3. Press your fist into the victim's abdomen with a quick upward thrust.
4. Repeat several times if necessary.

If the victim is lying down:

1. Position the victim on his or her back.
2. Face the victim, and kneel on his or her hips.
3. With one of your hands on top of the other, place the heel of your bottom hand on the abdomen, slightly above the navel and below the rib cage.
4. Press into the victim's abdomen with a quick upward thrust.
5. Repeat several times if necessary.

If you are alone and choking: Use anything that applies force just below your diaphragm. Press into a table or a sink, or use your own fist.

Persons saved from drowning and frequently shock victims experience apnea (cessation of breathing) and will soon die if not revived by someone performing artificial respiration. Figure 23.58 illustrates the accepted treatment for reviving a person who has stopped breathing.

Common Respiratory Disorders

A cough is the most common symptom of respiratory disorders. Acute problems may be accompanied by dyspnea or wheezing. Respiratory or circulatory problems may cause **cyanosis,** which is a blue discoloration of the skin caused by blood with a low oxygen content.

Pulmonary disorders are classified as **obstructive,** when there is increased resistance to air flow in the bronchioles, and as **restrictive,** when alveolar tissue is damaged. Asthma and acute bronchitis are usually just obstructive; chronic bronchitis and emphysema are both obstructive and restrictive. Pulmonary fibrosis, in contrast, is a purely restrictive disorder.

Asthma. The obstruction of air flow through the bronchioles may occur as a result of excessive mucus secretion, inflammation, and/or contraction of the smooth muscles in the bronchioles. *Asthma* results from bronchiolar constriction, which increases airway resistance and makes breathing difficult. Constriction of the bronchiolar smooth muscles is stimulated by leukotrienes and to a lesser degree by histamine, released by leukocytes and mast cells. This can be provoked by an allergic reaction or by the release of acetylcholine from parasympathetic nerve endings. There is evidence that some people with asthma may have a decreased sensitivity to the beta-adrenergic effects of epinephrine, which stimulate bronchodilation; the effects of chemicals that promote bronchoconstriction may, therefore, be enhanced.

Emphysema. Alveolar tissue is destroyed in emphysema, resulting in fewer but larger alveoli (fig. 23.59). This produces a decreased surface area for gas exchange and a decreased ability of the bronchioles to remain open during expiration. Collapse of the bronchioles as a result of the compression of the lungs during expiration produces *air trapping,* which further decreases the efficiency of gas exchange in the alveoli.

There are different types of emphysema. The most common type occurs almost exclusively in people who have smoked cigarettes heavily over a period of years. A component of cigarette smoke apparently stimulates the macrophages and leukocytes to secrete proteolytic (protein-digesting) enzymes that destroy lung tissues. A less common type of emphysema results from the genetic inability to produce a plasma protein called alpha-1-antitrypsin. This protein normally inhibits proteolytic enzymes such as trypsin and thus normally protects the lungs against the effects of enzymes that are released from alveolar macrophages.

Chronic bronchitis and emphysema, the most common causes of respiratory failure, together are called **chronic obstructive pulmonary disease (COPD).** In addition to the more direct obstructive and restrictive aspects of these conditions, other pathological changes may occur. These include edema, inflammation, hyperplasia (increased cell number), zones of pulmonary fibrosis, pneumonia, pulmonary emboli (traveling blood clots), and heart failure. Patients with severe emphysema may eventually develop *cor pulmonale*—pulmonary hypertension with hypertrophy and the eventual failure of the right ventricle.

Pulmonary Fibrosis. Under certain conditions, for reasons that are poorly understood, lung damage leads to pulmonary fibrosis instead of emphysema. In this condition the normal structure of the lungs is disrupted by the accumulation of fibrous connective tissue proteins. Fibrosis can result, for example, from the inhalation of particles less than 6 μm in size, which are able to accumulate in the respiratory zone of the lungs. Included in this category is *anthracosis,* or black lung, which is produced by the inhalation of carbon particles from coal dust (table 23.18).

cyanosis: Gk. *kyanosis,* dark-blue color
asthma: Gk. *asthma,* panting

emphysema: Gk. *emphysan,* blow up, inflate

Figure 23.58. Artificial respiration.

Mouth-to-Mouth Method

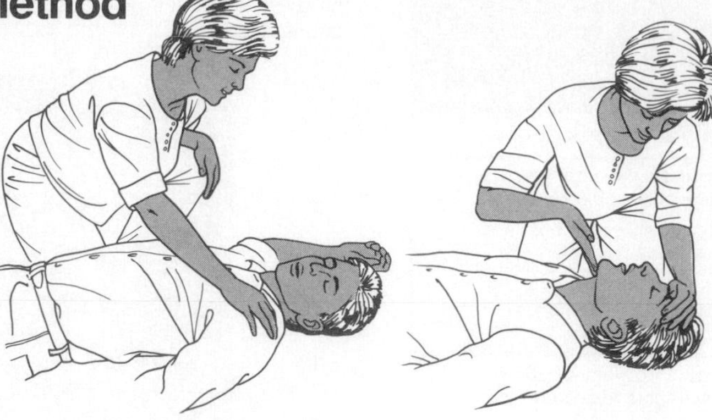

1. **Check the victim for unresponsiveness.** Gently shake him and shout, "Are you okay?" If no response, get the attention of someone who can phone for help. Make sure that the victim is on his back.

2. **Open the airway.** Tilt the victim's head back by pushing on his forehead with your hand and lifting his chin with your fingers under his jaw. This will open his airway by moving his tongue away from the back of his throat.

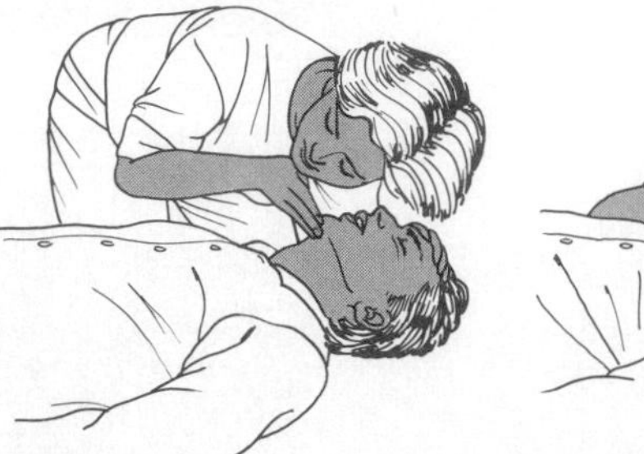

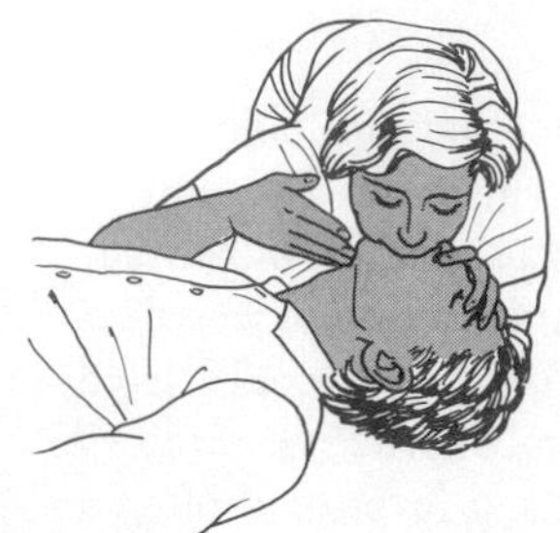

3. **Check for breathing.** Put your ear close to the victim's face to listen and feel for any return of air. At the same time, look to see if there is chest movement. Check for breathing for about five seconds.

4. **If no breathing, give two full breaths.** While maintaining the victim in the head-tilt position, pinch his nose to close off the nasal passageway. Take a deep breath, then seal your mouth around the victim's mouth and give two full breaths. (After the first breath, raise your head slightly to inhale quickly and then give the second breath.)

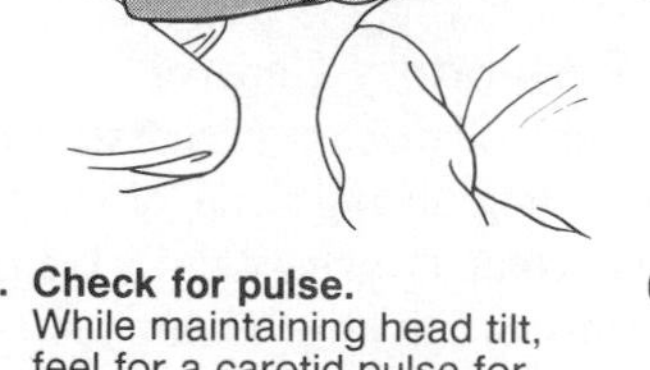

5. **Check for pulse.** While maintaining head tilt, feel for a carotid pulse for five to ten seconds on the side of the victim's neck.

6. **Continue rescue breathing.** With the victim in the head-tilt position and his nostrils pinched, give one breath every five seconds. Observe for signs of breathing between breaths. For an infant, give one gentle puff every three seconds.

7. **Recheck for pulse.** Feel for a carotid pulse at one-minute intervals. If the victim has a pulse but is not breathing, continue rescue breathing.

Mouth-to-Nose Method

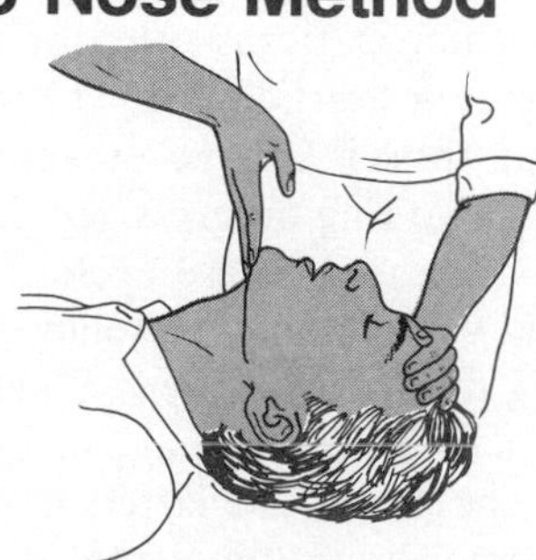

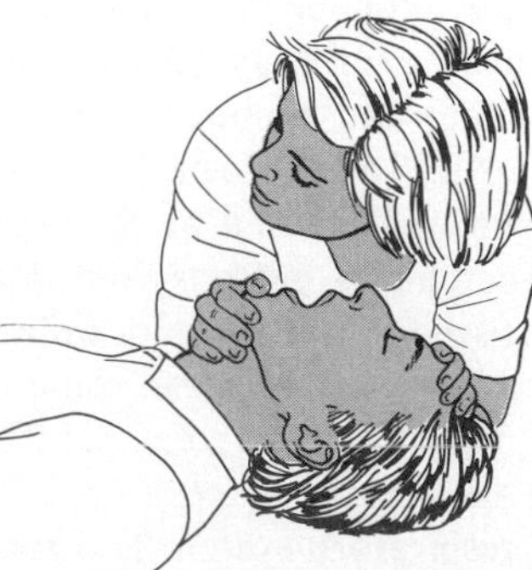

1. **Open the airway.** Place the victim in the head-tilt position as described above.

2. **Blow into the victim's nose.** Using the same sequence described above, blow into the victim's nose while holding his mouth closed.

3. **Feel and observe for breathing.** With the victim's mouth held open, detect for breathing between giving forced breaths.

Table 23.18 Pulmonary disorders caused by the inhalation of noxious agents containing particles less than 6 μm in size, which can lodge in the alveoli

Condition	Agent	Pulmonary disorders
Silicosis	Silica	Fibrosis, obstructive emphysema, tuberculosis, cor pulmonale
Anthracosis (black lung)	Coal dust	Similar to silicosis (most coal dust contains silica)
Asbestosis	Asbestos fibers	Diffuse fibrosis, cor pulmonale, pulmonary malignancy

From Stuart Ira Fox, *Human Physiology,* 2d ed.

Other Respiratory Disorders. The **common cold** is the most widespread of all respiratory diseases. No cure is presently available, only symptomatic relief. Colds occur repeatedly because acquired immunity to one virus does not protect against other viruses that cause colds. Cold viruses cause acute inflammation of the respiratory mucosa, resulting in a flow of mucus, a fever, and often a headache.

Nearly all the structure and regions of the respiratory tract can become infected and inflamed. **Influenza** is a viral disease that causes inflammation of the upper respiratory tract. Influenza can become epidemic, but fortunately vaccines are available. **Sinusitis** *(si″nŭ-si′tis)* is an inflammation of the paranasal sinuses. Sinusitis can be quite painful if the drainage ducts from the sinuses into the nasal cavity become blocked. **Tonsillitis** may involve one or all of the tonsils and frequently follows other lingering diseases of the oral or pharyngeal region. Chronic tonsillitis often requires a tonsillectomy. **Laryngitis** is inflammation of the larynx, which often produces a hoarse voice and limits the ability to talk. **Tracheobronchitis** and **bronchitis** are infections of the regions that give them their names. Severe inflammation of the bronchioles can cause significant airway resistance and dyspnea.

Diseases of the lungs are common and may be serious. **Pneumonia** is an acute infection and inflammation of lung tissue accompanied by exudation (the accumulation of fluid). It can have many causes, but the most common is a bacterial pneumonia. **Tuberculosis (TB)** is an inflammatory disease of the lungs caused by the presence of *tubercle bacilli.* Tuberculosis softens and leads to the ulceration of lung tissue. **Pleurisy** is an inflammation of the pleura and is usually secondary to some other respiratory disease. Inspiration may become painful, and fluid may collect within the pleural space.

Cancer in the respiratory system is often caused by repeated inhalation of irritating substances such as cigarette smoke. Cancers of the lips, larynx, and lungs are especially common in smokers over the age of fifty.

influenza: L. *influentia,* a flowing in
tuberculosis: L. *tuberculum,* diminutive of tuber, swelling

Figure 23.59. Photomicrographs of tissue from a normal lung (*a*) and from the lung of a person with emphysema (*b*). In emphysema, lung tissue is destroyed, resulting in the presence of fewer and larger alveoli.

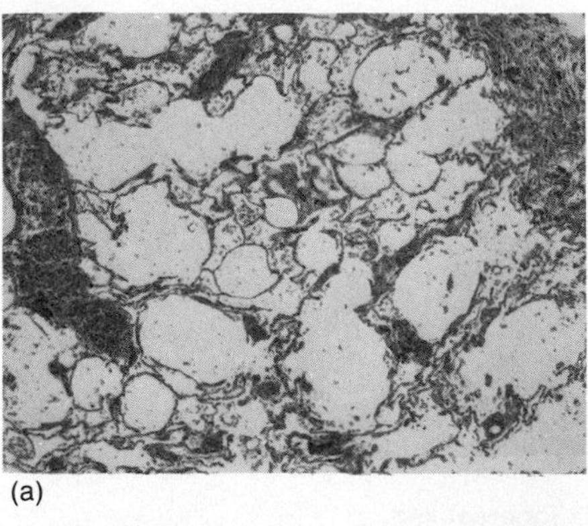
(a)

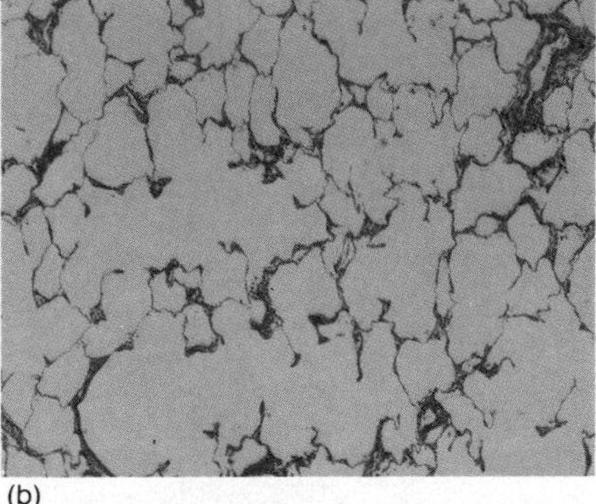
(b)

Disorders of Respiratory Control

There are a variety of disease processes that can produce cessation of breathing during sleep, or *sleep apnea.* **Sudden infant death syndrome** is an especially tragic form of sleep apnea that claims the lives of about ten thousand babies annually in the United States. Victims of this condition are apparently healthy two-to-five-month old babies who die in their sleep without apparent reason—hence, the layman's term of *crib death.* These deaths seem to be caused by failure of the respiratory control mechanisms in the brain stem and/or by failure of the carotid bodies to be stimulated by reduced arterial oxygen.

Abnormal breathing patterns often appear prior to death by brain damage or heart disease. The most common of these abnormal patterns is **Cheyne-Stokes breathing,** in which the depth of breathing progressively increases and then progressively decreases. These cycles of increasing and decreasing tidal volumes may be followed by periods of apnea of varying durations. Cheyne-Stokes breathing may be caused by neurological damage or by insufficient oxygen delivery to the brain. The latter may result from heart disease or from a brain tumor that diverts a large part of the vascular supply from the respiratory centers.

Cheyne-Stokes breathing: from John Cheyne, Scottish physician, 1777–1836, and William Stokes, Irish physician, 1804–1878

Chapter Summary

I. Functions and Development of the Respiratory System
  A. Respiration refers not only to simply breathing but also to the exchange of gases between the atmosphere, the blood, and individual cells.
  B. The functions of the respiratory system are gaseous exchange, sound production, assistance in abdominal compression, and coughing and sneezing.
  C. The upper respiratory pathway develops from ectoderm lining the oronasal cavity, whereas the lower respiratory system develops as an endodermal outpouching from the foregut.

II. Conducting Division
  A. The nose is supported by nasal bones and cartilages.
    1. The nasal epithelium warms, moistens, and cleanses the inspired air.
    2. Olfactory epithelium provides a sense of smell, and the nasal cavity acts as a resonating chamber for the voice.
  B. The paranasal sinuses are found in the maxillary, frontal, sphenoid, and ethmoid bones.
  C. The pharynx is a funnel-shaped passageway that connects the oral and nasal cavities with the larynx.
    1. The nasopharynx is connected by the auditory tubes to the middle ear cavities and contains the pharyngeal tonsils, or adenoids.
    2. The oropharynx is the middle portion, extending from the soft palate to the level of the hyoid bone, which contains the palatine and lingual tonsils.
    3. The laryngopharynx extends from the hyoid bone to the larynx.
  D. The larynx is composed of a number of cartilages that keep the passageway to the trachea open during breathing and close the respiratory passageway during swallowing.
    1. The epiglottis is a spoon-shaped structure that aids in closing the laryngeal opening, or glottis, during swallowing.
    2. The larynx contains vocal cords that are controlled by intrinsic muscles and used in sound production.
  E. The trachea is a rigid tube, supported by rings of cartilage, that leads from the larynx to the bronchial tree.
  F. The bronchial tree includes a right and left primary bronchus, which divides to produce secondary bronchi, tertiary bronchi, and bronchioles; the conducting division ends with the terminal bronchioles, which connect to the alveoli.

III. Alveoli, Lungs, and Pleura
  A. Alveoli are the functional units of the lungs where gas exchange occurs; they are small, numerous, thin-walled air sacs.
  B. The right and left lungs are separated by the mediastinum; each lung is divided into lobes and lobules.
  C. The lungs are covered by a visceral pleural membrane (or visceral pleura), and the thoracic cavity is lined by a parietal pleural membrane.
    1. There is a potential space between these two pleural membranes called the pleural cavity.
    2. The pleural membranes package each lung separately and exclude the structures located in the mediastinum.

IV. Physical Aspects of Ventilation
  A. The intrapleural and intrapulmonary pressures vary during ventilation.
    1. The intrapleural pressure is always less than the intrapulmonary pressure.
    2. The intrapulmonary pressure is subatmospheric during inspiration and greater than the atmospheric pressure during expiration.
  B. Pressure changes in the lungs are produced by variations in lung volume, in accordance with the inverse relationship between the volume and pressure of a gas described by Boyle's law.
  C. The mechanics of ventilation are influenced by the physical properties of the lungs.
    1. The compliance of the lungs is the change in lung volume per change in transpulmonary pressure.
    2. The elasticity of the lungs refers to their tendency to recoil after distension.
    3. The surface tension of the fluid in the alveoli exerts a force directed inward, which acts to resist distension.
  D. On first consideration, it seems that the surface tension in the alveoli would create a pressure causing smaller alveoli to collapse and empty their air into large alveoli.
    1. This would occur because the pressure caused by a given amount of surface tension would be greater in smaller alveoli than in larger alveoli, as described by the law of LaPlace.
    2. Collapse of alveoli due to surface tension does not normally occur, however, because the presence of pulmonary surfactant (a combination of phospholipid and protein) acts to lower surface tension.

V. Mechanics of Breathing
  A. Inspiration and expiration are accomplished by contraction and relaxation of striated muscles.
    1. During quiet inspiration, the diaphragm and external intercostal muscles contract and thus increase the volume of the thorax.
    2. During quiet expiration these muscles relax, and the elastic recoil of the lungs and thorax causes a decrease in thoracic volume.
    3. Forced inspiration and expiration are aided by contraction of the accessory respiratory muscles.
  B. Spirometry aids the diagnosis of a number of pulmonary disorders.
    1. In restrictive disease, such as pulmonary fibrosis, the vital capacity measurement is decreased below normal.
    2. In obstructive disease, such as asthma and bronchitis, the forced expiratory volume is reduced below normal because of increased airway resistance to air flow.

VI. Gas Exchange in the Lungs
  A. According to Dalton's law, the total pressure of a gas mixture is equal to the sum of the pressures that each gas in the mixture would exert independently.
    1. The partial pressure of a gas in a dry gas mixture is thus equal to the total pressure times the percent composition of that gas in the mixture.
    2. When the partial pressure of a gas in a wet gas mixture is calculated, the water vapor pressure must be taken into account.
  B. According to Henry's law, the amount of gas that can be dissolved in a fluid is directly proportional to the partial pressure of that gas in contact with the fluid.
    1. The concentrations of oxygen and carbon dioxide that are dissolved in plasma are proportional to an electric current generated by special electrodes that react with these gases.
    2. Normal arterial blood has a P_{O_2} of 100 mm Hg, indicating a concentration of dissolved oxygen of 0.3 ml per 100 ml of blood; the oxygen contained in red blood cells (about 19.7 ml per 100 ml blood) does not affect the P_{O_2} measurement.
  C. The P_{O_2} and P_{CO_2} measurements of arterial blood provide information about lung function.
  D. Abnormally high partial pressures of gases in blood can cause a variety of disorders, including oxygen toxicity, nitrogen narcosis, and decompression sickness.

VII. Regulation of Breathing
  A. The rhythmicity center in the medulla oblongata directly controls the muscles of respiration.
    1. Activity of the inspiratory and expiratory neurons varies in a reciprocal way to produce an automatic breathing cycle.
    2. Activity in the medulla is influenced by the apneustic and pneumotaxic centers in the pons, as well as by sensory feedback information.
    3. Conscious breathing involves direct control by the cerebral cortex via corticospinal tracts.
  B. Breathing is affected by chemoreceptors sensitive to the P_{O_2}, pH, and P_{CO_2} of the blood.
    1. The P_{CO_2} of the blood and consequent changes in pH are usually of greater importance than the blood P_{O_2} in the regulation of breathing.
    2. Central chemoreceptors in the medulla oblongata are sensitive to changes in blood P_{CO_2} because these changes cause the pH of cerebrospinal fluid to change.
    3. The peripheral chemoreceptors in the aortic and carotid bodies are sensitive to changes in blood P_{CO_2} indirectly, because of consequent changes in blood pH.
  C. Decreases in blood P_{O_2} directly stimulate breathing only when the blood P_{O_2} is less than 50 mm Hg; a drop in P_{O_2} also stimulates breathing indirectly, by making the chemoreceptors more sensitive to changes in P_{CO_2} and pH.
  D. At tidal volumes of one liter or more, inspiration is inhibited by stretch receptors in the lungs (the Hering-Breuer reflex); there is a similar deflation reflex.

VIII. Hemoglobin and Oxygen Transport
  A. Hemoglobin is composed of two alpha and two beta polypeptide chains and four heme groups that contain a central atom of iron.
  B. When the iron is in the reduced form and not attached to oxygen, the hemoglobin is called deoxyhemoglobin, or reduced hemoglobin; when it is attached to oxygen, it is called oxyhemoglobin.
  C. Deoxyhemoglobin combines with oxygen in the lungs (the loading reaction) and breaks its bonds with oxygen in the tissue capillaries (the unloading reaction); the extent of each reaction is determined by the P_{O_2} and the affinity of hemoglobin for oxygen.
  D. A graph of percent oxyhemoglobin saturation at different values of P_{O_2} is called an oxyhemoglobin dissociation curve.
    1. At rest, the difference between arterial and venous oxyhemoglobin saturations indicates that about 22% of the oxyhemoglobin unloads its oxygen to the tissues.
    2. During exercise, the venous P_{O_2} and percent oxyhemoglobin saturation are decreased, indicating that a higher percent of the oxyhemoglobin unloaded its oxygen to the tissues.
  E. The pH and temperature of the blood influence the affinity of hemoglobin for oxygen and thus the extent of loading and unloading.
    1. A fall in pH decreases the affinity, and a rise in pH increases the affinity of hemoglobin for oxygen; this is called the Bohr effect.
    2. A rise in temperature decreases the affinity of hemoglobin for oxygen.
    3. When the affinity is decreased, the oxyhemoglobin dissociation curve is shifted to the right; this indicates a greater percentage unloading of oxygen to the tissues.
  F. The affinity of hemoglobin for oxygen is also decreased by an organic molecule in the red blood cells called 2,3-diphosphoglyceric acid (2,3-DPG).
  G. Striated muscles have myoglobin, a pigment related to hemoglobin, which can combine with oxygen and deliver it to the muscle cell mitochondria at low values of P_{O_2}.

IX. Carbon Dioxide Transport and Acid-Base Balance
  A. Red blood cells contain an enzyme called carbonic anhydrase, which catalyzes the reversible reaction whereby carbon dioxide and water are used to form carbonic acid.
    1. This reaction is favored by the high P_{CO_2} in the tissue capillaries, and as a result, carbon dioxide produced by the tissues is converted into carbonic acid in the red blood cells.
    2. Carbonic acid then ionizes to form H^+ and HCO_3^- (bicarbonate).
    3. Since much of the H^+ is buffered by hemoglobin, but more bicarbonate is free to diffuse outward, an electrical gradient is established that draws Cl^- into the red blood cells; this is called the chloride shift.
    4. A reverse chloride shift occurs in the lungs; in this process, the low P_{CO_2} favors the conversion of carbonic acid to carbon dioxide, which can be exhaled.
  B. By adjusting the blood concentration of carbon dioxide and thus of carbonic acid, the process of ventilation helps to maintain proper acid-base balance of the blood.
    1. Normal arterial blood pH is 7.40; a pH less than 7.35 is termed acidosis, and a pH greater than 7.45 is termed alkalosis.
    2. Hyperventilation causes respiratory alkalosis, and hypoventilation causes respiratory acidosis.
    3. Metabolic acidosis stimulates hyperventilation, which can cause a respiratory alkalosis as a partial compensation.

X. Effect of Exercise and High Altitude on Respiratory Function
  A. During exercise there is increased ventilation, or hyperpnea, which can be matched to the increased metabolic rate so that the arterial blood P_{CO_2} remains normal.
    1. This hyperpnea may be caused by proprioceptor information, cerebral input, and/or changes in arterial P_{CO_2} and pH.
    2. During heavy exercise the anaerobic threshold may be reached at about 50%–60% of the maximal oxygen uptake; at this point, lactic acid is released into the blood by the muscles.
    3. Endurance training enables the muscles to utilize oxygen more effectively, so that greater levels of exercise can be performed before the anaerobic threshold is reached.
  B. Acclimatization to a high altitude involves changes that help to deliver oxygen more effectively to the tissues despite reduced arterial P_{O_2}.

Review Activities

Objective Questions

1. Which of the following statements about intrapulmonary and intrapleural pressure is true?
   (a) The intrapulmonary pressure is always subatmospheric.
   (b) The intrapleural pressure is always greater than the intrapulmonary pressure.
   (c) The intrapulmonary pressure is greater than the intrapleural pressure.
   (d) The intrapleural pressure equals the atmospheric pressure.
2. If the transpulmonary pressure equals zero,
   (a) a pneumothorax has probably occurred
   (b) the lungs cannot inflate
   (c) elastic recoil causes the lungs to collapse
   (d) all of the above
3. The maximum amount of air that can be expired after a maximum inspiration is the
   (a) tidal volume
   (b) forced expiratory volume
   (c) vital capacity
   (d) maximum expiratory flow rate
4. If the blood lacked red blood cells but the lungs were functioning normally,
   (a) the arterial P_{O_2} would be normal
   (b) the oxygen content of arterial blood would be normal
   (c) both a and b
   (d) neither a nor b
5. If a person were to dive with scuba equipment to a depth of sixty-six feet, which of the following statements would be *false?*
   (a) The arterial P_{O_2} would be three times normal.
   (b) The oxygen content of plasma would be three times normal.
   (c) The oxygen content of whole blood would be three times normal.
6. Which of the following would be most affected by a decrease in the affinity of hemoglobin for oxygen? The
   (a) arterial P_{O_2}
   (b) arterial percent oxyhemoglobin saturation
   (c) venous oxyhemoglobin saturation
   (d) arterial P_{CO_2}
7. If a person with normal lung function were to hyperventilate for several seconds, there would be a significant
   (a) increase in the arterial P_{O_2}
   (b) decrease in the arterial P_{CO_2}
   (c) increase in the arterial percent oxyhemoglobin saturation
   (d) decrease in the arterial pH
8. Erythropoietic factor is produced by the
   (a) kidneys
   (b) liver
   (c) lungs
   (d) bone marrow
9. The affinity of hemoglobin for oxygen is decreased under conditions of
   (a) acidosis
   (b) fever
   (c) anemia
   (d) acclimatization to a high altitude
   (e) all of the above
10. Most of the carbon dioxide in the blood is carried in the form of
    (a) dissolved CO_2
    (b) carbaminohemoglobin
    (c) carbonic acid and bicarbonate
    (d) carboxyhemoglobin
11. The bicarbonate concentration of the blood would be decreased during
    (a) metabolic acidosis
    (b) respiratory acidosis
    (c) metabolic alkalosis
    (d) respiratory alkalosis
12. The chemoreceptors in the medulla are directly stimulated by
    (a) CO_2 from the blood
    (b) H^+ from the blood
    (c) H^+ in cerebrospinal fluid that is derived from blood CO_2
    (d) decreased arterial P_{O_2}
13. The rhythmic control of breathing is produced by the activity of inspiratory and expiratory neurons in the
    (a) medulla oblongata
    (b) apneustic center of the pons
    (c) pneumotaxic center of the pons
    (d) cerebral cortex
14. Which of the following occurs during hypoxemia?
    (a) increased ventilation
    (b) increased production of 2,3-DPG
    (c) increased production of erythropoietin
    (d) all of the above
15. During exercise, which of the following statements is true?
    (a) The arterial percent oxyhemoglobin saturation is decreased.
    (b) The venous percent oxyhemoglobin saturation is decreased.
    (c) The arterial P_{CO_2} is measurably increased.
    (d) The arterial pH is measurably decreased.

Essay Questions

1. Using a flow diagram to show cause and effect, explain how contraction of the diaphragm produces inspiration.
2. Radiographic (X-ray) pictures show that the ribs of a person with a pneumothorax are expanded and farther apart. Explain why this occurs.
3. Explain, using a flowchart, how a rise in blood P_{CO_2} stimulates breathing. Include both the central and peripheral chemoreceptors in your answer.
4. A person with ketoacidosis may hyperventilate. Explain why this occurs, and explain why this hyperventilation can be stopped by an intravenous fluid containing bicarbonate.
5. What blood measurements can be performed to detect (*a*) anemia, (*b*) carbon monoxide poisoning, and (*c*) poor lung function?
6. Explain how measurements of blood P_{CO_2}, bicarbonate, and pH are affected by hypoventilation and hyperventilation.
7. Explain how blood pH and bicarbonate concentrations are affected by respiratory and metabolic acidosis.
8. How would an increase in the red blood cell content of 2,3-DPG affect the P_{O_2} of venous blood? Explain your answer.

Urinary System, Fluid and Electrolyte Balance

Outline

Concepts

Urine, which is produced by the kidneys as a blood filtrate, is modified as it passes through the nephrons, drains via the ureters to the urinary bladder, and is excreted through the urethra.

Humans have metanephric kidneys, which become functional after pronephric and mesonephric kidneys degenerate during embryonic development.

In gross structure, each kidney consists of a cortex, medulla, pyramids, papillae, columns, calyces, and a pelvis. In microscopic structure, each nephron consists of a glomerulus, glomerular (Bowman's) capsule, proximal convoluted tubule, nephron loop (loop of Henle), and a distal convoluted tubule connecting to a collecting duct.

The structures of the glomerulus and glomerular capsule allow large volumes of plasma filtrate with greatly reduced protein concentration to be produced. The glomerular filtration rate is controlled by the sympathetic nervous system and by autoregulatory mechanisms.

The proximal tubule reabsorbs most of the filtered salt and water from the filtrate. The transport of salt and water in the nephron loop creates a hypertonic renal medulla. Under the influence of ADH, this hypertonic medulla permits most of the remaining water to be reabsorbed.

Solutes in the plasma that are filtered or are secreted into the tubules will be excreted in the urine if they are not reabsorbed. Measurements of the renal plasma clearance of different compounds indicate how the kidneys dispose of these substances and provide clinically important information.

The reabsorption of Na^+ in exchange for the secretion of K^+ or H^+ in the distal convoluted tubules is regulated by aldosterone. Aldosterone secretion is controlled by negative feedback through the secretion of renin from the juxtaglomerular apparatus of the kidney and by a direct effect of blood potassium on the adrenal cortex.

The kidneys help regulate the acid-base balance of the blood through their excretion of H^+ and their reabsorption of HCO_3^-. By means of these processes, the kidneys are responsible for the metabolic component of acid-base balance.

Urine is channeled from the kidneys to the urinary bladder by the ureters and expelled from the body through the urethra. The mucosa of the urinary bladder permits distension, and the muscles of the urinary bladder are used in the control of micturition.

Introduction to the Urinary System

Urine, which is produced by the kidneys as a blood filtrate, is modified as it passes through the nephrons, drains via the ureters to the urinary bladder, and is excreted through the urethra.

Objective 1. Describe the path of flow of urine through the urinary system.
Objective 2. List the functions of the urinary system.

The urinary system consists of two kidneys, two ureters, the urinary bladder, and the urethra (fig. 24.1). The primary function of the urinary system is regulation of the extracellular fluid (plasma and tissue fluid) environment in the body. This function is accomplished through the formation of *urine,* which is a modified filtrate of plasma, within the kidneys. Urine is actually produced within the numerous microscopic tubules, called *nephrons,* within the kidneys. Urine passes out of the kidneys via the ureters to the urinary bladder, where it can be stored for a time but is eventually eliminated from the body through the urethra.

In the process of urine formation, the kidneys regulate (1) the volume of blood plasma and thus contribute significantly to the regulation of blood pressure; (2) the concentration of waste products in the blood; (3) the concentration of electrolytes (Na^+, K^+, HCO_3^-, and other ions) in the plasma; and (4) the pH of plasma.

nephron: Gk. *nephros,* kidney

Blood that is to be processed by a kidney enters through the large *renal artery* and after the filtration process exits through the *renal vein.* The importance of filtration of blood is shown by the fact that during normal resting conditions the kidneys receive approximately 20%–25% of the entire cardiac output. Every minute the kidneys process approximately 1,200 ml of blood.

1. List the organs of the urinary system in the order in which urine is processed through the system.
2. Explain the various ways in which the urinary system maintains the constancy of the blood within the circulatory system.

Development of the Urinary System

Humans have metanephric kidneys, which become functional after pronephric and mesonephric kidneys degenerate during embryonic development.

Objective 3. Describe the mesodermal site of origin of the urinary system.
Objective 4. Describe development of the pronephric, mesonephric, and metanephric kidneys.

The urinary and reproductive systems originate from a specialized elevation of mesodermal tissue called the **urogenital** *(u''ro-jen'i-tal)* **ridge.** The two systems share common structures during a portion of development, but by the time of birth two separate systems have formed.

Figure 24.1. The anatomical locations of the kidneys, ureters, and urinary bladder.

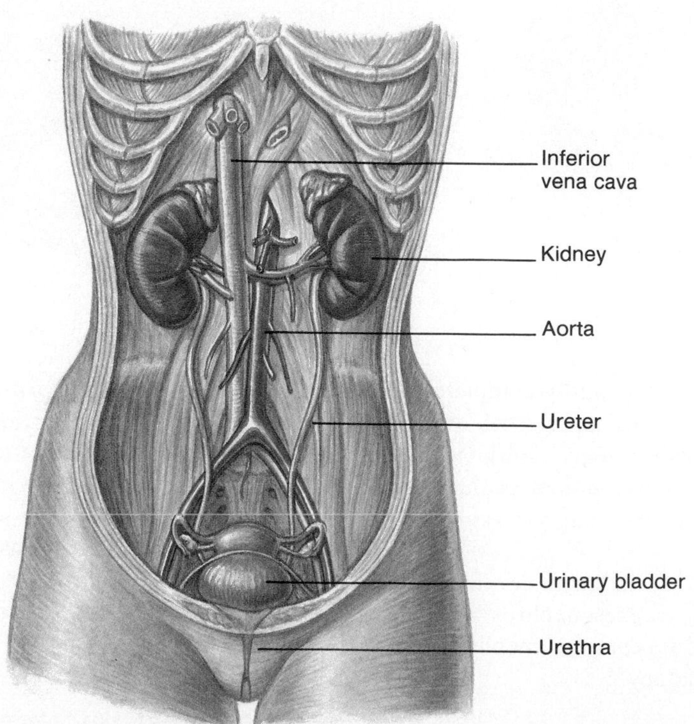

Figure 24.2. The embryonic development of the urinary system. (*a*) an anterolateral view of an embryo at five weeks showing the position of a transverse cut depicted in (a_1). During the sixth week, (*b* and b_1), the kidney is forming in the pelvic region and the mesonephric duct and gonadal ridge are prominent. The kidney begins migrating during the seventh week (b_2), and the urinary bladder and gonad are formed.

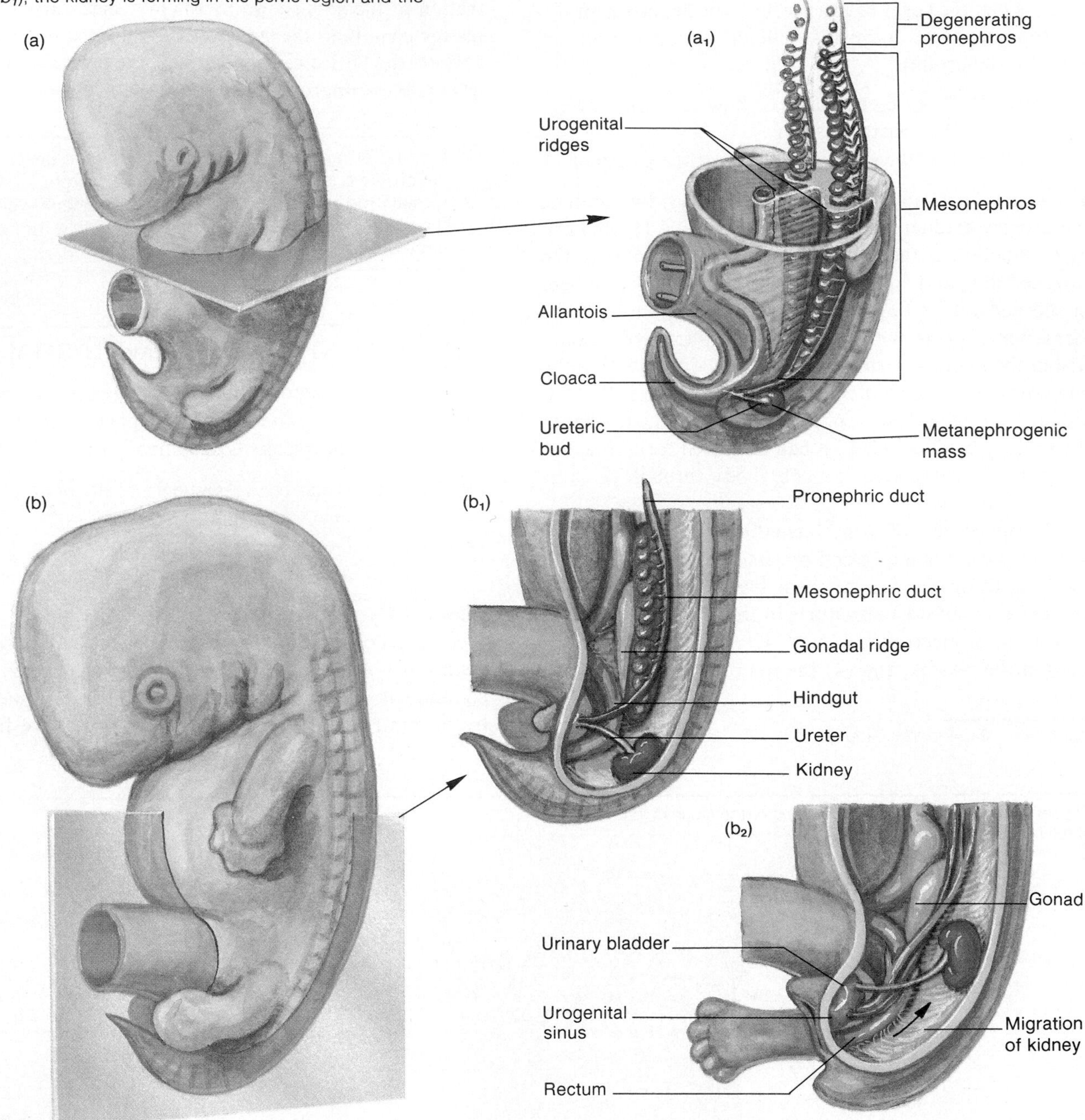

The separation in the male is not totally complete, however, since the urethra serves to transport both urine and semen. The development of both systems is initiated during the embryonic stage, but the development of the urinary system starts and finishes sooner than the reproductive system.

Three successive types of kidneys develop in the human embryo: the pronephros, mesonephros, and metanephros (fig. 24.2). The third type, or metanephric kidney, remains as the permanent kidney.

The **pronephros** *(pro-nef'ros)* develops during the fourth week after conception and persists only through the sixth week. It is the most superior in position on the urogenital ridge of the three kidneys and is connected to the embryonic **cloaca** by the **pronephric duct.** Although the pronephros is nonfunctional and degenerates in humans, most of its duct is used by the mesonephric kidney (fig. 24.2), and a portion of it is important in the formation of the metanephros.

pronephros: Gk. *pro,* before; *nephros,* kidney
cloaca: L. *cloaca,* sewer

Figure 24.3. The development of the metanephric kidney at (*a*) four weeks, (*b*) five weeks, (*c*) seven weeks, (*d*) birth. (d_1) magnified view of the arrangement of collecting ducts within a papilla.

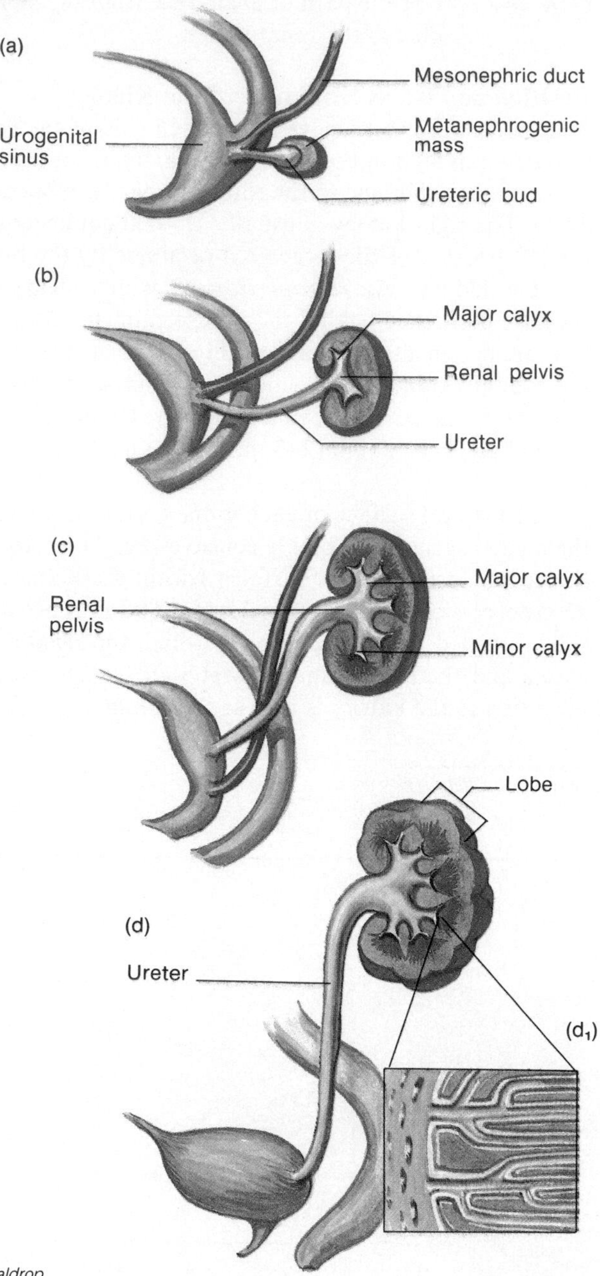

The **mesonephros** *(mes''o-nef'ros)* develops toward the end of the fourth week as the pronephros becomes vestigial. The mesonephros forms from an intermediate portion of the urogenital ridge and functions throughout the embryonic period of development.

Although the **metanephros** *(met''ah-nef'ros)* begins its formation during the fifth week, it does not become functional until immediately before the start of the fetal stage of development at the end of the eighth week. The paired metanephric kidneys continue to form urine throughout fetal development. The urine is expelled through the urinary system into the amniotic fluid.

The metanephros develops from two mesodermal sources (fig. 24.3). The meaty (glomerular) part of the kidney forms from a specialized caudal portion of the urogenital ridge called the **metanephrogenic mass.** The tubular drainage portion of the kidneys forms as a diverticulum emerges from the wall of the mesonephric duct near the cloaca. This diverticulum, known as the **ureteric** *(u''rĕ-ter'ik)* **bud,** expands into the metanephrogenic mass to form the drainage pathway for urine. The stalk of the ureteric bud develops into the ureter, whereas the expanded terminal portion forms the renal pelvis, calyces, and collecting tubules. A combination of the ureteric bud and metanephrogenic mass forms the other tubular channels within the kidney.

Once the metanephric kidneys are formed, they begin to migrate from the pelvis to the upper, posterior portion of the abdomen. The renal blood supply develops as the kidneys become positioned in the posterior body wall.

The urinary bladder develops from the urogenital sinus, which is connected to the embryonic umbilical cord by the fetal membrane called the allantois (fig. 24.2). By the twelfth week, the two ureters are emptying into the urinary bladder, the urethra is draining, and the connection of the urinary bladder to the allantois has been reduced to a supporting structure called the **urachus** *(u'rah-kus).*

Occasionally a *patent urachus* is present in a newborn and is discovered when urine is passed through the umbilicus, especially if there is a urethral obstruction. Usually, however, it remains undiscovered until old age, when an enlarged prostate obstructs the urethra and forces urine through the patent urachus and out the umbilicus (navel).

1. Discuss the basic developmental similarities between the urinary and the reproductive systems.
2. Which type of kidney is functional during the embryonic stage of development? When do the metanephric kidneys form and become functional?

KIDNEY STRUCTURE

In gross structure, each kidney consists of a cortex, medulla, pyramids, papillae, columns, calyces, and a pelvis. In microscopic structure, each nephron consists of a glomerulus, glomerular (Bowman's) capsule, proximal convoluted tubule, nephron loop (loop of Henle), and a distal convoluted tubule connecting to a collecting duct.

Objective 5. Describe the position and the gross structure of the kidney.

urachus: Gk. *ouron,* urine; *echein,* to hold

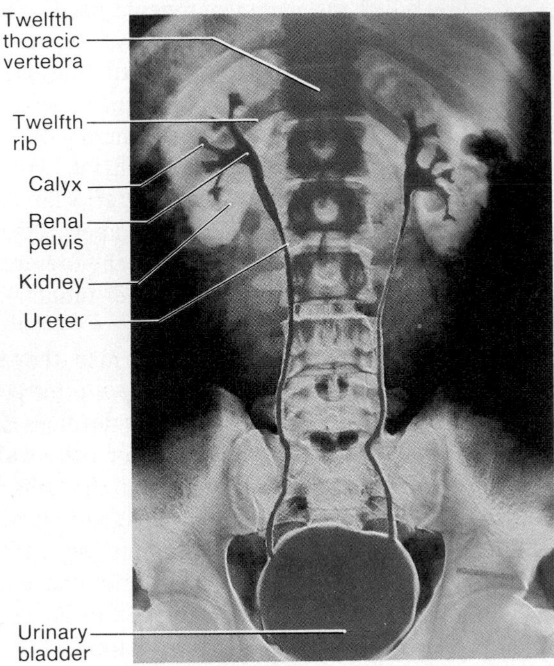

Figure 24.4. A color-enhanced X ray of the calyces and renal pelvises of the kidneys, the ureters, and the urinary bladder. Notice the position of the kidneys relative to the vertebral column and ribs.

Objective 6. Describe the structure of a nephron, and describe how the nephrons are oriented in the cortex and the medulla of the kidney.

Objective 7. List the path of urine flow from the glomerulus to the renal pelvis.

Position and Gross Structure of the Kidney

The reddish-brown **kidneys** lie on each side of the vertebral column, high in the abdominal cavity, from the level of the twelfth thoracic to the third lumbar vertebrae (fig. 24.4). The right kidney is usually 1.5–2.0 cm lower than the left because of the large area occupied by the liver.

The kidneys are ***retroperitoneal,*** which means that they are positioned behind the peritoneum (fig. 24.5) and thus are not, in a strict sense, within the peritoneal cavity. Each adult kidney is an anteroposteriorly compressed, lima-bean-shaped organ, which measures about 11.25 cm (4 in.) long, 5.5–7.7 cm (2–3 in.) wide, and 2.5 cm (1 in.) thick.

The lateral surface of each kidney is convex, whereas the medial surface is strongly concave (fig. 24.6). Toward the center of the medial surface, within the kidney, is a deep depression called the **renal sinus.** The entrance to the sinus is called the **hilum,** through which the **renal artery** enters and the **renal vein** and **ureter** *(u' re-ter)* exit. Innervation to the kidney is also at the hilum.

hilum: L. *hillum,* a trifle

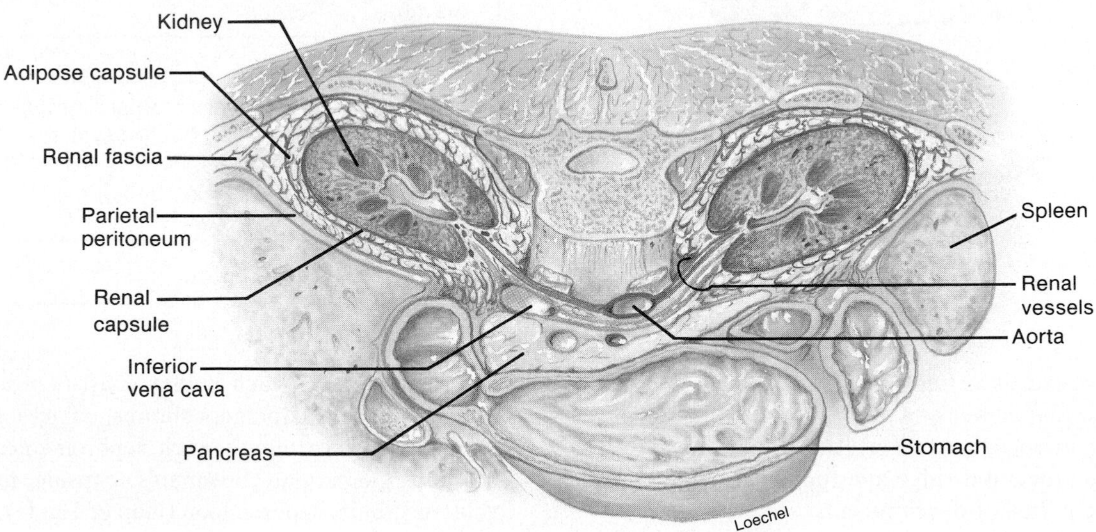

Figure 24.5. The positions of the kidneys as seen in transverse section through the upper abdominal cavity. The kidneys are embedded in adipose tissue behind the parietal peritoneum.

Each kidney is embedded in a fatty, fibrous pouch consisting of three layers. The **renal capsule** is the innermost layer that forms a strong, transparent fibrous attachment to the kidney. The renal capsule protects the kidney from trauma and the spread of infections. The second layer is formed by a firm, protective layer of adipose tissue called the **adipose capsule** (see fig. 24.5). The outermost layer, called the **renal fascia,** is a supportive layer that anchors the kidney to the peritoneum and the abdominal wall.

> Although the kidneys are firmly supported by the adipose capsule, renal fascia, and even the renal vessels, under certain conditions these structures may give in to the force of gravity and the kidneys may drop a bit in position. This condition is called *renal ptosis (to'sis)* and generally occurs in elderly persons who are extremely thin and have insufficient amounts of supportive fat in the adipose capsular layer. It may also affect victims of *anorexia nervosa,* who suffer from extreme weight loss. The potential danger of renal ptosis is that the ureter may kink, blocking the flow of urine from the affected kidney.

A coronal section of the kidney shows two distinct regions and a major cavity (fig. 24.7). The outer **cortex,** in contact with the capsule, is reddish brown and granular in appearance because of its many capillaries. The deeper region, or **medulla,** is darker in color and striped in appearance because of the presence of microscopic tubules and blood vessels. The medulla is composed of eight to fifteen conical **renal pyramids** separated by **renal columns.** The apexes of the renal pyramids are all directed into the renal sinus and are known as the **renal papillae.**

The cavity of the kidney collects and transports urine from the kidney to the ureter. It is divided into several portions. Each papilla of a pyramid projects into a small depression called the **minor calyx** *(ka'liks)*—the plural form is *calyces.* Several minor calyces unite to form a **major calyx.** In turn, the major calyces join to form the funnel-shaped **renal pelvis.** The renal pelvis is actually an expanded portion of the ureter in the kidney and serves to collect urine from the calyces and transport it to the ureter.

Figure 24.6. An anterior view of the kidneys, adrenal glands, and associated vessels. The right kidney is longitudinally dissected to show the internal structures.

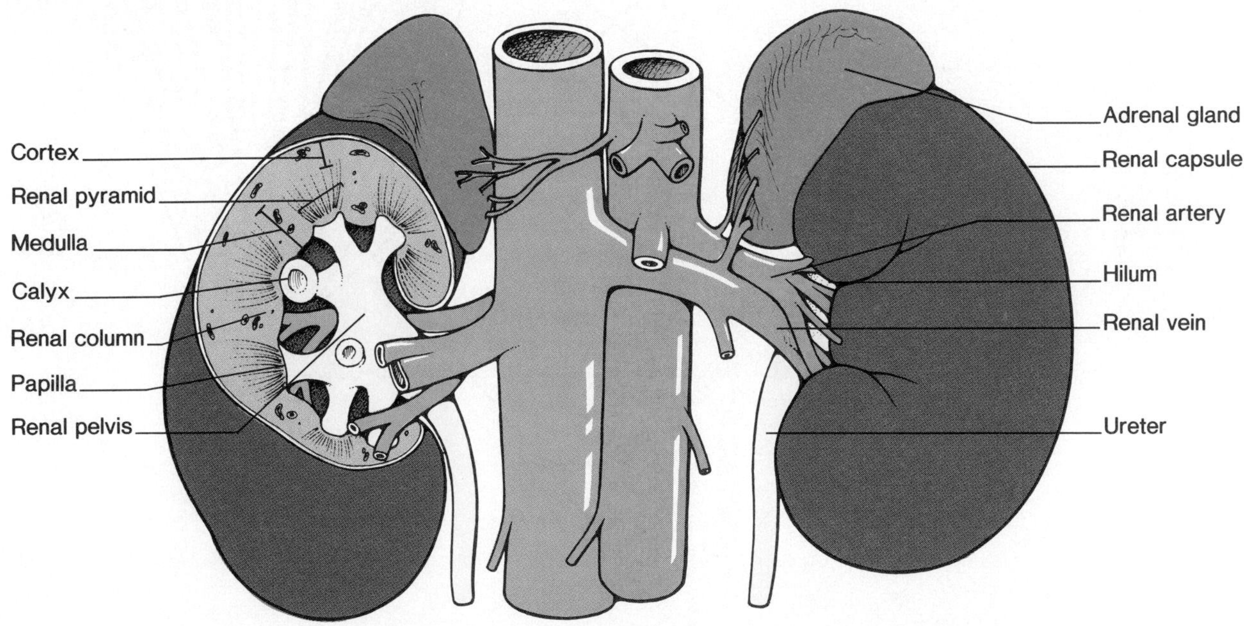

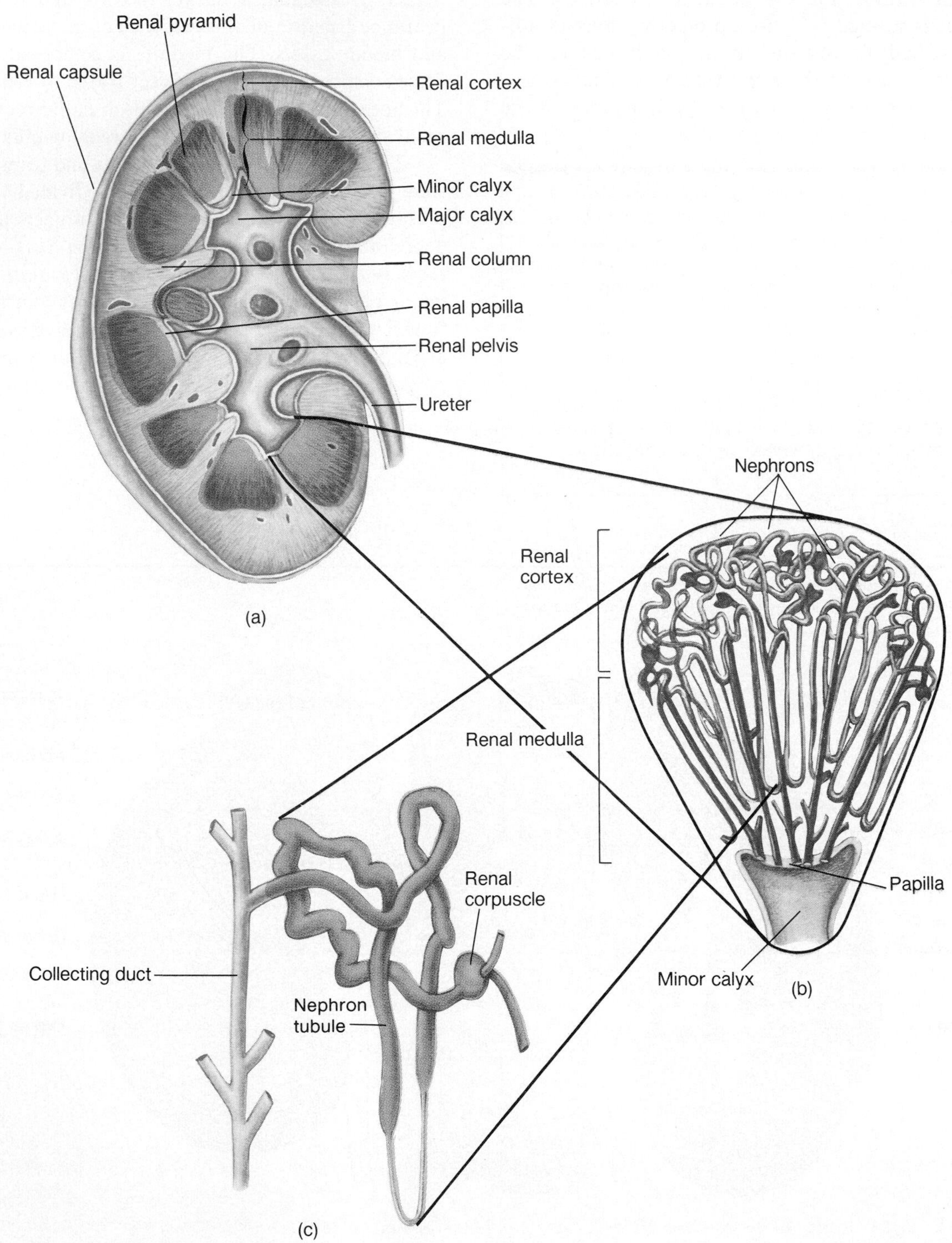

Figure 24.7. The internal structures of a kidney. (*a*) a coronal section showing the structure of the cortex, medulla, and renal pelvis. (*b*) a diagrammatic magnification of a renal pyramid and cortex to depict the tubules. (*c*) a diagrammatic view of a single nephron and a collecting duct.

Microscopic Structure of the Kidney

The **nephron** *(nef'ron)* is the functional unit of the kidney that is responsible for the formation of urine. Each kidney contains more than a million nephrons. A nephron consists of **urinary tubules** and associated small blood vessels. Fluid formed by capillary filtration enters the tubules and is subsequently modified by transport processes; the resulting fluid that leaves the tubules is *urine.*

Renal Blood Vessels. Arterial blood enters the kidney at the hilum through the **renal artery,** which divides into **interlobar** *(in''ter-lo'bar)* **arteries** (fig. 24.8), which pass between the pyramids through the renal columns. **Arcuate** *(ar'ku-āt)* **arteries** branch from the interlobar arteries at

arcuate: L. *arcuare,* to bend

Figure 24.8. The vascular structure of the kidneys. (*a*) an illustration of the major arterial supply and (*b*) a scanning electron micrograph of the glomeruli.

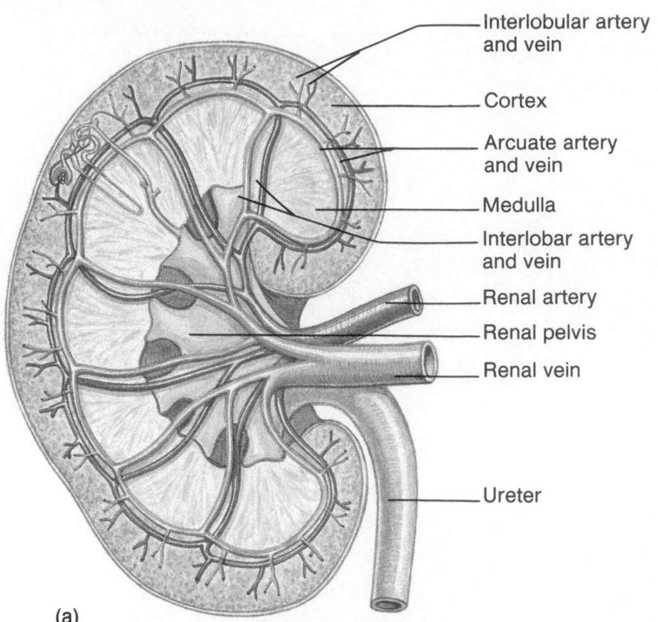

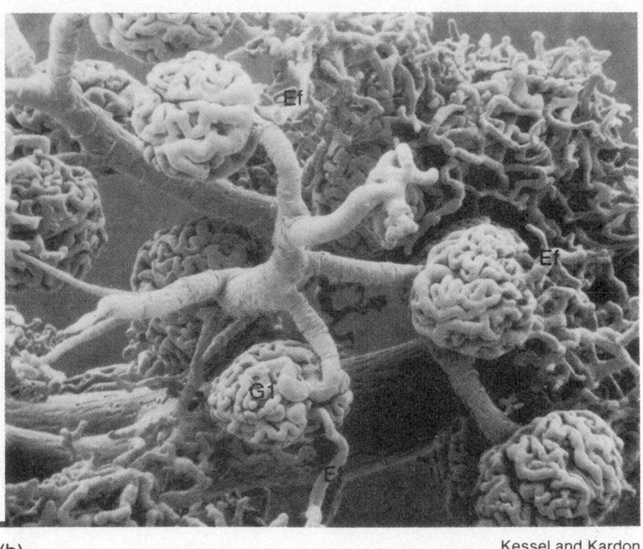

Kessel and Kardon

the boundary of the cortex and medulla. Many **interlobular arteries** radiate from the arcuate arteries and subdivide into numerous **afferent arterioles,** which are microscopic in size. The afferent arterioles deliver blood into capillary networks, called **glomeruli,** which produce a blood filtrate that enters the urinary tubules. The blood remaining in the glomerulus leaves through an **efferent arteriole,** which delivers the blood into another capillary network, the **peritubular capillaries,** that surround the tubules (fig. 24.9). This blood is drained into veins that parallel the course of the arteries in the kidney and are named the **interlobular veins, arcuate veins,** and **interlobar veins.** The interlobar veins descend between the pyramids, converge, and leave the kidney as a single **renal vein** that empties into the inferior vena cava.

Although the kidneys are generally well protected by being encapsulated retroperitoneally, they may be injured by a hard blow to the lumbar region. The immense vascularity of the kidney makes it highly susceptible to hemorrhage, and as a result such kidney damage can produce blood in the urine.

Nephron Tubules. The tubular portion of a nephron consists of a glomerular capsule, proximal convoluted tubule, descending limb of the nephron loop (loop of Henle), ascending limb of the nephron loop, and distal convoluted tubule (fig. 24.9).

The **glomerular (Bowman's) capsule** surrounds the glomerulus. The glomerular capsule and its associated glomerulus are located in the cortex of the kidney and together constitute the **renal corpuscle.** The glomerular capsule contains an inner visceral layer of epithelium around the glomerular capillaries and an outer parietal layer. The space between these two layers, called the **capsular space,** receives the glomerular filtrate.

Filtrate in the glomerular capsule passes into the lumen of the **proximal convoluted tubule.** The wall of the proximal convoluted tubule consists of a single layer of cuboidal cells containing millions of microvilli; these serve to increase the surface area for reabsorption. In the process of reabsorption, salt, water, and other molecules needed by the body are transported from the lumen, through the tubular cells, and into the surrounding peritubular capillaries.

The glomerulus, glomerular capsule, and proximal convoluted tubule are located in the renal cortex. Fluid passes from the proximal convoluted tubule to the **nephron loop (loop of Henle).** This fluid is carried into the medulla in the **descending limb** of the loop and returns to the cortex in the **ascending limb** of the loop. Back in the cortex, the

glomerulus: L. diminutive of *glomus,* ball
Bowman's capsule: from Sir William Bowman, English anatomist, 1816–92
loop of Henle: from Friedrich G. J. Henle, German anatomist, 1809–85

Figure 24.9. A simplified illustration of blood flow from a glomerulus to an efferent arteriole, to the peritubular capillaries, to the venous drainage of the kidneys.

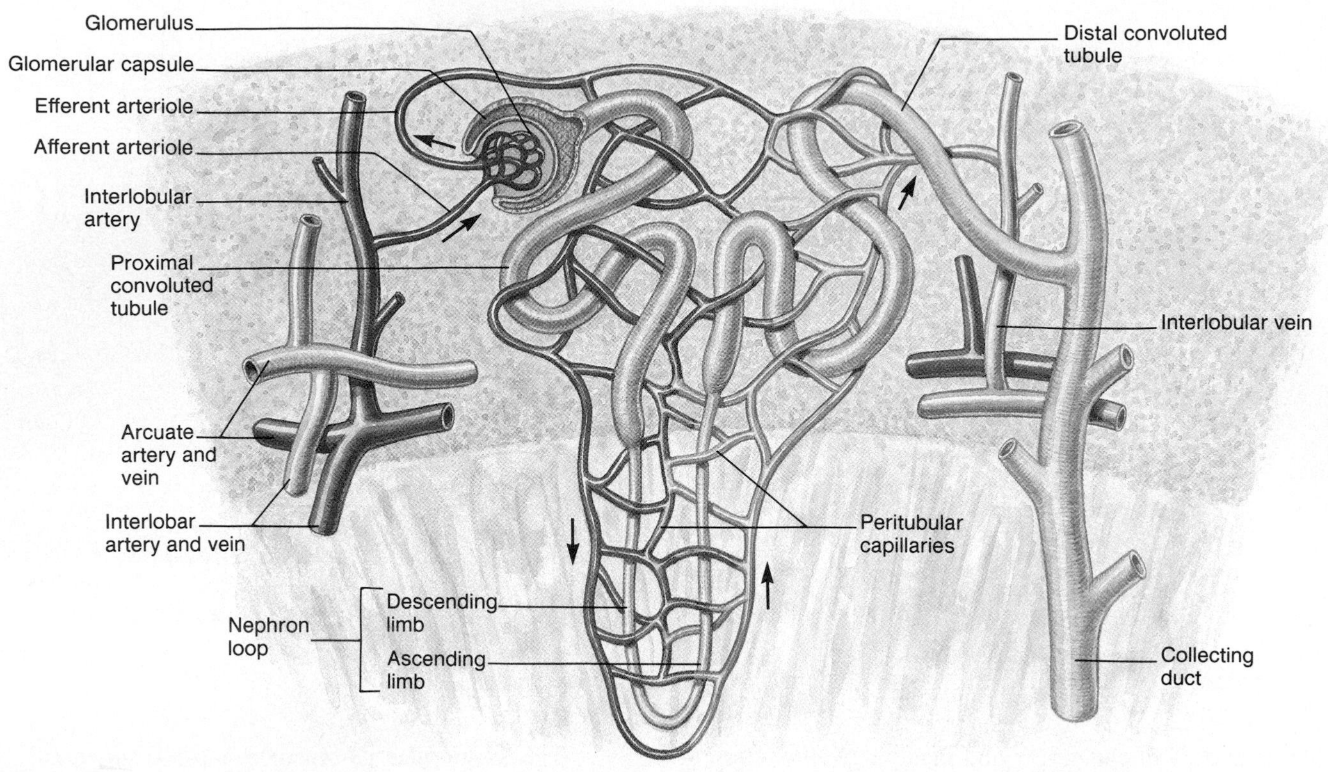

Figure 24.10. Cortical nephrons are located almost exclusively within the renal cortex, and juxtamedullary nephrons are located, for the most part, within the outer portion of the renal medulla.

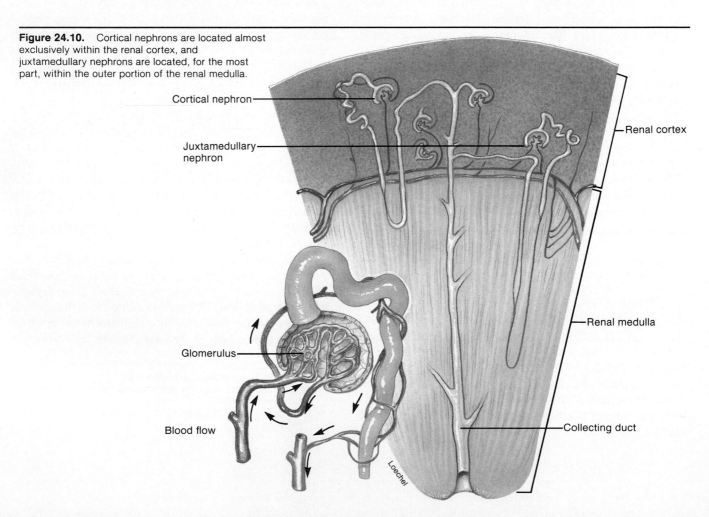

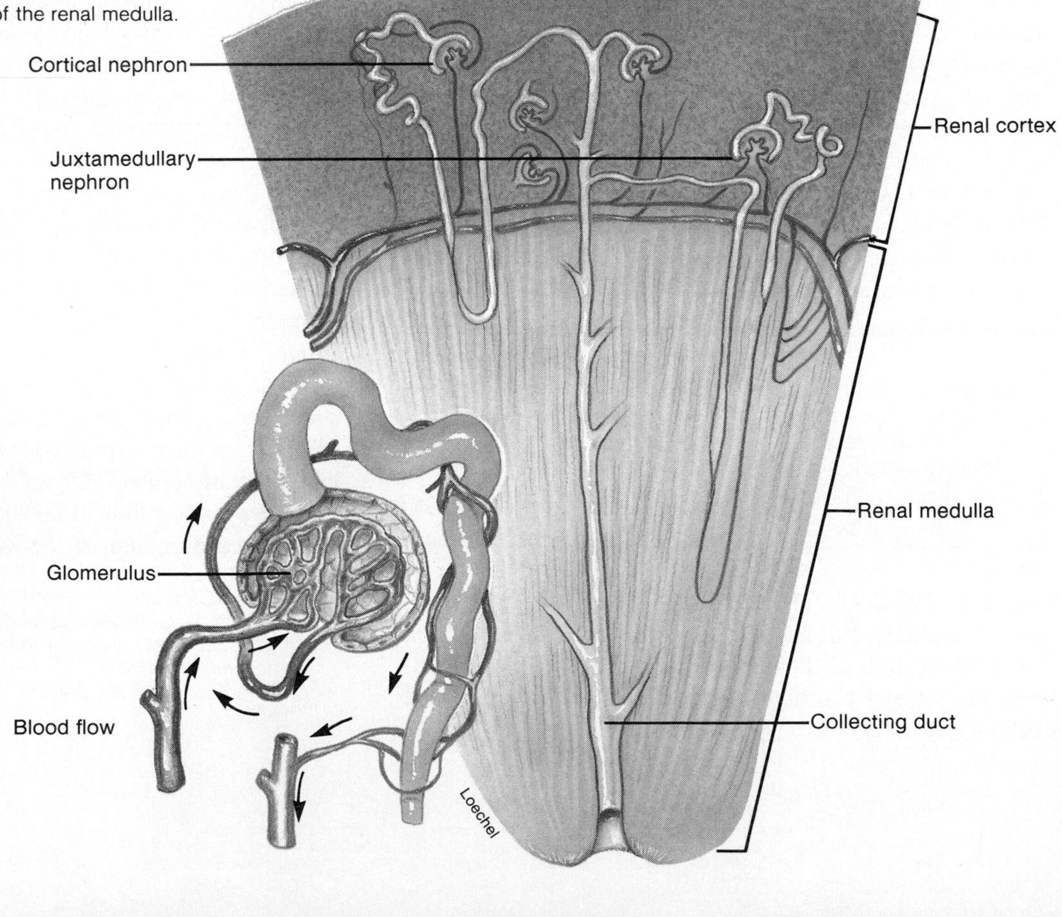

Cortical nephron
Juxtamedullary nephron
Renal cortex
Renal medulla
Glomerulus
Blood flow
Collecting duct
Loechel

tubule becomes coiled again and is called the **distal convoluted tubule.** In contrast to the proximal tubule, the distal convoluted tubule is shorter and has relatively few microvilli. The distal convoluted tubule is the last segment of the nephron and terminates as it empties into a collecting duct.

There are two types of nephrons, which are classified according to their position in the kidney and the lengths of their nephron loops. Nephrons that originate in the inner one-third of the cortex—called **juxtamedullary nephrons**—have longer loops than the **cortical nephrons** that originate in the outer two-thirds of the cortex (fig. 24.10).

The distal convoluted tubules of several nephrons drain into a **collecting duct.** Fluid is then drained by the collecting duct from the cortex into the medulla as the collecting duct passes through a renal pyramid. This fluid, now called urine, passes out of a renal papilla into a minor calyx. Urine is then funneled through the renal pelvis and out of the kidney into the ureter.

1. Describe the gross appearance of the renal cortex and medulla, and indicate the microscopic structures found in each layer.
2. Trace the course of blood flow through the kidney from the renal artery to the renal vein.
3. Trace the course of tubular fluid from the glomerular capsules to the ureter.
4. Draw a diagram of the tubular component of a nephron. Label the segments, and indicate which parts are in the cortex and which are in the medulla.

Glomerular Filtration

The structures of the glomerulus and glomerular capsule allow large volumes of plasma filtrate with greatly reduced protein concentration to be produced. The glomerular filtration rate is controlled by the sympathetic nervous system and by autoregulatory mechanisms.

Objective 8. Describe the fine structure of the glomerular capillaries and of the podocytes.

Objective 9. Describe the composition of glomerular ultrafiltrate, and explain why it contains a low concentration of proteins.

Objective 10. Define the glomerular filtration rate and explain how it is regulated.

The glomerular capillaries have extremely large pores (200–500 Å in diameter) and are thus said to be *fenestrated* (chapter 20). As a result of these large pores (fenestra), glomerular capillaries are one hundred to four hundred times more permeable to plasma water and dissolved solutes than are the capillaries of skeletal muscles. Though the pores of glomerular capillaries are large, they are still small enough to prevent the passage of red blood cells, white blood cells, and platelets into the filtrate.

Before the filtrate can enter the interior of the glomerular capsule, it must pass through the capillary pores, the basement membrane (a thin layer of glycoproteins immediately outside the endothelial cells), and the inner, visceral layer of the glomerular capsule. The inner layer of the glomerular capsule is composed of unique cells called **podocytes** *(pod'o-sīt),* or "foot processes," with numerous cytoplasmic extensions known as **pedicels** (fig. 24.11). Pedicels interdigitate, like fingers wrapped around the glomerular capillaries. The narrow slits between adjacent pedicels provide the passageways through which filtered molecules must pass to enter the interior of the glomerular capsule (fig. 24.12).

Although the glomerular capillary pores are apparently large enough to permit the passage of proteins, the fluid that enters the capsular space is almost completely free of plasma proteins. This exclusion of plasma proteins from the filtrate is partially a result of their negative charges, which hinder their passage through the negatively charged glycoproteins in the basement membrane of the capillaries. The large size and negative charges of plasma proteins may also restrict their movement through the filtration slits between pedicels.

Glomerular Ultrafiltrate

The fluid that enters the glomerular capsule is called *ultrafiltrate* (fig. 24.13) because it is formed under pressure (the hydrostatic pressure of the blood). This is similar to the formation of tissue fluid by other capillary beds in the body (described in chapter 21). The force favoring filtration is opposed by a counter force developed by the hydrostatic pressure of fluid in the glomerular capsule. Also, since the protein concentration of the tubular fluid is low (less than 2–5 mg per 100 ml) compared to that of plasma (6–8 g per 100 ml), the greater colloid osmotic pressure of plasma promotes the osmotic return of filtered water. When these opposing forces are subtracted from the hydrostatic pressure of the glomerular capillaries, a *net filtration pressure* of approximately 10 mm Hg is obtained.

Because glomerular capillaries are extremely permeable and have a high surface area, this modest net filtration pressure produces an extraordinarily large volume of filtrate. The **glomerular filtration rate** (**GFR**) averages 115 ml per minute in women and 125 ml per minute in men. This is equivalent to 7.5 L per hour or 180 L per day (about 45 gallons)! Since the total blood volume averages about 5 L, this means that the total blood volume is filtered into the urinary tubules every forty minutes. Most of the filtered water must obviously be returned immediately to the vascular system or people would literally urinate to death within several minutes.

podocyte: Gk. *pous,* foot; *kytos,* cell
pedicel: L. *pedicellus,* footplate

Figure 24.11. (*a*) the inner (visceral) layer of the glomerular (Bowman's) capsule is composed of podocytes, as shown in this scanning electron micrograph. Very fine extensions of these podocytes form foot processes, or pedicels, that interdigitate around the glomerular capillaries. Spaces between adjacent pedicels form the "filtration slits." (*b*) an illustration of the relationship between glomerular capillaries and the inner layer of the glomerular capsule.

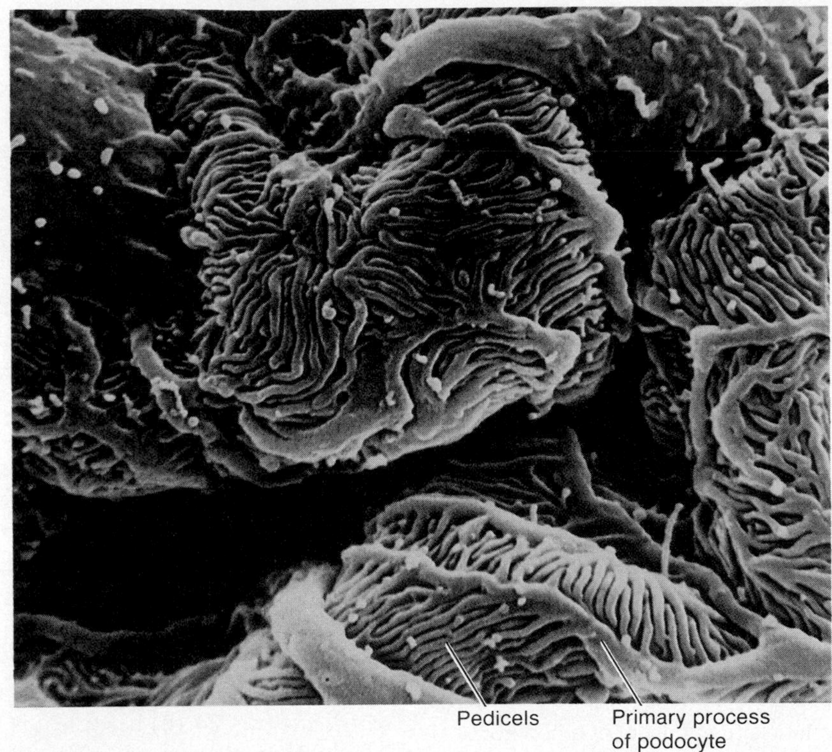

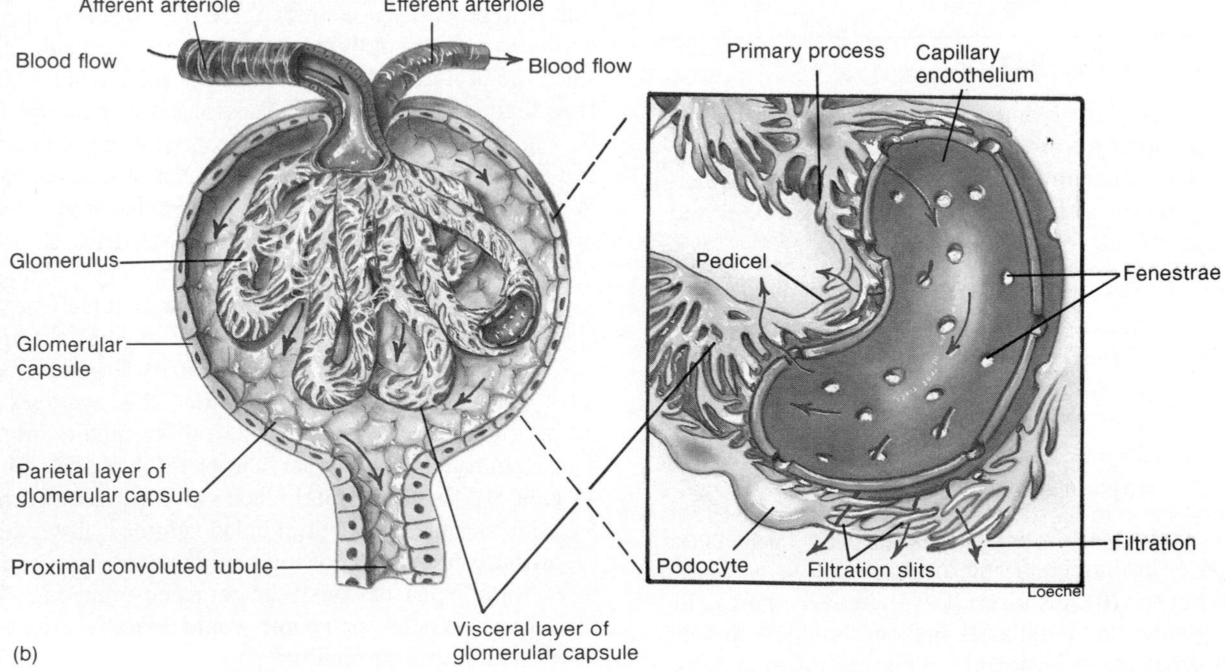

Figure 24.12. (*a*) an electron micrograph and (*b*) an illustration of the "filtration barrier" between the capillary lumen and the cavity of the glomerular (Bowman's) capsule.

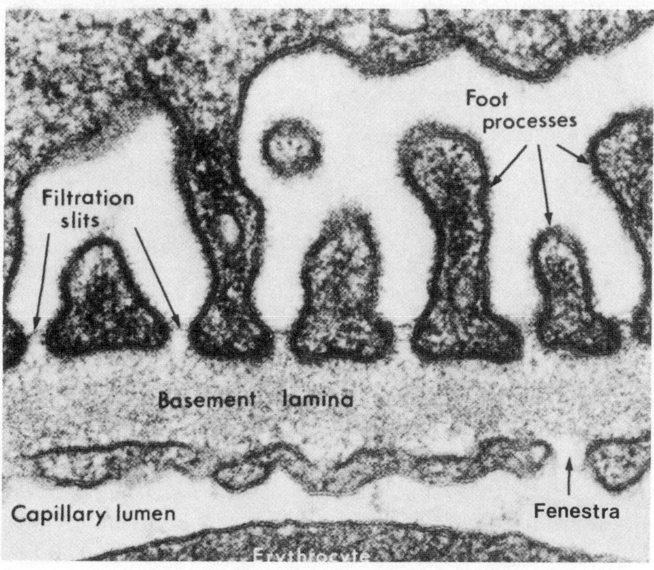

(a)

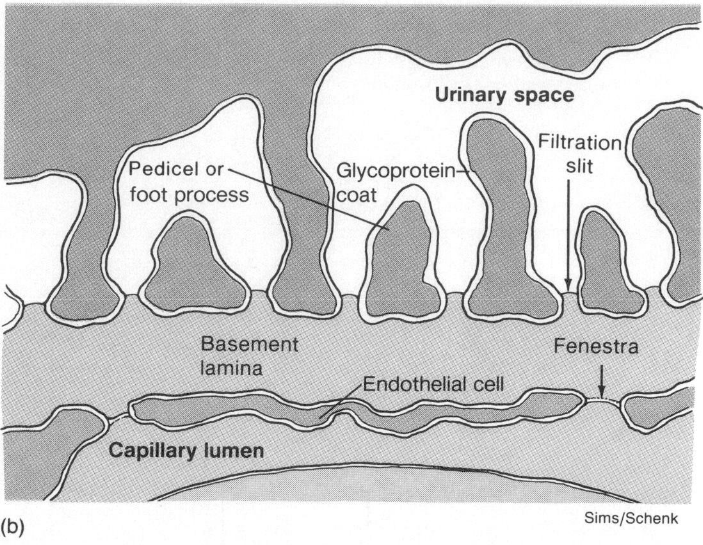

(b)

Figure 24.13. The formation of glomerular ultrafiltrate. Proteins (*large circles*) are not filtered, but smaller plasma solutes (*dots*) easily enter the glomerular ultrafiltrate. Arrows indicate the direction of filtration.

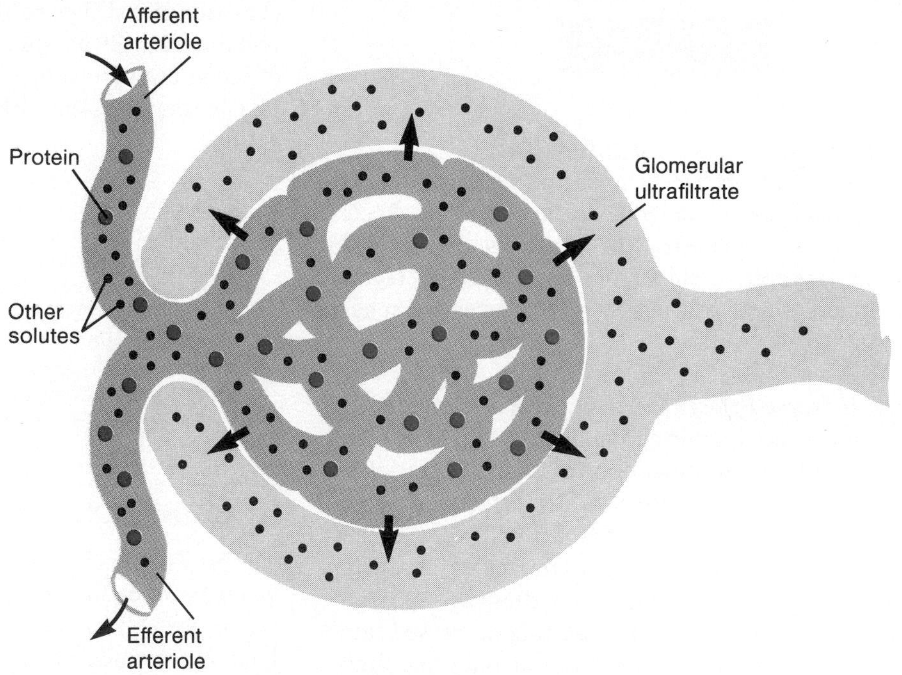

Figure 24.14. The effect of increased sympathetic nerve activity on kidney function and other physiological processes.

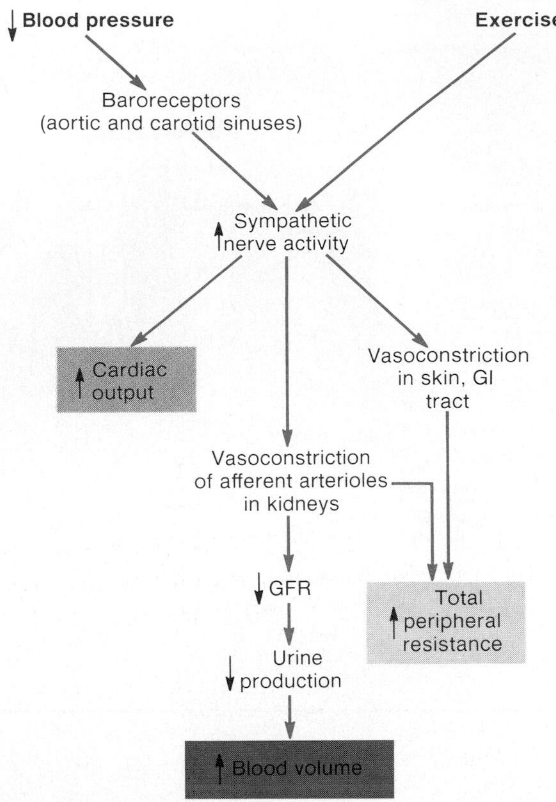

Regulation of Glomerular Filtration Rate

Vasoconstriction or dilation of afferent arterioles affects the rate of blood flow to the glomerulus and thus affects the glomerular filtration rate. Changes in the diameter of the afferent arterioles results from both extrinsic (sympathetic innervation) and intrinsic regulatory mechanisms.

Sympathetic Nerve Effects. An increase in sympathetic nerve activity, as occurs during the fight or flight reaction and exercise, stimulates constriction of afferent arterioles. This is an alpha-adrenergic effect, which helps to preserve blood volume and to divert blood to the muscles and heart. A similar effect occurs during cardiovascular shock in which sympathetic nerve activity stimulates vasoconstriction. The decreased GFR and the resulting decreased rate of urine formation help to compensate for the rapid drop of blood pressure under these circumstances (fig. 24.14).

Renal Autoregulation. When the direct effect of sympathetic stimulation is experimentally removed, the effect of systemic blood pressure on GFR can be observed. Under these conditions, surprisingly, the GFR remains relatively constant despite changes in mean arterial pressure within

Table 24.1 Regulation of glomerular filtration rate (GFR)

Regulation	Stimulus	Afferent arteriole	GFR
Sympathetic nerves	Activation by aortic and carotid baroreceptors or by higher brain centers	Constricts	Decreases
Autoregulation	Decreased blood pressure	Dilates	No change
Autoregulation	Increased blood pressure	Constricts	No change

From Stuart Ira Fox, *Human Physiology,* 2d ed. Copyright © 1987 Wm. C. Brown Publishers, Dubuque, Iowa. All Rights Reserved. Reprinted by permission.

a range of 70–180 mm Hg (normal mean arterial pressure is 100 mm Hg). The ability of the kidneys to maintain a relatively constant GFR in the face of fluctuating blood pressures is called **renal autoregulation.**

Renal autoregulation results from the effects of locally produced chemicals on the afferent arterioles (effects on the efferent arterioles are believed to be of secondary importance). When systemic arterial pressure falls toward a mean of 70 mm Hg, the afferent arterioles dilate, and when the pressure rises, the afferent arterioles constrict. Blood flow to the glomeruli and GFR can thus remain relatively constant within the autoregulatory range of blood pressure values. The effects of different regulatory mechanisms on GFR are summarized in table 24.1.

1. Describe the structures that plasma fluid must pass through to enter the capsular space. Explain how proteins are excluded from the filtrate.
2. Describe the forces that affect the formation of glomerular ultrafiltrate.
3. Describe the effect of sympathetic innervation on the glomerular filtrate rate, and explain the meaning of the term "renal autoregulation."

Reabsorption of Salt and Water

The proximal tubule reabsorbs most of the filtered salt and water from the filtrate. The transport of salt and water in the nephron loop creates a hypertonic renal medulla. Under the influence of ADH, this hypertonic medulla permits most of the remaining water to be reabsorbed.

Objective 11. Describe how salt and water are reabsorbed in the proximal tubule, and explain the significance of this process.

Objective 12. Describe the transport processes that occur in the descending and ascending limbs of the nephron loop, and explain how the countercurrent multiplier effect is produced.

Figure 24.15. Plasma water and its dissolved solutes (except proteins) enter the glomerular ultrafiltrate, but most of these filtered molecules are reabsorbed. The term *reabsorption* refers to the transport of molecules out of the tubular filtrate back into the blood.

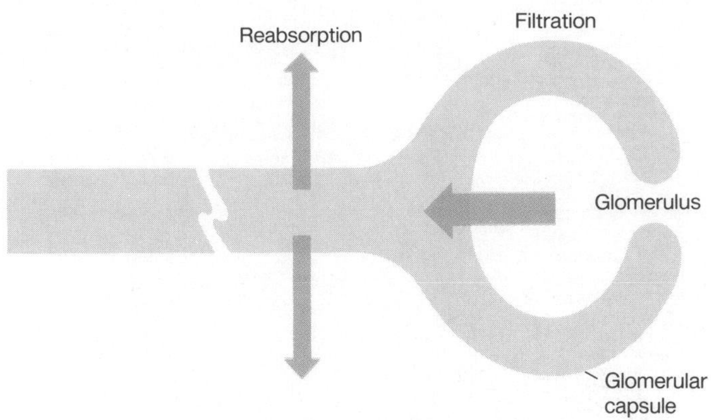

Objective 13. Describe the structure and function of the vasa recta.

Objective 14. Explain the significance of a hypertonic renal medulla, and describe the mechanism of ADH action.

Although about 180 L per day of glomerular ultrafiltrate are produced, the kidneys normally excrete only 1–2 L per day of urine. Approximately 99% of the filtrate must thus be returned to the vascular system, while 1% is excreted in the urine. The urine volume, however, varies according to the needs of the body. When a well-hydrated person drinks a liter or more of water, urine volume increases to 16 ml per minute (the equivalent of 23 L per day if this were to continue for twenty-four hours). In severe dehydration, when the body needs to conserve water, only 0.3 ml per minute, or 400 ml per day, of urine are produced. A volume of 400 ml per day of urine is needed to excrete the amount of metabolic wastes produced by the body; this is called the *obligatory water loss.* When water in excess of this amount is excreted, the urine volume is increased and its concentration is decreased.

Regardless of the body's state of hydration, it is clear that most of the filtered water must be returned to the vascular system to maintain blood volume and pressure. The return of filtered molecules from the tubules to the blood is called *reabsorption* (fig. 24.15). It is important to realize that the transport of water always occurs passively by *osmosis;* there is no such thing as active transport of water. A concentration gradient must thus be created between tubular fluid and blood that favors the osmotic return of water to the vascular system.

Reabsorption in the Proximal Tubule

Since all plasma solutes, with the exception of proteins, are able to freely enter the glomerular ultrafiltrate, the osmolality of the filtrate is essentially the same as that of the plasma (300 milliosmoles per liter, or 300 mOsm). The filtrate is therefore said to be isosmotic with the plasma. Osmosis thus cannot occur unless the concentration of plasma in the peritubular capillaries and the concentration of filtrate are altered by active transport processes. This is achieved by the active transport of Na^+ from the filtrate to the peritubular blood.

Active and Passive Transport. The epithelial cells that comprise the wall of the proximal tubule are joined together by gap junctions only on their apical sides—that is, the sides of each cell that are closest to the lumen of the tubule (fig. 24.16). Each cell therefore has four exposed surfaces: the apical side facing the lumen, which contains microvilli, the opposite, basal side facing the peritubular capillaries, and the lateral sides, facing the narrow clefts between adjacent epithelial cells.

The concentration of Na^+ in the glomerular ultrafiltrate—and thus in the fluid entering the proximal tubule—is the same as in plasma. The epithelial cells of the tubule, however, have a much lower Na^+ concentration. This lower Na^+ concentration is partially due to the low permeability of the cell membrane to Na^+ and partially due to the active transport of Na^+ out of the cell by Na^+/K^+ pumps, as described in chapter 5. In the cells of the proximal tubule, the Na^+/K^+ pumps are located in the basal and lateral sides of the cell membrane but not in the apical membrane. As a result of the action of these active transport

Figure 24.16. An illustration of the appearance of tubule cells in the electron microscope. Molecules that are reabsorbed pass through the tubule cells from the apical membrane (facing the filtrate) to the basolateral membrane (facing the blood).

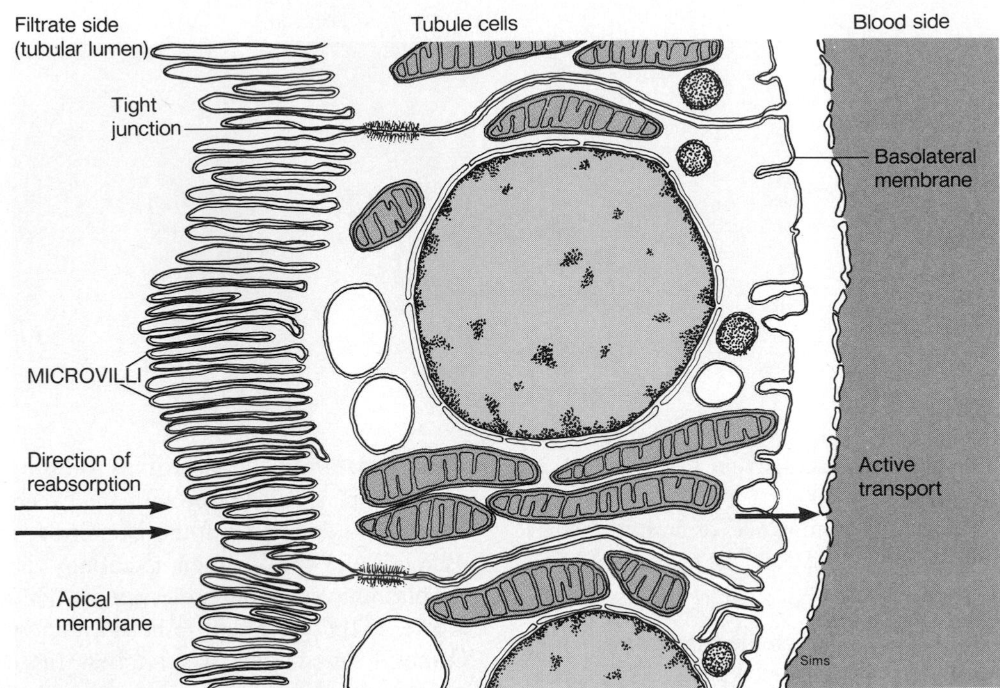

pumps, a concentration gradient is created that favors the diffusion of Na^+ from the tubular fluid, across the apical cell membranes, and into the epithelial cells of the proximal tubule. The Na^+ is then extruded into the surrounding tissue fluid by the Na^+/K^+ pumps.

The removal of Na^+ from the tubular fluid and its subsequent appearance in the tissue fluid surrounding the epithelial cells of the proximal tubule create an electrical gradient that favors the passive transport of Cl^- toward the higher Na^+ concentration in the tissue fluid. As a result of the accumulation of NaCl, the osmolality and osmotic pressure of the tissue fluid surrounding the epithelial cells is increased above that of the tubular fluid. This is particularly true of the tissue fluid between the lateral membranes of adjacent epithelial cells, where the narrow spaces permit the accumulated NaCl to become less diluted.

An osmotic gradient is thus created beween the tubular fluid and the tissue fluid surrounding the proximal tubule. Since the cells of the proximal tubule are permeable to water, water moves by osmosis from the tubular fluid into the epithelial cells and then across the basal and lateral sides of the epithelial cells into the tissue fluid. The salt and water which were reabsorbed from the tubular fluid can then move passively into the surrounding peritubular capillaries and in this way be returned to the blood (fig. 24.17).

Significance of Proximal Tubule Reabsorption. Approximately 65% of the salt and water in the original glomerular ultrafiltrate is reabsorbed across the proximal tubule and returned to the vascular system. The volume of tubular fluid remaining is reduced accordingly, but this fluid is still isosmotic with the blood (has a concentration of 300 mOsm). This results from the fact that the cell membranes in the proximal tubule are freely permeable to water so that water and salt are removed in proportionate amounts.

An additional smaller amount of salt and water is returned to the vascular system by reabsorption in the nephron loop. This reabsorption, like that in the proximal tubule, occurs constantly, regardless of the person's state of hydration. Unlike reabsorption in later regions of the nephron, it is not subject to hormonal regulation. Approximately 85% of the filtered salt and water is, therefore, reabsorbed in a constant, unregulated fashion in the early regions of the nephron (proximal tubule and nephron loop). This reabsorption is very costly in terms of energy expenditures, accounting for as much as 6% of the calories consumed by the body at rest.

Since 85% of the original glomerular ultrafiltrate is immediately reabsorbed in the early region of the nephron, only 15% of the initial filtrate remains to enter the distal

Figure 24.17. Mechanisms of salt and water reabsorption in the proximal tubule. Sodium is actively transported out of the filtrate, and chloride follows passively by electrical attraction. Water follows the salt out of the tubular filtrate into the peritubular capillaries by osmosis.

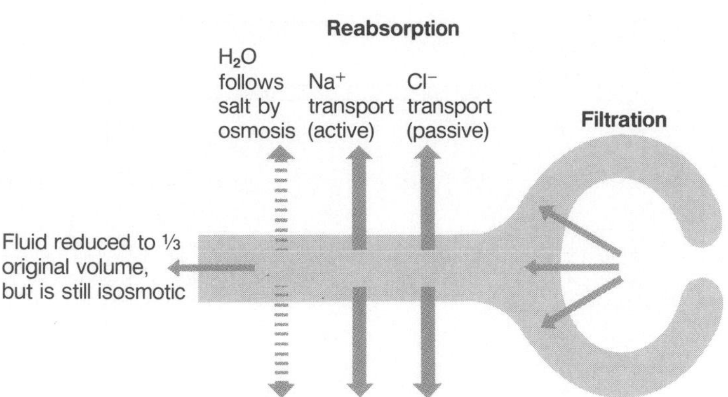

convoluted tubule and collecting duct. This is still a large volume of fluid—15% × GFR (180 L per day) = 27 L per day—that must be reabsorbed to varying degrees in accordance with the body's state of hydration. This "fine tuning" of the percent reabsorption and urine volume is accomplished by the action of hormones on the later regions of the nephron.

The Countercurrent Multiplier System

Water cannot be actively transported across the tubule wall, and osmosis of water cannot occur if the tubular fluid and surrounding tissue fluid are isotonic to each other. In order for water to be reabsorbed by osmosis, the surrounding tissue fluid must be hypertonic. The osmotic pressure of the tissue fluid in the renal medulla is, in fact, raised to over four times that of plasma. This results partly from the fact that the tubule bends; the geometry of the nephron loop allows interaction to occur between the descending and ascending limbs. Since the ascending limb is the active partner in this interaction, its properties will be described before those of the descending limb.

Ascending Limb of the Nephron Loop. Salt (NaCl) is actively extruded from the ascending limb into the surrounding tissue fluid. Until recently, it was thought that the cells of the ascending limb accomplish this by actively pumping out Cl^- and that Na^+ follows the Cl^- passively by electrical attraction (the opposite of the situation in the proximal tubule). Newer evidence, however, suggests that Na^+, K^+, and Cl^- move from the filtrate into the ascending limb cells, in a ratio of 1 Na^+ to 1 K^+ to 2 Cl^-, and that the Na^+ is then actively transported across the basolateral membrane by the Na^+/K^+ pump (fig. 24.18). The ion that is actively transported is Na^+; Cl^- follows the Na^+ passively because of electrical attraction, and K^+ probably diffuses back into the filtrate.

The ascending limb is structurally divisible into two regions: a *thin segment,* nearest to the tip of the nephron loop, and a *thick segment* of varying lengths, which carries the filtrate outward into the cortex and into the distal convoluted tubule. Some scientists believe that only the cells of the thick segments of the ascending limb are capable of actively transporting NaCl from the filtrate into the surrounding tissue fluid.

Regardless of the mechanism of active transport or the region of the ascending limb in which this transport occurs, the net effect is the same as in the proximal tubule: salt (NaCl) is extruded into the surrounding tissue fluid. Unlike the epithelial walls of the proximal tubule, however, the walls of the ascending limb of the nephron loop are *not permeable to water.* The tubular fluid thus becomes increasingly dilute as it ascends toward the cortex, whereas the tissue fluid around the nephron loops in the medulla becomes increasingly more concentrated. By means of these processes, the tubular fluid that enters the distal tubule in the cortex is made hypotonic (with a concentration of about 100 mOsm), whereas the tissue fluid in the medulla is made hypertonic.

Descending Limb of the Nephron Loop. The deeper regions of the medulla, around the tips of the loops of juxtamedullary nephrons, reach a concentration of 1200–1400 mOsm. In order to reach this high a concentration, the salt pumped out of the ascending limb must accumulate in the tissue fluid. This occurs as a result of the properties

Figure 24.18. In the thick segment of the ascending limb of the loop, Na^+ and K^+, together with two Cl^-, enter the tubule cells. Na^+ is then actively transported out into the interstitial space, and Cl^- follows passively. The K^+ diffuses back into the filtrate, and some also enters the interstitial space.

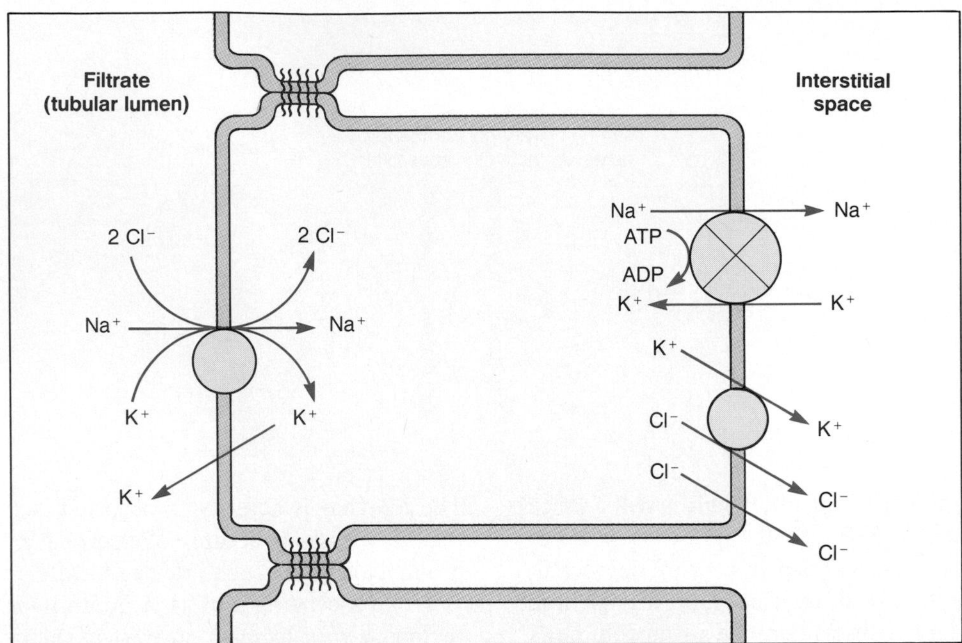

of the descending limb and because blood vessels around the loop do not carry back all of the extruded salt to the general circulation.

The descending limb does not actively transport salt. Instead, it is *passively permeable* to water and perhaps also to salt (this is currently controversial). The wall of the descending limb is like a porous plastic membrane; water and perhaps salt are free to diffuse according to their concentration gradients. Since the renal medulla is hypertonic to the fluid entering the descending limb, water moves by osmosis out of the descending limb and is removed by peritubular capillaries. At the same time, salt may diffuse into the descending tubule since it is present at a higher concentration in the surrounding tissue fluid. The concentration of tubular fluid is thus increased, and its volume is decreased, as it descends toward the tips of the nephron loops.

As a result of these passive transport processes in the descending limb, the fluid that "rounds the bend" at the tip of the nephron loop has the same osmolality as the surrounding tissue fluid (1200–1400 mOsm). There is, therefore, a higher salt concentration arriving in the ascending limb than there would be if the descending limb delivered isotonic fluid. Salt transport by the ascending limb is increased accordingly, so that the "saltiness" of the tissue fluid is multiplied (fig. 24.19).

Countercurrent Multiplication. Countercurrent flow (flow in opposite directions) in the ascending and descending limbs and the close proximity of the two limbs allow interaction to occur. Since the concentration of the tubular fluid in the descending limb reflects the concentration of surrounding tissue fluid, and since the concentration of this tissue fluid is raised by the active extrusion of salt from the ascending limb, a *positive feedback* mechanism is created. The more salt the ascending limb extrudes, the more concentrated will be the fluid that returns to it from the descending limb. This positive feedback mechanism multiplies the concentration of tissue fluid and descending limb fluid and is thus called the **countercurrent multiplier system.**

The countercurrent multiplier system recirculates salt and thus traps some of the salt that enters the nephron loop in the tissue fluid of the renal medulla. This system results in a gradually increasing concentration of renal tissue fluid from the cortex to the inner medulla; the osmolality of tissue fluid increases from 300 mOsm (isosmotic) in the cortex to 1200–1400 mOsm in the deepest part of the medulla.

Vasa Recta. In order for the countercurrent multiplier system to be effective, most of the salt that is extruded from the ascending limbs must remain in the tissue fluid

Figure 24.19. The countercurrent multiplier system. The active extrusion of Cl^- followed by Na^+ from the ascending limb makes the surrounding tissue fluid more concentrated. This concentration is multiplied by the fact that the descending limb is passively permeable so that its fluid increases in concentration as the surrounding tissue fluid becomes more concentrated. The transport properties of the loop and their effect on tubular fluid concentration is shown in (*a*). The values of these changes in osmolality, together with the effect on surrounding tissue fluid concentration, are shown in (*b*).

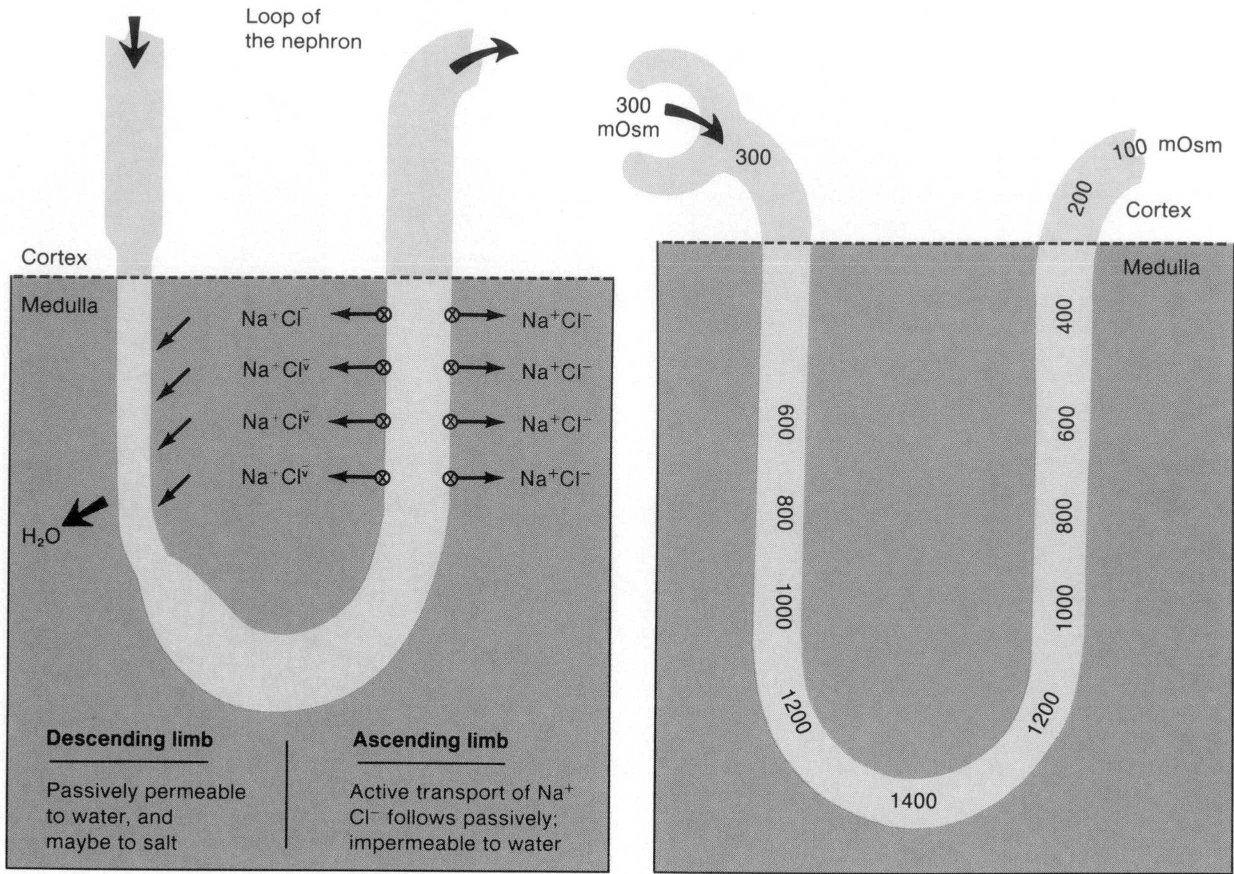

of the medulla, while most of the water that leaves the descending limbs must be removed by the blood. This is accomplished by vessels known as the **vasa recta,** which form long capillary loops that parallel the long nephron loops of the juxtamedullary nephrons (see fig. 24.22).

The vasa recta maintain the hypertonicity of the renal medulla by means of a mechanism known as **countercurrent exchange.** Salt and other solutes (such as urea) that are present at high concentrations in the medullary tissue fluid diffuse into the blood as the blood descends into the capillary loops of the vasa recta, but then passively diffuse out of the ascending vessels and back into the descending vessels (where the concentration is lower). Solutes are thus recirculated and trapped within the medulla. Water, in contrast, diffuses out of the descending vessels and into the ascending vessels (where the osmotic pressure is higher) and is thus transported out of the medulla (fig. 24.20).

Possible Effects of Urea. Experimental evidence suggests that the active transport of Na^+ may only occur in the thick segments of the ascending limbs. The thin segments of the ascending limbs, which are located in the deeper regions of the medulla, may not be able to extrude salt actively. Since salt does leave the thin segments, it must leave by diffusion even though the surrounding tissue fluid is at least as concentrated as the tubular fluid. Some investigators therefore conclude that molecules other than salt—specifically urea—contribute to the hypertonicity of the tissue fluid. Urea may accumulate in the medullary tissue fluid as a result of the recycling of urea between the collecting duct and the nephron loop (fig. 24.21). The transport properties of different tubule segments are summarized in table 24.2.

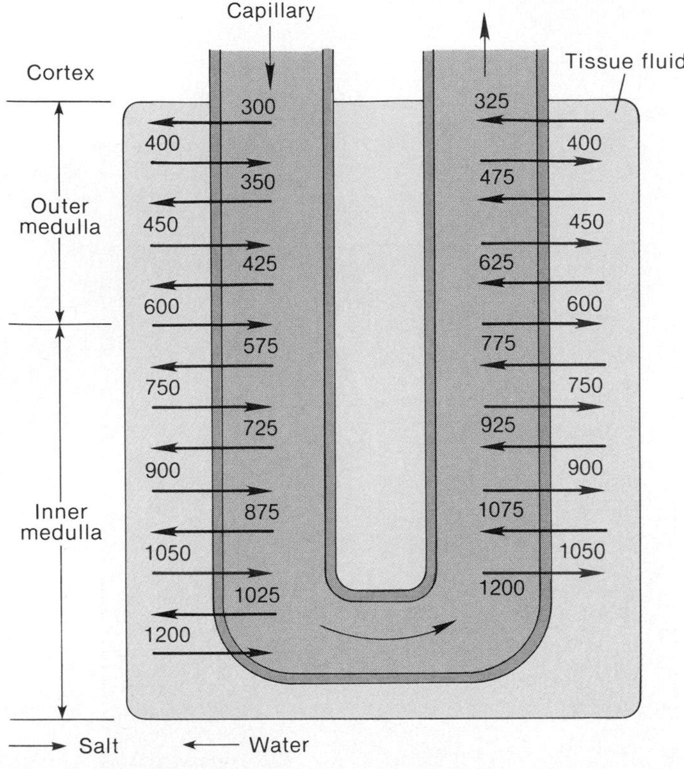

Figure 24.20. Countercurrent exchange in the vasa recta. The diffusion of salt and water first into and then out of these blood vessels helps to maintain the "saltiness" (hypertonicity) of the interstitial fluid in the renal medulla (numbers indicate osmolality).

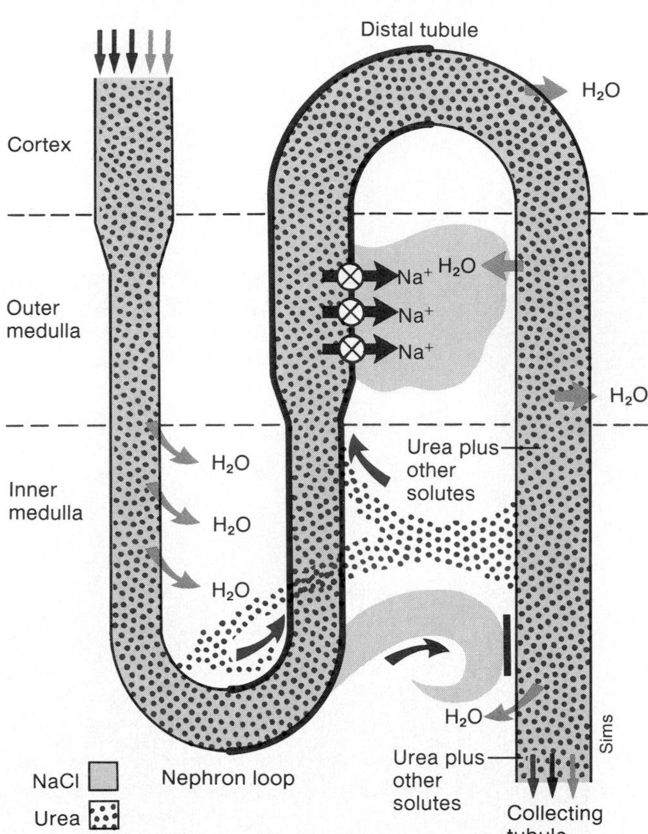

Figure 24.21. According to some authorities, urea diffuses out of the collecting duct and contributes significantly to the concentration of the interstitial fluid in the renal medulla. The active transport of Na^+ out of the thick segments of the ascending limbs also contributes to the hypertonicity of the medulla so that water is reabsorbed by osmosis from the collecting ducts.

Table 24.2 Properties of different nephron segments in regard to the concentrating-diluting mechanisms of the kidney

Nephron segment	Active transport	Passive transport		
		Salt	Water	Urea
Proximal tubule	Na^+	Cl^-	Yes	Yes
Descending limb of the nephron loop	None	Maybe	Yes	No
Thin segment of ascending limb*	Na^+ or none	Cl^- or NaCl	No	Yes
Thick segment of ascending limb	Na^+	Cl^-	No	No
Distal tubule	Na^+	No	No	No
Collecting duct**	slight Na^+	No	Yes (ADH) or slight (no ADH)	Yes

*The thin segment may actively transport Na^+ in the same manner as the thick segment, or it may only permit the passive diffusion of NaCl. Both possibilities are presented in this table.

**The permeability of the collecting duct to water depends on the presence of ADH.

From Stuart Ira Fox, *Human Physiology,* 2d ed. Copyright © 1987 Wm. C. Brown Publishers, Dubuque, Iowa. All Rights Reserved. Reprinted by permission.

Collecting Duct: Effect of Antidiuretic Hormone (ADH)

As a result of the recycling of salt between the ascending and descending limbs, and possibly of the recycling of urea between the collecting duct and the nephron loop, the medullary tissue fluid is made very hypertonic. The collecting ducts must transport their fluid through this hypertonic environment in order to empty their contents of urine into the calyces. Since the distal convoluted tubules are impermeable to water, the fluid that enters the collecting ducts in the cortex is hypotonic as a result of the active extrusion of salt by the ascending limbs of the nephron loops.

The walls of the collecting ducts are *permeable to water but not to salt.* Since the surrounding tissue fluid in the renal medulla is very hypertonic, as a result of the

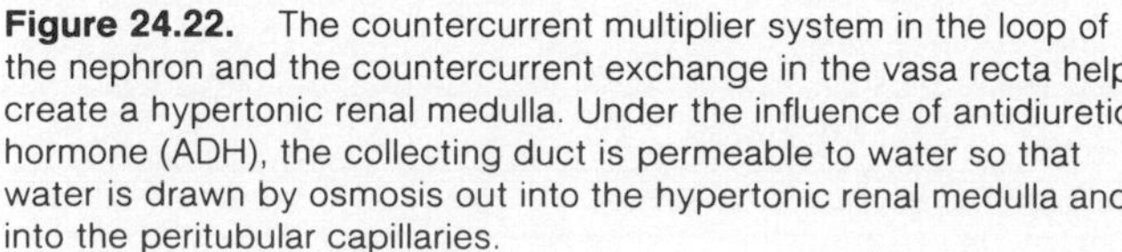

Figure 24.22. The countercurrent multiplier system in the loop of the nephron and the countercurrent exchange in the vasa recta help create a hypertonic renal medulla. Under the influence of antidiuretic hormone (ADH), the collecting duct is permeable to water so that water is drawn by osmosis out into the hypertonic renal medulla and into the peritubular capillaries.

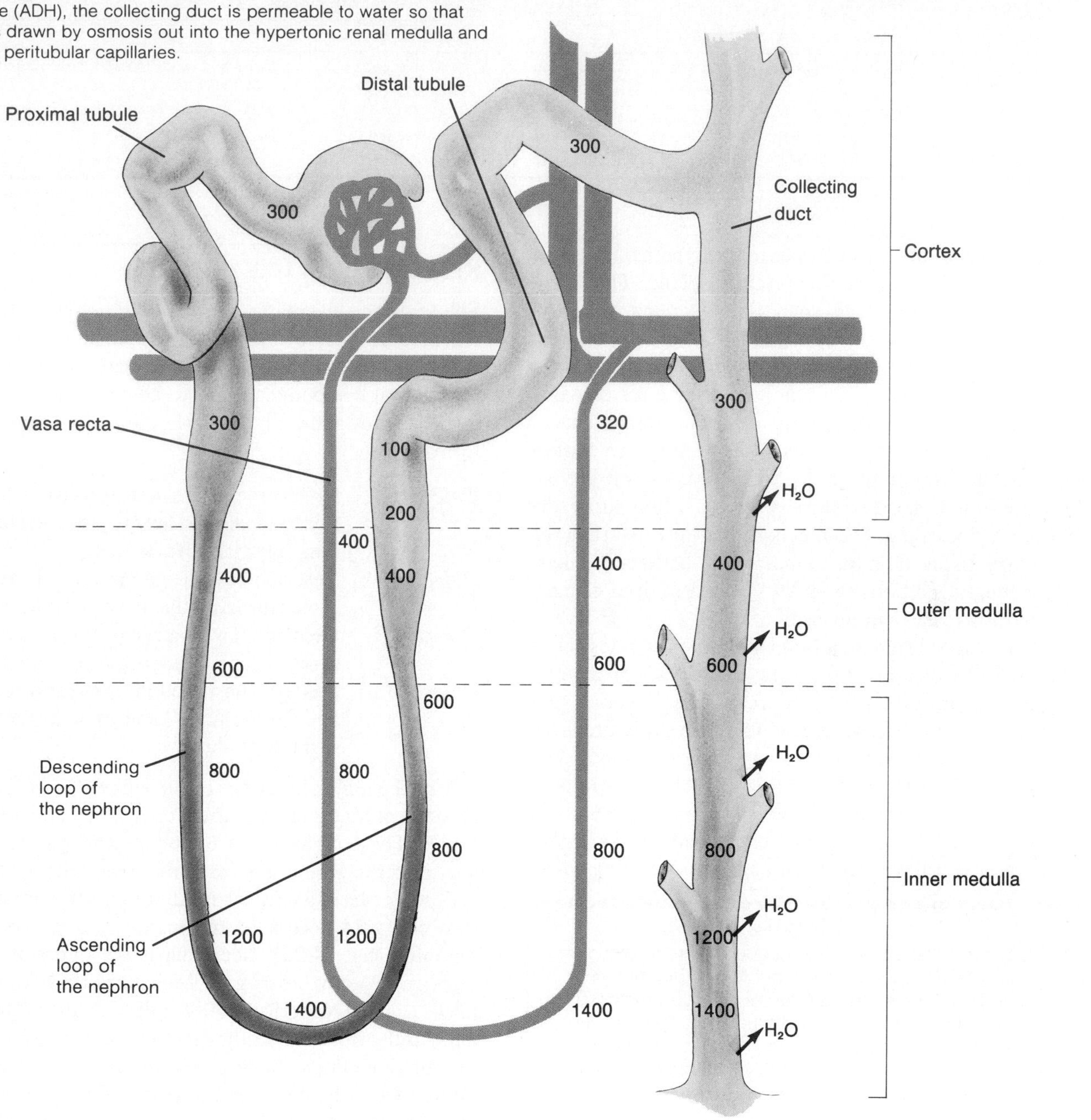

countercurrent multiplier system, water is drawn out of the collecting ducts by osmosis. This water does not dilute the surrounding tissue fluid because it is transported by capillaries to the general circulation. In this way, most of the water remaining in the tubules after reabsorption in the proximal tubules is returned to the vascular system (fig. 24.22).

The osmotic gradient created by the countercurrent multiplier system provides the force for water reabsorption through the collecting ducts. The rate of this reabsorption, however, is determined by the permeability of the collecting duct cell membranes to water. The permeability of the collecting duct to water, in turn, is determined by the concentration of **antidiuretic** *(an″tĭ-di″u-ret′ik)* **hormone (ADH)** in the blood. When the concentration of ADH is increased, the collecting ducts become more permeable to water, and more water is reabsorbed. A decrease in ADH, conversely, results in less reabsorption of water and thus in the excretion of a larger volume of more dilute urine.

Table 24.3 Antidiuretic hormone secretion and action

Stimulus	Receptors	Secretion of ADH	Effects on	
			Urine volume	Blood
↑Osmolality (dehydration)	Osmoreceptors in hypothalamus	Increased	Decreased	Increased water retention; decreased blood osmolality
↓Osmolality	Osmoreceptors in hypothalamus	Decreased	Increased	Water loss increases blood osmolality
↑Blood volume	Stretch receptors in left atrium	Decreased	Increased	Decreased blood volume
↓Blood volume	Stretch receptors in left atrium	Increased	Decreased	Increased blood volume

ADH is produced by neurons in the hypothalamus and is secreted from the posterior pituitary gland. The secretion of ADH is stimulated when osmoreceptors in the hypothalamus respond to an increase in blood osmotic pressure. During dehydration, therefore, when the plasma becomes more concentrated, increased secretion of ADH promotes increased permeability of the collecting ducts to water. In severe dehydration, only the minimal amount of water needed to eliminate the body's wastes is excreted. This minimum, about 400 ml per day, is limited by the fact that urine cannot become more concentrated than the medullary tissue fluid surrounding the collecting ducts. Under these conditions about 99.8% of the initial glomerular ultrafiltrate is reabsorbed.

A person in a state of normal hydration excretes about 1.5 L per day of urine, indicating that 99.2% of the glomerular ultrafiltrate volume is reabsorbed. Notice that small changes in percent reabsorption translate into large changes in urine volume. Increasing water ingestion—and thus decreasing ADH secretion (table 24.3)—results in correspondingly larger volumes of urine excretion. It should be noted that even in the complete absence of ADH some water is still reabsorbed through the collecting ducts.

Diabetes insipidus is a disease associated with the inadequate secretion or action of ADH. The collecting ducts are thus not very permeable to water and, therefore, a large volume (5–10 L per day) of dilute urine is produced. The dehydration that results causes intense thirst, but a person with this condition has difficulty drinking enough to compensate for the large volumes of water lost in the urine.

1. Describe the mechanisms for salt and water reabsorption in the proximal tubule.
2. Compare the transport of Na^+, Cl^-, and water across the walls of the proximal tubule, ascending and descending limbs of the nephron loop, and collecting duct.
3. Explain the interaction of the ascending and descending limbs of the nephron loop and how this interaction results in a hypertonic renal medulla.
4. Explain how ADH helps the body to conserve water, and describe how variations in ADH secretion affect the volume and concentration of urine.

Renal Plasma Clearance

Solutes in the plasma that are filtered or are secreted into the tubules will be excreted in the urine if they are not reabsorbed. Measurements of the renal plasma clearance of different compounds indicate how the kidneys dispose of these substances and provide clinically important information.

Objective 15. Define the terms *reabsorption* and *secretion,* and explain how these processes affect renal clearance rates.

Objective 16. Describe how the glomerular filtration rate is measured by the clearance rate of inulin.

Objective 17. Describe how the renal blood flow is measured by the clearance rate of PAH.

Objective 18. Describe how the kidneys reabsorb glucose and amino acids, and explain how glycosuria is produced.

One of the major functions of the kidneys is the excretion of waste products such as urea, creatinine, and other molecules. These molecules are filtered through the glomerulus into the glomerular capsule along with water, salt, and other plasma solutes. In addition, some waste products can gain access to the urine by a process called **secretion** (fig. 24.23). Secretion is the opposite of reabsorption by active transport. Molecules that are secreted move out of the peritubular capillaries and into the tubular cells, from which they are actively transported into the tubular lumen. In this way, molecules that were not filtered from the blood in the glomerulus can still be excreted in the urine.

Although most (about 99%) of the filtered water is returned to the vascular system by reabsorption, most of the wastes that are filtered or secreted are eliminated in the urine. The concentration of these substances in the renal vein leaving the kidneys is therefore lower than their concentrations in the blood entering the kidneys in the renal artery. Some of the blood that passes through the kidneys, in other words, is "cleared" of these waste products; this is known as the **renal plasma clearance.**

Figure 24.23. Secretion refers to the active transport of substances from the peritubular capillaries into the tubular fluid. This transport is in a direction opposite to that of reabsorption.

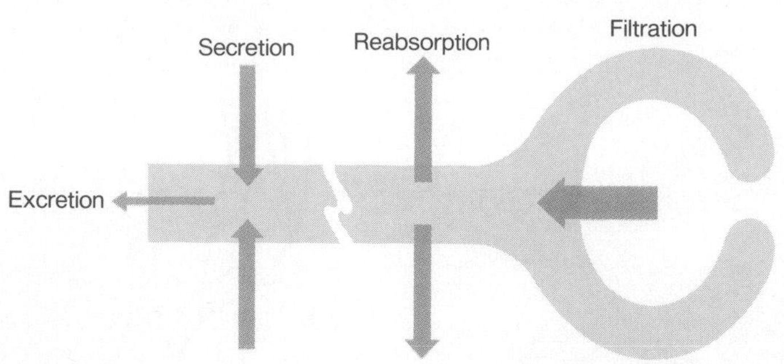

Table 24.4 The effects of filtration, reabsorption, and secretion on renal clearance rates

Term	Means	Effect on renal clearance
Filtered	A substance enters the glomerular ultrafiltrate	Some or all of a filtered substance may enter the urine and be "cleared" from the blood.
Reabsorbed	The transport of a substance from the filtrate, through tubular cells, and into the blood	Reabsorption decreases the rate at which a substance is cleared; clearance rate is less than the glomerular filtration rate (GFR).
Secreted	The transport of a substance from peritubular blood, through tubular cells, and into the filtrate	When a substance is secreted by the nephrons, its clearance rate is greater than the GFR.

The quantity of a substance excreted in the urine within a given period of time depends on (1) the quantity *filtered* through the glomeruli into the tubular fluid; (2) the quantity *secreted* from the unfiltered blood in peritubular capillaries by active transport into the tubules; and (3) the quantity *reabsorbed* by transport from the tubules into the peritubular blood. Filtration and secretion increase renal plasma clearance; reabsorption decreases the amount excreted and thus decreases the clearance (table 24.4).

Renal Clearance of Inulin: Measurement of GFR

If a substance is neither reabsorbed nor secreted by the tubules, the amount excreted per minute in the urine will be equal to the amount that is filtered out of the glomeruli. There does not seem to be a single substance produced by the body, however, that is not reabsorbed or secreted to some degree. Plants such as artichokes, dahlias, onions, and garlic, fortunately, do produce such a compound. This compound, a plant product which is a polymer of the monosaccharide fructose, is **inulin** *(in'u-lin)*. Once injected into the blood, inulin is filtered by the glomeruli, and the amount of inulin excreted per minute is exactly equal to the amount that was filtered per minute (fig. 24.24).

If the concentration of inulin in urine is measured and the rate of urine formation is determined, the rate of inulin excretion can easily be calculated:

$$\underset{(\text{mg/min})}{\textit{Quantity excreted per minute}} = \underset{\left(\frac{\text{ml}}{\text{min}}\right)}{V} \times \underset{\left(\frac{\text{mg}}{\text{ml}}\right)}{U}$$

Where V = rate of urine formation
U = inulin concentration in urine.

The rate at which a substance is filtered by the glomeruli (in mg per minute) can be calculated by multiplying the ml per minute of plasma that is filtered (the **glomerular filtration rate,** or **GFR**) by the concentration of that substance in the plasma. This is shown in the following equation:

$$\underset{(\text{mg/min})}{\textit{Quantity filtered per minute}} = \underset{\left(\frac{\text{ml}}{\text{min}}\right)}{GFR} \times \underset{\left(\frac{\text{mg}}{\text{ml}}\right)}{P}$$

Where P = inulin concentration in plasma.

Since inulin is neither reabsorbed nor secreted, the amount filtered equals the amount excreted:

$$\underset{(\text{amount filtered})}{GFR \times P} = \underset{(\text{amount excreted})}{V \times U.}$$

Figure 24.24. The renal clearance of inulin. (*a*) inulin is present in the blood entering the glomeruli, and (*b*) some of this blood, together with its dissolved inulin, is filtered. All of this filtered inulin enters the urine, whereas most of the filtered water is returned to the vascular system (is reabsorbed). (*c*) the blood leaving the kidneys in the renal vein, therefore, contains less inulin than the blood that entered the kidneys in the renal artery. Since inulin is filtered but neither reabsorbed nor secreted, the inulin clearance rate equals the glomerular filtration rate (GFR).

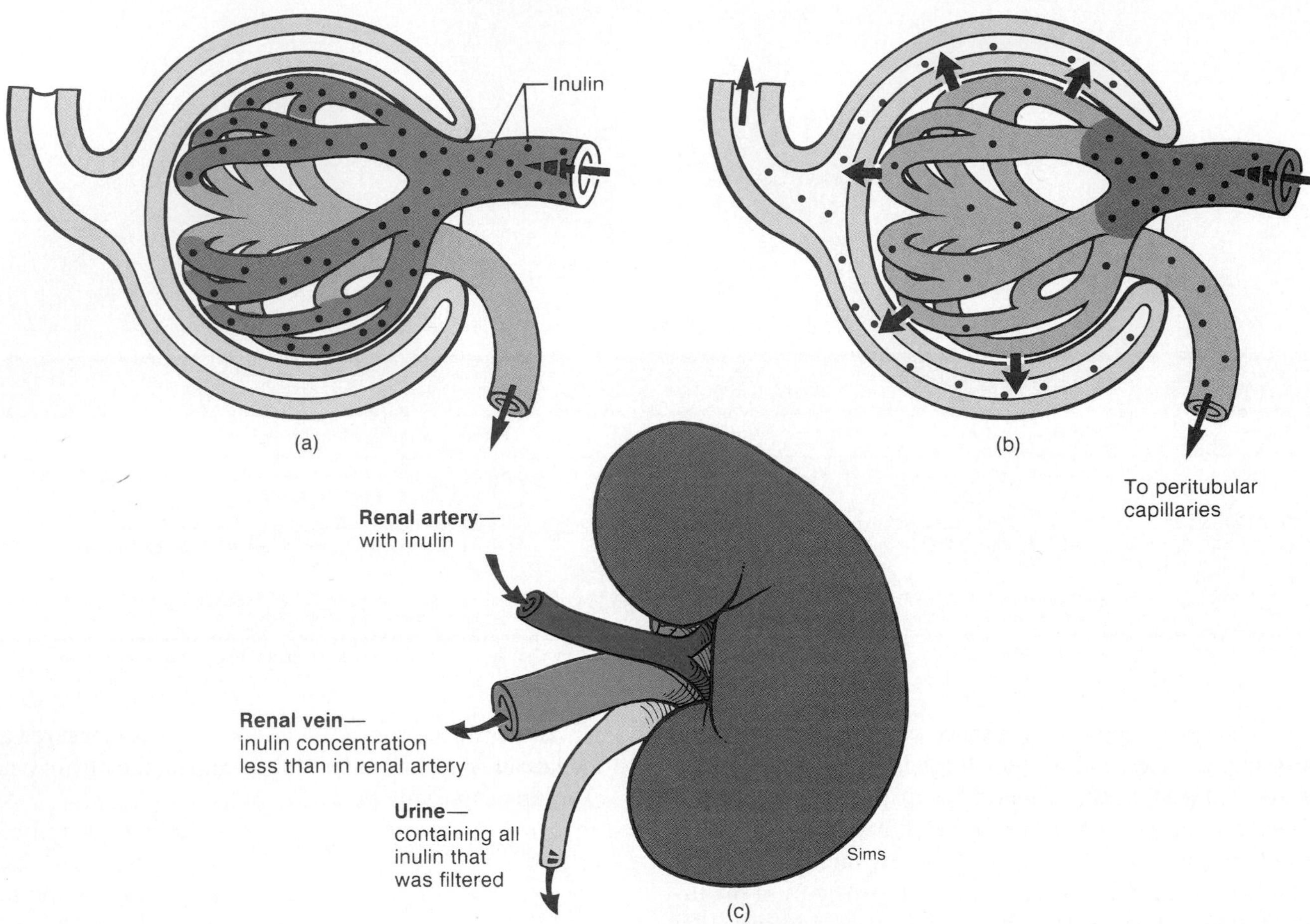

If the last equation is now solved for the glomerular filtration rate,

$$GFR_{(ml/min)} = \frac{V_{(ml/min)} \times U_{(mg/ml)}}{P_{(mg/ml)}}.$$

Suppose, for example, that inulin is infused into a vein and its concentration in the urine and plasma are found to be 30 mg per ml and 0.5 mg per ml, respectively. If the rate of urine formation is 2 ml per minute, the GFR can be calculated as follows:

$$GFR = \frac{2\ ml/min \times 30\ mg/ml}{0.5\ mg/ml} = 120\ ml/min.$$

This equation states that, at a plasma inulin concentration of 0.5 mg per ml, 120 ml of plasma must have been filtered in order to excrete the measured amount of 60 mg that appears in the urine per minute. The glomerular filtration rate is thus 120 ml per minute in this example.

Measurements of the plasma concentration of *creatinine (kre-at'i-nin)* are often used clinically as an index of kidney function. Creatinine, produced as a waste product of muscle creatine, is secreted to a slight degree by the renal tubules so that its excretion rate is a little above that of inulin. Since it is released into the blood at a constant rate and since its excretion is closely matched to the GFR, an abnormal decrease in GFR causes the plasma creatinine concentration to rise. A simple measurement of blood creatinine concentration can thus provide information about the health of the kidneys.

Table 24.5 Renal "handling" of different plasma molecules

If substance is	Example	Concentration in renal vein	Renal clearance
Not filtered	Proteins	Same as in renal artery	Zero
Filtered, not reabsorbed nor secreted	Inulin	Less than in renal artery	Equal to GFR (115–125 ml/min)
Filtered, partially reabsorbed	Urea	Less than in renal artery	Less than GFR
Filtered, completely reabsorbed	Glucose	Same as in renal artery	Zero
Filtered and secreted	PAH	Less than in renal artery; approaches zero	Greater than GFR; equal to total plasma flow rate (~625 ml/min)
Filtered, reabsorbed, and secreted	K^+	Variable	Variable

Clearance Calculations. If both the amount of inulin excreted per minute and the plasma inulin concentration are known, the volume of plasma that was filtered per minute (the GFR) can be calculated. Since 120 ml per minute are filtered (enter the glomerular capsules) in the previous example, the amount of inulin contained in 120 ml of plasma is excreted per minute. Since creatinine is filtered like inulin but is also secreted to a slight degree, the excretion rate of creatinine is slightly greater than the filtration rate. The amount of creatinine excreted per minute, in other words, is greater than the amount contained in the 120 ml of plasma that was filtered.

If a substance is reabsorbed to some degree, conversely, the excretion rate will be less than the amount contained in the 120 ml of plasma that was filtered. In order to compare the renal "handling" of various substances in terms of their reabsorption or secretion, the **renal plasma clearance** of these substances can be calculated using the same formula used for the determination of the GFR:

$$\textit{Renal Clearance} = \frac{V \times U}{P}$$

Where V = urine volume per minute
U = concentration of substance in urine
P = concentration of substance in plasma.

In the case of inulin, its renal clearance is equal to the glomerular filtration rate. If a substance is secreted by the tubules, its clearance is greater than the GFR; if a substance is reabsorbed, its clearance is less than the GFR (table 24.5). In each of these cases, the clearance, in ml per minute, indicates the amount of plasma that originally contained the amount of the substance excreted in a minute's time.

Clearance of Urea

Urea is a waste product of amino acid metabolism that is secreted by the liver into the blood. Despite the fact that it is a waste product, a significant proportion of the filtered urea (40%–60%) is reabsorbed passively, as a result of the high permeability of the tubule membranes to this compound. Not all of the urea contained in the filtered plasma, therefore, is cleared. Using the formula for renal clearance previously described,

If $V = 2$ ml/min
$U = 7.5$ mg/ml of urea
$P = 0.2$ mg/ml of urea

$$\text{urea clearance} = \frac{2 \text{ ml/min} \times 7.5 \text{ mg/ml}}{0.2 \text{ mg/ml}} = 75 \text{ ml/min.}$$

The amount of urea excreted per minute, in other words, was originally contained in 75 ml of plasma. Since 120 ml of plasma were filtered, as determined by the inulin clearance rate, this indicates that approximately 40% of the filtered urea must have been reabsorbed in this example.

Clearance of PAH: Measurement of Renal Blood Flow

Not all of the blood delivered to the glomeruli is filtered into the glomerular capsules; most of the glomerular blood passes through to the efferent arterioles and peritubular capillaries. The inulin and urea in this unfiltered blood is not excreted but instead returns to the general circulation. Blood must thus make many passes through the kidneys before it can be completely cleared of a given amount of inulin or urea.

In order for compounds in the unfiltered renal blood to be cleared, they must be *secreted* into the tubules by active transport from the peritubular capillaries. In this way all of the blood going to the kidneys can potentially be cleared of a secreted compound in a single pass. This is the case for a molecule called **para-aminohippuric acid, or PAH,** (fig. 24.25). The clearance (in ml/min) of PAH can be used to measure the *total renal blood flow* (in ml/min). The normal PAH clearance has been found to average 625 ml/min. Since the glomerular filtration rate averages about 120 ml/min, this indicates that only about 120/625, or roughly 20%, of the renal blood flow is filtered. The remaining 80% passes on to the efferent arterioles.

Figure 24.25. Some of the para-aminohippuric acid (PAH) in glomerular blood (*a*) is filtered into glomerular capsules (*b*). The PAH present in the unfiltered blood is secreted from peritubular capillaries into the nephron (*c*), so that all of the blood leaving the kidneys is free of PAH (*d*). The clearance rate of PAH therefore equals the total plasma flow to the glomeruli.

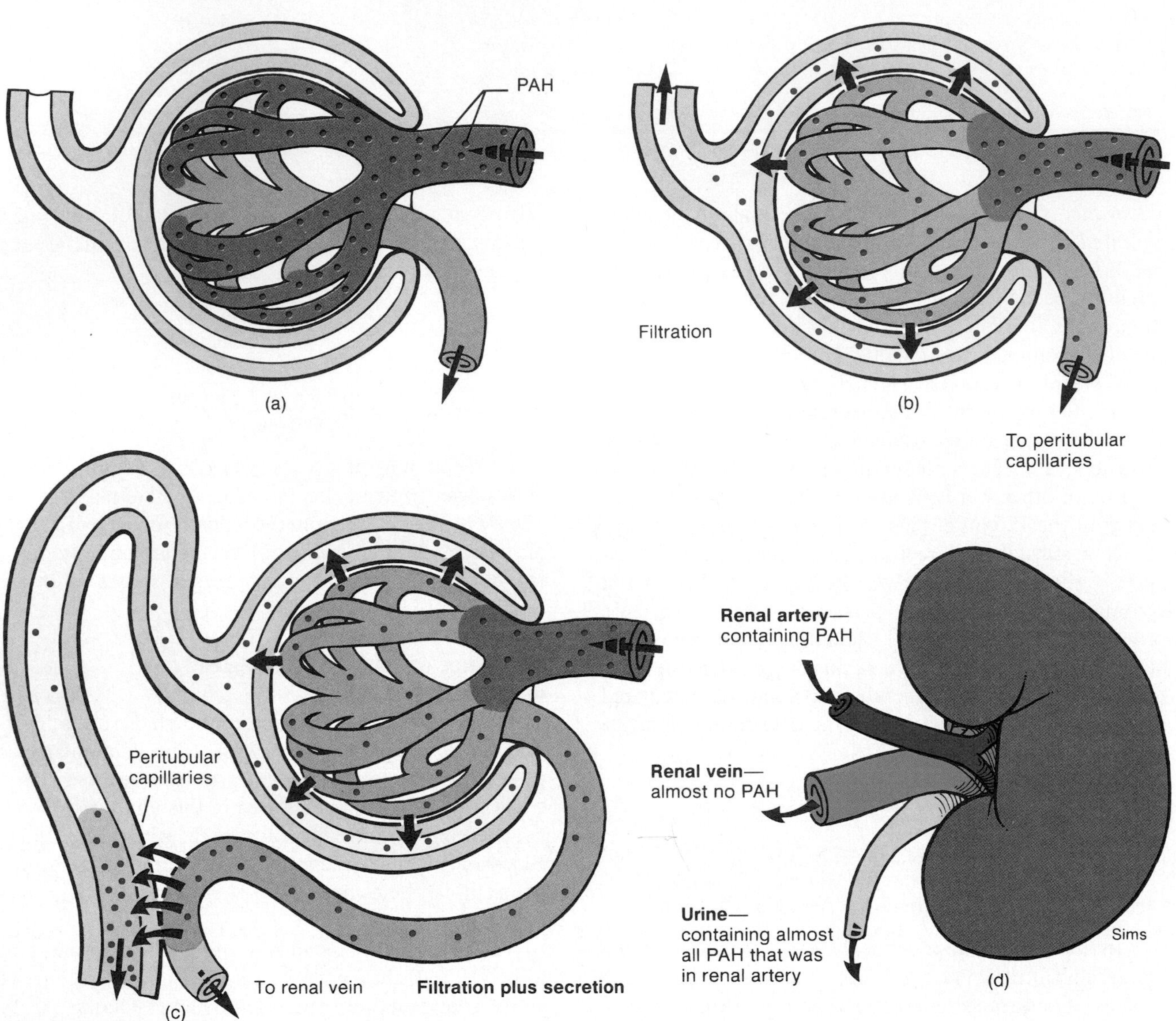

Since filtration and secretion clear only the molecules dissolved in plasma, the PAH clearance measures the renal plasma flow. In order to convert this to the total renal blood flow, the volume of blood occupied by erythrocytes must be taken into account. If the hematocrit is 45, for example, erythrocytes occupy 45% of the blood volume, and plasma accounts for the remaining 55% of the blood volume. The **total renal blood flow** is calculated by dividing the PAH clearance by the fractional blood volume occupied by plasma (0.55, in this example). The total renal blood flow in this example is thus 625 ml/min divided by 0.55, or 1.1 L/min.

Many antibiotics, like penicillin, are secreted by the renal tubules and thus have clearance rates greater than the glomerular filtration rate. Because penicillin is rapidly removed from the blood by renal clearance, large amounts must be administered to be effective. The ability of the kidneys to be visualized in radiographs is improved by the injection of Diodrast, a material that is secreted into the tubules and improves contrast by absorbing X rays. Many drugs and some hormones are inactivated in the liver by chemical transformations and are rapidly cleared from the blood by active secretion in the nephrons.

Figure 24.26. The reabsorption of glucose in the proximal tubule. Glucose is reabsorbed by cotransport with Na^+ from the tubular fluid and then transported by Na^+-independent carriers through the basolateral membrane into the peritubular blood. By this means normally all of the filtered glucose is reabsorbed.

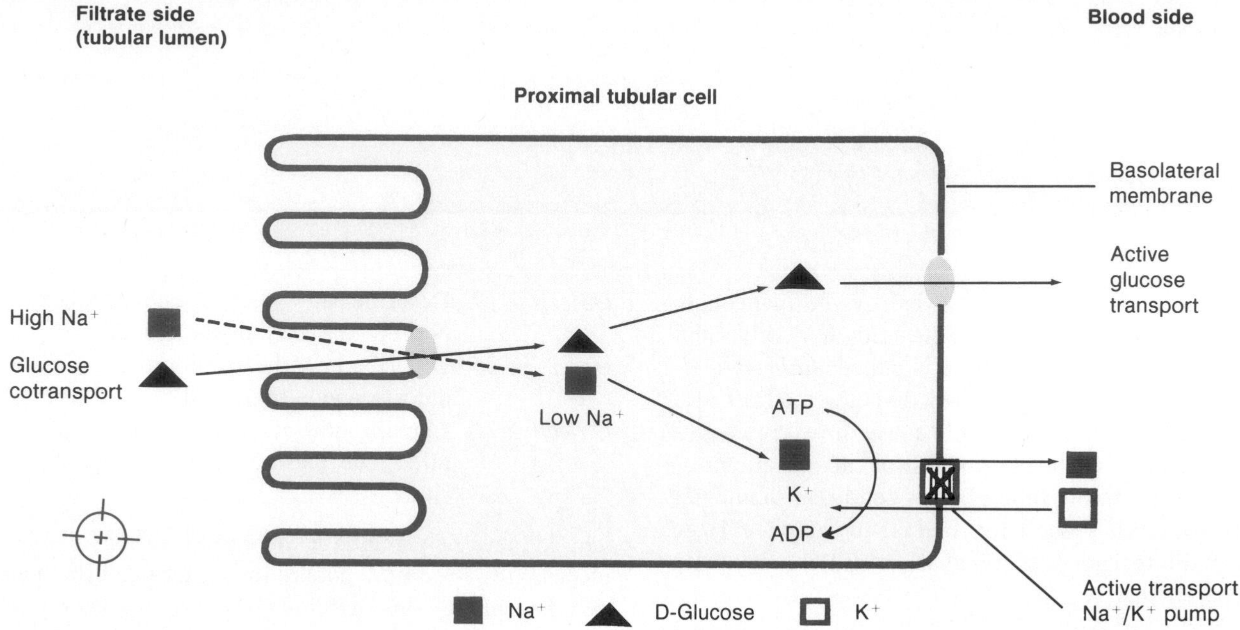

Reabsorption of Glucose and Amino Acids

Glucose and amino acids in the blood are easily filtered by the glomeruli into the renal tubules. These molecules, however, are usually not present in the urine. It can be concluded, therefore, that filtered glucose and amino acids are normally completely reabsorbed by the nephrons.

The reabsorption of glucose and amino acids is an energy-requiring process, which occurs primarily in the proximal convoluted tubules. The energy required for movement of these compounds from the filtrate into the tubule cells is provided by cotransport with Na^+, which diffuses down its electrochemical gradient when it enters the cells. The glucose and amino acids appear to share carriers with Na^+ in the apical membrane. The extrusion of glucose and amino acids from the other side of the cell (across the basolateral membranes) occurs by means of a Na^+-independent active transport carrier (fig. 24.26).

Carrier-mediated transport displays the property of *saturation.* This means that when the transported molecule (such as glucose) is present in sufficiently high concentrations, all of the carriers become "busy," and the transport rate reaches a maximal value. The concentration of transported molecules needed to just saturate the carriers and to just achieve the maximal transport rate is called the **transport maximum** (abbreviated T_m).

The carriers for glucose and amino acids in the renal tubules are not normally saturated and so are able to remove the filtered molecules completely. The T_m for glucose, for example, averages 375 mg per minute, which is well above the rate at which glucose is delivered to the tubules. The rate of glucose delivery can be calculated by multiplying the plasma glucose concentration (about 1 mg per ml) by the GFR (about 125 ml per minute). Approximately 125 mg per minute are thus delivered to the tubules, whereas a rate of 375 mg per minute is required to reach saturation.

Glycosuria. Glucose appears in the urine—a condition called glycosuria *(gli''ko-su're-ah)*—when more glucose passes through the tubules than can be reabsorbed. This occurs when the plasma glucose concentration reaches 180–200 mg per 100 ml. Since the rate of glucose delivery under these conditions is still below the average T_m for glucose, one must conclude that some nephrons have considerably lower T_m values than the average.

The **renal plasma threshold** is the minimum plasma concentration of a substance that results in the excretion of that substance in the urine. The renal plasma threshold for glucose, for example, is 180–200 mg per 100 ml. Glucose is normally absent from urine because plasma glucose concentrations normally remain below this threshold value. The appearance of glucose in the urine (glycosuria) thus occurs only when the plasma concentration of glucose is abnormally high (hyperglycemia).

Table 24.6 Inherited diseases associated with the presence of specific amino acids in the urine

Disease	Cause of disease	Effect of defect	Treatment
Cystinuria	Renal carriers for cystine and related amino acids are defective.	Kidney stones	Bicarbonate and diuretic administration
Hartnup disease	Renal carriers for tryptophane are defective.	Decreased NAD and NADP within body cells	Nicotinamide administration
Homocystinuria	Enzyme defect results in excessive blood levels of homocystine.	Speech defects, mental retardation	Diet low in methionine, high in cystine
Phenylketonuria	Enzyme defect results in excessive accumulation of phenylalanine and in urinary excretion of phenylpyruvic acid.	Severe mental retardation	Diet low in phenylalanine

Fasting hyperglycemia is caused by the inadequate secretion or action of insulin. When this hyperglycemia results in glycosuria, the disease is called *diabetes mellitus*. A person with uncontrolled diabetes mellitus also excretes a large volume of urine because the excreted glucose carries water with it as a result of the osmotic pressure it generates in the tubules. This condition should not be confused with diabetes insipidus, in which a large volume of dilute urine is excreted as a result of inadequate ADH secretion.

The excretion of amino acids in the urine is not usually due to the presence of high amino acid concentration in the blood. Rather, specific amino acids are excreted when their carriers are missing or defective due to a genetic disease. Since different classes of amino acids are reabsorbed by different carriers, the types of amino acids that "spill" into the urine are characteristic of the genetic defect (table 24.6).

1. Define *renal clearance,* and describe how it is measured. Explain why the glomerular filtration rate is equal to the clearance of inulin.
2. Define the terms *reabsorption* and *secretion.* Describe how the renal plasma clearance is affected by the processes of reabsorption and secretion, and give examples.
3. Explain why the total renal blood flow can be measured by the clearance of PAH.
4. Define transport maximum and renal plasma threshold. Explain why people with diabetes mellitus have glycosuria.

Renal Control of Electrolyte Balance

The reabsorption of Na^+ in exchange for the secretion of K^+ or H^+ in the distal convoluted tubules is regulated by aldosterone. Aldosterone secretion is controlled by negative feedback through the secretion of renin from the juxtaglomerular apparatus of the kidney and by a direct effect of blood potassium on the adrenal cortex.

Objective 19. Describe how aldosterone regulates the Na^+/K^+ balance of the blood.
Objective 20. Describe how the secretion of renin and of aldosterone is regulated.
Objective 21. Explain how changes in blood pH may affect the plasma concentration of potassium.

The kidneys help regulate the concentrations of plasma electrolytes—sodium, potassium, chloride, bicarbonate, and phosphate—by matching the urinary excretion of these compounds to the amounts ingested. The control of plasma Na^+ is important in the regulation of blood volume and pressure; the control of plasma K^+ is required to maintain proper function of cardiac and skeletal muscles. The regulation of Na^+/K^+ balance is also intimately related to renal control of acid-base balance.

Role of Aldosterone in Na^+/K^+ Balance

Approximately 90% of the filtered Na^+ and K^+ is reabsorbed in the early part of the nephron before the filtrate reaches the distal tubule. This reabsorption occurs at a constant rate and is not subject to hormonal regulation. The final concentration of Na^+ and K^+ in the urine is varied according to the needs of the body by processes that occur in the distal tubule. These processes are regulated by aldosterone *(al''do-ster'ōn),* a steroid hormone secreted by the adrenal cortex.

Sodium Reabsorption. Although 90% of the filtered sodium is reabsorbed in the early region of the nephron, the amount left in the filtrate delivered to the distal convoluted tubule is still quite large. In the absence of aldosterone, 80% of this amount is automatically reabsorbed through the wall of the distal tubule into the peritubular blood; this is 8% of the amount filtered. The amount of sodium excreted without aldosterone is thus 2% of the amount filtered. Although this percentage seems small, the actual amount of sodium this represents is an impressive 30 g per day excreted in the urine. When aldosterone is secreted in maximal amounts, in contrast, all of the sodium delivered to the distal tubule is reabsorbed. Under these conditions urine contains no Na^+ at all.

Potassium Secretion. About 90% of the filtered K^+ is reabsorbed in the early regions of the nephron (mainly in the proximal tubule). In the absence of aldosterone, all of the filtered K^+ that remains is reabsorbed in the distal tubule. In the absence of aldosterone, therefore, no K^+ is excreted in the urine. The presence of aldosterone stimulates the *secretion of* K^+ from the peritubular blood into the distal tubule (fig. 24.27). This secretion is the only means by which K^+ can be eliminated in the urine. When aldosterone secretion is maximal, as much as fifty times more K^+ is excreted in the urine, because of secretion into the distal tubule, than was originally filtered through the glomeruli.

In summary, aldosterone promotes sodium retention and potassium loss from the blood by stimulating the reabsorption of Na^+ and the secretion of K^+ across the wall of the distal convoluted tubules. Since aldosterone promotes the retention of Na^+, it contributes to an increased blood volume and pressure.

The body cannot get rid of excess K^+ in the absence of aldosterone-stimulated secretion of K^+ into the distal tubules. Indeed, when both adrenal glands are removed from an experimental animal, the *hyperkalemia* (high blood K^+) that results can produce fatal cardiac arrhythmias. Abnormally low plasma K^+ concentrations, as might result from excessive aldosterone secretion, can also produce arrhythmias as well as muscle weakness and cramps.

Figure 24.27. Potassium is almost completely reabsorbed in the proximal tubule, but under aldosterone stimulation, it is secreted into the distal tubule. All of the K^+ in urine is derived from secretion rather than from filtration.

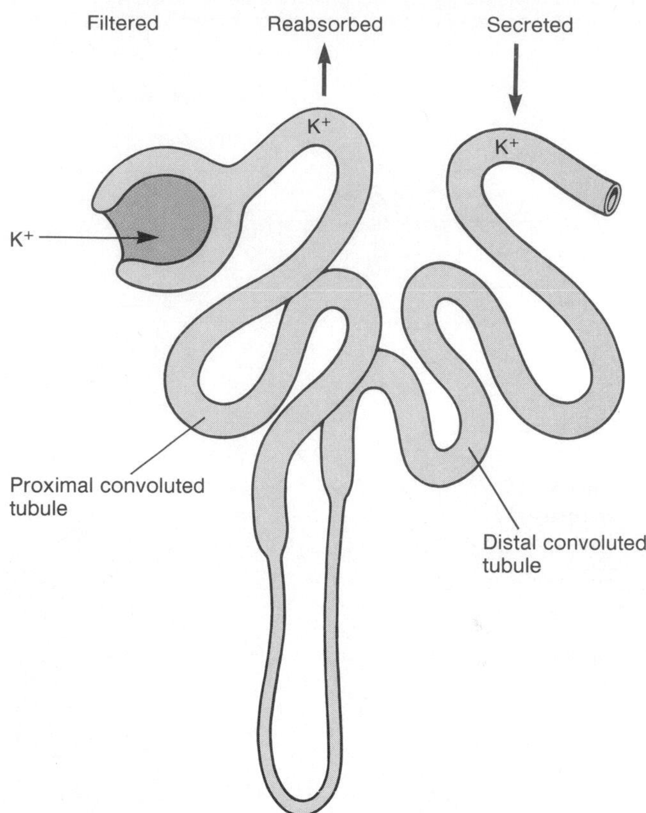

Control of Aldosterone Secretion

Since aldosterone promotes Na^+ retention and K^+ loss, one might predict (on the basis of negative feedback) that aldosterone secretion will be increased when there is a low Na^+ or a high K^+ concentration in the blood. This indeed is the case. A rise in blood K^+ *directly* stimulates the secretion of aldosterone from the adrenal cortex. Decreases in plasma Na^+ concentrations also promote aldosterone secretion, but they do so indirectly.

Juxtaglomerular Apparatus. The juxtaglomerular *(juks"tah-glo-mer'u-lar)* apparatus is the region in each nephron where the afferent arteriole and distal tubule come into contact (fig. 24.28). The microscopic appearance of the afferent arteriole and distal tubule in this small region differs from the appearance in other regions. **Granular cells** within the afferent arteriole secrete the enzyme **renin** *(re'nin)* into the blood; this enzyme catalyzes the conversion of angiotensinogen (a protein) into angiotensin I (a ten-amino-acid polypeptide).

Secretion of renin into the blood thus results in the formation of angiotensin I, which is then converted to **angiotensin** *(an"je-o-ten'sin)* **II** by a *converting enzyme* as blood passes through the lungs and other organs. Angiotensin II, in addition to other effects, stimulates the adrenal cortex to secrete aldosterone. Secretion of renin from the granular cells of the juxtaglomerular apparatus is thus said to initiate the **renin-angiotensin-aldosterone system.** Conditions that result in renin secretion thus cause increased aldosterone secretion and, by this means, promote the reabsorption of Na^+ in the distal convoluted tubules.

Regulation of Renin Secretion. A fall in plasma Na^+ concentration is always accompanied by a fall in blood volume. This is because ADH secretion is inhibited by the decreased plasma concentration (osmolality); with less ADH, less water is reabsorbed through the collecting ducts, and more is excreted in the urine. The fall in blood volume and the fall in renal blood flow that results causes increased renin secretion. Renin secretion is believed to be due in part to a direct effect of blood flow on the granular cells, which may function as baroreceptors in the afferent arterioles. Renin secretion is also stimulated by sympathetic nerve activity, which is increased when the blood volume and pressure fall.

Figure 24.28. The juxtaglomerular apparatus (*a*) includes the region of contact of the afferent arteriole with the distal tubule. The afferent arterioles in this region contain granular cells with renin, and the distal tubule cells in contact with the granular cells form an area called the macula densa (*b*). The granular cells of the afferent arteriole are innervated by renal sympathetic nerve fibers.

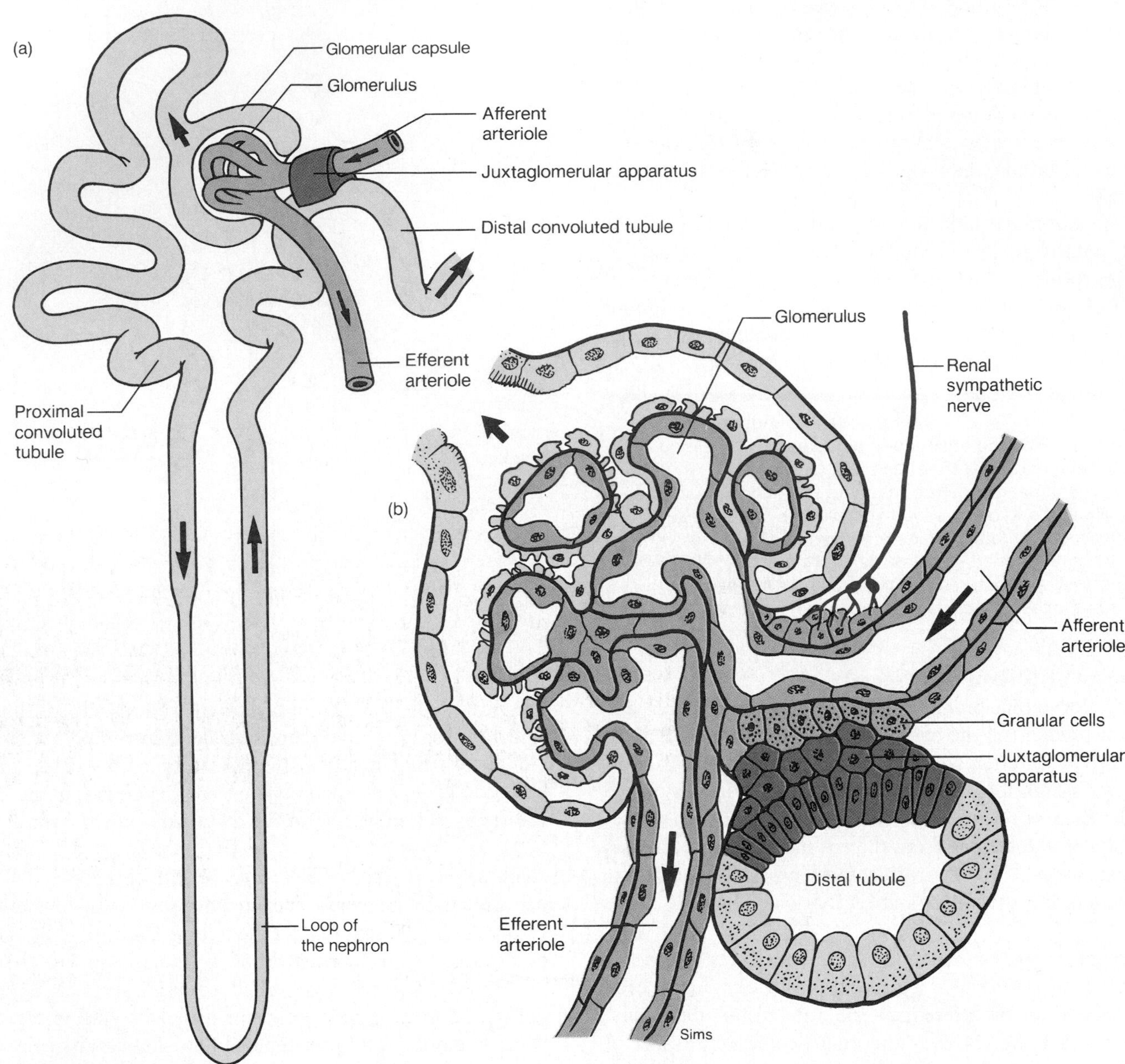

If there is inadequate sodium intake, therefore, the fall in plasma volume that results acts to increase renin secretion. The increased renin secretion acts, via the increased production of angiotensin II, to stimulate aldosterone secretion. In consequence, there is less Na^+ excreted in the urine. This negative feedback system is illustrated in figure 24.29.

Role of the Macula Densa. The region of the distal tubule in contact with the granular cells of the afferent arteriole is called the **macula densa** (see fig. 24.28). There is evidence that this region helps to inhibit renin secretion when the blood Na^+ concentration is raised.

According to the proposed mechanism, the cells of the macula densa respond to Na^+ within the filtrate delivered to the distal tubule. When the plasma Na^+ concentration

macula densa: L. *macula,* a spot; *densitas,* thick

Figure 24.29. The sequence of events by which a low sodium (salt) intake leads to increased sodium reabsorption by the kidneys. The dotted arrow and negative sign show the completion of the negative feedback loop.

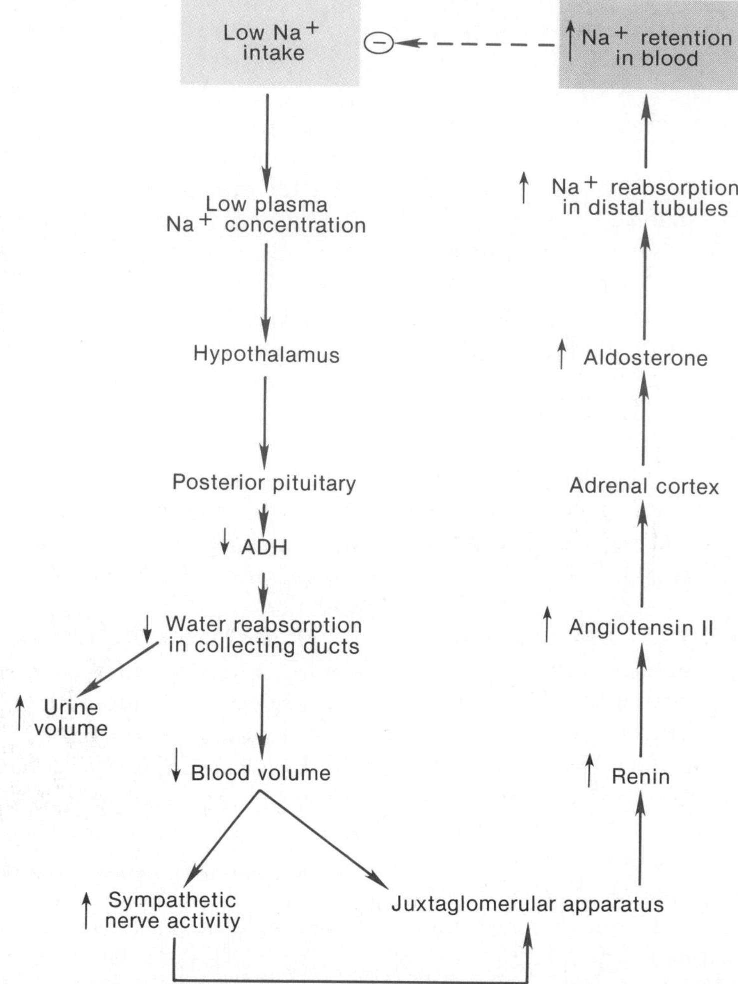

Table 24.7 Regulation of renin and aldosterone secretion

Stimulus	Effect on renin secretion	Mechanisms	Angiotensin II production	Aldosterone secretion
↓Na^+	Increased	Low blood volume stimulates renal baroreceptors; granular cells release renin.	Increased	Increased
↑Na^+	Decreased	Increased blood volume inhibits baroreceptors; increased Na^+ in distal tubule acts via macula densa to inhibit release of renin from granular cells.	Decreased	Decreased
↑K^+	None	Not applicable	Not changed	Increased
↑Sympathetic nerve activity	Increased	α-adrenergic effect stimulates constriction of afferent arterioles; β-adrenergic effect stimulates renin secretion directly.	Increased	Increased

is raised, the rate of Na^+ delivered to the distal tubule is also increased. Through an effect on the macula densa, this increase in filtered Na^+ may inhibit the granular cells from secreting renin. Aldosterone secretion thus decreases, and since less Na^+ is reabsorbed in the distal tubule, more Na^+ is excreted in the urine. The regulation of renin and aldosterone secretion is summarized in table 24.7.

Natriuretic Hormone. Expansion of the blood volume causes increased salt and water excretion in the urine. This is due in part to an inhibition of aldosterone secretion, as previously described. There is much experimental evidence, however, that the increased salt excretion that occurs under these conditions is due not only to the inhibition of aldosterone secretion, but also to the increased secretion of another substance with hormone properties.

This other substance is called **natriuretic** *(na''tre-u-ret'ik)* **hormone** and is so named because it stimulates salt excretion (the opposite of aldosterone's action). The source and chemical nature of natriuretic hormone remained elusive for many years, but recent evidence has shown that the atria of the heart produce a polypeptide that appears to fit the description of the natriuretic hormone proposed by renal physiologists. This polypeptide is currently known as *atrial natriuretic factor.*

Relationship between Na^+, K^+, and H^+

Hormones, as a general rule, alter the rate of already existing processes. In the absence of aldosterone, the distal tubule reabsorbs 80% of the Na^+ delivered to it; aldosterone can increase this reabsorption to 100%. At a given level of aldosterone, the amount of Na^+ reabsorbed in the distal tubule is a given proportion of the Na^+ delivered to it. Some diuretic drugs inhibit Na^+ reabsorption in the loop of the nephron and, therefore, increase the delivery of Na^+ to the distal tubule. As a result, there is an increased reabsorption of Na^+ in the distal convoluted tubule when a person takes these types of diuretics.

Relationship between Na^+ and K^+. The reabsorption of Na^+ in the distal convoluted tubules occurs together with K^+ secretion. This occurs because the aldosterone-stimulated reabsorption of Na^+ creates a large potential difference between the two sides of the tubular wall, with the lumen side very negative (-50 mV) in comparison to the basolateral side. The secretion of K^+ into the tubular fluid is driven by this electrical gradient. Because of the Na^+/K^+ exchange in the distal tubule, an increase in Na^+ reabsorption in the distal tubule results in an increase in K^+ secretion. People who take diuretics that inhibit Na^+ reabsorption in the loop of Henle, for these reasons, tend to have excessive K^+ secretion into the distal tubules and, therefore, excessive K^+ loss in the urine. The actions of different diuretics and their side-effects on blood K^+ are discussed in the last section of this chapter.

The K^+ loss that occurs with many diuretics may present serious side effects to these medications. If K^+ secretion into the distal convoluted tubules is significantly increased, a condition of *hypokalemia* (low blood K^+) may be produced, which must be compensated for by the increased ingestion of potassium. People who take diuretics for the treatment of high blood pressure are usually on a low-sodium diet and often must supplement their meals with potassium chloride (KCl).

natriuretic: L. *natrium,* sodium; *urina,* urine
hypokalemia: Gk. *hypo,* under; L. *kalium,* potassium; Gk. *haima,* blood

Relationship between K^+ and H^+. The plasma K^+ concentration indirectly affects the plasma H^+ concentration (pH). Changes in plasma pH likewise affect the K^+ concentration of the blood. These effects serve to stabilize the *ratio* of K^+ to H^+. When the extracellular H^+ concentration increases, for example, some of the H^+ moves into the tissue cells and causes cellular K^+ to diffuse outward into the extracellular fluid. The plasma concentration of H^+ is thus decreased while the K^+ increases, helping to restabilize the ratio of these ions in the extracellular fluid.

This exchange of H^+ for K^+ also occurs in the cells of the distal tubule when the plasma pH is lowered (fig. 24.30). Under these conditions the tubule cells contain more H^+ and less K^+ than before. Whenever the tubule cells reabsorb Na^+ from the filtrate, they can replace it by secreting either K^+ or H^+ (they usually secrete both ions). When the tubule cells contain increased amounts of H^+ as a result of the fall in plasma pH, however, they secrete increased amounts of H^+ at the expense of K^+, which helps to lower the plasma H^+ concentration and make the urine more acidic. As a side effect, less K^+ is secreted; the plasma concentration of K^+ may thus be increased under these conditions.

Hyperkalemia (high blood K^+) may thus occur in metabolic acidosis. Metabolic alkalosis can have the opposite effect, resulting in lowered plasma K^+ concentrations. This lowering of plasma K^+ is sometimes induced intentionally by infusing bicarbonate in patients with heart or renal failure, who are susceptible to hyperkalemia as a result of decreased glomerular filtration.

Aldosterone appears to stimulate the secretion of H^+ as well as K^+ into the distal tubules. Abnormally high aldosterone secretion, in *primary aldosteronism,* or *Conn's syndrome,* therefore results in both hypokalemia and metabolic alkalosis. Conversely, abnormally low aldosterone secretion, as occurs in *Addison's disease,* can produce hyperkalemia and metabolic acidosis.

1. Describe the effects of aldosterone on the renal nephrons, and explain how aldosterone secretion is regulated.
2. Describe how changes in plasma Na^+ concentrations regulate renin secretion, and explain how the secretion of renin acts to help regulate the plasma Na^+ concentration.
3. Explain why people who take some diuretic drugs may suffer from hypokalemia and why they must supplement their diets with potassium.
4. Explain the mechanisms involved in the lowering of plasma K^+ concentrations by the intravenous infusion of bicarbonate.

Addison's disease: from Christopher Addison, English anatomist, 1869–1951

Figure 24.30. In the distal tubule, K^+ and H^+ are secreted in exchange for Na^+. High concentrations of H^+ may therefore decrease K^+ secretion, and vice versa.

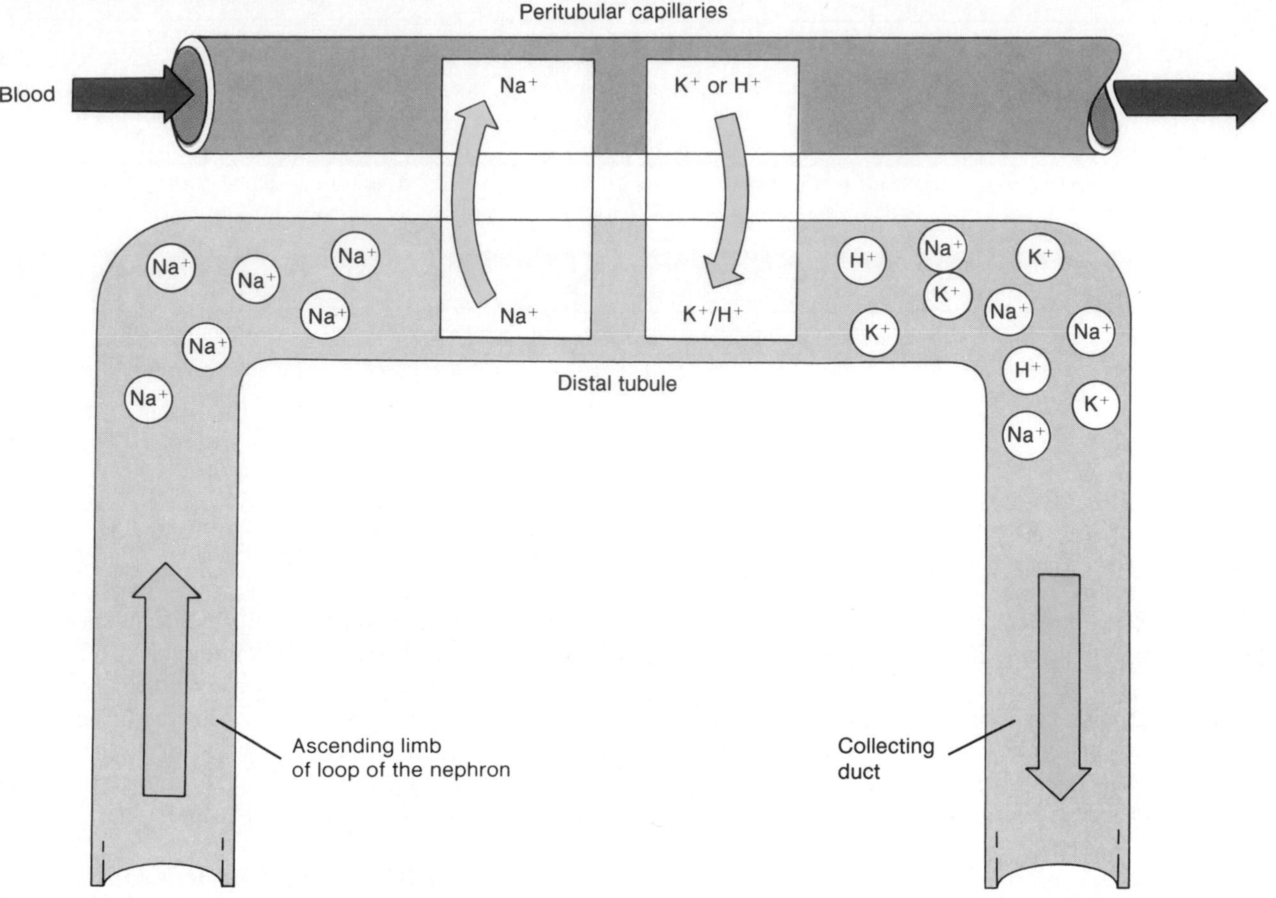

Renal Control of Acid-Base Balance

The kidneys help regulate the acid-base balance of the blood through their excretion of H^+ and their reabsorption of HCO_3^-. By means of these processes, the kidneys are responsible for the metabolic component of acid-base balance.

Objective 22. Describe the respiratory and metabolic components of acid-base balance, and explain how these relate to the bicarbonate and carbonic acid concentrations of the blood.

Objective 23. Describe how filtered bicarbonate is reabsorbed by the nephron.

Objective 24. Explain how the nephrons help compensate for acidosis and alkalosis, and describe how acid is excreted in the urine.

The kidneys help regulate the acid-base balance of the blood through their excretion of H^+ and their reabsorption of HCO_3^-. By means of these processes, the kidneys are responsible for the metabolic component of acid-base balance.

Normal arterial blood has a pH of 7.35–7.45. It should be recalled that the pH number is *inversely* related to the H^+ concentration. An increase in H^+, derived from carbonic acid or the nonvolatile metabolic acids, can lower blood pH. An increase in H^+ derived from metabolic acids, however, is normally prevented from changing the blood pH by **bicarbonate buffer,** which combines with and thus removes from solution the excess H^+.

Respiratory and Metabolic Components of Acid-Base Balance

Acid-base balance has both respiratory and metabolic components. The respiratory component describes the effect of ventilation on arterial P_{CO_2} and thus on the production of carbonic acid (H_2CO_3). The metabolic component describes the effect of nonvolatile metabolic acids—lactic acid, fatty acids, and ketone bodies—on blood pH. Since these acids are normally buffered by bicarbonate (HCO_3^-), the metabolic component can be described in terms of the free HCO_3^- concentration. An increase in metabolic acids "uses up" free bicarbonate, as HCO_3^- is converted to H_2CO_3, and is thus associated with a fall in plasma HCO_3^- concentrations. A decrease in metabolic acids, conversely, is associated with a rise in free HCO_3^-.

Table 24.8 Classification of metabolic and respiratory components of acidosis and alkalosis

P_{CO_2}	HCO_3^-	Condition	Causes
Normal	Low	Metabolic acidosis	Increased production of "nonvolatile" acids (lactic acid, ketone bodies, and others), or loss of HCO_3^- in diarrhea
Normal	High	Metabolic alkalosis	Vomiting of gastric acid; hypokalemia; excessive steroid administration
Low	Normal	Respiratory alkalosis	Hyperventilation
High	Normal	Respiratory acidosis	Hypoventilation

Table 24.9 The effect of changes in P_{CO_2} on the blood pH

P_{CO_2} (mm Hg)	H_2CO_3 (mEq/L)	HCO_3^- (mEq/L)	HCO_3^-/H_2CO_3 ratio	pH	Condition
20	0.6	24	40/1	7.70	Respiratory alkalosis
30	0.9	24	26.7/1	7.53	Respiratory alkalosis
40	1.2	24	20/1	7.40	Normal
50	1.5	24	16/1	7.30	Respiratory acidosis
60	1.8	24	13.3/1	7.22	Respiratory acidosis

Note: A blood pH of less than 7.35 due to high P_{CO_2} is called respiratory acidosis. A blood pH greater than 7.45 due to low P_{CO_2} is called respiratory alkalosis.

Since the respiratory component of acid-base balance is represented by the plasma P_{CO_2} and carbonic acid concentrations and the metabolic component is represented by the free bicarbonate concentrations, the study of acid-base balance can be considerably simplified. A normal plasma pH of 7.40 is obtained when the ratio of bicarbonate to carbonic acid concentrations is twenty to one:

$$\text{pH} \propto \log \frac{HCO_3^- \text{ concentration}}{H_2CO_3 \text{ concentration}} = \frac{20}{1} = 7.40.$$

($\propto$ means "is proportional to")

A change in this ratio results in an abnormal blood pH. Pure respiratory acidosis *(as''ĭ-do'sis)* or alkalosis *(al''kah-lo'sis)* occurs when the HCO_3^- concentration is normal but the P_{CO_2} and H_2CO_3 concentrations are altered. Pure metabolic acidosis or alkalosis occurs when the P_{CO_2} and H_2CO_3 are normal but the HCO_3^- concentration is abnormal. This classification is summarized in table 24.8.

A normal blood pH is produced when the P_{CO_2} is 40 mm Hg and the bicarbonate concentration is 24 milliequivalents (mEq) per liter (one mEq is equal to one millimole times the valence of the ion; in this case, since the valence is 1, mEq = mmole). Table 24.9 shows how changes in P_{CO_2} alter the ratio of HCO_3^- to H_2CO_3 and thus affect blood pH. For the sake of simplicity, this table shows the effect of changing P_{CO_2} when the bicarbonate concentration remains at a constant, normal value. Such changes in P_{CO_2} are produced by hyperventilation and hypoventilation and result in respiratory alkalosis and acidosis, respectively.

A normal plasma P_{CO_2} is maintained by proper lung function. A normal plasma bicarbonate concentration of 24 mEq per L is maintained, despite the continuous production of metabolic acids, by proper kidney function. The kidneys normally reabsorb all of the filtered bicarbonate. During alkalosis, however, the kidneys can excrete bicarbonate, and during acidosis they can reabsorb bicarbonate in excess of H^+, which is excreted in the urine. In summary, the plasma bicarbonate concentration is the responsibility of the kidneys, whereas the plasma carbonic acid concentration is regulated by the lungs.

Mechanisms of Renal Acid-Base Regulation

The kidneys help regulate the blood pH by excreting H^+ in the urine and by the reabsorption of bicarbonate. These two mechanisms are interdependent—the reabsorption of bicarbonate occurs as a result of the filtration and secretion of H^+. The kidneys normally reabsorb all of the filtered bicarbonate and excrete about 40–80 mEq of H^+ per day. Normal urine, therefore, is free of bicarbonate and is slightly acidic (with a pH range between 5 and 7).

The apical membranes of the tubule cells are impermeable to bicarbonate. The reabsorption of bicarbonate must therefore occur indirectly. When the urine is acidic, HCO_3^- combines with H^+ to form carbonic acid. Carbonic acid in the filtrate is then converted to CO_2 and H_2O by the action of **carbonic anhydrase** *(an-hi'dras)*. This enzyme is located in the apical cell membrane that faces the filtrate. Notice that the reaction that occurs in the tubule is the same one that occurs within the red blood cells in pulmonary capillaries (discussed in chapter 23).

Carbon dioxide, unlike bicarbonate, can easily pass from the filtrate, through the tubule cells, and enter the blood. Once inside the red blood cells, the CO_2 combines with water to form carbonic acid. This reaction is also catalyzed by carbonic anhydrase and results in the addition of equal amounts of H^+ and HCO_3^- to the blood (fig. 24.31).

Figure 24.31. Mechanisms of bicarbonate reabsorption.

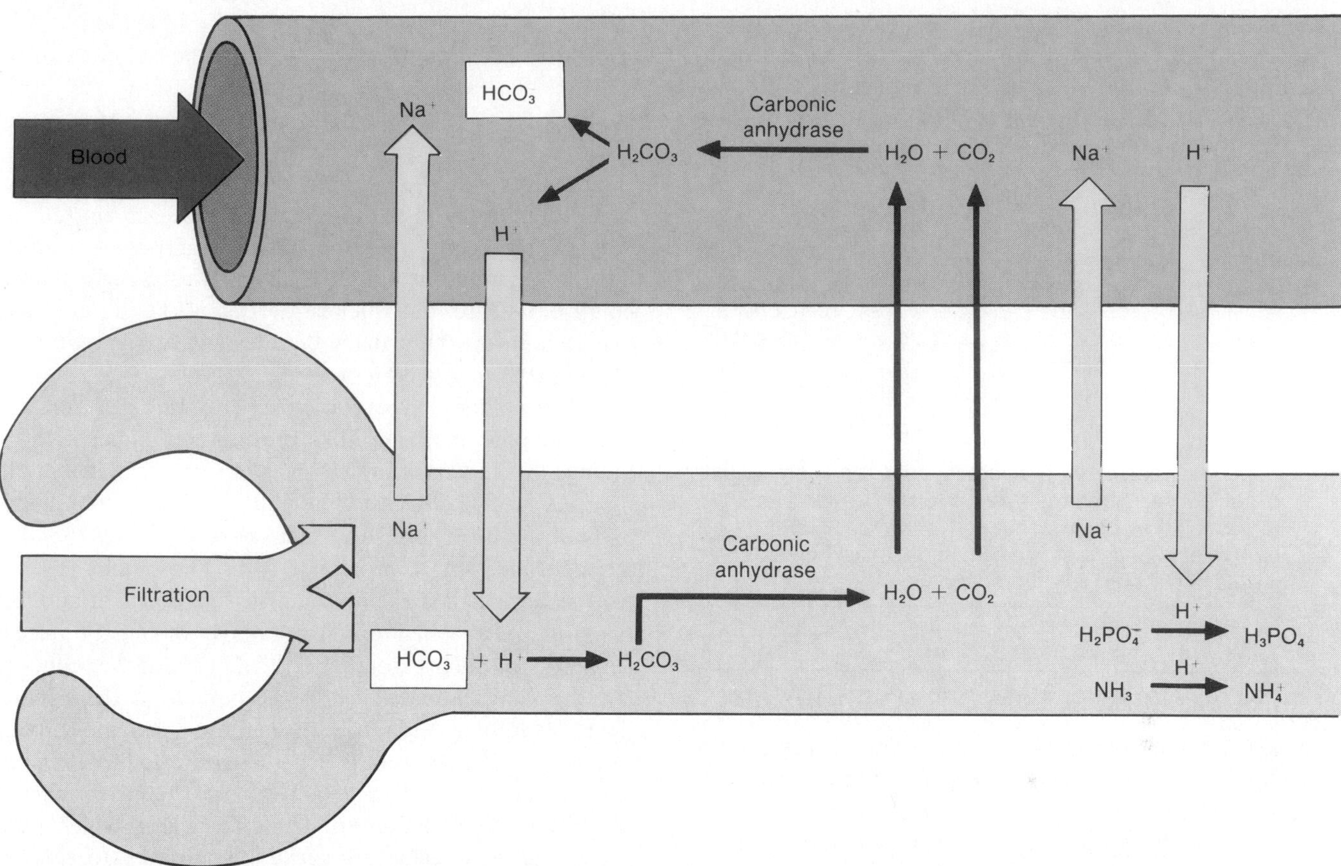

Table 24.10 Categories of disturbances in acid-base balance, including those that involve mixtures of respiratory and metabolic components

P_{CO_2} (mmHg)	Bicarbonate (mEq/L)		
	Less than 21	21–26	More than 26
More than 45	Combined metabolic and respiratory acidosis	Respiratory acidosis	Metabolic alkalosis and respiratory acidosis
35–45	Metabolic acidosis	Normal	Metabolic alkalosis
Less than 35	Metabolic acidosis and respiratory alkalosis	Respiratory alkalosis	Combined metabolic and respiratory alkalosis

Production of Bicarbonate in the Tubules. The tubule cell cytoplasm also contains carbonic anhydrase. Some of the CO_2 that enters the tubule cells can thus be converted to carbonic acid, which can in turn dissociate to HCO_3^- and H^+ within the tubule cells. Under acidotic conditions, the H^+ produced in this way is secreted back into the filtrate, whereas the bicarbonate diffuses across the basolateral membranes and enters the blood. More bicarbonate than H^+ is thus returned to the blood as a partial compensation for the state of acidosis.

During alkalosis there is less H^+ secreted into the filtrate. Since the reabsorption of filtered bicarbonate requires the combination of HCO_3^- with H^+ to form carbonic acid, less bicarbonate is reabsorbed. This results in urinary excretion of bicarbonate, which helps to partially compensate for the alkalosis.

In this way, disturbances in acid-base balance caused by respiratory problems can be partially compensated for by changes in plasma bicarbonate concentrations. Metabolic acidosis or alkalosis—in which changes in bicarbonate concentrations occur as the primary disturbance—can be similarly compensated for in part by changes in ventilation. These interactions of the respiratory and metabolic components of acid-base balance are shown in table 24.10.

Urinary Buffers. In order for H^+ to be excreted in the urine without making the urine damagingly (and perhaps painfully) acidic, the acid must be buffered. Bicarbonate cannot serve this function because it is normally completely reabsorbed. Instead, the urine pH is usually prevented from falling below a value of about 4.6 by the buffering action of phosphates (mainly HPO_4^{-2}) and ammonia (NH_3). Phosphate enters the urine by filtration. Ammonia (whose presence is strongly evident in a diaper pail or kitty litter box) is produced in the tubule cells by deamination of amino acids. These molecules buffer H^+ as described in the following equations:

$$NH_3 + H^+ \rightarrow NH_4^+ \text{ (ammonium ion)}$$

$$HPO_4^{-2} + H^+ \rightarrow H_2PO_4^-.$$

1. Describe the respiratory and metabolic components of acid-base balance, and explain how these components are represented by the carbonic acid and bicarbonate concentration of blood.
2. Explain how the kidneys reabsorb filtered bicarbonate and how this process is affected by acidosis and alkalosis.
3. Suppose a person with diabetes mellitus has an arterial pH of 7.30, an abnormally low arterial P_{CO_2}, and an abnormally low bicarbonate concentration. Identify the type of acid-base disturbance, and explain how these values might have been produced.

Ureters, Urinary Bladder, and Urethra

Urine is channeled from the kidneys to the urinary bladder by the ureters and expelled from the body through the urethra. The mucosa of the urinary bladder permits distension, and the muscles of the urinary bladder and urethra are used in the control of micturition.

Objective 25. Describe the location, structure, and function of the ureters.

Objective 26. Discuss the gross and histological structure of the urinary bladder and its innervation.

Objective 27. Explain the process of micturition.

Objective 28. Compare and contrast the structure of the male and female urethra.

Ureters

The **ureters,** like the kidneys, are retroperitoneal in location. Each ureter is a tubular organ about 25 cm (10 in.) long, which begins at the renal pelvis and courses inferiorly to enter the urinary bladder at the superior lateral angle of its base. The thickest portion of the ureter is near where it enters the urinary bladder and is approximately 1.7 cm (0.5 in.) in diameter.

The wall of the ureter consists of three layers, or tunicas. The inner **mucosa** *(mu-ko'sah)* is continuous with the linings of the renal tubules and the urinary bladder. The mucosa consists of transitional epithelium (fig. 24.32). The cells of this layer secrete a mucus that lubricates the walls of the ureter with a protective film. The middle layer of the ureter is called the **muscularis.** It consists of an inner longitudinal and an outer circular layer of smooth muscle. In addition, the proximal one-third of the ureter contains another longitudinal layer to the outside of the circular layer. Muscular peristaltic waves move the urine through the ureter. The peristaltic waves are initiated by the presence of urine in the renal pelvis, and their frequency is determined by the volume of urine. The waves, which occur from every few seconds to every few minutes, force urine through the ureter and cause it to spurt into the urinary bladder. The outer layer of the ureter is called the **fibrous**

Figure 24.32. A photomicrograph of the ureter in transverse section.

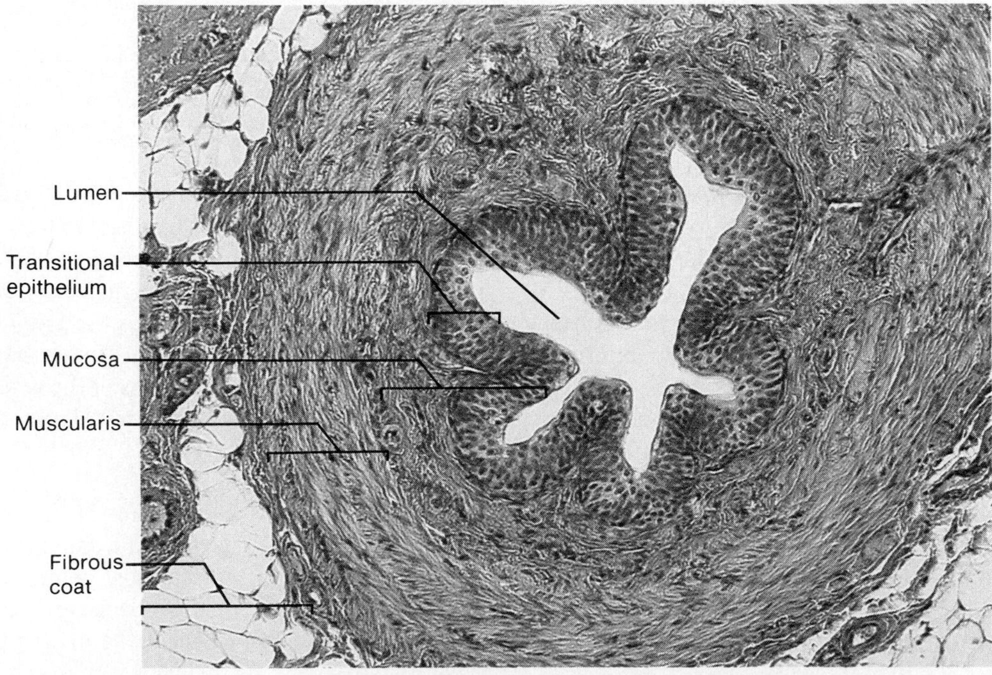

coat. The fibrous coat is composed of areolar connective tissue that not only covers the ureter but has extensions that anchor the ureter in place.

The arterial supply of the ureter comes from several sources. Branches from the renal artery serve the superior portion. The testicular (or ovarian) artery supplies the middle portion, and the superior vesicular artery serves the pelvic region. The venous return is through corresponding veins.

A *calculus,* or *kidney stone,* may obstruct the ureter and produce a tremendous amount of peristaltic waves in an attempt to pass the stone. The pain from a lodged calculus is extreme and extends throughout the pelvic area. A lodged calculus also causes a sympathetic ureterorenal reflex that results in constriction of renal arterioles, which reduces the production of urine in the kidney on the affected side.

Urinary Bladder

The **urinary bladder** is a storage sac for urine. It is located posterior to the symphysis pubis and anterior to the rectum. In females, the urinary bladder is in contact with the uterus and vagina. In males, the prostate gland is positioned below the urinary bladder (fig. 24.33).

The shape of the urinary bladder is determined by the volume of urine it contains. An empty urinary bladder is pyramidal in shape, having an anteroinferior apex, a superior surface, two inferolateral surfaces, a base (posterior surface), and a neck. The apex of the urinary bladder is superior to the symphysis pubis and is secured to the **median umbilical ligament** by a fibrous cord called the **urachus.** The base of the urinary bladder receives the ureters along the superolateral angles, and the urethra exits at the inferior angle. The urethra is a tubular continuation of the neck of the urinary bladder.

As the urinary bladder fills, it loses its pyramidal shape and becomes ovoid as the superior surface enlarges and bulges upward into the abdominal cavity (fig. 24.34). The urinary bladder is antiperitoneal when it is empty, but the peritoneal cover is retracted back along the superior surface when the bladder distends.

The wall of the urinary bladder consists of four layers: the mucosa, submucosa, muscularis, and serosa (adventitia). The **mucosa** is the innermost layer. It is composed of transitional epithelium, which decreases in thickness as the urinary bladder distends and the cells are stretched. Further distension is permitted by folds of the mucosa, called **rugae,** which can be seen when the urinary bladder is empty (see fig. 24.33). Fleshy flaps of mucosa located where the ureters pierce into the urinary bladder act as valves over the openings of the ureters to prevent a reverse flow of urine toward the kidneys as the bladder fills. A triangular area known as the **trigone** *(tri'gōn)* is formed on the mucosa between the two ureter openings and the single urethral opening. The internal trigone lacks rugae and is therefore smooth in appearance and remains relatively fixed in position as the urinary bladder changes shape during distension and contraction.

calculus: L. *calculus,* small stone

trigone: L. *trigonum,* triangle

Figure 24.33. The male urinary bladder and urethra. (*a*) a coronal view; (*b*) a posterior view.

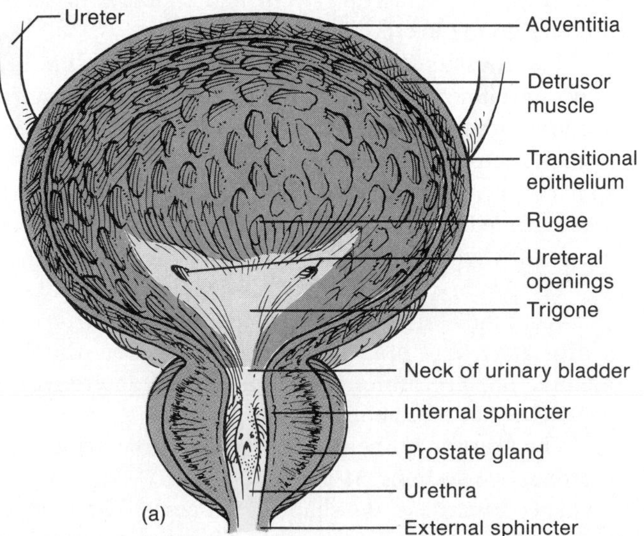

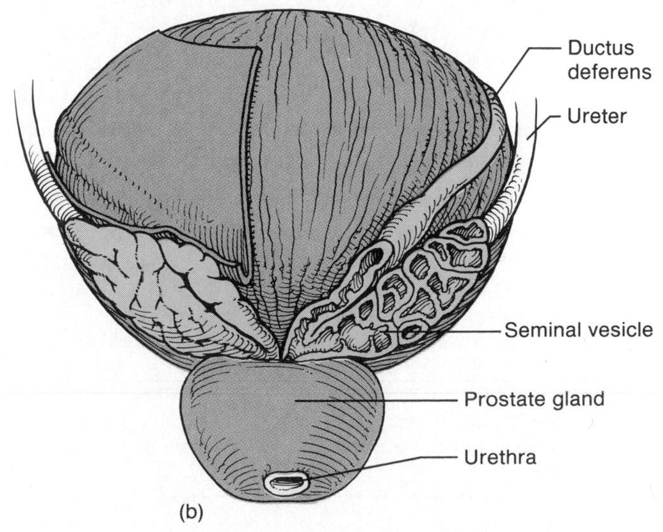

Figure 24.34. The position of an empty and a distended urinary bladder.

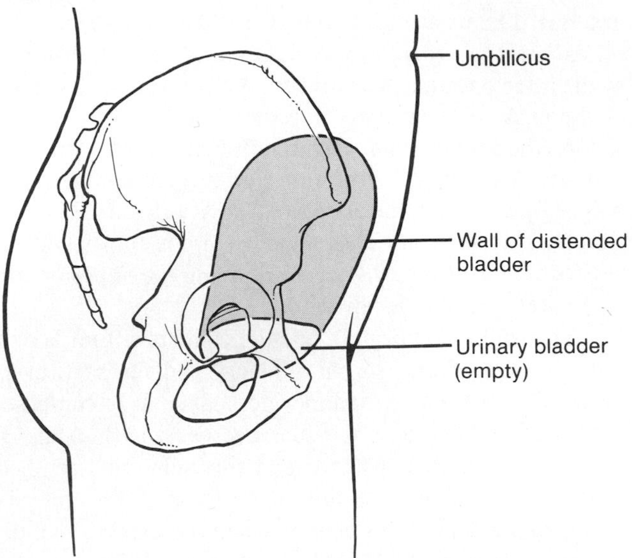

Figure 24.35. A longitudinal section of a male urethra.

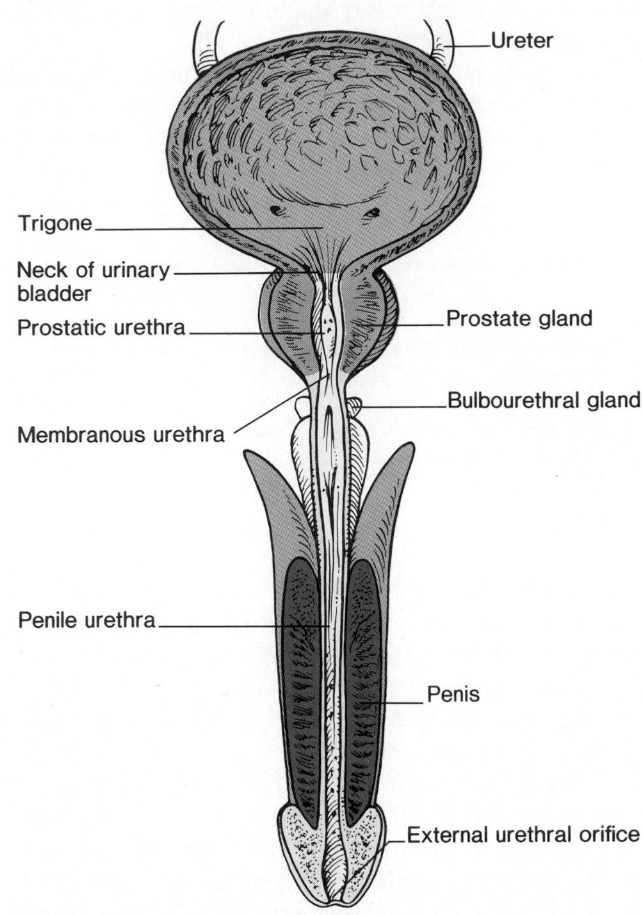

The second layer of the urinary bladder is the **submucosa,** which functions to support the mucosa. The **muscularis** consists of three interlaced smooth muscle layers and is referred to as the **detrusor muscle.** At the neck of the urinary bladder, the detrusor muscle is modified to form the upper of two muscular sphincters surrounding the urethra. The outer covering of the urinary bladder is the **serosa.** It appears only on the superior surface of the bladder and is actually a continuation of the peritoneum.

The arterial supply to the urinary bladder comes from the **superior** and **inferior vesicular arteries,** which arise from the internal iliac arteries. Blood draining from the urinary bladder enters from a **vesicular venous plexus** and empties into the internal iliac veins.

The nerves associated with the urinary bladder are derived from pelvic plexuses. The sympathetic fibers arise from the last thoracic and the first and second lumbar spinal nerves. They innervate the trigone, ureteral openings, and blood vessels of the urinary bladder. The parasympathetic innervation arises from the second, third, and fourth sacral nerves. These fibers serve the detrusor muscle. The afferent sensory fibers are specialized stretch receptors that respond to distension and relay impulses to the central nervous system via the pelvic spinal nerves.

The urinary bladder becomes infected easily, particularly in women because of their short urethra, which increases the possibility of contamination. A urinary bladder infection, called *cystitis,* may easily ascend from the bladder to the ureters since the mucous linings are continuous. An infection that involves the renal pelvis is called *pyelitis;* if it continues into the nephrons, it is known as *nephritis.*

detrusor: L. *detrudere,* thrust or forced down

Urethra

The tubular **urethra** *(u-re'thrah)* conveys urine from the urinary bladder to the outside of the body. The urethral wall has an inside lining of mucous membrane surrounded by a relatively thick layer of smooth muscle whose fibers are directed longitudinally. Specialized **urethral glands** are embedded in the urethral wall and function to secrete mucus into the urethral canal.

Two muscular sphincters surround the urethra. The upper, involuntary smooth muscle sphincter is the **internal urethral sphincter** (sphincter vesicae), which is formed from the detrusor muscle of the urinary bladder. The lower sphincter is composed of voluntary, striated muscle fibers and is called the **external urethral sphincter** (sphincter urethrae) (fig. 24.33).

The urethra of the female is a simple tube about 4 cm (1.5 in.) long, which empties urine through the **urethral orifice** *(or'ĭ-fis)* into the vestibule between the labia minora. The urethral orifice is positioned anterior to the vaginal orifice and about 2.5 cm posterior to the clitoris.

The urethra of the male serves both the urinary and reproductive systems. It is about 20 cm (8 in.) long and S-shaped because of the shape of the penis. Three regions can be identified in the male urethra: the prostatic urethra, membranous urethra, and penile urethra (fig. 24.35).

Figure 24.36. Innervation of the ureter, urinary bladder, and urethra.

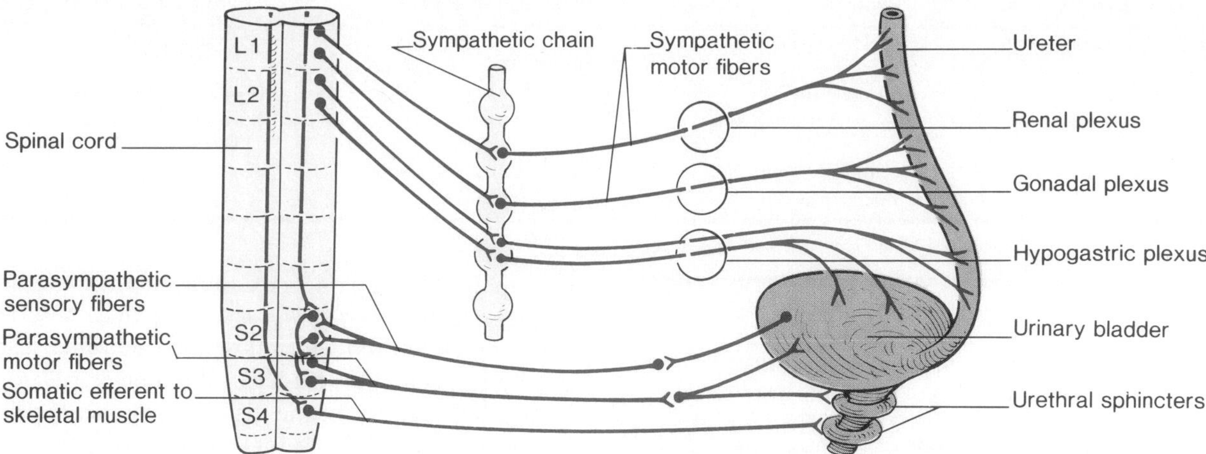

The **prostatic urethra** is the proximal portion, about 2.5 cm long, that passes through the **prostate gland** located near the neck of the urinary bladder. The prostatic urethra receives drainage from small ducts of the prostate gland and two **ejaculatory ducts** of the reproductive system.

The **membranous urethra** is the short (0.5 cm) portion of the urethra that passes through the urogenital diaphragm. The external urethral muscle is located in this region.

The **penile urethra** (cavernous urethra) is the longest portion (15 cm), extending from the outer edge of the urogenital diaphragm to the external urethral orifice on the glans penis. This portion is surrounded by erectile tissue as it passes through the corpus spongiosum of the penis. The paired ducts of the **bulbourethral glands** (Cowper's glands) of the reproductive system attach to the penile urethra near the urogenital diaphragm.

Micturition

Micturition *(mik''tu-rish'un),* commonly called urination or voiding, is a reflex action that expels urine from the urinary bladder. It is a complex function that requires a stimulus from the urinary bladder and a combination of involuntary and voluntary nerve impulses to the appropriate muscular structures of the bladder and urethra.

In young children, micturition is a simple reflex action that occurs when the urinary bladder becomes sufficiently distended. Voluntary control of micturition is normally developed at the time a child is two or three years old. Voluntary control requires the development of an inhibitory ability by the cerebral cortex and a maturing of various portions of the spinal cord.

The volume of urine produced by an adult averages about 1,200 ml per day, but it can vary from 600–2,500 ml. The average capacity of the urinary bladder is 700–800 ml. A volume of 200–300 ml will distend the bladder enough to stimulate stretch receptors and trigger the micturition reflex, creating a desire to urinate.

Table 24.11	Events of micturition
1.	The urinary bladder becomes distended as it fills with urine.
2.	Stretch receptors in the bladder wall are stimulated, and impulses are sent to the micturition center in the spinal cord.
3.	Parasympathetic nerve impulses travel to the detrusor muscle and the internal urethral sphincter.
4.	The detrusor muscle contracts rhythmically, and the internal urethral sphincter relaxes.
5.	The need to urinate is sensed as urgent.
6.	Urination is prevented by voluntary contraction of the external urethral sphincter and by inhibition of the micturition reflex by impulses from the midbrain and cerebral cortex.
7.	Following the decision to urinate, the external urethral sphincter is relaxed, and the micturition reflex is facilitated by impulses from the pons and the hypothalamus.
8.	The detrusor muscle contracts, and urine is expelled through the urethra.
9.	Neurons of the micturition reflex center are inactivated, the detrusor muscle relaxes, and the bladder begins to fill with urine.

The micturition reflex center is located in the second, third, and fouth sacral segments of the spinal cord. Following stimulation of this center by impulses arising from stretch receptors in the urinary bladder, parasympathetic nerves that stimulate the detrusor muscle and the internal urethral sphincter are activated. Stimulation of these muscles causes a rhythmic contraction of the bladder wall and a relaxation of the internal sphincter. At this point, a sensation of urgency is perceived in the brain, but there is still voluntary control over the external urethral sphincter. At the appropriate time, the conscious activity of the brain activates the motor nerve fibers (S4) to the external urethral sphincter via the pudendal nerve (S2, S3, and S4), causing the sphincter to relax and urination to occur.

The innervation of the ureter, urinary bladder, and urethra is shown in figure 24.36, and a summary of the process of micturition is presented in table 24.11.

micturition: L. *micturire,* to urinate

Table 24.12 Actions of different classes of diuretics

Category of diuretic	Example	Mechanism of action	Major site of action
Carbonic anhydrase inhibitors	Acetazolamide	Inhibits reabsorption of bicarbonate	Proximal tubule
Loop diuretics	Furosemide	Inhibits sodium transport	Thick segments of ascending limbs
Thiazides	Hydrochlorothiazide	Inhibits sodium transport	Last part of ascending limb and first part of distal tubule
Potassium-sparing diuretics	Spironolactone	Inhibits action of aldosterone	Distal convoluted tubule
	Triamterene	Inhibits Na^+/K^+ exchange	Distal convoluted tubule

Urinary retention, or the inability to void, may occur postoperatively, especially following surgery of the rectum, colon, or internal reproductive organs. The difficulty may be due to nervous tension, the effects of anesthetics, or pain and edema at the site of the operation. If urine is retained beyond six to eight hours, *catheterization* may become necessary. In this procedure, a tube or catheter is passed through the urethra into the urinary bladder so that urine can flow freely.

1. Describe the structure of a ureter, and indicate the function of its muscularis layer.
2. Describe the structure of the urinary bladder, and indicate the structures that allow it to distend.
3. Compare the male urinary system with that of the female.
4. Explain the structures and processes involved in the control of micturition.

Clinical Considerations

The importance of kidney function in maintaining homeostasis and the ease with which urine can be collected and used as a mirror of the plasma's chemical composition make the clinical study of renal function and urine composition particularly significant. **Urology** is the medical specialty concerned with dysfunctions of the urinary system. Urinary dysfunctions can be congenital, acquired, due to physical trauma, or the result of conditions that secondarily involve the urinary organs.

Use of Diuretics

People who need to lower their blood volume because of hypertension, congestive heart failure, or edema take medications that increase the volume of urine excreted. Such medications are called **diuretics.** There are a number of diuretic drugs in clinical use that act on the renal nephrons in different ways (table 24.12). Based on their chemical structure or aspects of their actions, the commonly used diuretics are categorized as carbonic acid inhibitors, loop diuretics, thiazides, osmotic diuretics, and potassium-sparing diuretics.

The most powerful diuretics, inhibiting salt and water reabsorption by as much as 25%, are the drugs that act to inhibit active salt transport out of the ascending limb of the nephron loop. Examples of these loop diuretics include *furosemide* and *ethacrynic acid.* The thiazide diuretics, like *hydrochlorothiazide,* inhibit salt and water reabsorption by as much as 8% through inhibition of salt transport by the first segment of the distal convoluted tubule. The carbonic anhydrase inhibitors (*acetazolamide*) are much weaker diuretics and act, primarily in the proximal tubule, to prevent the water reabsorption that occurs when bicarbonate is reabsorbed.

When extra solutes are present in the filtrate, they increase the osmotic pressure of the filtrate and in this way decrease the osmotic reabsorption of water throughout the nephron. *Mannitol* is sometimes used clinically for this purpose. Osmotic diuresis can occur in diabetes mellitus due to the presence of glucose in the filtrate and urine; this extra solute causes the excretion of excessive amounts of water in the urine and can result in severe dehydration of the person with uncontrolled diabetes.

Diuretics can result in the excessive secretion of K^+ into the filtrate and its excessive elimination in the urine. For this reason, potassium-sparing diuretics are sometimes used. *Spironolactones* are aldosterone antagonists, which compete with aldosterone for cytoplasmic receptor proteins in the cells of the distal tubule. These drugs, therefore, block the aldosterone stimulation of Na^+ reabsorption and K^+ secretion. *Triamterene* is a different type of potassium-sparing diuretic, which appears to act more directly on the Na^+/K^+ pumps in the distal tubule.

Symptoms and Diagnosis of Urinary Disorders

Normal micturition is painless. **Dysuria** *(dis-u're-ah),* or painful urination, is a sign of a urinary tract infection or obstruction of the urethra—as in an enlarged prostate gland in a male. **Hematuria** means blood in the urine and is usually associated with trauma. **Bacteriuria** means bacteria in the urine, and **pyuria** is the term for pus in the urine, which may result from a prolonged infection. **Oliguria** is an insufficient output of urine, whereas **polyuria** is an excessive output. Low blood pressure and kidney failure are two causes of oliguria. **Uremia** is a condition in which substances ordinarily excreted in the urine accumulate in the blood. **Enuresis** *(en''u-re'sis),* or **incontinence,** is the inability to control micturition. Its causes range from psychosomatic sources to actual physical impairment.

catheterization: L. *catheter,* to let down

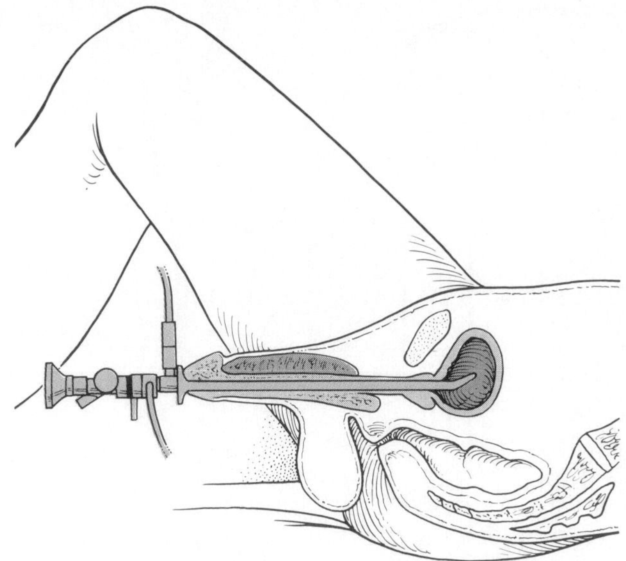

Figure 24.37. Cystoscopic examination of a male.

The palpation and inspection of urinary organs is an important aspect of physical assessment. The right kidney is palpable in the supine position; the left kidney usually is not. The distended urinary bladder is palpable along the superior pelvic rim.

The urinary system may be examined using X-ray techniques. An **intravenous pyelogram** (IVP) permits X-ray examination of the kidneys following the injection of radiopaque dye. In this procedure, the dye that has been injected intravenously is excreted by the kidneys so that the renal pelvises and the outlines of the ureters and urinary bladder can be observed in an X ray.

Cystoscopy *(sis-tos'ko-pe)* is the inspection of the inside of the urinary bladder by means of an instrument called a cystoscope (fig. 24.37). By using this technique, tissue samples can be obtained as well as urine samples from each kidney prior to mixing in the bladder. Once the cystoscope is in the bladder, the ureters and pelvis can be viewed through urethral catheterization and inspected for obstructions.

A **renal biopsy** is a diagnostic test for evaluating certain types and stages of kidney diseases. The biopsy is performed either through a skin puncture (closed biopsy) or through a surgical incision (open biopsy).

Urinalysis is a simple but important laboratory aspect of a physical examination. The voided urine specimen is tested for color, specific gravity, chemical composition, and for the presence of microscopic bacteria, crystals, and *casts.* Casts are accumulations of proteins that have leaked through the glomeruli and have been pushed through the tubules like toothpaste through a tube.

Infections of Urinary Organs

Urinary tract infections are a significant cause of illness and are also a major factor in the development of chronic renal failure. Females are more predisposed to urinary tract infections than are males, and the incidence of infection increases directly with sexual activity and aging. The higher infection rate in females has been attributed to their shorter urethra, which has a close proximity to the rectum, and to the lack of protection offered by prostatic secretions in males. To reduce the risk of urinary infections, a female should wipe her anal region in a posterior direction, away from the urethral orifice, after a bowel movement.

Infections of the urinary tract are named according to the infected organ. An infection of the urethra is called **urethritis,** and involvement of the urinary bladder is **cystitis.** Cystitis is frequently a secondary infection from some other part of the urinary tract.

Nephritis means inflammation of the kidney tissue. **Glomerulonephritis** *(glo-mer''u-lo-nĕ-fri'tis)* is inflammation of the glomeruli. Glomerulonephritis frequently occurs following an upper respiratory tract infection, because antibodies produced against streptococci bacteria can produce an autoimmune inflammation in the glomeruli. This inflammation may permanently change the glomeruli and figure significantly in the development of chronic renal disease and renal failure.

Any interference with the normal flow of urine, such as from a kidney stone or an enlarged prostate gland in a male, causes stagnation of urine in the renal pelvis and the development of pyelitis. **Pyelitis** is an inflammation of the renal pelvis and its calyces. **Pyelonephritis** is inflammation involving the renal pelvis, the calyces, and the tubules of the nephron within one or both kidneys. Bacterial invasion from the blood or from the lower urinary tract is another cause of both pyelitis and pyelonephritis.

Trauma to Urinary Organs

A sharp blow to a lumbar region of the back may cause a contusion or rupture of a kidney. Symptoms of kidney trauma include hematuria and pain of the upper abdominal quadrant and flank on the injured side.

Pelvic fractures from accidents may result in perforation of the urinary bladder and urethral tearing. When driving an automobile, it is advisable to stop periodically to urinate because an attached seat belt over the region of a full urinary bladder can cause it to rupture in even a relatively minor accident. Urethral injuries are more common in men than women because of the position of the urethra in the penis. "Straddle" injuries are those in which, for example, a man walking along a raised beam slips and compresses his urethra and penis between the hard surface and his pubic arch, rupturing the urethra.

Obstruction. The urinary system can become obstructed anywhere along the tract. Calculi (stones) are the most common cause, but blockage can also come from trauma, strictures, tumors or cysts, spasms or kinks of the ureters, or congenital anomaly. If not corrected, an obstruction causes urine to collect behind the blockage and generate pressure that may cause permanent functional and anatomic damage to one or both kidneys. Pressure

Figure 24.38. A diagram of a hemodialysis machine.

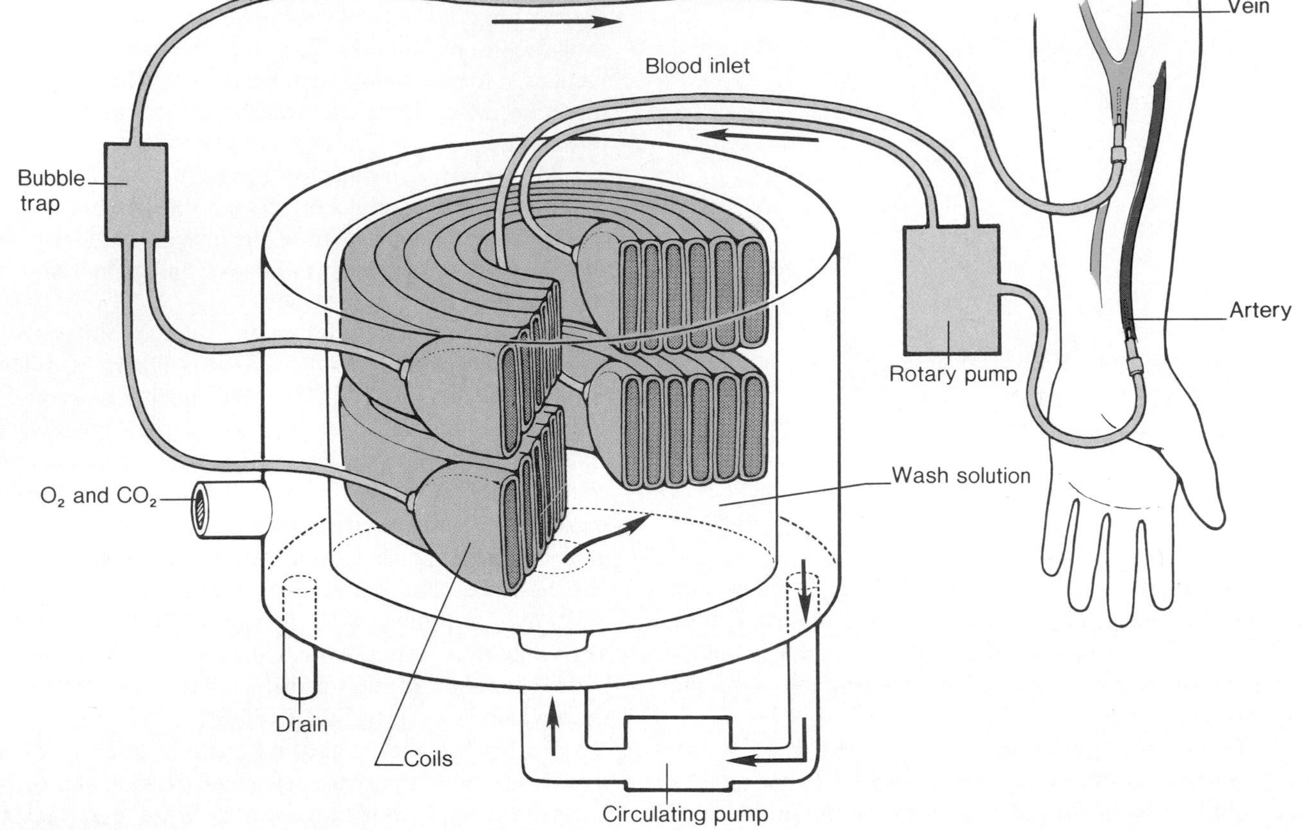

Figure 24.39. An artificial urethral sphincter. The entire device is implanted internally and consists of an inflation bulb that is manually pumped to inflate the urethral cuff from the fluid stored in a reservoir. Manual pumping of the deflation bulb releases the pressure of the urethral cuff and allows urine to flow from the bladder. The device for a female is similar except that the inflation and deflation bulbs are implanted in the tissue of the labia minora.

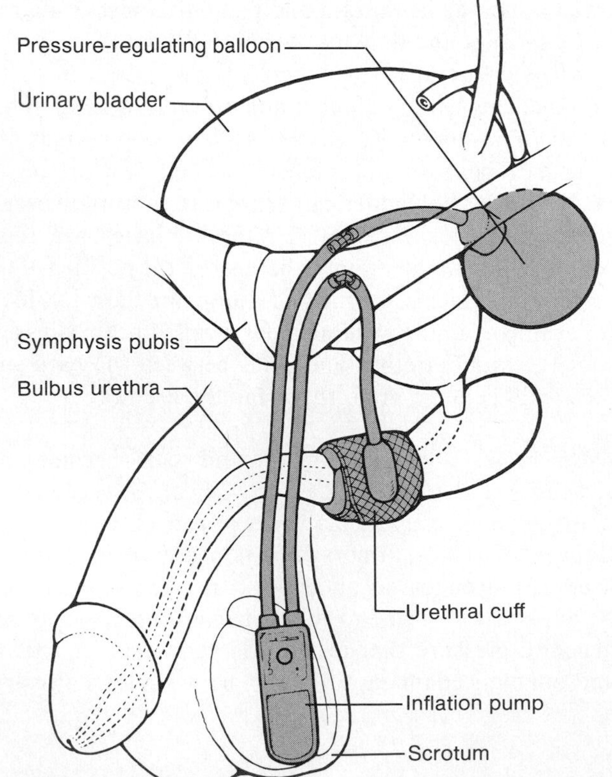

buildup in a ureter causes a distended ureter to develop, called a **hydroureter.** Dilation in the renal pelvis is called **hydronephrosis.**

Calculi, or kidney stones, are generally the result of infections or metabolic disorders that cause the excretion of large amounts of organic and inorganic substances. As the urine becomes concentrated, these substances may crystalize and form granules. The granule then serves as a core for further precipitation and development into a larger calculus. This becomes dangerous when the calculus is sufficiently large to cause an obstruction. It also causes intense pain when it passes through the urinary tract.

Renal Failure. An output of 50–60 cc of urine per hour is considered normal, and an output of less than 30 cc per hour may indicate renal failure. Renal failure is the loss of the kidney's ability to maintain fluid and electrolyte balance and to excrete waste products.

Renal failure can be either acute or chronic. **Acute renal failure** is the sudden loss of kidney function caused by shock and hemorrhage, thrombosis, or other physical trauma to the kidneys. The kidneys may sustain a 90% loss of their nephrons through tissue death and still not have an obvious loss of function. If a patient suffering acute renal failure is stabilized, the nephrons have an excellent capacity to regenerate.

A person with **chronic renal failure** cannot sustain life independently. Chronic renal failure is the end result of kidney disease in which the kidney tissue is progressively destroyed. As renal tissue continues to deteriorate, the options for sustaining life are hemodialysis or kidney transplantation.

Hemodialysis. Hemodialysis equipment is designed to filter the wastes from the blood of a patient who has chronic renal failure. During hemodialysis, the blood of a patient is pumped through a tube from the radial artery and passes through the machine, where it is cleansed, and then returned to the body through a vein (fig. 24.38). The cleaning process involves pumping the blood past a semipermeable cellophane membrane, which separates the blood from an isotonic solution containing molecules needed by the body (such as glucose). In a process called **dialysis,** waste products diffuse out of the blood through the membrane while glucose and other molecules needed by the body remain in the blood.

Urinary Incontinence. The inability to voluntarily retain urine in the urinary bladder is known as urinary incontinence. It has a number of temporary or permanent causes. Emotional stress is a cause of temporary incontinence in adults. Causes of permanent incontinence include neurological trauma, various urinary diseases, and tissue damage within the urinary bladder or urethra. Remarkable advances have been made in treating permanent urinary incontinence through the implantation of an artificial urethral sphincter (fig. 24.39).

Chapter Summary

- I. Introduction to the Urinary System
  - A. The urinary system consists of two kidneys, two ureters, the urinary bladder, and the urethra.
  - B. The urinary system maintains the composition and properties of the body fluid, which form the internal environment of the body cells. The end product of the urinary system is urine, which is voided from the body through the urethra.
- II. Development of the Urinary System
  - A. The urinary and reproductive systems both originate from mesodermal tissue called the urogenital ridge.
  - B. Three successive types of kidneys develop in the human embryo.
    1. The pronephric kidney persists only through the sixth week.
    2. The mesonephric kidney functions throughout the remainder of embryonic development.
    3. The metanephric kidney functions during fetal development and after birth.
- III. Kidney Structure
  - A. The gross structure of the kidney includes the renal sinus, medulla, and cortex.
    1. The renal sinus includes the renal pelvis and the major and minor calyces.
    2. The medulla is composed of the renal pyramids, which are separated by renal columns.
    3. The renal pyramids empty urine into the calyces, which drain into the renal pelvis and out the ureter.
  - B. Each kidney contains more than a million microscopic functional units called nephrons, which consist of vascular and tubular components.
    1. A capillary bed, called the glomerulus, produces a filtrate that enters the first part of the nephron tubule, known as the glomerular (Bowman's) capsule.
    2. Filtrate from the glomerular capsule enters, in turn, the proximal convoluted tubule, nephron loop (loop of Henle), distal convoluted tubule, and collecting duct.
    3. The glomerulus, proximal tubule, and distal tubule are located in the cortex; the nephron loop can descend into the medulla.
    4. The collecting ducts descend from the cortex through the medulla to empty their contents of urine into the calyces.
- IV. Glomerular Filtration
  - A. A filtrate derived from plasma in the glomerulus must pass through a basement membrane of the glomerular capillaries and through slits in the processes of the podocytes, which comprise the inner layer of the glomerular capsule.
    1. The glomerular ultrafiltrate is formed under the force of blood pressure and has a low protein concentration.
    2. The glomerular filtration rate is 115–125 ml/min.
  - B. The glomerular filtration rate can be regulated by constriction or dilation of the afferent arterioles.
    1. Sympathetic innervation causes constriction of the afferent arterioles.
    2. Intrinsic mechanisms help to autoregulate the rate of renal blood flow and the glomerular filtration rate.
- V. Reabsorption of Salt and Water
  - A. Approximately 65% of the filtered salt and water is reabsorbed across the proximal convoluted tubules.
    1. Sodium is actively transported, chloride follows passively by electrical attraction, and water follows the salt out of the proximal tubule.
    2. Salt transport in the proximal tubules is not under hormonal regulation.
  - B. The reabsorption of most of the remaining water occurs as a result of the action of the countercurrent multiplier system.
    1. Sodium is actively extruded from the ascending limb, followed passively by chloride.
    2. Since the ascending limb is impermeable to water, the remaining filtrate becomes hypotonic.
    3. Because of this salt transport and because of countercurrent exchange in the vasa recta, the tissue fluid of the medulla becomes hypertonic.
    4. The hypertonicity of the medulla is multiplied by a positive feedback mechanism involving the descending limb, which is passively permeable to water and perhaps to salt.
  - C. The collecting duct is permeable to water but not to salt.
    1. As the collecting ducts pass through the hypertonic renal medulla, water leaves by osmosis and is carried away in surrounding capillaries.
    2. The permeability of the collecting ducts to water is stimulated by antidiuretic hormone (ADH).
- VI. Renal Plasma Clearance
  - A. Inulin is filtered but neither reabsorbed nor secreted; its clearance is thus equal to the glomerular filtration rate.
  - B. Some of the filtered urea is reabsorbed; its clearance is therefore less than the glomerular filtration rate.
  - C. Since almost all the PAH in blood going through the kidneys is cleared by filtration and secretion, the PAH clearance is a measure of the total renal blood flow.
  - D. Normally all of the filtered glucose and amino acids are reabsorbed; glycosuria occurs when the transport carriers for glucose become saturated due to hyperglycemia.

VII. Renal Control of Electrolyte Balance
   A. Aldosterone stimulates sodium reabsorption and potassium secretion in the distal convoluted tubule.
   B. Aldosterone secretion is stimulated directly by a rise in blood potassium and indirectly by a fall in blood sodium.
      1. Decreased blood flow through the kidneys stimulates the secretion of the enzyme renin from the juxtaglomerular apparatus.
      2. Renin catalyzes the formation of angiotensin I, which is then converted to angiotensin II.
      3. Angiotensin II stimulates the adrenal cortex to secrete aldosterone.
   C. Aldosterone stimulates the secretion of H^+ as well as potassium into the filtrate in exchange for sodium.

VIII. Renal Control of Acid-Base Balance
   A. The lungs regulate the P_{CO_2} and carbonic acid concentration of the blood, whereas the kidneys regulate the bicarbonate concentration.
   B. Filtered bicarbonate combines with H^+ to form carbonic acid in the filtrate.
      1. Carbonic anhydrase in the membranes of microvilli in the tubules catalyzes the conversion of carbonic acid to carbon dioxide and water.
      2. Carbon dioxide is reabsorbed and converted in either the tubule cells or the red blood cells to carbonic acid, which dissociates to bicarbonate and H^+.
      3. In addition to reabsorbing bicarbonate, the kidneys excrete H^+, which is buffered by ammonium and phosphate buffers.

IX. Ureters, Urinary Bladder, and Urethra
   A. The ureters contain three layers: the mucosa, muscularis, and fibrous coat.
   B. The urinary bladder is lined by a transitional epithelium that is folded into rugae to permit distension.
   C. The urethra has an internal sphincter of smooth muscle and an external sphincter of skeletal muscle.
   D. Micturition is controlled by reflex centers in the second through fourth segments of the spinal cord.

Review Activities

Objective Questions

1. Which of the following statements about metanephric kidneys is *true?*
   (a) They become functional at the end of the eighth week.
   (b) They are active throughout fetal development.
   (c) They are the third pair of kidneys to develop.
   (d) All of the above are true.

Match the following:

2. Active transport of sodium; water follows passively
3. Active transport of sodium; impermeable to water
4. Passively permeable to water and maybe salt
5. Passively permeable to water only

   (a) proximal tubule
   (b) descending limb
   (c) ascending limb
   (d) distal tubule
   (e) collecting duct

6. Antidiuretic hormone promotes the retention of water by stimulating the
   (a) active transport of water
   (b) active transport of chloride
   (c) active transport of sodium
   (d) permeability of the collecting duct to water
7. Aldosterone stimulates sodium reabsorption and potassium secretion in the
   (a) proximal convoluted tubule
   (b) descending limb of the nephron loop
   (c) ascending limb of the nephron loop
   (d) distal convoluted tubule
   (e) collecting duct
8. Substance *X* has a clearance greater than zero but less than that of inulin. What can be concluded about substance *X?*
   (a) It is not filtered.
   (b) It is filtered, but neither reabsorbed nor secreted.
   (c) It is filtered and partially reabsorbed.
   (d) It is filtered and secreted.
9. Substance *Y* has a clearance greater than that of inulin. What can be concluded about *Y?*
   (a) It is not filtered.
   (b) It is filtered but neither reabsorbed nor secreted.
   (c) It is filtered and partially reabsorbed.
   (d) It is filtered and secreted.
10. About 65% of the glomerular ultrafiltrate is reabsorbed in the
   (a) proximal tubule
   (b) distal tubule
   (c) loop of the nephron
   (d) collecting duct
11. Which of the following statements about the renal pyramids is *false?*
   (a) They are located in the medulla.
   (b) They contain glomeruli.
   (c) They contain collecting ducts.
   (d) They open by renal papillae into the renal sinus.
12. The detrusor muscle is located in the
   (a) kidneys
   (b) ureters
   (c) urinary bladder
   (d) urethra
13. The internal urethral sphincter is innervated by
   (a) sympathetic nerve fibers
   (b) parasympathetic nerve fibers
   (c) somatic motor nerve fibers
   (d) all of the above
14. Diuretic drugs that act in the loop of the nephron
   (a) inhibit active sodium transport
   (b) result in the increased flow of filtrate to the distal convoluted tubule
   (c) cause the increased secretion of potassium into the tubule
   (d) promote the excretion of salt and water
   (e) all of the above
15. The appearance of glucose in the urine
   (a) occurs normally
   (b) indicates the presence of kidney disease
   (c) occurs only when the transport carriers for glucose become saturated
   (d) is a result of hypoglycemia
16. Reabsorption of water through the tubules occurs by
   (a) osmosis
   (b) active transport
   (c) facilitated diffusion
   (d) all of the above

Essay Questions

1. Explain how glomerular ultrafiltrate is produced and why it has a low protein concentration.
2. Explain how the countercurrent multiplier system works, and describe its functional significance.
3. Explain how countercurrent exchange occurs in the vasa recta, and describe its functional significance.
4. Explain the mechanisms whereby diuretic drugs may cause an excessive loss of potassium; also explain how the potassium-sparing diuretics work.
5. Explain how the structure of the epithelial wall of the proximal tubule and the distribution of Na^+/K^+ pumps in the epithelial cell membranes contribute to the ability of the proximal tubule to reabsorb salt and water.

25 Digestive System

Outline

Concepts

The digestive system, consisting of a tubular gastrointestinal tract and accessory digestive organs, is specialized for the digestion and absorption of food.

The primitive gut differentiates during the fourth week to give rise to the specific regions of the GI tract and the accessory digestive organs.

Serous membranes line the abdominal cavity and cover the visceral organs. The wall of the GI tract is composed of four tunics.

Ingested food is changed by the mechanical action of teeth and by the chemical activity of saliva into a bolus, which is swallowed in the process of deglutition.

A bolus of food is passed from the esophagus to the stomach, where it is mixed with gastric secretions. The chyme thus produced is sent past the pyloric sphincter to the duodenum.

The small intestine is adapted to provide a very high surface area for the exposure of chyme to brush border enzymes and for the absorption of nutrients.

The large intestine absorbs water and electrolytes from the chyme and passes undigested products out of the gastrointestinal tract.

The liver processes nutrients and secretes bile, which is stored and concentrated in the gallbladder prior to its discharge into the duodenum. The pancreas secretes important hormones into the blood and essential digestive enzymes into the duodenum.

Polysaccharides and polypeptides are digested into their subunits, which are absorbed and secreted into the the blood capillaries. Emulsified fat is digested, absorbed into the intestinal cells, and then resynthesized into triglycerides, which are secreted as small particles in the lymph.

Neural and hormonal mechanisms coordinate the activities of different regions of the GI tract and the actions of the liver, gallbladder, and pancreas.

Figure 25.1. The digestion of food molecules occurs by means of hydrolysis reactions.

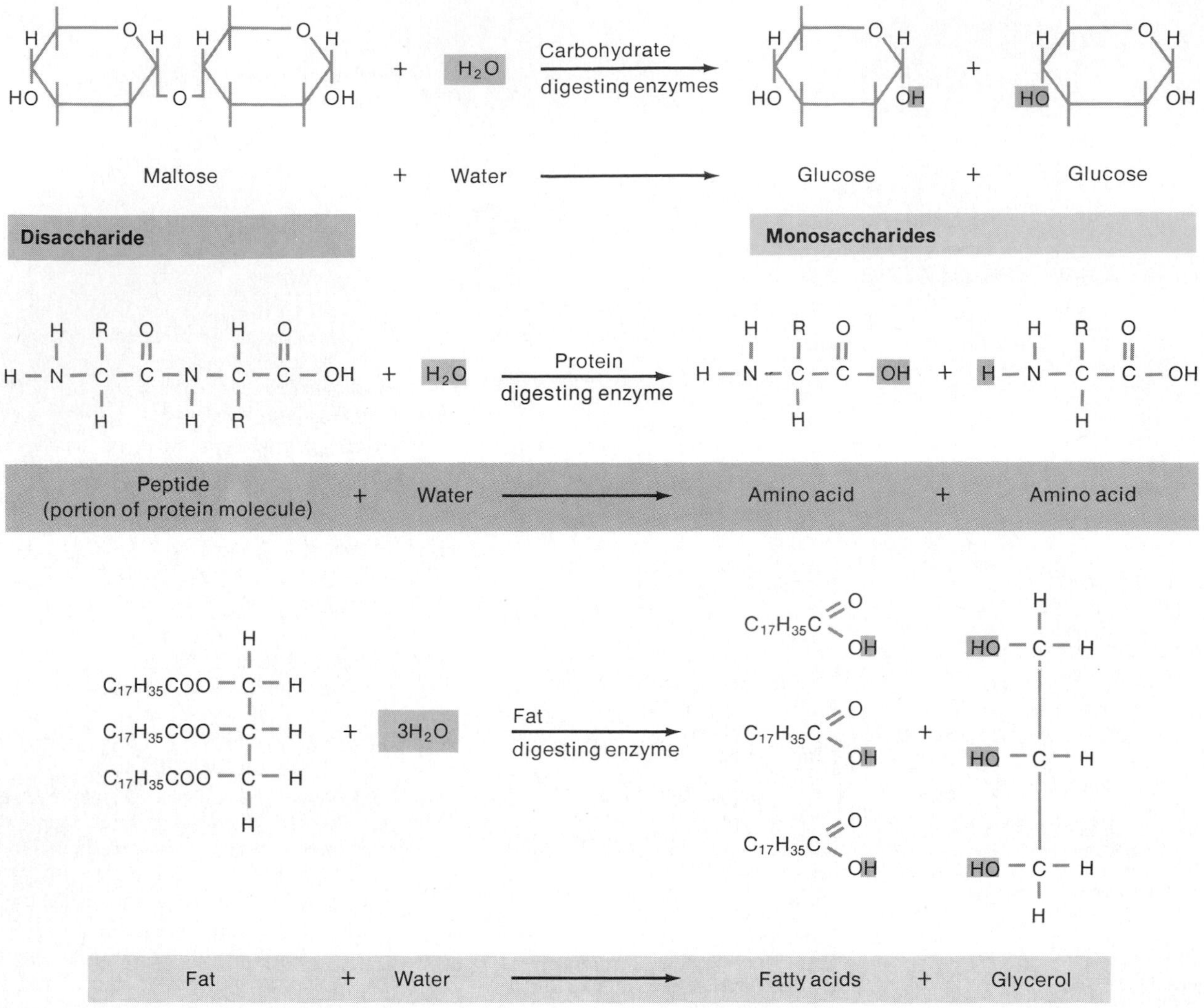

Introduction to the Digestive System

The digestive system, consisting of a tubular gastrointestinal tract and accessory digestive organs, is specialized for the digestion and absorption of food.

Objective 1. List the activities of the digestive system, and distinguish between digestion and absorption.

Objective 2. Identify the major structures and regions of the digestive system.

Objective 3. Define the terms *viscera* and *gut*.

Unlike plants, which can form organic molecules using inorganic compounds such as carbon dioxide, water, and ammonia, humans and other animals must obtain their basic organic molecules from food. Some of the ingested food molecules are needed for their energy (caloric) value—obtained by the reactions of cell respiration and used in the production of ATP—and the balance is used to make additional tissue.

Most of the organic molecules that are ingested are similar to the molecules that form the structure of human tissues. These are generally large molecules *(polymers)*, which are composed of subunits *(monomers)*. Within the gastrointestinal tract, these large molecules are **digested** into their monomers by *hydrolysis* reactions (reviewed in fig. 25.1). The monomers thus formed are transported across the wall of the intestine (**absorbed**) into the blood and lymph. Digestion and absorption are the two major functions of the digestive system.

Figure 25.2. The digestive system, including the gastrointestinal tract and the accessory digestive organs.

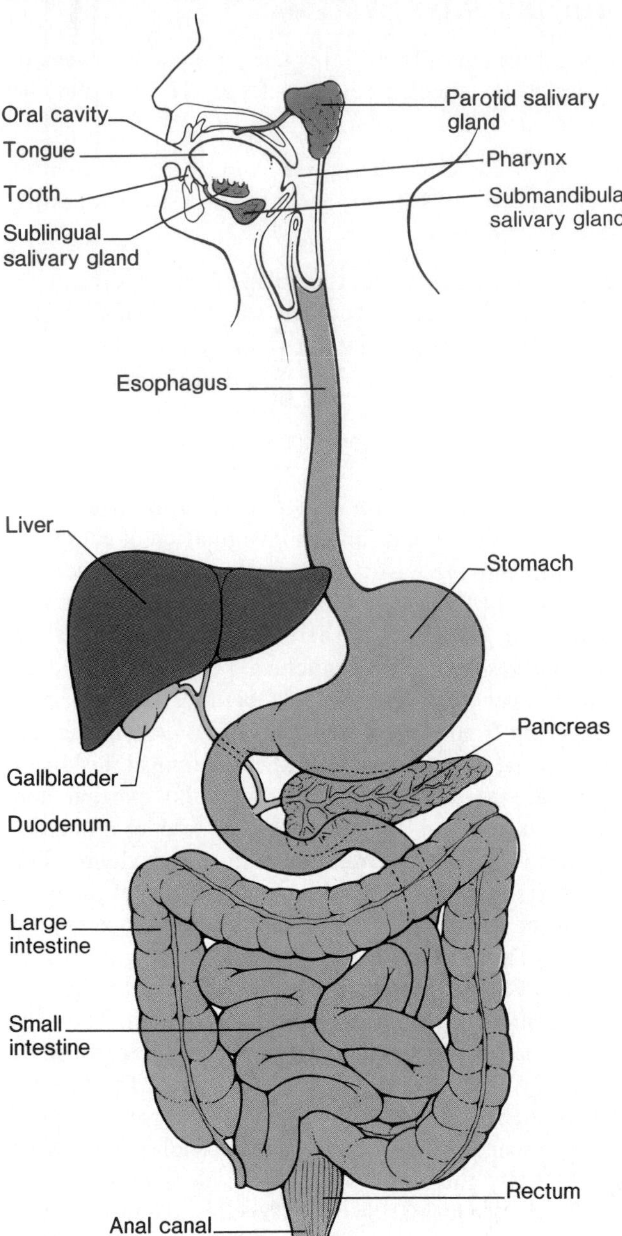

Since the composition of food is similar to the composition of body tissues, enzymes that digest food could also digest the person's own tissues. This does not normally occur because there is a variety of protective devices that inactivate digestive enzymes in the body and keep these enzymes separated from the cytoplasm of tissue cells. The fully active digestive enzymes are normally limited in their location to the lumen (cavity) of the gastrointestinal tract.

The lumen of the gastrointestinal tract is continuous with the environment because it is open at both ends (mouth and anus). Indigestible material, such as cellulose from plant walls, passes from one end to the other without crossing the epithelial lining of the digestive tract (that is, without being absorbed). In this sense, these indigestible materials never enter the body, and the harsh conditions required for digestion thus occur *outside* the body.

In *planaria* (a type of flatworm) the gastrointestinal tract has only one opening—the mouth is also the anus. Each cell that lines the gastrointestinal tract is therefore exposed to food, absorbable digestion products, and waste products. The two-ended digestive tract of higher organisms, in contrast, allows one-way transport, which is insured by wavelike muscle contractions and by the action of sphincter muscles. This one-way transport allows different regions of the gastrointestinal tract to be specialized for different functions, as a "dis-assembly line." The principal function of the digestive system is to prepare food for cellular utilization. This involves several functional activities listed as follows:

1. **Ingestion.** The taking of food into the digestive system by way of the mouth.
2. **Mastication.** Chewing movements to pulverize food and mix it with saliva.
3. **Deglutition.** The swallowing of food to move it from the mouth to the anus.
4. **Digestion.** Mechanical and chemical breakdown of food into absorbable molecules.
5. **Absorption.** The passage of monomers of food molecules through the mucous membrane of the intestine into blood or lymph.
6. **Peristalsis.** Rhythmic, wavelike contractions that move food through the digestive tract.
7. **Defecation** *(def''e-ka'shun).* The discharge of indigestible wastes, called feces, from the body.

Anatomically and functionally the digestive system can be divided into a tubular **gastrointestinal (GI) tract,** or an **alimentary canal,** and **accessory organs.** The GI tract is approximately 9 m (30 ft) long and extends from the mouth to the anus. It traverses the thoracic cavity and enters the abdominal cavity at the level of the diaphragm. The anus is located at the inferior portion of the pelvic cavity. The organs of the GI tract include the **oral (buccal) cavity, pharynx, esophagus, stomach, small intestine,** and **large intestine** (fig. 25.2). The accessory digestive organs

ingestion: L. *ingerere,* carry in
mastication: Gk. *mastichan,* gnash the teeth
deglutition: L. *deglutire,* swallow down
peristalsis: Gk. *peri,* around; *stellein,* compress
defecation: L. *de,* from, away; *faecare,* cleanse

Table 25.1 Regions of the GI tract and basic functions

Region	Function
Oral cavity	Ingests food; receives saliva; mastication; initiates digestion of carbohydrates; forms bolus; deglutition
Pharynx	Receives bolus from oral cavity; autonomically continues deglutition of bolus to esophagus
Esophagus	Transports bolus to stomach by peristalsis; esophageal sphincter restricts backflow of food
Stomach	Receives bolus from esophagus; churns bolus with gastric juice; initiates digestion of proteins; limited absorption; moves chyme into duodenum and prohibits backflow of chyme; regurgitates when necessary
Small intestine	Receives chyme from stomach and secretions from liver and pancreas; chemically and mechanically breaks down chyme; absorbs nutrients; transports wastes through peristalsis to large intestine; prohibits backflow of intestinal wastes from large intestine
Large intestine	Receives undigested wastes from small intestine; absorbs water and electrolytes; forms, stores, and expels feces through defecation reflex

include the **teeth, tongue, salivary glands, liver,** and **pancreas.** The term *viscera (vis'er-ah)* is frequently used to refer to the abdominal organs of digestion, but this term can actually be used to indicate any organ in the thoracic and abdominal cavities. **Gut** is an anatomical term that generally refers to the developing stomach and intestine in the embryo.

The regions of the GI tract and their functions are summarized in table 25.1.

Although there is an abundance of food in the United States, too many people eat the wrong kinds of foods, eat irregularly, or overeat, to the point that eating patterns have become a critical public health concern. Obesity is a major health problem; obese persons are at greater risk for cardiovascular disease, hypertension, osteoarthritis, and diabetes mellitus. People with good nutritional habits are better able to withstand trauma, are less likely to get sick, and are usually less seriously ill when they do become sick.

1. Which functional activities of the digestive system break down food? Define *absorption.* Which functional activities move the food through the GI tract?
2. List in order the regions of the GI tract through which ingested food would pass from the mouth to the anus.
3. List the abdominal visceral organs of the digestive system.
4. Write a sentence using correctly the term *gut.*

Embryological Development of the Digestive System

The primitive gut differentiates during the fourth week to give rise to the specific regions of the GI tract and the accessory digestive organs.

Objective 4. Describe the embryological development of the GI tract and the accessory digestive organs.

The entire digestive system develops from modifications of an elongated tubular structure called the **primitive gut.** These modifications are initiated during the fourth week of embryonic development. The primitive gut is composed solely of endoderm and for descriptive purposes can be divided into three regions: foregut, midgut, and hindgut.

Foregut. The **stomodeum** *(sto''moh-de'um),* or **oral pit,** is not part of the foregut but an invagination of ectoderm that breaks through a thin **oral membrane** to become continuous with the foregut and form the oral cavity, or mouth. Structures in the mouth, therefore, are ectodermal in origin. The esophagus, stomach, a portion of the duodenum, the pancreas, liver, and gallbladder are the organs that develop from the foregut (fig. 25.3). Along the alimentary canal, only the inside epithelial lining of the lumen is derived from endoderm. The vascular portion and smooth muscle layers are formed from mesoderm that develops from the surrounding splanchnic mesenchyme. The stomach first appears as an elongated dilation of the foregut. The dorsal border of the stomach undergoes more rapid growth than the ventral border, forming a distinct curvature. The caudal portion of the foregut and the cranial portion of the midgut form the duodenum. The liver and pancreas arise as small **hepatic** and **pancreatic buds,** respectively, from the wall of the duodenum. The hepatic bud experiences incredible growth to form the gallbladder, associated ducts, and the various lobes of the liver (fig. 25.3). By the sixth week, the liver is carrying out hemopoiesis (the formation of blood cells), and by the ninth week, the liver represents 10% of the total weight of the fetus.

The pancreas develops from dorsal and ventral pancreatic buds of endodermal cells. As the duodenum grows, it rotates clockwise and the two pancreatic buds fuse (fig. 25.3).

Midgut. During the fourth week of the embryonic stage, the midgut is continuous with the yolk sac. By the fifth week, the midgut forms a ventral U-shaped **midgut loop,**

stomodeum: Gk. *stoma,* mouth; *hodaios,* on the way to

Figure 25.3. Progressive stages of development of the foregut to form the stomach, duodenum, liver, gallbladder, and pancreas. (*a*) four weeks, (*b*) five weeks, (*c*) six weeks, and (*d*) seven weeks.

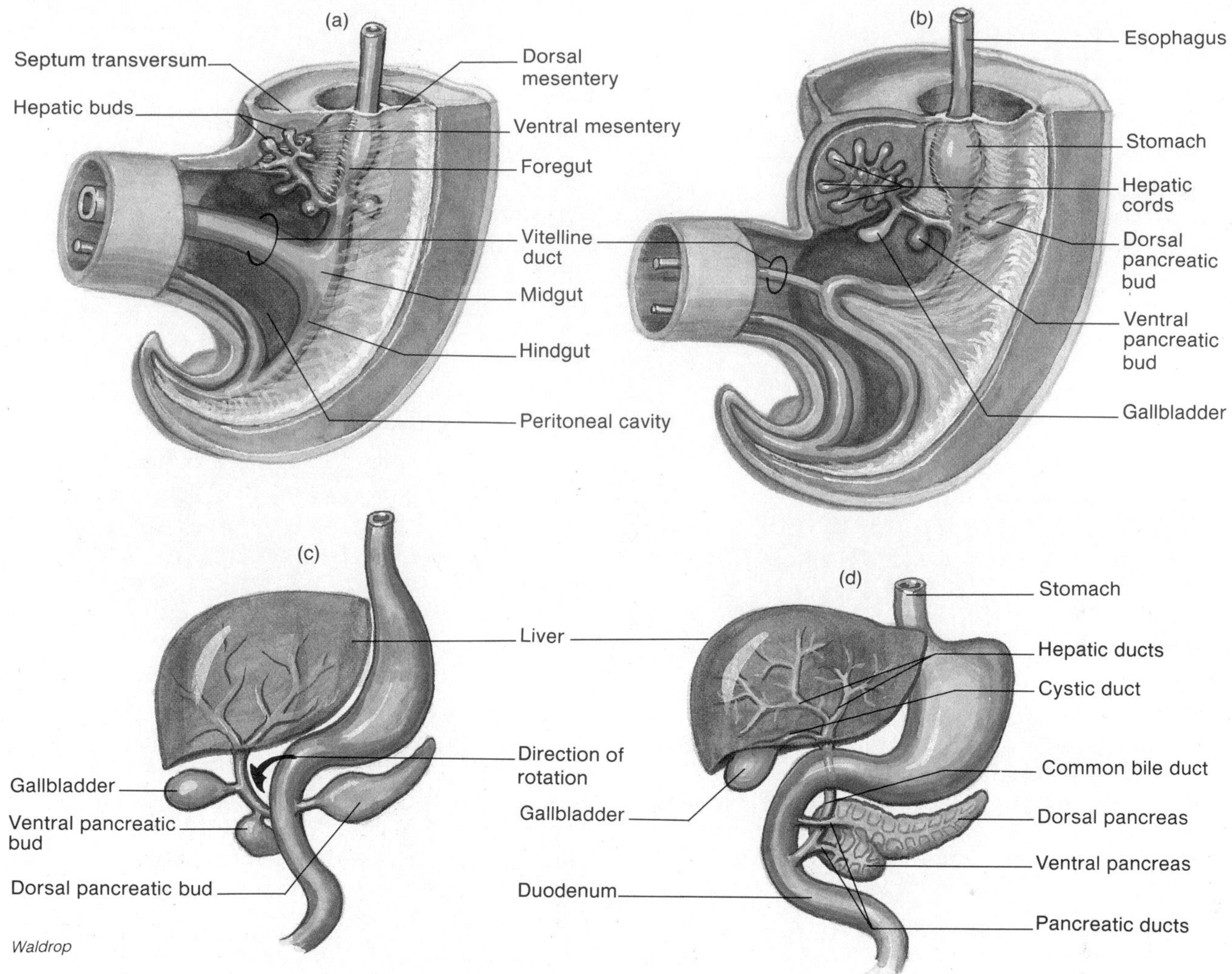

which projects into the umbilical cord (fig. 25.4). As development continues, the anterior limb of the midgut loop coils to form most of the small intestine. The posterior limb of the midgut loop expands to form a portion of the small and large intestines. A **cecal diverticulum** *(di''ver-tik'u-lum)* appears during the fifth week.

During the tenth week the intestines are withdrawn to the abdominal cavity, and further differentiation and rotation occur. The cecal diverticulum continues to develop, forming the cecum and appendix. The remainder of the midgut gives rise to the ascending colon and hepatic flexure (fig. 25.4).

Hindgut. The hindgut extends from the midgut to the **cloacal membrane.** The **proctodeum** *(prok''to-de'um),* or **anal pit,** is a depression in the anal region formed from an invagination of ectoderm that contributes to the cloacal membrane. The *allantois (ah-lan'to-is),* which receives urinary wastes from the fetus, connects to the hindgut at a region called the **cloaca** *(klo-a'kah),* as seen in figure 25.5. A band of mesenchymal cells called the **urorectal septum** grows caudally during the fourth through seventh weeks until a complete partition separates the cloaca into a dorsal **anal canal** and a ventral **urogenital sinus.** With

proctodeum: Gk. *proktos,* anus; *hodaios,* on the way to
cloaca: L. *cloaca,* sewer

Figure 25.4. Progressive stages of the development of the midgut to form the distal portion of the small intestine and the proximal portion of the large intestine. (*a*) five weeks, (*b*) six weeks, (*c*) seven weeks, (*d*) ten weeks, and (*e*) eighteen weeks.

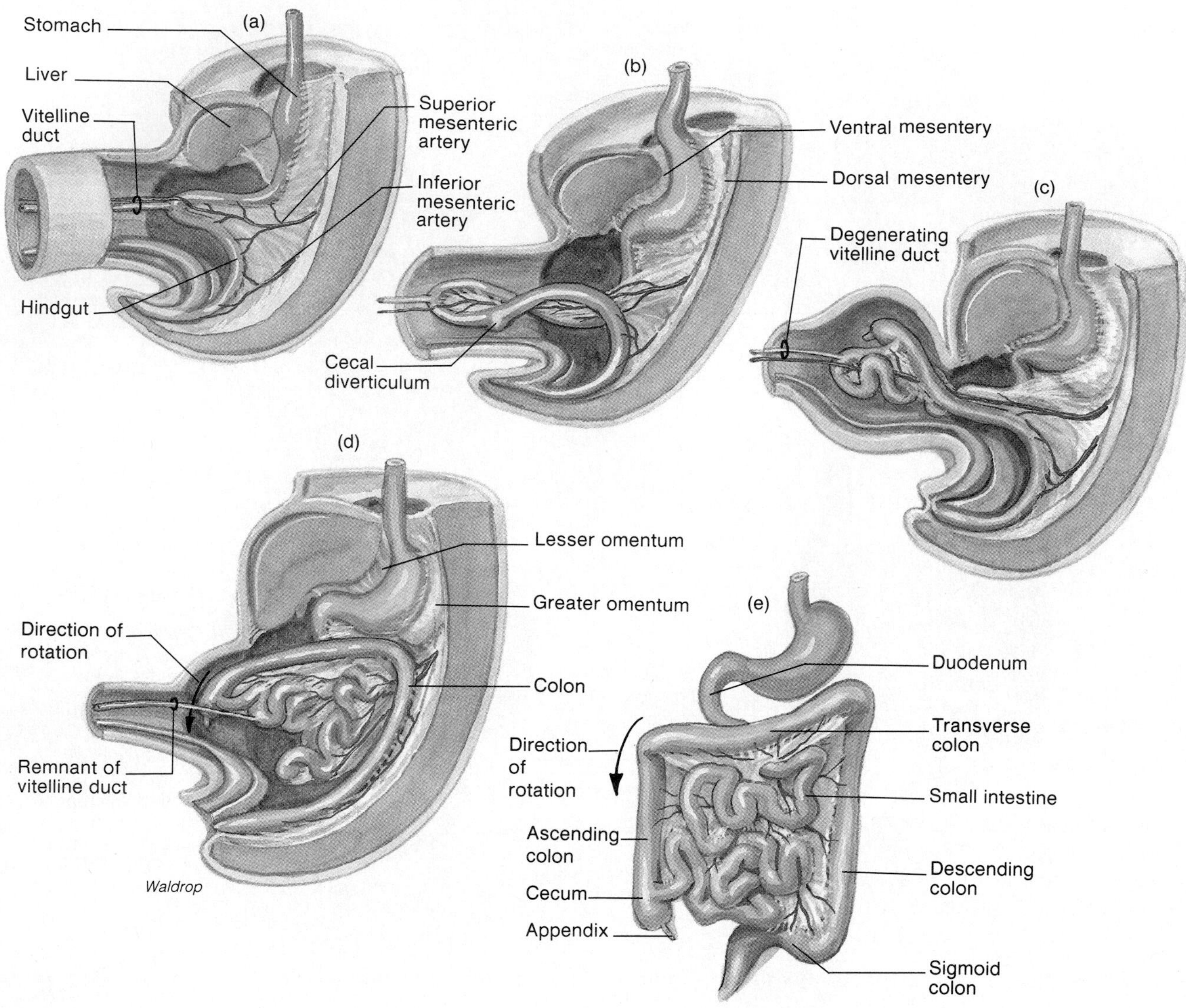

the completion of the urorectal septum, the cloacal membrane is divided into an anterior **urogenital membrane** and a posterior **anal membrane.** Toward the end of the seventh week, the anal membrane perforates and forms the anal opening, which is lined with ectodermal cells. About this time, the urogenital membrane ruptures to provide further development of the urinary and reproductive systems.

1. Describe the embryonic derivation of the mouth and anus.
2. Identify the portion of the gut that gives rise to the pancreas and liver.
3. Define the terms *stomodeum, proctodeum, allantois,* and *cloaca.*

Serous Membranes and Tunics of the Gastrointestinal Tract

Serous membranes line the abdominal cavity and cover the visceral organs. The wall of the GI tract is composed of four tunics.

Objective 5. Describe the arrangement of the serous membranes within the abdominal cavity.

Objective 6. Describe the generalized structure of the four tunics composing the wall of the GI tract.

Figure 25.5. The progressive development of the hindgut, illustrating the developmental separation of the digestive system from the urogenital system. (*a*) an anterolateral view of an embryo at four weeks showing the position of a sagittal cut depicted in (*b, c,* and *d*). (*b*) at four weeks, the hindgut, cloaca, and allantois are connected. (*c*) at six weeks, the connections between the gut and extraembryonic structures are greatly diminished. (*d*) by seven weeks, structural and functional separation between the digestive system and the urogenital system is nearly completed.

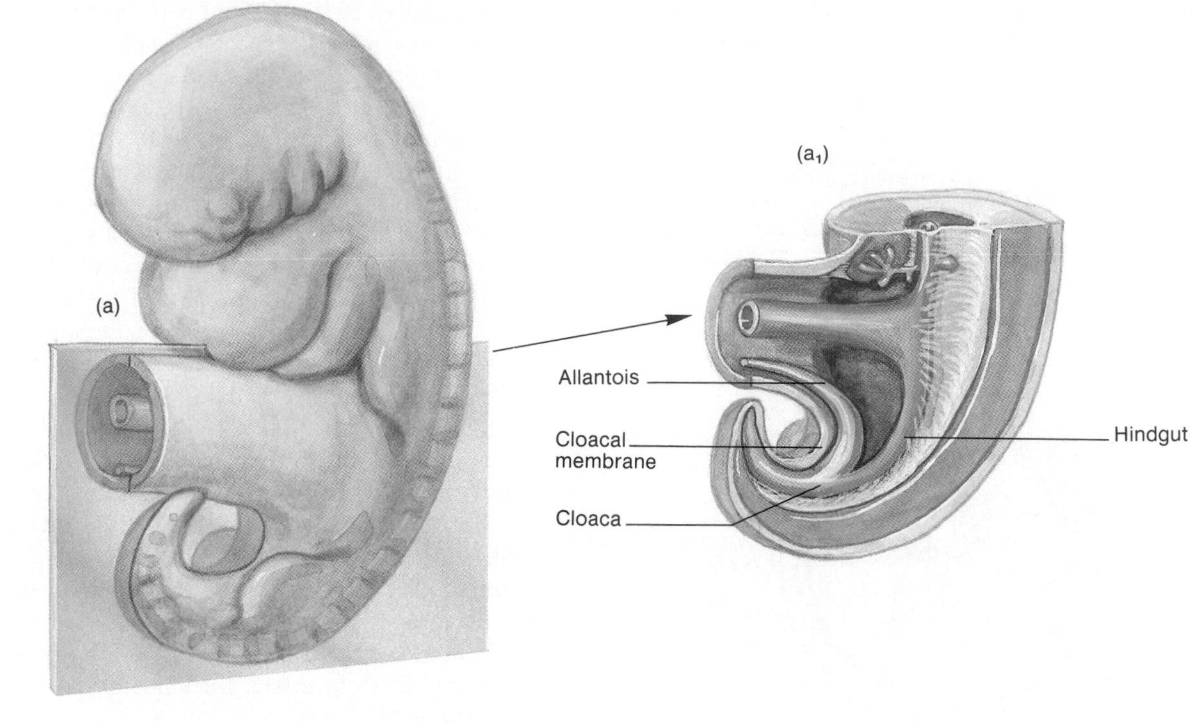

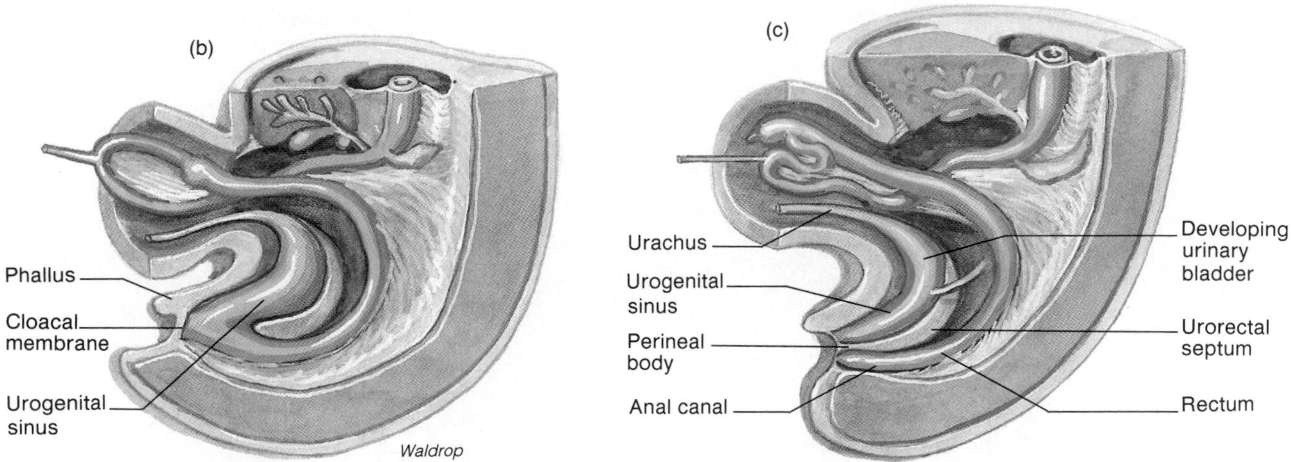

Serous Membranes

Most of the GI tract and abdominal accessory digestive organs are positioned within the **abdominal cavity.** These organs are not firmly embedded in solid tissue but are supported and covered by **serous membranes.** A serous membrane is an epithelial membrane that lines the thoracic and abdominal cavities and covers the organs that lie within these cavities. A serous membrane has a parietal portion lining the body wall and a visceral portion covering the internal organs. The serous membranes associated with the lungs are called pleurae. The serous membranes of the abdominal cavity are called **peritoneal membranes,** or **peritoneum** *(per''i-to-ne'um).* The peritoneum is the largest serous membrane of the body. It is composed of simple squamous epithelium with portions reinforced with connective tissue.

peritoneum: Gk. *peritonaion,* stretched over

The **parietal peritoneum** lines the wall of the abdominal cavity (fig. 25.6). Along the dorsal, or posterior, aspect of the abdominal cavity the parietal peritoneum comes together to form a double-layered peritoneal fold called the **mesentery,** which supports the GI tract. The **dorsal mesentery** gives the pendulous small intestine freedom for peristaltic movement and provides a structure through which intestinal nerves and vessels traverse. The **mesocolon** is a specific portion of the mesentery that supports the large intestine (fig. 25.7c and d). The peritoneal covering continues around the intestinal viscera as the **visceral peritoneum.** The **peritoneal cavity** is the space between the parietal and visceral portions of the peritoneum.

Extensions of the parietal peritoneum, located in the peritoneal cavity, serve specific functions (fig. 25.7). The **falciform** *(fal'si-form)* **ligament,** a serous membrane reinforced with connective tissue, attaches the liver to the diaphragm and anterior abdominal wall. The **lesser omentum** *(o-men'tum)* passes from the lesser curvature of the stomach and the upper duodenum to the inferior surface of the liver. The **greater omentum** extends from the greater curvature of the stomach to the transverse colon, forming an apronlike structure over most of the small intestine. Functions of the greater omentum include storing fat, cushioning visceral organs, supporting lymph nodes, and protecting against the spread of infections. In cases of localized inflammation, such as appendicitis, the greater omentum may compartmentalize the inflamed area, sealing it off from the rest of the peritoneal cavity.

Certain abdominal organs are not within the peritoneal cavity and are therefore covered by the parietal peritoneum. These organs are said to be *retroperitoneal* and include most of the pancreas, the kidneys, a portion of the duodenum, and the abdominal aorta. The urinary bladder is situated in front of the peritoneum and is considered *antiperitoneal.*

mesentery: Gk. *mesos,* middle; *enteron,* intestine
omentum: L. *omentum,* apron

Figure 25.6. A diagrammatic representation of the serous membranes.

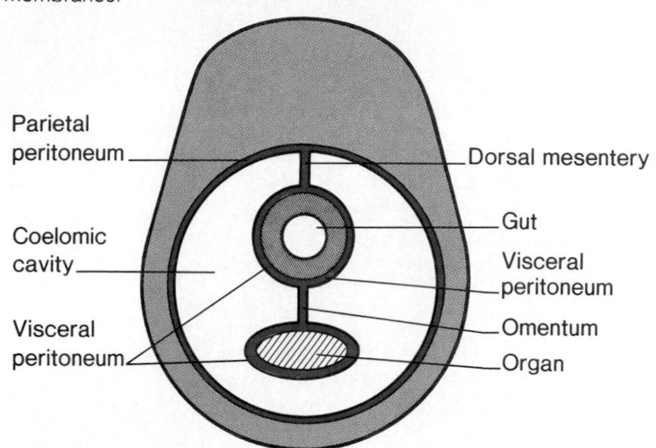

Figure 25.7. The structural arrangement of the abdominal organs and peritoneal membranes. (*a*) the greater omentum; (*b*) the lesser omentum with the liver lifted; (*c*) the mesentery with the greater omentum lifted; (*d*) the relationship of the peritoneal membranes to the visceral organs as shown in a sagittal view.

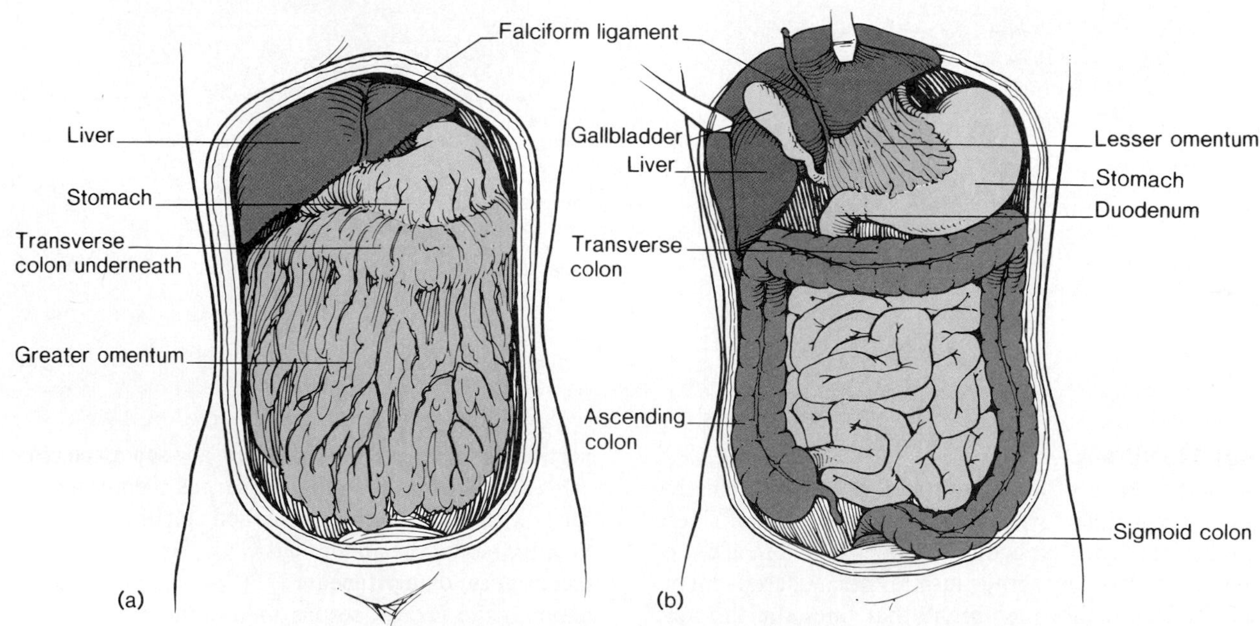

Peritonitis is a bacterial inflammation of the peritoneum. Peritonitis may be caused by trauma, rupture of a visceral organ, ectopic pregnancy, or postoperative complications. Peritonitis is usually extremely painful and serious. Treatment usually involves the injection of massive doses of antibiotics and perhaps peritoneal intubation to permit drainage.

Layers of the Gastrointestinal Tract

The GI tract from the stomach to the anal canal is composed of four layers, or **tunics.** Each tunic contains a dominant tissue type that performs specific functions in the digestive process. The four tunics of the GI tract, from the inside out, are the **mucosa, submucosa, muscularis,** and **serosa** (fig. 25.8).

Mucosa. The mucosa surrounds the lumen of the GI tract and is the absorptive and major secretory layer. It consists of a simple columnar epithelium supported by the **lamina** *(lam'ĭ-nah)* **propria,** which is a thin layer of connective tissue. The lamina propria contains numerous lymph nodules, which are important in protecting against disease (fig. 25.8). External to the lamina propria are thin layers of smooth muscle called the **muscularis mucosa.** This is the muscle layer that causes the small intestine portion of the GI tract to have numerous small folds, called plicae circulares (see fig. 25.25), which greatly increase the absorptive surface area. Specialized **goblet cells** in the mucosa secrete mucus throughout most of the GI tract.

Submucosa. The relatively thick submucosa is a highly vascular layer of connective tissue serving the mucosa. Absorbed molecules that pass through the columnar epithelial cells of the mucosa enter into blood vessels or lymph ductules of the submucosa. In addition to blood vessels, the submucosa contains glands and nerve plexuses. The *submucosal plexus (Meissner's plexus)* (fig. 25.8) provides autonomic innervation to the muscularis mucosa.

Muscularis. The muscularis (also called the muscularis externa) is responsible for segmental contractions and peristaltic movement through the GI tract. The muscularis has an inner circular and an outer longitudinal layer of smooth muscle. Contractions of these layers move the food peristaltically through the tract and physically pulverize and churn the food with digestive enzymes. The *myenteric plexus (Auerbach's plexus)* located between the

tunica: L. *tunica,* covering or coat

Meissner's plexus: from Georg Meissner, German histologist, 1829–1905

Auerbach's plexus: from Leopold Auerbach, German anatomist, 1828–97

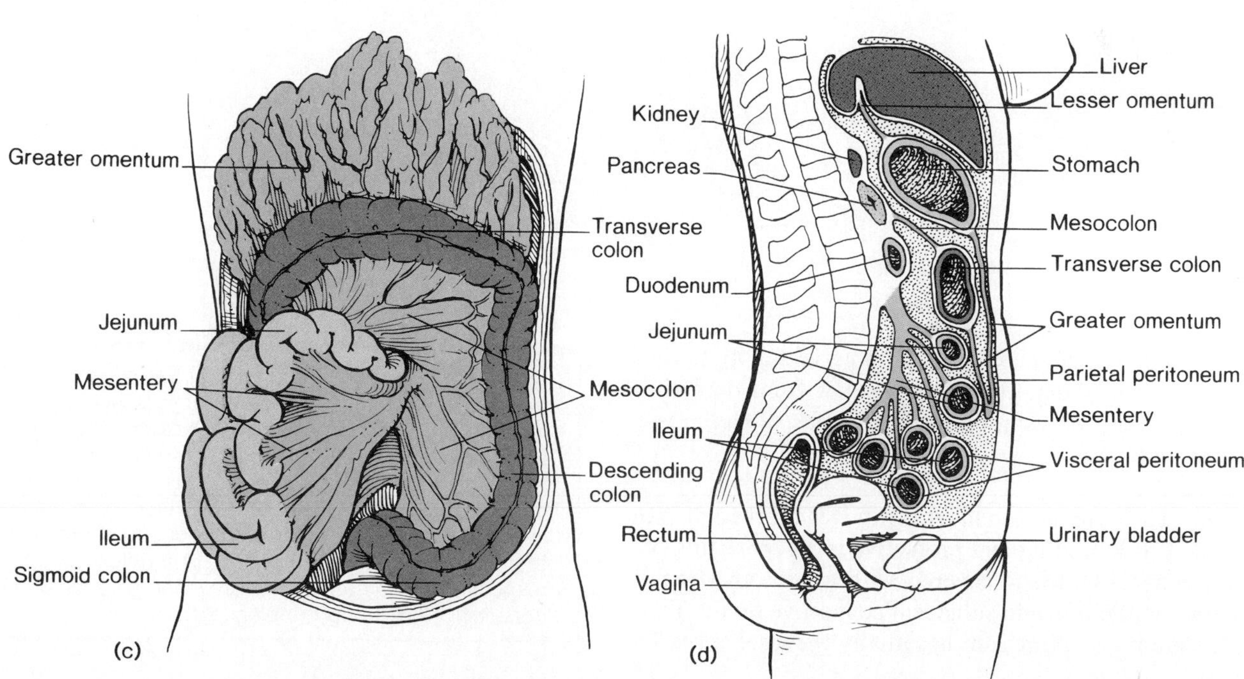

Figure 25.8. (*a*) an illustration of the major layers of the intestine. The insert shows how folds of mucosa form projections called *villi.* (*b*) an illustration of a cross section of the intestine showing layers and glands.

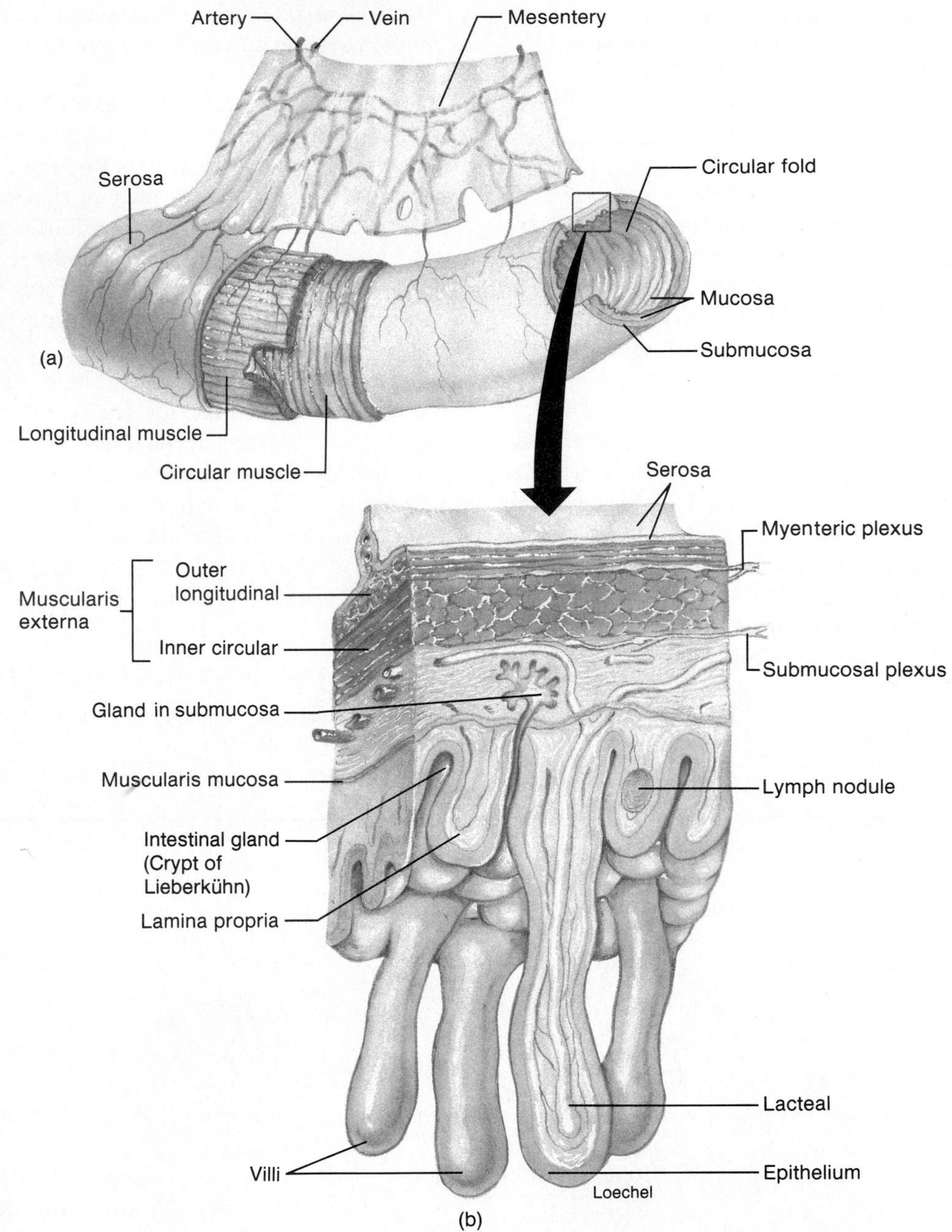

two muscle layers provides the major nerve supply to the GI tract and includes fibers and ganglia from both the sympathetic and parasympathetic divisions of the autonomic system.

Serosa. The outer serosa layer completes the wall of the GI tract. It is a binding and protective layer consisting of areolar connective tissue covered with a layer of simple squamous epithelium and subjacent connective tissue. The simple squamous epithelium is actually the visceral peritoneum.

The body has several defense mechanisms to protect against ingested material that may be harmful if absorbed. The acidic environment of the stomach and the lymphatic system kill many harmful bacteria. A mucous lining throughout the GI tract serves as a protective layer. Vomiting and in certain cases diarrhea are reactions to substances that irritate the GI tract. Vomiting is a reflexive response to many toxic chemicals and as such can be beneficial even though unpleasant.

Innervation. The GI tract is innervated by the sympathetic and parasympathetic divisions of the autonomic system. The vagus nerve is the source of parasympathetic activity in the esophagus, stomach, pancreas, gallbladder, small intestine, and upper portion of the large intestine. The lower portion of the large intestine receives parasympathetic innervation from spinal nerves in the sacral region. The submucosal plexus and myenteric plexus are the sites where preganglionic fibers synapse with postganglionic fibers that innervate the smooth muscle of the GI tract. Stimulation of the parasympathetic fibers increases peristalsis and the secretions of the GI tract.

Postganglionic sympathetic fibers pass through the submucosal and myenteric plexuses and innervate the GI tract. Sympathetic nerve stimulation acts antagonistically to the effects of parasympathetic nerves by reducing peristalsis and secretions and stimulating the contraction of sphincter muscles along the GI tract.

1. Describe the functions of the dorsal mesentery and greater omentum. Identify the location of the peritoneum, and list the organs that are retroperitoneal.
2. List the four tunics of the GI tract, and identify their major tissue types. Explain the functions of these four tunics.

Mouth, Pharynx, and Associated Structures

Ingested food is changed by the mechanical action of teeth and by the chemical activity of saliva into a bolus, which is swallowed in the process of deglutition.

Objective 7. Describe the anatomy of the oral cavity.
Objective 8. Contrast deciduous and permanent dentitions, and describe the structure of a typical tooth.
Objective 9. Describe the location and histological structures of the salivary glands, and list the functions of saliva.

The functions of the **mouth** and associated structures are to form a receptacle for food, to initiate digestion through mastication, to swallow food, and to form words in speech. The mouth can also assist the respiratory system in the passage of air. The **pharynx,** which is posterior to the mouth, serves as a common passageway for both the respiratory and digestive systems. Both the mouth and pharynx are lined with nonkeratinized, stratified squamous epithelium, which is constantly moistened by the secretion of saliva. The mouth is referred to as the **oral,** or **buccal** *(buk'al),* cavity (fig. 25.9). It is formed by the

pharynx: L. *pharynx,* throat
buccal: L. *bucca,* cheek

Figure 25.9. The superficial structures of the oral cavity.

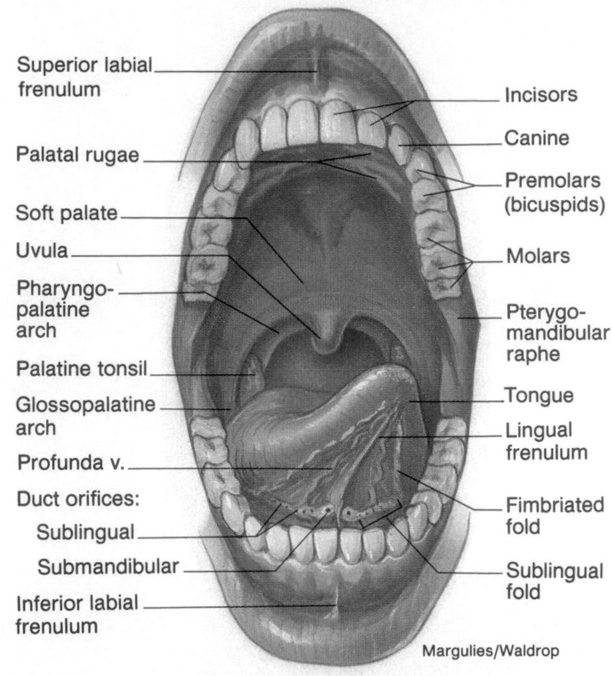

cheeks, lips, hard and **soft palates,** and **tongue.** The **vestibule** of the oral cavity is the depression between the cheeks and lips externally and the gums and teeth internally (fig. 25.10). The opening of the oral cavity is referred to as the **oral orifice,** and the opening between the oral cavity and the pharynx is called the **fauces.**

Cheeks and Lips

The **cheeks** form the lateral walls of the oral cavity and consist of outer layers of skin, subcutaneous fat, facial muscles that assist in manipulating food in the oral cavity, and inner linings of moistened, stratified squamous epithelium. The anterior portion of the cheeks terminates in the superior and inferior lips that surround the oral orifice.

The **lips** are fleshy, highly mobile organs whose principal function in humans is associated with speech. Lips also serve for suckling, manipulating food, and keeping food between the upper and lower teeth. Each lip is attached from its inner surface to the gum by a midline fold of mucous membrane called the **labial frenulum** *(fren'u-lum)* (fig. 25.9). The lips are formed from the orbicularis oris muscle and associated connective tissue, and they are covered with soft, pliable skin. Between the outer skin and the mucous membrane of the oral cavity is a transition zone called the **vermilion.** Lips are red to reddish-brown

fauces: L. *fauces,* throat
vermilion: O.E. *vermeylion,* red colored

Figure 25.10. A sagittal section of the facial region, showing the oral cavity, nasal cavity, and pharynx.

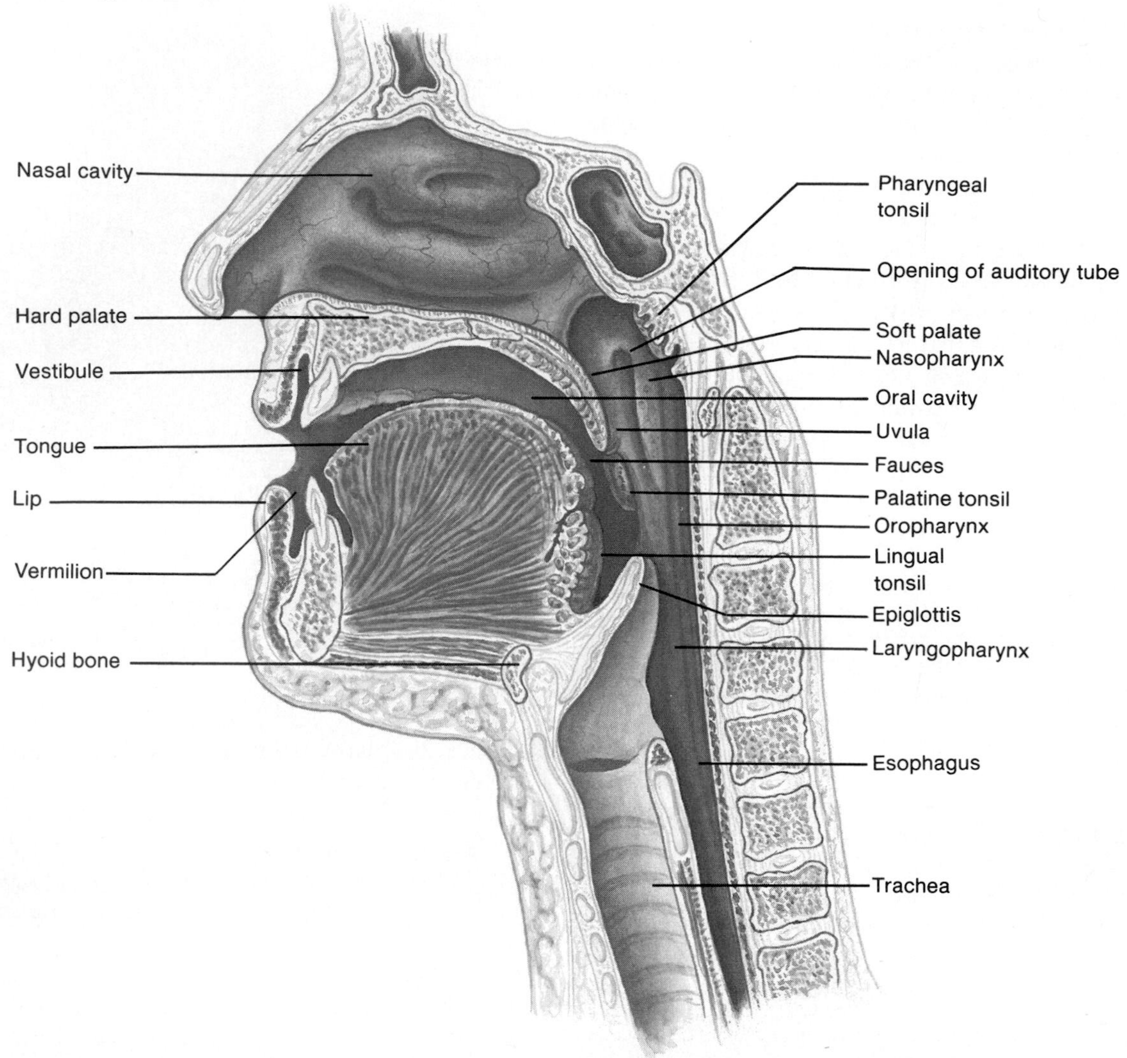

in color because of blood vessels close to the surface. The many sensory receptors in the lips aid in determining the temperature and texture of food.

Suckling is an innate characteristic of newborn infants. The lips are well formed for this activity and even contain blisterlike "milk pads" that aid in suckling. The wide nostrils and receding lower jaw of infants also facilitate suckling.

Tongue

As a digestive organ, the **tongue** functions to move food around in the mouth during mastication and to assist in swallowing food. It contains **taste buds,** which sense various food tastes. The tongue is also essential in producing articulated speech. The tongue is a mass of skeletal muscle covered with a mucous membrane. Extrinsic tongue muscles (those that insert upon the tongue) move the tongue from side to side and in and out. Only the anterior two-thirds of the tongue lie in the oral cavity; the remaining one-third lies in the pharynx (fig. 25.10) and is attached to the hyoid bone. Rounded masses of **lingual tonsils** are located on the dorsal surface of the base of the tongue (fig. 25.11). The undersurface of the tongue is connected along the midline anteriorly to the floor of the mouth by the vertically positioned **lingual frenulum** (fig. 25.9).

When a short lingual frenulum restricts tongue movements, the person is said to be *"tongue-tied."* If this developmental problem is severe, the infant may have difficulty suckling. Older children with this problem may have faulty speech. These functional problems can be easily corrected through surgery.

tonsil: L. *toles,* swelling
frenulum: L. diminutive of *frenum,* bridle

Figure 25.11. The dorsum of the tongue.

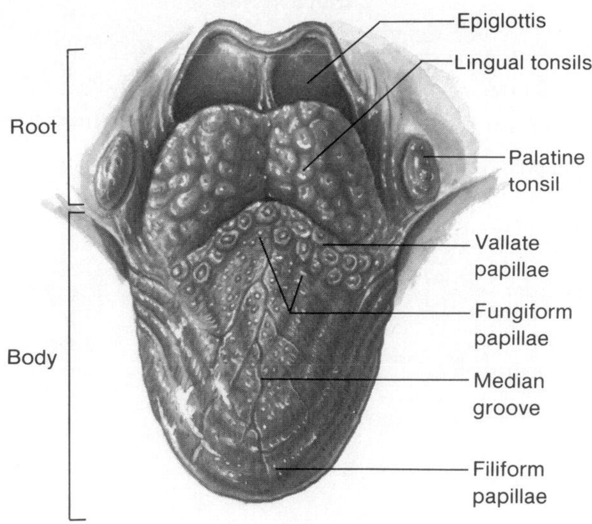

The dorsal surface of the tongue has numerous small elevations called **papillae,** which give the tongue a distinct roughened surface that aids the handling of food. These papillae also contain taste buds that can distinguish sweet, salty, sour, and bitter sensations. Three types of papillae are present on the dorsum of the tongue: **filiform, fungiform,** and **vallate** (**circumvallate**) (fig. 25.11). Filiform papillae are sensitive to touch, have tapered tips, and are by far the most numerous. The larger and rounded fungiform papillae are scattered among the filiform type. The few vallate papillae are arranged in a V-shape on the posterior surface of the tongue.

Palate

The **palate** is the roof of the oral cavity and consists of the bony hard palate anteriorly and the soft palate posteriorly (figs. 25.9, 25.10). The hard palate is formed by the palatine processes of the maxillae and the horizontal plates of the palatine bones and is covered with a mucous membrane. Transverse ridges called **palatal rugae** *(roo'je)* are located along the mucous membrane of the hard palate. These structures serve as friction ridges against which the tongue is placed during swallowing. The soft palate is a muscular arch covered with mucous membrane and is continuous anteriorly with the hard palate. Suspended from the middle lower border of the soft palate is a cone-shaped projection called the **uvula.** During swallowing, the soft palate and uvula are drawn upward, closing the nasopharynx and preventing food and fluid from entering the nasal cavity.

papilla: L. *papula,* swelling or pimple
filiform: L. *filum,* thread; *forma,* form
fungiform: L. *fungus,* fungus; *forma,* form
vallate: L. *vallatus,* surrounded with a rampart
uvula: L. *uvula,* small grapes

Two muscular folds extend downward from both sides of the base of the uvula (fig. 25.9). The anterior fold is called the **glossopalatine arch,** and the posterior fold is the **pharyngopalatine** *(fah-ring''go-pal'ah-tīn)* **arch.** Between these two arches, toward the posterior lateral portion of the oral cavity, is the **palatine tonsil.**

Teeth

Humans, being mammals, have *heterodont dentition,* which means **teeth,** or **dentes** *(den'tēz),* that differ in structure and are adapted to handle food in different ways (fig. 25.12). The four pairs (upper and lower jaws) of anteriormost teeth are the **incisors** *(in-si'zerz).* The chisel-shaped incisors are adapted for cutting and shearing food. The two pairs of cone-shaped **canines,** or **cuspids,** are located at the anterior corners of the mouth and are adapted for holding and tearing. Incisors and canines are further characterized by a single root on each tooth. Located behind the canines are the **premolars,** or **bicuspids,** and **molars.** These teeth have two or three roots and have somewhat rounded, irregular surfaces, called **cusps,** for crushing and grinding food.

Humans are *diphyodont (dif'ĭ-o-dont'');* that is, normally two sets of teeth develop in a person's lifetime. There are twenty **deciduous** (**milk**) **teeth,** which begin to erupt at about six months of age (fig. 25.13 and tables 25.2, 25.3), beginning with the incisors. All of the deciduous teeth have erupted by the age of two and a half. There are thirty-two **permanent teeth,** which replace the deciduous teeth in a predictable sequence. This process begins at about age six and continues until about age seventeen. The **third molars,** or **wisdom teeth,** are the last to erupt. Wisdom teeth are less predictable, and if they do erupt, it is between the ages of seventeen and twenty-five. Because the jaws are formed by this time and other teeth are in place, the eruption of wisdom teeth may cause serious problems of crowding or impaction.

A **dental formula** is a graphic representation of the types, number, and position of teeth in the oral cavity. Most mammals are *heterodont* and have a constant tooth count, so a dental formula can be written for each species of mammal that has heterodontia. Following are the deciduous and permanent dental formulae for humans:

Formula for deciduous dentition:
I 2/2, **C** 1/1, **DM** 2/2 = 10 × 2 = 20 teeth

Formula for permanent dentition:
I 2/2, **C** 1/1, **P** 2/2, **M** 3/3 = 16 × 2 = 32 teeth

(**I** = incisor; **C** = canine; **P** = premolar;
DM = deciduous molar; **M** = molar)

incisor: L. *incidere,* to cut
canine: L. *canis,* dog
molar: L. *mola,* millstone
deciduous: L. *deciduus,* to fall away
heterodont: Gk. *heteros,* other; *odous,* tooth

Figure 25.12. Dentitions and the sequence of eruptions. (*a*) deciduous teeth; (*b*) permanent teeth; (*c*) deciduous and permanent teeth viewed in a panoramic X ray of a female age ten years and four months.

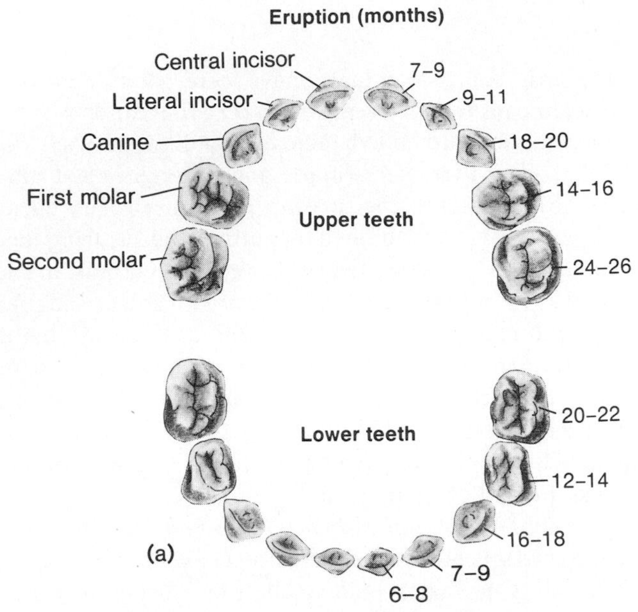

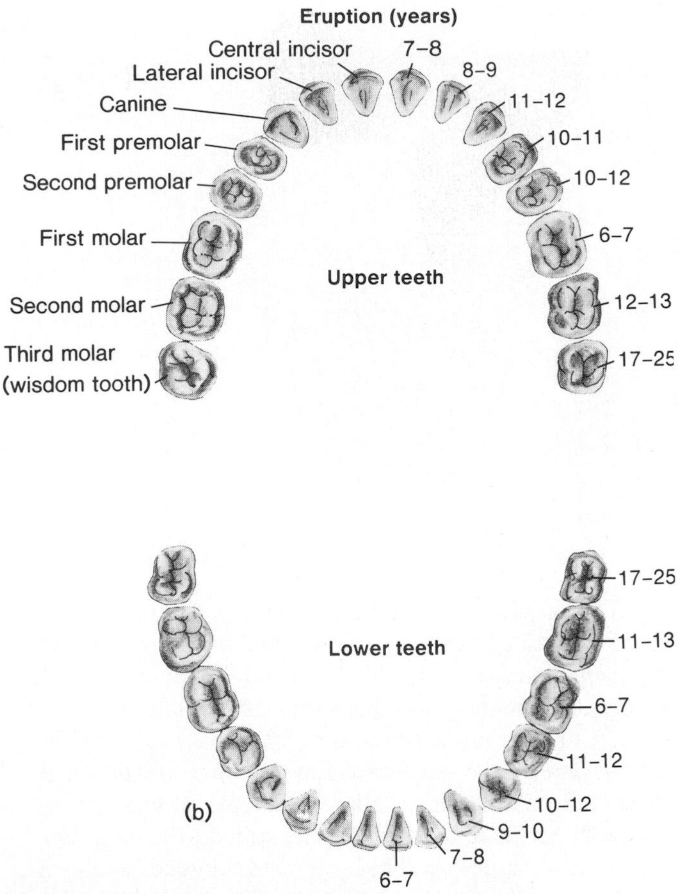

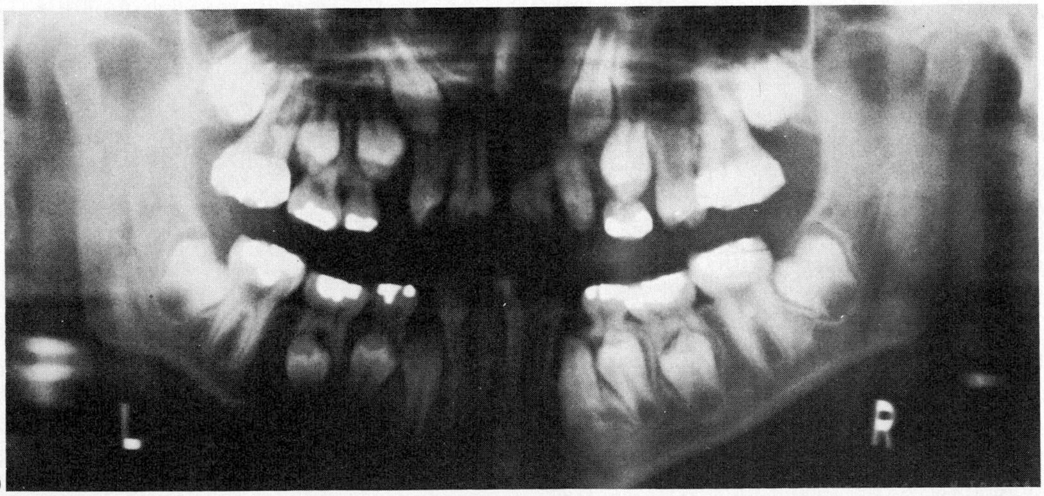

The cusps of the upper and lower premolar and molar teeth occlude for chewing food in a process called *mastication,* whereas the upper incisors normally form an overbite with the incisors of the lower jaw. An overbite of the upper incisors creates a shearing action as these teeth slide past one another. Masticated food is mixed with saliva, which initiates chemical digestion and facilitates swallowing. The soft, flexible mass of food that is swallowed is called a *bolus.*

A tooth consists of an exposed **crown,** supported by a **neck,** anchored firmly into the jaw by one or more **roots** (fig. 25.14). The roots of teeth fit into sockets, called **alveoli,** in the alveolar processes of the mandible and maxillae. Each socket is lined with a connective tissue

bolus: Gk. *bolos,* lump

Figure 25.13. The skull of a youth (nine to twelve years) showing the eruption of teeth.

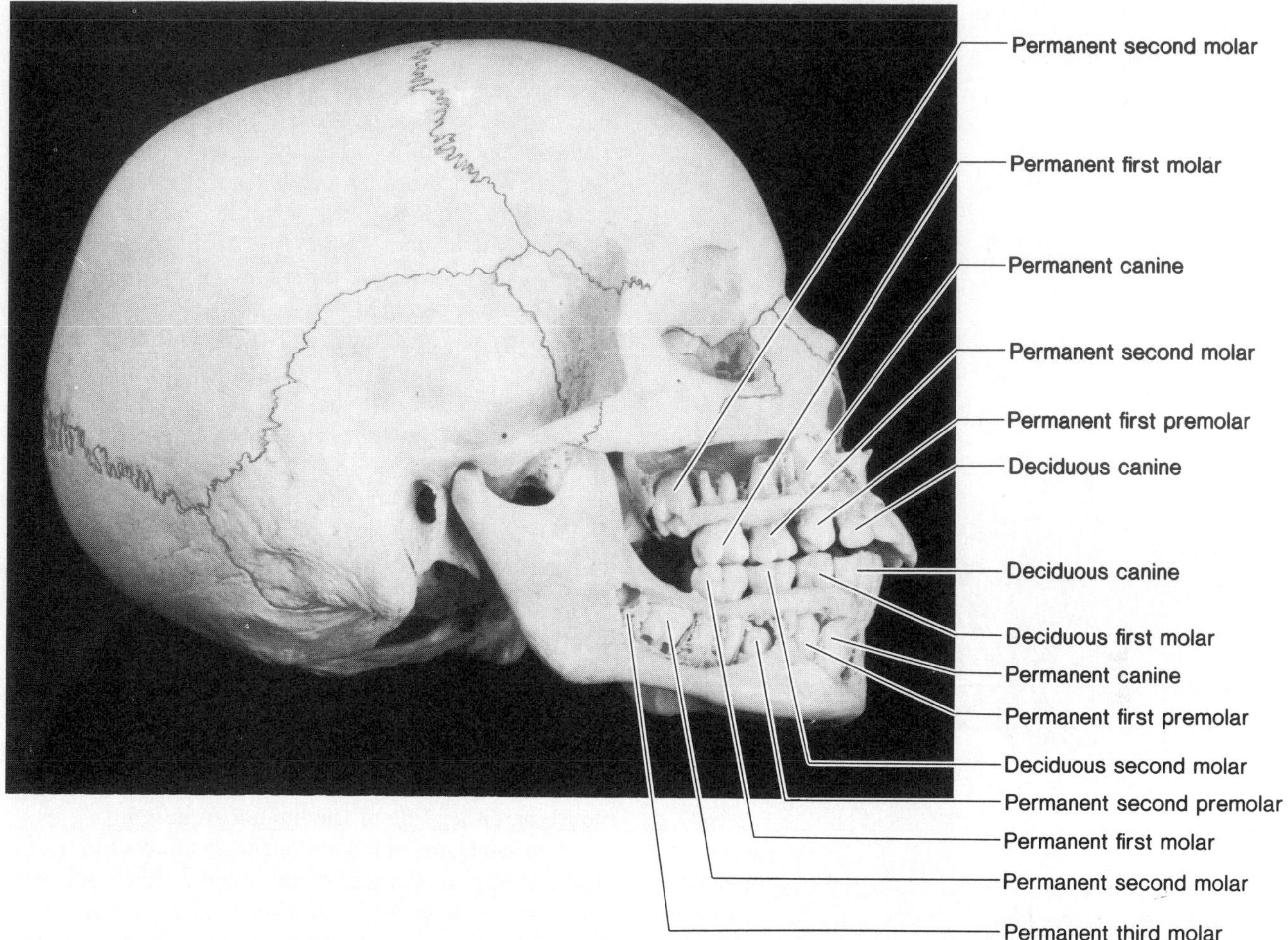

Table 25.2 Eruption sequence and loss of deciduous teeth

Type of tooth	Average age of eruption: Lower	Average age of eruption: Upper	Average age of loss
Central incisors	6–8 mos	7–9 mos	7 yrs
Lateral incisors	7–9 mos	9–11 mos	8 yrs
First molars	12–14 mos	14–16 mos	10 yrs
Canines (cuspids)	16–18 mos	18–20 mos	10 yrs
Second molars	20–22 mos	24–26 mos	11–12 yrs

From Kent M. Van De Graaff, *Human Anatomy,* 2d ed. Copyright © 1988 Wm. C. Brown Publishers, Dubuque, Iowa. All Rights Reserved. Reprinted by permission.

Table 25.3 Eruption sequence of permanent teeth

Type of tooth	Average age of eruption: Lower	Average age of eruption: Upper
First molars	6–7 yrs	6–7 yrs
Central incisors	6–7 yrs	7–8 yrs
Lateral incisors	7–8 yrs	8–9 yrs
Canines (cuspids)	9–10 yrs	11–12 yrs
First premolars (bicuspids)	10–12 yrs	10–11 yrs
Second premolars (bicuspids)	11–12 yrs	10–12 yrs
Second molars	11–13 yrs	12–13 yrs
Third molars (wisdom)	17–25 yrs	17–25 yrs

From Kent M. Van De Graaff, *Human Anatomy,* 2d ed. Copyright © 1988 Wm. C. Brown Publishers, Dubuque, Iowa. All Rights Reserved. Reprinted by permission.

periosteum, specifically called the **periodontal membrane.** The root of a tooth is covered with a bonelike material called the **cementum;** fibers in the periodontal membrane insert into the cementum and fasten the tooth in its socket. The **gingiva** *(jin-jī'vah),* or **gum,** is the mucous membrane surrounding the alveolar processes in the oral cavity.

The bulk of a tooth consists of **dentin** *(den'tin),* a substance similar to bone but harder. Covering the dentin on the outside and forming the crown is a tough, durable layer of **enamel.** Enamel is composed primarily of calcium phosphate and is the hardest substance in the body. The central region of the tooth contains the **pulp cavity.** The pulp

gingiva: L. *gingiva,* gum

dentin: L. *dens,* tooth

Figure 25.14. The structure of a tooth shown in a vertical section through a canine tooth.

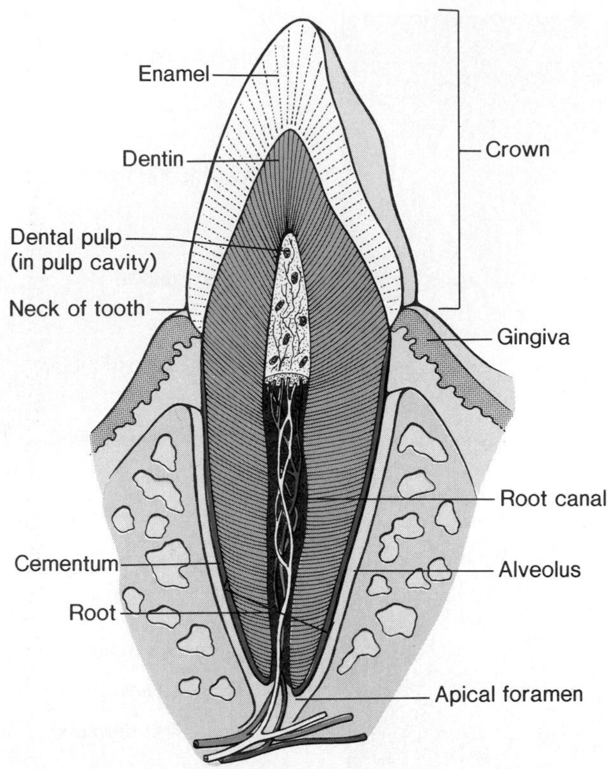

cavity contains the **pulp,** which is composed of connective tissue with blood vessels, lymph vessels, and nerves. A **root canal** is continuous with the pulp cavity through the root to an opening at the base called the **apical foramen.** The tooth receives nourishment through vessels traversing the apical foramen. Proper nourishment is particularly important during embryonic development. The diet of the mother should contain an abundance of calcium and vitamin D during pregnancy to insure the proper development of the baby's teeth.

Enamel is the hardest substance in the body, but it can be dissolved by bacterial activity that results in *dental caries (cavities).* These caries must be artificially filled because new enamel is not produced after a tooth erupts. The rate of tooth decay decreases after age thirty-five, but then periodontal diseases may develop. *Periodontal diseases* result from plaque or tartar buildup at the gum line, which wedges the gum away from the teeth, allowing bacterial infections to develop.

Salivary Glands

Salivary glands are accessory digestive glands that produce a fluid secretion called *saliva.* Saliva functions as a solvent in cleansing the teeth and dissolving food chemicals so they can be tasted. Saliva also contains enzymes, which digest starch, and mucus, which lubricates the pharynx to facilitate swallowing. Saliva is secreted continuously but usually only in sufficient amounts to keep the mucous membranes of the oral cavity moist. Numerous minor salivary glands, called **buccal glands,** are located in the mucous membranes of the palatal region of the oral cavity. But most of the saliva is produced by three pairs of salivary glands outside of the oral cavity and is transported to the mouth via **salivary ducts.** The three major pairs of salivary glands are the parotid, submandibular, and sublingual (fig. 25.15).

The **parotid** *(pah-rot'id)* **gland** is the largest and is positioned below and in front of the ear, between the skin and the masseter muscle. The **parotid (Stensen's) duct** parallels the zygomatic arch across the masseter muscle, pierces the buccinator muscle, and drains into the oral cavity opposite the second upper molar. It is the parotid gland that becomes infected and swollen with the mumps.

The **submandibular** *(sub''man-dib'u-lar)* **(submaxillary) gland** is inferior to the body of the mandible about midway along the inner side of the jaw. This gland is covered by the more superficial mylohyoid muscle. The **submandibular (Wharton's) duct** empties into the floor of the mouth on either side of the lingual frenulum.

The **sublingual gland** lies under the mucosa in the floor of the mouth on the side of the tongue. Each sublingual gland possesses several small ducts that empty into the floor of the mouth in an area posterior to the papilla of the submandibular duct.

Two types of secretory cells, called **serous** and **mucous cells,** are found in all salivary glands in various proportions (fig. 25.16). Serous cells produce a watery fluid containing digestive enzymes; mucous cells secrete a thick, stringy mucus. Cuboidal epithelial cells line the lumina of the salivary ducts.

The salivary glands are innervated by both divisions of the autonomic nervous system. Sympathetic impulses stimulate the secretion of small amounts of viscous saliva.

parotid: Gk. *para,* beside; *otos,* ear
Stensen's duct: from Nicholaus Stensen, Danish anatomist, 1638–86
Wharton's duct: from Thomas Wharton, English anatomist, 1614–73

Figure 25.15. The salivary glands.

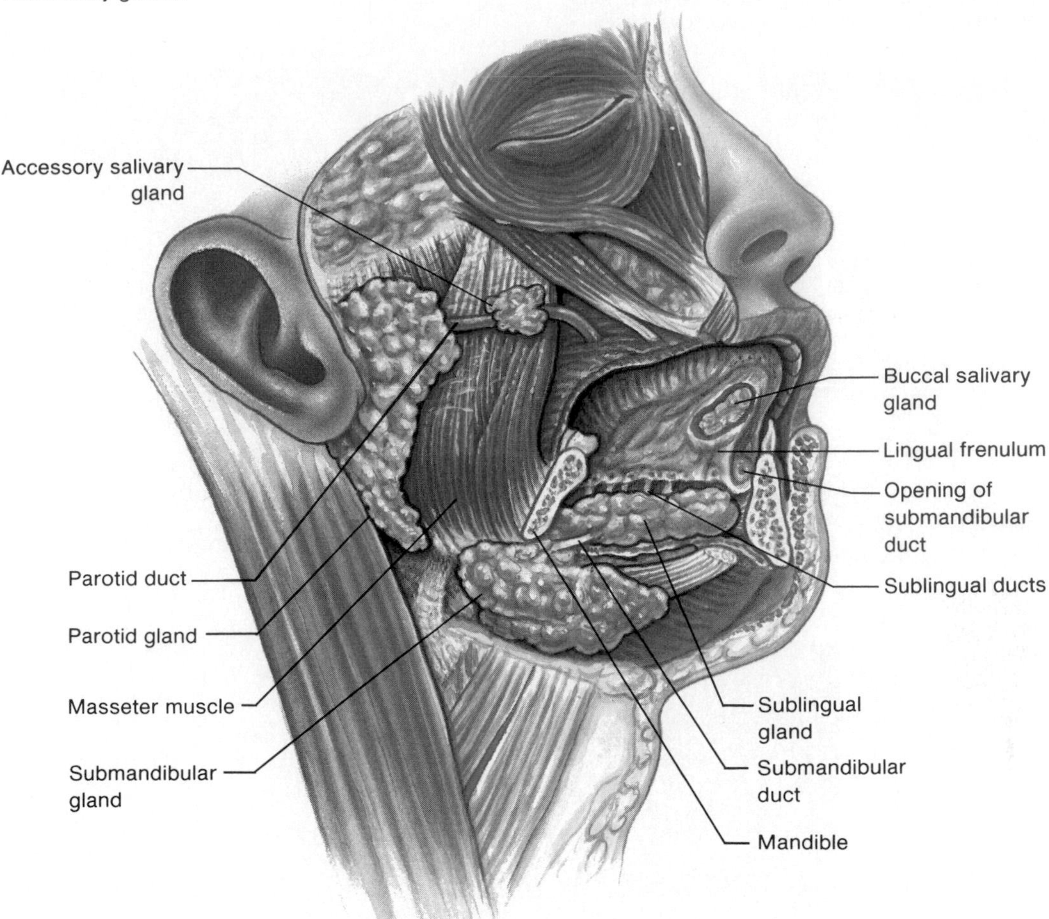

Parasympathetic stimulation causes the secretion of large volumes of watery saliva. Physiological responses of this type occur whenever a person sees, smells, tastes, or even thinks about desirable food. The amount of saliva secreted daily ranges from 1,000 to 1,500 ml. Information about the salivary glands is summarized in table 25.4.

Pharynx

The funnel-shaped **pharynx** *(far'inks)* is a passageway approximately 13 cm (5 in.) in length connecting the oral and nasal cavities to the esophagus and trachea. The pharynx has both digestive and respiratory functions. The supporting walls of the pharynx are composed of skeletal muscle, and the lumen is lined with a mucous membrane of stratified squamous epithelium.

The external, *circular layer* of pharyngeal muscles, called **constrictors** (fig. 25.17), serves to compress the lumen of the pharynx involuntarily during swallowing. The **superior constrictor muscle** attaches to bony processes of the skull and mandible and encircles the upper portion of the pharynx. The **middle constrictor muscle** arises from the hyoid bone and stylohyoid ligament and encircles the middle portion of the pharynx. The **inferior constrictor muscle** arises from the cartilages of the larynx and encircles the lower portion of the pharynx. During breathing, the lower portion of the inferior constrictor muscle is contracted, preventing air from entering the esophagus.

The motor and most of the sensory innervation to the pharynx is via the **pharyngeal plexus,** situated chiefly on the middle constrictor muscle. It is formed by the pharyngeal branches of the glossopharyngeal (ninth) and the vagus (tenth) cranial nerves, together with a deep sympathetic branch from the superior cervical ganglion.

The pharynx is served by the ascending pharyngeal and inferior thyroid arteries, both of which are branches of the external carotid arteries. Venous return is through the internal jugular veins.

Figure 25.16. The histology of the salivary glands. (*a*) the parotid gland; (*b*) the submandibular gland; and (*c*) the sublingual gland.

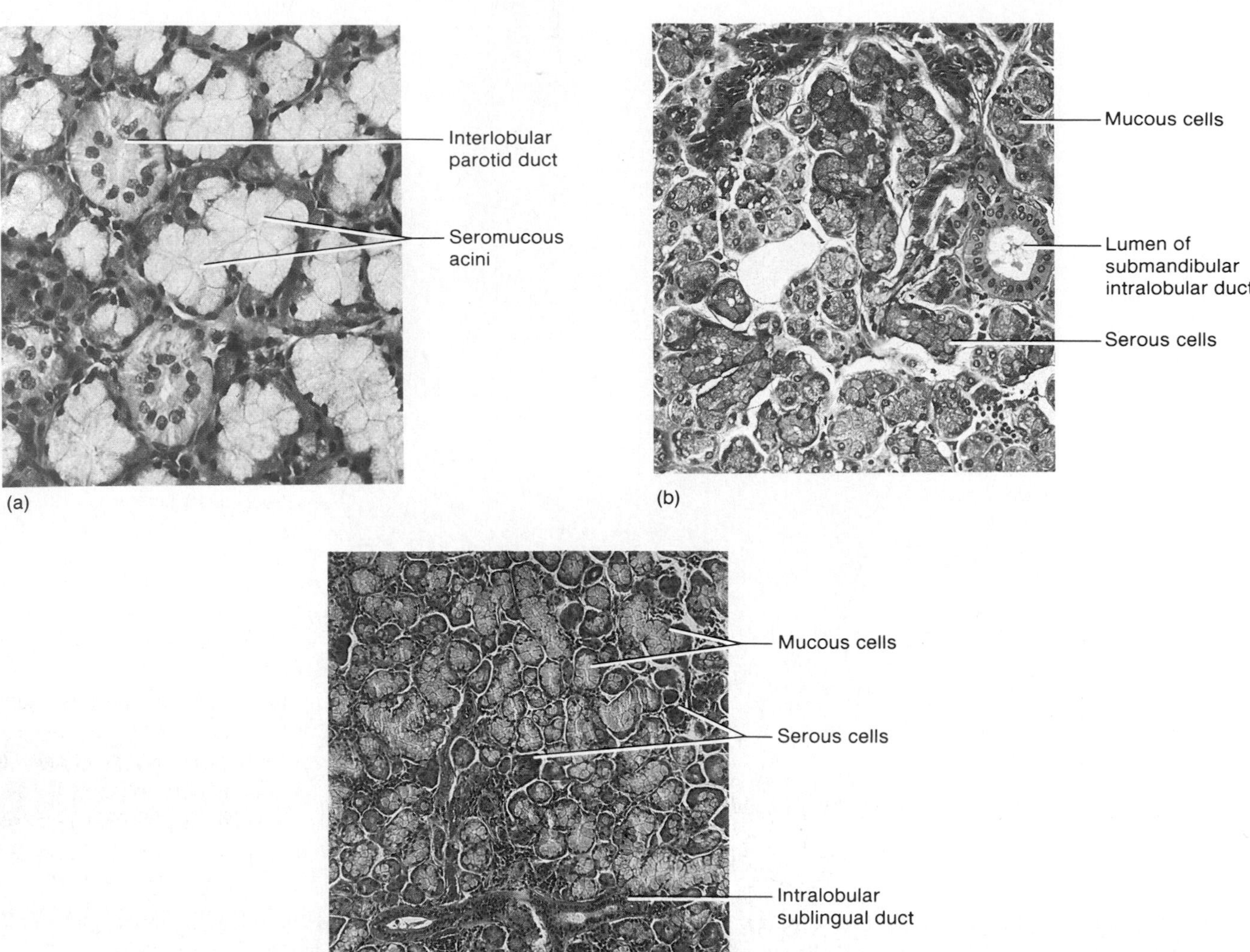

Table 25.4 Major salivary glands

Gland	Location	Duct	Entry into oral cavity	Type of secretion
Parotid	Anterior and inferior to auricle; subcutaneous over masseter muscle	Parotid (Stensen's duct)	Lateral to upper second molar	Watery serous fluid, salts, and enzyme
Submandibular	Inferior to the base of the tongue	Submandibular (Wharton's duct)	Papilla lateral to lingual frenulum	Watery serous fluid with some mucus
Sublingual	Anterior to submandibular; under tongue	Several small ducts (Rivinus's ducts)	Ducts along the base of the tongue	Mostly thick, stringy mucus, salts, and enzyme

Figure 25.17. A posterior view of the constrictor muscles of the pharynx. The right side has been cut away to illustrate the interior structures in the pharynx.

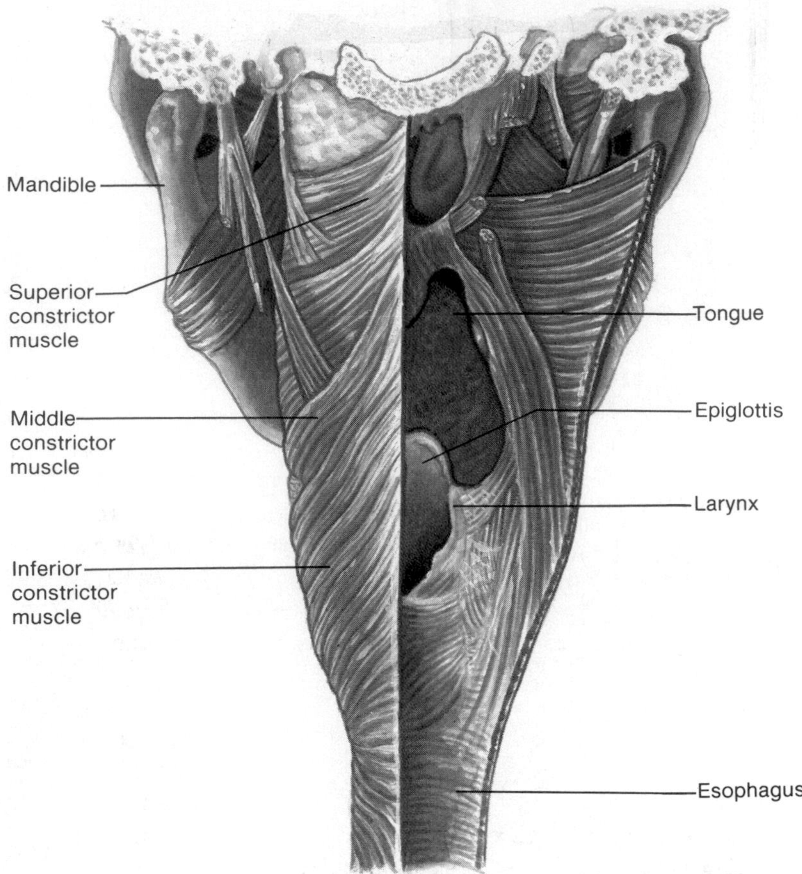

1. Define the terms *diphyodont* and *heterodont.* Which of the four kinds of teeth is absent in deciduous dentition?
2. Identify the locations of the enamel, dentin, cementum, and pulp of a tooth. Explain how a tooth is anchored into its socket.
3. Identify the location of the parotid and the submandibular ducts, and describe where they empty in the oral cavity.
4. Describe the effects of autonomic stimulation on salivary secretion.

Esophagus and Stomach

A bolus of food is passed from the esophagus to the stomach, where it is mixed with gastric secretions. The chyme thus produced is sent past the pyloric sphincter to the duodenum.

Objective 10. Describe the steps in deglutition.

Objective 11. Describe the location, gross structure, and functions of the stomach.

Objective 12. Describe the histological structure of the esophagus and stomach. List the cell types in the gastric mucosa and their secretory products.

Objective 13. Describe the functions of hydrochloric acid and pepsin.

Objective 14. Explain how gastric and duodenal peptic ulcers might be produced.

Esophagus

The **esophagus** *(ĕ-sof'ah-gus)* is that portion of the GI tract that connects the pharynx to the stomach. It is a collapsible muscular tube approximately 25 cm (10 in.) long, originating at the larynx and located posterior to the trachea.

The esophagus is located within the mediastinum of the thorax and passes through the diaphragm just above the opening into the stomach. The opening through the diaphragm is called the **esophageal hiatus** *(ĕ-sof''ah-je'al hi-a'tus).* The esophagus is lined with a nonkeratinized stratified squamous epithelium; its walls contain either skeletal or smooth muscle, depending on the location. The upper third of the esophagus contains skeletal muscle; the middle third contains a mixture of skeletal and smooth muscle, and the terminal portion contains only smooth muscle (fig. 25.18).

esophagus: Gk. *oisein,* to carry; *phagema,* food

Figure 25.18. The microscopic appearance of the esophagus.

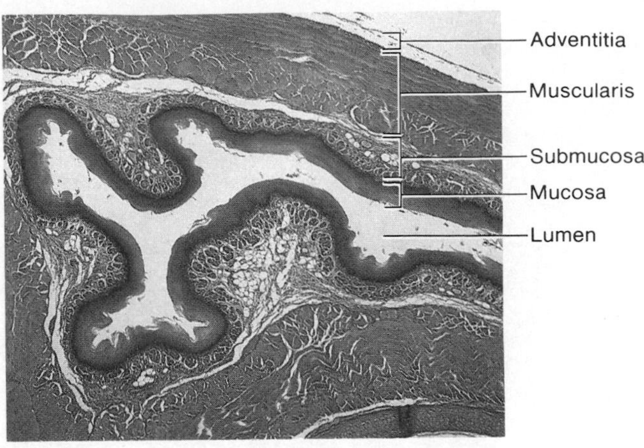

Figure 25.19. The esophagus. (*a-d*) the deglutition reflex and peristalsis of the esophagus, and (*e*) an anteroposterior X ray. (*a*) a bolus is formed as the tongue is elevated against the palate. (*b*) the bolus passes through the oropharynx as the nasopharynx is sealed by the elevated uvula. (*c*) the bolus enters the esophagus as the glottis is sealed against the epiglottis. (*d*) a peristaltic wave moves the bolus through the esophagus to the stomach.

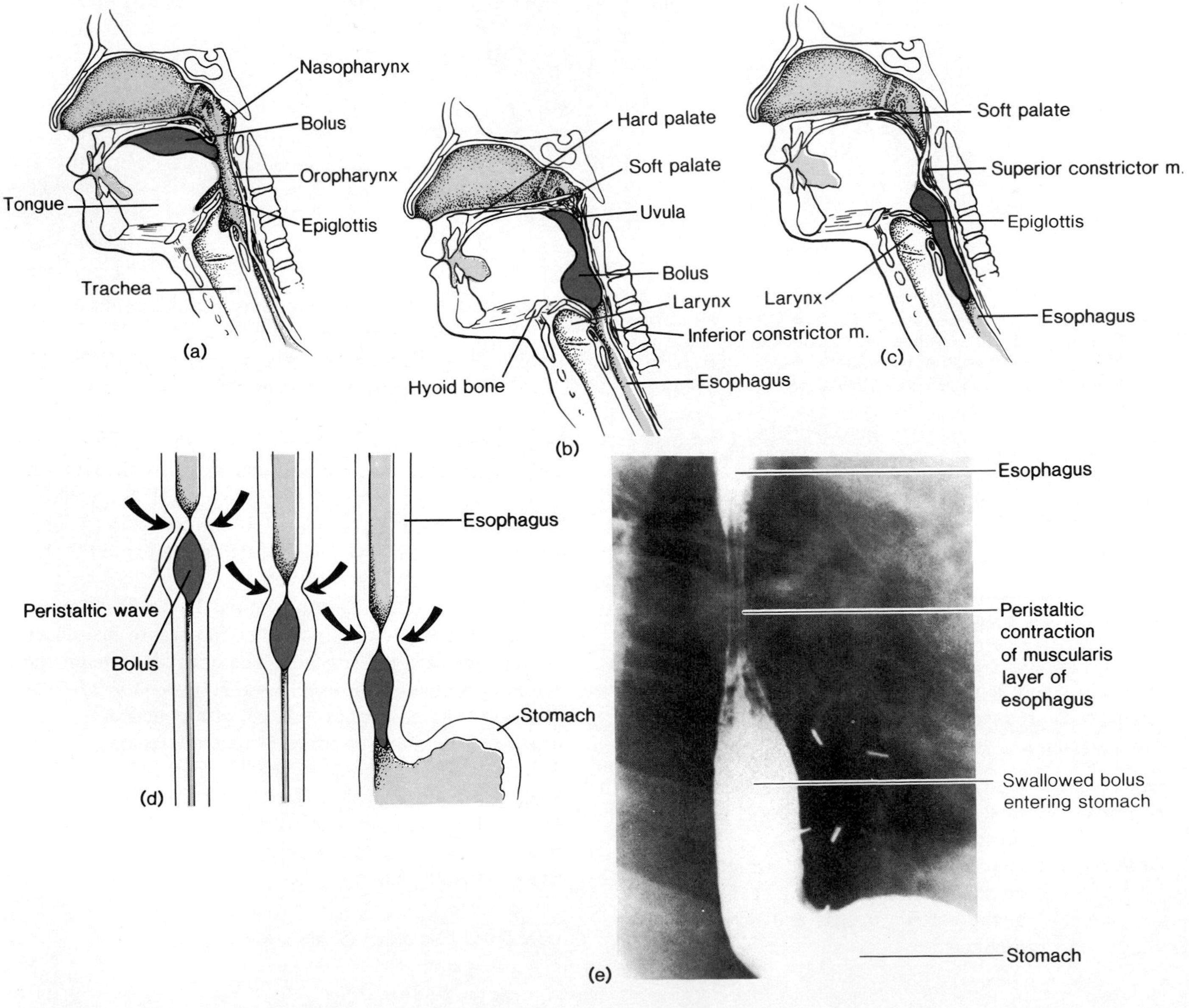

The lumen of the terminal portion of the esophagus is slightly narrowed because of a thickening of the circular muscle fibers in its wall. This portion is referred to as the **lower esophageal (gastroesophageal) sphincter.** The muscle fibers of this region constrict after food passes into the stomach to help prevent the stomach contents from regurgitating into the esophagus. Regurgitation would occur because the pressure in the abdominal cavity is greater than the pressure in the thoracic cavity as a result of respiratory movements. The lower esophageal sphincter must remain closed, therefore, until food is pushed through it by peristalsis into the stomach.

The lower esophageal sphincter is not a true sphincter muscle that can be identified histologically, and it does at times permit the acidic contents of the stomach to enter the esophagus. This can create a burning sensation commonly called *heartburn,* although the heart is not involved. Infants under one year old have difficulty controlling their lower esophageal sphincter and thus often "spit up" following meals. Certain mammals, such as rodents, have a true lower gastroesophageal sphincter and cannot regurgitate, which is why poison grains can kill mice and rats effectively.

Mechanism of Swallowing

Swallowing, or **deglutition** *(deg''loo-tish'un),* is the complex mechanical and physiological act of moving food or fluid from the oral cavity to the stomach. For descriptive purposes, deglutition of solid food is divided into three stages (fig. 25.19).

The first stage is voluntary and follows mastication, if food is involved. During this stage, the mouth is closed and breathing is temporarily interrupted. A bolus is formed as the tongue is elevated against the palate through contraction of the mylohyoid and styloglossus muscles and the intrinsic muscles of the tongue.

The second stage of deglutition is the passage of the bolus through the pharynx. The events of this stage are involuntary and are elicited by stimulation of sensory receptors located at the opening of the oral pharynx. Pressure of the tongue against the rugae of the hard palate seals the pharynx from the oral cavity, creates a pressure, and forces the bolus into the oral pharynx. The soft palate and pendulant uvula are elevated to close the nasal pharynx as the bolus passes. The hyoid bone and the larynx are also elevated. Elevation of the larynx against the epiglottis seals the glottis so that food or fluid is less likely to enter the trachea. Sequential contraction of the constrictor muscles of the pharynx moves the bolus through the pharynx to the esophagus. This stage is completed in one second or less.

The third stage, the entry and passage of food through the esophagus, is also involuntary. The bolus is moved through the esophagus by peristalsis. The entire time for deglutition varies, but it is slightly more than one second in the case of fluids and five to eight seconds with solid food material.

Stomach

The **stomach** is the most distensible part of the GI tract and is positioned in the upper left quadrant of the peritoneal cavity immediately below the diaphragm. It is a J-shaped pouch that is continuous with the esophagus superiorly and empties into the duodenal portion of the small intestine inferiorly. The functions of the stomach are to store food as it is mechanically churned with gastric secretions; to initiate the digestion of proteins; to carry on limited absorption; and to move food into the small intestine as a pasty material called **chyme** *(kīm).*

The stomach (figs. 25.20, 25.21), can be divided into four regions: the cardia, fundus, body, and pyloris. The **cardia** is the upper, narrow region immediately below the lower gastroesophageal sphincter. The **fundus** is the dome-shaped portion to the left of the cardia and in direct contact with the diaphragm. The **body** is the large central portion, and the **pylorus** is the funnel-shaped terminal portion. The pyloris communicates with the duodenal portion of the small intestine through a **sphincter.** *Pylorus* is a Greek word meaning "gatekeeper," and this junction is just that, regulating the movement of chyme into the small intestine and prohibiting backflow.

The stomach has two surfaces and two borders. The broadly rounded surfaces are referred to as the **anterior** and **posterior surfaces.** The medial concave border is the **lesser curvature** (fig. 25.20), and the lateral convex border is the **greater curvature.** The lesser omentum extends between the lesser curvature and the liver, and the greater omentum is attached to the greater curvature.

The wall of the stomach consists of the same four layers found in other regions of the GI tract, with certain modifications. The muscularis is composed of three layers of smooth muscle named according to the direction of fiber arrangement: an outer **longitudinal layer,** a middle **circular layer,** and an inner **oblique layer.** The circular muscle layer is further thickened at the gastroduodenal junction to form the pyloric sphincter.

chyme: L. *chymus,* juice
cardia: Gk. *kardia,* heart (upper portion, nearer the heart)
fundus: L. *fundus,* bottom
pylorus: Gk. *pyloros,* gatekeeper

Figure 25.20. The major regions and structures of the stomach.

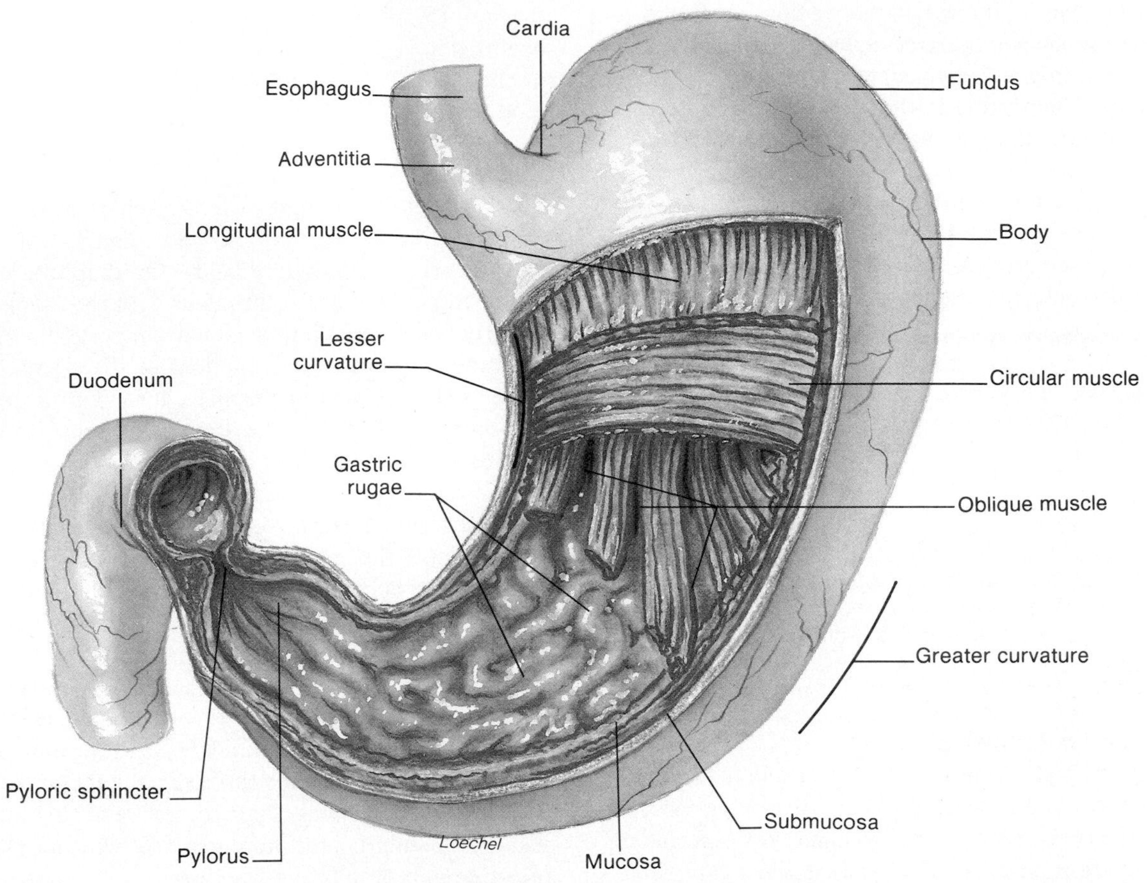

Figure 25.21. An X ray of a stomach.

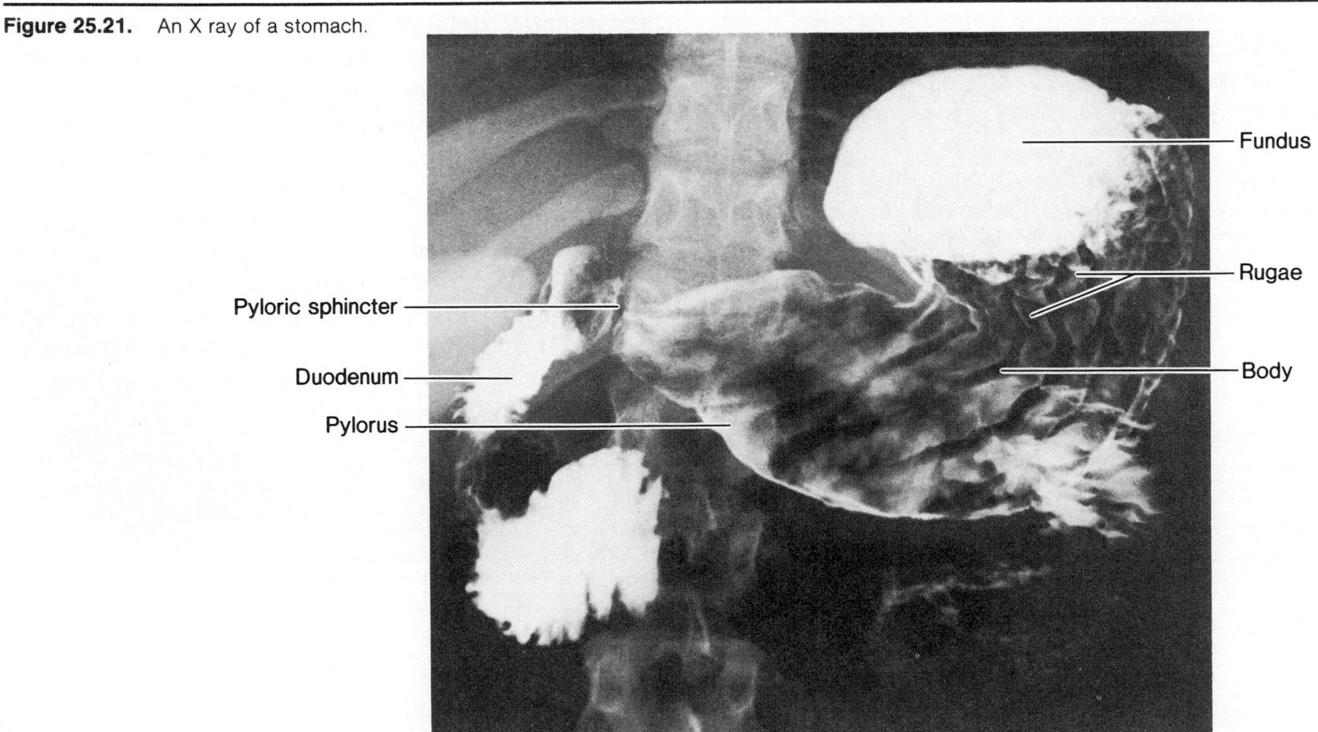

Figure 25.22. Microscopic structures of the mucosa of the stomach.

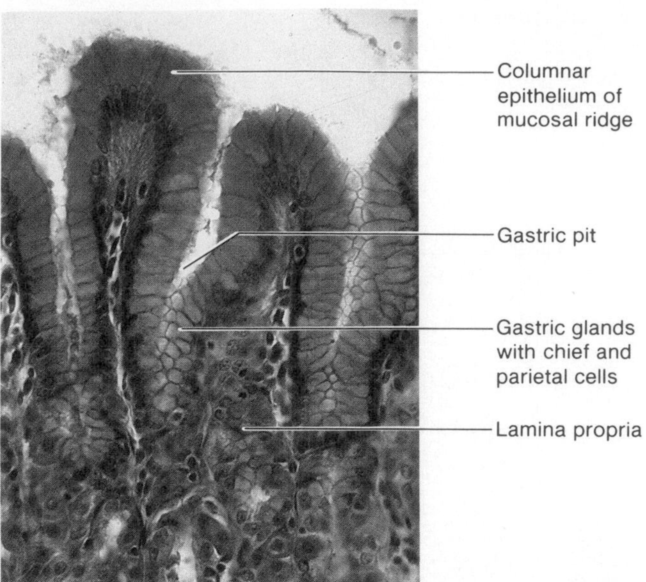

The mucosa is shaped into numerous longitudinal folds, called **gastric rugae,** which permit stomach distension. The gastric rugae gradually smooth out as the stomach fills. The mucosal layer contains a simple columnar epithelium. The entire mucosal layer is folded to form **gastric pits** and **gastric glands** (figs. 25.22 and 25.23).

There are several different types of cells in the gastric glands that secrete different products: (1) **goblet cells,** which secrete mucus; (2) **parietal cells,** which secrete hydrochloric acid; (3) **chief cells,** which secrete pepsinogen, an inactive form of the protein-digesting enzyme pepsin; (4) **argentaffen** *(ar-jen'tah-fin)* **cells,** which secrete serotonin and histamine; and (5) **G cells,** which secrete the hormone gastrin. In addition to these products (table 25.5), the gastric mucosa (probably the parietal cells) secretes a polypeptide called *intrinsic factor,* which is required for absorption of vitamin B_{12} in the intestine.

argentaffen cells: L. *argentum,* silver; *affinis,* attraction (become colored with silver stain)

Figure 25.23. Gastric pits and gastric glands of the mucosa. (*a*) gastric pits are the openings of the gastric glands. (*b*) gastric glands consist of mucous cells, chief cells, and parietal cells, each of which produces a specific secretion.

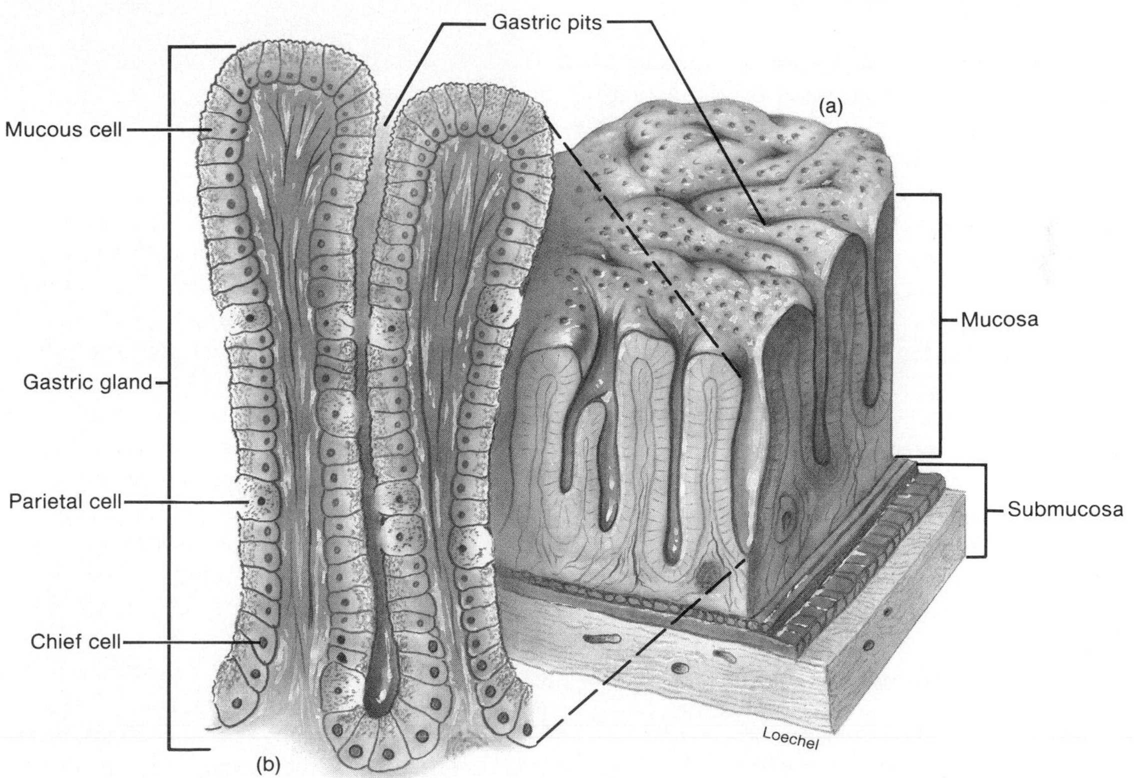

Table 25.5 Secretions of the fundus and pyloric regions of the stomach

Stomach region	Cell type	Secretions
Fundus	Parietal cells	Hydrochloric acid; intrinsic factor
	Chief cells	Pepsinogen
	Goblet cells	Mucus
	Argentaffen cells	Histamine, serotonin
Pyloric	G cells	Gastrin
	Chief cells	Pepsinogen
	Goblet cells	Mucus

From Stuart Ira Fox, *Human Physiology,* 2d ed. Copyright © 1987 Wm. C. Brown Publishers, Dubuque, Iowa. Reprinted by permission.

Pepsin and Hydrochloric Acid. The secretion of hydrochloric acid by parietal cells makes gastric juice very acidic, with a pH less than 2. This strong acidity serves three functions: (1) ingested proteins are denatured at low pH—that is, their tertiary structure is altered so that they become more digestible; (2) under acidic conditions, weak pepsinogen enzymes partially digest each other—this frees the active pepsin enzyme as small peptide fragments are removed (fig. 25.24); and (3) pepsin is more active under acidic conditions—it has a pH optimum of about 2.0. The peptide bonds of ingested protein are broken (through hydrolysis reactions) by pepsin under acidic conditions; the HCl itself does not directly digest proteins.

Digestion and Absorption in the Stomach. Proteins are only partially digested in the stomach by the action of pepsin. Fats are also partially digested, through the action of a lipase secreted by the lingual salivary glands and activated by low pH in the stomach. Carbohydrates are not digested at all in the stomach. The complete digestion of food molecules occurs later, when chyme enters the small intestine. Patients with partial gastric resections, therefore, and even those with complete gastrectomies (removal of the stomach), can still adequately digest and absorb their food.

Almost all of the products of digestion are absorbed through the wall of the intestine; the only commonly ingested substances that can be absorbed across the stomach wall are alcohol and aspirin. This occurs as a result of the lipid solubility of these molecules. The passage of aspirin through the gastric mucosa has been shown to cause bleeding, which may be significant if large amounts of aspirin are taken.

The only function of the stomach that appears to be essential for life is the secretion of *intrinsic factor.* This polypeptide is needed for the intestinal absorption of vitamin B_{12}, required for maturation of red blood cells in the bone marrow. A patient with a gastrectomy has to receive B_{12} orally (together with intrinsic factor) or through injections so that he or she will not develop *pernicious anemia.*

Figure 25.24. The gastric mucosa secretes the inactive enzyme pepsinogen and hydrochloric acid (HCl). In the presence of HCl, the active enzyme pepsin is produced. Pepsin digests proteins into shorter polypeptides.

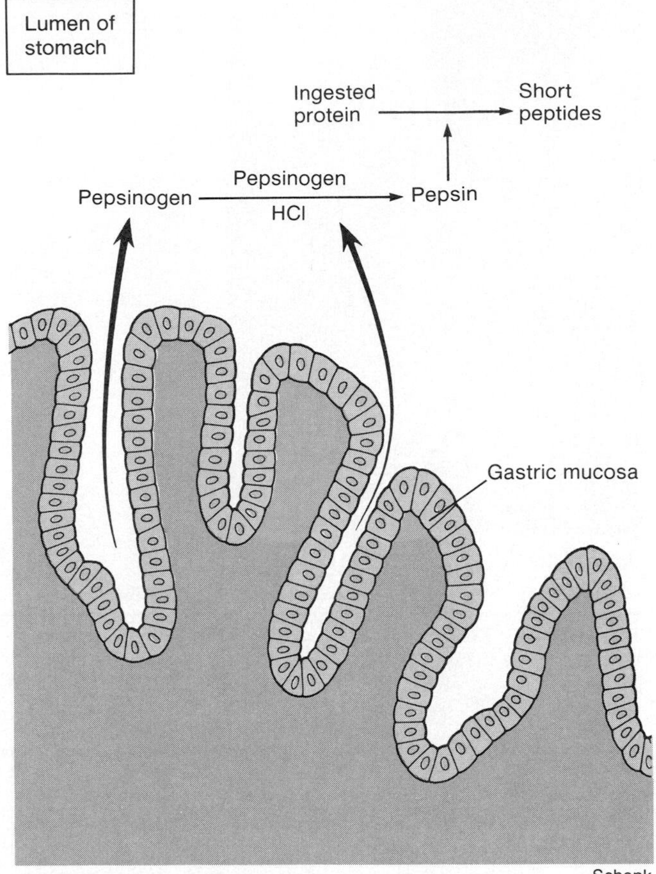

Gastritis and Peptic Ulcers. The gastric mucosa is normally resistant to the harsh conditions of low pH and pepsin activity in gastric juice. The reasons for its resistance to pepsin digestion are not well understood. Its resistance to hydrochloric acid appears to be due to three interrelated mechanisms: (1) the stomach lining is covered with a thin layer of alkaline mucus, which may offer some protection; (2) the epithelial cells of the mucosa are joined together by tight junctions, which prevent acid from leaking into the submucosa; and (3) the epithelial cells that are damaged are exfoliated (shed) and replaced by new cells. This latter process results in the loss of about one-half million cells a minute, so that the entire epithelial lining is replaced every three days.

These three mechanisms provide a barrier that prevents self-digestion of the stomach. Breakdown of the gastric mucosa barrier may occur at times, however, perhaps as a result of the detergent action of bile salts that are

Table 25.6 Gastric secretion of hydrochloric acid, measured in milliequivalents (mEq) (one milliequivalent is equal to one millimole of H^+)

Condition	Day (mEq/hr)	Night (mEq/12 hrs)	Maximal (mEq/hr)
Normal	2–3	18	16–20
Gastric ulcer	2–4	8	16–20
Duodenal ulcer	4–10	60	25–40
Zollinger-Ellison syndrome (G cell tumor)	30	120	45

regurgitated from the small intestine through the pyloric sphincter. When this occurs, acid can leak through the mucosal barrier to the submucosa, which causes direct damage and stimulates inflammation. The histamine released from mast cells during inflammation may stimulate further acid secretion and result in further damage to the mucosa. The inflammation that occurs during these events is called *acute gastritis.*

Once the mucosal barrier is broken, the acid can produce craterlike holes and even complete perforations of the stomach wall. These are called gastric ulcers. More commonly encountered, however, are ulcers of the first 3–4 cm of the duodenum. Ulcers of the stomach and duodenum are known as *peptic ulcers.* It has been estimated that about 10% of American men and 4% of American women will develop peptic ulcers.

Since 1983, evidence has been accumulating which suggests that acute gastritis, and perhaps peptic ulcers, may be the result of a bacterial infection. The bacterium, *Campylobacter pylori,* appears to penetrate between cells in the gastric and duodenal mucosa and has been found in 90% of patients with duodenal ulcers and 70% of patients with gastric ulcers. The association of this bacterium with the presence of ulcers does not prove that the ulcers were caused by the bacterium, but it does suggest that this might be the case. Further, it has been shown that treating ulcer patients with antibacterial drugs appears to reduce the chances of a relapse occurring.

The duodenum is normally protected against gastric acid by the buffering action of bicarbonate in alkaline pancreatic juice, as well as by other mechanisms (including secretion of bicarbonate by Brunner's glands in the duodenum). The presence of acid is required for the development of duodenal ulcers, and many people who develop duodenal ulcers produce excessive amounts of gastric acid (table 25.6), which cannot be neutralized by the bicarbonate.

The secretion of gastric acid is stimulated by *acetylcholine* (from parasympathetic nerve endings), *gastrin* (a hormone secreted by the stomach), and *histamine* (secreted by mast cells in the lamina propria). In response to these stimulants, the parietal cells secrete H^+ into the gastric lumen by active transport against a million-to-one concentration gradient, using a carrier that functions as a *H^+/K^+ ATPase pump.* People with ulcers, consequently, can be treated with drugs that block ACh (atropine; pirenzepine) or gastrin (proglumide) or histamine (cimetidine) action. Experimental drugs that may be important in the future treatment of ulcers include one that specifically blocks the H^+/K^+ pump (omeprazole), and various prostaglandin analogues that appear to block secretion of H^+.

1. Describe the structure and function of the lower esophageal sphincter.
2. List the secretory cells of the gastric mucosa and the products they secrete.
3. Explain the roles of hydrochloric acid in the stomach and how pepsinogen is activated.
4. Explain the cause of peptic ulcers and why they are more likely to be duodenal than gastric in location.
5. Explain how the parietal cells secrete H^+ and how various drugs function to inhibit acid secretion.

Small Intestine

The small intestine is adapted to provide a very high surface area for the exposure of chyme to brush border enzymes and for the absorption of nutrients.

Objective 15. Identify the regions of the small intestine, and describe how bile and pancreatic juice are delivered to the small intestine.

Objective 16. Describe the structure and function of the intestinal villi, the microvilli, and the intestinal glands (crypts of Lieberkühn).

Objective 17. Identify the location and action of brush border enzymes.

Objective 18. Describe the types of intestinal motility, and explain how intestinal smooth muscle contraction is produced.

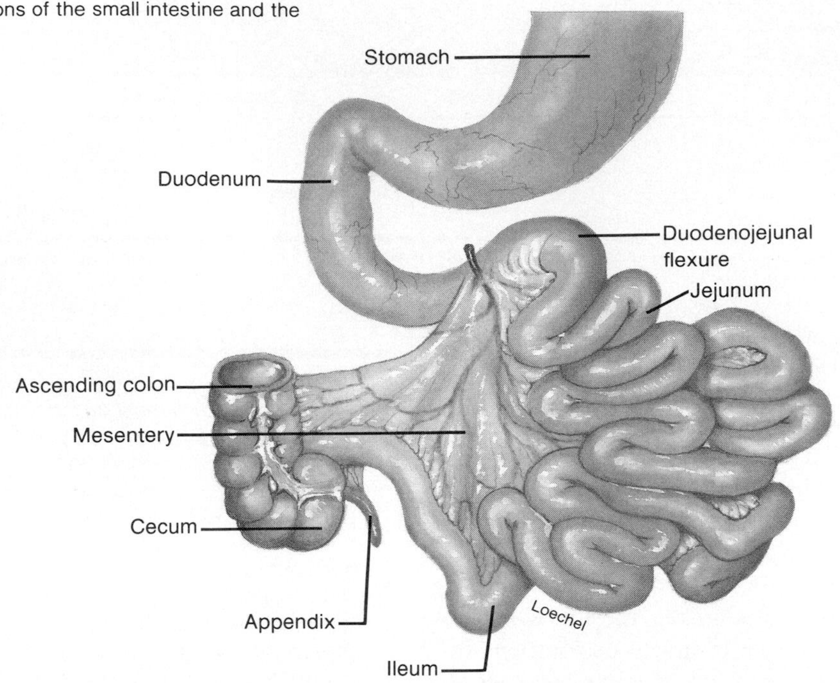

Figure 25.25. The regions of the small intestine and the mesenteric attachment.

The **small intestine** is that portion of the GI tract between the pyloric sphincter of the stomach and the ileocecal valve opening into the large intestine. It is positioned in the central and lower portions of the abdominal cavity and is supported, except for the first portion, by mesentery (fig. 25.25). The fan-shaped attachment of mesentery to the small intestine maintains mobility while allowing little chance of the intestine becoming twisted or kinked. Enclosed within the mesentery are blood vessels, nerves, and lymphatic vessels that supply the intestinal wall.

The small intestine is approximately 3 m (12 ft) long and 2.5 cm (1 in.) wide in a living person, but it will measure nearly twice this length in a cadaver when the muscle wall is relaxed. It is called the "small" intestine because of its relatively small diameter compared to that of the large intestine. The small intestine serves as the major site of digestion and absorption in the GI tract.

The small intestine is innervated by the **superior mesenteric** *(mes''en-ter'ik)* **plexus.** The branches of the plexus contain afferent fibers, postganglionic sympathetic fibers, and preganglionic parasympathetic fibers. The small intestine's arterial blood supply comes through the superior mesenteric artery and small branches from the celiac and inferior mesenteric arteries. Venous drainage is through the superior mesenteric vein, which unites with the splenic vein to form the hepatic portal vein carrying nutrient-rich blood to the liver.

Regions of the Small Intestine

The small intestine is divided into three regions on the basis of function and histological structure. These three regions are the duodenum, jejunum, and ileum.

The **duodenum** *(du''o-de'num* or *du-od'ĕ-num)* is a relatively fixed, C-shaped tube measuring approximately 25 cm (10 in.) from the pyloric sphincter of the stomach to the **duodenojejunal** *(du-od''ĕ-no''jĕ-joo'nal)* **flexure.** The concave surface of the duodenum faces to the left, where it receives bile secretions through the **common bile duct** from the liver and gallbladder and pancreatic secretions through the **pancreatic duct** of the pancreas (fig. 25.26). These two ducts unite to form a common entry into the duodenum called the **hepatopancreatic ampulla** *(hep''ah-to-pan''kre-at'ik am-pul'lah)* (or **ampulla of Vater**), which pierces the duodenal wall and drains into the duodenum from an elevation called the **duodenal papilla.** The duodenum is retroperitoneal except for a short portion near the stomach. The duodenum differs histologically from the rest of the small intestine by the presence of **duodenal (Brunner's) glands** in the submucosa (fig. 25.27). These compound tubuloalveolar glands secrete mucus and are most numerous near the superior end of the duodenum.

The **jejunum** *(je-joo'num)* is approximately 1 m (3 ft) long and extends from the duodenum to the ileum. The jejunum has a slightly larger lumen and more internal folds than does the ileum but lacks unique histological structures.

---

duodenum: L. *duodeni,* twelve each (length of twelve fingers' breadth)
ampulla of Vater: from Abraham Vater, German anatomist, 1684–1751
Brunner's glands: from Johann C. Brunner, Swiss anatomist, 1653–1727

Figure 25.26. The duodenum and associated structures.

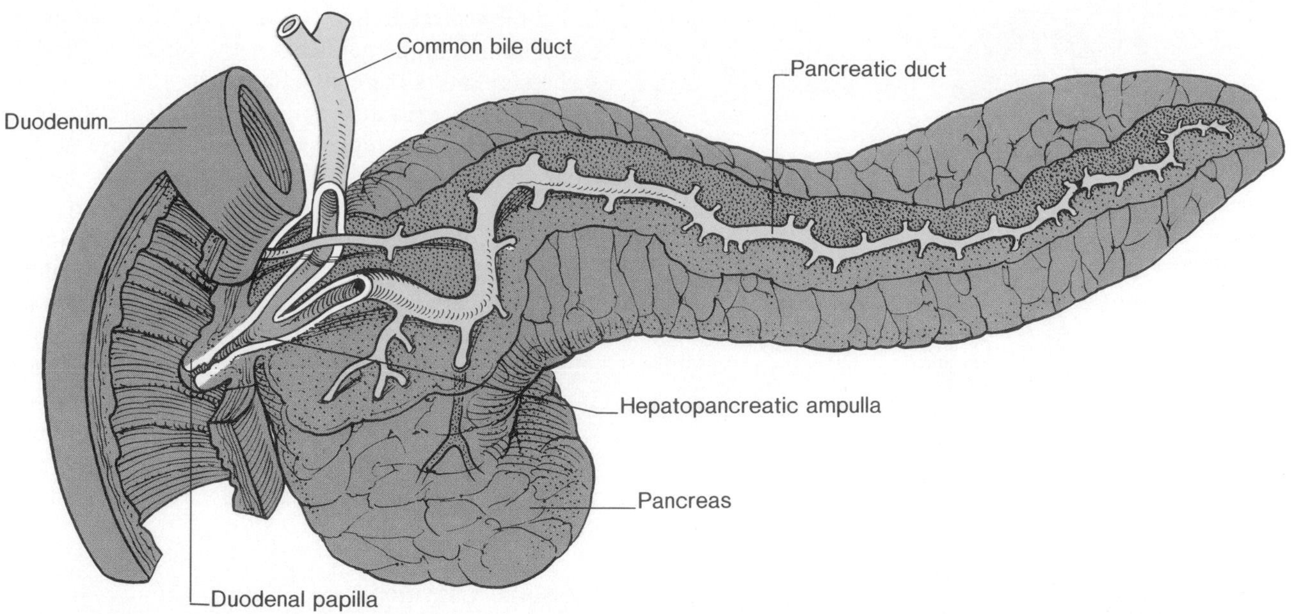

The **ileum** *(il'e-um)* makes up the remaining 2 m (6–7 ft) of the small intestine. The terminal portion of the ileum empties into the medial side of the cecum through the **ileocecal** *(il''e-o-se'kal)* **valve.** The walls of the ileum have an abundance of lymphatic tissue. Aggregates of lymph nodules, called **mesenteric (Peyer's) patches,** are characteristic of the ileum.

Structural Modifications of the Small Intestinal Wall

The products of digestion are absorbed across the epithelial lining of the intestinal mucosa. Absorption occurs primarily in the jejunum, although some also occurs in the duodenum and ileum. Absorption occurs at a rapid rate as a result of the high mucosal surface area in the small intestine, provided by folds. The mucosa and submucosa form large folds called the **plicae** *(pli'se)* **circularis,** which can be observed with the unaided eye. The surface area is further increased by the microscopic folds of mucosa called villi and by the foldings of the apical cell membrane of epithelial cells (which can only be seen with an electron microscope), called microvilli.

Each **villus** is a fingerlike fold of mucosa that projects into the intestinal lumen. The villi are covered with columnar epithelial cells, and interspersed among these are the mucus-secreting goblet cells. The lamina propria forms

Figure 25.27. The microscopic structure of the duodenum.

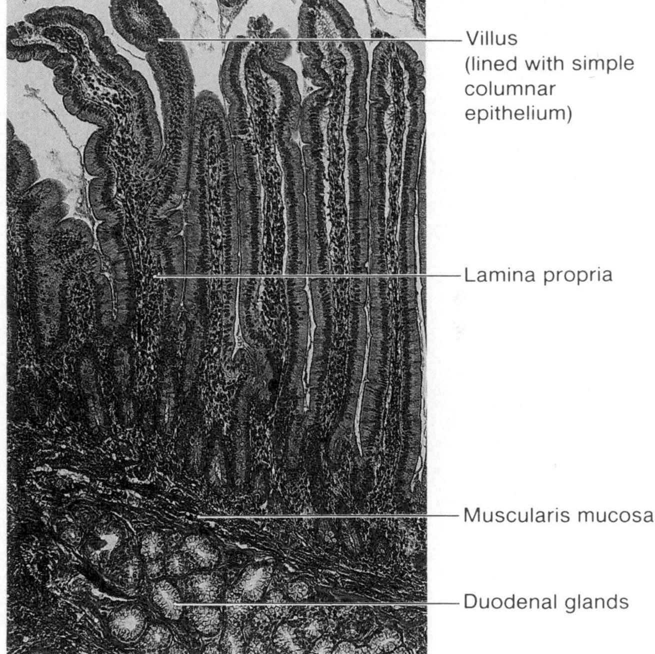

Peyer's patches: from Johann K. Peyer, Swiss anatomist, 1653–1712
plica: L. *plicatus,* folded
villus: L. *villosus,* shaggy

Figure 25.28. A diagram of the structure of an intestinal villus.

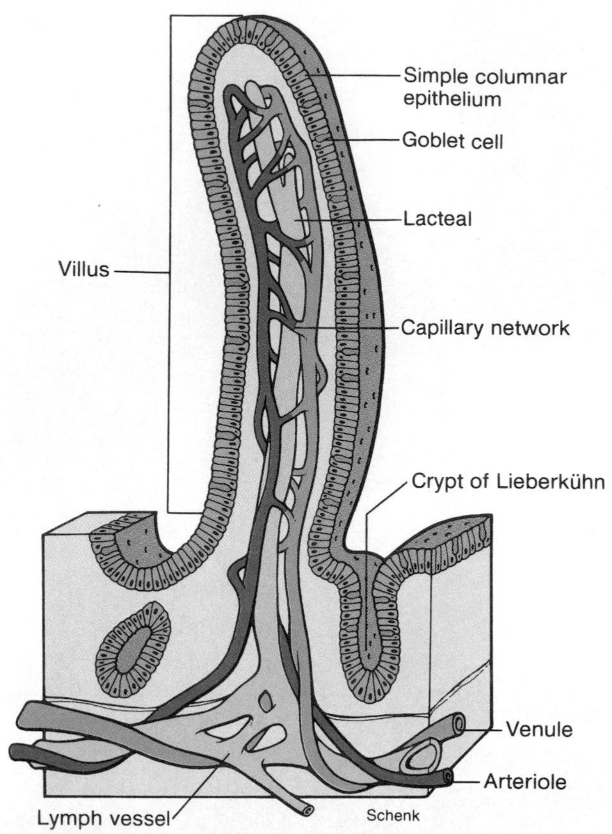

a connective core of each villus and contains numerous lymphocytes, blood capillaries, and a lymphatic vessel called the **central lacteal** (fig. 25.28). Absorbed monosaccharides and amino acids enter the blood capillaries; absorbed fat enters the central lacteals.

Epithelial cells at the tips of the villi are continuously exfoliated and are replaced by cells that are pushed up from the bases of the villi. The epithelium at the base of the villi invaginates downward at various points to form narrow pouches that open through pores to the intestinal lumen. These structures are called the **intestinal glands (crypts of Lieberkühn)** (fig. 25.29). Microvilli are fingerlike projections formed by foldings of the cell membrane, which can only be clearly seen in an electron microscope. In a light microscope, the microvilli display a somewhat vague **brush border** on the edges of the columnar epithelial cells. The term *brush border* is thus often used synonymously with microvilli in descriptions of the intestine (fig. 25.30).

Intestinal Enzymes

In addition to providing a large surface area for absorption, the cell membranes of the microvilli contain digestive enzymes. These enzymes are *not* secreted into the

lacteal: L. *lacteus,* milk
crypts of Lieberkühn: from Johann N. Lieberkühn, German anatomist, 1711–56

Figure 25.29. Intestinal villi and intestinal glands, or crypts of Lieberkühn are shown in (*a*). The crypts serve as sites for production of new epithelial cells. The time required for migration of these new cells to the tip of the villi is shown in (*b*). Epithelial cells are exfoliated from the tips of the villi.

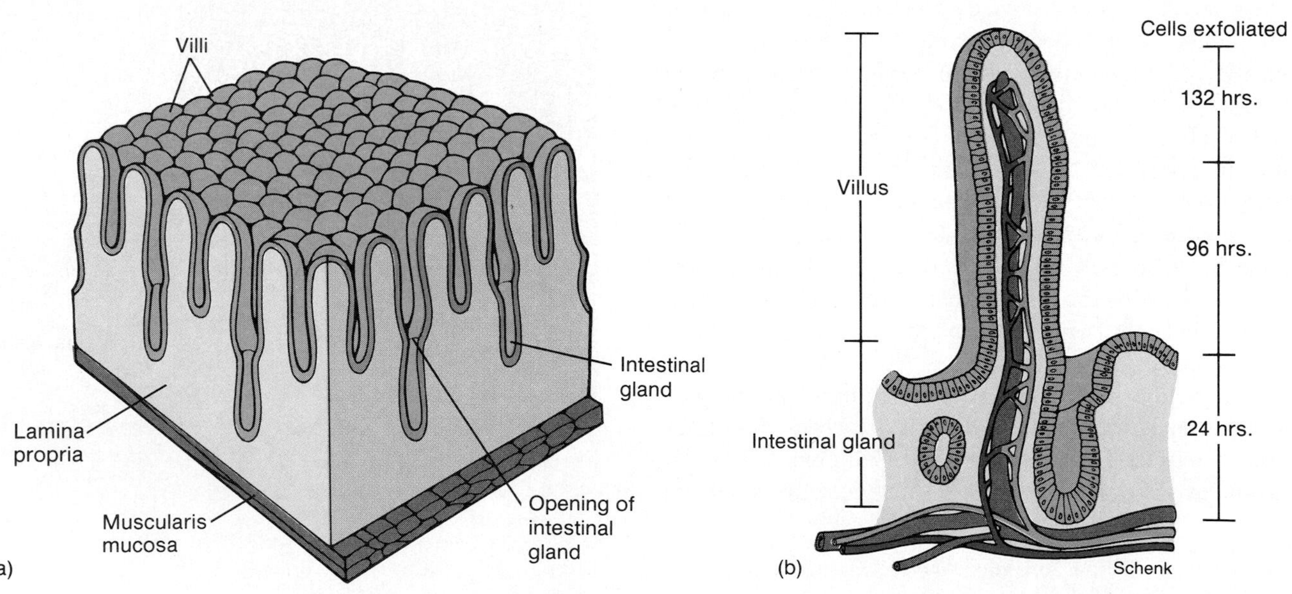

lumen (as implied by the name "intestinal glands"), but instead remain attached to the cell membrane with their active sites exposed to the chyme. These **brush border enzymes** hydrolyze disaccharides, polypeptides, and other substrates (table 25.7). One brush border enzyme, *enterokinase (en″ter-o-ki′nās),* is required for activation of the protein-digesting enzyme *trypsin,* which enters the intestine in pancreatic juice.

Intestinal Contractions and Motility

Like cardiac muscle, intestinal smooth muscle is capable of spontaneous electrical activity and automatic, rhythmic contractions. Spontaneous depolarization begins in the

The ability to digest milk sugar, or lactose, depends on the presence of a brush border enzyme called lactase. This enzyme is present in all children under the age of four but becomes inactive to some degree in most adults (with the exception of Scandinavians and some others). This can result in *lactose intolerance.* The presence of large amounts of undigested lactose in the intestine causes diarrhea, gas, cramps, and other unpleasant symptoms. Yogurt is better tolerated than milk because it contains lactase produced by the yogurt bacteria, which becomes activated in the duodenum and digests lactose.

Figure 25.30. (*a*) the cell membrane of epithelial cells that line the small intestine is folded into microvilli. (*b*) glycoproteins within the membranes of the microvilli contain polysaccharides that extend out into the lumen and form the glycocalyx, which covers the brush border epithelium. (*c*) a diagram of an intestinal epithelial cell. (*d*) the microvilli contain actin filaments attached to a terminal web of myosin filaments. This allows the microvilli to shorten.

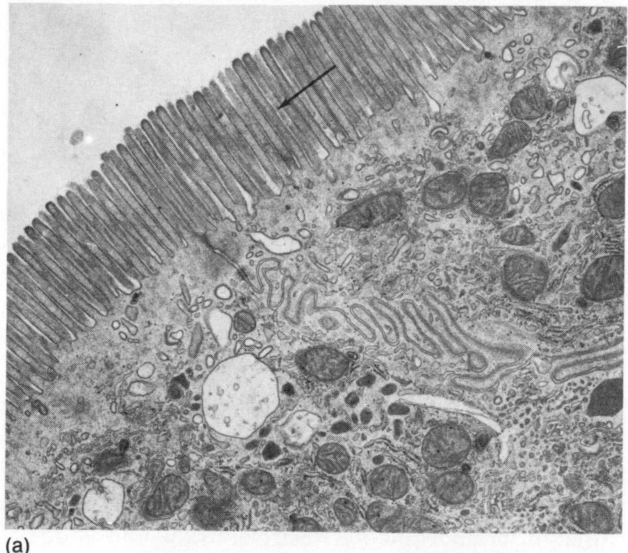

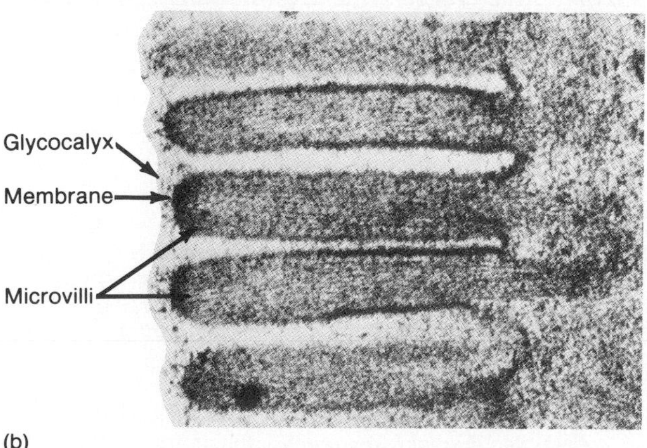

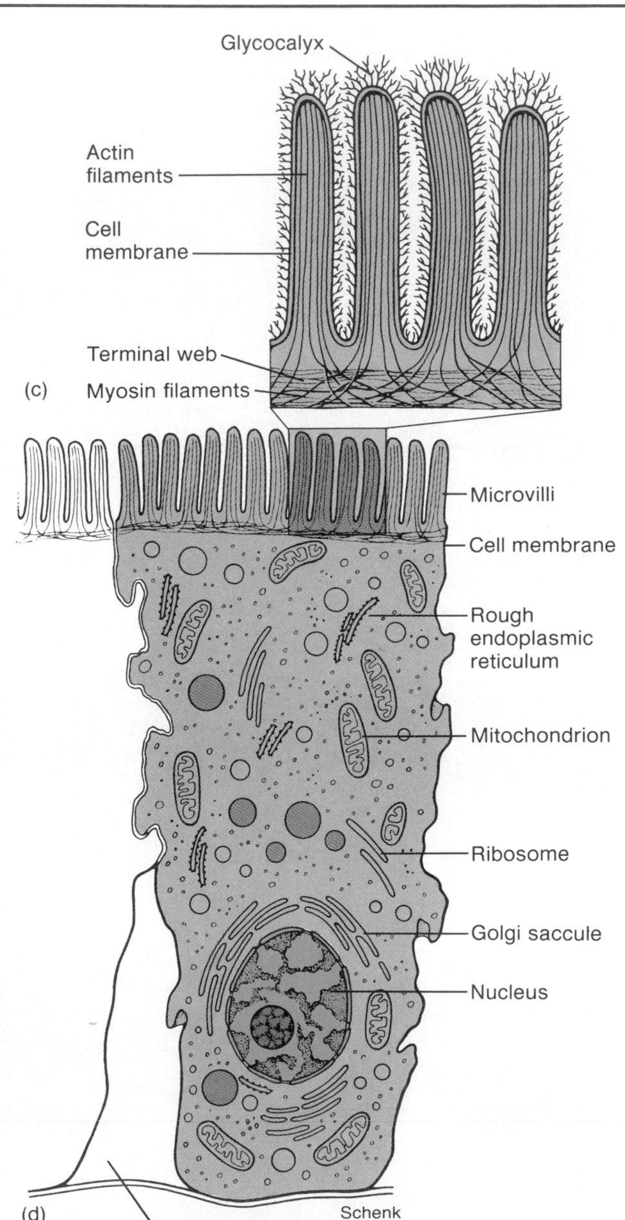

Table 25.7 Brush border enzymes attached to the cell membrane of microvilli in the small intestine

Category	Enzyme	Comments
Disaccharidase	Sucrase	Digests sucrose to glucose and fructose; deficiency produces gastrointestinal disturbances
	Maltase	Digests maltose to glucose
	Lactose	Digests lactose to glucose and galactose; deficiency produces gastrointestinal disturbances (lactose intolerance)
Peptidase	Aminopeptidase	Produces free amino acids, dipeptides, and tripeptides
	Enterokinase	Activates trypsin (and indirectly other pancreatic juice enzymes); deficiency results in protein malnutrition
Phosphatase	Ca^{++}, Mg^{++}—ATPase	Needed for absorption of dietary calcium; enzyme activity regulated by vitamin D
	Alkaline phosphatase	Removes phosphate groups from organic molecules; enzyme activity may be regulated by vitamin D

Figure 25.31. The smooth muscle of the gastrointestinal tract produces and conducts spontaneous pacesetter potentials. As these potential changes reach a threshold level of depolarization, they stimulate the production of action potentials, which in turn stimulate smooth muscle contraction.

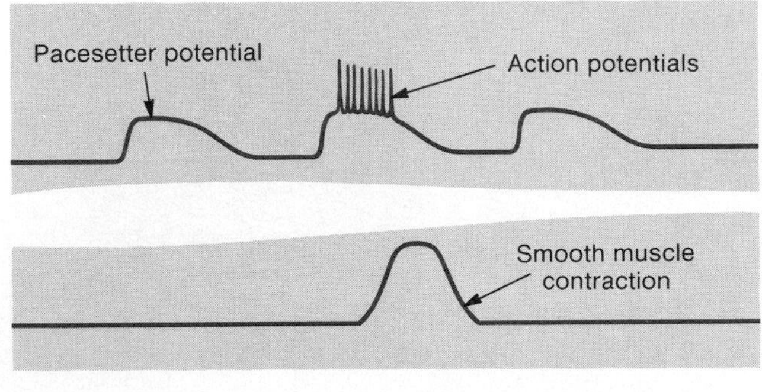

Figure 25.32. Segmentation of the small intestine. Simultaneous contractions of many segments of the intestine help to mix the chyme with digestive enzymes and mucus.

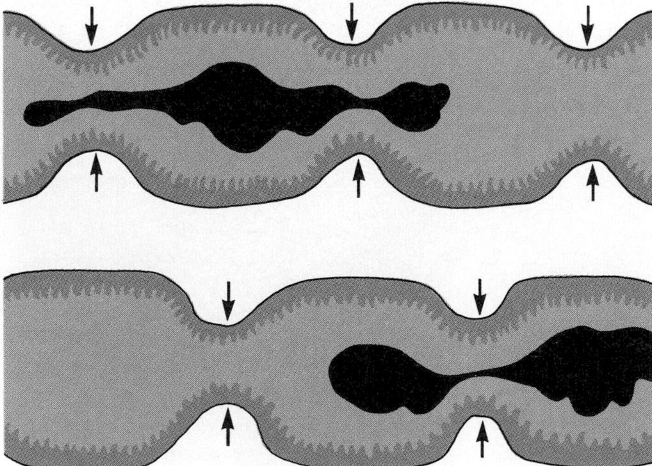

longitudinal smooth muscle and is conducted to the circular smooth muscle layer across *nexuses.* The term *nexus* is used here to indicate an electrical synapse between smooth muscle cells. The spontaneous depolarizations, called **pacesetter potentials,** decrease in amplitude as they are conducted from one muscle cell to another, much like excitatory postsynaptic potentials (EPSPs). Also like EPSPs, pacesetter potentials stimulate the production of action potentials in the smooth muscle cells through which they are conducted (fig. 25.31).

The nexuses conduct the pacesetter potentials, not the action potentials. Action potentials are, therefore, limited to those smooth muscle cells that are depolarized to threshold by the spreading pacesetter potentials. When action potentials are produced, they stimulate smooth muscle contraction. The rate at which this automatic activity occurs is influenced by autonomic nerves. Contraction is stimulated by parasympathetic (vagus nerve) innervation and is reduced by sympathetic nerve activity.

The small intestine has two major types of contractions: peristalsis and segmentation. Peristalsis is much weaker in the small intestine than in the esophagus and stomach. **Intestinal motility**—the movement of chyme through the intestine—is relatively slow and is due primarily to the fact that the pressure at the pyloric end of the small intestine is greater than at the distal end.

The major contractile activity of the small intestine is **segmentation.** This term refers to muscular constrictions of the lumen, which occur simultaneously at different intestinal segments (fig. 25.32). This action serves to mix the chyme more thoroughly.

1. Describe the structures involved in the delivery of pancreatic juice and bile to the small intestine.
2. Describe the intestinal structures that increase surface area, and explain the function of the intestinal glands (crypts of Lieberkühn).
3. Define the term *brush border enzymes,* and list examples. Explain why many adults cannot tolerate milk.
4. Describe how smooth muscle contraction in the small intestine is regulated, and explain the function of segmentation.

nexus: L. *nexus,* interconnection

Figure 25.33. The large intestine.

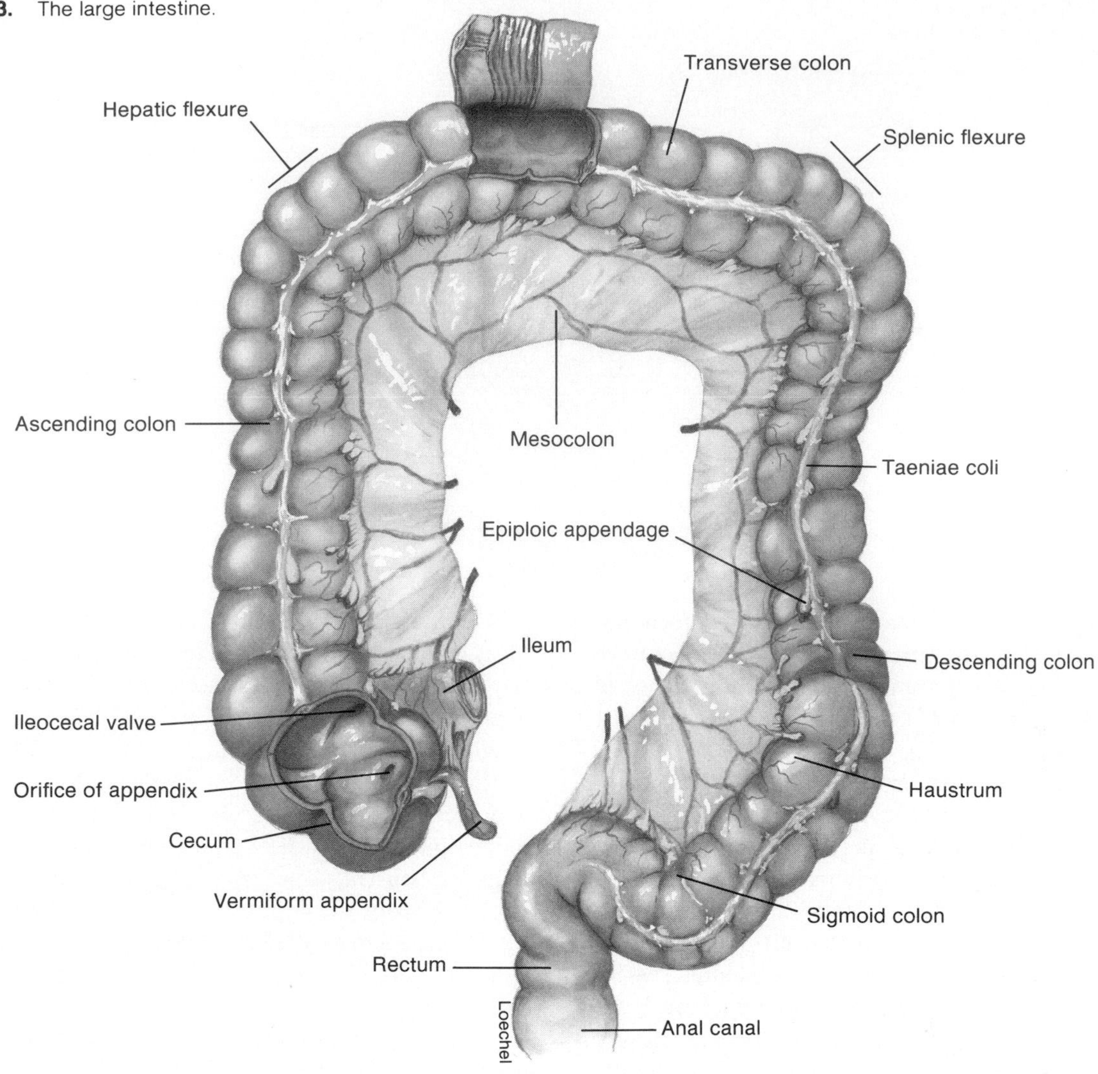

Large Intestine

The large intestine absorbs water and electrolytes from the chyme and passes undigested products out of the gastrointestinal tract.

Objective 19. Identify the regions of the large intestine, and describe its gross and histological structure.

Objective 20. Describe the functions of the large intestine, and explain how defecation is accomplished.

The **large intestine** is about 1.5 m (5 ft) long and 6.5 cm (2.5 in.) in diameter. It is named the "large" intestine because its diameter is larger than that of the small intestine. The large intestine begins at the terminal end of the ileum in the lower right quadrant of the abdominal cavity. From there, the large intestine leads superiorly on the right side to just below the liver, where it crosses to the left and descends into the pelvis and terminates at the anus. A specialized portion of the dorsal mesentery, called the **mesocolon,** supports the large intestine along the posterior abdominal wall.

The large intestine has little or no digestive function, but it does absorb water and electrolytes from the remaining chyme. In addition, the large intestine functions to form, store, and expel feces from the body.

Regions and Structures of the Large Intestine

The large intestine is structurally divided into the cecum, colon, rectum, and anal canal (fig. 25.33).

The **cecum** *(se'kum)* is a dilated pouch that hangs inferiorly, slightly below the ileocecal valve. The ileocecal valve is a fold of mucous membrane at the junction of the small and large intestine, which prohibits the backflow of

cecum: L. *caecum,* blind pouch

Figure 25.34. The microscopic appearance of a cross section of the human appendix.

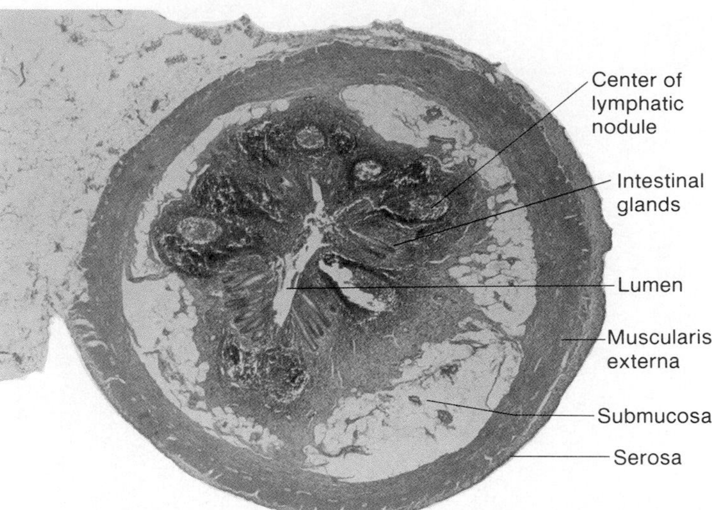

chyme. A fingerlike projection called the **vermiform appendix** is attached to the inferior medial margin of the cecum. The 8 cm (3 in.) appendix has an abundance of lymphatic tissue (fig. 25.34), which may serve to resist infection.

The open, superior portion of the cecum is continuous with the colon. The colon consists of ascending, transverse, descending, and sigmoid portions (fig. 25.33). The **ascending colon** extends superiorly from the cecum along the right abdominal wall to the inferior surface of the liver. Here the colon bends sharply to the left at the **hepatic flexure** and transversely crosses the upper abdominal cavity as the **transverse colon.** At the left abdominal wall, another right angle bend called the **splenic** *(splen'ik)* **flexure** denotes the beginning of the **descending colon.** The descending colon traverses inferiorly along the left abdominal wall to the pelvic region. The colon then angles medially from the brim of the pelvis to form an S-shaped bend known as the **sigmoid colon.**

The terminal 20 cm (7.5 in.) of the GI tract is the **rectum,** and the last 2 to 3 cm of the rectum is referred to as the **anal canal** (fig. 25.35). The rectum lies anterior to the sacrum, where it is firmly attached by peritoneum. The **anus** is the external opening of the anal canal. Two sphincter muscles guard the anal opening: the **internal anal sphincter,** which is composed of smooth muscle fibers, and the **external anal sphincter,** consisting of skeletal muscle. The mucous membrane of the anal canal is arranged into highly vascular, longitudinal folds called **anal columns.** Varicose veins in the anal area are known as *hemorrhoids (hem'o-roid),* or piles.

vermiform appendix: L. *vermiformis,* wormlike; *appendix,* attachment
colon: Gk. *kolon,* member of the whole
sigmoid: Gk. *sigmoeides,* shaped like the letter C
rectum: L. *rectum,* straight tube
anus: L. *anus,* ring

Although the large intestine consists of the same tunicas as the small intestine, there are some structural differences. The large intestine lacks villi but does have numerous goblet cells in the mucosal layer. The longitudinal muscle layer of the muscularis externa forms three distinct muscle bands, called **taeniae coli** *(te'ne-ah co'li),* which run the length of the large intestine. A series of bulges in the walls of the large intestine form sacculations, or **haustra** *(hows'tra),* along its entire length (fig. 25.36). Finally, the large intestine has small but numerous fat-filled pouches called **epiploic appendages** (fig. 25.33), which are attached superficially to the taeniae coli in the serous layer.

The large intestine has both types of autonomic innervation. The sympathetic innervation arises from superior and inferior mesenteric plexuses as well as from the celiac plexus. The parasympathetic innervation arises from the paired pelvic splanchnic and vagus nerves. Sensory fibers from the large intestine respond to bowel pressure and signal the need to defecate.

Branches from the superior mesenteric and inferior mesenteric arteries supply blood to the large intestine. Venous blood is returned through the superior and inferior mesenteric veins, which are tributaries to the hepatic portal vein that drains the liver.

A common disorder of the large intestine is inflammation of the appendix, or *appendicitis.* Wastes that accumulate in the appendix cannot be moved easily by peristalsis since the appendix has only one opening. The symptoms of appendicitis include muscular rigidity, localized pain in the lower right quadrant, and vomiting. The chief danger of appendicitis is that the appendix might rupture and produce peritonitis.

taenia: Gk. *teinein,* to stretch
haustrum: L. *haustrum,* bucket or scoop
epiploic: Gk. *epiplein,* to float on

Figure 25.35. The anal canal.

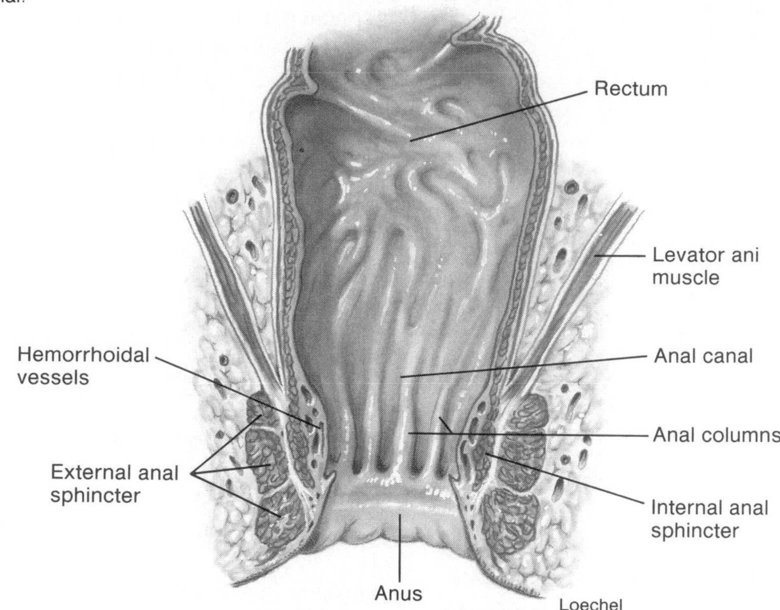

Fluid and Electrolyte Absorption in the Intestine

Most of the fluid and electrolytes in the lumen of the digestive tract are absorbed by the small intestine. Although a person may drink only about 1.5 L/day of water, the small intestine receives 7–9 L/day as a result of the fluid secreted into the digestive tract by the salivary glands, stomach, pancreas, liver, and gallbladder. The small intestine absorbs most of this fluid and passes about 1.5–2.0 L/day of fluid to the large intestine. The large intestine absorbs about 90% of this remaining volume, leaving less than 200 ml/day of fluid to be excreted in the feces.

Absorption of water in the intestine occurs passively as a result of the osmotic gradient created by the active transport of ions. The epithelial cells of the intestinal mucosa are joined together much like those of the kidney tubules and, like the kidney tubules, contain Na^+/K^+ pumps in the basolateral membrane. The analogy with kidney tubules is emphasized by the observation that aldosterone, which stimulates salt and water reabsorption in the renal tubules, also appears to stimulate salt and water absorption in the ileum.

The handling of salt and water transport in the large intestine is made more complex by the fact that the large intestine can secrete, as well as absorb, water. The secretion of water by the mucosa of the large intestine occurs passively as a result of the active transport of NaCl out of the epithelial cells into the intestinal lumen. Normally the absorption of NaCl and water far exceeds their secretion, but this balance may be altered in some disease states.

Figure 25.36. A radiograph after a barium enema showing the haustra of the large intestine.

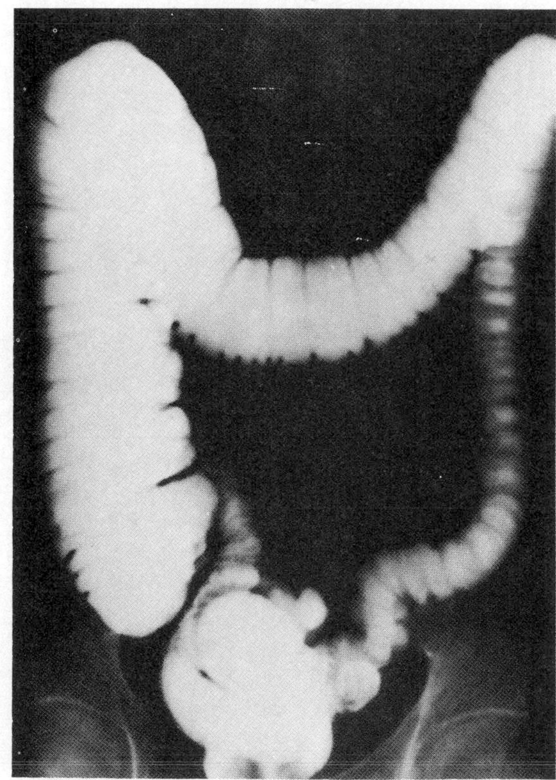

Diarrhea is characterized by excessive fluid excretion in the feces. There are three different mechanisms, illustrated by three different diseases, which can cause diarrhea. In *cholera,* severe diarrhea results from a chemical called *enterotoxin,* released from the infecting bacteria. Enterotoxin, acting via stimulation of adenylate cyclase and cAMP production, stimulates active NaCl transport followed by the osmotic movement of water into the lumen of the intestine. In *celiac sprue,* a disease produced in susceptible people by eating foods that contain gluten (proteins from grains such as wheat), diarrhea results from inadequate absorption of fluid due to damage to the intestinal mucosa. In *lactose intolerance,* diarrhea is produced by the increased osmolarity of the contents of the intestinal lumen as a result of the presence of undigested lactose.

Mechanical Activities of the Large Intestine

Chyme enters the large intestine through the ileocecal valve. About 15 ml of pasty material enters the cecum with each rhythmic opening of the valve. The ingestion of food intensifies peristalsis of the ileum and increases the frequency with which the ileocecal valve opens; this is called the **gastroileal reflex.** Material entering the large intestine accumulates in the cecum and ascending colon.

Three types of movements occur throughout the large intestine: peristalsis, haustral churning, and mass movement. **Peristaltic movements** of the colon are similar to those of the small intestine, though they are usually more sluggish. In **haustral churning,** a relaxed haustrum is filled until it reaches a certain point of distension, and then the muscularis layer is stimulated to contract. This contraction not only moves the material to the next haustrum but churns the contents and exposes it to the mucosa, where water and electrolytes are absorbed. **Mass movement** is a very strong peristaltic wave, involving the action of the taeniae coli, which moves the colonic contents toward the rectum. Mass movements generally occur only two or three times a day, usually during or shortly after a meal. This response to eating is called the **gastrocolic reflex** and can best be observed in infants who have a bowel movement during or shortly after feeding.

After electrolytes and water have been absorbed, the waste material that is left then passes to the rectum, leading to an increase in rectal pressure and the urge to defecate. If the urge to defecate is denied, the feces are prevented from entering the anal canal by the internal anal sphincter. In this case the feces remain in the rectum and may even back up into the sigmoid colon. The **defecation reflex** normally occurs when the rectal pressure rises to a particular level that is determined, to a large degree, by habit. At this point the internal anal sphincter relaxes to admit the feces into the anal canal.

During the act of defecation the longitudinal rectal muscles contract to increase rectal pressure, and the internal and external anal sphincter muscles relax. Excretion is aided by contractions of abdominal and pelvic skeletal muscles, which raise the intra-abdominal pressure and help push the feces from the rectum through the anal canal and out the anus.

Table 25.8 summarizes the various mechanical activities of the GI tract.

1. Identify the regions of the large intestine, the haustra, and the taeniae coli.
2. Describe how electrolytes and water are absorbed in the large intestine, and explain how diarrhea may be produced.
3. Describe the structures and mechanisms involved in defecation.

Table 25.8 Summary of the mechanical activity in the GI tract

Region	Type of motility	Frequency	Stimulus	Result
Oral cavity	Mastication	Variable	Initiated voluntarily, proceeds reflexly	Subdivision, mixing with saliva
Oral cavity and pharynx	Deglutition	Maximum of 20 per min	Initiated voluntarily, reflexly controlled by swallowing center	Clears oral cavity of food
Esophagus	Peristalsis	Depends on frequency of swallowing	Initiated by swallowing	Movement through the esophagus
Stomach	Receptive relaxation	Matches frequency of swallowing	Unknown	Permits filling of stomach
	Tonic contraction	15–20 per min	Autonomic plexuses	Mixing and churning
	Peristalsis	1–2 per min	Autonomic plexuses	Evacuation of stomach
	"Hunger contractions"	3 per min	Low blood sugar level	"Feeding"
Small intestine	Peristalsis	17–18 per min	Autonomic plexuses	Transfer through intestine
	Rhythmic segmentation	12–16 per min	Autonomic plexuses	Mixing
	Pendular movements	Variable	Autonomic plexuses	Mixing
Large intestine	Peristalsis	3–12 per min	Autonomic plexuses	Transport
	Mass movement	2–3 per day	Stretch	Fills pelvic colon
	Haustral churning	3–12 per min	Autonomic plexuses	Mixing
	Defecation	Variable: 1 per day to 3 per week	Reflex triggered by rectal distension	Defecation

Liver, Gallbladder, and Pancreas

The liver processes nutrients and secretes bile, which is stored and concentrated in the gallbladder prior to its discharge into the duodenum. The pancreas secretes important hormones into the blood and essential digestive enzymes into the duodenum.

Objective 21. Describe the structure of liver lobules, and trace the flow of blood and bile through a liver lobule.

Objective 22. List the categories of liver functions, and explain the anatomical basis for these functions.

Objective 23. Describe enterohepatic circulation, and give examples.

Objective 24. Describe the anatomical relationships among the liver, gallbladder, hepatic ducts, cystic duct, and common bile duct.

Objective 25. Describe the structure and functions of the pancreas.

Three accessory digestive organs in the abdominal cavity aid in the chemical breakdown of food. These are the liver, gallbladder, and pancreas. The liver and pancreas function as exocrine glands in this process because their secretions are transported to the lumen of the GI tract through ducts.

Structure of the Liver

The **liver** is the largest internal organ of the body, weighing about 1.3 kg (3.5–4.0 lbs) in an adult. It is positioned immediately beneath the diaphragm in the right hypochondrium of the abdominal cavity. The liver is reddish brown in color because of its great vascularity.

The liver has two major lobes and two minor lobes. Anteriorly, the **right lobe** is separated from the smaller **left lobe** by the **falciform ligament** (fig. 25.37). Inferiorly,

falciform: L. *falcis,* sickle; *forma,* form

Figure 25.37. The liver. (*a*) an anterior view of the gross structure; (*b*) a CT scan showing the relative position of the liver to other abdominal organs.

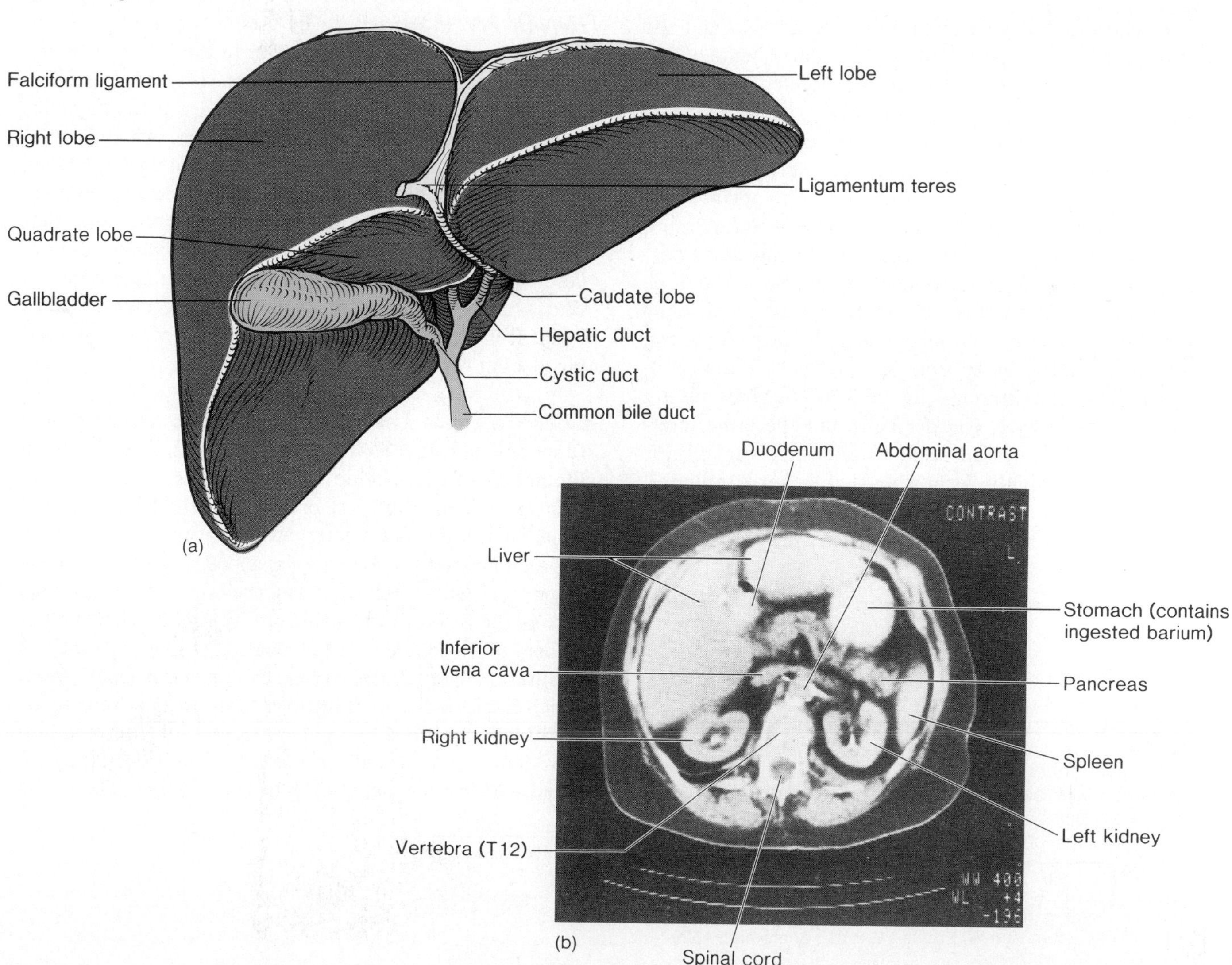

Figure 25.38. The structure of the liver. (*a*) a scanning electron micrograph of the liver. Hepatocytes are arranged in plates so that blood that passes through sinusoids (*b*) will be in contact with each liver cell. (Photo from: *Tissues and Organs: A Text Atlas of Scanning Electron Microscopy* by R. G. Kessel and R. Kardon. W. H. Freeman and Company. © 1979.)

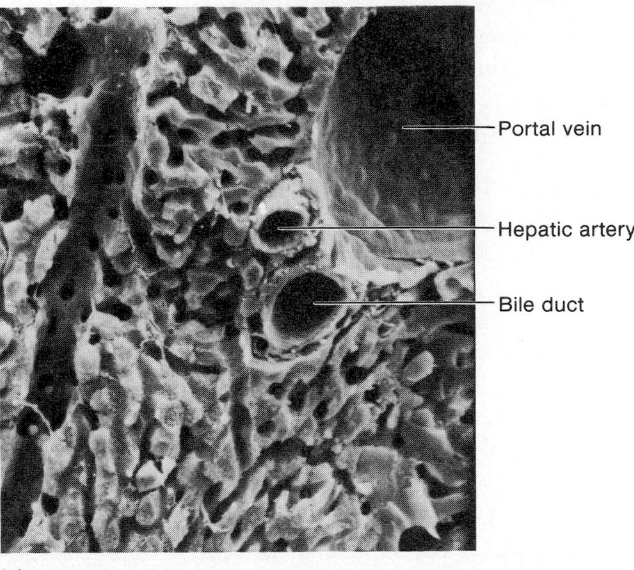

(a)

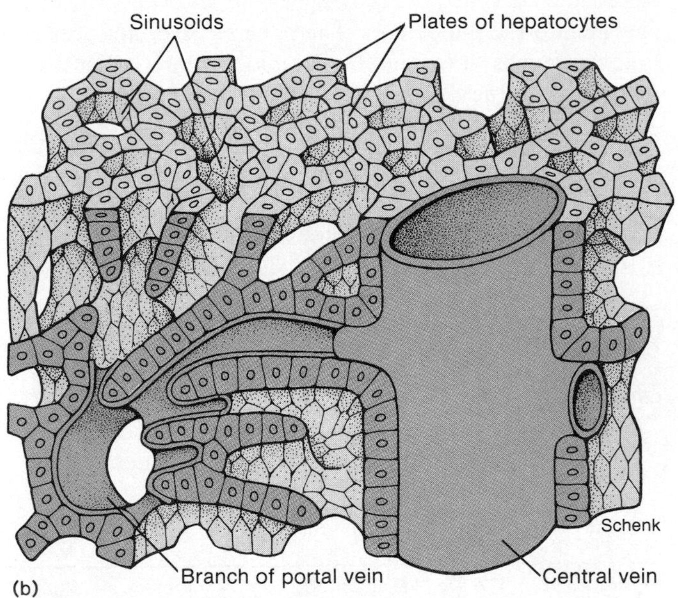

(b)

the **caudate lobe** is near the inferior vena cava, and the **quadrate lobe** is adjacent to the gallbladder. The falciform ligament attaches the liver to the anterior abdominal wall and the diaphragm. A **ligamentum teres** (round ligament) is continuous along the free border of the falciform ligament to the umbilicus. The ligamentum teres is the remnant of the umbilical vein of the fetus. The **porta** of the liver is where the hepatic artery, portal vein, lymphatics, and nerves enter the liver and where the **hepatic ducts** exit.

Although the liver is the largest internal organ, it is, in a sense, only one to two cells thick. This is because the liver cells, or **hepatocytes,** form **plates** that are one to two cells thick and separated from each other by large capillary spaces called **sinusoids** (fig. 25.38). The sinusoids are lined with phagocytic **Kupffer cells,** but the large intercellular gaps between adjacent Kupffer cells make these sinusoids more highly permeable than other capillaries. The plate structure of the liver and the high permeability of the sinusoids allow each hepatocyte to have direct contact with the blood.

In *cirrhosis,* large numbers of liver lobules are destroyed and replaced with permanent connective tissue and "regenerative nodules" of hepatocytes. These regenerative nodules don't have the platelike structure of normal liver tissue and are therefore less functional. One indication of this decreased function is the entry of ammonia from the hepatic portal blood into the general circulation. Cirrhosis may be caused by chronic alcohol abuse, viral hepatitis, and other agents that attack liver cells.

porta: L. *porta,* gate
Kupffer cells: from Karl W. Kupffer, German anatomist, 1829–1902

Hepatic Portal System. The products of digestion that are absorbed into blood capillaries in the intestine do not directly enter the general circulation. Instead, this blood is delivered first to the liver. Capillaries in the digestive tract drain into the *hepatic portal vein,* which carries this blood to capillaries in the liver; it is not until the blood has passed through this second capillary bed that it enters the general circulation through the *hepatic vein* that drains the liver. The term **hepatic portal system** is used to describe this unique pattern of circulation: capillaries → vein → capillaries → vein. In addition to receiving venous blood from the intestine, the liver also receives arterial blood from the *hepatic artery.*

Liver Lobules. The hepatic plates are arranged into functional units called **liver lobules** (figs. 25.39, 25.40). In the middle of each lobule is a **central vein,** and at the periphery of each lobule are branches of the hepatic portal vein and of the hepatic artery, which open into the spaces *between* hepatic plates. Arterial blood and portal venous blood, containing molecules absorbed in the GI tract, thus mix as the blood flows within the sinusoids from the periphery of the lobule to the central vein. The central veins of different liver lobules converge to form the hepatic vein, which carries blood from the liver to the inferior vena cava.

Bile is produced by the hepatocytes and secreted into thin channels called **bile canaliculi** *(kan"ah-lik'u-li)* located *within* each hepatic plate (fig. 25.39). These bile

hepatic: Gk. *hepatos,* liver

Figure 25.39. A liver lobule and the histology of the liver. (*a*) a liver lobule seen in cross section and (*b*) longitudinal section. Blood enters a liver lobule through the vessels in a hepatic triad, passes through hepatic sinusoids, and leaves the lobule through a central vein. The central veins converge to form hepatic veins that transport venous blood from the liver. (*c*) a photomicrograph of a liver lobule in transverse section.

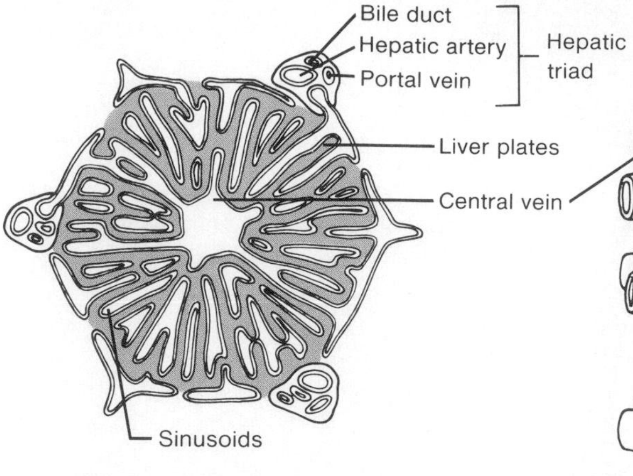

Bile duct
Hepatic artery
Portal vein
Hepatic triad
Liver plates
Central vein
Sinusoids
(a)

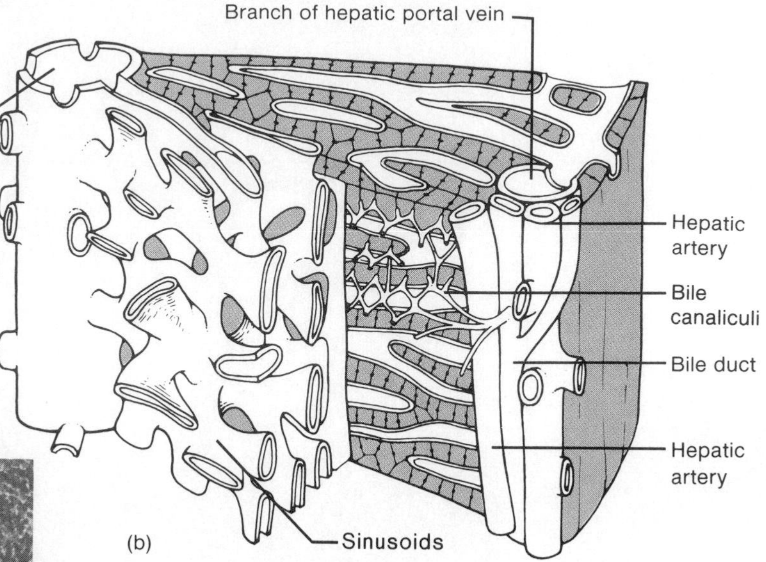

Branch of hepatic portal vein
Hepatic artery
Bile canaliculi
Bile duct
Hepatic artery
Sinusoids
(b)

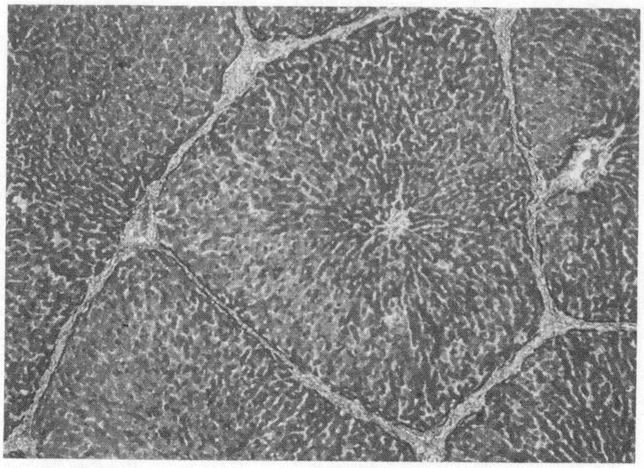

(c)

Figure 25.40. The flow of blood and bile in a liver lobule. Blood flows within sinusoids from a portal vein to the central vein (from the periphery to the center of a lobule). Bile flows within hepatic plates from the center to bile ducts at the periphery of a lobule.

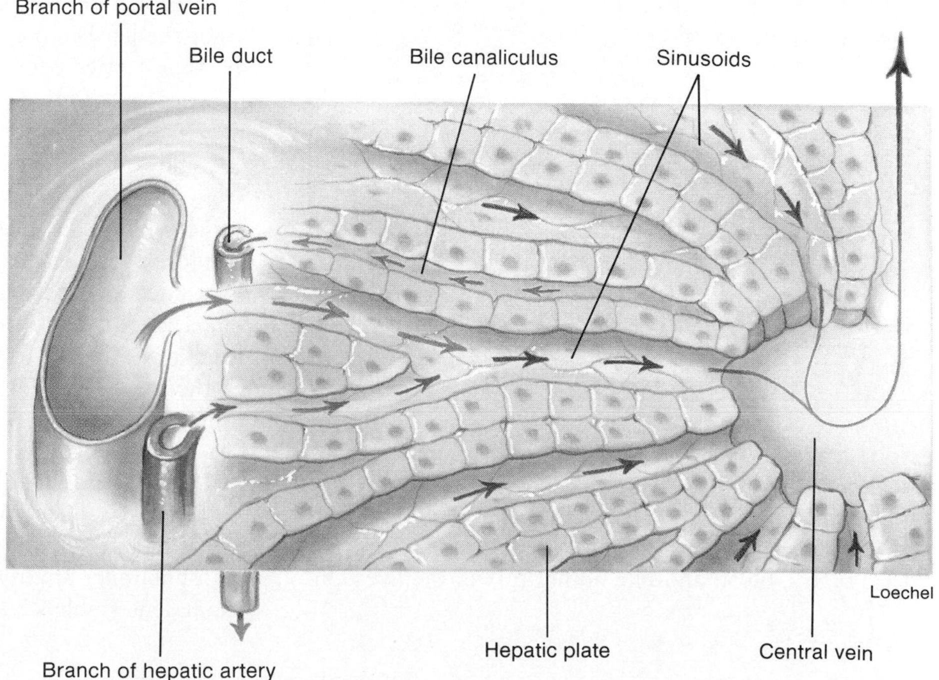

Figure 25.41. Enterohepatic circulation. Substances secreted in the bile may be absorbed by the intestinal epithelium and recycled to the liver via the hepatic portal vein.

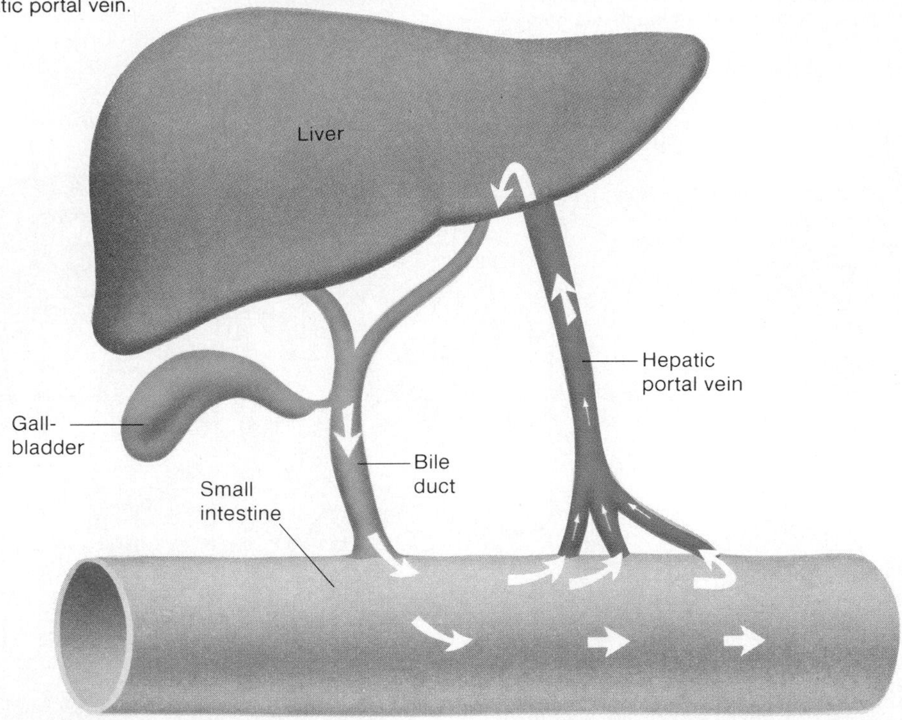

Table 25.9 Compounds that the liver excretes into the bile		
	Compound	**Comments**
Endogenous (naturally occurring)	Bile salts, urobilinogen, cholesterol	High percentage is absorbed and has an enterohepatic circulation[1]
	Lecithin	Small percentage is absorbed and has an enterohepatic circulation
	Bilirubin	No enterohepatic circulation
Exogenous (drugs)	Ampicillin, streptomycin, tetracycline	High percentage is absorbed and has an enterohepatic circulation
	Sulfonamides, penicillin	Small percentage is absorbed and has an enterohepatic circulation

[1]Compounds with an enterohepatic circulation are absorbed to some degree by the intestine and are returned to the liver in the hepatic portal vein.

canaliculi are drained at the periphery of each lobule by **bile ducts,** which in turn drain into **hepatic ducts** that carry bile away from the liver. Since blood travels in the sinusoids and bile travels in the opposite direction within the hepatic plates, blood and bile do not mix in the liver lobules.

Enterohepatic Circulation. In addition to the normal constituents of bile, a wide variety of exogenous compounds (drugs) are secreted by the liver into the bile ducts (table 25.9). The liver can thus "clear" the blood of particular compounds by removing them from the blood and excreting them into the intestine with the bile. (The liver can also clear the blood by other mechanisms that will be described in a later section.)

Many compounds that are released with the bile into the intestine are not excreted with the feces, however. Some of these can be absorbed through the small intestine and enter the hepatic portal blood. These absorbed molecules are thus carried back to the liver, where they can be again secreted by hepatocytes into the bile ducts. Compounds that recirculate between the liver and intestine in this way are said to have an **enterohepatic circulation** (fig. 25.41).

Functions of the Liver

Because of its very large and diverse enzymatic content, its unique structure, and the fact that it receives venous blood from the intestine, the liver has a wider variety of functions than any other organ in the body. A summary of the major categories of liver function is presented in table 25.10.

Bile Production and Secretion. The liver produces and secretes 250–1,500 ml of bile per day. The major constituents of bile include bile salts, bile pigment (bilirubin), phospholipids (mainly lecithin), cholesterol, and inorganic ions (table 25.11).

Table 25.10 Summary of the major categories of liver functions

Functional category	Actions
Detoxification of blood	Phagocytosis by Kupffer cells
	Chemical alteration of biologically active molecules (hormones and drugs)
	Production of urea, uric acid, and other molecules that are less toxic than parent compounds
	Excretion of molecules in bile
Carbohydrate metabolism	Conversion of blood glucose to glycogen and fat
	Production of glucose from liver glycogen and from other molecules (amino acids, lactic acid) by gluconeogenesis
	Secretion of glucose into the blood
Lipid metabolism	Synthesis of triglyceride and cholesterol
	Excretion of cholesterol in bile
	Production of ketone bodies from fatty acids
Protein synthesis	Production of albumin
	Production of plasma transport proteins
	Production of clotting factors (fibrinogen, prothrombin, and others)
Secretion of bile	Synthesis of bile salts
	Conjugation and excretion of bile pigment (bilirubin)

Table 25.11 Composition of the bile

Component	Concentration
pH	5.7–8.6
Bile salts	140–2230 mg/100 ml
Lecithin	140–810 mg/100 ml
Cholesterol	97–320 mg/100 ml
Bilirubin	12–70 mg/100 ml
Urobilinogen	5–45 mg/100 ml
Sodium	145–165 mEq/L
Potassium	2.7–4.9 mEq/L
Chloride	88–115 mEq/L
Bicarbonate	27–55 mEq/L

Bile pigment, or *bilirubin (bil''i-roo'bin),* is produced in the spleen, liver, and bone marrow from heme groups (minus the iron) derived from hemoglobin. Without the protein part of hemoglobin, the **free bilirubin** is not very water-soluble and thus must be carried in the blood attached to albumin proteins. This protein-bound bilirubin can neither be filtered by the kidneys into the urine, nor directly excreted by the liver into the bile.

The liver can take some of the free bilirubin out of the blood and conjugate (combine) it with glucuronic acid. This **conjugated bilirubin** is water-soluble and is secreted into the bile. Once the conjugated bilirubin enters the intestine it is converted by bacteria into another pigment—**urobilinogen** *(u''ro-bi-lin'o-jen)*—which is partially responsible for the color of the feces. About 30% to 50% of the urobilinogen, however, is absorbed by the intestine and enters the hepatic portal blood. Some of this is returned to the intestine in an enterohepatic circulation; the rest enters the general circulation (fig. 25.42). The urobilinogen in plasma, unlike free bilirubin, is not attached to albumin and therefore is easily filtered by the kidneys into the urine, giving urine its characteristic yellow color.

The **bile salts** are derivatives of cholesterol that have two to four polar groups on each molecule. The principal bile salts in humans are *cholic acid* and *chenodeoxycholic (ke''no-de-ok-sĭ-ko'lik) acid* (fig. 25.43). In aqueous solutions these molecules "huddle" together to form aggregates known as **micelles** *(mi-sel').* The nonpolar parts are located in the central region of the micelle (away from water), whereas the polar groups face water around the periphery of the micelle. Lecithin, cholesterol, and other lipids enter these micelles in a process that aids the digestion and absorption of fats.

Detoxification of the Blood. The liver can remove biologically active molecules such as hormones and drugs from the blood by (1) excretion of these compounds in the bile; (2) phagocytosis by Kupffer cells, which line the sinusoids; and (3) chemical alteration of these molecules within the hepatocytes.

Ammonia, for example, is a very toxic molecule produced by the action of bacteria in the intestine. The observation that hepatic portal blood has an ammonia concentration that is four to fifty times greater than in the hepatic vein means that this compound is removed by the liver. The liver has the enzymes needed to convert ammonia into less toxic **urea** molecules, which are secreted by the liver into the blood and excreted by the kidneys in the urine. Similarly, the liver converts toxic porphyrins into **bilirubin** and toxic purines into **uric acid.**

Figure 25.42. The enterohepatic circulation of uroglobin. Bacteria in the intestine convert bile pigment (bilirubin) into urobilinogen. Some of this pigment goes out in the feces; some is absorbed by the intestine and is recycled through the bile. A portion of the uroglobin that is absorbed goes into the general circulation and is therefore filtered by the kidneys into the urine.

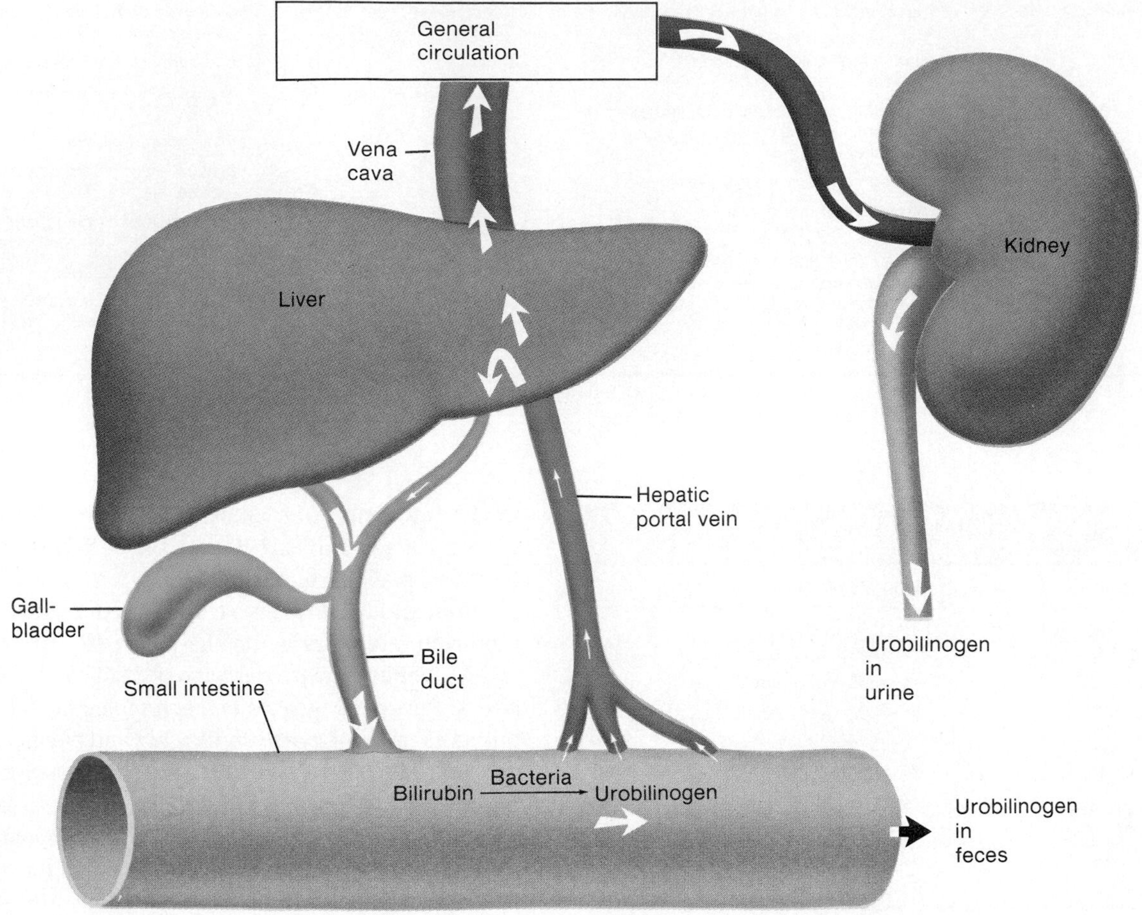

Figure 25.43. The two major bile acids (which form bile salts) in humans.

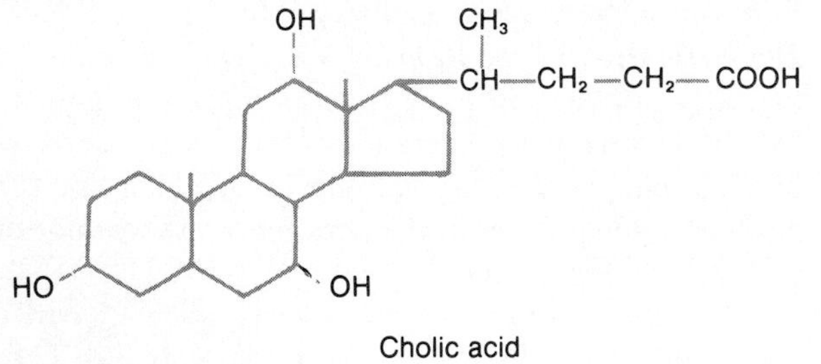

OH, CH_3, CH—CH_2—CH_2—COOH, HO, OH

Cholic acid

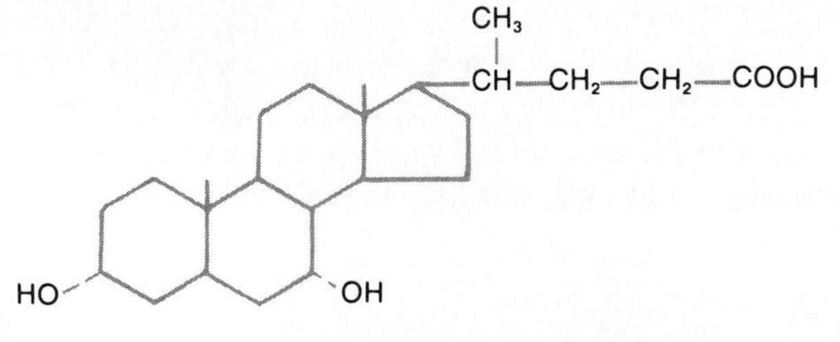

CH_3, CH—CH_2—CH_2—COOH, HO, OH

Chenodeoxycholic acid

Steroid hormones and other nonpolar compounds, such as many drugs, are inactivated in their passage through the liver by modifications of their chemical structures. The liver has enzymes that convert these molecules into more polar (more water-soluble) forms by **hydroxylation** (the addition of OH^- groups) and by **conjugation** with highly polar groups such as sulfate and glucuronic acid. Polar derivatives of steroid hormones and drugs are less biologically active and, because of their increased water solubility, are more easily excreted by the kidneys into the urine.

Secretion of Glucose, Triglycerides, and Ketone Bodies. The liver helps to regulate the blood glucose concentration by either removing glucose from or adding glucose to the blood, according to the needs of the body. After a carbohydrate-rich meal, the liver can remove some glucose from the hepatic portal blood and convert it into glycogen and triglycerides (**glycogenesis** *[gli''ko-jen'ĕ-sis]* and **lipogenesis,** respectively). During fasting, the liver secretes glucose into the blood. This glucose can be derived

Figure 25.44. The pancreatic duct joins the common bile duct to empty its secretions through the duodenal papilla into the duodenum. The release of bile and pancreatic juice into the duodenum is controlled by the sphincter of ampulla (Oddi).

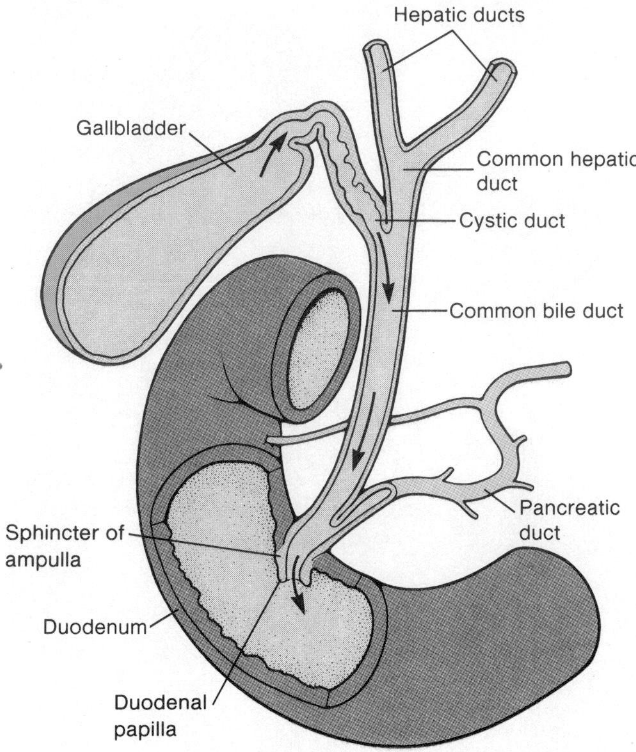

from the breakdown of stored glycogen in a process called **glycogenolysis,** or it can be produced by the conversion of noncarbohydrate molecules (such as amino acids) into glucose in a process called **gluconeogenesis** *(gloo''ko-ne''o-jen'ĕ-sis)*. The liver also contains the enzymes required to convert free fatty acids into ketone bodies (**ketogenesis),** which are secreted into the blood in large amounts during fasting. These processes are controlled by hormones (chapter 26).

Production of Plasma Proteins. Plasma albumin and most of the plasma globulins (with the exception of immunoglobulins) are produced by the liver. Albumin comprises about 70% of the total plasma protein and contributes most to the colloid osmotic pressure of the blood. The globulins produced by the liver have a wide variety of functions, including the transport of cholesterol and triglycerides, transport of steroid and thyroid hormones, inhibition of trypsin activity, and blood clotting. Clotting factors I (fibrinogen), II (prothrombin), III, V, VII, IX, and XI are all produced by the liver.

Gallbladder

The **gallbladder** is a saclike organ attached to the inferior surface of the liver. This organ stores and concentrates bile, which drains to it from the liver by way of the bile ducts, hepatic duct, and **cystic duct,** respectively. A sphincter valve at the neck of the gallbladder allows a storage capacity of about 35 to 50 ml. The inner mucosal layer of the gallbladder is arranged in rugae similar to those of the stomach. When the gallbladder fills with bile, it expands to the size and shape of a small pear. Bile is a yellowish-green fluid containing bile salts, bilirubin, cholesterol, and other compounds as previously discussed. Contraction of the muscularis ejects bile from the cystic duct into the **common bile duct,** which conveys bile into the duodenum (fig. 25.44).

Bile is continuously produced by the liver and drains through the hepatic and common bile ducts to the duodenum. When the small intestine is empty of food, the **sphincter of ampulla** (of Oddi) closes, and bile backs up in the cystic duct to the gallbladder for storage.

The gallbladder is supplied with blood from the cystic artery, which branches from the right hepatic artery. Venous blood is returned through the cystic vein, which empties into the hepatic portal vein. Autonomic innervation of the gallbladder is similar to the liver; both receive parasympathetic innervation from the vagus nerves and sympathetic innervation from thoracolumbar nerves through the celiac ganglia.

Approximately twenty million Americans have *gallstones,* which can produce painful symptoms by obstructing the cystic or common bile ducts. Gallstones commonly contain cholesterol as their major component. Cholesterol normally has an extremely low water solubility (20 μg/L), but it can be present in bile at two million times its water solubility (40 g/L) because it enters the hydrophobic centers of mixed micelles of bile salts and lecithin. In order for gallstones to be produced, the liver must secrete enough cholesterol to create a supersaturated solution, and some substance within the gallbladder must serve as a nucleus for the formation of cholesterol crystals. The gallstone is formed from cholesterol crystals that become hardened by the precipitation of inorganic salts (fig. 25.45). Gallstones may sometimes be dissolved by treatment with the bile salt chenodeoxycholic acid, or they may have to be removed surgically. A newer treatment involves fragmentation of the gallstones by high-energy shock waves delivered to a patient immersed in a water bath.

cystic: Gk. *kystis,* pouch
sphincter of Oddi: from Ruggero Oddi, Italian physician, 19th century

Figure 25.45. (*a*) an X ray of a gallbladder that contains gallstones. (*b*) a posterior view of a gallbladder that has been removed (cholecystectomy) and cut open to reveal its gallstones (biliary calculi). A dime is placed in the photo to show relative size.

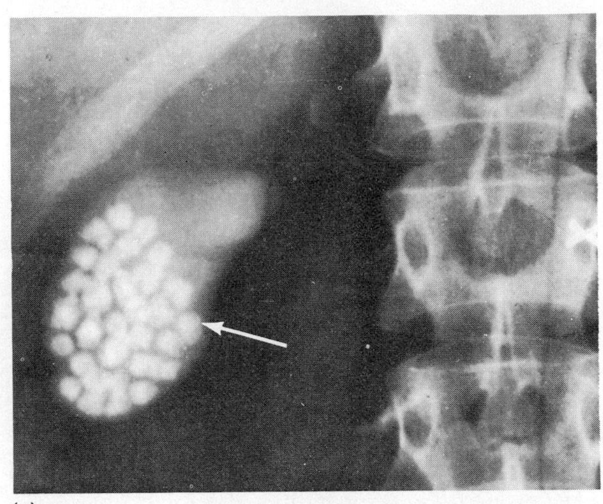

(a)

(b)

Pancreas

The **pancreas** is a soft, lobulated, glandular organ that has both exocrine and endocrine functions. The endocrine function is performed by clusters of cells called the **pancreatic islets (islets of Langerhans),** which secrete the hormones insulin and glucagon into the blood. As an exocrine gland, the pancreas secretes pancreatic juice through the pancreatic duct (fig. 25.46) into the duodenum. The pancreas is positioned horizontally along the posterior abdominal wall, adjacent to the greater curvature of the stomach.

pancreas: Gk. *pan,* all; *kreas,* flesh
islets of Langerhans: from Paul Langerhans, German anatomist, 1847–1888

Figure 25.46. The pancreas is both an exocrine and an endocrine gland. Pancreatic juice—the exocrine product—is secreted by acinar cells into the pancreatic duct. Scattered "islands" of cells, called the pancreatic islets (islets of Langerhans), secrete the hormones insulin and glucagon into the blood.

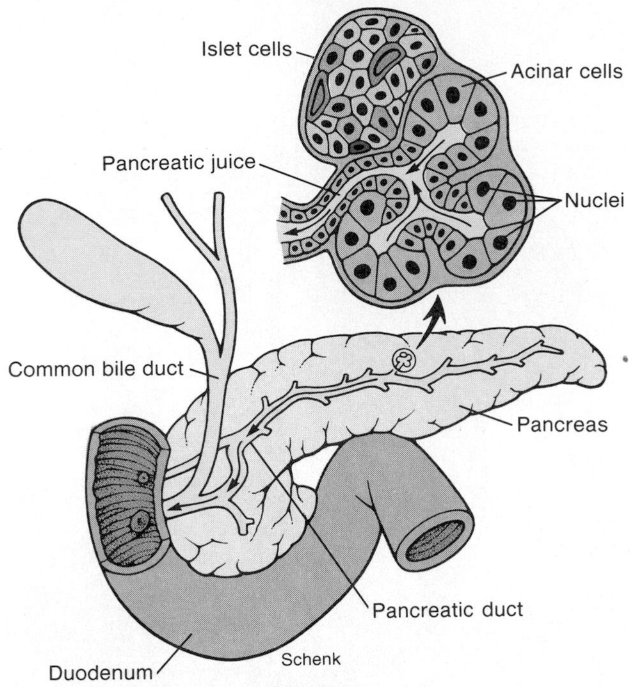

The pancreas is about 12.5 cm (6 in.) long and 2.5 cm (1 in.) thick. It has an expanded **head** near the duodenum, centrally located **body,** and a tapering **tail.** All but a portion of the head is positioned retroperitoneally. Within the lobules of the pancreas are the exocrine secretory units, called **acini** *(as'ĭ-ni).* Each acinus consists of a single layer of epithelial cells surrounding a lumen into which the constituents of pancreatic juice are secreted.

The pancreas is innervated by branches of the celiac plexus. The glandular portion of the pancreas receives parasympathetic innervation, whereas the pancreatic blood vessels receive sympathetic innervation. The pancreas is supplied with blood by the pancreatic branch of the splenic artery arising from the celiac artery and by the pancreatoduodenal branches from the superior mesenteric artery. Venous blood is returned through the splenic and superior mesenteric veins into the hepatic portal vein.

Pancreatic cancer has the worst prognosis of all types of cancer. This is probably because of the spongy, vascular nature of this organ and its vital exocrine and endocrine functions. Pancreatic surgery is a problem because the soft, spongy tissue is difficult to suture.

acinus: L. *acinus,* grape

Figure 25.47. The pancreatic protein-digesting enzyme *trypsin* is secreted in an inactive form known as trypsinogen. This inactive enzyme (zymogen) is activated by a brush border enzyme, enterokinase (*EN*), located in the cell membrane of microvilli. Active trypsin in turn activates other zymogens in pancreatic juice.

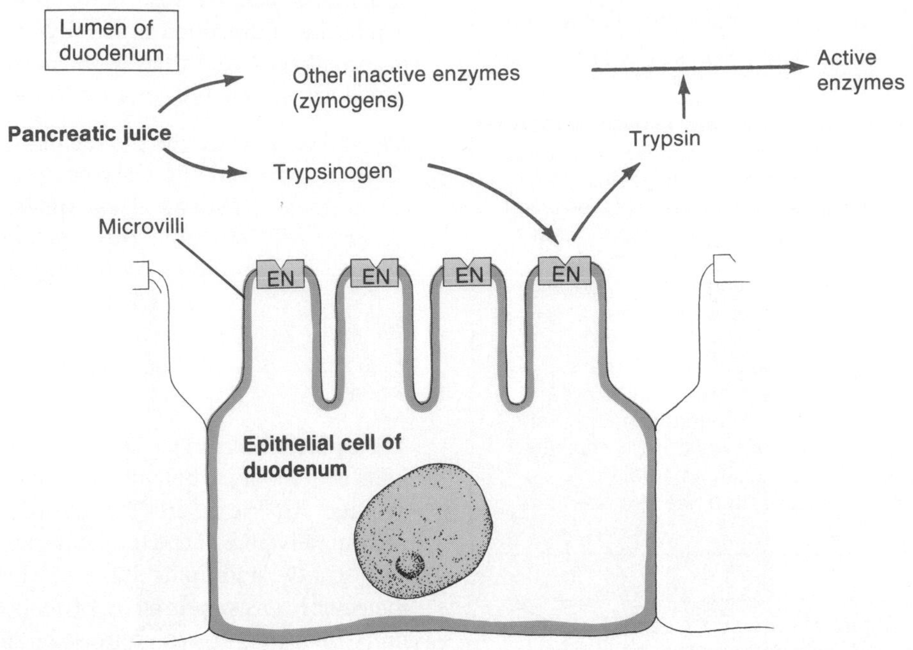

Table 25.12 Enzymes in pancreatic juice

Enzyme	Zymogen	Activator	Action
Trypsin	Trypsinogen	Enterokinase	Cleaves internal peptide bonds
Chymotrypsin	Chymotrypsinogen	Trypsin	Cleaves internal peptide bonds
Elastase	Proelastase	Trypsin	Cleaves internal peptide bonds
Carboxypeptidase	Procarboxypeptidase	Trypsin	Cleaves last amino acid from carboxyl-terminal end of polypeptide
Phospholipase	Prophospholipase	Trypsin	Cleaves fatty acids from phospholipids such as lecithin
Lipase	None	None	Cleaves fatty acids from glycerol
Amylase	None	None	Digests starch to maltose and short chains of glucose molecules
Cholesterolesterase	None	None	Releases cholesterol from its bonds with other molecules
Ribonuclease	None	None	Cleaves RNA to form short chains
Deoxyribonuclease	None	None	Cleaves DNA to form short chains

Pancreatic Juice. Pancreatic juice contains water, bicarbonate, and a wide variety of digestive enzymes that are secreted into the duodenum. These enzymes include (1) **amylase,** which digests starch; (2) **trypsin** *(trip'sin),* which digests protein; and (3) **lipase,** which digests triglycerides; other pancreatic enzymes are indicated in table 25.12. It should be noted that the complete digestion of food molecules in the small intestine requires the action of both pancreatic enzymes and brush border enzymes.

Most pancreatic enzymes are produced as inactive molecules, or *zymogens,* which help to minimize the risk of self-digestion within the pancreas. The inactive form of trypsin, called trypsinogen, is activated within the small intestine by the catalytic action of the brush border enzyme *enterokinase.* Enterokinase converts trypsinogen to active trypsin. Trypsin, in turn, activates the other zymogens of pancreatic juice (fig. 25.47) by cleaving off polypeptide sequences that inhibit the activity of these enzymes.

---

zymogen: Gk. *zyme,* leaven

The activation of trypsin is, therefore, the triggering event for the activation of other pancreatic enzymes. Actually, the pancreas does produce small amounts of active trypsin, yet the other enzymes don't become active until pancreatic juice enters the duodenum. This is because pancreatic juice also contains a small protein called *pancreatic trypsin inhibitor,* which attaches to trypsin and inactivates it in the pancreas.

Inflammation of the pancreas may result when the various safeguards against self-digestion are insufficient. *Acute pancreatitis* is believed to be caused by the reflux of pancreatic juice and bile from the duodenum into the pancreatic duct. The leakage of trypsin into the blood also occurs, but trypsin is inactive in the blood because of the inhibitory action of two plasma proteins, alpha-1-antitrypsin and alpha-2-macroglobulin. Pancreatic amylase may also leak into the blood, but it is not active because its substrate (glycogen) is not present in blood. Pancreatic amylase activity can be measured *in vitro,* however, and these measurements are commonly performed to assess the health of the pancreas.

1. Describe the structure of liver lobules, and trace the pathways for the flow of blood and bile in the lobules.
2. Describe the composition and function of bile, and trace the flow of bile from the liver and gallbladder to the duodenum.
3. Explain how the liver inactivates and excretes compounds such as hormones and drugs.
4. Describe the enterohepatic circulation of bilirubin and urobilinogen.
5. Explain how the liver helps maintain a constant blood glucose concentration and how the pattern of venous blood flow permits this task to be performed.
6. Describe the endocrine and exocrine structures and functions of the pancreas, and explain why the pancreas does not normally digest itsef.

Digestion and Absorption of Carbohydrates, Lipids, and Proteins

Polysaccharides and polypeptides are digested into their subunits, which are absorbed and secreted into the blood capillaries. Emulsified fat is digested, absorbed into the intestinal cells, and then resynthesized into triglycerides, which are secreted as small particles in the lymph.

Objective 26. Describe the regions of the digestive system and the enzymes involved with the digestion and absorption of carbohydrates.

Objective 27. Describe the digestion and absorption of proteins, indicating the enzymes involved and their location in the GI tract.

Objective 28. Describe the actions of bile and enzymes in lipid digestion, and explain how lipids are absorbed.

The caloric (energy) value of food is found predominantly in its content of carbohydrates, lipids, and proteins. In the average American diet, carbohydrates account for approximately 50% of the total calories, protein accounts for 11% to 14%, and lipids make up the balance. These food molecules consist primarily of long combinations of subunits (monomers), which must be digested by hydrolysis reactions into the free monomers before absorption can occur. The characteristics of the major digestive enzymes are summarized in table 25.13.

Digestion and Absorption of Carbohydrates

Most of the ingested carbohydrates are in the form of starch, which is a long polysaccharide of glucose in the form of straight chains with occasional branchings. The most commonly ingested sugars are the disaccharides sucrose (table sugar, consisting of glucose and fructose) and lactose (milk sugar, consisting of glucose and galactose).

Table 25.13 Summary of the sources and activities of the major digestive enzymes

Region or source					
Organ	**Source**	**Substrate**	**Enzymes**	**Optimum pH**	**Products**
Mouth	Saliva	Starch	Salivary amylase	6.7	Maltose
Stomach	Gastric glands	Protein	Pepsin	1.6–2.4	Shorter polypeptides
Duodenum	Pancreatic juice	Starch	Pancreatic amylase	6.7–7.0	Maltose, maltriose, and oligosaccharides
		Polypeptides	Trypsin, chymotrypsin, carboxypeptidase	8.0	Amino acids, dipeptides, and tripeptides
		Triglycerides	Pancreatic lipase	8.0	Fatty acids and monoglycerides
	Epithelial membranes	Maltose	Maltase	5.0–7.0	Glucose
		Sucrose	Sucrase	5.0–7.0	Glucose + fructose
		Lactose	Lactase	5.8–6.2	Glucose + galactose
		Polypeptides	Aminopeptidase	8.0	Amino acids, dipeptides, tripeptides

The digestion of starch begins in the mouth with the action of **salivary amylase,** or **ptyalin** *(ti'ah-lin).* This enzyme cleaves some of the bonds between adjacent glucose molecules, but most people don't chew their food long enough for sufficient digestion to occur in the mouth. The digestive action of salivary amylase stops when the bolus enters the stomach because this enzyme is inactivated at the low pH of gastric juice.

The digestion of starch, therefore, occurs mainly in the duodenum as a result of the action of **pancreatic amylase.** This enzyme cleaves the straight chains of starch to produce the disaccharide *maltose* and the trisaccharide *maltriose.* Pancreatic amylase, however, cannot hydrolyze the bond between glucose molecules at the branch points in the starch. As a result, short, branched chains of glucose molecules, called *oligosaccharides,* are released together with maltose and maltriose by the activity of this enzyme (fig. 25.48).

Maltose, maltriose, and oligosaccharides of glucose released from partially digested starch, together with the disaccharides sucrose and lactose, are hydrolyzed to their monosaccharides by brush border enzymes located on the microvilli of the epithelial cells in the small intestine. The absorption of these monosaccharides across the membrane of the microvilli occurs by means of *coupled transport.* In this process, glucose binds to the same carrier as Na^+ and enters the epithelial cell as Na^+ diffuses down its electrochemical gradient. This is a type of active transport, because energy from ATP is needed to maintain the Na^+ gradient. Glucose is then secreted from the epithelial cells into capillaries within the villi.

Digestion and Absorption of Proteins

Protein digestion begins in the stomach with the action of pepsin. Pepsin results in the liberation of some amino acids, but the major products of pepsin digestion are short-chain polypeptides. This activity helps to produce a more homogenous chyme, but it is not essential for the complete digestion of protein that occurs—even in people with total gastrectomies—in the small intestine.

Most protein digestion occurs in the duodenum and jejunum. The pancreatic juice enzymes **trypsin, chymotrypsin** *(ki''mo-trip'sin),* and **elastase** cleave peptide bonds within the interior of the polypeptide chains. These enzymes are thus grouped together as *endopeptidases (en''do-pep'tĭ-dās).* Enzymes that remove amino acids from the ends of polypeptide chains, in contrast, are *exopeptidases.* These include the pancreatic juice enzyme **carboxypeptidase,** which removes amino acids from the carboxyl-terminal end of polypeptide chains, and the brush border enzyme **aminopeptidase.** Aminopeptidase cleaves amino acids from the amino-terminal end of polypeptide chains.

As a result of the action of these enzymes, polypeptide chains are digested into free amino acids, dipeptides, and tripeptides. The free amino acids are absorbed through the epithelial cells of the intestinal mucosa and secreted into blood capillaries. This absorption is carrier-mediated, involving the coupled transport of free amino acids with Na^+, and uses four different carrier systems for different classes of amino acids. The dipeptides and tripeptides may enter epithelial cells by a different carrier system, but they are then digested within these cells into amino acids, which

oligosaccharide: Gk. *oligos,* little; *sakcharon,* sugar

Figure 25.48. Pancreatic amylase digests starch into maltrose, maltriose, and short oligosaccharides containing branch points in the chain of glucose molecules.

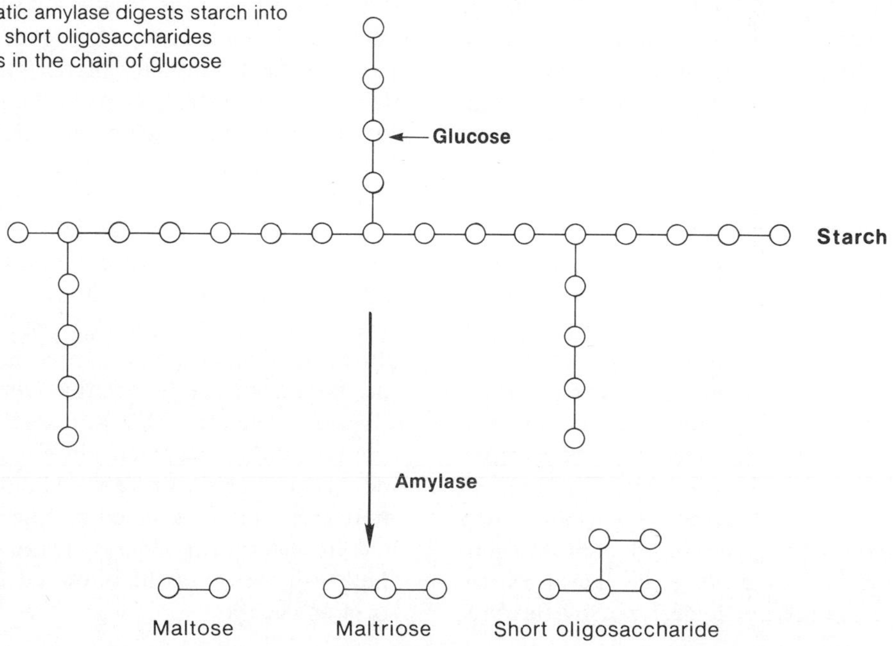

Figure 25.49. Polypeptide chains are digested into free amino acids, dipeptides, and tripeptides by the action of pancreatic juice enzymes and brush border enzymes. The amino acids, dipeptides, and tripeptides enter duodenal epithelial cells. Dipeptides and tripeptides are hydrolyzed into free amino acids within the epithelial cells, and these products are secreted into capillaries that carry them to the hepatic portal vein.

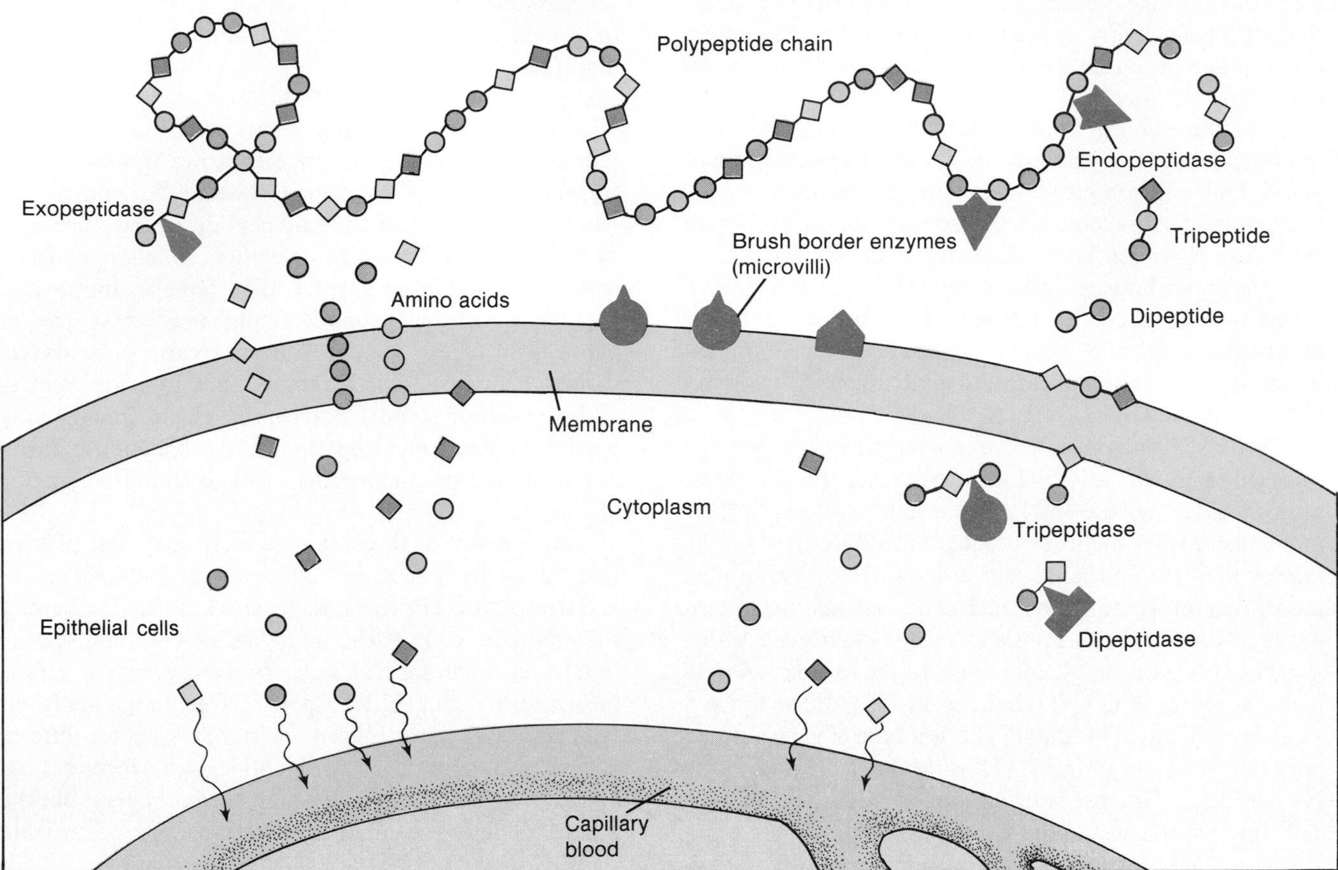

are secreted into the blood (fig. 25.49). Newborn babies appear to be capable of absorbing a substantial amount of undigested proteins (hence they can absorb antibodies from their mother's first milk); in adults, however, only the free amino acids enter the hepatic portal vein. Foreign food protein, which would be very antigenic, does not normally enter the blood.

Digestion and Absorption of Lipids

Although the salivary glands and stomach produce lipases, there is very little fat digestion until the fat in chyme arrives in the duodenum in the form of fat globules. Through mechanisms described in the next section, the arrival of fat in the duodenum serves as a stimulus for the secretion of bile. Mixed micelles of bile salts, lecithin, and cholesterol are secreted into the duodenum and act to break up the fat droplets into much finer droplets. This process, called **emulsification,** results in the formation of tiny *emulsification droplets* of triglycerides. Note that emulsification is not chemical digestion—the bonds joining glycerol and fatty acids are not hydrolyzed by this process.

Digestion of Lipids. The emulsification of fat aids digestion because the smaller and more numerous emulsification droplets present a greater surface area than the unemulsified fat droplets that originally entered the duodenum. Fat digestion occurs at the surface of the droplets through the enzymatic action of **pancreatic lipase,** which is aided in its action by a protein called *colipase,* also secreted by the pancreas, which coats the emulsification droplets and "anchors" the lipase enzyme to the droplets. Through hydrolysis, lipase removes two of the three fatty acids from each triglyceride molecule and thus liberates *free fatty acids* and *monoglycerides* (fig. 25.50). **Phospholipase A** likewise digests phospholipids such as lecithin into fatty acids and lysolecithin (the remainder of the lecithin molecule after two fatty acids are removed).

Free fatty acids, monoglycerides, and lysolecithin are more polar than the undigested lipids and are able to move more easily into the mixed micelles of bile salts, lecithin, and cholesterol (fig. 25.51). These micelles then move to the brush border of the intestinal epithelium, where absorption occurs.

Figure 25.50. Pancreatic lipase digests fat (triglycerides) by cleaving off the first and third fatty acids. This produces free fatty acids and monoglycerides. Sawtooth structures indicate hydrocarbon chains in the fatty acids.

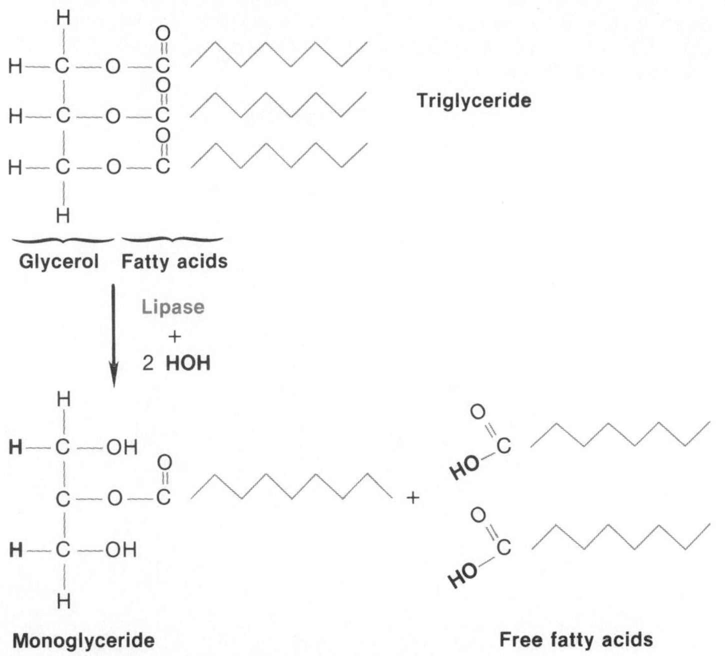

Triglyceride

Glycerol Fatty acids

Lipase + 2 HOH

Monoglyceride

Free fatty acids

Figure 25.51. Steps in the digestion of fat (triglycerides) and the entry of fat digestion products (fatty acids and monoglycerides) into micelles of bile salts secreted by the liver.

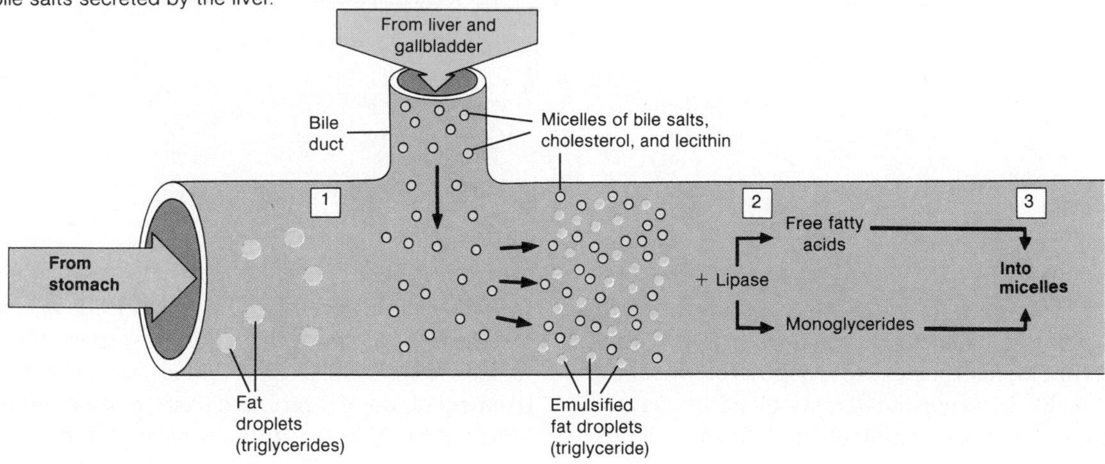

Step 1 Emulsification of fat droplets by bile salts

Step 2 Hydrolysis of triglycerides in emulsified fat droplets into fatty acid and monoglycerides

Step 3 Dissolving of fatty acids and monoglycerides into micelles to produce "mixed micelles"

Absorption of Lipids. Free fatty acids, monoglycerides, and lysolecithin can leave the micelles and pass through the membrane of the microvilli to enter the intestinal epithelial cells. There is also some evidence that the micelles may be transported intact into the epithelial cells and that the lipid digestion products may be removed intracellularly from the micelles. In either event, these products are used to *resynthesize* triglycerides and phospholipids within the epithelial cells. This is different from the absorption of amino acids and monosaccharides, which pass through the epithelial cells without being altered.

Triglycerides, phospholipids, and cholesterol are then combined with protein inside the epithelial cells to form small particles called **chylomicrons** *(ki″lo-mi′krons).*

Figure 25.52. Fatty acids and monoglycerides from the micelles within the small intestine are absorbed by epithelial cells and converted intracellularly into triglycerides. These are then combined with protein to form chylomicrons, which enter the lymphatic vessels (lacteals) of the villi. These lymphatic vessels transport the chylomicrons to the thoracic duct, which empties them into the venous blood.

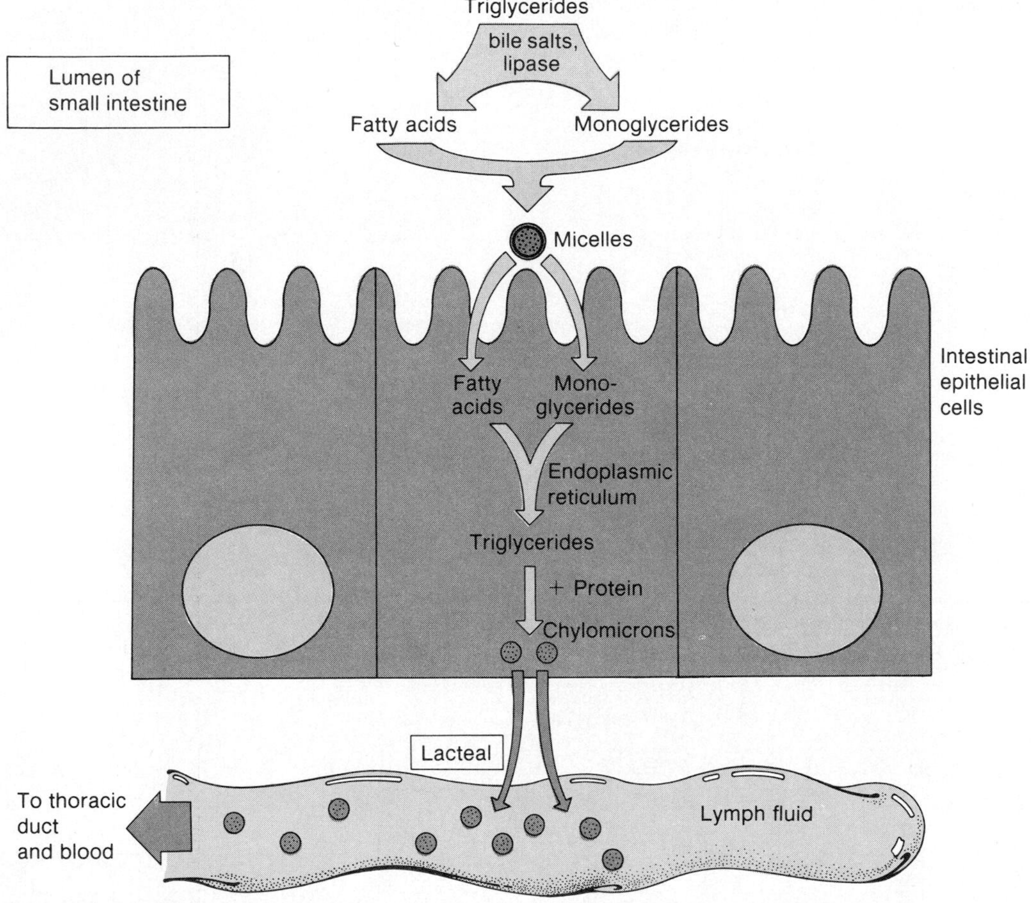

These tiny combinations of lipid and protein are secreted into the lymphatic capillaries of the intestinal villi (fig. 25.52). Absorbed lipids thus pass through the lymphatic system, eventually entering the venous blood by way of the thoracic duct. The absorption of lipids is thus significantly different from that of amino acids and monosaccharides, which enter the hepatic portal blood.

Once the chylomicrons are in the blood, their triglyceride content is removed by the enzyme **lipoprotein lipase,** which is attached to the endothelium of blood vessels. This enzyme hydrolyzes triglycerides and thus provides free fatty acids and glycerol for use by the tissue cells. The remaining *remnant particles,* containing cholesterol, are taken up by the liver; this is a process of endocytosis, which requires membrane receptors for the protein part (or *apoprotein*) of the remnant particle.

Cholesterol and triglycerides produced by liver cells are combined with other apoproteins and secreted into the blood as *very-low-density lipoproteins (VLDL),* which serve to deliver these triglycerides to different organs. Once the triglycerides are removed, the VLDL is converted to *low-density lipoproteins,* which transport cholesterol to various organs, including blood vessels (chapter 20). Excess cholesterol is returned from these organs to the liver attached to *high-density lipoproteins (HDL).* These various lipoproteins are summarized in table 25.14.

1. List the enzymes involved in carbohydrate digestion, indicating their origin, sites of action, substrates, and products.
2. List the enzymes involved in protein digestion, indicating their origins, sites of action, and mode of action (endopeptidase or exopeptidase). Compare the characteristics of pepsin and trypsin.
3. Describe how bile aids both the digestion and absorption of fats. Explain how the absorption of fat differs from the absorption of amino acids and monosaccharides.
4. Trace the pathway and fate of a molecule of triglyceride and cholesterol in a chylomicron within an intestinal epithelial cell.

Table 25.14 Summary of the characteristics of the lipid carrier proteins (lipoproteins) found in plasma

Lipoprotein class	Origin	Destination	Major lipids	Functions
Chylomicrons	Intestine	Many organs	Triglycerides, other lipids	Delivers lipids of dietary origin to body cells
Very-low density lipoproteins (VLDL)	Liver	Many organs	Triglycerides, cholesterol	Delivers endogenously produced triglycerides to body cells
Low-density lipoproteins (LDL)	Intravascular removal of triglycerides from VLDL	Blood vessels, liver	Cholesterol	Delivers endogenously produced cholesterol to various organs
High-density lipoproteins (HDL)	Liver and intestine	Liver and steroid hormone producing glands	Cholesterol	Transports cholesterol from various organs to liver

Neural and Endocrine Regulation of the Digestive System

Neural and hormonal mechanisms coordinate the activities of different regions of the GI tract and the actions of the liver, gallbladder, and pancreas.

Objective 29. Describe the mechanisms that regulate gastric juice secretion, and indicate those that belong to the cephalic, gastric, and intestinal phases of regulation.

Objective 30. Describe the neural and hormonal mechanisms that regulate the secretion of pancreatic juice and bile.

Objective 31. Explain the significance of the trophic effects of gastrointestinal hormones.

The motility and glandular secretions of the GI tract are, to a large degree, automatic. Neural and endocrine control mechanisms, however, can stimulate or inhibit these automatic functions to help coordinate the different stages of digestion. The sight, smell, or taste of food, for example, can stimulate salivary and gastric secretions by activation of the vagus nerve, which helps to "prime" the digestive system in preparation for a meal. Stimulation of the vagus, in this case, originates in the brain and is a conditioned reflex (as Pavlov demonstrated by training dogs to salivate in response to a bell). The vagus nerve is also involved in the reflex control of one part of the digestive system by another—these are "short reflexes," which don't involve the brain.

The GI tract is both an endocrine gland and a target for the action of various hormones. Indeed, the first hormones to be discovered were gastrointestinal hormones. In 1902, Bayliss and Starling discovered that the duodenum produced a chemical regulator, which they named **secretin** *(se-kre'tin);* in 1905, these scientists proposed that secretin was but one of many yet undiscovered chemical regulators produced by the body. They coined the term *hormones* for this new class of regulators. Other investigators in 1905 discovered that an extract from the stomach antrum (pyloric region) stimulated gastric secretion. The hormone **gastrin** was thus the second hormone to be discovered.

The chemical structures of gastrin, secretin, and the duodenal hormone **cholecystokinin** *(ko''le-sis''to-kin'in),* CCK, were determined in the 1960s. More recently, a fourth hormone produced by the small intestine, **gastric inhibitory peptide** *(GIP),* has been added to the list of proven GI tract hormones. The effects of these hormones are summarized in table 25.15.

Regulation of Gastric Function

Gastric motility and secretion are, to some extent, automatic. Waves of contraction that serve to push chyme through the pyloric sphincter, for example, are initiated spontaneously by pacesetter cells in the greater curvature of the stomach. The secretion of hydrochloric acid (HCl) and pepsinogen, likewise, can be stimulated in the absence of neural and hormonal influences by the presence of cooked or partially digested protein in the stomach. The effects of autonomic nerves and hormones are superimposed on this automatic activity. The extrinsic control of gastric function is conveniently divided into three phases: (1) the cephalic phase; (2) the gastric phase; and (3) the intestinal phase. These are summarized in table 25.16.

Cephalic Phase. The cephalic phase of gastric regulation refers to control by the brain via the vagus nerves. As previously discussed, various conditioned stimuli can evoke gastric secretion. Activation of the vagus nerves can stimulate HCl and pepsinogen secretion by three mechanisms: (1) direct vagal stimulation of the gastric parietal and chief cells (the primary mechanism); (2) vagal stimulation of gastrin secretion by the G cells, which in turn stimulates the parietal and chief cells to secrete HCl and pepsinogen, respectively; and (3) stimulation of secretion by increases in gastric blood flow.

Pavlov: Ivan Petrovich Pavlov, Russian physiologist, 1849–1936
Starling: Ernest Henry Starling, English physiologist, 1866–1927

Table 25.15 Summary of the physiological effects of gastrointestinal hormones

Secreted by	Hormone	Effects
Stomach	Gastrin	Stimulates parietal cells to secrete HCl Stimulates chief cells to secrete pepsinogen Maintains structure of gastric mucosa
Small intestine	Secretin	Stimulates water and bicarbonate secretion in pancreatic juice Potentiates actions of cholecystokinin on pancreas
Small intestine	Cholecystokinin (CCK)	Stimulates contraction of the gallbladder Stimulates secretion of pancreatic juice enzymes Potentiates action of secretin on pancreas Maintains structure of exocrine pancreas (acini)
Small intestine	Gastric inhibitory peptide (GIP)	Inhibits gastric emptying Inhibits gastric acid secretion Stimulates secretion of insulin from endocrine pancreas (pancreatic islets)

Table 25.16 The cephalic, gastric, and intestinal phases in the regulation of gastric acid secretion

Phase of Regulation	Description
Cephalic phase	1. Sight, smell, and taste of food cause stimulation of vagus nuclei in brain 2. Vagus stimulates acid secretion a) Direct stimulation of parietal cells (major effect) b) Stimulation of gastrin secretion; gastrin stimulates acid secretion (lesser effect)
Gastric phase	1. Distension of stomach stimulates vagus nerve; vagus stimulates acid secretion 2. Amino acids and peptides in stomach lumen stimulate acid secretion a) Direct stimulation of parietal cells (lesser effect) b) Stimulation of gastrin secretion; gastrin stimulates acid secretion (major effect) 3. Gastrin secretion inhibited when pH of gastric juice falls below 2.5
Intestinal phase	1. Neural inhibition of gastric emptying and acid secretion a) Arrival of chyme in duodenum causes distension, increase in osmotic pressure b) These stimuli activate a neural reflex that inhibits gastric activity 2. Gastric inhibitory peptide (GIP) secreted by duodenum in response to fat in chyme; GIP inhibits gastric acid secretion

Gastric Phase. The presence of short polypeptides and amino acids in the stomach stimulates the G cells to secrete gastrin and the parietal and chief cells to secrete HCl and pepsinogen; this is called the gastric phase of regulation. Since gastrin also stimulates HCl and pepsinogen secretion, a *positive feedback mechanism* develops: as more HCl and pepsinogen are secreted, more short polypeptides and amino acids are released from the ingested protein, thus stimulating more secretion of gastrin and, therefore, more secretion of HCl and pepsinogen (fig. 25.53).

Secretion of HCl during the gastric phase is also regulated by a *negative feedback mechanism.* As the pH of gastric juice drops, so does the secretion of gastrin: at a pH of 2.5 gastrin secretion is reduced, and at a pH of 1.0 gastrin secretion is totally abolished. The secretion of HCl thus declines accordingly. The presence of proteins and polypeptides in the stomach help to buffer the acid and thus to prevent a rapid fall in gastric pH; more acid can thus be secreted when proteins are present than when they are absent. The arrival of protein into the stomach thus stimulates acid secretion in two ways—by the positive feedback mechanism previously discussed and by lessening of the negative feedback control of acid secretion. The amount of acid secreted is, by means of these effects, closely matched to the amount of protein ingested. As the stomach is emptied, the protein buffers leave, the pH thus falls, and the secretion of gastrin and HCl is accordingly inhibited.

Intestinal Phase. The intestinal phase of gastric regulation refers to the inhibition of gastric activity when chyme enters the small intestine. Investigators in 1886 demonstrated that the addition of olive oil to a meal inhibits gastric emptying, and in 1929 it was shown that the presence of fat inhibits gastric juice secretion. This inhibitory intestinal phase of gastric regulation is due to both a neural reflex originating from the duodenum and to a hormone secreted by the duodenum.

The arrival of chyme into the duodenum increases its osmolality. This stimulus, together with stretch of the duodenum and possibly other stimuli, produces a neural reflex that results in the inhibition of gastric motility and secretion. The presence of fat in the chyme also stimulates the duodenum to secrete a hormone that inhibits gastric function. This inhibitory hormone is believed to be gastric inhibitory peptide (GIP).

Figure 25.53. The stimulation of gastric acid (HCl) secretion by the presence of proteins in the stomach lumen and by the hormone gastrin. The secretion of gastrin is inhibited by gastric acidity. This forms a negative feedback loop.

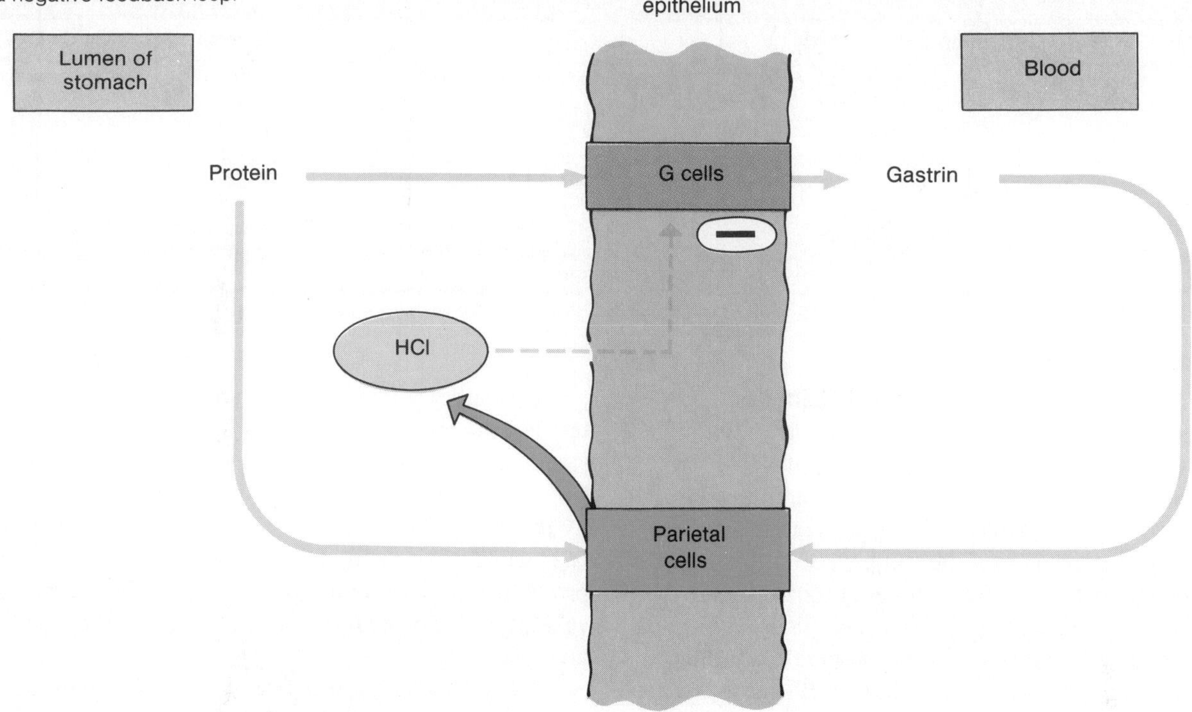

The inhibitory neural and endocrine mechanisms during the intestinal phase prevent the further passage of chyme from the stomach to the duodenum. This gives the duodenum time to process the load of chyme that it has previously received. Since GIP is stimulated by fat in the chyme, a breakfast of bacon and eggs takes longer to pass through the stomach—and makes one feel "fuller" for a longer time—than does a breakfast of pancakes and syrup.

Regulation of Pancreatic Juice and Bile Secretion

The arrival of chyme into the duodenum stimulates the intestinal phase of gastric regulation and, at the same time, stimulates reflex secretion of pancreatic juice and bile. The entry of new chyme is thus retarded as the previous load is digested. The secretion of pancreatic juice and bile is stimulated by both neural reflexes initiated in the duodenum and by secretion of the duodenal hormones cholecystokinin (CCK) and secretin.

Pancreatic Juice. The secretion of pancreatic juice is stimulated by both secretin and CCK. The release of secretin occurs in response to a fall in duodenal pH below 4.5; this pH fall occurs for only a short time, however, because the acidic chyme is rapidly neutralized by alkaline pancreatic juice. The secretion of CCK occurs in response to the fat content of chyme in the duodenum.

Secretin stimulates the production of bicarbonate by the pancreas. Since bicarbonate neutralizes the acidic chyme and since secretin is released in response to the low pH of chyme, this completes a negative feedback loop where the effects of secretin inhibit its secretion. Cholecystokinin, in contrast, stimulates the production of pancreatic enzymes such as trypsin, lipase, and amylase. Secretin and CCK can have different effects on the same cells (the pancreatic acinar cells) because their actions are mediated by different intracellular compounds that act as second messengers. The second messenger of secretin action is cyclic AMP, whereas the second messenger for CCK is Ca^{++} (table 25.17).

Secretion of Bile. The liver secretes bile continuously, but this secretion is greatly augmented following a meal. This increased secretion is due to the release of secretin and CCK from the duodenum. Secretin is the major stimulator of bile secretion by the liver, and CCK enhances this effect. The arrival of chyme in the duodenum also causes the gallbladder to contract and eject bile. Contraction of the gallbladder occurs in response to neural reflexes from the duodenum and in response to stimulation by CCK.

Table 25.17 Regulation of pancreatic juice and bile secretion by the hormones *secretin* and *cholecystokinin* (CCK) and by the neurotransmitter *acetylcholine,* released from parasympathetic nerve endings

Description	Secretin	CCK	Acetylcholine (vagus nerve)
Stimulus for release	Acidity of chyme decreases duodenal pH below 4.5	Fat and protein in chyme	Sight, smell of food; distension of stomach
Second messenger	Cyclic AMP	Ca^{++}	Ca^{++}
Effect on pancreatic juice	Stimulates water and bicarbonate secretion; potentiates action of CCK	Stimulates enzyme secretion; potentiates action of secretin	Stimulates enzyme secretion
Effect on bile	Stimulates secretion	Potentiates action of secretin; stimulates contraction of gallbladder	Stimulates contraction of gallbladder

Trophic Effects of Gastrointestinal Hormones

Patients with tumors of the stomach pylorus have high acid secretion and hyperplasia (growth) of the gastric mucosa. Surgical removal of the pylorus reduces gastric secretion and prevents growth of the gastric mucosa. Patients with peptic ulcers are sometimes treated by gastric vagotomy—cutting of gastric branches of the vagus nerves. A gastric vagotomy also reduces acid secretion but has no effect on the gastric mucosa. These observations suggest that the hormone gastrin, secreted by the pyloric mucosa, may exert stimulatory, or *trophic,* effects on the gastric mucosa. The structure of the gastric mucosa, in other words, is dependent on the effects of gastrin.

In the same way, the structure of the acinar (exocrine) cells of the pancreas is dependent upon the trophic effects of CCK. Perhaps this explains why the pancreas, as well as the stomach, atrophies during starvation. Since neural reflexes appear to be capable of regulating digestion, perhaps the primary function of the GI hormone is trophic—that is, maintenance of the structure of their target organs.

1. Describe the positive and negative feedback mechanisms that operate during the gastric phase of HCl and pepsinogen secretion.
2. Describe the mechanisms involved in the intestinal phase of gastric regulation, and explain why a high-fat meal takes longer to leave the stomach than a low-fat meal.
3. Explain the hormonal mechanisms involved in the production and release of pancreatic juice and bile.

Clinical Considerations

Pathogens and Poisons

The GI tract is a suitable environment for an array of microorganisms. Many of these are beneficial, but some bacteria and protozoa can cause diseases. Only a few examples of these will be discussed.

trophic: Gk. *trophikos,* pertaining to nutrition

Dysentery *(dis'en-ter''e)* is an inflammation of the intestinal mucosa, characterized by the discharge of loose stools that contain mucus, pus, and blood. The most common dysentery is **amoebic dysentery,** which is caused by the protozoan *Entamoeba histolytica.* Cysts from this organism are ingested in contaminated food, and after the protective coat is removed by HCl in the stomach, the vegetative form invades the mucosal walls of the ileum and colon.

Food poisoning comes from ingesting pathogenic bacteria or their toxins. *Salmonella* is a bacterium that commonly causes food poisoning. **Botulism** is the most serious type of food poisoning and is caused by ingesting food contaminated with the bacterium *Clostridium botulinum.* This organism is widely distributed in nature, and the spores from it are frequently on food being processed by canning. For this reason food must be heated to 120°C (248°F) before it is canned. It is the toxins produced by the bacterium growing in the food that are pathogenic, rather than the organisms themselves. The poison is a neurotoxin that is readily absorbed in the blood, where it affects the nervous system.

Clinical Problems of the Teeth and Salivary Glands

Dental caries, or tooth decay, is the gradual decalcification of tooth enamel (fig. 25.54) and underlying dentin, produced by the acid products of bacteria. These bacteria thrive between teeth where food particles accumulate and form part of the thin layer of bacteria, proteins, and other debris called *plaque* that covers teeth. The development of dental caries can thus be reduced by brushing at least once a day and by flossing between teeth at regular intervals.

People over the age of thirty-five are particularly susceptible to **periodontal diseases.** There are many types of periodontal disease, but basically they all cause inflammation and deterioration of the gingiva, alveolar sockets, periodontal membranes, and teeth cementum. Some of the

dysentery: Gk. *dys,* bad; *entera,* intestine

Figure 25.54. Clinical problems of the teeth. (*a*) trench mouth and dental caries. (*b*) severe alveolar bone destruction from periodontitis. (*c*) pyogenic granuloma and dental caries. (*d*) malposition of teeth.

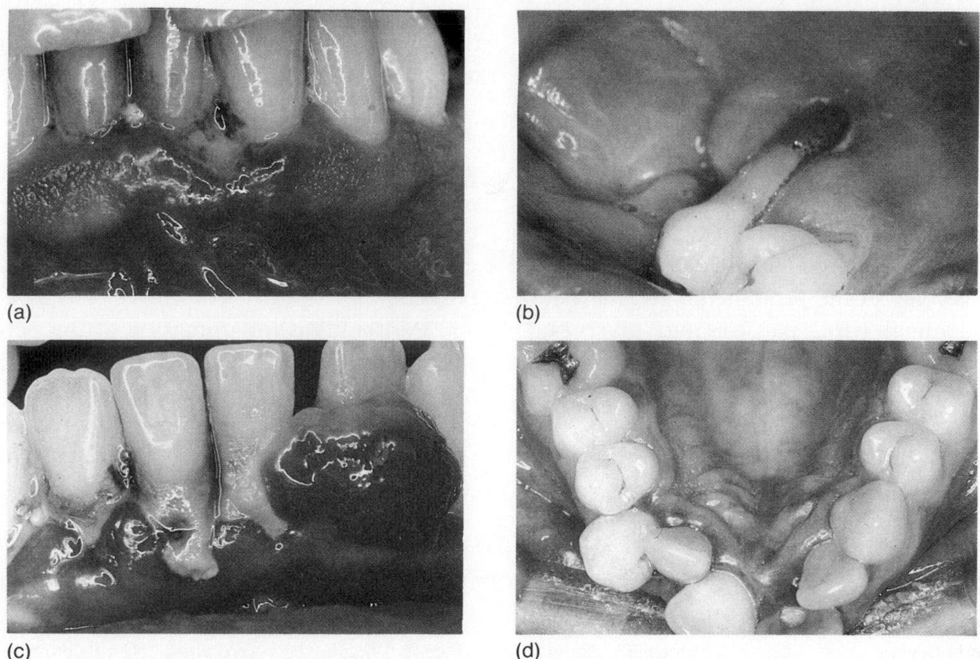

symptoms are loosening of the teeth, bad breath, bleeding gums when brushing, and some edema. Periodontal diseases are caused by many things, including impacted plaque, cigarette smoking, malocclusion, and poor diet.

Mumps is a viral disease of the parotid salivary glands and in advanced stages may involve the pancreas and testes. It is generally not serious in children but can be very serious in adults, in whom it may cause deafness and destroy the pancreatic islet tissue or testicular cells.

Disorders of the Liver

The liver is a remarkable organ that has the ability to regenerate itself even if up to 80 percent has been removed. The most serious diseases of the liver (hepatitis, cirrhosis, and hepatomas) affect the liver throughout so that it cannot repair itself. **Hepatitis** is inflammation of the liver. Certain chemicals may cause hepatitis, but generally it is caused by infectious viral agents. **Infectious hepatitis** is a viral disease transmitted through contaminated foods and liquids. **Serum hepatitis** is also caused by a virus and is transmitted in serum or plasma during transfusions or by improperly sterilized needles and syringes.

Jaundice is a yellow staining of the tissue produced by high blood concentrations of either free or conjugated bilirubin. Since free bilirubin is derived from heme, abnormally high concentrations of this pigment may result from an unusually high rate of red blood cell destruction. This can occur, for example, as a result of Rh disease (erythroblastosis fetalis) in an Rh positive baby born to a sensitized Rh negative mother. Jaundice may also occur in otherwise healthy infants because of the fact that red blood cells are normally destroyed at about the time of birth (hemoglobin concentrations decrease from 19 g per 100 ml to 14 g per 100 ml near the time of birth). This condition is called *physiological jaundice of the newborn* and is not indicative of disease. Premature infants may also develop jaundice because the hepatic enzymes that conjugate bilirubin (a reaction needed for the excretion of bilirubin in the bile) mature late in gestation. Jaundice due to high levels of conjugated bilirubin in the blood is commonly produced in adults when the excretion of bile is blocked by gallstones.

Newborn infants with jaundice are usually treated by *phototherapy,* in which they are placed under blue light in the 400–500 nm wavelength range. This light is absorbed by bilirubin in cutaneous vessels and results in the conversion of bilirubin to a more water-soluble isomer, which is soluble in plasma without having to be conjugated with glucuronic acid. The more water-soluble photoisomer of bilirubin can then be excreted in the bile.

Hepatomas *(hep''ah-to'mah)* are malignant tumors that originate in or secondarily invade the liver. Those that originate in the liver (primary hepatomas) are relatively rare, but those that secondarily metastasize to the liver from other organs (secondary hepatomas) are common. Carcinoma of the liver is usually fatal.

cirrhosis: Gk. *kirrhos,* yellow-orange
jaundice: L. *galbus,* yellow

Disorders of the GI Tract

Peptic ulcers are erosions of the mucous membranes of the stomach or duodenum produced by the action of HCl. Agents that weaken the mucosal lining of the stomach, such as alcohol and aspirin, and abnormally high secretions of HCl thus increase the likelihood of developing peptic ulcers. Many people subject to chronic stress produce too much gastric acid and develop a peptic ulcer as a result.

Enteritis is inflammation of the intestinal mucosa and is frequently referred to as intestinal flu. Causes of enteritis include bacterial or viral infections, irritating foods or fluids (including alcohol), and emotional stress. The symptoms are abdominal pain, nausea, and diarrhea. **Diarrhea** is the passage of watery, unformed stools. This condition is symptomatic of inflammation, stress, and other body dysfunctions.

A **hernia** is a protrusion of a portion of a visceral organ, usually the small intestine, through a weakened portion of the abdominal wall. Inguinal, femoral, umbilical, and hiatal hernias are the most common. A **hiatal hernia** occurs when a portion of the stomach pushes superiorly through the esophageal hiatus in the diaphragm and into the thorax. The potential dangers of a hernia are strangulation of the blood supply followed by gangrene, blockage of chyme, or rupture—each of which can threaten life.

Diverticulosis is a condition in which the intestinal wall weakens and an outpouching occurs. Recent studies suggest that suppressing the passage of flatus (intestinal gas) may contribute to diverticulosis, especially in the sigmoid colon. **Diverticulitis,** or inflammation of a diverticulum, can develop if fecal material becomes impacted in these pockets.

Peritonitis is inflammation of the peritoneum lining the abdominal cavity and covering the viscera. The causes of peritonitis include bacterial contamination of the abdominal cavity through accidental or surgical wounds in the abdominal wall or perforation of the intestinal wall (as with ulcers or a ruptured appendix).

Other Important Clinical Terminology

chilitis Inflammation of the lips.

colitis Inflammation of the colon and rectum.

colostomy The formation of an abdominal exit from the GI tract by bringing a loop of the colon to the surface of the abdomen. If the rectum is removed because of cancer, the colostomy provides a permanent outlet for the feces.

cystic fibrosis An inherited disease of the exocrine glands, particularly the pancreas. Pancreatic secretions are too thick to drain easily, causing the ducts to become inflamed and promoting connective tissue formation that occludes drainage from the ducts still further.

gingivitis Inflammation of the gums. It may result from improper hygiene, poorly fitted dentures, improper diet, or certain infections.

halitosis Offensive breath odor. It may result from dental caries, certain diseases, or eating particular foods.

heartburn A burning sensation of the esophagus and stomach. It may result from the regurgitation of gastric juice into the lower portion of the esophagus.

hemorrhoids Varicose veins of the rectum and anus.

jejunoileal bypass A surgical procedure for creating a bypass of a considerable portion of the small intestine. It reduces the absorptive capacity of the small intestine and is thus used to control extreme obesity.

nausea Gastric discomfort and sensations of illness with a tendency to vomit. This feeling is symptomatic of many conditions (e.g., motion sickness, foul odors or sights, pregnancy, and diseases).

pyorrhea The discharge of pus at the base of the teeth at the gum line.

regurgitation (vomiting) The forceful expulsion of gastric contents into the mouth. Nausea and vomiting are common symptoms of almost any dysfunction of the digestive system.

trench mouth A contagious bacterial infection that causes inflammation, ulceration, and painful swelling of the floor of the mouth. Generally it is contracted through direct contact by kissing an infected person. Trench mouth can be treated by penicillin and other medications.

vagotomy The surgical removal of sections of both of the vagus nerves where they enter the stomach to eliminate nerve impulses that stimulate gastric secretion. This procedure is used to treat ulcers.

Chapter Summary

I. Introduction to the Digestive System
   A. The digestive system prepares ingested food for cellular utilization through a series of mechanical and chemical processes that reduce food to forms that can be absorbed through the intestinal wall and transported by the blood and lymph.
   B. The digestive system consists of a gastrointestinal tract and accessory digestive organs.

II. Embryological Development of the Digestive System
   A. The digestive tract is derived embryologically from the primitive gut, which is composed of endoderm.
   B. The structures of the mouth are derived from ectoderm of the stomodeum, and the epithelium of the anal canal is derived from ectoderm of the proctodeum; these structures become continuous with the gut endoderm.
   C. The liver and pancreas develop from buds that arise from the wall of the small intestine.

III. Serous Membranes and Tunics of the Gastrointestinal Tract
   A. Peritoneal membranes line the abdominal wall and cover the visceral organs; the GI tract is supported by a double layer of peritoneum called the mesentery.
   B. The layers (tunics) of the abdominal GI tract are, from the inside outward, mucosa, submucosa, muscularis, and serosa.

IV. Mouth, Pharynx, and Associated Structures
   A. The oral cavity is formed by the cheeks, lips, hard and soft palates, and tongue.
   B. The four most anterior pairs of teeth are the incisors and canines, which have one root each; the bicuspids and molars have two or three roots.
      1. The tooth is anchored to its bony alveolar socket by a periodontal membrane, which contains fibers that insert into the cementum of the tooth root.

2. Enamel forms the outer layer of the tooth crown; beneath the enamel is dentin.
3. The interior of a tooth contains a pulp cavity, which is continuous through the apical foramen of the root with the connective tissue around the tooth.

C. The major salivary glands are the parotid glands and the submandibular glands.

D. The muscular pharynx is a passageway connecting the oral and nasal cavities to the esophagus and trachea.

V. Esophagus and Stomach

A. Peristaltic waves of contraction push food through the lower esophageal sphincter into the stomach.

B. Swallowing, or deglutition, occurs in three phases and involves structures of the buccal cavity, pharynx, and esophagus.

C. The stomach consists of a cardia, fundus, body, and pylorus (antrum), which ends with the pyloric sphincter.

1. The lining of the stomach is thrown into folds, or rugae, and the mucosa is formed into gastric pits and gastric glands.
2. The parietal cells of the gastric glands secrete HCl, and the chief cells secrete pepsinogen.
3. In the acidic environment of gastric juice, pepsinogen is converted into the active protein-digesting enzyme called pepsin.

VI. Small Intestine

A. Regions of the small intestine include the duodenum, jejunum, and ileum; the common bile duct and pancreatic duct empty into the duodenum.

B. Fingerlike extensions of mucosa, called villi, project into the lumen, and at the bases of the villi the mucosa forms narrow pouches called the intestinal glands.

C. Digestive enzymes, called brush border enzymes, are located in the membranes of the microvilli.

D. The small intestine exhibits two major types of movements—peristalsis and segmentation.

VII. Large Intestine

A. The large intestine is divided into the cecum, colon, rectum, and anal canal.

1. The appendix is attached to the inferior medial margin of the cecum.
2. The colon consists of ascending, transverse, descending, and sigmoid portions.
3. Bulges in the walls of the large intestine are called haustra.

B. Three types of movements occur in the large intestine: peristalsis, haustral churning, and mass movement.

C. The large intestine absorbs water and electrolytes.

D. Defecation occurs when the anal sphincters relax and contraction of other muscles raises the rectal pressure.

VIII. Liver, Gallbladder, and Pancreas

A. The liver, the largest internal organ, is composed of functional units called lobules.

1. Liver lobules consist of plates of hepatic cells separated by capillary sinusoids.
2. Blood flows from the periphery of each lobule, where the hepatic artery and portal vein empty through the sinusoids and out the central vein.
3. Bile flows within the hepatocyte plates, in canaliculi, to the bile ducts.
4. Substances excreted in the bile can be returned to the liver in the hepatic portal blood; this is called an enterophepatic circulation.
5. The liver detoxifies the blood by excreting substances in the bile, by phagocytosis, and by chemical inactivation.
6. The liver modifies the plasma concentrations of proteins, glucose, triglycerides, and ketone bodies.

B. The gallbladder serves to store and concentrate the bile, and it releases bile through the cystic duct and common bile duct to the duodenum.

C. The pancreas is both an exocrine and an endocrine gland.

1. The endocrine portion is known as the pancreatic islets and secretes the hormones insulin and glucagon.
2. The exocrine acini of the pancreas produce pancreatic juice, which contains various digestive enzymes and bicarbonate.

IX. Digestion and Absorption of Carbohydrates, Lipids, and Proteins

A. The digestion of starch begins in the mouth through the action of salivary amylase.

1. Pancreatic amylase digests starch into disaccharides and short-chain oligosaccharides.
2. Complete digestion into monosaccharides is accomplished by brush border enzymes.

B. Protein digestion begins in the stomach by the action of pepsin.

1. Pancreatic juice contains protein-digesting enzymes, including trypsin, chymotrypsin, and others.
2. The brush border contains digestive enzymes that help to complete the digestion of proteins into amino acids.
3. Amino acids, like monosaccharides, are absorbed and secreted into capillary blood entering the portal vein.

C. Lipids are digested in the small intestine after being emulsified by bile salts.

1. Free fatty acids and monoglycerides enter particles called micelles, formed in large part by bile salts, and in this form, or as free molecules, they are absorbed.
2. Once inside the mucosal epithelial cells, these subunits are used to resynthesize triglycerides.
3. Triglycerides in the epithelial cells, together with proteins, form chylomicrons, which are secreted into the central lacteals of the villi.

X. Neural and Endocrine Regulation of the Digestive System

A. The regulation of gastric function occurs in three phases.

1. In the cephalic phase, the activity of higher brain centers, acting via the vagus nerve, stimulate gastric juice secretion.
2. In the gastric phase, the secretion of HCl and pepsin is controlled by the gastric contents and by the hormone gastrin, secreted by the gastric mucosa.
3. In the intestinal phase, the activity of the stomach is inhibited by neural reflexes from the duodenum and by gastric inhibitory peptide (GIP), secreted by the duodenum.

B. The secretion of the hormones secretin and cholecystokinin (CCK) regulate pancreatic juice and bile secretion.

C. Gastrointestinal hormones may be needed for the maintenance of the GI tract and accessory digestive organs.

Review Activities

Objective Questions

1. Which of the following types of teeth are found in the permanent but not in the deciduous dentition?
   (a) incisors
   (b) canines
   (c) premolars
   (d) molars
2. The epithelium of the mouth is derived from the
   (a) ectoderm of the stomodeum
   (b) ectoderm of the proctodeum
   (c) endoderm of the foregut
   (d) all of the above
3. A double layer of peritoneum that supports the GI tract is called the
   (a) visceral peritoneum
   (b) dorsal mesentery
   (c) greater omentum
   (d) lesser omentum
4. Which of the following tissue layers in the small intestine contains the central lacteals?
   (a) submucosa
   (b) muscularis mucosa
   (c) lamina propria
   (d) muscularis externa
5. Which of the following statements about gastric secretion of HCl is *false?*
   (a) HCl is secreted by parietal cells.
   (b) HCl hydrolyzes peptide bonds.
   (c) HCl is needed for the conversion of pepsinogen to pepsin.
   (d) HCl is needed for maximum activity of pepsin.
6. Intrinsic factor
   (a) is secreted by the stomach
   (b) is a polypeptide
   (c) promotes absorption of vitamin B_{12} in the intestine
   (d) helps prevent pernicious anemia
   (e) all of the above
7. Intestinal enzymes such as lactase are
   (a) secreted by the intestine into the chyme
   (b) produced by the intestinal glands (crypts of Lieberkühn)
   (c) produced by the pancreas
   (d) attached to the cell membrane of microvilli in the epithelial cells of the mucosa
8. Most digestion occurs in the
   (a) mouth
   (b) stomach
   (c) small intestine
   (d) large intestine
9. Which of the following statements about trypsin is true?
   (a) Trypsin is derived from trypsinogen by the digestive action of pepsin.
   (b) Active trypsin is secreted into the pancreatic acini.
   (c) Trypsin is produced in the intestinal glands (crypts of Lieberkühn).
   (d) Trypsinogen is converted to trypsin by the brush border enzyme enterokinase.
10. During the gastric phase, the secretion of HCl and pepsinogen is stimulated by
    (a) vagus nerve stimulation that originates in the brain
    (b) polypeptides in the gastric lumen and by gastrin secretion
    (c) secretin and cholecystokinin from the duodenum
    (d) all of the above
11. The secretion of HCl by the stomach mucosa is inhibited by
    (a) neural reflexes from the duodenum
    (b) the secretion of gastric inhibitory peptide from the duodenum
    (c) the lowering of gastric pH
    (d) all of the above
12. The first organ to receive the blood-borne products of digestion is the
    (a) liver
    (b) pancreas
    (c) heart
    (d) brain
13. Which of the following statements about hepatic portal blood is true?
    (a) It contains absorbed fat.
    (b) It contains ingested proteins.
    (c) It is mixed with bile in the liver.
    (d) It is mixed with blood from the hepatic artery in the liver.

Essay Questions

1. Explain the embryological and anatomical relationships that the liver and pancreas have with the intestine.
2. Explain how the gastric secretion of HCl and pepsin is regulated during the cephalic, gastric, and intestinal phases.
3. Describe how pancreatic enzymes become activated in the lumen of the intestine, and explain the need for these mechanisms.
4. What is the function of bicarbonate in pancreatic juice? Explain why stress ulcers are more likely to be located in the duodenum than in the stomach.
5. Explain why the pancreas is considered to be both an exocrine and an endocrine gland. Given this information, predict what effects tying of the pancreatic duct would have on pancreatic structure and function.

26 Regulation of Metabolism

Outline

Concepts

Vitamins and minerals are needed in the diet for enzymatic reactions to proceed, and the essential amino acids and fatty acids are required for the metabolism of protein and fat. The energy needs of the body are met by the caloric value of food.

Food and stored energy reserves in the body can provide circulating energy substrates for the tissue cells. The utilization and storage of energy is regulated by a variety of hormones.

The secretion of insulin is increased and that of glucagon is decreased during intestinal absorption of a meal. These two hormones have antagonistic effects on energy balance in the body.

Epinephrine, glucocorticoids, thyroxine, and growth hormone stimulate the catabolism of carbohydrates and lipids. In addition, thyroxine and growth hormone stimulate protein synthesis and are needed for proper body growth and development.

The calcium and phosphate concentrations of plasma are affected by bone formation and resorption, intestinal absorption, and urinary excretion of these ions. These processes are regulated by parathyroid hormone, 1,25-dihydroxyvitamin D_3, and calcitonin.

Nutritional Requirements

Vitamins and minerals are needed in the diet for enzymatic reactions to proceed, and the essential amino acids and fatty acids are required for the metabolism of protein and fat. The energy needs of the body are met by the caloric value of food.

Objective 1. Distinguish between the water-soluble and fat-soluble vitamins, and describe some of the functions that vitamins serve in the body.

Objective 2. Explain the importance of the essential amino acids and essential fatty acids.

Objective 3. Explain the caloric needs of the body and some of the dangers of excessive caloric intake.

Living tissue is maintained by the constant expenditure of energy. This energy is obtained directly from ATP and indirectly from the cell respiration of glucose, fatty acids, ketone bodies, amino acids, and other organic molecules. These molecules are ultimately obtained from food, but they can also be obtained from the glycogen, fat, and protein stored in the body. The energy value of food is commonly measured in *kilocalories,* which are also called "big calories" and spelled with a capital letter (*C*alories).

In addition to its caloric value, food also supplies the essential amino acids and fatty acids. The eight **essential amino acids** are lysine, methionine, valine, leucine, isoleucine, tryptophane, phenylalanine, and threonine. The **essential fatty acids** are linoleic acid and linolenic acid. These molecules are termed *essential* because the body cannot make them and thus is forced to obtain them in the diet for proper protein and fat synthesis.

Vitamins and Elements

Vitamins are small organic molecules that cannot be made by the body and that serve as coenzymes in metabolic reactions. There are two groups of vitamins—the *fat-soluble vitamins* (A, D, E, and K) and the *water-soluble vitamins.* Water-soluble vitamins include thiamine *(thi'ah-min),* B_1, riboflavin (B_2), niacin (B_3), pyridoxine (B_6), pantothenic acid, biotin, folic acid, B_{12}, and vitamin C (ascorbic acid). Recommended daily allowances for these vitamins are shown in table 26.1.

Many of the water-soluble vitamins serve as coenzymes in the metabolism of carbohydrates, lipids, and proteins. Thiamine, for example, is needed for the activity of the enzyme that converts pyruvic acid to acetyl coenzyme A. Riboflavin and niacin are needed for the production of FAD and NAD, respectively; these latter compounds serve as coenzymes that transfer hydrogens during cell respiration. Pyridoxine is a cofactor for the enzymes involved in amino acid metabolism. Deficiencies of the water-soluble vitamins can, for obvious reasons, have widespread effects in the body.

Many fat-soluble vitamins have highly specialized functions. Vitamin K, for example, is required for the production of prothrombin and for clotting factors VII, IX, and X. Vitamin D is converted into a hormone that participates in the regulation of calcium balance. The visual

Table 26.1 Recommended daily allowances for vitamins and elements

	Infants	Children	Adolescents (15–18 yrs)		Adults (23–50 yrs)	
	0–6 mos	4–6 yrs	Males	Females	Males	Females
Weight, kg (lb)	6 (13)	20 (44)	66 (145)	55 (120)	70 (154)	55 (120)
Height, cm (in.)	60 (24)	112 (44)	176 (69)	163 (64)	178 (70)	163 (64)
Protein, g	kg × 2.2	30	56	46	56	44
Fat-soluble vitamins†						
Vitamin A, μg	420	500	1000	800	1000	800
Vitamin D, μg	10	10	10	10	5	5
Vitamin E activity, mg	3	6	10	8	10	8
Water-soluble vitamins						
Ascorbic acid, mg	35	45	60	60	60	60
Folic acid, μg	30	200	400	400	400	400
Niacin, mg	6	11	18	14	18	13
Riboflavin, mg	0.4	1.0	1.7	1.3	1.6	1.2
Thiamine, mg	0.3	0.9	1.4	1.1	1.4	1.0
Vitamin B_6, mg	0.3	1.3	2.0	2.0	2.2	2.0
Vitamin B_{12}, μg	0.5	2.5	3.0	3.0	3.0	3.0
Elements						
Calcium, mg	360	800	1200	1200	800	800
Phosphorus, mg	240	800	1200	1200	800	800
Iodine, μg	40	90	150	150	150	150
Iron, mg	10	10	18	18	10	18
Magnesium, mg	50	200	400	300	350	300
Zinc, mg	3	10	15	15	15	15

†Microgram

Source: Food and Nutrition Board, *Recommended Dietary Allowances,* 9th ed., National Academy of Sciences—National Research Council, Washington, D.C., 1980.

pigments in the rods and cones of the retina are derived from vitamin A. Vitamin A and related compounds, called *retinoids (ret'i-noid),* also have effects on genetic expression in epithelial cells; these compounds are now used clinically in the treatment of some skin conditions, and researchers are attempting to derive related compounds that may aid the treatment of some cancers.

Elements are needed as cofactors for specific enzymes and for a wide variety of other critical functions. Elements that are required in relatively large amounts per day include sodium, potassium, magnesium, calcium, phosphorous, and chlorine (table 26.1). In addition, the following **trace elements** are recognized as essential: iron, zinc, manganese, fluorine, copper, molybdenum *(mo-lib'dĕ-num),* chromium, and selenium. These must be ingested in amounts ranging from 50 micrograms to 18 milligrams per day (table 26.2).

Table 26.2 Recommended daily intake of trace elements

Element	Safe and adequate intake (mg/day)
Iron (males)	10
Iron (females)	18
Zinc	15
Manganese	2.5 to 5.0
Fluorine	1.5 to 4.0
Copper	2.0 to 3.0
Molybdenum	0.15 to 0.5
Chromium	0.05 to 0.2
Selenium	0.05 to 0.2
Iodine	0.15

Source: Food and Nutrition Board of the National Academy of Sciences, 1980.

Caloric Requirements

There are tremendous variations in the energy requirements of people. These are partly due to differences in "fuel efficiency," so that some people consume more calories than others do at comparable levels of physical activity. This "fuel efficiency" is regulated in part by the thyroid gland. The differences in daily energy requirements among people, however, are largely due to differences in physical activity.

Average daily energy expenditures may range from about 1,300 to 5,000 kilocalories per day. The average values for people not engaged in heavy manual labor but who are active during their leisure time are about 2,900 kilocalories per day for a man and 2,100 kilocalories per day for a woman. People engaged in office work, the professions, sales, and comparable occupations consume approximately 5 kilocalories per minute during work. More physically demanding occupations may require energy expenditures of 7.5 to 10 kilocalories per minute.

When the caloric intake is greater than the energy expenditures, excess calories are stored primarily as fat. This is true regardless of the source of the calories—carbohydrates, protein, or fat—because these molecules can be converted to fat by the metabolic pathways described in chapter 4. Appropriate body weights are indicated in table 26.3.

The degree of obesity can be most conveniently determined by the *body mass index.* This is equal to the nude body weight (in kilograms) divided by the square of the barefoot height (in meters). For example, a person who weighs 170 pounds (77.1 kg), and is five-feet-nine-inches tall (1.75 m), has a body mass index of $77.1 \div 1.75^2 = 25.1$. The normal body mass index is 20–25 for males and

Table 26.3 Desirable body weights, according to sex, age, height, and body frame

	Height: Feet	Height: Inches	Small frame	Medium frame	Large frame
Men	5	2	128–134	131–141	138–150
	5	3	130–136	133–143	140–153
	5	4	132–138	135–145	142–156
	5	5	134–140	137–148	144–160
	5	6	136–142	139–151	146–164
	5	7	138–145	142–154	149–168
	5	8	140–148	145–157	152–172
	5	9	142–151	148–160	155–176
	5	10	144–154	151–163	158–180
	5	11	146–157	154–166	161–184
	6	0	149–160	157–170	164–188
	6	1	152–164	160–174	168–192
	6	2	155–168	164–178	172–197
	6	3	158–172	167–182	176–202
	6	4	162–176	171–187	181–207

Weights at ages 25–59 based on lowest mortality. Weight in pounds according to frame (in indoor clothing weighing 5 lbs., shoes with 1" heels).

	Height: Feet	Height: Inches	Small frame	Medium frame	Large frame
Women	4	10	102–111	109–121	118–131
	4	11	103–113	111–123	120–134
	5	0	104–115	113–126	122–137
	5	1	106–118	115–129	125–140
	5	2	108–121	118–132	128–143
	5	3	111–124	121–135	131–147
	5	4	114–127	124–138	134–151
	5	5	117–130	127–141	137–155
	5	6	120–133	130–144	140–159
	5	7	123–136	133–147	143–163
	5	8	126–139	136–150	146–167
	5	9	129–142	139–153	149–170
	5	10	132–145	142–156	152–173
	5	11	135–148	145–159	155–176
	6	0	138–151	148–162	158–179

Weights at ages 25–59 based on lowest mortality. Weight in pounds according to frame (in indoor clothing weighing 3 lbs., shoes with 1" heels).

"Courtesy of the Metropolitan Life Insurance Company."

19–24 for females. Statistically, the morbidity and mortality rates are significantly increased when the body mass index is greater than 30. When the body mass index is over 40, the risk factor for cardiovascular disease due to obesity is comparable to risk factors due to smoking, hypertension, or hyperlipidemia.

Obesity is a risk factor for cardiovascular diseases, renal disease, diabetes mellitus, gallbladder disease, the development of kidney stones, and some malignancies (particularly endometrial and breast cancer). Obesity in childhood is due to an increase in both the size and number of adipose cells; weight gain in adulthood is due mainly to an increase in adipose cell size, although the number of these cells may also increase in extreme weight gains. When weight is lost, the size of the adipose cells decreases, but the number of adipose cells does not decrease. It is thus important to prevent further increases in weight in all overweight people but particularly so in children.

Weight is lost when the caloric value of the food ingested is less than the amount required in cell respiration over a period of time. Weight loss, therefore, can be achieved by dieting alone or in combination with an exercise program to raise the metabolic rate. A summary of the caloric expenditure produced by different types of exercises is provided in table 26.4. In addition to the calories expended directly during the exercise, additional calories may be consumed by a heightened resting metabolic rate that may result from a regular exercise program. This, however, is controversial and appears to occur only in people who exercise extensively rather than moderately. Since vigorous exercise in obese people may place dangerous demands on the cardiovascular system, an exercise program under these conditions should be attempted only under medical supervision.

Table 26.4 Energy consumed (in kilocalories per minute) by different types of activities

Activity	Weight in pounds 105–115	127–137	160–170	182–192
Bicycling				
10 mph	5.41	6.16	7.33	7.91
Stationary, 10 mph	5.50	6.25	7.41	8.16
Calisthenics	3.91	4.50	7.33	7.91
Dancing				
Aerobic	5.83	6.58	7.83	8.58
Square	5.50	6.25	7.41	8.00
Gardening, weeding, and digging	5.08	5.75	6.83	7.50
Jogging				
5.5 mph	8.58	9.75	11.50	12.66
6.5 mph	8.90	10.20	12.00	13.20
8.0 mph	10.40	11.90	14.10	15.50
9.0 mph	12.00	13.80	16.20	17.80
Rowing, machine				
Easily	3.91	4.50	5.25	5.83
Vigorously	8.58	9.75	11.50	12.66
Skiing				
Downhill	7.75	8.83	10.41	11.50
Cross-country, 5 mph	9.16	10.41	12.25	13.33
Cross-country, 9 mph	13.08	14.83	17.58	19.33
Swimming, crawl				
20 yards per minute	3.91	4.50	5.25	5.83
40 yards per minute	7.83	8.91	10.50	11.58
55 yards per minute	11.00	12.50	14.75	16.25
Walking				
2 mph	2.40	2.80	3.30	3.60
3 mph	3.90	4.50	5.30	5.80
4 mph	4.50	5.20	6.10	6.80

1. List the fat-soluble vitamins, and describe some of their functions.
2. Explain the roles of vitamins B_1, B_2, and B_3 in energy metabolism.
3. List the essential amino acids and fatty acids, and explain their significance.
4. In a single sentence, completely explain how a person can gain fat. Describe the effects of weight gain and loss on adipose cells.

Regulation of Energy Metabolism

Food and stored energy reserves in the body can provide circulating energy substrates for the tissue cells. The utilization and storage of energy is regulated by a variety of hormones.

Objective 4. List molecules that function as energy reserves and as circulating energy substrates, and explain why these terms are used to describe these molecules.

Objective 5. Identify the parts of the brain that appear to be involved in eating behavior, and describe how some chemicals affect hunger and satiety.

Objective 6. Describe the types of reactions that occur during anabolism and catabolism, and list the hormones that promote these processes.

The term **metabolism** refers to all of the chemical changes that occur within the cells of the body. On the basis of energy flow, these reactions comprise two categories: **anabolism** *(ah-nab'o-lizm)* and **catabolism** *(kah-tab'o-lizm).* Anabolic reactions require the input of energy (obtained from the hydrolysis of ATP) and result in the formation of large, energy-rich molecules such as triglycerides, glycogen, and protein. Catabolism refers to the hydrolysis of these molecules into their subunits and to the use of these subunits in cell respiration for the generation of energy used to make ATP.

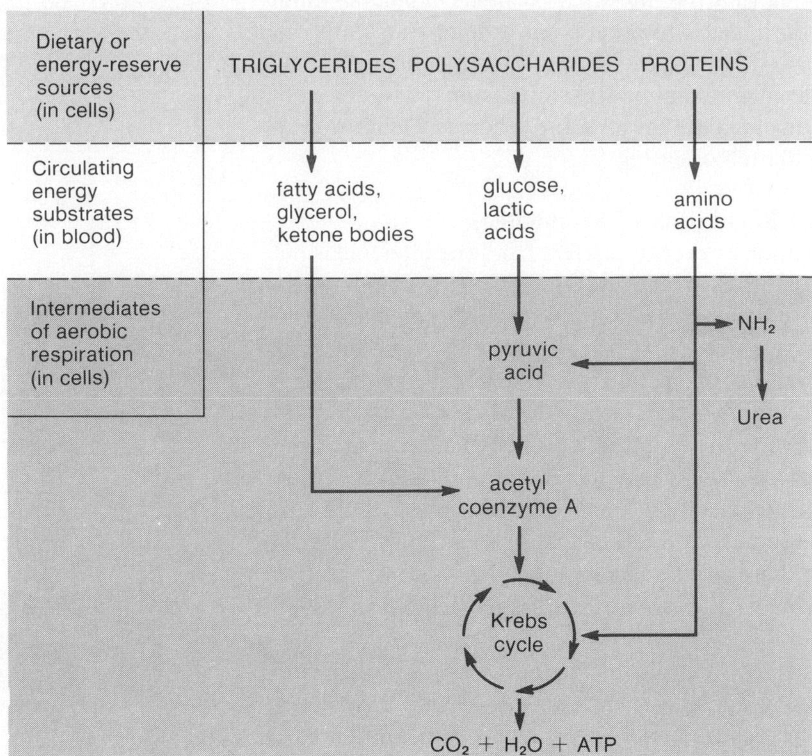

Figure 26.1. A schematic flowchart of energy pathways in the body.

The molecules that can be oxidized for energy by the processes of cell respiration may be derived from the **energy reserves** of glycogen, fat, or protein in the body, or they can be derived from the products of digestion that are absorbed through the small intestine. Since these molecules—glucose, fatty acids, amino acids, and others—are carried by the blood to the tissue cells for use in cell respiration, they can be called **circulating energy substrates** (fig. 26.1).

Because of differences in cellular enzyme content, different organs have different *preferred energy sources.* The brain has an almost absolute requirement for blood glucose as its energy source, for example. A fall in the plasma concentration of glucose below about 50 mg per 100 ml can thus "starve" the brain and have disastrous consequences. Resting skeletal muscles, in contrast, use fatty acids as their preferred energy source. Similarly, ketone bodies, lactic acid, and amino acids can be used to different degrees as energy sources by various organs. The plasma normally contains adequate concentrations of all of these circulating energy substrates to meet the energy needs of the body.

Eating

Ideally, one should eat the kinds and amounts of foods that provide adequate vitamins, minerals, essential amino acids and fatty acids, and calories. Proper caloric intake is that which maintains energy reserves (primarily fat and glycogen) and results in a body weight within an optimum range for health.

There is a tendency for body weight to be stable despite short-term changes in caloric intake. It has thus been proposed that there may be some mechanism that is sensitive to the amount of body fat. Although this mechanism is not known, it is clear that there is a relationship between body fat and endocrine function. The secretion of anterior pituitary hormones is affected in a variety of ways. Obese women, for example, may experience menstrual cycle abnormalities and hirsutism (hairiness), whereas the sudden weight loss seen in women with *anorexia nervosa* may produce amenorrhea (cessation of the menstrual cycle). Abnormalities in growth hormone, ACTH, and prolactin secretion have also been observed in obese people.

Eating behavior appears to be at least partially controlled by areas of the hypothalamus. Lesions (destruction) in the ventromedial area of the hypothalamus produce *hyperphagia (hi''per-fa'je-ah),* or overeating, and obesity in experimental animals. Lesions of the lateral hypothalamus, in contrast, produce *hypophagia* and weight loss.

The chemical neurotransmitters that may be involved in neural pathways mediating eating behavior are being investigated. There is evidence, for example, that endorphins may be involved, because injections of noloxone (a morphine-blocking drug) suppresses overeating in rats. There is also evidence that the neurotransmitters norepinephrine and serotonin may be involved; injections of norepinephrine into the brain of rats cause overeating, whereas injections of serotonin have the opposite effect. It is interesting that the intestinal hormone cholecystokinin (CCK) also appears to function as a neurotransmitter in the brain, and it has been shown that injections of CCK cause experimental animals to stop eating. More research is needed to elucidate the structure and processes involved in the regulation of eating.

Hormonal Regulation of Metabolism

The absorption of energy carriers from the intestine is not continuous; it rises to high levels following meals and tapers toward zero between meals. Despite this, the plasma concentration of glucose and other energy substrates does not remain high during periods of absorption and does not normally fall below a certain level during periods of fasting. During the absorption of digestion products from the intestine, energy substrates are removed from the blood and deposited as energy reserves from which withdrawals can be made during times of fasting (fig. 26.2). This assures that there will be an adequate plasma concentration of energy substrates to sustain tissue metabolism at all times.

The rate of deposit and withdrawal of energy substrates into and from the energy reserves and the conversion of one type of energy substrate into another are regulated by the actions of hormones. The balance between anabolism and catabolism is determined by the antagonistic effects of hormones such as insulin, glucagon, growth hormone, thyroxine, and others (fig. 26.2). The specific metabolic effects of these hormones are summarized in table 26.5, and some of these actions are illustrated in figure 26.3.

1. Define the terms *energy reserves* and *circulating energy carriers,* and give examples.
2. Describe the structures and neurotransmitters that may be involved in the regulation of eating.
3. Which hormones promote an increase in blood glucose? Which promote a decrease? List the hormones that stimulate fat synthesis (lipogenesis) and fat breakdown (lipolysis).

Figure 26.2. The balance of metabolism can be tilted toward anabolism (synthesis of energy reserves) or catabolism (utilization of energy reserves) by the combined actions of various hormones. Growth hormone and thyroxine have both anabolic and catabolic effects.

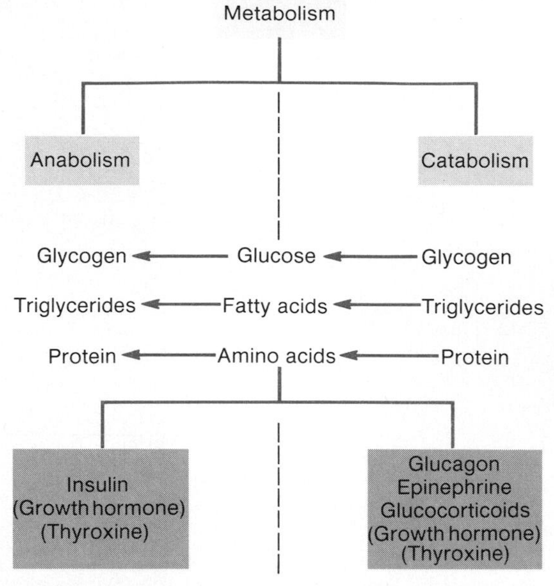

Energy Regulation by the Pancreatic Islets

The secretion of insulin is increased and that of glucagon is decreased during intestinal absorption of a meal. These two hormones have antagonistic effects on energy balance in the body.

Objective 7. Describe how the secretion of insulin and glucagon is affected by periods of absorption and fasting, and explain how the secretion of these hormones is regulated.

Objective 8. Describe the effects of insulin and glucagon on the metabolism of glycogen, fat, and protein. Explain the physiological significance of the effects of these two islet hormones.

Table 26.5 Summary of the endocrine regulation of metabolism

Hormone	Blood glucose	Carbohydrate metabolism	Protein metabolism	Lipid metabolism
Insulin	Decreased	↑Glycogen formation ↓Glycogenolysis ↓Gluconeogenesis	↑Amino acid transport	↑Lipogenesis ↓Lipolysis ↓Ketogenesis
Glucagon	Increased	↓Glycogen formation ↑Glycogenolysis ↑Gluconeogenesis	No direct effect	↑Lipolysis ↑Ketogenesis
Growth hormone	Increased	↑Glycogen formation ↑Gluconeogenesis ↓Glucose utilization	↑Protein synthesis	↓Lipogenesis ↑Lipolysis ↑Ketogenesis
Glucocorticoids	Increased	↑Glycogen formation ↑Gluconeogenesis	↓Protein synthesis	↓Lipogenesis ↑Lipolysis ↑Ketogenesis
Epinephrine	Increased	↓Glycogen formation ↑Glycogenolysis ↑Gluconeogenesis	No direct effect	↑Lipolysis ↑Ketogenesis
Thyroxine	No effect	↑Glucose utilization	↑Protein synthesis	No direct effect

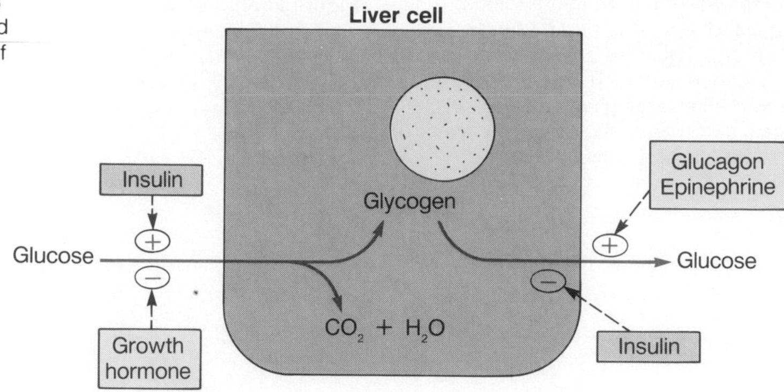

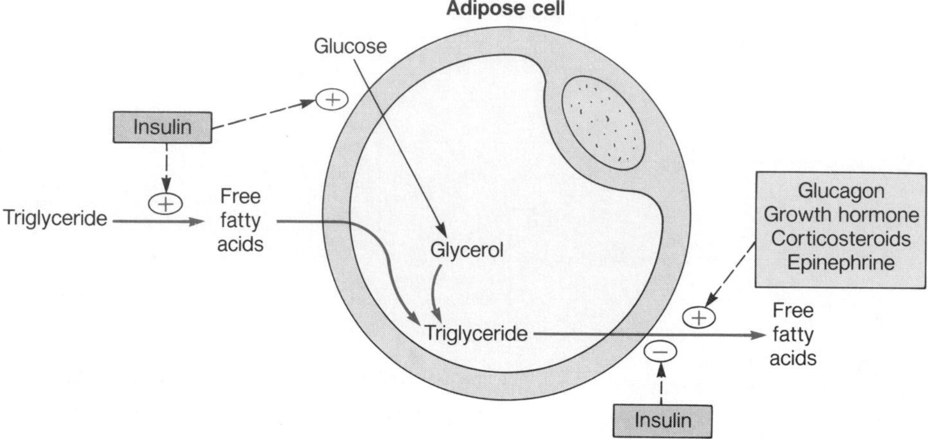

Figure 26.3. Different hormones participate both synergistically and antagonistically in the regulation of metabolism.

Scattered within a "sea" of pancreatic exocrine tissue (the acini) are islands of hormone-secreting cells. These **pancreatic islets (of Langerhans)** contain three distinct cell types that secrete different hormones (fig. 26.4). The most numerous are the **beta cells;** these cells comprise 60% of each islet and secrete the hormone **insulin.** The **alpha cells** comprise about 25% of each islet and secrete the hormone **glucagon.** The least numerous cell type, the **delta cells,** produce **somatostatin.** The latter hormone is identical to the somatostatin produced by the hypothalamus, which acts to inhibit growth-hormone secretion from the pituitary.

All three hormones are polypeptides. Insulin consists of two polypeptide chains—one that is twenty-one amino acids long and another that is thirty amino acids long—joined together by disulfide bonds. Glucagon is a twenty-one-amino-acid polypeptide, and somatostatin contains fourteen amino acids. Insulin was the first of these hormones to be discovered (in 1921). The importance of insulin in diabetes mellitus was immediately appreciated, and clinical use of insulin in the treatment of this disease began almost immediately after its discovery. The physiological role of glucagon was discovered later, and the importance of glucagon in the development of diabetes has only recently been suspected. The physiological significance of islet-secreted somatostatin is not currently known.

Figure 26.4. The cellular composition of a normal pancreatic islet.

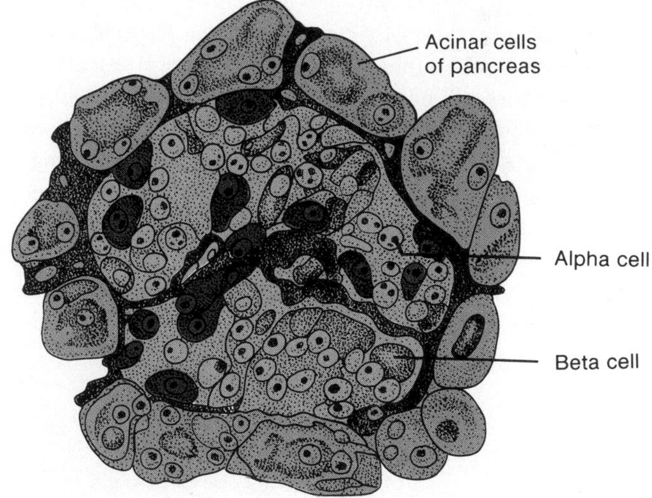

Figure 26.5. The secretion from the B (beta) cells and A (alpha) cells of the pancreatic islets is regulated to a large degree by the blood glucose concentration. A high blood glucose concentration (*a*) stimulates insulin and inhibits glucagon secretion. A low blood glucose concentration (*b*), conversely, stimulates glucagon and inhibits insulin secretion. The actions of insulin and glucagon provide negative feedback control of the blood glucose concentration.

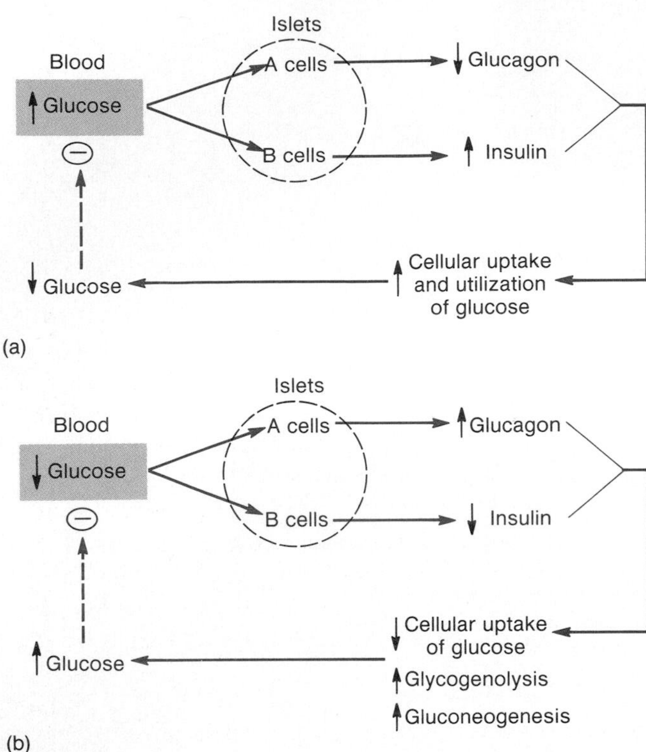

Figure 26.6. Changes in blood glucose and plasma insulin concentrations after the ingestion of 100 grams of glucose in an oral glucose tolerance test.

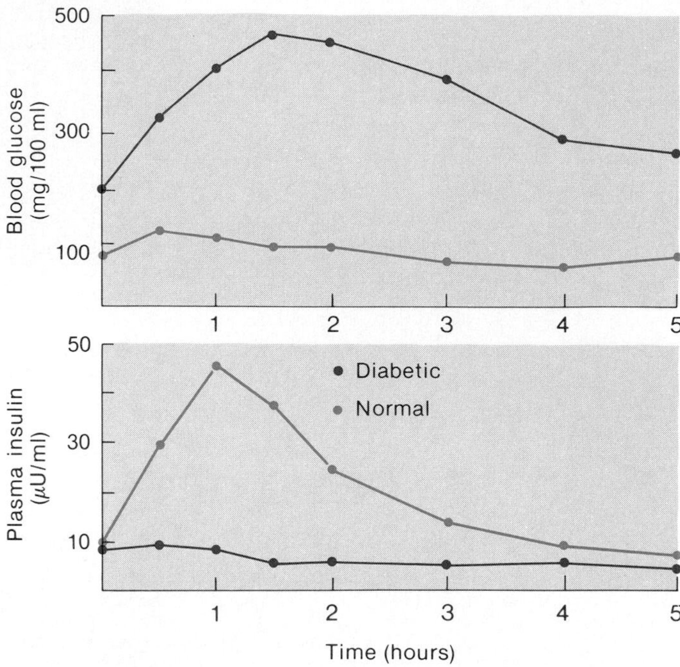

Regulation of Insulin and Glucagon Secretion

Insulin and glucagon secretion is largely regulated by the plasma concentrations of glucose and, to a lesser degree, of amino acids. The alpha and beta cells, therefore, act as both the sensors and effectors in this control system. Since the plasma concentration of glucose and amino acids rises during the absorption of a meal and falls during fasting, the secretion of insulin and glucagon likewise changes between periods of absorption and periods of fasting. These changes in insulin and glucagon secretion, in turn, cause changes in plasma glucose and amino acid concentrations and thus help to maintain homeostasis via negative feedback loops (fig. 26.5).

Effect of Glucose. During the absorption of a carbohydrate meal, the plasma glucose concentration rises. This rise in plasma glucose (1) stimulates the beta cells to secrete insulin and (2) inhibits the secretion of glucagon from the alpha cells. Insulin acts to *stimulate the cellular uptake of plasma glucose.* A rise in insulin secretion therefore lowers the plasma glucose concentration. Since glucagon has the antagonistic effect of raising the plasma glucose concentration, the inhibition of glucagon secretion complements the effect of increased insulin during the absorption of a carbohydrate meal. A rise in insulin and a fall in glucagon secretion thus help to lower the high plasma glucose concentration that occurs during periods of absorption.

During fasting, the plasma glucose concentration falls; therefore, (1) insulin secretion decreases and (2) glucagon secretion increases. These changes in hormone secretion prevent the cellular uptake of blood glucose into organs such as the muscles, liver, and adipose tissue and promote the release of glucose from the liver (through the actions of glucagon). A negative feedback loop is therefore completed (fig. 26.5), which helps to retard the fall in plasma glucose concentration that occurs during fasting.

The ability of the beta cells to secrete insulin, as well as the ability of insulin to lower blood glucose, is measured clinically by the *oral glucose tolerance test* (fig. 26.6). In this procedure, a person drinks a glucose solution and blood samples are taken periodically for plasma glucose measurements. In a normal person the rise in blood glucose produced by drinking this solution is reversed to normal levels within two hours following glucose ingestion.

People with *diabetes mellitus*—due to the inadequate secretion or action of insulin—maintain a state of high plasma glucose concentration (hyperglycemia) during the oral glucose tolerance test (fig. 26.6). People who have *reactive hypoglycemia* (low plasma glucose concentration due to excessive insulin secretion) have lower-than-normal blood glucose concentrations five hours following glucose ingestion. These conditions are discussed in more detail at the end of this chapter.

Effects of Ingested Protein. Although the ingestion of carbohydrates causes a rise in insulin and a fall in glucagon secretion, the ingestion of proteins stimulates the secretion of both hormones. This rise in insulin secretion helps to complete a negative feedback loop because insulin promotes the uptake of amino acids into the tissues and the incorporation of these amino acids into proteins. The importance of a simultaneous rise in glucagon and insulin secretions can be understood in terms of the need to maintain a constant blood glucose concentration. A rise in insulin secretion, if not accompanied by a simultaneous rise in glucagon secretion, would cause hypoglycemia in the absence of ingested carbohydrates. A rise in both insulin and glucagon secretions thus promotes lowering of the amino acid concentration of the blood, while the blood glucose concentration remains relatively constant due to the antagonistic effects of insulin and glucagon.

Since most meals contain both carbohydrates and proteins, insulin secretion is normally raised following meals, but the effects on glucagon secretion are variable. In an average meal that is rich in carbohydrates, the suppressive effects of high plasma glucose are more potent than the stimulatory effects of amino acids on glucagon secretion. In general, therefore, the insulin secretion is increased and the glucagon secretion is decreased during the period of absorption.

Effects of Autonomic Nerves. The pancreatic islets receive both parasympathetic and sympathetic innervation. The activation of the parasympathetic system during meals stimulates insulin secretion at the same time that gastrointestinal function is stimulated. The activation of the sympathetic system, in contrast, stimulates glucagon secretion and inhibits insulin secretion. The effects of glucagon, together with those of epinephrine, produce a "stress hyperglycemia" when the sympathoadrenal system is activated.

Effects of Gastric Inhibitory Peptide (GIP). Surprisingly, insulin secretion increases more rapidly following glucose ingestion than it does following an intravenous injection of glucose. This is due to the fact that the intestine, in response to glucose ingestion, secretes a hormone that stimulates insulin secretion before the glucose is absorbed. Insulin secretion thus begins to rise in anticipation of a rise in blood glucose. The intestinal hormone that mediates this effect is believed to be gastric inhibitory peptide (GIP).

Table 26.6 summarizes the effects of various factors that regulate insulin and glucagon secretion.

The mechanisms that regulate insulin and glucagon secretion and the actions of these hormones normally prevent the plasma glucose concentration from rising above 170 mg per 100 ml after a meal or from falling below about 50 mg per 100 ml between meals. This regulation is important because abnormally high blood glucose can damage tissue cells (as may occur in diabetes mellitus), and abnormally low blood glucose can damage the brain. The later effect results from the fact that glucose enters the brain by facilitated diffusion; when the rate of this diffusion is too low, due to low plasma glucose concentrations, the supply of metabolic energy for the brain may become inadequate. This can result in weakness, dizziness, personality changes, and ultimately in coma and death.

Table 26.6 Regulation of insulin and glucagon secretion

Regulator	Effect on insulin secretion	Effect on glucagon secretion
Hyperglycemia	Stimulates	Inhibits
Hypoglycemia	Inhibits	Stimulates
Gastric inhibitory peptide	Stimulates	Stimulates (?)
Sympathetic nerve impulses	Inhibits	Stimulates
Parasympathetic nerve impulses	Stimulates	Inhibits
Amino acids	Stimulates	Stimulates
Somatostatin	Inhibits*	Inhibits*

*Inhibitory effects of somatostatin on insulin and glucagon secretion may not occur under normal conditions.

Insulin and Glucagon: Period of Absorption

The lowering of plasma glucose by insulin is, in a sense, a side effect of the primary action of this hormone. Insulin is the major hormone that promotes anabolism in the body. During absorption of the products of digestion and the subsequent rise in the plasma concentrations of circulating energy substrates, insulin promotes the cellular uptake of plasma glucose and its incorporation into glycogen and fat. As previously described, insulin also promotes the cellular uptake of amino acids and their incorporation into proteins. The stores of large, energy reserve molecules are thus increased while the plasma concentrations of glucose and amino acids are decreased.

The synthesis of triglycerides (fat) within adipose cells depends upon insulin-stimulated glucose uptake from plasma. Once inside the adipose cells, glucose can be converted into α-glycerol phosphate and acetyl coenzyme A (acetyl CoA). The formation of fatty acids from acetyl CoA and condensation of three fatty acids with glycerol yields triglycerides. Although the adipose cells can also utilize free fatty acids from the blood, they cannot incorporate glycerol from the blood into triglycerides. This is because adipose cells lack the enzymes needed to convert glycerol to α-glycerol phosphate, which is the precursor needed in triglyceride synthesis (fig. 26.7). Entry of blood glucose into adipose cells—which is directly dependent on insulin—thus determines the rate at which fat is produced.

Figure 26.7. The synthesis of triglycerides (fat) within adipose cells. Notice that fat can be produced from glucose and from the fatty acids released by the hydrolysis of plasma lipoproteins.

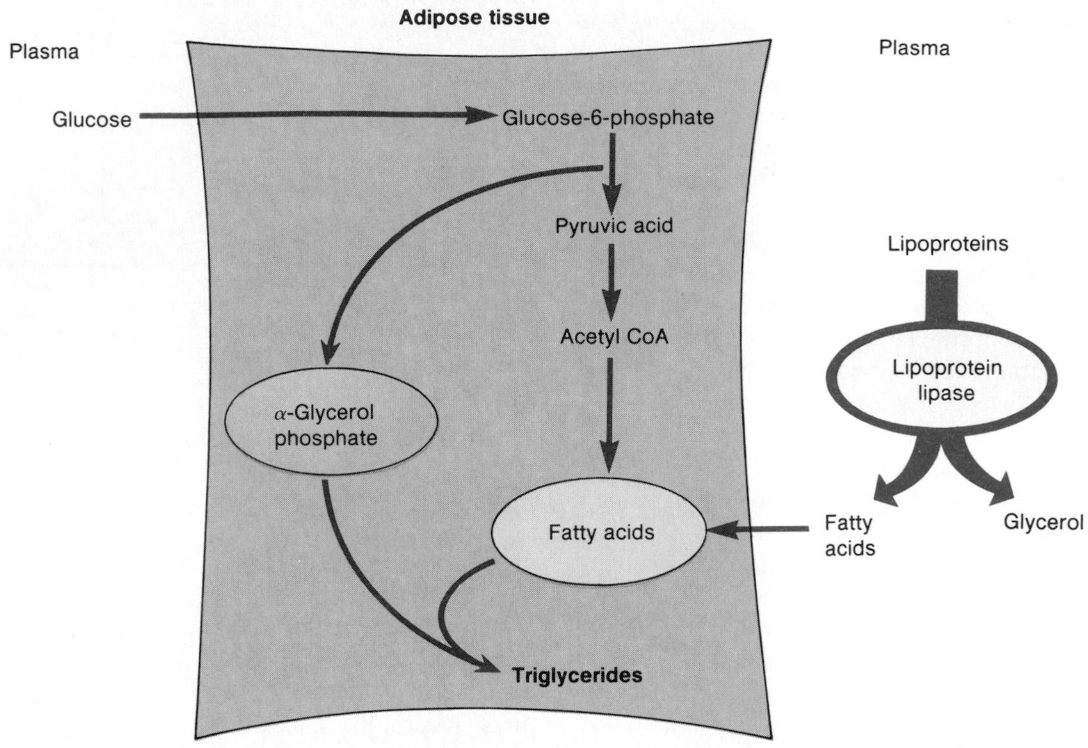

A nonobese 70 kg man has approximately 10 kg of stored fat. Since 250 g of fat can supply the energy requirements for one day (about 2300 K calories), this reserve fuel is sufficient for about forty days. Glycogen is less efficient as an energy reserve (4.1 K cal/g, versus 9.3 K cal/g for fat), and less is stored in the body; there is about 100 g of glycogen stored in the liver and about 200 g in skeletal muscles. Insulin promotes the cellular uptake of glucose into the liver and muscles and the conversion of glucose into glucose–6-phosphate. In the liver and muscles, this can be changed into glucose–1-phosphate, which is used as the precursor of glycogen. Once the stores of glycogen are filled, the continued ingestion of excess calories results in the continued production of fat rather than of glycogen.

Insulin and Glucagon: Period of Fasting

Glucagon stimulates and insulin suppresses the hydrolysis of liver glycogen, or **glycogenolysis** *(gli″ko-jĕ-nol′ĭ-sis)*. During times of fasting, when glucagon secretion is high and insulin secretion is low, therefore, liver glycogen is used as a source of additional blood glucose. This process is essentially the reverse of that which formed glycogen and results in the liberation of free glucose from glucose–6-phosphate by the action of an enzyme called *glucose–6-phosphatase*. Only the liver has this enzyme and therefore only the liver can use its stored glycogen as a source of additional blood glucose. Since skeletal muscles lack glucose–6-phosphatase, the glucose–6-phosphate produced from muscle glycogen can only be used for glycolysis by the muscle cells themselves.

Since there are only about 100 g of stored glycogen in the liver, adequate blood glucose levels could not be maintained for very long during fasting using this source alone. The low levels of insulin secretion during fasting, together with elevated glucagon secretion, however, promote **gluconeogenesis:** the formation of glucose from noncarbohydrate molecules. Low insulin allows the release of amino acids from skeletal muscles, while glucagon stimulates the production of enzymes in the liver that convert amino acids to pyruvic acid and pyruvic acid to glucose. During prolonged fasting and exercise, gluconeogenesis in the liver using amino acids from muscles may be the only source of blood glucose.

The secretion of glucose from the liver during fasting thus compensates for the low plasma glucose concentrations and helps to provide the glucose needed by the brain. In the presence of low insulin secretion, however, other organs cannot utilize blood glucose as an energy source. This helps to spare glucose for the brain, but alternative energy sources are needed for these other organs. The major alternative energy sources that are used by the liver and skeletal muscles are free fatty acids and ketone bodies.

Figure 26.8. Increased glucagon secretion and decreased insulin secretion during fasting favors catabolism. These hormonal changes result in elevated release of glucose, fatty acids, ketone bodies, and amino acids into the blood. Notice that the liver secretes glucose that is derived both from the breakdown of liver glycogen and from the conversion of amino acids in gluconeogenesis.

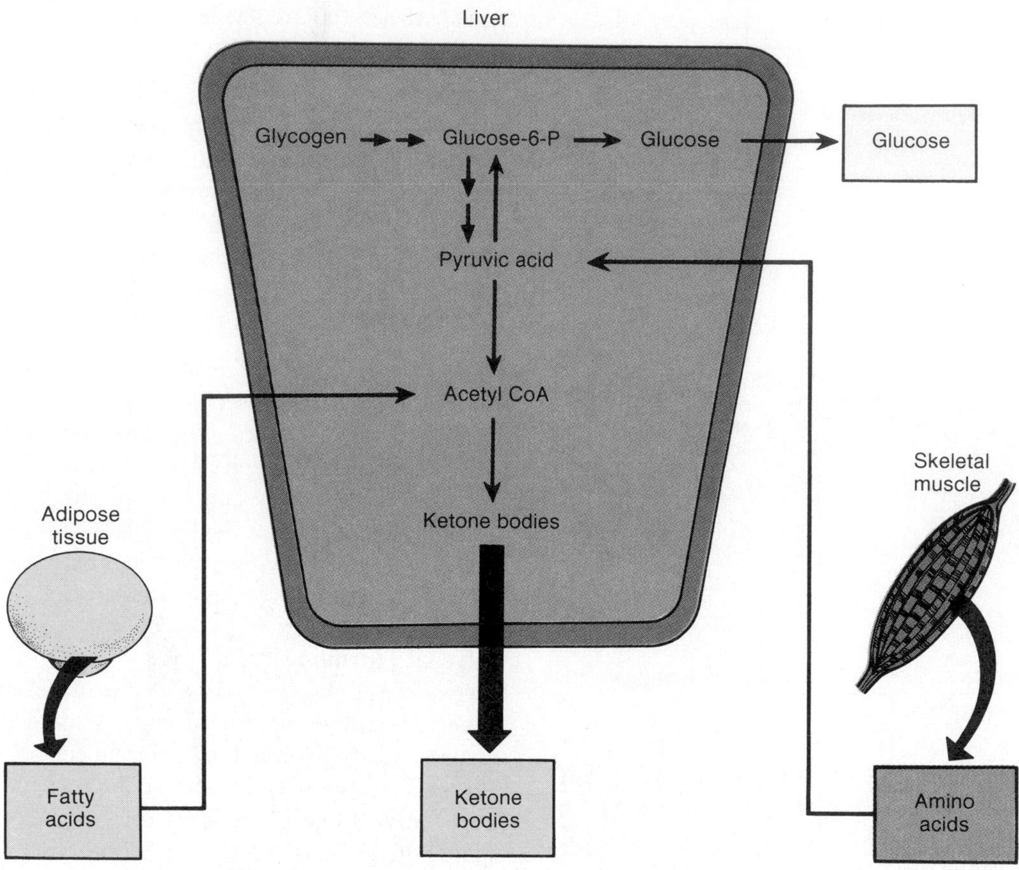

Glucagon, in the presence of low insulin levels, stimulates in adipose cells an enzyme called hormone-sensitive lipase. This enzyme catalyzes the hydrolysis of stored triglycerides and the release of free fatty acids and glycerol into the blood. Glucagon, in the presence of low insulin levels, also stimulates enzymes in the liver that convert some of these fatty acids into ketone bodies. These derivatives of fatty acids—including acetoacetic acid, β-hydroxybutyric acid, and acetone—can also be released into the blood (fig. 26.8). Several organs in the body can use ketone bodies, like fatty acids, as a source of acetyl CoA in aerobic respiration.

Through the stimulation of **lipolysis** *(lĭ-pol'ĭ-sis)* (the breakdown of fat), and **ketogenesis** (the formation of ketone bodies), the high glucagon and low insulin levels that are found during fasting provide circulating energy substrates for use by the muscles, liver, and other organs. Through liver glycogenolysis and gluconeogenesis, these hormonal changes help to provide adequate levels of blood glucose to sustain the metabolism of the brain. The antagonistic action of insulin and glucagon (fig. 26.9) thus promotes appropriate metabolic responses during periods of fasting and periods of absorption. The actions of insulin and glucagon are summarized in table 26.7.

1. Describe how the secretions of insulin and glucagon change during periods of absorption and periods of fasting. Explain how these changes in hormone secretion are produced.
2. Describe how the synthesis of fat in adipose cells is regulated by insulin. Explain how fat metabolism is regulated by insulin and glucagon during periods of absorption and fasting.
3. Define the following terms: *glycogenolysis, gluconeogenesis,* and *ketogenesis.* Explain how insulin and glucagon affect the processes these terms describe during periods of absorption and fasting.
4. Explain why the liver, but not skeletal muscles, can secrete glucose into the blood. Describe two pathways that contribute to the hepatic secretion of glucose.

Figure 26.9. The inverse relationship between insulin and glucagon secretion during the absorption of a meal and during fasting. Changes in the insulin : glucagon ratio tilts metabolism toward anabolism during the absorption of food and toward catabolism during fasting.

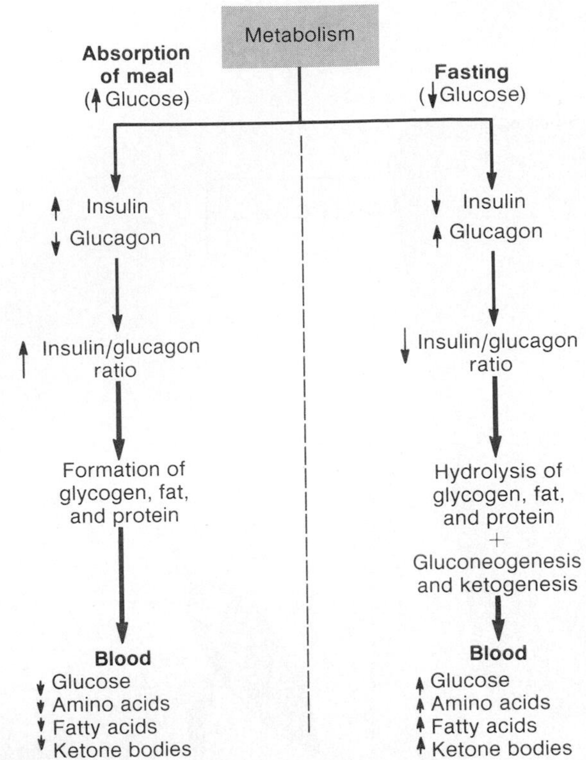

Table 26.7 Comparison of the metabolic effects of insulin and glucagon

Actions	Insulin	Glucagon
Cellular glucose transport	Increased	No effect
Glycogen synthesis	Increased	Decreased
Glycogenolysis in liver	Decreased	Increased
Gluconeogenesis	Decreased	Increased
Amino acid uptake; protein synthesis	Increased	No effect
Inhibition of amino acid release; protein degradation	Decreased	No effect
Lipogenesis	Increased	No effect
Lipolysis	Decreased	Increased
Ketogenesis	Decreased	Increased

Metabolic Regulation by the Adrenal Hormones, Thyroxine, and Growth Hormone

Epinephrine, glucocorticoids, thyroxine, and growth hormone stimulate the catabolism of carbohydrates and lipids. In addition, thyroxine and growth hormone stimulate protein synthesis and are needed for proper body growth and development.

Objective 9. Describe the metabolic effects of epinephrine and the glucocorticoids.

Objective 10. Describe the metabolic effects of thyroxine, and explain how these effects may be produced.

Objective 11. Describe the metabolic effects of growth hormone, and explain the role of somatomedins in these effects.

The anabolic effects of insulin are antagonized by glucagon, as previously described, and by the actions of a variety of other hormones. The hormones of the adrenals, thyroid, and anterior pituitary (specifically growth hormone) antagonize the action of insulin on carbohydrate and lipid metabolism. The actions of insulin, thyroxine, and growth hormone, however, can act synergistically in the stimulation of protein synthesis.

Adrenal Hormones

The adrenal gland consists of two different parts that have different embryonic origins, secrete different hormones, and are regulated by different control systems. The **adrenal medulla** secretes catecholamine hormones—epinephrine and lesser amounts of norepinephrine—in response to sympathetic nerve stimulation. The **adrenal cortex** secretes corticosteroid hormones. These are grouped into two functional categories: **mineralocorticoids** *(min''er-al-o-kor'ti-koid),* such as aldosterone, which regulate Na^+ and K^+ balance, and **glucocorticoids** *(gloo''ko-kor'ti-koids),* such as hydrocortisone (cortisol), which participate in metabolic regulation. The secretion of the glucocorticoids is stimulated by the anterior pituitary hormone ACTH.

Metabolic Effects of Epinephrine. The metabolic effects of epinephrine are similar to those of glucagon. Both stimulate glycogenolysis and the release of glucose from the liver and lipolysis and the release of fatty acids from adipose tissue. These actions occur in response to glucagon during fasting, when low blood glucose stimulates glucagon secretion, and in response to epinephrine during the "fight or flight" reaction to stress. The latter effect provides circulating energy substrates in anticipation of the need for intense physical activity. Glucagon and epinephrine have similar mechanisms of action; the actions of both are mediated by cyclic AMP (fig. 26.10).

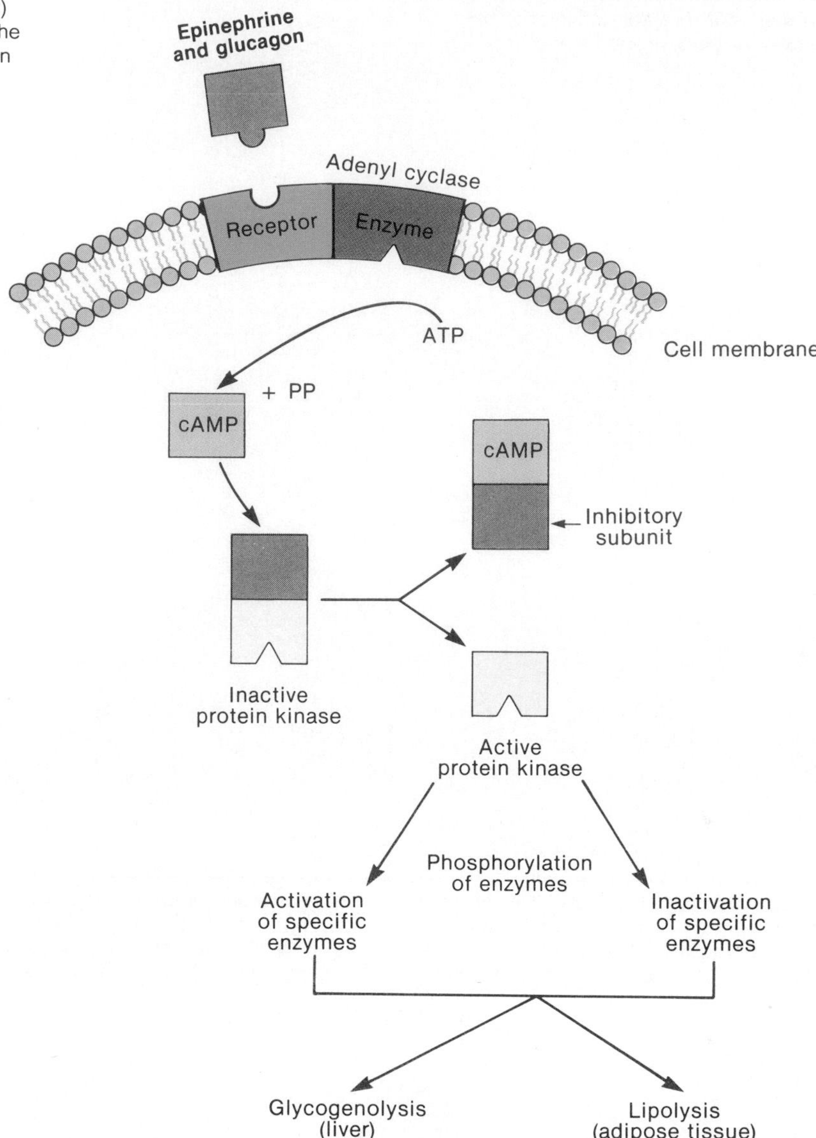

Figure 26.10. Cyclic AMP (cAMP) serves as a second messenger in the actions of epinephrine and glucagon on liver and adipose tissue metabolism.

Metabolic Effects of Glucocorticoids. Hydrocortisone and other glucocorticoids are secreted by the adrenal cortex in response to ACTH stimulation. The secretion of ACTH from the anterior pituitary occurs as part of the general adaptation syndrome in response to stress (chapter 19). Since prolonged fasting or prolonged exercise certainly qualify as stressors, ACTH—and thus glucocorticoid secretion—is stimulated under these conditions. The increased secretion of glucocorticoids during prolonged fasting or exercise supports the effects of increased glucagon and decreased insulin secretion from the islets.

Like glucagon, hydrocortisone promotes lipolysis and ketogenesis; it also stimulates the synthesis of hepatic enzymes that promote gluconeogenesis. Although it stimulates enzyme (protein) synthesis in the liver, hydrocortisone promotes protein breakdown in the muscles. This latter effect increases the blood levels of amino acids and thus provides the substrates needed by the liver for gluconeogenesis. The release of circulating energy substrates—amino acids, glucose, fatty acids, and ketone bodies—into the blood in response to hydrocortisone (fig. 26.11) helps to compensate for a state of prolonged fasting or exercise. Whether these metabolic responses are beneficial in other stressful states is open to question.

Thyroxine

The thyroid follicles secrete **thyroxine** *(thi-rok'sin),* also called **tetraiodothyronine** (T_4), in response to stimulation by TSH from the anterior pituitary. Almost all organs in the body are targets of thyroxine action. Thyroxine itself, however, is not the active form of the hormone within the target cells; thyroxine is a prehormone that must first be converted to triiodothyronine (T_3) within the target cells

Figure 26.11. The catabolic actions of glucocorticoids help raise the blood concentration of glucose and other energy-carrier molecules.

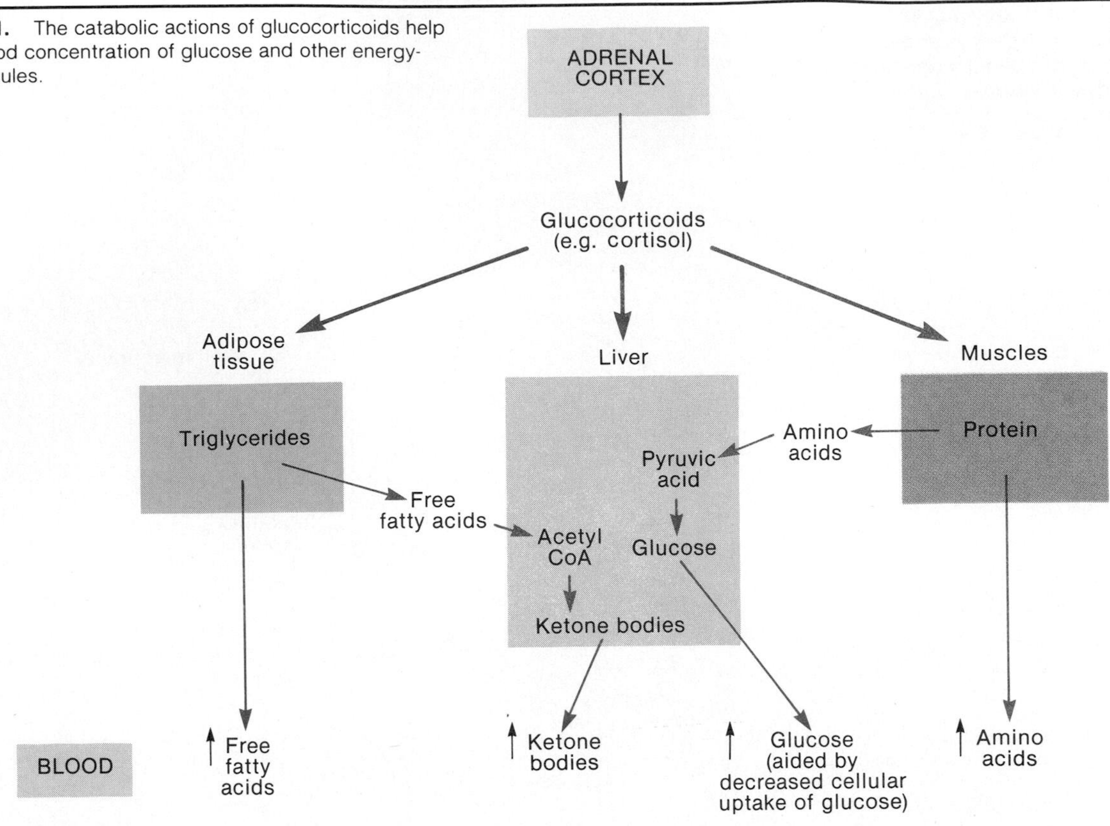

Figure 26.12. A mechanism proposed to explain the effects of thyroid hormones on basal metabolic rate. Through the activation of genes and the stimulation of protein synthesis, thyroid hormones increase the activity of the Na^+/K^+ pump. This active transport carrier accounts for a large percentage of the energy expenditures in the cell. The concentration of ATP, therefore, declines as a result of this increased energy usage, and the decreased ATP concentrations stimulate increased cellular respiration.

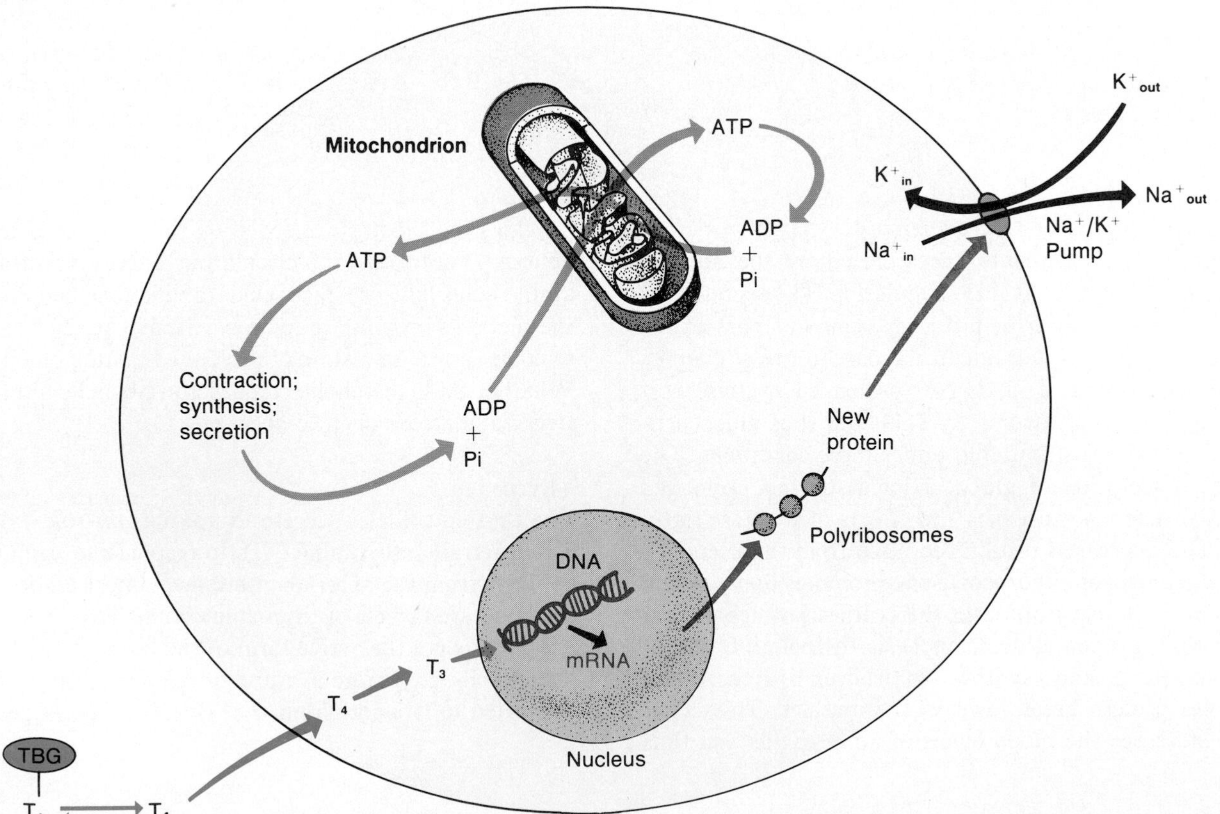

to be active. Acting via its conversion to T_3, thyroxine (1) regulates the rate of cell respiration and (2) contributes to proper growth and development, particularly during early childhood.

Thyroxine and Cell Respiration. Thyroxine stimulates the rate of cell respiration in almost all cells in the body. This is believed to be due to thyroxine-induced lowering of cellular ATP concentrations. ATP exerts an end-product inhibition of cell respiration so that when ATP concentrations increase, the rate of cell respiration decreases. Conversely, a lowering of ATP concentrations, as may occur in response to thyroxine, stimulates cell respiration.

The mechanisms of thyroxine action are not entirely understood. One effect of thyroxine, however—the stimulation of Na^+/K^+ pump activity in cell membranes—could account for the thyroxine-induced lowering of cell ATP concentrations. The active transport of Na^+ and K^+ represents a significant energy "sink" in the cell, accounting for about 12% of the calories consumed at rest. Through stimulation of Na^+/K^+ pumps, therefore, thyroxine could significantly decrease ATP concentrations and thus stimulate the rate of cell respiration (fig. 26.12).

When a person wakes up in the morning, the total energy consumption of the body is at its lowest or basal level, known as the **basal metabolic rate (BMR).** Most commonly measured by the rate of oxygen consumption, the BMR indicates the "idling speed" of the body. Since the activity of the Na^+/K^+ pumps contributes significantly to the energy consumed in the basal state and since the activity of these pumps is set by thyroxine secretion, the BMR can be used as an index of thyroid function. Indeed, such measurements were used clinically to evaluate thyroid function prior to the development of direct chemical determinations of T_4 and T_3 in the blood.

The coupling of energy-releasing reactions to energy-requiring reactions is never 100% efficient; a proportion of the energy is always lost as heat. Much of the energy liberated during cell respiration and much of the energy released by the hydrolysis of ATP escapes as heat. Since thyroxine stimulates both ATP consumption and cell respiration, the actions of thyroxine result in the production of metabolic heat.

The heat-producing, or *calorigenic (kah-lor"ĭ-jen'ik)* (calor = heat), *effects* of thyroxine are required for cold adaptation. This does not mean that people who are cold-adapted have high levels of thyroxine secretion. Rather, thyroxine levels in the normal range coupled with the increased activity of the sympathoadrenal system are responsible for cold adaptation. Thyroxine exerts a permissive effect on the ability of the sympathoadrenal system to increase heat production in response to cold stress.

Figure 26.13. Cretinism is a disease of infancy caused by an underactive thyroid gland.

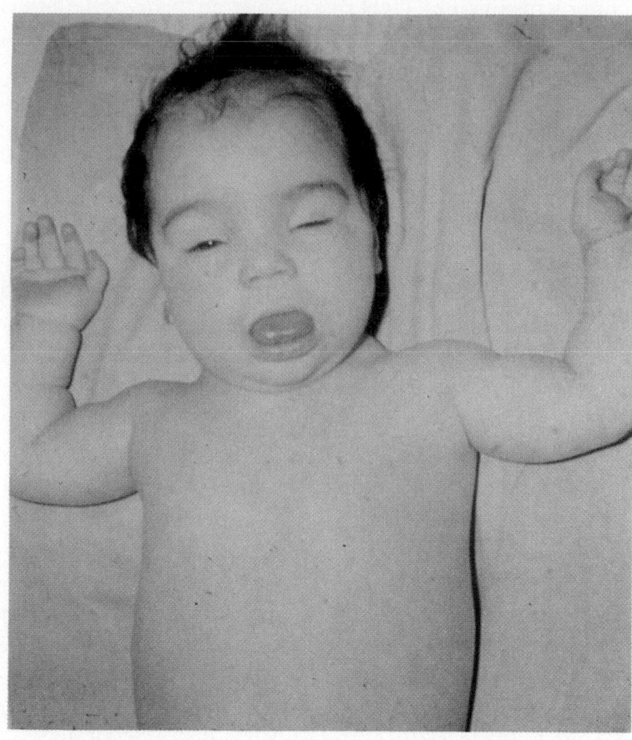

Thyroxine in Growth and Development. Through its stimulation of cell respiration, thyroxine stimulates the increased consumption of circulating energy substrates such as glucose, fatty acids, and other molecules. These effects, however, are mediated at least in part by the activation of genes; thyroxine thus stimulates both RNA and protein synthesis. As a result of its stimulation of protein synthesis throughout the body, thyroxine is considered to be an anabolic hormone like insulin and growth hormone.

Because of its stimulation of protein synthesis, thyroxine is needed for growth of the skeleton and, most importantly, for the proper development of the central nervous system. This latter effect is particularly significant during prenatal development and the first two years after birth. Hypothyroidism during this time may result in *cretinism (kre'tin-izm)* (fig. 26.13). Unlike dwarfs, who have normal thyroxine secretion but a low secretion of growth hormone, cretins suffer from severe mental retardation.

Hypothyroidism and Hyperthyroidism. As might be predicted from the effects of thyroxine, people who are hypothyroid have an abnormally low basal metabolic rate (BMR) and experience weight gain and lethargy. There is also a decreased ability to adapt to cold stress when there is a thyroxine deficiency. Another symptom of hypothyroidism is *myxedema (mik"sĕ-de'mah)*—accumulation of mucoproteins in subcutaneous connective tissues. Hypothyroidism can be produced by a variety of causes, including insufficient TRH secretion from the hypothalamus, insufficient TSH secretion from the pituitary, or insufficient iodine in the diet. Hypothyroidism due to lack of iodine is accompanied by excessive TSH secretion, which stimulates growth of the thyroid and the production of a goiter.

A goiter can also be produced by another mechanism. In *Graves' disease,* autoantibodies are produced that have TSH-like effects on the thyroid. Since the production of these autoantibodies is not controlled by negative feedback, this results in the excessive stimulation of the thyroid. A goiter is thus produced that is associated with a hyperthyroid state. Hyperthyroidism produces a high BMR accompanied by weight loss, nervousness, irritability, and an intolerance to heat. The symptoms of hypothyroidism and hyperthyroidism are compared in table 26.8.

Table 26.8 Comparison of hypothyroidism and hyperthyroidism

	Hypothyroid	Hyperthyroid
Growth and development	Impaired growth	Accelerated growth
Activity and sleep	Decreased activity; increased sleep	Increased activity; decreased sleep
Temperature tolerance	Intolerance to cold	Intolerance to heat
Skin characteristics	Coarse, dry skin	Smooth skin
Perspiration	Absent	Excessive
Pulse	Slow	Rapid
Gastrointestinal symptoms	Constipation; decreased appetite; increased weight	Frequent bowel movements; increased appetite; decreased weight
Reflexes	Slow	Rapid
Psychological aspects	Depression and apathy	Nervous, "emotional"
Plasma T_4 levels	Decreased	Increased

Growth Hormone

The anterior pituitary secretes **growth hormone,** also called **somatotropic hormone,** in larger amounts than any other of its hormones. As its name implies, growth hormone stimulates growth in children and adolescents. The continued high secretion of growth hormone in adults, particularly under the conditions of fasting and other forms of stress, implies that this hormone can have important metabolic effects even after the growing years have ended.

Regulation of Growth Hormone Secretion. The secretion of growth hormone is inhibited by somatostatin, which is produced by the hypothalamus and secreted into the hypothalamo-hypophyseal portal system. In addition, a newly discovered hypothalamic-releasing hormone appears to stimulate growth hormone secretion. Growth hormone thus appears to be unique among the anterior pituitary hormones in that its secretion is controlled by both a releasing and an inhibiting hormone from the hypothalamus. The secretion of growth hormone follows a circadian ("about a day") pattern, increasing during sleep and decreasing during periods of wakefulness.

Growth hormone secretion is stimulated by an increase in the plasma concentrations of amino acids and by a decrease in the plasma glucose concentration. The secretion of growth hormone is, therefore, stimulated during absorption of a high protein meal, when amino acids are absorbed. The secretion of growth hormone is also stimulated during prolonged fasting, when plasma glucose is low and plasma amino acid concentration is raised by the breakdown of muscle protein.

Effects of Growth Hormone on Metabolism. The fact that growth hormone secretion is increased during fasting and also during absorption of a protein meal reflects the complex nature of this hormone's action. Growth hormone has both anabolic and catabolic effects; it promotes protein synthesis (anabolism), and it also stimulates the catabolism of fat and release of fatty acids from adipose tissue. Thus, during times of fasting, growth hormone acts to help maintain skeletal muscle mass at the expense of fat, which is mobilized for energy.

In terms of its action on lipid and carbohydrate metabolism, growth hormone is said to have an anti-insulin effect. A rise in the plasma fatty acid concentration induced by growth hormone results in decreased rates of glycolysis in many organs. This inhibition of glycolysis by fatty acids, perhaps together with a more direct action of growth hormone, results in decreased glucose utilization by the tissues. Growth hormone thus acts to raise the blood glucose concentration.

Growth hormone stimulates the cellular uptake of amino acids and protein synthesis in many organs of the body. These actions are useful during a protein-rich meal; amino acids are removed from the blood and used to form proteins, and the plasma concentration of glucose and fatty acids is increased to provide alternate energy sources (fig. 26.14). The anabolic effect of growth hormone on protein synthesis is particularly important during the growing years, when it contributes to increases in bone length and in the mass of many soft tissues.

Graves' disease: from Robert James Graves, Irish physician, 1796–1853

Figure 26.14. The effects of growth hormone. The growth-promoting, or anabolic, effects of growth hormone are mediated indirectly via stimulation and somatomedin production by the liver.

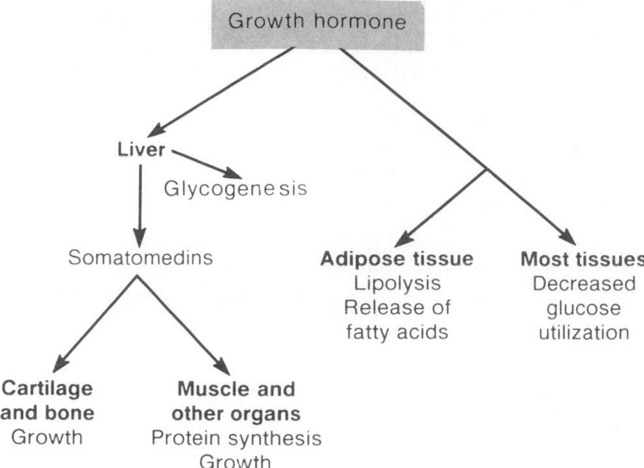

Figure 26.15. The hepatic production of polypeptides called *somatomedins* is stimulated by growth hormone. These compounds in turn produce the anabolic effects that are characteristic of the actions of growth hormone in the body.

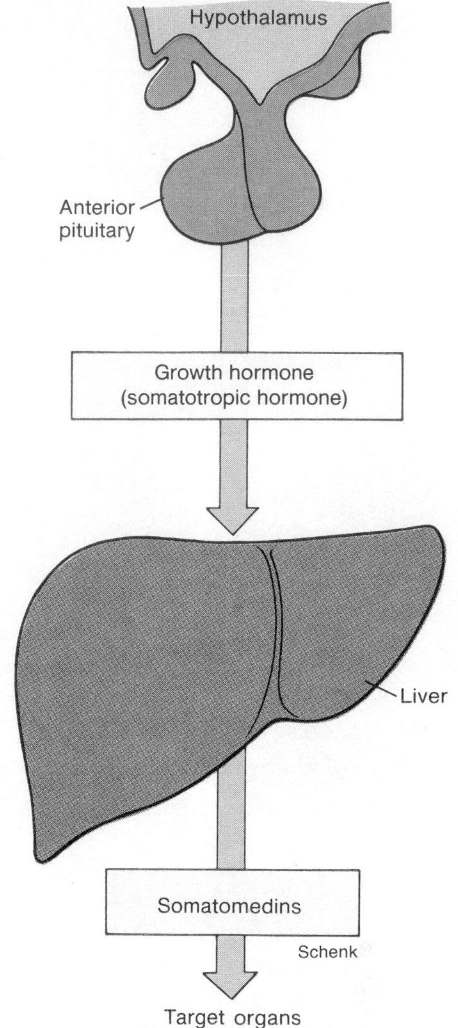

Effects of Growth Hormone on Body Growth. The stimulatory effects of growth hormone on skeletal growth result from stimulation of mitosis in the epiphyseal discs of cartilage within long bones (chapter 10). This effect, however, is believed to be indirect. Growth hormone stimulates the production of polypeptides called *somatomedins (so''mah-to-me'din),* which in turn stimulate the chondrocytes within the epiphyseal discs of growing children and adolescents. The liver is believed to be the primary source of somatomedins found in the plasma (fig. 26.15), but there is evidence that somatomedins may also be produced locally within the epiphyseal cartilages in response to growth hormone stimulation. In either case, the growth-promoting effects of growth hormone appear to be mediated indirectly via somatomedins. This skeletal growth stops when the epiphyseal discs are converted to bone after the growth spurt during puberty, despite the fact that growth hormone secretion normally continues throughout adulthood.

An excessive secretion of growth hormone in children can produce **gigantism,** in which people may grow up to eight feet tall. An excessive growth hormone secretion that occurs after the epiphyseal discs have converted to bone, however, cannot produce increases in height. The oversecretion of growth hormone in adults results in an elongation of the jaw and deformities in the bones of the face, hands, and feet. This condition, called **acromegaly,** is accompanied by the growth of soft tissues and coarsening of the skin (fig. 26.16).

An inadequate secretion of growth hormone during the growing years results in **dwarfism.** An interesting variant of this is *Laron dwarfism,* in which there is a genetic insensitivity to the effects of growth hormone. This insensitivity is associated with, but may not be caused by, a reduction in the number of growth hormone receptors in the target cells.

It is also believed that the short stature of adult African pygmies may be due to a genetically low sensitivity to the effects of growth hormone, since these people have normal amounts of growth hormone secretion. In a recent study, it was observed that pygmy children grow normally and have blood levels of somatomedins that are similar to those of other children. During puberty, however, the pygmy adolescents had only one-third the normal blood levels of somatomedins and lacked the normal pubertal growth spurt, even though the blood concentrations of sex steroids were within the normal range.

These observations suggest that the primary cause of the pubertal growth spurt in normal children may be increased production of somatomedins in response to growth

acromegaly: Gk. *akron,* extremity; *melos,* limb

Figure 26.16. The progression of acromegaly in one individual, from age nine (*a*), sixteen (*b*), thirty-three (*c*), and fifty-two (*d*) years. The coarsening of features and disfigurement are evident by age thirty-three and severe at age fifty-two.

hormone stimulation. The increased secretion of sex steroids during puberty, according to this hypothesis, promotes growth of nonsexual organs only indirectly: sex steroids stimulate increased growth hormone secretion, which results in more somatomedin production, which in turn directly stimulates growth. (The sex steroids eventually also cause the conversion of the epiphyseal discs of cartilage to bone, thus stopping growth.) Further research is required to test this exciting hypothesis and to learn the precise hormonal mechanisms responsible for the rapid growth that occurs during puberty.

An adequate diet, particularly of proteins, is required for the production of somatomedins. This helps to explain the common observation that many children are significantly taller than their immigrant parents, who may not have had an adequate diet in their youth. Children with protein malnutrition, a condition called *kwashiorkor (kwash-e-or'kor)*, have low growth rates and low somatomedin levels, despite the fact that their growth hormone levels are abnormally elevated. When these children eat an adequate diet, somatomedin levels and growth rates increase.

1. Describe the effects of epinephrine and of the glucocorticoids on the metabolism of carbohydrates and lipids, and explain the significance of these effects as a response to stress.
2. Explain the actions of thyroxine on the basal metabolic rate, and explain why people who are hypothyroid have a tendency to gain weight and are less resistant to cold stress.
3. Describe the effects of growth hormone on the metabolism of lipids, glucose, and amino acids.
4. Explain how growth hormone stimulates skeletal growth.

Regulation of Calcium and Phosphate Balance

The calcium and phosphate concentrations of plasma are affected by bone formation and resorption, intestinal absorption, and urinary excretion of these ions. These processes are regulated by parathyroid hormone, 1,25-dihydroxyvitamin D_3, and calcitonin.

Objective 12. Describe the relationships beween bone and plasma calcium levels and the importance of maintaining normal blood calcium concentrations.

Objective 13. Describe how parathyroid hormone and calcitonin secretions are regulated and how these hormones regulate the plasma calcium and phosphate concentrations.

Objective 14. Explain how vitamin D functions as a prehormone, and describe its effects on calcium and phosphate balance.

The skeleton, in addition to providing support for the body, serves as a large store of calcium and phosphate in the form of *hydroxyapatite (hi-drok''se-ap'ah-tīt)* crystals, which have the formula $Ca_{10}(PO_4)_6(OH)_2$. The calcium phosphate in these hydroxyapatite crystals is derived from the blood by the action of bone-forming cells, or *osteoblasts*. The osteoblasts secrete an organic matrix, composed largely of collagen fibers, which becomes hardened by deposits of hydroxyapatite. This process is called **bone deposition. Bone resorption** (dissolution), produced by the action of *osteoclasts* (fig. 26.17), results in the return of bone calcium and phosphate to the blood.

The formation and resorption of bone occur constantly at rates determined by the hormonal balance. Body growth during the first two decades of life occurs because bone formation proceeds at a faster rate than bone resorption. By age fifty or sixty, the rate of bone resorption often exceeds the rate of bone deposition. The constant activity of osteoblasts and osteoclasts allows bone to be remodeled throughout life. The position of the teeth, for example, can be changed by orthodontic appliances (braces), which cause bone resorption on the pressure-bearing side and bone formation on the opposite side of the alveolar sockets.

Figure 26.17. (*a*) the resorption of bone by osteoclasts and (*b*) the formation of new bone by osteoblasts. Both resorption and deposition (formation) occur simultaneously throughout the body.

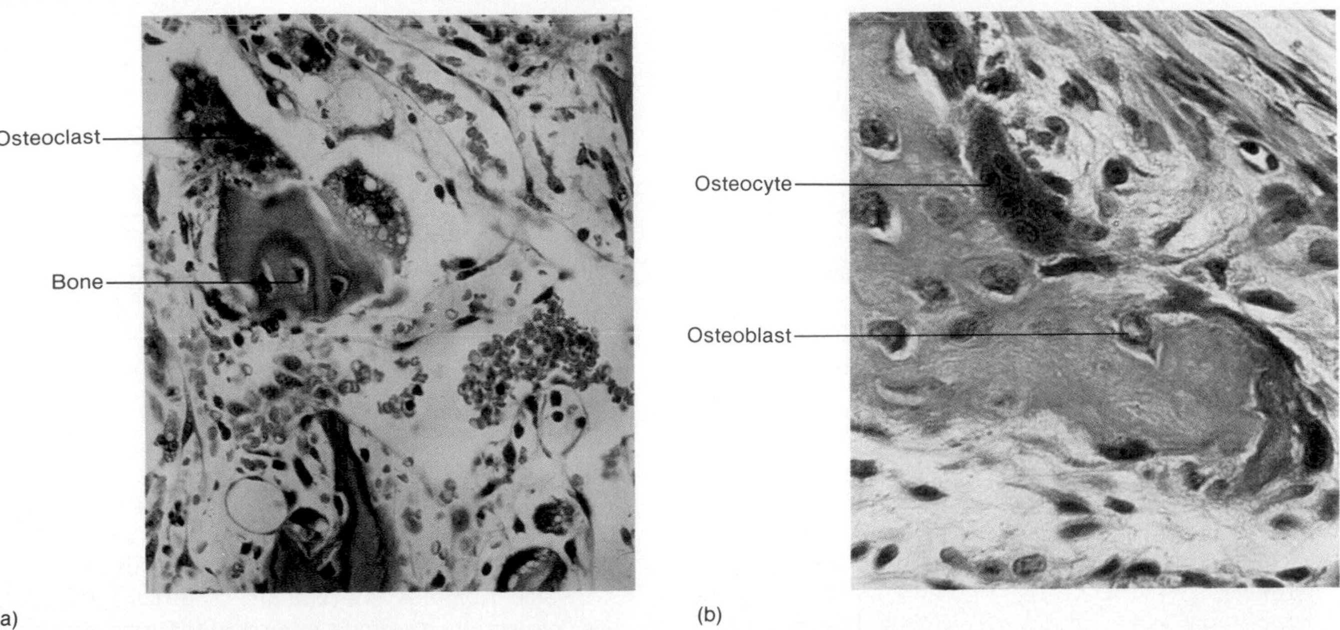

Table 26.9 The endocrine regulation of calcium and phosphate balance

Hormone	Effect on intestine	Effect on kidneys	Effect on bone	Associated diseases
Parathyroid hormone (PTH)	No direct effect	Stimulates Ca^{++} reabsorption Inhibits PO_4^{-3} reabsorption	Stimulates resorption	Osteitis fibrosa cystica with hypercalcemia due to excess PTH
1,25-dihydroxyvitamin D_3	Stimulates absorption of Ca^{++} and PO_4^{-3}	Stimulates reabsorption of Ca^{++} and PO_4^{-3}	Stimulates resorption	Osteomalacia (adults) and rickets (children) due to deficiency of 1,25-dihydroxyvitamin D_3
Calcitonin	None	Inhibits resorption of Ca^{++} and PO_4^{-3}	Stimulates deposition	None

The plasma concentrations of calcium and phosphate are maintained, despite the changing rates of bone formation and resorption, by hormonal control of the intestinal absorption and urinary excretion of these ions. These hormonal control mechanisms are very effective at maintaining the plasma calcium and phosphate concentrations within narrow limits. Plasma calcium, for example, is normally maintained at about 2.5 millimolar, or 5 milliequivalents per liter (a milliequivalent equals a millimole times the valence of the ion, in this case, times two).

The maintenance of normal plasma calcium concentrations is important because of the wide variety of effects that calcium has in the body. In addition to its role in bone formation and excitation-contraction coupling in muscles, and as a second messenger in the action of some hormones, calcium is needed to maintain proper membrane permeability. An abnormally low plasma calcium concentration increases the permeability of the cell membranes to Na^+ and other ions. Hypocalcemia, therefore, enhances the excitability of nerves and muscles and can result in muscle spasm (tetany).

A variety of bone disorders are associated with improper mineralization. In *osteomalacia* (in adults) and *rickets,* (in children), inadequate intake of vitamin D results in inadequate mineralization of the organic matrix of collagen. Excessive secretion of parathyroid hormone results in *osteitis fibrosa* (table 26.9), in which high osteoclast activity causes resorption of both the mineral and organic matrix components of bone, which is then replaced by fibrous tissue. The most common of the bone disorders is *osteoporosis,* in which there is a continuous loss of both the organic matrix and mineral components due to inadequate activity of osteoblasts. Although the causes of osteoporosis are not well understood, this condition is known to be promoted by the decreased estrogen secretion occurring in women following menopause (osteoporosis is almost ten times more common in women than men at comparable ages). Women prior to menopause should have a daily calcium intake of at least 800 mg, and after menopause should increase this to 1500 mg per day in order to minimize the progression of osteoporosis. Many women also benefit from supplementary estrogen treatments after menopause.

Parathyroid Hormone

Whenever the plasma concentration of Ca^{++} begins to fall, the parathyroid glands are stimulated to secrete increased amounts of **parathyroid hormone** *(PTH),* which acts to raise the blood Ca^{++} back to normal levels (fig. 26.18). As might be predicted from this action of PTH, people who have had their parathyroid glands removed (as may occur accidentally during thyroid surgery) will experience hypocalcemia. This can cause severe muscle tetany, for reasons previously discussed, and serves as a dramatic reminder of the importance of PTH in normal physiology.

Parathyroid hormone helps to raise the blood Ca^{++} concentrations primarily by stimulating the activity of osteoclasts to resorb bone. In addition, PTH stimulates the kidneys to reabsorb Ca^{++} from the glomerular filtrate while inhibiting the reabsorption of PO_4^{-3}. This raises blood Ca^{++} levels without promoting the deposition of calcium phosphate crystals in bone. Finally, PTH promotes the formation of 1,25-dihydroxyvitamin D_3 (as described in the next section), and so it also helps to raise the blood calcium levels indirectly through the effects of this other hormone.

Figure 26.18. The negative feedback control of parathyroid hormone secretion.

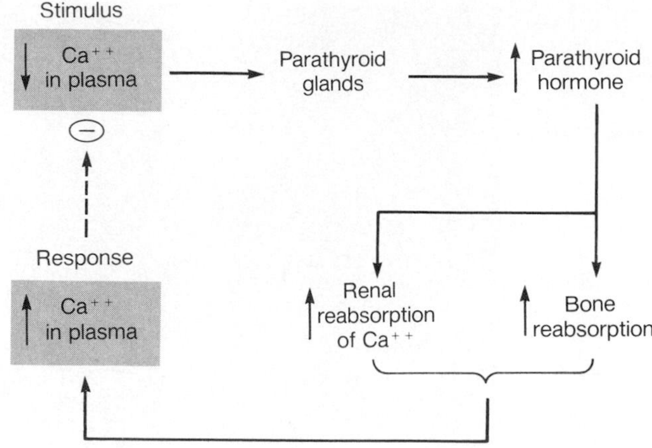

1,25-Dihydroxyvitamin D_3

The production of **1,25-dihydroxyvitamin D_3** begins in the skin, where vitamin D_3 is produced from its precursor molecule (7-dehydrocholesterol) under the influence of sunlight. When the skin does not make sufficient vitamin D_3 because of insufficient exposure to sunlight, this compound must be ingested in the diet—that is why it is called a vitamin. Whether this compound is secreted into the blood from the skin or enters the blood after being absorbed from the intestine, vitamin D_3 functions as a *prehormone* (chapter 19), which must be chemically changed in order to be biologically active.

An enzyme in the liver adds a hydroxyl group (OH) to carbon 25, which converts vitamin D_3 into 25-hydroxyvitamin D_3. In order to be active, however, another hydroxyl group must be added to the first carbon. Hydroxylation of the first carbon is accomplished by an enzyme in the kidneys, which converts the molecule to 1,25-dihydroxyvitamin D_3 (fig. 26.19). The activity of this enzyme in the kidneys is stimulated by parathyroid hormone (fig. 26.20). Increased secretion of PTH, stimulated by low blood Ca^{++}, is thus accompanied by the increased production of 1,25-dihydroxyvitamin D_3.

The hormone 1,25-dihydroxyvitamin D_3 helps to raise the plasma concentrations of calcium and phosphate by stimulating (1) the intestinal absorption of calcium and phosphate, (2) the resorption of bones, and (3) the renal reabsorption of phosphate so that less is excreted in the urine. Notice that 1,25-dihydroxyvitamin D_3, but not parathyroid hormone, directly stimulates intestinal absorption of calcium and phosphate and promotes the reabsorption of phosphate in the kidneys. The effect of simultaneously raising the blood concentrations of Ca^{++} and PO_4^{-3} results in the increased tendency of these two ions to precipitate as hydroxyapatite crystals in bone.

Since 1,25-dihydroxyvitamin D_3 directly stimulates bone resorption, it seems paradoxical that this hormone is needed for proper bone deposition and, in fact, that inadequate amounts of 1,25-dihydroxyvitamin D_3 result in the bone demineralization of osteomalacia and rickets. This apparent paradox may be explained logically by the fact that the primary function of 1,25-dihydroxyvitamin D_3 is stimulation of intestinal Ca^{++} and PO_4^{-3} absorption. When calcium intake is adequate, the major result of 1,25-dihydroxyvitamin D_3 action is the availability of Ca^{++} and PO_4^{-3} in sufficient amounts to promote bone deposition. Only when calcium intake is inadequate does the direct effect of 1,25-dihydroxyvitamin D_3 on bone resorption become significant, acting to assure proper blood Ca^{++} levels.

Calcitonin

Experiments in the 1960s revealed that high blood calcium in dogs may be lowered by a hormone secreted from the thyroid gland. This hormone thus has an effect opposite to that of parathyroid hormone and 1,25-dihydroxyvitamin D_3. The calcium-lowering hormone, called **calcitonin** *(kal"sĭ-to'nin),* was found to be a thirty-two-amino-acid polypeptide secreted by *parafollicular cells,* or *C cells,* in the thyroid, which are distinct from the follicular cells that secreted thyroxine.

The secretion of calcitonin is stimulated by high plasma calcium levels and acts to lower calcium levels by (1) inhibiting the activity of osteoclasts, thus reducing bone resorption and (2) stimulating the urinary excretion of calcium and phosphate by inhibiting their reabsorption in the kidneys (fig. 26.21).

Although it is attractive to think that calcium balance is regulated by the effects of antagonistic hormones, the significance of calcitonin in human physiology

Skin

Sunlight

HO

7-Dehydrocholesterol

CH_2

HO

Vitamin D_3

Blood

Liver

25-Hydroxylase

Kidney

1 α-Hydroxylase

OH

CH_2

HO

OH

1,25-Dihydroxy vitamin D_3

Blood

Blood

Target organs

(Bone, intestine, kidneys)

OH

CH_2

HO

25-Hydroxy vitamin D_3

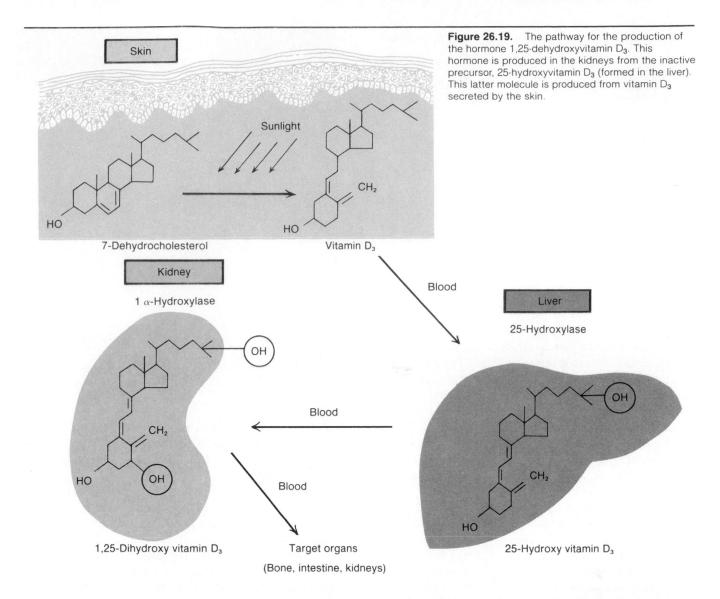

Figure 26.19. The pathway for the production of the hormone 1,25-dehydroxyvitamin D_3. This hormone is produced in the kidneys from the inactive precursor, 25-hydroxyvitamin D_3 (formed in the liver). This latter molecule is produced from vitamin D_3 secreted by the skin.

remains unclear. Patients with their thyroid gland surgically removed (as for thyroid cancer) are *not* hypercalcemic, as would have been expected if calcitonin were needed to lower blood calcium levels. Similarly, patients who receive injections of many times the normal amounts of calcitonin do not become hypocalcemic. The ability of very large, pharmacological doses of calcitonin to inhibit osteoclast activity and bone resorption, however, is clinically useful in the treatment of *Paget's disease,* in which osteoclast activity causes softening of bone.

Figure 26.20. A decrease in plasma Ca^{++} directly stimulates the secretion of parathyroid hormone (*PTH*). The production of 1,25-dehydroxyvitamin D_3 also rises when Ca^{++} is low because PTH stimulates the final hydroxylation step in the formation of this compound in the kidneys.

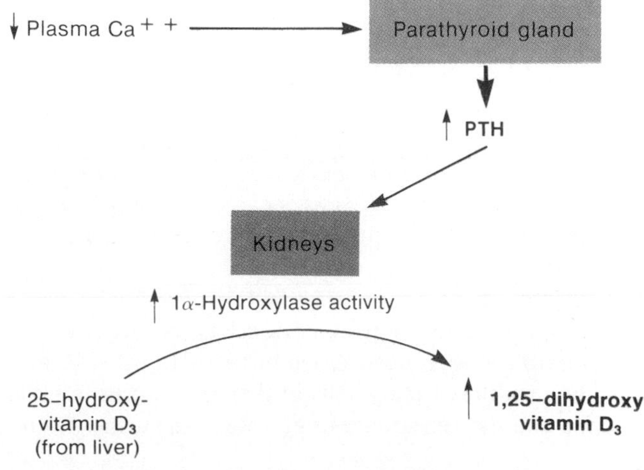

1. Describe how the secretion of parathyroid hormone and of calcitonin is regulated.
2. List the steps involved in the formation of 1,25-dihydroxyvitamin D_3, and indicate how this formation is influenced by parathyroid hormone.
3. Describe the actions of parathyroid hormone, 1,25-dihydroxyvitamin D_3, and calcitonin on the intestine, skeletal system, and kidneys, and explain how these actions affect the blood levels of calcium.
4. Explain how the *in vivo* effects of 1,25-dihydroxyvitamin D_3 differ when calcium intake is adequate or inadequate.

Figure 26.21. Negative feedback control of calcitonin secretion.

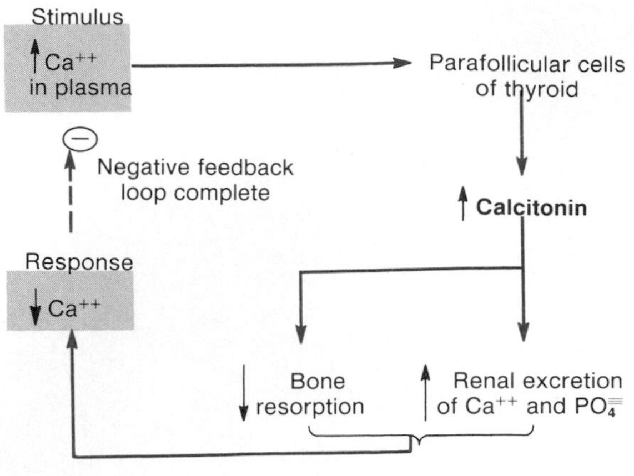

Table 26.10	Comparison of juvenile-onset and maturity-onset diabetes mellitus	
Characteristics	**Juvenile-onset (Type I)**	**Maturity-onset (Type II)**
Usual age at onset	Under 20 years	Over 40 years
Development of symptoms	Rapid	Slow
Percent of diabetic population	About 10%	About 90%
Development of ketoacidosis	Common	Rare
Association with obesity	Rare	Common
Beta cells of islets	Destroyed	Usually not destroyed
Insulin secretion	Decreased	Normal or increased
Autoantibodies to islet cells	Present	Absent
Associated with particular HLA antigens	Yes	No
Usual treatment	Insulin injections	Diet; oral stimulators of insulin secretion

From Stuart Ira Fox, *Human Physiology,* 2d ed. Copyright © 1987 Wm. C. Brown Publishers, Dubuque, Iowa. All Rights Reserved. Reprinted by permission.

Clinical Considerations

Chronic high blood glucose, or hyperglycemia, is the hallmark of the disease **diabetes mellitus.** The name of this disease is derived from the fact that glucose "spills over" into the urine when the blood glucose concentration is too high. The hyperglycemia of diabetes mellitus results from either the insufficient secretion of insulin by the beta cells of the pancreatic islets or the inability of secreted insulin to stimulate the cellular uptake of glucose from the blood. Diabetes mellitus, in short, results from the inadequate secretion or action of insulin.

There are two forms of diabetes mellitus. In **type I,** or **insulin-dependent** diabetes, the beta cells are destroyed and secrete little or no insulin. This form of the disease accounts for only about 10% of the cases of diabetes in the country. About 90% of the people who have diabetes have **type II,** or **non-insulin-dependent** diabetes mellitus. Type I diabetes is also called *juvenile-onset diabetes* because this condition is usually diagnosed in people under the age of thirty. Type II, or *maturity-onset diabetes,* is usually diagnosed in people over the age of thirty. Some comparisons of these two forms of diabetes mellitus are shown in table 26.10.

Type I Diabetes Mellitus

Type I diabetes mellitus is an autoimmune disease (chapter 22) triggered by a virus or other environmental agent in a genetically susceptible person. At least one of the genes determining susceptibility is located in the major histocompatibility region (described in chapter 22) and has been identified. Once triggered by the appropriate environmental agent, the immune system attacks the person's own beta cells in the pancreatic islets, eventually culminating in the total destruction of the beta cells. Removal of the insulin secreting cells in this way causes hyperglycemia and the appearance of glucose in the urine. Without insulin, glucose cannot enter the adipose cells; the rate of fat synthesis thus lags behind the rate of fat breakdown, and large amounts of free fatty acids are released from the adipose cells.

In a person with uncontrolled type I diabetes, many of the fatty acids released from adipose cells are converted into ketone bodies in the liver. This may result in an elevated ketone body concentration in the blood (ketosis), and if the buffer reserve of bicarbonate is neutralized, it may also result in *ketoacidosis.* During this time the glucose and excess ketone bodies that are excreted in the urine act as osmotic diuretics and cause the excessive excretion of water in the urine. This can produce severe dehydration, which together with ketoacidosis and associated disturbances in electrolyte balance may lead to coma and death (fig. 26.22).

In addition to the lack of insulin, people with type I diabetes have an abnormally high secretion of glucagon from the alpha cells of the islets. Glucagon stimulates glycogenolysis in the liver and thus helps to raise the blood glucose concentration. Glucagon also stimulates the production of enzymes in the liver that convert fatty acids to ketone bodies. Some researchers believe that the full symptoms of diabetes result from high glucagon secretion as well as from the absence of insulin. The lack of insulin may be largely responsible for hyperglycemia and for the release of large amounts of fatty acids into the blood. The high glucagon secretion may contribute to the hyperglycemia and be largely responsible for the development of ketoacidosis.

Figure 26.22. The sequence of events by which an insulin deficiency may lead to coma and death.

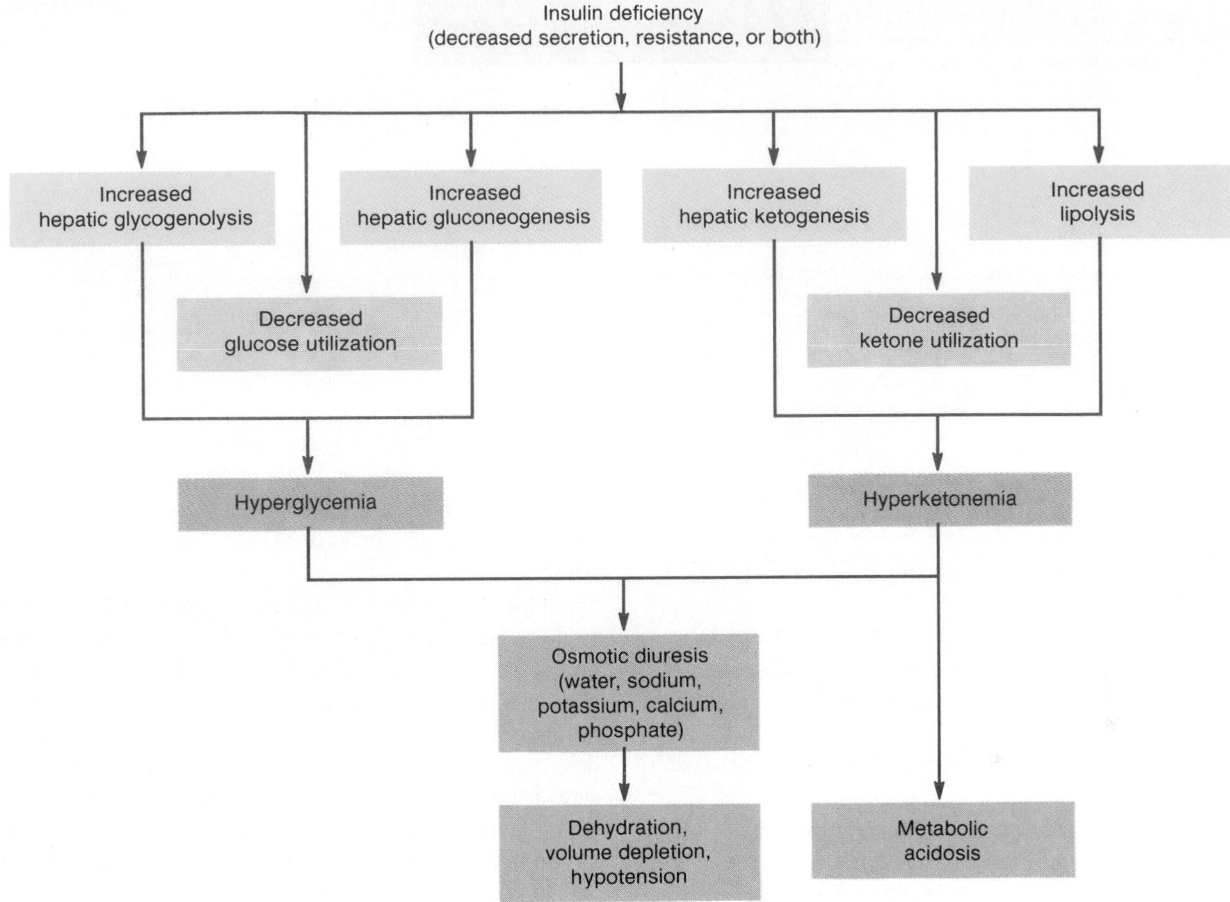

Insulin Resistance: Obesity and Type II Diabetes

The effects produced by insulin, or any hormone, depend on the concentration of that hormone in the blood and on the sensitivity of the target tissue to given amounts of the hormone. Tissue responsiveness to insulin, for example, varies under normal conditions. For reasons that are incompletely understood, exercise increases insulin sensitivity and obesity decreases insulin sensitivity of the target tissues. The pancreatic islets of a nondiabetic obese person, therefore, must secrete high amounts of insulin to maintain the blood glucose concentration in the normal range.

Type II diabetes is usually slow to develop, is hereditary, and occurs most often in people who are overweight. Unlike type I diabetes mellitus, people who have type II diabetes have normal or even elevated levels of insulin in their blood. Despite this, people with type II diabetes have hyperglycemia if untreated. This must mean that even though the insulin levels may be in the normal range, the amount of insulin secreted is inadequate because there is a decreased tissue sensitivity to the effects of insulin.

Since obesity decreases insulin sensitivity, people who are genetically predisposed to insulin resistance may develop symptoms of diabetes when they gain weight. Conversely, this type of diabetes mellitus can usually be controlled by increasing tissue sensitivity to insulin through diet and exercise. If this is not sufficient, oral drugs are available that increase insulin secretion and also stimulate tissue responsiveness to insulin.

People with type II diabetes don't usually develop ketoacidosis. The hyperglycemia itself, however, can be dangerous on a long-term basis. Diabetes is the second leading cause of blindness in the United States, and people with diabetes frequently have circulatory problems that can lead to kidney failure. These circulatory problems also increase the tendency to get gangrene and increase the risk of atherosclerosis. The causes of damage to the retina and lens of the eyes and to blood vessels are not well understood. It is believed, however, that these problems may result from a long-term exposure to high blood glucose.

Hypoglycemia

People with type I diabetes mellitus are dependent upon insulin injections to prevent hyperglycemia and ketoacidosis. If inadequate insulin is injected, the person may enter a coma as a result of the ketoacidosis, electrolyte imbalance, and dehydration that develop. An overdose of insulin, however, can also produce a coma as a result of

Table 26.11 Comparison of coma due to diabetic ketoacidosis and to hypoglycemia

	Diabetic ketoacidosis	Hypoglycemia
Onset	Hours to days	Minutes
Causes	Insufficient insulin; other diseases	Excess insulin; insufficient food; excessive exercise
Symptoms	Excessive urination and thirst; headache, nausea, and vomiting	Hunger, headache, confusion, stupor
Physical findings	Deep, labored breathing; breath has acetone odor; blood pressure decreased, pulse weak; skin is dry	Pulse, blood pressure, and respiration are normal; skin is pale and moist
Laboratory findings	Urine: glucose present, ketone bodies increased Plasma: glucose and ketone bodies increased, bicarbonate decreased	Urine: no glucose; ketone bodies at normal concentration Plasma: glucose concentration low, bicarbonate normal

From Stuart Ira Fox, *Human Physiology,* 2d ed. Copyright © 1987 Wm. C. Brown Publishers, Dubuque, Iowa. All Rights Reserved. Reprinted by permission.

the hypoglycemia (abnormally low blood glucose levels) produced. The physical signs and symptoms of diabetic and hypoglycemic coma are sufficiently different (table 26.11) to allow hospital personnel to distinguish between these two types.

Less severe symptoms of hypoglycemia are usually produced by an oversecretion of insulin from the pancreatic islets after a carbohydrate meal. This **reactive hypoglycemia** is caused by an exaggerated response of the beta cells to a rise in blood glucose and is most commonly

Figure 26.23. An idealized oral glucose tolerance test in a person with reactive hypoglycemia. The blood glucose concentration falls below the normal range within five hours of glucose ingestion as a result of excessive insulin secretion.

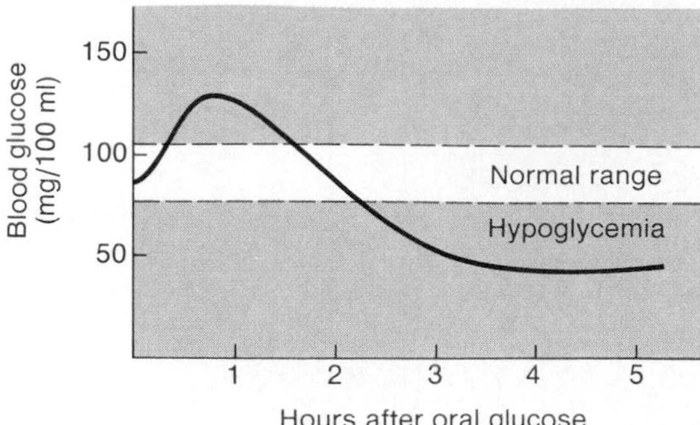

seen in adults who are genetically predisposed to type II diabetes. People with reactive hypoglycemia must, therefore, limit their intake of carbohydrates and eat small meals at frequent intervals, rather than two or three meals per day.

The symptoms of reactive hypoglycemia include tremor, hunger, weakness, hypoglycemia, blurred vision, and impaired mental ability. The appearance of some of these symptoms, however, does not necessarily indicate reactive hypoglycemia, and a given level of blood glucose does not always produce these symptoms. For these reasons a number of tests must be performed, including the oral glucose tolerance test, to confirm the diagnosis of reactive hypoglycemia. In the glucose tolerance test, reactive hypoglycemia is shown when the initial rise in blood glucose produced by the ingestion of a glucose solution triggers excessive insulin secretion, so that the blood glucose levels fall below normal within five hours (fig. 26.23).

Chapter Summary

I. Nutritional Requirements
   A. Vitamins and elements serve as cofactors and coenzymes.
      1. Vitamins are classified as fat-soluble (A, D, E, and K) or water-soluble.
      2. Many water-soluble vitamins are needed for the activity of the enzymes involved in cell respiration.
   B. The caloric intake must be sufficient to meet the energy expenditures of the body; an excessive caloric intake results in obesity.

II. Regulation of Energy Metabolism
   A. The body tissues can use circulating energy substrates, including glucose, fatty acids, ketone bodies, lactic acid, amino acid, and others, for cell respiration.
      1. Different organs have different preferred energy sources.
      2. Circulating energy substrates can be obtained from food or from the energy reserves of glycogen, fat, and protein in the body.
   B. Eating behavior is regulated, at least in part, by the hypothalamus.
   C. The control of energy balance in the body is regulated by the anabolic and catabolic effects of a variety of hormones.

III. Energy Regulation by the Pancreatic Islets
   A. A rise in plasma glucose concentration stimulates insulin and inhibits glucagon secretion.
      1. Amino acids stimulate the secretion of both insulin and glucagon.
      2. Insulin secretion is also stimulated by parasympathetic innervation of the islets and by the action of gastric inhibitory peptide (GIP), secreted by the intestine.
   B. During the intestinal absorption of a meal, insulin promotes the uptake of blood glucose into tissue cells.
      1. This lowers the blood glucose concentration and increases the energy reserves of glycogen, fat, and protein.
      2. Insulin is required for the production of fat by adipose cells.
   C. During periods of fasting, insulin secretion decreases and glucagon secretion increases.
      1. Glucagon stimulates glycogenolysis in the liver, gluconeogenesis, lipolysis, and ketogenesis.
      2. These effects help maintain adequate levels of blood glucose for the brain and provide alternate energy sources for other organs.

IV. Metabolic Regulation by the Adrenal Hormones, Thyroxine, and Growth Hormone
   A. The adrenal hormones involved in energy regulation include epinephrine from the adrenal medulla and glucocorticoids (mainly hydrocortisone) from the adrenal cortex.

1. The effects of epinephrine are similar to those of glucagon.
2. Glucocorticoids promote the breakdown of muscle protein and the conversion of amino acids to glucose in the liver.

B. Thyroxine stimulates the rate of cell respiration in almost all cells in the body.
1. Thyroxine thus sets the basal metabolic rate (BMR), which is the rate at which energy is consumed by the body under resting conditions.
2. Thyroxine also promotes protein synthesis and by this means is needed for proper body growth and development, particularly of the central nervous system.

C. The secretion of growth hormone is regulated by releasing and inhibiting hormones from the hypothalamus.
1. The secretion of growth hormone is stimulated by a protein meal and by a fall in glucose, as occurs during fasting.
2. Growth hormone stimulates catabolism of lipids and inhibits glucose utilization.
3. Growth hormone also stimulates protein synthesis and thus promotes body growth.
4. The anabolic effects of growth hormone, including the stimulation of bone growth in childhood, is believed to be produced indirectly via polypeptides called somatomedins.

V. Regulation of Calcium and Phosphate Balance

A. Bone contains calcium and phosphate in the form of hydroxyapatite crystals; this serves as a reserve supply of calcium and phosphate for the blood.
1. The formation and resorption of bone is produced by the action of osteoblasts and osteoclasts, respectively.
2. The plasma concentrations of calcium and phosphate are also affected by absorption from the intestine and by the urinary excretion of these ions.

B. Parathyroid hormone stimulates bone resorption and calcium reabsorption in the kidneys; this hormone thus acts to raise the blood calcium concentration.
1. The secretion of parathyroid hormone is stimulated by a fall in blood calcium levels.
2. Parathyroid hormone also inhibits reabsorption of phosphate in the kidneys, so that more phosphate is excreted in the urine.

C. 1,25-dihydroxyvitamin D_3 is derived from vitamin D by hydroxylation reactions in the liver and kidneys.
1. The last hydroxylation step is stimulated by parathyroid hormone.
2. 1,25-dihydroxyvitamin D_3 stimulates the intestinal absorption of calcium and phosphate, resorption of bone, and renal reabsorption of phosphate.

D. A rise in parathyroid hormone, accompanied by the increased production of 1,25-dihydroxyvitamin D_3, helps to maintain proper blood levels of calcium and phosphate in response to a fall in calcium levels.

E. Calcitonin is secreted by the parafollicular cells of the thyroid gland and may inhibit bone resorption.

Review Activities

Objective Questions:

Match the following:

1. Absorption of carbohydrate meal
2. Absorption of protein meal
3. Fasting

(a) rise in insulin, rise in glucagon
(b) fall in insulin, rise in glucagon
(c) rise in insulin, fall in glucagon
(d) fall in insulin, fall in glucagon

Match the following:

4. Growth hormone
5. Thyroxine
6. Hydrocortisone

(a) increased protein synthesis, increased cell respiration
(b) protein catabolism in muscles; gluconeogenesis in liver
(c) protein synthesis in muscles, decreased glucose utilization
(d) fall in blood glucose, increased fat synthesis

7. A lowering of blood glucose concentration promotes
(a) decreased lipogenesis
(b) increased lipolysis
(c) glycogenolysis
(d) all of the above
8. Glucose can be secreted into the blood by the
(a) liver
(b) muscles
(c) liver and muscles
(d) liver, muscles, and brain
9. The basal metabolic rate is determined primarily by
(a) hydrocortisone
(b) insulin
(c) growth hormone
(d) thyroxine
10. Somatomedins are required for the anabolic effects of
(a) hydrocortisone
(b) insulin
(c) growth hormone
(d) thyroxine
11. The increased intestinal absorption of calcium is stimulated directly by
(a) parathyroid hormone
(b) 1,25-dihydroxyvitamin D_3
(c) calcitonin
(d) all of the above
12. A rise in blood calcium levels directly stimulates
(a) parathyroid hormone secretion
(b) calcitonin secretion
(c) 1,25-dihydroxyvitamin D_3 formation
(d) all of the above

Essay Questions

1. Compare the metabolic effects of fasting to the state of uncontrolled type I diabetes. Explain the hormonal similarities of these conditions.
2. Glucocorticoids stimulate the breakdown of protein in muscles but the synthesis of protein in the liver. Explain the significance of these differences.
3. Describe how thyroxine affects cell respiration, and explain why a hypothyroid person has a tendency to gain weight and has a reduced tolerance to cold.
4. Compare and contrast the metabolic effects of thyroxine and growth hormone.
5. Why is vitamin D considered to be both a vitamin and a prehormone? Explain why people with osteoporosis might be helped by taking controlled amounts of vitamin D.

UNIT 5

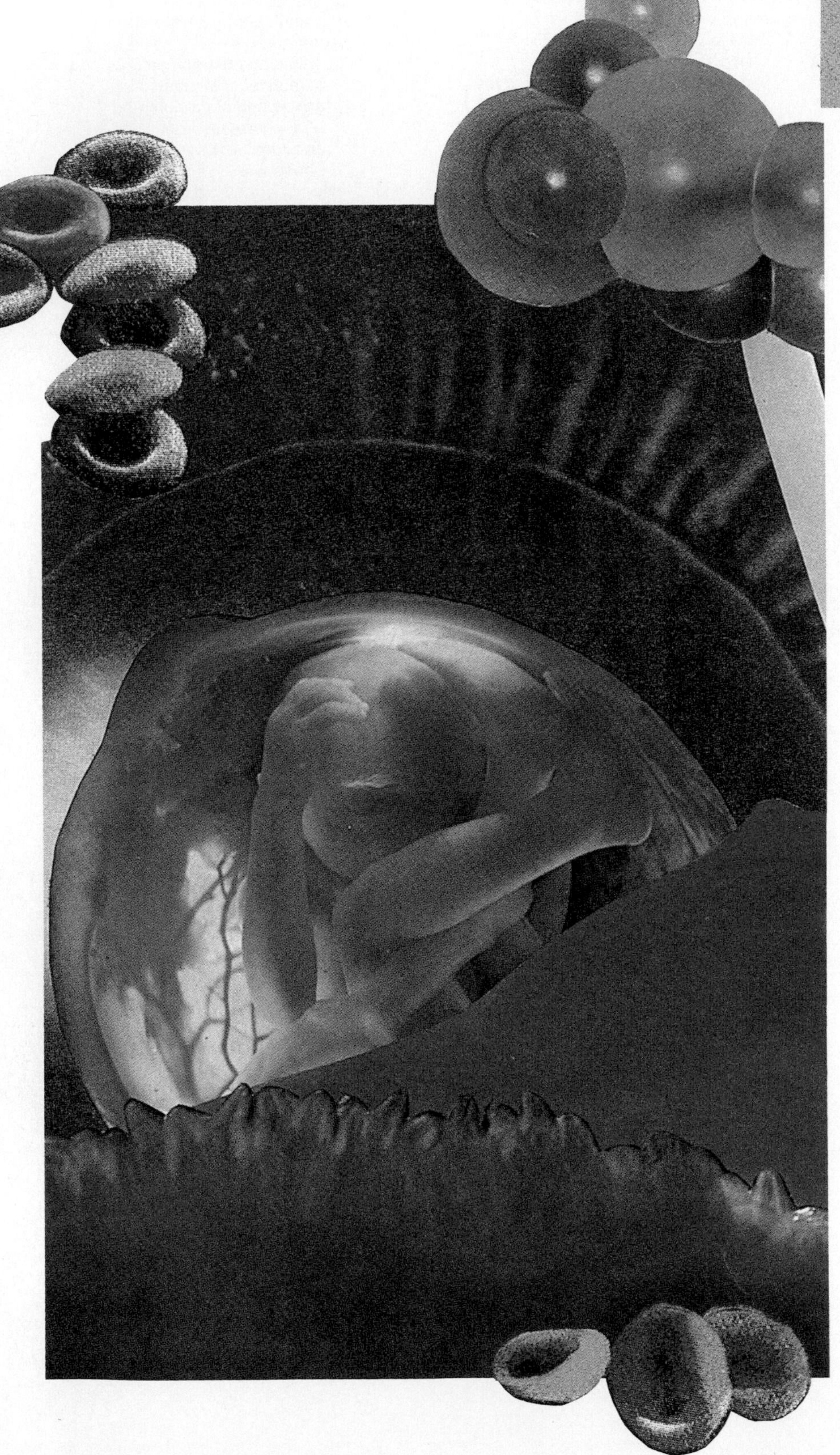

The chapters of unit 5 are concerned with the reproduction, genetics, development, growth, and aging of humans. They describe the anatomy and physiology of the male and female reproductive systems and how these systems produce a new and genetically unique individual. The fascinating process of prenatal development is described. The final chapter discusses the anatomical and physiological changes that occur in each body system as a person ages.

Continuance of the Human Species

27 Male Reproductive System
The organs of the male reproductive system produce and transport sperm to the female reproductive system. They also produce the male sex hormones, which have various regulatory functions within the body.

28 Female Reproductive System
The organs of the female reproductive system produce ova and provide the sites for fertilization, implantation, and fetal development. They are also adapted to deliver the baby and nurse it during infancy. Various sex hormones, which have regulatory functions within the body, are produced by the female reproductive organs.

29 Prenatal Development and Inheritance
Prenatal development, which is divided into pre-embryonic, embryonic, and fetal stages, permits precise genetic expressions during the crucial events of morphogenesis.

30 Postnatal Growth, Development, and Aging
The human life following birth is divided, on the basis of predictable physical and physiological changes, into the neonatal period, infancy, childhood, adolescence, and adulthood.

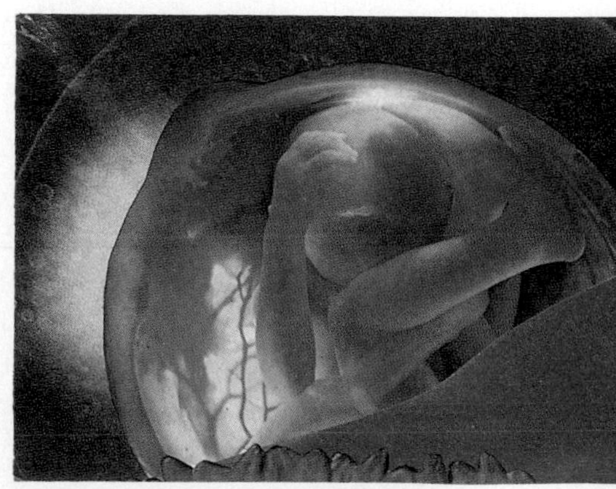

27

Male Reproductive System

Outline

Concepts

The organs of the male and female reproductive systems are adapted to produce and unite gametes that contain specific genes. A random combination of the genes during sexual reproduction permits the propagation of individuals with genetic differences.

Sexual distinction is determined upon fertilization of an ovum by a sperm containing either an X or a Y chromosome. Sexual differentiation under hormonal control is a progressive process through embryonic and early fetal development.

Gonadotropic hormones from the anterior pituitary and maturational changes in the central nervous system result in an increased secretion of sex steroids from the testes, which in turn stimulates a growth spurt and sexual maturation during puberty.

The testes within the scrotum produce spermatozoa and androgens. Androgens regulate spermatogenesis and the development and functioning of the secondary sex organs.

The spermatic ducts store spermatozoa and transport them from the testes to the urethra. The accessory reproductive glands provide additives to the spermatozoa in the formation of semen, which is discharged from the erect penis during ejaculation.

Erection of the penis results from parasympathetic induced vasodilation of arteries within the penis. Emission and ejaculation are stimulated by sympathetic impulses, resulting in the forceful expulsion of semen from the penis.

Introduction to the Reproductive System

The organs of the male and female reproductive systems are adapted to produce and unite gametes that contain specific genes. A random combination of the genes during sexual reproduction permits the propagation of individuals with genetic differences.

Objective 1. Explain why sexual reproduction is biologically advantageous.

Objective 2. List the functions of the male reproductive system, and compare them with those of the female.

Objective 3. Distinguish between primary and secondary sex organs.

Sexual reproduction is common in organisms. Through this process, individuals of a species are propagated and have a genetic diversity inherited from both parents. Sexual reproduction provides the important advantage of genetic recombination, whereby a genetic change becomes a part of the gene pool. Genetic differences in the offspring help some of them adapt to environmental changes that might otherwise cause the extinction of a population.

Because sexual reproduction requires the production of two types of **gametes** *(gam'ēts),* or sex cells, the species has a male and female, each with its own unique reproductive system. The functions of the female reproductive system are more complex than those of the male. The functions of the male reproductive system are to produce the male gametes, **spermatozoa** (sperm), 'and to transfer them to the female through the process of *coitus (sexual intercourse),* or *copulation.* The female not only produces her own gametes, called **ova,** and receives the sperm from the male, but her reproductive organs are specialized to provide a site for fertilization, implantation of the blastocyst, pregnancy, and delivery of a baby. The reproductive system of the female also provides a means of nourishing the baby through the secretions of the mammary glands in the breast.

The reproductive system of the male and female is a unique body system in three respects. First, whereas all the other body systems function to sustain the individual, the reproductive system is specialized to perpetuate the species and pass genetic material from generation to generation. Second, the anatomy and physiology of the reproductive organs are the major differences between the male and female. Other systems, such as the integumentary, skeletal, and urinary, have minor sexual differences, but none to the extent of the reproductive system. The third uniqueness of the reproductive system is its latent development under hormonal control. The other body systems are totally functional at birth or shortly thereafter, whereas the reproductive system does not become functional until it is acted on by hormones during puberty. *Puberty (pu'ber-te)* is the period of human development during which the reproductive organs become functional.

gamete: Gk. *gameta,* husband or wife
spermatozoa: Gk. *sperma,* seed; *zoon,* animal
puberty: L. *puberty,* grown up

Table 27.1 Functions of the organs of the male reproductive system

Organ	Function
Testes	
Seminiferous tubules	Produce spermatozoa
Interstitial cells	Produce and secrete male sex hormones
Epididymides	Storage and maturation of spermatozoa; convey spermatozoa to ductus deferentia
Ductus deferentia	Store spermatozoa; convey spermatozoa to ejaculatory ducts
Ejaculatory ducts	Receive spermatozoa and additives to produce seminal fluid
Seminal vesicles	Secrete alkaline fluid containing nutrients and prostaglandins
Prostate gland	Secretes alkaline fluid that helps neutralize acidic seminal fluid and enhances motility of spermatozoa
Bulbourethral glands	Secrete fluid that lubricates urethra and end of penis
Scrotum	Encloses and protects testes
Penis	Conveys urine and seminal fluid to outside of body; organ of copulation

From Kent M. Van De Graaff, *Human Anatomy,* 2d ed.

The structures of the male reproductive system can be divided into three categories based on function.

1. **Primary sex organs.** The primary sex organs are called **gonads,** specifically the **testes** in the male. Gonads produce the gametes, or sperm, and produce and secrete sex hormones. The appropriate amounts and timing of the production and secretion of male sex hormones cause the development of secondary sex organs and the expression of secondary sex characteristics.
2. **Secondary sex organs.** Secondary sex organs are those structures that are essential in caring for and transporting sperm. The three categories of secondary sex organs are the sperm-transporting ducts, the accessory glands, and the copulatory organ. The ducts that transport sperm include the **epididymides, ductus deferentia, ejaculatory ducts,** and **urethra.** The accessory glands are the **seminal vesicles,** the **prostate gland,** and the **bulbourethral** *(bul"bo-u-re'thral)* **glands.** The **penis,** which contains erectile tissue, is the copulatory organ. The **scrotum** is a pouch of skin that encloses the testes.
3. **Secondary sex characteristics.** Secondary sex characteristics are features that are not essential for the reproductive process but are generally considered sexually attractant features. Body physique, body hair, and voice pitch are examples.

Figure 27.1 depicts the external and internal organs of the male reproductive system, and table 27.1 summarizes their functions.

Figure 27.1. Organs of the male reproductive system. (*a*) a sagittal view. (*b*) a posterior view.

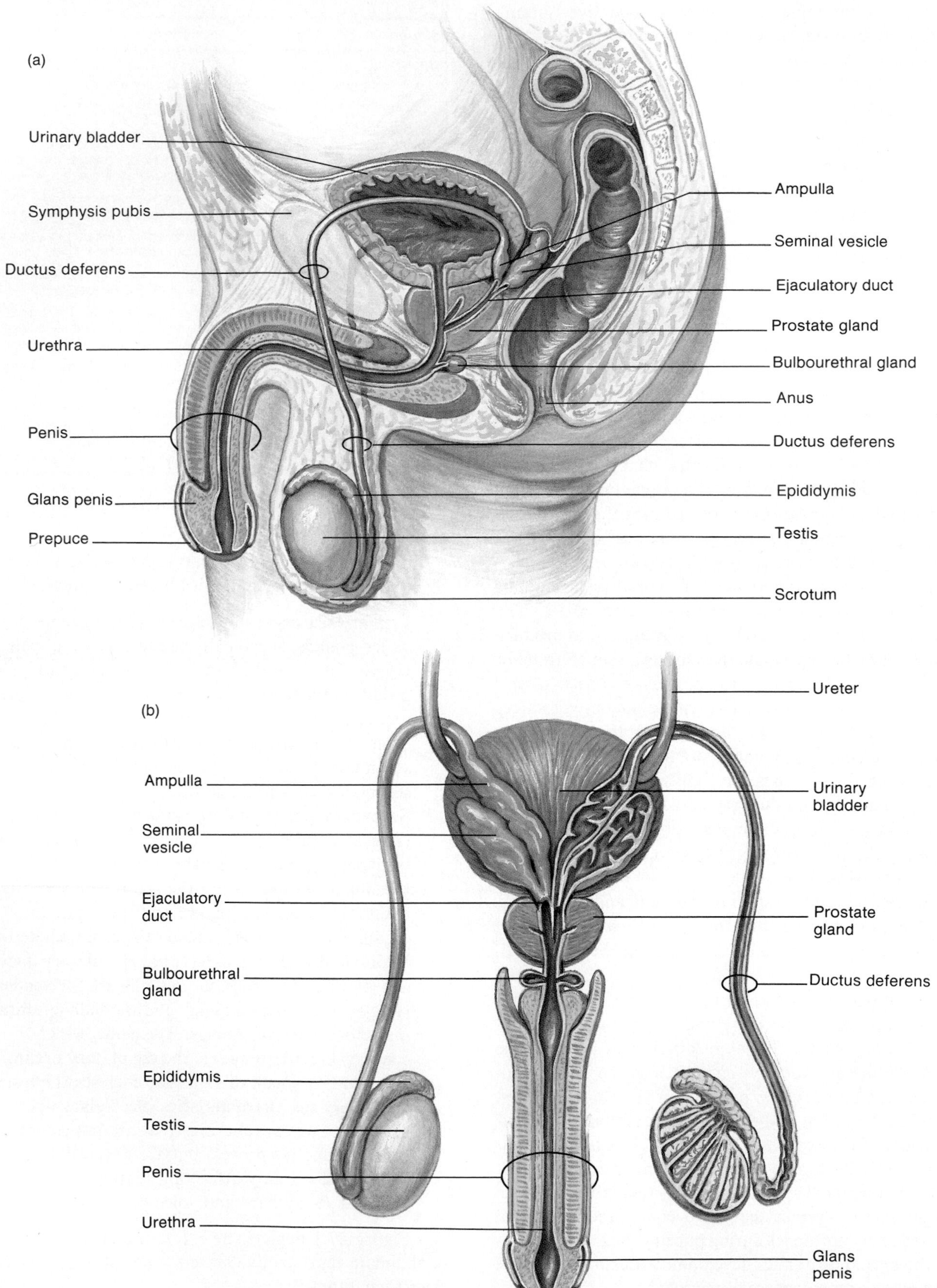

1. What is the principal value of sexual reproduction?
2. What are the functions of the male reproductive system?
3. List the organs of the male reproductive system, and indicate whether they are primary or secondary sex organs.
4. Define latent development with respect to the reproductive system.

Development of the Male Reproductive System

Sexual distinction is determined upon fertilization of an ovum by a sperm containing either an X or a Y chromosome. Sexual differentiation under hormonal control is a progressive process through embryonic and early fetal development.

Objective 4. Discuss how the increased production of testosterone affects the development of the secondary sex organs and secondary sex characteristics.

Objective 5. Explain how the genetic sex of a person is determined by the action of an XX or XY chromosome combination.

Objective 6. Trace the development of the reproductive organs to the indifferent stage.

Objective 7. Describe the sequence of events in the development of the male reproductive organs during the fetal period—including the descent of the testes into the scrotum.

Sex Determination

Sexual identity is initiated at the moment of conception when the *genetic sex* of the zygote (fertilized egg) is determined. The ovum is fertilized by a sperm containing either an X or a Y chromosome. If the sperm contains an X chromosome, it will pair with the X chromosome of the ovum and a female child will develop. A sperm carrying a Y chromosome results in an XY combination, and a male child will develop.

Genetic sex determines whether the gonads will be testes or ovaries. If testes develop, they will produce and secrete male sex hormones during late embryonic and early fetal development and cause the secondary sex organs of the male to develop.

It has long been recognized that male sex is determined by the presence of a Y chromosome and female sex by the absence of a Y chromosome. Occasionally, however, a male baby is born who has the genotype XX. Scientists have recently discovered that in these cases, one of the X chromosomes contains a segment of the Y chromosome, which was inserted into the X chromosome of the father's sperm. Rare XY females were found to be missing the same portion of the Y chromosome that was inserted into the chromosome of XX males. Through these and other studies, it has been learned that the gene for sex determination, called the *testis-determining factor (TDF),* is located on the short arm of the Y chromosome.

Embryonic Development

The first sign of development of either the male or the female reproductive organs occurs during the fifth week as the medial aspect of each mesonephros enlarges to form the **gonadal ridge.** The gonadal ridge continues to grow behind the developing peritoneal membrane. By the sixth week, stringlike masses called **primary sex cords** form within the enlarging gonadal ridge (fig. 27.2). Externally, a swelling called the **genital tubercle** appears cephalad to the cloacal membrane (see fig. 27.6).

Reproductive development is well progressed by the eighth week, but it is still in what is known as the *indifferent stage,* because although the sex organs are apparent, external distinction is not (see fig. 27.6). The gonads at six weeks are relatively large and have a distinct outer cortex composed of primary sex cords and an inner medulla (fig. 27.3). During this time, specialized **primordial germ cells** are forming and migrating from the yolk sac to the embryonic gonads. These primordial cells are more specifically called *spermatogonia* in the developing male and *oogonia* in the developing female. Prior to approximately seven weeks of development, the gonads have the potential to become either testes or ovaries (fig. 27.4). The sexual organs for both male and female are derived from the same developmental tissues and are considered *homologous* structures.

spermatogonia: Gk. *sperma,* seed; *gone,* generation
oogonia: Gk. *oon,* egg; *gone,* generation
homologous: Gk. *homos,* the same

Figure 27.2. The development of the gonadal ridge and the primary sex cords. (*a*) an embryo at five weeks showing the position of a transverse cut depicted in (*b*) and (*c*). (*b*) at five weeks, the primordia of the gonads are forming along the gonadal ridges. (*c*) at seven weeks, the germ cells are migrating into the developing gonads.

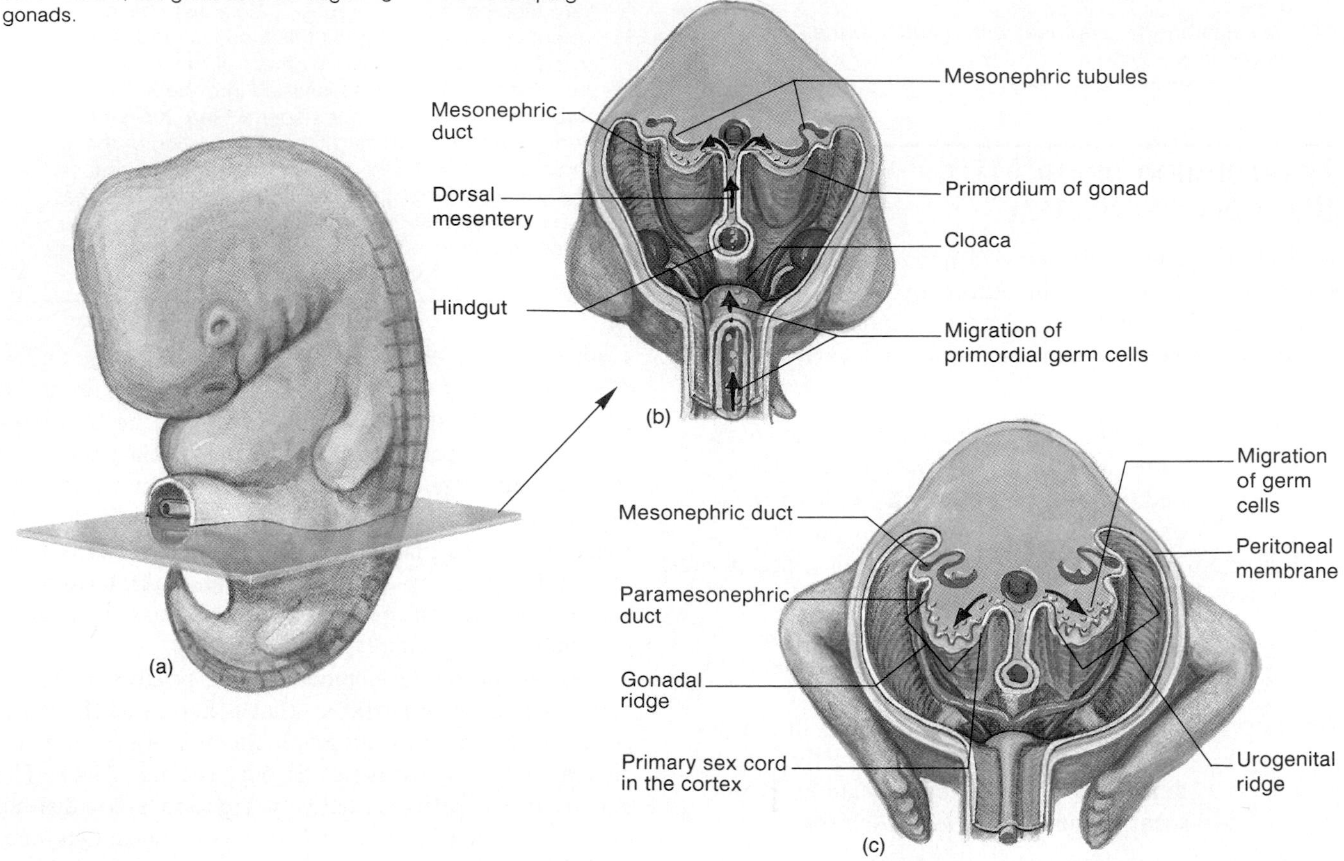

Waldrop

Notice that it is the presence or absence of the Y chromosome that determines whether the embryo will have testes or ovaries. This point is well illustrated by two genetic abnormalities. In *Klinefelter's syndrome* the affected person has forty-seven instead of forty-six chromosomes because of the presence of an extra X chromosome. These people, with XXY genotypes, develop testes despite the presence of two X chromosomes. Patients with *Turner's syndrome,* who have the genotype XO (and therefore have only forty-five chromosomes) develop ovaries.

Once the testes have differentiated, large amounts of male sex hormones, called *androgens (an'dro-jens),* are secreted from **interstitial cells (cells of Leydig).** The major androgen secreted by these cells is *testosterone (tes-tos'tĕ-rōn).* Testosterone secretion begins as early as eight to ten weeks after conception, reaches a peak at twelve to fourteen weeks, and thereafter declines to very low levels by the end of the second trimester (about twenty-one weeks). High levels of testosterone will not appear again in the life of the individual until the time of puberty.

The **seminiferous** *(se''mĭ-nif'er-us)* **tubules,** which will eventually produce sperm within the testes, appear about forty-five to fifty days following conception. Although spermatogenesis begins during embryonic life, it is arrested until the onset of puberty.

androgen: Gk. *andros,* male producing

Figure 27.3. Differentiation of the male and female gonads. (*a*) an embryo at six weeks showing the positions of a transverse cut depicted in (a_1, *b*, and *c*). (a_1) at six weeks, the developing gonads (primary sex cords) are still indifferent. By four months, the gonads have differentiated into male (*b*) or female (*c*). The oogonia are formed within the ovaries (c_1) by six months.

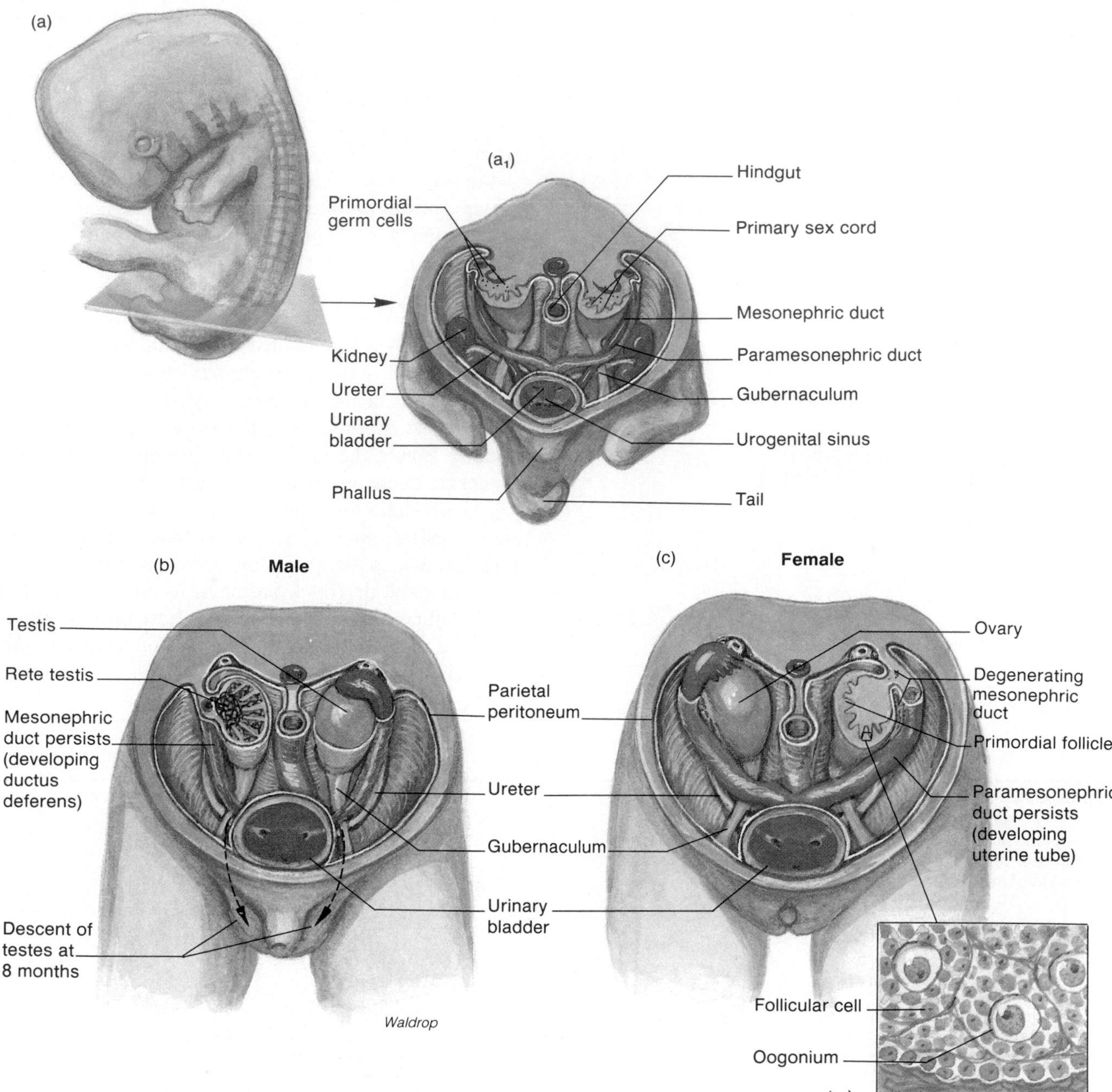

Figure 27.4. The formation of the chromosomal sex of the embryo and the development of the gonads. The very early embryo has "indifferent gonads" that can develop into either testes or ovaries. If the embryonic cells have Y chromosomes, which contain the testis-determining factor (TDF) gene, the gonads become testes. If no Y chromosomes and therefore no TDF gene is present, the gonads become ovaries. Embryonic testes develop quickly, forming seminiferous tubules (which will produce sperm later in life) and interstitial cells (which secrete testosterone during embryonic development).

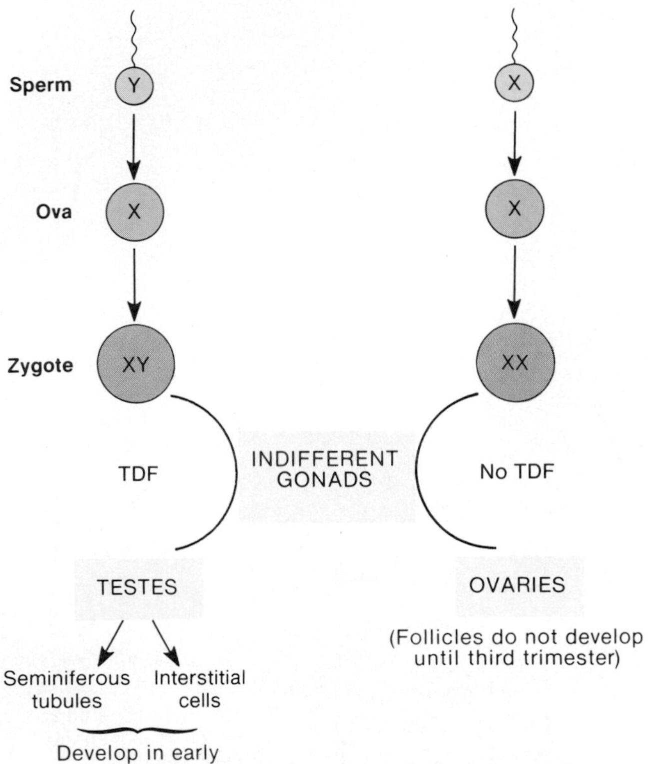

In addition to gonads, various internal accessory sexual organs are needed for reproductive functioning. Most of these sex accessory organs are derived from two systems of embryonic ducts. Male accessory organs are derived from mesonephric (wolffian) ducts, and female accessory organs are derived from paramesonephric (müllerian) ducts (fig. 27.5). Interestingly, both male and female embryos between day twenty-five and day fifty have both duct systems and, therefore, have the potential to form the accessory organs characteristic of either sex.

The experimental removal of the testes (castration) from male embryonic animals results in the regression of the mesonephric ducts and the development of the paramesonephric ducts into female accessory organs: the uterus and uterine tubes. Female sex accessory organs, therefore, develop as a result of the absence of testes rather than as a result of the presence of ovaries.

The developing seminiferous tubules within the testes secrete a polypeptide called *müllerian inhibition factor* (MIF), which causes the regression of the paramesonephric ducts beginning about day sixty. The secretion of testosterone by the interstitial cells of the testes subsequently causes the growth and development of the mesonephric ducts into male secondary sex organs.

Other embryonic reproductive structures (**urogenital sinus, genital tubercle, urogenital folds,** and **labioscrotal folds**) are also masculinized by secretions of the testes. The prostate gland derives from the urogenital sinus, and the other embryonic structures differentiate into the external genitalia (fig. 27.6). In the absence of testicular secretions, the female genitalia are formed.

Figure 27.5. The embryonic development of male and female accessory organs and external genitalia. In the presence of testosterone and müllerian inhibition factor (MIF) secreted by the testes, male structures develop. In the absence of these secretions, female structures develop.

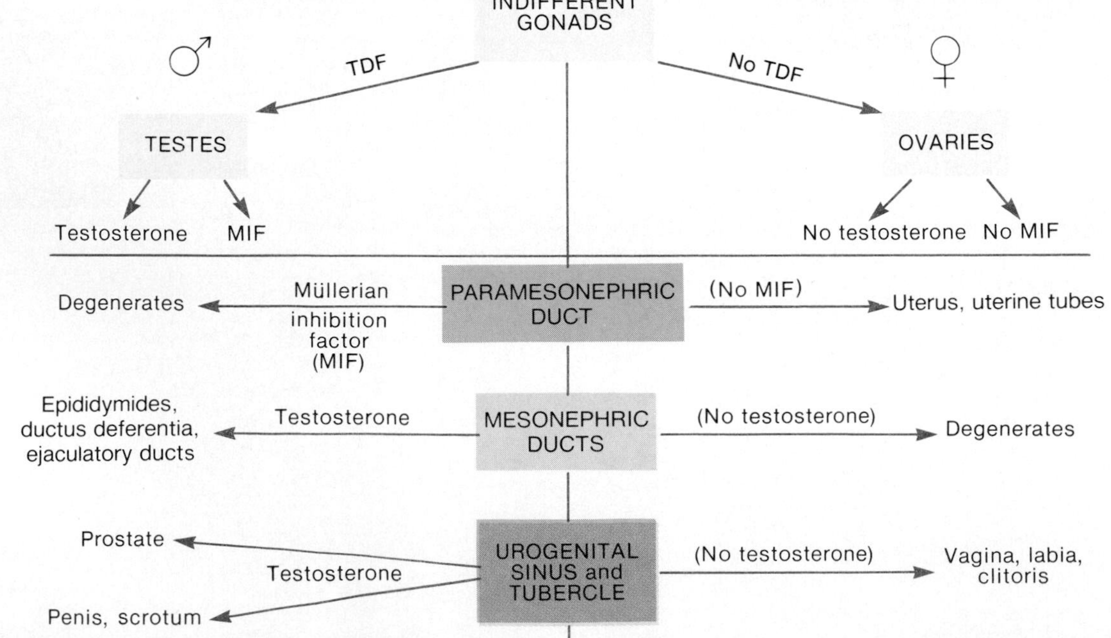

Figure 27.6. Differentiation of the external genitalia in the male and female. (*a*) and (a_1, sagittal view) at six weeks, the genital tubercle, urogenital fold, and labioscrotal swelling have differentiated from the genital tubercle. (*b*) at eight weeks, a distinct phallus is present during the indifferent stage. By the twelfth week, the genitalia are distinctly male (*c*) or female (*d*), being derived from homologous structures. (*e* and *f*) at sixteen weeks, the genitalia are formed. (*g* and *h*) photographs at week ten of male and female genitalia, respectively.

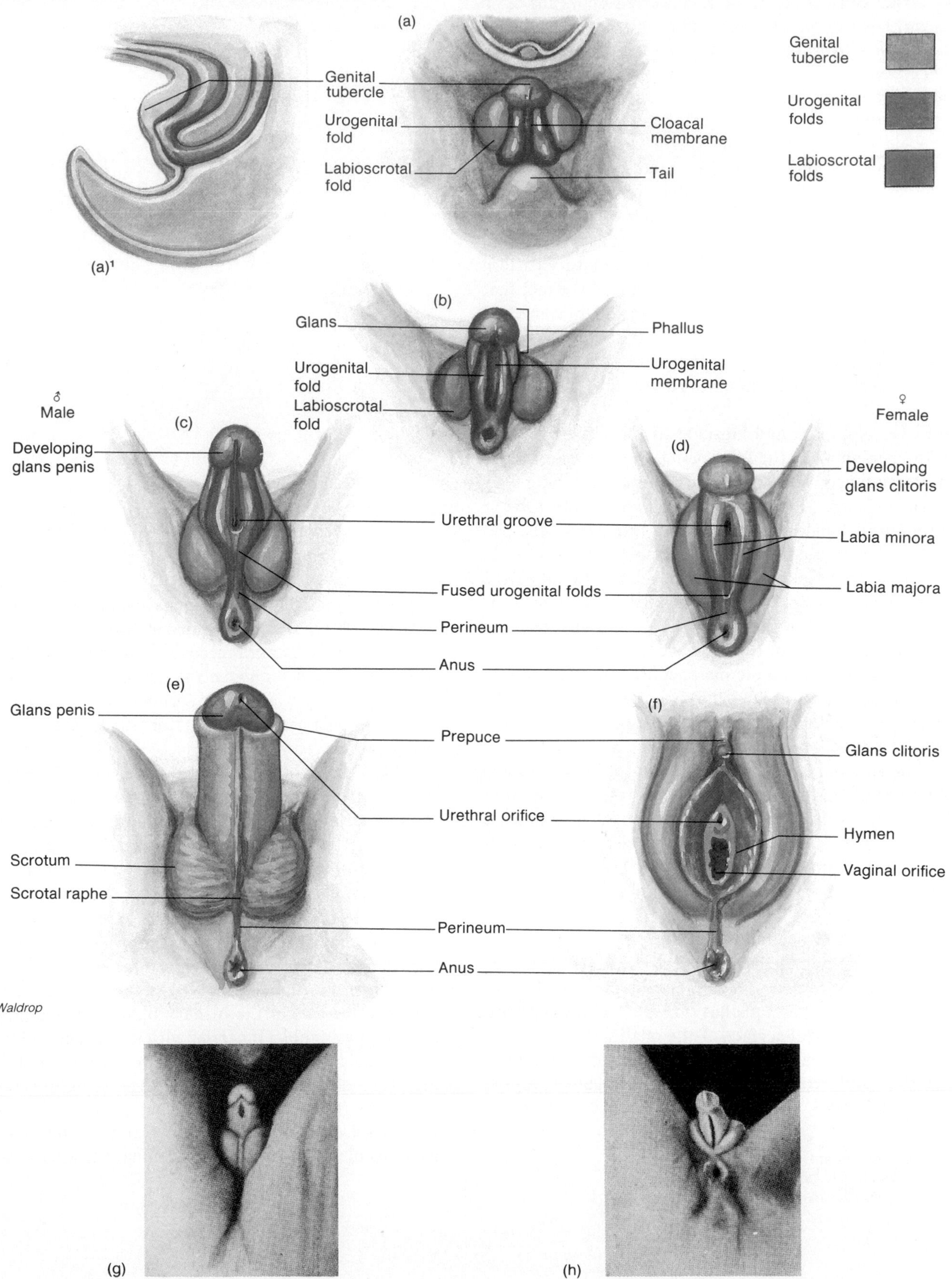

The masculinization of the embryonic reproductive structures occurs as a result of testosterone secreted by the embryonic testes. Testosterone itself, however, is not the active agent within these target organs. Once inside the target cells, testosterone is converted by means of an enzyme called *5α-reductase* into the active hormone known as **dihydrotestosterone** *(di-hi''dro-tes-tos'ter-ōn)* **(DHT),** which directly mediates the androgen effect in these organs (fig. 27.7).

In summary, the genetic sex is determined by whether a Y-bearing or an X-bearing sperm fertilizes the ovum; the presence or absence of a Y chromosome in turn determines whether the gonads of the embryo will be testes or ovaries; and the presence or absence of testes, finally, determines whether the sex accessory organs and external genitalia will be male or female. This regulatory pattern of sex determination makes sense in light of the fact that both male and female embryos develop within an environment high in estrogen, which is secreted by the mother's ovaries and the placenta. If estrogen determined the sex, all embryos would become feminized.

Figure 27.7. The conversion of testosterone, secreted by the interstitial cells of the testes, into dihydrotestosterone (DHT) within the target cells. This reaction involves the addition of a hydrogen (and the removal of the double carbon bond) in the first (*A*) ring of the steroid.

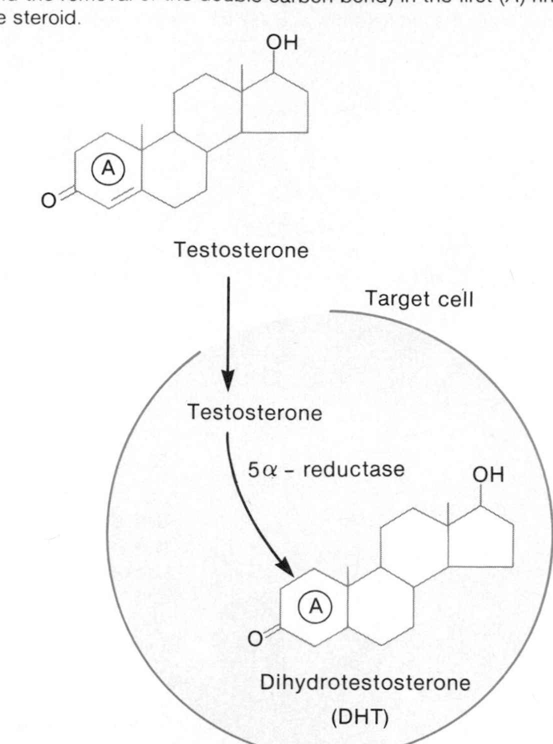

Fetal Development and Descent of the Testes

By the beginning of the fetal period at nine weeks, male differentiation of the gonads into testes is well under way. Internal changes include the formation of the tubular **seminiferous tubules** and **rete** *(re'te)* **testis** from the primary sex cord (see fig. 27.3). Developing on the outside surface of each testis is a fibromuscular cord called the **gubernaculum** *(gu''ber-nak'u-lum)* (fig. 27.8), which attaches to the inferior portion of the testis and extends to the labioscrotal fold of the same side. At the same time, a portion of the embryonic mesonephric duct adjacent to the testis becomes attached and convoluted and forms the **epididymis.** Another portion of the mesonephric duct becomes the **ductus deferens.**

The accessory reproductive glands in a male develop from two different sources. The **seminal vesicles** form from lateral outgrowths of the caudal ends of each mesonephric duct. The **prostate gland** arises from an endodermal outgrowth of the urogenital sinus. The **bulbourethral glands** develop from outgrowths in the membranous portion of the urethra.

The external genitalia of the male become distinct from those of a female by the end of the ninth week (see fig. 27.6). Prior to that, the genital tubercle in both sexes elongates and is called a **phallus** *(fal'us)*. A **urethral groove** forms on the ventral surface of the phallus.

The differentiation of the external genitalia of a male from the indifferent stage to discernible organs is caused by the androgens produced and secreted by the testes. These changes include the elongation and differentiation of the phallus into a **penis,** a fusion of the **urogenital folds** surrounding the urethral groove along the ventral surface of the penis, and a midventral fusion of the labioscrotal folds to form the wall of the **scrotum.** The external genitalia of a male are completely formed by the end of the twelfth week.

The descent of the testes from the site of development begins between the sixth and tenth week. Descent into the scrotal sac, however, does not occur until about the twenty-eighth week, when paired inguinal canals form in the abdominal wall to provide openings from the pelvic cavity to the scrotal sac. The process by which a testis descends is not well understood, but it seems to be associated with the shortening of the gubernaculum, which is attached to the testis and extends through the inguinal canal to the wall of the scrotum (fig. 27.8). As the testis descends, it passes to the side of the urinary bladder and anterior to the symphysis pubis. It carries with it the ductus deferens, the testicular vessels and nerve, a portion of the abdominal muscle, and lymph vessels. All of these structures remain attached to the testis and form what is known as the **spermatic cord.** By the time the testis is in the scrotal sac, the gubernaculum is no more than a remnant of scarlike tissue.

rete: L. *rete,* a net
gubernaculum: L. *gubernaculum,* helm
phallus: Gk. *phallus,* penis

Figure 27.8. The descent of the testes. (*a*) at ten weeks; (*b*) at eighteen weeks; and (*c*) at twenty-eight weeks. During development each testis descends through an inguinal canal in front of the symphysis pubis and enters the scrotum.

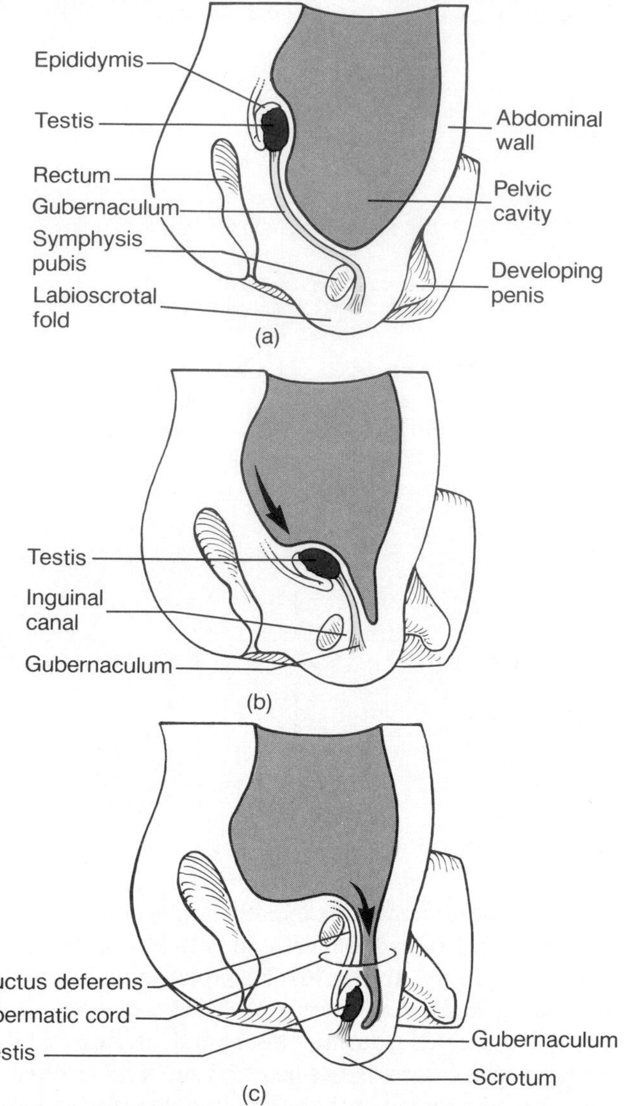

During the physical examination of a neonatal male child, a physician will palpate the scrotum to determine if the testes are in position. If one or both are not in the scrotal sac, it may be possible to induce descent by administering certain hormones. If this procedure does not work, surgery is necessary. The surgery is generally performed before the age of five. Failure to correct the situation may result in sterility and possibly the development of a tumorous testicle.

1. Describe the hormonal process of genital masculinization.
2. Explain how the chromosomal sex determines whether testes or ovaries will be formed.
3. What early embryonic structures give rise to the gonads, and what structures give rise to the external genitalia?
4. List the homologous structures of the male and female reproductive systems.
5. Describe the reproductive changes that occur during the fetal stage of development. How and why do the testes descend into the scrotum?

Puberty

Gonadotropic hormones from the anterior pituitary and maturational changes in the central nervous system result in an increased secretion of sex steroids from the testes, which in turn stimulate a growth spurt and sexual maturation during puberty.

Objective 8. Describe the hormonal changes and feedback mechanisms that result in puberty in males.

Objective 9. List the physical changes that occur in the male secondary sex organs during puberty.

The testes are formed and receive spermatogonia (sperm-forming cells) early in embryonic development. During the first trimester of pregnancy, the embryonic testes are active endocrine glands, secreting the high amounts of testosterone needed to masculinize the male embryo's secondary sex organs. During the second trimester of pregnancy, testosterone secretion in a male fetus declines so that at birth the testes are relatively inactive.

There is no difference in the blood concentrations of *sex steroids*—androgens and estrogens—in prepubertal boys and girls. This does not appear to be due to deficiencies in the ability of the gonads to produce these hormones, but rather to lack of sufficient stimulation. During *puberty,* the gonads secrete increased amounts of sex steroid hormones as a result of increased stimulation by **gonadotropic hormones** from the anterior pituitary gland.

Interactions between the Hypothalamus, Pituitary, and Gonads

The anterior pituitary gland produces and secretes **FSH (follicle-stimulating hormone)** and **LH (luteinizing hormone),** which is also called **ICSH (interstitial cell stimulating hormone)** in the male. Although these two hormones are named according to their actions in the female, the same hormones are secreted by the male's pituitary. The gonadotropic hormones of both sexes have three primary effects on the gonads: (1) stimulation of spermatogenesis or oogenesis (formation of sperm or ova); (2) stimulation of gonadal hormone secretion; and (3) maintenance of the structure of the gonads (gonads atrophy if the pituitary is removed).

There is a strict compartmentation in the testes with regard to gonadotropin action. Cellular receptor proteins for FSH are located exclusively in the seminiferous tubules, whereas LH receptor proteins are confined exclusively to the interstitial tissue. Secretion of testosterone by the interstitial cells is stimulated by LH but not by FSH. Spermatogenesis in the tubules is stimulated by FSH. The secretion of both LH and FSH from the anterior pituitary is stimulated by a hormone produced by the hypothalamus and secreted into the hypothalamo-hypophyseal portal vessels. This releasing hormone is called *LHRH (luteinizing hormone–releasing hormone).* Since attempts to find a separate FSH-releasing hormone have thus far failed and since LHRH stimulates FSH as well as LH secretion, LHRH is often referred to as **gonadotropin-releasing hormone** (and accordingly abbreviated **GnRH**).

If a male or female animal is castrated (has had its gonads surgically removed), the secretion of FSH and LH increases to much higher levels than those measured in the intact animal. This demonstrates that the gonads secrete products that have a **negative feedback inhibition** of gonadotropin secretion. This negative feedback is exerted in a large part by sex steroids: testosterone in the male and estrogen and progesterone in the female.

The negative feedback effects of steroid hormones are believed to occur by means of two mechanisms: (1) inhibition of GnRH secretion, and (2) inhibition of the pituitary's response to a given amount of GnRH secreted from the hypothalamus. In addition to steroid hormones, there is evidence that the testes (and perhaps the ovaries as well) may secrete a polypeptide hormone, called **inhibin** *(in-hib'in),* that specifically inhibits FSH secretion in response to GnRH, without affecting the secretion of LH.

Figure 27.9 illustrates the similarities in the gonadal regulation of males and females. Important differences in hypothalamus-pituitary-gonad interactions exist, however, between males and females. The secretion of gonadotropins and sex steroids are more or less constant in adult males. The secretion of gonadotropins and sex steroids in adult females, in contrast, shows cyclic variations (during the menstrual cycle). Also, during the normal female cycle, estrogen exerts a positive effect on LH secretion.

Figure 27.9. Interactions between the hypothalamus, anterior pituitary, and gonads. Sex steroids secreted by the gonads have a negative feedback effect on the secretion of GnRH (gonadotropin-releasing hormone) and on the secretion of gonadotropins. The gonads may also secrete a polypeptide hormone called *inhibin* that exerts negative feedback control of FSH secretion.

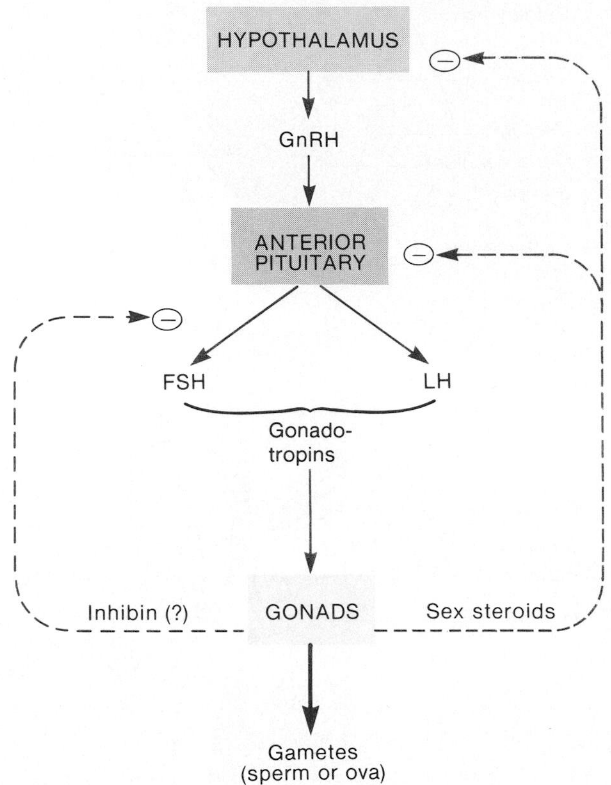

The Onset of Puberty

The secretion of FSH and LH is high in the newborn but falls to very low levels a few weeks after birth. Gonadotropin secretion remains low until the beginning of puberty, which is marked by rising levels of FSH followed by LH secretion. Experimental evidence suggests that this rise in gonadotropin secretion is the result of two processes: (1) maturational changes in the brain that result in increased GnRH secretion by the hypothalamus, and (2) a decreased sensitivity of gonadotropin secretion to the negative feedback effects of sex steroid hormones.

The maturation of the hypothalamus or other regions of the brain that lead to increased GnRH secretion at the time of puberty appear to be programmed—children without gonads show increased FSH secretion at the normal time. Also during this period of time, a given amount of sex steroids has less of a suppressive effect on gonadotropin secretion than the same dose would have if administered prior to puberty. This suggests that the sensitivity of the hypothalamus and the pituitary to negative feedback effects decreases at puberty, which would also help account for the rising gonadotropin secretion at this time.

During late puberty there is a "pulsatile" secretion of gonadotropins—FSH and LH secretion increases during periods of sleep and decreases during periods of wakefulness. These pulses of increased gonadotropin secretion during puberty stimulate a rise in sex steroid secretion from the gonads. The increased secretion of testosterone from the testes during puberty causes the male **secondary sexual characteristics** to be manifested (table 27.2).

Table 27.2 Development of secondary sexual characteristics and other changes that occur during puberty in males

Characteristic	Age of first appearance	Hormonal stimulation
Growth of testes	10–14	Testosterone, FSH, growth hormone
Pubic hair	10–15	Testosterone
Body growth	11–16	Testosterone, growth hormone
Growth of penis	11–15	Testosterone
Growth of larynx (voice lowers)	Same time as growth of penis	Testosterone
Facial and axillary (underarm) hair	About two years after appearance of pubic hair	Testosterone
Eccrine sweat glands and sebaceous glands; acne (from blocked sebaceous glands)	About same time as facial and axillary hair growth	Testosterone

1. Using a flow diagram, show the negative feedback control that the gonads exert on GnRH and gonadotropin secretion. Explain the effects of castration on FSH and LH secretion and the effects of the removal of the pituitary on the structure of the gonads and accessory sex organs.
2. Describe the two mechanisms that have been proposed to explain the rise in sex steroid secretion that occurs at puberty.
3. Contrast the relative levels of testosterone during early fetal development, at birth, and during puberty.
4. Describe the physical changes that occur in a male during puberty.

Structure and Function of the Testes

The testes within the scrotum produce spermatozoa and androgens. Androgens regulate spermatogenesis and the development and functioning of the secondary sex organs.

Objective 10. Describe the location, structure, and functions of the testes.

Objective 11. Explain the hormonal actions of testosterone, and specifically describe the interactions between the interstitial cells and the seminiferous tubules in each testis.

Objective 12. List the events of spermatogenesis, and distinguish between spermatogenesis and spermiogenesis.

Objective 13. Diagram the structure of a sperm, and explain the function of each of its parts.

Scrotum

The saclike **scrotum** is suspended immediately behind the base of the penis, anterior to the anal opening, in a region known as the **perineum** *(per''ĭ-ne'um)* (see fig. 1.17). The functions of the scrotum are to support and protect the testes and to regulate their position relative to the pelvic region of the body. The soft textured skin of the scrotum is covered with sparse hair in mature males and is darker in color than most of the other skin of the body. It also contains numerous sebaceous glands. The external appearance of the scrotum varies at different times in the same individual, depending on environmental conditions and the contraction of the dartos and cremaster muscles. The **dartos** *(dar'tos)* is a layer of smooth muscle fibers in the subcutaneous tissue of the scrotum, and the **cremaster** *(kre-mas'ter)* is a thin strand of skeletal muscle associated with the spermatic cord. Both muscles involuntarily contract to cold temperatures to move the testes closer to the heat of the body in the pelvic region. The cremaster muscle is a continuation of the internal oblique muscle of the abdominal wall, which is derived as the testes descend into the scrotum. Because it is a skeletal muscle, it can contract voluntarily as well. When these muscles are contracted, the scrotum appears tightly wrinkled. Warm temperatures cause the dartos and cremaster muscles to relax and the testes to become pendent in the flaccid scrotum. The temperature of the testes is maintained at about 35°C (95°F or about 3.6°F below normal body temperature) by the contraction or relaxation of the scrotal muscles. This temperature has been determined to be an optimum temperature for the production and storage of sperm.

The scrotum is subdivided into two longitudinal compartments by a fibrous **median septum.** The purpose of the median septum is to compartmentalize each testis so that the trauma or infections of one will generally not affect the other. Another protective feature is that the left testis is generally suspended lower in the scrotum than the right so that the two will not as likely be compressed forcefully together. The site of the median septum is apparent on the surface of the scrotum along a median longitudinal ridge called the **perineal raphe** *(ra'fe)*. The perineal raphe extends forward to the undersurface of the penis and backward to the anal opening. The blood supply and innervation of the scrotum are extensive. The arteries that serve the scrotum are the internal pudendal branch of the internal

dartos: Gk. *dartos,* skinned or flayed
cremaster: Gk. *cremaster,* a suspender, to hang
septum: L. *septum,* a partition
raphe: Gk. *raphe,* a seam
pudendal: L. *pudeo,* to feel ashamed

Figure 27.10. Structural features of the testis and epididymis. (*a*) a longitudinal view; (*b*) a transverse view.

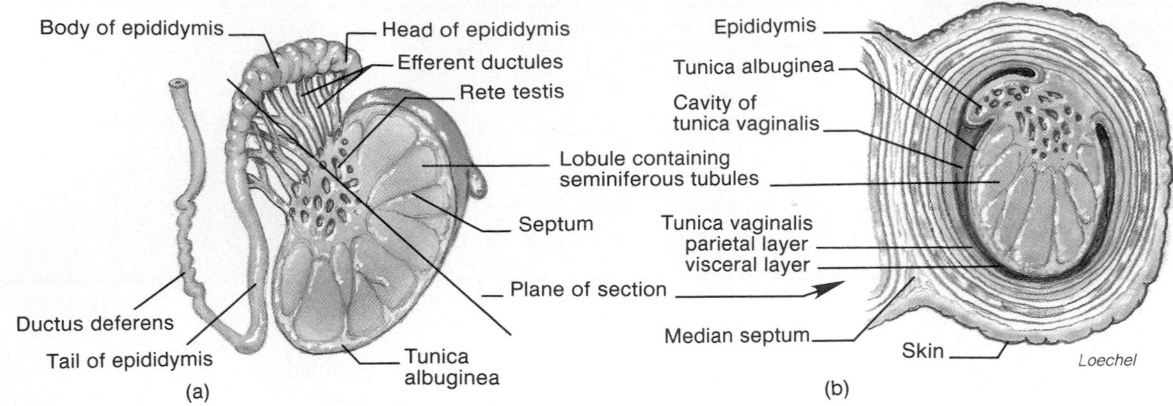

iliac artery, the external pudendal branch from the femoral artery, and the cremasteric branch of the inferior epigastric artery. The venous drainage follows a pattern similar to the arteries. The scrotal nerves are primarily sensory and include the pudendal nerves, ilioinguinal nerves, and posterior cutaneous nerves of the thigh.

Although uncommon, male infertility may result from an excessively high temperature of the testes over an extended period of time. Tight clothing that keeps the testes close to the body or frequent hot baths or saunas may destroy sperm to the extent that the sperm count of discharged semen is below that necessary to cause fertilization.

Testes

Structure of the Testis. The **testes** are paired, whitish, ovoid organs, which measure about 4 cm (1.5 in.) in length and 2.5 cm (1 in.) in diameter. Each testis weighs between 10 and 14 g.

Two tissue layers, or tunicas, cover the testis. The outer **tunica vaginalis** is a thin, serous sac derived from the peritoneum during the descent of the testis. The **tunica albuginea** *(al''bu-jin'e-ah)* is a tough, fibrous membrane that directly encapsules the testis (fig. 27.10). Fibrous, inward extensions of the tunica albuginea partition the testis into 250 to 300 wedge-shaped **lobules.**

Each lobule of the testis contains one to three tightly convoluted **seminiferous tubules,** which may exceed 70 cm (28 in.) in length if uncoiled. The seminiferous tubules are the functional units of the testis because it is here that *spermatogenesis* occurs. Sperm are produced at the rate of thousands per second throughout the life of a healthy, sexually mature male.

tunica: L. *tunica,* a coat
vaginalis: L. *vagina,* a sheath
albuginea: L. *albus,* white

Various stages of spermatogenesis can be observed in a histological section of seminiferous tubules (fig. 27.11). The germinal cells, called **spermatogonia** *(sper''mah-to-go'ne-ah),* are in contact with the basement membrane. Spermatogonia undergo meiosis (discussed in a later section), to produce, in order of advancing maturity, the primary spermatocytes, secondary spermatocytes, and spermatids (see fig. 27.14). Forming the walls of the seminiferous tubules are **nurse cells,** or **Sertoli cells,** which produce and secrete nutrients to the developing spermatozoa embedded between them. The spermatozoa are formed, but not fully matured, by the time they reach the lumina of the seminiferous tubules.

Between the seminiferous tubules are specialized endocrine cells called **interstitial cells (cells of Leydig).** The function of these cells is to produce and secrete the male sex hormones. The testes are considered mixed exocrine and endocrine glands because they produce both sperm and androgens.

Once the sperm are produced, they move through the seminiferous tubules and enter the **rete testis** for further maturation. Cilia are located on some of the cells of the rete testis, presumably for moving sperm. The sperm are transported out of the testis and into the epididymis through a series of **efferent ductules.**

The testes receive blood through the **testicular arteries,** which arise from the abdominal aorta immediately below the origin of the renal arteries. The **testicular veins** drain the testes. The testicular vein of the right side enters directly into the inferior vena cava, whereas the testicular vein of the left side drains into the left renal vein.

Testicular nerves innervate the testes with both efferent and afferent fibers arising from the tenth thoracic segment of the spinal cord. Innervation is primarily through sympathetic fibers, but there is limited parasympathetic stimulation as well.

Sertoli cells: from Enrico Sertoli, Italian histologist, 1842–1910
efferent ductules: L. *efferre,* to bring out; *ducere,* to lead

Figure 27.11. A diagrammatic representation of seminiferous tubules. (*a*) a sagittal section of a testis; (*b*) a transverse section of a seminiferous tubule.

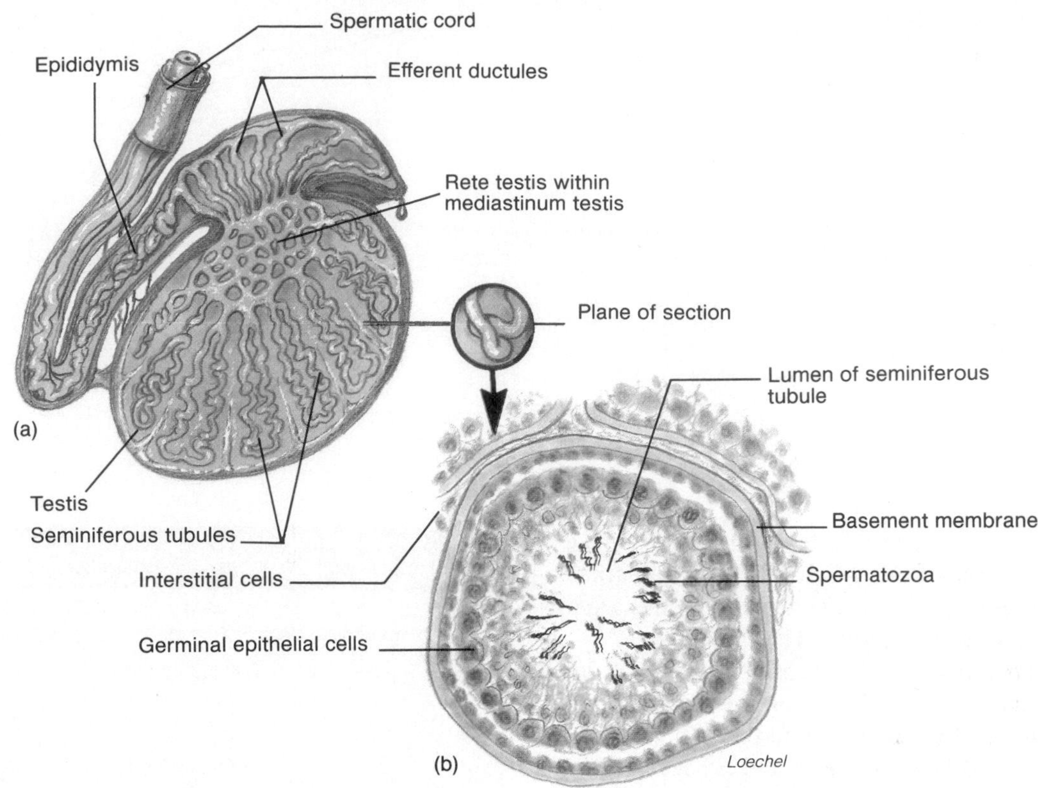

The primary cause of male infertility is a condition called *varicocele (var'ĭ-ko-sēl")*. Varicocele occurs when one or both of the testicular veins draining from the testes becomes swollen, resulting in poor vascular circulation in the testes. A varicocele generally occurs on the left side, because the left spermatic vein drains into the renal vein, where the blood pressure is higher than it is in the inferior vena cava, into which the right testicular vein empties.

Endocrine Functions of the Testis. Testosterone is by far the major androgen secreted by the adult testis. This hormone (or derivatives of it—the 5 α-reduced androgens and estradiol) is responsible for the initiation and maintenance of the body changes of puberty in males. Androgens are sometimes called *anabolic steroids* because they stimulate the growth of muscles and other structures (table 27.3). Increased testosterone secretion during puberty is also required for the growth of the sex accessory organs, primarily the seminal vesicles and the prostate gland. The removal of androgens by castration results in atrophy of these organs.

Androgens stimulate the growth of the larynx (causing lowering of the voice), increased hemoglobin synthesis (males have higher hemoglobin levels than females), and bone growth. The effect of androgens on bone growth is self-limiting, however, because androgens ultimately cause the conversion of cartilage to bone in the epiphyseal discs, thus "sealing" the discs and preventing further lengthening of the bones.

Although androgens are by far the major secretory product of the testes, the testes do produce and secrete small amounts of estradiol (fig. 27.12). The source of these

Table 27.3 Summary of some of the actions of androgens in the male

Category	Action
Sex determination	Growth and development of mesonephric ducts into epididymides, ductus deferentia, seminal vesicles, and ejaculatory ducts Development of the urogenital sinus and tubercle into prostate gland Development of male external genitalia
Spermatogenesis	At puberty: completion of meiotic division and early maturation of spermatids After puberty: maintenance of spermatogenesis
Secondary sex characteristics	Growth and maintenance of accessory sex organs Growth of the penis Growth of facial and axillary hair Body growth
Anabolic effects	Protein synthesis and muscle growth Growth of bones Growth of other organs (including the larynx) Erythropoiesis (red blood cell formation)

From Stuart Ira Fox, *Human Physiology,* 2d ed. Copyright © 1987 Wm. C. Brown Publishers, Dubuque, Iowa. All Rights Reserved. Reprinted by permission.

varicocele: L. *varico,* a dilated vein; Gk. *kele,* tumor or hernia

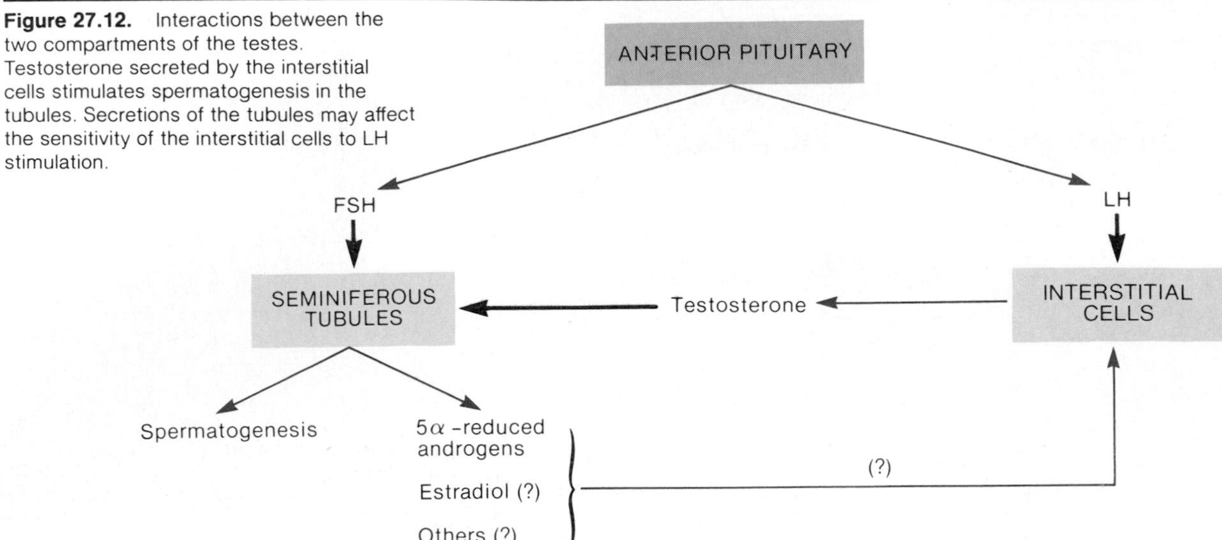

Figure 27.12. Interactions between the two compartments of the testes. Testosterone secreted by the interstitial cells stimulates spermatogenesis in the tubules. Secretions of the tubules may affect the sensitivity of the interstitial cells to LH stimulation.

estrogens is not known—some evidence suggests that the tubules produce estradiol as a result of FSH stimulation, whereas other evidence suggests that estradiol comes from the interstitial cells as a result of LH stimulation.

The testes contain estrogen receptors, which are confined to the interstitial tissue. Receptors for androgens, in contrast, are confined to the seminiferous tubules. This suggests the possibility that androgens and estrogens might have regulatory functions within the testes themselves. This possibility has proven correct for androgens (they are needed for spermatogenesis). The physiological role of estrogens in the testes remains to be demonstrated.

The two compartments of the testes interact with each other. Testosterone from the interstitial cells is metabolized by the tubules into other active androgens and serves to regulate tubular functions. The tubules also secrete products—5α-reduced androgens and possibly other hormones—that could conceivably influence interstitial cell function. Such interactions are suggested by evidence that in the pubertal male rat, exposure to FSH augments the responsiveness of the interstitial cells to LH. Since FSH can only directly affect the tubules, the FSH-induced enhancement of LH responsiveness must be mediated by products secreted from the tubules. The nature of these regulators is not currently known.

Table 27.4 Stages of meiosis

Stage	Events
First meiotic division	
Prophase I	Chromosomes appear double-stranded. Each strand, called a chromatid, contains duplicate DNA joined together by a structure known as a centromere. Homologous chromosomes pair up side by side.
Metaphase I	Homologous chromosome pairs line up at equator. Spindle apparatus is completed.
Anaphase I	Homologous chromosomes are separated; each member of a homologous pair moves to opposite poles.
Telophase I	Cytoplasm divides to produce two haploid cells.
Second meiotic division	
Prophase II	Chromosomes appear, each containing two chromatids.
Metaphase II	Chromosomes line up single file along equator as spindle formation is completed.
Anaphase II	Centromeres split and chromatids move to opposite poles.
Telophase II	Cytoplasm divides to produce two haploid cells from each of the haploid cells formed at telophase I.

Spermatogenesis. The germ cells that migrate from the yolk sac to the testes during early embryonic development become "stem cells" called **spermatogonia** within the outer region of the seminiferous tubules. Spermatogonia are diploid cells (with forty-six chromosomes) that ultimately give rise to mature haploid gametes by a process of cell division called *meiosis (mi-o'sis).*

Meiosis occurs in two parts as summarized in table 27.4. During the first part of this process, the DNA duplicates (prophase I), and homologous chromosomes are separated (during anaphase I) into two daughter cells (telophase I). Since each daughter cell contains only one of each homologous pair of chromosomes, the cells formed at the end of this *first meiotic division* contain twenty-three chromosomes each and are haploid (fig. 27.13). Each of the twenty-three chromosomes at this stage, however, consists of two strands (called *chromatids*) of identical DNA. During the *second meiotic division,* these duplicate chromatids are separated (at anaphase II) into daughter cells at telophase II. Meiosis of one diploid spermatogonia cell therefore produces four haploid daughter cells.

Figure 27.13. Meiosis, or reduction division. In the first meiotic division the homologous chromosomes of a diploid parent cell are separated into two haploid daughter cells. Each of these chromosomes contain duplicate strands, or chromatids. In the second meiotic division these chromatids are distributed to two new haploid daughter cells.

First meiotic division

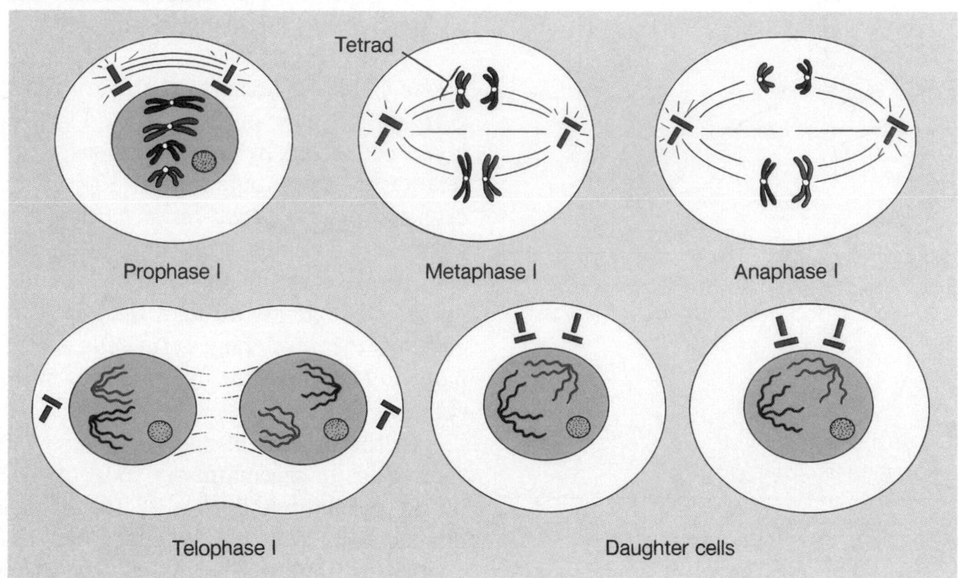

Second meiotic division

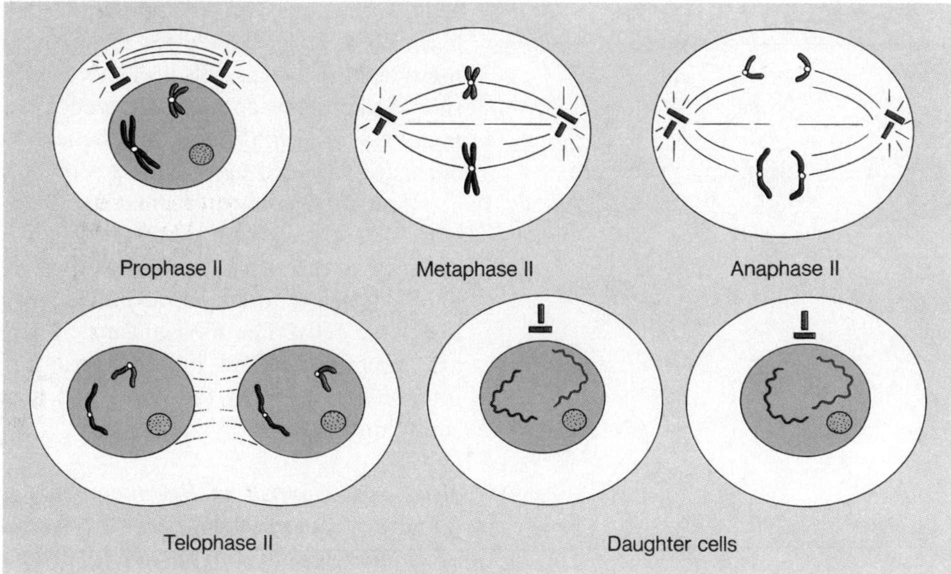

Actually, only about 1,000–2,000 stem cells migrate from the yolk sac into the embryonic testes. In order to produce many millions of sperm throughout adult life, these spermatogonia cells duplicate themselves by mitotic division, and only one of the two cells—now called a **primary spermatocyte**—undergoes meiotic division (fig. 27.14). In this way, spermatogenesis can occur continuously without exhausting the number of spermatogonia.

When a diploid primary spermatocyte completes the first meiotic division (at telophase I), the two haploid daughter cells thus produced are called **secondary spermatocytes.** At the end of the second meiotic division, each of the two secondary spermatocytes produces two haploid **spermatids.** One primary spermatocyte therefore produces four spermatids.

Figure 27.14. Spermatogonia undergo mitotic division to replace themselves and produce a daughter cell that will undergo meiotic division. This cell is called a primary spermatocyte. Upon completion of the first meiotic division, the daughter cells are called secondary spermatocytes. Each of these completes a second meiotic division to form spermatids. Notice that the four spermatids produced by the meiosis of a primary spermatocyte are interconnected. Each spermatid forms a mature spermatozoan.

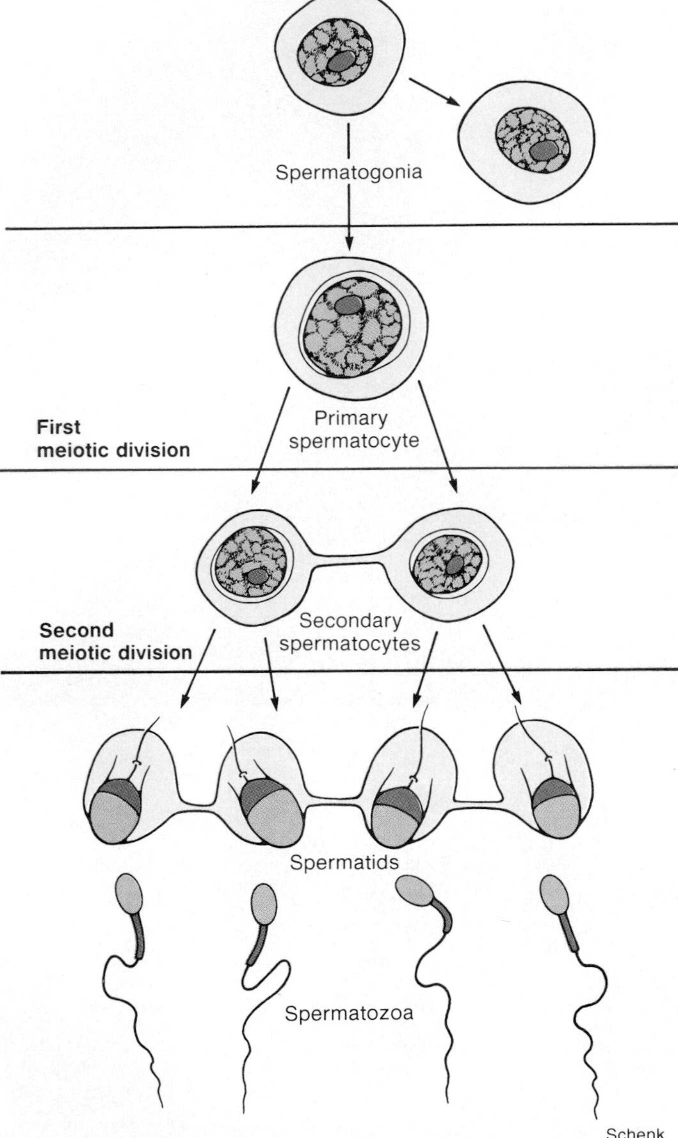

The stages of spermatogenesis are arranged sequentially in the wall of the seminiferous tubule. The epithelial wall of the tubule, called the **germinal epithelium,** is indeed composed of germ cells in different stages of spermatogenesis. The spermatogonia and primary spermatocytes are located toward the outer side of the tubule, whereas the spermatids and mature spermatozoa are located on the side of the tubule facing the lumen (fig. 27.15).

At the end of the second meiotic division, the four spermatids produced by meiosis of one primary spermatocyte are interconnected with each other—their cytoplasm does not completely pinch off at the end of each division. The development of these interconnected spermatids into separate, mature **spermatozoa** (a process called **spermiogenesis**) requires the participation of another type of cell in the tubules, the nurse cells.

Nurse Cells. The nurse (Sertoli) cells are the only nongerminal cell type in the tubules. They form a continuous layer, connected by tight junctions, around the circumference of each tubule (fig. 27.15). In this way the nurse cells comprise a *blood-testis barrier,* because molecules from the blood must pass through the cytoplasm of nurse cells before entering germinal cells. The cytoplasm of the nurse cells extends through the width of the tubule and envelops the developing germ cells, so that it is often difficult to tell where the cytoplasm of nurse cells and germ cells are separated.

In the process of *spermiogenesis* (the conversion of spermatids to spermatozoa), most of the spermatid cytoplasm is eliminated (fig. 27.15). This occurs through phagocytosis by the nurse cells of the "residual bodies" of cytoplasm from the spermatids. Many believe that phagocytosis of the residual bodies may transmit informational molecules from the germ cells to the nurse cells. The nurse cells, in turn, may provide many molecules needed by the germ cells. It is known, for example, that the X chromosome of the germ cells is inactive during meiosis. Since this chromosome contains the genes needed to produce many essential molecules, it is believed that these molecules are provided by the nurse cells during this time.

The importance of the nurse cells in tubular function is further evidenced by the fact that FSH receptors are confined to the nurse cells. Any effect of FSH on the tubules, therefore, must be mediated through the action of the nurse cells. The mechanisms by which these effects are accomplished and, indeed, the nature of all chemical interactions between the nurse and germ cells are currently unknown.

Hormonal Control of Spermatogenesis. The very beginning of spermatogenesis—the formation of primary spermatocytes and entry into early prophase I—is apparently somewhat independent of hormonal control and, in fact, starts during embryonic development. Spermatogenesis is arrested at this stage, however, until puberty, when testosterone secretion rises (fig. 27.16). Testosterone is required for the completion of meiotic division and for the early stages of spermatid maturation. This effect is probably not mediated directly by testosterone but, rather, by some of the molecules produced from testosterone by the tubules.

Figure 27.15. Seminiferous tubules. (*a*) a cross section with surrounding interstitial tissue; (*b*) the stages of spermatogenesis within the germinal epithelium of a seminiferous tubule, showing the relationship between nurse cells and developing spermatozoa. ([*a*] From: *Tissues and Organs: A Text Atlas of Scanning-Electron Microscopy* by R. G. Kessel and R. Kardon. W. H. Freeman and Company. © 1979.)

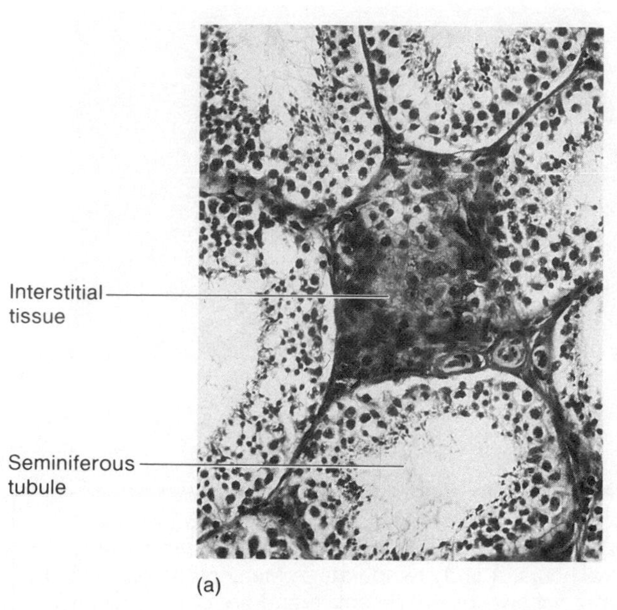

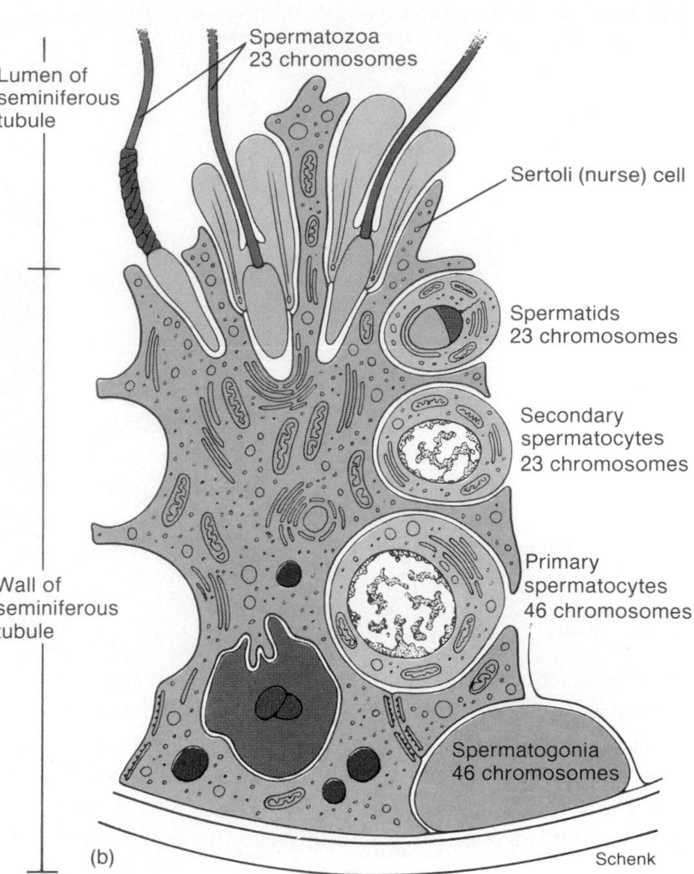

Figure 27.16. The endocrine control of spermatogenesis. During puberty both testosterone and FSH are required to initiate spermatogenesis. In the adult, however, testosterone alone can maintain spermatogenesis.

Spermatogonia
↓
Primary spermatocytes
↓ Meiosis (first division)
Secondary spermatocytes
↓ Meiosis (second division)
Spermatids
↓ Spermiogenesis
Spermatozoa

Testosterone Required at puberty →

FSH Required at puberty →

← **Testosterone** Maintains after puberty

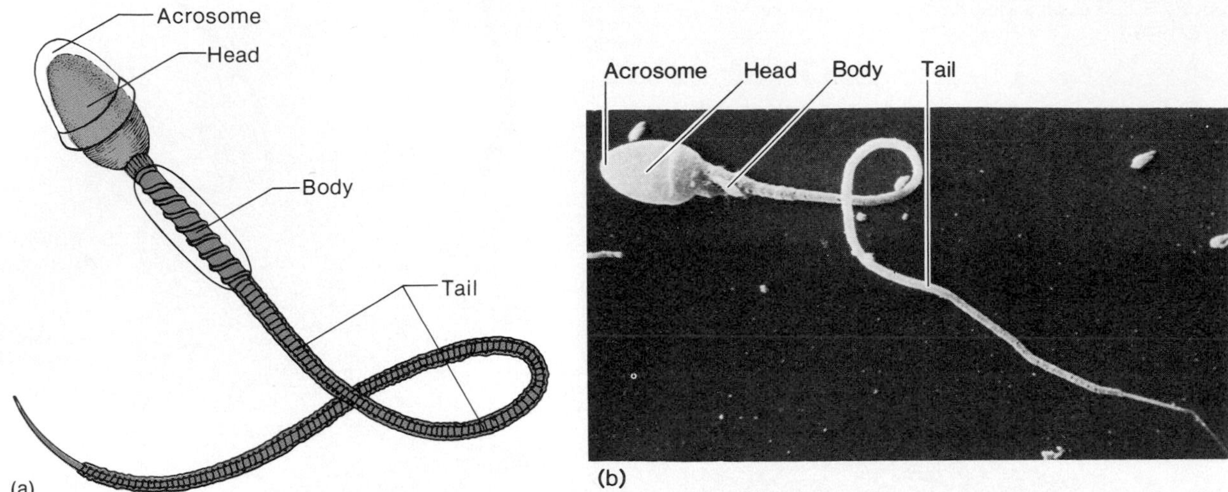

Figure 27.17. A human spermatozoan. (*a*) a diagrammatic representation; (*b*) a scanning electron micrograph. (From: *Tissues and Organs: A Text Atlas of Scanning Electron Microscopy* by R. G. Kessel and R. Kardon. W. H. Freeman and Company. © 1979.)

The later stages of spermatid maturation during puberty appear to require stimulation by FSH. This FSH effect is, as previously mentioned, mediated by the nurse cells, which contain FSH receptors and surround and interact with the spermatids. During puberty, therefore, both FSH and androgens are needed for the initiation of spermatogenesis.

Experiments in rats, and more recently in humans, have revealed that spermatogenesis within the adult testis can be maintained by androgens alone, in the absence of FSH. It appears, in other words, that FSH is needed to initiate spermatogenesis at puberty but that once spermatogenesis has begun, FSH may no longer be required. The physiological importance of FSH in the adult male is, as a result of these experiments, currently controversial.

Structure of Spermatozoa

A mature sperm cell, or **spermatazoon,** is a microscopic, tadpole-shaped structure about 0.06 mm long (fig. 27.17). It consists of an oval-shaped **head,** a cylindrical **body,** and an elongated **tail.** The head of a sperm contains a nucleus with twenty-three chromosomes.The tip of the head, called the *acrosome,* contains enzymes that help the sperm penetrate into the ovum. The body of the sperm contains numerous mitochondria spiraled around a filamentous core. The mitochondria provide the energy necessary for locomotion. The tail of the sperm is a flagellum that propels the sperm with a lashing movement. The maximum unassisted rate of sperm movement is about 3 mm per hour.

acrosome: Gk. *akros,* extremity; *soma* body

The life expectancy of ejaculated sperm is between forty-eight and seventy-two hours at body temperature. Many of the ejaculated sperm, however, are defective and are of no value. It is not uncommon for sperm to have enlarged heads, dwarfed and misshapen heads, two flagella, or a flagellum that is bent. Sperm such as these are unable to propel themselves adequately.

1. Describe the location and structure of the testes.
2. Describe the effects that castration has in the male on FSH and LH secretion, and explain the experimental evidence suggesting that the testes produce a polypeptide that specifically inhibits FSH secretion.
3. Describe the two compartments of the testes with respect to their structure, function, and response to gonadotropin stimulation. Describe two ways that these compartments appear to interact.
4. Using a diagram, describe the stages of spermatogenesis. Explain why spermatogenesis can continue throughout life without using up all of the spermatogonia.
5. Diagram a sperm and its adjacent nurse cell, and explain the functions of nurse cells in the seminiferous tubules.
6. Explain how FSH and androgens synergize to stimulate sperm production at puberty. Describe the hormonal requirements for spermatogenesis after puberty.

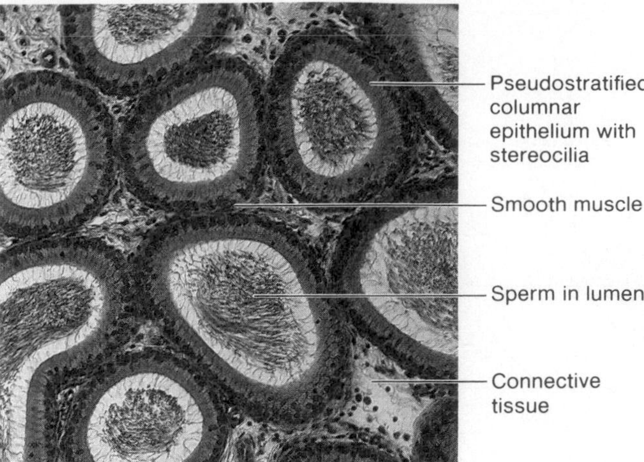

Figure 27.18. A photomicrograph of the epididymis showing sperm in the lumen (50×).

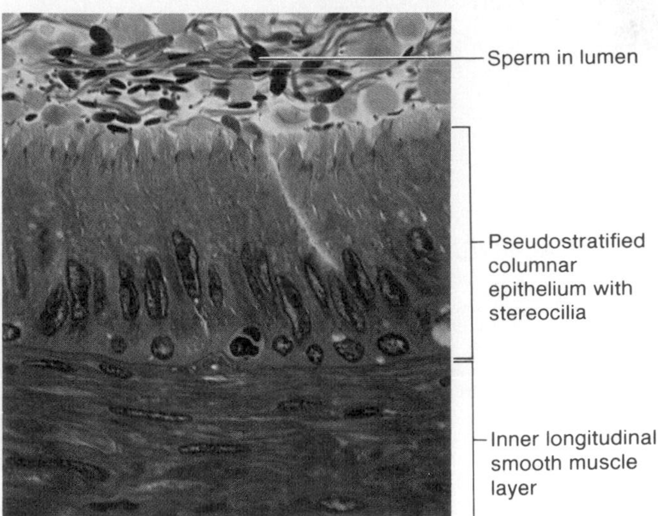

Figure 27.19. A photomicrograph of the ductus deferens (250×).

Spermatic Ducts, Accessory Glands, and the Penis

The spermatic ducts store spermatozoa and transport them from the testes to the urethra. The accessory reproductive glands provide additives to the spermatozoa in the formation of semen, which is discharged from the erect penis during ejaculation.

Objective 14. List the various spermatic ducts and describe the location and structure of each segment.

Objective 15. Describe the structure and contents of the spermatic cord.

Objective 16. Describe the location, structure, and function of the ejaculatory ducts, seminal vesicles, prostate gland, and bulbourethral gland.

Objective 17. Describe the structure and function of the penis.

Spermatic Ducts

The duct system, which stores and transports spermatozoa from the testes to the urethra, includes the epididymides, the ductus deferentia, and the ejaculatory ducts.

Epididymis. The **epididymis** *(ep''i-did'i-mis)*—pl., *epididymides*—is a long, flattened organ attached to the posterior surface of the testis (see fig. 27.10). The tubular portion of the epididymis is highly coiled and contains millions of sperm in their final stages of maturation. It is estimated that if the epididymis were uncoiled, it would measure 5–6 m (about 17 ft). The upper, expanded portion of the epididymis is the **head,** the tapering middle section is the **body,** and the lower, tubular portion is the **tail.** The tail of the epididymis is continuous with the ductus deferens. Both the tail of the epididymis and the ductus deferens store the sperm that is to be discharged during ejaculation (fig. 27.18). The time required to produce mature sperm—from meiosis in the seminiferous tubules to storage in the ductus deferens—is approximately two months.

Ductus Deferens. The **ductus deferens** *(duk'tus def'er-enz)*—pl., *ductus deferentia*—also called *vas deferens,* is a fibromuscular tube about 45 cm (18 in.) long and 2.5 mm thick (see fig. 27.1), that conveys sperm from the epididymis to the ejaculatory duct. The ductus deferens originates where the tail of the epididymis becomes less convoluted and is no longer attached to the testis. This first portion of the ductus deferens is important for the storage of sperm. The ductus deferens exits the scrotum as it ascends along the posterior border of the testis. From here, it penetrates the inguinal canal, enters the pelvic cavity, and passes to the side of the urinary bladder medial to the ureter. The **ampulla** of the ductus deferens is the terminal portion that joins the ejaculatory duct.

The histological structure of the ductus deferens includes a layer of pseudostratified epithelium in contact with the tubular lumen and surrounded by three layers of tightly packed smooth muscle (fig. 27.19). Sympathetic nerves from the pelvic plexus serve the ductus deferens. Stimulation through these nerves causes peristaltic contractions of the muscular layer, which forcefully ejects the stored sperm toward the ejaculatory duct.

deferens: L. *deferens,* conducting away
ampulla: L. *ampulla,* a two-handled bottle

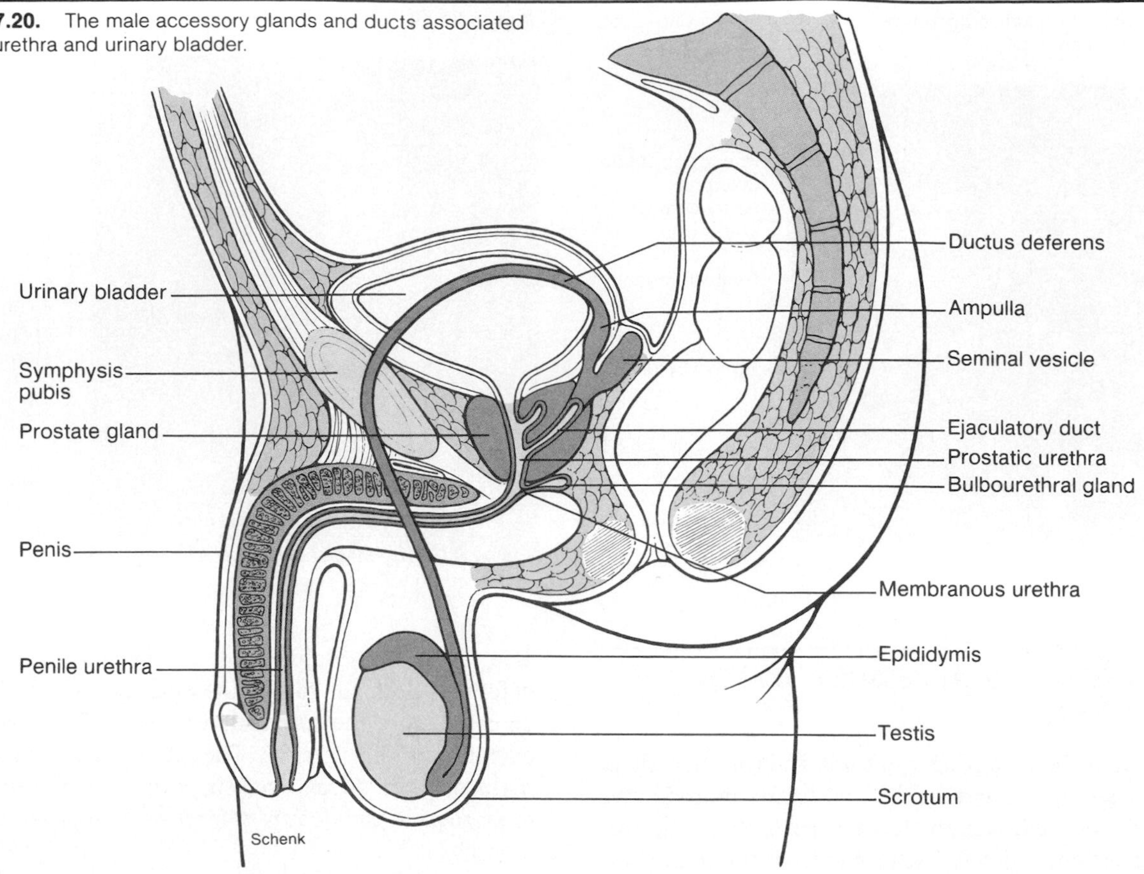

Figure 27.20. The male accessory glands and ducts associated with the urethra and urinary bladder.

Much of the ductus deferens is located within a structure known as the **spermatic cord.** The spermatic cord extends from the testis to the inguinal ring and consists of the ductus deferens, spermatic vessels, nerves, cremaster muscle, lymph vessels, and connective tissue. The portion of the spermatic cord positioned anterior to the pubic bone can be palpated on a male as it is compressed between the skin and the bone. The **inguinal** *(ing'gwi-nal)* **canal** is an opening for the spermatic cord to traverse the inguinal ligament and is a potentially weak area and common site for a hernia to develop.

Ejaculatory Duct. The **ejaculatory duct** is 2 cm (1 in.) long and is formed by the union of the ampulla of the ductus deferens and the duct of the seminal vesicle. The ejaculatory duct then pierces the capsule of the prostate gland on its posterior surface and continues through this gland (fig. 27.20). Both ejaculatory ducts receive secretions from the seminal vesicles and prostate gland and then eject the sperm with its additives into the prostatic urethra.

Accessory Glands

Accessory reproductive glands include the seminal vesicles, the prostate gland, and the bulbourethral glands (fig. 27.20). The contents of the seminal vesicles and the prostate gland are mixed with the sperm during ejaculation to form *semen (seminal fluid).* The fluid from the bulbourethral glands is released in response to sexual stimulation prior to ejaculation. Both the seminal vesicles and prostate are androgen-dependent organs. They atrophy if androgen is deprived by castration.

Seminal Vesicles. The **seminal vesicles** are convoluted, club-shaped glands about 5 cm (2 in.) long and are positioned immediately posterior to and at the base of the urinary bladder. They secrete a sticky, slightly alkaline, yellowish substance, which serves as a fluid medium to sperm movement and longevity. The secretion from the seminal vesicles contains a variety of nutrients, including fructose, a monosaccharide that provides sperm with an energy source. It also contains citric acid, coagulation proteins, and prostaglandins. The discharge from the seminal vesicles makes up about 60% of the volume of semen.

Histologically, the seminal vesicle appears as a mass of cuts embedded in connective tissue (fig. 27.21). The extensively coiled mucosal layer breaks the lumen into numerous intercommunicating spaces that are lined by pseudostratified columnar and cuboidal secretory epithelia (referred to as glandular epithelium).

Blood is supplied to the seminal vesicles by the branches from the middle rectal arteries. The seminal vesicles are innervated by both sympathetic and parasympathetic fibers. Sympathetic stimulation causes the contents of the seminal vesicles to empty into the ejaculatory ducts of their respective sides.

Figure 27.21. A photomicrograph of the seminal vesicle.

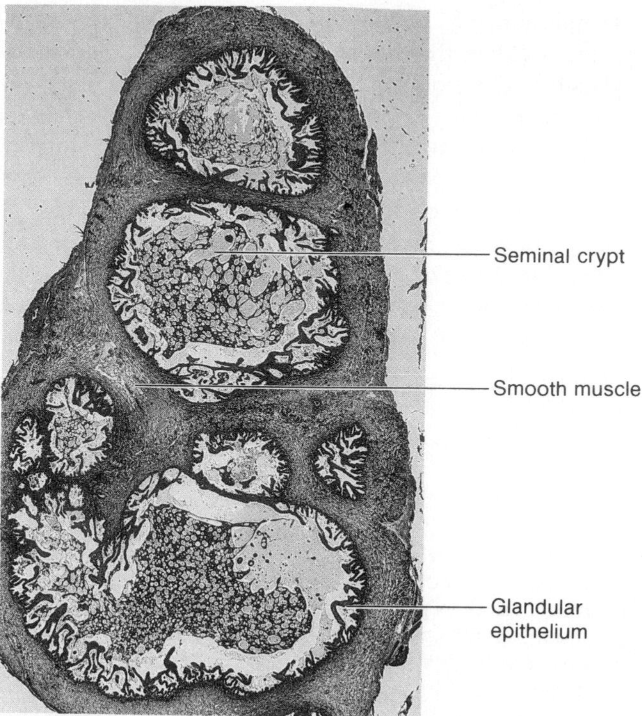

Figure 27.22. A photomicrograph of the prostate gland.

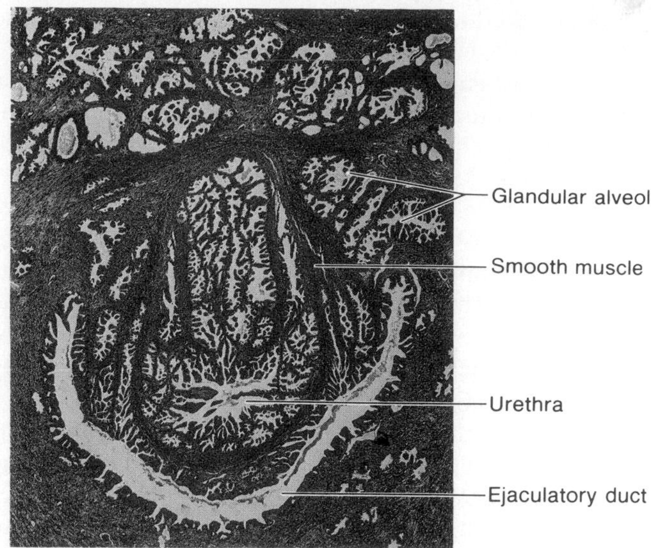

A routine physical examination of the male includes rectal palpation of the prostate gland. Enlargement or overgrowth of the glandular substance of the prostate, called *benign prostatic hypertrophy,* is relatively common in older men. This may constrict the urethra and cause difficult micturition. An enlarged prostate gland usually requires surgery. If the obstruction is slight, the surgery may be accomplished through the urethral canal using a technique called a *transurethral prostatic resection,* in which excessive tissue is cut and cauterized.

Prostate Gland. The firm **prostate gland** is the size and shape of a horse chestnut. It is about 4 cm (1.6 in.) across and 3 cm (1.2 in.) thick and positioned immediately below the urinary bladder surrounding the beginning of the urethra (see figs. 27.1, 27.20). It is enclosed by a fibrous capsule and divided into five distinct lobules. The lobules are formed by the urethra and the ejaculatory ducts that extend through the gland. The ducts from the lobules open into the urethra. Extensive bands of smooth muscular tissue course throughout the prostate to form a meshwork that supports the glandular tissue (fig. 27.22). Contraction of the smooth muscle within the prostate empties the contents from the gland and provides part of the propulsive force needed to ejaculate the semen. The thin, milky-colored, prostatic secretion assists sperm motility as a liquefying agent, and its alkalinity protects sperm in their passage through the acidic environment of the female vagina. The prostate also secretes the enzyme *acid phosphatase,* which is often measured clinically to assess prostate function.

Blood is supplied to the prostate from branches of the middle rectal and inferior vesical arteries. The venous return forms the prostatic venous plexus along with blood draining from the penis. The prostatic venous plexus drains into the internal iliac veins. The prostate gland has both sympathetic and parasympathetic innervation arising from the pelvic plexuses.

Bulbourethral Glands. The paired, pea-sized **bulbourethral** (Cowper's) **glands** are located inferior to the prostate gland. Each bulbourethral gland is brownish in color, about 1 cm in diameter, and drains by a 2.5 cm (1 in.) duct into the urethra (see fig. 27.20). Upon sexual excitement and prior to ejaculation, the bulbourethral glands are stimulated to secrete a mucoid substance, which coats the lining of the urethra to neutralize the pH of the urine residue and lubricates the tip of the penis in preparation for coitus.

Urethra

The **urethra** of the male serves as a common tube for both the urinary and reproductive systems. However, urine and semen cannot simultaneously pass through the urethra because the nervous reflex during ejaculation automatically inhibits micturition. The urethra of the male is about 20 cm (8 in.) long and S-shaped due to the shape of the penis. Three regions can be identified—the prostatic urethra, the membranous urethra, and the penile urethra (see fig. 27.20).

prostate: Gk. *prostate,* one standing before

cauterize: Gk. *kauterion,* a branding iron
Cowper's gland: from William Cowper, English anatomist, 1666–1709

The **prostatic urethra** is the proximal (2.5 cm) portion of the urethra that passes through the prostate gland. The prostatic urethra receives drainage from small ducts of the prostate gland and the two ejaculatory ducts.

The **membranous urethra** is the short (0.5 cm) portion of the urethra that passes through the urogenital diaphragm. The external urethral sphincter muscle is located in this region.

The **penile urethra** (cavernous urethra) is the longest portion (15 cm), extending from the outer edge of the urogenital diaphragm to the external urethral orifice on the glans penis. This portion is surrounded by erectile tissue as it passes through the corpus spongiosum of the penis. The paired ducts from the bulbourethral glands attach to the penile urethra near the urogenital diaphragm.

The wall of the urethra has an inside lining of mucous membrane, composed of transitional epithelium (fig. 27.23) and surrounded by a relatively thick layer of smooth muscle tissue, called *tunica muscularis*. Specialized **urethral glands** are embedded in the urethral wall and function to secrete mucus into the urethral canal.

Penis

The **penis** is composed mainly of erectile tissue and when distended serves as the intromittent, or copulatory, organ of the male reproductive system. **Erectile tissue** contains numerous vascular spaces that become engorged with blood. The penis is a pendent structure, positioned anterior to the scrotum and attached to the pubic arch. It is divided into a proximally attached root, an elongated tubular shaft, and a distal cone-shaped glans (fig. 27.24).

The **root of the penis** expands posteriorly to form the **bulb of the penis** and the **crus** *(krus)* **of the penis.** The bulb is positioned in the urogenital triangle of the perineum, where it is attached to the undersurface of the perineal membrane and enveloped by the bulbospongiosus muscle. The crus, in turn, attaches the root of the penis to the pubic arch (ischiopubic ramus) and to the perineal membrane. The crus is positioned superior to the bulb and is enveloped by the ischiocavernosus muscle.

The **shaft,** or body, **of the penis** is composed of three cylindrical columns of erectile tissue that are bound together by fibrous tissue and covered with skin (fig. 27.24). The paired dorsally positioned masses are named the **corpora cavernosa penis.** The fibrous tissue between the two corpora forms a **median septum.** The **corpus spongiosum penis (corpus cavernosum urethrae)** lies ventral to the other two and surrounds the penile urethra. The penis is flaccid and relaxed when the spongelike tissue is not engorged with blood but becomes firm and erect when the spaces are filled.

crus: L. *crus,* leg; resembling a leg
cavernous: L. *cavus,* hollow

The **glans penis** is the cone-shaped terminal portion of the penis that is formed from the expanded corpus spongiosum. The opening of the urethra at the tip of the glans is called the **external urinary meatus.** The **corona glandis** is the prominent posterior ridge of the glans. On the undersurface of the glans, a vertical fold of tissue, called the **frenulum** *(fren'u-lum),* attaches the skin covering the penis to the glans.

The skin covering the penis is hairless, contains no fat cells, and generally has a darker pigment than the other body skin. The skin of the shaft is loosely attached and is continuous over the glans as a retractable sheath called the **prepuce** *(pre'pūs),* or **foreskin.** The prepuce is commonly removed in a newborn by a surgical procedure called *circumcision.*

A *circumcision* is generally performed for hygienic purposes because the glans is easier to clean if exposed. A sebaceous secretion from the glans, called *smegma,* will accumulate along the border of the corona glandis if good hygiene is not practiced. Smegma can foster bacteria that may cause infections and therefore should be removed through washing. Cleaning the glans of an uncircumcised male requires retraction of the prepuce. Occasionally, a child is born with a prepuce that is too tight to permit retraction. This condition is called *phimosis* and necessitates circumcision.

The penis is supplied with blood through the superficial external pudendal branch of the femoral artery. The venous return is through a superficial median dorsal vein that drains into the great saphenous vein in the thigh and through the deep median vein that drains into the prostatic plexus.

The penis has many afferent tactile receptors, especially in the glans, that make it a highly sensitive organ. In addition, the penis has extensive efferent innervation from both parasympathetic and sympathetic fibers.

1. What are the functions of each of the regions of the spermatic duct?
2. Differentiate between the ductus deferens and the spermatic cord.
3. Describe the structure and location of each of the accessory glands.
4. Why is the position of the prostate gland of clinical importance?
5. What do each of the accessory glands secrete?
6. Describe the external structure of the penis and the internal arrangement of the erectile tissue within the penis.

glans: L. *glans,* acorn
corona: L. *corona,* garland, crown
frenulum: L. diminutive of *frenum,* a bridle
prepuce: L. *prae,* before; *putium,* penis
phimosis: Gk. *phimosis,* a muzzling

Figure 27.23. A photomicrograph of the urethra (10X).

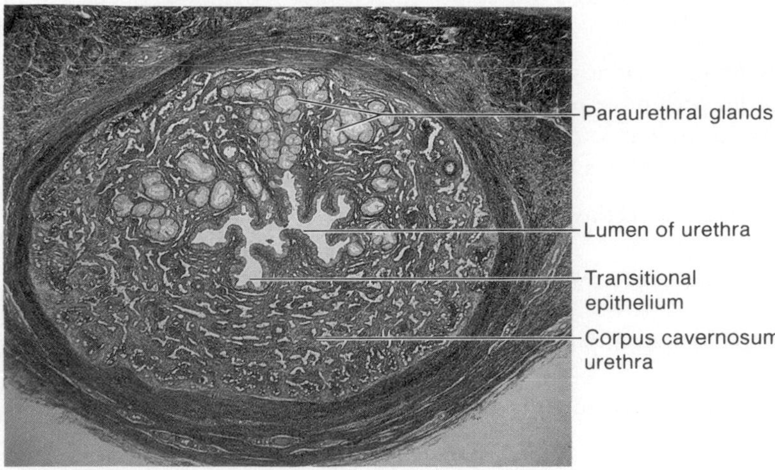

Figure 27.24. The structure of the penis showing the attachment, blood and nerve supply, and the arrangement of the erectile tissue.

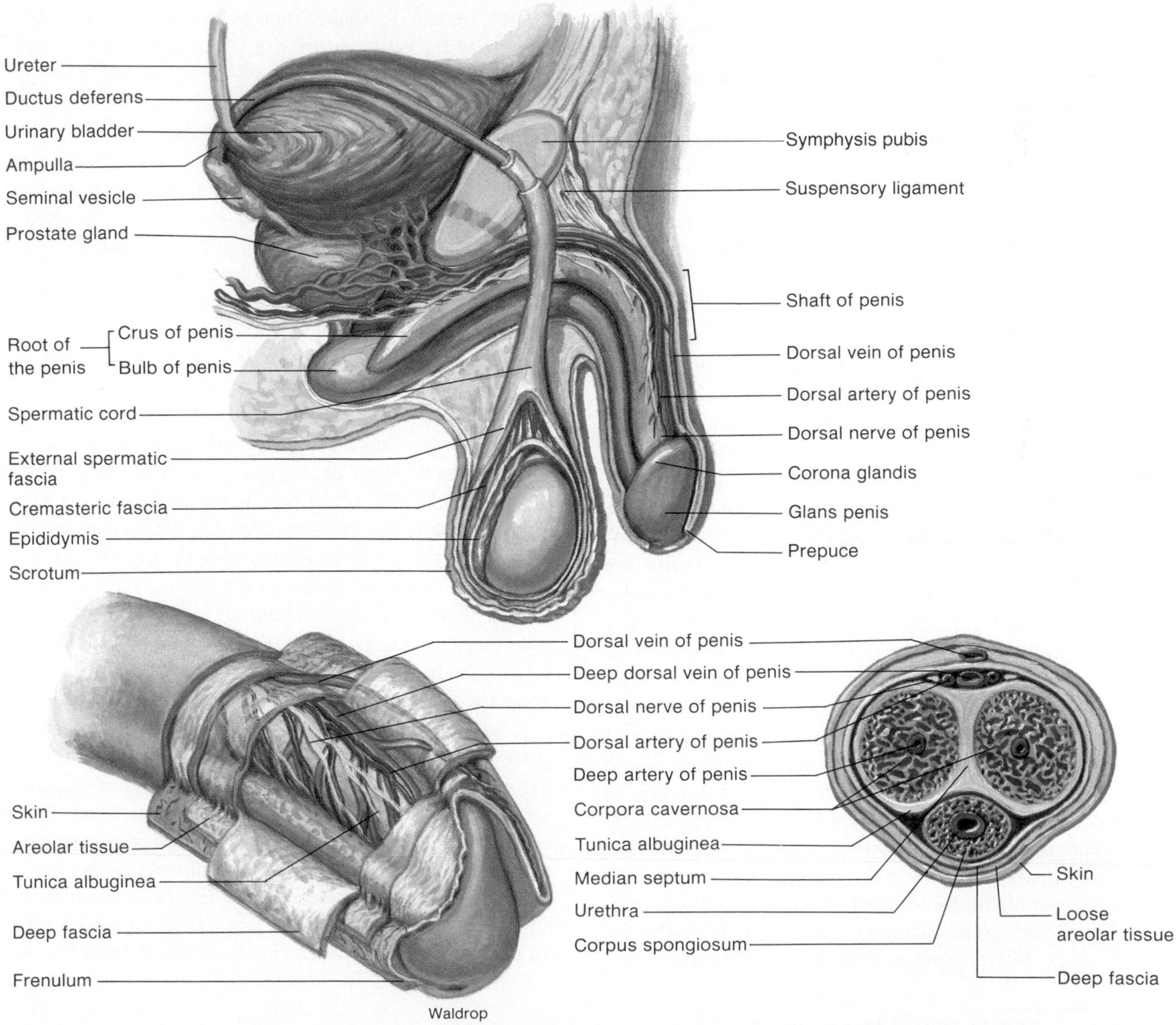

Mechanisms of Erection, Emission, and Ejaculation

Erection of the penis results from parasympathetic induced vasodilation of arteries within the penis. Emission and ejaculation are stimulated by sympathetic impulses, resulting in the forceful expulsion of semen from the penis.

Objective 18. Distinguish between erection, emission, and ejaculation.
Objective 19. Describe the events that result in erection of the penis.
Objective 20. Explain the physiological process of ejaculation.
Objective 21. Describe the characteristic properties of semen.

Erection, emission, and ejaculation are interrelated events that are necessary for the deposition of semen into the vagina by the natural means of coitus. *Erection* usually occurs as a male becomes sexually aroused; the erectile tissue of the penis becomes engorged with blood, causing the penis to become wider, longer, and firmer. *Emission (e-mish'un)* is the movement of spermatozoa from the epididymides to the ejaculatory ducts. *Ejaculation (e-jak"u-la'shun)* is the forceful expulsion of the ejaculate, or *semen* (seminal fluid), from the ejaculatory ducts and urethra of the penis. Emission and ejaculation do not have to follow erection of the penis and only occur if there is sufficient stimulation of the sensory receptors in the penis to elicit the ejaculatory response.

Erection of the Penis

Erection of the penis depends on the volume of blood that enters the arteries of the penis as compared to the volume that exits through venous drainage. Normally, constant sympathetic stimuli to the arterioles of the penis maintain a partial constriction of smooth muscles within the arteriole walls so that there is an even flow of blood throughout the penis. During sexual excitement, however, parasympathetic impulses cause marked vasodilation within the arterioles of the penis, resulting in more blood entering than venous blood draining. At the same time, there may be slight vasoconstriction of the dorsal vein of the penis and an increase in cardiac output. These combined events cause the spongy tissue of the corpora cavernosa and the corpus spongiosum to become distended with blood and the penis to become turgid. In this condition, the penis can be inserted into the vagina of the female and function as an intromittent, or copulatory, organ to discharge semen.

Erection is controlled by two portions of the central nervous system—the hypothalamus in the brain and the sacral portion of the spinal cord. The hypothalamus controls conscious sexual thoughts that originate in the cerebral cortex. Nerve impulses from the hypothalamus elicit parasympathetic responses from the sacral region, which cause vasodilation of the arterioles within the penis. Conscious thought is not required for an erection, however, and stimulation of the penis can cause an erection because of a reflex response in the spinal cord. This reflexive action makes possible an erection in a sleeping male or in an infant—perhaps from the stimulus of a diaper.

The mechanism of erection of the penis is summarized in figure 27.25.

Ejaculation is the expulsion of semen through the urethra of the penis. In contrast to erection, ejaculation is a response involving the sympathetic innervation of the accessory reproductive organs. Ejaculation is preceded by continued sexual stimulation, usually through activated tactile receptors in the glans penis and the skin of the shaft. Rhythmic friction of these structures during coitus causes afferent impulses to be transmitted to the thalamus and cerebral cortex. The first sympathetic response, which occurs prior to ejaculation, is the discharge of the contents from the bulbourethral glands. The fluids from these glands are usually discharged before penetration into the vagina and serve to lubricate the urethra and the glans penis.

Emission and Ejaculation of Semen

Emission. Continued sexual stimulation following erection of the penis causes emission. Emission is the movement of sperm from the epididymides to the ejaculatory ducts and the secretions of the accessory glands into the ejaculatory ducts and urethra in the formation of semen. Emission occurs as sympathetic impulses from the pelvic plexus cause a rhythmic contraction of the smooth muscle layers of the testes, epididymides, ductus deferentia, ejaculatory ducts, seminal vesicles, and prostate gland.

Ejaculation. Ejaculation immediately follows emission and is accompanied by *orgasm,* which is considered the climax of the sex act. Ejaculation occurs in a series of spurts of semen from the urethra. This takes place as parasympathetic impulses traveling through the pudendal nerves stimulate the bulbocavernosus muscles at the base of the penis and cause them to contract rhythmically. There is also sympathetic stimulation of the smooth muscles in the urethral wall that peristaltically contract to help eject the semen.

Sexual function in the male thus requires the synergistic action (rather than antagonistic action) of the parasympathetic and sympathetic systems. The mechanism of emission and ejaculation is summarized in figure 27.26.

Immediately following ejaculation or a cessation of sexual stimulus, sympathetic impulses cause vasoconstriction of the arterioles within the penis, reducing the inflow of blood. At the same time, cardiac output returns to normal, as does venous return of blood from the penis.

---

emission: L. *emittere,* expel or eject
turgid: L. *turgeo,* to swell

Figure 27.25. The mechanism of erection of the penis.

Sexual excitement; stimulation
Inhibition of sympathetic impulses to penis
Increase in parasympathetic impulses cause vasodilation of arterioles within penis
Reduction of venous flow from penis; increased cardiac output through sympathetic stimulation to heart
Accumulation of blood within erectile tissue of penis
Penis becomes turgid and erect

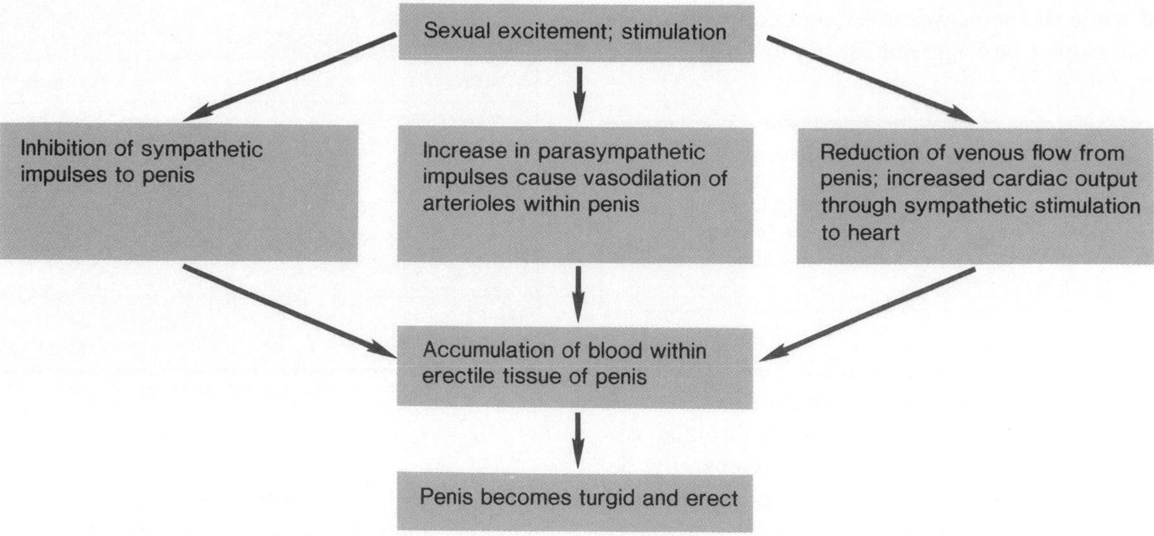

Figure 27.26. The mechanism of emission and ejaculation.

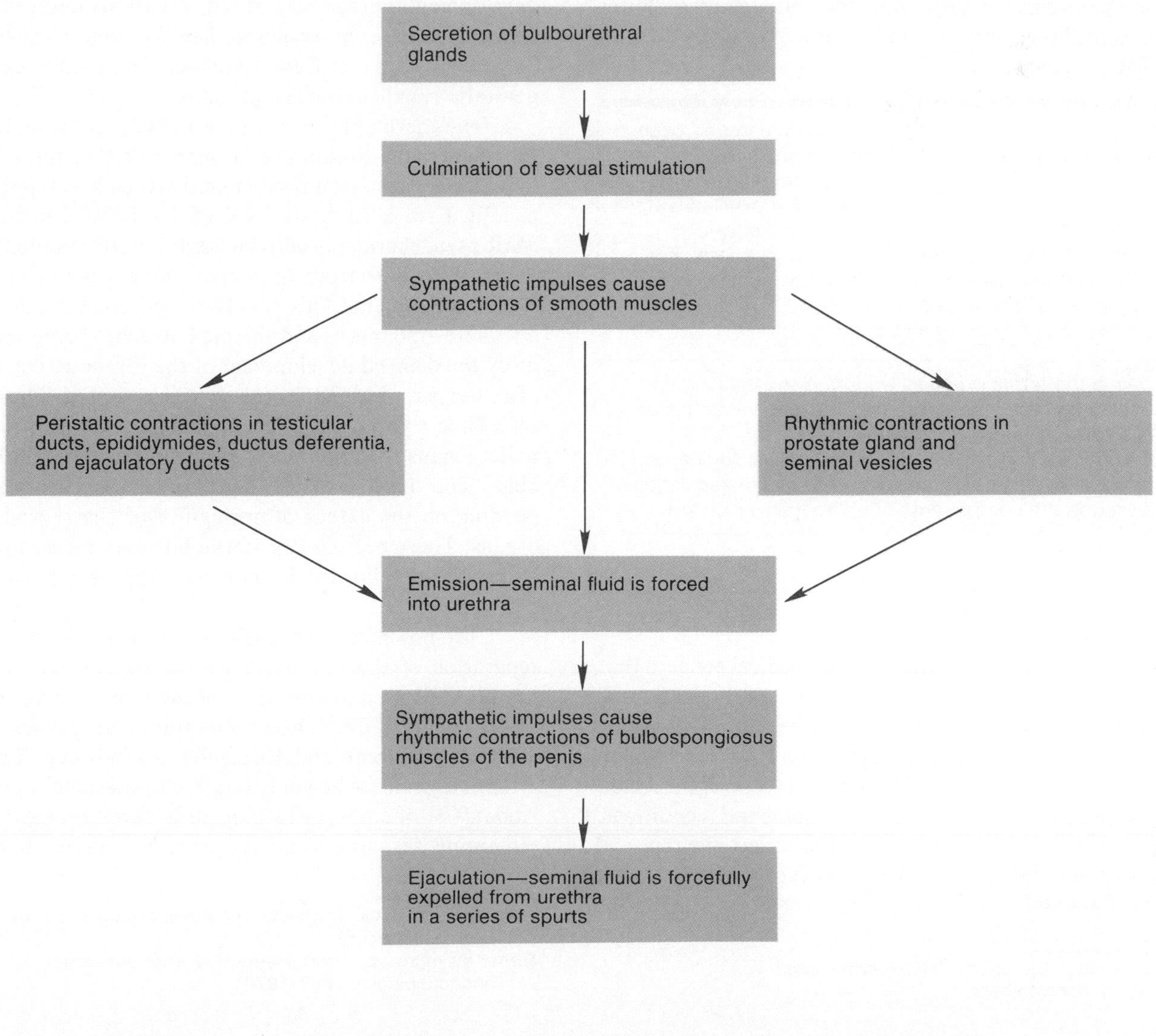

Secretion of bulbourethral glands
Culmination of sexual stimulation
Sympathetic impulses cause contractions of smooth muscles
Peristaltic contractions in testicular ducts, epididymides, ductus deferentia, and ejaculatory ducts
Rhythmic contractions in prostate gland and seminal vesicles
Emission—seminal fluid is forced into urethra
Sympathetic impulses cause rhythmic contractions of bulbospongiosus muscles of the penis
Ejaculation—seminal fluid is forcefully expelled from urethra in a series of spurts

With the normal flow of blood through the penis, it returns to its flaccid condition. If an ejaculation of semen occurred while the penis was erect, another erection and ejaculation cannot be triggered for ten to thirty minutes or longer.

Adolescent males may experience erection of the penis and spontaneous emission and ejaculation of semen during sleep. These *nocturnal emissions* are thought to be caused by changes in hormonal concentrations that accompany adolescent development.

Semen. Semen, also called seminal *(sem'ĭ-nal)* fluid, which consists of spermatozoa plus the additives from the accessory reproductive glands, is the substance discharged during ejaculation (table 27.5). Generally, between 1.5 and 5.0 ml of semen are ejected during ejaculation. The bulk of the fluid (about 60%) is produced by the seminal vesicles, and the remaining (about 40%) is contributed by the prostate gland. There are usually between 60 and 150 million sperm per milliliter of ejaculate. In the condition of *oligospermia,* the male ejaculates fewer than 10 million sperm per milliliter and is likely to have fertility problems.

Human semen can be frozen and stored in sperm banks for future artificial insemination. In this procedure, the semen is diluted with 10% glycerol, monosaccharide, and distilled water buffer and frozen in liquid nitrogen. The freezing process destroys defective and abnormal sperm. For some unknown reason, however, not all human sperm is suitable for freezing.

1. Define the terms *erection, emission* and *ejaculation.*
2. Explain the statement that male sexual function is an autonomic synergistic action.
3. Compose a flowchart to explain the physiological and physical events of erection, emission, and ejaculation.
4. Describe the components of a "normal" ejaculate.

Table 27.5 Some characteristics and reference values used in clinical examination of semen

Characteristic	Reference value
Volume of ejaculate	1.5–5.0 ml
Sperm count	40–250 million/ml
Sperm motility	
Percent of motile forms:	
1 hour after ejaculation	70% or more
3 hours after ejaculation	60% or more
Leukocyte count	0–2000/ml
pH	7.2–7.8
Fructose concentration	150–600 mg/100 ml

"Reprinted by permission from the July/August 1981 issue of *Diagnostic Medicine.*"

Clinical Considerations

Sexual dysfunction is a broad area of medical concern that includes developmental and psychogenic problems as well as conditions resulting from various diseases. Psychogenic problems of the reproductive system are extremely complex, poorly understood, and beyond the scope of this book. Only a few of the principal developmental conditions, functional disorders, and diseases that affect the physical structure and function of the male reproductive system will be discussed.

Developmental Abnormalities

The reproductive organs of both sexes develop from similar embryonic tissue that follows a consistent pattern of formation well into the fetal period. Because an embryo has the potential to differentiate into a male or a female, developmental errors can result in various degrees of intermediate sex, or **hermaphroditism** *(her-maf'ro-di-tizm'').* A person with undifferentiated or ambiguous external genitalia is called a **hermaphrodite.**

True hermaphroditism—in which both male and female gonadal tissues are present, in either the same or opposite gonads—is a rare anomaly. True hermaphrodites usually have a forty-six, XX chromosome constitution. **Male pseudohermaphroditism** occurs more commonly and generally results from hormonal influences during early fetal development. This condition is caused either by inadequate amounts of androgenic hormones being secreted or by the delayed development of the reproductive organs after the period of tissue sensitivity has passed. These persons have a forty-six, XY chromosome constitution and male gonads, but the genitalia are intersexual and variable. The treatment of hermaphroditism varies, depending on the extent of ambiguity of the reproductive organs. These persons are sterile but may marry and live a normal life following hormonal therapy and plastic surgery.

Chromosomal anomalies result from the improper separation of the chromosomes during meiosis and are usually expressed in deviations of the reproductive organs. The two most frequent chromosomal anomalies cause Turner's syndrome and Klinefelter's syndrome. **Turner's syndrome** occurs when only one X chromosome is present. About 97% of embryos lacking an X chromosome die; the remaining 3% survive and appear to be females, but their

oligospermia: Gk. *oligos,* few; *sperma,* seed
fertility: L. *fere,* to bear

hermaphrodite: Gk. (mythology) *Hermaphroditos,* son of Hermes (Mercury)
Turner's syndrome: from Henry H. Turner, American endocrinologist, 1892–1970

gonads are rudimentary or absent, and they do not mature at puberty. A person with **Klinefelter's syndrome** has an XXY chromosome constitution, develops breasts and male genitalia, but has underdeveloped seminiferous tubules and is generally retarded.

A more common developmental problem than genetic abnormalities, and fortunately less serious, is cryptorchidism. **Cryptorchidism** *(krip-tor'kĭ-dizm)* means "hidden testis" and is characterized by the failure of one or both testes to descend into the scrotum. A cryptorchid testis is usually located along the path of descent but can be anywhere in the pelvic cavity (fig. 27.27). It occurs in about 3% of male infants and should be treated before the infant is five years old to reduce the chance of infertility or other complications.

Functional Considerations

Functional disorders of the male reproductive system include impotence, infertility, and sterility. **Impotence** *(im'po-tens)* is the inability of a sexually mature male to maintain erection long enough to have an ejaculation. The causes of impotence may be physical, such as abnormalities of the penis, vascular irregularities, neurological disorders, or the result of diseases. Generally, however, the cause of impotence is psychological, and the patient requires skilled counseling by a sex therapist.

Infertility is the inability of the sperm to fertilize the ovum and may involve the male or female, or both. The term *impotence* should not be used when referring to infertility. Infertility in males may be caused by a number of things, the most common of which is the inadequate production of viable sperm. Some of the causes of infertility in males are alcoholism, dietary deficiencies, local injury, varicocele, excessive heat, or exposure to X rays. A hormonal imbalance may also contribute to infertility. Many of the causes of infertility can be treated through proper nutrition, gonadotropic hormone treatment, or microsurgery. If corrective treatment is not possible, however, it may be possible to concentrate the sperm obtained through *masturbation* (self-stimulation to the point of ejaculation) and use this concentrate to inseminate the woman artificially.

Sterility is similar to infertility except that it is a permanent condition. Sterility may be genetically caused, or it may be the result of degenerative changes in the seminiferous tubules (for example, mumps in a mature male may secondarily infect the testes and cause irreversible tissue damage).

Voluntary sterilization of the male in a procedure called a **vasectomy** is a common technique of birth control and can be performed on an out-patient basis. In this procedure, a small section of each ductus deferens near the epididymis is surgically removed, and the cut ends of the ducts are tied (fig. 27.28). A vasectomy interferes with sperm transport but does not directly affect the secretion of androgens from interstitial cells in the interstitial tissue. Since spermatogenesis continues, the sperm cannot be drained from the testes and instead accumulate in the

Figure 27.27. Cryptorchidism. (*a*) incomplete descent of a testis may involve four separate regions: (*1*) in the pelvic cavity, (*2*) in the inguinal canal, (*3*) at the superficial inguinal ring, (*4*) in the upper scrotum. (*b*) an ectopic testis may be (*1*) in the superficial fascia of the anterior pelvic wall, (*2*) at the root of the penis, (*3*) in the perineum, in the thigh alongside the femoral vessels.

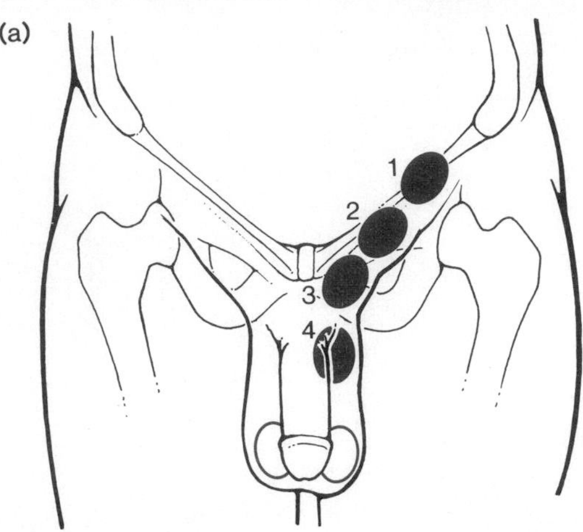

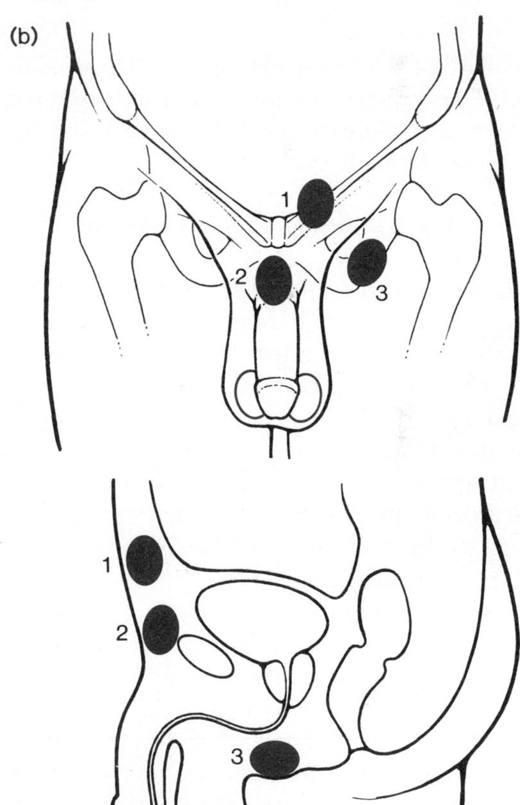

Klinefelter's syndrome: from Harry F. Klinefelter, Jr., American physician, b. 1912
cryptorchidism: Gk. *crypto,* hidden; *orchis,* testis
impotence: L. *im,* not; *potens,* potent
sterility: L. *sterilis,* barren
vasectomy: L. *vas,* vessel; Gk. *ektome,* excision

Figure 27.28. A simplified illustration of a vasectomy, in which segments of the ductus deferentia are removed through incisions in the scrotum.

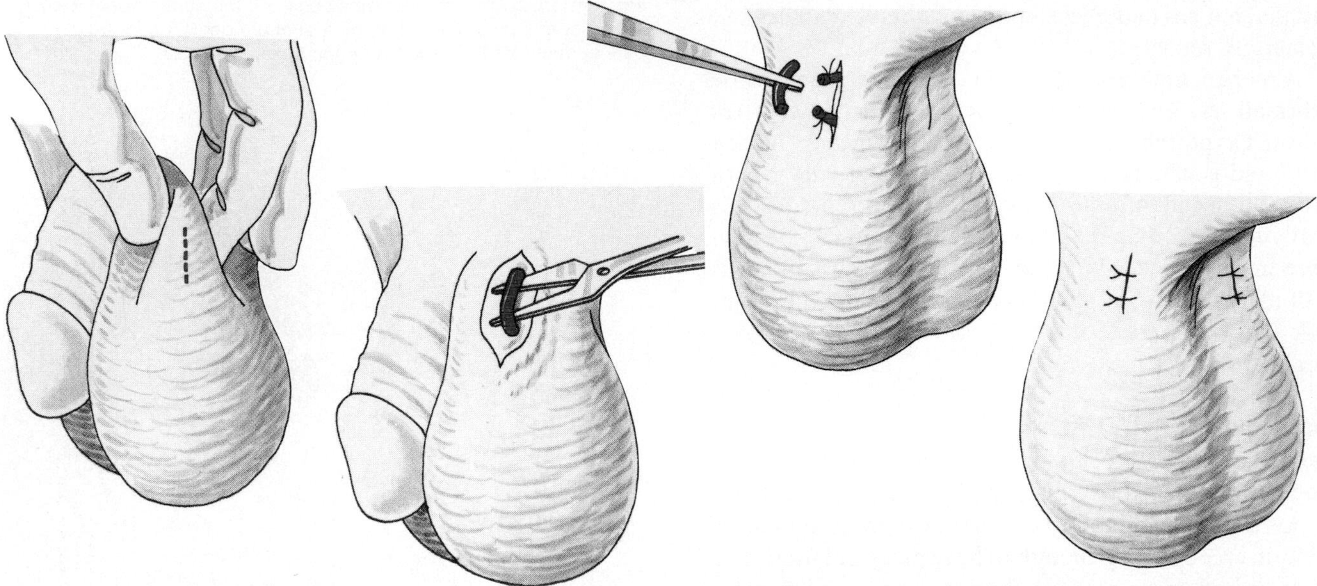

"crypts" that form in the seminiferous tubules and ductus deferens. These crypts present sites of inflammatory reactions in which spermatozoa are phagocytosed and destroyed by the immune system.

Diseases of the Male Reproductive System

Sexually Transmitted Diseases. Sexually transmitted diseases, frequently called *venereal diseases (VD),* are contagious diseases that affect the reproductive systems of both the male and the female (table 27.6) and are transmitted during sexual activity. The frequency of sexually transmitted diseases in the United States is regarded by health authorities as epidemic. These diseases have not been eradicated, mainly because humans cannot develop immunity to them and increased sexual promiscuity increases the chances of reinfection.

Gonorrhea *(gon''o-re'ah),* commonly called "clap," is caused by the bacterium gonococcus, or *Neisseria gonorrhoeae.* Males with this disease suffer inflammation of the urethra, accompanied by painful urination and frequently the discharge of pus. In females, the condition is usually asymptomatic, and therefore many women may be unsuspecting carriers of the disease. Advanced stages of gonorrhea in females may infect the uterus and the uterine tubes. A pregnant woman with gonorrhea that is not treated may transmit the disease to the eyes of her newborn, causing blindness.

venereal: L. (mythology) from *Venus,* the goddess of love
gonorrhea: L. *gonos,* seed; *rhoia,* a flow

Syphilis *(sif'ĭ-lis)* is caused by the bacterium *Treponema pallidum.* Syphilis is less common than gonorrhea but is the more serious of the two diseases. During the *primary stage* of syphilis, a lesion called a *chancre* develops at the point where contact was made with a similar sore from an infected person. The chancre is an ulcerated sore that has hard edges and endures for ten days to three months. It is only during the primary stage that syphilis can be spread to another sexual partner. The chancre will heal with time, but if not treated, it will be followed by secondary and tertiary stages of syphilis. During the initial contact, the bacteria enter the bloodstream and spread throughout the body. The *secondary stage* of syphilis is expressed by lesions or a rash of the skin and mucous membranes, accompanied by fever (fig. 27.29). This stage lasts from two weeks to six months, and the symptoms disappear of their own accord. The *tertiary stage* occurs ten to twenty years following the primary infection. The circulatory, integumentary, skeletal, and nervous systems are particularly vulnerable to the degenerative changes caused by this disease. The end result of untreated syphilis is blindness, insanity, and eventual death.

AIDS, or **acquired immune deficiency syndrome,** is believed to be a viral disease that is transmitted primarily through sexual intercourse and drug abuse (by sharing contaminated syringe needles). This fatal disease, for which there is currently no cure, is discussed in more detail in chapter 22.

chancre: Fr. *chancre,* indirectly from L. *cancer,* a crab

Table 27.6 Kinds of sexually transmitted diseases

Name	Organism	Resulting condition	Treatment
Gonorrhea	*Gonococcus* (bacterium)	Adult: sterility due to scarring of epididymides and tubes; rarely: septicemia; newborn: blindness	Penicillin injections; tetracycline tablets; eye drops (silver nitrate or penicillin)
Syphilis	*Treponema pallidum* (bacterium)	Adult: gummas, cardiovascular neurosyphilis; newborn: congenital syphilis (abnormalities, blindness)	Penicillin injections; tetracycline tablets
Chancroid (soft chancre)	*Hemophilus ducreyi* (bacterium)	Chancres, buboes	Tetracycline; sulfa drugs
Urethritis in men	Various microorganisms	Clear discharge	Tetracycline
Vaginitis	*Trichomonas* (protozoan)	Frothy white or yellow discharge	Metronidazole
	Candida albicans (yeast)	Thick, white, curdy discharge (moniliasis)	Nystatin
Acquired immunity deficiency syndrome (AIDS)	Human immunodeficiency virus (HIV)	Early symptoms include extreme fatigue, weight loss, fever, diarrhea; severe susceptibility to pneumonia, rare infections, and cancer	Azidothymidine (AZT, or Retrovir); no cure available
Chlamydia	*Chlamydia trachomatis* (bacterium)	Whitish discharge from penis or vagina; pain during urination	Tetracycline and sulfonamides
Lymphogranuloma venereum (LGV)	Microorganism	Ulcerating buboes; rectal stricture	Tetracycline; sulfa drugs
Granuloma venereum (inguinale)	*Donovania granulomatis*	Raw, open, extended sore	Tetracycline
Venereal warts	Virus	Warts	Podophyllin
Genital herpes	Herpes simplex virus	Sores	Palliative treatment
Crabs	Arthropod	Itching	Gamma benzene hexachloride

Disorders of the Prostate Gland. The prostate gland is subject to several disorders, most of which are common in older men. The four most frequent prostatic problems are acute prostatitis, chronic prostatitis, benign prostatic hyperplasia, and carcinoma of the prostate.

Acute prostatitis is common in sexually active young men through infections acquired from a gonococcus bacterium. The symptoms of acute prostatitis are a swollen and tender prostate gland, painful urination, and in extreme conditions, pus dripping from the penis. It is treated with penicillin, bed rest, and increased fluid intake.

Chronic prostatitis is one of the most common afflictions of middle-aged and elderly men. The symptoms of this condition vary considerably from irritation and slight difficulty in urination to extreme pain and urine blockage, which commonly causes secondary renal infections. In this disease, several kinds of infectious microorganisms are believed to be harbored in the prostate gland and are responsible for inflammations elsewhere in the body, such as in the nerves (neuritis), the joints (arthritis), the muscles (myositis), and the iris (iritis).

Prostatic hyperplasia, or an enlarged prostate gland, occurs in approximately one-third of all males over the age of sixty. In this condition, an overgrowth of granular material compresses the prostatic urethra. The cause of prostatic hyperplasia is not known. As the prostate enlarges, urination becomes painful and difficult. If the urinary bladder is not emptied completely, cystitis eventually occurs. Persons with cystitis may become incontinent and

Figure 27.29. The secondary stages of syphilis as expressed by lesions of the skin of this young woman.

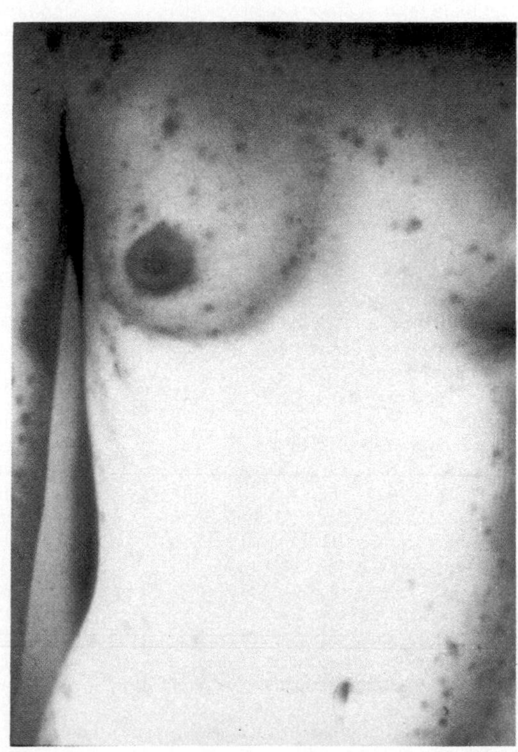

dribble urine continuously. Prostatic hypertrophy is usually treated by the surgical removal of portions of the gland through transurethral curetting (cutting and removal of a small section) or the removal of the entire prostate gland, called **prostatectomy.**

Prostatic carcinoma, or cancer of the prostate gland, is the second leading cause of death from cancer in males in the United States. It is common in males over sixty and accounts for 19,000 deaths annually. When prostatic cancer is confined to the prostate gland, it is generally small and asymptomatic. But as the cancer grows and invades surrounding nerve plexuses, it becomes extremely painful and easily detected. The metastases of this cancer to the spinal column and brain are generally what kills the patient.

As prostatic carcinoma develops, it has symptoms nearly identical to prostatic hyperplasia—painful urination and cystitis. When examined by rectal palpation with a gloved finger, however, a hard cancerous mass can be detected in contrast to the enlarged, soft, and tender prostate diagnostic of prostatic hypertrophy. Prostatic carcinoma is treated by prostatectomy and frequently by the removal of the testes (called **orchiectomy**) as well. An orchiectomy inhibits metastases by eliminating testosterone secretion.

Disorders of the Testes and Scrotum. A **hydrocele** *(hi'dro-sēl)* is a benign fluid mass within the tunica vaginalis that causes swelling of the scrotum. It is a frequent, minor disorder in infant boys as well as in adults. The cause is unknown.

An infection in the testes is called **orchitis.** Orchitis may develop from a primary infection from a tubercle bacterium or as a secondary complication of mumps contracted after puberty. If orchitis from mumps involves both testes, it usually causes sterility.

Trauma to the testes and scrotum is common because of their pendent position. The testes are extremely sensitive to pain, and a male responds reflexively to protect the groin area.

Chapter Summary

I. Introduction to the Reproductive System
   A. The functions of the male reproductive system are to produce sperm, secrete testosterone, and transfer sperm to the reproductive system of the female.
   B. The male reproductive system is divided into primary sex organs (the testes), secondary sex organs (those that are essential for sexual reproduction), and secondary sex characteristics (features that are sexual attractants, which are expressed after puberty).

II. Development of the Male Reproductive System
   A. An XY chromosome combination produces a male, and an XX chromosome combination produces a female.
   B. The appearance of the gonadal ridge during the fifth week is the first indication of sex organ formation.
      1. Primary sex cords and a genital tubercle develop during the sixth week.
      2. Reproductive development is well progressed by the eighth week but is in the indifferent stage because external sexual distinction is not apparent.
   C. The testes secrete testosterone, which stimulates the development of male accessory sex organs and male genitalia.
   D. The penis and scrotum are formed by the end of the twelfth week, and the descent of the testes occurs during the twenty-eighth week.

III. Puberty
   A. Androgens stimulate the development of secondary sex characteristics.
   B. The sensitivity of the hypothalamus and pituitary to negative feedback may decrease at puberty and thus contribute to the rise in gonadotropin secretion.

IV. Structure and Function of the Testes
   A. The saclike scrotum supports and protects the testes and regulates their position relative to the pelvic region of the body.
   B. The testes are separated into lobules composed of seminiferous tubules that produce sperm and interstitial tissue that produces androgens.
      1. LH stimulates the interstitial (Leydig) cells to secrete androgens; FSH stimulates the seminiferous tubules to produce sperm.
      2. Testosterone from the interstitial cells exerts negative feedback inhibition on LH secretion; inhibin, secreted from the tubules, is believed to exert negative feedback inhibition on FSH secretion.
   C. Spermatogenesis occurs by meiotic division of the cells that line the seminiferous tubules.
      1. At the end of the first meiotic division, two secondary spermatocytes are produced.
      2. At the end of the second meiotic division, four haploid spermatids are produced.
   D. The conversion of spermatids to spermatozoa is called spermiogenesis.
   E. A sperm consists of a head, body, and tail and matures in the epididymides prior to ejaculation.

V. Spermatic Ducts, Accessory Glands, and the Penis
   A. The epididymides, ductus (vas) deferentia, and ejaculatory ducts are the components of the spermatic ducts.
      1. The highly coiled epididymides are the tubular structures on the testes where sperm mature and are stored.
      2. The ductus deferentia convey sperm from the epididymides to the ejaculatory ducts during emission. Each ductus deferens forms a component of a spermatic cord.
   B. The seminal vesicles and prostate gland provide additives to the sperm in the formation of semen.
      1. The seminal vesicles are posterior to the base of the urinary bladder and secrete about 60% of the additive fluid of semen.
      2. The prostate gland surrounds the urethra just below the urinary bladder and secretes about 40% of the additive fluid of semen.
      3. The small bulbourethral glands secrete fluid that serves as a lubricant for the erect penis in preparation for coitus.

C. The male urethra, which serves both the urinary and reproductive systems, is divided into the prostatic, membranous, and penile portions.

D. The penis is specialized to become erect for insertion into the vagina during coitus.
   1. The body of the penis is composed of three columns of erectile tissue, the penile urethra, and associated vessels and nerves.
   2. The root of the penis is attached to the pubic arch and urogenital diaphragm.
   3. The glans penis is the terminal end, which is covered with the prepuce in an uncircumcised male.

VI. Mechanisms of Erection, Emission, and Ejaculation

A. Erection of the penis occurs as the erectile tissue becomes engorged with blood; emission is the movement of the spermatozoa from the epididymides to the ejaculatory ducts; and ejaculation is the forceful expulsion of semen from the ejaculatory ducts and urethra of the penis.

B. Parasympathetic stimuli to arteries in the penis cause the erectile tissue to engorge with blood as arteriole flow increases and venous drainage decreases.

C. Ejaculation is the result of sympathetic reflexes in the smooth muscles of the reproductive organs.

Review Activities

Objective Questions

1. An embryo with the genotype XY develops male accessory sex organs because of
   (a) androgens
   (b) estrogens
   (c) the absence of androgens
   (d) the absence of estrogens
2. Which of the following does *not* arise from the embryonic mesonephric duct? The
   (a) epididymis
   (b) ductus deferens
   (c) seminal vesicle
   (d) prostate gland
3. The external genitalia of a male are completely formed by the end of the
   (a) embryonic period
   (b) ninth week
   (c) tenth week
   (d) twelfth week
4. In the male, FSH
   (a) is not secreted by the pituitary
   (b) receptors are located in the interstitial cells
   (c) receptors are located in the spermatogonia
   (d) receptors are located in the nurse cells
5. The secretion of FSH in a male is inhibited by negative feedback effects of
   (a) inhibin secreted from the tubules
   (b) inhibin secreted from the interstitial cells
   (c) testosterone secreted from the tubules
   (d) testosterone secreted from the interstitial cells
6. Which of the following is *not* a spermatic duct? The
   (a) epididymis
   (b) spermatic cord
   (c) ejaculatory duct
   (d) ductus deferens
7. Spermatozoa are stored prior to emission and ejaculation in the
   (a) epididymides
   (b) seminal vesicles
   (c) penile urethra
   (d) prostate gland
8. Urethral glands function to
   (a) secrete mucus
   (b) produce nutrients
   (c) secrete hormones
   (d) regulate sperm production
9. Which statement is *false* regarding erection of the penis?
   (a) It is a parasympathetic response.
   (b) It may be both a voluntary and an involuntary response.
   (c) It has to be followed by emission and ejaculation.
   (d) It is controlled by the hypothalamus of the brain and sacral portion of the spinal cord.
10. The condition in which one or both testes fail to descend into the scrotum is
   (a) cryptorchidism
   (b) Turner's syndrome
   (c) hermaphroditism
   (d) Klinefelter's syndrome

Essay Questions

1. Explain what is meant by the latent development of the reproductive organs during puberty, and discuss the physiological mechanisms that cause puberty in a male.
2. Discuss the significance of the indifferent stage in reproductive development. When is the indifferent stage, and what are its implications in the potential development of abnormalities of the reproductive organs?
3. Describe the location and structure of the scrotum. Explain how the scrotal muscles regulate the position of the testes in the scrotum. Why is this important?
4. Describe the internal structure of a testis. Discuss the function of the nurse cells, interstitial cells, seminiferous tubules, rete testis, and efferent ductules.
5. List the structures that constitute the spermatic cord. Where is the inguinal canal? Why are the inguinal canal and inguinal ring clinically important?
6. Define the term *semen*. What amount of semen is ejected during ejaculation, and what are the properties and substances of semen?
7. Compare the seminal vesicles and the prostate gland in terms of location, structure, and function.
8. Describe the structure of the penis, and explain the mechanisms that result in erection, emission, and ejaculation.
9. Distinguish between impotence, infertility, and sterility.
10. Distinguish between gonorrhea and syphilis. Describe the stages through which syphilis will progress if untreated.

28 Female Reproductive System

Outline

Concepts

The female reproductive system consists of ovaries and the secondary sex organs, including the vagina, uterine tubes, uterus, and mammary glands. These structures are involved in coitus, fertilization, development of the embryo and fetus, and birth and nurturing of the baby.

The ovaries contain a large number of follicles, each of which encloses an ovum. During the ovarian cycle some follicles mature, and the ova they contain progress to the secondary oocyte stage of meiosis. At ovulation, the largest follicle breaks open to extrude a secondary oocyte from the ovary. The empty follicle then becomes a corpus luteum, which ultimately degenerates at the end of a nonfertile cycle.

Cyclic changes in the secretion of gonadotropic hormones from the anterior pituitary cause the ovarian changes during a monthly cycle. The ovarian cycle is accompanied by cyclic changes in the secretion of sex steroids, which interact with the hypothalamus and pituitary to regulate gonadotropin secretion. The cyclic changes in ovarian hormone secretion also cause the changes in the endometrium of the uterus during a menstrual cycle.

The structure and function of the mammary glands is dependent on the action of a number of hormones. The secretion of prolactin and oxytocin is directly required for the production and delivery of milk to the suckling infant.

Components and Structures of the Female Reproductive System

The female reproductive system consists of ovaries and the secondary sex organs, including the vagina, uterine tubes, uterus, and mammary glands. These structures are involved in coitus, fertilization, development of the embryo and fetus, and birth and nurturing of the baby.

Objective 1. Identify the primary and secondary female sex organs and the secondary sexual characteristics.

Objective 2. Describe the embryonic origin of the female genital tract, and explain why certain structures in the male and female reproductive systems are considered to be homologous.

Objective 3. Describe the structures of the uterine tubes and uterus.

Objective 4. Identify the components of the vulva and vagina.

Objective 5. Describe the changes that occur in the female genital system during sexual excitement and coitus.

The general term for the primary sex organs in both sexes is *gonads*. These are the organs that produce the *gametes*, which is a general term for sperm and ova. In females, the gonads are the **ovaries**, and the gametes are the **ova.** The structure of the ovaries, their production of ova and of sex steroid hormones, and the effects of these steroid hormones will be discussed in later sections.

Secondary Sex Organs and Characteristics

Secondary sex organs (fig. 28.1 and table 28.1) are those structures that are essential for successful fertilization, implantation of the embryo, and development of the embryo and fetus. The secondary sex organs include the **vagina** *(vah-ji'nah),* which receives the penis and ejaculated semen during coitus and through which the baby passes during delivery; the **uterine (fallopian) tubes,** also called the **oviducts,** through which an ovum is transported toward the uterus after ovulation, and in which fertilization normally occurs; and the **uterus** (womb), where implantation and development occur. The muscular walls of the uterus play an active role in *parturition* (labor and delivery). **Mammary glands** are also considered to be secondary sex organs because the milk secreted after parturition provides nourishment to the baby. The structure and function of mammary glands will be discussed in a separate section.

vagina: L. *vagina,* sheath or scabbard

Figure 28.1. Organs of the female reproductive system seen in sagittal section.

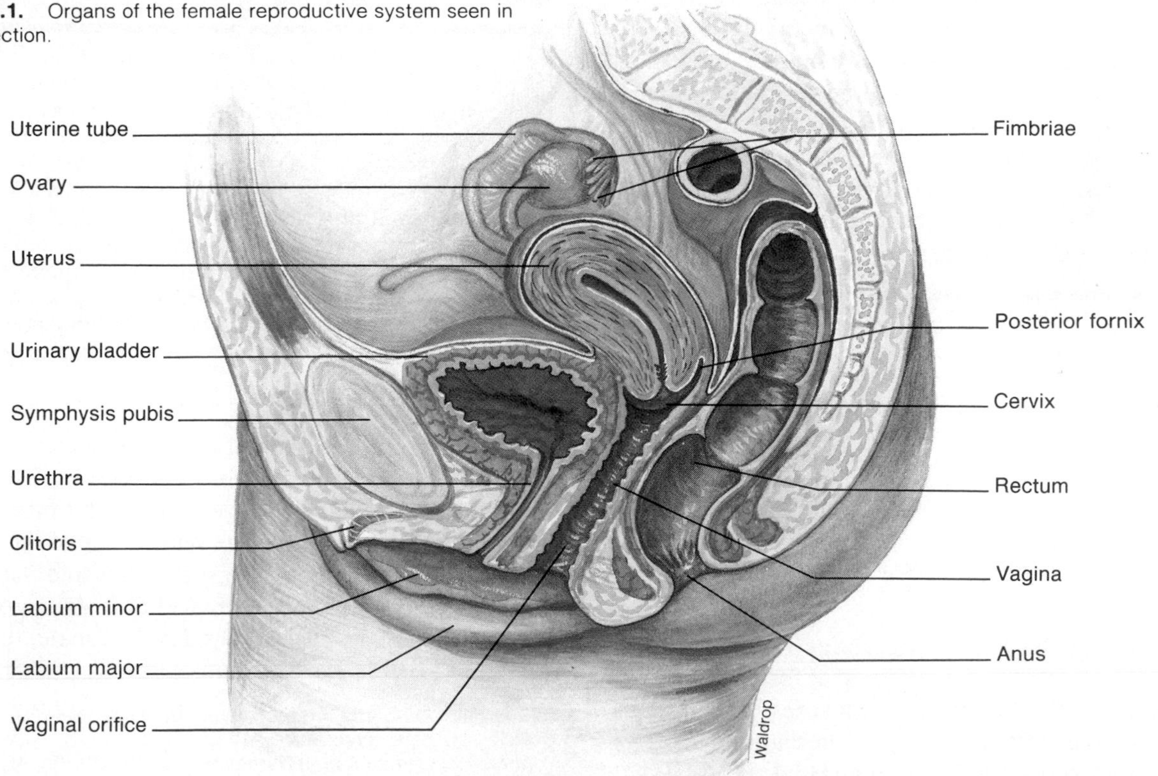

Table 28.1 Functions of the organs of the female reproductive system

Organ(s)	Function(s)
Ovaries	Produce ova and female sex hormones
Uterine tubes	Convey ovum toward uterus; site of fertilization; convey developing embryo to uterus
Uterus	Site of implantation; protects and sustains life of embryo and fetus during pregnancy; active role in parturition
Vagina	Conveys uterine secretions to outside of body; receives erect penis and semen during copulation and ejaculation; passage for fetus during parturition
Labia majora	Form margins of pudendal cleft; enclose and protect other external reproductive organs
Labia minora	Form margins of vestibule; protect openings of vagina and urethra
Clitoris	Glans of the clitoris is richly supplied with sensory nerve endings associated with feeling of pleasure during sexual stimulation
Vestibule	Cleft between labia minora that includes vaginal and urethral openings
Vestibular glands	Secrete fluid that moistens and lubricates the vestibule and vaginal opening during intercourse
Mammary glands	Produce and secrete milk for nourishment of an infant

Secondary sex characteristics are features that are not essential for the reproductive process but that are considered to be sexual attractants. Body physique, contributed largely by the female distribution of subcutaneous fat, the pattern of body hair, and breast development are examples. Although the breasts contain the mammary glands, large breasts are not essential for nursing the young. Indeed, while all mammals have mammary glands and are capable of nursing, only human females have protruding breasts; the sole purpose of this characteristic, it seems, is to function as a sexual attractant.

Puberty occurs at age twelve to fourteen, varying with the nutritional condition, genetic background, and even sexual exposure of the individual. Generally, girls attain puberty six months to one year earlier than boys, and the transition is more abrupt because of the onset of menstruation, or *menarche (mĕ-nar′ke).* Puberty results from the increased secretion of gonadotropins from the anterior pituitary, which stimulates the ovaries to begin their cycles of ova development and sex steroid secretion.

Development of the Female Reproductive System

Although the genetic sex is determined at fertilization (XX for females and XY for males), both sexes develop similarly through the *indifferent stage* of the eighth week. The gonads of both sexes develop from **gonadal ridges** medial to the functioning kidneys of that stage (the mesonephric kidneys, described in chapter 24). The **genital tubercle** also develops during the sixth week as an external swelling.

The ovaries develop more slowly than do the testes. Ovarian development begins at about the tenth week when **primordial follicles** begin to form within the medulla of the gonads. Each of the primordial follicles consists of an **oogonium** *(o″o-go′ne-um)* surrounded by a layer of **follicular cells.** Mitosis of the oogonia occurs during fetal development, so that thousands of germ cells are formed. Unlike the male, in which spermatogonia are formed by mitosis throughout life, all oogonia are formed prenatally and their number continuously decreases after birth.

The uterus and uterine tubes develop from a pair of embryonic tubes called the **paramesonephric (müllerian) ducts,** which are so called because they are located to the sides of the mesonephric ducts (which form temporary, embryonic kidneys). As the mesonephric kidneys degenerate (chapter 24), the lower portions of the paramesonephric ducts fuse to form the uterus, and the upper portions give rise to the uterine tubes. As described in chapter 27, if the embryo is male the paramesonephric ducts degenerate due to secretion of müllerian inhibition factor from the testes.

The epithelial lining of the vagina develops from the endoderm of the urogenital sinus. A thin membrane called the **hymen** forms to separate the lumen of the vagina from the urethral sinus. The hymen usually is perforated during later fetal development.

There is tremendous individual variation in the structure of the hymen. The hymen of a baby girl may be absent, or it may partially, or occasionally completely, cover the vaginal orifice, in which case it is called an *imperforate hymen.* An imperforate hymen is usually not detected until the first menstruation (menarche), when the discharge cannot be expelled. If the hymen is present, it may be ruptured during childhood in the course of normal exercise. On the other hand, a hymen may be so elastic that it persists even after sexual intercourse. The presence of a hymen is not, therefore, a reliable sign of virginity.

The external genitalia of both sexes appear the same during the indifferent stage of the eighth week. A prominent **phallus** *(fal′us)* forms from the genital tubercle, and a **urethral groove** forms on the ventral side of the phallus. The phallus becomes the penis in a male and the smaller **clitoris** in a female. Paired **urogenital folds** surround the urethral groove on the lateral sides. In a male, these fuse

---

primordial: L. *prima,* first; *ordior,* to begin
follicle: L. diminutive of *follis,* bag
hymen: Gk. (mythology) Hymen was god of marriage; *hymen,* thin skin or membrane

Table 28.2	Homologous reproductive organs in the male and female and the undifferentiated structures from which they develop	
Indifferent stage	**Male**	**Female**
Gonads	Testes	Ovaries
Urogenital groove	Membranous urethra	Vestibule
Genital tubercle	Glans penis	Clitoris
Urogenital folds	Penile urethra	Labia minora
Genital swelling	Scrotum Bulbourethral glands	Labia majora Vestibular glands

to form the urethra of the penis; in a female, the urogenital folds remain unfused and form the inner **labia minor** (table 28.2). Similarly, the labiosacral folds in a male fuse to form the scrotum; in a female, these remain unfused and form the prominent **labia majora.** The male and female structures that share a common embryological origin are said to be **homologous structures** (table 28.2).

Uterine Tubes

The paired **uterine tubes,** also known as the **fallopian** *(fal-lo'pe-an)* **tubes** or **oviducts,** transport ova from the ovaries to the uterus. Fertilization normally occurs within the uterine tube. Each uterine tube is approximately 10 cm (4 in.) long and 0.7 cm (0.3 in.) in diameter and is positioned between the folds of the broad ligament of the uterus.

The term *salpinx* is occasionally used to refer to the uterine tubes. It is a Greek word meaning "trumpet" or "tube" and is the root of such clinical terms as *salpingitis (sal"pin-ji'tis),* or inflammation of the uterine tubes, *salpingography* (radiography of the uterine tubes), and *salpingolysis* (the breaking up of adhesions of the uterine tube to correct female infertility).

The funnel-shaped, open-ended portion of the uterine tube is called the **infundibulum.** Although the infundibulum is close to the ovary, it is not attached. A number of fringed, fingerlike processes, called **fimbriae,** project from the margins of the infundibulum over the lateral surface of the ovary. The function of the fimbriae is to direct the ovum into the lumen of the uterine tube. From the infundibulum, the uterine tube extends medially and inferiorly to open into the superior-lateral cavity of the uterus at the **uterine opening.** The **ampulla** *(am-pul'ah)* is the longest and widest portion of the uterine tube.

fallopian tubes: from Gabriele Fallopius, Italian anatomist, 1523–62
oviduct: L. *ovum,* egg; *ductus,* a leading
fimbriae: L. *fimbria,* fringe

Figure 28.2. The histological structure of the uterine tube.

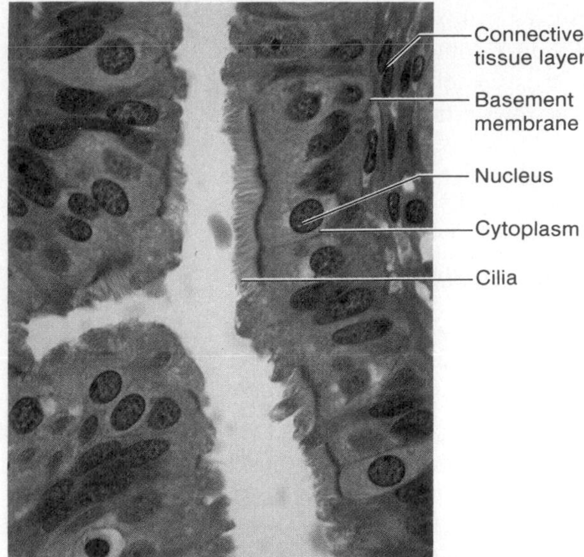

The wall of the uterine tube consists of three histological layers (fig. 28.2). The internal **mucosa** lines the lumen and is composed of ciliated columnar cells that are drawn into numerous folds. The **muscularis** is the middle layer, composed of a thick, circular layer of smooth muscle and a thin, outer layer of smooth muscle. Peristaltic contractions of the muscularis and ciliary action of the mucosa move the ovum through the lumen of the uterine tube. The outer, lubricative **serous layer** of the uterine tube is part of the visceral peritoneum.

The oocyte takes four to five days to move through the uterine tube. If enough viable sperm are ejaculated into the vagina during coitus and an oocyte is in the uterine tube, fertilization will occur within hours after discharge of the semen. The zygote will move toward the uterus, where implantation occurs. If the embryo (called a blastocyst) implants instead into the uterine tube, an *ectopic pregnancy* will be produced. An ectopic pregnancy is an implantation of the blastocyst in a site other than the uterus.

Since the infundibulum of the uterine tube is unattached, it provides a pathway for pathogens to enter the abdominopelvic cavity. The mucosa of the uterine tube is continuous with that of the uterus and vagina, and it is possible for infectious agents to enter the vagina and cause infections that may ultimately spread to the peritoneal linings, resulting in *pelvic inflammatory disease (PID).* There is no opening into the abdominopelvic cavity other than through the uterine tubes. The abdominopelvic cavity of a male is totally sealed from external contamination.

ectopic: Gk. *ex,* out; *topos,* place

Uterus

The **uterus** receives the blastocyst that develops from a fertilized ovum and provides a site for implantation. Prenatal development continues within the uterus until gestation is completed, at which time the uterus plays an active role in the delivery of the baby.

Structure. The uterus is a hollow, thick-walled, muscular organ that is shaped like an inverted pear. Although the shape and position of the uterus changes immensely during pregnancy (fig. 28.3), in its nonpregnant state it is about 7 cm (2.8 in.) long, 5 cm (2 in.) wide (through its broadest region), and 2.5 cm (1 in.) in diameter. The anatomical regions of the uterus include the uppermost, dome-shaped portion above the entrance of the uterine tubes, called the **fundus;** the enlarged main portion, called the **body;** and the inferior constricted portion opening into the vagina, called the **cervix** (fig. 28.4). The uterus is located between the urinary bladder anteriorly and the rectum and sigmoid colon posteriorly. The fundus projects anteriorly and slightly superiorly over the urinary bladder. The cervix projects posteriorly and inferiorly, joining the vagina at nearly a right angle.

The **uterine cavity** is the space within the regions of the fundus and body. The lumina of the uterine tubes open into the uterine cavity on the superior-lateral portions. The uterine cavity is continuous inferiorly with the **cervical canal,** which extends through the cervix and opens into the lumen of the vagina. The junction of the uterine cavity with the cervical canal is called the **internal os,** whereas the opening of the uterine cavity into the cavity of the vagina is called the **external os.**

Support of the Uterus. The uterus is maintained in position by muscular support and ligaments that extend from the pelvic girdle or body wall to the uterus. Muscles of the perineum, especially the levator ani muscle, provide the principal muscular support. The ligaments that support the uterus undergo marked hypertrophy during pregnancy, regress in size after parturition, and atrophy after menopause.

Four paired ligaments support the uterus in position within the pelvic cavity. The paired **broad ligaments** are folds of the peritoneum that extend from the pelvic walls and floor to the lateral walls of the uterus. The ovaries and uterine tubes are also supported by the broad ligaments. The paired **uterosacral ligaments** are also folds of peritoneum that curve along the lateral pelvic wall on either side of the rectum to connect the uterus to the sacrum.

fundus: L. *fundus,* bottom
cervix: L. *cervix,* neck
os: L. *os,* mouth

Figure 28.3. The size and position of the uterus in a full-term pregnant woman in sagittal section.

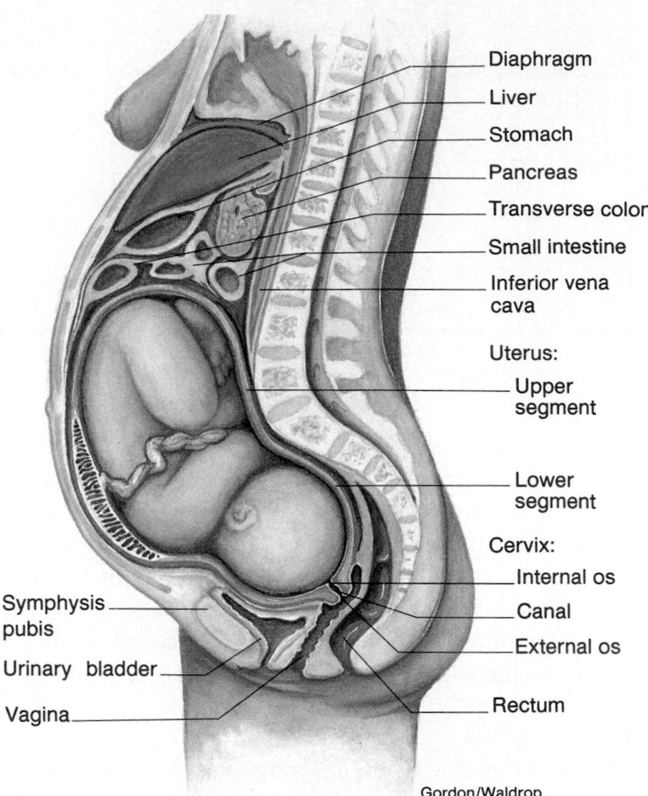

The **cardinal (lateral cervical) ligaments** are fibrous sheets of peritoneum that extend laterally from the cervix and vagina across the pelvic floor where they attach to the wall of the pelvis. The cardinal ligaments contain some smooth muscle as well as vessels and nerves to the cervix and vagina. The fourth paired ligaments are the **round ligaments.** Each round ligament extends from the lateral border of the uterus just below the position where the uterine tube attaches to the lateral pelvic wall. Similar to the course taken by the ductus deferens in the male, the round ligaments continue through the inguinal canal of the abdominal wall where they attach to the deep tissues of the labia majora.

Although the uterus has extensive support, considerable movement is possible. The uterus tilts slightly posteriorly as the urinary bladder fills and moves anteriorly during defecation. In some women, the uterus may go out of position and interfere with the normal progress of pregnancy. A posterior tilting of the uterus is called *retroflexion,* whereas an anterior tilting is called *anteflexion.*

Uterine Wall. The wall of the uterus is composed of three layers: the perimetrium, myometrium, and endometrium (fig. 28.4).

Figure 28.4. The organs of the female reproductive system and the supporting ligaments.

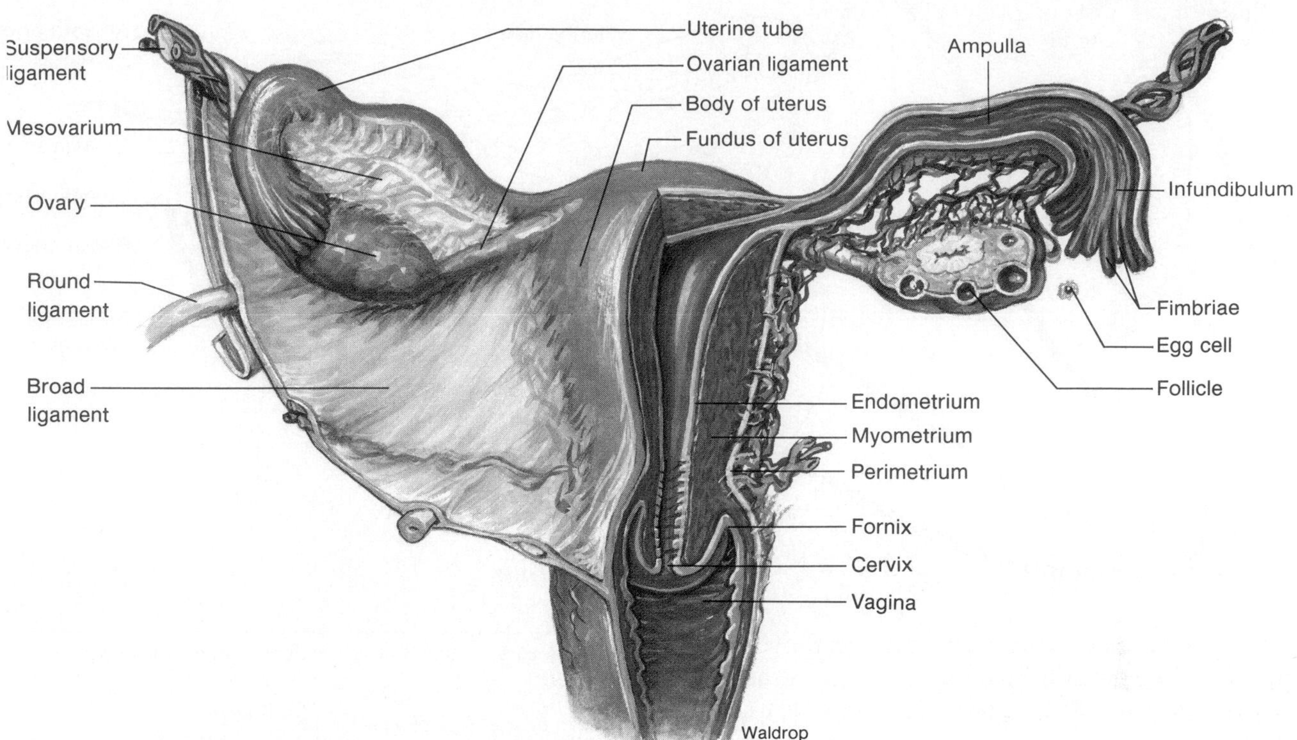

The **perimetrium** is the thin, outer serosal covering and a part of the peritoneum.The thick **myometrium** is composed of three poorly defined layers of smooth muscle arranged in longitudinal, circular, and spiral patterns. The myometrium is thickest in the fundus and thinnest in the cervix. During parturition, the muscles of this layer are stimulated to contract forcefully.

The **endometrium** *(en-do-me'tre-um)* is the inner mucosal lining of the uterus. The endometrium has two distinct layers. The superficial **stratum functionale** layer, composed of columnar epithelium and containing secretory glands, is shed as *menses* during menstruation and built up again under the stimulation of ovarian steroid hormones. The deeper **stratum basale** layer is highly vascular and serves to regenerate the stratum functionale after each menstruation.

The size of the uterus changes tremendously during pregnancy. Its weight increases more than sixteen times (from about 60 g to about 1,000 g), and its capacity increases from about 2.5 ml to over 5,000 ml. The principal change in the myometrium is a marked hypertrophy, or elongation, of the individual muscle cells to as much as ten times their original length. There is some atrophy of the muscle cells after parturition, but the uterus never returns to its original size.

perimetrium: Gk. *peri,* around; *metra,* uterus
myometrium: Gk. *mys,* muscle; *metra,* uterus
endometrium: Gk. *endon,* within; *metra,* uterus

Uterine Blood Supply and Innervation. The uterus is supplied with blood through the **uterine arteries** that arise from the internal iliac arteries and by the **uterine branches** of the **ovarian arteries** (fig. 28.5). Each pair of these two vessels anastomose on the upper lateral margin of the uterus. Branches from the uterine arteries penetrate the perimetrium to form a vascular arch, called the **arcuate artery,** along the outside of the myometrium. Numerous **radial arteries** arise from the arcuate artery to penetrate the myometrium and supply it with blood. **Spiral arterioles** arise from the radial arteries to serve the endometrium. The blood from the uterus returns through uterine veins that parallel the pattern of the arteries.

Vagina

The **vagina** is the organ that receives sperm through the urethra of the erect penis during coitus. It also serves as the birth canal during parturition and provides for the passage of menses to the outside. The musculomembranous vagina is a tubular organ about 9 cm (3.6 in.) in length, passing from the cervix of the uterus to the vestibule. The vagina is positioned between the urinary bladder and urethra anteriorly and the rectum posteriorly, where it is continuous with the cervical canal of the uterus.

menses: L. *menses,* plural of *mensis,* monthly

Figure 28.5. Vascular supply to the uterus.

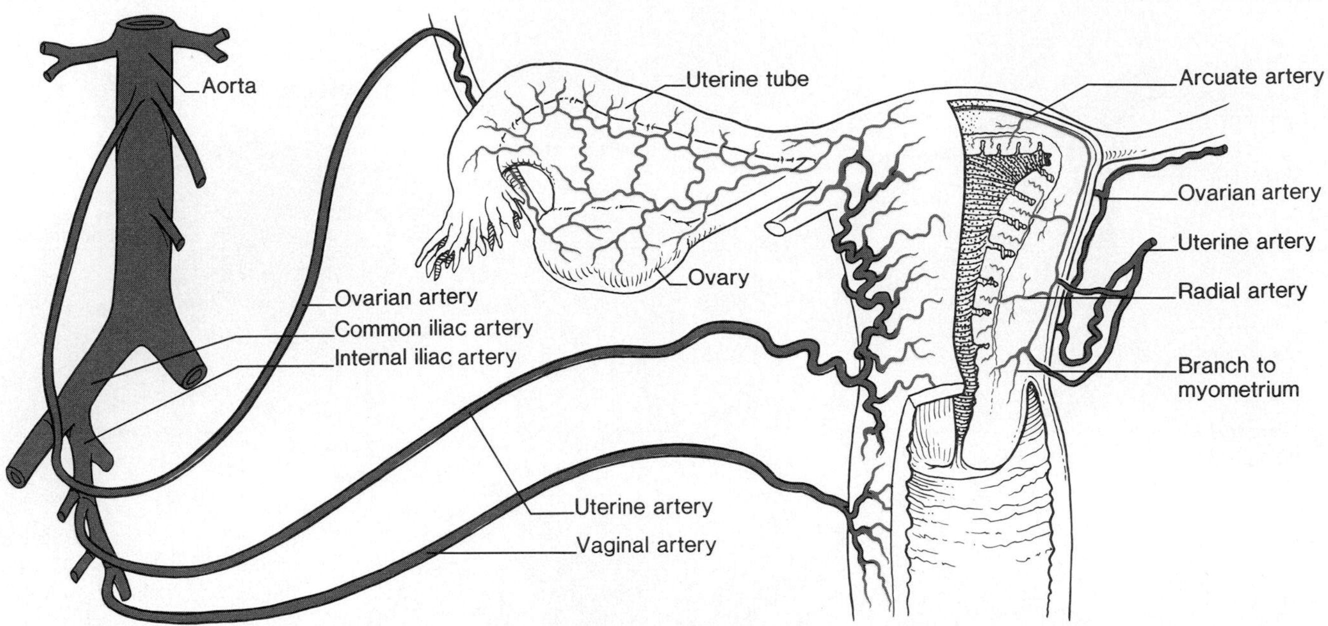

The cervix attaches to the vagina at a nearly 90 degree angle. The deep recess behind the protrusion of the cervix into the vagina is called the **posterior fornix.** The smaller recesses in front and to the sides are called the **anterior** and **lateral fornices.**

The fornices are of clinical importance since they permit the cervix to be palpated during a *gynecological examination.* Occasionally, the deep posterior fornix provides surgical access to the pelvic cavity through the vagina. In addition, the fornices are important in the placement of two forms of *birth-control devices*—the *cervical cap* and the *diaphragm.*

The exterior opening of the vagina, at its lower end, is called the **vaginal orifice.** A thin fold of mucous membrane, called the **hymen,** may partially cover the vaginal orifice.

The vaginal wall is composed of three layers: an inner mucosa, a middle muscularis, and an outer fibrous layer. The **mucosal layer** consists of stratified squamous epithelium, which forms a series of transverse folds called **vaginal rugae** (fig. 28.6). The vaginal rugae provide friction ridges for stimulation of the erect penis during sexual intercourse. They also permit considerable distension of the vagina to facilitate coitus. The mucosal layer contains few glands; the acidic mucus that is present in the vagina comes primarily from glands within the uterus. The acidic environment of the vagina retards microbial growth. The additives within semen, however, temporarily neutralize the acidity of the vagina to insure the survival of the ejaculated sperm deposited within the vagina.

Figure 28.6. The histological structure of a vaginal ruga.

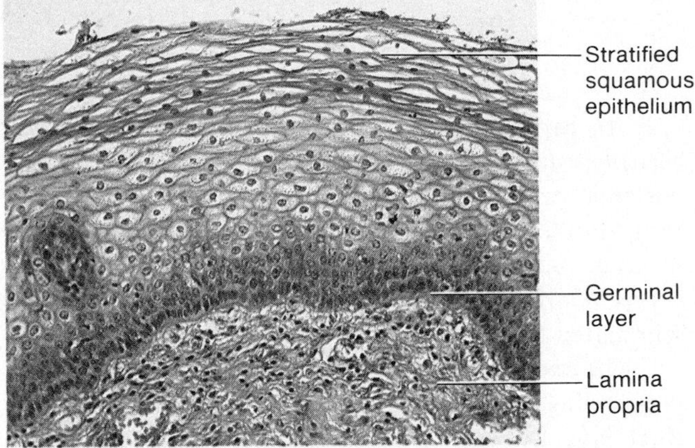

The **muscularis layer** consists of longitudinal and circular bands of smooth muscle interlaced with distensible connective tissue. The distension of this layer is especially important during parturition. Skeletal muscle strands, including the levator ani muscle, near the vaginal orifice partially constrict this opening.

The **fibrous layer** covers the vagina and attaches it to surrounding pelvic organs. This layer consists of dense fibrous connective tissue interlaced with strands of elastic fibers.

Figure 28.7. The external female genitalia.

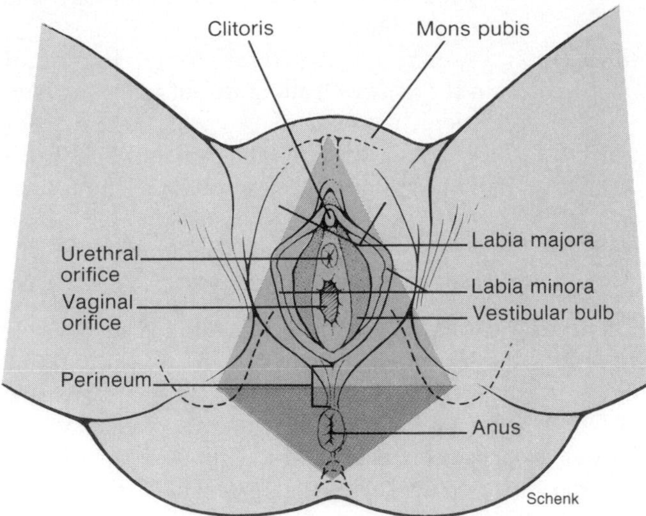

Vulva

The external genitalia of the female are referred to as the **vulva,** or **pudendum** *(pu-den'dum),* which is shown in figure 28.7. The structures of the vulva surround the vaginal orifice and include the mons pubis, labia majora, labia minora, clitoris, vestibule, and vestibular glands.

The **mons pubis** is the subcutaneous pad of adipose connective tissue covering the symphysis pubis. At puberty, the mons pubis becomes covered with a pattern of coarse pubic hair that is somewhat triangular and usually has a horizontal upper border. The elevated and padded mons pubis cushions the symphysis pubis and vulva during coitus.

The **labia majora** are two thickened longitudinal folds of skin that contain areolar and adipose tissue as well as some smooth muscle. The labia majora are continuous anteriorly with the mons pubis, are separated longitudinally by the **pudendal cleft,** and converge again posteriorly on the **perineum** *(per''ĭ-ne'um).* They are also covered with hair and contain numerous sebaceous and sweat glands. The labia majora are homologous to the scrotum of the male and function to enclose and protect the other organs of the vulva.

The **labia minora** are two smaller longitudinal folds positioned close together between the labia majora. The labia minora are hairless but do contain sebaceous glands. On the anterior side, the labia minora split to form the **prepuce** *(pre'pūs)* of the clitoris. These inner folds of skin further protect the vaginal and urethral openings.

The **clitoris** is a small rounded projection at the upper portion of the pudendal cleft. The clitoris corresponds in structure and origin to the penis in the male; it is, however, smaller and has no urethra. Although most of the clitoris is embedded, it does have an exposed **glans** of erectile tissue that is richly innervated with afferent endings. The clitoris is about 2 cm (0.8 in.) long and 0.5 cm (0.2 in.) in diameter. The unexposed portion of the clitoris is composed of two columns of erectile tissue called the **corpora cavernosa,** which diverge posteriorly to form the **crura** and attach to the sides of the pubic arch.

The **vestibule** is the longitudinal cleft enclosed by the labia minora. The openings for the urethra and vagina are located in the vestibule. The external opening of the urethra is about 2.5 cm (1 in.) behind the glans of the clitoris and immediately in front of the vaginal orifice. The vaginal orifice is lubricated during sexual excitement by secretions from a pair of **vestibular glands** (Bartholin's glands) located within the wall of the region immediately inside the vaginal orifice. The ducts from these glands open into the vestibule near the lateral margins of the vaginal orifice. Bodies of vascular erectile tissue, called **vestibular bulbs,** are located immediately below the skin forming the lateral walls of the vestibule. The vestibular bulbs are separated from each other by the vagina and urethra and extend from the level of the vaginal orifice to the clitoris.

The vulva is highly vascular and is supplied with arterial blood from internal pudendal branches of the internal iliac arteries and external pudendal branches from the femoral arteries. Extensive vascular networks exist within most of the organs of the vulva. The venous return is through vessels that correspond in name and position to the arteries.

The vulva has both sympathetic and parasympathetic innervation, as well as extensive somatic fibers that respond to sensory stimulation. Parasympathetic stimulation causes a response similar to that of the male: dilation of the arterioles of the genital erectile tissue and compression of the venous return.

Mechanism of Erection and Orgasm

The homologous structures of the male and female reproductive systems respond to sexual stimulation in a similar fashion. The erectile tissues of a female, like those of a male, become engorged with blood and swollen during sexual arousal. During sexual excitement, the hypothalamus of the brain sends parasympathetic nerve impulses through the sacral segments of the spinal cord, which cause dilation of arteries serving the clitoris and vestibular bulbs. This increased blood flow causes the erectile tissues to swell. In addition, the erectile tissues in the areola of the breasts become engorged.

vulva: L. *volvere,* to roll, wrapper
pudendum: L. *pudere,* to be ashamed
mons pubis: L. *mons,* mountain; *pubis,* genital area

vestibule: L. *vestibule,* an entrance, court
Bartholin's glands: from Casper Bartholin, Jr., Danish anatomist, 1655–1738

Simultaneous with the erection of the clitoris and vestibular bulbs, the vagina expands and elongates to accommodate the erect penis of the male, and parasympathetic impulses cause the vestibular glands to secrete mucus near the vaginal orifice. The vestibular secretion moistens and lubricates the tissues of the vestibule, thus facilitating the penetration of the erect penis into the vagina during coitus. Mucus continues to be secreted during coitus so that the male and female genitalia do not become irritated as they would if the vagina became dry.

The position of the sensitive clitoris usually permits its being stimulated during coitus. If stimulation of the clitoris is of sufficient intensity and duration, a woman will experience a culmination of pleasurable psychological and physiological release called *orgasm.*

If orgasm occurs, there is an associated rhythmic contraction of the muscles of the perineum and the muscular walls of the uterus and uterine tubes. These reflexive muscular actions are thought to aid the movement of sperm through the female reproductive tract toward the upper end of a uterine tube, where an ovum might be located.

Following an orgasm or completion of the sexual act, sympathetic impulses cause a reduction in arterial flow to the erectile tissues, and their size diminishes to that prior to sexual stimulation.

---

1. Distinguish between primary and secondary sex organs, and between secondary sex organs and secondary sexual characteristics.
2. Describe the location and fate of the paramesonephric (müllerian) ducts, and explain which structures of the male and female are homologous.
3. Describe the structure and position of the uterine tubes and of the uterus, and explain why the endometrium is subdivided into a stratum basale and a stratum functionale.
4. Describe the structures of the vagina and vulva, and explain how these structures change during sexual excitement and coitus.

Ovaries and the Ovarian Cycle

The ovaries contain a large number of follicles, each of which encloses an ovum. During the ovarian cycle some follicles mature, and the ova they contain progress to the secondary oocyte stage of meiosis. At ovulation, the largest follicle breaks open to extrude a secondary oocyte from the ovary. The empty follicle then becomes a corpus luteum, which ultimately degenerates at the end of a nonfertile cycle.

Objective 6. Describe the position of the ovaries and the ligaments associated with the ovaries and genital ducts.

Objective 7. Describe the structural changes that occur in the ovaries leading to and following ovulation.

Objective 8. Describe oogenesis, and explain why meiosis of one primary oocyte results in the formation of only one mature ovum.

Objective 9. Describe the hormonal secretions of the ovaries during an ovarian cycle.

The **ovaries** of sexually mature females are solid, ovoid structures that are about 3.5 cm (1.4 in.) long, 2 cm (0.8 in.) wide, and 1 cm (0.4 in.) thick. On the medial portion of each ovary is a **hilum,** which is the point of entrance for ovarian blood vessels and nerves. The lateral portion of the ovary is in contact with the open ends of the uterine tube (fig. 28.4).

Position and Structure of the Ovaries

The ovaries are positioned in the upper pelvic cavity on both sides of the uterus. Each ovary is situated in a shallow depression of the posterior body wall, called the **ovarian fossa,** and is secured by several membranous attachments. The principal supporting membrane of the female reproductive tract is the **broad ligament.** The broad ligament is the parietal peritoneum that supports the uterine tubes and uterus. The **mesovarium** *(mes''o-va're-um)* is a specialized posterior extension of the broad ligament that attaches to an ovary. Each ovary is additionally supported by an **ovarian ligament** anchored to the uterus and a **suspensory ligament** attached to the pelvic wall (fig. 28.4).

Each ovary consists of four layers. The **germinal epithelium** is the thin, outermost layer composed of cuboidal epithelial cells. A collagenous connective tissue layer called the **tunica albuginea** *(al''bu-jin'e-ah)* is immediately below the germinal epithelium. The principal substance of the ovary is divided into an outer **cortex** and an inner, vascular **medulla,** although the boundary between these layers is not distinct. The **stroma** is the material of the ovary in which follicles and blood vessels are embedded and lies in both cortical and medullary layers.

Blood is supplied by ovarian arteries that arise from the lateral sides of the abdominal aorta just below the origin of the renal arteries. An additional supply comes from the ovarian branches of the uterine arteries. Venous return is through the ovarian veins. The right ovarian vein empties into the inferior vena cava, whereas the left ovarian vein drains into the left renal vein.

---

orgasm: Gk. *orgasmos,* to swell; to become excited

ovaries: L. *ovaries,* egg holders; *ovum,* egg
hilum: L. *hilum,* a trifle (of little significance)
stroma: Gk. *stroma,* a couch or bed

Figure 28.8. (*a*) Photomicrographs of primordial and primary follicles; and (*b*) a secondary follicle.

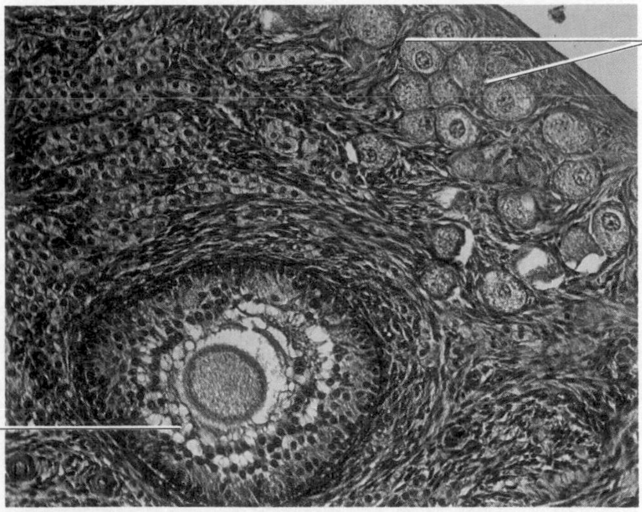

Ovarian Cycle

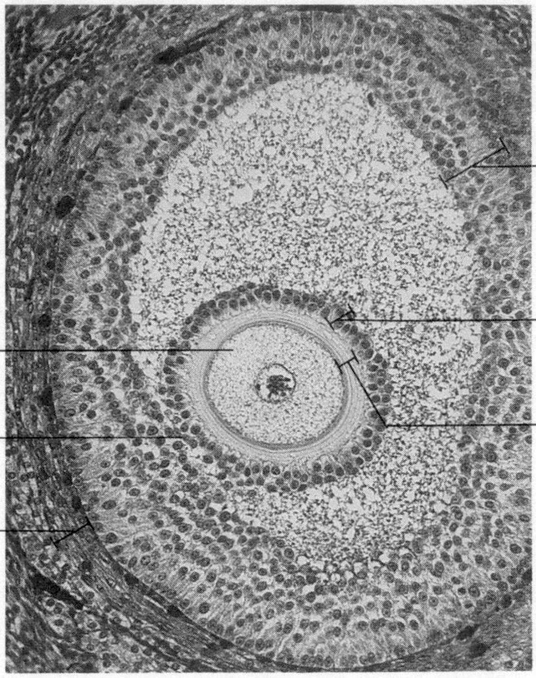

The germ cells that migrate into the ovaries during early embryonic development multiply, so that by about five months of gestation the ovaries contain approximately 6 to 7 million oogonia. Production of new oogonia stops at this point and never resumes again. Toward the end of gestation the oogonia begin meiosis, at which time they are called **primary oocytes** *(o'o-sīts)*. Like spermatogenesis in the male, oogenesis is arrested at prophase I of the first meiotic division. The number of primary oocytes decreases throughout a woman's reproductive years. The ovaries of a newborn girl contain about 2 million oocytes, but this number is reduced to about 300,000–400,000 by the time the girl enters puberty. Oogenesis ceases entirely at menopause (the time menstruation stops).

Primary oocytes that are not stimulated to complete the first meiotic division are contained within tiny follicles called **primordial follicles.** In response to gonadotropin stimulation, some of these oocytes and follicles get larger, and the follicular cells divide to produce numerous small **granulosa cells** that surround the oocyte and fill the follicle. A follicle at this stage in development is called a **primary follicle.**

Some primary follicles will be stimulated to grow still bigger and develop a fluid-filled cavity, called an *antrum,* at which time they are called **secondary follicles** (fig. 28.8). The granulosa cells of secondary follicles form a ring around the circumference of the follicle and form a mound

oogonium: Gk. *oion,* egg; *gonos,* generation

Figure 28.9. (*a*) a primary oocyte at metaphase I of meiosis. Note the alignment of chromosomes (*arrow*). (*b*) a secondary oocyte formed at the end of the first meiotic division and the first polar body (*arrow*).

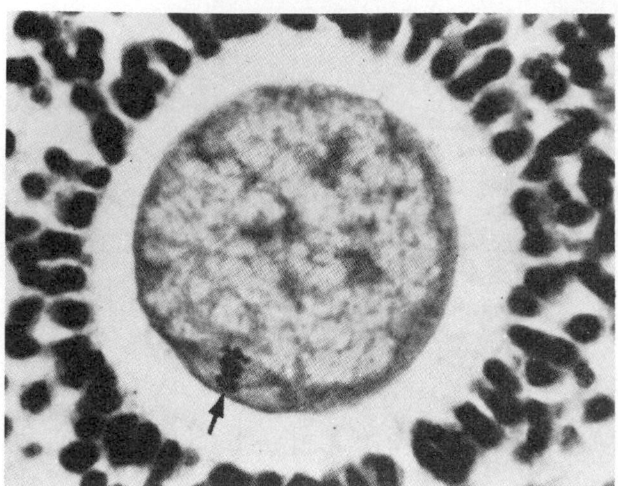

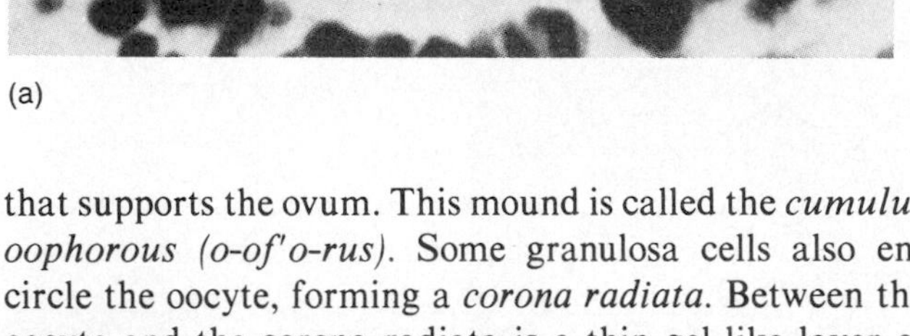

(a)

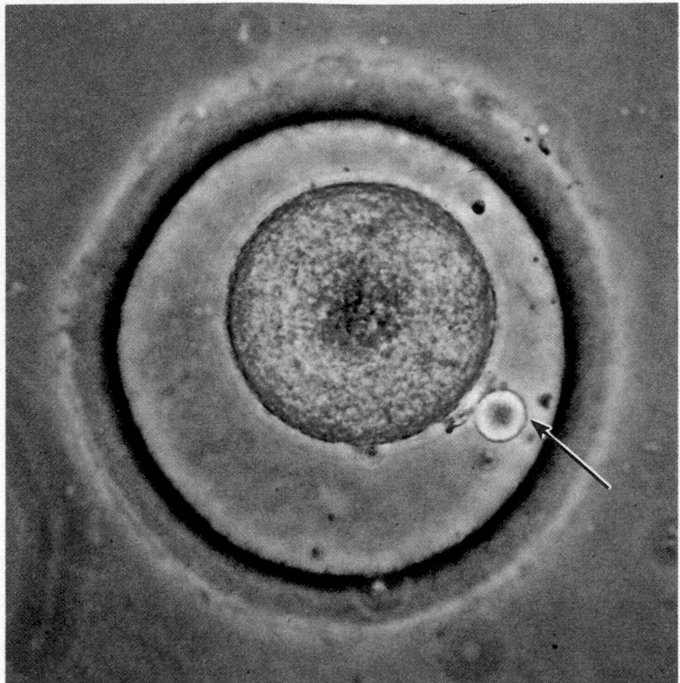

(b)

that supports the ovum. This mound is called the *cumulus oophorous (o-of'o-rus).* Some granulosa cells also encircle the oocyte, forming a *corona radiata.* Between the oocyte and the corona radiata is a thin gel-like layer of proteins and polysaccharides called the *zona pellucida (pel-lu'sĭ-dah).* Under stimulation of FSH from the anterior pituitary, the granulosa cells secrete increasing amounts of estrogen as the follicles grow. Interestingly, the granulosa cells produce estrogen from its precursor testosterone, which is supplied by cells of the *theca interna* layer immediately outside the follicle.

As the follicle develops, the primary oocyte completes its first meiotic division. This does not form two complete cells, however, because only one cell—the **secondary oocyte**—gets all the cytoplasm. The other cell formed at this time becomes a small *polar body* (fig. 28.9) which eventually fragments and disappears. The secondary oocyte enters the second meiotic division, but meiosis is arrested at metaphase II and is never completed unless fertilization occurs.

Ovulation. Usually, by about ten to fourteen days after the first day of menstruation, only one follicle has continued its growth to become a mature **vesicular ovarian (graafian) follicle** (fig. 28.10); other secondary follicles during that cycle regress and become *atretic (ah-tret'ik).* The ovarian follicle is so large that it forms a bulge on the surface of the ovary. Under proper hormonal stimulation this follicle will rupture—much like the popping of a blister—and extrude its oocyte into the uterine tube in the process of **ovulation** (fig. 28.11).

The released cell is a *secondary oocyte,* surrounded by the zona pellucida and corona radiata. If it is not fertilized, it disintegrates in a couple of days. If a sperm passes through the corona radiata and zona pellucida and enters the cytoplasm of the secondary oocyte, the oocyte completes the second meiotic division. In this process the cytoplasm is again not divided equally; most of the cytoplasm remains in the zygote (fertilized egg), leaving another polar body that, like the first, disintegrates (fig. 28.12).

Changes continue in the ovary following ovulation. The empty follicle, under the influence of luteinizing hormone from the anterior pituitary, undergoes structural and biochemical changes to become a **corpus luteum.** Unlike the ovarian follicles, which secrete only estrogen, the corpus luteum secretes two sex steroid hormones: estrogen

cumulus oophorous: L. *cumulus,* a mound; G. *oophoros,* egg-bearing
corona radiata: G. *korone,* crown; L. *radiata,* radiate
zona pellucida: G. *zone,* girdle; L. *pellis,* skin
theca interna: G. *theke,* a box; L. *internus,* interior

atretic: Gk. *atretos,* not perforated
graafian follicle: from Regnier de Graaf, Dutch anatomist and physician, 1641–73
corpus luteum: L. *corpus,* body; *luteum,* yellow

Figure 28.10. A vesicular ovarian (graafian) follicle within the ovary of a monkey.

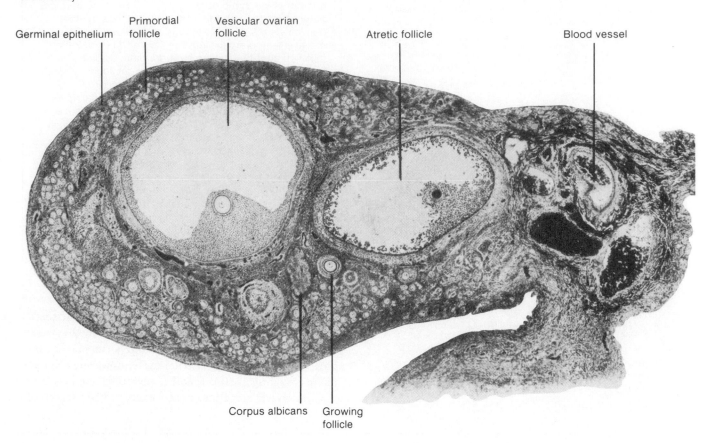

Germinal epithelium
Primordial follicle
Vesicular ovarian follicle
Atretic follicle
Blood vessel
Corpus albicans
Growing follicle

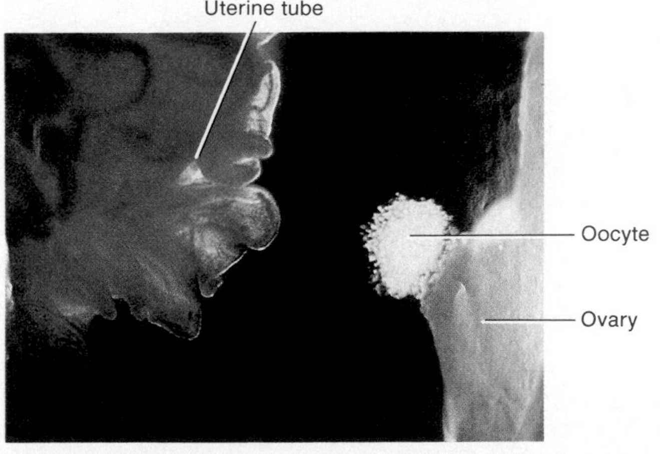

Figure 28.11. Ovulation from a human ovary.

and progesterone. Toward the end of a nonfertile cycle the corpus luteum regresses and is changed into a nonfunctional **corpus albicans.** These cyclic changes in the ovary are summarized in figure 28.13.

albicans: L. *albicare,* to whiten

Pituitary-Ovarian Axis

The term **"pituitary-ovarian axis"** refers to the hormonal interactions between the anterior pituitary and the ovaries. The anterior pituitary secretes two gonadotropic hormones—**FSH (follicle stimulating hormone)** and **LH (luteinizing hormone)**—that stimulate the ovaries. The secretion of both gonadotropic hormones from the anterior pituitary, in turn, is stimulated by a single releasing hormone—**gonadotropin releasing hormone (GnRH)**—which is secreted by the hypothalamus (chapter 19). Feedback loops are completed in this control system because ovarian sex steroid hormones modify the secretion of both GnRH and the gonadotropic hormones.

Studies have shown that secretion of GnRH from the hypothalamus is pulsatile rather than continuous, and thus the secretion of FSH and LH follows this pulsatile pattern. The pulsatile pattern of GnRh secretion is apparently required to prevent desensitization of the anterior pituitary, which has been shown to occur in experiments in which GnRH was maintained at a constant level. In addition, the frequency of GnRH pulses, as well as their amplitude (how much GnRH is secreted per pulse) is believed to influence the nature of the pituitary's response to this releasing hormone.

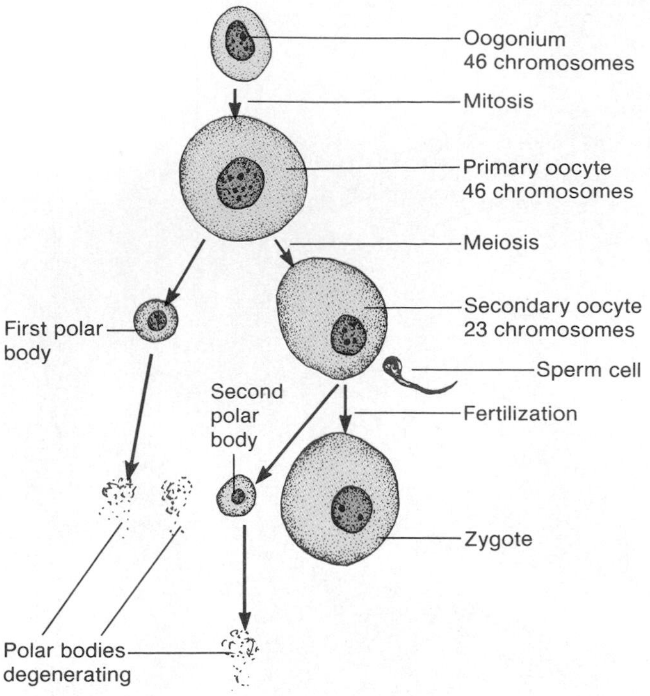

Figure 28.12. A schematic diagram of the process of oogenesis. During meiosis, each primary oocyte produces a single haploid gamete. If the secondary oocyte is fertilized, it forms a secondary polar body and becomes a zygote.

Scientists have produced a number of synthetic derivatives of GnRH that are more potent than the natural GnRH. When these synthetic derivatives are administered to men or women, they cause a temporary increase in FSH and LH secretion, followed after two weeks by a dramatic decline in the secretion of the gonadotropins. This decline is related to the fact that GnRH is normally secreted in a pulsatile fashion; when powerful synthetic analogues are present continuously in the blood, they produce desensitization of the pituitary. The consequent fall in LH secretion in men results in lower testosterone secretion, which is beneficial in the treatment of prostate disease. In women, the fall in LH results in a decrease in estrogen secretion by the ovaries. This has recently been exploited in the treatment of *endometriosis,* a condition in which endometrial tissue leaves the uterus through the fallopian tubes and enters the peritoneal cavity. Since growth of this ectopic tissue is dependent on estrogen, a reversible decrease in estrogen induced by synthetic GnRH analogues (such as *nafarelin*) provides an exciting new form of treatment for this condition.

Since one releasing hormone can stimulate the secretion of both FSH and LH, one might expect to always see parallel changes in the secretion of both hormones. During an early phase of the menstrual cycle, however, FSH secretion is slightly greater than LH secretion, and just prior to ovulation LH secretion greatly exceeds FSH secretion.

Figure 28.13. A schematic diagram of an ovary showing the various stages of ovum and follicle development.

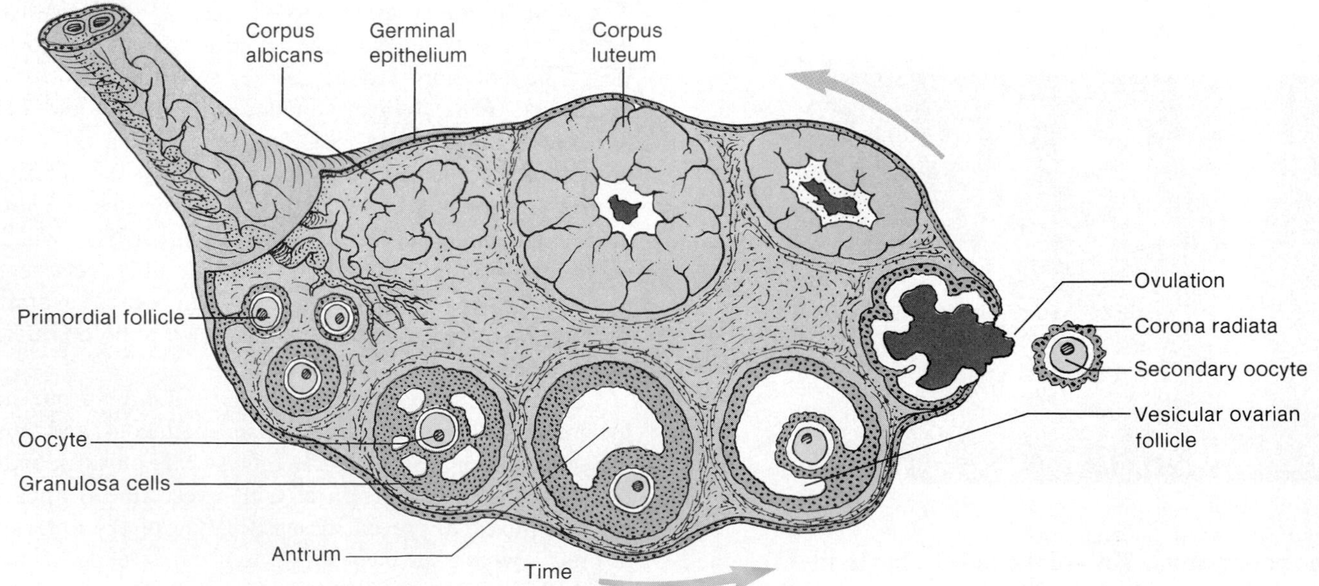

These and other changes in gonadotropin secretion are believed to be a result of the feedback effects of ovarian sex steroids, which can change the amount of GnRH secreted, the pulse frequency of GnRH secretion, and the ability of the pituitary to produce and secrete FSH and/or LH. These complex interactions result in a pattern of hormone secretion that is responsible for, and characteristic of, each phase of the menstrual cycle.

1. Describe the position of the ovaries relative to the uterine tubes, and describe the position and functions of the broad ligament and mesovarium.
2. Compare the structure and contents of a primordial follicle, primary follicle, secondary follicle, and vesicular ovarian follicle.
3. Define ovulation, and describe the changes that occur in the ovary following ovulation in a nonfertile cycle.
4. Describe oogenesis, and explain why only one mature ovum is produced by this process.
5. Compare the hormonal secretions of the ovarian follicles with those of a corpus luteum.

Menstrual Cycle

Cyclic changes in the secretion of gonadotropic hormones from the anterior pituitary cause the ovarian changes during a monthly cycle. The ovarian cycle is accompanied by cyclic changes in the secretion of sex steroids, which interact with the hypothalamus and pituitary to regulate gonadotropin secretion. The cyclic changes in ovarian hormone secretion also cause the changes in the endometrium of the uterus during a menstrual cycle.

Objective 10. Describe the hormonal changes that occur during the follicular phase of the cycle, and explain how these changes are regulated.

Objective 11. Describe the hormonal control of ovulation, and explain how the time for ovulation is regulated.

Objective 12. Describe the formation, function, and fate of the corpus luteum.

Objective 13. Describe the structural changes that occur in the endometrium during the cycle, and explain the hormonal control of these changes.

Humans, apes, and old-world monkeys have cycles of ovarian activity that repeat at approximately one-month intervals; hence the name *menstrual cycle.* The term *menstruation (men''stroo-a'shun)* is used to indicate the periodic shedding of the stratum functionale of the endometrium, which becomes thickened prior to menstruation under stimulation by ovarian steroid hormones. In primates (other than new-world monkeys) this shedding of the endometrium is accompanied by bleeding. There is no bleeding, in contrast, when other mammals shed the endometrium; their cycles, therefore, are not called menstrual cycles (even though the cycle of a cow is also about a month long).

Humans and other primates that have menstrual cycles may permit coitus at any time of the cycle. Nonprimate mammals, in contrast, are sexually receptive only at a particular time in their cycles (shortly before or shortly after ovulation). These animals are therefore said to have *estrous cycles.* In some animals (such as dogs and cats) that have estrous cycles bleeding occurs shortly before they permit coitus. This bleeding is a result of high estrogen secretion and is not associated with shedding of the endometrium.

The bleeding that accompanies menstruation, in contrast, can be experimentally induced by removing the ovaries. This demonstrates that menstrual bleeding is caused by the withdrawal of ovarian hormones. During the normal menstrual cyle, the secretion of ovarian hormones rises and falls in a regular fashion, causing cyclic changes in the endometrium and other sex-steroid-dependent tissues.

Phases of the Menstrual Cycle

The average menstrual cycle has a duration of about twenty-eight days. Since it is a cycle, there is no beginning or end, and the changes that occur are generally gradual. It is convenient, however, to call the first day of menstruation "day one" of the cycle, because menstrual blood flow is the most apparent of the changes that occur. It is also convenient to divide the cycle into four phases based on changes that occur in the ovary and in the endometrium. The ovaries are in the **follicular phase** starting on the first day of menstruation and ending on the day of ovulation. After ovulation, the ovaries are in the **luteal phase** until the first day of menstruation. The cyclic changes that occur in the endometrium are called the *menstrual, proliferative,* and *secretory phases* and will be discussed separately.

Menstrual and Follicular Phases. The menstrual phase lasts from day one to day four or five of the average cycle. During this time, the secretions of ovarian steroid hormones are at their lowest ebb, and the ovaries contain only primordial and primary follicles. During the *follicular phase,* which lasts from day one to about day thirteen of the cycle (this is highly variable), some of the primary follicles grow, form an antrum, and become secondary follicles. Toward the end of the follicular phase, one follicle in one ovary (the "dominant follicle") reaches maturity and becomes a vesicular ovarian follicle. As follicles grow, the granulosa cells secrete an increasing amount of **estradiol** (the principal estrogen), which reaches its highest concentration in the blood at about day twelve of the cycle (two days before ovulation; fig. 28.14).

The growth of the follicles and their secretion of estradiol are stimulated by, and dependent upon, FSH secreted from the anterior pituitary. The amount of FSH secreted during the early follicular phase is believed to be slightly greater than the amount secreted in the late follicular phase, although this can vary from cycle to cycle (a measure of variance is shown by vertical bars in figure 28.14). FSH stimulates the production of FSH receptors in the granulosa cells, so that the follicles become increasingly sensitive to a given amount of FSH. This increased sensitivity is augmented by estradiol, which also stimulates the production of new FSH receptors in the follicles. As a result, the stimulatory effect of FSH on the follicles, and later on the dominant follicle, increases despite the fact that FSH levels in the blood do not increase throughout the follicular phase. Toward the end of the follicular phase, FSH and estradiol also stimulate the production of LH receptors in the vesicular ovarian follicle. This prepares the vesicular follicle for the next major event in the cycle.

The rapid rise in estradiol secretion from the granulosa cells during the follicular phase acts on the hypothalamus to increase the frequency of GnRH pulses. In addition, estradiol augments the ability of the pituitary to respond to GnRH with an increase in LH secretion. As a result of this **positive feedback** effect of estradiol on the pituitary, there is an increase in LH secretion in the late follicular phase that culminates in an **LH surge** (fig. 28.14).

The LH surge begins about twenty-four hours before ovulation and reaches its peak about sixteen hours before ovulation. It is this LH surge that acts to trigger ovulation. Since GnRH stimulates the anterior pituitary to secrete both FSH and LH, there is a simultaneous, though smaller, surge in FSH secretion. Some investigators believe that this midcycle peak in FSH acts as a stimulus for the development of new follicles for the next month's cycle.

Ovulation. Under the influence of FSH stimulation, the vesicular follicle grows so large that it becomes a thin-walled "blister" on the surface of the ovary. The growth of the follicle is accompanied by a rapid rate of increase in estradiol secretion. This rapid increase in estradiol, in turn, triggers the LH surge. Finally, the surge in LH secretion causes the wall of the vesicular ovarian follicle to rupture (fig. 28.15). Ovulation occurs, therefore, as a result of the sequential effects of FSH followed by LH on the ovarian follicles.

Figure 28.14. Sample values for LH, FSH, progesterone, and estradiol during the menstrual cycle. The midcycle peak of LH is used as a reference day. (IU = international unit.) Vertical bars indicate the variance of the measurements.

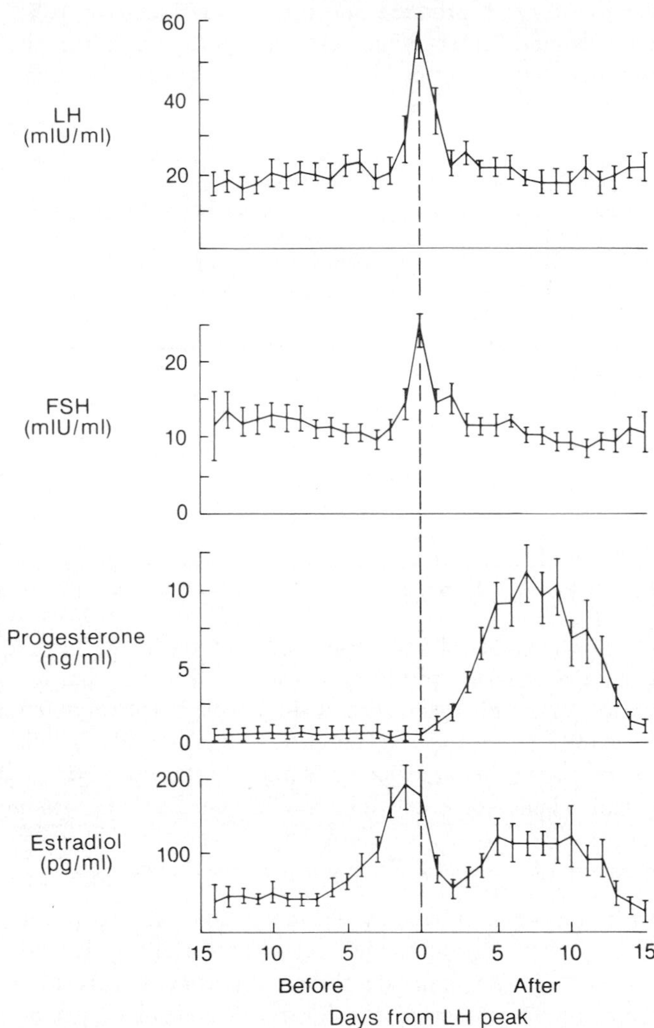

By means of the positive feedback effect of estradiol on LH secretion, the follicle in a sense sets the time for its own ovulation. This is because ovulation is triggered by an LH surge, and the LH surge is triggered by increased estradiol secretion, which occurs while the follicle grows. In this way the vesicular ovarian follicle does not normally ovulate until it has reached the proper size and degree of maturation.

In ovulation, a secondary oocyte, arrested at metaphase II of meiosis, is released into a uterine tube. This oocyte is still surrounded by a zona pellucida and corona radiata as it begins its journey to the uterus. Normally only one ovary ovulates per cycle, with the left and right ovary alternating in successive cycles. Interestingly, if one ovary is removed, the remaining ovary does not skip cycles but ovulates every month. The mechanisms by which this regulation is achieved are not understood.

Figure 28.15. Phases of the menstrual cycle in relation to ovarian changes and hormone secretion, and the relationship between changes in the ovaries and the endometrium of the uterus during different phases of the menstrual cycle.

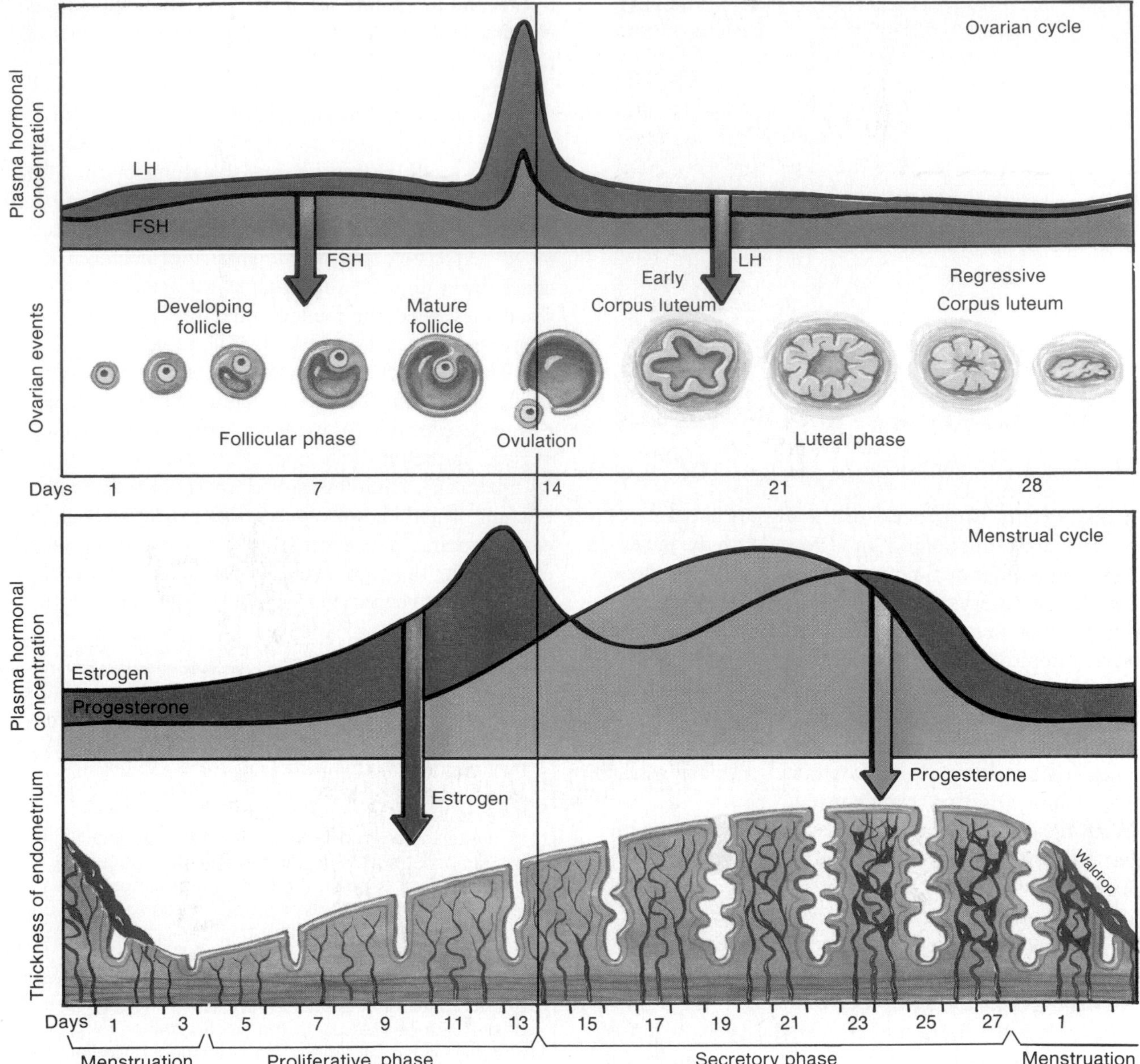

Luteal Phase. After ovulation, the empty follicle is stimulated by LH to become a new structure, the **corpus luteum** (fig. 28.16). This change in structure is accompanied by a change in function. Whereas the developing follicles secrete only estradiol, the corpus luteum secretes both estradiol and **progesterone.** Progesterone levels in the blood are negligible before ovulation but rise rapidly to reach a peak during the luteal phase at approximately one week after ovulation.

The combined high levels of estradiol and progesterone during the luteal phase exert a **negative feedback inhibition** of FSH and LH secretion. This serves to retard development of new follicles, so that further ovulation does not normally occur during that cycle. Although this might seem like locking the barn door after the horse (ovum) has escaped, it does prevent more horses (ova) from escaping. In this way multiple ovulations (and possible pregnancies) on succeeding days of the cycle are prevented.

High levels of estrogen and progesterone during the nonfertile cycle do not persist for very long, however, and new follicles do start to develop toward the end of one cycle, in preparation for the next cycle. Estrogen and progesterone levels fall during the late luteal phase (starting

Figure 28.16. A corpus luteum in a human ovary.

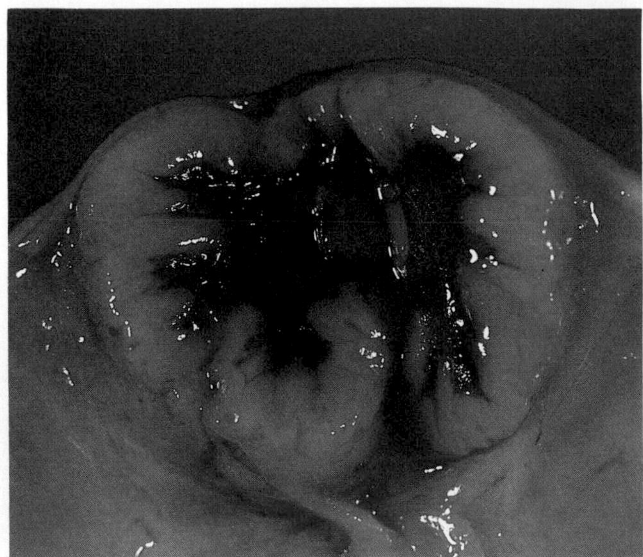

about day twenty-two), because the corpus luteum regresses and stops functioning. In lower mammals, the decline in corpus luteum function is caused by a hormone secreted by the uterus called *luteolysin.* A similar hormone has not yet been identified in humans, and the cause of corpus luteum regression in humans is not well understood. Luteolysis (breakdown of the corpus luteum) might be prevented by continuous LH secretion, but LH levels remain low during the luteal phase as a result of negative feedback inhibition by ovarian steroids. In this way, the corpus luteum causes its own demise.

With the declining function of the corpus luteum, estrogen and progesterone fall to very low levels by day twenty-eight of the cycle. The withdrawal of ovarian steroids causes menstruation and permits a new cycle of ovarian follicle development to progress.

Cyclic Changes in the Endometrium

In addition to a description of the female cycle in terms of the phases of ovarian function, the cycle can also be described in terms of the changes that occur in the endometrium. Three phases can be identified on this basis (fig. 28.15 *bottom*): (1) the proliferative phase; (2) the secretory phase; and (3) the menstrual phase.

The **proliferative phase** of the endometrium occurs while the ovary is in its follicular phase. The increasing amounts of estradiol secreted by the developing follicles stimulates growth (proliferation) of the stratum functionale of the endometrium. In humans and other primates, spiral arteries develop in the endometrium during this phase. Estradiol may also stimulate the production of receptor proteins for progesterone at this time, in preparation for the next phase of the cycle.

The **secretory phase** of the endometrium occurs when the ovary is in its luteal phase. In this phase, increased progesterone secretion stimulates the development of mucus glands. As a result of the combined actions of estradiol and progesterone, the endometrium becomes thick, vascular, and "spongy" in appearance during the time of the cycle following ovulation. It is therefore well prepared to accept and nourish an embryo if fertilization occurs.

The **menstrual phase** occurs as a result of the fall in ovarian hormone secretion during the late luteal phase. Necrosis (cellular death) and sloughing of the stratum functionale of the endometrium may be produced by constriction of the spiral arteries. The spiral arteries appear to be responsible for bleeding during menstruation because lower animals that lack spiral arteries don't bleed when they shed their endometrium. The phases of the menstrual cycle are summarized in table 28.3.

The cyclic changes in ovarian secretion cause other cyclic changes in the female genital ducts. High levels of estradiol secretion, for example, cause cornification of the vaginal epithelium (the upper cells die and become filled with keratin). High levels of estradiol also cause the production of a thin, watery cervical mucus, which can be easily penetrated by spermatozoa. During the luteal phase of the cycle, the high levels of progesterone cause the cervical mucus to become thick and sticky after ovulation has occurred.

Cyclic changes in ovarian hormone secretion also cause cyclic changes in *basal body temperature.* In the *rhythm method* of birth control, a woman measures her oral basal body temperature upon waking to determine when ovulation has occurred. On the day of the LH peak, when estradiol secretion begins to decline, there is a slight drop in basal body temperature. Starting about one day after the LH peak, the basal body temperature sharply rises as a result of progesterone secretion and remains elevated throughout the luteal phase of the cycle (fig. 28.17). The day of ovulation can be accurately determined by this method, making the method useful in increasing fertility if conception is desired. Since the day of the cycle in which ovulation occurs is quite variable in many women, however, the rhythm method is not very reliable for contraception by predicting when the next ovulation will occur. The contraceptive pill is a much more effective means of contraception.

Contraceptive Pill

About ten million women in the United States and sixty million women in the world are currently using **oral steroid contraceptives,** also called birth-control pills. These contraceptives usually consist of a synthetic estrogen combined with a synthetic progesterone in the form of pills that are taken once each day for three weeks after the last

Figure 28.17. Changes in basal body temperature during the menstrual cycle.

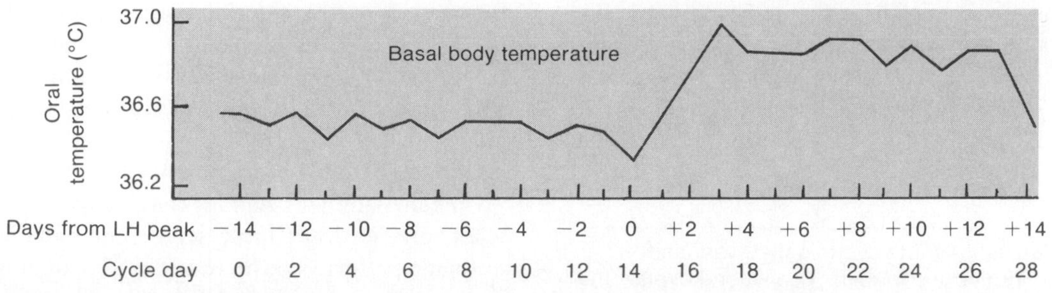

Table 28.3 Phases of the menstrual cycle

Phase of cycle		Hormonal changes		Tissue changes	
Ovarian	**Endometrial**	**Pituitary**	**Ovary**	**Ovarian**	**Endometrial**
Follicular (days 1–4)	Menstrual	FSH and LH secretion low	Estradiol and progesterone remain low	Primary follicles grow	Outer two-thirds of endometrium is shed with accompanying bleeding
Follicular (days 5–13)	Proliferative	FSH slightly higher than LH secretion in early follicular phase	Estradiol secretion rises (due to FSH stimulation of follicles)	Follicles grow; vesicular ovarian follicle develops (due to FSH stimulation)	Mitotic division increases thickness of endometrium; spiral arteries develop (due to estradiol stimulation)
Ovulatory (day 14)	Proliferative	LH surge (and increased FSH) stimulated by positive feedback from estradiol	Fall in estradiol secretion	Vesicular ovarian follicle is ruptured and ovum is extruded into fallopian tube	No change
Luteal (days 15–28)	Secretory	LH and FSH decrease (due to negative feedback of steroids)	Progesterone and estrogen secretion increase, then fall	Development of corpus luteum (due to LH stimulation); regression of corpus luteum	Glandular development in endometrium (due to progesterone stimulation)

day of a menstrual period. This procedure causes an immediate increase in blood levels of ovarian steroids (from the pill), which is maintained for the normal duration of a monthly cycle. As a result of *negative feedback inhibition* of gonadotropin secretion, *ovulation never occurs.* The entire cycle is like a false luteal phase, with high levels of progesterone and estrogen and low levels of gonadotropins.

Since the contraceptive pills contain ovarian steroid hormones, the endometrium proliferates and becomes secretory just as it does during a normal cycle. In order to prevent an abnormal growth of the endometrium, women stop taking the steroid pills after three weeks (placebo pills are taken during the fourth week). This causes estrogen and progesterone levels to fall, and permits menstruation to occur. The contraceptive pill is an extremely effective method of birth control, but it does have potentially serious side effects—including an increased incidence of thromboembolism, cardiovascular disorders, and endometrial and breast cancer. It has been pointed out, however, that the mortality risk of contraceptive pills is still much lower than the risk of death from the complications of pregnancy—or even from automobile accidents. Additionally, there is evidence that oral contraceptives may actually provide some protection against cancer of the uterus and ovary.

Menopause

The term *menopause* means literally "pause in the menses" and refers to the cessation of ovarian activity that occurs at about the age of fifty. During the postmenopausal years, which account for about a third of a woman's life span, no new ovarian follicles develop and the ovaries cease secreting estradiol. This fall in estradiol is due to defects in the ovaries, not in the pituitary; indeed, FSH and LH secretion by the pituitary is elevated due to the absence of negative feedback inhibition from estradiol. As in prepubertal boys and girls, the only estrogen found in the blood of postmenopausal women is that formed by conversion of the weak androgen androstenedione, secreted principally by the adrenal cortex, into a weak estrogen called estrone.

It is the withdrawal of estradiol secretion from the ovaries that is most responsible for the many debilitating symptoms of menopause. These include vasomotor disturbances (which produce "hot flashes"), urogenital atrophy, and the increased development of osteoporosis (see chapter 27). Estrogen replacement therapy, often in combination with progesterone may be prescribed to help alleviate these symptoms.

1. Describe the changes that occur in the ovary and endometrium during the follicular phase, and explain the hormonal control of these changes.
2. Describe the hormonal regulation of ovulation.
3. Describe the formation, function, and fate of the corpus luteum, and describe the changes that occur in the endometrium during the luteal phase.
4. Explain the significance of negative feedback inhibition during the luteal phase, and explain the hormonal control of menstruation.

Mammary Glands and Lactation

The structure and function of the mammary glands is dependent on the action of a number of hormones. The secretion of prolactin and oxytocin is directly required for the production and delivery of milk to a suckling infant.

Objective 14. Describe the structure of breasts and of the mammary glands.

Objective 15. Describe the interaction of hormones required for breast and mammary gland development.

Objective 16. Explain the actions of prolactin and oxytocin on milk production and delivery, and explain how the secretion of these hormones is regulated during lactation.

In structure, the mammary glands, located in the **breasts,** are modified sweat glands and part of the integumentary system. In function, however, these glands are associated with the reproductive system because they secrete milk for the nourishment of the young. The size and shape of the breasts vary considerably from person to person because of genetic differences, age, percentage of body fat, or pregnancy. At puberty, estrogen from the ovaries stimulates growth of the mammary glands and the deposition of adipose tissue within the breasts. Mammary glands hypertrophy in pregnant and lactating women and usually atrophy somewhat after menopause.

Structure of Breasts and Mammary Glands

Each breast is positioned over the second to the sixth rib and overlies the pectoralis major muscle and portions of the serratus anterior and external oblique muscles (fig. 28.18). The medial boundary of the breast is over the lateral margin of the sternum, and the lateral margin of the breast is along the anterior border of the axilla. The **axillary tail** of the breast extends upward and laterally toward the axilla, where it comes into close relationship with the axillary vessels. This region of the breast is clinically significant because of the high incidence of breast cancer within the lymphatic drainage of the axillary tail.

Each mammary gland is composed of fifteen to twenty **lobes,** divided by adipose tissue and each with its own drainage pathway to the outside. The amount of adipose tissue determines the size and shape of the breast but has nothing to do with the ability of a woman to nurse. Each lobe is subdivided into **lobules,** which contain the glandular **alveoli** (fig. 28.19) that secrete the milk of a lactating female. **Suspensory ligaments (of Cooper)** between the lobules extend from the skin to the deep fascia overlying the pectoralis muscle and support the breasts. The clustered alveoli secrete milk into a series of **secondary tubules.** The secondary tubules in each lobe converge to form a **lactiferous** *(lak-tif'er-us)* **duct.** The lumen of each lactiferous duct expands near the nipple to form an **ampulla** *(am-pul'lah),* where milk may be stored before it drains at the tip of the nipple.

The **nipple** is a cylindrical projection containing some erectile tissue. A circular pigmented **areola** *(ah-re'o-lah)* surrounds the nipple. The surface of the areola may appear rough because of the presence of sebaceous **areolar glands** close to the surface. The secretions of the areolar glands keep the nipple pliable. The color of the areola and nipple varies with the complexion of the woman and whether or not she is pregnant. During pregnancy, the areola becomes darker and enlarges somewhat, presumably to become more conspicuous to a nursing infant.

Blood is supplied to the mammary gland through the perforating branches of the internal thoracic artery, which enter the breast through the second, third, and fourth intercostal spaces just lateral to the sternum, and through the more superficial mammary artery, branching from the lateral thoracic artery. Venous return is through a series of veins that parallel the pattern of the arteries. A superficial venous plexus may be apparent through the skin of the breast, especially during pregnancy and lactation.

The breast is innervated primarily through afferent somatic fibers that are derived from the anterior and lateral cutaneous branches of the fourth, fifth, and sixth thoracic nerves. Sensory nerve endings in the nipple and areola are especially important in stimulating the release of milk from the mammary glands to a suckling infant.

ligaments of Cooper: from Sir Astley P. Cooper, English anatomist and surgeon, 1768–1841

Figure 28.18. The structure of the breast and mammary glands. (*a*) a sagittal section; (*b*) an anterior view partially sectioned.

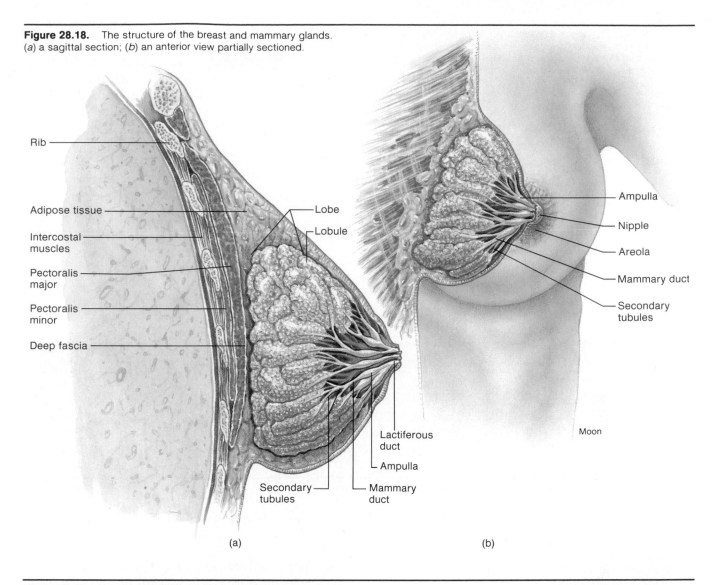

Rib
Adipose tissue
Intercostal muscles
Pectoralis major
Pectoralis minor
Deep fascia
Lobe
Lobule
Lactiferous duct
Ampulla
Secondary tubules
Mammary duct
(a)

Ampulla
Nipple
Areola
Mammary duct
Secondary tubules
Moon
(b)

Figure 28.19. The histology of the mammary gland. (*a*) nonlactating (63×); (*b*) lactating (63×).

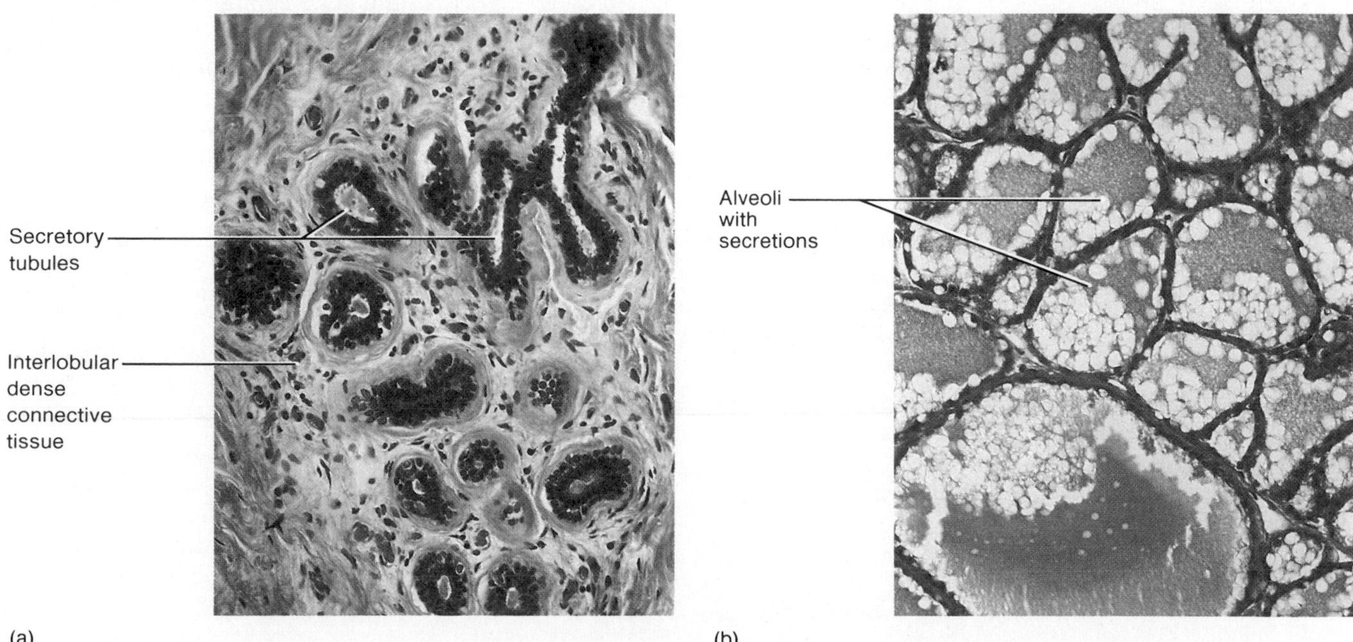

Figure 28.20. The hormonal control of mammary gland development during pregnancy and lactation. Note that milk production is prevented during pregnancy by estrogen inhibition of prolactin secretion. This inhibition is accomplished by the stimulation of PIH (prolactin-inhibiting hormone) secretion from the hypothalamus.

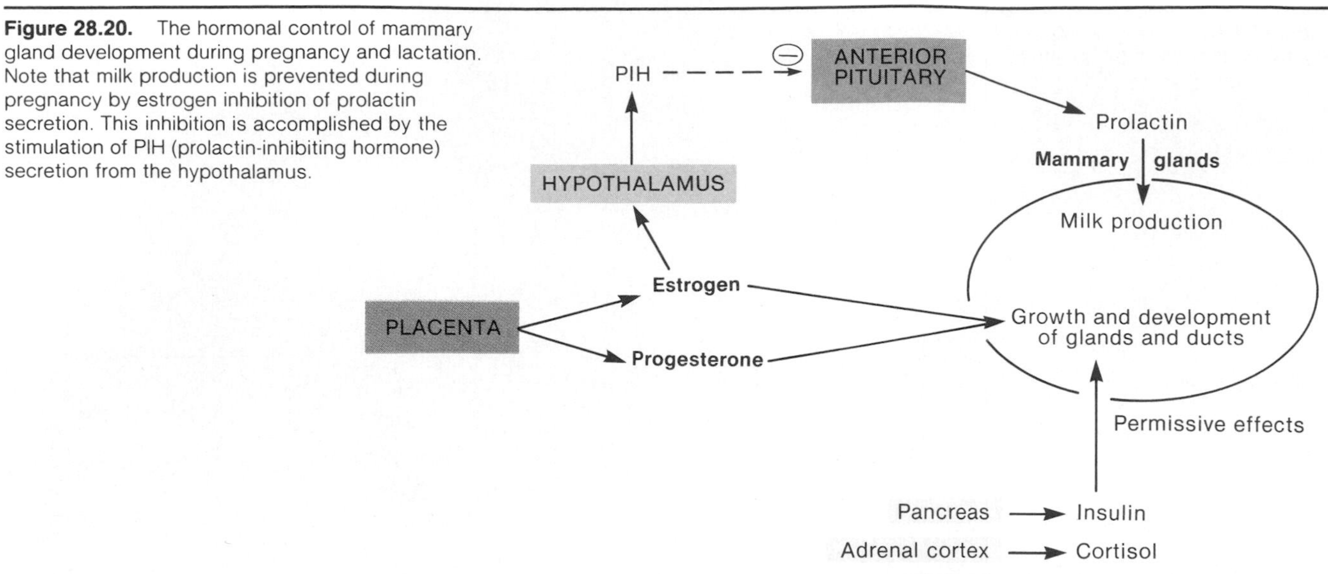

Table 28.4 Hormonal factors affecting lactation

Hormones	Major source	Effects
Insulin, cortisol, thyroid hormones	Pancreas, adrenal cortex, and thyroid	Permissive effects—adequate amounts of these must be present for other hormones to exert their effects on mammary glands
Estrogen and progesterone	Placenta	Growth and development of secretory units (alveoli) and ducts in mammary glands
Prolactin	Anterior pituitary	Production of milk proteins, including casein and lactalbumin
Oxytocin	Posterior pituitary	Stimulates milk-ejection reflex

Lymphatic drainage and the location of lymph nodes within the breast are of considerable clinical importance because of the frequency of *breast cancer* and the high incidence of metastases. About 75% of the lymph drains through the axillary tail of the breast into the axillary lymph nodes. Some 20% of the lymph passes toward the sternum to the internal thoracic lymph nodes. The remaining 5% of the lymph is subcutaneous and follows the lymph drainage pathway in the skin toward the back, where it reaches the intercostal nodes near the neck of the ribs.

Lactation

The changes that occur in the mammary glands during pregnancy and the regulation of lactation provide excellent examples of hormonal interactions and neuroendocrine regulation (table 28.4). Growth and development of the mammary glands during pregnancy requires the permissive actions of insulin, cortisol, and thyroid hormones; in the presence of adequate amounts of these hormones, high levels of estrogen stimulate the development of the mammary alveoli, and progesterone stimulates proliferation of the tubules and ducts (fig. 28.20).

The production of milk proteins, including casein and lactalbumin, is stimulated after parturition by **prolactin,** a hormone secreted by the anterior pituitary gland. The secretion of prolactin is controlled primarily by *prolactin-inhibiting hormone (PIH),* which is produced by the hypothalamus and secreted into the portal blood vessels. The secretion of PIH is stimulated by high levels of estrogen. During pregnancy, consequently, the high levels of estrogen prepare the breasts for lactation but prevent prolactin secretion.

After parturition, when the placenta is eliminated, declining levels of estrogen are accompanied by an increase in the secretion of prolactin. Lactation, therefore, commences. If a woman does not wish to breast-feed her baby, she may be injected with a powerful synthetic estrogen (diethylstilbestrol, or DES), which inhibits further prolactin secretion.

The act of nursing helps to maintain high levels of prolactin secretion via a *neuroendocrine reflex* (fig. 28.21). Sensory endings in the breast, activated by the stimulus of suckling, relay impulses to the hypothalamus and inhibit PIH secretion. The continued secretion of high levels of prolactin results in the secretion of milk from the alveoli into the ducts. In order for the baby to get the milk, however, the action of another hormone is needed.

Figure 28.21. Lactation occurs in two stages: milk production (stimulated by prolactin) and milk ejection (stimulated by oxytocin). The stimulus of suckling triggers a neuroendocrine reflex that results in increased secretion of oxytocin and prolactin.

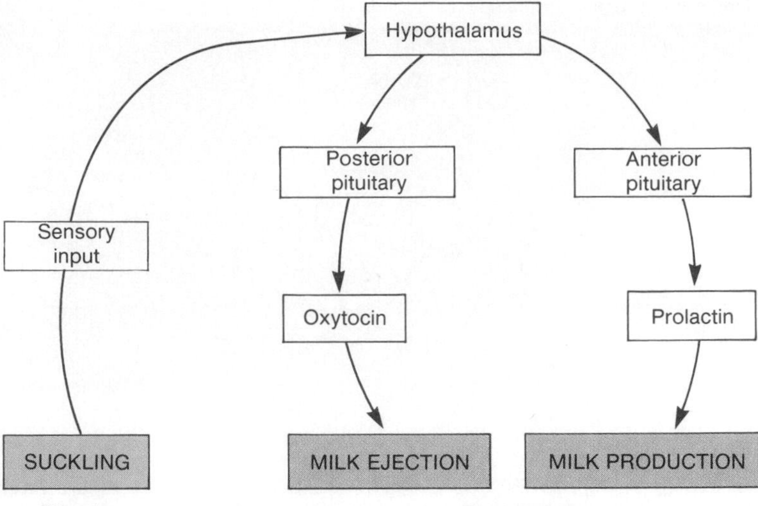

The stimulus of suckling also results in the reflex secretion of **oxytocin** from the posterior pituitary. This hormone is produced in the hypothalamus and stored in the posterior pituitary; its secretion results in the **milk-ejection reflex,** or milk let-down. Oxytocin stimulates contraction of the lactiferous ducts, as well as of the uterus (this is why women who breast-feed regain uterine muscle tone faster than women who bottle-feed).

For reasons that are poorly understood, breast-feeding often suppresses the release of FSH and LH from the hypothalamus. This inhibits ovulation for a period of months to a year or so, producing a natural spacing of births. In many women, however, suckling does not suppress FSH and LH secretion, and ovulation commences soon after parturition. For this reason, lactation is not regarded as a reliable means of birth control.

1. Describe the structure of the breasts and mammary glands, and explain why variations in breast size do not affect the ability to lactate.
2. Describe the structural changes in the mammary glands that occur during pregnancy, and explain how these changes are produced.
3. Describe the roles of prolactin and oxytocin in lactation, and explain how the secretion of these hormones is regulated.

Clinical Considerations

Females are more prone to dysfunctions and diseases of the reproductive organs than are males because of cyclic changes in reproductive events, problems associated with pregnancy, and the susceptibility of the breasts to infections and neoplasms. The termination of reproductive capabilities at menopause can also cause complications due to hormonal alterations. *Gynecology* is the specialty of medicine concerned with dysfunction and diseases of the female reproductive system, whereas *obstetrics* is the specialty dealing with pregnancy and childbirth. Frequently a physician will specialize in both obstetrics and gynecology (OBGYN).

There are numerous clinical aspects of the female reproductive system, but only the most important conditions will be discussed.

Diagnostic Procedures

A gynecological, or pelvic, examination is generally given in a thorough physical examination, especially prior to marriage, during pregnancy, or if problems involving the reproductive organs are suspected. In a gynecological examination, the physician inspects the vulva for irritations, lesions, or abnormal vaginal discharge, and palpates the vulva and internal organs. Most of the internal organs can be palpated through the vagina, especially if they are enlarged or tender. Inserting a lubricated *speculum* into the vagina allows visual examination of the cervix and vaginal walls. A speculum is an instrument for opening or distending a body opening to permit visual inspection.

In special cases, it may be necessary to examine the cavities of the uterus and uterine tubes by *hysterosalpingography (his''ter-o-sal''ping-gog'rah-fe)* (fig. 28.22). This technique involves injecting a radiopaque dye into the reproductive tract. The patency of the uterine tubes, irregular pregnancies, and various types of tumors may be detected using this technique. A *laparoscopy (lap''ah-ros'ko-pe)* permits *in vivo* visualization of the internal reproductive organs. The entrance for the laparoscope may be the umbilicus, a small incision in the lower abdominal

hysterosalpingograph: Gk. *hystera,* uterus; *salpinx,* trumpet (uterine tube); *graphein,* to record

Figure 28.22. A hysterosalpingogram showing the cavities of the uterus and uterine tubes.

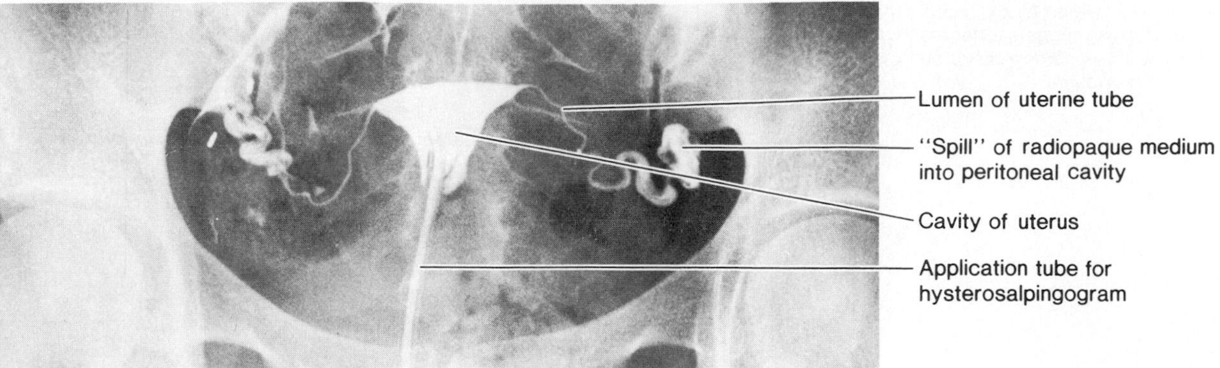

wall, or through the posterior fornix of the vagina into the rectouterine pouch. Although a laparoscope is used primarily in diagnosis, it can be used when performing a **tubal ligation** (fig. 28.23), which is a method of sterilizing a female.

One diagnostic procedure that should be routinely performed by a woman is a *breast self-examination* (BSE). The importance of a BSE is not to prevent diseases of the breast but to detect any problems before they become serious. A BSE should be performed monthly one week after the cessation of menstruation so that the breast will not be swollen or especially tender. The procedure for a BSE is presented in figure 28.24.

Another important diagnostic procedure is a *Papanicolaou (Pap) smear.* The Pap smear permits microscopic examination of cells obtained from near the external os of the cervix. Samples of cells are obtained by using a vaginal pipette or a specially designed wooden spatula. Women should have routine Pap smears for the early detection of cervical cancer.

Problems Involving the Ovaries and Uterine Tubes

Most **ovarian neoplasms** are nonmalignant ovarian cysts lined by cuboidal epithelium and filled with a serous albuminous fluid. These tumors may frequently be palpated during a gynecological examination and may require surgical removal if they exceed about 4 cm in diameter. They are generally removed as a precaution because it is impossible to determine by palpation whether the mass is malignant or benign.

Malignant ovarian tumors, which generally occur in women over the age of sixty, may reach massive size. Ovarian tumors of 5 kg (14 lbs) are not uncommon, and ovarian tumors of 110 kg (300 lbs) have been reported. Some ovarian tumors produce estrogen and thus cause feminization in elderly women, including recommencement of menstrual periods. The prognosis for women with ovarian tumors varies depending on the type of tumor and whether or not metastasis has occurred.

Two frequent problems involving the uterine tubes are salpingitis and ectopic pregnancies. **Salpingitis** is an inflammation of one or both uterine tubes. Infection of the uterine tubes is generally caused by venereal disease, although secondary bacterial infections from the vagina may also cause salpingitis. Salpingitis may cause sterility if the uterine tubes become occluded.

Figure 28.23. A simplified illustration of a tubal ligation of each uterine tube.

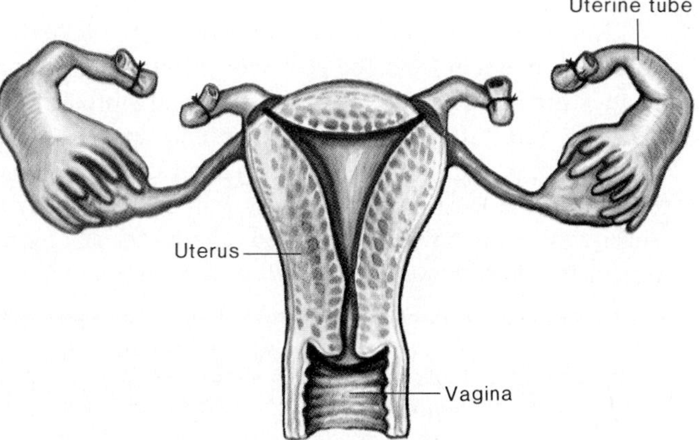

Ectopic pregnancy results from implantation of the blastocyst in a location other than the body or fundus of the uterus. The most frequent ectopic site is in the uterine tube, where an implanted blastocyst causes what is commonly called a **tubular pregnancy.** One danger of a tubular pregnancy is the enlargement, rupture, and subsequent hemorrhage of the uterine tube where implantation has occurred. A tubular pregnancy is frequently treated by removing the affected tube.

Infertility, or the inability to conceive, is a clinical problem that involves the male or female reproductive system and affects about 10% of couples. Generally, when a male is infertile, it is because of inadequate sperm counts. Female infertility is frequently caused by an obstruction of the uterine tubes or abnormal ovulation.

Problems Involving the Uterus

Abnormal menstruations are among the most common disorders of the female reproductive system. Abnormal menstruations may be directly related to problems of the reproductive organs and pituitary gland or associated with emotional and psychological stress.

Figure 28.24. The procedures for a breast self-examination (BSE).

How to examine your breasts

1

In the shower:

Examine your breasts during bath or shower; hands glide easier over wet skin. Fingers flat, move gently over every part of each breast. Use right hand to examine left breast, left hand for right breast. Check for any lump, hard knot or thickening.

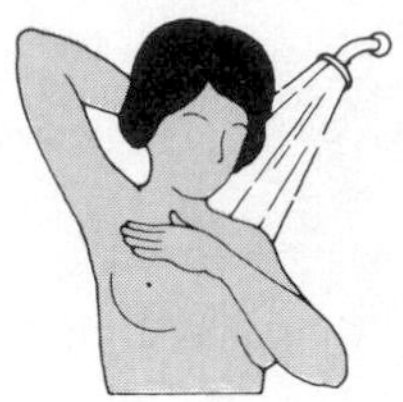

2

Before a mirror:

Inspect your breasts with arms at your sides. Next, raise your arms high overhead. Look for any changes in contour of each breast, a swelling, dimpling of skin or changes in the nipple.

Then, rest palms on hips and press down firmly to flex your chest muscles. Left and right breast will not exactly match—few women's breasts do.

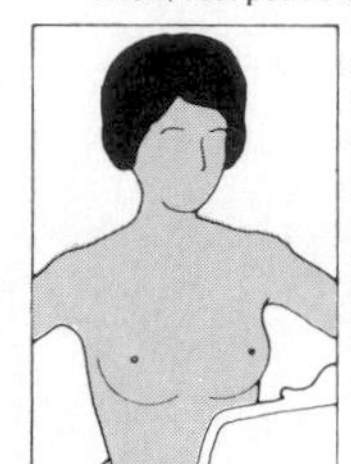

Regular inspection shows what is normal for you and will give you confidence in your examination.

3

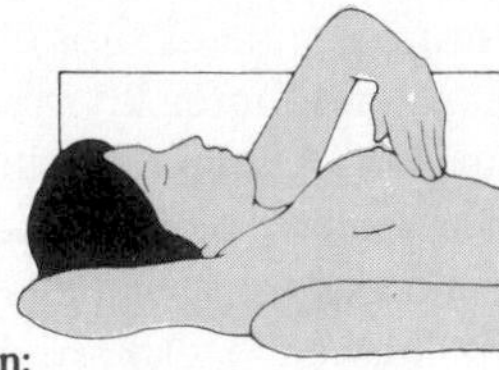

Lying down:

To examine your right breast, put a pillow or folded towel under your right shoulder. Place right hand behind your head—this distributes breast tissue more evenly on the chest. With left hand, fingers flat, press gently in small circular motions around an imaginary clock face. Begin at outermost top of your right breast for 12 o'clock, then move to 1 o'clock, and so on around the circle back to 12. A ridge of firm tissue in the lower curve of each breast is normal. Then move in an inch, toward the nipple, keep circling to examine *every part of your breast*, including nipple. This requires at least three more circles. Now slowly repeat procedure on your left breast with a pillow under your left shoulder and left hand behind head. Notice how your breast structure feels.

Finally, squeeze the nipple of each breast gently between thumb and index finger. Any discharge, clear or bloody, should be reported to your doctor immediately.

Amenorrhea *(ah-men''o-re'ah)* is the absence of menstruation and can be categorized as normal, primary, or secondary. **Normal amenorrhea** follows menopause, occurs during pregnancy, and in some women may occur during lactation. **Primary amenorrhea** means that a woman has never menstruated although she has passed the age when menstruation normally begins. Primary amenorrhea is generally accompanied by failure of the secondary sexual characteristics to develop. Endocrine disorders may cause primary amenorrhea and abnormal development of the ovaries or uterus.

Secondary amenorrhea is the cessation of menstruation in women who previously have had normal menstrual periods and are not pregnant and have not gone through menopause. Various endocrine disturbances as well as psychological factors cause secondary amenorrhea. It is not uncommon, for example, for young women who are in the process of making major changes or adjustments in their lives to miss menstrual periods. Secondary amenorrhea is also frequent in women athletes during periods of intense training. A low percentage of body fat may be a contributing factor. Sickness, fatigue, poor nutrition, or emotional stress also cause secondary amenorrhea.

Dysmenorrhea is painful or difficult menstruation accompanied by severe menstrual cramps. The causes of dysmenorrhea are not totally understood but may include endocrine disturbances (inadequate progesterone levels), a faulty position of the uterus, emotional stress, or some type of obstruction that prohibits menstrual discharge.

Abnormal uterine bleeding includes **menorrhagia** *(men''o-ra'je-ah),* or excessive bleeding during the menstrual period, and **metrorrhagia,** or spotting between menstrual periods. Other types of abnormal uterine bleeding are menstruations of excessive duration, too frequent menstruations, and postmenopausal bleeding. These abnormalities may be caused by hormonal irregularities, emotional factors, or various diseases and physical conditions.

Uterine neoplasms are an extremely common problem of the female reproductive tract. Most of the neoplasms are benign and include cysts, polyps, and smooth muscle tumors (leiomyoma). Any of these conditions may provoke irregular menstruations and may cause infertility if the neoplasms are massive.

Cancer of the uterus is the most common malignancy of the female reproductive tract. The most common site of uterine cancer is the cervix (fig. 28.25). Cervical cancer is second only to cancer of the breast in frequency of occurrence and is a disease of young women (ages thirty through fifty), especially those who have had frequent intercourse with multiple partners during their teens and onward. If detected early through regular Pap smears, the disease can be cured before it metastasizes. The treatment of cervical cancer depends on the stage of the malignancy and the age and health of the woman. In the case of women who are through having children, a **hysterectomy** *(his''tĕ-rek'to-me)* (surgical removal of the uterus) is usually performed.

neoplasm: Gk. *neos,* new; *plasma,* something formed

Endometriosis is a condition characterized by the presence of endometrial tissues at sites other than the inner lining of the uterus. Frequent sites of ectopic endometrial cells are on the ovaries, outer layer of the uterus, abdominal wall, and urinary bladder. Although it is not certain how endometrial cells become established outside the uterus, it is speculated that some discharged endometrial tissue might be flushed backward from the uterus and through the uterine tubes during menstruation. Women with endometriosis will bleed internally with each menstrual period because the ectopic endometrial cells are stimulated along with the normal endometrium by ovarian hormones. The most common symptoms of endometriosis are extreme dysmenorrhea and a feeling of fullness during each menstruation period. Endometriosis can cause infertility. It is most often treated by suppressing the endometrial tissues with oral contraceptive pills or by surgery. An oophorectomy, or removal of the ovaries, may be necessary in extreme cases.

Uterine displacements are relatively common in elderly women. When uterine displacements occur in younger women, they are important because of dysmenorrhea, fertility, or parturition problems. **Retroversion,** or **retroflexion,** is a displacement backward; **anteversion,** or **anteflexion,** is displacement forward. **Prolapse of the uterus** is a marked downward displacement into the vagina.

An **abortion** is defined as the termination of a pregnancy before the twenty-eighth week of gestation. A **spontaneous abortion,** or **miscarriage,** occurs without mechanical aid or medicinal intervention and may happen in as many as 10% of all pregnancies. Spontaneous abortions usually occur when there is abnormal development of the fetus or disease of the maternal reproductive system. An **induced abortion** is removal of the fetus from the uterus by mechanical means or drugs. Induced abortions are a major issue of controversy because of questions regarding individual rights—those of the mother and those of the fetus—the definition of life, and moral concerns.

Diseases of the Vagina and Vulva

Pelvic inflammatory disease (PID) is a general term for inflammation of the female reproductive organs within the pelvis. The infection may be confined to a single organ, or it may involve all the internal reproductive organs. The pathogens generally enter through the vagina during coitus, induced abortion, childbirth, or postpartum. Inflammation of the ovaries is called **oophoritis,** and inflammation of the uterine tube is **salpingitis** *(sal''pin-ji'tis).*

The vagina and vulva are generally resistant to infection because of the acidity of the vaginal secretions. Occasionally, however, localized infections and inflammations do occur; these are termed **vaginitis,** if confined to the vagina, or **vulvovaginitis,** if both the vagina and external genitalia are affected. The symptoms of vaginitis are a discharge of pus *(leukorrhea)* and itching *(pruritus).* The two most common organisms that cause vaginitis are the protozoan *Trichomonas vaginalis* and the fungus *Candida albicans.*

Figure 28.25. Sites of various conditions and diseases of the female reproductive tract, each of which could cause an abnormal discharge of blood.

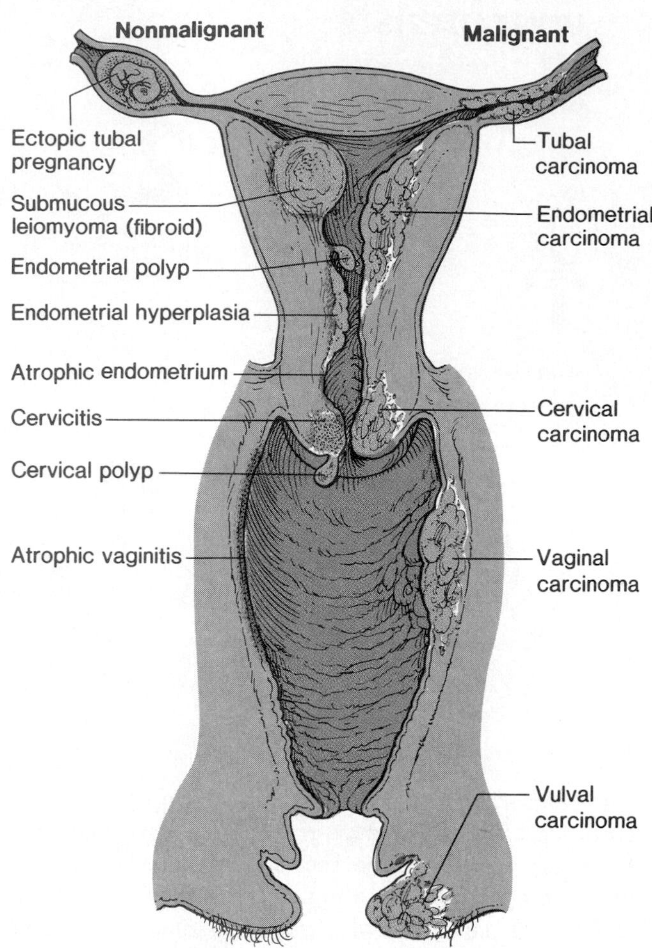

Diseases of the Breasts and Mammary Glands

The breasts and mammary glands of females are highly susceptible to infections, cysts, and tumors. Infections involving the mammary glands usually follow the development of a dry and cracked nipple during lactation. Bacteria enter the wound and establish an infection within the lobules of the gland. During an infection of the mammary gland, a blocked duct frequently causes a lobe to become engorged with milk. This localized swelling is usually accompanied by redness, pain, and an elevation of temperature. Administering specific antibiotics and applying heat are the usual treatments.

Nonmalignant cysts are the most frequent diseases of the breast. These masses are generally of two types, neither of which is life threatening.

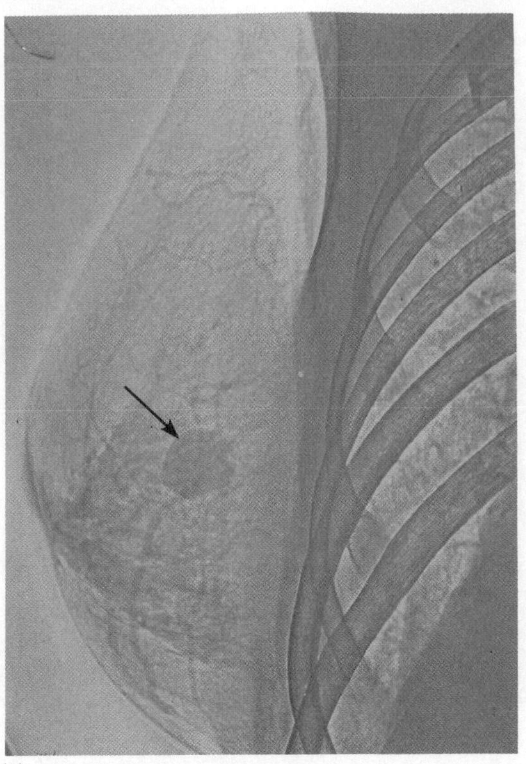

(a)

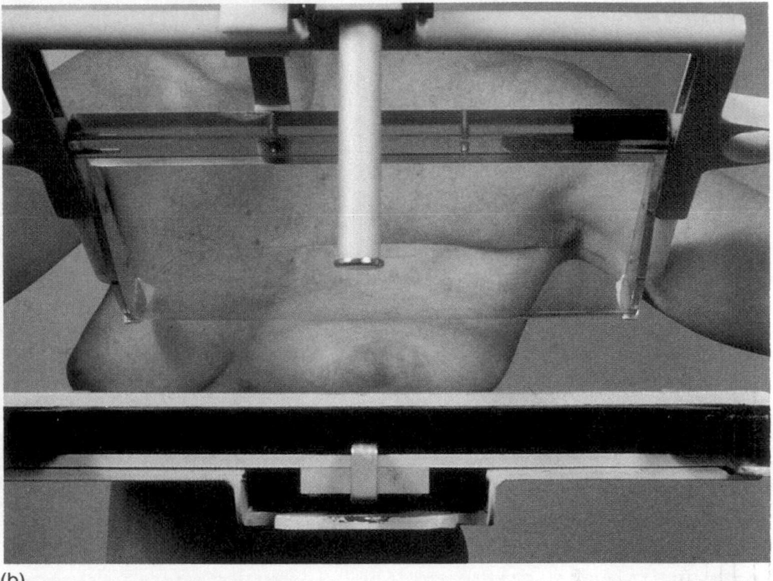

(b)

Figure 28.26. (*a*) a mammogram of a patient with carcinoma of the upper breast. Note the presence of a neoplasm indicated with an arrow. (*b*) in mammography, the breasts are placed alternately on a metal plate and x-rayed from the side and from above.

Dysplasia (fibrocystic disease) is a broad condition involving several nonmalignant diseases of the breast. All dysplasias are benign neoplasms of various sizes that may become painful during or prior to menstruation. Most of the masses are small and remain undetected. Dysplasia affects nearly one-half of women over thirty years of age prior to menopause.

Fibroadenoma *(fi''bro-ad''ĕ-no'mah)* is a benign tumor of the breast that frequently occurs in women under the age of thirty-five. Fibroadenomas are nontender, rubbery masses, which are easily moved about in the mammary tissue. A fibroadenoma can be excised in a physician's office under local anesthetics.

Carcinoma of the breast is the most common malignancy in women. One in thirteen women, or about 7%, will develop breast cancer, and one-third of these will die from the disease. Breast cancer is the leading cause of death in women between forty and fifty years of age. Men are also susceptible to breast cancer, but it is one hundred times more frequent in women. Breast cancer in men is usually fatal.

The causes of breast cancer are not known, but women who are most susceptible are those who are over age thirty-five, who have a family history of breast cancer, and who are nulliparus (no children). The early detection of breast cancer is important because the progressed state of the disease will determine the treatment and prognosis.

Confirming suspected breast cancer generally requires *mammography* (fig. 28.26). If the mammograph indicates breast cancer, surgery is performed so that a biopsy can be obtained and the tumor assessed. If the tumor is found to be malignant, extensive surgery is performed, depending on the size of the tumor and whether metastasis has occurred. The surgical treatment for breast cancer is generally some degree of *mastectomy.* A *simple mastectomy* is removal of the entire breast but not the underlying lymph nodes. A *modified radical mastectomy* is the complete removal of the breast, the lymphatic drainage, and perhaps the pectoralis major muscle. A *radical mastectomy* is similar to a modified except that the pectoralis major muscle is always removed as well as the axillary lymph nodes and adjacent connective tissue.

Methods of Contraception

In addition to the rhythm method and the contraceptive (birth-control) pill, which were discussed in a previous section, contraception may be accomplished by sterilization, intrauterine devices (IUDs), and barrier methods—including condoms for the male and diaphragms for the female (fig. 28.27).

Sterilization techniques include *vasectomy* for the male and *tubal ligation* for the female (fig. 28.23). In the latter technique (which currently accounts for over 60% of sterilization procedures performed in the United States),

Figure 28.27. Various types of birth control devices. (*a*) IUD; (*b*) contraceptive sponge; (*c*) diaphragm; (*d*) birth control pills; (*e*) vaginal spermicide; (*f*) condom.

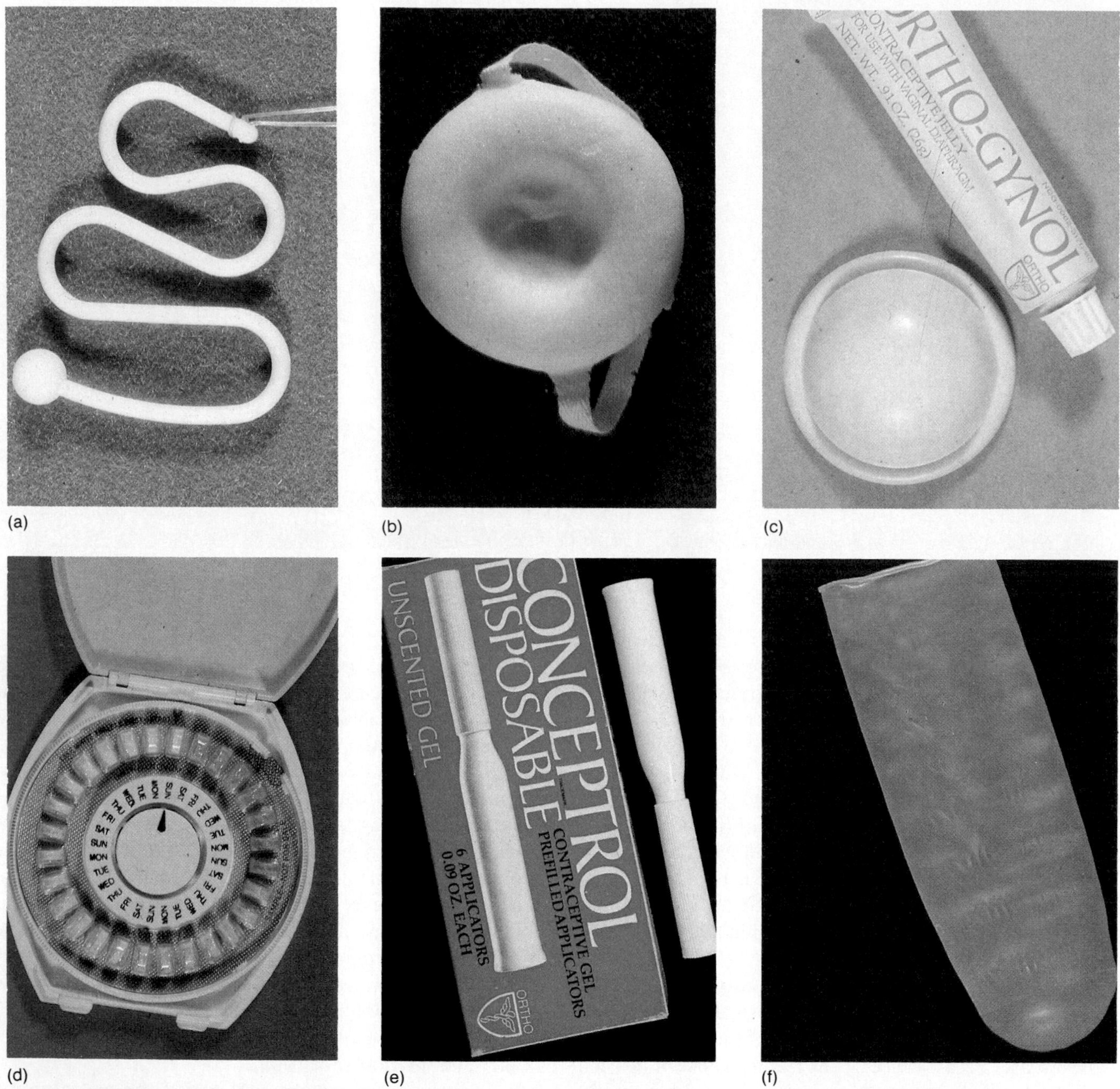

the uterine tubes are cut and tied. This is analogous to the procedure performed on the ductus (vas) deferens in a vasectomy and prevents fertilization of the ovulated ovum.

Studies on the long-term effects of these procedures have failed to show deleterious side effects. With current procedures, tubal ligations (as well as vasectomies) should be considered essentially irreversible.

Intrauterine devices (IUDs) don't prevent ovulation but instead prevent implantation of the embryo into the uterus in the event fertilization occurs. The mechanisms by which these contraceptive effects are produced are not well understood, but the efficiency of different IUDs appears to be related to their ability to cause inflammatory reactions in the uterus. Uterine perforations are the foremost complication of the use of IUDs.

Barrier contraceptives—condoms and diaphragms—are slightly less effective than the other methods of contraception, but they do not cause serious side effects. The failure rate of condoms—one of the oldest methods of contraception—is 12 to 20 pregnancies per 100 woman years of use, whereas the failure rate of the diaphragm is 12 to 18 pregnancies per 100 woman years. Latex condoms offer an additional benefit: they provide some protection against sexually transmitted diseases, including AIDS.

Chapter Summary

I. Components and Structures of the Female Reproductive System
  A. The female secondary sex organs include the vagina, uterine tubes, uterus, and mammary glands.
  B. In embryonic development, the paramesonephric (müllerian) ducts form the uterus and uterine tubes; the clitoris is homologous to the penis, and the labia majora are homologous to the scrotum.
  C. The uterine (fallopian) tubes end in fimbria, which project over the ovary and help direct a secondary oocyte into the infundibulum of the tube.
  D. The uterus contains a fundus, body, and cervix, and is supported by four pairs of ligaments.
    1. The wall of the uterus consists of a perimetrium, a muscular myometrium, and an epithelial lining called the endometrium.
    2. The endothelium is stratified, and the layers are divided into a stratum basale and a stratum functionale; the latter is shed during menstruation and rebuilt during the next cycle.
  E. The vagina opens to the cervix of the uterus; the vulva is the external genitalia.
  F. During sexual excitement the clitoris and erectile tissue of the areola of the breasts swell with blood; during orgasm there are rhythmic contractions of the muscles of the perineum, uterus, and uterine tubes.

II. Ovaries and the Ovarian Cycle
  A. The ovaries are supported by the mesovarium, which extends from the broad ligament, and by the ovarian and suspensory ligaments.
  B. Primary oocytes, arrested at prophase I of the first meiotic division, are contained within primordial follicles in the ovary.
    1. Upon stimulation by gonadotropic hormones, granulosa cells divide and fill the follicle, forming a primary follicle.
    2. Upon further stimulation, a fluid-filled cavity called the antrum forms to produce a secondary follicle.
    3. As a secondary follicle develops, the primary oocyte completes its first meiotic division to form a secondary oocyte, arrested at metaphase II, and a polar body.
    4. A single, fully mature follicle called the vesicular ovarian (graafian) follicle releases its secondary oocyte at ovulation; the empty follicle then becomes a corpus luteum.
  C. The pituitary secretes FSH and LH in response to the pulsatile secretion of GnRH from the hypothalamus; secretion of FSH and LH, as well as GnRH, is modified by feedback from sex steroids secreted by the ovaries.

III. Menstrual Cycle
  A. The ovarian cycle can be divided into follicular and luteal phases, which are separated by the event of ovulation.
    1. During the follicular phase, FSH stimulates the growth and development of follicles; this is accompanied by increasing secretion of estradiol from the granulosa cells of the follicles.
    2. Estradiol exerts a positive feedback effect on the hypothalamus and pituitary, resulting in an LH surge that triggers ovulation.
    3. At ovulation, the vesicular ovarian follicle breaks and its secondary oocyte is released out of the ovary.
    4. The empty vesicular ovarian follicle becomes a corpus luteum, which secretes estradiol and progesterone.
    5. Estradiol and progesterone inhibit FSH and LH secretion during the luteal phase.
    6. The corpus luteum regresses at the end of the cycle, and the resulting decline in estradiol and progesterone secretion causes menstruation.
  B. In terms of the changes that occur in the endometrium, the cycle can be described by three phases: proliferative phase (following menstruation), secretory phase (corresponding to the luteal phase of the ovaries), and menstrual phase.
  C. The contraceptive pill acts by duplicating the negative feedback inhibition of FSH and LH by estradiol and progesterone that normally occurs during the luteal phase.

IV. Mammary Glands and Lactation
  A. Mammary glands consist of secretory alveoli that drain into lactiferous ducts, which in turn open into the tip of the nipple of the breast.
  B. Prolactin stimulates the production of milk proteins; secretion of oxytocin in response to a neuroendocrine reflex stimulates contraction of the lactiferous ducts and ejection of milk from the nipple.

Review Activities

Objective Questions

Match the following:

1. High estradiol and progesterone; low FSH and LH
2. Low estradiol; no progesterone
3. LH surge
4. Estradiol and progesterone begin to decline
   (a) early follicular phase
   (b) late follicular phase
   (c) early luteal phase
   (d) late luteal phase
5. The secretory phase of the endometrium corresponds to which of the following ovarian phases?
   (a) follicular phase
   (b) ovulation
   (c) luteal phase
   (d) menstrual phase
6. Which of the following statements about oogenesis is *true?*
   (a) Oogonia form continuously in postnatal life.
   (b) Primary oocytes are haploid.
   (c) Meiosis is completed prior to ovulation.
   (d) A secondary oocyte is released.
7. The paramesonephric (müllerian) ducts give rise to the
   (a) uterine tubes (d) both (a) and (b)
   (b) uterus (e) both (b) and (c)
   (c) pudendum
8. In a female, the homologous structure to the male scrotum is the
   (a) labia majora (c) clitoris
   (b) labia minora (d) vestibule
9. The cervix is a portion of the
   (a) vulva (c) uterus
   (b) vagina (d) uterine tubes
10. Fertilization normally occurs in the
   (a) ovaries (c) uterus
   (b) uterine tubes (d) vagina
11. Contractions of the mammary ducts are stimulated by
   (a) prolactin (c) estrogen
   (b) oxytocin (d) progesterone
12. The suspensory ligaments (of Cooper) support the
   (a) ovaries
   (b) uterus
   (c) uterine tubes
   (d) breasts

Essay Questions

1. Explain how the genital ducts develop and why the external genitalia of males and females are considered homologous.
2. Describe the gross and histologic structure of the uterus, and explain the significance of the strata functionale and basale of the endometrium.
3. Explain the hormonal interactions that control ovulation and make it occur at the proper time.
4. Compare menstrual bleeding and bleeding that occurs during the estrus cycle of a dog, in terms of hormonal control mechanisms and the ovarian cycle.
5. "The contraceptive pill tricks the brain into thinking you're pregnant." Interpret this popularized explanation in terms of physiological mechanisms.

29

Prenatal Development and Inheritance

Outline

Concepts

The fertilization of a secondary oocyte by a sperm in the uterine tube promotes the completion of meiotic development and the formation of a diploid zygote.

The events of the two-week pre-embryonic stage include fertilization, transportation of the zygote through the uterine tube, mitotic divisions, implantation, and the formation of primordial embryonic tissue.

The events of the six-week embryonic stage include the differentiation of the germ layers into specific body organs and the formation of the placenta, the umbilical cord, and the extraembryonic membranes, which provide sustenance and protection to the embryo.

The fetal stage lasts from nine weeks until birth and is characterized by tremendous growth and the specialization of body structures.

Labor and parturition are the culmination of gestation and require the action of oxytocin secreted by the posterior pituitary, and prostaglandins, produced in the uterus.

Inheritance is the passage of hereditary traits carried on the genes of chromosomes from one generation to another.

Figure 29.1. The process of fertilization. (*a, b*) diagrammatic representations. As the head of the sperm encounters the gelatinous corona radiata of the egg, which has progressed in meiotic development to the secondary oocyte stage (*2*), the acrosomal vesicle ruptures and the sperm digests a path for itself by the action of enzymes released from the acrosome (*3, 4*). When the cell membrane of the sperm contacts the cell membrane of the egg (*5*), they become continuous, and the sperm nucleus and other contents move into the egg cytoplasm. (*c*) a scanning electron micrograph of sperm bound to the egg surface.

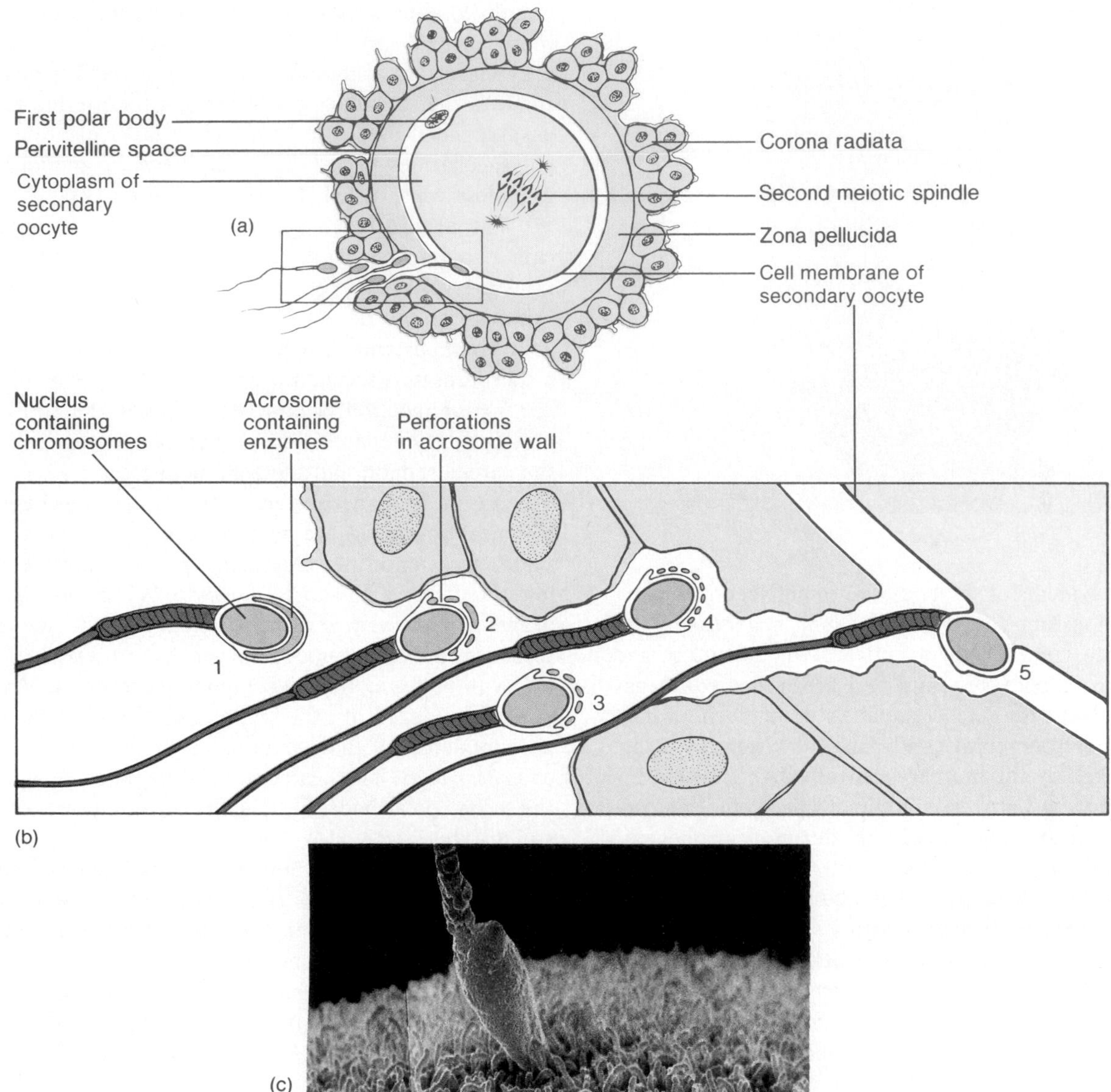

Fertilization

The fertilization of a secondary oocyte by a sperm in the uterine tube promotes the completion of meiotic development and the formation of a diploid zygote.

Objective 1. Define *capacitation, fertilization,* and *morphogenesis.*

Objective 2. Describe the changes that occur in the sperm and ovum during fertilization.

The structure of the human body forms before birth, or *prenatally,* through a process called **morphogenesis** *(mor''fo-jen'ĕ-sis).* Through morphogenic events the

morphogenesis: Gk. *morphe,* form; *genesis,* beginning

Figure 29.2. An electron micrograph showing the head of a human sperm with its nucleus and acrosomal cap.

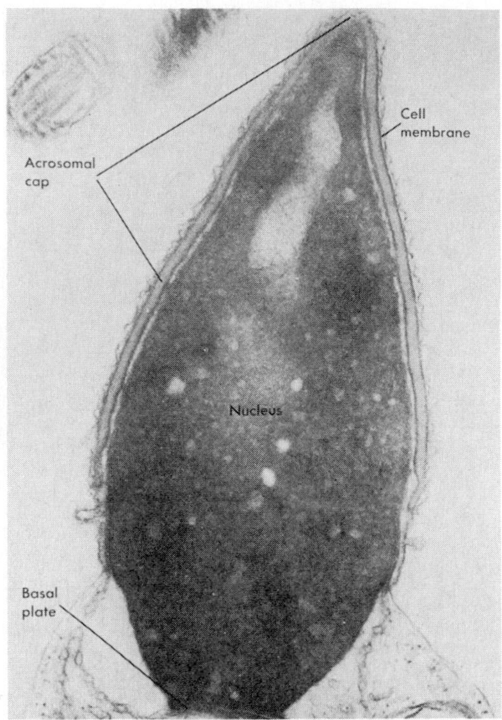

organs and systems of the body are established in a functional relationship. Understanding morphogenesis is important for a complete perspective of the structure and function of each body system. There are "sensitive" stages of morphogenesis for each organ and system, during which genetic or environmental conditions (nutrition, smoking, or drugs taken by the mother) may affect the normal development of the baby. Many clinical problems are congenital in nature and originate during morphogenic development.

The prenatal development of a human is a fascinating and awesome event. It begins with a single fertilized egg and culminates some thirty-eight weeks later with a complex organization of over 200 million cells. Prenatal development can be divided into a *pre-embryonic stage,* which is initiated by the fertilization of a secondary oocyte; an *embryonic stage,* during which the body's organ systems are formed; and a *fetal stage,* which culminates in parturition, or the birth of the baby.

During coitus (sexual intercourse), a male ejaculates between 100 million and 500 million sperm into the female's vagina. This tremendous number is needed because of the high rate of sperm fatality—only about 100 survive to contact the secondary oocyte in the uterine tube. During passage through the acidic female tract, sperm gain the ability to fertilize a secondary oocyte through a process called **capacitation.** The changes that occur in capacitation are not fully understood. Experiments have shown, however, that freshly ejaculated sperm are infertile and must be in the female tract for at least seven hours before they can fertilize a secondary oocyte.

A woman usually ovulates one secondary oocyte a month, totaling approximately four hundred during her reproductive years. Each ovulation releases a secondary oocyte arrested at metaphase of the second meiotic division. As the secondary oocyte enters the uterine tube, it is surrounded by a thin transparent layer of protein and polysaccharides, called the *zona pellucida,* and a layer of granulosa cells, called the *corona radiata* (fig. 29.1).

The head of each sperm is capped by an organelle called an **acrosome** (figs. 29.1, 29.2). The acrosome contains a trypsinlike protein-digesting enzyme and *hyaluronidase (hi''ah-lu-ron'i-das),* which digests hyaluronic acid, an important constituent of connective tissue. When a sperm meets a secondary oocyte in the uterine tube, an *acrosomal reaction* occurs that exposes the acrosome's digestive enzymes and allows a sperm to penetrate through the corona radiata and the zona pellucida.

As the first sperm penetrates the zona pellucida, a chemical change in the zona occurs that prevents other sperm from entering. Only one sperm, therefore, is allowed to fertilize a secondary oocyte. When the cell membranes of the sperm and the secondary oocyte merge, the secondary oocyte becomes an **ovum.** As fertilization occurs, the ovum is stimulated to complete its second meiotic division (fig. 29.3). Like the first meiotic division, the second produces one cell that contains all of the cytoplasm and one *polar body.* The healthy cell is the mature ovum, and the second polar body, like the first, ultimately fragments and disintegrates.

At fertilization, the entire sperm enters the cytoplasm of the much larger ovum (fig. 29.4). Within twelve hours the nuclear membrane in the ovum disappears, and the *haploid number* of chromosomes (twenty-three) in the ovum is joined by the haploid number of chromosomes from the sperm. A fertilized egg, or zygote *(zi'gōt),* containing the *diploid number* of chromosomes (forty-six) is thus formed.

---

capacitation: L. *capacitas,* capable of
zona pellucida: L. *zone,* a girdle; *pellis,* skin
corona radiata: Gk. *korone,* crown; *radiata,* radiate
acrosome: Gk. *akron,* extremity; *soma,* body
haploid: Gk. *haplous,* single; L. *ploideus,* multiple in form
diploid: Gk. *diplous,* double; L. *ploideus,* multiple in form

Figure 29.3. A secondary oocyte, arrested at metaphase II of meiosis, is released at ovulation. If this cell is fertilized, it becomes an ovum, completes its second meiotic division, and produces a second polar body.

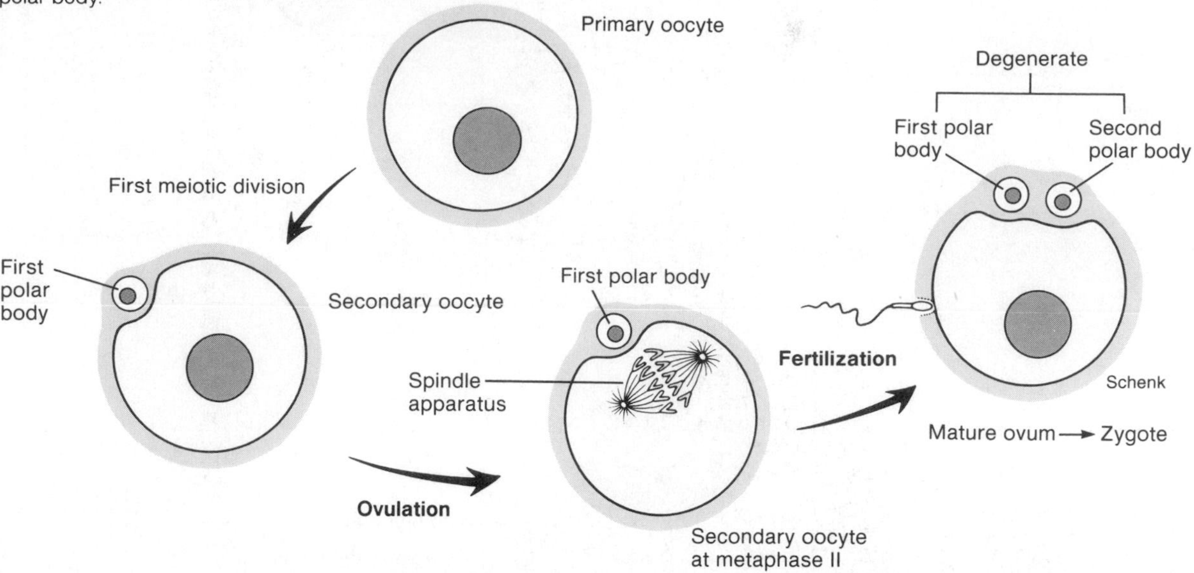

Figure 29.4. Fertilization and the union of chromosomes from the sperm and ovum to form the zygote. (*a*) sperm penetration; (*b*) haploid number of chromosomes within the nucleus of each sex cell; (*c*) degeneration of nuclear membranes; and (*d*) matching and alignment of chromosomes.

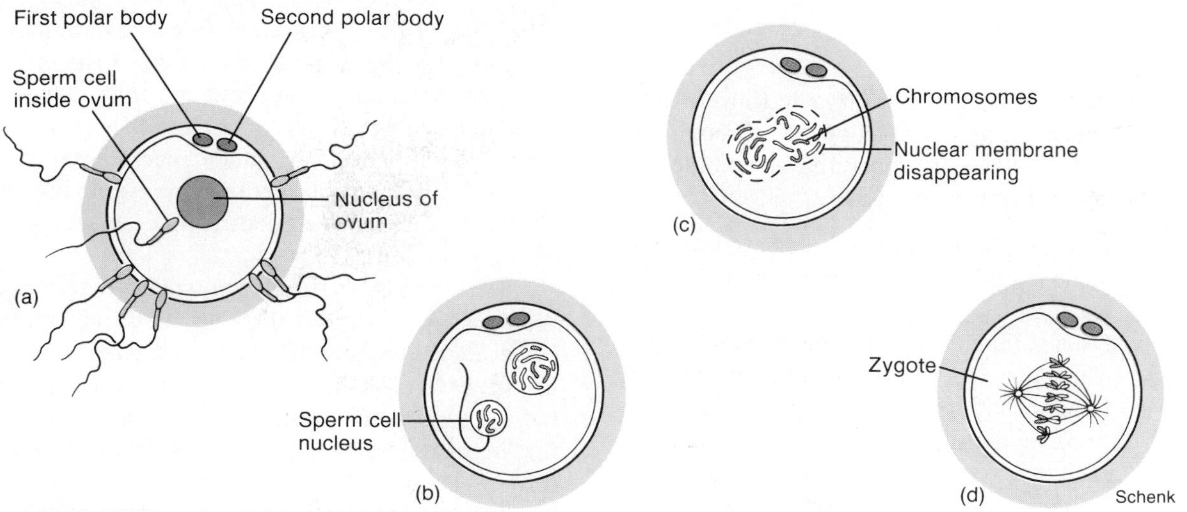

A secondary oocyte that is ovulated but not fertilized does not complete its second meiotic division, but instead it disintegrates twelve to twenty-four hours after ovulation. Fertilization therefore cannot occur if coitus takes place beyond one day following ovulation. Sperm, in contrast, can survive up to three days in the female reproductive tract. Fertilization therefore can occur if coitus is performed within three days prior to the day of ovulation.

1. Explain why capacitation and the acrosomal reaction of sperm are necessary to accomplish fertilization of a secondary oocyte.
2. Define the term *morphogenesis*, and discuss its importance in the development of a healthy baby.
3. Discuss the changes that occur in a sperm from the time of ejaculation to the time of fertilization. What changes occur in a secondary oocyte following ovulation to the time of fertilization?

Figure 29.5. Sequential illustrations from the first cleavage of the zygote to the formation of the blastocyst. Note the deterioration of the zona pellucida in the early blastocyst.

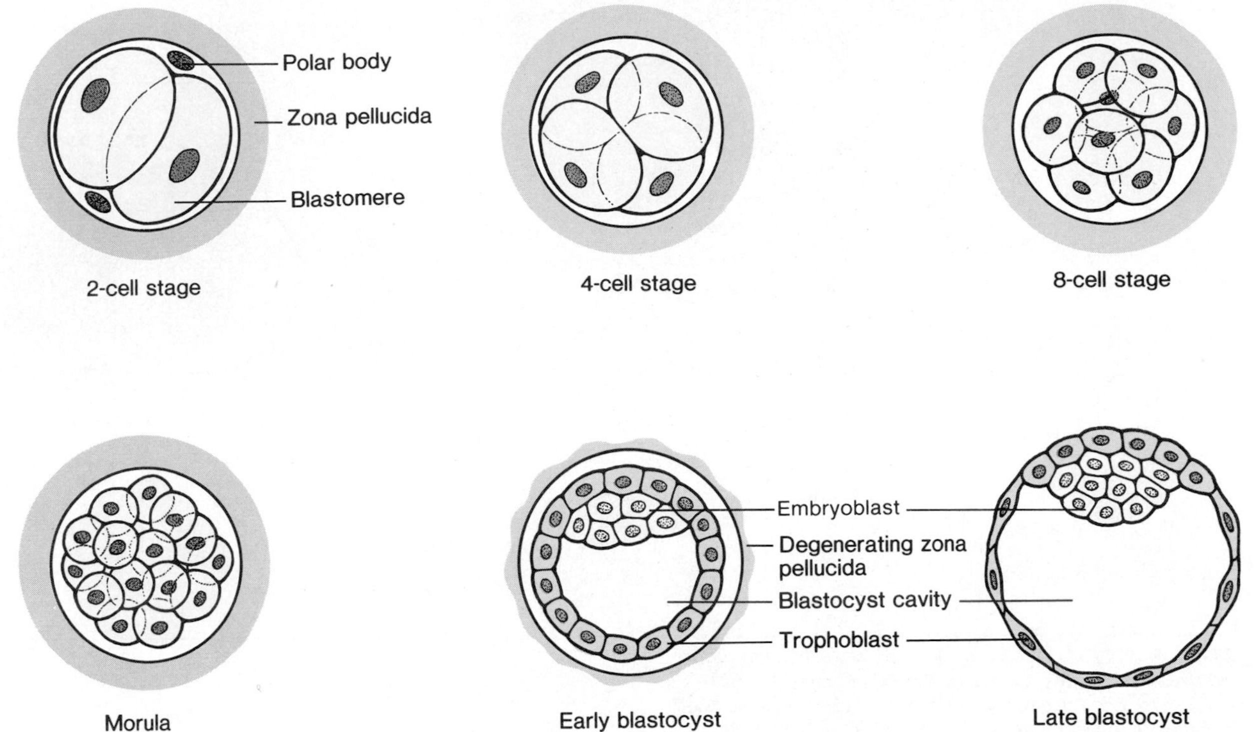

Pre-Embryonic Stage

The events of the two-week pre-embryonic stage include fertilization, transportation of the zygote through the uterine tube, mitotic divisions, implantation, and the formation of primordial embryonic tissue.

Objective 3. Describe the events of pre-embryonic development that result in the formation of the blastocyst.

Objective 4. Discuss the role of the trophoblast in the implantation and development of the placenta.

Objective 5. Explain how the primary germ layers develop, and list the structures produced by each layer.

Objective 6. Define *gestation,* and explain how the parturition date is determined.

Cleavage and Formation of the Blastocyst

Fertilization occurs within the uterine tube, usually about twelve to twenty-four hours following ovulation. The fertilized egg (ovum) is referred to as a **zygote.** Within thirty hours the **cleavage** process begins with a mitotic division that results in the formation of two identical daughter cells called *blastomeres* (figs. 29.5, 29.6). Several more cleavages occur as the structure passes down the uterine tube and enters the uterus on about the third day. It is now composed of a ball of sixteen or more cells called a **morula.** Although the morula has undergone several mitotic divisions, it is not much larger than the zygote because there have not been additional nutrients necessary for growth entering the cells.

The developing structure remains unattached in the uterine cavity for about three days. During this time, the center of the morula fills with fluid passing in from the uterine cavity. As the fluid-filled space develops inside the morula, two distinct groups of cells form, and the structure becomes known as a **blastocyst.** The single outer layer forming the wall of the blastocyst is known as the **trophoblast,** whereas the small, inner aggregation of cells is called the **embryoblast,** or *inner cell mass.* With further development the trophoblast differentiates into a structure called the **chorion,** which will become a portion of the placenta, and the embryoblast will become the embryo. The hollow, fluid-filled center of the blastocyst is called the **blastocyst cavity.** A diagrammatic summary of the ovarian cycle, fertilization, and the morphogenic events of the first week is presented in figure 29.7.

zygote: Gk. *zygotos,* yolked, joined

Figure 29.6. Stages of pre-embryonic development of a human ovum fertilized in a laboratory (*in vitro*) as seen in scanning electron micrographs. (*a*) four-cell stage; (*b*) cleavage at the sixteen-cell stage; (*c*) a morula; and (*d*) a blastocyst.

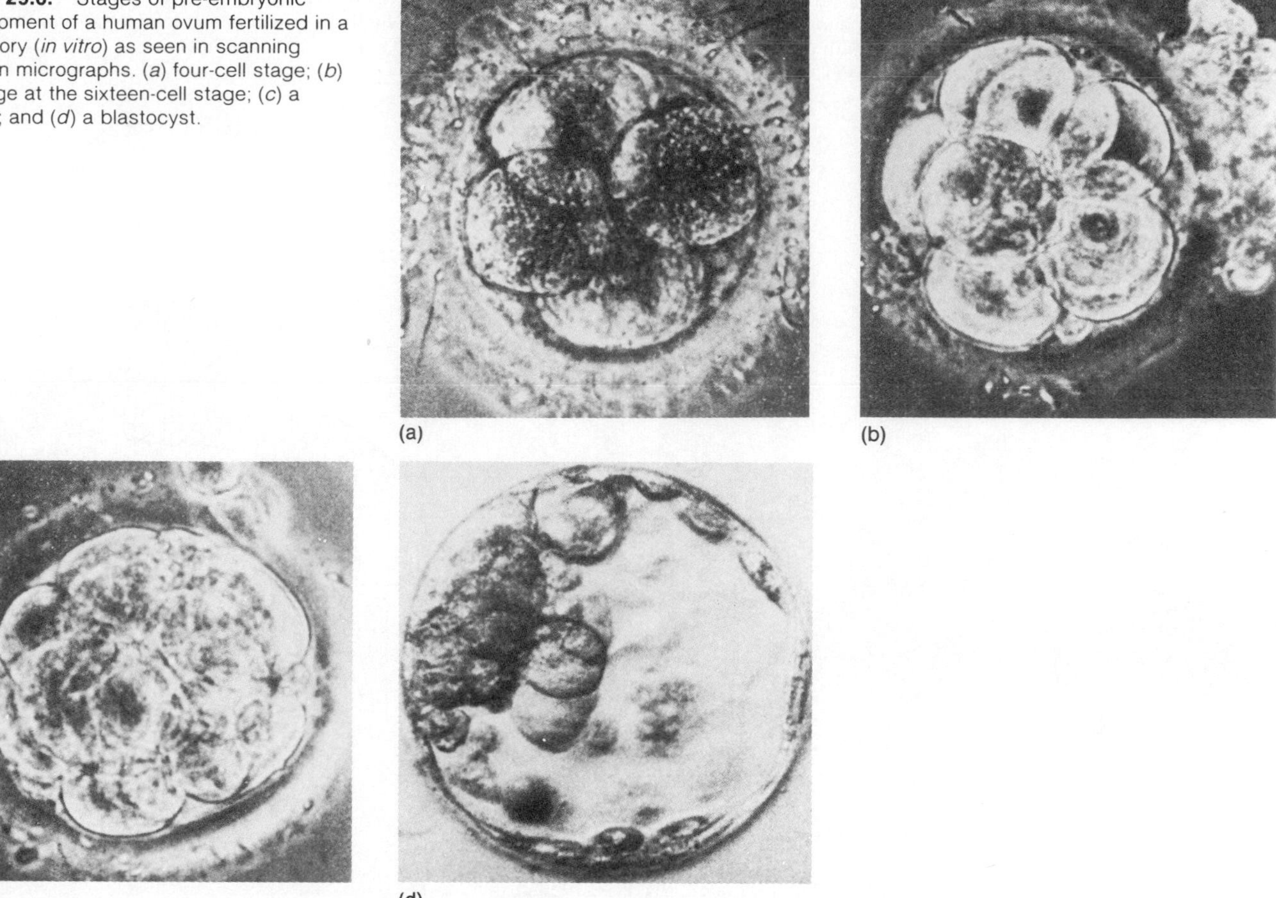

Figure 29.7. A diagrammatic representation of the ovarian cycle, fertilization, and the morphogenic events of the first week. The numbers indicate the days after fertilization. Implantation of the blastocyst begins between the fifth and seventh day and is generally completed by the tenth day.

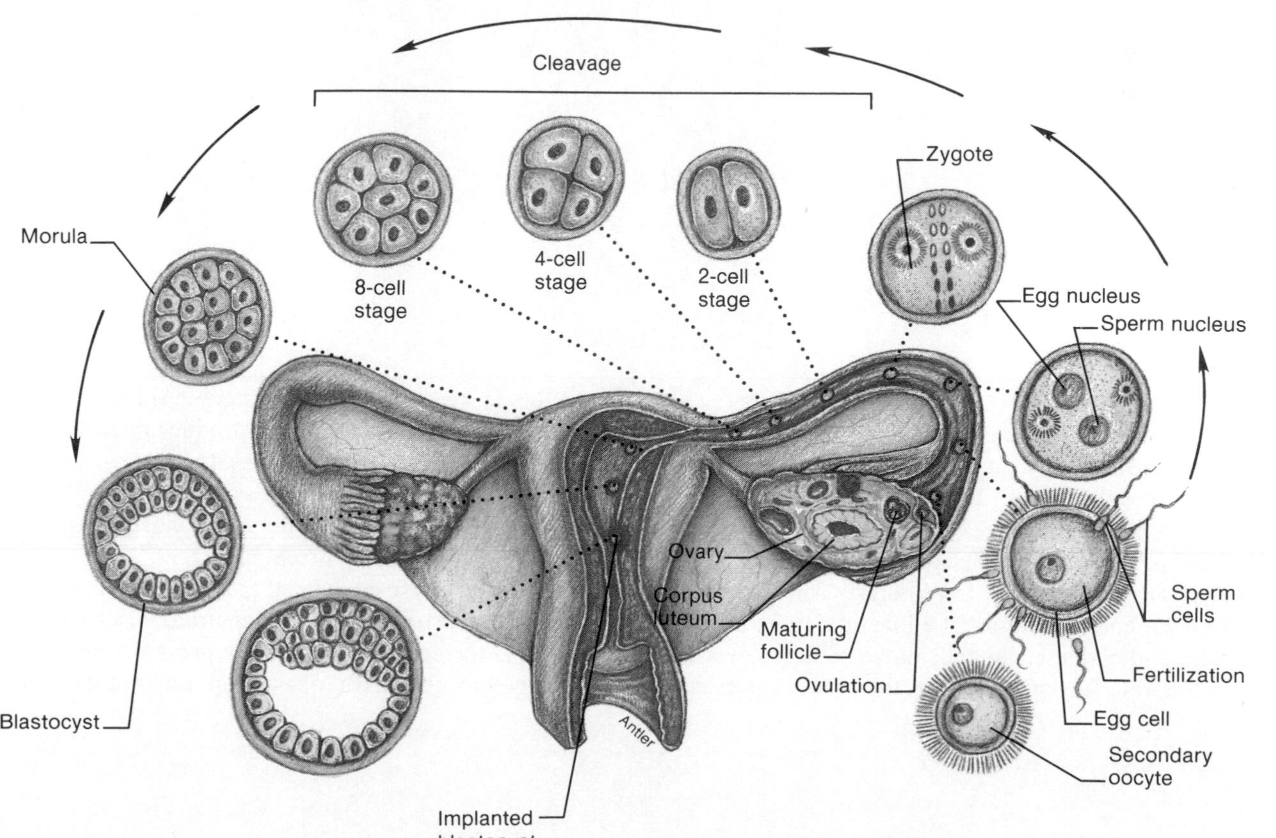

Figure 29.8. The blastocyst adheres to the endometrium on about the sixth day as seen in (*a*) a photomicrograph and (*b*) a diagram of the specific cells of the blastocyst. By the seventh day (*c*), specialized syncytiotrophoblasts from the trophoblast invade the endometrium. The syncytiotrophoblasts secrete hCG to sustain pregnancy and will eventually participate in the formation of the placenta for embryonic and fetal sustenance.

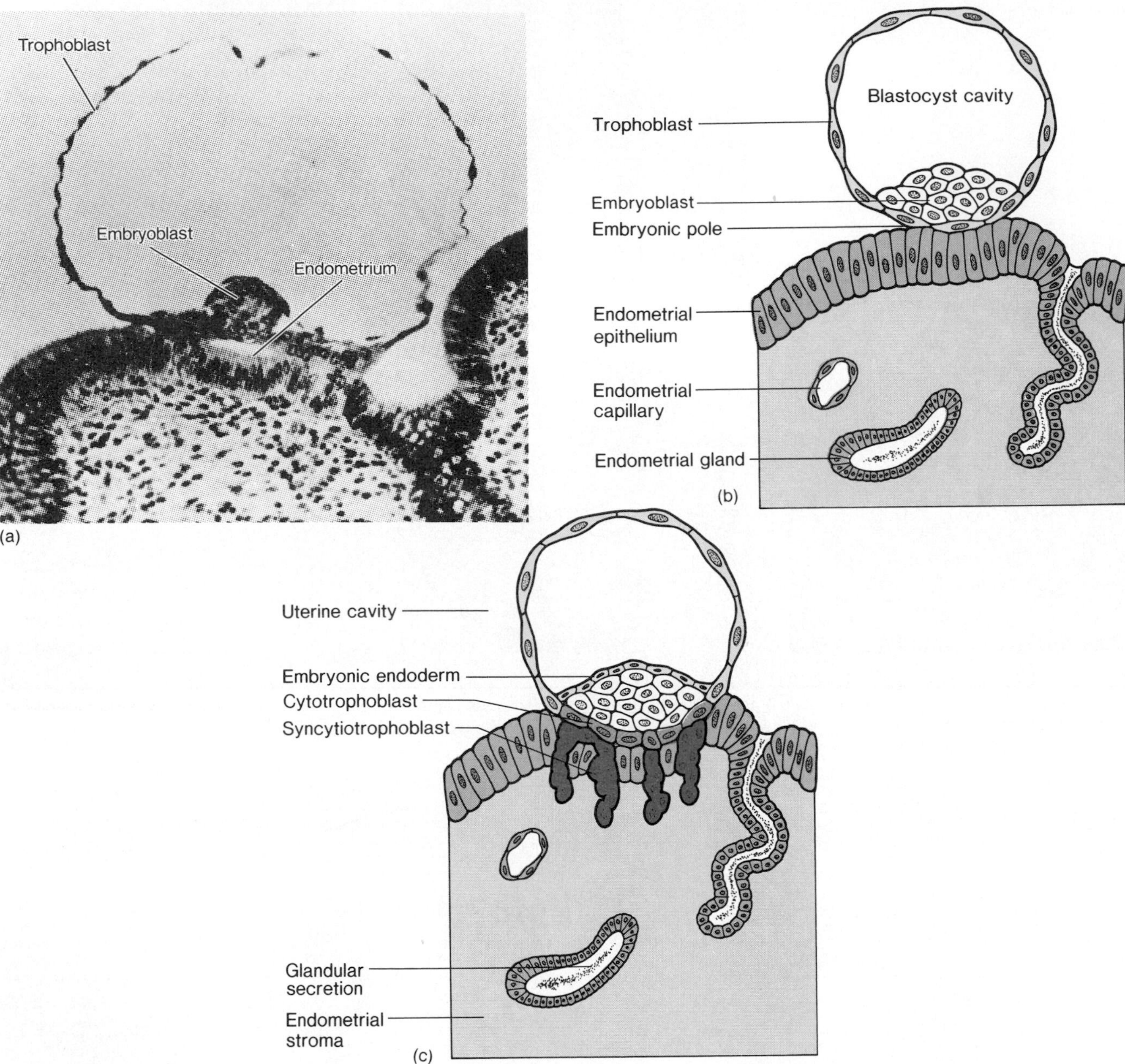

Implantation

The process of **implantation,** or *nidation,* begins between the fifth and seventh day. Attachment is usually upon the posterior wall of the body of the uterus, with the side containing the embryoblast against the endometrium (fig. 29.8). Implantation is made possible by the secretion of *proteolytic enzymes* by the trophoblast, which digest a portion of the endometrium. The blastula sinks into the depression, and endometrial cells move back to cover the defect in the wall. At the same time, the part of the uterine wall below the implanting blastocyst thickens, and specialized cells of the trophoblast produce fingerlike projections, called **syncytiotrophoblasts** *(sin-sit''e-o-trof'o-blast),* into the thickened area. The syncytiotrophoblasts arise from a specific portion of the trophoblast, at the embryonic pole, called the **cytotrophoblast.**

The blastocyst saves itself from being eliminated with the endometrium during a menstrual discharge by secreting a hormone that indirectly prevents menstruation. Even before the sixth day when implantation begins,

implantation: L. *im,* in; *planto,* to plant
nidation: L. *nidus,* nest

the syncytiotrophoblasts secrete **chorionic gonadotropin** *(ko''re-on'ik gon''ah-do-tro'pin),* or **hCG** (the h stands for *human*). This hormone is identical to LH in its effects and therefore is able to maintain the corpus luteum past the time when it would otherwise regress. The secretion of estrogen and progesterone is thus maintained and menstruation is normally prevented (fig. 29.9).

The secretion of hCG declines by the tenth week of pregnancy. Actually, this hormone is required only for the first five to six weeks of pregnancy, because the placenta itself becomes an active steroid-secreting gland by this time. At the fifth to sixth week, the mother's corpus luteum begins to regress (even in the presence of hCG), but the placenta secretes more than sufficient amounts of steroids to maintain the endometrium and prevent menstruation.

All *pregnancy tests* assay for the presence of hCG in the blood or urine, because this hormone is secreted by the blastocyst but not by the mother's endocrine glands. Modern pregnancy tests detect the presence of hCG by use of antibodies against hCG or by the use of cellular receptor proteins for hCG.

Figure 29.9. Human chorionic gonadotropin (hCG) is secreted by syncytiotrophoblasts during the first trimester of pregnancy. This hormone maintains the mother's corpus luteum for the first five and one-half weeks. After that time the placenta becomes the major sex-hormone-producing gland, secreting increasing amounts of estrogen and progesterone throughout pregnancy.

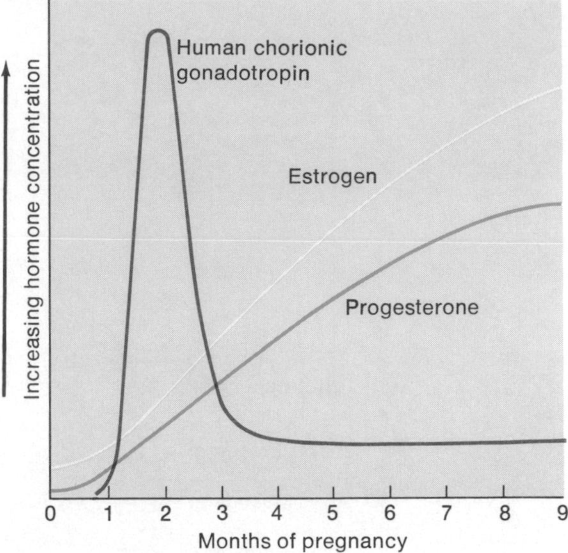

Formation of Germ Layers

As the blastocyst completes implantation during the second week of development, the embryoblast undergoes marked differentiation. A slitlike space, called the **amniotic cavity,** forms between the embryoblast and the invading trophoblast (fig. 29.10). The inner cell mass flattens into the **embryonic disc** (see fig. 29.12), which consists of two layers: an upper **ectoderm,** which is closer to the amniotic cavity, and a lower **endoderm,** which borders the blastocyst cavity. A short time later, a third layer called the **mesoderm** forms between the endoderm and ectoderm. These three layers constitute the **primary germ layers.** Once they are formed, at the end of the second week, the pre-embryonic stage is completed, and the embryonic stage begins.

The primary germ layers are important because various cells and tissues of the body are derived from them. Ectodermal cells form the nervous system; the outer layer of skin (epidermis), including hair, nails, and skin glands; and portions of the sensory organs. Mesodermal cells form the skeleton, muscles, blood, reproductive organs, dermis of the skin, and connective tissue. Endodermal cells produce the lining of the digestive tract, the digestive organs, the respiratory tract and lungs, and the urinary bladder and urethra.

ectoderm: Gk. *ecto,* outside; *derm,* skin
endoderm: Gk. *endo,* within; *derm,* skin
mesoderm: Gk. *meso,* middle; *derm,* skin

Figure 29.11 illustrates the organs and body systems that derive from each of the three primary germ layers. Table 29.1 summarizes the events of the pre-embryonic stage, and table 29.2 lists the derivatives of the primary germ layers.

The period of prenatal development is referred to as *gestation.* Normal gestation for humans is nine months. Knowing this and the pattern of menstruation make it possible to determine the delivery date of a baby. In a typical reproductive cycle, a woman ovulates fourteen days prior to the onset of the next menstruation and is fertile for approximately twenty to twenty-four hours following ovulation. Adding nine months, or thirty-eight weeks, to the time of ovulation gives one the estimated delivery date.

1. List the structural characteristics of a zygote, morula, and blastocyst. What is the general time of each of these stages of the pre-embryonic period of development?
2. Discuss the process of implantation, and describe the physiological events that ensure pregnancy.
3. Describe the development of the placenta.
4. List the major structures or organs that derive from each germ layer.
5. What is the length of time of a typical pregnancy? Define *gestation.* How is the parturition date determined?

gestation: L. *gestatus,* to bear

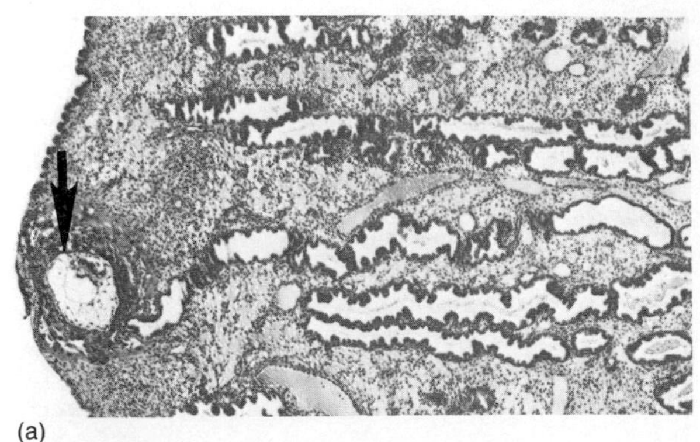

(a)

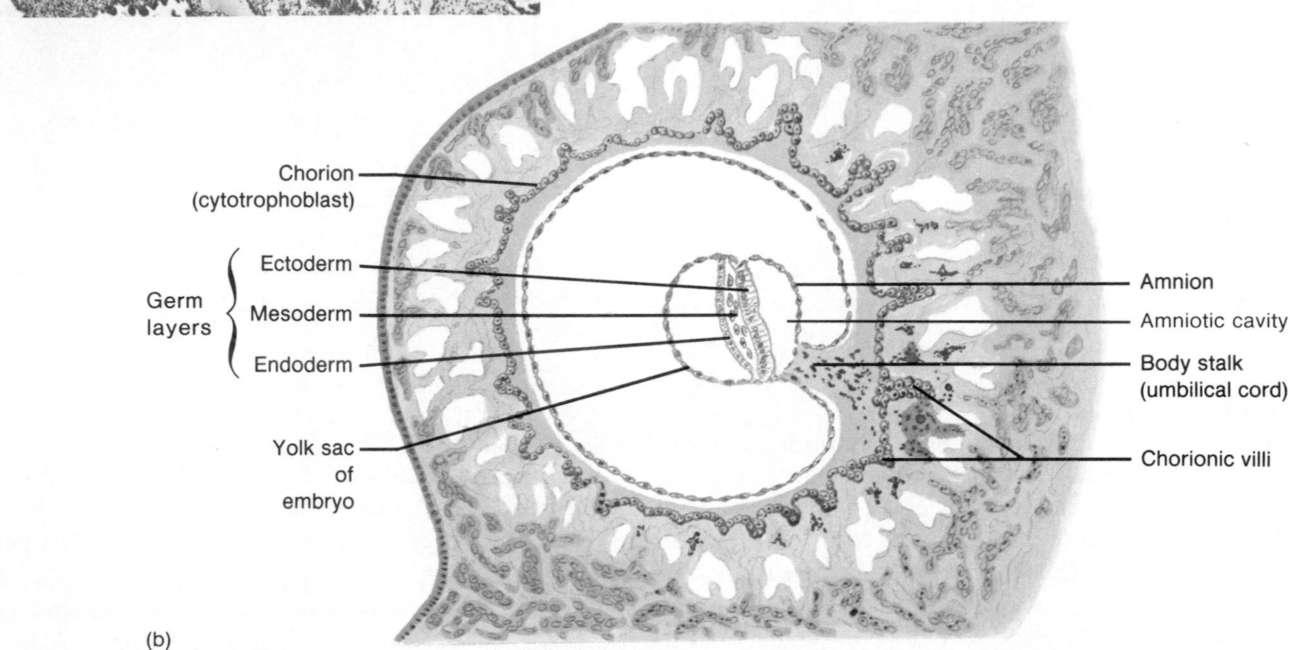

(b)

Figure 29.10. The completion of implantation. (*a*) a photomicrograph of an implanted blastocyst in the uterine wall. (*b*) the formation of the primary germ layers at the end of two weeks demarks the end of the pre-embryonic stage and the beginning of the embryonic stage. Note the formation of extraembryonic membranes at this early period of development.

Table 29.1 Morphogenic stages and principal events during pre-embryonic development

Stage	Time period	Principal events
Zygote	Twelve to twenty-four hours following ovulation	Egg is fertilized; zygote has twenty-three pairs of chromosomes (diploid) from haploid sperm and haploid egg; genetically unique
Cleavage	Thirty hours to third day	Mitotic divisions produce increased number of cells
Morula	Third to fourth day	Hollow ball-like structure forms; single layer thick
Blastocyst	Fifth day to end of second week	Inner cell mass and trophoblast form; implantation occurs; embryonic disc forms followed by primary germ layers

Table 29.2 Derivatives of germ layers

Ectoderm	Mesoderm	Endoderm
Epidermis of skin and epidermal derivatives: hair, nails, glands of the skin; linings of oral, nasal, anal, and vaginal cavities	Muscle: smooth, cardiac, and skeletal	Epithelium of pharynx, auditory canal, tonsils, thyroid, parathyroid, thymus, larynx, trachea, lungs, digestive tract, urinary bladder and urethra, and vagina
Nervous tissue; sense organs	Connective tissue: embryonic, connective tissue proper, cartilage, bone, blood	Liver and pancreas
Lens of eye; enamel of teeth	Dermis of skin; dentin of teeth	
Pituitary gland	Epithelium of blood vessels, lymphatic vessels, body cavities, joint cavities	
Adrenal medulla	Internal reproductive organs	
	Kidneys and ureters	
	Adrenal cortex	

Figure 29.11. The body systems and the primary germ layers from which they develop.

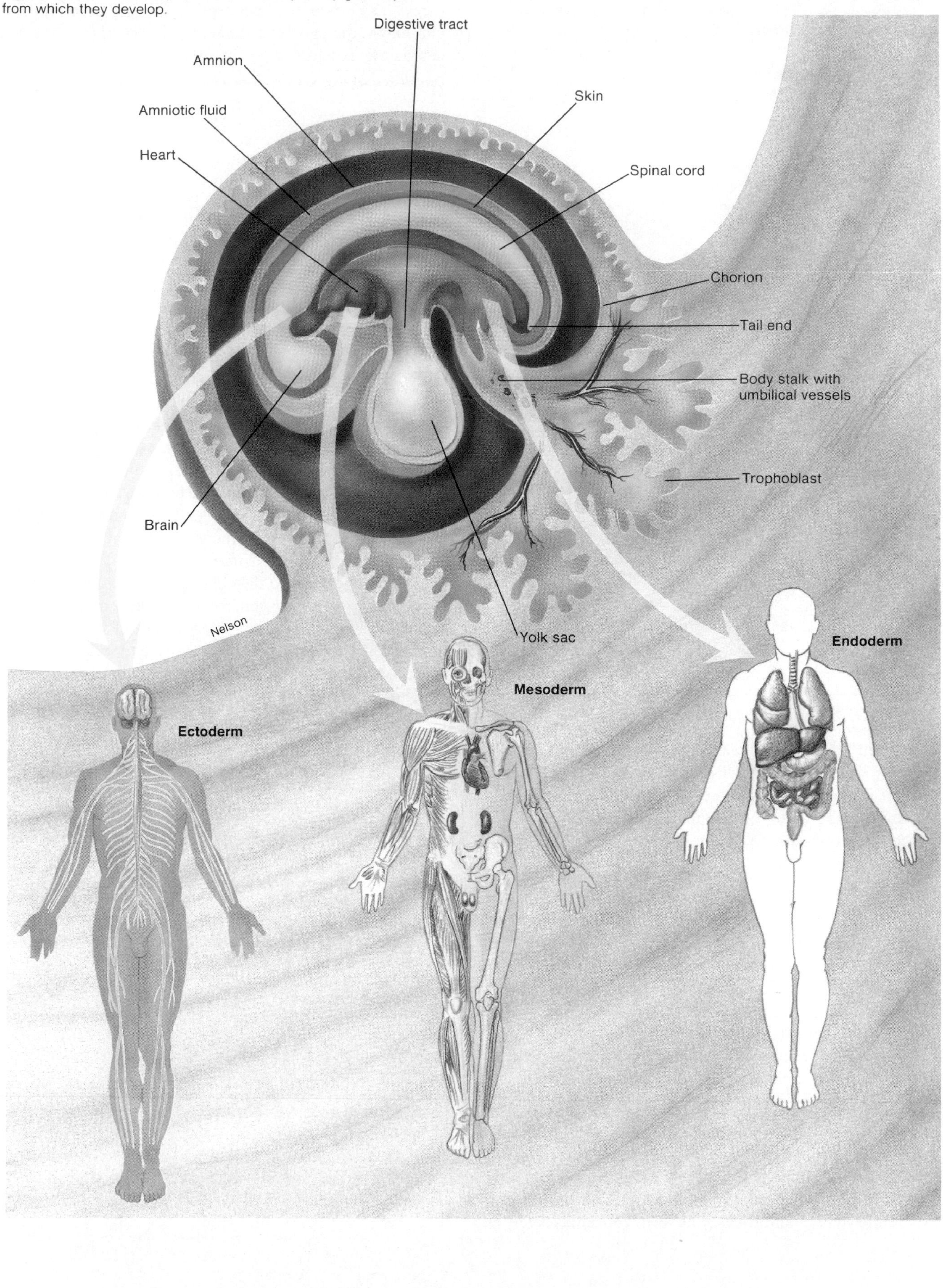

Figure 29.12. The formation of the extraembryonic membranes during a single week of rapid embryonic development. (*a*) three weeks; (*b*) three and one-half weeks; (*c*) four weeks.

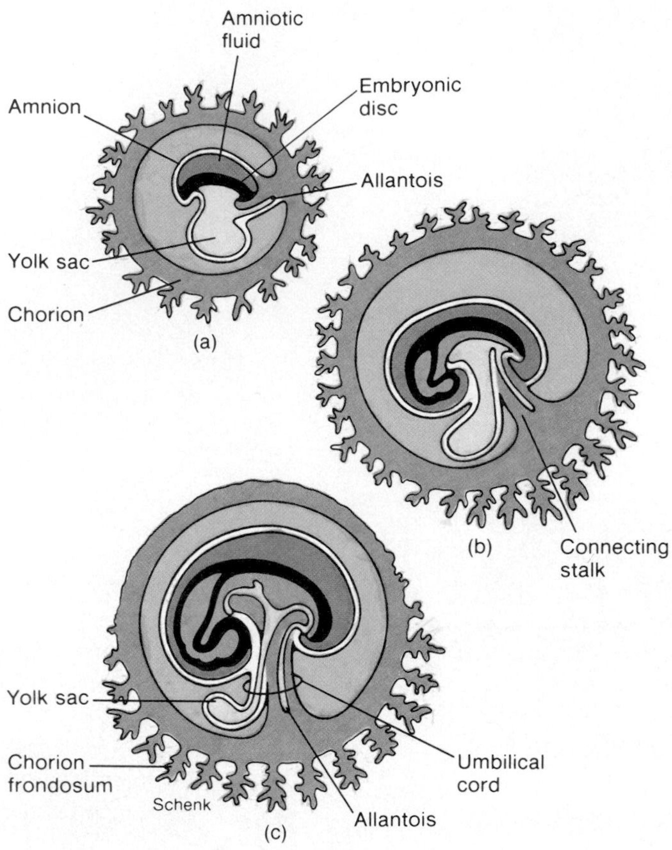

EMBRYONIC STAGE

The events of the six-week embryonic stage include the differentiation of the germ layers into specific body organs and the formation of the placenta, the umbilical cord, and the extraembryonic membranes, which provide sustenance and protection to the embryo.

Objective 7. Define the term *embryo,* give the time duration of the embryonic stage, and describe the major events of this period of development.

Objective 8. List the embryonic needs that must be met to avoid a spontaneous abortion.

Objective 9. Describe the structure and function of each of the extraembryonic membranes.

Objective 10. Describe the development and function of the placenta and the umbilical cord.

The embryonic stage lasts from the beginning of the third week to the end of the eighth week. At this stage, the developing organism can correctly be called an **embryo** *(em'bre-o).* During the embryonic stage all of the body organs form, as well as the placenta, umbilical cord, and extraembryonic membranes. The term **conceptus** refers to the embryo, or fetus, and all of the extraembryonic structures—the products of conception. During **parturition,** or childbirth, the baby and the remaining portion of the conceptus are expelled from the uterus.

Embryology is the study of the sequential changes in an organism as the various tissues, organs, and systems develop. Chick embryos are frequently studied because of the easy access through the shell and their rapid development. Mice and pig embryos are also extensively studied as mammalian models. Genetic manipulation, induction of drugs, exposure to disease, radioactive tagging or dyeing of developing tissues, and X-ray treatments are some of the commonly conducted experiments that provide information that can be applied to human development and birth defects.

During the pre-embryonic stage of cell divisions and differentiation, the developing structure is self-sustaining. The embryo, however, is not self-supporting and must derive sustenance from the mother. For morphogenesis to continue, certain immediate needs must be met, which include the (1) formation of a vascular connection between the uterus of the mother and the embryo so that nutrients and oxygen can be provided and metabolic wastes and carbon dioxide can be removed; (2) establishment of a constant, protective environment around the embryo, which is conducive to development; (3) establishment of a structural organization for embryonic morphogenesis along a longitudinal axis; (4) ensurance of structural support for the embryo both internally and externally; and (5) coordination of the morphogenic events through genetic expression. If these needs are not met, a spontaneous abortion will generally occur.

The first two of the aforementioned needs are provided by extraembryonic structures, whereas the other three are provided intraembryonically. The extraembryonic membranes, the placenta, and the umbilical cord will be discussed separately from the development of the embryo.

Most serious developmental defects cause the embryo to be naturally aborted. About 25% of early aborted embryos have chromosomal abnormalities. Other abortions may be caused by environmental factors (infectious agents and teratogenic drugs). In addition, an implanting, developing embryo is regarded as foreign tissue by the mother and is rejected and aborted unless maternal immune responses are suppressed.

Extraembryonic Membranes

While the many intraembryonic events are forming the body organs, a complex system of extraembryonic membranes is developing as well (fig. 29.12). The **extraembryonic membranes** are the amnion, yolk sac, allantois, and chorion. These membranes are responsible for the protection, respiration, excretion, and nutrition of the embryo and subsequent fetus. At parturition, the placenta, umbilical cord, and extraembryonic membranes separate from the fetus and are expelled from the uterus as the *afterbirth.*

Amnion. The amnion *(am'ne-on)* is a thin extraembryonic membrane, derived from ectoderm and mesoderm, which loosely envelops the embryo, forming an **amniotic sac,** filled with *amniotic fluid* (fig. 29.13). In later stages of fetal development, the amnion expands to come in contact with the chorion. Amniotic development is initiated during early embryonic development, at which time its margin is attached around the free edge of the embryonic disc (fig. 29.12). As the amniotic sac enlarges during the late embryonic period (about eight weeks), the amnion gradually sheaths the developing umbilical cord with an epithelial covering (fig. 29.14).

As a buoyant medium, amniotic fluid performs four functions for the embryo and subsequent fetus: (1) it permits symmetrical structural development and growth; (2) it cushions and protects by absorbing jolts that the mother may receive; (3) it helps to maintain consistent pressure and temperature; and (4) it enables freedom of fetal development, which is important for musculoskeletal development and blood flow.

Amniotic fluid is formed initially as an isotonic fluid absorbed from the maternal blood in the endometrium surrounding the developing embryo. Later, the volume is increased and the concentration changed by urine excreted from the fetus into the amniotic sac. Amniotic fluid also contains cells that are sloughed off from the fetus, placenta, and amniotic sac. Since all of these cells are derived from the same fertilized egg, all have the same genetic composition. Many genetic abnormalities can be detected by aspirating this fluid and examining the cells thus obtained in a procedure called *amniocentesis (am''ne-o-sen-te'sis).*

Amniotic fluid is normally swallowed by the fetus and absorbed in the gastrointestinal tract. Prior to delivery, the amnion is naturally or surgically ruptured, and the amniotic fluid ("bag of waters") is released.

> *Amniocentesis* (fig. 29.15) is usually performed at the fourteenth or fifteenth weeks of pregnancy, when the amniotic sac contains 175–225 ml of fluid. Genetic diseases such as *Down's syndrome* (in which there are three instead of two number-21 chromosomes) can be detected by examining chromosomes; diseases such as *Tay-Sachs disease,* in which there is a defective enzyme involved in formation of myelin sheaths, can be detected by biochemical techniques.

Down's syndrome: from John L. H. Down, English physician, 1828–1896

Yolk Sac. The yolk sac is established during the end of the second week as cells from the trophoblast form a thin *exocoelomic (ek''so-se-lo'mik) membrane.* Unlike many vertebrates, the human yolk sac contains no nutritive yolk but is an essential structure during early embryonic development. It is attached to the underside of the embryonic disc (figs. 29.12, 29.13), where it produces blood for the embryo until the liver forms during the sixth week. The dorsal portion of the yolk sac is involved in the formation of the primitive gut. In addition, primordial germ cells form in the wall of the yolk sac and migrate during the fourth week to the developing gonads, where they become the primitive germ cells (spermatogonia or oogonia).

The stalk of the yolk sac usually detaches from the gut by the sixth week. Following this, the yolk sac gradually shrinks as pregnancy advances. Eventually it becomes very small and serves no additional developmental functions.

Allantois. The allantois forms during the third week as a small outpouching, or diverticulum, from the caudal wall of the yolk sac (see fig. 29.12). The allantois remains small but is involved in the formation of blood cells and gives rise to the fetal umbilical arteries and vein. It also contributes to the development of the urinary bladder.

The extraembryonic portion of the allantois degenerates during the second month. The intraembryonic portion involutes to form a thick urinary tube called the **urachus.** After birth, the urachus becomes a fibrous cord, called the median umbilical ligament, which attaches to the urinary bladder.

Chorion. The chorion is a highly specialized extraembryonic membrane that participates in the formation of the placenta (see fig. 29.12). It is the outermost membrane and originates from the trophoblast of the blastocyst. Numerous, small, fingerlike extensions, called **villi** (see fig. 29.13), form from the chorion and penetrate deeply into the uterine tissue. Initially, the entire surface of the chorion is covered with villi. But those villi on the surface toward the uterine cavity gradually degenerate and produce a smooth, bare area known as the **smooth chorion.** As this occurs, the villi associated with the uterine wall rapidly increase in number and branch out. This portion of the chorion is known as the **villous chorion.** The villous chorion becomes highly vascular, and as the embryonic heart begins to function, blood is pumped in close proximity to the uterine wall.

allantois: Gk. *allanto,* sausage; *iodos,* resemblance
chorion: Gk. *chorion,* external fetal membrane
villous: L. *villus,* tuft of hair

Figure 29.13. An implanted embryo at approximately four and one-half weeks. (*a*) the interior of a uterus showing the implantation site and elevated *decidua capsularis* caused by the expanded chorion. (*b*) the developing embryo, extraembryonic membranes, and the formation of the placenta.

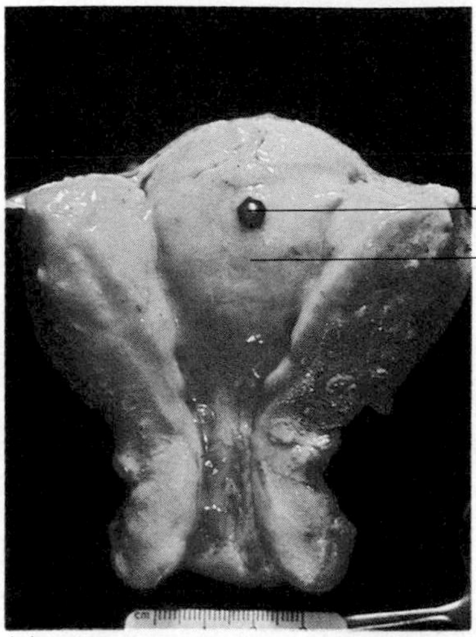

Decidua capsularis
Body of uterus
(a)

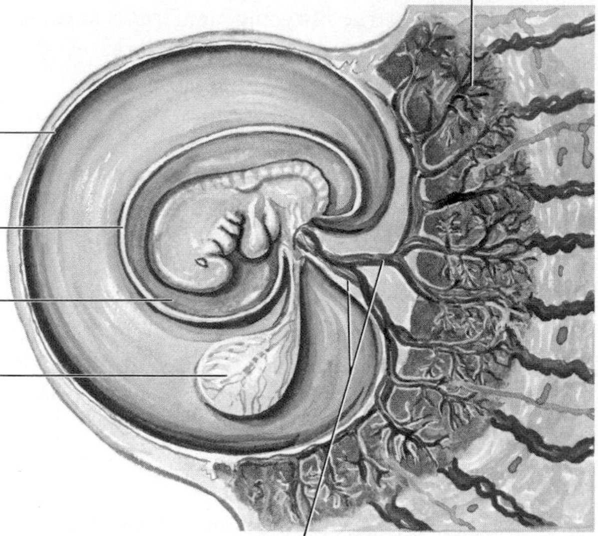

Villi of chorion frondosum
Chorion
Amnion
Amniotic sac
Yolk sac
Umbilical blood vessels
(b)

Figure 29.14. The embryo, extraembryonic membranes, and placenta at approximately seven weeks of development. Note that at this time, the amnion and chorion are adherent and are frequently referred to as the *amniochorionic membrane*. Blood from the embryo is carried to and from the chorion frondosum by umbilical arteries and vein. The maternal tissue between the chorionic villi is known as the decidua basalis, and this tissue, together with the villi, form the functioning placenta.

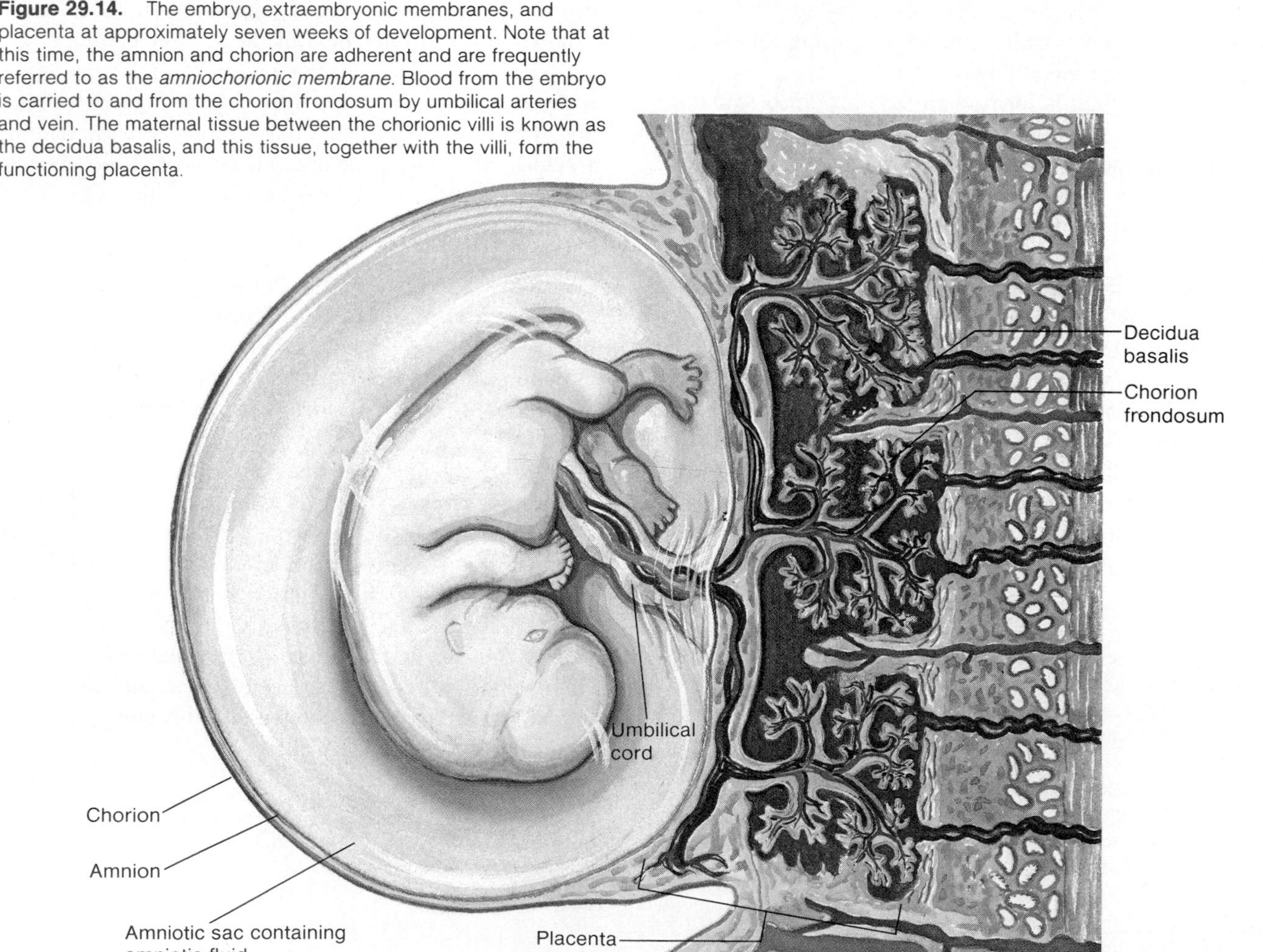

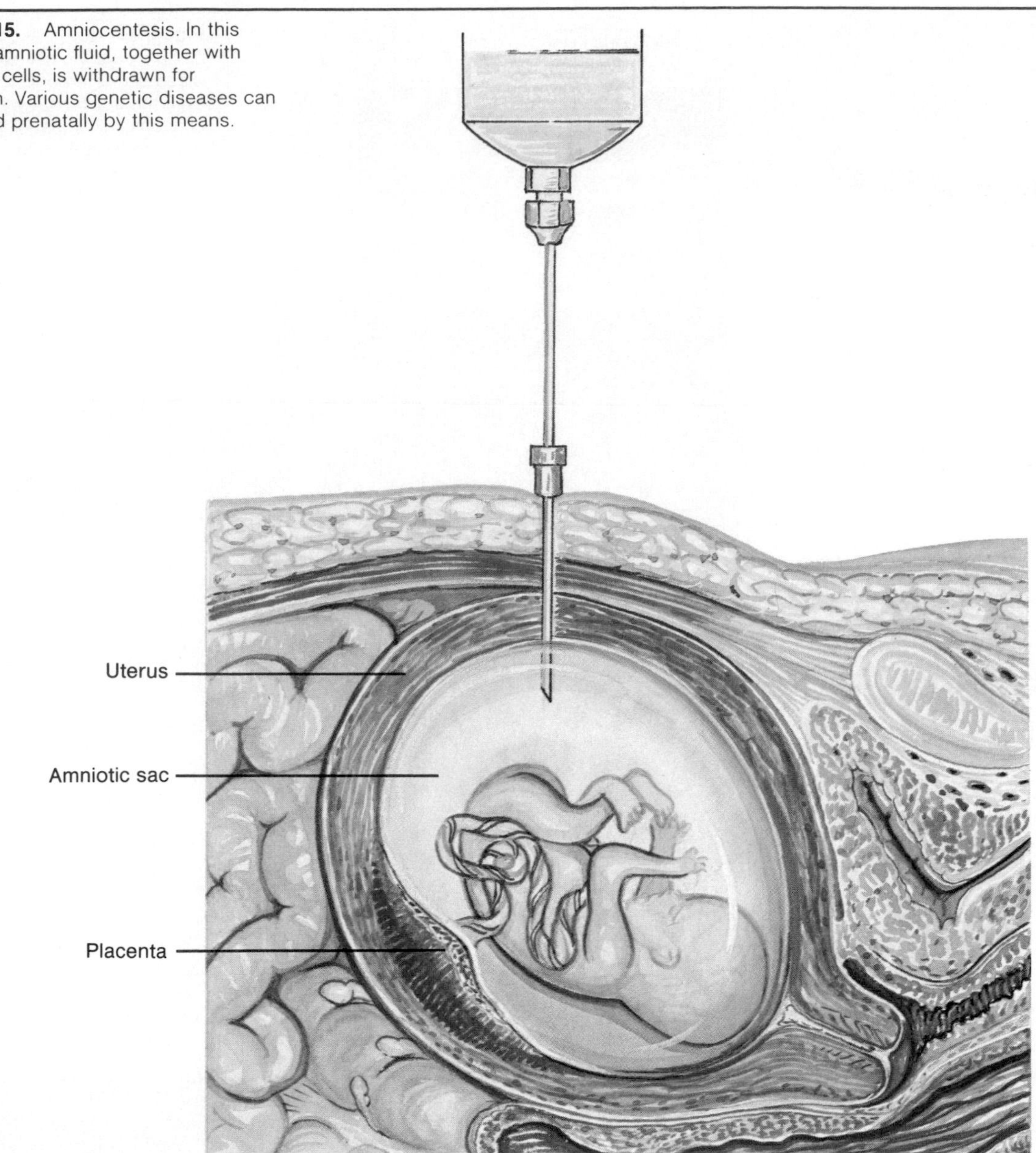

Figure 29.15. Amniocentesis. In this procedure amniotic fluid, together with suspended cells, is withdrawn for examination. Various genetic diseases can be detected prenatally by this means.

The amniotic fluid that is withdrawn contains fetal cells at a concentration that is too low to permit direct determination of genetic or chromosomal disorders. These cells must therefore be cultured *in vitro* for at least two weeks before they are present in sufficient numbers for the laboratory tests required. A newer method, called *chorionic villus biopsy,* is now available to detect genetic disorders much earlier than permitted by amniocentesis. In chorionic villus biopsy, a catheter is inserted through the cervix to the chorion, and a sample of a chorionic villus is obtained by suction or cutting. Genetic tests can be performed directly on the villus sample, since this sample contains much larger numbers of fetal cells than does a sample of amniotic fluid. Chorionic villus biopsy can provide genetic information at ten to twelve weeks gestation.

Placenta

The **placenta** *(plah-sen'tah)* is a vascular structure by which an unborn child is attached to its mother's uterine wall and through which metabolic exchange occurs (fig. 29.16). The placenta is formed in part from maternal tissue and in part from embryonic tissue. The embryonic portion of the placenta consists of the villi of the **chorion frondosum,** whereas the maternal portion is composed of the area of the uterine wall called the **decidua basalis** (see fig. 29.14), to which the villi penetrate. Blood does not flow directly between these two portions, but because of the close membranous proximity, certain substances diffuse readily.

placenta: L. *placenta,* a flat cake

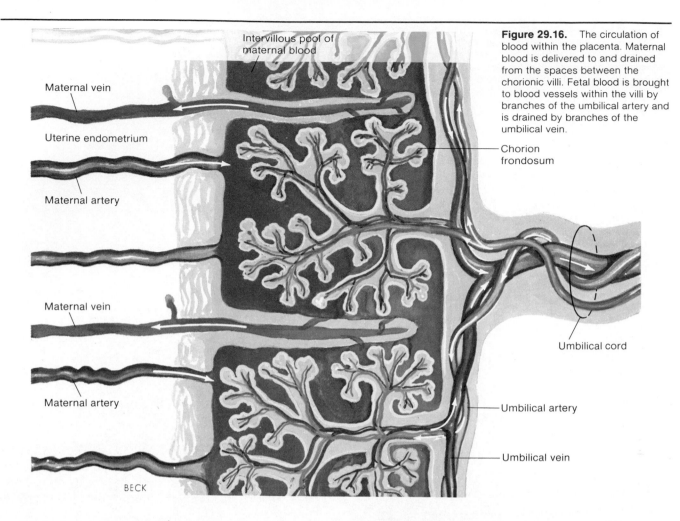

Intervillous pool of maternal blood
Maternal vein
Uterine endometrium
Maternal artery
Maternal vein
Maternal artery
BECK
Chorion frondosum
Umbilical cord
Umbilical artery
Umbilical vein

Figure 29.16. The circulation of blood within the placenta. Maternal blood is delivered to and drained from the spaces between the chorionic villi. Fetal blood is brought to blood vessels within the villi by branches of the umbilical artery and is drained by branches of the umbilical vein.

When fully formed, the placenta is a reddish brown oval disc (fig. 29.17) with a diameter of 15–20 cm (8 in.) and a thickness of 2.5 cm (1 in.). It weighs 500–600 gm, being about one-sixth the weight of the fetus.

Exchange of Molecules across the Placenta. The **umbilical artery** delivers fetal blood to vessels within the villi of the chorion frondosum of the placenta. This blood circulates within the villi and returns to the fetus via the **umbilical vein.** Maternal blood is delivered to and drained from the cavities within the decidua basalis, which are located between the chorionic villi. In this way, maternal and fetal blood are brought close together but never mix within the placenta.

The placenta serves as a site for the exchange of gases and other molecules between the maternal and fetal blood. Oxygen diffuses from mother to fetus, and carbon dioxide diffuses in the opposite direction. Nutrient molecules and waste products likewise pass between maternal and fetal blood.

The placenta is not merely a passive conduit for exchange between maternal and fetal blood, however. It has a very high metabolic rate, utilizing about one-third of all the oxygen and glucose supplied by the maternal blood.

Figure 29.17. The examination of a placenta. Following the birth of a baby, the placenta, umbilical cord, and extraembryonic membranes are expelled from the uterus as the *afterbirth*.

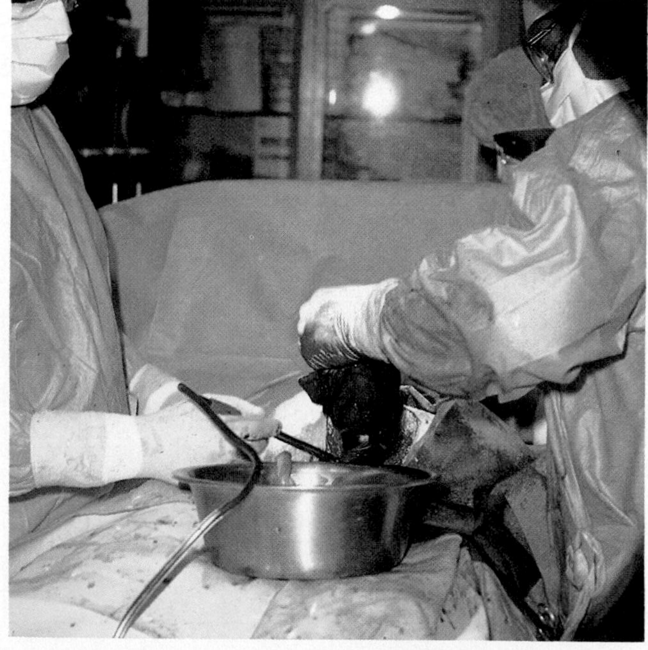

Table 29.3 Hormones secreted by placenta

Hormones	Effects
Pituitary-like hormones	
Chorionic gonadotropin (hCG)	Similar to LH; maintains mother's corpus luteum for first 5½ weeks of pregnancy; may be involved in suppressing immunological rejection of embryo; also has TSH-like activity
Chorionic somatomammotropin (hCS)	Similar to prolactin and growth hormone; in the mother, hCS acts to promote increased fat breakdown and fatty acid release from adipose tissue and to promote the sparing of glucose use by maternal tissues ("diabetic-like" effects)
Sex steroids	
Progesterone	Helps maintain endometrium during pregnancy; helps suppress gonadotropin secretion; promotes uterine sensitivity to oxytocin; helps stimulate mammary gland development
Estrogens	Help maintain endometrium during pregnancy; help suppress gonadotropin secretion; help stimulate mammary gland development; inhibit prolactin secretion

The rate of protein synthesis is, in fact, higher in the placenta than it is in the liver. Like the liver, the placenta produces a great variety of enzymes capable of converting biologically active molecules (such as hormones and drugs) into less active, more water-soluble forms. In this way, potentially dangerous molecules in the maternal blood are often prevented from harming the fetus.

Most drugs ingested by a pregnant woman can readily pass through the placenta and may be deleterious to the developing baby. For example, the nicotine taken in by a chain-smoking pregnant woman will stunt the growth of her fetus. Hard drugs, such as heroin, can lead to *fetal drug addiction.* Depressant drugs given to a mother during labor can readily cross the placenta and, if given in high dosages, can cause respiratory depression in the newborn infant.

Endocrine Functions of the Placenta

The placenta secretes both steroid hormones and protein hormones that have actions similar to those of some anterior pituitary hormones. This latter category of hormones includes **chorionic gonadotropin (hCG)** and **chorionic somatomammotropin (hCS)** (table 29.3). Chorionic gonadotropin has LH-like effects, as previously described; it also has thyroid-stimulating ability, like pituitary TSH. Chorionic somatomammotropin likewise has actions that are similar to two pituitary hormones: growth hormone and prolactin. The placental hormones hCG and hCS thus duplicate the actions of four anterior pituitary hormones.

Pituitary-Like Hormones from the Placenta. The importance of chorionic gonadotropin in maintaining the mother's corpus luteum for the first five and a half weeks of pregnancy has been previously discussed. There is also some evidence that hCG may in some way help to prevent immunological rejection of the implanting embryo. Chorionic somatomammotropin synergizes (acts together) with growth hormone from the mother's pituitary to produce a "diabetic-like" effect in the pregnant woman. The effects of these two hormones stimulate (1) lipolysis and increased plasma fatty acid concentration; (2) decreased maternal utilization of glucose and, therefore, increased blood glucose concentrations; and (3) polyuria (excretion of large volumes of urine), thereby producing a degree of dehydration and thirst. This "diabetic-like" effect in the mother helps to spare glucose for the placenta and fetus that, like the brain, use glucose as their primary energy source.

Steroid Hormones from the Placenta. After the first five and a half weeks of pregnancy, when the corpus luteum regresses, the placenta becomes the major sex steroid–producing gland. The blood concentration of estrogens, as a result of placental secretion, rises to levels more than 100 times greater than those existing at the beginning of pregnancy. The placenta also secretes large amounts of progesterone, changing the estrogen/progesterone ratio in the blood from 100:1 at the beginning of pregnancy to a ratio of close to 1:1 toward full-term.

The placenta, however, is an "incomplete endocrine gland" because it cannot produce estrogen and progesterone without the aid of precursors supplied to it by both the mother and the fetus. The placenta, for example, cannot produce cholesterol from acetate and so must be supplied with cholesterol from the mother's circulation. Cholesterol, which is a steroid containing twenty-seven carbons, can then be converted by enzymes in the placenta into steroids that contain twenty-one carbons—such as progesterone. The placenta, however, lacks the enzymes needed to convert progesterone into androgens (which have nineteen carbons) and estrogens (which have eighteen carbons).

In order for the placenta to produce estrogens, it needs to cooperate with steroid-producing tissues in the fetus. Fetus and placenta, therefore, form a single functioning system in terms of steroid hormone production. This system has been called the **fetal-placental unit** (fig. 29.18).

Figure 29.18. The secretion of progesterone and estrogen from the placenta requires a supply of cholesterol from the mother's blood and the cooperation of fetal enzymes that convert progesterone to androgens.

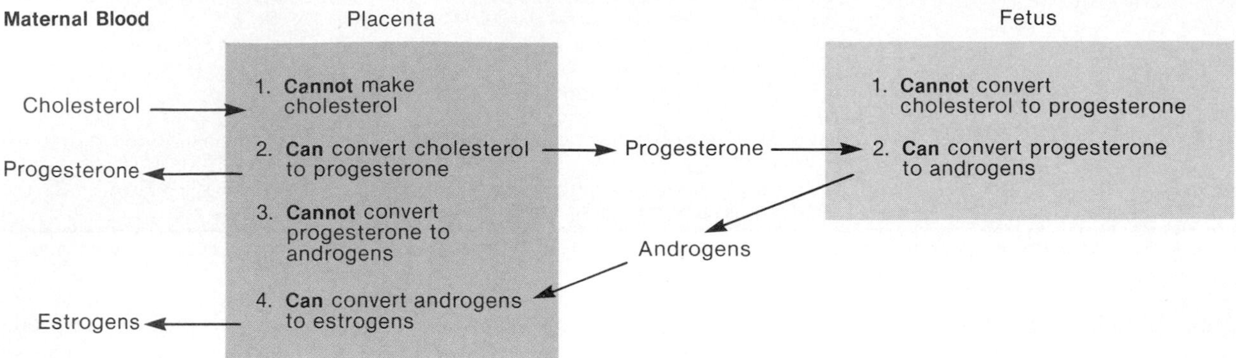

The ability of the placenta to convert androgens into estrogen helps to protect the female embryo from becoming masculinized by the androgens secreted from the mother's adrenal glands. In addition to producing estradiol, the placenta secretes large amounts of a weak estrogen called **estriol.** The production of estriol increases tenfold during pregnancy, so that by the third trimester estriol accounts for about 90% of the estrogens excreted in the mother's urine. Since almost all of this estriol comes from the placenta (rather than from maternal tissues), measurements of urinary estriol can be used clinically to assess the health of the placenta.

Umbilical Cord

The **umbilical** *(um-bil'i-kal)* **cord** forms as the yolk sac shrinks and the amnion expands to envelop the tissues on the underside of the embryo. Figure 29.19 illustrates the formation of the umbilical cord.

The umbilical cord usually attaches near the center of the placenta. When fully formed, the umbilical cord is about 1–2 cm (0.5–1 in.) in diameter and approximately 55 cm (2 ft) in length. The umbilical cord contains two **umbilical arteries,** which carry deoxygenated blood toward the placenta, and one **umbilical vein,** which carries oxygenated blood from the placenta to the embryo. These vessels are surrounded by embryonic connective tissue called *Wharton's jelly.*

The umbilical cord has natural twists because the umbilical vein is longer than the arteries. In about one-fifth of all deliveries, the cord is looped once around the baby's neck. If drawn tightly, the cord may cause death or serious perinatal problems.

Wharton's jelly: from Thomas Wharton, English anatomist, 1614–1673

Third Week. Early in the third week a thick linear band, known as the **primitive streak,** appears along the dorsal midline of the embryonic disc (see fig. 29.12). The primitive streak is derived from the mesodermal cells, which are specialized from the ectodermal layer of the embryonic mass. The primitive streak establishes a structural foundation for embryonic morphogenesis along a longitudinal axis. As the primitive streak elongates, a prominent thickening known as the **primitive knot** appears at its cranial end (fig. 29.20). The primitive knot later gives rise to the mesodermal structures of the head and the **notochord.** To give support to the embryo, the notochord forms a midline axis that is the basis of the embryonic skeleton. The primitive streak also gives rise to loose embryonic connective tissue called **mesenchyme,** which differentiates into all the various kinds of connective tissue found in the adult. One of the earliest formed organs is the skin, which develops to support and maintain homeostasis within the embryo.

A tremendous amount of change and specialization occurs during the embryonic stage. The factors that cause precise, sequential change from one cell or tissue type to another are not fully understood. It is known, however, that the potential for change is programmed into the genetics of each cell and that under conducive environmental conditions these characteristics become expressed. The process of developmental change is referred to as **induction.** Induction occurs when one tissue, called the **inductor tissue,** has a marked effect on an adjacent **induced tissue** and stimulates it to differentiate.

Fourth Week. During the fourth week of development, the embryo increases about 4 mm in length. A **connecting stalk,** which is later involved in the formation of the umbilical cord, is established from the body of the embryo to

induction: L. *inductus,* to lead in

Figure 29.19. The formation of the umbilical cord and other extraembryonic structures as seen in sagittal sections of the gravid uterus from the fourth to the twenty-second week. (*a*) a connecting stalk forms as the developing amnion expands around the embryo and finally meets ventrally. (*b*) the umbilical cord begins to take form as the amnion ensheathes the yolk sac. (*c*) a cross section of the umbilical cord showing the embryonic vessels, Wharton's jelly, and the tubular connection to the yolk sac. (*d*) by the twenty-second week, the amnion and chorion are fused, and the umbilical cord and placenta are well-developed structures. (*e*) sixteen-week-old fetus.

Uterine wall
Chorion
Uterine cavity
Amnion
Chorionic cavity
Connecting stalk
Yolk sac
Cervix
Vagina
(a)

Intervillous space
Villous chorion
Umbilical cord
Amnion
Smooth chorion
Internal os
(b)

Placenta
Villous chorion
Amnion
Uterine cavity
Smooth chorion
(d)

Umbilical vein
Wharton's jelly
Umbilical arteries
Yolk sac
Amnion
(c)

(e)

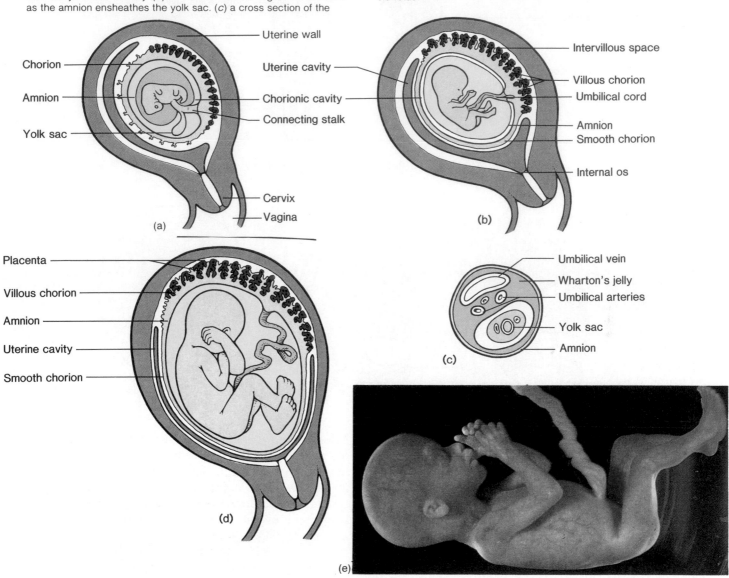

Figure 29.20. The appearance of the primitive streak and primitive knot along the embryonic disc. These progressive changes occur through the process of induction.

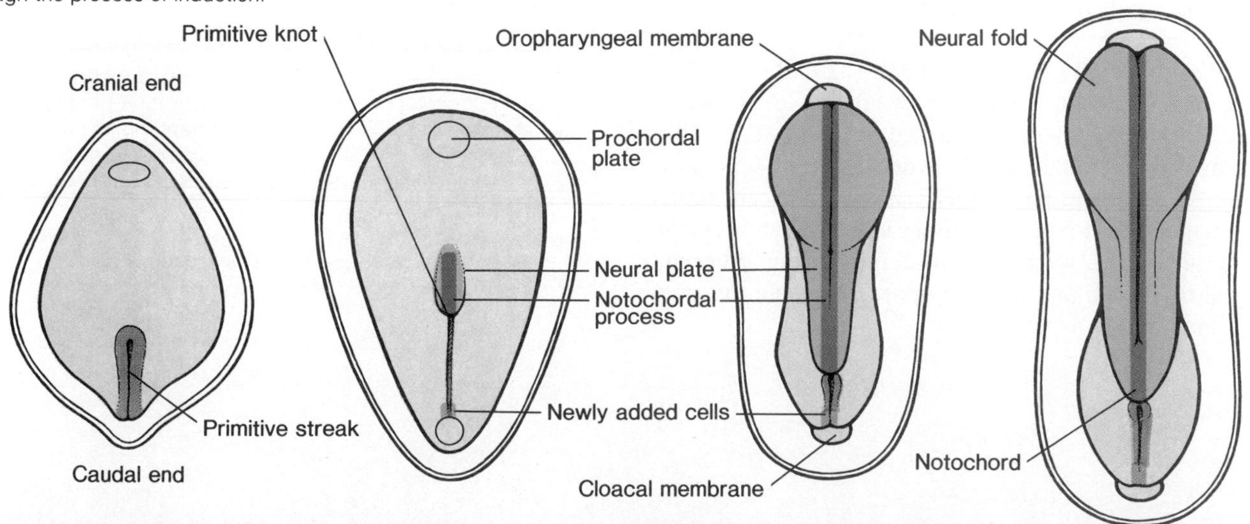

Figure 29.21. Progressive illustrations of four-week-old embryos.

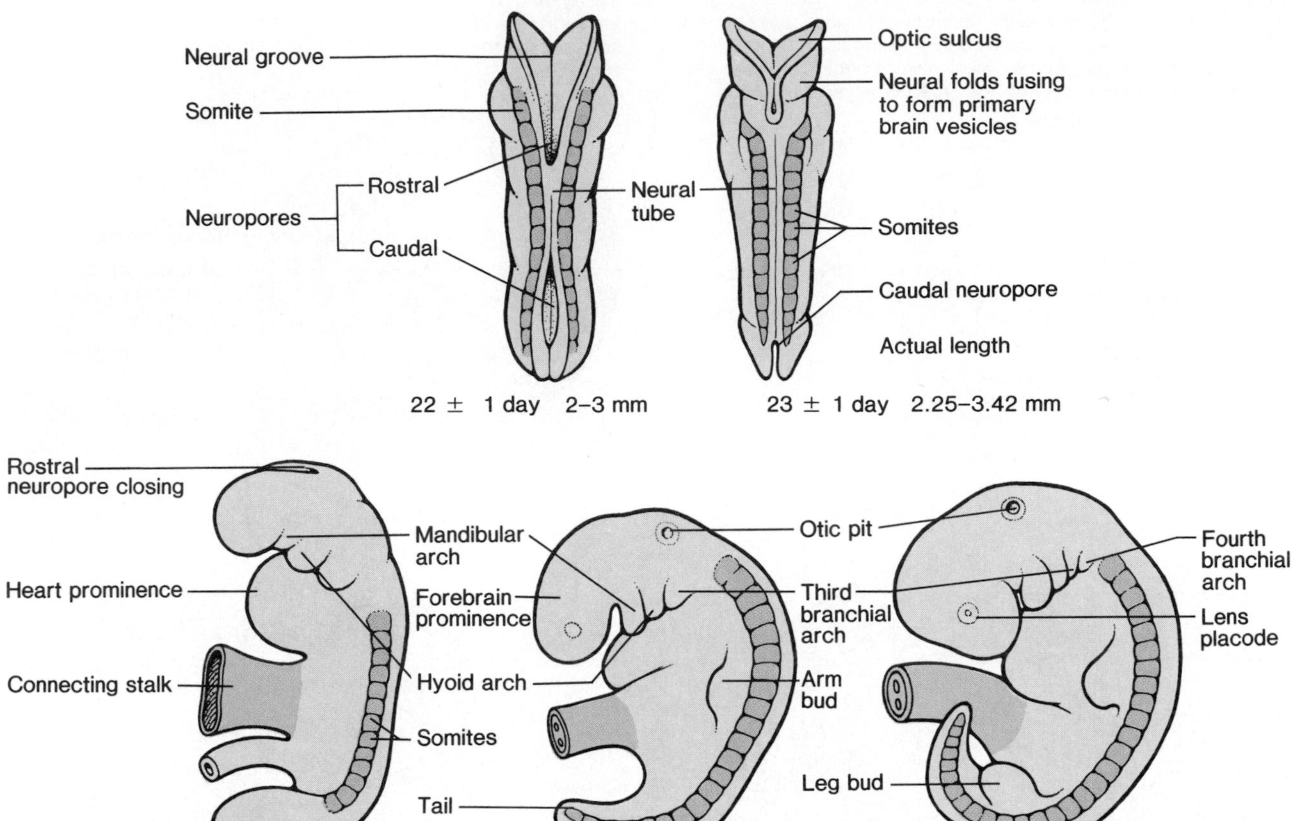

the developing placenta (fig. 29.21). By this time, the heart is beating firmly to pump blood to all parts of the embryo. The head and jaws are apparent, and the primordial tissue that will form the eyes, brain, spinal cord, lungs, and digestive organs has developed. The **arm** and **leg buds** are recognizable as small swellings on the lateral body walls.

Fifth Week. Embryonic changes during the fifth week are not as extensive as those during the fourth week. The head enlarges, and the developing eyes, ears, and nasal pit are obvious (fig. 29.22). The appendages have formed from the limb buds, and paddle-shaped hand plates develop digital ridges called **finger,** or **digital, rays.**

Sixth Week. During the sixth week, the embryo is 16–24 mm long. The head is much larger relative to the trunk, and the brain has undergone marked differentiation. This is the most vulnerable period of development for many organs. An interruption at this critical time can easily cause congenital damage. The limbs undergo considerable change during this week. The forelimbs are lengthened, and slightly flexed, and notches appear between the rays in the hand and foot plates.

Seventh and Eighth Weeks. During the last two weeks of the embryonic stage, the embryo, which is now 28–40 mm long, has distinct human characteristics (figs. 29.23, 29.24). The body organs are formed, and the nervous system begins coordinating body activity. The neck region is apparent, and the abdomen is less protuberant. The eyes are well developed, but the lids are stuck together to protect against probing fingers during muscular movement. The nostrils are developed but plugged with mucus. The external genitalia are forming but are still undifferentiated. The body systems are developed by the end of the eighth week, and from this time on the embryo is called a **fetus** *(fe'tus).*

The most precarious time of prenatal development is during the embryonic stage, when there is much tissue differentiation and organ formation. Frequently, however, a woman does not even realize that she is pregnant until she is well into this period of development. For this reason, a woman should consistently take good care of herself if there is even a chance that she might become pregnant.

Figure 29.22. Progressive illustrations of five-week-old embryos and a photograph of a five-week-old embryo. Note the digital rays and the developing eyes and ears.

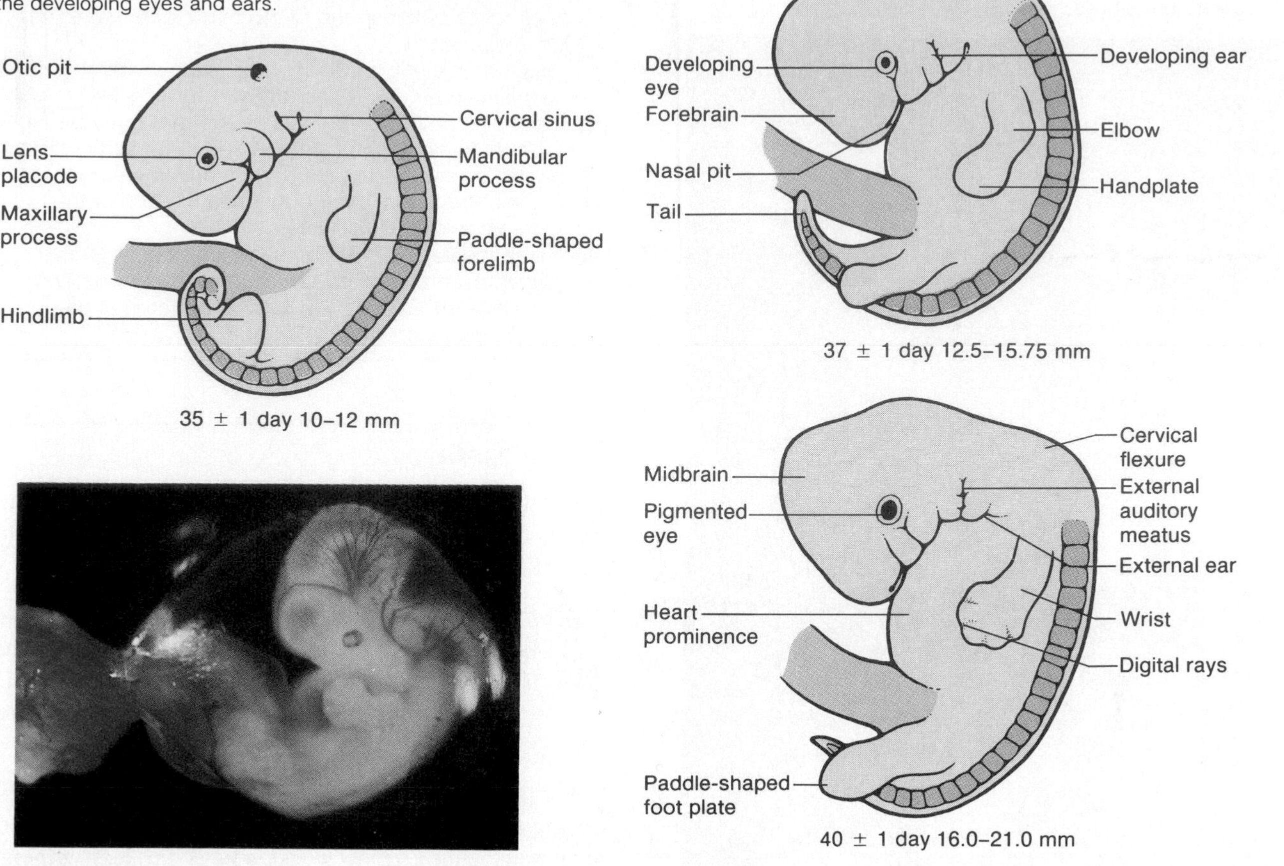

Figure 29.23. A photograph of a six-week-old embryo; progressive illustrations of six- and seven-week-old embryos.

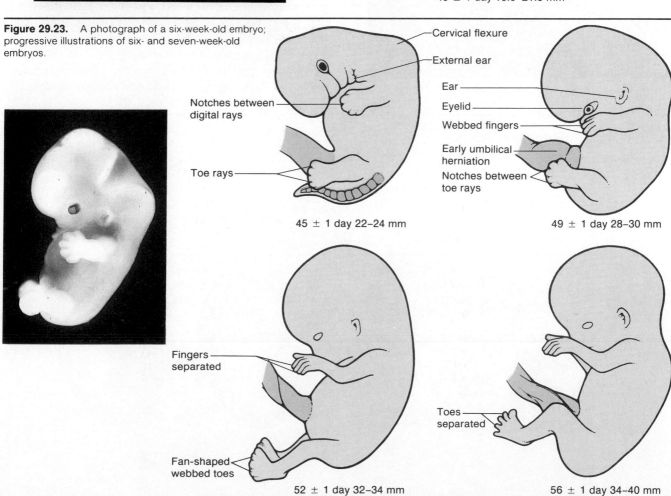

Cervical flexure
External ear
Notches between digital rays
Toe rays
45 ± 1 day 22–24 mm
Ear
Eyelid
Webbed fingers
Early umbilical herniation
Notches between toe rays
49 ± 1 day 28–30 mm
Fingers separated
Fan-shaped webbed toes
52 ± 1 day 32–34 mm
Toes separated
56 ± 1 day 34–40 mm

Figure 29.24. Photograph of an eight-week-old embryo. The body systems are developed by the end of the eighth week, and the embryo is recognizable as human.

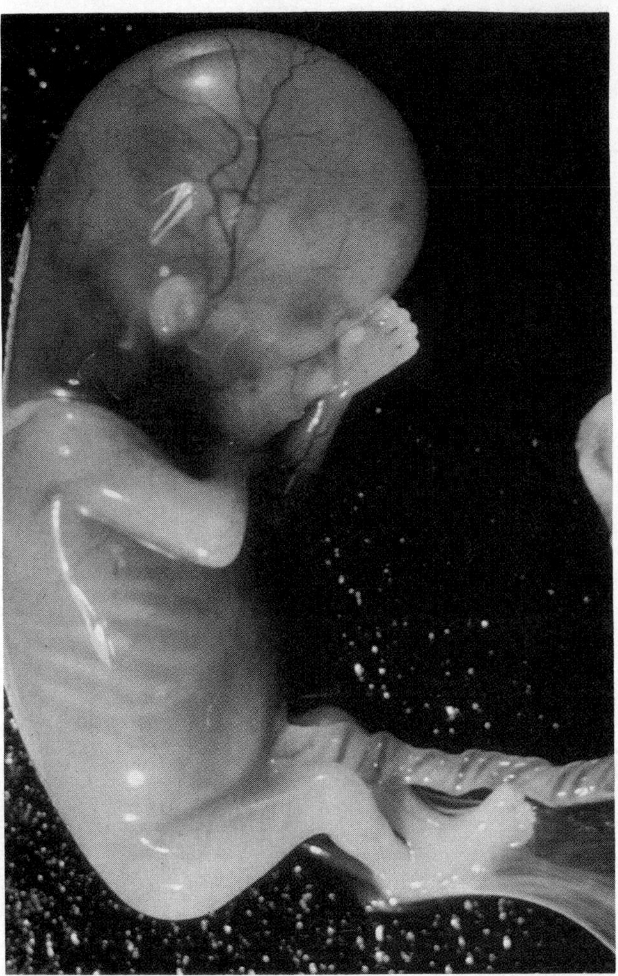

1. Distinguish between the terms *embryo* and *fetus.* Briefly summarize the structural changes that an embryo undergoes between the fourth and eighth weeks of development.
2. Describe the origin of hCG, and explain why it is needed to maintain pregnancy for the first ten weeks.
3. What are the five embryonic needs that must be met in order to avoid spontaneous abortion? Which of these needs are met by the extraembryonic membranes?
4. Explain the formation and function of each of the extraembryonic membranes.
5. Name the fetal and maternal components of the placenta, and describe the circulation in these two components. Explain how fetal and maternal gas exchange occurs.

Fetal Stage

The fetal stage lasts from nine weeks until birth and is characterized by tremendous growth and the specialization of body structures.

Objective 11. Define the term *fetus,* and discuss the major events of the fetal stage of development.

Objective 12. Describe the various techniques available for fetal examination or for monitoring fetal activity.

Since most of the tissues and organs of the body appear during the embryonic period, the **fetus** is recognizable as a human being at nine weeks and is far less vulnerable than the embryo to deformation from viruses, drugs, and radiation. A small amount of tissue differentiation and organ development still occurs during the fetal stage, but for the most part fetal development is primarily limited to body growth. Figure 29.25 depicts changes in external

Figure 29.25. Changes in the external appearance of the fetus from the ninth through the thirty-eighth week.

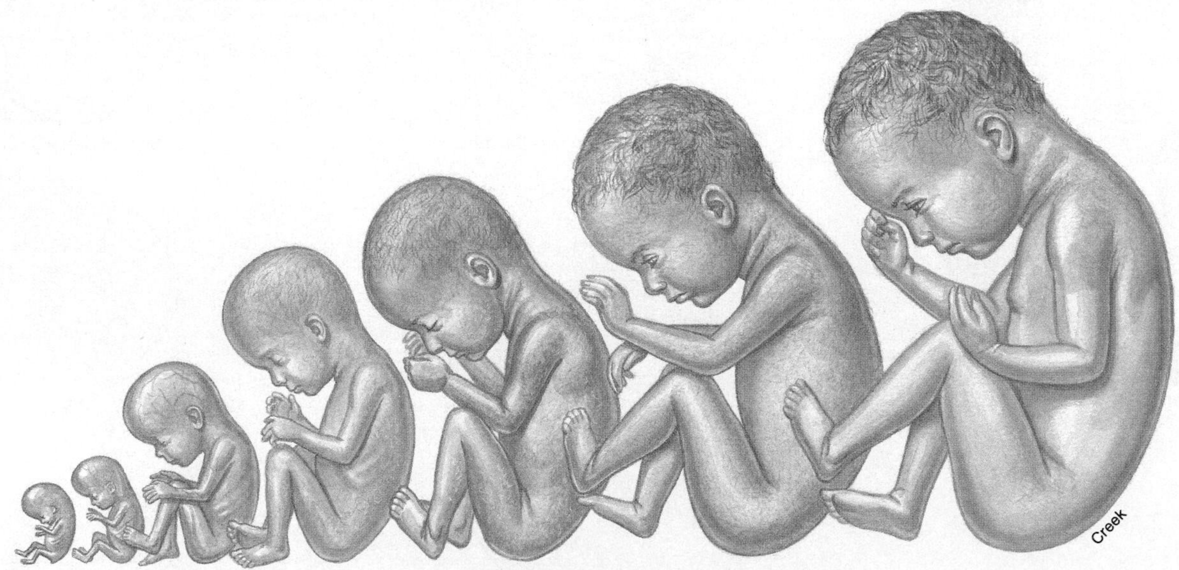

Figure 29.26. External appearances of fetuses at (*a*) ten weeks and (*b*) twelve weeks.

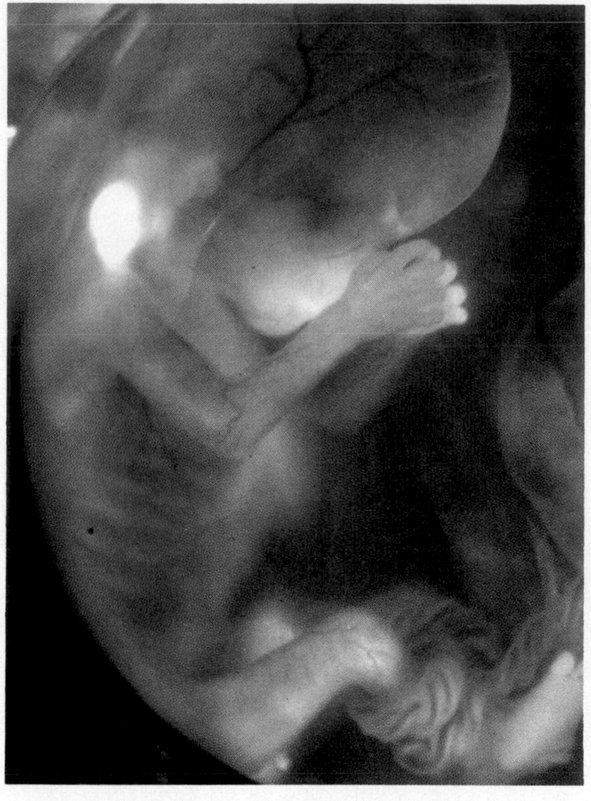

(a)

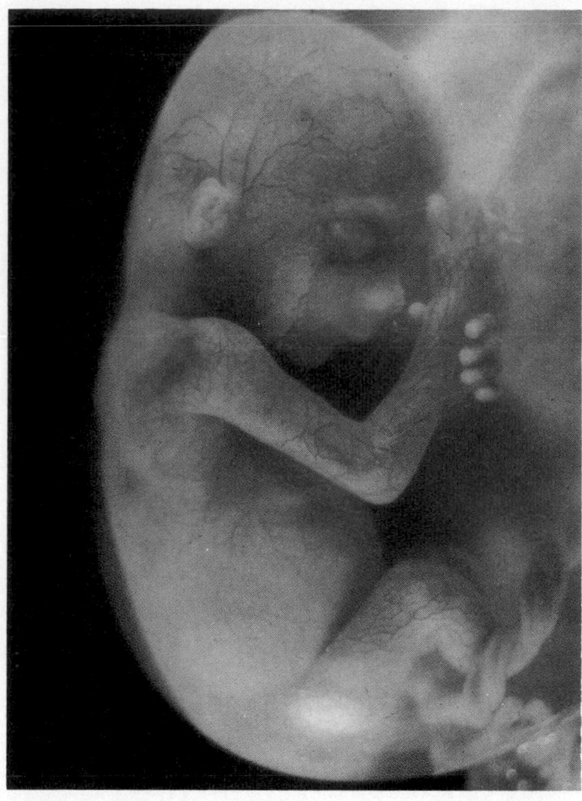

(b)

appearance of the fetus from the eleventh through the thirty-eighth week. A discussion of the structural changes of the fetus by weeks follows.

Nine to Twelve Weeks. At the beginning of the ninth week, the head is as large as the rest of the body. The eyes are widely spaced, and the ears are set low. Head growth slows during the next three weeks, whereas growth in body length accelerates. Ossification centers appear in most bones during the ninth week. Differentiation of the external genitalia becomes apparent at the end of the ninth week, but the genitalia are not developed to the point of sex determination until the twelfth week. By the end of the twelfth week, the fetus is 87 mm (3.5 in.) long and weighs about 45 g (1.6 oz). It can swallow, digest the fluid that passes through its system, and defecate and urinate into the amniotic fluid. The nervous system and muscle coordination are developed enough so that the fetus will withdraw its leg if tickled. The fetus begins inhaling through its nose but can take in only amniotic fluid. Figure 29.26 depicts the external appearance of fetuses at ten and twelve weeks.

Major structural abnormalities, which may not be predictable from genetic analysis, can often be detected by *ultrasound* (see fig. 29.27). Sound-wave vibrations are reflected from the interface of tissues with different densities—such as the interface between the fetus and amniotic fluid—and used to produce an image. This technique is so sensitive that it can be used to detect a fetal heartbeat several weeks before it can be detected by a stethoscope.

Thirteen to Sixteen Weeks. During the period from thirteen through sixteen weeks, the facial features are well formed, and epidermal structures such as eyelashes, eyebrows, hair on the head, fingernails, and nipples begin to develop. The appendages lengthen, and by the sixteenth week the skeleton is sufficiently developed so that it shows clearly on X-ray films. During the sixteenth week, the fetal heartbeat can be heard by applying a stethoscope to the mother's abdomen. By the end of the sixteenth week, the fetus is 140 mm in length (5.5 in.) and weighs about 200 g (7 oz).

Figure 29.27. Structures of the human fetus observed through an ultrasound scan.

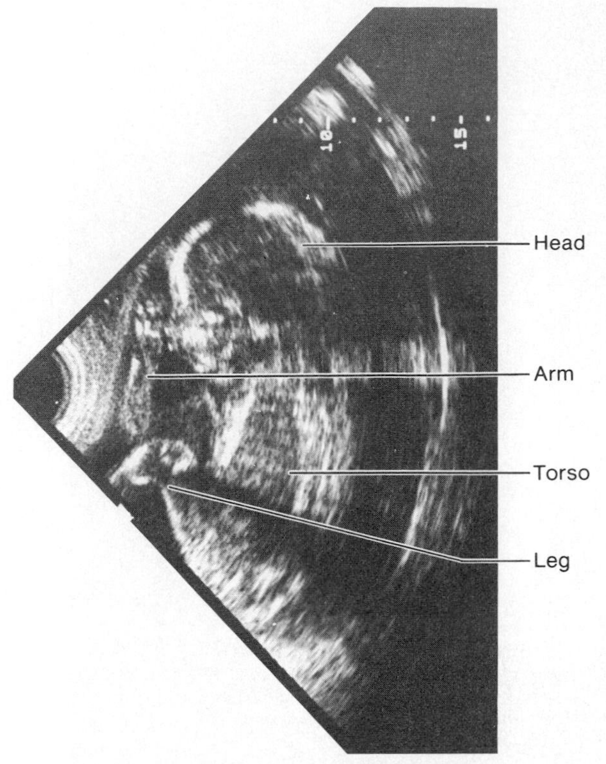

Figure 29.28. A fetus in vertex position. Toward the end of most pregnancies, the weight of the fetal head causes a rotation of its entire body such that the head is positioned in contact with the cervix of the uterus.

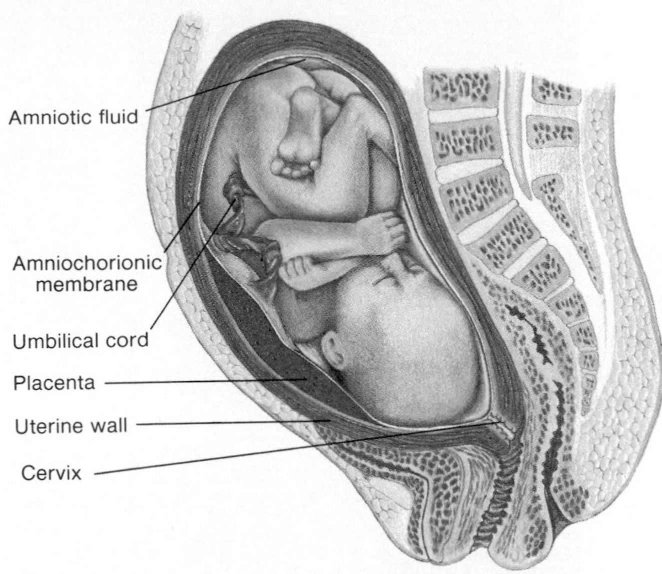

After the sixteenth week, fetal length can be determined using X-ray films. The reported length of a fetus is generally derived from a straight line measurement from the crown of the head to the developing ischium (crown-rump length). Measurements made on an embryo prior to the fetal stage, however, are not reported as crown-rump measurements but as total length.

Seventeen to Twenty Weeks. During the period from seventeen to twenty weeks, the legs achieve their final relative proportions, and fetal movements, known as **quickening,** are commonly felt by the mother. The skin is covered with a white, cheeselike material known as **vernix caseosa.** It consists of fatty secretions from the sebaceous glands and dead epidermal cells. The function of vernix caseosa is to protect the fetus while it is bathed in amniotic fluid. Twenty-week-old fetuses usually have fine, silklike fetal hair, called **lanugo** *(lah-nu'go),* covering the skin. Lanugo is thought to hold the vernix caseosa on the skin and produce a ciliarylike motion that moves amniotic fluid. The length of a twenty-week-old fetus is about 190 mm (7.5 in.) and the weight is 460 gm (16 oz). Because of cramped space, the fetus develops a marked spinal flexure and is in what is commonly called the "fetal posture," with the head bent down in contact with the flexed knees.

Twenty-one to Twenty-five Weeks. During the period from twenty-one to twenty-five weeks, the fetus increases its weight substantially to about 900 gm (32 oz). Body length increases only moderately (240 mm), however, so the weight is evenly proportioned. The skin is quite wrinkled and is translucent pinkish in color because the blood flowing in the capillaries is now visible.

Twenty-six to Twenty-nine Weeks. Toward the end of the period from twenty-six to twenty-nine weeks, the fetus will be 275 mm (about 11 in.) in length and will weigh 1,300 gm (46 oz). A fetus might now survive if born prematurely, but the mortality rate is high. Its body metabolism cannot yet maintain a constant temperature, and the respiratory muscles have not matured enough to provide a regular respiratory rate. If, however, the premature infant is put in an incubator and a respirator is used to maintain its breathing, it may survive. The eyes open during this period, and the body is well covered with lanugo. If the fetus is a male, the testes should have begun descent into the scrotum (see fig. 27.8). As the time of birth approaches, the fetus rotates to a **vertex position** (fig. 29.28). The head repositions toward the cervix because of the shape of the uterus and because the head is the heaviest part of the body.

Thirty to Thirty-eight Weeks. By the end of thirty-eight weeks, the fetus is considered "full-term." It has reached a crown-rump length of 360 mm (14 in.) and weighs 3,400 gm (7.5 lbs). The average total length from crown to heel

vernix caseosa: L. *vernix,* varnish; L. *caseus,* cheese
lanugo: L. *lana,* wool

vertex: L. *vertex,* summit

is 50 cm (20 in.). Most fetuses are plump with smooth skin because of the accumulation of subcutaneous fat. The skin is pinkish blue in color even on fetuses of dark-skinned parents, because melanocytes do not produce melanin until the skin is exposed to sunlight. Lanugo hair is sparse and is generally found on the head and back. The chest is prominent, and the mammary area protrudes in both sexes. The external genitalia are somewhat swollen.

1. Explain why the ninth week is designated as the beginning of the fetal stage of development.
2. List the approximate fetal age at which each of the following occur: first detection of fetal heartbeat; presence of vernix caseosa and lanugo; fetal rotation into vertex position.
3. Compare the techniques of ultrasound and X rays in examining fetal development and structure.

Labor and Parturition

Labor and parturition are the culmination of gestation and require the action of oxytocin, secreted by the posterior pituitary, and prostaglandins, produced in the uterus.

Objective 13. Describe the hormonal action that controls labor and parturition.

Objective 14. Describe the three stages of labor.

The time of prenatal development, or the time of pregnancy, is called **gestation** *(jes-ta'shun)*. The human gestational period is usually 266 days or about 280 days from the beginning of the last menstrual period to **parturition,** or birth. Most fetuses are born within 10 to 15 days before or after this time. Parturition is accompanied by a sequence of physiological and physical events called **labor.**

The *onset of labor* is denoted by rhythmic and forceful contractions of the myometrium layer of the uterus (see table 29.4). In *true labor,* the pains from uterine contractions occur at regular intervals and intensify as the time between contractions shorten. A reliable indication of true labor is dilation of the cervix and a "show," or discharge of blood-containing mucus in the cervical canal and vagina. In *false labor,* abdominal pain is felt at irregular intervals, and there is a lack of cervical dilations and cervical "show."

The uterine contractions of labor are stimulated by two agents: (1) **oxytocin** *(ok''si-to'sin),* a polypeptide hormone produced in the hypothalamus and secreted by the posterior pituitary, and (2) **prostaglandins** *(pros''tah-glan'dins),* a class of fatty acids produced within the uterus itself. Labor can indeed be induced artificially by injections of oxytocin or by the insertion of prostaglandins into the vagina as a suppository.

gestation: L. *gestatus,* to bear

Table 29.4 Possible sequence of events leading to the onset of labor in humans

Step	Event
1	High estrogen secretion from the placenta stimulates production of oxytocin receptors in the uterus.
2	Uterine muscle (myometrium) becomes increasingly sensitive to effects of oxytocin during pregnancy.
3	Oxytocin may stimulate production of prostaglandins in the uterus.
4	Prostaglandins may stimulate uterine contractions.
5	Contractions of the uterus stimulate oxytocin secretion from the posterior pituitary.
6	Increased oxytocin secretion stimulates increased uterine contractions, creating a positive feedback loop and resulting in labor.

From Stuart Ira Fox, *Human Physiology,* 2d ed.

The hormone *relaxin,* produced by the corpus luteum, may also be involved in labor and parturition. Relaxin is known to soften the symphysis pubis in preparation for parturition and is thought to also soften the cervix in preparation for dilation. It may be, however, that relaxin does not affect the uterus, but rather progesterone and estradiol may be responsible for this effect. Further research is necessary to understand the total physiological effect of these hormones.

Labor is divided into three stages (fig. 29.29).

1. **Dilation stage.** In this period the cervix dilates to a diameter of approximately 10 cm. There are regular contractions during this stage and usually a rupturing of the amniotic sac ("bag of waters"). If the amniotic sac does not rupture spontaneously, it is done surgically. The dilation stage may last eight to twenty-four hours, depending on whether it is occurring in the first or subsequent pregnancies.
2. **Expulsion stage.** This is the period of parturition, or actual childbirth. It consists of forceful uterine contractions and abdominal compressions to expel the fetus from the uterus and through the vagina. This stage may require thirty minutes in a first pregnancy, but only a few minutes in subsequent pregnancies.
3. **Placental stage.** Generally within ten to fifteen minutes after parturition, the placenta is separated from the uterine wall and expelled as the *"afterbirth."* Forceful uterine contractions characterize this stage, constricting uterine blood vessels to prevent hemorrhage. In a normal delivery, blood loss does not exceed 350 milliliters.

A *pudendal nerve block* may be administered during the early part of the expulsion stage to ease the trauma of delivery for the mother and make the performance of an episiotomy possible.

Figure 29.29. The stages of labor and parturition. (*a*) the position of the fetus prior to labor. (*b*) the ruptured amniotic sac and early dilation of the cervix. (*c*) expulsion stage, or the period of parturition. (*d*) the placental stage.

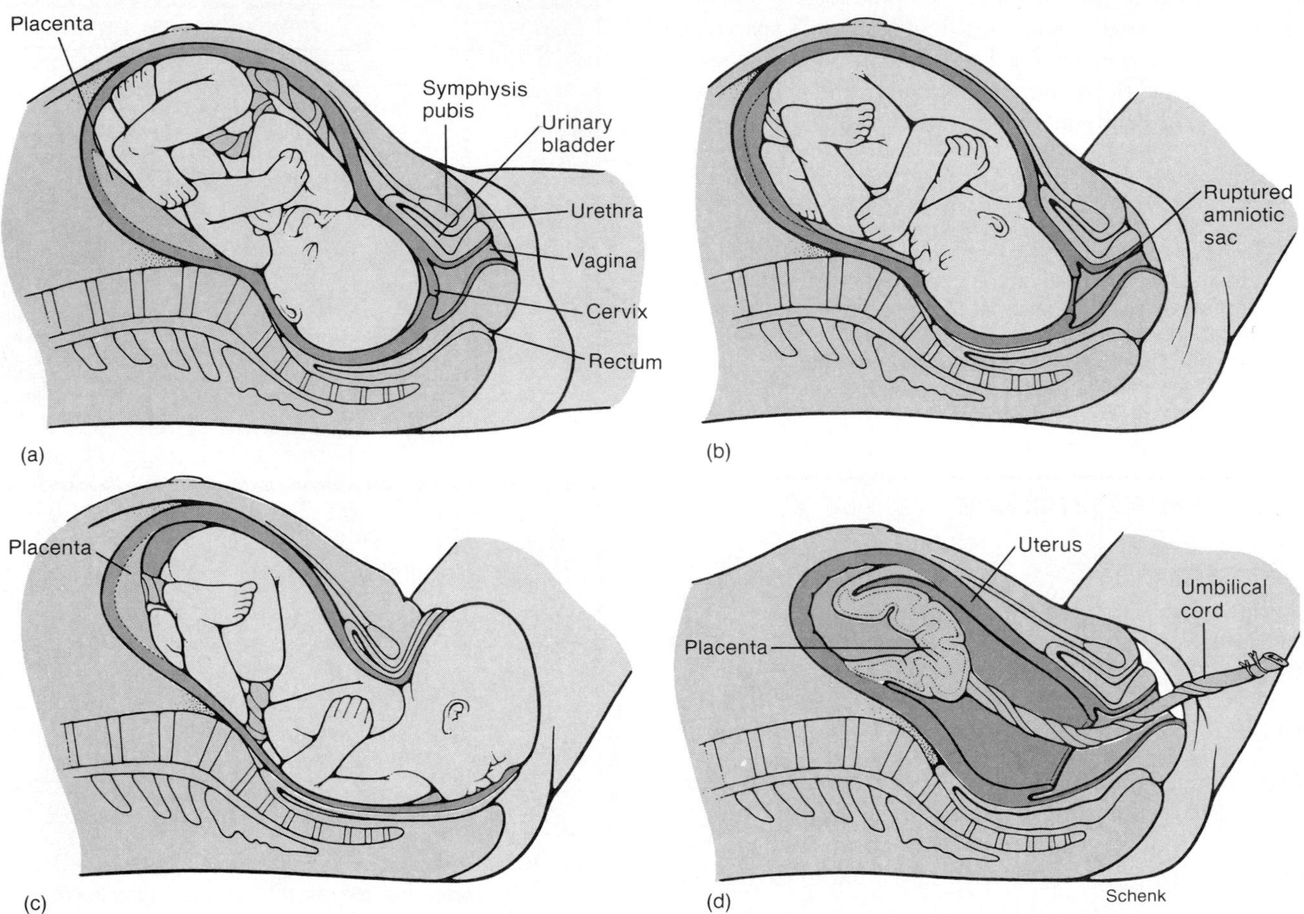

Five percent of newborns are born *breech.* In a breech birth, the fetus has not rotated and the buttocks are the presenting part. The principal concern of a breech birth is the increased time and difficulty of the expulsion stage of parturition. Attempts to rotate the fetus through the use of forceps may injure the infant. If an infant cannot be delivered breech, a *cesarean (se-sa're-an) section* must be performed. A cesarean section is delivery of the fetus through an incision made into the abdominal wall and the uterus.

INHERITANCE

Inheritance is the passage of hereditary traits carried on the genes of chromosomes from one generation to another.

Objective 15. Define the term *genetics.*

Objective 16. Discuss the variables that account for a person's phenotype.

Objective 17. Explain how probability is involved in predicting inheritance, and use a Punnett square to illustrate selected probabilities.

Genetics is the branch of biology that deals with inheritance. Genetics and inheritance are important in anatomy and physiology because of the numerous developmental and functional disorders that have a genetic basis. Genetic counseling is the practical application of knowing which disorders and diseases are inherited. The genetic inheritance of an individual begins with conception.

Each zygote inherits twenty-three chromosomes from its mother and twenty-three chromosomes from its father. This does not produce forty-six different chromosomes but, rather, twenty-three pairs of *homologous chromosomes.* Each member of a homologous pair, with the important exception of the sex chromosomes, looks like the other and contains similar genes (such as those coding for eye color, height, and so on). These homologous pairs of chromosomes can be **karyotyped** (photographed or illustrated) and identified (as shown in figures 29.30, 29.31). Each cell that contains forty-six chromosomes (that is *diploid*) has two chromosomes number 1, two chromosomes number 2, and so on through chromosomes number 22. The first twenty-two pairs of chromosomes are called **autosomal** *(aw'' to-so'-mal)* **chromosomes.** The twenty-third pair of chro-

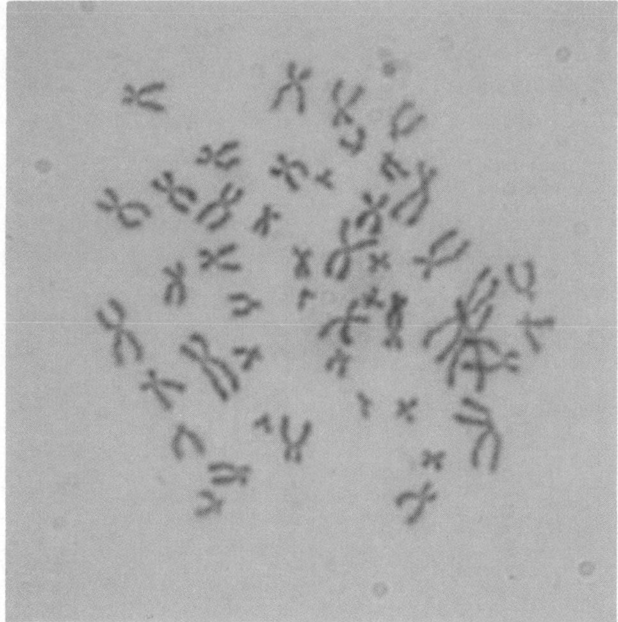

Figure 29.30. A karyotype photograph of the twenty-three pairs of homologous human chromosomes.

mosomes are the **sex chromosomes,** which may look different and may carry different genes. In a female these consist of two X chromosomes, whereas in a male there is only one X chromosome and one Y chromosome.

Genes and Alleles. A **gene** is a portion of the DNA of a chromosome that contains the information needed to synthesize a particular protein molecule. Although each diploid cell has a pair of gene locations for each characteristic, there may be a number of alternate forms of each gene. Those alternative forms of a gene that affect the same characteristic, but produce different expressions of that characteristic, are called **alleles** *(ah-lēls')*. One allele of each pair originates from the female parent and the other from the male. The shape of a person's ears, for example, is determined by the kind of allele received from each parent and how the alleles interact with one another. Alleles are always located on the same spot (called a **locus**), on homologous chromosomes (fig. 29.32).

For any particular pair of alleles in a person, the two alleles are either identical or not identical. If the alleles are identical, the person is said to be **homozygous** *(ho''mo-zi'gus)* for that particular characteristic. But if the two alleles are different, the person is **heterozygous** *(het''er-o-zi'gus)* for that particular trait.

Genotype and Phenotype. A person's DNA contains a catalog of genes known as the **genotype** *(je'no-tīp)* of that person. The expression of those genes results in certain observable characteristics referred to as the **phenotype** *(fe'no-tīp)*.

Figure 29.31. A karyotype of homologous pairs of chromosomes obtained from a human diploid cell. The first twenty-two pairs of chromosomes are called the autosomal chromosomes. The sex chromosomes are (*a*) XY for a male and (*b*) XX for a female.

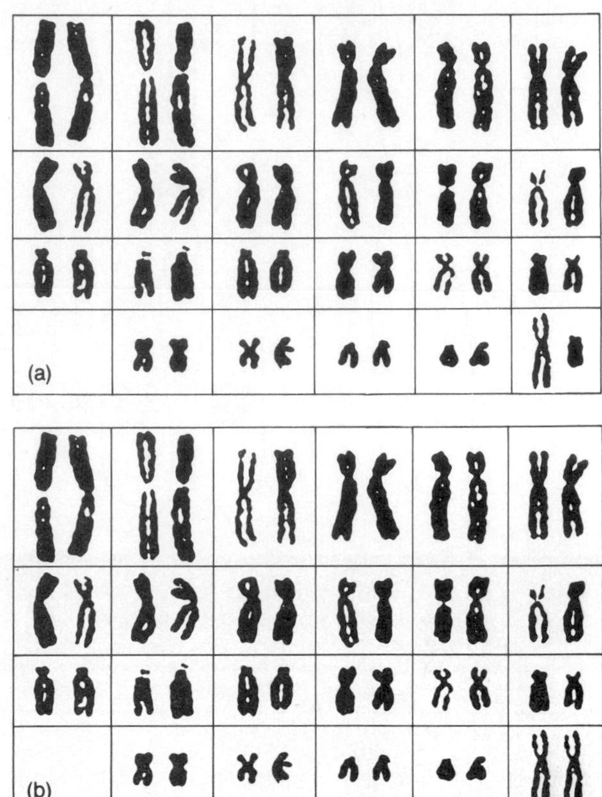

If the alleles for a particular trait are homozygous, the characteristic expresses itself in a specific manner (two alleles for attached earlobes, for example, results in a person with attached earlobes). If the alleles for a particular trait are heterozygous, however, the allele that expresses itself and how the genes for that trait interact will determine the phenotype. Often one of the alleles expresses itself as the **dominant allele,** whereas the other does not and is the **recessive allele.** The combinations of dominant and recessive alleles are responsible for a person's hereditary traits (table 29.5).

In describing genotypes, it is traditional to use letter symbols to refer to the alleles of an organism. The dominant alleles are symbolized by uppercase letters and the recessive alleles are symbolized by lowercase. Thus, the *genotype* of a person who is homozygous for free earlobes due to a dominant allele is symbolized *EE;* a heterozygous pair is symbolized *Ee.* In both of these instances, the *phenotypes* of the individuals would be free earlobes, since a dominant allele is present in each genotype. A person who inherited two recessive alleles for earlobes has the genotype *ee* and will have attached earlobes.

Figure 29.32. A pair of homologous chromosomes. Homologous chromosomes contain genes for the same characteristic at the same locus.

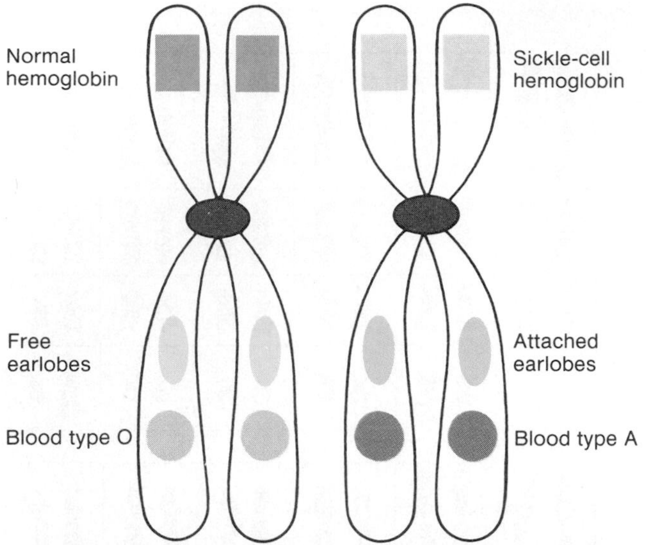

Figure 29.33. Inheritance of ear shape. Two heterozygous parents with free earlobes can have a child with attached earlobes.

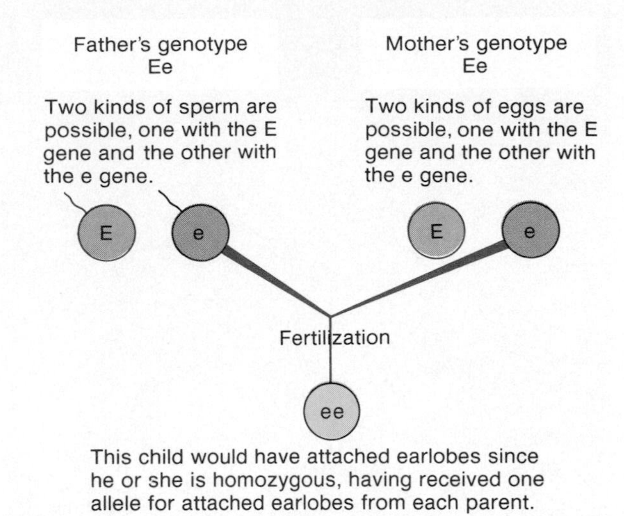

Table 29.5	Some hereditary traits in humans determined by single pairs of dominant and recessive alleles		
Dominant	**Recessive**	**Dominant**	**Recessive**
Free earlobes	Attached earlobes	Color vision	Color blindness
Dark brown hair	All other colors	Broad lips	Thin lips
Curly hair	Straight hair	Ability to roll tongue	Lack of this ability
Pattern baldness (♂♂)	Baldness (♀♀)	Arched feet	Flatfeet
Pigmented skin	Albinism	A or B blood factor	O blood factor
Brown eyes	Blue or green eyes	Rh blood factor	No Rh blood factor

Thus, three genotypes are possible when gene pairing involves dominant and recessive alleles. They are *homozygous dominant (EE)*, *heterozygous (Ee)*, and *homozygous recessive (ee)*. Only two phenotypes are possible, however, since the dominant allele is expressed in both the homozygous dominant (*EE*) and the heterozygous (*Ee*) individuals. The recessive allele is expressed only in the homozygous recessive (*ee*) condition. Figure 29.33 illustrates how a homozygous recessive trait may be expressed in a child of parents who are heterozygous dominant.

Figure 29.34. Inheritance of the shape of earlobes.

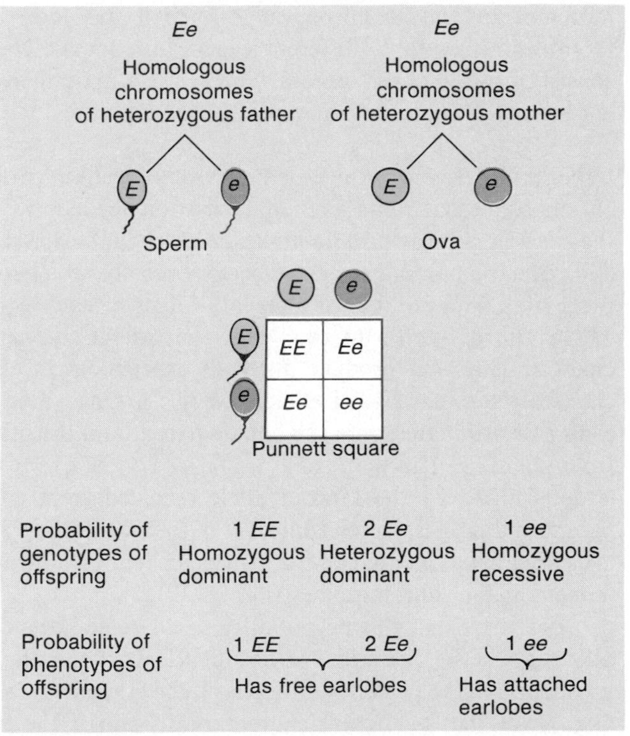

Probability. A **Punnett square** is a convenient way to express the probabilities of allele combinations for a particular inheritable trait. In constructing a Punnett square, the male gametes (spermatozoa) carrying a particular trait are placed at the side of the chart, and the female gametes (ova) at the top (as in figure 29.34). The four spaces on the chart represent the possible combinations of male and female gametes that could form zygotes. The probability of an offspring having a particular genotype is one in four (.25) for homozygous dominant and homozygous recessive, and one in two (.50) for heterozygous dominant.

Figure 29.35. A dihybrid cross studies the probability of two characteristics at the same time. Any of the combinations of genes that have a *D* and an *E* (nine possibilities) will have free earlobes and dark hair. These are indicated with an asterisk (*). Three of the possible combinations have two alleles for attached earlobes (*ee*) and at least one allele for dark hair. They are indicated with a dot (•). Three of the combinations have free earlobes and light hair. These are indicated with a square (■). The remaining possibility has the genotype (*eedd*) for attached earlobes and light hair.

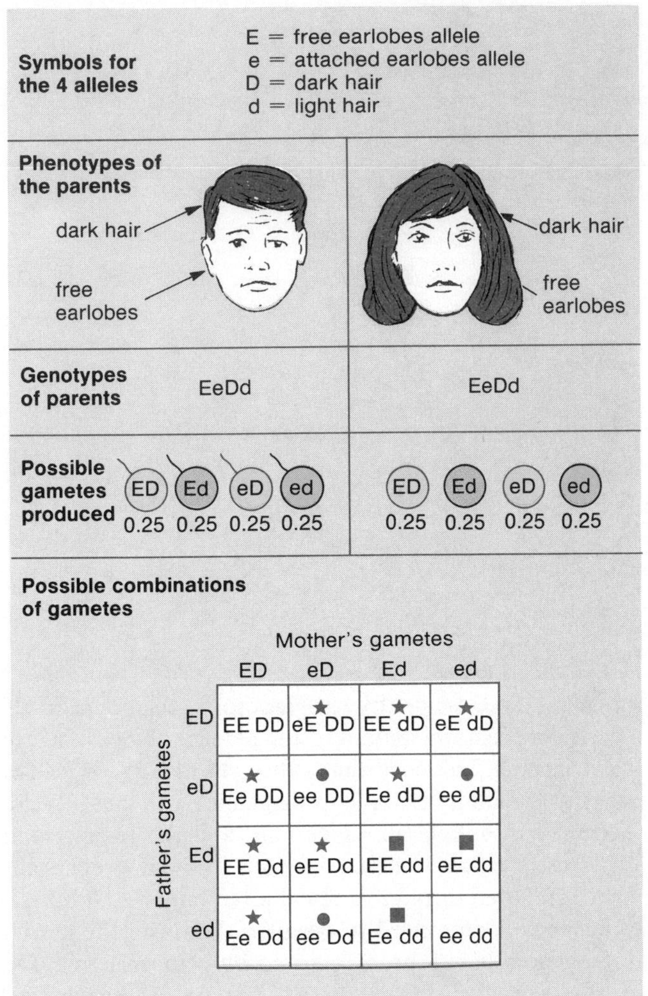

A genetic study in which a single characteristic (e.g., ear shape) is followed from parents to offspring is referred to as a **monohybrid cross.** A genetic study in which two characteristics are followed from parents to offspring is referred to as a **dihybrid cross** (fig. 29.35). The term **hybrid** refers to an offspring descended from parents that have different genotypes.

Sex-linked Inheritance. Certain inherited traits are located on a sex-determining chromosome and are called **sex-linked** characteristics. The allele for red-green *color blindness,* for example, is determined by a recessive allele (designated c) found on the X chromosome but not the Y chromosome. Normal color vision (designated C) dominates. The ability to discern red-green colors, therefore, depends entirely on the X chromosomes. The genotype possibilities are

X^CY	Normal male
X^cY	Color-blind male
X^CX^C	Normal female
X^CX^c	Normal female carrying the recessive allele
X^cX^c	Color-blind female

In order for a female to be red-green color-blind, she must have the recessive allele on both of her X chromosomes. Her father would have to be red-green color blind, and her mother would have to be a carrier for this condition. A male with only one such allele on his X chromosome, however, will show the characteristic. Since a male receives his X chromosome from his mother, the inheritance of sex-linked characteristics usually passes from mother to son.

Hemophilia is a sex-linked condition caused by a recessive allele. The blood in a person with hemophilia fails to clot or clots very slowly after an injury. If H represents normal clotting and h represents abnormal clotting, then males with X^HY will be normal and males with X^hY will be hemophiliac. Females with X^hX^h will have the disorder.

1. Define the following terms: *genetics, genotype, phenotype, allele, dominant, recessive, homozygous,* and *heterozygous.*
2. List several dominant and recessive traits inherited in humans. What are some variables that determine a person's phenotype?
3. Construct a Punnett square to show the possible genotypes for color blindness of an X^CY male and an X^CX^c female.

Clinical Considerations

Pregnancy and childbirth are natural events in human biology and generally progress smoothly without complications. Prenatal development is amazingly precise, and although traumatic, childbirth for most women in the world takes place without the aid of a physician. Occasionally, however, serious complications arise, and the knowledge of an obstetrician is required. The physician's knowledge of normal development and the causes of congenital malformations ensures the embryo and fetus every possible chance to develop normally. Many of the clinical aspects of prenatal development involve what might be referred to as applied developmental biology.

In clinical terms, gestation is frequently divided into three phases, or **trimesters,** each lasting three calendar months. By the end of the **first trimester** all of the major body systems are formed, the fetal heart can be detected,

Figure 29.36. The various sites of ectopic pregnancies. The normal site is indicated by an X, and the abnormal sites are indicated by letters in order of frequency of occurrence.

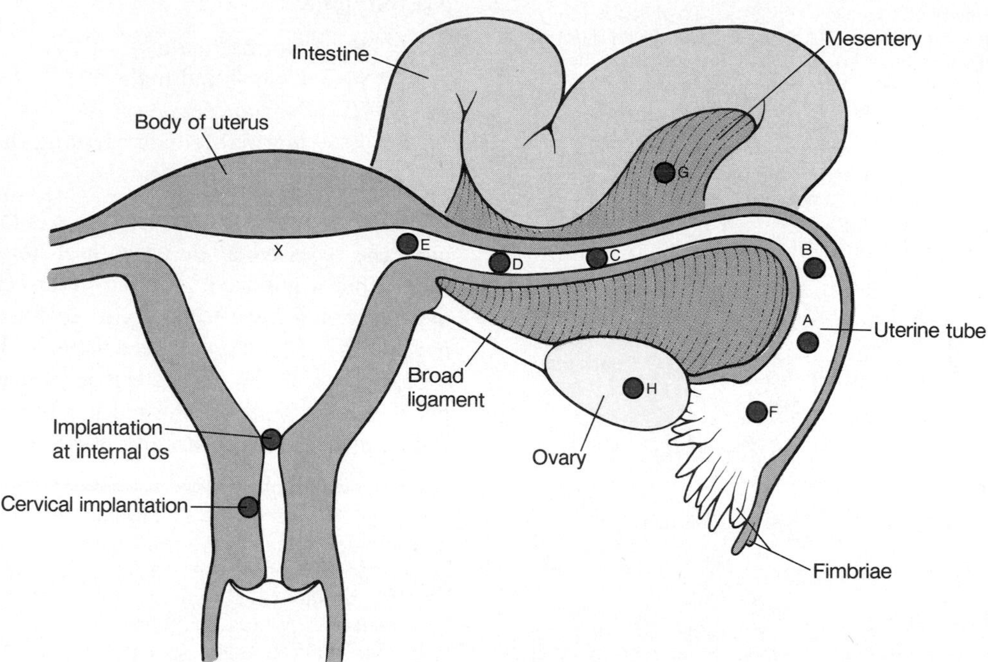

the external genitalia are developed, and the fetus is about the width of the palm of an adult's hand. During the **second trimester,** fetal quickening can be detected, epidermal features are formed, and the vital body systems are functioning. The fetus, however, is still unlikely to survive if birth were to occur. At the end of the second trimester, fetal length is about equal to the length of an adult's hand. The fetus experiences a tremendous amount of growth and refinement in system functioning during the **third trimester.** A fetus of this age may survive if born prematurely, and of course, the chances of survival improve as the length of pregnancy approaches the natural delivery date.

Many clinical considerations are associated with prenatal development, some of which relate directly to the female reproductive system. Other developmental problems are genetically related and will be mentioned only briefly. Of clinical concern for developmental anatomy are topics such as implantation sites, test-tube development, multiple pregnancy, fetal monitoring, and congenital defects.

Abnormal Implantation Sites

In an **ectopic pregnancy** the blastocyst implants outside the uterus or in an abnormal site within the uterus (fig. 29.36). The most common ectopic location (about 95%) is within the uterine tube and is referred to as a **tubal pregnancy.** Occasionally, implantation occurs near the cervix, where development of the placenta blocks the cervical opening. This condition, called **placenta previa,** causes serious bleeding. Ectopic pregnancies will not develop normally in unfavorable locations, and the fetus seldom survives beyond the first trimester. Tubular pregnancies are terminated through medical intervention. If a tubular pregnancy is permitted to progress, however, the uterine tube generally ruptures, followed by hemorrhaging. Depending on the location and the stage of development (hence vascularity) of a tubal pregnancy, it may not be serious or the hemorrhaging and shock may cause the death of the woman.

In Vitro Fertilization and Artificial Implantation

Reproductive biologists have been able to fertilize a human oocyte *in vitro* (outside the body), culture it to the blastocyst stage, and then perform artificial implantation, leading to a full-term development and delivery. This is the so-called test-tube baby. To obtain the oocyte, a specialized laparoscope (fig. 29.37) is used to aspirate the preovulatory egg from a vesicular ovarian (graafian) follicle. The oocyte is then placed in a suitable culture medium, where it is fertilized with sperm. After the zygote

Figure 29.37. A laparoscope, used for various abdominal operations including the extraction of a preovulatory ovum.

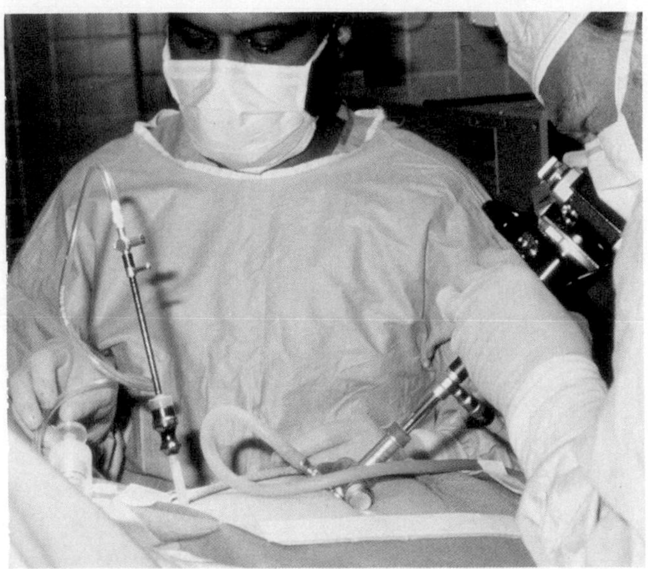

forms, the sequential pre-embryonic development continues until the blastocyst stage, at which time implantation is performed. *In vitro* fertilization with artificial implantation is a means of overcoming infertility problems due to blocked uterine tubes in females or low sperm counts in males.

Multiple Pregnancy

Twins occur about once in eighty-five pregnancies. They can develop in two ways. **Dizygotic** (fraternal) **twins** develop from two zygotes resulting from two spermatozoa fertilizing two oocytes in the same ovulatory cycle (fig. 29.38). **Monozygotic** (identical) **twins** form from a single zygote (fig. 29.39). Approximately two-thirds of twins are dizygotic.

Dizygotic twins may be of the same sex or different sexes and are not any more alike than brothers or sisters born at different times. Dizygotic twins always have two chorions and two amnions, but the chorions and the placentas may be fused.

Monozygotic twins are of the same sex and are genetically identical. Any physical differences in monozygotic twins are caused by environmental factors during

Figure 29.38. The formation of dizygotic twins. Twins of this type are not identical but fraternal, and may have (*a*) separate or (*b*) fused placentas. (*c*) photo of fraternal twins at eleven weeks.

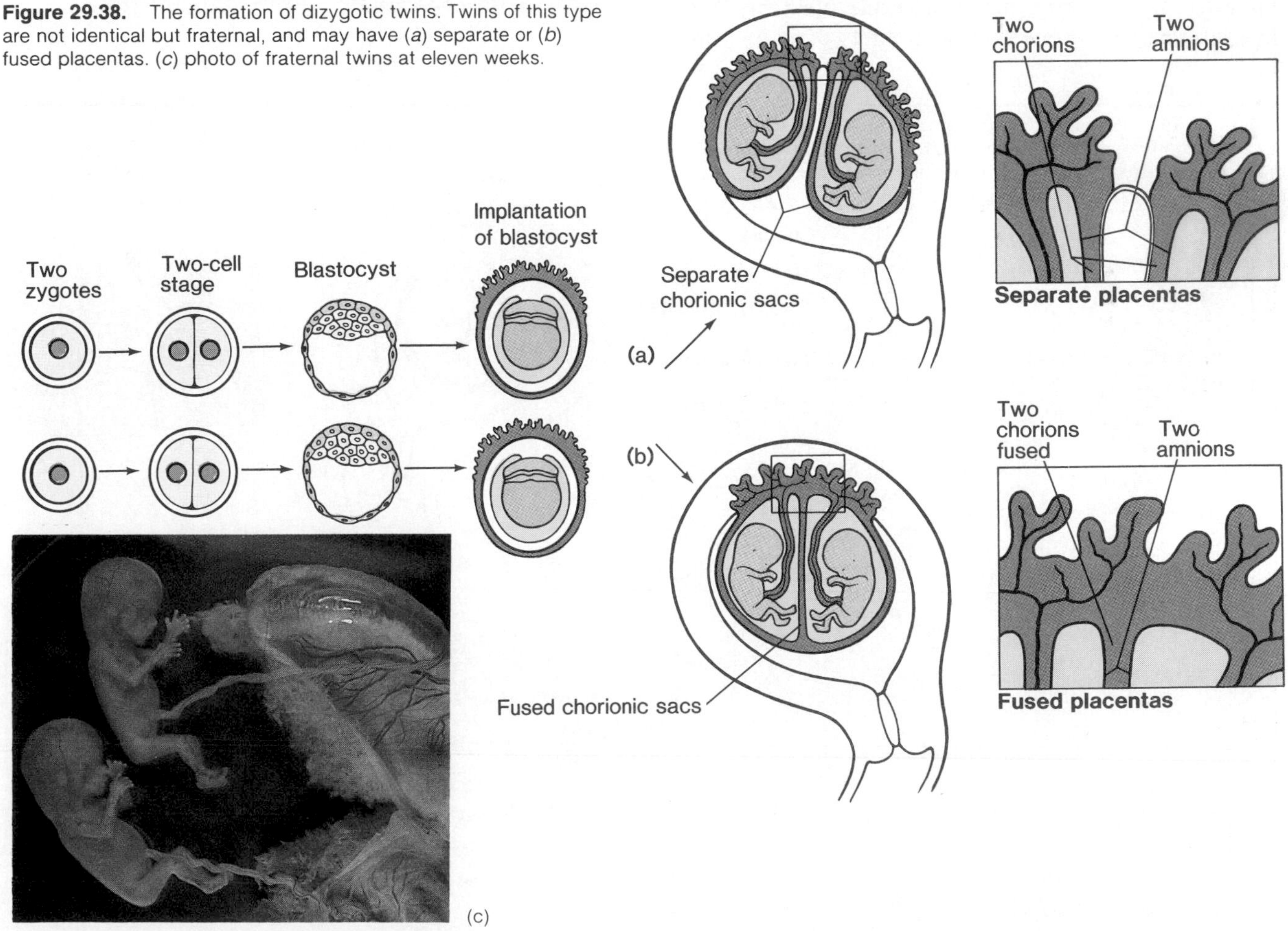

Two zygotes
Two-cell stage
Blastocyst
Implantation of blastocyst
Separate chorionic sacs
(a)
Two chorions
Two amnions
Separate placentas
(b)
Two chorions fused
Two amnions
Fused chorionic sacs
Fused placentas
(c)

morphogenic development (e.g., there might be a differential vascular supply that causes slight differences to be expressed). Monozygotic twinning is usually initiated toward the end of the first week when the inner cell mass divides to form two embryonic primordia. Monozygotic twins have two amnions but only one chorion and a common placenta. If the inner mass fails to completely divide, **conjoined twins** (Siamese twins) may form.

Triplets occur about once in 7,600 pregnancies and may be (1) all from the same ovum and identical, (2) two identical and the third from another ovum, or (3) three zygotes from three different ova. Similar combinations occur in quadruplets, quintuplets, and so on.

Fetal Monitoring

Obstetrics has benefited greatly from the advancements made in fetal monitoring in the last two decades. Before these techniques became available, physicians could determine the welfare of the unborn child only by auscultation of the fetal heart and palpation of the fetus. Currently, there are several tests that provide much information about the fetus during any stage of development. Fetal conditions that can now be diagnosed and evaluated include genetic disorders, hypoxia, blood disorders, growth retardation, placental functioning, prematurity, postmaturity, and intrauterine infections. These tests also help determine the advisability of an abortion.

X rays of the fetus were once commonly performed but were found harmful and have been replaced by other methods of evaluation that are safer and more informative. **Ultrasonography,** produced by a mechanical vibration of high frequency, produces a safe, high resolution (sharp image) of fetal structure (fig. 29.40). Ultrasonic

Figure 29.40. Color-enhanced ultrasonogram of a fetus during the third trimester. The left hand is raised, as if waving to the viewer.

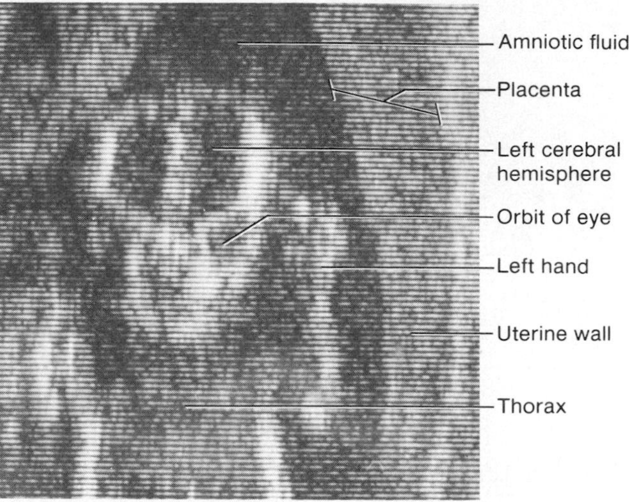

Figure 29.39. The formation of monozygotic twins. Twins of this type develop from a single zygote and are identical. Such twins have two amnions but one chorion and a common placenta.

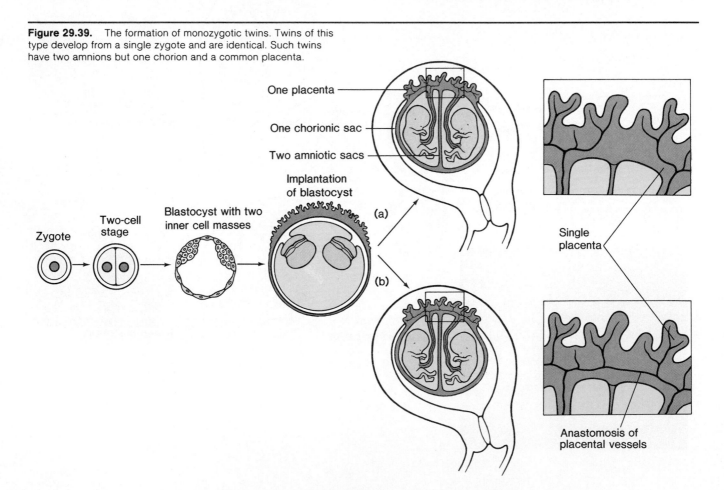

Figure 29.41. Fetoscopy.

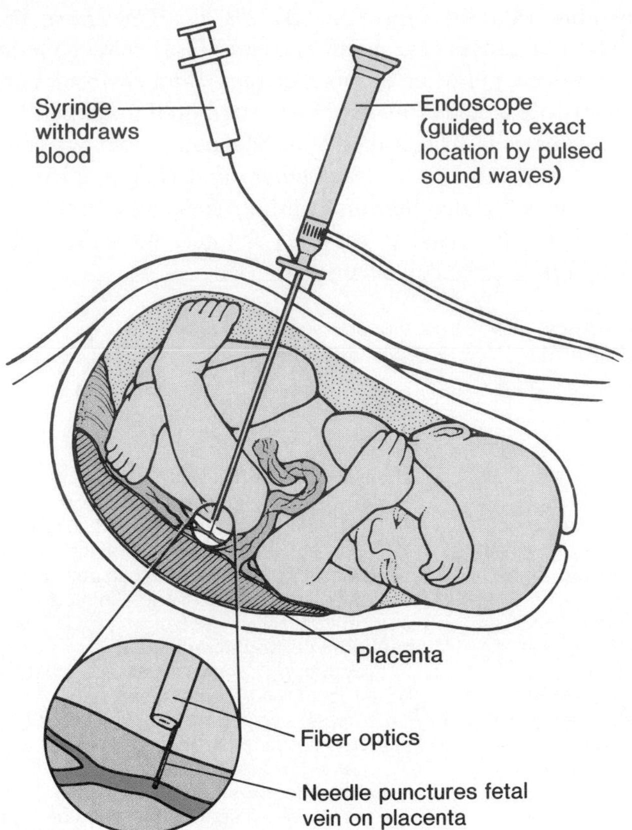

imaging is a reliable way to determine pregnancy as early as six weeks after ovulation. It can also be used to determine fetal weight, length, and position, as well as to diagnose multiple fetuses.

Amniocentesis is a technique used to obtain a small sample (5–10 ml) of amniotic fluid with a syringe so that the fluid can be assessed (see fig. 29.15). Amniocentesis is most often performed to determine fetal maturity, but it can also help predict serious disorders like *Down's syndrome (mongolism)* and *Gaucher's disease* (a metabolic disorder).

The new technique of **fetoscopy** (fig. 29.41) goes beyond amniocentesis by allowing direct examination of the fetus. Using fetoscopy, physicians scan the uterus with pulsed sound waves to locate fetal structures, the umbilical cord, and the placenta. Skin samples are taken from the head of the fetus and blood samples extracted from the placenta. The principal advantage of fetoscopy is that external features of the fetus (such as fingers, eyes, ears, mouth, and genitals) can be carefully observed. Fetoscopy is also used to determine several diseases, including hemophilia, thalassemia, and the 40% of sickle-cell anemia cases missed by amniocentesis.

amniocentesis: Gk. *amnion,* lamb (fetal membrane); *kentesis,* puncture
fetoscopy: L. *fetus,* offspring, *skopein,* to view

Figure 29.42. Monitoring fetal heart rate and uterine contractions using an FHR-UC device.

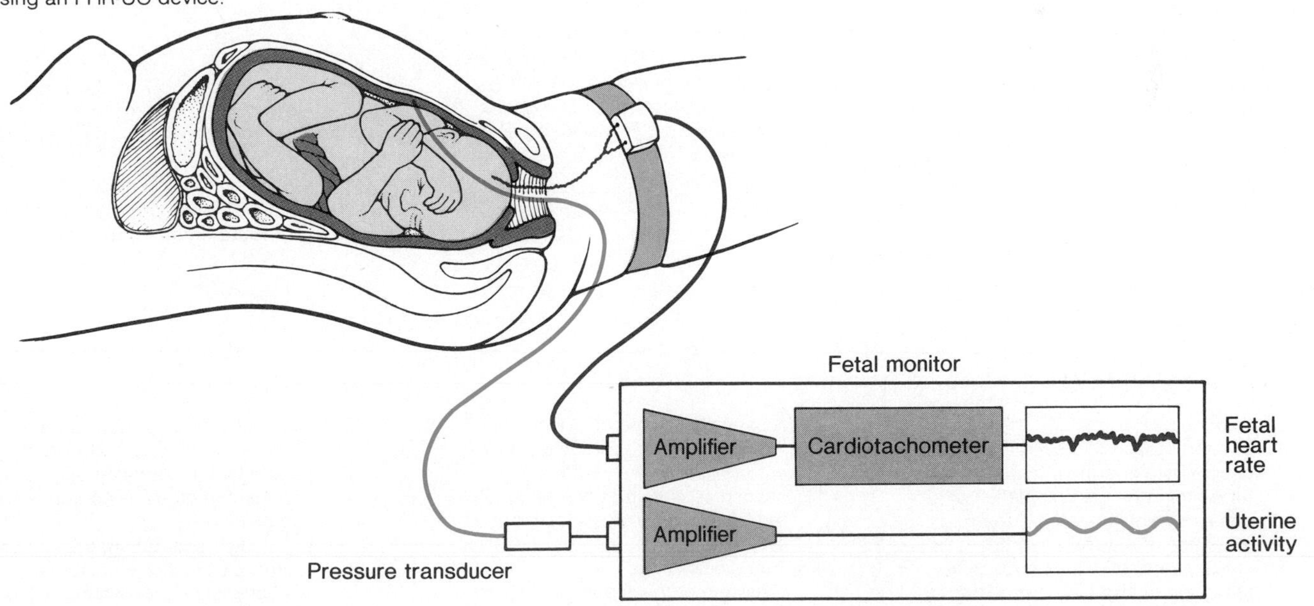

Most hospitals are now equipped with instruments that monitor fetal heart rate and uterine contractions during labor and can detect any complication that arises during the delivery. This procedure is called Electronic Monitoring of Fetal Heart Rate and Uterine Contractions (FHR-UC Monitoring). The stress to the fetus from uterine contractions can be determined through monitoring (fig. 29.42). Long, arduous deliveries can be taxing to both the mother and fetus. If the baby's health and vitality are diagnosed to be in danger because of a difficult delivery, the physician may decide to perform a cesarean section.

Congenital Defects

Major developmental problems called **congenital malformations** occur in approximately 2% of all newborn infants. The causes of congenital conditions include genetic inheritance, mutation (genetic change), and environment. About 15% of neonatal deaths are attributed to congenital malformations. The branch of developmental biology concerned with abnormal development and congenital malformations is called *teratology.* Many congenital problems have been discussed in previous chapters with the body system in which they occur.

teratology: Gk. *teras,* monster; *logos,* study of
congenital: L. *congenitus,* born with

Some Genetic Disorders of Clinical Importance

cystic fibrosis An autosomal recessive disorder that is characterized by the formation of thick mucus in the lungs and pancreas, which interferes with normal breathing and digestion.

familial cretinism An autosomal recessive disorder that is characterized by a lack of thyroid secretion, due to a defect in the iodine transport mechanism. Untreated children are dwarfed, sterile, and may be mentally retarded.

galactosemia An autosomal recessive disorder that is characterized by an inability to metabolize galactose, a component of milk sugar. Patients with this disorder have cataracts, damaged livers, and mental retardation.

gout An autosomal dominant disorder that is characterized by an accumulation of uric acid in the blood and tissue due to an abnormal metabolism of purines.

hepatic porphyria An autosomal dominant disorder that is characterized by painful gastrointestinal disorders and neurologic disturbances due to an abnormal metabolism of porphyrins.

hereditary hemochromatosis A sex-influenced, autosomal dominant disorder that is characterized by an accumulation of iron in the pancreas, liver, and heart, resulting in diabetes, cirrhosis, and heart failure.

hereditary leukomelanopathy An autosomal recessive disorder that is characterized by decreased pigmentation in the skin, hair, and eyes, and abnormal white blood cells. Patients with this condition are generally susceptible to infections and early deaths.

Huntington's chorea An autosomal dominant disorder that is characterized by uncontrolled twitching of skeletal muscles and the deterioration of mental capacities. A latent expression of this disorder allows the mutant gene to be passed to children before the symptoms develop.

Marfan's syndrome An autosomal dominant disorder that is characterized by tremendous growth of the extremities, dislocation of the lenses, and congenital cardiovascular defects.

phenylketonuria *(fen''il-ke''to-nu're-ah)* **(PKU)** An autosomal recessive disorder that is characterized by an inability to metabolize the amino acid phenylalanine. This is accompanied by brain and nerve damage and mental retardation.

pseudohypertrophic muscular dystrophy A sex-linked recessive disorder that is characterized by progressive muscle atrophy. It usually begins during childhood and causes death in adolescence.

retinitis pigmentosa A sex-linked recessive disorder that is characterized by progressive atrophy of the retina and eventual blindness.

Tay-Sachs disease An autosomal recessive disorder that is characterized by a deterioration of physical and mental abilities, early blindness, and death.

Huntington's chorea: from George Huntington, U.S. physician, 1850–1916
Marfan's syndrome: from Antoine Bernard-Jean Marfan, French physician, 1858–1942
Tay-Sachs disease: from Warren Tay, English physician, 1843–1927, and Bernard Sachs, U.S. neurologist, 1858–1944

Chapter Summary

I. Fertilization
   A. The fertilization of a secondary oocyte by a sperm in the uterine tube promotes the completion of meiotic development and the formation of a diploid zygote.
   B. Morphogenesis is the sequential formation of body structures during the prenatal period of human life. The prenatal period lasts for thirty-eight weeks and is divided into a pre-embryonic, an embryonic, and a fetal stage.
   C. A capacitated sperm digests its way through the zona pellucida and corona radiata layers of the secondary oocyte to complete the fertilization process and formation of a zygote.

II. Pre-Embryonic Stage
   A. Cleavage of the zygote is initiated within thirty hours and continues until a morula forms, which enters the uterine cavity on about the third day.
   B. A hollow, fluid-filled space forms within the morula, and it is then called a blastocyst.
   C. Implantation begins between the fifth and seventh day and is made possible by the secretion of enzymes that digest a portion of the endometrium.
      1. During implantation, the trophoblast secretes human chorionic gonadotropin (hCG), which prevents the breakdown of the endometrium and menstruation.
      2. The secretion of hCG declines by the tenth week as the developed placenta secretes steroids that maintain the endometrium.

D. The embryoblast of the implanted blastocyst flattens into the embryonic disc, from which the primary germ layers of the embryo develop.
   1. Ectoderm gives rise to the nervous system, the epidermis of the skin and epidermal derivatives, and portions of sensory organs.
   2. Mesoderm gives rise to bones, muscles, blood, reproductive organs, the dermis of the skin, and connective tissue.
   3. Endoderm gives rise to linings of the digestive tract, digestive organs, the respiratory tract and lungs, and the urinary bladder and urethra.

III. Embryonic Stage
   A. The events of the six-week embryonic stage include the differentiation of the germ layers into specific body organs and the formation of the placenta, the umbilical cord, and the extraembryonic membranes, which provide sustenance and protection to the embryo.
   B. The extraembryonic membranes include the amnion, yolk sac, allantois, and chorion.
      1. The amnion is a thin membrane surrounding the embryo and contains amniotic fluid, which cushions and protects the embryo.
      2. The yolk sac produces blood for the embryo.
      3. The allantois also produces blood for the embryo and gives rise to the umbilical arteries and vein.
      4. The chorion participates in the formation of the placenta.
   C. The placenta, formed from both maternal and embryonic tissue, has a transport role in providing for the metabolic needs of the fetus and in removing its wastes.
      1. The placenta produces steroids and hormones.
      2. Nicotine, drugs, alcohol, and viruses can cross the placenta to the fetus.
   D. The umbilical cord, containing two umbilical arteries and one umbilical vein, is formed as the amnion envelops the tissues on the underside of the embryo.
   E. During the third to the eighth week, the structure of all the body organs, except the genitalia, becomes apparent.
      1. During the third week, the primitive knot forms from the primitive streak, which later gives rise to the notochord and mesenchyme.
      2. By the end of the fourth week, the heart is beating; the primordial tissue of the eyes, brain, spinal cord, lungs, and digestive organs are in their proper place; and the arm and leg buds are recognizable.
      3. At the end of the fifth week, the sense organs are formed in the enlarged head, and the appendages have developed with finger rays present.
      4. The brain is well developed by the end of the sixth week, and the digits are separate on elongated appendages.
      5. During the seventh and eighth weeks the body organs, except the genitalia, are formed and the embryo appears distinctly human.

IV. Fetal Stage
   A. A small amount of tissue differentiation and organ development occurs during the fetal stage, but for the most part fetal development is primarily limited to body growth.
   B. Between weeks nine and twelve, ossification centers appear; the genitalia are formed; and the digestive, urinary, respiratory, and muscle systems show functional activity.
   C. Between weeks thirteen and sixteen, facial features are formed, and the fetal heart beat can be detected with a stethoscope.
   D. During the period from seventeen to twenty weeks, quickening can be felt by the mother, and vernix caseosa and lanugo cover the skin of the fetus.
   E. During the period from twenty-one to twenty-five weeks, substantial weight gain occurs, and the fetal skin becomes wrinkled and pinkish.
   F. Toward the end of the period from twenty-six to twenty-nine weeks, the eyes have opened, the gonads have descended in a male, and the weight and development of the fetus is such that it may survive if born prematurely.
   G. By thirty-eight weeks, the fetus is full-term; the normal gestation is 266 days.

V. Labor and Parturition
   A. Labor and parturition are the culmination of gestation and require the action of oxytocin, secreted by the posterior pituitary, and prostaglandins, produced in the uterus.
   B. Labor is divided into the dilation, expulsion, and placental stages.

VI. Inheritance
   A. Inheritance is the passage of hereditary traits carried on the genes of chromosomes from one generation to another.
   B. Each zygote contains twenty-two pairs of autosomal chromosomes and one pair of sex chromosomes; XX in a female and XY in a male.
   C. A gene is a portion of a DNA molecule that contains information for the production of one kind of protein molecule; alleles are different forms of genes that occupy corresponding positions on homologous chromosomes.
   D. The combination of genes present in a person's cells constitutes a genotype; the appearance of a person is a phenotype.
      1. Dominant alleles are symbolized by uppercase letters and recessive alleles are symbolized by lowercase letters.
      2. The three possible genotypes are homozygous dominant, heterozygous, and homozygous recessive.
   E. A Punnett square is a convenient way to express probability.
      1. The probability of a particular genotype is one in four (.25) for homozygous dominant and homozygous recessive, and one in two (.50) for heterozygous dominant.
      2. A single trait is studied in a monohybrid cross, and two traits are studied in a dihybrid cross.
   F. Sex-linked traits, such as color blindness or hemophilia, are carried on the sex-determining chromosome.

Review Activities

Objective Questions

1. The pre-embryonic stage is completed when the
   (a) blastocyst implants
   (b) placenta forms
   (c) blastocyst reaches the uterus
   (d) primary germ layers form
2. The yolk sac produces blood for the embryo until the
   (a) heart is functional
   (b) kidneys are functional
   (c) liver is functional
   (d) baby is delivered
3. Which of the following is a function of the placenta?
   (a) production of steroids and hormones
   (b) diffusion of nutrients and oxygen
   (c) production of enzymes
   (d) all of the above
4. The decidua basalis is
   (a) a component of the umbilical cord
   (b) the embryonic portion of the villous chorion
   (c) the maternal portion of the placenta
   (d) a vascular membrane derived from the trophoblast
5. Which of the following could diffuse across the placenta?
   (a) nicotine
   (b) alcohol
   (c) heroin
   (d) all of the above
6. During which week following conception does the embryonic heart begin pumping blood?
   (a) fourth
   (b) fifth
   (c) sixth
   (d) eighth
7. Twins that develop from two zygotes resulting from the fertilization of two ova by two sperm in the same ovulatory cycle are referred to as
   (a) monozygotic
   (b) conjoined
   (c) dizygotic
   (d) identical
8. Match the genotype descriptions with the correct symbols:
   homozygous recessive — Bb
   heterozygous — bb
   homozygous dominant — BB
9. An allele that is *not* expressed in a heterozygous genotype is called
   (a) recessive
   (b) dominant
   (c) genotypic
   (d) phenotypic
10. If the genotypes of both parents are Aa and Aa, the offspring probably will be
    (a) ½ AA and ½ aa
    (b) all Aa
    (c) ¼ AA, ½ Aa, ¼ aa
    (d) ¾ AA and ¼ aa

Essay Questions

1. Discuss the implantation of the trophoblast into the uterine wall and its involvement in the formation of the placenta.
2. Explain how the primary germ layers form. What major structures does each germ layer give rise to?
3. Explain why development during the embryonic stage is so critical, and list the embryonic needs that must be met during the embryonic stage for morphogenesis to continue.
4. Identify the approximate time period (in weeks) for the following occurrences:
   (a) The arm and leg buds appear.
   (b) The external genitalia differentiate.
   (c) Quickening is perceived by the mother.
   (d) The embryonic heart is functioning.
   (e) The ossification of bone is initiated.
   (f) Lanugo and vernix caseosa appear.
   (g) The fetus has a chance of survival if born prematurely.
   (h) All major body organs are formed.
5. State the features of a genetic disorder that would lead to the belief that it is a form of sex-linked inheritance.

30

Postnatal Growth, Development, and Aging

Outline

Concepts

Mitotic potential is an important aspect of cellular specialization and body growth. Growth is not a linear process, and growth rates vary for different organs and structures of the body.

The course of human life after birth is seen in terms of physical and physiological changes and the attainment of maturity in the neonatal, infant, childhood, adolescent, and adulthood periods.

Every living organism experiences chronological and biological aging. Senescence accompanies biological aging as each body system experiences a decrease in viability and an increase in susceptibility to injury and disease.

Biological aging and senescence are just as universal as chronological aging, but individuals, within limits, age biologically at different rates. Regardless of the rate of biological aging, each person has a life expectancy, and the individuals within the human population have a life span.

The many theories of aging are evidence that it is a complex, multifaceted process that is not well understood. Evidence suggests that there is some genetic control of the aging process.

Introduction to Postnatal Growth, Development, and Aging

Mitotic potential is an important aspect of cellular specialization and body growth. Growth is not a linear process, and growth rates vary for different organs and structures of the body.

Objective 1. List the factors that presumably influence mitosis.

Objective 2. Discuss how an organism increases in body size.

Although the body organs are formed prenatally, the body undergoes continual growth and development into postnatal life. Body growth, developmental changes, and death are fundamental principles of life. Understanding the biological importance of these principles will make the material in this chapter more meaningful. Growth and physical development are necessary for an organism to sustain itself and contribute to its own species population. The principal contribution of an individual is reproduction, which ensures the continuance of the species. In a biological sense, once an individual has completed its reproductive potential, it is a liability to the species population because of competition for sustaining resources. Death of the aged is beneficial to a population because it eliminates those members who can no longer reproduce and provides room for offspring with genetic differences to add to the gene pool of the population.

Although humans have the capability and opportunity long beyond the natural reproductive years to contribute far more to the population than simply producing and protecting offspring, humans are still subject to the biological phenomena of growth, development, aging, and death. Clinical aspects of anatomy and physiology have been stressed throughout this book, and many of these conditions are, in part, the result of changes in the body during postnatal growth, development, and aging.

Cell Division

A zygote is a single fertilized egg cell that has the potential to divide repeatedly, differentiate, and grow into a complex individual with a body that contains trillions of cells. As specialized cell types are formed, further mitotic divisions occur until distinct organ systems become functional. Embryonic development and early fetal development are characterized by phenomenal cellular mitotic activity, differentiation and specialization, and growth.

Apparently included within cellular specialization of structure and function is mitotic potential. Certain cells do not require further division once the organ to which they contribute becomes functional. Others, as part of their specialization, require continuous mitosis to keep an organ healthy. Thus, in the adult, it is found that some cells divide continually, some occasionally, and some not at all. For example, epidermal cells, hemopoietic cells, and cells that line the lumen of the gastrointestinal tract divide continually throughout life. Cells within specialized organs, such as the liver or kidneys, divide as the need becomes apparent. Disease or trauma from surgery or injury may necessitate mitosis in these organs. Still other cells, such as muscle or nerve cells, lose their mitotic ability as they become differentiated. Trauma to these cells frequently causes a permanent loss of function.

The factors regulating mitosis are unclear. Evidence suggests that mitotic ability is genetically controlled and, of those cells that do divide, even the number of divisions is predetermined. If this is true, it would certainly account for the aging process. Physical stress, nutrition, and hormones definitely have an effect on mitotic activity. It is thought that the reproductive activity of cells might be controlled through a feedback mechanism involving the release of a *growth-inhibiting substance.* Such a substance may slow or inhibit the divisions and growth of particular organs once the organ has a certain number of cells or reaches a certain size.

Growth

Growth is a normal process by which an organism increases in size as a result of the accretion of tissues similar to those already present. Growth is an integral part of development that continues until adulthood. There are three aspects to the growth process: (1) an increase in the number of constituent cells; (2) an increase in the size of the existing cells through the addition of protoplasmic substance; and (3) an increase in the amount of intercellular substance.

Human growth is not a steady, linear process. It varies with age and the individual and is somewhat sex-dependent (fig. 30.1). At birth, an average full-term baby has a weight of 3,400 g (7.5 lb) and a length of about 50 cm (20 in.). The head circumference of newborns averages 35.5 cm (14 in.). In the first days of life, most newborns lose nearly 100 g (about 4 oz) before their bodies adapt to ingested food. The head of the newborn is disproportionately large compared to the trunk and appendages.

Growth rates vary for different organs and structures of the body. For example, the brain at birth is about 24% of its adult weight, whereas the neonatal body is only 6% of its adult weight. Phenomenally rapid growth of the brain continues, so that by the time a child is four years old the brain has reached 90% of its adult weight, whereas its body weight is only 25% of its eventual adult weight. The reproductive organs experience latent development because they are under hormonal control. These organs remain at less than 10% of their final weight until the onset of puberty.

Figure 30.1. A chronological sequence of heights, growth rates, and physical changes from infancy through adolescence. Each chart (females above and males below) displays a growth curve of total average heights. The sharp peaks in the growth-rate curves are the adolescent growth spurts. Note that the adolescent growth spurt in males follows that of females by about two years. (From "Growing Up," by J. M. Tanner. Copyright ©1973 by Scientific American, Inc. All rights reserved.)

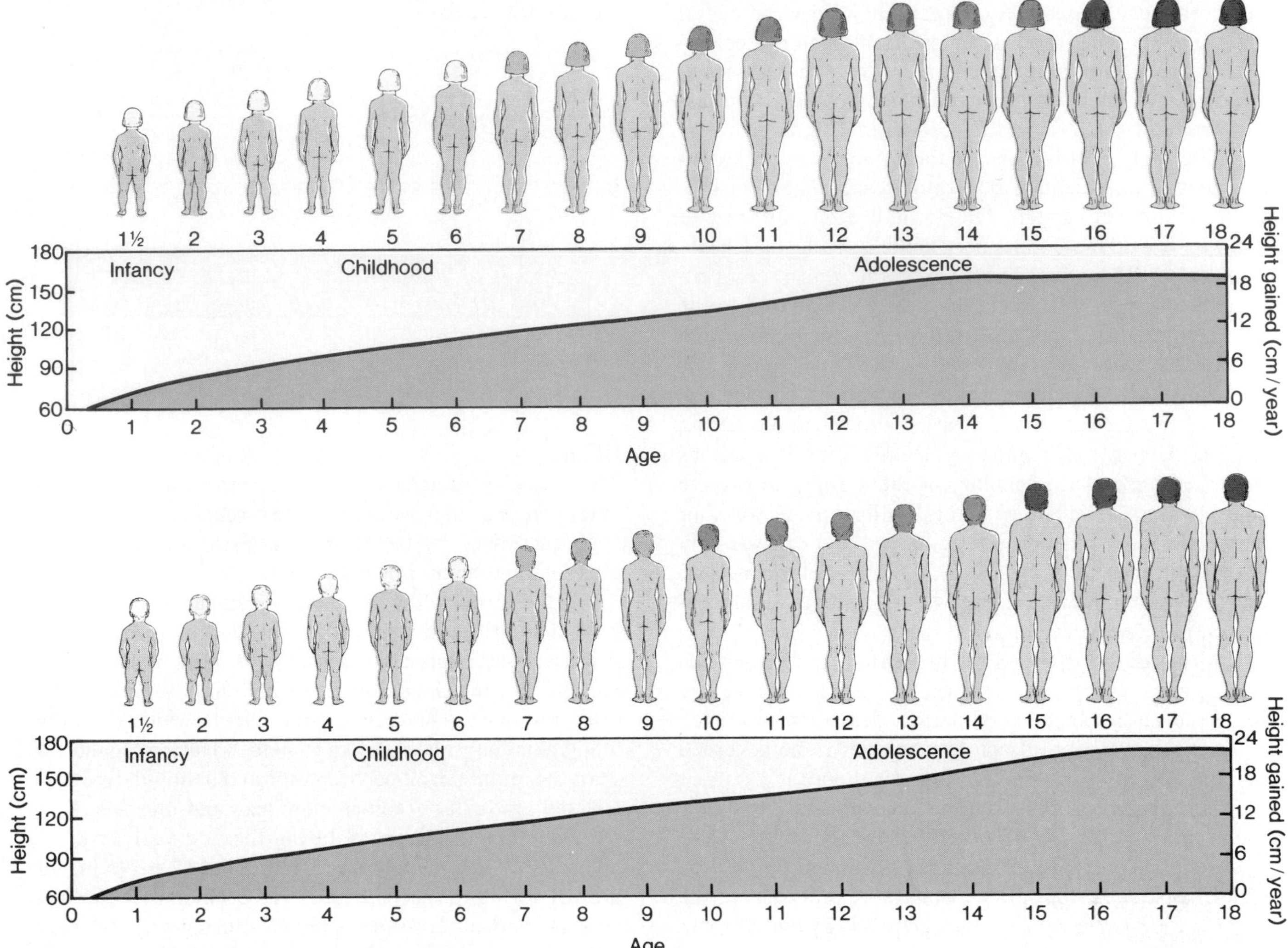

Normal growth depends not only on proper nutrition but on the concerted effect of several hormones, including insulin, growth hormone, thyroid hormone, and (during adolescence) the androgens. The hypothalamus, through the action of hormone-releasing factors, seems to play the governing role in growth, including the initiation of puberty and the adolescent growth spurt. Both hyper- and hypoactivity of the hypothalamus may have profound effects on the growth process.

1. List examples of cells that divide continuously, those that divide occasionally, and those that never divide in an adult person.
2. Diagram a cell, and construct a model that depicts the factors that may influence mitosis. Which factor is intrinsic to the cell, and which factors are extrinsic?
3. List the three aspects of the growth process that occur at the cellular level.

Stages of Postnatal Growth

The course of human life after birth is seen in terms of physical and physiological changes and the attainment of maturity in the neonatal, infant, childhood, adolescent, and adulthood periods.

Objective 3. Describe the growth and developmental events of the neonatal, infant, childhood, and adolescent periods.

Objective 4. Define *puberty,* and discuss what determines its onset in males and females.

Objective 5. Define the term *adulthood,* and discuss sexual dimorphism in adult humans.

Neonatal Period

The **neonatal period** extends from birth to the end of four weeks. Although growth is rapid during this period, the most drastic changes are physiological. The body of a newborn must immediately adapt to major environmental changes, including thermal stress; rapid bacterial colonization of the skin, oral cavity, and gastrointestinal tract; a barrage of sensory stimuli; and sudden demands on cardiorespiratory, gastrointestinal, and renal functions.

The most critical need of the newborn is the establishment of an adequate respiratory rate to ensure sufficient amounts of oxygen. The normal respiratory rate of a newborn is 30 to 40 times per minute. An adequate heart rate is also imperative. The heart of a newborn seems enlarged in respect to the thoracic cavity (compared to the heart of an adult) and has a rapid rate that ranges from 120 to 160 beats per minute.

Most full-term newborn babies appear chubby because of the deposition of fat within adipose tissue during the last trimester of pregnancy. Dehydration is a serious threat because of the inability of the kidneys to excrete concentrated urine; large volumes of dilute urine are eliminated. Immunity is not well developed and consists only of that obtained from the placental transfer of the mother. For this reason, newborns need to be guarded against exposure to infected persons.

Although virtually all of the neurons of the nervous system are present in a newborn, they are immature and the newborn has little coordination. Most behavior such as sleep, hunger, and discomfort appears to be governed by lower cerebral centers and the spinal cord.

A newborn has many reflexes, some indicative of neuromuscular maturity and others essential for life itself. Four reflexes critical to survival are (1) the *suckling reflex,* which causes a newborn to suck anything that touches the lips; (2) the *rooting reflex,* which helps a baby find a nipple by causing it to turn its head and start suckling whenever something brushes its cheek; (3) the *crying reflex,* when its stomach is empty or it is experiencing other discomforts; and (4) the *breathing reflex,* which is apparent in a normal newborn even before the umbilical cord, with its supply of oxygen, is cut.

Babies born more than three weeks before the due date are generally considered *premature,* but because errors are commonly made in calculating the conception date, prematurity is defined by neonatal body weight rather than due date. Newborns weighing less than 2,500 grams (5.5 lbs) are considered premature. By this definition, approximately 8% of newborns in the United States are premature.

Postmature babies are those born two or more weeks after the due date. They frequently weigh less than they would have if they had been born at term because the placenta often becomes less efficient after a full-term pregnancy. Approximately 10% of newborns in the United States are postmature.

neonatal: Gk. *neos,* new; *natus,* born
premature: L. *prae,* before; *maturus,* ripe

Figure 30.2. The relative proportions of the body from embryo to adult. The head of a newborn accounts for a quarter of the total body length, and the lower appendages make up about one-third. In an adult, the head accounts for about 13% of the total body length, whereas the length of the lower appendages constitutes approximately one-half.

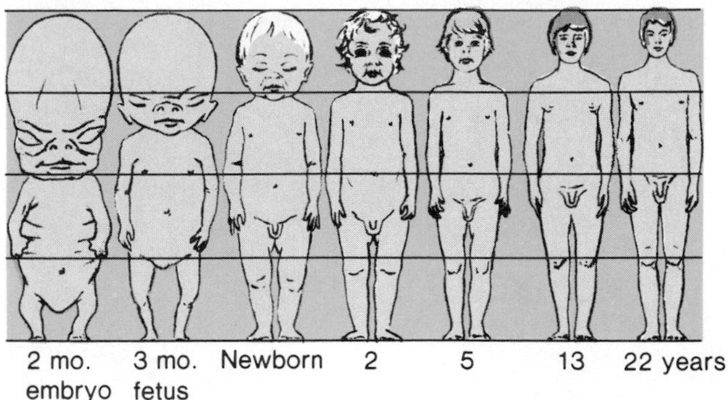

Infancy

The period of **infancy** extends from the end of the neonatal period at four weeks until two years of age. Infancy is characterized by tremendous growth, increased coordination, and mental development.

A full-term child will generally double its birth weight by five months and triple it in a year. The formation of subcutaneous adipose tissue reaches its peak at about nine months, causing the infant to appear chubby. Growth decelerates during the second year, during which time the infant gains only about 2.5 kg (5–6 lbs). During the second year, the infant develops locomotor and manipulative control and gradually becomes more lean and muscular.

Body length increases during the first year by 25 to 30 cm (10–12 in.). There is an additional 12 cm (5 in.) of growth during the second year. The brain and circumference of the head also grow rapidly during the first year and only moderately so during the second. Head circumference increases approximately 12 cm (5 in.) during the first year and only an additional 2 cm during the second. The anterior fontanel gradually diminishes in size after six months and becomes effectively closed at any time from nine to eighteen months. It is the last of the fontanels to close. The brain is two-thirds of its adult size at the end of the first year and four-fifths of its adult size by the end of the second year.

By two years, most infants weigh approximately four times their birth weight and average between 81 and 91 cm (32–36 in.) in length. The body proportions of a two-year-old are certainly not the same as an adult (fig. 30.2). Growth is a differential process, resulting in gradual changes in body proportions.

infancy: L. *in,* not; *fans,* speaking

Figure 30.3. X rays of the right hand of (*a*) a child, (*b*) an adolescent, and (*c*) an adult.

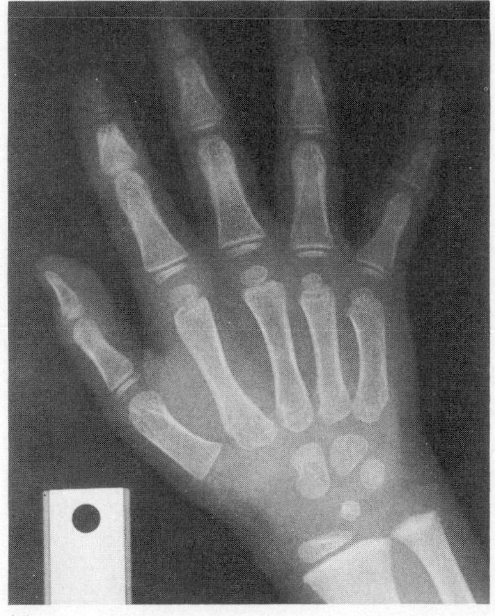

(a)

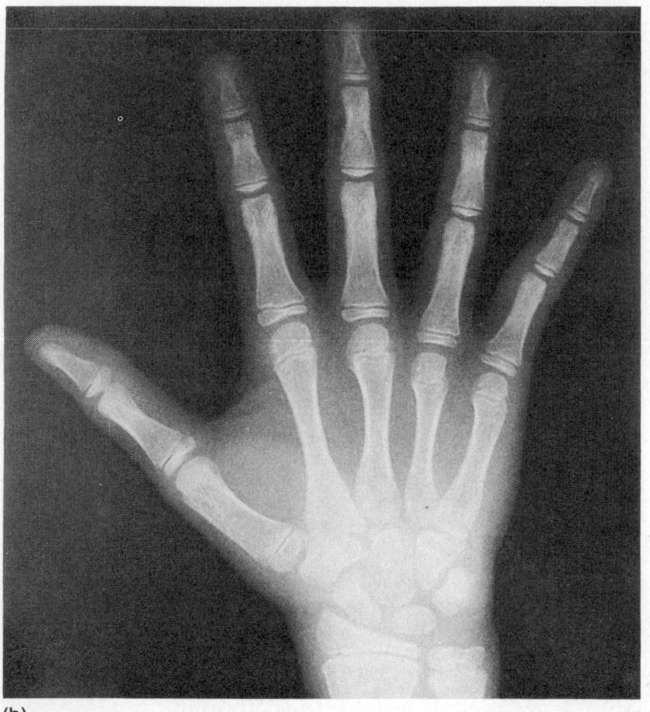

(b)

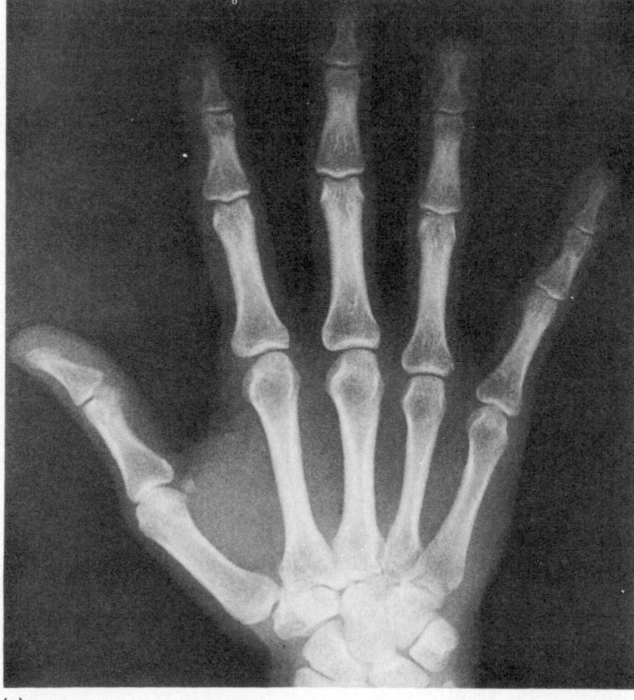

(c)

Deciduous teeth begin to erupt in most infants between five and nine months. By one year of age, most infants have six to eight teeth. Eight more teeth erupt during the second year, making a total of fourteen to sixteen, including the first deciduous molars and canine teeth.

The growth rates of children vary tremendously. Body lengths and weights are not always reliable indicators of normal growth and development. A more objective evaluation of a child's physical developmental progress is determined through X-ray analysis of skeletal ossification in the carpal region (fig. 30.3).

Childhood

Childhood is the period of growth and development extending from infancy to adolescence, at which time puberty begins. The chronological duration of childhood varies because puberty begins at different ages for different people.

Childhood years are a period of relatively steady growth until preadolescence, when there is a growth spurt. The average weight gain during childhood is about 3 to 3.5 kg (7 lb) per year. There is an average increase in height of 6 cm (2.5 in.) per year. The circumference of the head increases by only about 3 to 4 cm (1.5 in.) during childhood, and by adolescence the head and brain are virtually adult size.

The facial bones continue to develop during childhood (fig. 30.4). Especially significant is the enlargement of the sinuses. The first permanent teeth generally erupt during the seventh year, and then the deciduous teeth are shed approximately in the same sequence as they were acquired. Deciduous teeth are replaced at a rate of about four per year over the next seven years.

Figure 30.4. Growth of the skull. The height of the cranial vault (distance between planes *a* and *b*) is drawn the same in both the infant and adult skulls. Growth of the skull occurs almost exclusively within the bones of the facial region.

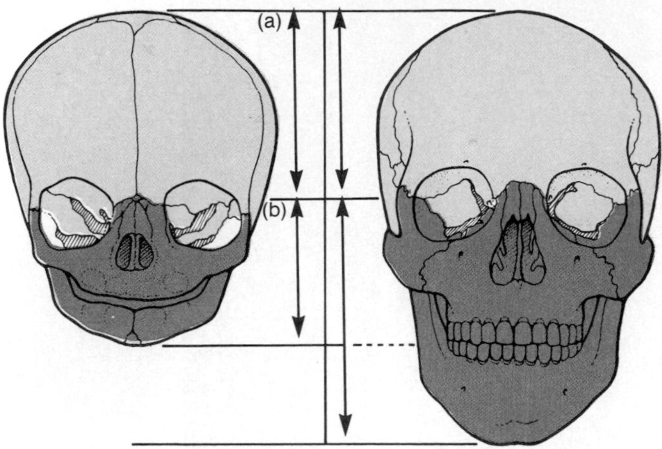

Although there is an average rate of growth during childhood due to genetics, there is a tremendous range in what is considered normal growth. If, for example, the eight-year-olds who are within the tallest and heaviest 10% of their age group were to stop growing for a year while their classmates grew normally, they would still be taller than half of their contemporaries and heavier than three-quarters of them.

During childhood, the average child becomes thinner and stronger each year as he or she grows taller. The average ten-year-old, for example, can throw a ball twice as far as the average six-year-old. Visceral organs, particularly the heart and lungs, develop tremendously during this period, enabling a child to run faster and exercise longer.

Lymphatic tissue is at its peak of development during midchildhood and generally exceeds the amount of such tissue in the normal adult. Children need the extra lymphatic tissue to combat childhood diseases, which take a tremendous toll, particularly in countries where nutrition is poor and health care is minimal.

Childhood *obesity* can become a serious physical and psychological problem if not corrected. Overweight children usually exercise less and run a greater risk of serious illnesses. Frequently, they are teased and rejected by classmates, which causes psychological stress and learning impairments. At least 5% of children in the United States can be classified as obese. There is generally a correlation between childhood obesity and adult obesity. Obesity in adults is a major health problem, in light of the fact that one of five adults is 30% or more over the ideal, healthy weight. The obvious help for obese children and adults is a controlled diet and regular exercise.

Adolescence

Adolescence *(ad''o-les'ens)* is the period of growth and development between childhood and adulthood. It begins around the age of ten years in girls and the age of twelve years in boys. The end of adolescence is frequently said to be the age of twenty years, but it is not clearly delineated and varies with the developmental, physical, emotional, mental, or cultural criteria that define an adult.

Puberty *(pu'ber-te)* is the stage of early adolescence when the secondary sexual characteristics become expressed and the sexual organs become functional. **Pubescence** *(pu-bes'ens)* refers to the continuum of physical changes during puberty, particularly in regard to body hair.

For both sexes there is wide individual variation in the onset and duration of puberty. Although puberty is under hormonal control, a complex interaction of other factors, including nutrition and socioeconomic forces, has a decisive influence on the onset and duration of puberty. The end result of puberty is that the sexes are **sexually dimorphic** *(di-mor'fik);* that is, they are structurally different. The average adult male, for example, has a deeper voice and more body hair and is taller than the average adult female. Prior to puberty, male and female children have few major structural differences aside from the general appearance of the external genitalia.

Puberty actually begins before it is physically expressed. In most instances, significant amounts of sex hormones appear in the blood of females by the age of ten years and in males by the age of eleven years. Sexual changes are usually thought of as the only features of puberty, but major musculoskeletal changes take place as well. During late childhood, the body proportions of the sexes are similar, males being slightly taller. Under the influence of hormones, females experience a growth spurt in early adolescence that precedes that of males by nearly two years (see fig. 30.1). During this time, females are temporarily taller. Once puberty begins in males, the heights of both sexes are soon the same, and at the culmination of puberty, males average about 10 cm (4 in.) taller than females. By the time growth is completed at the end of adolescence, males average 13 cm (5 in.) taller than females.

Other dimorphic differences involving skeletal structures include a broadening of the pelvic girdle in females. The muscles of males become more massive and stronger than those of females. Females acquire a thicker subcutaneous layer of the skin during adolescence, which gives them a softer appearance.

Sexual maturation during adolescence includes not only the development of the reproductive organs but the appearance of secondary sexual characteristics. The sequence of average sexual maturation and the expression of secondary sexual characteristics for both males and females is presented in table 30.1.

adolescence: L. *adolescere,* to grow up
puberty: L. *pubertas,* adult form
dimorphic: Gk. *di,* two; *morphe,* form

Table 30.1 Sequence of adolescent physical development

Females	Age span		Males
Growth spurt begins; breast buds appear; sparse pubic hair	10–11 yrs	11.5–13 yrs	Growth of testes and scrotum; sparse pubic hair; growth spurt begins; penis growth begins
Appearance of straight, pigmented pubic hair; some deepening of voice; rapid growth of ovaries, uterus, and vagina; acidic vaginal secretion; menarche; further enlargement of breasts; kinky pubic hair; age of maximum growth	11–14 yrs	13–16 yrs	Appearance of straight, pigmented pubic hair; deepening of voice; maturation of penis, testes, scrotum, and accessory reproductive glands; ejaculation of semen; axillary hair; kinky pubic hair; sparse facial hair; age of maximum growth
Appearance of axillary hair; breasts are adult size and shape; culmination of physical growth	14–16 yrs	16–18 yrs	Increased body hair; marked vocal change; culmination of physical growth

In females, the first physical indication of puberty is the appearance of **breast buds,** which are swellings of the breasts and slight enlargement and pigmentation of the areolar areas. The average age for breast buds to appear in healthy girls is about eleven years, but it ranges from nine to thirteen years. Approximately three years are required after the appearance of breast buds for the maturation of the breasts. Usually pubic hair begins to appear shortly after the breast buds, but in about one-third of all girls, sparse pubic hair appears before the breast buds. Axillary hair appears a year or two after pubic hair.

The first menstrual period, referred to as **menarche** *(mĕ-nar′ke),* is generally at the age of thirteen years but may range as much as from nine to seventeen years of age. During puberty, the vaginal secretions change from alkaline to acidic.

The onset of puberty in males varies just as much as in females but generally lags behind by about one and a half years. The first indication of puberty in boys is growth of the testes and the appearance of sparse pubic hair at the age of twelve years on the average. This is followed by growth of the penis, which continues for about two years, and the appearance of axillary hair. Vocal changes generally begin during early puberty but are not completed until midpuberty. Facial hair and chest hair (which may or may not be present) come toward the end of puberty. The mean age when semen can be ejaculated is 13.7 years, but sufficient mature sperm for fertility are generally not produced until fourteen to sixteen years of age.

Acne is an inflammatory disease of the integument, which is common during adolescence. The increase in hormonal activity that is responsible for the physical changes taking place during puberty also affects the activity of the sebaceous glands and promotes the formation of inflamed superficial pustules and comedones (blackheads). Tension and emotional stress may also promote acne. Acne in teenagers may cause serious psychological problems related to a concern about and need for peer acceptance.

Adulthood

Adulthood is the final stage of human physical change. It is the period of life beyond adolescence. An adult has reached maximum physical stature as determined by genetic, nutritional, and environmental factors. Although skeletal maturity is reached in early adulthood, anatomical and physiological changes continue throughout adulthood and are part of the aging process.

Sexual dimorphism exists in adults apart from the obvious primary and secondary sex differences. The sexes differ anatomically, physiologically, metabolically, and behaviorally (psychologically or socially). Some of these differences manifest themselves prenatally and during childhood. Others are characteristics of adolescence and adulthood. It is uncertain to what extent specific dimorphisms of the sexes are genetically determined through hormonal action or influenced by environmental (including cultural) factors. It is also unclear how these governing factors are expressed in observed physical characteristics.

The shape of the adult body is determined primarily by the skeleton and attached muscles as well as the subcutaneous connective tissue (especially adipose tissue) and extracellular body fluids. Although there is considerable variation in body proportions between adult males and females (fig. 30.5), in general, adult males have relatively longer appendages than females, their shoulders are relatively broader, and their pelvises are narrower. A male also has a relatively longer neck.

The general body composition of males and females can also be compared numerically. Mean data of body composition are summarized in table 30.2. These data show that total body fluid and skeletal weight is lower in adult females than in adult males. Females, however, have a greater percentage of body fat.

Other differences not shown in table 30.2 are that adult females have lower blood pressures, erythrocyte counts (hematocrits), basal metabolic rates, and respiratory rates than adult males. Females, however, have higher heart rates and oral temperatures.

Figure 30.5. Relative differences in the physiques of an adult male and female. A male has proportionately longer appendages than a female.

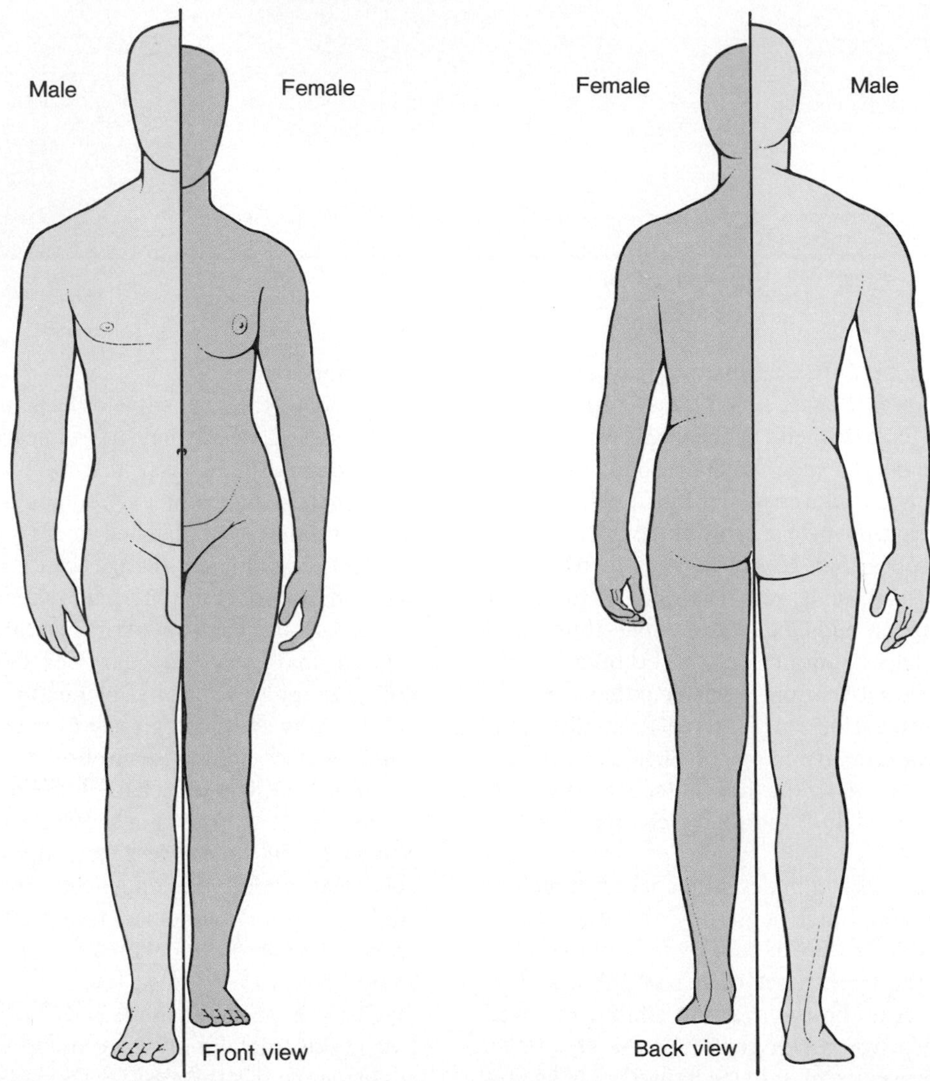

Table 30.2 Body composition in average adult males and females at age 25 and 65								
	Males				**Females**			
	Absolute		**Relative (% body weight)**		**Absolute**		**Relative (% body weight)**	
	Age 25	**Age 65**	**Age 25**	**Age 65**	**Age 25**	**Age 65**	**Age 25**	**Age 65**
Body weight	70.0 kg	70.0 kg			69.0 kg	60.0 kg		
Body fluids	41.0 liters	37.0 liters	58.9%	52.9%	30.8 liters	28.0 liters	51.3%	46.7%
Intracellular	24.0 liters	19.2 liters	34.3%	27.4%	16.6 liters	14.3 liters	27.7%	23.8%
Extracellular	17.2 liters	17.8 liters	24.6%	25.5%	14.2 liters	13.7 liters	23.6%	22.9%
Plasma volume	3,302 ml	2,940 ml	4.7%	4.2%	2,760 ml	2,462 ml	4.6%	4.1%
Lean body weight	56.3 kg	50.5 kg	80.4%	72.1%	42.0 kg	38.2 kg	70.2%	63.7%
Body fat	13.7 kg	19.5 kg	19.6%	27.9%	17.9 kg	21.8 kg	29.8%	36.3%
Skeletal weight	5.8 kg	5.7 kg	8.3%	8.1%	4.4 kg	4.2 kg	7.3%	7.0%

From K. H. Oleson, "Body Composition in Normal Adults," in *Human Body Composition, 7,* edited by J. Brozek, Symposium of the Society for the Study of Human Body Composition. Copyright © 1965 Pergamon Press Ltd., Oxford, England, Reprinted by permission.

Table 30.3 Summary of stages of postnatal periods

Stage	Time period	Some physical and behavioral characteristics
Neonatal period	Birth to end of fourth week	Stabilizing of body systems necessary to carry on respiration, obtain nutrients, digest nutrients, excrete wastes, regulate body temperature, and circulate blood
Infancy	End of fourth week to two years	Tremendous growth; teeth begin to erupt; muscular and nervous systems develop so that motor activities are possible; verbal communication begins
Childhood	One year to puberty	Consistent growth; deciduous teeth erupt and are replaced by permanent teeth; good motor control; urinary bladder and bowel controls are established; intellect greatly improved
Adolescence	Puberty to adulthood	Reproductive system matures; growth spurts in skeletal and muscular systems; intellect and emotional maturity increase
Adulthood	Adolescence to old age	Maximum physical stature and strength obtained; anatomical and physiological degenerative changes begin
Senescence	Old age to death	Senescence continues; body becomes less able to cope with diseases and physical demands; death—usually from physical disturbances in the cardiovascular system or disease processes in vital organs

Some of the physical and behavioral characteristics of each stage of human growth and development are summarized in table 30.3.

1. Construct a table that lists the stages from infancy through adolescence of postnatal growth, and indicate the events or characteristics of each.
2. List four reflexes in a newborn that are critical for survival.
3. Define the terms *puberty, pubescence, sexual dimorphism,* and *menarche.* What is the mean age of puberty, and what causes its occurrence?
4. Describe the physical characteristics of adulthood. Compare the body structure of an adult female to an adult male.

AGING AND SENESCENCE

Every living organism experiences chronological and biological aging. Senescence accompanies biological aging as each body system experiences a decrease in viability and an increase in susceptibility to injury and disease.

Objective 6. Define the terms *aging* and *senescence.*
Objective 7. Discuss the senescent changes that occur in each body system.

Aging

Aging is a term that has a broad definition. *Chronological aging* is an expression of time that can apply to prenatal or postnatal life. The age of an embryo is given chronologically in days or weeks from the date of conception, whereas the age of the fetus is expressed in weeks or months. The age during the first year is given in weeks and months from the date of birth and thereafter is indicated in years. *Biological aging,* which begins with conception and ends at death, is a process characterized by changes in the structure and function of organ systems. These changes are generally predictable and, at least during certain age periods, not always detrimental. The changes in biological aging that take place during adolescence, for example, permit humans to be at peak performance as they enter early adulthood. Aging during midadulthood, on the other hand, results in a gradual reduction of capabilities.

Senescence of Body Systems

Senescence *(se-nes'ens)* refers to the biological aging process, characterized by a gradual deterioration of body structure and function. As senescence progresses through adulthood and old age, the viability of the body decreases and vulnerability increases. The body becomes more susceptible to injury and disease.

The following paragraphs of this section briefly discuss senescence of the body systems that have been discussed in the previous chapters of the text.

Integumentary System. With senescence, the skin becomes thin, dry, and inelastic. Collagen fibers in the dermis become thicker and less elastic, and the amount of adipose tissue in the hypodermis diminishes, making it thinner. Skinfold measurements indicate that the diminution of the hypodermis begins with regularity at the age of forty-five years. With a loss of elasticity and a reduction in the thickness of the hypodermis, wrinkling, or the permanent infolding of the skin, becomes apparent (fig. 30.6).

During senescence of the integument, the number of hair follicles, sweat glands, and sebaceous glands also diminishes, as well as their activity. Consequently, there is a marked thinning of scalp hair and hair on the extremities, reduced sweating, and decreased sebum production.

senescence: L. *senis,* old

Figure 30.6. Senescence of the skin results in a loss of elasticity and the appearance of wrinkles.

Figure 30.7. The geriatric skull. Note the loss of teeth and the degeneration of bone, particularly in the facial region.

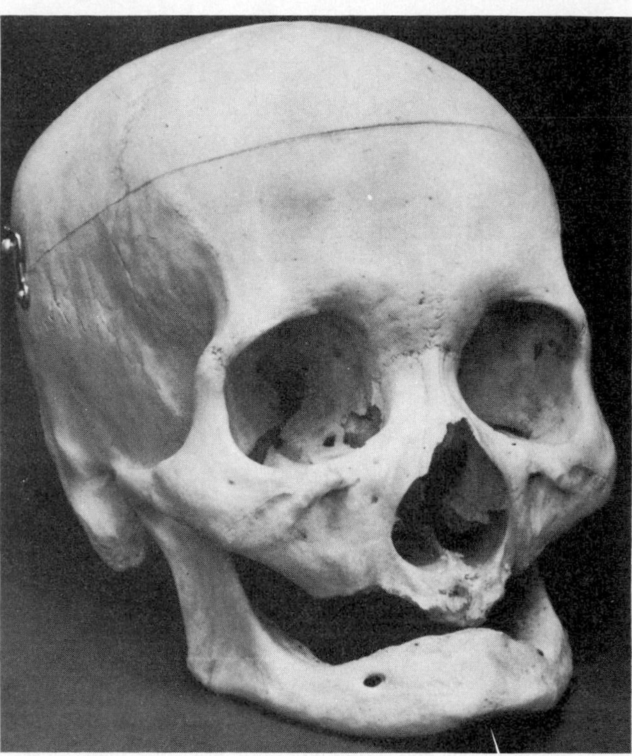

Since elderly people cannot perspire as freely, they are more likely to complain of heat and are more subject to heat exhaustion. They also become more sensitive to cold because of the loss of insulating adipose tissue and diminished circulation. A decrease in the production of sebum causes the skin to dry and crack frequently.

The integument is not as well protected from the sun because of thinning, and melanocytes that produce the brown pigment melanin gradually atrophy. The loss of melanocytes accounts for graying of the hair and pallor of the skin. After the age of fifty, brown, plaquelike growths, called *seborrheic (seb''o-re'ik) hyperkeratoses,* appear within the skin, particularly on exposed portions. Skin that has been exposed to excessive sunlight tends to develop more cutaneous carcinomas than less exposed skin. Caucasians are more susceptible to skin cancer than people of more darkly pigmented races.

Whereas a general loss of hair is characteristic of aging, females who have gone through menopause may develop more facial hair, particularly around the lips and on the chin. Changes in the androgen/estrogen ratio are responsible for the growth of this coarse, darkly pigmented hair. It can be removed through a procedure referred to as *depilation.*

Skeletal System. Senescence affects the skeletal system by decreasing skeletal mass and density and increasing porosity and erosion (fig. 30.7). Bones become more brittle and susceptible to fracture. Articulating surfaces also deteriorate, contributing to arthritic conditions. Arthritic diseases are second to heart disease as the most common debilitation in elderly persons. *Osteoporosis (os''te-o-po-ro'sis)* is the most prevalent metabolic disorder of bone and develops in the aged. It is characterized by a decrease in skeletal mass and density. Postmenopausal women are most susceptible to this condition and because of it frequently sustain fractures of the hip, vertebrae, or wrist. Osteoporosis is attributed to immobilization, decreased estrogen levels, high steroid levels, and environmental factors.

Distinct losses in height occur during middle and old age. Between the ages of fifty and fifty-five years, there is a decrease of 0.5–2 cm (0.25–0.75 in.) because of compression and shrinkage of the intervertebral discs. Elderly persons may suffer a further major loss of height because of osteoporosis.

Muscular System. Although the aged experience a general decrease in the strength, endurance, and agility of skeletal muscle (fig. 30.8), the extent of senescent changes varies considerably among individuals. Apparently the

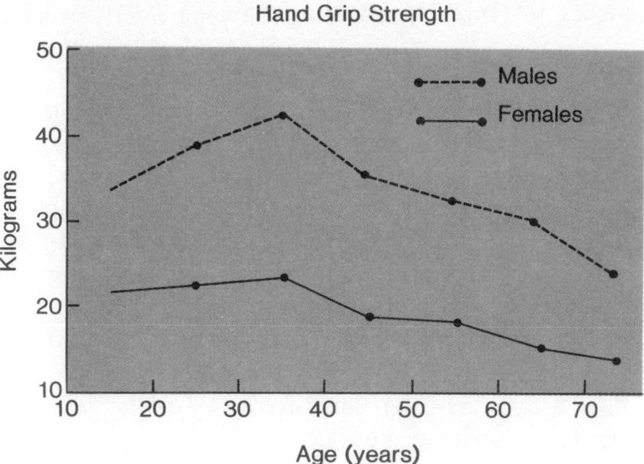

Figure 30.8. A gradual diminishing of muscle strength occurs after the age of thirty-five, as shown with a graph of hand-grip strength.

Figure 30.9. Competitive runners in late adulthood, recognizing the benefits of exercise.

muscular system is one of the body systems in which a person may actively slow senescent changes. A decrease in muscle mass, in part, is due to changes in connective and circulatory tissues. Atrophy of the muscles of the appendages causes the arms and legs to appear thin and bony. Degenerative changes in the nervous system decrease the effectiveness of motor activity. Muscle reflexes become less efficient, causing a marked reduction in physical capabilities.

Diminished muscular capabilities may affect the functioning of other body systems. A decrease in the strength of the respiratory muscles may limit the ability of the lungs to ventilate. Reduced muscularity of the urinary bladder causes difficult micturition and may cause urinary infections.

Exercise is important at all stages of life but is especially beneficial as one approaches old age (fig. 30.9). Exercise not only strengthens muscles, but it also contributes to a healthy circulatory system and thus ensures an adequate blood supply to all body tissues. If an elderly person does not maintain muscular strength through exercise, he or she will be more prone to debility. Learning to use a cane, crutches, or a walker would be extremely difficult for such a person.

Nervous System. The extent of senescent changes within the brain is not known. Previous textbooks have reported that perhaps 100,000 neurons die each day of our adult life. Recent studies, however, show that such claims are unfounded. It is believed that relatively few neural cells are lost during the normal aging process. Neurons are, however, extremely sensitive and susceptible to various drugs or interruptions of vascular supply such as those caused by a stroke or other cardiovascular diseases.

There is evidence that senescence alters neurotransmitters. Age-related conditions such as depression or specific diseases such as Alzheimer's disease may be caused by an imbalance of neurotransmitter chemicals. Changes in sleeping patterns in aged persons also probably result from neurotransmitter problems.

The slowing of the nervous system with age is most apparent in tests of reaction time. It is not certain whether this is a result of a slowing in the transmission of impulses along neurons or in neurotransmitter relays at the synapses.

Although the nervous system does deteriorate with age (in ways that are not well understood), in most people it functions effectively throughout life. Brain dysfunction is not a common characteristic of senescence.

Sensory Organs. Senescence is undoubtedly characterized by sensory impairments. An elderly person may require more intense levels of stimulation in vision, hearing, taste, and smell in order for the sensory receptors to perform with precise acuity. Such declines in sensory capabilities, especially in vision and hearing, are important not only because they affect a person's ability to function, but because they may socially isolate the person and cause serious psychological problems.

Many conditions of the eye are related to age, and some cause visual impairments. Most elderly persons develop lipid infiltrates of the cornea in a condition called *arcus senilis.* Visual acuity remains fairly constant until the ages of twenty-five to thirty in females and thirty-five to forty in males and then gradually declines (fig. 30.10). By the age of sixty-five, 40% of males and 60% of females have vision poorer than 20/70. The pupil of an elderly person's eye cannot dilate fully, and the amount of light reaching the retina may be only 50% of the amount that reaches the retina of a youth. In addition, the depth of the

Figure 30.10. The decline in visual acuity in males and females due to senescence.

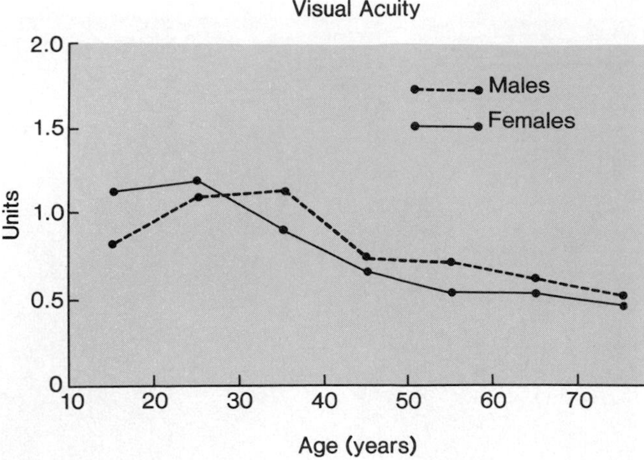

anterior chamber diminishes, creating poor drainage of aqueous humor, which results in increased intraocular pressure and a greater likelihood of glaucoma. Ptosis, or drooping, of the lids and diminished lacrimal secretion, contributing to dryness and irritation of the eyes, is typical in elderly persons.

The lens gradually loses its elasticity during senescence, and its ability to accommodate for viewing near objects diminishes. This change, called *presbyopia (pres″ be-o′pe-ah),* makes it necessary to hold reading material farther from the eye for the print to be focused on the retina. A person with presbyopia may require bifocals to restore visual abilities.

Cataract is the major cause of visual disability in elderly persons. In this condition, the transparent lens becomes opaque (cataractal), restricting the passage of light waves.

Retinal blindness is caused by a slow degeneration of the macula of the retina. Focusing for reading is gradually impaired, whereas perimeter retinal vision remains unaffected.

Actually, the number of legally blind persons in the United States is relatively low (about 225 per 100,000 population), but 60% of blind people are older than seventy years. Most cases of senescent blindness are preventable with early detection and treatment.

Senescence of hearing is a relatively frequent occurrence. About 5% of the population have hearing impairments by the age of fifty; after the age of sixty-five, 30% have hearing difficulties. Senescent hearing loss is referred to as *presbycusis (pres″be-ku-sis)* and is generally a loss of the ability to perceive high-frequency sounds. Changes that accompany presbycusis are so gradual that many elderly people fail to recognize them until the disability is extreme.

The discernment of taste and smell declines with age. About 30% of people over the age of eighty have difficulty identifying common substances by smell. The ability to recognize common foods by taste likewise diminishes with age. The loss in the sensory perception of taste and smell may perhaps be responsible for the clinically noted frequency of complaints about food among elderly people.

Endocrine System. Although the organs of the endocrine system do undergo senescence, the consequences on overall body function are not well understood. The alteration in responsiveness to hormones is likely a major factor in the deterioration of biochemical activity in aging. The ability to secrete insulin and glucocorticoids is impaired in aged persons, and there seem to be changes in hormone transport, action, and feedback systems.

The organs of the endocrine system are susceptible to disorders of the circulatory system and to the diseases of other body organs that have feedback mechanisms with specific endocrine glands.

Circulatory System. Cardiovascular senescence is the leading cause of death in the aged. As the heart and blood vessels age, a person develops or becomes more susceptible to life-threatening diseases. The pulse and electrocardiogram give some insight into general circulatory changes in aging. Changes in blood vessels include an increased stiffness of the arterial walls (*arteriosclerosis*) and the deposition of fatty, plaquelike material (*atherosclerosis*) along the lumina of arteries.

The senescent heart may exhibit three principal characteristics: (1) structural changes in the heart and valves; (2) a diminished blood flow through the heart; and (3) an altered conduction system. The heart generally shrinks with age and accumulates adipose tissue along the coronary vessels. The endocardium increases 25% in thickness from the age of thirty to the age of eighty. After the age of thirty, the valves gradually become more rigid, thickened, and distorted because of sclerosis. The heart in an aged person reacts poorly to stress and cannot utilize oxygen as well. The maximum blood flow through the coronary arteries at the age of sixty is about 35% lower than in a person at the age of thirty. Blood pressure in elderly people may vary from 120 to 160 systolic (fig. 30.11) and from 70 to 90 diastolic. Electrocardiograms indicate slower and more erratic depolarization within the conduction system of the senescent heart.

Respiratory System. Senescence significantly changes the structure of the lung and affects the capacity for ventilation. After the age of twenty-five, there is a gradual loss of elasticity within the alveoli and in the pulmonary capillaries so that by the age of seventy, vital capacity has decreased by about 40% (fig. 30.12). Elderly persons thus require increased muscular exertion to maintain an adequate level of ventilation. Vital lung capacity decreases at a much faster rate if the person is a cigarette smoker.

Figure 30.11. There is a marked increase in systolic blood pressure after the age of thirty-five as the heart and blood vessels undergo senescent changes.

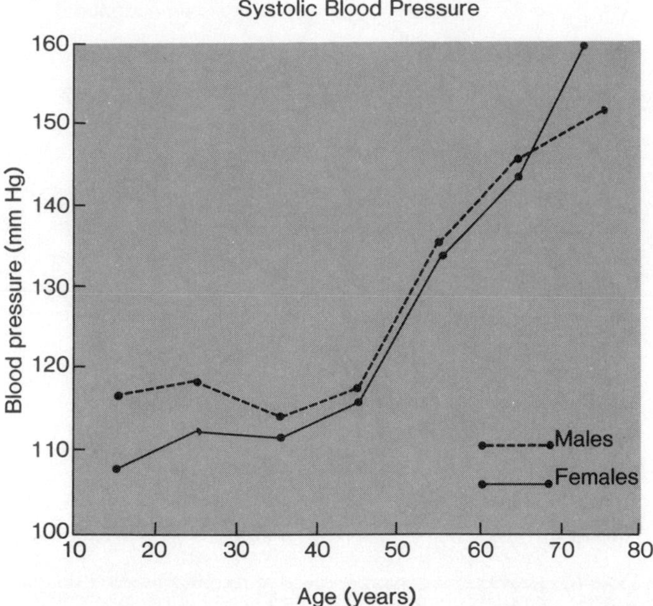

In addition to a loss of elasticity, pulmonary aging causes a change in gaseous exchange across pulmonary membranes. By the age of sixty-five, there is about a 48% decrease in diffusion capacity from that at the age of thirty.

Digestive System. It is estimated that 60% of persons over fifty years of age have senescent problems of the digestive system that are generally more of an irritation than an impairment of health. Common problems include constipation, difficulties in swallowing (dysphagia), increased amounts of intestinal gas, heartburn, and the formation of diverticula along the large intestine. In addition, persons over the age of thirty-five are more susceptible to periodontal disease than those under the age of thirty-five.

Although gastrointestinal problems such as those listed above are normal in elderly persons, they may be symptomatic of more serious conditions or diseases such as hypertension, ulcers, or cancer.

The *basal metabolism rate (BMR)* is the minimum amount of energy a person utilizes in a state of rest and is an index of caloric usage and requirements. The BMR's are usually higher for males than for females (fig. 30.13) and decline proportionally with age for both sexes. Young adults with a high BMR can eat almost anything and not get fat, whereas middle-aged people with a lower BMR must constantly control food intake to avoid obesity.

Urinary System. A progressive decline in kidney function accompanies senescence, due to general atrophy of the kidney and sclerosis of glomeruli. At the age of seventy the rate of glomerular filtration is only about 50% of the rate at the age of thirty. Renal blood flow diminishes

Figure 30.12. Senescence of the lungs and muscles of respiration reduces vital capacity. The graph represents data accumulated from healthy nonsmokers.

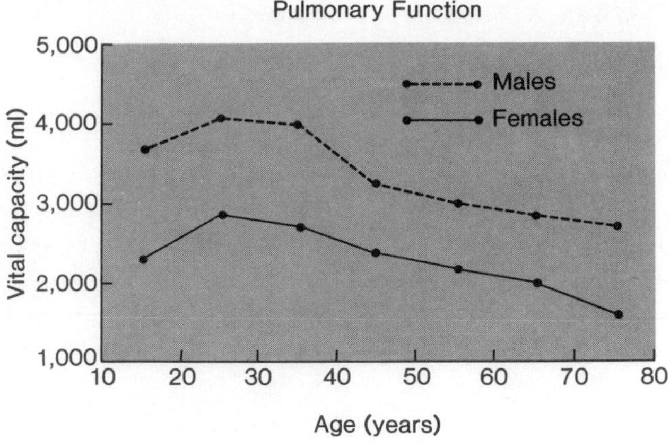

Figure 30.13. The basal metabolism rate.

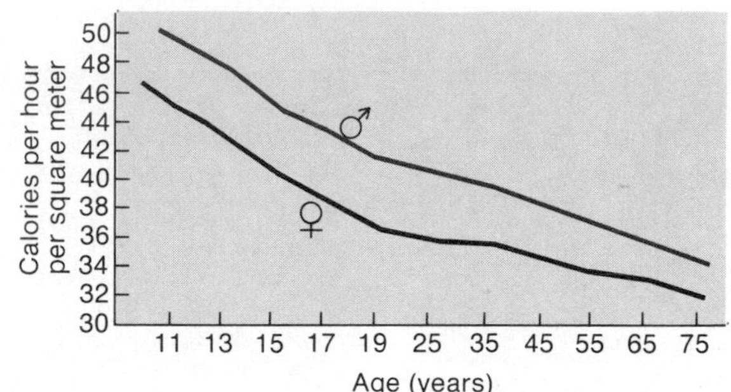

from approximately 1,100 ml per minute at the age of thirty to only about 475 ml per minute at the age of eighty. The kidneys are able to regulate acid-base balance during senescence, although they respond more slowly to a sudden, large acid load.

Reproductive System. The senescent changes in the reproductive system in both males and females are generally more important psychologically than pathologically. A decrease in sex hormones in elderly persons may decrease sexual desires or abilities. There is a decreased sensitivity of the genitalia, which may also affect sexual responsiveness.

A number of physical changes occur in aging males. The testes atrophy and the seminiferous tubules thicken and decrease in diameter. The prostate gland hypertrophies, which may cause pathological conditions. Although healthy males maintain the ability to ejaculate throughout old age, there is a reduction in the force of ejaculate, the volume, and the general quality of the seminal fluid.

Females experience major physiological changes at menopause, which generally occurs near the age of fifty. Hormonal changes bring about a cessation of menstruation and ovulation. Physical changes following menopause

Figure 30.14. Changes in the body and cervix of the uterus as a female ages. The uterus of a postmenopausal woman atrophies and becomes fibrous.

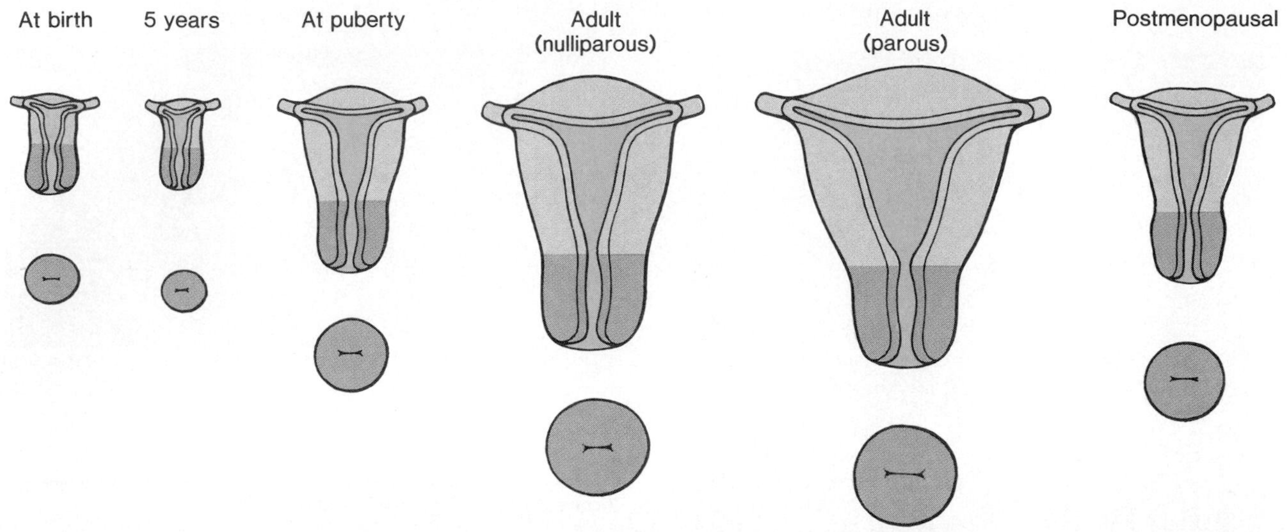

Table 30.4 Summary of aging in body systems

Organ system	Principal senescent changes
Integumentary system	Degenerative change in collagenous and elastic fibers in dermis; decreased production of pigment in skin and hair follicles; reduced activity of sweat and sebaceous glands Skin tends to become thinner, wrinkled, and dry with pigment spots; hair becomes gray and then white
Skeletal system	Degenerative loss of bone matrix; deteriorating articulations Bones become thinner, more brittle, and more likely to fracture; stature may shorten due to compression of intervertebral discs; susceptibility to joint diseases increases
Muscular system	Loss of skeletal muscle mass; degenerative changes in neuromuscular junctions Loss of muscular strength and motor response
Nervous system	Degenerative changes in neurons; loss of dendrites and synapses; decrease in sensory sensitivity Decreased efficiency in processing and recalling information; decreased ability to communicate; diminished senses of smell, taste, sight, hearing, and touch
Endocrine system	Slightly reduced hormonal secretions Decreased responsiveness to hormones; decreased metabolic rate; reduced ability to maintain homeostasis
Circulatory system	Degenerative changes in cardiac muscle; decreased diameters of lumina of arteries and arterioles; decreased efficiency of immune system Decreased cardiac output; increased resistance to blood flow; increased blood pressure; increased incidence of autoimmune diseases
Respiratory system	Degenerative loss of elastic fibers in lungs; reduced number of functional alveoli Reduced vital capacity; increased dead air space; reduced ability to clear airways by coughing
Digestive system	Decreased motility in GI tract; reduced secretion of digestive enzymes; increased occurrence of periodontal disease Reduced efficiency of digestion
Urinary system	Degenerative changes in kidneys; reduced number of functional nephrons Reduced filtration rate, tubular secretion, and reabsorption
Reproductive system	
Male	Reduced secretion of sex hormones; reduced production of spermatozoa; enlargement of prostate gland Decreased sexual capabilities
Female	Degenerative changes in ovaries; decreased secretion of sex hormones Menopause; regressed secondary sex organs and characteristics

include atrophy of the uterus (fig. 30.14), with a reduction in the size of the cervix and a thinning of the vaginal wall and walls of the uterine tubes. The ovaries become irregular in shape. The vulva develops less pronounced folds as the skin becomes thinner. Vascularity and elasticity decrease, causing the vulva to become more susceptible to tissue trauma and pruritus (itching).

The principal senescent changes that occur within each body system are summarized in table 30.4.

1. Using the term *senescence,* distinguish between chronological and biological aging.
2. Construct a table that lists the senescence of the principal organs of the body. Which body systems seem to age most rapidly? In which body systems does aging present a greater threat to life?
3. List three sensory impairments that result from senescence.

Life Epectancy and Life Span

Biological aging and senescence are just as universal as chronological aging, but individuals, within limits, age biologically at different rates. Regardless of the rate of biological aging, each person has a life expectancy, and the individuals within the human population have a life span.

Objective 8. Define the terms *life expectancy, life span,* and *gerontology.*

Life expectancy is best defined as the average number of years lived by persons born in a particular location and period of time. A person born today in the United States has a greater life expectancy than a person born a century ago or a person born today in a Third World country. Life expectancy is relative to age. For example, life expectancy at birth is greater than it is at adolescence.

Life span is the maximum length of life possible in a particular culture. The average life span in the United States has increased twenty-six years in this century, from forty-seven to seventy-three years. Owing to cultural and environmental factors, the current average life span of males is about six years less (seventy years) than females (seventy-six years). The potential human life span is between 120 and 150 years if unaffected by accidents or diseases.

Gerontology *(jer'' on-tol' o-je)* is the study of senescence. This science is concerned primarily with the physical and psychological changes that occur between middle and old age until death. Gerontological research is concentrated in two principal areas: (1) studying the process of aging to determine if and how rates of senescence can be altered; and (2) examining and improving the quality of old age. Because of the increasing percentage of elderly people in our population, gerontology is becoming a more popular and vital science.

1. Discuss what determines life expectancy and what determines life span. Can either life expectancy or life span be increased? If so, how?
2. Define gerontology. List the kinds of classes a person would probably take who is studying to become a gerontologist.

Theories of Aging

The many theories of aging are evidence that it is a complex, multifaceted process that is not well understood. Evidence suggests that there is some genetic control of the aging process.

Objective 9. Discuss the five principal theories of aging.

Current literature in gerontology presents nearly two hundred theories of the mechanism of aging. These theories can be synthesized, however, into several principal types. Most of these theories continue to be scrutinized scientifically and are subject to continued modification.

Genetic Mutation Theory. The foundation for the mutation *(mu-ta'shun)* theory is that cells with unusual or different characteristics are more frequently noted as a person ages. It is thought that senescence is related to the accumulation of mutational damage within the genetic mechanism of somatic cells. The processes that DNA undergoes during mitosis and protein synthesis increase the chance of damage or defect to the copying mechanism. As DNA becomes a modified or inaccurate blueprint, defective proteins are synthesized. Some of these proteins are essential enzymes whose loss will cause the death of a cell. Senescence occurs as cells mutate. When a critical number of cells is altered or killed, the organism dies.

Autoimmunity Theory. The autoimmunity *(au''to-im-mu'ni-te)* theory suggests that senescence results when various body systems begin to reject their own tissues. As a person ages, the production of autoantibodies in the body increases. The antibodies are produced in response to somatic mutations or post-DNA errors. This causes organs to behave self-destructively, and the immune mechanism operates against its own body cells as if they were foreign bodies. Autoimmune reactions are known to occur in various cardiovascular diseases (e.g., giant cell arteries and amyloidosis), diabetes, rheumatoid arthritis, and certain types of cancer. It is also suggested that immunological surveillance becomes impaired as a person ages, so that cells that might give rise to neoplasms are no longer suppressed.

Cross-Linking of Molecules Theory. Homogeneity is a characteristic of some organic molecules. That is, different molecules tend to attract each other and become uniform. Chemical substances adhere to, or cross-link with, adjacent protein molecules so that each loses its functional identity. During senescence, tendons, skin, and even blood vessels lose elasticity. As the functional characteristics of collagen and elastic fibers change, the skin wrinkles, muscles sag, and wounds in the skin heal more slowly.

Cellular Aging Theory. The cellular aging theory suggests that there is a genetically determined limit to the doubling potential of each normal mitotically active cell. This means that cells have an intrinsic finite life span. As cells approach their final divisions, the time interval for mitosis progressively increases, causing a gradual cessation of activity and an accumulation of cellular debris. Eventually there is a total loss of vigor, and cellular death occurs. Degenerative phenomena are manifest after about fifty cell-population doublings.

Programmed Aging Theory. The preceding theories fail to explain why the specific life span for each species (even species that are closely related biologically) is so consistently fixed. Another unexplained observation is that the offspring of long-lived parents tend to live longer than the offspring of short-lived parents. Furthermore, the various cells constituting an organism have different fixed life spans. Epidermal cells, for example, live only a relatively short time before they are replaced, whereas muscle and nerve cells endure the lifetime of the body.

Clearly, there must be some genetic control of the aging process that is not only species and individual specific but also cell specific. If there is an aging gene responsible for senescence, then someday it should be possible to identify and perhaps eventually to manipulate that gene.

Senescence is a complex process, and all of the theories are probably valid to some extent. The process almost certainly has numerous factors. A major challenge for gerontology is to explain how events that occur at the cellular level are transcribed into events at the organismic level.

1. Using hypothetical situations, describe how laboratory experiments could be conducted to test each of the five current theories of aging.
2. Which of the five theories of aging accounts for closely related species having markedly different life spans?

Chapter Summary

I. Introduction to Postnatal Growth, Development, and Aging
   A. Growth and development are necessary so that an individual organism can sustain itself and contribute to its own species population.
   B. Mitotic potential is apparently just as much of a cellular specialization as is any structural or functional feature of a cell.
   C. Growth is not a linear process, and growth rates vary for different organs and structures of the body.
   D. A growth-inhibiting substance released via a feedback mechanism may influence cellular reproductive activities to limit the number of cells or the size of specific organs.

II. Stages of Postnatal Growth
   A. The course of human life after birth is seen in terms of physical and physiological changes and the attainment of maturity in the neonatal, infant, childhood, adolescent, and adulthood periods.
   B. The neonatal period, extending from birth to the end of the fourth week, is characterized by major physiological changes.
   C. Infancy, extending from four weeks until the age of two years, is characterized by tremendous growth, increased coordination, and mental development.
   D. Childhood, extending from infancy to adolescence, is characterized by steady growth until preadolescence, when there is a marked growth spurt.
   E. Adolescence is the period of growth and development between childhood and adulthood.
      1. Puberty is the stage of early adolescence when the secondary sexual characteristics become expressed and the sexual organs become functional.
      2. The first physical indications of puberty are the appearance of breast buds in females and the growth of the testes and the appearance of sparse pubic hair in males.
   F. Adulthood, the final stage of human physical change, is characterized by gradual senescence as a person ages.
      1. Sexual dimorphism exists anatomically, physiologically, metabolically, and behaviorally between males and females.
      2. Differences in body statures, proportions, and compositions between males and females may become more expressed with age.

III. Aging and Senescence
   A. Aging can be expressed as chronological or as biological events. Senescence refers to the biological aging process, characterized by a gradual deterioration of body structure and function.
   B. Senescence affects each body system in a predictable but somewhat individual sequence.
      1. In the integumentary system, the skin thins, becomes dry, is less elastic, and is more fragile. There is an increased sensitivity to temperature extremes and an increased occurrence of cutaneous carcinomas in less-pigmented races.
      2. In the skeletal system, the skeletal mass decreases in density and increases in porosity and erosion, predisposing humans to osteoporosis and arthritic conditions. A decrease in height during middle and old age results from the compression or shrinkage of intervertebral discs.
      3. In the muscular system, there is a general decrease in strength, endurance, and agility that varies considerably among individuals. Senescent changes may be slowed through regular exercise.
      4. In the nervous system, there is evidence that senescence causes an alteration in neurotransmitters, which may result in depression or conditions such as Parkinson's disease. The nervous system slows with age but functions effectively in most people throughout life.
      5. Sense organs are impaired and become less acute in response to stimuli with aging. Vision may be decreased by corneal lipid infiltrates (arcus senilis), glaucoma, ptosis of the eyelids, presbyopia, cataracts, or retinal blindness. Presbycusis refers to the loss of high-frequency sound perception. The discernment of taste and smell also decline with age.
      6. In the endocrine system, the consequences of senescence on overall body functions are not well understood. Organs of this system are susceptible to circulatory disorders and to the diseases of other body organs through feedback mechanisms.
      7. Cardiovascular senescence is the leading cause of death in the aged. Blood vessels may develop arteriosclerosis or atherosclerosis, and the heart may develop structural changes in the valves, a decreased blood flow, and an altered conduction system.
      8. In the respiratory system, the structure of the lung changes significantly with senescence, decreasing its capacity for ventilation. Elasticity decreases, as does the effectiveness of gaseous diffusion in lung tissue.
      9. In the digestive system, senescence irritates rather than impairs health in most cases. Common problems include constipation, dysphagia, increased flatulence, heartburn, diverticuli formation along the colon, and an increased susceptibility to periodontal disease. These problems may be symptomatic of more serious diseases such as hypertension, ulcers, or cancer.

10. In the urinary system, kidney function progressively declines. The kidneys also respond slower in regulating acid-base balance.
11. In the reproductive system, senescent changes are more important psychologically in regard to maintaining sexuality than they are pathological. Although able to ejaculate, males experience a decrease in the force, volume, and general quality of the ejaculate. Females experience major physiological changes at menopause.

IV. Life Expectancy and Life Span
   A. Biological aging and senescence are just as universal as chronological aging, but individuals, within limits, age biologically at different rates.
   B. Life expectancy is the average number of years lived by persons born during a particular time period in a certain location.
   C. Life span is the maximum length of life attainable in a culture.

V. Theories of Aging
   A. The many theories of aging are evidence that it is a complex, multifaceted process that is not well understood.
   B. There are five principal theories of aging that are probably not mutually exclusive.
      1. Genetic mutation theory—with age, somatic cells with mutational damage in the genetic mechanism accumulate.
      2. Autoimmunity theory—senescence results when various body systems reject their own tissue.
      3. Cross-linking of molecules theory—chemical substances adhere to, or cross-link with, adjacent protein molecules so that each loses its functional identity.
      4. Cellular aging theory—cells have an intrinsic, finite life span due to the genetically determined, limited doubling potential characteristic of each normal mitotically active cell.
      5. Programmed aging theory—evidence indicates that the aging process is genetically species specific, individual specific, and cell specific.

Review Activities

Objective Questions

1. Which of the following is the period of growth from birth to the end of the fourth week? The
   (a) neonatal
   (b) fetal
   (c) infant
   (d) suckling
2. The most drastic changes in the first four weeks of life are
   (a) anatomical
   (b) environmental
   (c) psychological
   (d) physiological
3. The normal newborn heart rate is
   (a) 70–80 beats/min
   (b) 120–160 beats/min
   (c) 100–120 beats/min
   (d) 180–200 beats/min
4. The continuum of physical change occurring in adolescence that regulates the growth of body hair is known as
   (a) puberty
   (b) pubal progression
   (c) pubescence
   (d) dimorphism
5. At birth, a baby is regarded as premature if it weighs less than
   (a) 2,500 grams (5.5 lbs)
   (b) 1,350 grams (3.0 lbs)
   (c) 1,050 grams (2.5 lbs)
   (d) 3,200 grams (7.0 lbs)
6. Which is *not* characteristic of the female when compared to the male?
   (a) lower blood pressure
   (b) higher basal metabolic rate
   (c) lower red blood cell count
   (d) faster heart rate
7. The first physical indication of puberty in females is generally
   (a) alkaline vaginal secretions
   (b) a widening pelvis
   (c) breast buds
   (d) axillary hair
8. Which statement is *not* true of a senescent heart of an elderly person?
   (a) The heart enlarges to pump additional blood.
   (b) The endocardium increases in thickness.
   (c) Adipose tissue accumulates along the coronary vessels.
   (d) Slower and more erratic depolarization occurs within the conduction system.
9. Which statement is *not* true of life expectancy?
   (a) It can be changed with improved diet and health care.
   (b) It has an inverse ratio to chronological age.
   (c) It is the maximum length of life possible in a particular culture.
   (d) It varies from country to country.
10. The tendency that the offspring of long-lived parents are often long-lived themselves supports which theory of aging? The
   (a) cross-linked
   (b) autoimmunity
   (c) programmed aging
   (d) genetic mutation

Essay Questions

1. Explain the biological importance of growth, development, aging, and death.
2. Discuss the growth and developmental events characteristic of the neonatal period.
3. Define the term *infancy,* and discuss the growth and developmental events that are characteristic of this stage of life.
4. Define the term *adolescence,* and discuss the role that puberty has in this stage of life for both males and females.
5. Distinguish between puberty and pubescence.
6. List the sexual dimorphic structural features of adult males and females.
7. For each body system, list two senescent changes.
8. Distinguish between chronological aging, biological aging, and senescence.
9. In which body systems do senescent changes threaten life?
10. Discuss gerontology as a contemporary science.

Appendixes

Appendix A

Answers to Objective Questions

Chapter 1
1. c 2. a 3. c 4. b 5. b 6. c 7. b 8. c 9. c 10. d

Chapter 2
1. c 2. b 3. a 4. d 5. c 6. b 7. c 8. d 9. d 10. b 11. a 12. d

Chapter 3
1. d 2. b 3. a 4. c 5. d 6. b 7. a 8. c 9. a 10. b

Chapter 4
1. b 2. d 3. d 4. d 5. e 6. e 7. d 8. b 9. a 10. c 11. e 12. d 13. c 14. a 15. a 16. d 17. b 18. d

Chapter 5
1. c 2. b 3. a 4. c 5. b 6. d 7. a 8. a 9. b 10. d

Chapter 6
1. b 2. c 3. a 4. b 5. d 6. a 7. c 8. a 9. b 10. d

Chapter 7
1. b 2. b 3. a 4. c 5. d

Chapter 8
1. a 2. b 3. b 4. a 5. d 6. a 7. c 8. b 9. b 10. d

Chapter 9
1. c 2. a 3. b 4. b 5. a 6. d 7. a 8. b 9. d 10. c

Chapter 10
1. a 2. c 3. d 4. b 5. b 6. d 7. e 8. b 9. d 10. a

Chapter 11
1. b 2. a 3. d 4. d 5. a 6. d 7. c 8. b 9. b 10. d

Chapter 12
1. e 2. b 3. b 4. a 5. c 6. d 7. a 8. c 9. b 10. b

Chapter 13
1. d 2. b 3. d 4. d 5. d 6. a 7. a 8. b 9. c 10. a

Chapter 14
1. d 2. a 3. e 4. c 5. d 6. a 7. a 8. c 9. d 10. d 11. a 12. c 13. c 14. b 15. d

Chapter 15
1. b 2. a 3. d 4. b 5. c 6. d 7. c 8. a 9. b 10. d

Chapter 16
1. a 2. b 3. b 4. c 5. a 6. b 7. c 8. d 9. a 10. c

Chapter 17
1. d 2. d 3. d 4. d 5. a 6. c 7. b 8. c 9. c 10. a 11. c 12. b 13. b 14. e 15. c

Chapter 18
1. d 2. b 3. b 4. c 5. d 6. b 7. a 8. a 9. b

Chapter 19
1. e 2. b 3. d 4. c 5. a 6. d 7. c 8. e 9. e 10. d 11. a 12. b 13. e 14. d 15. a 16. d 17. c

Chapter 20
1. b 2. c 3. b 4. b 5. a 6. a 7. c 8. e 9. a 10. b 11. c 12. a 13. d 14. c 15. d

Chapter 21
1. a 2. d 3. c 4. e 5. b 6. c 7. a 8. c 9. d 10. b 11. c 12. d 13. c

Chapter 22
1. c 2. b 3. d 4. a 5. c 6. d 7. d 8. b 9. e 10. a 11. d 12. a 13. d 14. b 15. d 16. a

Chapter 23
1. c 2. d 3. c 4. a 5. c 6. c 7. b 8. a 9. e 10. c 11. a 12. c 13. a 14. d 15. b

Chapter 24
1. d 2. a 3. c 4. b 5. e 6. d 7. d 8. c 9. d 10. a 11. b 12. c 13. b 14. e 15. c 16. a

Chapter 25
1. c 2. a 3. b 4. c 5. b 6. e 7. d 8. c 9. d 10. b 11. d 12. a 13. d

Chapter 26
1. c 2. a 3. b 4. c 5. a 6. b 7. d 8. a 9. d 10. c 11. b 12. a

Chapter 27
1. a 2. d 3. b 4. d 5. a 6. b 7. a 8. a 9. c 10. a

Chapter 28
1. c 2. a 3. b 4. d 5. c 6. d 7. d 8. a 9. c 10. b 11. b 12. d

Chapter 29
1. d 2. c 3. d 4. c 5. d 6. a 7. c 8. bb Bb BB 9. a 10. c

Chapter 30
1. a 2. d 3. b 4. c 5. a 6. b 7. c 8. a 9. c 10. c

Appendix B

Scientific Journals of Anatomy and Physiology

Acta Anatomica (Basel)
Acta Biologica et Medica Germanica (Berlin)
Acta Cytologica (St. Louis)
Acta Embryologiae Experimentalis (Palermo)
Acta Morphologica Academiae Scientiarum Hungaricae (Budapest)
Acta Morphologica Academiae Scientiarum Hungaricae Supplementum (Budapest)
Acta Morphologica Neerlando-Scandinavica (Utrecht)
Acta Physiologica et Pharmacologica Bulgarica (Sofia)
Acta Physiologica Latino Americana (Buenos Aires)
Acta Physiologica Scandinavica (Stockholm)
Activitas Nervosa Superior (Prague)
Age and Ageing (London)
Agressologie (Paris)
American Journal of Anatomy (New York)
American Journal of Physiology (Bethesda)
Anatomical Record (New York)
Anatomischer Anzeiger (Jena)
Anatomy and Embryology (Berlin)
Andrologia (Berlin)
Archives d'Anatomie, d'Histologie et d'Embryologie (Strasbourg)
Archives d'Anatomie et de Cytologie Pathologiques (Paris)
Archives d'Anatomie Microscopique et de Morphologie Experimentale (Paris)
Archives Internationales de Physiologie et de Biochimie (Liege)
Archivio di Fisiologia (Florence)
Archivio Italiano di Anatomia e di Embriologia (Florence)
Archivum Histologicum Japonicum. Nihon Soshikigaku Kiroku (Niigata)
Arkhiv Anatomii, Gistologii i Embriologii (Moscow)
Biology of the Neonate (Basel)
Brain, Behavior, and Evolution (Basel)
Calcified Tissue International (New York)
Canadian Journal of Genetics and Cytology (Ottawa)
Canadian Journal of Physiology and Pharmacology (Ottawa)
Cell (Cambridge MA)
Cell Biology International Reports (London)
Cell Differentiation (Limerick)
Cell and Tissue Kinetics (Oxford)
Cell and Tissue Research (Berlin)
Cellule (Brussels)
Ceskoslovenska Fysiologie (Prague)
Chronobiologia (Milan)
Clinical and Experimental Pharmacology and Physiology (Carlton)
Comparative Biochemistry and Physiology. A. Comparative Physiology (Oxford)
Computers and Biomedical Research (New York)
Connective Tissue Research (London)
Contraception (Los Altos CA)
Cytobiologie (Stuttgart)
Cytobios (Cambridge ENG)
Cytogenetics and Cell Genetics (Basel)
Cytologia (Tokyo)
Developmental Biology (New York)
Differentiation (New York)
EEG/EMG (Stuttgart)
Electroencephalography and Clinical Neurophysiology (Limerick)
European Journal of Applied Physiology and Occupational Physiology (Berlin)
Experimental Cell Research (New York)
Fiziologicheskii Zhurnal (Leningrad)
Folia Morphologica (Prague)
Growth (Lakeland FL)
Human Development (Basel)
Human Physiology (New York)
Indian Journal of Physiology and Pharmacology (New Delhi)
International Journal of Aging and Human Development (Farmingdale NY)
In Vitro (Gaithersburg MD)
Japanese Journal of Physiology (Tokyo)
Journal de Physiologie (Paris)
Journal of Anatomy (Cambridge ENG)
Journal of Applied Physiology: Respiratory, Environmental and Exercise Physiology (Bethesda)
Journal of Cell Biology (New York)
Journal of Cell Science (London)
Journal of Cellular Physiology (New York)
Journal of Comparative and Physiological Psychology (Washington)
Journal of Electron Microscopy (Tokyo)
Journal of Embryology and Experimental Morphology (Colchester)
Journal of General Physiology (New York)
Journal of Human Ergology (Tokyo)
Journal of Insect Physiology (Elmsford)
Journal of Membrane Biology (New York)
Journal of Microscopy (Oxford)
Journal of Molecular Evolution (Berlin)
Journal of Morphology (New York)
Journal of Neurocytology (London)
Journal of Neurophysiology (Bethesda)
Journal of Physiology (London)
Journal of Ultrastructure Research (New York)
Kaibogaku Zasshi. Journal of Anatomy (Tokyo)
Life Sciences (Oxford)
Mechanisms of Ageing and Development (Limerick)
Okajima's Folia Anatomica Japonica (Tokyo)
Patologicheskaya Fiziologiya i Eksperimental'naya Terapiya (Moscow)
Pfluegers Archiv. European Journal of Physiology (Berlin)
Physiologia Bohemoslovaca (Prague)
Physiological Reviews (Bethesda)
Physiologist (Bethesda)
Physiology and Behavior (Elmsford NY)
Prostaglandins, Leukotrienes, and Medicine (Edinburgh)
Psychophysiology (Baltimore)
Quarterly Journal of Experimental Physiology and Cognate Medical Sciences (Edinburgh)
Respiration Physiology (Amsterdam)
Revista Espanola De Fisiologia (Barcelona)
Scanning Electron Microscopy (Chicago)
Stain Technology (Baltimore)
Teratology (New York)
Tissue and Cell (Harlow)
Tsitologiya (Leningrad)
Tsitologiya i Genetika (Kiev)
Ultramicroscopy (Amsterdam)
Undersea Biomedical Research (Bethesda)
Uspekhi Fizicheskikh Nauk (Moscow)
Virchows Archiv. A. Pathological Anatomy and Histology (New York)
Zeitschrift fuer Mikroskopisch-Anatomische Forschung (Leipzig)
Zeitschrift fuer Morphologie und Anthropologie (Stuttgart)
Zeitschrift fuer Tierphysiologie Tierernaehrung und Futtermittelkunde (Hamburg)
Zhurnal Evolyutsionnoi Biokhimii i Fiziologii (Leningrad)

APPENDIX C
Selected Readings

This appendix lists articles and textbooks that may be of reference value to students who wish to deepen their understanding of particular topics in anatomy and physiology. The references are divided into ten sections based on general topics. The sequence of these sections corresponds to the order in which the subjects are covered in the text.

Cell Structure and Function

Afzelius, B. 1986. Disorders of ciliary motility. *Hospital Practice* 21:73.
Allen, R. D. 1987 (February). The microtubule as an intracellular engine. *Scientific American.*
Bretscher, M. S. 1985 (October). The molecules of the cell membrane. *Scientific American.*
Brown, D. D. 1981. Gene expression in eukaryotes. *Science* 211:667.
Cech, T. R. 1986 (November). RNA as an enzyme. *Scientific American.*
Chambon, P. 1981 (May). Split genes. *Scientific American.*
Crick, F. 1962 (October). The genetic code. *Scientific American.*
Danielli, J. F. 1973. The bilayer hypothesis of membrane structure. *Hospital Practice* 8:63.
Darnell, J. E., Jr. 1985 (October). RNA. *Scientific American.*
Dautry-Varsal, A., and H. F. Lodish. 1984 (May). How receptors bring proteins and particles into cells. *Scientific American.*
DeDuve, C. 1983 (May). Microbodies in the living cell. *Scientific American.*
Doolittle, R. F. 1985 (October). Proteins. *Scientific American.*
Dustin, P. 1980 (August). Microtubules. *Scientific American.*
Felsenfled, G. 1985 (October). DNA. *Scientific American.*
Fox, C. F. 1972 (February). The structure of cell membranes. *Scientific American.*
Grivell, L. A. 1983 (March). Mitochondrial DNA. *Scientific American.*
Hayflick, H. 1980 (January). The cell biology of human aging. *Scientific American.*
Hinkle, P., and R. E. McCarty. 1979 (March). How cells make ATP. *Scientific American.*
Kornfeld, S., and W. S. Sly. 1985. Lysosomal storage defects. *Hospital Practice* 20:71
Lake, J. A. 1981 (August). The ribosome. *Scientific American.*
Lazarides, E., and J. P. Ravel. 1979 (May). The molecular basis of cell movement. *Scientific American.*
Lodish, H. F., and J. E. Rothman. 1979 (January). The assembly of cell membranes. *Scientific American.*
McKusick, V. A. 1981. The anatomy of the human genome. *Hospital Practice* 16:82.
Mazia, D. 1974 (January). The cell cycle. *Scientific American.*
Miller, O. L., Jr. 1973 (March). The visualization of genes in action. *Scientific American.*
Palade, G. 1975. Intracellular aspects of the process of protein synthesis. *Science* 189:347.
Rothman, J. E. 1985 (September). The compartmental organization of the Golgi apparatus. *Scientific American.*
Singer, S. J. 1973. Biological membranes. *Hospital Practice* 8:81.
Singer, S. J., and G. L. Nicolson. 1972. The fluid mosaic model of the structure of cell membranes. *Science* 175:720.
Sloboda, R. D. 1980. The role of microtubules in cell structure and cell division. *American Scientist* 68:290.
Stein, G., J. S. Stein, and L. J. Kleinsmith. 1975 (February). Chromosomal proteins and gene regulation. *Scientific American.*
Wallace, D. C. 1986. Mitochondrial genes and disease. *Hospital Practice* 21:77.
Weinberg, R. A. 1985 (October). The molecules of life. *Scientific American.*
Wheeler, T. J., and P. C. Hinkle. 1985. The glucose transporter of mammalian cells. *Annual Review of Physiology* 47:503.
White, R., and J M. Lalouel. 1988 (February). Chromosome mapping with DNA markers. *Scientific American.*

Integumentary, Skeletal, and Muscular Systems

Astrand, P. O., and K. Rodahl. 1977. *Textbook of Work Physiology: Physiological Basis of Exercise.* New York: McGraw-Hill Book Co.
Bluefarb, S. M. 1974. *Dermatology.* Kalamazoo, Mich.: The Upjohn Co.
Bourne, G. H., ed. 1973. *The Structure and Function of Muscle.* 2d ed. 4 vols. New York: Academic Press.
Cohen, C. 1975 (November). The protein switch of muscle contraction. *Scientific American.*
Edelson, R. L., and J. M. Fink. 1985 (June). The immunologic function of the skin. *Scientific American.*
Evans, F. G., ed. 1966. *Studies in the Anatomy and Function of Bones and Joints.* New York: Springer-Verlag.
Felig, P., and J. Wahren. 1975. Fuel homeostasis in exercise. *New England Journal of Medicine* 293:1078.
Grinnel, A. D., and M. A. B. Brazier, eds. 1981. *Regulation of Muscle Contraction: Excitation-Contraction Coupling.* New York: Academic Press.
Hall, B. K. 1988 (March-April). The embryonic development of bone. *American Scientist.*
Hoyle, G. 1970 (April). How is muscle turned on and off? *Scientific American.*
Huxley, H. E. 1969. The mechanism of muscle contraction. *Science* 164:1356.
Loomis, W. F. 1970 (December). Rickets. *Scientific American.*
Margaria, R. 1972 (March). The sources of muscular energy. *Scientific American.*
Marples, M. J. 1979 (January). Life on the human skin. *Scientific American.*
Merton, P. A. 1972 (May). How we control the contraction of our muscles. *Scientific American.*
Moncrief, J. A. 1973. Burns. *New England Journal of Medicine.* 228:444.
Montagna, W. 1969 (June). The skin. *Scientific American.*
Murray, J. H., and A. Weber. 1974 (February). The cooperative action of muscle proteins. *Scientific American.*
Nadel, E. R. 1985. Physiological adaptations to aerobic training. *American Scientist* 73:334.
Pawelek, J. M., and A. M. Korner. 1982 (March-April). The biosynthesis of mammalian melanin. *American Scientist.*
Rasche, P. J., and R. K. Burke. 1978. *Kinesiology and Applied Anatomy: The Science of Human Movement.* 6th ed. Philadelphia: Lea and Febiger.
Ross, R. 1969 (June). Wound healing. *Scientific American.*
Rosse, C., and D. K. Clawson. 1980. *The Musculoskeletal System in Human Health and Disease.* Philadelphia: Harper and Row Publishers.
Rushmer, R. L., et al. 1966. The skin. *Science* 154:343.
Sharpe, W. D. 1979. Age changes in human bones: An overview. *Bulletin of the New York Academy of Medicine* 55:757.
Sonstegard, D. A., L. S. Mathews, and H. Kaufer. 1979 (January). The surgical replacement of the human knee joint. *Scientific American.*
Vaughan, J. M. 1981. *The Physiology of Bone.* 3d ed. New York: Oxford University Press.

Nervous System

Andreasen, N. C. 1988. Brain imaging: Applications in psychiatry. *Science* 239:1381.
Angevine, J. B., Jr., and C. Cottman. 1981. *Principles of Neuroanatomy.* New York: Oxford University Press.
Axelrod, J. 1974 (June). Neurotransmitters. *Scientific American.*
Barchas, J. D., et al. 1978. Behavioral neurochemistry: Neuroregulators and behavioral states. *Science* 200:964.
Bartus, R. T., et al. 1982. The cholinergic hypothesis of geriatric memory dysfunction. *Science* 217:408.
Blusztain, J. K., and R. J. Wurtman. 1983. Choline and cholinergic neurons. *Science* 221:614.
Catteral, W. A. 1982. The molecular basis of neuronal excitability. *Science* 223:653.
Coyle, J. T., D. L. Prince, and M. R. DeLong. 1983. Alzheimer's disease: A disorder of cortical cholinergic innervation. *Science* 219:1184.
Dunant, Y., and M. Israel. 1985 (April). The release of acetylcholine. *Scientific American.*
Fine, A. 1986 (August). Transplantation in the central nervous system. *Scientific American.*
Frohman, L. A. 1975. Neurotransmitters as regulators of endocrine function. *Hospital Practice* 10:54.
Geschwind, N. 1972 (April). Language and the brain. *Scientific American.*
Goldstein, G. W., and A. L. Betz. 1986 (September). The blood-brain barrier. *Scientific American.*
Gottlieb, D. I. 1988 (February). GABAergic neurons. *Scientific American.*
Hubel, D. H. 1979 (September). The brain. *Scientific American.*

Kandel, E. R. 1979. Psychotherapy and the single synapse. *New England Journal of Medicine* 301:1028.

Kandel, E. R., and J. H. Schwartz, eds. 1981. *Principles of Neural Science.* New York: Elsevier North Holland.

Keynes, R. D. 1979 (March). Ion channels in the nerve cell membrane. *Scientific American.*

Krieger, D. T. 1983. Brain peptides: What, where, and why? *Science* 222:975.

Kuffler, S. W., and J. G. Nicholls. 1976. *From Neuron to Brain: A Cellular Approach to the Function of the Nervous System.* Sunderland, Mass.: Sinauer Associates, Inc. Publishers.

Lester, H. A. 1977 (February). The response of acetylcholine. *Scientific American.*

Mishkin, M., and T. Appenzeller. 1987 (June). The anatomy of memory. *Scientific American.*

Morrel, P., and W. Norton. 1980 (May). Myelin. *Scientific American.*

Motulski, J. H., and P. A. Insel. 1982. Adrenergic receptors in man. *New England Journal of Medicine* 307:18.

Nathason, J. A., and P. Greegard. 1977 (August). "Second messengers" in the brain. *Scientific American.*

Noback, C. E., and R. J. Demerest. 1975. *The Human Nervous System: Basic Principles of Neurobiology.* 2d ed. New York: McGraw-Hill Book Company.

Routtenberg, A. 1978 (November). The transport of substances in nerve cells. *Scientific American.*

Shashoua, V. E. 1985. The role of extracellular proteins in learning and memory. *American Scientist* 73:364.

Snyder, S. H. 1980. Brain peptides as neurotransmitters. Science 209:976.

Snyder, S. H. 1984. Drug and neurotransmitter receptors in the brain. *Science* 224:22.

Springer, S. P., and G. Deutch. 1985. *Left Brain, Right Brain.* Rev. ed. New York: W. H. Freeman and Company.

Squire, L. R. 1986. Mechanisms of memory. *Science* 232:1612.

Stevens, C.F. 1979 (September). The neuron. *Scientific American.*

Thompson, R. F. 1985. *The Brain.* New York: W. H. Freeman and Company.

Thompson, R. F. 1986. The neurobiology of learning and memory. *Science* 233:941.

Wagner, H. N. 1984. Imaging CNS receptors: The dopaminergic system. *Hospital Practice* 19:187.

Wurtman, R. J. 1982 (April). Nutrients that modify brain function. *Scientific American.*

Wurtman, R. J. 1985 (January). Alzheimer's disease. *Scientific American.*

Sensory Organs

Botstein, D. 1986. The molecular biology of color vision. *Science* 232:142.

Boynton, R. M. 1979. *Human Color Vision.* New York: Holt, Rinehart, and Winston.

Casey, K. L. 1973. Pain: A current view of neural mechanisms. *American Scientist* 61:194.

Fireman, P. 1987. Newer concepts in otitis media. *Hospital Practice* 22:85.

Freese, A. J. 1977. *The Miracle of Vision.* New York: Harper and Row Publishers.

Goldberg, J. M., and C. Fernandez. 1975. Vestibular mechanisms. *Annual Review of Physiology* 37:129.

Green, D. M. 1976. *An Introduction to Hearing.* New York: L. Erlbaum Assoc.

Hubel, D. H. 1979. The visual cortex of normal and deprived monkeys. *American Scientist* 67:532.

Hubel, D. H., and T. Wiesel. 1979 (September). Brain mechanisms of vision. *Scientific American.*

Hudspeth, A. J. 1983 (February). The hair cells of the inner ear. *Scientific American.*

Koretz, J. F., and G. H. Handelman. 1988 (July). How the human eye focuses. *Scientific American.*

Loeb, G. E. 1985 (February). The functional replacement of the ear. *Scientific American.*

Masland, R. H. 1987 (December). The functional architecture of the retina. *Scientific American.*

Nathans, J., D. Thomas, and D. S. Hogness. 1986. Molecular genetics of human vision: The genes encoding blue, green, and red pigments. *Science* 232:193.

O'Brian, D. F. 1982. The chemistry of vision. *Science* 218:961.

Parker, D. E. 1980 (November). The vestibular apparatus. *Scientific American.*

Pettigrew, J. D. 1972 (August). The neurophysiology of binocular vision. *Scientific American.*

Pfaffmann, C., M. Frank, and R. Norgren. 1979. Neural mechanisms and behavioral aspects of taste. *Annual Review of Physiology* 30:283

Rhode, W. S. 1984. Cochlear mechanics. *Annual Review of Physiology* 46:231.

Rushton, W. A. H. 1975 (March). Visual pigments and color blindness. *Scientific American.*

Schnapf, J. L., and D. A. Baylor. 1987 (April). How photoreceptor cells respond to light. *Scientific American.*

Stryer, L. 1987 (July). The molecules of visual excitation. *Scientific American.*

Van Essen, D. C. 1979. Visual areas of the mammalian cerebral cortex. *Annual Review of Neurosciences* 2:277.

Van Heyninger, R. 1975 (December). What happens to the human lens in cataract? *Scientific American.*

Von Bekesky, G. 1975 (August). The ear. *Scientific American.*

Endocrine System

Axelrod, J., and T. D. Reisine. 1984. Stress hormones: Their interaction and regulation. *Science* 224:452.

Baxter, J. D., and W. J. Funder. 1979. Hormone receptors. *New England Journal of Medicine* 300:117.

Brownstein, M. J., et al. 1980. Synthesis, transport, and release of posterior pituitary hormones. *Science* 207:373.

Carmichael, S. W., and H. Winkler. 1985 (August). The adrenal chromaffin cell. *Scientific American.*

Demers, L. M. 1984 (September). The effects of prostaglandins. *Diagnostic Medicine.*

Ganong, W. F., L. C. Alpert, and T. C. Lee. 1974. ACTH and the regulation of adrenocortical secretion. *New England Journal of Medicine* 290:1006.

Gelato, M. C., and G. R. Merriam. 1986. Growth hormone releasing hormone. *Annual Review of Physiology* 48:569.

Gillie, R. B. 1971 (June). Endemic goiter. *Scientific American.*

Katzenellenbogen, B. S. 1980. Dynamics of steroid hormone receptor action. *Annual Review of Physiology* 42:17.

McEwen, B. S. 1976 (July). Interactions between hormones and nerve tissue. *Scientific American.*

O'Malley, B., and W. T. Shrader. 1976 (February). The receptors of steroid hormones. *Scientific American.*

Quinn, S. J., and G. H. Williams. 1988. Regulation of aldosterone secretion. *Annual Review of Physiology* 50:409.

Rasmussen, H. 1986. The calcium messenger system. *New England Journal of Medicine* 314:1094, 1164.

Reisine, T. 1988. Neurohumoral aspects of ACTH release. *Hospital Practice* 23:77.

Roth, J., and S. I. Taylor. 1982. Receptors for peptide hormones: Alterations in diseases of humans. *Annual Review of Physiology* 44:639.

Schally, A. V. 1978. Aspects of the hypothalamic control of the pituitary gland. *Science* 202:18.

Selye, H. 1973. The evolution of the stress concept. *American Scientist* 61:693.

Thorner, M. O. 1986. Hypothalamic releasing hormones. *Hospital Practice* 21:63.

Circulatory System

Atlas, S. A. 1986. Atrial natriuretic factor: Renal and systemic effects. *Hospital Practice* 21:67.

Berne, R. M., and M. N. Levy. 1981. *Cardiovascular Physiology.* 4th ed. St. Louis: The C. V. Mosby Co.

Braunwald, E. 1974. Regulation of the circulation. *New England Journal of Medicine* 290:1124, 1420.

Brody, H. J., J. R. Haywood, and K. B. Toun. 1980. Neural mechanisms in hypertension. *Annual Review of Physiology* 42:441.

Brown, M. S., and J. L. Goldstein. 1984 (November). How LDL receptors influence cholesterol and atherosclerosis. *Scientific American.*

Brown, M. S., and J. L. Goldstein. 1986. A receptor-mediated pathway for cholesterol homeostasis. *Science* 232:34.

Cantin, M., and J. Genest. 1986. (February). The heart as an endocrine gland. *Scientific American.*

Conover, M. B. 1980. *Understanding Electrocariography.* 3d ed. St. Louis: The C. V. Mosby Co.

Del Zoppo, G. J., and L. A. Harker. 1984. Blood/vessel interaction in coronary disease. *Hospital Practice* 19:163.

Donald, D. E., and J. T. Shepard. 1980. Autonomic regulation of the peripheral circulation. *Annual Review of Physiology* 42:429.

Dublin, D. 1981. *Rapid Interpretation of EKG's.* 3d ed. Tampa, Fla.: Cover Publishing Co.

Fulkow, B., and E. Neill. 1971. *Circulation.* London: Oxford University Press.

Gerard, J. M. 1988. Platelet aggregation: Cellular regulation and physiologic role. *Hospital Practice* 23:89.

Glasser, S. P., and R. G. Zobie. 1985. Management of cardiac arrhythmias. *Hospital Practice* 20:127.

Herd, J. A. 1984. Cardiovascular response to stress in man. *Annual Review of Physiology* 46:177.

Hills, D., and E. Braunwald. 1977. Myocardial ischemia. *New England Journal of Medicine* 296:971, 1033, 1093.

Hilton, S. M., and K. M. Spyer. 1980. Central nervous regulation of vascular resistance. *Annual Review of Physiology* 42:399.

Katz, A. M. 1987. A physiologic approach to the treatment of heart failure. *Hospital Practice* 22:117.

Kontos, H. A. 1981. Regulation of the cerebral circulation. *Annual Review of Physiology* 43:397.

Laragh, J. H. 1985. Atrial natriuretic hormone, the renin-aldosterone axis, and blood pressure–electrolyte homeostasis. *New England Journal of Medicine* 313:1330.

Leon, A. S. 1983. Exercise and coronary heart disease. *Hospital Practice* 18:38.

Little, R. C. 1981. *Physiology of the heart and circulation.* 2d ed. Chicago: Year Book Medical Publishers.

Nadel, E. R. 1985. Physiological adaptations to aerobic training. *American Scientist* 73:334.
Needleman, P., and J. E. Greenwald. 1986. Atriopeptin: A cardiac hormone intimately involved in fluid, electrolyte, and blood pressure homeostasis. *New England Journal of Medicine* 314:828.
Olsson, R. A. 1981. Local factors regulating cardiac and skeletal muscle blood flow. *Annual Review of Physiology* 43:385.
Robinson, T. F., S. M. Factor, and E. H. Sonnenblick. 1986 (June). The heart as a suction pump. *Scientific American.*
Ross, R. 1986. The pathogenesis of atherosclerosis: An update. *New England Journal of Medicine* 314:488.
Smith, J. J., and J. P. Kampine. 1980. *Circulatory Physiology: The essentials.* Baltimore: Williams and Wilkins Co.
Spear, J. F., and E. N. Moore. 1982. Mechanisms of cardiac arrhythmias. *Annual Review of Physiology* 44:485.
Stephenson, R. B. 1984. Modification of reflex regulation of blood pressure by behavior. *Annual Review of Physiology* 46:133.
Vatner, S. F., and E. Braunwald. 1975. Cardiovascular control mechanisms in the conscious state. *New England Journal of Medicine* 293:970.
Weber, K. T., J. S. Janicki, and W. Laskey. 1983. The mechanics of ventricular function. *Hospital Practice* 18:113.
Zellis, R. S., S. F. Flaim, A. J. Liedke, and S. H. Nellis. 1981. Cardiovascular dynamics in the normal and failing heart. *Annual Review of Physiology* 43:455.
Zucker, M. B. 1980 (June). The function of blood platelets. *Scientific American.*

Lymphatic System and Immunity

Acuto, O., and E. Reinhertz. 1985. The human T cell receptor. Structure and function. *New England Journal of Medicine* 312:1100.
Ada, G. L., and G. Nossal. 1987 (August). The clonal selection theory. *Scientific American.*
Alt, F. W., T. K. Blackwell, and G. D. Yancopoulos. 1987. Development of the primary antibody repertoire. *Science* 238:1079.
Baglioni, C., and T. W. Nilsen. 1981. The action of interferon at the molecular level. *American Scientist* 69:392.
Barrett, J. T. 1978. *Textbook of Immunology.* 3d. ed. St. Louis: C. V. Mosby Co.
Biusseret, P. D. 1982 (August). Allergy. *Scientific American.*
Burnet, F. M. 1976. *Immunology: Readings from Scientific American.* San Francisco: W. H. Freeman and Co.
Capra, J. D., and A. B. Edmunson. 1977 (January). The antibody combining site. *Scientific American.*
Cohen, I. R. 1988 (April). The self, the world, and autoimmunity. *Scientific American.*
Cunningham, B. A. 1977 (October). The structure and function of histocompatibility antigens. *Scientific American.*
Dausset, J. 1981. The major histocompatibility complex in man: Past, present, and future concepts. *Science* 213:1469.
DiNome, M. A., and D. E. Young. 1987 (August). The clonal selection theory. *Scientific American.*
Geha, R. S. 1988. Regulation of IgE synthesis in atopic disease. *Hospital Practice* 23:91.
Gleich, G. J. 1988. Current understanding of eosinophil function. *Hospital Practice* 23:137.
Hamburger, R. N. 1976. Allergy and the immune system. *American Scientist* 64:157.
Herberman, R. B., and J. R. Ortaldo. 1981. Natural killer cells: Their role in defense against disease. *Science* 214:24.
Hirsch, M. S., and J. C. Kaplan. 1987 (April). Antiviral therapy. *Scientific American.*
Kapp, J. A., C. W. Pierce, and C. M. Sorensen. 1984. Antigen-specific suppressor T cell factors. *Hospital Practice* 19:85.
Koffler, D. 1980 (July). Systemic lupus erythematosus. *Scientific American.*
Laurence, J. 1985 (December). The immune system in AIDS. *Scientific American.*
Leder, P. 1982 (November). The genetics of antibody diversity. *Scientific American.*
McDevitt, H. O. 1985. The HLA system and its relation to disease. *Hospital Practice* 20:57.
Marrack, P., and J. Kappler. 1986 (February). The T cell and its receptor. *Scientific American.*
Milstein, C. 1980 (October). Monoclonal antibodies. *Scientific American.*
Milstein, C. 1986. From antibody structure to immunological diversification of immune response. *Science* 231:1261.
Nossal, G. J. V. 1987. The basic components of the immune system. *New England Journal of Medicine* 316:1320.
Oettgen, H. F. 1981. Immunological aspects of cancer. *Hospital Practice* 16:93.
Old, L. J. 1977 (May). Cancer immunology. *Scientific American.*
Old, L. J. 1988 (May). Tumor necrosis factor. *Scientific American.*
Rose, N. R. 1981 (February). Autoimmune diseases. *Scientific American.*
Sachs, L. 1986 (January). Growth, differentiation, and reversal of malignancy. *Scientific American.*
Samuelsson, B. 1983. Leukotrienes: Mediators of immediate hypersensitvity and inflammation. *Science* 220:568.
Tannock, I. F. 1983. Biology of tumor growth. *Hospital Practice* 18:81.
Tonegawa, S. 1985 (October). The molecules of the immune system. *Scientific American.*
Unanue, E. R., and P. M. Allen. 1987. The immunoregulatory role of the macrophase. *Hospital Practice* 22:87.
Vaghan, J. A. 1984. Rheumatoid arthritis: Evidence of a defect in T cell function. *Hospital Practice* 19:101.
Yelton, D. E., and M. D. Scharff. 1980. Monoclonal antibodies. *American Scientist* 63:510.
Young, J. D., and Z. A. Cohen. 1988 (January). How killer cells kill. *Scientific American.*

Respiratory and Urinary Systems

Alexander, E. 1986. Metabolic acidosis: Recognition and etiologic diagnosis. *Hospital Practice* 21:100E.
Anderson, E. 1977. Regulation of body fluids. *Annual Review of Physiology* 39:185.
Avery, M. E., S. S. Wang, and H. W. Taeusch. 1975 (March). The lung of the newborn infant. *Scientific American.*
Bauman, J. W., and F. P. Chinard. 1975. *Renal Function: Physiological and Medical Aspects.* St. Louis: C. V. Mosby Co.
Beeuwkes, R., III. 1980. The vascular organization of the kidney. *Annual Review of Physiology* 42:531.
Berger, A. J., R. A. Mitchel, and J. W. Severinghaus. 1977. Regulation of respiration. *New England Journal of Medicine* 297:92, 138, 194.
Bramble, D. M., and D. R. Carrier. 1983. Running and breathing in mammals. *Science* 219:251.
Brenner, B. M., and R. Beeuwkes, III. 1978. The renal circulation. *Hospital Practice* 13:35.
Brenner, B. M., T. H. Hostetter, and H. D. Humes. 1978. Molecular basis of proteinuria of glomerular origin. *New England Journal of Medicine* 298:826.
Browning, R. J. 1982 (January–February: 39; March–April: 59). Pulmonary disease. Part 1: Back to basics; Part 2: Putting blood gasses to work. *Diagnostic Medicine.*
Buckalew, V. M., Jr., and K. A. Gruber. 1984. Natriuretic hormone. *Annual Review of Physiology* 46:343.
Cherniak, N. S. 1986. Breathing disorders during sleep. *Hospital Practice* 21:81.
Dantzker, D. R. 1986. Physiology and pathophysiology of pulmonary gas exchange. *Hospital Practice* 121:135.
Epstein, F. H., and R. S. Brown. 1988. Acute renal failure: A collection of paradoxes. *Hospital Practice* 23:171.
Finch, C. A., and C. Lenfant. 1972. Oxygen transport in man. *New England Journal of Medicine* 286:407.
Flenley, D. C., and P. M. Warren. 1983. Ventilatory response to O_2 and CO_2 during exercise. *Annual Review of Physiology* 45:415.
Fraser, R. G., and J. A. P. Pare. 1977. *Structure and Function of the Lung.* 2d ed. Philadelphia: W. B. Saunders Co.
Galla, J. H., and R. G. Luke. 1987. Pathophysiology of metabolic alkalosis. *Hospital Practice* 22:123.
Giebisch, G. H., and B. Stanton. 1979. Potassium transport in the nephron. *Annual Review of Physiology* 41:241.
Glassock, R. J. 1987. Pathophysiology of acute glomerulonephritis. *Hospital Practice* 22:163.
Guz, A. 1975. Regulation of respiration in man. *Annual Review of Physiology* 37:303.
Haddad, G. G., and R. B. Mellins. 1984. Hypoxia and respiratory control in early life. *Annual Review of Physiology* 46:629.
Hays, R. M. 1978. Principles of ion and water transport in the kidneys. *Hospital Practice* 13:79.
Hollenberg, N. K. 1986. The kidney in heart failure. *Hospital Practice* 21:81.
Irsigler, G. B., and J. W. Severinghaus. 1980. Clinical problems of ventilatory control. *Annual Review of Medicine* 31:109.
Jacobson, H. R. 1987. Diuretics: Mechanisms of action and uses. *Hospital Practice* 22:129.
Kassirer, J. P., and N. E. Madias. 1980. Respiratory acid-base disorders. *Hospital Practice* 15:57.
Kokko, J. S. 1979. Renal concentrating and diluting mechanisms. *Hospital Practice* 14:110.
Macklem, P. T. 1986. Respiratory muscle dysfunction. *Hospital Practice* 21:83.
Murray, J. F. 1985. The lungs and heart failure. *Hospital Practice* 20:55.
Naeye, R. L. 1980 (April). Sudden infant death. *Scientific American.*
Peart, W. S. 1975. Renin-angiotensin system. *New England Journal of Medicine* 292:302.
Perutz, M. F. 1978 (December). Hemoglobin structure and respiratory transport. *Scientific American.*
Reid, I. A., B. J. Morris, and W. F. Ganong. 1978. The renin-angiotensin system. *Annual Review of Physiology* 40:377.
Renkin, E. M., and R. R. Robinson. 1974. Glomerular filtration. *New England Journal of Medicine* 290:70.
Rigatto, H. 1984. Control of ventilation in the newborn. *Annual Review of Physiology* 46:661.
Roussos, C., and P. T. Macklem. 1982. The respiratory muscles. *New England Journal of Medicine* 307:786.
Steinmetz, P. R., and B. M. Koeppen. 1984. Cellular mechanisms of diuretic action along the nephron. *Hospital Practice* 19:125.
Their, S. O. 1987. Diuretic mechanisms as a guide to therapy. *Hospital Practice* 22:81.

Tobin, M. J. 1986. Update on strategies in mechanical ventilation. *Hospital Practice* 21:69.

Vander, A. J. 1980. *Renal Physiology.* 2d ed. New York: McGraw-Hill Book Co.

Walker, D. W. 1984. Peripheral and central chemoreceptors in the fetus and newborn. *Annual Review of Physiology* 46:687.

Walker, L. A., and H. Vatlin. 1982. Biological importance of nephron heterogeneity. *Annual Review of Physiology* 44:203.

Warnock, D. G., and F. C. Rector, Jr. 1979. Proton secretion by the kidney. *Annual Review of Physiology* 41:197.

Whipp, B. J. 1983. Ventilatory control during exercise in humans. *Annual Review of Physiology* 45:393.

Digestive System and the Regulation of Metabolism

Austin, L. A., and H. Heath III. 1981. Calcitonin: Physiology and pathophysiology. *New England Journal of Medicine* 304:269.

Barret, E. J., and R. A. DeFronzo, 1984. Diabetic ketoacidosis: Diagnosis and treatment. *Hospital Practice* 19:89.

Binder, H. J. 1984. The pathophysiology of diarrhea. *Hospital Practice* 19:107.

Bleich, H. L., and E. S. Boro. 1979. Protein digestion and absorption. *New England Journal of Medicine* 300:659.

Cahill, G. F., and H. O. McDevitt. 1981. Insulin-dependent diabetes mellitus: The initial lesion. *New England Journal of Medicine* 304:454.

Carey, M. C., D. M. Small, and C. M. Bliss. 1983. Lipid digestion and absorption. *Annual Review of Physiology* 45:651.

Cheng, K., and J. Larner. 1985. Intracellular mediators of insulin action. *Annual Review of Physiology* 47:405.

Chou, C. C. 1982. Relationship between intestinal blood flow and motility. *Annual Review of Physiology* 44:29.

Cohen, S. 1983. Neuromuscular disorders of the gastrointestinal tract. *Hospital Practice* 18:121.

Davenport, H. W. 1982. *Physiology of the Digestive Tract.* 5th ed. Chicago: Year Book Medical Publishers.

DeLuca, H. F. 1980. The vitamin D hormonal system: Implications for bone disease. *Hospital Practice* 15:57.

Dockray, G. J. 1979. Comparative biochemistry and physiology of gut hormones. *Annual Review of Physiology* 41:83.

Eisenbarth, G. S. 1986. Type I diabetes mellitus: A chronic autoimmune disease. *New England Journal of Medicine* 314:1360.

Freeman, H. J., and Y. S. Kim. 1978. Digestion and absorption of proteins. *Annual Review of Physiology* 29:99.

Gardner, J. D., and R. T. Jensen. 1986. Receptors and cell activation associated with pancreatic enzyme secretion. *Annual Review of Physiology* 48:103.

Gardner, L. I. 1972 (July). Deprivation dwarfism. *Scientific American.*

Gollan, J. L., and A. B. Knapp. 1985. Bilirubin metabolism and congenital jaundice. *Hospital Practice* 20:83.

Goodman, D. S. 1984. Vitamin A and retinoids in health and disease. *New England Journal of Medicine* 310:1023.

Gray, G. M. 1975. Carbohydrate digestion and absorption: Role of the small intestine. *New England Journal of Medicine* 292:1225.

Grossman, M. I. 1979. Neural and hormonal regulation of gastrointestinal function: An overview. *Annual Review of Physiology* 41:27.

Habener, J. F., and J. E. Mahaffey. 1978. Osteomalacia and disorders of vitamin D metabolism. *Annual Review of Medicine* 29:327.

Hahn, T. J. 1986. Physiology of bone: Mechanisms of osteopenic disorders. *Hospital Practice* 21:73.

Hirsch, J. 1984. Hypothalamic control of appetite. *Hospital Practice* 19:131.

Holt, K. M., and J. I. Isenberg. 1985. Peptic ulcer disease: Physiology and pathophysiology. *Hospital Practice* 20:89.

Isaaksson, O. G., P. S. Eden, and J. O. Jansson. 1985. Mode of action of pituitary growth hormone on target cell. *Annual Review of Physiology* 47:483.

Kappas, A., and A. P. Alvarez. 1975 (June). How the liver metabolizes foreign substances. *Scientific American.*

McGuigan, J. E. 1978. Gastrointestinal hormones. *Annual Review of Physiology* 29:99.

Marshall, B. J. 1987. Peptic ulcer: An infectious disease? *Hospital Practice* 22:87.

Moog, F. 1981 (November). The lining of the small intestine. *Scientific American.*

Notkins, A. L. 1979 (November). The cause of diabetes. *Scientific American.*

Oppenheimer, J. H. 1979. Thyroid hormone action at the cellular level. *Science* 203:971.

Raisz, L. G., and B. E. Kream. 1981. Hormonal control of skeletal growth. *Annual Review of Physiology* 43:225.

Siperstein, M. D. 1985. Type II diabetes: Some problems in diagnosis and treatment. *Hospital Practice* 20:55.

Smith, B. F., and T. Lamont. 1984. The pathogenesis of gallstones. *Hospital Practice* 19:93.

Soll, A., and J. H. Walsh. 1979. Regulation of gastric acid secretion. *Annual Review of Physiology* 41:35.

Tepperman, J. 1980. *Metabolic and Endocrine Physiology.* 4th ed. Chicago: Year Book Medical Publishers.

Unger, R. H., and L. Orci. 1981. Glucagon and the A cell: Physiology and pathophysiology. *New England Journal of Medicine* 304:1518, 1575.

Unger, R. H., and L. Orci. 1981. Insulin, glucagon, and somatostatin secretion in the regulation of metabolism. *Annual Review of Physiology* 40:307.

Van De Graaff, K. M. 1986. Anatomy and physiology of the gastrointestinal tract. *Pediatric Infectious Diseases* 5:S11.

Van Wyk, J., and L. E. Underwood. 1978. Growth hormone, somatomedins, and growth failure. *Hospital Practice* 13:57.

Walsh, J. H., and M. I. Grossman. 1975. Gastrin. *New England Journal of Medicine* 292:1324, 1377.

Weinberg, R. H. 1987. Lipoprotein metabolism: Hormonal regulation. *Hospital Practice* 22:223.

Weisbrodt, N. W. 1981. Patterns of intestinal motility. *Annual Review of Physiology* 43:21.

Williams, J. A. 1984. Regulatory mechanisms in pancreas and salivary acini. *Annual Review of Physiology* 46:361.

Wood, J. D. 1981. Intrinsic neural control of intestinal motility. *Annual Review of Physiology* 43:33.

Wynder, E. L., and D. P. Rose. 1984. Diet and breast cancer. *Hospital Practice* 19:73.

Reproductive System, Development, and Aging

Balinsky, B. I. 1981. *An Introduction to Embryology.* 5th ed. Philadelphia: W. B. Saunders.

Bardin, C. W. 1979. The neuroendocrinology of male reproduction. *Hospital Practice* 14:65.

Bartke, A. A., et al. 1978. Hormonal interaction in the regulation of androgen secretion. *Biology of Reproduction* 18:44.

Beaconsfield, P., G. Birdwood, and R. Beaconsfield. 1980 (July). The placenta. *Scientific American.*

Birnholz, J. C., and E. E. Farrel. 1984. Ultrasound images of human fetal development. *American Scientist* 72:608.

Boyar, R. M. 1978. Control of the onset of puberty. *Annual Review of Medicine* 31:329.

Carter, N., ed. 1980. *Development, Growth, and Aging.* London: Croon Helm.

Chervenak, F. A., G. Isaacson, and M. J. Mahoney. 1986. Advances in the diagnosis of fetal defects. *New England Journal of Medicine* 315:305.

Comfort, A. 1979. *The Biology of Senescence.* 3d ed. London: Churchill Livingstone.

Diamond, M. C. 1978 (January–February). The aging brain. *American Scientist.*

Dufau, M. L. 1988. Endocrine regulation and communicating functions of the corpus luteum. *Annual Review of Physiology* 50:483.

England, M. A. 1983. *Color atlas of life before birth: Normal fetal development.* Chicago: Year Book Medical Publishers.

Epel, D. 1977 (November). The program of fertilization. *Scientific American.*

Fink, G. 1979. Feedback action of target hormones on hypothalamus and pituitary with special reference to gonadal steroids. *Annual Review of Physiology* 41:571.

Frantz, A. G. 1978. Prolactin. *New England Journal of Medicine* 298:112.

Goldzieher, J. W., and A. N. Poindexter. 1987. Medical aspects of contraception. *Hospital Practice* 22:93.

Grabowski, C. T. 1983. *Human Reproduction and Development.* Philadelphia: W. B. Saunders.

Grobstein, C. 1979 (March). External human fertilization. *Scientific American.*

Grumbach, M. M. 1979. The neuroendocrinology of puberty. *Hospital Practice* 14:65.

Hatcher, R. A., and A. K. Stewart. 1987. *Contraceptive Technology.* 13th rev. ed. New York: Wiley.

Hayflick, L. 1980 (January). The cell biology of human aging. *American Scientist.*

Jackson, L. G. 1985. First trimester diagnosis of fetal genetic disorders. *Hospital Practice* 20:39.

Jones, K. L., et al. 1985. *Dimensions of Human Sexuality.* Dubuque, Ia.: Wm. C. Brown Publishers.

Katchadourian, H. A., and D. T. Lunde. 1985. *Fundamentals of Human Sexuality.* 4th ed. New York: Holt, Rinehart, and Winston.

Keyes, P. L., and M. C. Wiltbank. 1988. Endocrine regulation of the corpus luteum. *Annual Review of Physiology* 50:465.

Lagerkrantz, H., and T. A. Slotkin. 1986 (April). The "stress" of being born. *Scientific American.*

Leong, D. S., L. S. Frawley, and J. D. Neill. 1983. Neuroendocrine control of prolactin secretion. *Annual Review of Physiology* 45:109.

Lipsett, M. B. 1980. Physiology and pathology of the Leydig cell. *New England Journal of Medicine* 303:682.

Marshall, J. C., and R. P. Kelch. 1986. Gonadotropin-releasing hormone: Role of pulsatile secretion in the regulation of reproduction. *New England Journal of Medicine* 315:1459.

Marx, J. L. 1978. The mating game: What happens when sperm meets egg. *Science* 200:1256.

Masters, W. H. 1986. Sex and aging—expectations and reality. *Hospital Practice* 21:175.

Means, A. R., et al. 1980. Regulation of the testis Sertoli cell by follicle stimulating hormone. *Annual Review of Physiology* 42:59.

Naftolin, F. 1981 (March). Understanding the basis of sex differences. *Science* 211:1263.

Nilsson, L., A. Ingelman-Sundbert, and C. Wirsen. 1977. *A Child is Born.* Rev. ed. New York: Dell Publishing Co.

Odell, W. D., and D. L. Moyer. 1971. *Physiology of Reproduction.* St. Louis: C. V. Mosby Co.

Oppenheimer, S. B., and G. Lefevere. 1984. *Introduction to Embryonic Development.* 2d ed. Boston: Allyn and Bacon.

Reiter, E. O., and M. M. Grumbach. 1982. Neuroendocrine control mechanisms and the onset of puberty. *Annual Review of Physiology* 44:595.

Santrock, J. W. 1985. *Adult Development and Aging.* Dubuque, Ia.: Wm. C. Brown Publishers.

Segal, S. J. 1974 (September). The physiology of human reproduction. *Scientific American.*

Short, R. V. 1984 (April). Breast feeding. *Scientific American.*

Simpson, E. R., and P. C. MacDonald. 1981. Endocrine physiology of the placenta. *Annual Review of Physiology* 43:163.

Tamarkin, L. C., C. J. Baird, and O. F. X. Almeida. 1985. Melatonin: A coordinating signal for mammalian reproduction? *Science* 227:714.

Tanner, J. M. 1973 (September). Growing up. *Scientific American.*

Tyson, J. E. 1984 (April). Reproductive endocrinology: New problems call for new solutions. *Diagnostic Medicine.*

Wilson, J. D. 1978. Sexual differentiation. *Annual Review of Physiology* 40:279.

Wilson, J. D., F. W. George, and J. E. Griffin. 1981. The hormonal control of sexual development. *Science* 211:1278.

Yen, S. S. C. 1979. Neuroendocrine regulation of the menstrual cycle. *Hospital Practice* 14:83.

Appendix D

Medical and Pharmacological Abbreviations

aa	of each
a.c.	before meals
A/G	albumin globulin ratio
ANS	autonomic nervous system
Bib	drink
b.i.d.	twice a day
bihor	during two hours
B.M.R.	basal metabolic rate
B.P.	blood pressure
BUN	blood urea nitrogen
b.v.	vapor bath
c̄	with
caps.	capsule
c.b.c.	complete blood count
cc.	cubic centimeter(s)
cm.	centimeter(s)
CNS	central nervous system
Co., Comp.	compound
cr	tomorrow
C.S.F.	cerebrospinal fluid
CVP	central venous pressure
d.	a day
D. & C.	dilatation and curettage
D.C.	discontinue
D, Det	give
de d. in d.	from day to day
Dieb, secund	every second day
Dieb, tert.	every third day
dim.	one-half
d. in dup.	give twice as much
D. in p. aeq.	divide into equal parts
dr.	dram
D.T.D.	give of such doses
Dur. dolor.	while pain lasts
e	out of, with
ECG, EKG	electrocardiogram
EEG	electroencephalogram
e.m.p.	in the manner prescribed
feb	fever
G.I.	gastrointestinal
gm.	gram
gr.	grain
Grad.	gradually
gtt.	drop(s)
h.	hour
HCT	hematocrit
Hg.	mercury
Hgb	hemoglobin
h.s.	at bedtime
ibid	in the same place
I.M.	intramuscular
incid.	cut
in d.	in a day
inj.	an injection
int. cib.	between meals
int. noct.	during the night
IPPB	intermittent positive pressure breathing
I.V.	intravenous
kg.	kilogram
Lat. dol.	to the painful side
M.	mix
man.	in the morning
mEq.	milliequivalent
mg.	milligram
ml.	milliliter
Noct.	at night
Noct. maneq.	night and morning
N.P.O.	nothing by mouth
O.D.	in the right eye
o.d.	every day
Omn. hor.	every hour
Omn. man.	every morning
Omn. noct.	every night
O.S.	in the left eye
O.U.	in each eye
oz.	ounce
Part. aeq.	equal parts
PBI	protein-bound iodine
p.c.	after meals
pCO_2	partial pressure of carbon dioxide
PNS	peripheral nervous system
P.O.	by mouth
pO_2	partial pressure of oxygen
p.p.a.	having first shaken the bottle
p.r.n.	as needed
pro. us. ext.	for external use
pt.	let it be continued
q.	each; every
q.d.	every day
q.h.	every hour
q. ____ h.	every ____ hours
q.i.d.	four times a day
q. noct.	every night
q.o.d.	every other day
q.q.	also
q.s.	sufficient quantity
RBC	red blood cell
s̄	without
Semih.	half an hour
Sig.	write, label
S.O.S.	if needed
sp. gr.	specific gravity
ss., s̄s̄	one-half
s.s.s.	layer on layer
stat.	immediately
sum.	take
s.v.r.	alcohol
t.	three times
tab.	tablet
t.i.d.	three times a day
ung.	ointment
Ut. dict.	as directed
vic.	times
WBC	white blood cell

Appendix E

Some Laboratory Tests of Clinical Importance

Common tests performed on blood		
Test	**Normal values (adult)**	**Clinical significance**
Acetone and acetoacetate (serum)	0.3–2.0 mg/100 ml	Values increase in diabetic acidosis, toxemia of pregnancy, fasting, and high-fat diet.
Albumin-globulin ratio or A/G ratio (serum)	1.5:1 to 2.5:1	Ratio of albumin to globulin is lowered in kidney diseases and malnutrition.
Albumin (serum)	3.2–5.5 gm/100 ml	Values increase in multiple myeloma and decrease with proteinuria and as a result of severe burns.
Ammonia (plasma)	50–170 μg/100 ml	Values increase in severe liver disease, pneumonia, shock, and congestive heart failure.
Amylase (serum)	80–160 Somogyi units/100 ml	Values increase in acute pancreatitis, intestinal obstructions, and mumps. They decrease in chronic pancreatitis, cirrhosis of the liver, and toxemia of pregnancy.
Bilirubin, total (serum)	0.3–1.1 mg/100 ml	Values increase in conditions causing red blood cell destruction or biliary obstruction.
Blood urea nitrogen or BUN (plasma or serum)	10–20 mg/100 ml	Values increase in various kidney disorders and decrease in liver failure and during pregnancy.
Calcium (serum)	9.0–11.0 mg/100 ml	Values increase in hyperparathyroidism, hypervitaminosis D, and respiratory conditions that cause a rise in CO_2 concentration. They decrease in hypoparathyroidism, malnutrition, and severe diarrhea.
Carbon dioxide (serum)	24–30 mEq/l	Values increase in respiratory diseases, intestinal obstruction, and vomiting. They decrease in acidosis, nephritis, and diarrhea.
Chloride (serum)	96–106 mEq/l	Values increase in nephritis, Cushing's syndrome, and hyperventilation. They decrease in diabetic acidosis, Addison's disease, diarrhea, and following severe burns.
Cholesterol, total (serum)	150–250 mg/100 ml	Values increase in diabetes mellitus and hypothyroidism. They decrease in pernicious anemia, hyperthyroidism, and acute infections.
Creatine phosphokinase or CPK (serum)	Men: 0–20 IU/l Women: 0–14 IU/l	Values increase in myocardial infarction and skeletal muscle diseases such as muscular dystrophy.
Creatine (serum)	0.2–0.8 mg/100 ml	Values increase in muscular dystrophy, nephritis, severe damage to muscle tissue, and during pregnancy.
Creatinine (serum)	0.7–1.5 mg/100 ml	Values increase in various kidney diseases.
Erythrocyte count or red cell count (whole blood)	Men: 4,600,000–6,200,000/cu mm Women: 4,200,000–5,400,000/cu mm Children: 4,500,000–5,100,000/cu mm (varies with age)	Values increase as a result of severe dehydration or diarrhea and decrease in anemia, leukemia, and following severe hemorrhage.
Fatty acids, total (serum)	190–420 mg/100 ml	Values increase in diabetes mellitus, anemia, kidney disease, and hypothyroidism. They decrease in hyperthyroidism.
Globulin (serum)	2.5–3.5 gm/100 ml	Values increase as a result of chronic infections.

Common tests performed on blood (cont.)		
Test	**Normal values (adult)**	**Clinical significance**
Glucose (plasma)	70–115 mg/100 ml	Values increase in diabetes mellitus, liver diseases, nephritis, hyperthyroidism, and pregnancy. They decrease in hyperinsulinism, hypothyroidism, and Addison's disease.
Hematocrit (whole blood)	Men: 40–54 ml/100 ml Women: 37–47 ml/100 ml Children: 35–49 ml/100 ml (varies with age)	Values increase in polycythemia due to dehydration or shock. They decrease in anemia and following severe hemorrhage.
Hemoglobin (whole blood)	Men: 14–18 gm/100 ml Women: 12–16 gm/100 ml Children: 11.2–16.5 gm/100 ml (varies with age)	Values increase in polycythemia, obstructive pulmonary diseases, congestive heart failure, and at high altitudes. They decrease in anemia, pregnancy, and as a result of severe hemorrhage or excessive fluid intake.
Iron (serum)	75–175 μg/100 ml	Values increase in various anemias and liver disease. They decrease in iron deficiency anemia.
Iron-binding capacity (serum)	250–410 μg/100 ml	Values increase in iron deficiency anemia and pregnancy. They decrease in pernicious anemia, liver disease, and chronic infections.
Lactic acid (whole blood)	6–16 mg/100 ml	Values increase with muscular activity and in congestive heart failure, severe hemorrhage, and shock.
Lactic dehydrogenase or LDH (serum)	90–200 milliunits/ml	Values increase in pernicious anemia, myocardial infarction, liver diseases, acute leukemia, and widespread carcinoma.
Lipids, total (serum)	450–850 mg/100 ml	Values increase in hypothyroidism, diabetes mellitus, and nephritis. They decrease in hyperthyroidism.
Oxygen saturation (whole blood)	Arterial: 94–100% Venous: 60–85%	Values increase in polycythemia and decrease in anemia and obstructive pulmonary diseases.
ph (whole blood)	7.35–7.45	Values increase due to vomiting, Cushing's syndrome, and hyperventilation. They decrease as a result of hypoventilation, severe diarrhea, Addison's disease, and diabetic acidosis.
Phosphatase, acid (serum)	1.0–5.0 King-Armstrong units/ml	Values increase in cancer of the prostate gland, hyperparathyroidism, certain liver diseases, myocardial infarction, and pulmonary embolism.
Phosphatase, alkaline (serum)	5–13 King-Armstrong units/ml	Values increase in hyperparathyroidism (and in other conditions that promote resorption of bone), liver diseases, and pregnancy.
Phospholipids (serum)	6–12 mg/100 ml as lipid phosphorus	Values increase in diabetes mellitus and nephritis.
Phosphorus (serum)	3.0–4.5 mg/100 ml	Values increase in kidney diseases, hypoparathyroidism, acromegaly, and hypervitaminosis D. They decrease in hyperparathyroidism.
Platelet count (whole blood)	150,000–350,000/cu mm	Values increase in polycythemia and certain anemias. They decrease in acute leukemia and aplastic anemia.
Potassium (serum)	3.5–5.0 mEq/l	Values increase in Addison's disease, hypoventilation, and conditions that cause severe cellular destruction. They decrease in diarrhea, vomiting, diabetic acidosis, and chronic kidney disease.
Protein, total (serum)	6.0–8.0 gm/100 ml	Values increase in severe dehydration and shock. They decrease in severe malnutrition and hemorrhage.
Protein-bound iodine or PBI (serum)	3.5–8.0 μg/100 ml	Values increase in hyperthyroidism and liver disease. They decrease in hypothyroidism.
Prothrombin time (serum)	12–14 sec (one stage)	Values increase in certain hemorrhagic diseases, liver disease, vitamin K deficiency, and following the use of various drugs.

Common tests performed on blood (cont.)		
Test	**Normal values (adult)**	**Clinical significance**
Sedimentation rate, Westergren (whole blood)	Men: 0–15 mm/hr Women: 0–20 mm/hr	Values increase in infectious diseases, menstruation, pregnancy, and as a result of severe tissue damage.
Sodium (serum)	136–145 mEq/l	Values increase in nephritis and severe dehydration. They decrease in Addison's disease, myxedema, kidney disease, and diarrhea.
Thyroxine or T_4 (serum)	2.9–6.4 μg/100 ml	Values increase in hyperthyroidism and pregnancy. They decrease in hypothyroidism.
Thromboplastin time, partial (plasma)	35–45 sec	Values increase in deficiencies of blood factors VIII, IX, and X.
Transaminases or SGOT (serum)	5–40 units/ml	Values increase in myocardial infarction, liver disease, and diseases of skeletal muscles.
Uric acid (serum)	Men: 2.5–8.0 mg/100 ml Women: 1.5–6.0 mg/100 ml	Values increase in gout, leukemia, pneumonia, toxemia of pregnancy, and as a result of severe tissue damage.
White blood cell count, differential (whole blood)	Neutrophils 54–62% Eosinophils 1–3% Basophils 0–1% Lymphocytes 25–33% Monocytes 3–7%	Neutrophils increase in bacterial diseases; lymphocytes and monocytes increase in viral diseases; eosinophils increase in collagen diseases, allergies, and in the presence of intestinal parasites.
White blood cell count, total (whole blood)	5,000–10,000/cu mm	Values increase in acute infections, acute leukemia, and following menstruation. They decrease in aplastic anemia and as a result of drug toxicity.

Common tests performed on urine		
Test	**Normal values**	**Clinical significance**
Acetone and acetoacetate	0	Values increase in diabetic acidosis.
Albumin, qualitative	0 to trace	Values increase in kidney disease, hypertension, and heart failure.
Ammonia	20–70 mEq/l	Values increase in diabetes mellitus and liver diseases.
Bacterial count	Under 10,000/ml	Values increase in urinary tract infection.
Bile and bilirubin	0	Values increase in melanoma and biliary tract obstruction.
Calcium	Under 250 mg/24 hr	Values increase in hyperparathyroidism and decrease in hypoparathyroidism.
Creatinine clearance	100–140 ml/min	Values increase in renal diseases.
Creatinine	1–2 gm/24 hr	Values increase in infections and decrease in muscular atrophy, anemia, leukemia, and kidney diseases.
Glucose	0	Values increase in diabetes mellitus and various pituitary gland disorders.
17-hydroxycorticosteroids	2–10 mg/24 hr	Values increase in Cushing's syndrome and decrease in Addison's disease.
Phenylpyruvic acid	0	Values increase in phenylketonuria.
Urea clearance	Over 40 ml blood cleared of urea/min	Values increase in renal diseases.
Urobilinogen	0–4 mg/24 hr	Values increase in liver diseases and hemolytic anemia. They decrease in complete biliary obstruction and severe diarrhea.
Urea	25–35 gm/24 hr	Values increase as a result of excessive protein breakdown. They decrease as a result of impaired renal function.
Uric acid	0.6–1.0 gm/24 hr. as urate	Values increase in gout and decrease in various kidney diseases.

Glossary

The words in this glossary are followed by a phonetic guide to pronunciation. This is a simplified system that is standard in medical usage and terminology.

Any unmarked vowel that ends a syllable or stands alone as a syllable is long. Any unmarked vowel that is followed by a consonant has the short sound.

If a long vowel appears in the middle of a syllable (followed by a consonant), it is marked with a macron (-). Similarly, if a vowel stands alone or ends a syllable, but should have a short sound, it is marked with a breve (˘).

a

abdomen (ab-do'men) A region of the body between the diaphragm and pelvis.

abduction (ab-duk'shun) The movement of a body part away from the axis or midline of the body.

ABO system The most common system of classification for red blood cell antigens. On the basis of antigens on the red blood cell surface, individuals can be type A, type B, type AB, or type O.

absorption (ab-sorp'shun) The transport of molecules across epithelial membranes into the body fluids.

accessory organs (ak-ses'o-re) Organs that assist the functioning of other organs within a system.

accommodation (ah-kom''o-da'shun) To fit or adjust; specifically, the ability of the eyes to adjust their curvature so that an image of an object is focused on the retina at different distances.

acetabulum (as''ĕ-tab'u-lum) A socket in the lateral surface of the hipbone (os coxa) into which the head of the femur articulates.

acetone (as'e-tōn) A ketone body produced as a result of the oxidation of fats.

acetylcholine (ACh) (as''ĕ-til-ko'lēn) A molecule—which is an acetic acid ester of choline—that functions as a neurotransmitter chemical in somatic motor nerve and parasympathetic nerve fibers.

acetylcholinesterase (as'' ĕ-til-ko''lin-es'ter-ās) An enzyme in the membrane of postsynaptic cells that catalyzes the conversion of ACh into choline and acetic acid. This enzymatic reaction inactivates the neurotransmitter.

Achilles tendon (ah-kil'ēz) Tendon that attaches the calf muscles to the calcaneus bone.

acid (as'id) A substance that releases hydrogen ions when ionized in water.

acidosis (as''i-do'sis) An abnormal increase in the H^+ concentration of the blood that lowers arterial pH below 7.35.

acromegaly (ak''ro-meg'ah-le) A condition caused by the hypersecretion of growth hormone from the pituitary after maturity and characterized by enlargement of the extremities, such as the nose, jaws, fingers, and toes.

actin (ak'tin) A protein in muscle fibers that together with myosin is responsible for contraction.

action potential An all-or-none electrical event in an axon or muscle fiber, in which the polarity of the membrane potential is rapidly reversed and reestablished.

active immunity (ĭ-mu'ni-te) Immunity involving sensitization, in which antibody production is stimulated by prior exposure to an antigen.

active transport The movement of molecules or ions across the cell membranes of epithelial cells by membrane carriers; an expenditure of cellular energy (ATP) is required.

adduction (ah'duk'shun) The movement of a body part toward the axis or midline of the body.

adenoids (ad'ĕ-noids) The tonsils located in the nasopharynx; pharyngeal tonsils.

adenylate cyclase (ah-den'i-lāt si'klās) An enzyme, found in cell membranes, that catalyzes the conversion of ATP to cyclic AMP and pyrophosphate (PP_1). This enzyme is activated by an interaction between a specific hormone and its membrane receptor protein.

ADH (antidiuretic hormone) Also known as *vasopressin.* A hormone produced by the hypothalamus and secreted by the posterior pituitary gland; it acts on the kidneys to promote water reabsorption.

ADP Adenosine diphosphate; a molecule that together with inorganic phosphate is used to make ATP (adenosine triphosphate).

adrenal cortex (ah-dre'nal kor'teks) The outer part of the adrenal gland. Derived from embryonic mesoderm, the adrenal cortex secretes corticosteroid hormones (such as aldosterone and hydrocortisone).

adrenal medulla (mĕ-dul'ah) The inner part of the adrenal gland. Derived from embryonic postganglionic sympathetic neurons, the adrenal medulla secretes catecholamine hormones—epinephrine and (to a lesser degree) norepinephrine.

adrenergic (ad''ren-er'jik) An adjective describing the actions of epinephrine, norepinephrine, or other molecules with similar activity (as in *adrenergic receptor* and *adrenergic stimulation*).

aerobic capacity (a-er-o'bik) The ability of an organ to utilize oxygen and respire aerobically to meet its energy needs.

afferent (af'er-ent) Conveying or transmitting to.

afferent arteriole (ar-te're-ōl) A blood vessel within the kidney that supplies blood to the glomerulus.

afferent neuron (nu'ron) A sensory nerve cell that transmits an impulse toward the central nervous system.

agglutinate (ah-gloo'ti-nāt) A clump of cells (usually erythrocytes) due to specific chemical interaction between surface antigens and antibodies.

agonist (ag'o-nist) The "prime mover" muscle, which is directly engaged in the contraction that produces the desired movement.

agranular leukocytes (ah-gran'u-ler loo'ko-sīts) White blood cells (leukocytes) that do not contain cytoplasmic granules; specifically, lymphocytes and monocytes.

albumin (al-bu'min) A water-soluble protein, produced in the liver, that is the major component of the plasma proteins.

aldosterone (al-dos'ter-ōn) The principal corticosteroid hormone involved in the regulation of electrolyte balance (mineralocorticoid).

alimentary canal (al''ĕ-men'tar-e) The tubular portion of the digestive tract.

allergens (al'er-jens) Antigens that evoke an allergic response rather than a normal immune response.

allergy (al'er-je) A state of hypersensitivity caused by exposure to allergens; it results in the liberation of histamine and other molecules with histamine-like effects.

all-or-none principle The statement of the fact that muscle fibers of a motor unit contract to their maximum extent when exposed to a stimulus of threshold strength.

allosteric (al''o-ster'ik) Denoting the alteration of an enzyme's activity by its combination with a regulator molecule; allosteric inhibition by an end product represents negative feedback control of an enzyme's activity.

alveolar sacs (al-ve'o-lar) A cluster of alveoli that share a common chamber or central atrium.

alveolus (al-ve'o-lus) An individual air capsule within the lung. The alveoli are the basic functional units of respiration.

amniocentesis (am''ne-o-sen-te'sis) A procedure to obtain amniotic fluid and fetal cells in this fluid through transabdominal perforation of the uterus.

amnion (am'ne-on) A developmental membrane that surrounds the fetus and contains amniotic fluid.

amphiarthrosis (am''fe-ar-thro'sis) A slightly movable articulation.

amphoteric (am-fo-ter'ik) Pertaining to having opposite characteristics; denoting a molecule that can be positively or negatively charged, depending on the pH of its environment.

ampulla (am-pul'lah) A saclike enlargement of a duct or tube.

ampulla of Vater (fah'ter) A small, elevated area within the duodenum where the combined pancreatic and common bile duct empties.

anabolic steroids (an″ah-bol′ik ste′roids) Steroids with androgenlike stimulatory effects on protein synthesis.

anabolism (ah′nab′o-lizm) Chemical reactions within cells that result in the production of larger molecules from smaller ones; specifically, the synthesis of protein, glycogen, and fat.

anaerobic respiration (an″a-er-o′bik res″pi-ra′shun) A form of cell respiration, involving the conversion of glucose to lactic acid, in which energy is obtained without the use of molecular oxygen.

anal canal (a′nal) The terminal tubular portion of the rectum that opens through the anus of the alimentary canal.

anal glands Enlarged and modified sweat glands that empty into the anal opening.

anaphylaxis (an″ah-fi-lak′sis) An unusually severe allergic reaction, which can result in cardiovascular shock and death.

anastomosis (ah-nas″to-mo′sis) An interconnecting aggregation of blood vessels or nerves that form a network plexus.

anatomical position (an″ah-tom′e-kal) An erect body stance with the eyes directed forward, the arms at the sides, and the palms of the hands facing forward.

anatomy (ah-nat′o-me) The branch of science concerned with the structure of the body and the relationship of its organs.

androgens (an′dro-jens) Steroids, containing eighteen carbons, that have masculinizing effects; primarily those hormones (such as testosterone) secreted by the testes, although weaker androgens are also secreted by the adrenal cortex.

anemia (ah-ne′me-ah) An abnormal reduction in the red blood cell count, hemoglobin concentration, or hematocrit, or any combination of these measurements. This condition is associated with a decreased ability of the blood to carry oxygen.

angina pectoris (an-ji′nah pek′tō-ris) A thoracic pain, often referred to the left pectoral and arm area, caused by myocardial ischemia.

angiotensin II (an″je-o-ten′sin) An eight-amino-acid polypeptide formed from angiotensin I (a ten-amino-acid precursor), which in turn is formed from cleavage of a protein (angiotensinogen) by the action of renin (which is an enzyme secreted by the kidneys). Angiotensin II is a powerful vasoconstrictor and a stimulator of aldosterone secretion from the adrenal cortex.

anions (an′i-ons) Ions that are negatively charged, such as chloride, bicarbonate, and phosphate.

antagonist (an-tag′-o-nist) A muscle that acts in opposition to a "prime mover," or an agonist.

antebrachium (an″te-bra′ke-um) The forearm.

anterior (ventral) (an-te′re-or) Toward the front; the opposite of *posterior,* or *dorsal.*

anterior pituitary (pi-tu′i-tār″e) Also called the *adenohypophysis;* this part of the pituitary gland secretes FSH (follicle-stimulating hormone), LH (luteinizing hormone), ACTH (adrenocorticotropic hormone), TSH (thyroid-stimulating hormone), GH (growth hormone), and prolactin. Secretions of the anterior pituitary are controlled by hormones secreted by the hypothalamus.

anterior root The anterior projection of the spinal cord, which is composed of axons of motor or efferent fibers.

antibodies (an″ti-bod″es) Immunoglobin proteins secreted by B lymphocytes that have transformed into plasma cells. Antibodies are responsible for humoral immunity. Their synthesis is induced by specific antigens, and they combine with these specific antigens but not with unrelated antigens.

anticodon (an″ti-ko′don) A base triplet provided by three nucleotides within a loop of transfer RNA, which is complementary in its base pairing properties to a triplet (the codon) in mRNA; the matching of codon to anticodon provides the mechanism for translating of genetic code into a specific sequence of amino acids.

antigen (an′ti-jen) A molecule able to induce the production of antibodies and to react in a specific manner with antibodies.

antigenic determinant site (an-ti-jen′ik) The region of an antigen molecule that specifically reacts with particular antibodies. A large antigen molecule may have a number of such sites.

antiserum (an″ti-se′rum) A serum that contains specific antibodies.

anus (a′nus) The terminal portion and outlet of the alimentary canal.

aorta (a-or′tah) The major systemic vessel of the arterial system of the body, emerging from the left ventricle.

apex (a′peks) The tip or pointed end of a conical structure.

aphasia (ah-fa′ze-ah) Defects in speech, writing, or the comprehension of spoken or written language; caused by brain damage or disease.

apneustic center (ap-nu′stik) A collection of neurons in the brain stem that participates in the rhythmic control of breathing.

apocrine gland (ap′o-krin) A type of sweat gland that functions in evaporative cooling. It may respond during periods of emotional stress.

aponeurosis (ap″o-nu-ro′sis) A fibrous or membranous sheetlike tendon.

aqueous humor (a′-kwe-us hum′or) The watery fluid that fills the anterior and posterior chambers of the eye.

arachnoid (ah-rak′noid) The weblike middle covering (one of the three meninges) of the central nervous system.

arbor vitae (ar′bor vi′tah) The branching arrangement of white matter within the cerebellum.

arch of aorta The superior left bend of the aorta between the ascending and descending portions.

arm (brachium) The portion of the upper extremity from the shoulder to the elbow.

arrector pili (ah-rek′tor pih′le) The smooth muscle attached to a hair follicle, which upon contraction, pulls the hair vertical, resulting in "goose bumps."

arteriole (ar-te′re-ōl) A minute arterial branch.

arteriosclerosis (ar-te″re-o-sklĕ-ro′sis) A group of diseases characterized by thickening and hardening of the artery wall and in the narrowing of its lumen.

arteriovenous anastomoses (ar-te″re-o-ve′nus ah-nas″to-mo′sis) Direct connections between arteries and veins that bypass capillary beds.

artery (ar′ter-e) A blood vessel that carries blood away from the heart.

arthrology (ar-throl′o-je) The scientific study of the structure and function of joints.

articular cartilage (ar-tik′u-lar kar′ti-lij) A hyaline cartilaginous covering over the articulating surface of the bones of synovial joints.

articulation (ar-tik″u-la′shun) A joint.

arytenoid cartilages (ar″ĕ-te′noid) A pair of small cartilages located on the superior aspect of the larynx.

ascending colon (ko′lon) The portion of the large intestine between the cecum and the hepatic flexure.

association neuron (nu′ron) A nerve cell located completely within the central nervous system. It conveys impulses in an arc from afferent to efferent neurons.

astigmatism (ah-stig′mah-tizm) Unequal curvature of the refractive surfaces of the eye (cornea and/or lens), so that light entering the eye along certain meridians does not focus on the retina.

atherosclerosis (ath″er-o″skle-ro′sis) A common type of arteriosclerosis found in medium and large arteries in which raised areas, or plaque, within the tunica intima are formed from smooth muscle cells, cholesterol, and other lipids. These plaques occlude arteries and serve as sites for the formation of thrombi.

atomic number (ah-tom′ik) A whole number representing the number of positively charged protons in the nucleus of an atom.

atopic dermatitis (ah-top′ik der″mah-ti′tis) An allergic skin reaction to agents such as poison ivy and poison oak; a type of delayed hypersensitivity.

ATP Adenosine triphosphate; the universal energy donor of the cell.

atretic (ah-tret′ik) Without an opening; atretic ovarian follicles are those that fail to ovulate.

atrioventricular bundle (a″tre-o-ven-trik′u-lar) A group of specialized cardiac fibers that conduct impulses from the atrioventricular node to the ventricular muscles or the heart; also called the *bundle of His* or *AV bundle.*

atrioventricular node A microscopic aggregation of specialized cardiac fibers located in the interatrial septum of the heart that are a part of the conduction system of the heart; *AV node.*

atrioventricular valve A cardiac valve located between an atrium and a ventricle of the heart; *AV valve.*

atrium (a′tre-um) Either of the two superior chambers of the heart that receive venous blood.

atrophy (at′ro-fe) A gradual wasting away or decrease in the size of a tissue or an organ.

atropine (at′ro-pēn) An alkaloid drug obtained from a plant of the species *Belladonna* that acts as an anticholinergic agent. Used medically to inhibit parasympathetic nerve effects, dilate the pupils of the eye, increase the heart rate, and inhibit movements of the intestine.

auditory (aw′di-to″re) Pertaining to the structures of the ear that are associated with hearing.

auricle (aw′rĕ-kl) 1. The fleshy pinna of the ear. 2. An ear-shaped appendage of each atrium of the heart.

autoantibodies (aw″to-an′ti-bod″e) Antibodies that are formed in response to, and which react with, molecules that are part of one's own body.

autonomic nervous system (aw″to-nom′ik) The sympathetic and parasympathetic portions of the nervous system that function to control the actions of the visceral organs and skin.

autosomal chromosomes (aw″to-so′mal kro′mo-sōms) The paired chromosomes; those other than the sex chromosomes.

axilla (ak-sil′ah) Pertaining to the depressed hollow commonly called the armpit.

axon (ak′son) The elongated process of a nerve cell that transmits an impulse away from the cell body.

b

ball-and-socket joint The most freely movable type of diarthrosis (e.g., the shoulder or hip joint).

Barr body (bahr) A microscopic structure in the cell nucleus produced from an inactive X chromosome in females.

basal ganglion (ba'sal gang'gle-on) A mass of nerve cell bodies located deep within a cerebral hemisphere of the brain.

basal metabolic rate (BMR) (ba'sal met''ah-bol'ik) The rate of metabolism (expressed as oxygen consumption or heat production) under resting or basal conditions (fourteen to eighteen hours after eating).

base (bās) A chemical substance that ionizes in water to release hydroxyl ions (OH^-) or other ions that combine with hydrogen ions.

basement membrane A thin sheet of extracellular substance to which the basal surfaces of epithelial cells are attached; also called the *basal lamina.*

basophil (ba'so-fil) A granular leukocyte that readily stains with basophilic dye.

B cell lymphocytes Lymphocytes that can be transformed by antigens into plasma cells that secrete antibodies (and are thus responsible for humoral immunity). The *B* stands for *bursa equivalent.*

belly The thickest circumference of a skeletal muscle.

benign (be-nīn') Not malignant.

bifurcation (bi''fur-ka'shun) Forked; divided into two branches.

bile (bīl) Fluid produced by the liver and stored in the gallbladder that contains bile salts, bile pigments, cholesterol, and other molecules. The bile is secreted into the small intestine.

bilirubin (bil''i-roo'bin) Bile pigment derived from the breakdown of the heme portion of hemoglobin.

bipennate (bi-pen'āt) Denoting muscles that have a fiber architecture coursing obliquely on both sides of a tendon.

blocking antibodies An antibody that is specific for antigens attacked by other antibodies or by T cells and, therefore, may block this attack.

blood The fluid connective tissue that circulates through the cardiovascular system to transport substances throughout the body.

blood-brain barrier A specialized mechanism that inhibits the passage of certain materials from the blood into brain tissue and cerebrospinal fluid.

bone A solid, rigid, ossified connective tissue forming the skeletal system.

bony labyrinth (lab'i-rinth) A series of chambers within the petrous portion of the temporal bone associated with the vestibular organs and the cochlea.

Bowman's capsule (bo'manz kap'sūl) The double-walled proximal portion of a renal tubule that encloses the glomerulus of a nephron.

brachial plexus (bra'ke-al plek'sus) A network of nerve fibers that arise from C5–C8 and T1. Nerves arising from the brachial plexuses supply the upper extremities.

bradycardia (brad''e-kar'de-ah) A slow cardiac rate; less than sixty beats per minute.

bradykinins (brad''e-ki'nins) Short polypeptides that stimulate vasodilation and other cardiovascular changes.

brain The enlarged superior portion of the central nervous system located in the cranial cavity of the skull.

brain stem The portion of the brain consisting of the medulla oblongata, pons, and midbrain.

bronchial tree (brong'ke-al) The bronchi and their branching bronchioles.

bronchiole (brong'ke-ōl) A small division of a bronchus within the lung.

bronchus (brong'kus) A branch of the trachea that leads to a lung.

buccal cavity (buk'al) The mouth, or oral cavity.

buffer (buf'er) A molecule that serves to prevent large changes in pH by either combining with H^+ or by releasing H^+ into solution.

bulbourethral glands (bul''bo-u-re'thral) A pair of glands that secrete a viscous fluid into the male urethra during sexual excitement; also called *Cowper's glands.*

bundle of His A band of rapidly conducting cardiac fibers that originate in the AV node and extend down the atrioventricular septum to the apex of the heart. This tissue conducts action potentials from the atria into the ventricles.

bursa (ber'sah) A saclike structure filled with synovial fluid, which occurs around joints and over which tendons can slide without contacting bone.

buttocks (but'oks) The rump or fleshy masses on the posterior aspect of the lower trunk, formed primarily by the gluteal muscles.

c

calcitonin (kal''si-to'nin) Also called *thyrocalcitonin.* A polypeptide hormone produced by the parafollicular cells of the thyroid and secreted in response to hypercalcemia. It acts to lower blood calcium and phosphate concentrations and may serve as an antagonist of parathyroid hormones.

calmodulin (kal'mod-u'lin) A receptor protein for Ca^{++} located within the cytoplasm of target cells. It appears to mediate the effects of this ion on cellular activities.

calorie (kal'o-re) A unit of heat equal to the amount of heat needed to raise the temperature of one gram of water by 1°C.

calyx (ka'liks) A cup-shaped portion of the renal pelvis that encircles renal papillae.

cAMP Cyclic adenosine monophosphate; a second messenger in the action of many hormones, such as catecholamines, polypeptides, and glycoproteins. It serves to mediate the effects of these hormones on their target cells.

canaliculus (kan''ah-lik'u-lus) A microscopic channel in bone tissue that connects lacunae.

canal of Schlemm (schlem) A circular venous drainage for the aqueous humor from the anterior chambers; located at the junction of the sclera and the cornea.

cancellous bone (kan'sĕ-lus) Spongy bone; bone tissue with a latticelike structure.

cancer A tumor characterized by abnormally rapid cell division and the loss of specialized tissue characteristics. This term usually refers to malignant tumors.

capacitation (kah-pas''i-ta'shun) Changes that occur within spermatozoa in the female reproductive tract that enable the sperm to fertilize ova; sperm that have not been capacitated in the female tract cannot fertilize ova.

capillary (cap'i-lar''e) A microscopic blood vessel that connects an arteriole and a venule; the functional unit of the circulatory system.

carbonic anhydrase (kar-bon'ik an-hi'drās) An enzyme that catalyzes the formation or breakdown of carbonic acid. When carbon dioxide concentrations are relatively high, this enzyme catalyzes the formation of carbonic acid from CO_2 and H_2O. When carbon dioxide concentrations are low, the breakdown of carbonic acid to CO_2 and H_2O is catalyzed. These reactions aid the transport of carbon dioxide from tissues to alveolar air.

cardiac muscle (kar'de-ak) Muscle of the heart, consisting of striated muscle cells. These cells are interconnected into a mass called the myocardium.

cardiac output The volume of blood pumped per minute by either the right or left ventricle.

cardiogenic shock (kar''de-o-jen'ik) Shock that results from low cardiac output in heart disease.

carpus (kar'pus) Pertaining to the wrist; collectively, the eight wrist bones.

carrier-mediated transport The transport of molecules or ions across a cell membrane by means of specific protein carriers. It includes both facilitated diffusion and active transport.

cartilage (kar'ti-lij) A type of connective tissue with a solid elastic matrix.

cartilaginous joint (kar''ti-laj'i-nus) A joint that lacks a joint cavity and permits little movement between the bones held together by cartilage.

cast An accumulation of proteins that produces molds of kidney tubules and that appears in urine sediment.

catabolism (kah-tab'o-lizm) Chemical reactions in a cell whereby larger, more complex molecules are converted into smaller molecules.

catecholamines (kat''ĕ-kol'ah-mēns) A group of molecules including epinephrine, norepinephrine, L-dopa, and related molecules with effects that are similar to those produced by activation of the sympathetic nervous system.

cations (kat'i-ons) Positively charged ions, such as sodium, potassium, calcium, and magnesium.

cauda equina (kaw'dah e-kwi'nah) The lower end of the spinal cord where the roots of spinal nerves have a tail-like appearance.

caudal (kaw'dal) Referring to a position more toward the tail.

cecum (se'kum) The pouchlike portion of the large intestine to which the ileum of the small intestine is attached.

cell The structural and functional unit of an organism; the smallest structure capable of performing all the functions necessary for life.

cell-mediated immunity (ĭ-mu'ni-te) Immunological defense provided by T cell lymphocytes, which come into close proximity to their victim cells (as opposed to humoral immunity provided by the secretion of antibodies by plasma cells).

cellular respiration (sel'u-lar res''pi-ra'shun) The energy-releasing metabolic pathways in a cell that oxidize organic molecules such as glucose, fatty acids, and others.

cementum (se-men'tum) Bonelike material that binds the root of a tooth into its bony socket.

central canal A small tube that runs the length of the spinal cord in the gray commissure and contains cerebrospinal fluid.

central nervous system The brain and the spinal cord; *CNS.*

centrioles (sen'tri-ōl) Cell organelles that form the spindle apparatus during cell division.

centromere (sen'tro-mēr) The central region of a chromosome to which the chromosomal arms are attached.

centrosome (sen'tro-sōm) A dense body near the nucleus of a cell that contains a pair of centrioles.

cerebellar peduncle (ser″ĕ-bel′ar pe-dung′-k′l) An aggregation of nerve fibers that connect the cerebellum with the brain stem.

cerebellum (ser″ĕ-bel′um) The portion of the brain concerned with the coordination of movements. Part of the metencephalon, it consists of two hemispheres and a central vermis.

cerebral aqueduct (ser′ĕ-bral ak′we-dukt″) The channel that connects the third and fourth ventricles of the brain; also called the *aqueduct of Sylvius.*

cerebral peduncles A paired bundle of nerve fibers along the ventral surface of the midbrain, conducting impulses between the pons and the cerebral hemispheres.

cerebrospinal fluid (ser″ĕ-bro-spi′nal) A fluid produced by the choroid plexus of the ventricles of the brain. It fills the ventricles and surrounds the central nervous system in association with the meninges.

cerebrum (ser′ĕ-brum) The largest portion of the brain, composed of the right and left hemispheres.

cervical (ser′vi-kal) Pertaining to the neck or a necklike portion of an organ.

cervical ganglion (gang′gle-on) A cluster of postganglionic sympathetic nerve cell bodies located in the neck, near the cervical vertebrae.

cervical plexus (plek′sus) A network of spinal nerves formed by the anterior branches of the first four cervical nerves.

cervix (ser′viks) 1. The narrow necklike portion of an organ. 2. The inferior end of the uterus that adjoins the vagina.

chemoreceptor (ke″mo-re-sep′tor) A neuroreceptor that is stimulated by the presence of chemical molecules.

chemotaxis (ke″mo-tak′sis) The movement of an organism or a cell, such as a leukocyte, toward a chemical stimulus.

Cheyne-Stokes respiration (chān′stōkes res″pi-ra′shun) Breathing characterized by rhythmic waxing and waning of the depth of respiration, with regularly occurring periods of apnea (lack of breathing).

cholesterol (ko-les′ter-ol) A twenty-seven-carbon steroid that serves as the precursor of steroid hormones.

cholinergic (ko″lin-er′jik) Denoting nerve endings that liberate acetylcholine as a neurotransmitter, such as those of the parasympathetic system.

chondrocranium (kon″dro-kra′ne-um) The portion of the skull that supports the brain and is derived from endochondral bone.

chondrocytes (kon′dro-sīts) Cartilage-forming cells.

chordae tendineae (kor′de ten-din′e-e) Chordlike tendinous bands that connect papillary muscles to the atrioventricular valves within the ventricles of the heart.

chorea (ko-re′ah) The occurrence of a wide variety of rapid, complex, jerky movements that appear to be well coordinated but are performed involuntarily.

choroid (ko′roid) The vascular, pigmented middle layer of the wall of the eye.

choroid plexus (ko′roid plek′sus) A mass of vascular capillaries from which cerebrospinal fluid is secreted into the ventricles of the brain.

chromatids (kro′mah-tids) Duplicated chromosomes that are joined together at the centromere and separate during cell division.

chromatin (kro′mah-tin) Threadlike structures in the cell nucleus consisting primarily of DNA and protein. They represent the extended form of chromosomes during interphase.

chromosome (kro′mo-sōm) Structures in the nucleus that contain the genes for genetic expression.

chyme (kīm) The mixture of partially digested food and digestive juices within the stomach and small intestine.

cilia (sil′e-ah) Plural of cilium; tiny hairlike processes that extend from the cell surface and beat in a coordinated fashion.

ciliary body (sil′e-er″e) A portion of the choroid layer of the eye that secretes aqueous humor and contains the ciliary muscle.

circadian rhythms (ser″kah-de′an) Physiological changes that repeat at about a twenty-four-hour period. These are often synchronized to changes in the external environment, such as the day-night cycles.

circle of Willis An arterial vessel located on the ventral surface of the brain around the pituitary gland.

circumduction (ser″kum-duk′shun) A conelike movement of a body part, such that the distal end moves in a circle while the proximal portion remains relatively stable.

circumvallate papilla (sir″kum-val′āt pah-pil′ah) The largest papillae on the dorsal surface of the tongue. They are arranged in an inverted V-shaped pattern at the posterior portion of the tongue.

cirrhosis (sir-ro′sis) Liver disease characterized by loss of normal microscopic structure, which is replaced by fibrosis and nodular regeneration.

clitoris (kli′to-ris) A small, erectile structure in the vulva of the female, homologous to the glans penis in the male.

clone (klōn) 1. A group of cells derived from a single parent cell by mitotic cell division; since reproduction is asexual, the descendants of the parent cell are genetically identical. 2. A term used when cells are separate individuals (as in white blood cells) rather than part of a growing organ.

CNS Central nervous system; part of the nervous system consisting of the brain and spinal cord.

coccygeal (kok-sij′e-al) Pertaining to the region of the coccyx; the caudal termination of the vertebral column.

cochlea (kok′le-ah) The organ of hearing in the inner ear where nerve impulses are generated in response to sound waves.

codon (ko′don) The sequence of three nucleotide bases in mRNA that specifies a given amino acid and determines the position of that amino acid in a polypeptide chain through complementary base pairing with an anticodon in RNA.

coelom (se′lom) The abdominal cavity.

coenzyme (ko-en′zīm) An organic molecule, usually derived from a water-soluble vitamin, that combines with and activates specific enzyme proteins.

cofactor (ko′fak-tor) A substance needed for the catalytic action of an enzyme; usually refers to inorganic ions such as Ca^{++} and Mg^{++}.

collateral (kŏ-lat′er-al) A small branch of a blood vessel or nerve fiber.

colloid osmotic pressure (kol′oid oz-mot′ik) Osmotic pressure exerted by plasma proteins, which are present as a colloidal suspension. Also called oncotic pressure.

colon (ko′lon) The large intestine.

common bile duct A tube that is formed by the union of the hepatic duct and cystic duct and transports bile to the duodenum.

compact (dense) bone Tightly packed bone that is superficial to spongy bone and covered by the periosteum.

compliance (kom-pli′ans) A measure of the ease with which a structure such as the lungs expands under pressure; a measure of the change in volume as a function of pressure changes.

condyle (kon′dīl) A rounded process at the end of a long bone that forms an articulation.

cone (kōn) A color receptor cell in the retina of the eye.

congenital (kon-jen′i-tal) Present at the time of birth.

congestive heart failure (kon-jes′tiv) The inability of the heart to deliver an adequate blood flow, due to heart disease or hypertension; it is associated with breathlessness, salt and water retention, and edema.

conjunctiva (kon″junk-ti′vah) The thin membranous covering on the anterior surface of the eyeball and lining the eyelids.

conjunctivitis (kon″junk″ti-vi′tis) Inflammation of the conjunctiva of the eye, which sometimes is called "pink eye."

connective tissue One of the four basic tissue types within the body. It is a binding and supportive tissue with abundant matrix.

Conn's syndrome (konz) Primary hyperaldosteronism; excessive secretion of aldosterone produces electrolyte imbalances.

contralateral (kon″trah-lat′er-al) Affecting the opposite side of the body.

conus medullaris (ko′nus med″u-lār′is) The caudal, tapering portion of the spinal cord.

convolution (kon-vo-lu′shun) An elevation on the surface of a structure and an infolding of the tissue upon itself.

cornea (kor′ne-ah) The transparent convex, anterior portion of the outer layer of the eyeball.

coronary circulation (kor′ŏ-na-re) The arterial and venous blood circulation to the wall of the heart.

coronary sinus A large venous channel on the posterior surface of the heart into which the cardiac veins drain.

corpora quadrigemina (kor′po-rah kwod″ri-jem′i-nah) Four dorsal lobes of the midbrain concerned with visual and auditory functions.

corpus callosum (kor′pus kah-lo′sum) A major tract within the brain that is composed of white matter and connects the right and left cerebral hemispheres.

cortex (kor′teks) The outer layer of an organ such as the convoluted cerebrum, adrenal gland, or kidney.

corticosteroids (kor″ti-ko-ste′roids) Steroid hormones of the adrenal cortex, consisting of glucocorticoids (such as hydrocortisone) and mineralocorticoids (such as aldosterone).

costal cartilage (kos′tal) The cartilage that connects the ribs to the sternum.

cranial (kra′ne-al) Pertaining to the cranium.

cranial nerve One of twelve pairs of nerves that arise from the ventral surface of the brain.

cranium (kra′ne-um) The endochondral bones of the skull that enclose or support the brain and the organs of sight, hearing, and balance.

creatine phosphate (kre′ah-tin fos′fāt) An organic phosphate molecule in muscle cells that serves as a source of high-energy phosphate for the synthesis of ATP. Also called *phosphocreatine.*

crenation (kre-na′shun) A notched or scalloped appearance of the red blood cell membrane caused by the osmotic loss of water from these cells.

crest A thickened ridge of bone for the attachment of muscle.
cretinism (kre'tin-izm) A condition caused by insufficient thyroid secretion during prenatal development or the years of early childhood; results in stunted growth and inadequate mental development.
cricoid cartilage (kri'koid) A ring-shaped cartilage that forms the inferior end of the larynx.
crista (kris'tă) A crest, such as the crista galli extending superiorly from the cribriform plate.
cryptorchidism (krip-tor'ki-dizm) A developmental defect in which the testes fail to descend into the scrotum and, instead, remain in the body cavity.
cubital (ku'bĭ-tal) Pertaining to the forearm. The cubital fossa is the anterior aspect of the elbow joint.
curare (koo-rah're) A chemical derived from plant sources that causes flaccid paralysis by blocking ACh receptor proteins in muscle cell membranes.
Cushing's syndrome (koosh'ingz) Symptoms caused by the hypersecretion of adrenal steroid hormones, due to tumors of the adrenal cortex or to ACTH-secreting tumors of the anterior pituitary.
cyanosis (si''ah-no'sis) A blue color given to the skin or mucous membranes by deoxyhemoglobin; indicates inadequate oxygen concentration in the blood.
cystic duct (sis'tik dukt) The tube that transports bile from the gallbladder to the common bile duct.
cytochrome (si'to-krōm) A pigment in mitochondria that transports electrons in the process of aerobic respiration.
cytokinesis (si''to-ki-ne'sis) The division of the cytoplasm that occurs in mitosis and meiosis, when a parent cell divides to produce two daughter cells.
cytology (si-tol'o-je) The science dealing with the study of cells.
cytoplasm (si'to-plazm'') In a cell, the protoplasm located outside of the nucleus.
cytoskeleton (si''to-skel'ĕ-ton) A latticework of structural proteins in the cytoplasm arranged in the form of microfilaments and microtubules.

d

deciduous (de-sid'u-us) Not permanent. Deciduous teeth are shed and replaced by permanent teeth during development.
defecation (def''ĕ-ka'shun) The elimination of feces from the rectum through the anus.
deglutition (deg''loo-tish'un) The process of swallowing.
delayed hypersensitivity An allergic response in which the onset of symptoms takes as long as two to three days after exposure to an antigen. Produced by T cells, it is a type of cell-mediated immunity.
dendrite (den'drīt) A nerve cell process that transmits impulses toward a neuron cell body.
dentin (den'tin) The main substance of a tooth, covered by enamel over the crown of the tooth and by cementum on the root.
dentition (den-tish'un) The number, arrangement, and shape of teeth.
depolarization (de-po''lar-i-za'shun) The loss of membrane polarity in which the inside of the cell membrane becomes less negative in comparison to the outside of the membrane. The term is also used to indicate the reversal of membrane polarity that occurs during the production of action potentials in nerve and muscle cells.
dermis (der'mis) The second, or deep, layer of skin beneath the epidermis.
descending colon The segment of the large intestine that descends on the left side from the level of the spleen to the level of the left iliac crest.
diabetes insipidus (di''ah-be'tēz in-sip'id-es) A condition in which inadequate amounts of antidiuretic hormone (ADH) are secreted by the posterior pituitary. It results in the inadequate reabsorption of water by the kidney tubules and, thus, in the excretion of a large volume of dilute urine.
diabetes mellitus (di''ah-be'tēz mel'li-tus) The appearance of glucose in the urine due to the presence of high plasma glucose concentrations, even in the fasting state. This disease is caused by either lack of sufficient insulin secretion or inadequate responsiveness of the target tissues to the effects of insulin.
diapedesis (di''ah-pĕ-de'sis) The migration of white blood cells through the endothelial walls of blood capillaries into the surrounding connective tissues.
diaphragm (di'ah-fram) A sheetlike dome of muscle and connective tissue that separates the thoracic and abdominal cavities.
diaphysis (di-af'ĭ-sis) The shaft of a long bone.
diarrhea (di''ah-re'ah) Abnormal frequence of defecation accompanied by abnormal liquidity of the feces.
diarthrosis (di''ar-thro'sis) A type of joint in which the articulating bones are freely movable; also called a *synovial joint.*
diastole (di-as'to-le) The sequence of the cardiac cycle during which a heart chamber wall is relaxed.
diencephalon (di''en-sef'ah-lon) A major region of the brain that includes the third ventricle, thalamus, hypothalamus, and pituitary.
diffusion (di-fu'zhun) The net movement of molecules or ions from regions of higher to regions of lower concentration.
digestion The process by which larger molecules of food substance are broken down mechanically and chemically into smaller molecules that can be absorbed.
diploid (dip'loid) Denoting cells having two of each chromosome, or twice the number of chromosomes that are present in sperm or ova.
disaccharide (di-sak'ah-rīd) A class of double sugars; carbohydrates that yield two simple sugars, or monosaccharides, upon hydrolysis.
distal (dis'tal) Away from the midline or origin; the opposite of *proximal.*
diuretic (di''u-ret'ik) An agent that promotes the excretion of urine, thereby lowering blood volume and pressure.
DNA Deoxyribonucleic acid; composed of nucleotide bases and deoxyribose sugar; contains the genetic code.
dopa (do'pah) A derivative of the amino acid tyrosine, L-dopa serves as the precursor for the neurotransmitter molecule dopamine. L-dopa is given to patients with Parkinson's disease to stimulate dopamine production.
dopamine (do'pah-mēn) A type of neurotransmitter in the central nervous system; also is the precursor of norepinephrine, another neurotransmitter molecule.
dorsal (dor'sal) Pertaining to the back or posterior portion of a body part; the opposite of *ventral.*
dorsal root ganglion Collections of cell bodies of sensory neurons that form swellings in the dorsal roots of spinal nerves.
ductus arteriosus (duk'tus ar-te''re-o'sus) The blood vessel that connects the pulmonary trunk and the aorta in a fetus.
ductus deferens (def'er-enz) A tube that carries spermatozoa from the epididymis to the ejaculatory duct; also called the *vas deferens* or *seminal duct.*
ductus venosus (ven-o'sus) A fetal blood vessel that connects the umbilical vein and the inferior vena cava.
duodenum (du''o-de'num) The first portion of the small intestine that leads from the pyloric sphincter of the stomach to the jejunum.
dura mater (du'rah ma'ter) The outermost meninx.
dwarfism A condition in which a person is undersized due to inadequate secretion of growth hormone.
dyspnea (disp'ne-ah) Subjective difficulty in breathing.

e

eccrine gland (ek'rin) A sweat gland that functions in thermoregulation.
ECG Electrocardiogram; EKG.
E. coli A species of bacteria normally found in the human intestine; full name is *Escherichia coli.*
ectoderm (ek'to-derm) The outermost of the three primary germ layers of an embryo.
ectopic (ek-top'ik) Foreign, out of place.
ectopic focus An area of the heart other than the SA node that assumes pacemaker activity.
ectopic pregnancy Embryonic development that occurs anywhere other than in the uterus (as in the uterine tubes or body cavity).
edema (ĕ-de'mah) An excessive accumulation of fluid in the body tissues.
EEG Electroencephalogram; a recording of the electrical activity of the brain from electrodes placed on the scalp.
effector (ef-fek'tor) An organ such as a gland or muscle that responds to a motor stimulation.
efferent (ef'er-ent) Conveying away from the center of an organ or structure.
efferent arteriole (ar-te're-ōl) An arteriole of the renal vascular system that conducts blood away from the glomerulus of a nephron.
efferent ductules (dukt'ūls) A series of coiled tubules that convey spermatozoa from the rete testis to the epididymis.
efferent neuron (nu'ron) A motor nerve cell that conducts impulses from the central nervous system to effector organs such as muscles or glands.
ejaculation (e-jak''u-la'shun) The discharge of semen from the male urethra during climax.
ejaculatory duct (e-jak'u-lah-to''re) A tube that transports spermatozoa from the ductus deferens to the prostatic urethra.
elastic fibers (e-las'tik) Protein strands that are found in certain connective tissue and have contractile properties.
elbow The diarthrotic joint between the brachium and the forearm.
electrocardiogram (e-lek''tro-kar'de-o-gram'') A recording of the electrical activity that accompanies the cardiac cycle; ECG or EKG.
electroencephalogram (e-lek''tro-en-sef'ah-lo-gram'') A recording of the brain wave patterns or electrical impulses of the brain; EEG.
electrolytes (e-lek'tro-līts) Ions and molecules that are able to ionize and thus carry an electric current. The most common electrolytes in the plasma are Na^+, HCO_3^-, and K^+.

electromyogram (e-lek″tro-mi′o-gram″) A recording of the electrical impulses or activity of a muscle; EMG.

electrophoresis (e-lek″tro-fo-re′sis) A biochemical technique in which different molecules can be separated and identified by their rate of movement in an electric field.

elephantiasis (el″ĕ-fan-ti′ah-sis) A disease caused by infection with a nematode worm, in which the larvae block lymphatic drainage and produce edema; the lower areas of the body can become enormously swollen as a result.

embryology (em′bre-ol″o-je) The study of prenatal development from conception through the eighth week in utero.

EMG Electromyogram; electrical recordings of the activity of skeletal muscles using surface electrodes.

emmetropia (em-ĕ-tro′pe-ah) A condition of normal vision in which the image of objects is focused on the retina, as opposed to nearsightedness (myopia) or farsightedness (hypermetropia).

emphysema (em″fī-se′mah, em″fī-ze′mah) A lung disease in which the alveoli are destroyed and the remaining alveoli become larger; results in decreased vital capacity and increased airways resistance.

emulsification (e-mul″si-fī-kā′shun) The process of producing an emulsion or fine suspension; in the intestine, fat globules are emulsified by the detergent action of bile.

enamel (en-am′el) The outer, dense substance covering the crown of a tooth.

endergonic (end″er-gon′ik) A chemical reaction that requires the input of energy from an external source in order to proceed.

endocardium (en″do-kar′de-um) The endothelial lining of the heart chambers and valves.

endocrine gland (en′do-krīn) A ductless gland that secretes hormones directly into the bloodstream or body fluids.

endocytosis (en″do-si-to′sis) The cellular uptake of particles that are too large to cross the cell membrane; occurs by invagination of the cell membrane until a membrane-enclosed vesicle is inched off within the cytoplasm.

endoderm (en′do-derm) The innermost of the three primary germ layers of an embryo.

endogenous (en-doj′ĕ-nus) A product or process arising from within the body; as opposed to exogenous products, or influences from external sources.

endolymph (en′do-limf) The fluid within the membranous labyrinth of the inner ear.

endometrium (en″do-me′tre-um) The inner lining of the uterus.

endomysium (en″do-mis′e-um) The connective tissue sheath surrounding each skeletal muscle fiber, separating the muscle cells from one another.

endoneurium (en″do-nu′re-um) The connective tissue sheath surrounding each nerve fiber, separating the nerve fibers one from another within a nerve.

endoplasmic reticulum (en-do-plas′mik rē-tik′u-lum) A cytoplasmic organelle composed of a network of canals running through the cytoplasm of a cell.

endorphins (en-dor′fins, en′dor-fins) A group of endogenous opiate molecules that may act as a natural analgesic.

endothelium (en″do-the′le-um) The layer of epithelial tissue that forms the thin inner lining of blood vessels and heart chambers.

endotoxin (en″do-tok′sin) A toxin contained within certain types of bacteria that is able to stimulate the release of endogenous pyrogen and produce a fever.

enkephalins (en-kef′ah-lins) Short polypeptides, containing five amino acids, that have analgesic effects; may function as neurotransmitters in the brain. The two known enkephalins (which differ in only one amino acid) are endorphins.

enteric (en-ter′ik) A term referring to the intestine.

entropy (en′tro-pe) The energy of a system that is not available to perform work; a measure of the degree of disorder in a system, entropy increases whenever energy is transformed.

enzyme (en′zīm) A protein catalyst that increases the rate of specific chemical reactions.

eosinophil (e″o-sin′o-fil) A type of white blood cell characterized by the presence of cytoplasmic granules that become stained by acidic eosin dye; eosinophils normally constitute about 2%–4% of the white blood cells.

epicardium (ep″ī-kar′de-um) A thin, outer layer of the heart; also called the *visceral pericardium.*

epicondyle (ep″ī-kon′dīl) A projection of bone above a condyle.

epidermis (ep″ī-der′mis) The outermost layer of the skin, composed of several stratified squamous epithelial layers.

epididymis (ep″ī-did′ī-mis) A highly coiled tube located along the posterior border of the testis. It stores spermatozoa and transports them from the seminiferous tubules of the testis to the vas deferens.

epidural space (ep″ī-du′ral) A space between the spinal dura mater and the bone of the vertebral canal.

epiglottis (ep″ī-glot′is) A cartilaginous leaflike structure positioned on top of the larynx that covers the glottis during swallowing.

epimysium (ep″ī-mis′e-um) A fibrous, outer sheath of connective tissue surrounding a skeletal muscle.

epinephrine (ep″ī-nef′rin) Also known as adrenalin; a catecholamine hormone secreted by the adrenal medulla in response to sympathetic nerve stimulation; acts together with norepinephrine released from sympathetic nerve endings to prepare the organism for "fight or flight."

epineurium (ep″ī-nu′re-um) A fibrous, outer sheath of connective tissue surrounding a nerve.

epiphyseal plate (ep″ī-fiz′e-al) A cartilaginous layer that is located between the epiphysis and diaphysis of a long bone and functions as a longitudinal growing region.

epiphysis (ĕ-pif′ī-sis) The end segment of a long bone, separated from the diaphysis early in life by an epiphyseal plate but later becoming part of the larger bone.

episiotomy (ĕ-piz″e-ot′o-me) An incision of the perineum at the end of the second stage of labor to facilitate delivery and avoid tearing the perineum.

epithelial tissue (ep″ī-the′le-al) One of the four basic tissue types; the type of tissue that covers or lines all exposed body surfaces.

eponychium (ep″o-nik′e-um) The thin layer of stratum corneum of the epidermis of the skin, which overlaps and protects the lunula of the nail.

EPSP Excitatory postsynaptic potential; a graded depolarization of a postsynaptic membrane in response to stimulation by a neutrotransmitter chemical. EPSPs can be summated but can only be transmitted short distances; EPSPs can stimulate the production of action potentials when a threshold level of depolarization is attained.

erythroblastosis fetalis (ĕ-rith″ro-blas-to′sis fī-tal′is) Hemolytic anemia in a newborn Rh positive baby caused by maternal antibodies against the Rh factor that have crossed the placenta.

erythrocyte (ĕ-rith′ro-sīt) A corpuscle or red blood cell.

esophagus (ĕ-sof′ah-gus) A tubular portion of the digestive tract that leads from the pharynx to the stomach as it passes through the thoracic cavity.

essential amino acids Those eight amino acids in adults or nine amino acids in children that cannot be made by the human body and therefore must be obtained in the diet.

estrus cycle (es′trus) Cyclic changes in the structure and function of the ovaries and female reproductive tract, accompanied by periods of "heat" (estrus) or sexual receptivity; the lower mammalian equivalent of the menstrual cycle, but differing from the menstrual cycle in that the endometrium is not shed with accompanying bleeding.

etiology (e″te-ol′o-je) The study of cause, especially of disease, including the origin and what pathogens, if any, are involved.

eustachian canal (u-sta′ke-an) A narrow tube that connects the middle ear chamber to the pharynx; also called the *auditory tube.*

eversion (e-ver′zhun) A movement of the foot in which the sole is turned outward.

exergonic (ek″ser-gon′ik) Chemical reactions that liberate energy.

exocrine gland (ek′so-krin) A gland that secretes its product to an epithelial surface, directly or through ducts.

exocytosis (eks″o-si-to′sis) The process of cellular secretion in which the secretory products are contained within a membrane-enclosed vesicle; the vesicle fuses with the cell membrane so that the lumen of the vesicle is open to the extracellular environment.

expiration (ek″spi-ra′shun) The process of expelling air from the lungs through breathing out; also called *exhalation.*

extension (ek-sten′shun) A movement that increases the angle between parts of a joint.

extensor (eks-ten′sor) A muscle that upon contraction increases the angle of a joint.

external (superficial) Located on or toward the surface.

external auditory meatus (aw′dī-to″re me-a′tus) An opening through the temporal bone that connects with the tympanum and the middle ear chamber and through which sound vibrations pass.

external ear The outer portion of the ear, consisting of the pinna, external auditory meatus, and tympanum.

external nares (na′rēz) The opening into the nasal cavity; also called the *nostrils.*

exteroceptors (eks″ter-o-sep′tors) Sensory receptors that are sensitive to changes in the external environment (as opposed to interoceptors).

extraocular muscles (eks″trah-ok′u-lar) The muscles that insert into the sclera of the eye and act to change the position of the eye in its orbit, as opposed to the intraocular muscles such as those of the iris and ciliary body within the eye.

extrinsic (eks-trin′sik) Pertaining to an outside or external origin.

f

face 1. The anterior aspect of the head not supporting or covering the brain. 2. The exposed surface of a structure.

facet (fas'et) A small, smooth surface of a bone where articulation occurs.

facilitated diffusion (fah-sil'ĭ-ta''tid) The carrier-mediated transport of molecules through the cell membrane along the direction of their concentration gradients; it does not require the expenditure of metabolic energy.

FAD Flavin adenine dinucleotide; a coenzyme derived from riboflavin that participates in electron transport within the mitochondria.

falciform ligament (fal'si-form lig'ah-ment) An extension of parietal peritoneum that separates the two major lobes of the liver.

fallopian tube (fal-lo'pe-an) The tube through which the ovum is transported to the uterus and the site of fertilization; also called the *oviduct* or *uterine tube*.

false vocal cords The supporting folds of tissue for the true vocal cords within the larynx.

fascia (fash'e-ah) A tough sheet of fibrous tissue binding the skin to underlying muscles or supporting and separating muscles.

fasciculus (fah-sik'u-lus) A small bundle of muscle or nerve fibers.

fauces (faw'sēz) The passageway between the mouth and the pharynx.

feces (fe'sēz) Material expelled from the digestive tract during defecation, composed of food residue, bacteria, and secretions; also called *stool*.

fertilization The fusion of an ovum and sperm.

fetus (fe'tus) A prenatal human after eight weeks of development.

fibrillation (fi-bri-la'shun) A condition of cardiac muscle characterized electrically by random and continuously changing patterns of electrical activity and resulting in the inability of the myocardium to contract as a unit and pump blood. It can be fatal if it occurs in the ventricles.

fibrin (fi'brin) The insoluble protein formed from fibrinogen by the enzymatic action of thrombin during the process of blood clot formation.

fibrinogen (fi-brin'o-jen) Also called *factor I;* a soluble plasma protein that serves as the precursor of fibrin.

fibrous joint (fi'brus) A type of articulation that allows little or no movement (e.g., syndesmosis).

filiform papillae (fil'ĭ-form pah-pil'e) The numerous small projections over the entire dorsal surface of the tongue that contain no taste buds.

filum terminale (fi'lum ter-mĭ-nal'e) A fibrous, threadlike continuation of the pia mater, extending inferiorly from the terminal end of the spinal cord to the coccyx.

fimbriae (fim'bre-e) Fringelike extensions from the open end of the uterine tube.

fissure (fish'ūr) A groove or narrow cleft that separates two parts, such as the cerebral hemispheres of the brain.

flagellum (flah-jel'um) A whiplike structure that provides motility for sperm.

flare-and-wheal reaction (hwēl, wēl) A cutaneous reaction to skin injury or the administration of antigens, produced by release of histamine and related molecules and characterized by local edema and a red flare.

flavoprotein (fla''vo-pro'te-in) A conjugated protein that contains a flavin pigment and involved in electron transport within the mitochondria.

flexion (flek'shun) A movement that decreases the angle between parts of a joint.

flexor (flek'sor) A muscle that decreases the angle of a joint when it contracts.

fontanel (fon''tah-nel) A membranous-covered region on the skull of a fetus or baby where ossification has not yet occurred; commonly called a *soft spot*.

foot The terminal portion of the lower extremity, consisting of the tarsus, metatarsus, and phalanges.

foramen (fo-ra'men), pl. *foramina* An opening, usually in a bone, for the passage of a blood vessel or a nerve.

foramen ovale (o-val'e) An opening through the interatrial septum of the fetal heart.

forearm (fōr'arm) The portion of the upper extremity between the elbow and the wrist; also called the *antebrachium*.

fornix (for'niks) 1. A recess around the cervix of the uterus where it protrudes into the vagina. 2. A tract within the brain connecting the hippocampus with the mammillary bodies.

fossa (fos'ah) A depressed area, usually on a bone.

fourth ventricle (ven'tri-k'l) A cavity within the brain, between the cerebellum and the medulla and the pons, containing cerebrospinal fluid.

fovea centralis (fo've-ah sen-tra'lis) A depression on the macula lutea of the eye, where only cones are located, which is the area of keenest vision.

frenulum (fren'u-lum) A membranous tissue that serves to anchor and limit the movement of a body part.

frontal 1. Pertaining to the region of the forehead. 2. A plane through the body, dividing the body into anterior and posterior portions; also called the *coronal plane*.

FSH Follicle-stimulating hormone; one of the two gonadotropic hormones secreted from the anterior pituitary. In females FSH stimulates the development of the ovarian follicles, whereas in males it stimulates the production of sperm in the seminiferous tubules.

fungiform papillae (fun'ji-form pah-pil'e) Flattened, mushroom-shaped projections that are interspersed over the dorsal surface of the tongue and contain taste buds.

g

GABA Gamma-aminobutyric acid; believed to function as an inhibitory neurotransmitter in the central nervous system.

gallbladder A pouchlike organ that is attached to the underside of the liver and stores and concentrates bile.

gamete (gam'ēt) A haploid sex cell; either an egg cell or a sperm cell.

ganglion (gang'gle-on) An aggregation of nerve cell bodies occurring outside the central nervous system.

gastric intrinsic factor (gas'trik) A glycoprotein secreted by the stomach and needed for the absorption of vitamin B_{12}.

gastrin (gas'trin) A hormone secreted by the stomach that stimulates the gastric secretion of hydrochloric acid and pepsin.

gastrointestinal tract (GI tract) (gas''tro-in-tes'tĭ-nal) The portion of the digestive tract that includes the stomach and the small and large intestines.

gates A term used to describe structures within the cell membrane that regulate the passage of ions through membrane channels. Such gates may be chemically regulated (by neurotransmitters) or voltage regulated (in which case they open in response to a threshold level of depolarization).

genetic recombination (jĕ-net'ik re''kom-bĭ-na'shun) The formation of new combinations of genes, as by crossing-over between homologous chromosomes.

genetic transcription (trans-krip'shun) The process by which RNA is produced with a sequence of nucleotide bases that is complementary to a region of DNA.

genetic translation (trans-la'shun) The process by which proteins are produced with amino acid sequences specified by the sequence of codons in messenger RNA.

gigantism (ji-gan'tism; ji'gan-tizm) Abnormal body growth due to the excessive secretion of growth hormone.

gingiva (jin-ji'vah) The fleshy covering over the mandible and maxilla through which the teeth protrude within the mouth; also called the *gum*.

gland An organ that produces a specific substance or secretion.

glans penis (glanz pe'nis) The enlarged, sensitive, distal end of the penis.

gliding joint A type of diarthrotic joint in which the articular surfaces are flat, permitting only side-to-side and back-and-forth movements.

glomerular ultrafiltrate (glo-mer'u-lar ul''trah-fil'trāt) Fluid filtered through the glomerular capillaries into Bowman's capsule of the kidney tubules.

glomerulonephritis (glo-mer''u-lo-nĕ-fri'tis) Inflammation of the renal glomeruli, associated with fluid retention, edema, hypertension, and the appearance of protein in the uterine.

glomerulus (glo-mer'u-lus) A coiled tuft of capillaries that is surrounded by Bowman's capsule and filtrates urine from the blood.

glottis (glot'is) A slitlike opening into the larynx, positioned between the true vocal cords.

glucagon (gloo'kah-gon) A polypeptide hormone that is secreted by the alpha cells of the islets of Langerhans in the pancreas and acts to promote glycogenolysis and raise the blood glucose levels.

glucocorticoids (gloo''ko-kor'ti-koids) Steroid hormones secreted by the adrenal cortex (corticosteroids) that affect the metabolism of glucose, protein, and fat. These hormones also have anti-inflammatory and immunosuppressive effects; the major glucocorticoid in humans is hydrocortisone (cortisol).

gluconeogenesis (gloo''ko-ne''o-jen'ĕ-sis) The formation of glucose from non-carbohydrate molecules such as amino acids and lactic acid.

glycogen (gli'ko-jen) A polysaccharide of glucose—also called *animal starch*—produced primarily in the liver and skeletal muscles. Similar to plant starch in composition, glycogen contains more highly branched chains of glucose subunits than does plant starch.

glycogenesis (gli''ko-jen'ĕ-sis) The formation of glycogen from glucose.

glycogenolysis (gli''ko-jĕ-nol'ĭ-sis) The hydrolysis of glycogen to glucose-1-phosphate, which can be converted to glucose-6-phosphate, which then may be oxidized via glycolysis or (in the liver) converted to free glucose.

glycolysis (gli''kol'ĭ-sis) The metabolic pathway that converts glucose to pyruvic acid and yields a net production of two ATP molecules and two molecules of reduced NAD. In anaerobic respiration the reduced NAD is oxidized by the conversion of pyruvic acid to lactic acid. In aerobic respiration pyruvic acid enters the Krebs cycle in mitochondria, and reduced NAD is ultimately oxidized by oxygen to yield water.

glycosuria (gli''ko-su're-ah) The excretion of an abnormal amount of glucose in the urine (urine normally only contains trace amounts of glucose).

goblet cell A unicellular gland that secretes mucus and is associated with columnar epithelia.

Golgi apparatus Stacks of flattened membranous sacs within the cytoplasm of cells that are believed to bud off vesicles containing secretory proteins.

Golgi tendon organ An afferent receptor found near the junction of tendons and muscles.

gonad (go'nad) A reproductive organ, testis or ovary, that produces gametes and sex hormones.

gonadotropin hormones (gon''ah-do-tro'pin) Hormones of the anterior pituitary gland that stimulate gonadal function—the formation of gametes and secretion of sex steroids. The two gonadotropins are FSH (follicle-stimulating hormone) and LH (luteinizing hormone), which are essentially the same in males and females.

graafian follicle (graf'e-an) A mature ovarian follicle, containing a single fluid-filled cavity, with the ovum located toward one side of the follicle and perched on top of a hill of granulosa cells.

granular leukocytes (gran'u-lar loo'ko-sīts) Leukocytes with granules in the cytoplasm; on the basis of the staining properties of the granules, these cells are of three types: neutrophils, eosinophils, and basophils.

Graves' disease A hyperthyroid condition believed to be caused by excessive stimulation of the thyroid gland by autoantibodies; it is associated with exophthalmos (bulging eyes), high pulse rate, high metabolic rate, and other symptoms of hyperthyroidism.

gray matter The region of the central nervous system that is composed of nonmyelinated nerve tissue.

greater omentum (o-men'tum) A double-layered serosa membrane that originates on the greater curvature of the stomach and hangs inferiorly like an apron over the contents of the abdominal cavity.

gross anatomy A branch of anatomy concerned with structures of the body that can be studied without a microscope.

growth hormone A hormone secreted by the anterior pituitary that stimulates growth of the skeleton and soft tissues during the growing years and that influences the metabolism of protein, carbohydrate, and fat throughout life.

gustatory (gus'tah-to''re) Pertaining to the sense of taste.

gut (gŭt) Pertaining to the intestines; generally a developmental term.

gyrus (jī'rus) A convoluted elevation or ridge.

h

hair A threadlike appendage of the epidermis, consisting of keratinized dead cells that have been pushed up from a dividing germinativum layer.

hair cells Specialized receptor nerve endings for detecting sensations, such as in the organ of Corti.

hair follicle (fol'lĭ-k'l) A tubular depression in the dermis of the skin in which a hair develops.

hand The terminal portion of the upper extremity, consisting of the carpus, metacarpus, and phalanges.

haploid (hap'loid) A cell that has one of each chromosome type and therefore half the number of chromosomes present in most other body cells; only the gametes (sperm and ova) are haploid.

haptens (hap'tens) Small molecules that are not antigenic by themselves, but which—when combined with proteins—become antigenic and thus able to stimulate the production of specific antibodies.

hard palate (pal'at) The bony partition between the oral and nasal cavities, formed by the maxillae and palatine bones and lined by mucous membrane.

haustra (haws'trah) Sacculations or pouches of the colon.

haversian canal (ha-ver'shan) An elongated, longitudinal channel in the center of a haversian system in bone, containing nutrient vessels and nerves.

haversian system A group of osteocytes and concentric lamellae surrounding a haversian canal, constituting the basic unit of structure in osseous tissue; also called an *osteon.*

hay fever A seasonal type of allergic rhinitis caused by pollen; it is characterized by itching and tearing of the eyes, swelling of the nasal mucosa, attacks of sneezing, and often by asthma.

head The superior portion of a human that contains the brain and major sense organs.

heart A four-chambered, muscular, pumping organ positioned in the thoracic cavity slightly to the left of midline.

heart murmur Abnormal heart sounds caused by an abnormal flow of blood in the heart due to structural defects, usually of the valves or septum.

helper T cells A subpopulation of T cells (lymphocytes), that help stimulate the antibody production of B lymphocytes by antigens.

hematocrit (he-mat'o-krit) The ratio of packed red blood cells to total blood volume in a centrifuged sample of blood, expressed as a percentage.

heme (hēm) The iron-containing red pigment that, together with the protein globin, forms hemoglobin.

hemoglobin (he''mo-glo'bin) The pigment of red blood cells that constitutes about 33% of the cell volume and transports oxygen and carbon dioxide.

hemopoiesis (he''mo-poi-e'sis) The production of red blood cells.

heparin (hep'ah-rin) A mucopolysaccharide, found in many tissues but most abundantly in the lungs and liver, that is used medically as an anticoagulant.

hepatic duct (hē-pat'ik) Tubules that drain bile from the liver and merge with the cystic duct from the gallbladder to form the common bile duct.

hepatic portal circulation (por'tal) The return of venous blood from the digestive organs through a capillary network within the liver before draining into the heart.

hepatitis (hep''ah-ti'tis) Inflammation of the liver.

Hering-Breuer reflex A reflex in which distension of the lungs stimulates stretch receptors, which in turn act to inhibit further distension of the lungs.

hermaphrodite (her-maf'ro-dīt) An organism having both testes and ovaries.

heterochromatin (het''er-o-kro'mah-tin) A condensed, inactive form of chromatin.

hiatal hernia (hi-a'tal her'ne-ah) A protrusion of an abdominal structure through the esophageal hiatus of the diaphragm into the thoracic cavity.

high-density lipoproteins (lip''o-pro'te-ins) Combinations of lipids and proteins that migrate rapidly to the bottom of a test tube during centrifugation; carrier proteins for lipids, such as cholesterol, which appear to offer some protection from atherosclerosis.

hilus (hi'lus) A concave or depressed area where vessels or nerves enter or exit an organ.

hinge joint A type of diarthrotic articulation characterized by a convex surface of one bone fitting into a concave surface of another so that movement is confined to one plane, such as in the knee or interphalangeal joint.

histamine (his''tah-mēn) A compound secreted by tissue mast cells and other connective tissue cells that stimulates vasodilation and increases capillary permeability; it is responsible for many of the symptoms of inflammation and allergy.

histology (his-tol'o-je) Microscopic anatomy of the structure and function of tissues.

histone (his'tōn) A basic protein associated with DNA that is believed to repress genetic expression.

homeostasis (ho''me-o-sta'sis) The dynamic constancy of the internal environment, which serves as the principal function of physiological regulatory mechanisms. The concept of homeostasis provides a framework for the understanding of most physiological processes.

homologous chromosomes (ho-mol'o-gus) The matching pairs of chromosomes in a diploid cell.

horizontal (transverse) A directional plane that divides the body, organ, or appendage into superior and inferior or proximal and distal portions.

hormone (hor'mōn) A chemical substance that is produced in an endocrine gland and secreted into the bloodstream to cause an effect in a specific target organ.

humoral immunity (hu'mor-al ĭ-mu'nĭ-te) The form of acquired immunity in which antibody molecules are secreted in response to antigenic stimulation (as opposed to cell-mediated immunity).

hyaline cartilage (hi'ah-līn) A cartilage with a homogeneous matrix. It is the most common type, occurring at the articular ends of bones, in the trachea, and within the nose, and forms the precursor to most of the bones in the body.

hyaline membrane disease A disease of some premature infants who lack pulmonary surfactant, it is characterized by collapse of the alveoli (atelectasis) and pulmonary edema. Also called *respiratory distress syndrome.*

hydrocortisone (hi''dro-kor'tĭ-son) Also called *cortisol;* the principal corticosteroid hormone secreted by the adrenal cortex, with glucocorticoid action.

hydrophilic (hy''dro-fil'ik) A substance that readily absorbs water, literally "water loving."

hydrophobic (hi''dro-fo'bik) A substance that repels, and that is repelled by, water; "water fearing."

hymen (hi'men) A developmental remnant of membranous tissue that partially covers the vaginal opening.

hyperbaric oxygen (hi''per-bār'ik) Oxygen gas present at greater than atmospheric pressure.

hypercapnia (hi″per-kap′ne-ah) Excessive concentration of carbon dioxide in the blood.
hyperextension (hi″per-ek-sten′shun) Extension beyond the normal anatomical position or 180°.
hyperglycemia (hi″per-gli-se′me-ah) Abnormally increased concentration of glucose in the blood.
hyperkalemia (hi″per-kal-le′me-ah) Abnormally high concentration of potassium in the blood.
hyperopia (hi″per-o′pe-ah) Also called *farsightedness:* a refractive disorder in which rays of light are brought to a focus behind the retina as a result of the eyeball being too short.
hyperplasia (hi″per-pla′ze-ah) An increase in organ size due to an increase in cell numbers as a result of mitotic cell division.
hyperpolarization (hi″per-po″lar-i-za′shun) An increase in the negativity of the inside of a cell membrane with respect to the resting membrane potential.
hypersensitivity (hi″per-sen′si-tiv″i-te) Another name for *allergy;* abnormal immune response that may be immediate (due to antibodies of the IgE class) or delayed (due to cell-mediated immunity).
hypertension (hi″per-ten′shun) Elevated or excessive blood pressure.
hypertonic (hi″per-ton′ik) A solution with a greater solute concentration and thus a greater osmotic pressure than plasma.
hypertrophy (hi-per′tro-fe) Growth of an organ due to an increase in the size of its cells.
hyperventilation (hi-per-ven″ti-la′shun) A high rate and depth of breathing that results in a decrease in the blood carbon dioxide concentration below normal.
hypodermis (hi″po-der′mis) A layer of fat beneath the dermis of the skin.
hyponychium (hi″po-nik′e-um) A thickened, supportive layer of stratum corneum at the distal end of a digit under the free edge of the nail.
hypothalamic hormones (hi″po-thah-lam′ik) Hormones produced by the hypothalamus; these include antidiuretic hormone and oxytocin, which are secreted by the posterior pituitary gland; and both releasing and inhibiting hormones that regulate the secretion of the anterior pituitary.
hypothalamo-hypophyseal portal system (hi″po-fiz′e-al) A vascular system that transports releasing and inhibiting hormones from the hypothalamus to the anterior pituitary.
hypothalamo-hypophyseal tract The tract of nerve fibers (axons) that transports antidiuretic hormone and oxytocin from the hypothalamus to the posterior pituitary gland.
hypothalamus (hi″po-thal′ah-mus) A portion of the forebrain within the diencephalonic region that is positioned below the thalamus and functions as an autonomic nerve center and regulates the pituitary gland.
hypovolemic shock (hi″po-vo-le′mik) A rapid fall in blood pressure due to diminished blood volume.
hypoxemia (hi″pok-se′me-ah) A low oxygen concentration of the arterial blood.

i

ileocecal valve (il″e-o-se′kal) A modification of the mucosa at the junction of the small and large intestine that forms a one-way passage and prevents the backflow of food materials.
ileum (il′e-um) The terminal portion of the small intestine between the jejunum and cecum.
immediate hypersensitivity (hi″per-sen′si-tiv″i-te) Hypersensitivity (allergy) that is mediated by antibodies of the IgE class and results in the release of histamine and related compounds from tissue cells.
immunization (im″u-ni-za′shun) The process of increasing one's resistance to pathogens. In active immunity a person is injected with antigens that stimulate the development of clones of specific B or T lymphocytes; in passive immunity a person is injected with the antibodies made by another organism.
immunoassay (im″u-no-as′sa) The detection or measurement of a molecule that acts as an antigen by its reaction with specific antibodies. The antigen-antibody reaction may be followed in a variety of ways, such as agglutination (if the antigen is attached to visible cells or particles), fluorescence, or radioactivity (a radioimmunoassay, or RIA).
immunoglobins (im″u-no-glob′u-lins) Subclasses of the gamma globulin fraction of plasma proteins that have antibody functions, providing humoral immunity.
immunosurveillance (im″u-no-ser-va′lens) The function of the immune system to recognize and attack malignant cells that produce antigens not recognized as "self." This function is believed to be cell mediated rather than humoral.
implantation (im″plan-ta′shun) The process by which a blastocyst attaches itself to and penetrates into the endometrium of the uterus.
incus (ing′kus) The middle of three ear ossicles within the middle ear chamber; commonly called the *anvil.*
infundibulum (in″fun-dib′u-lum) The flesh stalk that attaches the pituitary gland to the hypothalamus of the brain.
ingestion (in-jes′chun) The process of taking food or liquid into the body by way of the oral cavity.
inguinal (ing′gwi-nal) Pertaining to the groin region.
inguinal canal The circular passage through which a testis descends into the scrotum.
inhibin (in-hib′in) Believed to be a water-soluble hormone secreted by the seminiferous tubules of the testes that specifically exerts negative feedback inhibition of FSH secretion from the anterior pituitary gland.
inositol (in-o′si-tol) A sugarlike B-complex vitamin. Inositol triphosphate is believed to act as a second messenger in the action of some hormones.
insertion The more movable attachment of a muscle, usually more distal.
inspiration (in″spi-ra′shun) The act of breathing air into the alveoli of the lungs; also called *inhalation.*
insulin (in′su-lin) A polypeptide hormone, secreted by the beta cells of the islets of Langerhans in the pancreas, that promotes the anabolism of carbohydrates, fat, and protein. Insulin acts to promote the cellular uptake of blood glucose and, therefore, to lower the blood glucose concentration; insulin deficiency produces hyperglycemia and diabetes mellitus.
integument (in-teg′u-ment) Pertaining to the skin.
intercalated disc (in-ter′kah-lāt-ed) A thickened portion of the sarcolemma that extends across a cardiac muscle fiber and indicates the boundary between cells.
intercellular substance (in″ter-sel′u-lar) The matrix or material between cells that largely determines tissue types.
interferons (in″ter-fēr′ons) A group of small proteins that inhibit the multiplication of viruses inside host cells and also have antitumor properties.
internal (deep) Toward the center, away from the surface of the body.
internal ear The innermost portion or chamber of the ear, containing the cochlea and the vestibular organs.
internal nares (na′rēz) The two posterior openings from the nasal cavity into the nasopharynx; also called the *choanae.*
interneurons (in″ter-nu′rons) Also called *association neurons;* those neurons within the central nervous system that do not extend into the peripheral nervous system; they are interposed between sensory (afferent) and motor (efferent) neurons.
interoceptors (in″ter-o-sep′tors) Sensory receptors that respond to changes in the internal environment (as opposed to exteroceptors).
interphase The interval between successive cell divisions; during this time the chromosomes are in an extended state and are active in directing RNA synthesis.
intervertebral disc (in″ter-ver′tĕ-bral) A pad of fibrocartilage located between the bodies of adjacent vertebrae.
intrafusal fibers (in″trah-fu′zal) Modified muscle fibers that are encapsulated to form muscle spindle organs, which are muscle stretch receptors.
intramembranous ossification (in″trah-mem′brah-nus os″i-fi-ka′shun) A type of bone formation in which there is a membranous, rather than cartilaginous, precursor.
intrapleural space (in″trah-ploor′al) An actual or potential space between the visceral pleural membrane covering the lungs and the somatic pleural membrane lining the thoracic wall.
intrapulmonary space (in″trah-pul′mo-ner″e) The space within the air sacs and airways of the lungs.
intrinsic (in-trin′sik) Situated in or pertaining to internal origin.
inulin (in′u-lin) A polysaccharide of fructose, produced by certain plants, that is filtered by the human kidneys but neither reabsorbed nor secreted. The clearance rate of injected inulin is thus used to measure the glomerular filtration rate.
inversion (in-ver′zhun) A movement of the foot in which the sole is turned inward.
in vitro (in ve′tro) Occurring outside the body, in a test tube or other artificial environment.
in vivo (in ve′vo) Occurring within the body.
ion (i′on) An atom or group of atoms that has either lost or gained electrons and thus has a net positive or a net negative charge.
ionization (i″on-i-za′shun) The dissociation of a solute to form ions.
ipsilateral (ip″si-lat′er-al) On the same side (as opposed to contralateral).
IPSP Inhibitory postsynaptic potential; hyperpolarization of the postsynaptic membrane in response to a particular neurotransmitter chemical, which makes it more difficult for the postsynaptic cell to attain a threshold level of depolarization required to produce action potentials; it is responsible for postsynaptic inhibition.
iris (i′ris) The pigmented muscular portion of the eye that surrounds the pupil and regulates its diameter.
ischemia (is-ke′me-ah) A rate of blood flow to an organ that is inadequate to supply sufficient oxygen and maintain aerobic respiration in that organ.

islets of Langerhans (i'lets of lahng'er-hanz) A cluster of cells within the pancreas that forms the endocrine portion and secretes insulin and glucagon.
isoenzymes (i''so-en'zīms) Enzymes, usually produced by different organs, that catalyze the same reaction but that differ from each other in amino acid composition.
isometric contraction (i''so-met'rik) Muscle contraction in which there is no appreciable shortening of the muscle.
isotonic contraction (i''so-ton'ik) Muscle contraction in which the muscle shortens in length and maintains approximately the same amount of tension throughout the shortening process.
isotonic solution A solution having the same total solute concentration, osmolality, and osmotic pressure as the solution with which it is compared; a solution with the same solute concentration and osmotic pressure as plasma.
isthmus (is'mus) A narrow neck or portion of tissue connecting two structures.

j

jaundice (jawn'dis) A condition characterized by high blood bilirubin levels and staining of the tissues with bilirubin, which gives skin and mucous membranes a yellow color.
jejunum (je-joo'num) The middle portion of the small intestine, located between the duodenum and the ileum.
joint capsule (kap'sūl) A fibrous tissue cuff surrounding a diarthrotic joint.
junctional complexes (junk'shun-al) The structures that join adjacent epithelial cells together, including the zonula occludens, zonula adherens, and macula adherens (desmosome).

k

keratin (ker'ah-tin) An insoluble protein present in the epidermis and in epidermal derivatives such as hair and nails.
ketoacidosis (ke''to-ah''sī-do'sis) A type of metabolic acidosis resulting from the excessive production of ketone bodies, as in diabetes mellitus.
ketogenesis (ke''to-jen'ĕ-sis) The production of ketone bodies.
ketone bodies (ke'tōn) The substances acetone, acetoacetic acid, and β hydroxybutyric acid, which are derived from fatty acids via acetyl coenzyme A in the liver. Ketone bodies are oxidized by skeletal muscles for energy.
ketosis (ke-to'sis) An abnormal elevation in the blood concentration of ketone bodies that does not necessarily produce acidosis.
kidney (kid'ne) One of the paired organs of the urinary system that contains nephrons and filters urine from the blood.
kilocalorie (kil'o-kal''o-re) A unit of measurement equal to 1,000 calories, which are units of heat (a kilocalorie is the amount of heat required to raise the temperature of 1 kilogram of water by 1°C). In nutrition the kilocalorie is called a big calorie (Calorie).
Klinefelter's syndrome (klīn'fel-terz sin'drōm) The syndrome produced in a male by the presence of an extra X chromosome (genotype XXY).
knee A region in the lower extremity, between the thigh and the ankle, containing a diarthrotic hinge joint.
Krebs cycle (krebz) A cyclic metabolic pathway in the matrix of mitochondria by which the acetic acid part of acetyl CoA is oxidized and substrates are provided for reactions that are coupled to the formation of ATP.
Kupffer cells (koop'fer) Phagocytic cells that line the sinusoids of the liver, and which are part of the reticuloendothelial system.

l

labial frenulum (la'be-al fren'u-lum) A longitudinal fold of mucous membrane that attaches the lips to the gum along the midline of both the upper and lower lip.
labia majora (la'be-ah ma-jor'ah) A portion of the external genitalia of a female, consisting of two longitudinal folds of skin extending downward and backward from the mons pubis.
labia minora (ma-nor'-ah) Two small folds of skin devoid of hair and sweat glands, lying between the labia majora of the external genitalia of a female.
labyrinth (lab'ĭ-rinth) The complex system of interconnecting tubes within the inner ear, which includes the semicircular canals, cochlea, and vestibule.
lacrimal canal (lak'ri-mal) A drainage duct for tears, located at the medial corner of an eyelid and conveying the tears medially into the nasolacrimal sac.
lacrimal gland A tear-secreting gland, located on the superior lateral portion of the eyeball underneath the upper eyelid.
lactation (lak-ta'shun) The production and secretion of milk by the mammary glands.
lacteal (lak'te-al) A small lymphatic duct associated with a villus of the small intestine.
lactose (lak'tōs) Milk sugar; a disaccharide of glucose and galactose.
lactose intolerance The inability of many adults to digest lactose due to a loss of the ability of the intestine to produce lactase enzyme.
lacuna (lah-ku'nah) A small, hollow chamber that houses an osteocyte in mature bone tissue or a chondrocyte in cartilage tissue.
lamella (lah-mel'ah) A concentric ring of matrix surrounding the haversian canal in a haversian system of mature bone tissue.
lamina (lam'ĭ-nah) A thin plate of bone that extends superiorly from the body of a vertebra to form either side of the arch of a vertebra.
lanugo (lah-nu'go) Short, silky fetal hair, which may be present for a short time on a premature infant.
large intestine The last major portion of the alimentary canal, consisting of the cecum, colon, rectum, and anal canal.
laryngopharynx (lah-ring''go-far'inks) The inferior or lower portion of the pharynx in contact with the larynx.
larynx (lar'inks) The structure located between the pharynx and trachea that houses the vocal cords; commonly called the *voice box*.
lateral (lat'er-al) Pertaining to the side; farther from the midline.
lateral ventricle (ven'tri-k'l) A cavity located in the cerebral hemisphere of the brain and filled with cerebrospinal fluid.
leg The portion of the lower extremity between the knee and ankle.
lens (lenz) A transparent refractive organ of the eye, derived from ectoderm and positioned posterior to the pupil and iris.
lesion (le'zhun) A wounded or damaged area.
lesser omentum (o-men'tum) A peritoneal fold of tissue extending from the lesser curvature of the stomach to the liver.
leukocyte (loo'ko-sīt) A white blood cell; also spelled leucocyte.
ligament (lig'ah-ment) A tough chord or fibrous band of connective tissue that binds bone to bone to strengthen and provide flexibility to a joint; it also may support viscera.
limbic system (lim'bik) A portion of the brain concerned with emotions and autonomic activity.
linea alba (lin'e-ah al'bah) A vertical fibrous band extending down the anterior medial portion of the abdominal wall.
lingual frenulum (ling'gwal fren'u-lum) A longitudinal fold of mucous membrane that attaches the tongue to the floor of the mouth.
lipogenesis (lip''o-jen'ĕ-sis) The formation of fat or triglycerides.
lipolysis (li-pol'ĭ-sis) The hydrolysis of triglycerides into free fatty acids and glycerol.
low-density lipoproteins (lip''o-pro'te-ins) Plasma proteins that transport triglycerides and cholesterol; they are believed to contribute to arteriosclerosis.
lower extremity A lower appendage, including the thigh, leg, and foot.
lumbar (lum'ber) Pertaining to the region of the loins.
lumbar plexus (plek'sus) A network of nerves formed by the anterior branches of spinal nerves L1 through L4.
lumen (lu'men) The space within a tubular structure through which a substance passes.
lung One of the two major organs of respiration positioned within the thoracic cavity on either side of the mediastinum.
lung surfactant (sur-fak'tant) A mixture of lipoproteins (containing phospholipids) secreted by type II alveolar cells into the alveoli of the lungs; it lowers surface tension and prevents collapse of the lungs as occurs in hyaline membrane disease, in which surfactant is absent.
lunula (lu'nu-lah) The half-moon-shaped whitish area at the proximal portion of a nail.
luteinizing hormone (LH) (loo''te-in''i-zing) A gonadotropic hormone secreted by the anterior pituitary, which in a female stimulates ovulation and the development of a corpus luteum. In a male, LH stimulates the Leydig cells to secrete androgens.
lymph (limf) A clear, plasmalike fluid that flows through lymphatic vessels.
lymphatic system (lim-fat'ik) The lymphatic vessels and lymph nodes.
lymph node A small, oval mass of reticular tissue located along the course of lymph vessels.
lymphocyte (lim'fo-sīt) A type of white blood cell characterized by agranular cytoplasm. Lymphocytes usually constitute about 20%–25% of the white cell count.
lymphokines (lim'fo-kīns) A group of chemicals released from T cells that contribute to cell-mediated immunity.
lysosome (li'so-sōm) Organelles containing digestive enzymes and responsible for intracellular digestion.

m

macromolecules (mak''ro-mol'ĕ-kūl) Large molecules; a term that usually refers to protein, RNA, and DNA.
macrophage (mak'ro-fāj) A wandering phagocytic cell.

macula lutea (mak'u-lah lu'te-ah) A yellowish depression in the retina of the eye that contains the fovea centralis, the area of keenest vision.
malignant (mah-lig'nant) Tending to become worse and end in death.
malleus (mal'e-us) The first of three ear ossicles attached to the tympanum; commonly called the *hammer.*
mammary gland (mam'er-e) The gland of the female breast responsible for lactation and nourishment of the young.
marrow (mar'o) The soft connective tissue that occupies the inner cavity of certain bones and produces red blood cells.
mast cells A type of connective tissue cells that produce and secrete histamine and heparin.
mastication (mas''ti-ka'shun) Pertaining to the chewing of food.
matrix (ma'triks) The intercellular substance of a tissue.
maximal oxygen uptake The maximum amount of oxygen that can be consumed by the body per unit time during heavy exercise.
meatus (me-a'tus) A passageway or opening into a structure.
mechanoreceptor (mek''ah-no-re-sep'tor) A sensory receptor that responds to a mechanical stimulus.
medial (me'de-al) Toward or nearer the midline of the body.
mediastinum (me''de-as-ti'num) The space in the center of the thorax between the two pleural cavities.
medulla (mē-dul'ah) The center portion of an organ.
medulla oblongata (ob''long-ga'tah) A portion of the brain stem located between the spinal cord and the pons.
medullary (marrow) cavity (med'u-lār''e) The hollow core of the diaphysis of a long bone, occupied by marrow.
megakaryocyte (meg''ah-kar'e-o-sīt) A bone marrow cell that gives rise to blood platelets.
meiosis (mi-o'sis) A specialized type of cell division by which gametes or haploid sex cells are formed.
Meissner's corpuscle (mīs'nerz kor'pus'l) A touch sensory receptor found in the papillary layer of the dermis of the skin.
melanin (mel'ah-nin) A dark pigment found within the epidermis or epidermal derivatives of the skin.
melanocyte (mel'ah-no-sīt) A specialized melanin-producing cell found in the deepest layer of epidermis.
melatonin (mel''ah-to'nin) A hormone secreted by the pineal gland that produces lightening of the skin in lower animals and may contribute to the regulation of gonadal function in mammals.
membrane potential The potential difference or voltage that exists between the inner and outer sides of a cell membrane; it exists in all cells but is capable of being changed by excitable cells (neurons and muscle cells).
membranous bone (mem'brah-nus) Bone that forms from membranous connective tissue rather than from cartilage; also called *intramembranous bone.*
membranous labyrinth (lab'i-rinth) A system of communicating sacs and ducts within the bony labyrinth of the inner ear.
menarche (mē-nar'ke) The first menstrual discharge.
Meniere's disease (men''e-ārz') Deafness, tinnitus, and vertigo resulting from a disease of the labyrinth.
meninges (mē-nin'jēz) A group of three fibrous membranes that cover the central nervous system, composed of the dura, arachnoid, and pia maters.
menisci (mēn-is'si) Wedge-shaped fibrocartilages in certain movable joints; also called *semilunar cartilages.*
menopause (men'o-pawz) The cessation of menstrual periods in the human female.
menstrual cycle (men'stroo-al) The rhythmic female reproductive cycle, which is characterized by changes in hormone levels and physical changes in the uterine lining.
menstruation (men''stroo-a'shun) The discharge of blood and tissue from the uterus at the end of the female reproductive cycle.
mesentery (mes'en-ter''e) A fold of peritoneal membrane that attaches an abdominal organ to the abdominal wall.
mesoderm (mes'o-derm) The middle of the three germ layers.
mesovarium (mes''o-va're-um) The peritoneal fold that attaches an ovary to the broad ligament of the uterus.
messenger RNA (mRNA) A type of RNA that contains a base sequence complementary to a part of the DNA that specifies the synthesis of a particular protein.
metabolism (mē-tab'o-lizm) The sum total of the chemical changes that occur within a cell.
metacarpus (met''ah-kar'pus) The region of the hand between the wrist and the phalanges, including the five bones that compose the palm of the hand.
metastasis (mē-tas'tah-sis) The spread of a disease from one organ or body part to another.
metatarsus (met''ah-tar'sus) The region of the foot between the ankle and the phalanges, comprising five bones.
micelles (mi-sels') Colloidal particles formed by the aggregation of many molecules.
microglia (mi-krog'le-ah) Small phagocytic cells found in the central nervous system.
microvilli (mi''kro-vil'i) Microscopic, hairlike projections of cell membranes on certain epithelial cells.
micturition (mik''tu-rish'un) The process of voiding urine; also called *urination.*
midbrain The portion of the brain between the pons and the forebrain.
middle ear The middle of the three ear chambers, which contains the three ear ossicles.
midsagittal (mid-saj'i-tal) A plane that divides the body or an organ into right and left halves.
mineralocorticoids (min''er-al-o-kor'ti-koids) Steroid hormones of the adrenal cortex (corticosteroids) that regulate electrolyte balance.
mitosis (mi-to'sis) The process of cell division, in which the two daughter cells are identical and contain the same number of chromosomes.
mitral valve (mi'tral) The left atrioventricular heart valve; also called the *bicuspid valve.*
mixed nerve A nerve that contains both motor and sensory nerve fibers.
molal (mo'lal) Pertaining to the number of moles of solute per kilogram of solvent.
molar (mo'lar) Pertaining to the number of moles of solute per liter of solution.
mole (mōl) The number of grams of a chemical that is equal to its formula weight (atomic weight for an element or molecular weight for a compound).
monoclonal antibodies (mon''o-klōn'al an''ti-bod''es) Identical antibodies derived from a clone of genetically identical plasma cells.
monocyte (mon'o-sit) A phagocytic type of white blood cell, normally constituting about 3%–8% of the white blood cell count.
monomer (mon'o-mer) A single molecular unit of a longer, more complex molecule; monomers are joined together to form dimers, trimers, and polymers; the hydrolysis of polymers eventually yields separate monomers.
monosaccharide (mon''o-sak'ah-rīd) Also called a *simple sugar;* the monomers of the more complex carbohydrates. Examples include glucose, fructose, and galactose.
mons pubis (monz pu'bis) A fatty tissue pad covering the symphysis pubis, which is covered by pubic hair in the female.
morula (mor'u-lah) An early stage of embryonic development characterized by a solid ball of cells.
motile (mo'til; mō't'l, mo'tīl) Capable of self-propelled movement.
motor area A region of the cerebral cortex from which originate motor impulses to muscles or glands.
motor nerve A nerve composed of motor nerve fibers.
motor neuron (nu'ron) An efferent neuron that conducts action potentials away from the central nervous system and innervates effector organs (muscles and glands). It forms the ventral roots of the spinal nerves.
motor unit A single motor neuron and the muscle fibers it innervates.
mucosa (mu-ko'sah) A mucous membrane that lines cavities and tracts opening to the exterior.
mucous cell (mu'kus) A specialized unicellular gland that produces and secretes mucus; also called a *goblet cell.*
mucous membrane The layers of visceral organs that include the lining epithelium, submucosal connective tissue, and (in some cases) a thin layer of smooth muscle (the muscularis mucosa).
multipolar neuron (mul''ti-po'lar nu'ron) A nerve cell with many processes originating from the cell body.
muscle (mus'el) A major type of tissue that is adapted to contract. The three kinds of muscle are cardiac, smooth, and skeletal.
muscle spindles Sensory organs within skeletal muscles that are composed of intrafusal fibers and are sensitive to muscle stretch; they provide a length detector within muscles.
muscularis (mus''ku-la'ris) A muscular layer or tunic of an organ, composed of smooth muscle tissue.
myelin (mi'ĕ-lin) A lipoprotein material that forms a sheathlike covering around nerve fibers.
myelin sheath (shēth) A sheath surrounding axons, which is formed by successive wrappings of a neuroglial cell membrane. Myelin sheaths are formed by Schwann cells in the peripheral nervous system and by oligodendrocytes within the central nervous system.
myocardial infarction (mi''o-kar'de-al in-fark'shun) An area of necrotic tissue in the myocardium, which is filled in by scar (connective) tissue.
myocardium (mi''o-kar'de-um) The cardiac muscle layer of the heart.
myofibril (mi''o-fi'bril) A bundle of contractile fibers within muscle cells.
myogenic (mi''o-jen'ik) Originating within muscle cells; used to describe self-excitation by cardiac and smooth muscle cells.

myoglobin (mi″o-glo′bin) A molecule composed of globin protein and heme pigment; it is related to hemoglobin but contains only one subunit (instead of the four in hemoglobin) and is found in striated muscles. Myoglobin serves to store oxygen in skeletal and cardiac muscle cells.
myogram (mi′o-gram) A recording of electrical activity within a muscle.
myology (mi-ol′o-je) The science or study of muscle structure function.
myometrium (mi″o-me′tre-um) The layer or tunic of smooth muscle within the uterine wall.
myoneural junction (mi″o-nu′ral) The site of contact between an axon of a motor neuron and a muscle fiber.
myosin (mi′o-sin) A thick filament protein that together with actin causes muscle contraction.
myxedema (mik″sĕ-de′mah) A type of edema associated with hypothyroidism; it is characterized by the accumulation of mucoproteins in tissue fluid.

n

NAD Nicotinamide adenine dinucleotide; a coenzyme derived from niacin, which functions to transport electrons in oxidation-reduction reactions; it helps to transport electrons to the electron transport chain within mitochondria.
nail A hardened, keratinized plate that develops from the epidermis and forms a protective covering on the dorsal surface of the distal phalanges of fingers and toes.
naloxone (nal-oks′ōn) A drug that antagonizes the effects of morphine and endorphins.
nasal cavity (na′zal) A mucosa-lined space above the oral cavity that is divided by a nasal septum and is the first chamber of the respiratory system.
nasal concha (kong′kah) A scroll-like bone extending medially from the lateral wall of the nasal cavity; also called a *turbinate bone.*
nasal septum (sep′tum) A bony and cartilaginous partition that separates the nasal cavity into two portions.
nasopharynx (na″zo-far′inks) The first or uppermost chamber of the pharynx, positioned posterior to the nasal cavity and extending down to the soft palate.
natriuretic (na″tre-u-ret′ik) An agent that promotes the excretion of sodium in the urine. Atrial natriuretic hormone has this effect.
neck 1. Any constricted portion, such as the neck of an organ. 2. The cervical region of the body between the head and thorax.
necrosis (nĕ-kro′sis) Cellular death within tissues and organs.
negative feedback Mechanisms in the body that act to maintain a state of internal constancy, or homeostasis; effectors are activated by changes in the internal environment, and the actions of the effectors serve to counteract these changes and maintain a state of balance.
neonatal (ne″o-na′tal) The stage of life from birth to the end of four weeks.
neoplasm (ne′o-plazm) A new, abnormal growth of tissue, as in a tumor.
nephron (nef′ron) The functional unit of the kidney, consisting of a renal corpuscle, Bowman's capsule, convoluted tubules, and a loop of Henle.
nerve A bundle of nerve fibers outside the central nervous system.
neurilemma (nu″ri-lem′ah) A thin, membranous covering surrounding the myelin sheath of a nerve fiber.
neurofibril (nu″ro-fi′bril) One of many delicate threadlike structures within the cytoplasm of a cell body and the axon hillock of a neuron.
neuroglia (nu-rog′le-ah) Specialized supportive cells of the central nervous system.
neurohypophysis (nu″ro-hi-pof′ĭ-sis) The posterior lobe of the pituitary gland.
neuron (nu′ron) The structural and functional unit of the nervous system, composed of a cell body, dendrites, and an axon; also called a *nerve cell.*
neurotransmitter (nu″ro-trans′mit-er) A chemical contained in synaptic vesicles in nerve endings, which is released into the synaptic cleft and stimulates the production of either excitatory or inhibitory postsynaptic potentials.
neutrons (nu′trons) Electrically neutral particles that exist together with positively charged protons in the nucleus of atoms.
neutrophil (nu′tro-fil) A type of phagocytic white blood cell, normally constituting about 60%–70% of the white blood cell count.
nexus (nek′sus) A bond between members of a group; the type of bonds present in single-unit smooth muscles.
nidation (ni-da′shun) Implantation of the blastocyst into the endometrium of the uterus.
nipple A dark pigmented, rounded projection at the tip of the breast.
Nissl bodies (nis′zl) Clumps of rough endoplasmic reticulum in the cell bodies of neurons.
node of Ranvier (rah-ve-a′) A gap in the myelin sheath of a nerve fiber.
norepinephrine (nor″ep-i-ne′frin) A catecholamine released as a neurotransmitter from postganglionic sympathetic nerve endings and as a hormone (together with epinephrine) from the adrenal medulla.
nucleolus (nu-kle′o-lus) A dark-staining area within a cell nucleus; the site where ribosomal RNA is produced.
nucleoplasm (nu′kle-o-plazm″) The protoplasmic contents of the nucleus of a cell.
nucleotide (nu′kle-o-tīd) The subunit of DNA and RNA macromolecules; each nucleotide is composed of a nitrogenous base (adenine, guanine, cytosine, and thymine or uracil), a sugar (deoxyribose or ribose), and a phosphate group.
nucleus (nu′kle-us) A spheroid body within a cell that contains the genetic factors of the cell.
nucleus pulposus (pul-po′sus) The soft, pulpy core of an intervertebral disc; a remnant of the notochord.
nystagmus (nis-tag′mus) Involuntary, oscillatory movements of the eye.

o

obese (o-bēs′) Excessively fat.
olfactory (ol-fak′to-re) Pertaining to the sense of smell.
olfactory bulb An aggregation of sensory neurons of an olfactory nerve, lying inferior to the frontal lobe of the cerebrum on either side of the crista galli of the ethmoid bone.
olfactory tract The olfactory sensory tract of axons, which conveys impulses from the olfactory bulb to the olfactory portion of the cerebral cortex.
oligodendrocyte (ol″ĭ-go-den′dro-sīt) A type of neuroglial cell concerned with the formation of myelin of nerve fibers within the central nervous system.
oncology (ong-kol′o-je) The study of tumors.
oncotic pressure (ong-kot′ik) The colloid osmotic pressure of solutions produced by proteins; in plasma, it serves to counterbalance the outward filtration of fluid from capillaries due to hydrostatic pressure.
oocyte (o′o-sīt) A developing egg cell.
oogenesis (o″o-jen′ĕ-sis) The process of female gamete formation.
oogenesis (o″o-jen′ĕ-sis) The formation of ova in the ovaries.
opsonization (op″so-ni-za′shun) The role of antibodies in enhancing the ability of phagocytic cells to attack bacteria.
optic (op′tik) Pertaining to the eye.
optic chiasma (ki-as′mah) An X-shaped structure on the inferior aspect of the brain, anterior to the pituitary gland, where there is a partial crossing over of fibers in the optic nerves.
optic disc A small region of the retina where the fibers of the ganglion neurons exit from the eyeball to form the optic nerve; also called the *blind spot.*
optic tract A bundle of sensory axons located between the optic chiasma and the thalamus that functions to convey visual impulses from the photoreceptors within the eye.
oral Pertaining to the mouth; also called *buccal.*
organ A structure consisting of two or more tissues and which performs a specific function.
organ of Corti (kor′te) The spiral organ, or functional unit of hearing, consisting of a basilar membrane supporting receptor hair cells and a tectorial membrane within the endolymph of the cochlear duct.
organelle (or″gan-el′) A minute living structure of a cell with a specific function.
organism (or′gah-nizm) An individual living creature.
orifice (or′ĭ-fis) An opening into a body cavity or tube.
origin (or′ĭ-jin) The place of muscle attachment—usually the more stationary point or the proximal bone; opposite the insertion.
oropharynx (o″ro-far′inks) The second portion of the pharynx, located in a position posterior to the oral cavity and extending from the soft palate to the hyoid bone.
osmolality (oz″mo-lal′ĭ-te) A measure of the total concentration of a solution; the number of moles of solute per kilogram of solvent.
osmoreceptors (oz″mo-re-cep′tors) Sensory neurons that respond to changes in the osmotic pressure of the surrounding fluid.
osmosis (oz-mo′sis) The passage of solvent (water) from a more dilute to a more concentrated solution through a membrane that is more permeable to water than to the solute.
osmotic pressure (oz-mot′ik) A measure of the tendency for a solution to gain water by osmosis when separated by a membrane from pure water; directly related to the osmolality of the solution, it is the pressure required to just prevent osmosis.
osseous tissue (os′e-us) Bone tissue.
ossification (os″ĭ-fĭ-ka′shun) The process of bone tissue formation.
osteoblast (os′te-o-blast″) A bone-forming cell.
osteoclast (os′te-o-klast″) A cell that causes erosion and resorption of bone tissue.
osteocyte (os″te-o-sīt″) A mature bone cell.
osteology (os″te-ol′o-je) The study of the structure and function of bone and the entire skeleton.

osteomalacia (os'te-o-mah-la'she-ah) Softening of bones due to a deficiency of vitamin D and calcium.

osteoporosis (os''te-o-po-ro'sis) Demineralization of bone, seen most commonly in the elderly. It may be accompanied by pain, loss of stature, and other deformities and fractures.

otoliths (o'to-liths) Small, hardened particles of calcium carbonate in the saccule and utricle of the inner ear, associated with the receptors of equilibrium.

oval window An oval opening in the bony wall between the middle and inner ear, into which the footplate of the stapes fits.

ovarian follicle (o-va're-an fol'li-k'l) A developing ovum and its surrounding epithelial cells.

ovarian ligament (lig'ah-ment) A cordlike connective tissue that attaches the ovary to the uterus.

ovary (o'vah-re) The female gonad in which ova and certain sexual hormones are produced.

oviduct (o'vi-dukt) The tube that transports ova from the ovary to the uterus; also called the *uterine tube* or *fallopian tube.*

ovulation (o''vu-la'shun) The rupture of an ovarian follicle with the release of an ovum.

ovum (o'vum) A mature egg cell.

oxidative phosphorylation (ok''si-da'tiv fos''fōr-i-la'shun) The formation of ATP using energy derived from electron transport to oxygen; it occurs in the mitochondria.

oxidizing agent (ok''si-dīz-ing) An atom that accepts electrons in an oxidation-reduction reaction.

oxyhemoglobin (ok''se-he''mo-glo'bin) A compound formed by the bonding of molecular oxygen to hemoglobin.

oxyhemoglobin saturation The ratio, expressed as a percentage, of the amount of oxyhemoglobin compared to the total amount of hemoglobin in blood.

oxytocin (ok''si-to'sin) One of the two hormones produced in the hypothalamus and secreted by the posterior pituitary (the other hormone is vasopressin); oxytocin stimulates the contraction of uterine smooth muscles and promotes milk ejection in females.

P

pacemaker (pās'māk-er) A group of cells that has the fastest spontaneous rate of depolarization and contraction in a mass of electrically coupled cells; in the heart, this is the sinoatrial, or SA, node.

Pacinian corpuscle (pah-sin'e-an kor'pusl) A sensory receptor for pressure, found in tendons, around joints, and in visceral tissues.

PAH Para-aminohippuric acid; a substance used to measure total renal plasma flow because its clearance rate is equal to the total rate of plasma flow to the kidneys; PAH is filtered and secreted but not reabsorbed by the renal nephrons.

palate (pal'at) The roof of the mouth or oral cavity.

palatine (pal'ah-tīn) Pertaining to the palate.

palmar (pal'mar) Pertaining to the palm of the hand.

palpebra (pal'pĕ-brah) An eyelid.

pancreas (pan'kre-as) A mixed organ in the abdominal cavity that secretes gastric juices into the digestive tract and insulin and glucagon into the blood.

pancreatic duct (pan''kre-at'ik) A drainage tube that carries pancreatic juice from the pancreas into the duodenum of the ampulla of Vater.

papillae (pah-pil'e) Small, nipplelike projections.

papillary muscle (pap'ĭ-ler''e) Muscular projections from the ventricular walls of the heart to which the chordae tendineae are attached.

paranasal sinus (par''ah-na'zal si'nus) A mucous-lined air chamber that communicates with the nasal cavity.

parasympathetic (par''ah-sim''pah-thet'ik) Pertaining to the division of the autonomic nervous system concerned with activities that restore and conserve metabolic energy.

parathyroid hormone (PTH) A polypeptide hormone secreted by the parathyroid glands, PTH acts to raise the blood Ca^{++} levels primarily by stimulating reabsorption of bone.

parathyroids (par''ah-thi'roid) Small endocrine glands that are embedded on the posterior surface of the thyroid glands and are concerned with calcium metabolism.

parietal (pah-ri'ĕ-tal) Pertaining to a wall of an organ or cavity.

parietal pleura (ploor'ah) The thin serous membrane attached to the thoracic walls of the pleural cavity.

Parkinson's disease (par'kin-sunz) A tremor of the resting muscles and other symptoms caused by inadequate dopamine-producing neurons in the basal ganglia of the cerebrum. Also called *paralysis agitans.*

parotid gland (pah-rot'id) One of the paired salivary glands located on the sides of the face over the masseter muscle just anterior to the ear and connected to the oral cavity through a salivary duct.

parturition (par''tu-rish'un) The process of childbirth.

passive immunity (i-mu'ni-te) Specific immunity granted by the administration of antibodies made by another organism.

Pasteur effect (pas-ter) A decrease in the rate of glucose utilization and lactic acid production in tissues or microorganisms by their exposure to oxygen.

pathogen (path'o-jen) Any disease-producing microorganism or substance.

pectoral (pek'to-ral) Pertaining to the chest region.

pectoral girdle (ger'd'l) The portion of the skeleton that supports the upper extremities.

pedicle (ped'ĭ-k'l) The portion of a vertebra that connects and attaches the lamina to the body.

pelvic (pel'vik) Pertaining to the pelvis.

pelvic girdle (ger'd'l) The portion of the skeleton to which the lower extremities are attached.

pelvis (pel'vis) A basinlike bony structure formed by the sacrum and os coxae.

penis (pe'nis) The external male genital organ, through which urine passes during urination and which transports spermatozoa to the female during sexual intercourse.

pepsin The protein-digesting enzyme secreted in gastric juice.

peptic ulcer (pep'tik ul'ser) An injury to the mucosa of the esophagus, stomach, or small intestine, caused by acidic gastric juice.

pericardium (per''ĭ-kar'de-um) A protective serous membrane that surrounds the heart.

perichondrium (per''ĭ-kon'dre-um) A toughened connective sheet that covers cartilage.

perikaryon (per''ĭ-kar'e-on) The cell body of a neuron.

perilymph (per'ĭ-limf) The fluid between the membranous and bony labyrinth of the inner ear.

perimysium (per''ĭ-mis'e-um) Fascia or connective tissue surrounding a bundle (fascicle) of muscle fibers.

perineum (per''ĭ-ne'um) The floor of the pelvis, which is the region between the anus and the scrotum in the male and between the anus and the vulva in the female.

perineurium (per''ĭ-nu're-um) Connective tissue surrounding a bundle (fascicle) of nerve fibers.

periodontal membrane (per''e-o-don'tal) A fibrous connective tissue lining the sockets of teeth.

periosteum (per''e-os'te-um) A fibrous connective tissue covering the surface of bone.

peripheral nervous system (pĕ-rif'er-al) The nerves and ganglia of the nervous system that lie outside of the brain and spinal cord.

peristalsis (per''ĭ-stal'sis) Rhythmic contractions of smooth muscle in the walls of various tubular organs, which move the contents along.

peritoneum (per''ĭ-to-ne'um) The serous membrane that lines the abdominal cavity and covers the abdominal visceral organs.

permease (per'me-ās) A term used to indicate membrane transport carriers and to emphasize the similarity of specificity and other properties that transport carriers have with enzymes.

Peyer's patches (pi'erz) Clusters of lymph nodes on the walls of the small intestine.

pH The pH of a solution is equal to 1 over the logarithm of the hydrogen ion concentration. The pH scale goes from zero to 14; a pH of 7.0 is neutral, whereas solutions with lower pH are acidic and solutions with higher pH are basic.

phagocytosis (fag''o-si-to'sis) Cellular eating; the ability of some cells (such as white blood cells) to engulf large particles such as bacteria and digest these particles by merging the food vacuole containing these particles with a lysosome containing digestive enzymes.

phalanx (fa'lanks), pl. *phalanges* A bone of a finger or toe.

pharynx (far'inks) The region of the digestive system and respiratory system located at the back of the oral and nasal cavities and extending to the larynx anteriorly and the esophagus posteriorly; also called the *throat.*

phonocardiogram (fo''no-kar'de-o-gram) A visual display of the heart sounds.

photoreceptor (fo''to-re-sep'tor) A sensory nerve ending that responds to the stimulation light.

physiology (fiz''e-ol'o-je) The science that deals with the study of body functions.

pia mater (pi'ah ma'ter) The innermost meninx that is in direct contact with the brain and spinal cord.

pineal gland (pin'e-al) A small cone-shaped gland located in the root of the third ventricle.

pinna (pin'nah) The outer, fleshy portion of the external ear; also called the *auricle.*

pinocytosis (pi''no-, pin''-o-si-to'sis) Cell drinking; invagination of the cell membrane forming narrow channels that pinch off into vacuoles; this provides cellular intake of extracellular fluid and dissolved molecules.

pituitary gland (pĭ-tu'ĭ-tār''e) A small, pea-shaped endocrine gland situated on the interior surface of the diencephalonic region of the brain, consisting of anterior and posterior lobes; also called the *hypophysis* and commonly called the "master gland."

pivot joint (piv'ut) A diarthrotic joint in which a rounded head of one bone articulates with a depressed cup of another to permit a rotational type of movement.
placenta (plah-sen'tah) The organ of metabolic exchange between the mother and the fetus.
plantar (plan'tar) Pertaining to the sole of the foot.
plasma (plaz'mah) The fluid, extracellular portion of circulating blood.
plasma cells Cells derived from B lymphocytes that produce and secrete large amounts of antibodies; they are responsible for humoral immunity.
platelets (plāt'lets) Small fragments of specific bone marrow cells that function in blood coagulation; also called *thrombocytes.*
pleural (ploor'al) Pertaining to the serous membranes associated with the lungs.
pleural cavity The potential space between the visceral pleural and parietal pleural membranes.
pleural membranes Serous membranes that surround the lungs and provide protection and compartmentalization.
plexus (plek'sus) A network of interlaced nerves or vessels.
plexus of Auerbach (ow'er-bahk) A network of sympathetic and parasympathetic nerve fibers located in the muscularis tunic of the small intestine.
plexus of Meissner (mis'ner) A network of autonomic nerve fibers located in the submucosa of the small intestine.
plica circularis (pli'kah ser-ku-lar'is) A deep fold within the wall of the small intestine, which increases the absorptive surface area.
pluripotent (ploo-rip'o-tent) A property of early embryonic cells; the ability to specialize in a number of ways to produce tissues characteristic of different organs.
pneumotaxic area (nu''mo-tak'sik) The region of the respiratory control center, located in the pons of the brain.
polar body A small daughter cell formed by meiosis that degenerates in the process of oocyte production.
polar molecule A molecule in which the shared electrons are not evenly distributed, so that one side of the molecule is relatively negatively (or positively) charged in comparison with the other side; polar molecules are soluble in polar solvents such as water.
polydipsia (pol''e-dip'se-ah) Excessive thirst.
polymer (pol'i-mer) A large molecule formed by the combination of smaller subunits, or monomers.
polymorphonuclear leukocyte (pol''e-mor''fo-nu'kle-ar loo'ko-sīt) A granular leukocyte containing a nucleus with a number of lobes connected by thin, cytoplasmic strands; this type includes neutrophils, eosinophils, and basophils.
polypeptide (pol''e-pep'tīd) A chain of amino acids connected by covalent bonds called peptide bonds. A very large polypeptide is called a protein.
polyphagia (pol''e-fa'je-ah) Excessive eating.
polysaccharide (pol''e-sak'ah-rīd) A carbohydrate formed by covalent bonding of numerous monosaccharides; examples include glycogen and starch.
polyuria (pol''e-u're-ah) Excretion of an excessively large volume of urine in a given period.
pons (ponz) The portion of the brain stem just above the medulla oblongata and anterior to the cerebellum.
popliteal (pop-lit'e-al, pop''lī-te'al) Pertaining to the concave region on the posterior aspect of the knee joint.
posterior (pos-tēr'e-or) Toward the back; also called *dorsal.*
posterior pituitary (pi-tu'i-tār''e) The part of the pituitary gland that is derived from the brain; it secretes vasopressin (ADH) and oxytocin, produced in the hypothalamus. Also called the *neurohypophysis.*
posterior root An aggregation of sensory neuron fibers lying between a spinal nerve and the dorsolateral aspect of the spinal cord; also called the *dorsal root* or *sensory root.*
posterior root ganglion (gang'gle-on) A cluster of cell bodies of sensory neurons located along the posterior root of a spinal nerve.
postganglionic neuron (pōst''gang-gle-on'ik) The second neuron in an autonomic efferent pathway. Its cell body is outside the central nervous system, and it terminates at an effector organ.
postnatal (pōst-na'tal) After birth.
postsynaptic inhibition (pōst''si-nap'tik) The inhibition of a postsynaptic neuron by axon endings that release a neurotransmitter that induces hyperpolarization (inhibitory postsynaptic potentials).
preganglionic neuron (pre''gang-gle-on'ik) The first neuron in an autonomic efferent pathway. Its cell body is inside the central nervous system, and it terminates on a postganglionic neuron.
pregnancy A condition in which a female has a developing offspring in the uterus.
prenatal (pre-na'tal) The period of offspring development during pregnancy; before birth.
prepuce (pre'pūs) A fold of loose, retractable skin covering the glans of the penis or clitoris; also called the *foreskin.*
presynaptic inhibition (pre''si-nap'tik) Neural inhibition in which axoaxonic synapses inhibit the release of neurotransmitter chemicals from the presynaptic axon.
prime mover The muscle most directly responsible for a particular movement; an agonist.
prolactin (pro-lak'tin) A hormone secreted by the anterior pituitary that stimulates lactation (acting together with other hormones) in the postpartum female. It may also participate (along with the gonadotropins) in regulating gonadal function in some mammals.
pronation (pro-na'shun) A rotational movement of the forearm in which the palm of the hand is turned posteriorly.
prophylaxis (pro''fi-lak'sis) Prevention or protection.
proprioceptor (pro''pre-o-sep'tor) A sensory nerve ending that responds to changes in tension in a muscle or tendon.
prostaglandin (pros''tah-glan'din) A family of fatty acids that have numerous autocrine regulatory functions, which include stimulating contractions of the uterus, stimulating gastric acid secretion, and promoting inflammation.
prostate gland (pros'tāt) A walnut-shaped gland that surrounds the male urethra just below the urinary bladder and secretes an additive to seminal fluid during ejaculation.
proton A unit of positive charge in the nucleus of atoms.
protoplasm (pro'to-plazm) A general term that includes cytoplasm and nucleoplasm.
protraction (pro-trak'shun) The movement of a body part, such as the mandible, forward on a plane parallel with the ground; the opposite of *retraction.*
proximal (prok'si-mal) Closer to the midline of the body or origin of an appendage; the opposite of *distal.*
pseudohermaphrodite (soo''do-her-maf'ro-dīt) An individual with some of the physical characteristics of both sexes, but who lacks functioning gonads of both sexes; a true hermaphrodite has both testes and ovaries.
pseudopods (soo'do-pods) Footlike extensions of the cytoplasm that enable some cells (with amoeboid motion) to move across a substrate; pseudopods also are used to surround food particles in the process of phagocytosis.
ptyalin (ti'ah-lin) Also called *salivary amylase;* an enzyme in saliva that catalyzes the hydrolysis of starch into smaller molecules.
puberty (pu'ber-te) The period of development in which the reproductive organs become functional.
pulmonary (pul'mo-ner''e) Pertaining to the lungs.
pulmonary circulation The system of blood vessels from the right ventricle of the heart to the lungs, transporting deoxygenated blood and returning oxygenated blood from the lungs to the left atrium of the heart.
pulp cavity A cavity within the center of a tooth, containing blood vessels, nerves, and lymphatics.
pupil The opening through the iris that permits light to enter the posterior cavity of the eyeball and be refracted by the lens.
Purkinje fibers (pur-kin'je) Specialized cardiac muscle fibers that conduct electrical impulses from the AV bundle into the ventricular walls.
pyloric sphincter (pi-lor'ik sfingk'ter) A modification of the muscularis tunica between the stomach and the duodenum, which functions to regulate the food material leaving the stomach.
pyramid (pir'ah-mid) Any of several structures that have a pyramidal shape, including the renal pyramids in the kidney and the medullary pyramids on the ventral surface of the brain.
pyrogen (pi'ro-jen) A fever-producing substance.

q

QRS complex The part of an electrocardiogram that is produced by depolarization of the ventricles.

r

ramus (ra'mus) A branch of a bone, artery, or nerve.
raphe (ra'fe) A ridge or a seamlike structure.
receptor (re-sep'tor) A sense organ or a specialized distal end of a sensory neuron that receives stimuli from the environment.
rectouterine pouch (rek''to-u'ter-in) A pocket of parietal peritoneal membrane between the uterus and the rectum, forming the lowest point in the pelvic cavity; also called the *pouch of Douglas.*
rectum (rek'tum) The terminal portion of the gastrointestinal tract, from the sigmoid colon to the anus.
red marrow (mar'o) A tissue that forms blood cells, located in the medullary cavity of certain bones.
red nucleus (nu'kle-us) An aggregation of gray matter of a reddish color that is located in the upper portion of the midbrain and sends fibers to certain brain tracts.
reduced hemoglobin (he''mo-glo'bin) Hemoglobin with iron in the reduced ferrous state that is able to bond with

oxygen but is not combined with oxygen. Also called *deoxyhemoglobin.*

reducing agent An electron donor in a coupled oxidation-reduction reaction.

reflex (re'fleks) A rapid involuntary response to a stimulus.

reflex arc The basic conduction pathway through the nervous system, consisting of a sensory neuron, an interneuron, and a motor neuron.

regional anatomy The division of anatomy concerned with structural arrangement in specific areas of the body such as the head, neck, thorax, or abdomen.

REM sleep The stage of sleep in which dreaming occurs; associated with rapid eye movements (REM). REM sleep occurs three to four times each night and lasts from a few minutes to over an hour.

renal (re'nal) Pertaining to the kidney.

renal corpuscle (kor'pus'l) The portion of the nephron consisting of the glomerulus and a Bowman's capsule; also called the *Malpighian corpuscle.*

renal cortex The outer portion of the kidney, primarily vascular.

renal medulla (mĕ-dul'ah) The inner portion of the kidney, including the renal pyramids and renal columns.

renal pelvis The inner cavity of the kidney formed by the expanded ureter and into which the calyces open.

renal plasma clearance rate The milliliters of plasma that are cleared of a particular solute per minute by the excretion of that solute in the urine; if there is no reabsorption or secretion of that solute by the nephron tubules, this is equal to the glomerular filtration rate.

renal pyramid A triangular structure within the renal medulla, composed of the loops of Henle and the collecting ducts.

repolarization (re-po"lar-ĭ-za'shun) The reestablishment of the resting membrane potential after depolarization has occurred.

respiration (res"pĭ-ra'shun) The exchange of gases between the external environment and the cells of an organism.

respiratory acidosis (re-spi'rah-to"re as"ĭ-do'sis) A lowering of the blood pH below 7.35 due to accumulation of CO_2 as a result of hypoventilation.

respiratory alkalosis (al"kah-lo'sis) A rise in blood pH above 7.45 due to excessive elimination of blood CO_2 as a result of hyperventilation.

respiratory center The structure or portion of the brain stem that regulates the depth and rate of breathing.

respiratory distress syndrome Also called *hyaline membrane disease;* most frequently occurring in premature infants, this syndrome is caused by abnormally high alveolar surface tension as a result of a deficiency in lung surfactant.

respiratory membrane A thin, moistened membrane within the lungs, composed of an alveolar portion and a capillary portion, through which gaseous exchange occurs.

rete testis (re'te tes'tis) A network of ducts in the center of the testis, associated with the production of spermatozoa.

reticular formation (rĕ-tik'u-lar) A network of nervous tissue fibers in the brain stem that arouses the higher brain centers.

retina (ret'ĭ-nah) The inner layer of the eyeball that contains the photoreceptors.

retraction (re-trak'shun) The movement of a body part, such as the mandible, backward on a plane parallel with the ground; the opposite of *protraction.*

retroperitoneal (ret"ro-per"ĭ-to-ne'al) Positioned behind the parietal peritoneum.

rhodopsin (ro-dop'sin) Visual purple; a pigment in rod cells that undergoes a photochemical dissociation in response to light and in so doing, stimulates electrical activity in the photoreceptors.

rhythmicity area (rith-mis'ĭ-te) A portion of the respiratory control center that is located in the medulla and controls inspiratory and expiratory phases.

ribosome (ri'bo-sōm) A cytoplasmic organelle composed of protein and RNA and which is the site of protein synthesis.

rickets (rik'ets) A condition caused by a deficiency of vitamin D and associated with an interference of the normal ossification of bone.

right lymphatic duct (lim-fat'ik) A major vessel of the lymphatic system that drains lymph from the upper right portion of the body into the right subclavian vein.

rigor mortis The stiffening of a dead body, due to the depletion of ATP and the production of rigor complexes between actin and myosin in muscles.

RNA Ribonucleic acid; a nucleic acid consisting of the nitrogenous bases adenine, guanine, cytosine, and uracil; the sugar ribose; and phosphate groups. There are three types of RNA found in cytoplasm: messenger RNA (mRNA), transfer RNA (tRNA), and ribosomal RNA (rRNA).

rod A photoreceptor in the retina of the eye that is specialized for colorless, dim light vision.

root canal The hollow, tubular extension of the pulp cavity into the root of the tooth, containing vessels and nerves.

rotation (ro-ta'shun) The movement of a bone around its own longitudinal axis.

round window A round, membrane-covered opening between the middle and inner ear, directly below the oval window.

rugae (ru'je) The folds or ridges of the mucosa of an organ.

S

saccadic eye movements (sah-kad'ik) Very rapid eye movements that occur constantly and that change the focus on the retina from one point to another.

saccule (sak'ūl) A saclike cavity in the membranous labyrinth inside the vestibule of the inner ear, containing a vestibular organ for equilibrium.

sacral (sa'kral) Pertaining to the sacrum.

sacral plexus (plek'sus) A network of nerve fibers that arise from spinal nerves L4 through S3. Nerves arising from the sacral plexus merge with those from the lumbar plexus to form the lumbosacral plexus and supply the lower extremity.

saddle joint A diarthrotic joint in which the articular surfaces of both bones are concave in one plane and convex, or saddle shaped, in the other plane, such as in the distal carpal and proximal metacarpal joint of the thumb.

sagittal (saj'ĭ-tal) A vertical plane through the body that divides it into right and left portions.

salivary gland (sal'ĭ-ver-e) An accessory digestive gland that secretes saliva into the oral cavity.

saltatory conduction (sal'tah-to"re) The rapid passage of action potentials from one node of Ranvier to another in myelinated axons.

sarcolemma (sar"ko-lem'ah) The cell membrane of a muscle fiber.

sarcomere (sar'ko-mēr) The portion of a striated muscle fiber between the two adjacent Z lines that is considered the functional unit of a myofibril.

sarcoplasm (sar'ko-plazm) The cytoplasm within a muscle fiber.

sarcoplasmic reticulum (sar'ko-plaz'mik rĕ-tik'u-lum) The smooth or agranular endoplasmic reticulum of skeletal muscle cells; it surrounds each myofibril and serves to store Ca^{++} when the muscle is at rest.

scala tympani (ska'lah tim'pah-ne) The lower channel of a cochlea that is filled with perilymph.

scala vestibuli (ves-tib'u-le) The upper channel of the cochlea that is filled with perilymph.

Schwann cell (shwahn) A specialized neuroglia cell that surrounds an axon fiber of a peripheral nerve and forms the neurilemmal sheath.

sclera (skle'rah) The outer white layer of fibrous connective tissue that forms the protective covering of the eyeball.

scrotum (skro'tum) A pouch of skin that contains the testes and their accessory organs.

sebaceous gland (se-ba'shus) An exocrine gland of the skin that secretes sebum.

sebum (se'bum) An oily, waterproofing secretion of the sebaceous glands.

second messenger A molecule or ion whose concentration within a target cell is increased by the action of a regulator compound (e.g., a hormone or neurotransmitter) and which stimulates the metabolism of that target cell in a way characteristic of the actions of that regulator molecule—that is, in a way that mediates the intracellular effects of that regulatory compound.

secretin (se-kre'tin) A polypeptide hormone secreted by the small intestine in response to acidity of the intestinal lumen; along with cholecystokinin, secretin stimulates the secretion of pancreatic juice into the small intestine.

semen (se'men) The thick, whitish fluid secretion of the reproductive organs of the male, consisting of spermatozoa and additives from several other accessory organs.

semicircular canals Tubular channels within the inner ear that contain receptors for equilibrium.

semilunar valve (sem"e-lu'nar) Crescent or half-moon-shaped heart valves positioned at the entrances to the aorta and the pulmonary trunk.

seminal vesicles (sem'ĭ-nal ves'ĭ-k'lz) A pair of accessory male reproductive organs lying posterior and inferior to the urinary bladder, which secrete additives to spermatozoa into the ejaculatory ducts.

seminiferous tubule (se"mĭ-nif'er-us too'būl) Numerous small ducts in the testis where spermatozoa are produced.

semipermeable membrane (sem"e-per'me-ah-b'l) A membrane with pores of a size that permits the passage of solvent and some solute molecules but restricts the passage of other solute molecules.

senescence (sĕ-nes'ens) The process of aging.

sensory area A region of the cerebral cortex that receives and interprets sensory nerve impulses.

sensory neuron (nu'ron) A nerve cell that conducts an impulse from a receptor organ to the central nervous system.

septum (sep'tum) A membranous or fleshy wall dividing two cavities.

serosa (se-ro'sah, se-ro'zah) An outer epithelial membrane that covers the surface of a visceral organ.
serous membrane (se'rus) An epithelial and connective tissue membrane that lines body cavities and covers visceral organs within these cavities; also called *serosa*.
Sertoli cells (ser-to'le) Specialized cells within the testes that supply nutrients to developing spermatozoa; also called *nurse cells*.
serum (se'rum) Blood plasma with the clotting elements removed.
sesamoid bone (ses'ah-moid) A membranous bone formed in a tendon in response to joint stress (e.g., the patella).
sex chromosomes The X and Y chromosomes; the unequal pairs of chromosomes involved in sex determination (which is due to the presence or absence of a Y chromosome). Females lack a Y chromosome and normally have the genotype XX; males have a Y chromosome and normally have the genotype XY.
shock As it relates to the cardiovascular system, this term refers to a rapid, uncontrolled fall in blood pressure, which in some cases becomes irreversible and leads to death.
shoulder The region of the body where the humerus articulates with the scapula; also called the *omos*.
sickle-cell anemia A hereditary, autosomal recessive trait that occurs primarily in people of African ancestry, in which it evolved apparently as a protection (in the carrier state) against malaria. In the homozygous state, hemoglobin S is made instead of hemoglobin A; this leads to the characteristic sickling of red blood cells, hemolytic anemia, and organ damage.
sigmoid colon (sig'moid ko'lon) The S-shaped portion of the large intestine between the descending colon and the rectum.
sinoatrial node (sin"o-a'tre-al) A mass of specialized cardiac tissue in the wall of the right atrium that initiates the cardiac cycle; the SA node; also called the *pacemaker*.
sinus (si'nus) A cavity or hollow space within a body organ such as a bone.
sinusoid (si'nū-soid) A small, blood-filled space in certain organs such as the spleen or liver.
sleep apnea (ap-ne'ah) A temporary cessation of breathing during sleep, usually lasting for several seconds.
sliding filament theory The theory that the thick and thin filaments of a myofibril slide past each other, while maintaining their initial length, during muscle contraction.
small intestine The portion of the alimentary canal between the stomach and the cecum whose function is the absorption of food nutrients.
smooth muscle A specialized type of muscle tissue that is nonstriated, composed of fusiform, single-nucleated fibers, and contracts in an involuntary, rhythmic fashion with the walls of visceral organs.
sodium/potassium pump (so'de-um po-tas'e-um) An active transport carrier, with ATPase enzymatic activity, that acts to accumulate K^+ within cells and extrude Na^+ from cells, thus maintaining gradients for these ions across the cell membrane.
soft palate (pal'at) The fleshy, posterior portion of the roof of the mouth from the palatine bones to the uvula.
somatic (so-mat'ik) Pertaining to the nonvisceral parts of the body.
somatomedins (so"mah-to-me'dins) A group of small polypeptides that are believed to be produced in the liver in response to growth hormone stimulation and to mediate the actions of growth hormone on the skeleton and other tissues.
somatostatin (so"mah-to-stat'in) A polypeptide produced in the hypothalamus that acts to inhibit the secretion of growth hormone from the anterior pituitary; somatostatin is also produced in the islets of Langerhans of the pancreas, but its function there has not been established.
somatotropic hormone (so"mah-to-trōp'ik) Growth hormone; an anabolic hormone secreted by the anterior pituitary that stimulates skeletal growth and protein synthesis in many organs.
sounds of Korotkoff (ko-rot'kof) The sounds heard when blood pressure measurements are taken. These sounds are produced by the turbulent flow of blood through an artery that has been partially constricted by a pressure cuff.
spermatic cord (sper-mat'ik) The structure of the male reproductive system composed of the ductus deferens, spermatic vessels, nerves, cremasteric muscle, and connective tissue. The spermatic cord extends from a testis to the inguinal ring.
spermatogenesis (sper"mah-to-jen'ĕ-sis) The production of male sex gametes, or spermatozoa.
spermatozoon (sper"mah-to-zo'on) A mature male sperm cell, or gamete.
spermiogenesis (sper"me-o-jen'ĕ-sis) The maturational changes that transform spermatids into spermatozoa.
sphincter (sfingk'ter) A circular muscle that functions to constrict a body opening or the lumen of a tubular structure.
sphincter of Oddi (o'de) The muscular constriction at the opening of the common bile and pancreatic ducts.
sphygmomanometer (sfig"mo-mah-nom'ĕ-ter) A manometer (pressure transducer) used to measure the blood pressure.
spinal cord (spi'nal) The portion of the central nervous system that extends downward from the brain stem through the vertebral canal.
spinal nerve One of the thirty-one pairs of nerves that arise from the spinal cord.
spindle fibers Filaments that extend from the poles of a cell to its equator and attach to the chromosomes during the metaphase stage of cell division. Contraction of the spindle fibers pulls the chromosomes to opposite poles of the cell.
spinous process (spi'nus) A sharp projection of bone or a ridge of bone, such as on the scapula.
spironolactones (spi-ro"no-lak'tōns) Diuretic drugs that act as an aldosterone antagonist.
spleen (splēn) A large, blood-filled, glandular organ located in the upper left of the abdomen and attached by mesenteries to the stomach.
spongy bone (spun'je) A type of bone that contains many porous spaces; also called *cancellous bone*.
squamous (skwa'mus) Flat or scalelike.
stapes (sta'pēz) The innermost of the ossicles of the ear, which fits against the oval window of the inner ear; also called the *stirrup*.
steroid (ste'roid) A lipid, derived from cholesterol, that has three six-sided carbon rings and one five-sided carbon ring. These form the steroid hormones of the adrenal cortex and gonads.
stomach A pouchlike digestive organ located between the esophagus and the duodenum.
stratified (strat'i-fīd) Arranged in layers, or strata.
stratum corneum (stra'tum kor'ne-um) The outer, cornified layer of the epidermis of the skin.
stratum germinativum (stra'tum jer'mi-na"tiv-um) The deepest epidermal layer, where mitotic activity occurs.
striated muscle (stri'āt-ed) A specialized type of muscle tissue that is multinucleated, occurs in bundles, has crossbands of proteins, and contracts in either a voluntary or involuntary fashion.
stroke volume The amount of blood ejected from each ventricle at each heartbeat.
stroma (stro'mah) A connective tissue framework in an organ, gland, or other tissue.
subarachnoid space (sub"ah-rak'noid) The space within the meninges, between the arachnoid and pia maters, where cerebrospinal fluid flows.
subdural space (sub-du'ral) A space within the meninges between the dura and arachnoid of the brain and spinal cord.
sublingual gland (sub-ling'gwal) One of the three pairs of salivary glands; it is located below the tongue and secretes to the side of the lingual frenulum.
submandibular gland (sub-man-dib'u-lar) One of the three pairs of salivary glands; it is located below the mandible and secretes to the side of the lingual frenulum.
submucosa (sub"mu-ko'sah) A layer of supportive connective tissue that underlies a mucous membrane.
substrate (sub'strāt) In enzymatic reactions, the molecules that combine with the active sites of an enzyme and are converted to products by catalysis of the enzyme.
sulcus (sul'kus) A shallow impression or groove.
superficial (su"per-fish'al) Toward or near the surface.
superficial fascia (fash'e-ah) A binding layer of connective tissue between the dermis of the skin and the underlying muscle.
superior Toward the upper part of a structure or toward the head; also called *cephalic*.
supination (su"pi-na'shun) Rotation of the arm so that the palm is directed forward or anteriorly; the opposite of *pronation*.
suppressor T cells A subpopulation of T lymphocytes that acts to inhibit the production of antibodies against specific antigens by B lymphocytes.
surface anatomy The division of anatomy concerned with the structures that can be identified from the outside of the body.
surfactant (sur-fak'tant) A substance produced by the lungs that decreases the surface tension within the alveoli.
suspensory ligament (sus-pen'so-re) A portion of the peritoneum that extends laterally from the surface of the ovary to the wall of the pelvic cavity.
suture (su'chur) A type of immovable joint articulation found between bones of the skull.
sweat gland A skin gland that secretes a fluid substance for evaporative cooling.
sympathetic (sim"pah-thet'ik) Pertaining to that part of the autonomic nervous system concerned with processes involving the utilization of energy; also called the *thoracolumbar division*.
symphysis (sim'fĭ-sis) A type of articulation characterized by a fibrocartilaginous pad between the articulating bones, which provides slight movement.

synapse (sin′aps) A region where a nerve fiber comes into close or actual contact with another cell and across which nerve impulses are transmitted either directly or indirectly (via release of chemical neurotransmitters).

synarthrosis (sin″ar-thro′sis) An immovable joint, such as a synchondrosis or a suture.

synchondrosis (sin″kon-dro′sis) An immovable cartilaginous joint in which the articulating bones are separated by hyaline cartilage.

syndesmosis (sin″des-mo′sis) A type of slightly movable, fibrous joint in which two bones are united by an interosseous ligament.

synergist (sin′er-jist) A muscle that assists the action of the prime mover.

synergistic (sin″er-jis′tik) Pertaining to regulatory processes or molecules (such as hormones) that have complementary or additive effects.

synovial cavity (si-no′ve-al) A space between the two bones of a diarthrotic joint, filled with synovial fluid.

synovial joint A freely movable joint in which there is a synovial cavity between the articulating bones; also called a *diarthrotic joint.*

synovial membrane The inner membrane of a synovial capsule, which secretes synovial fluid into the joint cavity.

system A group of body organs that function together.

systemic (sis-tem′ik) Relating to the entire organism rather than individual parts.

systemic anatomy The division of anatomy concerned with the structure and function of the various systems.

systemic circulation The portion of the circulatory system concerned with blood flow from the left ventricle of the heart to the entire body and back to the heart via the right atrium; as opposed to the pulmonary system, which involves the lungs.

systole (sis′to-le) The muscular contraction of a heart chamber during the cardiac cycle.

systolic pressure (sis′tol′ik) Arterial blood pressure during the ventricular systolic phase of the cardiac cycle.

t

tachycardia (tak″e-kar′de-ah) Excessively rapid heart rate, usually applied to rates in excess of 100 beats per minute. In contrast to an excessively slow heart rate (below 60 beats per minute), which is termed bradycardia.

tactile (tak′til) Pertaining to the sense of touch.

taeniae coli (te′ne-e ko′li) The three longitudinal bands of muscle in the wall of the large intestine.

target organ The specific body organ that a particular hormone affects.

tarsus (tahr′sus) The seven bones that form the ankle.

taste bud An organ containing the chemoreceptors associated with the sense of taste.

T cell A type of lymphocyte that provides cell-mediated immunity, in contrast to B lymphocytes, which provide humoral immunity through the secretion of antibodies. There are three subpopulations of T cells: cytotoxic, helper, and suppressor.

tectorial membrane (tek-to′re-al) A gelatinous membrane positioned over the hair cells of the organ of Corti in the cochlea.

tendon (ten′dun) A band of dense fibrous connective tissue that attaches muscle to bone.

tendon sheath A covering of synovial membrane surrounding certain tendons.

tentorium cerebelli (ten-to′re-um ser″ĕ-bel′ē) An extension of dura mater that forms a partition between the cerebral hemispheres and the cerebellum and covers the cerebellum.

testis (tes′tis) The primary reproductive organ of a male, which produces spermatozoa and male sex hormones.

testosterone (tes-tos′tĕ-rōn) The major androgenic steroid secreted by the Leydig cells of the testes after puberty.

tetanus (tet′a-nus) A term used to mean either a smooth contraction of a muscle (as opposed to muscle twitching), or a state of maintained contracture of high tension.

thalamus (thal′ah-mus) An oval mass of gray matter within the diencephalon, which serves as a sensory relay area.

thalassemia (thal″ah-se′me-ah) A group of hemolytic anemias caused by the hereditary inability to produce either the alpha or beta chain of hemoglobin. It is found primarily among Mediterranean people.

thigh The proximal portion of the lower extremity between the hip and the knee, containing the femur bone.

third ventricle (ven′tri-k′l) A narrow cavity between the right and left halves of the thalamus and between the lateral ventricles, containing cerebrospinal fluid.

thoracic (tho-ras′ik) Pertaining to the chest region.

thoracic duct The major lymphatic vessel of the body, which drains lymph from the entire body except the upper right quadrant and returns it to the left subclavian vein.

thorax (tho′raks) The chest.

threshold The minimum stimulus that just produces a response.

thrombocyte (throm′bo-sīt) A blood platelet formed from a fragmented megakaryocyte.

thrombus A blood clot produced by the formation of fibrin threads around a platelet plug.

thymus gland (thi′mus) A bilobed lymphoid organ positioned in the upper mediastinum, posterior to the sternum and between the lungs.

thyroid cartilage (thi′roid kar′tĭ-lij) The largest cartilage in the larynx, which supports and protects the vocal cords; also called the *Adam's apple.*

thyroxine (thi-rok′sin) Also called tetraiodothyronine, or T_4. The major hormone secreted by the thyroid gland, which regulates the basal metabolic rate and stimulates protein synthesis in many organs; a deficiency of this hormone in early childhood produces cretinism.

tinnitus (ti-ni′tus) A ringing sound or other noise that is heard but is not related to external sounds.

tissue An aggregation of similar cells and their binding intercellular substance, joined to perform a specific function.

tongue A protrusible muscular organ on the floor of the oral cavity.

tonsil (ton′sil) A node of lymphoid tissue located in the mucous membrane of the pharynx.

toxin (tok′sin) A poison.

toxoid (tok′soid) A modified bacterial endotoxin that has lost toxicity but retains its ability to act as an antigen and stimulate antibody production.

trabeculae (trah-bek′u-le) A supporting framework of fibers crossing the substance of a structure, as in the lamellae of spongy bone.

trachea (tra′ke-ah) The airway leading from the larynx to the bronchi, composed of cartilaginous rings and a ciliated mucosal lining of the lumen; also called the *windpipe.*

tract A bundle of nerve fibers within the central nervous system.

transamination (trans″am-i-na′shun) The transfer of an amino group from an amino acid to an alpha-keto acid, forming a new keto acid and a new amino acid, without the appearance of free ammonia.

transection (tran-sek′shun) A cross-sectional cut.

transpulmonary pressure (trans″pul′mo-ner″e) The pressure difference across the wall of the lung, equal to the difference between intrapulmonary pressure and intrapleural pressure.

transverse colon (trans-vers′ ko′lon) A portion of the large intestine that extends from right to left across the abdomen between the hepatic and splenic flexures.

transverse fissure (fish′ūr) The prominent cleft that horizontally separates the cerebrum from the cerebellum.

tricuspid valve (tri-kus′pid) The heart valve located between the right atrium and the right ventricle.

trigone (tri′gōn) A triangular area in the urinary bladder between the openings of the ureters and the urethra.

triiodothyronine (tri″i-o″do-thi′ro-nēn) Abbreviated T_3; a hormone secreted in small amounts by the thyroid; the active hormone in target cells, formed from thyroxine.

trochanter (tro-kan′ter) A broad, prominent process on a bone, specifically on the femur.

trochlea (trok′le-ah) A pulley-shaped structure.

tropomyosin (tro″po-mi′o-sin) A filamentous protein that attaches to actin in the thin filaments and that acts, together with another protein called troponin, to inhibit and regulate the attachment of myosin cross-bridges to actin.

true vocal cords Folds of the mucous membrane in the larynx that produce sound as they are pulled taut and vibrated.

trunk The thorax and abdomen together.

trypsin (trip′sin) A protein-digesting enzyme in pancreatic juice that is released into the small intestine.

tubercle (tu′ber-k′l) A small, elevated process on a bone.

tuberosity (tu″bĕ-ros′ĭ-te) An elevation or protuberance on a bone.

tunica albuginea (tu′ni-kah al″bu-jin′e-ah) A tough, fibrous tissue surrounding the testis.

tympanic membrane (tim-pan′ik) The membranous eardrum positioned between the external and middle ear.

u

umbilical cord (um-bil′ĭ-kal) A cordlike structure containing the umbilical arteries and vein and connecting the fetus with the placenta.

umbilicus (um-bil′ĭ-kus) The site where the umbilical cord was attached to the fetus; commonly called the *navel.*

unipolar neuron (u″ni-po′lar nu′ron) A nerve cell that has a single nerve fiber extending from its cell body.

universal donor A person with blood type O, who is able to donate blood to people with other blood types in emergency blood transfusions.

universal recipient A person with blood type AB, who can receive blood of any type in emergency transfusions.

upper extremity The appendage attached to the pectoral girdle, consisting of the brachium, forearm, and hand.

urea (u-re'ah) The chief nitrogenous waste product of protein catabolism in the urine, formed in the liver from amino acids.

uremia (u-re'me-ah) The retention of urea and other products of protein catabolism due to inadequate kidney function.

ureter (u-re'ter) A tube that transports urine from the kidney to the urinary bladder.

urethra (u-re'thrah) A tube that transports urine from the urinary bladder to the outside of the body.

urinary bladder (u'ri-ner''e) A distensible sac that stores urine and is situated in the pelvic cavity posterior to the symphysis pubis.

urobilinogen (u''ro-bi-lin'o-jen) A compound formed from bilirubin in the intestine; some is excreted in the feces, and some is absorbed and enters the enterohepatic circulation, where it may be excreted either in the bile or in the urine.

urogenital triangle (u''ro-jen'i-tal) The region of the pelvic floor containing the external genitalia.

uterine tube (u'ter-in) A tube that leads from the ovary to the uterus; also called the *oviduct* or *fallopian tube.*

uterus (u''ter-us) A hollow, muscular organ in which a fetus develops. It is located within the female pelvis between the urinary bladder and the rectum; commonly called the *womb.*

utricle (u'tre-k'l) An enlarged portion of the membranous labyrinth, located within the vestibule of the inner ear.

uvula (u'vu-lah) A fleshy, pendulous portion of the soft palate that blocks the nasopharynx during swallowing.

V

vacuole (vak'u-ōl) Small spaces or cavities within the cytoplasm of a cell.

vagina (vah-ji'nah) A tubular organ that leads from the uterus to the vestibule of the female reproductive tract and receives the male penis during intercourse.

vasa vasorum (va'sah va-so'rum) Tiny blood vessels that form a network through the tunica of larger blood vessels to supply them with nutrients.

vasectomy (vah-sek'to-me) Surgical removal of a portion of the vas (ductus) deferens to induce infertility.

vasoconstriction (vas''o-kon-strik'shun) Narrowing of the lumen of blood vessels due to contraction of the smooth muscles in their walls.

vasodilation (vas''o-di-la'shun) Widening of the lumen of blood vessels due to relaxation of the smooth muscles in their walls.

vein A blood vessel that conveys blood toward the heart.

vena cava (ve'nah ka'vah) One of two large vessels that return deoxygenated blood to the right atrium of the heart.

ventilation (ven''ti-la'shun) Breathing; the process of moving air into and out of the lungs.

ventral (ven'tral) Toward the front or belly surface; the opposite of *dorsal.*

ventricle (ven'tri-k'l) A cavity within an organ; especially those in the brain that contain cerebrospinal fluid and those in the heart that contain blood to be pumped from the heart.

venule (ven'ūl) A small vessel that carries venous blood from capillaries to a vein.

vermiform appendix (ver'mi-form ah-pen'diks) A short, blind pouch that attaches to the cecum; commonly called the *appendix.*

vermis (ver'mis) The coiled, middle lobular structure that separates the two cerebellar hemispheres.

vertebral canal (ver'te-bral) The tubelike cavity that extends through the vertebral column and contains the spinal cord; also called the *spinal canal.*

vertigo (ver'ti-go, ver-ti'go) A feeling of movement or loss of equilibrium.

vestibule (ves'ti-būl) A space or cavity at the entrance to a canal, especially that of the nose, inner ear, and vagina.

villus (vil'lus) A minute projection that extends outward into the lumen from the mucosal layer of the small intestine.

virulent (vir'u-lent) Pathogenic, or able to cause disease.

viscera (vis'er-ah) The organs within the abdominal or thoracic cavities.

visceral (vis'er-al) Pertaining to the membranous covering of the viscera.

visceral peritoneum (per''i-to-ne'um) A serous membrane that covers the surfaces of abdominal viscera.

visceral pleura (ploor'ah) A serous membrane that covers the surfaces of the lungs.

visceroceptor (vis''er-o-sep'tor) A sensory receptor that is located within body organs and responds to information concerning the internal environment.

vitreous humor (vit're-us hu'mor) The transparent gel that occupies the space between the lens and retina of the eyeball.

Volkmann's canal (fōlk'mahnz) Minute ducts through compact bone by which blood vessels and nerves penetrate.

vulva (vul'vah) The external genitalia of the female that surround the opening of the vagina; also called the *pudendum.*

W

white matter Bundles of myelinated axons located in the central nervous system.

wormian bone (wer'me-an) A small bone positioned within a suture of certain cranial bones; also called a *sutural bone.*

Y

yellow marrow (mar'o) Specialized lipid storage tissue within bone cavities.

Z

zygote (zi'gōt) A fertilized egg cell formed by the union of a sperm and an egg.

zymogens (zi'mo-jens) Inactive enzymes that become active when part of their structure is removed by another enzyme or by some other means.

Credits

Illustrations

Samuel Collins

Figures: 7.6, 9.11, 14.26, 15.20, 16.20, 17.1, 18.33

Chris Creek

Figures: 1.21, 6.7, 6.9, 6.14, 6.20, 12.21, 14.35, 20.24, 20.54, 23.58, 24.1, 24.8, 24.9, 24.27, 29.25

Rob Gordon

Figures: 3.1, 3.10, 5.14, 20.47

Rob Gordon/Tom Waldrop

Figures: 16.1, 18.51, 28.3

Stephen Moon

Figures: 19.22, 20.22, 20.28

Michael Schenk

Figures: 9.41, 15.2, 15.18, 16.3, 23.11, 25.28, 27.20

Nancy Marshburn/Tom Waldrop

Figures: 8.16, 8.26, 11.27, 11.31

Robert Margulies/Tom Waldrop

Figures: 12.1, 12.2, 20.27, 20.30, 20.31, 20.33, 20.34, 20.35, 20.37, 20.40, 20.41, 20.42, 20.44, 25.9

Rolin Graphics

Figures: 2.1, 2.2, 2.3, 2.8, 2.13, 2.29, 3.15, 3.20, 4.2, 4.5, 7.1, 7.2, 7.7, 18.65, 19.2, 19.3, 19.7, 20.11, 22.21, 24.18, 26.3, 26.18, 27.13

Chapter 1

Figures 1.9, 1.13, 1.14, 1.15 bottom, 1.16 bottom, 1.17, 1.18, and 1.19: From Kent M. Van De Graaff, *Human Anatomy,* 2d ed. Copyright © 1988 Wm. C. Brown Publishers, Dubuque, Iowa. All Rights Reserved. Reprinted by permission.
Figure 1.20: From John W. Hole, Jr., *Human Anatomy and Physiology,* 4th ed. Copyright © 1987 Wm. C. Brown Publishers, Dubuque, Iowa. All Rights Reserved. Reprinted by permission.

Chapter 2

Figures 2.1, 2.2, 2.3, 2.4, 2.6, 2.7, 2.8, 2.9, 2.10, 2.11, 2.12, 2.13, 2.14, 2.15, 2.16, 2.17, 2.18, 2.19, 2.20, 2.21, 2.22, 2.23, 2.24, 2.25b, 2.27, 2.28, 2.30: From Stuart Ira Fox, *Human Physiology,* 2d ed. Copyright © 1987 Wm. C. Brown Publishers, Dubuque, Iowa. All Rights Reserved. Reprinted by permission.
Figure 2.5: From John W. Hole, Jr., *Human Anatomy and Physiology,* 4th ed. Copyright © 1987 Wm. C. Brown Publishers, Dubuque, Iowa. All Rights Reserved. Reprinted by permission.
Figure 2.25a: From Dr. P. Sheeler, *Cell Biology: Structure, Biochemistry and Function.* Copyright © 1983 John Wiley & Sons, Inc., New York, NY. All Rights Reserved. Reprinted by permission.

Chapter 3

Figures 3.2, 3.13 bottom, 3.16, 3.17, 3.18, 3.21: From Stuart Ira Fox, *Human Physiology,* 2d ed. Copyright © 1987 Wm. C. Brown Publishers, Dubuque, Iowa. All Rights Reserved. Reprinted by permission.
Figure 3.9b: From Leland G. Johnson, *Biology,* 2d ed. Copyright © 1987 Wm. C. Brown Publishers, Dubuque, Iowa. All Rights Reserved. Reprinted by permission.
Figure 3.22b: From John W. Hole, Jr., *Human Anatomy and Physiology,* 4th ed. Copyright © 1987 Wm. C. Brown Publishers, Dubuque, Iowa. All Rights Reserved. Reprinted by permission.

Chapter 4

Figures 4.1, 4.3, 4.4, 4.6, 4.7, 4.8, 4.9, 4.10, 4.11, 4.12, 4.13, 4.14, 4.15, 4.16, 4.17, 4.18, 4.19, 4.20, 4.21, 4.22, 4.23, 4.24, 4.26, 4.27, 4.28, 4.29, 4.30, 4.31, 4.32, 4.33, 4.34, 4.35, and 4.36: From Stuart Ira Fox, *Human Physiology,* 2d ed. Copyright © 1987 Wm. C. Brown Publishers, Dubuque, Iowa. All Rights Reserved. Reprinted by permission.

Chapter 5

Figures 5.1, 5.2, 5.3, 5.5, 5.6, 5.7, 5.8, 5.9, 5.10, 5.12, 5.13, 5.15, 5.16, 5.17, 5.18, 5.19, 5.20, and 5.21: From Stuart Ira Fox, *Human Physiology,* 2d ed. Copyright © 1987 Wm. C. Brown Publishers, Dubuque, Iowa. All Rights Reserved. Reprinted by permission.

Chapter 6

Figures 6.3, 6.5, 6.6a, 6.7c, 6.8c, 6.9c, 6.10c, 6.11b, 6.12a&c, 6.13b, 6.15, 6.16, 6.17b, 6.18c, 6.19b, 6.20c, 6.21c, 6.22c, 6.23c, 6.24a, 6.25c, 6.28 line art, 6.29b, and 6.30: From Kent M. Van De Graaff, *Human Anatomy,* 2d ed. Copyright © 1988 Wm. C. Brown Publishers, Dubuque, Iowa. All Rights Reserved. Reprinted by permission.
Figures 6.8a, 6.10a, 6.19a, 6.22a, 6.25a, and 6.27: From John W. Hole, Jr., *Human Anatomy and Physiology,* 4th ed. Copyright © 1987 Wm. C. Brown Publishers, Dubuque, Iowa. All Rights Reserved. Reprinted by permission.
Figure 6.23a: From John W. Hole, Jr., *Human Anatomy and Physiology,* 3d ed. Copyright © 1984 Wm. C. Brown Publishers, Dubuque, Iowa. All Rights Reserved. Reprinted by permission.

Chapter 7

Figures 7.1 bottom, 7.2 bottom, 7.3, 7.4, and 7.8: From Stuart Ira Fox, *Human Physiology,* 2d ed. Copyright © 1987 Wm. C. Brown Publishers, Dubuque, Iowa. All Rights Reserved. Reprinted by permission.
Figure 7.5: From John W. Hole, Jr., *Human Anatomy and Physiology,* 4th ed. Copyright © 1987 Wm. C. Brown Publishers, Dubuque, Iowa. All Rights Reserved. Reprinted by permission.

Reference Figures

Figures 1, 2, 3, 4, 5, 6, and 7: From John W. Hole, Jr., *Human Anatomy and Physiology,* 4th ed. Copyright © 1987 Wm. C. Brown Publishers, Dubuque, Iowa. All Rights Reserved. Reprinted by permission.

Chapter 8

Figures 8.1, 8.2, 8.3, 8.4, 8.17a, 8.20, 8.21, and 8.23 line art: From Kent M. Van De Graaff, *Human Anatomy,* 2d ed. Copyright © 1988 Wm. C. Brown Publishers, Dubuque, Iowa. All Rights Reserved. Reprinted by permission.
Figures 8.6, 8.7a, and 8.18: From John W. Hole, Jr., *Human Anatomy and Physiology,* 4th ed. Copyright © 1987 Wm. C. Brown Publishers, Dubuque, Iowa. All Rights Reserved. Reprinted by permission.

Chapter 9

Figures 9.1, 9.2a, 9.3, 9.4, 9.6, 9.9, 9.14, 9.15, 9.16, 9.17, 9.19, 9.22, 9.23, 9.24, 9.26, 9.27, 9.28, 9.29, 9.30, 9.32, 9.35b&c, 9.36, 9.37b&c, 9.38, 9.39, and 9.40: From Kent M. Van De Graaff, *Human Anatomy,* 2d ed. Copyright © 1988 Wm. C. Brown Publishers, Dubuque, Iowa. All Rights Reserved. Reprinted by permission.
Figures 9.7, 9.8, 9.18, 9.20, 9.25, and 9.31: From John W. Hole, *Human Anatomy and Physiology,* 4th ed. Copyright © 1987 Wm. C. Brown Publishers, Dubuque, Iowa. All Rights Reserved. Reprinted by permission.

Chapter 10

Figures 10.1, 10.2, 10.3, 10.5, 10.6, 10.7, 10.8a, 10.9, 10.11, 10.13, 10.14, 10.15, 10.17, 10.18c&d, and 10.19: From Kent M. Van De Graaff, *Human Anatomy,* 2d ed. Copyright © 1988 Wm. C. Brown Publishers, Dubuque, Iowa. All Rights Reserved. Reprinted by permission.
Figures 10.22 and 10.23a-d: From John W. Hole, Jr., *Human Anatomy and Physiology,* 4th ed. Copyright © 1987 Wm. C. Brown Publishers, Dubuque, Iowa. All Rights Reserved. Reprinted by permission.

Chapter 11

Figures 11.2, 11.4, 11.5, 11.6, 11.7, 11.8a, 11.9, 11.10, 11.11, 11.12, 11.13, 11.14, 11.15, 11.16, 11.17, 11.18, 11.19, 11.23, 11.24, 11.26, 11.29, 11.30, and 11.32: From Kent M. Van De Graaff, *Human Anatomy,* 2d ed. Copyright © 1988 Wm. C. Brown Publishers, Dubuque, Iowa. All Rights Reserved. Reprinted by permission.

Figure 11.8b: From John W. Hole, Jr., *Human Anatomy and Physiology,* 4th ed. Copyright © 1987 Wm. C. Brown Publishers, Dubuque, Iowa. All Rights Reserved. Reprinted by permission.

Chapter 12

Figures 12.6, 12.12, and 12.19a: From John W. Hole, Jr., *Human Anatomy and Physiology,* 4th ed. Copyright © 1987 Wm. C. Brown Publishers, Dubuque, Iowa. All Rights Reserved. Reprinted by permission.
Figures 12.7, 12.8, 12.14 bottom, 12.30, 12.32, 12.33b, 12.34, 12.36, 12.37, 12.38, 12.40, 12.41, 12.42, 12.43, 12.44, 12.45 left, 12.47, 12.48, 12.49, 12.50, 12.51, 12.52, 12.53 left, 12.54 left, 12.56 left, 12.57 left, 12.58 left, 12.59 left, 12.60, 12.61, and 12.62: From Kent M. Van De Graaff, *Human Anatomy,* 2d ed. Copyright © 1988 Wm. C. Brown Publishers, Dubuque, Iowa. All Rights Reserved. Reprinted by permission.
Figure 12.10a: From John W. Hole, Jr., *Human Anatomy and Physiology,* 3d ed. Copyright © 1984 Wm. C. Brown Publishers, Dubuque, Iowa. All Rights Reserved. Reprinted by permission.
Figures 12.11, 12.15, 12.17, 12.20, 12.25, 12.26, and 12.28: From Stuart Ira Fox, *Human Physiology,* 2d ed. Copyright © 1987 Wm. C. Brown Publishers, Dubuque, Iowa. All Rights Reserved. Reprinted by permission.
Figure 12.16: Reprinted with permission. Eugene Morkin. "Contractile Proteins of the Heart." *Hospital Practice* Volume 18, Issue 6. Pages 97–112. Illustration by Bunji Tagawa.
Figure 12.22: From Stuart Ira Fox, *A Laboratory Guide to Human Physiology: Concepts and Clinical Applications,* 4th ed. Copyright © 1987 Wm. C. Brown Publishers, Dubuque, Iowa. All Rights Reserved. Reprinted by permission.
Figure 12.33a: From *The Color Atlas of Human Anatomy,* Dr. Richard T. Jolly, translator. Copyright © 1980 Fabbri Representative Offices, Inc., New York, NY. Reprinted by permission.
Figure 12.39: From John W. Hole, Jr., *Human Anatomy and Physiology,* 2d ed. Copyright © 1981 Wm. C. Brown Publishers, Dubuque, Iowa. All Rights Reserved. Reprinted by permission.

Chapter 13

Figures 13.12, 13.17, and 13.24: From Kent M. Van De Graaff, *Human Anatomy,* 2d ed. Copyright © 1988 Wm. C. Brown Publishers, Dubuque, Iowa. All Rights Reserved. Reprinted by permission.

Chapter 14

Figures 14.2, 14.3, 14.4, 14.5, 14.6, 14.7, 14.10, 14.11, 14.12, and 14.13: From Kent M. Van De Graaff, *Human Anatomy,* 2d ed. Copyright © 1988 Wm. C. Brown Publishers, Dubuque, Iowa. All Rights Reserved. Reprinted by permission.
Figure 14.9: From John W. Hole, Jr., *Human Anatomy and Physiology,* 3d ed. Copyright © 1984 Wm. C. Brown Publishers, Dubuque, Iowa. All Rights Reserved. Reprinted by permission.
Figures 14.14 and 14.17: From John W. Hole, Jr., *Human Anatomy and Physiology,* 4th ed. Copyright © 1987 Wm. C. Brown Publishers, Dubuque, Iowa. All Rights Reserved. Reprinted by permission.
Figures 14.15, 14.19, 14.20, 14.21, 14.24, 14.25, 14.29, 14.30, 14.31, 14.32, 14.33, 14.34, 14.36, 14.37, and 14.38: From Stuart Ira Fox, *Human Anatomy,* 2d ed. Copyright © 1987 Wm. C. Brown Publishers, Dubuque, Iowa. All Rights Reserved. Reprinted by permission.
Figure 14.23: Reproduced by permission from Howard C. Kutchai, "Generation and Conduction of Action Potentials," Robert M. Berne and Matthew N. Levy, *Physiology,* 2d ed., St. Louis, 1988, The C. V. Mosby Co.
Figure 14.27c: From Leland G. Johnson, *Biology,* 2d ed. Copyright © 1987 Wm. C. Brown Publishers, Dubuque, Iowa. All Rights Reserved. Reprinted by permission.

Chapter 15

Figures 15.4a, 15.10a, and 15.21: From Kent M. Van De Graaff, *Human Anatomy Laboratory Textbook,* 2d ed. Copyright © 1984 Wm. C. Brown Publishers, Dubuque, Iowa. All Rights Reserved. Reprinted by permission.
Figures 15.4b, 15.8b, 15.9, 15.14, 15.15, 15.16, 15.17, 15.25, and 15.26a: From Kent M. Van De Graaff, *Human Anatomy,* 2d ed. Copyright © 1988 Wm. C. Brown Publishers, Dubuque, Iowa. All Rights Reserved. Reprinted by permission.
Figures 15.6, 15.12, 15.22, 15.24, 15.32, and 15.33: From John W. Hole, Jr., *Human Anatomy and Physiology,* 4th ed. Copyright © 1987 Wm. C. Brown Publishers, Dubuque, Iowa. All Rights Reserved. Reprinted by permission.
Figure 15.7: From T. L. Peele, *Neuroanatomical Basis for Clinical Neurology,* 2d ed. Copyright © 1961 McGraw-Hill Book Company, New York, NY. Reprinted by permission.
Figures 15.11 and 15.31: From Stuart Ira Fox, *Human Physiology,* 2d ed. Copyright © 1987 Wm. C. Brown Publishers, Dubuque, Iowa. All Rights Reserved. Reprinted by permission.
Figure 15.19: From John W. Hole, Jr., *Human Anatomy and Physiology,* 2d ed. Copyright © 1981 Wm. C. Brown Publishers, Dubuque, Iowa. All Rights Reserved. Reprinted by permission.
Figures 15.27, 15.28, and 15.29: From John W. Hole, Jr., *Human Anatomy and Physiology,* 3d ed. Copyright © 1984 Wm. C. Brown Publishers, Dubuque, Iowa. All Rights Reserved. Reprinted by permission.

Chapter 16

Figures 16.7, 16.8, 16.11, 16.14, 16.15, 16.16, 16.17, and 16.18: From Kent M. Van De Graaff, *Human Anatomy,* 2d ed. Copyright © 1988 Wm. C. Brown Publishers, Dubuque, Iowa. All Rights Reserved. Reprinted by permission.
Figures 16.9, 16.21, 16.22, and 16.23: From John W. Hole, Jr., *Human Anatomy and Physiology,* 4th ed. Copyright © 1987 Wm. C. Brown Publishers, Dubuque, Iowa. All Rights Reserved. Reprinted by permission.
Figure 16.13: From John W. Hole, Jr., *Human Anatomy and Physiology,* 3d ed. Copyright © 1984 Wm. C. Brown Publishers, Dubuque, Iowa. All Rights Reserved. Reprinted by permission.

Chapter 17

Figure 17.4: From Kent M. Van De Graaff, *Human Anatomy,* 2d ed. Copyright © 1988 Wm. C. Brown Publishers, Dubuque, Iowa. All Rights Reserved. Reprinted by permission.
Figure 17.7: From Kent M. Van De Graaff, *Human Anatomy.* Copyright © 1984 Wm. C. Brown Publishers, Dubuque, Iowa. All Rights Reserved. Reprinted by permission.
Figures 17.8, 17.9, and 17.10: From Stuart Ira Fox, *Human Physiology,* 2d ed. Copyright © 1987 Wm. C. Brown Publishers, Dubuque, Iowa. All Rights Reserved. Reprinted by permission.

Chapter 18

Figures 18.1, 18.3, 18.4, 18.8, 18.9, 18.10, 18.37, 18.39, 18.40, 18.53, 18.56, 18.57, 18.58, 18.59, 18.60, and 18.62: From Stuart Ira Fox, *Human Physiology,* 2d ed. Copyright © 1987 Wm. C. Brown Publishers, Dubuque, Iowa. All Rights Reserved. Reprinted by permission.
Figure 18.2: From Stuart Ira Fox, *Human Physiology,*. Copyright © 1984 Wm. C. Brown Publishers, Dubuque, Iowa. All Rights Reserved. Reprinted by permission.
Figures 18.5, 18.6, 18.13, 18.16, 18.17, 18.20, 18.23, 18.29a, 18.31, 18.41, 18.44, 18.45, 18.46, 18.47, and 18.50: From Kent M. Van De Graaff, *Human Anatomy,* 2d ed. Copyright © 1988 Wm. C. Brown Publishers, Dubuque, Iowa. All Rights Reserved. Reprinted by permission.
Figure 18.12: From T. L. Peele, *Neuroanatomical Basis for Clinical Neurology,* 2d ed. Copyright © 1961 McGraw-Hill Book Company, New York, NY. Reprinted by permission.
Figures 18.14a&b, 18.15, 18.18, 18.21, 18.26, 18.30 left, 18.36, 18.48, 18.52, 18.63, and 18.64a: From John W. Hole, Jr., *Human Anatomy and Physiology,* 4th ed. Copyright © 1987 Wm. C. Brown Publishers, Dubuque, Iowa. All Rights Reserved. Reprinted by permission.
Figure 18.30 right: From John W. Hole, Jr., *Human Anatomy and Physiology,* 3d ed. Copyright © 1984 Wm. C. Brown Publishers, Dubuque, Iowa. All Rights Reserved. Reprinted by permission.
Figure 18.34: From Stuart Ira Fox, *A Laboratory Guide to Human Physiology: Concepts and Clinical Applications,* 4th ed. Copyright © 1987 Wm. C. Brown Publishers, Dubuque, Iowa. All Rights Reserved. Reprinted by permission.
Figure 18.54a: From J. R. McClintic, *Physiology of the Human Body,* 2d ed. Copyright © 1978 John W. Wiley & Sons, Inc., New York, NY.
Figure 18.64b: From John W. Hole, Jr., *Human Anatomy and Physiology,* 2d ed. Copyright © 1981 Wm. C. Brown Publishers, Dubuque, Iowa. All Rights Reserved. Reprinted by permission.

Chapter 19

Figures 19.1a, 19.9, 19.10, and 19.20: From Kent M. Van De Graaff, *Human Anatomy,* 2d ed. Copyright © 1988 Wm. C. Brown Publishers, Dubuque, Iowa. All Rights Reserved. Reprinted by permission.
Figures 19.4, 19.5, 19.6, 19.13, 19.14, 19.16, 19.17, and 19.21: From Stuart Ira Fox, *Human Physiology,* 2d ed. Copyright © 1987 Wm. C. Brown Publishers, Dubuque, Iowa. All Rights Reserved. Reprinted by permission.
Figure 19.8: "Reprinted by permission from the September 1984 issue of *Diagnostic Medicine.*"
Figures 19.15, 19.23, and 19.26: From John W. Hole, Jr., *Human Anatomy and Physiology,* 4th ed. Copyright © 1987 Wm. C. Brown Publishers, Dubuque, Iowa. All Rights Reserved. Reprinted by permission.
Figure 19.24: From John W. Hole, Jr., *Essentials of Human Anatomy and Physiology.* Copyright © 1983 Wm. C. Brown Publishers, Dubuque, Iowa. All Rights Reserved. Reprinted by permission.

Chapter 20

Figures 20.1, 20.7, 20.9b, 20.16, 20.29a, 20.32, 20.38, 20.39, 20.46, and 20.48: From John W. Hole, Jr., *Human Anatomy and Physiology,* 4th ed. Copyright © 1987 Wm. C. Brown Publishers, Dubuque, Iowa. All Rights Reserved. Reprinted by permission.
Figures 20.2, 20.3, 20.6, 20.8, 20.10b&c, 20.13, 20.21, 20.36, 20.42a, 20.43, and 20.58: From Kent M. Van De Graaff, *Human Anatomy,* 2d ed. Copyright © 1988 Wm. C. Brown Publishers, Dubuque, Iowa. All Rights Reserved. Reprinted by permission.
Figures 20.15, 20.17, 20.20, 20.25, 20.45, 20.51, 20.56, 20.57, and 20.60: From Stuart Ira Fox, *Human Physiology,* 2d ed. Copyright © 1987 Wm. C. Brown Publishers, Dubuque, Iowa. All Rights Reserved. Reprinted by permission.
Figure 20.19: By permission of *Emergency Medicine* Magazine.
Figure 20.52: Adapted from: Marchand, A., Case of the Month, Circulating anticoagulants: Chasing the diagnosis. *Diagnostic Medicine,* June 1983, p 14.

Chapter 21

Figures 21.1, 21.2, 21.3, 21.5, 21.7, 21.10, 21.11, 21.12, 21.13, 21.14, 21.18, 21.21, 21.23, 21.25, 21.27, 21.28, 21.30, and 21.31: From Stuart Ira Fox, *Human Physiology,* 2d ed. Copyright © 1987 Wm. C. Brown Publishers, Dubuque, Iowa. All Rights Reserved. Reprinted by permission.
Figure 21.8: From John W. Hole, Jr., *Human Anatomy and Physiology,* 4th ed. Copyright © 1987 Wm. C. Brown Publishers, Dubuque, Iowa. All Rights Reserved. Reprinted by permission.
Figure 21.16: *Practical Cardiology,* Vol. 4, No. 9, September 1978. Reprinted by permission.
Figure 21.17: From P. Astrand and K. Rodahl, *Textbook of Work Physiology,* 2d ed. Copyright © 1977 McGraw-Hill Book Company, New York, NY. Reprinted by permission.
Figure 21.22: From E. O. Feigel, "Physics in the Cardiovascular System" in *Physiology and Biophysics,* Vol. II, 20th edition, edited by T. C. Ruch and H. D. Patton. Copyright © 1974 W. B. Saunders Company, Philadelphia, PA. Reprinted by permission.
Figure 21.26: From Stuart Ira Fox, *A Laboratory Guide to Human Physiology,* 4th ed. Copyright © 1987 Wm. C. Brown Publishers, Dubuque, Iowa. All Rights Reserved. Reprinted by permission.
Figure 21.29: "Modified and reproduced with permission from Geddes, L. A.: *The Direct and Indirect Measurement of Blood Pressure.* Copyright © 1970 by Year Book Medical Publishers, Inc., Chicago."

Chapter 22

Figures 22.1, 22.2, 22.3a, and 22.5: From Kent M. Van De Graaff, *Human Anatomy,* 2d ed. Copyright © 1988 Wm. C. Brown Publishers, Dubuque, Iowa. All Rights Reserved. Reprinted by permission.
Figure 22.4: From John W. Hole, Jr., *Human Anatomy and Physiology,* 4th ed. Copyright © 1987 Wm. C. Brown Publishers, Dubuque, Iowa. All Rights Reserved. Reprinted by permission.
Figures 22.7, 22.8, 22.9, 22.10, 22.12b, 22.14, 22.15, 22.17, 22.18, 22.19b, 22.20, 22.23, 22.24, 22.25, 22.26, 22.29, and 22.30: From Stuart Ira Fox, *Human Physiology,* 2d ed. Copyright © 1987 Wm. C. Brown Publishers, Dubuque, Iowa. All Rights Reserved. Reprinted by permission.
Figure 22.22: Reprinted with permission. Hugh O. McDevitt. "The HLA System and Its Relation to Disease." *Hospital Practice* Volume 20, Issue 7. Page 57. Illustration by Mr. Alan Iselin.

Chapter 23

Figures 23.1, 23.2, 23.3, 23.4, 23.5, 23.6, 23.7, 23.9, 23.10, 23.19, 23.21, 23.27, 23.28, 23.36, and 23.45: From Kent M. Van De Graaff, *Human Anatomy,* 2d ed. Copyright © 1988 Wm. C. Brown Publishers, Dubuque, Iowa. All Rights Reserved. Reprinted by permission.
Figure 23.12: From John W. Hole, Jr., *Human Anatomy and Physiology,* 3d ed. Copyright © 1984 Wm. C. Brown Publishers, Dubuque, Iowa. All Rights Reserved. Reprinted by permission.
Figures 23.14, 23.15, 23.22, 23.26 left, 23.35, 23.38, 23.54, and 23.55: From John W. Hole, Jr., *Human Anatomy and Physiology,* 4th ed. Copyright © 1987 Wm. C. Brown Publishers, Dubuque, Iowa. All Rights Reserved. Reprinted by permission.
Figures 23.23, 23.29, 23.31, 23.32, 23.39, 23.40, 23.41, 23.42, 23.43, 23.46, 23.47, 23.49, 23.50, 23.51, 23.52, 23.53, and 23.56: From Stuart Ira Fox, *Human Physiology,* 2d ed. Copyright © 1987 Wm. C. Brown Publishers, Dubuque, Iowa. All Rights Reserved. Reprinted by permission.
Figure 23.57: From Karl Wasserman, et al., *Journal of Applied Physiology*, vol. 22, pp. 71–85. Copyright © 1967 American Physiological Society, Bethesda, MD. Reprinted by permission.

Chapter 24

Figures 24.2, 24.3, 24.6, 24.33, 24.34, 24.35, 24.36, 24.37, and 24.38: From Kent M. Van De Graaff, *Human Anatomy,* 2d ed. Copyright © 1988 Wm. C. Brown Publishers, Dubuque, Iowa. All Rights Reserved. Reprinted by permission.
Figures 24.5, 24.7, 24.10, and 24.11b: From John W. Hole, Jr., *Human Anatomy and Physiology,* 4th ed. Copyright © 1987 Wm. C. Brown Publishers, Dubuque, Iowa. All Rights Reserved. Reprinted by permission.
Figures 24.13, 24.14, 24.15, 24.17, 24.19, 24.20, 24.22, 24.23, 24.26, 24.29, 24.30, and 24.31: From Stuart Ira Fox, *Human Physiology,* 2d ed. Copyright © 1987 Wm. C. Brown Publishers, Dubuque, Iowa. All Rights Reserved. Reprinted by permission.
Figure 24.39: Courtesy of American Medical Systems, Inc., Minnetonka, Minnesota.

Chapter 25

Figures 25.1, 25.11, 25.15, 25.17, 25.19, 25.23, and 25.25: From John W. Hole, Jr., *Human Anatomy and Physiology,* 4th ed. Copyright © 1987 Wm. C. Brown Publishers, Dubuque, Iowa. All Rights Reserved. Reprinted by permission.
Figures 25.2, 25.3, 25.4, 25.5, 25.6, 25.7, 25.10, 25.12a&b, 25.14, 25.20, 25.26, 25.37a, and 25.39a&b: From Kent M. Van De Graaff, *Human Anatomy,* 2d ed. Copyright © 1988 Wm. C. Brown Publishers, Dubuque, Iowa. All Rights Reserved. Reprinted by permission.
Figures 25.32, 25.41, 25.42, 25.43, 25.44, 25.47, 25.48, 25.49, 25.50, 25.51, 25.52, and 25.53: From Stuart Ira Fox, *Human Physiology,* 2d ed. Copyright © 1987 Wm. C. Brown Publishers, Dubuque, Iowa. All Rights Reserved. Reprinted by permission.

Chapter 26

Figures 26.4, 26.7, 26.8, 26.9, 26.10, 26.11, 26.12, 26.14, 26.19, 26.20, 26.21, and 26.23: From Stuart Ira Fox, *Human Physiology,* 2d ed. Copyright © 1987 Wm. C. Brown Publishers, Dubuque, Iowa. All Rights Reserved. Reprinted by permission.
Figures 26.5 and 26.6: From Mary L. Parker, et al., "Juvenile Diabetes Mellitus, A Defiency in Insulin" in *Diabetes, 17:*27–32, 1968. Copyright © 1968 American Diabetes Association, Inc., Alexandria, VA. Reproduced with permission from the American Diabetes Association, Inc.
Figure 26.22: Barrett, E. J. and R. A. DeFronao. "Diabetic Ketoacidosis: Diagnosis and Management." *Hospital Practice,* April, 1984, p. 89. Reprinted by permission.

Chapter 27

Figures 27.1 and 27.17a: From John W. Hole, Jr., *Human Anatomy and Physiology,* 4th ed. Copyright © 1987 Wm. C. Brown Publishers, Dubuque, Iowa. All Rights Reserved. Reprinted by permission.
Figures 27.2, 27.3, 27.6a-f, 27.8, 27.10, 27.11, 27.25, 27.26, and 27.27: From Kent M. Van De Graaff, *Human Anatomy,* 2d ed. Copyright © 1988 Wm. C. Brown Publishers, Dubuque, Iowa. All Rights Reserved. Reprinted by permission.
Figures 27.4, 27.5, 27.7, 27.9, 27.12, 27.16, and 27.28: From Stuart Ira Fox, *Human Physiology,* 2d ed. Copyright © 1987 Wm. C. Brown Publishers, Dubuque, Iowa. All Rights Reserved. Reprinted by permission.

Chapter 28

Figures 28.1, 28.12, 28.15, and 28.23: From John W. Hole, Jr., *Human Anatomy and Physiology,* 4th ed. Copyright © 1987 Wm. C. Brown Publishers, Dubuque, Iowa. All Rights Reserved. Reprinted by permission.
Figures 28.5, 28.7, 28.13, and 28.25: From Kent M. Van De Graaff, *Human Anatomy,* 2d ed. Copyright © 1988 Wm. C. Brown Publishers, Dubuque, Iowa. All Rights Reserved. Reprinted by permission.
Figure 28.14: From Ian H. Thorneycroft, et al., *American Journal of Obstetrics and Gynecology,* Volume III. Copyright © 1971 The C. V. Mosby Company, St. Louis, MO. Reprinted by permission.
Figures 28.17, 28.20, and 28.21: From Stuart Ira Fox, *Human Physiology,* 2d ed. Copyright © 1987 Wm. C. Brown Publishers, Dubuque, Iowa. All Rights Reserved. Reprinted by permission.
Figure 28.24: "Courtesy American Cancer Society."

Chapter 29

Figures 29.1a&b, 29.5, 29.8b&c, 29.19a-d, 29.20, 29.21, 29.22 line art, 29.23 line art, 29.36, 29.38a&b, 29.39, 29.41, and 29.42: From Kent M. Van De Graaff, *Human Anatomy,* 2d ed. Copyright © 1988 Wm. C. Brown Publishers, Dubuque, Iowa. All Rights Reserved. Reprinted by permission.
Figures 29.7, 29.9, and 29.28: From John W. Hole, Jr., *Human Anatomy and Physiology,* 4th ed. Copyright © 1987 Wm. C. Brown Publishers, Dubuque, Iowa. All Rights Reserved. Reprinted by permission.
Figures 29.10b, 29.13b, 29.14, 29.15, 29.16, and 29.18: From Stuart Ira Fox, *Human Physiology,* 2d ed. Copyright © 1987 Wm. C. Brown Publishers, Dubuque, Iowa. All Rights Reserved. Reprinted by permission.

Chapter 30

Figure 30.2: From L. Patten, *Human Embryology,* 2d ed. Copyright © 1933 McGraw-Hill Book Company, New York, NY. Reprinted by permission.

Figures 30.4, 30.5, and 30.14: From Kent M. Van De Graaff, *Human Anatomy,* 2d ed. Copyright © 1988 Wm. C. Brown Publishers, Dubuque, Iowa. All Rights Reserved. Reprinted by permission.
Figures 30.8, 30.10, 30.11, and 30.12: From J. W. Hollingsworth, A. Hasizume, and S. Jablon, "Correlations Between Tests of Aging in Hiroshima Subjects: An Attempt to Define 'Physiologic Age,' " in *Yale Journal of Biology and Medicine 38*:11–26. Copyright © 1965 *Yale Journal of Biology and Medicine,* New Haven, CT. Reprinted by permission.

Appendix

Appendix D: From Kent M. Van De Graaff, *Human Anatomy,* 2d ed. Copyright © 1988 Wm. C. Brown Publishers, Dubuque, Iowa. All Rights Reserved. Reprinted by permission.
Appendix E: From Kent M. Van De Graaff, *Human Anatomy.* Copyright © 1984 Wm. C. Brown Publishers, Dubuque, Iowa. All Rights Reserved. Reprinted by permission.

Photos

Chapter 1

Figure 1.1: Academy Florence; **Figure 1.2:** Bibliotheque Nationale; **Figure 1.3:** Fratelli Alinari; **Figure 1.4:** New York Academy of Medicine Library; **Figure 1.5:** Historical Picture Services, Inc.; **Figure 1.7:** Monte S. Buschbaum, M.D.; **Figures 1.10–1.22:** Dr. Sheril D. Burton

Chapter 2

Figure 2.26: © Edwin Reschke

Chapter 3

Figure 3.3a: © Keith R. Porter; **Figure 3.3b:** © Richard Chao; **Figure 3.4a,b:** © Kwang W. Jeon; **Figure 3.5(all):** © M. M. Perry and A. B. Gilbert; **Figure 3.6:** © Sandra L. Wolin; **Figure 3.8:** © Richard Chao; **Figure 3.9a, 3.10a:** © Keith R. Porter; **Figure 3.11:** © E. G. Pollack; **Figure 3.12:** © Richard Chao; **Figure 3.13(top):** Oscar L. Miller, Jr., *Journal of Cell Physiology,* 74(1969)225–232; **Figure 3.14:** © Alexander Rich; **Figure 3.19a:** © J. E. Rothman; **Figure 3.22a:** © David M. Phillips/Visuals Unlimited; **Figure 3.23a–e:** © Edwin Reschke; **Figure 3.24:** Courtesy of the Upjohn Company

Chapter 4

Figure 4.25: © H. Fernandes Móran

Chapter 5

Figure 5.4a: © Carolyn Chambers; **Figure 5.4b:** From TISSUES AND ORGANS: A TEXT ATLAS OF SCANNING ELECTRON MICROSCOPY by R. G. Kessel and R. Kardon. W. H. Freeman and Company © 1979; **Figure 5.11:** © Richard Chao

Chapter 6

Figure 6.1a–c: © Edwin Reschke; **Figure 6.1d:** © Richard A. Heckman and Wilford M. Hess; **Figures 6.6b–6.10b:** © Edwin Reschke; **Figure 6.11a:** © Edwin Reschke/Peter Arnold, Inc.; **Figures 6.12b–6.25b:** © Edwin Reschke; **Figure 6.26:** From TISSUES AND ORGANS: A TEXT ATLAS OF SCANNING ELECTRON MICROSCOPY by R. G. Kessel and R. Kardon. W. H. Freeman and Company © 1979; **Figures 6.28a–6.29a:** © Edwin Reschke

Chapter 8

Figure 8.7b: © Dr. Kerry L. Openshaw; **Figure 8.8:** © Victor B. Eichler; **Figure 8.9a–d:** American Institute of Applied Science; **Figure 8.10:** *Andrew's Diseases of the Skin,* A. N. Domonkos, © W. B. Saunders Company, 1971; **Figure 8.11:** © Dr. Sheril D. Burton; **Figure 8.12:** © James M. Clayton; **Figure 8.13:** © Lester V. Bergman and Associates, Inc.; **Figure 8.14a:** World Health Organizations; **Figure 8.14b:** © George P. Bogumill; **Figure 8.15:** © Dr. Sheril D. Burton; **Figure 8.16a:** © Michael Abbey/Science Source/Photo Researchers, Inc.; **Figures 8.16b–8.19:** © Dr. Kerry Openshaw; **Figure 8.22:** © Martin M. Rotker/Taurus Photos; **Figure 8.23a,b,c:** © Charles H. Stewart, M.D.; **Figure 8.25a,b:** © Robert T. Neisworth; **Figure 8.27:** © Jacques Jangoux/Peter Arnold, Inc.; **Figure 8.28:** © Norman R. Lightfoot/Photo Researchers, Inc.

Chapter 9

Figure 9.2b: © Ted Conde; **Figure 9.11g:** © Edwin Reschke; **Figures 9.12–9.37a:** Courtesy of Utah Valley Hospital, Dept. of Radiology; **Figure 9.42:** © Blayne L. Hirsche; **Figure 9.43:** S.I.U. School of Medicine

Chapter 10

Figure 10.8b: © Dr. Sheril D. Burton; **Figures 10.10–10.16:** Courtesy of Utah Valley Hospital, Dept. of Radiology; **Figure 10.18a:** © Dr. Sheril D. Burton; **Figure 10.18b:** Courtesy of Utah Valley Hospital, Dept. of Radiology; **Figure 10.20:** © Blayne L. Hirsche; **Figure 10.22e:** Courtesy of Kodak

Chapter 11

Figure 11.3: Courtesy of Utah Valley Hospital, Dept. of Radiology; **Figures 11.20a–11.22b:** © Dr. Sheril D. Burton; **Figures 11.33, 11.34:** © Lester V. Bergman and Associates, Inc.; **Figure 11.35a,d:** S.I.U. School of Medicine

Chapter 12

Figure 12.10b: © John D. Cunningham/Visuals Unlimited; **Figure 12.13a,b:** © Dr. H. E. Huxley; **Figure 12.13c:** From TISSUES AND ORGANS: A TEXT ATLAS OF SCANNING ELECTRON MICROSCOPY by R. G. Kessel and R. Kardon. W. H. Freeman and Company © 1979; **Figure 12.14:** © Dr. H. E. Huxley; **Figure 12.19b:** © John D. Cunningham/Visual Unlimited; **Figure 12.23:** International Biomedical, Inc.; **Figures 12.27, 12.29:** Hans Hoppler, *Respiratory Physiology* 44:94(1981); **Figure 12.31:** © Dr. Sheril D. Burton

Chapter 13

Figure 13.1 Historical Picture Service, Chicago; **Figure 13.4a,b:** © Dr. Sheril D. Burton; **Figure 13.5a–c:** © Dr. Douglas W. Hacking; **Figures 13.6a–13.31d:** © Dr. Sheril D. Burton; **Figure 13.32a,b:** © Dr. Kenneth A. Johnson; **Figure 13.32c:** © Lester V. Bergman and Associates

Reference Figures

Figures 8–13: Reproduced by permission from S. D. Vidic, F. R. Suarez, from *Photographic Atlas of the Human Body,* 1984. St. Louis, The C. V. Mosby Company; **Figure 14:** © Dr. Sheril D. Burton; **Figures 15–17:** Reproduced by permission from S. D. Vidic, F. R. Suarez, from *Photographic Atlas of the Human Body,* 1984, St. Louis, The C. V. Mosby Company; **Figure 18:** © Dr. Sheril D. Burton; **Figures 19–20:** Reproduced by permission from S. D. Vidic, F. R. Suarez, from *Photographic Atlas of the Human Body,* 1984, St. Louis, The C. V. Mosby Company; **Figures 21–23:** © Dr. Sheril D. Burton; **Figures 24–25:** Reproduced by permission from S. D. Vidic, F. R. Suarez, from *Photographic Atlas of the Human Body,* 1984. St. Louis, The C. V. Mosby Company; **Figures 26–30:** © Dr. Sheril D. Burton

Chapter 14

Figure 14.8: © Dr. John D. Cunningham/Visuals Unlimited; **Figure 14.16:** H. Webster, from Hubbard, John. *The Vetebrate Peripheral Nervous System,* Plenum Press, 1974; **Figure 14.18:** © Edwin Reschke; **Figure 14.22:** Bell et al. *Textbook of Physiology and Biochemistry,* 8th ed., Churchill Livingstone, Inc. 1972; **Figure 14.27a,b:** Gilula, Reeves, and Steinbach. *Nature,* 235:262–265. © Macmillan Journals Limited; **Figure 14.28:** Micrograph produced by Dr. John E. Heuser of Washington University, School of Medicine, St. Louis, MO

Chapter 15

Figure 15.3: © Monte S. Buchsbaum, M.D.; **Figure 15.8a:** © Martin M. Rotker/Taurus Photos; **Figure 15.10b:** Igaku Shoin, Ltd.; **Figure 15.23:** Courtesy of Utah Valley Hospital, Dept. of Radiology; **Figure 15.26b:** © Per H. Kjeldson, University of MIchigan at Ann Arbor

Chapter 16

Figure 16.2: From TISSUES AND ORGANS: A TEXT ATLAS OF SCANNING ELECTRON MICROSCOPY by R. G. Kessel and R. Kardon. W. H. Freeman and Company © 1979; **Figure 16.24:** Photographed by Dr. Sheril D. Burton assisted by Dr. Douglas W. Hacking

Chapter 17

Figure 17.2: © Avril V. Somylo, Ph.D; **Figure 17.11a:** © Edwin Reschke/Peter Arnold Inc.

Chapter 18

Figure 18.14c: © Victor B. Eichler; **Figures 18.19, 18.22:** © Dr. Sheril D. Burton; **Figure 18.24:** © Camera M.D. Studios; **Figure 18.29b:** © Per H. Kjeldson, University of Michigan at Ann Arbor; **Figure 18.32a:** © Thomas Sims; **Figure 18.42:** © David H. Hubel; **Figure 18.49:** © Dr. Sheril D. Burton; **Figure 18.55:** From Kessel, R. G., and Shih, C. Y. SCANNING ELECTRON MICROSCOPY IN BIOLOGY © 1976, Springer-Verlag; **Figure 18.61:** © Dean E. Hillman; **Figure 18.66:** © Dr. Stephen Clark

Chapter 19

Figure 19.1b: © Edwin Reschke; **Figure 19.18b:** Journalism Services; **Figure 19.19:** © Martin Rotker/Taurus Photos; **Figure 19.27–19.29:** © Lester V. Bergman and Associates

Chapter 20

Figure 20.4: Courtesy of Utah Valley Hospital, Dept. of Radiology; **Figure 20.9a:** © Igaku Shoin, Ltd.; **Figure 20.10a:** © Carroll Weiss, RBP; **Figure 20.23:** © Don Fawcett; **Figure 20.26:** Historical Pictures Service, Chicago; **Figures 20.29b, 20.31a:** From *Practische Ontleedkunde* from J. Dankmeijer, H. G. Lambers & J. M. F. Landsmeer. Bohn, Scheltema & Holkema; **Figure 20.33b:** Courtesy of Utah Valley Hospital, Dept. of Radiology; **Figure 20.49:** © Stuart I. Fox; **Figure 20.50:** © Manfred Kage/Peter Arnold, Inc.; **Figure 20.55a,b:** © Richard Menard; **Figure 20.59a,b:** American Heart Association; **Figure 20.59c:** © Lewis Lainey

Chapter 21

Figure 21.9: Markell, E. K., and Vogue, M., *Medical Parasitology,* 5th ed. W. B. Saunders Co.; **Figure 21.15a,b:** Donald S. Baum, from Hurst et al., *The Heart,* 5/e, McGraw/Hill Book Co.; **Figure 21.19a,b:** © Niels A. Lassen, Copenhagen, Denmark

Chapter 22

Figure 22.3b: © Dr. John D. Cunningham/Visuals Unlimited; **Figure 22.3c:** © Igaku Shoin, Ltd.; **Figure 22.12a:** Computer graphic by Arthur J. Olson, Ph.D. © 1985. All rights reserved. Scripps Clinic and Research Foundation, LaJolla, CA; **Figure 22.13a,b:** © S. E. V. Phillips, Biophysics, The University of Leeds, From *Science* 233(1986)749; **Figure 22.19a:** Rosenthal, Alan S. *New England Journal of Medicine,* 303(1980)1153; **Figure 22.27a,b:** © Dr. Andrejs Liepins; **Figure 22.28:** © Dr. Noel R. Rose, M.D.; **Figure 22.31(both):** Custom Medical Stock Photo; **Figure 22.32a:** Kessel, R. G., and Shih, C. Y. SCANNING ELECTRON MICROSCOPY IN BIOLOGY © 1976, Springer-Verlag; **Figure 22.32b:** Catherin Ellis/Science Photo Library/Photo Researchers, Inc.

Chapter 23

Figure 23.8: American Lung Association; **Figure 23.13:** Yokochi, C., & Rohen, J. W., Photographic Anatomy of Human Body 2/e. © Igaku Shoin Ltd.; **Figure 23.16a:** © John D. Cunningham/Visuals Unlimited; **Figure 23.16b:** © Edwin Reschke; **Figure 23.18:** © John Watney Photo Library; **Figure 23.20:** Courtesy of Kodak; **Figure 23.24:** David M. Philips, The Population Council/Visuals Unlimited; **Figure 23.25:** From TISSUES AND ORGANS: A TEXT ATLAS OF SCANNING ELECTRON MICROSCOPY by R. G. Kessel and R. Kardon. W. H. Freeman and Company © 1979; **Figure 23.30:** Edward C. Vasquez, R. T., C. R. T., Dept. of Radiologic Technology, Los Angeles City College; **Figure 23.34a,b:** Comroe, J. H., Jr. *Physiology of Respiration* © 1974 Year Book Medical Publishers Inc., Chicago; **Figure 23.37:** Courtesy of Warren E. Collins, Inc., Braintree, MA; **Figure 23.59a,b:** © Martin Rotker/Peter Arnold, Inc.

Chapter 24

Figure 24.4: CNRI/Science Photo Library/Photo Researchers, Inc.; **Figure 24.8b:** R. B. Wilson, University of Nebraska; **Figure 24.12a:** Daniel Friend, from Bloom, William, and Fawcett, Don W. A. *Textbook of Histology* 10/e, W. B. Saunders Co.; **Figure 24.32:** © Per H. Kjeldson, University of Michigan, Ann Arbor

Chapter 25

Figure 25.12c: © Dr. Karl M. Francis; **Figure 25.13:** © Dr. Sheril D. Burton; **Figure 25.16a–c:** Biophoto Associates, Science Source, Photo Researchers, Inc.; **Figure 25.18:** © Edwin Reschke; **Figures 25.19e, 25.21:** Courtesy of Utah Valley Hospital, Dept. of Radiology; **Figure 25.22:** © Edwin Reschke; **Figure 25.27:** © Manfred Kage/Peter Arnold, Inc.; **Figure 25.30a,b:** © Keith R. Porter/Alpers, D. H., & Seetharam, D. *New England Journal of Medicine,* 296(1977)1047; **Figure 25.34:** © Lester V. Bergman and Associates, Inc.; **Figure 25.36:** Sodeman, W. A., & Watson, T. M., *Pathologic Psychology* 6/e, W. B. Saunders Co. 1969; **Figure 25.38a:** From TISSUES AND ORGANS: A TEXT ATLAS OF SCANNING ELECTRON MICROSCOPY by R. G. Kessel and R. Kardon. W. H. Freeman and Company © 1979; **Figure 25.39c:** © Victor B. Eichler; **Figure 25.45a:** © Carroll H. Weiss/RBP; **Figure 25.45b:** © Dr. Sheril D. Burton; **Figure 25.54a–d:** John Van De Graaff, DDS, and Larry S. Pierce, DDS, both at Northwestern University, School of Dentistry, Chicago

Chapter 26

Figure 26.13: © Lester V. Bergman and Associates, Inc.; **Figure 26.16a–d:** *American Journal of Medicine,* 20(1956)133; **Figure 26.17a,b:** © BioPhoto Associates/Photo Researchers, Inc.

Chapter 27

Figure 27.6g,h: © Dr. Landrum Shettles; **Figure 27.15a:** © BioPhoto Associates, Science Source/Photo Researchers, Inc.; **Figures 27.15b, 27.17b:** From TISSUES AND ORGANS: A TEXT ATLAS OF SCANNING ELECTRON MICROSCOPY by R. G. Kessel and R. Kardon. W. H. Freeman and Company © 1979; **Figures 27.18, 27.19:** © Edwin Reschke; **Figures 27.21, 27.22:** © Manfred Kage/Peter Arnold, Inc.; **Figure 27.23:** © Edwin Reschke; **Figure 27.29:** Center for Disease Control, Atlanta, Georgia

Chapter 29

Figure 29.1c: © David Philips/Visuals Unlimited; **Figure 29.2:** Lucian Zamboni from Greep, Roy, and Weiss, Leon, HISTOLOGY, 3d ed., McGraw-Hill Book Company, 1973; **Figure 29.6a–d:** © R. G. Edwards; **Figure 29.8a:** © Ronan O'Rahilly/Carnegie Laboratories of Embryology, University of California; **Figure 29.10a:** Carnegie Institute of Washington, Dept. of Embryology, Davis Division; **Figure 29.13a:** © Donald Yeager/Carroll Weiss; **Figure 29.17:** © Lester V. Bergman and Associates, Inc.; **Figure 29.19e:** © Dr. Landrum Shettles; **Figure 29.22:** © Donald Yeager/Camera M.D. Studios; **Figure 29.23:** © Landrum Shettles; **Figures 29.24–29.26b:** © Donald Yeager/Camera M.D. Studios; **Figure 29.30:** Runk/Shoenberger/Grant Heilman Inc.; **Figure 29.31a,b:** March of Dimes; **Figure 29.37:** © Lester V. Bergman and Associates, Inc.; **Figure 29.38c:** © Dr. Landrum Shettles; **Figure 29.40:** Gregory Dellore M.D., and Steven L. Clark M.D.

Chapter 30

Figure 30.3: Courtesy of Utah Valley Hospital, Dept. of Radiology; **Figure 30.6(child):** from Science Year, *The World Book Science Annual.* © 1973 Field Enterprises Educational Corporation. By permission of World Book Inc.; **Figure 30.6(woman):** Black Star Publishing Company; **Figure 30.7:** Carolina Biological Supply; **Figure 30.9:** © John Bird

INDEX

B

C

E

F

G

H

J

K

L

N

O

P

S

U

V

W

X

Y

Z

Nobel Laureates in Physiology/Medicine

Nobel Laureate	Country*	Year of Award	Accomplishment
Emil A. von Behring	Germany	1901	Development of serum therapy for diphtheria
Sir Ronald Ross	England	1902	Studies on the cause and transmission of malaria
Niels Ryberg Finsen	Denmark	1903	Treatment of lupus vulgaris with concentrated light rays
Ivan Petrovich Pavlov	Russia	1904	Work on the physiology of digestion
Robert Koch	Germany	1905	Cultivation of the tubercle bacillus
Camillo Golgi	Italy	1906	Structure of the nervous system
Santiago Ramon Y Cajal	Italy		
Charles Louis Alphonse Laveran	France	1907	Role of protozoa in causing disease
Paul Ehrlich	Germany	1908	Theories on the development of immunity
Elie Metchnikoff	France (USSR)		Description of phagocytosis
Theodor Kocher	Germany (Swtz.)	1909	Work on the physiology, pathology, and surgery of the thyroid gland
Albrecht Kossel	Germany (Swtz.)	1910	Biochemistry of the cell
Allvar Gullstrand	Sweden	1911	Work on the dioptrics of the eye
Alexis Carrel	France	1912	Work on vascular ligature and the grafting of blood vessels and organs
Charles R. Richet	France	1913	Investigations on anaphylaxis
Robert Barany	Austria	1914	Understanding the physiology and pathology of vestibular organs
No award		1915–1918	
Jules Bordet	Belgium	1919	Discoveries in regard to immunity
August Krogh	Denmark	1920	Discovery of the motor mechanism of capillaries
No Award		1921	
Archibald Vivian Hill	England	1922	Physiology of muscle
Otto Meyerhof	Germany		
Frederick Grant Banting	Canada	1923	Discovery of insulin
John James Richard Macleod	Canada (Scotland)		
Willem Einthoven	Netherlands	1924	Mechanism of the electrocardiogram
No Award		1925	
Johannes Fibiger	Denmark	1926	Discovery of the spiroptera carcinoma
Julius Wagner-Jauregg	Austria	1927	Discovery of the therapeutic value of malaria inoculation in the treatment of dementia paralytica
Charles J. H. Nicolle	France	1928	Studies on the cause and transmission of epidemic typhus
Christiaan Eijkman	Netherlands	1929	Discovery of the antineuritic vitamin
Frederick Gowland Hopkins	England		Discovery of the growth-stimulating vitamins
Karl Landsteiner	United States (Aust.)	1930	Discovery of the human blood groups
Otto Warburg	Germany	1931	Discovery of the nature and mode of action of the respiratory enzyme
Charles Scott Sherrington	England	1932	Functions of neurons
Edgar Douglas Adrian	England		
Thomas Hunt Morgan	United States	1933	Function of chromosomes in the transmission of heredity
George Hoyt Whipple	United States	1934	Discoveries concerning liver therapy against anemias
George Richards Minot	United States		
William Parry Murphy	United States		
Hans Spemann	Germany	1935	Discovery of the organizer effect in embryonic development
Henry Dale	England	1936	Chemical transmission of nerve impulses
Otto Loewi	Germany		
Albert Szent-Györgyi von Nagyrapolt	Hungary	1937	Discoveries of biological combustion
Corneille Heymans	Belgium	1938	Discovery of the role played by the sinus and aortic mechanisms in the regulation of respiration
Gerhard Domagk	Germany	1939	Discovery of the chemotherapeutic effects of prontosil
No Award		1940–1942	
Henrik Dam	Denmark	1943	Discovery of vitamin K
Edward A. Doisy	United States		
Joseph Erlanger	United States	1944	Functions of nerve fibers
Herbert Spencer Gasser	United States		
Sir Alexander Fleming	England	1945	Discovery and development of penicillin
Ernst Boris Chain	England (Germany)		
Sir Howard W. Florey	England (Australia)		
Hermann Joseph Muller	United States	1946	Discovery of the production of mutations by means of X-ray irradiation
Bernardo Alberto Houssay	United States	1947	Discovery of how glycogen is catalytically converted
Carl F. Cori	United States (Hungary)		
Gerty T. Cori	United States (Hungary)		
Paul Muller	Swtz.	1948	Discovery of the insect-killing properties of DDT
Walter Rudolf Hess	Switzerland	1949	Research on brain control of body
Egas Moniz	Portugal		
Edward Calvin Kendall	United States	1950	Discoveries concerning the suprarenal cortex hormones
Philip Showalter Hench	United States		
Tadeus Reichstein	Swtz. (Poland)		
Max Theiler	United States (South Africa)	1951	Development of a vaccine for yellow fever
Selman A. Waksman	United States	1952	Discovery and development of streptomycin

*Country of birth is in parentheses.